Advanced Engineering Mathematics

Second Edition

Advanced Engineering Mathematics

Second Edition

Dennis G. Zill
Loyola Marymount University

Michael R. Cullen

JONES AND BARTLETT PUBLISHERS
Sudbury, Massachusetts
BOSTON TORONTO LONDON SINGAPORE

World Headquarters
Jones and Bartlett Publishers
40 Tall Pine Drive
Sudbury, MA 01776
978-443-5000
info@jbpub.com
www.jbpub.com

Jones and Bartlett Publishers Canada
2100 Bloor St. West
Suite 6-272
Toronto, ON M6S 5A5
CANADA

Jones and Bartlett Publishers International
Barb House, Barb Mews
London W6 7PA
UK

Production Credits
Chief Executive Officer: Clayton Jones
Chief Operating Officer: Don Jones, Jr.
President: Tom Walker
V.P., Sales and Marketing: Tom Manning
V.P., College Editorial Director: Brian L. McKean
V.P., Director of Interactive Technology: Mike Campbell
V.P., Managing Editor: Judith H. Hauck
Director of Design and Production: Anne Spencer
Senior Marketing Manager: Jennifer M. Jacobson
Senior Acquisitions Editor: J. Michael Stranz
Production Editor: Rebecca S. Marks
Director of Manufacturing and
 Inventory Control: Therese Bräuer
Developmental Editor: Barbara Lovenvirth
Cover Design: Anne Spencer
Design and Packaging: PageMasters & Company
Printing and Binding: World Color
Cover Printing: Lehigh Press

Portions of this book also appear in *Differential Equations with Boundary-Value Problems*, 4th Edition, by Dennis G. Zill and Michael R. Cullen, copyright © 1997 by Brooks/Cole Publishing Company. Used with permission of the publisher.

Cover images (from left to right) © Hulton Deutsch/Corbis; © Hulton Deutsch/Corbis; © Hulton Deutsch/Corbis; © Grant Smith/Corbis.

Photo credits: pg. 79, Figure 2.39 © Shelly Grossman/ Woodfin Camp and Associates; pg. 80, Figure 2.40 © 1978 Barrie M. Schwortz.

Library of Congress Cataloging-in-Publication Data
Zill, Dennis G., 1940- .
 Advanced engineering mathematics / Dennis G. Zill, Michael R. Cullen.—
2nd ed.
 p. cm.
 Includes index.
 ISBN 0-7637-1065-2 (hardcover : alk. paper).—ISBN 0-7637-1357-0
(pbk. : alk. paper)
 1. Engineering mathematics. I. Cullen, Michael R. II. Title.
TA330.Z55 1999
620′.001′51—dc21 99-5398
 CIP

Printed in the United States of America
03 02 01 00 99 10 9 8 7 6 5 4 3 2 1

About the Cover

The 984-foot high Eiffel Tower opened in 1889 in the International Exhibition of Paris to celebrate the centenary of the French Revolution. The design competition for this commemorative tower was won easily by the famous engineer and architect Alexandre Gustave Eiffel. Eiffel was also involved in the design and construction of the interior support structure of the Statue of Liberty.

With the help of architect Stephen Sauvestre, Gustave Eiffel set out to design a tower that could withstand the "violent action of hurricanes" while portraying "strength and beauty." On September 18, 1884, Eiffel filed a patent "for a new configuration allowing the construction of metal supports and pylons capable of exceeding a height of 300 metres."

The Tower is made up of a base that rests on four pillars, a two-legged second story that rests on the base, and the actual tower. Construction began in January of 1887, and the Tower was complete on March 31, 1889. The structure at the time of completion weighed 7,300 tons and stood 312.27 meters high.

The cultural and artistic elite of Paris never approved of the Tower, and almost succeeded in having it torn down in 1909. Time and utility prevailed, and with the advent of radio there was a realization that by installing an antenna 300 meters in the sky, there was a practical purpose for this magnificent engineering feat.

The history and facts of the Eiffel Tower can be found at several sites on the Internet. You are referred to URL:

www.users.imaginet.fr/~chouard/index.html

for links to other web sites, particularly to the web site of the "Eiffel Tower Equation." If $f(x)$ gives the width of the tower as a function of distance x from its top, then for a constant a,

$$af(x) \int_0^x [f(t)]^2 dt = x \int_0^x f(t)dt - \int_0^x tf(t)dt.$$

If you can solve this equation, you are asked to contact Christophe Chouard by e-mail at an address given at the website.

The French are justifiably proud of their many accomplishments in science, engineering, and mathematics. Surrounding the first level of the tower are plaques that list the names of 72 prominent scientists and mathematicians, including some mentioned in this text: Cauchy, Sturm, Fourier, Legendre, Poisson, and Laplace. For more information go to the URL:

www.tour-eiffel.fr/

This edition is dedicated to my coauthor, colleague,
and friend of 27 years, as well as an award-winning teacher,
Michael R. Cullen, who passed away during its production.

Michael Cullen
1943–1999

Contents

Part 1
Ordinary Differential Equations 1

CHAPTER

4 The Laplace Transform 189

CHAPTER

5 Series Solutions of Linear Equations 235

CHAPTER

6 Numerical Solutions of Ordinary Differential Equations 267

Part 2
Vectors, Matrices, and Vector Calculus

Part 3
Systems of Differential Equations 563

CHAPTER

10 Systems of Linear Differential Equations 567

CHAPTER

11 Systems of Nonlinear Differential Equations 601

Part 4
Fourier Series and Partial Differential Equations 647

Part 5
Complex Analysis 785

Preface

Unlike a course in "calculus" or "differential equations" in which the content of the course is fairly standardized, the content of a course titled "engineering mathematics" sometimes varies considerably between two different academic institutions. A text in Advanced Engineering Mathematics is, therefore, a compendium of many mathematical topics, all of which are loosely related by the expedient of either being needed or useful in courses and subsequent careers in science and engineering. There is literally no upper bound to the number of topics that could be included in a text such as this. Consequently, this book represents the authors' opinion, *at this time,* of what constitutes "engineering mathematics."

Content of the Text

For flexibility in topic selection the text is divided into five major parts or subdivisions. As can be seen from the titles of these various parts, it is clear that our belief is that the backbone of science/engineering-related mathematics is the theory and applications of ordinary and partial differential equations.

Part I: Ordinary Differential Equations

The six chapters in this part constitute a complete short course in ordinary differential equations. These chapters, with some modifications, correspond to Chapters 1, 2, 3, 4, 5, 6, 7, and 9 from the text *A First Course in Differential Equations with Modeling Applications,* 6th Edition, by Dennis G. Zill (Brooks/Cole Publishing Company, 1997).

Part II: Vectors, Matrices, and Vector Calculus

Chapter 7, "Vectors," and Chapter 9, "Vector Calculus," include many of the standard topics usually covered in the third semester of a calculus sequence: geometric vectors, vector functions, directional derivatives, line integrals, double and triple integrals, surface integrals, Green's theorem, Stokes' theorem, and the divergence theorem. Chapter 8, "Matrices," is an introduction to systems of algebraic equations, determinants, and matrix algebra with special emphasis on those types of matrices that are useful in solving systems of linear differential equations. Sections on cryptography, error correcting codes, the method of least squares, and discrete compartmental models are presented as applications of matrix algebra.

Part III: Systems of Differential Equations

There are two chapters in this part. Chapter 10, "Systems of Linear First-Order Differential Equations," and Chapter 11, "Systems of Nonlinear Differential Equations" draw heavily on the matrix material presented in Chapter 8 of Part II. In Chapter 10, systems of linear first-order equations are solved utilizing the concepts of eigenvalues and eigenvectors, diagonalization, and, albeit briefly, by means of a matrix exponential function. In Chapter 11, stability concepts are motivated using two applications: fluid flow in a plane and the motion of a bead on a wire.

Part IV: Fourier Series and Partial Differential Equations

The core material on Fourier series and boundary-value problems was drawn from *Differential Equations with Boundary-Value Problems,* 4th Edition, by Dennis G. Zill and Michael R. Cullen (Brooks/Cole Publishing Company, 1997). In Chapter 12, "Orthogonal Functions and Fourier Series," we present the fundamental topics of sets of orthogonal functions and expansions of functions in terms of an infinite series of orthogonal functions. These topics are then utilized in Chapters 13 and 14, where we solve boundary-value problems in rectangular, polar, cylindrical, and spherical coordinates using the method of separation of variables. In Chapter 15, "Integral Transforms," we solve boundary-value problems by means of the Laplace and Fourier integral transforms.

Part V: Complex Analysis

Chapters 17, 18, 19, and 20 cover basic complex numbers through applications of conformal mappings in the solution of Dirichlet's problem. This material by itself could easily serve as a one-quarter course introduction to complex variables.

■ Changes in This Edition

Changes by Chapter:

- Chapter 1. The notions of first- and second-order initial-value problems have been moved to this chapter. An entire section is now devoted to this important concept.

- Chapter 2. The chapter opens with a new introductory section on qualitative analysis of autonomous first-order differential equations—the notions of phase line,

critical points, attractors, and repellers are examined in a most elementary manner. This additional material stresses the geometry of the solution curves of autonomous first-order equations. The material on homogeneous equations and Bernoulli equations has been combined into one section on the use of substitutions in solving a differential equation. Euler's method has been added to the chapter as an elementary means of familiarizing the student with the notion of approximating solutions of differential equations, as well as a means of explaining the nature of commercial software packages, which we refer to generically as "numerical solvers." There is also new material on linear and nonlinear systems of first-order differential equations as mathematical models.

• Chapter 3. New material on nonlinear models—especially on nonlinear springs and the nonlinear pendulum—has been added to this chapter. The approach to undetermined coefficients using annihilator differential operators has been deleted. The notions of eigenvalues and eigenvectors are introduced in a new section on second-order boundary-value problems and their applications.

• Chapter 4. The material of the Laplace transform has been substantially rearranged. Solutions of differential equations are now considered earlier than in the previous edition. Solutions of linear differential equations are introduced in the second section rather than waiting until all the operational properties of the transform have been presented.

• Chapter 5. This chapter has been shortened somewhat. The discussion on the Cauchy–Euler equation has been moved to Chapter 3. The motivation for this move was the similarity between the method of solution of a Cauchy–Euler equation and the solution method of linear constant-coefficient equations. This chapter is now focused exclusively on series solutions of differential equations.

• Chapters 2, 6, 8, and 16. The various numerical methods, which were gathered into Part V in the 1st edition, have now been spread out into the appropriate chapters.

• Chapter 8. New sections on the rank of a matrix, powers of matrices, and discrete compartmental models have been added.

• Chapter 10. The discussion of the matrix exponential has been expanded.

• Chapter 12. A section on the complex Fourier series has been added. The concept of a frequency spectrum is considered.

• Chapter 15. The section on the fast Fourier transform (FFT) is new to this chapter.

Changes Throughout the Text:

• Each of the five major parts or subdivisions of this text opens with an essay by an acknowledged expert in a particular field of engineering.

• There is a greater emphasis on differential equations as mathematical models. The notion of a mathematical model is now interwoven throughout the text, and the constructions and pitfalls of various models are discussed.

• Graphing software, computer algebra systems, and numerical solvers are utilized wherever appropriate in applications and examples, as well as in the exercise sets. The problems labeled "Computer Lab Assignments" can be used for classroom demonstrations, discussion, or as single-person or group projects.

• Many new conceptual problems have also been added to the exercise sets.

■ Design of the Text

Each chapter has an opening page that includes a table of contents, an introduction, a list of important topics in that chapter, and a list of relevant topics from calculus and from other parts of the text (with cross-references to chapter sections in parentheses) that could productively be reviewed in preparation for the material introduced in that chapter.

Answers to selected odd-numbered problems are given in the the back of the text. Review exercises are given at the end of each chapter. Remarks and marginal annotations are scattered throughout the text.

■ Supplements

A complete Solutions Manual by Warren S. Wright is available for instructors. A Student Solutions Manual, in which every third problem is worked out, is also available. Contact your Jones and Bartlett Publisher's Representative for more information.

■ Acknowledgments

We would like to thank the following individuals for their criticisms and suggestions for improvement, both of the previous edition and of the preliminary versions of the new edition:

Noal Harbertson, California State University
Ronald Guenther, Oregon State University
Robert Hunt, Humboldt
Herman Gollwitzer, Drexel University
Stan Friedlander, Bronx Community College/CUNY
William Criminale, University of Washington
Cecilia Knoll, Florida Tech
Jeff Dodd, Jacksonville State University

In addition, a special word of thanks goes to the following individuals:

Sherif F. Ali, Arizona State University and the
 University of Cairo
Robert Briedenthal, University of Washington
John Ellison, Grove City College
Arpad Horvath, Carnegie Mellon University
Gilbert Lewis, Michigan Technological University

They were kind enough to take time out of their busy schedules to supply the essays. And thanks to Gary S. Stoudt, Indiana University of Pennsylvania for his invaluable input and help with the new material in Chapter 15. Finally, we would like to express our appreciation to Barbara Lovenvirth, developmental editor, for her unfailing and pointed sense of humor that enabled us to laugh instead of cry when bad things happened; for her patience; and, of course, for all the time and effort that she put into this project.

The task of compiling a text this size was, to say the least, time-consuming and difficult. Through the process of passing hundreds of manuscript pages through many hands, undoubtedly some errors have sifted through. We apologize for this in advance and we would certainly appreciate hearing about any errors (in care of the publisher) in order to expedite their correction.

Dennis G. Zill
Los Angeles

1

Ordinary Differential Equations

The Collapse of the Tacoma Narrows Suspension Bridge: A Modern Viewpoint

Gilbert N. Lewis
Michigan Technological University

In the summer of 1940, the Tacoma Narrows suspension bridge in the state of Washington was completed and opened to traffic. Almost immediately, observers noticed that the wind blowing across the roadway would sometimes set up large vertical vibrations in the roadbed. The bridge became a tourist attraction as people came to watch and perhaps ride the undulating bridge. Finally, on November 7, 1940, during a powerful storm, the oscillations increased beyond any previously observed, and the bridge was evacuated. Soon the vertical oscillations became rotational, as observed by looking down the roadway. The entire span was eventually shaken apart by the large vibrations, and the bridge collapsed. See [1] and [2] for interesting and sometimes humorous anecdotes associated with the bridge.

The noted engineer Theodore von Karman was asked to determine the cause of the collapse. He and his coauthors [3] claimed that the wind blowing perpendicularly across the roadway separated into vortices (wind swirls) alternately above and below the roadbed, thereby setting up a periodic, vertical force acting on the bridge. It was this force that caused the oscillations. Others further hypothesized that the frequency of this forcing function exactly matched the natural frequency of the bridge, thus leading to resonance, large oscillations, and destruction, as described in equation (31), Section 3.8. For almost fifty years, resonance was blamed as the cause of the collapse of the bridge, although the von Karman group denied this, stating that "it is very improbable that resonance with alternating vortices plays an important role in the oscillations of suspension bridges" [3].

As we can see from equation (31), Section 3.8, resonance is a *linear* phenomenon. In addition, for resonance to occur, there must be an exact match between the frequency of the forcing function and the natural frequency of the bridge. Furthermore, there must be absolutely no damping in the system. It should not be surprising, then, that resonance was not the culprit in the collapse.

If resonance did not cause the collapse of the bridge, what did? Recent research provides an alternative explanation for the collapse of the Tacoma Narrows bridge. Lazer and McKenna [4] contend that nonlinear effects, not linear resonance, were the main factors leading to the large oscillations of the bridge (see [5] for a good review article). The theory involves partial differential equations. However, a simplified model leading to a nonlinear ordinary differential equation can be constructed.

The development of the model below is not exactly the same as that of Lazer and McKenna, but it results in a similar differential equation. The example is a new one, and it shows another way that amplitudes of oscillation can increase.

Consider a single vertical cable of the suspension bridge. We assume that it acts like a spring but with different characteristics in tension and compression and with no damping. When stretched, the cable acts like a spring with Hooke's constant, *b,* while, when compressed, it acts like a spring with a different Hooke's constant, *a.* We assume that the cable in compression exerts a smaller force on the roadway than when stretched the same distance, so that $0 < a < b$. Let the vertical deflection (positive direction downward) of the slice of the roadbed attached to this cable be denoted by $y(t)$, where t represents time, and $y = 0$ represents the equilibrium position. As the roadbed oscillates under the influence of an applied vertical force (due to the von Karman vortices), the cable provides an upward restoring force equal to by when $y > 0$ and a downward restoring force equal to ay when $y < 0$. This change in the Hooke's law constant at $y = 0$ provides the nonlinearity to the differential equation. We are thus led to consider the differential equation derived from Newton's second law of motion ($F = ma$),

$$my'' + f(y) = g(t),$$

where $f(y)$ is the nonlinear function given by

$$f(y) = \begin{cases} by, & \text{if } y \geq 0 \\ ay, & \text{if } y < 0, \end{cases}$$

$g(t)$ is the applied force, and m is the mass of the section of the roadway. Note that the differential equation is linear on any interval on which y does not change sign.

Now, let us see what a typical solution of this problem would look like. We will assume that $m = 1$, $b = 4$, $a = 1$, and $g(t) = \sin 4t$. Note that the frequency of the forcing function is larger than the natural frequencies of the cable in both tension and compression, so that we do not expect resonance to occur. We also assign the following initial values to y: $y(0) = 0$, $y'(0) = \alpha$, where $\alpha > 0$, so that the roadbed starts in the equilibrium position with downward velocity.

Because of the downward initial velocity and the positive applied force, $y(t)$ will initially increase and become positive. Therefore, we first solve

$$y'' + 4y = \sin 4t, \qquad y(0) = 0, \ y'(0) = \alpha.$$

The solution is the sum of the complementary solution, $y_c(t)$, and the particular solution, $y_p(t)$. It is easy to see that $y_c(t) = c_1 \cos 2t + c_2 \sin 2t$ (see Section 3.3), and $y_p(t) = -\frac{1}{12} \sin 4t$ (see Section 3.4). Thus,

$$y(t) = c_1 \cos 2t + c_2 \sin 2t - \frac{1}{12} \sin 4t.$$

The initial conditions give

$$y(0) = 0 = c_1,$$

$$y'(0) = \alpha = 2c_2 - \frac{1}{3},$$

so that $c_2 = (\alpha + \frac{1}{3})/2$. Therefore,

$$y(t) = \frac{1}{2}\left(\alpha + \frac{1}{3}\right)\sin 2t - \frac{1}{12}\sin 4t$$

$$= \sin 2t\left[\frac{1}{2}\left(\alpha + \frac{1}{3}\right) - \frac{1}{6}\cos 2t\right]$$

where we have used $\sin 4t = 2 \sin 2t \cos 2t$. We note that the first positive value of t for which $y(t)$ is again equal to zero is $t = \pi/2$. At that point, $y'(\pi/2) = -(\alpha + \frac{2}{3})$.

After $t = \pi/2$, y becomes negative, so we must now solve the new problem

$$y'' + y = \sin 4t, \qquad y\left(\frac{\pi}{2}\right) = 0, \ y'\left(\frac{\pi}{2}\right) = -\left(\alpha + \frac{2}{3}\right).$$

The solution, found as above, is

$$y(t) = \left(\alpha + \frac{2}{5}\right)\cos t - \frac{1}{15}\sin 4t$$

$$= \cos t\left[\left(\alpha + \frac{2}{5}\right) - \frac{4}{15}\sin t \cos 2t\right].$$

The next positive value of t after $t = \pi/2$ at which $y(t) = 0$ is $t = 3\pi/2$, at which point $y'(\pi/2) = \alpha + \frac{2}{15}$.

At this point, the solution has gone through one cycle in the time interval $[0, 3\pi/2]$. During this cycle, the section of the roadway started at the equilibrium with positive velocity, became positive, came back to the equilibrium position with negative velocity, became negative, and finally returned to the equilibrium position with positive velocity. This pattern continues indefinitely, with each cycle covering $3\pi/2$ time units. The solution for the next cycle is

$$y(t) = \sin 2t\left[-\frac{1}{2}\left(\alpha + \frac{7}{15}\right) - \frac{1}{6}\cos 2t\right] \text{ on } \left[\frac{3\pi}{2}, 2\pi\right],$$

$$y(t) = \sin t\left[-\left(\alpha + \frac{8}{15}\right) - \frac{4}{15}\cos t \cos 2t\right] \text{ on } [2\pi, 3\pi].$$

It is instructive to note that the velocity at the beginning of the second cycle is $(\alpha + \frac{2}{15})$, while, at the beginning of the third cycle, it is $(\alpha + \frac{4}{15})$. In fact, the velocity at the beginning of each cycle is $\frac{2}{15}$ greater than at the beginning of the previous cycle. It is not surprising, then, that the amplitude of oscillations will increase over time, since the amplitude of (one term in) the solution during any one cycle is directly proportional to the velocity at the beginning of the cycle.

It must be remembered that the model presented here is a very simplified one-dimensional model that cannot take into account all of the intricate interactions of real bridges. The reader is referred to the account by Lazer and McKenna [4] for a more complete model. More recently, McKenna [6] has refined that model to provide a different viewpoint of the torsional oscillations observed in the Tacoma bridge.

Research on the behavior of bridges under forces continues. It is likely that the models will be refined over time and new insights will be gained from the research. However, it should be clear at this point that the large oscillations causing the destruction of the Tacoma Narrows suspension bridge were not the result of resonance.

References

[1] Lewis, G. N., "Tacoma Narrows Suspension Bridge Collapse." In *A First Course in Differential Equations,* by Dennis G. Zill, 253–256. Boston: PWS-Kent, 1993.

[2] Braun, M., *Differential Equations and Their Applications.* New York: Springer-Verlag, 1978.

[3] Amman, O. H., T. von Karman, and G. B. Woodruff, *The Failure of the Tacoma Narrows Bridge.* Washington, D.C.: Federal Works Agency, 1941.

[4] Lazer, A. C., and P. J. McKenna, "Large Amplitude Periodic Oscillations in Suspension Bridges: Some New Connections with Nonlinear Analysis." *SIAM Review* 32 (December 1990): 537–578.

[5] Peterson, I., "Rock and Roll Bridge." *Science News* 137 (1991): 344–346.

[6] McKenna, P. J., "Large Torsional Oscillations in Suspension Bridges Revisited: Fixing an Old Approximation." *American Mathematical Monthly* 106 (1999): 1–18.

..

Gilbert N. Lewis is Associate Professor of Mathematics at Michigan Technological University, where he has taught since 1977. He received a B.S. in Applied Mathematics from Brown University in 1969 and was awarded a Ph.D., also in Applied Mathematics, from the University of Wisconsin–Milwaukee in 1976. Dr. Lewis was a visiting professor at the University of Wisconsin–Parkside and has pursued research in the areas of ordinary differential equations, asymptotic analysis, perturbation theory, and cosmology.

1

Introduction to Differential Equations

INTRODUCTION

The purpose of this short chapter is two-fold: to introduce the basic terminology of differential equations and to briefly examine how differential equations arise in an attempt to describe or *model* physical phenomena in mathematical terms.

1.1 Definitions and Terminology

■ **A Differential Equation** In calculus you learned that the derivative dy/dx of a function $y = \phi(x)$ is itself another function $\phi'(x)$ found by an appropriate rule. The function $y = e^{0.1x^2}$ is differentiable on the interval $(-\infty, \infty)$, and its derivative is $dy/dx = 0.2xe^{0.1x^2}$. If we replace $e^{0.1x^2}$ on the right-hand side of the derivative by the symbol y, we obtain

$$\frac{dy}{dx} = 0.2xy. \tag{1}$$

Now imagine that a friend of yours simply hands you the *differential equation* in (1), and you have no idea how it was constructed. Your friend asks: What is the function represented by the symbol y? You are now face-to-face with one of the basic problems in a course in differential equations: How do you solve such an equation of the unknown function $y = \phi(x)$? The problem is loosely equivalent to the familiar reverse problem of differential calculus: Given a derivative, find an anti-derivative.

Before proceeding, let us formalize the notion of a differential equation.

DEFINITION 1.1 Differential Equation

An equation containing the derivatives of one or more dependent variables, with respect to one or more independent variables, is said to be a **differential equation** (**DE**).

In order to talk about them, we shall classify differential equations by **type, order,** and **linearity.**

■ **Classification by Type** If an equation contains only ordinary derivatives of one or more dependent variables with respect to a single independent variable, it is said to be an **ordinary differential equation** (**ODE**). For example,

$$\frac{dy}{dx} + 5y = e^x, \quad \frac{d^2y}{dx^2} - \frac{dy}{dx} + 6y = 0, \quad \text{and} \quad \frac{dx}{dt} + \frac{dy}{dt} = 2x + y \tag{2}$$

are ordinary differential equations. An equation involving the partial derivatives of one or more dependent variables of two or more independent variables is called a **partial differential equation** (**PDE**). For example,

$$\frac{\partial^2 u}{\partial x^2} = \frac{\partial^2 u}{\partial t^2} - 2\frac{\partial u}{\partial t} \quad \text{and} \quad \frac{\partial u}{\partial y} = -\frac{\partial v}{\partial x}$$

are partial differential equations.

Throughout this text, ordinary derivatives will be written using either the **Leibniz notation** dy/dx, d^2y/dx^2, ..., or the **prime notation** y', y'', Using the latter notation, the first two differential equations in (2) can be written a little more compactly as $y' + 5y = e^x$ and $y'' - y' + 6y = 0$. Although less convenient to write and to typeset, the Leibniz notation has an advantage over the prime notation in that it clearly displays both the dependent and independent variables. For example, in the equation $d^2x/dt^2 + x = 0$, it is immediately seen that the symbol x now represents

a dependent variable, whereas the independent variable is t. You should also be aware that in physical sciences and engineering Newton's dot notation (derogatively referred to by some as the "flyspeck" notation) is sometimes used to denote derivatives with respect to time t. Thus the differential equation $d^2s/dt^2 = -32$ becomes $\ddot{s} = -32$.

■ **Classification by Order** The **order of a differential equation** (ODE or PDE) is the order of the highest derivative in the equation. For example,

$$\overset{\overset{\text{second order}}{\downarrow}}{\frac{d^2y}{dx^2}} + 5\overset{\overset{\text{first order}}{\downarrow}}{\left(\frac{dy}{dx}\right)^3} - 4y = e^x$$

is a second-order ordinary differential equation. First-order ordinary differential equations are occasionally written in differential form $M(x, y)dy + N(x, y)dx = 0$. For example, if we assume that y denotes the dependent variable in $(y - x)dx + 4x\,dy = 0$, then $y' = dy/dx$, and so by dividing by the differential dx we get the alternative form $4xy' + y = x$. See the Remarks at the end of this section.

In symbols, we can express an nth-order ordinary differential equation in one dependent variable by the general form

$$F(x, y, y', \ldots, y^{(n)}) = 0, \tag{3}$$

where F is a real-valued function of $n + 2$ variables: $x, y, y', \ldots, y^{(n)}$, and where $y^{(n)} = d^ny/dx^n$. For both practical and theoretical reasons, we shall also make the assumption hereafter that it is possible to solve an ordinary differential equation in the form (3) uniquely for the highest derivative $y^{(n)}$ in terms of the remaining $n + 1$ variables. The differential equation

$$\frac{d^ny}{dx^n} = f(x, y, y', \ldots, y^{(n-1)}), \tag{4}$$

where f is a real-valued continuous function, is referred to as the **normal form** of (3). Thus, when it suits our purposes, we shall use the normal forms

$$\frac{dy}{dx} = f(x, y) \quad \text{and} \quad \frac{d^2y}{dx^2} = f(x, y, y')$$

to represent general first- and second-order ordinary differential equations. For example, the normal form of the first-order equation $4xy' + y = x$ is $y' = (x - y)/4x$. See the Remarks.

■ **Classification as Linear or Nonlinear** An nth-order ordinary differential equation (3) is said to be **linear** if F is linear in $y, y', \ldots, y^{(n-1)}$. This means that an nth-order ODE is linear when (3) is $a_n(x)y^{(n)} + a_{n-1}(x)y^{(n-1)} + \cdots + a_1(x)y' + a_0(x)y - g(x) = 0$ or

$$a_n(x)\frac{d^ny}{dx^n} + a_{n-1}(x)\frac{d^{n-1}y}{dx^{n-1}} + \cdots + a_1(x)\frac{dy}{dx} + a_0(x)y = g(x). \tag{5}$$

Remember these two characteristics of a linear ODE.

From (5) we see that the characteristic two properties of a linear differential equation are: The dependent variable and all its derivatives are of the first degree—that is, the power of each term involving y is 1—and each coefficient depends at most on the independent variable x. The equations, in turn,

$$(y - x)\,dx + 4x\,dy = 0, \quad y'' - 2y' + y = 0, \quad \frac{d^3y}{dx^3} + 3x\,\frac{dy}{dx} - 5y = e^x,$$

are linear first-, second-, and third-order ordinary differential equations. We have just demonstrated that the first equation is linear in the variable y by writing it in the alternative form $4xy' + y = x$. A **nonlinear** ordinary differential equation is simply one that is not linear. Nonlinear functions of the dependent variable or its derivatives, such as $\sin y$ or $e^{y'}$, cannot appear in a linear equation. Therefore,

nonlinear term: coefficient depends on y	nonlinear term: nonlinear function of y	nonlinear term: power not 1
↓	↓	↓
$(1 - y)y' + 2y = e^x,$	$\dfrac{d^2y}{dx^2} + \sin y = 0,$	$\dfrac{d^4y}{dx^4} + y^2 = 0,$

are examples of nonlinear first-, second-, and fourth-order ordinary differential equations, respectively.

■ **Solution** As stated before, one of our goals in this course is to solve—or find solutions of—differential equations. In the next definition we define the concept of a solution of an ordinary differential equation.

> **DEFINITION 1.2 Solution of an ODE**
>
> Any function ϕ, defined on an interval I and possessing at least n derivatives that are continuous on I, which when substituted into an nth-order ordinary differential equation reduces the equation to an identity, is said to be a **solution** of the equation on the interval.

In other words, a solution of an nth-order ordinary differential equation (3) is a function ϕ that possesses at least n derivatives and

$$F(x,\ \phi(x),\ \phi'(x),\ \dots,\ \phi^{(n)}(x)) = 0 \quad \text{for all } x \text{ in } I.$$

We say that ϕ *satisfies* the differential equation on I. For our purposes, we shall also assume that a solution ϕ is a real-valued function. In our initial discussion we have already seen that $y = e^{0.1x^2}$ is a solution of $dy/dx = 0.2xy$ on the interval $(-\infty, \infty)$.

Occasionally it will be convenient to denote a solution by the alternative symbol $y(x)$.

■ **Interval of Definition** You can't think *solution* of an ordinary differential equation without simultaneously thinking *interval*. The interval I in Definition 1.2 is variously called the **interval of definition**, the **interval of existence**, the **interval of validity**, or the **domain of the solution** and can be an open interval (a, b), a closed interval $[a, b]$, an infinite interval (a, ∞), and so on.

Example 1 Verification of a Solution

Verify that the indicated function is a solution of the given differential equation on the interval $(-\infty, \infty)$.

(a) $dy/dx = xy^{1/2}; \quad y = x^4/16$ (b) $y'' - 2y' + y = 0; \quad y = xe^x$

SOLUTION One way of verifying that the given function is a solution is to see, after substituting, whether each side of the equation is the same for every x in the interval.

(a) From *left-hand side:* $\dfrac{dy}{dx} = 4 \cdot \dfrac{x^3}{16} = \dfrac{x^3}{4}$

$\qquad\qquad$ *right-hand side:* $xy^{1/2} = x \cdot \left(\dfrac{x^4}{16}\right)^{1/2} = x \cdot \dfrac{x^2}{4} = \dfrac{x^3}{4}$,

we see that each side of the equation is the same for every real number x. Note that $y^{1/2} = x^2/4$ is, by definition, the nonnegative square root of $x^4/16$.

(b) From the derivatives $y' = xe^x + e^x$ and $y'' = xe^x + 2e^x$ we have for every real number x,

\qquad *left-hand side:* $y'' - 2y' + y = (xe^x + 2e^x) - 2(xe^x + e^x) + xe^x = 0$
\qquad *right-hand side:* 0. $\qquad\qquad\qquad\qquad\qquad\qquad\qquad$ ❏

Note, too, that in Example 1 each differential equation possesses the constant solution $y = 0$, $-\infty < x < \infty$. A solution of a differential equation that is identically zero on an interval I is said to be a **trivial solution.**

■ **Solution Curve** The graph of a solution ϕ of an ODE is called a **solution curve.** Since ϕ is a differentiable function, it is continuous on its interval I of definition. Thus there may be a difference between the graph of the *function* ϕ and the graph of the *solution* ϕ. Put another way, the domain of the function ϕ does not need to be the same as the interval I of definition (or domain) of the solution ϕ. Example 2 illustrates the difference.

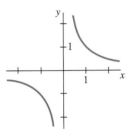

(a) Function $y = 1/x$ $x \neq 0$

(b) Solution $y = 1/x$, $(0, \infty)$

Figure 1.1

Example 2 **Domain vs. Interval I of Definition**

Considered simply as a *function,* the domain of $y = 1/x$ is the set of all real numbers x except 0. When we graph $y = 1/x$, we plot points in the xy-plane corresponding to a judicious sampling of numbers taken from its domain. The rational function $y = 1/x$ is discontinuous at 0, and its graph, in a neighborhood of the origin, is given in Figure 1.1(a). The function $y = 1/x$ is not differentiable at $x = 0$ since the y-axis (whose equation is $x = 0$) is a vertical asymptote of the graph.

Now $y = 1/x$ is also a solution of the linear first-order differential equation $xy' + y = 0$ (verify). But when we say $y = 1/x$ is a *solution* of this DE, we mean it is a function defined on an interval I on which it is differentiable and satisfies the equation. In other words, $y = 1/x$ is a solution of the DE on *any* interval not containing 0, such as $(-3, -1)$, $(\frac{1}{2}, 10)$, $(-\infty, 0)$, or $(0, \infty)$. Because the solution curves defined by $y = 1/x$ on the intervals $-3 < x < -1$ and on $\frac{1}{2} < x < 10$ are simply segments or pieces of the solution curves defined by $y = 1/x$ on $-\infty < x < 0$ and $0 < x < \infty$, respectively, it makes sense to take the interval I to be as large as possible. Thus we would take I to be either $(-\infty, 0)$ or $(0, \infty)$. The solution curve on $(0, \infty)$ is shown in Figure 1.1(b). $\qquad\qquad$ ❏

■ **Explicit and Implicit Solutions** You should be familiar with the terms *explicit* and *implicit functions* from your study of calculus. A solution in which the dependent variable is expressed solely in terms of the independent variable and constants is said to be an **explicit solution.** For our purposes let us think of an explicit solution as an explicit formula $y = \phi(x)$ that we can manipulate, evaluate, and differentiate using the standard rules. We have just seen in the last two examples that $y = x^4/16$, $y = xe^x$, and $y = 1/x$ are, in turn, explicit solutions of $dy/dx = xy^{1/2}$,

$y'' - 2y' + y = 0$, and $xy' + y = 0$. Moreover, the trivial solution $y = 0$ is an explicit solution of all three equations. We shall see when we get down to the business of actually solving some ordinary differential equations that methods of solution do not always lead directly to an explicit solution $y = \phi(x)$. This is particularly true when attempting to solve nonlinear first-order differential equations. Often we have to be content with a relation or expression $G(x, y) = 0$ that defines a solution ϕ implicitly.

DEFINITION 1.3 Implicit Solution of an ODE

A relation $G(x, y) = 0$ is said to be an **implicit solution** of an ordinary differential equation (3) on an interval I provided there exists at least one function ϕ that satisfies the relation as well as the differential equation on I.

It is beyond the scope of this course to investigate the conditions under which a relation $G(x, y) = 0$ defines a differentiable function ϕ. So we shall assume that if the formal implementation of a method of solution leads to a relation $G(x, y) = 0$, then there exists at least one function ϕ that satisfies both the relation (that is, $G(x, \phi(x)) = 0$) and the differential equation on an interval I. If the implicit solution $G(x, y) = 0$ is fairly simple, we may be able to solve for y in terms of x and obtain one or more explicit solutions. See the Remarks.

Example 3 **Verification of an Implicit Solution**

The relation $x^2 + y^2 = 25$ is an implicit solution of the differential equation

$$\frac{dy}{dx} = -\frac{x}{y} \tag{6}$$

on the interval $-5 < x < 5$. By implicit differentiation we obtain

$$\frac{d}{dx} x^2 + \frac{d}{dx} y^2 = \frac{d}{dx} 25 \quad \text{or} \quad 2x + 2y \frac{dy}{dx} = 0.$$

Solving the last equation for the symbol dy/dx gives (6). Moreover, solving $x^2 + y^2 = 25$ for y in terms of x yields $y = \pm\sqrt{25 - x^2}$. The two functions $y = \phi_1(x) = \sqrt{25 - x^2}$ and $y = \phi_2(x) = -\sqrt{25 - x^2}$ satisfy the relation (that is, $x^2 + \phi_1^2 = 25$ and $x^2 + \phi_2^2 = 25$) and are explicit solutions defined on the interval $-5 < x < 5$. The solution curves given in Figure 1.2(b) and 1.2(c) are segments of the graph of the implicit solution in Figure 1.2(a). ❑

Any relation of the form $x^2 + y^2 - c = 0$ *formally* satisfies (6) for any constant c. However, it is understood that the relation should always make sense in the real number system; thus, for example, we cannot say that $x^2 + y^2 + 25 = 0$ is an implicit solution of the equation. Why not?

Because the distinction between an explicit solution and an implicit solution should be intuitively clear, we will not belabor the issue by always saying, "Here is an explicit (implicit) solution."

■ **Families of Solutions** The study of differential equations is similar to that of integral calculus. In some texts, a solution ϕ is sometimes referred to as an **integral** of the equation, and its graph is called an **integral curve.** When evaluating an antiderivative or indefinite integral in calculus, we use a single constant c of inte-

(a) Implicit solution
$x^2 + y^2 = 25$

(b) Explicit solution
$y_1 = \sqrt{25-x^2}, -5 < x < 5$

(c) Explicit solution
$y_2 = -\sqrt{25-x^2}, -5 < x < 5$

Figure 1.2

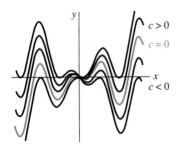

Figure 1.3 Some solutions of $xy' - y = x^2 \sin x$

gration. Analogously, when solving a first-order differential equation $F(x, y, y') = 0$, we *usually* obtain a solution containing a single arbitrary constant or parameter c. A solution containing an arbitrary constant represents a set $G(x, y, c) = 0$ of solutions called a **one-parameter family of solutions.** When solving an nth-order differential equation $F(x, y, y', \ldots, y^{(n)}) = 0$, we seek an ***n*-parameter family of solutions** $G(x, y, c_1, c_2, \ldots, c_n) = 0$. This means that a single differential equation can possess an infinite number of solutions corresponding to the unlimited number of choices for the parameter(s). A solution of a differential equation that is free of arbitrary parameters is called a **particular solution.** For example, the one-parameter family $y = cx - x \cos x$ is an explicit solution of the linear first-order equation $xy' - y = x^2 \sin x$ on the interval $(-\infty, \infty)$ (verify). Figure 1.3, obtained using graphing software, shows the graphs of some of the solutions in this family. The solution $y = -x \cos x$, the colored curve in the figure, is a particular solution corresponding to $c = 0$. Similarly, on the interval $(-\infty, \infty)$, $y = c_1 e^x + c_2 x e^x$ is a two-parameter family of solutions (verify) of the linear second-order equation $y'' - 2y' + y = 0$ in Example 1. Some particular solutions of the equation are the trivial solution $y = 0$ ($c_1 = c_2 = 0$), $y = xe^x$ ($c_1 = 0$, $c_2 = 1$), $y = 5e^x - 2xe^x$ ($c_1 = 5$, $c_2 = -2$), and so on.

In all the preceding examples we have used x and y to denote the independent and dependent variables, respectively. But you should become accustomed to seeing and working with other symbols to denote these variables. For example, we could denote the independent variable by t and the dependent variable by x.

Example 4 Using Different Symbols

The functions $x = c_1 \cos 4t$ and $x = c_2 \sin 4t$, where c_1 and c_2 are arbitrary constants or parameters, are both solutions of the linear differential equation

$$x'' + 16x = 0.$$

For $x = c_1 \cos 4t$ the first two derivatives with respect to t are $x' = -4c_1 \sin 4t$ and $x'' = -16c_1 \cos 4t$. Substituting x'' and x then gives

$$x'' + 16x = -16c_1 \cos 4t + 16(c_1 \cos 4t) = 0.$$

In like manner, for $x = c_2 \sin 4t$ we have $x'' = -16c_2 \sin 4t$, and so

$$x'' + 16x = -16c_2 \sin 4t + 16(c_2 \sin 4t) = 0.$$

Finally, it is straightforward to verify that the linear combination of solutions or two-parameter family $x = c_1 \cos 4t + c_2 \sin 4t$ is also a solution of the differential equation. ❏

The next example shows that a solution of a differential equation can be a piecewise defined function.

Example 5 A Piecewise Defined Solution

You should verify that the one-parameter family $y = cx^4$ is a one-parameter family of solutions of the differential equation $xy' - 4y = 0$ on the interval $(-\infty, \infty)$. See Figure 1.4(a). The piecewise defined differentiable function

$$y = \begin{cases} -x^4, & x < 0 \\ x^4, & x \geq 0 \end{cases}$$

is a particular solution of the equation but cannot be obtained from the family $y = cx^4$ by a single choice of c; the solution is constructed from the family by choosing $c = -1$ for $x < 0$ and $c = 1$ for $x \geq 0$. See Figure 1.4(b). ❏

(a)

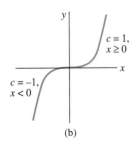

(b)

Figure 1.4 Some solutions of $xy' - 4y = 0$

Sometimes a differential equation possesses a solution that is not a member of a family of solutions of the equation, that is, a solution that cannot be obtained by specializing *any* of the parameters in the family of solutions. Such an extra solution is called a **singular solution.** For example, we have seen that $y = x^4/16$ and $y = 0$ are solutions of the differential equation $dy/dx = xy^{1/2}$ on $(-\infty, \infty)$. In Section 2.2 we shall demonstrate, by actually solving it, that the differential equation $dy/dx = xy^{1/2}$ possesses the one-parameter family of solutions $y = (x^2/4 + c)^2$. When $c = 0$, the resulting particular solution is $y = x^4/16$. But notice that the trivial solution $y = 0$ is a singular solution since it is not a member of the family $y = (x^2/4 + c)^2$; there is no way of assigning a value to the constant c to obtain $y = 0$.

■ **General Solution** If *every* solution of an nth-order ordinary differential equation $F(x, y, y', \ldots, y^{(n)}) = 0$ on an interval I can be obtained from an n-parameter family $G(x, y, c_1, c_2, \ldots, c_n) = 0$ by appropriate choices of the parameters $c_i, i = 1, 2, \ldots, n$, we then say that the family is the **general solution** of the differential equation. In solving linear differential equations, we shall impose relatively simple restrictions on the coefficients of the equation; with these restrictions one can be assured that not only does a solution exist on an interval but also that a family of solutions yields all possible solutions. Nonlinear equations, with the exception of some first-order equations, are usually difficult or impossible to solve in terms of familiar *elementary functions:* finite combinations of integer powers of x, roots, exponential and logarithmic functions, trigonometric and inverse trigonometric functions. Furthermore, if we happen to obtain a family of solutions for a nonlinear equation, it is not obvious whether this family contains all solutions. On a practical level, then, the designation "general solution" is applied only to linear differential equations. We will come back to this concept in Section 2.3 and in Chapter 3.

■ **Systems of Differential Equations** Up to this point we have been discussing single differential equations containing one unknown function. But often in theory, as well as in many applications, we must deal with systems of differential equations. A **system of ordinary differential equations** is two or more equations involving the derivatives of two or more unknown functions of a single independent variable. For example, if x and y denote dependent variables and t the independent variable, then a system of two first-order differential equations is given by

$$\frac{dx}{dt} = f(t, x, y)$$

$$\frac{dy}{dt} = g(t, x, y). \tag{7}$$

A **solution** of a system such as (7) is a pair of differentiable functions $x = \phi_1(t)$, $y = \phi_2(t)$ defined on a common interval I that satisfy each equation of the system on this interval.

Remarks

(*i*) A few last words about implicit solutions of differential equations are in order. In Example 3 we were able to solve the relation $x^2 + y^2 = 25$ for y in terms of x to get two explicit solutions, $\phi_1(x) = \sqrt{25 - x^2}$ and $\phi_2(x) = -\sqrt{25 - x^2}$, of the differential equation $dy/dx = -x/y$. But don't read too much into this one example.

Unless it is easy, obvious, convenient, important, or you are instructed to, there is usually no need to try to solve an implicit solution $G(x, y) = 0$ for y explicitly in terms of x. Also do not misinterpret the second sentence following Definition 1.3. An implicit solution $G(x, y) = 0$ can define a perfectly good differentiable function ϕ that is a solution of a DE, but yet we may not be able to solve $G(x, y) = 0$ using analytical methods such as algebra. The solution curve of ϕ may be a segment or piece of the graph of $G(x, y) = 0$. See Problems 41 and 42 in Exercises 1.1. Also, read the discussion following Example 4 in Section 2.2.

(*ii*) In the differential form $M(x, y)dx + N(x, y)dy = 0$ it may not be obvious whether a first-order ODE is linear or nonlinear since there is nothing in this form that tells us which symbol denotes the dependent variable. See Problems 3 and 4 in Exercises 1.1.

(*iii*) It may not seem like a big deal to assume that $F(x, y, y', \ldots, y^{(n)}) = 0$ can be solved for $y^{(n)}$, but one should be a little bit careful here. There are exceptions, and there certainly are some problems connected with this assumption. See Problems 34 and 35 in Exercises 1.1.

(*iv*) You will sometimes see in texts or hear instructors of courses in differential equations speak of "closed-form solutions." Translated, this phrase usually refers to explicit solutions that are expressible in terms of the elementary functions mentioned earlier in this section.

E X E R C I S E S 1 . 1

Answers to odd-numbered problems begin on page A-25.

In Problems 1–10, state whether the given differential is linear or nonlinear. Give the order of each equation.

1. $(1 - x)y'' - 4xy' + 5y = \cos x$ **2.** $x\dfrac{d^3y}{dx^3} - \left(\dfrac{dy}{dx}\right)^4 + y = 0$

3. $(y^2 - 1)dx + x\,dy = 0$ **4.** $u\,dv + (v + uv - ue^u)du = 0$

5. $t^5 y^{(4)} - t^3 y'' + 6y = 0$ **6.** $\dfrac{d^2u}{dr^2} + \dfrac{du}{dr} + u = \cos(r + u)$

7. $\dfrac{d^2y}{dx^2} = \sqrt{1 + \left(\dfrac{dy}{dx}\right)^2}$ **8.** $\dfrac{d^2R}{dt^2} = -\dfrac{k}{R^2}$

9. $(\sin\theta)y''' - (\cos\theta)y' = 2$ **10.** $\ddot{x} - \left(1 - \dfrac{\dot{x}^2}{3}\right)\dot{x} + x = 0$

In Problems 11–14, verify that the indicated function is an explicit solution of the given differential equation. Assume an appropriate interval I of definition.

11. $2y' + y = 0;$ $y = e^{-x/2}$

12. $\dfrac{dy}{dt} + 20y = 24;$ $y = \dfrac{6}{5} - \dfrac{6}{5}e^{-20t}$

13. $y'' - 6y' + 13y = 0;$ $y = e^{3x}\cos 2x$

14. $y'' + y = \tan x;$ $y = -(\cos x)\ln(\sec x + \tan x)$

⌨ In Problems 15 and 16, verify that the indicated expression is an implicit solution of the given differential equation. Find at least one explicit solution in each case. Use a graphing utility to obtain the graphs of the explicit solutions. Give the interval I of definition of each solution ϕ.

15. $\dfrac{dX}{dt} = (X - 1)(1 - 2X);$ $\ln\left(\dfrac{2X - 1}{X - 1}\right) = t$

16. $2xy\,dx + (x^2 - y)\,dy = 0;$ $-2x^2y + y^2 = 1$

In Problems 17–20, verify that the indicated family of functions is a solution of the given differential equation. Assume an appropriate interval I of definition.

17. $P' = P(1 - P);$ $P = \dfrac{c_1 e^t}{1 + c_1 e^t}$

18. $y' + 2xy = 1;$ $y = e^{-x^2}\displaystyle\int_0^x e^{t^2}\,dt + c_1 e^{-x^2}$

19. $\dfrac{d^2y}{dx^2} - 4\dfrac{dy}{dx} + 4y = 0;$ $y = c_1 e^{2x} + c_2 x e^{2x}$

20. $x^3\dfrac{d^3y}{dx^3} + 2x^2\dfrac{d^2y}{dx^2} - x\dfrac{dy}{dx} + y = 12x^2;$
$y = c_1 x^{-1} + c_2 x + c_3 x \ln x + 4x^2$

21. Verify that the piecewise defined function $y = \begin{cases} -x^2, & x < 0 \\ x^2, & x \geq 0 \end{cases}$ is
a solution of the differential equation $xy' - 2y = 0$ on $(-\infty, \infty)$.

22. In Example 3 we saw that $y = \phi_1(x) = \sqrt{25 - x^2}$ and $y = \phi_2(x) = -\sqrt{25 - x^2}$ are solutions of $dy/dx = -x/y$ on the interval $(-5, 5)$. Explain why

$$y = \begin{cases} \sqrt{25 - x^2}, & -5 < x < 0 \\ -\sqrt{25 - x^2}, & 0 \leq x < 5 \end{cases}$$

is not a solution of the differential equation on the interval $(-5, 5)$.

In Problems 23 and 24, verify that the given pair of functions is a solution of the given system of differential equations on $(-\infty, \infty)$.

23. $\dfrac{dx}{dt} = x + 3y$

$\dfrac{dy}{dt} = 5x + 3y;$

$x = e^{-2t} + 3e^{6t}, \quad y = -e^{-2t} + 5e^{6t}$

24. $\dfrac{d^2x}{dt^2} = 4y + e^t$

$\dfrac{d^2y}{dt^2} = 4x - e^t;$

$x = \cos 2t + \sin 2t + \dfrac{1}{5} e^t, \quad y = -\cos 2t - \sin 2t - \dfrac{1}{5} e^t$

In Problems 25 and 26, the indicated function is a solution of the given differential equation. Determine at least one such interval I of definition.

25. $y' = 25 + y^2; \quad y = 5 \tan 5x$

26. $2y' = y^3 \cos x; \quad y = (1 - \sin x)^{-1/2}$

27. Make up a differential equation that does not possess any real solutions.

28. Make up a differential equation that you feel confident possesses only the trivial solution $y = 0$. Explain your reasoning.

29. What function that you know from calculus is such that its first derivative is itself? What function that you know from calculus is such that its first derivative is a constant multiple k of itself? Write each of your answers in the form of a differential equation with a solution.

30. What function(s) that you know from calculus is such that its second derivative is itself? What function(s) is such that its second derivative is the negative of itself? Write each answer in the form of a differential equation with a solution.

31. Suppose $y = \phi(x)$ is a solution of an nth-order differential equation $F(x, y, y', \dots, y^{(n)}) = 0$ on an interval I. Explain why $\phi, \phi', \dots, \phi^{(n-1)}$ are necessarily continuous on I.

32. Suppose the two-parameter family $y(x) = c_1 y_1(x) + c_2 y_2(x)$ is the general solution of a linear second-order differential equation on an interval I. If $x = 0$ is in I, find a particular solution in the family that satisfies both conditions: $y(0) = 2$, $y'(0) = 0$. State any assumptions that you must make.

33. Discuss, and illustrate with examples, how to solve differential equations of the forms $dy/dx = f(x)$ and $d^2y/dx^2 = f(x)$.

34. The differential equation $x(y')^2 - 4y' - 12x^3 = 0$ has the form given in (3). Determine whether the equation can be put into the normal form $dy/dx = f(x, y)$.

35. The normal form (4) of an nth-order differential equation is equivalent to (3) whenever both forms have exactly the same solutions. Make up a first-order differential equation in which the equation $F(x, y, y') = 0$ is *not* equivalent to the normal form $dy/dx = f(x, y)$.

36. Find a linear second-order differential equation $F(x, y, y', y'') = 0$ for which $y = c_1 x + c_2 x^3$ is a two-parameter family of solutions. Make sure your equation is free of arbitrary parameters.

37. Find values of m so that the indicated function is a solution of the given differential equation. Explain your reasoning.

(a) $y' + 2y = 0; \quad y = e^{mx}$ (b) $y'' - 5y' + 6y = 0; \quad y = e^{mx}$

(c) $xy'' + 2y' = 0; \quad y = x^m$ (d) $x^2 y'' - 7xy' + 15y = 0; \quad y = x^m$

38. Examine the general form (5) of a linear differential equation. Discuss: Under what conditions is $y = 0$ a solution of a linear equation?

39. We know that $y' = 0$ if and only if y is a constant function. Use this idea to determine whether the given differential equation possesses constant solutions.

(a) $3xy' + 5y = 10$ (b) $y' = y^2 + 2y - 3$

(c) $(y - 1)y' = 1$ (d) $y'' + 4y' + 6y = 10$

40. Discuss: From its normal form can you predict a plausible interval I of definition for a solution ϕ of the differential equation $(x - 2)y' + y = 1$?

In Problems 41 and 42 the given figure represents the graph of an implicit solution $G(x, y) = 0$ of a differential equation $dy/dx = f(x, y)$. In each case the relation $G(x, y) = 0$ implicitly defines several solutions of the DE. Carefully reproduce each figure on a piece of paper. Use different colored pencils to mark off segments, or pieces, on each graph that correspond to graphs of solutions. Keep in mind that a solution ϕ must be a function and differentiable. Use the solution curve to estimate the interval I of definition of each solution ϕ.

41.

Figure 1.5

42.

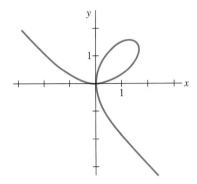

Figure 1.6

43. The graphs of the members of the one-parameter family $x^3 + y^3 = 3cxy$ are called **folia of Descartes.** Verify that this family is an implicit solution of the first-order differential equation

$$\frac{dy}{dx} = \frac{y(y^3 - 2x^3)}{x(2y^3 - x^3)}.$$

44. The graph in Figure 1.6 is the member of the family of folia in Problem 43 corresponding to $c = 1$. Discuss: How can the differential equation in Problem 43 help in finding points on the graph of $x^3 + y^3 = 3xy$ where the tangent line is vertical? How does knowing where a tangent line is vertical help in determining an interval I of definition of a solution ϕ of the DE? Carry out your ideas and compare with your estimates of the intervals in Problem 42.

Qualitative information about a solution of a differential equation can often be obtained from the equation itself. Before working Problems 45–47, review your calculus text on the geometric significance of the derivatives dy/dx and d^2y/dx^2.

45. Suppose $y = \phi(x)$ is a solution of $y' = y^2 + 4$ defined on an interval I. Describe the graph of $y = \phi(x)$ on I. For example, can a solution curve have any relative extrema?

46. (a) Verify that $y = 5$, $-\infty < x < \infty$, is a solution of $y' = 5 - y$. The graph of this constant solution is the horizontal black line in Figure 1.7.

(b) Why aren't the curves shown in color in Figure 1.7 plausible solution curves?

(c) Sketch several plausible solution curves in the regions defined by $y < 5$ and $y > 5$.

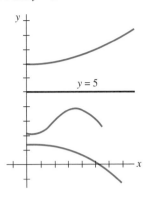

Figure 1.7

47. Suppose $y = \phi(x)$ is a solution of the differential equation $dy/dx = y(a - by)$, where a and b are positive constants.

(a) By inspection find two constant solutions of the equation.

(b) Using only the differential equation, find intervals on the y-axis on which a nonconstant solution $y = \phi(x)$ is increasing. On which $y = \phi(x)$ is decreasing.

(c) Using only the differential equation, explain why $y = a/2b$ is the y-coordinate of a point of inflection of the graph of a nonconstant solution $y = \phi(x)$.

(d) On the same coordinate axes, sketch the graphs of the two constant solutions found in part **(b).** These constant solutions partition the xy-plane into three regions. In each region, sketch the graph of a nonconstant solution $y = \phi(x)$ whose shape is suggested by the results in parts **(b)** and **(c).**

48. Computer Lab Assignment Use a CAS to compute all derivatives and to carry out the simplifications to verify that the indicated function is a particular solution of the given differential equation.

(a) $y^{(4)} - 20y''' + 158y'' - 580y' + 841y = 0$; $y = xe^{5x} \cos 2x$

(b) $x^3y''' + 2x^2y'' + 20xy' - 78y = 0$;

$$y = 20\,\frac{\cos(5 \ln x)}{x} - 3\,\frac{\sin(5 \ln x)}{x}$$

1.2 Initial-Value Problems

■ **Initial-Value Problem** We are often interested in solving a differential equation subject to prescribed side conditions—conditions that are imposed on the unknown solution $y = y(x)$ or its derivatives. On some interval I containing x_0, the problem

$$\textit{Solve:} \quad \frac{d^n y}{dx^n} = f(x, y, y', \cdots, y^{(n-1)})$$

$$\textit{Subject to:} \quad y(x_0) = y_0, \; y'(x_0) = y_1, \cdots, \; y^{(n-1)}(x_0) = y_{n-1}, \quad\quad (1)$$

where $y_0, y_1, \ldots, y_{n-1}$ are arbitrarily specified real constants, is called an **initial-value problem** (**IVP**). The values of $y(x)$ and its first $n-1$ derivatives at a single point x_0: $y(x_0) = y_0$, $y'(x_0) = y_1, \ldots, y^{(n-1)}(x_0) = y_{n-1}$, are called **initial conditions.**

■ **First- and Second-Order IVPs** The problem given in (1) is also called an **nth-order initial-value problem.** For example,

$$\text{Solve:} \qquad \frac{dy}{dx} = f(x, y) \tag{2}$$

$$\text{Subject to:} \qquad y(x_0) = y_0$$

and

$$\text{Solve:} \qquad \frac{d^2y}{dx^2} = f(x, y, y') \tag{3}$$

$$\text{Subject to:} \qquad y(x_0) = y_0, \ y'(x_0) = y_1$$

are first- and second-order initial-value problems, respectively. These two problems are easy to interpret in geometric terms. For (2) we are seeking a solution of the differential equation on an interval I containing x_0 so that a solution curve passes through the prescribed point (x_0, y_0). See Figure 1.8. For (3) we want to find a solution of the differential equation whose graph not only passes through (x_0, y_0) but passes through so that the slope of the curve at this point is y_1. See Figure 1.9. The term *initial condition* derives from physical systems where the independent variable is time t and where $y(t_0) = y_0$ and $y'(t_0) = y_1$ represent, respectively, the position and velocity of an object at some beginning, or initial, time t_0.

Solving an nth-order initial-value problem frequently entails using an n-parameter family of solutions of the given differential equation to find n specialized constants so that the resulting particular solution of the equation also "fits"—that is, satisfies—the n initial conditions.

solutions of the DE

Figure 1.8 First-order IVP

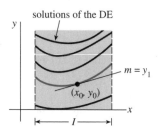

solutions of the DE

$m = y_1$

(x_0, y_0)

Figure 1.9 Second-order IVP

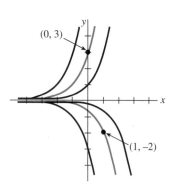

(0, 3)

(1, −2)

Figure 1.10 Solutions of IVPs

Example 1 First-Order IVPs

It is readily verified that $y = ce^x$ is a one-parameter family of solutions of the simple first-order equation $y' = y$ on the interval $(-\infty, \infty)$. If we specify an initial condition, say, $y(0) = 3$, then substituting $x = 0$, $y = 3$ in the family determines the constant $3 = ce^0 = c$. Thus the function $y = 3e^x$ is a solution of the initial-value problem

$$y' = y, \qquad y(0) = 3.$$

Now if we demand that a solution of the differential equation pass through the point $(1, -2)$ rather than $(0, 3)$, then $y(1) = -2$ will yield $-2 = ce$ or $c = -2e^{-1}$. The function $y = -2e^{x-1}$ is a solution of the initial-value problem

$$y' = y, \qquad y(1) = -2.$$

The graphs of these two functions are shown in color in Figure 1.10. ❏

Example 2 Second-Order IVP

In Example 4 of Section 1.1 we saw that $x = c_1 \cos 4t + c_2 \sin 4t$ is a two-parameter family of solutions of $x'' + 16x = 0$. Find a solution of the initial-value problem

$$x'' + 16x = 0, \qquad x\left(\frac{\pi}{2}\right) = -2, \quad x'\left(\frac{\pi}{2}\right) = 1. \tag{4}$$

SOLUTION We first apply $x(\pi/2) = -2$ to the given family of solutions: $c_1 \cos 2\pi + c_2 \sin 2\pi = -2$. Since $\cos 2\pi = 1$ and $\sin 2\pi = 0$, we find that $c_1 = -2$. We next apply $x'(\pi/2) = 1$ to the one-parameter family $x(t) = -2 \cos 4t + c_2 \sin 4t$. Differentiating and then setting $t = \pi/2$ and $x' = 1$ gives $8 \sin 2\pi + 4c_2 \cos 2\pi = 1$, from which we see that $c_2 = \frac{1}{4}$. Hence

$$x = -2 \cos 4t + \frac{1}{4} \sin 4t$$

is a solution of (4). ❏

■ **Existence and Uniqueness** Two fundamental questions arise in considering an initial-value problem:

> *Does a solution of the problem exist? If a solution exists, is it unique?*

For an initial-value problem such as (2), we ask:

Existence $\begin{cases} \textit{Does the differential equation } dy/dx = f(x, y) \textit{ possess solutions?} \\ \textit{Do any of the solution curves pass through the point } (x_0, y_0)? \end{cases}$

Uniqueness $\begin{cases} \textit{When can we be certain that there is precisely one solution} \\ \textit{curve passing through the point } (x_0, y_0)? \end{cases}$

Note that in Examples 1 and 2 the phrase "*a* solution" is used rather than "*the* solution" of the problem. The indefinite article "a" is used deliberately to suggest the possibility that other solutions may exist. At this point it has not been demonstrated that there is a single solution of each problem. The next example illustrates an initial-value problem with two solutions.

| *Example 3* **An IVP Can Have Several Solutions**
Each of the functions $y = 0$ and $y = x^4/16$ satisfies the differential equation $dy/dx = xy^{1/2}$ and the initial condition $y(0) = 0$, and so the initial-value problem $dy/dx = xy^{1/2}$, $y(0) = 0$, has at least two solutions. As illustrated in Figure 1.11, the graphs of both functions pass through the same point $(0, 0)$. ❏

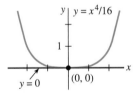

Figure 1.11 Two solutions of the same IVP

Within the safe confines of a formal course in differential equations one can be fairly confident that *most* differential equations will have solutions and that solutions of initial-value problems will *probably* be unique. Real life, however, is not so idyllic. Thus it is desirable to know in advance of trying to solve an initial-value problem whether a solution exists and, when it does, whether it is the only solution of the problem. Since we are going to consider first-order differential equations in the next two chapters, we state here without proof a straightforward theorem that gives conditions that are sufficient to guarantee the existence and uniqueness of a solution of a first-order initial-value problem of the form given in (2). We shall wait until Chapter 3 to address the question of existence and uniqueness of a second-order initial-value problem.

> ### THEOREM 1.1 Existence of a Unique Solution
>
> Let R be a rectangular region in the xy-plane defined by $a \le x \le b$, $c \le y \le d$, that contains the point (x_0, y_0) in its interior. If $f(x, y)$ and $\partial f/\partial y$ are continuous on R, then there exist some interval $I_0 : x_0 - h < x < x_0 + h$, $h > 0$, contained in $a \le x \le b$, and a unique function $y(x)$ defined on I_0 that is a solution of the initial-value problem (2).

The foregoing result is one of the most popular existence and uniqueness theorems for first-order differential equations, because the criteria of continuity of $f(x, y)$ and $\partial f/\partial y$ are relatively easy to check. The geometry of Theorem 1.1 is illustrated in Figure 1.12.

Figure 1.12 Rectangular region R

Example 4 Example 3 Revisited

We saw in Example 3 that the differential equation $dy/dx = xy^{1/2}$ possesses at least two solutions whose graphs pass through $(0, 0)$. Inspection of the functions

$$f(x, y) = xy^{1/2} \quad \text{and} \quad \frac{\partial f}{\partial y} = \frac{x}{2y^{1/2}}$$

shows that they are continuous in the upper half-plane defined by $y > 0$. Hence Theorem 1.1 enables us to conclude that through any point (x_0, y_0), $y_0 > 0$, in the upper half-plane there is some interval centered at x_0 on which the given differential equation has a unique solution. Thus, for example, even without solving it we know that there exists some interval centered at 2 on which the initial-value problem $dy/dx = xy^{1/2}$, $y(2) = 1$, has a unique solution. ❑

In Example 1, Theorem 1.1 guarantees that there are no other solutions of the initial-value problems $y' = y$, $y(0) = 3$, and $y' = y$, $y(1) = -2$, other than $y = 3e^x$ and $y = -2e^{x-1}$, respectively. This follows from the fact that $f(x, y) = y$ and $\partial f/\partial y = 1$ are continuous throughout the entire xy-plane. It can be further shown that the interval I on which each solution is defined is $(-\infty, \infty)$.

■ **Interval of Existence/Uniqueness** Suppose $y(x)$ represents a solution of the initial-value problem (2). The following three sets on the real x-axis may not be the same: the domain of the function $y(x)$, the interval I over which the solution $y(x)$ is defined or exists, and the interval I_0 of existence *and* uniqueness. In Example 2 of Section 1.1 we illustrated the difference between the domain of a function and the interval I of definition. Now suppose (x_0, y_0) is a point in the interior of the rectangular region R in Theorem 1.1. It turns out that the continuity of the function $f(x, y)$ on R by itself is sufficient to guarantee the existence of at least one solution of $dy/dx = f(x, y)$, $y(x_0) = y_0$, defined on some interval I. The interval I of definition for this initial-value problem is usually taken to be the largest interval containing x_0 over which the solution $y(x)$ is defined and differentiable. The interval I depends on both $f(x, y)$ and the initial condition $y(x_0) = y_0$. See Problems 27 and 28 in Exercises 1.2. The extra condition of continuity of the first partial derivative $\partial f/\partial y$ on R enables us to say that not only does a solution exist on some interval I_0 containing x_0 but it also is the *only* solution satisfying $y(x_0) = y_0$. However, Theorem 1.1 does not give any indication of the sizes of the intervals I and I_0; *the interval I of definition need not be as wide as the region R and the interval I_0 of existence and uniqueness may not be as*

large as I. The number $h > 0$ that defines the interval $I_0: x_0 - h < x < x_0 + h$, could be very small, and so it is best to think that the solution $y(x)$ is *unique in a local sense*, that is, a solution defined near the point (x_0, y_0). See Problem 32 in Exercises 1.2.

Remarks

(*i*) The conditions in Theorem 1.1 are sufficient but not necessary. When $f(x, y)$ and $\partial f/\partial y$ are continuous on a rectangular region R, it must always follow that a solution of (2) exists and is unique whenever (x_0, y_0) is a point interior to R. However, if the conditions stated in the hypothesis of Theorem 1.1 do not hold, then anything could happen: Problem (2) *may* still have a solution and this solution *may* be unique, or (2) may have several solutions, or it may have no solution at all. A rereading of Example 4 reveals that the hypotheses of Theorem 1.1 do not hold on the line $y = 0$ for the differential equation $dy/dx = xy^{1/2}$, and so it is not surprising, as we saw in Example 3 of this section, that there are two solutions defined on a common interval $-h < x < h$ satisfying $y(0) = 0$. On the other hand, the hypotheses of Theorem 1.1 do not hold on the line $y = 1$ for the differential equation $dy/dx = |y - 1|$. Nevertheless, it can be proved that the solution of the initial-value problem $dy/dx = |y - 1|$, $y(0) = 1$, is unique. Can you guess this solution?

(*ii*) You are encouraged to read, think about, work, and then keep in mind Problem 31 in Exercises 1.2.

EXERCISES 1.2

Answers to odd-numbered problems begin on page A-25.

In Problems 1 and 2, use the fact that $y = 1/(1 + c_1 e^{-x})$ is a one-parameter family of solutions of $y' = y - y^2$ to find a solution of the initial-value problem consisting of the differential equation and the given initial conditions.

1. $y(0) = -\frac{1}{3}$ **2.** $y(-1) = 2$

In Problems 3–6, use the fact that $x = c_1 \cos t + c_2 \sin t$ is a two-parameter family of solutions of $x'' + x = 0$ to find a solution of the initial-value problem consisting of the differential equation and the given initial conditions.

3. $x(0) = -1, \quad x'(0) = 8$

4. $x(\pi/2) = 0, \quad x'(\pi/2) = 1$

5. $x(\pi/6) = \frac{1}{2}, \quad x'(\pi/6) = 0$

6. $x(\pi/4) = \sqrt{2}, \quad x'(\pi/4) = 2\sqrt{2}$

In Problems 7–10, use the fact that $y = c_1 e^x + c_2 e^{-x}$ is a two-parameter family of solutions of $y'' - y = 0$ to find a solution of the initial-value problem consisting of the differential equation and the given initial conditions.

7. $y(0) = 1, \quad y'(0) = 2$ **8.** $y(1) = 0, \quad y'(1) = e$

9. $y(-1) = 5, \quad y'(-1) = -5$ **10.** $y(0) = 0, \quad y'(0) = 0$

In Problems 11 and 12, determine by inspection at least two solutions of the given initial-value problem.

11. $y' = 3y^{2/3}, \quad y(0) = 0$ **12.** $xy' = 2y, \quad y(0) = 0$

In Problems 13–20, determine a region of the xy-plane for which the given differential equation would have a unique solution whose graph passes through a point (x_0, y_0) in the region.

13. $\dfrac{dy}{dx} = y^{2/3}$ **14.** $\dfrac{dy}{dx} = \sqrt{xy}$

15. $x\dfrac{dy}{dx} = y$ **16.** $\dfrac{dy}{dx} - y = x$

17. $(4 - y^2)y' = x^2$ **18.** $(1 + y^3)y' = x^2$

19. $(x^2 + y^2)y' = y^2$ **20.** $(y - x)y' = y + x$

In Problems 21–24, determine whether Theorem 1.1 guarantees that the differential equation $y' = \sqrt{y^2 - 9}$ possesses a unique solution through the given point.

21. $(1, 4)$ **22.** $(5, 3)$ **23.** $(2, -3)$ **24.** $(-1, 1)$

25. (a) By inspection find a one-parameter family of solutions of the differential equation $xy' = y$. Verify that each member of the family is a solution of the initial-value problem $xy' = y$, $y(0) = 0$.

(b) Explain part **(a)** by determining a region R in the xy-plane for which the differential equation $xy' = y$ would have a unique solution through a point (x_0, y_0) in R.

(c) Verify that the piecewise-defined function

$$y = \begin{cases} 0, & x < 0 \\ x, & x \geq 0 \end{cases}$$

satisfies the condition $y(0) = 0$. Determine whether this function is also a solution of the initial-value problem in part **(a)**.

26. (a) Verify that $y = \tan(x + c)$ is a one-parameter family of solutions of the differential equation $y' = 1 + y^2$.

(b) Since $f(x, y) = 1 + y^2$ and $\partial f/\partial y = 2y$ are continuous everywhere, the region R in Theorem 1.1 can be taken to be the entire xy-plane. Use the family of solutions in part **(a)** to find an explicit solution of the initial-value problem $y' = 1 + y^2$, $y(0) = 0$. Even though $x_0 = 0$ is in the interval $-2 < x < 2$, explain why the solution is not defined on this interval.

(c) Determine the largest interval I of definition for the solution of the initial-value problem in part **(b)**

27. (a) Verify that $y = -1/(x + c)$ is a one-parameter family of solutions of the differential equation $y' = y^2$.

(b) Since $f(x, y) = y^2$ and $\partial f/\partial y = 2y$ are continuous everywhere, the region R in Theorem 1.1 can be taken to be the entire xy-plane. Find a solution from the family in part **(a)** that satisfies $y(0) = 1$. Find a solution from the family in part **(a)** that satisfies $y(0) = -1$. Determine the largest interval I of definition for the solution of each initial-value problem.

(c) Find a solution from the family in part **(a)** that satisfies $y' = y^2$, $y(0) = y_0$, $y_0 \neq 0$. Explain why the largest interval I of definition for this solution is either $-\infty < x < 1/y_0$ or $1/y_0 < x < \infty$.

(d) Determine the largest interval I of definition for the solution of the initial-value problem $y' = y^2$, $y(0) = 0$.

28. (a) Verify that $3x^2 - y^2 = c$ is a one-parameter family of solutions of the differential equation $y\,dy/dx = 3x$.

(b) By hand, sketch the graph of the implicit solution $3x^2 - y^2 = 3$. Find all explicit solutions $y = \phi(x)$ of the DE in part **(a)** defined by this relation. Give the interval I of definition of each explicit solution. The point $(-2, 3)$ is on the graph of $3x^2 - y^2 = 3$, but which of the explicit solutions satisfies $y(-2) = 3$?

(c) Use the family of solutions in part **(a)** to find an implicit solution of the initial-value problem $y\,dy/dx = 3x$, $y(2) = -4$. Then, by hand, sketch the graph of the explicit solution of this problem and give its interval I of definition.

(d) Are there any explicit solutions of $y\,dy/dx = 3x$ that pass through the origin?

29. Consider the initial-value problem: $y' = x - 2y$, $y(0) = \frac{1}{2}$. Determine which of the two curves shown in Figure 1.13 is the only plausible solution curve. Explain your reasoning.

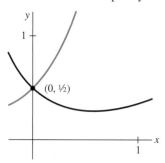

Figure 1.13

30. Determine a plausible value of x_0 for which the graph of the solution of the initial-value problem $y' + 2y = 3x - 6$, $y(x_0) = 0$, is tangent to the x-axis at $(x_0, 0)$. Explain your reasoning.

31. Suppose that the first-order differential equation $dy/dx = f(x, y)$ possesses a one-parameter family of solutions and that $f(x, y)$ satisfies the hypotheses of Theorem 1.1 in some rectangular region R of the xy-plane. Explain why two different solution curves cannot intersect or be tangent to each other at a point (x_0, y_0) in R.

32. The functions

$$y(x) = x^4/16, \quad -\infty < x < \infty, \quad \text{and} \quad y(x) = \begin{cases} 0, & x < 0 \\ x^4/16, & x \geq 0 \end{cases}$$

are clearly different. Show that both functions are solutions of the initial-value problem $dy/dx = xy^{1/2}$, $y(2) = 1$, on the interval $(-\infty, \infty)$. Resolve the apparent contradiction between this fact and the last sentence in Example 4. [*Hint:* Carefully examine the graph of each function.]

In Problems 33 and 34, use Problem 33 in Exercises 1.1 and equations (2) and (3) of this section.

33. Find a function $y = f(x)$ whose graph at each point (x, y) has the slope given by $8e^{2x} + 6x$ and has the y-intercept $(0, 9)$.

34. Find a function $y = f(x)$ whose second derivative is $y'' = 12x - 2$ at each point (x, y) on its graph and $y = -x + 5$ is tangent to the graph at the point corresponding to $x = 1$.

1.3 Differential Equations as Mathematical Models

■ **Mathematical Models** It is often desirable to describe the behavior of some real-life system or phenomenon, whether physical, sociological, or even economic, in mathematical terms. The mathematical description of a system or a phenomenon is called a **mathematical model** and is constructed with certain goals in mind. For example, we may wish to understand the mechanisms of a certain ecosystem by studying the growth of animal populations in that system, or we may wish to date fossils by means of analyzing the decay of a radioactive substance either in the fossil or in the stratum in which it was discovered.

Construction of a mathematical model of a system starts with *identification of the variables* that are responsible for changing the system. We may choose not to incorporate all these variables into the model at first. In this first step we are specifying the **level of resolution** of the model. Next, *we make a set of reasonable assumptions* or hypotheses about the system we are trying to describe. These assumptions will also include any empirical laws that may be applicable to the system.

For some purposes it may be perfectly within reason to be content with low-resolution models. For example, you may be already be aware that in modeling the motion of a body falling near the surface of the earth the retarding force of air friction is sometimes ignored in beginning physics courses; but if you are a scientist whose job it is to accurately predict the flight path of a long-range projectile, air resistance and other factors such as the curvature of the earth have to be taken into account.

Since the assumptions made about a system frequently involve *a rate of change* of one or more of the variables, the mathematical depiction of all these assumptions may be one or more equations involving *derivatives*. In other words, the mathematical model may be a differential equation or a system of differential equations.

Once we have formulated a mathematical model that is either a differential equation or a system of a differential equations, we are faced with the not insignificant problem of trying to solve it. *If* we can solve it, then we deem the model to be reasonable if its solution is consistent with either experimental data or known facts about the behavior of the system. But if the predictions produced by the solution are poor, we can either increase the level of resolution of the model or make alternative assumptions about the mechanisms for change in the system. The steps of the modeling process are then repeated as shown in the following diagram.

Of course, by increasing the resolution we add to the complexity of the mathematical model and increase the likelihood that we cannot obtain an explicit solution.

A mathematical model of a physical system will often involve the variable time t. A solution of the model then gives the **state of the system;** in other words, for appropriate values of t, the values of the dependent variable (or variables) describe the system in the past, present, and future.

■ **Population Dynamics** One of the earliest attempts to model human **population growth** by means of mathematics was by the English economist Thomas Malthus in 1798. Basically, the idea of the Malthusian model is the assumption that the rate at which a population of a country grows at a certain time is proportional* to the total population of the country at that time. In other words, the more people there are at time t, the more there are going to be in the future. In mathematical terms, if $P(t)$ denotes the total population at time t, then this assumption can be expressed as

$$\frac{dP}{dt} \propto P \quad \text{or} \quad \frac{dP}{dt} = kP, \tag{1}$$

where k is a constant of proportionality. This simple model, which fails to take into account many factors (immigration and emigration, for example) that can influence human populations to either grow or decline, nevertheless turned out to be fairly accurate in predicting the population of the United States during the years 1790–1860. Populations that grow at a rate described by (1) are rare; nevertheless, (1) is still used to model *growth of small populations over short intervals of time,* for example, bacteria growing in a petri dish.

■ **Radioactive Decay** The nucleus of an atom consists of combinations of protons and neutrons. Many of these combinations of protons and neutrons are unstable, that is, the atoms decay or transmute into the atoms of another substance. Such nuclei are said to be radioactive. For example, over time the highly radioactive radium, Ra-226, transmutes into the radioactive gas radon, Rn-222. In modeling the phenomenon of **radioactive decay,** it is assumed that the rate dA/dt at which the nuclei of a substance decays is proportional to the amount (more precisely, the number of nuclei) $A(t)$ of the substance remaining at time t:

$$\frac{dA}{dt} \propto A \quad \text{or} \quad \frac{dA}{dt} = kA. \tag{2}$$

Of course equations (1) and (2) are exactly the same; the difference is only in the interpretation of the symbols and the constants of proportionality. For growth, as we expect in (1), $k > 0$, and in the case of (2) and decay, $k < 0$.

The model (1) for growth can be seen as the equation $dS/dt = rS$, which describes the growth of capital S when an annual rate of interest r is compounded continuously. The model (2) for decay also occurs in a biological setting, such as determining the half-life of a drug—the time that it takes for 50% of a drug to be eliminated from a body by excretion or metabolism. In chemistry the decay model (2) appears as the mathematical description of a first-order chemical reaction. The point is this:

A single differential equation can serve as a mathematical model for many different phenomena.

* If two quantities u and v are proportional, we write $u \propto v$. This means one quantity is a constant multiple of the other: $u = kv$.

Mathematical models are often accompanied by certain side conditions. For example, in (1) and (2) we would expect to know, in turn, an initial population P_0 and an initial amount of radioactive substance A_0 that is on hand. If this initial point in time is taken to be $t = 0$, then we know that $P(0) = P_0$ and $A(0) = A_0$. In other words, a mathematical model can consist of either an initial-value problem or, as we shall see later on in Section 3.9, a boundary-value problem.

■ **Newton's Law of Cooling/Warming** According to Newton's empirical law of cooling—or warming—the rate at which the temperature of a body changes is proportional to the difference between the temperature of the body and the temperature of the surrounding medium, the so-called ambient temperature. If $T(t)$ represents the temperature of a body at time t, T_m the temperature of the surrounding medium, and dT/dt the rate at which the temperature of the body changes, then Newton's law of cooling/warming translates into the mathematical statement

$$\frac{dT}{dt} \propto T - T_m \quad \text{or} \quad \frac{dT}{dt} = k(T - T_m), \tag{3}$$

where k is a constant of proportionality. In either case, cooling or warming, if T_m is a constant, it stands to reason that $k < 0$.

■ **Spread of a Disease** A contagious disease—for example, a flu virus—is spread throughout a community by people coming into contact with other people. Let $x(t)$ denote the number of people who have contracted the disease and $y(t)$ the number of people who have not yet been exposed. It seems reasonable to assume that the rate dx/dt at which the disease spreads is proportional to the number of encounters or *interactions* between these two groups of people. If we assume that the number of interactions is jointly proportional to $x(t)$ and $y(t)$, that is, proportional to the product xy, then

$$\frac{dx}{dt} = kxy, \tag{4}$$

where k is the usual constant of proportionality. Suppose a small community has a fixed population of n people. If one infected person is introduced into this community, then it could be argued that $x(t)$ and $y(t)$ are related by $x + y = n + 1$. Using this last equation to eliminate y in (4) gives us the model

$$\frac{dx}{dt} = kx(n + 1 - x). \tag{5}$$

An obvious initial condition accompanying equation (5) is $x(0) = 1$.

■ **Chemical Reactions** The disintegration of a radioactive substance, governed by the differential equation (2), is said to be a **first-order reaction.** In chemistry a few reactions follow this same empirical law: If the molecules of substance A decompose into smaller molecules, it is a natural assumption that the rate at which this decomposition takes place is proportional to the amount of the first substance that has not undergone conversion; that is, if $X(t)$ is the amount of substance A remaining at any time, then $dX/dt = kX$, where k is negative constant since X is decreasing. An example of a first-order chemical reaction is the conversion of t-butyl chloride into t-butyl alcohol:

$$(CH_3)_3CCl + NaOH \rightarrow (CH_3)_3COH + NaCl.$$

Only the concentration of the *t*-butyl chloride controls the rate of reaction. But in the reaction

$$CH_3Cl + NaOH \rightarrow CH_3OH + NaCl,$$

for every molecule of methyl chloride one molecule of sodium hydroxide is consumed, thus forming one molecule of methyl alcohol and one molecule of sodium chloride. In this case the rate at which the reaction proceeds is proportional to the product of the remaining concentrations of CH_3Cl and of NaOH. If X denotes the amount of CH_3OH formed and α and β are the given amounts of the first two chemicals A and B, then the instantaneous amounts not converted to chemical C are $\alpha - X$ and $\beta - X$, respectively. Hence the rate of formation of C is given by

$$\frac{dX}{dt} = k(\alpha - X)(\beta - X), \tag{6}$$

where k is a constant of proportionality. A reaction whose model is equation (6) is said to be **second order.**

■ **Mixtures** The mixing of two salt solutions of differing concentrations gives rise to a first-order differential equation for the amount of salt contained in the mixture. Let us suppose that a large mixing tank holds 300 gallons of water in which a certain number of pounds of salt has been dissolved. Another brine solution is pumped into the large tank at a rate of 3 gallons per minute; the concentration of the solution entering is 2 pounds of salt per gallon. When the solution in the tank is well stirred, it is pumped out at the same rate as the entering solution. See Figure 1.14. If $A(t)$ denotes the amount of salt (measured in pounds) in the tank at time t, then the rate at which $A(t)$ changes is a net rate:

$$\frac{dA}{dt} = \left(\begin{array}{c} rate\ of \\ substance\ entering \end{array}\right) - \left(\begin{array}{c} rate\ of \\ substance\ leaving \end{array}\right) = R_1 - R_2. \tag{7}$$

The rate R_1 at which the salt enters the tank is, in pounds per minute,

$$R_1 = (3\ \text{gal/min}) \cdot (2\ \text{lb/gal}) = 6\ \text{lb/min}.$$

Now, since liquid is being pumped out of the tank at the same rate at which it is being pumped in, the number of gallons of solution in the tank at time t is 300 gallons, so that the concentration of the salt in the tank is $A(t)/300$ lb/gal. Hence the rate R_2 at which salt is leaving the tank is

$$R_2 = (3\ \text{gal/min}) \cdot \left(\frac{A}{300}\ \text{lb/gal}\right) = \frac{A}{100}\ \text{lb/min}.$$

The equation (7) becomes

$$\frac{dA}{dt} = 6 - \frac{A}{100}. \tag{8}$$

Figure 1.14

constant 300 gal

■ **Draining a Tank** In hydrodynamics, **Torricelli's law** states that the speed v of efflux of water through a sharp-edged hole at the bottom of a tank filled to a depth h is the same as the speed that a body (in this case a drop of water) would acquire in falling freely from a height h; that is, $v = \sqrt{2gh}$, where g is the acceleration due to

Figure 1.15

(a)

Inductor
inductance L: henrys (h)
voltage drop across: $L \dfrac{di}{dt}$

$i \rightarrow$ L

Resistor
resistance R: ohms (Ω)
voltage drop across: iR

$i \rightarrow$ R

Capacitor
capacitance C: farads (f)
voltage drop across: $\dfrac{1}{C} q$

$i \rightarrow$ C

(b)

Figure 1.16

gravity. This last expression comes from equating the kinetic energy $\frac{1}{2}mv^2$ with the potential energy mgh and solving for v. Suppose a tank filled with water is allowed to drain through a hole under the influence of gravity. We would like to find the depth h of water remaining in the tank at time t. Consider the tank shown in Figure 1.15. If the area of the hole is A_h (in ft^2) and the speed of the water leaving the tank is $v = \sqrt{2gh}$ (in ft/s), then the volume of water leaving the tank per second is $A_h\sqrt{2gh}$ (in ft^3/s). Thus if $V(t)$ denotes the volume of water in the tank at time t,

$$\frac{dV}{dt} = -A_h \sqrt{2gh}, \tag{9}$$

where the minus sign indicates that V is decreasing. Note here that we are ignoring the possibility of friction at the hole that might cause a reduction of the rate of flow there. Now if the tank is such that the volume of water in it at time t can be written $V(t) = A_w h$, where A_w (in ft^2) is the *constant* area of the upper surface of the water (see Figure 1.15), then $dV/dt = A_w dh/dt$. Substituting this last expression into (9) gives us the desired differential equation for the height of the water at time t:

$$\frac{dh}{dt} = -\frac{A_h}{A_w}\sqrt{2gh}. \tag{10}$$

It is interesting to note that (10) remains valid even when A_w is not constant. In this case we must express the upper surface area of the water as a function of h, that is, $A_w = A(h)$. See Problem 12 in Exercises 1.3.

■ **Series Circuits** Consider the single-loop series circuit containing an inductor, resistor, and capacitor shown in Figure 1.16(a). The current in a circuit after a switch is closed is denoted by $i(t)$; the charge on a capacitor at time t is denoted by $q(t)$. The letters L, C, and R are known as inductance, capacitance, and resistance, respectively, and are generally constants. Now according to **Kirchhoff's second law,** the impressed voltage $E(t)$ on a closed loop must equal the sum of the voltage drops in the loop. Figure 1.16(b) also shows the symbols and the formulas for the respective voltage drops across an inductor, a capacitor, and a resistor. Since current $i(t)$ is related to charge $q(t)$ on the capacitor by $i = dq/dt$, by adding the three voltage drops

Inductor Resistor Capacitor

$$L\frac{di}{dt} = L\frac{d^2q}{dt^2}, \qquad iR = R\frac{dq}{dt}, \qquad \frac{1}{C}q$$

and equating the sum to the impressed voltage, we obtain a second-order differential equation

$$L\frac{d^2q}{dt^2} + R\frac{dq}{dt} + \frac{1}{C}q = E(t). \tag{11}$$

We will examine a differential equation analogous to (11) in great detail in Section 3.8.

■ **Falling Bodies** In constructing a mathematical model of the motion of a body moving in a force field, one often starts with Newton's second law of motion. Recall from elementary physics that Newton's **first law of motion** states that a body will either remain at rest or will continue to move with a constant velocity unless acted upon by an external force. In each case this is equivalent to saying that when the sum of the forces $F = \sum F_k$—that is, the *net* or resultant force—acting on the body

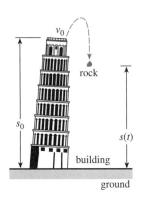

Figure 1.17

is zero, then the acceleration a of the body is zero. **Newton's second law of motion** indicates that when the net force acting on a body is not zero, then the net force is proportional to its acceleration a, or more precisely, $F = ma$, where m is the mass of the body.

Now suppose a rock is tossed upward from a roof of a building as illustrated in Figure 1.17. What is the position $s(t)$ of the rock relative to the ground at time t? The acceleration of the rock is the second derivative d^2s/dt^2. If we assume that the upward direction is positive and that no force acts on the rock other than the force of gravity, then Newton's second law gives

$$m \frac{d^2s}{dt^2} = -mg \quad \text{or} \quad \frac{d^2s}{dt^2} = -g. \tag{12}$$

In other words, the net force is simply the weight $F = F_1 = -W$ of the rock near the surface of the earth. Recall that the magnitude of the weight is $W = mg$, where m is mass of the body and g is the acceleration due to gravity. The minus sign in (12) is used because the weight of the rock is a force directed downward, which is opposite to the positive direction. If the height of the building is s_0 and the initial velocity of the rock is v_0, then s is determined from the second-order initial-value problem

$$\frac{d^2s}{dt^2} = -g, \qquad s(0) = s_0, s'(0) = v_0. \tag{13}$$

Although we have not stressed solutions of the equations we have constructed, we note that (13) can be solved by integrating the constant $-g$ twice with respect to t. The initial conditions determine the two constants of integration. You might recognize the solution of (13) from elementary physics as the formula $s(t) = -\frac{1}{2}gt^2 + v_0t + s_0$.

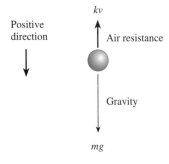

Figure 1.18 Falling body of mass m

■ **Falling Bodies and Air Resistance** Prior to Galileo's famous experiment from the Leaning Tower of Pisa, it was generally believed that heavier objects in free fall, such as a cannonball, fell with a greater acceleration than lighter objects, such as a feather. Obviously a cannonball and a feather, when dropped simultaneously from the same height, *do* fall at different rates, but it is not because a cannonball is heavier. The difference in rates is due to air resistance. The resistive force of air was ignored in the model given in (13). Under some circumstances a falling body of mass m—such as a feather with low density and irregular shape—encounters air resistance proportional to its instantaneous velocity v. If we take, in this circumstance, the positive direction to be oriented downward, then the net force acting on the mass is given by $F = F_1 + F_2 = mg - kv$, where the weight $F_1 = mg$ of the body is a force acting in the positive direction and air resistance $F_2 = -kv$ is a force, called **viscous damping,** acting in the opposite or upward direction. See Figure 1.18. Now since v is related to acceleration a by $a = dv/dt$, Newton's second law becomes $F = ma = m\,dv/dt$. By equating the net force to this form of Newton's second law, we obtain a first-order differential equation for the velocity $v(t)$ of the body at time t,

$$m \frac{dv}{dt} = mg - kv. \tag{14}$$

Here k is a positive constant of proportionality. If $s(t)$ is the distance the body falls in time t from its initial point of release, then $v = ds/dt$ and $a = dv/dt = d^2s/dt^2$. In terms of s, (14) is a second-order differential equation

$$m \frac{d^2s}{dt^2} = mg - k \frac{ds}{dt} \quad \text{or} \quad m \frac{d^2s}{dt^2} + k \frac{ds}{dt} = mg. \tag{15}$$

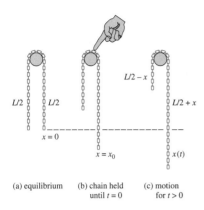

(a) equilibrium (b) chain held (c) motion
 until $t = 0$ for $t > 0$

Figure 1.19

■ **A Slipping Chain** Suppose a uniform chain of length L feet is draped over a metal peg anchored into a wall high above ground level. Let us assume that the peg is frictionless and that the chain weighs ρ lb/ft. Figure 1.19(a) illustrates the position of the chain where it hangs in equilibrium; if displaced a little to either the right or left, the chain would slip off the peg. Suppose the positive direction is taken to be downward, and suppose $x(t)$ denotes the distance that the right end of the chain would fall in time t. The equilibrium position corresponds to $x = 0$. In Figure 1.19(b) the chain is displaced an amount x_0 feet and is held on the peg until it is released at an initial time that is designated as $t = 0$. For the chain in motion, as shown in Figure 1.19(c), we have the following quantities:

weight of the chain: $W = (L \text{ ft}) \cdot (\rho \text{ lb/ft}) = L\rho$,

mass of the chain: $m = W/g = L\rho/32$

net force: $F = \left(\dfrac{L}{2} + x\right)\rho - \left(\dfrac{L}{2} - x\right)\rho = 2x\rho$.

Since $a = d^2x/dt^2$, $ma = F$ becomes

$$\frac{L\rho}{32}\frac{d^2x}{dt^2} = 2\rho x \quad \text{or} \quad \frac{d^2x}{dt^2} - \frac{64}{L}x = 0. \tag{16}$$

Remarks

Each example in this section has described a dynamical system: a system that changes or evolves with the flow of time t. Since the study of dynamical systems is a branch of mathematics currently in vogue, we shall occasionally relate the terminology of that field to the discussion at hand.

In more precise terms, a **dynamical system** consists of a set of time-dependent variables, called **state variables,** together with a rule that enables us to determine (without ambiguity) the state of the system (this may be past, present, or future states) in terms of a state prescribed at some time t_0. Dynamical systems are classified as either discrete-time systems or continuous-time systems. In this course we shall be concerned only with continuous-time dynamical systems—systems in which *all* variables are defined over a continuous range of time. The rule or the mathematical model in a continuous-time dynamical system is a differential equation or a system of differential equations. The **state of the system** at a time t is the value of the state variables at that time; the specified state of the system at a time t_0 is simply the initial conditions that accompany the mathematical model. The solution of the initial-value problem is referred to as the **response of the system.** For example, in the preceding case of radioactive decay, the rule is $dA/dt = kA$. Now if the quantity of a radioactive substance at some time t_0 is known, say $A(t_0) = A_0$, then by solving the rule the response of the system for $t \geq t_0$ is found to be $A(t) = A_0 e^{(t - t_0)}$ (see Section 2.7). The response $A(t)$ is the single-state variable for this system. In the case of the rock tossed from the roof of the building, the response of the system, the solution of the differential equation $d^2s/dt^2 = -g$ subject to the initial state $s(0) = s_0$, $s'(0) = v_0$, is the function $s(t) = -\frac{1}{2}gt^2 + v_0t + s_0$, $0 \leq t \leq T$, where the symbol T represents the time when the rock hits the ground. The state variables are $s(t)$ and $s'(t)$, which are, respectively, the vertical position of the rock above ground and its velocity at time t. The acceleration $s''(t)$ is *not* a state variable since we only have to know any initial position and initial velocity at a time t_0 to uniquely determine the rock's

position $s(t)$ and velocity $s'(t) = v(t)$ for any time in the interval $t_0 \leq t \leq T$. The acceleration $s''(t) = a(t)$ is, of course, given by the differential equation $s''(t) = -g$, $0 < t < T$.

One last point: Not every system studied in this text is a dynamical system. We shall also examine some static systems in which the model is a differential equation.

 Answers to odd-numbered problems begin on page A-26.

Population Dynamics

1. Using the same assumptions that underlie the model in (1), determine a differential equation governing the growing population $P(t)$ of a country when individuals are allowed to immigrate into the country at a constant rate r. What is the differential equation when individuals are allowed to emigrate at a constant rate r?

2. The model given in (1) fails to take death into consideration; that is, the growth rate equals the birth rate. In another model of a changing population of a community, it is assumed that the rate at which the population changes is a *net* rate—that is, the difference between the rate of births and the rate of deaths in the community. Determine a differential equation governing the population $P(t)$ if both the birth rate and the death rate are proportional to the population present at time t.

3. Using the concept of a net rate introduced in Problem 2, determine a differential equation governing a population $P(t)$ if the birth rate is proportional to the population present at time t but the death rate is proportional to the square of the population present at time t.

4. The number of field mice in a certain pasture is given by the function $200 - 10t$, where t is measured in years. Determine a differential equation governing a population of owls that feed on the mice if the rate at which the owl population grows is proportional to the difference between the number of owls at time t and the number of field mice at time t.

Newton's Law of Cooling/Warming

5. A cup of coffee cools according to Newton's law of cooling (3). Use data from the graph of the temperature $T(t)$ in Figure 1.20 to estimate the constants T_m, T_0, and k in a model of the form of the first-order initial-value problem: $dT/dt = k(T - T_m)$, $T(0) = T_0$.

Figure 1.20

6. The ambient temperature T_m in (3) could be a function of time t. Suppose in an artificially controlled environment, $T_m(t)$ is periodic with period 24 hr as illustrated in Figure 1.21. Devise a mathematical model for the temperature $T(t)$ of a body within this environment.

Figure 1.21

Spread of a Disease/Technology

7. Suppose a student carrying a flu virus returns to an isolated college campus of 1000 students. Determine a differential equation governing the number of people $x(t)$ who have

contracted the flu if the rate at which the disease spreads is proportional to the number of interactions between the number of students with the flu and the number of students who have not yet been exposed to it.

8. At a time $t = 0$ a technological innovation is introduced into a community with a fixed population of n people. Determine a differential equation governing the number of people $x(t)$ who have adopted the innovation at time t if it is assumed that the rate at which the innovation spread through the community is jointly proportional to the number of people who have adopted it and the number of people who have not adopted it.

Mixtures

9. Suppose that a large mixing tank initially holds 300 gallons of water in which 50 pounds of salt has been dissolved. Pure water is pumped into the tank at a rate of 3 gal/min, and when the solution is well stirred, it is pumped out at the same rate. Determine a differential equation for the amount $A(t)$ of salt in the tank at time t.

10. Suppose that a large mixing tank initially holds 300 gallons of water in which 50 pounds of salt has been dissolved. Another brine solution is pumped into the tank at a rate of 3 gal/min, and when the solution is well stirred, it is pumped out at a *slower* rate of 2 gal/min. If the concentration of the solution entering is 2 lb/gal, determine a differential equation for the amount $A(t)$ of salt in the tank at time t.

Draining a Tank

11. Suppose water is leaking from a tank through a circular hole of area A_h at its bottom. When water leaks through a hole, friction and contraction of the water stream near the hole reduce the volume of the water leaving the tank per second to $cA_h\sqrt{2gh}$, where c ($0 < c < 1$) is an empirical constant. Determine a differential equation for the height h of water at time t for the cubical tank in Figure 1.22. The radius of the hole is 2 in, $g = 32$ ft/s^2, and A_w is the constant area of the upper water surface.

Figure 1.22

12. The right-circular conical tank shown in Figure 1.23 loses water out of a circular hole at its bottom. Determine a differential equation for the height of the water h at time t. The radius of the hole is 2 in, $g = 32$ ft/s^2, and the friction/contraction factor introduced in Problem 11 is $c = 0.6$.

Figure 1.23

Series Circuits

13. A series circuit contains a resistor and an inductor as shown in Figure 1.24. Determine a differential equation for the current $i(t)$ if the resistance is R, the inductance is L, and the impressed voltage is $E(t)$.

Figure 1.24

14. A series circuit contains a resistor and a capacitor as shown in Figure 1.25. Determine a differential equation for the charge $q(t)$ on the capacitor if the resistance is R, the capacitance is C, and the impressed voltage is $E(t)$.

Figure 1.25

Falling Bodies and Air Resistance

15. For high-speed motion through the air—such as the skydiver shown in Figure 1.26 falling before the parachute is opened—air resistance is closer to a power of the instantaneous velocity. Determine a differential equation for the velocity $v(t)$ of a falling body of mass m if air resistance is proportional to the square of the instantaneous velocity.

Figure 1.26

Newton's Second Law and Archimedes' Principle

16. A cylindrical barrel s ft in diameter of weight w lb is floating in water. After an initial depression, the barrel exhibits an up-and-down bobbing motion along a vertical line. Using Figure 1.27(b), determine a differential equation for the vertical displacement $y(t)$ if the origin is taken to be on the vertical axis at the surface of the water when the barrel is at rest. Use **Archimedes' principle:** Buoyancy, or upward force of the water on the barrel, is equal to the weight of the water displaced. Assume that the downward direction is positive, that the density of water is 62.4 lb/ft³, and that there is no resistance between the barrel and the water.

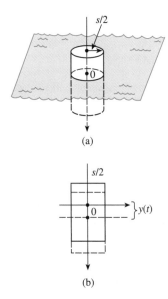

(a)

(b)

Figure 1.27

Newton's Second Law and Hooke's Law

17. After a mass m is attached to a spring, the spring stretches s units and then hangs at rest in the equilibrium position as shown in Figure 1.28(b). After the spring/mass system has been set in motion, let $x(t)$ denote the directed distance of the mass beyond the equilibrium position. Assume that the downward direction is positive, that the motion takes place in a vertical straight line through the center of gravity of the mass, and that the only forces acting on the system are the weight of the mass and the restoring force of the stretched spring. Use **Hooke's law:** The restoring force of a spring is proportional to its total elongation. Determine a differential equation for the displacement $x(t)$ at time t.

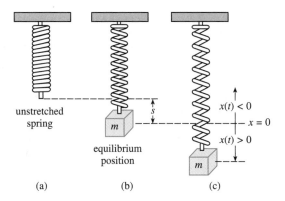

(a) (b) (c)

Figure 1.28

18. In Problem 17, determine a differential equation for the displacement $x(t)$ if the motion takes place in a medium that imparts a damping force on the spring/mass system that is proportional to the instantaneous velocity of the mass and acts in a direction opposite to that of motion.

Newton's Second Law and Variable Mass

When the mass m of a body moving through a force field is variable, **Newton's second law** takes on the form: When the net force acting on a body is not zero, then the net force F is equal to the time rate of change of momentum of the body. That is, $F = d/dt(mv),$* where mv is momentum. Use this formulation of Newton's second law in Problems 19 and 20.

19. A uniform 10-foot-long chain is coiled loosely on the ground. As shown in Figure 1.29, one end of the chain is pulled vertically upward by means of constant force of 5 lb. The chain weighs 1 lb/ft. Determine a differential equation for the height $x(t)$ of the end above ground level at time t. Assume that the positive direction is upward.

Figure 1.29

* Note that when m is constant, this is the same as $F = ma$.

20. A uniform chain of length L, measured in feet, is held vertically so that the lower end just touches the floor, as shown in Figure 1.30. The chain weighs 2 lb/ft. The upper end that is held is released from rest at $t = 0$, and the chain falls straight down. Ignore air resistance, assume that the positive direction is downward, and let $x(t)$ denote the length of the chain on the floor at time t. Show that a differential equation for $x(t)$ is

$$(L - x)\frac{d^2x}{dt^2} - \left(\frac{dx}{dt}\right)^2 = Lg.$$

Figure 1.30

Newton's Second Law and the Law of Universal Gravitation

21. By **Newton's universal law of gravitation** the free-fall acceleration a of a body, such as the satellite shown in Figure 1.31, falling a great distance to the surface is not the constant g. Rather, the acceleration a is inversely proportional to the square of the distance from the center of the earth, $a = k/r^2$, where k is the constant of proportionality. Use the fact that at the surface of the earth $r = R$ and $a = g$ to determine k. If the positive direction is upward, use Newton's second law and his universal law of gravitation to find a differential equation for the distance r. Let M denote the mass of the earth.

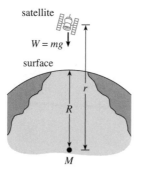

Figure 1.31

22. Suppose a hole is drilled through the center of the earth and a bowling ball of mass m is dropped into the hole as shown in Figure 1.32. Construct a mathematical model that describes the motion of the ball. Let the distance from the center of the earth to the mass at time t be denoted by r, let M denote the mass of the earth, M_r the mass of that portion of the earth within a sphere of radius r, and δ the constant density of the earth.

Figure 1.32

Miscellaneous Mathematical Models

23. In the theory of learning, the rate at which a subject is memorized is assumed to be proportional to the amount that is left to be memorized. Suppose M denotes the total amount of a subject to be memorized and $A(t)$ is the amount memorized in time t. Determine a differential equation for the amount $A(t)$.

24. In Problem 23 assume that the rate at which material is *forgotten* is proportional to the amount memorized in time t. Determine a differential equation for $A(t)$ when forgetfulness is taken into account.

25. A drug is infused into a patient's bloodstream at a constant rate of r grams per second. Simultaneously, the drug is removed at a rate proportional to the amount $x(t)$ of the drug present at time t. Determine a differential equation governing the amount $x(t)$.

26. A person P, starting at the origin, moves in the direction of the positive x-axis, pulling a weight along the curve C, called a **tractrix,** as shown in Figure 1.33. The weight, initially located on the y-axis at $(0, s)$, is pulled by a rope of constant length s, which is kept taut throughout the motion. Determine a differential equation of the path of motion. Assume that the rope is always tangent to C.

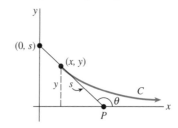

Figure 1.33

27. As illustrated in Figure 1.34, light rays strike a plane curve
C in such a manner that all rays *L* parallel to the *x*-axis are
reflected to a single point *O.* Assuming that the angle of
incidence is equal to the angle of reflection, determine a
differential equation that describes the shape of the curve
C. [*Hint:* Inspection of the figure shows that we can write
$\phi = 2\theta$. Why? Now use an appropriate trigonometric
identity.]

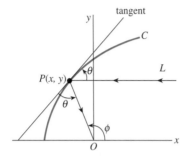

Figure 1.34

28. Reread Problem 29 in Exercises 1.1 and then give an
explicit solution *P*(*t*) for equation (1). Find a one-parameter
family of solutions of (1).

29. Reread the sentence following equation (3) and assume that
T_m is a positive constant. Supply the reasons why we would
expect $k < 0$ in (3) in both cases of cooling and warming.
You might start by interpreting, say, $T(t) > T_m$ in a graphi-
cal manner.

30. Reread the discussion leading up to equation (8). If we
assume that initially the tank holds, say, 50 lb of salt, it
stands to reason that since salt is being added to the tank
continuously for $t > 0$ that $A(t)$ should be an increasing
function. Discuss how you might determine from the DE,
without actually solving it, the number of pounds of salt in
the tank after a long period of time.

31. The differential equation $dP/dt = (k \cos t)P$, where *k* is a
positive constant, is a model of human population *P*(*t*) of a

certain community. Discuss an interpretation for the solu-
tion of this equation. In other words, what kind of popula-
tion do you think the differential equation describes?

32. Reread this section and classify each mathematical model
as linear or nonlinear.

33. In Problem 21, suppose $r = R + s$, where *s* is the distance
from the surface of the earth to the falling body. What does
the differential equation obtained in Problem 21 become
when *s* is very small compared to *R?*

34. In meteorology the term *virga* refers to falling raindrops or
ice particles that evaporate before they reach the ground.
Assume that a typical raindrop is spherical in shape.
Starting at some time, which we can designate as $t = 0$, the
raindrop of radius r_0 falls from rest from a cloud and begins
to evaporate. If it is assumed that a raindrop evaporates in
such a manner that its shape remains spherical, then it also
makes sense to assume that the rate at which the raindrop
evaporates, that is, the rate at which it loses mass, is
proportional to its surface area. Show that this latter
assumption implies that the rate at which the radius *r* of the
raindrop decreases is a constant. Find *r*(*t*). [*Hint:* See
Problem 33 in Exercises 1.1.] If the positive direction is
downward, construct a mathematical model for the velocity
v of the falling raindrop at time *t.* Ignore air resistance.
[*Hint:* See the introduction to Problems 19 and 20.]

35. Project The "snowplow problem" is a classic and appears
in many differential equations texts but was probably made
famous by Ralph Palmer Agnew:

> *"One day it started snowing at a heavy and steady rate. A
> snowplow started out at noon, going 2 miles the first hour
> and 1 mile the second hour. What time did it start
> snowing?"*

Find Agnew's text *Differential Equations* (McGraw-Hill
Book Company), and then discuss the construction and
solution of the mathematical model.

CHAPTER 1 REVIEW EXERCISES

Answers to odd-numbered problems begin on page A-26.

In Problems 1 and 2, fill in the blank and then write this result as a linear first-order differential equation that is free of the symbol c_1 and has the form $dy/dx = f(x, y)$. The symbol c_1 represents a constant.

1. $\dfrac{d}{dx}\dfrac{c_1}{x} = $ _____

2. $\dfrac{d}{dx}(5 + c_1 e^{-2x}) = $ _____

In Problems 3 and 4, fill in the blank and then write this result as a linear second-order differential equation that is free of the symbols c_1 and c_2 and has the form $F(y, y'') = 0$. The symbols c_1, c_2, and k represent constants.

3. $\dfrac{d^2}{dx^2}(c_1 \cos kx + c_2 \sin kx) = $ _____

4. $\dfrac{d^2}{dx^2}(c_1 \cosh kx + c_2 \sinh kx) = $ _____

In Problems 5 and 6, compute y' and y'' and then combine these results as a linear second-order differential equation that is free of the symbols c_1 and c_2 and has the form $F(y, y', y'') = 0$. The symbols c_1 and c_2 represent constants.

5. $y = c_1 e^x + c_2 x e^x$ **6.** $y = c_1 e^x \cos x + c_2 e^x \sin x$

In Problems 7–12, match each of the given differential equations with one or more of the solutions:
(a) $y = 0$, **(b)** $y = 2$, **(c)** $y = 2x$, **(d)** $y = 2x^2$.

7. $xy' = 2y$ **8.** $y' = 2$

9. $y' = 2y - 4$ **10.** $xy' = y$

11. $y'' + 9y = 18$ **12.** $xy'' - y' = 0$

In Problems 13 and 14, determine by inspection at least one solution of the given differential equation.

13. $y'' = y'$ **14.** $y' = y(y - 3)$

In Problems 15 and 16, interpret each statement as a differential equation.

15. On the graph of $y = \phi(x)$, the slope of the tangent line at a point P is the square of the distance from P to the origin.

16. On the graph of $y = \phi(x)$, the rate at which the slope changes with respect to x at a point P is the negative of the slope of the tangent line at P.

17. (a) Give the domain of the function $y = x^{2/3}$.

(b) Give an interval I of definition over which $y = x^{2/3}$ is a solution of $3xy' - 2y = 0$.

18. (a) Verify that the one-parameter family $y^2 - 2y = x^2 - x + c$ is an implicit solution of the differential equation $(2y - 2)y' = 2x - 1$.

(b) Find an implicit solution of the DE that satisfies the initial condition $y(0) = 1$.

(c) Use part **(b)** to find an explicit function $y = \phi(x)$ that satisfies $y(0) = 1$. Give the domain of ϕ. Is $y = \phi(x)$ a solution of the initial-value problem? If so, give its interval I of definition; if not, explain.

19. A differential equation may possess more than one family of solutions.

(a) Plot different members of the families $y = \phi_1(x) = x^2 + c_1$ and $y = \phi_2(x) = -x^2 + c_2$.

(b) Verify that $y = \phi_1(x)$ and $y = \phi_2(x)$ are two solutions of the nonlinear first-order differential equation $(y')^2 = 4x^2$.

(c) Construct a piecewise defined function that is a solution of the nonlinear DE in part **(b)** but is not a member of either family of solutions in part **(a).**

20. What is the slope of the tangent line to the graph of the solution of $y' = 6\sqrt{y} + 5x^3$ that passes through $(-1, 4)$?

In Problems 21 and 22, verify that the indicated function is a particular solution of the given differential equation. Give an interval of definition I for each solution.

21. $x^2 y'' + xy' + y = 0$; $y = \sin(\ln x)$

22. $x^2 y'' + xy' + y = \sec(\ln x)$;
$y = \cos(\ln x)\ln(\cos(\ln x)) + (\ln x)\sin(\ln x)$

23. The graph of a solution of a second-order initial-value problem $d^2y/dx^2 = f(x, y, y')$, $y(2) = y_0$, $y'(2) = y_1$, is given in Figure 1.35. Use the graph to estimate the values of y_0 and y_1.

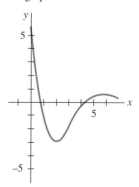

Figure 1.35

24. A tank in the form of a right circular cylinder of radius 2 ft and height 10 ft is standing on end. If the tank is initially full of water and water leaks from a circular hole of radius ½ inch at its bottom, determine a differential equation for the height *h* of the water at time *t*. Ignore friction and contraction of water at the hole.

25. A box weighing 96 lb slides down an incline that makes a 30° angle with the horizontal. If the coefficient of sliding friction is μ, determine a differential equation for the velocity *v*(*t*) of the box at time *t*. Use the fact that the force of friction opposing the motion is *μN,* where *N* is the normal component of the weight. See Figure 1.36.

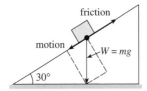

Figure 1.36

2

First-Order Differential Equations

INTRODUCTION

We are now in a position to solve *some* differential equations. We begin with first-order equations. The ability to find either explicit or implicit solutions usually depends on one's ability to recognize the *kind* of DE and then apply to it an equation-specific method of solution (Sections 2.2–2.5). As a rule, what works for one kind of first-order DE may not work on a different kind. We shall also see that a first-order DE can often yield information about the behavior of a solution even *before* we have attempted to find it (Section 2.1). But even in the case in which a DE cannot be solved analytically, the equation itself can be used to construct formulas that enable us to approximate a solution (Section 2.6). The chapter concludes with an introduction to systems of first-order differential equations as mathematical models.

2.1 Solution Curves Without the Solution

■ **What a First-Order DE Can Tell Us** Some differential equations do not possess *any* solutions. For example, there is no real function that can satisfy $(y')^2 + 1 = 0$. (Why?) Some differential equations possess solutions that can be found in explicit or implicit form by implementing an equation-specific method of solution (applying certain manipulations and analytical procedures such as integration to the DE). Some differential equations possess solutions but the equation cannot be solved analytically. In other words, when we say that a solution of a DE *exists,* we do not mean to imply that there also exists a method of solution that will produce an explicit or an implicit solution. Over a time span of centuries mathematicians have devised ingenious procedures to solve some very specialized equations, so there are, not surprisingly, a large number of differential equations that *can* be solved analytically. Although we will study some of these methods of solution for first-order equations in the subsequent sections of the chapter, let us imagine for the moment that we have in front of us a first-order differential equation in normal form $dy/dx = f(x, y)$, and let us imagine further that we can neither find nor invent a method for solving it analytically. This is not as bad a predicament as one might think, since it is often possible to glean useful information about the nature of solutions directly from the differential equation itself. For example, we have just seen that whenever $f(x, y)$ and $\partial f/\partial y$ satisfy certain continuity conditions, the qualitative questions about existence and uniqueness of solutions can be answered. We shall see in this section that other qualitative questions about properties of solutions—such as How does a solution behave near a certain point? and How does a solution behave as $x \to \infty$?—can often be answered when the function f depends solely on the variable y.

We begin, however, with a simple concept from calculus: A derivative dy/dx of a differential function $y = y(x)$ gives slopes of tangent lines at points on its graph.

■ **Slope** Because a solution $y = y(x)$ of a first-order differential equation $dy/dx = f(x, y)$ is necessarily a differentiable function on its interval I of definition it must also be continuous on I. Thus the corresponding solution curve on I must possess a tangent line at each point $(x, y(x))$ and must have no breaks. The slope of the tangent line at $(x, y(x))$ on a solution curve is the value of the first derivative dy/dx at this point, and this we know from the differential equation: $f(x, y(x))$. Now suppose that (x, y) represents any point in a region of the xy-plane over which the function f is defined. The value $f(x, y)$ that the function f assigns to the point represents the slope of a line, or as we shall envision it, a line segment called a **lineal element.** For example, consider the equation $dy/dx = 0.2xy$, where $f(x, y) = 0.2xy$. At, say, the point $(2, 3)$, the slope of a lineal element is $f(2, 3) = 1.2$. Figure 2.1(a) shows a line segment with positive slope passing though $(2, 3)$. As shown in Figure 2.1(b), *if* a solution curve also passes through $(2, 3)$, it does so tangent to this line segment; that is, the lineal element is a miniature tangent line at that point.

■ **Direction Fields** If we systematically evaluate f over a rectangular grid of points in the xy-plane and draw a lineal element at each point (x, y) of the grid with slope $f(x, y)$, then the collection of all these lineal elements is called a **direction field** or a **slope field** of the differential equation $dy/dx = f(x, y)$. Visually the direction field suggests the appearance or shape of a family of solution curves of the differential equation, and consequently it may be possible to see at a glance certain qualitative aspects of the solutions—regions in the plane, for example—in which a solution exhibits an unusual behavior. A single-solution curve that wends its way

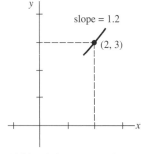

(a) Lineal element at a point

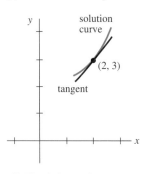

(b) Lineal element is tangent to solution curve that passes through the point

Figure 2.1

through a direction field must follow the flow pattern of the field; it is tangent to a lineal element when it intersects a point in the grid.

Sketching a direction field by hand is straightforward but time-consuming; it is probably one of those tasks about which an argument can be made for doing it once or twice in a lifetime, but overall it is most efficiently carried out with the aid of a computer.

Example 1 Direction Field

The direction field in Figure 2.2(a) was generated from computer software with $dy/dx = 0.2xy$ and using a 5×5 grid with points defined by (mh, nh), $h = 1$, m and n integers, $-5 \leq m \leq 5$, $-5 \leq n \leq 5$. In Figure 2.2(a), notice that at any point along the x-axis ($y = 0$) and the y-axis ($x = 0$) the slopes are $f(x, 0) = 0$ and $f(0, y) = 0$, respectively, so the lineal elements are horizontal. Moreover, observe in the first quadrant that for a fixed value of x the values of $f(x, y) = 0.2xy$ increase as y increases; similarly, for a fixed y the values of $f(x, y) = 0.2xy$ increase as x increases. This means that as both x and y increase the lineal elements almost become vertical and have positive slope $f(x, y) = 0.2xy > 0$ for $x > 0$, $y > 0$. In the second quadrant, $|f(x, y)|$ increases as $|x|$ and y increase, and so the lineal elements again become almost vertical but this time have negative slope $f(x, y) = 0.2xy < 0$ for $x < 0$, $y > 0$. Reading left to right, imagine that a solution curve starts at a point in the second quadrant, moves steeply downward, becomes flat at it passes through the y-axis, and then as it enters the first quadrant moves steeply upward—in other words, its shape would be concave upward and similar to a horseshoe. From this it could be surmised that $y \to \infty$ as $x \to \pm\infty$. Now in the third and fourth quadrants, since $f(x, y) = 0.2xy > 0$ and $f(x, y) = 0.2xy < 0$, respectively, the situation is reversed; a solution curve increases and then decreases as we move from left to right. We saw in (1) of Section 1.1 that $y = e^{0.1x^2}$ is an explicit solution of the differential equation $dy/dx = 0.2xy$; you should verify that a one-parameter family of solutions of the same equation is given by $y = ce^{0.1x^2}$. For purposes of comparison with Figure 2.2(a), some representative graphs of members of this family are shown in Figure 2.2(b).

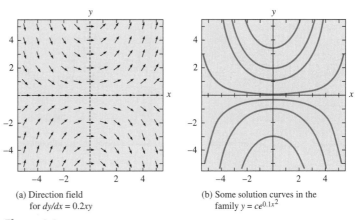

(a) Direction field
for $dy/dx = 0.2xy$

(b) Some solution curves in the
family $y = ce^{0.1x^2}$

Figure 2.2 ❑

Example 2 Direction Field

Use a direction field to sketch an approximate solution curve for the initial-value problem $dy/dx = \sin y$, $y(0) = -\frac{3}{2}$.

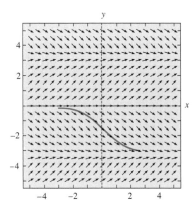

Figure 2.3 Direction field for $dy/dx = \sin y$

Autonomous DEs ☞
are very important
in applications.

SOLUTION Before proceeding, recall that from the continuity of $f(x, y) = \sin y$ and $\partial f/\partial y = \cos y$, Theorem 2.1 guarantees the existence of a unique solution curve passing through any specified point (x_0, y_0) in the plane. Now we set our computer software again for a 5×5 rectangular region and specify (because of the initial condition) points in that region with vertical and horizontal separation of $\frac{1}{2}$ unit, that is, at points (mh, nh), $h = \frac{1}{2}$, m and n integers such that $-10 \leq m \leq 10$, $-10 \leq n \leq 10$. The result is shown in Figure 2.3. Since the right-hand side of $dy/dx = \sin y$ is 0 at $y = 0$ and at $y = -\pi$, the lineal elements are horizontal at all points whose second coordinates are $y = 0$ or $y = -\pi$. It makes sense then that a solution curve passing through the initial point $(0, -\frac{3}{2})$ has the shape shown in color in the figure. ❑

■ **Increasing/Decreasing** Interpretation of the derivative dy/dx as a function that gives slope plays the key role in the construction of a direction field. Another telling property of the first derivative will be used next: namely, if $dy/dx > 0$ (or $dy/dx < 0$) for all x in an interval I, then a differentiable function $y = y(x)$ is increasing (or decreasing) on I.

■ **Autonomous First-Order DEs** In Section 1.1 we divided the class of ordinary differential equations into two types: linear and nonlinear. We now consider briefly another kind of classification of ordinary differential equations, a classification that is of particular importance in the qualitative investigation of differential equations. A differential equation in which the independent variable does not appear explicitly is said to be **autonomous.** If x is the independent variable, then an autonomous first-order differential equation can be written as $F(y, y') = 0$ or in normal form as

$$\frac{dy}{dx} = f(y) \tag{1}$$

We shall assume throughout that the function f in (1) and its derivative f' are continuous functions of y on some interval I. The first-order equations

$$\overset{f(y)}{\underset{\downarrow}{}} \qquad\qquad \overset{f(x,\, y)}{\underset{\downarrow}{}}$$

$$\frac{dy}{dx} = 1 + y^2 \qquad \text{and} \qquad \frac{dy}{dx} = 0.2xy$$

are autonomous and nonautonomous, respectively.

Many differential equations encountered in applications, or equations that are models of physical laws that do not change over time, are autonomous. As we have already seen in Section 1.3, in applied context, symbols other than y and x are routinely used to represent the dependent and independent variables. For example, if t represents time, then inspection of

$$\frac{dA}{dt} = kA, \quad \frac{dx}{dt} = kx(n + 1 - x), \quad \frac{dT}{dt} = k(T - T_m), \quad \frac{dA}{dt} = 6 - \frac{1}{100}A,$$

where k, n, and T_m are constants, shows that each equation is time-independent. Indeed, *all* of the first-order differential equations introduced in Section 1.3 are time-independent and so are autonomous.

■ **Critical Points** The zeros of the function f in (1) are of special importance. We say that a real number c is a **critical point** of the autonomous differential equation (1) if it is a zero of f, that is, $f(c) = 0$. A critical point is also called an **equilibrium point** or **stationary point.** Now observe that if we substitute $y(x) = c$ into (1), then both sides of the equation equal zero. This means:

If c is a critical point of (1), then y(x) = c is a constant solution of the autonomous equation.

A constant solution $y(x) = c$ of (1) is called an **equilibrium solution;** equilibria are the *only* constant solutions of (1).

As mentioned previously, we can tell when a nonconstant solution $y = y(x)$ of (1) is increasing or decreasing by determining the algebraic sign of the derivative dy/dx; this we do by identifying the intervals over which $f(y)$ is positive or negative.

Figure 2.4

> **Example 3** **An Autonomous Differential Equation**
>
> The differential equation
>
> $$\frac{dP}{dt} = P(a - bP), \qquad a > 0, \quad b > 0,$$
>
> has the form $dP/dt = f(P)$, which is (1) with t and P playing the parts of x and y, and hence is autonomous. From $f(P) = P(a - bP) = 0$, we see that 0 and a/b are critical points of the equation, and so the equilibrium solutions are $P(t) = 0$ and $P(t) = a/b$. By putting the critical points on a vertical line we divide the line into three intervals: $-\infty < P < 0$, $0 < P < a/b$, $a/b < P < \infty$. The arrows on the line shown in Figure 2.4 indicate the algebraic sign of $f(P) = P(a - bP)$ on these intervals and whether a nonconstant solution $P(t)$ is increasing or decreasing on an interval. The following table explains the figure.

Interval	Sign of $f(P)$	$P(t)$	Arrow
$(-\infty, 0)$	minus	decreasing	points down
$(0, a/b)$	plus	increasing	points up
$(a/b, \infty)$	minus	decreasing	points down

Figure 2.4 is called a **one-dimensional phase portrait,** or simply **phase portrait,** of the differential equation $dP/dt = P(a - bP)$. The vertical line is called a **phase line.**

■ **Solution Curves** Without solving an autonomous differential equation, we can usually say a great deal about its solution curves. Since the function f in (1) is independent of the variable x, we may consider f defined for $-\infty < x < \infty$ or for $0 \le x < \infty$. Also, since f and its derivative f' are continuous functions of x on some interval I, the fundamental results of Theorem 1.1 hold in some horizontal strip or region R in the xy-plane corresponding to I, and so through any point (x_0, y_0) in R there passes only one solution curve of (1). See Figure 2.5(a). For the sake of discussion let us suppose that (1) possesses exactly two critical points c_1 and c_2 and that $c_1 < c_2$. The graphs of the equilibrium solutions $y(x) = c_1$ and $y(x) = c_2$ are horizontal lines, and these lines partition the region R into three subregions R_1, R_2, and R_3 as illustrated in Figure 2.5(b). Without proof, here are some conclusions that we can draw about a nonconstant solution $y(x)$ of (1):

(a) Region R

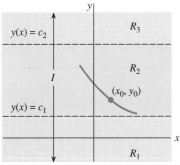

(b) Subregions R_1, R_2, and R_3

Figure 2.5

- If (x_0, y_0) is in a subregion R_i, $i = 1, 2, 3$, and $y(x)$ is a solution whose graph passes through this point, then $y(x)$ remains in the subregion for all x. As illustrated in Figure 2.5(b), the solution $y(x)$ in R_2 is bounded below by c_1 and above by c_2; that is, $c_1 < y(x) < c_2$ for all x. The solution curve stays within R_2 for all x because the graph of a nonconstant solution of (1) cannot cross the graph of either equilibrium solution $y(x) = c_1$ or $y(x) = c_2$. See Problem 31 in Exercises 2.1.

- By continuity of f we must then have either $f(y) > 0$ or $f(y) < 0$ for all x in a subregion R_i, $i = 1, 2, 3$. In other words, $f(y)$ cannot change signs in a subregion. See Problem 31 in Exercises 2.1.

- Since $dy/dx = f(y(x))$ is either positive or negative in a subregion, a solution $y(x)$ is strictly monotonic—that is, $y(x)$ is either increasing or decreasing in a subregion R_i, $i = 1, 2, 3$. Therefore $y(x)$ cannot be oscillatory, nor can it have a relative extremum (maximum or minimum). See Problem 31 in Exercises 2.1.

- If $y(x)$ is bounded above by a critical point (as in subregion R_1) or is bounded below by a critical point (as in subregion R_3), then $y(x)$ must approach this point either as $x \to \infty$ or as $x \to -\infty$. If $y(x)$ is bounded, that is, bounded above and below by two consecutive critical points (as in subregion R_2), then $y(x)$ must approach both critical points, one as $x \to \infty$ and the other as $x \to -\infty$. See Problem 32 in Exercises 2.1.

With the foregoing facts in mind, let us reexamine the differential equation in Example 3.

Example 4 **Example 3 Revisited**

The three intervals determined on the P-axis or phase line by the critical points 0 and a/b now correspond in the xy-plane to three subregions:

$$R_1: -\infty < P < 0, \quad R_2: 0 < P < a/b, \quad R_3: a/b < P < \infty,$$

where $-\infty < x < \infty$. The phase portrait in Figure 2.4 tells us that $P(t)$ is decreasing in R_1, increasing in R_2, and decreasing in R_3. If $P(0) = P_0$ is an initial value, then in R_1, R_2, and R_3, we have, respectively:

(*i*) For $P_0 < 0$, $P(t)$ is bounded above. Since $P(t)$ is decreasing, $P(t)$ decreases without bound for increasing t and $P(t) \to 0$ as $t \to -\infty$. This means the negative t-axis, $P = 0$, is a horizontal asymptote for a solution curve.

(*ii*) For $0 < P_0 < a/b$, $P(t)$ is bounded. Since $P(t)$ is increasing, $P(t) \to a/b$ as $t \to \infty$ and $P(t) \to 0$ as $t \to -\infty$. The two lines $P = 0$ and $P = a/b$ are horizontal asymptotes for any solution curve starting in this subregion.

(*iii*) For $P_0 > a/b$, $P(t)$ is bounded below. Since $P(t)$ is decreasing, $P(t) \to a/b$ as $t \to \infty$. This means $P = a/b$ is a horizontal asymptote for a solution curve.

In Figure 2.6 the original phase portrait is reproduced to the left of the tP-plane in which the subregions R_1, R_2, and R_3 are shaded. The graphs of the equilibrium solutions $P = a/b$ and $P = 0$ (the t-axis) are shown in the figure as colored dashed lines; the solid graphs represent typical graphs of $P(t)$ illustrating the three cases just discussed.

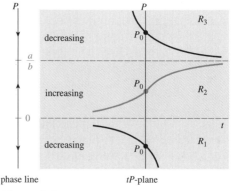

phase line tP-plane

Figure 2.6 ❏

In a subregion such as R_1 in Example 4, where $P(t)$ is decreasing and unbounded below, we must necessarily have $P(t) \to -\infty$. Do *not* interpret this last statement to mean $P(t) \to -\infty$ as $t \to \infty$; we could have $P(t) \to -\infty$ as $t \to T$, where $T > 0$ is a finite number that depends on the initial condition $P(t_0) = P_0$. Thinking in dynamic terms, $x(t)$ could "blow up" in finite time; thinking graphically, $P(t)$ could have a vertical asymptote at $t = T > 0$. A similar remark holds for the subregion R_3.

The differential equation $dy/dx = \sin y$ in Example 2 is autonomous and has an infinite number of critical points, since $\sin y = 0$ at $y = n\pi$, n an integer. Moreover, we now know that because the solution $y(x)$ that passes through $(0, -\frac{3}{2})$ is bounded above and below by two consecutive critical points $(-\pi < y(x) < 0)$ and is decreasing $(\sin y < 0$ for $-\pi < y < 0)$, $y(x)$ must approach the horizontal asymptotes $y = -\pi$ and $y = 0$ as $x \to \infty$ and as $x \to -\infty$, respectively.

Example 5 Solution Curves of an Autonomous DE

The autonomous equation $dy/dx = (y - 1)^2$ possesses the single critical point 1. From the phase portrait in Figure 2.7(a) we conclude that a solution $y(x)$ is an increasing function in the subregions $-\infty < y < 1$ and $1 < y < \infty$, where $-\infty < x < \infty$. For an initial condition $y(0) = y_0 < 1$ a solution $y(x)$ is increasing and bounded above by 1, and so $y(x) \to 1$ as $x \to \infty$; for $y(0) = y_0 > 1$ a solution $y(x)$ is increasing and unbounded.

Now $y(x) = 1 - 1/(x + c)$ is a one-parameter family of solutions of the differential equation. (See Problem 4 in Exercises 2.2.) A given initial condition determines a value for c. If we choose, say, $y(0) = -1 < 1$, we find $c = -\frac{1}{2}$, and so $y(x) = 1 - 1/(x + \frac{1}{2})$. Observe $x = -\frac{1}{2}$ is a vertical asymptote and $y(x) \to -\infty$ as $x \to -\frac{1}{2}$ from the right. See Figure 2.7(b). For an initial condition, say this time, $y(0) = 2 > 1$, we find $c = -1$ and $y(x) = 1 - 1/(x - 1)$. The last function has a vertical asymptote at $x = 1$, and thus $y(x) \to \infty$ as $x \to 1$ from the left. See Figure 2.7(c).

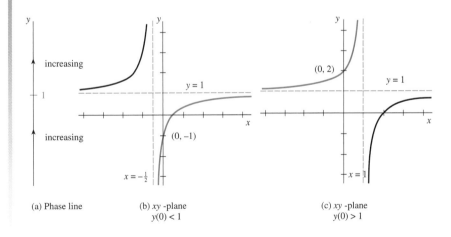

(a) Phase line

(b) *xy*-plane
$y(0) < 1$

(c) *xy*-plane
$y(0) > 1$

Figure 2.7

■ **Attractors and Repellers** There are basically three types of behavior of a solution of (1) near a critical point. When both arrowheads on either side of c point toward c, as in Figure 2.8(a), all solutions $y(x)$ of (1) that start from an initial point (x_0, y_0) sufficiently near c exhibit the asymptotic behavior $\lim_{x \to \infty} y(x) = c$. For this reason c is said to be **asymptotically stable.** Using a physical analogy, a solution that starts near c is like a charged particle that, over time, is drawn to the particle c

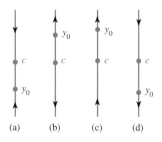

Figure 2.8

of opposite charge, and so c is also referred to as an **attractor.** The arrowheads in Figure 2.8(b) pointing away from c indicate that all solutions that start near c move away from c as x increases. In this case c is said to be an **unstable** critical point. An unstable critical point for obvious reasons is also called a **repeller.** The critical point c illustrated in Figures 2.8(c) and 2.8(d) is neither an attractor nor a repeller. But since c exhibits characteristics of both an attractor and a repeller—that is, a solution starting from an initial point (x_0, y_0) sufficiently near c is attracted to c from one side and repelled from the other side—we say that c is **semistable.** In Example 3 the critical points a/b and 0 are asymptotically stable (an attractor) and unstable (a repeller), respectively. The critical point 1 in Example 5 is semistable.

E X E R C I S E S 2 . 1

Answers to odd-numbered problems begin on page A-26.

In Problems 1–4, use the given computer-generated direction field to sketch, by hand, an approximate solution curve that passes through the indicated points.

1. $dy/dx = x/y$

(a) $y(0) = 5$

(b) $y(3) = 3$

(c) $y(4) = 2$

(d) $y(-5) = -3$

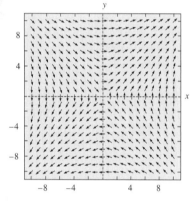

Figure 2.9

3. $dy/dx = 1 - xy$

(a) $y(0) = 0$

(b) $y(-1) = 0$

(c) $y(2) = 2$

(d) $y(0) = -4$

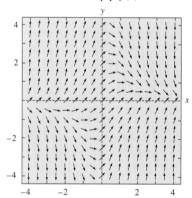

Figure 2.11

2. $dy/dx = e^{-0.01xy^2}$

(a) $y(-6) = 0$

(b) $y(0) = 1$

(c) $y(0) = -4$

(d) $y(8) = -4$

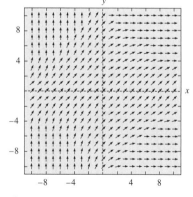

Figure 2.10

4. $dy/dx = (\sin x) \cos y$

(a) $y(0) = 1$

(b) $y(1) = 0$

(c) $y(3) = 3$

(d) $y(0) = -\frac{5}{2}$

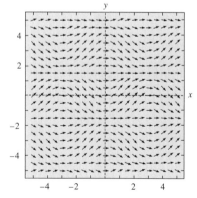

Figure 2.12

Computer Lab Assignment

In Problems 5–12, use computer software to obtain a direction field for the given differential equation. By hand, sketch an approximate solution curve passing through each of the given points.

5. $y' = x$

(a) $y(0) = 0$ (b) $y(0) = -3$

6. $y' = x + y$

(a) $y(-2) = 2$ (b) $y(1) = -3$

7. $y\dfrac{dy}{dx} = -x$

(a) $y(1) = 1$ (b) $y(0) = 4$

8. $\dfrac{dy}{dx} = \dfrac{1}{y}$

(a) $y(0) = 1$ (b) $y(-2) = -1$

9. $\dfrac{dy}{dx} = 0.2x^2 + y$

(a) $y(0) = \dfrac{1}{2}$ (b) $y(2) = -1$

10. $\dfrac{dy}{dx} = xe^y$

(a) $y(0) = -2$ (b) $y(1) = 2.5$

11. $y' = y - \cos\dfrac{\pi}{2}x$

(a) $y(2) = 2$ (b) $y(-1) = 0$

12. $\dfrac{dy}{dx} = 1 - \dfrac{y}{x}$

(a) $y\left(-\dfrac{1}{2}\right) = 2$ (b) $y\left(\dfrac{3}{2}\right) = 0$

In Problems 13 and 14, the given figures represent the graphs of $f(y)$ and $f(x)$, respectively. By hand, sketch a direction field over an appropriate grid for $dy/dx = f(y)$ (Problem 13) and then for $dy/dx = f(x)$ (Problem 14).

13.

14.

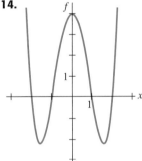

Figure 2.13 Figure 2.14

15. Consider the autonomous first-order differential equation $dy/dx = y - y^3$ and the initial condition $y(0) = y_0$. By hand, sketch the graph of a typical solution $y(x)$ when y_0 has the given values.

(a) $y_0 > 1$ (b) $0 < y_0 < 1$ (c) $-1 < y_0 < 0$ (d) $y_0 < -1$

16. Consider the autonomous first-order differential equation $dy/dx = y^2 - y^4$ and the initial condition $y(0) = y_0$. By hand, sketch a graph of a typical solution $y(x)$ when y_0 has the given values.

(a) $y_0 > 1$ (b) $0 < y_0 < 1$ (c) $-1 < y_0 < 0$ (d) $y_0 < -1$

In Problems 17–24, find the critical points and phase portrait of the given autonomous first-order differential equation. Classify each critical point as asymptotically stable, unstable, or semi-

stable. By hand, sketch typical solution curves in the regions in the xy-plane determined by the graphs of the equilibrium solutions.

17. $\dfrac{dy}{dx} = y^2 - 3y$

18. $\dfrac{dy}{dx} = y^2 - y^3$

19. $\dfrac{dy}{dx} = -(y - 2)^2$

20. $\dfrac{dy}{dx} = 10 + 3y - y^2$

21. $\dfrac{dy}{dx} = y^2(4 - y^2)$

22. $\dfrac{dy}{dx} = y(2 - y)(4 - y)$

23. $\dfrac{dy}{dx} = y\ln(y + 2)$

24. $\dfrac{dy}{dx} = \dfrac{ye^y - 9y}{e^y}$

25. The differential equation

$$m\frac{dv}{dt} = mg - kv,$$

where k is a positive constant of proportionality and g is the acceleration due to gravity, governs the velocity v of a body of mass m falling under the influence of gravity. Because the term $-kv$ represents air resistance, the velocity of a body falling from a great height does not increase without bound as time t increases.

(a) Use the fact that the equation is autonomous to find the limiting or terminal velocity of the body. Explain your reasoning.

(b) Find the terminal velocity of the body if air resistance is proportional to v^2. See pages 26 and 29.

26. When certain kinds of chemicals are combined, the rate at which a new compound is formed is governed by the differential equation

$$\frac{dX}{dt} = k(\alpha - X)(\beta - X),$$

where $k > 0$ is a constant of proportionality and $\beta > \alpha > 0$. Here $X(t)$ denotes the number of grams of the new compound formed in time t. See pages 23–24.

(a) Use the fact that the equation is autonomous to describe the behavior of X as $t \to \infty$.

(b) Consider the case when $\alpha = \beta$. What is the behavior of X as $t \to \infty$ if $X(0) < \alpha$? From the phase portrait of the differential equation, can you predict the behavior of X as $t \to \infty$ if $X(0) > \alpha$?

(c) Verify that an explicit solution of the differential equation in the case when $k = 1$ and $\alpha = \beta$ is $X(t) = \alpha - 1/(t + c)$. Find a solution satisfying $X(0) = \alpha/2$. Find a solution satisfying $X(0) = 2\alpha$. Graph these two solutions. Does the behavior of the solutions as $t \to \infty$ agree with your answers to part (b).

In Problems 27 and 28, consider the autonomous differential equation $dy/dx = f(y)$ where the graph of f is given. Use the graph to locate the critical points of each differential equation. Sketch a phase portrait of each differential equation. By hand,

sketch typical solution curves in the subregions in the xy-plane determined by the graphs of the equilibrium solutions.

27.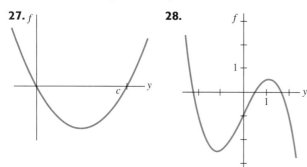

28.

Figure 2.15 **Figure 2.16**

29. Consider the differential equation $dy/dx = x^2 + y^2$. What can you say about the lineal elements at the points (x, y) on the graph of the circles $x^2 + y^2 = c^2$, $c > 0$?

30. For the first-order differential equation $dy/dx = f(x, y)$, a curve in the plane defined by $f(x, y) = 0$ is called a **nullcline,** since a lineal element at a point on the curve has zero slope. Use computer software to obtain a direction field over a rectangular grid of points for $dy/dx = x^2 - 2y$ and then to superimpose the graph of the nullcline $y = \frac{1}{2}x^2$ over the direction field. Describe the behavior of solution curves for $y < \frac{1}{2}x^2$ and for $y > \frac{1}{2}x^2$. Sketch some approximate solution curves. Try to generalize your observations.

31. Suppose that $y(x)$ is a nonconstant solution of the autonomous differential equation (1). Discuss: If c is a critical point of (1), why can't the graph of $y(x)$ cross the graph of the equilibrium solution $y = c$? Why can't $f(y)$ change signs in one of the subregions discussed on pages 39–40? What can't $y(x)$ be oscillatory or have a relative extremum (maximum or minimum)?

32. Suppose that $y(x)$ is a solution of the autonomous equation $dy/dx = f(y)$ and is bounded above and below by two consecutive critical points $c_1 < c_2$ as in region R_2 of Figure 2.5. If $f(y) > 0$ in the region, then $\lim_{x \to \infty} y(x) = c_2$. Why can't it happen that $\lim_{x \to \infty} y(x) = L$, where $L < c_2$? As part of your explanation, consider what happens to $y'(x)$ as $x \to \infty$.

33. (a) Using the autonomous equation (1), explain how it is possible to obtain information about the location of **points of inflection** of a solution curve.

(b) Consider the differential equation $dy/dx = y^2 - y - 6$. Use your ideas from part **(a)** to find intervals on the y-axis for which solution curves are concave up and intervals for which solution curves are concave down. Why does *each* solution curve of an initial-value problem of the form $dy/dx = y^2 - y - 6$, $y(0) = y_0$, where $-2 < y_0 < 3$, have a point of inflection with the same y-coordinate? What is that y-coordinate? By hand, carefully sketch the solution curve for which $y(0) = -1$. Repeat for $y(2) = 2$.

2.2 Separable Variables

It may be worthwhile to review techniques of integration.

■ **Solution by Integration** We begin our study of how to solve first-order equations $dy/dx = f(x, y)$ with the simplest of all differential equations. When f does not depend on the variable y, that is, $f(x, y) = g(x)$, the differential equation

$$\frac{dy}{dx} = g(x) \qquad (1)$$

can be solved by integration. If $g(x)$ is a continuous function, then integrating both sides of (1) gives the solution $y = \int g(x)\, dx = G(x) + c$, where $G(x)$ is an antiderivative (indefinite integral) of $g(x)$. For example, if $dy/dx = 1 + e^{2x}$, then $y = \int (1 + e^{2x}) dx = \frac{1}{2} e^{2x} + x + c$.

Equation (1), as well as its method of solution, is just a special case when f in $dy/dx = f(x, y)$ is a product of a function of x and a function of y.

> **DEFINITION 2.1 Separable Equation**
>
> A first-order differential equation of the form
>
> $$\frac{dy}{dx} = g(x)h(y)$$
>
> is said to be **separable** or to have **separable variables.**

Observe that by dividing by the function $h(y)$ a separable equation can be written as

$$p(y)\frac{dy}{dx} = g(x), \tag{2}$$

where, of convenience, we have denoted $1/h(y)$ by $p(y)$. From this last form we can see immediately that (2) reduces to (1) when $h(y) = 1$.

Now if $y = \phi(x)$ represents a solution of (2), we must have $p(\phi(x))\phi'(x) = g(x)$, and therefore,

$$\int p(\phi(x))\phi'(x)\,dx = \int g(x)\,dx. \tag{3}$$

But $dy = \phi'(x)\,dx$, and so (3) is the same as

$$\int p(y)\,dy = \int g(x)\,dx \qquad \text{or} \qquad H(y) = G(x) + c, \tag{4}$$

where $H(y)$ and $G(x)$ are antiderivatives of $p(y) = 1/h(y)$ and $g(x)$, respectively.

■ **Method of Solution** Equation (4) indicates the procedure for solving separable equations. A one-parameter family of solutions, usually given implicitly, is obtained by integrating both sides of $p(y)dy = g(x)\,dx$.

In solving first-order DEs use ☞ only one constant.

There is no need to use two constants in the integration of a separable equation, because if we write $H(y) + c_1 = G(x) + c_2$, then the difference $c_2 - c_1$ can be replaced by a single constant c, as in (4). In many instances throughout the chapters that follow, we will relabel constants in a manner convenient to a given equation. For example, multiples of constants or combinations of constants can sometimes by replaced by a single constant.

> ### *Example 1* Solving a Separable DE
> Solve $(1 + x)\,dy - y\,dx = 0$.
>
> **SOLUTION** Dividing by $(1 + x)y$, we can write $dy/y = dx/(1 + x)$, from which it follows that
>
> $$\int \frac{dy}{y} = \int \frac{dx}{1 + x}$$
>
> $$\ln|y| = \ln|1 + x| + c_1$$
>
> $$y = e^{\ln|1 + x| + c_1} = e^{\ln|1 + x|} \cdot e^{c_1} \qquad \leftarrow \text{laws of exponents}$$
>
> $$= |1 + x|e^{c_1}$$
>
> $$= \pm e^{c_1}(1 + x). \qquad \leftarrow \begin{cases} |1 + x| = 1 + x, & x \geq -1 \\ |1 + x| = -(1 + x), & x < -1 \end{cases}$$
>
> Relabeling $\pm e^{c_1}$ by c then gives $y = c(1 + x)$.
>
> **ALTERNATIVE SOLUTION** Since each integral results in a logarithm, a judicious choice for the constant of integration is $\ln|c|$ rather than c. Rewriting the second line of the solution as $\ln|y| = \ln|1 + x| + \ln|c|$ enables us to combine the terms on the right-hand side by the properties of logarithms. From $\ln|y| = \ln|c(1 + x)|$ we immediately get $y = c(1 + x)$. Even if the indefinite integrals are not *all* logarithms, it may still be advantageous to use $\ln|c|$. However, no firm rule can be given. ❏

In Section 1.1 we have already seen that a solution curve may be only a segment or an arc of the graph of an implicit solution $G(x, y) = 0$.

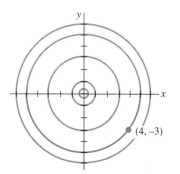

Figure 2.17

Example 2 Solution Curve

Solve the initial-value problem $\dfrac{dy}{dx} = -\dfrac{x}{y}$, $y(4) = -3$.

SOLUTION By rewriting the equation as $y\,dy = -x\,dx$ we get

$$\int y\,dy = -\int x\,dx \qquad \text{and} \qquad \frac{y^2}{2} = -\frac{x^2}{2} + c_1.$$

We can write the result of the integration as $x^2 + y^2 = c^2$ by replacing the constant $2c_1$ by c^2. This solution of the differential equation represents a family of concentric circles centered at the origin.

Now when $x = 4$, $y = -3$, so that $16 + 9 = 25 = c^2$. Thus the initial-value problem determines the circle $x^2 + y^2 = 25$ with radius 5. Because of its simplicity we can solve this implicit solution for an explicit solution that satisfies the initial condition. We have seen this solution as $y = \phi_2(x)$ or $y = -\sqrt{25 - x^2}$, $-5 < x < 5$ in Example 3 of Section 1.1. A solution curve is the graph of a differentiable *function*. In this case the solution curve is the lower semicircle, shown in color in Figure 2.17, that contains the point $(4, -3)$. ❏

■ **Losing a Solution** Some care should be exercised when separating variables, since the variable divisors could be zero at a point. Specifically, if r is a zero of the function $h(y)$, then substituting $y = r$ into $dy/dx = g(x)h(y)$ makes both sides zero; in other words, $y = r$ is a constant solution of the differential equation. But after separating variables, observe that the left side of $dy/h(y) = g(x)dx$ is undefined at r. As a consequence, $y = r$ may not show up in the family of solutions obtained after integration and simplification. Recall, such a solution is called a singular solution.

Example 3 Losing a Solution

Solve $dy/dx = y^2 - 4$.

SOLUTION We put the equation in the form

$$\frac{dy}{y^2 - 4} = dx \qquad \text{or} \qquad \left[\frac{1/4}{y-2} - \frac{1/4}{y+2}\right] dy = dx. \tag{5}$$

The second equation in (5) is the result of using partial fractions on the left side of the first equation. Integrating and using the laws of logarithms gives

$$\frac{1}{4}\ln|y - 2| - \frac{1}{4}|y + 2| = x + c_1 \quad \text{or} \quad \ln\left|\frac{y-2}{y+2}\right| = 4x + c_2 \quad \text{or} \quad \frac{y-2}{y+2} = e^{4x + c_2}.$$

Here we have replaced $4c_1$ by c_2. Finally, after replacing e^{c_2} by c and solving the last equation for y, we get the one-parameter family of solutions

$$y = 2\frac{1 + ce^{4x}}{1 - ce^{4x}} \tag{6}$$

Now if we factor the right side of the differential equation as $dy/dx = (y - 2)(y + 2)$ we know from the discussion in Section 2.1 that $y = 2$ and $y = -2$ are two constant (equilibrium) solutions. The solution $y = 2$ is a member of the family of solutions defined by (6) corresponding to the value $c = 0$. However, $y = -2$ is a

singular solution; it cannot be obtained from (6) for any choice of the parameter c. This latter solution was lost early on in the solution process. Inspection of (5) clearly indicates that we must preclude $y = \pm 2$ in these steps. ❑

Example 4 An Initial-Value Problem

Solve the initial-value problem

$$\cos x(e^{2y} - y)\frac{dy}{dx} = e^y \sin 2x, \qquad y(0) = 0.$$

SOLUTION Dividing by the equation by $e^y \cos x$ gives

$$\frac{e^{2y} - y}{e^y}dy = \frac{\sin 2x}{\cos x}dx.$$

Before integrating we use termwise division on the left side and the trigonometric identity $\sin 2x = 2 \sin x \cos x$ on the right side. Then

integration by parts → $$\int (e^y - ye^{-y}) \, dy = 2 \int \sin x \, dx$$

yields $$e^y + ye^{-y} + e^{-y} = -2 \cos x + c. \qquad (7)$$

The initial condition $y = 0$ when $x = 0$ implies $c = 4$. Thus a solution of the initial-value problem is

$$e^y + ye^{-y} + e^{-y} = 4 - 2 \cos x. \qquad (8) \ ❑$$

■ **Use of Computers** In the Remarks at the end of Section 1.1 we mentioned that it may be difficult to use an implicit solution $G(x, y) = 0$ to find an explicit solution $y = \phi(x)$. Equation (8) shows that the task of solving for y in terms of x may present more problems than just the drudgery of symbol pushing—it simply can't be done! Implicit solutions such as (8) are somewhat frustrating; neither the graph of the equation nor an interval over which a solution satisfying $y(0) = 0$ is defined is apparent. The problem of "seeing" what an implicit solution looks like can be overcome in some cases by means of technology. One way* of proceeding is to use the contour plot application of a CAS. Recall from multivariate calculus that for a function of two variables $z = G(x, y)$ the *two-dimensional* curves defined by $G(x, y) = c$, where c is constant, are called the *level curves* of the function. With the aid of a CAS we have illustrated in Figure 2.18 some of the level curves of the function $G(x, y) = e^y + ye^{-y} + e^{-y} + 2 \cos x$. The family of solutions defined by (7) are the level curves $G(x, y) = c$. Figure 2.19 illustrates, in color, the level curve $G(x, y) = 4$, which is the particular solution (8). The other curve in Figure 2.19 is the level curve $G(x, y) = 2$, which is the member of the family $G(x, y) = c$ that satisfies $y(\pi/2) = 0$.

If an initial condition leads to a particular solution by finding a specific value of the parameter c in a family of solutions for a first-order differential equation, it is a natural inclination for most students (and instructors) to relax and be content. However, a solution of an initial-value problem may not be unique. We saw in Example 3 of Section 1.2 that the initial-value problem

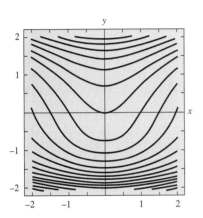

Figure 2.18 Level curves $G(x, y) = c$, where $G(x, y) = e^y + ye^{-y} + e^{-y} + 2 \cos x$

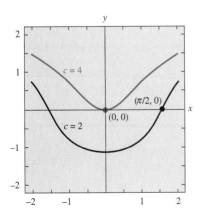

Figure 2.19 Level curves $c = 2$ and $c = 4$

* In Section 2.6 we discuss several other ways of proceeding that are based on the concept of a numerical solver.

$$\frac{dy}{dx} = xy^{1/2}, \qquad y(0) = 0, \tag{9}$$

has at least two solutions, $y = 0$ and $y = x^4/16$. We are now in a position to solve the equation. Separating variables and integrating $y^{-1/2}dy = x\,dx$ give

$$2y^{1/2} = \frac{x^2}{2} + c_1 \qquad \text{or} \qquad y = \left(\frac{x^2}{4} + c\right)^2.$$

When $x = 0$, then $y = 0$, and so necessarily $c = 0$. Therefore $y = x^4/16$. The trivial solution $y = 0$ was lost by dividing by $y^{1/2}$. In addition, the initial-value problem (9) possesses infinitely many more solutions, since for any choice of the parameter $a \geq 0$ the piecewise defined function

$$y = \begin{cases} 0, & x < a \\ (x^2 - a^2)^2/16, & x \geq a \end{cases}$$

satisfies both the differential equation and initial condition. See Figure 2.20.

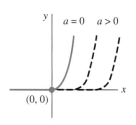

Figure 2.20

Remarks

(*i*) If g is a function continuous on an interval I containing x_0, then from the fundamental theorem of calculus we have

$$\frac{d}{dx}\int_{x_0}^{x} g(t)\,dt = g(x).$$

In other words, $\int_{x_0}^{x} g(t)\,dt$ is an antiderivative of the function g. There are times when this form is convenient. For example, if g is continuous on an interval I containing x_0, then a solution of the simple initial-value problem $dy/dx = g(x)$, $y(x_0) = y_0$, that is defined on I is given by

$$y(x) = y_0 + \int_{x_0}^{x} g(t)\,dt.$$

You should verify this. Since an antiderivative of a continuous function cannot always be expressed in terms of elementary functions, this may be the best we can do in obtaining an explicit solution of an IVP.

(*ii*) In some of the preceding examples we saw that the constant in the one-parameter family of solutions for a first-order differential equation can be relabeled when convenient. Also, it can easily happen that two individuals solving the same equation correctly arrive at dissimilar expressions for their answers. For example, by separation of variables we can show that one-parameter families of solutions for the DE $(1 + y^2)dx + (1 + x^2)dy = 0$ are

$$\arctan x + \arctan y = c \qquad \text{or} \qquad \frac{x + y}{1 - xy} = c.$$

As you work your way through the next several sections, keep in mind that families of solutions may be equivalent in the sense that one family may be obtained from another by either relabeling the constant or applying algebra and trigonometry. See Problems 27 and 28 in Exercises 2.2.

In Problems 1–22, solve the given differential equation by separation of variables.

1. $\dfrac{dy}{dx} = \sin 5x$

2. $\dfrac{dy}{dx} = (x + 1)^2$

3. $dx + e^{3x}\, dy = 0$

4. $dy - (y - 1)^2\, dx = 0$

5. $x\dfrac{dy}{dx} = 4y$

6. $\dfrac{dy}{dx} + 2xy = 0$

7. $\dfrac{dy}{dx} = e^{3x + 2y}$

8. $e^x y\dfrac{dy}{dx} = e^{-y} + e^{-2x - y}$

9. $y \ln x\, \dfrac{dx}{dy} = \left(\dfrac{y + 1}{x}\right)^2$

10. $\dfrac{dy}{dx} = \left(\dfrac{2y + 3}{4x + 5}\right)^2$

11. $\sec^2 x\, dy + \csc y\, dx = 0$

12. $\sin 3x\, dx + 2y \cos^3 3x\, dy = 0$

13. $(e^y + 1)^2 e^{-y}\, dx + (e^x + 1)^3 e^{-x}\, dy = 0$

14. $y\,dy = x(1 + x^2)^{-1/2}(1 + y^2)^{1/2}dx$

15. $\dfrac{dS}{dr} = kS$

16. $\dfrac{dQ}{dt} = k(Q - 70)$

17. $\dfrac{dP}{dt} = P - P^2$

18. $\dfrac{dN}{dt} + N = Nte^{t + 2}$

19. $\dfrac{dy}{dx} = \dfrac{xy + 3x - y - 3}{xy - 2x + 4y - 8}$

20. $\dfrac{dy}{dx} = \dfrac{xy + 2y - x - 2}{xy - 3y + x - 3}$

21. $\dfrac{dy}{dx} = x\sqrt{1 - y^2}$

22. $(e^x + e^{-x})\dfrac{dy}{dx} = y^2$

In Problems 23–26, solve the given initial-value problem.

23. $\dfrac{dx}{dt} = 4(x^2 + 1),\ x(\pi/4) = 1$ **24.** $\dfrac{dy}{dx} = \dfrac{y^2 - 1}{x^2 - 1},\ y(2) = 2$

25. $x^2\dfrac{dy}{dx} = y - xy,\ y(-1) = -1$ **26.** $\dfrac{dy}{dt} + 2y = 1,\ y(0) = \dfrac{5}{2}$

In Problems 27 and 28, find an implicit and an explicit solution of the given initial-value problem.

27. $\sqrt{1 - y^2}\, dx - \sqrt{1 - x^2}\, dy = 0,\quad y(0) = \dfrac{\sqrt{3}}{2}$

28. $(1 + x^4)\, dy + x(1 + 4y^2)\, dx = 0,\quad y(1) = 0$

29. (a) Find a solution of the initial-value problem consisting of the differential equation in Example 3 and the initial conditions: $y(0) = 2;\ y(0) = -2;\ y(\tfrac{1}{4}) = 1$.

(b) Find the solution of the differential equation in Example 3 when $\ln c_1$ is used as the constant of integration on the *left* side in the solution and $4 \ln c_1$ is replaced by $\ln c$. Then solve the same initial-value problems in part **(a).**

30. Find a solution of $x\dfrac{dy}{dx} = y^2 - y$ that passes through the indicated points.

(a) $(0, 1)$ **(b)** $(0, 0)$ **(c)** $(\tfrac{1}{2}, \tfrac{1}{2})$ **(d)** $(2, \tfrac{1}{4})$

31. Find a singular solution of Problem 21; of Problem 22.

32. Show that an implicit solution of

$$2x \sin^2 y\, dx - (x^2 + 10) \cos y\, dy = 0$$

is given by $\ln(x^2 + 10) + \csc y = c$. Find the constant solutions, if any, that were lost in the solution of the differential equation.

Often a radical change in the form of the solution of a differential equation corresponds to a very small change in either the initial condition or the equation itself. In Problems 33–36 find an explicit solution of the given initial-value problem. Use a graphing utility to plot the graph of each solution. Compare each solution curve in a neighborhood of $(0, 1)$.

33. $\dfrac{dy}{dx} = (y - 1)^2,\ y(0) = 1$

34. $\dfrac{dy}{dx} = (y - 1)^2,\ y(0) = 1.01$

35. $\dfrac{dy}{dx} = (y - 1)^2 + 0.01,\ y(0) = 1$

36. $\dfrac{dy}{dx} = (y - 1)^2 - 0.01,\ y(0) = 1$

37. Every autonomous first-order equation $dy/dx = f(y)$ is separable. Find explicit solutions $y_1(x),\ y_2(x),\ y_3(x),$ and $y_4(x)$ of the differential equation $dy/dx = y - y^3$ that satisfy, in turn, the initial conditions $y_1(0) = 2,\ y_2(0) = \tfrac{1}{2},$ $y_3(0) = -\tfrac{1}{2},$ and $y_4(0) = -2$. Use a graphing utility to plot the graphs of each solution. Compare these graphs with those predicted in of Problem 15 in Exercises 2.1. Give the exact interval of definition for each solution.

38. (a) The autonomous differential equation $dy/dx = 1/(y - 3)$ has no critical points. Nevertheless, place 3 on a phase line and obtain a phase portrait of the equation. Compute d^2y/dx^2 to determine where solution curves are concave up and concave down (see Problem 33 in Exercises 2.1). Use the phase portrait and concavity to sketch, by hand, some typical solution curves.

(b) Find explicit solutions $y_1(x),\ y_2(x),\ y_3(x),$ and $y_4(x)$ of the differential equation in part **(a)** that satisfy, in turn, the initial conditions $y_1(0) = 4,\ y_2(0) = 2,\ y_3(1) = 2,$ and $y_4(-1) = 4$. Graph each solution and compare with your sketches in part **(a).** Give the exact interval of definition for each solution.

39. (a) Find an explicit solution of the initial-value problem

$$\frac{dy}{dx} = \frac{2x+1}{2y}, \quad y(-2) = -1.$$

(b) Use a graphing utility to plot the graph of the solution in part **(a)**. Use the graph to estimate the interval I of definition of the solution.

(c) Determine the exact interval I of definition by analytical methods.

40. Repeat parts **(a)–(c)** of Problem 39 for the initial-value problem consisting of the differential equation in Problem 7 and the condition $y(0) = 0$.

41. (a) In Example 2 explain why the interval of definition of the explicit solution $y_2(x)$ of the initial-value problem in Example 2 is the *open* interval $-5 < x < 5$.

(b) Can any solution of the differential equation cross the *x*-axis? Do you think that $x^2 + y^2 = 1$ is an implicit solution of the initial-value problem $dy/dx = -x/y$, $y(1) = 0$?

42. (a) If $a > 0$, discuss the differences, if any, between the solutions of the initial-value problems consisting of the differential equation $dy/dx = x/y$ and each of the initial conditions: $y(a) = a$; $y(a) = -a$; $y(-a) = a$; $y(-a) = -a$.

(b) Does the initial-value problem $dy/dx = x/y$, $y(0) = 0$, have a solution?

(c) Solve $dy/dx = x/y$, $y(1) = 2$, and give the exact interval I of definition of its solution.

43. In Problems 37 and 38 we saw that every autonomous first-order differential equation $dy/dx = f(y)$ is separable. Does this fact help in the solution of the initial-value problem
$$\frac{dy}{dx} = \sqrt{1 + y^2}\, \sin^2 y, \quad y(0) = \tfrac{1}{2}?$$ Sketch, by hand, a plausible solution curve of the problem.

44. Without the use of technology, how would you solve

$$(\sqrt{x} + x)\frac{dy}{dx} = (\sqrt{y} + y)?$$

Carry out your ideas.

45. Find a function whose square plus the square of its derivative is 1.

46. (a) The differential equation in Problem 27 is equivalent to the normal form

$$\frac{dy}{dx} = \sqrt{\frac{1 - y^2}{1 - x^2}}$$

in the square region in the *xy*-plane defined by $|x| < 1$, $|y| < 1$.

But the quantity under the radical is nonnegative also in the regions defined by $|x| > 1$, $|y| > 1$. Sketch all regions in the *xy*-plane for which this differential equation possesses real solutions.

(b) Solve the DE in part **(a)** in the regions defined by $|x| > 1$, $|y| > 1$. Then find an implicit and an explicit solution of the differential equation subject to $y(2) = 2$.

47. Computer Lab Assignment (a) Use a CAS and the concept of level curves to plot representative graphs of members of the family of solutions of the differential equation $\frac{dy}{dx} = -\frac{8x + 5}{3y^2 + 1}$. Experiment with different numbers of level curves as well as various rectangular regions defined by $a \le x \le b$, $c \le y \le d$.

(b) On separate coordinate axes plot the graphs of the particular solutions corresponding to the initial conditions: $y(0) = -1$; $y(0) = 2$; $y(-1) = 4$; $y(-1) = -3$.

48. Computer Lab Assignment (a) Use a CAS and the concept of level curves to plot representative graphs of members of the family of solutions of the differential equation $\frac{dy}{dx} = \frac{x(1 - x)}{y(-2 + y)}$. Experiment with different numbers of level curves as well as various rectangular regions in the *xy*-plane until your result resembles Figure 2.21.

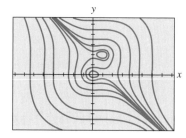

Figure 2.21

(b) On separate coordinate axes plot the graph of the implicit solution corresponding to the initial condition $y(0) = \tfrac{3}{2}$. Use a colored pencil to mark off that segment of the graph that corresponds to the solution curve of a solution ϕ that satisfies the initial condition. With the aid of a root-finding application, determine the approximate largest interval I of definition of the solution ϕ. [*Hint:* First find the points on the curve in part **(a)** where the tangent is vertical.]

(c) Repeat part **(b)** for the initial condition $y(0) = -2$.

2.3 Linear Equations

We continue our quest for solutions of first-order DEs by next examining linear equations. Linear differential equations are an especially "friendly" family of differential equations in that, given a linear equation, whether first-order or a higher-order kin, there is always a good possibility that we can find some sort of solution of the equation that we can look at.

Recall from Section 1.1 that a differential equation that is of the first degree in the dependent variable and all its derivatives is said to be linear. When $n = 1$ in (5) of Section 1.1 we obtain a linear first-order differential equation.

DEFINITION 2.2 Linear Equation

A first-order differential equation of the form

$$a_1(x)\frac{dy}{dx} + a_0(x)y = g(x) \tag{1}$$

is said to be a **linear equation.**

When $g(x) = 0$ the linear equation is said to be **homogeneous,** otherwise it is **nonhomogeneous.**

■ **Standard Form** By dividing both sides of (1) by the lead coefficient $a_1(x)$ we obtain a more useful form, the **standard form,** of a linear equation

$$\frac{dy}{dx} + P(x)y = f(x). \tag{2}$$

We seek a solution of (2) on an interval I for which both functions P and f are continuous.

In the discussion that follows we illustrate a property and a procedure and end up with a formula representing the form that every solution of (2) must have. But more than the formula, the property and the procedure are important, because these two concepts carry over to linear equations of higher order.

■ **The Property** The differential equation (2) has the property that its solution is the **sum** of the two solutions, $y = y_c + y_p$, where y_c is a solution of the associated homogeneous equation

$$\frac{dy}{dx} + P(x)y = 0 \tag{3}$$

and y_p is a particular solution of the nonhomogeneous equation (2). To see this, observe

$$\frac{d}{dx}[y_c + y_p] + P(x)[y_c + y_p] = \underbrace{\left[\frac{dy_c}{dx} + P(x)y_c\right]}_{0} + \underbrace{\left[\frac{dy_p}{dx} + P(x)y_p\right]}_{f(x)} = f(x).$$

Now the homogeneous equation (3) is also separable. This fact enables us to find y_c by writing (3) as

$$\frac{dy}{y} + P(x)dx = 0$$

and integrating. Solving for y gives $y_c = ce^{-\int P(x)dx}$. For convenience let us write $y_c = cy_1(x)$, where $y_1 = e^{-\int P(x)dx}$. The fact that $dy_1/dx + P(x)y_1 = 0$ will be used next to determine y_p.

■ **The Procedure** We can now find a particular solution of equation (2) by a procedure known as **variation of parameters.** The basic idea here is to find a function u so that $y_p = u(x)y_1(x) = u(x)e^{-\int P(x)dx}$ is a solution of (2). In other words, our assumption for y_p is the same as $y_c = cy_1(x)$ except that c is replaced by the "variable parameter" u. Substituting $y_p = uy_1$ into (2) gives

$$\overset{\underset{\text{product rule}}{\downarrow}}{u\frac{dy_1}{dx}} + y_1\frac{du}{dx} + P(x)uy_1 = f(x) \quad \text{or} \quad u\left[\frac{dy_1}{dx} + P(x)y_1\right]^{\overset{\underset{\text{zero}}{\downarrow}}{}} + y_1\frac{du}{dx} = f(x)$$

so that

$$y_1\frac{du}{dx} = f(x).$$

Separating variables and integrating then gives

$$du = \frac{f(x)}{y_1(x)}dx \quad \text{and} \quad u = \int \frac{f(x)}{y_1(x)}dx.$$

From the definition of $y_1(x)$ we see $1/y_1(x) = e^{\int P(x)dx}$. Therefore

$$y_p = uy_1 = \left(\int \frac{f(x)}{y_1(x)}\,dx\right)e^{-\int P(x)dx} = e^{-\int P(x)dx}\int e^{\int P(x)dx}f(x)dx,$$

and

$$y = y_c + y_p = ce^{-\int P(x)dx} + e^{-\int P(x)dx}\int e^{\int P(x)dx}f(x)dx. \tag{4}$$

Hence if (2) has a solution, it must be of form (4). Conversely, it is a straightforward exercise in differentiation to verify that (4) constitutes a one-parameter family of solutions of equation (2).

 You should not memorize the formula given in (4). There is an equivalent but easier way of solving (2). If (4) is multiplied by

$$e^{\int P(x)dx} \tag{5}$$

and then

$$e^{\int P(x)dx}\, y = c + \int e^{\int P(x)dx}f(x)dx \tag{6}$$

is differentiated,

$$\frac{d}{dx}[e^{\int P(x)dx}\,y] = e^{\int P(x)dx}f(x), \tag{7}$$

we get

$$e^{\int P(x)dx}\frac{dy}{dx} + P(x)\,e^{\int P(x)dx}\,y = e^{\int P(x)dx}f(x). \tag{8}$$

Dividing the last result by $e^{\int P(x)dx}$ gives (2).

■ **Method of Solution** The recommended method of solving (2) actually consists of (6)–(8) worked in reverse order. In other words, if (2) is multiplied by (5) we get (8). The left side of (8) is recognized as the derivative of the product of $e^{\int P(x)dx}$ and y. This gets us to (7). We then integrate both sides of (7) to get the solution (6). Because we can solve (2) by integration after multiplication by $e^{\int P(x)dx}$ we call this function an **integrating factor** for the differential equation. For convenience we summarize these results. We again emphasize that you should not memorize formula (4) but work through the following procedure each time.

> ### Solving a Linear First-Order Equation
>
> (*i*) Put a linear equation of form (1) into standard form (2) and then determine $P(x)$ and the integrating factor $e^{\int P(x)dx}$.
> (*ii*) Multiply (2) by the integrating factor. The left side of the resulting equation is automatically the derivative of the integrating factor and *y*: Write
>
> $$\frac{d}{dx}\left[e^{\int P(x)dx}\, y\right] = e^{\int P(x)dx}\, f(x)$$
>
> and then integrate both sides of this equation.

Example 1 Solving a Linear DE

Solve $\dfrac{dy}{dx} - 3y = 6$.

SOLUTION This linear equation can be solved by separation of variables. Alternatively, since the equation is already in the standard form (2) we see that the integrating factor is $e^{\int(-3)dx} = e^{-3x}$. We multiply the equation by this factor and recognize that

$$e^{-3x}\frac{dy}{dx} - 3e^{-3x}y = 6e^{-3x} \quad \text{is the same as} \quad \frac{d}{dx}[e^{-3x}y] = 6e^{-3x}.$$

Integrating both sides of the last equation gives $e^{-3x}y = -2e^{-3x} + c$. Thus a solution of the differential equation is $y = -2 + ce^{3x}$, $-\infty < x < \infty$. ❑

When a_1, a_0, and g in (1) are constants, the differential equation is autonomous. In Example 1, you can verify from the form $dy/dx = 3(y + 2)$ that -2 is a critical point and that it is unstable and a repeller. Thus a solution curve with an initial point either above or below the graph of the equilibrium solution $y = -2$ pushes away from this horizontal line as x increases.

■ **Constant of Integration** Notice in the general discussion and in Example 1 we disregarded a constant of integration in the evaluation of the indefinite integral in the exponent of $e^{\int P(x)dx}$. If you think about the laws of exponents and the fact that the integrating factor multiplies both sides of the differential equation, you should be able to answer why writing $\int P(x)dx + c$ is unnecessary. See Problem 47 in Exercises 2.3.

■ **General Solution** Suppose again that the functions P and f in (2) are continuous on a common interval I. In the steps leading to (4) we showed that *if* (2) has a solution on I, then it must be of the form given in (4). Conversely, it is a straightforward exercise in differentiation to verify that any function of the form given in (4) is a solution of the differential equation (2) on I. In other words, (4) is a one-parameter family of solutions of equation (2), and *every* solution of (2) defined on I is a member of this family. Consequently we are justified in calling (4) the **general solution** of the differential equation on the interval I. Now by writing (2) in the normal form $y' = F(x, y)$ we can identify $F(x, y) = -P(x)y + f(x)$ and $\partial F/\partial y = -P(x)$. From the continuity of P and f on the interval I, we see that F and $\partial F/\partial y$ are also continuous on I. With Theorem 1.1 as our justification, we conclude that there exists one and only one solution of the initial-value problem

$$\frac{dy}{dx} + P(x)y = f(x), \qquad y(x_0) = y_0 \tag{9}$$

defined on *some* interval I_0 containing x_0. But when x_0 is in I, finding a solution of (9) is just a matter of finding an appropriate value of c in (4); that is, for each x_0 in I there corresponds a distinct c. In other words, the interval I_0 of existence and uniqueness in Theorem 1.1 for the initial-value problem (9) is the entire interval I.

Example 2 General Solution

Solve $x\dfrac{dy}{dx} - 4y = x^6 e^x$.

SOLUTION By dividing by x we get the standard form

$$\frac{dy}{dx} - \frac{4}{x}y = x^5 e^x. \tag{10}$$

From this form we identify $P(x) = -4/x$ and $f(x) = x^5 e^x$ and observe that P and f are continuous on $(0, \infty)$. Hence the integrating factor is

we can use ln x instead of ln$|x|$ since $x > 0$
$$\downarrow$$
$$e^{-4\int dx/x} = e^{-4\ln x} = e^{\ln x^{-4}} = x^{-4}.$$

Here we have used the basic identity $b^{\log_b N} = N, N > 0$. Now we multiply (10) by x^{-4},

$$x^{-4}\frac{dy}{dx} - 4x^{-5}y = xe^x, \qquad \text{and obtain} \qquad \frac{d}{dx}[x^{-4}y] = xe^x.$$

It follows from integration by parts that the general solution defined on $(0, \infty)$ is $x^{-4}y = xe^x - e^x + c$ or $y = x^5 e^x - x^4 e^x + cx^4$. ❏

Except in the case when the lead coefficient is 1, the recasting of equation (1) into the standard form (2) requires division by $a_1(x)$. Values of x for which $a_1(x) = 0$ are called **singular points** of the equation. Singular points are potentially troublesome. Specifically in (2), if $P(x)$ (formed by dividing $a_0(x)$ by $a_1(x)$) is discontinuous at a point, the discontinuity may carry over to solutions of the differential equation.

Example 3 General Solution

Find the general solution of $(x^2 - 9)\dfrac{dy}{dx} + xy = 0$.

SOLUTION We write the differential equation in standard form

$$\frac{dy}{dx} + \frac{x}{x^2 - 9}y = 0 \tag{11}$$

and identify $P(x) = x/(x^2 - 9)$. Although P is continuous on $(-\infty, -3)$, on $(-3, 3)$, and on $(3, \infty)$, we shall solve the equation on the first and third intervals. On these intervals the integrating factor is

$$e^{\int x\,dx/(x^2 - 9)} = e^{\frac{1}{2}\int 2x\,dx/(x^2 - 9)} = e^{\frac{1}{2}\ln|x^2 - 9|} = \sqrt{x^2 - 9}.$$

After multiplying the standard form (11) by this factor, we get

$$\frac{d}{dx}[\sqrt{x^2 - 9}\,y] = 0, \qquad \text{and integrating gives} \qquad \sqrt{x^2 - 9}\,y = c.$$

Thus for either $x > 3$ or $x < -3$ the general solution of the equation is $y = c/\sqrt{x^2 - 9}$. ❏

Notice in the preceding example that $x = 3$ and $x = -3$ are singular points of the equation and that every function in the general solution $y = c/\sqrt{x^2 - 9}$ is discontinuous at these points. On the other hand, $x = 0$ is a singular point of the differential equation in Example 2, but the general solution $y = x^5 e^x - x^4 e^x + cx^4$ is noteworthy in that every function in this one-parameter family is continuous at $x = 0$ and is defined on the interval $(-\infty, \infty)$ and not just on $(0, \infty)$ as stated in the solution. However, the family $y = x^5 e^x - x^4 e^x + cx^4$ defined on $(-\infty, \infty)$ cannot be considered the general solution of the DE, since the singular point $x = 0$ still causes a problem. See Problem 43 in Exercises 2.3.

Example 4 An Initial-Value Problem

Solve the initial-value problem $\dfrac{dy}{dx} + y = x$, $y(0) = 4$.

SOLUTION The equation is in standard form, and $P(x) = 1$ and $f(x) = x$ are continuous on $(-\infty, \infty)$. The integrating factor is $e^{\int dx/x} = e^x$, and so integrating

$$\frac{d}{dx}[e^x y] = xe^x$$

gives $e^x y = xe^x - e^x + c$. Solving this last equation for y yields the general solution $y = x - 1 + ce^{-x}$. But from the initial condition we know that $y = 4$ when $x = 0$. Substituting these values in the general solution implies $c = 5$. Hence the solution of the problem is

$$y = x - 1 + 5e^{-x}, \qquad -\infty < x < \infty. \tag{12} \; \square$$

Recall that the general solution of every linear first-order differential equation is a sum of two special solutions: y_c, the general solution of the associated homogeneous equation (3), and y_p, a particular solution of the nonhomogeneous equation (2). In Example 4 we identify $y_c = ce^{-x}$ and $y_p = x - 1$. Figure 2.22, obtained with the aid of a graphing utility, shows (12) in solid color along with other representative solutions in the family $y = x - 1 + ce^{-x}$. It is interesting to observe that as x gets large the graphs of *all* members of the family are close to the graph of $y_p = x - 1$, which is shown in solid black in Figure 2.22. This is because the contribution of $y_c = ce^{-x}$ to the values of a solution becomes negligible for increasing values of x. We say that $y_c = ce^{-x}$ is a **transient term** since $y_c \to 0$ as $x \to \infty$. While this behavior is not a characteristic of all general solutions of linear equations (see Example 2), the notion of a transient is often important in applied problems.

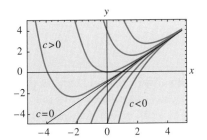

Figure 2.22 Some solutions of $y' + y = x$

Example 5 A Discontinuous $f(x)$

Find a continuous solution satisfying

$$\frac{dy}{dx} + y = f(x), \qquad \text{where} \qquad f(x) = \begin{cases} 1, & 0 \le x \le 1 \\ 0, & x > 1 \end{cases}$$

and the initial condition $y(0) = 0$.

SOLUTION From Figure 2.23 we see that f is piecewise continuous with a discontinuity at $x = 1$. Consequently, we solve the problem in two parts corresponding to the two intervals over which f is defined. For $0 \le x \le 1$ we have

$$\frac{dy}{dx} + y = 1 \qquad \text{or, equivalently,} \qquad \frac{d}{dx}[e^x y] = e^x.$$

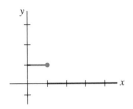

Figure 2.23

Integrating this last equation and solving for y gives $y = 1 + c_1 e^{-x}$. Since $y(0) = 0$, we must have $c_1 = -1$, and therefore $y = 1 - e^{-x}$, $0 \leq x \leq 1$. Then for $x > 1$, the equation

$$\frac{dy}{dx} + y = 0$$

leads to $y = c_2 e^{-x}$. Hence we can write

$$y = \begin{cases} 1 - e^{-x}, & 0 \leq x \leq 1 \\ c_2 e^{-x}, & x > 1. \end{cases}$$

Now in order for y to be a continuous function we certainly want $\lim_{x \to 1^+} y(x) = y(1)$. This latter requirement is equivalent to $c_2 e^{-1} = 1 - e^{-1}$ or $c_2 = e - 1$. As Figure 2.24 shows, the function

$$y = \begin{cases} 1 - e^{-x}, & 0 \leq x \leq 1 \\ (e - 1)e^{-x}, & x > 1 \end{cases} \tag{13}$$

is continuous on $(0, \infty)$. Although we shall refer to this function as a solution of the initial-value problem for $0 \leq x < \infty$, technically the interval is not correct. See Problem 45 in Exercises 2.3.

Figure 2.24

■ **Functions Defined by Integrals** Some simple functions do not possess antiderivatives that are elementary functions, and integrals of these kinds of functions are called **nonelementary.** For example, you may have seen in calculus that $\int e^{x^2} dx$ and $\int \sin x^2 dx$ are nonelementary integrals. In applied mathematics some important functions are *defined* in terms of nonelementary integrals. Two such functions are the **error function** and **complementary error function:**

$$\text{erf}(x) = \frac{2}{\sqrt{\pi}} \int_0^x e^{-t^2} \, dt \qquad \text{and} \qquad \text{erfc}(x) = \frac{2}{\sqrt{\pi}} \int_x^\infty e^{-t^2} \, dt. \tag{14}$$

Since $(2/\sqrt{\pi}) \int_0^\infty e^{-t^2} \, dt = 1$ it is seen from (14) that the error function $\text{erf}(x)$ and the complementary error function $\text{erfc}(x)$ are related by $\text{erf}(x) + \text{erfc}(x) = 1$. Because of its importance in areas such as probability and statistics, the error function has been extensively tabulated. Note that $\text{erf}(0) = 0$ is one obvious functional value. Values of $\text{erf}(x)$ can also be found using a CAS. Before working through the next example, you are urged to reread (*i*) of the Remarks at the end of Section 2.2.

Example 6 The Error Function

Solve the initial-value problem $\dfrac{dy}{dx} - 2xy = 2$, $y(0) = 1$.

SOLUTION Since the equation is already in standard form, we see that the integrating factor is e^{-x^2}, and so from

$$\frac{d}{dx}[e^{-x^2}] = 2e^{-x^2} \quad \text{we get} \quad y = 2e^{x^2} \int_0^x e^{-t^2} \, dt + c e^{x^2}. \tag{15}$$

Applying $y(0) = 1$ to the last expression then gives $c = 1$. Hence, the solution to the problem is

$$y = 2e^{x^2} \int_0^x e^{-t^2} \, dt + e^{x^2} = e^{x^2}\left[1 + \sqrt{\pi}\,\text{erf}(x)\right].$$

Figure 2.25 Some solutions of $y' - 2xy = 2$

The graph of this solution, shown in color in Figure 2.25 among other members of the family defined by (15), was obtained with the aid of a computer algebra system. ❏

■ **Use of Computers** Some computer algebra systems are capable of producing explicit solutions for some kinds of differential equations. For example, to solve the equation $y' + 2y = x$, the input commands

DSolve[y′[x] + 2 y[x] == x, y[x], x] (in *Mathematica*)

and **dsolve(diff(y(x), x) + 2*y = x, y(x));** (in *Maple*)

yield, in turn, the output

$$y[x] \; -> \; -(\frac{1}{4}) + \frac{x}{2} + \frac{C[1]}{E^{2x}}$$

and **y(x) = 1/2 x − 1/4 + exp(−2 x)_C1.**

Translated to standard symbols this means $y = -\frac{1}{4} + \frac{1}{2}x + ce^{-2x}$.

Remarks

(*i*) Occasionally a first-order differential equation is not linear in one variable but is linear in the other variable. For example, the differential equation

$$\frac{dy}{dx} = \frac{1}{x + y^2}$$

is not linear in the variable y. But its reciprocal

$$\frac{dx}{dy} = x + y^2 \quad \text{or} \quad \frac{dx}{dy} - x = y^2$$

is recognized as linear in the variable x. You should verify that the integrating factor $e^{\int (-1)dy} = e^{-y}$ and integration by parts yield an implicit solution of the first equation: $x = -y^2 - 2y - 2 + ce^y$.

(*ii*) Because mathematicians thought they were appropriately descriptive, certain words were "adopted" from engineering and made their own. The word *transient,* used earlier, is one of these terms. In future discussions the words *input* and *output* will occasionally pop up. The function f in (2) is called the **input** or **driving function;** a solution of the differential equation for a given input is called the **output** or **response.**

EXERCISES 2.3

Answers to odd-numbered problems begin on page A-26.

In Problems 1–24, find the general solution of the given differential equation. Give the largest interval over which the general solution is defined. Determine whether there are any transient terms in the general solution.

1. $\dfrac{dy}{dx} = 5y$

2. $\dfrac{dy}{dx} + 2y = 0$

3. $\dfrac{dy}{dx} + y = e^{3x}$

4. $3\dfrac{dy}{dx} + 12y = 4$

5. $y' + 3x^2 y = x^2$

6. $y' + 2xy = x^3$

7. $x^2 y' + xy = 1$

8. $y' = 2y + x^2 + 5$

9. $x\dfrac{dy}{dx} - y = x^2 \sin x$

10. $x\dfrac{dy}{dx} + 2y = 3$

11. $x\dfrac{dy}{dx} + 4y = x^3 - x$

12. $(1 + x)\dfrac{dy}{dx} - xy = x + x^2$

13. $x^2 y' + x(x + 2)y = e^x$

14. $xy' + (1 + x)y = e^{-x} \sin 2x$

15. $y\,dx - 4(x + y^6)dy = 0$

16. $y\,dx = (ye^y - 2x)\,dy$

17. $\cos x \dfrac{dy}{dx} + (\sin x)y = 1$

18. $\cos^2 x \sin x \, dy + (y \cos^3 x - 1)dx = 0$

19. $(x+1)\dfrac{dy}{dx} + (x+2)y = 2xe^{-x}$

20. $(x+2)^2 \dfrac{dy}{dx} = 5 - 8y - 4xy$

21. $\dfrac{dr}{d\theta} + r \sec \theta = \cos \theta$ **22.** $\dfrac{dP}{dt} + 2tP = P + 4t - 2$

23. $x\dfrac{dy}{dx} + (3x+1)y = e^{-3x}$ **24.** $(x^2 - 1)\dfrac{dy}{dx} + 2y = (x+1)^2$

In Problems 25–30, solve the given initial-value problem. Give the largest interval over which the solution is defined.

25. $xy' + y = e^x, \quad y(1) = 2$ **26.** $y\dfrac{dx}{dy} - x = 2y^2, \quad y(1) = 5$

27. $L\dfrac{di}{dt} + Ri = E; \quad i(0) = i_0, \, L, \, R, \, E, \text{ and } i_0 \text{ constants}$

28. $\dfrac{dT}{dt} = k(T - T_m); \quad T(0) = T_0, \, k, \, T_m, \text{ and } T_0 \text{ constants}$

29. $(x+1)\dfrac{dy}{dx} + y = \ln x, \quad y(1) = 10$

30. $y' + (\tan x)y = \cos^2 x, \quad y(0) = -1$

💻 In Problems 31–34, find a continuous solution satisfying the given differential equation and the indicated initial condition. Use a graphing utility to plot the solution curve for the IVP.

31. $\dfrac{dy}{dx} + 2y = f(x), \quad f(x) = \begin{cases} 1, & 0 \le x \le 3 \\ 0, & x > 3 \end{cases}, \quad y(0) = 0$

32. $\dfrac{dy}{dx} + y = f(x), \quad f(x) = \begin{cases} 1, & 0 \le x \le 1 \\ -1, & x > 1 \end{cases}, \quad y(0) = 1$

33. $\dfrac{dy}{dx} + 2xy = f(x), \quad f(x) = \begin{cases} x, & 0 \le x < 1 \\ 0, & x \ge 1 \end{cases}, \quad y(0) = 2$

34. $(1+x^2)\dfrac{dy}{dx} + 2xy = f(x), \quad f(x) = \begin{cases} x, & 0 \le x < 1 \\ -x, & x \ge 1 \end{cases}, \quad y(0) = 0$

💻 **35.** Find a continuous solution of $y' + P(x)y = 4x, \, y(0) = 3$, where

$$P(x) = \begin{cases} 2, & 0 \le x \le 1 \\ -2/x, & x > 1. \end{cases}$$

Use a graphing utility to plot the solution curve for the IVP.

36. Computer Lab Assignment (a) Express the solution of the initial-value problem $y' - 2xy = -1, y(0) = \sqrt{\pi}/2$, in terms of erfc$(x)$.

(b) Use a CAS to calculate $y(2)$. Then use a CAS to obtain the solution curve for the IVP for $x \ge 0$.

37. Express the solution of the initial-value problem $y' - 2xy = 1$, $y(1) = 1$, in terms of erf(x).

38. Computer Lab Assignment (a) The **sine integral function** is defined by Si$(x) = \displaystyle\int_0^x \dfrac{\sin t}{t} dt$, where the integrand is defined to be 1 at $t = 0$. Express the solution $y(x)$ of the initial-value problem $x^3 y' + 2x^2 y = 10 \sin x, \, y(1) = 0$, in terms of Si$(x)$.

(b) Use a CAS to calculate $y(2)$. Then use a CAS to plot the solution curve for the IVP for $x > 0$.

39. Computer Lab Assignment (a) The **Fresnel sine integral** is defined by $S(x) = \displaystyle\int_0^x \sin\left(\dfrac{\pi}{2}t^2\right)dt$. Express the solution $y(x)$ of the initial-value problem $y' - (\sin x^2)y = 0, \, y(0) = 5$, in terms of $S(x)$.

(b) Use a CAS to calculate $y(2)$. Then use a CAS to plot the solution curve for the IVP for $x \ge 0$. What is the behavior of the solution $y(x)$ as $x \to \infty$?

40. Consider the initial-value problem $y' + e^x y = f(x), y(0) = 1$. Express the solution of the IVP for $x \ge 0$ as a nonelementary integral when $f(x) = 1$. What is the solution when $f(x) = 0$? When $f(x) = e^x$?

41. Find the general solution of the differential equation in Example 3 on the interval $(-3, 3)$.

42. Reread the discussion following Example 1. Construct a linear first-order differential equation for which all nonconstant solutions approach the horizontal asymptote $y = 4$ as $x \to \infty$.

43. Reread Example 2 and then discuss, with reference to Theorem 1.1, the existence and uniqueness of a solution of the initial-value problem consisting of $xy' - 4y = x^6 e^x$ and the given initial condition.

(a) $y(0) = 0$ **(b)** $y(0) = y_0, \, y_0 \ne 0$ **(c)** $y(x_0) = y_0, \, x_0 \ne 0$

44. Reread the discussion following Example 4. Construct a linear first-order differential equation for which all solutions are asymptotic to the line $y = 3x - 5$ as $x \to \infty$.

45. Reread Example 5 and then discuss why it is technically incorrect to say that the function in (13) is a solution of the IVP on the interval $[0, \infty)$.

46. Construct a linear first-order differential equation for which $y_c = c/x^3$ and $y_p = x^3$.

47. In determining the integrating factor (5), we did not use a constant of integration in the evaluation of $\int P(x) \, dx$. Explain why using $\int P(x) \, dx + c$ has no effect on the solution of (2).

48. Suppose $P(x)$ is continuous on some interval I and a is a number in I. What can be said about the solution of the initial-value problem $y' + P(x)y = 0, \, y(a) = 0$?

2.4 Exact Equations

Although the simple equation $y\,dx + x\,dy = 0$ is separable, we can solve the equation in an alternative manner by recognizing that the left side is equivalent to the differential of the product of x and y; that is, $y\,dx + x\,dy = d(xy) = 0$. By integrating both sides we immediately obtain the implicit solution $xy = c$.

■ **Differential of a Function of Two Variables** If $z = f(x, y)$ is a function of two variables with continuous first partial derivatives in a region R of the xy-plane, then its differential (also called the total differential) is

$$dz = \frac{\partial f}{\partial x}\,dx + \frac{\partial f}{\partial y}\,dy. \tag{1}$$

Now if $f(x, y) = c$, it follows from (1) that

$$\frac{\partial f}{\partial x}\,dx + \frac{\partial f}{\partial y}\,dy = 0. \tag{2}$$

In other words, given a one-parameter family of curves $f(x, y) = c$, we can generate a first-order differential equation by computing the differential. For example, if $x^2 - 5xy + y^3 = c$, then (2) gives

$$(2x - 5y)dx + (-5x + 3y^2)dy = 0. \tag{3}$$

For our purposes it is more important to turn the problem around; namely, given a first-order DE such as (3), can we recognize that it is equivalent to the differential $d(x^2 - 5xy + y^3) = 0$?

DEFINITION 2.3 Exact Equation

A differential expression $M(x, y)dx + N(x, y)dy$ is an **exact differential** in a region R of the xy-plane if it corresponds to the differential of some function $f(x, y)$. A first-order differential equation of the form

$$M(x, y)\,dx + N(x, y)\,dy = 0$$

is said to be an **exact equation** if the expression on the left side is an exact differential.

For example, the equation $x^2y^3dx + x^3y^2dy = 0$ is exact, because the left side is $d(\frac{1}{3}x^3y^3) = x^2y^3dx + x^3y^2dy$. Notice that if $M(x, y) = x^2y^3$ and $N(x, y) = x^3y^2$ then $\partial M/\partial y = 3x^2y^2 = \partial N/\partial x$. Theorem 2.1 shows that the equality of these partial derivatives is no coincidence.

THEOREM 2.1 Criterion for an Exact Differential

Let $M(x, y)$ and $N(x, y)$ be continuous and have continuous first partial derivatives in a rectangular region R defined by $a < x < b$, $c < y < d$. Then a necessary and sufficient condition that $M(x, y)\,dx + N(x, y)\,dy$ be an exact differential is

$$\frac{\partial M}{\partial y} = \frac{\partial N}{\partial x}. \tag{4}$$

PROOF OF THE NECESSITY For simplicity let us assume that $M(x, y)$ and $N(x, y)$ have continuous first partial derivatives for all (x, y). Now if the expression $M(x, y)\,dx + N(x, y)\,dy$ is exact, there exists some function f such that for all x in R,

$$M(x, y)\,dx + N(x, y)\,dy = \frac{\partial f}{\partial x}\,dx + \frac{\partial f}{\partial y}\,dy.$$

Therefore,
$$M(x, y) = \frac{\partial f}{\partial x}, \qquad N(x, y) = \frac{\partial f}{\partial y},$$

and
$$\frac{\partial M}{\partial y} = \frac{\partial}{\partial y}\left(\frac{\partial f}{\partial x}\right) = \frac{\partial^2 f}{\partial y\,\partial x} = \frac{\partial}{\partial x}\left(\frac{\partial f}{\partial y}\right) = \frac{\partial N}{\partial x}.$$

The equality of the mixed partials is a consequence of the continuity of the first partial derivatives of $M(x, y)$ and $N(x, y)$. ❑

The sufficiency part of Theorem 2.1 consists of showing that there exists a function f for which $\partial f/\partial x = M(x, y)$ and $\partial f/\partial y = N(x, y)$ whenever (4) holds. The construction of the function f actually reflects a basic procedure for solving exact equations.

■ **Method of Solution** Given an equation of the form $M(x, y)dx + N(x, y)dy = 0$, determine whether the equality in (4) holds. If it does, then there exists a function f for which

$$\frac{\partial f}{\partial x} = M(x, y).$$

We can find f by integrating $M(x, y)$ with respect to x, while holding y constant:

$$f(x, y) = \int M(x, y)\,dx + g(y), \tag{5}$$

where the arbitrary function $g(y)$ is the "constant" of integration. Now differentiate (5) with respect to y and assume $\partial f/\partial y = N(x, y)$:

$$\frac{\partial f}{\partial y} = \frac{\partial}{\partial y}\int M(x, y)\,dx + g'(y) = N(x, y).$$

This gives
$$g'(y) = N(x, y) - \frac{\partial}{\partial y}\int M(x, y)\,dx. \tag{6}$$

Finally, integrate (6) with respect to y and substitute the result in (5). The implicit solution of the equation is $f(x, y) = c$.

Some observations are in order. First, it is important to realize that the expression $N(x, y) - (\partial/\partial y)\int M(x, y)\,dx$ in (6) is independent of x, because

$$\frac{\partial}{\partial x}\left[N(x, y) - \frac{\partial}{\partial y}\int M(x, y)\,dx\right] = \frac{\partial N}{\partial x} - \frac{\partial}{\partial y}\left(\frac{\partial}{\partial x}\int M(x, y)\,dx\right) = \frac{\partial N}{\partial x} - \frac{\partial M}{\partial y} = 0.$$

Second, we could just as well start the foregoing procedure with the assumption that $\partial f/\partial y = N(x, y)$. After integrating N with respect to y and then differentiating that result, we would find the analogues of (5) and (6) to be, respectively,

$$f(x, y) = \int N(x, y)\,dy + h(x) \quad \text{and} \quad h'(x) = M(x, y) - \frac{\partial}{\partial x}\int N(x, y)\,dy.$$

In either case *none of these formulas should be memorized.*

Example 1 **Solving an Exact DE**

Solve $2xy\,dx + (x^2 - 1)\,dy = 0$.

SOLUTION With $M(x, y) = 2xy$ and $N(x, y) = x^2 - 1$ we have

$$\frac{\partial M}{\partial y} = 2x = \frac{\partial N}{\partial x}.$$

Thus the equation is exact, and so, by Theorem 2.1, there exists a function $f(x, y)$ such that

$$\frac{\partial f}{\partial x} = 2xy \quad \text{and} \quad \frac{\partial f}{\partial y} = x^2 - 1.$$

From the first of these equations we obtain, after integrating,

$$f(x, y) = x^2 y + g(y).$$

Taking the partial derivative of the last expression with respect to y and setting the result equal to $N(x, y)$ gives

$$\frac{\partial f}{\partial y} = x^2 + g'(y) = x^2 - 1. \qquad \leftarrow N(x,\,y)$$

It follows that $\qquad g'(y) = -1 \quad \text{and} \quad g(y) = -y.$

Hence, $f(x, y) = x^2 y - y$, and so the solution of the equation in implicit form is $x^2 y - y = c$. The explicit form of the solution is easily seen to be $y = c/(1 - x^2)$ and is defined on any interval not containing either $x = 1$ or $x = -1$. ❑

Note the form of the solution. ☞
It is $f(x, y) = c$.

The solution of the DE in Example 1 is *not* $f(x, y) = x^2 - y$. Rather it is $f(x, y) = c$; or if a constant is used in the integration of $g'(y)$, we can then write the solution as $f(x, y) = 0$. Note, too, that the equation could be solved by separation of variables.

Example 2 **Solving an Exact DE**

Solve $(e^{2y} - y\cos xy)\,dx + (2xe^{2y} - x\cos xy + 2y)\,dy = 0$.

SOLUTION The equation is exact because

$$\frac{\partial M}{\partial y} = 2e^{2y} + xy\sin xy - \cos xy = \frac{\partial N}{\partial x}.$$

Hence a function $f(x, y)$ exists for which

$$M(x, y) = \frac{\partial f}{\partial x} \quad \text{and} \quad N(x, y) = \frac{\partial f}{\partial y}.$$

Now for variety we shall start with the assumption that $\partial f/\partial y = N(x, y)$;

that is, $\qquad\qquad \dfrac{\partial f}{\partial y} = 2xe^{2y} - x\cos xy + 2y$

$$f(x, y) = 2x \int e^{2y}\,dy - x \int \cos xy\,dy + 2 \int y\,dy.$$

Remember, the reason x can come out in front of the symbol \int is that in the integration with respect to y, x is treated as an ordinary constant. It follows that

$$f(x, y) = xe^{2y} - \sin xy + y^2 + h(x)$$

$$\frac{\partial f}{\partial x} = e^{2y} - y\cos xy + h'(x) = e^{2y} - y\cos xy, \qquad \leftarrow M(x,\,y)$$

and so $h'(x) = 0$ or $h(x) = c$. Hence a family of solutions is

$$xe^{2y} - \sin xy + y^2 + c = 0.$$ ❑

Example 3 An Initial-Value Problem

Solve the initial-value problem $\dfrac{dy}{dx} = \dfrac{xy^2 - \cos x \sin x}{y(1 - x^2)}$, $y(0) = 2$.

SOLUTION By writing the differential equation in the form

$$(\cos x \sin x - xy^2)\, dx + y(1 - x^2)\, dy = 0$$

we recognize that the equation is exact because

$$\frac{\partial M}{\partial y} = -2xy = \frac{\partial N}{\partial x}.$$

Now

$$\frac{\partial f}{\partial y} = y(1 - x^2)$$

$$f(x,\, y) = \frac{y^2}{2}(1 - x^2) + h(x)$$

$$\frac{\partial f}{\partial x} = -xy^2 + h'(x) = \cos x \sin x - xy^2.$$

The last equation implies that $h'(x) = \cos x \sin x$. Integrating gives

$$h(x) = -\int (\cos x)(-\sin x\, dx) = -\frac{1}{2}\cos^2 x.$$

Thus $\dfrac{y^2}{2}(1 - x^2) - \dfrac{1}{2}\cos^2 x = c_1$ or $y^2(1 - x^2) - \cos^2 x = c,$ (7)

where $2c_1$ has been replaced by c. The initial condition $y = 2$ when $x = 0$ demands that $4(1) - \cos^2(0) = c$ and so $c = 3$. An implicit solution of the problem is then $y^2(1 - x^2) - \cos^2 x = 3$.

 The solution curve of the IVP is part of an interesting family of curves and is the curve drawn in color in Figure 2.26. The graphs of the members of the one-parameter family of solutions given in (7) can be obtained in several ways, two of which are: using software to graph level curves as discussed in the last section, or using a graphing utility and carefully graphing the explicit functions obtained for various values of c by solving $y^2 = (c + \cos^2 x)/(1 - x^2)$ for y. ❑

Figure 2.26 Some solution curves in the family $y^2(1 - x^2) - \cos^2 x = c$

Remarks

 (*i*) When testing an equation for exactness, make sure it is of the precise form $M(x,\, y)\, dx + N(x,\, y)\, dy = 0$. Sometimes a differential equation is written $G(x,\, y)\, dx = H(x,\, y)\, dy$. In this case, first rewrite it as $G(x,\, y)\, dx - H(x,\, y)\, dy = 0$, and then identify $M(x,\, y) = G(x,\, y)$ and $N(x,\, y) = -H(x,\, y)$ before using (4).

 (*ii*) It is sometimes possible to transform a nonexact equation $M(x,\, y)\, dx + N(x,\, y)\, dy = 0$ into an equation that is exact by multiplying it by a function called an *integrating factor*. For example, verify that $(y^2 + 32x - 160)\, dx + xy\, dy = 0$ is not exact. However, if we multiply the equation by $\mu(x) = x$, the resulting equation $(xy^2 + 32x^2 - 160x)\, dx + x^2y\, dy = 0$ is now exact. Verify. See Problems 29–34 in Exercises 2.4.

Answers to odd-numbered problems begin on page A-27.

In Problems 1–20, determine whether the given differential equation is exact. If exact, solve it.

1. $(2x - 1) \, dx + (3y + 7) \, dy = 0$

2. $(2x + y) \, dx - (x + 6y) \, dy = 0$

3. $(5x + 4y) \, dx + (4x - 8y^3) \, dy = 0$

4. $(\sin y - y \sin x) \, dx + (\cos x + x \cos y - y) \, dy = 0$

5. $(2xy^2 - 3)dx + (2x^2y + 4)dy = 0$

6. $\left(2y - \dfrac{1}{x} + \cos 3x \right) \dfrac{dy}{dx} + \dfrac{y}{x^2} - 4x^3 + 3y \sin 3x = 0$

7. $(x^2 - y^2) \, dx + (x^2 - 2xy) \, dy = 0$

8. $\left(1 + \ln x + \dfrac{y}{x} \right) \, dx = (1 - \ln x) \, dy$

9. $(x - y^3 + y^2 \sin x) \, dx = (3xy^2 + 2y \cos x) \, dy$

10. $(x^3 + y^3) \, dx + 3xy^2 \, dy = 0$

11. $(y \ln y - e^{-xy}) \, dx + \left(\dfrac{1}{y} + x \ln y \right) dy = 0$

12. $(3x^2y + e^y) \, dx + (x^3 + xe^y - 2y) \, dy = 0$

13. $x \dfrac{dy}{dx} = 2xe^x - y + 6x^2$

14. $\left(1 - \dfrac{3}{y} + x \right) \dfrac{dy}{dx} + y = \dfrac{3}{x} - 1$

15. $\left(x^2y^3 - \dfrac{1}{1 + 9x^2} \right) \dfrac{dx}{dy} + x^3y^2 = 0$

16. $(5y - 2x)y' - 2y = 0$

17. $(\tan x - \sin x \sin y) \, dx + \cos x \cos y \, dy = 0$

18. $(2y \sin x \cos x - y + 2y^2 e^{xy^2}) \, dx = (x - \sin^2 x - 4xy e^{xy^2}) \, dy$

19. $(4t^3y - 15t^2 - y) \, dt + (t^4 + 3y^2 - t) \, dy = 0$

20. $\left(\dfrac{1}{t} + \dfrac{1}{t^2} - \dfrac{y}{t^2 + y^2} \right) dt + \left(ye^y + \dfrac{t}{t^2 + y^2} \right) dy = 0$

In Problems 21–26, solve the given initial-value problem.

21. $(x + y)^2 \, dx + (2xy + x^2 - 1) \, dy = 0, \quad y(1) = 1$

22. $(e^x + y) \, dx + (2 + x + ye^y) \, dy = 0, \quad y(0) = 1$

23. $(4y + 2t - 5) \, dt + (6y + 4t - 1) \, dy = 0, \quad y(-1) = 2$

24. $\left(\dfrac{3y^2 - t^2}{y^5} \right) \dfrac{dy}{dt} + \dfrac{t}{2y^4} = 0, \quad y(1) = 1$

25. $(y^2 \cos x - 3x^2y - 2x) \, dx + (2y \sin x - x^3 + \ln y) \, dy = 0,$
$y(0) = e$

26. $\left(\dfrac{1}{1 + y^2} + \cos x - 2xy \right) \dfrac{dy}{dx} = y(y + \sin x), \quad y(0) = 1$

In Problems 27 and 28, find the value of k so that the given differential equation is exact.

27. $(y^3 + kxy^4 - 2x) \, dx + (3xy^2 + 20x^2y^3) \, dy = 0$

28. $(6xy^3 + \cos y) \, dx + (2kx^2y^2 - x \sin y) \, dy = 0$

In Problems 29 and 30, verify that the given differential equation is not exact. Multiply the given differential equation by the indicated function $\mu(x, y)$ and verify that the new equation is exact. Solve. The function $\mu(x, y)$ is called an integrating factor.

29. $(-xy \sin x + 2y \cos x) \, dx + 2x \cos x \, dy = 0, \quad \mu(x, y) = xy$

30. $(x^2 + 2xy - y^2) \, dx + (y^2 + 2xy - x^2) \, dy = 0, \quad \mu(x, y) = (x + y)^{-2}$

Suppose $M(x, y) \, dx + N(x, y) \, dy = 0$ is a nonexact differential equation. Under certain conditions integrating factors (see Problems 29 and 30) can be found in a straightforward manner. If $(M_y - N_x)/N = p(x)$, that is, a function of x alone, then an integrating factor is $\mu(x) = e^{\int p(x)dx}$. If $(N_x - M_y)/M = q(y)$, that is, a function of y alone, then an integrating factor is $\mu(y) = e^{\int q(y)dy}$. In Problems 31–34, solve the given differential equation by finding an appropriate integrating factor.

31. $(2y^2 + 3x) \, dx + 2xy \, dy = 0$

32. $y(x + y + 1) \, dx + (x + 2y) \, dy = 0$

33. $6xy \, dx + (4y + 9x^2) \, dy = 0$

34. $\cos x \, dx + \left(1 + \dfrac{2}{y} \right) \sin x \, dy = 0$

35. (a) Show that every separable first-order equation $dy/dx = g(x)h(y)$ when written in the form $M(x) \, dx + N(y) \, dy = 0$, where $M(x) = -g(x)$ and $N(y) = 1/h(y)$, is exact.

(b) By any method find an implicit solution of the initial-value problem $dy/dx = -(\cos x)/\sin y, \; y(7\pi/6) = \pi/2$.

(c) Find an explicit solution $y(x)$ of the IVP in part **(b)**. Use a graphing utility to plot the graph of $y(x)$. Determine the exact interval I of definition of $y(x)$.

36. (a) Show that a one-parameter family of solutions of the equation

$$(4xy + 3x^2) \, dx + (2y + 2x^2) \, dy = 0$$

is $x^3 + 2x^2y + y^2 = c$.

(b) Show that the initial conditions $y(0) = -2$ and $y(1) = 1$ determine the same implicit solution.

(c) Find explicit solutions $y_1(x)$ and $y_2(x)$ of the differential equation in part **(a)** such that $y_1(0) = -2$ and $y_2(1) = 1$. Use a graphing utility to plot the graphs of $y_1(x)$ and $y_2(x)$.

37. Consider the concept of an integrating factor introduced in Problems 29–34. Are the two equations $M\,dx + N\,dy = 0$ and $\mu M \, dx + \mu N \, dy = 0$ necessarily equivalent in the sense that a solution of one is also a solution of the other?

38. Reread Example 3 and then discuss why we can conclude that the interval of definition of the explicit solution of the IVP (the colored curve in Figure 2.26) is $(-1, 1)$.

39. Find functions $M(x, y)$ and $N(x, y)$ so that each differential equation is exact.

 (a) $M(x, y)\, dx + \left(xe^{xy} + 2xy + \dfrac{1}{x}\right) dy = 0$

 (b) $\left(x^{-1/2}y^{1/2} + \dfrac{x}{x^2 + y}\right) dx + N(x, y)\, dy = 0$

40. Differential equations are sometimes solved by having a clever idea. Here is a little exercise in cleverness: The differential equation $(x - \sqrt{x^2 + y^2})\, dx + y\, dy = 0$ is not exact, but show how the rearrangement $x\, dx + y\, dy = \sqrt{x^2 + y^2}\, dx$ and the observation $\tfrac{1}{2}\, d(x^2 + y^2) = x\, dx + y\, dy$ leads to a solution.

2.5 Solutions by Substitutions

■ **Substitutions** We usually solve a differential equation by recognizing it as a certain kind of equation (say, separable) and then carrying out a procedure, consisting of equation-specific mathematical steps, that yields a function that satisfies the equation. Often the first step in solving a given differential equation consists of transforming it into another differential equation by means of a **substitution.** For example, suppose we wish to transform the first-order equation $dy/dx = f(x, y)$ by the substitution $y = g(x, u)$, where u is regarded as a function of the variable x.

If g possesses first-partial derivatives, then the Chain Rule gives

$$\frac{dy}{dx} = g_x(x, u) + g_u(x, u)\,\frac{du}{dx}.$$

By replacing dy/dx by $f(x, y)$ and y by $g(x, u)$ in the foregoing derivative, we get the new first-order differential equation

$$f(x,\, g(x, u)) = g_x(x, u) + g_u(x, u)\,\frac{du}{dx},$$

which, after solving for du/dx, has the form $du/dx = F(x, u)$. If we can determine a solution $u = \phi(x)$ of this second equation, then a solution of the original differential equation is $y = g(x,\, \phi(x))$.

■ **Homogeneous Equations** If a function f possesses the property $f(tx, ty) = t^{\alpha} f(x, y)$ for some real number α, then f is said to be a **homogeneous function** of degree α. For example, $f(x,y) = x^3 + y^3$ is a homogeneous function of degree 3 since

$$f(tx, ty) = (tx)^3 + (ty)^3 = t^3(x^3 + y^3) = t^3 f(x,y),$$

whereas $f(x,y) = x^3 + y^3 + 1$ is seen not to be homogeneous. A first-order DE in differential form

$$M(x, y)\, dx + N(x, y)\, dy = 0 \tag{1}$$

is said to be **homogeneous** if both coefficients M and N are homogeneous functions of the *same* degree. In other words, (1) is homogeneous if

$$M(tx, ty) = t^{\alpha} M(x, y) \quad \text{and} \quad N(tx, ty) = t^{\alpha} N(x, y).$$

Here the word "homogeneous" does not mean the same as it did in Section 2.3.

If M and N are homogeneous functions of degree α, we can also write

A linear first-order DE $a_1 y' + a_0 y = g(x)$ is homogeneous when $g(x) = 0$.

$$M(x, y) = x^{\alpha} M(1, u) \quad \text{and} \quad N(x,y) = x^{\alpha} N(1, u) \quad \text{where } u = y/x, \tag{2}$$

and

$$M(x, y) = y^{\alpha} M(v, 1) \quad \text{and} \quad N(x, y) = y^{\alpha} N(v, 1) \quad \text{where } v = x/y. \tag{3}$$

See Problem 31 in Exercises 2.5. Properties (2) and (3) suggest the substitutions that can used to solve a homogeneous differential equation. Specifically, *either* of the substitutions $y = ux$ or $x = vy$, where u and v are new dependent variables, will reduce a homogeneous equation to a *separable* first-order differential equation. To show this, observe that as a consequence of (2) a homogeneous equation $M(x, y)\, dx + N(x, y)\, dy = 0$ can be rewritten as

$$x^\alpha M(1, u)\, dx + x^\alpha N(1, u)\, dy = 0 \quad \text{or} \quad M(1, u)\, dx + N(1, u)\, dy = 0,$$

where $u = y/x$ or $y = ux$. By substituting the differential $dy = udx + xdu$ into the last equation and gathering terms, we obtain a separable DE in the variables u and x:

$$M(1, u)\, dx + N(1, u)[u\, dx + x\, du] = 0$$

$$[M(1, u) + uN(1, u)]\, dx + xN(1, u)\, du = 0$$

or

$$\frac{dx}{x} + \frac{N(1, u)\, du}{M(1, u) + uN(1, u)} = 0.$$

We hasten to point out that the preceding formula should not be memorized; rather, *the procedure should be worked through each time*. The proof that the substitutions $x = vy$ and $dx = vdy + ydv$ also lead to a separable equation follows in an analogous manner from (3).

Example 1 Solving a Homogeneous DE

Solve $(x^2 + y^2)\, dx + (x^2 - xy)\, dy = 0$.

SOLUTION Inspection of $M(x, y) = x^2 + y^2$ and $N(x, y) = x^2 - xy$ shows that these coefficients are homogeneous functions of degree 2. If we let $y = ux$, then $dy = u\, dx + x\, du$ so that, after substituting, the given equation becomes

$$(x^2 + u^2 x^2)\, dx + (x^2 - ux^2)[u\, dx + x\, du] = 0$$

$$x^2(1 + u)\, dx + x^3(1 - u)\, du = 0$$

$$\frac{1 - u}{1 + u}\, du + \frac{dx}{x} = 0$$

$$\left[-1 + \frac{2}{1 + u}\right] du + \frac{dx}{x} = 0. \qquad \leftarrow \text{long division}$$

After integration the last line gives

$$-u + 2\ln|1 + u| + \ln|x| = \ln|c|$$

$$-\frac{y}{x} + 2\ln\left|1 + \frac{y}{x}\right| + \ln|x| = \ln|c|. \qquad \leftarrow \text{resubstituting } u = y/x$$

Using the properties of logarithms, we can write the preceding solution as

$$\ln\left|\frac{(x + y)^2}{cx}\right| = \frac{y}{x} \quad \text{or} \quad (x + y)^2 = cxe^{y/x}. \qquad \square$$

Although either of the indicated substitutions can be used for every homogeneous differential equation, in practice we try $x = vy$ whenever the function $M(x, y)$ is simpler than $N(x, y)$. Also it could happen that after using one substitution, we may encounter integrals that are difficult or impossible to evaluate in closed form; switching substitutions may result in an easier problem.

■ **Bernoulli's Equation** The differential equation

$$\frac{dy}{dx} + P(x)y = f(x)y^n, \tag{4}$$

where n is any real number, is called **Bernoulli's equation.** Note that for $n = 0$ and $n = 1$, equation (4) is linear. For $n \neq 0$ and $n \neq 1$, the substitution $u = y^{1-n}$ reduces any equation of form (4) to a linear equation.

> ### *Example 2* Solving a Bernoulli DE
>
> Solve $x\dfrac{dy}{dx} + y = x^2y^2$.
>
> **SOLUTION** We first rewrite the equation as
>
> $$\frac{dy}{dx} + \frac{1}{x}y = xy^2$$
>
> by dividing by x. With $n = 2$, we next substitute $y = u^{-1}$ and
>
> $$\frac{dy}{dx} = -u^{-2}\frac{du}{dx} \qquad\qquad \leftarrow \text{Chain Rule}$$
>
> into the given equation and simplify. The result is
>
> $$\frac{du}{dx} - \frac{1}{x}u = -x.$$
>
> The integrating factor for this linear equation on, say, $(0, \infty)$ is
>
> $$e^{-\int dx/x} = e^{-\ln x} = e^{\ln x^{-1}} = x^{-1}.$$
>
> Integrating $$\frac{d}{dx}[x^{-1}u] = -1$$
>
> gives $x^{-1}u = -x + c$ or $u = -x^2 + cx$. Since $u = y^{-1}$ we have $y = 1/u$, and so a solution of the given equation is $y = 1/(-x^2 + cx)$. ❑

Note that we have not obtained the general solution of the original nonlinear differential equation in Example 2, since $y = 0$ is a singular solution of the equation.

■ **Reduction to Separation of Variables** A differential equation of the form

$$\frac{dy}{dx} = f(Ax + By + C) \tag{5}$$

can always be reduced to an equation with separable variables by means of the substitution $u = Ax + By + C$, $B \neq 0$. Example 3 illustrates the technique.

> ### *Example 3* An Initial-Value Problem
>
> Solve the initial-value problem $\dfrac{dy}{dx} = (-2x + y)^2 - 7$, $y(0) = 0$.
>
> **SOLUTION** If we let $u = -2x + y$, then $du/dx = -2 + dy/dx$, and so the differential equation is transformed into
>
> $$\frac{du}{dx} + 2 = u^2 - 7 \qquad \text{or} \qquad \frac{du}{dx} = u^2 - 9.$$

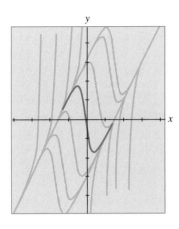

Figure 2.27 Some solutions of $y' = (-2x + y)^2 - 7$

The last equation is separable. Using partial fractions,

$$\frac{du}{(u-3)(u+3)} = dx \qquad \text{or} \qquad \frac{1}{6}\left[\frac{1}{u-3} - \frac{1}{u+3}\right]du = dx$$

and integrating, then yields

$$\frac{1}{6}\ln\left|\frac{u-3}{u+3}\right| = x + c_1 \qquad \text{or} \qquad \frac{u-3}{u+3} = e^{6x+6c_1} = ce^{6x}. \qquad \leftarrow \text{replace } e^{6c_1} \text{ by } c$$

Solving the last equation for u and then resubstituting gives the solution

$$u = \frac{3(1 + ce^{6x})}{1 - ce^{6x}} \qquad \text{or} \qquad y = 2x + \frac{3(1 + ce^{6x})}{1 - ce^{6x}}. \qquad (6)$$

Finally, applying the initial condition $y(0) = 0$ to the last equation in (6) gives $c = -1$. With the aid of a graphing utility we have shown in Figure 2.27 the graph of the particular solution

$$y = 2x + \frac{3(1 - e^{6x})}{1 + e^{6x}}$$

in solid color along with the graphs of some other members of the family solutions (6). ❑

Answers to odd-numbered problems begin on page A-27.

In Problems 1–10, solve the given homogeneous equation by using an appropriate substitution.

1. $(x - y)\,dx + x\,dy = 0$

2. $(x + y)\,dx + x\,dy = 0$

3. $x\,dx + (y - 2x)\,dy = 0$

4. $y\,dx = 2(x + y)\,dy$

5. $(y^2 + yx)\,dx - x^2\,dy = 0$

6. $(y^2 + yx)\,dx + x^2\,dy = 0$

7. $\dfrac{dy}{dx} = \dfrac{y - x}{y + x}$

8. $\dfrac{dy}{dx} = \dfrac{x + 3y}{3x + y}$

9. $-y\,dx + (x + \sqrt{xy})\,dy = 0$

10. $x\dfrac{dy}{dx} - y = \sqrt{x^2 + y^2}$

In Problems 11–14, solve the given initial-value problem.

11. $xy^2\dfrac{dy}{dx} = y^3 - x^3,\quad y(1) = 2$

12. $(x^2 + 2y^2)\dfrac{dx}{dy} = xy,\quad y(-1) = 1$

13. $(x + ye^{y/x})\,dx - xe^{y/x}\,dy = 0,\quad y(1) = 0$

14. $y\,dx + x(\ln x - \ln y - 1)\,dy = 0,\quad y(1) = e$

In Problems 15–20, solve the given Bernoulli equation by using an appropriate substitution.

15. $x\dfrac{dy}{dx} + y = \dfrac{1}{y^2}$

16. $\dfrac{dy}{dx} - y = e^x y^2$

17. $\dfrac{dy}{dx} = y(xy^3 - 1)$

18. $x\dfrac{dy}{dx} - (1 + x)y = xy^2$

19. $t^2\dfrac{dy}{dt} + y^2 = ty$

20. $3(1 + t^2)\dfrac{dy}{dt} = 2ty(y^3 - 1)$

In Problems 21 and 22, solve the given initial-value problem.

21. $x^2\dfrac{dy}{dx} - 2xy = 3y^4,\quad y(1) = \frac{1}{2}$

22. $y^{1/2}\dfrac{dy}{dx} + y^{3/2} = 1,\quad y(0) = 4$

In Problems 23–28, use the procedure illustrated in Example 3 to solve the given differential equation.

23. $\dfrac{dy}{dx} = (x + y + 1)^2$

24. $\dfrac{dy}{dx} = \dfrac{1 - x - y}{x + y}$

25. $\dfrac{dy}{dx} = \tan^2(x + y)$

26. $\dfrac{dy}{dx} = \sin(x + y)$

27. $\dfrac{dy}{dx} = 2 + \sqrt{y - 2x + 3}$

28. $\dfrac{dy}{dx} = 1 + e^{y - x + 5}$

In Problems 29 and 30, solve the given initial-value problem.

29. $\dfrac{dy}{dx} = \cos(x + y),\quad y(0) = \pi/4$

30. $\dfrac{dy}{dx} = \dfrac{3x + 2y}{3x + 2y + 2},\quad y(-1) = -1$

31. Explain why it is always possible to express any homogeneous differential equation $M(x, y)\,dx + N(x, y)\,dy = 0$ in the form

$$\frac{dy}{dx} = F\left(\frac{y}{x}\right) \qquad \text{or} \qquad \frac{dy}{dx} = G\left(\frac{x}{y}\right).$$

You might start by proving that $M(x, y) = x^\alpha M(1, y/x)$ and $N(x, y) = x^\alpha N(1, y/x)$.

32. In Example 3, the solution $y(x)$ becomes unbounded as $x \to \pm\infty$. Nevertheless $y(x)$ is asymptotic to a curve as $x \to -\infty$ and to a different curve as $x \to \infty$. What are the equations of these curves?

33. The differential equation $dy/dx = P(x) + Q(x)y + R(x)y^2$ is known as **Ricatti's equation.**

(a) A Ricatti equation can be solved by a succession of two substitutions *provided* we know a particular solution y_1 of the equation. Show that the substitution $y = y_1 + u$ reduces Ricatti's equation to a Bernoulli equation (4) with $n = 2$. The Bernoulli equation can then be reduced to a linear equation by the substitution $w = u^{-1}$.

(b) Find a one-parameter family of solutions for the differential equation

$$\frac{dy}{dx} = -\frac{4}{x^2} - \frac{1}{x}y + y^2,$$

where $y_1 = 2/x$ is a known solution of the equation.

34. Devise an appropriate substitution to solve $xy' = y \ln(xy)$.

2.6 A Numerical Solution

Note to the Instructor

In this section we are going to consider only the simplest of numerical methods, Euler's method, as a means of introducing the concept of a numerical solver. A more extensive treatment of numerical techniques for ODEs is found in Chapter 6.

■ **Using the Tangent Line** One of the simplest techniques for approximating a solution of the initial value problem

$$y' = f(x, y), \qquad y(x_0) = y_0 \tag{1}$$

is to use tangent lines. For example, let $y(x)$ denote the unknown solution of the initial-value problem $y' = 0.1\sqrt{y} + 0.4x^2$, $y(2) = 4$. The nonlinear differential equation cannot be solved directly by the methods considered in Sections 2.2, 2.4, and 2.5, but we can still find approximate numerical values of the unknown $y(x)$. Specifically, suppose we wish to know the value of $y(2.5)$. The IVP has a solution, and, as the flow of the direction field in Figure 2.28(a) suggests, a solution curve must have a shape similar to the curve shown in color.

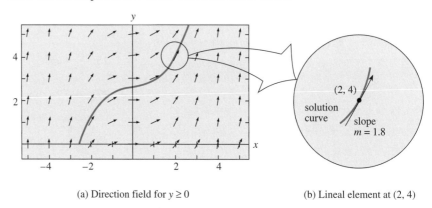

(a) Direction field for $y \geq 0$ (b) Lineal element at (2, 4)

Figure 2.28

The direction field was generated with lineal elements passing through points in a grid with integer coordinates. As the solution curve passes through the initial point (2, 4), the lineal element at this point is a tangent line with slope given by $f(2, 4) = 0.1\sqrt{4} + 0.4(2)^2 = 1.8$. As is obvious in Figure 2.28(a) and the "zoom in" in Figure 2.28(b), when x is close to 2 the points on the solution curve are close to the points on the tangent line (the lineal element). Using the point (2, 4), the slope $f(2, 4) = 1.8$, and the point-slope form of a line, we find that an equation of the tangent line is $y = L(x)$, where $L(x) = 1.8x + 0.4$. This last equation, called a **linearization** of $y(x)$ at $x = 2$, can be used to approximate $y(x)$ within a small neighborhood of $x = 2$. If $y_1 = L(x_1)$ denotes the value of the y-coordinate on the tangent line and $y(x_1)$ is the

y-coordinate on the solution curve corresponding to an x-coordinate x_1 that is close to $x = 2$, then $y(x_1) \approx y_1$. If we choose, say, $x_1 = 2.1$, then $y_1 = L(2.1) = 1.8(2.1) + 0.4 = 4.18$, and so $y(2.1) \approx 4.18$.

■ **Euler's Method** To generalize the procedure just illustrated, we use the linearization of the unknown solution $y(x)$ of (1) at $x = x_0$:

$$L(x) = f(x_0, y_0)(x - x_0) + y_0. \qquad (2)$$

The graph of this linearization is a straight line tangent to the graph of $y = y(x)$ at the point (x_0, y_0). We now let h be a positive increment of the x-axis, as shown in Figure 2.29. Then by replacing x by $x_1 = x_0 + h$ in (2) we get

$$L(x_1) = f(x_0, y_0)(x_0 + h - x_0) + y_0 \qquad \text{or} \qquad y_1 = y_0 + hf(x_0, y_0),$$

where $y_1 = L(x_1)$. The point (x_1, y_1) on the tangent line is an approximation to the point $(x_1, y(x_1))$ on the solution curve. Of course the accuracy of the approximation $y_1 \approx y(x_1)$ depends heavily on the size of the increment h. Usually we must choose this **step size** to be "reasonably small." We now repeat the process using a second "tangent line" at (x_1, y_1).* By identifying the new starting point as (x_1, y_1) with (x_0, y_0) in the above discussion, we obtain an approximation $y_2 \approx y(x_2)$ corresponding to two steps of length h from x_0, that is, $x_2 = x_1 + h = x_0 + 2h$ and

$$y(x_2) = y(x_0 + 2h) = y(x_1 + h) \approx y_2 = y_1 + hf(x_1, y_1).$$

Continuing in this manner, we see that y_1, y_2, y_3, \ldots, and so on, can be defined recursively by the general formula

$$y_{n+1} = y_n + hf(x_n, y_n), \qquad (3)$$

where $x_n = x_0 + nh$, $n = 0, 1, 2, \ldots$. This procedure of using successive "tangent lines" is called **Euler's method.**

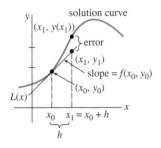

Figure 2.29

Example 1 Euler's Method

Consider the initial-value problem $y' = 0.1\sqrt{y} + 0.4x^2$, $y(2) = 4$. Use Euler's method to obtain an approximation to $y(2.5)$ using first $h = 0.1$ and then $h = 0.05$.

SOLUTION With the identification $f(x, y) = 0.1\sqrt{y} + 0.4x^2$, (3) becomes

$$y_{n+1} = y_n + h(0.1\sqrt{y_n} + 0.4x_n^2).$$

Then for $h = 0.1$, $x_0 = 2$, $y_0 = 4$, and $n = 0$, we find

$$y_1 = y_0 + h(0.1\sqrt{y_0} + 0.4x_0^2) = 4 + 0.1(0.1\sqrt{4} + 0.4(2)^2) = 4.18,$$

which, as we have already seen, is an estimate to the value of $y(2.1)$. However, if we use the smaller step size $h = 0.05$, it takes two steps to reach $x = 2.1$. From

$$y_1 = 4 + 0.05(0.1\sqrt{4} + 0.4(2)^2) = 4.09$$

$$y_2 = 4.09 + 0.05(0.1\sqrt{4.09} + 0.4(2.05)^2) = 4.18416187$$

we have $y_1 \approx y(2.05)$ and $y_2 \approx y(2.1)$. The remainder of the calculations were carried out using software; the results are summarized in Tables 2.1 and 2.2. We see in Tables 2.1 and 2.2 that it takes five steps with $h = 0.1$ and ten steps with $h = 0.05$, respectively, to get to $x = 2.5$. Also, each entry has been rounded to four decimal places.

*This is not an actual tangent line since (x_1, y_1) lies on the first tangent and not on the solution curve.

Table 2.1 *h* = 0.1

x_n	y_n
2.00	4.0000
2.10	4.1800
2.20	4.3768
2.30	4.5914
2.40	4.8244
2.50	5.0768

Table 2.2 *h* = 0.05

x_n	y_n
2.00	4.0000
2.05	4.0900
2.10	4.1842
2.15	4.2826
2.20	4.3854
2.25	4.4927
2.30	4.6045
2.35	4.7210
2.40	4.8423
2.45	4.9686
2.50	5.0997

In Example 2 we apply Euler's method to a differential equation for which we have already found a solution. We do this to compare the values of the approximations y_n at each step with the true values of the solution $y(x_n)$ of the initial-value problem.

Example 2 Comparison of Approximate and Exact Values

Consider the initial-value problem $y' = 0.2xy$, $y(1) = 1$. Use Euler's method to obtain an approximation to $y(1.5)$ using first $h = 0.1$ and then $h = 0.05$.

SOLUTION With the identification $f(x, y) = 0.2xy$, (3) becomes

$$y_{n+1} = y_n + h(0.2x_ny_n),$$

where $x_0 = 1$ and $y_0 = 1$. Again with the aid of computer software we obtain the values in Tables 2.3 and 2.4.

Table 2.3 *h* = 0.1

x_n	y_n	True Value	Abs. Error	% Rel. Error
1.00	1.0000	1.0000	0.0000	0.00
1.10	1.0200	1.0212	0.0012	0.12
1.20	1.0424	1.0450	0.0025	0.24
1.30	1.0675	1.0714	0.0040	0.37
1.40	1.0952	1.1008	0.0055	0.50
1.50	1.1259	1.1331	0.0073	0.64

Table 2.4 *h* = 0.05

x_n	y_n	True Value	Abs. Error	% Rel. Error
1.00	1.0000	1.0000	0.0000	0.00
1.05	1.0100	1.0103	0.0003	0.03
1.10	1.0206	1.0212	0.0006	0.06
1.15	1.0318	1.0328	0.0009	0.09
1.20	1.0437	1.0450	0.0013	0.12
1.25	1.0562	1.0579	0.0016	0.16
1.30	1.0694	1.0714	0.0020	0.19
1.35	1.0833	1.0857	0.0024	0.22
1.40	1.0980	1.1008	0.0028	0.25
1.45	1.1133	1.1166	0.0032	0.29
1.50	1.1295	1.1331	0.0037	0.32

In Example 1, the true values were calculated from the known solution $y = e^{0.1(x^2-1)}$ (verify). Also, the **absolute error** is defined to be

$$|true\ value - approximation|.$$

The **relative error** and **percentage relative error** are, in turn,

$$\frac{absolute\ error}{|true\ value|} \quad \text{and} \quad \frac{absolute\ error}{|true\ value|} \times 100.$$

It is apparent by comparing Tables 2.3 and 2.4 that the accuracy of the approximations improve as the step size h decreases. Also, we see that even though the percentage relative error is growing with each step, it does not appear to be that bad. But you should not be deceived by one example. If we simply change the coefficient of the right side of the DE in Example 2 from 0.2 to 2, then at $x_n = 1.5$ the percentage relative errors increase dramatically. See Problem 4 in Exercises 2.6.

 Note

Euler's method is just one of many different ways a solution of a differential equation can be approximated. Although attractive for its simplicity, *Euler's method is seldom used in serious calculations.* We have introduced this topic simply to give you a first taste of numerical methods. We will go into greater detail and discuss methods that give significantly greater accuracy, notably the **fourth-order Runge–Kutta method,** in Chapter 6.

■ **Numerical Solvers** Regardless of whether we can actually find an explicit or implicit solution, if a solution of a differential equation exists, it represents a smooth curve in the Cartesian plane. The basic idea behind *any* numerical method for ordinary differential equations is to somehow approximate the *y*-values of a solution for preselected values of *x*. We start at a specified initial point (x_0, y_0) on a solution curve and proceed to calculate in a step-by-step fashion a sequence of points $(x_1, y_1), (x_2, y_2), \ldots, (x_n, y_n)$ whose *y*-coordinates y_i approximate the *y*-coordinates $y(x_i)$ of points $(x_1, y(x_1)), (x_2, y(x_2)), \ldots, (x_n, y(x_n))$ that lie on the graph of the usually unknown solution $y(x)$. By taking the *x*-coordinates close together (that is, for small values of *h*) and by joining the points $(x_1, y_1), (x_2, y_2), \ldots, (x_n, y_n)$ with short line segments, we obtain a polygonal curve that appears smooth and whose qualitative characteristics we hope are close to those of an actual solution curve. Drawing curves is something well suited to a computer. A computer program written to either implement a numerical method or to render a visual representation of an approximate solution curve fitting the numerical data produced by this method is referred to as a **numerical solver.** There are many different numerical solvers commercially available, either embedded in a larger software package such as a computer algebra system or as a stand-alone package. Some software packages simply plot the generated numerical approximations, whereas others generate both hard numerical data and the corresponding approximate solution curves. As an illustration of the connect-the-dots nature of the graphs produced by a numerical solver, the two black polygonal graphs in Figure 2.30 are the "solution curves" for the initial-value problem $y' = 0.2xy$, $y(0) = 1$, on the interval $[0, 4]$ obtained from Euler's method and the Runge–Kutta method using the step size $h = 1$. The colored smooth curve is the graph of the exact solution $y = e^{0.1x^2}$ of the IVP. Notice in Figure 2.30 that even with the ridiculously large step size of $h = 1$, the Runge–Kutta method produces the more believable "solution curve." The approximation curve obtained from the Runge–Kutta method is indistinguishable from the actual solution curve on the interval $[0, 4]$ when a more typical step size of $h = 0.1$ is used.

■ **Using a Numerical Solver** Knowledge of the various numerical methods is not necessary in order to use a numerical solver. A solver usually requires that the differential equation be expressed in normal form $dy/dx = f(x, y)$. Numerical solvers that generate only curves usually require that you supply $f(x, y)$ and the initial data x_0 and y_0 and specify the desired numerical method. If the idea is to approximate the numerical value of $y(a)$, then a solver may additionally require that you state a value for h, or, equivalently, give the number of steps that you want to take to get from $x = x_0$ to $x = a$. For example, if we want to approximate $y(4)$ for the IVP illustrated in Figure 2.30, then, starting at $x = 0$, it takes four steps to reach $x = 4$ with a step size of $h = 1$; forty steps is equivalent to a step size of $h = 0.1$. Although it is not our intention to delve here into the many problems that one can encounter when attempting to approximate mathematical quantities, you should be at least aware of the fact that a numerical solver may break down near certain points or give an incomplete or misleading picture when applied to some first-order differential equations in the normal form. Figure 2.31 illustrates the graph obtained by applying Euler's method to a certain first-order initial value problem $dy/dx = f(x, y)$, $y(0) = 1$. Equivalent results were obtained using three different commercial numerical solvers, yet the graph is hardly a plausible solution curve. (Why?) There are several avenues of recourse when a numerical solver has difficulties; three of the more obvious are: decrease the step size, use another numerical method, or try a different numerical solver.

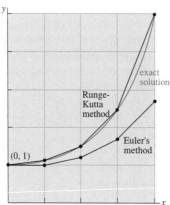

Figure 2.30 Comparison of numerical methods

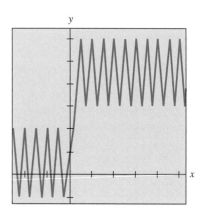

Figure 2.31

E X E R C I S E S 2 . 6 Answers to odd-numbered problems begin on page A-27.

In Problems 1 and 2, use Euler's method to obtain a four-decimal approximation of the indicated value. Carry out the recursion of (3) by hand first, using $h = 0.1$ and then $h = 0.05$.

1. $y' = 2x - 3y + 1$, $y(1) = 5$; $y(1.2)$

2. $y' = x + y^2$, $y(0) = 0$; $y(0.2)$

In Problems 3 and 4, use Euler's method to obtain a four-decimal approximation of the indicated value. First use $h = 0.1$ and then $h = 0.05$. Find an explicit solution for each initial-value problem and then construct tables similar to Tables 2.3 and 2.4.

3. $y' = y$, $y(0) = 1$; $y(1.0)$

4. $y' = 2xy$, $y(1) = 1$; $y(1.5)$

In Problems 5–10, use a numerical solver and Euler's method to obtain a four-decimal approximation to the indicated value. First use $h = 0.1$ and then $h = 0.05$.

5. $y' = e^{-y}$, $y(0) = 0$; $y(0.5)$

6. $y' = x^2 + y^2$, $y(0) = 1$; $y(0.5)$

7. $y' = (x - y)^2$, $y(0) = 0.5$; $y(0.5)$

8. $y' = xy + \sqrt{y}$, $y(0) = 1$; $y(0.5)$

9. $y' = xy^2 - \dfrac{y}{x}$, $y(1) = 1$; $y(1.5)$

10. $y' = y - y^2$, $y(0) = 0.5$; $y(0.5)$

In Problems 11 and 12, use a numerical solver to graph the solution curve for the given initial-value problem. First use Euler's method and then the fourth-order Runge–Kutta method. Use $h = 0.25$ in each case. Superimpose both solution curves on the same coordinate axes. If possible, use a different color for each curve. Repeat using $h = 0.1$ and $h = 0.05$.

11. $y' = 2(\cos x)y, \quad y(0) = 1$ **12.** $y' = y(10 - 2y), \quad y(0) = 1$

13. Use a numerical solver and Euler's method to approximate $y(1.0)$ where $y(x)$ is the solution to $y' = 2xy^2, \ y(0) = 1$. First use $h = 0.1$ and then $h = 0.05$. Repeat using the fourth-order Runge–Kutta method. Explain why the approximations to $y(1.0)$ differ so greatly.

2.7 Linear Models

■ **Growth and Decay** The initial-value problem

$$\frac{dx}{dt} = kx, \qquad x(t_0) = x_0, \tag{1}$$

where k is the constant of proportionality, serves as a model for diverse phenomena involving either **growth** or **decay.** We have seen in Section 1.3 that in biology, over short periods of time, the rate of growth of certain populations (bacteria, small animals) is observed to be proportional to the population present at time t. If a population at some arbitrary initial time t_0 is known, then the solution of (1) can be used to predict the population in the future—that is, at times $t > t_0$. The constant of proportionality k in (1) can be determined from the solution of the initial-value problem using a subsequent measurement of x at some time $t_1 > t_0$. In physics and chemistry (1) is seen in the form of a *first-order reaction,* that is, a reaction whose rate or velocity dx/dt is directly proportional to the first power of the reactant concentration x at time t. The decomposition or decay of U-238 (uranium) by radioactivity into Th-234 (thorium) is a first-order reaction.

Example 1 Bacterial Growth

A culture initially has P_0 number of bacteria. At $t = 1$ hr the number of bacteria is measured to be $\frac{3}{2} P_0$. If the rate of growth is proportional to the number of bacteria $P(t)$ present at time t, determine the time necessary for the number of bacteria to triple.

SOLUTION We first solve the differential equation in (1) with the symbol x replaced by P. With $t_0 = 0$ the initial condition is $P(0) = P_0$. We then use the empirical observation that $P(1) = \frac{3}{2} P_0$ to determine the constant of proportionality k.

Notice that the differential equation $dP/dt = kP$ is both separable and linear. When it is put in the standard form of a linear first-order DE,

$$\frac{dP}{dt} - kP = 0,$$

we can see by inspection that the integrating factor is e^{-kt}. Multiplying both sides of the equation by this term immediately gives

$$\frac{d}{dt}[e^{-kt} P] = 0.$$

Integrating both sides of the last equation yields $e^{-kt}P = c$ or $P(t) = ce^{kt}$. At $t = 0$ it follows that $P_0 = ce^0 = c$, and so $P(t) = P_0 e^{kt}$. At $t = 1$ we have $\frac{3}{2} P_0 = P_0 e^k$ or $e^k = \frac{3}{2}$. From the last equation we get $k = \ln \frac{3}{2} = 0.4055$. Thus $P(t) = P_0 e^{0.4055t}$. To find the time at which the number of bacteria have tripled, we solve

Figure 2.32

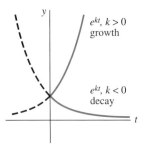

Figure 2.33

$3P_0 = P_0 e^{0.4055t}$ for t. It follows that $0.4055t = \ln 3$, and so

$$t = \frac{\ln 3}{0.4055} \approx 2.71 \text{ hrs.}$$

See Figure 2.32. ❏

Notice in Example 1 that the actual number P_0 of bacteria present at time $t = 0$ played no part in determining the time required for the number in the culture to triple. The time necessary for an initial population of, say, 100 or 1,000,000 bacteria to triple is still approximately 2.71 hours.

As shown in Figure 2.33, the exponential function e^{kt} increases as t increases for $k > 0$ and decreases as t increases for $k < 0$. Thus problems describing growth (whether of populations, bacteria, or even capital) are characterized by a positive value of k, whereas problems involving decay (as in radioactive disintegration) yield a negative k value. Accordingly, we say that k is either a **growth constant** ($k > 0$) or a **decay constant** ($k < 0$).

■ **Half-Life** In physics the **half-life** is a measure of the stability of a radioactive substance. The half-life is simply the time it takes for one-half of the atoms in an initial amount A_0 to disintegrate, or transmute, into the atoms of another element. The longer the half-life of a substance, the more stable it is. For example, the half-life of highly radioactive radium, Ra-226, is about 1700 years. In 1700 years one-half of a given quantity of Ra-226 is transmuted into radon, Rn-222. The most commonly occurring uranium isotope, U-238, has a half-life of approximately 4,500,000,000 years. In about 4.5 billion years, one-half of a quantity of U-238 is transmuted into lead, Pb-206.

Example 2 **Half-Life of Plutonium**

A breeder reactor converts relatively stable uranium-238 into the isotope pluto-nium-239. After 15 years it is determined that 0.043% of the initial amount A_0 of the plutonium has disintegrated. Find the half-life of this isotope if the rate of disintegration is proportional to the amount remaining.

SOLUTION Let $A(t)$ denote the amount of plutonium remaining at any time. As in Example 1, the solution of the initial-value problem

$$\frac{dA}{dt} = kA, \qquad A(0) = A_0,$$

is $A(t) = A_0 e^{kt}$. If 0.043% of the atoms of A_0 have disintegrated, then 99.957% of the substance remains. To find the decay constant k, we use $0.99957A_0 = A(15)$, that is, $0.99957A_0 = A_0 e^{15k}$. Solving for k then gives $k = \frac{1}{15} \ln 0.99957 = -0.00002867$. Hence $A(t) = A_0 e^{-0.00002867t}$. Now the half-life is the corresponding value of time at which $A(t) = A_0/2$. Solving for t gives $A_0/2 = A_0 e^{-0.00002867t}$ or $\frac{1}{2} = e^{-0.00002867t}$. The last equation yields

$$t = \frac{\ln 2}{0.00002867} \approx 24{,}180 \text{ years.}$$ ❏

■ **Carbon Dating** About 1950 the chemist Willard Libby devised a method of using radioactive carbon as a means of determining the approximate ages of fossils. The theory of **carbon dating** is based on the fact that the isotope carbon-14 is

produced in the atmosphere by the action of cosmic radiation on nitrogen. The ratio of the amount of C-14 to ordinary carbon in the atmosphere appears to be a constant, and as a consequence the proportionate amount of the isotope present in all living organisms is the same as that in the atmosphere. When an organism dies, the absorption of C-14, by either breathing or eating, ceases. Thus by comparing the proportionate amount of C-14 present, say, in a fossil with the constant ratio found in the atmosphere, it is possible to obtain a reasonable estimation of its age. The method is based on the knowledge that the half-life of the radioactive C-14 is approximately 5600 years. For his work Libby won the Nobel Prize for chemistry in 1960. Libby's method has been used to date wooden furniture in Egyptian tombs, the woven flax wrappings of the Dead Sea Scrolls, and the cloth of the enigmatic Shroud of Turin.

Example 3 Age of a Fossil

A fossilized bone is found to contain $\frac{1}{1000}$ the original amount of C-14. Determine the age of the fossil.

SOLUTION The starting point is again $A(t) = A_0 e^{kt}$. To determine the value of the decay constant k we use the fact that $A_0/2 = A(5600)$ or $A_0/2 = A_0 e^{5600k}$. From $5600k = \ln \frac{1}{2} = -\ln 2$ we then get $k = -(\ln 2)/5600 = -0.00012378$. Therefore $A(t) = A_0 e^{-0.00012378t}$. With $A(t) = A_0/1000$ we have $A_0/1000 = A_0 e^{-0.00012378t}$ so that $-0.00012378t = \ln \frac{1}{1000} = -\ln 1000$. Thus

$$t = \frac{\ln 1000}{0.00012378} \approx 55{,}800 \text{ years.} \qquad \square$$

The date found in Example 3 is really at the border of accuracy for this method. The usual carbon-14 technique is limited to about 9 half-lives of the isotope, or about 50,000 years. One reason is that the chemical analysis needed to obtain an accurate measurement of the remaining C-14 becomes somewhat formidable around the point of $A_0/1000$. Also, this analysis demands the destruction of a rather large sample of the specimen. If this measurement is accomplished indirectly, based on the actual radioactivity of the specimen, then it is very difficult to distinguish between the radiation from the fossil and the normal background radiation. But recently the use of a particle accelerator has enabled scientists to separate the C-14 from the stable C-12 directly. When the precise value of the ratio of C-14 to C-12 is computed, the accuracy of this method can be extended to 70,000–100,000 years. Other isotopic techniques, such as using potassium-40 and argon-40, can give dates of several million years. Nonisotopic methods based on the use of amino acids are also sometimes possible.

■ Newton's Law of Cooling

In equation (3) of Section 1.3 we saw that the mathematical formulation of Newton's empirical law of cooling of an object is given by the linear first-order differential equation

$$\frac{dT}{dt} = k(T - T_m), \tag{3}$$

where k is a constant of proportionality, $T(t)$ is the temperature of the object for $t > 0$, and T_m is the ambient temperature—that is, the temperature of the medium around the object. In Example 4 we assume that T_m is constant.

Example 4 Cooling of a Cake

When a cake is removed from an oven, its temperature is measured at 300°F. Three minutes later its temperature is 200°F. How long will it take for the cake to cool off to a room temperature of 70°F?

(a)

$T(t)$	t (min)
75°	20.1
74°	21.3
73°	22.8
72°	24.9
71°	28.6
70.5°	32.3

(b)

Figure 2.34

SOLUTION In (3) we make the identification $T_m = 70$. We must then solve the initial-value problem

$$\frac{dT}{dt} = k(T - 70), \qquad T(0) = 300 \tag{4}$$

and determine the value of k so that $T(3) = 200$.

Equation (4) is both linear and separable. Separating variables,

$$\frac{dT}{T - 70} = k\, dt,$$

yields $\ln|T - 70| = kt + c_1$, and so $T = 70 + c_2 e^{kt}$. When $t = 0$, $T = 300$, so that $300 = 70 + c_2$ gives $c_2 = 230$, and, therefore, $T = 70 + 230 e^{kt}$. Finally, the measurement $T(3) = 200$ leads to $e^{3k} = \frac{13}{23}$ or $k = \frac{1}{3} \ln \frac{13}{23} = -0.19018$. Thus

$$T(t) = 70 + 230 e^{-0.19018t}. \tag{5}$$

We note that (5) furnishes no finite solution to $T(t) = 70$ since $\lim_{t \to \infty} T(t) = 70$. Yet intuitively we expect the cake to reach the room temperature after a reasonably long period of time. How long is "long"? Of course, we should not be disturbed by the fact that the model (4) does not quite live up to our physical intuition. Parts (a) and (b) of Figure 2.34 clearly show that the cake will be approximately at room temperature in about one-half hour. ❏

■ **Mixtures** The mixing of two fluids sometimes gives rise to a linear first-order differential equation. When we discussed the mixing of two brine solutions in Section 1.3, we assumed that the rate $A'(t)$ at which the amount of salt in the mixing tank changes was a net rate:

$$\frac{dA}{dt} = \left(\begin{array}{c} rate\ of \\ substance\ entering \end{array} \right) - \left(\begin{array}{c} rate\ of \\ substance\ leaving \end{array} \right) = R_1 - R_2. \tag{6}$$

In Example 5 we solve equation (8) of Section 1.3.

Example 5 **Mixture of Two Salt Solutions**

Recall that the large tank considered in Section 1.3 initially held 300 gallons of a brine solution. Salt was entering and leaving the tank; a brine solution was being pumped into the tank at the rate of 3 gal/min, mixed with the solution there, and then pumped out of the tank at the rate of 3 gal/min. The concentration of the entering solution was 2 lb/gal, and so salt was entering the tank at the rate $R_1 = $ (2 lb/gal) \cdot (3 gal/min) = 6 lb/min and leaving the tank at the rate $R_2 = $ (3 gal/min) \cdot ($A/300$ lb/gal) = $A/100$ lb/min. From these data and (6) we get equation (8) of Section 1.3. Let us pose this question: If 50 pounds of salt was dissolved in the initial 300 gallons, how much salt would be in the tank after a long time?

SOLUTION To find $A(t)$ we solve the initial-value problem

$$\frac{dA}{dt} = 6 - \frac{A}{100}, \qquad A(0) = 50.$$

Note here that the side condition is the initial amount of salt, $A(0) = 50$, and not the initial amount of liquid. Now since the integrating factor of the linear differential equation is $e^{t/100}$, we can write the equation as

(a)

t (min)	A (lb)
50	266.41
100	397.67
150	477.27
200	525.57
300	572.62
400	589.93

(b)

Figure 2.35

$$\frac{d}{dt}[e^{t/100}\,A] = 6e^{t/100}.$$

Integrating the last equation and solving for A gives the general solution $A = 600 + ce^{-t/100}$. When $t = 0$, $A = 50$, so we find that $c = -550$. Thus the amount of salt in the tank at any time t is given by

$$A(t) = 600 - 550e^{-t/100}. \tag{7}$$

The solution (7) was used to construct the table in Figure 2.35(b). Also, it can be seen from (7) and Figure 2.35(a) that $A \to 600$ as $t \to \infty$. Of course, this is what we would expect in this case; over a long time the number of pounds of salt in the solution must be (300 gal)(2 lb/gal) = 600 lb. ❑

In Example 5 we assumed that the rate at which the solution was pumped in was the same as the rate at which the solution was pumped out. However, this need not be the situation; the mixed brine solution could be pumped out at a rate faster or slower than the rate at which the other solution was pumped in. For example, if the well-stirred solution in Example 5 is pumped out at a slower rate of, say, 2 gal/min, liquid will accumulate in the tank at the rate of (3 − 2) gal/min = 1 gal/min. After t minutes the tank will contain $300 + t$ gallons of brine. The rate at which the salt is leaving is then

$$R_2 = (2 \text{ gal/min})\left(\frac{A}{300 + t}\text{ lb/gal}\right).$$

Hence equation (6) becomes

$$\frac{dA}{dt} = 6 - \frac{2A}{300 + t} \quad \text{or} \quad \frac{dA}{dt} + \frac{2}{300 + t}A = 6.$$

You should verify that the solution of the last equation subject to $A(0) = 50$ is $A(t) = 600 + 2t - (4.95 \times 10^7)(300 + t)^{-2}$.

Figure 2.36 *LR* series circuit

■ **Series Circuits** For a series circuit containing only a resistor and an inductor, Kirchhoff's second law states that the sum of the voltage drop across the inductor $(L(di/dt))$ and the voltage drop across the resistor (iR) is the same as the impressed voltage $(E(t))$ on the circuit. See Figure 2.36.

Thus we obtain the linear differential equation for the current $i(t)$,

$$L\frac{di}{dt} + Ri = E(t), \tag{8}$$

where L and R are constants known as the inductance and the resistance, respectively. The current $i(t)$ is also called the **response** of the system.

The voltage drop across a capacitor with capacitance C is given by $q(t)/C$, where q is the charge on the capacitor. Hence, for the series circuit shown in Figure 2.37, Kirchhoff's second law gives

$$Ri + \frac{1}{C}q = E(t). \tag{9}$$

But current i and charge q are related by $i = dq/dt$, so (9) becomes the linear differential equation

$$R\frac{dq}{dt} + \frac{1}{C}q = E(t). \tag{10}$$

Figure 2.37 *RC* series circuit

Example 6 Series Circuit

A 12-volt battery is connected to a series circuit in which the inductance is ½ henry and the resistance is 10 ohms. Determine the current i if the initial current is zero.

SOLUTION From (8) we see that we must solve

$$\frac{1}{2}\frac{di}{dt} + 10i = 12$$

subject to $i(0) = 0$. First, we multiply the differential equation by 2 and read off the integrating factor e^{20t}. We then obtain

$$\frac{d}{dt}[e^{20t}i] = 24e^{20t}.$$

Integrating each side of the last equation and solving for i gives $i(t) = \frac{6}{5} + ce^{-20t}$. Now $i(0) = 0$ implies $0 = \frac{6}{5} + c$ or $c = -\frac{6}{5}$. Therefore the response is $i(t) = \frac{6}{5} - \frac{6}{5}e^{-20t}$. ❑

From (4) of Section 2.3 we can write a general solution of (8):

$$i(t) = \frac{e^{-(R/L)t}}{L}\int e^{(R/L)t} E(t)\, dt + ce^{-(R/L)t}. \tag{11}$$

In particular, when $E(t) = E_0$ is a constant, (11) becomes

$$i(t) = \frac{E_0}{R} + ce^{-(R/L)t}. \tag{12}$$

Note that as $t \to \infty$, the second term in (12) approaches zero. Such a term is usually called a **transient term;** any remaining terms are called the **steady-state** part of the solution. In this case E_0/R is also called the **steady-state current;** for large values of time it then appears that the current in the circuit is simply governed by Ohm's law ($E = iR$).

Remarks

The solution $P(t) = P_0 e^{0.4055t}$ of the initial-value problem in Example 1 described the population of a colony of bacteria at any time $t > 0$. Of course, $P(t)$ is a continuous function that takes on *all* real numbers in the interval $P_0 \le P < \infty$. But since we are talking about a population, common sense dictates that P can take on only positive integer values. Moreover, we would not expect the population to grow continuously—that is, every second, every microsecond, and so on—as predicted by our solution; there may be intervals of time $[t_1, t_2]$ over which there is no growth at all. Perhaps, then, the graph shown in Figure 2.38(a) is a more realistic description of P than is the graph of an exponential function. Using a continuous function to describe a discrete phenomenon is often more a matter of convenience than of accuracy. However, for some purposes we may be satisfied if our model describes the system fairly closely when viewed macroscopically in time, as in Figures 2.38(b) and (c), rather than microscopically, as in Figure 2.38(a). Keep firmly in mind, a mathematical model is not reality.

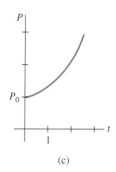

(a) (b) (c)

Figure 2.38

Growth and Decay

1. The population of a community is known to increase at a rate proportional to the number of people present at time t. If the population has doubled in 5 years, how long will it take to triple? To quadruple?

2. Suppose it is known that the population of the community in Problem 1 is 10,000 after 3 years. What was the initial population? What will be the population in 10 years?

3. The population of a town grows at a rate proportional to the population present at time t. Its initial population of 500 increases by 15% in 10 years. What will be the population in 30 years?

4. The population of bacteria in a culture grows at a rate proportional to the number of bacteria present at time t. After 3 hours it is observed that there are 400 bacteria present. After 10 hours there are 2000 bacteria present. What was the initial number of bacteria?

5. When a vertical beam of light passes through a transparent medium, the rate at which its intensity I decreases is proportional to $I(t)$, where t represents the thickness of the medium (in feet). In clear seawater, the intensity 3 feet below the surface is 25% of the initial intensity I_0 of the incident beam. What is the intensity of the beam 15 feet below the surface?

6. When interest is compounded continuously, the amount of money increases at a rate proportional to the amount S present at time t, that is, $dS/dt = rS$, where r is the annual rate of interest.

(a) Find the amount of money accrued at the end of 5 years when $5000 is deposited in a savings account drawing $5\frac{3}{4}$% annual interest compounded continuously.

(b) In how many years will the initial sum deposited have doubled?

(c) Use a calculator to compare the number obtained in part **(a)** with the amount $S = 5000(1 + \frac{1}{4}(0.0575))^{5(4)}$ that is accrued when interest is compounded quarterly.

7. The radioactive isotope of lead, Pb-209, decays at a rate proportional to the amount present at time t and has a half-life of 3.3 hours. If 1 gram of lead is present initially, how long will it take for 90% of the lead to decay?

8. Initially 100 milligrams of a radioactive substance were present. After 6 hours the mass had decreased by 3%. If the rate of decay is proportional to the amount of the substance present at time t, find the amount remaining after 24 hours.

9. Determine the half-life of the radioactive substance described in Problem 8.

10. (a) Consider the initial-value problem $dA/dt = kA$, $A(0) = A_0$, as the model for the decay of a radioactive substance. Show that in general the half-life T of the substance is $T = -(\ln 2)/k$.

(b) Show that the solution of the initial-value problem in part **(a)** can be written $A(t) = A_0 2^{-t/T}$.

(c) If a radioactive substance has a half-life T given in part **(a)**, how long will it take an initial amount A_0 of the substance to decay to $\frac{1}{8} A_0$?

Carbon Dating

11. Archaeologists used pieces of burned wood, or charcoal, to date prehistoric paintings and drawings on walls and ceilings in a cave in Lascaux, France. See Figure 2.39. Use the information in Example 3 to determine the approximate age of a piece of wood, if it was found that 85.5% of the C-14 had decayed.

Figure 2.39

12. The Shroud of Turin, which shows the negative image of a body of a crucified man, is believed by many to be the burial shroud of Jesus of Nazareth. See Figure 2.40. In 1988 the Vatican granted permission to have the shroud carbon dated. Three independent scientific laboratories analyzed the cloth and concluded that the shroud was approximately 660 years old,* an age consistent with its historical appearance. Using this age, determine what percentage of the original amount of C-14 remained in the cloth as of 1988.

Figure 2.40

Newton's Law of Cooling/Warming

13. A thermometer is removed from a room where the air temperature is 70°F to the outside, where the temperature is 10°F. After ½ minute the thermometer reads 50°F. What is the reading of the thermometer at $t = 1$ min? How long will it take for the thermometer to reach 15°F?

14. A thermometer is taken from an inside room to the outside, where the air temperature is 5°F. After 1 minute the thermometer reads 55°F, and after 5 minutes the reading is 30°F. What was the initial temperature of the room?

15. A small metal bar, whose initial temperature was 20°C, is dropped into a container of boiling water. How long will it take the bar to reach 90°C if it is known that its temperature increased 2° in 1 second? How long will it take the bar to reach 98°C?

16. A thermometer reading 70°F is placed in an oven preheated to a constant temperature. Through a glass window in the oven door an observer records that the thermometer read 110°F after ½ minute and 145°F after 1 minute. How hot is the oven?

17. Suppose a medical examiner arrives at the scene of a homicide and finds that the temperature of the corpse is 82°F; 2 hours later its temperature is 75°F. If the air temperature at the crime scene is a constant 65°F, determine an approximate time of death of the victim using Newton's law of cooling (3).

* Some scholars have disagreed with this finding. For more information on this fascinating mystery, see the Shroud of Turin Website home page at http://www.shroud.com/

18. Mr. Jones puts two cups of coffee on the breakfast table at the same time. He immediately pours cream into his coffee from a pitcher that was sitting on the table for a long time. He then reads the morning paper for 5 minutes before taking his first sip. Mrs. Jones arrives at the table 5 minutes after the cups were set down, adds cream to her coffee, and takes a sip. Assume that Mr. and Mrs. Jones add exactly the same amount of cream to their cups. Determine who drinks the hotter cup of coffee.

Mixtures

19. A tank contains 200 liters of fluid in which 30 grams of salt is dissolved. Brine containing 1 gram of salt per liter is then pumped into the tank at a rate of 4 L /min; the well-mixed solution is pumped out at the same rate. Find the number $A(t)$ of grams of salt in the tank at time t.

20. Solve Problem 19 assuming that pure water is pumped into the tank.

21. A large tank is filled with 500 gallons of pure water. Brine containing 2 pounds of salt per gallon is pumped into the tank at a rate of 5 gal/min. The well-mixed solution is pumped out at the same rate. Find the number $A(t)$ of pounds of salt in the tank at time t. What is the concentration of the solution in the tank at $t = 5$ min?

22. Solve Problem 21 under the assumption that the solution is pumped out at a faster rate of 10 gal/min. When is the tank empty?

23. A large tank is partially filled with 100 gallons of fluid in which 10 pounds of salt is dissolved. Brine containing ½ pound of salt per gallon is pumped into the tank at a rate of 6 gal/min. The well-mixed solution is then pumped out at a slower rate of 4 gal/min. Find the number of pounds of salt in the tank after 30 minutes.

24. In Example 5 the size of the tank containing the salt mixture was not given. Suppose, as discussed on page 77, that the rate at which brine is pumped into the tank is the same but that the mixed solution is pumped out a rate of 2 gal/min. It stands to reason that since brine is accumulating in the tank at the rate of 1 gal/min any finite tank must eventually overflow. Now suppose that the tank has an open top and has a total capacity of 400 gallons.

(a) When will the tank overflow?

(b) What is the number of pounds of salt in the tank at the instant it overflows?

(c) Assume that the tank is overflowing and that brine solution continues to be pumped in at a rate of 3 gal/min and the well-stirred solution continues to be pumped out at a rate of 2 gal/min. Devise a method for determining the number of pounds of salt in the tank at $t = 150$ min.

(d) Determine the number of pounds of salt in the tank as $t \to \infty$. Does your answer agree with your intuition?

(e) Use a graphing utility to obtain the graph $A(t)$ on the interval [0, 500).

Series Circuits

25. A 30-volt electromotive force is applied to an LR series circuit in which the inductance is 0.1 henry and the resistance is 50 ohms. Find the current $i(t)$ if $i(0) = 0$. Determine the current as $t \to \infty$.

26. Solve equation (8) under the assumption that $E(t) = E_0 \sin \omega t$ and $i(0) = i_0$.

27. A 100-volt electromotive force is applied to an RC series circuit in which the resistance is 200 ohms and the capacitance is 10^{-4} farad. Find the charge $q(t)$ on the capacitor if $q(0) = 0$. Find the current $i(t)$.

28. A 200-volt electromotive force is applied to an RC series circuit in which the resistance is 1000 ohms and the capacitance is 5×10^{-6} farad. Find the charge $q(t)$ on the capacitor if $i(0) = 0.4$. Determine the charge and current at $t = 0.005$ s. Determine the charge as $t \to \infty$.

29. An electromotive force

$$E(t) = \begin{cases} 120, & 0 \le t \le 20 \\ 0, & t > 20 \end{cases}$$

is applied to an LR series circuit in which the inductance is 20 henries and the resistance is 2 ohms. Find the current $i(t)$ if $i(0) = 0$.

30. Suppose an RC series circuit has a variable resistor. If the resistance at time t is given by $R = k_1 + k_2 t$ and k_1 and k_2 are known positive constants, then (10) becomes

$$(k_1 + k_2 t) \frac{dq}{dt} + \frac{1}{C} q = E(t).$$

If $E(t) = E_0$ and $q(0) = q_0$, show that

$$q(t) = E_0 C + (q_0 - E_0 C)\left(\frac{k_1}{k_1 + k_2 t}\right)^{1/Ck_2}.$$

Miscellaneous Linear Models

31. In (14) of Section 1.3 we saw that a differential equation governing the velocity v of a falling mass subject to air resistance proportional to the instantaneous velocity is

$$m\frac{dv}{dt} = mg - kv,$$

where $k > 0$ is a constant of proportionality. The positive direction is downward.

(a) Solve the equation subject to the initial condition $v(0) = v_0$.

(b) Use the solution in part **(a)** to determine the limiting, or terminal, velocity of the mass. We saw how to determine the terminal velocity without solving the DE in Problem 25 in Exercises 2.1.

(c) If the distance s, measured from the point where the mass was released above ground, is related to velocity by $ds/dt = v$, find an explicit expression for s if $s(0) = 0$.

32. Suppose a small cannonball weighing 16 lb is shot vertically upward with an initial velocity $v_0 = 300$ ft/s. The answer to the question, How high does the cannonball go? depends on whether we take air resistance into account.

(a) If air resistance is ignored and the positive direction is upward, then a model governing the state of the cannonball is given by $d^2s/dt^2 = -g$ (equation (12) of Section 1.3). Integrate this equation once to show that velocity is $v(t) = -32t + 300$, where we have taken $g = 32$ ft/s^2. Use the fact that $ds/dt = v(t)$ to determine the height $s(t)$ of the cannonball measured from ground level. Find the maximum height attained by the cannonball.

(b) Now if air resistance is assumed to be proportional to instantaneous velocity, then it stands to reason that the maximum height attained by the cannonball must be less than that in part **(a)**. Show this by supposing that the constant of proportionality is $k = 0.0025$. [*Hint:* You will have to slightly modify the DE in Problem 31.]

33. Computer Lab Assignment It is well known that the model in which air resistance is ignored, part **(a)** of Problem 32, predicts that the time t_a it takes the cannonball to attain its maximum height is the same as the time t_d it takes the cannonball to fall from the maximum height to the ground. Moreover, the magnitude of the impact velocity v_i will be the same as the initial velocity v_0 of the cannonball. Verify these results. Then use the model that takes air resistance into account, part **(b)** of Problem 32, to compare the value of t_a with t_d and the value of the magnitude of v_i with v_0. The root finding application of a CAS may be useful here.

34. A skydiver weighs 125 lb and her parachute and equipment combined weigh another 35 lb. After exiting from a plane at an altitude of 15,000 feet, she waits 15 seconds and opens her parachute. Assume the constant of proportionality in the model in Problem 31 has the value $k = 0.5$ during free fall and $k = 10$ after the parachute is opened. Assume her initial velocity upon leaving the plane is zero. What is her velocity and how far has she traveled 20 seconds after leaving the plane? How does her velocity at 20 seconds compare with her terminal velocity? How long does it take her to reach the ground? [*Hint:* Think in terms of two distinct IVPs.]

35. As a raindrop falls, it evaporates while retaining its spherical form. If the further assumptions are made that the rate at which the raindrop evaporates is proportional to its surface area and that air resistance is negligible, then velocity $v(t)$ of the raindrop is given by

$$\frac{dv}{dt} + \frac{3(k/\rho)}{(k/\rho)t + r_0} v = g.$$

Here ρ is the density of water, r_0 is the radius of the raindrop at $t = 0$, $k < 0$ is the constant of proportionality, and the downward direction is taken to be the positive direction.

(a) Solve for $v(t)$ if the raindrop falls from rest.

(b) Reread Problem 34 of Exercises 1.3 and then show that the radius of the raindrop at time t is $r(t) = (k/\rho)t + r_0$.

(c) If $r_0 = 0.01$ ft and $r = 0.007$ ft 10 seconds after the raindrop falls from a cloud, determine the time at which the raindrop has evaporated completely.

36. The differential equation $dP/dt = (k \cos t)P$, where k is a positive constant, is a mathematical model for a population $P(t)$ that undergoes yearly seasonal fluctuations. Solve the equation subject to $P(0) = P_0$. Use a graphing utility to obtain the graph of the solution for different choices of P_0.

37. In one model of the changing population $P(t)$ of a community, it is assumed that

$$\frac{dP}{dt} = \frac{dB}{dt} - \frac{dD}{dt},$$

where dB/dt and dD/dt are the birth and death rates, respectively.

(a) Solve for $P(t)$ if $dB/dt = k_1 P$ and $dD/dt = k_2 P$.

(b) Analyze the cases $k_1 > k_2$, $k_1 = k_2$, and $k_1 < k_2$.

38. The following system of differential equations is encountered in the study of a series of elements that are decaying by radioactivity:

$$\frac{dx}{dt} = -\lambda_1 x$$

$$\frac{dy}{dt} = \lambda_1 x - \lambda_2 y,$$

where λ_1 and λ_2 are constants. It does not take any specialized knowledge of systems to solve this particular system. Solve for $x(t)$ and $y(t)$ subject to $x(0) = x_0$ and $y(0) = y_0$.

39. When forgetfulness is taken into account, the rate of memorization of a subject is given by

$$\frac{dA}{dt} = k_1(M - A) - k_2 A,$$

where $k_1 > 0$, $k_2 > 0$, $A(t)$ is the amount to be memorized in time t, M is the total amount to be memorized, and $M - A$ is the amount remaining to be memorized. (See Problems 23 and 24 in Exercises 1.3.)

(a) Since the DE is autonomous, use the phase portrait concept of Section 2.1 to find the limiting value of $A(t)$ as $t \to \infty$. Interpret the result.

(b) Solve for $A(t)$ subject to $A(0) = 0$. Sketch the graph of $A(t)$ and verify your prediction in part **(a)**.

40. The rate at which a drug disseminates into the bloodstream is governed by $dx/dt = r - kx$, where r and k are positive constants. The function $x(t)$ describes the concentration of the drug in the bloodstream at time t.

(a) Since the DE is autonomous, use the phase portrait concept of Section 2.1 to find the limiting value of $x(t)$ as $t \to \infty$.

(b) Solve the DE subject to $x(0) = 0$. Sketch the graph of $x(t)$ and verify your prediction in part **(a).** At what time is the concentration one-half this limiting value?

2.8 Nonlinear Models

■ **Population Dynamics** If $P(t)$ denotes the size of a population at time t, the model for exponential growth begins with the assumption that $dP/dt = kP$ for some $k > 0$. In this model the **relative,** or **specific, growth rate** defined by

$$\frac{dP/dt}{P} \tag{1}$$

is assumed to be a constant k. True cases of exponential growth over long periods of time are hard to find, because the limited resources of the environment will at some time exert restrictions on the growth of a population. Thus (1) can be expected to decrease as P increases in size.

The assumption that the rate at which a population grows (or declines) is dependent only on the number present and not on any time-dependent mechanisms such as seasonal phenomena (see Problem 31 in Exercises 1.3) can be stated as

$$\frac{dP/dt}{P} = f(P) \quad \text{or} \quad \frac{dP}{dt} = Pf(P). \tag{2}$$

The differential equation in (2), which is widely assumed in models of animal populations, is called the **density-dependent hypothesis.**

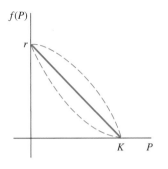

$f(P)$

r

K P

Figure 2.41

■ **Logistic Equation** Suppose an environment is capable of sustaining no more than a fixed number of K individuals in its population. The quantity K is called the **carrying capacity** of the environment. Hence, for the function f in (2) we have $f(K) = 0$, and we simply let $f(0) = r$. Figure 2.41 shows three functions f that satisfy these two conditions. The simplest assumption that we can make is that $f(P)$ is linear—that is, $f(P) = c_1 P + c_2$. If we use the conditions $f(0) = r$ and $f(K) = 0$, we find, in turn, $c_2 = r$, $c_1 = -r/K$, and so f takes on the form $f(P) = r - (r/K)P$. Equation (2) becomes

$$\frac{dP}{dt} = P\left(r - \frac{r}{K} P\right). \tag{3}$$

Relabeling constants, the nonlinear equation (3) is the same as

$$\frac{dP}{dt} = P(a - bP). \tag{4}$$

Around 1840 the Belgian mathematician/biologist P. F. Verhulst was concerned with mathematical models for predicting the human population of various countries. One of the equations he studied was (4), where $a > 0$, $b > 0$. Equation (4) came to be known as the **logistic equation,** and its solution is called the **logistic function.** The graph of a logistic function is called a **logistic curve.**

The linear differential equation $dP/dt = kP$ does not provide a very accurate model for population when the population itself is very large. Overcrowded conditions, with the resulting detrimental effects on the environment, such as pollution and excessive and competitive demands for food and fuel, can have an inhibiting effect on population growth. As we shall now see, the solution of (4) is bounded as $t \to \infty$. If we rewrite (4) as $dP/dt = aP - bP^2$, the nonlinear term $-bP^2$, $b > 0$, can be interpreted as an "inhibition" or "competition" term. Also, in most applications, the positive constant a is much larger than the constant b.

Logistic curves have proved to be quite accurate in predicting the growth patterns, in a limited space, of certain types of bacteria, protozoa, water fleas (*Daphnia*), and fruit flies (*Drosophila*).

■ **Solution of the Logistic Equation** One method of solving (4) is by separation of variables. Decomposing the left side of $dP/P(a - bP) = dt$ into partial fractions and integrating gives

$$\left(\frac{1/a}{P} + \frac{b/a}{a - bP}\right) dP = dt$$

$$\frac{1}{a}\ln|P| - \frac{1}{a}\ln|a - bP| = t + c$$

$$\ln\left|\frac{P}{a - bP}\right| = at + ac$$

$$\frac{P}{a - bP} = c_1 e^{at}.$$

It follows from the last equation that

$$P(t) = \frac{ac_1 e^{at}}{1 + bc_1 e^{at}} = \frac{ac_1}{bc_1 + e^{-at}}.$$

If $P(0) = P_0$, $P_0 \neq a/b$, we find $c_1 = P_0/(a - bP_0)$, and so, after substituting and simplifying, the solution becomes

$$P(t) = \frac{aP_0}{bP_0 + (a - bP_0)e^{-at}}. \tag{5}$$

■ **Graphs of** $P(t)$ The basic shape of the graph of the logistic function $P(t)$ can be obtained without too much effort. Although the variable t usually represents time and we are seldom concerned with applications in which $t < 0$, it is nonetheless of some interest to include this interval when displaying the various graphs of P. From (5) we see that

$$P(t) \rightarrow \frac{aP_0}{bP_0} = \frac{a}{b} \text{ as } t \rightarrow \infty \quad \text{and} \quad P(t) \rightarrow 0 \text{ as } t \rightarrow -\infty.$$

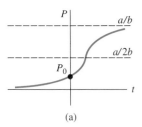

The dashed line $P = a/2b$ shown in Figure 2.42 corresponds to the ordinate of a point of inflection of the logistic curve. To show this, we differentiate (4) by the product rule:

$$\frac{d^2P}{dt^2} = P\left(-b\frac{dP}{dt}\right) + (a - bP)\frac{dP}{dt} = \frac{dP}{dt}(a - 2bP)$$

$$= P(a - bP)(a - 2bP)$$

$$= 2b^2 P\left(P - \frac{a}{b}\right)\left(P - \frac{a}{2b}\right).$$

From calculus recall that the points where $d^2P/dt^2 = 0$ are possible points of inflection, but $P = 0$ and $P = a/b$ can obviously be ruled out. Hence $P = a/2b$ is the only possible ordinate value at which the concavity of the graph can change. For $0 < P < a/2b$ it follows that $P'' > 0$, and $a/2b < P < a/b$ implies $P'' < 0$. Thus, as we read from left to right, the graph changes from concave up to concave down at the point corresponding to $P = a/2b$. When the initial value satisfies $0 < P_0 < a/2b$, the graph of $P(t)$ assumes the shape of an S, as we see in Figure 2.42(a). For $a/2b < P_0 < a/b$ the graph is still S-shaped, but the point of inflection occurs at a negative value of t, as shown in Figure 2.42(b).

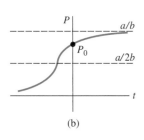

(b)

Figure 2.42

We have already seen equation (4) above in (5) of Section 1.3 in the form $dx/dt = kx(n + 1 - x)$, $k > 0$. This differential equation provides a reasonable model for describing the spread of an epidemic brought about initially by introducing an infected individual into a static population. The solution $x(t)$ represents the number of individuals infected with the disease at time t.

Example 1 Logistic Growth

Suppose a student carrying a flu virus returns to an isolated college campus of 1000 students. If it is assumed that the rate at which the virus spreads is proportional not only to the number x of infected students but also to the number of students not infected, determine the number of infected students after 6 days if it is further observed that after 4 days $x(4) = 50$.

SOLUTION Assuming that no one leaves the campus throughout the duration of the disease, we must solve the initial-value problem

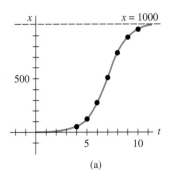

(a)

t (days)	x (number infected)
4	50 (observed)
5	124
6	276
7	507
8	735
9	882
10	953

(b)

Figure 2.43

$$\frac{dx}{dt} = kx(1000 - x), \qquad x(0) = 1.$$

By making the identifications $a = 1000k$ and $b = k$, we have immediately from (5) that

$$x(t) = \frac{1000k}{k + 999ke^{-1000kt}} = \frac{1000}{1 + 999e^{-1000kt}}.$$

Now, using the information $x(4) = 50$, we determine k from

$$50 = \frac{1000}{1 + 999e^{-4000k}}.$$

We find $-1000k = \frac{1}{4} \ln \frac{19}{999} = -0.9906$. Thus

$$x(t) = \frac{1000}{1 + 999e^{-0.9906t}}.$$

Finally, $\qquad x(6) = \dfrac{1000}{1 + 999e^{-5.9436}} = 276$ students.

Additional calculated values of $x(t)$ are given in the table in Figure 2.43(b). ❑

■ **Modifications of the Logistic Equation** There are many variations of the logistic equation. For example, the differential equations

$$\frac{dP}{dt} = P(a - bP) - h \quad \text{and} \quad \frac{dP}{dt} = P(a - bP) + h \tag{6}$$

could serve, in turn, as models for the population in a fishery where fish are **harvested** or are **restocked** at rate h. When $h > 0$ is a constant, the DEs in (6) can be readily analyzed qualitatively or solved by separation of variables. The equations in (6) could also serve as models of a human population either increased by immigration or decreased by emigration. The rate h in (6) could be a function of time t or may be population dependent; for example, harvesting might be done periodically in time or may be done at a rate proportional to the population P at time t. In the latter instance, the model would look like $P' = P(a - bP) - cP$, $c > 0$. A human population of a community might change due to immigration in such a manner that the contribution due to immigration is large when the population P of the community is itself small, but then the immigration contribution might be small when P is large; a reasonable model for the population of the community is then $P' = P(a - bP) + ce^{-kP}$, $c > 0$, $k > 0$. Another equation of the form given in (2),

$$\frac{dP}{dt} = P(a - b \ln P), \tag{7}$$

is a modification of the logistic equation known as the **Gompertz differential equation.** This DE is sometimes used as a model in the study of the growth or decline of population, in the growth of solid tumors, and in certain kinds of actuarial predictions. See Problems 5–8 in Exercises 2.8.

■ **Chemical Reactions** Suppose that a grams of chemical A are combined with b grams of chemical B. If there are M parts of A and N parts of B formed in the compound and $X(t)$ is the number of grams of chemical C formed, then the numbers

of grams of chemicals A and B remaining at any time are, respectively,

$$a - \frac{M}{M+N}X \quad \text{and} \quad b - \frac{N}{M+N}X.$$

By the law of mass action, the rate of the reaction satisfies

$$\frac{dX}{dt} \propto \left(a - \frac{M}{M+N}X\right)\left(b - \frac{N}{M+N}X\right). \tag{8}$$

If we factor out $M/(M+N)$ from the first factor and $N/(M+N)$ from the second and introduce a constant $k > 0$ of proportionality, (8) has the form

$$\frac{dX}{dt} = k(\alpha - X)(\beta - X), \tag{9}$$

where $\alpha = a(M+N)/M$ and $\beta = b(M+N)/N$. Recall from (6) of Section 1.3 that a chemical reaction governed by the nonlinear differential equation (9) is said to be a **second-order reaction.**

Example 2 Second-Order Chemical Reaction

A compound C is formed when two chemicals A and B are combined. The resulting reaction between the two chemicals is such that for each gram of A, 4 grams of B is used. It is observed that 30 grams of the compound C is formed in 10 minutes. Determine the amount of C at any time if the rate of the reaction is proportional to the amounts of A and B remaining and if initially there are 50 grams of A and 32 grams of B. How much of the compound C is present at 15 minutes? Interpret the solution as $t \to \infty$.

SOLUTION Let $X(t)$ denote the number of grams of the compound C present at any time t. Clearly $X(0) = 0$ g and $X(10) = 30$ g.

If, for example, 2 grams of compound C is present, we must have used, say, a grams of A and b grams of B so that $a + b = 2$ and $b = 4a$. Thus we must use $a = \tfrac{2}{5} = 2(\tfrac{1}{5})$ g of chemical A and $b = \tfrac{8}{5} = 2(\tfrac{4}{5})$ g of B. In general, for X grams of C we must use

$$\frac{X}{5} \text{ grams of } A \quad \text{and} \quad \frac{4}{5}X \text{ grams of } B.$$

The amounts of A and B remaining at any time are then

$$50 - \frac{X}{5} \quad \text{and} \quad 32 - \frac{4}{5}X,$$

respectively.

Now we know that the rate at which compound C is formed satisfies

$$\frac{dX}{dt} \propto \left(50 - \frac{X}{5}\right)\left(32 - \frac{4}{5}X\right).$$

To simplify the subsequent algebra, we factor $\tfrac{1}{5}$ from the first term and $\tfrac{4}{5}$ from the second and then introduce the constant of proportionality:

$$\frac{dX}{dt} = k(250 - X)(40 - X).$$

By separation of variables and partial fractions we can write

$$-\frac{1/210}{250 - X}\,dX + \frac{1/210}{40 - X}\,dX = k\,dt.$$

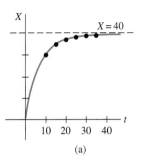

(a)

t(min)	x(g)
10	30 (measured)
15	34.78
20	37.25
25	38.54
30	39.22
35	39.59

(b)

Figure 2.44

Integrating gives

$$\ln\left|\frac{250 - X}{40 - X}\right| = 210kt + c_1 \quad \text{or} \quad \frac{250 - X}{40 - X} = c_2 e^{210kt}. \tag{10}$$

When $t = 0$, $X = 0$, so it follows at this point that $c_2 = \frac{25}{4}$. Using $X = 30$ g at $t = 10$, we find $210k = \frac{1}{10} \ln \frac{88}{25} = 0.1258$. With this information we solve the last equation in (10) for X:

$$X(t) = 1000 \frac{1 - e^{-0.1258t}}{25 - 4e^{-0.1258t}}. \tag{11}$$

The behavior of X as a function of time is displayed in Figure 2.44. It is clear from the accompanying table and (11) that $X \to 40$ as $t \to \infty$. This means that 40 grams of compound C is formed, leaving

$$50 - \frac{1}{5}(40) = 42 \text{ g of } A \quad \text{and} \quad 32 - \frac{4}{5}(40) = 0 \text{ g of } B. \qquad \square$$

Remarks

The indefinite integral $\int \dfrac{du}{a^2 - u^2}$ can be evaluated in terms of logarithms, the inverse hyperbolic tangent, or the inverse hyperbolic cotangent. For example, of the two results,

$$\int \frac{du}{a^2 - u^2} = \frac{1}{a} \tanh^{-1} \frac{u}{a} + c, \qquad |u| < a \tag{12}$$

$$\int \frac{du}{a^2 - u^2} = \frac{1}{2a} \ln\left|\frac{a + u}{a - u}\right| + c, \qquad |u| \neq a \tag{13}$$

(12) may be convenient for Problems 15 and 18 in Exercises 2.8, whereas (13) may be preferable in Problem 19.

EXERCISES 2.8

Answers to odd-numbered problems begin on page A-28.

Logistic Equation

1. The number $N(t)$ of supermarkets throughout the country that are using a computerized checkout system is described by the initial-value problem

$$\frac{dN}{dt} = N(1 - 0.0005N), \quad N(0) = 1.$$

(a) Use the phase portrait concept of Section 2.1 to predict how many supermarkets are expected to adopt the new procedure over a long period of time. By hand, sketch a solution curve of the given initial-value problem.

(b) Solve the initial-value problem and then use a graphing utility to verify the solution curve in part **(a)**. How many companies are expected to adopt the new technology when $t = 10$?

2. The number $N(t)$ of people in a community who are exposed to a particular advertisement is governed by the logistic equation. Initially $N(0) = 500$, and it is observed that $N(1) = 1000$. Solve for $N(t)$ if it is predicted that the limiting number of people in the community who will see the advertisement is 50,000.

3. A model for the population $P(t)$ in a suburb of a large city is given by the initial-value problem

$$\frac{dP}{dt} = P(10^{-1} - 10^{-7} P), \quad P(0) = 5000,$$

where t is measured in months. What is the limiting value of the population? At what time will the population be equal to one-half of this limiting value?

4. (a) Census data for the United States between 1790 and 1950 is given in the following table. Construct a logistic population model using the data from 1790, 1850, and 1910.

Year	Population (in millions)
1790	3.929
1800	5.308
1810	7.240
1820	9.638
1830	12.866
1840	17.069
1850	23.192
1860	31.433
1870	38.558
1880	50.156
1890	62.948
1900	75.996
1910	91.972
1920	105.711
1930	122.775
1940	131.669
1950	150.697

(b) Construct a table comparing actual census population with the population predicted by the model in part **(a)**. Compute the error and the percentage error for each entry pair.

Modifications of the Logistic Equation

5. (a) If a constant number h of animals is removed or harvested per unit time, then a model for the population $P(t)$ of animals at time t is given by

$$\frac{dP}{dt} = P(a - bP) - h, \quad P(0) = P_0,$$

where $a, b, h,$ and P_0 are positive constants. Suppose $a = 5$, $b = 1$, and $h = 4$. Since the DE is autonomous, use the phase portrait concept of Section 2.1 to sketch representative solution curves corresponding to the cases $P_0 > 4$, $1 < P_0 < 4$, and $0 < P_0 < 1$. Determine the long-term behavior of the population in each case.

(b) Solve the IVP in part **(a)**. Use a graphing utility to verify your results in part **(a)**.

(c) Use the information in parts **(a)** and **(b)** to determine whether the population becomes extinct in finite time. If so, find that time.

6. Investigate the harvesting model in Problem 5 both qualitatively and analytically in the following two additional cases to determine whether the population becomes extinct in finite time. If so, find that time.

(a) $a = 5, \ b = 1, \ h = \tfrac{25}{4}$ **(b)** $a = 5, \ b = 1, \ h = 7$

7. (a) Suppose $a = b = 1$ in the Gompertz differential equation (7). Since the DE is autonomous, use the phase portrait concept of Section 2.1 to sketch representative solution curves corresponding to the cases $P_0 > e^{a/b}$ and $0 < P_0 < e^{a/b}$.

(b) Suppose $a = 1, b = -1$ in (7). Use a new phase portrait to sketch representative solution curves corresponding to the cases $P_0 > e^{-a/b}$ and $0 < P_0 < e^{-a/b}$.

(c) Find an explicit solution of (7) subject to $P(0) = P_0$.

8. Computer Lab Assignment (a) In Examples 3 and 4 of Section 2.1 we saw that any solution $P(t)$ of (4) possessed the asymptotic behavior $P(t) \rightarrow a/b$ as $t \rightarrow \infty$ for $P_0 > a/b$ and for $0 < P_0 < a/b$, and as a consequence we called the equilibrium solution $P = a/b$ an attractor. Use a CAS or a graphic calculator with root-finding capabilities to approximate the equilibrium solution of the immigration model $P' = P(1 - P) + 0.3e^{-P}$.

(b) Use a graphing utility to graph the function $f(P) = P(1 - P) + 0.3e^{-P}$. Explain how this graph can be used to determine whether the number found in part **(a)** is an attractor.

(c) Use a numerical solver to compare the solution curves for the IVPs

$$P' = P(1 - P), \ P(0) = P_0 \text{ for } P_0 = 0.2 \text{ and } P_0 = 1.2$$

with the solution curves for the IVPs

$$P' = P(1 - P) + 0.3e^{-P}, \ P(0) = P_0 \text{ for } P_0 = 0.2 \text{ and } P_0 = 1.2.$$

Superimpose all curves on the same coordinate axes but, if possible, use a different color for the curves of the second initial-value problem. Over a long period of time, what percentage increase does the immigration model predict in the population compared to the logistic model?

Chemical Reactions

9. Two chemicals A and B are combined to form a chemical C. The rate, or velocity, of the reaction is proportional to the product of the instantaneous amounts of A and B not converted to chemical C. Initially there are 40 grams of A and 50 grams of B, and for each gram of B, 2 grams of A is used. It is observed that 10 grams of C is formed in 5 minutes. How much is formed in 20 minutes? What is the limiting amount of C after a long time? How much of chemicals A and B remains after a long time?

10. Solve Problem 9 if 100 grams of chemical A is present initially. At what time is chemical C half-formed?

Miscellaneous Nonlinear Models

11. A tank in the form of a right circular cylinder standing on end is leaking water through a circular hole in its bottom. When friction and contraction of water at the hole are ignored, we saw in (10) of Section 1.3 that the height h of water in the tank is described by

$$\frac{dh}{dt} = -\frac{A_h}{A_w}\sqrt{2gh},$$

where A_w and A_h are the cross-sectional areas of the water and the hole, respectively.

(a) Solve for $h(t)$ if the initial height of the water is H. By hand, draw the solution curve and give the interval I of its definition in terms of the symbols A_w, A_h, and H. Use $g = 32$ ft/s^2.

(b) Suppose the tank is 10 ft high and has radius 2 ft and the circular hole has radius ½ in. If the tank is initially full, how long will it take to empty?

12. When friction and contraction of the water at the hole is taken into account, the model in Problem 11 becomes

$$\frac{dh}{dt} = -c\frac{A_h}{A_w}\sqrt{2gh},$$

where $0 < c < 1$. How long will it take the tank in Problem 11(b) to empty if $c = 0.6$? See Problem 11 in Exercises 1.3.

13. A tank in the form of a rightcircular cone standing on end, vertex down, is leaking water through a circular hole in its bottom.

(a) Suppose the tank is 20 ft high and has radius 8 ft and the circular hole has radius 2 in. In Problem 12 in Exercises 1.3 you were asked to show that the differential equation governing the height h of water leaking from a tank is

$$\frac{dh}{dt} = -\frac{5}{6h^{3/2}}.$$

In this model friction and contraction of the water at the hole were taken into account with $c = 0.6$, and g was taken to be 32 ft/s^2. See Figure 1.26. If the tank is initially full, how long will it take the tank to empty?

(b) Suppose the tank has a vertex angle of 60°, and the circular hole has radius 2 in. Determine the differential equation governing the height h of water. Use $c = 0.6$ and $g = 32$ ft/s^2. If the height of the water is initially 9 ft, how long will it take the tank to empty?

14. Suppose that the conical tank in Problem 13(a) is inverted, as shown in Figure 2.45, and that water leaks out a circular hole of radius 2 in. in the center of the circular base. Is the time it takes to empty a full tank the same as for the tank with vertex down in Problem 13? Take the friction/contraction coefficient to be $c = 0.6$ and $g = 32$ ft/s^2.

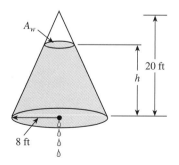

Figure 2.45

15. A differential equation governing the velocity v of a falling mass m subjected to air resistance proportional to the square of the instantaneous velocity is

$$m\frac{dv}{dt} = mg - kv^2,$$

where $k > 0$ is a constant of proportionality. The positive direction is downward.

(a) Solve this equation subject to the initial condition $v(0) = v_0$.

(b) Determine the limiting, or terminal, velocity of the mass.

(c) If distance s is related to velocity by $ds/dt = v$, find an explicit expression for s if it is further known that $s(0) = 0$.

16. Consider the 16 lb cannonball shot vertically upward in Problem 32 in Exercises 2.7 with an initial velocity of 300 ft/s. Determine the maximum height attained by the cannonball if air resistance is assumed to be proportional to the square of the instantaneous velocity. Assume the positive direction is upward and take $k = 0.0003$. [*Hint:* You will have to slightly modify the DE in Problem 15.]

17. Computer Lab Assignment In Problem 16 let t_a be the time it takes the cannonball to attain its maximum height and t_d the time it takes the cannonball to fall from the maximum height to the ground. Compare the value of t_a with t_d and the magnitude of the impact velocity v_i with v_0. See Problem 33 in Exercises 2.7. The root-finding application of a CAS may be useful here. [*Hint:* Use the model in Problem 15 when the cannonball is falling.]

18. Computer Lab Assignment A skydiver is equipped with a stop watch and an altimeter. After exiting a plane flying at an altitude of 20,000 feet, she waits 25 seconds, opens her parachute, and observes that her altitude is 14,800 ft. Assume that air resistance is proportional to the square of the instantaneous velocity, her initial velocity upon leaving the plane is zero, and $g = 32$ ft/s^2.

(a) Find the distance $s(t)$, measured from the plane, that the skydiver has traveled during free fall in time t. [*Hint:* The constant of proportionality k in the model given in Problem 15 is not specified. Use the expression for terminal velocity v_t obtained in part **(b)** of Problem 15 to eliminate k from the IVP. Then eventually solve for v_t.]

(b) How far does the skydiver fall and what is her velocity at $t = 15$ s?

19. Computer Lab Assignment A helicopter hovers 500 feet above a large open tank full of liquid (not water). A dense compact object weighing 160 pounds is dropped (released from rest) from the helicopter into the liquid. Assume that air resistance is proportional to v while the object is in the air and that viscous damping is proportional to v^2 after the object has entered the liquid. For air, take $k = \frac{1}{4}$, and for

the liquid, $k = 0.1$. Assume the positive direction is downward. If the tank is 75 ft high, determine the time and impact velocity when the object hits the bottom of the tank. [*Hint:* Think in terms of two distinct IVPs. Also, do not be discouraged by messy algebra. If you use (13), be careful in removing the absolute value sign. You might compare the velocity when the object hits the liquid (the initial velocity for the second problem) with the terminal velocity v_t of the object falling through the liquid.]

20. (a) Determine a differential equation for the velocity $v(t)$ of a mass m sinking in water that imparts a resistance proportional to the square of the instantaneous velocity and also exerts an upward buoyant force whose magnitude is given by Archimedes' principle. See Problem 16 in Exercises 1.3. Assume the positive direction is downward.

(b) Solve the differential equation in part **(a).**

(c) Determine the limiting, or terminal, velocity of the sinking mass.

21. The differential equation
$$\frac{dy}{dx} = \frac{-x + \sqrt{x^2 + y^2}}{y}$$
describes the shape of a plane curve C that will reflect all incoming light beams to the same point. See Problem 27 in Exercises 1.3. There are several ways of solving this equation.

(a) Verify that the differential equation is homogeneous (see Section 2.5). Show that the substitution $y = ux$ yields
$$\frac{u\ du}{\sqrt{1 + u^2}(1 - \sqrt{1 + u^2})} = \frac{dx}{x}.$$
Use a CAS (or another judicious substitution) to integrate the left side of the equation. Show that the curve C must be a parabola with focus at the origin and is symmetric with respect to the x-axis.

(b) Show that the first differential equation can also be solved by means of the substitution $u = x^2 + y^2$.

22. (a) A simple model for the shape of a tsunami, or tidal wave, is given by
$$\frac{dW}{dx} = W\sqrt{4 - 2W},$$
where $W(x) > 0$ is the height of the wave expressed as a function of its position relative to a point offshore. By inspection, find all constant solutions of the differential equation.

(b) Solve the differential equation in part **(a).** A CAS may be useful for integration.

(c) Use a graphing utility to obtain the graphs of all solutions that satisfy the initial condition $W(0) = 2$.

23. In Palms Springs, California, an outdoor decorative pond is in the shape of a hemispherical tank and is to be filled with water pumped into the tank through an inlet in its bottom. Suppose that the radius of the tank is $R = 10$ ft, that water is pumped in at a rate of π ft^3/min, and that the tank is initially empty. See Figure 2.46. As the tank fills, it loses water through evaporation. Assume that the rate of evaporation is proportional to the area A of the surface of the water and that the constant of proportionality is $k = 0.01$.

(a) Hemispherical tank

(b) Cross-section of tank

Figure 2.46

(a) The rate of change dV/dt of the volume of the water at time t is a net rate (see Example 5 on page 76). Use this net rate to determine a differential equation for the height h of the water at time t. The volume of the water shown in the figure is $V = \pi Rh^2 - \frac{1}{3}\pi h^3$, where $R = 10$. Express the area of the surface of the water $A = \pi r^2$ in terms of h.

(b) Solve the differential equation in part **(a).** Graph the solution.

(c) If there were no evaporation, how long would it take the tank to fill?

(d) With evaporation, what is the depth of the water at the time found in part **(c)**? Will the tank ever be filled? Prove your assertion.

24. Computer Lab Assignment Read the documentation for your CAS on *scatterplots* or *scatter diagrams* and *least-squares linear fit.* The straight line that best fits a set of data points is called a **regression line** or a **least squares line.** In this problem we wish to construct a logistic model for the population of the United States, defining $f(P)$ in (2) as an equation of a regression line based on the population data in the table in Problem 4. One way of doing this is to

approximate the left-hand side $\dfrac{1}{P}\dfrac{dP}{dt}$ of the first equation in (2) using the forward difference quotient in place of dP/dt:

$$Q(t) = \frac{1}{P(t)}\frac{P(t+h)-P(t)}{h}$$

(a) Make a table of the values t, $P(t)$, and $Q(t)$ using $t = 0$, $10, 20, \ldots, 160$, and $h = 10$. For example, the first line of the table should contain $t = 0$, $P(0)$, and $Q(0)$. With $P(0) = 3.929$ and $P(10) = 5.308$,

$$Q(0) = \frac{1}{P(0)}\frac{P(10)-P(0)}{10} = 0.035.$$

Note that $Q(160)$ depends on the 1960 census population $P(170)$. Look up this value.

(b) Use a CAS to obtain a scatterplot of the data $(P(t),$ $Q(t))$ computed in part **(a)**. Also use a CAS to find an equa-tion of the regression line and to superimpose its graph on the scatterplot.

(c) Construct a logistic model $dP/dt = Pf(P)$, where $f(P)$ is the equation of the regression line found in part **(b)**.

(d) Solve the model in part **(c)** using the initial condition $P(0) = 3.929$.

(e) Use a CAS to obtain another scatterplot, this time of the ordered pairs $(t, P(t))$ from your table in part **(a)**. Use your CAS to superimpose the graph of the solution in part **(d)** on the scatterplot.

(f) Look up the U.S. census data for 1970, 1980, and 1990. What population does the logistic model in part **(c)** predict for these years? What does the model predict for the U.S. population $P(t)$ as $t \to \infty$?

2.9 Systems: Linear and Nonlinear Models

Up to now all the mathematical models that we have considered were single differ-ential equations. A single differential equation could describe a single population in an environment; but if there are, say, two interacting and perhaps competing species living in the same environment (for example, rabbits and foxes), then a model for their populations $x(t)$ and $y(t)$ *might* be a system of two first-order differential equa-tions such as

$$\frac{dx}{dt} = g_1(t, x, y)$$

$$\frac{dy}{dt} = g_2(t, x, y).$$

(1)

When g_1 and g_2 are linear in the variables x and y, that is,

$$g_1(x,y) = c_1x + c_2y + f_1(t) \quad \text{and} \quad g_2(x,y) = c_3x + c_4y + f_2(t),$$

then (1) is said to be a **linear system.** A system of differential equations that is not linear is said to be **nonlinear.**

In this section we are going to discuss mathematical models based on some of the topics that we have already discussed in the preceding two sections. This section will be similar to Section 1.3 in that we are just going to discuss certain mathematical models that are systems of first-order differential equations; we are not going to develop any methods for solving these systems. There are reasons for not solving systems at this point: First, we do not as yet possess the necessary mathematical tools for solving systems, and second, some of the systems that we discuss simply cannot be solved analytically.

We shall examine solution methods for systems of linear first-order equations in Chapter 9 and for systems of linear higher-order differential equations in Chapters 3 and 4.

■ **Radioactive Series** In the discussion of radioactive decay in Sections 1.3 and 2.7, we assumed that the rate of decay was proportional to the number $A(t)$ of nuclei of the substance present at time t. When a substance decays by radioactivity, it usually doesn't just transmute into one stable substance and then the process stops, Rather, the first substance decays into another radioactive substance, this substance

in turn decays into yet a third substance, and so on. This process, called a **radioactive decay series,** continues until a stable element is reached. For example, the uranium decay series is U-238 → Th-234 → . . . → Pb-206, where Pb-206 is a stable isotope of lead. The half-lives of the various elements in a radioactive series can range from billions of years (4.5×10^9 years for U-238) to a fraction of a second. Suppose a radioactive series is described schematically by $X \xrightarrow{-\lambda_1} Y \xrightarrow{-\lambda_2} Z$, where $k_1 = -\lambda_1 < 0$ and $k_2 = -\lambda_2 < 0$ are the decay constants for substances X and Y, respectively, and Z is a stable element. Suppose too, that $x(t)$, $y(t)$, and $z(t)$ denote amounts of substances X, Y, and Z, respectively, remaining at time t. The decay of element X is described by

$$\frac{dx}{dt} = -\lambda_1 x,$$

whereas the rate at which the second element Y decays is the net rate,

$$\frac{dy}{dt} = \lambda_1 x - \lambda_2 y,$$

since it is *gaining* atoms from the decay of X and at the same time *losing* atoms due to its own decay. Since Z is a stable element, it is simply gaining atoms from the decay of element Y:

$$\frac{dz}{dt} = \lambda_2 y.$$

In other words, a model of the radioactive decay series for three elements is the linear system of three first-order differential equations

$$\frac{dx}{dt} = -\lambda_1 x$$

$$\frac{dy}{dt} = \lambda_1 x - \lambda_2 y \tag{2}$$

$$\frac{dz}{dt} = \lambda_2 y.$$

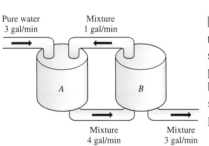

Pure water
3 gal/min

Mixture
1 gal/min

Mixture
4 gal/min

Mixture
3 gal/min

Figure 2.47

■ **Mixtures** Consider the two tanks shown in Figure 2.47. Let us suppose for the sake of discussion that tank A contains 50 gallons of water in which 25 pounds of salt is dissolved. Suppose tank B contains 50 gallons of pure water. Liquid is pumped in and out of the tanks as indicated in the figure; the mixture exchanged between the two tanks and the liquid pumped out of tank B is assumed to be well stirred. We wish to construct a mathematical model that describes the number of pounds $x_1(t)$ and $x_2(t)$ of salt in tanks A and B, respectively, at time t.

By an analysis similar to that on page 24 in Section 1.3 and Example 5 of Section 2.7, we see for tank A that the net rate of change of $x_1(t)$ is

$$\frac{dx_1}{dt} = \overbrace{(3 \text{ gal/min}) \cdot (0 \text{ lb/gal}) + (1 \text{ gal/min}) \cdot \left(\frac{x_2}{50} \text{ lb/gal}\right)}^{\text{input rate of salt}} - \overbrace{(4 \text{ gal/min}) \cdot \left(\frac{x_1}{50} \text{ lb/gal}\right)}^{\text{output rate of salt}}$$

$$= -\frac{2}{25}x_1 + \frac{1}{50}x_2.$$

Similarly, for tank B the net rate of change of $x_2(t)$ is

$$\frac{dx_2}{dt} = 4 \cdot \frac{x_1}{50} - 3 \cdot \frac{x_2}{50} - 1 \cdot \frac{x_2}{50} = \frac{2}{25}x_1 - \frac{2}{25}x_2.$$

Thus we obtain the linear system

$$\frac{dx_1}{dt} = -\frac{2}{25}x_1 + \frac{1}{50}x_2$$
$$\frac{dx_2}{dt} = \frac{2}{25}x_1 - \frac{2}{25}x_2. \tag{3}$$

Observe that the foregoing system is accompanied by the initial conditions $x_1(0) = 25$, $x_2(0) = 0$.

■ **A Predator-Prey Model** Suppose that two different species of animals interact within the same environment or ecosystem, and suppose further that the first species eats only vegetation and the second eats only the first species. In other words, one species is a predator and the other is a prey. For example, wolves hunt grass-eating caribou, sharks devour little fish, and the snowy owl pursues an arctic rodent called the lemming. For the sake of discussion, let us imagine that the predators are foxes and the prey are rabbits.

Let $x(t)$ and $y(t)$ denote, respectively, the fox and rabbit populations at time t. If there were no rabbits, then one might expect that the foxes, lacking an adequate food supply, would decline in number according to

$$\frac{dx}{dt} = -ax, \qquad a > 0. \tag{4}$$

When rabbits are present in the environment, however, it seems reasonable that the number of encounters or interactions between these two species per unit time is jointly proportional to their populations x and y, that is, proportional to the product xy. Thus when rabbits are present there is a supply of food, and so foxes are added to the system at a rate bxy, $b > 0$. Adding this last rate to (4) gives a model for the fox population:

$$\frac{dx}{dt} = -ax + bxy. \tag{5}$$

On the other hand, were there no foxes, then the rabbits would, with an added assumption of unlimited food supply, grow at a rate that is proportional to the number of rabbits present at time t:

$$\frac{dy}{dt} = dy, \qquad d > 0. \tag{6}$$

But when foxes are present, a model for the rabbit population is (6) decreased by cxy, $c > 0$, that is, decreased by the rate at which the rabbits are eaten during their encounters with the foxes:

$$\frac{dy}{dt} = dy - cxy. \tag{7}$$

Equations (5) and (7) constitute a system of nonlinear differential equations

$$\frac{dx}{dt} = -ax + bxy = x(-a + by)$$
$$\frac{dy}{dt} = dy - cxy = y(d - cx), \tag{8}$$

where a, b, c, and d are positive constants. This famous system of equations is known as the **Lotka–Volterra predator-prey model.**

Except for two constant solutions, $x(t) = 0$, $y(t) = 0$ and $x(t) = d/c$, $y(t) = a/b$, the nonlinear system (8) cannot be solved in terms of elementary functions. However, we can analyze such systems quantitatively and qualitatively. See Chapter 6 and Chapter 11.

Example 1 Predator-Prey Model

Suppose

$$\frac{dx}{dt} = -0.16x + 0.08xy$$

$$\frac{dy}{dt} = 4.5y - 0.9xy$$

represents a predator-prey model. Since we are dealing with populations, we have $x(t) \geq 0$, $y(t) \geq 0$. Figure 2.48, obtained with the aid of a numerical solver, shows typical population curves of the predators and prey for this model superimposed on the same coordinate axes. The initial conditions used were $x(0) = 4$, $y(0) = 4$. The curve in black represents the population $x(t)$ of the predator (foxes), and the colored curve is the population $y(t)$ of the prey (rabbits). Observe that the model seems to predict that both populations $x(t)$ and $y(t)$ are periodic in time. This makes intuitive sense since, as the number of prey decreases, the predator population eventually decreases because of a diminished food supply; but attendant to a decrease in the number of predators is an increase in the number of prey; this in turn gives rise to an increased number of predators, which ultimately brings about another decrease in the number of prey. ❏

Figure 2.48

■ **Competition Models** Now suppose two different species of animals occupy the same ecosystem, not as predator and prey but rather as competitors for the same resources (such as food and living space) in the system. In the absence of the other, let us assume that the rate at which each population grows is given by

$$\frac{dx}{dt} = ax \quad \text{and} \quad \frac{dy}{dt} = cy, \tag{9}$$

respectively.

Since the two species compete, another assumption might be that each of these rates is diminished simply by the influence, or existence, of the other population. Thus a model for the two populations is given by the linear system

$$\frac{dx}{dt} = ax - by$$

$$\frac{dy}{dt} = cy - dx, \tag{10}$$

where a, b, c, and d are positive constants.

On the other hand, we might assume, as we did in (5), that each growth rate in (9) should be reduced by a rate proportional to the number of interactions between the two species:

$$\frac{dx}{dt} = ax - bxy$$

$$\frac{dy}{dt} = cy - dxy. \tag{11}$$

Inspection shows that this nonlinear system is similar to the Lotka–Volterra predator-prey model. Lastly, it might be more realistic to replace the rates in (9), which indicate that the population of each species in isolation grows exponentially, with rates indicating that each population grows logistically (that is, over a long time the population is bounded):

$$\frac{dx}{dt} = a_1 x - b_1 x^2 \qquad \text{and} \qquad \frac{dy}{dt} = a_2 y - b_2 y^2. \tag{12}$$

When these new rates are decreased by rates proportional to the number of interactions, we obtain another nonlinear model

$$\frac{dx}{dt} = a_1 x - b_1 x^2 - c_1 xy = x(a_1 - b_1 x - c_1 y)$$

$$\tag{13}$$

$$\frac{dy}{dt} = a_2 y - b_2 y^2 - c_2 xy = y(a_2 - b_2 y - c_2 x).$$

where all coefficients are positive. The linear system (10) and the nonlinear systems (11) and (13) are, of course, called **competition models.**

■ **Networks** An electrical network having more than one loop also gives rise to simultaneous differential equations. As shown in Figure 2.49, the current $i_1(t)$ splits in the directions shown at point B_1, called a *branch point* of the network. By Kirchhoff's first law we can write

$$i_1(t) = i_2(t) + i_3(t). \tag{14}$$

In addition, we can also apply **Kirchhoff's second law** to each loop. For loop $A_1 B_1 B_2 A_2 A_1$, summing the voltage drops across each part of the loop gives

$$E(t) = i_1 R_1 + L_1 \frac{di_2}{dt} + i_2 R_2. \tag{15}$$

Similarly, for loop $A_1 B_1 C_1 C_2 B_2 A_2 A_1$ we find

$$E(t) = i_1 R_1 + L_2 \frac{di_3}{dt}. \tag{16}$$

Figure 2.49

Using (14) to eliminate i_1 in (15) and (16) yields two linear first-order equations for the currents $i_2(t)$ and $i_3(t)$:

$$L_1 \frac{di_2}{dt} + (R_1 + R_2) i_2 + R_1 i_3 = E(t)$$

$$\tag{17}$$

$$L_2 \frac{di_3}{dt} + R_1 i_2 + R_1 i_3 = E(t).$$

We leave it as an exercise (see Problem 14 in Exercises 2.9) to show that the system of differential equations describing the currents $i_1(t)$ and $i_2(t)$ in the network containing a resistor, an inductor, and a capacitor shown in Figure 2.50 is

$$L \frac{di_1}{dt} + R i_2 = E(t)$$

$$\tag{18}$$

$$RC \frac{di_2}{dt} + i_2 - i_1 = 0.$$

Figure 2.50

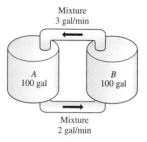

Figure 2.52

Radioactive Series

1. We have not discussed methods by which systems of first-order differential equations can be solved. Nevertheless, systems such as (2) can be solved with no other knowledge than how to solve a single linear first-order equation. Find a solution of (2) subject to the initial conditions $x(0) = x_0$, $y(0) = 0$, $z(0) = 0$.

2. In Problem 1, suppose time is measured in days, the decay constants are $k_1 = -0.138629$ and $k_2 = -0.004951$, and $x_0 = 20$. Use a graphing utility to obtain the graphs of the solutions $x(t)$, $y(t)$, and $z(t)$ on the same set of coordinate axes. Use the graphs to approximate the half-lives of substances X and Y.

3. Use the graphs in Problem 2 to approximate the times when the amounts $x(t)$ and $y(t)$ are the same, the times when the amounts $x(t)$ and $z(t)$ are the same, and the times when the amounts $y(t)$ and $z(t)$ are the same. Why does the time determined when the amounts $y(t)$ and $z(t)$ are the same make intuitive sense?

4. Construct a mathematical model for a radioactive series of four elements W, X, Y, and Z, where Z is a stable element.

Mixtures

5. Consider the two tanks A and B with liquid being pumped in and out at the same rates described by the system of equations (3). What is the system of differential equations if, instead of pure water, a brine solution containing 2 lb of salt per gallon is pumped into tank A?

6. Use the information given in Figure 2.51 to construct a mathematical model for the number of pounds of salt $x_1(t)$, $x_2(t)$, and $x_3(t)$ at any time in tanks A, B, and C, respectively.

Figure 2.51

7. Two large tanks A and B each hold 100 gallons of brine. Initially, 100 lb of salt is dissolved in the solution in tank A and 50 lb of salt is dissolved in the solution in tank B. The system is closed because the well-stirred liquid is pumped only between the tanks, as shown in Figure 2.52.

(a) Use the information given in the figure to construct a mathematical model for the number of pounds of salt $x_1(t)$ and $x_2(t)$ at time t in tanks A and B, respectively.

(b) Find a relationship between the variables $x_1(t)$ and $x_2(t)$ that holds at time t. Explain why this relationship makes intuitive sense. Use this relationship to help find the amount of salt in tank B at $t = 30$ min.

8. Three large tanks contain brine, as shown in Figure 2.53. Use the information in the figure to construct a mathematical model for the number of pounds of salt $x_1(t)$, $x_2(t)$, and $x_3(t)$ at time t in tanks A, B, and C, respectively. Without solving the system, predict limiting values of $x_1(t)$, $x_2(t)$, and $x_3(t)$ as $t \to \infty$.

Figure 2.53

Predator-Prey Models

9. Consider the Lotka–Volterra predator-prey model defined by

$$\frac{dx}{dt} = -0.1x + 0.02xy$$

$$\frac{dy}{dt} = 0.2y - 0.025xy,$$

where the populations $x(t)$ (predators) and $y(t)$ (prey) are measured in the thousands. Suppose $x(0) = 6$ and $y(0) = 6$. Use a numerical solver to graph $x(t)$ and $y(t)$. Use the graphs to approximate the time $t > 0$ when the two populations are first equal. Use the graphs to approximate the period of each population.

Competition Models

10. Consider the competition model defined by

$$\frac{dx}{dt} = x(2 - 0.4x - 0.3y)$$

$$\frac{dy}{dt} = y(1 - 0.1y - 0.3x),$$

where the populations $x(t)$ and $y(t)$ are measured in the thousands and t in years. Use a numerical solver to analyze the populations over a long period of time for each of the cases:

(a) $x(0) = 1.5$, $y(0) = 3.5$ **(b)** $x(0) = 1$, $y(0) = 1$

(c) $x(0) = 2$, $y(0) = 7$ **(d)** $x(0) = 4.5$, $y(0) = 0.5$

11. Consider the competition model defined by

$$\frac{dx}{dt} = x(1 - 0.1x - 0.05y)$$

$$\frac{dy}{dt} = y(1.7 - 0.1y - 0.15x),$$

where the populations $x(t)$ and $y(t)$ are measured in the thousands and t in years. Use a numerical solver to analyze the populations over a long period of time for each of the cases:

(a) $x(0) = 1$, $y(0) = 1$ **(b)** $x(0) = 4$, $y(0) = 10$

(c) $x(0) = 9$, $y(0) = 4$ **(d)** $x(0) = 5.5$, $y(0) = 3.5$

Networks

12. Show that a system of differential equations that describe the currents $i_2(t)$ and $i_3(t)$ in the electrical network shown in Figure 2.54 is

$$L\frac{di_2}{dt} + L\frac{di_3}{dt} + R_1 i_2 = E(t)$$

$$R_1\frac{di_2}{dt} + R_2\frac{di_3}{dt} + \frac{1}{C}i_3 = 0.$$

Figure 2.54

13. Determine a system of first-order differential equations that describe the currents $i_2(t)$ and $i_3(t)$ in the electrical network shown in Figure 2.55.

Figure 2.55

14. Show that the linear system given in (18) describes the currents $i_1(t)$ and $i_2(t)$ in the network shown in Figure 2.50. [*Hint*: $dq/dt = i_3$.]

Miscellaneous Models

15. A communicable disease is spread throughout a small community with a fixed population of n people by contact between infected persons and persons who are susceptible to the disease. Suppose initially that everyone is susceptible to the disease and that no one leaves the community while the epidemic is spreading. At time t, let $s(t)$, $i(t)$, and $r(t)$ denote, in turn, the number of people in the community (measured in hundreds) who are *susceptible* to the disease but not yet infected with it, the number of people who are *infected* with the disease, and the number of people who have *recovered* from the disease. Explain why the system of differential equations

$$\frac{ds}{dt} = -k_1 si$$

$$\frac{di}{dt} = -k_2 i + k_1 si$$

$$\frac{dr}{dt} = k_2 i,$$

where k_1 (called the *infection rate*) and k_2 (called the *removal rate*) are positive constants, is a reasonable mathematical model, commonly called a **SIR model**, for the spread of the epidemic throughout the community. Give plausible initial conditions associated with this system of equations.

16. Computer Lab Assignment **(a)** In Problem 15 explain why it is sufficient to analyze only

$$\frac{ds}{dt} = -k_1 si$$

$$\frac{di}{dt} = -k_2 i + k_1 si.$$

(b) Suppose $k_1 = 0.2$, $k_2 = 0.7$, and $n = 10$. Choose various values of $i(0) = i_0$, $0 < i_0 < 10$. Use a numerical solver to determine what the model predicts about the epidemic in the two cases $s_0 > k_2/k_1$ and $s_0 \leq k_2/k_1$. In the case of an epidemic, estimate the number of people who are eventually infected.

CHAPTER 2
REVIEW EXERCISES

Answers to odd-numbered problems begin on page A-28.

In Problems 1 and 2, fill in the blanks.

1. The DE $y' - ky = A$, where k and A are constants, is autonomous. The critical point _____ of the equation is a(n) _____ (attractor or repeller) for $k > 0$ and a(n) _____ (attractor or repeller) for $k < 0$.

2. The initial-value problem $x\dfrac{dy}{dx} - 4y = 0$, $y(0) = k$, has an infinite number of solutions for $k =$ _____ and no solution for $k =$ _____ .

In Problems 3 and 4, construct an autonomous first-order differential equation $dy/dx = f(y)$ whose phase portrait is consistent with the given figure.

3.

Figure 2.56

4.

Figure 2.57

5. The number 0 is a critical point of the autonomous differential equation $dx/dt = x^n$, where n is a positive integer. For what values of n is 0 asymptotically stable? Semistable? Unstable? Repeat for the equation $dx/dt = -x^n$.

6. Consider the differential equation

$$\frac{dP}{dt} = f(P), \quad \text{where} \quad f(P) = -0.5P^3 - 1.7P + 3.4.$$

The function $f(P)$ has one real zero, as shown in Figure 2.58. Without attempting to solve the differential equation, estimate the value of $\lim_{t \to \infty} P(t)$.

Figure 2.58

7. Figure 2.59 is a portion of the direction field of a differential equation $dy/dx = f(x, y)$. By hand, sketch two different solution curves, one that is tangent to the lineal element shown in black and the other tangent to the lineal element shown in color.

Figure 2.59

8. Classify each differential equation as to: separable, exact, linear, homogeneous, or Bernoulli. Some equations may be more than one kind. Do not solve.

(a) $\dfrac{dy}{dx} = \dfrac{x - y}{x}$ **(b)** $\dfrac{dy}{dx} = \dfrac{1}{y - x}$

(c) $(x + 1)\dfrac{dy}{dx} = -y + 10$ **(d)** $\dfrac{dy}{dx} = \dfrac{1}{x(x - y)}$

(e) $\dfrac{dy}{dx} = \dfrac{y^2 + y}{x^2 + x}$ **(f)** $\dfrac{dy}{dx} = 5y + y^2$

(g) $y\,dx = (y - xy^2)\,dy$ **(h)** $x\dfrac{dy}{dx} = ye^{x/y} - x$

(i) $xy\,y' + y^2 = 2x$ **(j)** $2xy\,y' + y^2 = 2x^2$

(k) $y\,dx + x\,dy = 0$ **(l)** $\left(x^2 + \dfrac{2y}{x}\right)dx = (3 - \ln x^2)\,dy$

(m) $\dfrac{dy}{dx} = \dfrac{x}{y} + \dfrac{y}{x} + 1$ **(n)** $\dfrac{y}{x^2}\dfrac{dy}{dx} + e^{2x^3 + y^2} = 0$

In Problems 9–16, solve the given differential equation.

9. $(y^2 + 1)\,dx = y\sec^2 x\,dy$

10. $y(\ln x - \ln y)\,dx = (x \ln x - x \ln y - y)\,dy$

11. $(6x + 1)y^2\dfrac{dy}{dx} + 3x^2 + 2y^3 = 0$

12. $\dfrac{dx}{dy} = -\dfrac{4y^2 + 6xy}{3y^2 + 2x}$

13. $t\dfrac{dQ}{dt} + Q = t^4 \ln t$

14. $(2x + y + 1)y' = 1$

15. $(x^2 + 4)\,dy = (2x - 8xy)\,dx$

16. $(2r^2 \cos\theta \sin\theta + r\cos\theta)\, d\theta + (4r + \sin\theta - 2r\cos^2\theta)\, dr = 0$

In Problems 17 and 18, solve the given initial-value problem and give the largest interval I on which the solution is defined.

17. $\sin x \dfrac{dy}{dx} + (\cos x)\, y = 0, \qquad y\!\left(\dfrac{7\pi}{6}\right) = -2$

18. $\dfrac{dy}{dt} + 2(t+1)y^2 = 0, \qquad y(0) = -\dfrac{1}{8}$

19. (a) Without solving, explain why the initial-value problem
$$\dfrac{dy}{dx} = \sqrt{y}, \quad y(x_0) = y_0,$$
has no solution for $y_0 < 0$.

(b) Solve the initial-value problem in part **(a)** for $y_0 > 0$ and find the largest interval I on which the solution is defined.

20. (a) Find an implicit solution of the initial-value problem
$$\dfrac{dy}{dx} = \dfrac{y^2 - x^2}{xy}, \quad y(1) = -\sqrt{2}.$$

(b) Find an explicit solution of the problem in part **(a)** and give the largest interval I over which the solution is defined. A graphing utility may be helpful here.

21. Graphs of some members of a family of solutions for a first-order differential equation $dy/dx = f(x, y)$ are shown in Figure 2.60. The graph of an implicit solution $G(x, y) = 0$ that passes through the points $(1, -1)$ and $(-1, 3)$ is shown in black. With colored pencils trace out the solution curves of the solutions $y = y_1(x)$ and $y = y_2(x)$ defined by the implicit solution such that $y_1(1) = -1$ and $y_2(-1) = 3$. Estimate the interval I on which each solution is defined.

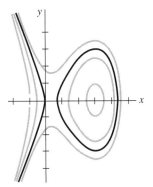

Figure 2.60

22. Use Euler's method with step size $h = 0.1$ to approximate $y(1.2)$ where $y(x)$ is a solution of the initial-value problem $y' = 1 + x\sqrt{y}, \ y(1) = 9$.

23. In March 1976 the world population reached 4 billion. A popular news magazine predicted that with an average yearly growth rate of 1.8%, the world population would be 8 billion in 45 years. How does this value compare with

that predicted by the model that says the rate of increase is proportional to the population at any time?

24. Air containing 0.06% carbon dioxide is pumped into a room whose volume is 8000 ft³. The air is pumped in at a rate of 2000 ft³/min, and the circulated air is then pumped out at the same rate. If there is an initial concentration of 0.2% carbon dioxide, determine the subsequent amount in the room at any time. What is the concentration at 10 minutes? What is the steady-state, or equilibrium, concentration of carbon dioxide?

25. A heart pacemaker, as shown in Figure 2.61, consists of a battery, a capacitor, and the heart as a resistor. When the switch S is at P the capacitor charges; when S is at Q the capacitor discharges, sending an electrical stimulus to the heart. During this time the voltage E applied to the heart is given by
$$\dfrac{dE}{dt} = -\dfrac{1}{RC}\, E, \quad t_1 < t < t_2,$$
where R and C are constants. Determine $E(t)$ if $E(t_1) = E_0$. (Of course, the opening and closing of the switch are periodic in time, to simulate the natural heartbeat.)

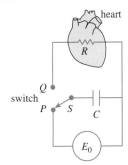

Figure 2.61

26. Suppose a cell is suspended in a solution containing a solute of constant concentration C_s. Suppose further that the cell has constant volume V and that the area of its permeable membrane is the constant A. By Fick's law the rate of change of its mass m is directly proportional to the area A and the difference $C_s - C(t)$, where $C(t)$ is the concentration of the solute inside the cell at any time t. Find $C(t)$ if $m = VC(t)$ and $C(0) = C_0$. See Figure 2.62.

Figure 2.62

27. Consider Newton's law of cooling $dT/dt = k(T - T_m)$, $k < 0$, where the temperature of the surrounding medium T_m changes with time. Suppose the initial temperature of a body is T_1 and the initial temperature of the surrounding medium is T_2 and $T_m = T_2 + B(T_1 - T)$, where $B > 0$ is a constant.

(a) Find the temperature of the body at any time t.

(b) What is the limiting value of the temperature as $t \to \infty$?

(c) What is the limiting value of T_m as $t \to \infty$?

28. An LR series circuit has a variable inductor with the inductance defined by

$$L = \begin{cases} 1 - \dfrac{t}{10}, & 0 \le t < 10 \\ 0, & t \ge 10 \end{cases}$$

Find the current $i(t)$ if the resistance is 0.2 ohm, the impressed voltage is $E(t) = 4$, and $i(0) = 0$. Graph $i(t)$.

29. A classical problem in the calculus of variations is to find the shape of a curve \mathscr{C} such that a bead, under the influence of gravity, will slide from point $A(0, 0)$ to point $B(x_1, y_1)$ in the least time. See Figure 2.63. It can be shown that a nonlinear differential equation for the shape $y(x)$ of the path is $y[1 + (y')^2] = k$, where k is a constant. First solve for dx in terms of y and dy, and then use the substitution $y = k \sin^2\theta$ to obtain the parametric form of the solution. The curve \mathscr{C} turns out to be a cycloid.

Figure 2.63

30. When all the curves in a family $G(x, y, c_1) = 0$ intersect orthogonally all the curves in another family $H(x, y, c_2) = 0$, the families are said to be **orthogonal trajectories** of each other. See Figure 2.64. If $dy/dx = f(x, y)$ is the differential equation of one family, then the differential equation for the orthogonal trajectories of this family is $dy/dx = -1/f(x, y)$.

(a) Find a differential equation for the family $y = -x - 1 + c_1 e^x$. Find the orthogonal trajectories of this family.

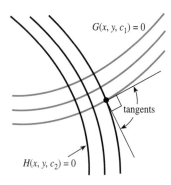

Figure 2.64

(b) Find the orthogonal trajectories for the family $y = 1/(x + c_1)$. Use a graphing utility to graph both families on the same set of coordinate axes.

31. The populations of two species of animals are described by the nonlinear system of first-order differential equations

$$\frac{dx}{dt} = k_1 x(\alpha - x)$$
$$\frac{dy}{dt} = k_2 xy.$$

Solve for x and y in terms of t.

32. Initially, two large tanks A and B each hold 100 gallons of brine. The well-stirred liquid is pumped between the tanks as shown in Figure 2.65. Use the information given in the figure to construct a mathematical model for the numbers $x_1(t)$ and $x_2(t)$ of pounds of salt at any time in tanks A and B, respectively.

Figure 2.65

**TOPICS TO REVIEW
FOR THIS CHAPTER**

solving polynomial equations
rational roots
determinants (8.4)
Cramer's rule (8.7)

**IMPORTANT CONCEPTS
IN THIS CHAPTER**

initial-value problem
boundary-value problem
linear dependence
linear independence
Wronskian
general solution
auxiliary equation
complementary function
particular solution
free motion
simple harmonic motion
damped motion
driven motion
aging spring
linear spring
deflection of beams
eigenvalues
eigenfunctions
buckling of thin columns
rotating string
nonlinear spring
pendulum motion
variable mass

INTRODUCTION

We turn now to the differential equations of order two and higher. In the first seven sections of this chapter we examine some of the underlying theory of *linear* DEs and methods for solving certain kinds of linear equations. The difficulties that surround higher-order *nonlinear* DEs and the few methods that yield analytic solutions of such equations are examined next (Section 3.7). The chapter concludes with higher-order linear and nonlinear mathematical models (Sections 3.8–3.10) and the first of several methods to be considered on solving systems of linear DEs (Section 3.11).

3.1 Preliminary Theory: Linear Equations

3.1.1 Initial-Value and Boundary-Value Problems

■ **Initial-Value Problem** In Section 1.2 we defined an initial-value problem for a general *n*th-order differential equation. For a linear differential equation, an **nth-order initial-value problem** is

$$\text{Solve: } a_n(x)\frac{d^n y}{dx^n} + a_{n-1}(x)\frac{d^{n-1}y}{dx^{n-1}} + \cdots + a_1(x)\frac{dy}{dx} + a_0(x)y = g(x)$$

$$\text{Subject to: } y(x_0) = y_0, \quad y'(x_0) = y_1, \quad \dots, \quad y^{(n-1)}(x_0) = y_{n-1}. \tag{1}$$

Recall that for a problem such as this we seek a function defined on some interval I containing x_0 that satisfies the differential equation and the n initial conditions specified at x_0: $y(x_0) = y_0$, $y'(x_0) = y_1$, \dots, $y^{(n-1)}(x_0) = y_{n-1}$. We have already seen that in the case of a second-order initial-value problem a solution curve must pass through the point (x_0, y_0) and have slope y_1 at this point.

■ **Existence and Uniqueness** In Section 1.2 we stated a theorem that gave conditions under which the existence and uniqueness of a solution of a first-order initial-value problem were guaranteed. The theorem that follows gives sufficient conditions for the existence of a unique solution of the problem in (1).

THEOREM 3.1 Existence of a Unique Solution

Let $a_n(x)$, $a_{n-1}(x)$, \dots, $a_1(x)$, $a_0(x)$, and $g(x)$ be continuous on an interval I, and let $a_n(x) \neq 0$ for every x in this interval. If $x = x_0$ is any point in this interval, then a solution $y(x)$ of the initial-value problem (1) exists on the interval and is unique.

Example 1 **Unique Solution of an IVP**

The initial-value problem

$$3y''' + 5y'' - y' + 7y = 0, \qquad y(1) = 0, \quad y'(1) = 0, \quad y''(1) = 0$$

possesses the trivial solution $y = 0$. Since the third-order equation is linear with constant coefficients, it follows that all the conditions of Theorem 3.1 are fulfilled. Hence $y = 0$ is the *only* solution on any interval containing $x = 1$. ❑

Example 2 Unique Solution of an IVP

You should verify that the function $y = 3e^{2x} + e^{-2x} - 3x$ is a solution of the initial-value problem $y'' - 4y = 12x$, $y(0) = 4$, $y'(0) = 1$. Now the differential equation is linear, the coefficients as well as $g(x) = 12x$ are continuous, and $a_2(x) = 1 \neq 0$ on any interval I containing $x = 0$. We conclude from Theorem 3.1 that the given function is the unique solution on I. ❏

The requirements in Theorem 3.1 that $a_i(x)$, $i = 0, 1, 2, \ldots, n$ be continuous and $a_n(x) \neq 0$ for every x in I are both important. Specifically, if $a_n(x) = 0$ for some x in the interval, then the solution of a linear initial-value problem may not be unique or even exist. For example, you should verify that the function $y = cx^2 + x + 3$ is a solution of the initial-value problem

$$x^2 y'' - 2xy' + 2y = 6, \qquad y(0) = 3, \quad y'(0) = 1$$

on the interval $(-\infty, \infty)$ for any choice of the parameter c. In other words, there is no unique solution of the problem. Although most of the conditions of Theorem 3.1 are satisfied, the obvious difficulties are that $a_2(x) = x^2$ is zero at $x = 0$ and that the initial conditions are also imposed at $x = 0$.

■ **Boundary-Value Problem** Another type of problem consists of solving a linear differential equation of order two or greater in which the dependent variable y or its derivatives are specified at *different points*. A problem such as

$$\textit{Solve:} \quad a_2(x) \frac{d^2y}{dx^2} + a_1(x) \frac{dy}{dx} + a_0(x)y = g(x)$$

$$\textit{Subject to:} \quad y(a) = y_0, \qquad y(b) = y_1$$

is called a **boundary-value problem** (**BVP**). The prescribed values $y(a) = y_0$ and $y(b) = y_1$ are called **boundary conditions.** A solution of the foregoing problem is a function satisfying the differential equation on some interval I, containing a and b, whose graph passes through the two points (a, y_0) and (b, y_1). See Figure 3.1.

For a second-order differential equation, other pairs of boundary conditions could be

$$y'(a) = y_0, \qquad y(b) = y_1$$

$$y(a) = y_0, \qquad y'(b) = y_1$$

$$y'(a) = y_0, \qquad y'(b) = y_1,$$

where y_0 and y_1 denote arbitrary constants. These three pairs of conditions are just special cases of the general boundary conditions

$$\alpha_1 y(a) + \beta_1 y'(a) = \gamma_1$$

$$\alpha_2 y(b) + \beta_2 y'(b) = \gamma_2.$$

The next example shows that even when the conditions of Theorem 3.1 are fulfilled, a boundary-value problem may have several solutions (as suggested in Figure 3.1), a unique solution, or no solution at all.

Example 3 A BVP Can Have Many, One, or No Solutions

In Example 4 of Section 1.1 we saw that the two-parameter family of solutions of the differential equation $x'' + 16x = 0$ is

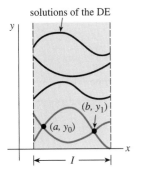

solutions of the DE

Figure 3.1

$$x = c_1 \cos 4t + c_2 \sin 4t. \qquad (2)$$

(a) Suppose we now wish to determine that solution of the equation that further satisfies the boundary conditions $x(0) = 0$, $x(\pi/2) = 0$. Observe that the first condition $0 = c_1 \cos 0 + c_2 \sin 0$ implies $c_1 = 0$, so that $x = c_2 \sin 4t$. But when $t = \pi/2$, $0 = c_2 \sin 2\pi$ is satisfied for any choice of c_2 since $\sin 2\pi = 0$. Hence the boundary-value problem

$$x'' + 16x = 0, \qquad x(0) = 0, \quad x\left(\frac{\pi}{2}\right) = 0 \qquad (3)$$

has infinitely many solutions. Figure 3.2 shows the graphs of some of the members of the one-parameter family $x = c_2 \sin 4t$ that pass through the two points $(0, 0)$ and $(\pi/2, 0)$.

(b) If the boundary-value problem in (3) is changed to

$$x'' + 16x = 0, \qquad x(0) = 0, \quad x\left(\frac{\pi}{8}\right) = 0, \qquad (4)$$

then $x(0) = 0$ still requires $c_1 = 0$ in the solution (2). But applying $x(\pi/8) = 0$ to $x = c_2 \sin 4t$ demands that $0 = c_2 \sin(\pi/2) = c_2 \cdot 1$. Hence $x = 0$ is a solution of this new boundary-value problem. Indeed, it can be proved that $x = 0$ is the *only* solution of (4).

(c) Finally, if we change the problem to

$$x'' + 16x = 0, \qquad x(0) = 0, \quad x\left(\frac{\pi}{2}\right) = 1, \qquad (5)$$

we find again that $c_1 = 0$ from $x(0) = 0$, but that applying $x(\pi/2) = 1$ to $x = c_2 \sin 4t$ leads to the contradiction $1 = c_2 \sin 2\pi = c_2 \cdot 0 = 0$. Hence the boundary-value problem (5) has no solution. ❏

Figure 3.2

3.1.2 Homogeneous Equations

A linear nth-order differential equation of the form

$$a_n(x) \frac{d^n y}{dx^n} + a_{n-1}(x) \frac{d^{n-1}y}{dx^{n-1}} + \cdots + a_1(x) \frac{dy}{dx} + a_0(x)y = 0 \qquad (6)$$

is said to be **homogeneous,** whereas an equation

$$a_n(x) \frac{d^n y}{dx^n} + a_{n-1}(x) \frac{d^{n-1}y}{dx^{n-1}} + \cdots + a_1(x) \frac{dy}{dx} + a_0(x)y = g(x). \qquad (7)$$

with $g(x)$ not identically zero, is said to be **nonhomogeneous.** For example, $2y'' + 3y' - 5y = 0$ is a homogeneous linear second-order differential equation, whereas $x^3 y''' + 6y' + 10y = e^x$ is a nonhomogeneous linear third-order differential equation. The word *homogeneous* in this context does not refer to coefficients that are homogeneous functions as in Section 2.5; rather, the word has exactly the same meaning as in Section 2.3.

We shall see that in order to solve a nonhomogeneous linear equation (7), we must first be able to solve the associated homogeneous equation (6).

To avoid needless repetition throughout the remainder of this text, we shall, as a matter of course, make the following important assumptions when stating definitions and theorems about the linear equations (6) and (7). On some common interval I

Remember these assumptions in the definitions and theorems of this chapter.

- the coefficients $a_i(x)$, $i = 0, 1, 2, \ldots, n$, are continuous;
- the right-hand member $g(x)$ is continuous; and
- $a_n(x) \neq 0$ for every x in the interval.

■ **Differential Operators** In calculus, differentiation is often denoted by the capital letter D, that is, $dy/dx = Dy$. The symbol D is called a **differential operator** because it transforms a differentiable function into another function. For example, $D(\cos 4x) = -4 \sin 4x$, and $D(5x^3 - 6x^2) = 15x^2 - 12x$. Higher-order derivatives can be expressed in terms of D in a natural manner:

$$\frac{d}{dx}\left(\frac{dy}{dx}\right) = \frac{d^2y}{dx^2} = D(Dy) = D^2y \qquad \text{and in general} \qquad \frac{d^ny}{dx^n} = D^ny,$$

where y represents a sufficiently differentiable function. Polynomial expressions involving D, such as $D + 3$, $D^2 + 3D - 4$, and $5x^3D^3 - 6x^2D^2 + 4xD + 9$, are also differential operators. In general, we define an **nth-order differential operator** to be

$$L = a_n(x)D^n + a_{n-1}(x)D^{n-1} + \cdots + a_1(x)D + a_0(x). \qquad (8)$$

As a consequence of two basic properties of differentiation, $D(cf(x)) = c\,Df(x)$, c a constant, and $D\{f(x) + g(x)\} = Df(x) + Dg(x)$, the differential operator L possesses a linearity property, that is, L operating on a linear combination of two differentiable functions is the same as the linear combination of L operating on the individual functions. In symbols, this means

$$L\{\alpha f(x) + \beta g(x)\} = \alpha L(f(x)) + \beta L(g(x)), \qquad (9)$$

where α and β are constants. Because of (9) we say that the nth-order differential operator L is a **linear operator.**

■ **Differential Equations** Any linear differential equation can be expressed in terms of the D notation. For example, the differential equation $y'' + 5y' + 6y = 5x - 3$ can be written as $D^2y + 5Dy + 6y = 5x - 3$ or $(D^2 + 5D + 6)y = 5x - 3$. Using (8), the nth-order linear differential equations (6) and (7) can be written compactly as

$$L(y) = 0 \qquad \text{and} \qquad L(y) = g(x),$$

respectively.

■ **Superposition Principle** In the next theorem we see that the sum, or **superposition,** of two or more solutions of a homogeneous linear differential equation is also a solution.

THEOREM 3.2 **Superposition Principle—Homogeneous Equations**

Let y_1, y_2, \ldots, y_k be solutions of the homogeneous nth-order differential equation (6) on an interval I. Then the linear combination

$$y = c_1 y_1(x) + c_2 y_2(x) + \cdots + c_k y_k(x),$$

where the c_i, $i = 1, 2, \ldots, k$ are arbitrary constants, is also a solution on the interval.

PROOF We prove the case $k = 2$. Let L be the differential operator defined in (8), and let $y_1(x)$ and $y_2(x)$ be solutions of the homogeneous equation $L(y) = 0$. If we define $y = c_1 y_1(x) + c_2 y_2(x)$, then by linearity of L we have

$$L(y) = L\{c_1 y_1(x) + c_2 y_2(x)\} = c_1 L(y_1) + c_2 L(y_2) = c_1 \cdot 0 + c_2 \cdot 0 = 0. \qquad \square$$

Corollaries to Theorem 3.2

(A) A constant multiple $y = c_1 y_1(x)$ of a solution $y_1(x)$ of a homogeneous linear differential equation is also a solution.

(B) A homogeneous linear differential equation always possesses the trivial solution $y = 0$.

> ### *Example 4* Superposition—Homogeneous DE
>
> The functions $y_1 = x^2$ and $y_2 = x^2 \ln x$ are both solutions of the homogeneous linear equation $x^3 y''' - 2xy' + 4y = 0$ on the interval $(0, \infty)$. By the superposition principle, the linear combination
>
> $$y = c_1 x^2 + c_2 x^2 \ln x$$
>
> is also a solution of the equation on the interval. ❑

The function $y = e^{7x}$ is a solution of $y'' - 9y' + 14y = 0$. Since the differential equation is linear and homogeneous, the constant multiple $y = ce^{7x}$ is also a solution. For various values of c we see that $y = 9e^{7x}$, $y = 0$, $y = -\sqrt{5}e^{7x}$, . . . , are all solutions of the equation.

■ **Linear Dependence and Linear Independence** The next two concepts are basic to the study of linear differential equations.

> ### DEFINITION 3.1 Linear Dependence/Independence
>
> A set of functions $f_1(x), f_2(x), \ldots, f_n(x)$ is said to be **linearly dependent** on an interval I if there exist constants c_1, c_2, \ldots, c_n, not all zero, such that
>
> $$c_1 f_1(x) + c_2 f_2(x) + \cdots + c_n f_n(x) = 0$$
>
> for every x in the interval. If the set of functions is not linearly dependent on the interval, it is said to be **linearly independent.**

In other words, a set of functions is linearly independent on an interval if the only constants for which

$$c_1 f_1(x) + c_2 f_2(x) + \cdots + c_n f_n(x) = 0$$

for every x in the interval are $c_1 = c_2 = \cdots = c_n = 0$.

It is easy to understand these definitions in the case of two functions $f_1(x)$ and $f_2(x)$. If the functions are linearly dependent on an interval, then there exist constants c_1 and c_2 that are not both zero such that for every x in the interval $c_1 f_1(x) + c_2 f_2(x) = 0$. Therefore, if we assume that $c_1 \neq 0$, it follows that $f_1(x) = (-c_2/c_1)f_2(x)$; that is, *if two functions are linearly dependent, then one is simply a constant multiple of the other.* Conversely, if $f_1(x) = c_2 f_2(x)$ for some constant c_2, then $(-1) \cdot f_1(x) + c_2 f_2(x) = 0$ for every x on some interval. Hence the functions are linearly dependent, since at least one of the constants (namely, $c_1 = -1$) is not zero. We conclude that *two functions are linearly independent when neither is a constant multiple of the other* on an interval. For example, the functions $f_1(x) = \sin 2x$ and $f_2(x) = \sin x \cos x$ are linearly dependent on $(-\infty, \infty)$ because $f_1(x)$ is a constant multiple of $f_2(x)$. Recall from the double angle formula for the sine that $\sin 2x = 2 \sin x \cos x$. On the other hand, the functions $f_1(x) = x$ and $f_2(x) = |x|$ are linearly independent on $(-\infty, \infty)$. Inspection of Figure 3.3 should convince you that neither function is a constant multiple of the other on the interval.

(a)

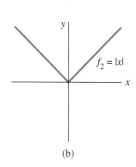

(b)

Figure 3.3

It follows from the preceding discussion that the ratio $f_2(x)/f_1(x)$ is not a constant on an interval on which $f_1(x)$ and $f_2(x)$ are linearly independent. This little fact will be used in the next section.

Example 5 Linearly Dependent Functions

The functions $f_1(x) = \cos^2 x$, $f_2(x) = \sin^2 x$, $f_3(x) = \sec^2 x$, $f_4(x) = \tan^2 x$ are linearly dependent on the interval $(-\pi/2, \pi/2)$ since

$$c_1 \cos^2 x + c_2 \sin^2 x + c_3 \sec^2 x + c_4 \tan^2 x = 0,$$

when $c_1 = c_2 = 1$, $c_3 = -1$, $c_4 = 1$. We used here $\cos^2 x + \sin^2 x = 1$ and $1 + \tan^2 x = \sec^2 x$. ❏

A set of functions $f_1(x), f_2(x), \ldots, f_n(x)$ is linearly dependent on an interval if at least one function can be expressed as a linear combination of the remaining functions.

Example 6 Linearly Dependent Functions

The functions $f_1(x) = \sqrt{x} + 5$, $f_2(x) = \sqrt{x} + 5x$, $f_3(x) = x - 1$, $f_4(x) = x^2$ are linearly dependent on the interval $(0, \infty)$ since f_2 can be written as a linear combination of f_1, f_3, and f_4. Observe that

$$f_2(x) = 1 \cdot f_1(x) + 5 \cdot f_3(x) + 0 \cdot f_4(x)$$

for every x in the interval $(0, \infty)$. ❏

■ **Solutions of Differential Equations** We are primarily interested in linearly independent functions or, more to the point, linearly independent solutions of a linear differential equation. Although we could always appeal directly to Definition 3.1, it turns out that the question of whether n solutions y_1, y_2, \ldots, y_n of a homogeneous linear nth-order differential equation (6) are linearly independent can be settled somewhat mechanically using a determinant.

DEFINITION 3.2 Wronskian

Suppose each of the functions $f_1(x), f_2(x), \ldots, f_n(x)$ possesses at least $n - 1$ derivatives. The determinant

$$W(f_1, f_2, \ldots, f_n) = \begin{vmatrix} f_1 & f_2 & \cdots & f_n \\ f_1' & f_2' & \cdots & f_n' \\ \vdots & \vdots & & \vdots \\ f_1^{(n-1)} & f_2^{(n-1)} & \cdots & f_n^{(n-1)} \end{vmatrix},$$

where the primes denote derivatives, is called the **Wronskian** of the functions.

THEOREM 3.3 Criterion for Linearly Independent Solutions

Let y_1, y_2, \ldots, y_n be n solutions of the homogeneous linear nth-order differential equation (6) on an interval I. Then the set of solutions is **linearly independent** on I if and only if $W(y_1, y_2, \ldots, y_n) \neq 0$ for every x in the interval.

It follows from Theorem 3.3 that when y_1, y_2, \ldots, y_n are n solutions of (6) on an interval I, the Wronskian $W(y_1, y_2, \ldots, y_n)$ is either identically zero or never zero on the interval.

A set of n linearly independent solutions of a homogeneous linear nth-order differential equation is given a special name.

DEFINITION 3.3 **Fundamental Set of Solutions**

Any set y_1, y_2, \ldots, y_n of n linearly independent solutions of the homogeneous linear nth-order differential equation (6) on an interval I is said to be a **fundamental set of solutions** on the interval.

The basic question of whether a fundamental set of solutions exists for a linear equation is answered in the next theorem.

THEOREM 3.4 **Existence of a Fundamental Set**

There exists a fundamental set of solutions for the homogeneous linear nth-order differential equation (6) on an interval I.

Analogous to the fact that any vector in three dimensions can be expressed as a linear combination of the *linearly independent* vectors **i, j, k,** any solution of an nth-order homogeneous linear differential equation on an interval I can be expressed as a linear combination of n linearly independent solutions on I. In other words, n linearly independent solutions y_1, y_2, \ldots, y_n are the basic building blocks for the general solution of the equation.

THEOREM 3.5 **General Solution—Homogeneous Equations**

Let y_1, y_2, \ldots, y_n be a fundamental set of solutions of the homogeneous linear nth-order differential equation (6) on an interval I. Then the **general solution** of the equation on the interval is

$$y = c_1 y_1(x) + c_2 y_2(x) + \cdots + c_n y_n(x),$$

where c_i, $i = 1, 2, \ldots, n$ are arbitrary constants.

Theorem 3.5 states that if $Y(x)$ is any solution of (6) on the interval, then constants C_1, C_2, \ldots, C_n can always be found so that

$$Y(x) = C_1 y_1(x) + C_2 y_2(x) + \cdots + C_n y_n(x).$$

We will prove the case when $n = 2$.

PROOF Let Y be a solution and y_1 and y_2 be linearly independent solutions of $a_2 y'' + a_1 y' + a_0 y = 0$ on an interval I. Suppose $x = t$ is a point in I for which $W(y_1(t), y_2(t)) \neq 0$. Suppose also that $Y(t) = k_1$ and $Y'(t) = k_2$. If we now examine the equations

$$C_1 y_1(t) + C_2 y_2(t) = k_1$$

$$C_1 y_1'(t) + C_2 y_2'(t) = k_2,$$

it follows that we can determine C_1 and C_2 uniquely, provided that the determinant of the coefficients satisfies

$$\begin{vmatrix} y_1(t) & y_2(t) \\ y_1'(t) & y_2'(t) \end{vmatrix} \neq 0.$$

But this determinant is simply the Wronskian evaluated at $x = t$, and, by assumption, $W \neq 0$. If we define $G(x) = C_1 y_1(x) + C_2 y_2(x)$, we observe that $G(x)$ satisfies the differential equation, since it is a superposition of two known solutions; $G(x)$ satisfies the initial conditions

$$G(t) = C_1 y_1(t) + C_2 y_2(t) = k_1 \quad \text{and} \quad G'(t) = C_1 y_1'(t) + C_2 y_2'(t) = k_2;$$

$Y(x)$ satisfies the *same* linear equation and the *same* initial conditions. Since the solution of this linear initial-value problem is unique (Theorem 3.1), we have $Y(x) = G(x)$ or $Y(x) = C_1 y_1(x) + C_2 y_2(x)$. ❏

Example 7 General Solution of a Homogeneous DE

The functions $y_1 = e^{3x}$ and $y_2 = e^{-3x}$ are both solutions of the homogeneous linear equation $y'' - 9y = 0$ on the interval $(-\infty, \infty)$. By inspection, the solutions are linearly independent on the x-axis. This fact can be corroborated by observing that the Wronskian

$$W(e^{3x}, e^{-3x}) = \begin{vmatrix} e^{3x} & e^{-3x} \\ 3e^{3x} & -3e^{-3x} \end{vmatrix} = -6 \neq 0$$

for every x. We conclude that y_1 and y_2 form a fundamental set of solutions, and consequently $y = c_1 e^{3x} + c_2 e^{-3x}$ is the general solution of the equation on the interval. ❏

Example 8 A Solution Obtained from a General Solution

The function $y = 4 \sinh 3x - 5e^{3x}$ is a solution of the differential equation in Example 7. (Verify this.) In view of Theorem 3.5, we must be able to obtain this solution from the general solution $y = c_1 e^{3x} + c_2 e^{-3x}$. Observe that if we choose $c_1 = 2$ and $c_2 = -7$, then $y = 2e^{3x} - 7e^{-3x}$ can be rewritten as

$$y = 2e^{3x} - 2e^{-3x} - 5e^{-3x} = 4\left(\frac{e^{3x} - e^{-3x}}{2}\right) - 5e^{-3x}.$$

The last expression is recognized as $y = 4 \sinh 3x - 5e^{-3x}$. ❏

Example 9 General Solution of a Homogeneous DE

The functions $y_1 = e^x$, $y_2 = e^{2x}$, and $y_3 = e^{3x}$ satisfy the third-order equation $y''' - 6y'' + 11y' - 6y = 0$. Since

$$W(e^x, e^{2x}, e^{3x}) = \begin{vmatrix} e^x & e^{2x} & e^{3x} \\ e^x & 2e^{2x} & 3e^{3x} \\ e^x & 4e^{2x} & 9e^{3x} \end{vmatrix} = 2e^{6x} \neq 0$$

for every real value of x, the functions y_1, y_2, and y_3 form a fundamental set of solutions on $(-\infty, \infty)$. We conclude that $y = c_1 e^x + c_2 e^{2x} + c_3 e^{3x}$ is the general solution of the differential equation on the interval. ❏

3.1.3 Nonhomogeneous Equations

Any function y_p free of arbitrary parameters that satisfies (7) is said to be a **particular solution** or **particular integral** of the equation. For example, it is a straightforward task to show that the constant function $y_p = 3$ is a particular solution of the nonhomogeneous equation $y'' + 9y = 27$.

Now if y_1, y_2, \ldots, y_k are solutions of (6) on an interval I and y_p is any particular solution of (7) on I, then the linear combination

$$y = c_1 y_1(x) + c_2 y_2(x) + \cdots + c_k y_k(x) + y_p \qquad (10)$$

is also a solution of the nonhomogeneous equation (7). If you think about it, this makes sense, because the linear combination $c_1 y_1(x) + c_2 y_2(x) + \cdots + c_k y_k(x)$ is mapped into 0 by the operator $L = a_n D^n + a_{n-1} D^{n-1} + \cdots + a_1 D + a_0$, whereas y_p is mapped into $g(x)$. If we use $k = n$ linearly independent solutions of the nth-order equation (6), then the expression in (10) becomes the general solution of (7).

> **THEOREM 3.6** General Solution—
> Nonhomogeneous Equations
>
> Let y_p be any particular solution of the nonhomogeneous linear nth-order differential equation (7) on an interval I, and let y_1, y_2, \ldots, y_n be a fundamental set of solutions of the associated homogeneous differential equation (6) on I. Then the **general solution** of the equation on the interval is
>
> $$y = c_1 y_1(x) + c_2 y_2(x) + \cdots + c_n y_n(x) + y_p,$$
>
> where the c_i, $i = 1, 2, \ldots, n$ are arbitrary constants.

PROOF Let L be the differential operator defined in (8), and let $Y(x)$ and $y_p(x)$ be particular solutions of the nonhomogeneous equation $L(y) = g(x)$. If we define $u(x) = Y(x) - y_p(x)$, then by linearity of L we have

$$L(u) = L\{Y(x) - y_p(x)\} = L(Y(x)) - L(y_p(x)) = g(x) - g(x) = 0.$$

This shows that $u(x)$ is a solution of the homogeneous equation $L(y) = 0$. Hence, by Theorem 3.5, $u(x) = c_1 y_1(x) + c_2 y_2(x) + \cdots + c_n y_n(x)$, and so

$$Y(x) - y_p(x) = c_1 y_1(x) + c_2 y_2(x) + \cdots + c_n y_n(x)$$

or $\qquad\qquad Y(x) = c_1 y_1(x) + c_2 y_2(x) + \cdots + c_n y_n(x) + y_p(x).$ ❏

■ **Complementary Function** We see in Theorem 3.6 that the general solution of a nonhomogeneous linear equation consists of the sum of two functions:

$$y = c_1 y_1(x) + c_2 y_2(x) + \cdots + c_n y_n(x) + y_p(x) = y_c(x) + y_p(x).$$

The linear combination $y_c(x) = c_1 y_1(x) + c_2 y_2(x) + \cdots + c_n y_n(x)$, which is the general solution of (6), is called the **complementary function** for equation (7). In other words, to solve a nonhomogeneous linear differential equation we first solve the associated homogeneous equation and then find any particular solution of the nonhomogeneous equation. The general solution of the nonhomogeneous equation is then

$$y = complementary\ function + any\ particular\ solution.$$

Example 10 **General Solution of a Nonhomogeneous DE**

By substitution, the function $y_p = -\frac{11}{12} - \frac{1}{2}x$ is readily shown to be a particular solution of the nonhomogeneous equation

$$y''' - 6y'' + 11y' - 6y = 3x. \tag{11}$$

In order to write the general solution of (11), we must also be able to solve the associated homogeneous equation

$$y''' - 6y'' + 11y' - 6y = 0.$$

But in Example 9 we saw that the general solution of this latter equation on the interval $(-\infty, \infty)$ was $y_c = c_1 e^x + c_2 e^{2x} + c_3 e^{3x}$. Hence the general solution of (11) on the interval is

$$y = y_c + y_p = c_1 e^x + c_2 e^{2x} + c_3 e^{3x} - \frac{11}{12} - \frac{1}{2}x. \qquad \square$$

■ **Another Superposition Principle** The last theorem of this discussion will be useful in Section 3.4, when we consider a method for finding particular solutions of nonhomogeneous equations.

THEOREM 3.7 Superposition Principle— Nonhomogeneous Equations

Let $y_{p_1}, y_{p_2}, \ldots, y_{p_k}$ be k particular solutions of the nonhomogeneous linear nth-order differential equation (7) on an interval I corresponding, in turn, to k distinct functions g_1, g_2, \ldots, g_k. That is, suppose y_{p_i} denotes a particular solution of the corresponding differential equation

$$a_n(x)y^{(n)} + a_{n-1}(x)y^{(n-1)} + \cdots + a_1(x)y' + a_0(x)y = g_i(x), \tag{12}$$

where $i = 1, 2, \ldots, k$. Then

$$y_p = y_{p_1}(x) + y_{p_2}(x) + \cdots + y_{p_k}(x) \tag{13}$$

is a particular solution of

$$a_n(x)y^{(n)} + a_{n-1}(x)y^{(n-1)} + \cdots + a_1(x)y' + a_0(x)y$$

$$= g_1(x) + g_2(x) + \cdots + g_k(x). \tag{14}$$

PROOF We prove the case $k = 2$. Let L be the differential operator defined in (8), and let $y_{p_1}(x)$ and $y_{p_2}(x)$ be particular solutions of the nonhomogeneous equations $L(y) = g_1(x)$ and $L(y) = g_2(x)$, respectively. If we define $y_p = y_{p_1}(x) + y_{p_2}(x)$, we want to show that y_p is a particular solution of $L(y) = g_1(x) + g_2(x)$. The result follows again by the linearity of the operator L:

$$L(y_p) = L\{y_{p_1}(x) + y_{p_2}(x)\} = L(y_{p_1}(x)) + L(y_{p_2}(x)) = g_1(x) + g_2(x). \qquad \square$$

Example 11 **Superposition—Nonhomogeneous DE**

You should verify that

$$y_{p_1} = -4x^2 \quad \text{is a particular solution of} \quad y'' - 3y' + 4y = -16x^2 + 24x - 8,$$

$$y_{p_2} = e^{2x} \quad \text{is a particular solution of} \quad y'' - 3y' + 4y = 2e^{2x},$$

$y_{p_3} = xe^x$ is a particular solution of $y'' - 3y' + 4y = 2xe^x - e^x$.

It follows from Theorem 3.7 that the superposition of y_{p_1}, y_{p_2}, and y_{p_3},

$$y = y_{p_1} + y_{p_2} + y_{p_3} = -4x^2 + e^{2x} + xe^x,$$

is a solution of

$$y'' - 3y' + 4y = \underbrace{-16x^2 + 24x - 8}_{g_1(x)} + \underbrace{2e^{2x}}_{g_2(x)} + \underbrace{2xe^x - e^x}_{g_3(x)}.$$ ❏

This sentence is a ☞
generalization of Theorem 3.7.

If the y_{p_i} are particular solutions of (12) for $i = 1, 2, \ldots, k,$ then the linear combination

$$y_p = c_1 y_{p_1} + c_2 y_{p_2} + \cdots + c_k y_{p_k},$$

where the c_i are constants, is also a particular solution of (14) when the right-hand member of the equation is the linear combination

$$c_1 g_1(x) + c_2 g_2(x) + \cdots + c_k g_k(x).$$

Before we actually start solving homogeneous and nonhomogeneous linear differential equations, we need one additional bit of theory presented in the next section.

Remarks

This remark is a continuation of the brief discussion of dynamical systems given at the end of Section 1.3.

A dynamical system whose rule or mathematical model is a linear nth-order differential equation

$$a_n(t)y^{(n)} + a_{n-1}(t)y^{(n-1)} + \cdots + a_1(t)y' + a_0(t)y = g(t)$$

is said to be a **linear system.** The set of n time-dependent functions $y(t)$, $y'(t)$, . . . , $y^{(n-1)}(t)$ are the **state variables** of the system. Recall, their values at some time t give the **state of the system.** The function g is variously called the **input function, forcing function,** or **excitation function.** A solution $y(t)$ of the differential equation is said to be the **output** or **response of the system.** Under the conditions stated in Theorem 3.1, the output or response $y(t)$ is uniquely determined by the input and the state of the system prescribed at a time t_0, that is, by the initial conditions $y(t_0)$, $y'(t_0)$, . . . , $y^{(n-1)}(t_0)$.

In order that a dynamical system be a linear system it is necessary that the superposition principle (Theorem 3.7) hold in the system; that is, the response of the system to a superposition of inputs is a superposition of outputs. We have already examined some simple linear systems in Section 2.7 (linear first-order equations); in Section 3.8 we examine linear systems in which the mathematical models are second-order differential equations.

3.1.1 Initial-Value and Boundary-Value Problems

In Problems 1–4, the given family of functions is the general solution of the differential equation on the indicated interval. Find a member of the family that is a solution of the initial-value problem.

1. $y = c_1 e^x + c_2 e^{-x}$, $(-\infty, \infty)$; $y'' - y = 0$, $y(0) = 0$, $y'(0) = 1$

2. $y = c_1 e^{4x} + c_2 e^{-x}$, $(-\infty, \infty)$; $y'' - 3y' - 4y = 0$, $y(0) = 1$, $y'(0) = 2$

3. $y = c_1 x + c_2 x \ln x$, $(0, \infty)$; $x^2 y'' - xy' + y = 0$, $y(1) = 3$, $y'(1) = -1$

4. $y = c_1 + c_2 \cos x + c_3 \sin x$, $(-\infty, \infty)$; $y''' + y' = 0$, $y(\pi) = 0$, $y'(\pi) = 2$, $y''(\pi) = -1$

5. Given that $y = c_1 + c_2 x^2$ is a two-parameter family of solutions of $xy'' - y' = 0$ on the interval $(-\infty, \infty)$, show that constants c_1 and c_2 cannot be found so that a member of the family satisfies the initial conditions $y(0) = 0$, $y'(0) = 1$. Explain why this does not violate Theorem 3.1.

6. Find two members of the family of solutions in Problem 5 that satisfy the initial conditions $y(0) = 0$, $y'(0) = 0$.

7. Given that $x(t) = c_1 \cos \omega t + c_2 \sin \omega t$ is the general solution of $x'' + \omega^2 x = 0$ on the interval $(-\infty, \infty)$, show that a solution satisfying the initial conditions $x(0) = x_0$, $x'(0) = x_1$, is given by

$$x(t) = x_0 \cos \omega t + \frac{x_1}{\omega} \sin \omega t.$$

8. Use the general solution of $x'' + \omega^2 x = 0$ given in Problem 7 to show that a solution satisfying the initial conditions $x(t_0) = x_0$, $x'(t_0) = x_1$, is the solution given in Problem 7 shifted by an amount t_0:

$$x(t) = x_0 \cos \omega(t - t_0) + \frac{x_1}{\omega} \sin \omega(t - t_0).$$

In Problems 9 and 10, find an interval centered about $x = 0$ for which the given initial-value problem has a unique solution.

9. $(x - 2)y'' + 3y = x$, $y(0) = 0$, $y'(0) = 1$

10. $y'' + (\tan x)y = e^x$, $y(0) = 1$, $y'(0) = 0$

11. Use the family in Problem 1 to find a solution of $y'' - y = 0$ that satisfies the boundary conditions $y(0) = 0$, $y(1) = 1$.

12. Use the family in Problem 5 to find a solution of $xy'' - y' = 0$ that satisfies the boundary conditions $y(0) = 1$, $y'(1) = 6$.

In Problems 13 and 14, the given two-parameter family is a solution of the indicated differential equation on the interval $(-\infty, \infty)$. Determine whether a member of the family can be found that satisfies the boundary conditions.

13. $y = c_1 e^x \cos x + c_2 e^x \sin x$; $y'' - 2y' + 2y = 0$

(a) $y(0) = 1$, $y'(\pi) = 0$ **(b)** $y(0) = 1$, $y(\pi) = -1$

(c) $y(0) = 1$, $y\left(\dfrac{\pi}{2}\right) = 1$ **(d)** $y(0) = 0$, $y(\pi) = 0$

14. $y = c_1 x^2 + c_2 x^4 + 3$; $x^2 y'' - 5xy' + 8y = 24$

(a) $y(-1) = 0$, $y(1) = 4$ **(b)** $y(0) = 1$, $y(1) = 2$

(c) $y(0) = 3$, $y(1) = 0$ **(d)** $y(1) = 3$, $y(2) = 15$

3.1.2 Homogeneous Equations

In Problems 15–22, determine whether the given set of functions is linearly dependent or linearly independent on the interval $(-\infty, \infty)$.

15. $f_1(x) = x$, $f_2(x) = x^2$, $f_3(x) = 4x - 3x^2$

16. $f_1(x) = 0$, $f_2(x) = x$, $f_3(x) = e^x$

17. $f_1(x) = 5$, $f_2(x) = \cos^2 x$, $f_3(x) = \sin^2 x$

18. $f_1(x) = \cos 2x$, $f_2(x) = 1$, $f_3(x) = \cos^2 x$

19. $f_1(x) = x$, $f_2(x) = x - 1$, $f_3(x) = x + 3$

20. $f_1(x) = 2 + x$, $f_2(x) = 2 + |x|$

21. $f_1(x) = 1 + x$, $f_2(x) = x$, $f_3(x) = x^2$

22. $f_1(x) = e^x$, $f_2(x) = e^{-x}$, $f_3(x) = \sinh x$

In Problems 23–30, verify that the given functions form a fundamental set of solutions of the differential equation on the indicated interval. Form the general solution.

23. $y'' - y' - 12y = 0$; e^{-3x}, e^{4x}, $(-\infty, \infty)$

24. $y'' - 4y = 0$; $\cosh 2x$, $\sinh 2x$, $(-\infty, \infty)$

25. $y'' - 2y' + 5y = 0$; $e^x \cos 2x$, $e^x \sin 2x$, $(-\infty, \infty)$

26. $4y'' - 4y' + y = 0$; $e^{x/2}$, $xe^{x/2}$, $(-\infty, \infty)$

27. $x^2 y'' - 6xy' + 12y = 0$; x^3, x^4, $(0, \infty)$

28. $x^2 y'' + xy' + y = 0$; $\cos(\ln x)$, $\sin(\ln x)$, $(0, \infty)$

29. $x^3 y''' + 6x^2 y'' + 4xy' - 4y = 0$; x, x^{-2}, $x^{-2} \ln x$, $(0, \infty)$

30. $y^{(4)} + y'' = 0$; 1, x, $\cos x$, $\sin x$, $(-\infty, \infty)$

31. (a) Verify that $y_1 = x^3$ and $y_2 = |x|^3$ are linearly independent solutions of the differential equation $x^2 y'' - 4xy' + 6y = 0$ on the interval $(-\infty, \infty)$.

(b) Show that $W(y_1, y_2) = 0$ for every real number x. Does this result violate Theorem 3.3? Explain.

(c) Verify that $Y_1 = x^3$ and $Y_2 = x^2$ are also linearly independent solutions of the differential equation in part **(a)** on the interval $(-\infty, \infty)$.

(d) Find a solution of the differential equation satisfying $y(0) = 0$, $y'(0) = 0$.

(e) By the superposition principle, Theorem 3.2, both linear combinations $y = c_1 y_1 + c_2 y_2$ and $Y = c_1 Y_1 + c_2 Y_2$ are solutions of the differential equation. Determine whether one, both, or neither of the linear combinations is a general solution of the differential equation on the interval $(-\infty, \infty)$.

32. Suppose that $y_1 = e^x$ and $y_2 = e^{-x}$ are two solutions of a homogeneous linear differential equation. Explain why $y_3 = \cosh x$ and $y_4 = \sinh x$ are also solutions of the equation.

33. Is the set of functions $f_1(x) = e^{x+2}$, $f_2(x) = e^{x-3}$ linearly dependent or linearly independent on the interval $(-\infty, \infty)$?

34. Suppose y_1, y_2, \ldots, y_k are k linearly independent solutions on $(-\infty, \infty)$ of a homogeneous linear nth-order differential equation with constant coefficients. By Theorem 3.2 it follows that $y_{k+1} = 0$ is also a solution of the differential equation. Is the set of solutions $y_1, y_2, \ldots, y_k, y_{k+1}$ linearly dependent or linearly independent on the interval $(-\infty, \infty)$?

3.1.3 Nonhomogeneous Equations

In Problems 35–38, verify that the given two-parameter family of functions is the general solution of the nonhomogeneous differential equation on the indicated interval.

35. $y'' - 7y' + 10y = 24e^x$;
$y = c_1 e^{2x} + c_2 e^{5x} + 6e^x$, $(-\infty, \infty)$

36. $y'' + y = \sec x$;
$y = c_1 \cos x + c_2 \sin x + x \sin x + (\cos x) \ln(\cos x)$,
$(-\pi/2, \pi/2)$

37. $y'' - 4y' + 4y = 2e^{2x} + 4x - 12$;
$y = c_1 e^{2x} + c_2 x e^{2x} + x^2 e^{2x} + x - 2$, $(-\infty, \infty)$

38. $2x^2 y'' + 5xy' + y = x^2 - x$;
$y = c_1 x^{-1/2} + c_2 x^{-1} + \frac{1}{15} x^2 - \frac{1}{6} x$, $(0, \infty)$

39. (a) Verify that $y_{p_1} = 3e^{2x}$ and $y_{p_2} = x^2 + 3x$ are, respectively, particular solutions of

$$y'' - 6y' + 5y = -9e^{2x}$$

and $y'' - 6y' + 5y = 5x^2 + 3x - 16$.

(b) Use part **(a)** to find particular solutions of

$$y'' - 6y' + 5y = 5x^2 + 3x - 16 - 9e^{2x}$$

and $y'' - 6y' + 5y = -10x^2 - 6x + 32 + e^{2x}$.

40. (a) By inspection find a particular solution of $y'' + 2y = 10$.

(b) By inspection find a particular solution of $y'' + 2y = -4x$.

(c) Find a particular solution of $y'' + 2y = -4x + 10$.

(d) Find a particular solution of $y'' + 2y = 8x + 5$.

3.2 Reduction of Order

It is one of the more interesting facts of mathematical life in the study of linear *second-order* differential equations that we can construct a second solution y_2 of

$$a_2(x)y'' + a_1(x)y' + a_0(x)y = 0 \tag{1}$$

on an interval I from a known nontrivial solution y_1. We seek a second solution $y_2(x)$ of (1) so that y_1 and y_2 are linearly independent on I. Recall that if y_1 and y_2 are linearly independent, then their ratio y_2/y_1 is nonconstant on I; that is, $y_2/y_1 = u(x)$ or $y_2(x) = u(x)y_1(x)$. The idea is to find the function $u(x)$ by substituting $y_2(x) = u(x)y_1(x)$ into the given differential equation. The method is called **reduction of order** since we must solve a linear *first-order* equation to find u.

> ### Example 1 A Second Solution by Reduction of Order
>
> Given that $y_1 = e^x$ is a solution of $y'' - y = 0$ on the interval $(-\infty, \infty)$, use reduction of order to find a second solution y_2.
>
> **SOLUTION** If $y = u(x)y_1(x) = u(x)e^x$, then the product rule gives
>
> $$y' = ue^x + e^x u', \qquad y'' = ue^x + 2e^x u' + e^x u'',$$
>
> and so $y'' - y = e^x(u'' + 2u') = 0.$

Since $e^x \neq 0$, the last equation requires $u'' + 2u' = 0$. If we make the substitution $w = u'$, this linear second-order equation in u becomes $w' + 2w = 0$, which is a linear first-order equation in w. Using the integrating factor e^{2x}, we can write $d/dx\,[e^{2x}w] = 0$. After integrating we get $w = c_1 e^{-2x}$ or $u' = c_1 e^{-2x}$. Integrating again then yields $u = -\frac{1}{2}\,c_1 e^{-2x} + c_2$. Thus

$$y = u(x)e^x = -\frac{c_1}{2}\,e^{-x} + c_2 e^x. \tag{2}$$

By picking $c_2 = 0$ and $c_1 = -2$ we obtain the desired second solution, $y_2 = e^{-x}$. Because $W(e^x, e^{-x}) \neq 0$ for every x, the solutions are linearly independent on $(-\infty, \infty)$. ❏

Since we have shown that $y_1 = e^x$ and $y_2 = e^{-x}$ are linearly independent solutions of a linear second-order equation, the expression in (2) is actually the general solution of $y'' - y = 0$ on $(-\infty, \infty)$.

■ **General Case** Suppose we divide by $a_2(x)$ in order to put equation (1) in the **standard form**

$$y'' + P(x)y' + Q(x)y = 0, \tag{3}$$

where $P(x)$ and $Q(x)$ are continuous on some interval I. Let us suppose further that $y_1(x)$ is a known solution of (3) on I and that $y_1(x) \neq 0$ for every x in the interval. If we define $y = u(x)y_1(x)$, it follows that

$$y' = uy_1' + y_1 u', \qquad y'' = uy_1'' + 2y_1' u' + y_1 u''$$

$$y'' + Py' + Qy = u[\underbrace{y_1'' + Py_1' + Qy_1}_{\text{zero}}] + y_1 u'' + (2y_1' + Py_1)u' = 0.$$

This implies that we must have

$$y_1 u'' + (2y_1' + Py_1)u' = 0 \qquad \text{or} \qquad y_1 w' + (2y_1' + Py_1)w = 0, \tag{4}$$

where we have let $w = u'$. Observe that the last equation in (4) is both linear and separable. Separating variables and integrating, we obtain

$$\frac{dw}{w} + 2\,\frac{y_1'}{y_1}\,dx + P\,dx = 0$$

$$\ln\left|wy_1^2\right| = -\int P\,dx + c \qquad \text{or} \qquad wy_1^2 = c_1 e^{-\int P\,dx}.$$

We solve the last equation for w, use $w = u'$, and integrate again:

$$u = c_1 \int \frac{e^{-\int P\,dx}}{y_1^2}\,dx + c_2.$$

By choosing $c_1 = 1$ and $c_2 = 0$, we find from $y = u(x)y_1(x)$ that a second solution of equation (3) is

$$y_2 = y_1(x) \int \frac{e^{-\int P(x)\,dx}}{y_1^2(x)}\,dx. \tag{5}$$

It makes a good review of differentiation to verify that the function $y_2(x)$ defined in (5) satisfies equation (3) and that y_1 and y_2 are linearly independent on any interval on which $y_1(x)$ is not zero.

Example 2 A Second Solution by Formula (5)

The function $y_1 = x^2$ is a solution of $x^2 y'' - 3xy' + 4y = 0$. Find the general solution on the interval $(0, \infty)$.

SOLUTION From the standard form of the equation,

$$y'' - \frac{3}{x} y' + \frac{4}{x^2} y = 0,$$

we find from (5)
$$y_2 = x^2 \int \frac{e^{3\int dx/x}}{x^4} \, dx \qquad \leftarrow e^{3\int dx/x} = e^{\ln x^3} = x^3$$

$$= x^2 \int \frac{dx}{x} = x^2 \ln x.$$

The general solution on $(0, \infty)$ is given by $y = c_1 y_1 + c_2 y_2$; that is,

$$y = c_1 x^2 + c_2 x^2 \ln x. \qquad \square$$

Remarks

We have derived and illustrated how to use (5) because this formula appears again in the next section and in Section 5.2. We use (5) simply to save time in obtaining a desired result. Your instructor will tell you whether you should memorize (5) or whether you should know the first principles of reduction of order.

EXERCISES 3.2

Answers to odd-numbered problems begin on page A-29.

In Problems 1–16, the indicated function $y_1(x)$ is a solution of the given equation. Use reduction of order or formula (5), as instructed, to find a second solution $y_2(x)$.

1. $y'' - 4y' + 4y = 0$; $y_1 = e^{2x}$

2. $y'' + 2y' + y = 0$; $y_1 = xe^{-x}$

3. $y'' + 16y = 0$; $y_1 = \cos 4x$

4. $y'' + 9y = 0$; $y_1 = \sin 3x$

5. $y'' - y = 0$; $y_1 = \cosh x$

6. $y'' - 25y = 0$; $y_1 = e^{5x}$

7. $9y'' - 12y' + 4y = 0$; $y_1 = e^{2x/3}$

8. $6y'' + y' - y = 0$; $y_1 = e^{x/3}$

9. $x^2 y'' - 7xy' + 16y = 0$; $y_1 = x^4$

10. $x^2 y'' + 2xy' - 6y = 0$; $y_1 = x^2$

11. $xy'' + y' = 0$; $y_1 = \ln x$

12. $4x^2 y'' + y = 0$; $y_1 = x^{1/2} \ln x$

13. $x^2 y'' - xy' + 2y = 0$; $y_1 = x \sin(\ln x)$

14. $x^2 y'' - 3xy' + 5y = 0$; $y_1 = x^2 \cos(\ln x)$

15. $(1 - 2x - x^2)y'' + 2(1 + x)y' - 2y = 0$; $y_1 = x + 1$

16. $(1 - x^2)y'' + 2xy' = 0$; $y_1 = 1$

In Problems 17–20, the indicated function $y_1(x)$ is a solution of the associated homogeneous equation. Use the method of reduction of order to find a second solution $y_2(x)$ of the homogeneous equation and a particular solution of the given nonhomogeneous equation.

17. $y'' - 4y = 2$; $y_1 = e^{-2x}$

18. $y'' + y' = 1$; $y_1 = 1$

19. $y'' - 3y' + 2y = 5e^{3x}$; $y_1 = e^x$

20. $y'' - 4y' + 3y = x$; $y_1 = e^x$

21. (a) Give a convincing demonstration that the second-order equation $ay'' + by' + cy = 0$, a, b, and c constants, always possesses at least one solution of the form $y_1 = e^{m_1 x}$, m_1 a constant.

(b) Explain why the differential equation in part **(a)** must then have a second solution, either of the form $y_2 = e^{m_2 x}$ or of the form $y_2 = xe^{m_1 x}$, m_1 and m_2 constants.

(c) Reexamine problems 1–8. Can you explain why the statements in parts **(a)** and **(b)** are not contradicted by the answers to Problems 3–5?

22. Verify that $y_1(x) = x$ is a solution of $xy'' - xy' + y = 0$. Use reduction of order to find a second solution $y_2(x)$ in the form of an infinite series. Conjecture an interval of definition for $y_2(x)$.

3.3 Homogeneous Linear Equations with Constant Coefficients

We have seen that the linear first-order DE $y' + ay = 0$, where a is a constant, possesses the exponential solution $y = c_1 e^{-ax}$ on the interval $(-\infty, \infty)$. Therefore, it is natural to ask whether exponential solutions exist on $(-\infty, \infty)$ for homogeneous linear higher-order DEs

$$a_n y^{(n)} + a_{n-1} y^{(n-1)} + \cdots + a_2 y'' + a_1 y' + a_0 y = 0, \qquad (1)$$

where the coefficients a_i, $i = 0, 1, \ldots, n$ are real constants and $a_n \neq 0$. The surprising fact is that *all* solutions of these higher-order equations are exponential functions or constructed out of exponential functions.

■ **Auxiliary Equation** We begin by considering the special case of a second-order equation

$$ay'' + by' + cy = 0. \qquad (2)$$

If we try a solution of the form $y = e^{mx}$, then after substituting $y' = me^{mx}$ and $y'' = m^2 e^{mx}$ equation (2) becomes

$$am^2 e^{mx} + bm e^{mx} + c e^{mx} = 0 \quad \text{or} \quad e^{mx}(am^2 + bm + c) = 0.$$

Since e^{mx} is never zero for real values of x, it is apparent that the only way that this exponential function can satisfy the differential equation (2) is to choose m as a root of the quadratic equation

$$am^2 + bm + c = 0. \qquad (3)$$

This last equation is called the **auxiliary equation** of the differential equation (2). Since the two roots of (3) are $m_1 = (-b + \sqrt{b^2 - 4ac})/2a$ and $m_2 = (-b - \sqrt{b^2 - 4ac})/2a$, there will be three forms of the general solution of (1) corresponding to the three cases:

- m_1 and m_2 are real and distinct ($b^2 - 4ac > 0$),
- m_1 and m_2 are real and equal ($b^2 - 4ac = 0$), and
- m_1 and m_2 are conjugate complex numbers ($b^2 - 4ac < 0$).

We discuss each of these cases in turn.

Case I: ***Distinct Real Roots*** Under the assumption that the auxiliary equation (3) has two unequal real roots m_1 and m_2, we find two solutions, $y_1 = e^{m_1 x}$ and $y_2 = e^{m_2 x}$. We see that these functions are linearly independent on $(-\infty, \infty)$ and hence form a fundamental set. It follows that the general solution of (2) on this interval is

$$y = c_1 e^{m_1 x} + c_2 e^{m_2 x}. \qquad (4)$$

Case II: ***Repeated Real Roots*** When $m_1 = m_2$ we necessarily obtain only one exponential solution, $y_1 = e^{m_1 x}$. From the quadratic formula we find that $m_1 = -b/2a$ since the only way to have $m_1 = m_2$ is to have $b^2 - 4ac = 0$. It follows from the discussion in Section 3.2 that a second solution of the equation is

$$y_2 = e^{m_1 x} \int \frac{e^{2m_1 x}}{e^{2m_1 x}} \, dx = e^{m_1 x} \int dx = x e^{m_1 x}. \qquad (5)$$

In (5) we have used the fact that $-b/a = 2m_1$. The general solution is then

$$y = c_1 e^{m_1 x} + c_2 x e^{m_1 x}. \tag{6}$$

Case III: Conjugate Complex Roots If m_1 and m_2 are complex, then we can write $m_1 = \alpha + i\beta$ and $m_2 = \alpha - i\beta$, where α and $\beta > 0$ are real and $i^2 = -1$. Formally, there is no difference between this case and Case I, hence

$$y = C_1 e^{(\alpha + i\beta)x} + C_2 e^{(\alpha - i\beta)x}.$$

However, in practice we prefer to work with real functions instead of complex exponentials. To this end we use Euler's formula:

$$e^{i\theta} = \cos \theta + i \sin \theta,$$

where θ is any real number.* It follows from this formula that

$$e^{i\beta x} = \cos \beta x + i \sin \beta x \quad \text{and} \quad e^{-i\beta x} = \cos \beta x - i \sin \beta x, \tag{7}$$

where we have used $\cos(-\beta x) = \cos \beta x$ and $\sin(-\beta x) = -\sin \beta x$. Note that by first adding and then subtracting the two equations in (7), we obtain, respectively,

$$e^{i\beta x} + e^{-i\beta x} = 2 \cos \beta x \quad \text{and} \quad e^{i\beta x} - e^{-i\beta x} = 2i \sin \beta x.$$

Since $y = C_1 e^{(\alpha + i\beta)x} + C_2 e^{(\alpha - i\beta)x}$ is a solution of (2) for any choice of the constants C_1 and C_2, the choices $C_1 = C_2 = 1$ and $C_1 = 1, C_2 = -1$ give, in turn, two solutions:

$$y_1 = e^{(\alpha + i\beta)x} + e^{(\alpha - i\beta)x} \quad \text{and} \quad y_2 = e^{(\alpha + i\beta)x} - e^{(\alpha - i\beta)x}.$$

But $y_1 = e^{\alpha x}(e^{i\beta x} + e^{-i\beta x}) = 2e^{\alpha x} \cos \beta x$

and $y_2 = e^{\alpha x}(e^{i\beta x} - e^{-i\beta x}) = 2i e^{\alpha x} \sin \beta x.$

Hence from Corollary (A) of Theorem 3.2 the last two results show that $e^{\alpha x} \cos \beta x$ and $e^{\alpha x} \sin \beta x$ are *real* solutions of (2). Moreover, these solutions form a fundamental set on $(-\infty, \infty)$. Consequently, the general solution is

$$y = c_1 e^{\alpha x} \cos \beta x + c_2 e^{\alpha x} \sin \beta x = e^{\alpha x}(c_1 \cos \beta x + c_2 \sin \beta x). \tag{8}$$

Example 1 Second-Order DEs

Solve the following differential equations.

(a) $2y'' - 5y' - 3y = 0$ (b) $y'' - 10y' + 25y = 0$ (c) $y'' + 4y' + 7y = 0$

SOLUTION We give the auxiliary equations, the roots, and the corresponding general solutions.

(a) $2m^2 - 5m - 3 = (2m + 1)(m - 3)$, $m_1 = -\tfrac{1}{2}$, $m_2 = 3$.
From (4), $y = c_1 e^{-x/2} + c_2 e^{3x}$.

(b) $m^2 - 10m + 25 = (m - 5)^2$, $m_1 = m_2 = 5$.
From (6), $y = c_1 e^{5x} + c_2 x e^{5x}$.

*A formal derivation of Euler's formula can be obtained from the Maclaurin series $e^x = \sum_{n=0}^{\infty} x^n/n!$ by substituting $x = i\theta$, using $i^2 = -1$, $i^3 = -i, \ldots$, and then separating the series into real and imaginary parts. The plausibility thus established, we can adopt $\cos \theta + i \sin \theta$ as the *definition* of $e^{i\theta}$.

(c) $m^2 + 4m + 7 = 0$, $m_1 = -2 + \sqrt{3}\, i$, $m_2 = -2 - \sqrt{3}\, i$.
From (8) with $\alpha = -2$, $\beta = \sqrt{3}$, $y = e^{-2x}(c_1 \cos \sqrt{3}\, x + c_2 \sin \sqrt{3}\, x)$. ❑

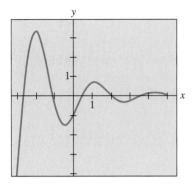

Figure 3.4

Example 2 **An Initial-Value Problem**

Solve the initial-value problem $4y'' + 4y' + 17y = 0$, $y(0) = -1$, $y'(0) = 2$.

SOLUTION By the quadratic formula we find that the roots of the auxiliary equation $4m^2 + 4m + 17 = 0$ are $m_1 = -\frac{1}{2} + 2i$ and $m_2 = -\frac{1}{2} - 2i$. Thus from (8) we have $y = e^{-x/2}(c_1 \cos 2x + c_2 \sin 2x)$. Applying the condition $y(0) = -1$, we see from $e^0(c_1 \cos 0 + c_2 \sin 0) = -1$ that $c_1 = -1$. Differentiating $y = e^{-x/2}(-\cos 2x + c_2 \sin 2x)$ and then using $y'(0) = 2$ gives $2c_2 + \frac{1}{2} = 2$ or $c_2 = \frac{3}{4}$. Hence the solution of the IVP is $y = e^{-x/2}(-\cos 2x + \frac{3}{4} \sin 2x)$. In Figure 3.4 we see that the solution is oscillatory but $y \to 0$ as $x \to \infty$ and $|y| \to \infty$ as $x \to -\infty$. ❑

■ **Two Equations Worth Knowing** The two differential equations

$$y'' + k^2 y = 0 \qquad \text{and} \qquad y'' - k^2 y = 0,$$

k real, are important in applied mathematics. For $y'' + k^2 y = 0$, the auxiliary equation $m^2 + k^2 = 0$ has imaginary roots $m_1 = ki$ and $m_2 = -ki$. With $\alpha = 0$ and $\beta = k$ in (8), the general solution of the DE is seen to be

$$y = c_1 \cos kx + c_2 \sin kx. \tag{9}$$

On the other hand, the auxiliary equation $m^2 - k^2 = 0$ for $y'' - k^2 y = 0$ has distinct real roots $m_1 = k$ and $m_2 = -k$ and so by (4) the general solution of the DE is

$$y = c_1 e^{kx} + c_2 e^{-kx}. \tag{10}$$

Notice that if we choose $c_1 = c_2 = \frac{1}{2}$ and $c_1 = \frac{1}{2}$, $c_2 = -\frac{1}{2}$ in (10) we get the particular solutions $y = \frac{1}{2}(e^{kx} + e^{-kx}) = \cosh kx$ and $y = \frac{1}{2}(e^{kx} - e^{-kx}) = \sinh kx$. Since $\cosh kx$ and $\sinh kx$ are linearly independent on any interval of the x-axis, an alternative form for the general solution of $y'' - k^2 y = 0$ is

$$y = c_1 \cosh kx + c_2 \sinh kx. \tag{11}$$

See Problems 43, 44, and 55 in Exercises 3.3.

■ **Higher-Order Equations** In general, to solve an nth-order differential equation

$$a_n y^{(n)} + a_{n-1} y^{(n-1)} + \cdots + a_2 y'' + a_1 y' + a_0 y = 0, \tag{12}$$

where the a_i, $i = 0, 1, \ldots, n$ are real constants, we must solve an nth-degree polynomial equation

$$a_n m^n + a_{n-1} m^{n-1} + \cdots + a_2 m^2 + a_1 m + a_0 = 0. \tag{13}$$

If all the roots of (13) are real and distinct, then the general solution of (12) is

$$y = c_1 e^{m_1 x} + c_2 e^{m_2 x} + \cdots + c_n e^{m_n x}.$$

It is somewhat harder to summarize the analogues of Cases II and III because the roots of an auxiliary equation of degree greater than two can occur in many combinations. For example, a fifth-degree equation could have five distinct real roots, or three distinct real and two complex roots, or one real and four complex roots, or five real but equal roots, or five real roots but with two of them equal, and so on. When

m_1 is a root of multiplicity k of an nth-degree auxiliary equation (that is, k roots are equal to m_1), it can be shown that the linearly independent solutions are

$$e^{m_1x}, \ xe^{m_1x}, \ x^2e^{m_1x}, \ \dots, \ x^{k-1}e^{m_1x}$$

and the general solution must contain the linear combination

$$c_1e^{m_1x} + c_2xe^{m_1x} + c_3x^2e^{m_1x} + \cdots + c_kx^{k-1}e^{m_1x}.$$

Lastly, it should be remembered that when the coefficients are real, complex roots of an auxiliary equation always appear in conjugate pairs. Thus, for example, a cubic polynomial equation can have at most two complex roots.

Example 3 Third-Order DE

Solve $y''' + 3y'' - 4y = 0$.

SOLUTION It should be apparent from inspection of $m^3 + 3m^2 - 4 = 0$ that one root is $m_1 = 1$ and so $m - 1$ is a factor of $m^3 + 3m^2 - 4$. By division we find

$$m^3 + 3m^2 - 4 = (m-1)(m^2 + 4m + 4) = (m-1)(m+2)^2,$$

and so the other roots are $m_2 = m_3 = -2$. Thus the general solution is

$$y = c_1e^x + c_2e^{-2x} + c_3xe^{-2x}.$$ ❏

Example 4 Fourth-Order DE

Solve $\dfrac{d^4y}{dx^4} + 2\dfrac{d^2y}{dx^2} + y = 0$.

SOLUTION The auxiliary equation $m^4 + 2m^2 + 1 = (m^2 + 1)^2 = 0$ has roots $m_1 = m_3 = i$ and $m_2 = m_4 = -i$. Thus from Case II the solution is

$$y = C_1e^{ix} + C_2e^{-ix} + C_3xe^{ix} + C_4xe^{-ix}.$$

By Euler's formula the grouping $C_1e^{ix} + C_2e^{-ix}$ can be rewritten as $c_1\cos x + c_2\sin x$ after a relabeling of constants. Similarly, $x(C_3e^{ix} + C_4e^{-ix})$ can be expressed as $x(c_3\cos x + c_4\sin x)$. Hence the general solution is

$$y = c_1\cos x + c_2\sin x + c_3x\cos x + c_4x\sin x.$$ ❏

Example 4 illustrates a special case when the auxiliary equation has repeated complex roots. In general, if $m_1 = \alpha + i\beta$, $\beta > 0$, is a complex root of multiplicity k of an auxiliary equation with real coefficients, then its conjugate $m_2 = \alpha - i\beta$ is also a root of multiplicity k. From the $2k$ complex-valued solutions

$$e^{(\alpha+i\beta)x}, \quad xe^{(\alpha+i\beta)x}, \quad x^2e^{(\alpha+i\beta)x}, \quad \dots, \quad x^{k-1}e^{(\alpha-i\beta)x}$$
$$e^{(\alpha-i\beta)x}, \quad xe^{(\alpha-i\beta)x}, \quad x^2e^{(\alpha-i\beta)x}, \quad \dots, \quad x^{k-1}e^{(\alpha-i\beta)x}$$

we conclude, with the aid of Euler's formula, that the general solution of the corresponding differential equation must then contain a linear combination of the $2k$ real linearly independent solutions

$$e^{\alpha x}\cos\beta x, \quad xe^{\alpha x}\cos\beta x, \quad x^2e^{\alpha x}\cos\beta x, \quad \dots, \quad x^{k-1}e^{\alpha x}\cos\beta x$$
$$e^{\alpha x}\sin\beta x, \quad xe^{\alpha x}\sin\beta x, \quad x^2e^{\alpha x}\sin\beta x, \quad \dots, \quad x^{k-1}e^{\alpha x}\sin\beta x.$$

In Example 4 we identify $k = 2$, $\alpha = 0$, and $\beta = 1$.

Of course the most difficult aspect of solving constant-coefficient differential equations is finding roots of auxiliary equations of degree greater than two. For

example, to solve $3y''' + 5y'' + 10y' - 4y = 0$ we must solve $3m^3 + 5m^2 + 10m - 4 = 0$. Something we can try is to test the auxiliary equation for rational roots. Recall, if $m_1 = p/q$ is a rational root (expressed in lowest terms) of an auxiliary equation $a_n m^n + \cdots + a_1 m + a_0 = 0$ with integer coefficients, then p is a factor of a_0 and q is a factor of a_n. For our specific cubic auxiliary equation, all the factors of $a_0 = -4$ and $a_n = 3$ are p: ± 1, ± 2, ± 4 and q: ± 1, ± 3, so the possible rational roots are p/q: ± 1, ± 2, ± 4, $\pm \frac{1}{3}$, $\pm \frac{2}{3}$, $\pm \frac{4}{3}$. Each of these numbers can then be tested, say, by synthetic division. In this way we discover both the root $m_1 = \frac{1}{3}$ and the factorization

$$3m^3 + 5m^2 + 10m - 4 = \left(m - \frac{1}{3} \right)(3m^2 + 6m + 12).$$

The quadratic formula then yields the remaining roots $m_2 = -1 + \sqrt{3}\, i$ and $m_3 = -1 - \sqrt{3}\, i$. Therefore the general solution of $3y''' + 5y'' + 10y' - 4y = 0$ is $y = c_1 e^{x/3} + e^{-x}(c_2 \cos \sqrt{3}\, x + c_3 \sin \sqrt{3}x)$.

■ **Use of Computers** Finding roots or approximations of roots of polynomial equations is a routine problem with an appropriate calculator or computer software. The computer algebra systems *Mathematica* and *Maple* can solve polynomial equations (in one variable) of degree less than five in terms of algebraic formulas. For the auxiliary equation in the preceding paragraph, the commands

Solve[3 m^3 + 5 m^2 + 10 m − 4 = = 0, m]　　(in *Mathematica*)

solve(3*m^3 +5*m^2 + 10*m − 4, m);　　(in *Maple*)

yield immediately their representations of the roots $\frac{1}{3}$, $-1 + \sqrt{3}\, i$, $-1 - \sqrt{3}\, i$. For auxiliary equations of higher degree it may be necessary to resort to numerical commands such as NSolve and FindRoot in *Mathematica*. Because of their capability of solving polynomial equations, it is not surprising that some computer algebra systems are also able to give explicit solutions of homogeneous linear constant-coefficient differential equations. For example, the inputs

DSolve [y″[x] + 2 y′[x] + 2 y[x] = = 0, y[x], x]　　(in *Mathematica*)

dsolve(diff(y(x),x\$2) + 2*diff(y(x),x) + 2*y(x) = 0, y(x));　　(in *Maple*)

give, respectively,

$$\textbf{y[x] − >} \frac{\textbf{C[2] Cos [x] − C[1] Sin [x]}}{\textbf{E}^{\textbf{x}}} \tag{14}$$

and　　　　**y(x) = _C1 exp(−x) sin(x) − _C2 exp(−x) cos(x)**

Translated, this means $y = c_2 e^{-x} \cos x + c_1 e^{-x} \sin x$ is a solution of $y'' + 2y' + 2y = 0$.

In the classic text *Differential Equations* by Ralph Palmer Agnew* (used by the author as a student), the following statement is made:

It is not reasonable to expect students in this course to have computing skill and equipment necessary for efficient solving of equations such as

$$4.317 \frac{d^4 y}{dx^4} + 2.179 \frac{d^3 y}{dx^3} + 1.416 \frac{d^2 y}{dx^2} + 1.295 \frac{dy}{dx} + 3.169 y = 0. \tag{15}$$

Although it is debatable whether computing skills have improved in the intervening years, it is a certainty that technology has. If one has access to a computer algebra

* McGraw-Hill, New York, 1960.

system, equation (15) could be considered reasonable. After simplification and some relabeling of the output, *Mathematica* yields the (approximate) general solution

$$y = c_1 e^{-0.728852x} \cos(0.618605x) + c_2 e^{-0.728852x} \sin(0.618605x)$$

$$+ c_3 e^{0.476478x} \cos(0.759081x) + c_4 e^{0.476478x} \sin(0.759081x).$$

We note in passing that the DSolve and dsolve commands in *Mathematica* and *Maple*, like most aspects of any CAS, have their limitations.

Finally, if we are faced with an initial-value problem consisting of, say, a fourth-order equation, then to fit the general solution of the DE to the four initial conditions we must solve a system of four linear equations in four unknowns (the c_1, c_2, c_3, c_4 in the general solution). Using a CAS to solve the system can save lots of time. See Problems 35–38 in Exercises 3.3 and Problem 39 in Chapter 3 Review Exercises.

EXERCISES 3.3

Answers to odd-numbered problems begin on page A-29.

In Problems 1–14, find the general solution of the given second-order differential equation.

1. $4y'' + y' = 0$

2. $y'' - 36y = 0$

3. $y'' - y' - 6y = 0$

4. $y'' - 3y' + 2y = 0$

5. $y'' + 8y' + 16y = 0$

6. $y'' - 10y' + 25y = 0$

7. $12y'' - 5y' - 2y = 0$

8. $y'' + 4y' - y = 0$

9. $y'' + 9y = 0$

10. $3y'' + y = 0$

11. $y'' - 4y' + 5y = 0$

12. $2y'' + 2y' + y = 0$

13. $3y'' + 2y' + y = 0$

14. $2y'' - 3y' + 4y = 0$

In Problems 15–28, find the general solution of the given higher-order differential equation.

15. $y''' - 4y'' - 5y' = 0$

16. $y''' - y = 0$

17. $y''' - 5y'' + 3y' + 9y = 0$

18. $y''' + 3y'' - 4y' - 12y = 0$

19. $\dfrac{d^3u}{dt^3} + \dfrac{d^2u}{dt^2} - 2u = 0$

20. $\dfrac{d^3x}{dt^3} - \dfrac{d^2x}{dt^2} - 4x = 0$

21. $y''' + 3y'' + 3y' + y = 0$

22. $y''' - 6y'' + 12y' - 8y = 0$

23. $y^{(4)} + y''' + y'' = 0$

24. $y^{(4)} - 2y'' + y = 0$

25. $16\dfrac{d^4y}{dx^4} + 24\dfrac{d^2y}{dx^2} + 9y = 0$

26. $\dfrac{d^4y}{dx^4} - 7\dfrac{d^2y}{dx^2} - 18y = 0$

27. $\dfrac{d^5u}{dr^5} + 5\dfrac{d^4u}{dr^4} - 2\dfrac{d^3u}{dr^3} - 10\dfrac{d^2u}{dr^2} + \dfrac{du}{dr} + 5u = 0$

28. $2\dfrac{d^5x}{ds^5} - 7\dfrac{d^4x}{ds^4} + 12\dfrac{d^3x}{ds^3} + 8\dfrac{d^2x}{ds^2} = 0$

In Problems 29–38, solve the given initial-value problem.

29. $y'' + 16y = 0$, $y(0) = 2$, $y'(0) = -2$

30. $\dfrac{d^2y}{d\theta^2} + y = 0$, $y\left(\dfrac{\pi}{3}\right) = 0$, $y'\left(\dfrac{\pi}{3}\right) = 2$

31. $\dfrac{d^2y}{dt^2} - 4\dfrac{dy}{dt} - 5y = 0$, $y(1) = 0$, $y'(1) = 2$

32. $4y'' - 4y' - 3y = 0$, $y(0) = 1$, $y'(0) = 5$

33. $y'' + y' + 2y = 0$, $y(0) = y'(0) = 0$

34. $y'' - 2y' + y = 0$, $y(0) = 5$, $y'(0) = 10$

35. $y''' + 12y'' + 36y' = 0$, $y(0) = 0$, $y'(0) = 1$, $y''(0) = -7$

36. $y''' + 2y'' - 5y' - 6y = 0$, $y(0) = y'(0) = 0$, $y''(0) = 1$

37. $y^{(4)} - 3y''' + 3y'' - y' = 0$, $y(0) = y'(0) = 0$, $y''(0) = y'''(0) = 1$

38. $y^{(4)} - y = 0$, $y(0) = y'(0) = y''(0) = 0$, $y'''(0) = 1$

In Problems 39–42, solve the given boundary-value problem.

39. $y'' - 10y' + 25y = 0$, $y(0) = 1$, $y(1) = 0$

40. $y'' + 4y = 0$, $y(0) = 0$, $y(\pi) = 0$

41. $y'' + y = 0$, $y'(0) = 0$, $y'\left(\dfrac{\pi}{2}\right) = 2$

42. $y'' - 2y' + 2y = 0$, $y(0) = 1$, $y(\pi) = 1$

In Problems 43 and 44, solve the given problem first using the form of the general solution given in (10). Solve again, this time using the form given in (11).

43. $y'' - 3y = 0$, $y(0) = 1$, $y'(0) = 5$

44. $y'' - y = 0$, $y(0) = 1$, $y'(1) = 0$.

Computer Lab Assignment

In Problems 45–48, use a computer either as an aid in solving the auxiliary equation or as a means of directly obtaining the general solution of the given differential equation. If you use a CAS to obtain the general solution, simplify the output and, if necessary, write the solution in terms of real functions.

45. $y''' - 6y'' + 2y' + y = 0$

46. $6.11y''' + 8.59y'' + 7.93y' + 0.778y = 0$

47. $3.15y^{(4)} - 5.34y'' + 6.33y' - 2.03y = 0$

48. $y^{(4)} + 2y'' - y' + 2y = 0$

In Problems 49–54, each figure represents the graph of a particular solution of one of the following differential equations.

(a) $y'' - 3y' - 4y = 0$ (b) $y'' + 4y = 0$

(c) $y'' + 2y' + y = 0$ (d) $y'' + y = 0$

(e) $y'' + 2y' + 2y = 0$ (f) $y'' - 3y' + 2y = 0$

Match a solution curve with one of the differential equations. Explain your reasoning.

49.

Figure 3.5

50.

Figure 3.6

51.

Figure 3.7

52.

Figure 3.8

53.

Figure 3.9

54.

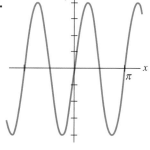

Figure 3.10

55. Reread the discussion on the slipping chain in Section 1.3 and illustrated in Figure 1.21 on page 27.

(a) Use the form of the solution given in (11) of this section to find the general solution of equation (16) of Section 1.3,

$$\frac{d^2x}{dt^2} - \frac{64}{L}x = 0.$$

(b) Find a particular solution that satisfies the initial conditions stated in the discussion on page 27.

(c) Suppose that the total length of the chain is $L = 20$ ft and that $x_0 = 1$. Find the velocity at which the slipping chain will leave the supporting peg.

56. Reexamine equation (14). Notice that when we rewrote this computer-generated solution, $-C[1]$ was changed to $+ c_1$. Why can we do that?

57. The roots of a cubic auxiliary equation are $m_1 = 4$ and $m_2 = m_3 = -5$. What is the corresponding homogeneous linear differential equation? Is your answer unique?

58. Two roots of a cubic auxiliary equation with real coefficients are $m_1 = -\frac{1}{2}$ and $m_2 = 3 + i$. What is the corresponding homogeneous linear differential equation?

59. Find the general solution of $y''' + 6y'' + y' - 34y = 0$ if it is known that $y_1 = e^{-4x} \cos x$ is one solution.

60. In order to solve $y^{(4)} + y = 0$ we must find the roots of $m^4 + 1 = 0$. This is a trivial problem using a CAS, but it can also be done by hand working with complex numbers. Observe that $m^4 + 1 = (m^2 + 1)^2 - 2m^2$. How does this help? Solve the differential equation.

61. Consider the boundary-value problem $y'' + \lambda y = 0$, $y(0) = 0$, $y(\pi/2) = 0$. Discuss: Is it possible to determine values of λ so that the problem possesses

(a) trivial solutions (b) nontrivial solutions

62. In the study of techniques of integration in calculus, certain indefinite integrals of the form $\int e^{ax} f(x)\, dx$ could be evaluated by applying integration by parts twice, recovering the original integral on the right side, solving for the original integral, and obtaining a constant multiple $k \int e^{ax} f(x)\, dx$ on the left side. The value of the integral is then found by dividing by k. For what kinds of functions f does the described procedure work? Your solution should lead to a differential equation. Carefully analyze this equation and solve for f.

3.4 Undetermined Coefficients

To solve a nonhomogeneous linear differential equation

$$a_n y^{(n)} + a_{n-1} y^{(n-1)} + \cdots + a_1 y' + a_0 y = g(x) \tag{1}$$

we must do two things:

(i) find the complementary function y_c;
(ii) find any particular solution y_p of the nonhomogeneous equation.

Then, as discussed in Section 3.1, the general solution of (1) on an interval I is $y = y_c + y_p$.

The complementary function y_c is the general solution of the associated homogeneous equation $a_n y^{(n)} + a_{n-1} y^{(n-1)} + \cdots + a_1 y' + a_0 y = 0$. In the last section we saw how to solve these kinds of equations when the coefficients were constants. The first of two ways we shall consider for obtaining a particular solution y_p is called the **method of undetermined coefficients.** The underlying idea in this method is a conjecture, an educated guess really, about the form of y_p, motivated by the kinds of functions that make up the input function $g(x)$. The general method is limited to nonhomogeneous linear DEs such as (1) where

- the coefficients, a_i, $i = 0, 1, \ldots, n$ are constants, and
- where $g(x)$ is a constant, a polynomial function, exponential function $e^{\alpha x}$, sine or cosine functions $\sin \beta x$ or $\cos \beta x$, or finite sums and products of these functions.

A constant k is a polynomial function. ☞

Strictly speaking, $g(x) = k$ (a constant) is a polynomial function. Since a constant function is probably not the first thing that comes to mind when you think of polynomial functions, for emphasis we shall continue to use the redundancy "constant functions, polynomial functions,. . . ."

The following functions are some examples of the types of inputs $g(x)$ that are appropriate for this discussion:

$$g(x) = 10, \qquad g(x) = x^2 - 5x, \qquad g(x) = 15x - 6 + 8e^{-x}$$

$$g(x) = \sin 3x - 5x \cos 2x, \qquad g(x) = xe^x \sin x + (3x^2 - 1)e^{-4x}.$$

That is, $g(x)$ is a linear combination of functions of the type

$$P(x) = a_n x^n + a_{n-1} x^{n-1} + \cdots + a_1 x + a_0, \quad P(x)e^{\alpha x}, \quad P(x)e^{\alpha x} \sin \beta x, \quad \text{and} \quad P(x)e^{\alpha x} \cos \beta x$$

where n is a nonnegative integer and α and β are real numbers. The method of undetermined coefficients is not applicable to equations of form (1) when

$$g(x) = \ln x, \quad g(x) = \frac{1}{x}, \quad g(x) = \tan x, \quad g(x) = \sin^{-1} x,$$

and so on. Differential equations in which the input $g(x)$ is a function of this last kind will be considered in Section 3.5.

The set of functions that consists of constants, polynomials, exponentials $e^{\alpha x}$, sines, and cosines has the remarkable property that derivatives of their sums and products are again sums and products of constants, polynomials, exponentials $e^{\alpha x}$, sines, and cosines. Since the linear combination of derivatives $a_n y_p^{(n)} + a_{n-1} y_p^{(n-1)} + \cdots + a_1 y_p' + a_0 y_p$ must be identical to $g(x)$, it seems reasonable to assume that y_p *has the same form as* $g(x)$.

The next two examples illustrate the basic method.

Example 1 General Solution Using Undetermined Coefficients

Solve $y'' + 4y' - 2y = 2x^2 - 3x + 6$. $\qquad\qquad$ (2)

SOLUTION **Step 1.** We first solve the associated homogeneous equation $y'' + 4y' - 2y = 0$. From the quadratic formula we find that the roots of the auxiliary equation $m^2 + 4m - 2 = 0$ are $m_1 = -2 - \sqrt{6}$ and $m_2 = -2 + \sqrt{6}$. Hence the complementary function is

$$y_c = c_1 e^{-(2 + \sqrt{6})x} + c_2 e^{(-2 + \sqrt{6})x}.$$

Step 2. Now, since the function $g(x)$ is a quadratic polynomial, let us assume a particular solution that is also in the form of a quadratic polynomial:

$$y_p = Ax^2 + Bx + C.$$

We seek to determine *specific* coefficients A, B, and C for which y_p is a solution of (2). Substituting y_p and the derivatives $y_p' = 2Ax + B$ and $y_p'' = 2A$ into the given differential equation (2), we get

$$y_p'' + 4y_p' - 2y_p = 2A + 8Ax + 4B - 2Ax^2 - 2Bx - 2C$$
$$= 2x^2 - 3x + 6.$$

Since the last equation is supposed to be an identity, the coefficients of like powers of x must be equal:

That is,

$$-2A = 2, \qquad 8A - 2B = -3, \qquad 2A + 4B - 2C = 6.$$

Solving this system of equations leads to the values $A = -1$, $B = -\frac{5}{2}$, and $C = -9$. Thus a particular solution is

$$y_p = -x^2 - \frac{5}{2}x - 9.$$

Step 3. The general solution of the given equation is

$$y = y_c + y_p = c_1 e^{-(2+\sqrt{6})x} + c_2 e^{(-2+\sqrt{6})x} - x^2 - \frac{5}{2}x - 9. \qquad \square$$

Example 2 Particular Solution Using Undetermined Coefficients

Find a particular solution of $y'' - y' + y = 2\sin 3x$.

SOLUTION A natural first guess for a particular solution would be $A\sin 3x$. But since successive differentiations of $\sin 3x$ produce $\sin 3x$ *and* $\cos 3x$, we are prompted instead to assume a particular solution that includes both of these terms:

$$y_p = A\cos 3x + B\sin 3x.$$

Differentiating y_p and substituting the results into the differential equation gives, after regrouping,

$$y_p'' - y_p' + y_p = (-8A - 3B)\cos 3x + (3A - 8B)\sin 3x = 2\sin 3x$$

or

From the resulting system of equations,

$$-8A - 3B = 0, \qquad 3A - 8B = 2,$$

we get $A = \frac{6}{73}$ and $B = -\frac{16}{73}$. A particular solution of the equation is

$$y_p = \frac{6}{73}\cos 3x - \frac{16}{73}\sin 3x. \qquad \square$$

As we mentioned, the form that we assume for the particular solution y_p is an educated guess; it is not a blind guess. This educated guess must take into consideration not only the types of functions that make up $g(x)$ but also, as we shall see in Example 4, the functions that make up the complementary function y_c.

Example 3 Forming y_p by Superposition

Solve $y'' - 2y' - 3y = 4x - 5 + 6xe^{2x}$. \hfill (3)

SOLUTION **Step 1.** First, the solution of the associated homogeneous equation $y'' - 2y' - 3y = 0$ is found to be $y_c = c_1 e^{-x} + c_2 e^{3x}$.

Step 2. Next, the presence of $4x - 5$ in $g(x)$ suggests that the particular solution includes a linear polynomial. Furthermore, since the derivative of the product xe^{2x} produces $2xe^{2x}$ and e^{2x}, we also assume that the particular solution includes both xe^{2x} and e^{2x}. In other words, g is the sum of two basic kinds of functions:

$$g(x) = g_1(x) + g_2(x) = polynomial + exponentials.$$

How to use Theorem 3.7 in the solution of Example 3. 👉

Correspondingly, the superposition principle for nonhomogeneous equations (Theorem 3.7) suggests that we seek a particular solution

$$y_p = y_{p_1} + y_{p_2},$$

where $y_{p_1} = Ax + B$ and $y_{p_2} = Cxe^{2x} + Ee^{2x}$. Substituting

$$y_p = Ax + B + Cxe^{2x} + Ee^{2x}$$

into the given equation (3) and grouping like terms gives

$$y_p'' - 2y_p' - 3y_p = -3Ax - 2A - 3B - 3Cxe^{2x} + (2C - 3E)e^{2x} = 4x - 5 + 6xe^{2x}. \quad (4)$$

From this identity we obtain the four equations

$$-3A = 4, \quad -2A - 3B = -5, \quad -3C = 6, \quad 2C - 3E = 0.$$

The last equation in this system results from the interpretation that the coefficient of e^{2x} in the right member of (4) is zero. Solving, we find $A = -\frac{4}{3}$, $B = \frac{23}{9}$, $C = -2$, and $E = -\frac{4}{3}$. Consequently,

$$y_p = -\frac{4}{3}x + \frac{23}{9} - 2xe^{2x} - \frac{4}{3}e^{2x}.$$

Step 3. The general solution of the equation is

$$y = c_1 e^{-x} + c_2 e^{3x} - \frac{4}{3}x + \frac{23}{9} - \left(2x + \frac{4}{3}\right)e^{2x}. \hfill \square$$

In light of the superposition principle (Theorem 3.7) we can also approach Example 3 from the viewpoint of solving two simpler problems. You should verify that substituting

$$y_{p_1} = Ax + B \qquad \text{into} \quad y'' - 2y' - 3y = 4x - 5$$

and $\qquad y_{p_2} = Cxe^{2x} + Ee^{2x} \quad \text{into} \quad y'' - 2y' - 3y = 6xe^{2x}$

yields, in turn, $y_{p_1} = \frac{4}{3}x + \frac{23}{9}$ and $y_{p_2} = -(2x + \frac{4}{3})e^{2x}$. A particular solution of (3) is then $y_p = y_{p_1} + y_{p_2}$.

The next example illustrates that sometimes the "obvious" assumption for the form of y_p is not a correct assumption.

Example 4 A Glitch in the Method

Find a particular solution of $y'' - 5y' + 4y = 8e^x$.

SOLUTION Differentiation of e^x produces no new functions. Thus, proceeding as we did in the earlier examples, we can reasonably assume a particular solution of the form $y_p = Ae^x$. But substitution of this expression into the differential equation yields the contradictory statement $0 = 8e^x$, and so we have clearly made the wrong guess for y_p.

The difficulty here is apparent upon examining the complementary function $y_c = c_1 e^x + c_2 e^{4x}$. Observe that our assumption Ae^x is already present in y_c. This means that e^x is a solution of the associated homogeneous differential equation, and a constant multiple Ae^x when substituted into the differential equation necessarily produces zero.

What then should be the form of y_p? Inspired by Case II of Section 3.3, let's see whether we can find a particular solution of the form

$$y_p = Axe^x.$$

Substituting $y_p' = Axe^x + Ae^x$ and $y_p'' = Axe^x + 2Ae^x$ into the differential equation and simplifying gives

$$y_p'' - 5y_p' + 4y_p = -3Ae^x = 8e^x.$$

From the last equality we see that the value of A is now determined as $A = -\frac{8}{3}$. Therefore a particular solution of the given equation is $y_p = -\frac{8}{3}\, xe^x$. ❏

The difference in the procedures used in Examples 1–3 and in Example 4 suggests that we consider two cases. The first case reflects the situation in Examples 1–3.

> ***Case I:*** No function in the assumed particular solution is a solution of the associated homogeneous differential equation.

In Table 3.1 we illustrate some specific examples of $g(x)$ in (1) along with the corresponding form of the particular solution. We are, of course, taking for granted that no function in the assumed particular solution y_p is duplicated by a function in the complementary function y_c.

Table 3.1 Trial Particular Solutions

$g(x)$	Form of y_p
1. 1 (any constant)	A
2. $5x + 7$	$Ax + B$
3. $3x^2 - 2$	$Ax^2 + Bx + C$
4. $x^3 - x + 1$	$Ax^3 + Bx^2 + Cx + E$
5. $\sin 4x$	$A \cos 4x + B \sin 4x$
6. $\cos 4x$	$A \cos 4x + B \sin 4x$
7. e^{5x}	Ae^{5x}
8. $(9x - 2)e^{5x}$	$(Ax + B)e^{5x}$
9. $x^2 e^{5x}$	$(Ax^2 + Bx + C)e^{5x}$
10. $e^{3x} \sin 4x$	$Ae^{3x} \cos 4x + Be^{3x} \sin 4x$
11. $5x^2 \sin 4x$	$(Ax^2 + Bx + C) \cos 4x + (Ex^2 + Fx + G) \sin 4x$
12. $xe^{3x} \cos 4x$	$(Ax + B)e^{3x} \cos 4x + (Cx + E)e^{3x} \sin 4x$

Example 5 Forms of Particular Solutions—Case I

Determine the form of a particular solution of

(a) $y'' - 8y' + 25y = 5x^3 e^{-x} - 7e^{-x}$ (b) $y'' + 4y = x\cos x.$

SOLUTION (a) We can write $g(x) = (5x^3 - 7)e^{-x}$. Using entry 9 in Table 3.1 as a model, we assume a particular solution of the form

$$y_p = (Ax^3 + Bx^2 + Cx + E)e^{-x}.$$

Note that there is no duplication between the terms in y_p and the terms in the complementary function $y_c = e^{4x}(c_1 \cos 3x + c_2 \sin 3x)$.

(b) The function $g(x) = x \cos x$ is similar to entry 11 in Table 3.1 except, of course, that we use a linear rather than a quadratic polynomial and $\cos x$ and $\sin x$ instead of $\cos 4x$ and $\sin 4x$ in the form of y_p:

$$y_p = (Ax + B)\cos x + (Cx + E)\sin x.$$

Again observe that there is no duplication of terms between y_p and $y_c = c_1 \cos 2x + c_2 \sin 2x.$ ❏

If $g(x)$ consists of a sum of, say, m terms of the kind listed in the table, then (as in Example 3) the assumption for a particular solution y_p consists of the sum of the trial forms $y_{p_1}, y_{p_2}, \ldots, y_{p_m}$ corresponding to these terms:

$$y_p = y_{p_1} + y_{p_2} + \cdots + y_{p_m}.$$

The foregoing sentence can be put another way.

Form Rule for Case I *The form of y_p is a linear combination of all linearly independent functions that are generated by repeated differentiations of $g(x)$.*

Example 6 Forming y_p by Superposition—Case I

Determine the form of a particular solution of

$$y'' - 9y' + 14y = 3x^2 - 5\sin 2x + 7xe^{6x}.$$

SOLUTION

Corresponding to $3x^2$ we assume $y_{p_1} = Ax^2 + Bx + C.$

Corresponding to $-5\sin 2x$ we assume $y_{p_2} = E\cos 2x + F\sin 2x.$

Corresponding to $7xe^{6x}$ we assume $y_{p_3} = (Gx + H)e^{6x}.$

The assumption for the particular solution is then

$$y_p = y_{p_1} + y_{p_2} + y_{p_3} = Ax^2 + Bx + C + E\cos 2x + F\sin 2x + (Gx + H)e^{6x}.$$

No term in this assumption duplicates a term in $y_c = c_1 e^{2x} + c_2 e^{7x}.$ ❏

Case II: A function in the assumed particular solution is also a solution of the associated homogeneous differential equation.

The next example is similar to Example 4.

Example 7 Particular Solution—Case II

Find a particular solution of $y'' - 2y' + y = e^x.$

SOLUTION The complementary function is $y_c = c_1 e^x + c_2 x e^x.$ As in Example 4, the assumption $y_p = A e^x$ will fail since it is apparent from y_c that e^x is a solution

of the associated homogeneous equation $y'' - 2y' + y = 0$. Moreover, we will not be able to find a particular solution of the form $y_p = Axe^x$ since the term xe^x is also duplicated in y_c. We next try

$$y_p = Ax^2 e^x.$$

Substituting into the given differential equation yields $2Ae^x = e^x$ and so $A = \frac{1}{2}$. Thus a particular solution is $y_p = \frac{1}{2}x^2 e^x$. ❏

Suppose again that $g(x)$ consists of m terms of the kind given in Table 3.1, and suppose further that the usual assumption for a particular solution is

$$y_p = y_{p_1} + y_{p_2} + \cdots + y_{p_m},$$

where the y_{p_i}, $i = 1, 2, \ldots, m$ are the trial particular solution forms corresponding to these terms. Under the circumstances described in Case II, we can make up the following general rule.

> **Multiplication Rule for Case II** *If any y_{p_i} contains terms that duplicate terms in y_c, then that y_{p_i} must be multiplied by x^n, where n is the smallest positive integer that eliminates that duplication.*

Example 8 An Initial-Value Problem

Solve the initial-value problem $y'' + y = 4x + 10 \sin x$, $y(\pi) = 0$, $y'(\pi) = 2$.

SOLUTION The solution of the associated homogeneous equation $y'' + y = 0$ is $y_c = c_1 \cos x + c_2 \sin x$. Since $g(x) = 4x + 10 \sin x$ is the sum of a linear polynomial and a sine function, our normal assumption for y_p, from entries 2 and 5 of Table 3.1, would be the sum of $y_{p_1} = Ax + B$ and $y_{p_2} = C \cos x + E \sin x$:

$$y_p = Ax + B + C \cos x + E \sin x. \tag{5}$$

But there is an obvious duplication of the terms $\cos x$ and $\sin x$ in this assumed form and two terms in the complementary function. This duplication can be eliminated by simply multiplying y_{p_2} by x. Instead of (5) we now use

$$y_p = Ax + B + Cx \cos x + Ex \sin x. \tag{6}$$

Differentiating this expression and substituting the results into the differential equation gives

$$y_p'' + y_p = Ax + B - 2C \sin x + 2E \cos x = 4x + 10 \sin x,$$

and so $A = 4$, $B = 0$, $-2C = 10$, $2E = 0$. The solutions of the system are immediate: $A = 4$, $B = 0$, $C = -5$, and $E = 0$. Therefore from (6) we obtain $y_p = 4x - 5x \cos x$. The general solution of the given equation is

$$y = y_c + y_p = c_1 \cos x + c_2 \sin x + 4x - 5x \cos x.$$

We now apply the prescribed initial conditions to the general solution of the equation. First, $y(\pi) = c_1 \cos \pi + c_2 \sin \pi + 4\pi - 5\pi \cos \pi = 0$ yields $c_1 = 9\pi$ since $\cos \pi = -1$ and $\sin \pi = 0$. Next, from the derivative

$$y' = -9\pi \sin x + c_2 \cos x + 4 + 5x \sin x - 5 \cos x$$

and $\qquad y'(\pi) = -9\pi \sin \pi + c_2 \cos \pi + 4 + 5\pi \sin \pi - 5 \cos \pi = 2$

we find $c_2 = 7$. The solution of the initial value is then

$$y = 9\pi \cos x + 7 \sin x + 4x - 5x \cos x. \qquad$$ ❏

Example 9 Using the Multiplication Rule

Solve $y'' - 6y' + 9y = 6x^2 + 2 - 12e^{3x}$.

SOLUTION The complementary function is $y_c = c_1 e^{3x} + c_2 x e^{3x}$. And so, based on entries 3 and 7 of Table 3.1, the usual assumption for a particular solution would be

$$y_p = \underbrace{Ax^2 + Bx + C}_{y_{p_1}} + \underbrace{Ee^{3x}}_{y_{p_2}}.$$

Inspection of these functions shows that the one term in y_{p_2} is duplicated in y_c. If we multiply y_{p_2} by x, we note that the term xe^{3x} is still part of y_c. But multiplying y_{p_2} by x^2 eliminates all duplications. Thus the operative form of a particular solution is

$$y_p = Ax^2 + Bx + C + Ex^2 e^{3x}.$$

Differentiating this last form, substituting into the differential equation, and collecting like terms gives

$$y_p'' - 6y_p' + 9y_p = 9Ax^2 + (-12A + 9B)x + 2A - 6B + 9C + 2Ee^{3x} = 6x^2 + 2 - 12e^{3x}.$$

It follows from this identity that $A = \tfrac{2}{3}$, $B = \tfrac{8}{9}$, $C = \tfrac{2}{3}$, and $E = -6$. Hence the general solution $y = y_c + y_p$ is

$$y = c_1 e^{3x} + c_2 x e^{3x} + \frac{2}{3}x^2 + \frac{8}{9}x + \frac{2}{3} - 6x^2 e^{3x}. \qquad \square$$

Example 10 Third-Order DE—Case I

Solve $y''' + y'' = e^x \cos x$.

SOLUTION From the characteristic equation $m^3 + m^2 = 0$ we find $m_1 = m_2 = 0$ and $m_3 = -1$. Hence the complementary function of the equation is $y_c = c_1 + c_2 x + c_3 e^{-x}$. With $g(x) = e^x \cos x$, we see from entry 10 of Table 3.1 that we should assume

$$y_p = Ae^x \cos x + Be^x \sin x.$$

Since there are no functions in y_p that duplicate functions in the complementary solution, we proceed in the usual manner. From

$$y_p''' + y_p'' = (-2A + 4B)e^x \cos x + (-4A - 2B)e^x \sin x = e^x \cos x$$

we get $-2A + 4B = 1$, $-4A - 2B = 0$. This system gives $A = -\tfrac{1}{10}$ and $B = \tfrac{1}{5}$, so that a particular solution is $y_p = -\tfrac{1}{10}e^x \cos x + \tfrac{1}{5}e^x \sin x$. The general solution of the equation is

$$y = y_c + y_p = c_1 + c_2 x + c_3 e^{-x} - \frac{1}{10}e^x \cos x + \frac{1}{5}e^x \sin x. \qquad \square$$

Example 11 Fourth-Order DE—Case II

Determine the form of a particular solution of $y^{(4)} + y''' = 1 - x^2 e^{-x}$.

SOLUTION Comparing $y_c = c_1 + c_2 x + c_3 x^2 + c_4 e^{-x}$ with our normal assumption for a particular solution

$$y_p = \underbrace{A}_{y_{p_1}} + \underbrace{Bx^2 e^{-x} + Cxe^{-x} + Ee^{-x}}_{y_{p_2}},$$

we see that the duplications between y_c and y_p are eliminated when y_{p_1} is multiplied by x^3 and y_{p_2} is multiplied by x. Thus the correct assumption for a particular solution is

$$y_p = Ax^3 + Bx^3 e^{-x} + Cx^2 e^{-x} + Exe^{-x}. \qquad \square$$

Remarks

In Problems 27–36 of Exercises 3.4, you are asked to solve initial-value problems, and in Problems 37 and 38, boundary-value problems. As illustrated in Example 8, be sure to apply the initial conditions or the boundary conditions to the general solution $y = y_c + y_p$. Students often make the mistake of applying these conditions only to the complementary function y_c since it is that part of the solution that contains the constants.

EXERCISES 3.4

Answers to odd-numbered problems begin on page A-29.

In Problems 1–26, solve the given differential equation by undetermined coefficients.

1. $y'' + 3y' + 2y = 6$

2. $4y'' + 9y = 15$

3. $y'' - 10y' + 25y = 30x + 3$

4. $y'' + y' - 6y = 2x$

5. $\frac{1}{4}y'' + y' + y = x^2 - 2x$

6. $y'' - 8y' + 20y = 100x^2 - 26xe^x$

7. $y'' + 3y = -48x^2 e^{3x}$

8. $4y'' - 4y' - 3y = \cos 2x$

9. $y'' - y' = -3$

10. $y'' + 2y' = 2x + 5 - e^{-2x}$

11. $y'' - y' + \frac{1}{4}y = 3 + e^{x/2}$

12. $y'' - 16y = 2e^{4x}$

13. $y'' + 4y = 3\sin 2x$

14. $y'' + 4y = (x^2 - 3)\sin 2x$

15. $y'' + y = 2x \sin x$

16. $y'' - 5y' = 2x^3 - 4x^2 - x + 6$

17. $y'' - 2y' + 5y = e^x \cos 2x$

18. $y'' - 2y' + 2y = e^{2x}(\cos x - 3\sin x)$

19. $y'' + 2y' + y = \sin x + 3\cos 2x$

20. $y'' + 2y' - 24y = 16 - (x + 2)e^{4x}$

21. $y''' - 6y'' = 3 - \cos x$

22. $y''' - 2y'' - 4y' + 8y = 6xe^{2x}$

23. $y''' - 3y'' + 3y' - y = x - 4e^x$

24. $y''' - y'' - 4y' + 4y = 5 - e^x + e^{2x}$

25. $y^{(4)} + 2y'' + y = (x - 1)^2$

26. $y^{(4)} - y'' = 4x + 2xe^{-x}$

In Problems 27–36, solve the given initial-value problem.

27. $y'' + 4y = -2$, $y\left(\dfrac{\pi}{8}\right) = \dfrac{1}{2}$, $y'\left(\dfrac{\pi}{8}\right) = 2$

28. $2y'' + 3y' - 2y = 14x^2 - 4x - 11$, $y(0) = 0$, $y'(0) = 0$

29. $5y'' + y' = -6x$, $y(0) = 0$, $y'(0) = -10$

30. $y'' + 4y' + 4y = (3 + x)e^{-2x}$, $y(0) = 2$, $y'(0) = 5$

31. $y'' + 4y' + 5y = 35e^{-4x}$, $y(0) = -3$, $y'(0) = 1$

32. $y'' - y = \cosh x$, $y(0) = 2$, $y'(0) = 12$

33. $\dfrac{d^2x}{dt^2} + \omega^2 x = F_0 \sin \omega t$, $x(0) = 0$, $x'(0) = 0$

34. $\dfrac{d^2x}{dt^2} + \omega^2 x = F_0 \cos \gamma t$, $x(0) = 0$, $x'(0) = 0$

35. $y''' - 2y'' + y' = 2 - 24e^x + 40e^{5x}$, $y(0) = \dfrac{1}{2}$, $y'(0) = \dfrac{5}{2}$, $y''(0) = -\dfrac{9}{2}$

36. $y''' + 8y = 2x - 5 + 8e^{-2x}$, $y(0) = -5$, $y'(0) = 3$, $y''(0) = -4$

In Problems 37 and 38, solve the given boundary-value problem.

37. $y'' + y = x^2 + 1$, $y(0) = 5$, $y(1) = 0$

38. $y'' - 2y' + 2y = 2x - 2$, $y(0) = 0$, $y(\pi) = \pi$

39. In applications the input function $g(x)$ is often discontinuous. Solve the initial-value problem

$$y'' + 4y = g(x), \qquad y(0) = 1, \quad y'(0) = 2,$$

where
$$g(x) = \begin{cases} \sin x, & 0 \le x \le \dfrac{\pi}{2} \\ 0, & x > \dfrac{\pi}{2} \end{cases}$$

[*Hint:* Solve the problem on the two intervals, and then find a solution so that y and y' are continuous at $x = \pi/2$.]

40. Computer Lab Assignment Find a particular solution of $y'' - 4y' + 8y = (2x^2 - 3x)e^{2x}\cos 2x + (10x^2 - x - 1)e^{2x}\sin 2x$. Use a CAS as an aid in carrying out differentiations, simplifications, and algebra.

In Problems 41–44, without solving, match the solution curve of $y'' + y = f(x)$ shown in the given figure exactly with one of the input functions

(i) $f(x) = 1,$ **(ii)** $f(x) = e^{-x},$ **(iii)** $f(x) = e^{x},$

(iv) $f(x) = \sin 2x,$ **(v)** $f(x) = e^{x} \sin x,$ **(vi)** $f(x) = \sin x.$

Briefly explain your reasoning.

41.

Figure 3.11

42.

Figure 3.12

43.

Figure 3.13

44.

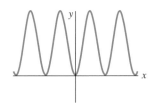

Figure 3.14

3.5 Variation of Parameters

The procedure that we used in Section 2.3 to find a particular solution of a linear first-order differential equation on an interval is applicable to linear higher-order equations as well. To adapt the method of **variation of parameters** to a linear second-order differential equation

$$a_2(x)y'' + a_1(x)y' + a_0(x)y = g(x), \tag{1}$$

we begin as we did in Section 3.2—we put the differential equation in the standard form

$$y'' + P(x)y' + Q(x)y = f(x) \tag{2}$$

by dividing through by the lead coefficient $a_2(x)$. Equation (2) is the second-order analogue of the linear first-order equation $dy/dx + P(x)y = f(x)$. In (2) we suppose that $P(x)$, $Q(x)$, and $f(x)$ are continuous on some common interval I. As we have already seen in Section 3.3, there is no difficulty in obtaining the complimentary function y_c of (2) when the coefficients are constants.

■ **The Assumptions** Corresponding to the assumption $y_p = u_1(x)y_1(x)$ that we used in Section 2.3 to find a particular solution y_p of $dy/dx + P(x)y = f(x)$, for the linear second-order equation (2) we seek a solution of the form

$$y_p = u_1(x)y_1(x) + u_2(x)y_2(x), \tag{3}$$

where y_1 and y_2 form a fundamental set of solutions on I of the associated homogeneous form of (1). Using the product rule to differentiate y_p twice, we get

$$y_p' = u_1 y_1' + y_1 u_1' + u_2 y_2' + y_2 u_2'$$

$$y_p'' = u_1 y_1'' + y_1' u_1' + y_1 u_1'' + u_1' y_1' + u_2 y_2'' + y_2' u_2' + y_2 u_2'' + u_2' y_2'.$$

Substituting (3) and the foregoing derivatives into (2) and grouping terms yields

$$y_p'' + P(x)y_p' + Q(x)y_p = u_1[\overset{\text{zero}}{\overbrace{y_1'' + Py_1' + Qy_1}}] + u_2[\overset{\text{zero}}{\overbrace{y_2'' + Py_2' + Qy_2}}]$$

$$+ y_1 u_1'' + u_1' y_1' + y_2 u_2'' + u_2' y_2' + P[y_1 u_1' + y_2 u_2'] + y_1' u_1' + y_2' u_2'$$

$$= \frac{d}{dx}[y_1 u_1'] + \frac{d}{dx}[y_2 u_2'] + P[y_1 u_1' + y_2 u_2'] + y_1' u_1' + y_2' u_2'$$

$$= \frac{d}{dx}[y_1 u_1' + y_2 u_2'] + P[y_1 u_1' + y_2 u_2'] + y_1' u_1' + y_2' u_2' = f(x). \quad (4)$$

Because we seek to determine two unknown functions u_1 and u_2, reason dictates that we need two equations. We can obtain these equations by making the further assumption that the functions u_1 and u_2 satisfy $y_1 u_1' + y_2 u_2' = 0$. This assumption does not come out of the blue but is prompted by the first two terms in (4), since, if we demand that $y_1 u_1' + y_2 u_2' = 0$, then (4) reduces to $y_1' u_1' + y_2' u_2' = f(x)$. We now have our desired two equations, albeit two equations for determining the derivatives u_1' and u_2'. By Cramer's rule, the solution of the system

$$y_1 u_1' + y_2 u_2' = 0$$

$$y_1' u_1' + y_2' u_2' = f(x)$$

can be expressed in terms of determinants:

$$u_1' = \frac{W_1}{W} = -\frac{y_2 f(x)}{W} \quad \text{and} \quad u_2' = \frac{W_2}{W} = \frac{y_1 f(x)}{W} \quad (5)$$

where

$$W = \begin{vmatrix} y_1 & y_2 \\ y_1' & y_2' \end{vmatrix}, \quad W_1 = \begin{vmatrix} 0 & y_2 \\ f(x) & y_2' \end{vmatrix}, \quad W_2 = \begin{vmatrix} y_1 & 0 \\ y_1' & f(x) \end{vmatrix}. \quad (6)$$

The functions u_1 and u_2 are found by integrating the results in (5). The determinant W is recognized as the Wronskian of y_1 and y_2. By linear independence of y_1 and y_2 on I, we know that $W(y_1(x), y_2(x)) \neq 0$ for every x in the interval.

■ **Summary of the Method** Usually it is not a good idea to memorize formulas in lieu of understanding a procedure. However, the foregoing procedure is too long and complicated to use each time we wish to solve a differential equation. In this case it is more efficient to simply use the formulas in (5). Thus to solve $a_2 y'' + a_1 y' + a_0 y = g(x)$, first find the complementary function $y_c = c_1 y_1 + c_2 y_2$ and then compute the Wronskian $W(y_1(x), y_2(x))$. By dividing by a_2, we put the equation into the standard form $y'' + Py' + Qy = f(x)$ to determine $f(x)$. We find u_1 and u_2 by integrating $u_1' = W_1/W$ and $u_2' = W_2/W$, where W_1 and W_2 are defined as in (6). A particular solution is $y_p = u_1 y_1 + u_2 y_2$. The general solution of the equation is then $y = y_c + y_p$.

Example 1 **General Solution Using Variation of Parameters**

Solve $y'' - 4y' + 4y = (x + 1)e^{2x}$.

SOLUTION From auxiliary equation $m^2 - 4m + 4 = (m - 2)^2 = 0$ we have $y_c = c_1 e^{2x} + c_2 x e^{2x}$. With the identifications $y_1 = e^{2x}$ and $y_2 = xe^{2x}$, we next compute the Wronskian:

$$W(e^{2x}, xe^{2x}) = \begin{vmatrix} e^{2x} & xe^{2x} \\ 2e^{2x} & 2xe^{2x} + e^{2x} \end{vmatrix} = e^{4x}.$$

Since the given differential equation is already in form (2) (that is, the coefficient of y'' is 1), we identify $f(x) = (x + 1)e^{2x}$. From (6) we obtain

$$W_1 = \begin{vmatrix} 0 & xe^{2x} \\ (x+1)e^{2x} & 2xe^{2x} + e^{2x} \end{vmatrix} = -(x+1)xe^{4x}, \qquad W_2 = \begin{vmatrix} e^{2x} & 0 \\ 2e^{2x} & (x+1)e^{2x} \end{vmatrix} = (x+1)e^{4x},$$

and so from (5)

$$u_1' = -\frac{(x+1)xe^{4x}}{e^{4x}} = -x^2 - x, \qquad u_2' = \frac{(x+1)e^{4x}}{e^{4x}} = x + 1.$$

It follows that $u_1 = -\frac{1}{3}x^3 - \frac{1}{2}x^2$ and $u_2 = \frac{1}{2}x^2 + x$. Hence

$$y_p = \left(-\frac{1}{3}x^3 - \frac{1}{2}x^2\right)e^{2x} + \left(\frac{1}{2}x^2 + x\right)xe^{2x} = \frac{1}{6}x^3e^{2x} + \frac{1}{2}x^2e^{2x}$$

and $$y = y_c + y_p = c_1 e^{2x} + c_2 x e^{2x} + \frac{1}{6}x^3 e^{2x} + \frac{1}{2}x^2 e^{2x}. \qquad \square$$

Example 2 General Solution Using Variation of Parameters

Solve $4y'' + 36y = \csc 3x$.

SOLUTION We first put the equation in the standard form (2) by dividing by 4:

$$y'' + 9y = \frac{1}{4}\csc 3x.$$

Since the roots of the auxiliary equation $m^2 + 9 = 0$ are $m_1 = 3i$ and $m_2 = -3i$, the complementary function is $y_c = c_1 \cos 3x + c_2 \sin 3x$. Using $y_1 = \cos 3x$, $y_2 = \sin 3x$, and $f(x) = \frac{1}{4}\csc 3x$, we obtain

$$W(\cos 3x, \sin 3x) = \begin{vmatrix} \cos 3x & \sin 3x \\ -3\sin 3x & 3\cos 3x \end{vmatrix} = 3$$

$$W_1 = \begin{vmatrix} 0 & \sin 3x \\ \frac{1}{4}\csc 3x & 3\cos 3x \end{vmatrix} = -\frac{1}{4}, \quad W_2 = \begin{vmatrix} \cos 3x & 0 \\ -3\sin 3x & \frac{1}{4}\csc 3x \end{vmatrix} = \frac{1}{4}\frac{\cos 3x}{\sin 3x}.$$

Integrating

$$u_1' = \frac{W_1}{W} = -\frac{1}{12} \quad \text{and} \quad u_2' = \frac{W_2}{W} = \frac{1}{12}\frac{\cos 3x}{\sin 3x}$$

gives $u_1 = -\frac{1}{12}x$ and $u_2 = \frac{1}{36}\ln|\sin 3x|$. Thus a particular solution is

$$y_p = -\frac{1}{12}x\cos 3x + \frac{1}{36}(\sin 3x)\ln|\sin 3x|.$$

The general solution of the equation is

$$y = y_c + y_p = c_1 \cos 3x + c_2 \sin 3x - \frac{1}{12}x\cos 3x + \frac{1}{36}(\sin 3x)\ln|\sin 3x|. \quad (7) \; \square$$

Equation (7) represents the general solution of the differential equation on, say, the interval $(0, \pi/6)$.

■ **Constants of Integration** When computing the indefinite integrals of u_1' and u_2', we need not introduce any constants. This is because

$$y = y_c + y_p = c_1 y_1 + c_2 y_2 + (u_1 + a_1)y_1 + (u_2 + b_1)y_2$$

$$= (c_1 + a_1)y_1 + (c_2 + b_1)y_2 + u_1 y_1 + u_2 y_2$$

$$= C_1 y_1 + C_2 y_2 + u_1 y_1 + u_2 y_2.$$

Example 3 General Solution Using Variation of Parameters

Solve $y'' - y = \dfrac{1}{x}$.

SOLUTION The auxiliary equation $m^2 - 1 = 0$ yields $m_1 = -1$ and $m_2 = 1$. Therefore $y_c = c_1 e^x + c_2 e^{-x}$. Now $W(e^x, e^{-x}) = -2$ and

$$u_1' = -\frac{e^{-x}(1/x)}{-2}, \qquad u_1 = \frac{1}{2} \int_{x_0}^{x} \frac{e^{-t}}{t}\, dt,$$

$$u_2' = \frac{e^{x}(1/x)}{-2}, \qquad u_2 = -\frac{1}{2} \int_{x_0}^{x} \frac{e^{t}}{t}\, dt.$$

Since the foregoing integrals are nonelementary we are forced to write

$$y_p = \frac{1}{2} e^x \int_{x_0}^{x} \frac{e^{-t}}{t}\, dt - \frac{1}{2} e^{-x} \int_{x_0}^{x} \frac{e^{t}}{t}\, dt,$$

and so

$$y = y_c + y_p = c_1 e^x + c_2 e^{-x} + \frac{1}{2} e^x \int_{x_0}^{x} \frac{e^{-t}}{t}\, dt - \frac{1}{2} e^{-x} \int_{x_0}^{x} \frac{e^{t}}{t}\, dt. \qquad \square$$

In Example 3 we can integrate on any interval $x_0 \le t \le x$ not containing the origin.

■ Higher-Order Equations

The method we have just examined for nonhomogeneous second-order differential equations can be generalized to linear nth-order equations that have been put into the standard form

$$y^{(n)} + P_{n-1}(x)y^{(n-1)} + \cdots + P_1(x)y' + P_0(x)y = f(x). \tag{8}$$

If $y_c = c_1 y_1 + c_2 y_2 + \cdots + c_n y_n$ is the complementary function for (8), then a particular solution is

$$y_p = u_1(x)y_1(x) + u_2(x)y_2(x) + \cdots + u_n(x)y_n(x),$$

where the u'_k, $k = 1, 2, \ldots, n$, are determined by the n equations

$$
\begin{aligned}
y_1 u_1' + y_2 u_2' + \cdots + \quad & y_n u_n' = 0 \\
y_1' u_1' + y_2' u_2' + \cdots + \quad & y_n' u_n' = 0 \\
\vdots \qquad\qquad\qquad & \vdots \\
y_1^{(n-1)} u_1' + y_2^{(n-1)} u_2' + \cdots + y_n^{(n-1)} u_n' &= f(x).
\end{aligned}
\tag{9}
$$

The first $n - 1$ equations in this system, like $y_1 u_1' + y_2 u_2' = 0$ in (4), are assumptions made to simplify the resulting equation after $y_p = u_1(x)y_1(x) + \cdots + u_n(x)y_n(x)$ is substituted in (8). In this case Cramer's rule gives

$$u_k' = \frac{W_k}{W}, \qquad k = 1, 2, \ldots, n,$$

where W is the Wronskian of y_1, y_2, \ldots, y_n and W_k is the determinant obtained by replacing the kth column of the Wronskian by the column consisting of the right-hand side of (9), that is, the column $(0, 0, \ldots, f(x))$. When $n = 2$ we get (5).

Remarks

(*i*) Variation of parameters has a distinct advantage over the method of undetermined coefficients in that it will *always* yield a particular solution y_p provided the related homogeneous equation can be solved. The present method is not limited to a function $f(x)$, which is a combination of the four types of functions listed on page 124. Also, variation of parameters, unlike undetermined coefficients, is applicable to differential equations with variable coefficients.

(*ii*) In the problems that follow do not hesitate to simplify the form of y_p. Depending on how the antiderivatives of u_1' and u_2' are found, you may not obtain the same y_p as given in the answer section. For example, in Problem 3 in Exercises 3.5 both $y_p = \frac{1}{2}\sin x - \frac{1}{2}x\cos x$ and $y_p = \frac{1}{4}\sin x - \frac{1}{2}x\cos x$ are valid answers. In either case the general solution $y = y_c + y_p$ simplifies to $y = c_1\cos x + c_2\sin x - \frac{1}{2}x\cos x$. Why?

In Problems 1–18, solve each differential equation by variation of parameters.

1. $y'' + y = \sec x$

2. $y'' + y = \tan x$

3. $y'' + y = \sin x$

4. $y'' + y = \sec\theta\tan\theta$

5. $y'' + y = \cos^2 x$

6. $y'' + y = \sec^2 x$

7. $y'' - y = \cosh x$

8. $y'' - y = \sinh 2x$

9. $y'' - 4y = \dfrac{e^{2x}}{x}$

10. $y'' - 9y = \dfrac{9x}{e^{3x}}$

11. $y'' + 3y' + 2y = \dfrac{1}{1+e^x}$

12. $y'' - 2y' + y = \dfrac{e^x}{1+x^2}$

13. $y'' + 3y' + 2y = \sin e^x$

14. $y'' - 2y' + y = e^t\arctan t$

15. $y'' + 2y' + y = e^{-t}\ln t$

16. $2y'' + 2y' + y = 4\sqrt{x}$

17. $3y'' - 6y' + 6y = e^x\sec x$

18. $4y'' - 4y' + y = e^{x/2}\sqrt{1-x^2}$

In Problems 19–20, solve each differential equation by variation of parameters subject to the initial conditions $y(0) = 1$, $y'(0) = 0$.

19. $4y'' - y = xe^{x/2}$

20. $2y'' + y' - y = x + 1$

21. $y'' + 2y' - 8y = 2e^{-2x} - e^{-x}$

22. $y'' - 4y' + 4y = (12x^2 - 6x)e^{2x}$

In Problems 23 and 24, the indicated functions are known linearly independent solutions of the associated homogeneous differential equation on $(0, \infty)$. Find the general solution of the given nonhomogeneous equation.

23. $x^2 y'' + xy' + (x^2 - \frac{1}{4})y = x^{3/2}$; $\quad y_1 = x^{-1/2}\cos x$, $\quad y_2 = x^{-1/2}\sin x$

24. $x^2 y'' + xy' + y = \sec(\ln x)$; $\quad y_1 = \cos(\ln x)$, $\quad y_2 = \sin(\ln x)$

In Problems 25 and 26, solve the given third-order differential equation by variation of parameters.

25. $y''' + y' = \tan x$

26. $y''' + 4y' = \sec 2x$

In Problems 27 and 28, discuss how the methods of undetermined coefficients and variation of parameters can be combined to solve the given differential equation. Carry out your ideas.

27. $3y'' - 6y' + 30y = 15\sin x + e^x\tan 3x$

28. $y'' - 2y' + y = 4x^2 - 3 + x^{-1}e^x$

29. What are the intervals of definition of the general solutions in Problems 1, 7, 9, and 18? Discuss why the interval of definition of the general solution in Problem 24 is *not* $(0, \infty)$.

30. Find the general solution of $x^4 y'' + x^3 y' - 4x^2 y = 1$ given that $y_1 = x^2$ is a solution of the associated homogeneous equation.

3.6 Cauchy–Euler Equation

The relative ease with which we were able to find explicit solutions of linear higher-order differential equations with constant coefficients in the preceding sections does not, in general, carry over to linear equations with variable coefficients. We shall see in Chapter 6 that when a linear differential equation has variable coefficients, the best that we can *usually* expect is to find a solution in the form of an infinite series. However, the type of differential equation considered in this section is an

exception to this rule; it is an equation with variable coefficients whose general solution can always be expressed in terms of powers of x, sines, cosines, and logarithmic and exponential functions. Moreover, its method of solution is quite similar to that for constant-coefficient equations.

■ **Cauchy–Euler Equation** Any linear differential equation of the form

$$a_n x^n \frac{d^n y}{dx^n} + a_{n-1} x^{n-1} \frac{d^{n-1} y}{dx^{n-1}} + \cdots + a_1 x \frac{dy}{dx} + a_0 y = g(x),$$

where the coefficients $a_n, a_{n-1}, \ldots, a_0$ are constants, is known diversely as a **Cauchy–Euler equation,** an **Euler–Cauchy equation,** an **Euler equation,** or an **equidimensional equation.** The observable characteristic of this type of equation is that the degree $k = n, n - 1, \ldots, 1, 0$ of the monomial coefficients x^k matches the order k of differentiation $d^k y/dx^k$:

$$\overset{\text{same}}{\underset{\downarrow}{a_n x^n}} \frac{d^n y}{dx^n} + \overset{\text{same}}{\underset{\downarrow}{a_{n-1} x^{n-1}}} \frac{d^{n-1} y}{dx^{n-1}} + \cdots.$$

As in Section 4.3, we start the discussion with a detailed examination of the forms of the general solutions of the homogeneous second-order equation

$$ax^2 \frac{d^2 y}{dx^2} + bx \frac{dy}{dx} + cy = 0.$$

The solution of higher-order equations follows analogously. Also, we can solve the nonhomogeneous equation $ax^2 y'' + bxy' + cy = g(x)$ by variation of parameters, once we have determined the complementary function $y_c(x)$.

Lead coefficient being zero at ☞ *$x = 0$ could cause a problem.* The coefficient of $d^2 y/dx^2$ is zero at $x = 0$. Hence in order to guarantee that the fundamental results of Theorem 3.1 are applicable to the Cauchy-Euler equation, we confine our attention to finding the general solution on the interval $(0, \infty)$. Solutions on the interval $(-\infty, 0)$ can be obtained by substituting $t = -x$ into the differential equation. See Problems 37 and 38 in Exercises 3.6.

■ **Method of Solution** We try a solution of the form $y = x^m$, where m is to be determined. Analogous to what happened when we substituted e^{mx} into a linear equation with constant coefficients, after substituting x^m each term of a Cauchy–Euler equation becomes a polynomial in m times x^m since

$$a_k x^k \frac{d^k y}{dx^k} = a_k x^k m(m-1)(m-2) \cdots (m-k+1)x^{m-k}$$
$$= a_k m(m-1)(m-2) \cdots (m-k+1)x^m.$$

For example, by substituting $y = x^m$ the second-order equation becomes

$$ax^2 \frac{d^2 y}{dx^2} + bx \frac{dy}{dx} + cy = am(m-1)x^m + bmx^m + cx^m = (am(m-1) + bm + c)x^m.$$

Thus $y = x^m$ is a solution of the differential equation whenever m is a solution of the **auxiliary equation**

$$am(m-1) + bm + c = 0 \quad \text{or} \quad am^2 + (b-a)m + c = 0. \tag{1}$$

There are three different cases to be considered, depending on whether the roots of this quadratic equation are real and distinct, real and equal, or complex. In the last case the roots appear as a conjugate pair.

Case I: **Distinct Real Roots** Let m_1 and m_2 denote the real roots of (1) such that $m_1 \neq m_2$. Then $y_1 = x^{m_1}$ and $y_2 = x^{m_2}$ form a fundamental set of solutions. Hence the general solution is

$$y = c_1 x^{m_1} + c_2 x^{m_2}. \tag{2}$$

Example 1 **Distinct Roots**

Solve $x^2 \dfrac{d^2 y}{dx^2} - 2x \dfrac{dy}{dx} - 4y = 0$.

SOLUTION Rather than just memorizing equation (1), it is preferable to assume $y = x^m$ as the solution a few times in order to understand the origin and the difference between this new form of the auxiliary equation and that obtained in Section 3.3. Differentiate twice,

$$\frac{dy}{dx} = mx^{m-1}, \qquad \frac{d^2 y}{dx^2} = m(m-1)x^{m-2},$$

and substitute back into the differential equation:

$$x^2 \frac{d^2 y}{dx^2} - 2x \frac{dy}{dx} - 4y = x^2 \cdot m(m-1)x^{m-2} - 2x \cdot mx^{m-1} - 4x^m$$
$$= x^m(m(m-1) - 2m - 4) = x^m(m^2 - 3m - 4) = 0$$

if $m^2 - 3m - 4 = 0$. Now $(m+1)(m-4) = 0$ implies $m_1 = -1$, $m_2 = 4$ so that $y = c_1 x^{-1} + c_2 x^4$. ❑

Case II: **Repeated Real Roots** If the roots of (1) are repeated (that is, $m_1 = m_2$), then we obtain only one solution, namely, $y = x^{m_1}$. When the roots of the quadratic equation $am^2 + (b-a)m + c = 0$ are equal, the discriminant of the coefficients is necessarily zero. It follows from the quadratic formula that the root must be $m_1 = -(b-a)/2a$.

Now we can construct a second solution y_2, using (5) of Section 3.2. We first write the Cauchy-Euler equation in the standard form

$$\frac{d^2 y}{dx^2} + \frac{b}{ax} \frac{dy}{dx} + \frac{c}{ax^2} y = 0$$

and make the identifications $P(x) = b/ax$ and $\int (b/ax)\, dx = (b/a) \ln x$. Thus

$$y_2 = x^{m_1} \int \frac{e^{-(b/a) \ln x}}{x^{2m_1}}\, dx$$

$$= x^{m_1} \int x^{-b/a} \cdot x^{-2m_1}\, dx \qquad \leftarrow e^{-(b/a) \ln x} = e^{\ln x^{-b/a}} = x^{-b/a}$$

$$= x^{m_1} \int x^{-b/a} \cdot x^{(b-a)/a}\, dx \qquad \leftarrow -2m_1 = (b-a)/a$$

$$= x^{m_1} \int \frac{dx}{x} = x^{m_1} \ln x.$$

The general solution is then

$$y = c_1 x^{m_1} + c_2 x^{m_1} \ln x. \tag{3}$$

Example 2 **Repeated Roots**

Solve $4x^2 \dfrac{d^2y}{dx^2} + 8x \dfrac{dy}{dx} + y = 0$.

SOLUTION The substitution $y = x^m$ yields

$$4x^2 \frac{d^2y}{dx^2} + 8x \frac{dy}{dx} + y = x^m(4m(m-1) + 8m + 1) = x^m(4m^2 + 4m + 1) = 0$$

when $4m^2 + 4m + 1 = 0$ or $(2m + 1)^2 = 0$. Since $m_1 = -\frac{1}{2}$, the general solution is
$y = c_1 x^{-1/2} + c_2 x^{-1/2} \ln x$. ❑

For higher-order equations, if m_1 is a root of multiplicity k, then it can be shown that

$$x^{m_1}, \quad x^{m_1} \ln x, \quad x^{m_1}(\ln x)^2, \quad \ldots, \quad x^{m_1}(\ln x)^{k-1}$$

are k linearly independent solutions. Correspondingly, the general solution of the differential equation must then contain a linear combination of these k solutions.

Case III: *Conjugate Complex Roots* If the roots of (1) are the conjugate pair $m_1 = \alpha + i\beta$, $m_2 = \alpha - i\beta$, where α and $\beta > 0$ are real, then a solution is $y = C_1 x^{\alpha + i\beta} + C_2 x^{\alpha - i\beta}$. But when the roots of the auxiliary equation are complex, as in the case of equations with constant coefficients, we wish to write the solution in terms of real functions only. We note the identity

$$x^{i\beta} = (e^{\ln x})^{i\beta} = e^{i\beta \ln x},$$

which, by Euler's formula, is the same as

$$x^{i\beta} = \cos(\beta \ln x) + i \sin(\beta \ln x).$$

Similarly, $x^{-i\beta} = \cos(\beta \ln x) - i \sin(\beta \ln x).$

Adding and subtracting the last two results yields

$$x^{i\beta} + x^{-i\beta} = 2\cos(\beta \ln x) \quad \text{and} \quad x^{i\beta} - x^{-i\beta} = 2i\sin(\beta \ln x),$$

respectively. From the fact that $y = C_1 x^{\alpha + i\beta} + C_2 x^{\alpha - i\beta}$ is a solution for any values of the constants, we see, in turn, for $C_1 = C_2 = 1$ and $C_1 = 1$, $C_2 = -1$ that

$$y_1 = x^\alpha(x^{i\beta} + x^{-i\beta}) \qquad \text{and} \quad y_2 = x^\alpha(x^{i\beta} - x^{-i\beta})$$

or $y_1 = 2x^\alpha \cos(\beta \ln x) \quad \text{and} \quad y_2 = 2ix^\alpha \sin(\beta \ln x)$

are also solutions. Since $W(x^\alpha \cos(\beta \ln x), x^\alpha \sin(\beta \ln x)) = \beta x^{2\alpha - 1} \neq 0$, $\beta > 0$, on the interval $(0, \infty)$, we conclude that

$$y_1 = x^\alpha \cos(\beta \ln x) \quad \text{and} \quad y_2 = x^\alpha \sin(\beta \ln x)$$

constitute a fundamental set of real solutions of the differential equation. Hence the general solution is

$$y = x^\alpha[c_1 \cos(\beta \ln x) + c_2 \sin(\beta \ln x)]. \tag{4}$$

Example 3 **An Initial-Value Problem**

Solve the initial-value problem $4x^2 y'' + 17y = 0$, $y(1) = -1$, $y'(1) = 0$.

SOLUTION The y' term is missing in the given Cauchy–Euler equation; nevertheless, the substitution $y = x^m$ yields

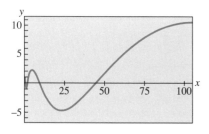

$$4x^2y'' + 17y = x^m(4m(m-1) + 17) = x^m(4m^2 - 4m + 17) = 0$$

when $4m^2 - 4m + 17 = 0$. From the quadratic formula we find that the roots are $m_1 = \frac{1}{2} + 2i$ and $m_2 = \frac{1}{2} - 2i$. With the identifications $\alpha = \frac{1}{2}$ and $\beta = 2$, we see from (4) that the general solution of the differential equation is

$$y = x^{1/2}[c_1 \cos(2 \ln x) + c_2 \sin(2 \ln x)].$$

By applying the initial conditions $y(1) = -1$, $y'(1) = 0$ to the foregoing solution and using $\ln 1 = 0$ we then find, in turn, that $c_1 = -1$ and $c_2 = 0$. Hence the solution of the initial-value problem is $y = -x^{1/2}\cos(2 \ln x)$. The graph of this function, obtained with the aid of computer software, is given in Figure 3.15. The particular solution is seen to be oscillatory and unbounded as $x \to \infty$. ❏

The next example illustrates the solution of a third-order Cauchy–Euler equation.

Example 4 **Third-Order Equation**

Solve $x^3 \dfrac{d^3y}{dx^3} + 5x^2 \dfrac{d^2y}{dx^2} + 7x \dfrac{dy}{dx} + 8y = 0$.

SOLUTION The first three derivatives of $y = x^m$ are

$$\frac{dy}{dx} = mx^{m-1}, \quad \frac{d^2y}{dx^2} = m(m-1)x^{m-2}, \quad \frac{d^3y}{dx^3} = m(m-1)(m-2)x^{m-3}$$

so that the given differential equation becomes

$$x^3 \frac{d^3y}{dx^3} + 5x^2 \frac{d^2y}{dx^2} + 7x \frac{dy}{dx} + 8y = x^3 m(m-1)(m-2)x^{m-3} + 5x^2 m(m-1)x^{m-2} + 7xmx^{m-1} + 8x^m$$

$$= x^m(m(m-1)(m-2) + 5m(m-1) + 7m + 8)$$

$$= x^m(m^3 + 2m^2 + 4m + 8) = x^m(m+2)(m^2 + 4) = 0.$$

In this case we see that $y = x^m$ will be a solution of the differential equation for $m_1 = -2$, $m_2 = 2i$, and $m_3 = -2i$. Hence the general solution is

$$y = c_1 x^{-2} + c_2 \cos(2 \ln x) + c_3 \sin(2 \ln x).$$ ❏

The method of undetermined coefficients as described in Section 3.4 does not carry over, *in general,* to linear differential equations with variable coefficients. Consequently, in the following example the method of variation of parameters is employed.

Example 5 **Variation of Parameters**

Solve $x^2y'' - 3xy' + 3y = 2x^4 e^x$.

SOLUTION Since the equation is nonhomogeneous we first solve the associated homogeneous equation. From the auxiliary equation $(m-1)(m-3) = 0$ we find $y_c = c_1 x + c_2 x^3$. Now before using variation of parameters to find a particular solution $y_p = u_1 y_1 + u_2 y_2$, recall that the formulas $u_1' = W_1/W$ and $u_2' = W_2/W$, where W_1, W_2, and W are the determinants defined on page 133, were derived under the assumption that the differential equation has been put into the standard form $y'' + P(x)y' + Q(x)y = f(x)$. Therefore we divide the given equation by x^2, and from

$$y'' - \frac{3}{x}y' + \frac{3}{x^2}y = 2x^2e^x$$

we make the identification $f(x) = 2x^2e^x$. Now with $y_1 = x$, $y_2 = x^3$ and

$$W = \begin{vmatrix} x & x^3 \\ 1 & 3x^2 \end{vmatrix} = 2x^3, \quad W_1 = \begin{vmatrix} 0 & x^3 \\ 2x^2e^x & 3x^2 \end{vmatrix} = -2x^5e^x, \quad W_2 = \begin{vmatrix} x & 0 \\ 1 & 2x^2e^x \end{vmatrix} = 2x^3e^x$$

we find $\quad u_1' = -\dfrac{2x^5e^x}{2x^3} = -x^2e^x \quad$ and $\quad u_2' = \dfrac{2x^5e^x}{2x^3} = e^x.$

The integral of the latter function is immediate, but in the case of u_1' we integrate by parts twice. The results are $u_1 = -x^2e^x + 2xe^x - 2e^x$ and $u_2 = e^x$. Hence

$$y_p = u_1y_1 + u_2y_2 = (-x^2e^x + 2xe^x - 2e^x)x + e^xx^3 = 2x^2e^x - 2xe^x.$$

Finally we have $y = y_c + y_p = c_1x + c_2x^3 + 2x^2e^x - 2xe^x.$ ❏

Remarks

The similarity between the forms of solutions of Cauchy–Euler equations and solutions of linear equations with constant coefficients is not just a coincidence. For example, when the roots of the auxiliary equations for $ay'' + by' + cy = 0$ and $ax^2y'' + bxy' + cy = 0$ are distinct and real, the respective general solutions are

$$y = c_1e^{m_1x} + c_2e^{m_2x} \quad \text{and} \quad y = c_1x^{m_1} + c_2x^{m_2}, \quad x > 0. \tag{5}$$

In view of the identity $e^{\ln x} = x$, $x > 0$, the second solution given in (5) can be expressed in the same form as the first solution:

$$y = c_1e^{m_1 \ln x} + c_2e^{m_2 \ln x} = c_1e^{m_1t} + c_2e^{m_2t},$$

where $t = \ln x$. This last result illustrates another fact of mathematical life: Any Cauchy–Euler equation can *always* be rewritten as a linear differential equation with constant coefficients by means of the substitution $x = e^t$. The idea is to solve the new differential equation in terms of the variable t, using the methods of the previous sections, and once the general solution is obtained, resubstitute $t = \ln x$. Since this procedure provides a good review of the Chain Rule of differentiation, you are urged to work Problems 29–32 in Exercises 3.6.

Answers to odd-numbered problems begin on page A-30.

In Problems 1–18, solve the given differential equation.

1. $x^2y'' - 2y = 0$

2. $4x^2y'' + y = 0$

3. $xy'' + y' = 0$

4. $xy'' - 3y' = 0$

5. $x^2y'' + xy' + 4y = 0$

6. $x^2y'' + 5xy' + 3y = 0$

7. $x^2y'' - 3xy' - 2y = 0$

8. $x^2y'' + 3xy' - 4y = 0$

9. $25x^2y'' + 25xy' + y = 0$

10. $4x^2y'' + 4xy' - y = 0$

11. $x^2y'' + 5xy' + 4y = 0$

12. $x^2y'' + 8xy' + 6y = 0$

13. $3x^2y'' + 6xy' + y = 0$

14. $x^2y'' - 7xy' + 41y = 0$

15. $x^3y''' - 6y = 0$

16. $x^3y''' + xy' - y = 0$

17. $xy^{(4)} + 6y''' = 0$

18. $x^4y^{(4)} + 6x^3y''' + 9x^2y'' + 3xy' + y = 0$

In Problems 19–22, solve the given differential equation by variation of parameters.

19. $xy'' - 4y' = x^4$

20. $2x^2y'' + 5xy' + y = x^2 - x$

21. $x^2y'' - xy' + y = 2x$

22. $x^2y'' - 2xy' + 2y = x^4e^x$

In Problems 23–28, solve the given initial-value problem.

23. $x^2y'' + 3xy' = 0$, $\quad y(1) = 0$, $\quad y'(1) = 4$

24. $x^2y'' - 5xy' + 8y = 0$, $\quad y(2) = 32$, $\quad y'(2) = 0$

25. $x^2y'' + xy' + y = 0$, $\quad y(1) = 1$, $\quad y'(1) = 2$

26. $x^2y'' - 3xy' + 4y = 0$, $\quad y(1) = 5$, $\quad y'(1) = 3$

27. $xy'' + y' = x$, $\quad y(1) = 1$, $\quad y'(1) = -\frac{1}{2}$

28. $x^2y'' - 5xy' + 8y = 8x^6$, $\quad y(\frac{1}{2}) = 0$, $\quad y'(\frac{1}{2}) = 0$

In Problems 29–32, use the substitution $x = e^t$ to transform the given Cauchy–Euler equation to a differential equation with constant coefficients. Solve the original equation by solving the new equation using the procedure in Section 3.4.

29. $x^2y'' + 10xy' + 8y = x^2$

30. $x^2y'' - 4xy' + 6y = \ln x^2$

31. $x^2y'' - 3xy' + 13y = 4 + 3x$

32. $x^3y''' - 3x^2y'' + 6xy' - 6y = 3 + \ln x^3$

Computer Lab Assignment In Problems 33–36, solve the given differential equation by using a CAS to find the (approximate) roots of the auxiliary equation.

33. $2x^3y''' - 10.98x^2y'' + 8.5xy' + 1.3y = 0$

34. $x^3y''' + 4x^2y'' + 5xy' - 9y = 0$

35. $x^4y^{(4)} + 6x^3y''' + 3x^2y'' - 3xy' + 4y = 0$

36. $x^4y^{(4)} - 6x^3y''' + 33x^2y'' - 105xy' + 169y = 0$

In Problems 37 and 38, solve the given initial-value problem on the interval $(-\infty, 0)$.

37. $4x^2y'' + y = 0$, $\quad y(-1) = 2$, $\quad y'(-1) = 4$

38. $x^2y'' - 4xy' + 6y = 0$, $\quad y(-2) = 8$, $\quad y'(-2) = 0$

39. How would you use the method of this section to solve

$$(x + 2)^2y'' + (x + 2)y' + y = 0?$$

Carry out your ideas. State an interval over which the solution is defined.

40. Find a Cauchy–Euler differential equation of lowest order with real coefficients if it is known that 2 and $1 - i$ are two roots of its auxiliary equation.

41. The initial conditions $y(0) = y_0$, $y'(0) = y_1$, apply to each of the following differential equations:

$$x^2y'' = 0, \quad x^2y'' - 2xy' + 2y = 0, \quad x^2y'' - 4xy' + 6y = 0.$$

For what values of y_0 and y_1 does each initial-value problem have a solution?

42. What are the x-intercepts of the solution curve shown in Figure 3.15? How many x-intercepts are there in the interval $0 < x < 0.5$?

3.7 Nonlinear Equations

■ **Some Differences** There are several significant differences between linear and nonlinear differential equations. We saw in Section 3.1 that homogeneous linear equations of order two or higher have the property that a linear combination of solutions is also a solution (Theorem 3.2). Nonlinear equations do not possess this property of superposability. For example, on the interval $(-\infty, \infty)$, $y_1 = e^x$, $y_2 = e^{-x}$, $y_3 = \cos x$, and $y_4 = \sin x$ are four linearly independent solutions of the nonlinear second-order differential equation $(y'')^2 - y^2 = 0$. But linear combinations such as $y = c_1e^x + c_3 \cos x$, $y = c_2e^{-x} + c_4 \sin x$, $y = c_1e^x + c_2e^{-x} + c_3 \cos x + c_4 \sin x$ are not solutions of the equation for arbitrary nonzero constants c_i. See Problem 1 in Exercises 3.7.

In Chapter 2 we saw that we could solve a few nonlinear first-order differential equations by recognizing them as separable, exact, homogeneous, or perhaps Bernoulli equations. Even though the solutions of these equations were in the form of a one-parameter family, this family did not, as a rule, represent the general solution of the differential equation. On the other hand, by paying attention to certain continuity conditions, we obtained general solutions of linear first-order equations. Stated another way, nonlinear first-order differential equations can possess singular solutions whereas linear equations cannot. But the major difference between linear and nonlinear equations of order two or higher lies in the realm of solvability. Given a linear equation there is a chance that we can find some form of a solution that we can look at, an explicit solution or perhaps a solution in the form of an infinite series. On the other hand, nonlinear higher-order differential equations virtually defy solution. This does not mean that a nonlinear higher-order differential equation

has no solution but rather that there are no analytical methods whereby either an explicit or implicit solution can be found.

Although this sounds disheartening, there are still things that can be done; we can always analyze a nonlinear DE qualitatively and numerically.

Let us make it clear at the outset that nonlinear higher-order differential equations are important—dare we say even more important than linear equations?—because as we fine-tune the mathematical model of, say, a physical system, we also increase the likelihood that this higher-resolution model will be nonlinear.

We begin by illustrating an analytical method that occasionally enables us to find explicit/implicit solutions of special kinds of nonlinear second-order differential equations.

■ **Reduction of Order** Nonlinear second-order differential equations $F(x, y', y'') = 0$, where the dependent variable y is missing, and $F(y, y', y'') = 0$, where the independent variable x is missing, can sometimes be solved using first-order methods. Each equation can be reduced to a first-order equation by means of the substitution $u = y'$.

The next example illustrates the substitution technique for an equation of the form $F(x, y', y'') = 0$. If $u = y'$, then the differential equation becomes $F(x, u, u') = 0$. If we can solve this last equation for u, we can find y by integration. Note that since we are solving a second-order equation, its solution will contain two arbitrary constants.

Example 1 Dependent Variable y Is Missing

Solve $y'' = 2x(y')^2$.

SOLUTION If we let $u = y'$ then $du/dx = y''$. After substituting, the second-order equation reduces to a first-order equation with separable variables; the independent variable is x and the dependent variable is u:

$$\frac{du}{dx} = 2xu^2 \qquad \text{or} \qquad \frac{du}{u^2} = 2x\, dx$$

$$\int u^{-2}\, du = \int 2x\, dx$$

$$-u^{-1} = x^2 + c_1^2.$$

The constant of integration is written as c_1^2 for convenience. The reason should be obvious in the next few steps. Since $u^{-1} = 1/y'$, it follows that

$$\frac{dy}{dx} = -\frac{1}{x^2 + c_1^2}$$

and so $\qquad y = -\int \dfrac{dx}{x^2 + c_1^2} \qquad$ or $\qquad y = -\dfrac{1}{c_1} \tan^{-1} \dfrac{x}{c_1} + c_2.$ ❑

Next we show how to solve an equation that has the form $F(y, y', y'') = 0$. Once more we let $u = y'$, but since the independent variable x is missing we use this substitution to transform the differential equation into one in which the independent variable is y and the dependent variable is u. To this end we use the Chain Rule to compute the second derivative of y:

$$y'' = \frac{du}{dx} = \frac{du}{dy}\frac{dy}{dx} = u\frac{du}{dy}.$$

In this case the first-order equation that we must now solve is $F(y, u, u\, du/dy) = 0$.

Example 2 **Independent Variable *x* Is Missing**

Solve $yy'' = (y')^2$.

SOLUTION With the aid of $u = y'$, the Chain Rule shown above, and separation of variables, the given differential equation becomes

$$y\left(u\frac{du}{dy}\right) = u^2 \quad\text{or}\quad \frac{du}{u} = \frac{dy}{y}.$$

Integrating the last equation then yields $\ln|u| = \ln|y| + c_1$, which, in turn, gives $u = c_2 y$, where the constant $\pm e^{c_1}$ has been relabeled as c_2. We now resubstitute $u = dy/dx$, separate variables once again, integrate, and relabel constants a second time:

$$\int \frac{dy}{y} = c_2 \int dx \quad\text{or}\quad \ln|y| = c_2 x + c_3 \quad\text{or}\quad y = c_4 e^{c_2 x}. \qquad\square$$

■ **Use of Taylor Series** In some instances a solution of a nonlinear initial-value problem, in which the initial conditions are specified at x_0, can be approximated by a Taylor series centered at x_0.

Example 3 **Taylor Series Solution of an IVP**

Let us assume that a solution of the initial-value problem

$$y'' = x + y - y^2, \quad y(0) = -1, \quad y'(0) = 1 \tag{1}$$

exists. If we further assume that the solution $y(x)$ of the problem is analytic at 0, then $y(x)$ possesses a Taylor series expansion centered at 0:

$$y(x) = y(0) + \frac{y'(0)}{1!}x + \frac{y''(0)}{2!}x^2 + \frac{y'''(0)}{3!}x^3 + \frac{y^{(4)}(0)}{4!}x^4 + \frac{y^{(5)}(0)}{5!}x^5 + \cdots. \tag{2}$$

Note that the value of the first and second terms in the series (2) are known since those values are the specified initial conditions $y(0) = -1$, $y'(0) = 1$. Moreover, the differential equation itself defines the value of the second derivative at 0: $y''(0) = 0 + y(0) - y(0)^2 = 0 + (-1) - (-1)^2 = -2$. We can then find expressions for the higher derivatives y''', $y^{(4)}$, . . . , by calculating the successive derivatives of the differential equation:

$$y'''(x) = \frac{d}{dx}(x + y - y^2) = 1 + y' - 2yy' \tag{3}$$

$$y^{(4)}(x) = \frac{d}{dx}(1 + y' - 2yy') = y'' - 2yy'' - 2(y')^2 \tag{4}$$

$$y^{(5)}(x) = \frac{d}{dx}(y'' - 2yy'' - 2(y')^2) = y''' - 2yy''' - 6y'y'' \tag{5}$$

and so on. Now using $y(0) = -1$ and $y'(0) = 1$ we find from (3) that $y'''(0) = 4$. From the values $y(0) = -1$, $y'(0) = 1$, and $y''(0) = -2$, we find $y^{(4)}(0) = -8$ from (4). With the additional information that $y'''(0) = 4$, we then see from (5) that $y^{(5)}(0) = 24$. Hence from (2), the first six terms of a series solution of the initial-value problem (1) are

$$y(x) = -1 + x - x^2 + \frac{2}{3}x^3 - \frac{1}{3}x^4 + \frac{1}{5}x^5 + \cdots. \qquad\square$$

■ **Use of a Numerical Solver** Numerical methods, such as Euler's method or a Runge–Kutta method, are developed solely for first-order differential equations and then are extended to systems of first-order equations. In order to analyze an *n*th-order initial-value problem numerically, we express the *n*th-order ODE as a system of *n* first-order equations. In brief, here is how it is done for a second-order initial-value problem: First, solve for y'', that is, put the DE into normal form $y'' = f(x, y, y')$, and then let $y' = u$. For example, if we substitute $y' = u$ in

$$\frac{d^2y}{dx^2} = f(x,\ y,\ y'), \quad y(x_0) = y_0, \quad y'(x_0) = u_0, \tag{6}$$

then $y'' = u'$ and $y'(x_0) = u(x_0)$ so that the initial-value problem (6) becomes

$$Solve: \begin{cases} y' = u \\ u' = f(x,\ y,\ u) \end{cases}$$

$$Subject\ to:\ y(x_0) = y_0, \quad u(x_0) = u_0.$$

However, it should be noted that a commercial numerical solver *may not* require* that you supply the system.

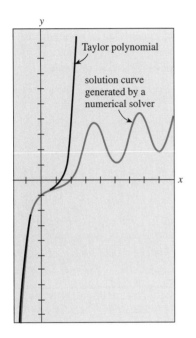

Figure 3.16 Comparison of two approximate solutions

Example 4 Graphical Analysis of Example 3

Following the foregoing procedure, the second-order initial-value problem in Example 3 is equivalent to

$$\frac{dy}{dx} = u$$

$$\frac{du}{dx} = x + y - y^2$$

with initial conditions $y(0) = -1$, $u(0) = 1$. With the aid of a numerical solver we get the solution curve shown in color in Figure 3.16. For comparison, the curve shown in black is the graph of the fifth-degree Taylor polynomial $T_5(x) = -1 + x - x^2 + \frac{2}{3}x^3 - \frac{1}{3}x^4 + \frac{1}{5}x^5$. Although we do not know the interval of convergence of the Taylor series obtained in Example 3, the closeness of the two curves in a neighborhood of the origin suggests that the power series may converge on the interval $(-1, 1)$. ❏

Figure 3.17

■ **Qualitative Questions** The colored graph in Figure 3.16 raises some questions of a qualitative nature: Is the solution of the original initial-value problem oscillatory as $x \to \infty$? The graph generated by a numerical solver on the larger interval shown in Figure 3.17 would seem to *suggest* that the answer is yes. But this single example, or even an assortment of examples, does not answer the basic question of whether *all* solutions of the differential equation $y'' = x + y - y^2$ are oscillatory in nature. Also, what is happening to the solution curves in Figure 3.17 when x is near -1? What is the behavior of solutions of the differential equation as $x \to -\infty$? Are solutions bounded as $x \to \infty$? Questions such as these are not easily answered, in general, for nonlinear second-order differential equations. But certain kinds of

* Some numerical solvers only require that a second-order differential equation be expressed in normal form $y'' = f(x,\ y,\ y')$. The translation of the single equation into a system of two equations is then built into the computer program, since the first equation of the system is always $y' = u$ and the second equation is $u' = f(x\ , y, u)$.

second-order equations lend themselves to a systematic qualitative analysis, and these, like their first-order relatives encountered in Section 2.1, are the kind that have no explicit dependence on the independent variable. Second-order ODEs of the form,

$$F(y, y', y'') = 0 \quad \text{or} \quad \frac{d^2y}{dx^2} = f(y, y'),$$

equations free of the independent variable x, are called **autonomous.** The differential equation in Example 2 is autonomous, and because of the presence of the x term on its right side, the equation in Example 3 is nonautonomous. For an in-depth treatment of the topic of stability of autonomous second-order differential equations and autonomous systems of differential equations, the reader is referred to Chapter 11.

E X E R C I S E S 3 . 7

Answers to odd-numbered problems begin on page A-30.

In Problems 1 and 2, verify that y_1 and y_2 are solutions of the given differential equation but that $y = c_1 y_1 + c_2 y_2$ is, in general, not a solution.

1. $(y'')^2 = y^2; \quad y_1 = e^x, \quad y_2 = \cos x$

2. $yy'' = \frac{1}{2}(y')^2; \quad y_1 = 1, \quad y_2 = x^2$

In Problems 3–8, solve the given differential equation by using the substitution $u = y'$.

3. $y'' + (y')^2 + 1 = 0$

4. $y'' = 1 + (y')^2$

5. $x^2 y'' + (y')^2 = 0$

6. $(y + 1)y'' = (y')^2$

7. $y'' + 2y(y')^3 = 0$

8. $y^2 y'' = y'$

9. Computer Lab Assignment Solve the initial-value problem $y'' + yy' = 0$, $y(0) = 1$, $y'(0) = -1$. Use the DE and a numerical solver to graph the solution curve. Then use a graphing utility to graph the explicit solution. Find an interval of validity for the solution of the problem.

10. Computer Lab Assignment Find two solutions of the initial-value problem

$$(y'')^2 + (y')^2 = 1, \quad y\left(\frac{\pi}{2}\right) = \frac{1}{2}, \quad y'\left(\frac{\pi}{2}\right) = \frac{\sqrt{3}}{2}.$$

Use a numerical solver to graph the solution curves.

In Problems 11 and 12, show that the substitution $u = y'$ leads to a Bernoulli equation. Solve this equation (see Section 2.5).

11. $xy'' = y' + (y')^3$

12. $xy'' = y' + x(y')^2$

Computer Lab Assignment In Problems 13–16, proceed as in Example 3 and obtain the first six nonzero terms of a Taylor series solution centered at 0 of the given initial-value problem. Use a numerical solver and a graphing utility to compare the solution curve and the graph of the Taylor polynomial.

13. $y'' = x + y^2, \quad y(0) = 1, \quad y'(0) = 1$

14. $y'' + y^2 = 1, \quad y(0) = 2, \quad y'(0) = 3$

15. $y'' = x^2 + y^2 - 2y', \quad y(0) = 1, \quad y'(0) = 1$

16. $y'' = e^y, \quad y(0) = 0, \quad y'(0) = -1$

17. In calculus the curvature of a curve that is defined by a function $y = f(x)$ is defined as

$$\kappa = \frac{y''}{[1 + (y')^2]^{3/2}}.$$

Find $y = f(x)$ for which $\kappa = 1$. [*Hint:* For simplicity ignore constants of integration.]

18. Computer Lab Assignment A mathematical model for the position $x(t)$ of a body moving rectilinearly on the x-axis in an inverse-square force field is given by

$$\frac{d^2x}{dt^2} = -\frac{k^2}{x^2}.$$

Suppose at $t = 0$ the body starts from rest from the position $x = x_0$, $x_0 > 0$. Show that the velocity of the body at any time is given by $v^2 = 2k^2(1/x - 1/x_0)$. Use the last equation and a CAS to carry out the integration to express time t in terms of x.

19. Computer Lab Assignment A mathematical model for the position $x(t)$ of a moving object is

$$\frac{d^2x}{dt^2} + \sin x = 0.$$

Use a numerical solver to graphically investigate the solutions of the equation subject to $x(0) = 0$, $x'(0) = x_1$, $x_1 \geq 0$. Discuss the motion of the object for $t \geq 0$ and for various choices of x_1. Investigate the equation

$$\frac{d^2x}{dt^2} + \frac{dx}{dt} + \sin x = 0$$

in the same manner. Give a possible physical interpretation of the dx/dt term.

20. We saw that $\sin x$, $\cos x$, e^x, and e^{-x} were four solutions of the nonlinear equation $(y'')^2 - y^2 = 0$. Without attempting to solve the differential equation, explain how these explicit solutions can be found using knowledge about linear equations. Without attempting to verify, discuss why the two special linear combinations $y = c_1 e^x + c_2 e^{-x}$ and $y = c_3 \cos x + c_4 \sin x$ must satisfy the differential equation.

21. Use the method of reduction of order considered in this section to solve the third-order differential equation

$$y''' = \sqrt{1 + (y'')^2}.$$

3.8 Linear Models: Initial-Value Problems

In this section we are going to consider several linear dynamical systems in which each mathematical model is a second-order differential equation with constant coefficients

$$a_2 \frac{d^2y}{dt^2} + a_1 \frac{dy}{dt} + a_0 y = g(t).$$

Recall, the function g is the **input** or **forcing function** of the system. The **output** or **response** of the system is a solution $y(t)$ of the differential equation that satisfies prescribed initial conditions $y(t_0) = y_0$, $y'(t_0) = y_1$ on an interval containing t_0.

3.8.1 Spring/Mass Systems: Free Undamped Motion

■ **Hooke's Law** Suppose a flexible spring is suspended vertically from a rigid support and then a mass m is attached to its free end. The amount of stretch, or elongation, of the spring will of course depend on the mass; masses with different weights stretch the spring by differing amounts. By Hooke's law, the spring itself exerts a restoring force F opposite to the direction of elongation and proportional to the amount of elongation s. Simply stated, $F = ks$, where k is a constant of proportionality called the **spring constant.** The spring is essentially characterized by the number k. For example, if a mass weighing 10 lb stretches a spring ½ ft, then $10 = k(½)$ implies $k = 20$ lb/ft. Necessarily then, a mass weighing, say, 8 lb stretches the same spring only ½ ft.

unstretched

equilibrium
position
$mg - ks = 0$

motion

(a) (b) (c)

Figure 3.18

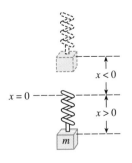

$x < 0$

$x = 0$

$x > 0$

Figure 3.19

■ **Newton's Second Law** After a mass m is attached to a spring, it stretches the spring by an amount s and attains a position of equilibrium at which its weight W is balanced by the restoring force ks. Recall that weight is defined by $W = mg$, where mass is measured in slugs, kilograms, or grams and $g = 32$ ft/s², 9.8 m/s², or 980 cm/s², respectively. As indicated in Figure 3.18(b), the condition of equilibrium is $mg = ks$ or $mg - ks = 0$. If the mass is displaced by an amount x from its equilibrium position, the restoring force of the spring is then $k(x + s)$. Assuming that there are no retarding forces acting on the system and assuming that the mass vibrates free of other external forces—**free motion**—we can equate Newton's second law with the net, or resultant, force of the restoring force and the weight:

$$m \frac{d^2x}{dt^2} = -k(s + x) + mg = -kx + \underbrace{mg - ks}_{\text{zero}} = -kx. \tag{1}$$

The negative sign in (1) indicates that the restoring force of the spring acts opposite to the direction of motion. Furthermore, we can adopt the convention that displacements measured *below* the equilibrium position are positive. See Figure 3.19.

■ **DE of Free Undamped Motion** By dividing (1) by the mass m we obtain the second-order differential equation $d^2x/dt^2 + (k/m)x = 0$ or

$$\frac{d^2x}{dt^2} + \omega^2 x = 0, \tag{2}$$

where $\omega^2 = k/m$. Equation (2) is said to describe **simple harmonic motion** or **free undamped motion.** Two obvious initial conditions associated with (2) are $x(0) = x_0$, the amount of initial displacement, and $x'(0) = x_1$, the initial velocity of the mass. For

example, if $x_0 > 0$, $x_1 < 0$, the mass starts from a point *below* the equilibrium position with an imparted *upward* velocity. When $x_1 = 0$ the mass is said to be released from *rest*. For example, if $x_0 < 0$, $x_1 = 0$, the mass is released from rest from a point $|x_0|$ units *above* the equilibrium position.

■ **Solution and Equation of Motion** To solve equation (2) we note that the solutions of the auxiliary equation $m^2 + \omega^2 = 0$ are the complex numbers $m_1 = \omega i$, $m_2 = -\omega i$. Thus from (8) of Section 3.3 we find the general solution of (2) to be

$$x(t) = c_1 \cos \omega t + c_2 \sin \omega t. \tag{3}$$

The **period** of free vibrations described by (3) is $T = 2\pi/\omega$, and the **frequency** is $f = 1/T = \omega/2\pi$. For example, for $x(t) = 2 \cos 3t - 4 \sin 3t$ the period is $2\pi/3$ and the frequency is $3/2\pi$. The former number means that the graph of $x(t)$ repeats every $2\pi/3$ units; the latter number means that there are 3 cycles of the graph every 2π units or, equivalently, that the mass undergoes $3/2\pi$ complete vibrations per unit time. In addition, it can be shown that the period $2\pi/\omega$ is the time interval between two successive maxima of $x(t)$. Keep in mind that a maximum of $x(t)$ is a positive displacement corresponding to the mass's attaining a maximum distance *below* the equilibrium position, whereas a minimum of $x(t)$ is a negative displacement corresponding to the mass's attaining a maximum height *above* the equilibrium position. We refer to either case as an **extreme displacement** of the mass. Finally, when the initial conditions are used to determine the constants c_1 and c_2 in (3), we say that the resulting particular solution or response is the **equation of motion.**

Example 1 **Free Undamped Motion**

A mass weighing 2 pounds stretches a spring 6 inches. At $t = 0$ the mass is released from a point 8 inches below the equilibrium position with an upward velocity of ⅓ ft/s. Determine the equation of free motion.

SOLUTION Because we are using the engineering system of units, the measurements given in terms of inches must be converted into feet: 6 in. = ½ ft; 8 in. = ⅔ ft. In addition, we must convert the units of weight given in pounds into units of mass. From $m = W/g$ we have $m = \tfrac{2}{32} = \tfrac{1}{16}$ slug. Also, from Hooke's law, $2 = k(\tfrac{1}{2})$ implies that the spring constant is $k = 4$ lb/ft. Hence (1) gives

$$\frac{1}{16} \frac{d^2x}{dt^2} = -4x \qquad \text{or} \qquad \frac{d^2x}{dt^2} + 64x = 0.$$

The initial displacement and initial velocity are $x(0) = \tfrac{2}{3}$, $x'(0) = -\tfrac{1}{3}$, where the negative sign in the last condition is a consequence of the fact that the mass is given an initial velocity in the negative, or upward, direction.

Now $\omega^2 = 64$ or $\omega = 8$, so that the general solution of the differential equation is

$$x(t) = c_1 \cos 8t + c_2 \sin 8t. \tag{4}$$

Applying the initial conditions to $x(t)$ and $x'(t)$ gives $c_1 = \tfrac{2}{3}$ and $c_2 = -\tfrac{1}{16}$. Thus the equation of motion is

$$x(t) = \frac{2}{3} \cos 8t - \frac{1}{6} \sin 8t. \tag{5}$$ □

■ **Alternative Form of $x(t)$** When $c_1 \neq 0$ and $c_2 \neq 0$, the actual **amplitude** A of free vibrations is not obvious from inspection of equation (3). For example,

although the mass in Example 1 is initially displaced ⅔ foot beyond the equilibrium position, the amplitude of vibrations is a number larger than ⅔. Hence it is often convenient to convert a solution of form (3) to the simpler form

$$x(t) = A \sin(\omega t + \phi),\tag{6}$$

where $A = \sqrt{c_1^2 + c_2^2}$ and ϕ is a **phase angle** defined by

$$\left.\begin{array}{l} \sin \phi = \dfrac{c_1}{A} \\[2ex] \cos \phi = \dfrac{c_2}{A} \end{array}\right\} \quad \tan \phi = \dfrac{c_1}{c_2}.\tag{7}$$

To verify this we expand (6) by the addition formula for the sine function:

$$A \sin \omega t \cos \phi + A \cos \omega t \sin \phi = (A \sin \phi) \cos \omega t + (A \cos \phi) \sin \omega t.\tag{8}$$

It follows from Figure 3.20 that if ϕ is defined by

$$\sin \phi = \frac{c_2}{\sqrt{c_1^2 + c_2^2}} = \frac{c_1}{A}, \qquad \cos \phi = \frac{c_2}{\sqrt{c_1^2 + c_2^2}} = \frac{c_2}{A},$$

then (8) becomes

$$A \frac{c_1}{A} \cos \omega t + A \frac{c_2}{A} \sin \omega t = c_1 \cos \omega t + c_2 \sin \omega t = x(t).$$

Figure 3.20

Be careful in the computation of the phase angle ϕ. ☞

Example 2 **Alternative Form of Solution (5)**

In view of the foregoing discussion, we can write the solution (5), $x(t) = ⅔ \cos 8t - ⅙ \sin 8t$, in the alternative form $x(t) = A \sin(8t + \phi)$. Computation of the amplitude is straightforward, $A = \sqrt{(⅔)^2 + (-⅙)^2} = \sqrt{^{17}/_{36}} \approx 0.69$ ft, but some care should be exercised when computing the phase angle ϕ defined by (7). With $c_1 = ⅔$ and $c_2 = -⅙$ we find $\tan \phi = -4$, and a calculator then gives $\tan^{-1}(-4) = -1.326$ rad. This is *not* the phase angle, since $\tan^{-1}(-4)$ is located in the *fourth quadrant* and therefore contradicts the fact that $\sin \phi > 0$ and $\cos \phi < 0$ because $c_1 > 0$ and $c_2 < 0$. Hence we must take ϕ to be the *second-quadrant* angle $\phi = \pi + (-1.326) = 1.816$ rad. Thus we have

$$x(t) = \frac{\sqrt{17}}{6} \sin(8t + 1.816).\tag{9}$$

The period of this function is $T = 2\pi/8 = \pi/4$. ❑

Figure 3.21(a) illustrates the mass in Example 2 going through approximately two complete cycles of motion. Reading left to right, the first five positions marked with black dots correspond to: the initial position of the mass below the equilibrium position ($x = ⅔$), the mass passing through the equilibrium position for the first time heading upward ($x = 0$), the mass at its extreme displacement above the equilibrium position ($x = -\sqrt{17}/6$), the mass at the equilibrium position for the second time heading downward ($x = 0$), and the mass at its extreme displacement below the equilibrium position ($x = \sqrt{17}/6$). The dots on the graph of (9) given in Figure 3.21(b) also agree with the five positions just given. Note, however, that in Figure 3.21(b) the positive direction in the *tx*-plane is the usual upward direction and so is opposite to the positive direction indicated in Figure 3.21(a). Hence the solid-colored graph

representing the motion of the mass in Figure 3.21(b) is the mirror image through the t-axis of the black dashed curve in Figure 3.21(a).

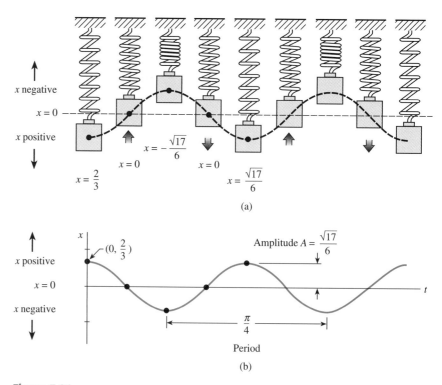

(a)

(b)

Figure 3.21

Form (6) is very useful, since it is easy to find values of time for which the graph of $x(t)$ crosses the positive t-axis (the line $x = 0$). We observe that $\sin(\omega t + \phi) = 0$ when $\omega t + \phi = n\pi$, where n is a nonnegative integer.

■ **Systems with Variable Spring Constants** In the model discussed above, we assumed an ideal world, a world in which the physical characteristics of the spring do not change over time. In the nonideal world, however, it seems reasonable to expect that when a spring/mass system is in motion for a long period the spring would weaken; in other words, the "spring constant" would vary, or, more specifically, decay with time. In one model for the **aging spring** the spring constant k in (1) is replaced by the decreasing function $K(t) = ke^{-\alpha t}$, $k > 0$, $\alpha > 0$. The linear differential equation $mx'' + ke^{-\alpha t}x = 0$ cannot be solved by the methods considered in this chapter. Nevertheless, we can obtain two linearly independent solutions using the methods in Chapter 5. See Problem 15 in Exercises 3.8, Example 3 in Section 5.3, and Problems 25 and 26 in Exercises 5.3.

When a spring/mass system is subjected to environment in which the temperature is rapidly decreasing, it might make sense to replace the constant k with $K(t) = kt$, $k > 0$, a function that increases with time. The resulting model, $mx'' + ktx = 0$, is a form of **Airy's differential equation.** Like the equation for an aging spring, Airy's equation can be solved by the methods of Chapter 5. See Problem 16 in Exercises 3.8, Example 2 in Section 5.1, and Problems 27–29 in Exercises 5.3.

(a)

(b)

Figure 3.22

3.8.2 Spring/Mass Systems: Free Damped Motion

The concept of free harmonic motion is somewhat unrealistic, since the motion described by equation (1) assumes that there are no retarding forces acting on the moving mass. Unless the mass is suspended in a perfect vacuum, there will be at least a resisting force due to the surrounding medium. As Figure 3.22 shows, the mass could be suspended in a viscous medium or connected to a dashpot damping device.

■ **DE of Free Damped Motion** In the study of mechanics, damping forces acting on a body are considered to be proportional to a power of the instantaneous velocity. In particular, we shall assume throughout the subsequent discussion that this force is given by a constant multiple of dx/dt. When no other external forces are impressed on the system, it follows from Newton's second law that

$$m\frac{d^2x}{dt^2} = -kx - \beta\frac{dx}{dt}, \tag{10}$$

where β is a positive *damping constant* and the negative sign is a consequence of the fact that the damping force acts in a direction opposite to the motion.

Dividing (10) by the mass m, we find the differential equation of **free damped motion** is $d^2x/dt^2 + (\beta/m)dx/dt + (k/m)x = 0$ or

$$\frac{d^2x}{dt^2} + 2\lambda\frac{dx}{dt} + \omega^2 x = 0, \tag{11}$$

where

$$2\lambda = \frac{\beta}{m}, \qquad \omega^2 = \frac{k}{m}. \tag{12}$$

The symbol 2λ is used only for algebraic convenience, since the auxiliary equation is $m^2 + 2\lambda m + \omega^2 = 0$ and the corresponding roots are then

$$m_1 = -\lambda + \sqrt{\lambda^2 - \omega^2}, \qquad m_2 = -\lambda - \sqrt{\lambda^2 - \omega^2}.$$

We can now distinguish three possible cases depending on the algebraic sign of $\lambda^2 - \omega^2$. Since each solution contains the *damping factor* $e^{-\lambda t}$, $\lambda > 0$, the displacements of the mass become negligible for a long period of time.

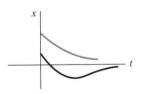

x

t

Figure 3.23

Case I: $\lambda^2 - \omega^2 > 0$. In this situation the system is said to be **overdamped** because the damping coefficient β is large when compared to the spring constant k. The corresponding solution of (11) is $x(t) = c_1 e^{m_1 t} + c_2 e^{m_2 t}$ or

$$x(t) = e^{-\lambda t}(c_1 e^{\sqrt{\lambda^2 - \omega^2}\, t} + c_2 e^{-\sqrt{\lambda^2 - \omega^2}\, t}). \tag{13}$$

This equation represents a smooth and nonoscillatory motion. Figure 3.23 shows two possible graphs of $x(t)$.

Case II: $\lambda^2 - \omega^2 = 0$. The system is said to be **critically damped** because any slight decrease in the damping force would result in oscillatory motion. The general solution of (11) is $x(t) = c_1 e^{m_1 t} + c_2 t e^{m_1 t}$ or

$$x(t) = e^{-\lambda t}(c_1 + c_2 t). \tag{14}$$

Some graphs of typical motion are given in Figure 3.24. Notice that the motion is quite similar to that of an overdamped system. It is also apparent from (14) that the mass can pass through the equilibrium position at most one time.

Figure 3.24

Figure 3.25

(a)

t	$x(t)$
1	0.601
1.5	0.370
2	0.225
2.5	0.137
3	0.083

(b)

Figure 3.26

Case III: $\lambda^2 - \omega^2 < 0$. In this case the system is said to be **underdamped** because the damping coefficient is small compared to the spring constant. The roots m_1 and m_2 are now complex:

$$m_1 = -\lambda + \sqrt{\omega^2 - \lambda^2}\, i, \qquad m_2 = -\lambda - \sqrt{\omega^2 - \lambda^2}\, i.$$

Thus the general solution of equation (11) is

$$x(t) = e^{-\lambda t}\,(c_1 \cos \sqrt{\omega^2 - \lambda^2}\, t + c_2 \sin \sqrt{\omega^2 - \lambda^2}\, t). \tag{15}$$

As indicated in Figure 3.25, the motion described by (15) is oscillatory, but because of the coefficient $e^{-\lambda t}$, the amplitudes of vibration $\to 0$ as $t \to \infty$.

Example 3 Overdamped Motion

It is readily verified that the solution of the initial-value problem

$$\frac{d^2x}{dt^2} + 5\frac{dx}{dt} + 4x = 0, \qquad x(0) = 1, \quad x'(0) = 1$$

is

$$x(t) = \frac{5}{3}e^{-t} - \frac{2}{3}e^{-4t}. \tag{16}$$

The problem can be interpreted as representing the overdamped motion of a mass on a spring. The mass starts from a position 1 unit *below* the equilibrium position with a *downward* velocity of 1 ft/s.

To graph $x(t)$ we find the value of t for which the function has an extremum—that is, the value of time for which the first derivative (velocity) is zero. Differentiating (16) gives $x'(t) = -\frac{5}{3}e^{-t} + \frac{8}{3}e^{-4t}$ so that $x'(t) = 0$ implies $e^{3t} = \frac{8}{5}$ or $t = \frac{1}{3}\ln\frac{8}{5} = 0.157$. It follows from the first derivative test, as well as our physical intuition, that $x(0.157) = 1.069$ ft is actually a maximum. In other words, the mass attains an extreme displacement of 1.069 feet below the equilibrium position.

We should also check to see whether the graph crosses the t-axis, that is, whether the mass passes through the equilibrium position. This cannot happen in this instance since the equation $x(t) = 0$, or $e^{3t} = \frac{2}{5}$, has the physically irrelevant solution $t = \frac{1}{3}\ln\frac{2}{5} = -0.305$.

The graph of $x(t)$, along with some other pertinent data, is given in Figure 3.26. ❑

Example 4 Critically Damped Motion

An 8-pound weight stretches a spring 2 feet. Assuming that a damping force numerically equal to 2 times the instantaneous velocity acts on the system, determine the equation of motion if the weight is released from the equilibrium position with an upward velocity of 3 ft/s.

SOLUTION From Hooke's law we see that $8 = k(2)$ gives $k = 4$ lb/ft and that $W = mg$ gives $m = \frac{8}{32} = \frac{1}{4}$ slug. The differential equation of motion is then

$$\frac{1}{4}\frac{d^2x}{dt^2} = -4x - 2\frac{dx}{dt} \qquad \text{or} \qquad \frac{d^2x}{dt^2} + 8\frac{dx}{dt} + 16x = 0. \tag{17}$$

The auxiliary equation for (17) is $m^2 + 8m + 16 = (m + 4)^2 = 0$ so that $m_1 = m_2 = -4$. Hence the system is critically damped and

$$x(t) = c_1 e^{-4t} + c_2 t e^{-4t}. \tag{18}$$

Figure 3.27

Applying the initial conditions $x(0) = 0$ and $x'(0) = -3$, we find, in turn, that $c_1 = 0$ and $c_2 = -3$. Thus the equation of motion is

$$x(t) = -3te^{-4t}. \qquad (19)$$

To graph $x(t)$ we proceed as in Example 4. From $x'(t) = -3e^{-4t}(1 - 4t)$ we see that $x'(t) = 0$ when $t = \frac{1}{4}$. The corresponding extreme displacement is $x(\frac{1}{4}) = -3(\frac{1}{4})e^{-1} = -0.276$ ft. As shown in Figure 3.27, we interpret this value to mean that the weight reaches a maximum height of 0.276 foot above the equilibrium position. ❑

Example 5 Underdamped Motion

A 16-pound weight is attached to a 5-foot-long spring. At equilibrium the spring measures 8.2 feet. If the weight is pushed up and released from rest at a point 2 feet above the equilibrium position, find the displacements $x(t)$ if it is further known that the surrounding medium offers a resistance numerically equal to the instantaneous velocity.

SOLUTION The elongation of the spring after the weight is attached is $8.2 - 5 = 3.2$ ft, so it follows from Hooke's law that $16 = k(3.2)$ or $k = 5$ lb/ft. In addition, $m = \frac{16}{32} = \frac{1}{2}$ slug so that the differential equation is given by

$$\frac{1}{2}\frac{d^2x}{dt^2} = -5x - \frac{dx}{dt} \qquad \text{or} \qquad \frac{d^2x}{dt^2} + 2\frac{dx}{dt} + 10x = 0. \qquad (20)$$

Proceeding, we find that the roots of $m^2 + 2m + 10 = 0$ are $m_1 = -1 + 3i$ and $m_2 = -1 - 3i$, which then implies the system is underdamped and

$$x(t) = e^{-t}(c_1 \cos 3t + c_2 \sin 3t). \qquad (21)$$

Finally, the initial conditions $x(0) = -2$ and $x'(0) = 0$ yield $c_1 = -2$ and $c_2 = -\frac{2}{3}$, so the equation of motion is

$$x(t) = e^{-t}\left(-2 \cos 3t - \frac{2}{3} \sin 3t\right). \qquad (22) \quad ❑$$

■ **Alternative Form of *x(t)*** In a manner identical to the procedure used on page 149, we can write any solution

$$x(t) = e^{-\lambda t}(c_1 \cos \sqrt{\omega^2 - \lambda^2}\, t + c_2 \sin \sqrt{\omega^2 - \lambda^2}\, t)$$

in the alternative form

$$x(t) = Ae^{-\lambda t}\sin(\sqrt{\omega^2 - \lambda^2}\, t + \phi), \qquad (23)$$

where $A = \sqrt{c_1^2 + c_2^2}$ and the phase angle ϕ is determined from the equations

$$\sin \phi = \frac{c_1}{A}, \quad \cos \phi = \frac{c_2}{A}, \quad \tan \phi = \frac{c_1}{c_2}.$$

The coefficient $Ae^{-\lambda t}$ is sometimes called the **damped amplitude** of vibrations. Because (23) is not a periodic function, the number $2\pi/\sqrt{\omega^2 - \lambda^2}$ is called the **quasi period** and $\sqrt{\omega^2 - \lambda^2}/2\pi$ is the **quasi frequency**. The quasi period is the time interval between two successive maxima of $x(t)$. You should verify, for the equation of motion in Example 5, that $A = 2\sqrt{10}/3$ and $\phi = 4.391$. Therefore an equivalent form of (22) is

$$x(t) = \frac{2\sqrt{10}}{3}\, e^{-t} \sin(3t + 4.391).$$

Figure 3.28

3.8.3 Spring/Mass Systems: Driven Motion

■ **DE of Driven Motion with Damping** Suppose we now take into consideration an external force $f(t)$ acting on a vibrating mass on a spring. For example, $f(t)$ could represent a driving force causing an oscillatory vertical motion of the support of the spring. See Figure 3.28. The inclusion of $f(t)$ in the formulation of Newton's second law gives the differential equation of **driven** or **forced motion:**

$$m\frac{d^2x}{dt^2} = -kx - \beta\frac{dx}{dt} + f(t). \tag{24}$$

Dividing (24) by m gives

$$\frac{d^2x}{dt^2} + 2\lambda\frac{dx}{dt} + \omega^2x = F(t), \tag{25}$$

where $F(t) = f(t)/m$ and, as in the preceding section, $2\lambda = \beta/m$, $\omega^2 = k/m$. To solve the latter nonhomogeneous equation we can use either the method of undetermined coefficients or variation of parameters.

Example 6 **Interpretation of an Initial-Value Problem**

Interpret and solve the initial-value problem

$$\frac{1}{5}\frac{d^2x}{dt^2} + 1.2\frac{dx}{dt} + 2x = 5\cos 4t, \qquad x(0) = \frac{1}{2}, \quad x'(0) = 0. \tag{26}$$

SOLUTION We can interpret the problem to represent a vibrational system consisting of a mass ($m = \frac{1}{5}$ slug or kilogram) attached to a spring ($k = 2$ lb/ft or N/m). The mass is released from rest $\frac{1}{2}$ unit (foot or meter) below the equilibrium position. The motion is damped ($\beta = 1.2$) and is being driven by an external periodic ($T = \pi/2$ s) force beginning at $t = 0$. Intuitively we would expect that even with damping the system would remain in motion until such time as the forcing function was "turned off," in which case the amplitudes would diminish. However, as the problem is given, $f(t) = 5\cos 4t$ will remain "on" forever.

We first multiply the differential equation in (26) by 5 and solve

$$\frac{dx^2}{dt^2} + 6\frac{dx}{dt} + 10x = 0$$

by the usual methods. Since $m_1 = -3 + i$, $m_2 = -3 - i$, it follows that

$$x_c(t) = e^{-3t}(c_1\cos t + c_2\sin t).$$

Using the method of undetermined coefficients, we assume a particular solution of the form $x_p(t) = A\cos 4t + B\sin 4t$. Differentiating $x_p(t)$ and substituting into the DE gives

$$x_p'' + 6x_p' + 10x_p = (-6A + 24B)\cos 4t + (-24A - 6B)\sin 4t = 25\cos 4t.$$

The resulting system of equations

$$-6A + 24B = 25, \qquad -24A - 6B = 0$$

yields $A = -\frac{25}{102}$ and $B = \frac{50}{51}$. It follows that

$$x(t) = e^{-3t}(c_1\cos t + c_2\sin t) - \frac{25}{102}\cos 4t + \frac{50}{51}\sin 4t. \tag{27}$$

When we set $t = 0$ in the above equation, we obtain $c_1 = \frac{38}{51}$. By differentiating the expression and then setting $t = 0$, we also find that $c_2 = -\frac{86}{51}$. Therefore the equation of motion is

$$x(t) = e^{-3t}\left(\frac{38}{51}\cos t - \frac{86}{51}\sin t\right) - \frac{25}{102}\cos 4t + \frac{50}{51}\sin 4t. \qquad (28) \quad \square$$

■ **Transient and Steady-State Terms** When F is a periodic function, such as $F(t) = F_0 \sin \gamma t$ or $F(t) = F_0 \cos \gamma t$, the general solution of (25) for $\lambda > 0$ is the sum of a nonperiodic function $x_c(t)$ and a periodic function $x_p(t)$. Moreover, $x_c(t)$ dies off as time increases, that is, $\lim_{t \to \infty} x_c(t) = 0$. Thus for long period of time, the displacements of the mass are closely approximated by the particular solution $x_p(t)$. The complementary function $x_c(t)$ is said to be a **transient term** or **transient solution,** and the function $x_p(t)$, the part of the solution that remains after an interval of time, is called a **steady-state term** or **steady-state solution.** Note therefore that the effect of the initial conditions on a spring/mass system driven by F is transient. In the particular solution (28), $e^{-3t}(\frac{38}{51}\cos t - \frac{86}{51}\sin t)$ is a transient term and $x_p(t) = -\frac{25}{102}\cos 4t + \frac{50}{51}\sin 4t$ is a steady-state term. The graphs of these two terms and the solution (28) are given in Figures 3.29(a) and 3.29(b), respectively.

Example 7 Transient/Steady-State Solutions

The solution of the initial-value problem

$$\frac{d^2x}{dt^2} + 2\frac{dx}{dt} + 2x = 4\cos t + 2\sin t, \qquad x(0) = 0, \quad x'(0) = x_1,$$

where x_1 is constant, is given by

$$x(t) = \underbrace{(x_1 - 2)e^{-t}\sin t}_{\text{transient}} + \underbrace{2\sin t}_{\text{steady-state}}.$$

Solution curves for selected values of the initial velocity x_1 are shown in Figure 3.30. The graphs show that the influence of the transient term is negligible for about $t > 3\pi/2$. $\qquad \square$

■ **DE of Driven Motion Without Damping** With a periodic impressed force and no damping force, there is no transient term in the solution of a problem. Also, we shall see that a periodic impressed force with a frequency near or the same as the frequency of free undamped vibrations can cause a severe problem in any oscillatory mechanical system.

Example 8 Undamped Forced Motion

Solve the initial-value problem

$$\frac{d^2x}{dt^2} + \omega^2 x = F_0 \sin \gamma t, \qquad x(0) = 0, \quad x'(0) = 0, \qquad (29)$$

where F_0 is a constant and $\gamma \neq \omega$.

SOLUTION The complementary function is $x_c(t) = c_1 \cos \omega t + c_2 \sin \omega t$. To obtain a particular solution we assume $x_p(t) = A \cos \gamma t + B \sin \gamma t$ so that

$$x_p'' + \omega^2 x_p = A(\omega^2 - \gamma^2)\cos \gamma t + B(\omega^2 - \gamma^2)\sin \gamma t = F_0 \sin \gamma t.$$

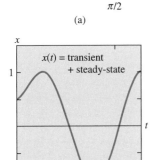

steady-state $x_p(t)$

transient

(a)

$x(t) = $ transient
+ steady-state

(b)

Figure 3.29

$x_1 = 7$
$x_1 = 3$
$x_1 = 0$
$x_1 = -3$

Figure 3.30

Equating coefficients immediately gives $A = 0$ and $B = F_0/(\omega^2 - \gamma^2)$. Therefore

$$x_p(t) = \frac{F_0}{\omega^2 - \gamma^2} \sin \gamma t.$$

Applying the given initial conditions to the general solution

$$x(t) = c_1 \cos \omega t + c_2 \sin \omega t + \frac{F_0}{\omega^2 - \gamma^2} \sin \gamma t$$

yields $c_1 = 0$ and $c_2 = -\gamma F_0/\omega(\omega^2 - \gamma^2)$. Thus the solution is

$$x(t) = \frac{F_0}{\omega(\omega^2 - \gamma^2)} (-\gamma \sin \omega t + \omega \sin \gamma t), \qquad \gamma \neq \omega. \qquad (30) \;\square$$

■ **Pure Resonance** Although equation (30) is not defined for $\gamma = \omega$, it is interesting to observe that its limiting value as $\gamma \to \omega$ can be obtained by applying L'Hôpital's rule. This limiting process is analogous to "tuning in" the frequency of the driving force ($\gamma/2\pi$) to the frequency of free vibrations ($\omega/2\pi$). Intuitively we expect that over a length of time we should be able to substantially increase the amplitudes of vibration. For $\gamma = \omega$ we define the solution to be

$$x(t) = \lim_{\gamma \to \omega} F_0 \frac{-\gamma \sin \omega t + \omega \sin \gamma t}{\omega(\omega^2 - \gamma^2)} = F_0 \lim_{\gamma \to \omega} \frac{\dfrac{d}{d\gamma}(-\gamma \sin \omega t + \omega \sin \gamma t)}{\dfrac{d}{d\gamma}(\omega^3 - \omega\gamma^2)}$$

$$= F_0 \lim_{\gamma \to \omega} \frac{-\sin \omega t + \omega t \cos \gamma t}{-2\omega\gamma}$$

$$= F_0 \frac{-\sin \omega t + \omega t \cos \omega t}{-2\omega^2}$$

$$= \frac{F_0}{2\omega^2} \sin \omega t - \frac{F_0}{2\omega} t \cos \omega t. \qquad (31)$$

As suspected, when $t \to \infty$ the displacements become large; in fact, $|x(t_n)| \to \infty$ when $t_n = n\pi/\omega$, $n = 1, 2, \ldots$. The phenomenon we have just described is known as **pure resonance.** The graph given in Figure 3.31 shows typical motion in this case.

In conclusion, it should be noted that there is no actual need to use a limiting process on (30) to obtain the solution for $\gamma = \omega$. Alternatively, equation (31) follows by solving the initial-value problem

$$\frac{d^2x}{dt^2} + \omega^2 x = F_0 \sin \omega t, \qquad x(0) = 0, \quad x'(0) = 0$$

directly by conventional methods.

If the displacements of a spring/mass system were actually described by a function such as (31), the system would necessarily fail. Large oscillations of the mass would eventually force the spring beyond its elastic limit. One might argue too that the resonating model presented in Figure 3.31 is completely unrealistic, because it ignores the retarding effects of ever-present damping forces. Although it is true that pure resonance cannot occur when the smallest amount of damping is taken into consideration, large and equally destructive amplitudes of vibration (although bounded as $t \to \infty$) can occur. See Problem 43 in Exercises 3.8.

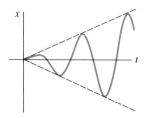

Figure 3.31

3.8.4 Series Circuit Analogue

■ *LRC* **Series Circuits** As mentioned in the introduction to this chapter, many different physical systems can be described by a linear second-order differential equation similar to the differential equation of forced motion with damping:

$$m\frac{d^2x}{dt^2} + \beta\frac{dx}{dt} + kx = f(t). \tag{32}$$

If $i(t)$ denotes current in the *LRC* **series electrical circuit** shown in Figure 3.32, then the voltage drops across the inductor, resistor, and capacitor are as shown in Figure 1.18. By Kirchhoff's second law, the sum of these voltages equals the voltage $E(t)$ impressed on the circuit; that is,

$$L\frac{di}{dt} + Ri + \frac{1}{C}q = E(t). \tag{33}$$

But the charge $q(t)$ on the capacitor is related to the current $i(t)$ by $i = dq/dt$, and so (33) becomes the linear second-order differential equation

$$L\frac{d^2q}{dt^2} + R\frac{dq}{dt} + \frac{1}{C}q = E(t). \tag{34}$$

The nomenclature used in the analysis of circuits is similar to that used to describe spring-mass systems.

If $E(t) = 0$, the **electrical vibrations** of the circuit are said to be **free.** Since the auxiliary equation for (34) is $Lm^2 + Rm + 1/C = 0$, there will be three forms of the solution with $R \neq 0$, depending on the value of the discriminant $R^2 - 4L/C$. We say that the circuit is

<div style="margin-left:2em">

overdamped if $R^2 - 4L/C > 0$,

critically damped if $R^2 - 4L/C = 0$,

and **underdamped** if $R^2 - 4L/C < 0$.

</div>

In each of these three cases the general solution of (34) contains the factor $e^{-Rt/2L}$, and so $q(t) \to 0$ as $t \to \infty$. In the underdamped case when $q(0) = q_0$, the charge on the capacitor oscillates as it decays; in other words, the capacitor is charging and discharging as $t \to \infty$. When $E(t) = 0$ and $R = 0$, the circuit is said to be undamped and the electrical vibrations do not approach zero as t increases without bound; the response of the circuit is **simple harmonic.**

Figure 3.32

Example 9 Underdamped Series Circuit

Find the charge $q(t)$ on the capacitor in an *LRC* series circuit when $L = 0.25$ henry (h), $R = 10$ ohms (Ω), $C = 0.001$ farad (f), $E(t) = 0$ volts (V), $q(0) = q_0$ coulombs (C), and $i(0) = 0$ amperes (A).

SOLUTION Since $1/C = 1000$, equation (34) becomes

$$\frac{1}{4}q'' + 10q' + 1000q = 0 \quad \text{or} \quad q'' + 40q' + 4000q = 0.$$

Solving this homogeneous equation in the usual manner, we find that the circuit is underdamped and $q(t) = e^{-20t}(c_1 \cos 60t + c_2 \sin 60t)$. Applying the initial conditions, we find $c_1 = q_0$ and $c_2 = q_0/3$. Thus $q(t) = q_0 e^{-20t}(\cos 60t + \frac{1}{3} \sin 60t)$. Using (23), we can write the foregoing solution as

$$q(t) = \frac{q_0\sqrt{10}}{3}e^{-20t}\sin(60t + 1.249). \qquad \square$$

When there is an impressed voltage $E(t)$ on the circuit, the electrical vibrations are said to be **forced.** In the case when $R \neq 0$, the complementary function $q_c(t)$ of (34) is called a **transient solution.** If $E(t)$ is periodic or a constant, then the particular solution $q_p(t)$ of (34) is a **steady-state solution.**

Example 10 Steady-State Current

Find the steady-state solution $q_p(t)$ and the **steady-state current** in an *LRC* series circuit when the impressed voltage is $E(t) = E_0 \sin \gamma t$.

SOLUTION The steady-state solution $q_p(t)$ is a particular solution of the differential equation

$$L \frac{d^2q}{dt^2} + R \frac{dq}{dt} + \frac{1}{C} q = E_0 \sin \gamma t.$$

Using the method of undetermined coefficients, we assume a particular solution of the form $q_p(t) = A \sin \gamma t + B \cos \gamma t$. Substituting this expression into the differential equation, simplifying, and equating coefficients gives

$$A = \frac{E_0 \left(L\gamma - \frac{1}{C\gamma} \right)}{-\gamma \left(L^2\gamma^2 - \frac{2L}{C} + \frac{1}{C^2\gamma^2} + R^2 \right)}, \qquad B = \frac{E_0 R}{-\gamma \left(L^2\gamma^2 - \frac{2L}{C} + \frac{1}{C^2\gamma^2} + R^2 \right)}.$$

It is convenient to express A and B in terms of some new symbols.

$$\text{If} \quad X = L\gamma - \frac{1}{C\gamma}, \quad \text{then} \quad X^2 = L^2\gamma^2 - \frac{2L}{C} + \frac{1}{C^2\gamma^2}.$$

$$\text{If} \quad Z = \sqrt{X^2 + R^2}, \quad \text{then} \quad Z^2 = L^2\gamma^2 - \frac{2L}{C} + \frac{1}{C^2\gamma^2} + R^2.$$

Therefore $A = E_0 X/(-\gamma Z^2)$ and $B = E_0 R/(-\gamma Z^2)$, so the steady-state charge is

$$q_p(t) = -\frac{E_0 X}{\gamma Z^2} \sin \gamma t - \frac{E_0 R}{\gamma Z^2} \cos \gamma t.$$

Now the steady-state current is given by $i_p(t) = q_p'(t)$:

$$i_p(t) = \frac{E_0}{Z} \left(\frac{R}{Z} \sin \gamma t - \frac{X}{Z} \cos \gamma t \right). \tag{35} \ \square$$

The quantities $X = L\gamma - 1/C\gamma$ and $Z = \sqrt{X^2 + R^2}$ defined in Example 10 are called, respectively, the **reactance** and **impedance** of the circuit. Both the reactance and the impedance are measured in ohms.

EXERCISES 3.8

Answers to odd-numbered problems begin on page A-30.

3.8.1 Spring/Mass Systems: Free Undamped Motion

1. A 4-pound weight is attached to a spring whose spring constant is 16 lb/ft. What is the period of simple harmonic motion?

2. A 20-kilogram mass is attached to a spring. If the frequency of simple harmonic motion is $2/\pi$ vibrations/second, what is the spring constant k? What is the frequency of simple harmonic motion if the original mass is replaced with an 80-kilogram mass?

3. A 24-pound weight, attached to the end of a spring, stretches it 4 inches. Find the equation of motion if the weight is released from rest from a point 3 inches above the equilibrium position.

4. Determine the equation of motion if the weight in Problem 3 is released from the equilibrium position with an initial downward velocity of 2 ft/s.

5. A 20-pound weight stretches a spring 6 inches. The weight is released from rest 6 inches below the equilibrium position.

(a) Find the position of the weight at $t = \pi/12$, $\pi/8$, $\pi/6$, $\pi/4$, $9\pi/32$ s.

(b) What is the velocity of the weight when $t = 3\pi/16$ s? In which direction is the weight heading at this instant?

(c) At what times does the weight pass through the equilibrium position?

6. A force of 400 newtons stretches a spring 2 meters. A mass of 50 kilograms is attached to the end of the spring and released from the equilibrium position with an upward velocity of 10 m/s. Find the equation of motion.

7. Another spring whose constant is 20 N/m is suspended from the same rigid support but parallel to the spring/mass system in Problem 6. A mass of 20 kilograms is attached to the second spring, and both masses are released from the equilibrium position with an upward velocity of 10 m/s.

(a) Which mass exhibits the greater amplitude of motion?

(b) Which mass is moving faster at $t = \pi/4$ s? at $\pi/2$ s?

(c) At what times are the two masses in the same position? Where are the masses at these times? In which directions are they moving?

8. A 32-pound weight stretches a spring 2 feet. Determine the amplitude and period of motion if the weight is released 1 foot above the equilibrium position with an initial upward velocity of 2 ft/s. How many complete vibrations will the weight have completed at the end of 4π seconds?

9. An 8-pound weight attached to a spring exhibits simple harmonic motion. Determine the equation of motion if the spring constant is 1 lb/ft and if the weight is released 6 inches below the equilibrium position with a downward velocity of ³⁄₂ ft/s. Express the solution in form (6).

10. A mass weighing 10 pounds stretches a spring ¼ foot. This mass is removed and replaced with a mass of 1.6 slugs, which is released ⅓ foot above the equilibrium position with a downward velocity of ¾ ft/s. Express the solution in form (6). At what times does the mass attain a displacement below the equilibrium position numerically equal to ½ the amplitude?

11. A 64-pound weight attached to the end of a spring stretches it 0.32 foot. From a position 8 inches above the equilibrium position the weight is given a downward velocity of 5 ft/s.

(a) Find the equation of motion.

(b) What are the amplitude and period of motion?

(c) How many complete vibrations will the weight have completed at the end of 3π seconds?

(d) At what time does the weight pass through the equilibrium position heading downward for the second time?

(e) At what time does the weight attain its extreme displacement on either side of the equilibrium position?

(f) What is the position of the weight at $t = 3$ s?

(g) What is the instantaneous velocity at $t = 3$ s?

(h) What is the acceleration at $t = 3$ s?

(i) What is the instantaneous velocity at the times when the weight passes through the equilibrium position?

(j) At what times is the weight 5 inches below the equilibrium position?

(k) At what times is the weight 5 inches below the equilibrium position heading in the upward direction?

12. A mass of 1 slug is suspended from a spring whose characteristic spring constant is 9 lb/ft. Initially the mass starts from a point 1 foot above the equilibrium position with an upward velocity of $\sqrt{3}$ ft/s. Find the times for which the mass is heading downward at a velocity of 3 ft/s.

13. Under some circumstances when two parallel springs, with constants k_1 and k_2, support a single weight W, the **effective spring constant** of the system is given by $k = 4k_1k_2/(k_1 + k_2)$. A 20-pound weight stretches one spring 6 inches and another spring 2 inches. The springs are attached to a common rigid support and then to a metal plate. As shown in Figure 3.33, the 20-pound weight is attached to the center of the plate in the double-spring arrangement. Determine the effective spring constant of this system. Find the equation of motion if the weight is released from the equilibrium position with a downward velocity of 2 ft/s.

Figure 3.33

14. A certain weight stretches one spring ⅓ foot and another spring ½ foot. The two springs are attached to a common rigid support in the manner indicated in Problem 13 and Figure 3.33. The first weight is set aside, an 8-pound weight is attached to the double-spring arrangement, and the system is set in motion. If the period of motion is $\pi/15$ second, determine the numerical value of the first weight.

15. By inspection of the differential equation only, describe the behavior of a spring/mass system described by $4x'' + e^{-0.1t}x = 0$ over a long period of time.

16. By inspection of the differential equation only, describe the behavior of a spring/mass system described by $4x'' + tx = 0$ over a long period of time.

3.8.2 Spring/Mass Systems: Free Damped Motion

In Problems 17–20, the given figure represents the graph of an equation of motion for a mass on a spring. The spring/mass system is damped. Use the graph to determine

(a) whether the initial displacement of the mass is above or below the equilibrium position and

(b) whether the mass is initially released from rest, heading downward, or heading upward.

17.

Figure 3.34

18.

Figure 3.35

19.

Figure 3.36

20.
Figure 3.37

21. A 4-pound weight is attached to a spring whose constant is 2 lb/ft. The medium offers a resistance to the motion of the weight numerically equal to the instantaneous velocity. If the weight is released from a point 1 foot above the equilibrium position with a downward velocity of 8 ft/s, determine the time at which the weight passes through the equilibrium position. Find the time at which the weight attains its extreme displacement from the equilibrium position. What is the position of the weight at this instant?

22. A 4-foot spring measures 8 feet long after an 8-pound weight is attached to it. The medium through which the weight moves offers a resistance numerically equal to $\sqrt{2}$ times the instantaneous velocity. Find the equation of motion if the weight is released from the equilibrium position with a downward velocity of 5 ft/s. Find the time at which the weight attains its extreme displacement from the equilibrium position. What is the position of the weight at this instant?

23. A 1-kilogram mass is attached to a spring whose constant is 16 N/m, and the entire system is then submerged in a liquid that imparts a damping force numerically equal to 10 times the instantaneous velocity. Determine the equations of motion if

(a) the weight is released from rest 1 meter below the equilibrium position and

(b) he weight is released 1 meter below the equilibrium position with an upward velocity of 12 m/s.

24. In parts **(a)** and **(b)** of Problem 23 determine whether the weight passes through the equilibrium position. In each case find the time at which the weight attains its extreme displacement from the equilibrium position. What is the position of the weight at this instant?

25. A force of 2 pounds stretches a spring 1 foot. A 3.2-pound weight is attached to the spring, and the system is then immersed in a medium that imparts a damping force numerically equal to 0.4 times the instantaneous velocity.

(a) Find the equation of motion if the weight is released from rest 1 foot above the equilibrium position.

(b) Express the equation of motion in the form given in (23).

(c) Find the first time at which the weight passes through the equilibrium position heading upward.

26. After a 10-pound weight is attached to a 5-foot spring, the spring measures 7 feet long. The 10-pound weight is removed and replaced with an 8-pound weight, and the entire system is placed in a medium offering a resistance numerically equal to the instantaneous velocity.

(a) Find the equation of motion if the weight is released ½ foot below the equilibrium position with a downward velocity of 1 ft/s.

(b) Express the equation of motion in the form given in (23).

(c) Find the times at which the weight passes through the equilibrium position heading downward.

(d) Graph the equation of motion.

27. A 10-pound weight attached to a spring stretches it 2 feet. The weight is attached to a dashpot damping device that offers a resistance numerically equal to β ($\beta > 0$) times the instantaneous velocity. Determine the values of the damping constant β so that the subsequent motion is **(a)** overdamped, **(b)** critically damped, and **(c)** underdamped.

28. A 24-pound weight stretches a spring 4 feet. The subsequent motion takes place in a medium offering a resistance numerically equal to β ($\beta > 0$) times the instantaneous velocity. If the weight starts from the equilibrium position

with an upward velocity of 2 ft/s, show that if $\beta > 3\sqrt{2}$ the equation of motion is

$$x(t) = \frac{-3}{\sqrt{\beta^2 - 18}} e^{-2\beta t/3} \sinh\frac{2}{3} \sqrt{\beta^2 - 18}\, t.$$

3.8.3 Spring/Mass Systems: Driven Motion

29. A 16-pound weight stretches a spring ⅔ feet. Initially the weight starts from rest 2 feet below the equilibrium position, and the subsequent motion takes place in a medium that offers a damping force numerically equal to ½ the instantaneous velocity. Find the equation of motion if the weight is driven by an external force equal to $f(t) = 10 \cos 3t$.

30. A mass of 1 slug is attached to a spring whose constant is 5 lb/ft. Initially the mass is released 1 foot below the equilibrium position with a downward velocity of 5 ft/s, and the subsequent motion takes place in a medium that offers a damping force numerically equal to 2 times the instantaneous velocity.

 (a) Find the equation of motion if the mass is driven by an external force equal to $f(t) = 12 \cos 2t + 3 \sin 2t$.

 (b) Graph the transient and steady-state solutions on the same coordinate axes.

 (c) Graph the equation of motion.

31. A mass of 1 slug, when attached to a spring, stretches it 2 feet and then comes to rest in the equilibrium position. Starting at $t = 0$, an external force equal to $f(t) = 8 \sin 4t$ is applied to the system. Find the equation of motion if the surrounding medium offers a damping force numerically equal to 8 times the instantaneous velocity.

32. In Problem 31 determine the equation of motion if the external force is $f(t) = e^{-t} \sin 4t$. Analyze the displacements for $t \to \infty$.

33. When a mass of 2 kilograms is attached to a spring whose constant is 32 N/m, it comes to rest in the equilibrium position. Starting at $t = 0$, a force equal to $f(t) = 68e^{-2t} \cos 4t$ is applied to the system. Find the equation of motion in the absence of damping.

34. In Problem 33 write the equation of motion in the form $x(t) = A \sin(\omega t + \phi) + Be^{-2t} \sin(4t + \theta)$. What is the amplitude of vibrations after a very long time?

35. A mass m is attached to the end of a spring whose constant is k. After the mass reaches equilibrium, its support begins to oscillate vertically about a horizontal line L according to a formula $h(t)$. The value of h represents the distance in feet measured from L. See Figure 3.38.

 (a) Determine the differential equation of motion if the entire system moves through a medium offering a damping force numerically equal to $\beta(dx/dt)$.

 (b) Solve the differential equation in part **(a)** if the spring is stretched 4 feet by a weight of 16 pounds and $\beta = 2$, $h(t) = 5 \cos t$, $x(0) = x'(0) = 0$.

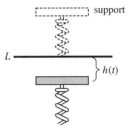

Figure 3.38

36. A mass of 100 grams is attached to a spring whose constant is 1600 dynes/cm. After the mass reaches equilibrium, its support oscillates according to the formula $h(t) = \sin 8t$, where h represents displacement from its original position. See Problem 35 and Figure 3.38.

 (a) In the absence of damping, determine the equation of motion if the mass starts from rest from the equilibrium position.

 (b) At what times does the mass pass through the equilibrium position?

 (c) At what times does the mass attain its extreme displacements?

 (d) What are the maximum and minimum displacements?

 (e) Graph the equation of motion.

In Problems 37 and 38, solve the given initial-value problem.

37. $\dfrac{d^2x}{dt^2} + 4x = -5 \sin 2t + 3 \cos 2t,$ $x(0) = -1,$ $x'(0) = 1$

38. $\dfrac{d^2x}{dt^2} + 9x = 5 \sin 3t,$ $x(0) = 2,$ $x'(0) = 0$

39. (a) Show that the solution of the initial-value problem

$$\frac{d^2x}{dt^2} + \omega^2 x = F_0 \cos \gamma t, \quad x(0) = 0, \quad x'(0) = 0$$

 is $x(t) = \dfrac{F_0}{\omega^2 - \gamma^2} (\cos \gamma t - \cos \omega t).$

 (b) Evaluate $\displaystyle\lim_{\gamma \to \omega} \frac{F_0}{\omega^2 - \gamma^2} (\cos \gamma t - \cos \omega t).$

40. Compare the result obtained in part **(b)** of Problem 39 with the solution obtained using variation of parameters when the external force is $F_0 \cos \omega t$.

41. Computer Lab Assignment (a) Show that $x(t)$ given in part **(a)** of Problem 39 can be written in the form

$$x(t) = \frac{-2F_0}{\omega^2 - \gamma^2} \sin\frac{1}{2} (\gamma - \omega)t \sin\frac{1}{2} (\gamma + \omega)t.$$

(b) If we define $\varepsilon = \frac{1}{2}(\gamma - \omega)$, show that when ε is small an *approximate* solution is

$$x(t) = \frac{F_0}{2\varepsilon\gamma} \sin \varepsilon t \sin \gamma t.$$

When ε is small the frequency $\gamma/2\pi$ of the impressed force is close to the frequency $\omega/2\pi$ of free vibrations. When this occurs, the motion is as indicated in Figure 3.39. Oscill-ations of this kind are called *beats* and are due to the fact that the frequency of $\sin \varepsilon t$ is quite small in comparison to the frequency of $\sin \gamma t$. The dashed curves, or *envelope* of the graph of $x(t)$, are obtained from the graphs of $\pm(F_0/2\varepsilon\gamma) \sin \varepsilon t$. Use a graphing utility with various values of F_0, ε, and γ to verify the graph in Figure 3.39.

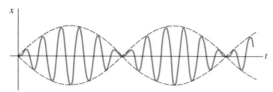

Figure 3.39

42. Computer Lab Assignment Can there be *beats* when a damping force is added to the model in part **(a)** of Problem 39? Defend your position with graphs obtained either from the explicit solution of the problem

$$\frac{d^2x}{dt^2} + 2\lambda\frac{dx}{dt} + \omega^2 x = F_0 \cos \gamma t, \qquad x(0) = 0, \quad x'(0) = 0$$

or from solution curves obtained using a numerical solver.

43. Computer Lab Assignment (a) Show that the general solution of

$$\frac{d^2x}{dt^2} + 2\lambda\frac{dx}{dt} + \omega^2 x = F_0 \sin \gamma t$$

is $x(t) = Ae^{-\lambda t}\sin(\sqrt{\omega^2 - \lambda^2}\, t + \phi)$

$$+ \frac{F_0}{\sqrt{(\omega^2 - \gamma^2)^2 + 4\lambda^2\gamma^2}}\sin(\gamma t + \theta),$$

where $A = \sqrt{c_1^2 + c_2^2}$ and the phase angles ϕ and θ are, respectively, defined by $\sin \phi = c_1/A$, $\cos \phi = c_2/A$ and

$$\sin \theta = \frac{-2\lambda\gamma}{\sqrt{(\omega^2 - \gamma^2)^2 + 4\lambda^2\gamma^2}}, \quad \cos \theta = \frac{\omega^2 - \gamma^2}{\sqrt{(\omega^2 - \gamma^2)^2 + 4\lambda^2\gamma^2}}.$$

(b) The solution in part **(a)** has the form $x(t) = x_c(t) + x_p(t)$. Inspection shows that $x_c(t)$ is transient, hence for large values of time, the solution is approximated by $x_p(t) = g(\gamma) \sin(\gamma t + \theta)$, where

$$g(\gamma) = \frac{F_0}{\sqrt{(\omega^2 - \gamma^2)^2 + 4\lambda^2\gamma^2}}.$$

Although the amplitude $g(\gamma)$ of $x_p(t)$ is bounded as $t \to \infty$, show that the maximum oscillations will occur at the value $\gamma_1 = \sqrt{\omega^2 - 2\lambda^2}$. What is the maximum value of g? The number $\sqrt{\omega^2 - 2\lambda^2}/2\pi$ is said to be the **resonance frequency** of the system.

(c) When $F_0 = 2$, $m = 1$, and $k = 4$, g becomes

$$g(\gamma) = \frac{2}{\sqrt{(4 - \gamma^2)^2 + \beta^2\gamma^2}}.$$

Construct a table of the values of γ_1 and $g(\gamma_1)$ corresponding to the damping coefficients $\beta = 2$, $\beta = 1$, $\beta = \frac{3}{4}$, $\beta = \frac{1}{2}$, and $\beta = \frac{1}{4}$. Use a graphing utility to obtain the graphs of g corresponding to these damping coefficients. Use the same coordinate axes. This family of graphs is called the **reso-nance curve** or **frequency response curve** of the system. What is γ_1 approaching as $\beta \to 0$? What is happening to the resonance curve as $\beta \to 0$?

44. Consider a driven undamped spring/mass system described by the initial-value problem

$$\frac{d^2x}{dt^2} + \omega^2 x = F_0 \sin^n \gamma t, \qquad x(0) = 0, \quad x'(0) = 0.$$

(a) For $n = 2$, discuss why there is a single frequency $\gamma_1/2\pi$ at which the system is in pure resonance.

(b) For $n = 3$, discuss why there are two frequencies $\gamma_1/2\pi$ and $\gamma_2/2\pi$ at which the system is in pure resonance.

(c) Suppose $\omega = 1$ and $F_0 = 1$. Use a numerical solver to obtain the graph of the solution of the initial-value problem for $n = 2$ and $\gamma = \gamma_1$ in part **(a)**. Obtain the graph of the solution of the initial-value problem for $n = 3$ correspond-ing, in turn, to $\gamma = \gamma_1$ and $\gamma = \gamma_2$ in part **(b)**.

3.8.4 Series Circuit Analogue

45. Find the charge on the capacitor in an *LRC* series circuit at $t = 0.01$ s when $L = 0.05$ h, $R = 2\ \Omega$, $C = 0.01$ f, $E(t) = 0$ V, $q(0) = 5$ C, and $i(0) = 0$ A. Determine the first time at which the charge on the capacitor is equal to zero.

46. Find the charge on the capacitor in an *LRC* series circuit when $L = \frac{1}{4}$ h, $R = 20\ \Omega$, $C = \frac{1}{300}$ f, $E(t) = 0$ V, $q(0) = 4$ C, and $i(0) = 0$ A. Is the charge on the capacitor ever equal to zero?

In Problems 47 and 48, find the charge on the capacitor and the current in the given *LRC* series circuit. Find the maximum charge on the capacitor.

47. $L = \dfrac{5}{3}$ h, $R = 10\ \Omega$, $C = \dfrac{1}{30}$ f, $E(t) = 300$ V,
$q(0) = 0$ C, $i(0) = 0$ A

48. $L = 1$ h, $R = 100\ \Omega$, $C = 0.0004$ f, $E(t) = 30$ V,
$q(0) = 0$ C, $i(0) = 2$ A

49. Find the steady-state charge and the steady-state current in an *LRC* series circuit when $L = 1$ h, $R = 2\ \Omega$, $C = 0.25$ f, and $E(t) = 50 \cos t$ V.

50. Show that the amplitude of the steady-state current in the *LRC* series circuit in Example 11 is given by E_0/Z, where Z is the impedance of the circuit.

51. Show that the steady-state current in an *LRC* series circuit when $L = \frac{1}{2}$ h, $R = 20\ \Omega$, $C = 0.001$ f, and $E(t) = 100 \sin 60t$ V is given by $i_p(t) = (4.160) \sin(60t - 0.588)$. [*Hint:* Use Problem 50.]

52. Find the steady-state current in an *LRC* series circuit when $L = \frac{1}{2}$ h, $R = 20\ \Omega$, $C = 0.001$ f, and $E(t) = 100 \sin 60t + 200 \cos 40t$ V.

53. Find the charge on the capacitor in an *LRC* series circuit when $L = \frac{1}{2}$ h, $R = 10\ \Omega$, $C = 0.01$ f, $E(t) = 150$ V, $q(0) = 1$ C, and $i(0) = 0$ A. What is the charge on the capacitor after a long time?

54. Show that if *L*, *R*, *C*, and E_0 are constant, then the amplitude of the steady-state current in Example 11 is a maximum when $\gamma = 1/\sqrt{LC}$. What is the maximum amplitude?

55. Show that if *L*, *R*, E_0, and γ are constant, then the amplitude of the steady-state current in Example 11 is a maximum when the capacitance is $C = 1/L\gamma^2$.

56. Find the charge on the capacitor and the current in an *LC* circuit when $L = 0.1$ h, $C = 0.1$ f, $E(t) = 100 \sin \gamma t$ V, $q(0) = 0$ C, and $i(0) = 0$ A.

57. Find the charge on the capacitor and the current in an *LC* circuit when $E(t) = E_0 \cos \gamma t$ V, $q(0) = q_0$ C, and $i(0) = i_0$ A.

58. In Problem 57 find the current when the circuit is in resonance.

3.9 Linear Models: Boundary-Value Problems

The preceding section was devoted to systems in which a second-order mathematical model was accompanied by prescribed initial conditions—that is, side conditions that are specified on the unknown function and its first derivative at a single point. But often the mathematical description of a physical system demands that we solve a differential equation subject to boundary conditions—that is, conditions specified on the unknown function, or on one of its derivatives, or even on a linear combination of the unknown function and one of its derivatives, at two (or more) different points.

■ **Deflection of a Beam** Many structures are constructed using girders, or beams, and these beams deflect or distort under their own weight or under the influence of some external force. As we shall now see, this deflection $y(x)$ is governed by a relatively simple linear fourth-order differential equation.

To begin, let us assume that a beam of length L is homogeneous and has uniform cross-sections along its length. In the absence of any load on the beam (including its weight), a curve joining the centroids of all its cross-sections is a straight line called the **axis of symmetry.** See Figure 3.40(a). If a load is applied to the beam in a vertical plane containing the axis of symmetry, the beam, as shown in Figure 3.40(b), undergoes a distortion, and the curve connecting the centroids of all cross-sections is called the **deflection curve** or **elastic curve.** The deflection curve approximates the shape of the beam. Now suppose that the *x*-axis coincides with the axis of symmetry and that the deflection $y(x)$, measured from this axis, is positive if downward. In the theory of elasticity it is shown that the bending moment $M(x)$ at a point x along the beam is related to the load per unit length $w(x)$ by the equation

$$\frac{d^2M}{dx^2} = w(x). \tag{1}$$

In addition, the bending moment $M(x)$ is proportional to the curvature κ of the elastic curve

$$M(x) = EI\kappa, \tag{2}$$

where E and I are constants; E is Young's modulus of elasticity of the material of the beam, and I is the moment of inertia of a cross-section of the beam (about an axis

axis of symmetry
(a)

deflection curve
(b)

Figure 3.40

known as the neutral axis). The product EI is called the **flexural rigidity** of the beam.

Now, from calculus, curvature is given by $\kappa = y''/[1 + (y')^2]^{3/2}$. When the deflection $y(x)$ is small, the slope $y' \approx 0$ and so $[1 + (y')^2]^{3/2} \approx 1$. If we let $\kappa = y''$, equation (2) becomes $M = EIy''$. The second derivative of this last expression is

$$\frac{d^2M}{dx^2} = EI\frac{d^2}{dx^2}y'' = EI\frac{d^4y}{dx^4}. \qquad (3)$$

Using the given result in (1) to replace d^2M/dx^2 in (3), we see that the deflection $y(x)$ satisfies the fourth-order differential equation

$$EI\frac{d^4y}{dx^4} = w(x). \qquad (4)$$

Boundary conditions associated with equation (4) depend on how the ends of the beam are supported. A cantilever beam is **embedded** or **clamped** at one end and **free** at the other. A diving board, an outstretched arm, an airplane wing, and a balcony are common examples of such beams, but even trees, flagpoles, skyscrapers, and the George Washington monument can act as cantilever beams, because they are embedded at one end and are subject to the bending force of the wind. For a cantilever beam, the deflection $y(x)$ must satisfy the following two conditions at the embedded end $x = 0$:

- $y(0) = 0$ since there is no deflection, and
- $y'(0) = 0$ since the deflection curve is tangent to the x-axis (in other words, the slope of the deflection curve is zero at this point).

At $x = L$ the free-end conditions are

- $y''(L) = 0$ since the bending moment is zero, and
- $y'''(L) = 0$ since the shear force is zero.

The function $F(x) = dM/dx = EI\,d^3y/dx^3$ is called the shear force. If an end of a beam is **simply supported** (also called **pin supported, fulcrum supported,** and **hinged**), then we must have $y = 0$ and $y'' = 0$ at that end. The following table summarizes the boundary conditions that are associated with (4). See Figure 3.41.

$x = 0$ $x = L$

(a) Embedded at both ends.

$x = 0$ $x = L$

(b) Cantilever beam: embedded at the left end, free at the right end.

$x = 0$ $x = L$

(c) Simply supported at both ends.

Figure 3.41

Ends of the Beam	Boundary Conditions
embedded	$y = 0, \quad y' = 0$
free	$y'' = 0, \quad y''' = 0$
simply supported	$y = 0, \quad y'' = 0$

Example 1 Embedded Beam

A beam of length L is embedded at both ends. Find the deflection of the beam if a constant load w_0 is uniformly distributed along its length—that is, $w(x) = w_0$, $0 < x < L$.

SOLUTION From the preceding discussion we see that the deflection $y(x)$ satisfies

$$EI\frac{d^4y}{dx^4} = w_0.$$

Because the beam is embedded at both its left end ($x = 0$) and right end ($x = L$), there is no vertical deflection and the line of deflection is horizontal at these points. Thus the boundary conditions are

$$y(0) = 0, \quad y'(0) = 0, \quad y(L) = 0, \quad y'(L) = 0.$$

We can solve the nonhomogeneous differential equation in the usual manner (find y_c by observing that $m = 0$ is a root of multiplicity four of the auxiliary equation $m^4 = 0$, and then find a particular solution y_p by undetermined coefficients), or we can simply integrate the equation $d^4y/dx^4 = w_0/EI$ four times in succession. Either way, we find the general solution of the equation to be

$$y(x) = c_1 + c_2 x + c_3 x^2 + c_4 x^3 + \frac{w_0}{24EI} x^4.$$

Now the conditions $y(0) = 0$ and $y'(0) = 0$ give, in turn, $c_1 = 0$ and $c_2 = 0$, whereas the remaining conditions $y(L) = 0$ and $y'(L) = 0$ applied to $y(x) = c_3 x^2 + c_4 x^3 + \frac{w_0}{24EI} x^4$ yield the equations

$$c_3 L^2 + c_4 L^3 + \frac{w_0}{24EI} L^4 = 0$$

$$2c_3 L + 3c_4 L^2 + \frac{w_0}{6EI} L^3 = 0.$$

Solving this system gives $c_3 = w_0 L^2/24EI$ and $c_4 = -w_0 L/12EI$. Thus the deflection is

$$y(x) = \frac{w_0 L^2}{24EI} x^2 - \frac{w_0 L}{12EI} x^3 + \frac{w_0}{24EI} x^4 = \frac{w_0}{24EI} x^2 (x - L)^2.$$

By choosing $w_0 = 24EI$ and $L = 1$, we obtain the graph of the deflection curve in Figure 3.42. ❑

Figure 3.42

■ **Eigenvalues and Eigenfunctions** Many applied problems demand that we solve a two-point boundary-value problem involving a linear differential equation that contains a parameter λ. We seek the values of λ for which the boundary-value problem has nontrivial solutions.

Example 2 **Nontrivial Solutions of a BVP**

Solve the boundary-value problem

$$y'' + \lambda y = 0, \quad y(0) = 0, \quad y(L) = 0.$$

SOLUTION We consider three cases: $\lambda = 0$, $\lambda < 0$, and $\lambda > 0$.

Case I: For $\lambda = 0$ the solution of $y'' = 0$ is $y = c_1 x + c_2$. The conditions $y(0) = 0$ and $y(L) = 0$ imply, in turn, $c_2 = 0$ and $c_1 = 0$. Hence for $\lambda = 0$ the only solution of the boundary-value problem is the trivial solution $y = 0$.

Case II: For $\lambda < 0$ we have $y = c_1 \cosh \sqrt{-\lambda}\, x + c_2 \sinh \sqrt{-\lambda}\, x$.* Again, $y(0) = 0$ gives $c_1 = 0$, and so $y = c_2 \sinh \sqrt{-\lambda}\, x$. The second condition $y(L) = 0$ dictates that $c_2 \sinh \sqrt{-\lambda}\, L = 0$. Since $\sinh \sqrt{-\lambda}\, L \neq 0$, we must have $c_2 = 0$. Thus $y = 0$.

Case III: For $\lambda > 0$ the general solution of $y'' + \lambda y = 0$ is given by $y = c_1 \cos \sqrt{\lambda}\, x + c_2 \sin \sqrt{\lambda}\, x$. As before, $y(0) = 0$ yields $c_1 = 0$, but $y(L) = 0$ implies

$$c_2 \sin \sqrt{\lambda}\, L = 0.$$

* $\sqrt{-\lambda}$ looks a little strange, but bear in mind that $\lambda < 0$ is equivalent to $-\lambda > 0$.

If $c_2 = 0$, then necessarily $y = 0$. However, if $c_2 \neq 0$, then $\sin \sqrt{\lambda}\, L = 0$. The last condition implies that the argument of the sine function must be an integer multiple of π:

$$\sqrt{\lambda}\, L = n\pi \qquad \text{or} \qquad \lambda = \frac{n^2 \pi^2}{L^2}, \quad n = 1, 2, 3, \dots.$$

Therefore for any real nonzero c_2, $y = c_2 \sin(n\pi x/L)$ is a solution of the problem for each n. Since the differential equation is homogeneous, we may, if desired, not write c_2. In other words, for a given number in the sequence

$$\frac{\pi^2}{L^2}, \quad \frac{4\pi^2}{L^2}, \quad \frac{9\pi^2}{L^2}, \dots,$$

the *corresponding* function in the sequence

$$\sin \frac{\pi}{L}\, x, \quad \sin \frac{2\pi}{L}\, x, \quad \sin \frac{3\pi}{L}\, x, \dots$$

is a nontrivial solution of the original problem. ❑

The numbers $\lambda_n = n^2 \pi^2/L^2$, $n = 1, 2, 3, \dots$ for which the boundary-value problem in Example 2 has a nontrivial solution are known as **characteristic values** or, more commonly, **eigenvalues.** The solutions depending on these values of λ_n, $y_n = c_2 \sin(n\pi x/L)$ or simply $y_n = \sin(n\pi x/L)$, are called **characteristic functions** or **eigenfunctions.**

■ **Buckling of a Thin Vertical Column** In the eighteenth century Leonhard Euler was one of the first mathematicians to study an eigenvalue problem in analyzing how a thin elastic column buckles under a compressive axial force.

Consider a long slender vertical column of uniform cross-section and length L. Let $y(x)$ denote the deflection of the column when a constant vertical compressive force, or load, P is applied to its top, as shown in Figure 3.43. By comparing bending moments at any point along the column we obtain

$$EI \frac{d^2 y}{dx^2} = -Py \quad \text{or} \quad EI \frac{d^2 y}{dx^2} + Py = 0, \tag{5}$$

where E is Young's modulus of elasticity and I is the moment of inertia of a cross-section about a vertical line through its centroid.

Example 3 The Euler Load

Find the deflection of a thin vertical homogeneous column of length L subjected to a constant axial load P if the column is hinged at both ends.

SOLUTION The boundary-value problem to be solved is

$$EI \frac{d^2 y}{dx^2} + Py = 0, \qquad y(0) = 0, \quad y(L) = 0.$$

First note that $y = 0$ is a perfectly good solution of this problem. This solution has a simple intuitive interpretation: If the load P is not great enough, there is no deflection. The question then is this: For what values of P will the column bend? In mathematical terms: For what values of P does the given boundary-value problem possess nontrivial solutions?

By writing $\lambda = P/EI$ we see that

$$y'' + \lambda y = 0, \qquad y(0) = 0, \quad y(L) = 0$$

Figure 3.43

(a) (b)

Figure 3.44

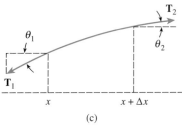

Figure 3.45

is identical to the problem in Example 2. From Case III of that discussion we see that the deflection curves are $y_n(x) = c_2 \sin(n\pi x/L)$, corresponding to the eigenvalues $\lambda_n = P_n/EI = n^2\pi^2/L^2$, $n = 1, 2, 3, \ldots$. Physically this means that the column will buckle or deflect only when the compressive force is one of the values $P_n = n^2\pi^2 EI/L^2$, $n = 1, 2, 3, \ldots$. These different forces are called **critical loads.** The deflection curve corresponding to the smallest critical load $P_1 = \pi^2 EI/L^2$, called the **Euler load,** is $y_1(x) = c_2 \sin(\pi x/L)$ and is known as the **first buckling mode.** ❏

The deflection curves in Example 3 corresponding to $n = 1$, $n = 2$, and $n = 3$ are shown in Figure 3.44. Note that if the original column has some sort of physical restraint put on it at $x = L/2$, then the smallest critical load will be $P_2 = 4\pi^2 EI/L^2$ and the deflection curve will be as shown in Figure 3.44(b). If restraints are put on the column at $x = L/3$ and at $x = 2L/3$, then the column will not buckle until the critical load $P_3 = 9\pi^2 EI/L^2$ is applied and the deflection curve will be as shown in Figure 3.44(c). See Problem 25 in Exercises 3.9.

■ **Rotating String** The simple linear second-order differential equation

$$y'' + \lambda y = 0 \tag{6}$$

occurs again and again as a mathematical model. In Section 3.8 we saw (6) in the forms $d^2x/dt^2 + (k/m)x = 0$ and $d^2q/dt^2 + (1/LC)q = 0$ as models for, respectively, the simple harmonic motion of a spring/mass system and the simple harmonic response of a series circuit. It is apparent when the model for the deflection of a thin column in (5) is written as $d^2y/dx^2 + (P/EI)y = 0$ that it is the same as (6). We encounter the basic equation (6) one more time in this section: as a model that defines the deflection curve or the shape $y(x)$ assumed by a rotating string. The physical situation is analogous to when two persons hold a jump rope and twirl it in a synchronous manner. See Figure 3.45 parts (a) and (b).

Suppose a string of length L with constant linear density ρ (mass per unit length) is stretched along the x-axis and fixed at $x = 0$ and $x = L$. Suppose the string is then rotated about that axis at a constant angular speed ω. Consider a portion of the string on the interval $[x, x + \Delta x]$, where Δx is small. If the magnitude T of the tension \mathbf{T}, acting tangential to the string, is constant along the string, then the desired differential equation can be obtained by equating two different formulations of the net force acting on the string on the interval $[x, x + \Delta x]$. First, we see from Figure 3.45(c) that the net vertical force is

$$F = T \sin \theta_2 - T \sin \theta_1. \tag{7}$$

When angles θ_1 and θ_2 (measured in radians) are small, we have $\sin \theta_2 \approx \tan \theta_2$ and $\sin \theta_1 \approx \tan \theta_1$. Moreover, since $\tan \theta_2$ and $\tan \theta_1$ are, in turn, slopes of the lines containing the vectors \mathbf{T}_2 and \mathbf{T}_1, we can also write

$$\tan \theta_2 = y'(x + \Delta x) \quad \text{and} \quad \tan \theta_1 = y'(x).$$

Thus (7) becomes

$$F \approx T[y'(x + \Delta x) - y'(x)]. \tag{8}$$

Second, we can obtain a different form of this same net force using Newton's second law, $F = ma$. Here the mass of string on the interval is $m = \rho\Delta x$; the centripetal acceleration of a body rotating with angular speed ω in a circle of radius r is $a = r\omega^2$. With Δx small we take $r = y$. Thus the net vertical force is also approximated by

$$F \approx -(\rho\Delta x)y\omega^2, \tag{9}$$

where the minus sign comes from the fact that the acceleration points in the direction opposite to the positive y-direction. Now by equating (8) and (9) we have

$$T[y'(x + \Delta x) - y'(x)] \approx -(\rho\Delta x)y\omega^2 \quad \text{or} \quad T\frac{y'(x + \Delta x) - y'(x)}{\Delta x} \approx -\rho\omega^2 y. \quad (10)$$

For Δx close to zero the difference quotient $[y'(x + \Delta x) - y'(x)]/\Delta x$ in (10) is approximated by the second derivative d^2y/dx^2. Finally we arrive at the model

$$T\frac{d^2y}{dx^2} = -\rho\omega^2 y \quad \text{or} \quad T\frac{d^2y}{dx^2} + \rho\omega^2 y = 0. \quad (11)$$

Since the string is anchored at its ends $x = 0$ and $y = L$, we expect that the solution $y(x)$ of the last equation in (11) should also satisfy the boundary conditions $y(0) = 0$ and $y(L) = 0$.

EXERCISES 3.9

Answers to odd-numbered problems begin on page A-31.

Deflection of a Beam

In Problems 1–4, solve equation (4) subject to the appropriate boundary conditions. The beam is of length L and w_0 is a constant.

1. (a) The beam is embedded at its left end and free at its right end and $w(x) = w_0$, $0 < x < L$.

(b) Use a graphing utility to obtain the graph of the deflection curve of the beam when $w_0 = 24EI$ and $L = 1$.

2. (a) The beam is simply supported at both ends and $w(x) = w_0$, $0 < x < L$.

(b) Use a graphing utility to obtain the graph of the deflection curve of the beam when $w_0 = 24EI$ and $L = 1$.

3. (a) The beam is embedded at its left end and simply supported at its right end and $w(x) = w_0$, $0 < x < L$.

(b) Use a graphing utility to obtain the graph of the deflection curve of the beam when $w_0 = 48EI$ and $L = 1$.

4. (a) The beam is embedded at its left end and simply supported at its right end and $w(x) = w_0 \sin(\pi x/L)$, $0 < x < L$.

(b) Use a graphing utility to obtain the graph of the deflection curve of the beam when $w_0 = 2\pi^3 EI$ and $L = 1$.

5. (a) Find the maximum deflection of the cantilever beam in Problem 1.

(b) How does the maximum deflection of a beam that is half as long compare with the value in part **(a)**?

6. (a) Find the maximum deflection of the simply supported beam in Problem 2.

(b) How does the maximum deflection of the simply supported beam compare with the value of maximum deflection of the embedded beam in Example 1?

7. A cantilever beam of length L is embedded at its right end and a horizontal tensile force of P pounds is applied to its free left end. When the origin O is taken at its free end as shown in Figure 3.46 the deflection $y(x)$ of the beam can be shown to satisfy the differential equation

$$EIy'' = Py - w(x)\frac{x}{2}.$$

Find the deflection of the cantilever beam if $w(x) = w_0 x$, $0 < x < L$, and $y(0) = 0$, $y'(L) = 0$.

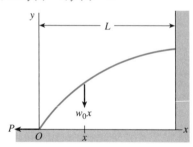

Figure 3.46

8. When a compressive instead of a tensile force is applied at the free end of the beam in Problem 7 the differential equation of the deflection is

$$EIy'' = -Py - w(x)\frac{x}{2}.$$

Solve this equation if $w(x) = w_0 x$, $0 < x < L$, and $y(0) = 0$, $y'(L) = 0$.

Eigenvalues and Eigenfunctions

In Problems 9–22, find the eigenvalues and eigenfunctions for the given boundary-value problem.

9. $y'' + \lambda y = 0$, $y(0) = 0$, $y(\pi) = 0$

10. $y'' + \lambda y = 0$, $y(0) = 0$, $y(\pi/4) = 0$

11. $y'' + \lambda y = 0$, $y'(0) = 0$, $y(L) = 0$

12. $y'' + \lambda y = 0$, $y(0) = 0$, $y'(\pi/2) = 0$

13. $y'' + \lambda y = 0$, $y'(0) = 0$, $y'(\pi) = 0$

14. $y'' + \lambda y = 0$, $y(-\pi) = 0$, $y(\pi) = 0$

15. $y'' + 2y' + (\lambda + 1)y = 0$, $y(0) = 0$, $y(5) = 0$

16. $y'' + (\lambda + 1)y = 0$, $y'(0) = 0$, $y'(1) = 0$

17. $y'' + \lambda^2 y = 0$, $y(0) = 0$, $y(L) = 0$

18. $y'' + \lambda^2 y = 0$, $y(0) = 0$, $y'(3\pi) = 0$

In Problems 19–22, the DE is a Cauchy–Euler equation.

19. $x^2 y'' + xy' + \lambda y = 0$, $y(1) = 0$, $y(e^\pi) = 0$

20. $x^2 y'' + xy' + \lambda y = 0$, $y'(e^{-1}) = 0$, $y(1) = 0$

21. $x^2 y'' + xy' + \lambda y = 0$, $y'(1) = 0$, $y'(e^2) = 0$

22. $x^2 y'' + 2xy' + \lambda y = 0$, $y(1) = 0$, $y(e^2) = 0$

23. Show that the eigenvalues and eigenfunctions of the boundary-value problem

$$y'' + \lambda y = 0, \qquad y(0) = 0, \qquad y(1) + y'(1) = 0,$$

are, respectively, $\lambda_n = x_n^2$, where x_n, $n = 1, 2, 3, \ldots$ are the consecutive *positive* roots of the equation $\tan\sqrt{\lambda} = -\sqrt{\lambda}$, and $y_n = \sin\sqrt{\lambda_n}\, x$.

24. Computer Lab Assignment (a) Use a CAS to plot graphs to convince yourself that the equation $\tan x = -x$ has an infinite number of roots. Explain why the negative roots of the equation can be ignored. Explain why $\lambda = 0$ is not an eigenvalue in Problem 23 even though $\lambda = 0$ is an obvious root of the equation $\tan\sqrt{\lambda} = -\sqrt{\lambda}$.

(b) Use a root-finding application of a CAS to approximate the first four eigenvalues λ_1, λ_2, λ_3, and λ_4 in Problem 23.

Buckling of a Thin Column

25. Consider Figure 3.44. Where should physical restraints be placed on the column if we want the critical load to be P_4? Give a sketch of the deflection curve corresponding to this load.

26. The critical loads of thin columns depend on the end conditions of the column. The value of the Euler load P_1 in Example 3 was derived under the assumption that the column was hinged at both ends.

Suppose a thin vertical homogeneous column is embedded at its base ($x = 0$) and free at its top ($x = L$) and that an axial load P is applied to its free end. Either this load causes a small deflection δ as shown in Figure 3.47 or it does not cause such a deflection. In either case the differential equation for the deflection $y(x)$ is

$$EI\frac{d^2 y}{dx^2} + Py = P\delta.$$

(a) What is the predicted deflection when $\delta = 0$?

(b) When $\delta \neq 0$, show that the Euler load for this column is one-fourth of the Euler load for the hinged column in Example 3.

Figure 3.47

Rotating String

27. Consider the boundary-value problem introduced in the construction of the mathematical model for the shape of a rotating string:

$$T\frac{d^2 y}{dx^2} + \rho\omega^2 y = 0, \qquad y(0) = 0, \quad y(L) = 0.$$

For constant T and ρ, define the critical speeds of angular rotation ω_n as the values of ω for which the boundary-value problem has nontrivial solutions. Find the critical speeds ω_n and the corresponding deflection curves $y_n(x)$.

28. When the magnitude of tension T is not constant, then a model for the deflection curve or shape $y(x)$ assumed by a rotating string is given by

$$\frac{d}{dx}\left[T(x)\frac{dy}{dx}\right] + \rho\omega^2 y = 0.$$

Suppose $1 < x < e$ and that $T(x) = x^2$.

(a) If $y(1) = 0$, $y(e) = 0$, and $\rho\omega^2 > 0.25$, show that the critical speeds of angular rotation are $\omega_n = \frac{1}{2}\sqrt{(4n^2\pi^2 + 1)/\rho}$ and the corresponding deflection curves are

$$y_n(x) = c_2 x^{-1/2}\sin(n\pi\ln x), \quad n = 1, 2, 3, \ldots.$$

(b) Use a graphing utility to graph the deflection curves on the interval $[1, e]$ for $n = 1, 2, 3$. Choose $c_2 = 1$.

Miscellaneous Boundary-Value Problems

29. Consider two concentric spheres of radius $r = a$ and $r = b$, $a < b$, as shown in Figure 3.48. The temperature $u(r)$ in the region between the spheres is determined from the boundary-value problem

$$r\frac{d^2 u}{dr^2} + 2\frac{du}{dr} = 0, \qquad u(a) = u_0, \quad u(b) = u_1,$$

where u_0 and u_1 are constants. Solve for $u(r)$.

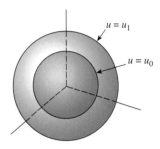

Figure 3.48

30. The temperature $u(r)$ in the circular ring shown in Figure 3.49 is determined from the boundary-value problem

$$r\frac{d^2u}{dr^2} + \frac{du}{dr} = 0, \qquad u(a) = u_0, \quad u(b) = u_1,$$

where u_0 and u_1 are constants. Show that

$$u(r) = \frac{u_0 \ln(r/b) - u_1 \ln(r/a)}{\ln(a/b)}.$$

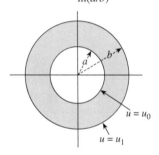

Figure 3.49

31. Consider the boundary-value problem $y'' + 16y = 0$, $y(0) = y_0$, $y(\pi/2) = y_1$. Determine whether it is possible to find values of y_0 and y_1 so that the problem possesses **(a)** precisely one nontrivial solution, **(b)** more than one solution, **(c)** no solution, **(d)** the trivial solution.

32. Consider the boundary-value problem $y'' + 16y = 0$, $y(0) = 1$, $y(L) = 1$. Determine whether it is possible to find values of $L > 0$ so that the problem possesses **(a)** precisely one nontrivial solution, **(b)** more than one solution, **(c)** no solution, **(d)** the trivial solution.

33. Consider the boundary-value problem

$$y'' + \lambda^2 y = 0, \quad y(-\pi) = y(\pi), \quad y'(-\pi) = y'(\pi).$$

(a) Give a geometric interpretation of the boundary conditions.

(b) Find the eigenvalues and eigenfunctions of the problem.

(c) Use a graphing utility to graph some of the eigenfunctions. Verify the geometric interpretation of the boundary conditions.

3.10 Nonlinear Models

■ **Nonlinear Springs** The mathematical model in (1) of Section 3.8 has the form

$$m\frac{d^2x}{dt^2} + F(x) = 0, \tag{1}$$

where $F(x) = kx$. Since x denotes the displacement of the mass from its equilibrium position, $F(x) = kx$ is Hooke's law, that is, the force exerted by the spring that tends to restore the mass to the equilibrium position. A spring acting under a linear restoring force $F(x) = kx$ is naturally referred to as a **linear spring.** But springs are seldom perfectly linear. Depending on how it is constructed and the material used, a spring can range from "mushy" or soft to "stiff" or hard, so that its restorative force may vary from something below to something above that given by the linear law. In the case of free motion, if we assume that a nonaging spring possesses some nonlinear characteristics, then it might be reasonable to assume that the restorative force $F(x)$ of a spring is proportional to, say, the cube of the displacement x of the mass beyond its equilibrium position or that $F(x)$ is a linear combination of powers of the displacement such as that given by the nonlinear function $F(x) = kx + k_1 x^3$. A spring whose mathematical model incorporates a nonlinear restorative force, such as

$$m\frac{d^2x}{dt^2} + kx^3 = 0 \qquad \text{or} \qquad m\frac{d^2x}{dt^2} + kx + k_1x^3 = 0, \qquad (2)$$

is called a **nonlinear spring.** In addition, we examined mathematical models in which damping imparted to the motion was proportional to the instantaneous velocity dx/dt, and the restoring force of a spring was given by the linear function $F(x) = kx$. But these were simply assumptions; in more realistic situations damping could be proportional to some power of the instantaneous velocity dx/dt. The nonlinear differential equation

$$m\frac{d^2x}{dt^2} + \beta\left|\frac{dx}{dt}\right|\frac{dx}{dt} + kx = 0 \qquad (3)$$

is one model of a free spring/mass system with damping proportional to the square of the velocity. One can then envision other kinds of models: linear damping and nonlinear restoring force, nonlinear damping and nonlinear restoring force, and so on. The point is, nonlinear characteristics of a physical system lead to a mathematical model that is nonlinear.

Notice in (2) that both $F(x) = kx^3$ and $F(x) = kx + k_1x^3$ are odd functions of x. To see why a polynomial function containing only odd powers of x provides a reasonable model for the restoring force, let us express F as a power series centered at the equilibrium position $x = 0$:

$$F(x) = c_0 + c_1x + c_2x^2 + c_3x^3 + \cdots.$$

When the displacements x are small, the values of x^n are negligible for n sufficiently large. If we truncate the power series with, say, the fourth term, then

$$F(x) = c_0 + c_1x + c_2x^2 + c_3x^3.$$

In order for the force at $x > 0$ ($F(x) = c_0 + c_1x + c_2x^2 + c_3x^3$) and the force at $-x < 0$ ($F(-x) = c_0 - c_1x + c_2x^2 - c_3x^3$) to have the same magnitude but act in the opposite directions, we must have $F(-x) = -F(x)$. Since this means F is an odd function, we must have $c_0 = 0$ and $c_2 = 0$, and so $F(x) = c_1x + c_3x^3$. Had we used only the first two terms in the series, the same argument yields the linear function $F(x) = c_1x$. For discussion purposes we shall write $c_1 = k$ and $c_2 = k_1$. A restoring force with mixed powers such as $F(x) = kx + k_1x^2$, and the corresponding vibrations, are said to be unsymmetrical.

■ **Hard and Soft Springs** Let us take a closer look at the equation in (1) in the case where the restoring force is given by $F(x) = kx + k_1x^3$, $k > 0$. The spring is said to be **hard** if $k_1 > 0$ and **soft** if $k_1 < 0$. Graphs of three types of restoring forces are illustrated in Figure 3.50. The next example illustrates these two special cases of the differential equation $m\,d^2x/dt^2 + kx + k_1x^3 = 0$, $m > 0$, $k > 0$.

Example 1 **Comparison of Hard and Soft Springs**

The differential equations

$$\frac{d^2x}{dt^2} + x + x^3 = 0 \qquad (4)$$

and

$$\frac{d^2x}{dt^2} + x - x^3 = 0 \qquad (5)$$

are special cases of (2) and are models of a hard spring and soft spring, respectively. Figure 3.51(a) shows two solutions of (4) and Figure 3.51(b) shows two

Figure 3.50

(a) Hard spring

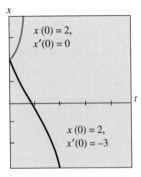

(b) Soft spring

Figure 3.51

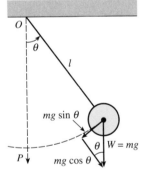

Figure 3.52

solutions of (5) obtained from a numerical solver. The curves shown in black are solutions satisfying the initial conditions $x(0) = 2$, $x'(0) = -3$; the two curves in color are solutions satisfying $x(0) = 2$, $x'(0) = 0$. These solution curves certainly suggest that the motion of a mass on the hard spring is oscillatory, whereas motion of a mass on the soft spring is not oscillatory. But we must be careful about drawing conclusions based on a couple of solution curves. A more complete picture of the nature of the solutions of both of these equations can be obtained from the qualitative analysis discussed in Chapter 11. ❑

■ **Nonlinear Pendulum** Any object that swings back and forth is called a **physical pendulum.** The **simple pendulum** is a special case of the physical pendulum and consists of a rod of length l to which a mass m is attached at one end. In describing the motion of a simple pendulum in a vertical plane, we make the simplifying assumptions that the mass of the rod is negligible and that no external damping or driving forces act on the system. The displacement angle θ of the pendulum, measured from the vertical as shown in Figure 3.52, is considered positive when measured to the right of OP and negative to the left of OP. Now recall that the arc s of a circle of radius l is related to the central angle θ by the formula $s = l\theta$. Hence angular acceleration is

$$a = \frac{d^2s}{dt^2} = l\frac{d^2\theta}{dt^2}.$$

From Newton's second law we then have

$$F = ma = ml\frac{d^2\theta}{dt^2}.$$

From Figure 3.52 we see that the magnitude of the tangential component of the force due to the weight W is $mg \sin \theta$. In direction this force is $-mg \sin \theta$, since it points to the left for $\theta > 0$ and to the right for $\theta < 0$. We equate the two different versions of the tangential force to obtain $ml\, d^2\theta/dt^2 = -mg \sin \theta$ or

$$\frac{d^2\theta}{dt^2} + \frac{g}{l} \sin \theta = 0. \tag{6}$$

■ **Linearization** Because of the presence of $\sin \theta$, the model in (6) is nonlinear. In an attempt to understand the behavior of the solutions of nonlinear higher-order differential equations, one sometimes tries to simplify the problem by replacing nonlinear terms by certain approximations. For example, the Maclaurin series for $\sin \theta$ is given by

$$\sin \theta = \theta - \frac{\theta^3}{3!} + \frac{\theta^5}{5!} - \cdots,$$

and so if we use the approximation $\sin \theta \approx \theta - \theta^3/6$ equation (6) becomes $d^2\theta/dt^2 + (g/l)\theta + (g/6l)\theta^3 = 0$. Observe that this last equation is the same as the second nonlinear equation in (2) with $m = 1$, $k = g/l$, and $k_1 = -g/6l$. However, if we assume that the displacements θ are small enough to justify using the replacement $\sin \theta \approx \theta$, then (6) becomes

$$\frac{d^2\theta}{dt^2} + \frac{g}{l}\,\theta = 0. \tag{7}$$

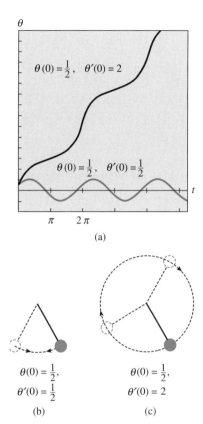

(a)

$\theta(0) = \frac{1}{2},$ $\theta(0) = \frac{1}{2},$
$\theta'(0) = \frac{1}{2}$ $\theta'(0) = 2$
(b) (c)

Figure 3.53

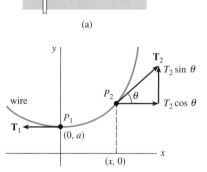

(a)

(b)

Figure 3.54

See Problem 14 in Exercises 3.10. If we set $\omega^2 = g/l$, we recognize (7) as the differential equation (2) of Section 3.8 that is a model for the free undamped vibrations of a linear spring/mass system. In other words, (7) is again the basic linear equation $y'' + \lambda y = 0$ discussed on page 165 of Section 3.9. As a consequence, we say that equation (7) is a **linearization** of equation (6). Since the general solution of (7) is $\theta(t) = c_1 \cos \omega t + c_2 \sin \omega t$, this linearization suggests that for initial conditions amenable to small oscillations the motion of the pendulum described by (6) will be periodic.

Example 2 Two Initial-Value Problems

The graphs in Figure 3.53(a) were obtained with the aid of a numerical solver and represent solution curves of equation (6) when $\omega^2 = 1$. The colored curve depicts the solution of (6) that satisfies the initial conditions $\theta(0) = \frac{1}{2}$, $\theta'(0) = \frac{1}{2}$ whereas the black curve is the solution of (6) that satisfies $\theta(0) = \frac{1}{2}$, $\theta'(0) = 2$. The colored curve represents a periodic solution—the pendulum oscillating back and forth as shown in Figure 3.53(b) with an apparent amplitude $A \leq 1$. The black curve shows that θ increases without bound as time increases—the pendulum, starting from the same initial displacement, is given an initial velocity of magnitude great enough to send it over the top; in other words, the pendulum is whirling about its pivot as shown in Figure 3.53(c). In the absence of damping the motion in each case is continued indefinitely. ❑

■ **Suspended Wire** Suppose a suspended wire hangs under its own weight. As Figure 3.54(a) shows, a physical model for this could be a long telephone wire strung between two posts. Our goal is to construct a mathematical model that describes the shape that the hanging wire assumes.

To begin, let us suppose that the y-axis in Figure 3.54(b) is chosen to pass through the lowest point P_1 on the curve and that the x-axis is a units below P_1. Let's also agree to examine only a portion of the wire between the lowest point P_1 and any arbitrary point P_2. Three forces are acting on the wire: the weight of the segment $P_1 P_2$ and the two tensions \mathbf{T}_1 and \mathbf{T}_2 in the wire at P_1 and P_2, respectively. If w is the linear density of the wire (measured, say, in lb/ft) and s is the length of the segment $P_1 P_2$, then its weight is ws. Now the tension \mathbf{T}_2 resolves into horizontal and vertical components (scalar quantities) $T_2 \cos \theta$ and $T_2 \sin \theta$. Because of equilibrium we can write

$$|\mathbf{T}_1| = T_1 = T_2 \cos \theta \quad \text{and} \quad ws = T_2 \sin \theta.$$

Dividing the last two equations we find $\tan \theta = ws/T_1$. That is,

$$\frac{dy}{dx} = \frac{ws}{T_1}. \tag{8}$$

Now since the arc length between points P_1 and P_2 is given by

$$s = \int_0^x \sqrt{1 + \left(\frac{dy}{dx}\right)^2} \, dx, \tag{9}$$

it follows from the fundamental theorem of calculus that the derivative of (9) is

$$\frac{ds}{dx} = \sqrt{1 + \left(\frac{dy}{dx}\right)^2}. \tag{10}$$

Differentiating (8) with respect to x and using (10) lead to

$$\frac{d^2y}{dx^2} = \frac{w}{T_1} \frac{ds}{dx} \qquad \text{or} \qquad \frac{d^2y}{dx^2} = \frac{w}{T_1}\sqrt{1 + \left(\frac{dy}{dx}\right)^2}. \qquad (11)$$

One might conclude from Figure 3.54 that the shape the hanging wire assumes is parabolic. The next example shows that this is not the case—the curve assumed by the suspended cable is called a **catenary**. Before proceeding, observe that the nonlinear second-order differential equation (11) is one of those equations having the form $F(x, y', y'') = 0$ discussed in Section 3.7. Recall, we have a chance of solving this type of equation by reducing the order of the equation by means of the substitution $u = y'$.

Example 3 An Initial-Value Problem

From the position of the y-axis in Figure 3.54(b) it is apparent that initial conditions associated with the second differential equation in (11) are $y(0) = a$ and $y'(0) = 0$. If we substitute $u = y'$, the last equation in (11) becomes $\dfrac{du}{dx} = \dfrac{w}{T_1}\sqrt{1 + u^2}$. Separating variables,

$$\int \frac{du}{\sqrt{1 + u^2}} = \frac{w}{T_1}\int dx \qquad \text{gives} \qquad \sinh^{-1} u = \frac{w}{T_1} x + c_1.$$

Now, $y'(0) = 0$ is equivalent to $u(0) = 0$. Since $\sinh^{-1} 0 = 0$, we find $c_1 = 0$ and so $u = \sinh (wx/T_1)$. Finally, by integrating both sides of

$$\frac{dy}{dx} = \sinh \frac{w}{T_1}x \qquad \text{we get} \qquad y = \frac{T_1}{w} \cosh \frac{w}{T_1}x + c_2.$$

Using $y(0) = a$, $\cosh 0 = 1$, the last equation implies that $c_2 = a - T_1/w$. Thus we see that the shape of the hanging wire is defined by

$$y = \frac{T_1}{w} \cosh \frac{w}{T_1}x + a - \frac{T_1}{w}. \qquad \qquad \square$$

In Example 3, had we been clever enough at the start to choose $a = T_1/w$, then the solution of the problem would have been simply the hyperbolic cosine $y = (T_1/w) \cosh (wx/T_1)$.

■ **Rocket Motion** In Section 1.3 we saw that the differential equation of a free-falling body of mass m near the surface of the earth is given by

$$m \frac{d^2s}{dt^2} = -mg \qquad \text{or simply} \qquad \frac{d^2s}{dt^2} = -g,$$

where s represents the distance from the surface of the earth to the object and the positive direction is considered to be upward. In other words, the underlying assumption here is that the distance s to the object is small when compared with the radius R of the earth; put yet another way, the distance y from the center of the earth to the object is approximately the same as R. If, on the other hand, the distance y to an object, such as a rocket or a space probe, is large compared to R, then we combine Newton's second law of motion and his universal law of gravitation to derive a differential equation in the variable y.

Suppose a rocket is launched vertically upward from the ground as shown in Figure 3.55. If the positive direction is upward and air resistance is ignored, then the differential equation of motion after fuel burnout is

$$m \frac{d^2y}{dt^2} = -k \frac{Mm}{y^2} \qquad \text{or} \qquad \frac{d^2y}{dt^2} = -k \frac{M}{y^2}, \qquad (12)$$

Figure 3.55

where k a constant of proportionality, y is the distance from the center of the earth to the rocket, M is the mass of the earth, and m is the mass of the rocket. To determine the constant k, we use the fact that when $y = R$, $kMm/R^2 = mg$ or $k = gR^2/M$. Thus the last equation in (12) becomes

$$\frac{d^2y}{dt^2} = -g\,\frac{R^2}{y^2}.$$ (13)

See Problem 18 in Exercises 3.10.

■ **Variable Mass** Notice in the preceding discussion that we described the motion of the rocket after it has burned all its fuel, when presumably its mass m is constant. Of course during its powered ascent the total mass of the rocket varies as its fuel is being expended. The second law of motion, as originally advanced by Newton, states that when a body of mass m moves through a force field with velocity v, the time rate of change of the momentum mv of the body is equal to applied or net force F acting on the body:

$$F = \frac{d}{dt}(mv).$$ (14)

If m is constant, then (14) yields the more familiar form $F = m\,dv/dt = ma$, where a is acceleration. We use the form of Newton's second law given in (14) in the next example, in which the mass m of the body is variable.

Example 4 Chain Pulled Upward by a Constant Force

A uniform 10-foot-long chain is coiled loosely on the ground. One end of the chain is pulled vertically upward by means of constant force of 5 lb. The chain weighs 1 lb per foot. Determine the height of the end above ground level at time t. See Figure 1.32 and Problem 19 in Exercises 1.3.

SOLUTION Let us suppose that $x = x(t)$ denotes the height of the end of the chain in the air at time t, $v = dx/dt$, and that the positive direction is upward. For that portion of the chain in the air at time t we have the following variable quantities:

weight: $W = (x \text{ ft}) \cdot (1 \text{ lb/ft}) = x$

mass: $m = W/g = x/32$,

net force: $F = 5 - W = 5 - x$.

Thus from (14) we have

product rule
↓

$$\frac{d}{dt}\left(\frac{x}{32}\,v\right) = 5 - x \qquad \text{or} \qquad x\frac{dv}{dt} + v\frac{dx}{dt} = 160 - 32x.$$ (15)

Since $v = dx/dt$ the last equation becomes

$$x\frac{d^2x}{dt^2} + \left(\frac{dx}{dt}\right)^2 + 32x = 160.$$ (16)

The nonlinear second-order differential equation (16) has the form $F(x, x', x'') = 0$, which is the second of the two forms considered in Section 3.7 that can possibly be solved by reduction of order. In order to solve (16), we revert back to (15) and

use $v = x'$ along with the Chain Rule. From $\dfrac{dv}{dt} = \dfrac{dv}{dx}\dfrac{dx}{dt} = v\dfrac{dv}{dx}$ the second equation in (15) can be rewritten as

$$xv\frac{dv}{dx} + v^2 = 160 - 32x. \tag{17}$$

On inspection (17) might appear intractable, since it cannot be characterized as any of the first-order equations that were solved in Chapter 2. However, by rewriting (17) in differential form $M(x, v)dx + N(x, v)dv = 0$, we observe that the nonexact equation

$$(v^2 + 32x - 160)dx + xvdv = 0 \tag{18}$$

can be transformed into an exact equation by multiplying it by an *integrating factor*.* When (18) is multiplied by $\mu(x) = x$, the resulting equation is exact (verify). If we identify $\partial f/\partial x = xv^2 + 32x^2 - 160x$, $\partial f/\partial v = x^2 v$, and then proceed as in Section 2.4, we arrive at

$$\frac{1}{2}x^2v^2 + \frac{32}{3}x^3 - 80x^2 = c_1. \tag{19}$$

From the initial condition $x(0) = 0$ it follows that $c_1 = 0$. Now by solving $\frac{1}{2}x^2v^2 + \frac{32}{3}x^3 - 80x^2 = 0$ for $v = dx/dt > 0$ we get another differential equation,

$$\frac{dx}{dt} = \sqrt{160 - \frac{64}{3}x},$$

which can be solved by separation of variables. You should verify that

$$-\frac{3}{32}\left(160 - \frac{64}{3}x\right)^{1/2} = t + c_2. \tag{20}$$

This time the initial condition $x(0) = 0$ implies $c_2 = -3\sqrt{10}/8$. Finally, by squaring both sides of (20) and solving for x we arrive at the desired result,

$$x(t) = \frac{15}{2} - \frac{15}{2}\left(1 - \frac{4\sqrt{10}}{15}t\right)^2. \tag{21} \ \square$$

See Problem 19 in Exercises 3.10.

* See the Remarks on page 62 as well as Problems 29–34 in Exercises 2.4.

To the Instructor

In addition to Problems 16 and 17, all or portions of Problems 1–6, 8–12, 14, 15, 19, and 21 could serve as Computer Lab Assignments.

Nonlinear Springs

In Problems 1–4, the given differential equation is a model of an undamped spring/mass system in which the restoring force $F(x)$ in (1) is nonlinear. For each equation use a numerical solver to plot the solution curves satisfying the given initial conditions. If the solutions appear to be periodic use the solution curve to estimate the period T of oscillations.

1. $\dfrac{d^2x}{dt^2} + x^3 = 0$, $\ x(0) = 1$, $\ x'(0) = 1$; $\ x(0) = \dfrac{1}{2}$, $\ x'(0) = -1$

2. $\dfrac{d^2x}{dt^2} + 4x - 16x^3 = 0$, $\ x(0) = 1$, $\ x'(0) = 1$;

$x(0) = -2$, $\ x'(0) = 2$

3. $\dfrac{d^2x}{dt^2} + 2x - x^2 = 0$, $\ x(0) = 1$, $\ x'(0) = 1$; $\ x(0) = \dfrac{3}{2}$, $\ x'(0) = -1$

4. $\dfrac{d^2x}{dt^2} + xe^{0.01x} = 0$, $\ x(0) = 1$, $\ x'(0) = 1$; $\ x(0) = 3$, $\ x'(0) = -1$

5. In Problem 3, suppose the mass is released from the initial position $x(0) = 1$ with an initial velocity $x'(0) = x_1$. Use a numerical solver to estimate the smallest value of $|x_1|$ at which the motion of the mass is nonperiodic.

6. In Problem 3, suppose the mass is released from an initial position $x(0) = x_0$ with the initial velocity $x'(0) = 1$. Use a numerical solver to estimate an interval $a \leq x_0 \leq b$ for which the motion is oscillatory.

7. Find a linearization of the differential equation in Problem 4.

8. Consider the model of an undamped nonlinear spring/mass system given by

$$\frac{d^2x}{dt^2} + 8x - 6x^3 + x^5 = 0.$$

Use a numerical solver to discuss the nature of the oscillations of the system corresponding to the initial conditions:

$x(0) = 1, x'(0) = 1;$ $x(0) = -2, x'(0) = 0.5;$

$x(0) = \sqrt{2}, x'(0) = 1;$ $x(0) = 2, x'(0) = 0.5;$

$x(0) = 2, x'(0) = 0;$ $x(0) = -\sqrt{2}, x'(0) = -1.$

In Problems 9 and 10, the given differential equation is a model of a damped nonlinear spring/mass system. Predict the behavior of each system as $t \to \infty$. For each equation use a numerical solver to obtain the solution curves satisfying the given initial conditions.

9. $\dfrac{d^2x}{dt^2} + \dfrac{dx}{dt} + x + x^3 = 0, \quad x(0) = -3, \quad x'(0) = 4;$

$x(0) = 0, \quad x'(0) = -8$

10. $\dfrac{d^2x}{dt^2} + \dfrac{dx}{dt} + x - x^3 = 0, \quad x(0) = 0, \quad x'(0) = \dfrac{3}{2};$

$x(0) = -1, \quad x'(0) = 1$

11. The model $mx'' + kx + k_1x^3 = F_0 \cos \omega t$ of an undamped periodically driven spring/mass system is called **Duffing's differential equation.** Consider the initial-value problem $x'' + x + k_1x^3 = 5 \cos t, x(0) = 1, x'(0) = 0.$ Use a numerical solver to investigate the behavior of the system for values of $k_1 > 0$ ranging from $k_1 = 0.01$ to $k_1 = 100$. State your conclusions.

12. (a) In Problem 11, find values of $k_1 < 0$ for which the system in part **(a)** is oscillatory.

(b) Consider the initial-value problem $x'' + x + k_1x^3 = \cos \frac{3}{2}t, x(0) = 0, x'(0) = 0.$ Find values for $k_1 < 0$ for which the system is oscillatory.

13. Explain why the damping term in equation (3)

$$\text{written as } \beta \left| \frac{dx}{dt} \right| \frac{dx}{dt} \quad \text{instead of} \quad \beta \left(\frac{dx}{dt} \right)^2.$$

Nonlinear Pendulum

14. (a) Experiment with a calculator and determine an interval $0 \le \theta < \theta_1$, where θ is measured in radians, for which you think $\sin \theta \approx \theta$ is a fairly good estimate. Then use a graphing utility to plot the graphs of $y = x$ and $y = \sin x$ on the same coordinate axes for $0 \le x \le \pi/2$. Do the graphs confirm your observations with the calculator?

(b) Use a numerical solver to plot the solutions curves of the initial-value problems

$$\frac{d^2\theta}{dt^2} + \sin \theta = 0, \quad \theta(0) = \theta_0, \theta'(0) = 0$$

$$\text{and} \quad \frac{d^2\theta}{dt^2} + \theta = 0, \quad \theta(0) = \theta_0, \theta'(0) = 0$$

for several values of θ_0 in the interval $0 \le \theta < \theta_1$ found in part **(a)**. Then plot solution curves of the initial-value problems for several values of θ_0 for which $\theta_0 > \theta_1$.

15. (a) Consider the nonlinear pendulum whose oscillations are defined by (6). Use a numerical solver as an aid to determine whether a pendulum of length l will oscillate faster on the earth or on the moon. Use the same initial conditions but choose these initial conditions so that the pendulum oscillates back and forth.

(b) Which pendulum in part **(a)** has the greater amplitude?

(c) Are the conclusions in parts **(a)** and **(b)** the same when the linear model (7) is used?

16. Computer Lab Assignment Consider the initial-value problem

$$\frac{d^2\theta}{dt^2} + \sin \theta = 0, \quad \theta(0) = \frac{\pi}{12}, \quad \theta'(0) = -\frac{1}{3}$$

for the nonlinear pendulum. Since we cannot solve the differential equation we can find no explicit solution of this problem. But suppose we wish to determine the first time $t_1 > 0$ for which the pendulum, starting from its initial position to the right, reaches the position OP in Figure 3.52, that is, find the first positive root of $\theta(t) = 0$. In this problem and the next we examine several ways to proceed.

(a) Approximate t_1 by solving the linear problem

$$d^2\theta/dt^2 + \theta = 0, \theta(0) = \pi/12, \theta'(0) = -\frac{1}{3}.$$

(b) Use the method illustrated in Example 3 of Section 3.7 to find the first four nonzero terms of a Taylor series solution $\theta(t)$ centered at 0 for the nonlinear initial-value problem. Give the exact values of all coefficients.

(c) Use the first two terms of the Taylor series in part **(b)** to approximate t_1.

(d) Use the first three terms of the Taylor series in part **(b)** to approximate t_1.

(e) Use a calculator with root-finding capability or a CAS and the first four terms of the Taylor series in part **(b)** to approximate t_1.

(f) In this part of the problem you are led through the commands in *Mathematica* that enable you to approximate the root t_1. The procedure is easily modified so that any root of $\theta(t) = 0$ can be approximated. (*If you do not have Mathematica, adapt the given procedure by finding the corresponding syntax for the CAS you have on hand.*) Precisely reproduce and then, in turn, execute each line in the given sequence of commands.

```
sol = NDSolve[{y0[t] + Sin[y[t]] == 0, y[0] == Pi/12,
    y'[0] == -1/3}, y, {t, 0, 5}]//Flatten
solution = y[t]/.sol
Clear[y]
y[t_]:= Evaluate[solution]
y[t]
gr1 = Plot[y[t], {t, 0, 5}]
root = FindRoot[y[t] == 0, {t, 1}]
```

(g) Appropriately modify the syntax in part **(f)** and find the next two positive roots of $\theta(t) = 0$.

17. **Computer Lab Assignment** Consider a pendulum that is released from rest from an initial displacement of θ_0 radians. Solving the linear model (7) subject to the initial conditions $\theta(0) = \theta_0$, $\theta'(0) = 0$ gives $\theta(t) = \theta_0 \cos \sqrt{g/l}\, t$. The period of oscillations predicted by this model is given by the familiar formula $T = 2\pi/\sqrt{g/l} = 2\pi\sqrt{l/g}$. The interesting thing about this formula for T is that it does not depend on the magnitude of the initial displacement θ_0; in other words, the linear model would predict that the time that it takes the pendulum to swing from an initial displacement of, say, $\theta_0 = \pi/2$ $(= 90°)$ to $-\pi/2$ and back again would be exactly the same time to cycle from, say, $\theta_0 = \pi/360$ $(= 0.5°)$ to $-\pi/360$. This is intuitively unreasonable; the actual period must depend on θ_0.

If we assume that $g = 32$ ft/s^2 and $l = 32$ ft, then the period of oscillation of the linear model is $T = 2\pi$ seconds. Let us compare this last number with the period predicted by the nonlinear model when $\theta_0 = \pi/4$. Use a numerical solver that is capable of generating hard data. Use the solver to approximate the solution of

$$\frac{d^2\theta}{dt^2} + \sin\theta = 0, \quad \theta(0) = \frac{\pi}{4}, \quad \theta'(0) = 0$$

on the interval $0 \leq t \leq 2$. As in Problem 16, if t_1 denotes the first time the pendulum reaches the position OP in Figure 3.52, then the period of the nonlinear pendulum is $4t_1$. Here is another way of solving the equation $\theta(t) = 0$. Experiment with small step sizes and advance time starting at $t = 0$ and ending at $t = 2$. Observe from your hard data the time t_1 when $\theta(t)$ changes, for the first time, from positive to negative. Use the value t_1 to determine the true value of the period of the nonlinear pendulum. Compute the percentage relative error in the period estimated by $T = 2\pi$.

Rocket Motion

18. **(a)** Use the substitution $v = dy/dt$ to solve (13) for v in terms of y. Assume that the velocity of the rocket is $v = v_0$ at burnout and that $y \approx R$ at that instant to show that the approximate value of the constant c of integration is $c = -gR + \frac{1}{2}v_0^2$.

(b) Use the solution for v in part **(a)** to show that the escape velocity of the rocket is given by $v_0 = \sqrt{2gR}$. [*Hint:* Take $y \to \infty$ and assume $v > 0$ for all time t.]

(c) The result in part **(b)** holds for any body in the solar system. Use the values $g = 32$ ft/s^2 and $R = 4000$ miles to show that the escape velocity from the earth is (approximately) $v_0 = 25{,}000$ mi/h.

(d) Find the escape velocity from the moon if the acceleration of gravity is $0.165g$ and $R = 1080$ miles.

Variable Mass

19. **(a)** In Example 4, how much of the chain would you intuitively expect the constant 5-lb force to be able to lift?

(b) What is the initial velocity of the chain?

(c) Use a graphing utility to plot the solution $x(t)$ given in (21) of Example 4. What is the interval I of definition of the solution? How much chain is actually lifted? Explain any difference between this answer and that predicted by your intuition in part **(a)**.

20. A uniform chain of length L, measured in feet, is held vertically so that the lower end just touches the floor. The chain weighs 2 pounds per foot. The upper end that is held is released from rest at $t = 0$ and the chain falls straight down. See Figure 1.33. If $x(t)$ denotes the length of the chain on the floor at time t, we saw in Problem 20 in Exercises 1.3 that, when air resistance is ignored and the positive direction is taken to be downward,

$$(L - x)\frac{d^2x}{dt^2} - \left(\frac{dx}{dt}\right)^2 = Lg.$$

(a) Solve for v in terms of x. Solve for x in terms of t. Express v in terms of t.

(b) Determine how long it takes for the chain to fall completely to the ground.

(c) What velocity does the model in part **(a)** predict for the upper end of the chain as it hits the ground?

21. A portion of a uniform chain of length 8 ft is loosely coiled around a peg at the edge of a high horizontal platform, and the remaining portion of the chain hangs at rest over the edge of the platform. Suppose the length of the overhang is 3 ft and that the chain weighs 2 lb/ft. Starting at $t = 0$, the weight of the overhanging portion causes the chain on the table to uncoil smoothly and to fall to the floor.

(a) Ignore any resistive forces and assume that the positive direction is downward. If $x(t)$ denotes the length of the chain overhanging the table at time $t > 0$ and $v = dx/dt$, find a differential equation that relates v to x.

(b) Proceed as in Example 4 and solve for v in terms of x by finding an appropriate integrating factor.

(c) Express time t in terms of x. Use a CAS as an aid in determining the time it takes for a 7-ft segment of chain to uncoil completely, that is, fall from the table.

22. A portion of a uniform chain of length 8 ft lies stretched out on a high horizontal platform and the remaining portion of the chain hangs over the edge of the platform as shown in Figure 3.56. Suppose the length of the overhang is 3 ft and that the chain weighs 2 lb/ft. The end of the chain on the platform is held until at $t = 0$ it is released from rest and the chain begins to slide off the platform due to the weight of the overhanging portion.

Figure 3.56

(a) Ignore any resistive forces and assume that the positive direction is downward. If $x(t)$ denotes the length of the chain overhanging the platform at time $t > 0$ and $v = dx/dt$, show that v is related to x by the differential equation $v\dfrac{dv}{dx} = 4x$.

(b) Solve for v in terms of x. Solve for x in terms of t. Express v in terms of t.

(c) Approximate the time it takes for the rest of the chain to slide off the platform. Find the velocity at which the end of the chain leaves the edge of the platform.

(d) Suppose the chain is L ft long and weighs a total of W lb. If the overhang at $t = 0$ is x_0 ft, show that the velocity at which the end of the chain leaves the edge of the platform is

$$v(L) = \sqrt{\frac{g}{L}\,(L^2 - x_0^2)}.$$

Miscellaneous Nonlinear Models

23. In a naval exercise a ship S_1 is pursued by a submarine S_2, as shown in Figure 3.57. Ship S_1 departs point $(0, 0)$ at $t = 0$ and proceeds along a straight-line course (the y-axis) at a constant speed v_1. The submarine S_2 keeps ship S_1 in visual contact, indicated by the straight dashed line L in the figure, while traveling at a constant speed v_2 along a curve C. Assume that S_2 starts at the point $(a, 0)$, $a > 0$, at $t = 0$ and that L is tangent to C.

Figure 3.57

(a) Determine a mathematical model that describes the curve C.

[*Hint:* $\dfrac{dt}{dx} = \dfrac{dt}{ds}\dfrac{ds}{dx}$, where s is arc length measured along C.]

(b) Find an explicit solution of the differential equation. For convenience define $r = v_1/v_2$.

(c) Determine whether the paths of S_1 and S_2 will ever intersect by considering the cases $r > 1$, $r < 1$, and $r = 1$.

24. In another naval exercise a destroyer S_1 pursues a submerged submarine S_2. Suppose that S_1 at $(9, 0)$ on the x-axis detects S_2 at $(0, 0)$ and that S_2 simultaneously detects S_1. The captain of the destroyer S_1 assumes that the submarine will take immediate evasive action and conjectures its likely new course is the straight line indicated in Figure 3.58. When S_1 is at $(3, 0)$ it changes from its straight-line course toward the origin to a pursuit curve C. Assume that the speed of the destroyer is, at all times, a constant 30 mi/h and the submarine's speed is a constant 15 mi/h.

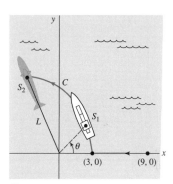

Figure 3.58

(a) Explain why the captain waits until S_1 reaches $(3, 0)$ before ordering a course change to C.

(b) Using polar coordinates, find an equation $r = f(\theta)$ for the curve C.

(c) Explain why the time, measured from the initial detection, at which the destroyer intercepts the submarine must be less than $\frac{1}{5}\,(1 + e^{2\pi/\sqrt{3}}\,)$.

25. Consider the model of the free damped nonlinear pendulum given by

$$\frac{d^2\theta}{dt^2} + 2\lambda\,\frac{d\theta}{dt} + \omega^2 \sin\,\theta = 0.$$

Use a numerical solver to investigate whether the motion in the two cases $\lambda^2 - \omega^2 > 0$ and $\lambda^2 - \omega^2 < 0$ correspond, respectively, to the overdamped and underdamped cases discussed in Section 3.8 for spring/mass systems. Choose appropriate initial conditions and values of λ and ω.

3.11 Solving Systems of Linear Equations

We conclude this chapter as we did Chapter 2, with systems of differential equations.

■ **Coupled Systems/Coupled Equations** In Section 2.9 we briefly examined some mathematical models that were systems of linear and nonlinear first-order ODEs. In Section 3.8 we saw that the mathematical model describing the displacement of a mass on a single spring, current in a series circuit, and charge on a capacitor in a series circuit consisted of a *single* differential equation. When physical systems are coupled—for example, when two or more mixing tanks are connected, when two or more spring/mass systems are attached, or when circuits are joined to form a network—the mathematical model of the system usually consists of a set of coupled differential equations, in other words, a system of differential equations.

We did not attempt to solve any of these systems considered in Section 2.9. The same remarks made in Sections 3.7 and 3.10 pertain as well to systems of nonlinear ODEs; that is, it is nearly impossible to solve such systems analytically. However, linear systems with constant coefficients can be solved. The method that we shall examine in this section for solving linear systems with constant coefficients simply uncouples the system into distinct linear ODEs in each dependent variable. Thus, this section gives you an opportunity to practice what you learned earlier in the chapter.

Before proceeding, let us continue in the same vein as Section 3.8 by considering a spring/mass system, but this time we derive a mathematical model that describes the vertical displacements of two masses in a coupled spring/mass system.

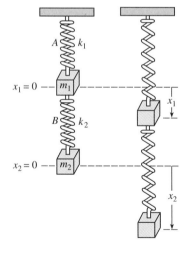

Figure 3.59

■ **Coupled Spring/Mass System** Suppose two masses m_1 and m_2 are connected to two springs A and B of negligible mass having spring constants k_1 and k_2, respectively. In turn, the two springs are attached as shown in Figure 3.59. Let $x_1(t)$ and $x_2(t)$ denote the vertical displacements of the masses from their equilibrium positions. When the system is in motion, spring B is subject to both an elongation and a compression; hence its net elongation is $x_2 - x_1$. Therefore it follows from Hooke's law that springs A and B exert forces $-k_1x_1$ and $k_2(x_2 - x_1)$, respectively, on m_1. If no damping is present and no external force is impressed on the system, then the net force on m_1 is $-k_1x_1 + k_2(x_2 - x_1)$. By Newton's second law we can write

$$m_1 \frac{d^2x_1}{dt^2} = -k_1x_1 + k_2(x_2 - x_1).$$

Similarly, the net force exerted on mass m_2 is due solely to the net elongation of spring B; that is, $-k_2(x_2 - x_1)$. Hence we have

$$m_2 \frac{d^2x_2}{dt^2} = -k_2(x_2 - x_1).$$

In other words, the motion of the coupled system is represented by the system of linear second-order equations

$$
\begin{aligned}
m_1 x_1'' &= -k_1 x_1 + k_2(x_2 - x_1) \\
m_2 x_2'' &= -k_2(x_2 - x_1).
\end{aligned}
\tag{1}
$$

After we have illustrated the main idea of this section, we will return to system (1).

■ **Systematic Elimination** The method of **systematic elimination** for solving systems of linear equations with constant coefficients is based on the algebraic principle of elimination of variables. The analogue of *multiplying* an algebraic equation by a constant is *operating* on an ODE with some combination of derivatives. The elimination process is expedited by rewriting each equation in a system using differential operator notation. Recall from Section 3.1 that a single linear equation

$$a_n y^{(n)} + a_{n-1} y^{(n-1)} + \cdots + a_1 y' + a_0 y = g(t),$$

where the a_i, $i = 0, 1, \ldots, n$ are constants, can be written as

$$(a_n D^n + a_{n-1} D^{(n-1)} + \cdots + a_1 D + a_0) y = g(t).$$

If an nth-order differential operator $a_n D^n + a_{n-1} D^{(n-1)} + \cdots + a_1 D + a_0$ factors into differential operators of lower order, then the factors commute. Now, for example, to rewrite the system

$$x'' + 2x' + y'' = x + 3y + \sin t$$

$$x' + y' = -4x + 2y + e^{-t}$$

in terms of the operator D, we first bring all terms involving the dependent variables to one side and group the same variables:

$$x'' + 2x' - x + y'' - 3y = \sin t \qquad\qquad (D^2 + 2D - 1)x + (D^2 - 3)y = \sin t$$
$$\text{so that}$$
$$x' - 4x + y' - 2y = e^{-t} \qquad\qquad (D - 4)x + (D - 2)y = e^{-t}$$

■ **Solution of a System** A **solution** of a system of differential equations is a set of sufficiently differentiable functions $x = \phi_1(t)$, $y = \phi_2(t)$, $z = \phi_3(t)$, and so on, that satisfies each equation in the system on some common interval I.

■ **Method of Solution** Consider the simple system of linear first-order equations

$$\begin{array}{ll} \dfrac{dx}{dt} = 3y & \\ & \text{or, equivalently,} \\ \dfrac{dy}{dt} = 2x & \end{array} \qquad \begin{array}{l} Dx - 3y = 0 \\ \\ 2x - Dy = 0. \end{array} \qquad (2)$$

Operating on the first equation in (2) by D while multiplying the second by -3 and then adding eliminates y from the system and gives $D^2 x - 6x = 0$. Since the roots of the auxiliary equation of the last DE are $m_1 = \sqrt{6}$ and $m_2 = -\sqrt{6}$, we obtain

$$x(t) = c_1 e^{-\sqrt{6}\, t} + c_2 e^{\sqrt{6}\, t}. \qquad (3)$$

Multiplying the first equation in (2) by 2 while operating on the second by D and then subtracting gives the differential equation for y, $D^2 y - 6y = 0$. It follows immediately that

$$y(t) = c_3 e^{-\sqrt{6}\, t} + c_4 e^{\sqrt{6}\, t}. \qquad (4)$$

This is important. 🖝 Now, (3) and (4) do not satisfy the system (2) for every choice of c_1, c_2, c_3, and c_4 because the system itself puts a constraint on the number of parameters in a solution that can be chosen arbitrarily. To see this, observe that after substituting $x(t)$ and $y(t)$ into the first equation of the original system, (2) gives, after simplification,

$$(-\sqrt{6}c_1 - 3c_3)e^{-\sqrt{6}\, t} + (\sqrt{6}c_2 - 3c_4)e^{\sqrt{6}\, t} = 0.$$

Since the latter expression is to be zero for all values of t, we must have $-\sqrt{6}c_1 - 3c_3 = 0$

and $\sqrt{6}c_2 - 3c_4 = 0$. Thus we can write c_3 and a multiple of c_1 and c_4 as a multiple of c_2:

$$c_3 = -\frac{\sqrt{6}}{3}c_1 \quad \text{and} \quad c_4 = \frac{\sqrt{6}}{3}c_2. \tag{5}$$

Hence we conclude that a solution of the system must be

$$x(t) = c_1 e^{-\sqrt{6}\,t} + c_2 e^{\sqrt{6}\,t}, \qquad y(t) = -\frac{\sqrt{6}}{3}c_1 e^{-\sqrt{6}\,t} + \frac{\sqrt{6}}{3}c_2 e^{\sqrt{6}\,t}.$$

You are urged to substitute (3) and (4) into the second equation of (2) and verify that the same relationship (5) holds between the constants.

Example 1 Solution by Elimination

Solve
$$Dx + (D + 2)y = 0$$
$$(D - 3)x - 2y = 0. \tag{6}$$

SOLUTION Operating on the first equation by $D - 3$ and on the second by D and then subtracting eliminates x from the system. It follows that the differential equation for y is

$$[(D - 3)(D + 2) + 2D]y = 0 \quad \text{or} \quad (D^2 + D - 6)y = 0.$$

Since the characteristic equation of this last differential equation is $m^2 + m - 6 = (m - 2)(m + 3) = 0$, we obtain the solution

$$y(t) = c_1 e^{2t} + c_2 e^{-3t}. \tag{7}$$

Eliminating y in a similar manner yields $(D^2 + D - 6)x = 0$, from which we find

$$x(t) = c_3 e^{2t} + c_4 e^{-3t}. \tag{8}$$

As we noted in the foregoing discussion, a solution of (6) does not contain four independent constants. Substituting (7) and (8) into the first equation of (6) gives

$$(4c_1 + 2c_3)e^{2t} + (-c_2 - 3c_4)e^{-3t} = 0.$$

From $4c_1 + 2c_3 = 0$ and $-c_2 - 3c_4 = 0$ we get $c_3 = -2c_1$ and $c_4 = -\frac{1}{3}c_2$. Accordingly, a solution of the system is

$$x(t) = -2c_1 e^{2t} - \frac{1}{3}c_2 e^{-3t}, \qquad y(t) = c_1 e^{2t} + c_2 e^{-3t}. \qquad \Box$$

Since we could just as easily solve for c_3 and c_4 in terms of c_1 and c_2, the solution in Example 1 can be written in the alternative form

$$x(t) = c_3 e^{2t} + c_4 e^{-3t}, \qquad y(t) = -\frac{1}{2}c_3 e^{2t} - 3c_4 e^{-3t}.$$

Watch for a shortcut.

It sometimes pays to keep one's eyes open when solving systems. Had we solved for x first, then y could be found, along with the relationship between the constants, by using the last equation in (6). You should verify that substituting $x(t)$ into $y = \frac{1}{2}(Dx - 3x)$ yields $y = -\frac{1}{2}c_3 e^{2t} - 3c_4 e^{-3t}$.

Example 2 Solution by Elimination

Solve
$$x' - 4x + y'' = t^2$$
$$x' + x + y' = 0. \tag{9}$$

SOLUTION First we write the system in differential operator notation:

$$(D - 4)x + D^2y = t^2$$
$$(D + 1)x + Dy = 0. \tag{10}$$

Then, by eliminating x, we obtain

$$[(D + 1)D^2 - (D - 4)D]y = (D + 1)t^2 - (D - 4)0$$

or

$$(D^3 + 4D)y = t^2 + 2t.$$

Since the roots of the auxiliary equation $m(m^2 + 4) = 0$ are $m_1 = 0$, $m_2 = 2i$, and $m_3 = -2i$, the complementary function is

$$y_c = c_1 + c_2 \cos 2t + c_3 \sin 2t.$$

To determine the particular solution y_p we use undetermined coefficients by assuming $y_p = At^3 + Bt^2 + Ct$. Therefore

$$y_p' = 3At^2 + 2Bt + C, \quad y_p'' = 6At + 2B, \quad y_p''' = 6A,$$

$$y_p''' + 4y_p' = 12At^2 + 8Bt + 6A + 4C = t^2 + 2t.$$

The last equality implies $12A = 1$, $8B = 2$, $6A + 4C = 0$, and hence $A = \frac{1}{12}$, $B = \frac{1}{4}$, $C = -\frac{1}{8}$. Thus

$$y = y_c + y_p = c_1 + c_2 \cos 2t + c_3 \sin 2t + \frac{1}{12}t^3 + \frac{1}{4}t^2 - \frac{1}{8}t. \tag{11}$$

Eliminating y from the system (9) leads to

$$[(D - 4) - D(D + 1)]x = t^2 \quad \text{or} \quad (D^2 + 4)x = -t^2.$$

It should be obvious that

$$x_c = c_4 \cos 2t + c_5 \sin 2t$$

and that undetermined coefficients can be applied to obtain a particular solution of the form $x_p = At^2 + Bt + C$. In this case the usual differentiations and algebra yield $x_p = -\frac{1}{4}t^2 + \frac{1}{8}$, and so

$$x = x_c + x_p = c_4 \cos 2t + c_5 \sin 2t - \frac{1}{4}t^2 + \frac{1}{8}. \tag{12}$$

Now c_4 and c_5 can be expressed in terms of c_2 and c_3 by substituting (11) and (12) into either equation of (9). By using the second equation, we find, after combining terms,

$$(c_5 - 2c_4 - 2c_2) \sin 2t + (2c_5 + c_4 + 2c_3) \cos 2t = 0$$

so that $c_5 - 2c_4 - 2c_2 = 0$ and $2c_5 + c_4 + 2c_3 = 0$. Solving for c_4 and c_5 in terms of c_2 and c_3 gives $c_4 = -\frac{1}{5}(4c_2 + 2c_3)$ and $c_5 = \frac{1}{5}(2c_2 - 4c_3)$. Finally, a solution of (9) is found to be

$$x(t) = -\frac{1}{5}(4c_2 + 2c_3) \cos 2t + \frac{1}{5}(2c_2 - 4c_3) \sin 2t - \frac{1}{4}t^2 + \frac{1}{8},$$

$$y(t) = c_1 + c_2 \cos 2t + c_3 \sin 2t + \frac{1}{12}t^3 + \frac{1}{4}t^2 - \frac{1}{8}t. \quad \square$$

Example 3 **A Mathematical Model Revisited**

In (3) of Section 2.9 we saw that a system of linear first-order differential equations described the number of pounds of salt $x_1(t)$ and $x_2(t)$ of a brine mixture that

flows between two tanks. At that time we were not able to solve the system. But now, in terms of differential operators, the system is

$$\left(D + \frac{2}{25}\right)x_1 - \frac{1}{50}x_2 = 0$$

$$-\frac{2}{25}x_1 + \left(D + \frac{2}{25}\right)x_2 = 0.$$

Operating on the first equation by $D + \frac{2}{25}$, multiplying the second equation by $\frac{1}{50}$, adding, and then simplifying, give

$$(625D^2 + 100D + 3)x_1 = 0.$$

From the auxiliary equation $625m^2 + 100m + 3 = (25m + 1)(25m + 3) = 0$ we see immediately that

$$x_1(t) = c_1 e^{-t/25} + c_2 e^{-3t/25}.$$

In like manner we find $(625D^2 + 100D + 3)x_2 = 0$ and so

$$x_2(t) = c_3 e^{-t/25} + c_4 e^{-3t/25}.$$

Substituting $x_1(t)$ and $x_2(t)$ into, say, the first equation of the system then gives

$$(2c_1 - c_3)e^{-t/25} + (-2c_2 - c_4)e^{-3t/25} = 0.$$

From this last equation we find $c_3 = 2c_1$ and $c_4 = -2c_2$. Thus a solution of the system is

$$x_1(t) = c_1 e^{-t/25} + c_2 e^{-3t/25}, \quad x_2(t) = 2c_1 e^{-t/25} - 2c_2 e^{-3t/25}.$$

In the original discussion we assumed that initial conditions were $x_1(0) = 25$ and $x_2(0) = 0$. Applying these conditions to the solution yields $c_1 + c_2 = 25$ and $2c_1 - 2c_2 = 0$. Solving these equations simultaneously gives $c_1 = c_2 = \frac{25}{2}$. Finally, a solution of the initial-value problem is

$$x_1(t) = \frac{25}{2}e^{-t/25} + \frac{25}{2}e^{-3t/25}, \quad x_2(t) = 25e^{-t/25} - 25e^{-3t/25}. \qquad \square$$

In our example we solve (1) under the assumption that $k_1 = 6$, $k_2 = 4$, $m_1 = 1$, and $m_2 = 1$.

Example 4 A Special Case of System (1)

Solve
$$x_1'' + 10x_1 \qquad - 4x_2 = 0$$
$$4x_1 + x_2'' + 4x_2 = 0 \tag{13}$$

subject to $x_1(0) = 0$, $x_1'(0) = 1$, $x_2(0) = 0$, $x_2'(0) = -1$.

SOLUTION Using elimination on the equivalent form of the system

$$(D^2 + 10)x_1 \qquad - 4x_2 = 0$$
$$4x_1 + (D^2 + 4)x_2 = 0$$

we find that x_1 and x_2 satisfy, respectively,

$$(D^2 + 2)(D^2 + 12)x_1 = 0 \quad \text{and} \quad (D^2 + 2)(D^2 + 12)x_2 = 0.$$

Thus we find

$$x_1(t) = c_1 \cos \sqrt{2}\,t + c_2 \sin \sqrt{2}\,t + c_3 \cos 2\sqrt{3}\,t + c_4 \sin 2\sqrt{3}t$$
$$x_2(t) = c_5 \cos \sqrt{2}\,t + c_6 \sin \sqrt{2}\,t + c_7 \cos 2\sqrt{3}\,t + c_8 \sin 2\sqrt{3}t$$

(a) $x_1(t)$

(b) $x_2(t)$

Figure 3.60

Substituting both expressions into the first equation of (13) and simplifying eventually yields $c_5 = 2c_1$, $c_6 = 2c_2$, $c_7 = -\frac{1}{2}c_3$, $c_8 = -\frac{1}{2}c_4$. Thus, a solution of (13) is

$$x_1(t) = c_1 \cos \sqrt{2}\,t + c_2 \sin \sqrt{2}\,t + c_3 \cos 2\sqrt{3}\,t + c_4 \sin 2\sqrt{3}\,t$$

$$x_2(t) = 2c_1 \cos \sqrt{2}\,t + 2c_2 \sin \sqrt{2}\,t - \frac{1}{2}c_3 \cos 2\sqrt{3}\,t - \frac{1}{2}c_4 \sin 2\sqrt{3}\,t.$$

The stipulated initial conditions then imply $c_1 = 0$, $c_2 = -\sqrt{2}/10$, $c_3 = 0$, $c_4 = \sqrt{3}/5$. And so the solution of the initial-value problem is

$$x_1(t) = -\frac{\sqrt{2}}{10} \sin \sqrt{2}t + \frac{\sqrt{3}}{5} \sin 2\sqrt{3}t,$$

$$x_2(t) = -\frac{\sqrt{2}}{5} \sin \sqrt{2}t - \frac{\sqrt{3}}{10} \sin 2\sqrt{3}t \qquad (14)$$

The graphs of x_1 and x_2 in Figure 3.60 reveal the complicated oscillatory motion of each mass. ❑

We will revisit Example 4 in Section 4.6, where we will solve (13) by means of the Laplace transform.

E X E R C I S E S 3 . 1 1

Answers to odd-numbered problems begin on page A-32.

In Problems 1–20, solve the given system of differential equations by systematic elimination.

1. $\dfrac{dx}{dt} = 2x - y$

$\dfrac{dy}{dt} = x$

2. $\dfrac{dx}{dt} = 4x + 7y$

$\dfrac{dy}{dt} = x - 2y$

3. $\dfrac{dx}{dt} = -y + t$

$\dfrac{dy}{dt} = x - t$

4. $\dfrac{dx}{dt} - 4y = 1$

$\dfrac{dy}{dt} + x = 2$

5. $(D^2 + 5)x \qquad -2y = 0$

$-2x + (D^2 + 2)y = 0$

6. $(D + 1)x + (D - 1)y = 2$

$3x + (D + 2)y = -1$

7. $\dfrac{d^2x}{dt^2} = 4y + e^t$

$\dfrac{d^2y}{dt^2} = 4x - e^t$

8. $\dfrac{d^2x}{dt^2} + \dfrac{dy}{dt} = -5x$

$\dfrac{dx}{dt} + \dfrac{dy}{dt} = -x + 4y$

9. $Dx + \qquad D^2y = e^{3t}$

$(D + 1)x + (D - 1)y = 4e^{3t}$

10. $D^2x - \qquad Dy = t$

$(D + 3)x + (D + 3)y = 2$

11. $(D^2 - 1)x - \quad y = 0$

$(D - 1)\,x + Dy = 0$

12. $(2D^2 - D - 1)x - (2D + 1)y = 1$

$(D - 1)x + \qquad Dy = -1$

13. $2\dfrac{dx}{dt} - 5x + \dfrac{dy}{dt} = e^t$

$\dfrac{dx}{dt} - \quad x + \dfrac{dy}{dt} = 5e^t$

14. $\dfrac{dx}{dt} + \dfrac{dy}{dt} \qquad = e^t$

$-\dfrac{d^2x}{dt^2} + \dfrac{dx}{dt} + x + y = 0$

15. $(D - 1)x + (D^2 + 1)y = 1$

$(D^2 - 1)x + (D + 1)y = 2$

16. $D^2x - 2(D^2 + D)y = \sin t$

$x + \qquad Dy = 0$

17. $Dx = y$

$Dy = z$

$Dz = x$

18. $\qquad Dx + \qquad z = e^t$

$(D - 1)x + Dy + Dz = 0$

$x + 2y + Dz = e^t$

19. $\dfrac{dx}{dt} = 6y$

$\dfrac{dy}{dt} = x + z$

$\dfrac{dz}{dt} = x + y$

20. $\dfrac{dx}{dt} = -x + z$

$\dfrac{dy}{dt} = -y + z$

$\dfrac{dz}{dt} = -x + y$

In Problems 21 and 22 solve the given initial-value problem.

21. $\dfrac{dx}{dt} = -5x - y$

$\dfrac{dy}{dt} = 4x - y$

$x(1) = 0, \quad y(1) = 1$

22. $\dfrac{dx}{dt} = y - 1$

$\dfrac{dy}{dt} = -3x + 2y$

$x(0) = 0, \quad y(0) = 0$

23. Computer Lab Assignment (a) Use systematic elimination to solve the system (1) for the coupled spring/mass system when $k_1 = 4$, $k_2 = 2$, $m_1 = 2$, and $m_2 = 1$ and with initial conditions $x_1(0) = 2$, $x_1'(0) = 1$, $x_2(0) = -1$, $x_2'(0) = 1$.

(b) Use a CAS to plot the graphs of $x_1(t)$ and $x_2(t)$ in the tx-plane. What is the fundamental difference in the motions of the masses m_1 and m_2 in this problem and that of the masses illustrated in Figure 3.60?

(c) As parametric equations, plot $x_1(t)$ and $x_2(t)$ in the x_1x_2-plane. The curve defined by these parametric equations is called a **Lissajous curve.**

24. A system of equations for the currents $i_1(t)$ and $i_2(t)$ in the network illustrated in Figure 2.49 is

$$L_1 \frac{di_2}{dt} + (R_1 + R_2)i_2 + R_1 i_3 = E(t)$$

$$L_2 \frac{di_3}{dt} + R_1 i_2 + R_1 i_3 = E(t).$$

Use systematic elimination to solve the system when $R_1 = 10\ \Omega$, $R_2 = 20\ \Omega$, $L_1 = 0.04$ h, $L_2 = 0.02$ h, $E = 50$ V, and with initial conditions $i_2(0) = 0$, $i_3(0) = 0$.

25. A projectile shot from a gun has weight $w = mg$ and velocity \mathbf{v} tangent to its path of motion. Ignoring air resistance and all other forces except its weight, determine a system of differential equations that describes the motion. See Figure 3.61. Use systematic elimination to solve the system. [*Hint:* Use Newton's second law of motion in the x and y directions.]

Figure 3.61

26. Determine a system of differential equations that describes the motion in Problem 25 if the projectile encounters a retarding force \mathbf{k} (of magnitude k) acting tangent to the path but opposite to the motion. See Figure 3.62. Use systematic elimination to solve the system. [*Hint:* \mathbf{k} is a multiple of velocity, say $c\mathbf{v}$.]

Figure 3.62

CHAPTER 3 REVIEW EXERCISES

Answers to odd-numbered problems begin on page A-32.

Answer Problems 1–8 without referring back to the text. Fill in the blank or answer true/false.

1. The only solution of the initial-value problem $y'' + x^2 y = 0$, $y(0) = 0$, $y'(0)$, $= 0$ is _____.

2. For the method of undetermined coefficients, the assumed form of the particular solution y_p for $y'' - y = 1 + e^x$ is _____.

3. A constant multiple of a solution of a linear differential equation is also a solution.

4. If f_1 and f_2 are linearly independent functions on an interval I, then their Wronskian $W(f_1, f_2) \neq 0$ for all x in I.

5. If a 10-pound weight stretches a spring 2.5 feet, a 32-pound weight will stretch it _____ feet.

6. The period of simple harmonic motion of an 8-pound weight attached to a spring whose constant is 6.25 lb/ft is _____ seconds.

7. The differential equation describing the motion of a mass attached to a spring is $x'' + 16x = 0$. If the mass is released at $t = 0$ from 1 meter above the equilibrium position with a downward velocity of 3 m/s, the amplitude of vibrations is _____ meters.

8. If simple harmonic motion is described by $x(t) = (\sqrt{2}/2) \sin(2t + \phi)$, the phase angle ϕ is _____ when $x(0) = -\frac{1}{2}$ and $x'(0) = 1$.

9. Give an interval over which $f_1(x) = x^2$ and $f_2(x) = x\,|x|$ are linearly independent. Then give an interval on which f_1 and f_2 are linearly dependent.

10. Without the aid of the Wronskian determine whether the given set of functions is linearly independent or linearly dependent on the indicated interval.

(a) $f_1(x) = \ln x$, $\quad f_2(x) = \ln x^2$, $\quad (0, \infty)$

(b) $f_1(x) = x^n$, $\quad f_2(x) = x^{n+1}$, $\quad n = 1, 2, \ldots, (-\infty, \infty)$

(c) $f_1(x) = x$, $\quad f_2(x) = x + 1$, $\quad (-\infty, \infty)$

(d) $f_1(x) = \cos\left(x + \frac{\pi}{2}\right)$, $\quad f_2(x) = \sin x$, $\quad (-\infty, \infty)$

(e) $f_1(x) = 0$, $\quad f_2(x) = x$, $\quad (-5, 5)$

(f) $f_1(x) = 2$, $\quad f_2(x) = 2x$, $\quad (-\infty, \infty)$

(g) $f_1(x) = x^2$, $\quad f_2(x) = 1 - x^2$, $\quad f_3(x) = 2 + x^2$, $\quad (-\infty, \infty)$

(h) $f_1(x) = xe^{x+1}$, $f_2(x) = (4x - 5)e^x$, $f_3(x) = xe^x$, $(-\infty, \infty)$

11. Suppose $m_1 = 3$, $m_2 = -5$, and $m_3 = 1$ are roots of multiplicity one, two, and three, respectively, of an auxiliary equation. Write down the general solution of the corresponding homogeneous linear DE if it is

(a) an equation with constant coefficients,

(b) a Cauchy–Euler equation.

12. Find a Cauchy–Euler differential equation $ax^2y'' + bxy' + cy = 0$, where a, b, and c are real constants, if it is known that:

(a) $m_1 = 3$ and $m_2 = -1$ are roots of its auxiliary equation,

(b) $m_1 = i$ is a complex root of its auxiliary equation.

In Problems 13–28, use the procedures developed in this chapter to find the general solution of each differential equation.

13. $y'' - 2y' - 2y = 0$

14. $2y'' + 2y' + 3y = 0$

15. $y''' + 10y'' + 25y' = 0$

16. $2y''' + 9y'' + 12y' + 5y = 0$

17. $3y''' + 10y'' + 15y' + 4y = 0$

18. $2y^{(4)} + 3y''' + 2y'' + 6y' - 4y = 0$

19. $y'' - 3y' + 5y = 4x^3 - 2x$

20. $y'' - 2y' + y = x^2e^x$

21. $y''' - 5y'' + 6y' = 8 + 2 \sin x$

22. $y''' - y'' = 6$

23. $y'' - 2y' + 2y = e^x \tan x$

24. $y'' - y = \dfrac{2e^x}{e^x + e^{-x}}$

25. $6x^2y'' + 5xy' - y = 0$

26. $2x^3y''' + 19x^2y'' + 39xy' + 9y = 0$

27. $x^2y'' - 4xy' + 6y = 2x^4 + x^2$

28. $x^2y'' - xy' + y = x^3$

29. Write down the form of the general solution $y = y_c + y_p$ of the given differential equation in the two cases $\omega \neq \alpha$ and $\omega = \alpha$. Do not determine the coefficients in y_p.

(a) $y'' + \omega^2 y = \sin \alpha x$

(b) $y'' - \omega^2 y = e^{\alpha x}$

30. (a) Given that $y = \sin x$ is a solution of $y^{(4)} + 2y''' + 11y'' + 2y' + 10y = 0$, find the general solution of the DE *without the aid of a calculator or a computer.*

(b) Find a linear second-order differential equation with constant coefficients for which $y_1 = 1$ and $y_2 = e^{-x}$ are solutions of the associated homogeneous equation and $y_p = \frac{1}{2}x^2 - x$ is a particular solution of the nonhomogeneous equation.

31. (a) Write the general solution of the fourth-order DE $y^{(4)} - 2y'' + y = 0$ entirely in terms of hyperbolic functions.

(b) Write down the form of a particular solution of $y^{(4)} - 2y'' + y = \sinh x$.

32. Consider the differential equation $x^2y'' - (x^2 + 2x)y' + (x + 2)y = x^3$. Verify that $y_1 = x$ is one solution of the associated homogeneous equation. Then show that the method of reduction of order discussed in Section 3.2 leads both to a second solution y_2 of the homogeneous equation and to a particular solution y_p of the nonhomogeneous equation. Form the general solution of the DE on the interval $(0, \infty)$.

In Problems 33–38, solve the given differential equation subject to the indicated conditions.

33. $y'' - 2y' + 2y = 0$, $y\left(\dfrac{\pi}{2}\right) = 0$, $y(\pi) = -1$

34. $y'' + 2y' + y = 0$, $y(-1) = 0$, $y'(0) = 0$

35. $y'' - y = x + \sin x$, $y(0) = 2$, $y'(0) = 3$

36. $y'' + y = \sec^3 x$, $y(0) = 1$, $y'(0) = \dfrac{1}{2}$

37. $y'y'' = 4x$, $y(1) = 5$, $y'(1) = 2$

38. $2y'' = 3y^2$, $y(0) = 1$, $y'(0) = 1$

39. Computer Lab Assignment (a) Use a CAS as an aid in finding the roots of the auxiliary equation for $12y^{(4)} + 64y''' + 59y'' - 23y' - 12y = 0$. Give the general solution of the equation.

(b) Solve the DE in part **(a)** subject to the initial conditions $y(0) = -1$, $y'(0) = 2$, $y''(0) = 5$, $y'''(0) = 0$. Use a CAS as an aid in solving the resulting systems of four equations in four unknowns.

40. Find a member of the family of solutions of $xy'' + y' + \sqrt{x} = 0$ whose graph is tangent to the x-axis at $x = 1$. Use a graphing utility to obtain the solution curve.

In Problems 41–44, use systematic elimination to solve the given system.

41. $\dfrac{dx}{dt} + \dfrac{dy}{dt} = 2x + 2y + 1$

$\dfrac{dx}{dt} + 2\dfrac{dy}{dt} = y + 3$

42. $\dfrac{dx}{dt} = 2x + y + t - 2$

$\dfrac{dy}{dt} = 3x + 4y - 4t$

43. $(D - 2)x \qquad - y = -e^t$

$\qquad -3x + (D - 4)y = -7e^t$

44. $(D + 2)x + (D + 1)y = \sin 2t$
 $5x + (D + 3)y = \cos 2t$

45. A free undamped spring/mass system oscillates with a period of 3 s. When 8 lb is removed from the spring, the system then has a period of 2 s. What was the weight of the original mass on the spring?

46. A 12-pound weight stretches a spring 2 feet. The weight is released from a point 1 foot below the equilibrium position with an upward velocity of 4 ft/s.

 (a) Find the equation describing the resulting simple harmonic motion.

 (b) What are the amplitude, period, and frequency of motion?

 (c) At what times does the weight return to the point 1 foot below the equilibrium position?

 (d) At what times does the weight pass through the equilibrium position moving upward? moving downward?

 (e) What is the velocity of the weight at $t = 3\pi/16$ s?

 (f) At what times is the velocity zero?

47. A spring with constant $k = 2$ is suspended in a liquid that offers a damping force numerically equal to 4 times the instantaneous velocity. If a mass m is suspended from the spring, determine the values of m for which the subsequent free motion is nonoscillatory.

48. A 32-pound weight stretches a spring 6 inches. The weight moves through a medium offering a damping force numerically equal to β times the instantaneous velocity. Determine the values of β for which the system will exhibit oscillatory motion.

49. A series circuit contains an inductance of $L = 1$ h, a capacitance of $C = 10^{-4}$ f, and an electromotive force of $E(t) = 100 \sin 50t$ V. Initially the charge q and current i are zero.

 (a) Find the equation for the charge at time t.

 (b) Find the equation for the current at time t.

 (c) Find the times for which the charge on the capacitor is zero.

50. Show that the current $i(t)$ in an *LRC* series circuit satisfies the differential equation

$$L\frac{d^2i}{dt^2} + R\frac{di}{dt} + \frac{1}{C}i = E'(t),$$

where $E'(t)$ denotes the derivative of $E(t)$.

51. Consider the boundary-value problem

$$y'' + \lambda y = 0, \quad y(0) = y(2\pi), \quad y'(0) = y'(2\pi).$$

Show that except for the case $\lambda = 0$, there are two independent eigenfunctions corresponding to each eigenvalue.

52. A bead is constrained to slide along a frictionless rod of length L. The rod is rotating in a vertical plane with a constant angular velocity ω about a pivot P fixed at the midpoint of the rod, but the design of the pivot allows the bead to move along the entire length of the rod. Let $r(t)$ denote the position of the bead relative to this rotating coordinate system as shown in Figure 3.63. In order to apply Newton's second law of motion to this rotating frame of reference it is necessary to use the fact that the net force acting on the bead is the sum of the real forces (in this case, the force due to gravity) and the inertial forces (coriolis, transverse, and centrifugal). The mathematics is a little complicated, so we just give the resulting differential equation for r,

$$m\frac{d^2r}{dt^2} = m\omega^2 r - mg \sin(\omega t).$$

 (a) Solve the foregoing DE subject to the initial conditions $r(0) = r_0,\ r'(0) = v_0$.

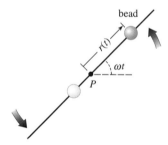

Figure 3.63

 (b) Determine initial conditions for which the bead exhibits simple harmonic motion. What is the minimum length L of the rod for which it can accommodate simple harmonic motion of the bead?

 (c) For initial conditions other than those obtained in part **(b)**, the bead must eventually fly off the rod. Explain using the solution $r(t)$ in part **(a)**.

 (d) Suppose $\omega = 1$ rad/s. Use a graphing utility to plot the graph of the solution $r(t)$ for the initial conditions $r(0) = 0$, $r'(0) = v_0$, where v_0 is 0, 10, 15, 16, 16.1, and 17.

 (e) Suppose the length of the rod is $L = 40$ ft. For each pair of initial conditions in part **(d)**, use a root-finding application to find the total time that the bead stays on the rod.

The Laplace Transform

INTRODUCTION

In the linear mathematical models for physical systems such as a spring/mass system or a series electrical circuit, the right member, or input, of the ODEs

$$m\,\frac{d^2x}{dt^2} + \beta\,\frac{dx}{dt} + kx = f(t) \qquad L\,\frac{d^2q}{dt^2} + R\,\frac{dq}{dt} + \frac{1}{C}q = E(t)$$

is a driving function and represents either an external force $f(t)$ or an impressed voltage $E(t)$. In Section 3.8 we considered problems in which f and E were *continuous*. However, *discontinuous* driving functions are not uncommon. For example, the impressed voltage on a circuit could be

piecewise continuous and periodic such as the function shown in the following figure. Solving

the differential equation of the circuit in this case is difficult but not impossible. The Laplace transform studied in this chapter is an invaluable tool in solving problems such as these (Section 4.4).

4.1 Definition of the Laplace Transform

■ **Linearity Property** In elementary calculus you learned that differentiation and integration transform a function into another function. For example, the function $f(x) = x^2$ is transformed, in turn, into a linear function, a family of cubic polynomial functions, and a constant by the operations of differentiation, indefinite integration, and definite integration:

$$\frac{d}{dx} x^2 = 2x, \qquad \int x^2 \, dx = \frac{x^3}{3} + c, \qquad \int_0^3 x^2 \, dx = 9.$$

Moreover, these three operations possess the **linearity property.** This means that for any constants α and β

$$\frac{d}{dx} [\alpha f(x) + \beta g(x)] = \alpha \frac{d}{dx} f(x) + \beta \frac{d}{dx} g(x)$$

$$\int [\alpha f(x) + \beta g(x)] \, dx = \alpha \int f(x) \, dx + \beta \int g(x) \, dx \qquad (1)$$

$$\int_a^b [\alpha f(x) + \beta g(x)] \, dx = \alpha \int_a^b f(x) \, dx + \beta \int_a^b g(x) \, dx$$

provided each derivative and integral exists.

If $f(x, y)$ is a function of two variables, then a definite integral of f with respect to one of the variables leads to a function of the other variable. For example, by holding y constant we see that $\int_1^2 2xy^2 \, dx = 3y^2$. Similarly, a definite integral such as $\int_a^b K(s, t)f(t)dt$ transforms a function $f(t)$ into a function of the variable s. We are particularly interested in **integral transforms** of this last kind, where the interval of integration is the unbounded interval $[0, \infty)$.

■ **Basic Definition** If $f(t)$ is defined for $t \geq 0$, then the improper integral $\int_0^\infty K(s, t)f(t) \, dt$ is defined as a limit:

$$\int_0^\infty K(s, t)f(t) \, dt = \lim_{b \to \infty} \int_0^b K(s, t)f(t) \, dt.$$

If the limit exists, the integral is said to exist or to be convergent; if the limit does not exist, the integral does not exist and is said to be divergent. The foregoing limit will, in general, exist for only certain values of the variable s. The choice $K(s, t) = e^{-st}$ gives us an especially important integral transform.

> **DEFINITION 4.1** Laplace Transform
>
> Let f be a function defined for $t \geq 0$. Then the integral
>
> $$\mathcal{L}\{f(t)\} = \int_0^\infty e^{-st} f(t)\, dt \qquad (2)$$
>
> is said to be the **Laplace transform** of f, provided the integral converges.

When the defining integral (2) converges, the result is a function of s. In general discussion we shall use a lowercase letter to denote the function being transformed and the corresponding capital letter to denote its Laplace transform; for example,

$$\mathcal{L}\{f(t)\} = F(s), \qquad \mathcal{L}\{g(t)\} = G(s), \qquad \mathcal{L}\{y(t)\} = Y(s).$$

Example 1 Using Definition 4.1

Evaluate $\mathcal{L}\{1\}$.

SOLUTION From (2),

$$\mathcal{L}\{1\} = \int_0^\infty e^{-st}(1)\, dt = \lim_{b \to \infty} \int_0^b e^{-st}\, dt$$

$$= \lim_{b \to \infty} \frac{-e^{-st}}{s} \bigg|_0^b = \lim_{b \to \infty} \frac{-e^{-sb} + 1}{s} = \frac{1}{s}$$

provided $s > 0$. In other words, when $s > 0$, the exponent $-sb$ is negative and $e^{-sb} \to 0$ as $b \to \infty$. The integral diverges for $s < 0$. ❑

The use of the limit sign becomes somewhat tedious, so we shall adopt the notation $\big|_0^\infty$ as a shorthand to writing $\lim_{b \to \infty} (\)\big|_0^b$. For example,

$$\mathcal{L}\{1\} = \int_0^\infty e^{-st}(1)\, dt = \frac{-e^{-st}}{s} \bigg|_0^\infty = \frac{1}{s}, \qquad s > 0$$

At the upper limit, it is understood we mean $e^{-st} \to 0$ as $t \to \infty$ for $s > 0$.

Example 2 Using Definition 4.1

Evaluate $\mathcal{L}\{t\}$.

SOLUTION From Definition 4.1, we have $\mathcal{L}\{t\} = \int_0^\infty e^{-st} t\, dt$. Integrating by parts and using $\lim_{t \to \infty} te^{-st} = 0$, $s > 0$, along with the result from Example 1, we obtain

$$\mathcal{L}\{t\} = \frac{-te^{-st}}{s} \bigg|_0^\infty + \frac{1}{s} \int_0^\infty e^{-st}\, dt = \frac{1}{s}\, \mathcal{L}\{1\} = \frac{1}{s}\left(\frac{1}{s}\right) = \frac{1}{s^2}.$$ ❑

Example 3 Using Definition 4.1

Evaluate $\mathcal{L}\{e^{-3t}\}$.

SOLUTION From Definition 4.1, we have

$$\mathscr{L}\{e^{-3t}\} = \int_0^\infty e^{-st}e^{-3t}\,dt = \int_0^\infty e^{-(s+3)t}\,dt$$

$$= \frac{-e^{(s+3)t}}{s+3}\bigg|_0^\infty$$

$$= \frac{1}{s+3}, \qquad s > -3.$$

The result follows from the fact that $\lim_{t\to\infty} e^{-(s+3)t} = 0$ for $s+3 > 0$ or $s > -3$. ❑

Example 4 **Using Definition 4.1**

Evaluate $\mathscr{L}\{\sin 2t\}$.

SOLUTION From Definition 4.1 and integration by parts we have

$$\mathscr{L}\{\sin 2t\} = \int_0^\infty e^{-st}\sin 2t\,dt = \frac{-e^{-st}\sin 2t}{s}\bigg|_0^\infty + \frac{2}{s}\int_0^\infty e^{-st}\cos 2t\,dt$$

$$= \frac{2}{s}\int_0^\infty e^{-st}\cos 2t\,dt, \qquad s > 0$$

$$\lim_{t\to\infty} e^{-st}\cos 2t = 0,\ s > 0 \qquad\qquad\qquad \text{Laplace transform of } \sin 2t$$
$$\downarrow \qquad\qquad\qquad\qquad\qquad\qquad\qquad \downarrow$$

$$= \frac{2}{s}\left[\frac{-e^{-st}\cos 2t}{s}\bigg|_0^\infty - \frac{2}{s}\int_0^\infty e^{-st}\sin 2t\,dt\right]$$

$$= \frac{2}{s^2} - \frac{4}{s^2}\mathscr{L}\{\sin 2t\}.$$

At this point we have an equation with $\mathscr{L}\{\sin 2t\}$ on both sides of the equality. Solving for that quantity yields the result

$$\mathscr{L}\{\sin 2t\} = \frac{2}{s^2+4}, \qquad s > 0.$$ ❑

■ **\mathscr{L} Is a Linear Transform** For a sum of functions, we can write

$$\int_0^\infty e^{-st}[\alpha f(t) + \beta g(t)]\,dt = \alpha \int_0^\infty e^{-st}f(t)\,dt + \beta \int_0^\infty e^{-st}g(t)\,dt$$

whenever both integrals converge for $s > c$. Hence it follows that

$$\mathscr{L}\{\alpha f(t) + \beta g(t)\} = \alpha\mathscr{L}\{f(t)\} + \beta\mathscr{L}\{g(t)\} = \alpha F(s) + \beta G(s). \tag{3}$$

Because of the property given in (3), \mathscr{L} is said to be a **linear transform.** For example, from Examples 1 and 2,

$$\mathscr{L}\{1 + 5t\} = \mathscr{L}\{1\} + 5\mathscr{L}\{t\} = \frac{1}{s} + \frac{1}{s^2},$$

and from Examples 3 and 4,

$$\mathscr{L}\{4e^{-3t} - 10\sin 2t\} = 4\mathscr{L}\{e^{-3t}\} - 10\mathscr{L}\{\sin 2t\} = \frac{4}{s+3} - \frac{20}{s^2+4}.$$

We state the generalization of some of the preceding examples by means of the next theorem. From this point on we shall also refrain from stating any restrictions on s; it is understood that s is sufficiently restricted to guarantee the convergence of the appropriate Laplace transform.

A more extensive list of 👉 *transforms is given in Appendix III.*

Figure 4.1

Figure 4.2

(a)

(b)

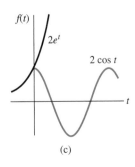

(c)

Figure 4.3

> ### THEOREM 4.1 Transforms of Some Basic Functions
>
> $$\text{(a) } \mathscr{L}\{1\} = \frac{1}{s}$$
>
> $$\text{(b) } \mathscr{L}\{t^n\} = \frac{n!}{s^{n+1}}, \quad n = 1, 2, 3, \cdots \qquad \text{(c) } \mathscr{L}\{e^{at}\} = \frac{1}{s-a}$$
>
> $$\text{(d) } \mathscr{L}\{\sin kt\} = \frac{k}{s^2 + k^2} \qquad\qquad \text{(e) } \mathscr{L}\{\cos kt\} = \frac{s}{s^2 + k^2}$$
>
> $$\text{(f) } \mathscr{L}\{\sinh kt\} = \frac{k}{s^2 - k^2} \qquad\qquad \text{(g) } \mathscr{L}\{\cosh kt\} = \frac{s}{s^2 - k^2}$$

■ **Sufficient Conditions for Existence of $\mathscr{L}\{f(t)\}$** The integral that defines the Laplace transform does not have to converge. For example, neither $\mathscr{L}\{1/t\}$ nor $\mathscr{L}\{e^{t^2}\}$ exists. Sufficient conditions guaranteeing the existence of $\mathscr{L}\{f(t)\}$ are that f be piecewise continuous on $[0, \infty)$ and that f be of exponential order for $t > T$. Recall that a function f is **piecewise continuous** on $[0, \infty)$ if, in any interval $0 \le a \le t \le b$, there are at most a finite number of points t_k, $k = 1, 2, \ldots, n$ $(t_{k-1} < t_k)$, at which f has finite discontinuities and is continuous on each open interval $t_{k-1} < t < t_k$. See Figure 4.1. The concept of **exponential order** is defined in the following manner.

> ### DEFINITION 4.2 Exponential Order
>
> A function f is said to be of **exponential order** c if there exist constants c, $M > 0$, and $T > 0$ such that $|f(t)| \le Me^{ct}$ for all $t > T$.

If f is an *increasing* function, then the condition $|f(t)| \le Me^{ct}$, $t > T$, simply states that the graph of f on the interval (T, ∞) does not grow faster than the graph of the exponential function Me^{ct}, where c is a positive constant. See Figure 4.2. The functions $f(t) = t$, $f(t) = e^{-t}$, and $f(t) = 2 \cos t$ are all of exponential order $c = 1$ for $t > 0$ since we have, respectively,

$$|t| \le e^t, \qquad |e^{-t}| \le e^t, \qquad |2 \cos t| \le 2e^t.$$

A comparison of the graphs on the interval $[0, \infty)$ is given in Figure 4.3.

A function such as $f(t) = e^{t^2}$ is not of exponential order since, as shown in Figure 4.4, its graph grows faster than any positive linear power of e for $t > c > 0$.

A positive integral power of t is always of exponential order since, for $c > 0$,

$$|t^n| \le Me^{ct} \quad \text{or} \quad \left|\frac{t^n}{e^{ct}}\right| \le M \qquad \text{for } t > T$$

is equivalent to showing that $\lim_{t \to \infty} t^n/e^{ct}$ is finite for $n = 1, 2, 3, \ldots$. The result follows by n applications of L'Hôpital's rule.

Figure 4.4

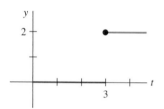

Figure 4.5

> **THEOREM 4.2 Sufficient Conditions for Existence**
>
> If $f(t)$ is piecewise continuous on the interval $[0, \infty)$ and of exponential order c for $t > T$, then $\mathcal{L}\{f(t)\}$ exists for $s > c$.

PROOF By the additive interval property of definite integrals,

$$\mathcal{L}\{f(t)\} = \int_0^T e^{-st}f(t)dt + \int_T^\infty e^{-st}f(t)\,dt = I_1 + I_2.$$

The integral I_1 exists because it can be written as a sum of integrals over intervals on which $e^{-st}f(t)$ is continuous. Now

$$|I_2| \le \int_T^\infty |e^{-st}f(t)|\,dt \le M \int_T^\infty e^{-st}e^{ct}\,dt$$

$$= M \int_T^\infty e^{-(s-c)t}\,dt = -M\,\frac{e^{-(s-c)t}}{s-c}\bigg|_T^\infty = M\,\frac{e^{-(s-c)T}}{s-c}$$

for $s > c$. Since $\int_T^\infty Me^{-(s-c)t}\,dt$ converges, the integral $\int_T^\infty |e^{-st}f(t)|\,dt$ converges by the comparison test for improper integrals. This, in turn, implies that I_2 exists for $s > c$. The existence of I_1 and I_2 implies that $\mathcal{L}\{f(t)\} = \int_0^\infty e^{-st}f(t)\,dt$ exists for $s > c$. ❑

Example 5 **Transform of a Piecewise Defined Function**

Evaluate $\mathcal{L}\{f(t)\}$ for $f(t) = \begin{cases} 0, & 0 \le t < 3 \\ 2, & t \ge 3. \end{cases}$

SOLUTION This piecewise continuous function appears in Figure 4.5. Since f is defined in two pieces, $\mathcal{L}\{f(t)\}$ is expressed as the sum of two integrals:

$$\mathcal{L}\{f(t)\} = \int_0^\infty e^{-st}f(t)\,dt = \int_0^3 e^{-st}(0)\,dt + \int_3^\infty e^{-st}(2)\,dt$$

$$= -\frac{2e^{-st}}{s}\bigg|_3^\infty$$

$$= \frac{2e^{-3s}}{s}, \qquad s > 0. ❑$$

Throughout this entire chapter we shall be concerned primarily with functions that are both piecewise continuous and of exponential order. We note, however, that these conditions are sufficient but not necessary for the existence of a Laplace transform. The function $f(t) = t^{-1/2}$ is not piecewise continuous on the interval $[0, \infty)$, but its Laplace transform exists. See Problem 40 in Exercises 4.1.

In Problems 1–18, use Definition 4.1 to find $\mathscr{L}\{f(t)\}$.

1. $f(t) = \begin{cases} -1, & 0 \le t < 1 \\ 1, & t \ge 1 \end{cases}$

2. $f(t) = \begin{cases} 4, & 0 \le t < 2 \\ 0, & t \ge 2 \end{cases}$

3. $f(t) = \begin{cases} t, & 0 \le t < 1 \\ 1, & t \ge 1 \end{cases}$

4. $f(t) = \begin{cases} 2t + 1, & 0 \le t < 1 \\ 0, & t \ge 1 \end{cases}$

5. $f(t) = \begin{cases} \sin t, & 0 \le t < \pi \\ 0, & t \ge \pi \end{cases}$

6. $f(t) = \begin{cases} 0, & 0 \le t < \pi/2 \\ \cos t, & t \ge \pi/2 \end{cases}$

7.

Figure 4.6

8.

Figure 4.7

9.

Figure 4.8

10.

Figure 4.9

11. $f(t) = e^{t+7}$

12. $f(t) = e^{-2t-5}$

13. $f(t) = te^{4t}$

14. $f(t) = t^2 e^{-2t}$

15. $f(t) = e^{-t} \sin t$

16. $f(t) = e^t \cos t$

17. $f(t) = t \cos t$

18. $f(t) = t \sin t$

In Problems 19–36, use Theorem 4.1 to find $\mathscr{L}\{f(t)\}$.

19. $f(t) = 2t^4$

20. $f(t) = t^5$

21. $f(t) = 4t - 10$

22. $f(t) = 7t + 3$

23. $f(t) = t^2 + 6t - 3$

24. $f(t) = -4t^2 + 16t + 9$

25. $f(t) = (t + 1)^3$

26. $f(t) = (2t - 1)^3$

27. $f(t) = 1 + e^{4t}$

28. $f(t) = t^2 - e^{-9t} + 5$

29. $f(t) = (1 + e^{2t})^2$

30. $f(t) = (e^t - e^{-t})^2$

31. $f(t) = 4t^2 - 5 \sin 3t$

32. $f(t) = \cos 5t + \sin 2t$

33. $f(t) = \sinh kt$

34. $f(t) = \cosh kt$

35. $f(t) = e^t \sinh t$

36. $f(t) = e^{-t} \cosh t$

In Problems 37 and 38, find $\mathscr{L}\{f(t)\}$ by first using an appropriate trigonometric identity.

37. $f(t) = \sin 2t \cos 2t$

38. $f(t) = \cos^2 t$

39. One definition of the **gamma function** is given by the improper integral

$$\Gamma(\alpha) = \int_0^\infty t^{\alpha-1} e^{-t}\, dt, \quad \alpha > 0.$$

(a) Show that $\Gamma(\alpha + 1) = \alpha \Gamma(\alpha)$.

(b) Show that $\mathscr{L}\{t^\alpha\} = \dfrac{\Gamma(\alpha+1)}{s^{\alpha+1}}$, $\quad \alpha > -1$.

40. Use the fact that $\Gamma(\frac12) = \sqrt{\pi}$ and Problem 39 to find the Laplace transform of:

(a) $f(t) = t^{-1/2}$ **(b)** $f(t) = t^{1/2}$ **(c)** $f(t) = t^{3/2}$.

41. Although the proof of part **(b)** of Theorem 4.1 requires the use of mathematical induction, the result can be justified informally. Show that $\mathscr{L}\{t^n\} = \dfrac{n}{s}\,\mathscr{L}\{t^{n-1}\}$. See what happens when you let n take on the successive values $1, 2, 3, \ldots$.

42. Make up a function $F(t)$ that is of exponential order, but $f(t) = F'(t)$ is not of exponential order. Make up a function f that is not of exponential order but whose Laplace transform exists.

43. Suppose that $\mathscr{L}\{f_1(t)\} = F_1(s)$ for $s > c_1$ and that $\mathscr{L}\{f_2(t)\} = F_2(s)$ for $s > c_2$. When does $\mathscr{L}\{f_1(t) + f_2(t)\} = F_1(s) + F_2(s)$?

44. Figure 4.4 suggests, but does not prove, that the function $f(t) = e^{t^2}$ is not of exponential order. How does the observation that $t^2 > \ln M + ct$, for $M > 0$ and t sufficiently large, show that $e^{t^2} > Me^{ct}$ for any c?

45. Use part **(c)** of Theorem 4.1 to show that $\mathscr{L}\{e^{(a+ib)t}\} = \dfrac{s - a + ib}{(s-a)^2 + b^2}$, where a and b are real and $i^2 = -1$. Show how Euler's formula (page 118) can then be used to deduce the results

$$\mathscr{L}\{e^{at} \cos bt\} = \frac{s - a}{(s-a)^2 + b^2}$$

and

$$\mathscr{L}\{e^{at} \sin bt\} = \frac{b}{(s-a)^2 + b^2}.$$

46. Under what conditions is a linear function $f(x) = mx + b$, $m \ne 0$, a linear transformation?

4.2 The Inverse Transform and Transforms of Derivatives

■ **Inverse Transform** If $F(s)$ represents the Laplace transform of a function $f(t)$, that is, $\mathscr{L}\{f(t)\} = F(s)$, we then say $f(t)$ is the **inverse Laplace transform** of $F(s)$ and write $f(t) = \mathscr{L}^{-1}\{F(s)\}$. For example, from Examples 1, 2, and 3 in Section 4.1 we have, respectively,

$$1 = \mathscr{L}^{-1}\left\{\frac{1}{s}\right\}, \quad t = \mathscr{L}^{-1}\left\{\frac{1}{s^2}\right\}, \quad \text{and} \quad e^{-3t} = \mathscr{L}^{-1}\left\{\frac{1}{s+3}\right\}.$$

The analogue of Theorem 4.1 for the inverse transform is presented next.

THEOREM 4.3 Some Inverse Transforms

$$\textbf{(a)}\; 1 = \mathscr{L}^{-1}\left\{\frac{1}{s}\right\}$$

$$\textbf{(b)}\; t^n = \mathscr{L}^{-1}\left\{\frac{n!}{s^{n+1}}\right\}, \quad n = 1, 2, 3, \ldots \qquad \textbf{(c)}\; e^{at} = \mathscr{L}^{-1}\left\{\frac{1}{s-a}\right\}$$

$$\textbf{(d)}\; \sin kt = \mathscr{L}^{-1}\left\{\frac{k}{s^2+k^2}\right\} \qquad\qquad \textbf{(e)}\; \cos kt = \mathscr{L}^{-1}\left\{\frac{s}{s^2+k^2}\right\}$$

$$\textbf{(f)}\; \sinh kt = \mathscr{L}^{-1}\left\{\frac{k}{s^2-k^2}\right\} \qquad\qquad \textbf{(g)}\; \cosh kt = \mathscr{L}^{-1}\left\{\frac{s}{s^2-k^2}\right\}$$

When evaluating inverse transforms, it often happens that a function of s under consideration does not match *exactly* the form of a Laplace transform $F(s)$ given in a table. It may be necessary to "fix up" the function of s by multiplying and dividing by an appropriate constant.

Example 1 Applying Theorem 4.3

Evaluate (a) $\mathscr{L}^{-1}\left\{\dfrac{1}{s^5}\right\}$ (b) $\mathscr{L}^{-1}\left\{\dfrac{1}{s^2+7}\right\}$.

SOLUTION (a) To match the form given in part **(b)** of Theorem 4.3, we identify $n + 1 = 5$ or $n = 4$ and then multiply and divide by 4!:

$$\mathscr{L}^{-1}\left\{\frac{1}{s^5}\right\} = \frac{1}{4!}\,\mathscr{L}^{-1}\left\{\frac{4!}{s^5}\right\} = \frac{1}{24}\,t^4.$$

(b) To match the form given in part **(d)** of Theorem 4.3 we identify $k^2 = 7$ and so $k = \sqrt{7}$. We fix up the expression by multiplying and dividing by $\sqrt{7}$:

$$\mathscr{L}^{-1}\left\{\frac{1}{s^2+7}\right\} = \frac{1}{\sqrt{7}}\,\mathscr{L}^{-1}\left\{\frac{\sqrt{7}}{s^2+7}\right\} = \frac{1}{\sqrt{7}}\,\sin\sqrt{7}\,t. \qquad \square$$

■ **\mathscr{L}^{-1} Is a Linear Transform** The inverse Laplace transform is also a linear transform; that is, for constants α and β,

$$\mathscr{L}^{-1}\{\alpha F(s) + \beta G(s)\} = \alpha\mathscr{L}^{-1}\{F(s)\} + \beta\mathscr{L}^{-1}\{G(s)\}, \qquad (1)$$

where F and G are the transforms of some functions f and g. Like (2) of Section 4.1, (1) extends to any finite linear combination of Laplace transforms.

Example 2 Termwise Division and Linearity

Evaluate $\mathscr{L}^{-1}\left\{\dfrac{-2s+6}{s^2+4}\right\}$.

SOLUTION We first rewrite the given function of s as two expressions by means of termwise division and then use (1):

$$
\mathscr{L}^{-1}\left\{\frac{-2s+6}{s^2+4}\right\} = \overset{\substack{\text{termwise division}\\\downarrow}}{\mathscr{L}^{-1}\left\{\frac{-2s}{s^2+4}+\frac{6}{s^2+4}\right\}} = \overset{\substack{\text{linearity and fixing up constants}\\\downarrow}}{-2\,\mathscr{L}^{-1}\left\{\frac{s}{s^2+4}\right\}+\frac{6}{2}\,\mathscr{L}^{-1}\left\{\frac{2}{s^2+4}\right\}} \qquad (2)
$$

$$
= -2\cos 2t + 3\sin 2t. \qquad \leftarrow \text{parts (e) and (d) of Theorem 4.3 with } k=2 \qquad \square
$$

■ **Partial Fractions** Partial fractions play an important role in finding inverse Laplace transforms. As mentioned in Section 2.2, the decomposition of a rational expression into component fractions can be done quickly by means of a single command on most computer algebra systems. Indeed, some CASs have packages that implement Laplace transform and inverse Laplace transform commands. But for those of you without access to such software, we will review in this and subsequent sections some of the basic algebra in the important cases in which the denominator of a Laplace transform $F(s)$ contains distinct linear factors, repeated linear factors, and quadratic polynomials with no real factors. Although we shall examine each of these cases as this chapter develops, it still might be a good idea for you to consult either a calculus text or a current precalculus text for a more comprehensive review of this theory.

Example 3 Partial Fractions and Linearity

Partial fractions: ☞
distinct linear factors

Evaluate $\mathscr{L}^{-1}\left\{\dfrac{s^2+6s+9}{(s-1)(s-2)(s+4)}\right\}$.

SOLUTION There exist unique constants A, B, and C such that

$$
\frac{s^2+6s+9}{(s-1)(s-2)(s+4)} = \frac{A}{s-1}+\frac{B}{s-2}+\frac{C}{s+4}
$$

$$
= \frac{A(s-2)(s+4)+B(s-1)(s+4)+C(s-1)(s-2)}{(s-1)(s-2)(s+4)}.
$$

Since the denominators are identical, the numerators are identical:

$$
s^2+6s+9 = A(s-2)(s+4)+B(s-1)(s+4)+C(s-1)(s-2). \qquad (3)
$$

By comparing coefficients of powers of s on both sides of the equality, we know that (3) is equivalent to a system of three equations in the three unknowns A, B, and C. However, recall that there is a shortcut for determining these unknowns. If we set $s=1$, $s=2$, and $s=-4$ in (3) we obtain, respectively,*

$$
16 = A(-1)(5), \quad 25 = B(1)(6), \quad 1 = C(-5)(-6),
$$

* The numbers 1, 2, and -4 are the zeros of the common denominator $(s-1)(s-2)(s+4)$.

and so $A = -16/5$, $B = 25/6$, $C = 1/30$. Hence the partial fraction decomposition is

$$\frac{s^2 + 6s + 9}{(s-1)(s-2)(s+4)} = -\frac{16/5}{s-1} + \frac{25/6}{s-2} + \frac{1/30}{s+4}, \tag{4}$$

and thus, from the linearity of \mathscr{L}^{-1} and part (c) of Theorem 4.3,

$$\mathscr{L}^{-1}\left\{\frac{s^2 + 6s + 9}{(s-1)(s-2)(s+4)}\right\} = -\frac{16}{5}\,\mathscr{L}^{-1}\left\{\frac{1}{s-1}\right\} + \frac{25}{6}\,\mathscr{L}^{-1}\left\{\frac{1}{s-2}\right\} + \frac{1}{30}\,\mathscr{L}^{-1}\left\{\frac{1}{s+4}\right\}$$

$$= -\frac{16}{5}\,e^t + \frac{25}{6}\,e^{2t} + \frac{1}{30}\,e^{-4t}. \tag{5} \quad \square$$

■ **Transforming a Derivative** As pointed out in the introduction to this chapter, our immediate goal is to use the Laplace transform to solve differential equations. To that end we need to evaluate quantities such as $\mathscr{L}\{dy/dt\}$ and $\mathscr{L}\{d^2y/dt^2\}$. For example, if f' is continuous for $t \geq 0$, then integration by parts gives

$$\mathscr{L}\{f'(t)\} = \int_0^\infty e^{-st}f'(t)\,dt = e^{-st}f(t)\Big|_0^\infty + s\int_0^\infty e^{-st}f(t)\,dt$$

$$= -f(0) + s\mathscr{L}\{f(t)\}$$

or $\qquad\qquad \mathscr{L}\{f'(t)\} = sF(s) - f(0). \tag{6}$

Here we have assumed that $e^{-st}f(t) \to 0$ as $t \to \infty$. Similarly, with the aid of (6),

$$\mathscr{L}\{f''(t)\} = \int_0^\infty e^{-st}f''(t)\,dt = e^{-st}f'(t)\Big|_0^\infty + s\int_0^\infty e^{-st}f'(t)\,dt$$

$$= -f'(0) + s\mathscr{L}\{f'(t)\}$$

$$= s[sF(s) - f(0)] - f'(0) \qquad\qquad\qquad \leftarrow \text{From (6)}$$

or $\qquad\qquad \mathscr{L}\{f''(t)\} = s^2F(s) - sf(0) - f'(0). \tag{7}$

In like manner it can be shown that

$$\mathscr{L}\{f'''(t)\} = s^3F(s) - s^2f(0) - sf'(0) - f''(0). \tag{8}$$

The recursive nature of the Laplace transform of the derivatives of a function f should be apparent from the results in (6), (7), and (8). The next theorem gives the Laplace transform of the nth derivative of f. The proof is omitted.

THEOREM 4.4 Transform of a Derivative

If $f, f', \ldots, f^{(n-1)}$ are continuous on $[0, \infty)$ and are of exponential order and if $f^{(n)}(t)$ is piecewise continuous on $[0, \infty)$, then

$$\mathscr{L}\{f^{(n)}(t)\} = s^nF(s) - s^{n-1}f(0) - s^{n-2}f'(0) - \cdots - f^{(n-1)}(0),$$

where $F(s) = \mathscr{L}\{f(t)\}$.

■ **Solving Linear ODEs** It is apparent from the general result given in Theorem 4.4 that $\mathscr{L}\{d^ny/dt^n\}$ depends on $Y(s) = \mathscr{L}\{y(t)\}$ and the $n-1$ derivatives of $y(t)$ evaluated at $t = 0$. This property makes the Laplace transform ideally suited for

solving linear initial-value problems in which the differential equation has *constant coefficients*. Such a differential equation is simply a linear combination of terms $y, y', y'', \ldots, y^{(n)}$:

$$a_n \frac{d^n y}{dt^n} + a_{n-1} \frac{d^{n-1} y}{dt^{n-1}} + \cdots + a_0 y = g(t),$$

$$y(0) = y_0, \qquad y'(0) = y_1, \cdots, \qquad y^{(n-1)}(0) = y_{n-1},$$

where the a_i, $i = 0, 1, \ldots, n$ and $y_0, y_1, \ldots, y_{n-1}$ are constants. By the linearity property, the Laplace transform of this linear combination is a linear combination of Laplace transforms:

$$a_n \mathcal{L}\left\{\frac{d^n y}{dt^n}\right\} + a_{n-1} \mathcal{L}\left\{\frac{d^{n-1} y}{dt^{n-1}}\right\} + \cdots + a_0 \mathcal{L}\{y\} = \mathcal{L}\{g(t)\}. \tag{9}$$

From Theorem 4.4, (9) becomes

$$a_n[s^n Y(s) - s^{n-1} y(0) - \cdots - y^{(n-1)}(0)]$$
$$+ a_{n-1}[s^{n-1} Y(s) - s^{n-2} y(0) - \cdots - y^{(n-2)}(0)] + \cdots + a_0 Y(s) = G(s), \tag{10}$$

where $\mathcal{L}\{y(t)\} = Y(s)$ and $\mathcal{L}\{g(t)\} = G(s)$. In other words, *the Laplace transform of a linear differential equation with constant coefficients becomes an algebraic equation in $Y(s)$.* If we solve the general transformed equation (10) for the symbol $Y(s)$, we first obtain $P(s)Y(s) = Q(s) + G(s)$, and then write

$$Y(s) = \frac{Q(s)}{P(s)} + \frac{G(s)}{P(s)}, \tag{11}$$

where $P(s) = a_n s^n + a_{n-1} s^{n-1} + \ldots + a_0$, $Q(s)$ is a polynomial in s of degree less than or equal to $n - 1$ consisting of the various products of the coefficients a_i, $i = 1, \ldots, n$, and the prescribed initial conditions $y_0, y_1, \ldots, y_{n-1}$, and $G(s)$ is the Laplace transform of $g(t)$.* Typically we put the two terms in (11) over the least common denominator and then decompose the expression into two or more partial fractions. Finally, the solution $y(t)$ of the original initial-value problem is $y(t) = \mathcal{L}^{-1}\{Y(s)\}$, where the inverse transform is done term by term.

The procedure is summarized in the following diagram.

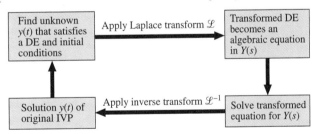

The next example illustrates the foregoing method of solving DEs.

Example 4 Solving a First-Order IVP

Use the Laplace transform to solve the initial-value problem

$$\frac{dy}{dt} + 3y = 13 \sin 2t, \qquad y(0) = 6.$$

* The polynomial $P(s)$ is the same as the nth degree auxiliary polynomial in (13) in Section 3.3, with the usual symbol m replaced by s.

SOLUTION We first take the transform of each member of the differential equation:

$$\mathscr{L}\left\{\frac{dy}{dt}\right\} + 3\mathscr{L}\{y\} = 13\mathscr{L}\{\sin 2t\}. \tag{12}$$

But from (6), $\mathscr{L}\{dy/dt\} = sY(s) - y(0) = sY(s) - 6$, and from part **(d)** of Theorem 4.1, $\mathscr{L}\{\sin 2t\} = 2/(s^2 + 4)$, and so (12) is the same as

$$sY(s) - 6 + 3Y(s) = \frac{26}{s^2 + 4} \quad \text{or} \quad (s + 3)Y(s) = 6 + \frac{26}{s^2 + 4}.$$

Solving the last equation for $Y(s)$, we get

Partial fractions: quadratic ☞
polynomial with no real factors

$$Y(s) = \frac{6}{s + 3} + \frac{26}{(s + 3)(s^2 + 4)} = \frac{6s^2 + 50}{(s + 3)(s^2 + 4)}. \tag{13}$$

Since the quadratic polynomial $s^2 + 4$ does not factor using real numbers, its assumed numerator in the partial fraction decomposition is a linear polynomial in s:

$$\frac{6s^2 + 50}{(s + 3)(s^2 + 4)} = \frac{A}{s + 3} + \frac{Bs + C}{s^2 + 4}.$$

Putting the right side of the equality over a common denominator and equating numerators gives $6s^2 + 50 = A(s^2 + 4) + (Bs + C)(s + 3)$. Setting $s = -3$ then yields immediately $A = 8$. Since the denominator has no more real zeros, we equate the coefficients of s^2 and s: $6 = A + B$ and $0 = 3B + C$. Using the value of A in the first equation gives $B = -2$, and then using this last value in the second equation gives $C = 6$. Thus

$$Y(s) = \frac{6s^2 + 50}{(s + 3)(s^2 + 4)} = \frac{8}{s + 3} + \frac{-2s + 6}{s^2 + 4}.$$

We are not quite finished because the last rational expression still has to be written as two fractions. But this was done by termwise division in Example 2. From (2) of that example,

$$y(t) = 8\mathscr{L}^{-1}\left\{\frac{1}{s + 3}\right\} - 2\mathscr{L}^{-1}\left\{\frac{s}{s^2 + 4}\right\} + 3\mathscr{L}^{-1}\left\{\frac{2}{s^2 + 4}\right\}.$$

If follows from parts **(c)**, **(d)**, and **(e)** of Theorem 4.3 that the solution of the initial value problem is $y(t) = 8e^{-3t} - 2\cos 2t + 3\sin 2t$. ❑

Example 5 **Solving a Second-Order IVP**

Solve $y'' - 3y' + 2y = e^{-4t}$, $y(0) = 1$, $y'(0) = 5$.

SOLUTION Proceeding as in Example 4, we transform the DE by taking the sum of the transforms of each term, use (6) and (7), use the given initial conditions, use **(c)** of Theorem 4.3, and then solve for $Y(s)$:

$$\mathscr{L}\left\{\frac{d^2y}{dt^2}\right\} - 3\mathscr{L}\left\{\frac{dy}{dt}\right\} + 2\mathscr{L}\{y\} = \mathscr{L}\{e^{-4t}\}$$

$$s^2Y(s) - sy(0) - y'(0) - 3[sY(s) - y(0)] + 2Y(s) = \frac{1}{s + 4}$$

$$(s^2 - 3s + 2)Y(s) = s + 2 + \frac{1}{s + 4}$$

$$Y(s) = \frac{s+2}{s^2 - 3s + 2} + \frac{1}{(s^2 - 3s + 2)(s+4)} = \frac{s^2 + 6s + 9}{(s-1)(s-2)(s+4)}. \qquad (14)$$

The details of the partial fraction decomposition of $Y(s)$ have already been carried out in Example 3. In view of results in (4) and (5) we have the solution of the initial-value problem:

$$y(t) = \mathscr{L}^{-1}\{Y(s)\} = -\frac{16}{5} e^t + \frac{25}{6} e^{2t} + \frac{1}{30} e^{-4t}. \qquad \square$$

Examples 4 and 5 illustrate the basic procedure for using the Laplace transform to solve a linear initial-value problem, but these examples may appear to demonstrate a method that is not much better than the approach to such problems outlined in Sections 2.4 and 3.3–3.6. Don't draw any negative conclusions from two examples. Yes, there is a lot of algebra inherent in the use of the Laplace transform, *but* observe that we do not have to use variation of parameters or worry about the cases and algebra in the method of undetermined coefficients. Moreover, since the method incorporates the prescribed initial conditions directly into the solution, there is no need for the separate operations of applying the initial conditions to the general solution $y = c_1 y_1 + c_2 y_2 + \ldots + c_n y_n + y_p$ of the DE to find specific constants in a particular solution of the IVP.

The Laplace transform has many operational properties. We will examine some of these properties and illustrate how they enable us to solve problems of greater complexity in the sections that follow.

We conclude this section with a little bit of additional theory related to the types of functions of s that we will generally be working with. The next theorem indicates that not every arbitrary function of s is a Laplace transform of a piecewise continuous function of exponential order.

THEOREM 4.5 **Behavior of $F(s)$ as $s \to \infty$**

If f is piecewise continuous on $[0, \infty)$ and of exponential order for $t > T$, then $\lim\limits_{s \to \infty} \mathscr{L}\{f(t)\} = 0$.

PROOF Since $f(t)$ is piecewise continuous on $0 \le t \le T$, it is necessarily bounded on the interval. That is, $|f(t)| \le M_1 = M_1 e^{0t}$. Also, $|f(t)| \le M_2 e^{\gamma t}$ for $t > T$. If M denotes the maximum of $\{M_1, M_2\}$ and c denotes the maximum of $\{0, \gamma\}$, then

$$\mathscr{L}\{f(t)\} \le \int_0^\infty e^{-st}|f(t)|\, dt \le M \int_0^\infty e^{-st} \cdot e^{ct}\, dt = -M \left. \frac{e^{-(s-c)t}}{s-c} \right|_0^\infty = \frac{M}{s-c}$$

for $s > c$. As $s \to \infty$, we have $\left|\mathscr{L}\{f(t)\}\right| \to 0$, and so $\to \mathscr{L}\{f(t)\} \to 0$. $\qquad \square$

As a consequence of Theorem 4.5 we can say that functions of s such as $F_1(s) = 1$ and $F_2(s) = s/(s+1)$ are not the Laplace transforms of piecewise continuous functions of exponential order since $F_1(s) \nrightarrow 0$ and $F_2(s) \nrightarrow 0$ as $s \infty$. But you should not conclude from this that $F_1(s)$ and $F_2(s)$ are *not* Laplace transforms. There are other kinds of functions.

Remarks

(*i*) The inverse Laplace of a function $F(s)$ may not be unique; in other words, it is possible that $\mathscr{L}\{f_1(t)\} = \mathscr{L}\{f_2(t)\}$ and yet $f_1 \ne f_2$. For our purposes this is not

anything to be concerned about. If f_1 and f_2 are piecewise continuous on $[0, \infty)$ and of exponential order, then f_1 and f_2 are *essentially* the same. See Problem 44 in Exercises 4.2. However, if f_1 and f_2 are continuous on $[0, \infty)$ and $\mathscr{L}\{f_1(t)\} = \mathscr{L}\{f_2(t)\}$, then $f_1 = f_2$ on the interval.

(*ii*) This remark is for those of you who will be required to do partial fraction decompositions by hand. There is another way of determining the coefficients in a partial fraction decomposition in the special case when $\mathscr{L}\{f(t)\} = F(s)$ is a rational function of s and the denominator of F is a product of *distinct* linear factors. Let us illustrate by reexamining Example 3. Suppose we multiply both sides of the assumed decomposition

$$\frac{s^2 + 6s + 9}{(s - 1)(s - 2)(s + 4)} = \frac{A}{s - 1} + \frac{B}{s - 2} + \frac{C}{s + 4} \tag{15}$$

by, say, $s - 1$, simplify, and then set $s = 1$. Since the coefficients of B and C on the right side of the equality are zero, we get

$$\left.\frac{s^2 + 6s + 9}{(s - 2)(s + 4)}\right|_{s=1} = A \quad \text{or} \quad A = -\frac{16}{5}.$$

Written another way,

$$\left.\frac{s^2 + 6s + 9}{(s - 1)(s - 2)(s + 4)}\right|_{s=1} = -\frac{16}{5} = A,$$

where we have colored or *covered up* the factor that canceled when the left side was multiplied by $s - 1$. Now to obtain B and C we simply evaluate the left-hand side of (15) while covering up, in turn, $s - 2$ and $s + 4$:

$$\left.\frac{s^2 + 6s + 9}{(s - 1)(s - 2)(s + 4)}\right|_{s=2} = \frac{25}{6} = B \quad \text{and} \quad \left.\frac{s^2 + 6s + 9}{(s - 1)(s - 2)(s + 4)}\right|_{s=-4} = \frac{1}{30} = C.$$

The desired decomposition (15) is given in (4). This special technique for determining coefficients is naturally known as the **cover-up method.**

(*iii*) In this remark we continue our introduction to the terminology of dynamical systems. Because of (9) and (10) the Laplace transform is well adapted to *linear* dynamical systems. In (11) the polynomial $P(s) = a_n s^n + a_{n-1} s^{n-1} + \ldots + a_0$ is the total coefficient of $Y(s)$ in (10) and is simply the left-hand side of the DE with the derivatives $d^k y/dt^k$ replaced by powers s^k, $k = 0, 1, \ldots, n$. It is usual practice to call the reciprocal of $P(s)$, namely $W(s) = 1/P(s)$, the **transfer function** of the system and write (11) as

$$Y(s) = W(s)Q(s) + W(s)G(s). \tag{16}$$

In this manner we have separated, in an additive sense, the effects on the response that are due to the initial conditions (that is, $W(s)Q(s)$) and to the input function g (that is, $W(s)G(s)$). See (13) and (14). Hence the response $y(t)$ of the system is a superposition of two responses

$$y(t) = \mathscr{L}^{-1}\{W(s)Q(s)\} + \mathscr{L}^{-1}\{W(s)G(s)\} = y_0(t) + y_1(t).$$

If the input is $g(t) = 0$, then the solution of the problem is $y_0(t) = \mathscr{L}^{-1}\{W(s)Q(s)\}$. This solution is called the **zero-input response** of the system. On the other hand, the function $y_1(t) = \mathscr{L}^{-1}\{W(s)G(s)\}$ is the output due to the input $g(t)$. Now if the initial state of the system is the zero state (all the initial conditions are zero), then $Q(s) = 0$, and so the only solution of the initial-value problem is $y_1(t)$. The latter solution is called the **zero-state response** of the system. Both $y_0(t)$ and $y_1(t)$ are

particular solutions: $y_0(t)$ is a solution of the IVP consisting of the associated homogeneous equation with the given initial conditions, and $y_1(t)$ is a solution of the IVP consisting of the nonhomogeneous equation with zero initial conditions. In Example 5, we see from (14) that the transfer function is $W(s) = 1/(s^2 - 3s + 2)$, the zero-input response is

$$y_0(t) = \mathscr{L}^{-1}\left\{\frac{s+2}{(s-1)(s-2)}\right\} = -3e^t + 4e^{2t},$$

and the zero-state response is

$$y_1(t) = \mathscr{L}^{-1}\left\{\frac{1}{(s-1)(s-2)(s+4)}\right\} = -\frac{1}{5}e^t + \frac{1}{6}e^{2t} + \frac{1}{30}e^{-4t}.$$

Verify that the sum of $y_0(t)$ and $y_1(t)$ is the solution $y(t)$ in that example and that $y_0(0) = 1$, $y_0'(0) = 5$, whereas $y_1(0) = 0$, $y_1'(0) = 0$.

E X E R C I S E S 4 . 2

Answers to odd-numbered problems begin on page A-33.

In Exercises 1–30, use Theorem 4.3 to find the given inverse transform.

1. $\mathscr{L}^{-1}\left\{\dfrac{1}{s^3}\right\}$

2. $\mathscr{L}^{-1}\left\{\dfrac{1}{s^4}\right\}$

3. $\mathscr{L}^{-1}\left\{\dfrac{1}{s^2} - \dfrac{48}{s^5}\right\}$

4. $\mathscr{L}^{-1}\left\{\left(\dfrac{2}{s} - \dfrac{1}{s^3}\right)^2\right\}$

5. $\mathscr{L}^{-1}\left\{\dfrac{(s+1)^3}{s^4}\right\}$

6. $\mathscr{L}^{-1}\left\{\dfrac{(s+2)^2}{s^3}\right\}$

7. $\mathscr{L}^{-1}\left\{\dfrac{1}{s^2} - \dfrac{1}{s} + \dfrac{1}{s-2}\right\}$

8. $\mathscr{L}^{-1}\left\{\dfrac{4}{s} + \dfrac{6}{s^5} - \dfrac{1}{s+8}\right\}$

9. $\mathscr{L}^{-1}\left\{\dfrac{1}{4s+1}\right\}$

10. $\mathscr{L}^{-1}\left\{\dfrac{1}{5s-2}\right\}$

11. $\mathscr{L}^{-1}\left\{\dfrac{5}{s^2+49}\right\}$

12. $\mathscr{L}^{-1}\left\{\dfrac{10s}{s^2+16}\right\}$

13. $\mathscr{L}^{-1}\left\{\dfrac{4s}{4s^2+1}\right\}$

14. $\mathscr{L}^{-1}\left\{\dfrac{1}{4s^2+1}\right\}$

15. $\mathscr{L}^{-1}\left\{\dfrac{2s-6}{s^2+9}\right\}$

16. $\mathscr{L}^{-1}\left\{\dfrac{s+1}{s^2+2}\right\}$

17. $\mathscr{L}^{-1}\left\{\dfrac{1}{s^2+3s}\right\}$

18. $\mathscr{L}^{-1}\left\{\dfrac{s+1}{s^2-4s}\right\}$

19. $\mathscr{L}^{-1}\left\{\dfrac{s}{s^2+2s-3}\right\}$

20. $\mathscr{L}^{-1}\left\{\dfrac{1}{s^2+s-20}\right\}$

21. $\mathscr{L}^{-1}\left\{\dfrac{0.9s}{(s-0.1)(s+0.2)}\right\}$

22. $\mathscr{L}^{-1}\left\{\dfrac{s-3}{(s-\sqrt{3})(s+\sqrt{3})}\right\}$

23. $\mathscr{L}^{-1}\left\{\dfrac{s}{(s-2)(s-3)(s-6)}\right\}$

24. $\mathscr{L}^{-1}\left\{\dfrac{s^2+1}{s(s-1)(s+1)(s-2)}\right\}$

25. $\mathscr{L}^{-1}\left\{\dfrac{1}{s^3+5s}\right\}$

26. $\mathscr{L}^{-1}\left\{\dfrac{s}{(s+2)(s^2+4)}\right\}$

27. $\mathscr{L}^{-1}\left\{\dfrac{2s-4}{(s^2+s)(s^2+1)}\right\}$

28. $\mathscr{L}^{-1}\left\{\dfrac{1}{s^4-9}\right\}$

29. $\mathscr{L}^{-1}\left\{\dfrac{1}{(s^2+1)(s^2+4)}\right\}$

30. $\mathscr{L}^{-1}\left\{\dfrac{6s+3}{(s^4+5s^2+4)}\right\}$

In Problems 31–40, use the Laplace transform to solve the given initial-value problem.

31. $\dfrac{dy}{dt} - y = 1$, $y(0) = 0$

32. $2\dfrac{dy}{dt} + y = 0$, $y(0) = -3$

33. $y' + 6y = e^{4t}$, $y(0) = 2$

34. $y' - y = 2\cos 5t$, $y(0) = 0$

35. $y'' + 5y' + 4y = 0$, $y(0) = 1$, $y'(0) = 0$

36. $y'' - 4y' = 6e^{3t} - 3e^{-t}$, $y(0) = 1$, $y'(0) = -1$

37. $y'' + y = \sqrt{2}\sin\sqrt{2}t$, $y(0) = 10$, $y'(0) = 0$

38. $y'' + 9y = e^t$, $y(0) = 0$, $y'(0) = 0$

39. $2y''' + 3y'' - 3y' - 2y = e^{-t}$, $y(0) = 0$, $y'(0) = 0$, $y''(0) = 1$

40. $y''' + 2y'' - y' - 2y = \sin 3t$, $y(0) = 0$, $y'(0) = 0$, $y''(0) = 1$

In Problems 41 and 42, find the zero-input and the zero-state responses for the indicated IVP. Reread Remark (*iii*) on page 202.

41. Problem 33

42. Problem 36

43. The inverse forms of the results in Problem 45 in Exercises 4.1 are

$$\mathscr{L}^{-1}\left\{\frac{s-a}{(s-a)^2+b^2}\right\} = e^{at}\cos bt$$

and

$$\mathscr{L}^{-1}\left\{\frac{b}{(s-a)^2+b^2}\right\} = e^{at}\sin bt.$$

Use these inverses and the Laplace transform to solve $y' + y = e^{-3t}\cos 2t$, $y(0) = 0$.

44. Make up two distinct functions f and g that have the same Laplace transform.

4.3 Translation Theorems

It is not convenient to use Definition 4.1 each time we wish to find the Laplace transform of a function $f(t)$. For example, the integration by parts involved in evaluating, say, $\mathscr{L}\{e^t t^2 \sin 3t\}$ is formidable to say the least. In this section and the next we present several labor-saving theorems that enable us to build up a more extensive list of transforms (see the table, Appendix III) without the necessity of using the definition of the Laplace transform.

4.3.1 Translation on the *s*-axis

Evaluating transforms such as $\mathscr{L}\{e^{5t} t^3\}$ and $\mathscr{L}\{e^{-2t} \cos 4t\}$ is straightforward provided we know $\mathscr{L}\{t^3\}$ and $\mathscr{L}\{\cos 4t\}$, which we do. In general, if we know $\mathscr{L}\{f(t)\} = F(s)$ it is possible to compute the Laplace transform of an exponential multiple of the function f, that is, $\mathscr{L}\{e^{at} f(t)\}$, with no additional effort other than *translating*, or *shifting*, $F(s)$ to $F(s-a)$. This result is known as the **first translation theorem** or **first shifting theorem.**

> **THEOREM 4.6 First Translation Theorem**
>
> If $\mathscr{L}\{f(t)\} = F(s)$ and a is any real number, then
>
> $$\mathscr{L}\{e^{at} f(t)\} = F(s-a).$$

PROOF The proof is immediate, since by Definition 4.1

$$\mathscr{L}\{e^{at} f(t)\} = \int_0^\infty e^{-st} e^{at} f(t)\, dt = \int_0^\infty e^{-(s-a)t} f(t)\, dt = F(s-a). \qquad \square$$

If we consider s a real variable, then the graph of $F(s-a)$ is the graph of $F(s)$ shifted on the s-axis by the amount $|a|$. If $a > 0$, the graph of $F(s)$ is shifted a units to the right, whereas if $a < 0$, the graph is shifted $|a|$ units to the left. See Figure 4.10.

For emphasis it is sometimes useful to use the symbolism

$$\mathscr{L}\{e^{at} f(t)\} = \mathscr{L}\{f(t)\}_{s \to s-a},$$

where $s \to s - a$ means that in the Laplace transform $F(s)$ of $f(t)$ we replace the symbol s wherever it appears by $s - a$.

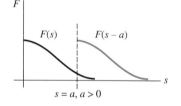

Figure 4.10 Shift on s-axis

Example 1 Using the First Translation Theorem

Evaluate (a) $\mathscr{L}\{e^{5t} t^3\}$ (b) $\mathscr{L}\{e^{-2t} \cos 4t\}$.

SOLUTION The results follow from Theorems 4.1 and 4.6.

(a) $\mathscr{L}\{e^{5t} t^3\} = \mathscr{L}\{t^3\}_{s \to s-5} = \left.\dfrac{3!}{s^4}\right|_{s \to s-5} = \dfrac{6}{(s-5)^4}$

(b) $\mathscr{L}\{e^{-2t} \cos 4t\} = \mathscr{L}\{\cos 4t\}_{s \to s-(-2)} = \left.\dfrac{s}{s^2 + 16}\right|_{s \to s+2} = \dfrac{s+2}{(s+2)^2 + 16}$. \square

■ **Inverse Form of Theorem 4.6** To compute the inverse of $F(s-a)$ we must recognize $F(s)$, find $f(t)$ by taking the inverse Laplace transform of $F(s)$, and then multiply $f(t)$ by the exponential function e^{at}. This procedure can be summarized symbolically in the following manner:

$$\mathscr{L}^{-1}\{F(s-a)\} = \mathscr{L}^{-1}\{F(s)|_{s \to s-a}\} = e^{at}f(t) \tag{1}$$

where $f(t) = \mathscr{L}^{-1}\{F(s)\}$.

Example 2 Partial Fractions and Completing the Square

Partial fractions: 👉
repeated linear factors

Evaluate (a) $\mathscr{L}^{-1}\left\{\dfrac{2s+5}{(s-3)^2}\right\}$ (b) $\mathscr{L}^{-1}\left\{\dfrac{s/2+5/3}{s^2+4s+6}\right\}$.

SOLUTION (a) A repeated linear factor is a term $(s-a)^n$, where a is a real number and n is a positive integer ≥ 2. Recall that if $(s-a)^n$ appears in the denominator of a rational expression, then the assumed decomposition contains n partial fractions with constant numerators and denominators $s-a$, $(s-a)^2$, ..., $(s-a)^n$. Hence with $a=3$ and $n=2$ we write

$$\frac{2s+5}{(s-3)^2} = \frac{A}{s-3} + \frac{B}{(s-3)^2}.$$

The numerator obtained by putting the two terms on the right over a common denominator is $2s+6 = A(s-3) + B$, and this identity yields $A=2$ and $B=11$. Therefore,

$$\frac{2s+5}{(s-3)^2} = \frac{2}{s-3} + \frac{11}{(s-3)^2} \tag{2}$$

and

$$\mathscr{L}^{-1}\left\{\frac{2s+5}{(s-3)^2}\right\} = 2\mathscr{L}^{-1}\left\{\frac{1}{s-3}\right\} + 11\mathscr{L}^{-1}\left\{\frac{1}{(s-3)^2}\right\}. \tag{3}$$

Now $1/(s-3)^2$ is $F(s) = 1/s^2$ shifted 3 units to the right. Since $\mathscr{L}^{-1}\{1/s^2\} = t$, it follows from (1) that

$$\mathscr{L}^{-1}\left\{\frac{1}{(s-3)^2}\right\} = \mathscr{L}^{-1}\left\{\frac{1}{s^2}\bigg|_{s \to s-3}\right\} = e^{-3t}t.$$

Finally, (3) is $\mathscr{L}^{-1}\left\{\dfrac{2s+5}{(s-3)^2}\right\} = 2e^{-3t} + 11e^{-3t}t.$ \hfill (4)

(b) To start, observe that the quadratic polynomial $s^2 + 4s + 6$ does not have real zeros and so has no real linear factors. In this situation we complete the square:

$$\frac{s/2+5/3}{s^2+4s+6} = \frac{s/2+5/3}{(s+2)^2+2}. \tag{5}$$

Our goal here is to recognize the expression on the right as some Laplace transform $F(s)$ in which s has been replaced throughout by $s+2$. What we are trying to do here is analogous to working part **(b)** of Example 1 backwards. The denominator in (5) is already in the correct form, that is, $s^2 + 2$ with s replaced by $s+2$. However, we must fix up the numerator by manipulating the constants: $\frac{1}{2}s + \frac{5}{3} = \frac{1}{2}(s+2) + \frac{5}{3} - \frac{2}{2} = \frac{1}{2}(s+2) + \frac{2}{3}$.

Now by termwise division, the linearity of \mathscr{L}^{-1}, parts **(e)** and **(d)** of Theorem 4.3, and finally from (1),

$$\frac{s/2+5/3}{(s+2)^2+2} = \frac{\frac{1}{2}(s+2)+\frac{2}{3}}{(s+2)^2+2} = \frac{1}{2}\frac{s+2}{(s+2)^2+2} + \frac{2}{3}\frac{1}{(s+2)^2+2}$$

$$\mathscr{L}^{-1}\left\{\frac{s/2+5/3}{s^2+4s+6}\right\} = \frac{1}{2}\mathscr{L}^{-1}\left\{\frac{s+2}{(s+2)^2+2}\right\} + \frac{2}{3}\mathscr{L}^{-1}\left\{\frac{1}{(s+2)^2+2}\right\}$$

$$= \frac{1}{2} \mathscr{L}^{-1}\left\{\frac{s}{s^2+2}\Big|_{s\to s+2}\right\} + \frac{2}{3\sqrt{2}} \mathscr{L}^{-1}\left\{\frac{\sqrt{2}}{s^2+2}\Big|_{s\to s+2}\right\} \tag{6}$$

$$= \frac{1}{2} e^{-2t} \cos \sqrt{2}\,t + \frac{3}{\sqrt{2}} e^{-2t} \sin \sqrt{2}\,t. \tag{7} \ \square$$

Example 3 Initial-Value Problem

Solve $y'' - 6y' + 9y = t^2 e^{3t}$, $y(0) = 2$, $y'(0) = 6$.

SOLUTION Before transforming the DE note that its right-hand side is similar to the function in part **(a)** of Example 1. We use Theorem 4.5, the initial conditions, simplify, and then solve for $Y(s) = \mathscr{L}\{f(t)\}$:

$$\mathscr{L}\{y''\} - 6\mathscr{L}\{y'\} + 9\mathscr{L}\{y\} = \mathscr{L}\{t^2 e^{3t}\}$$

$$s^2 Y(s) - sy(0) - y'(0) - 6[sY(s) - y(0)] + 9Y(s) = \frac{2}{(s-3)^3}$$

$$(s^2 - 6s + 9)Y(s) = 2s + 5 + \frac{2}{(s-3)^3}$$

$$(s-3)^2 Y(s) = 2s + 5 + \frac{2}{(s-3)^3}$$

$$Y(s) = \frac{2s+5}{(s-3)^2} + \frac{2}{(s-3)^5}.$$

The first term on the right has already been decomposed into individual partial fractions in (2) in part **(a)** of Example 2:

$$Y(s) = \frac{2}{s-3} + \frac{11}{(s-3)^2} + \frac{2}{(s-3)^5}.$$

Thus $y(t) = 2\mathscr{L}^{-1}\left\{\frac{1}{s-3}\right\} + 11\mathscr{L}^{-1}\left\{\frac{1}{(s-3)^2}\right\} + \frac{2}{4!} \mathscr{L}^{-1}\left\{\frac{4!}{(s-3)^5}\right\}. \tag{8}$

From the inverse form (1) of Theorem 4.6, the last two terms are

$$\mathscr{L}^{-1}\left\{\frac{1}{s^2}\Big|_{s\to s-3}\right\} = te^{3t} \quad \text{and} \quad \mathscr{L}^{-1}\left\{\frac{4!}{s^5}\Big|_{s\to s-3}\right\} = t^4 e^{3t},$$

and so (8) is $y(t) = 2e^{3t} + 11te^{3t} + \frac{1}{12} t^4 e^{3t}.$ \square

Example 4 An Initial-Value Problem

Solve $y'' + 4y' + 6y = 1 + e^{-t}$, $y(0) = 0$, $y'(0) = 0$.

SOLUTION $\mathscr{L}\{y''\} + 4\mathscr{L}\{y'\} + 6\mathscr{L}\{y\} = \mathscr{L}\{1\} + \mathscr{L}\{e^{-t}\}$

$$s^2 Y(s) - sy(0) - y'(0) + 4[sY(s) - y(0)] + 6Y(s) = \frac{1}{s} + \frac{1}{s+1}$$

$$(s^2 + 4s + 6)Y(s) = \frac{2s+1}{s(s+1)}$$

$$Y(s) = \frac{2s+1}{s(s+1)(s^2+4s+6)}.$$

Since the quadratic term in the denominator does not factor into real linear factors, the partial fraction decomposition for $Y(s)$ is found to be

$$Y(s) = \frac{1/6}{s} + \frac{1/3}{s+1} - \frac{s/2 + 5/3}{s^2 + 4s + 6}.$$

Moreover, in preparation for taking the inverse transform, we have already manipulated the last term into the necessary form in part **(b)** of Example 2. So in view of the results in (6) and (7) we have the solution

$$y(t) = \frac{1}{6} \mathcal{L}^{-1}\left\{\frac{1}{s}\right\} + \frac{1}{3} \mathcal{L}^{-1}\left\{\frac{1}{s+1}\right\} - \frac{1}{2} \mathcal{L}^{-1}\left\{\frac{s+2}{(s+2)^2 + 2}\right\} - \frac{2}{3\sqrt{2}} \mathcal{L}^{-1}\left\{\frac{\sqrt{2}}{(s+2)^2 + 2}\right\}$$

$$= \frac{1}{6} + \frac{1}{3} e^{-t} - \frac{1}{2} e^{-2t} \cos \sqrt{2}\,t - \frac{\sqrt{2}}{3} e^{-2t} \sin \sqrt{2}\,t. \qquad \square$$

4.3.2 Translation on the *t*-axis

■ **Unit Step Function** In engineering, one frequently encounters functions that are either "off" or "on." For example, an external force acting on a mechanical system or a voltage impressed on a circuit can be turned off after a period of time. It is convenient, then, to define a special function that is the number 0 (off) up to a certain time $t = a$ and then the number 1 (on) after that time. This function is called the **unit step function** or the **Heaviside function.**

Figure 4.11

> **DEFINITION 4.3 Unit Step Function**
>
> The **unit step function** $\mathcal{U}(t - a)$ is defined to be
>
> $$\mathcal{U}(t - a) = \begin{cases} 0, & 0 \le t < a \\ 1, & t \ge a. \end{cases}$$

Notice that we define $\mathcal{U}(t - a)$ only on the nonnegative *t*-axis since this is all that we are concerned with in the study of the Laplace transform. In a broader sense $\mathcal{U}(t - a) = 0$ for $t < a$. The graph of $\mathcal{U}(t - a)$ is given in Figure 4.11.

When a function f defined for $t \ge 0$ is multiplied by $\mathcal{U}(t - a)$, the unit step function "turns off" a portion of the graph of that function. For example, consider the function $f(t) = 2t - 3$. To "turn off" the portion of the graph of f on, say, the interval $0 \le t < 1$, we simply form the product $(2t - 3)\mathcal{U}(t - 1)$. See Figure 4.12. In general, the graph of $f(t)\mathcal{U}(t - a)$ is 0 (off) for $0 \le t < a$ and is the portion of the graph of f (on) for $t \ge a$.

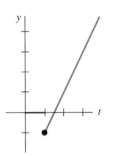

Figure 4.12

The unit step function can also be used to write piecewise defined functions in a compact form. For example, by considering the intervals $0 \le t < 2$, $2 \le t < 3$, $t \ge 3$, and the corresponding values of $\mathcal{U}(t - 2)$ and $\mathcal{U}(t - 3)$, it should be apparent that the piecewise defined function shown in Figure 4.13 is the same as $f(t) = 2 - 3\mathcal{U}(t - 2) + \mathcal{U}(t - 3)$. Also, a general piecewise defined function of the type

$$f(t) = \begin{cases} g(t), & 0 \le t < a \\ h(t), & t \ge a \end{cases} \qquad (9)$$

is the same as

$$f(t) = g(t) - g(t)\mathcal{U}(t - a) + h(t)\mathcal{U}(t - a). \qquad (10)$$

Figure 4.13

Similarly, a function of the type

$$f(t) = \begin{cases} 0, & 0 \le t < a \\ g(t), & a \le t < b \\ 0, & t \ge b \end{cases} \tag{11}$$

can be written

$$f(t) = g(t)[\mathcal{U}(t - a) - \mathcal{U}(t - a)]. \tag{12}$$

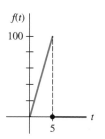

Figure 4.14

Example 5 A Piecewise Defined Function

Express $f(t) = \begin{cases} 20t, & 0 \le t < 5 \\ 0, & t \ge 5 \end{cases}$ in terms of unit step functions. Graph.

SOLUTION The graph of f is given in Figure 4.14. Now from (9) and (10) with $a = 5$, $g(t) = 20t$, and $h(t) = 0$, we get $f(t) = 20t - 20t\mathcal{U}(t - 5)$. ❑

Consider a general function $y = f(t)$ defined for $t \ge 0$. The piecewise defined function

$$f(t - a)\mathcal{U}(t - a) = \begin{cases} 0, & 0 \le t < a \\ f(t - a), & t \ge a \end{cases} \tag{13}$$

plays a significant role in the discussion that follows. As shown in Figure 4.15, for $a > 0$ the graph of the function $y = f(t - a)\mathcal{U}(t - a)$ coincides with the graph of $y = f(t - a)$ for $t \ge a$ (which is the *entire* graph of $y = f(t)$, $t \ge 0$, shifted a units to the right on the t-axis) but is identically zero for $0 \le t < a$.

We saw in Theorem 4.6 that an exponential multiple of $f(t)$ results in a translation of the transform $F(s)$ on the s-axis. As a consequence of the next theorem we see that whenever $F(s)$ is multiplied by an exponential function e^{-as}, $a > 0$, the inverse transform of the product $e^{-as}F(s)$ is the function f shifted along the t-axis in the manner illustrated in Figure 4.15(b). This result, presented next in its direct transform version, is called the **second translation theorem** or **second shifting theorem.**

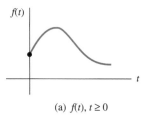

(a) $f(t), t \ge 0$

(b) $f(t - a)\,\mathcal{U}(t - a)$

shift on t-axis

Figure 4.15

THEOREM 4.7 Second Translation Theorem

If $F(s) = \mathcal{L}\{f(t)\}$ and $a > 0$, then

$$\mathcal{L}\{f(t - a)\mathcal{U}(t - a)\} = e^{-as}F(s).$$

PROOF By the additive interval property of integrals, $\int_0^\infty e^{-st}f(t - a)\mathcal{U}(t - a)\,dt$ can be written as two integrals:

$$\mathcal{L}\{f(t - a)\mathcal{U}(t - a)\} = \underbrace{\int_0^a e^{-st}f(t - a)\mathcal{U}(t - a)\,dt}_{\text{zero for } 0 \le t < a} + \underbrace{\int_a^\infty e^{-st}f(t - a)\mathcal{U}(t - a)\,dt}_{\text{one for } t \ge a}$$

$$= \int_a^\infty e^{-st}f(t - a)\,dt.$$

Now if we let $v = t - a$, $dv = dt$ in the last integral, then

$$\mathcal{L}\{f(t - a)\mathcal{U}(t - a)\} = \int_0^\infty e^{-s(v + a)}f(v)\,dv$$

$$= e^{-as}\int_0^\infty e^{-sv}f(v)\,dv = e^{-as}\mathcal{L}\{f(t)\}. \quad ❑$$

We often wish to find the Laplace transform of just a unit step function. This can be found from either Definition 4.1 or Theorem 4.6. If we identify $f(t) = 1$ in Theorem 4.6, then $f(t - a) = 1$, $F(s) = \mathscr{L}\{1\} = 1/s$, and so

$$\mathscr{L}\{\mathscr{U}(t - a)\} = \frac{e^{-as}}{s}. \tag{14}$$

For example, using (14), the Laplace transform of the function in Figure 4.13 is

$$\mathscr{L}\{f(t)\} = 2\mathscr{L}\{1\} - 3\mathscr{L}\{\mathscr{U}(t - 2)\} + \mathscr{L}\{\mathscr{U}(t - 3)\}$$

$$= 2\frac{1}{s} - 3\frac{e^{-2s}}{s} + \frac{e^{-3s}}{s}.$$

■ **Inverse Form of Theorem 4.7** If $f(t) = \mathscr{L}^{-1}\{F(s)\}$, the inverse form of Theorem 4.7, $a > 0$, is

$$\mathscr{L}^{-1}\{e^{-as} F(s)\} = f(t - a)\mathscr{U}(t - a). \tag{15}$$

Example 6 **Using Formula (15)**

Evaluate (a) $\mathscr{L}^{-1}\left\{\dfrac{1}{s - 4}\, e^{-2s}\right\}$ (b) $\mathscr{L}^{-1}\left\{\dfrac{s}{s^2 + 9}\, e^{-\pi s/2}\right\}$.

SOLUTION (a) With the identifications $a = 2$, $F(s) = 1/(s - 4)$, $\mathscr{L}^{-1}\{F(s)\} = e^{4t}$, we have from (15)

$$\mathscr{L}^{-1}\left\{\frac{1}{s - 4}\, e^{-2s}\right\} = e^{4(t - 2)}\mathscr{U}(t - 2).$$

(b) With $a = \pi/2$, $F(s) = s/(s^2 + 9)$, $\mathscr{L}^{-1}\{F(s)\} = \cos 3t$, (15) yields

$$\mathscr{L}^{-1}\left\{\frac{s}{s^2 + 9}\, e^{-\pi s/2}\right\} = \cos 3\left(t - \frac{\pi}{2}\right)\mathscr{U}\left(t - \frac{\pi}{2}\right).$$

The last expression can be simplified somewhat using the addition formula for the cosine. Verify that the result is the same as $-\sin 3t\,\mathscr{U}(t - \pi/2)$. ❏

■ **Alternative Form of Theorem 4.7** We are frequently confronted with the problem of finding the Laplace transform of a product of a function g and a unit step function $\mathscr{U}(t - a)$ where the function g lacks the precise shifted form $f(t - a)$ in Theorem 4.7. To find the Laplace transform of $g(t)\mathscr{U}(t - a)$, it is possible to fix up $g(t)$ into the required form $f(t - a)$ by algebraic manipulations. For example, if we want to use Theorem 4.7 to find the Laplace transform of $t^2\mathscr{U}(t - 2)$, we would have force to $g(t) = t^2$ into the form $f(t - 2)$. You should work through the details and verify that $t^2 = (t - 2)^2 + 4(t - 2) + 4$ is an identity. Therefore,

$$\mathscr{L}\{t^2\mathscr{U}(t - 2)\} = \mathscr{L}\{(t - 2)^2\mathscr{U}(t - 2) + 4(t - 2)\mathscr{U}(t - 2) + 4\mathscr{U}(t - 2)\},$$

where each term on the right can now be evaluated by Theorem 4.7. But since these manipulations are time-consuming and often not obvious, it is simpler to devise an alternative version of Theorem 4.7. Using Definition 4.1, the definition of $\mathscr{U}(t - a)$, and the substitution $u = t - a$, we obtain

$$\mathscr{L}\{g(t)\mathscr{U}(t - a)\} = \int_a^\infty e^{-st}g(t)\, dt = \int_0^\infty e^{-s(u + a)}g(u + a)\, du.$$

That is, $\qquad \mathscr{L}\{g(t)\mathscr{U}(t - a)\} = e^{-as}\mathscr{L}\{g(t + a)\}. \tag{16}$

Example 7 Second Translation Theorem—Alternative Form

Evaluate $\mathscr{L}\{\sin t\, \mathscr{U}(t - \pi)\}$.

SOLUTION With $g(t) = \sin t$, $a = \pi$, then $g(t + \pi) = \sin(t + \pi) = -\cos t$ by the addition formula for the sine function. Hence by (16),

$$\mathscr{L}\{\sin t\, \mathscr{U}(t - \pi)\} = -e^{-\pi s}\mathscr{L}\{\cos t\} = -\frac{s}{s^2 + 1}\, e^{-\pi s}. \qquad \square$$

Example 8 An Initial-Value Problem

Solve $y' + y = f(t)$, $\quad y(0) = 5$, \quad where $\quad f(t) = \begin{cases} 0, & 0 \le t < \pi, \\ 3\sin t, & t \ge \pi. \end{cases}$

SOLUTION The function f can be written as $f(t) = 3\sin t\, \mathscr{U}(t - \pi)$ and so by linearity, the results of Example 7, and the usual partial fractions, we have

$$\mathscr{L}\{y'\} + \mathscr{L}\{y\} = 3\mathscr{L}\{\sin t\, \mathscr{U}(t - \pi)\}$$

$$sY(s) - y(0) + Y(s) = -3\,\frac{s}{s^2 + 1}\, e^{-\pi s}$$

$$(s + 1)Y(s) = 5 - \frac{3s}{s^2 + 1}\, e^{-\pi s}$$

$$Y(s) = \frac{5}{s + 1} - \frac{3}{2}\left[-\frac{1}{s + 1}\, e^{-\pi s} + \frac{1}{s^2 + 1}\, e^{-\pi s} + \frac{s}{s^2 + 1}\, e^{-\pi s}\right]. \qquad (17)$$

Now proceeding as we did in Example 6, it follows from (15) with $a = \pi$ that the inverses of the terms in the bracket are

$$\mathscr{L}^{-1}\left\{\frac{1}{s + 1}\, e^{-\pi s}\right\} = e^{-(t - \pi)}\mathscr{U}(t - \pi), \qquad \mathscr{L}^{-1}\left\{\frac{1}{s^2 + 1}\, e^{-\pi s}\right\} = \sin(t - \pi)\mathscr{U}(t - \pi),$$

and

$$\mathscr{L}^{-1}\left\{\frac{s}{s^2 + 1}\, e^{-\pi s}\right\} = \cos(t - \pi)\mathscr{U}(t - \pi).$$

Thus the inverse of (17) is

$$y(t) = 5e^{-t} + \frac{3}{2}\, e^{-(t - \pi)}\mathscr{U}(t - \pi) - \frac{3}{2}\sin(t - \pi)\mathscr{U}(t - \pi) - \frac{3}{2}\cos(t - \pi)\mathscr{U}(t - \pi)$$

$$= 5e^{-t} + \frac{3}{2}\left[e^{-(t - \pi)} + \sin t + \cos t\right]\mathscr{U}(t - \pi) \qquad \leftarrow \text{trigonometric identities}$$

$$= \begin{cases} 5e^{-t}, & 0 \le t < \pi \\ 5e^{-t} + \frac{3}{2}\, e^{-(t - \pi)} + \frac{3}{2}\sin t + \frac{3}{2}\cos t, & t \ge \pi. \end{cases} \qquad (18)$$

With the aid of a graphing utility we get the graph of (18), shown in Figure 4.16. $\qquad \square$

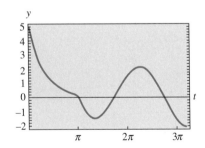

Figure 4.16

■ **Beams** In Section 3.9 we saw that the static deflection $y(x)$ of a uniform beam of length L carrying load $w(x)$ per unit length is found from the linear fourth-order differential equation

$$EI\,\frac{d^4 y}{dx^4} = w(x), \qquad (19)$$

where E is Young's modulus of elasticity and I is a moment of inertia of a cross-section of the beam. The Laplace transform is particularly useful when $w(x)$ is piecewise defined, but in order to use the transform, we must tacitly assume that $y(x)$ and $w(x)$ are defined on $(0, \infty)$ rather than on $(0, L)$. Note, too, that the next example is a boundary-value problem rather than an initial-value problem.

Example 9　A Boundary-Value Problem

A beam of length L is embedded at both ends as shown in Figure 4.17. Find the deflection of the beam when the load is given by

$$w(x) = \begin{cases} w_0\left(1 - \dfrac{2}{L}x\right), & 0 < x < L/2 \\ 0, & L/2 < x < L. \end{cases}$$

SOLUTION　Recall that, since the beam is embedded at both ends, the boundary conditions are $y(0) = 0$, $y'(0) = 0$, $y(L) = 0$, $y'(L) = 0$. Now by (10) we can express $w(x)$ in terms of the unit step function:

$$w(x) = w_0\left(1 - \frac{2}{L}x\right) - w_0\left(1 - \frac{2}{L}x\right)\mathcal{U}\left(x - \frac{L}{2}\right)$$

$$= \frac{2w_0}{L}\left[\frac{L}{2} - x + \left(x - \frac{L}{2}\right)\mathcal{U}\left(x - \frac{L}{2}\right)\right].$$

Transforming (9) with respect to the variable x gives

$$EI(s^4 Y(s) - s^3 y(0) - s^2 y'(0) - s y''(0) - y'''(0)) = \frac{2w_0}{EIL}\left[\frac{L/2}{s} - \frac{1}{s^2} + \frac{1}{s^2}e^{-Ls/2}\right]$$

or

$$s^4 Y(s) - s y''(0) - y'''(0) = \frac{2w_0}{EIL}\left[\frac{L/2}{s} - \frac{1}{s^2} + \frac{1}{s^2}e^{-Ls/2}\right].$$

If we let $c_1 = y''(0)$ and $c_2 = y'''(0)$, then

$$Y(s) = \frac{c_1}{s^3} + \frac{c_2}{s^4} + \frac{2w_0}{EIL}\left[\frac{L/2}{s^5} - \frac{1}{s^6} + \frac{1}{s^6}e^{-Ls/2}\right],$$

and consequently

$$y(x) = \frac{c_1}{2!}\mathcal{L}^{-1}\left\{\frac{2!}{s^3}\right\} + \frac{c_2}{3!}\mathcal{L}^{-1}\left\{\frac{3!}{s^4}\right\}$$

$$+ \frac{2w_0}{EIL}\left[\frac{L/2}{4!}\mathcal{L}^{-1}\left\{\frac{4!}{s^5}\right\} - \frac{1}{5!}\mathcal{L}^{-1}\left\{\frac{5!}{s^6}\right\} + \frac{1}{5!}\mathcal{L}^{-1}\left\{\frac{5!}{s^6}e^{-Ls/2}\right\}\right]$$

$$= \frac{c_1}{2}x^2 + \frac{c_2}{6}x^3 + \frac{w_0}{60EIL}\left[\frac{5L}{2}x^4 - x^5 + \left(x - \frac{L}{2}\right)^5\mathcal{U}\left(x - \frac{L}{2}\right)\right].$$

Applying the conditions $y(L) = 0$ and $y'(L) = 0$ to the last result yields a system of equations for c_1 and c_2:

$$c_1\frac{L^2}{2} + c_2\frac{L^3}{6} + \frac{49w_0 L^4}{1920EI} = 0$$

$$c_1 L + c_2\frac{L^2}{2} + \frac{85w_0 L^3}{960EI} = 0.$$

$w(x)$

wall

x

L

y

Figure 4.17

Solving, we find $c_1 = 23w_0 L^2/960EI$ and $c_2 = -9w_0 L/40EI$. Thus the deflection is given by

$$y(x) = \frac{23w_0 L^2}{1920EI}x^2 - \frac{3w_0 L}{80EI}x^3 + \frac{w_0}{60EIL}\left[\frac{5L}{2}x^4 - x^5 + \left(x - \frac{L}{2}\right)^5 \mathcal{U}\left(x - \frac{L}{2}\right)\right]. \qquad \square$$

4.3.1 Translation on the *s*-axis

In Problems 1–20, find either $F(s)$ or $f(t)$, as indicated.

1. $\mathcal{L}\{te^{10t}\}$

2. $\mathcal{L}\{te^{-6t}\}$

3. $\mathcal{L}\{t^3 e^{-2t}\}$

4. $\mathcal{L}\{t^{10} e^{-7t}\}$

5. $\mathcal{L}\{t(e^t + e^{2t})^2\}$

6. $\mathcal{L}\{e^{2t}(t-1)^2\}$

7. $\mathcal{L}\{e^t \sin 3t\}$

8. $\mathcal{L}\{e^{-2t} \cos 4t\}$

9. $\mathcal{L}\{(1 - e^t + 3e^{-4t}) \cos 5t\}$

10. $\mathcal{L}\left\{e^{3t}\left(9 - 4t + 10 \sin \dfrac{t}{2}\right)\right\}$

11. $\mathcal{L}^{-1}\left\{\dfrac{1}{(s+2)^3}\right\}$

12. $\mathcal{L}^{-1}\left\{\dfrac{1}{(s-1)^4}\right\}$

13. $\mathcal{L}^{-1}\left\{\dfrac{1}{s^2 - 6s + 10}\right\}$

14. $\mathcal{L}^{-1}\left\{\dfrac{1}{s^2 + 2s + 5}\right\}$

15. $\mathcal{L}^{-1}\left\{\dfrac{s}{s^2 + 4s + 5}\right\}$

16. $\mathcal{L}^{-1}\left\{\dfrac{2s+5}{s^2 + 6s + 34}\right\}$

17. $\mathcal{L}^{-1}\left\{\dfrac{s}{(s+1)^2}\right\}$

18. $\mathcal{L}^{-1}\left\{\dfrac{5s}{(s-2)^2}\right\}$

19. $\mathcal{L}^{-1}\left\{\dfrac{2s-1}{s^2(s+1)^3}\right\}$

20. $\mathcal{L}^{-1}\left\{\dfrac{(s+1)^2}{(s+2)^4}\right\}$

In Problems 21–30, use the Laplace transform to solve the given initial-value problem.

21. $y' + 4y = e^{-4t}$, $y(0) = 2$

22. $y' - y = 1 + te^t$, $y(0) = 0$

23. $y'' + 2y' + y = 0$, $y(0) = 1$, $y'(0) = 1$

24. $y'' - 4y' + 4y = t^3 e^{2t}$, $y(0) = 0$, $y'(0) = 0$

25. $y'' - 6y' + 9y = t$, $y(0) = 0$, $y'(0) = 1$

26. $y'' - 4y' + 4y = t^3$, $y(0) = 1$, $y'(0) = 0$

27. $y'' - 6y' + 13y = 0$, $y(0) = 0$, $y'(0) = -3$

28. $2y'' + 20y' + 51y = 0$, $y(0) = 2$, $y'(0) = 0$

29. $y'' - y' = e^t \cos t$, $y(0) = 0$, $y'(0) = 0$

30. $y'' - 2y' + 5y = 1 + t$, $y(0) = 0$, $y'(0) = 4$

In Problems 31 and 32, use the Laplace transform and the procedure outlined in Example 9 to solve the given boundary-value problem.

31. $y'' + 2y' + y = 0$, $y'(0) = 2$, $y(1) = 2$

32. $y'' + 8y' + 20y = 0$, $y(0) = 0$, $y'(\pi) = 0$

33. A 4-lb weight stretches a spring 2 ft. The weight is released from rest 18 in. above the equilibrium position, and the resulting motion takes place in a medium offering a damping force numerically equal to ⅞ times the instantaneous velocity. Use the Laplace transform to find the equation of motion $x(t)$.

34. Recall that the differential equation for the instantaneous charge $q(t)$ on the capacitor in an *LRC* series circuit is given by

$$L\frac{d^2 q}{dt^2} + R\frac{dq}{dt} + \frac{1}{C}q = E(t). \qquad (20)$$

See Section 3.8. Use the Laplace transform to find $q(t)$ when $L = 1$ h, $R = 20\ \Omega$, $C = 0.005$ f, $E(t) = 150$ V, $t > 0$, $q(0) = 0$, and $i(0) = 0$. What is the current $i(t)$?

35. Consider the battery of constant voltage E_0 that charges the capacitor shown in Figure 4.18. Divide equation (20) by L and define $2\lambda = R/L$ and $\omega^2 = 1/LC$. Use the Laplace transform to shown that the solution $q(t)$ of $q'' + 2\lambda q' + \omega^2 q = E_0/L$, subject to $q(0) = 0$, $i(0) = 0$, is

$$q(t) = \begin{cases} E_0 C\left[1 - e^{-\lambda t}\left(\cosh \sqrt{\lambda^2 - \omega^2}\,t + \dfrac{\lambda}{\sqrt{\lambda^2 - \omega^2}} \sinh \sqrt{\lambda^2 - \omega^2}\,t\right)\right], & \lambda > \omega \\[3mm] E_0 C\left[1 - e^{-\lambda t}(1 + \lambda t)\right], & \lambda = \omega \\[3mm] E_0 C\left[1 - e^{-\lambda t}\left(\cos \sqrt{\omega^2 - \lambda^2}\,t + \dfrac{\lambda}{\sqrt{\omega^2 - \lambda^2}} \sin \sqrt{\omega^2 - \lambda^2}\,t\right)\right], & \lambda < \omega \end{cases}$$

Figure 4.18

36. Use the Laplace transform to find the charge $q(t)$ in an *RC* series when $q(0) = 0$ and $E(t) = E_0 e^{-kt}$, $k > 0$. Consider two cases: $k \neq 1/RC$ and $k = 1/RC$.

4.3.2 Translation on the *t*-axis

In Problems 37–48, find either $F(s)$ or $f(t)$, as indicated.

37. $\mathcal{L}\{(t-1)\mathcal{U}(t-1)\}$

38. $\mathcal{L}\{e^{2-t}\,\mathcal{U}(t-2)\}$

39. $\mathcal{L}\{t\,\mathcal{U}(t-2)\}$

40. $\mathcal{L}\{(3t+1)\,\mathcal{U}(t-1)\}$

41. $\mathcal{L}\{\cos 2t\,\mathcal{U}(t-\pi)\}$

42. $\mathcal{L}\left\{\sin t\,\mathcal{U}\left(t-\dfrac{\pi}{2}\right)\right\}$

43. $\mathcal{L}^{-1}\left\{\dfrac{e^{-2s}}{s^3}\right\}$

44. $\mathcal{L}^{-1}\left\{\dfrac{(1+e^{-2s})^2}{s+2}\right\}$

45. $\mathcal{L}^{-1}\left\{\dfrac{e^{-\pi s}}{s^2+1}\right\}$

46. $\mathcal{L}^{-1}\left\{\dfrac{se^{-\pi s/2}}{s^2+4}\right\}$

47. $\mathcal{L}^{-1}\left\{\dfrac{e^{-s}}{s(s+1)}\right\}$

48. $\mathcal{L}^{-1}\left\{\dfrac{e^{-2s}}{s^2(s-1)}\right\}$

In Problems 49–54, match the given graph with one of the given functions in **(a)–(f)**. The graph of $f(t)$ is given in Figure 4.19.

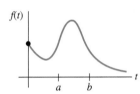

Figure 4.19

(a) $f(t) - f(t)\mathcal{U}(t-a)$

(b) $f(t-b)\mathcal{U}(t-b)$

(c) $f(t)\mathcal{U}(t-a)$

(d) $f(t) - f(t)\mathcal{U}(t-b)$

(e) $f(t)\mathcal{U}(t-a) - f(t)\mathcal{U}(t-b)$

(f) $f(t-a)\mathcal{U}(t-a) - f(t-a)\mathcal{U}(t-b)$

49.

Figure 4.20

50.

Figure 4.21

51.

Figure 4.22

52.

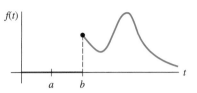

Figure 4.23

53. $f(t)$ **54.** $f(t)$

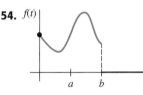

Figure 4.24 **Figure 4.25**

In Problems 55–62, write each function in terms of unit step functions. Find the Laplace transform of the given function.

55. $f(t) = \begin{cases} 2, & 0 \le t < 3 \\ -2, & t \ge 3 \end{cases}$

56. $f(t) = \begin{cases} 1, & 0 \le t < 4 \\ 0, & 4 \le t < 5 \\ 1, & t \ge 5 \end{cases}$

57. $f(t) = \begin{cases} 0, & 0 \le t < 1 \\ t^2, & t \ge 1 \end{cases}$

58. $f(t) = \begin{cases} 0, & 0 \le t < 3\pi/2 \\ \sin t, & t \ge 3\pi/2 \end{cases}$

59. $f(t) = \begin{cases} t, & 0 \le t < 2 \\ 0, & t \ge 2 \end{cases}$

60. $f(t) = \begin{cases} \sin t, & 0 \le t < 2\pi \\ 0, & t \ge 2\pi \end{cases}$

61. $f(t)$ **62.** $f(t)$

rectangular pulse

Figure 4.26

staircase function

Figure 4.27

In Problems 63–70, use the Laplace transform to solve the given initial-value problem.

63. $y' + y = f(t)$, $\quad y(0) = 0$, \quad where $f(t) = \begin{cases} 0, & 0 \le t < 1 \\ 5, & t \ge 1 \end{cases}$

64. $y' + y = f(t)$, $\quad y(0) = 0$, \quad where $f(t) = \begin{cases} 1, & 0 \le t < 1 \\ -1, & t \ge 1 \end{cases}$

65. $y' + 2y = f(t)$, $\quad y(0) = 0$, \quad where $f(t) = \begin{cases} t, & 0 \le t < 1 \\ 0, & t \ge 1 \end{cases}$

66. $y'' + 4y = f(t)$, $y(0) = 0$, $y'(0) = -1$,

\quad where $f(t) = \begin{cases} 1, & 0 \le t < 1 \\ 0, & t \ge 1 \end{cases}$

67. $y'' + 4y = \sin t\,\mathcal{U}(t-2\pi)$, $\quad y(0) = 1$, $\quad y'(0) = 0$

68. $y'' - 5y' + 6y = \mathcal{U}(t - 1)$, $y(0) = 0$, $y'(0) = 1$

69. $y'' + y = f(t)$, $y(0) = 0$, $y'(0) = 1$,

where $f(t) = \begin{cases} 0, & 0 \le t < \pi \\ 1, & \pi \le t < 2\pi \\ 0, & t \ge 2\pi \end{cases}$

70. $y'' + 4y' + 3y = 1 - \mathcal{U}(t - 2) - \mathcal{U}(t - 4) + \mathcal{U}(t - 6)$,
$y(0) = 0$, $y'(0) = 0$

71. Suppose a 32-lb weight stretches a spring 2 ft. If the weight is released from rest at the equilibrium position, find the equation of motion $x(t)$ if an impressed force $f(t) = 20t$ acts on the system for $0 \le t < 5$ and is then removed (see Example 5). Ignore any damping forces. Use a graphing utility to obtain the graph $x(t)$ on the interval [0, 10].

72. Solve Problem 71 if the impressed force $f(t) = \sin t$ acts on the system for $0 \le t < 2\pi$ and is then removed.

In Problems 73 and 74, use the Laplace transform to find the charge $q(t)$ on the capacitor in an RC series circuit subject to the given conditions.

73. $q(0) = 0$, $R = 2.5\ \Omega$,
$C = 0.08$ f, $E(t)$ given
in Figure 4.28

74. $q(0) = q_0$, $R = 10\ \Omega$,
$C = 0.1$ f, $E(t)$ given
in Figure 4.29

Figure 4.28

Figure 4.29

75. (a) Use the Laplace transform to find the current $i(t)$ in a single-loop LR series circuit when $i(0) = 0$, $L = 1$ h, $R = 10\ \Omega$, and $E(t)$ is as given in Figure 4.30.

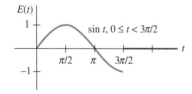

Figure 4.30

(b) Use a computer graphing program to graph $i(t)$ on the interval $0 \le t \le 6$. Use the graph to estimate i_{max} and i_{min}, the maximum and minimum values of the current.

76. (a) Use the Laplace transform to find the charge $q(t)$ on the capacitor in an RC series circuit when $q(0) = 0$, $R = 50\ \Omega$, $C = 0.01$ f, and $E(t)$ is as given in Figure 4.31.

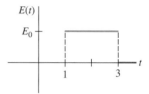

Figure 4.31

(b) Assume $E_0 = 100$ V. Use a computer graphing program to graph $q(t)$ on the interval $0 \le t \le 6$. Use the graph to estimate q_{max}, the maximum value of the charge.

77. A cantilever beam is embedded at its left end and free at its right end. Use the Laplace transform to find the deflection $y(x)$ when the load is given by

$$w(x) = \begin{cases} w_0, & 0 < x < L/2 \\ 0, & L/2 \le x < L. \end{cases}$$

78. Solve Problem 77 when the load is given by

$$w(x) = \begin{cases} 0, & 0 < x < L/3 \\ w_0, & L/3 < x < 2L/3 \\ 0, & 2L/3 < x < L. \end{cases}$$

79. Find the deflection $y(x)$ of a cantilever beam embedded at its left end and free at its right end when the load is as given in Example 9.

80. A beam is embedded at its left end and simply supported at its right end. Find the deflection $y(x)$ when the load is as given in Problem 77.

81. Assume Theorem 4.6 holds when the symbol a is replaced by ki, where k is a real number and $i^2 = -1$. Show that $\mathscr{L}\{te^{kti}\}$ can be used to deduce

$$\mathscr{L}\{t \cos kt\} = \frac{s^2 - k^2}{(s^2 + k^2)^2} \quad \text{and} \quad \mathscr{L}\{t \sin kt\} = \frac{2ks}{(s^2 + k^2)^2}.$$

82. Use the Laplace transform and Problem 81 to solve the initial-value problem $x'' + \omega^2 x = \cos \omega t$, $x(0) = 0$, $x'(0) = 0$.

4.4 Additional Operational Properties

■ **Multiplying a Function by t^n** The Laplace transform of the product of a function $f(t)$ with t can be found by differentiating the Laplace transform of $f(t)$. If $F(s) = \mathscr{L}\{f(t)\}$ and if we assume that interchanging of differentiation and integration is possible, then

$$\frac{d}{ds} F(s) = \frac{d}{ds} \int_0^\infty e^{-st} f(t)\, dt = \int_0^\infty \frac{\partial}{\partial s} [e^{-st} f(t)]\, dt = -\int_0^\infty e^{-st}\, tf(t)\, dt = -\mathscr{L}\{tf(t)\};$$

that is,
$$\mathscr{L}\{tf(t)\} = -\frac{d}{ds} \mathscr{L}\{f(t)\}.$$

Similarly,
$$\mathscr{L}\{t^2 f(t)\} = \mathscr{L}\{t \cdot tf(t)\} = -\frac{d}{ds} \mathscr{L}\{tf(t)\}$$

$$= -\frac{d}{ds}\left(-\frac{d}{ds} \mathscr{L}\{f(t)\}\right) = \frac{d^2}{ds^2} \mathscr{L}\{f(t)\}.$$

The preceding two cases suggest the general result for $\mathscr{L}\{t^n f(t)\}$.

THEOREM 4.8 Derivatives of Transforms

If $F(s) = \mathscr{L}\{f(t)\}$ and $n = 1, 2, 3, \ldots$, then

$$\mathscr{L}\{t^n f(t)\} = (-1)^n \frac{d^n}{ds^n} F(s).$$

Example 1 Using Theorem 4.8
Evaluate $\mathscr{L}\{t \sin kt\}$.

SOLUTION With $f(t) = \sin kt$, $F(s) = k/(s^2 + k^2)$, and $n = 1$, Theorem 4.8 gives

$$\mathscr{L}\{t \sin kt\} = -\frac{d}{ds} \mathscr{L}\{\sin kt\} = -\frac{d}{ds}\left(\frac{k}{s^2 + k^2}\right) = \frac{2ks}{(s^2 + k^2)^2}. \qquad \square$$

If we wanted to evaluate $\mathscr{L}\{t^2 \sin kt\}$ and $\mathscr{L}\{t^3 \sin kt\}$, all we need do, in turn, is take the negative of the derivative with respect to s of the result in Example 1 and then take the negative of the derivative with s of $\mathscr{L}\{t^2 \sin kt\}$.

 Note that either theorem could be used.

To find transforms of functions $t^n e^{at}$ we can use either the first translation theorem or Theorem 4.8. For example,

Theorem 4.6: $\mathscr{L}\{te^{3t}\} = \mathscr{L}\{t\}_{s \to s-3} = \left.\frac{1}{s^2}\right|_{s \to s-3} = \frac{1}{(s-3)^2}$

Theorem 4.8: $\mathscr{L}\{te^{3t}\} = -\dfrac{d}{ds} \mathscr{L}\{e^{3t}\} = -\dfrac{d}{ds} \dfrac{1}{s-3} = (s-3)^{-2} = \dfrac{1}{(s-3)^2}$.

Example 2 An Initial-Value Problem
Solve $x'' + 16x = \cos 4t$, $x(0) = 0$, $x'(0) = 1$.

SOLUTION The initial-value problem could describe the forced, undamped, and resonant motion of a mass on a spring. The mass starts with an initial velocity

of 1 foot per second in the downward direction from the equilibrium position. Transforming the differential equation gives

$$(s^2 + 16)X(s) = 1 + \frac{s}{s^2 + 16} \quad \text{or} \quad X(s) = \frac{1}{s^2 + 16} + \frac{s}{(s^2 + 16)^2}.$$

Now we have just learned in Example 1 that

$$\mathscr{L}^{-1}\left\{\frac{2ks}{(s^2 + k^2)^2}\right\} = t \sin kt, \tag{1}$$

and so with the identification $k = 4$ in (1) and in part **(d)** of Theorem 4.3, we obtain

$$x(t) = \frac{1}{4}\mathscr{L}^{-1}\left\{\frac{4}{s^2 + 16}\right\} + \frac{1}{8}\mathscr{L}^{-1}\left\{\frac{8s}{(s^2 + 16)^2}\right\}$$

$$= \frac{1}{4}\sin 4t + \frac{1}{8}t \sin 4t. \qquad\qquad ❑$$

■ **Convolution** If functions f and g are piecewise continuous on $[0, \infty)$, then a special product, denoted by $f * g$, is defined by the integral

$$f * g = \int_0^t f(\tau)g(t - \tau)\,d\tau \tag{2}$$

and is called the **convolution** of f and g. The convolution $f * g$ is a function of t. For example,

$$e^t * \sin t = \int_0^t e^\tau \sin(t - \tau)\,d\tau = \frac{1}{2}(-\sin t - \cos t + e^t). \tag{3}$$

It is left as an exercise to show that

$$\int_0^t f(\tau)g(t - \tau)\,d\tau = \int_0^t f(t - \tau)g(\tau)\,d\tau,$$

that is, $f * g = g * f$. This means that the convolution of two functions is commutative. See Problem 53 in Exercises 4.4.

It *is not* true that the integral of a product of functions is the product of the integrals. However, it *is* true that the Laplace transform of the special product (2) is the product of the Laplace transforms of f and g. This means it is possible to find the Laplace transform of the convolution of two functions without actually evaluating the integral as we did in (3). The result that follows is known as the **convolution theorem.**

THEOREM 4.9 Convolution Theorem

If $f(t)$ and $g(t)$ are piecewise continuous on $[0, \infty)$ and of exponential order, then

$$\mathscr{L}\{f * g\} = \mathscr{L}\{f(t)\}\mathscr{L}\{g(t)\} = F(s)G(s).$$

PROOF

Let

$$F(s) = \mathscr{L}\{f(t)\} = \int_0^\infty e^{-s\tau}f(\tau)\,d\tau$$

and
$$G(s) = \mathscr{L}\{g(t)\} = \int_0^\infty e^{-s\beta} g(\beta) \, d\beta.$$

Proceeding formally, we have

$$F(s)G(s) = \left(\int_0^\infty e^{-s\tau} f(\tau) \, d\tau \right) \left(\int_0^\infty e^{-s\beta} g(\beta) \, d\beta \right)$$

$$= \int_0^\infty \int_0^\infty e^{-s(\tau + \beta)} f(\tau) g(\beta) \, d\tau \, d\beta$$

$$= \int_0^\infty f(\tau) \, d\tau \int_0^\infty e^{-s(\tau + \beta)} g(\beta) \, d\beta.$$

Holding τ fixed, we let $t = \tau + \beta$, $dt = d\beta$, so that

$$F(s)G(s) = \int_0^\infty f(\tau) \, d\tau \int_\tau^\infty e^{-st} g(t - \tau) \, dt.$$

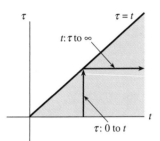

In the $t\tau$-plane we are integrating over the shaded region in Figure 4.32. Since f and g are piecewise continuous on $[0, \infty)$ and of exponential order, it is possible to interchange the order of integration:

$$F(s)G(s) = \int_0^\infty e^{-st} \, dt \int_0^t f(\tau) g(t - \tau) \, d\tau = \int_0^\infty e^{-st} \left\{ \int_0^t f(\tau) g(t - \tau) \, d\tau \right\} dt = \mathscr{L}\{f * g\}. \quad \square$$

Figure 4.32

Example 3 Transform of a Convolution

Evaluate $\mathscr{L}\left\{ \int_0^t e^\tau \sin(t - \tau) \, d\tau \right\}$.

SOLUTION With $f(t) = e^t$ and $g(t) = \sin t$ the convolution theorem states that the Laplace transform of the convolution of f and g is the product of their Laplace transforms:

$$\mathscr{L}\left\{ \int_0^t e^\tau \sin(t - \tau) \, d\tau \right\} = \mathscr{L}\{e^t\} \cdot \mathscr{L}\{\sin t\} = \frac{1}{s - 1} \cdot \frac{1}{s^2 + 1} = \frac{1}{(s - 1)(s^2 + 1)}. \quad \square$$

■ **Inverse Form of Theorem 4.9** The convolution theorem is sometimes useful in finding the inverse Laplace transforms of the product of two Laplace transforms. From Theorem 4.9 we have

$$\mathscr{L}^{-1}\{F(s)G(s)\} = f * g. \tag{4}$$

Many of the results in the table of Laplace transforms in Appendix III can be derived using (4). For instance, in the next example we obtain entry 25 of the table:

$$\mathscr{L}\{\sin kt - kt \cos kt\} = \frac{2k^3}{(s^2 + k^2)^2}. \tag{5}$$

Example 4 Inverse Transform as a Convolution

Evaluate $\mathscr{L}^{-1}\left\{ \frac{1}{(s^2 + k^2)^2} \right\}$.

SOLUTION Let $\qquad F(s) = G(s) = \frac{1}{s^2 + k^2}$

so that
$$f(t) = g(t) = \frac{1}{k} \mathcal{L}^{-1}\left\{ \frac{k}{s^2 + k^2} \right\} = \frac{1}{k} \sin kt.$$

In this case (4) gives

$$\mathcal{L}^{-1}\left\{ \frac{1}{(s^2 + k^2)^2} \right\} = \frac{1}{k^2} \int_0^t \sin k\tau \sin k(t - \tau) \, d\tau. \qquad (6)$$

Now recall from trigonometry that

$$\cos(A + B) = \cos A \cos B - \sin A \sin B$$

and
$$\cos(A - B) = \cos A \cos B + \sin A \sin B.$$

Subtracting the first from the second gives the identity

$$\sin A \sin B = \frac{1}{2}[\cos(A - B) - \cos(A + B)].$$

If we set $A = k\tau$ and $B = k(t - \tau)$, we can carry out the integration in (6):

$$\mathcal{L}^{-1}\left\{ \frac{1}{(s^2 + k^2)^2} \right\} = \frac{1}{2k^2} \int_0^t [\cos k(2\tau - t) - \cos kt] \, d\tau$$

$$= \frac{1}{2k^2}\left[\frac{1}{2k} \sin k(2\tau - t) - \tau \cos kt \right]_0^t$$

$$= \frac{\sin kt - kt \cos kt}{2k^3}.$$

Multiplying both sides by $2k^3$ gives the inverse form of (5). ❏

■ **Transform of an Integral** When $g(t) = 1$ and $\mathcal{L}\{g(t)\} = G(s) = 1/s$, the convolution theorem implies that the Laplace transform of the integral of f is

$$\mathcal{L}\left\{ \int_0^t f(\tau) \, d\tau \right\} = \frac{F(s)}{s}. \qquad (7)$$

The inverse form of (7),

$$\int_0^t f(\tau) \, d\tau = \mathcal{L}^{-1}\left\{ \frac{F(s)}{s} \right\}, \qquad (8)$$

can be used in lieu of partial fractions when s^n is a factor of the denominator and $f(t) = \mathcal{L}^{-1}\{F(s)\}$ is easy to integrate. For example, we know for $f(t) = \sin t$ that $F(s) = 1/(s^2 + 1)$, and so by (8)

$$\mathcal{L}^{-1}\left\{ \frac{1}{s(s^2 + 1)} \right\} = \int_0^t \sin \tau \, d\tau = 1 - \cos t$$

$$\mathcal{L}^{-1}\left\{ \frac{1}{s^2(s^2 + 1)} \right\} = \int_0^t (1 - \cos \tau) \, d\tau = t - \sin t$$

$$\mathcal{L}^{-1}\left\{ \frac{1}{s^3(s^2 + 1)} \right\} = \int_0^t (\tau - \sin \tau) \, d\tau = \frac{1}{2}t^2 - 1 + \cos t$$

and so on.

■ **Volterra Integral Equation** The convolution theorem and the result in (7) are useful in solving other types of equations in which an unknown function appears under an integral sign. In the next example, we solve a **Volterra integral equation** for $f(t)$,

$$f(t) = g(t) + \int_0^t f(\tau)h(t - \tau)\, d\tau. \tag{9}$$

The functions $g(t)$ and $h(t)$ are known. Notice that the integral in (9) has the convolution form (2) with the symbol h playing the part of g.

Example 5 An Integral Equation

Solve $f(t) = 3t^2 - e^{-t} - \int_0^t f(\tau)e^{t-\tau}\, d\tau$ for $f(t)$.

SOLUTION In the integral we identify $h(t - \tau) = e^{t-\tau}$ so that $h(t) = e^t$. We take the Laplace transform of each term; in particular, by Theorem 4.9 the transform of the integral is the product of $\mathscr{L}\{f(t)\} = F(s)$ and $\mathscr{L}\{e^t\} = 1/(s - 1)$:

$$F(s) = 3 \cdot \frac{2}{s^3} - \frac{1}{s + 1} - F(s) \cdot \frac{1}{s - 1}.$$

After solving the last equation for $F(s)$ and carrying out the partial fraction decomposition, we find

$$F(s) = \frac{6}{s^3} - \frac{6}{s^4} + \frac{1}{s} - \frac{2}{s + 1}.$$

The inverse transform then gives

$$f(t) = 3\mathscr{L}^{-1}\left\{\frac{2!}{s^3}\right\} - \mathscr{L}^{-1}\left\{\frac{3!}{s^4}\right\} + \mathscr{L}^{-1}\left\{\frac{1}{s}\right\} - 2\mathscr{L}^{-1}\left\{\frac{1}{s + 1}\right\}$$

$$= 3t^2 - t^3 + 1 - 2e^{-t}. \qquad \square$$

■ **Series Circuits** In a single-loop or series circuit, Kirchhoff's second law states that the sum of the voltage drops across an inductor, resistor, and capacitor is equal to the impressed voltage $E(t)$. Now it is known that the voltage drops across an inductor, resistor, and capacitor are, respectively,

$$L\frac{di}{dt}, \quad Ri(t), \quad \text{and} \quad \frac{1}{C}\int_0^t i(\tau)\, d\tau,$$

where $i(t)$ is the current and L, R, and C are constants. It follows that the current in a circuit, such as that shown in Figure 4.33, is governed by the **integrodifferential equation**

$$L\frac{di}{dt} + Ri(t) + \frac{1}{C}\int_0^t i(\tau)\, d\tau = E(t). \tag{10}$$

Figure 4.33

Example 6 An Integrodifferential Equation

Determine the current $i(t)$ in a single-loop LRC circuit when $L = 0.1$ h, $R = 2\ \Omega$, $C = 0.1$ f, $i(0) = 0$, and the impressed voltage is

$$E(t) = 120t - 120t\,\mathscr{U}(t - 1).$$

SOLUTION Using the given data, equation (10) becomes

$$0.1\frac{di}{dt} + 2i + 10\int_0^t i(\tau)\,d\tau = 120t - 120t\,\mathscr{U}(t-1).$$

Now by (7), $\mathscr{L}\{\int_0^t i(\tau)\,d\tau\} = I(s)/s$, where $I(s) = \mathscr{L}\{i(t)\}$. Thus the Laplace transform of the integrodifferential equation is

$$0.1sI(s) + 2I(s) + 10\frac{I(s)}{s} = 120\left[\frac{1}{s^2} - \frac{1}{s^2}e^{-s} - \frac{1}{s}e^{-s}\right].\quad \leftarrow \text{by (16) of Section 4.3}$$

Multiplying this equation by $10s$, using $s^2 + 20s + 100 = (s+10)^2$, and then solving for $I(s)$ gives

$$I(s) = 1200\left[\frac{1}{s(s+10)^2} - \frac{1}{s(s+10)^2}e^{-s} - \frac{1}{(s+10)^2}e^{-s}\right].$$

By partial fractions,

$$I(s) = 1200\left[\frac{1/100}{s} - \frac{1/100}{s+10} - \frac{1/10}{(s+10)^2} - \frac{1/100}{s}e^{-s}\right.$$

$$\left. + \frac{1/100}{s+10}e^{-s} + \frac{1/10}{(s+10)^2}e^{-s} - \frac{1}{(s+10)^2}e^{-s}\right].$$

From the inverse form of the second translation theorem, equation (15) of Section 4.3, we finally obtain

$$i(t) = 12[1 - \mathscr{U}(t-1)] - 12[e^{-10t} - e^{-10(t-1)}\mathscr{U}(t-1)]$$

$$- 120te^{-10t} - 1080(t-1)e^{-10(t-1)}\mathscr{U}(t-1).$$

Written as a piecewise defined function, the foregoing solution is

$$i(t) = \begin{cases} 12 - 12e^{-10t} - 120te^{-10t}, & 0 \le t < 1 \\ -12e^{-10t} + 12e^{-10(t-1)} - 120te^{-10t} - 1080(t-1)e^{-10(t-1)}, & t \ge 1. \end{cases}$$

Using the last form of the solution and a CAS, we graph $i(t)$ on each of the two intervals and then combine the graphs. Note in Figure 4.34 that even though the input $E(t)$ is discontinuous, the output or response $i(t)$ is a continuous function. ❏

Figure 4.34

■ **Transform of a Periodic Function** If a periodic function f has period T, $T > 0$, then $f(t + T) = f(t)$. The Laplace transform of a periodic function can be obtained by integration over one period.

THEOREM 4.10 Transform of a Periodic Function

If $f(t)$ is piecewise continuous on $[0, \infty)$, of exponential order, and periodic with period T, then

$$\mathscr{L}\{f(t)\} = \frac{1}{1 - e^{sT}}\int_0^T e^{-st} f(t)\,dt.$$

PROOF Write the Laplace transform of f as two integrals:

$$\mathscr{L}\{f(t)\} = \int_0^T e^{-st}f(t)\,dt + \int_T^\infty e^{-st}f(t)\,dt.$$

When we let $t = u + T$, the last integral becomes

$$\int_T^\infty e^{-st} f(t) \, dt = \int_0^\infty e^{-s(u+T)} f(u+T) \, du = e^{-sT} \int_0^\infty e^{-su} f(u) \, du = e^{-sT} \mathscr{L}\{f(t)\}.$$

Therefore

$$\mathscr{L}\{f(t)\} = \int_0^T e^{-st} f(t) \, dt + e^{-sT} \mathscr{L}\{f(t)\}.$$

Solving the equation in the last line for $\mathscr{L}\{f(t)\}$ proves the theorem. ❏

Figure 4.35

Example 7 Transform of a Periodic Function

Find the Laplace transform of the periodic function shown in Figure 4.35.

SOLUTION The function $E(t)$ is called a square wave and has period $T = 2$. On the interval $0 \le t < 2$, $E(t)$ can be defined by

$$E(t) = \begin{cases} 1, & 0 \le t < 1 \\ 0, & 1 \le t < 2, \end{cases}$$

and outside the interval by $f(t+2) = f(t)$. Now from Theorem 4.10,

$$\mathscr{L}\{E(t)\} = \frac{1}{1-e^{-2s}} \int_0^2 e^{-st} E(t) \, dt = \frac{1}{1-e^{-2s}} \left[\int_0^1 e^{-st} \cdot 1 \, dt + \int_1^2 e^{-st} \cdot 0 \, dt \right]$$

$$= \frac{1}{1-e^{-2s}} \frac{1-e^{-s}}{s} \qquad \leftarrow 1 - e^{-2s} = (1 + e^{-s})(1 - e^{-s})$$

$$= \frac{1}{s(1+e^{-s})}. \tag{11} ❏$$

Example 8 A Periodic Impressed Voltage

The differential equation for the current $i(t)$ in a single-loop LR series circuit is

$$L \frac{di}{dt} + Ri = E(t). \tag{12}$$

Determine the current $i(t)$ when $i(0) = 0$ and $E(t)$ is the square-wave function shown in Figure 4.35.

SOLUTION Using the result in (11) of the preceding example, the Laplace transform of the DE is

$$LsI(s) + RI(s) = \frac{1}{s(1+e^{-s})} \quad \text{or} \quad I(s) = \frac{1/L}{s(s+R/L)} \cdot \frac{1}{1+e^{-s}}. \tag{13}$$

To find the inverse Laplace transform of the last function, we first make use of geometric series. With the identification $x = e^{-s}$, $s > 0$, the geometric series

$$\frac{1}{1+x} = 1 - x + x^2 - x^3 + \dots \quad \text{becomes} \quad \frac{1}{1+e^{-s}} = 1 - e^{-s} + e^{-2s} - e^{-3s} + \dots.$$

From

$$\frac{1}{s(s+R/L)} = \frac{L/R}{s} - \frac{L/R}{s+R/L},$$

we can then rewrite (13) as

$$I(s) = \frac{1}{R}\left(\frac{1}{s} - \frac{1}{s + R/L}\right)(1 - e^{-s} + e^{-2s} - e^{-3s} + \cdots)$$

$$= \frac{1}{R}\left(\frac{1}{s} - \frac{e^{-s}}{s} + \frac{e^{-2s}}{s} - \frac{e^{-3s}}{s} + \cdots\right) - \frac{1}{R}\left(\frac{1}{s + R/L} - \frac{1}{s + R/L}e^{-s} + \frac{e^{-2s}}{s + R/L} - \frac{e^{-3s}}{s + R/L} + \cdots\right).$$

By applying the form of the second translation theorem to each term of both series we obtain

$$i(t) = \frac{1}{R}\left(1 - \mathcal{U}(t - 1) + \mathcal{U}(t - 2) - \mathcal{U}(t - 3) + \cdots\right)$$

$$- \frac{1}{R}\left(e^{-Rt/L} - e^{-R(t - 1)/L}\mathcal{U}(t - 1) + e^{-R(t - 2)/L}\mathcal{U}(t - 2) - e^{-R(t - 3)/L}\mathcal{U}(t - 3) + \cdots\right)$$

or, equivalently,

$$i(t) = \frac{1}{R}\left(1 - e^{-Rt/L}\right) + \frac{1}{R}\sum_{n=1}^{\infty}(-1)^n(1 - e^{-R(t - n)/L})\mathcal{U}(t - n).$$

To interpret the solution, let us assume for the sake of illustration that $R = 1$, $L = 1$, and $0 \le t < 4$. In this case

$$i(t) = 1 - e^{-t} - (1 - e^{t - 1})\mathcal{U}(t - 1) + (1 - e^{-(t - 2)})\mathcal{U}(t - 2) - (1 - e^{-(t - 3)})\mathcal{U}(t - 3);$$

in other words,

$$i(t) = \begin{cases} 1 - e^{-t}, & 0 \le t < 1 \\ -e^{-t} + e^{-(t - 1)}, & 1 \le t < 2 \\ 1 - e^{-t} + e^{-(t - 1)} - e^{-(t - 2)}, & 2 \le t < 3 \\ -e^{-t} + e^{-(t - 1)} - e^{-(t - 2)} + e^{-(t - 3)}, & 3 \le t < 4. \end{cases}$$

The graph of $i(t)$ on the interval $0 \le t < 4$, given in Figure 4.36, was obtained with the help of a CAS. ❑

Figure 4.36

Answers to odd-numbered problems begin on page A-34.

In Problems 1–10, find $F(s)$.

1. $\mathcal{L}\{t \cos 2t\}$

2. $\mathcal{L}\{t \sinh 3t\}$

3. $\mathcal{L}\{t^2 \sinh t\}$

4. $\mathcal{L}\{t^2 \cos t\}$

5. $\mathcal{L}\{te^{2t} \sin 6t\}$

6. $\mathcal{L}\{te^{-3t} \cos 3t\}$

7. $\mathcal{L}\{1 * t^3\}$

8. $\mathcal{L}\{t^2 * te^t\}$

9. $\mathcal{L}\{e^{-t} * e^t \cos t\}$

10. $\mathcal{L}\{e^{2t} * \sin t\}$

In Problems 11–18, evaluate the given Laplace transform without evaluating the integral.

11. $\mathcal{L}\left\{\int_0^t e^\tau \, d\tau\right\}$

12. $\mathcal{L}\left\{\int_0^t \cos \tau \, d\tau\right\}$

13. $\mathcal{L}\left\{\int_0^t e^{-\tau} \cos \tau \, d\tau\right\}$

14. $\mathcal{L}\left\{\int_0^t \tau \sin \tau \, d\tau\right\}$

15. $\mathcal{L}\left\{\int_0^t \tau e^{t - \tau} \, d\tau\right\}$

16. $\mathcal{L}\left\{\int_0^t \sin \tau \cos(t - \tau) \, d\tau\right\}$

17. $\mathcal{L}\left\{t\int_0^t \sin \tau \, d\tau\right\}$

18. $\mathcal{L}\left\{t\int_0^t \tau e^{-\tau} \, d\tau\right\}$

19. Use (8) to evaluate the given inverse transform.

(a) $\mathcal{L}^{-1}\left\{\dfrac{1}{s(s - 1)}\right\}$

(b) $\mathcal{L}^{-1}\left\{\dfrac{1}{s^2(s - 1)}\right\}$

(c) $\mathcal{L}^{-1}\left\{\dfrac{1}{s^3(s - 1)}\right\}$

20. The table in Appendix III does not contain an entry for

$$\mathcal{L}^{-1}\left\{\frac{8k^3 s}{(s^2 + k^2)^3}\right\}.$$

(a) Use (4), along with the results in (5), to evaluate this inverse transform. Use a CAS as an aid in evaluating the convolution integral.

(b) Reexamine your answer to part **(a)**. Could you have obtained the result in a different manner?

In Problems 21 and 22, use Theorem 4.8 in the form ($n = 1$)

$$f(t) = -\frac{1}{t}\,\mathcal{L}^{-1}\left\{\frac{d}{ds}\,F(s)\right\}$$

to evaluate the given inverse Laplace transform.

21. $\mathcal{L}^{-1}\left\{\ln\dfrac{s-3}{s+1}\right\}$ **22.** $\mathcal{L}^{-1}\left\{\ln\dfrac{s^2+1}{s^2+4}\right\}$

In Problems 23–28, use Theorem 4.10 to find the Laplace transform of the given periodic function.

23.

Meander function

Figure 4.37

24.

Square wave

Figure 4.38

25.

Sawtooth function

Figure 4.39

26.

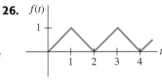

Triangular wave

Figure 4.40

27.

Full-wave rectification of sint

Figure 4.41

28.

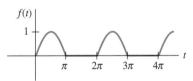

Half-wave rectification of sint

Figure 4.42

In the remaining problems in this exercise set use the table of Laplace transforms in Appendix III as needed.

In Problems 29–34, use the Laplace transform to solve the given initial-value problem.

29. $y' + y = t \sin t$, $y(0) = 0$

30. $y' - y = te^t \sin t$, $y(0) = 0$

31. $y'' + 9y = \cos 3t$, $y(0) = 2$, $y'(0) = 5$

32. $y'' + y = \sin t$, $y(0) = 1$, $y'(0) = -1$

33. $y'' + 16y = f(t)$, $y(0) = 0$, $y'(0) = 1$, where

$$f(t) = \begin{cases} \cos 4t, & 0 \le t < \pi \\ 0, & t \ge \pi \end{cases}$$

34. $y'' + y = f(t)$, $y(0) = 1$, $y'(0) = 0$, where

$$f(t) = \begin{cases} 1, & 0 \le t < \pi/2 \\ \sin t, & t \ge \pi/2 \end{cases}$$

35. Use the Laplace transform and the results obtained in Problem 20 to solve the initial-value problem
$y'' + y = \sin t + t \sin t$, $y(0) = 0$, $y'(0) = 0$.

36. Use a graphing utility to obtain the graph of the indicated solution on $0 \le t < 2\pi$.

(a) $y(t)$ of Problem 33 **(b)** $y(t)$ of Problem 34

In Problems 37–46, use the Laplace transform to solve the given integral equation or integrodifferential equation.

37. $f(t) + \displaystyle\int_0^t (t - \tau)f(\tau)\, d\tau = t$

38. $f(t) = 2t - 4\displaystyle\int_0^t \sin \tau\, f(t - \tau)\, d\tau$

39. $f(t) = te^t + \displaystyle\int_0^t \tau f(t - \tau)\, d\tau$

40. $f(t) + 2\displaystyle\int_0^t f(\tau) \cos(t - \tau)\, d\tau = 4e^{-t} + \sin t$

41. $f(t) + \displaystyle\int_0^t f(\tau)\, d\tau = 1$

42. $f(t) = \cos t + \displaystyle\int_0^t e^{-\tau} f(t - \tau)\, d\tau$

43. $f(t) = 1 + t - \dfrac{8}{3}\displaystyle\int_0^t (t - \tau)^3 f(\tau)\, d\tau$

44. $t - 2f(t) = \displaystyle\int_0^t (e^\tau - e^{-\tau})f(t - \tau)\, d\tau$

45. $y'(t) = 1 - \sin t - \displaystyle\int_0^t y(\tau)\, d\tau$, $y(0) = 0$

46. $\dfrac{dy}{dt} + 6y(t) + 9\displaystyle\int_0^t y(\tau)\, d\tau = 1$, $y(0) = 0$

In Problems 47 and 48, solve equation (10) subject to $i(0) = 0$ with L, R, C, and $E(t)$ as given. Use a graphing utility to obtain the graph of the solution of the interval $0 \le t \le 3$.

47. $L = 0.1$ h, $R = 3\ \Omega$, $C = 0.05$ f,
$E(t) = 100[\mathcal{U}(t - 1) - \mathcal{U}(t - 2)]$

48. $L = 0.005$ h, $R = 1\ \Omega$, $C = 0.02$ f,
$E(t) = 100[t - (t - 1)\mathcal{U}(t - 1)]$

In Problems 49 and 50, solve equation (12) subject to $i(0) = 0$ with $E(t)$ as given. Use a graphing utility to obtain the graph of the solution on the interval $0 \le t < 4$ in the case when $L = 1$ and $R = 1$.

49. $E(t)$ is the meander function in Problem 23 with amplitude 1 and $a = 1$.

50. $E(t)$ is the sawtooth function in Problem 25 with amplitude 1 and $b = 1$.

In Problems 51 and 52, solve the model for a driven spring/mass system with damping

$$m\frac{d^2x}{dt^2} + \beta\frac{dx}{dt} + kx = f(t), \quad x(0) = 0, \quad x'(0) = 0,$$

where the driving function f is as specified. Use a graphing utility to obtain the graph of $x(t)$ on the indicated interval.

51. $m = \frac{1}{2}$, $\beta = 1$, $k = 5$, f is the meander function in Problem 23 with amplitude 10, and $a = \pi$, $0 \le t < 2\pi$.

52. $m = 1$, $\beta = 2$, $k = 1$, f is the square wave in Problem 24 with amplitude 5, and $a = \pi$, $0 \le t < 4\pi$.

53. Prove the commutative property $f * g = g * f$ of the convolution integral.

54. Prove the distributive property $f * (g + h) = f * g + f * h$ of the convolution integral.

55. Computer Lab Assignment In part **(a)** of this problem, you are led through the commands in *Mathematica* that enable you to obtain the symbolic Laplace transform of a differential equation and the solution of the initial-value problem by finding the inverse transform. In *Mathematica* the Laplace transform of a function $y(t)$ is obtained using **LaplaceTransform** [y[t], t, s]. In line two of the syntax, we replace **LaplaceTransform** [y[t], t, s] by the symbol **Y**. (*If you do not have* Mathematica, *then adapt the given procedure by finding the corresponding syntax for the CAS you have on hand.*)

(a) Consider the initial-value problem

$$y'' + 6y' + 9y = t \sin t, \quad y(0) = 2, \quad y'(0) = -1.$$

Load the Laplace transform package. Precisely reproduce and then, in turn, execute each line in the given sequence of commands. Either copy the output by hand or print out the results.

```
diffequat = y''[t] + 6y'[t] + 9y[t] == t Sin[t]
transformdeq = LaplaceTransform [diffequat, t, s] /.
  { y[0] -> 2,
    y'[0] -> -1, LaplaceTransform [y[t], t, s] -> Y}
soln = Solve[transformdeq, Y] // Flatten
Y = Y/. soln
InverseLaplaceTransform[Y, s, t]
```

(b) Appropriately modify the procedure of part **(a)** to find a solution of

$$y''' + 3y' - 4y = 0, \quad y(0) = 0, \quad y'(0) = 0, \quad y''(0) = 1.$$

(c) The charge $q(t)$ on a capacitor in an *L-C* series circuit is given by

$$\frac{d^2q}{dt^2} + q = 1 - 4\mathcal{U}(t - \pi) + 6\mathcal{U}(t - 3\pi), \quad q(0) = 0, \quad q'(0) = 0.$$

Appropriately modify the procedure of part **(a)** to find $q(t)$. (In *Mathematica* the unit step function $\mathcal{U}(t - a)$ is written **UnitStep**[t − a].) Graph your solution.

4.5 Dirac Delta Function

■ **Unit Impulse** Mechanical systems are often acted on by an external force (or emf in an electrical circuit) of large magnitude that acts only for a very short period of time. For example, a vibrating airplane wing could be struck by lightning, a mass on a spring could be given a sharp blow by a ballpeen hammer, a ball (baseball, golf ball, tennis ball) could be sent soaring when struck violently by some kind of club (baseball bat, golf club, tennis racket). The function

$$\delta_a(t - t_0) = \begin{cases} 0, & 0 \le t < t_0 - a \\ \dfrac{1}{2a}, & t_0 - a \le t < t_0 + a \\ 0, & t \ge t_0 + a, \end{cases} \tag{1}$$

$a > 0$, $t_0 > 0$, shown in Figure 4.43(a), could serve as a mathematical model for such a force. For a small value of a, $\delta_a(t - t_0)$ is essentially a constant function of large magnitude that is "on" for just a very short period of time, around t_0. The behavior of $\delta_a(t - t_0)$ as $a \to 0$ is illustrated in Figure 4.43(b). The function $\delta_a(t - t_0)$ is called a **unit impulse** since it possesses the integration property $\int_0^\infty \delta_a(t - t_0) \, dt = 1$.

(a)

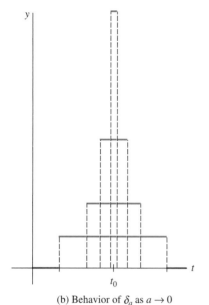

(b) Behavior of δ_a as $a \to 0$

Figure 4.43

■ **Dirac Delta Function** In practice it is convenient to work with another type of unit impulse, a "function" that approximates $\delta_a(t - t_0)$ and is defined by the limit

$$\delta(t - t_0) = \lim_{a \to 0} \delta_a(t - t_0). \tag{2}$$

The latter expression, which is not a function at all, can be characterized by the two properties

$$(i)\ \delta(t - t_0) = \begin{cases} \infty, & t = t_0 \\ 0, & t \neq t_0 \end{cases} \quad \text{and} \quad (ii) \int_0^x \delta(t - t_0)\,dt = 1.$$

The unit impulse $\delta(t - t_0)$ is called the **Dirac delta function.**

It is possible to obtain the Laplace transform of the Dirac delta function by the formal assumption that $\mathscr{L}\{\delta(t - t_0)\} = \lim_{a \to 0} \mathscr{L}\{\delta_a(t - t_0)\}$.

THEOREM 4.11 Transform of the Dirac Delta Function

For $t_0 > 0$, $\qquad\qquad \mathscr{L}\{\delta(t - t_0)\} = e^{-st_0}. \tag{3}$

PROOF To begin, we can write $\delta_a(t - t_0)$ in terms of the unit step function by virtue of (11) and (12) of Section 4.3:

$$\delta_a(t - t_0) = \frac{1}{2a}\left[\mathscr{U}(t - (t_0 - a)) - \mathscr{U}(t - (t_0 + a))\right].$$

By linearity and (14) of Section 4.3, the Laplace transform of this last expression is

$$\mathscr{L}\{\delta_a(t - t_0)\} = \frac{1}{2a}\left[\frac{e^{-s(t_0 - a)}}{s} - \frac{e^{-s(t_0 + a)}}{s}\right] = e^{-st_0}\left(\frac{e^{sa} - e^{-sa}}{2sa}\right). \tag{4}$$

Since (4) has the indeterminate form 0/0 as $a \to 0$, we apply L'Hôpital's rule:

$$\mathscr{L}\{\delta(t - t_0)\} = \lim_{a \to 0} \mathscr{L}\{\delta_a(t - t_0)\} = e^{-st_0}\lim_{a \to 0}\left(\frac{e^{sa} - e^{-sa}}{2sa}\right) = e^{-st_0}. \qquad \square$$

Now when $t_0 = 0$, it seems plausible to conclude from (3) that

$$\mathscr{L}\{\delta(t)\} = 1.$$

The last result emphasizes the fact that $\delta(t)$ is not the usual type of function that we have been considering, since we expect from Theorem 4.5 that $\mathscr{L}\{f(t)\} \to 0$ as $s \to \infty$.

Example 1 Two Initial-Value Problems

Solve $y'' + y = 4\delta(t - 2\pi)$ subject to

(a) $y(0) = 1, \quad y'(0) = 0$ (b) $y(0) = 0, \quad y'(0) = 0$.

The two initial-value problems could serve as models for describing the motion of a mass on a spring moving in a medium in which damping is negligible. At $t = 2\pi$ the mass is given a sharp blow. In **(a)** the mass is released from rest 1 unit below the equilibrium position. In (b) the mass is at rest in the equilibrium position.

SOLUTION (a) From (3) the Laplace transform of the differential equation is

$$s^2Y(s) - s + Y(s) = 4e^{-2\pi s} \quad \text{or} \quad Y(s) = \frac{s}{s^2 + 1} + \frac{4e^{-2\pi s}}{s^2 + 1}.$$

Figure 4.44

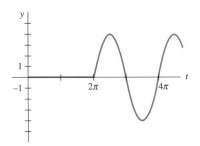

Figure 4.45

Using the inverse form of the second translation theorem, we find

$$y(t) = \cos t + 4 \sin(t - 2\pi)\,\mathcal{U}(t - 2\pi).$$

Since $\sin(t - 2\pi) = \sin t$, the foregoing solution can be written as

$$y(t) = \begin{cases} \cos t, & 0 \le t < 2\pi \\ \cos t + 4 \sin t, & t \ge 2\pi. \end{cases} \tag{5}$$

In Figure 4.44 we see from the graph of (5) that the mass is exhibiting simple harmonic motion until it is struck at $t = 2\pi$. The influence of the unit impulse is to increase the amplitude of vibration to $\sqrt{17}$ for $t > 2\pi$.

(b) In this case the transform of the equation is simply

$$Y(s) = \frac{4e^{-2\pi s}}{s^2 + 1},$$

and so

$$y(t) = 4 \sin(t - 2\pi)\,\mathcal{U}(t - 2\pi)$$

$$= \begin{cases} 0, & 0 \le t < 2\pi \\ 4 \sin t, & t \ge 2\pi. \end{cases} \tag{6}$$

The graph of (6) in Figure 4.45 shows, as we would expect from the initial conditions, that the mass exhibits no motion until it is struck at $t = 2\pi$. ❏

Remarks

(*i*) If $\delta(t - t_0)$ were a function in the usual sense, then property (*i*) on page 225 would imply $\int_0^\infty \delta(t - t_0)\, dt = 0$ rather than $\int_0^\infty \delta(t - t_0)\, dt = 1$. Since the Dirac delta function did not "behave" like an ordinary function, even though its users produced correct results, it was met initially with great scorn by mathematicians. However, in the 1940s Dirac's controversial function was put on a rigorous footing by the French mathematician Laurent Schwartz in his book *La Théorie de distribution*, and this, in turn, led to an entirely new branch of mathematics known as the **theory of distributions** or **generalized functions.** In this theory, (2) is not an accepted definition of $\delta(t - t_0)$, nor does one speak of a function whose values are either ∞ or 0. Although we shall not pursue this topic any further, suffice it to say that the Dirac delta function is best characterized by its effect on other functions. If f is a continuous function, then

$$\int_0^\infty f(t)\,\delta(t - t_0)\, dt = f(t_0) \tag{7}$$

can be taken as the *definition* of $\delta(t - t_0)$. This result is known as the **sifting property** since $\delta(t - t_0)$ has the effect of sifting the value $f(t_0)$ out of the set of values of f on $[0, \infty)$. Note that property (*ii*) (with $f(t) = 1$) and (3) (with $f(t) = e^{-st}$) are consistent with (7).

(*ii*) In the remark in Section 4.2 we indicated that the transfer function of a general linear nth-order differential equation with constant coefficients is $W(s) = 1/P(s)$, where $P(s) = a_n s^n + a_{n-1} s^{n-1} + \ldots + a_0$. The transfer function is the Laplace transform of function $w(t)$, called the **weight function** of a linear system. But $w(t)$ can be characterized in terms of the discussion at hand. For simplicity let us consider a second-order linear system in which the input is a unit impulse at $t = 0$:

$$a_2 y'' + a_1 y' + a_0 y = \delta(t), \qquad y(0) = 0, \quad y'(0) = 0.$$

Applying the Laplace transform and using $\mathscr{L}\{\delta(t)\} = 1$ shows that the transform of the response y in this case is the transfer function

$$Y(s) = \frac{1}{a_2 s^2 + a_1 s + a_0} = \frac{1}{P(s)} = W(s) \quad \text{and so} \quad y = \mathscr{L}^{-1}\left\{\frac{1}{P(s)}\right\} = w(t).$$

From this we can see, in general, that the weight function $y = w(t)$ of an nth-order linear system is the zero-state response of the system to a unit impulse. For this reason $w(t)$ is called as well the **impulse response** of the system.

In Problems 1–12, use the Laplace transform to solve the given differential equation subject to the indicated initial conditions.

1. $y' - 3y = \delta(t - 2), \quad y(0) = 0$

2. $y' + y = \delta(t - 1), \quad y(0) = 2$

3. $y'' + y = \delta(t - 2\pi), \quad y(0) = 0, \quad y'(0) = 1$

4. $y'' + 16y = \delta(t - 2\pi), \quad y(0) = 0, \quad y'(0) = 0$

5. $y'' + y = \delta\left(t - \frac{\pi}{2}\right) + \delta\left(t - \frac{3\pi}{2}\right), \quad y(0) = 0, \quad y'(0) = 0$

6. $y'' + y = \delta(t - 2\pi) + \delta(t - 4\pi), \quad y(0) = 1, \quad y'(0) = 0$

7. $y'' + 2y' = \delta(t - 1), \quad y(0) = 0, \quad y'(0) = 1$

8. $y'' - 2y' = 1 + \delta(t - 2), \quad y(0) = 0, \quad y'(0) = 1$

9. $y'' + 4y' + 5y = \delta(t - 2\pi), \quad y(0) = 0, \quad y'(0) = 0$

10. $y'' + 2y' + y = \delta(t - 1), \quad y(0) = 0, \quad y'(0) = 0$

11. $y'' + 4y' + 13y = \delta(t - \pi) + \delta(t - 3\pi), \quad y(0) = 1, \quad y'(0) = 0$

12. $y'' - 7y' + 6y = e^t + \delta(t - 2) + \delta(t - 4), \quad y(0) = 0, \quad y'(0) = 0$

13. A uniform beam of length L carries a concentrated load w_0 at $x = L/2$. The beam is embedded at its left end and is free at its right end. Use the Laplace transform to determine the deflection $y(x)$ from

$$EI \frac{d^4 y}{dx^4} = w_0 \delta\left(x - \frac{L}{2}\right),$$

where $y(0) = 0$, $y'(0) = 0$, $y''(L) = 0$, and $y'''(L) = 0$.

14. Solve the differential equation in Problem 13 subject to $y(0) = 0$, $y'(0) = 0$, $y(L) = 0$, $y'(L) = 0$. In this case the beam is embedded at both ends. See Figure 4.46.

Figure 4.46

15. Use the Laplace transform to solve the initial-value problem $y'' + \omega^2 y = \delta(t)$, $y(0) = 0$, $y'(0) = 0$. Do you see anything unusual about the solution?

4.6 Solving Systems of Linear Equations

When initial conditions are specified, the Laplace transform reduces a system of linear differential equations with constant coefficients to a set of simultaneous algebraic equations in the transformed functions.

■ **Coupled Springs** In our first example we solve the model

$$m_1 x_1'' = -k_1 x_1 + k_2(x_2 - x_1)$$

$$m_2 x_2'' = -k_2(x_2 - x_1)$$

(1)

that describes the motion of two masses m_1 and m_2 in the coupled spring system shown in Figure 3.59 of Section 3.11.

Example 1 Example 4 of Section 3.11 Revisited

Use the Laplace transform to solve

$$x_1'' + 10x_1 \qquad - 4x_2 = 0$$
$$4x_1 + x_2'' + 4x_2 = 0 \tag{2}$$

subject to $x_1(0) = 0$, $x_1'(0) = 1$, $x_2(0) = 0$, $x_2'(0) = -1$. This is system (1) with $k_1 = 6$, $k_2 = 4$, $m_1 = 1$, and $m_2 = 1$.

SOLUTION The transform of each equation is

$$s^2 X_1(s) - s x_1(0) - x_1'(0) + 10 X_1(s) - 4 X_2(s) = 0$$
$$-4 X_1(s) + s^2 X_2(s) - s x_2(0) - x_2'(0) + 4 X_2(s) = 0,$$

where $X_1(s) = \mathscr{L}\{x_1(t)\}$ and $X_2(s) = \mathscr{L}\{x_2(t)\}$. The preceding system is the same as

$$(s^2 + 10)X_1(s) - \qquad 4X_2(s) = 1$$
$$-4X_1(s) + (s^2 + 4)X_2(s) = -1. \tag{3}$$

Solving (3) for $X_1(s)$ and using partial fractions on the result yields

$$X_1(s) = \frac{s^2}{(s^2 + 2)(s^2 + 12)} = -\frac{1/5}{s^2 + 2} + \frac{6/5}{s^2 + 12},$$

and therefore

$$x_1(t) = -\frac{1}{5\sqrt{2}} \mathscr{L}^{-1}\left\{\frac{\sqrt{2}}{s^2 + 2}\right\} + \frac{6}{5\sqrt{12}} \mathscr{L}^{-1}\left\{\frac{\sqrt{12}}{s^2 + 12}\right\}$$

$$= -\frac{\sqrt{2}}{10} \sin \sqrt{2}\, t + \frac{\sqrt{3}}{5} \sin 2\sqrt{3}\, t.$$

Substituting the expression for $X_1(s)$ into the first equation of (3) gives us

$$X_2(s) = -\frac{s^2 + 6}{(s^2 + 2)(s^2 + 12)} = -\frac{2/5}{s^2 + 2} - \frac{3/5}{s^2 + 12}$$

and

$$x_2(t) = -\frac{2}{5\sqrt{2}} \mathscr{L}^{-1}\left\{\frac{\sqrt{2}}{s^2 + 2}\right\} - \frac{3}{5\sqrt{12}} \mathscr{L}^{-1}\left\{\frac{\sqrt{12}}{s^2 + 12}\right\}$$

$$= -\frac{\sqrt{2}}{10} \sin \sqrt{2}\, t - \frac{\sqrt{3}}{10} \sin 2\sqrt{3}\, t.$$

Finally the solution to the given system (2) is

$$x_1(t) = -\frac{\sqrt{2}}{10} \sin \sqrt{2}\, t + \frac{\sqrt{3}}{5} \sin 2\sqrt{3}\, t$$

$$x_2(t) = -\frac{\sqrt{2}}{5} \sin \sqrt{2}\, t - \frac{\sqrt{3}}{10} \sin 2\sqrt{3}\, t. \tag{4}$$

The solution (4) is the same as (14) of Section 3.11. ❑

■ **Networks** In (18) of Section 2.9 we saw that currents $i_1(t)$ and $i_2(t)$ in the network containing an inductor, a resistor, and a capacitor shown in Figure 4.47 were governed by the system of first-order differential equations

Figure 4.47

$$L\frac{di_1}{dt} + Ri_2 = E(t)$$

$$RC\frac{di_2}{dt} + i_2 - i_1 = 0. \tag{5}$$

We solve this system by the Laplace transform in the next example.

Example 2 An Electrical Network

Solve the system in (5) under the conditions $E(t) = 60$ V, $L = 1$ h, $R = 50$ Ω, $C = 10^{-4}$ f, and the currents i_1 and i_2 are initially zero.

SOLUTION We must solve

$$\frac{di_1}{dt} + 50i_2 = 60$$

$$50(10^{-4})\frac{di_2}{dt} + i_2 - i_1 = 0$$

subject to $i_1(0) = 0$, $i_2(0) = 0$.

Applying the Laplace transform to each equation of the system and simplifying gives

$$sI_1(s) + \qquad 50I_2(s) = \frac{60}{s}$$

$$-200I_1(s) + (s + 200)I_2(s) = 0,$$

where $I_1(s) = \mathscr{L}\{i_1(t)\}$ and $I_2(s) = \mathscr{L}\{i_2(t)\}$. Solving the system for I_1 and I_2 and decomposing the results into partial fractions gives

$$I_1(s) = \frac{60s + 12{,}000}{s(s + 100)^2} = \frac{6/5}{s} - \frac{6/5}{s + 100} - \frac{60}{(s + 100)^2}$$

$$I_2(s) = \frac{12{,}000}{s(s + 100)^2} = \frac{6/5}{s} - \frac{6/5}{s + 100} - \frac{120}{(s + 100)^2}.$$

Taking the inverse Laplace transform, we find the currents to be

$$i_1(t) = \frac{6}{5} - \frac{6}{5}e^{-100t} - 60te^{-100t}$$

$$i_2(t) = \frac{6}{5} - \frac{6}{5}e^{-100t} - 120te^{-100t}. \qquad \square$$

Note that both $i_1(t)$ and $i_2(t)$ in Example 2 tend toward the value $E/R = \tfrac{6}{5}$ as $t \to \infty$. Furthermore, since the current through the capacitor is $i_3(t) = i_1(t) - i_2(t) = 60te^{-100t}$, we observe that $i_3(t) \to 0$ as $t \to \infty$.

■ **Double Pendulum** As shown in Figure 4.48, a double pendulum oscillates in a vertical plane under the influence of gravity. For small displacements $\theta_1(t)$ and $\theta_2(t)$, it can be shown that the system of differential equations describing the motion is

$$(m_1 + m_2)l_1^2\theta_1'' + m_2l_1l_2\theta_2'' + (m_1 + m_2)l_1g\theta_1 = 0$$

$$m_2l_2^2\theta_2'' + m_2l_1l_2\theta_1'' + m_2l_2g\theta_2 = 0. \tag{6}$$

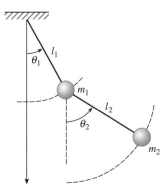

Figure 4.48

As indicated in Figure 4.48, θ_1 is measured (in radians) from a vertical line extending downward from the pivot of the system and θ_2 is measured from a vertical line extending downward from the center of mass m_1. The positive direction is to the right, negative direction to the left.

Example 3 **Double Pendulum**

It is left as an exercise to fill in the details of using the Laplace transform to solve system (6) when $m_1 = 3$, $m_2 = 1$, $l_1 = l_2 = 16$, $\theta_1(0) = 1$, $\theta_2(0) = -1$, $\theta_1'(0) = 0$, and $\theta_2'(0) = 0$. You should find that

$$\theta_1(t) = \frac{1}{4} \cos \frac{2}{\sqrt{3}} t + \frac{3}{4} \cos 2t$$

$$\theta_2(t) = \frac{1}{2} \cos \frac{2}{\sqrt{3}} t - \frac{3}{2} \cos 2t. \tag{7}$$

With the aid of a CAS, the positions of the two masses at $t = 0$ and at subsequent times are shown in Figure 4.49. See Problem 21 in Exercises 4.6.

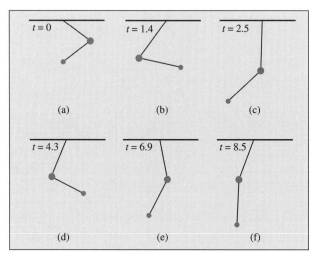

Figure 4.49

Answers to odd-numbered problems begin on page A-34.

In Problems 1–12, use the Laplace transform to solve the given system of differential equations.

1. $\dfrac{dx}{dt} = -x + y$

$\dfrac{dy}{dt} = 2x$

$x(0) = 0, \quad y(0) = 1$

2. $\dfrac{dx}{dt} = 2y + e^t$

$\dfrac{dy}{dt} = 8x - t$

$x(0) = 1, \quad y(0) = 1$

3. $\dfrac{dx}{dt} = x - 2y$

$\dfrac{dy}{dt} = 5x - y$

$x(0) = -1, \quad y(0) = 2$

4. $\dfrac{dx}{dt} + 3x + \dfrac{dy}{dt} = 1$

$\dfrac{dx}{dt} - x + \dfrac{dy}{dt} - y = e^t$

$x(0) = 0, \quad y(0) = 0$

5. $2\dfrac{dx}{dt} + \dfrac{dy}{dt} - 2x = 1$

$\dfrac{dx}{dt} + \dfrac{dy}{dt} - 3x - 3y = 2$

$x(0) = 0, \quad y(0) = 0$

6. $\dfrac{dx}{dt} + x - \dfrac{dy}{dt} + y = 0$

$\dfrac{dx}{dt} + \dfrac{dy}{dt} + 2y = 0$

$x(0) = 0, \quad y(0) = 1$

7. $\dfrac{d^2x}{dt^2} + x - y = 0$

$\dfrac{d^2y}{dt^2} + y - x = 0$

$x(0) = 0, \quad x'(0) = -2$

$y(0) = 0, \quad y'(0) = 1$

8. $\dfrac{d^2x}{dt^2} + \dfrac{dx}{dt} + \dfrac{dy}{dt} = 0$

$\dfrac{d^2y}{dt^2} + \dfrac{dy}{dt} - 4\dfrac{dx}{dt} = 0$

$x(0) = 1, \quad x'(0) = 0,$

$y(0) = -1, \quad y'(0) = 5$

9. $\dfrac{d^2x}{dt^2} + \dfrac{d^2y}{dt^2} = t^2$ **10.** $\dfrac{dx}{dt} - 4x + \dfrac{d^3y}{dt^3} = 6 \sin t$

$\dfrac{d^2x}{dt^2} - \dfrac{d^2y}{dt^2} = 4t$ $\dfrac{dx}{dt} + 2x - 2\dfrac{d^3y}{dt^3} = 0$

$x(0) = 8, \quad x'(0) = 0,$ $x(0) = 0, \quad y(0) = 0,$

$y(0) = 0, \quad y'(0) = 0$ $y'(0) = 0, \quad y''(0) = 0$

11. $\dfrac{d^2x}{dt^2} + 3\dfrac{dy}{dt} + 3y = 0$ **12.** $\dfrac{dx}{dt} = 4x - 2y + 2\,\mathcal{U}(t - 1)$

$\dfrac{d^2x}{dt^2} \qquad\quad + 3y = te^{-t}$ $\dfrac{dy}{dt} = 3x - y + \mathcal{U}(t - 1)$

$x(0) = 0, \quad x'(0) = 2, y(0) = 0$ $x(0) = 0, \quad y(0) = \dfrac{1}{2}$

13. Solve system (1) when $k_1 = 3, k_2 = 2, m_1 = 1, m_2 = 1$ and $x_1(0) = 0, x_1'(0) = 1, x_2(0) = 1, x_2'(0) = 0$.

14. Derive the system of differential equations describing the straight-line vertical motion of the coupled springs shown in Figure 4.50. Use the Laplace transform to solve the system when $k_1 = 1, k_2 = 1, k_3 = 1, m_1 = 1, m_2 = 1$ and $x_1(0) = 0, x_1'(0) = -1, x_2(0) = 0, x_2'(0) = 1$.

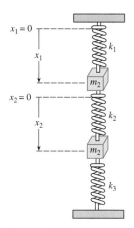

Figure 4.50

15. (a) Show that the system of differential equations for the currents $i_2(t)$ and $i_3(t)$ in the electrical network shown in Figure 4.51 is

$$L_1 \frac{di_2}{dt} + Ri_2 + Ri_3 = E(t)$$

$$L_2 \frac{di_3}{dt} + Ri_2 + Ri_3 = E(t).$$

Figure 4.51

(b) Solve the system in part **(a)** if $R = 5 \ \Omega, L_1 = 0.01$ h, $L_2 = 0.0125$ h, $E = 100$ V, $i_2(0) = 0$, and $i_3(0) = 0$.

(c) Determine the current $i_1(t)$.

16. (a) In Problem 12 in Exercises 2.9 you were asked to show that the currents $i_2(t)$ and $i_3(t)$ in the electrical network shown in Figure 4.52 satisfy

Figure 4.52

$$L\frac{di_2}{dt} + L\frac{di_3}{dt} + R_1 i_2 = E(t)$$

$$-R_1 \frac{di_2}{dt} + R_2 \frac{di_3}{dt} + \frac{1}{C} i_3 = 0.$$

Solve the system if $R_1 = 10 \ \Omega, R_2 = 5 \ \Omega, L = 1$ h, $C = 0.2$ f,

$$E(t) = \begin{cases} 120, & 0 \le t < 2 \\ 0, & t \ge 2, \end{cases} \quad i_2(0) = 0, \text{ and } i_3(0) = 0.$$

(b) Determine the current $i_1(t)$.

17. Solve the system given in (17) of Section 2.9 when $R_1 = 6 \ \Omega, R_2 = 5 \ \Omega, L_1 = 1$ h, $L_2 = 1$ h, $E(t) = 50 \sin t$ V, $i_2(0) = 0$, and $i_3(0) = 0$.

18. Solve (5) when $E = 60$ V, $L = \dfrac{1}{2}$ h, $R = 50 \ \Omega, C = 10^{-4}$ f, $i_1(0) = 0$, and $i_2(0) = 0$.

19. Solve (5) when $E = 60$ V, $L = 2$ h, $R = 50 \ \Omega, C = 10^{-4}$ f, $i_1(0) = 0$, and $i_2(0) = 0$.

20. (a) Show that the system of differential equations for the charge on the capacitor $q(t)$ and the current $i_3(t)$ in the electrical network shown in Figure 4.53 is

$$R_1 \frac{dq}{dt} + \frac{1}{C} q + R_1 i_3 = E(t)$$

$$L\frac{di_3}{dt} + R_2 i_3 - \frac{1}{C} q = 0.$$

Figure 4.53

(b) Find the charge on the capacitor when $L = 1$ h, $R_1 = 1\ \Omega$, $R_2 = 1\ \Omega$, $C = 1$ f,

$$E(t) = \begin{cases} 0, & 0 < t < 1 \\ 50e^{-t}, & t \geq 1, \end{cases} \quad i_3(0) = 0, \text{ and } q(0) = 0.$$

21. Computer Lab Assignment (a) Use the Laplace transform and the information given in Example 3 to obtain the solution (7) of the system given in (6).

(b) Use a graphing utility to plot the graphs of $\theta_1(t)$ and $\theta_2(t)$ in the $t\theta$-plane. Which mass has extreme displacements of greater magnitude? Use the graphs to estimate the first time that each mass passes through its equilibrium position. Discuss whether the motion of the pendulums is periodic.

(c) As parametric equations, graph $\theta_1(t)$ and $\theta_2(t)$ in the $\theta_1\theta_2$-plane. The curve defined by these parametric equations is called a **Lissajous curve.**

(d) The position of the masses at $t = 0$ is given in Figure 4.49(a). Note that we have used 1 radian $\approx 57.3°$. Use a calculator or a table application in a CAS to construct a table of values of the angles θ_1 and θ_2 for $t = 1, 2, \ldots, 10$ seconds. Then plot the positions of the two masses at these times.

(e) Use a CAS to find the first time that $\theta_1(t) = \theta_2(t)$ and compute the corresponding angular value. Plot the positions of the two masses at these times.

(f) Utilize the CAS to also draw appropriate lines to simulate the pendulum rods as in Figure 4.49. Use the animation capability of your CAS to make a "movie" of the motion of the double pendulum from $t = 0$ to $t = 10$ using a time increment of 0.1. [*Hint:* Express the coordinates $(x_1(t), y_1(t))$ and $(x_2(t), y_2(t))$ of the masses m_1 and m_2, respectively, in terms of $\theta_1(t)$ and $\theta_2(t)$.]

C H A P T E R 4
R E V I E W E X E R C I S E S Answers to odd-numbered problems begin on page A-35.

In Problems 1 and 2, use the definition of the Laplace transform to find $\mathscr{L}\{f(t)\}$.

1. $f(t) = \begin{cases} t, & 0 \leq t < 1 \\ 2 - t, & t \geq 1 \end{cases}$

2. $f(t) = \begin{cases} 0, & 0 \leq t < 2 \\ 1, & 2 \leq t < 4 \\ 0, & t \geq 4 \end{cases}$

In Problems 3–24, fill in the blanks or answer true/false.

3. If f is not piecewise continuous on $[0, \infty)$, then $\mathscr{L}\{f(t)\}$ will not exist. _____

4. The function $f(t) = (e^t)^{10}$ is not of exponential order. _____

5. $F(s) = s^2/(s^2 + 4)$ is not the Laplace transform of a function that is piecewise continuous and of exponential order. _____

6. If $\mathscr{L}\{f(t)\} = F(s)$ and $\mathscr{L}\{g(t)\} = G(s)$, then $\mathscr{L}^{-1}\{F(s)G(s)\} = f(t)g(t)$. _____

7. $\mathscr{L}\{e^{-7t}\} =$ _____

8. $\mathscr{L}\{te^{-7t}\} =$ _____

9. $\mathscr{L}\{\sin 2t\} =$ _____

10. $\mathscr{L}\{e^{-3t} \sin 2t\} =$ _____

11. $\mathscr{L}\{t \sin 2t\} =$ _____

12. $\mathscr{L}\{\sin 2t\, \mathcal{U}(t - \pi)\} =$ _____

13. $\mathscr{L}^{-1}\left\{\dfrac{20}{s^6}\right\} =$ _____

14. $\mathscr{L}^{-1}\left\{\dfrac{1}{3s - 1}\right\} =$ _____

15. $\mathscr{L}^{-1}\left\{\dfrac{1}{(s - 5)^3}\right\} =$ _____

16. $\mathscr{L}^{-1}\left\{\dfrac{1}{s^2 - 5}\right\} =$ _____

17. $\mathscr{L}^{-1}\left\{\dfrac{s}{s^2 - 10s + 29}\right\} =$ _____

18. $\mathscr{L}^{-1}\left\{\dfrac{e^{-5s}}{s^2}\right\} =$ _____

19. $\mathscr{L}^{-1}\left\{\dfrac{s + \pi}{s^2 + \pi^2}\, e^{-s}\right\} =$ _____

20. $\mathscr{L}^{-1}\left\{\dfrac{1}{L^2s^2 + n^2\pi^2}\right\} =$ _____

21. $\mathscr{L}\{e^{-5t}\}$ exists for $s >$ _____.

22. If $\mathscr{L}\{f(t)\} = F(s)$, then $\mathscr{L}\{te^{8t}f(t)\} =$ _____.

23. If $\mathscr{L}\{f(t)\} = F(s)$ and $k > 0$, then $\mathscr{L}\{e^{at}f(t - k)\mathcal{U}(t - k)\} =$ _____.

24. $\mathscr{L}\left\{\displaystyle\int_0^t e^{a\tau} f(\tau)\, d\tau\right\} =$ _____ whereas $\mathscr{L}\left\{e^{at} \displaystyle\int_0^t f(\tau)\, d\tau\right\} =$ _____.

In Problems 25–28, use the unit step function to write down an equation for each graph in terms of the function $y = f(t)$ whose graph is given in Figure 4.54.

Figure 4.54

25.

Figure 4.55

26.

Figure 4.56

27.

Figure 4.57

28.

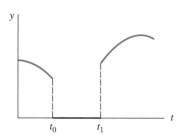

Figure 4.58

In Problems 29–32, express f in terms of unit step functions. Find $\mathscr{L}\{f(t)\}$ and $\mathscr{L}\{e^t f(t)\}$.

29.

Figure 4.59

30.

Figure 4.60

31.

Figure 4.61

32.

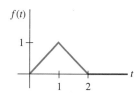

Figure 4.62

In Problems 33–38, use the Laplace transform to solve the given equation.

33. $y'' - 2y' + y = e^t$, $\quad y(0) = 0$, $\quad y'(0) = 5$

34. $y'' - 8y' + 20y = te^t$, $\quad y(0) = 0$, $\quad y'(0) = 0$

35. $y'' + 6y' + 5y = t - t\,\mathscr{U}(t - 2)$, $\quad y(0) = 1$, $\quad y'(0) = 0$

36. $y' - 5y = f(t)$, \quad where $f(t) = \begin{cases} t^2, & 0 \le t < 1 \\ 0, & t \ge 1 \end{cases}$, $\quad y(0) = 1$

37. $y'(t) = \cos t + \displaystyle\int_0^t y(\tau) \cos(t - \tau)\, d\tau$, $\quad y(0) = 1$

38. $\displaystyle\int_0^t f(\tau) f(t - \tau)\, d\tau = 6t^3$

In Problems 39 and 40, use the Laplace transform to solve each system.

39. $x' + y = t$
$\quad\;\; 4x + y' = 0$
$\quad\;\; x(0) = 1$, $\quad y(0) = 2$

40. $x'' + y'' = e^{2t}$
$\quad\;\; 2x' + y'' = -e^{2t}$
$\quad\;\; x(0) = 0$, $\quad y(0) = 0$
$\quad\;\; x'(0) = 0$, $\quad y'(0) = 0$

41. The current $i(t)$ in an RC series circuit can be determined from the integral equation

$$Ri + \frac{1}{C} \int_0^t i(\tau)\, d\tau = E(t),$$

where $E(t)$ is the impressed voltage. Determine $i(t)$ when $R = 10\ \Omega$, $C = 0.5$ f, and $E(t) = 2(t^2 + t)$.

42. A series circuit contains an inductor, a resistor, and a capacitor for which $L = \frac{1}{2}$ h, $R = 10\ \Omega$, and $C = 0.01$ f, respectively. The voltage

$$E(t) = \begin{cases} 10, & 0 \le t < 5 \\ 0, & t \ge 5 \end{cases}$$

is applied to the circuit. Determine the instantaneous charge $q(t)$ on the capacitor for $t > 0$ if $q(0) = 0$ and $q'(0) = 0$.

43. A uniform cantilever beam of length L is embedded at its left end $(x = 0)$ and is free at its right end. Find the deflection $y(x)$ if the load per unit length is given by

$$w(x) = \frac{2w_0}{L}\left[\frac{L}{2} - x + \left(x - \frac{L}{2}\right)\mathcal{U}\left(x - \frac{L}{2}\right)\right].$$

44. When a uniform beam is supported by an elastic foundation, the differential equation for its deflection $y(x)$ is

$$\frac{d^4y}{dx^4} + 4a^4y = \frac{w(x)}{EI},$$

where a is a constant. In the case when $a = 1$, find the deflection $y(x)$ of an elastically supported beam of length π that is embedded in concrete at both ends when a concentrated load w_0 is applied at $x = \pi/2$. [*Hint:* Use the table of Laplace transforms in Appendix III.]

5

Series Solutions of Linear Equations

INTRODUCTION

Up to this point we have primarily solved differential equations of order two or higher when the equation was linear and had constant coefficients. In applications, linear second-order equations with variable coefficients are just as important as, if not more so than, differential equations with constant coefficients. The only linear DE with variable coefficients considered so far was the Cauchy-Euler equation. In this chapter we shall see that the same ease with which we solved second-order Cauchy-Euler equations does not carry over to even an unpretentious linear second-order equation with variable coefficients such as $y'' + xy = 0$.

5.1 Solutions about Ordinary Points

5.1.1 Review of Power Series

Section 3.6 notwithstanding, most higher-order linear differential equations with variable coefficients cannot be solved in terms of elementary functions. A standard technique for equations of this sort is to try to find a solution defined in terms of an infinite series, and often these series solutions are power series. Because of this, it is appropriate to list some of the more important facts about power series. For a more in-depth review of this theory, you are encouraged to consult a calculus text.

- **Definition.** A power series in $x - a$ is an infinite series of the form $\sum_{n=0}^{\infty} c_n(x-a)^n$. Such a series is also said to be a power series centered at a. For example, the power series $\sum_{n=0}^{\infty}(x+1)^n$ is centered at $a = -1$. In this section we are concerned principally with power series in x; in other words, power series centered at $a = 0$. For example, $\sum_{n=1}^{\infty} 2^n x^n$ is a power series in x.

- **Convergence.** A power series $\sum_{n=0}^{\infty} c_n(x-a)^n$ is convergent at a specified value of x if its sequence of partial sums $\{S_N(x)\}$ converges; that is, if $\lim_{N \to \infty} S_N(x) = \lim_{N \to \infty} \sum_{n=0}^{N} c_n(x-a)^n$ exists. If the limit does not exist at x, the series is said to be divergent.

- **Interval of Convergence.** Every power series has an interval of convergence. The interval of convergence is the set of all real numbers x for which the series converges.

- **Radius of Convergence.** Every power series has a radius of convergence R. If $R > 0$, then a power series $\sum_{n=0}^{\infty} c_n(x-a)^n$ converges for $|x - a| < R$ and diverges for $|x - a| > R$. If the series converges only at its center a, then $R = 0$. If the series converges for all x, then we write $R = \infty$. Recall that $|x - a| < R$ is equivalent to $a - R < x < a + R$. A power series may or may not converge at the endpoints $a - R$ and $a + R$ of this interval.

- **Absolute Convergence.** Within its interval of convergence a power series converges absolutely. In other words, if x is in the interval of convergence and is not an endpoint of the interval, then the series of absolute values $\sum_{n=0}^{\infty} |c_n(x-a)^n|$ converges.

- **Ratio Test.** Convergence of power series $\sum_{n=0}^{\infty} c_n(x-a)^n$ can often be determined by the ratio test. Suppose $c_n \neq 0$ for all n, and that

$$\lim_{n \to \infty} \left| \frac{c_{n+1}(x-a)^{n+1}}{c_n(x-a)^n} \right| = |x - a| \lim_{n \to \infty} \left| \frac{c_{n+1}}{c_n} \right| = L.$$

If $L < 1$ the series converges absolutely, if $L > 1$ the series diverges, and if $L = 1$ the test is inconclusive. For example, for the power series $\sum_{n=1}^{\infty} (x-3)^n/2^n n$ the ratio test gives

$$\lim_{n \to \infty} \left| \frac{(x-3)^{n+1}/2^{n+1}(n+1)}{(x-3)^n/2^n n} \right| = |x - 3| \lim_{n \to \infty} \frac{n+1}{2n} = \frac{1}{2}|x - 3|.$$

The series converges absolutely for $\frac{1}{2}|x - 3| < 1$ or $|x - 3| < 2$ or $1 < x < 5$. This last interval is referred to as the *open* interval of convergence. The series diverges for $|x - 3| > 2$, that is, for $x > 5$ or $x < 1$. At the left endpoint $x = 1$ of the open interval of convergence, the series of constants $\sum_{n=1}^{\infty} ((-1)^n/n)$ is convergent by the alternating series test. At the right endpoint $x = 5$, the series $\sum_{n=1}^{\infty} (1/n)$ is the diver-

gent harmonic series. The interval of convergence of the series is $[1, 5)$ and the radius of convergence is $R = 2$.

- **A Power Series Defines a Function.** A power series defines a function $f(x) = \sum_{n=0}^{\infty} c_n(x - a)^n$ whose domain is the interval of convergence of the series. If the radius of convergence is $R > 0$, then f is continuous, differentiable, and integrable on the interval $(a - R, a + R)$. Moreover, $f'(x)$ and $\int f(x)dx$ can be found by term-by-term differentiation and integration. Convergence at an endpoint may be either lost by differentiation or gained through integration. If $y = \sum_{n=0}^{\infty} c_n x^n$ is a power series in x, then the first two derivatives are $y' = \sum_{n=0}^{\infty} nx^{n-1}$ and $y'' = \sum_{n=0}^{\infty} n(n - 1)x^{n-2}$. Notice that the first term in the first derivative and the first two terms in the second derivative are zero. We omit these zero terms and write

$$y' = \sum_{n=1}^{\infty} c_n n x^{n-1} \quad \text{and} \quad y'' = \sum_{n=2}^{\infty} c_n n(n - 1)x^{n-2}. \tag{1}$$

These results are important and will be used shortly.

- **Identity Property.** If $\sum_{n=0}^{\infty} c_n(x - a)^n = 0$, $R > 0$, for all numbers x in the interval of convergence, then $c_n = 0$ for all n.

- **Analytic at a Point.** A function f is analytic at a point a if it can be represented by a power series in $x - a$ with a positive radius of convergence. In calculus it is seen that functions such as e^x, $\cos x$, $\sin x$, $\ln(x - 1)$, and so on can be represented by Taylor series. Recall, for example, that

$$e^x = 1 + \frac{x}{1!} + \frac{x^2}{2!} + \cdots, \quad \sin x = x - \frac{x^3}{3!} + \frac{x^5}{5!} - \cdots, \quad \cos x = 1 - \frac{x^2}{2!} + \frac{x^4}{4!} - \frac{x^6}{6!} + \cdots, \tag{2}$$

for $|x| < \infty$. These Taylor series centered at 0, called Maclaurin series, show that e^x, $\sin x$, and $\cos x$ are analytic at $x = 0$.

- **Arithmetic of Power Series.** Power series can be combined through the operations of addition, multiplication, and division. The procedures for power series are similar to the way in which two polynomials are added, multiplied, and divided—that is, we add coefficients of like powers of x, use the distributive law and collect like terms, and perform long division. For example, using the series in (2),

$$e^x \sin x = \left(1 + x + \frac{x^2}{2} + \frac{x^3}{6} + \frac{x^4}{24} + \cdots\right)\left(x - \frac{x^3}{6} + \frac{x^5}{120} - \frac{x^7}{5040} + \cdots\right)$$

$$= (1)x + (1)x^2 + \left(-\frac{1}{6} + \frac{1}{2}\right)x^3 + \left(-\frac{1}{6} + \frac{1}{6}\right)x^4 + \left(\frac{1}{120} - \frac{1}{12} + \frac{1}{24}\right)x^5 + \cdots$$

$$= x + x^2 + \frac{x^3}{3} - \frac{x^5}{30} - \cdots.$$

Since the power series for e^x and $\sin x$ converge for $|x| < \infty$, the product series converges on the same interval. Problems involving multiplication or division of power series can be done with minimal fuss using a computer algebra system.

■ **Shifting the Summation Index** For the remainder of this section, as well as this chapter, it is important that you become adept at simplifying the sum of two or more power series, each series expressed in summation (sigma) notation, to

an expression with a single \sum. As the next example illustrates, combining two or more summations as a single summation often requires a reindexing, that is, a shift in the index of summation.

Example 1 Adding Two Power Series

Write $\sum_{n=2}^{\infty} n(n-1)c_n x^{n-2} + \sum_{n=0}^{\infty} c_n x^{n+1}$ as one power series.

SOLUTION In order to add the two series, it is necessary that *both summation indices start with the same number and that the powers of x in each series be "in phase,"* that is, if one series starts with a multiple of, say, x to the first power, then we want the other series to start with the same power. Note that in the given problem, the first series starts with x^0, whereas the second series starts with x^1. By writing the first term of the first series outside of the summation notation,

Important

$$\underset{n=2}{\overset{\infty}{\sum}} n(n-1)c_n x^{n-2} + \underset{n=0}{\overset{\infty}{\sum}} c_n x^{n+1} = 2 \cdot 1 c_2 x^0 + \underset{\substack{\downarrow \\ n=3 \\ \text{series starts with} \\ x \text{ for } n=3}}{\overset{\infty}{\sum}} n(n-1)c_n x^{n-2} + \underset{\substack{\downarrow \\ n=0 \\ \text{series starts with} \\ x \text{ for } n=0}}{\overset{\infty}{\sum}} c_n x^{n+1},$$

we see that both series on the right side start with the same power of x, namely, x^1. Now to get the same summation index we are inspired by the exponents of x; we let $k = n - 2$ in the first series and at the same time let $k = n + 1$ in the second series. The right side becomes

$$2c_2 + \overset{\text{same}}{\underset{k=1}{\overset{\infty}{\sum}} (k+2)(k+1)c_{k+2} x^k} + \underset{k=1}{\overset{\infty}{\sum}} c_{k-1} x^k. \tag{3}$$

$$\uparrow\underline{\qquad\text{same}\qquad\qquad}\uparrow$$

Remember, the summation index is a "dummy" variable; the fact that $k = n - 1$ in one case and $k = n + 1$ in the other should cause no confusion if you keep in mind that it is the *value* of the summation index that is important. In both cases k takes on the same successive values $k = 1, 2, 3, \ldots$ when n takes on the values $n = 2, 3, 4, \ldots$ for $k = n - 1$ and $n = 0, 1, 2, \ldots$ for $k = n + 1$. We are now in a position to add the series in (3) term by term:

$$\underset{n=2}{\overset{\infty}{\sum}} n(n-1)c_n x^{n-2} + \underset{n=0}{\overset{\infty}{\sum}} c_n x^{n+1} = 2c_2 + \underset{k=1}{\overset{\infty}{\sum}} [(k+2)(k+1)c_{k+2} + c_{k-1}]x^k. \tag{4}$$

If you are not convinced of the result in (4), then write out a few terms on both sides of the equality.

5.1.2 Power Series Solutions

Suppose the linear second-order differential equation

$$a_2(x)y'' + a_1(x)y' + a_0(x)y = 0 \tag{5}$$

is put into standard form

$$y'' + P(x)y' + Q(x)y = 0 \tag{6}$$

by dividing by the leading coefficient $a_2(x)$. We make the following definition.

> **DEFINITION 5.1** **Ordinary and Singular Points**
>
> A point x_0 is said to be an **ordinary point** of the differential equation (5) if both $P(x)$ and $Q(x)$ in the standard form (6) are analytic at x_0. A point that is not an ordinary point is said to be a **singular point** of the equation.

Every finite value of x is an ordinary point of $y'' + (e^x)y' + (\sin x)y = 0$. In particular, $x = 0$ is an ordinary point, since, as we have already seen in (2), both e^x and sin x are analytic at this point. The negation in the second sentence of Definition 5.1 stipulates that if at least one of the functions $P(x)$ and $Q(x)$ in (6) fails to be analytic at x_0, then x_0 is a singular point. Note that $x = 0$ is a singular point of the differential equation $y'' + (e^x)y' + (\ln x)y = 0$, since $Q(x) = \ln x$ is discontinuous at $x = 0$ and so cannot be represented by a power series in x.

■ **Polynomial Coefficients** We shall be interested primarily in the case in which (5) has polynomial coefficients. A polynomial is analytic at any value x, and a rational function is analytic *except* at points where its denominator is zero. Thus, if $a_2(x)$, $a_1(x)$, and $a_0(x)$ are polynomials with no common factors, then both rational functions $P(x) = a_1(x)/a_2(x)$ and $Q(x) = a_0(x)/a_2(x)$ are analytic except where $a_2(x) = 0$. It follows, then, that $x = x_0$ is an ordinary point of (5) if $a_2(x_0) \neq 0$, whereas $x = x_0$ is a singular point of (5) if $a_2(x_0) = 0$. For example, the only singular points of the equation $(x^2 - 1)y'' + 2xy' + 6y = 0$ are solutions of $x^2 - 1 = 0$ or $x = \pm 1$. All other finite values* of x are ordinary points. Inspection of the Cauchy–Euler equation $ax^2y'' + bxy' + cy = 0$ shows that it has a singular point at $x = 0$. Singular points need not be real numbers. The equation $(x^2 + 1)y'' + xy' - y = 0$ has singular points at the solutions of $x^2 + 1 = 0$, namely, $x = \pm i$. All other values of x, real or complex, are ordinary points.

We state the following theorem about the existence of power series solutions without proof.

> **THEOREM 5.1** **Existence of Power Series Solutions**
>
> If $x = x_0$ is an ordinary point of the differential equation (5), we can always find two linearly independent solutions in the form of a power series centered at x_0, that is, $y = \sum_{n=0}^{\infty} c_n(x - x_0)^n$. A series solution converges at least on some interval defined by $|x - x_0| < R$, where R is the distance from x_0 to the closest singular point.

A solution of the form $y = \sum_{n=0}^{\infty} c_n(x - x_0)^n$ is said to be a solution *about* the ordinary point x_0. The distance R in Theorem 5.1 is the *minimum* value for the radius of convergence. For example, the complex numbers $1 \pm 2i$ are singular points of $(x^2 - 2x + 5)y'' + xy' - y = 0$, but since $x = 0$ is an ordinary point of the equation, Theorem 5.1 guarantees that we can find two power series solutions centered at 0. That is, the solutions look like $y = \sum_{n=0}^{\infty} c_n x^n$, and, moreover, we know without actually finding these solutions that each series must converge *at least* for $|x| < \sqrt{5}$, where $R = \sqrt{5}$ is the distance in the complex plane from 0 to either of the numbers $1 + 2i$ or $1 - 2i$. However, the differential equation has a solution that is

*For our purposes ordinary points and singular points will always be finite points. It is possible for an ODE to have, say, a singular point at infinity.

valid for much larger values of x; indeed, this solution is valid on $(-\infty, \infty)$ because it can be shown that one of the two solutions is a polynomial.

In the examples that follow, as well as in Exercises 5.1, we shall, for the sake of simplicity, only find power series solutions about the ordinary point $x = 0$. If it is necessary to find a power series solution of an ODE about an ordinary point $x_0 \neq 0$, we can simply make the change of variable $t = x - x_0$ in the equation (this translates $x = x_0$ to $t = 0$), find solutions of the new equation of the form $y = \sum_{n=0}^{\infty} c_n t^n$, and then resubstitute $t = x - x_0$.

All the power series solutions ☞
will be centered at 0.

Finding a power series solution of a homogeneous linear second-order ODE has been accurately described as "the method of undetermined *series* coefficients," since the procedure is quite analogous to what we did in Section 3.4. In brief, here is the idea: We substitute $y = \sum_{n=0}^{\infty} c_n x^n$ into the differential equation, combine series as we did in Example 1, and then equate all coefficients to the right side of the equation to determine the coefficients c_n. But since the right side is zero, the last step requires, by the identity property in the preceding bulleted list, that all coefficients of x must be equated to zero. No, this does *not* mean that all coefficients *are* zero; this would not make sense, because after all, Theorem 5.1 guarantees that we can find two solutions. Example 2 illustrates how the single assumption that $y = \sum_{n=0}^{\infty} c_n x^n = c_0 + c_1 x + c_2 x^2 + \ldots$ leads to two sets of coefficients so that we have two distinct power series $y_1(x)$ and $y_2(x)$, both expanded about the ordinary point $x = 0$. The general solution of the differential equation is $y = C_1 y_1(x) + C_2 y_2(x)$; indeed, it can be shown that $C_1 = c_0$ and $C_2 = c_1$.

Example 2 **Power Series Solutions**

Solve $y'' + xy = 0$.

SOLUTION Since there are no finite singular points, Theorem 5.1 guarantees two power series solutions, centered at 0, convergent for $|x| < \infty$. Substituting $y = \sum_{n=0}^{\infty} c_n x^n$ and the second derivative $y'' = \sum_{n=2}^{\infty} n(n-1) c_n x^{n-2}$ (see (1)) into the differential equation gives

$$y'' + xy = \sum_{n=2}^{\infty} c_n n(n-1) x^{n-2} + x \sum_{n=0}^{\infty} c_n x^n = \sum_{n=2}^{\infty} c_n n(n-1) x^{n-2} + \sum_{n=0}^{\infty} c_n x^{n+1}. \quad (7)$$

Now we have already added the last two series on the right side of the equality in (7) by shifting the summation index in Example 1. From the result given in (4),

$$y'' + xy = 2c_2 + \sum_{k=1}^{\infty} [(k+1)(k+2)c_{k+2} + c_{k-1}]x^k = 0. \quad (8)$$

At this point we invoke the identity property. Since (8) is identically zero, it is necessary that the coefficient of each power of x be set equal to zero—that is, $2c_2 = 0$ (it is the coefficient of x^0)—and

$$(k+1)(k+2)c_{k+2} + c_{k-1} = 0, \qquad k = 1, 2, 3, \ldots. \quad (9)$$

Now $2c_2 = 0$ obviously dictates that $c_2 = 0$. But the expression in (9), called a **recurrence relation,** determines the c_k in such a manner that we can choose a certain subset of the set of coefficients to be *nonzero*. Since $(k+1)(k+2) \neq 0$ for all values of k, we can solve (9) for c_{k+2} in terms of c_{k-1}:

This formula is called a ☞
two-term recurrence relation.

$$c_{k+2} = -\frac{c_{k-1}}{(k+1)(k+2)}, \qquad k = 1, 2, 3, \ldots. \quad (10)$$

This relation generates consecutive coefficients of the assumed solution one at a time as we let k take on the successive integers indicated in (10):

$$k = 1, \qquad c_3 = -\frac{c_0}{2 \cdot 3}$$

$$k = 2, \qquad c_4 = -\frac{c_1}{3 \cdot 4}$$

$$k = 3, \qquad c_5 = -\frac{c_2}{4 \cdot 5} = 0 \qquad\qquad \leftarrow c_2 \text{ is zero}$$

$$k = 4, \qquad c_6 = -\frac{c_3}{5 \cdot 6} = \frac{1}{2 \cdot 3 \cdot 5 \cdot 6} c_0$$

$$k = 5, \qquad c_7 = -\frac{c_4}{6 \cdot 7} = \frac{1}{3 \cdot 4 \cdot 6 \cdot 7} c_1$$

$$k = 6, \qquad c_8 = -\frac{c_5}{7 \cdot 8} = 0 \qquad\qquad \leftarrow c_5 \text{ is zero}$$

$$k = 7, \qquad c_9 = -\frac{c_6}{8 \cdot 9} = -\frac{1}{2 \cdot 3 \cdot 5 \cdot 6 \cdot 8 \cdot 9} c_0$$

$$k = 8, \qquad c_{10} = -\frac{c_7}{9 \cdot 10} = -\frac{1}{3 \cdot 4 \cdot 6 \cdot 7 \cdot 9 \cdot 10} c_1$$

$$k = 9, \qquad c_{11} = -\frac{c_8}{10 \cdot 11} = 0 \qquad\qquad \leftarrow c_8 \text{ is zero}$$

and so on. Now substituting the coefficients just obtained into the original assumption

$$y = c_0 + c_1 x + c_2 x^2 + c_3 x^3 + c_4 x^4 + c_5 x^5 + c_6 x^6 + c_7 x^7 + c_8 x^8 + c_9 x^9 + c_{10} x^{10} + c_{11} x^{11} + \cdots$$

we get

$$y = c_0 + c_1 x + 0 - \frac{c_0}{2 \cdot 3} x^3 - \frac{c_1}{3 \cdot 4} x^4 + 0 + \frac{c_0}{2 \cdot 3 \cdot 5 \cdot 6} x^6$$

$$+ \frac{c_1}{3 \cdot 4 \cdot 6 \cdot 7} x^7 + 0 - \frac{c_0}{2 \cdot 3 \cdot 5 \cdot 6 \cdot 8 \cdot 9} x^9 - \frac{c_1}{3 \cdot 4 \cdot 6 \cdot 7 \cdot 9 \cdot 10} x^{10} + 0 + \cdots.$$

After grouping the terms containing c_0 and the terms containing c_1, we obtain $y = c_0 y_1(x) + c_1 y_2(x)$, where

$$y_1(x) = 1 - \frac{1}{2 \cdot 3} x^3 + \frac{1}{2 \cdot 3 \cdot 5 \cdot 6} x^6 - \frac{1}{2 \cdot 3 \cdot 5 \cdot 6 \cdot 8 \cdot 9} x^9 + \cdots = 1 + \sum_{k=1}^{\infty} \frac{(-1)^k}{2 \cdot 3 \cdots (3n-1)(3n)} x^{3k}$$

$$y_2(x) = x - \frac{1}{3 \cdot 4} x^4 + \frac{1}{3 \cdot 4 \cdot 6 \cdot 7} x^7 - \frac{1}{3 \cdot 4 \cdot 6 \cdot 7 \cdot 9 \cdot 10} x^{10} + \cdots = x + \sum_{k=1}^{\infty} \frac{(-1)^k}{3 \cdot 4 \cdots (3n)(3n+1)} x^{3k+1}.$$

Since the recursive use of (10) leaves c_0 and c_1 completely undetermined, they can be chosen arbitrarily. As mentioned prior to this example, the linear combination $y = c_0 y_1(x) + c_1 y_2(x)$ actually represents the general solution of the differential equation. Although we know from Theorem 5.1 that each series solution converges for $|x| < \infty$, this fact can also be verified by the ratio test. ❏

The differential equation in Example 2 is called **Airy's equation** and is encountered in the study of diffraction of light, diffraction of radio waves around the surface of the earth, aerodynamics, and the deflection of a uniform thin vertical column that bends under its own weight. Other common forms of Airy's equation are $y'' - xy = 0$ and $y'' + \alpha^2 xy = 0$. See Problem 31 in Exercises 5.3 for an application of the last equation.

Example 3 Power Series Solution

Solve $(x^2 + 1)y'' + xy' - y = 0$.

SOLUTION As we have already seen on page 239, the given differential equation has singular points at $x = \pm i$, and so a power series solution centered at 0 will converge at least for $|x| < 1$, where 1 is the distance in the complex plane from 0 to either i or $-i$. The assumption $y = \sum_{n=0}^{\infty} c_n x^n$ and its first two derivatives (see (1)) lead to

$$(x^2 + 1) \sum_{n=2}^{\infty} n(n-1)c_n x^{n-2} + x \sum_{n=1}^{\infty} n c_n x^{n-1} - \sum_{n=0}^{\infty} c_n x^n$$

$$= \sum_{n=2}^{\infty} n(n-1)c_n x^n + \sum_{n=2}^{\infty} n(n-1)c_n x^{n-2} + \sum_{n=1}^{\infty} n c_n x^n - \sum_{n=0}^{\infty} c_n x^n$$

$$= 2c_2 x^0 - c_0 x^0 + 6c_3 x + c_1 x - c_1 x + \underbrace{\sum_{n=2}^{\infty} n(n-1)c_n x^n}_{k=n}$$

$$+ \underbrace{\sum_{n=4}^{\infty} n(n-1)c_n x^{n-2}}_{k=n-2} + \underbrace{\sum_{n=2}^{\infty} n c_n x^n}_{k=n} - \underbrace{\sum_{n=2}^{\infty} c_n x^n}_{k=n}$$

$$= 2c_2 - c_0 + 6c_3 x + \sum_{n=2}^{\infty} [k(k-1)c_k + (k+2)(k+1)c_{k+2} + kc_k - c_k]x^k$$

$$= 2c_2 - c_0 + 6c_3 x + \sum_{n=2}^{\infty} [(k+1)(k-1)c_k + (k+2)(k+1)c_{k+2}]x^k = 0.$$

From this last identity we conclude that $2c_2 - c_0 = 0$, $6c_3 = 0$, and

$$(k+1)(k-1)c_k + (k+2)(k+1)c_{k+2} = 0.$$

Thus,

$$c_2 = \frac{1}{2}c_0, \qquad c_3 = 0$$

$$c_{k+2} = \frac{1-k}{k+2}c_k, \qquad k = 2, 3, 4, \ldots.$$

Substituting $k = 2, 3, 4, \ldots$ into the last formula gives

$$c_4 = -\frac{1}{4}c_2 = -\frac{1}{2 \cdot 4}c_0 = -\frac{1}{2^2\, 2!}c_0$$

$$c_5 = -\frac{2}{5}c_3 = 0 \qquad\qquad\qquad \leftarrow c_3 \text{ is zero}$$

$$c_6 = -\frac{3}{6}c_4 = \frac{3}{2 \cdot 4 \cdot 6}c_0 = \frac{1 \cdot 3}{2^3\, 3!}c_0$$

$$c_7 = -\frac{4}{7} c_5 = 0 \qquad\qquad \leftarrow c_5 \text{ is zero}$$

$$c_8 = -\frac{5}{8} c_6 = -\frac{3 \cdot 5}{2 \cdot 4 \cdot 6 \cdot 8} c_0 = -\frac{1 \cdot 3 \cdot 5}{2^4\, 4!} c_0$$

$$c_9 = -\frac{6}{9} c_7 = 0 \qquad\qquad \leftarrow c_7 \text{ is zero}$$

$$c_{10} = -\frac{7}{10} c_8 = \frac{3 \cdot 5 \cdot 7}{2 \cdot 4 \cdot 6 \cdot 8 \cdot 10} c_0 = -\frac{1 \cdot 3 \cdot 5 \cdot 7}{2^5\, 5!} c_0$$

and so on. Therefore,

$$y = c_0 + c_1 x + c_2 x^2 + c_3 x^3 + c_4 x^4 + c_5 x^5 + c_6 x^6 + c_7 x^7 + c_8 x^8 + c_9 x^9 + c_{10} x^{10} + \cdots$$

$$= c_0\left[1 + \frac{1}{2} x^2 - \frac{1}{2^2\, 2!} x^4 + \frac{1 \cdot 3}{2^3\, 3!} x^6 - \frac{1 \cdot 3 \cdot 5}{2^4\, 4!} x^2 + \frac{1 \cdot 3 \cdot 5 \cdot 7}{2^5\, 5!} x^{10} - \cdots \right] + c_1 x$$

$$= c_0 y_1(x) + c_1 y_2(x).$$

The solutions are the polynomial $y_2(x) = x$ and the power series

$$y_1(x) = 1 + \frac{1}{2} x^2 + \sum_{n=2}^{\infty} (-1)^{n-1} \frac{1 \cdot 3 \cdot 5 \cdots (2n-3)}{2^n\, n!} x^{2n}, \qquad |x| < 1. \qquad \square$$

Example 4 **Three-Term Recurrence Relation**

If we seek a power series solution $y = \sum_{n=0}^{\infty} c_n x^n$ for the differential equation

$$y'' - (1 + x)y = 0,$$

we obtain $c_2 = c_0/2$ and the recurrence relation

This formula is called a 👉
three-term recurrence relation.

$$c_{k+2} = \frac{c_k + c_{k-1}}{(k+1)(k+2)}, \qquad k = 1, 2, 3, \dots.$$

Examination of the formula shows that the coefficients c_3, c_4, c_5, ... are expressed in terms of both c_1 and c_0, and, moreover, the algebra required to do this becomes a little messy. To simplify life we can first choose $c_0 \neq 0$, $c_1 = 0$; this yields consecutive coefficients for one solution that are expressed entirely in terms of c_0:

$$c_2 = \frac{1}{2} c_0$$

$$c_3 = \frac{c_1 + c_0}{2 \cdot 3} = \frac{c_0}{2 \cdot 3} = \frac{1}{6} c_0$$

$$c_4 = \frac{c_2 + c_1}{3 \cdot 4} = \frac{c_0}{2 \cdot 3 \cdot 4} = \frac{1}{24} c_0$$

$$c_5 = \frac{c_3 + c_2}{4 \cdot 5} = \frac{c_0}{4 \cdot 5}\left[\frac{1}{6} + \frac{1}{2} \right] = \frac{1}{30} c_0$$

and so on. Next, by choosing $c_0 = 0$, $c_1 \neq 0$, the coefficients for the other solution are expressed in terms of c_1:

$$c_2 = \frac{1}{2} \, c_0 = 0$$

$$c_3 = \frac{c_1 + c_0}{2 \cdot 3} = \frac{c_1}{2 \cdot 3} = \frac{1}{6} \, c_1$$

$$c_4 = \frac{c_2 + c_1}{3 \cdot 4} = \frac{c_1}{3 \cdot 4} = \frac{1}{12} \, c_1$$

$$c_5 = \frac{c_3 + c_2}{4 \cdot 5} = \frac{c_1}{4 \cdot 5 \cdot 6} = \frac{1}{120} \, c_1$$

and so on. Finally, we see that the general solution of the equation is $y = c_0 y_1(x) + c_1 y_2(x)$, where

$$y_1(x) = 1 + \frac{1}{2} x^2 + \frac{1}{6} x^3 + \frac{1}{24} x^4 + \frac{1}{30} x^5 + \cdots$$

and

$$y_2(x) = x + \frac{1}{6} x^3 + \frac{1}{12} x^4 + \frac{1}{120} x^5 + \cdots.$$

Each series converges for all finite values of x. ❑

■ **Nonpolynomial Coefficients** The next example illustrates how to find a power series solution about the ordinary point $x_0 = 0$ of a differential equation when its coefficients are not polynomials. In this example we see an application of multiplication of two power series.

Example 5 ODE with Nonpolynomial Coefficients

Solve $y'' + (\cos x)y = 0$.

SOLUTION We see $x = 0$ is an ordinary point of the equation because, as we have already seen, $\cos x$ is analytic at that point. Using the Maclaurin series for $\cos x$ given in (2), along with the usual assumption $y = \sum_{n=0}^{\infty} c_n x^n$ and the results in (1), we find

$$y'' + (\cos x)y = \sum_{n=2}^{\infty} n(n-1)c_n x^{m-2} + \left(1 - \frac{x^2}{2!} + \frac{x^4}{4!} - \frac{x^6}{6!} + \cdots \right) \sum_{n=0}^{\infty} c_n x^n$$

$$= 2c_2 + 6c_3 x + 12c_4 x^2 + 20c_5 x^3 + \cdots + \left(1 - \frac{x^2}{2!} + \frac{x^4}{4!} + \cdots \right)(c_0 + c_1 x + c_2 x^2 + c_3 x^3 + \cdots)$$

$$= 2c_2 + c_0 + (6c_3 + c_1)x + \left(12c_4 + c_2 - \frac{1}{2} c_0 \right)x^2 + \left(20c_5 + c_3 - \frac{1}{2} c_1 \right)x^3 + \cdots = 0.$$

It follows that

$$2c_2 + c_0 = 0, \quad 6c_3 + c_1 = 0, \quad 12c_4 + c_2 - \frac{1}{2} c_0 = 0, \quad 20c_5 + c_3 - \frac{1}{2} c_1 = 0,$$

and so on. This gives $c_2 = -\frac{1}{2} c_0$, $c_3 = -\frac{1}{6} c_1$, $c_4 = \frac{1}{12} c_0$, $c_5 = \frac{1}{30} c_1$, By grouping terms we arrive at the general solution $y = c_0 y_1(x) + c_1 y_2(x)$, where

$$y_1(x) = 1 - \frac{1}{2}x^2 + \frac{1}{12}x^4 - \cdots \quad \text{and} \quad y_2(x) = x - \frac{1}{6}x^3 + \frac{1}{30}x^5 - \cdots .$$

Since the differential equation has no finite singular points, both power series converge for $|x| < \infty$. ❏

■ **Solution Curves** The approximate graph of a power series solution $y(x) = \sum_{n=0}^{\infty} c_n x^n$ can be obtained in several ways. We can always resort to graphing the terms in the sequence of partial sums of the series; in other words, the graphs of the polynomials $S_N(x) = \sum_{n=0}^{N} c_n x^n$. For large values of N, $S_N(x)$ should give us an indication of the behavior of $y(x)$ near the ordinary point $x = 0$. We can also obtain an approximate solution curve by using a numerical solver as we did in Section 3.8. For example, if you carefully scrutinize the series solutions of Airy's equation in Example 2, you should see that $y_1(x)$ and $y_2(x)$ are, in turn, the solutions of the initial-value problems

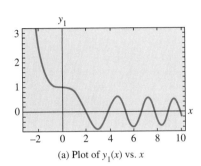

(a) Plot of $y_1(x)$ vs. x

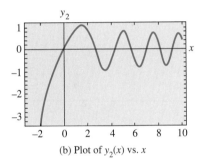

(b) Plot of $y_2(x)$ vs. x

Figure 5.1 Solutions of Airy's equation

$$y'' + xy = 0, \qquad y(0) = 1, \quad y'(0) = 0,$$
$$y'' + xy = 0, \qquad y(0) = 0, \quad y'(0) = 1. \tag{11}$$

The specified initial conditions "pick out" the solutions $y_1(x)$ and $y_2(x)$ from $y = c_0 y_1(x) + c_1 y_2(x)$, since it should be apparent from our basic series assumption $y = \sum_{n=0}^{\infty} c_n x^n$ that $y(0) = c_0$ and $y'(0) = c_1$. Now if your numerical solver requires a system of equations, the substitution $y' = u$ in $y'' + xy = 0$ gives $y'' = u' = -xy$, and so a system of two first-order equations equivalent to Airy's equation is

$$\begin{aligned} y' &= u \\ u' &= -xy. \end{aligned} \tag{12}$$

Initial conditions for the system in (12) are the two sets of initial conditions in (11) but rewritten as $y(0) = 1$, $u(0) = 0$, and $y(0) = 0$, $u(0) = 1$. The graphs of $y_1(x)$ and $y_2(x)$ shown in Figure 5.1 were obtained with the aid of a numerical solver using the fourth-order Runge–Kutta method with a step size of $h = 0.1$.

Remarks

In the problems that follow do not expect, in each case, to be able to write a solution in terms of summation notation. Even though we can obtain as many terms as desired in a series solution, either through iteration of a recurrence relation or, as in Example 4, by multiplication, it may not be possible to deduce any general term for the c_n. We may have to settle, as in Examples 4 and 5, for just writing out the first few terms of the series.

Answers to odd-numbered problems begin on page A-35.

5.1.1 Review of Power Series

In Problems 1–4, find the radius of convergence and interval of convergence for the given power series.

1. $\displaystyle\sum_{n=1}^{\infty} \frac{2^n}{n} x^n$

2. $\displaystyle\sum_{n=0}^{\infty} \frac{(100)^n}{n!} (x+7)^n$

3. $\displaystyle\sum_{k=1}^{\infty} \frac{(-1)^k}{10^k} (x-5)^k$

4. $\displaystyle\sum_{k=0}^{\infty} k!(x-1)^k$

💻 In Problems 5 and 6, the given function is analytic at $x = 0$. Find the first four terms of a power series in x. Perform the multiplication by hand or use a CAS, as instructed.

5. $\sin x \cos x$

6. $e^{-x} \cos x$

In Problems 7 and 8, the given function is analytic at x = 0. Find the first four terms of a power series in x. Perform the long division by hand or use a CAS, as instructed. Give the open interval of convergence.

7. $\dfrac{1}{\cos x}$ **8.** $\dfrac{1-x}{2+x}$

In Problems 9 and 10, rewrite the given expression as a single power series.

9. $\displaystyle\sum_{n=1}^{\infty} 2nc_n x^{n-1} + \sum_{n=0}^{\infty} 6c_n x^{n+1}$

10. $\displaystyle\sum_{n=2}^{\infty} n(n-1)c_n x^n + 2\sum_{n=2}^{\infty} n(n-1)c_n x^{n-2} + 3\sum_{n=1}^{\infty} nc_n x^n$

In Problems 11 and 12, verify by direct substitution that the given power series is a particular solution of the indicated differential equation.

11. $y = \displaystyle\sum_{n=1}^{\infty} \dfrac{(-1)^{n+1}}{n} x^n$, $(x+1)y'' + y' = 0$

12. $y = \displaystyle\sum_{n=0}^{\infty} \dfrac{(-1)^n}{2^{2n}(n!)^2} x^{2n}$, $xy'' + y' + xy = 0$

5.1.2 Power Series Solutions

In Problems 13–24, find two power series solutions of the given differential equation about the ordinary point x = 0.

13. $y'' - xy = 0$ **14.** $y'' + x^2 y = 0$

15. $y'' - 2xy' + y = 0$ **16.** $y'' - xy' + 2y = 0$

17. $y'' + x^2 y' + xy = 0$ **18.** $y'' + 2xy' + 2y = 0$

19. $(x-1)y'' + y' = 0$ **20.** $(x+2)y'' + xy' - y = 0$

21. $y'' - (x+1)y' - y = 0$ **22.** $(x^2+1)y'' - 6y = 0$

23. $(x^2+2)y'' + 3xy' - y = 0$ **24.** $(x^2-1)y'' + xy' - y = 0$

In Problems 25–28, use the power series method to solve the given initial-value problem.

25. $(x-1)y'' - xy' + y = 0$, $y(0) = -2$, $y'(0) = 6$

26. $(x+1)y'' - (2-x)y' + y = 0$, $y(0) = 2$, $y'(0) = -1$

27. $y'' - 2xy' + 8y = 0$, $y(0) = 3$, $y'(0) = 0$

28. $(x^2+1)y'' + 2xy' = 0$, $y(0) = 0$, $y'(0) = 1$

In Problems 29 and 30, use the procedure in Example 5 to find two power series solutions of the given differential equation about the ordinary point x = 0.

29. $y'' + (\sin x)y = 0$ **30.** $y'' + e^x y' - y = 0$

31. How can the power series method be used to solve the nonhomogeneous equations such as $y'' - xy = 1$ and $y'' - 4xy' - 4y = e^x$? Carry out your ideas by solving both equations.

32. Is $x = 0$ an ordinary or a singular point of the differential equation $xy'' + (\sin x)y = 0$? Defend your answer with sound mathematics.

33. Computer Lab Assignment (a) Find two power series solutions for $y'' + xy' + y = 0$ and express the solutions $y_1(x)$ and $y_2(x)$ in terms of summation notation.

(b) Use a graphing utility to plot the graphs of the partial sums $S_N(x)$ for $y_1(x)$. Use $N = 2, 3, 5, 6, 8, 10$. Repeat, using the partial sums $S_N(x)$ for $y_2(x)$.

(c) Compare the graphs obtained in part **(b)** with the curve obtained using a numerical solver. Use the initial conditions $y_1(0) = 1$, $y_1'(0) = 0$, and $y_2(0) = 0$, $y_2'(0) = 1$.

34. Computer Lab Assignment (a) Find one more nonzero term for each of the solutions $y_1(x)$ and $y_2(x)$ in Example 5.

(b) Find a series solution $y(x)$ of the initial-value problem $y'' + (\cos x)y = 0$, $y(0) = 1$, $y'(0) = 1$.

(c) Use a graphing utility to plot the graphs of the partial sums $S_N(x)$ for the solution $y(x)$ in part **(b)**. Use $N = 2, 3, 4, 5, 6, 7$.

(d) Compare the graphs obtained in part **(c)** with the curve obtained using a numerical solver for the initial-value problem in part **(b)**.

35. Reexamine the solution $y_1(x)$ in part **(a)** of Problem 33. Express this series as an elementary function. Then use (5) of Section 3.2 to find a second solution of the equation. Verify that this second solution is the same as the power series solution $y_2(x)$.

36. For purposes of this problem ignore the graphs given in Figure 5.1. If Airy's equation is written as $y'' = -xy$, what can we say about the shape of a solution curve if $x > 0$ and $y > 0$? If $x > 0$ and $y < 0$?

5.2 Solutions about Singular Points

The two differential equations $y'' + xy = 0$ and $xy'' + y = 0$ are similar only in that they are both examples of simple linear second-order equations with variable coefficients. That is all they have in common. We saw in the preceding section that since $x = 0$ is an *ordinary point* of the first equation, there was no problem in finding two linearly independent power series solutions centered at that point. In contrast,

because $x = 0$ is a *singular point* of the second ODE, finding two infinite series solutions of the equation about that point becomes a more difficult task. See Examples 4 and 5.

■ **Regular and Irregular Singular Points**　A singular point $x = x_0$ of a linear differential equation

$$a_2(x)y'' + a_1(x)y' + a_0(x)y = 0 \qquad (1)$$

is further classified as either regular or irregular. The classification again depends on the functions P and Q in the standard form

$$y'' + P(x)y' + Q(x)y = 0. \qquad (2)$$

DEFINITION 5.2　Regular/Irregular Singular Points

A singular point x_0 is said to be an **regular singular point** of the differential equation (1) if the functions $p(x) = (x - x_0)p(x)$ and $q(x) = (x - x_0)^2Q(x)$ are both analytic at x_0. A singular point that is not regular is said to be an **irregular singular point** of the equation.

The second sentence in Definition 5.2 indicates that if one or both of the functions $p(x) = (x - x_0)P(x)$ and $q(x) = (x - x_0)^2Q(x)$ fails to be analytic at x_0, then x_0 is an irregular singular point.

■ **Polynomial Coefficients**　As in Section 5.1, we are mainly interested in linear equations (1) where the coefficients $a_2(x)$, $a_1(x)$, and $a_0(x)$ are polynomials with no common factors. We have already seen that if $a_2(x_0) = 0$, then $x = x_0$ is a singular point of (1) since at least one of the rational functions $P(x) = a_1(x)/a_2(x)$ and $Q(x) = a_0(x)/a_2(x)$ in the standard form (2) fails to be analytic at that point. But since $a_2(x)$ is a polynomial and x_0 is one of its zeros, it follows from the Factor Theorem of algebra that $x - x_0$ is a factor of $a_2(x)$. This means that after $a_1(x)/a_2(x)$ and $a_0(x)/a_2(x)$ are reduced to lowest terms, the factor $x - x_0$ must remain, to some positive integer power, in one or both denominators. Now suppose that $x = x_0$ is a singular point of (1) but that both the functions defined by the products $p(x) = (x - x_0)P(x)$ and $q(x) = (x - x_0)^2Q(x)$ are analytic at x_0. We are led to the conclusion that multiplying $P(x)$ by $x - x_0$ and $Q(x)$ by $(x - x_0)^2$ has the effect (through cancellation) that $x - x_0$ no longer appears in either denominator. We can now determine whether x_0 is regular by a quick visual check of denominators: If $x - x_0$ appears *at most* to the first power in the denominator of $P(x)$ and *at most* to the second power in the denominator of $Q(x)$, then $x = x_0$ is a regular singular point. Moreover, observe that if $x = x_0$ is a regular singular point and we multiply (2) by $(x - x_0)^2$, then the original DE can be put into the form

$$(x - x_0)^2y'' + (x - x_0)p(x)y' + q(x)y = 0, \qquad (3)$$

where p and q are analytic at $x = x_0$.

Example 1　**Classification of Singular Points**

It should be clear that $x = 2$ and $x = -2$ are singular points of

$$(x^2 - 4)^2y'' + 3(x - 2)y' + 5y = 0.$$

After dividing the equation by $(x^2 - 4)^2 = (x - 2)^2(x + 2)^2$ and reducing the coefficients to lowest terms, we find that

$$P(x) = \frac{3}{(x - 2)(x + 2)^2} \quad \text{and} \quad Q(x) = \frac{5}{(x - 2)^2(x + 2)^2}.$$

We now test $P(x)$ and $Q(x)$ at each singular point.

In order for $x = 2$ to be a regular singular point, the factor $x - 2$ can appear at most to the first power in the denominator of $P(x)$ and at most to the second power in the denominator of $Q(x)$. A check of the denominators of $P(x)$ and $Q(x)$ shows that both these conditions are satisfied, so $x = 2$ is a regular singular point. Alternatively, we are led to the same conclusion by noting that both rational functions

$$p(x) = (x - 2)P(x) = \frac{3}{(x + 2)^2} \quad \text{and} \quad q(x) = (x - 2)^2 Q(x) = \frac{5}{(x + 2)^2}$$

are analytic at $x = 2$.

Now since the factor $x - (-2) = x + 2$ appears to the second power in the denominator of $P(x)$, we can conclude immediately that $x = -2$ is an irregular singular point of the equation. This also follows from the fact that $p(x) = (x + 2)P(x) = 3/[(x - 2)(x + 2)]$ is not analytic at $x = -2$. ☐

In Example 1, notice that since $x = 2$ is a regular singular point, the original equation can be written as

$$\overset{p(x) \text{ analytic at } x = 2}{\underset{\downarrow}{}} \qquad \overset{q(x) \text{ analytic at } x = 2}{\underset{\downarrow}{}}$$

$$(x - 2)^2 y'' + (x - 2)\frac{3}{(x + 2)^2} y' + \frac{5}{(x + 2)^2} y = 0.$$

As another example, we can see that $x = 0$ is an irregular singular point of $x^3 y'' - 2xy' + 8y = 0$ by inspection of the denominators of $P(x) = -2/x^2$ and $Q(x) = 8/x^3$. On the other hand, $x = 0$ is a regular singular point of $xy'' - 2xy' + 8y = 0$ since $x - 0$ and $(x - 0)^2$ do not even appear in the respective denominators of $P(x) = -2$ and $Q(x) = 8/x$. For a singular point $x = x_0$, any nonnegative power of $x - x_0$ less than one (namely zero) and any nonnegative power less than two (namely, zero and one) in the denominators of $P(x)$ and $Q(x)$, respectively, imply that x_0 is a regular singular point. A singular point can be a complex number. You should verify that $x = 3i$ and $x = -3i$ are two regular singular points of $(x^2 + 9)y'' - 3xy' + (1 - x)y = 0$.

Any second-order Cauchy–Euler equation $ax^2 y'' + bxy' + cy = 0$, with a, b, and c real constants, has a regular singular point at $x = 0$. You should verify that two solutions of the Cauchy–Euler equation $x^2 y'' - 3xy' + 4y = 0$ on the interval $(0, \infty)$ are $y_1 = x^2$ and $y_2 = x^2 \ln x$. If we attempt to find a power series solution about the regular singular point $x = 0$, namely $y = \sum_{n=0}^{\infty} c_n x^n$, we would succeed in obtaining only the polynomial solution $y_1 = x^2$. The fact that we would not obtain the second solution is not surprising since $\ln x$, and consequently $y_2 = x^2 \ln x$, is not analytic at $x = 0$; that is, y_2 does not possess a Taylor series expansion centered at $x = 0$.

■ Method of Frobenius

To solve a differential equation (1) about a regular singular point, we employ the following theorem due to Frobenius.

> ### THEOREM 5.2 Frobenius' Theorem
>
> If $x = x_0$ is a regular singular point of the differential equation (1), then there exists at least one solution of the form
>
> $$y = (x - x_0)^r \sum_{n=0}^{\infty} c_n(x - x_0)^n = \sum_{n=0}^{\infty} c_n(x - x_0)^{n+r}, \qquad (4)$$
>
> where the number r is a constant to be determined. The series will converge at least on some interval $0 < x - x_0 < R$.

Notice the words *at least* in the first sentence of Theorem 5.2. This means that, in contrast to Theorem 5.1, we have no assurance that two series solutions of the type indicated in (4) can be found. The **method of Frobenius,** finding series solutions about a regular singular point x_0, is similar to the "method of undetermined series coefficients" of the preceding section in that we substitute $y = \sum_{n=0}^{\infty} c_n(x - x_0)^{n+r}$ into the given differential equation and determine the unknown coefficients c_n by a recursion relation. However, we have an additional task in this procedure; before determining the coefficients, we must first find the unknown exponent r. If r is found to be a number that is not a nonnegative integer, then the corresponding solution $y = \sum_{n=0}^{\infty} c_n(x - x_0)^{n+r}$ is not a power series.

All solutions will be about ☞
the regular singular point 0.

As we did in the discussion of solutions about ordinary points, we shall always assume, for the sake of simplicity in solving differential equations, that the regular singular point is $x = 0$.

Example 2 Two Series Solutions

Because $x = 0$ is a regular singular point of the differential equation

$$3xy'' + y' - y = 0, \qquad (5)$$

we try to find a solution of the form $y = \sum_{n=0}^{\infty} c_n x^{n+r}$. Now

$$y' = \sum_{n=0}^{\infty} (n + r)c_n x^{n+r-1} \quad \text{and} \quad y'' = \sum_{n=0}^{\infty} (n + r)(n + r - 1)c_n x^{n+r-2}$$

so that

$$3xy'' + y' - y = 3\sum_{n=0}^{\infty} (n + r)(n + r - 1)c_n x^{n+r-1} + \sum_{n=0}^{\infty} (n + r)c_n x^{n+r-1} - \sum_{n=0}^{\infty} c_n x^{n+r}$$

$$= \sum_{n=0}^{\infty} (n + r)(3n + 3r - 2)c_n x^{n+r-1} - \sum_{n=0}^{\infty} c_n x^{n+r}$$

$$= x^r \left[r(3r - 2)c_0 x^{-1} + \underbrace{\sum_{n=1}^{\infty} (n + r)(3n + 3r - 2)c_n x^{n-1}}_{k = n-1} - \underbrace{\sum_{n=0}^{\infty} c_n x^n}_{k = n} \right]$$

$$= x^r \left[r(3r - 2)c_0 x^{-1} + \sum_{k=0}^{\infty} [(k + r + 1)(3k + 3r + 1)c_{k+1} - c_k]x^k \right] = 0,$$

which implies
$$r(3r - 2)c_0 = 0$$

$$(k + r + 1)(3k + 3r + 1)c_{k+1} - c_k = 0, \qquad k = 0, 1, 2, \ldots.$$

Since nothing is gained by taking $c_0 = 0$, we must then have

$$r(3r - 2) = 0, \tag{6}$$

and
$$c_{k+1} = \frac{c_k}{(k + r + 1)(3k + 3r + 1)}, \qquad k = 0, 1, 2, \dots. \tag{7}$$

The two values of r that satisfy the quadratic equation (6), $r_1 = \frac{2}{3}$ and $r_2 = 0$, when substituted in (7), give two different recurrence relations:

$$r_1 = \tfrac{2}{3}, \qquad c_{k+1} = \frac{c_k}{(3k + 5)(k + 1)}, \qquad k = 0, 1, 2, \dots \tag{8}$$

$$r_2 = 0, \qquad c_{k+1} = \frac{c_k}{(k + 1)(3k + 1)}, \qquad k = 0, 1, 2, \dots. \tag{9}$$

From (8) we find:

$$c_1 = \frac{c_0}{5 \cdot 1}$$

$$c_2 = \frac{c_1}{8 \cdot 2} = \frac{c_0}{2! \, 5 \cdot 8}$$

$$c_3 = \frac{c_2}{11 \cdot 3} = \frac{c_0}{3! \, 5 \cdot 8 \cdot 11}$$

$$c_4 = \frac{c_3}{14 \cdot 4} = \frac{c_0}{4! \, 5 \cdot 8 \cdot 11 \cdot 14}$$

$$\vdots$$

$$c_n = \frac{c_0}{n! \, 5 \cdot 8 \cdot 11 \cdots (3n + 2)}.$$

From (9) we find:

$$c_1 = \frac{c_0}{1 \cdot 1}$$

$$c_2 = \frac{c_1}{2 \cdot 4} = \frac{c_0}{2! \, 1 \cdot 4}$$

$$c_3 = \frac{c_2}{3 \cdot 7} = \frac{c_0}{3! \, 1 \cdot 4 \cdot 7}$$

$$c_4 = \frac{c_3}{4 \cdot 10} = \frac{c_0}{4! \, 1 \cdot 4 \cdot 7 \cdot 10}$$

$$\vdots$$

$$c_n = \frac{(-1)^n c_0}{n! \, 1 \cdot 4 \cdot 7 \cdots (3n - 2)}.$$

Here we encounter something that did not happen when we obtained solutions about an ordinary point; we have what looks to be two different sets of coefficients, but each set contains the *same* multiple c_0. Omitting this term, the series solutions are

$$y_1(x) = x^{2/3}\left[1 + \sum_{n=1}^{\infty} \frac{1}{n! \, 5 \cdot 8 \cdot 11 \cdots (3n + 2)} x^n\right] \tag{10}$$

$$y_2(x) = x^0\left[1 + \sum_{n=1}^{\infty} \frac{1}{n! \, 1 \cdot 4 \cdot 7 \cdots (3n - 2)} x^n\right]. \tag{11}$$

By the ratio test it can be demonstrated that both (10) and (11) converge for all finite values of x, that is, $|x| < \infty$. Also, it should be apparent from the form of these solutions that neither series is a constant multiple of the other and, therefore, $y_1(x)$ and $y_2(x)$ are linearly independent on the entire x-axis. Hence by the superposition principle $y = C_1 y_1(x) + C_2 y_2(x)$ is another solution of (5). On any interval not containing the origin, such as $(0, \infty)$, this linear combination represents the general solution of the differential equation. ❏

■ **Indicial Equation** Equation (6) is called the **indicial equation** of the problem, and the values $r_1 = \frac{2}{3}$ and $r_2 = 0$ are called the **indicial roots,** or **exponents,** of the singularity $x = 0$. In general, after substituting $y = \sum_{n=0}^{\infty} c_n x^{n+r}$ into the given

differential equation and simplifying, the indicial equation is a quadratic equation in r that results from equating the *total coefficient of the lowest power of x to zero.* We solve for the two values of r and substitute these values into a recurrence relation such as (7). Theorem 5.2 guarantees that at least one solution of the assumed series form can be found.

It is possible to obtain the indicial equation in advance of substituting $y = \sum_{n=0}^{\infty} c_n x^{n+r}$ into the differential equation. If $x = 0$ is a regular singular point of (1), then by Definition 5.2 both functions $p(x) = xP(x)$ and $q(x) = x^2 Q(x)$, where P and Q are defined by the standard form (2), are analytic at $x = 0$; that is, the power series expansions

$$p(x) = xP(x) = a_0 + a_1 x + a_2 x^2 + \cdots \quad \text{and} \quad q(x) = x^2 Q(x) = b_0 + b_1 x + b_2 x^2 + \cdots \quad (12)$$

are valid on intervals that have a positive radius of convergence. By multiplying (2) by x^2, we get the form given in (3):

$$x^2 y'' + x[xP(x)]y' + [x^2 Q(x)]y = 0. \quad (13)$$

After substituting $y = \sum_{n=0}^{\infty} c_n x^{n+r}$ and the two series in (12) into (13) and carrying out the multiplication of series, we find the general indicial equation to be

$$r(r - 1) + a_0 r + b_0 = 0, \quad (14)$$

where a_0 and b_0 are defined in (12). See Problems 13 and 14 in Exercises 5.2.

Example 3 Two Series Solutions

Solve $2xy'' + (1 + x)y' + y = 0$.

SOLUTION Substituting $y = \sum_{n=0}^{\infty} c_n x^{n+r}$ gives

$$2xy'' + (1 + x)y' + y = 2 \sum_{n=0}^{\infty} (n + r)(n + r - 1)c_n x^{n+r-1} + \sum_{n=0}^{\infty} (n + r)c_n x^{n+r-1}$$

$$+ \sum_{n=0}^{\infty} (n + r)c_n x^{n+r} + \sum_{n=0}^{\infty} c_n x^{n+r}$$

$$= \sum_{n=0}^{\infty} (n + r)(2n + 2r - 1)c_n x^{n+r-1} + \sum_{n=0}^{\infty} (n + r + 1)c_n x^{n+r}$$

$$= x^r \left[r(2r - 1)c_0 x^{-1} + \underbrace{\sum_{n=1}^{\infty} (n + r)(2n + 2r - 1)c_n x^{n-1}}_{k = n - 1} + \underbrace{\sum_{n=0}^{\infty} (n + r + 1)c_n x^{n}}_{k = n} \right]$$

$$= x^r \left[r(2r - 1)c_0 x^{-1} + \sum_{k=0}^{\infty} [(k + r + 1)(2k + 2r + 1)c_{k+1} + (k + r + 1)c_k]x^k \right]$$

which implies
$$r(2r - 1) = 0 \quad (15)$$

$$(k + r + 1)(2k + 2r + 1)c_{k+1} + (k + r + 1)c_k = 0, \qquad k = 0, 1, 2, \ldots . \quad (16)$$

From (15) we see that the indicial roots are $r_1 = \frac{1}{2}$ and $r_2 = 0$.

For $r_1 = \frac{1}{2}$, we can divide by $k + \frac{3}{2}$ in (16) to obtain

$$c_{k+1} = \frac{-c_k}{2(k + 1)}, \qquad k = 0, 1, 2, \ldots, \quad (17)$$

whereas for $r_2 = 0$, (16) becomes

$$c_{k+1} = \frac{-c_k}{2k + 1}, \qquad k = 0, 1, 2, \ldots . \quad (18)$$

From (17):

$$c_1 = \frac{-c_0}{2 \cdot 1}$$

$$c_2 = \frac{-c_1}{2 \cdot 2} = \frac{c_0}{2^2 \cdot 2!}$$

$$c_3 = \frac{-c_2}{2 \cdot 3} = \frac{-c_0}{2^3 \cdot 3!}$$

$$c_4 = \frac{-c_3}{2 \cdot 4} = \frac{c_0}{2^4 \cdot 4!}$$

$$\vdots$$

$$c_n = \frac{(-1)^n c_0}{2^n \, n!}.$$

From (18):

$$c_1 = \frac{-c_0}{1}$$

$$c_2 = \frac{-c_1}{3} = \frac{c_0}{1 \cdot 3}$$

$$c_3 = \frac{-c_2}{5} = \frac{-c_0}{1 \cdot 3 \cdot 5}$$

$$c_4 = \frac{-c_3}{7} = \frac{c_0}{1 \cdot 3 \cdot 5 \cdot 7}$$

$$\vdots$$

$$c_n = \frac{(-1)^n c_0}{1 \cdot 3 \cdot 5 \cdot 7 \cdots (2n-1)}.$$

Thus for the indicial root $r_1 = \frac{1}{2}$ we obtain the solution

$$y_1(x) = x^{1/2}\left[1 + \sum_{n=1}^{\infty} \frac{(-1)^n}{2^n \, n!} x^n\right] = \sum_{n=0}^{\infty} \frac{(-1)^n}{2^n \, n!} x^{n+1/2},$$

where we have again omitted c_0. The series converges for $x \geq 0$; as given, the series is not defined for negative values of x because of the presence of $x^{1/2}$. For $r_2 = 0$, a second solution is

$$y_2(x) = 1 + \sum_{n=1}^{\infty} \frac{(-1)^n}{1 \cdot 3 \cdot 5 \cdot 7 \cdots (2n-1)} x^n, \qquad |x| < \infty.$$

On the interval $(0, \infty)$, the general solution is $y = C_1 y_1(x) + C_2 y_2(x)$. ❏

Example 4 **Only One Series Solution**

Solve $xy'' + y = 0$.

SOLUTION From $xP(x) = 0$ and $x^2 Q(x) = x$ and the fact that 0 and x are their own power series centered at 0, we conclude $a_0 = 0$ and $b_0 = 0$ and so from (14) the indicial equation is $r(r-1) = 0$. You should verify that the two recurrence relations corresponding to the indicial roots $r_1 = 1$ and $r_2 = 0$ yield exactly the same set of coefficients. In other words, in this case the method of Frobenius produces only a single series solution

$$y_1(x) = \sum_{n=0}^{\infty} \frac{(-1)^n}{n!(n+1)!} x^{n+1} = x + \frac{1}{2} x^2 + \frac{1}{12} x^3 + \frac{1}{144} x^4 + \cdots. \qquad ❏$$

■ **Three Cases** For the sake of discussion, let us suppose again that $x = 0$ is a regular singular point of (1) and that indicial roots r_1 and r_2 of the singularity are real, with r_1 denoting the largest root. When using the method of Frobenius, we usually distinguish three cases corresponding to the nature of the indicial roots r_1 and r_2.

Case I: If r_1 and r_2 are distinct and do not differ by an integer, there exist two linearly independent solutions $y_1(x)$ and $y_2(x)$ of (1) of the form $y = \sum_{n=0}^{\infty} c_n x^{n+r}$.

This is the case illustrated in Examples 2 and 3.

In the next case, we see that when the difference of indicial roots $r_1 - r_2$ is a positive integer, the second solution *may* contain a logarithm.

> **Case II:** If $r_1 - r_2 = N$, where N is a positive integer, then there exist two linearly independent solutions of equation (1) of the form
>
> $$y_1(x) = \sum_{n=0}^{\infty} c_n x^{n+r_1}, \qquad c_0 \neq 0, \tag{19}$$
>
> $$y_2(x) = Cy_1(x) \ln x + \sum_{n=0}^{\infty} b_n x^{n+r_2}, \qquad b_0 \neq 0, \tag{20}$$
>
> where C is a constant that could be zero.

When the difference $r_1 - r_2$ is a positive integer, we *may* or *may not* be able to find *two* solutions having the form $y = \sum_{n=0}^{\infty} c_n x^{n+r}$. This is something we do not know in advance but that is determined after we have found the indicial roots and have carefully examined the recurrence relation that defines the coefficients c_n. We just may be lucky enough to find two solutions that involve only powers of x, that is, $y_1 = \sum_{n=0}^{\infty} c_n x^{n+r_1}$ (equation (19)) and $y_2 = \sum_{n=0}^{\infty} b_n x^{n+r_2}$ (equation (20) with $C = 0$). See Problem 31 in Exercises 5.2. On the other hand, in Example 4 we see that for $r_1 - r_2 = 1$, an integer and the method of Frobenius failed to give a second series solution. In this situation, equation (20), with $C \neq 0$, indicates what the second solution looks like. One way to obtain this second solution with the logarithmic term is to use the fact that

$$y_2(x) = y_1(x) \int \frac{e^{-\int P(x)\,dx}}{y_1^2(x)}\,dx \tag{21}$$

is also a solution of $y'' + P(x)y' + Q(x)y = 0$ whenever $y_1(x)$ is the known solution (see (5) in Section 3.2). We will illustrate how to use this formula in Example 5.

Finally, in the last case, the case when the indicial roots are equal, a second solution will *always* contain a logarithm. This situation is analogous to the solution of a Cauchy–Euler equation when the roots of the auxiliary equation are equal.

> **Case III:** If $r_1 = r_2$, then there always exist two linearly independent solutions of the equation (1) of the form
>
> $$y_1(x) = \sum_{n=0}^{\infty} c_n x^{n+r_1}, \qquad c_0 \neq 0, \tag{22}$$
>
> $$y_2(x) = y_1(x) \ln x + \sum_{n=1}^{\infty} b_n x^{n+r_2}. \tag{23}$$

The solution $y_2(x)$ in (23) can also be obtained using (21).

Example 5 Example 4 Revisited—Using a CAS

Find the general solution of $xy'' + y = 0$.

SOLUTION From the known solution given in Example 4,

$$y_1(x) = x + \frac{1}{2}x^2 + \frac{1}{12}x^3 + \frac{1}{144}x^4 + \cdots,$$

we can construct a second solution $y_2(x)$ using formula (21). For those with the time, energy, and patience, the drudgery of squaring a series, long division, and

integration of the quotient can be carried out by hand. But all these operations can be done with relative ease with the help of a CAS. We give the results:

Here is a good place to use a computer algebra system. ☞

$$y_2(x) = y_1(x) \int \frac{e^{-\int 0\, dx}}{[y_1(x)]^2}\, dx = y_1(x) \int \frac{dx}{\left[x + \dfrac{1}{2}x^2 + \dfrac{1}{12}x^3 + \dfrac{1}{144}x^4 + \cdots\right]^2}$$

$$= y_1(x) \int \frac{dx}{\left[x^2 - x^3 + \dfrac{5}{12}x^4 - \dfrac{7}{12}x^5 + \cdots\right]} \qquad \leftarrow \text{after squaring}$$

$$= y_1(x) \int \left[\dfrac{1}{x^2} + \dfrac{1}{x} + \dfrac{7}{12} + \dfrac{19}{72}x + \cdots\right] dx \qquad \leftarrow \text{after long division}$$

$$= y_1(x)\left[-\dfrac{1}{x} + \ln x + \dfrac{7}{12}x + \dfrac{19}{144}x^2 + \cdots\right] \qquad \leftarrow \text{after integrating}$$

or $\quad y_2(x) = y_1(x) \ln x + y_1(x)\left[-\dfrac{1}{x} + \dfrac{7}{12}x + \dfrac{19}{144}x^2 + \cdots\right].$

On the interval $(0, \infty)$, the general solution is $y = C_1 y_1(x) + C_2 y_2(x)$. ❏

R e m a r k s

(*i*) The three different forms of a linear second-order differential equation in (1), (2), and (3) were used to discuss various theoretical concepts. But on a practical level, when it comes to actually solving a differential equation using the method of Frobenius, it is advisable to work with the form of the DE given in (1).

(*ii*) When the difference of indicial roots $r_1 - r_2$ is a positive integer $(r_1 > r_2)$, it sometimes pays to iterate the recurrence relation using the smaller root r_2 first. See Problems 31 and 32 in Exercises 5.2.

(*ii*) Since r is a root of a quadratic equation, it could be complex. We shall not, however, investigate this case.

(*iii*) If $x = 0$ is an irregular singular point, we may not be able to find *any* solution of the form $y = \sum_{n=0}^{\infty} c_n x^{n+r}$.

E X E R C I S E S 5 . 2

Answers to odd-numbered problems begin on page A-35.

In Problems 1–10, determine the singular points of the given differential equation. Classify each singular point as regular or irregular.

1. $x^3 y'' + 4x^2 y' + 3y = 0$

2. $x(x + 3)^2 y'' - y = 0$

3. $(x^2 - 9)^2 y'' + (x + 3)y' + 2y = 0$

4. $y'' - \dfrac{1}{x}y' + \dfrac{1}{(x-1)^3}y = 0$

5. $(x^3 + 4x)y'' - 2xy' + 6y = 0$

6. $x^2(x - 5)^2 y'' + 4xy' + (x^2 - 25)y = 0$

7. $(x^2 + x - 6)y'' + (x + 3)y' + (x - 2)y = 0$

8. $x(x^2 + 1)^2 y'' + y = 0$

9. $x^3(x^2 - 25)(x - 2)^2 y'' + 3x(x - 2)y' + 7(x + 5)y = 0$

10. $(x^3 - 2x^2 + 3x)^2 y'' + x(x - 3)^2 y' - (x + 1)y = 0$

In Problems 11 and 12, put the given differential equation into the form (3) for each regular singular point of the equation. Identify the functions $p(x)$ and $q(x)$.

11. $(x^2 - 1)y'' + 5(x + 1)y' + (x^2 - x)y = 0$

12. $xy'' + (x + 3)y' + 7x^2 y = 0$

In Problems 13 and 14, $x = 0$ is a regular singular point of the given differential equation. Use the general form of the indicial equation in (14) to find the indicial roots of the singularity. Without solving, discuss the number of series solutions you would expect to find using the method of Frobenius.

13. $x^2y'' + (\frac{5}{3}x + x^2)y' - \frac{1}{3}y = 0$ **14.** $xy'' + y' + 10y = 0$

In Problems 15–24, $x = 0$ is a regular singular point of the given differential equation. Show that the indicial roots of the singularity do not differ by an integer. Use the method of Frobenius to obtain two linearly independent series solutions about $x = 0$. Form the general solution on $(0, \infty)$.

15. $2xy'' - y' + 2y = 0$ **16.** $2xy'' + 5y' + xy = 0$

17. $4xy'' + \frac{1}{2}y' + y = 0$ **18.** $2x^2y'' - xy' + (x^2 + 1)y = 0$

19. $3xy'' + (2 - x)y' - y = 0$ **20.** $x^2y'' - (x - \frac{2}{9})y = 0$

21. $2xy'' - (3 + 2x)y' + y = 0$ **22.** $x^2y'' + xy' + (x^2 - \frac{4}{9})y = 0$

23. $9x^2y'' + 9x^2y' + 2y = 0$

24. $2x^2y'' + 3xy' + (2x - 1)y = 0$

In Problems 25–30, $x = 0$ is a regular singular point of the given differential equation. Show that the indicial roots of the singularity differ by an integer. Use the method of Frobenius to obtain at least one series solution about $x = 0$. Use (21) where necessary and a CAS, if instructed, to find a second solution. Form the general solution on $(0, \infty)$.

25. $xy'' + 2y' - xy = 0$ **26.** $x^2y'' + xy' + (x^2 - \frac{1}{4})y = 0$

27. $xy'' - xy' + y = 0$ **28.** $y'' + \dfrac{3}{x}y' - 2y = 0$

29. $xy'' + (1 - x)y' - y = 0$ **30.** $xy'' + y' + y = 0$

In Problems 31 and 32, $x = 0$ is a regular singular point of the given differential equation. Show that the indicial roots of the singularity differ by an integer. Use the recurrence relation found by the method of Frobenius first with the largest root r_1. How many solutions did you find? Next use the recurrence relation with the smaller root r_2. How many solutions did you find?

31. $xy'' + (x - 6)y' - 3y = 0$ **32.** $x(x - 1)y'' + 3y' - 2y = 0$

33. (a) The differential equation $x^4y'' + \lambda y = 0$ has an irregular singular point at $x = 0$. Show that the substitution $t = 1/x$ yields the differential equation

$$\frac{d^2y}{dt^2} + \frac{2}{t}\frac{dy}{dt} + \lambda y = 0,$$

which now has a regular singular point at $t = 0$.

(b) Use the method of this section to find two series solutions of the second equation in part **(a)** about the singular point $t = 0$.

(c) Express each series solution of the original equation in terms of elementary functions.

34. In Example 3 of Section 3.9, we saw that when a constant vertical compressive force or load P was applied to a thin column of uniform cross section, the deflection $y(x)$ satisfied the boundary-value problem

$$EI\frac{d^2y}{dx^2} + Py = 0, \quad y(0) = 0, \quad y(L) = 0. \tag{24}$$

The column will buckle or deflect only when the compressive force is a critical load P_n.

(a) Now suppose the column is again of length L and has circular cross-sections but is tapered as shown in Figure 5.2(a). If the column, a truncated cone, has a linear taper $y = cx$ as shown in cross-section in Figure 5.2(b), the moment of inertia of a cross-section with respect to an axis perpendicular to the xy-plane is $I = \frac{1}{4}\pi r^4$, where $r = y$ and $y = cx$. Hence we can write $I(x) = I_0(x/b)^4$, where $I_0 = I(b) = \frac{1}{4}\pi(cb)^4$.

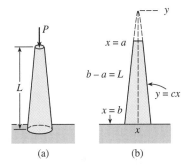

Figure 5.2

Substituting $I(x)$ into the differential equation (24), we see that the deflection in this case is determined from the boundary-value problem

$$x^4\frac{d^2y}{dx^2} + \lambda y = 0, \quad y(a) = 0, \quad y(b) = 0,$$

where $\lambda = Pb^4/EI_0$. Use the results of Problem 33 to find the critical loads P_n for the tapered column. Use an appropriate identity to express the buckling modes $y_n(x)$ as a single function.

(b) Use a CAS to plot the graph of the first buckling mode $y_1(x)$ corresponding to the Euler load P_1 when $b = 11$ and $a = 1$.

5.3 Two Special Equations

The two differential equations

$$x^2 y'' + xy' + (x^2 - v^2)y = 0 \tag{1}$$

$$(1 - x^2)y'' - 2xy' + n(n + 1)y = 0 \tag{2}$$

occur frequently in advanced studies in applied mathematics, physics, and engineering. They are called **Bessel's equation** and **Legendre's equation,** respectively. In solving (1) we shall assume $v \geq 0$, whereas in (2) we shall consider only the case when n is a nonnegative integer. Since we seek series solutions of each equation about $x = 0$, we observe that the origin is a regular singular point of Bessel's equation but is an ordinary point of Legendre's equation.

■ **Solution of Bessel's Equation** Because $x = 0$ is a regular singular point of Bessel's equation, we know that there exists at least one solution of the form $y = \sum_{n=0}^{\infty} c_n x^{n+r}$. Substituting the last expression into (1) then gives

$$x^2 y'' + xy' + (x^2 - v^2)y = \sum_{n=0}^{\infty} c_n (n+r)(n+r-1)x^{n+r} + \sum_{n=0}^{\infty} c_n (n+r)x^{n+r} + \sum_{n=0}^{\infty} c_n x^{n+r+2} - v^2 \sum_{n=0}^{\infty} c_n x^{n+r}$$

$$= c_0(r^2 - r + r - v^2)x^r$$

$$+ x^r \sum_{n=1}^{\infty} c_n[(n+r)(n+r-1) + (n+r) - v^2]x^n + x^r \sum_{n=0}^{\infty} c_n x^{n+2}$$

$$= c_0(r^2 - v^2)x^r + x^r \sum_{n=1}^{\infty} c_n[(n+r)^2 - v^2]x^n + x^r \sum_{n=0}^{\infty} c_n x^{n+2}. \tag{3}$$

From (3) we see that the indicial equation is $r^2 - v^2 = 0$ so that the indicial roots are $r_1 = v$ and $r_2 = -v$. When $r_1 = v$, (3) becomes

$$x^v \sum_{n=1}^{\infty} c_n n(n + 2v)x^n + x^v \sum_{n=0}^{\infty} c_n x^{n+2}$$

$$= x^v \left[(1 + 2v)c_1 x + \underbrace{\sum_{n=2}^{\infty} c_n n(n + 2v)x^n}_{k=n-2} + \underbrace{\sum_{n=0}^{\infty} c_n x^{n+2}}_{k=n} \right]$$

$$= x^v \left[(1 + 2v)c_1 x + \sum_{k=0}^{\infty} [(k+2)(k+2+2v)c_{k+2} + c_k]x^{k+2} \right] = 0.$$

Therefore, by the usual argument we can write $(1 + 2v)c_1 = 0$ and

$$(k + 2)(k + 2 + 2v)c_{k+2} + c_k = 0$$

or

$$c_{k+2} = \frac{-c_k}{(k+2)(k+2+2v)}, \qquad k = 0, 1, 2, \cdots. \tag{4}$$

The choice $c_1 = 0$ in (4) implies $c_3 = c_5 = c_7 = \cdots = 0$, so for $k = 0, 2, 4, \ldots$ we find, after letting $k + 2 = 2n$, $n = 1, 2, 3, \ldots$, that

$$c_{2n} = -\frac{c_{2n-2}}{2^2 n(n + v)}. \tag{5}$$

Thus $c_2 = -\dfrac{c_0}{2^2 \cdot 1 \cdot (1 + v)}$

$$c_4 = -\dfrac{c_2}{2^2 \cdot 2(2 + v)} = \dfrac{c_0}{2^4 \cdot 1 \cdot 2(1 + v)(2 + v)}$$

$$c_6 = -\dfrac{c_4}{2^2 \cdot 3(3 + v)} = -\dfrac{c_0}{2^6 \cdot 1 \cdot 2 \cdot 3(1 + v)(2 + v)(3 + v)}$$

$$\vdots$$

$$c_{2n} = \dfrac{(-1)^n c_0}{2^{2n} n! (1 + v)(2 + v) \cdots (n + v)}, \qquad n = 1, 2, 3, \cdots \qquad (6)$$

It is standard practice to choose c_0 to be a specific value—namely,

$$c_0 = \dfrac{1}{2^v \Gamma(1 + v)},$$

where $\Gamma(1 + v)$ is the gamma function. See Appendix II. Since this latter function possesses the convenient property $\Gamma(1 + \alpha) = \alpha \Gamma(\alpha)$, we can reduce the indicated product in the denominator of (6) to one term. For example,

$$\Gamma(1 + v + 1) = (1 + v)\Gamma(1 + v)$$

$$\Gamma(1 + v + 2) = (2 + v)\Gamma(2 + v) = (2 + v)(1 + v)\Gamma(1 + v).$$

Hence we can write (6) as

$$c_{2n} = \dfrac{(-1)^n}{2^{2n + v} n! (1 + v)(2 + v) \cdots (n + v)\Gamma(1 + v)} = \dfrac{(-1)^n}{2^{2n + v} n! \Gamma(1 + v + n)}$$

for $n = 0, 1, 2, \ldots$.

■ Bessel Functions of the First Kind The series solution
$y = \sum_{n=0}^{\infty} c_{2n} x^{2n + v}$ is usually denoted by $J_v(x)$:

$$J_v(x) = \sum_{n=0}^{\infty} \dfrac{(-1)^n}{n! \Gamma(1 + v + n)} \left(\dfrac{x}{2}\right)^{2n + v}. \qquad (7)$$

If $v \geq 0$, the series converges at least on the interval $[0, \infty)$. Also, for the second exponent $r_2 = -v$ we obtain, in exactly the same manner,

$$J_{-v}(x) = \sum_{n=0}^{\infty} \dfrac{(-1)^n}{n! \Gamma(1 - v + n)} \left(\dfrac{x}{2}\right)^{2n - v}. \qquad (8)$$

The functions $J_v(x)$ and $J_{-v}(x)$ are called **Bessel functions of the first kind** of order v and $-v$, respectively. Depending on the value of v, (8) may contain negative powers of x and hence converge on $(0, \infty)$.*

Now some care must be taken in writing the general solution of (1). When $v = 0$, it is apparent that (7) and (8) are the same. If $v > 0$ and $r_1 - r_2 = v - (-v) = 2v$ is not a positive integer, it follows from Case I of Section 5.2 that $J_v(x)$ and $J_{-v}(x)$ are linearly independent solutions of (1) on $(0, \infty)$, and so the general solution of the interval is $y = c_1 J_v(x) + c_2 J_{-v}(x)$. But we also know from Case II of Section 5.2 that when $r_1 - r_2 = 2v$ is a positive integer, a second series solution of (1) *may* exist. In this second case we distinguish two possibilities. When $v = m =$ positive integer, $J_{-m}(x)$ defined by (8) and $J_m(x)$ are not linearly independent solutions. It can be

* When we replace x by $|x|$, the series given in (7) and (8) converge for $0 < |x| < \infty$.

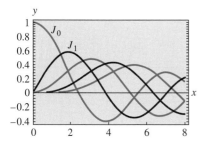

Figure 5.3 Bessel functions of the first kind for $n = 0, 1, 2, 3, 4$

shown that J_{-m} is a constant multiple of J_m (see Property (*i*) on page 259). In addition, $r_1 - r_2 = 2v$ can be a positive integer when v is half an odd positive integer. It can be shown in this latter event that $J_v(x)$ and $J_{-v}(x)$ are linearly independent. In other words, the general solution of (1) on $(0, \infty)$ is

$$y = c_1 J_v(x) + c_2 J_{-v}(x), \qquad v \neq \text{integer}. \tag{9}$$

The graphs of $y = J_0(x)$ and $y = J_1(x)$ are given in Figure 5.3.

Example 1 General Solution: *v* Not an Integer

By identifying $v^2 = \frac{1}{4}$ and $v = \frac{1}{2}$ we can see from (9) that the general solution of the equation $x^2 y'' + xy' + (x^2 - \frac{1}{4})y = 0$ on $(0, \infty)$ is $y = c_1 J_{1/2}(x) + c_2 J_{-1/2}(x)$. ❏

■ Bessel Functions of the Second Kind If $v \neq$ integer, the function defined by the linear combination

$$Y_v(x) = \frac{\cos v\pi J_v(x) - J_{-v}(x)}{\sin v\pi} \tag{10}$$

and the function $J_v(x)$ are linearly independent solutions of (1). Thus another form of the general solution of (1) is $y = c_1 J_v(x) + c_2 Y_v(x)$, provided $v \neq$ integer. As $v \rightarrow m$, m an integer, (10) has the indeterminate form 0/0. However, it can be shown by L'Hôpital's rule that $\lim_{v \to m} Y_v(x)$ exists. Moreover, the function

$$Y_m(x) = \lim_{v \to m} Y_v(x)$$

and $J_m(x)$ are linearly independent solutions of $x^2 y'' + xy' + (x^2 - m^2)y = 0$. Hence for *any* value of v the general solution of (1) on $(0, \infty)$ can be written as

$$y = c_1 J_v(x) + c_2 Y_v(x). \tag{11}$$

$Y_v(x)$ is called the **Bessel function of the second kind** of order v. Figure 5.4 shows the graphs of $Y_0(x)$ and $Y_1(x)$.

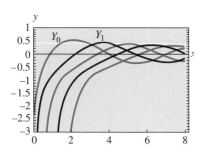

Figure 5.4 Bessel functions of the second kind for $n = 0, 1, 2, 3, 4$

Example 2 General Solution: *v* an Integer

By identifying $v^2 = 9$ and $v = 3$ we see from (11) that the general solution of the equation $x^2 y'' + xy' + (x^2 - 9)y = 0$ on $(0, \infty)$ is $y = c_1 J_3(x) + c_2 Y_3(x)$. ❏

Sometimes it is possible to transform a given differential equation into equation (1) by means of a change of variable. We can then express the general solution of the original equation in terms of Bessel functions. Example 3 illustrates this technique.

Example 3 The Aging Spring Revisited

Recall that in Section 3.8 we saw that one mathematical model for the free undamped motion of a mass on an aging spring is given by $mx'' + ke^{-\alpha t}x = 0$, $\alpha > 0$. We are now in a position to find the general solution of the equation. It is left as a problem to show that the change of variables $s = \dfrac{2}{\alpha}\sqrt{\dfrac{k}{m}}\,e^{-\alpha t/2}$ transforms the differential equation of the aging spring into

$$s^2 \frac{d^2 x}{ds^2} + s \frac{dx}{ds} + s^2 x = 0.$$

The last equation is recognized as (1) with $v = 0$ and with the symbols x and s playing the roles of y and x, respectively. The general solution of the new equation is $x = c_1 J_0(s) + c_2 Y_0(s)$. If we resubstitute s, the general solution of $mx'' + ke^{-\alpha t}x = 0$ is then seen to be

$$x(t) = c_1 J_0\left(\frac{2}{\alpha}\sqrt{\frac{k}{m}}\,e^{-\alpha t/2}\right) + c_2 Y_0\left(\frac{2}{\alpha}\sqrt{\frac{k}{m}}\,e^{-\alpha t/2}\right).$$

See Problems 27 and 28 in Exercises 5.3. ❑

The other model discussed in Section 3.9 of a spring whose characteristics change with time was $mx'' + ktx = 0$. By dividing through by m we see that this equation is Airy's equation, $y'' + \alpha^2 xy = 0$. See Example 2 in Section 5.1. The general solution of Airy's differential equation can also be written in terms of Bessel functions. See Problems 29–31 in Exercises 5.3.

■ **Parametric Bessel Equation** By replacing x by λx in (1) and using the Chain Rule we obtain an alternative form of Bessel's equation known as the **parametric Bessel equation:**

$$x^2 y'' + xy' + (\lambda^2 x^2 - v^2)y = 0. \tag{12}$$

The general solution of (12) is

$$y = c_1 J_v(\lambda x) + c_2 Y_v(\lambda x). \tag{13}$$

■ **Properties** We list below a few of the more useful properties of Bessel functions of order m, $m = 0, 1, 2, \ldots$:

$(i)\ J_{-m}(x) = (-1)^m J_m(x)$ $(ii)\ J_m(-x) = (-1)^m J_m(x)$

$(iii)\ J_m(0) = \begin{cases} 0, & m > 0 \\ 1, & m = 0 \end{cases}$ $(iv)\ \lim_{x \to 0^+} Y_m(x) = -\infty.$

Note that Property (ii) indicates that $J_m(x)$ is an even function if m is an even integer and an odd function if m is an odd integer. The graphs of $Y_0(x)$ and $Y_1(x)$ in Figure 5.4 illustrate Property (iv): $Y_m(x)$ is unbounded at the origin. This last fact is not obvious from (10). It can be shown either from (10) or by the methods of Section 5.2 that for $x > 0$,

$$Y_0(x) = \frac{2}{\pi} J_0(x)\left[\gamma + \ln \frac{x}{2}\right] - \frac{2}{\pi}\sum_{k=1}^{\infty} \frac{(-1)^k}{(k!)^2}\left(1 + \frac{1}{2} + \cdots + \frac{1}{k}\right)\left(\frac{x}{2}\right)^{2k},$$

where $\gamma = 0.57721566\ldots$ is **Euler's constant.** Because of the presence of the logarithmic term, $Y_0(x)$ is discontinuous at $x = 0$.

■ **Numerical Values** Some functional values of $J_0(x)$, $J_1(x)$, $Y_0(x)$, and $Y_1(x)$ for selected values of x are given in Table 5.1. The first five nonnegative zeros of $J_0(x)$, $J_1(x)$, $Y_0(x)$, and $Y_1(x)$ are given in Table 5.2.

Table 5.1 Numerical Values of J_0, J_1, Y_0, and Y_1

x	$J_0(x)$	$J_1(x)$	$Y_0(x)$	$Y_1(x)$
0	1.0000	0.0000	—	—
1	0.7652	0.4401	0.0883	−0.7812
2	0.2239	0.5767	0.5104	−0.1070
3	−0.2601	0.3391	0.3769	0.3247
4	−0.3971	−0.0660	−0.0169	0.3979
5	−0.1776	−0.3276	−0.3085	0.1479
6	0.1506	−0.2767	−0.2882	−0.1750
7	0.3001	−0.0047	−0.0259	−0.3027
8	0.1717	0.2346	0.2235	−0.1581
9	−0.0903	0.2453	0.2499	0.1043
10	−0.2459	0.0435	0.0557	0.2490
11	−0.1712	−0.1768	−0.1688	0.1637
12	0.0477	−0.2234	−0.2252	−0.0571
13	0.2069	−0.0703	−0.0782	−0.2101
14	0.1711	0.1334	0.1272	−0.1666
15	−0.0142	0.2051	0.2055	0.0211

Table 5.2 Zeros of J_0, J_1, Y_0, and Y_1

$J_0(x)$	$J_1(x)$	$Y_0(x)$	$Y_1(x)$
2.4048	0.0000	0.8936	2.1971
5.5201	3.8317	3.9577	5.4297
8.6537	7.0156	7.0861	8.5960
11.7915	10.1735	10.2223	11.7492
14.9309	13.3237	13.3611	14.8974

■ **Differential Recurrence Relation** Recurrence formulas that relate Bessel functions of different orders are important in theory and in applications. In the next example we derive a **differential recurrence relation.**

> ### *Example 4* Derivation Using the Series Definition
>
> Derive the formula $xJ_v'(x) = vJ_v(x) - xJ_{v+1}(x)$.
>
> **SOLUTION** It follows from (7) that
>
> $$xJ_v'(x) = \sum_{n=0}^{\infty} \frac{(-1)^n(2n+v)}{n!\,\Gamma(1+v+n)} \left(\frac{x}{2}\right)^{2n+v}$$
>
> $$= v \sum_{n=0}^{\infty} \frac{(-1)^n}{n!\,\Gamma(1+v+n)} \left(\frac{x}{2}\right)^{2n+v} + 2 \sum_{n=0}^{\infty} \frac{(-1)^n n}{n!\,\Gamma(1+v+n)} \left(\frac{x}{2}\right)^{2n+v}$$
>
> $$= vJ_v(x) + x \underbrace{\sum_{n=1}^{\infty} \frac{(-1)^n}{(n-1)!\,\Gamma(1+v+n)} \left(\frac{x}{2}\right)^{2n+v-1}}_{k\,=\,n\,-\,1}$$
>
> $$= vJ_v(x) - x \sum_{k=0}^{\infty} \frac{(-1)^k}{k!\,\Gamma(2+v+k)} \left(\frac{x}{2}\right)^{2k+v+1} = vJ_v(x) - xJ_{v+1}(x).$$ ❑

The result in Example 4 can be written in an alternative form. Dividing $xJ_v'(x) - vJ_v(x) = -xJ_{v+1}(x)$ by x gives

$$J_v'(x) - \frac{v}{x} J_v(x) = -J_{v+1}(x).$$

This last expression is recognized as a linear first-order differential equation in $J_v(x)$. Multiplying both sides of the equality by the integrating factor x^{-v} then yields

$$\frac{d}{dx}[x^{-v}J_v(x)] = -x^{-v}J_{v+1}(x). \tag{14}$$

It can be shown in a similar manner that

$$\frac{d}{dx}[x^vJ_v(x)] = x^vJ_{v-1}(x). \tag{15}$$

See Problem 19 in Exercises 5.3. The differential recurrence relations (14) and (15) are also valid for the Bessel function of the second kind $Y_v(x)$. Observe that when $v = 0$ it follows from (14) that

$$J_0'(x) = -J_1(x) \quad \text{and} \quad Y_0'(x) = -Y_1(x). \tag{16}$$

An application of these results is given in Problem 28 in Exercises 5.3.

■ **Spherical Bessel Functions** When the order v is half an odd integer, that is, $\pm\frac{1}{2}, \pm\frac{3}{2}, \pm\frac{5}{2}, \ldots$, the Bessel function of the first kind $J_v(x)$ can be expressed in terms of the elementary functions $\sin x$, $\cos x$, and powers of x. Such Bessel functions are called **spherical Bessel functions.** Let's consider the case when $v = \frac{1}{2}$. From (7),

$$J_{1/2}(x) = \sum_{n=0}^{\infty} \frac{(-1)^n}{n!\Gamma(1+\frac{1}{2}+n)} \left(\frac{x}{2}\right)^{2n+\frac{1}{2}}.$$

Now in view of the property $\Gamma(1+\alpha) = \alpha\Gamma(\alpha)$ and the fact that $\Gamma(\frac{1}{2}) = \sqrt{\pi}$ we obtain

$n = 0$: $\quad \Gamma(1+\frac{1}{2}) = \frac{1}{2}\Gamma(\frac{1}{2}) = \frac{1}{2}\sqrt{\pi}$

$n = 1$: $\quad \Gamma(1+\frac{3}{2}) = \frac{3}{2}\Gamma(\frac{3}{2}) = \frac{3}{2^2}\sqrt{\pi}$

$n = 2$: $\quad \Gamma(1+\frac{5}{2}) = \frac{5}{2}\Gamma(\frac{5}{2}) = \frac{5\cdot3}{2^3}\sqrt{\pi} = \frac{5\cdot4\cdot3\cdot2\cdot1}{2^34\cdot2}\sqrt{\pi} = \frac{5!}{2^52!}\sqrt{\pi}$

$n = 3$: $\quad \Gamma(1+\frac{7}{2}) = \frac{7}{2}\Gamma(\frac{7}{2}) = \frac{7\cdot5!}{2^62!}\sqrt{\pi} = \frac{7\cdot6\cdot5!}{2^6\cdot6\cdot2!}\sqrt{\pi} = \frac{7!}{2^73!}\sqrt{\pi}.$

In general, $\quad \Gamma\left(1+\frac{1}{2}+n\right) = \frac{(2n+1)!}{2^{2n+1}n!}\sqrt{\pi}.$

Hence

$$J_{1/2}(x) = \sum_{n=0}^{\infty} \frac{(-1)^n}{n!\frac{(2n+1)!\sqrt{\pi}}{2^{2n+1}n!}} \left(\frac{x}{2}\right)^{2n+1/2} = \sqrt{\frac{2}{\pi x}}\sum_{n=0}^{\infty}\frac{(-1)^n}{(2n+1)!}x^{2n+1}.$$

Since the series in the last line is the Maclaurin series for sin x, we have shown that

$$J_{1/2}(x) = \sqrt{\frac{2}{\pi x}} \sin x.$$

See Problems 23 and 24 in Exercises 5.3.

■ **Solution of Legendre's Equation** Since $x = 0$ is an ordinary point of the equation, we substitute the power series $y = \sum_{n=0}^{\infty} c_n x^n$, shift summation indices, and combine series to get

$$(1 - x^2)y'' - 2xy' + n(n + 1)y = [n(n + 1)c_0 + 2c_2] + [(n - 1)(n + 2)c_1 + 6c_3]x$$

$$+ \sum_{j=2}^{\infty} [(j + 2)(j + 1)c_{j+2} + (n - j)(n + j + 1)c_j]x^j = 0$$

implies that

$$n(n + 1)c_0 + 2c_2 = 0$$

$$(n - 1)(n + 2)c_1 + 6c_3 = 0$$

$$(j + 2)(j + 1)c_{j+2} + (n - j)(n + j + 1)c_j = 0$$

or

$$c_2 = -\frac{n(n + 1)}{2!} c_0$$

$$c_3 = -\frac{(n - 1)(n + 2)}{3!} c_1$$

$$c_{j+2} = -\frac{(n - j)(n + j + 1)}{(j + 2)(j + 1)} c_j, \qquad j = 2, 3, 4, \dots . \qquad (17)$$

Letting j take on the values 2, 3, 4, . . . , recurrence relation (17) yields.

$$c_4 = -\frac{(n - 2)(n + 3)}{4 \cdot 3} c_2 = \frac{(n - 2)n(n + 1)(n + 3)}{4!} c_0$$

$$c_5 = -\frac{(n - 3)(n + 4)}{5 \cdot 4} c_3 = \frac{(n - 3)(n - 1)(n + 2)(n + 4)}{5!} c_1$$

$$c_6 = -\frac{(n - 4)(n + 5)}{6 \cdot 5} c_4 = -\frac{(n - 4)(n - 2)n(n + 1)(n + 3)(n + 5)}{6!} c_0$$

$$c_7 = -\frac{(n - 5)(n + 6)}{7 \cdot 6} c_5$$

$$= -\frac{(n - 5)(n - 3)(n - 1)(n + 2)(n + 4)(n + 6)}{7!} c_1$$

and so on. Thus for at least $|x| < 1$ we obtain two linearly independent power series solutions:

$$y_1(x) = c_0 \left[1 - \frac{n(n + 1)}{2!} x^2 + \frac{(n - 2)n(n + 1)(n + 3)}{4!} x^4 \right.$$

$$\left. - \frac{(n - 4)(n - 2)n(n + 1)(n + 3)(n + 5)}{6!} x^6 + \cdots \right]$$

$$y_2(x) = c_1 \left[x - \frac{(n - 1)(n + 2)}{3!} x^3 + \frac{(n - 3)(n - 1)(n + 2)(n + 4)}{5!} x^5 \right.$$

$$\left. - \frac{(n - 5)(n - 3)(n - 1)(n + 2)(n + 4)(n + 6)}{7!} x^7 + \cdots \right]. \qquad (18)$$

Notice that if n is an even integer, the first series terminates, whereas $y_2(x)$ is an infinite series. For example, if $n = 4$, then

$$y_1(x) = c_0 \left[1 - \frac{4 \cdot 5}{2!} x^2 + \frac{2 \cdot 4 \cdot 5 \cdot 7}{4!} x^4 \right] = c_0 \left[1 - 10x^2 + \frac{35}{3} x^4 \right].$$

Similarly, when n is an odd integer, the series for $y_2(x)$ terminates with x^n; that is, *when n is a nonnegative integer, we obtain an nth-degree polynomial solution* of Legendre's equation.

Since we know that a constant multiple of a solution of Legendre's equation is also a solution, it is traditional to choose specific values for c_0 or c_1, depending on whether n is an even or odd positive integer, respectively. For $n = 0$ we choose $c_0 = 1$, and for $n = 2, 4, 6, \ldots,$

$$c_0 = (-1)^{n/2} \frac{1 \cdot 3 \cdots (n-1)}{2 \cdot 4 \cdots n};$$

whereas for $n = 1$ we choose $c_1 = 1$, and for $n = 3, 5, 7, \ldots,$

$$c_1 = (-1)^{(n-1)/2} \frac{1 \cdot 3 \cdots n}{2 \cdot 4 \cdots (n-1)}.$$

For example, when $n = 4$ we have

$$y_1(x) = (-1)^{4/2} \frac{1 \cdot 3}{2 \cdot 4} \left[1 - 10x^2 + \frac{35}{3} x^4 \right] = \frac{1}{8} (35x^4 - 30x^2 + 3).$$

■ **Legendre Polynomials** These specific nth-degree polynomial solutions are called **Legendre polynomials** and are denoted by $P_n(x)$. From the series for $y_1(x)$ and $y_2(x)$ and from the above choices of c_0 and c_1 we find that the first several Legendre polynomials are

$$P_0(x) = 1, \qquad\qquad P_1(x) = x,$$

$$P_2(x) = \frac{1}{2} (3x^2 - 1), \qquad\qquad P_3(x) = \frac{1}{2} (5x^3 - 3x), \qquad\qquad (19)$$

$$P_4(x) = \frac{1}{8} (35x^4 - 30x^2 + 3), \qquad P_5(x) = \frac{1}{8} (63x^5 - 70x^3 + 15x).$$

Remember, $P_0(x)$, $P_1(x)$, $P_2(x)$, $P_3(x)$, ..., are, in turn, particular solutions of the differential equations

$$\begin{array}{ll} n = 0: & (1 - x^2)y'' - 2xy' = 0 \\[4pt] n = 1: & (1 - x^2)y'' - 2xy' + 2y = 0 \\[4pt] n = 2: & (1 - x^2)y'' - 2xy' + 6y = 0 \\[4pt] n = 3: & (1 - x^2)y'' - 2xy' + 12y = 0 \end{array} \qquad (20)$$

$$\vdots \qquad\qquad \vdots$$

The graphs, on the interval $-1 \le x \le 1$, of the six Legendre polynomials in (19) are given in Figure 5.5.

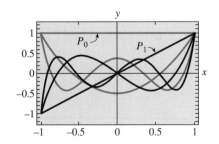

Figure 5.5 Legendre polynomials for $n = 0, 1, 2, 3, 4, 5$

■ **Properties** You are encouraged to verify the following properties using the Legendre polynomials in (19).

$$(i)\ P_n(-x) = (-1)^n P_n(x)$$

$$(ii)\ P_n(1) = 1 \qquad\qquad (iii)\ P_n(-1) = (-1)^n$$

$$(iv)\ P_n(0) = 0, \quad n \text{ odd} \qquad (v)\ P'_n(0) = 0, \quad n \text{ even}$$

Property (*i*) indicates, as is apparent in Figure 5.5, that $P_n(x)$ is an even or odd function according to whether n is even or odd.

■ **Recurrence Relation** Recurrence relations that relate Legendre polynomials of different degrees are also important in some aspects of their applications. We state, without proof, the following three-term recurrence relation

$$(k + 1)P_{k+1}(x) - (2k + 1)xP_k(x) + kP_{k-1}(x) = 0, \tag{21}$$

which is valid for $k = 1, 2, 3, \ldots$. In (19) we listed the first six Legendre polynomials. If, say, we wish to find $P_6(x)$, we can use (21) with $k = 5$. This relation expresses $P_6(x)$ in terms of the known $P_4(x)$ and $P_5(x)$. See Problem 35 in Exercises 5.3.

EXERCISES 5.3

Answers to odd-numbered problems begin on page A-36.

In Problems 1–8, find the general solution of the given differential equation on $(0, \infty)$.

1. $x^2y'' + xy' + (x^2 - \frac{1}{9})y = 0$

2. $x^2y'' + xy' + (x^2 - 1)y = 0$

3. $4x^2y'' + 4xy' + (4x^2 - 25)y = 0$

4. $16x^2y'' + 16xy' + (16x^2 - 1)y = 0$

5. $xy'' + y' + xy = 0$

6. $\dfrac{d}{dx}[xy'] + \left(x - \dfrac{4}{x}\right)y = 0$

7. $x^2y'' + xy' + (9x^2 - 4)y = 0$

8. $x^2y'' + xy' + (36x^2 - \frac{1}{4})y = 0$

9. Use the change of variables $y = x^{-1/2}v(x)$ to find the general solution of

$$x^2y'' + 2xy' + \lambda^2 x^2 y = 0, \qquad x > 0.$$

10. Verify that $y = x^n J_n(x)$ is a particular solution of

$$xy'' + (1 - 2n)y' + xy = 0, \qquad x > 0.$$

11. Verify that $y = x^{-n} J_n(x)$ is a particular solution of

$$xy'' + (1 + 2n)y' + xy = 0, \qquad x > 0.$$

12. Verify that $y = \sqrt{x} J_n(\lambda x)$, $\lambda > 0$, is a particular solution of

$$x^2y'' + (\lambda^2 x^2 - v^2 + \frac{1}{4})y = 0, \qquad x > 0.$$

In Problems 13–18, use the results of Problems 10, 11, and 12 to find a particular solution of the given differential equation on $(0, \infty)$.

13. $y'' + y = 0$

14. $xy'' - y' + xy = 0$

15. $xy'' + 3y' + xy = 0$

16. $4x^2y'' + (16x^2 + 1)y = 0$

17. $x^2y'' + (x^2 - 2)y = 0$

18. $xy'' - 5y' + xy = 0$

19. (a) Proceed as in Example 4 to show that

$$xJ'_v(x) = -vJ_v(x) + xJ_{v-1}(x).$$

[*Hint:* Write $2n + v = 2(n + v) - v$.]

(b) Use the result in part (a) to derive (15).

20. Use the formula obtained in Example 4 along with part (a) of Problem 19 to derive the recurrence relation

$$2vJ_v(x) = xJ_{v+1}(x) + xJ_{v-1}(x).$$

In Problems 21 and 22, use (14) or (15) to obtain the given result.

21. $\displaystyle\int_0^x rJ_0(r)\,dr = xJ_1(x)$

22. $J'_0(x) = J_{-1}(x) = -J_1(x)$

23. Proceed as on page 261 to express $J_{-1/2}(x)$ in terms of $\cos x$ and a power of x.

24. (a) Use Problem 23, the recurrence relation in Problem 20, and $J_{-1/2}(x)$ on page 262 to express $J_{3/2}(x)$, $J_{-3/2}(x)$, and $J_{5/2}(x)$ in terms of $\sin x$, $\cos x$, and powers of x.

(b) Use a graphing utility to plot the graphs of $J_{1/2}(x)$, $J_{-1/2}(x)$, $J_{3/2}(x)$, $J_{-3/2}(x)$, and $J_{5/2}(x)$.

25. Explain how $J'_0(x) = -J_1(x)$ (Problem 22) and Rolle's theorem from calculus can be used to demonstrate that the zeros of $J_0(x)$ and $J_1(x)$ interlace; that is, between any two consecutive zeros of $J_0(x)$, there exists a zero of $J_1(x)$. See Figure 5.3 and Table 5.2.

26. When $v = n$, n a nonnegative integer, the Bessel function of the first kind $J_n(x)$ can be represented by the definite integral: $J_n(x) = \dfrac{1}{\pi} \displaystyle\int_0^\pi \cos\,(x \sin t - nt)\,dt$. How does it follow from this integral that $\left|J_n(x)\right| \le 1$ for all x?

27. Use the change of variables $s = \dfrac{2}{\alpha}\sqrt{\dfrac{k}{m}}\,e^{-\alpha t/2}$ to show that the differential equation of an aging spring $mx'' + ke^{-\alpha t}x = 0$, $\alpha > 0$, becomes

$$s^2 \frac{d^2x}{ds^2} + s\frac{dx}{ds} + s^2 x = 0.$$

28. Computer Lab Assignment (a) Use the general solution given in Example 3 to solve the initial-value problem

$$4x'' + e^{-0.1t}x = 0, \quad x(0) = 1, \quad x'(0) = -\tfrac{1}{2}.$$

Also use $J_0'(x) = -J_1(x)$ and $Y_0'(x) = -Y_1(x)$ along with Table 5.1, or a CAS, to evaluate coefficients.

(b) Use a CAS to plot the graph of the solution obtained in part **(a)** over the interval $0 \le t \le 200$.

29. Show that $y = x^{1/2}w(\tfrac{2}{3}\,\alpha x^{3/2})$ is a solution of Airy's differential equation $y'' + \alpha^2 xy = 0$, $x > 0$, whenever w is a solution of Bessel's equation $t^2 w'' + tw' + (t^2 - \tfrac{1}{9})w = 0$, $t > 0$. [*Hint:* After differentiating, substituting, and simplifying, then let $t = \tfrac{2}{3}\,\alpha x^{3/2}$.]

30. Use the result of Problem 29 to express the general solution of Airy's equation for $x > 0$ in terms of Bessel functions.

31. Computer Lab Assignment (a) Use the general solution obtained in Problem 30 to solve the initial-value problem

$$4x'' + tx = 0, \quad x(0.1) = 1, \quad x'(0.1) = -\tfrac{1}{2}.$$

Use a CAS to evaluate coefficients.

(b) Use a CAS to plot the graph of the solution obtained in part **(a)** over the interval $0 \le t \le 200$.

32. Computer Lab Assignment A uniform thin column, positioned vertically with one end embedded in the ground, will deflect, or bend away, from the vertical under the influence of its own weight when its length is greater than a certain critical height. It can be shown that the angular deflection $\theta(x)$ of the column from the vertical at a point $P(x)$ is a solution of the boundary-value problem:

$$EI\frac{d^2\theta}{dx^2} + \delta g(L - x)\theta = 0, \quad \theta(0) = 0, \quad \theta'(0) = 0,$$

where E is Young's modulus, I is the cross-sectional moment of inertia, δ is the constant linear density, and x is distance along the column measured from its base. See Figure 5.6. The column will bend only if this boundary-value problem has a nontrivial solution.

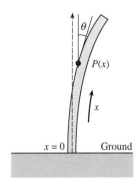

Figure 5.6

(a) First make the change of variables $t = L - x$ and state the resulting boundary-value problem. Then use the result of Problem 30 to express the general solution of the differential equation in terms of Bessel functions.

(b) With the aid of a CAS, find the critical length L of a solid steel rod of radius $r = 0.05$ in, $\delta g = 0.28\,A$ lb/in, $E = 2.6 \times 10^7$ lb/in^2, $A = \pi r^2$, and $I = \tfrac{1}{4}\,\pi r^4$.

33. It can be shown by a change of variables that the general solution of the differential equation $xy'' + \lambda y = 0$ on the interval $(0, \infty)$ is

$$y = c_1\sqrt{x}J_1(2\sqrt{\lambda x}) + c_2\sqrt{x}Y_1(2\sqrt{\lambda x}).$$

Verify by direct substitution that $y = \sqrt{x}J_1(2\sqrt{x})$ is a particular solution of the equation in the case $\lambda = 1$.

34. Computer Lab Assignment In Example 3 of Section 3.9, we saw that when a constant vertical compressive force or load P was applied to a thin column of uniform cross section, the deflection $y(x)$ satisfied the boundary-value problem

$$EI\frac{d^2y}{dx^2} + Py = 0, \quad y(0) = 0, \quad y(L) = 0.$$

(a) If the bending stiffness factor EI is proportional to x, then $EI(x) = kx$, where k is a constant of proportionality. If $EI(L) = kL = M$ is the maximum stiffness factor, then $k = M/L$ and so $EI(x) = Mx/L$. Use the information in Problem 33 to find the solution of

$$M\frac{x}{L}\frac{d^2y}{dx^2} + Py = 0, \quad y(0) = 0, \quad y(L) = 0$$

if it is known that $\sqrt{x}Y_1(2\sqrt{\lambda x})$ is *not* zero at $x = 0$.

(b) Use Table 5.2 to find the Euler load P_1 for the column.

(c) Use a CAS to plot the graph of the first buckling mode $y_1(x)$ corresponding to the Euler load P_1. For simplicity, assume that $c_1 = 1$ and $L = 1$.

35. (a) Use the explicit solutions $y_1(x)$ and $y_2(x)$ of Legendre's equation and the appropriate choice of c_0 and c_1 to find the Legendre polynomials $P_6(x)$ and $P_7(x)$.

(b) Write the differential equations for which $P_6(x)$ and $P_7(x)$ are particular solutions.

36. Computer Lab Assignment (a) For purposes of this problem, ignore the graphs given in Figure 5.5. Use a CAS to plot the graphs of $P_1(x), P_2(x), \ldots, P_7(x)$ on the interval $[-1, 1]$.

(b) Use a root-finding application to find the zeros of $P_1(x), P_2(x), \ldots, P_7(x)$. Check your CAS to see whether the Legendre polynomials are built-in functions. If so, find zeros of Legendre polynomials of higher degree. Form a conjecture about the location of the zeros of *any* Legendre polynomial $P_n(x)$, and then investigate to see whether it is true.

37. Use the recurrence relation (21) and $P_0(x) = 1$, $P_1(x) = x$, to generate the next six Legendre polynomials.

38. Show that the differential equation

$$\sin \theta \frac{d^2y}{d\theta^2} + \cos \theta \frac{dy}{d\theta} + n(n+1)(\sin \theta)y = 0$$

can be transformed into Legendre's equation by means of the substitution $x = \cos \theta$.

1. Suppose the power series $\sum_{k=0}^{\infty} c_k(x-4)^k$ is known to converge at -2 and diverge at 13. Discuss whether the series converges at 10, 7, -7, and 11. Possible answers are: *does, does not, might.*

2. Use the Maclaurin series for $\sin x$ and $\cos x$, along with long division, to find the first three nonzero terms of a power series in x for the function $f(x) = \dfrac{\sin x}{\cos x}$.

In Problems 3 and 4, give the minimum value of the radius R of convergence of a power series solution that is centered at 0. Do not solve.

3. $(x^2 - 2x + 10)y'' + y = 0$ **4.** $(1 - \sin x)y'' + xy = 0$

In Problems 5 and 6, construct a linear second-order differential equation that has the given properties.

5. a regular singular point at $x = 1$ and an irregular singular point at $x = 0$

6. regular singular points at $x = 1$ and at $x = -3$

In Problems 7–12, use an appropriate infinite series method about $x = 0$ to find two solutions of the given differential equation.

7. $2xy'' + y' + y = 0$ **8.** $y'' - xy' - y = 0$

9. $(x-1)y'' + 3y = 0$ **10.** $y'' - x^2y' + xy = 0$

11. $xy'' + (x-6)y' - 3y = 0$ **12.** $(\cos x)y'' + y = 0$

In Problems 13 and 14, solve the given initial-value problem.

13. $y'' + xy' + 2y = 0$, $y(0) = 3$, $y'(0) = -2$

14. $(x+2)y'' + 3y = 0$, $y(0) = 0$, $y'(0) = 1$

In Problems 15 and 16, investigate whether $x = 0$ is an ordinary point, singular point, or irregular singular point of the given differential equation. [*Hint:* Recall the Maclaurin series for $\cos x$ and e^x.]

15. $xy'' + (1 - \cos x)y' + x^2y = 0$ **16.** $(e^x - 1 - x)y'' + xy = 0$

17. Note that $x = 0$ is an ordinary point of the differential equation $y'' + x^2y' + 2xy = 5 - 2x + 10x^3$. Use the assumption $y = \sum_{n=0}^{\infty} c_nx^n$ to find the general solution $y = y_c + y_p$ that consists of three power series centered at $x = 0$.

18. The first-order differential equation $dy/dx = x^2 + y^2$ cannot be solved in terms of elementary functions. However, a solution can be expressed in terms of Bessel functions.

(a) Show that the substitution $y = -\dfrac{1}{u}\dfrac{du}{dx}$ leads to the equation $u'' + x^2u = 0$.

(b) Show that $u = x^{1/2}w(\tfrac{1}{2}x^2)$ is a solution of $u'' + x^2u = 0$ whenever w is a solution of Bessel's equation

$$x^2w'' + xw' + (x^2 - \tfrac{1}{16})w = 0.$$

(c) Use (14) and (15) in Section 5.3 in the forms

$$J_v'(x) = \frac{v}{x}J_v(x) - J_{v+1}(x) \quad \text{and} \quad J_v'(x) = \frac{v}{x}J_v(x) + J_{v-1}(x)$$

as an aid to show that a one-parameter family of solutions of $dy/dx = x^2 + y^2$ is given by

$$y = x\frac{J_{3/4}(\tfrac{1}{2}x^2) - cJ_{-3/4}(\tfrac{1}{2}x^2)}{cJ_{1/4}(\tfrac{1}{2}x^2) + J_{-1/4}(\tfrac{1}{2}x^2)}.$$

19. We know that when $n = 1$, $y_1(x) = x$ is a solution to Legendre's differential equation $(1 - x^2)y'' - 2xy' + 2y = 0$. Show that a second linearly independent solution on the interval $-1 < x < 1$ is

$$y_2(x) = \frac{x}{2} \ln\left(\frac{1+x}{1-x}\right) - 1.$$

20. The Legendre polynomials can be generated by means of **Rodrigues' formula**

$$P_n(x) = \frac{1}{2^n n!}\frac{d^n}{dx^n}(x^2 - 1)^n.$$

Verify the results for $n = 0, 1, 2, 3$.

6

Numerical Solutions of Ordinary Differential Equations

INTRODUCTION

A differential equation does not have to have a solution. But even when a solution exists, we may not be able to exhibit it in explicit or implicit form. In many instances we may have to be content with an approximation to the solution. We have already discussed in Section 2.6 that if a solution exists, it represents a set of points in the Cartesian plane. In this chapter we continue the basic idea of Section 2.6, that is, using the differential equation to construct algorithms to approximate the coordinates of the points on the actual solution curve.

Our concentration in this chapter is primarily on first-order initial-value problems $dy/dx = f(x, y)$, $y(x_0) = y_0$. We have already seen in Section 3.7 the numerical procedures developed for first-order DEs extend in a natural way to systems of first-order DEs, and so we can approximate solutions of higher-order equations by reducing the equation to a system of first-order equations. The chapter concludes with methods for approximating solutions of linear second-order boundary-value problems.

6.1 Euler Methods and Error Analysis

■ **Euler's Method Revisited** In Section 2.6 we examined one of the simplest numerical methods for approximating solutions of first-order initial-value problems $y' = f(x, y)$, $y(x_0) = y_0$. Recall that the backbone of Euler's method was the formula

$$y_{n+1} = y_n + hf(x_n, y_n), \tag{1}$$

where f is the function obtained from the differential equation $y' = f(x, y)$. The recursive use of (1) for $n = 0, 1, 2, \ldots$, yields the y-coordinates y_1, y_2, y_3, \ldots, of points on successive "tangent lines" to the solution curve at x_1, x_2, x_3, \ldots, or $x_n = x_0 + nh$, where h is a constant and is the size of the step between x_n and x_{n+1}. The values y_1, y_2, y_3, \ldots, approximate the values of a solution $y(x)$ of the IVP at x_1, x_2, x_3, \ldots. But whatever advantage (1) has in its simplicity is lost in the crudeness of its approximations. In Problem 4 in Exercise 2.6 you were asked to use Euler's method to obtain the approximate value of $y(1.5)$ for the solution of the initial-value problem $y' = 2xy$, $y(1) = 1$. You should have obtained the analytic solution $y = e^{x^2 - 1}$ and results similar to those given in Tables 6.1 and 6.2.

Table 6.1　$h = 0.1$

x_n	y_n	True value	Abs. error	% rel. error
1.00	1.0000	1.0000	0.0000	0.00
1.10	1.2000	1.2337	0.0337	2.73
1.20	1.4640	1.5527	0.0887	5.71
1.30	1.8154	1.9937	0.1784	8.95
1.40	2.2874	2.6117	0.3244	12.42
1.50	2.9278	3.4903	0.5625	16.12

Table 6.2　$h = 0.05$

x_n	y_n	True value	Abs. error	% rel. error
1.00	1.0000	1.0000	0.0000	0.00
1.05	1.1000	1.1079	0.0079	0.72
1.10	1.2155	1.2337	0.0182	1.47
1.15	1.3492	1.3806	0.0314	2.27
1.20	1.5044	1.5527	0.0483	3.11
1.25	1.6849	1.7551	0.0702	4.00
1.30	1.8955	1.9937	0.0982	4.93
1.35	2.1419	2.2762	0.1343	5.90
1.40	2.4311	2.6117	0.1806	6.92
1.45	2.7714	3.0117	0.2403	7.98
1.50	3.1733	3.4903	0.3171	9.08

In this case, with a step size $h = 0.1$, a 16% relative error in the calculation of the approximation to $y(1.5)$ is totally unacceptable. At the expense of doubling the number of calculations, some improvement in accuracy is obtained by halving the step size to $h = 0.05$.

■ **Errors in Numerical Methods** In choosing and using a numerical method for the solution of an initial-value problem, we must be aware of the various sources of errors. For some kinds of computation the accumulation of errors might reduce the accuracy of an approximation to the point of making the computation useless. On the other hand, depending on the use to which a numerical solution may be put, extreme accuracy may not be worth the added expense and complication.

One source of error always present in calculations is **round-off error.** This error results from the fact that any calculator or computer can represent numbers using only a finite number of digits. Suppose, for the sake of illustration, that we have a calculator that uses base 10 arithmetic and carries four digits, so that ⅓ is represented in the calculator as 0.3333 and ⅑ is represented as 0.1111. If we use this calculator to compute $(x^2 - ⅑)/(x - ⅓)$ for $x = 0.3334$, we obtain

$$\frac{(0.3334)^2 - 0.1111}{0.3334 - 0.3333} = \frac{0.1112 - 0.1111}{0.3334 - 0.3333} = 1.$$

With the help of a little algebra, however, we see that

$$\frac{x^2 - 1/9}{x - 1/3} = \frac{(x - 1/3)(x + 1/3)}{x - 1/3} = x + \frac{1}{3},$$

so that when $x = 0.3334$, $(x^2 - ⅑)/(x - ⅓) \approx 0.3334 + 0.3333 = 0.6667$. This example shows that the effects of round-off error can be quite serious unless some care is taken. One way to reduce the effect of round-off error is to minimize the number of calculations. Another technique on a computer is to use double-precision arithmetic to check the results. In general, round-off error is unpredictable and difficult to analyze, and we will neglect it in the error analysis that follows. We will concentrate on investigating the error introduced by using a formula or algorithm to approximate the values of the solution.

■ **Truncation Errors for Euler's Method** In the sequence of values y_1, y_2, y_3, . . . , generated from (1), usually the value of y_1 will not agree with the actual solution evaluated at x_1, namely, $y(x_1)$, because the algorithm gives only a straight-line approximation to the solution. See Figure 2.29. The error is called the **local truncation error, formula error,** or **discretization error.** It occurs at each step; that is, if we assume that y_n is accurate, then y_{n+1} will contain a local truncation error.

To derive a formula for the local truncation error for Euler's method, we use Taylor's formula with remainder. If a function $y(x)$ possesses $k + 1$ derivatives that are continuous on an open interval containing a and x, then

$$y(x) = y(a) + y'(a)\frac{x - a}{1!} + \cdots + y^{(k)}(a)\frac{(x - a)^k}{k!} + y^{(k+1)}(c)\frac{(x - a)^{k+1}}{(k+1)!},$$

where c is some point between a and x. Setting $k = 1$, $a = x_n$, and $x = x_{n+1} = x_n + h$, we get

$$y(x_{n+1}) = y(x_n) + y'(x_n)\frac{h}{1!} + y''(c)\frac{h^2}{2!} \quad \text{or} \quad y(x_{n+1}) = \underbrace{y_n + hf(x_n, y_n)}_{y_{n+1}} + y''(c)\frac{h^2}{2!}.$$

Euler's method (1) is the last formula without the last term; hence the local truncation error in y_{n+1} is

$$y''(c)\frac{h^2}{2!} \quad \text{where} \quad x_n < c < x_{n+1}.$$

The value of c is usually unknown (it exists theoretically), and so the *exact* error cannot be calculated, but an upper bound on the absolute value of the error is

$$M \frac{h^2}{2!} \quad \text{where} \quad M = \max_{x_n < x < x_{n+1}} |y''(x)|.$$

In discussing errors arising from the use of numerical methods, it is helpful to use the notation $O(h^n)$. To define this concept we let $e(h)$ denote the error in a numerical calculation depending on h. Then $e(h)$ is said to be of order h^n, denoted by $O(h^n)$, if there is a constant C and a positive integer n such that $|e(h)| \leq Ch^n$ for h sufficiently small. Thus the local truncation error for Euler's method is $O(h^2)$. We note that, in general, if $e(h)$ in a numerical method is of order h^n, and h is halved, the new error is approximately $C(h/2)^n = Ch^n/2^n$; that is, the error is reduced by a factor of $1/2^n$.

Example 1 **Bound for Local Truncation Error**

Find a bound for the local truncation errors for Euler's method applied to $y' = 2xy$, $y(1) = 1$.

SOLUTION From the solution $y = e^{x^2-1}$ we get $y'' = (2 + 4x^2)e^{x^2-1}$, and so the local truncation error is

$$y''(c) \frac{h^2}{2} = (2 + 4c^2)e^{(c^2-1)} \frac{h^2}{2},$$

where c is between x_n and $x_n + h$. In particular, for $h = 0.1$ we can get an upper bound on the local truncation error for y_1 by replacing c by 1.1:

$$[2 + (4)(1.1)^2]e^{((1.1)^2-1)} \frac{(0.1)^2}{2} = 0.0422.$$

From Table 6.1 we see that the error after the first step is 0.0337, less than the value given by the bound.

Similarly, we can get a bound for the local truncation error for any of the five steps given in Table 6.1 by replacing c by 1.5 (this value of c gives the largest value of $y''(c)$ for any of the steps and may be too generous for the first few steps). Doing this gives

$$[2 + (4)(1.5)^2]e^{((1.5)^2-1)} \frac{(0.1)^2}{2} = 0.1920 \tag{2}$$

as an upper bound for the local truncation error in each step. ❑

Note in Example 1 that if h is halved to 0.05, then the error bound is 0.0480, about one-fourth as much as shown in (2). This is expected because the local truncation error for Euler's method is $O(h^2)$.

In the above analysis we assumed that the value of y_n was exact in the calculation of y_{n+1}, but it is not because it contains local truncation errors from previous steps. The total error in y_{n+1} is an accumulation of the errors in each of the previous steps. This total error is called the **global truncation error.** A complete analysis of the global truncation error is beyond the scope of this text, but it can be shown that the global truncation error for Euler's method is $O(h)$.

We expect that, for Euler's method, if the step size is halved the error will be approximately halved as well. This is borne out in Tables 6.1 and 6.2, where the absolute error at $x = 1.50$ with $h = 0.1$ is 0.5625 and with $h = 0.05$ is 0.3171, approximately half as large.

In general it can be shown that if a method for the numerical solution of a differential equation has local truncation error $O(h^{\alpha+1})$, then the global truncation error is $O(h^{\alpha})$.

For the remainder of this section and in the subsequent sections we study methods that give significantly greater accuracy than Euler's method.

■ **Improved Euler's Method** The numerical method defined by the formula

$$y_{n+1} = y_n + h\,\frac{f(x_n,\,y_n) + f(x_{n+1},\,y^*_{n+1})}{2}, \tag{3}$$

where

$$y^*_{n+1} = y_n + hf(x_n, y_n), \tag{4}$$

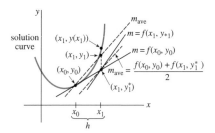

solution curve

Figure 6.1

is commonly known as the **improved Euler's method.** In order to compute y_{n+1} for $n = 0, 1, 2, \ldots$, from (3) we must, at each step, first use Euler's method (4) to obtain an initial estimate y^*_{n+1}. For example, with $n = 0$, (4) would give $y^*_1 = y_0 + hf(x_0, y_0)$, and then knowing this value we use (3) to get $y_1 = y_0 + h(f(x_0, y_0) + f(x_1, y^*_1))/2$, where $x_1 = x_0 + h$. These equations can be readily visualized. In Figure 6.1 observe that $m_0 = f(x_0, y_0)$ and $m_1 = f(x_1, y^*_1)$ are slopes of the solid straight lines shown passing through the points (x_0, y_0) and (x_1, y^*_1), respectively. By taking an average of these slopes, that is, $m_{ave} = (f(x_0, y_0) + f(x_1, y^*_1))/2$, we obtain the slope of the parallel dashed skew lines. With the first step, rather than advancing along the line through (x_0, y_0) with slope $f(x_0, y_0)$ to the point with y-coordinate y^*_1 obtained by Euler's method, we advance instead along the colored dashed line through (x_0, y_0) with slope m_{ave} until we reach x_1. It seems plausible from inspection of the figure that y_1 is an improvement over y^*_1.

In general, the improved Euler's method is an example of a **predictor-corrector method.** The value of y^*_{n+1} given by (4) *predicts* a value of $y(x_n)$, whereas the value of y_{n+1} defined by formula (3) *corrects* this estimate.

Example 2 Improved Euler's Method

Use the improved Euler's method to obtain the approximate value of $y(1.5)$ for the solution of the initial-value problem $y' = 2xy$, $y(1) = 1$. Compare the results for $h = 0.1$ and $h = 0.05$.

SOLUTION With $x_0 = 1$, $y_0 = 1$, $f(x_n, y_n) = 2x_ny_n$, $n = 0$, and $h = 0.1$ we first compute (4):

$$y^*_1 = y_0 + (0.1)(2x_0y_0) = 1 + (0.1)2(1)(1) = 1.2.$$

We use this last value in (3) along with $x_1 = 1 + h = 1 + 0.1 = 1.1$:

$$y_1 = y_0 + (0.1)\,\frac{2x_0y_0 + 2x_1y^*_1}{2} = 1 + (0.1)\,\frac{2(1)(1) + 2(1.1)(1.2)}{2} = 1.232.$$

The comparative values of the calculations for $h = 0.1$ and $h = 0.05$ are given in Tables 6.3 and 6.4, respectively.

272 Chapter 6 Numerical Solutions of Ordinary Differential Equations

Table 6.3 $h = 0.1$

x_n	y_n	True value	Abs. error	% rel. error
1.00	1.0000	1.0000	0.0000	0.00
1.10	1.2320	1.2337	0.0017	0.14
1.20	1.5479	1.5527	0.0048	0.31
1.30	1.9832	1.9937	0.0106	0.53
1.40	2.5908	2.6117	0.0209	0.80
1.50	3.4509	3.4904	0.0394	1.13

Table 6.4 $h = 0.05$

x_n	y_n	True value	Abs. error	% rel. error
1.00	1.0000	1.0000	0.0000	0.00
1.05	1.1077	1.1079	0.0002	0.02
1.10	1.2332	1.2337	0.0004	0.04
1.15	1.3798	1.3806	0.0008	0.06
1.20	1.5514	1.5527	0.0013	0.08
1.25	1.7531	1.7551	0.0020	0.11
1.30	1.9909	1.9937	0.0029	0.14
1.35	2.2721	2.2762	0.0041	0.18
1.40	2.6060	2.6117	0.0057	0.22
1.45	3.0038	3.0117	0.0079	0.26
1.50	3.4795	3.4904	0.0108	0.31

Note that we can't compute all the y_n^ first.* ☞ A brief word of caution is in order here. We cannot compute all the values of y_n^* first and then substitute these values into formula (3). In other words, we cannot use the data in Table 6.1 to help construct the values in Table 6.3. Why not?

■ **Truncation Errors for the Improved Euler's Method** The local truncation error for the improved Euler's method is $O(h^3)$. The derivation of this result is similar to the derivation of the local truncation error for Euler's method. Since the local truncation error for the improved Euler's method is $O(h^3)$, the global truncation error is $O(h^2)$. This can be seen in Example 2; when the step size is halved from $h = 0.1$ to $h = 0.05$, the absolute error at $x = 1.50$ is reduced from 0.0394 to 0.0108, a reduction of approximately $(\frac{1}{2})^2 = \frac{1}{4}$.

EXERCISES 6.1

Answers to odd-numbered problems begin on page A-37.

Given the initial-value problems in Problems 1–10, use the improved Euler's method to obtain a four-decimal approximation to the indicated value. First use $h = 0.1$ and then use $h = 0.05$.

1. $y' = 2x - 3y + 1$, $y(1) = 5$; $y(1.5)$

2. $y' = 4x - 2y$, $y(0) = 2$; $y(0.5)$

3. $y' = 1 + y^2$, $y(0) = 0$; $y(0.5)$

4. $y' = x^2 + y^2$, $y(0) = 1$; $y(0.5)$

5. $y' = e^{-y}$, $y(0) = 0$; $y(0.5)$

6. $y' = x + y^2$, $y(0) = 0$; $y(0.5)$

7. $y' = (x - y)^2$, $y(0) = 0.5$; $y(0.5)$

8. $y' = xy + \sqrt{y}$, $y(0) = 1$; $y(0.5)$

9. $y' = xy^2 - \dfrac{y}{x}$, $y(1) = 1$; $y(1.5)$

10. $y' = y - y^2$, $y(0) = 0.5$; $y(0.5)$

11. Consider the initial-value problem $y' = (x + y - 1)^2$, $y(0) = 2$. Use the improved Euler's method with $h = 0.1$ and $h = 0.05$ to obtain approximate values of the solution at $x = 0.5$. At each step compare the approximate value with the exact value of the analytic solution.

12. Although it may not be obvious from the differential equation, its solution could "behave badly" near a point x at which we wish to approximate $y(x)$. Numerical procedures may give widely differing results near this point. Let $y(x)$ be the solution of the initial-value problem $y' = x^2 + y^3$, $y(1) = 1$.

(a) Use a numerical solver to obtain the graph of the solution on the interval $[1, 1.4]$.

(b) Using the step size $h = 0.1$, compare the results obtained from Euler's method with the results from the improved Euler's method in the approximation of $y(1.4)$.

13. Consider the initial-value problem $y' = 2y$, $y(0) = 1$. The analytic solution is $y = e^{2x}$.

(a) Approximate $y(0.1)$ using one step and Euler's method.

(b) Find a bound for the local truncation error in y_1.

(c) Compare the actual error in y_1 with your error bound.

(d) Approximate $y(0.1)$ using two steps and Euler's method.

(e) Verify that the global truncation error for Euler's method is $O(h)$ by comparing the errors in parts **(a)** and **(d)**.

14. Repeat Problem 13 using the improved Euler's method. Its global truncation error is $O(h^2)$.

15. Repeat Problem 13 using the initial-value problem $y' = x - 2y$, $y(0) = 1$. The analytic solution is $y = \frac{1}{2}x - \frac{1}{4} + \frac{5}{4}e^{-2x}$.

16. Repeat Problem 15 using the improved Euler's method. Its global truncation error is $O(h^2)$.

17. Consider the initial-value problem $y' = 2x - 3y + 1$, $y(1) = 5$. The analytic solution is $y(x) = \frac{1}{9} + \frac{2}{3}x + \frac{38}{9}e^{-3(x-1)}$.

(a) Find a formula involving c and h for the local truncation error in the nth step if Euler's method is used.

(b) Find a bound for the local truncation error in each step if $h = 0.1$ is used to approximate $y(1.5)$.

(c) Approximate $y(1.5)$ using $h = 0.1$ and $h = 0.05$ with Euler's method. See Problem 1 in Exercises 2.6.

(d) Calculate the errors in part **(c)** and verify that the global truncation error of Euler's method is $O(h)$.

18. Repeat Problem 17 using the improved Euler's method, which has a global truncation error $O(h^2)$. See Problem 1. You may need to keep more than four decimal places to see the effect of reducing the order of error.

19. Repeat Problem 17 for the initial-value problem $y' = e^{-y}$, $y(0) = 0$. The analytic solution is $y(x) = \ln(x + 1)$. Approximate $y(0.5)$. See Problem 5 in Exercises 2.6.

20. Repeat Problem 19 using the improved Euler's method, which has global truncation error $O(h^2)$. See Problem 5. You may need to keep more than four decimal places to see the effect of reducing the order of error.

21. Answer the question: Why not? following the three sentences after Example 2 on page 272.

6.2 Runge–Kutta Methods

Probably one of the popular, as well as most accurate, numerical procedures used in obtaining approximate solutions to the initial-value problem $y' = f(x, y)$, $y(x_0) = y_0$, is the **fourth-order Runge–Kutta method.** As the name suggests, there are Runge–Kutta methods of different orders. These methods are derived using the Taylor series expansion with remainder of function $y(x)$.

■ **First-Order Runge–Kutta Method** We have seen in the last section that if a function $y(x)$ possesses $k + 1$ derivatives that are continuous on an open interval containing a and x, then we can write

$$y(x) = y(a) + y'(a) \frac{x - a}{1!} + y''(a) \frac{(x - a)^2}{2!} + \cdots + y^{(k+1)}(c) \frac{(x - a)^{k+1}}{(k+1)!},$$

where c is some number between a and x. If we replace a by x_n and x by $x_{n+1} = x_n + h$, then the foregoing formula becomes

$$y(x_{n+1}) = y(x_n + h) = y(x_n) + hy'(x_n) + \frac{h^2}{2!} y''(x_n) + \cdots + \frac{h^{k+1}}{(k+1)!} y^{(k+1)}(c),$$

where c is now some number between x_n and x_{n+1}. In the case when $k = 1$ and the remainder $(h^2/2)y''(c)$ is small, we obtain the familiar formula

$$y_{n+1} = y_n + hy'_n = y_n + hf(x_n, y_n).$$

In other words, the basic Euler method is a **first-order Runge–Kutta method.**

■ **Second-Order Runge–Kutta Method** We consider now the **second-order Runge–Kutta method.** This consists of finding constants a, b, α, and β so that the formula

$$y_{n+1} = y_n + ak_1 + bk_2, \tag{1}$$

where
$$k_1 = hf(x_n, y_n),$$

$$k_2 = hf(x_n + \alpha h, y_n + \beta k_1)$$

agrees with a Taylor polynomial of degree 2. It can be shown that this can be done whenever the constants satisfy

$$a + b = 1, \quad b\alpha = \frac{1}{2}, \quad \text{and} \quad b\beta = \frac{1}{2}. \tag{2}$$

This is a system of three equations in four unknowns and has infinitely many solutions. Observe that when $a = b = \frac{1}{2}$ and $\alpha = \beta = 1$, (1) reduces to the improved Euler formula. Since the formula agrees with a Taylor polynomial of degree 2, the local truncation error for this method is $O(h^3)$ and the global truncation error is $O(h^2)$.

Notice that the sum $ak_1 + bk_2$ in (1) is a weighted average of k_1 and k_2 since $a + b = 1$. The numbers k_1 and k_2 are multiples of approximations to the slope of the solution curve $y(x)$ at two different points in the interval from x_n to x_{n+1}.

■ **Fourth-Order Runge–Kutta Method** The **fourth-order Runge–Kutta method** consists of finding appropriate constants so that the formula

$$y_{n+1} = y_n + ak_1 + bk_2 + ck_3 + dk_4,$$

where
$$k_1 = hf(x_n, y_n)$$

$$k_2 = hf(x_n + \alpha_1 h, y_n + \beta_1 k_1)$$

$$k_3 = hf(x_n + \alpha_2 h, y_n + \beta_2 k_1 + \beta_3 k_2)$$

$$k_4 = hf(x_n + \alpha_3 h, y_n + \beta_4 k_1 + \beta_5 k_2 + \beta_6 k_3)$$

agrees with a Taylor polynomial of degree 4. This results in 11 equations in 13 unknowns. The most commonly used set of values for the constants yields the following result:

$$y_{n+1} = y_n + \frac{1}{6}(k_1 + 2k_2 + 2k_3 + k_4),$$

$$k_1 = hf(x_n, y_n)$$

$$k_2 = hf(x_n + \tfrac{1}{2}h, y_n + \tfrac{1}{2}k_1)$$

$$k_3 = hf(x_n + \tfrac{1}{2}h, y_n + \tfrac{1}{2}k_2)$$

$$k_4 = hf(x_n + h, y_n + k_3). \tag{3}$$

You are advised to look carefully at the formulas in (3); note that k_2 depends on k_1, k_3 depends on k_2, and k_4 depends on k_3. Also k_2 and k_3 involve approximations to the slope at the midpoint of the interval between x_n and x_{n+1}.

Example 1 Runge–Kutta Method

Use the Runge–Kutta method with $h = 0.1$ to obtain an approximation to $y(1.5)$ for the solution of

$$y' = 2xy, \qquad y(1) = 1.$$

SOLUTION For the sake of illustration let us compute the case when $n = 0$. From (3) we find

$$k_1 = (0.1)f(x_0, y_0) = (0.1)(2x_0y_0) = 0.2$$

$$k_2 = (0.1)f(x_0 + \tfrac{1}{2}(0.1), y_0 + \tfrac{1}{2}(0.2))$$

$$= (0.1)2(x_0 + \tfrac{1}{2}(0.1))(y_0 + \tfrac{1}{2}(0.2)) = 0.231$$

$$k_3 = (0.1)f(x_0 + \tfrac{1}{2}(0.1), y_0 + \tfrac{1}{2}(0.231))$$

$$= (0.1)2(x_0 + \tfrac{1}{2}(0.1))(y_0 + \tfrac{1}{2}(0.231)) = 0.234255$$

$$k_4 = (0.1)f(x_0 + 0.1, y_0 + 0.234255)$$

$$= (0.1)2(x_0 + 0.1)(y_0 + 0.234255) = 0.2715361$$

and therefore

$$y_1 = y_0 + \frac{1}{6}(k_1 + 2k_2 + 2k_3 + k_4)$$

$$= 1 + \frac{1}{6}(0.2 + 2(0.231) + 2(0.234255) + 0.2715361) = 1.23367435.$$

The remaining calculations are summarized in Table 6.7, whose entries are rounded to four decimal places.

Table 6.7 Runge–Kutta Method with $h = 0.1$

x_n	y_n	True value	Abs. error	% rel. error
1.00	1.0000	1.0000	0.0000	0.00
1.10	1.2337	1.2337	0.0000	0.00
1.20	1.5527	1.5527	0.0000	0.00
1.30	1.9937	1.9937	0.0000	0.00
1.40	2.6116	2.6117	0.0001	0.00
1.50	3.4902	3.4904	0.0001	0.00

Inspection of Table 6.7 shows why the fourth-order Runge–Kutta method is so popular. If four-decimal-place accuracy is all that we desire, there is no need to use a smaller step size. Table 6.8 compares the results of applying the Euler, improved Euler, and fourth-order Runge–Kutta methods to the initial-value problem $y' = 2xy$, $y(1) = 1$. (See Tables 6.1 and 6.3.)

Table 6.8 $y' = 2xy,$ $y(1) = 1$

	Comparison of numerical methods with $h = 0.1$					Comparison of numerical methods with $h = 0.05$			
x_n	Euler	Improved Euler	Runge–Kutta	True value	x_n	Euler	Improved Euler	Runge–Kutta	True value
1.00	1.0000	1.0000	1.0000	1.0000	1.00	1.0000	1.0000	1.0000	1.0000
1.10	1.2000	1.2320	1.2337	1.2337	1.05	1.1000	1.1077	1.1079	1.1079
1.20	1.4640	1.5479	1.5527	1.5527	1.10	1.2155	1.2332	1.2337	1.2337
1.30	1.8154	1.9832	1.9937	1.9937	1.15	1.3492	1.3798	1.3806	1.3806
1.40	2.2874	2.5908	2.6116	2.6117	1.20	1.5044	1.5514	1.5527	1.5527
1.50	2.9278	3.4509	3.4902	3.4904	1.25	1.6849	1.7531	1.7551	1.7551
					1.30	1.8955	1.9909	1.9937	1.9937
					1.35	2.1419	2.2721	2.2762	2.2762
					1.40	2.4311	2.6060	2.6117	2.6117
					1.45	2.7714	3.0038	3.0117	3.0117
					1.50	3.1733	3.4795	3.4903	3.4904

■ **Truncation Errors for the Runge–Kutta Method** Since the first equation in (3) agrees with a Taylor polynomial of degree 4, the local truncation error for this method is

$$y^{(5)}(c) \frac{h^5}{5!} \quad \text{or} \quad O(h^5),$$

and the global truncation error is thus $O(h^4)$. It is now obvious why this is called the *fourth-order* Runge–Kutta method.

Example 2 Bound for Local/Global Truncation Errors

Analyze the local and global truncation errors for the fourth-order Runge–Kutta method applied to $y' = 2xy$, $y(1) = 1$.

SOLUTION By differentiating the known solution $y(x) = e^{x^2 - 1}$ we get

$$y^{(5)}(c) \frac{h^5}{5!} = (120c + 160c^3 + 32c^5)e^{c^2 - 1} \frac{h^5}{5!}. \tag{4}$$

Thus with $c = 1.5$, (4) yields a bound of 0.00028 on the local truncation error for each of the five steps when $h = 0.1$. Note that in Table 6.7 the actual error in y_1 is much less than this bound.

 Table 6.9 gives the approximations to the solution of the initial-value problem at $x = 1.5$ that are obtained from the fourth-order Runge–Kutta method. By computing the value of the exact solution at $x = 1.5$ we can find the error in these approximations. Because the method is so accurate, many decimal places must be used in the numerical solution to see the effect of halving the step size. Note that when h is halved, from $h = 0.1$ to $h = 0.05$, the error is divided by a factor of about $2^4 = 16$, as expected.

Table 6.9 Runge–Kutta Method

h	Approximation	Error
0.1	3.49021064	$1.323210889 \times 10^{-4}$
0.05	3.49033382	$9.137760898 \times 10^{-6}$

Adaptive Methods We have seen that the accuracy of a numerical method can be improved by decreasing the step size h. Of course, this enhanced accuracy is usually obtained at a cost—namely, increased computation time and greater possibility of round-off error. In general, over the interval of approximation there may be subintervals where a relatively large step size suffices and other subintervals where a smaller step size is necessary in order to keep the truncation error within a desired limit. Numerical methods that use a variable step size are called **adaptive methods.** One of the more popular of these methods for approximating solutions of differential equations is called the **Runge–Kutta–Fehlberg algorithm.**

E X E R C I S E S 6 . 2 Answers to odd-numbered problems begin on page A-38.

1. Use the fourth-order Runge–Kutta method with $h = 0.1$ to obtain a four-decimal approximation to the solution of the initial-value problem

$$y' = (x + y - 1)^2, \quad y(0) = 2,$$

at $x = 0.5$. Compare the approximate values with the exact values obtained in Problem 11 in Exercises 6.1.

2. Solve the equations in (2) using the assumption $a = \frac{1}{4}$. Use the resulting second-order Runge–Kutta method to obtain a four-decimal place approximation to the solution of the initial-value problem in Problem 1 at $x = 0.5$. Compare the approximate values with the approximate values obtained in Problem 11 in Exercises 6.1.

Given the initial-value problems in Problems 3–12, use the Runge–Kutta method with $h = 0.1$ to obtain a four-decimal approximation to the indicated value.

3. $y' = 2x - 3y + 1, \quad y(1) = 5; \quad y(1.5)$

4. $y' = 4x - 2y, \quad y(0) = 2; \quad y(0.5)$

5. $y' = 1 + y^2, \quad y(0) = 0; \quad y(0.5)$

6. $y' = x^2 + y^2, \quad y(0) = 1; \quad y(0.5)$

7. $y' = e^{-y}, \quad y(0) = 0; \quad y(0.5)$

8. $y' = x + y^2, \quad y(0) = 0; \quad y(0.5)$

9. $y' = (x - y)^2, \quad y(0) = 0.5; \quad y(0.5)$

10. $y' = xy + \sqrt{y}, \quad y(0) = 1; \quad y(0.5)$

11. $y' = xy^2 - \dfrac{y}{x}, \quad y(1) = 1; \quad y(1.5)$

12. $y' = y - y^2, \quad y(0) = 0.5; \quad y(0.5)$

13. If air resistance is proportional to the square of the instantaneous velocity, then the velocity v of a mass m dropped from a height h is determined from

$$m \frac{dv}{dt} = mg - kv^2, \quad k > 0.$$

Let $v(0) = 0$, $k = 0.125$, $m = 5$ slugs, and $g = 32$ ft/s^2.

(a) Use the Runge–Kutta method with $h = 1$ to find an approximation to the velocity of the falling mass at $t = 5$ s.

(b) Use a numerical solver to graph the solution of the initial-value problem.

(c) Use separation of variables to solve the initial-value problem and find the true value $v(5)$.

14. A mathematical model for the area A (in cm^2) that a colony of bacteria (*B. dendroides*) occupies is given by

$$\frac{dA}{dt} = A(2.128 - 0.0432A).*$$

Suppose that the initial area is 0.24 cm^2.

(a) Use the Runge–Kutta method with $h = 0.5$ to complete the following table.

t (days)	1	2	3	4	5
A(observed)	2.78	13.53	36.30	47.50	49.40
A(approximated)					

(b) Use a numerical solver to graph the solution of the initial-value problem. Estimate the values $A(1)$, $A(2)$, $A(3)$, $A(4)$, and $A(5)$ from the graph.

(c) Use separation of variables to solve the initial-value problem and compute the values $A(1)$, $A(2)$, $A(3)$, $A(4)$, and $A(5)$.

15. Consider the initial-value problem

$$y' = x^2 + y^3, \quad y(1) = 1.$$

(See Problem 12 in Exercises 6.1.)

(a) Compare the results obtained from using the Runge–Kutta formula over the interval $[1, 1.4]$ with step sizes $h = 0.1$ and $h = 0.05$.

(b) Use a numerical solver to graph the solution of the initial-value problem on the interval $[1, 1.4]$.

16. Consider the initial-value problem $y' = 2y$, $y(0) = 1$. The analytic solution is $y(x) = e^{2x}$.

* See V. A. Kostitzin, *Mathematical Biology* (London: Harrap, 1939).

(a) Approximate $y(0.1)$ using one step and the fourth-order Runge–Kutta method.

(b) Find a bound for the local truncation error in y_1.

(c) Compare the actual error in y_1 with your error bound.

(d) Approximate $y(0.1)$ using two steps and the fourth-order Runge–Kutta method.

(e) Verify that the global truncation error for the fourth-order Runge–Kutta method is $O(h^4)$ by comparing the errors in parts **(a)** and **(d)**.

17. Repeat Problem 16 using the initial-value problem $y' = -2y + x$, $y(0) = 1$. The analytic solution is $y(x) = \frac{1}{2}x - \frac{1}{4} + \frac{5}{4}e^{-2x}$.

18. Consider the initial-value problem $y' = 2x - 3y + 1$, $y(1) = 5$. The analytic solution is $y(x) = \frac{1}{9} + \frac{2}{3}x + \frac{38}{9}e^{-3(x-1)}$.

(a) Find a formula involving c and h for the local truncation error in the nth step if the fourth-order Runge–Kutta method is used.

(b) Find a bound for the local truncation error in each step if $h = 0.1$ is used to approximate $y(1.5)$.

(c) Approximate $y(1.5)$ using the fourth-order Runge–Kutta method with $h = 0.1$ and $h = 0.05$. See Problem 3. You will need to carry more than six decimal places to see the effect of reducing the step size.

19. Repeat Problem 18 for the initial-value problem $y' = e^{-y}$, $y(0) = 0$. The analytic solution is $y(x) = \ln(x + 1)$. Approximate $y(0.5)$. See Problem 7.

20. Computer Lab Assignment The Runge–Kutta method for solving an initial-value problem over an interval $[a, b]$ results in a finite set of points that are supposed to approximate points on the graph of the exact solution. In order to expand this set of discrete points to an approximate solution defined at *all* points on the interval $[a, b]$, we can use an **interpolating function.** This is a function, supported by most computer algebra systems, that agrees with the given data exactly and assumes a smooth transition between data points. These interpolating functions may be polynomials or sets of polynomials joined together smoothly. In *Mathematica* the command **y = Interpolation[data]** can be used to obtain an interpolating function through the points **data = {{x_0, y_0}, {x_1, y_1}, ..., {x_n, y_n}}**. The interpolating function **y[x]** can now be treated like any other function built into the computer algebra system.

(a) Find the exact solution of the initial-value problem $y' = -y + 10 \sin x$, $y(0) = 0$ on the interval $[0, 2]$. Graph this solution and find its positive roots.

(b) Use the fourth-order Runge–Kutta method with $h = 0.1$ to approximate a solution of the initial-value problem in part **(a)**. Obtain an interpolating function and graph it. Find the positive roots of the interpolating function on the interval $[0, 2]$.

6.3 Multistep Methods

The Euler and Runge–Kutta methods are examples of **single-step** methods. In these methods each successive value y_{n+1} is computed based only on information about the immediately preceding value y_n. A **multistep** or **continuing method,** on the other hand, uses the values from several previously computed steps to obtain the value of y_{n+1}. There are a large number of multistep formulas that can be used to approximate solutions of DEs. Since it is not our intention to survey the vast field of numerical procedures, we will consider only one such method here. This, like the improved Euler's method, is a **predictor-corrector method.** That is, one formula is used to predict a value y^*_{n+1}, which in turn is used to obtain a corrected value y_{n+1}.

■ **Adams–Bashforth/Adams–Moulton Method** One of the most popular multistep methods is the fourth-order **Adams–Bashforth/Adams–Moulton method.** The predictor in this method is the Adams–Bashforth formula

$$y^*_{n+1} = y_n + \frac{h}{24}(55y'_n - 59y'_{n-1} + 37y'_{n-2} - 9y'_{n-3}), \qquad (1)$$

$$y'_n = f(x_n, y_n)$$
$$y'_{n-1} = f(x_{n-1}, y_{n-1})$$
$$y'_{n-2} = f(x_{n-2}, y_{n-2})$$
$$y'_{n-3} = f(x_{n-3}, y_{n-3})$$

for $n \geq 3$. The value of y_{n+1}^* is then substituted into the Adams–Moulton corrector

$$y_{n+1} = y_n + \frac{h}{24}(9y_{n+1}' + 19y_n' - 5y_{n-1}' + y_{n-2}'),$$

$$y_{n+1}' = f(x_{n+1}, y_{n+1}^*). \tag{2}$$

Notice that formula (1) requires that we know the values of y_0, y_1, y_2, and y_3 in order to obtain y_4. The value of y_0 is, of course, the given initial condition. Since the local truncation error of the Adams–Bashforth/Adams–Moulton method is $O(h^5)$, the values of y_1, y_2, and y_3 are generally computed by a method with the same error property, such as the fourth-order Runge–Kutta formula.

Example 1 Adams–Bashforth/Adams–Moulton Method

Use the Adams–Bashforth/Adams–Moulton method with $h = 0.2$ to obtain an approximation to $y(0.8)$ for the solution of

$$y' = x + y - 1, \qquad y(0) = 1.$$

SOLUTION With a step size of $h = 0.2$, $y(0.8)$ will be approximated by y_4. To get started we use the Runge–Kutta method with $x_0 = 0$, $y_0 = 1$, and $h = 0.2$ to obtain

$$y_1 = 1.02140000, \quad y_2 = 1.09181796, \quad y_3 = 1.22210646.$$

Now with the identifications $x_0 = 0$, $x_1 = 0.2$, $x_2 = 0.4$, $x_3 = 0.6$, and $f(x, y) = x + y - 1$, we find

$$y_0' = f(x_0, y_0) = (0) + (1) - 1 = 0$$

$$y_1' = f(x_1, y_1) = (0.2) + (1.02140000) - 1 = 0.22140000$$

$$y_2' = f(x_2, y_2) = (0.4) + (1.09181796) - 1 = 0.49181796$$

$$y_3' = f(x_3, y_3) = (0.6) + (1.22210646) - 1 = 0.82210646.$$

With the foregoing values, the predictor (1) then gives

$$y_4^* = y_3 + \frac{0.2}{24}(55y_3' - 59y_2' + 37y_1' - 9y_0') = 1.42535975.$$

To use the corrector (2) we first need

$$y_4' = f(x_4, y_4^*) = 0.8 + 1.42535975 - 1 = 1.22535975.$$

Finally, (2) yields

$$y_4 = y_3 + \frac{0.2}{24}(9y_4' + 19y_3' - 5y_2' + y_1') = 1.42552788. \qquad \square$$

You should verify that the exact value of $y(0.8)$ in Example 1 is $y(0.8) = 1.42554093$.

■ **Stability of Numerical Methods** An important consideration in using numerical methods to approximate the solution of an initial-value problem is the stability of the method. Simply stated, a numerical method is **stable** if small changes in the initial condition result in only small changes in the computed solution. A numerical method is said to be **unstable** if it is not stable. The reason that stability considerations are important is that in each step after the first step of a numerical

technique we are essentially starting over again with a new initial-value problem, where the initial condition is the approximate solution value computed in the preceding step. Because of the presence of round-off error, this value will almost certainly vary at least slightly from the true value of the solution. Besides round-off error, another common source of error occurs in the initial condition itself; in physical applications the data are often obtained by imprecise measurements.

One possible method for detecting instability in the numerical solution of a specific initial-value problem is to compare the approximate solutions obtained when decreasing step sizes are used. If the numerical method is unstable, the error may actually increase with smaller step sizes. Another way of checking stability is to observe what happens to solutions when the initial condition is slightly perturbed (for example, change $y(0) = 1$ to $y(0) = 0.999$).

For a more detailed and precise discussion of stability, consult a numerical analysis text. In general, all of the methods we have discussed in this chapter have good stability characteristics.

■ **Advantages/Disadvantages of Multistep Methods** Many considerations enter into the choice of a method to solve a differential equation numerically. Single-step methods, particularly the Runge–Kutta method, are often chosen because of their accuracy and the fact that they are easy to program. However, a major drawback is that the right-hand side of the differential equation must be evaluated many times at each step. For instance, the fourth-order Runge–Kutta method requires four function evaluations for each step. On the other hand, if the function evaluations in the previous step have been calculated and stored, a multistep method requires only one new function evaluation for each step. This can lead to great savings in time and expense.

As an example, to solve $y' = f(x, y)$, $y(x_0) = y_0$ numerically using n steps by the fourth-order Runge–Kutta method requires $4n$ function evaluations. The Adams–Bashforth multistep method requires 16 function evaluations for the Runge–Kutta fourth-order starter and $n - 4$ for the n Adams–Bashforth steps, giving a total of $n + 12$ function evaluations for this method. In general the Adams–Bashforth multistep method requires slightly more than a quarter of the number of function evaluations required for the fourth-order Runge–Kutta method. If the evaluation of $f(x, y)$ is complicated, the multistep method will be more efficient.

Another issue involved with multistep methods is how many times the Adams–Moulton corrector formula should be repeated in each step. Each time the corrector is used, another function evaluation is done, and so the accuracy is increased at the expense of losing an advantage of the multistep method. In practice, the corrector is calculated once, and if the value of y_{n+1} is changed by a large amount, the entire problem is restarted using a smaller step size. This is often the basis of the variable step size methods, whose discussion is beyond the scope of this text.

E X E R C I S E S 6 . 3 Answers to odd-numbered problems begin on page A-39.

1. Find the exact solution of the initial-value problem in Example 1. Compare the exact value of $y(0.2)$, $y(0.4)$, $y(0.6)$, and $y(0.8)$ with the approximations y_1, y_2, y_3, and y_4.

2. Write a computer program to implement the Adams–Bashforth/Adams–Moulton method.

In Problems 3 and 4, use the Adams–Bashforth/Adams–Moulton method to approximate $y(0.8)$, where $y(x)$ is the solution of the given initial-value problem. Use $h = 0.2$ and the Runge–Kutta method to compute y_1, y_2, and y_3.

3. $y' = 2x - 3y + 1$, $y(0) = 1$ **4.** $y' = 4x - 2y$, $y(0) = 2$

In Problems 5–8, use the Adams–Bashforth/Adams–Moulton method to approximate $y(1.0)$, where $y(x)$ is the solution of the given initial-value problem. Use $h = 0.2$ and $h = 0.1$ and the Runge–Kutta method to compute y_1, y_2, and y_3.

5. $y' = 1 + y^2$, $\quad y(0) = 0$

6. $y' = y + \cos x$, $\quad y(0) = 1$

7. $y' = (x - y)^2$, $\quad y(0) = 0$

8. $y' = xy + \sqrt{y}$, $\quad y(0) = 1$

6.4 Higher-Order Equations and Systems

In Section 2.6 and the first three sections of the present chapter we focused on numerical techniques that can be used to approximate the solution of a first-order initial-value problem $dy/dx = f(x, y)$, $y(x_0) = y_0$. To approximate the solution of a second-order initial-value problem, we must, as already discussed in Section 3.7, express a second-order DE as a system of two first-order DEs. To do this we write the second-order DE in normal form (see Section 1.1) by solving for y'' in terms of x, y, and y'.

■ **Second-Order Initial-Value Problems** A second-order initial-value problem

$$y'' = f(x, y, y'), \qquad y(x_0) = y_0, \quad y'(x_0) = u_0, \tag{1}$$

can be expressed as an initial-value problem for a system of first-order differential equations. If we let $y' = u$, the differential equation in (1) becomes the system

$$\begin{aligned} y' &= u \\ u' &= f(x, y, u). \end{aligned} \tag{2}$$

Since $y'(x_0) = u(x_0)$, the corresponding initial conditions for (2) are then $y(x_0) = y_0$, $u(x_0) = u_0$. This system of DEs (2) can now be solved numerically by simply applying a particular numerical method to each first-order differential equation in the system. For example, **Euler's method** applied to the system (2) would be

$$\begin{aligned} y_{n+1} &= y_n + hu_n \\ u_{n+1} &= u_n + hf(x_n, y_n, u_n), \end{aligned} \tag{3}$$

whereas the **fourth-order Runge–Kutta method** would be

$$\begin{aligned} y_{n+1} &= y_n + \frac{1}{6}(m_1 + 2m_2 + 2m_3 + m_4) \\ u_{n+1} &= u_n + \frac{1}{6}(k_1 + 2k_2 + 2k_3 + k_4), \end{aligned} \tag{4}$$

where

$$\begin{aligned} m_1 &= hu_n & k_1 &= hf(x_n, y_n, u_n) \\ m_2 &= h(u_n + \tfrac{1}{2}k_1) & k_2 &= hf(x_n + \tfrac{1}{2}h,\ y_n + \tfrac{1}{2}m_1,\ u_n + \tfrac{1}{2}k_1) \\ m_3 &= h(u_n + \tfrac{1}{2}k_2) & k_3 &= hf(x_n + \tfrac{1}{2}h,\ y_n + \tfrac{1}{2}m_2,\ u_n + \tfrac{1}{2}k_2) \\ m_4 &= h(u_n + k_3) & k_4 &= hf(x_n + h,\ y_n + m_3,\ u_n + k_3). \end{aligned}$$

In general, we can express every nth-order ODE $y^{(n)} = f(x, y, y', \ldots, y^{(n-1)})$ as a system of n first-order equations using the substitutions $y = u_1$, $y' = u_2$, $y'' = u_3, \ldots,$ $y^{(n-1)} = u_n$.

(a) Euler's method (color)
Runge–Kutta (black)

(b) Runge–Kutta

Figure 6.2

Example 1 Euler's Method

Use Euler's method to obtain the approximate value of $y(0.2)$, where $y(x)$ is the solution of the initial-value problem

$$y'' + xy' + y = 0, \qquad y(0) = 1, \quad y'(0) = 2. \tag{5}$$

SOLUTION In terms of the substitution $y' = u$, the equation is equivalent to the system

$$y' = u$$
$$u' = -xu - y.$$

Thus from (3) we obtain

$$y_{n+1} = y_n + hu_n$$
$$u_{n+1} = u_n + h[-x_n u_n - y_n].$$

Using the step size $h = 0.1$ and $y_0 = 1$, $u_0 = 2$, we find

$$y_1 = y_0 + (0.1)u_0 = 1 + (0.1)2 = 1.2$$
$$u_1 = u_0 + (0.1)[-x_0 u_0 - y_0] = 2 + (0.1)[-(0)(2) - 1] = 1.9$$
$$y_2 = y_1 + (0.1)u_1 = 1.2 + (0.1)(1.9) = 1.39$$
$$u_2 = u_1 + (0.1)[-x_1 u_1 - y_1] = 1.9 + (0.1)[-(0.1)(1.9) - 1.2] = 1.761.$$

In other words, $y(0.2) \approx 1.39$ and $y'(0.2) \approx 1.761$. ❏

With the aid of the graphing feature of a numerical solver we have compared in Figure 6.2(a) the solution curve of (5) generated by Euler's method ($h = 0.1$) on the interval $[0, 3]$ with the solution curve generated by the fourth-order Runge–Kutta method ($h = 0.1$). From Figure 6.2(b), it would appear that the solution $y(x)$ of (4) has the property that $y(x) \to 0$ as $x \to \infty$.

■ Systems Reduced to First-Order Systems Using a procedure similar to that just discussed, we can often reduce a system of higher-order differential equations to a system of first-order equations by first solving for the highest-order derivative of each dependent variable and then making appropriate substitutions for the lower-order derivatives.

Example 2 A System Rewritten as a First-Order System

Write
$$x'' - x' + 5x + 2y'' = e^t$$
$$-2x + y'' + 2y = 3t^2$$

as a system of first-order differential equations.

SOLUTION Write the system as

$$x'' + 2y'' = e^t - 5x + x'$$
$$y'' = 3t^2 + 2x - 2y$$

and then eliminate y'' by multiplying the second equation by 2 and subtracting. This gives

$$x'' = -9x + 4y + x' + e^t - 6t^2.$$

Since the second equation of the system already expresses the highest-order derivative of y in terms of the remaining functions, we are now in a position to introduce new variables. If we let $x' = u$ and $y' = v$, the expressions for x'' and y'' become, respectively,

$$u' = x'' = -9x + 4y + u + e^t - 6t^2$$

$$v' = y'' = 2x - 2y + 3t^2.$$

The original system can then be written in the form

$$x' = u$$

$$y' = v$$

$$u' = -9x + 4y + u + e^t - 6t^2$$

$$v' = 2x - 2y + 3t^2.$$ ❑

It may not always be possible to carry out the reductions illustrated in Example 2.

■ **Numerical Solution of a System** The solution of a system of the form

$$\frac{dx_1}{dt} = f_1(t, x_1, x_2, \cdots, x_n)$$

$$\frac{dx_2}{dt} = f_2(t, x_1, x_2, \cdots, x_n)$$

$$\vdots \qquad \vdots$$

$$\frac{dx_n}{dt} = f_n(t, x_1, x_2, \cdots, x_n)$$

can be approximated by a version of the Euler, Runge–Kutta, or Adams–Bashforth/Adams–Moulton method adapted to the system. For example, the **fourth-order Runge–Kutta method** applied to the system

$$x' = f(t, x, y)$$
$$y' = g(t, x, y) \tag{6}$$
$$x(t_0) = x_0, \qquad y(t_0) = y_0$$

looks like this:

$$x_{n+1} = x_n + \frac{1}{6}(m_1 + 2m_2 + 2m_3 + m_4)$$

$$\tag{7}$$

$$y_{n+1} = y_n + \frac{1}{6}(k_1 + 2k_2 + 2k_3 + k_4),$$

where

$$
\begin{aligned}
m_1 &= hf(t_n, x_n, y_n) & k_1 &= hg(t_n, x_n, y_n) \\
m_2 &= hf(t_n + \tfrac{1}{2}h, x_n + \tfrac{1}{2}m_1, y_n + \tfrac{1}{2}k_1) & k_2 &= hg(t_n + \tfrac{1}{2}h, x_n + \tfrac{1}{2}m_1, y_n + \tfrac{1}{2}k_1) \\
m_3 &= hf(t_n + \tfrac{1}{2}h, x_n + \tfrac{1}{2}m_2, y_n + \tfrac{1}{2}k_2) & k_3 &= hg(t_n + \tfrac{1}{2}h, x_n + \tfrac{1}{2}m_2, y_n + \tfrac{1}{2}k_2) \\
m_4 &= hf(t_n + h, x_n + m_3, y_n + k_3) & k_4 &= hg(t_n + h, x_n + m_3, y_n + k_3).
\end{aligned}
\tag{8}
$$

Example 3 Runge–Kutta Method

Consider the initial-value problem

$$x' = 2x + 4y$$

$$y' = -x + 6y$$

$$x(0) = -1, \qquad y(0) = 6.$$

Use the fourth-order Runge–Kutta method to approximate $x(0.6)$ and $y(0.6)$. Compare the results for $h = 0.2$ and $h = 0.1$.

SOLUTION We illustrate the computations of x_1 and y_1 with the step size $h = 0.2$. With the identifications $f(t, x, y) = 2x + 4y$, $g(t, x, y) = -x + 6y$, $t_0 = 0$, $x_0 = -1$, and $y_0 = 6$, we see from (8) that

$$m_1 = hf(t_0, x_0, y_0) = 0.2f(0, -1, 6) = 0.2[2(-1) + 4(6)] = 4.4000$$

$$k_1 = hg(t_0, x_0, y_0) = 0.2g(0, -1, 6) = 0.2[-1(-1) + 6(6)] = 7.4000$$

$$m_2 = hf(t_0 + \tfrac{1}{2}h, x_0 + \tfrac{1}{2}m_1, y_0 + \tfrac{1}{2}k_1) = 0.2f(0.1, 1.2, 9.7) = 8.2400$$

$$k_2 = hg(t_0 + \tfrac{1}{2}h, x_0 + \tfrac{1}{2}m_1, y_0 + \tfrac{1}{2}k_1) = 0.2g(0.1, 1.2, 9.7) = 11.4000$$

$$m_3 = hf(t_0 + \tfrac{1}{2}h, x_0 + \tfrac{1}{2}m_2, y_0 + \tfrac{1}{2}k_2) = 0.2f(0.1, 3.12, 11.7) = 10.6080$$

$$k_3 = hg(t_0 + \tfrac{1}{2}h, x_0 + \tfrac{1}{2}m_2, y_0 + \tfrac{1}{2}k_2) = 0.2g(0.1, 3.12, 11.7) = 13.4160$$

$$m_4 = hf(t_0 + h, x_0 + m_3, y_0 + k_3) = 0.2f(0.2, 8, 20.216) = 19.3760$$

$$k_4 = hg(t_0 + h, x_0 + m_3, y_0 + k_3) = 0.2g(0.2, 8, 20.216) = 21.3776.$$

Therefore from (7) we get

$$x_1 = x_0 + \frac{1}{6}(m_1 + 2m_2 + 2m_3 + m_4)$$

$$= -1 + \frac{1}{6}(4.4 + 2(8.24) + 2(10.608) + 19.3760) = 9.2453$$

$$y_1 = y_0 + \frac{1}{6}(k_1 + 2k_2 + 2k_3 + k_4)$$

$$= 6 + \frac{1}{6}(7.4 + 2(11.4) + 2(13.416) + 21.3776) = 19.0683,$$

where, as usual, the computed values are rounded to four decimal places. These numbers give us the approximations $x_1 \approx x(0.2)$ and $y_1 \approx y(0.2)$. The subsequent values, obtained with the aid of a computer, are summarized in Tables 6.10 and 6.11.

Table 6.10 Runge–Kutta Method with $h = 0.2$

m_1	m_2	m_3	m_4	k_1	k_2	k_3	k_4	t_n	x_n	y_n
								0.00	-1.0000	6.0000
4.4000	8.2400	10.6080	19.3760	7.4000	11.4000	13.4160	21.3776	0.20	9.2453	19.0683
18.9527	31.1564	37.8870	63.6848	21.0329	31.7573	36.9716	57.8214	0.40	46.0327	55.1203
62.5093	97.7863	116.0063	187.3669	56.9378	84.8495	98.0688	151.4191	0.60	158.9430	150.8192

Table 6.11 Runge–Kutta Method with $h = 0.1$

m_1	m_2	m_3	m_4	k_1	k_2	k_3	k_4	t_n	x_n	y_n
								0.00	−1.0000	6.0000
2.2000	3.1600	3.4560	4.8720	3.7000	4.7000	4.9520	6.3256	0.10	2.3840	10.8883
4.8321	6.5742	7.0778	9.5870	6.2946	7.9413	8.3482	10.5957	0.20	9.3379	19.1332
9.5208	12.5821	13.4258	17.7609	10.5461	13.2339	13.8872	17.5358	0.30	22.5541	32.8539
17.6524	22.9090	24.3055	31.6554	17.4569	21.8114	22.8549	28.7393	0.40	46.5103	55.4420
31.4788	40.3496	42.6387	54.9202	28.6141	35.6245	37.2840	46.7207	0.50	88.5729	93.3006
54.6348	69.4029	73.1247	93.4107	46.5231	57.7482	60.3774	75.4370	0.60	160.7563	152.0025

Figure 6.3

You should verify that the solution of the initial-value problem in Example 3 is given by $x(t) = (26t - 1)e^{4t}$, $y(t) = (13t + 6)e^{4t}$. From these equations we see that the exact values are $x(0.6) = 160.9384$ and $y(0.6) = 152.1198$. The graph of the solution in a neighborhood of $t = 0$, shown in Figure 6.3, was obtained from a numerical solver using the Runge–Kutta method with $h = 0.1$.

In conclusion, we state **Euler's method** for the general system (4):

$$x_{n+1} = x_n + hf(t_n, x_n, y_n)$$

$$y_{n+1} = y_n + hg(t_n, x_n, y_n).$$

E X E R C I S E S 6 . 4

Answers to odd-numbered problems begin on page A-39.

1. Use Euler's method to approximate $y(0.2)$, where $y(x)$ is the solution of the initial-value problem

$$y'' - 4y' + 4y = 0, \quad y(0) = -2, \quad y'(0) = 1.$$

Use $h = 0.1$. Find the exact solution of the problem, and compare the exact value of $y(0.2)$ with y_2.

2. Use Euler's method to approximate $y(1.2)$, where $y(x)$ is the solution of the initial-value problem

$$x^2 y'' - 2xy' + 2y = 0, \quad y(1) = 4, \quad y'(1) = 9,$$

where $x > 0$. Use $h = 0.1$. Find the exact solution of the problem, and compare the exact value of $y(1.2)$ with y_2.

3. Repeat Problem 1 using the Runge–Kutta method with $h = 0.2$ and $h = 0.1$.

4. Repeat Problem 2 using the Runge–Kutta method with $h = 0.2$ and $h = 0.1$.

5. Use the Runge–Kutta method to obtain the approximate value of $y(0.2)$, where $y(x)$ is a solution of the initial-value problem

$$y'' - 2y' + 2y = e^t \cos t, \quad y(0) = 1, \quad y'(0) = 2.$$

Use $h = 0.2$ and $h = 0.1$.

6. When $E = 100$ V, $R = 10$ Ω, and $L = 1$ h, the system of differential equations for the currents $i_1(t)$ and $i_3(t)$ in the electrical network given in Figure 6.4 is

$$\frac{di_1}{dt} = -20i_1 + 10i_3 + 100$$

$$\frac{di_3}{dt} = 10i_1 - 20i_3,$$

Figure 6.4

where $i_1(0) = 0$ and $i_3(0) = 0$. Use the Runge–Kutta method to approximate $i_1(t)$ and $i_3(t)$ at $t = 0.1, 0.2, 0.3, 0.4$, and 0.5. Use $h = 0.1$. Use a numerical solver to obtain the graph

of the solution $i_1(t)$, $i_3(t)$ on the interval $0 \le t \le 5$. Use their graphs to predict the behavior of $i_1(t)$ and $i_3(t)$ as $t \to \infty$.

In Problems 7–12, use the Runge–Kutta method to approximate $x(0.2)$ and $y(0.2)$. Compare the results for $h = 0.2$ and $h = 0.1$. Use a numerical solver and $h = 0.1$ to obtain the graph of the solution in a neighborhood of $t = 0$.

7. $x' = 2x - y$

$y' = x$

$x(0) = 6, \quad y(0) = 2$

8. $x' = x + 2y$

$y' = 4x + 3y$

$x(0) = 1, \quad y(0) = 1$

9. $x' = -y + t$

$y' = x - t$

$x(0) = -3, \quad y(0) = 5$

10. $x' = 6x + y + 6t$

$y' = 4x + 3y - 10t + 4$

$x(0) = 0.5, \quad y(0) = 0.2$

11. $x' + 4x - y' = 7t$

$x' + y' - 2y = 3t$

$x(0) = 1, \quad y(0) = -2$

12. $x' + y' = 4t$

$-x' + y' + y = 6t^2 + 10$

$x(0) = 3, \quad y(0) = -1$

6.5 Second-Order Boundary-Value Problems

In Sections 6.1–6.3 we considered techniques that yield an approximation to a solution of a first-order initial-value problem $y' = f(x, y)$, $y(x_0) = y_0$. In addition, we just saw in Section 6.4 that we can adapt these approximation techniques to a *second-order initial-value problem* $y'' = f(x, y, y')$, $y(x_0) = y_0$, $y'(x_0) = u_0$ by expressing the second-order DE as a system of first-order DEs. We are now going to examine two methods for approximating a solution to a *second-order boundary-value problem* $y'' = f(x, y, y')$, $y(a) = \alpha$, $y(b) = \beta$. Unlike second-order initial-value problems, these methods for second-order boundary-value problems do not require writing the differential equation as a system.

■ **Finite Difference Approximations** The Taylor series expansion centered at a point a of a function $y(x)$ is

$$y(x) = y(a) + y'(a)\frac{x-a}{1!} + y''(a)\frac{(x-a)^2}{2!} + y'''(a)\frac{(x-a)^3}{3!} + \cdots.$$

If we set $h = x - a$, then the preceding line is the same as

$$y(x) = y(a) + y'(a)\frac{h}{1!} + y''(a)\frac{h^2}{2!} + y'''(a)\frac{h^3}{3!} + \cdots.$$

For the subsequent discussion it is convenient, then, to rewrite this last expression in two alternative forms:

$$y(x + h) = y(x) + y'(x)h + y''(x)\frac{h^2}{2} + y'''(x)\frac{h^3}{6} + \cdots \tag{1}$$

and

$$y(x - h) = y(x) - y'(x)h + y''(x)\frac{h^2}{2} - y'''(x)\frac{h^3}{6} + \cdots. \tag{2}$$

If h is small, we can ignore terms involving h^4, h^5, . . . , since these values are negligible. Indeed, if we ignore all terms involving h^2 and higher, then (1) and (2) yield, in turn, the following approximations for the first derivative $y'(x)$:

$$y'(x) \approx \frac{1}{h}\,[y(x+h) - y(x)] \tag{3}$$

$$y'(x) \approx \frac{1}{h}\,[y(x) - y(x-h)]. \tag{4}$$

Subtracting (1) and (2) also gives

$$y'(x) \approx \frac{1}{2h}\,[y(x+h) - y(x-h)]. \tag{5}$$

On the other hand, if we ignore terms involving h^3 and higher, then by adding (1) and (2) we obtain an approximation for the second derivative $y''(x)$:

$$y''(x) \approx \frac{1}{h^2}\,[y(x+h) - 2y(x) + y(x-h)]. \tag{6}$$

The right sides of (3), (4), (5), and (6) are called **difference quotients.** The expressions

$$y(x+h) - y(x), \quad y(x) - y(x-h), \quad y(x+h) - y(x-h)$$

and

$$y(x+h) - 2y(x) + y(x-h)$$

are called **finite differences.** Specifically, $y(x+h) - y(x)$ is called a **forward difference,** $y(x) - y(x-h)$ is a **backward difference,** and both $y(x+h) - y(x-h)$ and $y(x+h) - 2y(x) + y(x-h)$ are called **central differences.** The results given in (5) and (6) are referred to as **central difference approximations** for the derivatives y' and y''.

■ **Finite Difference Method** Consider now a linear second-order boundary-value problem

$$y'' + P(x)y' + Q(x)y = f(x), \qquad y(a) = \alpha, \quad y(b) = \beta. \tag{7}$$

Suppose $a = x_0 < x_1 < x_2 < \cdots < x_{n-1} < x_n = b$ represents a regular partition of the interval $[a, b]$; that is, $x_i = a + ih$, where $i = 0, 1, 2, \ldots, n$ and $h = (b - a)/n$. The points

$$x_1 = a + h, \quad x_2 = a + 2h, \ldots, \quad x_{n-1} = a + (n-1)h,$$

are called **interior mesh points** of the interval $[a, b]$. If we let

$$y_i = y(x_i), \quad P_i = P(x_i), \quad Q_i = Q(x_i), \quad \text{and} \quad f_i = f(x_i)$$

and if y'' and y' in (7) are replaced by the central difference approximations (5) and (6), we get

$$\frac{y_{i+1} - 2y_i + y_{i-1}}{h^2} + P_i\,\frac{y_{i+1} - y_{i-1}}{2h} + Q_i y_i = f_i$$

or, after simplifying,

$$\left(1 + \frac{h}{2}\,P_i\right)y_{i+1} + (-2 + h^2 Q_i)y_i + \left(1 - \frac{h}{2}\,P_i\right)y_{i-1} = h^2 f_i. \tag{8}$$

The last equation, known as a **finite difference equation,** is an approximation to the differential equation. It enables us to approximate the solution $y(x)$ of (7) at the

interior mesh points $x_1, x_2, \ldots, x_{n-1}$ of the interval $[a, b]$. By letting i take on the values $1, 2, \ldots, n-1$ in (8), we obtain $n-1$ equations in the $n-1$ unknowns y_1, y_2, \ldots, y_{n-1}. Bear in mind that we know y_0 and y_n, since these are the prescribed boundary conditions $y_0 = y(x_0) = y(a) = \alpha$ and $y_n = y(x_n) = y(b) = \beta$.

In Example 1 we consider a boundary-value problem for which we can compare the approximate values found with the exact values of an explicit solution.

Example 1 Using the Finite Difference Method

Use the difference equation (8) with $n = 4$ to approximate the solution of the boundary-value problem

$$y'' - 4y = 0, \qquad y(0) = 0, \quad y(1) = 5.$$

SOLUTION To use (8) we identify $P(x) = 0$, $Q(x) = -4$, $f(x) = 0$, and $h = (1 - 0)/4 = \frac{1}{4}$. Hence the difference equation is

$$y_{i+1} - 2.25y_i + y_{i-1} = 0. \tag{9}$$

Now the interior points are $x_1 = 0 + \frac{1}{4}$, $x_2 = 0 + \frac{2}{4}$, $x_3 = 0 + \frac{3}{4}$, and so for $i = 1, 2$, and 3, (9) yields the following system for the corresponding y_1, y_2, and y_3:

$$y_2 - 2.25y_1 + y_0 = 0$$
$$y_3 - 2.25y_2 + y_1 = 0$$
$$y_4 - 2.25y_3 + y_2 = 0.$$

With the boundary conditions $y_0 = 0$ and $y_4 = 5$, the foregoing system becomes

$$-2.25y_1 + \quad y_2 \qquad\qquad = 0$$
$$y_1 - 2.25y_2 + \quad y_3 = 0$$
$$y_2 - 2.25y_3 = -5.$$

Solving the system gives $y_1 = 0.7256$, $y_2 = 1.6327$, and $y_3 = 2.9479$.

Now the general solution of the given differential equation is $y = c_1 \cosh 2x + c_2 \sinh 2x$. The condition $y(0) = 0$ implies $c_1 = 0$. The other boundary condition gives c_2. In this way we see that an explicit solution of the boundary-value problem is $y(x) = (5 \sinh 2x)/\sinh 2$.

Thus the exact values (rounded to four decimal places) of this solution at the interior points are as follows: $y(0.25) = 0.7184$, $y(0.5) = 1.6201$, and $y(0.75) = 2.9354$. ❑

The accuracy of the approximations in Example 1 can be improved by using a smaller value of h. Of course, the trade-off here is that a smaller value of h necessitates solving a larger system of equations. It is left as an exercise to show that with $h = \frac{1}{8}$, approximations to $y(0.25)$, $y(0.5)$, and $y(0.75)$ are 0.7202, 1.6233, and 2.9386, respectively. See Problem 11 in Exercises 6.5.

Example 2 Using the Finite Difference Method

Use the difference equation (8) with $n = 10$ to approximate the solution of

$$y'' + 3y' + 2y = 4x^2, \qquad y(1) = 1, \quad y(2) = 6.$$

SOLUTION In this case we identify $P(x) = 3$, $Q(x) = 2$, $f(x) = 4x^2$, and $h = (2 - 1)/10 = 0.1$, and so (8) becomes

$$1.15y_{i+1} - 1.98y_i + 0.85y_{i-1} = 0.04x_i^2. \tag{10}$$

Now the interior points are $x_1 = 1.1$, $x_2 = 1.2$, $x_3 = 1.3$, $x_4 = 1.4$, $x_5 = 1.5$, $x_6 = 1.6$, $x_7 = 1.7$, $x_8 = 1.8$, and $x_9 = 1.9$. For $i = 1, 2, \ldots, 9$ and $y_0 = 1$, $y_{10} = 6$, (10) gives a system of nine equations and nine unknowns:

$$1.15y_2 - 1.98y_1 \qquad\qquad = -0.8016$$
$$1.15y_3 - 1.98y_2 + 0.85y_1 = 0.0576$$
$$1.15y_4 - 1.98y_3 + 0.85y_2 = 0.0676$$
$$1.15y_5 - 1.98y_4 + 0.85y_3 = 0.0784$$
$$1.15y_6 - 1.98y_5 + 0.85y_4 = 0.0900$$
$$1.15y_7 - 1.98y_6 + 0.85y_5 = 0.1024$$
$$1.15y_8 - 1.98y_7 + 0.85y_6 = 0.1156$$
$$1.15y_9 - 1.98y_8 + 0.85y_7 = 0.1296$$
$$-1.98y_9 + 0.85y_8 = -6.7556.$$

We can solve this large system using Gaussian elimination or, with relative ease, by means of a computer algebra system. The result is found to be $y_1 = 2.4047$, $y_2 = 3.4432$, $y_3 = 4.2010$, $y_4 = 4.7469$, $y_5 = 5.1359$, $y_6 = 5.4124$, $y_7 = 5.6117$, $y_8 = 5.7620$, and $y_9 = 5.8855$. ❏

■ **Shooting Method** Another way of approximating a solution of a boundary-value problem $y'' = f(x, y, y')$, $y(a) = \alpha$, $y(b) = \beta$, is called the **shooting method.** The starting point in this method is the replacement of the boundary-value problem by an initial-value problem

$$y'' = f(x, y, y'), \qquad y(a) = \alpha, \quad y'(a) = m_1. \tag{11}$$

The number m_1 in (11) is simply a guess for the unknown slope of the solution curve at the known point $(a, y(a))$. We then apply one of the step-by-step numerical techniques to the second-order equation in (11) to find an approximation β_1 for the value of $y(b)$. If β_1 agrees with the given value $y(b) = \beta$ to some preassigned tolerance, we stop; otherwise the calculations are repeated, starting with a different guess $y'(a) = m_2$ to obtain a second approximation β_2 for $y(b)$. This method can be continued in a trial-and-error manner or the subsequent slopes m_3, m_4, . . . , can be adjusted in some systematic way; linear interpolation is particularly successful when the differential equation in (11) is linear. The procedure is analogous to shooting (the "aim" is the choice of the initial slope) at a target until the bull's-eye $y(b)$ is hit. See Problem 14 in Exercises 6.5.

Of course, underlying the use of these numerical methods is the assumption, which we know is not always warranted, that a solution of the boundary-value problem exists.

Remarks

The approximation method using finite differences can be extended to boundary-value problems in which the first derivative is specified at a boundary—for example, a problem such as $y'' = f(x, y, y')$, $y'(a) = \alpha$, $y(b) = \beta$. See Problem 13 in Exercises 6.5.

Answers to odd-numbered problems begin on page A-39.

In Problems 1–10, use the finite difference method and the indicated value of n to approximate the solution of the given boundary-value problem.

1. $y'' + 9y = 0$, $y(0) = 4$, $y(2) = 1$; $n = 4$

2. $y'' - y = x^2$, $y(0) = 0$, $y(1) = 0$; $n = 4$

3. $y'' + 2y' + y = 5x$, $y(0) = 0$, $y(1) = 0$; $n = 5$

4. $y'' - 10y' + 25y = 1$, $y(0) = 1$, $y(1) = 0$; $n = 5$

5. $y'' - 4y' + 4y = (x + 1)e^{2x}$, $y(0) = 3$, $y(1) = 0$; $n = 6$

6. $y'' + 5y' = 4\sqrt{x}$, $y(1) = 1$, $y(2) = -1$; $n = 6$

7. $x^2 y'' + 3xy' + 3y = 0$, $y(1) = 5$, $y(2) = 0$; $n = 8$

8. $x^2 y'' - xy' + y = \ln x$, $y(1) = 0$, $y(2) = -2$; $n = 8$

9. $y'' + (1 - x)y' + xy = x$, $y(0) = 0$, $y(1) = 2$; $n = 10$

10. $y'' + xy' + y = x$, $y(0) = 1$, $y(1) = 0$; $n = 10$

11. Rework Example 1 using $n = 8$.

12. The electrostatic potential u between two concentric spheres of radius $r = 1$ and $r = 4$ is determined from

$$\frac{d^2 u}{dr^2} + \frac{2}{r}\frac{du}{dr} = 0, \quad u(1) = 50, \quad u(4) = 100.$$

Use the method of this section with $n = 6$ to approximate the solution of this boundary-value problem.

13. Consider the boundary-value problem

$$y'' + xy = 0, \quad y'(0) = 1, \quad y(1) = -1.$$

(a) Find the difference equation corresponding to the differential equation. Show that for $i = 0, 1, 2, \ldots, n - 1$, the difference equation yields n equations in $n + 1$ unknowns $y_{-1}, y_0, y_1, y_2, \ldots, y_{n-1}$. Here y_{-1} and y_0 are unknowns since y_{-1} represents an approximation to y at the exterior point $x = -h$ and y_0 is not specified at $x = 0$.

(b) Use the central difference approximation (5) to show that $y_1 - y_{-1} = 2h$. Use this equation to eliminate y_{-1} from the system in part **(a)**.

(c) Use $n = 5$ and the system of equations found in parts **(a)** and **(b)** to approximate the solution of the original boundary-value problem.

14. Computer Lab Assignment Consider the boundary-value problem $y'' = y' - \sin(xy)$, $y(0) = 1$, $y(1) = 1.5$. Use the shooting method to approximate the solution of this problem. (The actual approximation can be obtained using a numerical technique, say, the fourth-order Runge–Kutta method with $h = 0.1$; even better, if you have access to a CAS, such as *Mathematica* or *Maple*, the **NDSolve** function can be used.)

Answers to odd-numbered problems begin on page A-40.

In Problems 1–4, construct a table comparing the indicated values of $y(x)$ using Euler's method, the improved Euler's method, and the Runge–Kutta method. Compute to four rounded decimal places. Use $h = 0.1$ and $h = 0.05$.

1. $y' = 2 \ln xy$, $y(1) = 2$;

$y(1.1), y(1.2), y(1.3), y(1.4), y(1.5)$

2. $y' = \sin x^2 + \cos y^2$, $y(0) = 0$;

$y(0.1), y(0.2), y(0.3), y(0.4), y(0.5)$

3. $y' = \sqrt{x + y}$, $y(0.5) = 0.5$;

$y(0.6), y(0.7), y(0.8), y(0.9), y(1.0)$

4. $y' = xy + y^2$, $y(1) = 1$;

$y(1.1), y(1.2), y(1.3), y(1.4), y(1.5)$

5. Use Euler's method to obtain the approximate value of $y(0.2)$, where $y(x)$ is the solution of the initial-value problem $y'' - (2x + 1)y = 1$, $y(0) = 3$, $y'(0) = 1$. First use one

step with $h = 0.2$, and then repeat the calculations using two steps with $h = 0.1$.

6. Use the Adams–Bashforth/Adams–Moulton method to approximate the value of $y(0.4)$, where $y(x)$ is the solution of the initial-value problem $y' = 4x - 2y$, $y(0) = 2$. Use the Runge–Kutta method and $h = 0.1$ to obtain the values of y_1, y_2, and y_3.

7. Use Euler's method with $h = 0.1$ to approximate values of $x(0.2)$ and $y(0.2)$, where $x(t)$, $y(t)$ is the solution of the initial-value problem

$$x' = x + y$$
$$y' = x - y$$
$$x(0) = 1, \quad y(0) = 2.$$

8. Use the finite difference method with $n = 10$ to approximate the solution of the boundary-value problem

$$y'' + 6.55(1 + x)y = 1, \quad y(0) = 0, \quad y(1) = 0.$$

PART

2

Vectors, Matrices, and Vector Calculus

Application *Essay*

A Matrix Model for Environmental Life Cycle Assessment

Arpad Horvath
Carnegie Mellon University

Life cycle assessment (LCA) is an important tool for designing with environmental impact in mind, or designing for environment. Selecting product designs, materials, processes, reuse or recycle strategies, and options for final disposal should rely on a careful examination of the energy, resources, and environmental discharges associated with each alternative. Process and product models are commonly used for LCA in which materials and products are analyzed through different stages of fabrication, use, or end-of-life (reuse, recycling, disposal) options. LCA models have been developed as flow sheets for individual processes or as matrices based on economic input–output (EIO) data. In this example, we will demonstrate the use of matrix approaches to LCA.

In analyzing the environmental effects of products, note that products typically require more than one material input, so that multiple material process chains are required to capture all environmental impacts. For example, an LCA model for concrete includes the resource inputs of aggregates mining, cement production, chemical additives manufacturing, and electricity generation, and the environmental effects, or outputs, such as toxic emissions or hazardous wastes.

Consider a case in which reinforced concrete used to pave a 1-kilometer-long roadway costs $150,000 and requires 3,680 metric tons of concrete (costing $104,000) and 78 metric tons of reinforcing steel (costing $46,000). This reinforced concrete product LCA is illustrated in a tabular, or matrix, form, shown in Table 1. For reinforced concrete, both iron and steel production (for reinforcing bars) and concrete are required. A column in this matrix represents the required direct inputs (from the rows) for a unit of output. For example, $1,000 of concrete production requires $120 of aggregate and $200 of cement production, and $1,000 of iron and steel production requires $40 of iron and ferroalloy ores mining.

$$\mathbf{X}_{\text{direct suppliers}} = (\mathbf{I} + \mathbf{D})\mathbf{F}, \qquad (1)$$

The direct supplier inputs for concrete production can be obtained by multiplying the matrix in Table 1 by the desired output of reinforced concrete:
where $\mathbf{X}_{\text{direct suppliers}}$ are the direct supplier inputs to reinforced concrete (in dollars), \mathbf{I} is an identity matrix (to include the output of the concrete production stage

Table 1 A Process-Model-Requirements Matrix for $1 of Reinforced Concrete Product

Inputs	Transportation	Aggregates Mining	Iron and Ferroalloy Ores Mining	Lime	Electricity	Coal Mining	Chemical Additives	Cement Production	Iron and Steel Production	Concrete Production
Transportation	0	0	0	0	0	0	0	0	0.03	0.10
Aggregates mining	0	0	0	0	0	0	0	0	0	0.12
Iron and ferroalloy ores mining	0	0	0	0	0	0	0	0	0.04	0
Lime	0	0	0	0	0	0	0	0	0.005	0
Electricity	0	0	0	0	0	0	0	0	0.07	0.04
Coal mining	0	0	0	0	0	0	0	0	0.05	0.02
Chemical additives	0	0	0	0	0	0	0	0	0.05	0.03
Cement production	0	0	0	0	0	0	0	0	0	0.20
Iron and steel production	0	0	0	0	0	0	0	0	0	0
Concrete production	0	0	0	0	0	0	0	0	0	0

Note. Rows represent inputs into the sectors named at column heads.

itself), **D** is the requirements matrix (shown in Table 1), and **F** is a vector of desired output. For $150,000 of reinforced concrete roadway, **F** consists of $104,000 of concrete and $46,000 of reinforcing steel.

For the most complete environmental assessment, LCA should take into account the entire supply chain for a product, including suppliers to a supplier, called indirect suppliers. Consider, for example, that iron and steel production contributes to aggregates mining, iron and ferroalloy ores mining, cement production, concrete production, even to iron and steel production. Therefore, each sector of the economy named in the column heads has inputs represented by rows. In Table 1, we only have nonzero values for the direct suppliers to the concrete and reinforcing steel processes, so the supply chain is only one level deep. The second level supplier requirements could be calculated as **D*D*F,** but in some cases third or fourth levels of suppliers exist also. The total output for the various stages of LCA can be calculated as the series:

$$X = (I + D + D*D + D*D*D + \ldots)F, \qquad (2)$$

where **X** is the change in total output (in dollars), **I** is an identity matrix (to include the output of the concrete production stage itself), **D** is the requirements matrix

(shown in Table 1), and **F** is a vector representing the desired final demand (of concrete in this case). The supplier requirements series $(I + D + D*D + D*D*D + \ldots)$ can be approximated by $(I - D)^{-1}$, so the total output including indirect suppliers is:

$$X = (I - D)^{-1} F. \qquad (3)$$

The $(I - D)^{-1}$ table is illustrated in Table 2.

Once the economic output for each stage is calculated, then a vector of direct environmental outputs can be obtained by multiplying the output at each stage by the environmental impact per dollar of output:

$$B_i = R_iX = R_i(I - D)^{-1}F \qquad (4)$$

where B_i is the vector of environmental burdens (such as toxic emissions or electricity use), and R_i is a matrix with diagonal elements representing the impact per dollar of output for each stage. Illustrated below is an **R** matrix in which the elements are given in kilograms of Resource Conservation and Recovery Act (RCRA) Subtitle C hazardous waste generated per $1,000 of output. For example, 0.027 kg of hazardous waste are generated per $1,000 of output from the coal-mining sector.

Table 2 Ten-sector Total Requirements for $1 of Each of the Inputs to the Example Reinforced Concrete Product

Inputs	Transpor-tation	Aggregates Mining	Iron and Ferroalloy Ores Mining	Lime	Electricity	Coal Mining	Chemical Additives	Cement Production	Iron and Steel Production	Concrete Production
Transportation	1.19700	0.02556	0.02408	0.05055	0.00984	0.01426	0.02865	0.05523	0.03322	0.10076
Aggregates mining	0.00010	1.04037	0.00062	0.00124	0.00059	0.00178	0.00022	0.00168	0.00070	0.12402
Iron and ferroalloy ores mining	0.00013	0.00046	1.00745	0.00035	0.00021	0.00052	0.00139	0.00215	0.04099	0.00063
Lime	0.00005	0.00008	0.00056	1.00332	0.00013	0.00032	0.00082	0.00017	0.00488	0.00009
Electricity	0.01683	0.05801	0.20138	0.08399	1.00861	0.03367	0.05604	0.12515	0.07199	0.04477
Coal mining	0.00212	0.00889	0.02507	0.07788	0.11145	1.12447	0.01201	0.10313	0.04676	0.02265
Chemical additives	0.00583	0.00892	0.02152	0.01837	0.00582	0.00830	1.25858	0.01621	0.04903	0.03182
Cement production	0.00026	0.01429	0.00052	0.00208	0.00074	0.00046	0.00059	1.03467	0.00049	0.19756
Iron and steel production	0.00323	0.00963	0.01261	0.00540	0.00491	0.01154	0.00936	0.00601	1.16573	0.00464
Concrete production	0.00029	0.00122	0.00071	0.00077	0.00160	0.00057	0.00047	0.00091	0.00092	1.01104

Note. Rows represent inputs into the sectors named at column heads.

$$\begin{pmatrix} 2.096 & 0 & 0 & 0 & 0 & 0 & 0 & 0 & 0 & 0 \\ 0 & 0.002 & 0 & 0 & 0 & 0 & 0 & 0 & 0 & 0 \\ 0 & 0 & 0.162 & 0 & 0 & 0 & 0 & 0 & 0 & 0 \\ 0 & 0 & 0 & 0.025 & 0 & 0 & 0 & 0 & 0 & 0 \\ 0 & 0 & 0 & 0 & 2.745 & 0 & 0 & 0 & 0 & 0 \\ 0 & 0 & 0 & 0 & 0 & 0.027 & 0 & 0 & 0 & 0 \\ 0 & 0 & 0 & 0 & 0 & 0 & 94.158 & 0 & 0 & 0 \\ 0 & 0 & 0 & 0 & 0 & 0 & 0 & 13.866 & 0 & 0 \\ 0 & 0 & 0 & 0 & 0 & 0 & 0 & 0 & 42.454 & 0 \\ 0 & 0 & 0 & 0 & 0 & 0 & 0 & 0 & 0 & 0.0002 \end{pmatrix}$$

EXERCISES

1. Calculate the direct supplier inputs to reinforced concrete ($\mathbf{X}_{\text{direct suppliers}}$) in thousands of dollars, corresponding to the example presented above.

Solution:

$\mathbf{X}_{\text{direct suppliers}} = (\mathbf{I} + \mathbf{D})\mathbf{F} =$

$$\begin{pmatrix} 1 & 0 & 0 & 0 & 0 & 0 & 0 & 0 & 0.03 & 0.10 \\ 0 & 1 & 0 & 0 & 0 & 0 & 0 & 0 & 0 & 0.12 \\ 0 & 0 & 1 & 0 & 0 & 0 & 0 & 0 & 0.04 & 0 \\ 0 & 0 & 0 & 1 & 0 & 0 & 0 & 0 & 0.005 & 0 \\ 0 & 0 & 0 & 0 & 1 & 0 & 0 & 0 & 0.07 & 0.04 \\ 0 & 0 & 0 & 0 & 0 & 1 & 0 & 0 & 0.05 & 0.02 \\ 0 & 0 & 0 & 0 & 0 & 0 & 1 & 0 & 0.05 & 0.03 \\ 0 & 0 & 0 & 0 & 0 & 0 & 0 & 1 & 0 & 0.20 \\ 0 & 0 & 0 & 0 & 0 & 0 & 0 & 0 & 1 & 0 \\ 0 & 0 & 0 & 0 & 0 & 0 & 0 & 0 & 0 & 1 \end{pmatrix} \begin{pmatrix} 0 \\ 0 \\ 0 \\ 0 \\ 0 \\ 0 \\ 0 \\ 0 \\ 46 \\ 104 \end{pmatrix} = \begin{pmatrix} 12 \\ 12 \\ 2 \\ 0.2 \\ 7 \\ 4 \\ 5 \\ 21 \\ 46 \\ 104 \end{pmatrix}$$

2. Interpret the results from Exercise 1. How much aggregate mining is required for $150,000 of concrete?

Solution: $12,000

3. How much RCRA hazardous waste is generated for the direct suppliers in the example presented above, in kilograms, using (4)? (Use $\mathbf{X}_{\text{direct suppliers}}$ from Exercise 1.)

Solution:

$\mathbf{B} = \mathbf{R} \ \mathbf{X}_{\text{direct suppliers}} =$

$$
\begin{pmatrix}
2.096 & 0 & 0 & 0 & 0 & 0 & 0 & 0 & 0 & 0 \\
0 & 0.002 & 0 & 0 & 0 & 0 & 0 & 0 & 0 & 0 \\
0 & 0 & 0.162 & 0 & 0 & 0 & 0 & 0 & 0 & 0 \\
0 & 0 & 0 & 0.025 & 0 & 0 & 0 & 0 & 0 & 0 \\
0 & 0 & 0 & 0 & 2.745 & 0 & 0 & 0 & 0 & 0 \\
0 & 0 & 0 & 0 & 0 & 0.027 & 0 & 0 & 0 & 0 \\
0 & 0 & 0 & 0 & 0 & 0 & 94.158 & 0 & 0 & 0 \\
0 & 0 & 0 & 0 & 0 & 0 & 0 & 13.866 & 0 & 0 \\
0 & 0 & 0 & 0 & 0 & 0 & 0 & 0 & 42.454 & 0 \\
0 & 0 & 0 & 0 & 0 & 0 & 0 & 0 & 0 & 0.0002
\end{pmatrix}
\begin{pmatrix}
12 \\
12 \\
2 \\
0.2 \\
7 \\
4 \\
5 \\
21 \\
46 \\
104
\end{pmatrix}
=
\begin{pmatrix}
25 \\
0.03 \\
0.3 \\
0.005 \\
19 \\
0.1 \\
471 \\
291 \\
1,953 \\
0.02
\end{pmatrix}
$$

4. Interpret the results from Exercise 3. How much RCRA Subtitle C hazardous waste is generated for $4,000 of coal mining?

Solution: 0.027 kg of hazardous waste is generated per $1,000 of output in the coal-mining sector, thus for $4,000 of coal mining, 0.1 kg of hazardous waste is generated.

5. Overall, how much RCRA hazardous waste is generated for the direct suppliers, as calculated in Exercise 3?

Solution: 2,759 kg of RCRA Subtitle C hazardous waste are generated (obtained by summing the elements of the **B** vector).

Application Essay

Reference

[1] Hendrickson, C. T., A. Horvath, S. Joshi, and L. B. Lave, "Economic Input–Output Models for Environmental Life Cycle Assessment."
Environmental Science and Technology, ACS, 32, no. 4 (1998): 184A–191A.

...

Arpad Horvath is a research faculty member in the Department of Civil and Environmental Engineering, Green Design Initiative, at Carnegie Mellon University. His current research focuses on life cycle engineering, infrastructure systems, and environmental management issues. Dr. Horvath holds Master of Science degrees in civil engineering from both the Technical University of Budapest and Carnegie Mellon and a Ph.D. from Carnegie Mellon, where he presently teaches courses in project management for construction.

CHAPTER

7

Vectors

INTRODUCTION

You have undoubtedly encountered the notion of vectors in your study of calculus, as well as in physics and engineering courses. For most of you, then, this chapter is a review of familiar topics such as the dot and cross products. However, in Section 7.6 we shall consider an abstraction of the vector concept.

7.1 Vectors in 2-Space

■ **Geometric Vectors** In science, mathematics, and engineering, we distinguish two important quantities: **scalars** and **vectors.** A **scalar** is simply a real number or a quantity that has *magnitude.* For example, length, temperature, and blood pressure are represented by numbers such as 80 m, 20°C, and the systolic–diastolic ratio $^{120}/_{80}$. A **vector,** on the other hand, is usually described as a quantity that has both *magnitude* and *direction.* Geometrically, a vector is represented by a directed line segment—that is, by an arrow—and is written either as a boldface symbol **v** or as \vec{v} or \overrightarrow{AB}. Examples of vector quantities shown in Figure 7.1 are weight **w**, velocity **v**, and the retarding force of friction \mathbf{F}_f.

Figure 7.2

Figure 7.1

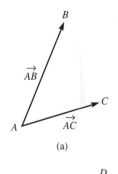

Figure 7.3

■ **Notation and Terminology** A vector whose initial point (or end) is A and whose terminal point (or tip) is B is written \overrightarrow{AB}. The magnitude of a vector is written $\|\overrightarrow{AB}\|$. Two vectors that have the same magnitude and same direction are said to be **equal.** Thus, in Figure 7.2, we have $\overrightarrow{AB} = \overrightarrow{CD}$. Vectors are said to be **free,** which means that a vector can be moved from one position to another provided its magnitude and direction are not changed. The **negative** of a vector \overrightarrow{AB}, written $-\overrightarrow{AB}$, is a vector that has the same magnitude as \overrightarrow{AB} but is opposite in direction. If $k \neq 0$ is a scalar, the **scalar multiple** of a vector, $k\overrightarrow{AB}$, is a vector that is $|k|$ times as long as \overrightarrow{AB}. If $k > 0$, then $k\overrightarrow{AB}$ has the same direction as the vector \overrightarrow{AB}; if $k < 0$, then $k\overrightarrow{AB}$ has the direction opposite that of \overrightarrow{AB}. When $k = 0$, we say $0\overrightarrow{AB} = \mathbf{0}$ is the **zero vector.*** Two vectors are **parallel** if and only if they are nonzero scalar multiples of each other. See Figure 7.3.

■ **Addition and Subtraction** Two vectors can be considered as having a common initial point, such as A in Figure 7.4(a). Thus, if nonparallel vectors \overrightarrow{AB} and \overrightarrow{AC} are the sides of a parallelogram as in Figure 7.4(b), we say the vector that is the main diagonal, or \overrightarrow{AD}, is the **sum** of \overrightarrow{AB} and \overrightarrow{AC}. We write

$$\overrightarrow{AD} = \overrightarrow{AB} + \overrightarrow{AC}.$$

The **difference** of two vectors \overrightarrow{AB} and \overrightarrow{AC} is defined by

Figure 7.4

* The question of what is the direction of **0** is usually answered by saying that the zero vector can be assigned *any* direction. More to the point, **0** is needed in order to have a vector algebra.

(a)

(b)

Figure 7.5

Figure 7.6

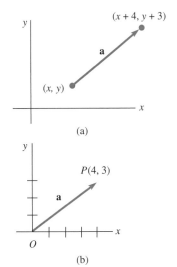

(a)

(b)

Figure 7.7

$$\overrightarrow{AB} - \overrightarrow{AC} = \overrightarrow{AB} + (-\overrightarrow{AC}).$$

As seen in Figure 7.5(a), the difference $\overrightarrow{AB} - \overrightarrow{AC}$ can be interpreted as the main diagonal of the parallelogram with sides \overrightarrow{AB} and $-\overrightarrow{AC}$. However, as shown in Figure 7.5(b), we can also interpret the same vector difference as the third side of a triangle with sides \overrightarrow{AB} and \overrightarrow{AC}. In this second interpretation, observe that the vector difference $\overrightarrow{CB} = \overrightarrow{AB} - \overrightarrow{AC}$ points toward the terminal point of the vector *from* which we are subtracting the second vector. If $\overrightarrow{AB} = \overrightarrow{AC}$, then

$$\overrightarrow{AB} - \overrightarrow{AC} = \mathbf{0}.$$

■ **Vectors in a Coordinate Plane** To describe a vector analytically, let us suppose for the remainder of this section that the vectors we are considering lie in a two-dimensional coordinate plane or **2-space.** We shall denote the set of all vectors in the plane by R^2. The vector shown in Figure 7.6, with initial point the origin O and terminal point $P(x_1, y_1)$, is called the **position vector** of the point P and is written

$$\overrightarrow{OP} = \langle x_1, y_1 \rangle.$$

■ **Components** In general, a vector \mathbf{a} in R^2 is any ordered pair of real numbers,

$$\mathbf{a} = \langle a_1, a_2 \rangle.$$

The numbers a_1 and a_2 are said to be the **components** of the vector \mathbf{a}.

As we shall see in the first example, the vector \mathbf{a} is not necessarily a position vector.

Example 1 Position Vector

The displacement between the point (x, y) and $(x + 4, y + 3)$ in Figure 7.7(a) is written $\langle 4, 3 \rangle$. As seen in Figure 7.7(b), the position vector of $\langle 4, 3 \rangle$ is the vector emanating from the origin and terminating at the point $P(4, 3)$. ❑

Addition and subtraction of vectors, multiplication of vectors by scalars, and so on, are defined in terms of their components.

DEFINITION 7.1 Addition, Scalar Multiplication, Equality

Let $\mathbf{a} = \langle a_1, a_2 \rangle$ and $\mathbf{b} = \langle b_1, b_2 \rangle$ be vectors in R^2.
(*i*) **Addition: $\mathbf{a} + \mathbf{b} = \langle a_1 + b_1, a_2 + b_2 \rangle$** (1)
(*ii*) **Scalar multiplication:** $k\mathbf{a} = \langle ka_1, ka_2 \rangle$ (2)
(*iii*) **Equality: $\mathbf{a} = \mathbf{b}$** if and only if $a_1 = b_1, a_2 = b_2$ (3)

■ **Subtraction** Using (2), we define the **negative** of a vector \mathbf{b} by

$$-\mathbf{b} = (-1)\mathbf{b} = \langle -b_1, -b_2 \rangle.$$

We can then define the **subtraction,** or the difference, of two vectors as

$$\mathbf{a} - \mathbf{b} = \mathbf{a} + (-\mathbf{b}) = \langle a_1 - b_1, a_2 - b_2 \rangle. \qquad (4)$$

(a)

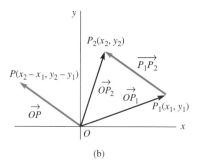

(b)

Figure 7.8

In Figure 7.8(a), we have illustrated the sum of two vectors $\overrightarrow{OP_1}$ and $\overrightarrow{OP_2}$. In Figure 7.8(b), the vector $\overrightarrow{P_1P_2}$, with initial point P_1 and terminal point P_2, is the difference of position vectors

$$\overrightarrow{P_1P_2} = \overrightarrow{OP_2} - \overrightarrow{OP_1} = \langle x_2 - x_1, y_2 - y_1 \rangle.$$

As shown in Figure 7.8(b), the vector $\overrightarrow{P_1P_2}$ can be drawn either starting from the terminal point of $\overrightarrow{OP_1}$ and ending at the terminal point of $\overrightarrow{OP_2}$ or as the position vector \overrightarrow{OP} whose terminal point has coordinates $(x_2 - x_1, y_2 - y_1)$. Remember, \overrightarrow{OP} and $\overrightarrow{P_1P_2}$ are considered equal, since they have the same magnitude and direction.

Example 2 **Addition and Subtraction of Two Vectors**

If $\mathbf{a} = \langle 1, 4 \rangle$ and $\mathbf{b} = \langle -6, 3 \rangle$, find $\mathbf{a} + \mathbf{b}$, $\mathbf{a} - \mathbf{b}$, and $2\mathbf{a} + 3\mathbf{b}$.

SOLUTION We use (1), (2), and (4).

$$\mathbf{a} + \mathbf{b} = \langle 1 + (-6), 4 + 3 \rangle = \langle -5, 7 \rangle$$

$$\mathbf{a} - \mathbf{b} = \langle 1 - (-6), 4 - 3 \rangle = \langle 7, 1 \rangle$$

$$2\mathbf{a} + 3\mathbf{b} = \langle 2, 8 \rangle + \langle -18, 9 \rangle = \langle -16, 17 \rangle. \qquad \square$$

■ **Properties** The component definition of a vector can be used to verify each of the following **properties** of vectors in R^2:

Properties of Vectors

(*i*) $\mathbf{a} + \mathbf{b} = \mathbf{b} + \mathbf{a}$ (commutative law)

(*ii*) $\mathbf{a} + (\mathbf{b} + \mathbf{c}) = (\mathbf{a} + \mathbf{b}) + \mathbf{c}$ (associative law)

(*iii*) $\mathbf{a} + \mathbf{0} = \mathbf{a}$ (additive identity)

(*iv*) $\mathbf{a} + (-\mathbf{a}) = \mathbf{0}$ (additive inverse)

(*v*) $k(\mathbf{a} + \mathbf{b}) = k\mathbf{a} + k\mathbf{b}$, k a scalar

(*vi*) $(k_1 + k_2)\mathbf{a} = k_1\mathbf{a} + k_2\mathbf{a}$, k_1 and k_2 scalars

(*vii*) $k_1(k_2\mathbf{a}) = (k_1k_2)\mathbf{a}$, k_1 and k_2 scalars

(*viii*) $1\mathbf{a} = \mathbf{a}$

(*ix*) $0\mathbf{a} = \mathbf{0}$ (zero vector)

The **zero vector 0** in properties (*iii*), (*iv*), and (*ix*) is defined as

$$\mathbf{0} = \langle 0, 0 \rangle.$$

■ **Magnitude** The **magnitude, length,** or **norm** of a vector \mathbf{a} is denoted by $\|\mathbf{a}\|$. Motivated by the Pythagorean Theorem and Figure 7.9, we define the magnitude of a vector

$$\mathbf{a} = \langle a_1, a_2 \rangle \quad \text{to be} \quad \|\mathbf{a}\| = \sqrt{a_1^2 + a_2^2}.$$

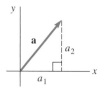

Figure 7.9

Clearly, $\|\mathbf{a}\| \geq 0$ for any vector \mathbf{a}, and $\|\mathbf{a}\| = 0$ if and only if $\mathbf{a} = \mathbf{0}$. For example, if $\mathbf{a} = \langle 6, -2 \rangle$, then $\|\mathbf{a}\| = \sqrt{6^2 + (-2)^2} = \sqrt{40} = 2\sqrt{10}$.

■ **Unit Vectors** A vector that has magnitude 1 is called a **unit vector.** We can obtain a unit vector **u** in the same direction as a nonzero vector **a** by multiplying **a** by the reciprocal of its magnitude. The vector $\mathbf{u} = (1/\|\mathbf{a}\|)\mathbf{a}$ is a unit vector, since

$$\|\mathbf{u}\| = \left\| \frac{1}{\|\mathbf{a}\|}\mathbf{a} \right\| = \frac{1}{\|\mathbf{a}\|}\|\mathbf{a}\| = 1.$$

Example 3 Unit Vectors

Given $\mathbf{a} = \langle 2, -1 \rangle$. Form a unit vector in the same direction as **a.** In the opposite direction of **a**.

SOLUTION The magnitude of the vector **a** is $\|\mathbf{a}\| = \sqrt{4 + (-1)^2} = \sqrt{5}$. Thus a unit vector in the same direction as **a** is the scalar multiple

$$\mathbf{u} = \frac{1}{\sqrt{5}}\,\mathbf{a} = \frac{1}{\sqrt{5}}\,\langle 2, -1 \rangle = \left\langle \frac{2}{\sqrt{5}}, \frac{-1}{\sqrt{5}} \right\rangle.$$

A unit vector in the opposite direction of **a** is the negative of **u**:

$$-\mathbf{u} = \left\langle -\frac{2}{\sqrt{5}}, \frac{1}{\sqrt{5}} \right\rangle. \qquad \square$$

If **a** and **b** are vectors and c_1 and c_2 are scalars, then the expression $c_1\mathbf{a} + c_2\mathbf{b}$ is called a **linear combination** of **a** and **b**. As we see next, any vector in R^2 can be written as a linear combination of two special vectors.

■ **The i, j Vectors** In view of (1) and (2), any vector $\mathbf{a} = \langle a_1, a_2 \rangle$ can be written as a sum:

$$\langle a_1, a_2 \rangle = \langle a_1, 0 \rangle + \langle 0, a_2 \rangle = a_1\langle 1, 0 \rangle + a_2\langle 0, 1 \rangle. \qquad (5)$$

The unit vectors $\langle 1, 0 \rangle$ and $\langle 0, 1 \rangle$ are usually given the special symbols **i** and **j**. See Figure 7.10(a). Thus, if

$$\mathbf{i} = \langle 1, 0 \rangle \quad \text{and} \quad \mathbf{j} = \langle 0, 1 \rangle,$$

then (5) becomes $\mathbf{a} = a_1\mathbf{i} + a_2\mathbf{j}. \qquad (6)$

The unit vectors **i** and **j** are said to form a **basis** for the system of two-dimensional vectors, since any vector **a** can be written uniquely as a linear combination of **i** and **j**. If $\mathbf{a} = a_1\mathbf{i} + a_2\mathbf{j}$ is a position vector, then Figure 7.10(b) shows that **a** is the sum of the vectors $a_1\mathbf{i}$ and $a_2\mathbf{j}$, which have the origin as a common initial point and which lie on the x- and y-axes, respectively. The scalar a_1 is called the **horizontal component** of **a,** and the scalar a_2 is called the **vertical component** of **a**.

(a)

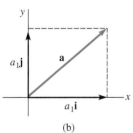

(b)

Figure 7.10

Example 4 Vector Operations Using i and j

(a) $\langle 4, 7 \rangle = 4\mathbf{i} + 7\mathbf{j}$

(b) $(2\mathbf{i} - 5\mathbf{j}) + (8\mathbf{i} + 13\mathbf{j}) = 10\mathbf{i} + 8\mathbf{j}$

(c) $\|\mathbf{i} + \mathbf{j}\| = \sqrt{2}$

(d) $10(3\mathbf{i} - \mathbf{j}) = 30\mathbf{i} - 10\mathbf{j}$

(e) $\mathbf{a} = 6\mathbf{i} + 4\mathbf{j}$ and $\mathbf{b} = 9\mathbf{i} + 6\mathbf{j}$ are parallel, since **b** is a scalar multiple of **a**. We see that $\mathbf{b} = \frac{3}{2}\mathbf{a}$. $\qquad \square$

Example 5 **Graphs of Vector Sum/Vector Difference**

Let $\mathbf{a} = 4\mathbf{i} + 2\mathbf{j}$ and $\mathbf{b} = -2\mathbf{i} + 5\mathbf{j}$. Graph $\mathbf{a} + \mathbf{b}$ and $\mathbf{a} - \mathbf{b}$.

SOLUTION The graphs of $\mathbf{a} + \mathbf{b} = 2\mathbf{i} + 7\mathbf{j}$ and $\mathbf{a} - \mathbf{b} = 6\mathbf{i} - 3\mathbf{j}$ are given in Figures 7.11(a) and 7.11(b), respectively.

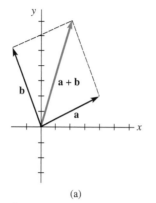

(a) (b)

Figure 7.11

Answers to odd-numbered problems begin on page A-40.

In Problems 1–8, find **(a)** $3\mathbf{a}$, **(b)** $\mathbf{a} + \mathbf{b}$, **(c)** $\mathbf{a} - \mathbf{b}$, **(d)** $\|\mathbf{a} + \mathbf{b}\|$, and **(e)** $\|\mathbf{a} - \mathbf{b}\|$.

1. $\mathbf{a} = 2\mathbf{i} + 4\mathbf{j}$, $\mathbf{b} = -\mathbf{i} + 4\mathbf{j}$

2. $\mathbf{a} = \langle 1, 1 \rangle$, $\mathbf{b} = \langle 2, 3 \rangle$

3. $\mathbf{a} = \langle 4, 0 \rangle$, $\mathbf{b} = \langle 0, -5 \rangle$

4. $\mathbf{a} = \tfrac{1}{6}\mathbf{i} - \tfrac{1}{6}\mathbf{j}$, $\mathbf{b} = \tfrac{1}{2}\mathbf{i} + \tfrac{5}{6}\mathbf{j}$

5. $\mathbf{a} = -3\mathbf{i} + 2\mathbf{j}$, $\mathbf{b} = 7\mathbf{j}$

6. $\mathbf{a} = \langle 1, 3 \rangle$, $\mathbf{b} = -5\mathbf{a}$

7. $\mathbf{a} = -\mathbf{b}$, $\mathbf{b} = 2\mathbf{i} - 9\mathbf{j}$

8. $\mathbf{a} = \langle 7, 10 \rangle$, $\mathbf{b} = \langle 1, 2 \rangle$

In Problems 9–14, find **(a)** $4\mathbf{a} - 2\mathbf{b}$ and **(b)** $-3\mathbf{a} - 5\mathbf{b}$.

9. $\mathbf{a} = \langle 1, -3 \rangle$, $\mathbf{b} = \langle -1, 1 \rangle$

10. $\mathbf{a} = \mathbf{i} + \mathbf{j}$, $\mathbf{b} = 3\mathbf{i} - 2\mathbf{j}$ **11.** $\mathbf{a} = \mathbf{i} - \mathbf{j}$, $\mathbf{b} = -3\mathbf{i} + 4\mathbf{j}$

12. $\mathbf{a} = \langle 2, 0 \rangle$, $\mathbf{b} = \langle 0, -3 \rangle$ **13.** $\mathbf{a} = \langle 4, 10 \rangle$, $\mathbf{b} = -2\langle 1, 3 \rangle$

14. $\mathbf{a} = \langle 3, 1 \rangle + \langle -1, 2 \rangle$, $\mathbf{b} = \langle 6, 5 \rangle - \langle 1, 2 \rangle$

In Problems 15–18, find the vector $\overrightarrow{P_1P_2}$. Graph $\overrightarrow{P_1P_2}$ and its corresponding position vector.

15. $P_1(3, 2)$, $P_2(5, 7)$ **16.** $P_1(-2, -1)$, $P_2(4, -5)$

17. $P_1(3, 3)$, $P_2(5, 5)$ **18.** $P_1(0, 3)$, $P_2(2, 0)$

19. Find the terminal point of the vector $\overrightarrow{P_1P_2} = 4\mathbf{i} + 8\mathbf{j}$ if its initial point is $(-3, 10)$.

20. Find the initial point of the vector $\overrightarrow{P_1P_2} = \langle -5, -1 \rangle$ if its terminal point is $\langle 4, 7 \rangle$.

21. Determine which of the following vectors are parallel to $\mathbf{a} = 4\mathbf{i} + 6\mathbf{j}$.

 (a) $-4\mathbf{i} - 6\mathbf{j}$ **(b)** $-\mathbf{i} - \tfrac{3}{2}\mathbf{j}$

 (c) $10\mathbf{i} + 15\mathbf{j}$ **(d)** $2(\mathbf{i} - \mathbf{j}) - 3(\tfrac{1}{2}\mathbf{i} - \tfrac{5}{12}\mathbf{j})$

 (e) $8\mathbf{i} + 12\mathbf{j}$ **(f)** $(5\mathbf{i} + \mathbf{j}) - (7\mathbf{i} + 4\mathbf{j})$

22. Determine a scalar c so that $\mathbf{a} = 3\mathbf{i} + c\mathbf{j}$ and $\mathbf{b} = -\mathbf{i} + 9\mathbf{j}$ are parallel.

In Problems 23 and 24, find $\mathbf{a} + (\mathbf{b} + \mathbf{c})$ for the given vectors.

23. $\mathbf{a} = \langle 5, 1 \rangle$, $\mathbf{b} = \langle -2, 4 \rangle$, $\mathbf{c} = \langle 3, 10 \rangle$

24. $\mathbf{a} = \langle 1, 1 \rangle$, $\mathbf{b} = \langle 4, 3 \rangle$, $\mathbf{c} = \langle 0, -2 \rangle$

In Problems 25–28, find a unit vector **(a)** in the same direction as **a**, and **(b)** in the opposite direction of **a**.

25. $\mathbf{a} = \langle 2, 2 \rangle$ **26.** $\mathbf{a} = \langle -3, 4 \rangle$

27. $\mathbf{a} = \langle 0, -5 \rangle$ **28.** $\mathbf{a} = \langle 1, -\sqrt{3} \rangle$

In Problems 29 and 30, $\mathbf{a} = \langle 2, 8 \rangle$ and $\mathbf{b} = \langle 3, 4 \rangle$. Find a unit vector in the same direction as the given vector.

29. $\mathbf{a} + \mathbf{b}$ **30.** $2\mathbf{a} - 3\mathbf{b}$

In Problems 31 and 32, find a vector **b** that is parallel to the given vector and has the indicated magnitude.

31. $\mathbf{a} = 3\mathbf{i} + 7\mathbf{j}$, $\|\mathbf{b}\| = 2$ **32.** $\mathbf{a} = \tfrac{1}{2}\mathbf{i} - \tfrac{1}{2}\mathbf{j}$, $\|\mathbf{b}\| = 3$

33. Find a vector in the opposite direction of $\mathbf{a} = \langle 4, 10 \rangle$ but ¾ as long.

34. Given that $\mathbf{a} = \langle 1, 1 \rangle$ and $\mathbf{b} = \langle -1, 0 \rangle$, find a vector in the same direction as $\mathbf{a} + \mathbf{b}$ but 5 times as long.

In Problems 35 and 36, use the given figure to draw the indicated vector.

35. $3\mathbf{b} - \mathbf{a}$

Figure 7.12

36. $\mathbf{a} + (\mathbf{b} + \mathbf{c})$

Figure 7.13

In Problems 37 and 38, express the vector \mathbf{x} in terms of the vectors \mathbf{a} and \mathbf{b}.

37.

Figure 7.14

38.

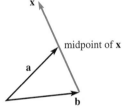

Figure 7.15

In Problems 39 and 40, use the given figure to prove the given result.

39. $\mathbf{a} + \mathbf{b} + \mathbf{c} = 0$

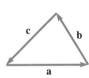

Figure 7.16

40. $\mathbf{a} + \mathbf{b} + \mathbf{c} + \mathbf{d} = 0$

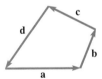

Figure 7.17

In Problems 41 and 42, express the vector $\mathbf{a} = 2\mathbf{i} + 3\mathbf{j}$ as a linear combination of the given vectors \mathbf{b} and \mathbf{c}.

41. $\mathbf{b} = \mathbf{i} + \mathbf{j}, \quad \mathbf{c} = \mathbf{i} - \mathbf{j}$

42. $\mathbf{b} = -2\mathbf{i} + 4\mathbf{j}, \quad \mathbf{c} = 5\mathbf{i} + 7\mathbf{j}$

A vector is said to be tangent to a curve at a point if it is parallel to the tangent line at the point. In Problems 43 and 44, find a unit tangent vector to the given curve at the indicated point.

43. $y = \frac14 x^2 + 1, \quad (2, 2)$ **44.** $y = -x^2 + 3x, \quad (0, 0)$

45. When walking, a person's foot strikes the ground with a force \mathbf{F} at an angle θ from the vertical. In Figure 7.18 the vector \mathbf{F} is resolved into vector components \mathbf{F}_g, which is parallel to the ground, and \mathbf{F}_n, which is perpendicular to the

ground. In order that the foot does not slip, the force \mathbf{F}_g must be offset by the opposing force \mathbf{F}_f of friction; that is, $\mathbf{F}_f = -\mathbf{F}_g$.

(a) Use the fact that $\|\mathbf{F}_f\| = \mu\|\mathbf{F}_n\|$, where μ is the coefficient of friction, to show that $\tan \theta = \mu$. The foot will not slip for angles less than or equal to θ.

(b) Given that $\mu = 0.6$ for a rubber heel striking an asphalt sidewalk, find the "no-slip" angle.

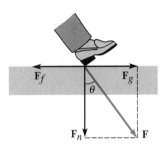

Figure 7.18

46. A 200-lb traffic light supported by two cables hangs in equilibrium. As shown in Figure 7.19(b), let the weight of the light be represented by \mathbf{w} and the forces in the two cables by \mathbf{F}_1 and \mathbf{F}_2. From Figure 7.19(c), we see that a condition of equilibrium is

$$\mathbf{w} + \mathbf{F}_1 + \mathbf{F}_2 = \mathbf{0}. \tag{7}$$

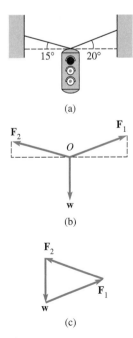

Figure 7.19

See Problem 39. If

$$\mathbf{w} = -200\mathbf{j}$$
$$\mathbf{F}_1 = (\|\mathbf{F}_1\| \cos 20°)\mathbf{i} + (\|\mathbf{F}_1\| \sin 20°)\mathbf{j}$$
$$\mathbf{F}_2 = -(\|\mathbf{F}_2\| \cos 15°)\mathbf{i} + (\|\mathbf{F}_2\| \sin 15°)\mathbf{j},$$

use (7) to determine the magnitudes of \mathbf{F}_1 and \mathbf{F}_2.
[*Hint:* Reread (*iii*) of Definition 7.1.]

47. An electric charge Q is uniformly distributed along the
y-axis between $y = -a$ and $y = a$. See Figure 7.20. The total
force exerted on the charge q on the x-axis by

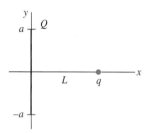

Figure 7.20

the charge Q is $\mathbf{F} = F_x \mathbf{i} + F_y \mathbf{j}$, where

$$F_x = \frac{qQ}{4\pi\varepsilon_0} \int_{-a}^{a} \frac{L\,dy}{2a(L^2 + y^2)^{3/2}}$$

and

$$F_y = -\frac{qQ}{4\pi\varepsilon_0} \int_{-a}^{a} \frac{y\,dy}{2a(L^2 + y^2)^{3/2}}.$$

Determine \mathbf{F}.

48. Using vectors, show that the diagonals of a parallelogram
bisect each other. [*Hint:* Let M be the midpoint of one diag-
onal and N the midpoint of the other.]

49. Using vectors, show that the line segment between the
midpoints of two sides of a triangle is parallel to the third
side and half as long.

50. An airplane starts from an airport located at the origin O
and flies 150 miles in the direction 20° north of east to city
A. From A the airplane then flies 200 miles in the direction
23° west of north to city B. From B the airplane flies 240
miles in the direction 10° south of west to city C. Express
the location of city C as a vector \mathbf{r} as shown in Figure 7.21.
Find the distance from O to C.

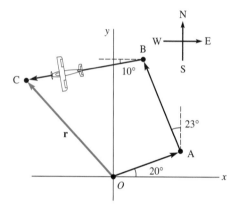

Figure 7.21

7.2 Vectors in 3-Space

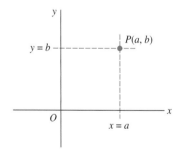

Figure 7.22

■ **Rectangular Coordinate System in Three Dimensions** In the
plane, or 2-space, one way of describing the position of a point P is to assign to it
coordinates relative to two mutually orthogonal axes called the x- and y-axes. If P is
the point of intersection of the line $x = a$ (perpendicular to the x-axis) and the line
$y = b$ (perpendicular to the y-axis), then the **ordered pair** (a, b) is said to be the
rectangular or **Cartesian coordinates** of the point. See Figure 7.22.

In three dimensions, or **3-space,** a rectangular coordinate system is constructed
using three mutually orthogonal axes. The point at which these axes intersect is
called the **origin** O. These axes, shown in Figure 7.23(a), are labeled in accordance
with the so-called **right-hand rule:** If the fingers of the right hand, pointing in the
direction of the positive x-axis, are curled toward the positive y-axis, then the thumb
will point in the direction of a new axis perpendicular to the plane of the x- and
y-axes. This new axis is labeled the z-axis.* The dashed lines in Figure 7.23(a) repre-
sent the negative axes. Now, if

$$x = a, \quad y = b, \quad z = c$$

* If the x- and y-axes are interchanged in Figure 7.23(a), the coordinate system is said to be **left-
handed.**

are planes perpendicular to the x-axis, y-axis, and z-axis, respectively, then the point P at which these planes intersect can be represented by an **ordered triple** of numbers (a, b, c) said to be the **rectangular** or **Cartesian coordinates** of the point. The numbers a, b, and c are, in turn, called the **x-, y-,** and **z-coordinates** of $P(a, b, c)$. See Figure 7.23(b).

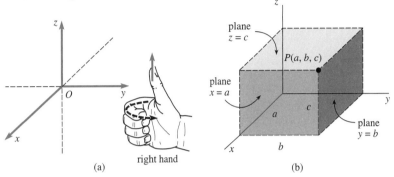

(a) right hand (b)

Figure 7.23

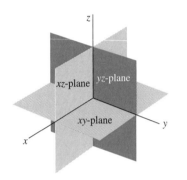

Figure 7.24

■ **Octants** Each pair of coordinate axes determines a **coordinate plane.** As shown in Figure 7.24, the x- and y-axes determine the xy-plane, the x- and z-axes determine the xz-plane, and so on. The coordinate planes divide 3-space into eight parts known as **octants.** The octant in which all three coordinates of a point are *positive* is called the **first octant.** There is no agreement for naming the other seven octants.

The following table summarizes the coordinates of a point either on a coordinate axis or in a coordinate plane. As seen in the table, we can also describe, say, the xy-plane by the simple equation $z = 0$. Similarly, the xz-plane is $y = 0$ and the yz-plane is $x = 0$.

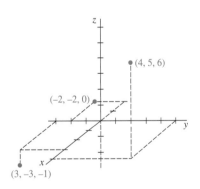

Figure 7.25

Axes	Coordinates	Plane	Coordinates
x	$(a, 0, 0)$	xy	$(a, b, 0)$
y	$(0, b, 0)$	xz	$(a, 0, c)$
z	$(0, 0, c)$	yz	$(0, b, c)$

Example 1 Graphs of Three Points

Graph the points $(4, 5, 6)$, $(3, -3, -1)$, and $(-2, -2, 0)$.

SOLUTION Of the three points shown in Figure 7.25, only $(4, 5, 6)$ is in the first octant. The point $(-2, -2, 0)$ is in the xy-plane. ❏

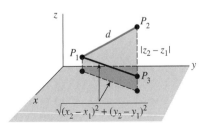

Figure 7.26

■ **Distance Formula** To find the **distance** between two points $P_1(x_1, y_1, z_1)$ and $P_2(x_2, y_2, z_2)$ in 3-space, let us first consider their projection onto the xy-plane. As seen in Figure 7.26, the distance between $(x_1, y_1, 0)$ and $(x_2, y_2, 0)$ follows from the usual distance formula in the plane and is $\sqrt{(x_2 - x_1)^2 + (y_2 - y_1)^2}$. If the coordinates of P_3 are (x_2, y_2, z_1), then the Pythagorean Theorem applied to the right triangle $P_1P_2P_3$ yields

$$[d(P_1, P_2)]^2 = [\sqrt{(x_2 - x_1)^2 + (y_2 - y_1)^2}]^2 + |z_2 - z_1|^2$$

or
$$d(P_1, P_2) = \sqrt{(x_2 - x_1)^2 + (y_2 - y_1)^2 + (z_2 - z_1)^2}. \quad (1)$$

Example 2 **Distance Between Two Points**

Find the distance between $(2, -3, 6)$ and $(-1, -7, 4)$.

SOLUTION Choosing P_2 as $(2, -3, 6)$ and P_1 as $(-1, -7, 4)$, formula (1) gives

$$d = \sqrt{(2 - (-1))^2 + (-3 - (-7))^2 + (6 - 4)^2} = \sqrt{29}. \qquad \square$$

■ **Midpoint Formula** The formula for finding the midpoint of a line segment between two points in 2-space carries over in an analogous fashion to 3-space. If $P_1(x_1, y_1, z_1)$ and $P_2(x_2, y_2, z_2)$ are two distinct points, then the **coordinates of the midpoint** of the line segment between them are

$$\left(\frac{x_1 + x_2}{2}, \frac{y_1 + y_2}{2}, \frac{z_1 + z_2}{2} \right). \qquad (2)$$

Example 3 **Coordinates of a Midpoint**

Find the coordinates of the midpoint of the line segment between the two points in Example 2.

SOLUTION From (2) we obtain

$$\left(\frac{2 + (-1)}{2}, \frac{-3 + (-7)}{2}, \frac{6 + 4}{2} \right) \quad \text{or} \quad (\tfrac{1}{2}, -5, 5). \qquad \square$$

■ **Vectors in 3-Space** A **vector a in 3-space** is any ordered triple of real numbers

$$\mathbf{a} = \langle a_1, a_2, a_3 \rangle,$$

where a_1, a_2, and a_3 are the **components** of the vector. The set of all vectors in 3-space will be denoted by the symbol R^3. The **position vector** of a point $P(x_1, y_1, z_1)$ in space is the vector $\overrightarrow{OP} = \langle x_1, y_1, z_1 \rangle$ whose initial point is the origin O and whose terminal point is P. See Figure 7.27.

The component definitions of addition, subtraction, scalar multiplication, and so on are natural generalizations of those given for vectors in R^2.

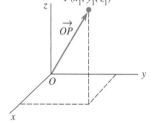

Figure 7.27

> **DEFINITION 7.2 Component Definitions in 3-Space**
>
> Let $\mathbf{a} = \langle a_1, a_2, a_3 \rangle$ and $\mathbf{b} = \langle b_1, b_2, b_3 \rangle$ be vectors in R^3.
> (*i*) Addition: $\mathbf{a} + \mathbf{b} = \langle a_1 + b_1, a_2 + b_2, a_3 + b_3 \rangle$
> (*ii*) Scalar multiplication: $k\mathbf{a} = \langle ka_1, ka_2, ka_3 \rangle$
> (*iii*) Equality: $\mathbf{a} = \mathbf{b}$ if and only if $a_1 = b_1, a_2 = b_2, a_3 = b_3$
> (*iv*) Negative: $-\mathbf{b} = (-1)\mathbf{b} = \langle -b_1, -b_2, -b_3 \rangle$
> (*v*) Subtraction: $\mathbf{a} - \mathbf{b} = \mathbf{a} + (-\mathbf{b}) = \langle a_1 - b_1, a_2 - b_2, a_3 - b_3 \rangle$
> (*vi*) Zero vector: $\mathbf{0} = \langle 0, 0, 0 \rangle$
> (*vii*) Magnitude: $\|\mathbf{a}\| = \sqrt{a_1^2 + a_2^2 + a_3^2}$

If $\overrightarrow{OP_1}$ and $\overrightarrow{OP_2}$ are the position vectors of the points $P_1(x_1, y_1, z_1)$ and $P_2(x_2, y_2, z_2)$, then the vector $\overrightarrow{P_1P_2}$ is given by

$$\overrightarrow{P_1P_2} = \overrightarrow{OP_2} - \overrightarrow{OP_1} = \langle x_2 - x_1, y_2 - y_1, z_2 - z_1 \rangle. \qquad (3)$$

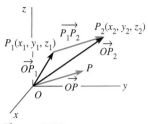

Figure 7.28

As in 2-space, $\overrightarrow{P_1P_2}$ can be drawn either as a vector whose initial point is P_1 and whose terminal point is P_2 or as position vector \overrightarrow{OP} with terminal point $P(x_2 - x_1, y_2 - y_1, z_2 - z_1)$. See Figure 7.28.

Example 4 Vector Between Two Points

Find the vector $\overrightarrow{P_1P_2}$ if the points P_1 and P_2 are given by $P_1(4, 6, -2)$ and $P_2(1, 8, 3)$.

SOLUTION If the position vectors of the points are $\overrightarrow{OP_1} = \langle 4, 6, -2 \rangle$ and $\overrightarrow{OP_2} = \langle 1, 8, 3 \rangle$, then from (3) we have

$$\overrightarrow{P_1P_2} = \overrightarrow{OP_2} - \overrightarrow{OP_1} = \langle 1 - 4, 8 - 6, 3 - (-2) \rangle = \langle -3, 2, 5 \rangle. \qquad \square$$

Example 5 Magnitude of a Vector

From part (*vii*) of Definition 7.2, we see that $\mathbf{a} = \langle -\frac{2}{7}, \frac{3}{7}, \frac{6}{7} \rangle$ is a unit vector, since

$$\|\mathbf{a}\| = \sqrt{\left(-\frac{2}{7}\right)^2 + \left(\frac{3}{7}\right)^2 + \left(\frac{6}{7}\right)^2} = \sqrt{\frac{4 + 9 + 36}{49}} = 1. \qquad \square$$

■ **The i, j, k Vectors** We saw in the preceding section that the unit vectors $\mathbf{i} = \langle 1, 0 \rangle$ and $\mathbf{j} = \langle 0, 1 \rangle$ are a basis for the system of two-dimensional vectors in that any vector \mathbf{a} in 2-space can be written as a linear combination of \mathbf{i} and \mathbf{j}: $\mathbf{a} = a_1\mathbf{i} + a_2\mathbf{j}$. A basis for the system of three-dimensional vectors is given by the set of unit vectors

$$\mathbf{i} = \langle 1, 0, 0 \rangle, \quad \mathbf{j} = \langle 0, 1, 0 \rangle, \quad \mathbf{k} = \langle 0, 0, 1 \rangle.$$

Any vector $\mathbf{a} = \langle a_1, a_2, a_3 \rangle$ in 3-space can be expressed as a linear combination of \mathbf{i}, \mathbf{j}, and \mathbf{k}:

$$\langle a_1, a_2, a_3 \rangle = \langle a_1, 0, 0 \rangle + \langle 0, a_2, 0 \rangle + \langle 0, 0, a_3 \rangle$$

$$= a_1\langle 1, 0, 0 \rangle + a_2\langle 0, 1, 0 \rangle + a_3\langle 0, 0, 1 \rangle,$$

that is, $\qquad\qquad \mathbf{a} = a_1\mathbf{i} + a_2\mathbf{j} + a_3\mathbf{k}.$

The vectors \mathbf{i}, \mathbf{j}, and \mathbf{k} are illustrated in Figure 7.29(a). In Figure 7.29(b) we see that a position vector $\mathbf{a} = a_1\mathbf{i} + a_2\mathbf{j} + a_3\mathbf{k}$ is the sum of the vectors $a_1\mathbf{i}$, $a_2\mathbf{j}$, and $a_3\mathbf{k}$, which lie along the coordinate axes and have the origin as a common initial point.

(a)

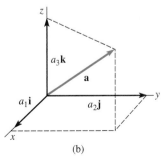

(b)

Figure 7.29

Example 6 Vector Expressed in Terms of i, j, k

The vector $\mathbf{a} = \langle 7, -5, 13 \rangle$ is the same as $\mathbf{a} = 7\mathbf{i} - 5\mathbf{j} + 13\mathbf{k}.$ $\qquad \square$

When the third dimension is taken into consideration, any vector in the xy-plane is equivalently described as a three-dimensional vector that lies in the coordinate plane $z = 0$. Although the vectors $\langle a_1, a_2 \rangle$ and $\langle a_1, a_2, 0 \rangle$ are technically not equal, we shall ignore the distinction. That is why, for example, we denoted $\langle 1, 0 \rangle$ and $\langle 1, 0, 0 \rangle$ by the same symbol \mathbf{i}. But to avoid any possible confusion, hereafter we shall always consider a vector a three-dimensional vector, and the symbols \mathbf{i} and \mathbf{j} will represent only $\langle 1, 0, 0 \rangle$ and $\langle 0, 1, 0 \rangle$, respectively. Similarly, a vector in either the xy-plane or the xz-plane must have one zero component. In the yz-plane a vector

$$\mathbf{b} = \langle 0, b_2, b_3 \rangle \quad \text{is written} \quad \mathbf{b} = b_2\mathbf{j} + b_3\mathbf{k}.$$

In the xz-plane a vector

$$\mathbf{c} = \langle c_1, 0, c_3 \rangle \quad \text{is the same as} \quad \mathbf{c} = c_1\mathbf{i} + c_3\mathbf{k}.$$

Example 7 Vector in *xz*-Plane

(a) The vector $\mathbf{a} = 5\mathbf{i} + 3\mathbf{k}$ is in the *xz*-coordinate plane.

(b) $\|5\mathbf{i} + 3\mathbf{k}\| = \sqrt{5^2 + 3^2} = \sqrt{34}$. ❏

Example 8 Linear Combination

If $\mathbf{a} = 3\mathbf{i} - 4\mathbf{j} + 8\mathbf{k}$ and $\mathbf{b} = \mathbf{i} - 4\mathbf{k}$, find $5\mathbf{a} - 2\mathbf{b}$.

SOLUTION We treat \mathbf{b} as a three-dimensional vector and write, for emphasis, $\mathbf{b} = \mathbf{i} + 0\mathbf{j} - 4\mathbf{k}$. From

$$5\mathbf{a} = 15\mathbf{i} - 20\mathbf{j} + 40\mathbf{k} \quad \text{and} \quad 2\mathbf{b} = 2\mathbf{i} + 0\mathbf{j} - 8\mathbf{k}$$

we get
$$5\mathbf{a} - 2\mathbf{b} = (15\mathbf{i} - 20\mathbf{j} + 40\mathbf{k}) - (2\mathbf{i} + 0\mathbf{j} - 8\mathbf{k})$$
$$= 13\mathbf{i} - 20\mathbf{j} + 48\mathbf{k}.$$ ❏

EXERCISES 7.2

Answers to odd-numbered problems begin on page A-41.

In Problems 1–6, graph the given point. Use the same coordinate axes.

1. $(1, 1, 5)$ **2.** $(0, 0, 4)$

3. $(3, 4, 0)$ **4.** $(6, 0, 0)$

5. $(6, -2, 0)$ **6.** $(5, -4, 3)$

In Problems 7–10, describe geometrically all points $P(x, y, z)$ that satisfy the given condition.

7. $z = 5$ **8.** $x = 1$

9. $x = 2, \quad y = 3$ **10.** $x = 4, \quad y = -1, \quad z = 7$

11. Give the coordinates of the vertices of the rectangular parallelepiped whose sides are the coordinate planes and the planes $x = 2$, $y = 5$, $z = 8$.

12. In Figure 7.30, two vertices are shown of a rectangular parallelepiped having sides parallel to the coordinate planes. Find the coordinates of the remaining six vertices.

Figure 7.30

13. Consider the point $P(-2, 5, 4)$.

(a) If lines are drawn from P perpendicular to the coordinate planes, what are the coordinates of the point at the base of each perpendicular?

(b) If a line is drawn from P to the plane $z = -2$, what are the coordinates of the point at the base of the perpendicular?

(c) Find the point in the plane $x = 3$ that is closest to P.

14. Determine an equation of a plane parallel to a coordinate plane that contains the given pair of points.

(a) $(3, 4, -5), (-2, 8, -5)$

(b) $(1, -1, 1), (1, -1, -1)$

(c) $(-2, 1, 2), (2, 4, 2)$

In Problems 15–20, describe the locus of points $P(x, y, z)$ that satisfy the given equation(s).

15. $xyz = 0$ **16.** $x^2 + y^2 + z^2 = 0$

17. $(x + 1)^2 + (y - 2)^2 + (z + 3)^2 = 0$

18. $(x - 2)(z - 8) = 0$

19. $z^2 - 25 = 0$ **20.** $x = y = z$

In Problems 21 and 22, find the distance between the given points.

21. $(3, -1, 2), (6, 4, 8)$ **22.** $(-1, -3, 5), (0, 4, 3)$

23. Find the distance from the point $(7, -3, -4)$ to (a) the *yz*-plane and (b) the *x*-axis.

24. Find the distance from the point $(-6, 2, -3)$ to (a) the *xz*-plane and (b) the origin.

In Problems 25–28, the given three points form a triangle. Determine which triangles are isosceles and which are right triangles.

25. $(0, 0, 0), (3, 6, -6), (2, 1, 2)$

26. $(0, 0, 0), (1, 2, 4), (3, 2, 2\sqrt{2})$

27. $(1, 2, 3)$, $(4, 1, 3)$, $(4, 6, 4)$

28. $(1, 1, -1)$, $(1, 1, 1)$, $(0, -1, 1)$

In Problems 29 and 30, use the distance formula to prove that the given points are collinear.

29. $P_1(1, 2, 0)$, $P_2(-2, -2, -3)$, $P_3(7, 10, 6)$

30. $P_1(2, 3, 2)$, $P_2(1, 4, 4)$, $P_3(5, 0, -4)$

In Problems 31 and 32, solve for the unknown.

31. $P_1(x, 2, 3)$, $P_2(2, 1, 1)$; $d(P_1, P_2) = \sqrt{21}$

32. $P_1(x, x, 1)$, $P_2(0, 3, 5)$; $d(P_1, P_2) = 5$

In Problems 33 and 34, find the coordinates of the midpoint of the line segment between the given points.

33. $(1, 3, \frac{1}{2})$, $(7, -2, \frac{5}{2})$ **34.** $(0, 5, -8)$, $(4, 1, -6)$

35. The coordinates of the midpoint of the line segment between $P_1(x_1, y_1, z_1)$ and $P_2(2, 3, 6)$ are $(-1, -4, 8)$. Find the coordinates of P_1.

36. Let P_3 be the midpoint of the line segment between $P_1(-3, 4, 1)$ and $P_2(-5, 8, 3)$. Find the coordinates of the midpoint of the line segment **(a)** between P_1 and P_3 and **(b)** between P_3 and P_2.

In Problems 37–40, find the vector $\overrightarrow{P_1P_2}$.

37. $P_1(3, 4, 5)$, $P_2(0, -2, 6)$

38. $P_1(-2, 4, 0)$, $P_2(6, \frac{3}{2}, 8)$

39. $P_1(0, -1, 0)$, $P_2(2, 0, 1)$

40. $P_1(\frac{1}{2}, \frac{3}{4}, 5)$, $P_2(-\frac{5}{2}, -\frac{9}{4}, 12)$

In Problems 41–48, $\mathbf{a} = \langle 1, -3, 2 \rangle$, $\mathbf{b} = \langle -1, 1, 1 \rangle$, and $\mathbf{c} = \langle 2, 6, 9 \rangle$. Find the indicated vector or scalar.

41. $\mathbf{a} + (\mathbf{b} + \mathbf{c})$ **42.** $2\mathbf{a} - (\mathbf{b} - \mathbf{c})$

43. $\mathbf{b} + 2(\mathbf{a} - 3\mathbf{c})$ **44.** $4(\mathbf{a} + 2\mathbf{c}) - 6\mathbf{b}$

45. $\|\mathbf{a} + \mathbf{c}\|$ **46.** $\|\mathbf{c}\| \|2\mathbf{b}\|$

47. $\left\| \dfrac{\mathbf{a}}{\|\mathbf{a}\|} \right\| + 5 \left\| \dfrac{\mathbf{b}}{\|\mathbf{b}\|} \right\|$ **48.** $\|\mathbf{b}\| \mathbf{a} + \|\mathbf{a}\| \mathbf{b}$

49. Find a unit vector in the opposite direction of $\mathbf{a} = \langle 10, -5, 10 \rangle$.

50. Find a unit vector in the same direction as $\mathbf{a} = \mathbf{i} - 3\mathbf{j} + 2\mathbf{k}$.

51. Find a vector \mathbf{b} that is 4 times as long as $\mathbf{a} = \mathbf{i} - \mathbf{j} + \mathbf{k}$ in the same direction as \mathbf{a}.

52. Find a vector \mathbf{b} for which $\|\mathbf{b}\| = \frac{1}{2}$ that is parallel to $\mathbf{a} = \langle -6, 3, -2 \rangle$ but has the opposite direction.

53. Using the vectors \mathbf{a} and \mathbf{b} shown in Figure 7.31, sketch the "average vector" $\frac{1}{2}(\mathbf{a} + \mathbf{b})$.

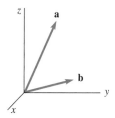

Figure 7.31

7.3 The Dot Product

(a)

(b)

(c)

Figure 7.32

In this and the following section, we shall consider two kinds of products between vectors that originated in the study of mechanics and electricity and magnetism. The first of these products, known as the **dot product, inner product,** or **scalar product,** yields a scalar. The dot product of two vectors \mathbf{a} and \mathbf{b} is denoted by $\mathbf{a} \cdot \mathbf{b}$.

DEFINITION 7.3 Dot Product of Two Vectors

The **dot product** of two vectors \mathbf{a} and \mathbf{b} is the scalar

$$\mathbf{a} \cdot \mathbf{b} = \|\mathbf{a}\| \|\mathbf{b}\| \cos \theta, \tag{1}$$

where θ is the angle between the vectors such that $0 \le \theta \le \pi$.

Figure 7.32 illustrates the angle θ in three cases. If the vectors \mathbf{a} and \mathbf{b} are not parallel, then θ is the *smaller* of the two possible angles between them.

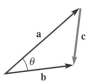

Figure 7.33

Example 1 **Dot Product Using (1)**

From (1) we obtain

$$\mathbf{i} \cdot \mathbf{i} = 1, \quad \mathbf{j} \cdot \mathbf{j} = 1, \quad \mathbf{k} \cdot \mathbf{k} = 1, \tag{2}$$

since $\|\mathbf{i}\| = \|\mathbf{j}\| = \|\mathbf{k}\| = 1$, and in each case $\cos \theta = 1$. ❑

■ **Component Form of Dot Product** The dot product can be expressed in terms of the components of two vectors. Suppose θ is the angle between the vectors $\mathbf{a} = a_1\mathbf{i} + a_2\mathbf{j} + a_3\mathbf{k}$ and $\mathbf{b} = b_1\mathbf{i} + b_2\mathbf{j} + b_3\mathbf{k}$. Then the vector

$$\mathbf{c} = \mathbf{b} - \mathbf{a} = (b_1 - a_1)\mathbf{i} + (b_2 - a_2)\mathbf{j} + (b_3 - a_3)\mathbf{k}$$

is the third side of the triangle indicated in Figure 7.33. By the law of cosines we can write

$$\|\mathbf{c}\|^2 = \|\mathbf{b}\|^2 + \|\mathbf{a}\|^2 - 2\|\mathbf{a}\|\|\mathbf{b}\| \cos \theta \quad \text{or} \quad \|\mathbf{a}\|\|\mathbf{b}\| \cos \theta = \tfrac{1}{2}(\|\mathbf{b}\|^2 + \|\mathbf{a}\|^2 - \|\mathbf{c}\|^2). \tag{3}$$

Using $\|\mathbf{a}\|^2 = a_1^2 + a_2^2 + a_3^2$, $\|\mathbf{b}\|^2 = b_1^2 + b_2^2 + b_3^2$, $\|\mathbf{b} - \mathbf{a}\|^2 = (b_1 - a_1)^2 + (b_2 - a_2)^2 + (b_3 - a_3)^2$, we can simplify the right side of the second equation in (3) to $a_1b_1 + a_2b_2 + a_3b_3$. Since the left side of this equation is the definition of the dot product, we have derived an alternative form of the dot product:

$$\mathbf{a} \cdot \mathbf{b} = a_1b_1 + a_2b_2 + a_3b_3. \tag{4}$$

In other words, the dot product of two vectors is the *sum of the products of their corresponding components*.

Example 2 **Dot Product Using (4)**

If $\mathbf{a} = 10\mathbf{i} + 2\mathbf{j} - 6\mathbf{k}$ and $\mathbf{b} = -\tfrac{1}{2}\mathbf{i} + 4\mathbf{j} - 3\mathbf{k}$, then it follows from (4) that

$$\mathbf{a} \cdot \mathbf{b} = (10)\left(-\frac{1}{2}\right) + (2)(4) + (-6)(-3) = 21.$$ ❑

■ **Properties** The dot product possesses the following properties.

Properties of the Dot Product

(*i*) $\mathbf{a} \cdot \mathbf{b} = 0$ if $\mathbf{a} = \mathbf{0}$ or $\mathbf{b} = \mathbf{0}$

(*ii*) $\mathbf{a} \cdot \mathbf{b} = \mathbf{b} \cdot \mathbf{a}$ (commutative law)

(*iii*) $\mathbf{a} \cdot (\mathbf{b} + \mathbf{c}) = \mathbf{a} \cdot \mathbf{b} + \mathbf{a} \cdot \mathbf{c}$ (distributive law)

(*iv*) $\mathbf{a} \cdot (k\mathbf{b}) = (k\mathbf{a}) \cdot \mathbf{b} = k(\mathbf{a} \cdot \mathbf{b})$, k a scalar

(*v*) $\mathbf{a} \cdot \mathbf{a} \geq 0$

(*vi*) $\mathbf{a} \cdot \mathbf{a} = \|\mathbf{a}\|^2$

Each of these properties, with the possible exception of (*iii*), should be apparent from (1). Notice that (*vi*) states that the magnitude of a vector

$$\mathbf{a} = a_1\mathbf{i} + a_2\mathbf{j} + a_3\mathbf{k}$$

can be written in terms of the dot product:

$$\|\mathbf{a}\| = \sqrt{\mathbf{a} \cdot \mathbf{a}} = \sqrt{a_1^2 + a_2^2 + a_3^2}.$$

We can use (4) to prove (*iii*). If $\mathbf{a} = a_1\mathbf{i} + a_2\mathbf{j} + a_3\mathbf{k}$, $\mathbf{b} = b_1\mathbf{i} + b_2\mathbf{j} + b_3\mathbf{k}$, and $\mathbf{c} = c_1\mathbf{i} + c_2\mathbf{j} + c_3\mathbf{k}$, then from (4) we have

$$\mathbf{a} \cdot (\mathbf{b} + \mathbf{c}) = a_1(b_1 + c_1) + a_2(b_2 + c_2) + a_3(b_3 + c_3)$$

$$= (a_1b_1 + a_2b_2 + a_3b_3) + (a_1c_1 + a_2c_2 + a_3c_3)$$

$$= \mathbf{a} \cdot \mathbf{b} + \mathbf{a} \cdot \mathbf{c}.$$

■ **Orthogonal Vectors** If \mathbf{a} and \mathbf{b} are nonzero vectors, then Definition 7.3 implies that

(*i*) $\mathbf{a} \cdot \mathbf{b} > 0$ if and only if θ is acute,

(*ii*) $\mathbf{a} \cdot \mathbf{b} < 0$ if and only if θ is obtuse, and

(*iii*) $\mathbf{a} \cdot \mathbf{b} = 0$ if and only if $\cos \theta = 0$.

But in the last case the only number in $[0, \pi]$ for which $\cos \theta = 0$ is $\theta = \pi/2$. When $\theta = \pi/2$, we say that the vectors are **perpendicular** or **orthogonal.** Thus we are led to the following result:

> **THEOREM 7.1 Criterion for Orthogonal Vectors**
>
> Two nonzero vectors \mathbf{a} and \mathbf{b} are orthogonal if and only if $\mathbf{a} \cdot \mathbf{b} = 0$.

Since $\mathbf{0} \cdot \mathbf{b} = 0$ for every vector \mathbf{b}, the zero vector is regarded to be orthogonal to every vector.

Example 3 i, j, k Are Orthogonal Vectors

It follows immediately from Theorem 7.1 and the fact that the dot product is commutative that

$$\mathbf{i} \cdot \mathbf{j} = \mathbf{j} \cdot \mathbf{i} = 0, \qquad \mathbf{j} \cdot \mathbf{k} = \mathbf{k} \cdot \mathbf{j} = 0, \qquad \mathbf{k} \cdot \mathbf{i} = \mathbf{i} \cdot \mathbf{k} = 0. \qquad (5) \quad ❑$$

Example 4 Orthogonal Vectors

If $\mathbf{a} = -3\mathbf{i} - \mathbf{j} + 4\mathbf{k}$ and $\mathbf{b} = 2\mathbf{i} + 14\mathbf{j} + 5\mathbf{k}$, then

$$\mathbf{a} \cdot \mathbf{b} = (-3)(2) + (-1)(14) + (4)(5) = 0.$$

From Theorem 7.1, we conclude that \mathbf{a} and \mathbf{b} are orthogonal. ❑

■ **Angle Between Two Vectors** By equating the two forms of the dot product, (1) and (4), we can determine **the angle between two vectors** from

$$\cos \theta = \frac{a_1b_1 + a_2b_2 + a_3b_3}{\|\mathbf{a}\|\|\mathbf{b}\|}. \qquad (6)$$

Example 5 Angles Between Two Vectors

Find the angle between $\mathbf{a} = 2\mathbf{i} + 3\mathbf{j} + \mathbf{k}$ and $\mathbf{b} = -\mathbf{i} + 5\mathbf{j} + \mathbf{k}$.

SOLUTION From $\|\mathbf{a}\| = \sqrt{14}$, $\|\mathbf{b}\| = \sqrt{27}$, $\mathbf{a} \cdot \mathbf{b} = 14$, we see from (6) that

$$\cos \theta = \frac{14}{\sqrt{14}\sqrt{27}} = \frac{\sqrt{42}}{9}$$

and so $\theta = \cos^{-1}\left(\dfrac{\sqrt{42}}{9}\right) \approx 0.77$ radian or $\theta \approx 44.9°$. ❏

Figure 7.34

■ **Direction Cosines** For a nonzero vector $\mathbf{a} = a_1\mathbf{i} + a_2\mathbf{j} + a_3\mathbf{k}$ in 3-space, the angles α, β, and γ between \mathbf{a} and the unit vectors \mathbf{i}, \mathbf{j}, and \mathbf{k}, respectively, are called **direction angles of a**. See Figure 7.34. Now, by (6),

$$\cos \alpha = \frac{\mathbf{a} \cdot \mathbf{i}}{\|\mathbf{a}\|\|\mathbf{i}\|}, \quad \cos \beta = \frac{\mathbf{a} \cdot \mathbf{j}}{\|\mathbf{a}\|\|\mathbf{j}\|}, \quad \cos \gamma = \frac{\mathbf{a} \cdot \mathbf{k}}{\|\mathbf{a}\|\|\mathbf{k}\|},$$

which simplify to

$$\cos \alpha = \frac{a_1}{\|\mathbf{a}\|}, \qquad \cos \beta = \frac{a_2}{\|\mathbf{a}\|}, \qquad \cos \gamma = \frac{a_3}{\|\mathbf{a}\|}.$$

We say that $\cos \alpha$, $\cos \beta$, and $\cos \gamma$ are the **direction cosines** of \mathbf{a}. The direction cosines of a nonzero vector \mathbf{a} are simply the components of the unit vector $(1/\|\mathbf{a}\|)\mathbf{a}$:

$$\frac{1}{\|\mathbf{a}\|}\,\mathbf{a} = \frac{a_1}{\|\mathbf{a}\|}\,\mathbf{i} + \frac{a_2}{\|\mathbf{a}\|}\,\mathbf{j} + \frac{a_3}{\|\mathbf{a}\|}\,\mathbf{k} = (\cos \alpha)\mathbf{i} + (\cos \beta)\mathbf{j} + (\cos \gamma)\mathbf{k}$$

Since the magnitude of $(1/\|\mathbf{a}\|)\mathbf{a}$ is 1, it follows from the last equation that

$$\cos^2\alpha + \cos^2\beta + \cos^2\gamma = 1.$$

> ### *Example 6* Direction Cosines/Angles
> Find the direction cosines and direction angles of the vector $\mathbf{a} = 2\mathbf{i} + 5\mathbf{j} + 4\mathbf{k}$.
> **SOLUTION** From $\|\mathbf{a}\| = \sqrt{2^2 + 5^2 + 4^2} = \sqrt{45} = 3\sqrt{5}$, we see that the direction cosines are
>
> $$\cos \alpha = \frac{2}{3\sqrt{5}}, \quad \cos \beta = \frac{5}{3\sqrt{5}}, \quad \cos \gamma = \frac{4}{3\sqrt{5}}.$$
>
> The direction angles are
>
> $$\alpha = \cos^{-1}\left(\frac{2}{3\sqrt{5}}\right) \approx 1.27 \text{ radians} \quad \text{or} \quad \alpha \approx 72.7°$$
>
> $$\beta = \cos^{-1}\left(\frac{5}{3\sqrt{5}}\right) \approx 0.73 \text{ radian} \quad \text{or} \quad \beta \approx 41.8°$$
>
> $$\gamma = \cos^{-1}\left(\frac{4}{3\sqrt{5}}\right) \approx 0.93 \text{ radian} \quad \text{or} \quad \gamma \approx 53.4°.$$ ❏

Observe in Example 6 that

$$\cos^2\alpha + \cos^2\beta + \cos^2\gamma = \frac{4}{45} + \frac{25}{45} + \frac{16}{45} = 1.$$

■ **Component of a on b** The distributive law and (5) enable us to express the components of a vector $\mathbf{a} = a_1\mathbf{i} + a_2\mathbf{j} + a_3\mathbf{k}$ in terms of the dot product:

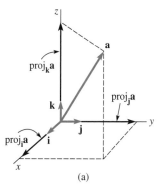

Figure 7.35

$$a_1 = \mathbf{a} \cdot \mathbf{i}, \quad a_2 = \mathbf{a} \cdot \mathbf{j}, \quad a_3 = \mathbf{a} \cdot \mathbf{k}. \tag{7}$$

Symbolically, we write the components of **a** as

$$\text{comp}_i \mathbf{a} = \mathbf{a} \cdot \mathbf{i}, \quad \text{comp}_j \mathbf{a} = \mathbf{a} \cdot \mathbf{j}, \quad \text{comp}_k \mathbf{a} = \mathbf{a} \cdot \mathbf{k}. \tag{8}$$

We shall now see that the results indicated in (8) carry over to finding the **component of a on an arbitrary vector b.** Note that in either of the two cases shown in Figure 7.35,

$$\text{comp}_\mathbf{b} \mathbf{a} = \|\mathbf{a}\| \cos \theta. \tag{9}$$

In Figure 7.35(b), $\text{comp}_\mathbf{b} \mathbf{a} < 0$ since $\pi/2 < \theta \le \pi$. Now, by writing (9) as

$$\text{comp}_\mathbf{b} \mathbf{a} = \frac{\|\mathbf{a}\|\|\mathbf{b}\| \cos \theta}{\|\mathbf{b}\|} = \frac{\mathbf{a} \cdot \mathbf{b}}{\|\mathbf{b}\|},$$

we see that

$$\text{comp}_\mathbf{b} \mathbf{a} = \mathbf{a} \cdot \left(\frac{1}{\|\mathbf{b}\|} \mathbf{b} \right). \tag{10}$$

In other words, *to find the component of **a** on a vector **b**, we dot **a** with a unit vector in the direction of **b**.*

Example 7 Component of a Vector on Another Vector

Let $\mathbf{a} = 2\mathbf{i} + 3\mathbf{j} - 4\mathbf{k}$ and $\mathbf{b} = \mathbf{i} + \mathbf{j} + 2\mathbf{k}$. Find $\text{comp}_\mathbf{b} \mathbf{a}$ and $\text{comp}_\mathbf{a} \mathbf{b}$.

SOLUTION We first form a unit vector in the direction of **b**:

$$\|\mathbf{b}\| = \sqrt{6}, \qquad \frac{1}{\|\mathbf{b}\|} \mathbf{b} = \frac{1}{\sqrt{6}} (\mathbf{i} + \mathbf{j} + 2\mathbf{k}).$$

Then from (10) we have

$$\text{comp}_\mathbf{b} \mathbf{a} = (2\mathbf{i} + 3\mathbf{j} - 4\mathbf{k}) \cdot \frac{1}{\sqrt{6}} (\mathbf{i} + \mathbf{j} + 2\mathbf{k}) = -\frac{3}{\sqrt{6}}.$$

By modifying (10) accordingly, we have

$$\text{comp}_\mathbf{a} \mathbf{b} = \mathbf{b} \cdot \left(\frac{1}{\|\mathbf{a}\|} \mathbf{a} \right).$$

Therefore, $\|\mathbf{a}\| = \sqrt{29}, \qquad \frac{1}{\|\mathbf{a}\|} \mathbf{a} = \frac{1}{\sqrt{29}} (2\mathbf{i} + 3\mathbf{j} - 4\mathbf{k})$

and $\text{comp}_\mathbf{a} \mathbf{b} = (\mathbf{i} + \mathbf{j} + 2\mathbf{k}) \cdot \frac{1}{\sqrt{29}} (2\mathbf{i} + 3\mathbf{j} - 4\mathbf{k}) = -\frac{3}{\sqrt{29}}.$ ❑

■ **Projection of a onto b** As illustrated in Figure 7.36(a), the **projection** of a vector **a** in any of the directions determined by **i**, **j**, **k** is simply the *vector* formed by multiplying the component of **a** in the specified direction with a unit vector in that direction; for example,

$$\text{proj}_i \mathbf{a} = (\text{comp}_i \mathbf{a})\mathbf{i} = (\mathbf{a} \cdot \mathbf{i})\mathbf{i} = a_1 \mathbf{i}$$

and so on. Figure 7.36(b) shows the general case of the **projection of a onto b**:

$$\text{proj}_\mathbf{b} \mathbf{a} = (\text{comp}_\mathbf{b} \mathbf{a}) \left(\frac{1}{\|\mathbf{b}\|} \mathbf{b} \right). \tag{11}$$

Figure 7.36

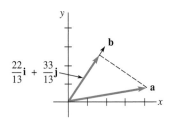

Figure 7.37

Example 8 Projection of a Vector on Another Vector

Find the projection of $\mathbf{a} = 4\mathbf{i} + \mathbf{j}$ on the vector $\mathbf{b} = 2\mathbf{i} + 3\mathbf{j}$. Graph.

SOLUTION First, we find the component of \mathbf{a} and \mathbf{b}. Since $\|\mathbf{b}\| = \sqrt{13}$, we find from (10) that

$$\text{comp}_{\mathbf{b}}\mathbf{a} = (4\mathbf{i} + \mathbf{j}) \cdot \frac{1}{\sqrt{13}} (2\mathbf{i} + 3\mathbf{j}) = \frac{11}{\sqrt{13}}.$$

Thus, from (11),

$$\text{proj}_{\mathbf{b}}\mathbf{a} = \left(\frac{11}{\sqrt{13}} \right)\left(\frac{1}{\sqrt{13}} \right)(2\mathbf{i} + 3\mathbf{j}) = \frac{22}{13}\mathbf{i} + \frac{33}{13}\mathbf{j}.$$

The graph of this vector is shown in Figure 7.37. ❏

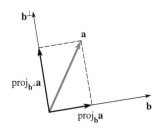

Figure 7.38

■ **Projection of a onto b$^{\perp}$** If $\mathbf{b} \neq \mathbf{0}$, any vector \mathbf{a} can be projected onto \mathbf{b}, as well as onto a vector \mathbf{b}^{\perp} of magnitude $\|\mathbf{b}\|$, that is orthogonal to \mathbf{b}. From Figure 7.38, we see that \mathbf{a} can be written as the sum of two projections:

$$\text{proj}_{\mathbf{b}}\mathbf{a} + \text{proj}_{\mathbf{b}^{\perp}}\mathbf{a} = \mathbf{a}. \tag{12}$$

Equation (12) enables us to define the projection of \mathbf{a} onto \mathbf{b}^{\perp}:

$$\text{proj}_{\mathbf{b}^{\perp}}\mathbf{a} = \mathbf{a} - \text{proj}_{\mathbf{b}}\mathbf{a}. \tag{13}$$

Example 9 Projection of a Vector on Another Vector

Let $\mathbf{a} = 3\mathbf{i} - \mathbf{j} + 5\mathbf{k}$ and $\mathbf{b} = 2\mathbf{i} + \mathbf{j} + 2\mathbf{k}$. Find $\text{proj}_{\mathbf{b}}\mathbf{a}$ and $\text{proj}_{\mathbf{b}^{\perp}}\mathbf{a}$.

SOLUTION Since $\|\mathbf{b}\| = 3$, we have

$$\text{comp}_{\mathbf{b}}\mathbf{a} = (3\mathbf{i} - \mathbf{j} + 5\mathbf{k}) \cdot \frac{1}{3} (2\mathbf{i} + \mathbf{j} + 2\mathbf{k}) = \frac{15}{3} = 5$$

$$\text{proj}_{\mathbf{b}}\mathbf{a} = (5)\left(\frac{1}{3} \right)(2\mathbf{i} + \mathbf{j} + 2\mathbf{k}) = \frac{10}{3}\mathbf{i} + \frac{5}{3}\mathbf{j} + \frac{10}{3}\mathbf{k}$$

$$\text{proj}_{\mathbf{b}^{\perp}}\mathbf{a} = \mathbf{a} - \text{proj}_{\mathbf{b}}\mathbf{a}$$

$$= (3\mathbf{i} - \mathbf{j} + 5\mathbf{k}) - \left(\frac{10}{3}\mathbf{i} + \frac{5}{3}\mathbf{j} + \frac{10}{3}\mathbf{k} \right) = -\frac{1}{3}\mathbf{i} - \frac{8}{3}\mathbf{j} + \frac{5}{3}\mathbf{k}. \quad ❏$$

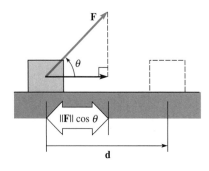

Figure 7.39

Note that (14) reduces to 👉
$W = Fd$ *when* $\theta = 0$.

■ **Physical Interpretation of the Dot Product** When a constant force of magnitude F moves an object a distance d in the same direction of the force, the work done is simply $W = Fd$. However, if a constant force \mathbf{F} applied to a body acts at an angle θ to the direction of motion, then the work done by \mathbf{F} is defined to be the product of the component of \mathbf{F} in the direction of the displacement and the distance $\|\mathbf{d}\|$ that the body moves:

$$W = (\|\mathbf{F}\| \cos \theta)\|\mathbf{d}\| = \|\mathbf{F}\|\|\mathbf{d}\| \cos \theta.$$

See Figure 7.39. It follows from Definition 7.3 that if \mathbf{F} causes a displacement \mathbf{d} of a body, then the work done is

$$W = \mathbf{F} \cdot \mathbf{d}. \tag{14}$$

Example 10 **Work Done by a Constant Force**

Find the work done by a constant force $\mathbf{F} = 2\mathbf{i} + 4\mathbf{j}$ if its point of application to a block moves from $P_1(1, 1)$ to $P_2(4, 6)$. Assume that $\|\mathbf{F}\|$ is measured in newtons and $\|\mathbf{d}\|$ is measured in meters.

SOLUTION The displacement of the block is given by

$$\mathbf{d} = \overrightarrow{P_1P_2} = \overrightarrow{OP_2} - \overrightarrow{OP_1} = 3\mathbf{i} + 5\mathbf{j}.$$

It follows from (14) that the work done is

$$W = (2\mathbf{i} + 4\mathbf{j}) \cdot (3\mathbf{i} + 5\mathbf{j}) = 26 \text{ N-m}. \qquad \square$$

EXERCISES 7.3

Answers to odd-numbered problems begin on page A-41.

In Problems 1 and 2, find $\mathbf{a} \cdot \mathbf{b}$ if the smaller angle between \mathbf{a} and \mathbf{b} is as given.

1. $\|\mathbf{a}\| = 10$, $\|\mathbf{b}\| = 5$, $\theta = \pi/4$

2. $\|\mathbf{a}\| = 6$, $\|\mathbf{b}\| = 12$, $\theta = \pi/6$

In Problems 3–14, $\mathbf{a} = \langle 2, -3, 4 \rangle$, $\mathbf{b} = \langle -1, 2, 5 \rangle$, and $\mathbf{c} = \langle 3, 6, -1 \rangle$. Find the indicated scalar or vector.

3. $\mathbf{a} \cdot \mathbf{b}$

4. $\mathbf{b} \cdot \mathbf{c}$

5. $\mathbf{a} \cdot \mathbf{c}$

6. $\mathbf{a} \cdot (\mathbf{b} + \mathbf{c})$

7. $\mathbf{a} \cdot (4\mathbf{b})$

8. $\mathbf{b} \cdot (\mathbf{a} - \mathbf{c})$

9. $\mathbf{a} \cdot \mathbf{a}$

10. $(2\mathbf{b}) \cdot (3\mathbf{c})$

11. $\mathbf{a} \cdot (\mathbf{a} + \mathbf{b} + \mathbf{c})$

12. $(2\mathbf{a}) \cdot (\mathbf{a} - 2\mathbf{b})$

13. $\left(\dfrac{\mathbf{a} \cdot \mathbf{b}}{\mathbf{b} \cdot \mathbf{b}} \right) \mathbf{b}$

14. $(\mathbf{c} \cdot \mathbf{b})\mathbf{a}$

15. Determine which pairs of the following vectors are orthogonal:

(a) $\langle 2, 0, 1 \rangle$

(b) $3\mathbf{i} + 2\mathbf{j} - \mathbf{k}$

(c) $2\mathbf{i} - \mathbf{j} - \mathbf{k}$

(d) $\mathbf{i} - 4\mathbf{j} + 6\mathbf{k}$

(e) $\langle 1, -1, 1 \rangle$

(f) $\langle -4, 3, 8 \rangle$

16. Determine a scalar c so that the given vectors are orthogonal.

(a) $\mathbf{a} = 2\mathbf{i} - c\mathbf{j} + 3\mathbf{k}$, $\mathbf{b} = 3\mathbf{i} + 2\mathbf{j} + 4\mathbf{k}$

(b) $\mathbf{a} = \langle c, \frac{1}{2}, c \rangle$, $\mathbf{b} = \langle -3, 4, c \rangle$

17. Find a vector $\mathbf{v} = \langle x_1, y_1, 1 \rangle$ that is orthogonal to both $\mathbf{a} = \langle 3, 1, -1 \rangle$ and $\mathbf{b} = \langle -3, 2, 2 \rangle$.

18. A rhombus is an oblique-angled parallelogram with all four sides equal. Use the dot product to show that the diagonals of a rhombus are perpendicular.

19. Verify that the vector

$$\mathbf{c} = \mathbf{b} - \frac{\mathbf{a} \cdot \mathbf{b}}{\|\mathbf{a}\|^2} \mathbf{a}$$

is orthogonal to the vector \mathbf{a}.

20. Determine a scalar c so that the angle between $\mathbf{a} = \mathbf{i} + c\mathbf{j}$ and $\mathbf{b} = \mathbf{i} + \mathbf{j}$ is $45°$.

In Problems 21–24, find the angle θ between the given vectors.

21. $\mathbf{a} = 3\mathbf{i} - \mathbf{k}$, $\mathbf{b} = 2\mathbf{i} + 2\mathbf{k}$

22. $\mathbf{a} = 2\mathbf{i} + \mathbf{j}$, $\mathbf{b} = -3\mathbf{i} - 4\mathbf{j}$

23. $\mathbf{a} = \langle 2, 4, 0 \rangle$, $\mathbf{b} = \langle -1, -1, 4 \rangle$

24. $\mathbf{a} = \langle \frac{1}{2}, \frac{1}{2}, \frac{3}{2} \rangle$, $\mathbf{b} = \langle 2, -4, 6 \rangle$

In Problems 25–28, find the direction cosines and direction angles of the given vector.

25. $\mathbf{a} = \mathbf{i} + 2\mathbf{j} + 3\mathbf{k}$

26. $\mathbf{a} = 6\mathbf{i} + 6\mathbf{j} - 3\mathbf{k}$

27. $\mathbf{a} = \langle 1, 0, -\sqrt{3} \rangle$

28. $\mathbf{a} = \langle 5, 7, 2 \rangle$

29. Find the angle between the diagonal \overrightarrow{AD} of the cube shown in Figure 7.40 and the edge AB. Find the angle between the diagonal AD of the cube and the diagonal \overrightarrow{AC}.

30. Show that if two nonzero vectors \mathbf{a} and \mathbf{b} are orthogonal, then their direction cosines satisfy

$$\cos \alpha_1 \cos \alpha_2 + \cos \beta_1 \cos \beta_2 + \cos \gamma_1 \cos \gamma_2 = 0.$$

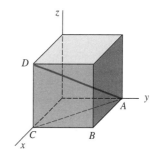

Figure 7.40

31. An airplane is 4 km high, 5 km south, and 7 km east of an airport. See Figure 7.41. Find the direction angles of the plane.

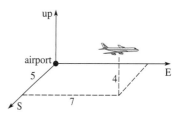

Figure 7.41

32. Determine a unit vector whose direction angles, relative to the three coordinate axes, are equal.

In Problems 33–36, $\mathbf{a} = \langle 1, -1, 3 \rangle$ and $\mathbf{b} = \langle 2, 6, 3 \rangle$. Find the indicated number.

33. $\text{comp}_\mathbf{b}\mathbf{a}$

34. $\text{comp}_\mathbf{a}\mathbf{b}$

35. $\text{comp}_\mathbf{a}(\mathbf{b} - \mathbf{a})$

36. $\text{comp}_{2\mathbf{b}}(\mathbf{a} + \mathbf{b})$

In Problems 37 and 38, find the component of the given vector in the direction from the origin to the indicated point.

37. $\mathbf{a} = 4\mathbf{i} + 6\mathbf{j}$, $P(3, 10)$

38. $\mathbf{a} = \langle 2, 1, -1 \rangle$, $P(1, -1, 1)$

In Problems 39–42, find **(a)** $\text{proj}_\mathbf{b}\mathbf{a}$ and **(b)** $\text{proj}_{\mathbf{b}^\perp}\mathbf{a}$.

39. $\mathbf{a} = -5\mathbf{i} + 5\mathbf{j}$, $\mathbf{b} = -3\mathbf{i} + 4\mathbf{j}$

40. $\mathbf{a} = 4\mathbf{i} + 2\mathbf{j}$, $\mathbf{b} = -3\mathbf{i} + \mathbf{j}$

41. $\mathbf{a} = -\mathbf{i} - 2\mathbf{j} + 7\mathbf{k}$, $\mathbf{b} = 6\mathbf{i} - 3\mathbf{j} - 2\mathbf{k}$

42. $\mathbf{a} = \langle 1, 1, 1 \rangle$, $\mathbf{b} = \langle -2, 2, -1 \rangle$

In Problems 43 and 44, $\mathbf{a} = 4\mathbf{i} + 3\mathbf{j}$ and $\mathbf{b} = -\mathbf{i} + \mathbf{j}$. Find the indicated vector.

43. $\text{proj}_{(\mathbf{a} + \mathbf{b})}\mathbf{a}$

44. $\text{proj}_{(\mathbf{a} - \mathbf{b})^\perp}\mathbf{b}$

45. A sled is pulled horizontally over ice by a rope attached to its front. A 20-lb force acting at an angle of 60° with the horizontal moves the sled 100 ft. Find the work done.

46. Find the work done if the point at which the constant force $\mathbf{F} = 4\mathbf{i} + 3\mathbf{j} + 5\mathbf{k}$ is applied to an object moves from $P_1(3, 1, -2)$ to $P_2(2, 4, 6)$. Assume $\|\mathbf{F}\|$ is measured in newtons and $\|\mathbf{d}\|$ is measured in meters.

47. A block with weight \mathbf{w} is pulled along a frictionless horizontal surface by a constant force \mathbf{F} of magnitude 30 newtons in the direction given by a vector \mathbf{d}. See Figure 7.42. Assume $\|\mathbf{d}\|$ is measured in meters.

Figure 7.42

(a) What is the work done by the weight \mathbf{w}?

(b) What is the work done by the force \mathbf{F} if $\mathbf{d} = 4\mathbf{i} + 3\mathbf{j}$?

48. A constant force \mathbf{F} of magnitude 3 lb is applied to the block shown in Figure 7.43. \mathbf{F} has the same direction as the vector $\mathbf{a} = 3\mathbf{i} + 4\mathbf{j}$. Find the work done in the direction of motion if the block moves from $P_1(3, 1)$ to $P_2(9, 3)$. Assume distance is measured in feet.

Figure 7.43

49. In the methane molecule CH_4, the hydrogen atoms are located at the four vertices of a regular tetrahedron. See Figure 7.44. The distance between the center of a hydrogen atom and the center of a carbon atom is 1.10 angstroms (1 angstrom $= 10^{-10}$ meter), and the hydrogen–carbon–hydrogen bond angle is $\theta = 109.5°$. Using vector methods only, find the distance between two hydrogen atoms.

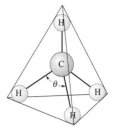

Figure 7.44

50. Use the dot product to prove the Cauchy–Schwarz inequality: $|\mathbf{a} \cdot \mathbf{b}| \le \|\mathbf{a}\|\|\mathbf{b}\|$.

51. Use the dot product to prove the triangle inequality, $\|\mathbf{a} + \mathbf{b}\| \le \|\mathbf{a}\| + \|\mathbf{b}\|$. [*Hint:* Consider property (*vi*) of the dot product.]

52. Prove that the vector $\mathbf{n} = a\mathbf{i} + b\mathbf{j}$ is perpendicular to the line whose equation is $ax + by + c = 0$. [*Hint:* Let $P_1(x_1, y_1)$ and $P_2(x_2, y_2)$ be distinct points on the line.]

53. Use the result of Problem 52 and Figure 7.45 to show that the distance d from a point $P_1(x_1, y_1)$ to a line $ax + by + c = 0$ is $d = |ax_1 + by_1 + c| / \sqrt{a^2 + b^2}$.

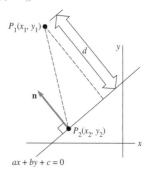

Figure 7.45

7.4 The Cross Product

In contrast to the dot product, which is a number or a scalar, the next special product of two vectors, called the **cross product,** is a vector.

> **DEFINITION 7.4 Cross Product of Two Vectors**
>
> The **cross product** of two vectors **a** and **b** in R^3 is the vector
>
> $$\mathbf{a} \times \mathbf{b} = (\|\mathbf{a}\|\|\mathbf{b}\| \sin \theta)\mathbf{n}, \tag{1}$$
>
> where θ is the angle between the vectors such that $0 \le \theta \le \pi$ and **n** is a unit vector perpendicular to the plane of **a** and **b** with direction given by the right-hand rule.

As seen in Figure 7.46(a), if the fingers of the right hand point along the vector **a** and then curl toward the vector **b,** the thumb will give the direction of **n** and hence **a** × **b**. In Figure 7.46(b) the right-hand rule shows the direction of **b** × **a**.

(a) (b)

Figure 7.46

Figure 7.47

Example 1 Torque as Cross Product

In physics a force **F** acting at the end of a position vector **r,** as shown in Figure 7.47, is said to produce a **torque** $\boldsymbol{\tau}$ defined by $\boldsymbol{\tau} = \mathbf{r} \times \mathbf{F}$. For example, if $\|\mathbf{F}\| = 20$ N, $\|\mathbf{r}\| = 3.5$ m, and $\theta = 30°$, then from (1) $\|\boldsymbol{\tau}\| = (3.5)(20)\sin 30° = 35$ N-m. If **F** and **r** are in the plane of the page, the right-hand rule implies that the direction of $\boldsymbol{\tau}$ is outward from, and perpendicular to, the page (toward the reader).

As we see in Figure 7.48, when a force **F** is applied to a wrench, the magnitude of the torque $\boldsymbol{\tau}$ is a measure of the turning effect about the pivot point P, and the vector $\boldsymbol{\tau}$ is directed along the axis of the bolt. In this case $\boldsymbol{\tau}$ points inward from the page. ❑

Figure 7.48

■ **Properties** The cross product has the following properties.

Properties of the Cross Product

(*i*) $\mathbf{a} \times \mathbf{b} = \mathbf{0}$ if $\mathbf{a} = \mathbf{0}$ or $\mathbf{b} = \mathbf{0}$

(*ii*) $\mathbf{a} \times \mathbf{b} = -\mathbf{b} \times \mathbf{a}$

(*iii*) $\mathbf{a} \times (\mathbf{b} + \mathbf{c}) = (\mathbf{a} \times \mathbf{b}) + (\mathbf{a} \times \mathbf{c})$

(*iv*) $(\mathbf{a} + \mathbf{b}) \times \mathbf{c} = (\mathbf{a} \times \mathbf{c}) + (\mathbf{b} \times \mathbf{c})$

(distributive laws)

(v) $\mathbf{a} \times (k\mathbf{b}) = (k\mathbf{a}) \times \mathbf{b} = k(\mathbf{a} \times \mathbf{b})$, k a scalar

(vi) $\mathbf{a} \times \mathbf{a} = \mathbf{0}$

(vii) $\mathbf{a} \cdot (\mathbf{a} \times \mathbf{b}) = 0$

(viii) $\mathbf{b} \cdot (\mathbf{a} \times \mathbf{b}) = 0$

Property (vi) follows from (1) because $\theta = 0$. Properties (vii) and (viii) are simply statements of the fact that $\mathbf{a} \times \mathbf{b}$ is perpendicular to the plane containing \mathbf{a} and \mathbf{b}. Property (ii) should be intuitively clear from Figure 7.46.

■ **Parallel Vectors** When the angle between two nonzero vectors is either $\theta = 0$ or $\theta = \pi$, then sin $\theta = 0$, and so we must have $\mathbf{a} \times \mathbf{b} = \mathbf{0}$. This is stated formally in the next theorem.

THEOREM 7.2 Criterion for Parallel Vectors

Two nonzero vectors \mathbf{a} and \mathbf{b} are **parallel** if and only if $\mathbf{a} \times \mathbf{b} = \mathbf{0}$.

Example 2 **Parallel Vectors**

(a) From property (vi) we have

$$\mathbf{i} \times \mathbf{i} = \mathbf{0}, \quad \mathbf{j} \times \mathbf{j} = \mathbf{0}, \quad \mathbf{k} \times \mathbf{k} = \mathbf{0}. \tag{2}$$

(b) If $\mathbf{a} = 2\mathbf{i} + \mathbf{j} - \mathbf{k}$ and $\mathbf{b} = -6\mathbf{i} - 3\mathbf{j} + 3\mathbf{k} = -3\mathbf{a}$, then \mathbf{a} and \mathbf{b} are parallel. Hence, from Theorem 7.2, $\mathbf{a} \times \mathbf{b} = \mathbf{0}$. Note that this result also follows by combining properties (v) and (vi). ❏

From (1), if $\mathbf{a} = \mathbf{i}$, $\mathbf{b} = \mathbf{j}$, then

$$\mathbf{i} \times \mathbf{j} = \left(\|\mathbf{i}\| \|\mathbf{j}\| \sin \frac{\pi}{2} \right) \mathbf{n} = \mathbf{n}. \tag{3}$$

But, since a unit vector perpendicular to the plane that contains \mathbf{i} and \mathbf{j} with the direction given by the right-hand rule is \mathbf{k}, it follows from (3) that $\mathbf{n} = \mathbf{k}$. In other words, $\mathbf{i} \times \mathbf{j} = \mathbf{k}$.

Example 3 **A Mnemonic**

The cross products of any pair of vectors in the set \mathbf{i}, \mathbf{j}, \mathbf{k} can be obtained by the circular mnemonic illustrated in Figure 7.49, that is,

$$\left. \begin{array}{l} \mathbf{i} \times \mathbf{j} = \mathbf{k} \\ \mathbf{j} \times \mathbf{k} = \mathbf{i} \\ \mathbf{k} \times \mathbf{i} = \mathbf{j} \end{array} \right\} \quad \text{and from property (ii)} \quad \left\{ \begin{array}{l} \mathbf{j} \times \mathbf{i} = -\mathbf{k} \\ \mathbf{k} \times \mathbf{j} = -\mathbf{i} . \\ \mathbf{i} \times \mathbf{k} = -\mathbf{j} \end{array} \right. \tag{4} ❏$$

Figure 7.49

■ **Alternative Definition of the Cross Product** As we did for the dot product, we can use the distributive law (iii) to arrive at an alternative form of the cross product:

$$\mathbf{a} \times \mathbf{b} = (a_1\mathbf{i} + a_2\mathbf{j} + a_3\mathbf{k}) \times (b_1\mathbf{i} + b_2\mathbf{j} + b_3\mathbf{k})$$

$$= a_1\mathbf{i} \times (b_1\mathbf{i} + b_2\mathbf{j} + b_3\mathbf{k}) + a_2\mathbf{j} \times (b_1\mathbf{i} + b_2\mathbf{j} + b_3\mathbf{k})$$

$$+ a_3\mathbf{k} \times (b_1\mathbf{i} + b_2\mathbf{j} + b_3\mathbf{k})$$

$$= a_1 b_1 (\mathbf{i} \times \mathbf{i}) + a_1 b_2 (\mathbf{i} \times \mathbf{j}) + a_1 b_3 (\mathbf{i} \times \mathbf{k})$$
$$+ a_2 b_1 (\mathbf{j} \times \mathbf{i}) + a_2 b_2 (\mathbf{j} \times \mathbf{j}) + a_2 b_3 (\mathbf{j} \times \mathbf{k})$$
$$+ a_3 b_1 (\mathbf{k} \times \mathbf{i}) + a_3 b_2 (\mathbf{k} \times \mathbf{j}) + a_3 b_3 (\mathbf{k} \times \mathbf{k}). \tag{5}$$

From the results in (2) and (4), (5) simplifies to

$$\mathbf{a} \times \mathbf{b} = (a_2 b_3 - a_3 b_2)\mathbf{i} - (a_1 b_3 - a_3 b_1)\mathbf{j} + (a_1 b_2 - a_2 b_1)\mathbf{k}. \tag{6}$$

We note that the components of the vector in (6) can be written as determinants of order 2:

$$\mathbf{a} \times \mathbf{b} = \begin{vmatrix} a_2 & a_3 \\ b_2 & b_3 \end{vmatrix} \mathbf{i} - \begin{vmatrix} a_1 & a_3 \\ b_1 & b_3 \end{vmatrix} \mathbf{j} + \begin{vmatrix} a_1 & a_2 \\ b_1 & b_2 \end{vmatrix} \mathbf{k}. \tag{7}$$

In turn, (7) can be written as a determinant of order 3:

$$\mathbf{a} \times \mathbf{b} = \begin{vmatrix} \mathbf{i} & \mathbf{j} & \mathbf{k} \\ a_1 & a_2 & a_3 \\ b_1 & b_2 & b_3 \end{vmatrix}. \tag{8}$$

The expression on the right-hand side of (8) is not an actual determinant since its entries are not all scalars; (8) is simply a way of remembering the complicated expression in (6).

> ### Example 4 Cross Product
> Let $\mathbf{a} = 4\mathbf{i} - 2\mathbf{j} + 5\mathbf{k}$ and $\mathbf{b} = 3\mathbf{i} + \mathbf{j} - \mathbf{k}$. Find $\mathbf{a} \times \mathbf{b}$.
> **SOLUTION** From (8) we have
> $$\mathbf{a} \times \mathbf{b} = \begin{vmatrix} \mathbf{i} & \mathbf{j} & \mathbf{k} \\ 4 & -2 & 5 \\ 3 & 1 & -1 \end{vmatrix} = \begin{vmatrix} -2 & 5 \\ 1 & -1 \end{vmatrix} \mathbf{i} - \begin{vmatrix} 4 & 5 \\ 3 & -1 \end{vmatrix} \mathbf{j} + \begin{vmatrix} 4 & -2 \\ 3 & 1 \end{vmatrix} \mathbf{k}$$
> $$= -3\mathbf{i} + 19\mathbf{j} + 10\mathbf{k}. \qquad \square$$

The form of the cross product given in (7) enables us to prove some of the properties (*i*)–(*viii*). For example, to prove (*ii*) we write

$$\mathbf{a} \times \mathbf{b} = \begin{vmatrix} a_2 & a_3 \\ b_2 & b_3 \end{vmatrix} \mathbf{i} - \begin{vmatrix} a_1 & a_3 \\ b_1 & b_3 \end{vmatrix} \mathbf{j} + \begin{vmatrix} a_1 & a_2 \\ b_1 & b_2 \end{vmatrix} \mathbf{k}$$

$$= -\begin{vmatrix} b_2 & b_3 \\ a_2 & a_3 \end{vmatrix} \mathbf{i} + \begin{vmatrix} b_1 & b_3 \\ a_1 & a_3 \end{vmatrix} \mathbf{j} - \begin{vmatrix} b_1 & b_2 \\ a_1 & a_2 \end{vmatrix} \mathbf{k}$$

$$= -\left(\begin{vmatrix} b_2 & b_3 \\ a_2 & a_3 \end{vmatrix} \mathbf{i} - \begin{vmatrix} b_1 & b_3 \\ a_1 & a_3 \end{vmatrix} \mathbf{j} + \begin{vmatrix} b_1 & b_2 \\ a_1 & a_2 \end{vmatrix} \mathbf{k} \right) = -\mathbf{b} \times \mathbf{a}.$$

We leave the proof of property (*iii*) as an exercise.

■ **Special Products** The so-called **triple scalar product** of vectors \mathbf{a}, \mathbf{b}, and \mathbf{c} is $\mathbf{a} \cdot (\mathbf{b} \times \mathbf{c})$. Now,

$$\mathbf{a} \cdot (\mathbf{b} \times \mathbf{c}) = (a_1 \mathbf{i} + a_2 \mathbf{j} + a_3 \mathbf{k}) \cdot \left[\begin{vmatrix} b_2 & b_3 \\ c_2 & c_3 \end{vmatrix} \mathbf{i} - \begin{vmatrix} b_1 & b_3 \\ c_1 & c_3 \end{vmatrix} \mathbf{j} + \begin{vmatrix} b_1 & b_2 \\ c_1 & c_2 \end{vmatrix} \mathbf{k} \right]$$

$$= a_1 \begin{vmatrix} b_2 & b_3 \\ c_2 & c_3 \end{vmatrix} - a_2 \begin{vmatrix} b_1 & b_3 \\ c_1 & c_3 \end{vmatrix} + a_3 \begin{vmatrix} b_1 & b_2 \\ c_1 & c_2 \end{vmatrix}.$$

Thus, we see

$$\mathbf{a} \cdot (\mathbf{b} \times \mathbf{c}) = \begin{vmatrix} a_1 & a_2 & a_3 \\ b_1 & b_2 & b_3 \\ c_1 & c_2 & c_3 \end{vmatrix}. \tag{9}$$

Furthermore, from the properties of determinants, we have

$$\mathbf{a} \cdot (\mathbf{b} \times \mathbf{c}) = (\mathbf{a} \times \mathbf{b}) \cdot \mathbf{c}.$$

The **triple vector product** of three vectors **a, b**, and **c** is $\mathbf{a} \times (\mathbf{b} \times \mathbf{c})$. It is left as an exercise to show that

$$\mathbf{a} \times (\mathbf{b} \times \mathbf{c}) = (\mathbf{a} \cdot \mathbf{c})\mathbf{b} - (\mathbf{a} \cdot \mathbf{b})\mathbf{c}. \tag{10}$$

■ **Areas and Volume** Two nonzero and nonparallel vectors **a** and **b** can be considered to be the sides of a parallelogram. The **area A of a parallelogram** is A = (base)(altitude). From Figure 7.50(a), we see that $A = \|\mathbf{b}\|(\|\mathbf{a}\| \sin \theta) = \|\mathbf{a}\|\|\mathbf{b}\| \sin \theta$

or

$$A = \|\mathbf{a} \times \mathbf{b}\|. \tag{11}$$

Likewise from Figure 7.50(b), we see that the **area of a triangle** with sides **a** and **b** is

$$A = \frac{1}{2}\|\mathbf{a} \times \mathbf{b}\|. \tag{12}$$

Similarly, if the vectors **a, b**, and **c** do not lie in the same plane, then the **volume of the parallelepiped** with edges **a, b**, and **c** shown in Figure 7.51 is

$$V = \text{(area of base)(height)}$$

$$= \|\mathbf{b} \times \mathbf{c}\| \, |\text{comp}_{\mathbf{b} \times \mathbf{c}} \mathbf{a}|$$

$$= \|\mathbf{b} \times \mathbf{c}\| \left| \mathbf{a} \cdot \left(\frac{1}{\|\mathbf{b} \times \mathbf{c}\|} \mathbf{b} \times \mathbf{c} \right) \right|$$

or

$$V = |\mathbf{a} \cdot (\mathbf{b} \times \mathbf{c})|. \tag{13}$$

Because of the last result, the triple scalar product is sometimes referred to as the **box product** of **a, b**, and **c**.

Example 5 Area of a Triangle

Find the area of the triangle determined by the points $P_1(1, 1, 1)$, $P_2(2, 3, 4)$, and $P_3(3, 0, -1)$.

SOLUTION The vectors $\overrightarrow{P_1P_2}$ and $\overrightarrow{P_1P_3}$ can be taken as two sides of the triangle. Since $\overrightarrow{P_1P_2} = \mathbf{i} + 2\mathbf{j} + 3\mathbf{k}$ and $\overrightarrow{P_1P_3} = \mathbf{i} - 3\mathbf{j} - 5\mathbf{k}$, we have

$$\overrightarrow{P_1P_2} \times \overrightarrow{P_2P_3} = \begin{vmatrix} \mathbf{i} & \mathbf{j} & \mathbf{k} \\ 1 & 2 & 3 \\ 1 & -3 & -5 \end{vmatrix} = \begin{vmatrix} 2 & 3 \\ -3 & -5 \end{vmatrix}\mathbf{i} - \begin{vmatrix} 1 & 3 \\ 1 & -5 \end{vmatrix}\mathbf{j} + \begin{vmatrix} 1 & 2 \\ 1 & -3 \end{vmatrix}\mathbf{k}$$

$$= -\mathbf{i} + 8\mathbf{j} - 5\mathbf{k}.$$

(a)

(b)

Figure 7.50

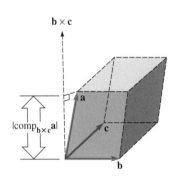

Figure 7.51

From (12) we see that the area is

$$A = \frac{1}{2} \|-\mathbf{i} + 8\mathbf{j} - 5\mathbf{k}\| = \frac{3}{2} \sqrt{10} \text{ square units.}$$ ❑

■ **Coplanar Vectors** Vectors that lie in the same plane are said to be **coplanar**. We have just seen that if the vectors **a**, **b**, and **c** are not coplanar, then necessarily $\mathbf{a} \cdot (\mathbf{b} \times \mathbf{c}) \neq 0$, since the volume of a parallelepiped with edges **a**, **b**, and **c** has nonzero volume. Equivalently stated, this means that if $\mathbf{a} \cdot (\mathbf{b} \times \mathbf{c}) = 0$, then the vectors **a**, **b**, and **c** are coplanar. Since the converse of this last statement is also true, we have

$$\mathbf{a} \cdot (\mathbf{b} \times \mathbf{c}) = 0 \quad \textit{if and only if } \mathbf{a}, \mathbf{b}, \textit{ and } \mathbf{c} \textit{ are coplanar.}$$

Remarks

When working with vectors, one should be careful not to mix the symbols · and ×
with the symbols for ordinary multiplication, and to be especially careful in the
use, or lack of use, of parentheses. For example, expressions such as

$$\mathbf{a} \times \mathbf{b} \times \mathbf{c} \qquad \mathbf{a} \cdot \mathbf{b} \times \mathbf{c} \qquad \mathbf{a} \cdot \mathbf{b} \cdot \mathbf{c} \qquad \mathbf{a} \cdot \mathbf{bc}$$

are not meaningful or well defined.

E X E R C I S E S 7 . 4

Answers to odd-numbered problems begin on page A-41.

In Problems 1–10, find $\mathbf{a} \times \mathbf{b}$.

1. $\mathbf{a} = \mathbf{i} - \mathbf{j}, \quad \mathbf{b} = 3\mathbf{j} + 5\mathbf{k}$

2. $\mathbf{a} = 2\mathbf{i} + \mathbf{j}, \quad \mathbf{b} = 4\mathbf{i} - \mathbf{k}$

3. $\mathbf{a} = \langle 1, -3, 1 \rangle, \quad \mathbf{b} = \langle 2, 0, 4 \rangle$

4. $\mathbf{a} = \langle 1, 1, 1 \rangle, \quad \mathbf{b} = \langle -5, 2, 3 \rangle$

5. $\mathbf{a} = 2\mathbf{i} - \mathbf{j} + 2\mathbf{k}, \quad \mathbf{b} = -\mathbf{i} + 3\mathbf{j} - \mathbf{k}$

6. $\mathbf{a} = 4\mathbf{i} + \mathbf{j} - 5\mathbf{k}, \quad \mathbf{b} = 2\mathbf{i} + 3\mathbf{j} - \mathbf{k}$

7. $\mathbf{a} = \langle \frac{1}{2}, 0, \frac{1}{2} \rangle, \quad \mathbf{b} = \langle 4, 6, 0 \rangle$

8. $\mathbf{a} = \langle 0, 5, 0 \rangle, \quad \mathbf{b} = \langle 2, -3, 4 \rangle$

9. $\mathbf{a} = \langle 2, 2, -4 \rangle, \quad \mathbf{b} = \langle -3, -3, 6 \rangle$

10. $\mathbf{a} = \langle 8, 1, -6 \rangle, \quad \mathbf{b} = \langle 1, -2, 10 \rangle$

In Problems 11 and 12, find $\overrightarrow{P_1P_2} \times \overrightarrow{P_1P_3}$.

11. $P_1(2, 1, 3), \quad P_2(0, 3, -1), \quad P_3(-1, 2, 4)$

12. $P_1(0, 0, 1), \quad P_2(0, 1, 2), \quad P_3(1, 2, 3)$

In Problems 13 and 14, find a vector that is perpendicular to both **a** and **b**.

13. $\mathbf{a} = 2\mathbf{i} + 7\mathbf{j} - 4\mathbf{k}, \quad \mathbf{b} = \mathbf{i} + \mathbf{j} - \mathbf{k}$

14. $\mathbf{a} = \langle -1, -2, 4 \rangle, \quad \mathbf{b} = \langle 4, -1, 0 \rangle$

In Problems 15 and 16, verify that $\mathbf{a} \cdot (\mathbf{a} \times \mathbf{b}) = 0$ and $\mathbf{b} \cdot (\mathbf{a} \times \mathbf{b}) = 0$.

15. $\mathbf{a} = \langle 5, -2, 1 \rangle, \quad \mathbf{b} = \langle 2, 0, -7 \rangle$

16. $\mathbf{a} = \frac{1}{2}\mathbf{i} - \frac{1}{4}\mathbf{j}, \quad \mathbf{b} = 2\mathbf{i} - 2\mathbf{j} + 6\mathbf{k}$

In Problems 17 and 18, **(a)** calculate $\mathbf{b} \times \mathbf{c}$ followed by $\mathbf{a} \times (\mathbf{b} \times \mathbf{c})$.
(b) Verify the results in part **(a)** by (10) of this section.

17. $\mathbf{a} = \mathbf{i} - \mathbf{j} + 2\mathbf{k}$

　　$\mathbf{b} = 2\mathbf{i} + \mathbf{j} + \mathbf{k}$

　　$\mathbf{c} = 3\mathbf{i} + \mathbf{j} + \mathbf{k}$

18. $\mathbf{a} = 3\mathbf{i} - 4\mathbf{k}$

　　$\mathbf{b} = \mathbf{i} + 2\mathbf{j} - \mathbf{k}$

　　$\mathbf{c} = -\mathbf{i} + 5\mathbf{j} + 8\mathbf{k}$

In Problems 19–36, find the indicated scalar or vector *without* using (8), (9), or (10).

19. $(2\mathbf{i}) \times \mathbf{j}$

20. $\mathbf{i} \times (-3\mathbf{k})$

21. $\mathbf{k} \times (2\mathbf{i} - \mathbf{j})$

22. $\mathbf{i} \times (\mathbf{j} \times \mathbf{k})$

23. $[(2\mathbf{k}) \times (3\mathbf{j})] \times (4\mathbf{j})$

24. $(2\mathbf{i} - \mathbf{j} + 5\mathbf{k}) \times \mathbf{i}$

25. $(\mathbf{i} + \mathbf{j}) \times (\mathbf{i} + 5\mathbf{k})$

26. $\mathbf{i} \times \mathbf{k} - 2(\mathbf{j} \times \mathbf{i})$

27. $\mathbf{k} \cdot (\mathbf{j} \times \mathbf{k})$

28. $\mathbf{i} \cdot [\mathbf{j} \times (-\mathbf{k})]$

29. $\|4\mathbf{j} - 5(\mathbf{i} \times \mathbf{j})\|$

30. $(\mathbf{i} \times \mathbf{j}) \cdot (3\mathbf{j} \times \mathbf{i})$

31. $\mathbf{i} \times (\mathbf{i} \times \mathbf{j})$

32. $(\mathbf{i} \times \mathbf{j}) \times \mathbf{i}$

33. $(\mathbf{i} \times \mathbf{i}) \times \mathbf{j}$

34. $(\mathbf{i} \cdot \mathbf{i})(\mathbf{i} \times \mathbf{j})$

35. $2\mathbf{j} \cdot [\mathbf{i} \times (\mathbf{j} - 3\mathbf{k})]$

36. $(\mathbf{i} \times \mathbf{k}) \times (\mathbf{j} \times \mathbf{i})$

In Problems 37–44, $\mathbf{a} \times \mathbf{b} = 4\mathbf{i} - 3\mathbf{j} + 6\mathbf{k}$ and $\mathbf{c} = 2\mathbf{i} + 4\mathbf{j} - \mathbf{k}$. Find the indicated scalar or vector.

37. $\mathbf{a} \times (3\mathbf{b})$

38. $\mathbf{b} \times \mathbf{a}$

39. $(-\mathbf{a}) \times \mathbf{b}$

40. $\|\mathbf{a} \times \mathbf{b}\|$

41. $(\mathbf{a} \times \mathbf{b}) \times \mathbf{c}$

42. $(\mathbf{a} \times \mathbf{b}) \cdot \mathbf{c}$

43. $\mathbf{a} \cdot (\mathbf{b} \times \mathbf{c})$.

44. $(4\mathbf{a}) \cdot (\mathbf{b} \times \mathbf{c})$

In Problems 45 and 46, **(a)** verify that the given quadrilateral is a parallelogram, and **(b)** find the area of the parallelogram.

45.

Figure 7.52

46.

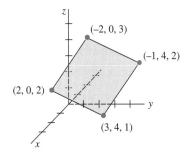

Figure 7.53

In Problems 47–50, find the area of the triangle determined by the given points.

47. $P_1(1, 1, 1)$, $P_2(1, 2, 1)$, $P_3(1, 1, 2)$

48. $P_1(0, 0, 0)$, $P_2(0, 1, 2)$, $P_3(2, 2, 0)$

49. $P_1(1, 2, 4)$, $P_2(1, -1, 3)$, $P_3(-1, -1, 2)$

50. $P_1(1, 0, 3)$, $P_2(0, 0, 6)$, $P_3(2, 4, 5)$

In Problems 51 and 52, find the volume of the parallelepiped for which the given vectors are three edges.

51. $\mathbf{a} = \mathbf{i} + \mathbf{j}$, $\mathbf{b} = -\mathbf{i} + 4\mathbf{j}$, $\mathbf{c} = 2\mathbf{i} + 2\mathbf{j} + 2\mathbf{k}$

52. $\mathbf{a} = 3\mathbf{i} + \mathbf{j} + \mathbf{k}$, $\mathbf{b} = \mathbf{i} + 4\mathbf{j} + \mathbf{k}$, $\mathbf{c} = \mathbf{i} + \mathbf{j} + 5\mathbf{k}$

53. Determine whether the vectors $\mathbf{a} = 4\mathbf{i} + 6\mathbf{j}$, $\mathbf{b} = -2\mathbf{i} + 6\mathbf{j} - 6\mathbf{k}$, and $\mathbf{c} = \frac{5}{3}\mathbf{i} + 3\mathbf{j} + \frac{1}{2}\mathbf{k}$ are coplanar.

54. Determine whether the four points $P_1(1, 1, -2)$, $P_2(4, 0, -3)$, $P_3(1, -5, 10)$, and $P_4(-7, 2, 4)$ lie in the same plane.

55. As shown in Figure 7.54, the vector **a** lies in the xy-plane and the vector **b** lies along the positive z-axis. Their magnitudes are $\|\mathbf{a}\| = 6.4$ and $\|\mathbf{b}\| = 5$.

(a) Use Definition 7.4 to find $\|\mathbf{a} \times \mathbf{b}\|$.

(b) Use the right-hand rule to find the direction of $\mathbf{a} \times \mathbf{b}$.

(c) Use part **(b)** to express $\mathbf{a} \times \mathbf{b}$ in terms of the unit vectors **i**, **j**, **k**.

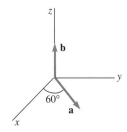

Figure 7.54

56. Two vectors **a** and **b** lie in the xz-plane so that the angle between them is $120°$. If $\|\mathbf{a}\| = \sqrt{27}$ and $\|\mathbf{b}\| = 8$, find all possible values of $\mathbf{a} \times \mathbf{b}$.

57. A three-dimensional lattice is a collection of integer combinations of three non-coplanar basis vectors **a**, **b**, and **c**. In crystallography, a lattice can specify the locations of atoms in a crystal. X-ray diffraction studies of crystals use the "reciprocal lattice" that has basis

$$\mathbf{A} = \frac{\mathbf{b} \times \mathbf{c}}{\mathbf{a} \cdot (\mathbf{b} \times \mathbf{c})} \qquad \mathbf{B} = \frac{\mathbf{c} \times \mathbf{a}}{\mathbf{b} \cdot (\mathbf{c} \times \mathbf{a})} \qquad \mathbf{C} = \frac{\mathbf{a} \times \mathbf{b}}{\mathbf{c} \cdot (\mathbf{a} \times \mathbf{b})}.$$

(a) A certain lattice has basis vectors $\mathbf{a} = \mathbf{i}$, $\mathbf{b} = \mathbf{j}$, and $\mathbf{c} = \frac{1}{2}(\mathbf{i} + \mathbf{j} + \mathbf{k})$. Find basis vectors for the reciprocal lattice.

(b) The unit cell of the reciprocal lattice is the parallelepiped with edges **A**, **B**, and **C**, while the unit cell of the original lattice is the parallelepiped with edges **a**, **b**, and **c**. Show that the volume of the unit cell of the reciprocal lattice is the reciprocal of the volume of the unit cell of the original lattice. [*Hint:* Start with $\mathbf{B} \times \mathbf{C}$ and use (10).]

58. Use (7) to prove property (*iii*) of the cross product.

59. Prove $\mathbf{a} \times (\mathbf{b} \times \mathbf{c}) = (\mathbf{a} \cdot \mathbf{c})\mathbf{b} - (\mathbf{a} \cdot \mathbf{b})\mathbf{c}$.

60. Prove or disprove $\mathbf{a} \times (\mathbf{b} \times \mathbf{c}) = (\mathbf{a} \times \mathbf{b}) \times \mathbf{c}$.

61. Prove $\mathbf{a} \cdot (\mathbf{b} \times \mathbf{c}) = (\mathbf{a} \times \mathbf{b}) \cdot \mathbf{c}$.

62. Prove $\mathbf{a} \times (\mathbf{b} \times \mathbf{c}) + \mathbf{b} \times (\mathbf{c} \times \mathbf{a}) + \mathbf{c} \times (\mathbf{a} \times \mathbf{b}) = \mathbf{0}$.

63. Prove **Lagrange's identity:**

$$\|\mathbf{a} \times \mathbf{b}\|^2 = \|\mathbf{a}\|^2 \|\mathbf{b}\|^2 - (\mathbf{a} \cdot \mathbf{b})^2$$

64. Does $\mathbf{a} \times \mathbf{b} = \mathbf{a} \times \mathbf{c}$ imply that $\mathbf{b} = \mathbf{c}$?

65. Show that $(\mathbf{a} + \mathbf{b}) \times (\mathbf{a} - \mathbf{b}) = 2\mathbf{b} \times \mathbf{a}$.

7.5 Lines and Planes in 3-Space

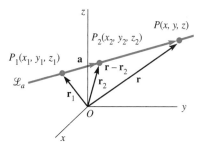

Figure 7.55

■ **Lines: Vector Equation** As in the plane, any two distinct points in 3-space determine only one line between them. To find an equation of the line through $P_1(x_1, y_1, z_1)$ and $P_2(x_2, y_2, z_2)$, let us assume that $P(x, y, z)$ is *any* point on the line. In Figure 7.55, if $\mathbf{r} = \overrightarrow{OP}$, $\mathbf{r}_1 = \overrightarrow{OP_1}$, and $\mathbf{r}_2 = \overrightarrow{OP_2}$, we see that vector $\mathbf{a} = \mathbf{r}_2 - \mathbf{r}_1$ is parallel to vector $\mathbf{r} - \mathbf{r}_2$. Thus,

$$\mathbf{r} - \mathbf{r}_2 = t(\mathbf{r}_2 - \mathbf{r}_1). \tag{1}$$

If we write

$$\mathbf{a} = \mathbf{r}_2 - \mathbf{r}_1 = \langle x_2 - x_1, y_2 - y_1, z_2 - z_1 \rangle = \langle a_1, a_2, a_3 \rangle,$$

then (1) implies that a **vector equation** for the line \mathscr{L}_a is

$$\mathbf{r} = \mathbf{r}_2 + t\mathbf{a}. \tag{2}$$

The vector \mathbf{a} is called a **direction vector** of the line.

Alternative form of the vector 👉 *equation.*

Since $\mathbf{r} - \mathbf{r}_1$ is also parallel to \mathscr{L}_a, an alternative vector equation for the line is $\mathbf{r} = \mathbf{r}_1 + t\mathbf{a}$. Indeed, $\mathbf{r} = \mathbf{r}_1 + t(-\mathbf{a})$ and $\mathbf{r} = \mathbf{r}_1 + t(k\mathbf{a})$, k a nonzero scalar, are also equations for \mathscr{L}_a.

> ### Example 1 Vector Equation of a Line
> Find a vector equation for the line through $(2, -1, 8)$ and $(5, 6, -3)$.
> **SOLUTION** Define $\mathbf{a} = \langle 2 - 5, -1 - 6, 8 - (-3) \rangle = \langle -3, -7, 11 \rangle$. The following are three possible vector equations for the line:
>
> $$\langle x, y, z \rangle = \langle 2, -1, 8 \rangle + t\langle -3, -7, 11 \rangle \tag{3}$$
>
> $$\langle x, y, z \rangle = \langle 5, 6, -3 \rangle + t\langle -3, -7, 11 \rangle \tag{4}$$
>
> $$\langle x, y, z \rangle = \langle 5, 6, -3 \rangle + t\langle 3, 7, -11 \rangle. \tag{5}$$ ❑

■ **Parametric Equations** By writing (2) as

$$\langle x, y, z \rangle = \langle x_2 + t(x_2 - x_1), y_2 + t(y_2 - y_1), z_2 + t(z_2 - z_1) \rangle$$

$$= \langle x_2 + a_1 t, y_2 + a_2 t, z_2 + a_3 t \rangle$$

and equating components, we obtain

$$x = x_2 + a_1 t, \quad y = y_2 + a_2 t, \quad z = z_2 + a_3 t. \tag{6}$$

The equations in (6) are called **parametric equations** for the line through P_1 and P_2. As the parameter t increases from $-\infty$ to ∞, we can think of the point $P(x, y, z)$ tracing out the entire line. If the parameter t is restricted to a closed interval $[t_0, t_1]$, then $P(x, y, z)$ traces out a **line segment** starting at the point corresponding to t_0 and ending at the point corresponding to t_1. For example, in Figure 7.55 if $-1 \leq t \leq 0$,

then $P(x, y, z)$ traces out the line segment starting at $P_1(x_1, y_1, z_1)$ and ending at $P_2(x_2, y_2, z_2)$.

Example 2 Parametric Equations of a Line

Find parametric equations for the line in Example 1.

SOLUTION From (3) it follows that

$$x = 2 - 3t, \quad y = -1 - 7t, \quad z = 8 + 11t. \tag{7}$$

An alternative set of parametric equations can be obtained from (5):

$$x = 5 + 3t, \quad y = 6 + 7t, \quad z = -3 - 11t. \tag{8} \ \Box$$

Note that the value $t = 0$ in (7) gives $(2, -1, 8)$, whereas in (8), $t = -1$ must be used to obtain the same point.

Example 3 Vector Parallel to a Line

Find a vector \mathbf{a} that is parallel to the line \mathscr{L}_a whose parametric equations are $x = 4 + 9t$, $y = -14 + 5t$, $z = 1 - 3t$.

SOLUTION The coefficients (or a nonzero constant multiple of the coefficients) of the parameter in each equation are the components of a vector that is parallel to the line. Thus, $\mathbf{a} = 9\mathbf{i} + 5\mathbf{j} - 3\mathbf{k}$ is parallel to \mathscr{L}_a and hence is a direction vector of the line. $\qquad \Box$

■ Symmetric Equations From (6) observe that we can clear the parameter by writing

$$t = \frac{x - x_2}{a_1} = \frac{y - y_2}{a_2} = \frac{z - z_2}{a_3}$$

provided that the three numbers a_1, a_2, and a_3 are nonzero. The resulting equations

$$\frac{x - x_2}{a_1} = \frac{y - y_2}{a_2} = \frac{z - z_2}{a_3} \tag{9}$$

are said to be **symmetric equations** for the line through P_1 and P_2.

Example 4 Symmetric Equations of a Line

Find symmetric equations for the line through $(4, 10, -6)$ and $(7, 9, 2)$.

SOLUTION Define $a_1 = 7 - 4 = 3$, $a_2 = 9 - 10 = -1$, and $a_3 = 2 - (-6) = 8$. It follows from (9) that symmetric equations for the line are

$$\frac{x - 7}{3} = \frac{y - 9}{-1} = \frac{z - 2}{8}. \qquad \Box$$

If one of the numbers a_1, a_2, or a_3 is zero in (6), we use the remaining two equations to eliminate the parameter t. For example, if $a_1 = 0$, $a_2 \neq 0$, $a_3 \neq 0$, then (6) yields

$$x = x_2 \quad \text{and} \quad t = \frac{y - y_2}{a_2} = \frac{z - z_2}{a_3}.$$

In this case, $x = x_2, \dfrac{y - y_2}{a_2} = \dfrac{z - z_2}{a_3}$

are symmetric equations for the line.

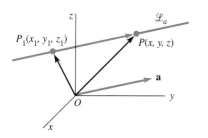

Figure 7.56

Example 5 Symmetric Equations of a Line

Find symmetric equations for the line through (5, 3, 1) and (2, 1, 1).

SOLUTION Define $a_1 = 5 - 2 = 3$, $a_2 = 3 - 1 = 2$, and $a_3 = 1 - 1 = 0$. From the preceding discussion it follows that symmetric equations for the line are

$$\frac{x-5}{3} = \frac{y-3}{2}, \qquad z = 1.$$

In other words, the symmetric equations describe a line in the plane $z = 1$. ❏

A line in space is also determined by specifying a point $P_1(x_1, y_1, z_1)$ and a nonzero direction vector **a**. Through the point P_1 there passes only one line \mathscr{L}_a parallel to the given vector. If $P(x, y, z)$ is a point on the line \mathscr{L}_a, shown in Figure 7.56, then, as before,

$$\overrightarrow{OP} - \overrightarrow{OP_1} = t\mathbf{a} \quad \text{or} \quad \mathbf{r} = \mathbf{r}_1 + t\mathbf{a}.$$

Example 6 Line Parallel to a Vector

Write vector, parametric, and symmetric equations for the line through $(4, 6, -3)$ and parallel to $\mathbf{a} = 5\mathbf{i} - 10\mathbf{j} + 2\mathbf{k}$.

SOLUTION With $a_1 = 5$, $a_2 = -10$, and $a_3 = 2$ we have immediately

Vector: $\langle x, y, z \rangle = \langle 4, 6, -3 \rangle + t\langle 5, -10, 2 \rangle$

Parametric: $x = 4 + 5t, \quad y = 6 - 10t, \quad z = -3 + 2t$

Symmetric: $\dfrac{x-4}{5} = \dfrac{y-6}{-10} = \dfrac{z+3}{2}.$ ❏

■ **Planes: Vector Equation** Figure 7.57(a) illustrates the fact that through a given point $P_1(x_1, y_1, z_1)$ there pass an infinite number of planes. However, as shown in Figure 7.57(b), if a point P_1 and a vector **n** are specified, there is only *one* plane \mathscr{P} containing P_1 with **n normal,** or perpendicular, to the plane. Moreover, if $P(x, y, z)$ is any point on \mathscr{P}, and $\mathbf{r} = \overrightarrow{OP}$, $\mathbf{r}_1 = OP_1$, then, as shown in Figure 7.57(c), $\mathbf{r} - \mathbf{r}_1$ is in the plane. It follows that a **vector equation** of the plane is

$$\mathbf{n} \cdot (\mathbf{r} - \mathbf{r}_1) = 0. \tag{10}$$

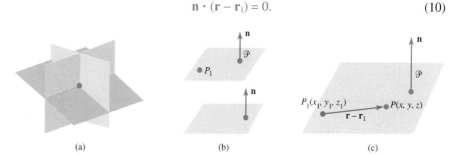

(a) (b) (c)

Figure 7.57

■ **Cartesian Equation** Specifically, if the normal vector is $\mathbf{n} = a\mathbf{i} + b\mathbf{j} + c\mathbf{k}$, then (10) yields a **Cartesian equation** of the plane containing $P_1(x_1, y_1, z_1)$:

$$a(x - x_1) + b(y - y_1) + c(z - z_1) = 0. \tag{11}$$

Example 7 Plane Perpendicular to a Vector

Find an equation of the plane that contains the point $(4, -1, 3)$ and is perpendicular to the vector $\mathbf{n} = 2\mathbf{i} + 8\mathbf{j} - 5\mathbf{k}$.

SOLUTION It follows immediately from (11) that the equation is

$$2(x - 4) + 8(y + 1) - 5(z - 3) = 0 \quad \text{or} \quad 2x + 8y - 5z + 15 = 0. \qquad \square$$

The equation (11) can always be written as $ax + by + cz + d = 0$ by identifying $d = -ax_1 - by_1 - cz_1$. Conversely, we shall now prove that any linear equation

$$ax + by + cz + d = 0, \qquad a, b, c \text{ not all zero} \tag{12}$$

is a plane.

> **THEOREM 7.3 Plane with Normal Vector**
>
> The graph of any equation $ax + by + cz + d = 0$, a, b, c not all zero, is a plane with the normal vector $\mathbf{n} = a\mathbf{i} + b\mathbf{j} + c\mathbf{k}$.

PROOF Suppose x_0, y_0, and z_0 are numbers that satisfy the given equation. Then, $ax_0 + by_0 + cz_0 + d = 0$ implies that $d = -ax_0 - by_0 - cz_0$. Replacing this latter value of d in the original equation gives, after simplifying, $a(x - x_0) + b(y - y_0) + c(z - z_0) = 0$, or, in terms of vectors,

$$[a\mathbf{i} + b\mathbf{j} + c\mathbf{k}] \cdot [(x - x_0)\mathbf{i} + (y - y_0)\mathbf{j} + (z - z_0)\mathbf{k}] = 0.$$

This last equation implies that $a\mathbf{i} + b\mathbf{j} + c\mathbf{k}$ is normal to the plane containing the point (x_0, y_0, z_0) and the vector $(x - x_0)\mathbf{i} + (y - y_0)\mathbf{j} + (z - z_0)\mathbf{k}$. $\qquad \square$

Example 8 A Vector Normal to a Plane

A vector normal to the plane $3x - 4y + 10z - 8 = 0$ is $\mathbf{n} = 3\mathbf{i} - 4\mathbf{j} + 10\mathbf{k}$. $\qquad \square$

Of course, a nonzero scalar multiple of a normal vector is still perpendicular to the plane.

Three noncollinear points P_1, P_2, and P_3 also determine a plane.* To obtain an equation of the plane, we need only form two vectors between two pairs of points. As shown in Figure 7.58, their cross product is a vector normal to the plane containing these vectors. If $P(x, y, z)$ represents any point on the plane, and $\mathbf{r} = \overrightarrow{OP}$, $\mathbf{r}_1 = \overrightarrow{OP_1}$, $\mathbf{r}_2 = \overrightarrow{OP_2}$, $\mathbf{r}_3 = \overrightarrow{OP_3}$, then $\mathbf{r} - \mathbf{r}_1$ (or, for that matter, $\mathbf{r} - \mathbf{r}_2$ or $\mathbf{r} - \mathbf{r}_3$) is in the plane. Hence,

$$[(\mathbf{r}_2 - \mathbf{r}_1) \times (\mathbf{r}_3 - \mathbf{r}_1)] \cdot (\mathbf{r} - \mathbf{r}_1) = 0 \tag{13}$$

is a vector equation of the plane. Do not memorize the last formula. The procedure is the same as (10) with the exception that the vector \mathbf{n} normal to the plane is obtained by means of the cross product.

$(\mathbf{r}_2 - \mathbf{r}_1) \times (\mathbf{r}_3 - \mathbf{r}_1)$

$\mathbf{r}_3 - \mathbf{r}_1$ P_3

P_1

P_2

$\mathbf{r} - \mathbf{r}_1$ $\mathbf{r}_2 - \mathbf{r}_1$

P

Figure 7.58

* If you have ever sat at a four-legged table that rocks, you might consider replacing it with a three-legged table.

Example 9 Three Points That Determine a Plane

Find an equation of the plane that contains $(1, 0, -1)$, $(3, 1, 4)$, and $(2, -2, 0)$.

SOLUTION We need three vectors. Pairing the points on the left as shown yields the vectors on the right. The order in which we subtract is irrelevant.

$$\left.\begin{matrix}(1,0,-1)\\(3,1,4)\end{matrix}\right\}\mathbf{u} = 2\mathbf{i} + \mathbf{j} + 5\mathbf{k}, \quad \left.\begin{matrix}(3,1,4)\\(2,-2,0)\end{matrix}\right\}\mathbf{v} = \mathbf{i} + 3\mathbf{j} + 4\mathbf{k}, \quad \left.\begin{matrix}(2,-2,0)\\(x,y,z)\end{matrix}\right\}\mathbf{w} = (x-2)\mathbf{i} + (y+2)\mathbf{j} + z\mathbf{k}.$$

Now,
$$\mathbf{u} \times \mathbf{v} = \begin{vmatrix} \mathbf{i} & \mathbf{j} & \mathbf{k} \\ 2 & 1 & 5 \\ 1 & 3 & 4 \end{vmatrix} = -11\mathbf{i} - 3\mathbf{j} + 5\mathbf{k}$$

is a vector normal to the plane containing the given points. Hence, a vector equation of the plane is $(\mathbf{u} \times \mathbf{v}) \cdot \mathbf{w} = 0$. The latter equation yields

$$-11(x-2) - 3(y+2) + 5z = 0 \quad \text{or} \quad -11x - 3y + 5z + 16 = 0. \qquad \square$$

■ **Graphs** The graph of (12) with one or even two variables missing is still a plane. For example, we saw in Section 7.2 that the graphs of

$$x = x_0, \quad y = y_0, \quad z = z_0,$$

where x_0, y_0, z_0 are constants, are planes perpendicular to the x-, y-, and z-axes, respectively. In general, to graph a plane, we should try to find

 (*i*) the x-, y-, and z-intercepts and, if necessary,
 (*ii*) the trace of the plane in each coordinate plane.

A **trace** of a plane in a coordinate plane is the line of intersection of the plane with a coordinate plane.

Example 10 Graph of a Plane

Graph the equation $2x + 3y + 6z = 18$.

SOLUTION Setting: $y = 0, z = 0$ gives $x = 9$
 $x = 0, z = 0$ gives $y = 6$
 $x = 0, y = 0$ gives $z = 3$.

The x-, y-, and z-intercepts are 9, 6, and 3, respectively. As shown in Figure 7.59, we use the points $(9, 0, 0)$, $(0, 6, 0)$, and $(0, 0, 3)$ to draw the graph of the plane in the first octant. $\qquad \square$

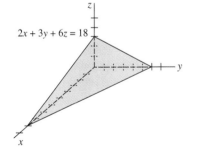

Figure 7.59

Example 11 Graph of a Plane

Graph the equation $6x + 4y = 12$.

SOLUTION In two dimensions the graph of the equation is a line with x-intercept 2 and y-intercept 3. However, in three dimensions this line is the trace of a plane in the xy-coordinate plane. Since z is not specified, it can be any real number. In other words, (x, y, z) is a point on the plane provided that x and y are related by the given equation. As shown in Figure 7.60, the graph is a plane parallel to the z-axis. $\qquad \square$

Figure 7.60

Figure 7.61

Figure 7.62

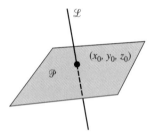

Figure 7.63

Example 12 Graph of a Plane

Graph the equation $x + y - z = 0$.

SOLUTION First observe that the plane passes through the origin $(0, 0, 0)$. Now, the trace of the plane in the xz-plane ($y = 0$) is $z = x$, whereas its trace in the yz-plane ($x = 0$) is $z = y$. Drawing these two lines leads to the graph given in Figure 7.61. ❏

Two planes \mathscr{P}_1 and \mathscr{P}_2 that are not parallel must intersect in a line \mathscr{L}. See Figure 7.62. Example 13 will illustrate one way of finding parametric equations for the line of intersection. In Example 14 we shall see how to find a point of intersection (x_0, y_0, z_0) of a plane \mathscr{P} and a line \mathscr{L}. See Figure 7.63.

Example 13 Line of Intersection of Two Planes

Find parametric equations for the line of intersection of

$$2x - 3y + 4z = 1$$
$$x - y - z = 5.$$

SOLUTION In a system of two equations and three unknowns, we choose one variable arbitrarily, say $z = t$, and solve for x and y from

$$2x - 3y = 1 - 4t$$
$$x - y = 5 + t.$$

Proceeding, we find $x = 14 + 7t$, $y = 9 + 6t$, $z = t$. These are parametric equations for the line of intersection of the given planes. ❏

Example 14 Point of Intersection of a Line and a Plane

Find the point of intersection of the plane $3x - 2y + z = -5$ and the line $x = 1 + t$, $y = -2 + 2t$, $z = 4t$.

SOLUTION If (x_0, y_0, z_0) denotes the point of intersection, then we must have $3x_0 - 2y_0 + z_0 = -5$ and $x_0 = 1 + t_0$, $y_0 = -2 + 2t_0$, $z_0 = 4t_0$, for some number t_0. Substituting the latter equations into the equation of the plane gives

$$3(1 + t_0) - 2(-2 + 2t_0) + 4t_0 = -5 \quad \text{or} \quad t_0 = -4.$$

From the parametric equations for the line we then obtain $x_0. = -3$, $y_0 = -10$, and $z_0 = -16$. The point of intersection is $(-3, -10, -16)$. ❏

E X E R C I S E S 7 . 5

Answers to odd-numbered problems begin on page A-41.

In Problems 1–6, find a vector equation for the line through the given points.

1. $(1, 2, 1)$, $(3, 5, -2)$ **2.** $(0, 4, 5)$, $(-2, 6, 3)$

3. $(\frac{1}{2}, -\frac{1}{2}, 1)$, $(-\frac{3}{2}, \frac{5}{2}, -\frac{1}{2})$ **4.** $(10, 2, -10)$, $(5, -3, 5)$

5. $(1, 1, -1)$, $(-4, 1, -1)$ **6.** $(3, 2, 1)$, $(\frac{5}{2}, 1, -2)$

In Problems 7–12, find parametric equations for the line through the given points.

7. $(2, 3, 5)$, $(6, -1, 8)$ **8.** $(2, 0, 0)$, $(0, 4, 9)$

9. $(1, 0, 0)$, $(3, -2, -7)$ **10.** $(0, 0, 5)$, $(-2, 4, 0)$

11. $(4, \frac{1}{2}, \frac{1}{3})$, $(-6, -\frac{1}{4}, \frac{1}{6})$ **12.** $(-3, 7, 9)$, $(4, -8, -1)$

In Problems 13–18, find symmetric equations for the line through the given points.

13. $(1, 4, -9)$, $(10, 14, -2)$ **14.** $(\tfrac{2}{3}, 0, -\tfrac{1}{4})$, $(1, 3, \tfrac{1}{4})$

15. $(4, 2, 1)$, $(-7, 2, 5)$ **16.** $(-5, -2, -4)$, $(1, 1, 2)$

17. $(5, 10, -2)$, $(5, 1, -14)$

18. $(\tfrac{5}{6}, -\tfrac{1}{4}, \tfrac{1}{5})$, $(\tfrac{1}{3}, \tfrac{3}{8}, -\tfrac{1}{10})$

In Problems 19–22, find parametric and symmetric equations for the line through the given point parallel to the given vector.

19. $(4, 6, -7)$, $\mathbf{a} = \langle 3, \tfrac{1}{2}, -\tfrac{3}{2} \rangle$

20. $(1, 8, -2)$, $\mathbf{a} = -7\mathbf{i} - 8\mathbf{j}$

21. $(0, 0, 0)$, $\mathbf{a} = 5\mathbf{i} + 9\mathbf{j} + 4\mathbf{k}$

22. $(0, -3, 10)$, $\mathbf{a} = \langle 12, -5, -6 \rangle$

23. Find parametric equations for the line through $(6, 4, -2)$ that is parallel to the line $x/2 = (1 - y)/3 = (z - 5)/6$.

24. Find symmetric equations for the line through $(4, -11, -7)$ that is parallel to the line $x = 2 + 5t$, $y = -1 + \tfrac{1}{3}t$, $z = 9 - 2t$.

25. Find parametric equations for the line through $(2, -2, 15)$ that is parallel to the xz-plane and the xy-plane.

26. Find parametric equations for the line through $(1, 2, 8)$ that is **(a)** parallel to the y-axis and **(b)** perpendicular to the xy-plane.

27. Show that the lines given by $\mathbf{r} = t\langle 1, 1, 1 \rangle$ and $\mathbf{r} = \langle 6, 6, 6 \rangle + t\langle -3, -3, -3 \rangle$ are the same.

28. Let \mathcal{L}_a and \mathcal{L}_b be lines with direction vectors \mathbf{a} and \mathbf{b}, respectively. \mathcal{L}_a and \mathcal{L}_b are orthogonal if \mathbf{a} and \mathbf{b} are orthogonal and parallel if \mathbf{a} and \mathbf{b} are parallel. Determine which of the following lines are orthogonal and which are parallel.

(a) $\mathbf{r} = \langle 1, 0, 2 \rangle + t\langle 9, -12, 6 \rangle$

(b) $x = 1 + 9t$, $y = 12t$, $z = 2 - 6t$

(c) $x = 2t$, $y = -3t$, $z = 4t$

(d) $x = 5 + t$, $y = 4t$, $z = 3 + \tfrac{5}{2}t$

(e) $x = 1 + t$, $y = \tfrac{3}{2}t$, $z = 2 - \tfrac{3}{2}t$

(f) $\dfrac{x+1}{-3} = \dfrac{y+6}{4} = \dfrac{z-3}{-2}$

In Problems 29 and 30, determine the points of intersection of the given line and the three coordinate planes.

29. $x = 4 - 2t$, $y = 1 + 2t$, $z = 9 + 3t$

30. $\dfrac{x-1}{2} = \dfrac{y+2}{3} = \dfrac{z-4}{2}$

In Problems 31–34, determine whether the given lines intersect. If so, find the point of intersection.

31. $x = 4 + t$, $y = 5 + t$, $z = -1 + 2t$

$x = 6 + 2s$, $y = 11 + 4s$, $z = -3 + s$

32. $x = 1 + t$, $y = 2 - t$, $z = 3t$

$x = 2 - s$, $y = 1 + s$, $z = 6s$

33. $x = 2 - t$, $y = 3 + t$, $z = 1 + t$

$x = 4 + s$, $y = 1 + s$, $z = 1 - s$

34. $x = 3 - t$, $y = 2 + t$, $z = 8 + 2t$

$x = 2 + 2s$, $y = -2 + 3s$, $z = -2 + 8s$

The angle between two lines \mathcal{L}_a and \mathcal{L}_b is the angle between their direction vectors \mathbf{a} and \mathbf{b}. In Problems 35 and 36, find the angle between the given lines.

35. $x = 4 - t$, $y = 3 + 2t$, $z = -2t$

$x = 5 + 2s$, $y = 1 + 3s$, $z = 5 - 6s$

36. $\dfrac{x-1}{2} = \dfrac{y+5}{7} = \dfrac{z-1}{-1}$; $\dfrac{x+3}{-2} = y - 9 = \dfrac{z}{4}$

In Problems 37 and 38, the given lines lie in the same plane. Find parametric equations for the line through the indicated point that is perpendicular to this plane.

37. $x = 3 + t$, $y = -2 + t$, $z = 9 + t$

$x = 1 - 2s$, $y = 5 + s$, $z = -2 - 5s$; $(4, 1, 6)$

38. $\dfrac{x-1}{3} = \dfrac{y+1}{2} = \dfrac{z}{4}$

$\dfrac{x+4}{6} = \dfrac{y-6}{4} = \dfrac{z-10}{8}$; $(1, -1, 0)$

In Problems 39–44, find an equation of the plane that contains the given point and is perpendicular to the indicated vector.

39. $(5, 1, 3)$; $2\mathbf{i} - 3\mathbf{j} + 4\mathbf{k}$

40. $(1, 2, 5)$; $4\mathbf{i} - 2\mathbf{j}$

41. $(6, 10, -7)$; $-5\mathbf{i} + 3\mathbf{k}$

42. $(0, 0, 0)$; $6\mathbf{i} - \mathbf{j} + 3\mathbf{k}$

43. $(\tfrac{1}{2}, \tfrac{3}{4}, -\tfrac{1}{2})$; $6\mathbf{i} + 8\mathbf{j} - 4\mathbf{k}$

44. $(-1, 1, 0)$; $-\mathbf{i} + \mathbf{j} - \mathbf{k}$

In Problems 45–50, find, if possible, an equation of a plane that contains the given points.

45. $(3, 5, 2)$, $(2, 3, 1)$, $(-1, -1, 4)$

46. $(0, 1, 0)$, $(0, 1, 1)$, $(1, 3, -1)$

47. $(0, 0, 0)$, $(1, 1, 1)$, $(3, 2, -1)$

48. $(0, 0, 3)$, $(0, -1, 0)$, $(0, 0, 6)$

49. $(1, 2, -1)$, $(4, 3, 1)$, $(7, 4, 3)$

50. $(2, 1, 2)$, $(4, 1, 0)$, $(5, 0, -5)$

In Problems 51–60, find an equation of the plane that satisfies the given conditions.

51. Contains $(2, 3, -5)$ and is parallel to $x + y - 4z = 1$

52. Contains the origin and is parallel to $5x - y + z = 6$

53. Contains $(3, 6, 12)$ and is parallel to the xy-plane

54. Contains $(-7, -5, 18)$ and is perpendicular to the y-axis

55. Contains the lines $x = 1 + 3t$, $y = 1 - t$, $z = 2 + t$; $x = 4 + 4s$, $y = 2s$, $z = 3 + s$

56. Contains the lines $\dfrac{x-1}{2} = \dfrac{y+1}{-1} = \dfrac{z-5}{6}$;
$\mathbf{r} = \langle 1, -1, 5 \rangle + t\langle 1, 1, -3 \rangle$

57. Contains the parallel lines $x = 1 + t$, $y = 1 + 2t$, $z = 3 + t$; $x = 3 + s$, $y = 2s$, $z = -2 + s$

58. Contains the point $(4, 0, -6)$ and the line $x = 3t$, $y = 2t$, $z = -2t$

59. Contains $(2, 4, 8)$ and is perpendicular to the line $x = 10 - 3t$, $y = 5 + t$, $z = 6 - \frac{1}{2}t$

60. Contains $(1, 1, 1)$ and is perpendicular to the line through $(2, 6, -3)$ and $(1, 0, -2)$

61. Let \mathscr{P}_1 and \mathscr{P}_2 be planes with normal vectors \mathbf{n}_1 and \mathbf{n}_2, respectively. \mathscr{P}_1 and \mathscr{P}_2 are orthogonal if \mathbf{n}_1 and \mathbf{n}_2 are orthogonal and parallel if \mathbf{n}_1 and \mathbf{n}_2 are parallel. Determine which of the following planes are orthogonal and which are parallel.

(a) $2x - y + 3z = 1$

(b) $x + 2y + 2z = 9$

(c) $x + y - \frac{3}{2}z = 2$

(d) $-5x + 2y + 4z = 0$

(e) $-8x - 8y + 12z = 1$

(f) $-2x + y - 3z = 5$

62. Find parametric equations for the line that contains $(-4, 1, 7)$ and is perpendicular to the plane $-7x + 2y + 3z = 1$.

63. Determine which of the following planes are perpendicular to the line $x = 4 - 6t$, $y = 1 + 9t$, $z = 2 + 3t$.

(a) $4x + y + 2z = 1$

(b) $2x - 3y + z = 4$

(c) $10x - 15y - 5z = 2$

(d) $-4x + 6y + 2z = 9$

64. Determine which of the following planes are parallel to the line $(1 - x)/2 = (y + 2)/4 = z - 5$.

(a) $x - y + 3z = 1$

(b) $6x - 3y = 1$

(c) $x - 2y + 5z = 0$

(d) $-2x + y - 2z = 7$

In Problems 65–68, find parametric equations for the line of intersection of the given planes.

65. $5x - 4y - 9z = 8$
$x + 4y + 3z = 4$

66. $x + 2y - z = 2$
$3x - y + 2z = 1$

67. $4x - 2y - z = 1$
$x + y + 2z = 1$

68. $2x - 5y + z = 0$
$y = 0$

In Problems 69–72, find the point of intersection of the given plane and line.

69. $2x - 3y + 2z = -7$; $x = 1 + 2t$, $y = 2 - t$, $z = -3t$

70. $x + y + 4z = 12$; $x = 3 - 2t$, $y = 1 + 6t$, $z = 2 - \frac{1}{2}t$

71. $x + y - z = 8$; $x = 1$, $y = 2$, $z = 1 + t$

72. $x - 3y + 2z = 0$; $x = 4 + t$, $y = 2 + t$, $z = 1 + 5t$

In Problems 73 and 74, find parametric equations for the line through the indicated point that is parallel to the given planes.

73. $x + y - 4z = 2$
$2x - y + z = 10$; $(5, 6, -12)$

74. $2x + z = 0$
$-x + 3y + z = 1$; $(-3, 5, -1)$

In Problems 75 and 76, find an equation of the plane that contains the given line and is orthogonal to the indicated plane.

75. $x = 4 + 3t$, $y = -t$, $z = 1 + 5t$; $x + y + z = 7$

76. $\dfrac{2-x}{3} = \dfrac{y+2}{5} = \dfrac{z-8}{2}$; $2x - 4y - z + 16 = 0$

In Problems 77–82, graph the given equation.

77. $5x + 2y + z = 10$

78. $3x + 2z = 9$

79. $-y - 3z + 6 = 0$

80. $3x + 4y - 2z - 12 = 0$

81. $-x + 2y + z = 4$

82. $x - y - 1 = 0$

7.6 Vector Spaces

In the preceding sections we were dealing with points and vectors in 2- and 3-space. Mathematicians in the nineteenth century, notably the English mathematicians Arthur Cayley (1821–1895) and James Joseph Sylvester (1814–1897) and the Irish mathematician William Rowan Hamilton (1805–1865), realized that the concepts of point and vector could be generalized, that vectors could be described, or defined, by analytic rather than geometric properties. This was a truly significant breakthrough in the history of mathematics. There is no need to stop with three dimensions. Ordered quadruples $\langle a_1, a_2, a_3, a_4 \rangle$, quintuples $\langle a_1, a_2, a_3, a_4, a_5 \rangle$, and n-tuples $\langle a_1, a_2, \ldots, a_n \rangle$ of real numbers can be thought of as vectors just as well as ordered pairs $\langle a_1, a_2 \rangle$ and ordered triples $\langle a_1, a_2, a_3 \rangle$, the only difference being that we lose our visualization of directed line segments or arrows in four-dimensional, five-dimensional, or n-dimensional space.

In formal terms, a **vector in n-space** is any ordered n-tuple $\langle a_1, a_2, \ldots, a_n \rangle$ of real numbers called the **components** of **a**. The set of all vectors in n-space is denoted by R^n. The concepts of vector addition, scalar multiplication, equality, and so on listed in Definition 7.2 carry over to R^n in a natural way. For example, if $\mathbf{a} = \langle a_1, a_2, \ldots, a_n \rangle$ and $\mathbf{b} = \langle b_1, b_2, \ldots, b_n \rangle$, then addition and scalar multiplication in n-space are defined by

$$\mathbf{a} + \mathbf{b} = \langle a_1 + b_1, a_2 + b_2, \cdots, a_n + b_n \rangle \quad \text{and} \quad k\mathbf{a} = \langle ka_1, ka_2, \cdots, ka_n \rangle. \quad (1)$$

The zero vector in R^n is $\langle 0, 0, \ldots, 0 \rangle$. The notion of **magnitude** or **length** of a vector $\mathbf{a} = \langle a_1, a_2, \ldots, a_n \rangle$ in n-space is just an extension of that concept in 2- and 3-space:

$$\|\mathbf{a}\| = \sqrt{a_1^2 + a_2^2 + \cdots + a_n^2}.$$

A vector **a** in R^n is a **unit vector** if $\|\mathbf{a}\| = 1$. The **dot product** of two n-vectors $\mathbf{a} = \langle a_1, a_2, \ldots, a_n \rangle$ and $\mathbf{b} = \langle b_1, b_2, \ldots, b_n \rangle$ is the real number defined by

$$\mathbf{a} \cdot \mathbf{b} = \langle a_1, a_2, \ldots, a_n \rangle \cdot \langle b_1, b_2, \ldots, b_n \rangle = a_1 b_1 + a_2 b_2 + \cdots + a_n b_n. \quad (2)$$

Two nonzero vectors **a** and **b** in R^n are **orthogonal** if and only if $\mathbf{a} \cdot \mathbf{b} = 0$.

■ **Vector Space** We can even go beyond the notion of a vector as an ordered n-tuple in R^n. A **vector** can be defined as anything we want it to be: an ordered n-tuple, a number, or even a function. But we are particularly interested in vectors that are elements in a special kind of set called a **vector space.** Fundamental to this notion of vector space are two kinds of objects, vectors and scalars, and two algebraic operations analogous to those given in (1). For a set of vectors we want to be able to add two vectors in this set and get another vector in the same set, and we want to multiply a vector by a scalar and obtain a vector in the same set. Whether a set of objects is a vector space depends on whether the set possesses these two algebraic operations along with certain other properties. These properties, the axioms of a vector space, are given next.

> ## DEFINITION 7.5 Vector Space
>
> Let V be a set of elements on which two operations called **vector addition** and **scalar multiplication** are defined. Then V is said to be a **vector space** if the following ten properties are satisfied.
>
> **Axioms for Vector Addition**
>
> (*i*) If \mathbf{x} and \mathbf{y} are in V, then $\mathbf{x} + \mathbf{y}$ is in V.
> (*ii*) For all \mathbf{x}, \mathbf{y} in V, $\mathbf{x} + \mathbf{y} = \mathbf{y} + \mathbf{x}$. (commutative law)
> (*iii*) For all \mathbf{x}, \mathbf{y}, \mathbf{z} in V, $\mathbf{x} + (\mathbf{y} + \mathbf{z}) = (\mathbf{x} + \mathbf{y}) + \mathbf{z}$. (associative law)
> (*iv*) There is a unique vector $\mathbf{0}$ in V such that
> $\mathbf{0} + \mathbf{x} = \mathbf{x} + \mathbf{0} = \mathbf{0}$. (zero vector)
> (*v*) For each \mathbf{x} in V, there exists a vector $-\mathbf{x}$ such that
> $\mathbf{x} + (-\mathbf{x}) = (-\mathbf{x}) + \mathbf{x} = \mathbf{0}$. (negative of a vector)
>
> **Axioms for Scalar Multiplication**
>
> (*vi*) If k is any scalar and \mathbf{x} is in V, then $k\mathbf{x}$ is in V.
> (*vii*) $k(\mathbf{x} + \mathbf{y}) = k\mathbf{x} + k\mathbf{y}$
> (*viii*) $(k_1 + k_2)\mathbf{x} = k_1\mathbf{x} + k_2\mathbf{x}$ (distributive laws)
> (*ix*) $k_1(k_2\mathbf{x}) = (k_1 k_2)\mathbf{x}$
> (*x*) $1\mathbf{x} = \mathbf{x}$

In this abbreviated introduction to abstract vectors, we shall take the scalars in Definition 7.5 to be real numbers. In this case V is referred to as a *real* vector space, although we shall not belabor this term. When the scalars are allowed to be complex numbers, we obtain a *complex* vector space. Since properties (*i*)–(*viii*) on page 300 are the prototypes for the axioms in Definition 7.5, it is clear that R^2 is a vector space. Moreover, since vectors in R^3 and R^n have these same properties, we conclude that R^3 and R^n are also vector spaces. Axioms (*i*) and (*vi*) are called the **closure** axioms; we say that a vector space V is closed under vector addition and scalar multiplication. Note too that concepts such as length and the dot product are not part of the axiomatic structure of a vector space.

Example 1 Checking the Closure Axioms

Determine whether the sets (a) $V = \{1\}$ and (b) $V = \{0\}$ under ordinary addition and multiplication by real numbers are vector spaces.

SOLUTION (a) For this system consisting of one element, many axioms are violated. In particular, axioms (*i*) and (*vi*) of closure are not satisfied. Neither the sum $1 + 1 = 2$ nor the scalar multiple $k \cdot 1 = k$, for $k \neq 1$, is in V. Hence V is not a vector space.

(b) In this case the closure axioms are satisfied, since $0 + 0 = 0$ and $k \cdot 0 = 0$ for any real number k. The commutative and associative axioms are satisfied, since $0 + 0 = 0 + 0$ and $0 + (0 + 0) = (0 + 0) + 0$. In this manner it is easy to verify that the remaining axioms are also satisfied. Hence V is a vector space. ❑

The vector space $V = \{\mathbf{0}\}$ is often called the **trivial** or **zero vector space**.

Readers should not take the names vector *addition* and scalar *multiplication* too literally. These operations are *defined*, and the student must accept them at face value even though these operations may not bear any resemblance to the usual

understanding of ordinary addition and multiplication in, say, R, R^2, R^3, or R^n. For example, the addition of two vectors \mathbf{x} and \mathbf{y} could be $\mathbf{x} - \mathbf{y}$. With this forewarning consider the next example.

> ### *Example 2* **An Example of a Vector Space**
>
> Consider the set V of positive real numbers where addition is defined by
>
> $$\mathbf{x} + \mathbf{y} = xy$$
>
> and scalar multiplication is defined by
>
> $$k\mathbf{x} = x^k.$$
>
> Determine whether V is a vector space.
>
> **SOLUTION** We shall go through all ten axioms.
>
> (i) For $\mathbf{x} = x > 0$ and $\mathbf{y} = y > 0$, $\mathbf{x} + \mathbf{y} = xy > 0$. Thus, the sum $\mathbf{x} + \mathbf{y}$ is in V; V is closed under addition.
>
> (ii) Since multiplication of positive real numbers is commutative, we have for all $\mathbf{x} = x$ and $\mathbf{y} = y$ in V, $\mathbf{x} + \mathbf{y} = xy = yx = \mathbf{y} + \mathbf{x}$. Thus, addition is commutative.
>
> (iii) For all $\mathbf{x} = x$, $\mathbf{y} = y$, $\mathbf{z} = z$ in V, $\mathbf{x} + (\mathbf{y} + \mathbf{z}) = x(yz) = (xy)z = (\mathbf{x} + \mathbf{y}) + \mathbf{z}$. Thus, addition is associative.
>
> (iv) Since $\mathbf{1} + \mathbf{x} = 1x = x = \mathbf{x}$ and $\mathbf{x} + \mathbf{1} = x1 = x = \mathbf{x}$, the zero vector $\mathbf{0}$ is $\mathbf{1} = 1$.
>
> (v) If we define $-\mathbf{x} = \dfrac{1}{x}$, then $\mathbf{x} + (-\mathbf{x}) = x\,\dfrac{1}{x} = 1 = \mathbf{1} = \mathbf{0}$ and $(-\mathbf{x}) + \mathbf{x} = \dfrac{1}{x}\,x = 1 = \mathbf{1} = \mathbf{0}$. The negative of a vector is its reciprocal.
>
> (vi) If k is any scalar and $\mathbf{x} = x > 0$ is any vector, then $k\mathbf{x} = x^k > 0$. Hence V is closed under scalar multiplication.
>
> (vii) If k is any scalar, then $k(\mathbf{x} + \mathbf{y}) = (xy)^k = x^k y^k = k\mathbf{x} + k\mathbf{y}$.
>
> (viii) For scalars k_1 and k_2, $(k_1 + k_2)\mathbf{x} = x^{(k_1 + k_2)} = x^{k_1} x^{k_2} = k_1\mathbf{x} + k_2\mathbf{x}$.
>
> (ix) For scalars k_1 and k_2, $k_1(k_2\mathbf{x}) = (x^{k_2})^{k_1} = x^{k_1 k_2} = (k_1 k_2)\mathbf{x}$.
>
> (x) $1\mathbf{x} = x^1 = x = \mathbf{x}$.
>
> Since all the axioms of Definition 7.5 are satisfied, we conclude that V is a vector space. ❏

Following are some important vector spaces. The operations of vector addition and scalar multiplication are the usual operations associated with the set.

- The set R of real numbers
- The set P_n of polynomials of degree less than or equal to n
- The set of real-valued functions f defined on the entire real line
- The set $C[a, b]$ of real-valued functions f continuous on the closed interval $a \leq x \leq b$
- The set $C(-\infty, \infty)$ of real-valued functions f continuous on the entire real line
- The set $C^n[a, b]$ of all real-valued functions f for which $f, f', \ldots, f^{(n)}$ exist and are continuous on the interval $[a, b]$

■ **Subspace** It may happen that a subset of vectors W of a vector space V is itself a vector space.

> **DEFINITION 7.6** **Subspace of a Vector Space**
>
> If a subset W of a vector space V is itself a vector space under the operations of vector addition and scalar multiplication defined on V, then W is called a **subspace** of V.

Every vector space V has at least two subspaces: V itself and the zero subspace $\{\mathbf{0}\}$; $\{\mathbf{0}\}$ is a subspace since the zero vector must be an element in every vector space.

To show that a subset W of a vector space V is a subspace it is not necessary to demonstrate that all ten axioms are satisfied. Since all the vectors in W are also in V, these vectors must satisfy axioms such as (*ii*) and (*iii*). In other words, W inherits most of the properties of a vector space from V. Indeed, to show that a subset W is a subspace we need only check the two closure axioms.

> **THEOREM 7.4** **Criteria for a Subspace**
>
> A nonempty subset W of a vector space V is a subspace of V if and only if W is closed under vector addition and scalar multiplication defined on V:
> - (*i*) If \mathbf{x} and \mathbf{y} are in W, then $\mathbf{x} + \mathbf{y}$ is in W.
> - (*ii*) If \mathbf{x} is in W and k is any scalar, then $k\mathbf{x}$ is W.

Example 3 A Subspace

Suppose \mathbf{f} and \mathbf{g} are continuous real-valued functions defined on the entire real line. Then we know from calculus that $\mathbf{f} + \mathbf{g}$ and $k\mathbf{f}$, for any real number k, are continuous and real-valued functions. From this we can conclude that $C(-\infty, \infty)$ is a subspace of the vector space of real-valued functions defined on the entire real line. ❏

It is always a good idea to have concrete visualizations of vector spaces and subspaces. The subspaces of the vector space R^3 of three-dimensional vectors can be easily visualized by thinking of a vector as point (a_1, a_2, a_3). Of course, R^3 and $\{\mathbf{0}\}$ are subspaces; other subspaces are all lines passing through the origin and all planes passing through the origin. The lines and planes must pass through the origin since the zero vector $\mathbf{0} = (0, 0, 0)$ must be an element in each subspace.

Similar to Definition 3.1, we can define linearly independent vectors:

> **DEFINITION 7.7** **Linear Independence**
>
> A set of vectors $\mathbf{x}_1, \mathbf{x}_2, \ldots, \mathbf{x}_n$ is said to be **linearly independent** if the only constants satisfying the equation
>
> $$k_1\mathbf{x}_1 + k_2\mathbf{x}_2 + \cdots + k_n\mathbf{x}_n = \mathbf{0} \qquad (3)$$
>
> are $k_1 = k_2 = \cdots = k_n = 0$. If the set of vectors is not linearly independent, then it is said to be **linearly dependent.**

In R^3, the vectors $\mathbf{i} = \langle 1, 0, 0 \rangle$, $\mathbf{j} = \langle 0, 1, 0 \rangle$, and $\mathbf{k} = \langle 0, 0, 1 \rangle$ are linearly independent, since the equation $k_1\mathbf{i} + k_2\mathbf{j} + k_3\mathbf{k} = 0$ is the same as

$$k_1\langle 1, 0, 0 \rangle + k_2\langle 0, 1, 0 \rangle + k_3\langle 0, 0, 1 \rangle = \langle 0, 0, 0 \rangle \quad \text{or} \quad \langle k_1, k_2, k_3 \rangle = \langle 0, 0, 0 \rangle.$$

By equality of vectors, (*ii*) of Definition 7.2, we conclude that $k_1 = 0$, $k_2 = 0$, and $k_3 = 0$. In Definition 7.7, linear dependence means that there are constants k_1, k_2, \ldots, k_n *not all zero* such that $k_1\mathbf{x}_1 + k_2\mathbf{x}_2 + \cdots + k_n\mathbf{x}_n = \mathbf{0}$. Hence the vectors $\mathbf{a} = \langle 1, 1, 1 \rangle$, $\mathbf{b} = \langle 2, -1, 4 \rangle$, and $\mathbf{c} = \langle 3, 0, 5 \rangle$ are linearly dependent since $\mathbf{a} + \mathbf{b} - \mathbf{c} = \mathbf{0}$; that is, (3) is satisfied when $k_1 = k_2 = 1$ and $k_3 = -1$. We observe that two vectors are linearly independent if neither is a constant multiple of the other.

■ **Basis** Any vector in R^3 can be written as a linear combination of the linearly independent vectors \mathbf{i}, \mathbf{j}, and \mathbf{k}. In Section 7.2 we said that these vectors form a **basis** for the system of three-dimensional vectors.

> ### DEFINITION 7.8 Basis for a Vector Space
>
> Consider a set of vectors $\mathbf{x}_1, \mathbf{x}_2, \ldots, \mathbf{x}_n$ in a vector space V. If this set of vectors is linearly independent and if every vector in V can be expressed as a linear combination of these vectors, then $\mathbf{x}_1, \mathbf{x}_2, \ldots, \mathbf{x}_n$ is said to be a **basis** for V.

Although we cannot prove it in this course, every vector space has a basis. The vector space P_n of all polynomials of degree less than or equal to n has the basis $1, x, x^2, \ldots, x^n$, since any vector (polynomial) $p(x)$ of degree n or less can be written as the linear combination $p(x) = c_n x^n + \cdots + c_2 x^2 + c_1 x + c_0$. A vector space may have many bases. We mentioned previously that the vectors $\mathbf{i}, \mathbf{j}, \mathbf{k}$ are a basis for R^3. But it can be proved that $\mathbf{u}_1 = \langle 1, 0, 0 \rangle$, $\mathbf{u}_2 = \langle 1, 1, 0 \rangle$, and $\mathbf{u}_3 = \langle 1, 1, 1 \rangle$ are linearly independent (see Problem 23 in Exercises 7.6) and, furthermore, every vector $\mathbf{a} = \langle a_1, a_2, a_3 \rangle$ can be expressed as a linear combination of $\mathbf{u}_1, \mathbf{u}_2, \mathbf{u}_3$:

$$\mathbf{a} = c_1\mathbf{u}_1 + c_2\mathbf{u}_2 + c_3\mathbf{u}_3.$$

Hence, the vectors $\mathbf{u}_1, \mathbf{u}_2, \mathbf{u}_3$ are another basis for R^3. Indeed, any set of *three* linearly independent vectors in R^3 forms a basis for that space. However, the vectors $\mathbf{i}, \mathbf{j}, \mathbf{k}$ are referred to as the **standard basis.** For the space P_n the standard basis is $1, x, x^2, \ldots, x^n$; for the vector space R^n the standard basis consists of the n vectors $\mathbf{e}_1 = \langle 1, 0, 0, \ldots, 0 \rangle$, $\mathbf{e}_2 = \langle 0, 1, 0, \ldots, 0 \rangle, \ldots, \mathbf{e}_n = \langle 0, 0, 0, \ldots, 1 \rangle$.

> ### DEFINITION 7.9 Dimension of a Vector Space
>
> The number of vectors in a basis for a vector space is said to be the **dimension** of the space.

Example 4 Dimensions of Some Vector Spaces

In agreement with our intuition, the dimensions of the vector spaces R, R^2, R^3, and R^n are, in turn, 1, 2, 3, and n. Since there are $n + 1$ vectors in the basis $1, x, x^2, \ldots, x^n$, the dimension of the vector space P_n of polynomials of degree less than or equal to n is $n + 1$. ❑

If a vector space has dimension n, then every basis for that space must contain n vectors. If the basis of a vector space contains a finite number of vectors, then we say that the vector space is **finite dimensional;** otherwise it is **infinite dimensional.** The function space $C^n(I)$ of n times continuously differentiable functions on an interval I is an example of an infinite-dimensional vector space. The zero vector space $\{\mathbf{0}\}$ is given special consideration. This space contains only $\mathbf{0}$ and since $\mathbf{0}$ is linearly dependent, it is not a basis. In this case it is customary to take the empty set as the basis and to define the dimension of $\{\mathbf{0}\}$ as zero.

■ **Linear Differential Equations** Consider the homogeneous linear nth-order differential equation

$$a_n(x)\frac{d^n y}{dx^n} + a_{n-1}(x)\frac{d^{n-1}y}{dx^{n-1}} + \cdots + a_1(x)\frac{dy}{dx} + a_0(x)y = 0 \qquad (4)$$

on an interval I on which the coefficients are continuous and $a_n(x) \neq 0$ for every x in the interval. A solution y_1 of (4) is necessarily a vector in the vector space $C^n(I)$. In addition, we know from the theory examined in Section 3.1 that if y_1 and y_2 are solutions of (4), then the sum $y_1 + y_2$ and any constant multiple ky_1 are also solutions. Since the solution set is closed under addition and scalar multiplication, it follows from Theorem 7.4 that the solution set of (4) is a subspace of $C^n(I)$. Hence the solution set of (4) deserves to be called the **solution space** of the differential equation. We also know that if y_1, y_2, \ldots, y_n are linearly independent solutions of (4), then its general solution is

$$y = c_1 y_1(x) + c_2 y_2(x) + \cdots + c_n y_n(x).$$

Recall that any solution of the equation can be found from this general solution by specialization of the constants c_1, c_2, \ldots, c_n. Hence the set of linearly independent solutions y_1, y_2, \ldots, y_n comprise a basis for the solution space. The dimension of this solution space is n.

> ### *Example 5* Dimension of a Solution Space
>
> The general solution of the homogeneous linear second-order differential equation $y'' + 25y = 0$ is $y = c_1 \cos 5x + c_2 \sin 5x$. A basis for the solution space consists of the linearly independent vectors $\cos 5x$ and $\sin 5x$. The solution space is two-dimensional. ❏

The set of solutions of a nonhomogeneous linear differential equation is not a vector space. Several axioms of a vector space are not satisfied; most notably the set of solutions does not contain a zero vector. In other words, $y = 0$ is not a solution of a nonhomogeneous linear differential equation.

■ **Span** If S denotes any set of vectors $\mathbf{x}_1, \mathbf{x}_2, \ldots, \mathbf{x}_n$ in a vector space V, then a sum of the form $k_1\mathbf{x}_1 + k_2\mathbf{x}_2 + \ldots + k_n\mathbf{x}_n$, where the k_i, $i = 1, \ldots, n$ are scalars, is called a **linear combination** of $\mathbf{x}_1, \mathbf{x}_2, \ldots, \mathbf{x}_n$. The set of *all* linear combinations of the vectors $\mathbf{x}_1, \mathbf{x}_2, \ldots, \mathbf{x}_n$ is called the **span** of the vectors and written Span(S) or Span($\mathbf{x}_1, \mathbf{x}_2, \ldots, \mathbf{x}_n$). It is left as an exercise to show that Span(S) is a subspace of the vector space V. See Problem 33 in Exercises 7.6. Span(S) is said to be a subspace *spanned* by the vectors $\mathbf{x}_1, \mathbf{x}_2, \ldots, \mathbf{x}_n$. If every vector in the vector space V can be written as a linear combination of the vectors in S, then S is called a **spanning set** for V. For example, each of the three sets

$$\mathbf{i}, \mathbf{j}, \mathbf{k}, \quad \text{and} \quad \mathbf{i}, \mathbf{i} + \mathbf{j}, \mathbf{i} + \mathbf{j} + \mathbf{k}, \quad \text{and} \quad \mathbf{i}, \mathbf{j}, \mathbf{k}, \mathbf{i} + \mathbf{j}, \mathbf{i} + \mathbf{j} + \mathbf{k}$$

is a spanning set for the vector space R^3. But note that the first two sets are linearly independent, whereas the third set is dependent. With these new concepts we can rephrase Definitions 7.8 and 7.9 in the following manner:

A set S of vectors $\mathbf{x}_1, \mathbf{x}_2, \ldots, \mathbf{x}_n$ in a vector space V is a **basis** *for V if S is a spanning set for V and S is linearly independent. The number of vectors in this spanning set S is the* **dimension** *of the space V.*

Answers to odd-numbered problems begin on page A-42.

In Problems 1–10, determine whether the given set is a vector space. If not, give at least one axiom that is not satisfied. Unless stated to the contrary, assume that vector addition and scalar multiplication are the ordinary operations defined on that set.

1. The set of vectors $\langle a_1, a_2 \rangle$, where $a_1 \geq 0$, $a_2 \geq 0$

2. The set of vectors $\langle a_1, a_2 \rangle$, where $a_2 = 3a_1 + 1$

3. The set of vectors $\langle a_1, a_2 \rangle$, scalar multiplication defined by $k\langle a_1, a_2 \rangle = \langle ka_1, 0 \rangle$

4. The set of vectors $\langle a_1, a_2 \rangle$, where $a_1 + a_2 = 0$

5. The set of vectors $\langle a_1, a_2, 0 \rangle$

6. The set of vectors $\langle a_1, a_2 \rangle$, addition and scalar multiplication defined by

$$\langle a_1, a_2 \rangle + \langle b_1, b_2 \rangle = \langle a_1 + b_1 + 1, a_2 + b_2 + 1 \rangle$$

$$k\langle a_1, a_2 \rangle = \langle ka_1 + k - 1, ka_2 + k - 1 \rangle$$

7. The set of real numbers, addition defined by $\mathbf{x} + \mathbf{y} = x - y$

8. The set of complex numbers $a + bi$, where $i^2 = -1$, addition and scalar multiplication defined by

$$(a_1 + b_1 i) + (a_2 + b_2 i) = (a_1 + a_2) + (b_1 + b_2)i$$

$$k(a + bi) = ka + kbi, \quad k \text{ a real number}$$

9. The set of arrays of real numbers

$$\begin{pmatrix} a_{11} & a_{12} \\ a_{21} & a_{22} \end{pmatrix}, \text{ addition and scalar multiplication defined by}$$

$$\begin{pmatrix} a_{11} & a_{12} \\ a_{21} & a_{22} \end{pmatrix} + \begin{pmatrix} b_{11} & b_{12} \\ b_{21} & b_{22} \end{pmatrix} = \begin{pmatrix} a_{12} + b_{12} & a_{11} + b_{11} \\ a_{22} + b_{22} & a_{21} + b_{21} \end{pmatrix}$$

$$k\begin{pmatrix} a_{11} & a_{12} \\ a_{21} & a_{22} \end{pmatrix} = \begin{pmatrix} ka_{11} & ka_{12} \\ ka_{21} & ka_{22} \end{pmatrix}$$

10. The set of all polynomials of degree 2

In Problems 11–16, determine whether or not the given set is a subspace of the vector space $C(-\infty, \infty)$.

11. All functions f such that $f(1) = 0$

12. All functions f such that $f(0) = 1$

13. All nonnegative functions f

14. All functions f such that $f(-x) = f(x)$

15. All differentiable functions f

16. All functions f of the form $f(x) = c_1 e^x + c_2 x e^x$

In Problems 17–20, determine whether or not the given set is a subspace of the indicated vector space.

17. Polynomials of the form $p(x) = c_3 x^3 + c_1 x$; P_3

18. Polynomials p that are divisible by $x - 2$; P_2

19. All unit vectors; R^3

20. Functions f such that $\int_a^b f(x)\, dx = 0$; $C[a, b]$

21. In 3-space a line through the origin can be written as $S = \{(x, y, z)\,|\, x = at,\, y = bt,\, z = ct,\, a, b, c \text{ real numbers}\}$. With addition and scalar multiplication the same as for vectors $\langle x, y, z \rangle$ show that S is a subspace of R^3.

22. In 3-space a plane through the origin can be written as $S = \{(x, y, z)\,|\, ax + by + cz = 0,\, a, b, c \text{ real numbers}\}$. Show that S is a subspace of R^3.

23. The vectors $\mathbf{u}_1 = \langle 1, 0, 0 \rangle$, $\mathbf{u}_2 = \langle 1, 1, 0 \rangle$, and $\mathbf{u}_3 = \langle 1, 1, 1 \rangle$ form a basis for the vector space R^3.

 (a) Show that \mathbf{u}_1, \mathbf{u}_2, and \mathbf{u}_3 are linearly independent.

 (b) Express the vector $\mathbf{a} = \langle 3, -4, 8 \rangle$ as a linear combination of \mathbf{u}_1, \mathbf{u}_2, and \mathbf{u}_3.

24. The vectors $p_1(x) = x + 1$, $p_2(x) = x - 1$ form a basis for the vector space P_1.

 (a) Show that $p_1(x)$ and $p_2(x)$ are linearly independent.

 (b) Express the vector $p(x) = 5x + 2$ as a linear combination of $p_1(x)$ and $p_2(x)$.

In Problems 25–28, determine whether the given vectors are linearly independent or linearly dependent.

25. $\langle 4, -8 \rangle, \langle -6, 12 \rangle$ in R^2

26. $\langle 1, 1 \rangle, \langle 0, 1 \rangle, \langle 2, 5 \rangle$ in R^2

27. $1, (x + 1), (x + 1)^2$ in P_2

28. $1, (x + 1), (x + 1)^2, x^2$ in P_2

29. Explain why $f(x) = \dfrac{x}{x^2 + 4x + 3}$ is a vector in $C[0, 3]$ but not a vector in $C[-3, 0]$.

30. A vector space V on which a dot or inner product has been defined is called an **inner product space.** An inner product for the vector space $C[a, b]$ is given by

$$(f, g) = \int_a^b f(x)g(x)\, dx.$$

In $C[0, 2\pi]$ compute $(x, \sin x)$.

31. The **norm** of a vector in an inner product space is defined in terms of the inner product. For the inner product given in Problem 30 the norm of a vector is given by $\|f\| = \sqrt{(f, f)}$. In $C[0, 2\pi]$ compute $\|x\|$ and $\|\sin x\|$.

32. Find a basis for the solution space of

$$\frac{d^4y}{dx^4} - 2\frac{d^3y}{dx^3} + 10\frac{d^2y}{dx^2} = 0.$$

33. Let $\mathbf{x}_1, \mathbf{x}_2, \ldots, \mathbf{x}_n$ be any set of vectors in a vector space V. Show that $\mathrm{Span}(\mathbf{x}_1, \mathbf{x}_2, \ldots, \mathbf{x}_n)$ is a subspace of V.

CHAPTER 7 REVIEW EXERCISES

Answers to odd-numbered problems begin on page A-42.

Answer Problems 1–30 without referring back to the text. Fill in the blank or answer true/false.

1. The vectors $\langle -4, -6, 10 \rangle$ and $\langle -10, -15, 25 \rangle$ are parallel. _____

2. In 3-space any three distinct points determine a plane. _____

3. The line $x = 1 + 5t, \ y = 1 - 2t, \ z = 4 + t$ and the plane $2x + 3y - 4z = 1$ are perpendicular. _____

4. Nonzero vectors \mathbf{a} and \mathbf{b} are parallel if $\mathbf{a} \times \mathbf{b} = \mathbf{0}$. _____

5. If $\mathbf{a} \cdot \mathbf{b} < 0$, the angle between \mathbf{a} and \mathbf{b} is obtuse. _____

6. If \mathbf{a} is a unit vector, then $\mathbf{a} \cdot \mathbf{a} = 1$. _____

7. The cross product of two vectors is not commutative. _____

8. The terminal point of the vector $\mathbf{a} - \mathbf{b}$ is at the terminal point of \mathbf{a}. _____

9. $(\mathbf{a} \times \mathbf{b}) \cdot \mathbf{c} = \mathbf{a} \cdot (\mathbf{b} \times \mathbf{c})$ _____

10. If $\mathbf{a}, \mathbf{b}, \mathbf{c}$, and \mathbf{d} are nonzero coplanar vectors, then $(\mathbf{a} \times \mathbf{b}) \times (\mathbf{c} \times \mathbf{d}) = \mathbf{0}$. _____

11. The sum of $3\mathbf{i} + 4\mathbf{j} + 5\mathbf{k}$ and $6\mathbf{i} - 2\mathbf{j} - 3\mathbf{k}$ is _____.

12. If $\mathbf{a} \cdot \mathbf{b} = 0$, the nonzero vectors \mathbf{a} and \mathbf{b} are _____.

13. $(-\mathbf{k}) \times (5\mathbf{j}) = $ _____

14. $\mathbf{i} \cdot (\mathbf{i} \times \mathbf{j}) = $ _____

15. $\| -12\mathbf{i} + 4\mathbf{j} + 6\mathbf{k} \| = $ _____

16. $\begin{vmatrix} \mathbf{i} & \mathbf{j} & \mathbf{k} \\ 2 & 1 & 5 \\ 0 & 4 & -1 \end{vmatrix} = $ _____

17. A vector that is normal to the plane $-6x + y - 7z + 10 = 0$ is _____.

18. The plane $x + 3y - z = 5$ contains the point $(1, -2, \underline{\hspace{1em}})$.

19. The point of intersection of the line $x - 1 = (y + 2)/3 = (z + 1)/2$ and the plane $x + 2y - z = 13$ is _____.

20. A unit vector that has the opposite direction of $\mathbf{a} = 4\mathbf{i} + 3\mathbf{j} - 5\mathbf{k}$ is _____.

21. If $\overrightarrow{P_1P_2} = \langle 3, 5, -4 \rangle$ and P_1 has coordinates $(2, 1, 7)$, then the coordinates of P_2 are _____.

22. The midpoint of the line segment between $P_1(4, 3, 10)$ and $P_2(6, -2, -5)$ has coordinates _____.

23. If $|\mathbf{a}| = 7.2, |\mathbf{b}| = 10$, and the angle between \mathbf{a} and \mathbf{b} is $135°$, then $\mathbf{a} \cdot \mathbf{b} = $ _____.

24. If $\mathbf{a} = \langle 3, 1, 0 \rangle, \mathbf{b} = \langle -1, 2, 1 \rangle$, and $\mathbf{c} = \langle 0, -2, 2 \rangle$, then $\mathbf{a} \cdot (2\mathbf{b} + 4\mathbf{c}) = $ _____.

25. The x-, y-, and z-intercepts of the plane $2x - 3y + 4z = 24$ are, respectively, _____.

26. The angle θ between the vectors $\mathbf{a} = \mathbf{i} + \mathbf{j}$ and $\mathbf{b} = \mathbf{i} - \mathbf{k}$ is _____.

27. The area of a triangle with two sides given by $\mathbf{a} = \langle 1, 3, -1 \rangle$ and $\mathbf{b} = \langle 2, -1, 2 \rangle$ is _____.

28. An equation of the plane containing $(3, 6, -2)$ and with normal vector $\mathbf{n} = 3\mathbf{i} + \mathbf{k}$ is _____.

29. The distance from the plane $y = -5$ to the point $(4, -3, 1)$ is _____.

30. The vectors $\langle 1, 3, c \rangle$ and $\langle -2, -6, 5 \rangle$ are parallel for $c =$ ____ and orthogonal for $c =$ ____.

31. Find a unit vector that is perpendicular to both $\mathbf{a} = \mathbf{i} + \mathbf{j}$ and $\mathbf{b} = \mathbf{i} - 2\mathbf{j} + \mathbf{k}$.

32. Find the direction cosines and direction angles of the vector $\mathbf{a} = \frac{1}{2}\mathbf{i} + \frac{1}{2}\mathbf{j} - \frac{1}{4}\mathbf{k}$.

In Problems 33–36, let $\mathbf{a} = \langle 1, 2, -2 \rangle$ and $\mathbf{b} = \langle 4, 3, 0 \rangle$. Find the indicated number or vector.

33. $\text{comp}_b \mathbf{a}$

34. $\text{proj}_a \mathbf{b}$

35. $\text{proj}_a \cdot \mathbf{b}$

36. $\text{proj}_{b^\perp}(\mathbf{a} - \mathbf{b})$

37. Let \mathbf{r} be the position vector of a variable point $P(x, y, z)$ in space and let \mathbf{a} be a constant vector. Determine the surface described by **(a)** $(\mathbf{r} - \mathbf{a}) \cdot \mathbf{r} = 0$ and **(b)** $(\mathbf{r} - \mathbf{a}) \cdot \mathbf{a} = 0$.

38. Use the dot product to determine whether the points $(4, 2, -2)$, $(2, 4, -3)$, and $(6, 7, -5)$ are vertices of a right triangle.

39. Find symmetric equations for the line through the point $(7, 3, -5)$ that is parallel to $(x - 3)/4 = (y + 4)/(-2) = (z - 9)/6$.

40. Find parametric equations for the line through the point $(5, -9, 3)$ that is perpendicular to the plane $8x + 3y - 4z = 13$.

41. Show that the lines $x = 1 - 2t$, $y = 3t$, $z = 1 + t$ and $x = 1 + 2s$, $y = -4 + s$, $z = -1 + s$ intersect orthogonally.

42. Find an equation of the plane containing the points $(0, 0, 0)$, $(2, 3, 1)$, $(1, 0, 2)$.

43. Find an equation of the plane containing the lines $x = t$, $y = 4t$, $z = -2t$, and $x = 1 + t$, $y = 1 + 4t$, $z = 3 - 2t$.

44. Find an equation of the plane containing $(1, 7, -1)$ that is perpendicular to the line of intersection of $-x + y - 8z = 4$ and $3x - y + 2z = 0$.

45. A constant force of 10 N in the direction of $\mathbf{a} = \mathbf{i} + \mathbf{j}$ moves a block on a frictionless surface from $P_1(4, 1, 0)$ to $P_2(7, 4, 0)$. Suppose distance is measured in meters. Find the work done.

46. In Problem 45 find the work done in moving the block between the same points if another constant force of 50 N in the direction of $\mathbf{b} = \mathbf{i}$ acts simultaneously with the original force.

47. Water rushing from a fire hose exerts a horizontal force \mathbf{F}_1 of magnitude 200 lb. See Figure 7.64. What is the magnitude of the force \mathbf{F}_3 that a firefighter must exert to hold the hose at an angle of $45°$ from the horizontal?

Figure 7.64

48. A uniform ball of weight 50 lb is supported by two frictionless planes as shown in Figure 7.65. Let the force exerted by the supporting plane \mathscr{P}_1 on the ball be \mathbf{F}_1 and the force exerted by the plane \mathscr{P}_2 be \mathbf{F}_2. Since the ball is held in equilibrium, we must have $\mathbf{w} + \mathbf{F}_1 + \mathbf{F}_2 = \mathbf{0}$, where $\mathbf{w} = -50\mathbf{j}$. Find the magnitudes of the forces \mathbf{F}_1 and \mathbf{F}_2. [*Hint:* Assume the forces \mathbf{F}_1 and \mathbf{F}_2 are normal to the planes \mathscr{P}_1 and \mathscr{P}_2, respectively, and act along lines through the center C of the ball. Place the origin of a two-dimensional coordinate system at C.]

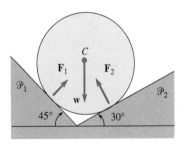

Figure 7.65

49. Determine whether the set of vectors $\langle a_1, 0, a_3 \rangle$ under addition and scalar multiplication defined by

$$\langle a_1, 0, a_3 \rangle + \langle b_1, 0, b_3 \rangle = \langle a_1 + b_1, 0, a_3 + b_3 \rangle$$

$$k\langle a_1, 0, a_3 \rangle = \langle ka_1, 0, a_3 \rangle$$

is a vector space.

50. Determine whether the vectors $\langle 1, 1, 2 \rangle$, $\langle 0, 2, 3 \rangle$, and $\langle 0, 1, -1 \rangle$ are linearly independent in R^3.

51. Determine whether the set of polynomials in P_n satisfying the condition $d^2p/dx^2 = 0$ is a subspace of P_n. If it is, find a basis for the subspace.

52. Recall that the intersection of two sets W_1 and W_2 is the set of all elements common to both sets, and the union of W_1 and W_2 is the set of elements that are in either W_1 or W_2. Suppose W_1 and W_2 are subspaces of a vector space V. Prove or disprove by counterexample the following propositions:

(a) $W_1 \cap W_2$ is a subspace of V.

(b) $W_1 \cup W_2$ is a subspace of V.

CHAPTER

Matrices

8

INTRODUCTION

In mathematics one is often faced with the task of handling arrays of numbers or functions. For example, we shall see that to solve a system of equations such as

$$2x - y = 1$$
$$3x + y = 4$$

it is irrelevant what symbols are used to denote variables; the solution of the system depends only on the numbers that appear in the system. We can solve a system by appropriate manipulations on the array of numbers

$$\begin{pmatrix} 2 & -1 & 1 \\ 3 & 1 & 4 \end{pmatrix}.$$

The array is called a matrix. The theory of matrices was the invention of the eminent English mathematician Arthur Cayley (1821–1895).

8.1 Matrix Algebra

■ **Matrix** In the last section of Chapter 7 we saw that a vector in R^n is an ordered n-tuple (x_1, x_2, \ldots, x_n). Vectors are often written as a horizontal or vertical array with no commas:

$$(x_1 \ x_2 \ \cdots \ x_n) \quad \text{or} \quad \begin{pmatrix} x_1 \\ x_2 \\ \vdots \\ x_n \end{pmatrix}. \tag{1}$$

Each array in (1) is called a matrix. Our objective in this section is to consider the algebra of such arrays. We begin with a definition.

DEFINITION 8.1 Matrix

A **matrix** is any rectangular array of numbers or functions:

$$\begin{pmatrix} a_{11} & a_{12} & \cdots & a_{1n} \\ a_{21} & a_{22} & \cdots & a_{2n} \\ \vdots & & & \vdots \\ a_{m1} & a_{m2} & \cdots & a_{mn} \end{pmatrix}. \tag{2}$$

The numbers or functions in the array are called **entries** or **elements** of the matrix. If a matrix has m rows and n columns, we say that its **size** is m by n (written $m \times n$). An $n \times n$ matrix is called a **square** matrix or a matrix of **order n**. A 1×1 matrix is simply a constant or a single function. For example, $\mathbf{A} = \begin{pmatrix} 1 & 2 & 5 \\ 6 & 9 & 3 \end{pmatrix}$ is a 2×3 matrix, whereas

$$\mathbf{B} = \begin{pmatrix} 9 & 7 & 0 & 8 \\ \frac{1}{2} & -2 & 6 & 1 \\ 0 & 0 & -1 & 6 \\ 5 & \sqrt{3} & \pi & -4 \end{pmatrix} \tag{3}$$

is a 4×4 square matrix or a matrix of order 4. Throughout this text we shall denote a matrix by a capital boldfaced letter such as **A, B, C, X.**

The entry in the ith row and jth column of an $m \times n$ matrix **A** is written a_{ij}. An $m \times n$ matrix **A** is then abbreviated as $\mathbf{A} = (a_{ij})_{m \times n}$. For an $n \times n$ square matrix, the entries $a_{11}, a_{22}, \ldots, a_{nn}$ are called the **main diagonal entries.** The main diagonal entries for the matrix **B** in (3) are 9, −2, −1, and −4.

DEFINITION 8.2 Column and Row Vectors

An $n \times 1$ matrix,

$$\begin{pmatrix} a_1 \\ a_2 \\ \vdots \\ a_n \end{pmatrix},$$

is called a **column vector.** A $1 \times n$ matrix,

$$(a_1 \quad a_2 \quad \cdots \quad a_n),$$

is called a **row vector.**

DEFINITION 8.3 Equality of Matrices

Two $m \times n$ matrices **A** and **B** are **equal** if $a_{ij} = b_{ij}$ for each i and j.

In other words, two matrices are equal if and only if they have the same size and their corresponding entries are equal.

Example 1 Equality

(a) The matrices $\begin{pmatrix} 1 & 1 \\ 1 & 1 \end{pmatrix}$ and $\begin{pmatrix} 1 & 1 & 1 \\ 1 & 1 & 1 \end{pmatrix}$ are not equal, since the size of the first

matrix is 2×2 and the size of the second matrix is 2×3.

(b) The matrices $\begin{pmatrix} 1 & 2 \\ 3 & 4 \end{pmatrix}$ and $\begin{pmatrix} 1 & 2 \\ 4 & 3 \end{pmatrix}$ are not equal, since the corresponding

entries in the second rows of the matrices are not the same. ❏

■ **Matrix Addition** When two matrices **A** and **B** are of the same size we can add them by adding their corresponding entries.

DEFINITION 8.4 Matrix Addition

If **A** and **B** are $m \times n$ matrices, then their **sum** is

$$\mathbf{A} + \mathbf{B} = (a_{ij} + b_{ij})_{m \times n}.$$

Example 2 **Addition of Two Matrices**

(a) The sum of $\mathbf{A} = \begin{pmatrix} 2 & -1 & 3 \\ 0 & 4 & 6 \\ -6 & 10 & -5 \end{pmatrix}$ and $\mathbf{B} = \begin{pmatrix} 4 & 7 & -8 \\ 9 & 3 & 5 \\ 1 & -1 & 2 \end{pmatrix}$ is

$$\mathbf{A} + \mathbf{B} = \begin{pmatrix} 2+4 & -1+7 & 3+(-8) \\ 0+9 & 4+3 & 6+5 \\ -6+1 & 10+(-1) & -5+2 \end{pmatrix} = \begin{pmatrix} 6 & 6 & -5 \\ 9 & 7 & 11 \\ -5 & 9 & -3 \end{pmatrix}.$$

(b) The sum of $\mathbf{A} = \begin{pmatrix} 1 & 2 & 3 \\ 4 & 5 & 6 \end{pmatrix}$ and $\mathbf{B} = \begin{pmatrix} 1 & 0 \\ 1 & 0 \end{pmatrix}$ is not defined, since \mathbf{A} and \mathbf{B} are of different sizes. ❏

DEFINITION 8.5 Scalar Multiple of a Matrix

If k is a real number, then the **scalar multiple** of a matrix \mathbf{A} is

$$k\mathbf{A} = \begin{pmatrix} ka_{11} & ka_{12} & \cdots & ka_{1n} \\ ka_{21} & ka_{22} & \cdots & ka_{2n} \\ \vdots & & & \vdots \\ ka_{m1} & ka_{m2} & \cdots & ka_{mn} \end{pmatrix} = (ka_{ij})_{m \times n}.$$

In other words, to compute $k\mathbf{A}$, we simply multiply each entry of \mathbf{A} by k. For example, from Definition 8.5, $5\begin{pmatrix} 2 & -3 \\ 4 & -1 \end{pmatrix} = \begin{pmatrix} 5\cdot 2 & 5\cdot(-3) \\ 5\cdot 4 & 5\cdot(-1) \end{pmatrix} = \begin{pmatrix} 10 & -15 \\ 20 & -5 \end{pmatrix}$. We note in passing that for any matrix \mathbf{A} the scalar multiple $k\mathbf{A}$ is the same as $\mathbf{A}k$.

The **difference** of two $m \times n$ matrices is defined in the usual manner: $\mathbf{A} - \mathbf{B} = \mathbf{A} + (-\mathbf{B})$ where $-\mathbf{B} = (-1)\mathbf{B}$.

The following theorem summarizes some of the properties of matrix addition and scalar multiplication. Each of the six parts of the theorem can be proved using Definitions 8.4 and 8.5.

THEOREM 8.1 Properties of Matrix Addition and Scalar Multiplication

Suppose \mathbf{A}, \mathbf{B}, and \mathbf{C} are $m \times n$ matrices and k_1 and k_2 are scalars. Then

(i) $\mathbf{A} + \mathbf{B} = \mathbf{B} + \mathbf{A}$ Commutative Law of Addition

(ii) $\mathbf{A} + (\mathbf{B} + \mathbf{C}) = (\mathbf{A} + \mathbf{B}) + \mathbf{C}$ Associative Law of Addition

(iii) $(k_1 k_2)\mathbf{A} = k_1(k_2\mathbf{A})$

(iv) $1\mathbf{A} = \mathbf{A}$

(v) $k_1(\mathbf{A} + \mathbf{B}) = k_1\mathbf{A} + k_1\mathbf{B}$ Distributive Law

(vi) $(k_1 + k_2)\mathbf{A} = k_1\mathbf{A} + k_2\mathbf{A}$ Distributive Law.

■ **Matrix Multiplication** We have just seen that any matrix \mathbf{A} can be multiplied by a scalar. But can two matrices be multiplied together? The next definition gives the answer.

> **DEFINITION 8.6 Matrix Multiplication**
>
> Let **A** be a matrix having m rows and p columns, and let **B** be a matrix having p rows and n columns. The **product AB** is the $m \times n$ matrix
>
> $$\mathbf{AB} = \begin{pmatrix} a_{11} & a_{12} & \cdots & a_{1p} \\ a_{21} & a_{22} & \cdots & a_{2p} \\ \vdots & & & \vdots \\ a_{m1} & a_{m2} & \cdots & a_{mp} \end{pmatrix} \begin{pmatrix} b_{11} & b_{12} & \cdots & b_{1n} \\ b_{21} & b_{22} & \cdots & b_{2n} \\ \vdots & & & \vdots \\ b_{p1} & b_{p2} & \cdots & b_{pn} \end{pmatrix}$$
>
> $$= \begin{pmatrix} a_{11}b_{11} + a_{12}b_{21} + \cdots + a_{1p}b_{p1} & \cdots & a_{11}b_{1n} + a_{12}b_{2n} + \cdots + a_{1p}b_{pn} \\ a_{21}b_{11} + a_{22}b_{21} + \cdots + a_{2p}b_{p1} & \cdots & a_{21}b_{1n} + a_{22}b_{2n} + \cdots + a_{2p}b_{pn} \\ \vdots & & \vdots \\ a_{m1}b_{11} + a_{m2}b_{21} + \cdots + a_{mp}b_{p1} & \cdots & a_{m1}b_{1n} + a_{m2}b_{2n} + \cdots + a_{mp}b_{pn} \end{pmatrix}$$
>
> $$= \left(\sum_{k=1}^{p} a_{ik}b_{kj} \right)_{m \times n}.$$

 Note: The number of columns of **A** *must equal the number of rows of* **B.**

Definition 8.6 states that the product $\mathbf{C} = \mathbf{AB}$ is defined only when the number of columns in the matrix **A** is the same as the number of rows in **B.** The dimension of the product can be determined from

$$\mathbf{A}_{m \times p} \mathbf{B}_{p \times n} = \mathbf{C}_{m \times n}.$$

Also, you might recognize that the entries in, say, the ith row of the final matrix $\mathbf{C} = \mathbf{AB}$ are formed by using the component definition of the inner or dot product of the ith row (vector) of **A** with each of the columns (vectors) of **B**.

Example 3 Matrix Multiplication

Find the product **AB** for the following matrices:

(a) $\mathbf{A} = \begin{pmatrix} 4 & 7 \\ 3 & 5 \end{pmatrix}$, $\mathbf{B} = \begin{pmatrix} 9 & -2 \\ 6 & 8 \end{pmatrix}$ (b) $\mathbf{A} = \begin{pmatrix} 5 & 8 \\ 1 & 0 \\ 2 & 7 \end{pmatrix}$, $\mathbf{B} = \begin{pmatrix} -4 & -3 \\ 2 & 0 \end{pmatrix}$

SOLUTION From Definition 8.6 we have:

(a) $\mathbf{AB} = \begin{pmatrix} 4 \cdot 9 + 7 \cdot 6 & 4 \cdot (-2) + 7 \cdot 8 \\ 3 \cdot 9 + 5 \cdot 6 & 3 \cdot (-2) + 5 \cdot 8 \end{pmatrix} = \begin{pmatrix} 78 & 48 \\ 57 & 34 \end{pmatrix}$

(b) $\mathbf{AB} = \begin{pmatrix} 5 \cdot (-4) + 8 \cdot 2 & 5 \cdot (-3) + 8 \cdot 0 \\ 1 \cdot (-4) + 0 \cdot 2 & 1 \cdot (-3) + 0 \cdot 0 \\ 2 \cdot (-4) + 7 \cdot 2 & 2 \cdot (-3) + 7 \cdot 0 \end{pmatrix} = \begin{pmatrix} -4 & -15 \\ -4 & -3 \\ 6 & -6 \end{pmatrix}$ ❑

Unlike matrix addition, matrix multiplication is, in general, not commutative. That is, $\mathbf{BA} \neq \mathbf{AB}$. Observe that in part **(a)** of Example 3 $\mathbf{BA} = \begin{pmatrix} 30 & 53 \\ 48 & 82 \end{pmatrix}$, whereas in part **(b)** the product **BA** is *not defined* because the first matrix (in this case **B**) does not have the same number of columns as the second matrix has rows.

The product of an $m \times n$ matrix with an $n \times 1$ column vector is an $m \times 1$ column vector. For example,

$$\begin{pmatrix} -4 & 2 \\ 3 & 8 \end{pmatrix} \begin{pmatrix} x_1 \\ x_2 \end{pmatrix} = \begin{pmatrix} -4x_1 + 2x_2 \\ 3x_1 + 8x_2 \end{pmatrix}. \tag{3}$$

It is often convenient to write a column vector as a sum of two or more column vectors. In view of Definitions 8.4 and 8.5, the result in (3) can be written

$$\begin{pmatrix} -4x_1 + 2x_2 \\ 3x_1 + 8x_2 \end{pmatrix} = x_1 \begin{pmatrix} -4 \\ 3 \end{pmatrix} + x_2 \begin{pmatrix} 2 \\ 8 \end{pmatrix}.$$

■ **Associative Law** Although we shall not prove it, matrix multiplication is **associative.** If \mathbf{A} is an $m \times p$ matrix, \mathbf{B} a $p \times r$ matrix, and \mathbf{C} an $r \times n$ matrix, then the product

$$\mathbf{A}(\mathbf{BC}) = (\mathbf{AB})\mathbf{C}$$

is an $m \times n$ matrix.

■ **Distributive Law** If \mathbf{B} and \mathbf{C} are both $r \times n$ matrices and \mathbf{A} is an $m \times r$ matrix, then the **distributive law** is

$$\mathbf{A}(\mathbf{B} + \mathbf{C}) = \mathbf{AB} + \mathbf{AC}.$$

Furthermore, if the product $(\mathbf{B} + \mathbf{C})\mathbf{A}$ is defined, then

$$(\mathbf{B} + \mathbf{C})\mathbf{A} = \mathbf{BA} + \mathbf{CA}.$$

DEFINITION 8.7 Transpose of a Matrix

The **transpose** of the $m \times n$ matrix (2) is the $n \times m$ matrix \mathbf{A}^T given by

$$\mathbf{A}^T = \begin{pmatrix} a_{11} & a_{21} & \cdots & a_{m1} \\ a_{12} & a_{22} & \cdots & a_{m2} \\ \vdots & & & \vdots \\ a_{1n} & a_{2n} & \cdots & a_{mn} \end{pmatrix}.$$

In other words, the rows of a matrix \mathbf{A} become the columns of its transpose \mathbf{A}^T. For example, if

$$\mathbf{A} = \begin{pmatrix} 3 & 2 & -1 \\ 6 & 5 & 2 \\ 2 & 1 & 4 \end{pmatrix}, \text{ then } \mathbf{A}^T = \begin{pmatrix} 3 & 6 & 2 \\ 2 & 5 & 1 \\ -1 & 2 & 4 \end{pmatrix}. \text{ If } \mathbf{B} = (5 \ 3), \text{ then } \mathbf{B}^T = \begin{pmatrix} 5 \\ 3 \end{pmatrix}.$$

In the next theorem we give some important properties of the transpose.

THEOREM 8.2 Properties of Transpose

Suppose \mathbf{A} and \mathbf{B} are matrices and k a scalar. Then

 (*i*) $(\mathbf{A}^T)^T = \mathbf{A}$ Transpose of a transpose

 (*ii*) $(\mathbf{A} + \mathbf{B})^T = \mathbf{A}^T + \mathbf{B}^T$ Transpose of a sum

 (*iii*) $(\mathbf{AB})^T = \mathbf{B}^T\mathbf{A}^T$ Transpose of a product

 (*iv*) $(k\mathbf{A})^T = k\mathbf{A}^T$ Transpose of a scalar multiple

Of course, in properties (*ii*) and (*iii*) of Theorem 8.2 we assume that the sum and product of **A** and **B** are defined. Note carefully that part (*iii*) of the theorem indicates that the transpose of the product is the product of the transposes *in reverse order.* Moreover, (*ii*) and (*iii*) extend to any finite sum or product of matrices. For example, in the case of three matrices we have

$$(\mathbf{A} + \mathbf{B} + \mathbf{C})^T = \mathbf{A}^T + \mathbf{B}^T + \mathbf{C}^T \quad \text{and} \quad (\mathbf{ABC})^T = \mathbf{C}^T\mathbf{B}^T\mathbf{A}^T.$$

■ **Special Matrices** In matrix theory there are many special kinds of matrices that are important because they possess certain properties. The following is a list of some of these matrices:

• A matrix that consists of all zero entries is called a **zero matrix** and is denoted by **0**. For example,

$$\mathbf{0} = \begin{pmatrix} 0 \\ 0 \end{pmatrix}, \quad \mathbf{0} = \begin{pmatrix} 0 & 0 \\ 0 & 0 \end{pmatrix}, \quad \mathbf{0} = \begin{pmatrix} 0 & 0 \\ 0 & 0 \\ 0 & 0 \end{pmatrix}$$

are zero matrices. If **A** and **0** are $m \times n$ matrices, then

$$\mathbf{A} + \mathbf{0} = \mathbf{A}. \tag{4}$$

In addition, $\qquad\qquad \mathbf{A} + (-\mathbf{A}) = \mathbf{0}. \tag{5}$

• An $n \times n$ matrix **A** is said to be a **triangular matrix** if all its entries below the main diagonal are zeros or if all its entries above the main diagonal are zeros. In other words, a square matrix **A** is triangular if $a_{ij} = 0$ for $i < j$ or $a_{ij} = 0$ for $i > j$. More specifically, in the first case the matrix is called **upper triangular,** and in the second case the matrix is called **lower triangular.** The following matrices are triangular:

$$\begin{pmatrix} 1 & 2 & 3 & 4 \\ 0 & 5 & 6 & 7 \\ 0 & 0 & 8 & 9 \\ 0 & 0 & 0 & 1 \end{pmatrix} \qquad \begin{pmatrix} -2 & 0 & 0 & 0 & 0 \\ 1 & 6 & 0 & 0 & 0 \\ 8 & 9 & 3 & 0 & 0 \\ 1 & 1 & 1 & 2 & 0 \\ 15 & 2 & 3 & 4 & 1 \end{pmatrix}$$

upper triangular matrix lower triangular matrix

• An $n \times n$ matrix **A** is said to be a **diagonal matrix** if all its entries not on the main diagonal are zeros. In terms of the symbolism $\mathbf{A} = (a_{ij})_{n \times n}$, **A** is a diagonal matrix if $a_{ij} = 0$ for $i \neq j$. The following is a diagonal matrix:

$$\begin{pmatrix} 7 & 0 & 0 \\ 0 & \frac{1}{2} & 0 \\ 0 & 0 & 1 \end{pmatrix}.$$

• When the entries a_{ii} of a diagonal matrix **A** are all equal, it is called a **scalar matrix.** For example, $\begin{pmatrix} 5 & 0 \\ 0 & 5 \end{pmatrix}$ is a scalar matrix. An $n \times n$ scalar matrix is simply a scalar multiple of a diagonal matrix in which the main diagonal entries are all equal to 1. For example, $\begin{pmatrix} 5 & 0 \\ 0 & 5 \end{pmatrix} = 5\begin{pmatrix} 1 & 0 \\ 0 & 1 \end{pmatrix}$. In general, the $n \times n$ matrix

$$\begin{pmatrix} 1 & 0 & 0 & \cdots & 0 \\ 0 & 1 & 0 & \cdots & 0 \\ \vdots & & & & \vdots \\ 0 & 0 & 0 & \cdots & 1 \end{pmatrix}$$

is denoted by the symbol \mathbf{I} (or \mathbf{I}_n when there is a need to emphasize the order of the matrix). For any $m \times n$ matrix \mathbf{A} it is readily verified that $\mathbf{I}_m \mathbf{A} = \mathbf{A} \mathbf{I}_n = \mathbf{A}$. Because this last property is analogous to $1 \cdot a = a \cdot 1 = a$, for any real number a, the matrix \mathbf{I} is called the **identity matrix.**

- An $n \times n$ matrix \mathbf{A} is said to be **symmetric** if $\mathbf{A}^T = \mathbf{A}$; that is, \mathbf{A} is symmetric if $a_{ij} = a_{ji}$ for all i and j. This means that the entries in a symmetric matrix are symmetric with respect to the main diagonal of the matrix. For example, a quick inspection of the matrix

$$\mathbf{A} = \begin{pmatrix} 1 & 2 & 7 \\ 2 & 5 & 6 \\ 7 & 6 & 4 \end{pmatrix}$$

shows that it is symmetric. In addition, by taking the transpose of \mathbf{A} we see that

$$\mathbf{A}^T = \begin{pmatrix} 1 & 2 & 7 \\ 2 & 5 & 6 \\ 7 & 6 & 4 \end{pmatrix} = \mathbf{A}.$$

Remarks

Suppose the symbol $M_{m,n}$ denotes the set of all $m \times n$ matrices on which the operations of matrix addition and scalar multiplication are defined. Then

$$\mathbf{A} + \mathbf{B} \text{ is in } M_{m,n} \quad \text{and} \quad k\mathbf{A} \text{ is in } M_{m,n} \qquad (6)$$

for every \mathbf{A} and \mathbf{B} in $M_{m,n}$ and for every scalar k. That is to say, $M_{m,n}$ is closed under matrix addition and scalar multiplication. When we combine (5) with properties (3) and (4) and the properties listed in Theorem 8.1, it follows immediately that $M_{m,n}$ is a vector space. For practical purposes, the vector spaces $M_{1,n}$ (row vectors) and $M_{n,1}$ (column vectors) are indistinguishable from the vector space R^n.

E X E R C I S E S 8 . 1 Answers to odd-numbered problems begin on page A-42.

In Problems 1–6, state the size of the given matrix.

1. $\begin{pmatrix} 1 & 2 & 3 & 9 \\ 5 & 6 & 0 & 1 \end{pmatrix}$

2. $\begin{pmatrix} 0 & 2 \\ 8 & 4 \\ 5 & 6 \end{pmatrix}$

3. $\begin{pmatrix} 1 & 2 & -1 \\ 0 & 7 & -2 \\ 0 & 0 & 5 \end{pmatrix}$

4. $(5 \quad 7 \quad -15)$

5. $\begin{pmatrix} 1 & 5 & -6 & 0 \\ 7 & -10 & 2 & 12 \\ 0 & 9 & 2 & -1 \end{pmatrix}$

6. $\begin{pmatrix} 1 \\ 5 \\ -6 \\ 0 \\ 7 \\ -10 \\ 2 \\ 12 \end{pmatrix}$

In Problems 7–10, determine whether the given matrices are equal.

7. $\begin{pmatrix} 1 & 2 & 3 \\ 4 & 5 & 6 \end{pmatrix}$, $\begin{pmatrix} 1 & 2 \\ 3 & 4 \\ 5 & 6 \end{pmatrix}$ **8.** $\begin{pmatrix} 1 & 2 \\ 0 & 1 \end{pmatrix}$, $\begin{pmatrix} 1 & 0 \\ 2 & 1 \end{pmatrix}$

9. $\begin{pmatrix} \sqrt{(-2)^2} & 1 \\ 2 & \frac{2}{8} \end{pmatrix}$, $\begin{pmatrix} -2 & 1 \\ 2 & \frac{1}{4} \end{pmatrix}$

10. $\begin{pmatrix} \frac{1}{8} & \frac{1}{5} \\ \sqrt{2} & 1 \end{pmatrix}$, $\begin{pmatrix} 0.125 & 0.2 \\ 1.414 & 1 \end{pmatrix}$

In Problems 11 and 12, determine the values of x and y for which the matrices are equal.

11. $\begin{pmatrix} 1 & x \\ y & -3 \end{pmatrix}$, $\begin{pmatrix} 1 & y-2 \\ 3x-2 & -3 \end{pmatrix}$ **12.** $\begin{pmatrix} x^2 & 1 \\ y & 5 \end{pmatrix}$, $\begin{pmatrix} 9 & 1 \\ 4x & 5 \end{pmatrix}$

In Problems 13 and 14, find the entries c_{23} and c_{12} for the matrix $\mathbf{C} = 2\mathbf{A} - 3\mathbf{B}$.

13. $\mathbf{A} = \begin{pmatrix} 2 & 3 & -1 \\ -1 & 6 & 0 \end{pmatrix}$, $\mathbf{B} = \begin{pmatrix} 4 & -2 & 6 \\ 1 & 3 & -3 \end{pmatrix}$

14. $\mathbf{A} = \begin{pmatrix} 1 & -1 & 1 \\ 2 & 2 & 1 \\ 0 & -4 & 1 \end{pmatrix}$, $\mathbf{B} = \begin{pmatrix} 2 & 0 & 5 \\ 0 & 4 & 0 \\ 3 & 0 & 7 \end{pmatrix}$

15. If $\mathbf{A} = \begin{pmatrix} 4 & 5 \\ -6 & 9 \end{pmatrix}$ and $\mathbf{B} = \begin{pmatrix} -2 & 6 \\ 8 & -10 \end{pmatrix}$, find **(a)** $\mathbf{A} + \mathbf{B}$,

(b) $\mathbf{B} - \mathbf{A}$, **(c)** $2\mathbf{A} + 3\mathbf{B}$.

16. If $\mathbf{A} = \begin{pmatrix} -2 & 0 \\ 4 & 1 \\ 7 & 3 \end{pmatrix}$ and $\mathbf{B} = \begin{pmatrix} 3 & -1 \\ 0 & 2 \\ -4 & -2 \end{pmatrix}$, find **(a)** $\mathbf{A} - \mathbf{B}$,

(b) $\mathbf{B} - \mathbf{A}$, **(c)** $2(\mathbf{A} + \mathbf{B})$.

17. If $\mathbf{A} = \begin{pmatrix} 2 & -3 \\ -5 & 4 \end{pmatrix}$ and $\mathbf{B} = \begin{pmatrix} -1 & 6 \\ 3 & 2 \end{pmatrix}$, find **(a)** \mathbf{AB},

(b) \mathbf{BA}, **(c)** $\mathbf{A}^2 = \mathbf{AA}$, **(d)** $\mathbf{B}^2 = \mathbf{BB}$.

18. If $\mathbf{A} = \begin{pmatrix} 1 & 4 \\ 5 & 10 \\ 8 & 12 \end{pmatrix}$ and $\mathbf{B} = \begin{pmatrix} -4 & 6 & -3 \\ 1 & -3 & 2 \end{pmatrix}$, find **(a)** \mathbf{AB},

(b) \mathbf{BA}.

19. If $\mathbf{A} = \begin{pmatrix} 1 & -2 \\ -2 & 4 \end{pmatrix}$, $\mathbf{B} = \begin{pmatrix} 6 & 3 \\ 2 & 1 \end{pmatrix}$, and $\mathbf{C} = \begin{pmatrix} 0 & 2 \\ 3 & 4 \end{pmatrix}$, find

(a) \mathbf{BC}, **(b)** $\mathbf{A(BC)}$, **(c)** $\mathbf{C(BA)}$, **(d)** $\mathbf{A(B + C)}$.

20. If $\mathbf{A} = (5 \quad -6 \quad 7)$, $\mathbf{B} = \begin{pmatrix} 3 \\ 4 \\ -1 \end{pmatrix}$, and $\mathbf{C} = \begin{pmatrix} 1 & 2 & 4 \\ 0 & 1 & -1 \\ 3 & 2 & 1 \end{pmatrix}$,

find **(a)** \mathbf{AB}, **(b)** \mathbf{BA}, **(c)** $(\mathbf{BA})\mathbf{C}$, **(d)** $(\mathbf{AB})\mathbf{C}$.

21. If $\mathbf{A} = \begin{pmatrix} 4 \\ 8 \\ -10 \end{pmatrix}$ and $\mathbf{B} = (2 \quad 4 \quad 5)$, find **(a)** $\mathbf{A}^T\mathbf{A}$, **(b)** $\mathbf{B}^T\mathbf{B}$,

(c) $\mathbf{A} + \mathbf{B}^T$.

22. If $\mathbf{A} = \begin{pmatrix} 1 & 2 \\ 2 & 4 \end{pmatrix}$ and $\mathbf{B} = \begin{pmatrix} -2 & 3 \\ 5 & 7 \end{pmatrix}$, find **(a)** $\mathbf{A} + \mathbf{B}^T$,

(b) $2\mathbf{A}^T - \mathbf{B}^T$, **(c)** $\mathbf{A}^T(\mathbf{A} - \mathbf{B})$.

23. If $\mathbf{A} = \begin{pmatrix} 3 & 4 \\ 8 & 1 \end{pmatrix}$ and $\mathbf{B} = \begin{pmatrix} 5 & 10 \\ -2 & -5 \end{pmatrix}$, find **(a)** $(\mathbf{AB})^T$,

(b) $\mathbf{B}^T\mathbf{A}^T$.

24. If $\mathbf{A} = \begin{pmatrix} 5 & 9 \\ -4 & 6 \end{pmatrix}$ and $\mathbf{B} = \begin{pmatrix} -3 & 11 \\ -7 & 2 \end{pmatrix}$, find **(a)** $\mathbf{A}^T + \mathbf{B}$,

(b) $2\mathbf{A} + \mathbf{B}^T$.

In Problems 25–28, write the given sum as a single column matrix.

25. $4\begin{pmatrix} -1 \\ 2 \end{pmatrix} - 2\begin{pmatrix} 2 \\ 8 \end{pmatrix} + 3\begin{pmatrix} -2 \\ 3 \end{pmatrix}$ **26.** $3\begin{pmatrix} 2 \\ 1 \\ -1 \end{pmatrix} + 5\begin{pmatrix} -1 \\ -1 \\ 3 \end{pmatrix} - 2\begin{pmatrix} 3 \\ 4 \\ -5 \end{pmatrix}$

27. $\begin{pmatrix} 2 & -3 \\ 1 & 4 \end{pmatrix}\begin{pmatrix} -2 \\ 5 \end{pmatrix} - \begin{pmatrix} -1 & 6 \\ -2 & 3 \end{pmatrix}\begin{pmatrix} -7 \\ 2 \end{pmatrix}$

28. $\begin{pmatrix} 1 & -3 & 4 \\ 2 & 5 & -1 \\ 0 & -4 & -2 \end{pmatrix}\begin{pmatrix} 3 \\ 2 \\ -1 \end{pmatrix} + \begin{pmatrix} -1 \\ 1 \\ 4 \end{pmatrix} - \begin{pmatrix} 2 \\ 8 \\ -6 \end{pmatrix}$

In Problems 29 and 30, determine the size of the matrix \mathbf{A} such that the given product is defined.

29. $\begin{pmatrix} 2 & 1 & 3 & 3 \\ 9 & 6 & 7 & 0 \end{pmatrix}\mathbf{A}\begin{pmatrix} 0 \\ 5 \\ 7 \\ 9 \\ 2 \end{pmatrix}$ **30.** $\begin{pmatrix} 2 & 1 & 3 \\ 3 & 9 & 6 \\ 7 & 0 & -1 \end{pmatrix}\mathbf{A}\begin{pmatrix} 0 & 1 \\ 7 & 4 \end{pmatrix}$

In Problems 31–34, suppose $\mathbf{A} = \begin{pmatrix} 2 & 4 \\ -3 & 2 \end{pmatrix}$ and $\mathbf{B} = \begin{pmatrix} 4 & 10 \\ 2 & 5 \end{pmatrix}$. Verify the given property by computing the left and right members of the given equality.

31. $(\mathbf{A}^T)^T = \mathbf{A}$ **32.** $(\mathbf{A} + \mathbf{B})^T = \mathbf{A}^T + \mathbf{B}^T$

33. $(\mathbf{AB})^T = \mathbf{B}^T\mathbf{A}^T$ **34.** $(6\mathbf{A})^T = 6\mathbf{A}^T$

35. Suppose $\mathbf{A} = \begin{pmatrix} 2 & 1 \\ 6 & 3 \\ 2 & 5 \end{pmatrix}$. Verify that the matrix $\mathbf{B} = \mathbf{AA}^T$ is symmetric.

36. Show that if \mathbf{A} is an $m \times n$ matrix, then \mathbf{AA}^T is symmetric.

37. In matrix theory many of the familiar properties of the real number system are not valid. If a and b are real numbers, then $ab = 0$ implies that $a = 0$ or $b = 0$. Find two matrices such that $\mathbf{AB} = \mathbf{0}$ but $\mathbf{A} \neq \mathbf{0}$ and $\mathbf{B} \neq \mathbf{0}$.

38. If a, b, and c are real numbers and $c \neq 0$, then $ac = bc$ implies $a = b$. For matrices, $\mathbf{AC} = \mathbf{BC}$, $\mathbf{C} \neq \mathbf{0}$, does not necessarily imply $\mathbf{A} = \mathbf{B}$. Verify this for

$$\mathbf{A} = \begin{pmatrix} 2 & 1 & 4 \\ 3 & 2 & 1 \\ 1 & 3 & 2 \end{pmatrix}, \mathbf{B} = \begin{pmatrix} 5 & 1 & 6 \\ 9 & 2 & -3 \\ -1 & 3 & 7 \end{pmatrix}, \text{ and } \mathbf{C} = \begin{pmatrix} 0 & 0 & 0 \\ 2 & 3 & 4 \\ 0 & 0 & 0 \end{pmatrix}$$

In Problems 39 and 40, let \mathbf{A} and \mathbf{B} be $n \times n$ matrices. Explain why, in general, the given formula is not valid.

39. $(\mathbf{A} + \mathbf{B})^2 = \mathbf{A}^2 + 2\mathbf{AB} + \mathbf{B}^2$

40. $(\mathbf{A} + \mathbf{B})(\mathbf{A} - \mathbf{B}) = \mathbf{A}^2 - \mathbf{B}^2$

41. Write $\begin{pmatrix} a_{11} & a_{12} \\ a_{21} & a_{22} \end{pmatrix} \begin{pmatrix} x_1 \\ x_2 \end{pmatrix} = \begin{pmatrix} b_1 \\ b_2 \end{pmatrix}$ without matrices.

42. Write the system of equations

$$2x_1 + 6x_2 + x_3 = 7$$
$$x_1 + 2x_2 - x_3 = -1$$
$$5x_1 + 7x_2 - 4x_3 = 9$$

as a matrix equation $\mathbf{AX} = \mathbf{B}$, where \mathbf{X} and \mathbf{B} are column vectors.

43. Verify that the quadratic form $ax^2 + bxy + cy^2$ is the same as

$$(x \ \ y) \begin{pmatrix} a & \tfrac{1}{2}b \\ \tfrac{1}{2}b & c \end{pmatrix} \begin{pmatrix} x \\ y \end{pmatrix}$$

44. Verify that the curl of the vector field $\mathbf{F} = P\mathbf{i} + Q\mathbf{j} + R\mathbf{k}$ can be written

$$\text{curl } \mathbf{F} = \begin{pmatrix} 0 & -\partial/\partial x & \partial/\partial x \\ \partial/\partial x & 0 & -\partial/\partial x \\ -\partial/\partial y & \partial/\partial x & 0 \end{pmatrix} \begin{pmatrix} P \\ Q \\ R \end{pmatrix}$$

(Readers who are not familiar with the concept of the curl of a vector field should see Section 9.7.)

45. As shown in Figure 8.1(a), a spacecraft can perform rotations called **pitch, roll,** and **yaw** about three distinct axes. To describe the coordinates of a point P we use two coordinate systems: a fixed three-dimensional Cartesian coordinate system in which the coordinates of P are (x, y, z) and a spacecraft coordinate system that moves with the particular rotation. In Figure 8.1(b) we have illustrated a yaw—that is, a rotation around the z-axis (which is perpendicular to the plane of the paper). The coordinates (x_Y, y_Y, z_Y) of the point P in the spacecraft system after the yaw are related to the coordinates (x, y, z) of P in the fixed coordinate system by the equations

$$x_Y = x \cos \gamma + y \sin \gamma$$
$$y_Y = -x \sin \gamma + y \cos \gamma$$
$$z_Y = z$$

where γ is the angle of rotation.

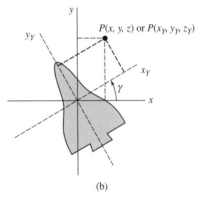

Figure 8.1

(a) Verify that the foregoing system of equations can be written as the matrix equation

$$\begin{pmatrix} x_Y \\ y_Y \\ z_Y \end{pmatrix} = \mathbf{M}_Y \begin{pmatrix} x \\ y \\ z \end{pmatrix}$$

where $\mathbf{M}_Y = \begin{pmatrix} \cos \gamma & \sin \gamma & 0 \\ -\sin \gamma & \cos \gamma & 0 \\ 0 & 0 & 1 \end{pmatrix}$

(b) When the spacecraft performs a pitch, roll, and yaw *in sequence* through the angles α, β, and γ, respectively, the final coordinates of the point P in the spacecraft system (x_S, y_S, z_S) are obtained from the sequence of transformations

$$x_P = x \qquad\qquad x_R = x_P \cos \beta - z_P \sin \beta$$
$$y_P = y \cos \alpha + z \sin \alpha \qquad y_R = y_P$$
$$z_P = -y \sin \alpha + z \cos \alpha \qquad z_R = x_P \sin \beta + z_P \cos \beta$$

$$x_S = x_R \cos \gamma + y_R \sin \gamma$$
$$y_S = -x_R \sin \gamma + y_R \cos \gamma$$
$$z_S = z_R$$

Write this sequence of transformations as a matrix equation

$$\begin{pmatrix} x_S \\ y_S \\ z_S \end{pmatrix} = \mathbf{M}_Y \mathbf{M}_R \mathbf{M}_P \begin{pmatrix} x \\ y \\ z \end{pmatrix}$$

The matrix \mathbf{M}_Y is the same as in part **(a)**. Identify the matrices \mathbf{M}_R and \mathbf{M}_P.

(c) Suppose the coordinates of a point are $(1, 1, 1)$ in the fixed coordinate system. Determine the coordinates of the point in the spacecraft system if the spacecraft performs a pitch, roll, and yaw in sequence through the angles $\alpha = 30°$, $\beta = 45°$, $\gamma = 60°$.

46. If an $n \times n$ matrix A can be written as a product of a lower triangular matrix L and an upper triangular matrix U, then $A = LU$ is said to be an **LU-factorization** of A. Verify that the given matrix A can be written as the product of the indicated matrices L and U.

(a) $A = \begin{pmatrix} 2 & -2 \\ 1 & 2 \end{pmatrix}$; $L = \begin{pmatrix} 1 & 0 \\ \frac{1}{2} & 1 \end{pmatrix}$, $U = \begin{pmatrix} 2 & -2 \\ 0 & 3 \end{pmatrix}$

(b) $A = \begin{pmatrix} 6 & 2 \\ 4 & 1 \end{pmatrix}$; $L = \begin{pmatrix} 1 & 0 \\ \frac{2}{3} & 1 \end{pmatrix}$, $U = \begin{pmatrix} 6 & 2 \\ 0 & -\frac{1}{3} \end{pmatrix}$

(c) $A = \begin{pmatrix} 1 & -2 & 1 \\ 0 & 1 & 2 \\ 2 & 6 & 1 \end{pmatrix}$; $L = \begin{pmatrix} 1 & 0 & 0 \\ 0 & 1 & 0 \\ 2 & 10 & 1 \end{pmatrix}$, $U = \begin{pmatrix} 1 & -2 & 1 \\ 0 & 1 & 2 \\ 0 & 0 & -21 \end{pmatrix}$

(d) $A = \begin{pmatrix} 1 & 1 & 1 \\ 3 & 1 & 2 \\ 1 & -1 & 1 \end{pmatrix}$; $L = \begin{pmatrix} 1 & 0 & 0 \\ 3 & 1 & 0 \\ 1 & 1 & 1 \end{pmatrix}$, $U = \begin{pmatrix} 1 & 1 & 1 \\ 0 & -2 & -1 \\ 0 & 0 & 1 \end{pmatrix}$

47. Project (a) A matrix A can be **partitioned** into submatrices. For example the 3×5 and 5×2 matrices

$$A = \left(\begin{array}{ccc:cc} 3 & 2 & -1 & 2 & 4 \\ \hdashline 1 & 6 & 3 & -1 & 5 \\ 0 & 4 & 6 & -2 & 3 \end{array} \right), \quad B = \begin{pmatrix} 3 & 4 \\ 0 & 7 \\ -4 & 1 \\ -2 & -1 \\ 2 & 5 \end{pmatrix}$$

can be written

$$A = \begin{pmatrix} A_{11} & A_{12} \\ A_{21} & A_{22} \end{pmatrix}, \quad B = \begin{pmatrix} B_1 \\ B_2 \end{pmatrix},$$

where A_{11} is the upper left-hand block, or submatrix, indicated in color in A, A_{12} is the upper right-hand block, and so on. Compute the product AB using the partitioned matrices.

(b) Investigate how partitioned matrices can be useful when using a computer to perform matrix calculations involving large matrices.

8.2 Systems of Linear Algebraic Equations

■ **Linear Equations** Recall that any equation of the form $ax + by = c$, where a, b, and c are real numbers, is said to be a **linear equation** in the variables x and y. The graph of a linear equation in two variables is a straight line. For real numbers a, b, c, and d, $ax + by + cz = d$ is a linear equation in the variables x, y, and z and is the equation of a plane. In general, an equation of the form

$$a_1 x_1 + a_2 x_2 + \cdots + a_n x_n = b_n,$$

where a_1, a_2, \ldots, a_n, and b_n are real numbers is a linear equation in the n variables x_1, x_2, \ldots, x_n.

In this section we will study **systems of linear equations.** Systems of linear equations are also called **linear systems.**

■ **General Form** A system of m linear equations in n unknowns has the general form

$$a_{11} x_1 + a_{12} x_2 + \cdots + a_{1n} x_n = b_1$$
$$a_{21} x_1 + a_{22} x_2 + \cdots + a_{2n} x_n = b_2 \tag{1}$$
$$\vdots \qquad\qquad \vdots$$
$$a_{m1} x_1 + a_{m2} x_2 + \cdots + a_{mn} x_n = b_m.$$

The **coefficients** of the unknowns in the linear system (1) can be abbreviated as a_{ij}, where i denotes the row and j denotes the column in which the coefficient appears. For example, a_{23} is the coefficient of the unknown in the second row and third column

(that is, x_3). Thus, $i = 1, 2, 3, \ldots, m$ and $j = 1, 2, 3, \ldots, n$. The numbers b_1, b_2, \ldots, b_m are called the **constants** of the system. If all the constants are zero the system (1) is said to be **homogeneous,** otherwise it is **nonhomogeneous.** For example,

This system is homogeneous
\downarrow

$$5x_1 - 9x_2 + x_3 = 0$$

$$x_1 + 3x_2 \qquad = 0$$

$$4x_1 + 6x_2 - x_3 = 0$$

This system is nonhomogeneous
\downarrow

$$2x_1 + 5x_2 + 6x_3 = 1$$

$$4x_1 + 3x_2 - x_3 = 9$$

■ **Solution** A **solution** of a linear system (1) is a set of n numbers x_1, x_2, \ldots, x_n that satisfies each equation in the system. For example, $x_1 = 3$, $x_2 = -1$ is a solution of the system

$$3x_1 + 6x_2 = 3$$

$$x_1 - 4x_2 = 7.$$

To see this we replace x_1 by 3 and x_2 by -1 in each equation:

$$3(3) + 6(-1)9 - 6 = 3 \quad \text{and} \quad 3 - 4(-1) = 3 + 4 = 7.$$

A linear system of equations is said to be **consistent** if it has at least one solution and **inconsistent** if it has no solutions. If a linear system is consistent, it has either

- a unique solution (that is, precisely one solution), or
- infinitely many solutions.

Thus a system of linear equations cannot have, say, exactly three solutions. For a linear system with two equations and two unknowns, the lines intersect at one point as in Figure 8.2(a) (unique solution), are identical as in Figure 8.2(b) (infinitely many solutions), or are parallel as in Figure 8.2(c) (inconsistent).

Example 1 Verification of a Solution

Verify that $x_1 = 14 + 7t$, $x_2 = 9 + 6t$, $x_3 = t$, where t is any real number, is a solution of the system

$$2x_1 - 3x_2 + 4x_3 = 1$$

$$x_1 - x_2 - x_3 = 5.$$

SOLUTION Replacing x_1, x_2, and x_3 in turn by $14 + 7t$, $9 + 6t$, and t, we have

$$2(14 + 7t) - 3(9 + 6t) + 4t = 1$$

$$14 + 7t - (9 + 6t) - t = 5.$$

For each real number t we obtain a different solution of the system; in other words, the system has an infinite number of solutions. For instance, $t = 0$, $t = 4$, and $t = -2$ give the three solutions

$$x_1 = 14, \qquad x_2 = 9, \qquad x_3 = 0,$$

$$x_1 = 42, \qquad x_2 = 33, \qquad x_3 = 4,$$

and
$$x_1 = 0, \qquad x_2 = -3, \qquad x_3 = -2,$$

respectively. Geometrically, each equation in the system represents a plane in R^3. In this case the planes intersect in a line; parametric equations of the line are $x_1 = 14 + 7t$, $x_2 = 9 + 6t$, $x_3 = t$. ❏

(a)

(b)

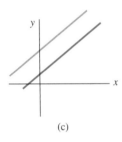

(c)

Figure 8.2

■ **Solving Systems** We can transform a system of linear equations into an equivalent system (that is, one having the same solutions) by the following **elementary operations:**

(*i*) Multiply an equation by a nonzero constant.
(*ii*) Interchange the positions of equations in the system.
(*iii*) Add a nonzero multiple of one equation to any other equation.

As the next example will show, these elementary operations enable us to systematically eliminate variables from the equations of the system.

> ### *Example 2* Solving a Linear System
>
> Solve
> $$2x_1 + 6x_2 + \ x_3 = 7$$
> $$x_1 + 2x_2 - \ x_3 = -1$$
> $$5x_1 + 7x_2 - 4x_3 = 9.$$
>
> **SOLUTION** We begin by interchanging the first and second rows:
> $$x_1 + 2x_2 - \ x_3 = -1$$
> $$2x_1 + 6x_2 + \ x_3 = 7$$
> $$5x_1 + 7x_2 - 4x_3 = 9.$$

Our goal now is to eliminate x_1 from the second and third equations. If we add to the second equation -2 times the first equation, we obtain the equivalent system

$$x_1 + 2x_2 - \ x_3 = -1$$
$$2x_2 + 3x_3 = 9$$
$$5x_1 + 7x_2 - 4x_3 = 9.$$

By adding to the third equation -5 times the first equation, we get a new equivalent system:

$$x_1 + 2x_2 - \ x_3 = -1$$
$$2x_2 + 3x_3 \quad = 9$$
$$-3x_2 + \ x_3 = 14.$$

We are now going to use the second equation to eliminate the variable x_2 from the first and third equations. To simplify matters let us multiply the second equation by ½:

$$x_1 + 2x_2 - \ x_3 = -1$$
$$x_2 + \frac{3}{2}x_3 = \frac{9}{2}$$
$$-3x_2 + \ x_3 = 14.$$

Adding to the first equation -2 times the second equation yields

$$x_1 \qquad - \ 4x_3 = -10$$
$$x_2 + \frac{3}{2}x_3 = \frac{9}{2}$$
$$-3x_2 + \ x_3 = 14.$$

Next, by adding 3 times the second equation to the third equation we get

$$x_1 \qquad - 4x_3 = -10$$
$$x_2 + \frac{3}{2}x_3 = \frac{9}{2}$$
$$\frac{11}{2}x_3 = \frac{55}{2}.$$

We shall use the last equation to eliminate the variable x_3 from the first and second equations. To this end we multiply the third equation by $\frac{2}{11}$:

$$x_1 \qquad - 4x_3 = -10$$
$$x_2 + \frac{3}{2}x_3 = \frac{9}{2}$$
$$x_3 = 5.$$

At this point we could use **back-substitution;** that is, substitute the value $x_3 = 5$ back into the remaining equations to determine x_1 and x_2. However, by continuing with our systematic elimination, we add to the second equation $-\frac{3}{2}$ times the third equation:

$$x_1 \qquad -4x_3 = -10$$
$$x_2 \qquad = -3$$
$$x_3 = 5.$$

Finally, by adding to the first equation 4 times the third equation, we obtain

$$x_1 \qquad = 10$$
$$x_2 \qquad = -3$$
$$x_3 = 5.$$

It is now apparent that $x_1 = 10$, $x_2 = -3$, $x_3 = 5$ is the solution of the original system. ❏

■ **Augmented Matrix** Reflecting on the solution of the linear system in Example 2 should convince you that the solution of the system does not depend on what symbols are used as variables. Thus, the systems

$$\begin{aligned} 2x + 6y + z &= 7 \\ x + 2y - z &= -1 \\ 5x + 7y - 4z &= 9 \end{aligned} \qquad \text{and} \qquad \begin{aligned} 2u + 6v + w &= 7 \\ u + 2v - w &= -1 \\ 5u + 7v - 4w &= 9 \end{aligned}$$

have the same solution as the system in Example 2. In other words, in the solution of a linear system the symbols used to denote the variables are immaterial; it is the coefficients of the variables and the constants that determine the solution of the system. In fact, we can solve a system of form (1) by dropping the variables entirely and performing operations on the rows of the array of coefficients and constants:

$$\begin{pmatrix} a_{11} & a_{12} & \cdots & a_{1n} & \bigg| & b_1 \\ a_{21} & a_{22} & \cdots & a_{2n} & \bigg| & b_2 \\ \vdots & & & \vdots & \bigg| & \vdots \\ a_{m1} & a_{m2} & \cdots & a_{mn} & \bigg| & b_m \end{pmatrix}. \tag{2}$$

This array is called the **augmented matrix** of the system or simply the **matrix of the system** (1).

Example 3 Augmented Matrices

(a) The augmented matrix $\begin{pmatrix} 1 & -3 & 5 & \big| & 2 \\ 4 & 7 & -1 & \big| & 8 \end{pmatrix}$ represents the linear system

$$x_1 - 3x_2 + 5x_3 = 2$$
$$4x_1 + 7x_2 - x_3 = 8.$$

(b) The linear system

$$x_1 - 5x_3 = -1 \qquad\qquad x_1 + 0x_2 - 5x_3 = -1$$
$$2x_1 + 8x_2 = 7 \quad \text{is the same as} \quad 2x_1 + 8x_2 + 0x_3 = 7$$
$$x_2 + 9x_3 = 1 \qquad\qquad 0x_1 + x_2 + 9x_3 = 1.$$

Thus the matrix of the system is

$$\begin{pmatrix} 1 & 0 & -5 & \big| & -1 \\ 2 & 8 & 0 & \big| & 7 \\ 0 & 1 & 9 & \big| & 1 \end{pmatrix}. \qquad \square$$

■ **Elementary Row Operations** Since the rows of an augmented matrix represent the equations in a linear system, the three elementary operations on a linear system listed above are equivalent to the following **elementary row operations** on a matrix:

 (*i*) Multiply a row by a nonzero constant.
 (*ii*) Interchange any two rows.
 (*iii*) Add a nonzero multiple of one row to any other row.

Of course, when we add a multiple of one row to another, we add the corresponding entries in the rows. We say that two matrices are **row equivalent** if one can be obtained from the other through a sequence of elementary row operations. The procedure of carrying out elementary row operations on a matrix to obtain a row equivalent matrix is called **row reduction.**

■ **Elimination Methods** To solve a system such as (1) using an augmented matrix we shall use either **Gaussian elimination** or the **Gauss–Jordan elimination method.** In the former method we row reduce the augmented matrix of the system until we arrive at a row-equivalent augmented matrix in **row-echelon form:**

 (*i*) The first nonzero entry in a nonzero row is a 1.
 (*ii*) In consecutive nonzero rows, the first entry 1 in the lower row appears to the right of the 1 in the higher row.

(*iii*) Rows consisting of all zeros are at the bottom of the matrix.
In the Gauss–Jordan method the row operations are continued until we obtain an augmented matrix that is in **reduced row-echelon form.** A reduced row-echelon matrix has the same three properties listed above, but in addition:

(*iv* A column containing a first entry 1 has zeros everywhere else.

Example 4 Echelon Forms

(a) The augmented matrices

$$\begin{pmatrix} 1 & 5 & 0 & | & 2 \\ 0 & 1 & 0 & | & -1 \\ 0 & 0 & 0 & | & 0 \end{pmatrix} \quad \text{and} \quad \begin{pmatrix} 0 & 0 & 1 & -6 & 2 & | & 2 \\ 0 & 0 & 0 & 0 & 1 & | & 4 \end{pmatrix}$$

are in row-echelon form. The reader should verify that the three criteria for this form are satisfied.

(b) The augmented matrices

$$\begin{pmatrix} 1 & 0 & 0 & | & 7 \\ 0 & 1 & 0 & | & -1 \\ 0 & 0 & 0 & | & 0 \end{pmatrix} \quad \text{and} \quad \begin{pmatrix} 0 & 0 & 1 & -6 & 0 & | & -6 \\ 0 & 0 & 0 & 0 & 1 & | & 4 \end{pmatrix}$$

are in reduced row-echelon form. Note that the remaining entries in the columns that contain a leading entry 1 are all zeros. ❏

Note: Row operations can lead to different row-echelon forms.

It should be noted that in Gaussian elimination, we stop when we have obtained *an* augmented matrix in row-echelon form. In other words, by using different sequences of row operations, we may arrive at different row-echelon forms. This method then requires the use of back-substitution. In Gauss–Jordan elimination we stop when we have obtained *the* augmented matrix in reduced row-echelon form. Any sequence of row operations will lead to the same augmented matrix in reduced row-echelon form. This method does not require back-substitution; the solution of the system will be apparent by inspection of the final matrix. In terms of the equations of the original system, our goal in both methods is simply to make the coefficient of x_1 in the first equation[*] equal to one and then use multiples of that equation to eliminate x_1 from other equations. The process is repeated for the other variables.

To keep track of the row operations used on an augmented matrix, we shall utilize the following notation:

Symbol	Meaning
R_{ij}	Interchange rows i and j
cR_i	Multiply the ith row by the nonzero constant c
$cR_i + R_j$	Multiply the ith row by c and add to the jth row

Example 5 Elimination Methods and Augmented Matrices

Solve the linear system in Example 2 using (a) Gaussian elimination and (b) Gauss–Jordan elimination.

SOLUTION (a) Using row operations on the augmented matrix of the system, we obtain:

[*] We can always interchange equations so that the first equation contains the variable x_1.

$$\begin{pmatrix} 2 & 6 & 1 & | & 7 \\ 1 & 2 & -1 & | & -1 \\ 5 & 7 & -4 & | & 9 \end{pmatrix} \xRightarrow{R_{12}} \begin{pmatrix} 1 & 2 & -1 & | & -1 \\ 2 & 6 & 1 & | & 7 \\ 5 & 7 & -4 & | & 9 \end{pmatrix}$$

$$\xRightarrow[-5R_1 + R_3]{-2R_1 + R_2} \begin{pmatrix} 1 & 2 & -1 & | & -1 \\ 0 & 2 & 3 & | & 9 \\ 0 & -3 & 1 & | & 14 \end{pmatrix} \xRightarrow{\frac{1}{2}R_2} \begin{pmatrix} 1 & 2 & -1 & | & -1 \\ 0 & 1 & \frac{3}{2} & | & \frac{9}{2} \\ 0 & -3 & 1 & | & 14 \end{pmatrix}$$

$$\xRightarrow{3R_2 + R_3} \begin{pmatrix} 1 & 2 & -1 & | & -1 \\ 0 & 1 & \frac{3}{2} & | & \frac{9}{2} \\ 0 & 0 & \frac{11}{2} & | & \frac{55}{2} \end{pmatrix} \xRightarrow{\frac{2}{11}R_3} \begin{pmatrix} 1 & 2 & -1 & | & -1 \\ 0 & 1 & \frac{3}{2} & | & \frac{9}{2} \\ 0 & 0 & 1 & | & 5 \end{pmatrix}.$$

The last matrix is in row-echelon form and represents the system

$$x_1 + 2x_2 - x_3 = -1$$

$$x_2 + \frac{3}{2}x_3 = \frac{9}{2}$$

$$x_3 = 5.$$

Substituting $x_3 = 5$ into the second equation gives $x_2 = -3$. Substituting both these values back into the first equation finally yields $x_1 = 10$.

(b) We start with the last matrix above. Since the first entries in the second and third rows are ones, we must, in turn, make the remaining entries in the second and third columns zeros:

$$\begin{pmatrix} 1 & 2 & -1 & | & -1 \\ 0 & 1 & \frac{3}{2} & | & \frac{9}{2} \\ 0 & 0 & 1 & | & 5 \end{pmatrix} \xRightarrow{-2R_2 + R_1} \begin{pmatrix} 1 & 0 & -4 & | & -10 \\ 0 & 1 & \frac{3}{2} & | & \frac{9}{2} \\ 0 & 0 & 1 & | & 5 \end{pmatrix} \xRightarrow[-\frac{3}{2}R_3 + R_2]{4R_3 + R_1} \begin{pmatrix} 1 & 0 & 0 & | & 10 \\ 0 & 1 & 0 & | & -3 \\ 0 & 0 & 1 & | & 5 \end{pmatrix}$$

The last matrix is in reduced row-echelon form. Bearing in mind what the matrix means in terms of equations, we see that the solution of the system is $x_1 = 10$, $x_2 = -3$, $x_3 = 5$. ❑

Example 6 Gauss–Jordan Elimination

Use Gauss–Jordan elimination to solve

$$x_1 + 3x_2 - 2x_3 = -7$$

$$4x_1 + x_2 + 3x_3 = 5$$

$$2x_1 - 5x_2 + 7x_3 = 19.$$

SOLUTION

$$\begin{pmatrix} 1 & 3 & -2 & | & -7 \\ 4 & 1 & 3 & | & 5 \\ 2 & -5 & 7 & | & 19 \end{pmatrix} \xRightarrow[-2R_1 + R_3]{-4R_1 + R_2} \begin{pmatrix} 1 & 3 & -2 & | & -7 \\ 0 & -11 & 11 & | & 33 \\ 0 & -11 & 11 & | & 33 \end{pmatrix}$$

$$\xRightarrow[-\frac{1}{11}R_3]{-\frac{1}{11}R_2} \begin{pmatrix} 1 & 3 & -2 & | & -7 \\ 0 & 1 & -1 & | & -3 \\ 0 & 1 & -1 & | & -3 \end{pmatrix} \xRightarrow[-R_2 + R_3]{-3R_2 + R_1} \begin{pmatrix} 1 & 0 & 1 & | & 2 \\ 0 & 1 & -1 & | & -3 \\ 0 & 0 & 0 & | & 0 \end{pmatrix}.$$

In this case the last matrix in reduced row-echelon form implies that the original system of three equations in three unknowns is really equivalent to two equations

in the unknowns. Since only x_3 is common to both equations (the nonzero rows), we can assign its values arbitrarily. If we let $x_3 = t$, where t represents any real number, then we see that the system has infinitely many solutions: $x_1 = 2 - t$, $x_2 = -3 + t$, $x_3 = t$. Geometrically, these equations are the parametric equations for the line of intersection of the planes $x_1 + 0x_2 + x_3 = 2$ and $0x_1 + x_2 - x_3 = -3$. ❏

Example 7 Inconsistent System

Solve
$$x_1 + \ x_2 = 1$$
$$4x_1 - \ x_2 = -6$$
$$2x_1 - 3x_2 = 8.$$

SOLUTION In the process of applying Gauss–Jordan elimination to the matrix of the system, we stop at

$$\begin{pmatrix} 1 & 1 & | & 1 \\ 4 & -1 & | & -6 \\ 2 & -3 & | & 8 \end{pmatrix} \stackrel{\text{row operations}}{\Longrightarrow} \begin{pmatrix} 1 & 0 & | & -1 \\ 0 & 1 & | & 2 \\ 0 & 0 & | & 16 \end{pmatrix}.$$

The third row of the last matrix means $0x_1 + 0x_2 = 16$ (or $0 = 16$). Since no numbers x_1 and x_2 can satisfy this equation, we conclude that the system has no solution. ❏

Worth remembering.

Inconsistent systems of m linear equations in n unknowns will always yield the situation illustrated in Example 7; that is, there will be a row in the reduced row-echelon form of the augmented matrix in which the first n entries are zero and the $(n + 1)$st entry is nonzero.

■ Networks
The currents in the branches of an electrical network can be determined by using Kirchhoff's point and loop rules:

Point rule: The algebraic sum of the currents toward any branch point is 0.
Loop rule: The algebraic sum of the potential differences in any loop is 0.

When a loop is traversed in a chosen direction (clockwise or counterclockwise), an emf is taken to be *positive* when it is traversed from − to + and *negative* when traversed from + to −. An iR product is taken to be *positive* if the chosen direction through the resistor is opposite that of the assumed current, and *negative* if the chosen direction is in the same direction as the assumed current.

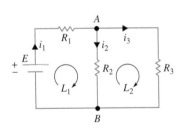

Figure 8.3

In Figure 8.3 the branch points of the network are labeled A and B, the loops are labeled L_1 and L_2, and the chosen direction in each loop is clockwise. Now applying the foregoing rules to the network yields the nonhomogeneous system of linear equations

$$i_1 - i_2 - i_3 = 0 \qquad\qquad i_1 \ \ - i_2 \ \ - i_3 \ \ = 0$$
$$E - i_1 R_1 - i_2 R_2 = 0 \quad\text{or}\quad i_1 R_1 + i_2 R_2 \qquad = E \qquad\qquad (3)$$
$$i_2 R_2 - i_3 R_3 = 0 \qquad\qquad i_2 R_2 - i_3 R_3 = 0.$$

Example 8 Currents in a Network

Use Gauss–Jordan elimination to solve the system (3) when $R_1 = 10$ ohms, $R_2 = 20$ ohms, $R_3 = 10$ ohms, and $E = 12$ volts.

SOLUTION The system to be solved is

$$i_1 - i_2 - i_3 = 0$$
$$10i_1 + 20i_2 \qquad = 12$$
$$20i_2 - 10i_3 = 0.$$

In this case Gauss–Jordan elimination yields

$$\begin{pmatrix} 1 & -1 & -1 & | & 0 \\ 10 & 20 & 0 & | & 12 \\ 0 & 20 & -10 & | & 0 \end{pmatrix} \overset{\overset{\text{row}}{\text{operations}}}{\Longrightarrow} \begin{pmatrix} 1 & 0 & 0 & | & {}^{18}\!/_{25} \\ 0 & 1 & 0 & | & {}^{6}\!/_{25} \\ 0 & 0 & 1 & | & {}^{12}\!/_{25} \end{pmatrix}.$$

Hence, we see that the currents in the three branches are $i_1 = {}^{18}\!/_{25} = 0.72$ ampere, $i_2 = {}^{6}\!/_{25} = 0.24$ ampere, and $i_3 = {}^{12}\!/_{25} = 0.48$ ampere. ❏

■ **Homogeneous Systems** All the systems in the preceding examples are nonhomogeneous systems. As we have seen, a nonhomogeneous system can be consistent or inconsistent. By contrast, a homogeneous system of linear equations

$$a_{11}x_1 + a_{12}x_2 + \cdots + a_{1n}x_n = 0$$
$$a_{12}x_1 + a_{22}x_2 + \cdots + a_{2n}x_n = 0 \qquad (4)$$
$$\vdots \qquad\qquad \vdots$$
$$a_{m1}x_1 + a_{m2}x_2 + \cdots + a_{mn}x_n = 0$$

is *always consistent,* since $x_1 = 0$, $x_2 = 0, \ldots, x_n = 0$ will satisfy each equation in the system. The solution consisting of all zeros is called the **trivial solution.** But naturally we are interested in whether a system of form (4) has any solutions for which *some* of the x_i, $i = 1, 2, \ldots, n$, are not zero. Such a solution is called a **nontrivial solution.** A homogeneous system either possesses only the trivial solution or possesses the trivial solution along with infinitely many nontrivial solutions. The next theorem, presented without proof, will give us a sufficient condition for the existence of nontrivial solutions.

THEOREM 8.3 Existence of Nontrivial Solutions

A homogeneous system of form (4) possesses nontrivial solutions if the number m of equations is less than the number n of unknowns ($m < n$).

Example 9 **Solving a Homogeneous System**

Solve
$$2x_1 - 4x_2 + 3x_3 = 0$$
$$x_1 + x_2 - 2x_3 = 0.$$

SOLUTION Since the number of equations is less than the number of unknowns, we know from Theorem 8.3 that the given system has nontrivial solutions. Using Gauss–Jordan elimination, we find

$$\begin{pmatrix} 2 & -4 & 3 & | & 0 \\ 1 & 1 & -2 & | & 0 \end{pmatrix} \overset{R_{12}}{\Longrightarrow} \begin{pmatrix} 1 & 1 & -2 & | & 0 \\ 2 & -4 & 3 & | & 0 \end{pmatrix} \overset{-2R_1 + R_2}{\Longrightarrow} \begin{pmatrix} 1 & 1 & -2 & | & 0 \\ 0 & -6 & 7 & | & 0 \end{pmatrix}$$

$$\overset{-\frac{1}{6}R_2}{\Longrightarrow} \begin{pmatrix} 1 & 1 & -2 & | & 0 \\ 0 & 1 & -\frac{7}{6} & | & 0 \end{pmatrix} \overset{-R_2 + R_1}{\Longrightarrow} \begin{pmatrix} 1 & 0 & -\frac{5}{6} & | & 0 \\ 0 & 1 & -\frac{7}{6} & | & 0 \end{pmatrix}.$$

As in Example 6, if $x_3 = t$, then the solution of the system is $x_1 = \frac{5}{6}t$, $x_2 = \frac{7}{6}t$, $x_3 = t$. Note we obtain the trivial solution $x_1 = 0$, $x_2 = 0$, $x_3 = 0$ for this system by choosing $t = 0$. For $t \neq 0$ we get nontrivial solutions. For example, the solutions corresponding to $t = 6$, $t = -12$, and $t = 3$ are, in turn, $x_1 = 5$, $x_2 = 7$, $x_3 = 6$; $x_1 = -10$, $x_2 = -14$, $x_3 = -12$; and $x_1 = \frac{5}{2}$, $x_2 = \frac{7}{2}$, $x_3 = 3$. ❏

■ **Chemical Equations** The next example will give an application of homogeneous systems in chemistry.

Example 10 **Balancing a Chemical Equation**

Balance the chemical equation $C_2H_6 + O_2 \rightarrow CO_2 + H_2O$.

SOLUTION We want to find positive integers x_1, x_2, x_3, and x_4 so that

$$x_1 C_2H_6 + x_2 O_2 \rightarrow x_3 CO_2 + x_4 H_2O.$$

Because the number of atoms of each element must be the same on each side of the last equation, we have:

carbon (C): $2x_1 = x_3$ $\qquad\qquad\qquad\qquad$ $2x_1 + 0x_2 - x_3 + 0x_4 = 0$

hydrogen (H): $6x_1 = 2x_4$ \qquad or \qquad $6x_1 + 0x_2 + 0x_3 - 2x_4 = 0$

oxygen (O): $2x_2 = 2x_3 + x_4$ $\qquad\qquad$ $0x_1 + 2x_2 - 2x_3 - x_4 = 0.$

Gauss–Jordan elimination gives

$$\begin{pmatrix} 2 & 0 & -1 & 0 & | & 0 \\ 6 & 0 & 0 & -2 & | & 0 \\ 0 & 2 & -2 & -1 & | & 0 \end{pmatrix} \overset{\text{row}}{\underset{\text{operations}}{\Longrightarrow}} \begin{pmatrix} 1 & 0 & 0 & -\frac{1}{3} & | & 0 \\ 0 & 1 & 0 & -\frac{7}{6} & | & 0 \\ 0 & 0 & 1 & -\frac{2}{3} & | & 0 \end{pmatrix}$$

and so $x_1 = \frac{1}{3}t$, $x_2 = \frac{7}{6}t$, $x_3 = \frac{2}{3}t$, $x_4 = t$. In this case t must be a positive integer chosen in such a manner so that x_1, x_2, and x_3 are positive integers. To accomplish this we pick $t = 6$. This gives $x_1 = 2$, $x_2 = 7$, $x_3 = 4$, $x_4 = 6$. The balanced chemical equation is then

$$2C_2H_6 + 7O_2 \rightarrow 4CO_2 + 6H_2O.$$ ❏

■ **Notation** In view of matrix multiplication and equality of matrices defined in Section 8.1, observe that the linear system (1) can be written compactly as a matrix equation $\mathbf{AX} = \mathbf{B}$, where

$$\mathbf{A} = \begin{pmatrix} a_{11} & a_{12} & \cdots & a_{1n} \\ a_{21} & a_{22} & \cdots & a_{2n} \\ \vdots & & & \vdots \\ a_{m1} & a_{m2} & \cdots & a_{mn} \end{pmatrix}, \quad \mathbf{X} = \begin{pmatrix} x_1 \\ x_2 \\ \vdots \\ x_n \end{pmatrix}, \quad \mathbf{B} = \begin{pmatrix} b_1 \\ b_2 \\ \vdots \\ b_m \end{pmatrix}.$$

The matrix \mathbf{A} is, naturally, called the coefficient matrix. The augmented matrix of a system $\mathbf{AX} = \mathbf{B}$ is often denoted by $(\mathbf{A}|\mathbf{B})$.

A similarity with linear DEs. ☞ A consistent nonhomogeneous linear system $\mathbf{AX} = \mathbf{B}$, $\mathbf{B} \neq \mathbf{0}$, shares a property with nonhomogeneous linear differential equations. If \mathbf{X}_h is a solution of the associated homogeneous system $\mathbf{AX} = \mathbf{0}$, and \mathbf{X}_p is a particular solution of the nonhomogeneous system $\mathbf{AX} = \mathbf{B}$, then the superposition $\mathbf{X}_h + \mathbf{X}_p$ is also a solution of the

nonhomogeneous system. This is easy to verify: $\mathbf{A}(\mathbf{X}_h + \mathbf{X}_p) = \mathbf{A}\mathbf{X}_h + \mathbf{A}\mathbf{X}_p = \mathbf{0} + \mathbf{B} = \mathbf{B}$. Moreover, analogous to the notion of a general solution of a linear DE, *every* solution of the nonhomogeneous system can be obtained from $\mathbf{X}_h + \mathbf{X}_p$.

Remarks

(*i*) To solve large systems of linear equations, we obviously need the help of a computer. Since the two methods in this section are so systematic, they can be programmed easily. However, the requirement that each nonzero row start with a one may at times demand division by a very small number. Problems can then occur. Large systems are often solved indirectly—that is, by an approximation technique such as **Gauss–Seidel iteration.** See Section 16.1.

(*ii*) Since Gauss–Jordan elimination avoids the necessity of back-substitution, it would appear that it is the more efficient of the two methods we have considered. Actually, this is not the case. It has been shown that for large systems, Gauss–Jordan elimination can require about 50% more operations than Gaussian elimination.

(*iii*) A system of linear equations that has more equations than unknowns is said to be **overdetermined,** whereas a system that has fewer equations than unknowns is called **underdetermined.** As a rule, an overdetermined system is usually—not always—inconsistent, and an underdetermined system is usually—not always—consistent. (See Examples 7 and 9.) It should be noted that it is impossible for a consistent underdetermined system to possess a single or unique solution. To see this, suppose we have m equations and n unknowns where $m < n$. If Gaussian elimination is used to solve such a system, then the row-echelon form that is row equivalent to the matrix of the system will contain $r \leq m$ nonzero rows. Thus we can solve for r of the variables in terms of $n - r > 0$ variables. If the underdetermined system is consistent, then those remaining $n - r$ variables can be chosen arbitrarily, and so the system has an infinite number of solutions.

EXERCISES 8.2

Answers to odd-numbered problems begin on page A-43.

In Problems 1–20, use either Gaussian elimination or Gauss–Jordan elimination to solve the given system or show that no solution exists.

1. $\begin{aligned} x_1 - x_2 &= 11 \\ 4x_1 + 3x_2 &= -5 \end{aligned}$

2. $\begin{aligned} 3x_1 - 2x_2 &= 4 \\ x_1 - x_2 &= -2 \end{aligned}$

3. $\begin{aligned} 9x_1 + 3x_2 &= -5 \\ 2x_1 + x_2 &= -1 \end{aligned}$

4. $\begin{aligned} 10x_1 + 15x_2 &= 1 \\ 3x_1 + 2x_2 &= -1 \end{aligned}$

5. $\begin{aligned} x_1 - x_2 - x_3 &= -3 \\ 2x_1 + 3x_2 + 5x_3 &= 7 \\ x_1 - 2x_2 + 3x_3 &= -11 \end{aligned}$

6. $\begin{aligned} x_1 + 2x_2 - x_3 &= 0 \\ 2x_1 + x_2 + 2x_3 &= 9 \\ x_1 - x_2 + x_3 &= 3 \end{aligned}$

7. $\begin{aligned} x_1 + x_2 + x_3 &= 0 \\ x_1 + x_2 + 3x_3 &= 0 \end{aligned}$

8. $\begin{aligned} x_1 + 2x_2 - 4x_3 &= 9 \\ 5x_1 - x_2 + 2x_3 &= 1 \end{aligned}$

9. $\begin{aligned} x_1 - x_2 - x_3 &= 8 \\ x_1 - x_2 + x_3 &= 3 \\ -x_1 + x_2 + x_3 &= 4 \end{aligned}$

10. $\begin{aligned} 3x_1 + x_2 &= 4 \\ 4x_1 + 3x_2 &= -3 \\ 2x_1 - x_2 &= 11 \end{aligned}$

11. $\begin{aligned} 2x_1 + 2x_2 &= 0 \\ -2x_1 + x_2 + x_3 &= 0 \\ 3x_1 + x_3 &= 0 \end{aligned}$

12. $\begin{aligned} x_1 - x_2 - 2x_3 &= 0 \\ 2x_1 + 4x_2 + 5x_3 &= 0 \\ 6x_1 - 3x_3 &= 0 \end{aligned}$

13. $\begin{aligned} x_1 + 2x_2 + 2x_3 &= 2 \\ x_1 + x_2 + x_3 &= 0 \\ x_1 - 3x_2 - x_3 &= 0 \end{aligned}$

14. $\begin{aligned} x_1 - 2x_2 + x_3 &= 2 \\ 3x_1 - x_2 + 2x_3 &= 5 \\ 2x_1 + x_2 + x_3 &= 1 \end{aligned}$

15. $\begin{aligned} x_1 + x_2 + x_3 &= 3 \\ x_1 - x_2 - x_3 &= -1 \\ 3x_1 + x_2 + x_3 &= 5 \end{aligned}$

16. $x_1 - x_2 - 2x_3 = -1$
$-3x_1 - 2x_2 + x_3 = -7$
$2x_1 + 3x_2 + x_3 = 8$

17. $x_1 + x_3 - x_4 = 1$
$2x_2 + x_3 + x_4 = 3$
$x_1 - x_2 + x_4 = -1$
$x_1 + x_2 + x_3 + x_4 = 2$

18. $2x_1 + x_2 + x_3 = 3$
$3x_1 + x_2 + x_3 + x_4 = 4$
$x_1 + 2x_2 + 2x_3 + 3x_4 = 3$
$4x_1 + 5x_2 - 2x_3 + x_4 = 16$

19. $x_2 + x_3 - x_4 = 4$
$x_1 + 3x_2 + 5x_3 - x_4 = 1$
$x_1 + 2x_2 + 5x_3 - 4x_4 = -2$
$x_1 + 4x_2 + 6x_3 - 2x_4 = 6$

20. $x_1 + 2x_2 + x_4 = 0$
$4x_1 + 9x_2 + x_3 + 12x_4 = 0$
$3x_1 + 9x_2 + 6x_3 + 21x_4 = 0$
$x_1 + 3x_2 + x_3 + 9x_4 = 0$

In Problems 21 and 22, use a calculator to solve the given system.

21. $x_1 + x_2 + x_3 = 4.280$
$0.2x_1 - 0.1x_2 - 0.5x_3 = -1.978$
$4.1x_1 + 0.3x_2 + 0.12x_3 = 1.686$

22. $2.5x_1 + 1.4x_2 + 4.5x_3 = 2.6170$
$1.35x_1 + 0.95x_2 + 1.2x_3 = 0.7545$
$2.7x_1 + 3.05x_2 - 1.44x_3 = -1.4292$

In Problems 23–28, use the procedures illustrated in Example 10 to balance the given chemical equation.

23. $Na + H_2O \rightarrow NaOH + H_2$

24. $KClO_3 \rightarrow KCl + O_2$

25. $Fe_3O_4 + C \rightarrow Fe + CO$

26. $C_5H_8 + O_2 \rightarrow CO_2 + H_2O$

27. $Cu + HNO_3 \rightarrow Cu(NO_3)_2 + H_2O + NO$

28. $Ca_3(PO_4)_2 + H_3PO_4 \rightarrow Ca(H_2PO_4)_2$

In Problems 29 and 30, set up and solve the system of equations for the currents in the branches of the given network.

29.

Figure 8.4

30.

Figure 8.5

An **elementary matrix E** is one obtained by performing a single row operation on the identity matrix **I**. In Problems 31–34, verify that the given matrix is an elementary matrix.

31. $\begin{pmatrix} 0 & 1 & 0 \\ 1 & 0 & 0 \\ 0 & 0 & 1 \end{pmatrix}$ **32.** $\begin{pmatrix} 1 & 0 & 0 \\ 0 & 1 & 0 \\ 0 & 0 & c \end{pmatrix}$

33. $\begin{pmatrix} 1 & 0 & 0 \\ 0 & 1 & 0 \\ 0 & c & 1 \end{pmatrix}$ **34.** $\begin{pmatrix} 1 & 0 & 0 & 1 \\ 0 & 1 & 0 & 0 \\ 0 & 0 & 1 & 0 \\ 0 & 0 & 0 & 1 \end{pmatrix}$

If a matrix **A** is premultiplied by an elementary matrix **E**, the product **EA** will be that matrix obtained from **A** by performing the elementary row operation symbolized by **E**. In Problems 35–38, compute the given product for an arbitrary 3×3 matrix **A**.

35. $\begin{pmatrix} 0 & 1 & 0 \\ 1 & 0 & 0 \\ 0 & 0 & 1 \end{pmatrix}\mathbf{A}$ **36.** $\begin{pmatrix} 1 & 0 & 0 \\ 0 & 1 & 0 \\ 0 & 0 & c \end{pmatrix}\mathbf{A}$

37. $\begin{pmatrix} 1 & 0 & 0 \\ 0 & 1 & 0 \\ 0 & c & 1 \end{pmatrix}\mathbf{A}$ **38.** $\begin{pmatrix} 0 & 1 & 0 \\ 1 & 0 & 0 \\ 0 & 0 & 1 \end{pmatrix}\begin{pmatrix} 1 & 0 & 0 \\ 0 & 1 & 0 \\ 0 & c & 1 \end{pmatrix}\mathbf{A}$

The linear system (1) can be written as the matrix equation **AX = B**. Suppose $m = n$. If the $n \times n$ coefficient matrix **A** in the system has an LU-factorization **A = LU** (see page 351), then the system **AX = B**, or **LUX = B**, can be solved efficiently in two steps *without* Gaussian or Gauss–Jordan elimination:

(*i*) First, let **Y = UX** and solve **LY = B** for **Y** by forward-substitution.

(*ii*) Then solve **UX = Y** for **X** using back-substitution.

In Problems 39–42, use the results of Problem 46 in Exercises 8.1 to solve the given system.

39. $\begin{pmatrix} 2 & -2 \\ 1 & 2 \end{pmatrix}\mathbf{X} = \begin{pmatrix} 2 \\ 6 \end{pmatrix}$ **40.** $\begin{pmatrix} 6 & 2 \\ 4 & 1 \end{pmatrix}\mathbf{X} = \begin{pmatrix} 1 \\ -1 \end{pmatrix}$

41. $\begin{pmatrix} 1 & -2 & 1 \\ 0 & 1 & 2 \\ 2 & 6 & 1 \end{pmatrix}\mathbf{X} = \begin{pmatrix} 2 \\ -1 \\ 1 \end{pmatrix}$ **42.** $\begin{pmatrix} 1 & 1 & 1 \\ 3 & 1 & 2 \\ 1 & -1 & 1 \end{pmatrix}\mathbf{X} = \begin{pmatrix} 0 \\ 1 \\ 4 \end{pmatrix}$

43. Computer Lab Assignment Use a CAS to solve the given system.

(a) $1.567x_1 - 3.48x_2 + 5.22x_3 = 1.045$
$3.56x_1 + 4.118x_2 + 1.57x_3 = -1.625$

(b) $x_1 + 2x_2 - 2x_3 = 0$
$2x_1 - 2x_2 + x_3 = 0$
$3x_1 - 6x_2 + 4x_3 = 0$
$4x_1 + 14x_2 - 13x_3 = 0$

(c) $1.2x_1 + 3.5x_2 - 4.4x_3 + 3.1x_4 = 1.8$
$0.2x_1 - 6.1x_2 - 2.3x_3 + 5.4x_4 = -0.6$
$3.3x_1 - 3.5x_2 - 2.4x_3 - 0.1x_4 = 2.5$
$5.2x_1 + 8.5x_2 - 4.4x_3 - 2.9x_4 = 0$

(d) $x_1 - x_2 - x_3 + 2x_4 - x_5 = 5$
$6x_1 + 9x_2 - 6x_3 + 17x_4 - x_5 = 40$
$2x_1 + x_2 - 2x_3 + 5x_4 - x_5 = 12$
$x_1 + 2x_2 - x_3 + 3x_4 = 7$
$x_1 + 2x_2 + x_3 + 3x_4 = 1$

8.3 Rank of a Matrix

■ **Row and Column Vectors** In a general $m \times n$ matrix

$$\mathbf{A} = \begin{pmatrix} a_{11} & a_{12} & \cdots & a_{1n} \\ a_{21} & a_{22} & \cdots & a_{2n} \\ \vdots & & & \vdots \\ a_{m1} & a_{m2} & \cdots & a_{mn} \end{pmatrix},$$

the rows

$$\mathbf{u}_1 = (a_{11} \quad a_{12} \cdots a_{1n}), \quad \mathbf{u}_2 = (a_{21} \quad a_{22} \cdots a_{2n}), \quad \ldots, \quad \mathbf{u}_m = (a_{m1} \quad a_{m2} \cdots a_{mn})$$

and columns

$$\mathbf{v}_1 = \begin{pmatrix} a_{11} \\ a_{21} \\ \vdots \\ a_{m1} \end{pmatrix}, \quad \mathbf{v}_2 = \begin{pmatrix} a_{12} \\ a_{22} \\ \vdots \\ a_{m2} \end{pmatrix}, \quad \ldots, \quad \mathbf{v}_n = \begin{pmatrix} a_{1n} \\ a_{2n} \\ \vdots \\ a_{mn} \end{pmatrix}$$

are called the **row vectors** of **A** and the **column vectors** of **A**, respectively. As vectors the set $\mathbf{u}_1, \mathbf{u}_2, \ldots, \mathbf{u}_m$ is either linearly independent or linearly dependent. We have the following definition.

> **DEFINITION 8.8** Rank of a Matrix
>
> The rank of an $m \times n$ matrix **A**, denoted by rank(**A**), is the maximum number of linearly independent row vectors in **A**.

Example 1 Rank of a 3×4 Matrix

Consider the 3×4 matrix

$$\mathbf{A} = \begin{pmatrix} 1 & 1 & -1 & 3 \\ 2 & -2 & 6 & 8 \\ 3 & 5 & -7 & 8 \end{pmatrix}. \tag{1}$$

With $\mathbf{u}_1 = (1 \quad 1 \quad -1 \quad 3)$, $\mathbf{u}_2 = (2 \quad -2 \quad 6 \quad 8)$, and $\mathbf{u}_3 = (3 \quad 5 \quad -7 \quad 8)$, we see that $4\mathbf{u}_1 - \frac{1}{2}\mathbf{u}_2 + \mathbf{u}_3 = \mathbf{0}$. In view of Definition 7.7 and the discussion following it, we conclude that the set $\mathbf{u}_1, \mathbf{u}_2, \mathbf{u}_3$ is linearly dependent. On the other hand, since neither \mathbf{u}_1 nor \mathbf{u}_2 is a constant multiple of the other, the set of row vectors $\mathbf{u}_1, \mathbf{u}_2$ is linearly independent. Hence by Definition 8.8, rank(**A**) = 2. ❑

See page 331 in Section 7.6.

■ **Row Space** In the terminology of the preceding chapter, the row vectors $\mathbf{u}_1, \mathbf{u}_2, \mathbf{u}_3$ of the matrix (1) are a set of vectors in the *vector space* R^4. Since $R_A = \text{Span}(\mathbf{u}_1, \mathbf{u}_2, \mathbf{u}_3)$ (the set of all linear combinations of the vectors $\mathbf{u}_1, \mathbf{u}_2, \mathbf{u}_3$) is a subspace of R^4, we are justified in calling R_A the **row space** of the matrix \mathbf{A}. Now the set of vectors $\mathbf{u}_1, \mathbf{u}_2$ is linearly independent and also spans R_A; in other words, the set $\mathbf{u}_1, \mathbf{u}_2$ is a *basis* for R_A. The dimension (the number of vectors in a basis) of the row space R_A is 2, which is rank(\mathbf{A}).

■ **Rank by Row Reduction** Example 1 notwithstanding, it is generally not easy to determine the rank of a matrix by inspection. Although there are several mechanical ways of finding rank(\mathbf{A}), we examine one way which uses the elementary row operations introduced in the preceding section. *Specifically, the rank of* \mathbf{A} *can be found by row reducing* \mathbf{A} *to a row-echelon matrix* \mathbf{B}. To understand this, first recall that an $m \times n$ matrix \mathbf{B} is row equivalent to an $m \times n$ matrix \mathbf{A} if the rows of \mathbf{B} were obtained from the rows of \mathbf{A} by applying elementary row operations. If we simply interchange two rows in \mathbf{A} to obtain \mathbf{B}, then the row space R_A of \mathbf{A} and the row space R_B of \mathbf{B} are equal because the row vectors of \mathbf{A} and \mathbf{B} are the same. When the row vectors of \mathbf{B} are linear combinations of the rows of \mathbf{A}, it follows that the row vectors of \mathbf{B} are in the row space R_A, and so R_B is a subset of R_A (written $R_B \subseteq R_A$). Conversely, \mathbf{A} is row equivalent to \mathbf{B}, since we can obtain \mathbf{A} by applying row operations to \mathbf{B}. Hence the rows of \mathbf{A} are linear combinations of the rows of \mathbf{B}, and so it follows that R_A is a subset of R_B ($R_A \subseteq R_B$). From $R_B \subseteq R_A$ and $R_A \subseteq R_B$, we conclude that $R_A = R_B$. Finally, if we row reduce \mathbf{A} into a row-echelon matrix \mathbf{B}, then the rows of \mathbf{B} are linearly independent. (Why?) The rows of \mathbf{B} form a basis for the row space R_A, and so we have the result that rank(\mathbf{A}) = dimension of R_A.

We summarize these conclusions in the next theorem.

> **THEOREM 8.4 Rank of a Matrix by Row Reduction**
>
> If a matrix \mathbf{A} is row equivalent to a row-echelon form \mathbf{B}, then
>
> (*i*) the row space of \mathbf{A} = the row space of \mathbf{B},
>
> (*ii*) the nonzero rows of \mathbf{B} form a basis for the row space of \mathbf{A}, and
>
> (*iii*) rank(\mathbf{A}) = the number of nonzero rows in \mathbf{B}.

Example 2 **Rank by Row Reduction—Example 1 Revisited**

We row reduce a matrix \mathbf{A} to a row-echelon form \mathbf{B} in exactly the same manner as we row reduced the augmented matrix of a system of linear equations to an echelon form when we solved the system using Gaussian elimination. Using the matrix (1) in Example 1, elementary row operations give

$$\mathbf{A} = \begin{pmatrix} 1 & 1 & -1 & 3 \\ 2 & -2 & 6 & 8 \\ 3 & 5 & -7 & 8 \end{pmatrix} \overset{-2R_1 + R_2}{\underset{-3R_1 + R_3}{\Longrightarrow}} \begin{pmatrix} 1 & 1 & -1 & 3 \\ 0 & -4 & 8 & 2 \\ 0 & 2 & -4 & -1 \end{pmatrix} \overset{\frac{1}{2}R_2 + R_3}{\underset{-\frac{1}{4}R_2}{\Longrightarrow}} \begin{pmatrix} 1 & 1 & -1 & 3 \\ 0 & 1 & -2 & -\frac{1}{2} \\ 0 & 0 & 0 & 0 \end{pmatrix}.$$

Since the last matrix is in row-echelon form, and since the last matrix has two nonzero rows, we conclude from (*iii*) of Theorem 8.4 that rank(\mathbf{A}) = 2. ❏

Example 3 **Linear Independence/Dependence**

Determine whether the set of vectors $\mathbf{u}_1 = \langle 2, 1, 1 \rangle$, $\mathbf{u}_2 = \langle 0, 3, 0 \rangle$, $\mathbf{u}_3 = \langle 3, 1, 2 \rangle$, in R^3 is linearly dependent or linearly independent.

SOLUTION　It should be clear from the discussion above that if we form a matrix **A** with the given vectors as rows, and if we row reduce **A** to a row-echelon form **B** with rank 3, then the set of vectors is linearly independent. If rank(**A**) < 3, then the set of vectors is linearly dependent. In this case it is easy to carry the row reduction all the way to a reduced row-echelon form

$$\mathbf{A} = \begin{pmatrix} 2 & 1 & 1 \\ 0 & 3 & 0 \\ 3 & 1 & 2 \end{pmatrix} \overset{\text{row operations}}{\Longrightarrow} \begin{pmatrix} 1 & 0 & 0 \\ 0 & 1 & 0 \\ 0 & 0 & 1 \end{pmatrix}.$$

Thus rank(**A**) = 3 and the set of vectors \mathbf{u}_1, \mathbf{u}_2, \mathbf{u}_3 is linearly independent.　❏

As mentioned previously, the vectors in the row-echelon form of a matrix **A** can serve as a basis for the row space of the matrix **A.** In Example 3 we see that a basis for the row space of **A** is the standard basis $\langle 1, 0, 0 \rangle$, $\langle 0, 1, 0 \rangle$, $\langle 0, 0, 1 \rangle$ of R^3.

■ **Rank and Linear Systems**　The concept of rank can be related back to solvability of linear systems of algebraic equations. Suppose **AX** = **B** is a linear system and that $(\mathbf{A}|\mathbf{B})$ denotes the augmented matrix of the system. In Example 7 of Section 8.2, we saw that the system

$$x_1 + x_2 = 1$$
$$4x_1 - x_2 = -6$$
$$2x_1 - 3x_2 = 8$$

was inconsistent. The inconsistency of the system is seen in the fact that, after row reduction of the augmented matrix $(\mathbf{A}|\mathbf{B})$,

$$\begin{pmatrix} 1 & 1 & | & 1 \\ 4 & -1 & | & -6 \\ 2 & -3 & | & 8 \end{pmatrix} \overset{\text{row operations}}{\Longrightarrow} \begin{pmatrix} 1 & 0 & | & 1 \\ 0 & 1 & | & 2 \\ 0 & 0 & | & 16 \end{pmatrix} \Longrightarrow \begin{pmatrix} 1 & 0 & | & 0 \\ 0 & 1 & | & 0 \\ 0 & 0 & | & 1 \end{pmatrix} \qquad (2)$$

the last row of the reduced row-echelon form is nonzero. Of course, this reduction shows that rank$(\mathbf{A}|\mathbf{B})$ = 3. But note, too, that the result in (2) indicates that rank(**A**) = 2 because

$$\begin{pmatrix} 1 & 1 \\ 4 & -1 \\ 2 & -3 \end{pmatrix} \overset{\text{row operations}}{\Longrightarrow} \begin{pmatrix} 1 & 0 \\ 0 & 1 \\ 0 & 0 \end{pmatrix}.$$

We have illustrated a special case of the next theorem.

THEOREM 8.5　Consistency of AX = B

A linear system of equations **AX** = **B** is consistent if and only if the rank of the coefficient matrix **A** is the same as the rank of the augmented matrix of the system $(\mathbf{A}|\mathbf{B})$.

In Example 6 of Section 8.2, we saw that the system

$$x_1 + 3x_2 - 2x_3 = -7$$
$$4x_1 + x_2 + 3x_3 = 5 \qquad (3)$$
$$2x_1 - 5x_2 + 7x_3 = 19$$

was consistent and had an infinite number of solutions. We solved for two of the unknowns, x_1 and x_2, in terms of the remaining unknown x_3, which we relabeled as a parameter t. The number of parameters in a solution of a system is related to the rank of the coefficient matrix \mathbf{A}.

> **THEOREM 8.6 Number of Parameters in a Solution**
>
> Suppose a linear system $\mathbf{AX} = \mathbf{B}$ with m equations and n unknowns is consistent. If the coefficient matrix \mathbf{A} has rank r, then the solution of the system contains $n - r$ parameters.

For the system (3), we can see from the row reduction

$$\begin{pmatrix} 1 & 3 & -2 & -7 \\ 4 & 1 & 3 & 5 \\ 2 & -5 & 7 & 19 \end{pmatrix} \xrightarrow[\text{operations}]{\text{row}} \begin{pmatrix} 1 & 0 & 1 & 2 \\ 0 & 1 & -1 & -3 \\ 0 & 0 & 0 & 0 \end{pmatrix}$$

that $\text{rank}(\mathbf{A}) = \text{rank}(\mathbf{A}|\mathbf{B}) = 2$, and so the system is consistent by Theorem 8.5. With $n = 3$, we see from Theorem 8.6 the number of parameters in the solution is $3 - 2 = 1$.

The following diagram outlines the connection between the concept of rank of a matrix and the solution of a linear system.

For m linear equations in n unknowns $\mathbf{AX} = \mathbf{B}$.

Two cases: $\mathbf{B} = \mathbf{0}$, $\mathbf{B} \neq \mathbf{0}$. Let $\text{rank}(\mathbf{A}) = r$.

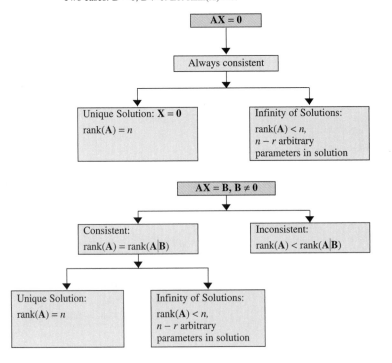

Remarks

We have not mentioned the connection between the columns of a matrix \mathbf{A} and the rank of \mathbf{A}. It turns out that the maximum number of independent columns that a matrix \mathbf{A} can have must equal the maximum number of independent rows. In the terminology of vector spaces, the row space $R_{\mathbf{A}}$ of matrix \mathbf{A} has the same dimension as its column space $C_{\mathbf{A}}$. For example, if we take the transpose of the matrix in (1) and reduce it to a row-echelon form:

$$\mathbf{A}^T = \begin{pmatrix} 1 & 2 & 3 \\ 1 & -2 & 5 \\ -1 & 6 & -7 \\ 3 & 8 & 8 \end{pmatrix} \xRightarrow[\text{operations}]{\text{row}} \begin{pmatrix} 1 & 2 & 3 \\ 0 & 1 & -\frac{1}{2} \\ 0 & 0 & 0 \\ 0 & 0 & 0 \end{pmatrix}$$

we see that the maximum number of rows of \mathbf{A}^T is 2, and so the maximum number of linearly independent columns of \mathbf{A} is 2.

EXERCISES 8.3

Answers to odd-numbered problems begin on page A-43.

In Problems 1–10, use (*iii*) of Theorem 8.4 to find the rank of the given matrix.

1. $\begin{pmatrix} 3 & -1 \\ 1 & 3 \end{pmatrix}$

2. $\begin{pmatrix} 2 & -2 \\ 0 & 0 \end{pmatrix}$

3. $\begin{pmatrix} 2 & 1 & 3 \\ 6 & 3 & 9 \\ -1 & -\frac{1}{2} & -\frac{3}{2} \end{pmatrix}$

4. $\begin{pmatrix} 1 & 1 & 2 \\ -1 & 2 & 4 \\ -1 & 0 & 3 \end{pmatrix}$

5. $\begin{pmatrix} 1 & 1 & 1 \\ 1 & 0 & 4 \\ 1 & 4 & 1 \end{pmatrix}$

6. $\begin{pmatrix} 3 & -1 & 2 & 0 \\ 6 & 2 & 4 & 5 \end{pmatrix}$

7. $\begin{pmatrix} 1 & -2 \\ 3 & -6 \\ 7 & -1 \\ 4 & 5 \end{pmatrix}$

8. $\begin{pmatrix} 1 & -2 & 3 & 4 \\ 1 & 4 & 6 & 8 \\ 0 & 1 & 0 & 0 \\ 2 & 5 & 6 & 8 \end{pmatrix}$

9. $\begin{pmatrix} 0 & 2 & 4 & 2 & 2 \\ 4 & 1 & 0 & 5 & 1 \\ 2 & 1 & \frac{2}{3} & 3 & \frac{1}{3} \\ 6 & 6 & 6 & 12 & 0 \end{pmatrix}$

10. $\begin{pmatrix} 1 & -2 & 1 & 8 & -1 & 1 & 1 & 6 \\ 0 & 0 & 1 & 3 & -1 & 1 & 1 & 5 \\ 0 & 0 & 1 & 3 & -1 & 2 & 10 & 8 \\ 0 & 0 & 0 & 0 & 0 & 1 & 1 & 3 \\ 1 & -2 & 1 & 8 & -1 & 1 & 2 & 6 \end{pmatrix}$

In Problems 11–14, determine whether the given set of vectors is linearly dependent or linearly independent.

11. $\mathbf{u}_1 = \langle 1, 2, 3 \rangle$, $\mathbf{u}_2 = \langle 1, 0, 1 \rangle$, $\mathbf{u}_3 = \langle 1, -1, 5 \rangle$

12. $\mathbf{u}_1 = \langle 2, 6, 3 \rangle$, $\mathbf{u}_2 = \langle 1, -1, 4 \rangle$, $\mathbf{u}_3 = \langle 3, 2, 1 \rangle$, $\mathbf{u}_4 = \langle 2, 5, 4 \rangle$

13. $\mathbf{u}_1 = \langle 1, -1, 3, -1 \rangle$, $\mathbf{u}_2 = \langle 1, -1, 4, 2 \rangle$, $\mathbf{u}_3 = \langle 1, -1, 5, 7 \rangle$

14. $\mathbf{u}_1 = \langle 2, 1, 1, 5 \rangle$, $\mathbf{u}_2 = \langle 2, 2, 1, 1 \rangle$, $\mathbf{u}_3 = \langle 3, -1, 6, 1 \rangle$, $\mathbf{u}_4 = \langle 1, 1, 1, -1 \rangle$

15. Suppose the system $\mathbf{AX} = \mathbf{B}$ is consistent and \mathbf{A} is a 5×8 matrix and rank(\mathbf{A}) = 3. How many parameters does the solution of the system have?

16. Let \mathbf{A} be a nonzero 4×6 matrix.

 (a) What is the maximum rank that \mathbf{A} can have?

 (b) If rank($\mathbf{A}|\mathbf{B}$) = 2, then for what value(s) of rank(\mathbf{A}) is the system $\mathbf{AX} = \mathbf{B}$, $\mathbf{B} \neq \mathbf{0}$, inconsistent? Consistent?

 (c) If rank(\mathbf{A}) = 3, then how many parameters does the solution of the system $\mathbf{AX} = \mathbf{0}$ have?

17. Let \mathbf{v}_1, \mathbf{v}_2, and \mathbf{v}_3 be the first, second, and third column vectors, respectively, of the matrix

$$\mathbf{A} = \begin{pmatrix} 2 & 1 & 7 \\ 1 & 0 & 2 \\ -1 & 5 & 13 \end{pmatrix}.$$

What can we conclude about rank(\mathbf{A}) from the observation $2\mathbf{v}_1 + 3\mathbf{v}_2 - \mathbf{v}_3 = \mathbf{0}$?

[*Hint:* Read the Remarks at the end of this section.]

18. Suppose the system $\mathbf{AX} = \mathbf{B}$ is consistent and \mathbf{A} is a 6×3 matrix. Suppose the maximum number of linearly independent rows in \mathbf{A} is 3. Discuss: Is the solution of the system unique?

19. Suppose we wish to determine whether the set of column vectors

$$\mathbf{v}_1 = \begin{pmatrix} 4 \\ 3 \\ 2 \\ 1 \end{pmatrix}, \ \mathbf{v}_2 = \begin{pmatrix} 1 \\ 2 \\ 2 \\ 1 \end{pmatrix}, \ \mathbf{v}_3 = \begin{pmatrix} -1 \\ 1 \\ 1 \\ 1 \end{pmatrix}, \ \mathbf{v}_4 = \begin{pmatrix} 2 \\ 3 \\ 4 \\ 1 \end{pmatrix}, \ \mathbf{v}_5 = \begin{pmatrix} 1 \\ 7 \\ -5 \\ 1 \end{pmatrix}$$

is linearly dependent or linearly independent. By Definition 7.7, if

$$c_1\mathbf{v}_1 + c_2\mathbf{v}_2 + c_3\mathbf{v}_3 + c_4\mathbf{v}_4 + c_5\mathbf{v}_5 = \mathbf{0} \qquad (4)$$

only for $c_1 = 0$, $c_2 = 0$, $c_3 = 0$, $c_4 = 0$, $c_5 = 0$, then the set of vectors is linearly independent; otherwise the set is linearly dependent. But (4) is equivalent to the linear system

$$4c_1 + c_2 - c_3 + 2c_4 + c_5 = 0$$
$$3c_1 + 2c_2 + c_3 + 3c_4 + 7c_5 = 0$$
$$2c_1 + 2c_2 + c_3 + 4c_4 - 5c_5 = 0$$
$$c_1 + c_2 + c_3 + c_4 + c_5 = 0.$$

Without doing any further work, explain why we can now conclude that the set of vectors is linearly dependent.

20. Computer Lab Assignment A CAS can be used to row reduce a matrix to a row-echelon form. Use a CAS to determine the ranks of the augmented matrix $(\mathbf{A}|\mathbf{B})$ and the coefficient matrix \mathbf{A} for

$$x_1 + 2x_2 - 6x_3 + x_4 + x_5 + x_6 = 2$$
$$5x_1 + 2x_2 - 2x_3 + 5x_4 + 4x_5 + 2x_6 = 3$$
$$6x_1 + 2x_2 - 2x_3 + x_4 + x_5 + 3x_6 = -1$$
$$-x_1 + 2x_2 + 3x_3 + x_4 - x_5 + 6x_6 = 0$$
$$9x_1 + 7x_2 - 2x_3 + x_4 + 4x_5 \qquad = 5.$$

Is the system consistent or inconsistent? If consistent, solve the system.

8.4 Determinants

■ **Determinant of a Square Matrix** Suppose \mathbf{A} is an $n \times n$ matrix. Associated with \mathbf{A} is a *number* called the **determinant of A** and is denoted by det \mathbf{A}. Symbolically, we distinguish a matrix \mathbf{A} from the determinant of \mathbf{A} by replacing the parentheses by vertical bars:

$$\mathbf{A} = \begin{pmatrix} a_{11} & a_{12} & \cdots & a_{1n} \\ a_{21} & a_{22} & \cdots & a_{2n} \\ \vdots & & & \vdots \\ a_{n1} & a_{n2} & \cdots & a_{nn} \end{pmatrix} \quad \text{and} \quad \det \mathbf{A} = \begin{vmatrix} a_{11} & a_{12} & \cdots & a_{1n} \\ a_{21} & a_{22} & \cdots & a_{2n} \\ \vdots & & & \vdots \\ a_{n1} & a_{n2} & \cdots & a_{nn} \end{vmatrix}.$$

A determinant of an $n \times n$ matrix is said to be **a determinant of order n.** We begin by defining the determinants of 1×1, 2×2, and 3×3 matrices.

For a 1×1 matrix $\mathbf{A} = (a)$ we have det $\mathbf{A} = |a| = a$. For example, if $\mathbf{A} = (-5)$, then det $\mathbf{A} = |-5| = -5$. In this case the vertical bars $|\ |$ around a number *do not* mean absolute value of the number.

DEFINITION 8.9 Determinant of a 2×2 Matrix

The determinant of $\mathbf{A} = \begin{pmatrix} a_{11} & a_{12} \\ a_{21} & a_{22} \end{pmatrix}$ is

$$\det \mathbf{A} = \begin{vmatrix} a_{11} & a_{12} \\ a_{21} & a_{22} \end{vmatrix} = a_{11}a_{22} - a_{12}a_{21}. \qquad (1)$$

As a mnemonic, a determinant of order 2 is thought to be the product of the main diagonal entries of \mathbf{A} minus the product of the other diagonal entries:

$$\overset{\substack{\text{multiply} \quad\quad \text{multiply} \quad\quad \text{subtract}\\ \qquad\qquad\qquad\qquad\qquad \downarrow}}{\begin{vmatrix} a_{11} & a_{12} \\ a_{21} & a_{22} \end{vmatrix} = a_{11}a_{22} - a_{12}a_{21}.} \tag{2}$$

For example, if $\mathbf{A} = \begin{pmatrix} 6 & -3 \\ 5 & 9 \end{pmatrix}$, then $\det \mathbf{A} = \begin{vmatrix} 6 & -3 \\ 5 & 9 \end{vmatrix} = 6(9) - (-3)(5) = 69.$

DEFINITION 8.10 Determinant of a 3×3 Matrix

The determinant of $\mathbf{A} = \begin{pmatrix} a_{11} & a_{12} & a_{13} \\ a_{21} & a_{22} & a_{23} \\ a_{31} & a_{32} & a_{33} \end{pmatrix}$ is

$$\det \mathbf{A} = \begin{vmatrix} a_{11} & a_{12} & a_{13} \\ a_{21} & a_{22} & a_{23} \\ a_{31} & a_{32} & a_{33} \end{vmatrix} = \begin{aligned} &a_{11}a_{22}a_{33} + a_{12}a_{23}a_{31} + a_{13}a_{21}a_{32} - a_{13}a_{22}a_{31} \\ &- a_{11}a_{23}a_{32} - a_{12}a_{21}a_{33}. \end{aligned} \tag{3}$$

The expression in (3) can be written in a more tractable form. By factoring, we have

$$\det \mathbf{A} = a_{11}(a_{22}a_{33} - a_{23}a_{32}) + a_{12}(-a_{21}a_{33} + a_{23}a_{31}) + a_{13}(a_{21}a_{32} - a_{22}a_{31}).$$

But in view of (1), each term in parentheses is recognized as the determinant of a 2×2 matrix:

$$\det \mathbf{A} = a_{11} \begin{vmatrix} a_{22} & a_{23} \\ a_{32} & a_{33} \end{vmatrix} + a_{12}\left(- \begin{vmatrix} a_{21} & a_{23} \\ a_{31} & a_{33} \end{vmatrix} \right) + a_{13} \begin{vmatrix} a_{21} & a_{22} \\ a_{31} & a_{32} \end{vmatrix}. \tag{4}$$

Observe that each determinant in (4) is a determinant of a *submatrix* of the matrix \mathbf{A} and corresponds to its coefficient in the following manner: a_{11} is the coefficient of the determinant of the submatrix obtained by deleting the *first row and first column* of \mathbf{A}; a_{12} is the coefficient of the negative of the determinant of the submatrix obtained by deleting the *first row and second column* of \mathbf{A}; and finally, a_{13} is the coefficient of the determinant of the submatrix obtained by deleting the *first row and third column* of \mathbf{A}. In other words, the coefficients in (4) are simply the entries of the first row of \mathbf{A}. We say that $\det \mathbf{A}$ has been **expanded by cofactors along the first row,** with the cofactors of a_{11}, a_{12}, and a_{13} being the determinants

$$C_{11} = \begin{vmatrix} a_{22} & a_{23} \\ a_{32} & a_{33} \end{vmatrix} \quad C_{12} = - \begin{vmatrix} a_{21} & a_{23} \\ a_{31} & a_{33} \end{vmatrix} \quad C_{13} = \begin{vmatrix} a_{21} & a_{22} \\ a_{31} & a_{32} \end{vmatrix}.$$

Thus (4) is

$$\det \mathbf{A} = a_{11}C_{11} + a_{12}C_{12} + a_{13}C_{13}. \tag{5}$$

In general, the **cofactor of a_{ij}** is the determinant

$$C_{ij} = (-1)^{i+j}M_{ij}, \tag{6}$$

where M_{ij} is the determinant of the submatrix obtained by deleting the ith row and the jth column of \mathbf{A}. The determinant M_{ij} is called a **minor determinant.** A cofactor is a *signed* minor determinant; that is, $C_{ij} = M_{ij}$ when $i + j$ is even and $C_{ij} = -M_{ij}$ when $i + j$ is odd.

A 3×3 matrix has nine cofactors:

$$C_{11} = M_{11} \qquad C_{12} = -M_{12} \qquad C_{13} = M_{13}$$

$$C_{21} = -M_{21} \qquad C_{22} = M_{22} \qquad C_{23} = -M_{23}$$

$$C_{31} = M_{31} \qquad C_{32} = -M_{32} \qquad C_{33} = M_{33}.$$

Inspection of the above array shows that the sign factor $+1$ or -1 associated with a cofactor can be obtained from the checkerboard pattern:

$$
\begin{matrix}
+ & - & + \\
- & + & - \\
+ & - & +
\end{matrix}
\tag{7}
$$

3×3 matrix

Now observe that (3) can be rearranged and factored again as

$$\det \mathbf{A} = -a_{12}(a_{21}a_{33} - a_{23}a_{31}) + a_{22}(a_{11}a_{33} - a_{13}a_{31}) - a_{32}(a_{11}a_{23} - a_{13}a_{21})$$

$$= a_{12}\left(-\begin{vmatrix} a_{21} & a_{23} \\ a_{31} & a_{33} \end{vmatrix}\right) + a_{22}\begin{vmatrix} a_{11} & a_{13} \\ a_{31} & a_{33} \end{vmatrix} + a_{32}\left(-\begin{vmatrix} a_{11} & a_{13} \\ a_{21} & a_{23} \end{vmatrix}\right) \tag{8}$$

$$= a_{12}C_{12} + a_{22}C_{22} + a_{32}C_{32},$$

which is the cofactor expansion of $\det \mathbf{A}$ along the second column. It is left as an exercise to show from (3) that $\det \mathbf{A}$ can also be expanded by cofactors along the third row:

$$\det \mathbf{A} = a_{31}C_{31} + a_{32}C_{32} + a_{33}C_{33}. \tag{9}$$

We are, of course, suggesting in (5), (8), and (9) the following general result: *The determinant of a 3×3 matrix can be evaluated by expanding* $\det \mathbf{A}$ *by cofactors along any row or along any column.*

Example 1 **Cofactor Expansion Along the First Row**

Evaluate the determinant of $\mathbf{A} = \begin{pmatrix} 2 & 4 & 7 \\ 6 & 0 & 3 \\ 1 & 5 & 3 \end{pmatrix}$.

SOLUTION Using cofactor expansion along the first row gives

$$\det \mathbf{A} = \begin{vmatrix} 2 & 4 & 7 \\ 6 & 0 & 3 \\ 1 & 5 & 3 \end{vmatrix} = 2C_{11} + 4C_{12} + 7C_{13}.$$

Now, the cofactors of the entries in the first row of \mathbf{A} are

$$C_{11} = (-1)^{1+1} \begin{vmatrix} 2 & 4 & 7 \\ 6 & 0 & 3 \\ 1 & 5 & 3 \end{vmatrix} = (-1)^{1+1} \begin{vmatrix} 0 & 3 \\ 5 & 3 \end{vmatrix}$$

$$C_{12} = (-1)^{1+2} \begin{vmatrix} 2 & 4 & 7 \\ 6 & 0 & 3 \\ 1 & 5 & 3 \end{vmatrix} = (-1)^{1+2} \begin{vmatrix} 6 & 3 \\ 1 & 3 \end{vmatrix}$$

$$C_{13} = (-1)^{1+3} \begin{vmatrix} 2 & 4 & 7 \\ 6 & 0 & 3 \\ 1 & 5 & 3 \end{vmatrix} = (-1)^{1+3} \begin{vmatrix} 6 & 0 \\ 1 & 5 \end{vmatrix},$$

where the dashed lines indicate the row and column that are deleted. Thus,

$$\det \mathbf{A} = 2(-1)^{1+1} \begin{vmatrix} 0 & 3 \\ 5 & 3 \end{vmatrix} + 4(-1)^{1+2} \begin{vmatrix} 6 & 3 \\ 1 & 3 \end{vmatrix} + 7(-1)^{1+3} \begin{vmatrix} 6 & 0 \\ 1 & 5 \end{vmatrix}$$

$$= 2[0(3) - 3(5)] - 4[6(3) - 3(1)] + 7[6(5) - 0(1)] = 120. \qquad \square$$

If a matrix has a row (or a column) containing many zero entries, then wisdom dictates that we evaluate the determinant of the matrix using cofactor expansion along that row (or column). Thus, in Example 1, had we expanded the determinant of **A** using cofactors along, say, the second row, then

$$\det \mathbf{A} = 6C_{21} + 0C_{22} + 3C_{23} = 6C_{21} + 3C_{23}$$

$$= 6(-1)^{1+2} \begin{vmatrix} 4 & 7 \\ 5 & 3 \end{vmatrix} + 3(-1)^{2+3} \begin{vmatrix} 2 & 4 \\ 1 & 5 \end{vmatrix}$$

$$= -6(-23) - 3(6) = 120.$$

Example 2 **Cofactor Expansion Along the Third Column**

Evaluate the determinant of $\mathbf{A} = \begin{pmatrix} 6 & 5 & 0 \\ -1 & 8 & -7 \\ -2 & 4 & 0 \end{pmatrix}$.

SOLUTION Since there are two zeros in the third column, we expand by cofactors of that column:

$$\det \mathbf{A} = \begin{vmatrix} 6 & 5 & 0 \\ -1 & 8 & -7 \\ -2 & 4 & 0 \end{vmatrix} = 0C_{13} + (-7)C_{23} + 0C_{33}$$

$$= (-7)(-1)^{2+3} \begin{vmatrix} 6 & 5 & 0 \\ -1 & 8 & -7 \\ -2 & 4 & 0 \end{vmatrix} = (-7)(-1)^{2+3} \begin{vmatrix} 6 & 5 \\ -2 & 4 \end{vmatrix}$$

$$= 7[6(4) - 5(-2)] = 238. \qquad \square$$

Carrying the above ideas one step further, we can evaluate the determinant of a 4×4 matrix by multiplying the entries in a row (or column) by their corresponding cofactors and adding the products. In this case each cofactor is a signed minor determinant of an appropriate 3×3 submatrix. The following theorem, which we shall give without proof, states that the determinant of any $n \times n$ matrix **A** can be evaluated by means of cofactors.

THEOREM 8.7 Cofactor Expansion of a Determinant

Let $\mathbf{A} = (a_{ij})_{n \times n}$ be an $n \times n$ matrix. For each $1 \leq i \leq n$, **the cofactor expansion of det A along the *i*th row** is

$$\det \mathbf{A} = a_{i1}C_{i1} + a_{i2}C_{i2} + \cdots + a_{in}C_{in}.$$

For each $1 \leq j \leq n$, **the cofactor expansion of det A along the *j*th column** is

$$\det \mathbf{A} = a_{1j}C_{1j} + a_{2j}C_{2j} + \cdots + a_{nj}C_{nj}.$$

The sign factor pattern for the cofactors illustrated in (7) extends to matrices of order greater than 3:

$$
\begin{matrix}
+ & - & + & - \\
- & + & - & + \\
+ & - & + & - \\
- & + & - & +
\end{matrix}
\qquad
\begin{matrix}
+ & - & + & - & + & \cdots \\
- & + & - & + & - & \cdots \\
+ & - & + & - & + & \cdots \\
- & + & - & + & - & \cdots \\
+ & - & + & - & + & \cdots \\
\vdots & \vdots & \vdots & \vdots & \vdots &
\end{matrix}
$$

4×4 matrix $n \times n$ matrix

Example 3 Cofactor Expansion Along the Fourth Row

Evaluate the determinant of the matrix

$$\mathbf{A} = \begin{pmatrix} 5 & 1 & 2 & 4 \\ -1 & 0 & 2 & 3 \\ 1 & 1 & 6 & 1 \\ 1 & 0 & 0 & -4 \end{pmatrix}.$$

SOLUTION Since the matrix has two zero entries in its fourth row, we choose to expand det \mathbf{A} by cofactors along that row:

$$\det \mathbf{A} = \begin{vmatrix} 5 & 1 & 2 & 4 \\ -1 & 0 & 2 & 3 \\ 1 & 1 & 6 & 1 \\ 1 & 0 & 0 & -4 \end{vmatrix} = (1)C_{41} + 0C_{42} + 0C_{43} + (-4)C_{44}, \qquad (10)$$

where $\quad C_{41} = (-1)^{4+1}\begin{vmatrix} 1 & 2 & 4 \\ 0 & 2 & 3 \\ 1 & 6 & 1 \end{vmatrix}$ and $\quad C_{44} = (-1)^{4+4}\begin{vmatrix} 5 & 1 & 2 \\ -1 & 0 & 2 \\ 1 & 1 & 6 \end{vmatrix}.$

We then expand both these determinants by cofactors along the second row:

$$C_{41} = (-1)\begin{vmatrix} 1 & 2 & 4 \\ 0 & 2 & 3 \\ 1 & 6 & 1 \end{vmatrix} = -\left(0(-1)^{2+1}\begin{vmatrix} 2 & 4 \\ 6 & 1 \end{vmatrix} + 2(-1)^{2+2}\begin{vmatrix} 1 & 4 \\ 1 & 1 \end{vmatrix} + 3(-1)^{2+3}\begin{vmatrix} 1 & 2 \\ 1 & 6 \end{vmatrix} \right)$$

$$= 18$$

$$C_{44} = \begin{vmatrix} 5 & 1 & 2 \\ -1 & 0 & 2 \\ 1 & 1 & 6 \end{vmatrix} = \left((-1)(-1)^{2+1}\begin{vmatrix} 1 & 2 \\ 1 & 6 \end{vmatrix} + 0(-1)^{2+2}\begin{vmatrix} 5 & 2 \\ 1 & 6 \end{vmatrix} + 2(-1)^{2+3}\begin{vmatrix} 5 & 1 \\ 1 & 1 \end{vmatrix} \right)$$

$$= -4.$$

Therefore (10) becomes

$$\det \mathbf{A} = \begin{vmatrix} 5 & 1 & 2 & 4 \\ -1 & 0 & 2 & 3 \\ 1 & 1 & 6 & 1 \\ 1 & 0 & 0 & -4 \end{vmatrix} = (1)(18) + (-4)(-4) = 34.$$

You should verify this result by expanding det **A** by cofactors along the second column. ❏

Remarks

In previous mathematics courses you may have seen the following memory device, analogous to (2), for evaluating a determinant of order 3:

<div align="center">
multiply multiply

$$\begin{vmatrix} a_{11} & a_{12} & a_{13} & a_{11} & a_{12} \\ a_{21} & a_{22} & a_{23} & a_{21} & a_{22} \\ a_{31} & a_{32} & a_{33} & a_{31} & a_{32} \end{vmatrix}$$ (11)
</div>

 (*i*) Add the products of the entries on the arrows that go from left to right.
 (*ii*) Subtract from the number in (*i*) the sum of the products of the entries on the arrows that go from right to left.

 Note: Method illustrated in (11) does not work for determinants of order n > 3.

A word of caution is in order here. The memory device given in (11), though easily adapted to matrices larger than 3×3, does not give the correct results. There are no mnemonic devices for evaluating the determinants of order 4 or greater.

EXERCISES 8.4

Answers to odd-numbered problems begin on page A-43.

In Problems 1–4, suppose

$$\mathbf{A} = \begin{pmatrix} 2 & 3 & 4 \\ 1 & -1 & 2 \\ -2 & 3 & 5 \end{pmatrix}.$$

Evaluate the indicated minor determinant or cofactor.

1. M_{12} **2.** M_{32} **3.** C_{13} **4.** C_{22}

In Problems 5–8, suppose

$$\mathbf{A} = \begin{pmatrix} 0 & 2 & 4 & 0 \\ 1 & 2 & -2 & 3 \\ 5 & 1 & 0 & -1 \\ 1 & 1 & 1 & 2 \end{pmatrix}.$$

Evaluate the indicated minor determinant or cofactor.

5. M_{33} **6.** M_{41} **7.** C_{34} **8.** C_{23}

In Problems 9–14, evaluate the determinant of the given matrix.

9. (-7) **10.** (2)

11. $\begin{pmatrix} 3 & 5 \\ -1 & 4 \end{pmatrix}$ **12.** $\begin{pmatrix} \frac{1}{4} & \frac{1}{2} \\ \frac{1}{3} & -\frac{4}{3} \end{pmatrix}$

13. $\begin{pmatrix} 1 - \lambda & 3 \\ 2 & 2 - \lambda \end{pmatrix}$ **14.** $\begin{pmatrix} -3 - \lambda & -4 \\ -2 & 5 - \lambda \end{pmatrix}$

In Problems 15–28, evaluate the determinant of the given matrix by cofactor expansion.

15. $\begin{pmatrix} 0 & 2 & 0 \\ 3 & 0 & 1 \\ 0 & 5 & 8 \end{pmatrix}$ **16.** $\begin{pmatrix} 5 & 0 & 0 \\ 0 & -3 & 0 \\ 0 & 0 & 2 \end{pmatrix}$

17. $\begin{pmatrix} 3 & 0 & 2 \\ 2 & 7 & 1 \\ 2 & 6 & 4 \end{pmatrix}$ **18.** $\begin{pmatrix} 1 & -1 & -1 \\ 2 & 2 & -2 \\ 1 & 1 & 9 \end{pmatrix}$

19. $\begin{pmatrix} 4 & 5 & 3 \\ 1 & 2 & 3 \\ 1 & 2 & 3 \end{pmatrix}$ **20.** $\begin{pmatrix} \frac{1}{4} & 6 & 0 \\ \frac{1}{3} & 8 & 0 \\ \frac{1}{2} & 9 & 0 \end{pmatrix}$

21. $\begin{pmatrix} -2 & -1 & 4 \\ -3 & 6 & 1 \\ -3 & 4 & 8 \end{pmatrix}$ **22.** $\begin{pmatrix} 3 & 5 & 1 \\ -1 & 2 & 5 \\ 7 & -4 & 10 \end{pmatrix}$

23. $\begin{pmatrix} 1 & 1 & 1 \\ x & y & z \\ 2 & 3 & 4 \end{pmatrix}$

24. $\begin{pmatrix} 1 & 1 & 1 \\ x & y & z \\ 2+x & 3+y & 4+z \end{pmatrix}$

25. $\begin{pmatrix} 1 & 1 & -3 & 0 \\ 1 & 5 & 3 & 2 \\ 1 & -2 & 1 & 0 \\ 4 & 8 & 0 & 0 \end{pmatrix}$

26. $\begin{pmatrix} 2 & 1 & -2 & 1 \\ 0 & 5 & 0 & 4 \\ 1 & 6 & 1 & 0 \\ 5 & -1 & 1 & 1 \end{pmatrix}$

27. $\begin{pmatrix} 3 & 2 & 0 & 1 & -1 \\ 0 & 1 & 4 & 2 & 3 \\ 0 & 0 & 2 & -1 & 1 \\ 0 & 0 & 0 & 4 & 3 \\ 0 & 0 & 0 & 0 & 2 \end{pmatrix}$

28. $\begin{pmatrix} 2 & 2 & 0 & 0 & -2 \\ 1 & 1 & 6 & 0 & 5 \\ 1 & 0 & 2 & -1 & -1 \\ 2 & 0 & 1 & -2 & 3 \\ 0 & 1 & 0 & 0 & 1 \end{pmatrix}$

In Problems 29 and 30, find the values of λ that satisfy the given equation.

29. $\begin{vmatrix} -3-\lambda & 10 \\ 2 & 5-\lambda \end{vmatrix} = 0$

30. $\begin{vmatrix} 1-\lambda & 0 & -1 \\ 1 & 2-\lambda & 1 \\ 3 & 3 & -\lambda \end{vmatrix} = 0$

8.5 Properties of Determinants

In this section we are going to consider some of the many properties of determinants. Our goal in the discussion is to use these properties to develop a means of evaluating a determinant that is an alternative to cofactor expansion.

The first property states that the determinant of an $n \times n$ matrix and its transpose are the same.

> **THEOREM 8.8 Determinant of a Transpose**
>
> If \mathbf{A}^T is the transpose of the $n \times n$ matrix \mathbf{A}, then det $\mathbf{A}^T =$ det \mathbf{A}.

For example, the transpose of the matrix $\mathbf{A} = \begin{pmatrix} 5 & 7 \\ 3 & -4 \end{pmatrix}$, is $\mathbf{A}^T = \begin{pmatrix} 5 & 3 \\ 7 & -4 \end{pmatrix}$. Observe that

$$\det \mathbf{A} = \begin{vmatrix} 5 & 7 \\ 3 & -4 \end{vmatrix} = -41 \quad \text{and} \quad \det \mathbf{A}^T = \begin{vmatrix} 5 & 3 \\ 7 & -4 \end{vmatrix} = -41.$$

Since transposing a matrix interchanges its rows and columns, the significance of Theorem 8.8 is that statements concerning determinants and the rows of a matrix also hold when the word "row" is replaced by the word "column."

> **THEOREM 8.9 Two Identical Rows**
>
> If any two rows (columns) of an $n \times n$ matrix \mathbf{A} are the same, then det $\mathbf{A} = 0$.

Example 1 Matrix with Two Identical Rows

Since the first and second columns in the matrix $\mathbf{A} = \begin{pmatrix} 6 & 2 & 2 \\ 4 & 2 & 2 \\ 9 & 2 & 2 \end{pmatrix}$ are the same, it follows from Theorem 8.9 that

$$\det \mathbf{A} = \begin{vmatrix} 6 & 2 & 2 \\ 4 & 2 & 2 \\ 9 & 2 & 2 \end{vmatrix} = 0.$$

You should verify this by expanding the determinant by cofactors. ❏

THEOREM 8.10 Zero Row or Column

If all the entries in a row (column) of an $n \times n$ matrix \mathbf{A} are zero, then $\det \mathbf{A} = 0$.

PROOF Suppose the ith row of \mathbf{A} consists of all zeros. Hence all the products in the expansion of $\det \mathbf{A}$ by cofactors along the ith row are zero and consequently $\det \mathbf{A} = 0$. ❏

For example, it follows immediately from Theorem 8.10 that

zero column ↓

zero row → $\begin{vmatrix} 0 & 0 \\ 7 & -6 \end{vmatrix} = 0$ and $\begin{vmatrix} 4 & 6 & 0 \\ 1 & 5 & 0 \\ 8 & -1 & 0 \end{vmatrix} = 0.$

THEOREM 8.11 Interchanging Rows

If \mathbf{B} is the matrix obtained by interchanging any two rows (columns) of an $n \times n$ matrix \mathbf{A}, then $\det \mathbf{B} = -\det \mathbf{A}$.

For example, if \mathbf{B} is the matrix obtained by interchanging the first and third rows of $\mathbf{A} = \begin{pmatrix} 4 & -1 & 9 \\ 6 & 0 & 7 \\ 2 & 1 & 3 \end{pmatrix}$, then from Theorem 8.11 we have

$$\det \mathbf{B} = \begin{vmatrix} 2 & 1 & 3 \\ 6 & 0 & 7 \\ 4 & -1 & 9 \end{vmatrix} = -\begin{vmatrix} 4 & -1 & 9 \\ 6 & 0 & 7 \\ 2 & 1 & 3 \end{vmatrix} = -\det \mathbf{A}.$$

You should verify this by computing both determinants.

THEOREM 8.12 Constant Multiple of a Row

If \mathbf{B} is the matrix obtained from an $n \times n$ matrix \mathbf{A} by multiplying a row (column) by a nonzero real number k, then $\det \mathbf{B} = k \det \mathbf{A}$.

PROOF Suppose the entries in the ith row of \mathbf{A} are multiplied by the number k. Call the resulting matrix \mathbf{B}. Expanding $\det \mathbf{B}$ by cofactors along the ith row then gives

$$\det \mathbf{B} = k a_{i1} C_{i1} + k a_{i2} C_{i2} + \cdots + k a_{in} C_{in}$$

$$= k \underbrace{(a_{i1} C_{i1} + a_{i2} C_{i2} + \cdots + a_{in} C_{in})}_{\text{expansion of } \det \mathbf{A} \text{ by cofactors along the } i\text{th row}} = k \det \mathbf{A}.$$

❏

Example 2 **Theorems 8.12 and 8.9**

from first column from second column from second row

(a) $\begin{vmatrix} 5 & 8 \\ 20 & 16 \end{vmatrix} = 5\begin{vmatrix} 1 & 8 \\ 4 & 16 \end{vmatrix} = 5 \cdot 8 \begin{vmatrix} 1 & 1 \\ 4 & 2 \end{vmatrix} = 5 \cdot 8 \cdot 2 \begin{vmatrix} 1 & 1 \\ 2 & 1 \end{vmatrix} = 80(1 - 2) = -80$

from second column from Theorem 8.9

(b) $\begin{vmatrix} 4 & 2 & -1 \\ 5 & -2 & 1 \\ 7 & 4 & -2 \end{vmatrix} = (-2)\begin{vmatrix} 4 & -1 & -1 \\ 5 & 1 & 1 \\ 7 & -2 & -2 \end{vmatrix} = (-2) \cdot 0 = 0$ ❏

THEOREM 8.13 Determinant of a Matrix Product

If **A** and **B** are both $n \times n$ matrices, then det **AB** = det **A** \cdot det **B**.

In other words, the determinant of a product of two $n \times n$ matrices is the same as the product of the determinants.

Example 3 **Determinant of a Matrix Product**

Suppose $\mathbf{A} = \begin{pmatrix} 2 & 6 \\ 1 & -1 \end{pmatrix}$ and $\mathbf{B} = \begin{pmatrix} 3 & -4 \\ -3 & 5 \end{pmatrix}$. Then $\mathbf{AB} = \begin{pmatrix} -12 & 22 \\ 6 & -9 \end{pmatrix}$. Now

det **AB** = −24, det **A** = −8, det **B** = 3, and so we see that

$$\det \mathbf{A} \cdot \det \mathbf{B} = (-8)(3) = -24 = \det \mathbf{AB}.$$ ❏

THEOREM 8.14 Determinant is Unchanged

Suppose **B** is the matrix obtained from an $n \times n$ matrix **A** by multiplying the entries in a row (column) by a nonzero real number k and adding the result to the corresponding entries in another row (column). Then det **B** = det **A**.

Example 4 **A Multiple of a Row Added to Another**

Suppose $\mathbf{A} = \begin{pmatrix} 5 & 1 & 2 \\ 3 & 0 & 7 \\ 4 & -1 & 4 \end{pmatrix}$ and suppose the matrix **B** is defined as that matrix

obtained from **A** by the elementary row operation

$$\mathbf{A} = \begin{pmatrix} 5 & 1 & 2 \\ 3 & 0 & 7 \\ 4 & -1 & 4 \end{pmatrix} \overset{-3R_1 + R_3}{\Longrightarrow} \begin{pmatrix} 5 & 1 & 2 \\ 3 & 0 & 7 \\ -11 & -4 & -2 \end{pmatrix} = \mathbf{B}.$$

Expanding by cofactors along, say, the second column, we find det **A** = 45 and det **B** = 45. The student should verify this result. ❏

> **THEOREM 8.15 Determinant of a Triangular Matrix**
>
> Suppose \mathbf{A} is an $n \times n$ triangular matrix (upper or lower). Then
>
> $$\det \mathbf{A} = a_{11}a_{22} \cdots a_{nn},$$
>
> where $a_{11}, a_{22}, \ldots, a_{nn}$ are the entries on the main diagonal of \mathbf{A}.

PROOF We prove the result for a 3×3 lower triangular matrix

$$\mathbf{A} = \begin{pmatrix} a_{11} & 0 & 0 \\ a_{21} & a_{22} & 0 \\ a_{31} & a_{32} & a_{33} \end{pmatrix}.$$

Expanding $\det \mathbf{A}$ by cofactors along the first row gives

$$\det \mathbf{A} = a_{11} \begin{vmatrix} a_{22} & 0 \\ a_{32} & a_{33} \end{vmatrix} = a_{11}(a_{22}a_{33} - 0 \cdot a_{32}) = a_{11}a_{22}a_{33}. \qquad \square$$

Example 5 Determinant of a Triangular Matrix

(a) The determinant of the lower triangular matrix

$$\mathbf{A} = \begin{pmatrix} 3 & 0 & 0 & 0 \\ 2 & 6 & 0 & 0 \\ 5 & 9 & -4 & 0 \\ 7 & 2 & 4 & -2 \end{pmatrix}$$

is

$$\det \mathbf{A} = \begin{vmatrix} 3 & 0 & 0 & 0 \\ 2 & 6 & 0 & 0 \\ 5 & 9 & -4 & 0 \\ 7 & 2 & 4 & -2 \end{vmatrix} = 3 \cdot 6 \cdot (-4) \cdot (-2) = 144.$$

(b) The determinant of the diagonal matrix $\mathbf{A} = \begin{pmatrix} -3 & 0 & 0 \\ 0 & 6 & 0 \\ 0 & 0 & 4 \end{pmatrix}$ is

$$\det \mathbf{A} = \begin{vmatrix} -3 & 0 & 0 \\ 0 & 6 & 0 \\ 0 & 0 & 4 \end{vmatrix} = (-3) \cdot 6 \cdot 4 = -72. \qquad \square$$

■ **Row Reduction** Evaluating the determinant of an $n \times n$ matrix by the method of cofactor expansion requires a herculean effort when the order of the matrix is large. To expand the determinant of, say, a 5×5 matrix with nonzero entries requires evaluating five cofactors that are determinants of 4×4 submatrices, each of these in turn requires four additional cofactors that are determinants of 3×3 submatrices, and so on. There is a more practical (and programmable) method for evaluating the determinant of a matrix. This method is based on **reducing** the matrix to a triangular form by row operations and the fact that determinants of triangular matrices are easy to evaluate (see Theorem 8.15).

Example 6 Reducing a Determinant to Triangular Form

Evaluate the determinant of $\mathbf{A} = \begin{pmatrix} 6 & 2 & 7 \\ -4 & -3 & 2 \\ 2 & 4 & 8 \end{pmatrix}$.

SOLUTION

$$\det \mathbf{A} = \begin{vmatrix} 6 & 2 & 7 \\ -4 & -3 & 2 \\ 2 & 4 & 8 \end{vmatrix}$$

$$= 2 \begin{vmatrix} 6 & 2 & 7 \\ -4 & -3 & 2 \\ 1 & 2 & 4 \end{vmatrix} \quad \text{(2 is a common factor in third row: Theorem 8.12)}$$

$$= -2 \begin{vmatrix} 1 & 2 & 4 \\ -4 & -3 & 2 \\ 6 & 2 & 7 \end{vmatrix} \quad \text{(Interchange first and third rows: Theorem 8.11)}$$

$$= -2 \begin{vmatrix} 1 & 2 & 4 \\ 0 & 5 & 18 \\ 6 & 2 & 7 \end{vmatrix} \quad \text{(4 times first row added to second row: Theorem 8.14)}$$

$$= -2 \begin{vmatrix} 1 & 2 & 4 \\ 0 & 5 & 18 \\ 0 & -10 & -17 \end{vmatrix} \quad \text{(−6 times first row added to third row: Theorem 8.14)}$$

$$= -2 \begin{vmatrix} 1 & 2 & 4 \\ 0 & 5 & 18 \\ 0 & 0 & 19 \end{vmatrix} \quad \text{(2 times second row added to third row: Theorem 8.14)}$$

$$= (-2)(1)(5)(19) = -190 \quad \text{(Theorem 8.15)}$$

Our final theorem concerns cofactors. We saw in Section 8.4 that a determinant $\det \mathbf{A}$ of an $n \times n$ matrix \mathbf{A} could be evaluated by cofactor expansion along any row (column). This means that the n entries a_{ij} of a row (column) are multiplied by the corresponding cofactors C_{ij} and the n products are added. If, however, the entries a_{ij} of a row (a_{ij} of a column) of \mathbf{A} are multiplied by the corresponding cofactors C_{kj} of a different row (C_{ik} of a different column), the sum of the n products is zero.

THEOREM 8.16 A Property of Cofactors

Suppose \mathbf{A} is an $n \times n$ matrix. If $a_{i1}, a_{i2}, \ldots, a_{in}$ are the entries in the ith row and $C_{k1}, C_{k2}, \ldots, C_{kn}$ are the cofactors of the entries in the kth row, then

$$a_{i1}C_{k1} + a_{i2}C_{k2} + \cdots + a_{in}C_{kn} = 0 \quad \text{for } i \neq k.$$

If $a_{1j}, a_{2j}, \ldots, a_{nj}$ are the entries in the jth column and $C_{1k}, C_{2k}, \ldots, C_{nk}$ are the cofactors of the entries in the kth column, then

$$a_{1j}C_{1k} + a_{2j}C_{2k} + \cdots + a_{nj}C_{nk} = 0 \quad \text{for } j \neq k.$$

PROOF We shall prove the result for rows. Let \mathbf{B} be the matrix obtained from \mathbf{A} by letting the entries in the ith row of \mathbf{A} be the same as the entries in the kth row—that is, $a_{i1} = a_{k1}, a_{i2} = a_{k2}, \ldots, a_{in} = a_{kn}$. Since \mathbf{B} has two rows that are the same, it follows from Theorem 8.9 that det $\mathbf{B} = 0$. Cofactor expansion along the kth row then gives the desired result:

$$0 = \det \mathbf{B} = a_{k1}C_{k1} + a_{k2}C_{k2} + \cdots + a_{kn}C_{kn}$$

$$= a_{i1}C_{k1} + a_{i2}C_{k2} + \cdots + a_{in}C_{kn}. \qquad \square$$

Example 7 **Cofactors of Third Row/Entries of First Row**

Consider the matrix $\mathbf{A} = \begin{pmatrix} 6 & 2 & 7 \\ -4 & -3 & 2 \\ 2 & 4 & 8 \end{pmatrix}$. Suppose we then multiply the entries of the first row by the cofactors of the third row and add the results; that is,

$$a_{11}C_{31} + a_{12}C_{32} + a_{13}C_{33} = 6 \begin{vmatrix} 2 & 7 \\ -3 & 2 \end{vmatrix} + 2\left(-\begin{vmatrix} 6 & 7 \\ -4 & 2 \end{vmatrix}\right) + 7 \begin{vmatrix} 6 & 2 \\ -4 & -3 \end{vmatrix}$$

$$= 6(25) + 2(-40) + 7(-10) = 0. \qquad \square$$

EXERCISES 8.5

Answers to odd-numbered problems begin on page A-43.

In Problems 1–10, state the appropriate theorem(s) in this section that justifies the given equality. Do not expand the determinants by cofactors.

1. $\begin{vmatrix} 1 & 2 \\ 3 & 4 \end{vmatrix} = -\begin{vmatrix} 3 & 4 \\ 1 & 2 \end{vmatrix}$

2. $\begin{vmatrix} 1 & 2 \\ 3 & 4 \end{vmatrix} = \begin{vmatrix} 1 & 2 \\ 4 & 6 \end{vmatrix}$

3. $\begin{vmatrix} -5 & 6 \\ 2 & -8 \end{vmatrix} = \begin{vmatrix} 1 & 6 \\ -6 & -8 \end{vmatrix}$

4. $\begin{vmatrix} 1 & 0 & 0 \\ 0 & 0 & 2 \\ 0 & 1 & 0 \end{vmatrix} = -2\begin{vmatrix} 1 & 0 & 0 \\ 0 & 1 & 0 \\ 0 & 0 & 1 \end{vmatrix}$

5. $\begin{vmatrix} 1 & 2 & 3 \\ 4 & 2 & 18 \\ 5 & 9 & -12 \end{vmatrix} = 6\begin{vmatrix} 1 & 2 & 1 \\ 2 & 1 & 3 \\ 5 & 9 & -4 \end{vmatrix}$

6. $\begin{vmatrix} 1 & 2 & 3 \\ 4 & 2 & 18 \\ 5 & 9 & -12 \end{vmatrix} = \begin{vmatrix} 4 & 2 & 18 \\ 5 & 9 & -12 \\ 1 & 2 & 3 \end{vmatrix}$

7. $\begin{vmatrix} 0 & 5 & 0 & 6 \\ 2 & 1 & 0 & 8 \\ 0 & 2 & 0 & -9 \\ 0 & 6 & 0 & 4 \end{vmatrix} = 0$

8. $\begin{vmatrix} 3 & 2 & 1 \\ 2 & 6 & 3 \\ 5 & -8 & -4 \end{vmatrix} = 0$

9. $\begin{vmatrix} 1 & 2 & 3 \\ 4 & 5 & 6 \\ 7 & 8 & 9 \end{vmatrix} = \begin{vmatrix} 1 & 4 & 7 \\ 2 & 5 & 8 \\ 3 & 6 & 9 \end{vmatrix}$

10. $\begin{vmatrix} 1 & 0 & 0 & 0 \\ 0 & 2 & 0 & 0 \\ 0 & 0 & 3 & 0 \\ 0 & 0 & 0 & 4 \end{vmatrix} = \begin{vmatrix} 0 & 0 & 0 & 1 \\ 0 & 0 & 2 & 0 \\ 0 & 3 & 0 & 0 \\ 4 & 0 & 0 & 0 \end{vmatrix}$

In Problems 11–14, evaluate the determinant of the given matrix using the result

$$\begin{vmatrix} a_1 & a_2 & a_3 \\ b_1 & b_2 & b_3 \\ c_1 & c_2 & c_3 \end{vmatrix} = 5.$$

11. $\mathbf{A} = \begin{pmatrix} a_3 & a_2 & a_1 \\ b_3 & b_2 & b_1 \\ c_3 & c_2 & c_1 \end{pmatrix}$

12. $\mathbf{B} = \begin{pmatrix} 2a_1 & a_2 & a_3 \\ 6b_1 & 3b_2 & 3b_3 \\ 2c_1 & c_2 & c_3 \end{pmatrix}$

13. $\mathbf{C} = \begin{pmatrix} -a_1 & -a_2 & -a_3 \\ b_1 & b_2 & b_3 \\ c_1 - a_1 & c_2 - a_2 & c_3 - a_3 \end{pmatrix}$

14. $\mathbf{D} = \begin{pmatrix} a_1 & b_1 & c_1 \\ a_2 & b_2 & c_2 \\ a_3 & b_3 & c_3 \end{pmatrix}$

In Problems 15–18, evaluate the determinant of the given matrix without expanding by cofactors.

15. A $= \begin{pmatrix} 6 & 1 & 8 & 10 \\ 0 & \frac{2}{3} & 7 & 2 \\ 0 & 0 & - & 9 \\ 0 & 0 & 40 & -5 \end{pmatrix}$

16. B $= \begin{pmatrix} 0 & 0 & a_{13} \\ 0 & a_{22} & a_{23} \\ a_{31} & a_{32} & a_{33} \end{pmatrix}$

17. C $= \begin{pmatrix} -5 & 0 & 0 \\ 0 & 7 & 0 \\ 0 & 0 & 3 \end{pmatrix}$ **18. D** $= \begin{pmatrix} 0 & 7 & 0 \\ 4 & 0 & 0 \\ 0 & 0 & -2 \end{pmatrix}$

In Problems 19 and 20, verify that $\det \mathbf{A} = \det \mathbf{A}^T$ for the given matrix **A**.

19. A $= \begin{pmatrix} 1 & 2 & 1 \\ 4 & 1 & -1 \\ 1 & 2 & -1 \end{pmatrix}$ **20. A** $= \begin{pmatrix} 2 & 3 & 4 \\ 1 & 0 & 5 \\ 7 & 2 & -1 \end{pmatrix}$

21. Consider the matrices

$$\mathbf{A} = \begin{pmatrix} 2 & -1 & 1 \\ 3 & 1 & -1 \\ 0 & 2 & 2 \end{pmatrix} \text{ and } \mathbf{B} = \begin{pmatrix} 2 & 1 & 5 \\ 4 & 3 & 8 \\ 0 & -1 & 0 \end{pmatrix}.$$

Verify that $\det \mathbf{AB} = \det \mathbf{A} \det \mathbf{B}$.

22. Suppose **A** is an $n \times n$ matrix such that $\mathbf{A}^2 = \mathbf{I}$, where $\mathbf{A}^2 = \mathbf{AA}$. Show that $\det \mathbf{A} = \pm 1$.

23. Consider the matrix

$$\mathbf{A} = \begin{pmatrix} a & a+1 & a+2 \\ b & b+1 & b+2 \\ c & c+1 & c+2 \end{pmatrix}.$$

Without expanding, evaluate $\det \mathbf{A}$.

24. Consider the matrix

$$\mathbf{A} = \begin{pmatrix} 1 & 1 & 1 \\ x & y & z \\ y+z & x+z & x+y \end{pmatrix}.$$

Without expanding, show that $\det \mathbf{A} = 0$.

In Problems 25–32, use the procedure illustrated in Example 6 to evaluate the determinant of the given matrix.

25. $\begin{pmatrix} 1 & 1 & 5 \\ 4 & 3 & 6 \\ 0 & -1 & 1 \end{pmatrix}$ **26.** $\begin{pmatrix} 2 & 4 & 5 \\ 4 & 2 & 0 \\ 8 & 7 & -2 \end{pmatrix}$

27. $\begin{pmatrix} -1 & 2 & 3 \\ 4 & -5 & -2 \\ 9 & -9 & 6 \end{pmatrix}$ **28.** $\begin{pmatrix} -2 & 2 & -6 \\ 5 & 0 & 1 \\ 1 & -2 & 2 \end{pmatrix}$

29. $\begin{vmatrix} 1 & -2 & 2 & 1 \\ 2 & 1 & -2 & 3 \\ 3 & 4 & -8 & 1 \\ 3 & -11 & 12 & 2 \end{vmatrix}$ **30.** $\begin{vmatrix} 0 & 1 & 4 & 5 \\ 2 & 5 & 0 & 1 \\ 1 & 2 & 2 & 0 \\ 3 & 1 & 3 & 2 \end{vmatrix}$

31. $\begin{vmatrix} 1 & 2 & 3 & 4 \\ 1 & 3 & 5 & 7 \\ 2 & 3 & 6 & 7 \\ 1 & 5 & 8 & 20 \end{vmatrix}$ **32.** $\begin{vmatrix} 2 & 9 & 1 & 8 \\ 1 & 3 & 7 & 4 \\ 0 & 1 & 6 & 5 \\ 3 & 1 & 4 & 2 \end{vmatrix}$

33. By proceeding as in Example 6, show that

$$\begin{vmatrix} 1 & 1 & 1 \\ a & b & c \\ a^2 & b^2 & c^2 \end{vmatrix} = (b-a)(c-a)(c-b)$$

34. Evaluate $\begin{vmatrix} 1 & 1 & 1 & 1 \\ a & b & c & d \\ a^2 & b^2 & c^2 & d^2 \\ a^3 & b^3 & c^3 & d^3 \end{vmatrix}$. [*Hint:* See Problem 33.]

In Problems 35 and 36, verify Theorem 8.16 by evaluating $a_{21}C_{11} + a_{22}C_{12} + a_{23}C_{13}$ and $a_{13}C_{12} + a_{23}C_{22} + a_{33}C_{32}$ for the given matrix.

35. A $= \begin{pmatrix} 1 & 1 & 2 \\ -1 & 2 & 1 \\ 4 & -2 & 1 \end{pmatrix}$

36. A $= \begin{pmatrix} 3 & 0 & 5 \\ -2 & 3 & -1 \\ 2 & 2 & -3 \end{pmatrix}$

37. Let $\mathbf{A} = \begin{pmatrix} 3 & -4 \\ 1 & 2 \end{pmatrix}$ and $\mathbf{B} = \begin{pmatrix} 7 & 4 \\ -1 & -5 \end{pmatrix}$.
Verify that $\det(\mathbf{A} + \mathbf{B}) \neq \det \mathbf{A} + \det \mathbf{B}$.

38. Suppose **A** is a 5×5 matrix for which $\det \mathbf{A} = -7$. What is the value of $\det(2\mathbf{A})$?

39. An $n \times n$ matrix **A** is said to be **skew-symmetric** if $\mathbf{A}^T = -\mathbf{A}$. If **A** is a 5×5 skew-symmetric matrix, show that $\det \mathbf{A} = 0$.

40. It takes about $n!$ multiplications to evaluate the determinant of an $n \times n$ matrix using expansion by cofactors, whereas it takes about $n^3/3$ arithmetic operations using the row reduction method illustrated in Example 6.

(a) Compare the number of operations for both methods using a 25×25 matrix.

(b) If a computer can do 50,000 operations per second, compare the lengths of time it would take the computer to evaluate the determinant of a 25×25 matrix using cofactor expansion and row reduction.

8.6 Inverse of a Matrix

The concept of the determinant of a matrix will play an important role in this and the following section.

8.6.1 Finding the Inverse

In the real number system, if a is a nonzero number, then there exists a number b such that $ab = ba = 1$. The number b is called the multiplicative inverse of the number a and is denoted by a^{-1}. For a *square* matrix \mathbf{A} it is also important to know whether we can find another square matrix \mathbf{B} of the same order such that $\mathbf{AB} = \mathbf{BA} = \mathbf{I}$. We have the following definition.

> **DEFINITION 8.11 Inverse of a Matrix**
>
> Let \mathbf{A} be an $n \times n$ matrix. If there exists an $n \times n$ matrix \mathbf{B} such that
>
> $$\mathbf{AB} = \mathbf{BA} = \mathbf{I}, \tag{1}$$
>
> where \mathbf{I} is the $n \times n$ identity, then the matrix \mathbf{A} is said to be **nonsingular** or **invertible.** The matrix \mathbf{B} is said to be the **inverse** of \mathbf{A}.

For example, the matrix $\mathbf{A} = \begin{pmatrix} 2 & 1 \\ 1 & 1 \end{pmatrix}$ is nonsingular or invertible since the matrix

$\mathbf{B} = \begin{pmatrix} 1 & -1 \\ -1 & 2 \end{pmatrix}$ is its inverse. To verify this observe that

$$\mathbf{AB} = \begin{pmatrix} 2 & 1 \\ 1 & 1 \end{pmatrix}\begin{pmatrix} 1 & -1 \\ -1 & 2 \end{pmatrix} = \begin{pmatrix} 1 & 0 \\ 0 & 1 \end{pmatrix} = \mathbf{I}$$

and

$$\mathbf{BA} = \begin{pmatrix} 1 & -1 \\ -1 & 2 \end{pmatrix}\begin{pmatrix} 2 & 1 \\ 1 & 1 \end{pmatrix} = \begin{pmatrix} 1 & 0 \\ 0 & 1 \end{pmatrix} = \mathbf{I}.$$

Unlike the real number system, where *every* nonzero number a has a multiplicative inverse, not every nonzero $n \times n$ matrix \mathbf{A} has an inverse.

For example, if $\mathbf{A} = \begin{pmatrix} 1 & 1 \\ 0 & 0 \end{pmatrix}$ and $\mathbf{B} = \begin{pmatrix} b_{11} & b_{12} \\ b_{21} & b_{22} \end{pmatrix}$, then

$$\mathbf{AB} = \begin{pmatrix} 1 & 1 \\ 0 & 0 \end{pmatrix}\begin{pmatrix} b_{11} & b_{12} \\ b_{21} & b_{22} \end{pmatrix} = \begin{pmatrix} b_{11} + b_{21} & b_{12} + b_{22} \\ 0 & 0 \end{pmatrix}.$$

Inspection of this result shows that it is impossible to obtain the 2×2 identity matrix \mathbf{I}, since there is no way of selecting b_{11}, b_{12}, b_{21}, and b_{22} to get 1 as the entry in the second row and second column. We have shown that the matrix $\mathbf{A} = \begin{pmatrix} 1 & 1 \\ 0 & 0 \end{pmatrix}$ has no inverse. An $n \times n$ matrix that has no inverse is called **singular.** If \mathbf{A} is nonsingular, its inverse is denoted by $\mathbf{B} = \mathbf{A}^{-1}$.

 Important

Note that symbol -1 in the notation \mathbf{A}^{-1} is not an exponent; in other words, \mathbf{A}^{-1} is not a reciprocal. Also, if \mathbf{A} is nonsingular, its inverse is unique.

■ **Properties** The following theorem lists some properties of the inverse of a matrix.

> **THEOREM 8.17 Properties of the Inverse**
>
> Let \mathbf{A} and \mathbf{B} be nonsingular matrices.
>
> (i) $(\mathbf{A}^{-1})^{-1} = \mathbf{A}$
>
> (ii) $(\mathbf{AB})^{-1} = \mathbf{B}^{-1}\mathbf{A}^{-1}$
>
> (iii) $(\mathbf{A}^T)^{-1} = (\mathbf{A}^{-1})^T$

PROOF OF *(i)* This part of the theorem states that if \mathbf{A} is nonsingular, then its \mathbf{A}^{-1} is also nonsingular and *its* inverse is \mathbf{A}. To prove that \mathbf{A}^{-1} is nonsingular, we have to show that a matrix \mathbf{B} can be found such that $\mathbf{A}^{-1}\mathbf{B} = \mathbf{B}\mathbf{A}^{-1} = \mathbf{I}$. But since \mathbf{A} is assumed to be nonsingular, we know from (1) that $\mathbf{AA}^{-1} = \mathbf{A}^{-1}\mathbf{A} = \mathbf{I}$ and, equivalently, $\mathbf{A}^{-1}\mathbf{A} = \mathbf{AA}^{-1} = \mathbf{I}$. The last matrix equation indicates that the required matrix, the inverse of \mathbf{A}^{-1}, is $\mathbf{B} = \mathbf{A}$. Consequently, $(\mathbf{A}^{-1})^{-1} = \mathbf{A}$. ❑

Theorem 8.17(*ii*) extends to any finite number of nonsingular matrices:

$$(\mathbf{A}_1\mathbf{A}_2 \cdots \mathbf{A}_k)^{-1} = \mathbf{A}_k^{-1}\mathbf{A}_{k-1}^{-1} \cdots \mathbf{A}_1^{-1},$$

that is, the inverse of a product of nonsingular matrices is the product of the inverses in reverse order.

In the discussion that follows we are going to consider two different ways of finding \mathbf{A}^{-1} for a nonsingular matrix \mathbf{A}. The first method utilizes determinants, whereas the second employs the elementary row operations introduced in Section 8.2.

■ **Adjoint Method** Recall from (6) of Section 8.4 that the cofactor C_{ij} of the entry a_{ij} of an $n \times n$ matrix \mathbf{A} is $C_{ij} = (-1)^{i+j}M_{ij}$, where M_{ij} is the minor of a_{ij}; that is, the determinant of the $(n-1) \times (n-1)$ submatrix obtained by deleting the ith row and the jth column of \mathbf{A}.

> **DEFINITION 8.12 Adjoint Matrix**
>
> Let \mathbf{A} be an $n \times n$ matrix. The matrix that is the transpose of the matrix of cofactors corresponding to the entries of \mathbf{A}:
>
> $$\begin{pmatrix} C_{11} & C_{12} & \cdots & C_{1n} \\ C_{21} & C_{22} & \cdots & C_{2n} \\ \vdots & & & \vdots \\ C_{n1} & C_{n2} & \cdots & C_{nn} \end{pmatrix}^T = \begin{pmatrix} C_{11} & C_{21} & \cdots & C_{n1} \\ C_{12} & C_{22} & \cdots & C_{n2} \\ \vdots & & & \vdots \\ C_{1n} & C_{2n} & \cdots & C_{nn} \end{pmatrix}$$
>
> is called the **adjoint** of \mathbf{A} and is denoted by adj \mathbf{A}.

The next theorem will give a compact formula for the inverse of a nonsingular matrix in terms of the adjoint of the matrix. However, because of the determinants involved, this method becomes unwieldy for $n \geq 4$.

> **THEOREM 8.18 Finding the Inverse**
>
> Let \mathbf{A} be an $n \times n$ matrix. If $\det \mathbf{A} \neq 0$, then
>
> $$\mathbf{A}^{-1} = \left(\frac{1}{\det \mathbf{A}}\right) \text{adj } \mathbf{A}. \tag{2}$$

PROOF For brevity we prove the case when $n = 3$. Observe that

$$\mathbf{A}(\text{adj } \mathbf{A}) = \begin{pmatrix} a_{11} & a_{12} & a_{13} \\ a_{21} & a_{22} & a_{23} \\ a_{31} & a_{32} & a_{33} \end{pmatrix} \begin{pmatrix} C_{11} & C_{21} & C_{31} \\ C_{12} & C_{22} & C_{32} \\ C_{13} & C_{23} & C_{33} \end{pmatrix}$$

$$= \begin{pmatrix} \det \mathbf{A} & 0 & 0 \\ 0 & \det \mathbf{A} & 0 \\ 0 & 0 & \det \mathbf{A} \end{pmatrix} \tag{3}$$

since $\det \mathbf{A} = a_{i1}C_{i1} + a_{i2}C_{i2} + a_{i3}C_{i3}$, for $i = 1, 2, 3$, are the expansions of $\det \mathbf{A}$ by cofactors along the first, second, and third rows, and

$$a_{11}C_{21} + a_{12}C_{22} + a_{13}C_{23} = 0 \qquad a_{11}C_{31} + a_{12}C_{32} + a_{13}C_{33} = 0$$

$$a_{21}C_{11} + a_{22}C_{12} + a_{23}C_{13} = 0 \qquad a_{21}C_{31} + a_{22}C_{32} + a_{23}C_{33} = 0$$

$$a_{31}C_{11} + a_{32}C_{12} + a_{33}C_{13} = 0 \qquad a_{31}C_{21} + a_{32}C_{22} + a_{33}C_{23} = 0$$

in view of Theorem 8.16. Thus, (3) is the same as

$$\mathbf{A}(\text{adj } \mathbf{A}) = (\det \mathbf{A}) \begin{pmatrix} 1 & 0 & 0 \\ 0 & 1 & 0 \\ 0 & 0 & 1 \end{pmatrix} = (\det \mathbf{A})\mathbf{I}$$

or $\mathbf{A}(1/\det \mathbf{A}) \text{ adj } \mathbf{A} = \mathbf{I}$. Similarly, it can be shown in exactly the same manner that $((1/\det \mathbf{A}) \text{ adj } \mathbf{A})\mathbf{A} = \mathbf{I}$. Hence, by definition, $\mathbf{A}^{-1} = (1/\det \mathbf{A})\text{adj } \mathbf{A}$. ❑

For future reference we observe in the case of a 2×2 nonsingular matrix

$$\mathbf{A} = \begin{pmatrix} a_{11} & a_{12} \\ a_{21} & a_{22} \end{pmatrix}$$

that the cofactors are $C_{11} = a_{22}$, $C_{12} = -a_{21}$, $C_{21} = -a_{12}$, and $C_{22} = a_{11}$. In this case,

$$\text{adj } \mathbf{A} = \begin{pmatrix} C_{11} & C_{12} \\ C_{21} & C_{22} \end{pmatrix}^T = \begin{pmatrix} a_{22} & -a_{21} \\ -a_{12} & -a_{11} \end{pmatrix}^T = \begin{pmatrix} a_{22} & -a_{12} \\ -a_{21} & a_{11} \end{pmatrix}.$$

It follows from (2) that

$$\mathbf{A}^{-1} = \frac{1}{\det \mathbf{A}} \begin{pmatrix} a_{22} & -a_{12} \\ -a_{21} & a_{11} \end{pmatrix}. \tag{4}$$

For a 3×3 nonsingular matrix

$$\mathbf{A} = \begin{pmatrix} a_{11} & a_{12} & a_{13} \\ a_{21} & a_{22} & a_{23} \\ a_{31} & a_{32} & a_{33} \end{pmatrix}$$

$$C_{11} = \begin{vmatrix} a_{22} & a_{23} \\ a_{32} & a_{33} \end{vmatrix} \qquad C_{12} = -\begin{vmatrix} a_{21} & a_{23} \\ a_{31} & a_{33} \end{vmatrix} \qquad C_{13} = \begin{vmatrix} a_{21} & a_{22} \\ a_{31} & a_{32} \end{vmatrix},$$

and so on. After the adjoint of \mathbf{A} is formed, (2) gives

$$\mathbf{A}^{-1} = \frac{1}{\det \mathbf{A}} \begin{pmatrix} C_{11} & C_{21} & C_{31} \\ C_{12} & C_{22} & C_{32} \\ C_{13} & C_{23} & C_{33} \end{pmatrix}. \qquad (5)$$

Example 1 Inverse of a Matrix

Find the inverse of $\mathbf{A} = \begin{pmatrix} 1 & 4 \\ 2 & 10 \end{pmatrix}$.

SOLUTION Since $\det \mathbf{A} = 10 - 8 = 2$, it follows from (4) that

$$\mathbf{A}^{-1} = \frac{1}{2} \begin{pmatrix} 10 & -4 \\ -2 & 1 \end{pmatrix} = \begin{pmatrix} 5 & -2 \\ -1 & \frac{1}{2} \end{pmatrix}.$$

CHECK $\mathbf{A}\mathbf{A}^{-1} = \begin{pmatrix} 1 & 4 \\ 2 & 10 \end{pmatrix} \begin{pmatrix} 5 & -2 \\ -1 & \frac{1}{2} \end{pmatrix} = \begin{pmatrix} 5-4 & -2+2 \\ 10-10 & -4+5 \end{pmatrix} = \begin{pmatrix} 1 & 0 \\ 0 & 1 \end{pmatrix}$

$$\mathbf{A}^{-1}\mathbf{A} = \begin{pmatrix} 5 & -2 \\ -1 & \frac{1}{2} \end{pmatrix} \begin{pmatrix} 1 & 4 \\ 2 & 10 \end{pmatrix} = \begin{pmatrix} 5-4 & 20-20 \\ -1+1 & -4+5 \end{pmatrix} = \begin{pmatrix} 1 & 0 \\ 0 & 1 \end{pmatrix}. \qquad \square$$

Example 2 Inverse of a Matrix

Find the inverse of $\mathbf{A} = \begin{pmatrix} 2 & 2 & 0 \\ -2 & 1 & 1 \\ 3 & 0 & 1 \end{pmatrix}$.

SOLUTION Since $\det \mathbf{A} = 12$, we can find \mathbf{A}^{-1} from (5). The cofactors corresponding to the entries in \mathbf{A} are

$$C_{11} = \begin{vmatrix} 1 & 1 \\ 0 & 1 \end{vmatrix} = 1 \qquad C_{12} = -\begin{vmatrix} -2 & 1 \\ 3 & 1 \end{vmatrix} = 5 \qquad C_{13} = \begin{vmatrix} -2 & 1 \\ 3 & 0 \end{vmatrix} = -3$$

$$C_{21} = -\begin{vmatrix} 2 & 0 \\ 0 & 1 \end{vmatrix} = -2 \qquad C_{22} = \begin{vmatrix} 2 & 0 \\ 3 & 1 \end{vmatrix} = 2 \qquad C_{23} = -\begin{vmatrix} 2 & 2 \\ 3 & 0 \end{vmatrix} = 6$$

$$C_{31} = \begin{vmatrix} 2 & 0 \\ 1 & 1 \end{vmatrix} = 2 \qquad C_{32} = -\begin{vmatrix} 2 & 0 \\ -2 & 1 \end{vmatrix} = -2 \qquad C_{33} = \begin{vmatrix} 2 & 2 \\ -2 & 1 \end{vmatrix} = 6$$

From (5) we then obtain

$$\mathbf{A}^{-1} = \frac{1}{12} \begin{pmatrix} 1 & -2 & 2 \\ 5 & 2 & -2 \\ -3 & 6 & 6 \end{pmatrix} = \begin{pmatrix} \frac{1}{12} & -\frac{1}{6} & \frac{1}{6} \\ \frac{5}{12} & \frac{1}{6} & -\frac{1}{6} \\ -\frac{1}{4} & \frac{1}{2} & \frac{1}{2} \end{pmatrix}.$$

The reader is urged to verify that $\mathbf{A}\mathbf{A}^{-1} = \mathbf{A}^{-1}\mathbf{A} = \mathbf{I}$. \square

We are now in a position to prove a necessary and sufficient condition for an $n \times n$ matrix \mathbf{A} to have an inverse.

> **THEOREM 8.19 Nonsingular Matrices and det A**
>
> An $n \times n$ matrix \mathbf{A} is nonsingular if and only if $\det \mathbf{A} \neq 0$.

PROOF We shall first prove the sufficiency. Assume $\det \mathbf{A} \neq 0$. Then \mathbf{A} is nonsingular, since \mathbf{A}^{-1} can be found from Theorem 8.18.

To prove the necessity, we must assume that \mathbf{A} is nonsingular and prove that $\det \mathbf{A} \neq 0$. Now from Theorem 8.13, $\mathbf{A}\mathbf{A}^{-1} = \mathbf{A}^{-1}\mathbf{A} = \mathbf{I}$ implies

$$(\det \mathbf{A})(\det \mathbf{A}^{-1}) = (\det \mathbf{A}^{-1})(\det \mathbf{A}) = \det \mathbf{I}.$$

But since $\det \mathbf{I} = 1$ (why?), the product $(\det \mathbf{A})(\det \mathbf{A}^{-1}) = 1 \neq 0$ shows that we must have $\det \mathbf{A} \neq 0$. ❏

Example 3 **A Singular Matrix**

The 2×2 matrix $\mathbf{A} = \begin{pmatrix} 2 & 2 \\ 3 & 3 \end{pmatrix}$ has no inverse; that is, \mathbf{A} is singular, since $\det \mathbf{A} = 6 - 6 = 0$. ❏

Because of the number of determinants that must be evaluated, the foregoing method for calculating the inverse of a matrix is tedious when the order of the matrix is large. In the case of 3×3 or larger matrices, the next method is a particularly efficient means for finding \mathbf{A}^{-1}.

■ **Row Operations Method** Although it would take us beyond the scope of this book to prove it, we shall nonetheless use the following results:

> **THEOREM 8.20 Finding the Inverse**
>
> If an $n \times n$ matrix \mathbf{A} can be transformed into the $n \times n$ identity \mathbf{I} by a sequence of elementary row operations, then \mathbf{A} is nonsingular. The same sequence of operations that transforms \mathbf{A} into the identity \mathbf{I} will also transform \mathbf{I} into \mathbf{A}^{-1}.

It is convenient to carry out these row operations on \mathbf{A} and \mathbf{I} simultaneously by means of an $n \times 2n$ matrix obtained by augmenting \mathbf{A} with the identity \mathbf{I} as shown here:

$$(\mathbf{A}|\mathbf{I}) = \begin{pmatrix} a_{11} & a_{12} & \cdots & a_{1n} & 1 & 0 & \cdots & 0 \\ a_{21} & a_{22} & \cdots & a_{2n} & 1 & 0 & \cdots & 0 \\ \vdots & & & \vdots & \vdots & & & \vdots \\ a_{n1} & a_{n2} & \cdots & a_{nn} & 0 & 0 & \cdots & 1 \end{pmatrix}.$$

The procedure for finding \mathbf{A}^{-1} is outlined in the following diagram:

Perform row operations on **A** until **I** is obtained. This means **A** is nonsingular.

$(\ \mathbf{A}\ |\ \mathbf{I}\)$ $(\mathbf{I}\ |\ \mathbf{A}^{-1})$

By simultaneously applying the same row operations to **I** we get \mathbf{A}^{-1}.

Example 4 **Inverse by Elementary Row Operations**

Find the inverse of $\mathbf{A} = \begin{pmatrix} 2 & 0 & 1 \\ -2 & 3 & 4 \\ -5 & 5 & 6 \end{pmatrix}$.

SOLUTION We shall use the same notation as we did in Section 8.2 when we reduced an augmented matrix to reduced row-echelon form:

$$\left(\begin{array}{ccc|ccc} 2 & 0 & 1 & 1 & 0 & 0 \\ -2 & 3 & 4 & 0 & 1 & 0 \\ -5 & 5 & 6 & 0 & 0 & 1 \end{array}\right) \quad \overset{\frac{1}{2}R_1}{\Longrightarrow} \quad \left(\begin{array}{ccc|ccc} 1 & 0 & \frac{1}{2} & \frac{1}{2} & 0 & 0 \\ -2 & 3 & 4 & 0 & 1 & 0 \\ -5 & 5 & 6 & 0 & 0 & 1 \end{array}\right)$$

$$\overset{\substack{2R_1 + R_2 \\ 5R_1 + R_3}}{\Longrightarrow} \left(\begin{array}{ccc|ccc} 1 & 0 & \frac{1}{2} & \frac{1}{2} & 0 & 0 \\ 0 & 3 & 5 & 1 & 1 & 0 \\ 0 & 5 & \frac{17}{2} & \frac{5}{2} & 0 & 1 \end{array}\right)$$

$$\overset{\substack{\frac{1}{3}R_2 \\ \frac{1}{5}R_3}}{\Longrightarrow} \left(\begin{array}{ccc|ccc} 1 & 0 & \frac{1}{2} & \frac{1}{2} & 0 & 0 \\ 0 & 1 & \frac{5}{3} & \frac{1}{3} & \frac{1}{3} & 0 \\ 0 & 1 & \frac{17}{10} & \frac{1}{2} & 0 & \frac{1}{5} \end{array}\right)$$

$$\overset{-R_2 + R_3}{\Longrightarrow} \left(\begin{array}{ccc|ccc} 1 & 0 & \frac{1}{2} & \frac{1}{2} & 0 & 0 \\ 0 & 1 & \frac{5}{3} & \frac{1}{3} & \frac{1}{3} & 0 \\ 0 & 0 & \frac{1}{30} & \frac{1}{6} & -\frac{1}{3} & \frac{1}{5} \end{array}\right)$$

$$\overset{30R_3}{\Longrightarrow} \left(\begin{array}{ccc|ccc} 1 & 0 & \frac{1}{2} & \frac{1}{2} & 0 & 0 \\ 0 & 1 & \frac{5}{3} & \frac{1}{3} & \frac{1}{3} & 0 \\ 0 & 0 & 1 & 5 & -10 & 6 \end{array}\right)$$

$$\overset{\substack{-\frac{1}{2}R_3 + R_1 \\ -\frac{5}{3}R_3 + R_2}}{\Longrightarrow} \left(\begin{array}{ccc|ccc} 1 & 0 & 0 & -2 & 5 & -3 \\ 0 & 1 & 0 & -8 & 17 & -10 \\ 0 & 0 & 1 & 5 & -10 & 6 \end{array}\right).$$

Since \mathbf{I} appears to the left of the vertical line, we conclude that the matrix to the right of the line is

$$\mathbf{A}^{-1} = \begin{pmatrix} -2 & 5 & -3 \\ -8 & 17 & -10 \\ 5 & -10 & 6 \end{pmatrix}. \qquad \square$$

If row reduction of $(\mathbf{A} \mid \mathbf{I})$ leads to the situation

$$(\mathbf{A} \mid \mathbf{I}) \overset{\substack{\text{row} \\ \text{operations}}}{\Longrightarrow} (\mathbf{B} \mid \mathbf{C}),$$

where the matrix \mathbf{B} contains a row of zeros, then necessarily \mathbf{A} is singular. Since further reduction of \mathbf{B} always yields another matrix with a row of zeros, we can never transform \mathbf{A} into \mathbf{I}.

Example 5 **A Singular Matrix**

The matrix $\mathbf{A} = \begin{pmatrix} 1 & -1 & -2 \\ 2 & 4 & 5 \\ 6 & 0 & -3 \end{pmatrix}$ has no inverse, since

$$\left(\begin{array}{ccc|ccc} 1 & -1 & -2 & 1 & 0 & 0 \\ 2 & 4 & 5 & 0 & 1 & 0 \\ 6 & 0 & -3 & 0 & 0 & 1 \end{array}\right) \overset{-2R_1 + R_2}{\Longrightarrow} \left(\begin{array}{ccc|ccc} 1 & -1 & -2 & 1 & 0 & 0 \\ 0 & 6 & 9 & -2 & 1 & 0 \\ 6 & 0 & -3 & 0 & 0 & 1 \end{array}\right)$$

$$\stackrel{-6R_1 + R_3}{\Longrightarrow} \begin{pmatrix} 1 & -1 & -2 & | & 1 & 0 & 0 \\ 0 & 6 & 9 & | & -2 & 1 & 0 \\ 0 & 6 & 9 & | & -6 & 0 & 1 \end{pmatrix}$$

$$\stackrel{-R_2 + R_3}{\Longrightarrow} \begin{pmatrix} 1 & -1 & -2 & | & 1 & 0 & 0 \\ 0 & 6 & 9 & | & -2 & 1 & 0 \\ 0 & 0 & 0 & | & -4 & -1 & 1 \end{pmatrix}.$$

Since the matrix to the left of the vertical bar has a row of zeros, we can stop at this point and conclude that **A** is singular. ❏

8.6.2 Using the Inverse to Solve Systems

A system of m linear equations in n unknowns x_1, x_2, \ldots, x_n,

$$a_{11}x_1 + a_{12}x_2 + \cdots + a_{1n}x_n = b_1$$
$$a_{21}x_1 + a_{22}x_2 + \cdots + a_{2n}x_n = b_2 \tag{6}$$
$$\vdots \qquad\qquad \vdots$$
$$a_{m1}x_1 + a_{m2}x_2 + \cdots + a_{mn}x_n = b_m$$

can be written compactly as a matrix equation $\mathbf{AX} = \mathbf{B}$, where

$$\mathbf{A} = \begin{pmatrix} a_{11} & a_{12} & \cdots & a_{1n} \\ a_{21} & a_{22} & \cdots & a_{2n} \\ \vdots & & & \vdots \\ a_{m1} & a_{m2} & \cdots & a_{mn} \end{pmatrix}, \qquad \mathbf{X} = \begin{pmatrix} x_1 \\ x_2 \\ \vdots \\ x_n \end{pmatrix}, \qquad \mathbf{B} = \begin{pmatrix} b_1 \\ b_2 \\ \vdots \\ b_m \end{pmatrix}.$$

■ **Special Case** Let us suppose now that $m = n$ in (6) so that the coefficient matrix **A** is $n \times n$. In particular, if **A** is nonsingular, then the system $\mathbf{AX} = \mathbf{B}$ can be solved by multiplying both of the equations by \mathbf{A}^{-1}. From $\mathbf{A}^{-1}(\mathbf{AX}) = \mathbf{A}^{-1}\mathbf{B}$, we get $(\mathbf{A}^{-1}\mathbf{A})\mathbf{X} = \mathbf{A}^{-1}\mathbf{B}$. Since $\mathbf{A}^{-1}\mathbf{A} = \mathbf{I}$ and $\mathbf{IX} = \mathbf{X}$, we have

$$\mathbf{X} = \mathbf{A}^{-1}\mathbf{B}. \tag{7}$$

Example 6 Using (7) to Solve a System
Use the inverse of the coefficient matrix to solve the system

$$2x_1 - 9x_2 = 15$$
$$3x_1 + 6x_2 = 16.$$

SOLUTION The given system can be written as

$$\begin{pmatrix} 2 & -9 \\ 3 & 6 \end{pmatrix} \begin{pmatrix} x_1 \\ x_2 \end{pmatrix} = \begin{pmatrix} 15 \\ 16 \end{pmatrix}.$$

Since $\begin{vmatrix} 3 & -9 \\ 3 & 6 \end{vmatrix} = 39 \neq 0$, the coefficient matrix is nonsingular. Consequently, from (4) we get

$$\begin{pmatrix} 2 & -9 \\ 3 & 6 \end{pmatrix}^{-1} = \frac{1}{39} \begin{pmatrix} 6 & 9 \\ -3 & 2 \end{pmatrix}.$$

Using (7) it follows that

$$\begin{pmatrix} x_1 \\ x_2 \end{pmatrix} = \frac{1}{39} \begin{pmatrix} 6 & 9 \\ -3 & 2 \end{pmatrix} \begin{pmatrix} 15 \\ 16 \end{pmatrix} = \frac{1}{39} \begin{pmatrix} 234 \\ -13 \end{pmatrix} = \begin{pmatrix} 6 \\ -\dfrac{1}{3} \end{pmatrix},$$

and so $x_1 = 6$ and $x_2 = -\frac{1}{3}$. ❏

Example 7 **Using (7) to Solve a System**

Use the inverse of the coefficient matrix to solve the system

$$2x_1 + x_3 = 2$$
$$5x_1 + 5x_2 + 6x_3 = -1$$
$$-2x_1 + 3x_2 + 4x_3 = 4.$$

SOLUTION We found the inverse of the coefficient matrix

$$\mathbf{A} = \begin{pmatrix} 2 & 0 & 1 \\ -2 & 3 & 4 \\ -5 & 5 & 6 \end{pmatrix}$$

in Example 4. Thus, (7) gives

$$\begin{pmatrix} x_1 \\ x_2 \\ x_3 \end{pmatrix} = \begin{pmatrix} 2 & 0 & 1 \\ -2 & 3 & 4 \\ -5 & 5 & 6 \end{pmatrix}^{-1} \begin{pmatrix} 2 \\ 4 \\ -1 \end{pmatrix} = \begin{pmatrix} -2 & 5 & -3 \\ -8 & 17 & -10 \\ 5 & -10 & 6 \end{pmatrix} \begin{pmatrix} 2 \\ 4 \\ -1 \end{pmatrix} = \begin{pmatrix} 19 \\ 62 \\ -36 \end{pmatrix}.$$

Consequently, $x_1 = 19$, $x_2 = 62$, and $x_3 = -36$. ❏

■ **Uniqueness** When det $\mathbf{A} \neq 0$ the solution of the system $\mathbf{AX} = \mathbf{B}$ is unique. Suppose not—that is, suppose det $\mathbf{A} \neq 0$ and \mathbf{X}_1 and \mathbf{X}_2 are two *different* solution vectors. Then $\mathbf{AX}_1 = \mathbf{B}$ and $\mathbf{AX}_2 = \mathbf{B}$ imply $\mathbf{AX}_1 = \mathbf{AX}_2$. Since \mathbf{A} is nonsingular, \mathbf{A}^{-1} exists, and so $\mathbf{A}^{-1}(\mathbf{AX}_1) = \mathbf{A}^{-1}(\mathbf{AX}_2)$ and $(\mathbf{A}^{-1}\mathbf{A})\mathbf{X}_1 = (\mathbf{A}^{-1}\mathbf{A})\mathbf{X}_2$. This gives $\mathbf{IX}_1 = \mathbf{IX}_2$ or $\mathbf{X}_1 = \mathbf{X}_2$, which contradicts our assumption that \mathbf{X}_1 and \mathbf{X}_2 were different solution vectors.

■ **Homogeneous Systems** A homogeneous system of equations can be written $\mathbf{AX} = \mathbf{0}$. Recall that a homogeneous system always possesses the trivial solution $\mathbf{X} = \mathbf{0}$ and possibly an infinite number of solutions. In the next theorem we shall see that a homogeneous system of n equations in n unknowns possesses *only* the trivial solution when \mathbf{A} is nonsingular.

THEOREM 8.21 Trivial Solution Only

A homogeneous system of n linear equations in n unknowns $\mathbf{AX} = \mathbf{0}$ has only the trivial solution if and only if \mathbf{A} is nonsingular.

PROOF We prove the sufficiency part of the theorem. Suppose \mathbf{A} is nonsingular. Then by (7) we have the unique solution $\mathbf{X} = \mathbf{A}^{-1}\mathbf{0} = \mathbf{0}$. ❏

The next theorem will answer the question: When does a homogeneous system of n linear equations in n unknowns possess a nontrivial solution? Bear in mind that if a homogeneous system has one nontrivial solution, it must have an infinite number of solutions.

THEOREM 8.22 Existence of Nontrivial Solutions

A homogeneous system of n linear equations in n unknowns $\mathbf{AX} = \mathbf{0}$ has a nontrivial solution if and only if \mathbf{A} is singular.

In view of Theorem 8.22 we can conclude that a homogeneous system of n linear equations in n unknowns $\mathbf{AX} = \mathbf{0}$ possesses

- only the trivial solution if and only if det $\mathbf{A} \neq 0$, and
- a nontrivial solution if and only if det $\mathbf{A} = 0$.

The last result will be put to use in Section 8.8.

Remarks

(*i*) As a practical means of solving n linear equations in n unknowns, the use of an inverse matrix offers few advantages over the method of Section 8.2. However, in applications we sometimes need to solve a system $\mathbf{AX} = \mathbf{B}$ several times; that is, we need to examine the solutions of the system corresponding to the same coefficient matrix \mathbf{A} but different input vectors \mathbf{B}. In this case, the single calculation of \mathbf{A}^{-1} enables us to obtain these solutions quickly through the matrix multiplication $\mathbf{A}^{-1}\mathbf{B}$.

(*ii*) In Definition 8.11 we saw that if \mathbf{A} is an $n \times n$ matrix and there exists another $n \times n$ matrix \mathbf{B} that commutes with \mathbf{A} such that

$$\mathbf{AB} = \mathbf{I} \quad \text{and} \quad \mathbf{BA} = \mathbf{I}, \tag{8}$$

then \mathbf{B} is the inverse of \mathbf{A}. Although matrix multiplication is in general not commutative, the condition in (8) can be relaxed somewhat in this sense: If we find an $n \times n$ matrix \mathbf{B} for which $\mathbf{AB} = \mathbf{I}$, then it can be proved that $\mathbf{BA} = \mathbf{I}$ as well, and so \mathbf{B} is the inverse of \mathbf{A}. As a consequence of this result, in the subsequent sections of this chapter if we wish to prove that a certain matrix \mathbf{B} is the inverse of a given matrix \mathbf{A}, it will suffice to show only that $\mathbf{AB} = \mathbf{I}$. We need not demonstrate that \mathbf{B} commutes with \mathbf{A} to give \mathbf{I}.

EXERCISES 8.6

Answers to odd-numbered problems begin on page A-43.

8.6.1 Finding the Inverse

In Problems 1 and 2, verify that the matrix \mathbf{B} is the inverse of the matrix \mathbf{A}.

1. $\mathbf{A} = \begin{pmatrix} 1 & \frac{1}{2} \\ 2 & \frac{3}{2} \end{pmatrix}$, $\quad \mathbf{B} = \begin{pmatrix} 3 & -1 \\ -4 & 2 \end{pmatrix}$

2. $\mathbf{A} = \begin{pmatrix} 1 & -1 & 0 \\ 3 & 0 & 2 \\ 1 & 1 & 1 \end{pmatrix}$, $\quad \mathbf{B} = \begin{pmatrix} 2 & -1 & 2 \\ 1 & -1 & 2 \\ -3 & 2 & -3 \end{pmatrix}$

In Problems 3–14, use Theorem 8.19 to determine whether the given matrix is singular or nonsingular. If it is nonsingular, use Theorem 8.18 to find the inverse.

3. $\begin{pmatrix} 5 & -1 \\ 4 & 1 \end{pmatrix}$

4. $\begin{pmatrix} \frac{1}{3} & -1 \\ 4 & 3 \end{pmatrix}$

5. $\begin{pmatrix} 6 & 0 \\ -3 & 2 \end{pmatrix}$

6. $\begin{pmatrix} -2\pi & -\pi \\ -\pi & \pi \end{pmatrix}$

7. $\begin{pmatrix} 1 & 3 & 5 \\ 2 & 4 & 4 \\ 1 & -1 & 1 \end{pmatrix}$

8. $\begin{pmatrix} 2 & 3 & 0 \\ 0 & 11 & 14 \\ -1 & 4 & 7 \end{pmatrix}$

9. $\begin{pmatrix} 1 & 2 & 3 \\ 0 & -4 & 2 \\ -1 & 5 & 1 \end{pmatrix}$

10. $\begin{pmatrix} 2 & -1 & 5 \\ 3 & 0 & -2 \\ 1 & 4 & 0 \end{pmatrix}$

11. $\begin{pmatrix} 3 & 0 & 0 \\ 0 & 6 & 0 \\ 0 & 0 & -2 \end{pmatrix}$

12. $\begin{pmatrix} 0 & 2 & 0 \\ 0 & 0 & 1 \\ 8 & 0 & 0 \end{pmatrix}$

13. $\begin{pmatrix} 0 & -1 & 1 & 4 \\ 3 & 2 & -2 & 1 \\ 0 & 4 & 0 & 1 \\ 1 & 0 & -1 & 1 \end{pmatrix}$

14. $\begin{pmatrix} 1 & 2 & 1 & 1 \\ 0 & 0 & 3 & 0 \\ 3 & 1 & 2 & 0 \\ 1 & 1 & 1 & 0 \end{pmatrix}$

In Problems 15–26, use Theorem 8.20 to find the inverse of the given matrix or show that no inverse exists.

15. $\begin{pmatrix} 6 & -2 \\ 0 & 4 \end{pmatrix}$

16. $\begin{pmatrix} 8 & 0 \\ 0 & \frac{1}{2} \end{pmatrix}$

17. $\begin{pmatrix} 1 & 3 \\ 5 & 3 \end{pmatrix}$

18. $\begin{pmatrix} 2 & -3 \\ -2 & 4 \end{pmatrix}$

19. $\begin{pmatrix} 1 & 2 & 3 \\ 4 & 5 & 6 \\ 7 & 8 & 9 \end{pmatrix}$

20. $\begin{pmatrix} 1 & 0 & -1 \\ 0 & -2 & 1 \\ 2 & -1 & 3 \end{pmatrix}$

21. $\begin{pmatrix} 4 & 2 & 3 \\ 2 & 1 & 0 \\ -1 & -2 & 0 \end{pmatrix}$

22. $\begin{pmatrix} 2 & 4 & -2 \\ 4 & 2 & -2 \\ 8 & 10 & -6 \end{pmatrix}$

23. $\begin{pmatrix} -1 & 3 & 0 \\ 1 & -2 & 1 \\ 0 & 1 & 2 \end{pmatrix}$

24. $\begin{pmatrix} 1 & 2 & 3 \\ 0 & 1 & 4 \\ 0 & 0 & 8 \end{pmatrix}$

25. $\begin{pmatrix} 1 & 2 & 3 & 1 \\ -1 & 0 & 2 & 1 \\ 2 & 1 & -3 & 0 \\ 1 & 1 & 2 & 1 \end{pmatrix}$

26. $\begin{pmatrix} 1 & 0 & 0 & 0 \\ 0 & 0 & 1 & 0 \\ 0 & 0 & 0 & 1 \\ 0 & 1 & 0 & 0 \end{pmatrix}$

In Problems 27 and 28, use the given matrices to find $(\mathbf{AB})^{-1}$.

27. $\mathbf{A}^{-1} = \begin{pmatrix} \frac{1}{2} & -\frac{5}{2} \\ -\frac{1}{2} & \frac{3}{2} \end{pmatrix}$, $\mathbf{B}^{-1} = \begin{pmatrix} \frac{2}{3} & \frac{4}{3} \\ -\frac{1}{3} & \frac{5}{3} \end{pmatrix}$

28. $\mathbf{A}^{-1} = \begin{pmatrix} 1 & 3 & -15 \\ 0 & -1 & 5 \\ -1 & -2 & 11 \end{pmatrix}$, $\mathbf{B}^{-1} = \begin{pmatrix} -1 & 1 & 0 \\ 2 & 0 & 0 \\ 1 & 1 & -2 \end{pmatrix}$

29. If $\mathbf{A}^{-1} = \begin{pmatrix} 4 & 3 \\ 3 & 2 \end{pmatrix}$, what is \mathbf{A}?

30. If \mathbf{A} is nonsingular, then $(\mathbf{A}^T)^{-1} = (\mathbf{A}^{-1})^T$. Verify this for
$\mathbf{A} = \begin{pmatrix} 1 & 4 \\ 2 & 10 \end{pmatrix}$.

31. Find a value of x such that the matrix $\mathbf{A} = \begin{pmatrix} 4 & -3 \\ x & -4 \end{pmatrix}$ is its own inverse.

32. Find the inverse of $\mathbf{A} = \begin{pmatrix} \sin\theta & \cos\theta \\ -\cos\theta & \sin\theta \end{pmatrix}$.

33. A nonsingular matrix \mathbf{A} is said to be **orthogonal** if $\mathbf{A}^{-1} = \mathbf{A}^T$.

(a) Verify that the matrix in Problem 32 is orthogonal.

(b) Verify that $\mathbf{A} = \begin{pmatrix} 1/\sqrt{3} & 0 & -2/\sqrt{6} \\ 1/\sqrt{3} & 1/\sqrt{2} & 1/\sqrt{6} \\ 1/\sqrt{3} & -1/\sqrt{2} & 1/\sqrt{6} \end{pmatrix}$ is an orthogonal matrix.

34. Show that if \mathbf{A} is an orthogonal matrix (see Problem 33), then $\det \mathbf{A} = \pm 1$.

35. If \mathbf{A} and \mathbf{B} are nonsingular $n \times n$ matrices, use Theorem 8.19 to show that \mathbf{AB} is nonsingular.

36. Suppose \mathbf{A} and \mathbf{B} are $n \times n$ matrices. Show that if either \mathbf{A} or \mathbf{B} is singular, then \mathbf{AB} is singular.

37. Show that if \mathbf{A} is a nonsingular matrix, then $\det \mathbf{A}^{-1} = 1/\det \mathbf{A}$.

38. Show that if $\mathbf{A}^2 = \mathbf{A}$, then either $\mathbf{A} = \mathbf{I}$ or \mathbf{A} is singular.

39. Suppose \mathbf{A} and \mathbf{B} are $n \times n$ matrices and \mathbf{A} is nonsingular. Show that if $\mathbf{AB} = \mathbf{0}$, then $\mathbf{B} = \mathbf{0}$.

40. Suppose \mathbf{A} and \mathbf{B} are $n \times n$ matrices and \mathbf{A} is nonsingular. Show that if $\mathbf{AB} = \mathbf{AC}$, then $\mathbf{B} = \mathbf{C}$.

41. If \mathbf{A} and \mathbf{B} are nonsingular $n \times n$ matrices, is $\mathbf{A} + \mathbf{B}$ necessarily nonsingular?

42. Consider the 3×3 diagonal matrix $\mathbf{A} = \begin{pmatrix} a_{11} & 0 & 0 \\ 0 & a_{22} & 0 \\ 0 & 0 & a_{33} \end{pmatrix}$. Determine conditions such that \mathbf{A} is nonsingular. If \mathbf{A} is nonsingular, find \mathbf{A}^{-1}. Generalize your results to an $n \times n$ diagonal matrix.

8.6.2 Using the Inverse to Solve Systems

In Problems 43–50, use an inverse matrix to solve the given system of equations.

43. $x_1 + x_2 = 4$
$2x_1 - x_2 = 14$

44. $x_1 - x_2 = 2$
$2x_1 + 4x_2 = -5$

45. $4x_1 - 6x_2 = 6$
$2x_1 + x_2 = 1$

46. $x_1 + 2x_2 = 4$
$3x_1 + 4x_2 = -3$

47. $x_1 + \quad x_3 = -4$
$x_1 + x_2 + x_3 = 0$
$5x_1 - x_2 \quad = 6$

48. $x_1 - x_2 + x_3 = 1$
$2x_1 + x_2 + 2x_3 = 2$
$3x_1 + 2x_2 - x_3 = -3$

49. $x_1 + 2x_2 + 2x_3 = 1$
$x_1 - 2x_2 + 2x_3 = -3$
$3x_1 - x_2 + 5x_3 = 7$

50. $x_1 - \quad x_3 \quad = 2$
$x_2 + x_3 \quad = 1$
$-x_1 + x_2 + 2x_3 + x_4 = -5$
$x_3 - x_4 = 3$

In Problems 51 and 52, write the system in the form $\mathbf{AX} = \mathbf{B}$. Use $\mathbf{X} = \mathbf{A}^{-1}\mathbf{B}$ to solve the system for each matrix \mathbf{B}.

51. $7x_1 - 2x_2 = b_1$,
$3x_1 - 2x_2 = b_2$

$\mathbf{B} = \begin{pmatrix} 5 \\ 4 \end{pmatrix}$, $\mathbf{B} = \begin{pmatrix} 10 \\ 50 \end{pmatrix}$, $\mathbf{B} = \begin{pmatrix} 0 \\ -20 \end{pmatrix}$

52. $x_1 + 2x_2 + 5x_3 = b_1$
$2x_1 + 3x_2 + 8x_3 = b_2$,
$-x_1 + x_2 + 2x_3 = b_3$

$\mathbf{B} = \begin{pmatrix} -1 \\ 4 \\ 6 \end{pmatrix}$, $\mathbf{B} = \begin{pmatrix} 3 \\ 3 \\ 3 \end{pmatrix}$, $\mathbf{B} = \begin{pmatrix} 0 \\ -5 \\ 4 \end{pmatrix}$

In Problems 53–56, without solving, determine whether the given homogeneous system of equations has only the trivial solution or a nontrivial solution.

53. $x_1 + 2x_2 - x_3 = 0$
$4x_1 - x_2 + x_3 = 0$
$5x_1 + x_2 - 2x_3 = 0$

54. $x_1 + x_2 + x_3 = 0$
$x_1 - 2x_2 + x_3 = 0$
$-2x_1 + x_2 - 2x_3 = 0$

55. $x_1 + x_2 - x_3 + x_4 = 0$
$5x_2 + 2x_4 = 0$
$x_1 + x_3 - x_4 = 0$
$3x_1 + 2x_2 - x_3 + x_4 = 0$

56. $x_1 + x_2 - x_3 + x_4 = 0$
$x_1 + x_2 + x_3 - x_4 = 0$
$2x_2 + x_3 + x_4 = 0$
$x_2 - x_3 - x_4 = 0$

57. The system of equations for the currents i_1, i_2, and i_3 in the circuit shown in Figure 8.6 is

$$i_1 + i_2 + i_3 = 0$$
$$-R_1 i_1 + R_2 i_2 = E_2 - E_1$$
$$-R_2 i_2 + R_3 i_3 = E_3 - E_2$$

where R_k and E_k, $k = 1, 2, 3$, are constants.

(a) Express the system as a matrix equation $\mathbf{AX} = \mathbf{B}$.

(b) Show that the coefficient matrix \mathbf{A} is nonsingular.

(c) Use $\mathbf{X} = \mathbf{A}^{-1}\mathbf{B}$ to solve for the currents.

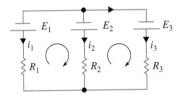

Figure 8.6

58. Consider the square plate shown in Figure 8.7, with the temperatures as indicated on each side. Under some circumstances it can be shown that the approximate temperatures u_1, u_2, u_3, and u_4 at the points P_1, P_2, P_3, and P_4, respectively, are given by

$$u_1 = \frac{u_2 + u_4 + 100 + 100}{4}$$

$$u_2 = \frac{200 + u_3 + u_1 + 100}{4}$$

$$u_3 = \frac{200 + 100 + u_4 + u_2}{4}$$

$$u_4 = \frac{u_3 + 100 + 100 + u_1}{4}$$

(a) Show that the above system can be written as the matrix equation

$$\begin{pmatrix} -4 & 1 & 0 & 1 \\ 1 & -4 & 1 & 0 \\ 0 & 1 & -4 & 1 \\ 1 & 0 & 1 & -4 \end{pmatrix} \begin{pmatrix} u_1 \\ u_2 \\ u_3 \\ u_4 \end{pmatrix} = \begin{pmatrix} -200 \\ -300 \\ -300 \\ -200 \end{pmatrix}$$

(b) Solve the system in part **(a)** by finding the inverse of the coefficient matrix.

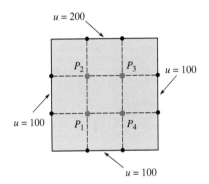

Figure 8.7

8.7 Cramer's Rule

■ **Using Determinants to Solve Systems** We saw at the end of the preceding section that a system of n linear equations in n unknowns $\mathbf{AX} = \mathbf{B}$ has precisely one solution if det $\mathbf{A} \neq 0$. This solution, as we shall now see, can be expressed in terms of determinants. For example, the system of two equations in two unknowns

$$a_{11}x_1 + a_{12}x_2 = b_1$$
$$a_{21}x_1 + a_{22}x_2 = b_2 \tag{1}$$

possesses the solution

$$x_1 = \frac{b_1 a_{22} - a_{12} b_2}{a_{11} a_{22} - a_{12} a_{21}} \quad \text{and} \quad x_2 = \frac{a_{11} b_2 - b_1 a_{21}}{a_{11} a_{22} - a_{12} a_{21}} \tag{2}$$

provided that $a_{11}a_{22} - a_{12}a_{21} \neq 0$. The numerators and denominators in (2) are recognized as determinants. That is, the system (1) has the unique solution

$$x_1 = \frac{\begin{vmatrix} b_1 & a_{12} \\ b_2 & a_{22} \end{vmatrix}}{\begin{vmatrix} a_{11} & a_{12} \\ a_{21} & a_{22} \end{vmatrix}}, \qquad x_2 = \frac{\begin{vmatrix} a_{11} & b_1 \\ a_{21} & b_2 \end{vmatrix}}{\begin{vmatrix} a_{11} & a_{12} \\ a_{21} & a_{22} \end{vmatrix}} \tag{3}$$

provided that the determinant of the matrix of coefficients $\begin{vmatrix} a_{11} & a_{12} \\ a_{21} & a_{22} \end{vmatrix} \neq 0$.

For a system of n linear equations in n unknowns

$$\begin{aligned} a_{11}x_1 + a_{12}x_2 + \cdots + a_{1n}x_n &= b_1 \\ a_{21}x_1 + a_{22}x_2 + \cdots + a_{2n}x_n &= b_2 \\ &\vdots \\ a_{n1}x_1 + a_{n2}x_2 + \cdots + a_{nn}x_n &= b_n, \end{aligned} \tag{4}$$

it is convenient to define a special matrix

$$\mathbf{A}_k = \begin{pmatrix} a_{11} & a_{12} & \cdots & a_{1\,k-1} & b_1 & a_{1\,k+1} & \cdots & a_{1n} \\ a_{21} & a_{22} & \cdots & a_{2\,k-1} & b_2 & a_{2\,k+1} & \cdots & a_{2n} \\ \vdots & & & \vdots & & \vdots & & \vdots \\ a_{n1} & a_{n2} & \cdots & a_{n\,k-1} & b_n & a_{n\,k+1} & \cdots & a_{nn} \end{pmatrix}. \tag{5}$$

kth column \downarrow

In other words, \mathbf{A}_k is the same as the matrix \mathbf{A} except that the kth column of \mathbf{A} has been replaced by the entries of the column matrix

$$\mathbf{B} = \begin{pmatrix} b_1 \\ b_2 \\ \vdots \\ b_n \end{pmatrix}.$$

The generalization of (3), known as **Cramer's Rule,** is given in the next theorem.

THEOREM 8.23 Cramer's Rule

Let \mathbf{A} be the coefficient matrix of the system (1). If $\det \mathbf{A} \neq 0$, then the solution of (1) is given by

$$x_1 = \frac{\det \mathbf{A}_1}{\det \mathbf{A}}, \quad x_2 = \frac{\det \mathbf{A}_2}{\det \mathbf{A}}, \quad \ldots, \quad x_n = \frac{\det \mathbf{A}_n}{\det \mathbf{A}}, \tag{6}$$

where \mathbf{A}_k, $k = 1, 2, \ldots, n$, is defined in (5).

PROOF We first write the system (1) as $\mathbf{AX} = \mathbf{B}$. Since $\det \mathbf{A} \neq 0$, \mathbf{A}^{-1} exists, and so

$$\mathbf{X} = \mathbf{A}^{-1}\mathbf{B} = \frac{1}{\det \mathbf{A}} \begin{pmatrix} C_{11} & C_{21} & \cdots & C_{n1} \\ C_{12} & C_{22} & \cdots & C_{n2} \\ \vdots & & & \vdots \\ C_{1n} & C_{2n} & \cdots & C_{nn} \end{pmatrix} \begin{pmatrix} b_1 \\ b_2 \\ \vdots \\ b_n \end{pmatrix}$$

$$= \frac{1}{\det \mathbf{A}} \begin{pmatrix} b_1 C_{11} + b_2 C_{21} + \cdots + b_n C_{n1} \\ b_1 C_{12} + b_2 C_{22} + \cdots + b_n C_{n2} \\ \vdots \\ b_1 C_{1n} + b_2 C_{2n} + \cdots + b_n C_{nn} \end{pmatrix}.$$

Now the entry in the kth row of the last matrix is

$$x_k = \frac{b_1 C_{1k} + b_2 C_{2k} + \cdots + b_n C_{nk}}{\det \mathbf{A}}. \tag{4}$$

But $b_1 C_{1k} + b_2 C_{2k} + \cdots + b_n C_{nk}$ is the cofactor expansion of $\det \mathbf{A}_k$, where \mathbf{A}_k is the matrix given in (5), along the kth column. Hence, we have $x_k = \det \mathbf{A}_k / \det \mathbf{A}$ for $k = 1, 2, \ldots, n$. ❏

Example 1 Using Cramer's Rule to Solve a System

Use Cramer's rule to solve the system

$$3x_1 + 2x_2 + \ x_3 = 7$$
$$x_1 - \ x_2 + 3x_3 = 3$$
$$5x_1 + 4x_2 - 2x_3 = 1.$$

SOLUTION The solution requires the evaluation of four determinants:

$$\det \mathbf{A} = \begin{vmatrix} 3 & 2 & 1 \\ 1 & -1 & 3 \\ 5 & 4 & -2 \end{vmatrix} = 13, \qquad \det \mathbf{A}_1 = \begin{vmatrix} 7 & 2 & 1 \\ 3 & -1 & 3 \\ 1 & 4 & -2 \end{vmatrix} = -39,$$

$$\det \mathbf{A}_2 = \begin{vmatrix} 3 & 7 & 1 \\ 1 & 3 & 3 \\ 5 & 1 & -2 \end{vmatrix} = 78, \qquad \det \mathbf{A}_3 = \begin{vmatrix} 3 & 2 & 7 \\ 1 & -1 & 3 \\ 5 & 4 & 1 \end{vmatrix} = 52.$$

Thus, (3) gives

$$x_1 = \frac{\det \mathbf{A}_1}{\det \mathbf{A}} = -3, \qquad x_2 = \frac{\det \mathbf{A}_2}{\det \mathbf{A}} = 6, \qquad x_3 = \frac{\det \mathbf{A}_3}{\det \mathbf{A}} = 4. \qquad ❏$$

Remarks

Like the method of the preceding section, Cramer's rule is not a very practical means for solving systems of n linear equations in n unknowns. For $n \geq 4$, the work required to evaluate the determinants becomes formidable. Nonetheless, as we saw in Section 3.5, Cramer's rule is used *sometimes,* and it is of theoretical importance.

In the application of Cramer's rule, some shortcuts can be taken. In Example 1, for instance, we really did not have to compute $\det \mathbf{A}_3$ since once we have found the values of x_1 and x_2 the value of x_3 can be found by using one of the equations in the system.

Answers to odd-numbered problems begin on page A-44.

In Problems 1–10, solve the given system of equations by Cramer's rule.

1. $-3x_1 + x_2 = 3$
$2x_1 - 4x_2 = -6$

2. $x_1 + x_2 = 4$
$2x_1 - x_2 = 2$

3. $0.1x_1 - 0.4x_2 = 0.13$
$x_1 - x_2 = 0.4$

4. $0.21x_1 + 0.57x_2 = 0.369$
$0.1x_1 + 0.2 \, x_2 = 0.135$

5. $2x + y = 1$
$3x + 2y = -2$

6. $5r + 4s = -1$
$10r - 6s = 5$

7. $x_1 - 2x_2 - 3x_3 = 3$
$x_1 + x_2 - x_3 = 5$
$3x_1 + 2x_2 = -4$

8. $x_1 - x_2 + 6x_3 = -2$
$-x_1 + 2x_2 + 4x_3 = 9$
$2x_1 + 3x_2 - x_3 = \frac{1}{2}$

9. $u + 2v + w = 8$
$2u - 2v + 2w = 7$
$u - 4v + 3w = 1$

10. $4x + 3y + 2z = 8$
$-x + 2z = 12$
$3x + 2y + z = 3$

11. Use Cramer's rule to determine the solution of the system

$$(2 - k)x_1 + kx_2 = 4$$
$$kx_1 + (3 - k)x_2 = 3.$$

For what value(s) of k is the system inconsistent?

12. Consider the system

$$x_1 + x_2 = 1$$
$$x_1 + \varepsilon x_2 = 2.$$

When ε is close to 1, the lines that make up the system are almost parallel.

(a) Use Cramer's rule to show that a solution of the system is $x_1 = 1 - \dfrac{1}{\varepsilon - 1}, x_2 = \dfrac{1}{\varepsilon - 1}$.

(b) The system is said to be **ill-conditioned** since small changes in the input data (for example, the coefficients) causes a significant or large change in the output or solution. Verify this by finding the solution of the system for $\varepsilon = 1.01$ and then for $\varepsilon = 0.99$.

13. The magnitudes T_1 and T_2 of the tensions in the wires shown in Figure 8.8 satisfy the equations

$$(\cos 25°)T_1 - (\cos 15°)T_2 = 0$$
$$(\sin 25°)T_1 + (\sin 15°)T_2 = 300.$$

Use Cramer's rule to solve for T_1 and T_2.

Figure 8.8

14. The 400-lb block shown in Figure 8.9 is kept from sliding down the inclined plane by friction and a force of smallest magnitude F. If the coefficient of friction between the block and the inclined plane is 0.5, then the magnitude of the frictional force is $0.5N$, where N is the magnitude of the normal force exerted on the block by the plane. Use the fact that the system is in equilibrium to set up a system of equations for F and N. Use Cramer's rule to solve for F and N.

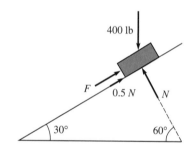

Figure 8.9

15. As shown in Figure 8.10, two batteries with internal resistances r_1 and r_2 are connected in parallel with a resistor. Use Cramer's rule to show that the current i through the resistor is given by

$$i = \frac{r_1E_2 + r_2E_1}{r_1R + r_2R + r_1r_2}.$$

Figure 8.10

8.8 The Eigenvalue Problem

■ **Eigenvalues and Eigenvectors** Gauss–Jordan elimination introduced in Section 8.2 can be used to find the **eigenvectors** of a square matrix **A**.

DEFINITION 8.13 Eigenvalues and Eigenvectors

Let **A** be an $n \times n$ matrix. A number λ is said to be an **eigenvalue** of **A** if there exists a nonzero solution vector **K** of the linear system

$$\mathbf{AK} = \lambda\mathbf{K}. \tag{1}$$

The solution vector **K** is said to be an **eigenvector** corresponding to the eigenvalue λ.

The word "eigenvalue" is a combination of German and English terms adapted from the German word *eigenwert,* which, translated literally, is "proper value." Eigenvalues and eigenvectors are also called **characteristic values** and **characteristic vectors,** respectively.

Example 1 **Verification of an Eigenvector**

Verify that $\mathbf{K} = \begin{pmatrix} 1 \\ -1 \\ 1 \end{pmatrix}$ is an eigenvector of the matrix

$$\mathbf{A} = \begin{pmatrix} 0 & -1 & -3 \\ 2 & 3 & 3 \\ -2 & 1 & 1 \end{pmatrix}.$$

SOLUTION By carrying out the multiplication **AK** we see

$$\mathbf{AK} = \begin{pmatrix} 0 & -1 & -3 \\ 2 & 3 & 3 \\ -2 & 1 & 1 \end{pmatrix}\begin{pmatrix} 1 \\ -1 \\ 1 \end{pmatrix} = \begin{pmatrix} -2 \\ 2 \\ -2 \end{pmatrix} = (-2)\begin{pmatrix} 1 \\ -1 \\ 1 \end{pmatrix} = \overset{\text{eigenvalue}}{(-2)}\mathbf{K}.$$

We see from the preceding line and Definition 8.13 that $\lambda = -2$ is an eigenvalue of **A**. ❑

Using properties of matrix algebra, we can write (1) in the alternative form

$$(\mathbf{A} - \lambda\mathbf{I})\mathbf{K} = \mathbf{0}, \tag{2}$$

where **I** is the multiplicative identity. If we let

$$\mathbf{K} = \begin{pmatrix} k_1 \\ k_2 \\ \vdots \\ k_n \end{pmatrix},$$

then (2) is the same as

$$\begin{aligned}
(a_{11} - \lambda)k_1 + \quad a_{12}k_2 + \cdots + \quad a_{1n}k_n &= 0 \\
a_{21}k_1 + (a_{22} - \lambda)k_2 + \cdots + \quad a_{2n}k_n &= 0 \\
\vdots \qquad\qquad\qquad\qquad \vdots & \\
a_{n1}k_1 + \quad a_{n2}k_2 + \cdots + (a_{nn} - \lambda)k_n &= 0.
\end{aligned} \tag{3}$$

Although an obvious solution of (3) is $k_1 = 0$, $k_2 = 0$, . . . , $k_n = 0$, we are seeking only nontrivial solutions. Now we know that a homogeneous system of n linear equations in n unknowns has a nontrivial solution if and only if the determinant of the coefficient matrix is equal to zero. Thus to find a nonzero solution \mathbf{K} for (2), we must have

$$\det(\mathbf{A} - \lambda \mathbf{I}) = 0. \tag{4}$$

Inspection of (4) shows that expansion of $\det(\mathbf{A} - \lambda \mathbf{I})$ by cofactors results in an nth-degree polynomial in λ. The equation (4) is called the **characteristic equation** of \mathbf{A}. Thus, *the eigenvalues of* \mathbf{A} *are the roots of the characteristic equation.* To find an eigenvector corresponding to an eigenvalue λ we simply solve the system of equations $(\mathbf{A} - \lambda \mathbf{I})\mathbf{K} = \mathbf{0}$ by applying Gauss–Jordan elimination to the augmented matrix $(\mathbf{A} - \lambda \mathbf{I} | \mathbf{0})$.

Example 2 Finding Eigenvalues and Eigenvectors

Find the eigenvalues and eigenvectors of

$$\mathbf{A} = \begin{pmatrix} 1 & 2 & 1 \\ 6 & -1 & 0 \\ -1 & -2 & -1 \end{pmatrix}.$$

SOLUTION To expand the determinant in the characteristic equation

$$\det(\mathbf{A} - \lambda \mathbf{I}) = \begin{vmatrix} 1 - \lambda & 2 & 1 \\ 6 & -1 - \lambda & 0 \\ -1 & -2 & -1 - \lambda \end{vmatrix} = 0,$$

we use the cofactors of the second row. It follows that the characteristic equation is

$$-\lambda^3 - \lambda^2 + 12\lambda = 0 \quad \text{or} \quad \lambda(\lambda + 4)(\lambda - 3) = 0.$$

Hence the eigenvalues are $\lambda_1 = 0$, $\lambda_2 = -4$, $\lambda_3 = 3$. To find the eigenvectors we must now reduce $(\mathbf{A} - \lambda \mathbf{I} | \mathbf{0})$ three times corresponding to the three distinct eigenvalues.

For $\lambda_1 = 0$, we have

$$(\mathbf{A} - 0\mathbf{I} | \mathbf{0}) = \begin{pmatrix} 1 & 2 & 1 & | & 0 \\ 6 & -1 & 0 & | & 0 \\ -1 & -2 & -1 & | & 0 \end{pmatrix} \xrightarrow[R_1 + R_3]{-6R_1 + R_2} \begin{pmatrix} 1 & 2 & 1 & | & 0 \\ 0 & -13 & -6 & | & 0 \\ 0 & 0 & 0 & | & 0 \end{pmatrix}$$

$$\xrightarrow{-\frac{1}{13}R_2} \begin{pmatrix} 1 & 2 & 1 & | & 0 \\ 0 & 1 & \frac{6}{13} & | & 0 \\ 0 & 0 & 0 & | & 0 \end{pmatrix} \xrightarrow{-2R_2 + R_1} \begin{pmatrix} 1 & 0 & \frac{1}{13} & | & 0 \\ 0 & 1 & \frac{6}{13} & | & 0 \\ 0 & 0 & 0 & | & 0 \end{pmatrix}.$$

Thus we see that $k_1 = -\frac{1}{13}k_3$ and $k_2 = -\frac{6}{13}k_3$. Choosing $k_3 = -13$ gives the eigenvector*

$$\mathbf{K}_1 = \begin{pmatrix} 1 \\ 6 \\ -13 \end{pmatrix}.$$

For $\lambda_2 = -4$,

* Of course k_3 could be chosen as any nonzero number. In other words, a nonzero constant multiple of an eigenvector is also an eigenvector.

$$(\mathbf{A} + 4\mathbf{I}\,|\,\mathbf{0}) = \begin{pmatrix} 5 & 2 & 1 & | & 0 \\ 6 & 3 & 0 & | & 0 \\ -1 & -2 & 3 & | & 0 \end{pmatrix} \overset{\substack{-R_3 \\ R_{13}}}{\Longrightarrow} \begin{pmatrix} 1 & 2 & -3 & | & 0 \\ 6 & 3 & 0 & | & 0 \\ 5 & 2 & 1 & | & 0 \end{pmatrix}$$

$$\overset{\substack{-6R_1 + R_2 \\ -5R_1 + R_3}}{\Longrightarrow} \begin{pmatrix} 1 & 2 & -3 & | & 0 \\ 0 & -9 & 18 & | & 0 \\ 0 & -8 & 16 & | & 0 \end{pmatrix} \overset{\substack{-\frac{1}{9}R_2 \\ -\frac{1}{8}R_3}}{\Longrightarrow} \begin{pmatrix} 1 & 2 & -3 & | & 0 \\ 0 & 1 & -2 & | & 0 \\ 0 & 1 & -2 & | & 0 \end{pmatrix}$$

$$\overset{\substack{-2R_2 + R_1 \\ -2R_2 + R_3}}{\Longrightarrow} \begin{pmatrix} 1 & 0 & 1 & | & 0 \\ 0 & 1 & -2 & | & 0 \\ 0 & 0 & 0 & | & 0 \end{pmatrix}$$

implies $k_1 = -k_3$ and $k_2 = 2k_3$. Choosing $k_3 = 1$ then yields a second eigenvector

$$\mathbf{K}_2 = \begin{pmatrix} -1 \\ 2 \\ 1 \end{pmatrix}.$$

Finally, for $\lambda_3 = 3$, Gauss–Jordan elimination gives

$$(\mathbf{A} - 3\mathbf{I}\,|\,\mathbf{0}) = \begin{pmatrix} -2 & 2 & 1 & | & 0 \\ 6 & -4 & 0 & | & 0 \\ -1 & -2 & -4 & | & 0 \end{pmatrix} \overset{\substack{\text{row} \\ \text{operations}}}{\Longrightarrow} \begin{pmatrix} 1 & 0 & 1 & | & 0 \\ 0 & 1 & \frac{3}{2} & | & 0 \\ 0 & 0 & 0 & | & 0 \end{pmatrix},$$

and so $k_1 = -k_3$ and $k_2 = -\frac{3}{2}k_3$. The choice of $k_3 = -2$ leads to a third eigenvector,

$$\mathbf{K}_3 = \begin{pmatrix} 2 \\ 3 \\ -2 \end{pmatrix}. \qquad \square$$

When an $n \times n$ matrix \mathbf{A} possesses n distinct eigenvalues $\lambda_1, \lambda_2, \ldots, \lambda_n$, it can be proved that a set of n linearly independent eigenvectors $\mathbf{K}_1, \mathbf{K}_2, \ldots, \mathbf{K}_n$ can be found. However, when the characteristic equation has repeated roots, it may not be possible to find n linearly independent eigenvectors for \mathbf{A}.

Example 3 **Finding Eigenvalues and Eigenvectors**

Find the eigenvalues and eigenvectors of $\mathbf{A} = \begin{pmatrix} 3 & 4 \\ -1 & 7 \end{pmatrix}$.

SOLUTION From the characteristic equation

$$\det(\mathbf{A} - \lambda\mathbf{I}) = \begin{vmatrix} 3 - \lambda & 4 \\ -1 & 7 - \lambda \end{vmatrix} = (\lambda - 5)^2 = 0,$$

we see $\lambda_1 = \lambda_2 = 5$ is an eigenvalue of multiplicity 2. In the case of a 2×2 matrix there is no need to use Gauss–Jordan elimination. To find the eigenvector(s) corresponding to $\lambda_1 = 5$, we resort to the system $(\mathbf{A} - 5\mathbf{I}\,|\,\mathbf{0})$ in its equivalent form

$$-2k_1 + 4k_2 = 0$$
$$-k_1 + 2k_2 = 0.$$

It is apparent from this system that $k_1 = 2k_2$. Thus, if we choose $k_2 = 1$, we find the single eigenvector $\mathbf{K}_1 = \begin{pmatrix} 2 \\ 1 \end{pmatrix}$. $\qquad \square$

Example 4 **Finding Eigenvalues and Eigenvectors**

Find the eigenvalues and eigenvectors of $\mathbf{A} = \begin{pmatrix} 9 & 1 & 1 \\ 1 & 9 & 1 \\ 1 & 1 & 9 \end{pmatrix}$.

SOLUTION The characteristic equation

$$\det(\mathbf{A} - \lambda\mathbf{I}) = \begin{vmatrix} 9-\lambda & 1 & 1 \\ 1 & 9-\lambda & 1 \\ 1 & 1 & 9-\lambda \end{vmatrix} = -(\lambda - 11)(\lambda - 8)^2 = 0$$

shows that $\lambda_1 = 11$ and that $\lambda_2 = \lambda_3 = 8$ is an eigenvalue of multiplicity 2.

For $\lambda_1 = 11$, Gauss–Jordan elimination gives

$$(\mathbf{A} - 11\mathbf{I}|\mathbf{0}) = \begin{pmatrix} -2 & 1 & 1 & | & 0 \\ 1 & -2 & 1 & | & 0 \\ 1 & 1 & -2 & | & 0 \end{pmatrix} \implies \begin{pmatrix} 1 & 0 & -1 & | & 0 \\ 0 & 1 & -1 & | & 0 \\ 0 & 0 & 0 & | & 0 \end{pmatrix}.$$

Hence $k_1 = k_3$ and $k_2 = k_3$. If $k_3 = 1$, then

$$\mathbf{K}_1 = \begin{pmatrix} 1 \\ 1 \\ 1 \end{pmatrix}.$$

Now for $\lambda_2 = 8$ we have

$$(\mathbf{A} - 8\mathbf{I}|\mathbf{0}) = \begin{pmatrix} 1 & 1 & 1 & | & 0 \\ 1 & 1 & 1 & | & 0 \\ 1 & 1 & 1 & | & 0 \end{pmatrix} \implies \begin{pmatrix} 1 & 1 & 1 & | & 0 \\ 0 & 0 & 0 & | & 0 \\ 0 & 0 & 0 & | & 0 \end{pmatrix}.$$

In the equation $k_1 + k_2 + k_3 = 0$ we are free to select two of the variables arbitrarily. Choosing, on the one hand, $k_2 = 1$, $k_3 = 0$, and on the other, $k_2 = 0$, $k_3 = 1$, we obtain two linearly independent eigenvectors:

$$\mathbf{K}_2 = \begin{pmatrix} -1 \\ 1 \\ 0 \end{pmatrix} \quad \text{and} \quad \mathbf{K}_3 = \begin{pmatrix} -1 \\ 0 \\ 1 \end{pmatrix}$$

corresponding to a single eigenvalue. ❑

■ Complex Eigenvalues A matrix \mathbf{A} may have **complex eigenvalues.**

THEOREM 8.24 Complex Eigenvalues and Eigenvectors

Let \mathbf{A} be a square matrix with real entries. If $\lambda = \alpha + i\beta$, $\beta \neq 0$, is a complex eigenvalue of \mathbf{A}, then its conjugate $\bar{\lambda} = \alpha - i\beta$ is also an eigenvalue of \mathbf{A}. If \mathbf{K} is an eigenvector corresponding to λ, then its conjugate $\bar{\mathbf{K}}$ is an eigenvector corresponding to $\bar{\lambda}$.

PROOF Since \mathbf{A} is a matrix with real entries, the characteristic equation $\det(\mathbf{A} - \lambda\mathbf{I}) = 0$ is a polynomial equation with real coefficients. From algebra we know that complex roots of such equations appear in conjugate pairs. In other words, if $\lambda = \alpha + i\beta$ is a root, then $\bar{\lambda} = \alpha - i\beta$ is also a root. Now let \mathbf{K} be an eigenvector of \mathbf{A} corresponding to λ. By definition, $\mathbf{AK} = \lambda\mathbf{K}$. Taking complex conjugates of the latter equation gives

$$\overline{\mathbf{AK}} = \overline{\lambda\mathbf{K}} \quad \text{or} \quad \mathbf{A}\overline{\mathbf{K}} = \bar{\lambda}\overline{\mathbf{K}},$$

since \mathbf{A} is a real matrix. The last equation indicates that $\overline{\mathbf{K}}$ is an eigenvector corresponding to $\bar{\lambda}$. ❑

Example 5 Complex Eigenvalues and Eigenvectors

Find the eigenvalues and eigenvectors of $\mathbf{A} = \begin{pmatrix} 6 & -1 \\ 5 & 4 \end{pmatrix}$.

SOLUTION The characteristic equation is

$$\det(\mathbf{A} - \lambda\mathbf{I}) = \begin{vmatrix} 6 - \lambda & -1 \\ 5 & 4 - \lambda \end{vmatrix} = \lambda^2 - 10\lambda + 29 = 0.$$

From the quadratic formula we find $\lambda_1 = 5 + 2i$ and $\lambda_2 = \bar{\lambda}_1 = 5 - 2i$.
 Now for $\lambda_1 = 5 + 2i$ we must solve

$$(1 - 2i)k_1 - \qquad k_2 = 0$$
$$5k_1 - (1 + 2i)k_2 = 0.$$

Since $k_2 = (1 - 2i)k_1$,* it follows, after choosing $k_1 = 1$, that one eigenvector is

$$\mathbf{K}_1 = \begin{pmatrix} 1 \\ 1 - 2i \end{pmatrix}.$$

From Theorem 8.24 we see that an eigenvector corresponding to $\lambda_2 = \bar{\lambda}_1 = 5 - 2i$ is

$$\mathbf{K}_2 = \bar{\mathbf{K}}_1 = \begin{pmatrix} 1 \\ 1 + 2i \end{pmatrix}. \qquad \square$$

Our last theorem follows immediately from the fact that the determinant of an upper triangular, lower triangular, or diagonal matrix is the product of the diagonal entries.

THEOREM 8.25 Triangular and Diagonal Matrices

The eigenvalues of an upper triangular, lower triangular, or diagonal matrix are the main diagonal entries.

EXERCISES 8.8

Answers to odd-numbered problems begin on page A-44.

In Problems 1–6, determine which of the indicated column vectors are eigenvectors of the given matrix \mathbf{A}. Give the corresponding eigenvalue.

1. $\mathbf{A} = \begin{pmatrix} 4 & 2 \\ 5 & 1 \end{pmatrix}$; $\mathbf{K}_1 = \begin{pmatrix} 5 \\ -2 \end{pmatrix}$, $\mathbf{K}_2 = \begin{pmatrix} 2 \\ 5 \end{pmatrix}$, $\mathbf{K}_3 = \begin{pmatrix} -2 \\ 5 \end{pmatrix}$

2. $\mathbf{A} = \begin{pmatrix} 2 & -1 \\ 2 & -2 \end{pmatrix}$; $\mathbf{K}_1 = \begin{pmatrix} 1 \\ 2 - \sqrt{2} \end{pmatrix}$, $\mathbf{K}_2 = \begin{pmatrix} 2 + \sqrt{2} \\ 2 \end{pmatrix}$,

$\mathbf{K}_3 = \begin{pmatrix} \sqrt{2} \\ -\sqrt{2} \end{pmatrix}$

3. $\mathbf{A} = \begin{pmatrix} 6 & 3 \\ 2 & 1 \end{pmatrix}$; $\mathbf{K}_1 = \begin{pmatrix} 3 \\ -2 \end{pmatrix}$, $\mathbf{K}_2 = \begin{pmatrix} 1 \\ 0 \end{pmatrix}$, $\mathbf{K}_3 = \begin{pmatrix} -5 \\ 10 \end{pmatrix}$

4. $\mathbf{A} = \begin{pmatrix} 2 & 8 \\ -1 & -2 \end{pmatrix}$; $\mathbf{K}_1 = \begin{pmatrix} 0 \\ 0 \end{pmatrix}$, $\mathbf{K}_2 = \begin{pmatrix} 2 + 2i \\ -1 \end{pmatrix}$,

$\mathbf{K}_3 = \begin{pmatrix} 2 + 2i \\ 1 \end{pmatrix}$

5. $\mathbf{A} = \begin{pmatrix} 1 & -2 & 2 \\ -2 & 1 & -2 \\ 2 & 2 & 1 \end{pmatrix}$; $\mathbf{K}_1 = \begin{pmatrix} 0 \\ 1 \\ 1 \end{pmatrix}$, $\mathbf{K}_2 = \begin{pmatrix} 4 \\ -4 \\ 0 \end{pmatrix}$,

$\mathbf{K}_3 = \begin{pmatrix} -1 \\ 1 \\ 1 \end{pmatrix}$

6. $\mathbf{A} = \begin{pmatrix} -1 & 1 & 0 \\ 1 & 2 & 1 \\ 0 & 3 & -1 \end{pmatrix}$; $\mathbf{K}_1 = \begin{pmatrix} -1 \\ 4 \\ 3 \end{pmatrix}$, $\mathbf{K}_2 = \begin{pmatrix} -1 \\ 4 \\ 3 \end{pmatrix}$, $\mathbf{K}_3 = \begin{pmatrix} 3 \\ 1 \\ 4 \end{pmatrix}$

*Note that the second equation is simply $1 + 2i$ times the first.

In Problems 7–22, find the eigenvalues and eigenvectors of the given matrix.

7. $\begin{pmatrix} -1 & 2 \\ -7 & 8 \end{pmatrix}$

8. $\begin{pmatrix} 2 & 1 \\ 2 & 1 \end{pmatrix}$

9. $\begin{pmatrix} -8 & -1 \\ 16 & 0 \end{pmatrix}$

10. $\begin{pmatrix} 1 & 1 \\ \frac{1}{4} & 1 \end{pmatrix}$

11. $\begin{pmatrix} -1 & 2 \\ -5 & 1 \end{pmatrix}$

12. $\begin{pmatrix} 1 & -1 \\ 1 & 1 \end{pmatrix}$

13. $\begin{pmatrix} 4 & 8 \\ 0 & -5 \end{pmatrix}$

14. $\begin{pmatrix} 7 & 0 \\ 0 & 13 \end{pmatrix}$

15. $\begin{pmatrix} 5 & -1 & 0 \\ 0 & -5 & 9 \\ 5 & -1 & 0 \end{pmatrix}$

16. $\begin{pmatrix} 3 & 0 & 0 \\ 0 & 2 & 0 \\ 4 & 0 & 1 \end{pmatrix}$

17. $\begin{pmatrix} 0 & 4 & 0 \\ -1 & - & 0 \\ 0 & 4 & -2 \end{pmatrix}$

18. $\begin{pmatrix} 1 & 6 & 0 \\ 0 & 2 & 1 \\ 0 & 1 & 2 \end{pmatrix}$

19. $\begin{pmatrix} 0 & 0 & -1 \\ 1 & 0 & 0 \\ 1 & 1 & -1 \end{pmatrix}$

20. $\begin{pmatrix} 2 & -1 & 0 \\ 5 & 2 & 4 \\ 0 & 1 & 2 \end{pmatrix}$

21. $\begin{pmatrix} 1 & 2 & 3 \\ 0 & 5 & 6 \\ 0 & 0 & -7 \end{pmatrix}$

22. $\begin{pmatrix} 0 & 0 & 0 \\ 0 & 0 & 0 \\ 0 & 0 & 1 \end{pmatrix}$

The eigenvalues of \mathbf{A}^{-1} are the reciprocals of the eigenvalues of a nonsingular matrix \mathbf{A}. Furthermore, the eigenvectors for \mathbf{A} and \mathbf{A}^{-1} are the same. In Problems 23 and 24, verify these facts for the given matrix.

23. $\mathbf{A} = \begin{pmatrix} 5 & 1 \\ 1 & 5 \end{pmatrix}$

24. $\mathbf{A} = \begin{pmatrix} 1 & 2 & -1 \\ 1 & 0 & 1 \\ 4 & -4 & 5 \end{pmatrix}$

A matrix \mathbf{A} is singular if and only if $\lambda = 0$ is an eigenvalue. In Problems 25 and 26, verify that the given matrix \mathbf{A} is singular. Find the characteristic equation for \mathbf{A} and verify that $\lambda = 0$ is an eigenvalue.

25. $\mathbf{A} = \begin{pmatrix} 6 & 0 \\ 3 & 0 \end{pmatrix}$

26. $\mathbf{A} = \begin{pmatrix} 1 & 0 & 1 \\ 4 & -4 & 5 \end{pmatrix}$

27. Computer Lab Assignment A square matrix \mathbf{A} is said to be a **stochastic matrix** if all its entries are nonnegative and the sum of the entries in each row (or the sum of the entries in each column) add up to 1. Stochastic matrices are important in probability theory.

(a) Verify that

$$\mathbf{A} = \begin{pmatrix} p & 1-p \\ q & 1-q \end{pmatrix}, \quad 0 \le p \le 1, 0 \le q \le 1,$$

and

$$\mathbf{A} = \begin{pmatrix} \frac{1}{2} & \frac{1}{4} & \frac{1}{4} \\ \frac{1}{3} & \frac{1}{3} & \frac{1}{3} \\ \frac{1}{6} & \frac{1}{3} & \frac{1}{2} \end{pmatrix}$$

are stochastic matrices.

(b) Use linear algebra software or a CAS to find the eigenvalues and eigenvectors of the 3×3 matrix \mathbf{A} in part **(a)**. Make up at least six more stochastic matrices of various sizes, 2×2, 3×3, 4×4, and 5×5. Find the eigenvalues and eigenvectors of each matrix. If you discern a pattern, form a conjecture and then try to prove it.

(c) For the 3×3 matrix \mathbf{A} in part **(a)**, use the software to find \mathbf{A}^2, \mathbf{A}^3, \mathbf{A}^4, . . . Repeat for the matrices that you constructed in part **(b)**. If you discern a pattern, form a conjecture and then try to prove it.

8.9 Powers of Matrices

■ **Computation of \mathbf{A}^m** It is sometimes important to be able to quickly compute a power \mathbf{A}^m, m a positive integer, of an $n \times n$ matrix \mathbf{A}. Of course, computation of \mathbf{A}^m could be done with the appropriate software or by writing a short computer program, but even then, you should be aware that it is inefficient to simply use brute force to carry out repeated multiplications: $\mathbf{A}^2 = \mathbf{AA}$, $\mathbf{A}^3 = \mathbf{AA}^2$, and so on. We are going to sketch an alternative method for computing \mathbf{A}^m by means of the following theorem known as the Cayley–Hamilton theorem.

> **THEOREM 8.26 Cayley–Hamilton Theorem**
>
> An $n \times n$ matrix \mathbf{A} satisfies its own characteristic equation.

If $(-1)^n\lambda^n + c_{n-1}\lambda^{n-1} + \cdots + c_1\lambda + c_0 = 0$ is the characteristic equation of \mathbf{A}, then Theorem 8.26 states that

$$(-1)^n\mathbf{A}^n + c_{n-1}\mathbf{A}^{n-1} + \cdots + c_1\mathbf{A} + c_0\mathbf{I} = \mathbf{0}. \tag{1}$$

■ **Matrices of Order 2** The characteristic equation of the 2×2 matrix $\mathbf{A} = \begin{pmatrix} -2 & 4 \\ -1 & 3 \end{pmatrix}$ is $\lambda^2 - \lambda - 2 = 0$, and the eigenvalues of \mathbf{A} are $\lambda_1 = -1$ and $\lambda_2 = 2$. Theorem 8.26 implies $\mathbf{A}^2 - \mathbf{A} - 2\mathbf{I} = 0$, or, solving for the highest power of \mathbf{A},

$$\mathbf{A}^2 = 2\mathbf{I} + \mathbf{A}. \tag{2}$$

Now if we multiply (2) by \mathbf{A}, we get $\mathbf{A}^3 = 2\mathbf{A} + \mathbf{A}^2$, and if we use (2) again to eliminate \mathbf{A}^2 on the right side of this new equation, then

$$\mathbf{A}^3 = 2\mathbf{A} + \mathbf{A}^2 = 2\mathbf{A} + (2\mathbf{I} + \mathbf{A}) = 2\mathbf{I} + 3\mathbf{A}.$$

Continuing in this manner—in other words, multiplying the last result by \mathbf{A} and using (2) to eliminate \mathbf{A}^2—we obtain in succession powers of \mathbf{A} expressed solely in terms of the identity matrix \mathbf{I} and \mathbf{A}:

$$\mathbf{A}^4 = 6\mathbf{I} + 5\mathbf{A}$$
$$\mathbf{A}^5 = 10\mathbf{I} + 11\mathbf{A} \tag{3}$$
$$\mathbf{A}^6 = 22\mathbf{I} + 21\mathbf{A}$$

and so on (verify). Thus, for example,

$$\mathbf{A}^6 = 22\begin{pmatrix} 1 & 0 \\ 0 & 1 \end{pmatrix} + 21\begin{pmatrix} -2 & 4 \\ -1 & 3 \end{pmatrix} = \begin{pmatrix} -20 & 84 \\ -21 & 85 \end{pmatrix}. \tag{4}$$

Now we can determine the c_k without actually carrying out the repeated multiplications and resubstitutions as we did in (3). First, note that since the characteristic equation of the matrix $\mathbf{A} = \begin{pmatrix} -2 & 4 \\ -1 & 3 \end{pmatrix}$ can be written $\lambda^2 = 2 + \lambda$, results analogous to (3) must also hold for the eigenvalues $\lambda_1 = -1$ and $\lambda_2 = 2$, that is, $\lambda^3 = 2 + 3\lambda$, $\lambda^4 = 6 + 5\lambda$, $\lambda^5 = 10 + 11\lambda$, $\lambda^6 = 22 + 21\lambda$, It follows then that the equations

$$\mathbf{A}^m = c_0\mathbf{I} + c_1\mathbf{A} \quad \text{and} \quad \lambda^m = c_0 + c_1\lambda \tag{5}$$

hold *for the same pair of constants* c_0 and c_1. We can determine the constants c_0 and c_1 by simply setting $\lambda = -1$ and $\lambda = 2$ in the last equation in (5) and solving the resulting system of two equations in two unknowns. The solution of the system

$$(-1)^m = c_0 + c_1(-1)$$
$$2^m = c_0 + c_1(2)$$

is $c_0 = \frac{1}{3}[2^m + 2(-1)^m]$, $c_1 = \frac{1}{3}[2^m - (-1)^m]$. Now by substituting these coefficients in the first equation in (5), adding the two matrices and simplifying each entry, we obtain

$$\mathbf{A}^m = \begin{pmatrix} \frac{1}{3}[-2^m + 4(-1)^m] & \frac{4}{3}[2^m - (-1)^m] \\ -\frac{1}{3}[2^m - (-1)^m] & \frac{1}{3}[2^{m+2} - (-1)^m] \end{pmatrix}. \tag{6}$$

You should verify the result in (4) by setting $m = 6$ in (6). Note that (5) and (6) are valid for $m \geq 0$ since $\mathbf{A}^0 = \mathbf{I}$ and $\mathbf{A}^1 = \mathbf{A}$.

■ **Matrices of Order** n If the matrix \mathbf{A} were 3×3, then the characteristic equation (1) is a cubic polynomial equation, and the analogue of (2) would enable us to express \mathbf{A}^3 in terms of \mathbf{I}, \mathbf{A}, and \mathbf{A}^2. We could proceed as just illustrated to write any power \mathbf{A}^m in terms of \mathbf{I}, \mathbf{A}, and \mathbf{A}^2. In general, for an $n \times n$ matrix \mathbf{A}, we can write

$$\mathbf{A}^m = c_0\mathbf{I} + c_1\mathbf{A} + c_2\mathbf{A}^2 + \cdots + c_{n-1}\mathbf{A}^{n-1},$$

and where each of the coefficients c_k, $k = 0, 1, \ldots, n-1$, depend on the value of m.

Example 1 \mathbf{A}^m For a 3×3 Matrix

Compute \mathbf{A}^m for $\mathbf{A} = \begin{pmatrix} 1 & 1 & -2 \\ -1 & 2 & 1 \\ 0 & 1 & -1 \end{pmatrix}$.

SOLUTION The characteristic equation of \mathbf{A} is $-\lambda^3 + 2\lambda^2 + \lambda - 2 = 0$ or $\lambda^3 = -2 + \lambda + 2\lambda^2$, and the eigenvalues are $\lambda_1 = -1$, $\lambda_2 = 1$, and $\lambda_3 = 2$. From the preceding discussion we know that the same coefficients hold in both of the following equations:

$$\mathbf{A}^m = c_0\mathbf{I} + c_1\mathbf{A} + c_2\mathbf{A}^2 \quad \text{and} \quad \lambda^m = c_0 + c_1\lambda + c_2\lambda^2. \tag{7}$$

Setting, in turn, $\lambda = -1$, $\lambda = 1$, $\lambda = 2$ in the last equation generates three equations in three unknowns:

$$\begin{aligned} (-1)^m &= c_0 - c_1 + c_2 \\ 1 &= c_0 + c_1 + c_2 \\ 2^m &= c_0 + 2c_1 + 4c_2. \end{aligned} \tag{8}$$

Solving (8) gives

$$\begin{aligned} c_0 &= \tfrac{1}{3}[3 + (-1)^m - 2^m], \\ c_1 &= 2[1 - (-1)^m], \\ c_2 &= \tfrac{1}{6}[-3 + (-1)^m + 2^{m+1}]. \end{aligned}$$

After computing \mathbf{A}^2, we substitute these coefficients into the first equation of (7) and simplify the entries of the resulting matrix. The result is

$$\mathbf{A}^m = \begin{pmatrix} \tfrac{1}{6}[9 - 2^{m+1} - (-1)^m] & \tfrac{1}{3}[2^m - (-1)^m] & \tfrac{1}{6}[-9 + 2^{m+1} + 7(-1)^m] \\ 1 - 2^m & 2^m & 2^m - 1 \\ \tfrac{1}{6}[3 - 2^{m+1} - (-1)^m] & \tfrac{1}{3}[2^m - (-1)^m] & \tfrac{1}{6}[-3 + 2^{m+1} + 7(-1)^m] \end{pmatrix}.$$

For example, with $m = 10$,

$$\mathbf{A}^{10} = \begin{pmatrix} -340 & 341 & 341 \\ -1023 & 1024 & 1023 \\ -341 & 341 & 342 \end{pmatrix}. \qquad \square$$

■ **Finding the Inverse** Suppose \mathbf{A} is a nonsingular matrix. The fact that \mathbf{A} satisfies its own characteristic equation can be used to compute \mathbf{A}^{-1} as a linear combination of powers of \mathbf{A}. For example, we have just seen that the nonsingular matrix $\mathbf{A} = \begin{pmatrix} -2 & 4 \\ -1 & 3 \end{pmatrix}$ satisfies $\mathbf{A}^2 - \mathbf{A} - 2\mathbf{I} = 0$. Solving for the identity matrix gives

$I = \frac{1}{2}A^2 - \frac{1}{2}A$. By multiplying the last result by A^{-1}, we find $A^{-1} = \frac{1}{2}A - \frac{1}{2}I$. In other words,

$$\begin{pmatrix} -2 & 4 \\ -1 & 3 \end{pmatrix}^{-1} = \frac{1}{2}\begin{pmatrix} -2 & 4 \\ -1 & 3 \end{pmatrix} - \frac{1}{2}\begin{pmatrix} 1 & 0 \\ 0 & 1 \end{pmatrix} = \begin{pmatrix} -\frac{3}{2} & 2 \\ -\frac{1}{2} & 1 \end{pmatrix}. \tag{9}$$

Remarks

There are some obvious problems with the method just illustrated for finding A^m. If, for example, the matrix in Example 1 had an eigenvalue of multiplicity two, then instead of three equations and three unknowns as in (8) we would have only two equations in three unknowns. How do we then find *unique* coefficients c_0, c_1, and c_2? See Problems 11–14 in Exercises 8.9. Also, for larger matrices, even with distinct eigenvalues, solving a large system of equations for $c_0, c_1, c_2, \ldots, c_{n-1}$ is far too tedious to do by hand.

EXERCISES 8.9

Answers to odd-numbered problems begin on page A-44.

In Problems 1 and 2, verify that the given matrix satisfies its own characteristic equation.

1. $A = \begin{pmatrix} 1 & -2 \\ 4 & 5 \end{pmatrix}$

2. $A = \begin{pmatrix} 0 & 1 & 2 \\ 1 & 0 & 3 \\ 0 & 1 & 1 \end{pmatrix}$

In Problems 3–10, use the method of section to compute A^m. Use this result to compute the indicated power of the matrix A.

3. $A = \begin{pmatrix} -1 & 3 \\ 2 & 4 \end{pmatrix}$; $m = 3$

4. $A = \begin{pmatrix} 5 & -3 \\ -3 & 5 \end{pmatrix}$; $m = 4$

5. $A = \begin{pmatrix} 8 & 5 \\ 4 & 0 \end{pmatrix}$; $m = 5$

6. $A = \begin{pmatrix} -1 & 2 \\ 0 & -3 \end{pmatrix}$; $m = 6$

7. $A = \begin{pmatrix} 1 & 1 & 1 \\ 0 & 1 & 2 \\ 0 & 1 & 0 \end{pmatrix}$; $m = 10$

8. $A = \begin{pmatrix} 0 & 1 & 1 \\ 0 & -1 & 1 \\ 1 & 1 & 0 \end{pmatrix}$; $m = 6$

9. $A = \begin{pmatrix} 2 & 2 & 0 \\ 4 & 0 & 0 \\ 1 & 2 & 1 \end{pmatrix}$; $m = 10$

10. $A = \begin{pmatrix} 0 & -\frac{1}{2} & 0 \\ -1 & \frac{1}{2} & 0 \\ 2 & -\frac{1}{2} & -2 \end{pmatrix}$; $m = 8$

In Problems 11 and 12, show that the given matrix has an eigenvalue λ_1 of multiplicity two. As a consequence, the equations $\lambda^m = c_0 + c_1\lambda$ (Problem 11) and $\lambda^m = c_0 + c_1\lambda + c_2\lambda^2$ (Problem 12) do not yield enough independent equations to form a system for determining the coefficients c_i. Use the derivative (with respect to λ) of each of these equations evaluated at λ_1 as the extra needed equation to form a system. Compute A^m and use this result to compute the indicated power of the matrix A.

11. $A = \begin{pmatrix} 7 & 3 \\ -3 & 1 \end{pmatrix}$; $m = 6$

12. $A = \begin{pmatrix} -2 & 2 & -1 \\ 2 & 1 & -2 \\ -3 & -6 & 0 \end{pmatrix}$; $m = 5$

13. Show that $\lambda = 0$ is an eigenvalue of each matrix. In this case the coefficient c_0 in the characteristic equation (1) is 0. Compute A^m in each case. In parts **(a)** and **(b)**, explain why we do not have to solve any system for the coefficients c_i in determining A^m.

(a) $A = \begin{pmatrix} 1 & 1 \\ 3 & 3 \end{pmatrix}$ **(b)** $A = \begin{pmatrix} 1 & -1 \\ 1 & -1 \end{pmatrix}$ **(c)** $A = \begin{pmatrix} 2 & 1 & 1 \\ 1 & 0 & -2 \\ 1 & 1 & 3 \end{pmatrix}$

14. In his work *Liber Abbaci*, published in 1202, Leonardo Fibonacci of Pisa speculated on the reproduction of rabbits:

How many pairs of rabbits will be produced in a year beginning with a single pair, if every month each pair bears a new pair which become productive from the second month on?

The answer to his question is contained in a sequence known as a Fibonacci sequence.

	After Each Month												
Start n =	**0**	**1**	**2**	**3**	**4**	**5**	**6**	**7**	**8**	**9**	**10**	**11**	**12**
Adult pairs	1	1	2	3	5	8	13	21	...				
Baby pairs	0	1	1	2	3	5	8	13	...				
Total pairs	1	2	3	5	8	13	21	34	...				

Each of the three rows describing rabbit pairs is a Fibonacci sequence and can be defined recursively by a **second-order difference equation** $x_n = x_{n-2} + x_{n-1}$, $n = 2, 3, \ldots$, where x_0 and x_1 depend on the row. For example, for the first row designating adult pairs of rabbits, $x_0 = 1$, $x_1 = 1$.

(a) If we let $y_{n-1} = x_{n-2}$, then $y_n = x_{n-1}$, and the difference equation can be written as a system of first-order difference equations

$$x_n = x_{n-1} + y_{n-1}$$

$$y_n = x_{n-1}.$$

Write this system in the matrix form $\mathbf{X}_n = \mathbf{A}\mathbf{X}_{n-1}$, $n = 2, 3, \ldots$.

(b) Show that

$$\mathbf{A}^m = \begin{pmatrix} \dfrac{\lambda_2\lambda_1^m - \lambda_1\lambda_2^m + \lambda_2^m - \lambda_1^m}{\lambda_2 - \lambda_1} & \dfrac{\lambda_2^m - \lambda_1^m}{\lambda_2 - \lambda_1} \\[3mm] \dfrac{\lambda_2^m - \lambda_1^m}{\lambda_2 - \lambda_1} & \dfrac{\lambda_2\lambda_1^m - \lambda_1\lambda_2^m}{\lambda_2 - \lambda_1} \end{pmatrix}$$

or

$$\mathbf{A}^m = \frac{1}{2^{m+1}\sqrt{5}} \begin{pmatrix} (1+\sqrt{5})^{m+1} - (1-\sqrt{5})^{m+1} & 2(1+\sqrt{5})^m - 2(1-\sqrt{5})^m \\ 2(1+\sqrt{5})^m - 2(1-\sqrt{5})^m & (1+\sqrt{5})(1-\sqrt{5})^m - (1-\sqrt{5})(1+\sqrt{5})^m \end{pmatrix},$$

where $\lambda_1 = \frac{1}{2}(1 - \sqrt{5})$ and $\lambda_2 = \frac{1}{2}(1 + \sqrt{5})$ are the distinct eigenvalues of \mathbf{A}.

(c) Use the result in part **(a)** to show $\mathbf{X}_n = \mathbf{A}^{n-1}\mathbf{X}_1$. Use the last result and the result in part **(b)** to find the number of adult pairs, baby pairs, and total pairs of rabbits after the twelfth month.

In Problems 15 and 16, use the procedure illustrated in (9) to find \mathbf{A}^{-1}.

15. $\mathbf{A} = \begin{pmatrix} 2 & -4 \\ 1 & 3 \end{pmatrix}$ **16.** $\mathbf{A} = \begin{pmatrix} 1 & 1 & -2 \\ -1 & 2 & 1 \\ 0 & 1 & -1 \end{pmatrix}$

17. A nonzero $n \times n$ matrix \mathbf{A} is said to be **nilpotent of index m** if m is the smallest positive integer for which $\mathbf{A}^m = \mathbf{0}$. Which of the following matrices are nilpotent? If nilpotent, what is its index?

(a) $\begin{pmatrix} 1 & 0 \\ -1 & 0 \end{pmatrix}$ **(b)** $\begin{pmatrix} 2 & 2 \\ -2 & -2 \end{pmatrix}$

(c) $\begin{pmatrix} 0 & 0 & 0 \\ 1 & 0 & 0 \\ 2 & 3 & 0 \end{pmatrix}$ **(d)** $\begin{pmatrix} 0 & 0 & 5 \\ 0 & 0 & 0 \\ 0 & 0 & 0 \end{pmatrix}$

(e) $\begin{pmatrix} 0 & 0 & 0 & 0 \\ -1 & 0 & 0 & 1 \\ 0 & 1 & 0 & 1 \\ -1 & 0 & 0 & 0 \end{pmatrix}$ **(f)** $\begin{pmatrix} 0 & 0 & 0 & 0 \\ 1 & 0 & 0 & 0 \\ 3 & 1 & 0 & 0 \\ 2 & 2 & 1 & 0 \end{pmatrix}$

18. (a) Explain why any nilpotent matrix \mathbf{A} is singular. [*Hint:* Review Section 8.5.] **(b)** Show that all the eigenvalues of a nilpotent matrix \mathbf{A} are 0. [Hint: Use (1) of Section 8.8.]

8.10 Orthogonal Matrices

■ **Complex Numbers** In the discussion that follows, we will use some elementary properties of complex numbers. Suppose $z = a + ib$ denotes a complex number, where a and b are real and $i^2 = -1$. If $\bar{z} = a - ib$ is the **conjugate** of z, then the equality $z = \bar{z}$ or $a + ib = a - ib$ implies that $b = 0$. In other words, if $z = \bar{z}$, then z is a real number. In addition, it is easily verified that the product of a complex number z and its conjugate \bar{z} is a real number: $z\bar{z} = a^2 + b^2$. The **magnitude** of z is defined to be the real number $|z| = \sqrt{a^2 + b^2}$. The magnitude of z can be expressed in terms of the product $z\bar{z}$: $|z| = \sqrt{a^2 + b^2} = \sqrt{z\bar{z}}$, or $|z|^2 = z\bar{z}$. A detailed discussion of complex numbers can be found in Section 17.1.

There are many types of special matrices, but two types occur again and again in applications: symmetric matrices (page 348) and orthogonal matrices (page 390). In this section we are going to examine both these matrices in further detail.

■ **Symmetric Matrices** We begin by recalling, in formal terms, the definition of a symmetric matrix.

DEFINITION 8.14 Symmetric Matrix

An $n \times n$ matrix \mathbf{A} is **symmetric** if $\mathbf{A} = \mathbf{A}^T$, where \mathbf{A}^T is the transpose of \mathbf{A}.

The proof of the next theorem depends on the properties of complex numbers discussed in the review at the start of this section.

THEOREM 8.27 Real Eigenvalues

Let \mathbf{A} be a symmetric matrix with real entries. Then the eigenvalues of \mathbf{A} are real.

PROOF If \mathbf{K} is an eigenvector corresponding to an eigenvalue λ of \mathbf{A}, then $\mathbf{AK} = \lambda\mathbf{K}$. The conjugate of the last equation is

$$\overline{\mathbf{AK}} = \overline{\lambda}\overline{\mathbf{K}}. \tag{1}$$

Since the entries of \mathbf{A} are real, we have $\mathbf{A} = \overline{\mathbf{A}}$, and so (1) is

$$\mathbf{A}\overline{\mathbf{K}} = \overline{\lambda}\overline{\mathbf{K}}. \tag{2}$$

We now take the transpose of (2), use the fact that \mathbf{A} is symmetric, and multiply the resulting equation on the right by \mathbf{K}:

$$\overline{\mathbf{K}}^T\mathbf{AK} = \overline{\lambda}\overline{\mathbf{K}}^T\mathbf{K}. \tag{3}$$

But if we multiply $\mathbf{AK} = \lambda\mathbf{K}$ on the left by $\overline{\mathbf{K}}^T$, we obtain

$$\overline{\mathbf{K}}^T\mathbf{AK} = \lambda\overline{\mathbf{K}}^T\mathbf{K}. \tag{4}$$

Subtracting (4) from (3) then gives

$$0 = (\overline{\lambda} - \lambda)\overline{\mathbf{K}}^T\mathbf{K}. \tag{5}$$

Now $\overline{\mathbf{K}}^T$ is a $1 \times n$ matrix and \mathbf{K} is an $n \times 1$ matrix, so the product $\overline{\mathbf{K}}^T\mathbf{K}$ is the 1×1 matrix $\overline{\mathbf{K}}^T\mathbf{K} = \left(|k_1|^2 + |k_2|^2 + \cdots + |k_n|^2\right)$. Since by definition, $\mathbf{K} \neq \mathbf{0}$, the last expression is a positive quantity. Therefore we conclude from (5) that $\overline{\lambda} - \lambda = 0$ or $\overline{\lambda} = \lambda$. This implies that λ is a real number. ❑

In R^n the **dot** or **inner product** of two vectors $\mathbf{x} = (x_1, x_2, \ldots, x_n)$ and $\mathbf{y} = (y_1, y_2, \ldots, y_n)$ is given by

$$\mathbf{x} \cdot \mathbf{y} = x_1y_1 + x_2y_2 + \cdots + x_ny_n. \tag{6}$$

Now if \mathbf{X} and \mathbf{Y} are $n \times 1$ column vectors, $\mathbf{X} = \begin{pmatrix} x_1 \\ x_2 \\ \vdots \\ x_n \end{pmatrix}$ and $\mathbf{Y} = \begin{pmatrix} y_1 \\ y_2 \\ \vdots \\ y_n \end{pmatrix}$, then the matrix analogue of (6) is

$$\mathbf{X} \cdot \mathbf{Y} = \mathbf{X}^T\mathbf{Y} = (x_1y_1 + x_2y_2 + \cdots + x_ny_n).^* \tag{7}$$

Of course, for the column vectors given, $\mathbf{Y}^T\mathbf{X} = \mathbf{X}^T\mathbf{Y}$. The **norm** of a column vector \mathbf{X} is given by

$$\|\mathbf{X}\| = \sqrt{\mathbf{X} \cdot \mathbf{X}} = \sqrt{\mathbf{X}^T\mathbf{X}} = \sqrt{x_1^2 + x_2^2 + \cdots + x_n^2}.$$

*Since a 1×1 matrix is simply a scalar, we will hereafter drop the parentheses and write $\mathbf{X}^T\mathbf{Y} = x_1y_1 + x_2y_2 + \cdots + x_ny_n$.

> **THEOREM 8.28** **Orthogonal Eigenvectors**
>
> Let \mathbf{A} be an $n \times n$ symmetric matrix. Then eigenvectors corresponding to distinct (different) eigenvalues are orthogonal.

PROOF Let λ_1 and λ_2 by two distinct eigenvalues of \mathbf{A} corresponding to the eigenvectors \mathbf{K}_1 and \mathbf{K}_2, respectively. We wish to show that $\mathbf{K}_1 \cdot \mathbf{K}_2 = \mathbf{K}_1^T \mathbf{K}_2 = 0$.
Now by definition we must have

$$\mathbf{AK}_1 = \lambda_1 \mathbf{K}_1 \quad \text{and} \quad \mathbf{AK}_2 = \lambda_2 \mathbf{K}_2. \tag{8}$$

We form the transpose of the first of these equations, use $\mathbf{A}^T = \mathbf{A}$, and then multiply the result on the right by \mathbf{K}_2:

$$\mathbf{K}_1^T \mathbf{A} \mathbf{K}_2 = \lambda_1 \mathbf{K}_1^T \mathbf{K}_2. \tag{9}$$

The second equation in (8) is multiplied on the left by \mathbf{K}_1^T:

$$\mathbf{K}_1^T \mathbf{A} \mathbf{K}_2 = \lambda_2 \mathbf{K}_1^T \mathbf{K}_2. \tag{10}$$

Subtracting (10) from (9) yields

$$0 = \lambda_1 \mathbf{K}_1^T \mathbf{K}_2 - \lambda_2 \mathbf{K}_1^T \mathbf{K}_2 \quad \text{or} \quad 0 = (\lambda_1 - \lambda_2)\mathbf{K}_1^T \mathbf{K}_2.$$

Since $\lambda_1 \neq \lambda_2$, it follows that $\mathbf{K}_1^T \mathbf{K}_2 = 0$. ❑

Example 1 **Orthogonal Eigenvectors**

The eigenvalues of the symmetric matrix $\mathbf{A} = \begin{pmatrix} 0 & -1 & 0 \\ -1 & -1 & 1 \\ 0 & 1 & 0 \end{pmatrix}$ are $\lambda_1 = 0$, $\lambda_2 = 1$, and $\lambda_3 = -2$. In turn, the corresponding eigenvectors are

$$\mathbf{K}_1 = \begin{pmatrix} 1 \\ 0 \\ 1 \end{pmatrix}, \qquad \mathbf{K}_2 = \begin{pmatrix} -1 \\ 1 \\ 1 \end{pmatrix}, \qquad \mathbf{K}_3 = \begin{pmatrix} 1 \\ 2 \\ -1 \end{pmatrix}.$$

Since all the eigenvalues are distinct, we have

$$\mathbf{K}_1^T \mathbf{K}_2 = (1 \quad 0 \quad 1)\begin{pmatrix} -1 \\ 1 \\ 1 \end{pmatrix} = 1 \cdot (-1) + 0 \cdot 1 + 1 \cdot 1 = 0$$

$$\mathbf{K}_1^T \mathbf{K}_3 = (1 \quad 0 \quad 1)\begin{pmatrix} 1 \\ 2 \\ -1 \end{pmatrix} = 1 \cdot 1 + 0 \cdot 2 + 1(-1) = 0$$

$$\mathbf{K}_2^T \mathbf{K}_3 = (-1 \quad 1 \quad 1)\begin{pmatrix} 1 \\ 2 \\ -1 \end{pmatrix} = (-1) \cdot 1 + 1 \cdot 2 + 1 \cdot (-1) = 0. \qquad ❑$$

We saw in Example 3 of Section 8.8 that it may not be possible to find n linearly independent eigenvectors for an $n \times n$ matrix \mathbf{A} when some of the eigenvalues are repeated. But a symmetric matrix is an exception. It can be proved that a set of n linearly independent eigenvectors can always be found for an $n \times n$ symmetric matrix \mathbf{A} even when there is some repetition of the eigenvalues. (See Example 4 of Section 8.8.)

It follows from (2) of Section 7.6 that a set of vectors $\mathbf{x}_1, \mathbf{x}_2, \ldots, \mathbf{x}_n$ in R^n is **orthonormal** if every pair of distinct vectors is orthogonal and each vector in the set is a unit vector. In terms of the inner product of vectors, the set is orthonormal if

$$\mathbf{x}_i \cdot \mathbf{x}_j = 0, \quad i \neq j, \quad i, j = 1, 2, \ldots, n \qquad \text{and} \qquad \mathbf{x}_i \cdot \mathbf{x}_i = 1, \quad i = 1, 2, \ldots, n.$$

The last condition simply states that $\|\mathbf{x}_i\| = \sqrt{\mathbf{x}_i \cdot \mathbf{x}_i} = 1, i = 1, 2, \ldots, n$.

■ **Orthogonal Matrix** The concept of an orthonormal set of vectors plays an important role in the consideration of the next type of matrix.

DEFINITION 8.15 Orthogonal Matrix

An $n \times n$ nonsingular matrix \mathbf{A} is **orthogonal** if $\mathbf{A}^{-1} = \mathbf{A}^T$.

In other words, \mathbf{A} is orthogonal if $\mathbf{A}^T\mathbf{A} = \mathbf{I}$.

Example 2 **Orthogonal Matrices**

(a) The $n \times n$ identity matrix \mathbf{I} is an orthogonal matrix. For example, in the case of the 3×3 identity

$$\mathbf{I} = \begin{pmatrix} 1 & 0 & 0 \\ 0 & 1 & 0 \\ 0 & 0 & 1 \end{pmatrix},$$

it is readily seen that $\mathbf{I}^T = \mathbf{I}$ and $\mathbf{I}^T\mathbf{I} = \mathbf{II} = \mathbf{I}$.

(b) The matrix

$$\mathbf{A} = \begin{pmatrix} \frac{1}{3} & -\frac{2}{3} & \frac{2}{3} \\ \frac{2}{3} & \frac{2}{3} & \frac{1}{3} \\ -\frac{2}{3} & \frac{1}{3} & \frac{2}{3} \end{pmatrix}$$

is orthogonal. To see this, we need only verify that $\mathbf{A}^T\mathbf{A} = \mathbf{I}$:

$$\mathbf{A}^T\mathbf{A} = \begin{pmatrix} \frac{1}{3} & \frac{2}{3} & -\frac{2}{3} \\ -\frac{2}{3} & \frac{2}{3} & \frac{1}{3} \\ \frac{2}{3} & \frac{1}{3} & \frac{2}{3} \end{pmatrix} \begin{pmatrix} \frac{1}{3} & -\frac{2}{3} & \frac{2}{3} \\ \frac{2}{3} & \frac{2}{3} & \frac{1}{3} \\ -\frac{2}{3} & \frac{1}{3} & \frac{2}{3} \end{pmatrix} = \begin{pmatrix} 1 & 0 & 0 \\ 0 & 1 & 0 \\ 0 & 0 & 1 \end{pmatrix}. \qquad \square$$

THEOREM 8.29 Criterion for an Orthogonal Matrix

An $n \times n$ matrix \mathbf{A} is orthogonal if and only if its columns $\mathbf{X}_1, \mathbf{X}_2, \ldots, \mathbf{X}_n$ form an orthonormal set.

PARTIAL PROOF Let us suppose that \mathbf{A} is an $n \times n$ orthogonal matrix with columns $\mathbf{X}_1, \mathbf{X}_2, \ldots, \mathbf{X}_n$. Hence, the rows of \mathbf{A}^T are $\mathbf{X}_1^T, \mathbf{X}_2^T, \ldots, \mathbf{X}_n^T$. But since \mathbf{A} is orthogonal, $\mathbf{A}^T\mathbf{A} = \mathbf{I}$; that is,

$$\mathbf{A}^T\mathbf{A} = \begin{pmatrix} \mathbf{X}_1^T\mathbf{X}_1 & \mathbf{X}_1^T\mathbf{X}_2 & \cdots & \mathbf{X}_1^T\mathbf{X}_n \\ \mathbf{X}_2^T\mathbf{X}_1 & \mathbf{X}_2^T\mathbf{X}_2 & \cdots & \mathbf{X}_2^T\mathbf{X}_n \\ \vdots & & & \vdots \\ \mathbf{X}_n^T\mathbf{X}_1 & \mathbf{X}_n^T\mathbf{X}_2 & \cdots & \mathbf{X}_n^T\mathbf{X}_n \end{pmatrix} = \begin{pmatrix} 1 & 0 & \cdots & 0 \\ 0 & 1 & \cdots & 0 \\ \vdots & & & \vdots \\ 0 & 0 & \cdots & 1 \end{pmatrix}.$$

It follows from the definition of equality of matrices that

$$\mathbf{X}_i^T \mathbf{X}_j = 0, \quad i \neq j, \quad i, j = 1, 2, \ldots, n \quad \text{and} \quad \mathbf{X}_i^T \mathbf{X}_i = 1, \quad i = 1, 2, \ldots, n.$$

This means that the columns of the orthogonal matrix form an orthonormal set of n vectors. ❏

If we write the columns of the matrix in part **(b)** of Example 2 as

$$\mathbf{X}_1 = \begin{pmatrix} 1/3 \\ 2/3 \\ -2/3 \end{pmatrix}, \qquad \mathbf{X}_2 = \begin{pmatrix} -2/3 \\ 2/3 \\ 1/3 \end{pmatrix}, \qquad \mathbf{X}_3 = \begin{pmatrix} 2/3 \\ 1/3 \\ 2/3 \end{pmatrix},$$

then the vectors are orthogonal:

$$\mathbf{X}_1^T \mathbf{X}_2 = (1/3 \quad 2/3 \quad -2/3) \begin{pmatrix} -2/3 \\ 2/3 \\ 1/3 \end{pmatrix} = -\frac{2}{9} + \frac{4}{9} - \frac{2}{9} = 0$$

$$\mathbf{X}_1^T \mathbf{X}_3 = (1/3 \quad 2/3 \quad -2/3) \begin{pmatrix} 2/3 \\ 1/3 \\ 2/3 \end{pmatrix} = \frac{2}{9} + \frac{2}{9} - \frac{4}{9} = 0$$

$$\mathbf{X}_2^T \mathbf{X}_3 = (-2/3 \quad 2/3 \quad 1/3) \begin{pmatrix} 2/3 \\ 1/3 \\ 2/3 \end{pmatrix} = -\frac{4}{9} + \frac{2}{9} + \frac{2}{9} = 0$$

and are unit vectors:

$$\mathbf{X}_1^T \mathbf{X}_1 = (1/3 \quad 2/3 \quad -2/3) \begin{pmatrix} 1/3 \\ 2/3 \\ -2/3 \end{pmatrix} = \frac{1}{9} + \frac{4}{9} + \frac{4}{9} = 1$$

$$\mathbf{X}_2^T \mathbf{X}_2 = (-2/3 \quad 2/3 \quad 1/3) \begin{pmatrix} -2/3 \\ 2/3 \\ 1/3 \end{pmatrix} = \frac{4}{9} + \frac{4}{9} + \frac{1}{9} = 1$$

$$\mathbf{X}_3^T \mathbf{X}_3 = (2/3 \quad 1/3 \quad 2/3) \begin{pmatrix} 2/3 \\ 1/3 \\ 2/3 \end{pmatrix} = \frac{4}{9} + \frac{1}{9} + \frac{4}{9} = 1.$$

■ **Constructing an Orthogonal Matrix** If an $n \times n$ real symmetric matrix **A** possesses n distinct eigenvalues $\lambda_1, \lambda_2, \ldots, \lambda_n$, it follows from Theorem 8.28 that the corresponding eigenvectors $\mathbf{K}_1, \mathbf{K}_2, \ldots, \mathbf{K}_n$ are mutually orthogonal. By multiplying each vector by the reciprocal of its norm, we obtain a set of mutually orthogonal unit vectors—that is, an orthonormal set. We can then construct an orthogonal matrix by forming an $n \times n$ matrix **P** whose columns are these normalized eigenvectors of **A**.

Example 3 **Constructing an Orthogonal Matrix**

In Example 1 we verified that the eigenvectors

$$\mathbf{K}_1 = \begin{pmatrix} 1 \\ 0 \\ 1 \end{pmatrix}, \qquad \mathbf{K}_2 = \begin{pmatrix} -1 \\ 1 \\ 1 \end{pmatrix}, \qquad \mathbf{K}_3 = \begin{pmatrix} 1 \\ 2 \\ -1 \end{pmatrix}$$

of the given symmetric matrix **A** are orthogonal. Now the norms of the eigenvectors are

$$\|\mathbf{K}_1\| = \sqrt{\mathbf{K}_1^T \mathbf{K}_1} = \sqrt{2}, \qquad \|\mathbf{K}_2\| = \sqrt{\mathbf{K}_2^T \mathbf{K}_2} = \sqrt{3}, \qquad \|\mathbf{K}_3\| = \sqrt{\mathbf{K}_3^T \mathbf{K}_3} = \sqrt{6}.$$

Thus, an orthonormal set of vectors is

$$\begin{pmatrix} \dfrac{1}{\sqrt{2}} \\[2mm] 0 \\[2mm] \dfrac{1}{\sqrt{2}} \end{pmatrix}, \qquad \begin{pmatrix} -\dfrac{1}{\sqrt{3}} \\[2mm] \dfrac{1}{\sqrt{3}} \\[2mm] \dfrac{1}{\sqrt{3}} \end{pmatrix}, \qquad \begin{pmatrix} \dfrac{1}{\sqrt{6}} \\[2mm] \dfrac{2}{\sqrt{6}} \\[2mm] -\dfrac{1}{\sqrt{6}} \end{pmatrix}.$$

Using these vectors as columns, we obtain the orthogonal matrix

$$\mathbf{P} = \begin{pmatrix} \dfrac{1}{\sqrt{2}} & -\dfrac{1}{\sqrt{3}} & \dfrac{1}{\sqrt{6}} \\[2mm] 0 & \dfrac{1}{\sqrt{3}} & \dfrac{2}{\sqrt{6}} \\[2mm] \dfrac{1}{\sqrt{2}} & \dfrac{1}{\sqrt{3}} & -\dfrac{1}{\sqrt{6}} \end{pmatrix}.$$

The student should verify that $\mathbf{P}^T = \mathbf{P}^{-1}$. ❑

We will use the technique of constructing an orthogonal matrix from the eigenvectors of a symmetric matrix in the next section.

Remarks

Do not misinterpret Theorem 8.28. We can always find n linearly independent eigenvectors for an $n \times n$ real symmetric matrix \mathbf{A}. The theorem does not state that all the eigenvectors are mutually orthogonal. The eigenvectors corresponding to distinct eigenvalues are orthogonal, but the different eigenvectors corresponding to a repeated eigenvalue may not be orthogonal. Consider the symmetric matrix

$$\mathbf{A} = \begin{pmatrix} 7 & 4 & -4 \\ 4 & -8 & -1 \\ -4 & -1 & -8 \end{pmatrix}.$$

The eigenvalues are $\lambda_1 = 9$ and $\lambda_2 = \lambda_3 = -9$. We leave it as an exercise to verify that corresponding eigenvectors are

$$\mathbf{K}_1 = \begin{pmatrix} 4 \\ 1 \\ -1 \end{pmatrix}, \qquad \mathbf{K}_2 = \begin{pmatrix} 1 \\ 0 \\ 4 \end{pmatrix}, \qquad \mathbf{K}_3 = \begin{pmatrix} 1 \\ -4 \\ 0 \end{pmatrix}.$$

Note that in compliance with Theorem 8.28, \mathbf{K}_1 is orthogonal to \mathbf{K}_2 and to \mathbf{K}_3, but \mathbf{K}_2 and \mathbf{K}_3, eigenvectors corresponding to the repeated eigenvalue $\lambda_2 = -9$, are not orthogonal. However, for a real $n \times n$ symmetric matrix it is always possible to find, or construct, a set of n mutually orthogonal eigenvectors (see Problem 21 in Exercises 8.10).

In Problems 1–4, **(a)** verify that the indicated column vectors are eigenvectors of the given symmetric matrix, **(b)** identify the corresponding eigenvalues, and **(c)** verify that the column vectors are orthogonal.

1. $\begin{pmatrix} 0 & 0 & -4 \\ 0 & -4 & 0 \\ -4 & 0 & 15 \end{pmatrix}$; $\begin{pmatrix} 0 \\ 1 \\ 0 \end{pmatrix}$, $\begin{pmatrix} 4 \\ 0 \\ 1 \end{pmatrix}$, $\begin{pmatrix} 1 \\ 0 \\ -4 \end{pmatrix}$

2. $\begin{pmatrix} 1 & -1 & -1 \\ -1 & 1 & -1 \\ -1 & -1 & 1 \end{pmatrix}$; $\begin{pmatrix} -2 \\ 1 \\ 1 \end{pmatrix}$, $\begin{pmatrix} 0 \\ 1 \\ -1 \end{pmatrix}$, $\begin{pmatrix} 1 \\ 1 \\ 1 \end{pmatrix}$

3. $\begin{pmatrix} 5 & 13 & 0 \\ 13 & 5 & 0 \\ 0 & 0 & -8 \end{pmatrix}$;

$\begin{pmatrix} \dfrac{\sqrt{2}}{2} \\[2mm] \dfrac{\sqrt{2}}{2} \\[2mm] 0 \end{pmatrix}$, $\begin{pmatrix} \dfrac{\sqrt{3}}{3} \\[2mm] -\dfrac{\sqrt{3}}{3} \\[2mm] \dfrac{\sqrt{3}}{3} \end{pmatrix}$, $\begin{pmatrix} \dfrac{\sqrt{6}}{6} \\[2mm] -\dfrac{\sqrt{6}}{6} \\[2mm] -\dfrac{\sqrt{6}}{3} \end{pmatrix}$,

4. $\begin{pmatrix} 3 & 2 & 2 \\ 2 & 2 & 0 \\ 2 & 0 & 4 \end{pmatrix}$; $\begin{pmatrix} -2 \\ 2 \\ 1 \end{pmatrix}$, $\begin{pmatrix} 1 \\ 2 \\ -2 \end{pmatrix}$, $\begin{pmatrix} 2 \\ 1 \\ 2 \end{pmatrix}$

In Problems 5–10, determine whether the given matrix is orthogonal.

5. $\begin{pmatrix} 0 & 1 & 0 \\ 1 & 0 & 0 \\ 0 & 0 & 1 \end{pmatrix}$ **6.** $\begin{pmatrix} \frac{1}{2} & 0 & \frac{1}{2} \\ 0 & 1 & 0 \\ \frac{1}{2} & 0 & -\frac{1}{2} \end{pmatrix}$

7. $\begin{pmatrix} 0 & 0 & 1 \\ -\frac{12}{13} & \frac{5}{13} & 0 \\ \frac{5}{13} & \frac{12}{13} & 0 \end{pmatrix}$ **8.** $\begin{pmatrix} 0 & 0 & 0 \\ 0 & 0 & 0 \\ 0 & 0 & 0 \end{pmatrix}$

9. $\begin{pmatrix} 1 & -1 & 1 \\ 1 & -1 & -1 \\ 1 & 2 & 0 \end{pmatrix}$

10. $\begin{pmatrix} 0 & \frac{8}{17} & 0 & -\frac{15}{17} \\ 0 & 0 & 1 & 0 \\ 1 & 0 & 0 & 0 \\ 0 & \frac{15}{17} & 0 & \frac{8}{17} \end{pmatrix}$

In Problems 11–18, proceed as in Example 3 to construct an orthogonal matrix from the eigenvectors of the given symmetric matrix. (The answers are not unique.)

11. $\begin{pmatrix} 1 & 9 \\ 9 & 1 \end{pmatrix}$ **12.** $\begin{pmatrix} 7 & 0 \\ 0 & 4 \end{pmatrix}$

13. $\begin{pmatrix} 1 & 3 \\ 3 & 9 \end{pmatrix}$ **14.** $\begin{pmatrix} 1 & 1 \\ 1 & 0 \end{pmatrix}$

15. $\begin{pmatrix} 1 & 0 & 1 \\ 0 & 1 & 0 \\ 1 & 0 & 1 \end{pmatrix}$ **16.** $\begin{pmatrix} 0 & 1 & 1 \\ 1 & 1 & 1 \\ 1 & 1 & 0 \end{pmatrix}$

17. $\begin{pmatrix} -8 & 5 & 4 \\ 5 & 3 & 1 \\ 4 & 1 & 0 \end{pmatrix}$ **18.** $\begin{pmatrix} 2 & 8 & -2 \\ 8 & -4 & 10 \\ -2 & 10 & -7 \end{pmatrix}$

In Problems 19 and 20, use Theorem 8.29 to find values of a and b so that the given matrix is orthogonal.

19. $\begin{pmatrix} \frac{3}{5} & a \\ \frac{4}{5} & b \end{pmatrix}$ **20.** $\begin{pmatrix} 1/\sqrt{5} & b \\ a & 1/\sqrt{5} \end{pmatrix}$

21. (a) Verify that

$$\mathbf{K}_1 = \begin{pmatrix} 4 \\ 1 \\ -1 \end{pmatrix}, \mathbf{K}_2 = \begin{pmatrix} 1 \\ 0 \\ 4 \end{pmatrix}, \text{ and } \mathbf{K}_3 = \begin{pmatrix} 1 \\ -4 \\ 0 \end{pmatrix}$$

are eigenvectors for the symmetric matrix

$$\mathbf{A} = \begin{pmatrix} 7 & 4 & -4 \\ 4 & -8 & -1 \\ -4 & -1 & -8 \end{pmatrix}$$

corresponding to the eigenvalues $\lambda_1 = 9$ and $\lambda_2 = \lambda_3 = -9$, respectively.

(b) Find a set of three mutually orthogonal eigenvectors for the matrix \mathbf{A} in part **(a)**.

22. Construct an orthogonal matrix from the eigenvectors of

$$\mathbf{A} = \begin{pmatrix} 1 & 2 & 0 & 0 \\ 2 & 1 & 0 & 0 \\ 0 & 0 & 1 & 2 \\ 0 & 0 & 2 & 1 \end{pmatrix}.$$

23. Show that if \mathbf{A} and \mathbf{B} are $n \times n$ orthogonal matrices, then \mathbf{AB} is orthogonal.

8.11 Approximation of Eigenvalues

Recall that to find the eigenvalues for a matrix \mathbf{A} we must find the roots of the polynomial equation $p(\lambda) = \det(\mathbf{A} - \lambda\mathbf{I}) = 0$. If \mathbf{A} is a large matrix, the computations involved in obtaining this characteristic equation can be overwhelming. Moreover, even if we can find the exact characteristic equation, it is likely that we would have to use a numerical procedure to approximate its roots. There are alternative numerical procedures for approximating eigenvalues and the corresponding eigenvectors. The procedure that we shall consider in this section deals with matrices that possess a **dominant eigenvalue.**

DEFINITION 8.16　Dominant Eigenvalue

Let $\lambda_1, \lambda_2, \ldots, \lambda_k, \ldots, \lambda_n$ denote the eigenvalues of an $n \times n$ matrix \mathbf{A}. The eigenvalue λ_k is said to be the **dominant eigenvalue** of \mathbf{A} if

$$|\lambda_k| > \lambda_i, \quad i = 1, 2, \ldots, n, \quad \text{but } i \neq k.$$

An eigenvector corresponding to λ_k is called the **dominant eigenvector** of \mathbf{A}.

In other words, a dominant eigenvalue is one whose absolute value is greater than the absolute value of each of the remaining eigenvalues. For example, in Example 2 of Section 8.8 we saw that the eigenvalues of the matrix

$$\mathbf{A} = \begin{pmatrix} 1 & 2 & 1 \\ 6 & -1 & 0 \\ -1 & -2 & -1 \end{pmatrix}$$

are $\lambda_1 = 0$, $\lambda_2 = -4$, and $\lambda_3 = 3$. Since $|-4| > 0$ and $|-4| > 3$, we see that $\lambda_2 = -4$ is the dominant eigenvalue of \mathbf{A}.

Example 1　Matrices With No Dominant Eigenvalue

(a)　The matrix $\mathbf{A} = \begin{pmatrix} 4 & 0 \\ 0 & -4 \end{pmatrix}$ has eigenvalues $\lambda_1 = -2$ and $\lambda_2 = 2$. Since $|\lambda_1| = |\lambda_2| = 2$, it follows that there is no dominant eigenvalue.

(b)　The eigenvalues of the matrix

$$\mathbf{A} = \begin{pmatrix} 2 & 0 & 0 \\ 0 & 5 & 1 \\ 0 & 0 & 5 \end{pmatrix} \quad \text{are} \quad \lambda_1 = 2, \lambda_2 = \lambda_3 = 5.$$

Again, the matrix has no dominant eigenvalue. ❏

■ **Power Method**　Suppose the $n \times n$ matrix \mathbf{A} has a dominant eigenvalue λ_1. The iterative technique for approximating a corresponding dominant eigenvector is due to the German mathematician Richard Von Mises (1883–1953) and is called the **power method.** The basic idea of this procedure is first to compute an approxima-

tion to a dominant eigenvector by means of the sequence

$$\mathbf{X}_i = \mathbf{A}\mathbf{X}_{i-1}, \qquad i = 1, 2, 3, \dots,$$ (1)

where \mathbf{X}_0 represents a nonzero $n \times 1$ vector that is an initial guess or approximation for the eigenvector we seek. Iterating (1) gives

$$\mathbf{X}_1 = \mathbf{A}\mathbf{X}_0$$

$$\mathbf{X}_2 = \mathbf{A}\mathbf{X}_1 = \mathbf{A}^2\mathbf{X}_0$$

$$\vdots$$ (2)

$$\mathbf{X}_m = \mathbf{A}\mathbf{X}_{m-1} = \mathbf{A}^m\mathbf{X}_0.$$

Under certain circumstances, for large values of m the vector defined by $\mathbf{X}_m = \mathbf{A}^m\mathbf{X}_0$ is an approximation to a dominant eigenvector. To see this, let us make some further assumptions about the matrix \mathbf{A}. Let us assume that the eigenvalues of \mathbf{A} are such that

$$|\lambda_1| > |\lambda_2| \geq |\lambda_3| \geq \cdots \geq |\lambda_n|$$

and that the corresponding n eigenvectors $\mathbf{K}_1, \mathbf{K}_2, \dots, \mathbf{K}_n$ are linearly independent. Because of this last assumption, $\mathbf{K}_1, \mathbf{K}_2, \dots, \mathbf{K}_n$ can serve as a basis for R^n (see Section 7.6). Thus, for any nonzero $n \times 1$ vector \mathbf{X}_0, constants c_1, c_2, \dots, c_n can be found such that

$$\mathbf{X}_0 = c_1\mathbf{K}_1 + c_2\mathbf{K}_2 + \cdots + c_n\mathbf{K}_n.$$ (3)

We shall also assume \mathbf{X}_0 is chosen so that $c_1 \neq 0$. Multiplying (3) by \mathbf{A} then gives

$$\mathbf{A}\mathbf{X}_0 = c_1\mathbf{A}\mathbf{K}_1 + c_2\mathbf{A}\mathbf{K}_2 + \cdots + c_n\mathbf{A}\mathbf{K}_n.$$

Since $\mathbf{A}\mathbf{K}_1 = \lambda_1\mathbf{K}_1$, $\mathbf{A}\mathbf{K}_2 = \lambda_2\mathbf{K}_2$, \dots, $\mathbf{A}\mathbf{K}_n = \lambda_n\mathbf{K}_n$, the last line becomes

$$\mathbf{A}\mathbf{X}_0 = c_1\lambda_1\mathbf{K}_1 + c_2\lambda_2\mathbf{K}_2 + \cdots + c_n\lambda_n\mathbf{A}\mathbf{K}_n.$$ (4)

Multiplying (4) by \mathbf{A} gives

$$\mathbf{A}^2\mathbf{X}_0 = c_1\lambda_1\mathbf{A}\mathbf{K}_1 + c_2\lambda_2\mathbf{A}\mathbf{K}_2 + \cdots + c_n\lambda_n\mathbf{A}\mathbf{K}_n$$

$$= c_1\lambda_1^2\mathbf{K}_1 + c_2\lambda_2^2\mathbf{K}_2 + \cdots + c_n\lambda_n^2\mathbf{K}_n.$$

Continuing in this manner, we find that

$$\mathbf{A}^m\mathbf{X}_0 = c_1\lambda_1^m\mathbf{K}_1 + c_2\lambda_2^m\mathbf{K}_2 + \cdots + c_n\lambda_n^m\mathbf{K}_n$$ (5)

$$= \lambda_1^m\left(c_1\mathbf{K}_1 + c_2\left(\frac{\lambda_2}{\lambda_1}\right)^m\mathbf{K}_2 + \cdots + c_n\left(\frac{\lambda_n}{\lambda_1}\right)^m\mathbf{K}_n\right).$$ (6)

Since $|\lambda_1| > |\lambda_i|$ for $i = 2, 3, \dots, n$, we have $|\lambda_i/\lambda_1| < 1$ and consequently $\lim_{m \to \infty}(\lambda_i/\lambda_1)^m = 0$. Therefore, as $m \to \infty$, we see from (6) that

$$\mathbf{A}^m\mathbf{X}_0 \approx \lambda_1^m c_1\mathbf{K}_1.$$ (7)

Since a nonzero constant multiple of an eigenvector is an eigenvector, it follows from (7) that for large values of m and under all the assumptions that were made, the $n \times 1$ matrix $\mathbf{X}_m = \mathbf{A}^m\mathbf{X}_0$ is an approximation to a dominant eigenvector associated with the dominant eigenvalue λ_1. The rate at which this method converges depends on the quotient λ_2/λ_1: If $|\lambda_2/\lambda_1|$ is very small, then convergence is fast, whereas if $|\lambda_2/\lambda_1|$ is close to one, convergence is slow. Of course, this information is not as useful as it seems because we generally do not know the eigenvalues in advance.

It remains, then, to approximate the dominant eigenvalue itself. This can be done by means of the inner product. If \mathbf{K} is an eigenvector of a matrix \mathbf{A} corresponding to the eigenvalue λ, we have $\mathbf{AK} = \lambda\mathbf{K}$, and so we have $\mathbf{AK} \cdot \mathbf{K} = \lambda\mathbf{K} \cdot \mathbf{K}$. Since $\mathbf{AK} \cdot \mathbf{K}$ and $\mathbf{K} \cdot \mathbf{K}$ are scalars, we can solve this last equation for λ:

$$\lambda = \frac{\mathbf{AK} \cdot \mathbf{K}}{\mathbf{K} \cdot \mathbf{K}}.$$

Hence, if $\mathbf{X}_m = \mathbf{A}^m\mathbf{X}_0$ is an approximation to a dominant eigenvector obtained by the iteration of (1), then the dominant eigenvalue λ_1 can be approximated by the quotient

$$\lambda_1 \approx \frac{\mathbf{AX}_m \cdot \mathbf{X}_m}{\mathbf{X}_m \cdot \mathbf{X}_m}. \tag{8}$$

The quotient in (8) is sometimes referred to as the **Rayleigh quotient.**

Example 2 Using the Power Method

Use the power method to approximate the dominant eigenvalue and a corresponding dominant eigenvector of $\mathbf{A} = \begin{pmatrix} 4 & 2 \\ 3 & -1 \end{pmatrix}$.

SOLUTION Since we do not know the eigenvalues and eigenvectors, we may as well take $\mathbf{X}_0 = \begin{pmatrix} 1 \\ 1 \end{pmatrix}$. Now the first two terms of the sequence of vectors defined by (1) are

$$\mathbf{X}_1 = \mathbf{AX}_0 = \begin{pmatrix} 4 & 2 \\ 3 & -1 \end{pmatrix}\begin{pmatrix} 1 \\ 1 \end{pmatrix} = \begin{pmatrix} 6 \\ 2 \end{pmatrix}$$

$$\mathbf{X}_2 = \mathbf{AX}_1 = \begin{pmatrix} 4 & 2 \\ 3 & -1 \end{pmatrix}\begin{pmatrix} 6 \\ 2 \end{pmatrix} = \begin{pmatrix} 28 \\ 16 \end{pmatrix}.$$

The next five vectors obtained in this manner are given in the following table:

i	3	4	5	6	7
\mathbf{X}_i	$\begin{pmatrix} 144 \\ 68 \end{pmatrix}$	$\begin{pmatrix} 712 \\ 364 \end{pmatrix}$	$\begin{pmatrix} 3576 \\ 1772 \end{pmatrix}$	$\begin{pmatrix} 17{,}848 \\ 8956 \end{pmatrix}$	$\begin{pmatrix} 89{,}304 \\ 44{,}588 \end{pmatrix}$

At this point it may not appear that we are getting anywhere, since the entries of the vectors in the table appear to be growing without bound. But bear in mind that (7) indicates we are obtaining a constant multiple of a vector. If the power method converges, then by factoring the entry with the largest absolute value from \mathbf{X}_m (for a large value of m), we should obtain a reasonable approximation to a dominant eigenvector. From the table,

$$\mathbf{X}_7 = 89{,}304\begin{pmatrix} 1 \\ 0.4933 \end{pmatrix}. \tag{9}$$

It appears then that the vectors are approaching scalar multiples of $\begin{pmatrix} 1 \\ 0.5 \end{pmatrix}$.

We now use (8) to approximate the dominant eigenvalue λ_1. First we have

$$\mathbf{AX}_7 = \begin{pmatrix} 4 & 2 \\ 3 & -1 \end{pmatrix} \begin{pmatrix} 1 \\ 0.4993 \end{pmatrix} = \begin{pmatrix} 4.9986 \\ 2.5007 \end{pmatrix}$$

$$\mathbf{AX}_7 \cdot \mathbf{X}_7 = \begin{pmatrix} 4.9986 \\ 2.5007 \end{pmatrix}^T \begin{pmatrix} 1 \\ 0.4993 \end{pmatrix} = 6.2472$$

$$\mathbf{X}_7 \cdot \mathbf{X}_7 = \begin{pmatrix} 1 \\ 0.4993 \end{pmatrix}^T \begin{pmatrix} 1 \\ 0.4993 \end{pmatrix} = 1.2493.$$

Finally, we have

$$\lambda_1 \approx \frac{\mathbf{AX}_7 \cdot \mathbf{X}_7}{\mathbf{X}_7 \cdot \mathbf{X}_7} = \frac{6.2472}{1.2493} = 5.0006.$$

The reader should use the procedure of Section 8.7 to verify that the eigenvalues and corresponding eigenvectors of **A** are $\lambda_1 = 5$, $\lambda_2 = -2$, $\mathbf{K}_1 = \begin{pmatrix} 1 \\ 0.5 \end{pmatrix}$, and $\mathbf{K}_2 = \begin{pmatrix} 1 \\ -3 \end{pmatrix}$. ❏

■ **Scaling** As we have just seen, iteration of (1) often results in vectors whose entries become very large in absolute value. Large numbers can, of course, cause a problem if a computer is used to carry out a great number of iterations. The result in (9) suggests that one way around this difficulty is to use a **scaled-down** vector at each step of the iteration. To do this we simply multiply the vector \mathbf{AX}_0 by the reciprocal of the entry with the largest absolute value. That is, we multiply

$$\mathbf{AX}_0 = \begin{pmatrix} x_1 \\ \vdots \\ x_n \end{pmatrix} \quad \text{by} \quad \frac{1}{\max\{|x_1|, |x_2|, \ldots, |x_n|\}}.$$

This resulting matrix, whose entries are now less than or equal to one, we label \mathbf{X}_1. We repeat the process with the vector \mathbf{AX}_1 to obtain the scaled-down vector \mathbf{X}_2, and so on.

Example 3 **Example 2 Revisited**

Repeat the iterations of Example 2 using scaled-down vectors.

SOLUTION From $\mathbf{AX}_0 = \begin{pmatrix} 4 & 2 \\ 3 & -1 \end{pmatrix} \begin{pmatrix} 1 \\ 1 \end{pmatrix} = \begin{pmatrix} 6 \\ 2 \end{pmatrix}$ we define

$$\mathbf{X}_1 = \frac{1}{6} \begin{pmatrix} 6 \\ 2 \end{pmatrix} = \begin{pmatrix} 1 \\ 0.3333 \end{pmatrix}.$$

From $\mathbf{AX}_1 = \begin{pmatrix} 4 & 2 \\ 3 & -1 \end{pmatrix} \begin{pmatrix} 1 \\ 0.3333 \end{pmatrix} = \begin{pmatrix} 4.6666 \\ 2.6667 \end{pmatrix}$ we define

$$\mathbf{X}_2 = \frac{1}{4.6666} \begin{pmatrix} 4.6666 \\ 2.6667 \end{pmatrix} = \begin{pmatrix} 1 \\ 0.5714 \end{pmatrix}.$$

We continue in this manner to construct the following table:

i	3	4	5	6	7
\mathbf{X}_i	$\begin{pmatrix} 1 \\ 0.4722 \end{pmatrix}$	$\begin{pmatrix} 1 \\ 0.5112 \end{pmatrix}$	$\begin{pmatrix} 1 \\ 0.4955 \end{pmatrix}$	$\begin{pmatrix} 1 \\ 0.5018 \end{pmatrix}$	$\begin{pmatrix} 1 \\ 0.4993 \end{pmatrix}$

In contrast to the table in Example 3, it is apparent from this table that the vectors are approaching $\begin{pmatrix} 1 \\ 0.5 \end{pmatrix}$. ❑

■ **Method of Deflation** After we have found the dominant eigenvalue λ_1 of a matrix \mathbf{A} it may still be necessary to find nondominant eigenvalues. The procedure we shall consider next is a modification of the power method and is called the **method of deflation.** We will limit the discussion to the case where \mathbf{A} is a *symmetric* matrix.

Suppose λ_1 and \mathbf{K}_1 are, respectively, the dominant eigenvalue and a corresponding *normalized* eigenvector[*] (that is, $\|\mathbf{K}_1\| = 1$) of a symmetric matrix \mathbf{A}. Furthermore, suppose the eigenvalues of \mathbf{A} are such that

$$|\lambda_1| > |\lambda_2| > |\lambda_3| \geq \cdots \geq |\lambda_n|.$$

It can be proved that the matrix

$$\mathbf{B} = \mathbf{A} - \lambda_1 \mathbf{K}_1 \mathbf{K}_1^T \tag{10}$$

has eigenvalues $0, \lambda_2, \lambda_3, \ldots, \lambda_n$ and that eigenvectors of \mathbf{B} are also eigenvectors of \mathbf{A}. Note that λ_2 is now the dominant eigenvalue of \mathbf{B}. We apply the power method to \mathbf{B} to approximate λ_2 and a corresponding eigenvector.

Example 4 **Using the Method of Deflation**

Use the method of deflation to approximate the eigenvalues of

$$\mathbf{A} = \begin{pmatrix} 1 & 2 & -1 \\ 2 & 1 & 1 \\ -1 & 1 & 0 \end{pmatrix}.$$

SOLUTION We begin by using the power method with scaling to find the dominant eigenvalue and a corresponding eigenvector of \mathbf{A}. Choosing $\mathbf{X}_0 = \begin{pmatrix} 1 \\ 1 \\ 1 \end{pmatrix}$, we see that

$$\mathbf{AX}_0 = \begin{pmatrix} 1 & 2 & -1 \\ 2 & 1 & 1 \\ -1 & 1 & 0 \end{pmatrix} \begin{pmatrix} 1 \\ 1 \\ 1 \end{pmatrix} = \begin{pmatrix} 2 \\ 4 \\ 0 \end{pmatrix} \qquad \text{so} \quad \mathbf{X}_1 = \frac{1}{4} \begin{pmatrix} 2 \\ 4 \\ 0 \end{pmatrix} = \begin{pmatrix} 0.5 \\ 1 \\ 0 \end{pmatrix}$$

$$\mathbf{AX}_1 = \begin{pmatrix} 1 & 2 & -1 \\ 2 & 1 & 1 \\ -1 & 1 & 0 \end{pmatrix} \begin{pmatrix} 0.5 \\ 1 \\ 0 \end{pmatrix} = \begin{pmatrix} 2.5 \\ 2 \\ 0.5 \end{pmatrix} \qquad \text{so} \quad \mathbf{X}_2 = \frac{1}{2.5} \begin{pmatrix} 2.5 \\ 2 \\ 0.5 \end{pmatrix} = \begin{pmatrix} 1 \\ 0.8 \\ 0.2 \end{pmatrix}.$$

The scaled vectors \mathbf{X}_3 to \mathbf{X}_{10} are given in the following table:

[*] See Example 3 in Section 8.10.

i	3	4	5	6	7	8	9	10
\mathbf{X}_i	$\begin{pmatrix} 0.8 \\ 1 \\ -0.0667 \end{pmatrix}$	$\begin{pmatrix} 1 \\ 0.8837 \\ 0.0698 \end{pmatrix}$	$\begin{pmatrix} 0.9134 \\ 1 \\ -0.0394 \end{pmatrix}$	$\begin{pmatrix} 1 \\ 0.9440 \\ 0.0293 \end{pmatrix}$	$\begin{pmatrix} 0.9614 \\ 1 \\ -0.0188 \end{pmatrix}$	$\begin{pmatrix} 1 \\ 0.9744 \\ 0.0129 \end{pmatrix}$	$\begin{pmatrix} 0.9828 \\ 1 \\ -0.0086 \end{pmatrix}$	$\begin{pmatrix} 1 \\ 0.9885 \\ 0.0058 \end{pmatrix}$

Utilizing \mathbf{X}_{10} and (8), we find

$$\lambda_1 \approx \frac{\mathbf{AX}_{10} \cdot \mathbf{X}_{10}}{\mathbf{X}_{10} \cdot \mathbf{X}_{10}} = 2.9997.$$

It appears that the dominant eigenvalue and a corresponding eigenvector are, respectively, $\lambda_1 = 3$ and $\mathbf{K} = \begin{pmatrix} 1 \\ 1 \\ 0 \end{pmatrix}$.

Our next task is to construct the matrix \mathbf{B} defined by (10). With $\|\mathbf{K}\| = \sqrt{2}$, the normalized eigenvector is $\mathbf{K}_1 = \begin{pmatrix} 1/\sqrt{2} \\ 1/\sqrt{2} \\ 0 \end{pmatrix}$. Thus,

$$\mathbf{B} = \begin{pmatrix} 1 & 2 & -1 \\ 2 & 1 & 1 \\ -1 & 1 & 0 \end{pmatrix} - 3\begin{pmatrix} 1/\sqrt{2} \\ 1/\sqrt{2} \\ 0 \end{pmatrix}(1/\sqrt{2} \quad 1/\sqrt{2} \quad 0)$$

$$= \begin{pmatrix} 1 & 2 & -1 \\ 2 & 1 & 1 \\ -1 & 1 & 0 \end{pmatrix} - \begin{pmatrix} \tfrac{3}{2} & \tfrac{3}{2} & 0 \\ \tfrac{3}{2} & \tfrac{3}{2} & 0 \\ 0 & 0 & 0 \end{pmatrix} = \begin{pmatrix} -0.5 & 0.5 & -1 \\ 0.5 & -0.5 & 1 \\ -1 & 1 & 0 \end{pmatrix}.$$

We now use the power method with scaling to find the dominant eigenvalue of \mathbf{B}. With $\mathbf{X}_0 = \begin{pmatrix} 1 \\ 1 \\ 1 \end{pmatrix}$ again, the results are summarized in the following table:

i	1	2	3	4	5	6	7
\mathbf{X}_i	$\begin{pmatrix} -1 \\ 1 \\ 0 \end{pmatrix}$	$\begin{pmatrix} 0.5 \\ -0.5 \\ 1 \end{pmatrix}$	$\begin{pmatrix} -1 \\ 1 \\ -0.6667 \end{pmatrix}$	$\begin{pmatrix} 0.8333 \\ -0.8333 \\ 1 \end{pmatrix}$	$\begin{pmatrix} -1 \\ 1 \\ -0.9091 \end{pmatrix}$	$\begin{pmatrix} 0.9545 \\ -0.9545 \\ 1 \end{pmatrix}$	$\begin{pmatrix} -1 \\ 1 \\ -0.9767 \end{pmatrix}$

Utilizing \mathbf{X}_7 and (8), we find

$$\lambda_2 \approx \frac{\mathbf{AX}_7 \cdot \mathbf{X}_7}{\mathbf{X}_7 \cdot \mathbf{X}_7} = -1.9996.$$

From these computations it seems apparent that the dominant eigenvalue of \mathbf{B} and a corresponding eigenvector are $\lambda_2 = -2$ and $\mathbf{K} = \begin{pmatrix} -1 \\ 1 \\ -1 \end{pmatrix}$.

To find the last eigenvalue of \mathbf{A}, we repeat the deflation process to find the dominant eigenvalue and a corresponding eigenvector of the matrix

$$\mathbf{C} = \mathbf{B} - \lambda_2 \mathbf{K}_2 \mathbf{K}_2^T = \begin{pmatrix} 0.1667 & -0.1667 & -0.3333 \\ -0.1667 & 0.1667 & 0.3333 \\ -0.3333 & 0.3333 & 0.6667 \end{pmatrix}$$

where we have used $\mathbf{K}_2 = \begin{pmatrix} -1/\sqrt{3} \\ 1/\sqrt{3} \\ -1/\sqrt{3} \end{pmatrix}$. The student is encouraged to verify that $\lambda_3 = 1$. ❑

Example 5 is somewhat artificial since eigenvalues of a matrix need not be "nice" numbers such as 3, −2, and 1. Moreover, we used the exact values of the dominant eigenvalues λ_1 and λ_2 in the formation of the matrices \mathbf{B} and \mathbf{C}. In practice, of course, we may have to be content with working with approximations to the dominant eigenvalue λ_1 and a corresponding normalized dominant eigenvector \mathbf{K}_1 of \mathbf{A}. If these approximations are used in (10), an error is introduced in the computation of the matrix \mathbf{B}, and so more errors may be introduced in the computation of its dominant eigenvalue λ_2 and dominant eigenvector \mathbf{K}_2. If λ_2 and \mathbf{K}_2 are now used to construct the matrix \mathbf{C}, it seems reasonable to conclude that errors are being compounded. In other words, the method of deflation can become increasingly inaccurate as more eigenvalues are computed.

■ **Inverse Power Method** In some applied problems we are more interested in approximating the eigenvalue of a matrix \mathbf{A} of smallest absolute value than the dominant eigenvalue. If \mathbf{A} is nonsingular, then the eigenvalues of \mathbf{A} are nonzero (prove this), and if $\lambda_1, \lambda_2, \ldots, \lambda_n$ are the eigenvalues of \mathbf{A}, then $1/\lambda_1, 1/\lambda_2, \ldots, 1/\lambda_n$ are the eigenvalues of \mathbf{A}^{-1}. The last fact can be seen by multiplying the equation $\mathbf{AK} = \lambda\mathbf{K}$, $\lambda \neq 0$, by \mathbf{A}^{-1} and $1/\lambda$ to get $\mathbf{A}^{-1}\mathbf{K} = (1/\lambda)\mathbf{K}$. Now if the eigenvalues of \mathbf{A} can be arranged in the order

$$|\lambda_1| \geq |\lambda_2| \geq |\lambda_3| \geq \cdots \geq |\lambda_{n-1}| > |\lambda_n|,$$

then we see that $1/\lambda_n$ is the dominant eigenvalue of \mathbf{A}^{-1}. By applying the power method to \mathbf{A}^{-1}, we approximate the eigenvalue of largest magnitude and, by taking its reciprocal, we find the eigenvalue of \mathbf{A} of least magnitude. This is called the **inverse power method.** See Problems 11–13 in Exercises 8.11.

E X E R C I S E S 8 . 1 1 Answers to odd-numbered problems begin on page A-45.

To the Instructor/Student: A calculator with matrix capabilities or a CAS would be useful in the following problems.

Each matrix in Problems 1–10 has a dominant eigenvalue.

In Problems 1 and 2, use the power method as illustrated in Example 3 to find the dominant eigenvalue and a corresponding dominant eigenvector of the given matrix.

1. $\begin{pmatrix} 1 & 1 \\ 2 & 0 \end{pmatrix}$

2. $\begin{pmatrix} -7 & 2 \\ 8 & -1 \end{pmatrix}$

In Problems 2–6, use the power method with scaling to find the dominant eigenvalue and a corresponding eigenvector of the given matrix.

3. $\begin{pmatrix} 2 & 4 \\ 3 & 13 \end{pmatrix}$

4. $\begin{pmatrix} -1 & 2 \\ -2 & 7 \end{pmatrix}$

5. $\begin{pmatrix} 5 & 4 & 2 \\ 4 & 5 & 2 \\ 2 & 2 & 2 \end{pmatrix}$

6. $\begin{pmatrix} 3 & 1 & 1 \\ 0 & 1 & 1 \\ 0 & 0 & 2 \end{pmatrix}$

In Problems 7–10, use the method of deflation to find the eigenvalues of the given matrix.

7. $\begin{pmatrix} 3 & 2 \\ 2 & 6 \end{pmatrix}$

8. $\begin{pmatrix} 1 & 3 \\ 3 & 9 \end{pmatrix}$

9. $\begin{pmatrix} 3 & -1 & 0 \\ -1 & 2 & -1 \\ 0 & -1 & 3 \end{pmatrix}$

10. $\begin{pmatrix} 0 & 0 & -4 \\ 0 & -4 & 0 \\ -4 & 0 & 15 \end{pmatrix}$

In Problems 11 and 12, use the inverse power method to find the eigenvalue of least magnitude for the given matrix.

11. $\begin{pmatrix} 1 & 1 \\ 3 & 4 \end{pmatrix}$

12. $\begin{pmatrix} -0.2 & 0.3 \\ 0.4 & -0.1 \end{pmatrix}$

13. In Example 3 of Section 3.9 we saw that the deflection curve of a thin column under an applied load P was defined by the boundary-value problem

$$EI \frac{d^2y}{dx^2} + Py = 0, \quad y(0) = 0, \quad y(L) = 0.$$

In this problem we show how to apply matrix techniques to compute the smallest critical load.

 Let the interval $[0, L]$ be divided into n subintervals of length $h = L/n$, and let $x_i = ih$, $i = 0, 1, \ldots, n$. For small values of h it follows from (6) of Section 6.5 that

$$\frac{d^2y}{dx^2} \approx \frac{y_{i+1} - 2y_i + y_{i-1}}{h^2},$$

where $y_i = y(x_i)$.

(a) Show that the differential equation can be replaced by the difference equation

$$EI(y_{i+1} - 2y_i + y_{i-1}) + Ph^2y_i = 0, \quad i = 1, 2, \ldots, n-1$$

(b) Show that for $n = 4$ the difference equation in part **(a)** yields the system of linear equations

$$\begin{pmatrix} 2 & -1 & 0 \\ -1 & 2 & -1 \\ 0 & -1 & 2 \end{pmatrix} \begin{pmatrix} y_1 \\ y_2 \\ y_3 \end{pmatrix} = \frac{PL^2}{16EI} \begin{pmatrix} y_1 \\ y_2 \\ y_3 \end{pmatrix}$$

Note that this system has the form of the eigenvalue problem $\mathbf{AY} = \lambda\mathbf{Y}$, where $\lambda = PL^2/16EI$.

(c) Find \mathbf{A}^{-1}.

(d) Use the inverse power method to find, to two decimal places, the eigenvalue of \mathbf{A} of least magnitude.

(e) Use the result of part **(d)** to compute the approximate smallest critical load. Compare your answer with that given in Section 3.9.

14. Suppose the column in Problem 13 is tapered so that the moment of inertia of a cross-section I varies linearly from $I(0) = I_0 = 0.002$ to $I(L) = I_L = 0.001$.

(a) Use the difference equation in part **(a)** of Problem 13 with $n = 4$ to set up a system of equations analogous to that given in part **(b)**.

(b) Proceed as in Problem 13 to find an approximation to the smallest critical load.

15. Computer Lab Assignment In Section 8.9 we saw how to compute a power \mathbf{A}^m for an $n \times n$ matrix \mathbf{A}. Consult the documentation for the CAS you have on hand for the command to compute the power \mathbf{A}^m. (In *Mathematica* the command is **MatrixPower[A, m]**.) The matrix

$$\mathbf{A} = \begin{pmatrix} 5 & -2 & 0 \\ -2 & 3 & -1 \\ 0 & -1 & 1 \end{pmatrix}$$

possesses a dominant eigenvalue.

(a) Use a CAS to compute \mathbf{A}^{10}.

(b) Now use (2), $\mathbf{X}_m = \mathbf{A}^m\mathbf{X}_0$, with $m = 10$ and $\mathbf{X}_0 = \begin{pmatrix} 1 \\ 0 \\ 0 \end{pmatrix}$, to compute \mathbf{X}_{10}. In the same manner compute \mathbf{X}_{12}. Then proceed as in (9) to find the approximate dominant eigenvector \mathbf{K}.

(c) If \mathbf{K} is an eigenvector of \mathbf{A}, then $\mathbf{AK} = \lambda\mathbf{K}$. Use this definition and the result in part **(b)** to find the dominant eigenvalue.

8.12 Diagonalization

■ **A Special Notation** We begin by introducing a shorthand notation for the product of two $n \times n$ matrices. This notation will be useful in proving the principal theorem of this section. To illustrate, suppose \mathbf{A} and \mathbf{B} are 2×2 matrices. Then

$$\mathbf{AB} = \begin{pmatrix} a_{11} & a_{12} \\ a_{21} & a_{22} \end{pmatrix}\begin{pmatrix} b_{11} & b_{12} \\ b_{21} & b_{22} \end{pmatrix} = \begin{pmatrix} \underbrace{a_{11}b_{11} + a_{12}b_{21}}_{\text{column 1}} & \underbrace{a_{11}b_{12} + a_{12}b_{22}}_{\text{column 2}} \\ a_{21}b_{11} + a_{22}b_{21} & a_{21}b_{12} + a_{22}b_{22} \end{pmatrix}. \quad (1)$$

If we write the columns of the matrix \mathbf{B} as vectors $\mathbf{X}_1 = \begin{pmatrix} b_{11} \\ b_{21} \end{pmatrix}$ and $\mathbf{X}_2 = \begin{pmatrix} b_{12} \\ b_{22} \end{pmatrix}$, then column 1 and column 2 in the product (1) can be expressed as the products \mathbf{AX}_1 and \mathbf{AX}_2. That is,

$$\mathbf{AB} = (\underbrace{\mathbf{AX}_1}_{\text{column 1}} \quad \underbrace{\mathbf{AX}_2}_{\text{column 2}}).$$

In general, for two $n \times n$ matrices

$$\mathbf{AB} = \mathbf{A}(\mathbf{X}_1 \quad \mathbf{X}_2 \quad \cdots \quad \mathbf{X}_n) = (\mathbf{AX}_1 \quad \mathbf{AX}_2 \quad \cdots \quad \mathbf{AX}_n), \tag{2}$$

where $\mathbf{X}_1, \mathbf{X}_2, \ldots, \mathbf{X}_n$, are the columns of \mathbf{B}.

■ **Diagonalizable Matrix** In Chapter 10 we shall see that eigenvalues, eigenvectors, orthogonal matrices, and the topic of this present section, diagonalization, are important tools in the solution of systems of linear first-order differential equations. The basic question that we shall consider in this section is:

> *For an $n \times n$ matrix \mathbf{A}, can we find an $n \times n$ nonsingular matrix \mathbf{P} such that $\mathbf{P}^{-1}\mathbf{AP} = \mathbf{D}$ is a diagonal matrix?*

If such a matrix \mathbf{P} can be found, then we say that the matrix \mathbf{A} can be **diagonalized,** or is **diagonalizable,** and that **P diagonalizes A**.

To discover how to diagonalize a matrix, let us assume for the sake of discussion that \mathbf{A} is a 3×3 diagonalizable matrix. Then there exists a 3×3 nonsingular matrix \mathbf{P} such that $\mathbf{P}^{-1}\mathbf{AP} = \mathbf{D}$ or $\mathbf{AP} = \mathbf{PD}$, where \mathbf{D} is a diagonal matrix

$$\mathbf{D} = \begin{pmatrix} d_{11} & 0 & 0 \\ 0 & d_{22} & 0 \\ 0 & 0 & d_{33} \end{pmatrix}.$$

If \mathbf{P}_1, \mathbf{P}_2, and \mathbf{P}_3 denote the columns of \mathbf{P}, then it follows from (2) that the equation $\mathbf{AP} = \mathbf{PD}$ is the same as

$$(\mathbf{AP}_1 \quad \mathbf{AP}_2 \quad \mathbf{AP}_3) = (d_{11}\mathbf{P}_1 \quad d_{22}\mathbf{P}_2 \quad d_{33}\mathbf{P}_3)$$

or

$$\mathbf{AP}_1 = d_{11}\mathbf{P}_1, \qquad \mathbf{AP}_2 = d_{22}\mathbf{P}_2, \qquad \mathbf{AP}_3 = d_{33}\mathbf{P}_3.$$

But by Definition 8.13 we see that d_{11}, d_{22}, and d_{33} are eigenvalues of \mathbf{A} associated with the eigenvectors \mathbf{P}_1, \mathbf{P}_2, and \mathbf{P}_3. These eigenvectors are linearly independent, since \mathbf{P} was assumed to be nonsingular.

We have just discovered, in a particular case, that if \mathbf{A} is diagonalizable, then the columns of the diagonalizing matrix \mathbf{P} consist of linearly independent eigenvectors of \mathbf{A}. Since we wish to diagonalize a matrix, we are really concerned with the validity of the converse of the last sentence. In other words, if we can find n linearly independent eigenvectors of an $n \times n$ matrix \mathbf{A} and form an $n \times n$ matrix \mathbf{P} whose columns consist of these eigenvectors, then does \mathbf{P} diagonalize \mathbf{A}? The answer is yes and will be proved in the next theorem.

THEOREM 8.30 Sufficient Condition for Diagonalizability

If an $n \times n$ matrix \mathbf{A} has n linearly independent eigenvectors $\mathbf{K}_1, \mathbf{K}_2, \ldots, \mathbf{K}_n$, then \mathbf{A} is diagonalizable.

PROOF We shall prove the theorem in the case when \mathbf{A} is a 3×3 matrix. Let \mathbf{K}_1, \mathbf{K}_2, and \mathbf{K}_3 be linearly independent eigenvectors corresponding to eigenvalues λ_1, λ_2, and λ_3; that is,

$$\mathbf{AK}_1 = \lambda_1\mathbf{K}_1, \quad \mathbf{AK}_2 = \lambda_2\mathbf{K}_2, \quad \text{and} \quad \mathbf{AK}_3 = \lambda_3\mathbf{K}_3. \tag{3}$$

Next form the 3×3 matrix \mathbf{P} with column vectors \mathbf{K}_1, \mathbf{K}_2, and \mathbf{K}_3: $\mathbf{P} = (\mathbf{K}_1 \ \mathbf{K}_2 \ \mathbf{K}_3)$. \mathbf{P} is nonsingular since, by hypothesis, the eigenvectors are linearly independent. Now using (2) and (3) we can write the product \mathbf{AP} as

$$\mathbf{AP} = (\mathbf{AK}_1 \quad \mathbf{AK}_2 \quad \mathbf{AK}_3) = (\lambda_1\mathbf{K}_1 \quad \lambda_2\mathbf{K}_2 \quad \lambda_3\mathbf{K}_3)$$

$$= (\mathbf{K}_1 \quad \mathbf{K}_2 \quad \mathbf{K}_3)\begin{pmatrix} \lambda_1 & 0 & 0 \\ 0 & \lambda_2 & 0 \\ 0 & 0 & \lambda_3 \end{pmatrix} = \mathbf{PD}.$$

Multiplying the last equation on the left by \mathbf{P}^{-1} then gives $\mathbf{P}^{-1}\mathbf{AP} = \mathbf{D}$. ❏

Note carefully in the proof of Theorem 8.30 that the entries in the diagonalized matrix are the eigenvalues of \mathbf{A} and the order in which these numbers appear on the diagonal of \mathbf{D} corresponds to the order in which the eigenvectors are used as columns in the matrix \mathbf{P}.

In view of the motivational discussion preceding Theorem 8.30, we can state the general result:

THEOREM 8.31 Criterion for Diagonalizability

An $n \times n$ matrix \mathbf{A} is diagonalizable if and only if \mathbf{A} has n linearly independent eigenvectors.

We saw in Section 8.8 that an $n \times n$ matrix \mathbf{A} has n linearly independent eigenvectors whenever it possesses n distinct eigenvalues.

THEOREM 8.32 Sufficient Condition for Diagonalizability

If an $n \times n$ matrix \mathbf{A} has n distinct eigenvalues, it is diagonalizable.

Example 1 Diagonalizing a Matrix

Diagonalize $\mathbf{A} = \begin{pmatrix} -5 & 9 \\ -6 & 10 \end{pmatrix}$ if possible.

SOLUTION First we find the eigenvalues of \mathbf{A}. The characteristic equation is $\det(\mathbf{A} - \lambda\mathbf{I}) = \begin{vmatrix} -5 - \lambda & 9 \\ -6 & 10 - \lambda \end{vmatrix} = \lambda^2 - 5\lambda + 4 = (\lambda - 1)(\lambda - 4) = 0$. The eigenvalues are $\lambda_1 = 1$ and $\lambda_2 = 4$. Since the eigenvalues are distinct, we know from Theorem 8.32 that \mathbf{A} is diagonalizable.

Next the eigenvectors of \mathbf{A} corresponding to $\lambda_1 = 1$ and $\lambda_2 = 4$ are, respectively,

$$\mathbf{K}_1 = \begin{pmatrix} 3 \\ 2 \end{pmatrix} \quad \text{and} \quad \mathbf{K}_2 = \begin{pmatrix} 1 \\ 1 \end{pmatrix}.$$

Using these vectors as columns, we find that the nonsingular matrix \mathbf{P} that diagonalizes \mathbf{A} is

$$\mathbf{P} = (\mathbf{K}_1 \quad \mathbf{K}_2) = \begin{pmatrix} 3 & 1 \\ 2 & 1 \end{pmatrix}.$$

Now
$$\mathbf{P}^{-1} = \begin{pmatrix} 1 & -1 \\ -2 & 3 \end{pmatrix},$$

and so carrying out the multiplication gives

$$\mathbf{P}^{-1}\mathbf{AP} = \begin{pmatrix} 1 & -1 \\ -2 & 3 \end{pmatrix}\begin{pmatrix} -5 & 9 \\ -6 & 10 \end{pmatrix}\begin{pmatrix} 3 & 1 \\ 2 & 1 \end{pmatrix} = \begin{pmatrix} 1 & 0 \\ 0 & 4 \end{pmatrix} = \mathbf{D}.$$ ❏

In Example 1, had we reversed the columns in \mathbf{P}, that is, $\mathbf{P} = \begin{pmatrix} 1 & 3 \\ 1 & 2 \end{pmatrix}$, then the diagonal matrix would have been $\mathbf{D} = \begin{pmatrix} 4 & 0 \\ 0 & 1 \end{pmatrix}$.

Example 2 Diagonalizing a Matrix

Consider the matrix $\mathbf{A} = \begin{pmatrix} 1 & 2 & 1 \\ 6 & -1 & 0 \\ -1 & -2 & -1 \end{pmatrix}$. We saw in Example 2 of Section 8.8 that the eigenvalues and corresponding eigenvectors are

$$\lambda_1 = 0, \quad \lambda_2 = -4, \quad \lambda_3 = 3, \quad \mathbf{K}_1 = \begin{pmatrix} 1 \\ 6 \\ -13 \end{pmatrix}, \quad \mathbf{K}_2 = \begin{pmatrix} -1 \\ 2 \\ 1 \end{pmatrix}, \quad \mathbf{K}_3 = \begin{pmatrix} 2 \\ 3 \\ -2 \end{pmatrix}.$$

Since the eigenvalues are distinct, \mathbf{A} is diagonalizable. We form the matrix

$$\mathbf{P} = (\mathbf{K}_1 \quad \mathbf{K}_2 \quad \mathbf{K}_3) = \begin{pmatrix} 1 & -1 & 2 \\ 6 & 2 & 3 \\ -13 & 1 & -2 \end{pmatrix}.$$

Matching the eigenvalues with the order in which the eigenvectors appear in \mathbf{P}, we know that the diagonal matrix will be

$$\mathbf{D} = \begin{pmatrix} 0 & 0 & 0 \\ 0 & -4 & 0 \\ 0 & 0 & 3 \end{pmatrix}.$$

Now from either of the methods of Section 8.6 we find

$$\mathbf{P}^{-1} = \begin{pmatrix} -1/12 & 0 & -1/12 \\ -9/28 & 2/7 & 3/28 \\ 8/21 & 1/7 & 2/21 \end{pmatrix},$$

and so
$$\mathbf{P}^{-1}\mathbf{AP} = \begin{pmatrix} -1/12 & 0 & -1/12 \\ -9/28 & 2/7 & 3/28 \\ 8/21 & 1/7 & 2/21 \end{pmatrix}\begin{pmatrix} 1 & 2 & 1 \\ 6 & -1 & 0 \\ -1 & -2 & -1 \end{pmatrix}\begin{pmatrix} 1 & -1 & 2 \\ 6 & 2 & 3 \\ -13 & 1 & -2 \end{pmatrix}$$

$$= \begin{pmatrix} 0 & 0 & 0 \\ 0 & -4 & 0 \\ 0 & 0 & 3 \end{pmatrix} = \mathbf{D}.$$ ❏

A matrix with repeated ☞
eigenvalues could be diagonalizable.

The condition that an $n \times n$ matrix \mathbf{A} have n distinct eigenvalues is sufficient—that is, a guarantee—that \mathbf{A} is diagonalizable. The condition that there be n distinct eigenvalues is not a necessary condition for the diagonalization of \mathbf{A}. In other words, if the matrix \mathbf{A} does not have n distinct eigenvalues, then it may or may not be diagonalizable.

Example 3 **A Matrix That Is Not Diagonalizable**

In Example 3 of Section 8.8 we saw that the matrix $\mathbf{A} = \begin{pmatrix} 3 & 4 \\ -1 & 7 \end{pmatrix}$ has a repeated eigenvalue $\lambda_1 = \lambda_2 = 5$. Correspondingly we were able to find only a *single* eigenvector $\mathbf{K}_1 = \begin{pmatrix} 2 \\ 1 \end{pmatrix}$. We conclude from Theorem 8.31 that \mathbf{A} is not diagonalizable. ❑

Example 4 **Repeated Eigenvalues Yet Diagonalizable**

The eigenvalues of the matrix $\mathbf{A} = \begin{pmatrix} 0 & 1 & 0 \\ 1 & 0 & 0 \\ 0 & 0 & 1 \end{pmatrix}$ are $\lambda_1 = -1$ and $\lambda_2 = \lambda_3 = 1$.

For $\lambda_1 = -1$ we find $\mathbf{K}_1 = \begin{pmatrix} 1 \\ -1 \\ 0 \end{pmatrix}$. For the repeated eigenvalue $\lambda_2 = \lambda_3 = 1$, Gauss–Jordan elimination gives

$$(\mathbf{A} - \mathbf{I}|\mathbf{0}) = \begin{pmatrix} -1 & 1 & 0 \\ 1 & -1 & 0 \\ 0 & 0 & 0 \end{pmatrix} \implies \begin{pmatrix} 1 & -1 & 0 \\ 0 & 0 & 0 \\ 0 & 0 & 0 \end{pmatrix}.$$

From the last matrix we see that $k_1 - k_2 = 0$. Since k_3 is not determined from the last matrix, we can choose its value arbitrarily. The choice $k_2 = 1$ gives $k_1 = 1$. If we then pick $k_3 = 0$, we get the eigenvector

$$\mathbf{K}_2 = \begin{pmatrix} 1 \\ 1 \\ 0 \end{pmatrix}.$$

The alternative choice $k_2 = 0$ gives $k_1 = 0$. If $k_3 = 1$, we get another eigenvector corresponding to $\lambda_2 = \lambda_3 = 1$:

$$\mathbf{K}_3 = \begin{pmatrix} 0 \\ 0 \\ 1 \end{pmatrix}.$$

Since the eigenvectors \mathbf{K}_1, \mathbf{K}_2, and \mathbf{K}_3 are linearly independent, a matrix that diagonalizes \mathbf{A} is

$$\mathbf{P} = \begin{pmatrix} 1 & 1 & 0 \\ -1 & 1 & 0 \\ 0 & 0 & 1 \end{pmatrix}.$$

Matching the eigenvalues with the eigenvectors in \mathbf{P}, we have $\mathbf{P}^{-1}\mathbf{A}\mathbf{P} = \mathbf{D}$, where

$$\mathbf{D} = \begin{pmatrix} -1 & 0 & 0 \\ 0 & 1 & 0 \\ 0 & 0 & 1 \end{pmatrix}.$$
❑

■ **Symmetric Matrices** An $n \times n$ symmetric matrix \mathbf{A} with real entries can always be diagonalized. This is a consequence of the fact that we can always find n linearly independent eigenvectors for such a matrix. Moreover, since we can find n mutually orthogonal eigenvectors, we can use an orthogonal matrix \mathbf{P} to diagonalize \mathbf{A}. A symmetric matrix is said to be **orthogonally diagonalizable**.

> **THEOREM 8.32 Criterion for Orthogonal Diagonalizability**
>
> An $n \times n$ matrix \mathbf{A} can be orthogonally diagonalized if and only if \mathbf{A} is symmetric.

PARTIAL PROOF We shall prove the necessity part (that is, the "only if" part) of the theorem. Assume an $n \times n$ matrix \mathbf{A} is orthogonally diagonalizable. Then there exists an orthogonal matrix \mathbf{P} such that $\mathbf{P}^{-1}\mathbf{A}\mathbf{P} = \mathbf{D}$ or $\mathbf{A} = \mathbf{P}\mathbf{D}\mathbf{P}^{-1}$. Since \mathbf{P} is orthogonal, $\mathbf{P}^{-1} = \mathbf{P}^T$ and consequently $\mathbf{A} = \mathbf{P}\mathbf{D}\mathbf{P}^T$. But from (*i*) and (*iii*) of Theorem 8.2 and the fact that a diagonal matrix is symmetric, we have

$$\mathbf{A}^T = (\mathbf{P}\mathbf{D}\mathbf{P}^T)^T = (\mathbf{P}^T)^T\mathbf{D}^T\mathbf{P}^T = \mathbf{P}\mathbf{D}\mathbf{P}^T = \mathbf{A}.$$

Thus, \mathbf{A} is symmetric. ❏

Example 5 Diagonalizing a Symmetric Matrix

Consider the symmetric matrix $\mathbf{A} = \begin{pmatrix} 9 & 1 & 1 \\ 1 & 9 & 1 \\ 1 & 1 & 9 \end{pmatrix}$. We saw in Example 4 of Section 8.8 that the eigenvalues and corresponding eigenvectors are

$$\lambda_1 = 11, \quad \lambda_2 = \lambda_3 = 8, \quad \mathbf{K}_1 = \begin{pmatrix} 1 \\ 1 \\ 1 \end{pmatrix}, \quad \mathbf{K}_2 = \begin{pmatrix} -1 \\ 1 \\ 0 \end{pmatrix}, \quad \mathbf{K}_3 = \begin{pmatrix} -1 \\ 0 \\ 1 \end{pmatrix}.$$

The eigenvectors \mathbf{K}_1, \mathbf{K}_2, and \mathbf{K}_3 are linearly independent, but note that they are not mutually orthogonal since \mathbf{K}_2 and \mathbf{K}_3, the eigenvectors corresponding to the repeated eigenvalue $\lambda_2 = \lambda_3 = 8$, are not orthogonal. For $\lambda_2 = \lambda_3 = 8$, we found the eigenvectors from Gauss–Jordan elimination

$$(\mathbf{A} - 8\mathbf{I}|\mathbf{0}) = \begin{pmatrix} 1 & 1 & 1 \\ 1 & 1 & 1 \\ 1 & 1 & 1 \end{pmatrix} \implies \begin{pmatrix} 1 & 1 & 1 \\ 0 & 0 & 0 \\ 0 & 0 & 0 \end{pmatrix},$$

which implies that $k_1 + k_2 + k_3 = 0$. Since two of the variables are arbitrary, we selected $k_2 = 1$, $k_3 = 0$ to obtain \mathbf{K}_2, and $k_2 = 0$, $k_3 = 1$ to obtain \mathbf{K}_3. Now if instead we choose $k_2 = 1$, $k_3 = 1$ and then $k_2 = 1$, $k_3 = -1$, we obtain, respectively, two entirely different but orthogonal eigenvectors:

$$\mathbf{K}_2 = \begin{pmatrix} -2 \\ 1 \\ 1 \end{pmatrix} \quad \text{and} \quad \mathbf{K}_3 = \begin{pmatrix} 0 \\ 1 \\ -1 \end{pmatrix}.$$

Thus, a new set of mutually orthogonal eigenvectors is

$$\mathbf{K}_1 = \begin{pmatrix} 1 \\ 1 \\ 1 \end{pmatrix}, \quad \mathbf{K}_2 = \begin{pmatrix} -2 \\ 1 \\ 1 \end{pmatrix}, \quad \mathbf{K}_3 = \begin{pmatrix} 0 \\ 1 \\ -1 \end{pmatrix}.$$

Multiplying these vectors, in turn, by the reciprocals of $\|\mathbf{K}_1\| = \sqrt{3}$, $\|\mathbf{K}_2\| = \sqrt{6}$, and $\|\mathbf{K}_1\| = \sqrt{2}$, we obtain an orthonormal set

$$\begin{pmatrix} \dfrac{1}{\sqrt{3}} \\[3mm] \dfrac{1}{\sqrt{3}} \\[3mm] \dfrac{1}{\sqrt{3}} \end{pmatrix}, \qquad \begin{pmatrix} -\dfrac{2}{\sqrt{6}} \\[3mm] \dfrac{1}{\sqrt{6}} \\[3mm] \dfrac{1}{\sqrt{6}} \end{pmatrix}, \qquad \begin{pmatrix} 0 \\[3mm] \dfrac{1}{\sqrt{2}} \\[3mm] -\dfrac{1}{\sqrt{2}} \end{pmatrix}.$$

We then use these vectors as columns to construct an orthogonal matrix that diagonalizes **A**:

$$\mathbf{P} = \begin{pmatrix} \dfrac{1}{\sqrt{3}} & -\dfrac{2}{\sqrt{6}} & 0 \\[3mm] \dfrac{1}{\sqrt{3}} & \dfrac{1}{\sqrt{6}} & \dfrac{1}{\sqrt{2}} \\[3mm] \dfrac{1}{\sqrt{3}} & \dfrac{1}{\sqrt{6}} & -\dfrac{1}{\sqrt{2}} \end{pmatrix}.$$

The diagonal matrix whose entries are the eigenvalues of **A** corresponding to the order in which the eigenvectors appear in **P** is then

$$\mathbf{D} = \begin{pmatrix} 11 & 0 & 0 \\ 0 & 8 & 0 \\ 0 & 0 & 8 \end{pmatrix}.$$

This is verified from

$$\mathbf{P}^{-1}\mathbf{AP} = \mathbf{P}^{T}\mathbf{AP} = \begin{pmatrix} \dfrac{1}{\sqrt{3}} & \dfrac{1}{\sqrt{3}} & \dfrac{1}{\sqrt{3}} \\[3mm] -\dfrac{2}{\sqrt{6}} & \dfrac{1}{\sqrt{6}} & \dfrac{1}{\sqrt{6}} \\[3mm] 0 & \dfrac{1}{\sqrt{2}} & -\dfrac{1}{\sqrt{2}} \end{pmatrix} \begin{pmatrix} 9 & 1 & 1 \\ 1 & 9 & 1 \\ 1 & 1 & 9 \end{pmatrix} \begin{pmatrix} \dfrac{1}{\sqrt{3}} & -\dfrac{2}{\sqrt{6}} & 0 \\[3mm] \dfrac{1}{\sqrt{3}} & \dfrac{1}{\sqrt{6}} & \dfrac{1}{\sqrt{2}} \\[3mm] \dfrac{1}{\sqrt{3}} & \dfrac{1}{\sqrt{6}} & -\dfrac{1}{\sqrt{2}} \end{pmatrix}$$

$$= \begin{pmatrix} 11 & 0 & 0 \\ 0 & 8 & 0 \\ 0 & 0 & 8 \end{pmatrix} = \mathbf{D}. \qquad \square$$

■ **Quadratic Forms** An algebraic expression of the form

$$ax^2 + bxy + cy^2 \tag{4}$$

is said to be a **quadratic form.** If we let $\mathbf{X} = \begin{pmatrix} x \\ y \end{pmatrix}$, then (4) can be written as the matrix product

$$\mathbf{X}^{T}\mathbf{AX} = (x \quad y) \begin{pmatrix} a & \tfrac{1}{2}b \\ \tfrac{1}{2}b & c \end{pmatrix} \begin{pmatrix} x \\ y \end{pmatrix}. \tag{5}$$

Observe that the matrix $\begin{pmatrix} a & \tfrac{1}{2}b \\ \tfrac{1}{2}b & c \end{pmatrix}$ is symmetric.

In calculus you may have seen that an appropriate rotation of axes enables us to eliminate the xy-term in an equation

$$ax^2 + bxy + cy^2 + dx + ey + f = 0.$$

As the next example will illustrate, we can eliminate the xy-term by means of an orthogonal matrix and diagonalization rather than by using trigonometry.

Example 6 Identifying a Conic Section

Identify the conic section whose equation is $2x^2 + 4xy - y^2 = 1$.

SOLUTION From (5) we can write the given equation as

$$(x \quad y)\begin{pmatrix} 2 & 2 \\ 2 & -1 \end{pmatrix}\begin{pmatrix} x \\ y \end{pmatrix} = 1 \quad \text{or} \quad \mathbf{X}^T\mathbf{A}\mathbf{X} = 1, \tag{6}$$

where $\mathbf{A} = \begin{pmatrix} 2 & 2 \\ 2 & -1 \end{pmatrix}$ and $\mathbf{X} = \begin{pmatrix} x \\ y \end{pmatrix}$. Now the eigenvalues and corresponding eigenvectors of \mathbf{A} are found to be

$$\lambda_1 = -2, \quad \lambda_2 = 3, \quad \mathbf{K}_1 = \begin{pmatrix} 1 \\ -2 \end{pmatrix}, \quad \mathbf{K}_2 = \begin{pmatrix} 2 \\ 1 \end{pmatrix}.$$

Observe that \mathbf{K}_1 and \mathbf{K}_2 are orthogonal. Moreover, $\|\mathbf{K}_1\| = \|\mathbf{K}_2\| = \sqrt{5}$, and so the vectors

$$\begin{pmatrix} \dfrac{1}{\sqrt{5}} \\ -\dfrac{2}{\sqrt{5}} \end{pmatrix} \quad \text{and} \quad \begin{pmatrix} \dfrac{2}{\sqrt{5}} \\ \dfrac{1}{\sqrt{5}} \end{pmatrix}$$

are orthonormal. Hence, the matrix

$$\mathbf{P} = \begin{pmatrix} \dfrac{1}{\sqrt{5}} & \dfrac{2}{\sqrt{5}} \\ -\dfrac{2}{\sqrt{5}} & \dfrac{1}{\sqrt{5}} \end{pmatrix}$$

is orthogonal. If we define the change of variables $\mathbf{X} = \mathbf{P}\mathbf{X}'$ where $\mathbf{X}' = \begin{pmatrix} X \\ Y \end{pmatrix}$ then the quadratic form $2x^2 + 4xy - y^2$ can be written

$$\mathbf{X}^T\mathbf{A}\mathbf{X} = (\mathbf{X}')^T\mathbf{P}^T\mathbf{A}\mathbf{P}\mathbf{X}' = (\mathbf{X}')^T(\mathbf{P}^T\mathbf{A}\mathbf{P})\mathbf{X}'.$$

Since \mathbf{P} orthogonally diagonalizes the symmetric matrix \mathbf{A}, the last equation is the same as

$$\mathbf{X}^T\mathbf{A}\mathbf{X} = (\mathbf{X}')^T\mathbf{D}\mathbf{X}'. \tag{7}$$

Using (7), we see that (6) becomes

$$(X \quad Y)\begin{pmatrix} -2 & 0 \\ 0 & 3 \end{pmatrix}\begin{pmatrix} X \\ Y \end{pmatrix} = 1 \quad \text{or} \quad -2X^2 + 3Y^2 = 1.$$

This last equation is recognized as the standard form of a hyperbola. The xy-coordinates of the eigenvectors are $(1, -2)$ and $(2, 1)$. Using the substitution $\mathbf{X} = \mathbf{P}\mathbf{X}'$ in the form $\mathbf{X}' = \mathbf{P}^{-1}\mathbf{X} = \mathbf{P}^T\mathbf{X}$, we find that the XY-coordinates of these two points are $(\sqrt{5}, 0)$ and $(0, \sqrt{5})$, respectively. From this we conclude that the X-axis and Y-axis are as shown in Figure 8.11. The eigenvectors, shown in black in the figure, lie along the new axes. The X- and Y-axes are called the **principal axes** of the conic. ❑

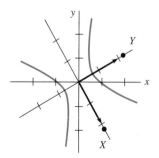

Figure 8.11

Answers to odd-numbered problems begin on page A-45.

In Problems 1–20, determine whether the given matrix \mathbf{A} is diagonalizable. If so, find the matrix \mathbf{P} that diagonalizes \mathbf{A} and the diagonal matrix \mathbf{D} such that $\mathbf{D} = \mathbf{P}^{-1}\mathbf{A}\mathbf{P}$.

1. $\begin{pmatrix} 2 & 3 \\ 1 & 4 \end{pmatrix}$

2. $\begin{pmatrix} -4 & -5 \\ 8 & 10 \end{pmatrix}$

3. $\begin{pmatrix} 0 & 1 \\ -1 & 2 \end{pmatrix}$

4. $\begin{pmatrix} 0 & 5 \\ 1 & 0 \end{pmatrix}$

5. $\begin{pmatrix} -9 & 13 \\ -2 & 6 \end{pmatrix}$

6. $\begin{pmatrix} -5 & -3 \\ 5 & 11 \end{pmatrix}$

7. $\begin{pmatrix} \frac{1}{2} & \frac{1}{6} \\ \frac{1}{6} & \frac{1}{2} \end{pmatrix}$

8. $\begin{pmatrix} -2 & -1 \\ 1 & -4 \end{pmatrix}$

9. $\begin{pmatrix} 0 & 1 \\ -1 & 0 \end{pmatrix}$

10. $\begin{pmatrix} 1 & 2 \\ -\frac{1}{2} & 1 \end{pmatrix}$

11. $\begin{pmatrix} 1 & 0 & 1 \\ 0 & -1 & 3 \\ 0 & 0 & 2 \end{pmatrix}$

12. $\begin{pmatrix} 1 & 2 & 2 \\ 2 & 3 & -2 \\ -5 & 3 & 8 \end{pmatrix}$

13. $\begin{pmatrix} 1 & -1 & 1 \\ 0 & 1 & 0 \\ 1 & -1 & 1 \end{pmatrix}$

14. $\begin{pmatrix} 0 & -9 & 0 \\ 1 & 0 & 0 \\ 0 & 0 & 1 \end{pmatrix}$

15. $\begin{pmatrix} 1 & 3 & -1 \\ 0 & 2 & 4 \\ 0 & 0 & 1 \end{pmatrix}$

16. $\begin{pmatrix} 1 & 1 & 0 \\ 0 & 2 & 0 \\ 0 & 0 & 3 \end{pmatrix}$

17. $\begin{pmatrix} 1 & 2 & 0 \\ 2 & -1 & 0 \\ 0 & 0 & 1 \end{pmatrix}$

18. $\begin{pmatrix} 0 & 0 & 1 \\ 1 & 0 & -3 \\ 0 & 1 & 3 \end{pmatrix}$

19. $\begin{pmatrix} -8 & -10 & 7 & -9 \\ 0 & 2 & 0 & 0 \\ -9 & -9 & 8 & -9 \\ 1 & 1 & -1 & 2 \end{pmatrix}$

20. $\begin{pmatrix} 4 & 2 & -1 & 4 \\ 0 & 2 & 0 & 0 \\ 1 & 3 & 2 & 1 \\ 0 & 0 & 0 & 2 \end{pmatrix}$

In Problems 21–30, the given matrix \mathbf{A} is symmetric. Find an orthogonal matrix \mathbf{P} that diagonalizes \mathbf{A} and the diagonal matrix \mathbf{D} such that $\mathbf{D} = \mathbf{P}^{T}\mathbf{A}\mathbf{P}$.

21. $\begin{pmatrix} 1 & 1 \\ 1 & 1 \end{pmatrix}$

22. $\begin{pmatrix} 3 & 2 \\ 2 & 0 \end{pmatrix}$

23. $\begin{pmatrix} 5 & \sqrt{10} \\ \sqrt{10} & 8 \end{pmatrix}$

24. $\begin{pmatrix} 1 & -2 \\ -2 & 1 \end{pmatrix}$

25. $\begin{pmatrix} 0 & 1 & 0 \\ 1 & 0 & 0 \\ 0 & 0 & 1 \end{pmatrix}$

26. $\begin{pmatrix} 1 & -2 & 2 \\ -2 & 1 & -2 \\ 2 & -2 & 1 \end{pmatrix}$

27. $\begin{pmatrix} 5 & -2 & 0 \\ -2 & 6 & -2 \\ 0 & -2 & 7 \end{pmatrix}$

28. $\begin{pmatrix} 3 & 0 & 1 \\ 0 & 1 & 0 \\ 1 & 0 & 1 \end{pmatrix}$

29. $\begin{pmatrix} 1 & 0 & 7 \\ 0 & 1 & 0 \\ 7 & 0 & 1 \end{pmatrix}$

30. $\begin{pmatrix} 0 & 1 & 0 & 1 \\ 1 & 0 & 1 & 0 \\ 0 & 1 & 0 & 1 \\ 1 & 0 & 1 & 0 \end{pmatrix}$

In Problems 31–34, use the procedure illustrated in Example 6 to identify the given conic section. Graph.

31. $5x^2 - 2xy + 5y^2 = 24$

32. $13x^2 - 10xy + 13y^2 = 288$

33. $-3x^2 + 8xy + 3y^2 = 20$

34. $16x^2 + 24xy + 9y^2 - 3x + 4y = 0$

35. Find a 2×2 matrix \mathbf{A} that has eigenvalues $\lambda_1 = 2$ and $\lambda_2 = 3$ and corresponding eigenvectors $\mathbf{K}_1 = \begin{pmatrix} 1 \\ 2 \end{pmatrix}$ and $\mathbf{K}_2 = \begin{pmatrix} 1 \\ 1 \end{pmatrix}$.

36. Find a 3×3 symmetric matrix that has eigenvalues $\lambda_1 = 1$, $\lambda_2 = 3$, and $\lambda_3 = 5$ and corresponding eigenvectors

$$\mathbf{K}_1 = \begin{pmatrix} 1 \\ -1 \\ 1 \end{pmatrix}, \ \mathbf{K}_2 = \begin{pmatrix} 1 \\ 0 \\ -1 \end{pmatrix}, \ \text{and } \mathbf{K}_3 = \begin{pmatrix} 1 \\ 2 \\ 1 \end{pmatrix}.$$

37. If \mathbf{A} is an $n \times n$ diagonalizable matrix, then $\mathbf{D} = \mathbf{P}^{-1}\mathbf{A}\mathbf{P}$, where \mathbf{D} is a diagonal matrix. Show that if m is a positive integer, then $\mathbf{A}^m = \mathbf{P}\mathbf{D}^m\mathbf{P}^{-1}$.

38. The mth power of a diagonal matrix

$$\mathbf{D} = \begin{pmatrix} a_{11} & 0 & \cdots & 0 \\ 0 & a_{22} & \cdots & 0 \\ \vdots & & & \vdots \\ 0 & 0 & \cdots & a_{nn} \end{pmatrix}$$

is $\quad \mathbf{D}^m = \begin{pmatrix} a_{11}^m & 0 & \cdots & 0 \\ 0 & a_{22}^m & \cdots & 0 \\ \vdots & & & \vdots \\ 0 & 0 & \cdots & a_{nn}^m \end{pmatrix}.$

Use this result to compute

$$\begin{pmatrix} 2 & 0 & 0 & 0 \\ 0 & 3 & 0 & 0 \\ 0 & 0 & -1 & 0 \\ 0 & 0 & 0 & 5 \end{pmatrix}^4.$$

In Problems 39 and 40, use the results of Problems 37 and 38 to find the indicated power of the given matrix.

39. $\mathbf{A} = \begin{pmatrix} 1 & 1 \\ 2 & 0 \end{pmatrix}$, $\quad \mathbf{A}^5$

40. $\mathbf{A} = \begin{pmatrix} 6 & -10 \\ 3 & -5 \end{pmatrix}$, $\quad \mathbf{A}^{10}$

8.13 Cryptography

■ **Codes** The word *cryptography* is a combination of two Greek words: *crypto,* meaning "hidden" or "secret," and *grapho,* which means "writing." Cryptography then is the study of making "secret writings" or **codes.**

In this section we will consider a system of encoding and decoding messages that requires both the sender of the message and the receiver of the message to know:

- a specified rule of correspondence between a set of symbols (such as letters of the alphabet and punctuation marks from which messages are composed) and a set of integers; and
- a specified *nonsingular* matrix **A.**

A natural correspondence between the first 27 nonnegative integers and the letters of the alphabet and a blank space (to separate words) is given by

$$
\begin{array}{ccccccccccccccccccccccccccc}
0 & & 1 & 2 & 3 & 4 & 5 & 6 & 7 & 8 & 9 & 10 & 11 & 12 & 13 & 14 & 15 & 16 & 17 & 18 & 19 & 20 & 21 & 22 & 23 & 24 & 25 & 26 \\
\text{space} & & a & b & c & d & e & f & g & h & i & j & k & l & m & n & o & p & q & r & s & t & u & v & w & x & y & z
\end{array} \tag{1}
$$

From (1) the numerical equivalent of the message

<div align="center">SEND THE DOCUMENT TODAY</div>

is

$$19\ 5\ 14\ 4\ 0\ 20\ 8\ 5\ 0\ 4\ 15\ 3\ 21\ 13\ 5\ 14\ 20\ 0\ 20\ 15\ 4\ 1\ 25 \tag{2}$$

The sender will **encode** the message by means of the nonsingular matrix **A** and, as we shall see, the receiver of the encoded message will **decode** the message by means of the (unique) matrix \mathbf{A}^{-1}. The numerical message (2) is now written as a matrix. Since there are 23 symbols in the message, we need a matrix that will hold a minimum of 24 entries (an $m \times n$ matrix has mn entries). We choose to write (2) as the 3×8 matrix

$$
\mathbf{M} = \begin{pmatrix} 19 & 5 & 14 & 4 & 0 & 20 & 8 & 5 \\ 0 & 4 & 15 & 3 & 21 & 13 & 5 & 14 \\ 20 & 0 & 20 & 15 & 4 & 1 & 25 & 0 \end{pmatrix}. \tag{3}
$$

Note that the last entry (a_{38}) in the message matrix **M** has been simply padded with a space represented by the number 0. Of course, we could have written (2) as a 6×4 or a 4×6 matrix but that would require a larger encoding matrix. A 3×8 matrix allows us to encode the message by means of a 3×3 matrix. The size of the matrices used is only a concern when the encoding and decoding are done by hand rather than by a computer.

The encoding matrix **A** is chosen, or rather constructed, so that

- **A** is nonsingular,
- **A** has only integer entries, and
- \mathbf{A}^{-1} has only integer entries.

The last criterion is not particularly difficult to accomplish. We need only select the integer entries of **A** in such a manner that $\det \mathbf{A} = \pm 1$. For a 2×2 or a 3×3 matrix we can then find \mathbf{A}^{-1} by the formulas in (4) and (5) of Section 8.6. If **A** has integer

entries, then all the cofactors C_{11}, C_{12}, and so on are also integers. For the discussion on hand we choose

$$\mathbf{A} = \begin{pmatrix} -1 & 0 & -1 \\ 2 & 3 & 4 \\ 2 & 4 & 5 \end{pmatrix}. \tag{4}$$

You should verify that det $\mathbf{A} = -1$.

The original message is encoded by premultiplying the message matrix \mathbf{M} by \mathbf{A}; that is, the message is sent as the matrix:

$$\mathbf{B} = \mathbf{AM} = \begin{pmatrix} -1 & 0 & -1 \\ 2 & 3 & 4 \\ 2 & 4 & 5 \end{pmatrix} \begin{pmatrix} 19 & 5 & 14 & 4 & 0 & 20 & 8 & 5 \\ 0 & 4 & 15 & 3 & 21 & 13 & 5 & 14 \\ 20 & 0 & 20 & 15 & 4 & 1 & 25 & 0 \end{pmatrix}$$

$$= \begin{pmatrix} -39 & -5 & -34 & -19 & -4 & -21 & -33 & -5 \\ 118 & 22 & 153 & 77 & 79 & 83 & 131 & 52 \\ 138 & 26 & 188 & 95 & 104 & 97 & 161 & 66 \end{pmatrix}. \tag{5}$$

You should try to imagine the difficulty of decoding (5) without knowledge of \mathbf{A}. But the receiver of the encoded message \mathbf{B} knows \mathbf{A} and its inverse, and so decoding is the straightforward computation of premultiplying \mathbf{B} by \mathbf{A}^{-1}:

$$\mathbf{AM} = \mathbf{B} \quad \text{implies} \quad \mathbf{M} = \mathbf{A}^{-1}\mathbf{B}.$$

For the matrix (4) we find from (5) of Section 8.5 that

$$\mathbf{A}^{-1} = \begin{pmatrix} 1 & 4 & -3 \\ 2 & 3 & -2 \\ -2 & -4 & 3 \end{pmatrix}.$$

Thus, the decoded message is

$$\mathbf{M} = \begin{pmatrix} 1 & 4 & -3 \\ 2 & 3 & -2 \\ -2 & -4 & 3 \end{pmatrix} \begin{pmatrix} -39 & -5 & -34 & -19 & -4 & -21 & -33 & -5 \\ 118 & 22 & 153 & 77 & 79 & 83 & 131 & 52 \\ 138 & 26 & 188 & 95 & 104 & 97 & 161 & 66 \end{pmatrix}$$

$$= \begin{pmatrix} 19 & 5 & 14 & 4 & 0 & 20 & 8 & 5 \\ 0 & 4 & 15 & 3 & 21 & 13 & 5 & 14 \\ 20 & 0 & 20 & 15 & 4 & 1 & 25 & 0 \end{pmatrix}$$

or

$$19\ 5\ 14\ 4\ 0\ 20\ 8\ 5\ 0\ 4\ 15\ 3\ 21\ 13\ 5\ 14\ 20\ 0\ 20\ 15\ 4\ 1\ 25\ 0.$$

By also knowing the original correspondence (1), the receiver translates the numbers into

$$\text{SEND__THE__DOCUMENT__TODAY__}$$

where we have indicated the blank spaces by lines.

Some observations are in order. The correspondence or mapping (1) is one of many such correspondences that can be set up between the letters of the alphabet (we could even include punctuation symbols such as the comma and period) and integers. Using the 26 letters of the alphabet and the blank space, we can set up 27! of these correspondences. (Why?) Furthermore, we could have used a 2×2 matrix to encode (2). The size of the message matrix \mathbf{M} would then have to be at least 2×12 in order to contain the 23 entries of the message. For example, if

$$A = \begin{pmatrix} 1 & 2 \\ 0 & 1 \end{pmatrix} \quad \text{and} \quad M = \begin{pmatrix} 19 & 5 & 14 & 4 & 0 & 20 & 8 & 5 & 0 & 4 & 15 & 3 \\ 21 & 13 & 5 & 14 & 20 & 0 & 20 & 15 & 4 & 1 & 25 & 0 \end{pmatrix},$$

then

$$B = AM = \begin{pmatrix} 61 & 31 & 24 & 32 & 40 & 20 & 48 & 35 & 8 & 6 & 65 & 3 \\ 21 & 13 & 5 & 14 & 20 & 0 & 20 & 15 & 4 & 1 & 25 & 0 \end{pmatrix}.$$

Using $A^{-1} = \begin{pmatrix} 1 & -2 \\ 0 & 1 \end{pmatrix}$, we obtain as before

$$M = A^{-1}B = \begin{pmatrix} 1 & -2 \\ 0 & 1 \end{pmatrix}\begin{pmatrix} 61 & 31 & 24 & 32 & 40 & 20 & 48 & 35 & 8 & 6 & 65 & 3 \\ 21 & 13 & 5 & 14 & 20 & 0 & 20 & 15 & 4 & 1 & 25 & 0 \end{pmatrix}$$

$$= \begin{pmatrix} 19 & 5 & 14 & 4 & 0 & 20 & 8 & 5 & 0 & 4 & 15 & 3 \\ 21 & 13 & 5 & 14 & 20 & 0 & 20 & 15 & 4 & 1 & 25 & 0 \end{pmatrix}.$$

There is no particular reason why the numerical message (2) has to be broken down as rows (1×8 vectors) as in the matrix (3). Alternatively, (2) could be broken down as columns (3×1 vectors) as shown in the matrix

$$\begin{pmatrix} 19 & 4 & 8 & 4 & 21 & 14 & 20 & 1 \\ 5 & 0 & 5 & 15 & 13 & 20 & 15 & 25 \\ 14 & 20 & 0 & 3 & 5 & 0 & 4 & 0 \end{pmatrix}.$$

Finally, it may be desirable to send the encoded message as letters of the alphabet rather than as numbers. In Problem 13 in Exercises 8.13, we shall show how to transmit SEND THE DOCUMENT TODAY encoded as

<div align="center">OVTHWFUVJVRWYBWYCZZNWPZL</div>

EXERCISES 8.13 Answers to odd-numbered problems begin on page A-46.

In Problems 1–6, use the matrix **A** and the correspondence (1) to encode the given message. Verify your work by decoding the encoded message.

1. $A = \begin{pmatrix} 1 & 2 \\ 1 & 1 \end{pmatrix}$; SEND HELP

2. $A = \begin{pmatrix} 3 & 5 \\ 1 & 2 \end{pmatrix}$; THE MONEY IS HERE

3. $A = \begin{pmatrix} 3 & 5 \\ 2 & 3 \end{pmatrix}$; PHONE HOME

4. $A = \begin{pmatrix} 1 & 2 & 3 \\ 1 & 1 & 2 \\ 0 & 1 & 2 \end{pmatrix}$; MADAME X HAS THE PLANS

5. $A = \begin{pmatrix} 2 & 1 & 1 \\ 1 & 1 & 1 \\ -1 & 1 & 0 \end{pmatrix}$; GO NORTH ON MAIN ST

6. $A = \begin{pmatrix} 5 & 3 & 0 \\ 4 & 3 & -1 \\ 5 & 2 & 2 \end{pmatrix}$; DR JOHN IS THE SPY

In Problems 7–10, use the matrix **A** and the correspondence (1) to decode the given message.

7. $A = \begin{pmatrix} 8 & 3 \\ 5 & 2 \end{pmatrix}$; $B = \begin{pmatrix} 152 & 184 & 171 & 86 & 212 \\ 95 & 116 & 107 & 56 & 133 \end{pmatrix}$

8. $A = \begin{pmatrix} 2 & -1 \\ 1 & -1 \end{pmatrix}$;

$$B = \begin{pmatrix} 46 & -7 & -13 & 22 & -18 & 1 & 10 \\ 23 & -15 & -14 & 2 & -18 & -12 & 5 \end{pmatrix}$$

9. $A = \begin{pmatrix} 1 & 0 & 1 \\ 0 & 1 & 0 \\ 1 & 0 & 0 \end{pmatrix}$; $B = \begin{pmatrix} 31 & 21 & 21 & 22 & 20 & 9 \\ 19 & 0 & 9 & 13 & 16 & 15 \\ 13 & 1 & 20 & 8 & 0 & 9 \end{pmatrix}$

10. $A = \begin{pmatrix} 2 & 1 & 1 \\ 0 & 0 & -1 \\ 1 & 1 & 1 \end{pmatrix}$;

$$B = \begin{pmatrix} 36 & 32 & 28 & 61 & 26 & 56 & 10 & 12 \\ -9 & -2 & -18 & -1 & -18 & -25 & 0 & 0 \\ 23 & 27 & 23 & 41 & 26 & 43 & 5 & 12 \end{pmatrix}$$

11. Using the correspondence (1), we encoded the following message by a 2×2 matrix:

$$\begin{pmatrix} 17 & 16 & 18 & 5 & 34 & 0 & 34 & 20 & 9 & 5 & 25 \\ -30 & -31 & -32 & -10 & -59 & 0 & -54 & -35 & -13 & -6 & -50 \end{pmatrix}$$

Decode the message if its first two letters are DA and its last two letters are AY.

12. (a) Using the correspondence

1 2 3 4 5 6 7 8 9 10 11 12 13 14 15 16 17 18 19 20 21 22 23 24 25 26 27

j k l n m s t u w x g h i o p q r v y z a b c d e f space

find the numerical equivalent of the message

BUY ALL AVAILABLE STOCK AT MARKET

(b) Encode the message by *postmultiplying* the message matrix **M** by

$$\mathbf{A} = \begin{pmatrix} 1 & 1 & 0 \\ 1 & 0 & 1 \\ 1 & 1 & -1 \end{pmatrix}.$$

(c) Verify your work by decoding the encoded message in part **(b)**.

13. Consider the matrices **A** and **B** defined in (4) and (5), respectively.

(a) Rewrite **B** as **B'** using integers modulo 27.*

(b) Verify that the encoded message to be sent in letters is

OVTHWFUVJVRWYBWYCZZNWPZL

(c) Decode the encoded message by computing $\mathbf{A}^{-1}\mathbf{B}'$ and rewriting the result using integers modulo 27.

8.14 An Error-Correcting Code

■ **Arithmetic Modulo 2** In this section we will deal only with matrices whose entries are binary digits, namely, zeros and ones. When adding and multiplying such matrices, we will use **arithmetic modulo 2**. This arithmetic is defined by the addition and multiplication tables

+	0	1		×	0	1
0	0	1		0	0	0
1	1	0		1	0	1

Fundamental properties such as the commutative and associative laws hold in this system. The only notable exception here is that $1 + 1 = 0$.

■ **Binary Strings** In contrast to the last section, there is no connotation of secretiveness to the word "code" as used in this section. We are going to examine briefly the concept of digital communication, say, communication between a satellite and a computer. In such communication the **messages** or **words** are binary n-tuples, that is, n-tuples consisting of only zeros and ones or **bits**. An n-bit word is also said to be a binary **string** of **length** n.

> ### *Example 1* Binary Strings
>
> (a) The ordered 4-tuple (0, 1, 0, 1) is a 4-bit word or a string of length four.
> (b) The binary (that is, base 2) representation of the number 39 is 1 0 0 1 1 1 or as a 6-tuple (1, 0, 0, 1, 1, 1).
> (c) The ASCII** word for the letter Z is the string of length 8: (1, 0, 0, 1, 1, 0, 1, 0). ❑

* For integers a and b, we write $a = b \pmod{27}$ if b is the remainder $(0 \le b < 27)$ when a is divided by 27. For example, $33 = 6 \pmod{27}$, $28 = 1 \pmod{27}$, and so on. Negative integers are handled in the following manner. If $27 = 0 \pmod{27}$, then, for example, $25 + 2 = 0 \pmod{27}$ so that $-25 = 2 \pmod{27}$ and $-2 = 25 \pmod{27}$. Also, $-30 = 24 \pmod{27}$, since $30 + 24 \ (= 54) = 0 \pmod{27}$.

** American Standard Code for Information Interchange.

For convenience a word of length n will be written as a $1 \times n$ matrix, that is, as a row vector. For example, the 4-bit word in Example 1 would be written as the 1×4 matrix $\mathbf{W} = (0 \quad 1 \quad 0 \quad 1)$.

■ **Codes** By **encoding** a message, we mean a process whereby we transform a word \mathbf{W} of length n into another word \mathbf{C} of length $n + m$ by augmenting \mathbf{W} with m additional bits, called **parity check bits.** An encoded word is said to be a **code word.** By **decoding** a received message we mean another process which gives either the decoded message or an indication that an error has occurred during transmission. An encoding/decoding scheme is called a **code.**

One of the simplest codes is the parity check code. In this code a word is encoded according to the rule:

If the number of ones in the word is even: Add 0 to the word. (1)
odd: Add 1 to the word.

The word *parity* refers to whether the number of ones in a word is even or odd. The encoding rule given in (1) makes the parity of the code word always even.

> ### *Example 2* Encoding Words
>
> Use the parity check code to encode the words (a) $\mathbf{W} = (1 \quad 0 \quad 0 \quad 0 \quad 1 \quad 1)$ and (b) $\mathbf{W} = (1 \quad 1 \quad 1 \quad 0 \quad 0 \quad 1)$.
>
> **SOLUTION** (a) Since the number of ones in \mathbf{W} is odd, we add the extra bit 1 to the end of the word \mathbf{W}. The code word is then $\mathbf{C} = (1 \quad 0 \quad 0 \quad 0 \quad 1 \quad 1 \quad 1)$.
> (b) In this case the number of ones is even, so the extra bit added to the word is 0. The encoded word is $\mathbf{C} = (1 \quad 1 \quad 1 \quad 0 \quad 0 \quad 1 \quad 0)$. ❑

In digital communication it is the encoded word \mathbf{C} that is transmitted. But due to some kind of interference or noise in the transmission channel, one or more of the bits of \mathbf{C} may be changed. Thus, the message transmitted is not always the message received. See Figure 8.12.

The parity check code enables the decoder to detect single errors. Suppose \mathbf{R} is the message received. A single error in \mathbf{R} means that one bit has been changed; either a zero has been changed to a one or a one has been changed to a zero. In either circumstance the parity of the word \mathbf{R} is odd.

Figure 8.12

> ### *Example 3* Decoding Words
>
> Use the parity check code to decode the words
> (a) $\mathbf{R} = (1 \quad 1 \quad 0 \quad 0 \quad 1 \quad 0 \quad 1)$ and
> (b) $\mathbf{R} = (1 \quad 0 \quad 1 \quad 1 \quad 0 \quad 0 \quad 0)$.
>
> **SOLUTION** (a) The parity of \mathbf{R} is even. We drop the last bit and take the decoded message to be $(1 \quad 1 \quad 0 \quad 0 \quad 1 \quad 0)$.
>
> (b) The parity of \mathbf{R} is odd. The decoding is simply: a parity error. ❑

For some types of digital communication, such as communication internal to a computer, the parity check code is adequate. But Example 2 clearly indicates one major pitfall of the code: If an error occurs, we do not know how to correct it since we do not know which bit is incorrect. Moreover, multiple errors could occur in transmission. If, say, two ones were changed to zeros in transmission, the message received still has even parity and the decoding takes place by dropping the last bit. In this case at least one of the bits in the decoded message is in error.

■ **Hamming Codes** The parity check code is an example of an *error-detecting,* but not an *error-correcting,* code. For the remainder of this discussion we will consider an error-detecting/correcting code called the **Hamming (7, 4) code.** This code, one of a widely used set of codes invented by the mathematician Richard W. Hamming at Bell Laboratories in the 1950s, is an encoding/decoding scheme that can detect the presence of a single error in a received message and can tell which bit must be corrected. In the (7, 4) code the encoding process consists of transforming a 4-bit word $\mathbf{W} = (w_1 \quad w_2 \quad w_3 \quad w_4)$ into a 7-bit code word

$$\mathbf{C} = (c_1 \quad c_2 \quad w_1 \quad c_3 \quad w_2 \quad w_3 \quad w_4),$$

where c_1, c_2, and c_3 denote the parity check bits. (Words longer than four bits can be broken up into sequences of words of length four.)

■ **Encoding** In the Hamming (7, 4) code the parity check bits c_1, c_2, and c_3 are defined in terms of the information bits w_1, w_2, w_3, and w_4:

$$c_1 = w_1 + w_2 + w_4$$

$$c_2 = w_1 + w_3 + w_4 \tag{2}$$

$$c_3 = w_2 + w_3 + w_4,$$

where the arithmetic is carried out modulo 2. Using matrices, we can write (2) as the product

$$\begin{pmatrix} c_1 \\ c_2 \\ c_3 \end{pmatrix} = \begin{pmatrix} 1 & 1 & 0 & 1 \\ 1 & 0 & 1 & 1 \\ 0 & 1 & 1 & 1 \end{pmatrix} \begin{pmatrix} w_1 \\ w_2 \\ w_3 \\ w_4 \end{pmatrix}. \tag{3}$$

Example 4 **Encoding a Word**

Encode the word $\mathbf{W} = (1 \quad 0 \quad 1 \quad 1)$.

SOLUTION From (3) we have, with $w_1 = 1$, $w_2 = 0$, $w_3 = 1$, and $w_4 = 1$:

$$\begin{pmatrix} c_1 \\ c_2 \\ c_3 \end{pmatrix} = \begin{pmatrix} 1 & 1 & 0 & 1 \\ 1 & 0 & 1 & 1 \\ 0 & 1 & 1 & 1 \end{pmatrix} \begin{pmatrix} 1 \\ 0 \\ 1 \\ 1 \end{pmatrix} = \begin{pmatrix} 1 \cdot 1 + 1 \cdot 0 + 0 \cdot 1 + 1 \cdot 1 \\ 1 \cdot 1 + 0 \cdot 0 + 1 \cdot 1 + 1 \cdot 1 \\ 0 \cdot 1 + 1 \cdot 0 + 1 \cdot 1 + 1 \cdot 1 \end{pmatrix} = \begin{pmatrix} 0 \\ 1 \\ 0 \end{pmatrix}$$

That is, $c_1 = 0$, $c_2 = 1$, $c_3 = 0$, and so the corresponding code word is $\mathbf{C} = (0 \quad 1 \quad 1 \quad 0 \quad 0 \quad 1 \quad 1)$. ❑

Before launching into the details of how to decode a message we need to introduce a special matrix. We first observe that in modulo 2 arithmetic there are no negative numbers; the additive inverse of 1 is 1, not -1. With this in mind, we can write the system (2) in the equivalent form

$$c_3 + w_2 + w_3 + w_4 = 0$$
$$c_2 + w_1 + w_3 + w_4 = 0 \qquad (4)$$
$$c_1 + w_1 + w_2 + w_4 = 0.$$

These are called **parity check equations.** This means that each c_i is a parity check on three of the digits in the original word. For instance, if the number of ones in the three digits w_2, w_3, and w_4 is odd, then, as in the parity check code discussed previously, we would take $c_1 = 1$ and so on. As a matrix product, (4) can be written

$$\begin{pmatrix} 0 & 0 & 0 & 1 & 1 & 1 & 1 \\ 0 & 1 & 1 & 0 & 0 & 1 & 1 \\ 1 & 0 & 1 & 0 & 1 & 0 & 1 \end{pmatrix} \begin{pmatrix} c_1 \\ c_2 \\ w_1 \\ c_3 \\ w_2 \\ w_3 \\ w_4 \end{pmatrix} = \begin{pmatrix} 0 \\ 0 \\ 0 \end{pmatrix}. \qquad (5)$$

The 3×7 matrix in (5),

$$\mathbf{H} = \begin{pmatrix} 0 & 0 & 0 & 1 & 1 & 1 & 1 \\ 0 & 1 & 1 & 0 & 0 & 1 & 1 \\ 1 & 0 & 1 & 0 & 1 & 0 & 1 \end{pmatrix}$$

is called the **parity check matrix**. We have shown in (5) that the binary digits of a code word $\mathbf{C} = (c_1 \quad c_2 \quad w_1 \quad c_3 \quad w_2 \quad w_3 \quad w_4)$ satisfy the matrix equation

$$\mathbf{H}\mathbf{C}^T = \mathbf{0}. \qquad (6)$$

A closer inspection of \mathbf{H} shows a surprising fact: The columns of \mathbf{H}, left to right, are the numbers 1 through 7 written in binary. For example, by writing the column $\begin{pmatrix} 1 \\ 1 \\ 0 \end{pmatrix}$ as 1 1 0, we recognize the binary representation of the number 6.

Let \mathbf{R} be a 1×7 matrix representing the received message. Since \mathbf{H} is a 3×7 matrix and \mathbf{R}^T is a 7×1 matrix, the product

$$\mathbf{S} = \mathbf{H}\mathbf{R}^T \qquad (7)$$

is a 3×1 matrix called the **syndrome** of \mathbf{R}.

■ **Decoding** If the syndrome of the received message \mathbf{R} is

$$\mathbf{S} = \mathbf{H}\mathbf{R}^T = \mathbf{0},$$

then, in view of the result in (6), we conclude that \mathbf{R} is a code word, and it is assumed that the transmission is correct and that \mathbf{R} is the same as the original encoded message \mathbf{C}. The decoding of the message is accomplished by simply dropping the three check bits in \mathbf{R}.

| *Example 5* **Syndromes**

Compute the syndrome of (a) $\mathbf{R} = (1 \quad 1 \quad 0 \quad 1 \quad 0 \quad 0 \quad 1)$ and (b) $\mathbf{R} = (1 \quad 0 \quad 0 \quad 1 \quad 0 \quad 1 \quad 0)$.

SOLUTION (a) From (7) we have

$$\mathbf{S} = \begin{pmatrix} 0 & 0 & 0 & 1 & 1 & 1 & 1 \\ 0 & 1 & 1 & 0 & 0 & 1 & 1 \\ 1 & 0 & 1 & 0 & 1 & 0 & 1 \end{pmatrix} \begin{pmatrix} 1 \\ 1 \\ 0 \\ 1 \\ 0 \\ 0 \\ 1 \end{pmatrix} = \begin{pmatrix} 0 \\ 0 \\ 0 \end{pmatrix}.$$

We conclude that \mathbf{R} is a code word. By dropping the colored check bits in $(1 \quad 1 \quad 0 \quad 1 \quad 0 \quad 0 \quad 1)$, we get the decoded message $(0 \quad 0 \quad 0 \quad 1)$.

(b) From (7), $\mathbf{S} = \begin{pmatrix} 0 & 0 & 0 & 1 & 1 & 1 & 1 \\ 0 & 1 & 1 & 0 & 0 & 1 & 1 \\ 1 & 0 & 1 & 0 & 1 & 0 & 1 \end{pmatrix} \begin{pmatrix} 1 \\ 0 \\ 0 \\ 1 \\ 0 \\ 1 \\ 0 \end{pmatrix} = \begin{pmatrix} 0 \\ 1 \\ 1 \end{pmatrix}.$

Since $\mathbf{S} \neq \mathbf{0}$, the received message \mathbf{R} is not a code word. ❏

As mentioned earlier, the Hamming (7, 4) code enables us to detect and also to correct a single error in the message \mathbf{R}. Now let \mathbf{C} be a code word and let $\mathbf{E} = [e_1 \ e_2 \ e_3 \ e_4 \ e_5 \ e_6 \ e_7]$ be a single-error noise word added to \mathbf{C} during its transmission. The entries of \mathbf{E} are defined by

$$e_i = \begin{cases} 1, & \text{if noise changes the } i\text{th bit} \\ 0, & \text{if noise does not change the } i\text{th bit.} \end{cases}$$

The received message is then $\mathbf{R} = \mathbf{C} + \mathbf{E}$. From the property $\mathbf{R}^T = \mathbf{C}^T + \mathbf{E}^T$ and the distributive law, we see that the syndrome of \mathbf{R} is the same as the syndrome of \mathbf{E}:

$$\mathbf{HR}^T = \mathbf{H}(\mathbf{C}^T + \mathbf{E}^T) = \mathbf{HC}^T + \mathbf{HE}^T = \mathbf{0} + \mathbf{HE}^T = \mathbf{HE}^T.$$

From the definition of matrix addition, it is a straightforward matter to verify that the syndrome of \mathbf{E}

$$\mathbf{HE}^T = \begin{pmatrix} e_4 + e_5 + e_6 + e_7 \\ e_2 + e_3 + e_6 + e_7 \\ e_1 + e_3 + e_5 + e_7 \end{pmatrix}$$

can be written as the sum of the column vectors of \mathbf{H} with coefficients the symbols that denote the bits where the error might occur:

$$\mathbf{HE}^T = e_1 \begin{pmatrix} 0 \\ 0 \\ 1 \end{pmatrix} + e_2 \begin{pmatrix} 0 \\ 1 \\ 0 \end{pmatrix} + e_3 \begin{pmatrix} 0 \\ 1 \\ 1 \end{pmatrix} + e_4 \begin{pmatrix} 1 \\ 0 \\ 0 \end{pmatrix} + e_5 \begin{pmatrix} 1 \\ 0 \\ 1 \end{pmatrix} + e_6 \begin{pmatrix} 1 \\ 1 \\ 0 \end{pmatrix} + e_7 \begin{pmatrix} 1 \\ 1 \\ 1 \end{pmatrix}. \tag{8}$$

Now consider the set of 3×1 column vectors whose entries are binary digits. Since there are only two ways to select each of the three entries, there are $2^3 = 8$ such vectors. The seven *nonzero* vectors are the columns of \mathbf{H} or the column vectors displayed in (8). The syndrome \mathbf{S} of the received message \mathbf{R} is a 3×1 column vector with binary entries; hence, if $\mathbf{S} \neq \mathbf{0}$, then \mathbf{S} must be one of the columns of \mathbf{H}. If \mathbf{R} contains a single error, then $\mathbf{S} \neq \mathbf{0}$ and, since the entries of \mathbf{E} are all zero except for one entry of one, we see from (8) that the syndrome itself indicates which bit is in error. In practice there is no need to write out (8); just compute the syndrome \mathbf{S} of the received message \mathbf{R}. \mathbf{S} is a column of \mathbf{H} and consequently is the binary number of that bit which is in error.

> ### *Example 6* Decoding a Word
>
> In part **(b)** of Example 5 we saw that the syndrome of the message $\mathbf{R} = (1 \quad 0 \quad 0 \quad 1 \quad 0 \quad 1 \quad 0)$ was $\mathbf{S} = \begin{pmatrix} 0 \\ 1 \\ 1 \end{pmatrix}$. This is the third column of \mathbf{H} (or the binary representation of the number 3) and so we conclude that the third bit in \mathbf{R} is incorrect. Changing zero to one gives the code word $\mathbf{C} = (1 \quad 0 \quad 1 \quad 1 \quad 0 \quad 1 \quad 0)$. Hence by dropping the first, second, and fourth bits from \mathbf{C} we arrive at the decoded message $(1 \quad 0 \quad 1 \quad 0)$. ❏

In these brief descriptions of cryptography and coding theory we have not even begun to scratch the surface of these fascinating subjects. Our goal was a modest one: to show how matrix theory is a natural working tool in various areas of mathematics and computer science.

Remarks

The Hamming (7, 4) code can detect but not correct any pair of errors. Students interested in how this is done or in further details of coding theory should check their library for more specialized texts.

EXERCISES 8.14

Answers to odd-numbered problems begin on page A-46.

In Problems 1–6, encode the given word using the parity check code.

1. $(0 \quad 1 \quad 1)$

2. $(1 \quad 1 \quad 1)$

3. $(0 \quad 0 \quad 0 \quad 1)$

4. $(1 \quad 0 \quad 1 \quad 0)$

5. $(1 \quad 0 \quad 1 \quad 0 \quad 1 \quad 0 \quad 0)$

6. $(0 \quad 1 \quad 1 \quad 0 \quad 1 \quad 0 \quad 1)$

In Problems 7–12, decode the given message using the parity check code.

7. $(1 \quad 0 \quad 0 \quad 1)$

8. $(0 \quad 0 \quad 1 \quad 1)$

9. $(1 \quad 1 \quad 1 \quad 0 \quad 0)$

10. $(1 \quad 0 \quad 1 \quad 0 \quad 0)$

11. $(1 \quad 0 \quad 0 \quad 1 \quad 1 \quad 1)$

12. $(1 \quad 0 \quad 0 \quad 1 \quad 0 \quad 1)$

In Problems 13–18, encode the given word using the Hamming (7, 4) code.

13. $(1 \quad 1 \quad 1 \quad 0)$

14. $(0 \quad 0 \quad 1 \quad 1)$

15. $(0 \quad 1 \quad 0 \quad 1)$

16. $(0 \quad 0 \quad 0 \quad 1)$

17. $(0 \quad 1 \quad 1 \quad 0)$

18. $(1 \quad 1 \quad 0 \quad 0)$

In Problems 19–28, determine whether the given message is a code word in the Hamming (7, 4) code. If it is, decode it. If it is not, correct the single error and decode the corrected message.

19. $(0 \quad 0 \quad 0 \quad 0 \quad 0 \quad 0 \quad 0)$

20. $(1 \quad 1 \quad 0 \quad 0 \quad 0 \quad 0 \quad 0)$

21. $(1 \quad 1 \quad 0 \quad 1 \quad 1 \quad 0 \quad 1)$

22. $(0 \quad 1 \quad 0 \quad 1 \quad 0 \quad 1 \quad 0)$

23. $(1 \quad 1 \quad 1 \quad 1 \quad 1 \quad 1 \quad 1)$

24. $(1 \quad 1 \quad 0 \quad 0 \quad 1 \quad 1 \quad 0)$

25. $(0 \quad 1 \quad 1 \quad 1 \quad 0 \quad 0 \quad 1)$

26. $(1 \quad 0 \quad 0 \quad 1 \quad 0 \quad 0 \quad 1)$

27. $(1 \quad 0 \quad 1 \quad 1 \quad 0 \quad 1 \quad 1)$

28. $(0 \quad 0 \quad 1 \quad 0 \quad 0 \quad 1 \quad 1)$

29. (a) Determine the total number of 7-tuples with binary entries.

(b) How many 7-tuple code words are there in the Hamming (7, 4) code?

(c) List all code words in the Hamming (7, 4) code.

30. (a) In the Hamming (8, 4) code a word

$$\mathbf{W} = (w_1 \quad w_2 \quad w_3 \quad w_4)$$

of length 4 is transformed into a code word of length 8:

$$\mathbf{C} = (c_1 \quad c_2 \quad c_3 \quad w_1 \quad c_4 \quad w_2 \quad w_3 \quad w_4),$$

where the parity check equations are

$$c_4 + w_2 + w_3 + w_4 = 0$$
$$c_3 + w_1 + w_3 + w_4 = 0$$
$$c_2 + w_1 + w_2 + w_4 = 0$$
$$c_1 + c_2 + c_3 + w_1 + c_4 + w_2 + w_3 + w_4 = 0.$$

Encode the word $(0 \quad 1 \quad 1 \quad 0)$.

(b) From the system in part **(a)**, determine the parity check matrix \mathbf{H}.

(c) Using the matrix \mathbf{H} from part **(b)**, compute the syndrome \mathbf{S} of the received message

$$\mathbf{R} = (0 \quad 0 \quad 1 \quad 1 \quad 1 \quad 1 \quad 0 \quad 0).$$

8.15 Method of Least Squares

When performing experiments we often tabulate data in the form of ordered pairs (x_1, y_1), (x_2, y_2), . . . , (x_n, y_n), with each x_i distinct. Given the data, it is then often desirable to be able to extrapolate or predict y from x by finding a mathematical model, that is, a function that approximates or "fits" the data. In other words, we want a function $f(x)$ such that

$$f(x_1) \approx y_1, \qquad f(x_2) \approx y_2, \cdots, \quad f(x_n) \approx y_n.$$

But naturally we do not want just any function but a function that fits the data as closely as possible.

In the discussion that follows we shall confine our attention to the problem of finding a linear polynomial $f(x) = ax + b$ or straight line that "best fits" the data (x_1, y_1), (x_2, y_2), . . . , (x_n, y_n). The procedure for finding this linear function is known as **the method of least squares.**

Example 1 Line of Best Fit

Consider the data (1, 1), (2, 3), (3, 4), (4, 6), (5, 5) shown in Figure 8.13(a). Visually, and the fact that the line $y = x + 1$, shown in Figure 8.13(b), passes through two of the data points, we might take this line as that best fitting the data. ❑

Obviously we need something better than a visual guess to determine the linear function $y = f(x)$, as in the last example. We need a criterion that defines the concept of "best fit" or as it is sometimes called "the goodness of fit."

If we try match the data points with the function $f(x) = ax + b$, then we wish to find a and b that satisfy the system of equations

$$
\begin{aligned}
y_1 &= ax_1 + b \\
y_2 &= ax_2 + b \\
&\ \ \vdots \\
y_n &= ax_n + b
\end{aligned}
\tag{1}
$$

or $\quad \mathbf{Y} = \mathbf{AX} \quad$ where $\quad \mathbf{Y} = \begin{pmatrix} y_1 \\ y_2 \\ \vdots \\ y_n \end{pmatrix}, \quad \mathbf{A} = \begin{pmatrix} x_1 & 1 \\ x_2 & 1 \\ \vdots & \vdots \\ x_n & 1 \end{pmatrix}, \quad \mathbf{X} = \begin{pmatrix} a \\ b \end{pmatrix}. \tag{2}$

Figure 8.13

Unfortunately (1) is an overdetermined system and, unless the data points lie on the same line, has no solution. Thus we shall be content to find a vector $\mathbf{X} = \begin{pmatrix} a \\ b \end{pmatrix}$ so that the right side \mathbf{AX} is close to the left side \mathbf{Y}.

■ **Least Squares Line** If the data points are (x_1, y_1), (x_2, y_2), . . . , (x_n, y_n), then one way to determine how well the linear function $f(x) = ax + b$ fits the data is to measure the vertical distances between the points and the graphs of f:

$$e_i = |y_i - f(x_i)|, \qquad i = 1, 2, \ldots, n.$$

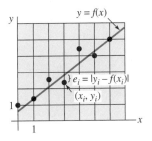

Figure 8.14

We can think of each e_i as the error in approximating the data value y_i by the functional value $f(x_i)$. See Figure 8.14. Intuitively, we know that the function f will fit the data well if the sum of all the e_i values is a minimum. Actually, a more convenient approach to the problem is to find a linear function f so that the sum of the *squares* of all the e_i values is a minimum. We shall define the solution of the system (1) to be those coefficients a and b that minimize the expression $E = e_1^2 + e_2^2 + \cdots + e_n^2$, that is,

$$E = [y_1 - f(x_1)]^2 + [y_2 - f(x_2)]^2 + \cdots + [y_n - f(x_n)]^2$$
$$= [y_1 - (ax_1 + b)]^2 + [y_2 - (ax_2 + b)]^2 + \cdots + [y_n - (ax_n + b)]^2$$

or

$$E = \sum_{i=1}^{n} [y_i - ax_i - b]^2. \tag{3}$$

The expression E is called the **sum of the square errors.** The line $y = ax + b$ that minimizes the sum of the square errors (3) is defined to be the **line of best fit** and is called the **least squares line** for the data $(x_1, y_1), (x_2, y_2), \ldots, (x_n, y_n)$.

The problem remains: How does one find a and b so that (3) is a minimum? The answer can be found from calculus. If we think of (3) as a function of two variables a and b, then to find the minimum value of E we set the first partial derivatives equal to zero:

$$\frac{\partial E}{\partial a} = 0 \quad \text{and} \quad \frac{\partial E}{\partial b} = 0.$$

The last two conditions yield, in turn,

$$-2 \sum_{i=1}^{n} x_i[y_i - ax_i - b] = 0$$
$$-2 \sum_{i=1}^{n} [y_i - ax_i - b] = 0. \tag{4}$$

Expanding the sums and using $\sum_{i=1}^{n} b = nb$, we find the system (4) is the same as

$$\left(\sum_{i=1}^{n} x_i^2 \right)a + \left(\sum_{i=1}^{n} x_i \right)b = \sum_{i=1}^{n} x_i y_i$$
$$\left(\sum_{i=1}^{n} x_i \right)a + \qquad nb = \sum_{i=1}^{n} y_i. \tag{5}$$

Although we shall not give the details, the values of a and b that satisfy the system (5) yield the minimum value of E.

In terms of matrices it can be shown that (5) is equivalent to

$$\mathbf{A}^T \mathbf{A} \mathbf{X} = \mathbf{A}^T \mathbf{Y}, \tag{6}$$

where \mathbf{A}, \mathbf{Y}, and \mathbf{X} are defined in (2). Since \mathbf{A} is an $n \times 2$ matrix and \mathbf{A}^T is a $2 \times n$ matrix, the matrix $\mathbf{A}^T \mathbf{A}$ is a 2×2 matrix. Moreover, unless the data points all lie on the same vertical line, the matrix $\mathbf{A}^T \mathbf{A}$ is nonsingular. Thus, (6) has the unique solution

$$\mathbf{X} = (\mathbf{A}^T \mathbf{A})^{-1} \mathbf{A}^T \mathbf{Y}. \tag{7}$$

We say that \mathbf{X} is the least squares solution of the overdetermined system (1).

Example 2 Least Squares Line

Find the least squares line for the data in Example 1. Calculate the sum of the square errors E for this line and the line $y = x + 1$.

SOLUTION For the function $f(x) = ax + b$ the data $(1, 1)$, $(2, 3)$, $(3, 4)$, $(4, 6)$, $(5, 5)$ lead to the overdetermined system

$$a + b = 1$$
$$2a + b = 3$$
$$3a + b = 4 \qquad\qquad (8)$$
$$4a + b = 6$$
$$5a + b = 5.$$

Now by identifying

$$\mathbf{Y} = \begin{pmatrix} 1 \\ 3 \\ 4 \\ 6 \\ 5 \end{pmatrix} \quad \text{and} \quad \mathbf{A} = \begin{pmatrix} 1 & 1 \\ 2 & 1 \\ 3 & 1 \\ 4 & 1 \\ 5 & 1 \end{pmatrix} \quad \text{we have} \quad \mathbf{A}^T\mathbf{A} = \begin{pmatrix} 55 & 15 \\ 15 & 5 \end{pmatrix},$$

and so (7) gives

$$\mathbf{X} = \begin{pmatrix} 55 & 15 \\ 15 & 5 \end{pmatrix}^{-1} \begin{pmatrix} 1 & 1 \\ 2 & 1 \\ 3 & 1 \\ 4 & 1 \\ 5 & 1 \end{pmatrix}^T \begin{pmatrix} 1 \\ 3 \\ 4 \\ 6 \\ 5 \end{pmatrix} = \frac{1}{50} \begin{pmatrix} 5 & -15 \\ -15 & 55 \end{pmatrix} \begin{pmatrix} 1 & 2 & 3 & 4 & 5 \\ 1 & 1 & 1 & 1 & 1 \end{pmatrix} \begin{pmatrix} 1 \\ 3 \\ 4 \\ 6 \\ 5 \end{pmatrix}$$

$$= \frac{1}{50} \begin{pmatrix} 5 & -15 \\ -15 & 55 \end{pmatrix} \begin{pmatrix} 68 \\ 19 \end{pmatrix} = \begin{pmatrix} 1.1 \\ 0.5 \end{pmatrix}.$$

Thus the least squares solution of (8) is $a = 1.1$ and $b = 0.5$, and the least squares line is $y = 1.1x + 0.5$. For this line the sum of the square errors is

$$E = [1 - f(1)]^2 + [3 - f(2)]^2 + [4 - f(3)]^2 + [6 - f(4)]^2 + [5 - f(5)]^2$$

$$= [1 - 1.6]^2 + [3 - 2.7]^2 + [4 - 3.8]^2 + [6 - 4.9]^2 + [5 - 6]^2 = 2.7.$$

For the line $y = x + 1$ that we guessed and that also passed through two of the data points, we find $E = 3.0$.

By way of comparison, Figure 8.15 shows the data points, the line $y = x + 1$, and the least squares line $y = 1.1x + 0.5$. ❑

Figure 8.15

Answers to odd-numbered problems begin on page A-46.

In Problems 1–6, find the least squares line for the given data.

1. $(2, 1)$, $(3, 2)$, $(4, 3)$, $(5, 2)$

2. $(0, -1)$, $(1, 3)$, $(2, 5)$, $(3, 7)$

3. $(1, 1)$, $(2, 1.5)$, $(3, 3)$, $(4, 4.5)$, $(5, 5)$

4. $(0, 0)$, $(2, 1.5)$, $(3, 3)$, $(4, 4.5)$, $(5, 5)$

5. $(0, 2)$, $(1, 3)$, $(2, 5)$, $(3, 5)$, $(4, 9)$, $(5, 8)$, $(6, 10)$

6. $(1, 2)$, $(2, 2.5)$, $(3, 1)$, $(4, 1.5)$, $(5, 2)$, $(6, 3.2)$, $(7, 5)$

7. In an experiment, the following correspondence was found between temperature T (in °C) and kinematic viscosity v (in Centistokes) of an oil with a certain additive:

T	20	40	60	80	100	120
v	220	200	180	170	150	135

Find the least squares line $v = aT + b$. Use this line to estimate the viscosity of the oil at $T = 140$ and $T = 160$.

8. In an experiment the following correspondence was found between temperature T (in °C) and electrical resistance R (in MΩ):

T	400	450	500	550	600	650
R	0.47	0.90	2.0	3.7	7.5	15

Find the least squares line $R = aT + b$. Use this line to estimate the resistance at $T = 700$.

8.16 Discrete Compartmental Models

The construction of the mathematical model (3) in Section 2.9, describing the number of pounds of salt in two connected tanks in which brine is flowing into and out of the tanks, is an example of **compartmental analysis.**

compartment 1 compartment 2

Figure 8.16

■ **The General Two-Compartment Model** Suppose material flows between two tanks with volumes V_1 and V_2. In the diagram shown in Figure 8.16, F_{01}, F_{12}, F_{21}, F_{10}, and F_{20} denote flow rates. Note that the double subscripted symbol F_{ij} denotes the flow rate from tank i to tank j. Next, suppose a second substance, called the **tracer,** is infused into compartment 1 at a known rate $I(t)$. As we did in Section 2.9, we will assume that the tracer is thoroughly mixed in both compartments at all times t. If $x(t)$ denotes the amount of tracer in compartment 1 and $y(t)$ the amount of tracer in compartment 2, then the concentrations are $c_1(t) = x(t)/V_1$ and $c_2(t) = y(t)/V_2$, respectively. It follows that the general two-compartment model is

$$\frac{dx}{dt} = -(F_{12} + F_{10})c_1(t) + F_{21}c_2(t) + I(t)$$

$$\frac{dy}{dt} = F_{21}c_1(t) - (F_{21} + F_{20})c_2(t).$$

(1)

The model in (1) keeps track of the amount of tracer that flows between the compartments. The material consisting of, say, a fluid and a tracer is *continually interchanged.* We present next a model that keeps track of compartmental contents every Δt units of time and assumes that the system changes only at times Δt, $2\Delta t, \ldots, n\Delta t, \ldots$. Of course, by selecting Δt very small, we can approximate the continuous case.

■ **Discrete Compartmental Models** In constructing a compartmental model of a physical system, we conceptually separate the system into a distinct number of small components between which material is transported. Compartments need not be spatially distinct (like the tanks used in Section 2.9) but must be distinguishable on some basis. The following are a few examples:

- Acid rain (containing strontium 90, for example) is deposited onto pastureland. Compartments might be grasses, soil, streams, and litter.

- In studying the flow of energy through an aquatic ecosystem, we might separate the system into phytoplankton, zooplankton, plankton predators, seaweed, small carnivores, large carnivores, and decay organisms.

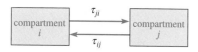

Figure 8.17

• A tracer is infused into the bloodstream and is lost to the body by the metabolism of a particular organ and by excretion. Appropriate compartments might be arterial blood, venous blood, the organ, and urine.

Suppose then that a system is divided into n compartments and, after each Δt units of time, material is interchanged between compartments. We will assume that a fixed fraction τ_{ij} of the contents of compartment j are passed to compartment i every Δt units of time, as depicted in Figure 8.17. This hypothesis is known as the **linear donor-controlled hypothesis.**

Let the entries x_i in the $n \times 1$ matrix \mathbf{X},

$$\mathbf{X} = \begin{pmatrix} x_1 \\ x_2 \\ \vdots \\ x_n \end{pmatrix}, \qquad \mathbf{Y} = \begin{pmatrix} y_1 \\ y_2 \\ \vdots \\ y_n \end{pmatrix}, \tag{2}$$

represent the amount of tracer in compartment i. We say that \mathbf{X} specifies the **state of the system.** The $n \times 1$ matrix \mathbf{Y} is the state of the system Δt units of time later. We will show that \mathbf{X} and \mathbf{Y} are related by the matrix equation $\mathbf{Y} = \mathbf{TX}$, where \mathbf{T} is an $n \times n$ matrix determined by the **transfer coefficients** τ_{ij}. To find \mathbf{T}, observe, for example, that

$$y_1 = x_1 + (\text{amount of tracer entering 1}) - (\text{amount of tracer leaving 1})$$

$$= x_1 + (\tau_{12}x_2 + \tau_{13}x_3 + \cdots + \tau_{1n}x_n) - (\tau_{21} + \tau_{31} + \cdots + \tau_{n1})x_1$$

$$= (1 - \tau_{21} - \tau_{31} - \cdots - \tau_{n1})x_1 + \tau_{12}x_2 + \cdots + \tau_{1n}x_n.$$

If we let $\tau_{11} = 1 - \tau_{21} - \tau_{31} - \cdots - \tau_{n1}$, then τ_{11} is just the fraction of the contents of compartment 1 that remains in 1.

Letting $\tau_{ii} = 1 - \sum_{j \neq i}\tau_{ji}$ we have, in general,

$$\begin{matrix} y_1 = \tau_{11}x_1 + \tau_{12}x_2 + \cdots + \tau_{1n}x_n \\ y_2 = \tau_{21}x_1 + \tau_{22}x_2 + \cdots + \tau_{2n}x_n \\ \vdots \qquad \vdots \\ y_n = \tau_{n1}x_1 + \tau_{n2}x_2 + \cdots + \tau_{nn}x_n \end{matrix} \quad \text{or} \quad \begin{pmatrix} y_1 \\ y_2 \\ \vdots \\ y_n \end{pmatrix} = \begin{pmatrix} \tau_{11} & \tau_{12} & \cdots & \tau_{1n} \\ \tau_{21} & \tau_{22} & \cdots & \tau_{2n} \\ \vdots & & & \vdots \\ \tau_{n1} & \tau_{n2} & \cdots & \tau_{nn} \end{pmatrix}\begin{pmatrix} x_1 \\ x_2 \\ \vdots \\ x_n \end{pmatrix}. \tag{3}$$

Note: A transfer matrix is an ☞ *example of a stochastic matrix. See Problem 27 in Exercises 8.8.*

The matrix equation in (3) is the desired equation $\mathbf{Y} = \mathbf{TX}$. The matrix $\mathbf{T} = (\tau_{ij})_{n \times n}$ is called the **transfer matrix.** Note that the sum of the entries in any column, the transfer coefficients, is equal to 1.

Discrete compartmental models are illustrated in the next two examples.

Example 1 Transfer Matrix

In Figure 8.18 the three boxes represent three compartments. The content of each compartment at time t is indicated in each box. The transfer coefficients are shown along the arrows connecting the compartments.
(a) Find the transfer matrix \mathbf{T}.
(b) Suppose $\Delta t = 1$ day. Find the state of the system \mathbf{Y} 1 day later.

SOLUTION (a) The state of the system at time $t = 0$ is $\mathbf{X} = \begin{pmatrix} 100 \\ 250 \\ 80 \end{pmatrix}$.

Remember that τ_{ij} specifies the rate of transfer to compartment i from compartment j. Hence we are given that $\tau_{21} = 0.2$, $\tau_{12} = 0.05$, $\tau_{32} = 0.3$, $\tau_{23} = 0$, $\tau_{13} = 0.25$, and $\tau_{31} = 0$. From these numbers we see that the matrix \mathbf{T} is

Figure 8.18

$$\mathbf{T} = \begin{pmatrix} - & 0.05 & 0.25 \\ 0.2 & - & 0 \\ 0 & 0.3 & - \end{pmatrix}. \tag{4}$$

But since the column entries must sum to 1, we can fill in the blanks in (4):

$$\mathbf{T} = \begin{pmatrix} 0.8 & 0.05 & 0.25 \\ 0.2 & 0.65 & 0 \\ 0 & 0.3 & 0.75 \end{pmatrix}.$$

(b) The state of the system 1 day later is then

$$\mathbf{Y} = \mathbf{TX} = \begin{pmatrix} 0.8 & 0.05 & 0.25 \\ 0.2 & 0.65 & 0 \\ 0 & 0.3 & 0.75 \end{pmatrix} \begin{pmatrix} 100 \\ 250 \\ 80 \end{pmatrix} = \begin{pmatrix} 112.5 \\ 182.5 \\ 135 \end{pmatrix}. \qquad \square$$

If \mathbf{X}_0 denotes the initial state of the system, and \mathbf{X}_n is the state after $n(\Delta t)$ units of time, then

$$\mathbf{X}_1 = \mathbf{TX}_0, \quad \mathbf{X}_2 = \mathbf{TX}_1, \quad \mathbf{X}_3 = \mathbf{TX}_2, \dots, \quad \mathbf{X}_{n+1} = \mathbf{TX}_n.$$

Because $\qquad\qquad \mathbf{X}_2 = \mathbf{T}(\mathbf{TX}_0) = \mathbf{T}^2\mathbf{X}_0, \quad \mathbf{X}_3 = \mathbf{T}(\mathbf{T}^2\mathbf{X}_0) = \mathbf{T}^3\mathbf{X}_0, \dots,$

we have in general $\qquad\qquad \mathbf{X}_n = \mathbf{T}^n\mathbf{X}_0, \quad n = 1, 2, \dots. \tag{5}$

We could, of course, use the method illustrated in Section 8.9 to compute \mathbf{T}^n, but with the aid of a calculator or a CAS it is just as easy to use the recursion formula $\mathbf{X}_{n+1} = \mathbf{TX}_n$ by letting $n = 0, 1, \dots$

Example 2 States of an Ecosystem

Strontium 90 is deposited into pastureland by rainfall. To study how this material is cycled through the ecosystem, we divide the system into the compartments shown in Figure 8.19. Suppose that $\Delta t = 1$ month and the transfer coefficients (which have been estimated experimentally) shown in the figure are measured in fraction/month. (We will ignore that some strontium 90 is lost due to radioactive decay.) Suppose that rainfall has deposited the strontium 90 into the compartments so that $\mathbf{X}_0 = \begin{pmatrix} 20 \\ 60 \\ 15 \\ 20 \end{pmatrix}$. (Units might be grams per hectare.) Compute the states of the ecosystem over the next 12 months.

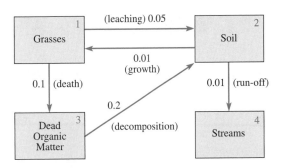

Figure 8.19

SOLUTION From the data in Figure 8.19 we see that transfer matrix \mathbf{T} is

$$\mathbf{T} = \begin{pmatrix} 0.85 & 0.01 & 0 & 0 \\ 0.05 & 0.98 & 0.2 & 0 \\ 0.1 & 0 & 0.8 & 0 \\ 0 & 0.01 & 0 & 1 \end{pmatrix}.$$

We must compute $\mathbf{X}_1, \mathbf{X}_2, \ldots, \mathbf{X}_{12}$. The state of the ecosystem after the first month is

$$\mathbf{X}_1 = \mathbf{T}\mathbf{X}_0 = \begin{pmatrix} 0.85 & 0.01 & 0 & 0 \\ 0.05 & 0.98 & 0.2 & 0 \\ 0.1 & 0 & 0.8 & 0 \\ 0 & 0.01 & 0 & 1 \end{pmatrix} \begin{pmatrix} 20 \\ 60 \\ 15 \\ 20 \end{pmatrix} = \begin{pmatrix} 17.6 \\ 62.8 \\ 14.0 \\ 20.6 \end{pmatrix}.$$

The remaining states, computed with the aid of a CAS and the recursion formula $\mathbf{X}_{n+1} = \mathbf{T}\mathbf{X}_n$ with $n = 1, 2, \ldots, 11$, are given in Table 8.1.

Table 8.1

Month	Grasses	Soil	Dead Organic Matter	Streams
0	20.00	60.00	15.00	20.00
1	17.60	62.80	14.00	20.60
2	15.59	65.22	12.96	21.23
3	13.90	67.29	11.93	21.88
4	12.49	69.03	10.93	22.55
5	11.31	70.46	9.99	23.24
6	10.32	71.61	9.13	23.95
7	9.48	72.52	8.33	24.66
8	8.79	73.21	7.61	25.39
9	8.20	73.71	6.97	26.12
10	7.71	74.04	6.40	26.86
11	7.29	74.22	5.89	27.60
12	6.94	74.28	5.44	28.34

EXERCISES 8.16

Answers to odd-numbered problems begin on page A-46.

1. (a) Use the data in the compartmental diagram in Figure 8.20 to determine the appropriate transfer matrix \mathbf{T} and the initial state of the system \mathbf{X}_0.

(b) Find the state of the system after 1 day. After 2 days.

(c) Eventually, the system will reach an equilibrium state $\hat{\mathbf{X}} = \begin{pmatrix} x_1 \\ x_2 \end{pmatrix}$ that satisfies $\mathbf{T}\hat{\mathbf{X}} = \hat{\mathbf{X}}$. Find $\hat{\mathbf{X}}$. [*Hint:* $x_1 + x_2 = 150$.]

Figure 8.20

2. (a) Use the data in the compartmental diagram in Figure 8.21 to determine the appropriate transfer matrix \mathbf{T} and the initial state of the system \mathbf{X}_0.

(b) Find the state of the system after 1 day. After 2 days.

(c) Find the equilibrium state $\hat{\mathbf{X}} = \begin{pmatrix} x_1 \\ x_2 \\ x_3 \end{pmatrix}$ that satisfies

$\mathbf{T}\hat{\mathbf{X}} = \hat{\mathbf{X}}$. [*Hint:* What is the analogue of the hint in part **(c)** of Problem 1?]

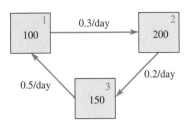

Figure 8.21

3. (a) Use the data in the compartmental diagram in Figure 8.22 to determine the appropriate transfer matrix \mathbf{T} and the initial state of the system \mathbf{X}_0.

(b) Find the state of the system after 1 day. After 2 days.

(c) Find the equilibrium state $\hat{\mathbf{X}} = \begin{pmatrix} x_1 \\ x_2 \\ x_3 \end{pmatrix}$ that satisfies $\mathbf{T}\hat{\mathbf{X}} = \hat{\mathbf{X}}$.

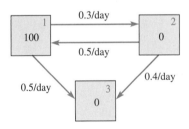

Figure 8.22

4. A field has been completely devastated by fire. Two types of vegetation, grasses and small shrubs, will first begin to grow, but the small shrubs can take over an area only if preceded by the grasses. In Figure 8.23, the transfer coefficient of 0.3 indicates that, by the end of the summer, 30% of the prior bare space in the field becomes occupied by grasses.

(a) Find the transfer matrix \mathbf{T}.

(b) Suppose $\mathbf{X} = \begin{pmatrix} 10 \\ 0 \\ 0 \end{pmatrix}$ and that area is measured in acres.

Use the recursion formula $\mathbf{X}_{n+1} = \mathbf{T}\mathbf{X}_n$, along with a calculator or a CAS, to determine the ground cover in each of the next 6 years.

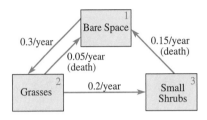

Figure 8.23

5. Computer Lab Assignment Radioisotopes (such as phosphorous 32 and carbon 14) have been used to study the transfer of nutrients in food chains. Figure 8.24 is a compartmental representation of a simple aquatic food chain. One hundred units (for example, microcuries) of tracer are dissolved in the water of an aquarium containing a species of phytoplankton and a species of zooplankton.

(a) Find the transfer matrix \mathbf{T} and the initial state of the system \mathbf{X}_0.

(b) Instead of the recursion formula, use $\mathbf{X}_n = \mathbf{T}^n\mathbf{X}_0$, $n = 1, 2, \ldots, 12$, to predict the state of the system for the next 12 hours. Use a CAS and the command to compute powers of matrices (in *Mathematica* it is **MatrixPower[, n]**) to find $\mathbf{T}^2, \mathbf{T}^3, \ldots, \mathbf{T}^{12}$.

Figure 8.24

6. Characterize the vector $\hat{\mathbf{X}}$ in part **(c)** of Problems 1–3 in terms of one of the principal concepts in Section 8.8.

CHAPTER 8
REVIEW EXERCISES
Answers to odd-numbered problems begin on page A-47.

In Problems 1–20, fill in the blank or answer true/false.

1. A matrix $\mathbf{A} = (a_{ij})_{4 \times 3}$ such that $a_{ij} = i + j$ is given by _____.

2. If \mathbf{A} is a 4×7 matrix and \mathbf{B} is a 7×3 matrix, then the size of \mathbf{AB} is _____.

3. If $\mathbf{A} = \begin{pmatrix} 1 \\ 2 \end{pmatrix}$ and $\mathbf{B} = (3 \quad 4)$, then $\mathbf{AB} =$ _____ and $\mathbf{BA} =$ _____.

4. If $\mathbf{A} = \begin{pmatrix} 1 & 2 \\ 3 & 4 \end{pmatrix}$, then $\mathbf{A}^{-1} =$ _____.

5. If \mathbf{A} and \mathbf{B} are $n \times n$ nonsingular matrices, then $\mathbf{A} + \mathbf{B}$ is necessarily nonsingular. _____

6. If \mathbf{A} is a nonsingular matrix for which $\mathbf{AB} = \mathbf{AC}$, then $\mathbf{B} = \mathbf{C}$. _____

7. If \mathbf{A} is a 3×3 matrix such that $\det \mathbf{A} = 5$, then $\det(\tfrac{1}{2}\mathbf{A}) =$ _____ and $\det(-\mathbf{A}^T) =$ _____.

8. If $\det \mathbf{A} = 6$ and $\det \mathbf{B} = 2$, then $\det \mathbf{AB}^{-1} =$ _____.

9. If \mathbf{A} and \mathbf{B} are $n \times n$ matrices whose corresponding entries in the third column are the same, then $\det(\mathbf{A} - \mathbf{B}) =$ _____.

10. Suppose \mathbf{A} is a 3×3 matrix such that $\det \mathbf{A} = 2$. If $\mathbf{B} = 10\mathbf{A}$ and $\mathbf{C} = -\mathbf{B}^{-1}$, then $\det \mathbf{C} =$ _____.

11. Let \mathbf{A} be an $n \times n$ matrix. The eigenvalues of \mathbf{A} are the nonzero solutions of $\det(\mathbf{A} - \lambda \mathbf{I}) = 0$. _____

12. A nonzero scalar multiple of an eigenvector is also an eigenvector corresponding to the same eigenvalue. _____

13. An $n \times 1$ column vector \mathbf{K} with all zero entries is never an eigenvector of an $n \times n$ matrix \mathbf{A}. _____

14. Let \mathbf{A} be an $n \times n$ matrix with real entries. If λ is a complex eigenvalue, then $\bar{\lambda}$ is also an eigenvalue of \mathbf{A}. _____

15. An $n \times n$ matrix \mathbf{A} always possesses n linearly independent eigenvectors. _____

16. The augmented matrix $\begin{pmatrix} 1 & 1 & 1 & 2 \\ 0 & 1 & 0 & 3 \\ 0 & 0 & 0 & 0 \end{pmatrix}$ is in reduced row-echelon form. _____

17. If a 3×3 matrix \mathbf{A} is diagonalizable, then it possesses three linearly independent eigenvectors. _____

18. The only matrices that are orthogonally diagonalizable are symmetric matrices. _____

19. The matrix $\mathbf{A} = \begin{pmatrix} 1 & -1 \\ -1 & 1 \end{pmatrix}$ is orthogonal since its columns are orthogonal vectors. _____

20. The eigenvalues of a symmetric matrix with real entries are always real numbers. _____

21. An $n \times n$ matrix \mathbf{B} is symmetric if $\mathbf{B}^T = \mathbf{B}$, and an $n \times n$ matrix \mathbf{C} is skew-symmetric if $\mathbf{C}^T = -\mathbf{C}$. By noting the identity $2\mathbf{A} = \mathbf{A} + \mathbf{A}^T + \mathbf{A} - \mathbf{A}^T$, show that any $n \times n$ matrix \mathbf{A} can be written as the sum of a symmetric matrix and a skew-symmetric matrix.

22. Show that there exists no 2×2 matrix with real entries such that $\mathbf{A}^2 = \begin{pmatrix} 0 & 1 \\ 1 & 0 \end{pmatrix}$.

23. An $n \times n$ matrix \mathbf{A} is said to be **nilpotent** if, for some positive integer n, $\mathbf{A}^n = \mathbf{0}$.

(a) Find a 2×2 nilpotent matrix $\mathbf{A} \neq \mathbf{0}$.

(b) Show that a nilpotent matrix is necessarily singular.

24. (a) Two $n \times n$ matrices \mathbf{A} and \mathbf{B} are said to **anticommute** if $\mathbf{AB} = -\mathbf{BA}$. Show that each of the **Pauli spin matrices**

$$\sigma_x = \begin{pmatrix} 0 & 1 \\ 1 & 0 \end{pmatrix} \quad \sigma_y = \begin{pmatrix} 0 & -i \\ i & 0 \end{pmatrix} \quad \sigma_z = \begin{pmatrix} 1 & 0 \\ 0 & -1 \end{pmatrix},$$

where $i^2 = -1$, anticommutes with the others. Pauli spin matrices are used in quantum mechanics.

(b) The matrix $\mathbf{C} = \mathbf{AB} - \mathbf{BA}$ is said to be the **commutator** of the $n \times n$ matrices \mathbf{A} and \mathbf{B}. Find the commutators of σ_x and σ_y, σ_y and σ_z, σ_z and σ_x.

In Problems 25 and 26, solve the given system by Gauss–Jordan elimination.

25. $\begin{pmatrix} 5 & -1 & 1 \\ 2 & 4 & 0 \\ 1 & 1 & 5 \end{pmatrix} \mathbf{X} = \begin{pmatrix} -9 \\ 27 \\ 9 \end{pmatrix}$

26. $\begin{aligned} x_1 + x_2 + x_3 &= 6 \\ x_1 - 2x_2 + 3x_3 &= 2 \\ 2x_1 \qquad - 3x_3 &= 3 \end{aligned}$

27. Without expanding, show that $\begin{vmatrix} 1 & 1 & 1 \\ 1/a & 1/b & 1/c \\ bc & ac & ab \end{vmatrix} = 0$.

28. Show that $\begin{vmatrix} y & x^2 & x & 1 \\ 2 & 1 & 1 & 1 \\ 3 & 4 & 2 & 1 \\ 5 & 9 & 3 & 1 \end{vmatrix} = 0$ is the equation of a parabola passing through the three points $(1, 2)$, $(2, 3)$, and $(3, 5)$.

In Problems 29 and 30, evaluate the determinant of the given matrix by inspection.

29.
$$\begin{vmatrix} 4 & 0 & 0 & 0 & 0 & 0 \\ 0 & -2 & 0 & 0 & 0 & 0 \\ 0 & 0 & 3 & 0 & 0 & 0 \\ 0 & 0 & 0 & -1 & 0 & 0 \\ 0 & 0 & 0 & 0 & 2 & 0 \\ 0 & 0 & 0 & 0 & 0 & 5 \end{vmatrix}$$

30.
$$\begin{pmatrix} -3 & 0 & 0 & 0 \\ 4 & 6 & 0 & 0 \\ 1 & 3 & 9 & 0 \\ 6 & 4 & 2 & 1 \end{pmatrix}$$

In Problems 31 and 32, without solving, state whether the given homogeneous system has only the trivial solution or has infinitely many solutions.

31. $x_1 - x_2 + x_3 = 0$
$5x_1 + x_2 - x_3 = 0$
$x_1 + 2x_2 + x_3 = 0$

32. $x_1 - x_2 - x_3 = 0$
$5x_1 + x_2 - x_3 = 0$
$x_1 + 2x_2 + x_3 = 0$

In Problems 33 and 34, balance the given chemical equation.

33. $I_2 + HNO_3 \rightarrow HIO_3 + NO_2 + H_2O$

34. $Ca + H_3PO_4 \rightarrow Ca_3P_2O_8 + H_2$

In Problems 35 and 36, solve the given system by Cramer's rule.

35. $x_1 + 2x_2 - 3x_3 = -2$
$2x_1 - 4x_2 + 3x_3 = 0$
$4x_2 + 6x_3 = 5$

36. $x_1 + x_3 = 4$
$2x_1 + 3x_2 + 4x_3 = 5$
$x_1 + 4x_2 + 5x_3 = 0$

37. Use Cramer's rule to solve the system

$$X = x \cos \theta + y \sin \theta$$
$$Y = -x \sin \theta + y \cos \theta$$

for x and y.

38. (a) Set up the system of equations for the currents in the branches of the network given in Figure 8.25.

Figure 8.25

(b) Use Cramer's rule to show that

$$i_1 = E\left(\frac{1}{R_1} + \frac{1}{R_2} + \frac{1}{R_3}\right).$$

39. Solve the system

$$2x_1 + 3x_2 - x_3 = 6$$
$$x_1 - 2x_2 = -3$$
$$-2x_1 + x_3 = 9$$

by writing it as a matrix equation and finding the inverse of the coefficient matrix.

40. Use the inverse of the matrix A to solve the system $AX = B$, where

$$A = \begin{pmatrix} 1 & 2 & 3 \\ 2 & 3 & 0 \\ 0 & 1 & 2 \end{pmatrix}$$

and the vector B is given by **(a)** $\begin{pmatrix} 1 \\ 1 \\ 1 \end{pmatrix}$ and **(b)** $\begin{pmatrix} -2 \\ 1 \\ 3 \end{pmatrix}$.

In Problems 41–46, find the eigenvalues and corresponding eigenvectors of the given matrix.

41. $\begin{pmatrix} 1 & 2 \\ 4 & 3 \end{pmatrix}$

42. $\begin{pmatrix} 0 & 0 \\ 4 & 0 \end{pmatrix}$

43. $\begin{pmatrix} 3 & 2 & 4 \\ 2 & 0 & 2 \\ 4 & 2 & 3 \end{pmatrix}$

44. $\begin{pmatrix} 7 & -2 & 0 \\ -2 & 6 & 2 \\ 0 & 2 & 5 \end{pmatrix}$

45. $\begin{pmatrix} -2 & 2 & -3 \\ 2 & 1 & -6 \\ -1 & -2 & 0 \end{pmatrix}$

46. $\begin{pmatrix} 0 & 0 & 0 \\ 0 & 0 & 1 \\ 2 & 2 & 1 \end{pmatrix}$

47. Supply a first column so that the matrix is orthogonal:

$$\begin{pmatrix} \Box & -\dfrac{1}{\sqrt{2}} & \dfrac{1}{\sqrt{3}} \\ \Box & 0 & \dfrac{1}{\sqrt{3}} \\ \Box & \dfrac{1}{\sqrt{2}} & \dfrac{1}{\sqrt{3}} \end{pmatrix}$$

48. Consider the symmetric matrix $A = \begin{pmatrix} 1 & 0 & -2 \\ 0 & 0 & 0 \\ -2 & 0 & 4 \end{pmatrix}$.

(a) Find matrices P and P^{-1} that orthogonally diagonalize the matrix A.

(b) Find the diagonal matrix D by actually carrying out the multiplication $P^{-1}AP$.

49. Identify the conic section $x^2 + 3xy + y^2 = 1$.

50. Consider the following population data:

Year	1890	1900	1910	1920	1930
Population (in millions)	63	76	92	106	123

The actual population in 1940 was 132 million. Compare this amount with the population predicted from the least squares line for the given data.

In Problems 51 and 52, use the matrix $\mathbf{A} = \begin{pmatrix} 10 & 1 \\ 9 & 1 \end{pmatrix}$ to encode the given message. Use the correspondence (1) of Section 8.13.

51. SATELLITE LAUNCHED ON FRI

52. SEC AGNT ARRVS TUES AM

In Problems 53 and 54, use the matrix $\mathbf{A} = \begin{pmatrix} 0 & 1 & 0 \\ 1 & 1 & 1 \\ 1 & -1 & 2 \end{pmatrix}$ to decode the given message. Use the correspondence (1) of Section 8.13.

53. $\mathbf{B} = \begin{pmatrix} 19 & 0 & 15 & 14 & 0 & 20 \\ 35 & 10 & 27 & 53 & 1 & 54 \\ 5 & 15 & -3 & 48 & 2 & 39 \end{pmatrix}$

54. $\mathbf{B} = \begin{pmatrix} 5 & 2 & 21 \\ 27 & 17 & 40 \\ 21 & 13 & -2 \end{pmatrix}$

55. Decode the following messages using the parity check code.

 (a) $(1 \quad 1 \quad 0 \quad 0 \quad 1 \quad 1)$

 (b) $(0 \quad 1 \quad 1 \quad 0 \quad 1 \quad 1 \quad 1 \quad 0)$

56. Encode the word $(1 \quad 0 \quad 0 \quad 1)$ using the Hamming (7, 4) code.

CHAPTER

Vector Calculus

INTRODUCTION

In Chapter 7 we studied the properties of vectors in 2- and 3-space. In this chapter we shall combine vector concepts with those from differential and integral calculus.

9.1 Vector Functions

(a) 2-space

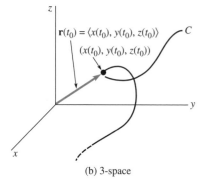

(b) 3-space

Figure 9.1

■ **Vector-Valued Functions** Recall that a curve C in the xy-plane can be parametrized by $x = f(t)$, $y = g(t)$, $a \leq t \leq b$. It is often convenient in science and engineering to introduce a vector \mathbf{r} with the functions f and g as components:

$$\mathbf{r}(t) = \langle f(t), g(t) \rangle = f(t)\mathbf{i} + g(t)\mathbf{j}.$$

We say that \mathbf{r} is a **vector-valued function** or simply a **vector function**. Similarly, a **space curve** is parametrized by three equations

$$x = f(t), \quad y = g(t), \quad z = h(t), \qquad a \leq t \leq b. \tag{1}$$

Correspondingly, a vector function is given by

$$\mathbf{r}(t) = \langle f(t), g(t), h(t) \rangle = f(t)\mathbf{i} + g(t)\mathbf{j} + h(t)\mathbf{k}.$$

As shown in Figure 9.1, for a given number t_0, the vector $\mathbf{r}(t_0)$ is the *position vector* of a point P on the curve C. In other words, as t varies, we can envision the curve C being traced out by the moving arrowhead of $\mathbf{r}(t)$.

We have already seen an example of parametric equations, as well as the vector function of a space curve, in Section 7.5, when we discussed the line in 3-space.

Example 1 Circular Helix

Graph the curve traced by the vector function

$$\mathbf{r}(t) = 2 \cos t\,\mathbf{i} + 2 \sin t\,\mathbf{j} + t\mathbf{k}, \qquad t \geq 0.$$

SOLUTION The parametric equations of the curve are $x = 2 \cos t$, $y = 2 \sin t$, $z = t$. By eliminating the parameter t from the first two equations:

$$x^2 + y^2 = (2 \cos t)^2 + (2 \sin t)^2 = 2^2$$

we see that a point on the curve lies on the circular cylinder $x^2 + y^2 = 4$. As seen in Figure 9.2 and the accompanying table, as the value of t increases, the curve winds upward in a spiral or *circular helix*.

t	x	y	z
0	2	0	0
$\pi/2$	0	2	$\pi/2$
π	-2	0	π
$3\pi/2$	0	-2	$3\pi/2$
2π	2	0	2π
$5\pi/2$	0	2	$5\pi/2$
3π	-2	0	3π
$7\pi/2$	0	-2	$7\pi/2$
4π	2	0	4π
$9\pi/2$	0	2	$9\pi/2$

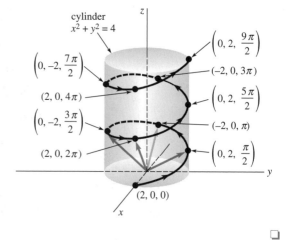

Figure 9.2

The curve in Example 1 is a special case of the vector function

$$\mathbf{r}(t) = a \cos t\,\mathbf{i} + b \sin t\,\mathbf{j} + ct\mathbf{k}, \qquad a > 0, \quad b > 0, \quad c > 0,$$

which describes an *elliptical helix*. When $a = b$, the helix is circular. The *pitch* of a helix is defined to be the number $2\pi c$. Problems 9 and 10 in Exercises 9.1 illustrate two other kinds of helixes.

Example 2 Circle in a Plane

Graph the curve traced by the vector function

$$\mathbf{r}(t) = 2\cos t\mathbf{i} + 2\sin t\mathbf{j} + 3\mathbf{k}.$$

SOLUTION The parametric equations of this curve are $x = 2\cos t$, $y = 2\sin t$, $z = 3$. As in Example 1, we see that a point on the curve must also lie on the cylinder $x^2 + y^2 = 4$. However, since the z-coordinate of any point has the constant value $z = 3$, the vector function $\mathbf{r}(t)$ traces out a circle 3 units above the xy-plane. See Figure 9.3. ❑

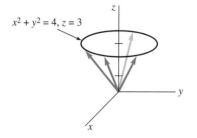

$x^2 + y^2 = 4$, $z = 3$

Figure 9.3

Example 3 Curve of Intersection

Find the vector function that describes the curve C of intersection of the plane $y = 2x$ and the paraboloid $z = 9 - x^2 - y^2$.

SOLUTION We first parametrize the curve C of intersection by letting $x = t$. It follows that $y = 2t$ and $z = 9 - t^2 - (2t)^2 = 9 - 5t^2$. From the parametric equations $x = t$, $y = 2t$, $z = 9 - 5t^2$, we see that a vector function describing the trace of the paraboloid in the plane $y = 2x$ is given by $\mathbf{r}(t) = t\mathbf{i} + 2t\mathbf{j} + (9 - 5t^2)\mathbf{k}$. See Figure 9.4. ❑

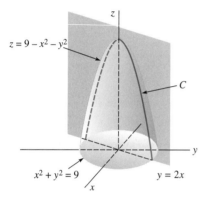

$z = 9 - x^2 - y^2$

$x^2 + y^2 = 9$

$y = 2x$

Figure 9.4

■ **Limits, Continuity, and Derivatives** The fundamental notion of the **limit** of a vector function $\mathbf{r}(t) = \langle f(t), g(t), h(t) \rangle$ is defined in terms of the limits of the component functions.

DEFINITION 9.1 Limit of a Vector Function

If $\lim_{t \to a} f(t)$, $\lim_{t \to a} g(t)$, and $\lim_{t \to a} h(t)$ exist, then

$$\lim_{t \to a} \mathbf{r}(t) = \left\langle \lim_{t \to a} f(t), \lim_{t \to a} g(t), \lim_{t \to a} h(t) \right\rangle.$$

The notation $t \to a$ in Definition 9.1 can, of course, be replaced by $t \to a^+$, $t \to a^-$, $t \to \infty$, or $t \to -\infty$.

As an immediate consequence of Definition 9.1, we have the following result.

THEOREM 9.1 Properties of Limits

If $\lim_{t \to a} \mathbf{r}_1(t) = \mathbf{L}_1$ and $\lim_{t \to a} \mathbf{r}_2(t) = \mathbf{L}_2$, then

(*i*) $\lim_{t \to a} c\mathbf{r}_1(t) = c\mathbf{L}_1$, c a scalar

(*ii*) $\lim_{t \to a} [\mathbf{r}_1(t) + \mathbf{r}_2(t)] = \mathbf{L}_1 + \mathbf{L}_2$

(*iii*) $\lim_{t \to a} \mathbf{r}_1(t) \cdot \mathbf{r}_2(t) = \mathbf{L}_1 \cdot \mathbf{L}_2$

DEFINITION 9.2 Continuity

A vector function \mathbf{r} is said to be **continuous** at $t = a$ if

 (*i*) $\mathbf{r}(a)$ is defined, (*ii*) $\lim_{t \to a} \mathbf{r}(t)$ exists, and (*iii*) $\lim_{t \to a} \mathbf{r}(t) = \mathbf{r}(a)$.

Equivalently, $\mathbf{r}(t)$ is continuous at $t = a$ if and only if the component functions f, g, and h are continuous there.

DEFINITION 9.3 Derivative of Vector Function

The **derivative** of a vector function \mathbf{r} is

$$\mathbf{r}'(t) = \lim_{\Delta t \to 0} \frac{1}{\Delta t}\left[\mathbf{r}(t + \Delta t) - \mathbf{r}(t)\right] \qquad (2)$$

for all t for which the limit exists.

The derivative of \mathbf{r} is also written $d\mathbf{r}/dt$. The next theorem will show that on a practical level the derivative of a vector function is obtained by simply differentiating its component functions.

THEOREM 9.2 Differentiation of Components

If $\mathbf{r}(t) = \langle f(t), g(t), h(t)\rangle$, where f, g, and h are differentiable, then

$$\mathbf{r}'(t) = \langle f'(t), g'(t), h'(t)\rangle.$$

PROOF From (2) we have

$$\mathbf{r}'(t) = \lim_{\Delta t \to 0} \frac{1}{\Delta t}\left[\langle f(t + \Delta t), g(t + \Delta t), h(t + \Delta t)\rangle - \langle f(t), g(t), h(t)\rangle\right]$$

$$= \lim_{\Delta t \to 0} \left\langle \frac{f(t + \Delta t) - f(t)}{\Delta t}, \frac{g(t + \Delta t) - g(t)}{\Delta t}, \frac{h(t + \Delta t) - h(t)}{\Delta t}\right\rangle.$$

Taking the limit of each component yields the desired result. ❑

■ **Smooth Curves** When the component functions of a vector function \mathbf{r} have continuous first derivatives and $\mathbf{r}'(t) \neq \mathbf{0}$ for all t in the open interval (a, b), then \mathbf{r} is said to be a **smooth function** and the curve C traced by \mathbf{r} is called a **smooth curve.**

■ **Geometric Interpretation of $\mathbf{r}'(t)$** If the vector $\mathbf{r}'(t)$ is not $\mathbf{0}$ at a point P, then it may be drawn *tangent to the curve* at P. As seen in Figure 9.5, the vectors

$$\Delta\mathbf{r} = \mathbf{r}(t + \Delta t) - \mathbf{r}(t) \quad \text{and} \quad \frac{\Delta\mathbf{r}}{\Delta t} = \frac{1}{\Delta t}\left[\mathbf{r}(t + \Delta t) - \mathbf{r}(t)\right]$$

are parallel. If we assume $\lim_{\Delta t \to 0} \Delta\mathbf{r}/\Delta t$ exists, it seems reasonable to conclude that as $\Delta t \to 0$, $\mathbf{r}(t)$ and $\mathbf{r}(t + \Delta t)$ become close and, as a consequence, the limiting position of the vector $\Delta\mathbf{r}/\Delta t$ is the tangent line at P. Indeed, the tangent line at P is *defined* as that line through P parallel to $\mathbf{r}'(t)$.

(a)

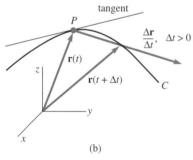

(b)

Figure 9.5

Example 4 Tangent Vectors

Graph the curve C that is traced by a point P whose position is given by $\mathbf{r}(t) = \cos 2t\,\mathbf{i} + \sin t\,\mathbf{j}$, $0 \leq t \leq 2\pi$. Graph $\mathbf{r}'(0)$ and $\mathbf{r}'(\pi/6)$.

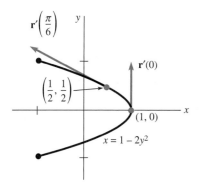

Figure 9.6

SOLUTION By clearing the parameter from the parametric equations $x = \cos 2t$, $y = \sin t$, $0 \le t \le 2\pi$, we find that C is the parabola $x = 1 - 2y^2$, $-1 \le x \le 1$. From $\mathbf{r}'(t) = -2 \sin 2t\mathbf{i} + \cos t\mathbf{j}$ we find

$$\mathbf{r}'(0) = \mathbf{j} \quad \text{and} \quad \mathbf{r}'\left(\frac{\pi}{6}\right) = -\sqrt{3}\mathbf{i} + \frac{\sqrt{3}}{2}\mathbf{j}.$$

In Figure 9.6 these vectors are drawn tangent to the curve C at $(1, 0)$ and $(½, ½)$, respectively. ❏

Example 5 Tangent Line

Find parametric equations of the tangent line to the graph of the curve C whose parametric equations are $x = t^2$, $y = t^2 - t$, $z = -7t$ at $t = 3$.

SOLUTION The vector function that gives the position of a point P on the curve is given by $\mathbf{r}(t) = t^2\mathbf{i} + (t^2 - t)\mathbf{j} - 7t\mathbf{k}$. Now

$$\mathbf{r}'(t) = 2t\mathbf{i} + (2t - 1)\mathbf{j} - 7\mathbf{k} \quad \text{and so} \quad \mathbf{r}'(3) = 6\mathbf{i} + 5\mathbf{j} - 7\mathbf{k},$$

which is tangent to C at the point whose position vector is

$$\mathbf{r}(3) = 9\mathbf{i} + 6\mathbf{j} - 21\mathbf{k},$$

that is, $P(9, 6, -21)$. Using the components of $\mathbf{r}'(3)$, we see that parametric equations of the tangent line are $x = 9 + 6t$, $y = 6 + 5t$, $z = -21 - 7t$. ❏

■ **Higher-Order Derivatives** Higher-order derivatives of a vector function are also obtained by differentiating its components. In the case of the **second derivative,** we have

$$\mathbf{r}''(t) = \langle f''(t), g''(t), h''(t) \rangle = f''(t)\mathbf{i} + g''(t)\mathbf{j} + h''(t)\mathbf{k}.$$

Example 6 Derivative of a Vector Function

If $\mathbf{r}(t) = (t^3 - 2t^2)\mathbf{i} + 4t\mathbf{j} + e^{-t}\mathbf{k}$, then

$$\mathbf{r}'(t) = (3t^2 - 4t)\mathbf{i} + 4\mathbf{j} - e^{-t}\mathbf{k} \quad \text{and} \quad \mathbf{r}''(t) = (6t - 4)\mathbf{i} + e^{-t}\mathbf{k}.$$ ❏

THEOREM 9.3 **Chain Rule**

If \mathbf{r} is a differentiable vector function and $s = u(t)$ is a differentiable scalar function, then the derivative of $\mathbf{r}(s)$ with respect to t is

$$\frac{d\mathbf{r}}{dt} = \frac{d\mathbf{r}}{ds}\frac{ds}{dt} = \mathbf{r}'(s)u'(t).$$

Example 7 Chain Rule

If $\mathbf{r}(s) = \cos 2s\mathbf{i} + \sin 2s\mathbf{j} + e^{-3s}\mathbf{k}$, where $s = t^4$, then

$$\frac{d\mathbf{r}}{dt} = [-2 \sin 2s\mathbf{i} + 2 \cos 2s\mathbf{j} - 3e^{-3s}\mathbf{k}]4t^3$$

$$= -8t^3 \sin(2t^4)\mathbf{i} + 8t^3 \cos(2t^4)\mathbf{j} - 12t^3 e^{-3t^4}\mathbf{k}.$$ ❏

Details of the proof of the next theorem are left as exercises.

THEOREM 9.4 Rules of Differentiation

Let \mathbf{r}_1 and \mathbf{r}_2 be differentiable vector functions and $u(t)$ a differentiable scalar function.

$(i) \quad \dfrac{d}{dt}[\mathbf{r}_1(t) + \mathbf{r}_2(t)] = \mathbf{r}_1'(t) + \mathbf{r}_2'(t)$

$(ii) \quad \dfrac{d}{dt}[u(t)\mathbf{r}_1(t)] = u(t)\mathbf{r}_1'(t) + u'(t)\mathbf{r}_1(t)$

$(iii) \quad \dfrac{d}{dt}[\mathbf{r}_1(t) \cdot \mathbf{r}_2(t)] = \mathbf{r}_1(t) \cdot \mathbf{r}_2'(t) + \mathbf{r}_1'(t) \cdot \mathbf{r}_2(t)$

$(iv) \quad \dfrac{d}{dt}[\mathbf{r}_1(t) \times \mathbf{r}_2(t)] = \mathbf{r}_1(t) \times \mathbf{r}_2'(t) + \mathbf{r}_1'(t) \times \mathbf{r}_2(t).$

 Note of caution.

Since the cross product of two vectors is not commutative, the order in which \mathbf{r}_1 and \mathbf{r}_2 appear in part (iv) of Theorem 9.4 must be strictly observed.

■ **Integrals of Vector Functions** If f, g, and h are integrable, then the indefinite and definite integrals of a vector function $\mathbf{r}(t) = f(t)\mathbf{i} + g(t)\mathbf{j} + h(t)\mathbf{k}$ are defined, respectively, by

$$\int \mathbf{r}(t)\, dt = \left[\int f(t)\, dt\right]\mathbf{i} + \left[\int g(t)\, dt\right]\mathbf{j} + \left[\int h(t)\, dt\right]\mathbf{k}$$

$$\int_a^b \mathbf{r}(t)\, dt = \left[\int_a^b f(t)\, dt\right]\mathbf{i} + \left[\int_a^b g(t)\, dt\right]\mathbf{j} + \left[\int_a^b h(t)\, dt\right]\mathbf{k}.$$

The indefinite integral of \mathbf{r} is another vector function $\mathbf{R} + \mathbf{c}$ such that $\mathbf{R}'(t) = \mathbf{r}(t)$.

Example 8 **Integral of a Vector Function**

If $\mathbf{r}(t) = 6t^2\mathbf{i} + 4e^{-2t}\mathbf{j} + 8\cos 4t\,\mathbf{k}$

then
$$\int \mathbf{r}(t)\, dt = \left[\int 6t^2\, dt\right]\mathbf{i} + \left[\int 4e^{-2t}\, dt\right]\mathbf{j} + \left[\int 8\cos 4t\, dt\right]\mathbf{k}$$

$$= [2t^3 + c_1]\mathbf{i} + [-2e^{-2t} + c_2]\mathbf{j} + [2\sin 4t + c_3]\mathbf{k}$$

$$= 2t^3\mathbf{i} - 2e^{-2t}\mathbf{j} + 2\sin 4t\,\mathbf{k} + \mathbf{c},$$

where $\mathbf{c} = c_1\mathbf{i} + c_2\mathbf{j} + c_3\mathbf{k}$. ❑

■ **Length of a Space Curve** If $\mathbf{r}(t) = f(t)\mathbf{i} + g(t)\mathbf{j} + h(t)\mathbf{k}$ is a smooth function, then it can be shown that the **length** of the smooth curve traced by \mathbf{r} is given by

$$s = \int_a^b \sqrt{[f'(t)]^2 + [g'(t)]^2 + [h'(t)]^2}\, dt = \int_a^b \|\mathbf{r}'(t)\|\, dt. \tag{3}$$

■ **Arc Length As a Parameter** A curve in the plane or in space can be parametrized in terms of the arc length s.

Example 9 **Example 1 Revisited**

Consider the helix of Example 1. Since $\|\mathbf{r}'(t)\| = \sqrt{5}$, it follows from (3) that the length of the curve from $\mathbf{r}(0)$ to an arbitrary point $\mathbf{r}(t)$ is

$$s = \int_0^t \sqrt{5}\, du = \sqrt{5}t,$$

where we have used u as a dummy variable of integration. Using $t = s/\sqrt{5}$, we obtain a vector equation of the helix as a function of arc length:

$$\mathbf{r}(s) = 2\cos\frac{s}{\sqrt{5}}\mathbf{i} + 2\sin\frac{s}{\sqrt{5}}\mathbf{j} + \frac{s}{\sqrt{5}}\mathbf{k}. \tag{4}$$

Parametric equations of the helix are then

$$f(s) = 2\cos\frac{s}{\sqrt{5}}, \qquad g(s) = 2\sin\frac{s}{\sqrt{5}}, \qquad h(s) = \frac{s}{\sqrt{5}}. \qquad \square$$

The derivative of a vector function $\mathbf{r}(t)$ with respect to the parameter t is a tangent vector to the curve traced by \mathbf{r}. However, if the curve is parametrized in terms of arc length s, then $\mathbf{r}'(s)$ *is a unit tangent vector*. To see this, let a curve be described by $\mathbf{r}(s)$, where s is arc length. From (3), the length of the curve from $\mathbf{r}(0)$ to $\mathbf{r}(s)$ is $s = \int_0^s \|\mathbf{r}'(u)\|\, du$. Differentiation of this last equation with respect to s then yields $\|\mathbf{r}'(s)\| = 1$.

EXERCISES 9.1

Answers to odd-numbered problems begin on page A-47.

In Problems 1–10, graph the curve traced by the given vector function.

1. $\mathbf{r}(t) = 2\sin t\mathbf{i} + 4\cos t\mathbf{j} + t\mathbf{k}$; $t \geq 0$

2. $\mathbf{r}(t) = \cos t\mathbf{i} + t\mathbf{j} + \sin t\mathbf{k}$; $t \geq 0$

3. $\mathbf{r}(t) = t\mathbf{i} + 2t\mathbf{j} + \cos t\mathbf{k}$; $t \geq 0$

4. $\mathbf{r}(t) = 4\mathbf{i} + 2\cos t\mathbf{j} + 3\sin t\mathbf{k}$

5. $\mathbf{r}(t) = \langle e^t, e^{2t} \rangle$

6. $\mathbf{r}(t) = \cosh t\mathbf{i} + 3\sinh t\mathbf{j}$

7. $\mathbf{r}(t) = \langle \sqrt{2}\sin t, \sqrt{2}\sin t, 2\cos t \rangle$; $0 \leq t \leq \pi/2$

8. $\mathbf{r}(t) = t\mathbf{i} + t^3\mathbf{j} + t\mathbf{k}$

9. $\mathbf{r}(t) = e^t\cos t\mathbf{i} + e^t\sin t\mathbf{j} + e^t\mathbf{k}$

10. $\mathbf{r}(t) = \langle t\cos t, t\sin t, t^2 \rangle$

In Problems 11–14, find the vector function that describes the curve C of intersection between the given surfaces. Sketch the curve C. Use the indicated parameter.

11. $z = x^2 + y^2$, $y = x$; $x = t$

12. $x^2 + y^2 - z^2 = 1$, $y = 2x$; $x = t$

13. $x^2 + y^2 = 9$, $z = 9 - x^2$; $x = 3\cos t$

14. $z = x^2 + y^2$, $z = 1$; $x = \sin t$

15. Given that $\mathbf{r}(t) = \dfrac{\sin 2t}{t}\mathbf{i} + (t-2)^5\mathbf{j} + t\ln t\mathbf{k}$, find $\lim_{t \to 0^+} \mathbf{r}(t)$.

16. Given that $\lim_{t \to a} \mathbf{r}_1(t) = \mathbf{i} - 2\mathbf{j} + \mathbf{k}$ and $\lim_{t \to a} \mathbf{r}_2(t) = 2\mathbf{i} + 5\mathbf{j} + 7\mathbf{k}$, find:

(a) $\lim_{t \to a} [-4\mathbf{r}_1(t) + 3\mathbf{r}_2(t)]$

(b) $\lim_{t \to a} \mathbf{r}_1(t) \cdot \mathbf{r}_2(t)$

In Problems 17–20, find $\mathbf{r}'(t)$ and $\mathbf{r}''(t)$ for the given vector function.

17. $\mathbf{r}(t) = \ln t\mathbf{i} + \dfrac{1}{t}\mathbf{j}$, $t > 0$

18. $\mathbf{r}(t) = \langle t\cos t - \sin t, t + \cos t \rangle$

19. $\mathbf{r}(t) = \langle te^{2t}, t^3, 4t^2 - t \rangle$

20. $\mathbf{r}(t) = t^2\mathbf{i} + t^3\mathbf{j} + \tan^{-1}t\mathbf{k}$

In Problems 21–24, graph the curve C that is described by \mathbf{r} and graph \mathbf{r}' at the indicated value of t.

21. $\mathbf{r}(t) = 2\cos t\mathbf{i} + 6\sin t\mathbf{j}$; $t = \pi/6$

22. $\mathbf{r}(t) = t^3\mathbf{i} + t^2\mathbf{j}$; $t = -1$

23. $\mathbf{r}(t) = 2\mathbf{i} + t\mathbf{j} + \dfrac{4}{1+t^2}\mathbf{k}$; $t = 1$

24. $\mathbf{r}(t) = 3\cos t\mathbf{i} + 3\sin t\mathbf{j} + 2t\mathbf{k}$; $t = \pi/4$

In Problems 25 and 26, find parametric equations of the tangent line to the given curve at the indicated value of t.

25. $x = t$, $y = \frac{1}{2}t^2$, $z = \frac{1}{3}t^3$; $t = 2$

26. $x = t^3 - t$, $y = \dfrac{6t}{t+1}$, $z = (2t + 1)^2$; $t = 1$

In Problems 27–32, find the indicated derivative. Assume that all vector functions are differentiable.

27. $\dfrac{d}{dt}[\mathbf{r}(t) \times \mathbf{r}'(t)]$

28. $\dfrac{d}{dt}[\mathbf{r}(t) \cdot (t\mathbf{r}(t))]$

29. $\dfrac{d}{dt}[\mathbf{r}(t) \cdot (\mathbf{r}'(t) \times \mathbf{r}''(t))]$

30. $\dfrac{d}{dt}[\mathbf{r}_1(t) \times (\mathbf{r}_2(t) \times \mathbf{r}_3(t))]$

31. $\dfrac{d}{dt}\left[\mathbf{r}_1(2t) + \mathbf{r}_2\left(\dfrac{1}{t}\right)\right]$

32. $\dfrac{d}{dt}[t^3\mathbf{r}(t^2)]$

In Problems 33–36, evaluate the given integral.

33. $\displaystyle\int_{-1}^{2} (t\mathbf{i} + 3t^2\mathbf{j} + 4t^3\mathbf{k})\, dt$

34. $\displaystyle\int_{0}^{4} (\sqrt{2t + 1}\,\mathbf{i} - \sqrt{t}\,\mathbf{j} + \sin \pi t\mathbf{k})\, dt$

35. $\displaystyle\int (te^t\mathbf{i} - e^{-2t}\mathbf{j} + te^{t^2}\mathbf{k})\, dt$

36. $\displaystyle\int \dfrac{1}{1 + t^2}\,(\mathbf{i} + t\mathbf{j} + t^2\mathbf{k})\, dt$

In Problems 37–40, find a vector function \mathbf{r} that satisfies the indicated conditions.

37. $\mathbf{r}'(t) = 6\mathbf{i} + 6t\mathbf{j} + 3t^2\mathbf{k}$; $\mathbf{r}(0) = \mathbf{i} - 2\mathbf{j} + \mathbf{k}$

38. $\mathbf{r}'(t) = t \sin t^2\mathbf{i} - \cos 2t\mathbf{j}$; $\mathbf{r}(0) = \frac{3}{2}\mathbf{i}$

39. $\mathbf{r}''(t) = 12t\mathbf{i} - 3t^{-1/2}\mathbf{j} + 2\mathbf{k}$; $\mathbf{r}'(1) = \mathbf{j}$, $\mathbf{r}(1) = 2\mathbf{i} - \mathbf{k}$

40. $\mathbf{r}''(t) = \sec^2 t\mathbf{i} + \cos t\mathbf{j} - \sin t\mathbf{k}$;

 $\mathbf{r}'(0) = \mathbf{i} + \mathbf{j} + \mathbf{k}$, $\mathbf{r}(0) = -\mathbf{j} + 5\mathbf{k}$

In Problems 41–44, find the length of the curve traced by the given vector function on the indicated interval.

41. $\mathbf{r}(t) = a \cos t\mathbf{i} + a \sin t\mathbf{j} + ct\mathbf{k}$; $0 \le t \le 2\pi$

42. $\mathbf{r}(t) = t\mathbf{i} + t \cos t\mathbf{j} + t \sin t\mathbf{k}$; $0 \le t \le \pi$

43. $\mathbf{r}(t) = e^t \cos 2t\mathbf{i} + e^t \sin 2t\mathbf{j} + e^t\mathbf{k}$; $0 \le t \le 3\pi$

44. $\mathbf{r}(t) = 3t\mathbf{i} + \sqrt{3}t^2\mathbf{j} + \frac{2}{3}t^3\mathbf{k}$; $0 \le t \le 1$

45. Express the vector equation of a circle $\mathbf{r}(t) = a \cos t\mathbf{i} + a \sin t\mathbf{j}$ as a function of arc length s. Verify that $\mathbf{r}'(s)$ is a unit vector.

46. If $\mathbf{r}(s)$ is the vector function given in (4), verify that $\mathbf{r}'(s)$ is a unit vector.

47. Suppose \mathbf{r} is a differentiable vector function for which $\|\mathbf{r}(t)\| = c$ for all t. Show that the tangent vector $\mathbf{r}'(t)$ is perpendicular to the position vector $\mathbf{r}(t)$ for all t.

48. In Problem 47, describe geometrically the kind of curve C for which $\|\mathbf{r}(t)\| = c$.

■ Miscellaneous Problems

49. Prove Theorem 9.4(*ii*).

50. Prove Theorem 9.4(*iii*).

51. Prove Theorem 9.4(*iv*).

52. If \mathbf{v} is a constant vector and \mathbf{r} is integrable on $[a, b]$, prove that $\int_a^b \mathbf{v} \cdot \mathbf{r}(t)\, dt = \mathbf{v} \cdot \int_a^b \mathbf{r}(t)\, dt$.

9.2 Motion on a Curve

■ Velocity and Acceleration Suppose a body or particle moves along a curve C so that its position at time t is given by the vector function $\mathbf{r}(t) = f(t)\mathbf{i} + g(t)\mathbf{j} + h(t)\mathbf{k}$. If f, g, and h have second derivatives, then the vectors

$$\mathbf{v}(t) = \mathbf{r}'(t) = f'(t)\mathbf{i} + g'(t)\mathbf{j} + h'(t)\mathbf{k}$$

$$\mathbf{a}(t) = \mathbf{r}''(t) = f''(t)\mathbf{i} + g''(t)\mathbf{j} + h''(t)\mathbf{k}$$

are called the **velocity** and **acceleration** of the particle, respectively. The scalar function $\|\mathbf{v}(t)\|$ is the **speed** of the particle. Since

$$\|\mathbf{v}(t)\| = \left\|\frac{d\mathbf{r}}{dt}\right\| = \sqrt{\left(\frac{dx}{dt}\right)^2 + \left(\frac{dy}{dt}\right)^2 + \left(\frac{dz}{dt}\right)^2}$$

speed is related to arc length s by $s'(t) = \|\mathbf{v}(t)\|$. In other words, arc length is given by $s = \int_{t_0}^{t_1} \|\mathbf{v}(t)\|\, dt$. It also follows from the discussion of Section 9.1 that if $P(x_1, y_1, z_1)$ is the position of the particle on C at time t_1, then we may draw $\mathbf{v}(t_1)$ *tangent to C at P*. Similar remarks hold for curves traced by the vector function $\mathbf{r}(t) = f(t)\mathbf{i} + g(t)\mathbf{j}$.

Figure 9.7

Figure 9.8

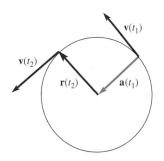

Figure 9.9

Example 1 Velocity and Acceleration Vectors

The position of a moving particle is given by $\mathbf{r}(t) = t^2\mathbf{i} + t\mathbf{j} + \frac{5}{2}t\mathbf{k}$. Graph the curve defined by $\mathbf{r}(t)$ and the vectors $\mathbf{v}(2)$ and $\mathbf{a}(2)$.

SOLUTION Since $x = t^2$, $y = t$, the path of the particle is above the parabola $x = y^2$. When $t = 2$ the position vector $\mathbf{r}(2) = 4\mathbf{i} + 2\mathbf{j} + 5\mathbf{k}$ indicates that the particle is at the point $P(4, 2, 5)$. Now,

$$\mathbf{v}(t) = \mathbf{r}'(t) = 2t\mathbf{i} + \mathbf{j} + \frac{5}{2}\mathbf{k} \quad \text{and} \quad \mathbf{a}(t) = \mathbf{r}''(t) = 2\mathbf{i}$$

so that $\mathbf{v}(2) = 4\mathbf{i} + \mathbf{j} + \frac{5}{2}\mathbf{k}$ and $\mathbf{a}(2) = 2\mathbf{i}$. These vectors are shown in Figure 9.7. ❑

If a particle moves with a constant speed c, then its acceleration vector is perpendicular to the velocity vector \mathbf{v}. To see this, note that $\|\mathbf{v}\|^2 = c^2$ or $\mathbf{v} \cdot \mathbf{v} = c^2$. We differentiate both sides with respect to t and obtain, with the aid of Theorem 9.4(iii):

$$\frac{d}{dt}(\mathbf{v} \cdot \mathbf{v}) = \mathbf{v} \cdot \frac{d\mathbf{v}}{dt} + \frac{d\mathbf{v}}{dt} \cdot \mathbf{v} = 2\mathbf{v} \cdot \frac{d\mathbf{v}}{dt} = 0.$$

Thus, $\qquad \dfrac{d\mathbf{v}}{dt} \cdot \mathbf{v} = 0 \quad \text{or} \quad \mathbf{a}(t) \cdot \mathbf{v}(t) = 0 \quad \text{for all } t.$

Example 2 Velocity and Acceleration Vectors

Suppose the vector function in Example 2 of Section 9.1 represents the position of a particle moving in a circular orbit. Graph the velocity and acceleration vector at $t = \pi/4$.

SOLUTION Recall that $\mathbf{r}(t) = 2\cos t\mathbf{i} + 2\sin t\mathbf{j} + 3\mathbf{k}$ is the position vector of a particle moving in a circular orbit of radius 2 in the plane $z = 3$. When $t = \pi/4$ the particle is at the point $P(\sqrt{2}, \sqrt{2}, 3)$. Now,

$$\mathbf{v}(t) = \mathbf{r}'(t) = -2\sin t\mathbf{i} + 2\cos t\mathbf{j}$$

$$\mathbf{a}(t) = \mathbf{r}''(t) = -2\cos t\mathbf{i} - 2\sin t\mathbf{j}.$$

Since the speed is $\|\mathbf{v}(t)\| = 2$ for all time t, it follows from the discussion preceding this example that $\mathbf{a}(t)$ is perpendicular to $\mathbf{v}(t)$. (Verify this.) As shown in Figure 9.8, the vectors

$$\mathbf{v}\left(\frac{\pi}{4}\right) = -2\sin\frac{\pi}{4}\mathbf{i} + 2\cos\frac{\pi}{4}\mathbf{j} = -\sqrt{2}\mathbf{i} + \sqrt{2}\mathbf{j}$$

$$\mathbf{a}\left(\frac{\pi}{4}\right) = -2\cos\frac{\pi}{4}\mathbf{i} - 2\sin\frac{\pi}{4}\mathbf{j} = -\sqrt{2}\mathbf{i} - \sqrt{2}\mathbf{j}$$

are drawn at the point P. The vector $\mathbf{v}(\pi/4)$ is tangent to the circular path and $\mathbf{a}(\pi/4)$ points along a radius toward the center of the circle. ❑

■ **Centripetal Acceleration** For circular motion in the plane, described by $\mathbf{r}(t) = r_0\cos\omega t\mathbf{i} + r_0\sin\omega t\mathbf{j}$, r_0 and ω constants, it is evident that $\mathbf{r}'' = -\omega^2\mathbf{r}$. This means that the acceleration vector $\mathbf{a}(t) = \mathbf{r}''(t)$ points in the direction opposite to that of the position vector $\mathbf{r}(t)$. We then say $\mathbf{a}(t)$ is **centripetal acceleration.** See Figure 9.9. If $v = \|\mathbf{v}(t)\|$ and $a = \|\mathbf{a}(t)\|$, we leave it as an exercise to show that $a = v^2/r_0$.

■ **Curvilinear Motion in the Plane** Many important applications of vector functions occur in regard to curvilinear motion in a plane. For example, planetary and projectile motion take place in a plane.

Figure 9.10

(a) **Maximum height H:**
Find t_1 for which $y'(t_1) = 0$;
$H = y_{max} = y(t_1)$

(b) **Range R:**
Find $t_1 > 0$ for which $y(t_1) = 0$;
$R = x_{max} = x(t_1)$

Figure 9.11

In analyzing the motion of short-range ballistic projectiles,* we begin with the acceleration of gravity written in vector form: $\mathbf{a}(t) = -g\mathbf{j}$.

If, as shown in Figure 9.10, a projectile is launched with an initial velocity $\mathbf{v}_0 = v_0 \cos \theta \mathbf{i} + v_0 \sin \theta \mathbf{j}$ from an initial height $\mathbf{s}_0 = s_0\mathbf{j}$, then

$$\mathbf{v}(t) = \int (-g\mathbf{j})\, dt = -gt\mathbf{j} + \mathbf{c}_1,$$

where $\mathbf{v}(0) = \mathbf{v}_0$ implies that $\mathbf{c}_1 = \mathbf{v}_0$. Therefore,

$$\mathbf{v}(t) = (v_0 \cos \theta)\mathbf{i} + (-gt + v_0 \sin \theta)\mathbf{j}.$$

Integrating again and using $\mathbf{r}(0) = \mathbf{s}_0$ yield

$$\mathbf{r}(t) = (v_0 \cos \theta)t\mathbf{i} + \left[-\frac{1}{2} gt^2 + (v_0 \sin \theta)t + s_0 \right]\mathbf{j}.$$

Hence, parametric equations for the trajectory of the projectile are

$$x(t) = (v_0 \cos \theta)t, \qquad y(t) = -\frac{1}{2} gt^2 + (v_0 \sin \theta)t + s_0. \tag{1}$$

We are naturally interested in finding the maximum height H and the range R attained by a projectile. As shown in Figure 9.11, these quantities are the maximum values of $y(t)$ and $x(t)$, respectively.

Example 3 Trajectory of a Shell

A shell is fired from ground level with an initial speed of 768 ft/s at an angle of elevation of 30°. Find (a) the vector function and parametric equations of the shell's trajectory, (b) the maximum altitude attained, (c) the range of the shell, and (d) the speed at impact.

SOLUTION

(a) Initially, we have $\mathbf{s}_0 = \mathbf{0}$ and

$$\mathbf{v}_0 = (768 \cos 30°)\mathbf{i} + (768 \sin 30°)\mathbf{j} = 384\sqrt{3}\mathbf{i} + 384\mathbf{j}. \tag{2}$$

Integrating $\mathbf{a}(t) = -32\mathbf{j}$ and using (2) give

$$\mathbf{v}(t) = (384\sqrt{3})\mathbf{i} + (-32t + 384)\mathbf{j}. \tag{3}$$

Integrating again gives

$$\mathbf{r}(t) = (384\sqrt{3}t)\mathbf{i} + (-16t^2 + 384t)\mathbf{j}.$$

Hence, the parametric equations of the shell's trajectory are

$$x(t) = 384\sqrt{3}t, \qquad y(t) = -16t^2 + 384t. \tag{4}$$

(b) From (4) we see that $dy/dt = 0$ when $-32t + 384 = 0$ or $t = 12$. Thus, the maximum height H attained by the shell is

$$H = y(12) = -16(12)^2 + 384(12) = 2304 \text{ ft}.$$

(c) From (4) we see that $y(t) = 0$ when $-16t(t - 24) = 0$ or $t = 0$, $t = 24$. The range R is then

$$R = x(24) = 384\sqrt{3}(24) \approx 15{,}963 \text{ ft}.$$

* The projectile is shot or hurled rather than self-propelled. In the analysis of *long-range* ballistic motion, the curvature of the earth must be taken into consideration.

(d) From (3) we obtain the impact speed of the shell:

$$\|\mathbf{v}(24)\| = \sqrt{(-384)^2 + (384\sqrt{3})^2} = 768 \text{ ft/s.}$$ ❏

Remarks

We have seen that the rate of change of arc length ds/dt is the same as the speed $\|\mathbf{v}(t)\| = \|\mathbf{r}'(t)\|$. However, as we shall see in the next section, it does *not* follow that the *scalar acceleration* d^2s/dt^2 is the same as $\|\mathbf{a}(t)\| = \|\mathbf{r}''(t)\|$. See Problem 20 in Exercises 9.2.

EXERCISES 9.2

Answers to odd-numbered problems begin on page A-48.

In Problems 1–8, $\mathbf{r}(t)$ is the position vector of a moving particle. Graph the curve and the velocity and acceleration vectors at the indicated time. Find the speed at that time.

1. $\mathbf{r}(t) = t^2\mathbf{i} + \tfrac{1}{4}t^4\mathbf{j}$; $t = 1$

2. $\mathbf{r}(t) = t^2\mathbf{i} + \dfrac{1}{t^2}\mathbf{j}$; $t = 1$

3. $\mathbf{r}(t) = -\cosh 2t\mathbf{i} + \sinh 2t\mathbf{j}$; $t = 0$

4. $\mathbf{r}(t) = 2\cos t\mathbf{i} + (1 + \sin t)\mathbf{j}$; $t = \pi/3$

5. $\mathbf{r}(t) = 2\mathbf{i} + (t - 1)^2\mathbf{j} + t\mathbf{k}$; $t = 2$

6. $\mathbf{r}(t) = t\mathbf{i} + t\mathbf{j} + t^3\mathbf{k}$; $t = 2$

7. $\mathbf{r}(t) = t\mathbf{i} + t^2\mathbf{j} + t^3\mathbf{k}$; $t = 1$

8. $\mathbf{r}(t) = t\mathbf{i} + t^3\mathbf{j} + t\mathbf{k}$; $t = 1$

9. Suppose $\mathbf{r}(t) = t^2\mathbf{i} + (t^3 - 2t)\mathbf{j} + (t^2 - 5t)\mathbf{k}$ is the position vector of a moving particle. At what points does the particle pass through the xy-plane? What are its velocity and acceleration at these points?

10. Suppose a particle moves in space so that $\mathbf{a}(t) = \mathbf{0}$ for all time t. Describe its path.

11. A shell is fired from ground level with an initial speed of 480 ft/s at an angle of elevation of 30°. Find:

 (a) a vector function and parametric equations of the shell's trajectory,

 (b) the maximum altitude attained,

 (c) the range of the shell, and

 (d) the speed at impact.

12. Rework Problem 11 if the shell is fired with the same initial speed and the same angle of elevation but from a cliff 1600 ft high.

13. A used car is pushed off an 81-ft-high sheer seaside cliff with a speed of 4 ft/s. Find the speed at which the car hits the water.

14. A small projectile is launched from ground level with an initial speed of 98 m/s. Find the possible angles of elevation so that its range is 490 m.

15. A football quarterback throws a 100-yd "bomb" at an angle of 45° from the horizontal. What is the initial speed of the football at the point of release?

16. A quarterback throws a football with the same initial speed at an angle of 60° from the horizontal and then at an angle of 30° from the horizontal. Show that the range of the football is the same in each case. Generalize this result for any release angle $0 < \theta < \pi/2$.

17. A projectile is fired from a cannon directly at a target that is dropped from rest simultaneously as the cannon is fired. Show that the projectile will strike the target in midair. See Figure 9.12. [*Hint:* Assume that the origin is at the muzzle of the cannon and that the angle of elevation is θ. If \mathbf{r}_p and \mathbf{r}_t are position vectors of the projectile and target, respectively, is there a time at which $\mathbf{r}_p = \mathbf{r}_t$?]

Figure 9.12

18. In army field maneuvers sturdy equipment and supply packs are simply dropped from planes that fly horizontally at a slow speed and a low altitude. A supply plane flies horizontally over a target at an altitude of 1024 ft at a constant speed of 180 mph. Use (1) to determine the horizontal distance a supply pack travels relative to the point from which it was dropped. At what line-of-sight angle α should the supply pack be released in order to hit the target indicated in Figure 9.13?

Figure 9.13

19. Suppose that $\mathbf{r}(t) = r_0 \cos \omega t \mathbf{i} + r_0 \sin \omega t \mathbf{j}$ is the position vector of an object that is moving in a circle of radius r_0 in the xy-plane. If $\|\mathbf{v}(t)\| = v$, show that the magnitude of the centripetal acceleration is $a = \|\mathbf{a}(t)\| = v^2/r_0$.

20. The motion of a particle in space is described by

$$\mathbf{r}(t) = b \cos t \mathbf{i} + b \sin t \mathbf{j} + ct \mathbf{k}, \quad t \geq 0$$

(a) Compute $\|\mathbf{v}(t)\|$.

(b) Compute $s = \int_0^t \|\mathbf{v}(t)\|\, dt$ and verify that ds/dt is the same as the result of part **(a)**.

(c) Verify that $d^2s/dt^2 \neq \|\mathbf{a}(t)\|$.

21. The **effective weight** w_e of mass m at the equator of the earth is defined by $w_e = mg - ma$, where a is the magnitude of the centripetal acceleration given in Problem 19. Determine the effective weight of a 192-lb person if the radius of the earth is 4000 mi, $g = 32$ ft/s^2, and $v = 1530$ ft/s.

22. Consider a bicyclist riding on a flat circular track of radius r_0. If m is the combined mass of the rider and bicycle, fill in the blanks in Figure 9.14. [*Hint:* Use Problem 19 and force = mass × acceleration. Assume that the directions are upward and to the left.] The resultant vector \mathbf{U} gives the direction the bicyclist must be tipped to avoid falling. Find the angle ϕ from the vertical at which the bicyclist must be tipped if her speed is 44 ft/s and the radius of the track is 60 ft.

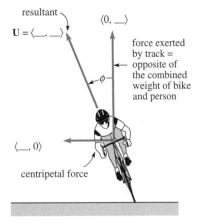

Figure 9.14

23. Use the results given in (1) to show that the trajectory of a ballistic projectile is parabolic.

24. A projectile is launched with an initial speed v_0 from ground level at an angle of elevation θ. Use (1) to show that the maximum height and range of the projectile are, respectively,

$$H = \frac{v_0^2 \sin^2 \theta}{2g} \quad \text{and} \quad R = \frac{v_0^2 \sin 2\theta}{g}.$$

25. The velocity of a particle moving in a fluid is described by means of a **velocity field** $\mathbf{v} = v_1 \mathbf{i} + v_2 \mathbf{j} + v_3 \mathbf{k}$, where the components v_1, v_2, and v_3 are functions of $x, y, z,$ and time t. If the velocity of the particle is $\mathbf{v}(t) = 6t^2 x \mathbf{i} - 4ty^2 \mathbf{j} + 2t(z + 1)\mathbf{k}$, find $\mathbf{r}(t)$. [*Hint:* Use separation of variables.]

26. Suppose m is the mass of a moving particle. Newton's second law of motion can be written in vector form as

$$\mathbf{F} = m\mathbf{a} = \frac{d}{dt}(m\mathbf{v}) = \frac{d\mathbf{p}}{dt},$$

where $\mathbf{p} = m\mathbf{v}$ is called **linear momentum.** The angular momentum of the particle with respect to the origin is defined to be $\mathbf{L} = \mathbf{r} \times \mathbf{p}$, where \mathbf{r} is its position vector. If the torque of the particle about the origin is $\boldsymbol{\tau} = \mathbf{r} \times \mathbf{F} = \mathbf{r} \times d\mathbf{p}/dt$, show that $\boldsymbol{\tau}$ is the time rate of change of angular momentum.

27. Suppose the sun is located at the origin. The gravitational force \mathbf{F} exerted on a planet of mass m by the sun of mass M is

$$\mathbf{F} = -k\frac{Mm}{r^2}\mathbf{u}.$$

\mathbf{F} is a **central force**—that is, a force directed along the position vector \mathbf{r} of the planet. Here k is the gravitational constant, $r = \|\mathbf{r}\|$, $\mathbf{u} = \mathbf{r}/r$ is a unit vector in the direction of \mathbf{r}, and the minus sign indicates that \mathbf{F} is an attractive force—that is, a force directed toward the sun. See Figure 9.15.

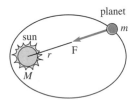

Figure 9.15

(a) Use Problem 26 to show that the torque acting on the planet due to this central force is $\mathbf{0}$.

(b) Explain why the angular momentum \mathbf{L} of a planet is constant.

28. (*This problem could present a challenge.*) In this problem the student will use the properties in Sections 7.4 and 9.1 to prove **Kepler's first law of planetary motion:** The orbit of a planet is an ellipse with the sun at one focus. We assume that the sun has mass M and is located at the origin, \mathbf{r} is the position vector of a body of mass m moving under the gravitational attraction of the sun, and $\mathbf{u} = \mathbf{r}/r$ is a unit vector in the direction of \mathbf{r}.

(a) Use Problem 27 and Newton's second law of motion $\mathbf{F} = m\mathbf{a}$ to show that

$$\frac{d^2\mathbf{r}}{dt^2} = -kM\,\frac{\mathbf{u}}{r^2}.$$

(b) Use part **(a)** to show that $\mathbf{r} \times \mathbf{r}'' = \mathbf{0}$.

(c) Use part **(b)** to show that $\dfrac{d}{dt}(\mathbf{r} \times \mathbf{v}) = \mathbf{0}$.

(d) It follows from part **(c)** that $\mathbf{r} \times \mathbf{v} = \mathbf{c}$, where \mathbf{c} is a constant vector. Show that $\mathbf{c} = r^2(\mathbf{u} \times \mathbf{u}')$.

(e) Show that $\dfrac{d}{dt}(\mathbf{u} \cdot \mathbf{u}) = 0$ and consequently $\mathbf{u} \cdot \mathbf{u}' = 0$.

(f) Use parts **(a)**, **(e)**, and **(d)** to show that

$$\frac{d}{dt}(\mathbf{v} \times \mathbf{c}) = kM\,\frac{d\mathbf{u}}{dt}.$$

(g) By integrating the result in part **(f)** with respect to t, we get $\mathbf{v} \times \mathbf{c} = kM\mathbf{u} + \mathbf{d}$, where \mathbf{d} is another constant vector.

Dot both sides of this last expression by the vector $\mathbf{r} = r\mathbf{u}$ and use Problem 61 in Exercises 7.4 to show that

$$r = \frac{c^2/kM}{1 + (d/kM)\cos\theta},$$

where $c = \|\mathbf{c}\|$, $d = \|\mathbf{d}\|$, and θ is the angle between \mathbf{d} and \mathbf{r}.

(h) Explain why the result in part **(g)** proves Kepler's first law.

(i) At perihelion* the vectors \mathbf{r} and \mathbf{v} are perpendicular and have magnitudes r_0 and v_0, respectively. Use this information and parts **(d)** and **(g)** to show that $c = r_0 v_0$ and $d = r_0 v_0^2 - kM$.

* This is the point in the orbit where the body is closest to the sun.

9.3 Curvature and Components of Acceleration

■ **Curvature** Let C be a smooth curve in either 2- or 3-space traced out by a vector function $\mathbf{r}(t)$. In this section we are going to consider in greater detail the acceleration vector $\mathbf{a}(t) = \mathbf{r}''(t)$ introduced in the last section. But before doing this, we need to examine a scalar quantity called the **curvature** of a curve. We know that $\mathbf{r}'(t)$ is a tangent vector to the curve C, and consequently

$$\mathbf{T} = \frac{\mathbf{r}'(t)}{\|\mathbf{r}'(t)\|} \tag{1}$$

is a **unit tangent.** But recall from the end of Section 9.1 that if C is parameterized by arc length s, then a unit tangent to the curve is also given by $d\mathbf{r}/ds$. The quantity $\|\mathbf{r}'(t)\|$ in (1) is related to arc length s by $ds/dt = \|\mathbf{r}'(t)\|$. Since the curve C is smooth, we know from page 454 that $ds/dt > 0$. Hence by the Chain Rule,

$$\frac{d\mathbf{r}}{dt} = \frac{d\mathbf{r}}{ds}\frac{ds}{dt} \quad \text{and so} \quad \frac{d\mathbf{r}}{ds} = \frac{d\mathbf{r}/dt}{ds/dt} = \frac{\mathbf{r}'(t)}{\|\mathbf{r}'(t)\|} = \mathbf{T}. \tag{2}$$

Now suppose C is as shown in Figure 9.16. As s increases, \mathbf{T} moves along C, changing direction but not length (it is always of unit length). Along the portion of the curve between P_1 and P_2 the vector \mathbf{T} varies little in direction; along the curve between P_2 and P_3, where C obviously bends more sharply, the change in the direction of the tangent \mathbf{T} is more pronounced. We use the *rate* at which the unit vector \mathbf{T} changes direction with respect to arc length as an indicator of the curvature of a smooth curve C.

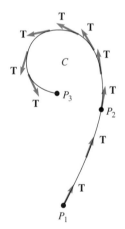

Figure 9.16

> **DEFINITION 9.4** Curvature
>
> Let $\mathbf{r}(t)$ be a vector function defining a smooth curve C. If s is the arc length parameter and $\mathbf{T} = d\mathbf{r}/ds$ is the unit tangent vector, then the **curvature** of C at a point is
>
> $$\kappa = \left\|\frac{d\mathbf{T}}{ds}\right\|. \tag{3}$$

The symbol κ in (3) is the Greek letter kappa. Now since curves are generally not parametrized by arc length, it is convenient to express (3) in terms of a general parameter t. Using the Chain Rule again, we can write

$$\frac{d\mathbf{T}}{dt} = \frac{d\mathbf{T}}{ds}\frac{ds}{dt} \quad \text{and consequently} \quad \frac{d\mathbf{T}}{ds} = \frac{d\mathbf{T}/dt}{ds/dt}.$$

In other words, curvature is given by

$$\kappa = \frac{\|\mathbf{T}'(t)\|}{\|\mathbf{r}'(t)\|}. \tag{4}$$

Example 1 **Curvature of a Circle**

Find the curvature of a circle of radius a.

SOLUTION A circle can be described by the vector function $\mathbf{r}(t) = a\cos t\mathbf{i} + a\sin t\mathbf{j}$. Now from $\mathbf{r}'(t) = -a\sin t\mathbf{i} + a\cos t\mathbf{j}$ and $\|\mathbf{r}'(t)\| = a$, we get

$$\mathbf{T}(t) = \frac{\mathbf{r}'(t)}{\|\mathbf{r}'(t)\|} = -\sin t\mathbf{i} + \cos t\mathbf{j} \quad \text{and} \quad \mathbf{T}'(t) = -\cos t\mathbf{i} - \sin t\mathbf{j}.$$

Hence, from (4) the curvature is

$$\kappa = \frac{\|\mathbf{T}'(t)\|}{\|\mathbf{r}'(t)\|} = \frac{\sqrt{\cos^2 t + \sin^2 t}}{a} = \frac{1}{a}. \tag{5} \ \square$$

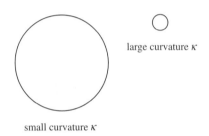

large curvature κ

small curvature κ

Figure 9.17

The result in (5) shows that the curvature at a point on a circle is the reciprocal of the radius of the circle and indicates a fact that is in keeping with our intuition: A circle with a small radius curves more than one with a large radius. See Figure 9.17.

■ **Tangential and Normal Components of Acceleration** Suppose a particle moves in 2- or 3-space on a smooth curve C described by the vector function $\mathbf{r}(t)$. Then the velocity of the particle on C is $\mathbf{v}(t) = \mathbf{r}'(t)$, whereas its speed is $ds/dt = v = \|\mathbf{v}(t)\|$. Thus, (1) implies $\mathbf{v}(t) = v\mathbf{T}$. Differentiating this last expression with respect to t gives acceleration:

$$\mathbf{a}(t) = v\frac{d\mathbf{T}}{dt} + \frac{dv}{dt}\mathbf{T}. \tag{6}$$

Furthermore, with the help of Theorem 9.4(*iii*), it follows from the differentiation of $\mathbf{T} \cdot \mathbf{T} = 1$ that $\mathbf{T} \cdot d\mathbf{T}/dt = 0$. Hence, at a point P on C the vectors \mathbf{T} and $d\mathbf{T}/dt$ are orthogonal. If $\|d\mathbf{T}/dt\| \neq 0$, the vector

$$\mathbf{N} = \frac{d\mathbf{T}/dt}{\|d\mathbf{T}/dt\|} \tag{7}$$

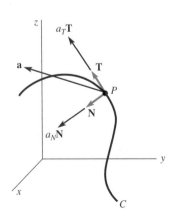

Figure 9.18

is a unit normal to the curve C at P with direction given by $d\mathbf{T}/dt$. The vector \mathbf{N} is also called the **principal normal.** But since curvature is $\kappa = \|d\mathbf{T}/dt\|/v$, it follows from (7) that $d\mathbf{T}/dt = \kappa v\mathbf{N}$. Thus, (6) becomes

$$\mathbf{a}(t) = \kappa v^2\mathbf{N} + \frac{dv}{dt}\mathbf{T}. \tag{8}$$

By writing (8) as $\qquad \mathbf{a}(t) = a_N\mathbf{N} + a_T\mathbf{T}, \tag{9}$

we see that the acceleration vector \mathbf{a} of the moving particle is the sum of two orthogonal vectors $a_N\mathbf{N}$ and $a_T\mathbf{T}$. See Figure 9.18. The scalar functions $a_T = dv/dt$ and $a_N = \kappa v^2$ are called the **tangential** and **normal components of the accelera-**

tion, respectively. Note that the tangential component of the acceleration results from a change in the *magnitude* of the velocity **v**, whereas the normal component of the acceleration results from a change in the *direction* of **v**.

■ **The Binormal** A third unit vector defined by

$$\mathbf{B} = \mathbf{T} \times \mathbf{N}$$

is called the **binormal.** The three unit vectors **T**, **N**, and **B** form a right-handed set of mutually orthogonal vectors called the **moving trihedral.** The plane of **T** and **N** is called the **osculating plane,*** the plane of **N** and **B** is said to be the **normal plane,** and the plane of **T** and **B** is the **rectifying plane.** See Figure 9.19.

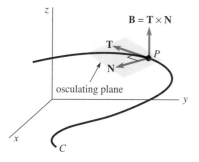

Figure 9.19

Example 2 Tangent, Normal, and Binormal Vectors

The position of a moving particle is given by $\mathbf{r}(t) = 2\cos t\mathbf{i} + 2\sin t\mathbf{j} + 3t\mathbf{k}$. Find the vectors **T**, **N**, and **B**. Find the curvature.

SOLUTION Since $\mathbf{r}'(t) = -2\sin t\mathbf{i} + 2\cos t\mathbf{j} + 3\mathbf{k}$, $\|\mathbf{r}'(t)\| = \sqrt{13}$, and so from (1) we see that a unit tangent is

$$\mathbf{T} = -\frac{2}{\sqrt{13}}\sin t\mathbf{i} + \frac{2}{\sqrt{13}}\cos t\mathbf{j} + \frac{3}{\sqrt{13}}\mathbf{k}.$$

Next, we have

$$\frac{d\mathbf{T}}{dt} = -\frac{2}{\sqrt{13}}\cos t\mathbf{i} - \frac{2}{\sqrt{13}}\sin t\mathbf{j} \quad \text{and} \quad \left\|\frac{d\mathbf{T}}{dt}\right\| = \frac{2}{\sqrt{13}}.$$

Hence, (3) gives the principal normal

$$\mathbf{N} = -\cos t\mathbf{i} - \sin t\mathbf{j}.$$

Now, the binormal is

$$\mathbf{B} = \mathbf{T} \times \mathbf{N} = \begin{vmatrix} \mathbf{i} & \mathbf{j} & \mathbf{k} \\ -\dfrac{2}{\sqrt{13}}\sin t & \dfrac{2}{\sqrt{13}}\cos t & \dfrac{3}{\sqrt{13}} \\ -\cos t & -\sin t & 0 \end{vmatrix}$$

$$= \frac{3}{\sqrt{13}}\sin t\mathbf{i} - \frac{3}{\sqrt{13}}\cos t\mathbf{j} + \frac{2}{\sqrt{13}}\mathbf{k}.$$

Finally, using $\|d\mathbf{T}/dt\| = 2/\sqrt{13}$ and $\|\mathbf{r}'(t)\| = \sqrt{13}$, we obtain from (4) that the curvature at any point is the constant

$$\kappa = \frac{2/\sqrt{13}}{\sqrt{13}} = \frac{2}{13}. \qquad \qquad ❏$$

The fact that the curvature in Example 2 is constant is not surprising, since the curve defined by $\mathbf{r}(t)$ is a circular helix.

■ **Formulas for a_T, a_N, and Curvature** By dotting, and in turn crossing, the vector $\mathbf{v} = v\mathbf{T}$ with (9), it is possible to obtain explicit formulas involving \mathbf{r}, \mathbf{r}', and \mathbf{r}'' for the tangential and normal components of the acceleration and the curvature. Observe that

$$\mathbf{v} \cdot \mathbf{a} = a_N(v\underbrace{\mathbf{T} \cdot \mathbf{N}}_{0}) + a_T(v\underbrace{\mathbf{T} \cdot \mathbf{T}}_{1}) = a_T v$$

* Literally, this means the "kissing" plane.

yields the tangential component of acceleration

$$a_T = \frac{dv}{dt} = \frac{\mathbf{v} \cdot \mathbf{a}}{\|\mathbf{v}\|} = \frac{\mathbf{r}'(t) \cdot \mathbf{r}''(t)}{\|\mathbf{r}'(t)\|}. \tag{10}$$

On the other hand,

$$\mathbf{v} \times \mathbf{a} = a_N \underbrace{(v\mathbf{T} \times \mathbf{N})}_{\mathbf{B}} + a_T \underbrace{(v\mathbf{T} \times \mathbf{T})}_{\mathbf{0}} = a_N v \mathbf{B}.$$

Since $\|\mathbf{B}\| = 1$, it follows that the normal component of acceleration is

$$a_N = \kappa v^2 = \frac{\|\mathbf{v} \times \mathbf{a}\|}{\|\mathbf{v}\|} = \frac{\|\mathbf{r}'(t) \times \mathbf{r}''(t)\|}{\|\mathbf{r}'(t)\|}. \tag{11}$$

Solving (11) for the curvature gives

$$\kappa = \frac{\|\mathbf{v} \times \mathbf{a}\|}{\|\mathbf{v}\|^3} = \frac{\|\mathbf{r}'(t) \times \mathbf{r}''(t)\|}{\|\mathbf{r}'(t)\|^3}. \tag{12}$$

Example 3 Curvature of Twisted Cubic

The curve traced by $\mathbf{r}(t) = t\mathbf{i} + \frac{1}{2} t^2\mathbf{j} + \frac{1}{3} t^3\mathbf{k}$ is said to be a "twisted cubic." If $\mathbf{r}(t)$ is the position vector of a moving particle, find the tangential and normal components of the acceleration at any t. Find the curvature.

SOLUTION $\mathbf{v}(t) = \mathbf{r}'(t) = \mathbf{i} + t\mathbf{j} + t^2\mathbf{k}$, $\mathbf{a}(t) = \mathbf{r}''(t) = \mathbf{j} + 2t\mathbf{k}$.

Since $\mathbf{v} \cdot \mathbf{a} = t + 2t^3$ and $\|\mathbf{v}\| = \sqrt{1 + t^2 + t^4}$, it follows from (10) that

$$a_T = \frac{dv}{dt} = \frac{t + 2t^3}{\sqrt{1 + t^2 + t^4}}.$$

Now,

$$\mathbf{v} \times \mathbf{a} = \begin{vmatrix} \mathbf{i} & \mathbf{j} & \mathbf{k} \\ 1 & t & t^2 \\ 0 & 1 & 2t \end{vmatrix} = t^2\mathbf{i} - 2t\mathbf{j} + \mathbf{k}$$

and $\|\mathbf{v} \times \mathbf{a}\| = \sqrt{t^4 + 4t^2 + 1}$. Thus, (11) gives

$$a_N = \kappa v^2 = \frac{\sqrt{t^4 + 4t^2 + 1}}{\sqrt{1 + t^2 + t^4}} = \sqrt{\frac{t^4 + 4t^2 + 1}{t^4 + t^2 + 1}}.$$

From (12) we find that the curvature of the twisted cubic is given by

$$\kappa = \frac{(t^4 + 4t^2 + 1)^{1/2}}{(t^4 + t^2 + 1)^{3/2}}. \qquad \square$$

■ **Radius of Curvature** The reciprocal of the curvature, $\rho = 1/\kappa$, is called the **radius of curvature.** The radius of curvature at a point P on a curve C is the radius of a circle that "fits" the curve there better than any other circle. The circle at P is called the **circle of curvature** and its center is the **center of curvature.** The circle of curvature has the same tangent line at P as the curve C, and its center lies on the concave side of C. For example, a car moving on a curved track, as shown in Figure 9.20, can, at any instant, be thought to be moving on a circle of radius ρ. Hence, the normal component of its acceleration $a_N = \kappa v^2$ must be the same as the magnitude of its centripetal acceleration $a = v^2/\rho$. Therefore, $\kappa = 1/\rho$ and $\rho = 1/\kappa$. Knowing the radius of curvature, we can determine the speed v at

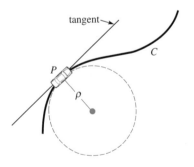

Figure 9.20

which a car can negotiate a banked curve without skidding. (This is essentially the idea in Problem 22 in Exercises 9.2.)

Remarks

By writing (6) as

$$\mathbf{a}(t) = \frac{ds}{dt}\frac{d\mathbf{T}}{dt} + \frac{d^2s}{dt^2}\,\mathbf{T},$$

we note that the so-called scalar acceleration d^2s/dt^2, referred to in the last remark, is now seen to be the tangential component of the acceleration a_T.

EXERCISES 9.3

Answers to odd-numbered problems begin on page A-48.

In Problems 1 and 2, for the given position function, find the unit tangent.

1. $\mathbf{r}(t) = (t \cos t - \sin t)\mathbf{i} + (t \sin t + \cos t)\mathbf{j} + t^2\mathbf{k}$; $t > 0$

2. $\mathbf{r}(t) = e^t \cos t\mathbf{i} + e^t \sin t\mathbf{j} + \sqrt{2}e^t\mathbf{k}$

3. Use the procedure outlined in Example 2 to find $\mathbf{T}, \mathbf{N}, \mathbf{B}$, and κ for motion on a general circular helix that is described by $\mathbf{r}(t) = a \cos t\mathbf{i} + a \sin t\mathbf{j} + ct\mathbf{k}$.

4. Use the procedure outlined in Example 2 to show on the twisted cubic of Example 3 that at $t = 1$:

$$\mathbf{T} = \frac{1}{\sqrt{3}}(\mathbf{i}+\mathbf{j}+\mathbf{k}), \quad \mathbf{N} = -\frac{1}{\sqrt{2}}(\mathbf{i}-\mathbf{k}),$$

$$\mathbf{B} = -\frac{1}{\sqrt{6}}(-\mathbf{i}+2\mathbf{j}-\mathbf{k}), \quad \kappa = \frac{\sqrt{2}}{3}.$$

In Problems 5 and 6, find an equation of the osculating plane to the given space curve at the point that corresponds to the indicated value of t.

5. The circular helix of Example 2; $t = \pi/4$

6. The twisted cubic of Example 3; $t = 1$

In Problems 7–16, $\mathbf{r}(t)$ is the position vector of a moving particle. Find the tangential and normal components of the acceleration at any t.

7. $\mathbf{r}(t) = \mathbf{i} + t\mathbf{j} + t^2\mathbf{k}$

8. $\mathbf{r}(t) = 3 \cos t\mathbf{i} + 2 \sin t\mathbf{j} + t\mathbf{k}$

9. $\mathbf{r}(t) = t^2\mathbf{i} + (t^2 - 1)\mathbf{j} + 2t^2\mathbf{k}$

10. $\mathbf{r}(t) = t^2\mathbf{i} - t^3\mathbf{j} + t^4\mathbf{k}$

11. $\mathbf{r}(t) = 2t\mathbf{i} + t^2\mathbf{j}$

12. $\mathbf{r}(t) = \tan^{-1} t\mathbf{i} + \frac{1}{2} \ln(1 + t^2)\mathbf{j}$

13. $\mathbf{r}(t) = 5 \cos t\mathbf{i} + 5 \sin t\mathbf{j}$

14. $\mathbf{r}(t) = \cosh t\mathbf{i} + \sinh t\mathbf{j}$

15. $\mathbf{r}(t) = e^{-t}(\mathbf{i} + \mathbf{j} + \mathbf{k})$

16. $\mathbf{r}(t) = t\mathbf{i} + (2t - 1)\mathbf{j} + (4t + 2)\mathbf{k}$

17. Find the curvature of an elliptical helix that is described by $\mathbf{r}(t) = a \cos t\mathbf{i} + b \sin t\mathbf{j} + ct\mathbf{k}$, $a > 0$, $b > 0$, $c > 0$.

18. (a) Find the curvature of an elliptical orbit that is described by $\mathbf{r}(t) = a \cos t\mathbf{i} + b \sin t\mathbf{j} + c\mathbf{k}$, $a > 0$, $b > 0$, $c > 0$.

(b) Show that when $a = b$, the curvature of a circular orbit is the constant $\kappa = 1/a$.

19. Show that the curvature of a straight line is the constant $\kappa = 0$. [*Hint:* Use (2) in Section 7.5.]

20. Find the curvature at $t = \pi$ of the cycloid that is described by

$$\mathbf{r}(t) = a(t - \sin t)\mathbf{i} + a(1 - \cos t)\mathbf{j}, \qquad a > 0$$

21. Let C be a plane curve traced by $\mathbf{r}(t) = f(t)\mathbf{i} + g(t)\mathbf{j}$, where f and g have second derivatives. Show that the curvature at a point is given by

$$\kappa = \frac{|f'(t)g''(t) - g'(t)f''(t)|}{([f'(t)]^2 + [g'(t)]^2)^{3/2}}$$

22. Show that if $y = F(x)$, the formula for κ in Problem 21 reduces to

$$\kappa = \frac{|F''(x)|}{[1 + (F'(x))^2]^{3/2}}$$

In Problems 23 and 24, use the result of Problem 22 to find the curvature and radius of curvature of the curve at the indicated points. Decide at which point the curve is "sharper."

23. $y = x^2$; $(0, 0)$, $(1, 1)$

24. $y = x^3$; $(-1, -1)$, $(\frac{1}{2}, \frac{1}{8})$

25. Discuss the curvature near a point of inflection of $y = F(x)$.

26. Show that $\|\mathbf{a}(t)\|^2 = a_N^2 + a_T^2$.

9.4 Functions of Several Variables

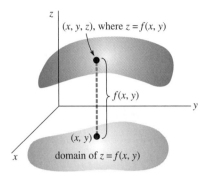

Figure 9.21

■ **Functions of Two Variables** Recall from calculus that a **function of two variables** is a rule of correspondence that assigns to each ordered pair of real numbers (x, y) of a subset of the xy-plane one and only one number z in the set R of real numbers. The set of ordered pairs (x, y) is called the **domain** of the function and the set of corresponding values of z is called the **range.** A function of two variables is usually written $z = f(x, y)$. The variables x and y are called the **independent variables** of the function, and z is called the **dependent variable.** The **graph** of a function $z = f(x, y)$ is a surface in 3-space. See Figure 9.21.

■ **Level Curves** For a function $z = f(x, y)$, the curves defined by $f(x, y) = c$, for suitable c, are called the **level curves** of f. The word *level* arises from the fact that we can interpret the equation $f(x, y) = c$ as the projection onto the xy-plane of the curve of intersection, or trace, of $z = f(x, y)$ and the (horizontal or level) plane $z = c$. See Figure 9.22

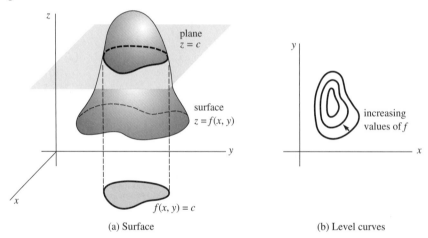

(a) Surface (b) Level curves

Figure 9.22

Example 1 Level Curves

The level curves of the function $f(x, y) = y^2 - x^2$ are defined by $y^2 - x^2 = c$. As shown in Figure 9.23, when $c > 0$ or $c < 0$, a member of this family of curves is a hyperbola. For $c = 0$, we obtain the lines $y = x$ and $y = -x$.

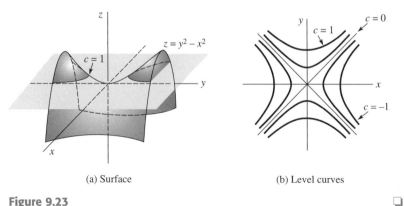

(a) Surface (b) Level curves

Figure 9.23

■ **Functions of Three or More Variables**　Functions of three or more variables are defined analogously to functions of two variables. For example, a **function of three variables** is a rule of correspondence that assigns to each ordered triple of real numbers (x, y, z) of a subset of 3-space one and only one number w in the set R of real numbers. We write $w = F(x, y, z)$.

■ **Level Surfaces**　Although we cannot draw a graph of a function of three variables $w = F(x, y, z)$, we can draw the surfaces defined by $F(x, y, z) = c$ for suitable values of c. These surfaces are called **level surfaces.** This is an unfortunate, though standard, choice of words, since *level* surfaces are usually not level.

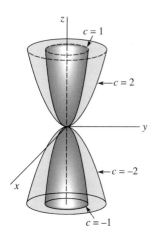

Figure 9.24

Example 2　Level Surfaces

Describe the level surfaces of the function $F(x, y, z) = (x^2 + y^2)/z$.

SOLUTION　For $c \neq 0$ the level surfaces are given by

$$\frac{x^2 + y^2}{z} = c \quad \text{or} \quad x^2 + y^2 = cz.$$

A few members of this family of paraboloids are shown in Figure 9.24.　❑

■ **Partial Derivatives**　The derivative of a function of one variable $y = f(x)$ is given by

$$\frac{dy}{dx} = \lim_{\Delta x \to 0} \frac{f(x + \Delta x) - f(x)}{\Delta x}.$$

In exactly the same manner, we can define a derivative of a function of two variables with respect to *each* variable. If $z = f(x, y)$, then the **partial derivative with respect to x** is

$$\frac{\partial z}{\partial x} = \lim_{\Delta x \to 0} \frac{f(x + \Delta x, y) - f(x, y)}{\Delta x} \tag{1}$$

and the **partial derivative with respect to y** is

$$\frac{\partial z}{\partial y} = \lim_{\Delta y \to 0} \frac{f(x, y + \Delta y) - f(x, y)}{\Delta y}, \tag{2}$$

provided each limit exists.

　In (1) the variable y does not change in the limiting process; that is, y is held fixed. Similarly, in (2) the variable x is held fixed. The two partial derivatives (1) and (2) then represent the *rates of change* of f with respect to x and y, respectively. On a practical level:

To compute $\partial z/\partial x$, use the laws of ordinary differentiation while treating y as a constant.

To compute $\partial z/\partial y$, use the laws of ordinary differentiation while treating x as a constant.

Example 3　Partial Derivatives

If $z = 4x^3 y^2 - 4x^2 + y^6 + 1$, find $\dfrac{\partial z}{\partial x}$ and $\dfrac{\partial z}{\partial y}$.

SOLUTION We hold y fixed and treat constants in the usual manner. Thus,

$$\frac{\partial z}{\partial x} = 12x^2y^2 - 8x.$$

By treating x as a constant, we obtain

$$\frac{\partial z}{\partial y} = 8x^3y + 6y^5.$$ ❏

■ **Alternative Symbols** The partial derivatives $\partial z/\partial x$ and $\partial z/\partial y$ are often represented by alternative symbols. If $z = f(x, y)$, then

$$\frac{\partial z}{\partial x} = \frac{\partial f}{\partial x} = z_x = f_x \quad \text{and} \quad \frac{\partial z}{\partial y} = \frac{\partial f}{\partial y} = z_y = f_y.$$

■ **Higher-Order and Mixed Derivatives** For a function of two variables $z = f(x, y)$, the partial derivatives $\partial z/\partial x$ and $\partial z/\partial y$ are themselves functions of x and y. Consequently, we can compute **second** and higher **partial derivatives.** Indeed, we can find the partial derivative of $\partial z/\partial x$ with respect to y, and the partial derivative of $\partial z/\partial y$ with respect to x. The latter types of partial derivatives are called **mixed partial derivatives.** In summary, for $z = f(x, y)$:

Second-order partial derivatives:

$$\frac{\partial^2 z}{\partial x^2} = \frac{\partial}{\partial x}\left(\frac{\partial z}{\partial x}\right) \quad \text{and} \quad \frac{\partial^2 z}{\partial y^2} = \frac{\partial}{\partial y}\left(\frac{\partial z}{\partial y}\right)$$

Third-order partial derivatives:

$$\frac{\partial^3 z}{\partial x^3} = \frac{\partial}{\partial x}\left(\frac{\partial^2 z}{\partial x^2}\right) \quad \text{and} \quad \frac{\partial^3 z}{\partial y^3} = \frac{\partial}{\partial y}\left(\frac{\partial^2 z}{\partial y^2}\right)$$

Mixed second-order partial derivatives:

$$\frac{\partial^2 z}{\partial x\,\partial y} = \frac{\partial}{\partial x}\left(\frac{\partial z}{\partial y}\right) \quad \text{and} \quad \frac{\partial^2 z}{\partial y\,\partial x} = \frac{\partial}{\partial y}\left(\frac{\partial z}{\partial x}\right)$$

■ **Alternative Symbols** The second- and third-order partial derivatives are denoted by f_{xx}, f_{yy}, f_{xxx}, and so on. The subscript notation for mixed second partial derivatives is f_{xy} or f_{yx}. Note that

$$f_{xy} = (f_x)_y = \frac{\partial}{\partial y}\left(\frac{\partial z}{\partial x}\right) = \frac{\partial^2 z}{\partial y\,\partial x} \quad \text{and} \quad f_{yx} = \frac{\partial^2 z}{\partial x\,\partial y}.$$

Although we shall not prove it, if a function f has continuous second partial derivatives, then the order in which a mixed second partial derivative is done is irrelevant; that is,

$$f_{xy} = f_{yx}. \tag{3}$$

■ **Functions of Three or More Variables** The rates of change of a function of three variables $w = F(x, y, z)$ in the x, y, and z directions are $\partial w/\partial x$, $\partial w/\partial y$, and $\partial w/\partial z$, respectively. To compute, say, $\partial w/\partial x$, we differentiate with respect to x in the usual manner while holding *both* y and z constant. In this manner we extend the process of partial differentiation to functions of any number of variables.

Example 4 **Partial Derivatives**

If $F(x, y, t) = e^{-3\pi t} \cos 4x \sin 6y$, then the partial derivatives with respect to x, y, and t are, in turn,

$$F_x(x, y, t) = -4e^{-3\pi t} \sin 4x \sin 6y$$

$$F_y(x, y, t) = 6e^{-3\pi t} \cos 4x \cos 6y$$

$$F_t(x, y, t) = -3\pi e^{-3\pi t} \cos 4x \sin 6y.$$ ❑

■ **Chain Rule** The Chain Rule for functions of one variable states that if $y = f(u)$ is a differentiable function of u, and $u = g(x)$ is a differentiable function of x, then the derivative of the composite function is

$$\frac{dy}{dx} = \frac{dy}{du}\frac{du}{dx}. \tag{4}$$

For a composite function of two variables $z = f(u, v)$, where $u = g(x, y)$ and $v = h(x, y)$, we would naturally expect *two* formulas analogous to (4), since we can compute both $\partial z/\partial x$ and $\partial z/\partial y$. The **Chain Rule** for functions of two variables is summarized as follows:

THEOREM 9.5 Chain Rule

If $z = f(u, v)$ is differentiable and $u = g(x, y)$ and $v = h(x, y)$ have continuous first partial derivatives, then

$$\frac{\partial z}{\partial x} = \frac{\partial z}{\partial u}\frac{\partial u}{\partial x} + \frac{\partial z}{\partial v}\frac{\partial v}{\partial x}, \qquad \frac{\partial z}{\partial y} = \frac{\partial z}{\partial u}\frac{\partial u}{\partial y} + \frac{\partial z}{\partial v}\frac{\partial v}{\partial y}. \tag{5}$$

Example 5 **Chain Rule**

If $z = u^2 - v^3$ and $u = e^{2x-3y}$, $v = \sin(x^2 - y^2)$, find $\partial z/\partial x$ and $\partial z/\partial y$.

SOLUTION Since $\partial z/\partial u = 2u$ and $\partial z/\partial v = -3v^2$, it follows from (5) that

$$\frac{\partial z}{\partial x} = 2u(2e^{2x-3y}) - 3v^2[2x\cos(x^2 - y^2)] = 4ue^{2x-3y} - 6xv^2\cos(x^2 - y^2) \tag{6}$$

$$\frac{\partial z}{\partial y} = 2u(-3e^{2x-3y}) - 3v^2[(-2y)\cos(x^2 - y^2)] = -6ue^{2x-3y} + 6yv^2\cos(x^2 - y^2). \tag{7} ❑$$

Of course, in Example 5 we could substitute the expressions for u and v in the original function and then find the partial derivatives directly. In the same manner, the answers (6) and (7) can be expressed in terms of x and y.

■ **Special Case** If $z = f(u, v)$ is differentiable and $u = g(t)$ and $v = h(t)$ are differentiable functions of a single variable t, then Theorem 9.5 implies that the ordinary derivative dz/dt is

$$\frac{dz}{dt} = \frac{\partial z}{\partial u}\frac{du}{dt} + \frac{\partial z}{\partial v}\frac{dv}{dt}. \tag{8}$$

■ **Generalizations** The results given in (4) and (8) immediately generalize to any number of variables. If $z = f(u_1, u_2, \ldots, u_n)$ and each of the variables u_1, u_2,

u_3, \ldots, u_n are functions of x_1, x_2, \ldots, x_k, then under the same assumptions as in Theorem 9.5 we have

$$\frac{\partial z}{\partial x_i} = \frac{\partial z}{\partial u_1}\frac{\partial u_1}{\partial x_i} + \frac{\partial z}{\partial u_2}\frac{\partial u_2}{\partial x_i} + \cdots + \frac{\partial z}{\partial u_n}\frac{\partial u_n}{\partial x_i}, \qquad (9)$$

where $i = 1, 2, \ldots, k$. Similarly, if the u_i, $i = 1, \ldots, n$, are differentiable functions of a single variable t, then

$$\frac{dz}{dt} = \frac{\partial z}{\partial u_1}\frac{du_1}{dt} + \frac{\partial z}{\partial u_2}\frac{du_2}{dt} + \cdots + \frac{\partial z}{\partial u_n}\frac{du_n}{dt}. \qquad (10)$$

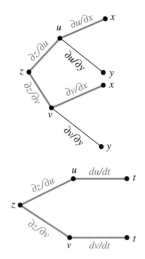

■ **Tree Diagrams** The results in (4) can be memorized in terms of a **tree diagram.** The dots in the first diagram that follows indicate the fact that z depends on u and v; u and v depend, in turn, on x and y. To compute $\partial z/\partial x$, for example, we read from left to right and follow the *two* gray polygonal paths leading from z to x, multiply the partial derivatives on each path, and then add the products. The result given in (8) is represented by the second tree diagram.

We shall use tree diagrams in the next two examples to illustrate special cases of (9) and (10).

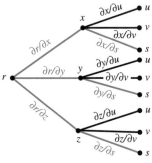

Example 6 Using Tree Diagrams

If $r = x^2 + y^5 z^3$ and $x = uve^{2s}$, $y = u^2 - v^2 s$, $z = \sin(uvs^2)$, find $\partial r/\partial s$.

SOLUTION From the gray paths of the accompanying tree diagram we obtain

$$\frac{\partial r}{\partial s} = \frac{\partial r}{\partial x}\frac{\partial x}{\partial s} + \frac{\partial r}{\partial y}\frac{\partial y}{\partial s} + \frac{\partial r}{\partial z}\frac{\partial z}{\partial s}$$

$$= 2x(2uve^{2s}) + 5y^4 z^3(-v^2) + 3y^5 z^2(2uvs\cos(uvs^2)). \qquad \square$$

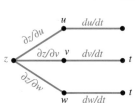

Example 7 Using Tree Diagrams

If $z = u^2 v^3 w^4$ and $u = t^2$, $v = 5t - 8$, $w = t^3 + t$, find dz/dt.

SOLUTION In this case the tree diagram indicates that

$$\frac{dz}{dt} = \frac{\partial z}{\partial u}\frac{du}{dt} + \frac{\partial z}{\partial v}\frac{dv}{dt} + \frac{\partial z}{\partial w}\frac{dw}{dt}$$

$$= 2uv^3 w^4(2t) + 3u^2 v^2 w^4(5) + 4u^2 v^3 w^3(3t^2 + 1).$$

ALTERNATIVE SOLUTION Differentiate $z = t^4(5t - 8)^3(t^3 + t)^4$ by the product rule. \square

Remarks

If $w = F(x, y, z)$ has continuous partial derivatives of any order, then analogous to (3), the mixed partial derivatives are equal:

$$F_{xyz} = F_{yzx} = F_{zyx}, \qquad F_{xxy} = F_{yxx} = F_{xyx}$$

and so on.

Answers to odd-numbered problems begin on page A-48.

In Problems 1–6, sketch some of the level curves associated with the given function.

1. $f(x, y) = x + 2y$

2. $f(x, y) = y^2 - x$

3. $f(x, y) = \sqrt{x^2 - y^2 - 1}$

4. $f(x, y) = \sqrt{36 - 4x^2 - 9y^2}$

5. $f(x, y) = e^{y - x^2}$

6. $f(x, y) = \tan^{-1}(y - x)$

In Problems 7–10, describe the level surfaces but do not graph.

7. $F(x, y, z) = \dfrac{x^2}{9} + \dfrac{z^2}{4}$

8. $F(x, y, z) = x^2 + y^2 + z^2$

9. $F(x, y, z) = x^2 + 3y^2 + 6z^2$

10. $F(x, y, z) = 4y - 2z + 1$

11. Graph some of the level surfaces associated with $F(x, y, z) = x^2 + y^2 - z^2$ for $c = 0$, $c > 0$, and $c < 0$.

12. Given that

$$F(x, y, z) = \frac{x^2}{16} + \frac{y^2}{4} + \frac{z^2}{9},$$

find the x-, y-, and z-intercepts of the level surface that passes through $(-4, 2, -3)$.

In Problems 13–32, find the first partial derivatives of the given function.

13. $z = x^2 - xy^2 + 4y^5$

14. $z = -x^3 + 6x^2y^3 + 5y^2$

15. $z = 5x^4y^3 - x^2y^6 + 6x^5 - 4y$

16. $z = \tan(x^3y^2)$

17. $z = \dfrac{4\sqrt{x}}{3y^2 + 1}$

18. $z = 4x^3 - 5x^2 + 8x$

19. $z = (x^3 - y^2)^{-1}$

20. $z = (-x^4 + 7y^2 + 3y)^6$

21. $z = \cos^2 5x + \sin^2 5y$

22. $z = e^{x^2 \tan^{-1} y^2}$

23. $f(x, y) = xe^{x^3 y}$

24. $f(\theta, \phi) = \phi^2 \sin \dfrac{\theta}{\phi}$

25. $f(x, y) = \dfrac{3x - y}{x + 2y}$

26. $f(x, y) = \dfrac{xy}{(x^2 - y^2)^2}$

27. $g(u, v) = \ln(4u^2 + 5v^3)$

28. $h(r, s) = \dfrac{\sqrt{r}}{s} - \dfrac{\sqrt{s}}{r}$

29. $w = 2\sqrt{xy} - ye^{y/z}$

30. $w = xy \ln xz$

31. $F(u, v, x, t) = u^2w^2 - uv^3 + vw \cos(ut^2) + (2x^2t)^4$

32. $G(p, q, r, s) = (p^2q^3)^{r^4 s^5}$

In Problems 33 and 34, verify that the given function satisfies **Laplace's equation:**

$$\frac{\partial^2 z}{\partial x^2} + \frac{\partial^2 z}{\partial y^2} = 0$$

33. $z = \ln(x^2 + y^2)$

34. $z = e^{x^2 - y^2} \cos 2xy$

In Problems 35 and 36 verify that the given function satisfies the **wave equation:**

$$a^2 \frac{\partial^2 u}{\partial x^2} = \frac{\partial^2 u}{\partial t^2}$$

35. $u = \cos at \sin x$

36. $u = \cos(x + at) + \sin(x - at)$

37. The molecular concentration $C(x, t)$ of a liquid is given by $C(x, t) = t^{-1/2} e^{-x^2/kt}$. Verify that this function satisfies the **diffusion equation:**

$$\frac{k}{4} \frac{\partial^2 C}{\partial x^2} = \frac{\partial C}{\partial t}$$

38. The pressure P exerted by an enclosed ideal gas is given by $P = k(T/V)$, where k is a constant, T is temperature, and V is volume. Find:

(a) the rate of change of P with respect to V,

(b) the rate of change of V with respect to T, and

(c) the rate of change of T with respect to P.

In Problems 39–48, use the Chain Rule to find the indicated partial derivatives.

39. $z = e^{uv^2}$; $u = x^3$, $v = x - y^2$; $\dfrac{\partial z}{\partial x}$, $\dfrac{\partial z}{\partial y}$

40. $z = u^2 \cos 4v$; $u = x^2y^3$, $v = x^3 + y^3$; $\dfrac{\partial z}{\partial x}$, $\dfrac{\partial z}{\partial y}$

41. $z = 4x - 5y^2$; $x = u^4 - 8v^3$, $y = (2u - v)^2$; $\dfrac{\partial z}{\partial u}$, $\dfrac{\partial z}{\partial v}$

42. $z = \dfrac{x - y}{x + y}$; $x = \dfrac{u}{v}$, $y = \dfrac{v^2}{u}$; $\dfrac{\partial z}{\partial u}$, $\dfrac{\partial z}{\partial v}$

43. $w = (u^2 + v^2)^{3/2}$; $u = e^{-t} \sin \theta$, $v = e^{-t} \cos \theta$; $\dfrac{\partial w}{\partial t}$, $\dfrac{\partial w}{\partial \theta}$

44. $w = \tan^{-1}\sqrt{uv}$; $u = r^2 - s^2$, $v = r^2s^2$; $\dfrac{\partial w}{\partial r}, \dfrac{\partial w}{\partial s}$

45. $R = rs^2t^4$; $r = ue^{v^2}$, $s = ve^{-u^2}$, $t = e^{u^2v^2}$; $\dfrac{\partial R}{\partial u}, \dfrac{\partial R}{\partial v}$

46. $Q = \ln(pqr)$; $p = t^2\sin^{-1}x$, $q = \dfrac{x}{t^2}$, $r = \tan^{-1}\dfrac{x}{t}$; $\dfrac{\partial Q}{\partial x}, \dfrac{\partial Q}{\partial t}$

47. $w = \sqrt{x^2 + y^2}$; $x = \ln(rs + tu)$, $y = \dfrac{t}{u}\cosh rs$; $\dfrac{\partial w}{\partial t}, \dfrac{\partial w}{\partial r}, \dfrac{\partial w}{\partial u}$

48. $s = p^2 + q^2 - r^2 + 4t$; $p = \phi e^{3\theta}$, $q = \cos(\phi + \theta)$, $r = \phi\theta^2$, $t = 2\phi + 8\theta$; $\dfrac{\partial s}{\partial \phi}, \dfrac{\partial s}{\partial \theta}$

In Problems 49–52, use (8) to find the indicated derivative.

49. $z = \ln(u^2 + v^2)$; $u = t^2$, $v = t^{-2}$; $\dfrac{dz}{dt}$

50. $z = u^3v - uv^4$; $u = e^{-5t}$, $v = \sec 5t$; $\dfrac{dz}{dt}$

51. $w = \cos(3u + 4v)$; $u = 2t + \dfrac{\pi}{2}$, $v = -t - \dfrac{\pi}{4}$; $\dfrac{dw}{dt}\Big|_{t = \pi}$

52. $w = e^{xy}$; $x = \dfrac{4}{2t + 1}$, $y = 3t + 5$; $\dfrac{dw}{dt}\Big|_{t = 0}$

53. If $u = f(x, y)$ and $x = r\cos\theta$, $y = r\sin\theta$, show that Laplace's equation $\partial^2 u/\partial x^2 + \partial^2 u/\partial y^2 = 0$ becomes

$$\frac{\partial^2 u}{\partial r^2} + \frac{1}{r}\frac{\partial u}{\partial r} + \frac{1}{r^2}\frac{\partial^2 u}{\partial \theta^2} = 0$$

54. Van der Waals' equation of state for the real gas CO_2 is

$$P = \frac{0.08T}{V - 0.0427} - \frac{3.6}{V^2}$$

If dT/dt and dV/dt are rates at which the temperature and volume change, respectively, use the Chain Rule to find dP/dt.

55. The equation of state for a thermodynamic system is $F(P, V, T) = 0$, where P, V, and T are pressure, volume, and temperature, respectively. If the equation defines V as a function of P and T, and also defines T as a function of V and P, show that

$$\frac{\partial V}{\partial T} = -\frac{\dfrac{\partial F}{\partial T}}{\dfrac{\partial F}{\partial V}} = \frac{1}{\dfrac{\partial T}{\partial V}}.$$

56. The voltage across a conductor is increasing at a rate of 2 volts/min and the resistance is decreasing at a rate of 1 ohm/min. Use $I = E/R$ and the Chain Rule to find the rate at which the current passing through the conductor is changing when $R = 50$ ohms and $E = 60$ volts.

57. The length of the side labeled x of the triangle in Figure 9.25 increases at a rate of 0.3 cm/s, the side labeled y increases at a rate of 0.5 cm/s, and the included angle θ increases at a rate of 0.1 rad/s. Use the Chain Rule to find the rate at which the area of the triangle is changing at the instant $x = 10$ cm, $y = 8$ cm, and $\theta = \pi/6$.

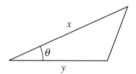

Figure 9.25

58. A particle moves in 3-space so that its coordinates at any time are $x = 4\cos t$, $y = 4\sin t$, $z = 5t$, $t \geq 0$. Use the Chain Rule to find the rate at which its distance

$$w = \sqrt{x^2 + y^2 + z^2}$$

from the origin is changing at $t = 5\pi/2$ seconds.

9.5 The Directional Derivative

■ **The Gradient of a Function** In this and the next sections it is convenient to introduce a new vector based on partial differentiation. When the **vector differential operator**

$$\nabla = \mathbf{i}\frac{\partial}{\partial x} + \mathbf{j}\frac{\partial}{\partial y} \quad \text{or} \quad \nabla = \mathbf{i}\frac{\partial}{\partial x} + \mathbf{j}\frac{\partial}{\partial y} + \mathbf{k}\frac{\partial}{\partial z}$$

is applied to a differentiable function $z = f(x, y)$ or $w = F(x, y, z)$, we say that the vectors

$$\nabla f(x, y) = \frac{\partial f}{\partial x}\mathbf{i} + \frac{\partial f}{\partial y}\mathbf{j} \tag{1}$$

$$\nabla F(x, y, z) = \frac{\partial F}{\partial x}\mathbf{i} + \frac{\partial F}{\partial y}\mathbf{j} + \frac{\partial F}{\partial z}\mathbf{k} \tag{2}$$

are the **gradients** of the respective functions. The symbol ∇, an inverted capital Greek delta, is called "del" or "nabla." The vector ∇f is usually read "grad f."

Example 1 Gradient

Compute $\nabla f(x, y)$ for $f(x, y) = 5y - x^3y^2$.

SOLUTION From (1), $\nabla f(x, y) = \dfrac{\partial}{\partial x}(5y - x^3y^2)\mathbf{i} + \dfrac{\partial}{\partial y}(5y - x^3y^2)\mathbf{j}$, therefore

$$\nabla f(x, y) = -3x^2y^2\mathbf{i} + (5 - 2x^3y)\mathbf{j}.$$ ❏

Example 2 Gradient at a Point

If $F(x, y, z) = xy^2 + 3x^2 - z^3$, find $\nabla F(x, y, z)$ at $(2, -1, 4)$.

SOLUTION From (2), $\nabla F(x, y, z) = (y^2 + 6x)\mathbf{i} + 2xy\mathbf{j} - 3z^2\mathbf{k}$ and so

$$\nabla F(2, -1, 4) = 13\mathbf{i} - 4\mathbf{j} - 48\mathbf{k}.$$ ❏

■ **A Generalization of Partial Differentiation** The partial derivatives $\partial z/\partial x$ and $\partial z/\partial y$ give the slope of a tangent to the trace, or curve of intersection, of the surface given by $z = f(x, y)$ and vertical planes which are, respectively, parallel to the x- and y-coordinate axes. Equivalently, $\partial z/\partial x$ is the rate of change of the function in the direction given by the vector \mathbf{i}, and $\partial z/\partial y$ is the rate of change of $z = f(x, y)$ in the \mathbf{j}-direction. There is no reason to confine our attention to just two directions; we can find the rate of change of a differentiable function in *any* direction. See Figure 9.26.

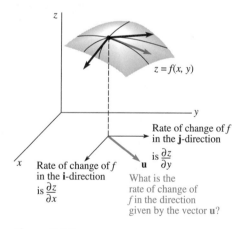

Figure 9.26

Suppose $\mathbf{u} = \cos\theta\,\mathbf{i} + \sin\theta\,\mathbf{j}$ is a unit vector in the xy-plane that makes an angle θ with the positive x-axis and is parallel to the vector \mathbf{v} from $(x, y, 0)$ to $(x + \Delta x, y + \Delta y, 0)$. If $h = \sqrt{(\Delta x)^2 + (\Delta y)^2} > 0$, then $\mathbf{v} = h\mathbf{u}$. Furthermore, let the plane perpendicular to the xy-plane that contains these points slice the surface $z = f(x, y)$ in a curve C. We ask: What is the slope of the tangent line to C at a point P with coordinates $(x, y, f(x, y))$ in the direction given by \mathbf{v}? See Figure 9.27.

From the figure we see that $\Delta x = h\cos\theta$ and $\Delta y = h\sin\theta$ so that the slope of the indicated secant line is

$$\frac{f(x + \Delta x, y + \Delta y) - f(x, y)}{h} = \frac{f(x + h\cos\theta, y + h\sin\theta) - f(x, y)}{h}. \tag{3}$$

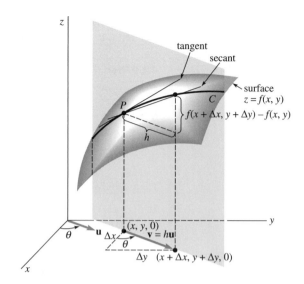

Figure 9.27

We expect the slope of the tangent at P to be the limit of (3) as $h \to 0$. This slope is the rate of change of f at P in the direction specified by the unit vector \mathbf{u}. This leads us to the following definition:

DEFINITION 9.5 Directional Derivative

The **directional derivative** of $z = f(x, y)$ in the direction of a unit vector $\mathbf{u} = \cos \theta \mathbf{i} + \sin \theta \mathbf{j}$ is

$$D_{\mathbf{u}}f(x, y) = \lim_{h \to 0} \frac{f(x + h \cos \theta, y + h \sin \theta) - f(x, y)}{h} \tag{4}$$

provided the limit exists.

Observe that (4) is truly a generalization of partial differentiation, since

$$\theta = 0 \quad \text{implies that} \quad D_{\mathbf{i}}f(x, y) = \lim_{h \to 0} \frac{f(x + h, y) - f(x, y)}{h} = \frac{\partial z}{\partial x}$$

and
$$\theta = \frac{\pi}{2} \quad \text{implies that} \quad D_{\mathbf{j}}f(x, y) = \lim_{h \to 0} \frac{f(x, y + h) - f(x, y)}{h} = \frac{\partial z}{\partial y}.$$

■ **Method for Computing the Directional Derivative** While (4) could be used to find $D_{\mathbf{u}}f(x, y)$ for a given function, as usual we seek a more efficient procedure. The next theorem will show how the concept of the gradient of a function plays a key role in computing a directional derivative.

THEOREM 9.6 Computing a Directional Derivative

If $z = f(x, y)$ is a differentiable function of x and y and $\mathbf{u} = \cos \theta \mathbf{i} + \sin \theta \mathbf{j}$, then

$$D_{\mathbf{u}}f(x, y) = \nabla f(x, y) \cdot \mathbf{u}. \tag{5}$$

PROOF Let x, y, and θ be fixed so that $g(t) = f(x + t \cos \theta, y + t \sin \theta)$ is a function of one variable. We wish to compare the value of $g'(0)$, which is found by two different methods. First, by the definition of a derivative,

$$g'(0) = \lim_{h \to 0} \frac{g(0 + h) - g(0)}{h} = \lim_{h \to 0} \frac{f(x + h \cos \theta, y + h \sin \theta) - f(x, y)}{h}. \quad (6)$$

Second, by the Chain Rule,

$$g'(t) = f_1(x + t \cos \theta, y + t \sin \theta) \frac{d}{dt}(x + t \cos \theta) + f_2(x + t \cos \theta, y + t \sin \theta) \frac{d}{dt}(x + t \sin \theta)$$
$$(7)$$
$$= f_1(x + t \cos \theta, y + t \sin \theta) \cos \theta + f_2(x + t \cos \theta, y + t \sin \theta) \sin \theta$$

Here the subscripts 1 and 2 refer to the partial derivatives of $f(x + t \cos \theta, y + t \sin \theta)$ with respect to $x + t \cos \theta$ and $y + t \sin \theta$, respectively. When $t = 0$, we note that $x + t \cos \theta$ and $y + t \sin \theta$ are simply x and y, and therefore (7) becomes

$$g'(0) = f_x(x, y) \cos \theta + f_y(x, y) \sin \theta. \quad (8)$$

Comparing (4), (6), and (8) then gives

$$D_\mathbf{u}f(x, y) = f_x(x, y) \cos \theta + f_y(x, y) \sin \theta$$
$$= [f_x(x, y)\mathbf{i} + f_y(x, y)\mathbf{j}] \cdot (\cos \theta \mathbf{i} + \sin \theta \mathbf{j})$$
$$= \nabla f(x, y) \cdot \mathbf{u}. \qquad \square$$

Example 3 **Directional Derivative**

Find the directional derivative of $f(x, y) = 2x^2y^3 + 6xy$ at $(1, 1)$ in the direction of a unit vector whose angle with the positive x-axis is $\pi/6$.

SOLUTION Since $\dfrac{\partial f}{\partial x} = 4xy^3 + 6y$ and $\dfrac{\partial f}{\partial y} = 6x^2y^2 + 6x$, we have

$$\nabla f(x, y) = (4xy^3 + 6y)\mathbf{i} + (6x^2y^2 + 6x)\mathbf{j} \quad \text{and} \quad \nabla f(1, 1) = 10\mathbf{i} + 12\mathbf{j}.$$

Now, at $\theta = \pi/6$, $\mathbf{u} = \cos \theta \mathbf{i} + \sin \theta \mathbf{j}$ becomes $\mathbf{u} = \dfrac{\sqrt{3}}{2}\mathbf{i} + \dfrac{1}{2}\mathbf{j}$. Therefore,

$$D_\mathbf{u}f(1, 1) = \nabla f(1, 1) \cdot \mathbf{u} = (10\mathbf{i} + 12\mathbf{j}) \cdot \left(\frac{\sqrt{3}}{2}\mathbf{i} + \frac{1}{2}\mathbf{j}\right) = 5\sqrt{3} + 6. \qquad \square$$

Example 4 **Directional Derivative**

Consider the plane that is perpendicular to the xy-plane and passes through the points $P(2, 1)$ and $Q(3, 2)$. What is the slope of the tangent line to the curve of intersection of this plane with the surface $f(x, y) = 4x^2 + y^2$ at $(2, 1, 17)$ in the direction of Q?

SOLUTION We want $D_\mathbf{u}f(2, 1)$ in the direction given by the vector $\overrightarrow{PQ} = \mathbf{i} + \mathbf{j}$. But since \overrightarrow{PQ} is not a unit vector, we form $\mathbf{u} = (1/\sqrt{2})\mathbf{i} + (1/\sqrt{2})\mathbf{j}$. Now,

$$\nabla f(x, y) = 8x\mathbf{i} + 2y\mathbf{j} \quad \text{and} \quad \nabla f(2, 1) = 16\mathbf{i} + 2\mathbf{j}.$$

Therefore, the required slope is

$$D_\mathbf{u}f(2, 1) = (16\mathbf{i} + 2\mathbf{j}) \cdot \left(\frac{1}{\sqrt{2}}\mathbf{i} + \frac{1}{\sqrt{2}}\mathbf{j}\right) = 9\sqrt{2}. \qquad \square$$

■ **Functions of Three Variables** For a function $w = F(x, y, z)$ the directional derivative is defined by

$$D_\mathbf{u}f(x, y, z) = \lim_{h \to 0} \frac{F(x + h \cos \alpha, y + h \cos \beta, z + h \cos \gamma) - F(x, y, z)}{h},$$

where α, β, and γ are the direction angles of the vector \mathbf{u} measured relative to the positive x-, y-, and z-axes, respectively.* But in the same manner as before, we can show that

$$D_{\mathbf{u}}F(x, y, z) = \nabla F(x, y, z) \cdot \mathbf{u}. \tag{9}$$

Notice that since \mathbf{u} is a unit vector, it follows from (10) of Section 7.3 that

$$D_{\mathbf{u}}f(x, y) = \text{comp}_{\mathbf{u}}\nabla f(x, y) \quad \text{and} \quad D_{\mathbf{u}}F(x, y, z) = \text{comp}_{\mathbf{u}}\nabla F(x, y, z).$$

In addition, (9) reveals that

$$D_{\mathbf{k}}F(x, y, z) = \frac{\partial w}{\partial z}.$$

Example 5 Directional Derivative

Find the directional derivative of $F(x, y, z) = xy^2 - 4x^2y + z^2$ at $(1, -1, 2)$ in the direction of $6\mathbf{i} + 2\mathbf{j} + 3\mathbf{k}$.

SOLUTION We have $\dfrac{\partial F}{\partial x} = y^2 - 8xy$, $\dfrac{\partial F}{\partial y} = 2xy - 4x^2$, and $\dfrac{\partial F}{\partial z} = 2z$ so that

$$\nabla F(x, y, z) = (y^2 - 8xy)\mathbf{i} + (2xy - 4x^2)\mathbf{j} + 2z\mathbf{k}$$

$$\nabla F(1, -1, 2) = 9\mathbf{i} - 6\mathbf{j} + 4\mathbf{k}.$$

Since $\|6\mathbf{i} + 2\mathbf{j} + 3\mathbf{k}\| = 7$ then $\mathbf{u} = \frac{6}{7}\mathbf{i} + \frac{2}{7}\mathbf{j} + \frac{3}{7}\mathbf{k}$ is a unit vector in the indicated direction. It follows from (9) that

$$D_{\mathbf{u}}F(1, -1, 2) = (9\mathbf{i} - 6\mathbf{j} + 4\mathbf{k}) \cdot \left(\frac{6}{7}\mathbf{i} + \frac{2}{7}\mathbf{j} + \frac{3}{7}\mathbf{k}\right) = \frac{54}{7}. \qquad \square$$

■ **Maximum Value of the Directional Derivative** Let f represent a function of either two or three variables. Since (5) and (9) express the directional derivative as a dot product, we see from Definition 7.3 that

$$D_{\mathbf{u}}f = \|\nabla f\| \, \|\mathbf{u}\| \cos \phi = \|\nabla f\| \cos \phi, \qquad (\|\mathbf{u}\| = 1),$$

where ϕ is the angle between ∇f and \mathbf{u}. Because $0 \le \phi \le \pi$, we have $-1 \le \cos \phi \le 1$ and, consequently, $-\|\nabla f\| \le D_{\mathbf{u}}f \le \|\nabla f\|$. In other words:

> *The maximum value of the directional derivative is $\|\nabla f\|$ and it occurs when \mathbf{u} has the same direction as ∇f (when $\cos \phi = 1$),* $\tag{10}$

and:

> *The minimum value of the directional derivative is $-\|\nabla f\|$ and it occurs when \mathbf{u} and ∇f have opposite directions (when $\cos \phi = -1$).* $\tag{11}$

Example 6 Max/Min of Directional Derivative

In Example 5 the maximum value of the directional derivative at F at $(1, -1, 2)$ is $\|\nabla F(1, -1, 2)\| = \sqrt{133}$. The minimum value of $D_{\mathbf{u}}F(1, -1, 2)$ is then $-\sqrt{133}$. \square

■ **Gradient Points in Direction of Most Rapid Increase of f** Put yet another way, (10) and (11) state:

* Note that the numerator of (4) can be written

$$f(x + h \cos \alpha, y + h \cos \beta) - f(x, y),$$

where $\beta = (\pi/2) - \alpha$.

The gradient vector ∇f points in the direction in which f increases most rapidly, whereas $-\nabla f$ points in the direction of the most rapid decrease of f.

(a) (b)

Figure 9.28

Example 7 Direction of Steepest Ascent

Each year in Los Angeles there is a bicycle race up to the top of a hill by a road known to be the steepest in the city. To understand why a bicyclist with a modicum of sanity will zigzag up the road, let us suppose the graph of $f(x, y) = 4 - \frac{2}{3}\sqrt{x^2 + y^2}$, $0 \leq z \leq 4$, shown in Figure 9.28(a) is a mathematical model of the hill. The gradient of f is

$$\nabla f(x, y) = \frac{2}{3}\left[\frac{-x}{\sqrt{x^2 + y^2}}\mathbf{i} + \frac{-y}{\sqrt{x^2 + y^2}}\mathbf{j}\right] = \frac{2/3}{\sqrt{x^2 + y^2}}\mathbf{r},$$

where $\mathbf{r} = -x\mathbf{i} - y\mathbf{j}$ is a vector pointing to the center of the circular base.

Thus the steepest ascent up the hill is a straight road whose projection in the xy-plane is a radius of the circular base. Since $D_\mathbf{u}f = \text{comp}_\mathbf{u}\nabla f$, a bicyclist will zigzag, or seek a direction \mathbf{u} other than ∇f, in order to reduce this component. ❏

Example 8 Direction to Cool Off Fastest

The temperature in a rectangular box is approximated by

$$T(x, y, z) = xyz(1 - x)(2 - y)(3 - z), \qquad 0 \leq x \leq 1, \quad 0 \leq y \leq 2, \quad 0 \leq z \leq 3.$$

If a mosquito is located at $(\frac{1}{2}, 1, 1)$, in which direction should it fly to cool off as rapidly as possible?

SOLUTION The gradient of T is

$$\nabla T(x, y, z) = yz(2 - y)(3 - z)(1 - 2x)\mathbf{i} + xz(1 - x)(3 - z)(2 - 2y)\mathbf{j}$$
$$+ xy(1 - x)(2 - y)(3 - 2z)\mathbf{k}.$$

Therefore, $\nabla T(\frac{1}{2}, 1, 1) = \frac{1}{4}\mathbf{k}$. To cool off most rapidly, the mosquito should fly in the direction of $-\frac{1}{4}\mathbf{k}$; that is, it should dive for the floor of the box, where the temperature is $T(x, y, 0) = 0$. ❏

EXERCISES 9.5

Answers to odd-numbered problems begin on page A-49.

In Problems 1–4, compute the gradient for the given function.

1. $f(x, y) = x^2 - x^3y^2 + y^4$

2. $f(x, y) = y - e^{-2x^2y}$

3. $F(x, y, z) = \dfrac{xy^2}{z^3}$

4. $F(x, y, z) = xy \cos yz$

In Problems 5–8, find the gradient of the given function at the indicated point.

5. $f(x, y) = x^2 - 4y^2$; $(2, 4)$

6. $f(x, y) = \sqrt{x^3y - y^4}$; $(3, 2)$

7. $F(x, y, z) = x^2z^2 \sin 4y$; $(-2, \pi/3, 1)$

8. $F(x, y, z) = \ln(x^2 + y^2 + z^2)$; $(-4, 3, 5)$

In Problems 9 and 10, use Definition 9.5 to find $D_\mathbf{u}f(x, y)$ given that \mathbf{u} makes the indicated angle with the positive x-axis.

9. $f(x, y) = x^2 + y^2$; $\theta = 30°$

10. $f(x, y) = 3x - y^2$; $\theta = 45°$

In Problems 11–20, find the directional derivative of the given function at the given point in the indicated direction.

11. $f(x, y) = 5x^3y^6$; $(-1, 1)$, $\theta = \pi/6$

12. $f(x, y) = 4x + xy^2 - 5y$; $(3, -1)$, $\theta = \pi/4$

13. $f(x, y) = \tan^{-1}\dfrac{y}{x}$; $(2, -2)$, $\mathbf{i} - 3\mathbf{j}$

14. $f(x, y) = \dfrac{xy}{x + y}$; $(2, -1)$, $6\mathbf{i} + 8\mathbf{j}$

15. $f(x, y) = (xy + 1)^2$; $(3, 2)$, in the direction of $(5, 3)$

16. $f(x, y) = x^2 \tan y$; $\left(\dfrac{1}{2}, \dfrac{\pi}{3}\right)$, in the direction of the negative

x-axis

17. $F(x, y, z) = x^2y^2(2z + 1)^2$; $(1, -1, 1)$, $\langle 0, 3, 3\rangle$

18. $F(x, y, z) = \dfrac{x^2 - y^2}{z^2}$; $(2, 4, -1)$, $\mathbf{i} - 2\mathbf{j} + \mathbf{k}$

19. $F(x, y, z) = \sqrt{x^2y + 2y^2z}$; $(-2, 2, 1)$, in the direction of the negative z-axis

20. $F(x, y, z) = 2x - y^2 + z^2$; $(4, -4, 2)$, in the direction of the origin

In Problems 21 and 22, consider the plane through the points P and Q that is perpendicular to the xy-plane. Find the slope of the tangent at the indicated point to the curve of intersection of this plane and the graph of the given function in the direction of Q.

21. $f(x, y) = (x - y)^2$; $P(4, 2)$, $Q(0, 1)$; $(4, 2, 4)$

22. $f(x, y) = x^3 - 5xy + y^2$; $P(1, 1)$, $Q(-1, 6)$; $(1, 1, -3)$

In Problems 23–26, find a vector that gives the direction in which the given function increases most rapidly at the indicated point. Find the maximum rate.

23. $f(x, y) = e^{2x} \sin y$; $(0, \pi/4)$

24. $f(x, y) = xye^{x-y}$; $(5, 5)$

25. $F(x, y, z) = x^2 + 4xz + 2yz^2$; $(1, 2, -1)$

26. $F(x, y, z) = xyz$; $(3, 1, -5)$

In Problems 27–30, find a vector that gives the direction in which the given function decreases most rapidly at the indicated point. Find the minimum rate.

27. $f(x, y) = \tan(x^2 + y^2)$; $(\sqrt{\pi/6}, \sqrt{\pi/6})$

28. $f(x, y) = x^3 - y^3$; $(2, -2)$

29. $F(x, y, z) = \sqrt{xz}e^y$; $(16, 0, 9)$

30. $F(x, y, z) = \ln\dfrac{xy}{z}$; $\left(\dfrac{1}{2}, \dfrac{1}{6}, \dfrac{1}{3}\right)$

31. Find the directional derivative(s) of $f(x, y) = x + y^2$ at $(3, 4)$ in the direction of a tangent vector to the graph of $2x^2 + y^2 = 9$ at $(2, 1)$.

32. If $f(x, y) = x^2 + xy + y^2 - x$, find all points where $D_{\mathbf{u}}f(x, y)$ in the direction of $\mathbf{u} = (1/\sqrt{2})(\mathbf{i} + \mathbf{j})$ is zero.

33. Suppose $\nabla f(a, b) = 4\mathbf{i} + 3\mathbf{j}$. Find a unit vector \mathbf{u} so that:

(a) $D_{\mathbf{u}}f(a, b) = 0$,

(b) $D_{\mathbf{u}}f(a, b)$ is a maximum, and

(c) $D_{\mathbf{u}}f(a, b)$ is a minimum.

34. Suppose $D_{\mathbf{u}}f(a, b) = 6$. What is the value of $D_{-\mathbf{u}}f(a, b)$?

35. (a) If $f(x, y) = x^3 - 3x^2y^2 + y^3$, find the directional derivative of f at a point (x, y) in the direction of $\mathbf{u} = (1/\sqrt{10})(3\mathbf{i} + \mathbf{j})$.

(b) If $F(x, y) = D_{\mathbf{u}}f(x, y)$ of part **(a)**, find $D_{\mathbf{u}}F(x, y)$.

36. Consider the gravitational potential

$$U(x, y) = \dfrac{-Gm}{\sqrt{x^2 + y^2}},$$

where G and m are constants. Show that U increases or decreases most rapidly along a line through the origin.

37. If $f(x, y) = x^3 - 12x + y^2 - 10y$, find all points at which $\|\nabla f\| = 0$.

38. Suppose

$$D_{\mathbf{u}}f(a, b) = 7, \quad D_{\mathbf{v}}f(a, b) = 3$$

$$\mathbf{u} = \dfrac{5}{13}\mathbf{i} - \dfrac{12}{13}\mathbf{j}, \quad \mathbf{v} = \dfrac{5}{13}\mathbf{i} + \dfrac{12}{13}\mathbf{j}.$$

Find $\nabla f(a, b)$.

39. Consider the rectangular plate shown in Figure 9.29. The temperature at a point (x, y) on the plate is given by $T(x, y) = 5 + 2x^2 + y^2$. Determine the direction an insect should take, starting at $(4, 2)$, in order to cool off as rapidly as possible.

Figure 9.29

40. In Problem 39, observe that $(0, 0)$ is the coolest point of the plate. Find the path the cold-seeking insect, starting at $(4, 2)$, will take to the origin. If $\langle x(t), y(t)\rangle$ is the vector equation of the path, then use the fact that $-\nabla T(x, y) = \langle x'(t), y'(t)\rangle$. Why is this? [*Hint:* Remember separation of variables?]

41. The temperature at a point (x, y) on a rectangular metal plate is given by $T(x, y) = 100 - 2x^2 - y^2$. Find the path a heat-seeking particle will take, starting at $(3, 4)$, as it moves in the direction in which the temperature increases most rapidly.

42. The temperature T at a point (x, y, z) in space is inversely proportional to the square of the distance from (x, y, z) to the origin. It is known that $T(0, 0, 1) = 500$. Find the rate of change of T at $(2, 3, 3)$ in the direction of $(3, 1, 1)$. In which direction from $(2, 3, 3)$ does the temperature T increase most rapidly? At $(2, 3, 3)$ what is the maximum rate of change of T?

43. Find a function f such that

$$\nabla f = (3x^2 + y^3 + ye^{xy})\mathbf{i} + (-2y^2 + 3xy^2 + xe^{xy})\mathbf{j}$$

44. Let f_x, f_y, f_{xy}, f_{yx} be continuous and \mathbf{u} and \mathbf{v} be unit vectors. Show that $D_{\mathbf{u}}D_{\mathbf{v}}f = D_{\mathbf{v}}D_{\mathbf{u}}f$.

In Problems 45–48, assume that f and g are differentiable functions of two variables. Prove the given identity.

45. $\nabla(cf) = c\,\nabla f$

46. $\nabla(f + g) = \nabla f + \nabla g$

47. $\nabla(fg) = f\,\nabla g + g\,\nabla f$

48. $\nabla\left(\dfrac{f}{g}\right) = \dfrac{g\nabla f - f\nabla g}{g^2}$

49. If $\mathbf{F}(x, y, z) = f_1(x, y, z)\mathbf{i} + f_2(x, y, z)\mathbf{j} + f_3(x, y, z)\mathbf{k}$ and

$$\nabla = \mathbf{i}\frac{\partial}{\partial x} + \mathbf{j}\frac{\partial}{\partial y} + \mathbf{k}\frac{\partial}{\partial z}$$

show that

$$\nabla \times \mathbf{F} = \left(\frac{\partial f_3}{\partial y} - \frac{\partial f_2}{\partial z}\right)\mathbf{i} + \left(\frac{\partial f_1}{\partial z} - \frac{\partial f_3}{\partial x}\right)\mathbf{j} + \left(\frac{\partial f_2}{\partial x} - \frac{\partial f_1}{\partial y}\right)\mathbf{k}$$

9.6 Tangent Planes and Normal Lines

■ **Geometric Interpretation of the Gradient—Functions of Two Variables** Suppose $f(x, y) = c$ is the level curve of the differential function $z = f(x, y)$ that passes through a specified point $P(x_0, y_0)$; that is, $f(x_0, y_0) = c$.

If this level curve is parameterized by the differentiable functions

$$x = g(t), \; y = h(t) \quad \text{such that} \quad x_0 = g(t_0), \; y_0 = h(t_0),$$

then the derivative of $f(g(t), h(t)) = c$ with respect to t is

$$\frac{\partial f}{\partial x}\frac{dx}{dt} + \frac{\partial f}{\partial y}\frac{dy}{dt} = 0. \tag{1}$$

When we introduce the vectors

$$\nabla f(x, y) = \frac{\partial f}{\partial x}\mathbf{i} + \frac{\partial f}{\partial y}\mathbf{j} \quad \text{and} \quad \mathbf{r}'(t) = \frac{dx}{dt}\mathbf{i} + \frac{dy}{dt}\mathbf{j},$$

(1) becomes $\nabla f \cdot \mathbf{r}' = 0$. Specifically, at $t = t_0$, we have

$$\nabla f(x_0, y_0) \cdot \mathbf{r}'(t_0) = 0. \tag{2}$$

Thus, if $\mathbf{r}'(t_0) \neq 0$, the vector $\nabla f(x_0, y_0)$ is orthogonal to the tangent vector $\mathbf{r}'(t_0)$ at $P(x_0, y_0)$. We interpret this to mean ∇f *is perpendicular to the level curve at P.* See Figure 9.30

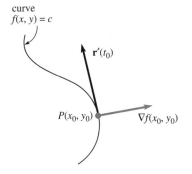

curve
$f(x, y) = c$

Figure 9.30

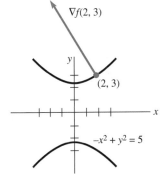

Figure 9.31

Example 1 **Gradient at a Point**

Find the level curve of $f(x, y) = -x^2 + y^2$ passing through $(2, 3)$. Graph the gradient at the point.

SOLUTION Since $f(2, 3) = -4 + 9 = 5$, the level curve is the hyperbola $-x^2 + y^2 = 5$. Now,

$$\nabla f(x, y) = -2x\mathbf{i} + 2y\mathbf{j} \quad \text{and} \quad \nabla f(2, 3) = -4\mathbf{i} + 6\mathbf{j}.$$

Figure 9.31 shows the level curve and $\nabla f(2, 3)$. ❑

■ **Geometric Interpretation of the Gradient—Functions of Three Variables** Proceeding as before, let $F(x, y, z) = c$ be the *level surface* of a differentiable function $w = F(x, y, z)$ that passes through $P(x_0, y_0, z_0)$. If the differentiable functions $x = f(t), y = g(t), z = h(t)$ are the parametric equations of a curve C on the surface for which $x_0 = f(t_0), y_0 = g(t_0), z_0 = h(t_0)$, then the derivative of $F(f(t), g(t), h(t)) = 0$ implies that

$$\frac{\partial F}{\partial x}\frac{dx}{dt} + \frac{\partial F}{\partial y}\frac{dy}{dt} + \frac{\partial F}{\partial z}\frac{dz}{dt} = 0$$

Figure 9.32

Figure 9.33

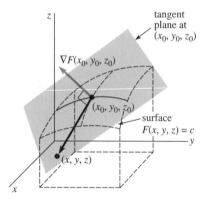

Figure 9.34

or

$$\left(\frac{\partial F}{\partial x}\mathbf{i} + \frac{\partial F}{\partial y}\mathbf{j} + \frac{\partial F}{\partial z}\mathbf{k}\right) \cdot \left(\frac{dx}{dt}\mathbf{i} + \frac{dy}{dt}\mathbf{j} + \frac{dz}{dt}\mathbf{k}\right) = 0. \qquad (3)$$

In particular, at $t = t_0$, (3) is

$$\nabla F(x_0, y_0, z_0) \cdot \mathbf{r}'(t_0) = 0. \qquad (4)$$

Thus, when $\mathbf{r}'(t_0) \neq 0$, the vector $\nabla F(x_0, y_0, z_0)$ is orthogonal to the tangent vector $\mathbf{r}'(t_0)$. Since this argument holds for any differentiable curve through $P(x_0, y_0, z_0)$ on the surface, we conclude that ∇F *is perpendicular (normal) to the level surface at P.* See Figure 9.32.

Example 2 Gradient at a Point

Find the level surface of $F(x, y, z) = x^2 + y^2 + z^2$ passing through $(1, 1, 1)$. Graph the gradient at the point.

SOLUTION Since $F(1, 1, 1) = 3$, the level surface passing through $(1, 1, 1)$ is the sphere $x^2 + y^2 + z^2 = 3$. The gradient of the function is

$$\nabla F(x, y, z) = 2x\mathbf{i} + 2y\mathbf{j} + 2z\mathbf{k},$$

and so, at the given point, $\nabla F(1, 1, 1) = 2\mathbf{i} + 2\mathbf{j} + 2\mathbf{k}$. The level surface and $\nabla F(1, 1, 1)$ are illustrated in Figure 9.33. ❑

■ **Tangent Plane** In the study of differential calculus a basic problem was finding an equation of a tangent line to the graph of a function. In 3-space the analogous problem is finding an equation of a **tangent plane** to a surface. We assume again that $w = F(x, y, z)$ is a differentiable function and that a surface is given by $F(x, y, z) = c$.

> **DEFINITION 9.6 Tangent Plane**
>
> Let $P(x_0, y_0, z_0)$ be a point on the graph of $F(x, y, z) = c$, where ∇F is not $\mathbf{0}$. The tangent plane at P is that plane through P that is perpendicular to ∇F evaluated at P.

Thus, if $P(x, y, z)$ and $P(x_0, y_0, z_0)$ are points on the tangent plane and \mathbf{r} and \mathbf{r}_0 are their respective position vectors, then a vector equation of the tangent plane is $\nabla F(x_0, y_0, z_0) \cdot (\mathbf{r} - \mathbf{r}_0) = 0$. See Figure 9.34. We summarize this last result:

> **THEOREM 9.7 Equation of Tangent Plane**
>
> Let $P(x_0, y_0, z_0)$ be a point on the graph of $F(x, y, z) = c$, where ∇F is not $\mathbf{0}$. Then an equation of the tangent plane at P is
>
> $$F_x(x_0, y_0, z_0)(x - x_0) + F_y(x_0, y_0, z_0)(y - y_0) + F_z(x_0, y_0, z_0)(z - z_0) = 0. \qquad (5)$$

Example 3 Equation of Tangent Plane

Find an equation of the tangent plane to the graph of $x^2 - 4y^2 + z^2 = 16$ at $(2, 1, 4)$.

SOLUTION By defining $F(x, y, z) = x^2 - 4y^2 + z^2$, the given surface is the level surface $F(x, y, z) = F(2, 1, 4) = 16$ passing through $(2, 1, 4)$. Now, $F_x(x, y, z) = 2x$, $F_y(x, y, z) = -8y$, and $F_z(x, y, z) = 2z$, so that

$$\nabla F(x, y, z) = 2x\mathbf{i} - 8y\mathbf{j} + 2z\mathbf{k} \quad \text{and} \quad \nabla F(2, 1, 4) = 4\mathbf{i} - 8\mathbf{j} + 8\mathbf{k}.$$

It follows from (5) that an equation of the tangent plane is

$$4(x - 2) - 8(y - 1) + 8(z - 4) = 0 \quad \text{or} \quad x - 2y + 2z = 8.$$ ❑

■ **Surfaces Given by** $z = f(x, y)$ For a surface given explicitly by a differentiable function $z = f(x, y)$, we define $F(x, y, z) = f(x, y) - z$ or $F(x, y, z) = z - f(x, y)$. Thus, a point (x_0, y_0, z_0) is on the graph of $z = f(x, y)$ if and only if it is also on the level surface $F(x, y, z) = 0$. This follows from $F(x_0, y_0, z_0) = f(x_0, y_0) - z_0 = 0$.

Example 4 Equation of Tangent Plane

Find an equation of the tangent plane to the graph of $z = \frac{1}{2}x^2 + \frac{1}{2}y^2 + 4$ at $(1, -1, 5)$.

SOLUTION Define $F(x, y, z) = \frac{1}{2}x^2 + \frac{1}{2}y^2 - z + 4$ so that the level surface of F passing through the given point is $F(x, y, z) = F(1, -1, 5)$ or $F(x, y, z) = 0$. Now, $F_x = x$, $F_y = y$, and $F_z = -1$, so that

$$\nabla F(x, y, z) = x\mathbf{i} + y\mathbf{j} - \mathbf{k} \quad \text{and} \quad \nabla F(1, -1, 5) = \mathbf{i} - \mathbf{j} - \mathbf{k}.$$

Thus, from (5) the desired equation is

$$(x + 1) - (y - 1) - (z - 5) = 0 \quad \text{or} \quad -x + y + z = 7.$$

See Figure 9.35. ❑

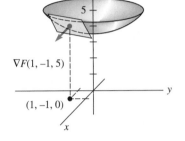

$\nabla F(1, -1, 5)$

$(1, -1, 0)$

Figure 9.35

■ **Normal Line** Let $P(x_0, y_0, z_0)$ be a point on the graph of $F(x, y, z) = c$, where ∇F is not **0**. The line containing $P(x_0, y_0, z_0)$ that is parallel to $\nabla F(x_0, y_0, z_0)$ is called the **normal line** to the surface at P. The normal line is perpendicular to the tangent plane to the surface at P.

Example 5 Normal Line to a Surface

Find parametric equations for the normal line to the surface in Example 4 at $(1, -1, 5)$.

SOLUTION A direction vector for the normal line at $(1, -1, 5)$ is $\nabla F(1, -1, 5) = \mathbf{i} - \mathbf{j} - \mathbf{k}$. It follows that parametric equations for the normal line are $x = 1 + t$, $y = -1 - t$, $z = 5 - t$. ❑

Remarks

Water flowing down a hill chooses a path in the direction of the greatest change in altitude. Figure 9.36 shows the contours, or level curves, of a hill. As shown in the figure, a stream starting at point P will take a path that is perpendicular to the contours. After reading Sections 9.5 and 9.6, the student should be able to explain why.

stream

contours
of a hill

Figure 9.36

 Answers to odd-numbered problems begin on page A-49.

In Problems 1–12, sketch the level curve or surface passing through the indicated point. Sketch the gradient at the point.

1. $f(x, y) = x - 2y$; $(6, 1)$

2. $f(x, y) = \dfrac{y + 2x}{x}$; $(1, 3)$

3. $f(x, y) = y - x^2$; $(2, 5)$

4. $f(x, y) = x^2 + y^2$; $(-1, 3)$

5. $f(x, y) = \dfrac{x^2}{4} + \dfrac{y^2}{9}$; $(-2, -3)$

6. $f(x, y) = \dfrac{y^2}{x}$; $(2, 2)$

7. $f(x, y) = (x - 1)^2 - y^2$; $(1, 1)$

8. $f(x, y) = \dfrac{y - 1}{\sin x}$; $\left(\dfrac{\pi}{6}, \dfrac{3}{2}\right)$

9. $F(x, y, z) = y + z$; $(3, 1, 1)$

10. $F(x, y, z) = x^2 + y^2 - z$; $(1, 1, 3)$

11. $F(x, y, z) = \sqrt{x^2 + y^2 + z^2}$; $(3, 4, 0)$

12. $F(x, y, z) = x^2 - y^2 + z$; $(0, -1, 1)$

In Problems 13 and 14, find the points on the given surface at which the gradient is parallel to the indicated vector.

13. $z = x^2 + y^2$; $4\mathbf{i} + \mathbf{j} + \frac{1}{2}\mathbf{k}$

14. $x^3 + y^2 + z = 15$; $27\mathbf{i} + 8\mathbf{j} + \mathbf{k}$

In Problems 15–24, find an equation of the tangent plane to the graph of the given equation at the indicated point.

15. $x^2 + y^2 + z^2 = 9$; $(-2, 2, 1)$

16. $5x^2 - y^2 + 4z^2 = 8$; $(2, 4, 1)$

17. $x^2 - y^2 - 3z^2 = 5$; $(6, 2, 3)$

18. $xy + yz + zx = 7$; $(1, -3, -5)$

19. $z = 25 - x^2 - y^2$; $(3, -4, 0)$

20. $xz = 6$; $(2, 0, 3)$

21. $z = \cos(2x + y)$; $\left(\dfrac{\pi}{2}, \dfrac{\pi}{4}, -\dfrac{1}{\sqrt{2}}\right)$

22. $x^2 y^3 + 6z = 10$; $(2, 1, 1)$

23. $z = \ln(x^2 + y^2)$; $\left(\dfrac{1}{\sqrt{2}}, \dfrac{1}{\sqrt{2}}, 0\right)$

24. $z = 8e^{-2y}\sin 4x$; $\left(\dfrac{\pi}{24}, 0, 4\right)$

In Problems 25 and 26, find the points on the given surface at which the tangent plane is parallel to the indicated plane.

25. $x^2 + y^2 + z^2 = 7$; $2x + 4y + 6z = 1$

26. $x^2 - 2y^2 - 3z^2 = 33$; $8x + 4y + 6z = 5$

27. Find points on the surface $x^2 + 4x + y^2 + z^2 - 2z = 11$ at which the tangent plane is horizontal.

28. Find points on the surface $x^2 + 3y^2 + 4z^2 - 2xy = 16$ at which the tangent plane is parallel to (a) the xz-plane (b) the yz-plane, and (c) the xy-plane.

In Problems 29 and 30, show that the second equation is an equation of the tangent plane to the graph of the first equation at (x_0, y_0, z_0).

29. $\dfrac{x^2}{a^2} + \dfrac{y^2}{b^2} + \dfrac{z^2}{c^2} = 1$; $\dfrac{xx_0}{a^2} + \dfrac{yy_0}{b^2} + \dfrac{zz_0}{c^2} = 1$

30. $\dfrac{x^2}{a^2} - \dfrac{y^2}{b^2} + \dfrac{z^2}{c^2} = 1$; $\dfrac{xx_0}{a^2} - \dfrac{yy_0}{b^2} + \dfrac{zz_0}{c^2} = 1$

31. Show that every tangent plane to the graph of $z^2 = x^2 + y^2$ passes through the origin.

32. Show that the sum of the x-, y-, and z-intercepts of every tangent plane to the graph of $\sqrt{x} + \sqrt{y} + \sqrt{z} = \sqrt{a}$, $a > 0$, is the number a.

In Problems 33 and 34, find parametric equations for the normal line at the indicated point. In Problems 35 and 36, find symmetric equations for the normal line.

33. $x^2 + 2y^2 + z^2 = 4$; $(1, -1, 1)$

34. $z = 2x^2 - 4y^2$; $(3, -2, 2)$

35. $z = 4x^2 + 9y^2 + 1$; $(\frac{1}{2}, \frac{1}{3}, 3)$

36. $x^2 + y^2 - z^2 = 0$; $(3, 4, 5)$

37. Show that every normal line to the graph $x^2 + y^2 + z^2 = a^2$ passes through the origin.

38. Two surfaces are said to be **orthogonal** at a point P of intersection if their normal lines at P are orthogonal. Prove that the surfaces given by $F(x, y, z) = 0$ and $G(x, y, z) = 0$ are orthogonal at P if and only if $F_x G_x + F_y G_y + F_z G_z = 0$.

In Problems 39 and 40, use the result of Problem 38 to show that the given surfaces are orthogonal at a point of intersection.

39. $x^2 + y^2 + z^2 = 25$; $-x^2 + y^2 + z^2 = 0$

40. $x^2 - y^2 + z^2 = 4$; $z = 1/xy^2$

9.7 Divergence and Curl

■ **Vector Fields** In Section 9.1 we introduced the concept of vector function of one variable. Vector functions of two and three variables:

$$\mathbf{F}(x, y) = P(x, y)\mathbf{i} + Q(x, y)\mathbf{j}$$

$$\mathbf{F}(x, y, z) = P(x, y, z)\mathbf{i} + Q(x, y, z)\mathbf{j} + R(x, y, z)\mathbf{k}$$

are also called **vector fields.** For example, the motion of a wind or a fluid can be described by a *velocity field* because a vector can be assigned at each point repre-

senting the velocity of a particle at the point. See Figures 9.37(a) and 9.37(b). The concept of a *force field* plays an important role in mechanics, electricity, and magnetism. See Figures 9.37(c) and 9.37(d).

(a) Airflow around an airplane wing

(b) Laminar flow of blood in an artery; cylindrical layers of blood flow faster near the center of the artery

(c) Inverse square force field; magnitude of the attractive force is large near the particle

(d) Lines of force around two equal positive charges

Figure 9.37

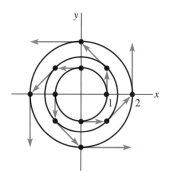

Figure 9.38

Example 1 Two-dimensional Vector Field

Graph the two-dimensional vector field $\mathbf{F}(x, y) = -y\mathbf{i} + x\mathbf{j}$.

SOLUTION One manner of proceeding is simply to choose points in the xy-plane and then graph the vector \mathbf{F} at that point. For example, at $(1, 1)$ we would draw the vector $\mathbf{F}(1, 1) = -\mathbf{i} + \mathbf{j}$.

For the given vector field it is possible to systematically draw vectors of the same length. Observe that $\|\mathbf{F}\| = \sqrt{x^2 + y^2}$, and so vectors of the same length k must lie along the curve defined by $\sqrt{x^2 + y^2} = k$; that is, at any point on the circle $x^2 + y^2 = k^2$ a vector would have length k. For simplicity let us choose circles that have some points on them with integer coordinates. For example, for $k = 1$, $k = \sqrt{2}$, and $k = 2$, we have:

$x^2 + y^2 = 1$: At the points $(1, 0)$, $(0, 1)$, $(-1, 0)$, $(0, -1)$, the corresponding vectors \mathbf{j}, $-\mathbf{i}$, $-\mathbf{j}$, \mathbf{i} have the same length 1.

$x^2 + y^2 = 2$: At the points $(1, 1)$, $(-1, 1)$, $(-1, -1)$, $(1, -1)$ the corresponding vectors $-\mathbf{i} + \mathbf{j}$, $-\mathbf{i} - \mathbf{j}$, $\mathbf{i} - \mathbf{j}$, $\mathbf{i} + \mathbf{j}$ have the same length $\sqrt{2}$.

$x^2 + y^2 = 4$: At the points $(2, 0)$, $(0, 2)$, $(-2, 0)$, $(0, -2)$ the corresponding vectors $2\mathbf{j}$, $-2\mathbf{i}$, $-2\mathbf{j}$, $2\mathbf{i}$ have the same length 2.

The vectors at these points are shown in Figure 9.38. ❑

In the previous section we saw that the del operator

$$\nabla = \frac{\partial}{\partial x}\mathbf{i} + \frac{\partial}{\partial y}\mathbf{j} + \frac{\partial}{\partial z}\mathbf{k}$$

combined with a scalar function $\phi(x, y, z)$ produces a vector field

$$\mathbf{F}(x, y, z) = \nabla\phi = \frac{\partial\phi}{\partial x}\mathbf{i} + \frac{\partial\phi}{\partial y}\mathbf{j} + \frac{\partial\phi}{\partial z}\mathbf{k}$$

called the gradient of ϕ. The del operator can also be combined with a vector field $\mathbf{F}(x, y, z) = P(x, y, z)\mathbf{i} + Q(x, y, z)\mathbf{j} + R(x, y, z)\mathbf{k}$ in two different ways: in one case producing another vector field and in the other producing a scalar function. We will assume hereafter that P, Q, and R have continuous partial derivatives.

> ### DEFINITION 9.7 Curl
>
> The **curl** of a vector field $\mathbf{F} = P\mathbf{i} + Q\mathbf{j} + R\mathbf{k}$ is the vector field
>
> $$\text{curl } \mathbf{F} = \left(\frac{\partial R}{\partial y} - \frac{\partial Q}{\partial z}\right)\mathbf{i} + \left(\frac{\partial P}{\partial z} - \frac{\partial R}{\partial x}\right)\mathbf{j} + \left(\frac{\partial Q}{\partial x} - \frac{\partial P}{\partial y}\right)\mathbf{k}.$$

Figure 9.39

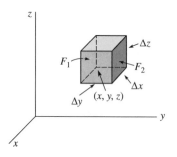

Figure 9.40

In practice, curl \mathbf{F} can be computed from the cross product of the del operator and the vector \mathbf{F}:

$$\text{curl } \mathbf{F} = \nabla \times \mathbf{F} = \begin{vmatrix} \mathbf{i} & \mathbf{j} & \mathbf{k} \\ \dfrac{\partial}{\partial x} & \dfrac{\partial}{\partial y} & \dfrac{\partial}{\partial z} \\ P & Q & R \end{vmatrix}. \tag{1}$$

There is another combination of partial derivatives of the component functions of a vector field that occurs frequently in science and engineering. Before we state the next definition, consider the following motivation.

If $\mathbf{F}(x, y, z) = P(x, y, z)\mathbf{i} + Q(x, y, z)\mathbf{j} + R(x, y, z)\mathbf{k}$ is the velocity field of a fluid, then as shown in Figure 9.39, the volume of the fluid flowing through an element of surface area ΔS per unit time—that is, the **flux** of the vector field \mathbf{F} through the area ΔS—is approximated by

$$(\text{height})(\text{area of base}) = (\text{comp}_{\mathbf{n}}\mathbf{F})\,\Delta S = (\mathbf{F} \cdot \mathbf{n})\,\Delta S, \tag{2}$$

where \mathbf{n} is a unit vector normal to the surface. Now consider the rectangular parallelepiped shown in Figure 9.40. To compute the total flux of \mathbf{F} through its six sides in the outward direction we first compute the total flux out of parallel faces. The area of face F_1 is $\Delta x\,\Delta z$ and its outward unit normal is $-\mathbf{j}$, and so by (2) the flux of \mathbf{F} through F_1 is approximately

$$\mathbf{F} \cdot (-\mathbf{j})\,\Delta x\,\Delta z = -Q(x, y, z)\,\Delta x\,\Delta z.$$

The flux out of face F_2, whose outward normal is \mathbf{j}, is approximated by

$$(\mathbf{F} \cdot \mathbf{j})\,\Delta x\,\Delta z = Q(x, y + \Delta y, z)\,\Delta x\,\Delta z.$$

Consequently the total flux out of these parallel faces is

$$Q(x, y + \Delta y, z)\,\Delta x\,\Delta z + (-Q(x, y, z)\,\Delta x\,\Delta z) = [Q(x, y + \Delta y, z) - Q(x, y, z)]\,\Delta x\,\Delta z. \tag{3}$$

By multiplying (3) by $\Delta y/\Delta y$ and recalling the definition of a partial derivative, we get for small Δy,

$$\frac{[Q(x, y + \Delta y, z) - Q(x, y, z)]}{\Delta y}\,\Delta x\,\Delta y\,\Delta z \approx \frac{\partial Q}{\partial y}\,\Delta x\,\Delta y\,\Delta z.$$

Arguing in exactly the same manner, we see that the contributions to the total flux out of the parallelepiped from the two faces parallel to the yz-plane, and from the two faces parallel to the xy-plane, are, in turn,

$$\frac{\partial P}{\partial x}\,\Delta x\,\Delta y\,\Delta z \quad \text{and} \quad \frac{\partial R}{\partial z}\,\Delta x\,\Delta y\,\Delta z.$$

Adding the results, we see that the net flux of \mathbf{F} out of the parallelepiped is approximately

$$\left(\frac{\partial P}{\partial x} + \frac{\partial Q}{\partial y} + \frac{\partial R}{\partial z}\right)\Delta x\,\Delta y\,\Delta z.$$

By dividing the last expression by $\Delta x \, \Delta y \, \Delta z$, we get the outward flux of **F** per unit volume:

$$\frac{\partial P}{\partial x} + \frac{\partial Q}{\partial y} + \frac{\partial R}{\partial z}.$$

It is this combination of partial derivatives that is given a special name.

DEFINITION 9.8 Divergence

The **divergence** of a vector field $\mathbf{F} = P\mathbf{i} + Q\mathbf{j} + R\mathbf{k}$ is the scalar function

$$\text{div } \mathbf{F} = \frac{\partial P}{\partial x} + \frac{\partial Q}{\partial y} + \frac{\partial R}{\partial z}.$$

Observe that div **F** can also be written in terms of the del operator as:

$$\text{div } \mathbf{F} = \nabla \cdot \mathbf{F} = \frac{\partial}{\partial x} P(x, y, z) + \frac{\partial}{\partial y} Q(x, y, z) + \frac{\partial}{\partial z} R(x, y, z). \tag{4}$$

Example 2 Curl and Divergence

If $\mathbf{F} = (x^2y^3 - z^4)\mathbf{i} + 4x^5y^2z\mathbf{j} - y^4z^6\mathbf{k}$, find curl **F** and div **F**.

SOLUTION From (1),

$$\text{curl } \mathbf{F} = \nabla \times \mathbf{F} = \begin{vmatrix} \mathbf{i} & \mathbf{j} & \mathbf{k} \\ \dfrac{\partial}{\partial x} & \dfrac{\partial}{\partial y} & \dfrac{\partial}{\partial z} \\ x^2y^3 - z^4 & 4x^5y^2z & -y^4z^6 \end{vmatrix}$$

$$= (-4y^3z^6 - 4x^5y^2)\mathbf{i} - 4z^3\mathbf{j} + (20x^4y^2z - 3x^2y^2)\mathbf{k}.$$

From (4),

$$\text{div } \mathbf{F} = \nabla \cdot \mathbf{F} = \frac{\partial}{\partial x}(x^2y^3 - z^4) + \frac{\partial}{\partial y}(4x^5y^2z) + \frac{\partial}{\partial z}(-y^4z^6)$$

$$= 2xy^3 + 8x^5yz - 6y^4z^5. \qquad \square$$

We ask you to prove the following two important properties. If f is a scalar function with continuous second partial derivatives, then

$$\text{curl}(\text{grad } f) = \nabla \times \nabla f = \mathbf{0}. \tag{5}$$

Also, if **F** is a vector field having continuous second partial derivatives, then

$$\text{div}(\text{curl } \mathbf{F}) = \nabla \cdot (\nabla \times \mathbf{F}) = 0. \tag{6}$$

See Problems 29 and 30 in Exercises 9.7.

■ **Physical Interpretations** The word *curl* was introduced by Maxwell* in his studies of electromagnetic fields. However, the curl is easily understood in connection with the flow of fluids. If a paddle device, such as shown in Figure 9.41, is inserted in a flowing fluid, then the curl of the velocity field **F** is a measure of the tendency of the fluid to turn the device about its vertical axis w. If curl $\mathbf{F} = \mathbf{0}$, then

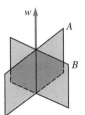

Figure 9.41

* James Clerk Maxwell (1831–1879), a Scottish physicist.

the flow of the fluid is said to be **irrotational,** which means that it is free of vortices or whirlpools that would cause the paddle to rotate.[*] In Figure 9.42 the axis w of the paddle points straight out of the page. Note from Figure 9.42(b) that "irrotational" does *not* mean that the fluid does not rotate.

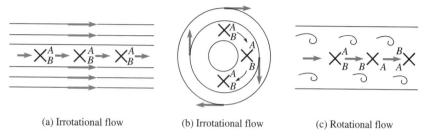

(a) Irrotational flow (b) Irrotational flow (c) Rotational flow

Figure 9.42

In the motivational discussion leading to Definition 9.8 we saw that the divergence of a velocity field \mathbf{F} near a point $P(x, y, z)$ is the flux per unit volume. If div $\mathbf{F}(P) > 0$, then P is said to be a **source** for \mathbf{F}, since there is a net outward flow of fluid near P; if div $\mathbf{F}(P) < 0$, then P is said to be a **sink** for \mathbf{F}, since there is a net inward flow of fluid near P; if div $\mathbf{F}(P) = 0$, there are no sources or sinks near P. See Figure 9.43.

The divergence of a vector field can also be interpreted as a measure of the rate of change of the density of the fluid at a point. In other words, div \mathbf{F} is a measure of the fluid's compressibility. If $\nabla \cdot \mathbf{F} = 0$, the fluid is said to be **incompressible.** In electromagnetic theory, if $\nabla \cdot \mathbf{F} = 0$, the vector field \mathbf{F} is said to be **solenoidal.**

(a) div $\mathbf{F}(P) > 0$; P a source

(b) div $\mathbf{F}(P) < 0$; P a sink

Figure 9.43

[*] In science texts the word *rotation* is sometimes used instead of *curl*. The symbol curl \mathbf{F} is then replaced by rot \mathbf{F}.

EXERCISES 9.7

Answers to odd-numbered problems begin on page A-50.

In Problems 1–6, graph some representative vectors in the given vector field.

1. $\mathbf{F}(x, y) = x\mathbf{i} + y\mathbf{j}$

2. $\mathbf{F}(x, y) = -x\mathbf{i} + y\mathbf{j}$

3. $\mathbf{F}(x, y) = y\mathbf{i} + x\mathbf{j}$

4. $\mathbf{F}(x, y) = x\mathbf{i} + 2y\mathbf{j}$

5. $\mathbf{F}(x, y) = y\mathbf{j}$

6. $\mathbf{F}(x, y) = x\mathbf{j}$

In Problems 7–16, find the curl and the divergence of the given vector field.

7. $\mathbf{F}(x, y, z) = xz\mathbf{i} + yz\mathbf{j} + xy\mathbf{k}$

8. $\mathbf{F}(x, y, z) = 10yz\mathbf{i} + 2x^2z\mathbf{j} + 6x^3\mathbf{k}$

9. $\mathbf{F}(x, y, z) = 4xy\mathbf{i} + (2x^2 + 2yz)\mathbf{j} + (3z^2 + y^2)\mathbf{k}$

10. $\mathbf{F}(x, y, z) = (x - y)^3\mathbf{i} + e^{-yz}\mathbf{j} + xye^{2y}\mathbf{k}$

11. $\mathbf{F}(x, y, z) = 3x^2y\mathbf{i} + 2xz^3\mathbf{j} + y^4\mathbf{k}$

12. $\mathbf{F}(x, y, z) = 5y^3\mathbf{i} + (\tfrac{1}{2}x^3y^2 - xy)\mathbf{j} - (x^3yz - xz)\mathbf{k}$

13. $\mathbf{F}(x, y, z) = xe^{-z}\mathbf{i} + 4yz^2\mathbf{j} + 3ye^{-z}\mathbf{k}$

14. $\mathbf{F}(x, y, z) = yz \ln x\mathbf{i} + (2x - 3yz)\mathbf{j} + xy^2z^3\mathbf{k}$

15. $\mathbf{F}(x, y, z) = xye^x\mathbf{i} - x^3yze^z\mathbf{j} + xy^2e^y\mathbf{k}$

16. $\mathbf{F}(x, y, z) = x^2 \sin yz\mathbf{i} + z \cos xz^3\mathbf{j} + ye^{5xy}\mathbf{k}$

In Problems 17–24, let \mathbf{a} be a constant vector and $\mathbf{r} = x\mathbf{i} + y\mathbf{j} + z\mathbf{k}$. Verify the given identity.

17. div $\mathbf{r} = 3$

18. curl $\mathbf{r} = \mathbf{0}$

19. $(\mathbf{a} \times \nabla) \times \mathbf{r} = -2\mathbf{a}$

20. $\nabla \times (\mathbf{a} \times \mathbf{r}) = 2\mathbf{a}$

21. $\nabla \cdot (\mathbf{a} \times \mathbf{r}) = 0$

22. $\mathbf{a} \times (\nabla \times \mathbf{r}) = \mathbf{0}$

23. $\nabla \times [(\mathbf{r} \cdot \mathbf{r})\mathbf{a}] = 2(\mathbf{r} \times \mathbf{a})$

24. $\nabla \cdot [(\mathbf{r} \cdot \mathbf{r})\mathbf{a}] = 2(\mathbf{r} \cdot \mathbf{a})$

In Problems 25–32, verify the given identity. Assume continuity of all partial derivatives.

25. $\nabla \cdot (\mathbf{F} + \mathbf{G}) = \nabla \cdot \mathbf{F} + \nabla \cdot \mathbf{G}$

26. $\nabla \times (\mathbf{F} + \mathbf{G}) = \nabla \times \mathbf{F} + \nabla \times \mathbf{G}$

27. $\nabla \cdot (f\mathbf{F}) = f(\nabla \cdot \mathbf{F}) + \mathbf{F} \cdot \nabla f$

28. $\nabla \times (f\mathbf{F}) = f(\nabla \times \mathbf{F}) + (\nabla f) \times \mathbf{F}$

29. curl(grad f) = $\mathbf{0}$

30. div(curl \mathbf{F}) = 0

31. div($\mathbf{F} \times \mathbf{G}$) = $\mathbf{G} \cdot$ curl $\mathbf{F} - \mathbf{F} \cdot$ curl \mathbf{G}

32. curl(curl \mathbf{F} + grad f) = curl(curl \mathbf{F})

33. Show that

$$\nabla \cdot \nabla f = \frac{\partial^2 f}{\partial x^2} + \frac{\partial^2 f}{\partial y^2} + \frac{\partial^2 f}{\partial z^2}$$

This is known as the **Laplacian** and is also written $\nabla^2 f$.

34. Show that $\nabla \cdot (f\nabla f) = f\nabla^2 f + \|\nabla f\|^2$, where $\nabla^2 f$ is the Laplacian defined in Problem 33. [*Hint:* See Problem 27.]

35. Find curl(curl \mathbf{F}) for the vector field $\mathbf{F} = xy\mathbf{i} + 4yz^2\mathbf{j} + 2xz\mathbf{k}$.

36. (a) Assuming continuity of all partial derivatives, show that curl(curl \mathbf{F}) = $-\nabla^2\mathbf{F}$ + grad(div \mathbf{F}), where

$$\nabla^2\mathbf{F} = \nabla^2(P\mathbf{i} + Q\mathbf{j} + R\mathbf{k}) = \nabla^2 P\mathbf{i} + \nabla^2 Q\mathbf{j} + \nabla^2 R\mathbf{k}$$

(b) Use the identity in part **(a)** to obtain the result in Problem 35.

37. Any scalar function f for which $\nabla^2 f = 0$ is said to be **harmonic.** Verify that $f(x, y, z) = (x^2 + y^2 + z^2)^{-1/2}$ is harmonic except at the origin. $\nabla^2 f = 0$ is called **Laplace's equation.**

38. Verify that

$$f(x, y) = \arctan\left(\frac{2y}{x^2 + y^2 - 1}\right), \qquad x^2 + y^2 \neq 1$$

satisfies Laplace's equation in two variables

$$\nabla^2 f = \frac{\partial^2 f}{\partial x^2} + \frac{\partial^2 f}{\partial y^2} = 0$$

39. Let $\mathbf{r} = x\mathbf{i} + y\mathbf{j} + z\mathbf{k}$ be the position vector of a mass m_1 and let the mass m_2 be located at the origin. If the force of gravitational attraction is

$$\mathbf{F} = -\frac{Gm_1 m_2}{\|\mathbf{r}\|^3}\,\mathbf{r}$$

verify that curl $\mathbf{F} = \mathbf{0}$ and div $\mathbf{F} = 0$, $\mathbf{r} \neq \mathbf{0}$.

40. Suppose a body rotates with a constant angular velocity ω about an axis. If \mathbf{r} is the position vector of a point P on the body measured from the origin, then the linear velocity vector \mathbf{v} of rotation is $\mathbf{v} = \omega \times \mathbf{r}$. See Figure 9.44. If $\mathbf{r} = x\mathbf{i} + y\mathbf{j} + z\mathbf{k}$ and $\omega = \omega_1\mathbf{i} + \omega_2\mathbf{j} + \omega_3\mathbf{k}$, show that $\omega = \frac{1}{2}$ curl \mathbf{v}.

Figure 9.44

In Problems 41 and 42, assume that f and g have continuous second partial derivatives. Show that the given vector field is solenoidal. [*Hint:* See Problem 31.]

41. $\mathbf{F} = \nabla f \times \nabla g$ **42.** $\mathbf{F} = \nabla f \times (f\nabla g)$

43. The velocity vector field for the two-dimensional flow of an ideal fluid around a cylinder is given by

$$\mathbf{F}(x, y) = A\left[\left(1 - \frac{x^2 - y^2}{(x^2 + y^2)^2}\right)\mathbf{i} - \frac{2xy}{(x^2 + y^2)^2}\,\mathbf{j}\right]$$

for some positive constant A. See Figure 9.45.

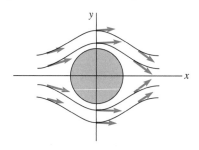

Figure 9.45

(a) Show that when the point (x, y) is far from the origin, $\mathbf{F}(x, y) \approx A\mathbf{i}$.

(b) Show that \mathbf{F} is irrotational.

(c) Show that \mathbf{F} is incompressible.

44. If $\mathbf{E} = \mathbf{E}(x, y, z, t)$ and $\mathbf{H} = \mathbf{H}(x, y, z, t)$ represent electric and magnetic fields in empty space, then Maxwell's equations are

$$\text{div } \mathbf{E} = 0, \quad \text{curl } \mathbf{E} = -\frac{1}{c}\frac{\partial \mathbf{H}}{\partial t},$$

$$\text{div } \mathbf{H} = 0, \quad \text{curl } \mathbf{H} = \frac{1}{c}\frac{\partial \mathbf{E}}{\partial t},$$

where c is the speed of light. Use the identity in Problem 36(a) to show that \mathbf{E} and \mathbf{H} satisfy

$$\nabla^2\mathbf{E} = \frac{1}{c^2}\frac{\partial^2 \mathbf{E}}{\partial t^2}, \quad \nabla^2\mathbf{H} = \frac{1}{c^2}\frac{\partial^2 \mathbf{H}}{\partial t^2}.$$

45. Consider the vector field $\mathbf{F} = x^2yz\mathbf{i} - xy^2z\mathbf{j} + (z + 5x)\mathbf{k}$. Explain why \mathbf{F} is not the curl of another vector field \mathbf{G}.

9.8 Line Integrals

(a) Smooth
curve

(c) Closed
but not
simple

Figure 9.46

■ **Terminology** The notion of the definite integral $\int_a^b f(x)\,dx$, that is, *integration of a function defined over an interval*, can be generalized to *integration of a function defined along a curve*. To this end we need to introduce some terminology about curves. Suppose C is a curve parameterized by $x = f(t)$, $y = g(t)$, $a \le t \le b$, and A and B are the points $(f(a), g(a))$ and $(f(b), g(b))$, respectively. We say that:

(*i*) C is a **smooth curve** if f' and g' are continuous on the closed interval $[a, b]$ and not simultaneously zero on the open interval (a, b).

(*ii*) C is **piecewise smooth** if it consists of a finite number of smooth curves C_1, C_2, \ldots, C_n joined end to end—that is, $C = C_1 \cup C_2 \cup \cdots \cup C_n$.

(*iii*) C is a **closed curve** if $A = B$.

(*iv*) C is a **simple closed curve** if $A = B$ and the curve does not cross itself.

(*v*) If C is not a closed curve, then the **positive direction** on C is the direction corresponding to increasing values of t.

Figure 9.46 illustrates each type of curve defined in (*i*)–(*iv*).

This same terminology carries over in a natural manner to curves in space. For example, a curve C defined by $x = f(t)$, $y = g(t)$, $z = h(t)$, $a \le t \le b$, is smooth if f', g', and h' are continuous on $[a, b]$ and not simultaneously zero on (a, b).

■ **Definite Integral** Before defining integration along a curve, let us review the five steps leading to the definition of the definite integral.

$y = f(x)$

1. Let f be defined on a closed interval $[a, b]$.
2. Partition the interval $[a, b]$ into n subintervals $[x_{k-1}, x_k]$ of length $\Delta x_k = x_k - x_{k-1}$. Let P denote the partition

$$a = x_0 < x_1 < x_2 < \cdots < x_{n-1} < x_n = b$$

$a = x_0 \quad x_1 \qquad x_{k-1} \quad x_k \qquad\qquad x_n = b$

3. Let $\|P\|$ be the length of the longest subinterval. The number $\|P\|$ is called the **norm** of the partition P.
4. Choose a number x_k^* in each subinterval.

x_k^*

$a = x_0 \qquad x_{k-1}\ x_k \qquad\quad x_n = b$

5. Form the sum $\displaystyle\sum_{k=1}^{n} f(x_k^*)\,\Delta x_k$.

The definite integral of a function of a single variable is given by the limit of a sum:

$$\int_a^b f(x)\,dx = \lim_{\|P\| \to 0} \sum_{k=1}^{n} f(x_k^*)\,\Delta x_k.$$

■ **Line Integrals in the Plane** The following analogous five steps lead to the definitions of three **line integrals*** in the plane.

* An unfortunate choice of names. **Curve integrals** would be more appropriate.

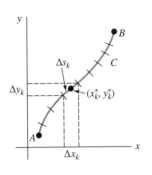

1. Let G be defined in some region that contains the smooth curve C defined by $x = f(t)$, $y = g(t)$, $a \le t \le b$.
2. Divide C into n subarcs of lengths Δs_k according to the partition $a = t_0 < t_1 < t_2 < \cdots < t_n = b$ of $[a, b]$. Let the projection of each subarc onto the x- and y-axes have lengths Δx_k and Δy_k, respectively.
3. Let $\|P\|$ be the **norm** of the partition or the length of the longest subarc.
4. Choose a point (x_k^*, y_k^*) on each subarc.
5. Form the sums

$$\sum_{k=1}^{n} G(x_k^*, y_k^*) \, \Delta x_k, \quad \sum_{k=1}^{n} G(x_k^*, y_k^*) \, \Delta y_k, \quad \sum_{k=1}^{n} G(x_k^*, y_k^*) \, \Delta s_k.$$

DEFINITION 9.9 Line Integrals in the Plane

Let G be a function of two variables x and y defined on a region of the plane containing a smooth curve C.

(*i*) The **line integral of G along C from A to B with respect to x** is

$$\int_C G(x, y) \, dx = \lim_{\|P\| \to 0} \sum_{k=1}^{n} G(x_k^*, y_k^*) \, \Delta x_k.$$

(*ii*) The **line integral of G along C from A to B with respect to y** is

$$\int_C G(x, y) \, dy = \lim_{\|P\| \to 0} \sum_{k=1}^{n} G(x_k^*, y_k^*) \, \Delta y_k.$$

(*iii*) The **line integral of G along C from A to B with respect to arc length** is

$$\int_C G(x, y) \, ds = \lim_{\|P\| \to 0} \sum_{k=1}^{n} G(x_k^*, y_k^*) \, \Delta s_k.$$

It can be proved that if $G(x, y)$ is continuous on C, then the integrals defined in (*i*), (*ii*), and (*iii*) exist. We shall assume continuity of G as a matter of course.

■ **Method of Evaluation—Curve Defined Parametrically** The line integrals in Definition 9.9 can be evaluated in two ways, depending on whether the curve C is defined parametrically or by an explicit function. In either case the basic idea is to convert the line integral to a definite integral in a single variable. If C is a smooth curve parameterized by $x = f(t)$, $y = g(t)$, $a \le t \le b$, then we simply replace x and y in the integral by the functions $f(t)$ and $g(t)$, and the appropriate differential dx, dy, or ds by $f'(t) \, dt$, $g'(t) \, dt$, or $\sqrt{[f'(t)]^2 + [g'(t)]^2} \, dt$. The expression $ds = \sqrt{[f'(t)]^2 + [g'(t)]^2} \, dt$ is called the differential of arc length. The integration is carried out with respect to the variable t in the usual manner:

$$\int_C G(x, y) \, dx = \int_a^b G(f(t), g(t)) f'(t) \, dt, \tag{1}$$

$$\int_C G(x, y) \, dy = \int_a^b G(f(t), g(t)) g'(t) \, dt, \tag{2}$$

$$\int_C G(x, y) \, ds = \int_a^b G(f(t), g(t)) \sqrt{[f'(t)]^2 + [g'(t)]^2} \, dt. \tag{3}$$

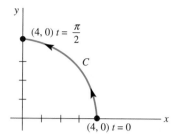

Figure 9.47

Example 1 Evaluation of Line Integrals

Evaluate (a) $\int_C xy^2\,dx$, (b) $\int_C xy^2\,dy$, and (c) $\int_C xy^2\,ds$ on the quarter-circle C defined by $x = 4\cos t$, $y = 4\sin t$, $0 \le t \le \pi/2$. See Figure 9.47.

SOLUTION (a) From (1),

$$\int_C xy^2\,dx = \int_0^{\pi/2} \overbrace{(4\cos t)}^{x}\overbrace{(16\sin^2 t)}^{y^2}\overbrace{(-4\sin t\,dt)}^{dx}$$

$$= -256\int_0^{\pi/2} \sin^3 t\cos t\,dt$$

$$= -256\left[\frac{1}{4}\sin^4 t\right]_0^{\pi/2} = -64.$$

(b) From (2),

$$\int_C xy^2\,dy = \int_0^{\pi/2} \overbrace{(4\cos t)}^{x}\overbrace{(16\sin^2 t)}^{y^2}\overbrace{(4\cos t\,dt)}^{dy}$$

$$= 256\int_0^{\pi/2} \sin^2 t\cos^2 t\,dt \quad \leftarrow \text{trig identities}$$

$$= 256\int_0^{\pi/2} \frac{1}{4}\sin^2 2t\,dt$$

$$= 64\int_0^{\pi/2} \frac{1}{2}(1-\cos 4t)\,dt$$

$$= 32\left[t - \frac{1}{4}\sin 4t\right]_0^{\pi/2} = 16\pi.$$

(c) From (3),

$$\int_C xy^2\,ds = \int_0^{\pi/2} \overbrace{(4\cos t)}^{x}\overbrace{(16\sin^2 t)}^{y^2}\overbrace{\sqrt{16(\cos^2 t + \sin^2 t)}\,dt}^{ds}$$

$$= 256\int_0^{\pi/2} \sin^2 t\cos t\,dt$$

$$= 256\left[\frac{1}{3}\sin^3 t\right]_0^{\pi/2} = \frac{256}{3}. \qquad \square$$

■ Method of Evaluation—Curve Defined by an Explicit Function

If the curve C is defined by an explicit function $y = f(x)$, $a \le x \le b$, we can use x as a parameter. With $dy = f'(x)\,dx$ and $ds = \sqrt{1 + [f'(x)]^2}\,dx$, the foregoing line integrals become, in turn,

$$\int_C G(x, y)\,dx = \int_a^b G(x, f(x))\,dx, \tag{4}$$

$$\int_C G(x, y)\,dy = \int_a^b G(x, f(x))f'(x)\,dx, \tag{5}$$

$$\int_C G(x, y)\,ds = \int_a^b G(x, f(x))\sqrt{1 + [f'(x)]^2}\,dx. \tag{6}$$

A line integral along a *piecewise smooth* curve C is defined as the *sum* of the integrals over the various smooth curves whose union comprises C. For example, if C is composed of smooth curves C_1 and C_2, then

$$\int_C G(x, y)\, ds = \int_{C_1} G(x, y)\, ds + \int_{C_2} G(x, y)\, ds.$$

■ **Notation** In many applications, line integrals appear as a sum

$$\int_C P(x, y)\, dx + \int_C Q(x, y)\, dy.$$

It is common practice to write this sum as one integral without parentheses as

$$\int_C P(x, y)\, dx + Q(x, y)\, dy \quad \text{or simply} \quad \int_C P\, dx + Q\, dy. \tag{7}$$

A line integral along a *closed* curve C is very often denoted by

$$\oint_C P\, dx + Q\, dy.$$

Example 2 Curve Defined by an Explicit Function

Evaluate $\int_C xy\, dx + x^2\, dy$, where C is given by $y = x^3, -1 \le x \le 2$.

SOLUTION The curve C is illustrated in Figure 9.48 and is defined by the explicit function $y = x^3$. Hence we can use x as the parameter. Using $dy = 3x^2\, dx$, we get

$$\int_C xy\, dx + x^2\, dy = \int_{-1}^{2} x\overbrace{(x^3)}^{y}\, dx + x^2\overbrace{(3x^2\, dx)}^{dy}$$

$$= \int_{-1}^{2} 4x^4\, dx$$

$$= \frac{4}{5} x^5 \Big]_{-1}^{2} = \frac{132}{5}. \qquad \square$$

Example 3 Curve Defined Parametrically

Evaluate $\oint_C x\, dx$, where C is the circle $x = \cos t$, $y = \sin t$, $0 \le t \le 2\pi$.

SOLUTION From (1),

$$\oint_C x\, dx = \int_0^{2\pi} \cos t(-\sin t\, dt) = \frac{1}{2} \cos^2 t \Big]_0^{2\pi} = \frac{1}{2}[1 - 1] = 0. \qquad \square$$

Example 4 Closed Curve

Evaluate $\oint_C y^2\, dx - x^2\, dy$ on the closed curve C that is shown in Figure 9.49(a).

SOLUTION Since C is piecewise smooth, we express the integral as a sum of integrals. Symbolically, we write

$$\oint_C = \int_{C_1} + \int_{C_2} + \int_{C_3},$$

where C_1, C_2, and C_3 are the curves shown in Figure 9.49(b). On C_1, we use x as a parameter. Since $y = 0$, $dy = 0$; therefore,

$$\int_{C_1} y^2\, dx - x^2\, dy = \int_0^2 0\, dx - x^2(0) = 0.$$

Figure 9.48

(a)

(b)

Figure 9.49

On C_2, we use y as a parameter. From $x = 2$, $dx = 0$, we have

$$\int_{C_2} y^2 \, dx - x^2 \, dy = \int_0^4 y^2(0) - 4 \, dy$$

$$= -\int_0^4 4 \, dy = -4y \Big]_0^4 = -16.$$

Finally, on C_3, we again use x as a parameter. From $y = x^2$, we get $dy = 2x \, dx$ and so

$$\int_{C_3} y^2 \, dx - x^2 \, dy = \int_2^0 x^4 \, dx - x^2(2x \, dx)$$

$$= \int_2^0 (x^4 - 2x^3) \, dx$$

$$= \left(\frac{1}{5} x^5 - \frac{1}{2} x^4 \right) \Big]_2^0 = \frac{8}{5}.$$

Hence,

$$\int_C y^2 \, dx - x^2 \, dy = 0 - 16 + \frac{8}{5} = -\frac{72}{5}. \qquad \square$$

 Note

It is important to be aware that a line integral is independent of the parameterization of the curve C provided C is given the same orientation by all sets of parametric equations defining the curve. See Problem 37 in Exercises 9.8. Also, recall for definite integrals that $\int_b^a f(x) \, dx = -\int_a^b f(x) \, dx$. Line integrals possess a similar property. Suppose, as shown in Figure 9.50, that $-C$ denotes the curve having the opposite orientation of C. Then it can be shown that

$$\int_{-C} P \, dx + Q \, dy = -\int_C P \, dx + Q \, dy,$$

or equivalently,

$$\int_{-C} P \, dx + Q \, dy + \int_C P \, dx + Q \, dy = 0. \qquad (8)$$

For example, in part (a) of Example 1, $\int_{-C} xy^2 \, dx = 64$.

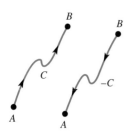

Figure 9.50

■ **Line Integrals in Space** Line integrals of a function G of three variables, $\int_C G(x, y, z) \, dx$, $\int_C G(x, y, z) \, dy$, and $\int_C G(x, y, z) \, ds$, are defined in a manner analogous to Definition 9.9. However, to that list we add a fourth **line integral along a space curve C with respect to z:**

$$\int_C G(x, y, z) \, dz = \lim_{\|P\| \to 0} \sum_{k=1}^n G(x_k^*, y_k^*, z_k^*) \, \Delta z_k. \qquad (9)$$

■ **Method of Evaluation** If C is a smooth curve in 3-space defined by the parametric equations $x = f(t)$, $y = g(t)$, $z = h(t)$, $a \le t \le b$, then the integral in (9) can be evaluated by using

$$\int_C G(x, y, z) \, dz = \int_a^b G(f(t), g(t), h(t)) h'(t) \, dt.$$

The integrals $\int_C G(x, y, z) \, dx$ and $\int_C G(x, y, z) \, dy$ are evaluated in a similar fashion. The line integral with respect to arc length is

$$\int_C G(x, y, z) \, ds = \int_a^b G(f(t), g(t), h(t)) \sqrt{[f'(t)]^2 + [g'(t)]^2 + [h'(t)]^2} \, dt.$$

As in (7), in 3-space we are often concerned with line integrals in the form of a sum:

$$\int_C P(x, y, z)\, dx + Q(x, y, z)\, dy + R(x, y, z)\, dz.$$

Example 5 Line Integral on a Curve in 3-Space

Evaluate $\int_C y\, dx + x\, dy + z\, dz$, where C is the helix $x = 2 \cos t$, $y = 2 \sin t$, $z = t$, $0 \le t \le 2\pi$.

SOLUTION Substituting the expressions for x, y, and z along with $dx = -2 \sin t\, dt$, $dy = 2 \cos t\, dt$, $dz = dt$, we get

$$\int_C y\, dx + x\, dy + z\, dz = \int_0^{2\pi} \underbrace{(-4 \sin^2 t + 4 \cos^2 t)}_{\text{double-angle formula}}\, dt + t\, dt$$

$$= \int_0^{2\pi} (4 \cos 2t + t)\, dt$$

$$= \left(2 \sin 2t + \frac{t^2}{2} \right)\Bigg]_0^{2\pi} = 2\pi^2. \qquad \square$$

We can use the concept of a vector function of several variables to write a general line integral in a compact fashion. For example, suppose the vector-valued function $\mathbf{F}(x, y) = P(x, y)\mathbf{i} + Q(x, y)\mathbf{j}$ is defined along a curve C: $x = f(t)$, $y = g(t)$, $a \le t \le b$, and suppose $\mathbf{r}(t) = f(t)\mathbf{i} + g(t)\mathbf{j}$ is the position vector of points on C. Then the derivative of $\mathbf{r}(t)$,

$$\frac{d\mathbf{r}}{dt} = f'(t)\mathbf{i} + g'(t)\mathbf{j} = \frac{dx}{dt}\mathbf{i} + \frac{dy}{dt}\mathbf{j},$$

prompts us to define $d\mathbf{r} = \dfrac{d\mathbf{r}}{dt}\, dt = dx\mathbf{i} + dy\mathbf{j}$. Since $\mathbf{F}(x, y) \cdot d\mathbf{r} = P(x, y)\, dx + Q(x, y)\, dy$ we can write

$$\int_C P(x, y)\, dx + Q(x, y)\, dy = \int_C \mathbf{F} \cdot d\mathbf{r}. \qquad (10)$$

Similarly, for a line integral on a space curve,

$$\int_C P(x, y, z)\, dx + Q(x, y, z)\, dy + R(x, y, z)\, dz = \int_C \mathbf{F} \cdot d\mathbf{r}, \qquad (11)$$

where $\mathbf{F}(x, y, z) = P(x, y, z)\mathbf{i} + Q(x, y, z)\mathbf{j} + R(x, y, z)\mathbf{j}$ and $d\mathbf{r} = dx\mathbf{i} + dy\mathbf{j} + dz\mathbf{k}$.

■ **Work** In Section 7.3 we saw that the work W done by a constant force \mathbf{F} that causes a straight-line displacement \mathbf{d} of an object is $W = \mathbf{F} \cdot \mathbf{d}$. In beginning courses in calculus or physics it is then shown that the work done in moving an object from $x = a$ to $x = b$ by a force $F(x)$, which varies in magnitude but not in direction, is given by the definite integral $W = \int_a^b F(x)\, dx$. In general, a force field $\mathbf{F}(x, y) = P(x, y)\mathbf{i} + Q(x, y)\mathbf{j}$ acting at each point on a smooth curve C: $x = f(t)$, $y = g(t)$, $a \le t \le b$, varies in both magnitude and direction. See Figure 9.51(a). If A and B are the points $(f(a), g(a))$ and $(f(b), g(b))$, respectively, we ask: What is the work done by \mathbf{F} as its point of application moves along C from A to B? To answer this question, suppose C is divided into n subarcs of lengths Δs_k. On each subarc $\mathbf{F}(x_k^*, y_k^*)$ is a constant force. If, as shown in Figure 9.51(b), the length of the vector $\Delta\mathbf{r}_k = (x_k - x_{k-1})\mathbf{i} + (y_k - y_{k-1})\mathbf{j} = \Delta x_k\mathbf{i} + \Delta y_k\mathbf{j}$ is

(a)

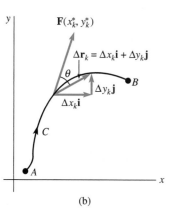

(b)

Figure 9.51

an approximation to the length of the kth subarc, then the approximate work done by \mathbf{F} over the subarc is

$$(\|\mathbf{F}(x_k^*, y_k^*)\| \cos \theta)\|\Delta \mathbf{r}_k\| = \mathbf{F}(x_k^*, y_k^*) \cdot \Delta \mathbf{r}_k$$

$$= P(x_k^*, y_k^*) \, \Delta x_k + Q(x_k^*, y_k^*) \, \Delta y_k.$$

By summing these elements of work and passing to the limit, we naturally define the **work done by F along *C*** as the line integral

$$W = \int_C P(x, y) \, dx + Q(x, y) \, dy \quad \text{or} \quad W = \int_C \mathbf{F} \cdot d\mathbf{r}. \tag{12}$$

Of course, (12) extends to force fields acting at points on a space curve. In this case, work $\int_C \mathbf{F} \cdot d\mathbf{r}$ is defined as in (11).

Now, since

$$\frac{d\mathbf{r}}{dt} = \frac{d\mathbf{r}}{ds} \frac{ds}{dt},$$

we let $d\mathbf{r} = \mathbf{T} \, ds$, where $\mathbf{T} = d\mathbf{r}/ds$ is a unit tangent to C. Hence,

$$W = \int_C \mathbf{F} \cdot d\mathbf{r} = \int_C \mathbf{F} \cdot \mathbf{T} \, ds = \int_C \text{comp}_{\mathbf{T}} \mathbf{F} \, ds. \tag{13}$$

In other words, *the work done by a force* \mathbf{F} *along a curve C is due entirely to the tangential component of* \mathbf{F}.

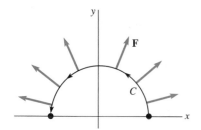

Figure 9.52

Example 6 Work Done by a Force

Find the work done by (a) $\mathbf{F} = x\mathbf{i} + y\mathbf{j}$ and (b) $\mathbf{F} = \frac{3}{4}\mathbf{i} + \frac{1}{2}\mathbf{j}$ along the curve C traced by $\mathbf{r}(t) = \cos t\mathbf{i} + \sin t\mathbf{j}$ from $t = 0$ to $t = \pi$.

SOLUTION (a) The vector function $\mathbf{r}(t)$ gives the parametric equations $x = \cos t$, $y = \sin t$, $0 \leq t \leq \pi$, which we recognize as a half-circle. As seen in Figure 9.52, the force field \mathbf{F} is perpendicular to C at every point. Since the tangential components of \mathbf{F} are zero, the work done along C is zero. To see this we use (12):

$$W = \int_C \mathbf{F} \cdot d\mathbf{r} = \int_C (x\mathbf{i} + y\mathbf{j}) \cdot d\mathbf{r}$$

$$= \int_0^\pi (\cos t\mathbf{i} + \sin t\mathbf{j}) \cdot (-\sin t\mathbf{i} + \cos t\mathbf{j}) \, dt$$

$$= \int_0^\pi (-\cos t \sin t + \sin t \cos t) \, dt = 0.$$

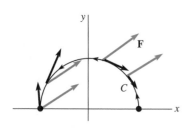

Figure 9.53

(b) In Figure 9.53 the vectors in black are the projections of \mathbf{F} on the unit tangent vectors. The work done by \mathbf{F} is

$$W = \int_C \mathbf{F} \cdot d\mathbf{r} = \int_C \left(\frac{3}{4}\mathbf{i} + \frac{1}{2}\mathbf{j}\right) \cdot d\mathbf{r}$$

$$= \int_0^\pi \left(\frac{3}{4}\mathbf{i} + \frac{1}{2}\mathbf{j}\right) \cdot (-\sin t\mathbf{i} + \cos t\mathbf{j}) \, dt$$

$$= \int_0^\pi \left(-\frac{3}{4} \sin t + \frac{1}{2} \cos t\right) dt$$

$$= \left(\frac{3}{4} \cos t + \frac{1}{2} \sin t\right)\Bigg]_0^\pi = -\frac{3}{2}.$$

The units of work depend on the units of $\|\mathbf{F}\|$ and on the units of distance. ❏

Figure 9.54

(a) Vertical rectangle

(b) "Fence" or "curtain" of varying
height $G(x, y)$ with base C

Figure 9.55

■ **Circulation** A line integral of a vector field **F** around a simple closed curve C is said to be the **circulation** of **F** around C; that is,

$$\text{circulation} = \oint_C \mathbf{F} \cdot d\mathbf{r} = \oint_C \mathbf{F} \cdot \mathbf{T}\, ds.$$

In particular, if **F** is the velocity field of a fluid, then the circulation is a measure of the amount by which the fluid tends to turn the curve C by rotating, or circulating, around it. For example, if **F** is perpendicular to **T** for every (x, y) on C, then $\int_C \mathbf{F} \cdot \mathbf{T}\, ds = 0$, and the curve does not move at all. On the other hand, $\int_C \mathbf{F} \cdot \mathbf{T}\, ds > 0$ and $\int_C \mathbf{F} \cdot \mathbf{T}\, ds < 0$ mean that the fluid tends to rotate C in the counterclockwise and clockwise directions, respectively. See Figure 9.54.

> ### Remarks
>
> In the case of two variables, the line integral with respect to arc length $\int_C G(x, y)\, ds$ can be interpreted in a geometric manner when $G(x, y) \geq 0$ on C. In Definition 9.9 the symbol Δs_k represents the length of the kth subarc on the curve C. But from the figure accompanying that definition, we have the approximation $\Delta s_k = \sqrt{(\Delta x_k)^2 + (\Delta y_k)^2}$. With this interpretation of Δs_k we see from Figure 9.55(a) that the product $G(x_k^*, y_k^*)\, \Delta s_k$ is the area of a vertical rectangle of height $G(x_k^*, y_k^*)$ and width Δs_k. The integral $\int_C G(x, y)\, ds$ then represents the area of one side of a "fence" or "curtain" extending from the curve C in the xy-plane up to the graph of $G(x, y)$ that corresponds to points (x, y) on C. See Figure 9.55(b).

EXERCISES 9.8

Answers to odd-numbered problems begin on page A-50.

In Problems 1–4, evaluate $\int_C G(x, y)\, dx$, $\int_C G(x, y)\, dy$, and $\int_C G(x, y)\, ds$ on the indicated curve C.

1. $G(x, y) = 2xy$; $x = 5 \cos t$, $y = 5 \sin t$, $0 \leq t \leq \pi/4$

2. $G(x, y) = x^3 + 2xy^2 + 2x$; $x = 2t$, $y = t^2$, $0 \leq t \leq 1$

3. $G(x, y) = 3x^2 + 6y^2$; $y = 2x + 1$, $-1 \leq x \leq 0$

4. $G(x, y) = x^2/y^3$; $2y = 3x^{2/3}$, $1 \leq x \leq 8$

In Problems 5 and 6, evaluate $\int_C G(x, y, z)\, dx$, $\int_C G(x, y, z)\, dy$, $\int_C G(x, y, z)\, dz$, and $\int_C G(x, y, z)\, ds$ on the indicated curve C.

5. $G(x, y, z) = z$; $x = \cos t$, $y = \sin t$, $z = t$, $0 \leq t \leq \pi/2$

6. $G(x, y, z) = 4xyz$; $x = \tfrac{1}{3}t^3$, $y = t^2$, $z = 2t$, $0 \leq t \leq 1$

In Problems 7–10, evaluate $\int_C (2x + y)\, dx + xy\, dy$ on the given curve C between $(-1, 2)$ and $(2, 5)$.

7. $y = x + 3$ **8.** $y = x^2 + 1$

9.

Figure 9.56

10.

Figure 9.57

In Problems 11–14, evaluate $\int_C y\, dx + x\, dy$ on the given curve C between $(0, 0)$ and $(1, 1)$.

11. $y = x^2$ **12.** $y = x$

13. C consists of the line segments from $(0, 0)$ to $(0, 1)$ and from $(0, 1)$ to $(1, 1)$.

14. *C* consists of the line segments from $(0, 0)$ to $(1, 0)$ and from $(1, 0)$ to $(1, 1)$.

15. Evaluate $\int_C (6x^2 + 2y^2)\, dx + 4xy\, dy$, where *C* is given by $x = \sqrt{t}$, $y = t$, $4 \leq t \leq 9$.

16. Evaluate $\int_C -y^2\, dx + xy\, dy$, where *C* is given by $x = 2t$, $y = t^3$, $0 \leq t \leq 2$.

17. Evaluate $\int_C 2x^3y\, dx + (3x + y)\, dy$, where *C* is given by $x = y^2$ from $(1, -1)$ to $(1, 1)$.

18. Evaluate $\int_C 4x\, dx + 2y\, dy$, where *C* is given by $x = y^3 + 1$ from $(0, -1)$ to $(9, 2)$.

In Problems 19 and 20, evaluate $\oint_C (x^2 + y^2)\, dx - 2xy\, dy$ on the given closed curve *C*.

19.

$x^2 + y^2 = 4$

Figure 9.58

20.

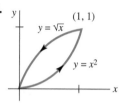

$y = \sqrt{x}$ $(1, 1)$

$y = x^2$

Figure 9.59

In Problems 21 and 22, evaluate $\oint_C x^2y^3\, dx - xy^2\, dy$ on the given closed curve *C*.

21.

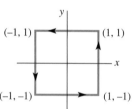

$(-1, 1)$ $(1, 1)$

$(-1, -1)$ $(1, -1)$

Figure 9.60

22.

$(2, 4)$

Figure 9.61

23. Evaluate $\oint_C (x^2 - y^2)\, ds$, where *C* is given by
$$x = 5\cos t, \quad y = 5\sin t, \quad 0 \leq t \leq 2\pi$$

24. Evaluate $\int_{-C} y\, dx - x\, dy$, where *C* is given by
$$x = 2\cos t, \quad y = 3\sin t, \quad 0 \leq t \leq \pi$$

In Problems 25–28, evaluate $\int_C y\, dx + z\, dy + x\, dz$ on the given curve *C* between $(0, 0, 0)$ and $(6, 8, 5)$.

25. *C* consists of the line segments from $(0, 0, 0)$ to $(2, 3, 4)$ and from $(2, 3, 4)$ to $(6, 8, 5)$.

26. $x = 3t$, $y = t^3$, $z = \frac{5}{4}t^2$, $0 \leq t \leq 2$

27.

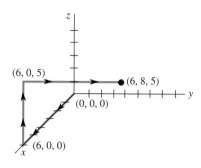

$(6, 0, 5)$ $(6, 8, 5)$

$(0, 0, 0)$

$(6, 0, 0)$

Figure 9.62

28.

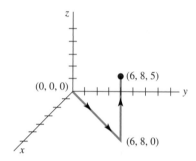

$(0, 0, 0)$ $(6, 8, 5)$

$(6, 8, 0)$

Figure 9.63

In Problems 29 and 30, evaluate $\int_C \mathbf{F} \cdot d\mathbf{r}$.

29. $\mathbf{F}(x, y) = y^3\mathbf{i} - x^2y\mathbf{j}$; $\mathbf{r}(t) = e^{-2t}\mathbf{i} + e^t\mathbf{j}$, $0 \leq t \leq \ln 2$

30. $\mathbf{F}(x, y, z) = e^x\mathbf{i} + xe^{xy}\mathbf{j} + xye^{xyz}\mathbf{k}$; $\mathbf{r}(t) = t\mathbf{i} + t^2\mathbf{j} + t^3\mathbf{k}$, $0 \leq t \leq 1$

31. Find the work done by the force $\mathbf{F}(x, y) = y\mathbf{i} + x\mathbf{j}$ acting along $y = \ln x$ from $(1, 0)$ to $(e, 1)$.

32. Find the work done by the force $\mathbf{F}(x, y) = 2xy\mathbf{i} + 4y^2\mathbf{j}$ acting along the piecewise smooth curve consisting of the line segments from $(-2, 2)$ to $(0, 0)$ and from $(0, 0)$ to $(2, 3)$.

33. Find the work done by the force $\mathbf{F}(x, y) = (x + 2y)\mathbf{i} + (6y - 2x)\mathbf{j}$ acting counterclockwise once around the triangle with vertices $(1, 1)$, $(3, 1)$, and $(3, 2)$.

34. Find the work done by the force $\mathbf{F}(x, y, z) = yz\mathbf{i} + xz\mathbf{j} + xy\mathbf{k}$ acting along the curve given by $\mathbf{r}(t) = t^3\mathbf{i} + t^2\mathbf{j} + t\mathbf{k}$ from $t = 1$ to $t = 3$.

35. Find the work done by a constant force $\mathbf{F}(x, y) = a\mathbf{i} + b\mathbf{j}$ acting counterclockwise once around the circle $x^2 + y^2 = 9$.

36. In an inverse square force field $\mathbf{F} = c\mathbf{r}/\|\mathbf{r}\|^3$, where *c* is a constant and $\mathbf{r} = x\mathbf{i} + y\mathbf{j} + z\mathbf{k}$,* find the work done in moving a particle along the line from $(1, 1, 1)$ to $(3, 3, 3)$.

* Note that the magnitude of \mathbf{F} is inversely proportional to $\|\mathbf{r}\|^2$.

37. Verify that the line integral $\int_C y^2\,dx + xy\,dy$ has the same value on C for each of the following parameterizations:

$$C:\ x = 2t + 1, \quad y = 4t + 2, \quad 0 \le t \le 1$$
$$C:\ x = t^2, \qquad y = 2t^2, \qquad 1 \le t \le \sqrt{3}$$
$$C:\ x = \ln t, \qquad y = 2\ln t, \qquad e \le t \le e^3$$

38. Consider the three curves between $(0, 0)$ and $(2, 4)$:

$$C_1:\ x = t, \qquad y = 2t, \qquad 0 \le t \le 2$$
$$C_2:\ x = t, \qquad y = t^2, \qquad 0 \le t \le 2$$
$$C_3:\ x = 2t - 4, \quad y = 4t - 8, \quad 2 \le t \le 3$$

Show that $\int_{C_1} xy\,ds = \int_{C_3} xy\,ds$, but $\int_{C_1} xy\,ds \ne \int_{C_2} xy\,ds$. Explain.

39. Assume a smooth curve C is described by the vector function $\mathbf{r}(t)$ for $a \le t \le b$. Let acceleration, velocity, and speed be given by $\mathbf{a} = d\mathbf{v}/dt$, $\mathbf{v} = d\mathbf{r}/dt$, and $v = \|\mathbf{v}\|$, respectively. Using Newton's second law $\mathbf{F} = m\mathbf{a}$, show that, in the absence of friction, the work done by \mathbf{F} in moving a particle of constant mass m from point A at $t = a$ to point B at $t = b$ is the same as the change in kinetic energy:

$$K(B) - K(A) = \frac{1}{2} m[v(b)]^2 - \frac{1}{2} m[v(a)]^2$$

$$\left[\textit{Hint: Consider } \frac{d}{dt}\, v^2 = \frac{d}{dt}\, \mathbf{v} \cdot \mathbf{v}. \right]$$

40. If $\rho(x, y)$ is the density of a wire (mass per unit length), then $m = \int_C \rho(x, y)\,ds$ is the mass of the wire. Find the mass of a wire having the shape of the semicircle $x = 1 + \cos t$, $y = \sin t$,

$0 \le t \le \pi$, if the density at a point P is directly proportional to distance from the y-axis.

41. The coordinates of the center of mass of a wire with variable density are given by $\bar{x} = M_y/m$, $\bar{y} = M_x/m$, where

$$m = \int_C \rho(x, y)\,ds, \quad M_x = \int_C y\rho(x, y)\,ds$$

and

$$M_y = \int_C x\rho(x, y)\,ds.$$

Find the center of mass of the wire in Problem 40.

42. A force field $\mathbf{F}(x, y)$ acts at each point on the curve C, which is the union of C_1, C_2, and C_3 shown in Figure 9.64. $\|\mathbf{F}\|$ is measured in pounds and distance is measured in feet using the scale given in the figure. Use the representative vectors shown to approximate the work done by \mathbf{F} along C. [*Hint:* Use $W = \int_C \mathbf{F} \cdot \mathbf{T}\,ds$.]

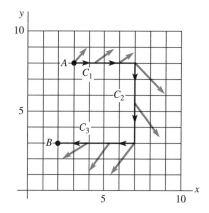

Figure 9.64

9.9 Line Integrals Independent of the Path

The following concepts play an important role in the discussion that follows.

■ **Differential—Functions of Two Variables** The differential of a function of two variables $\phi(x, y)$ is

$$d\phi = \frac{\partial \phi}{\partial x}\,dx + \frac{\partial \phi}{\partial y}\,dy.$$

A differential expression $P(x, y)dx + Q(x, y)dy$ is said to be an **exact differential** if there exists a function $\phi(x, y)$ such that

$$d\phi = P(x, y)dx + Q(x, y)dy.$$

For example, the expression $x^2y^3dx + x^3y^2dy$ is an exact differential since it is the differential of $\phi(x, y) = \frac{1}{3}x^3y^3$. Verify this. On the other hand, $(2y^2 - 2y)dx + (2xy - x)dy$ is not an exact differential.

■ **Differential—Functions of Three Variables** The differential of a function of three variables $\phi(x, y, z)$ is

$$d\phi = \frac{\partial \phi}{\partial x}\, dx + \frac{\partial \phi}{\partial y}\, dy + \frac{\partial \phi}{\partial z}\, dz.$$

A differential expression $P(x, y)dx + Q(x, y)dy + R(x, y, z)dz$ is an exact differential if there exists a function $\phi(x, y, z)$ such that

$$d\phi = P(x, y, z)dx + Q(x, y, z)dy + R(x, y, z)dz.$$

■ **Path Independence** The value of a line integral generally depends on the curve or **path** between two points A and B. However, there are exceptions. A line integral whose value is the same for *every* curve connecting A and B is said to be **independent of the path.**

As in the last section, we begin with a discussion of line integrals in the plane.

Example 1 An Integral Independent of the Path

The integral $\int_C y\, dx + x\, dy$ has the same value on each path C between $(0, 0)$ and $(1, 1)$ shown in Figure 9.65. You may recall from Problems 11–14 of Exercises 9.8 that on these paths

$$\int_C y\, dx + x\, dy = 1.$$

In Example 2 we shall prove that the given integral is independent of the path.

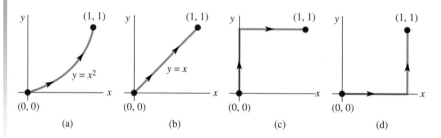

Figure 9.65

■ **A Fundamental Theorem** The next theorem establishes an important relationship between the seemingly disparate notions of path independence and exact differentials. In addition, it provides a means of evaluating path-independent line integrals in a manner analogous to the fundamental theorem of calculus: $\int_a^b f'(x)\, dx = f(b) - f(a)$.

THEOREM 9.8 Fundamental Theorem for Line Integrals

Suppose there exists a function $\phi(x, y)$ such that $d\phi = P\, dx + Q\, dy$; that is, $P\, dx + Q\, dy$ is an exact differential. Then $\int_C P\, dx + Q\, dy$ depends on only the endpoints A and B of the path C and

$$\int_C P\, dx + Q\, dy = \phi(B) - \phi(A).$$

PROOF Let C be a smooth path defined parametrically by $x = f(t)$, $y = g(t)$, $a \le t \le b$, and let the coordinates of A and B be $(f(a), g(a))$ and $(f(b), g(b))$, respectively. Then, by the Chain Rule,

$$\int_C P\,dx + Q\,dy = \int_a^b \left(\frac{\partial \phi}{\partial x} \frac{dx}{dt} + \frac{\partial \phi}{\partial y} \frac{dy}{dt} \right) dt$$

$$= \int_a^b \frac{d\phi}{dt}\,dt = \phi(f(t), g(t)) \Big]_a^b$$

$$= \phi(f(b), g(b)) - \phi(f(a), g(a))$$

$$= \phi(B) - \phi(A). \qquad \square$$

Two observations are in order. The theorem is also valid for piecewise smooth curves, but the foregoing proof would have to be modified by considering each smooth arc of the curve C. Furthermore, the converse of the theorem is also true. Hence,

$$\int_C P\,dx + Q\,dy \text{ is independent of the path if and only if } P\,dx + Q\,dy \text{ is an exact differential.} \qquad (1)$$

■ **Notation** A line integral $\int_C P\,dx + Q\,dy$, which is independent of the path between the endpoints A and B, is often written

$$\int_A^B P\,dx + Q\,dy.$$

Example 2 Using Theorem 9.8

In Example 1 note that $d(xy) = y\,dx + x\,dy$; that is, $y\,dx + x\,dy$ is an exact differential. Hence, $\int_C y\,dx + x\,dy$ is independent of the path between any two points A and B. Specifically, if A and B are, respectively, $(0, 0)$, and $(1, 1)$, we then have, from Theorem 9.8,

$$\int_{(0,0)}^{(1,1)} y\,dx + x\,dy = \int_{(0,0)}^{(1,1)} d(xy) = xy \Big]_{(0,0)}^{(1,1)} = 1. \qquad \square$$

In order to be able to state the next result we need to consider a particular kind of region in the plane. We say that a region R of the plane is **simply connected** if:

- R is connected; that is, every pair of points in the region can be joined by a piecewise smooth curve that lies entirely in R, and
- every simple closed curve C lying entirely within R can be shrunk, or contracted, to a point without leaving R.

The last statement means that if C is any simple closed curve lying entirely in R, then the region in the interior of C also lies entirely in R. Roughly put, a simply connected region has no holes in it. In Figure 9.66(a) we have illustrated a simply connected region. The representative simple closed curve C could be shrunk to a point without leaving the region. In Figure 9.66(b) the region shown has three holes in it; since the representative curve C surrounds one of the holes, it could not be shrunk to a point without leaving the region. This last region is said to be **multiply connected.** A simply connected region is said to be **open** if it contains no boundary points.

(a)

(b)

Figure 9.66

■ **Test for Path Independence in the Plane** In view of the statement given in (1), the same criteria for determining an exact differential become the criteria, or test, for path independence in the *xy*-plane.

THEOREM 9.9 Test for Path Independence

Let P and Q have continuous first partial derivatives in an open simply connected region. Then $\int_C P\, dx + Q\, dy$ is independent of the path C if and only if

$$\frac{\partial P}{\partial y} = \frac{\partial Q}{\partial x}$$

for all (x, y) in the region.

Example 3 **A Path-Dependent Integral**

Show that the integral $\int_C (x^2 - 2y^3)\, dx + (x + 5y)\, dy$ is not independent of the path C.

SOLUTION From $P = x^2 - 2y^3$ and $Q = x + 5y$, we find

$$\frac{\partial P}{\partial y} = -6y^2 \quad \text{and} \quad \frac{\partial Q}{\partial x} = 1.$$

Since $\partial P/\partial y \neq \partial Q/\partial x$, it follows from Theorem 9.9 that the integral is not independent of the path. In other words, the differential expression $(x^2 - 2y^3)\, dx + (x + 5y)\, dy$ is not an exact differential. ❑

Example 4 **An Integral Independent of the Path**

Show that $\int_C (y^2 - 6xy + 6)\, dx + (2xy - 3x^2)\, dy$ is independent of any path C between $(-1, 0)$ and $(3, 4)$. Evaluate.

SOLUTION Identifying $P = y^2 - 6xy + 6$ and $Q = 2xy - 3x^2$ yields

$$\frac{\partial P}{\partial y} = 2y - 6x \quad \text{and} \quad \frac{\partial Q}{\partial x} = 2y - 6x.$$

Since $\partial P/\partial y = \partial Q/\partial x$, the integral is independent of the path and so there exists a function ϕ such that $\partial\phi/\partial x = y^2 - 6xy + 6$ and $\partial\phi/\partial y = 2xy - 3x^2$. To find the function ϕ we can integrate either $\partial\phi/\partial x$ or $\partial\phi/\partial y$. Integrating $\partial\phi/\partial x$ with respect to x gives $\phi = y^2x - 3x^2y + 6x + g(y)$, where $g(y)$ is the "constant" of integration. Taking the partial derivative of this last expression with respect to y and setting the result equal to Q (that is, $\partial\phi/\partial y$) then give

$$\frac{\partial\phi}{\partial y} = 2yx - 3x^2 + g'(y) = 2yx - 3x^2,$$

which implies $g'(y) = 0$ and so $g(y) = C$, a constant. But since the differentials

$$d(y^2x - 3x^2y + 6x + C) \quad \text{and} \quad d(y^2x - 3x^2y + 6x)$$

both yield $(y^2 - 6xy + 6)\,dx + (2xy - 3x^2)\,dx$, we can drop the constant C and take $\phi = xy^2 - 3x^2y + 6x$. It follows from the fundamental theorem for line integrals that

$$\int_{(-1,\,0)}^{(3,\,4)} (y^2 - 6xy + 6)\,dx + (2xy - 3x^2)\,dy = \int_{(-1,\,0)}^{(3,\,4)} d(xy^2 - 3x^2y + 6x)$$

$$= (xy^2 - 3x^2y + 6x)\Big]_{(-1,\,0)}^{(3,\,4)}$$

$$= (48 - 108 + 18) - (-6) = -36.$$

ALTERNATIVE SOLUTION Since the integral is independent of the path, we can integrate along any convenient curve connecting the given points. In particular, $y = x + 1$ is such a curve. Using x as a parameter then gives

$$\int_C (y^2 - 6xy + 6)\,dx + (2xy - 3x^2)\,dy$$

$$= \int_{-1}^{3} [(x + 1)^2 - 6x(x + 1) + 6]\,dx + [2x(x + 1) - 3x^2]\,dx$$

$$= \int_{-1}^{3} (-6x^2 - 2x + 7)\,dx = -36. \qquad \square$$

■ **Conservative Vector Fields** If $\int_C P\,dx + Q\,dy$ is independent of the path C, we know there exists a function ϕ such that

$$d\phi = \frac{\partial \phi}{\partial x}\,dx + \frac{\partial \phi}{\partial y}\,dy = P\,dx + Q\,dy$$

$$= (P\mathbf{i} + Q\mathbf{j}) \cdot (dx\mathbf{i} + dy\mathbf{j}) = \mathbf{F} \cdot d\mathbf{r},$$

where $\mathbf{F} = P\mathbf{i} + Q\mathbf{j}$ is a vector field and $P = \partial\phi/\partial x$, $Q = \partial\phi/\partial y$. In other words, the vector field \mathbf{F} is a gradient of the function ϕ. Since $\mathbf{F} = \nabla\phi$, \mathbf{F} is said to be a **gradient field** and the function ϕ is then said to be a **potential function** for \mathbf{F}. In a gradient force field \mathbf{F}, the work done by the force upon a particle moving from position A to position B is the same for all paths between these points. Moreover, the work done by the force along a closed path is *zero*. See Problem 29 in Exercises 9.9. For this reason, such a force field is also said to be **conservative.** In a conservative field \mathbf{F} the *law of conservation of mechanical energy* holds: For a particle moving along a path in a conservative field,

kinetic energy + potential energy = constant

See Problem 31 in Exercises 9.9. In a simply connected region the hypotheses of Theorem 9.9 imply that a force field $\mathbf{F}(x, y) = P(x, y)\mathbf{i} + Q(x, y)\mathbf{j}$ is a gradient field (that is, conservative) if and only if

$$\frac{\partial P}{\partial y} = \frac{\partial Q}{\partial x}.$$

Example 5 Gradient Field

Show that the vector field $\mathbf{F} = (y^2 + 5)\mathbf{i} + (2xy - 8)\mathbf{j}$ is a gradient field. Find a potential function for \mathbf{F}.

SOLUTION By identifying $P = y^2 + 5$ and $Q = 2xy - 8$, we see that

$$\frac{\partial P}{\partial y} = \frac{\partial Q}{\partial x} = 2y.$$

Hence, **F** is a gradient field and so there exists a potential function ϕ satisfying

$$\frac{\partial \phi}{\partial x} = y^2 + 5 \quad \text{and} \quad \frac{\partial \phi}{\partial y} = 2xy - 8.$$

Proceeding as in Example 4, we find that $\phi = xy^2 - 8y + 5x$.

CHECK $\nabla \phi = \dfrac{\partial \phi}{\partial x}\,\mathbf{i} + \dfrac{\partial \phi}{\partial y}\,\mathbf{j} = (y^2 + 5)\mathbf{i} + (2xy - 8)\mathbf{j}.$ ❑

■ **Test for Path Independence in Space** If C is a space curve, then a line integral $\int_C \mathbf{F} \cdot d\mathbf{r}$ is independent of the path whenever the differential expression $P(x, y, z)\, dx + Q(x, y, z)\, dy + R(x, y, z)\, dz$ is an exact differential. The three-dimensional analogue of Theorem 9.9 is given next.

THEOREM 9.10 Test for Path Independence

Let P, Q, and R have continuous first partial derivatives in an open simply connected region of space. Then $\int_C P\, dx + Q\, dy + R\, dz$ is independent of the path C if and only if

$$\frac{\partial P}{\partial y} = \frac{\partial Q}{\partial x}, \quad \frac{\partial P}{\partial z} = \frac{\partial R}{\partial x}, \quad \text{and} \quad \frac{\partial Q}{\partial z} = \frac{\partial R}{\partial y}.$$

Example 6 **An Integral Independent of the Path**

Show that $\int_C (y + yz)\, dx + (x + 3z^3 + xz)\, dy + (9yz^2 + xy - 1)\, dz$ is independent of any path C between $(1, 1, 1)$ and $(2, 1, 4)$. Evaluate.

SOLUTION With the identifications

$$P = y + yz, \quad Q = x + 3z^3 + xz, \quad R = 9yz^2 + xy - 1,$$

we see that

$$\frac{\partial P}{\partial y} = 1 + z = \frac{\partial Q}{\partial x}, \quad \frac{\partial P}{\partial z} = y = \frac{\partial R}{\partial x}, \quad \frac{\partial Q}{\partial z} = 9z^2 + x = \frac{\partial R}{\partial y}.$$

From Theorem 9.10 we conclude that the integral is independent of the path. Moreover, $(y + yz)\, dx + (x + 3z^3 + xz)\, dy + (9yz^2 + xy - 1)\, dz$ is an exact differential and so there exists a function $\phi(x, y, z)$ such that

$$\frac{\partial \phi}{\partial x} = P, \quad \frac{\partial \phi}{\partial y} = Q, \quad \frac{\partial \phi}{\partial z} = R.$$

Integrating the first of these three equations with respect to x gives

$$\phi = xy + xyz + g(y, z).$$

The derivative of this last expression with respect to y must then be equal to Q:

$$\frac{\partial \phi}{\partial y} = x + xz + \frac{\partial g}{\partial y} = x + 3z^3 + xz.$$

Thus, $$\frac{\partial g}{\partial y} = 3z^3 \quad \text{and so} \quad g = 3yz^3 + h(z)$$

Consequently, $\phi = xy + xyz + 3yz^3 + h(z)$. The partial derivative of this last expression with respect to z must now be equal to R:

$$\frac{\partial \phi}{\partial z} = xy + 9yz^2 + h'(z) = 9yz^2 + xy - 1.$$

From this we get $h'(z) = -1$ and $h(z) = -z + C$. Disregarding C, we can write

$$\phi = xy + xyz + 3yz^3 - z. \tag{2}$$

Finally, we obtain

$$\int_{(1,1,1)}^{(2,1,4)} (y + yz)\, dx + (x + 3z^3 + xz)\, dy + (9yz^2 + xy - 1)\, dz$$

$$= \int_{(1,1,1)}^{(2,1,4)} d(xy + xyz + 3yz^3 - z)$$

$$= (xy + xyz + 3yz^3 - z)\Big|_{(1,1,1)}^{(2,1,4)} = 198 - 4 = 194. \qquad \square$$

In Example 6 the force field defined by

$$\mathbf{F}(x,\ y,\ z) = (y + yz)\mathbf{i} + (x + 3z^3 + xz)\mathbf{j} + (9yz^2 + xy - 1)\mathbf{k}$$

is a conservative field or a gradient field, since the proof that the differential form $(y + yz)\, dx + (x + 3z^3 + xz)\, dy + (9yz^2 + xy - 1)\, dz$ is exact also proves that $\mathbf{F} = \nabla\phi$, where ϕ is the potential function given in (2). In general, it follows from Theorem 9.10 and (1) of Section 9.7 that in a simply connected region of space, a force field $\mathbf{F}(x,\ y,\ z) = P(x,\ y,\ z)\mathbf{i} + Q(x,\ y,\ z)\mathbf{j} + R(x,\ y,\ z)\mathbf{k}$ is conservative if and only if curl $\mathbf{F} = \mathbf{0}$ (that is, $\nabla \times \mathbf{F} = \mathbf{0}$).

Remarks

A frictional force such as air resistance is *nonconservative*. Nonconservative forces are *dissipative* in that their action reduces kinetic energy without a corresponding increase in potential energy. In other words, if the work done $\int_C \mathbf{F} \cdot d\mathbf{r}$ depends on the path, then \mathbf{F} is nonconservative.

EXERCISES 9.9

Answers to odd-numbered problems begin on page A-50.

In Problems 1–10, show that the given integral is independent of the path. Evaluate in two ways: (a) find a function ϕ such that $d\phi = P\, dx + Q\, dy$, and (b) integrate along any convenient path between the points.

1. $\displaystyle\int_{(0,0)}^{(2,2)} x^2\, dx + y^2\, dy$

2. $\displaystyle\int_{(1,1)}^{(2,4)} 2xy\, dx + x^2\, dy$

3. $\displaystyle\int_{(1,0)}^{(3,2)} (x + 2y)\, dx + (2x - y)\, dy$

4. $\displaystyle\int_{(0,0)}^{(\pi/2,\,0)} \cos x \cos y\, dx + (1 - \sin x \sin y)\, dy$

5. $\displaystyle\int_{(4,1)}^{(4,4)} \frac{-y\, dx + x\, dy}{y^2}$ on any path not crossing the x-axis

6. $\displaystyle\int_{(1,0)}^{(3,4)} \frac{x\, dx + y\, dy}{\sqrt{x^2 + y^2}}$ on any path not through the origin

7. $\displaystyle\int_{(1,2)}^{(3,6)} (2y^2x - 3)\, dx + (2yx^2 + 4)\, dy$

8. $\displaystyle\int_{(-1,1)}^{(0,0)} (5x + 4y)\, dx + (4x - 8y^3)\, dy$

9. $\int_{(0,0)}^{(2,8)} (y^3 + 3x^2y)\, dx + (x^3 + 3y^2x + 1)\, dy$

10. $\int_{(-2,0)}^{(1,0)} (2x - y \sin xy - 5y^4)\, dx - (20xy^3 + x \sin xy)\, dy$

In Problems 11–16, determine whether the given vector field is a gradient field. If so, find a potential function ϕ for **F**.

11. $\mathbf{F}(x, y) = (4x^3y^3 + 3)\mathbf{i} + (3x^4y^2 + 1)\mathbf{j}$

12. $\mathbf{F}(x, y) = 2xy^3\mathbf{i} + 3y^2(x^2 + 1)\mathbf{j}$

13. $\mathbf{F}(x, y) = y^2 \cos xy^2\mathbf{i} - 2xy \sin xy^2\mathbf{j}$

14. $\mathbf{F}(x, y) = (x^2 + y^2 + 1)^{-2}(x\mathbf{i} + y\mathbf{j})$

15. $\mathbf{F}(x, y) = (x^3 + y)\mathbf{i} + (x + y^3)\mathbf{j}$

16. $\mathbf{F}(x, y) = 2e^{2y}\mathbf{i} + xe^{2y}\mathbf{j}$

In Problems 17 and 18, find the work done by the force $\mathbf{F}(x, y) = (2x + e^{-y})\mathbf{i} + (4y - xe^{-y})\mathbf{j}$ along the indicated curve.

17.

Figure 9.67

18.

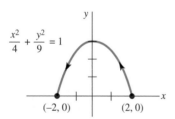

Figure 9.68

In Problems 19–24, show that the given integral is independent of the path. Evaluate.

19. $\int_{(1,1,1)}^{(2,4,8)} yz\, dx + xz\, dy + xy\, dz$

20. $\int_{(0,0,0)}^{(1,1,1)} 2x\, dx + 3y^2\, dy + 4z^3\, dz$

21. $\int_{(1,0,0)}^{(2,\pi/2,1)} (2x \sin y + e^{3z})\, dx + x^2 \cos y\, dy + (3xe^{3z} + 5)\, dz$

22. $\int_{(1,2,1)}^{(3,4,1)} (2x + 1)\, dx + 3y^2\, dy + \dfrac{1}{z}\, dz$

23. $\int_{(1,1,\ln 3)}^{(2,2,\ln 3)} e^{2z}\, dx + 3y^2\, dy + 2xe^{2z}\, dz$

24. $\int_{(-2,3,1)}^{(0,0,0)} 2xz\, dx + 2yz\, dy + (x^2 + y^2)\, dz$

In Problems 25 and 26, evaluate $\int_C \mathbf{F} \cdot d\mathbf{r}$.

25. $\mathbf{F}(x, y, z) = (y - yz \sin x)\mathbf{i} + (x + z \cos x)\mathbf{j} + y \cos x\mathbf{k}$; $\mathbf{r}(t) = 2t\mathbf{i} + (1 + \cos t)^2\mathbf{j} + 4 \sin^3 t\mathbf{k}, \quad 0 \le t \le \pi/2$

26. $\mathbf{F}(x, y, z) = (2 - e^z)\mathbf{i} + (2y - 1)\mathbf{j} + (2 - xe^z)\mathbf{k}$; $\mathbf{r}(t) = t\mathbf{i} + t^2\mathbf{j} + t^3\mathbf{k}, \quad (-1, 1, -1)$ to $(2, 4, 8)$

27. The inverse square law of gravitational attraction between two masses m_1 and m_2 is given by $\mathbf{F} = -Gm_1m_2\mathbf{r}/\|\mathbf{r}\|^3$, where $\mathbf{r} = x\mathbf{i} + y\mathbf{j} + z\mathbf{k}$. Show that \mathbf{F} is conservative. Find a potential function for \mathbf{F}.

28. Find the work done by the force $\mathbf{F}(x, y, z) = 8xy^3z\mathbf{i} + 12x^2y^2z\mathbf{j} + 4x^2y^3\mathbf{k}$ acting along the helix $\mathbf{r}(t) = 2 \cos t\mathbf{i} + 2 \sin t\mathbf{j} + t\mathbf{k}$ from $(2, 0, 0)$ to $(1, \sqrt{3}, \pi/3)$. From $(2, 0, 0)$ to $(0, 2, \pi/2)$. [*Hint:* Show that \mathbf{F} is conservative.]

29. If \mathbf{F} is a conservative force field, show that the work done along any simple closed path is zero.

30. A particle in the plane is attracted to the origin with a force $\mathbf{F} = \|\mathbf{r}\|^n\mathbf{r}$, where n is a positive integer and $\mathbf{r} = x\mathbf{i} + y\mathbf{j}$ is the position vector of the particle. Show that \mathbf{F} is conservative. Find the work done in moving the particle between (x_1, y_1) and (x_2, y_2).

31. Suppose \mathbf{F} is a conservative force field with potential function ϕ. In physics the function $p = -\phi$ is called **potential energy**. Since $\mathbf{F} = -\nabla p$, Newton's second law becomes

$$m\mathbf{r}'' = -\nabla p \quad \text{or} \quad m\frac{d\mathbf{v}}{dt} + \nabla p = \mathbf{0}.$$

By integrating $m\dfrac{d\mathbf{v}}{dt} \cdot \dfrac{d\mathbf{r}}{dt} + \nabla p \cdot \dfrac{d\mathbf{r}}{dt} = 0$ with respect to t, derive the law of conservation of mechanical energy: $\frac{1}{2}mv^2 + p = $ constant. [*Hint:* See Problem 39 in Exercises 9.8.]

32. Suppose that C is a smooth curve between points A (at $t = a$) and B (at $t = b$) and that p is potential energy, defined in Problem 31. If \mathbf{F} is a conservative force field and $K = \frac{1}{2}mv^2$ is kinetic energy, show that $p(B) + K(B) = p(A) + K(A)$.

9.10 Review of Double Integrals

In Section 9.8 we reviewed the five steps leading to the definition of the definite integral $\int_a^b f(x)\, dx$. The analogous five steps that lead to the concept of the *two-dimensional definite integral*, known simply as the **double integral** of a function f of two variables, will be given next.

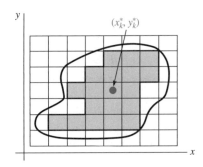

<thinking_since_2023>
Wait, let me place images appropriately.
</thinking_since_2023>

$z = f(x, y)$

1. Let f be defined in a closed and bounded region R.
2. By means of a grid of vertical and horizontal lines parallel to the coordinate axes, form a partition P of R into n rectangular subregions R_k of areas ΔA_k that lie entirely in R.
3. Let $\|P\|$ be the **norm** of the partition or the length of the longest diagonal of the R_k.
4. Choose a point (x_k^*, y_k^*) in each subregion R_k.
5. Form the sum $\sum_{k=1}^{n} f(x_k^*, y_k^*)\, \Delta A_k$.

Thus we have the following definition:

DEFINITION 9.10 The Double Integral

Let f be a function of two variables defined on a closed region R. Then the **double integral of f over R** is given by

$$\iint_R f(x, y)\, dA = \lim_{\|P\| \to 0} \sum_{k=1}^{n} f(x_k^*, y_k^*)\, \Delta A_k. \tag{1}$$

■ **Integrability** If the limit in (1) exists, we say that f is **integrable** over R and that R is the **region of integration.** When f is continuous on R, then f is necessarily integrable over R.

■ **Area** When $f(x, y) = 1$ on R, then $\lim_{\|P\| \to 0} \sum_{k=1}^{n} \Delta A_k$ simply gives the **area** A of the region; that is,

$$A = \iint_R dA. \tag{2}$$

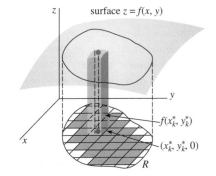

Figure 9.69

surface $z = f(x, y)$

$f(x_k^*, y_k^*)$

$(x_k^*, y_k^*, 0)$

R

■ **Volume** If $f(x, y) \geq 0$ on R, then, as shown in Figure 9.69, the product $f(x_k^*, y_k^*)\, \Delta A_k$ can be interpreted as the volume of a rectangular prism of height $f(x_k^*, y_k^*)$ and base of area ΔA_k. The summation of volumes $\sum_{k=1}^{n} f(x_k^*, y_k^*)\, \Delta A_k$ is an approximation to the **volume** V of the solid *above* the region R and *below* the surface $z = f(x, y)$. The limit of this sum as $\|P\| \to 0$, if it exists, gives the exact volume of this solid; that is, if f is non-negative on R, then

$$V = \iint_R f(x, y)\, dA. \tag{3}$$

■ **Properties** The double integral possesses the following properties:

THEOREM 9.11 Properties of Double Integrals

Let f and g be functions of two variables that are integrable over a region R. Then

(i) $\displaystyle \iint_R k f(x, y)\, dA = k \iint_R f(x, y)\, dA$, where k is any constant

(ii) $\displaystyle \iint_R [f(x, y) \pm g(x, y)]\, dA = \iint_R f(x, y)\, dA \pm \iint_R g(x, y)\, dA$

continued

$R = R_1 \cup R_2$

Figure 9.70

(a) Type I region

(b) Type II region

Figure 9.71

$$(iii) \iint_R f(x, y) \, dA = \iint_{R_1} f(x, y) \, dA + \iint_{R_2} f(x, y) \, dA, \text{ where } R_1 \text{ and } R_2$$

are subregions of R that do not overlap and $R = R_1 \cup R_2$

Part (*iii*) of Theorem 9.11 is the two-dimensional equivalent of $\int_a^b f(x) \, dx = \int_a^c f(x) \, dx + \int_c^b f(x) \, dx$. Figure 9.70 illustrates the division of a region into subregions R_1 and R_2 for which $R = R_1 \cup R_2$. R_1 and R_2 can have no points in common except possibly on their common border. Furthermore, Theorem 9.11(*iii*) extends to any finite number of nonoverlapping subregions whose union is R.

■ **Regions of Type I and II** The region shown in Figure 9.71(a),

$$R: a \leq x \leq b, \qquad g_1(x) \leq y \leq g_2(x),$$

where the boundary functions g_1 and g_2 are continuous, is called a **region of Type I.** In Figure 9.71(b), the region

$$R: c \leq y \leq d, \qquad h_1(y) \leq x \leq h_2(y),$$

where h_1 and h_2 are continuous, is called a **region of Type II.**

■ **Iterated Integrals** Since the partial integral $\int_{g_1(x)}^{g_2(x)} f(x, y) \, dy$ is a function of x alone, we may in turn integrate the resulting function now with respect to x. If f is continuous on a region of Type I, we define an **iterated integral** of f over the region by

$$\int_a^b \int_{g_1(x)}^{g_2(x)} f(x, y) \, dy \, dx = \int_a^b \left[\int_{g_1(x)}^{g_2(x)} f(x, y) \, dy \right] dx. \tag{4}$$

The basic idea in (4) is to carry out *successive integrations*. The partial integral gives a function of x, which is then integrated in the usual manner from $x = a$ to $x = b$. The end result of both integrations will be a real number. In a similar manner, we define an iterated integral of a continuous function f over a region of Type II by

$$\int_c^d \int_{h_1(y)}^{h_2(y)} f(x, y) \, dx \, dy = \int_c^d \left[\int_{h_1(y)}^{h_2(y)} f(x, y) \, dx \right] dy. \tag{5}$$

■ **Evaluation of Double Integrals** Iterated integrals provide the means for evaluating a double integral $\iint_R f(x, y) \, dA$ over a region of Type I or Type II or a region that can be expressed as a union of a finite number of these regions.

THEOREM 9.12 Evaluation of Double Integrals

Let f be continuous on a region R.
 (*i*) If R is of Type I, then

$$\iint_R f(x, y) \, dA = \int_a^b \int_{g_1(x)}^{g_2(x)} f(x, y) \, dy \, dx. \tag{6}$$

 (*ii*) If R is of Type II, then

$$\iint_R f(x, y) \, dA = \int_c^d \int_{h_1(y)}^{h_2(y)} f(x, y) \, dx \, dy. \tag{7}$$

Theorem 9.12 is the double integral analogue of the fundamental theorem of calculus. While Theorem 9.12 is difficult to prove, we can get some intuitive feeling for its significance by considering volumes. Let R be a Type I region and $z = f(x, y)$ be continuous and nonnegative on R. The area A of the vertical plane, as shown in Figure 9.72, is the area under the trace of the surface $z = f(x, y)$ in the plane $x =$ constant and hence is given by the partial integral

$$A(x) = \int_{g_1(x)}^{g_2(x)} f(x, y) \, dy.$$

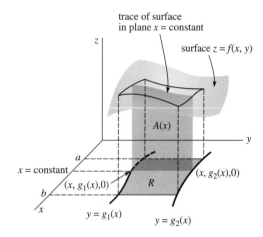

Figure 9.72

By summing all these areas from $x = a$ to $x = b$, we obtain the volume V of the solid above R and below the surface:

$$V = \int_a^b A(x) \, dx = \int_a^b \int_{g_1(x)}^{g_2(x)} f(x, y) \, dy \, dx.$$

But, as we have already seen in (3), this volume is also given by the double integral

$$V = \iint_R f(x, y) \, dA.$$

Example 1 Evaluation of a Double Integral

Evaluate the double integral $\iint_R e^{x+3y} \, dA$ over the region bounded by the graphs of $y = 1$, $y = 2$, $y = x$, and $y = -x + 5$.

SOLUTION As seen in Figure 9.73, the region is of Type II; hence, by (7) we integrate first with respect to x from the left boundary $x = y$ to the right boundary $x = 5 - y$:

$$\iint_R e^{x+3y} \, dA = \int_1^2 \int_y^{5-y} e^{x+3y} \, dx \, dy$$

$$= \int_1^2 e^{x+3y} \Big]_y^{5-y} \, dy$$

$$= \int_1^2 (e^{5+2y} - e^{4y}) \, dy = \left[\frac{1}{2} e^{5+2y} - \frac{1}{4} e^{4y} \right]_1^2$$

$$= \frac{1}{2} e^9 - \frac{1}{4} e^8 - \frac{1}{2} e^7 + \frac{1}{4} e^4 \approx 2771.64. \qquad \square$$

Figure 9.73

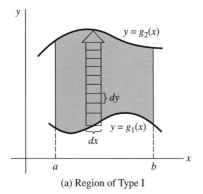

(a) Region of Type I

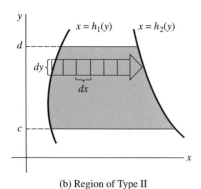

(b) Region of Type II

Figure 9.74

(a) Type I region

(b) Type II region

Figure 9.75

As an aid in reducing a double integral to an iterated integral with correct limits of integration, it is useful to visualize, as suggested in the foregoing discussion, the double integral as a double summation process. Over a Type I region the iterated integral $\int_a^b \int_{g_1(x)}^{g_2(x)} f(x, y)\, dy\, dx$ is first a summation in the y-direction. Pictorially, this is indicated by the vertical arrow in Figure 9.74(a); the typical rectangle in the arrow has area $dy\, dx$. The dy placed before the dx signifies that the "volumes" $f(x, y)\, dy\, dx$ of prisms built up on the rectangles are summed vertically with respect to y from the lower boundary curve g_1 to the upper boundary curve g_2. The dx following the dy signifies that the result of each vertical summation is then summed horizontally with respect to x from left ($x = a$) to right ($x = b$). Similar remarks hold for double integrals over regions of Type II. See Figure 9.74(b). Recall from (2) that when $f(x, y) = 1$, the double integral $A = \iint_R dA$ gives the area of the region. Thus, Figure 9.74(a) shows that $\int_a^b \int_{g_1(x)}^{g_2(x)} dy\, dx$ adds the rectangular *areas* vertically and then horizontally, whereas Figure 9.74(b) shows that $\int_c^d \int_{h_1(y)}^{h_2(y)} dx\, dy$ adds the rectangular areas horizontally and then vertically.

■ **Reversing the Order of Integration** A problem may become easier when the order of integration is **changed** or **reversed.** Also, some iterated integrals that may be impossible to evaluate using one order of integration can perhaps be evaluated using the reverse order of integration.

Example 2 **Reversing the Order of Integration**

Evaluate $\iint_R xe^{y^2}\, dA$ over the region R in the first quadrant bounded by the graphs of $y = x^2$, $x = 0$, $y = 4$.

SOLUTION When the region is viewed as Type I, we have from Figure 9.75(a), $0 \le x \le 2$, $x^2 \le y \le 4$, and so

$$\iint_R xe^{y^2}\, dA = \int_0^2 \int_{x^2}^4 xe^{y^2}\, dy\, dx.$$

The difficulty here is that the partial integral $\int_{x^2}^4 xe^{y^2}\, dy$ cannot be evaluated, since e^{y^2} has no elementary antiderivative with respect to y. However, as we see in Figure 9.75(b), we can interpret the same region as a Type II region defined by $0 \le y \le 4$, $0 \le x \le \sqrt{y}$. Hence, from (7)

$$\iint_R xe^{y^2}\, dA = \int_0^4 \int_0^{\sqrt{y}} xe^{y^2}\, dx\, dy$$

$$= \int_0^4 \frac{x^2}{2} e^{y^2} \Big]_0^{\sqrt{y}} dy$$

$$= \int_0^4 \frac{1}{2} ye^{y^2}\, dy = \frac{1}{4} e^{y^2} \Big]_0^4 = \frac{1}{4}(e^{16} - 1). \qquad \square$$

■ **Laminas with Variable Density—Center of Mass** If ρ is a constant density (mass per unit area), then the mass of the lamina coinciding with a region bounded by the graphs of $y = f(x)$, the x-axis, and the lines $x = a$ and $x = b$ is

$$m = \lim_{\|P\| \to 0} \sum_{k=1}^n \rho f(x_k^*)\, \Delta x_k = \int_a^b \rho f(x)\, dx. \qquad (8)$$

If a lamina corresponding to a region R has a *variable density* $\rho(x, y)$, where ρ is nonnegative and continuous on R, then analogous to (8) we define its **mass** m by the double integral

$$m = \lim_{\|P\| \to 0} \sum_{k=1}^n \rho(x_k^*, y_k^*)\, \Delta A_k = \iint_R \rho(x, y)\, dA. \qquad (9)$$

The coordinates of the **center of mass** of the lamina are then

$$\bar{x} = \frac{M_y}{m}, \qquad \bar{y} = \frac{M_x}{m}, \tag{10}$$

where

$$M_y = \iint_R x\rho(x, y)\, dA \quad \text{and} \quad M_x = \iint_R y\rho(x, y)\, dA \tag{11}$$

are the **moments** of the lamina about the y- and x-axes, respectively. The center of mass is the point where we consider all the mass of the lamina to be concentrated. If $\rho(x, y) = $ a constant, the center of mass is called the **centroid** of the lamina.

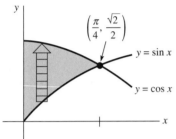

Figure 9.76

Example 3 Center of Mass

A lamina has the shape of the region in the first quadrant that is bounded by the graphs of $y = \sin x$, $y = \cos x$, between $x = 0$ and $x = \pi/4$. Find its center of mass if the density is $\rho(x, y) = y$.

SOLUTION From Figure 9.76 we see that

$$m = \iint_R y\, dA = \int_0^{\pi/4}\int_{\sin x}^{\cos x} y\, dy\, dx$$

$$= \int_0^{\pi/4} \frac{y^2}{2}\Bigg]_{\sin x}^{\cos x} dx$$

$$= \frac{1}{2}\int_0^{\pi/4} (\cos^2 x - \sin^2 x)\, dx \qquad \leftarrow \text{double-angle formula}$$

$$= \frac{1}{2}\int_0^{\pi/4} \cos 2x\, dx = \frac{1}{4}\sin 2x\Bigg]_0^{\pi/4} = \frac{1}{4}.$$

Now,

$$M_y = \iint_R xy\, dA = \int_0^{\pi/4}\int_{\sin x}^{\cos x} xy\, dy\, dx$$

$$= \int_0^{\pi/4} \frac{1}{2} xy^2\Bigg]_{\sin x}^{\cos x} dx$$

$$= \frac{1}{2}\int_0^{\pi/4} x \cos 2x\, dx \qquad \leftarrow \text{integration by parts}$$

$$= \left[\frac{1}{4} x \sin 2x + \frac{1}{8}\cos 2x\right]_0^{\pi/4} = \frac{\pi - 2}{16}.$$

Similarly,

$$M_x = \iint_R y^2\, dA = \int_0^{\pi/4}\int_{\sin x}^{\cos x} y^2\, dy\, dx$$

$$= \frac{1}{3}\int_0^{\pi/4} (\cos^3 x - \sin^3 x)\, dx$$

$$= \frac{1}{3}\int_0^{\pi/4} [\cos x(1 - \sin^2 x) - \sin x(1 - \cos^2 x)]\, dx$$

$$= \frac{1}{3}\left[\sin x - \frac{1}{3}\sin^3 x + \cos x - \frac{1}{3}\cos^3 x\right]_0^{\pi/4} = \frac{5\sqrt{2} - 4}{18}.$$

Hence, from (10),

$$\bar{x} = \frac{M_y}{m} = \frac{(\pi - 2)/16}{1/4} \approx 0.29, \quad \text{and} \quad \bar{y} = \frac{M_x}{m} = \frac{(5\sqrt{2} - 4)/18}{1/4} \approx 0.68.$$

Thus the center of mass has the approximate coordinates $(0.29, 0.68)$. ❑

■ **Moments of Inertia** The integrals M_x and M_y in (11) are also called the **first moments** of a lamina about the x-axis and y-axis, respectively. The so-called **second moments** of a lamina or **moments of inertia** about the x- and y-axes are, in turn, defined by the double integrals

$$I_x = \iint_R y^2 \rho(x, y) \, dA \quad \text{and} \quad I_y = \iint_R x^2 \rho(x, y) \, dA. \qquad (12)$$

A moment of inertia is the rotational equivalent of mass. For translational motion, kinetic energy is given by $K = \frac{1}{2}mv^2$, where m is mass and v is linear speed. The kinetic energy of a particle of mass m rotating at a distance r from an axis is $K = \frac{1}{2}mv^2 = \frac{1}{2}m(r\omega)^2 = \frac{1}{2}(mr^2)\omega^2 = \frac{1}{2}I\omega^2$, where $I = mr^2$ is its moment of inertia about the axis of rotation and ω is angular speed.

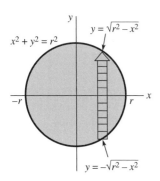

$x^2 + y^2 = r^2$

$y = \sqrt{r^2 - x^2}$

$y = -\sqrt{r^2 - x^2}$

Figure 9.77

Example 4 Moment of Inertia

Find the moment of inertia about the y-axis of the thin homogeneous disk of mass m shown in Figure 9.77.

SOLUTION Since the disk is homogeneous, its density is the constant $\rho(x, y) = m/\pi r^2$. Hence, from (12),

$$I_y = \iint_R x^2 \left(\frac{m}{\pi r^2}\right) dA = \frac{m}{\pi r^2} \int_{-r}^{r} \int_{-\sqrt{r^2 - x^2}}^{\sqrt{r^2 - x^2}} x^2 \, dy \, dx$$

$$= \frac{2m}{\pi r^2} \int_{-r}^{r} x^2 \sqrt{r^2 - x^2} \, dx \qquad \leftarrow \text{trig substitution}$$

$$= \frac{2mr^2}{\pi} \int_{-\pi/2}^{\pi/2} \sin^2\theta \cos^2\theta \, d\theta$$

$$= \frac{mr^2}{2\pi} \int_{-\pi/2}^{\pi/2} \sin^2 2\theta \, d\theta$$

$$= \frac{mr^2}{4\pi} \int_{-\pi/2}^{\pi/2} (1 - \cos 4\theta) \, d\theta = \frac{1}{4} mr^2. \qquad ❑$$

■ **Radius of Gyration** The **radius of gyration** of a lamina of mass m and the moment of inertia I about an axis is defined by

$$R_g = \sqrt{\frac{I}{m}}. \qquad (13)$$

Since (13) implies that $I = mR_g^2$, the radius of gyration is interpreted as the radial distance the lamina, considered as a point mass, can rotate about the axis without changing the rotational inertia of the body. In Example 4 the radius of gyration is $R_g = \sqrt{I_y/m} = \sqrt{(mr^2/4)/m} = r/2$.

Answers to odd-numbered problems begin on page A-50.

In Problems 1–8, evaluate the given partial integral.

1. $\displaystyle\int_{-1}^{3} (6xy - 5e^y)\, dx$

2. $\displaystyle\int_{1}^{2} \tan xy\, dy$

3. $\displaystyle\int_{1}^{3x} x^3 e^{xy}\, dy$

4. $\displaystyle\int_{\sqrt{y}}^{y^3} (8x^3y - 4xy^2)\, dx$

5. $\displaystyle\int_{0}^{2x} \frac{xy}{x^2 + y^2}\, dy$

6. $\displaystyle\int_{x^3}^{x} e^{2y/x}\, dy$

7. $\displaystyle\int_{\tan y}^{\sec y} (2x + \cos y)\, dx$

8. $\displaystyle\int_{\sqrt{y}}^{1} y \ln x\, dx$

In Problems 9–12, sketch the region of integration for the given iterated integral.

9. $\displaystyle\int_{0}^{2} \int_{1}^{2x+1} f(x, y)\, dy\, dx$

10. $\displaystyle\int_{1}^{4} \int_{-\sqrt{y}}^{\sqrt{y}} f(x, y)\, dx\, dy$

11. $\displaystyle\int_{-1}^{3} \int_{0}^{\sqrt{16-y^2}} f(x, y)\, dx\, dy$

12. $\displaystyle\int_{-1}^{2} \int_{-x^2}^{x^2+1} f(x, y)\, dy\, dx$

In Problems 13–22, evaluate the double integral over the region R that is bounded by the graphs of the given equations. Choose the most convenient order of integration.

13. $\displaystyle\iint_R x^3 y^2\, dA;\ y = x,\ y = 0,\ x = 1$

14. $\displaystyle\iint_R (x + 1)\, dA;\ y = x,\ x + y = 4,\ x = 0$

15. $\displaystyle\iint_R (2x + 4y + 1)\, dA;\ y = x^2,\ y = x^3$

16. $\displaystyle\iint_R xe^y\, dA;\ R$ the same as in Problem 13

17. $\displaystyle\iint_R 2xy\, dA;\ y = x^3,\ y = 8,\ x = 0$

18. $\displaystyle\iint_R \frac{x}{\sqrt{y}}\, dA;\ y = x^2 + 1,\ y = 3 - x^2$

19. $\displaystyle\iint_R \frac{y}{1 + xy}\, dA;\ y = 0,\ y = 1,\ x = 0,\ x = 1$

20. $\displaystyle\iint_R \sin \frac{\pi x}{y}\, dA;\ x = y^2,\ x = 0,\ y = 1,\ y = 2$

21. $\displaystyle\iint_R \sqrt{x^2 + 1}\, dA;\ x = y,\ x = -y,\ x = \sqrt{3}$

22. $\displaystyle\iint_R x\, dA;\ y = \tan^{-1}x,\ y = 0,\ x = 1$

23. Consider the solid bounded by the graphs of $x^2 + y^2 = 4$, $z = 4 - y$, and $z = 0$ shown in Figure 9.78. Choose and evaluate the correct integral representing the volume V of the solid.

(a) $\displaystyle 4\int_{0}^{2} \int_{0}^{\sqrt{4-x^2}} (4 - y)\, dy\, dx$

(b) $\displaystyle 2\int_{-2}^{2} \int_{0}^{\sqrt{4-x^2}} (4 - y)\, dy\, dx$

(c) $\displaystyle 2\int_{-2}^{2} \int_{0}^{\sqrt{4-y^2}} (4 - y)\, dx\, dy$

Figure 9.78

24. Consider the solid bounded by the graphs of $x^2 + y^2 = 4$ and $y^2 + z^2 = 4$. An eighth of the solid is shown in Figure 9.79. Choose and evaluate the correct integral representing the volume V of the solid.

(a) $\displaystyle 4\int_{-2}^{2} \int_{-\sqrt{4-x^2}}^{\sqrt{4-x^2}} (4 - y^2)^{1/2}\, dy\, dx$

(b) $\displaystyle 8\int_{0}^{2} \int_{0}^{\sqrt{4-y^2}} (4 - y^2)^{1/2}\, dx\, dy$

(c) $\displaystyle 8\int_{0}^{2} \int_{0}^{\sqrt{4-x^2}} (4 - x^2)^{1/2}\, dy\, dx$

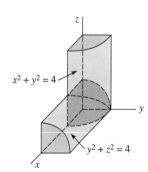

Figure 9.79

In Problems 25–34, find the volume of the solid bounded by the graphs of the given equations.

25. $2x + y + z = 6$, $x = 0$, $y = 0$, $z = 0$, first octant

26. $z = 4 - y^2$, $x = 3$, $x = 0$, $y = 0$, $z = 0$, first octant

27. $x^2 + y^2 = 4$, $x - y + 2z = 4$, $x = 0$, $y = 0$, $z = 0$, first octant

28. $y = x^2$, $y + z = 3$, $z = 0$

29. $z = 1 + x^2 + y^2$, $3x + y = 3$, $x = 0$, $y = 0$, $z = 0$, first octant

30. $z = x + y$, $x^2 + y^2 = 9$, $x = 0$, $y = 0$, $z = 0$, first octant

31. $yz = 6$, $x = 0$, $x = 5$, $y = 1$, $y = 6$, $z = 0$

32. $z = 4 - x^2 - \frac{1}{4}y^2$, $z = 0$

33. $z = 4 - y^2$, $x^2 + y^2 = 2x$, $z = 0$

34. $z = 1 - x^2$, $z = 1 - y^2$, $x = 0$, $y = 0$, $z = 0$, first octant

In Problems 35–40, evaluate the given iterated integral by reversing the order of integration.

35. $\displaystyle\int_0^1 \int_x^1 x^2\sqrt{1 + y^4}\ dy\ dx$

36. $\displaystyle\int_0^1 \int_{2y}^2 e^{-y/x}\ dx\ dy$

37. $\displaystyle\int_0^2 \int_{y^2}^4 \cos \sqrt{x^3}\ dx\ dy$

38. $\displaystyle\int_{-1}^1 \int_{-\sqrt{1-x^2}}^{\sqrt{1-x^2}} x\sqrt{1 - x^2 - y^2}\ dy\ dx$

39. $\displaystyle\int_0^1 \int_x^1 \frac{1}{1 + y^4}\ dy\ dx$

40. $\displaystyle\int_0^4 \int_{\sqrt{y}}^2 \sqrt{x^3 + 1}\ dx\ dy$

In Problems 41–50, find the center of mass of the lamina that has the given shape and density.

41. $x = 0$, $x = 4$, $y = 0$, $y = 3$; $\rho(x, y) = xy$

42. $x = 0$, $y = 0$, $2x + y = 4$; $\rho(x, y) = x^2$

43. $y = x$, $x + y = 6$, $y = 0$; $\rho(x, y) = 2y$

44. $y = |x|$, $y = 3$; $\rho(x, y) = x^2 + y^2$

45. $y = x^2$, $x = 1$, $y = 0$; $\rho(x, y) = x + y$

46. $x = y^2$, $x = 4$; $\rho(x, y) = y + 5$

47. $y = 1 - x^2$, $y = 0$; density at a point P directly proportional to the distance from the x-axis

48. $y = \sin x$, $0 \le x \le \pi$, $y = 0$; density at a point P directly proportional to the distance from the y-axis

49. $y = e^x$, $x = 0$, $x = 1$, $y = 0$; $\rho(x, y) = y^3$

50. $y = \sqrt{9 - x^2}$, $y = 0$; $\rho(x, y) = x^2$

In Problems 51–54, find the moment of inertia about the x-axis of the lamina that has the given shape and density.

51. $x = y - y^2$, $x = 0$; $\rho(x, y) = 2x$

52. $y = x^2$, $y = \sqrt{x}$; $\rho(x, y) = x^2$

53. $y = \cos x$, $-\pi/2 \le x \le \pi/2$, $y = 0$; $\rho(x, y) = k$ (constant)

54. $y = \sqrt{4 - x^2}$, $x = 0$, $y = 0$, first quadrant; $\rho(x, y) = y$

In Problems 55–58, find the moment of inertia about the y-axis of the lamina that has the given shape and density.

55. $y = x^2$, $x = 0$, $y = 4$, first quadrant; $\rho(x, y) = y$

56. $y = x^2$, $y = \sqrt{x}$; $\rho(x, y) = x^2$

57. $y = x$, $y = 0$, $y = 1$, $x = 3$; $\rho(x, y) = 4x + 3y$

58. Same R and density as in Problem 47

In Problems 59 and 60, find the radius of gyration about the indicated axis of the lamina that has the given shape and density.

59. $x = \sqrt{a^2 - y^2}$, $x = 0$; $\rho(x, y) = x$; y-axis

60. $x + y = a$, $a > 0$, $x = 0$, $y = 0$; $\rho(x, y) = k$ (constant); x-axis

61. A lamina has the shape of the region bounded by the graph of the ellipse $x^2/a^2 + y^2/b^2 = 1$. If its density is $\rho(x, y) = 1$, find:

 (a) the moment of inertia about the x-axis of the lamina,

 (b) the moment of inertia about the y-axis of the lamina,

 (c) the radius of gyration about the x-axis [*Hint:* The area of the ellipse is πab], and

 (d) the radius of gyration about the y-axis.

62. A cross-section of an experimental airfoil is the lamina shown in Figure 9.80. The arc ABC is elliptical, whereas the two arcs AD and CD are parabolic. Find the moment of inertia about the x-axis of the lamina under the assumption that the density is $\rho(x, y) = 1$.

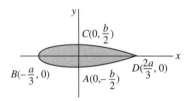

Figure 9.80

The **polar moment of inertia** of a lamina with respect to the origin is defined to be

$$I_0 = \iint_R (x^2 + y^2)\rho(x, y)\ dA = I_x + I_y.$$

In Problems 63–66, find the polar moment of inertia of the lamina that has the given shape and density.

63. $x + y = a$, $a > 0$, $x = 0$, $y = 0$; $\rho(x, y) = k$ (constant)

64. $y = x^2$, $y = \sqrt{x}$; $\rho(x, y) = x^2$ [*Hint:* See Problems 52 and 56.]

65. $x = y^2 + 2$, $x = 6 - y^2$; density at a point P inversely proportional to the square of the distance from the origin

66. $y = x$, $y = 0$, $y = 3$, $x = 4$; $\rho(x, y) = k$ (constant)

67. Find the radius of gyration in Problem 63.

68. Show that the polar moment of inertia about the center of a thin homogeneous rectangular plate of mass m, width w, and length l is $I_0 = m(l^2 + w^2)/12$.

9.11 Double Integrals in Polar Coordinates

Suppose R is a region bounded by the graphs of the polar equations $r = g_1(\theta)$, $r = g_2(\theta)$, and the rays $\theta = \alpha$, $\theta = \beta$, and f is a function of r and θ that is continuous on R. In order to define the double integral of f over R, we use rays and concentric circles to partition the region into a grid of "polar rectangles" or subregions R_k. See Figure 9.81(a) and (b). The area ΔA_k of typical subregion R_k, shown in Figure 9.81(c), is the difference of areas of two circular sectors: $\Delta A_k = \frac{1}{2}r_{k+1}^2\Delta\theta_k - \frac{1}{2}r_k^2\Delta\theta_k$. Now ΔA_k can be written as

$$\Delta A_k = \frac{1}{2}(r_{k+1}^2 - r_k^2)\Delta\theta_k = \frac{1}{2}(r_{k+1} + r_k)(r_{k+1} - r_k)\Delta\theta_k = r_k^*\Delta r_k\Delta\theta_k$$

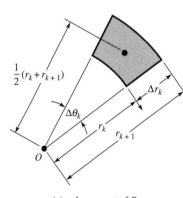

(a) region R (b) subregion R_k (c) enlargement of R_k

Figure 9.81

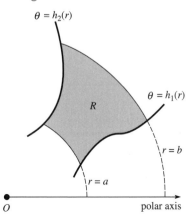

Figure 9.82

where $\Delta r_k = (r_{k+1} - r_k)$ and $r_k^* = \frac{1}{2}(r_{k+1} + r_k)$ denotes the average radius. By choosing (r_k^*, θ_k^*) on each R_k, the double integral of f over R is

$$\lim_{\|P\| \to 0} \sum_{k=1}^{n} f(r_k^*, \theta_k^*)r_k^*\,\Delta r_k\,\Delta\theta_k = \iint_R f(r, \theta)\,dA.$$

The double integral is then evaluated by means of the iterated integral:

$$\iint_R f(r, \theta)\,dA = \int_\alpha^\beta \int_{g_1(\theta)}^{g_2(\theta)} f(r, \theta)r\,dr\,d\theta. \tag{1}$$

On the other hand, if the region R is as given in Figure 9.82, the double integral of f over R is then

$$\iint_R f(r, \theta)\,dA = \int_a^b \int_{h_1(r)}^{h_2(r)} f(r, \theta)r\,d\theta\,dr. \tag{2}$$

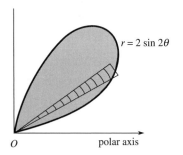

$r = 2 \sin 2\theta$

O polar axis

Figure 9.83

Example 1 Center of Mass

Find the center of mass of the lamina that corresponds to the region bounded by one leaf of the rose $r = 2 \sin 2\theta$ in the first quadrant if the density at a point P in the lamina is directly proportional to the distance from the pole.

SOLUTION By varying θ from 0 to $\pi/2$, we obtain the graph in Figure 9.83. Now, $d(O, P) = |r|$. Hence, $\rho(r, \theta) = k|r|$, where k is a constant of proportionality. From (9) of Section 9.10, we have

$$
m = \iint_R k|r| \, dA
$$

$$
= k \int_0^{\pi/2} \int_0^{2 \sin 2\theta} (r) r \, dr \, d\theta
$$

$$
= k \int_0^{\pi/2} \frac{r^3}{3} \Big]_0^{2 \sin 2\theta} d\theta
$$

$$
= \frac{8}{3} k \int_0^{\pi/2} \sin^3 2\theta \, d\theta \qquad \leftarrow \; \sin^2 2\theta = 1 - \cos^2 2\theta
$$

$$
= \frac{8}{3} k \int_0^{\pi/2} (1 - \cos^2 2\theta) \sin 2\theta \, d\theta
$$

$$
= \frac{8}{3} k \left[-\frac{1}{2} \cos 2\theta + \frac{1}{6} \cos^3 2\theta \right]_0^{\pi/2} = \frac{16}{9} k.
$$

Since $x = r \cos \theta$, we can write

$$
M_y = k \iint_R x|r| \, dA
$$

as

$$
M_y = \int_0^{\pi/2} \int_0^{2 \sin 2\theta} r^3 \cos \theta \, dr \, d\theta
$$

$$
= k \int_0^{\pi/2} \frac{r^4}{4} \cos \theta \Big]_0^{2 \sin 2\theta} d\theta
$$

$$
= 4k \int_0^{\pi/2} \sin^4 2\theta \cos \theta \, d\theta \qquad \leftarrow \; \text{double-angle formula}
$$

$$
= 4k \int_0^{\pi/2} 16 \sin^4 \theta \cos^4 \theta \cos \theta \, d\theta
$$

$$
= 64k \int_0^{\pi/2} \sin^4 \theta \cos^5 \theta \, d\theta
$$

$$
= 64k \int_0^{\pi/2} \sin^4 \theta (1 - \sin^2 \theta)^2 \cos \theta \, d\theta
$$

$$
= 64k \int_0^{\pi/2} (\sin^4 \theta - 2 \sin^6 \theta + \sin^8 \theta) \cos \theta \, d\theta
$$

$$
= 64k \left[\frac{1}{5} \sin^5 \theta - \frac{2}{7} \sin^7 \theta + \frac{1}{9} \sin^9 \theta \right]_0^{\pi/2} = \frac{512}{315} k.
$$

Similarly, by using $x = r \sin \theta$, we find*

$$
M_x = k \int_0^{\pi/2} \int_0^{2 \sin 2\theta} r^2 \sin \theta \, dr \, d\theta = \frac{512}{315} k.
$$

* We could have argued to the fact that $M_x = M_y$ and hence $\bar{x} = \bar{y}$ from the fact that the lamina and the density function are symmetric about the ray $\theta = \pi/4$.

Here the rectangular coordinates of the center of mass are

$$\bar{x} = \bar{y} = \frac{512k/315}{16k/9} = \frac{32}{35}. \qquad \square$$

■ **Change of Variables: Rectangular to Polar Coordinates** In some instances a double integral $\iint_R f(x, y)\, dA$ that is difficult or even impossible to evaluate using rectangular coordinates may be readily evaluated when a change of variables is used. If we assume that f is continuous on the region R and if R can be described in polar coordinates as $0 \le g_1(\theta) \le r \le g_2(\theta)$, $\alpha \le \theta \le \beta$, $0 < \beta - \alpha \le 2\pi$, then

$$\iint_R f(x, y)\, dA = \int_\alpha^\beta \int_{g_1(\theta)}^{g_2(\theta)} f(r\cos\theta, r\sin\theta) r\, dr\, d\theta. \qquad (3)$$

Equation (3) is particularly useful when f contains the expression $x^2 + y^2$ since, in polar coordinates, we can now write

$$x^2 + y^2 = r^2 \quad \text{and} \quad \sqrt{x^2 + y^2} = r.$$

Example 2 Changing an Integral to Polar Coordinates

Use polar coordinates to evaluate

$$\int_0^2 \int_x^{\sqrt{8-x^2}} \frac{1}{5 + x^2 + y^2}\, dy\, dx.$$

SOLUTION From $x \le y \le \sqrt{8 - x^2}$, $0 \le x \le 2$, we have sketched the region R of integration in Figure 9.84. Since $x^2 + y^2 = r^2$, the polar description of the circle $x^2 + y^2 = 8$ is $r = \sqrt{8}$. Hence, in polar coordinates, the region of R is given by $0 \le r \le \sqrt{8}$, $\pi/4 \le \theta \le \pi/2$. With $1/(5 + x^2 + y^2) = 1/(5 + r^2)$, the original integral becomes

$$\int_0^2 \int_x^{\sqrt{8-x^2}} \frac{1}{5 + x^2 + y^2}\, dy\, dx = \int_{\pi/4}^{\pi/2} \int_0^{\sqrt{8}} \frac{1}{5 + r^2}\, r\, dr\, d\theta$$

$$= \frac{1}{2} \int_{\pi/4}^{\pi/2} \int_0^{\sqrt{8}} \frac{2r\, dr}{5 + r^2}\, d\theta$$

$$= \frac{1}{2} \int_{\pi/4}^{\pi/2} \ln(5 + r^2)\Big]_0^{\sqrt{8}}\, d\theta$$

$$= \frac{1}{2} (\ln 13 - \ln 5) \int_{\pi/4}^{\pi/2} d\theta$$

$$= \frac{1}{2} (\ln 13 - \ln 5) \left(\frac{\pi}{2} - \frac{\pi}{4}\right) = \frac{\pi}{8} \ln \frac{13}{5}. \qquad \square$$

Example 3 Volume

Find the volume of the solid that is under the hemisphere $z = \sqrt{1 - x^2 - y^2}$ and above the region bounded by the graph of the circle $x^2 + y^2 - y = 0$.

SOLUTION From Figure 9.85, we see that $V = \iint_R \sqrt{1 - x^2 - y^2}\, dA$. In polar coordinates the equations of the hemisphere and the circle become, respectively, $z = \sqrt{1 - r^2}$ and $r = \sin\theta$. Now, using symmetry, we have

$$V = \iint_R \sqrt{1 - r^2}\, dA = 2 \int_0^{\pi/2} \int_0^{\sin\theta} (1 - r^2)^{1/2} r\, dr\, d\theta$$

$$= 2 \int_0^{\pi/2} \left[-\frac{1}{3}(1 - r^2)^{3/2}\right]_0^{\sin\theta} d\theta$$

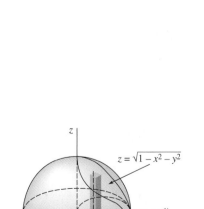

Figure 9.84

Figure 9.85

$$= \frac{2}{3} \int_0^{\pi/2} [1 - (1 - \sin^2\theta)^{3/2}] \, d\theta$$

$$= \frac{2}{3} \int_0^{\pi/2} [1 - (\cos^2\theta)^{3/2}] \, d\theta$$

$$= \frac{2}{3} \int_0^{\pi/2} [1 - \cos^3\theta] \, d\theta$$

$$= \frac{2}{3} \int_0^{\pi/2} [1 - (1 - \sin^2\theta) \cos\theta] \, d\theta$$

$$= \frac{2}{3} \left[\theta - \sin\theta + \frac{1}{3} \sin^3\theta \right]_0^{\pi/2} = \frac{\pi}{3} - \frac{4}{9} \approx 0.60 \text{ cubic unit.} \quad \square$$

■ **Area** Note that in (1) if $f(r, \theta) = 1$, then the **area** of the region R in Figure 9.81(a) is given by

$$A = \iint_R dA = \int_\alpha^\beta \int_{g_1(\theta)}^{g_2(\theta)} r \, dr \, d\theta.$$

The same observation holds for (2) and Figure 9.82 when $f(r, \theta) = 1$.

Remarks

The reader is invited to reexamine Example 3. The graph of the circle $r = \sin\theta$ is obtained by varying θ from 0 to π. However, carry out the iterated integration

$$V = \int_0^\pi \int_0^{\sin\theta} (1 - r^2)^{1/2} r \, dr \, d\theta$$

and see if the result is the *incorrect* answer $\pi/3$. What goes wrong?

In Problems 1–4, use a double integral in polar coordinates to find the area of the region bounded by the graphs of the given polar equations.

1. $r = 3 + 3 \sin\theta$

2. $r = 2 + \cos\theta$

3. $r = 2 \sin\theta, r = 1$, common area

4. $r = 8 \sin 4\theta$, one petal

In Problems 5–10, find the volume of the solid bounded by the graphs of the given equations.

5. One petal of $r = 5 \cos 3\theta$, $z = 0$, $z = 4$

6. $x^2 + y^2 = 4$, $z = \sqrt{9 - x^2 - y^2}$, $z = 0$

7. Between $x^2 + y^2 = 1$ and $x^2 + y^2 = 9$, $z = \sqrt{16 - x^2 - y^2}$, $z = 0$

8. $z = \sqrt{x^2 + y^2}$, $x^2 + y^2 = 25$, $z = 0$

9. $r = 1 + \cos\theta$, $z = y$, $z = 0$, first octant

10. $r = \cos\theta$, $z = 2 + x^2 + y^2$, $z = 0$

In Problems 11–16, find the center of mass of the lamina that has the given shape and density.

11. $r = 1$, $r = 3$, $x = 0$, $y = 0$, first quadrant; $\rho(r, \theta) = k$ (constant)

12. $r = \cos\theta$; density at point P directly proportional to the distance from the pole

13. $y = \sqrt{3}x$, $y = 0$, $x = 3$; $\rho(r, \theta) = r^2$

14. $r = 4 \cos 2\theta$, petal on the polar axis; $\rho(r, \theta) = k$ (constant)

15. Outside $r = 2$ and inside $r = 2 + 2 \cos\theta$, $y = 0$, first quadrant; density at a point P inversely proportional to the distance from the pole

16. $r = 2 + 2\cos\theta$, $y = 0$, first and second quadrants; $\rho(r, \theta) = k$ (constant)

In Problems 17–20, find the indicated moment of inertia of the lamina that has the given shape and density.

17. $r = a$; $\rho(r, \theta) = k$ (constant); I_x

18. $r = a$; $\rho(r, \theta) = \dfrac{1}{1 + r^4}$; I_x

19. Outside $r = a$ and inside $r = 2a\cos\theta$; density at a point P inversely proportional to the cube of the distance from the pole; I_y

20. Outside $r = 1$ and inside $r = 2\sin 2\theta$, first quadrant; $\rho(r, \theta) = \sec^2\theta$; I_y

In Problems 21–24, find the **polar moment of inertia** $I_0 = \iint_R r^2\rho(r, \theta)\, dA = I_x + I_y$ of the lamina that has the given shape and density.

21. $r = a$; $\rho(r, \theta) = k$ (constant) [*Hint:* Use Problem 17 and the fact that $I_x = I_y$.]

22. $r = \theta$, $0 \le \theta \le \pi$, $y = 0$; density at a point P proportional to the distance from the pole

23. $r\theta = 1$, $\frac{1}{3} \le \theta \le 1$, $r = 1$, $r = 3$, $y = 0$; density at a point P inversely proportional to the distance from the pole [*Hint:* Integrate first with respect to θ.]

24. $r = 2a\cos\theta$; $\rho(r, \theta) = k$ (constant)

In Problems 25–32, evaluate the given iterated integral by changing to polar coordinates.

25. $\displaystyle\int_{-3}^{3}\int_{0}^{\sqrt{9-x^2}} \sqrt{x^2 + y^2}\, dy\, dx$

26. $\displaystyle\int_{0}^{\sqrt{2}/2}\int_{y}^{\sqrt{1-y^2}} \frac{y^2}{\sqrt{x^2 + y^2}}\, dx\, dy$

27. $\displaystyle\int_{0}^{1}\int_{0}^{\sqrt{1-y^2}} e^{x^2 + y^2}\, dx\, dy$

28. $\displaystyle\int_{-\sqrt{\pi}}^{\sqrt{\pi}}\int_{0}^{\sqrt{\pi - x^2}} \sin(x^2 + y^2)\, dy\, dx$

29. $\displaystyle\int_{0}^{1}\int_{\sqrt{1-x^2}}^{\sqrt{4-x^2}} \frac{x^2}{x^2 + y^2}\, dy\, dx + \int_{1}^{2}\int_{0}^{\sqrt{4-x^2}} \frac{x^2}{x^2 + y^2}\, dy\, dx$

30. $\displaystyle\int_{0}^{1}\int_{0}^{\sqrt{2y-y^2}} (1 - x^2 - y^2)\, dx\, dy$

31. $\displaystyle\int_{-5}^{5}\int_{0}^{\sqrt{25-x^2}} (4x + 3y)\, dy\, dx$

32. $\displaystyle\int_{0}^{1}\int_{0}^{\sqrt{1-y^2}} \frac{1}{1 + \sqrt{x^2 + y^2}}\, dx\, dy$

33. The liquid hydrogen tank in the space shuttle has the form of a right circular cylinder with a semi-ellipsoidal cap at each end. The radius of the cylindrical part of the tank is 4.2 m. Find the volume of the tank shown in Figure 9.86.

5.15 m

19.3 m

5.15 m

Figure 9.86

34. Evaluate $\iint_R (x + y)\, dA$ over the region shown in Figure 9.87.

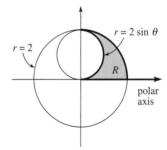

$r = 2\sin\theta$

$r = 2$

R

polar axis

Figure 9.87

35. The improper integral $\int_0^\infty e^{-x^2}\, dx$ is important in the theory of probability, statistics, and other areas of applied mathematics. If I denotes the integral, then

$$I = \int_0^\infty e^{-x^2}\, dx \quad\text{and}\quad I = \int_0^\infty e^{-y^2}\, dy$$

and consequently

$$I^2 = \left(\int_0^\infty e^{-x^2}\, dx\right)\left(\int_0^\infty e^{-y^2}\, dy\right) = \int_0^\infty\int_0^\infty e^{-(x^2 + y^2)}\, dx\, dy$$

Use polar coordinates to evaluate the last integral. Find the value of I.

9.12 Green's Theorem

(a) Positive direction

(b) Positive direction

(c) Negative direction

Figure 9.88

One of the most important theorems in vector integral calculus relates a line integral around a piecewise smooth *simple closed curve* to a double integral over the region bounded by the curve.

■ **Line Integrals Along Simple Closed Curves** We say the **positive direction** around a simple closed curve C is that direction a point on the curve must move, or the direction a person must walk on C, in order to keep the region R bounded by C to the left. See Figure 9.88(a). Roughly, as shown in Figure 9.88(b) and 9.88(c), the *positive* and *negative* directions correspond to the *counterclockwise* and *clockwise directions,* respectively. Line integrals on simple closed curves are written

$$\oint_C P(x, y)\, dx + Q(x, y)\, dy, \quad \oint_C P(x, y)\, dx + Q(x, y)\, dy, \quad \oint_C F(x, y)\, ds,$$

and so on. The symbols \oint_C and \oint_C refer, in turn, to integrations in the positive and negative directions.

THEOREM 9.13 Green's Theorem in the Plane*

Suppose that C is a piecewise smooth simple closed curve bounding a region R. If P, Q, $\partial P/\partial y$, and $\partial Q/\partial x$ are continuous on R, then

$$\oint_C P\, dx + Q\, dy = \iint_R \left(\frac{\partial Q}{\partial x} - \frac{\partial P}{\partial y} \right) dA. \tag{1}$$

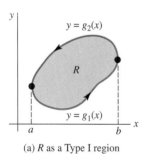

(a) R as a Type I region

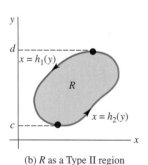

(b) R as a Type II region

Figure 9.89

PARTIAL PROOF We shall prove (1) only for a region R that is simultaneously of Type I and Type II:

$$R: g_1(x) \leq y \leq g_2(x), \qquad a \leq x \leq b$$
$$R: h_1(y) \leq x \leq h_2(y), \qquad c \leq y \leq d.$$

Using Figure 9.89(a), we have

$$-\iint_R \frac{\partial P}{\partial y}\, dA = -\int_a^b \int_{g_1(x)}^{g_2(x)} \frac{\partial P}{\partial y}\, dy\, dx$$

$$= -\int_a^b [P(x, g_2(x)) - P(x, g_1(x))]\, dx$$

$$= \int_a^b P(x, g_1(x))\, dx + \int_b^a P(x, g_2(x))\, dx$$

$$= \oint_C P(x, y)\, dx.$$

$$\tag{2}$$

*Named after George Green (1793–1841), an English mathematician and physicist. The words *in the plane* suggest that the theorem generalizes to 3-space. It does—read on.

Similarly, from Figure 9.89(b),

$$\iint_R \frac{\partial Q}{\partial x}\, dA = \int_c^d \int_{h_2(y)}^{h_2(y)} \frac{\partial Q}{\partial x}\, dx\, dy$$

$$= \int_c^d [Q(h_2(y), y) - Q(h_1(y), y)]\, dy \qquad (3)$$

$$= \int_c^d Q(h_2(y), y)\, dy + \int_d^c Q(h_1(y), y)\, dy$$

$$= \oint_C Q(x, y)\, dy.$$

Adding the results in (2) and (3) yields (1). ❏

Although the foregoing proof is not valid, the theorem is applicable to more complicated regions, such as those shown in Figure 9.90. The proof consists of decomposing R into a finite number of subregions to which (1) can be applied and then adding the results.

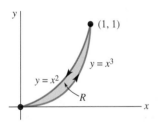

Figure 9.90

Example 1 Using Green's Theorem

Evaluate $\oint_C (x^2 - y^2)\, dx + (2y - x)\, dy$, where C consists of the boundary of the region in the first quadrant that is bounded by the graphs of $y = x^2$ and $y = x^3$.

SOLUTION If $P(x, y) = x^2 - y^2$ and $Q(x, y) = 2y - x$, then $\partial P/\partial y = -2y$ and $\partial Q/\partial x = -1$. From (1) and Figure 9.91, we have

$$\oint_C (x^2 - y^2)\, dx + (2y - x)\, dy = \iint_R (-1 + 2y)\, dA$$

$$= \int_0^1 \int_{x^3}^{x^2} (-1 + 2y)\, dy\, dx$$

$$= \int_0^1 \left. (-y + y^2) \right]_{x^3}^{x^2} dx$$

$$= \int_0^1 (-x^6 + x^4 + x^3 - x^2)\, dx = -\frac{11}{420}. ❏$$

Figure 9.91

We note that the line integral in Example 1 could have been evaluated in a straightforward manner using the variable x as a parameter. However, in the next example, ponder the problem of evaluating the given line integral in the usual manner.

Example 2 Using Green's Theorem

Evaluate $\oint_C (x^5 + 3y)\, dx + (2x - e^{y^3})\, dy$, where C is the circle $(x - 1)^2 + (y - 5)^2 = 4$.

SOLUTION Identifying $P(x, y) = x^5 + 3y$ and $Q(x, y) = 2x - e^{y^3}$, we have $\partial P/\partial y = 3$ and $\partial Q/\partial x = 2$. Hence, (1) gives

$$\oint_C (x^5 + 3y)\, dx + (2x - e^{y^3})\, dy = \iint_R (2 - 3)\, dA = -\iint_R dA.$$

Now the double integral $\iint_R dA$ gives the area of the region R bounded by the circle of radius 2 shown in Figure 9.92. Since the area of the circle is $\pi 2^2 = 4\pi$, it follows that

$$\oint_C (x^5 + 3y)\, dx + (2x - e^{y^3})\, dy = -4\pi. ❏$$

Figure 9.92

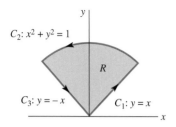

$C_2: x^2 + y^2 = 1$

R

$C_3: y = -x$ $C_1: y = x$

Figure 9.93

Example 3 Work Done by a Force

Find the work done by the force $\mathbf{F} = (-16y + \sin x^2)\mathbf{i} + (4e^y + 3x^2)\mathbf{j}$ acting along the simple closed curve C shown in Figure 9.93.

SOLUTION From (12) of Section 9.8 the work done by \mathbf{F} is given by

$$W = \oint_C \mathbf{F} \cdot d\mathbf{r} = \oint_C (-16y + \sin x^2)\, dx + (4e^y + 3x^2)\, dy$$

and so by Green's theorem $W = \iint_R (6x + 16)\, dA$. In view of the region R, the last integral is best handled in polar coordinates. Since R is defined by $0 \le r \le 1$, $\pi/4 \le \theta \le 3\pi/4$,

$$W = \int_{\pi/4}^{3\pi/4} \int_0^1 (6r \cos\theta + 16)r\, dr\, d\theta$$

$$= \int_{\pi/4}^{3\pi/4} (2r^3 \cos\theta + 8r^2) \Big]_0^1 d\theta$$

$$= \int_{\pi/4}^{3\pi/4} (2\cos\theta + 8)\, d\theta = 4\pi. \qquad \square$$

Example 4 Green's Theorem Not Applicable

Let C be the closed curve consisting of the four straight line segments C_1, C_2, C_3, C_4 shown in Figure 9.94. Green's theorem is *not* applicable to the line integral

$$\oint_C \frac{-y}{x^2 + y^2}\, dx + \frac{x}{x^2 + y^2}\, dy$$

since P, Q, $\partial P/\partial y$, and $\partial Q/\partial x$ are not continuous at the origin. \square

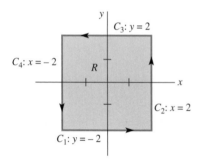

$C_3: y = 2$

$C_4: x = -2$ R

$C_2: x = 2$

$C_1: y = -2$

Figure 9.94

■ **Region with Holes** Green's theorem can also be extended to a region R with "holes," that is, a region bounded between two or more piecewise smooth simple closed curves. In Figure 9.95(a) we have shown a region R bounded by a curve C that consists of two simple closed curves C_1 and C_2, that is, $C = C_1 \cup C_2$. The curve C is positively oriented, since if we traverse C_1 in a counterclockwise direction and C_2 and a clockwise direction, the region R is always to the left. If we now introduce crosscuts as shown in Figure 9.95(b), the region R is divided into two subregions, R_1 and R_2. By applying Green's theorem to R_1 and R_2, we obtain

$$\iint_R \left(\frac{\partial Q}{\partial x} - \frac{\partial P}{\partial y} \right) dA = \iint_{R_1} \left(\frac{\partial Q}{\partial x} - \frac{\partial P}{\partial y} \right) dA + \iint_{R_2} \left(\frac{\partial Q}{\partial x} - \frac{\partial P}{\partial y} \right) dA$$

$$= \oint_{C_1} P\, dx + Q\, dy + \oint_{C_2} P\, dx + Q\, dy \qquad (4)$$

$$= \oint_C P\, dx + Q\, dy.$$

The last result follows from the fact that the line integrals on the crosscuts (paths with opposite orientations) cancel each other. See (8) of Section 9.8.

C_2

R

C_1

(a)

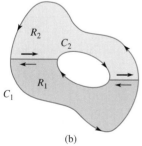

R_2

C_2

R_1

C_1

(b)

Figure 9.95

Example 5 Region with a Hole in It

Evaluate $\displaystyle\oint_C \frac{-y}{x^2 + y^2}\, dx + \frac{x}{x^2 + y^2}\, dy$, where $C = C_1 \cup C_2$ is the boundary of the shaded region R shown in Figure 9.96.

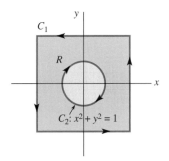

C_1

R

$C_2: x^2 + y^2 = 1$

Figure 9.96

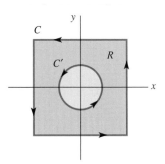

C_1

R

C_2

Figure 9.97

C

C'

R

Figure 9.98

SOLUTION Because $P(x, y) = \dfrac{-y}{x^2 + y^2}$, $\qquad Q(x, y) = \dfrac{x}{x^2 + y^2}$,

$$\frac{\partial P}{\partial y} = \frac{y^2 - x^2}{(x^2 + y^2)^2}, \qquad \frac{\partial Q}{\partial x} = \frac{y^2 - x^2}{(x^2 + y^2)^2},$$

are continuous on the region R bounded by C, it follows from the above discussion that

$$\oint_C \frac{-y}{x^2 + y^2}\, dx + \frac{x}{x^2 + y^2}\, dy = \iint_R \left[\frac{y^2 - x^2}{(x^2 + y^2)^2} - \frac{y^2 - x^2}{(x^2 + y^2)^2} \right] dA = 0. \qquad \Box$$

As a consequence of the discussion preceding Example 5 we can establish a result for line integrals that enables us, under certain circumstances, to replace a complicated closed path with a path that is simpler. Suppose, as shown in Figure 9.97, that C_1 and C_2 are two nonintersecting piecewise smooth simple closed paths that have the same counterclockwise orientation. Suppose further that P and Q have continuous first partial derivatives such that

$$\frac{\partial P}{\partial y} = \frac{\partial Q}{\partial x}$$

in the region R bounded between C_1 and C_2. Then from (4) above and (8) of Section 9.8 we have

$$\oint_{C_1} P\, dx + Q\, dy + \oint_{-C_2} P\, dx + Q\, dy = 0$$

or $\qquad\qquad \oint_{C_1} P\, dx + Q\, dy = \oint_{C_2} P\, dx + Q\, dy. \qquad (5)$

Example 6 **Example 4 Revisited**

Evaluate the line integral in Example 4.

SOLUTION One method of evaluating the line integral is to write

$$\oint_C = \int_{C_1} + \int_{C_2} + \int_{C_3} + \int_{C_4}$$

and then evaluate the four integrals on the line segments C_1, C_2, C_3, and C_4. Alternatively, if we note that the circle C': $x^2 + y^2 = 1$ lies entirely within C (see Figure 9.98), then from Example 5 it is apparent that $P = -y/(x^2 + y^2)$ and $Q = x/(x^2 + y^2)$ have continuous first partial derivatives in the region R bounded between C and C'. Moreover,

$$\frac{\partial P}{\partial y} = \frac{y^2 - x^2}{(x^2 + y^2)^2} = \frac{\partial Q}{\partial x}$$

in R. Hence, it follows from (5) that

$$\oint_C \frac{-y}{x^2 + y^2}\, dx + \frac{x}{x^2 + y^2}\, dy = \oint_{C'} \frac{-y}{x^2 + y^2}\, dx + \frac{x}{x^2 + y^2}\, dy.$$

Using the parametrization $x = \cos t$, $y = \sin t$, $0 \le t \le 2\pi$ for C' we obtain

$$\oint_C \frac{-y}{x^2 + y^2}\, dx + \frac{x}{x^2 + y^2}\, dy = \int_0^{2\pi} [-\sin t(-\sin t) + \cos t(\cos t)]\, dt$$

$$= \int_0^{2\pi} (\sin^2 t + \cos^2 t)\, dt \qquad (6)$$

$$= \int_0^{2\pi} dt = 2\pi. \qquad \Box$$

It is interesting to note that the result in (6):

$$\oint_C \frac{-y}{x^2 + y^2}\,dx + \frac{x}{x^2 + y^2}\,dy = 2\pi$$

is true for every piecewise smooth simple closed curve C with the origin in its interior. We need only choose C' to be $x^2 + y^2 = a^2$, where a is small enough so that the circle lies entirely within C.

E X E R C I S E S 9 . 1 2 Answers to odd-numbered problems begin on page A-51.

In Problems 1–4, verify Green's theorem by evaluating both integrals.

1. $\oint_C (x - y)\,dx + xy\,dy = \iint_R (y + 1)\,dA$, where C is the triangle with vertices $(0, 0)$, $(1, 0)$, $(1, 3)$

2. $\oint_C 3x^2y\,dx + (x^2 - 5y)\,dy = \iint_R (2x - 3x^2)\,dA$, where C is the rectangle with vertices $(-1, 0)$, $(1, 0)$, $(1, 1)$, $(-1, 1)$

3. $\oint_C -y^2\,dx + x^2\,dy = \iint_R (2x + 2y)\,dA$, where C is the circle $x = 3\cos t$, $y = 3\sin t$, $0 \le t \le 2\pi$

4. $\oint_C -2y^2\,dx + 4xy\,dy = \iint_R 8y\,dA$, where C is the boundary of the region in the first quadrant determined by the graphs of $y = 0$, $y = \sqrt{x}$, $y = -x + 2$

In Problems 5–14, use Green's theorem to evaluate the given line integral.

5. $\oint_C 2y\,dx + 5x\,dy$, where C is the circle $(x - 1)^2 + (y + 3)^2 = 25$

6. $\oint_C (x + y^2)\,dx + (2x^2 - y)\,dy$, where C is the boundary of the region determined by the graphs of $y = x^2$, $y = 4$

7. $\oint_C (x^4 - 2y^3)\,dx + (2x^3 - y^4)\,dy$, where C is the circle $x^2 + y^2 = 4$

8. $\oint_C (x - 3y)\,dx + (4x + y)\,dy$, where C is the rectangle with vertices $(-2, 0)$, $(3, 0)$, $(3, 2)$, $(-2, 2)$

9. $\oint_C 2xy\,dx + 3xy^2\,dy$, where C is the triangle with vertices $(1, 2)$, $(2, 2)$, $(2, 4)$

10. $\oint_C e^{2x}\sin 2y\,dx + e^{2x}\cos 2y\,dy$, where C is the ellipse $9(x - 1)^2 + 4(y - 3)^2 = 36$

11. $\oint_C xy\,dx + x^2\,dy$, where C is the boundary of the region determined by the graphs of $x = 0$, $x^2 + y^2 = 1$, $x \ge 0$

12. $\oint_C e^{x^2}\,dx + 2\tan^{-1}x\,dy$, where C is the triangle with vertices $(0, 0)$, $(0, 1)$, $(-1, 1)$

13. $\oint_C \frac{1}{3}y^3\,dx + (xy + xy^2)\,dy$, where C is the boundary of the region in the first quadrant determined by the graphs of $y = 0$, $x = y^2$, $x = 1 - y^2$

14. $\oint_C xy^2\,dx + 3\cos y\,dy$, where C is the boundary of the region in the first quadrant determined by the graphs of $y = x^2$, $y = x^3$

In Problems 15 and 16, evaluate the given integral on any piecewise smooth simple closed curve C.

15. $\oint_C ay\,dx + bx\,dy$ **16.** $\oint_C P(x)\,dx + Q(y)\,dy$

In Problems 17 and 18, let R be the region bounded by a piecewise smooth simple closed curve C. Prove the given result.

17. $\oint_C x\,dy = -\oint_C y\,dx = \text{area of } R$

18. $\frac{1}{2}\oint_C -y\,dx + x\,dy = \text{area of } R$

In Problems 19 and 20, use the results of Problems 17 and 18 to find the area of the region bounded by the given closed curve.

19. The hypocycloid $x = a\cos^3 t$, $y = a\sin^3 t$, $a > 0$, $0 \le t \le 2\pi$

20. The ellipse $x = a\cos t$, $y = b\sin t$, $a > 0$, $b > 0$, $0 \le t \le 2\pi$

21. (a) Show that

$$\int_C -y\,dx + x\,dy = x_1 y_2 - x_2 y_1,$$

where C is the line segment from the point (x_1, y_1) to (x_2, y_2).

(b) Use part **(a)** and Problem 18 to show that the area A of a polygon with vertices (x_1, y_1), (x_2, y_2), . . . , (x_n, y_n), labeled counterclockwise, is

$$A = \frac{1}{2}(x_1 y_2 - x_2 y_1) + \frac{1}{2}(x_2 y_3 - x_3 y_2) + \cdots$$
$$+ \frac{1}{2}(x_{n-1} y_n - x_n y_{n-1}) + \frac{1}{2}(x_n y_1 - x_1 y_n).$$

22. Use part **(b)** of Problem 21 to find the area of the quadrilateral with vertices $(-1, 3)$, $(1, 1)$, $(4, 2)$, and $(3, 5)$.

In Problems 23 and 24, evaluate the given line integral where $C = C_1 \cup C_2$ is the boundary of the shaded region R.

23. $\oint_C (4x^2 - y^3)\,dx + (x^3 + y^2)\,dy$

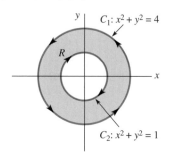

Figure 9.99

24. $\oint_C (\cos x^2 - y)\,dx + \sqrt{y^3 + 1}\,dy$

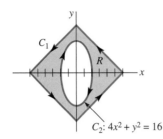

C_2: $4x^2 + y^2 = 16$

Figure 9.100

In Problems 25 and 26, proceed as in Example 6 to evaluate the given line integral.

25. $\oint_C \dfrac{-y^3\,dx + xy^2\,dy}{(x^2 + y^2)^2}$, where C is the ellipse $x^2 + 4y^2 = 4$

26. $\oint_C \dfrac{-y}{(x+1)^2 + 4y^2}\,dx + \dfrac{x+1}{(x+1)^2 + 4y^2}\,dy$, where C is the circle $x^2 + y^2 = 16$

In Problems 27 and 28, use Green's theorem to evaluate the given double integral by means of a line integral. [*Hint:* Find appropriate functions P and Q.]

27. $\iint_R x^2\,dA$; R is the region bounded by the ellipse $x^2/9 + y^2/4 = 1$

28. $\iint_R [1 - 2(y - 1)]\,dA$; R is the region in the first quadrant bounded by the circle $x^2 + (y - 1)^2 = 1$ and $x = 0$

In Problems 29 and 30, use Green's theorem to find the work done by the given force \mathbf{F} around the closed curve in Figure 9.101.

29. $\mathbf{F} = (x - y)\mathbf{i} + (x + y)\mathbf{j}$ **30.** $\mathbf{F} = -xy^2\mathbf{i} + x^2y\mathbf{j}$

31. Let P and Q be continuous and have continuous first partial derivatives in a simply connected region of the xy-plane. If $\int_A^B P\,dx + Q\,dy$ is independent of the path, show that $\oint_C P\,dx + Q\,dy = 0$ on every piecewise smooth simple closed curve C in the region.

32. Let R be a region bounded by a piecewise smooth simple closed curve C. Show that the coordinates of the centroid of the region are given by

$$\bar{x} = \frac{1}{2A} \oint_C x^2\,dy, \quad \bar{y} = -\frac{1}{2A} \oint_C y^2\,dx.$$

33. Find the work done by the force $\mathbf{F} = -y\mathbf{i} + x\mathbf{j}$ acting along the cardioid $r = 1 + \cos\theta$.

Figure 9.101

9.13 Surface Integrals

■ **Surface Area** In the plane the length of an arc of the graph of $y = f(x)$ from $x = a$ to $x = b$ is given by

$$s = \int_a^b \sqrt{1 + \left(\frac{dy}{dx}\right)^2}\,dx. \tag{1}$$

The problem in three dimensions, which is the counterpart of the arc length problem, is to find the area $A(S)$ of that portion of the surface S given by a function $z = f(x, y)$ having continuous first partial derivatives on a closed region R in the xy-plane. Such a surface is said to be **smooth.**

Suppose, as shown in Figure 9.102(a), that an inner partition P of R is formed using lines parallel to the x- and y-axes. P then consists of n rectangular elements R_k of area $\Delta A_k = \Delta x_k \Delta y_k$ that lie entirely within R. Let $(x_k, y_k, 0)$ denote any point in an R_k. As we see in Figure 9.102(a), by projecting the sides of R_k upward, we determine two quantities: a portion S_k of the surface and a portion T_k of a tangent plane at $(x_k, y_k, f(x_k, y_k))$. It seems reasonable to assume that when R_k is small, the area ΔT_k of T_k is approximately the same as the area ΔS_k of S_k.

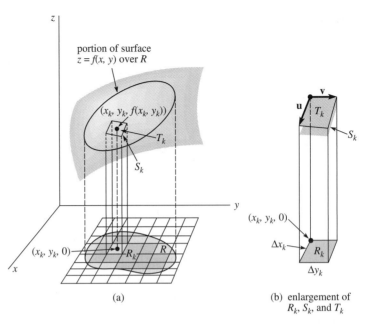

(a)

(b) enlargement of
R_k, S_k, and T_k

Figure 9.102

To find the area of T_k let us choose $(x_k, y_k, 0)$ at a corner of R_k as shown in Figure 9.102(b). The indicated vectors **u** and **v**, which form two sides of T_k, are given by

$$\mathbf{u} = \Delta x_k \mathbf{i} + f_x(x_k, y_k)\, \Delta x_k \mathbf{k}$$

$$\mathbf{v} = \Delta y_k \mathbf{j} + f_y(x_k, y_k)\, \Delta y_k \mathbf{k},$$

where $f_x(x_k, y_k)$ and $f_y(x_k, y_k)$ are the slopes of the lines containing **u** and **v**, respectively. Now from (11) of Section 7.4 we know that $\Delta T_k = |\mathbf{u} \times \mathbf{v}|$ where

$$\mathbf{u} \times \mathbf{v} = \begin{vmatrix} \mathbf{i} & \mathbf{j} & \mathbf{k} \\ \Delta x_k & 0 & f_x(x_k, y_k)\, \Delta x_k \\ 0 & \Delta y_k & f_y(x_k, y_k)\, \Delta y_k \end{vmatrix}$$

$$= [-f_x(x_k, y_k)\mathbf{i} - f_y(x_k, y_k)\mathbf{j} + \mathbf{k}]\, \Delta x_k\, \Delta y_k.$$

In other words,

$$\Delta T_k = \sqrt{[f_x(x_k, y_k)]^2 + [f_y(x_k, y_k)]^2 + 1}\; \Delta x_k\, \Delta y_k.$$

Consequently, the area A is approximately

$$\sum_{k=1}^{n} \sqrt{1 + [f_x(x_k, y_k)]^2 + [f_y(x_k, y_k)]^2}\; \Delta x_k\, \Delta y_k.$$

Taking the limit of the foregoing sum as $\|P\| \to 0$ leads us to the next definition.

DEFINITION 9.11 Surface Area

Let f be a function for which the first partial derivatives f_x and f_y are continuous on a closed region R. Then the **area of the surface over R** is given by

$$A(S) = \iint\limits_{R} \sqrt{1 + [f_x(x, y)]^2 + [f_y(x, y)]^2}\; dA. \qquad (2)$$

One could have almost guessed the form of (2) by naturally extending the one-variable structure of (1) to two variables.

Example 1 Surface Area

Find the surface area of that portion of the sphere $x^2 + y^2 + z^2 = a^2$ that is above the xy-plane and within the cylinder $x^2 + y^2 = b^2$, $0 < b < a$.

SOLUTION If we define $z = f(x, y)$ by $f(x, y) = \sqrt{a^2 - x^2 - y^2}$, then

$$f_x(x, y) = \frac{-x}{\sqrt{a^2 - x^2 - y^2}} \quad \text{and} \quad f_y(x, y) = \frac{-y}{\sqrt{a^2 - x^2 - y^2}}$$

and so

$$1 + [f_x(x, y)]^2 + [f_y(x, y)]^2 = \frac{a^2}{a^2 - x^2 - y^2}.$$

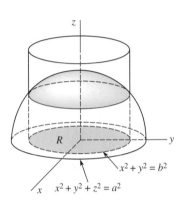

Hence, (2) is

$$A(S) = \iint_R \frac{a}{\sqrt{a^2 - x^2 - y^2}} \, dA,$$

where R is indicated in Figure 9.103. To evaluate this double integral, we change to polar coordinates:

$$A(S) = a \int_0^{2\pi} \int_0^b (a^2 - r^2)^{-1/2} r \, dr \, d\theta$$

$$= a \int_0^{2\pi} \left[-(a^2 - r^2)^{1/2} \right]_0^b \, d\theta = a(a - \sqrt{a^2 - b^2}) \int_0^{2\pi} d\theta$$

$$= 2\pi a(a - \sqrt{a^2 - b^2}) \text{ square units.} \qquad \square$$

Figure 9.103

■ **Differential of Surface Area** The function

$$dS = \sqrt{1 + [f_x(x, y)]^2 + [f_y(x, y)]^2} \, dA \qquad (3)$$

is called the differential of the surface area. We will use this function in the discussion that follows.

■ **Surface Integral** As we have seen, double and triple integrals,

$$\iint_R f(x, y) \, dA \quad \text{and} \quad \iiint_D f(x, y, z) \, dV,$$

are generalizations of the definite integral $\int_a^b f(x) \, dx$. The surface area integral (2) is a generalization of the arc length integral (1). We are now going to consider a generalization of the line integral $\int_C G(x, y) \, ds$. This generalization is called a **surface integral.**

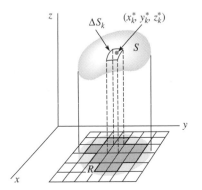

$w = G(x, y, z)$

1. Let G be defined in a region of 3-space that contains a surface S, which is the graph of a function $z = f(x, y)$. Let the projection R of the surface onto the xy-plane be either a Type I or a Type II region.
2. Divide the surface into n pieces of areas ΔS_k corresponding to a partition P of R into n rectangles R_k of areas ΔA_k.
3. Let $\|P\|$ be the **norm** of the partition or the length of the longest diagonal of the R_k.
4. Choose a point (x_k^*, y_k^*, z_k^*) on each element of surface area.
5. Form the sum $\sum_{k=1}^n G(x_k^*, y_k^*, z_k^*) \, \Delta S_k$.

> **DEFINITION 9.12** Surface Integral
>
> Let G be a function of three variables defined over a region of space containing the surface S. Then the **surface integral of G over S** is given by
>
> $$\iint_S G(x, y, z)\, dS = \lim_{\|P\| \to 0} \sum_{k=1}^{n} G(x_k^*, y_k^*, z_k^*)\, \Delta S_k. \qquad (4)$$

■ **Method of Evaluation** If G, f, f_x, and f_y are continuous throughout a region containing S, we can evaluate (4) by means of a double integral. From (3) the left side of (4) becomes

$$\iint_S G(x, y, z)\, dS = \iint_R G(x, y, f(x, y))\sqrt{1 + [f_x(x, y)]^2 + [f_y(x, y)]^2}\, dA. \qquad (5)$$

Note that when $G = 1$, (5) reduces to formula (2) for surface area.

■ **Projection of S Into Other Planes** If $y = g(x, z)$ is the equation of a surface S that projects onto a region R of the xz-plane, then

$$\iint_S G(x, y, z)\, dS = \iint_R G(x, g(x, z), z)\sqrt{1 + [g_x(x, z)]^2 + [g_z(x, z)]^2}\, dA. \qquad (6)$$

Similarly, if $x = h(y, z)$ is the equation of a surface that projects onto the yz-plane, then the analogue of (5) is

$$\iint_S G(x, y, z)\, dS = \iint_R G(h(y, z), y, z)\sqrt{1 + [h_y(y, z)]^2 + [h_z(y, z)]^2}\, dA. \qquad (7)$$

■ **Mass of a Surface** Suppose $\rho(x, y, z)$ represents the density of a surface at any point, or mass per unit surface area; then the **mass** m of the surface is

$$m = \iint_S \rho(x, y, z)\, dS. \qquad (8)$$

Example 2 Mass of a Surface

Find the mass of the surface of the paraboloid $z = 1 + x^2 + y^2$ in the first octant for $1 \le z \le 5$ if the density at a point P on the surface is directly proportional to its distance from the xy-plane.

SOLUTION The surface in question and its projection onto the xy-plane are shown in Figure 9.104. Now, since $\rho(x, y, z) = kz$ and $z = 1 + x^2 + y^2$, (8) and (5) give

$$m = \iint_S kz\, dS = k \iint_R (1 + x^2 + y^2)\sqrt{1 + 4x^2 + 4y^2}\, dA.$$

By changing to polar coordinates, we obtain

$$m = k \int_0^{\pi/2} \int_0^2 (1 + r^2)\sqrt{1 + 4r^2}\, r\, dr\, d\theta$$

$$= k \int_0^{\pi/2} \int_0^2 [r(1 + 4r^2)^{1/2} + r^3(1 + 4r^2)^{1/2}]\, dr\, d\theta \quad \leftarrow \text{integration by parts}$$

$$= k \int_0^{\pi/2} \left[\frac{1}{12}(1 + 4r^2)^{3/2} + \frac{1}{12}r^2(1 + 4r^2)^{3/2} - \frac{1}{120}(1 + 4r^2)^{5/2} \right]_0^2 d\theta$$

$$= \frac{k\pi}{2}\left[\frac{5(17)^{3/2}}{12} - \frac{17^{5/2}}{120} - \frac{3}{40} \right] \approx 19.2k \qquad \Box$$

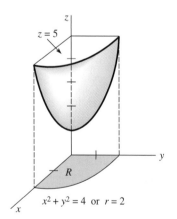

$z = 5$

R

$x^2 + y^2 = 4$ or $r = 2$

Figure 9.104

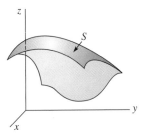

$y = 2x^2 + 1$

Figure 9.105

Example 3 Evaluating a Surface Integral

Evaluate $\iint_S xz^2 \, dS$, where S is that portion of the cylinder $y = 2x^2 + 1$ in the first octant bounded by $x = 0$, $x = 2$, $z = 4$, and $z = 8$.

SOLUTION We shall use (6) with $g(x, z) = 2x^2 + 1$ and R the rectangular region in the xz-plane shown in Figure 9.105. Since $g_x(x, z) = 4x$ and $g_z(x, z) = 0$, it follows that

$$\iint_S xz^2 \, dS = \int_0^2 \int_4^8 xz^2 \sqrt{1 + 16x^2} \, dz \, dx$$

$$= \int_0^2 \frac{z^3}{3} x\sqrt{1 + 16x^2} \Big]_4^8 \, dx = \frac{448}{3} \int_0^2 x(1 + 16x^2)^{1/2} \, dx$$

$$= \frac{28}{9} (1 + 16x^2)^{3/2} \Big]_0^2 = \frac{28}{9} [65^{3/2} - 1] \approx 1627.3. \qquad \square$$

(a) Two-sided surface

(b) One-sided surface

Figure 9.106

■ **Orientable Surfaces** In Example 5, we are going to evaluate a surface integral of a vector field. In order to do this we need to examine the concept of an **orientable surface.** Roughly, an orientable surface S, such as that given in Figure 9.106(a), has two sides that could be painted different colors. The Möbius strip* shown in Figure 9.106(b) is not an orientable surface and is one-sided. A person who starts to paint the surface of a Möbius strip at a point will paint the entire surface and return to the starting point.

Specifically, we say a smooth surface S is **orientable** or is an **oriented surface** if there exists a continuous unit normal vector function \mathbf{n} defined at each point (x, y, z) on the surface. The vector field $\mathbf{n}(x, y, z)$ is called the **orientation** of S. But since a unit normal to the surface S at (x, y, z) can be either $\mathbf{n}(x, y, z)$ or $-\mathbf{n}(x, y, z)$, an orientable surface has two orientations. See Figure 9.107(a)–(c). The Möbius strip shown again in Figure 9.107(d) is not an oriented surface, since if a unit normal \mathbf{n} starts at P on the surface and moves *once* around the strip on the curve C, it ends up on the "opposite side" of the strip at P and so points in the opposite direction. A surface S defined by $z = f(x, y)$ has an **upward orientation** (Figure 9.107(b)) when the unit normals are directed upward—that is, have positive \mathbf{k} components, and it has a **downward orientation** (Figure 9.107(c)) when the unit normals are directed downward—that is, have negative \mathbf{k} components.

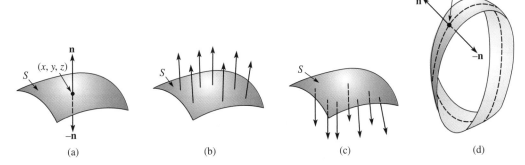

(a)　　　(b)　　　(c)　　　(d)

Figure 9.107

* To construct a Möbius strip cut out a long strip of paper, give one end a half-twist, and then attach the ends with tape.

(a)

(b)

Figure 9.108

Figure 9.109

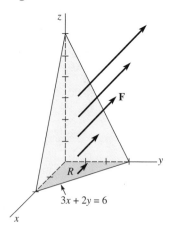

$3x + 2y = 6$

Figure 9.110

If a smooth surface S is defined by $g(x, y, z) = 0$, then recall that a unit normal is

$$\mathbf{n} = \frac{1}{\|\nabla g\|} \nabla g, \tag{9}$$

where $\nabla g = \dfrac{\partial g}{\partial x}\mathbf{i} + \dfrac{\partial g}{\partial y}\mathbf{j} + \dfrac{\partial g}{\partial z}\mathbf{k}$ is the gradient of g. If S is defined by $z = f(x, y)$, then we can use $g(x, y, z) = z - f(x, y) = 0$ or $g(x, y, z) = f(x, y) - z = 0$ depending on the orientation of S.

As we shall see in the next example, the two orientations of an orientable closed surface are **outward** and **inward.**

Example 4 Orientations of a Surface

Consider the sphere of radius $a > 0$: $x^2 + y^2 + z^2 = a^2$. If we define $g(x, y, z) = x^2 + y^2 + z^2 - a^2$, then

$$\nabla g = 2x\mathbf{i} + 2y\mathbf{j} + 2z\mathbf{k} \quad \text{and} \quad \|\nabla g\| = \sqrt{4x^2 + 4y^2 + 4z^2} = 2a.$$

Then the two orientations of the surface are

$$\mathbf{n} = \frac{x}{a}\mathbf{i} + \frac{y}{a}\mathbf{j} + \frac{z}{a}\mathbf{k} \quad \text{and} \quad \mathbf{n}_1 = -\mathbf{n} = -\frac{x}{a}\mathbf{i} - \frac{y}{a}\mathbf{j} - \frac{z}{a}\mathbf{k}.$$

The vector field \mathbf{n} defines an outward orientation, whereas $\mathbf{n}_1 = -\mathbf{n}$ defines an inward orientation. See Figure 9.108. ❑

■ Integrals of Vector Fields

If $\mathbf{F}(x, y, z) = P(x, y, z)\mathbf{i} + Q(x, y, z)\mathbf{j} + R(x, y, z)\mathbf{k}$ is the velocity field of a fluid, then, as we saw in Figure 9.39, the volume of the fluid flowing through an element of surface area ΔS per unit time is approximated by

$$(\text{height})(\text{area of base}) = (\text{comp}_\mathbf{n}\mathbf{F}) \, \Delta S = (\mathbf{F} \cdot \mathbf{n}) \, \Delta S,$$

where \mathbf{n} is a unit normal to the surface. See Figure 9.109. The total volume of a fluid passing through S per unit time is called the **flux of F through S** and is given by

$$\text{flux} = \iint_S (\mathbf{F} \cdot \mathbf{n}) \, dS. \tag{10}$$

In the case of a closed surface S, if \mathbf{n} is the outer (inner) normal, then (10) gives the volume of fluid flowing out (in) through S per unit time.

Example 5 Flux Through a Surface

Let $\mathbf{F}(x, y, z) = z\mathbf{j} + z\mathbf{k}$ represent the flow of a liquid. Find the flux of \mathbf{F} through the surface S given by that portion of the plane $z = 6 - 3x - 2y$ in the first octant oriented upward.

SOLUTION The vector field and the surface are illustrated in Figure 9.110. By defining the plane by $g(x, y, z) = 3x + 2y + z - 6 = 0$, we see that a unit normal with a positive \mathbf{k} component is

$$\mathbf{n} = \frac{\nabla g}{\|\nabla g\|} = \frac{3}{\sqrt{14}}\mathbf{i} + \frac{2}{\sqrt{14}}\mathbf{j} + \frac{1}{\sqrt{14}}\mathbf{k}.$$

Hence, $$\text{flux} = \iint_S (\mathbf{F} \cdot \mathbf{n}) \, dS = \frac{1}{\sqrt{14}} \iint_S 3z \, dS.$$

With R the projection of the surface onto the xy-plane, we find from (10) that

$$\text{flux} = \frac{1}{\sqrt{14}} \iint_R 3(6 - 3x - 2y)(\sqrt{14}\, dA)$$

$$= 3 \int_0^2 \int_0^{3-3x/2} (6 - 3x - 2y)\, dy\, dx = 18. \qquad \square$$

Depending on the nature of the vector field, the integral in (10) can represent other kinds of flux. For example, (10) could also give electric flux, magnetic flux, flux of heat, and so on.

R e m a r k s

If the surface S is piecewise defined, we express a surface integral over S as the sum of the surface integrals over the various pieces of the surface. For example, suppose S is the orientable piecewise smooth closed surface bounded by the paraboloid $z = x^2 + y^2$ (S_1) and the plane $z = 1$ (S_2). Then the flux of a vector field \mathbf{F} out of the surface S is

$$\iint_S \mathbf{F} \cdot \mathbf{n}\, dS = \iint_{S_1} \mathbf{F} \cdot \mathbf{n}\, dS + \iint_{S_2} \mathbf{F} \cdot \mathbf{n}\, dS,$$

where we take S_1 oriented upward and S_2 oriented downward. See Figure 9.111 and Problem 35 in Exercises 9.13.

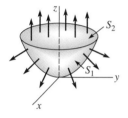

Figure 9.111

Answers to odd-numbered problems begin on page A-51.

1. Find the surface area of that portion of the plane $2x + 3y + 4z = 12$ that is bounded by the coordinate planes in the first octant.

2. Find the surface area of that portion of the plane $2x + 3y + 4z = 12$ that is above the region in the first quadrant bounded by the graph $r = \sin 2\theta$.

3. Find the surface area of that portion of the cylinder $x^2 + z^2 = 16$ that is above the region in the first quadrant bounded on the graphs of $x = 0$, $x = 2$, $y = 0$, $y = 5$.

4. Find the surface area of that portion of the paraboloid $z = x^2 + y^2$ that is below the plane $z = 2$.

5. Find the surface area of that portion of the paraboloid $z = 4 - x^2 - y^2$ that is above the xy-plane.

6. Find the surface area of those portions of the sphere $x^2 + y^2 + z^2 = 2$ that are within the cone $z^2 = x^2 + y^2$.

7. Find the surface area of the portion of the sphere $x^2 + y^2 + z^2 = 25$ that is above the region in the first quadrant bounded by the graphs of $x = 0$, $y = 0$, $4x^2 + y^2 = 25$. [*Hint:* Integrate first with respect to x.]

8. Find the surface area of that portion of the graph of $z = x^2 - y^2$ that is in the first octant within the cylinder $x^2 + y^2 = 4$.

9. Find the surface area of the portions of the sphere $x^2 + y^2 + z^2 = a^2$ that are within the cylinder $x^2 + y^2 = ay$.

10. Find the surface area of the portions of the cone $z^2 = \frac{1}{4}(x^2 + y^2)$ that are within the cylinder $(x - 1)^2 + y^2 = 1$.

11. Find the surface area of the portions of the cylinder $y^2 + z^2 = a^2$ that are within the cylinder $x^2 + y^2 = a^2$. [*Hint:* See Figure 9.79.]

12. Use the result given in Example 1 to prove that the surface area of a sphere of radius a is $4\pi a^2$. [*Hint:* Consider a limit as $b \to a$.]

13. Find the surface area of that portion of the sphere $x^2 + y^2 + z^2 = a^2$ that is bounded between $y = c_1$ and $y = c_2$, $0 < c_1 < c_2 < a$. [*Hint:* Use polar coordinates in the xz-plane.]

14. Show that the area found in Problem 13 is the same as the surface area of the cylinder $x^2 + z^2 = a^2$ between $y = c_1$ and $y = c_2$.

In Problems 15–24, evaluate the surface integral $\iint_S G(x, y, z)\, dS$.

15. $G(x, y, z) = x$; S the portion of the cylinder $z = 2 - x^2$ in the first octant bounded by $x = 0$, $y = 0$, $y = 4$, $z = 0$

16. $G(x, y, z) = xy(9 - 4z)$; same surface as in Problem 15

17. $G(x, y, z) = xz^3$; S the cone $z = \sqrt{x^2 + y^2}$ inside the cylinder $x^2 + y^2 = 1$

18. $G(x, y, z) = x + y + z$; S the cone $z = \sqrt{x^2 + y^2}$ between $z = 1$ and $z = 4$

19. $G(x, y, z) = (x^2 + y^2)z$; S that portion of the sphere $x^2 + y^2 + z^2 = 36$ in the first octant

20. $G(x, y, z) = z^2$; S that portion of the plane $z = x + 1$ within the cylinder $y = 1 - x^2$, $0 \le y \le 1$

21. $G(x, y, z) = xy$; S that portion of the paraboloid $2z = 4 - x^2 - y^2$ within $0 \le x \le 1$, $0 \le y \le 1$

22. $G(x, y, z) = 2z$; S that portion of the paraboloid $2z = 1 + x^2 + y^2$ in the first octant bounded by $x = 0$, $y = \sqrt{3}x$, $z = 1$

23. $G(x, y, z) = 24\sqrt{yz}$; S that portion of the cylinder $y = x^2$ in the first octant bounded by $y = 0$, $y = 4$, $z = 0$, $z = 3$

24. $G(x, y, z) = (1 + 4y^2 + 4z^2)^{1/2}$; S that portion of the paraboloid $x = 4 - y^2 - z^2$ in the first octant outside the cylinder $y^2 + z^2 = 1$

In Problems 25 and 26, evaluate $\iint_S (3z^2 + 4yz)\, dS$, where S is that portion of the plane $x + 2y + 3z = 6$ in the first octant. Use the projection of S onto the coordinate plane indicated in the given figure.

25.

26.

Figure 9.112 **Figure 9.113**

In Problems 27 and 28, find the mass of the given surface with the indicated density function.

27. S that portion of the plane $x + y + z = 1$ in the first octant; density at a point P directly proportional to the square of the distance from the yz-plane

28. S the hemisphere $z = \sqrt{4 - x^2 - y^2}$; $\rho(x, y, z) = |xy|$

In Problems 29–34, let \mathbf{F} be a vector field. Find the flux of \mathbf{F} through the given surface. Assume the surface S is oriented upward.

29. $\mathbf{F} = x\mathbf{i} + 2z\mathbf{j} + y\mathbf{k}$; S that portion of the cylinder $y^2 + z^2 = 4$ in the first octant bounded by $x = 0$, $x = 3$, $y = 0$, $z = 0$

30. $\mathbf{F} = z\mathbf{k}$; S that part of the paraboloid $z = 5 - x^2 - y^2$ inside the cylinder $x^2 + y^2 = 4$

31. $\mathbf{F} = x\mathbf{i} + y\mathbf{j} + z\mathbf{k}$; same surface S as in Problem 30

32. $\mathbf{F} = -x^3y\mathbf{i} + yz^3\mathbf{j} + xy^3\mathbf{k}$; S that portion of the plane $z = x + 3$ in the first octant within the cylinder $x^2 + y^2 = 2x$

33. $\mathbf{F} = \frac{1}{2}x^2\mathbf{i} + \frac{1}{2}y^2\mathbf{j} + z\mathbf{k}$; S that portion of the paraboloid $z = 4 - x^2 - y^2$ for $0 \le z \le 4$

34. $\mathbf{F} = e^y\mathbf{i} + e^x\mathbf{j} + 18y\mathbf{k}$; S that portion of the plane $x + y + z = 6$ in the first octant

35. Find the flux of $\mathbf{F} = y^2\mathbf{i} + x^2\mathbf{j} + 5z\mathbf{k}$ out of the closed surface S given in Figure 9.111.

36. Find the flux of $\mathbf{F} = -y\mathbf{i} + x\mathbf{j} + 6z^2\mathbf{k}$ out of the closed surface S bounded by the paraboloids $z = 4 - x^2 - y^2$ and $z = x^2 + y^2$.

37. Let $T(x, y, z) = x^2 + y^2 + z^2$ represent temperature and let the "flow" of heat be given by the vector field $\mathbf{F} = -\nabla T$. Find the flux of heat out of the sphere $x^2 + y^2 + z^2 = a^2$. [*Hint:* The surface area of a sphere of radius a is $4\pi a^2$.]

38. Find the flux of $\mathbf{F} = x\mathbf{i} + y\mathbf{j} + z\mathbf{k}$ out of the unit cube $0 \le x \le 1$, $0 \le y \le 1$, $0 \le z \le 1$. See Figure 9.114. Use the fact that the flux out of the cube is the sum of the fluxes out of the sides.

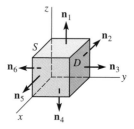

Figure 9.114

39. Coulomb's law states that the electric field \mathbf{E} due to a point charge q at the origin is given by $\mathbf{E} = kq\mathbf{r}/\|\mathbf{r}\|^3$, where k is a constant and $\mathbf{r} = x\mathbf{i} + y\mathbf{j} + z\mathbf{k}$. Determine the flux out of a sphere $x^2 + y^2 + z^2 = a^2$.

40. If $\sigma(x, y, z)$ is charge density in an electrostatic field, then the total charge on a surface S is $Q = \iint_S \sigma(x, y, z)\, dS$. Find the total charge on that part of the hemisphere $z = \sqrt{16 - x^2 - y^2}$ that is inside the cylinder $x^2 + y^2 = 9$ if the charge density at a point P on the surface is directly proportional to distance from the xy-plane.

41. The coordinates of the centroid of a surface are given by

$$\bar{x} = \frac{\iint_S x\, dS}{A(S)}, \quad \bar{y} = \frac{\iint_S y\, dS}{A(S)}, \quad \bar{z} = \frac{\iint_S z\, dS}{A(S)},$$

where $A(S)$ is the area of the surface. Find the centroid of that portion of the plane $2x + 3y + z = 6$ in the first octant.

42. Use the information in Problem 41 to find the centroid of the hemisphere $z = \sqrt{a^2 - x^2 - y^2}$.

43. Let $z = f(x, y)$ be the equation of a surface S and \mathbf{F} be the vector field $\mathbf{F}(x, y, z) = P(x, y, z)\mathbf{i} + Q(x, y, z)\mathbf{j} + R(x, y, z)\mathbf{k}$. Show that $\iint_S (\mathbf{F} \cdot \mathbf{n})\, dS$ equals

$$\iint_R \left[-P(x, y, z)\frac{\partial z}{\partial x} - Q(x, y, z)\frac{\partial z}{\partial y} + R(x, y, z) \right] dA$$

9.14 Stokes' Theorem

■ **A Vector Form of Green's Theorem** Green's theorem of the preceding section has two vector forms. In this and the next sections we shall generalize these forms to three dimensions. If $\mathbf{F}(x, y) = P(x, y)\mathbf{i} + Q(x, y)\mathbf{j}$ is a two-dimensional vector field, then

$$\text{curl } \mathbf{F} = \nabla \times \mathbf{F} = \begin{vmatrix} \mathbf{i} & \mathbf{j} & \mathbf{k} \\ \dfrac{\partial}{\partial x} & \dfrac{\partial}{\partial y} & \dfrac{\partial}{\partial z} \\ P & 0 & 0 \end{vmatrix} = \left(\dfrac{\partial Q}{\partial x} - \dfrac{\partial P}{\partial y} \right)\mathbf{k}.$$

From (12) and (13) of Section 9.8, Green's theorem can be written in vector notation as

$$\oint_C \mathbf{F} \cdot d\mathbf{r} = \oint_C \mathbf{F} \cdot \mathbf{T} \, ds = \iint_R (\text{curl } \mathbf{F}) \cdot \mathbf{k} \, dA, \tag{1}$$

that is, the line integral of the tangential component of \mathbf{F} is the double integral of the normal component of curl \mathbf{F}.

■ **Green's Theorem in 3-Space** The vector form of Green's theorem given in (1) relates a line integral around a piecewise smooth simple closed curve C forming the boundary of a plane region R to a double integral over R. Green's theorem in 3-space relates a line integral around a piecewise smooth simple closed curve C forming the boundary of a surface S with a surface integral over S. Suppose $z = f(x, y)$ is a continuous function whose graph is a piecewise smooth orientable surface over a region R on the xy-plane. Let C form the boundary of S and let the projection of C onto the xy-plane form the boundary of R. The positive direction on C is induced by the orientation of the surface S; the positive direction on C corresponds to the direction a person would have to walk on C to have his or her head point in the direction of the orientation of the surface while keeping the surface to the left. See Figure 9.115. More precisely, the positive orientation of C is in accordance with the right-hand rule: If the thumb of the right hand points in the direction of the orientation of the surface, then roughly the fingers of the right hand wrap around the surface in the positive direction. Finally, let \mathbf{T} be a unit tangent vector to C that points in the positive direction. The three-dimensional form of Green's theorem, which we now give, is called **Stokes' theorem.**

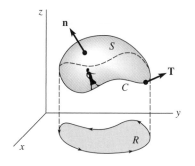

Figure 9.115

THEOREM 9.14 Stokes' Theorem

Let S be a piecewise smooth orientable surface bounded by a piecewise smooth simple closed curve C. Let $\mathbf{F}(x, y, z) = P(x, y, z)\mathbf{i} + Q(x, y, z)\mathbf{j} + R(x, y, z)\mathbf{k}$ be a vector field for which P, Q, and R are continuous and have continuous first partial derivatives in a region of 3-space containing S. If C is traversed in the positive direction, then

$$\oint_C \mathbf{F} \cdot d\mathbf{r} = \oint_C (\mathbf{F} \cdot \mathbf{T}) \, dS = \iint_S (\text{curl } \mathbf{F}) \cdot \mathbf{n} \, dS, \tag{2}$$

where \mathbf{n} is a unit normal to S in the direction of the orientation of S.

PARTIAL PROOF Suppose the surface S is oriented upward and is defined by a function $z = f(x, y)$ that has continuous second partial derivatives. From Definition 9.7 we have

$$\text{curl } \mathbf{F} = \left(\frac{\partial R}{\partial y} - \frac{\partial Q}{\partial z}\right)\mathbf{i} + \left(\frac{\partial P}{\partial z} - \frac{\partial R}{\partial x}\right)\mathbf{j} + \left(\frac{\partial Q}{\partial x} - \frac{\partial P}{\partial y}\right)\mathbf{k}.$$

Furthermore, if we write $g(x, y, z) = z - f(x, y) = 0$, then

$$\mathbf{n} = \frac{-\dfrac{\partial f}{\partial x}\,\mathbf{i} - \dfrac{\partial f}{\partial y}\,\mathbf{j} + \mathbf{k}}{\sqrt{1 + \left(\dfrac{\partial f}{\partial x}\right)^2 + \left(\dfrac{\partial f}{\partial y}\right)^2}}.$$

Hence,

$$\iint_S (\text{curl } \mathbf{F}) \cdot \mathbf{n}\, dS = \iint_R \left[-\left(\frac{\partial R}{\partial y} - \frac{\partial Q}{\partial z}\right)\frac{\partial f}{\partial x} - \left(\frac{\partial P}{\partial z} - \frac{\partial R}{\partial x}\right)\frac{\partial f}{\partial y} + \left(\frac{\partial Q}{\partial x} - \frac{\partial P}{\partial y}\right)\right] dA. \quad (3)$$

Our goal is now to show that $\oint_C \mathbf{F} \cdot d\mathbf{r}$ reduces to (3).

If C_{xy} is the projection of C onto the xy-plane and has the parametric equations $x = x(t)$, $y = y(t)$, $a \le t \le b$, then parametric equations for C are $x = x(t)$, $y = y(t)$, $z = f(x(t), y(t))$, $a \le t \le b$. Thus,

$$\oint_C \mathbf{F} \cdot d\mathbf{r} = \oint_C P\, dx + Q\, dy + R\, dz$$

$$= \int_a^b \left[P\frac{dx}{dt} + Q\frac{dy}{dt} + R\left(\frac{\partial f}{\partial x}\frac{dx}{dt} + \frac{\partial f}{\partial y}\frac{dy}{dt}\right)\right] dt \quad \text{Chain rule}$$

$$= \oint_{C_{xy}} \left(P + R\frac{\partial f}{\partial x}\right) dx + \left(Q + R\frac{\partial f}{\partial y}\right) dy \quad (4)$$

$$= \iint_R \left[\frac{\partial}{\partial x}\left(Q + R\frac{\partial f}{\partial y}\right) - \frac{\partial}{\partial y}\left(P + R\frac{\partial f}{\partial x}\right)\right] dA. \quad \text{Green's theorem}$$

Now,

$$\frac{\partial}{\partial x}\left(Q + R\frac{\partial f}{\partial y}\right) = \frac{\partial}{\partial x}\left[Q(x, y, f(x, y)) + R(x, y, f(x, y))\frac{\partial f}{\partial y}\right]$$

$$= \frac{\partial Q}{\partial x} + \frac{\partial Q}{\partial z}\frac{\partial f}{\partial x} + R\frac{\partial^2 f}{\partial x\, \partial y} + \frac{\partial f}{\partial y}\left(\frac{\partial R}{\partial x} + \frac{\partial R}{\partial z}\frac{\partial f}{\partial x}\right) \quad \substack{\text{Chain and} \\ \text{product rules}} \quad (5)$$

$$= \frac{\partial Q}{\partial x} + \frac{\partial Q}{\partial z}\frac{\partial f}{\partial x} + R\frac{\partial^2 f}{\partial x\, \partial y} + \frac{\partial R}{\partial x}\frac{\partial f}{\partial y} + \frac{\partial R}{\partial z}\frac{\partial f}{\partial y}\frac{\partial f}{\partial x}.$$

Similarly,

$$\frac{\partial}{\partial y}\left(P + R\frac{\partial f}{\partial x}\right) = \frac{\partial P}{\partial y} + \frac{\partial P}{\partial z}\frac{\partial f}{\partial y} + R\frac{\partial^2 f}{\partial y\, \partial x} + \frac{\partial R}{\partial y}\frac{\partial f}{\partial x} + \frac{\partial R}{\partial z}\frac{\partial f}{\partial x}\frac{\partial f}{\partial y}. \quad (6)$$

Subtracting (6) from (5) and using the fact that $\partial^2 f / \partial x\, \partial y = \partial^2 f / \partial y\, \partial x$, we see that (4) becomes, after rearranging,

$$\iint_R \left[-\left(\frac{\partial R}{\partial y} - \frac{\partial Q}{\partial z}\right)\frac{\partial f}{\partial x} - \left(\frac{\partial P}{\partial z} - \frac{\partial R}{\partial x}\right)\frac{\partial f}{\partial y} + \left(\frac{\partial Q}{\partial x} - \frac{\partial P}{\partial y}\right)\right] dA.$$

This last expression is the same as the right side of (3), which was to be shown. ❏

Example 1 Verifying Stokes' Theorem

Let S be the part of the cylinder $z = 1 - x^2$ for $0 \le x \le 1$, $-2 \le y \le 2$. Verify Stokes' theorem if $\mathbf{F} = xy\mathbf{i} + yz\mathbf{j} + xz\mathbf{k}$.

SOLUTION The surface S, the curve C (which is composed of the union of C_1, C_2, C_3, and C_4), and the region R are shown in Figure 9.116.

Surface Integral: From $\mathbf{F} = xy\mathbf{i} + yz\mathbf{j} + xz\mathbf{k}$, we find

$$\text{curl } \mathbf{F} = \begin{vmatrix} \mathbf{i} & \mathbf{j} & \mathbf{k} \\ \dfrac{\partial}{\partial x} & \dfrac{\partial}{\partial y} & \dfrac{\partial}{\partial z} \\ xy & yz & xz \end{vmatrix} = -y\mathbf{i} - z\mathbf{j} - x\mathbf{k}.$$

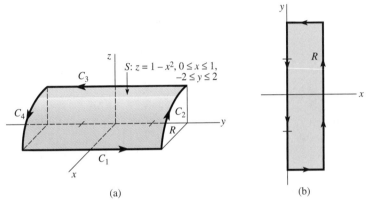

(a) (b)

Figure 9.116

Now, if $g(x, y, z) = z + x^2 - 1 = 0$ defines the cylinder, then the upper normal is

$$\mathbf{n} = \frac{\nabla g}{\|\nabla g\|} = \frac{2x\mathbf{i} + \mathbf{k}}{\sqrt{4x^2 + 1}}.$$

Therefore,
$$\iint\limits_{S} (\text{curl } \mathbf{F} \cdot \mathbf{n}) \, dS = \iint\limits_{S} \frac{-2xy - x}{\sqrt{4x^2 + 1}} \, dS.$$

To evaluate the latter surface integral, we use (5) of Section 9.13:

$$\iint\limits_{S} \frac{-2xy - x}{\sqrt{4x^2 + 1}} \, dS = \iint\limits_{R} (-2xy - x) \, dA$$

$$= \int_0^1 \int_{-2}^2 (-2xy - x) \, dy \, dx$$

$$= \int_0^1 \left[-xy^2 - xy \right]_{-2}^2 \, dx \tag{7}$$

$$= \int_0^1 (-4x) \, dx = -2.$$

Line Integral: We write $\displaystyle\oint_C = \int_{C_1} + \int_{C_2} + \int_{C_3} + \int_{C_4}$. On C_1: $x = 1$, $z = 0$, $dx = 0$, $dz = 0$, so

$$\int_{C_1} y(0) + y(0) \, dy + 0 = 0.$$

On C_2: $y = 2$, $z = 1 - x^2$, $dy = 0$, $dz = -2x \, dx$, so

$$\int_{C_2} 2x \, dx + 2(1 - x^2)0 + x(1 - x^2)(-2x \, dx) = \int_1^0 (2x - 2x^2 + 2x^4) \, dx = -\frac{11}{15}.$$

On C_3: $x = 0$, $z = 1$, $dx = 0$, $dz = 0$, so

$$\int_{C_3} 0 + y \, dy + 0 = \int_2^{-2} y \, dy = 0.$$

On C_4: $y = -2$, $z = 1 - x^2$, $dy = 0$, $dz = -2x \, dx$, so

$$\int_{C_4} -2x \, dx - 2(1 - x^2)0 + x(1 - x^2)(-2x \, dx) = \int_0^1 (-2x - 2x^2 + 2x^4) \, dx = -\frac{19}{15}.$$

Hence, $\displaystyle\oint_C xy \, dx + yz \, dy + xz \, dz = 0 - \frac{11}{15} + 0 - \frac{19}{15} = -2,$

which, of course, agrees with (7). ☐

Example 2 Using Stokes' Theorem

Evaluate $\oint_C z \, dx + x \, dy + y \, dz$, where C is the trace of the cylinder $x^2 + y^2 = 1$ in the plane $y + z = 2$. Orient C counterclockwise as viewed from above. See Figure 9.117.

SOLUTION If $\mathbf{F} = z\mathbf{i} + x\mathbf{j} + y\mathbf{k}$, then

$$\operatorname{curl} \mathbf{F} = \begin{vmatrix} \mathbf{i} & \mathbf{j} & \mathbf{k} \\ \dfrac{\partial}{\partial x} & \dfrac{\partial}{\partial y} & \dfrac{\partial}{\partial z} \\ z & x & y \end{vmatrix} = \mathbf{i} + \mathbf{j} + \mathbf{k}.$$

The given orientation of C corresponds to an upward orientation of the surface S. Thus, if $g(x, y, z) = y + z - 2 = 0$ defines the plane, then the upper normal is

$$\mathbf{n} = \frac{\nabla g}{\|\nabla g\|} = \frac{1}{\sqrt{2}}\mathbf{j} + \frac{1}{\sqrt{2}}\mathbf{k}.$$

Hence, from (2),

$$\oint_C \mathbf{F} \cdot d\mathbf{r} = \iint_S \left[(\mathbf{i} + \mathbf{j} + \mathbf{k}) \cdot \left(\frac{1}{\sqrt{2}}\mathbf{j} + \frac{1}{\sqrt{2}}\mathbf{k} \right) \right] dS$$

$$= \sqrt{2} \iint_S dS = \sqrt{2} \iint_R \sqrt{2} \, dA = 2\pi.$$ ☐

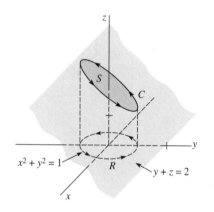

Figure 9.117

Note that if \mathbf{F} is the gradient of a scalar function, then, in view of (5) in Section 9.7, (2) implies that the circulation $\oint_C \mathbf{F} \cdot d\mathbf{r}$ is zero. Conversely, it can be shown that if the circulation is zero for every simple closed curve, then \mathbf{F} is the gradient of a scalar function. In other words, \mathbf{F} is irrotational if and only if $\mathbf{F} = \nabla \phi$, where ϕ is a potential for \mathbf{F}. Equivalently, this gives a test for a conservative vector field: \mathbf{F} *is a conservative vector field if and only if* $\operatorname{curl} \mathbf{F} = \mathbf{0}$.

■ **Physical Interpretation of Curl** In Section 9.8 we saw that if \mathbf{F} is a velocity field of a fluid, then the circulation $\oint_C \mathbf{F} \cdot d\mathbf{r}$ of \mathbf{F} around C is a measure of the amount by which the fluid tends to turn the curve C by circulating around it. The circulation of \mathbf{F} is closely related to the curl of \mathbf{F}. To see this, suppose $P_0(x_0, y_0, z_0)$ is any

Figure 9.118

Figure 9.119

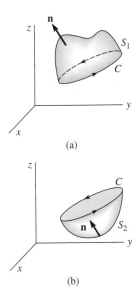

(a)

(b)

Figure 9.120

point in the fluid and C_r is a small circle of radius r centered at P_0. See Figure 9.118. Then by Stokes' theorem,

$$\oint_{C_r} \mathbf{F} \cdot d\mathbf{r} = \iint_{S_r} (\text{curl } \mathbf{F}) \cdot \mathbf{n} \, dS. \tag{8}$$

Now at all points $P(x, y, z)$ within the small circle C_r, if we take curl $\mathbf{F}(P) \approx$ curl $\mathbf{F}(P_0)$, then (8) gives the approximation

$$\oint_{C_r} \mathbf{F} \cdot d\mathbf{r} \approx \iint_{S_r} (\text{curl } \mathbf{F}(P_0)) \cdot \mathbf{n}(P_0) \, dS$$

$$= (\text{curl } \mathbf{F}(P_0)) \cdot \mathbf{n}(P_0) \iint_{S_r} dS \tag{9}$$

$$= (\text{curl } \mathbf{F}(P_0)) \cdot \mathbf{n}(P_0) A_r,$$

where A_r is the area (πr^2) of the circular surface S_r. As we let $r \to 0$, the approximation curl $\mathbf{F}(P) \approx$ curl $\mathbf{F}(P_0)$ becomes better, and so (9) yields

$$(\text{curl } \mathbf{F}(P_0)) \cdot \mathbf{n}(P_0) = \lim_{r \to 0} \frac{1}{A_r} \oint_{C_r} \mathbf{F} \cdot d\mathbf{r}. \tag{10}$$

Thus we see that the normal component of curl \mathbf{F} is the limiting value of the ratio of the circulation of \mathbf{F} to the area of the circular surface. For a small but fixed value of r, we have

$$(\text{curl } \mathbf{F}(P_0)) \cdot \mathbf{n}(P_0) \approx \frac{1}{A_r} \oint_{C_r} \mathbf{F} \cdot d\mathbf{r}. \tag{11}$$

Roughly then, the curl of \mathbf{F} is the circulation of \mathbf{F} per unit area. If curl $\mathbf{F}(P_0) \neq \mathbf{0}$, then the left-hand side of (11) is a maximum when the circle C_r is situated in a manner so that $\mathbf{n}(P_0)$ points in the same direction as curl $\mathbf{F}(P_0)$. In this case, the circulation on the right side of (11) will also be a maximum. Thus, a paddle wheel inserted into the fluid at P_0 will rotate fastest when its axis points in the direction of curl $\mathbf{F}(P_0)$. See Figure 9.119. Note, too, that the paddle wheel will not rotate if its axis is perpendicular to curl $\mathbf{F}(P_0)$.

Remarks

The value of the surface integral in (2) is determined solely by the integral around its boundary C. This basically means that the shape of the surface S is irrelevant. Assuming that the hypotheses of Theorem 9.14 are satisfied, then for two different surfaces S_1 and S_2 with the same orientation and with the same boundary C, we have

$$\oint_C \mathbf{F} \cdot d\mathbf{r} = \iint_{S_1} (\text{curl } \mathbf{F}) \cdot \mathbf{n} \, dS = \iint_{S_2} (\text{curl } \mathbf{F}) \cdot \mathbf{n} \, dS.$$

See Figure 9.120 and Problems 17 and 18 in Exercises 9.14.

E X E R C I S E S 9 . 1 4 Answers to odd-numbered problems begin on page A-51.

In Problems 1–4, verify Stokes' theorem. Assume that the surface S is oriented upward.

1. $\mathbf{F} = 5y\mathbf{i} - 5x\mathbf{j} + 3\mathbf{k}$; S that portion of the plane $z = 1$ within the cylinder $x^2 + y^2 = 4$

2. $\mathbf{F} = 2z\mathbf{i} - 3x\mathbf{j} + 4y\mathbf{k}$; S that portion of the paraboloid $z = 16 - x^2 - y^2$ for $z \geq 0$

3. $\mathbf{F} = z\mathbf{i} + x\mathbf{j} + y\mathbf{k}$; S that portion of the plane $2x + y + 2z = 6$ in the first octant

continued

4. $\mathbf{F} = x\mathbf{i} + y\mathbf{j} + z\mathbf{k}$; S that portion of the sphere $x^2 + y^2 + z^2 = 1$ for $z \geq 0$

In Problems 5–12, use Stokes' theorem to evaluate $\oint_C \mathbf{F} \cdot d\mathbf{r}$. Assume C is oriented counterclockwise as viewed from above.

5. $\mathbf{F} = (2z + x)\mathbf{i} + (y - z)\mathbf{j} + (x + y)\mathbf{k}$; C the triangle with vertices $(1, 0, 0)$, $(0, 1, 0)$, $(0, 0, 1)$

6. $\mathbf{F} = z^2y \cos xy\mathbf{i} + z^2x(1 + \cos xy)\mathbf{j} + 2z \sin xy\mathbf{k}$; C the boundary of the plane $z = 1 - y$ shown in Figure 9.121.

Figure 9.121

7. $\mathbf{F} = xy\mathbf{i} + 2yz\mathbf{j} + xz\mathbf{k}$; C the boundary given in Problem 6

8. $\mathbf{F} = (x + 2z)\mathbf{i} + (3x + y)\mathbf{j} + (2y - z)\mathbf{k}$; C the curve of intersection of the plane $x + 2y + z = 4$ with the coordinate planes

9. $\mathbf{F} = y^3\mathbf{i} - x^3\mathbf{j} + z^3\mathbf{k}$; C the trace of the cylinder $x^2 + y^2 = 1$ in the plane $x + y + z = 1$

10. $\mathbf{F} = x^2y\mathbf{i} + (x + y^2)\mathbf{j} + xy^2z\mathbf{k}$; C the boundary of the surface shown in Figure 9.122.

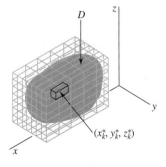

Figure 9.122

11. $\mathbf{F} = x\mathbf{i} + x^3y^2\mathbf{j} + z\mathbf{k}$; C the boundary of the semi-ellipsoid $z = \sqrt{4 - 4x^2 - y^2}$ in the plane $z = 0$

12. $\mathbf{F} = z\mathbf{i} + x\mathbf{j} + y\mathbf{k}$; C the curve of intersection of the plane $x + y + z = 0$ and the sphere $x^2 + y^2 + z^2 = 1$ [*Hint:* Recall that the area of an ellipse $x^2/a^2 + y^2/b^2 = 1$ is πab.]

In Problems 13–16, use Stokes' theorem to evaluate $\iint_S (\text{curl } \mathbf{F}) \cdot \mathbf{n} \, dS$. Assume that the surface S is oriented upward.

13. $\mathbf{F} = 6yz\mathbf{i} + 5x\mathbf{j} + yze^{x^2}\mathbf{k}$; S that portion of the paraboloid $z = \frac{1}{4}x^2 + y^2$ for $0 \leq z \leq 4$

14. $\mathbf{F} = y\mathbf{i} + (y - x)\mathbf{j} + z^2\mathbf{k}$; S that portion of the sphere $x^2 + y^2 + (z - 4)^2 = 25$ for $z \geq 0$

15. $\mathbf{F} = 3x^2\mathbf{i} + 8x^3y\mathbf{j} + 3x^2y\mathbf{k}$; S that portion of the plane $z = x$ that lies inside the rectangular cylinder defined by the planes $x = 0$, $y = 0$, $x = 2$, $y = 2$

16. $\mathbf{F} = 2xy^2z\mathbf{i} + 2x^2yz\mathbf{j} + (x^2y^2 - 6x)\mathbf{k}$; S that portion of the plane $z = y$ that lies inside the cylinder $x^2 + y^2 = 1$

17. Use Stokes' theorem to evaluate

$$\oint_C z^2e^{x^2} \, dx + xy^2 \, dy + \tan^{-1} y \, dz$$

where C is the circle $x^2 + y^2 = 9$, by finding a surface S with C as its boundary and such that the orientation of C is counterclockwise as viewed from above.

18. Consider the surface integral $\iint_S (\text{curl } \mathbf{F}) \cdot \mathbf{n} \, dS$, where $\mathbf{F} = xyz\mathbf{k}$ and S is that portion of the paraboloid $z = 1 - x^2 - y^2$ for $z \geq 0$ oriented upward.

(a) Evaluate the surface integral by the method of Section 9.13; that is, do not use Stokes' theorem.

(b) Evaluate the surface integral by finding a simpler surface that is oriented upward and has the same boundary as the paraboloid.

(c) Use Stokes' theorem to verify the result in part **(b)**.

9.15 Review of Triple Integrals

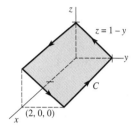

The steps leading to the definition of the *three-dimensional definite integral* or **triple integral** $\iiint_D F(x, y, z) \, dV$ are quite similar to those of the double integral.

> **$w = F(x, y, z)$**
>
> 1. Let F be defined over a closed and bounded region D of space.
> 2. By means of a three-dimensional grid of vertical and horizontal planes parallel to the coordinate planes, form a partition P of D into n subregions (boxes) D_k of volumes ΔV_k that lie entirely in D.

continued

3. Let $\|P\|$ be the **norm** of the partition or the length of the longest diagonal of the D_k.
4. Choose a point (x_k^*, y_k^*, z_k^*) in each subregion D_k.
5. Form the sum $\sum_{k=1}^{n} F(x_k^*, y_k^*, z_k^*) \, \Delta V_k$.

A sum of the form $\sum_{k=1}^{n} F(x_k^*, y_k^*, z_k^*) \, \Delta V_k$, where (x_k^*, y_k^*, z_k^*) is an arbitrary point within each D_k and ΔV_k denotes the volume of each D_k, is called a **Riemann sum.** The type of partition used in step 2, where all the D_k lie completely within D, is called an **inner partition** of D.

> **DEFINITION 9.13 The Triple Integral**
>
> Let F be a function of three variables defined over a closed region D of space. Then the **triple integral of F over D** is given by
>
> $$\iiint\limits_{D} F(x, y, z) \, dV = \lim_{\|P\| \to 0} \sum_{k=1}^{n} F(x_k^*, y_k^*, z_k^*) \, \Delta V_k. \qquad (1)$$

As in our previous discussions on the integral, when F is continuous over D, the limit in (1) exists; that is, F is **integrable** over D.

■ **Evaluation by Iterated Integrals** If the region D is bounded above by the graph of $z = f_2(x, y)$ and bounded below by the graph of $z = f_1(x, y)$, then it can be shown that the triple integral (1) can be expressed as a double integral of the partial integral $\int_{f_1(x, y)}^{f_2(x, y)} F(x, y, z) \, dz$; that is,

$$\iiint\limits_{D} F(x, y, z) \, dV = \iint\limits_{R} \left[\int_{f_1(x, y)}^{f_2(x, y)} F(x, y, z) \, dz \right] dA,$$

where R is the orthogonal projection of D onto the xy-plane. In particular, if R is a Type I region, then, as shown in Figure 9.123, the triple integral of F over D can be written as an iterated integral:

$$\iiint\limits_{D} F(x, y, z) \, dV = \int_{a}^{b} \int_{g_1(x)}^{g_2(x)} \int_{f_1(x, y)}^{f_2(x, y)} F(x, y, z) \, dz \, dy \, dx. \qquad (2)$$

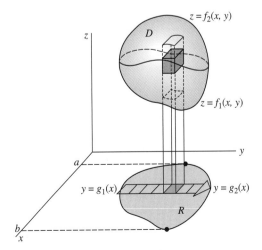

Figure 9.123

To evaluate the iterated integral in (2) we begin by evaluating the partial integral

$$\int_{f_1(x,\,y)}^{f_2(x,\,y)} F(x,\,y,\,z)\,dz,$$

in which *both* x and y are held fixed.

In a double integral there are only two possible orders of integration, *dy dx* and *dx dy*. The triple integral in (2) illustrates one of *six* possible orders of integration:

$$dz\,dy\,dx, \qquad dz\,dx\,dy, \qquad dy\,dx\,dz,$$

$$dx\,dy\,dz, \qquad dx\,dz\,dy, \qquad dy\,dz\,dx.$$

The last two differentials tell the coordinate plane in which the region *R* is situated. For example, the iterated integral corresponding to the order of integration *dx dz dy* must have the form

$$\int_c^d \int_{k_1(y)}^{k_2(y)} \int_{h_1(y,\,z)}^{h_2(y,\,z)} F(x,\,y,\,z)\,dx\,dz\,dy.$$

The geometric interpretation of this integral and the region *R* of integration in the *yz*-plane are shown in Figure 9.124.

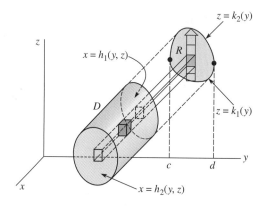

Figure 9.124

■ Applications A list of some of the standard applications of the triple integral follows:

Volume

If $F(x,\,y,\,z) = 1$, then the **volume** of the solid *D* is

$$V = \iiint_D dV.$$

Mass

If $\rho(x,\,y,\,z)$ is density, then the **mass** of the solid *D* is given by

$$m = \iiint_D \rho(x,\,y,\,z)\,dV.$$

First Moments

The **first moments** of the solid about the coordinate planes indicated by the subscripts are given by

$$M_{xy} = \iiint_D z\rho(x,\,y,\,z)\,dV, \qquad M_{xz} = \iiint_D y\rho(x,\,y,\,z)\,dV,$$

$$M_{yz} = \iiint_D x\rho(x,\,y,\,z)\,dV.$$

Center of Mass

The coordinates of the **center of mass** of D are given by

$$\bar{x} = \frac{M_{yz}}{m}, \qquad \bar{y} = \frac{M_{xz}}{m}, \qquad \bar{z} = \frac{M_{xy}}{m}.$$

Centroid

If $\rho(x, y, z) = $ a constant, the center of mass is called the **centroid** of the solid.

Second Moments

The **second moments**, or **moments of inertia** of D about the coordinate axes indicated by the subscripts, are given by

$$I_x = \iiint_D (y^2 + z^2)\rho(x, y, z) \, dV, \qquad I_y = \iiint_D (x^2 + z^2)\rho(x, y, z) \, dV,$$

$$I_z = \iiint_D (x^2 + y^2)\rho(x, y, z) \, dV.$$

Radius of Gyration

As in Section 9.10, if I is a moment of inertia of the solid about a given axis, then the **radius of gyration** is

$$R_g = \sqrt{\frac{I}{m}}.$$

Example 1 Volume of a Solid

Find the volume of the solid in the first octant bounded by the graphs of $z = 1 - y^2$, $y = 2x$, and $x = 3$.

SOLUTION As indicated in Figure 9.125(a), the first integration with respect to z is from 0 to $1 - y^2$. Furthermore, from Figure 9.125(b) we see that the projection of the solid D in the xy-plane is a region of Type II. Hence, we next integrate,

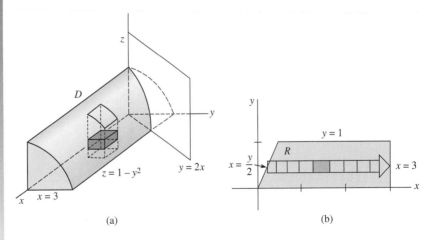

(a) (b)

Figure 9.125

with respect to x, from $y/2$ to 3. The last integration is with respect to y from 0 to 1. Thus,

$$V = \iiint_D dV = \int_0^1 \int_{y/2}^3 \int_0^{1-y^2} dz \, dx \, dy$$

$$= \int_0^1 \int_{y/2}^3 (1 - y^2) \, dx \, dy$$

$$= \int_0^1 \left[x - xy^2 \right]_{y/2}^3 dy$$

$$= \int_0^1 \left(3 - 3y^2 - \frac{1}{2}y + \frac{1}{2}y^3 \right) dy$$

$$= \left[3y - y^3 - \frac{1}{4}y^2 + \frac{1}{8}y^4 \right]_0^1 = \frac{15}{8} \text{ cubic units} \qquad \square$$

Example 2 Changing the Order of Integration

Change the order of integration in

$$\int_0^6 \int_0^{4 - 2x/3} \int_0^{3 - x/2 - 3y/4} F(x, y, z) \, dz \, dy \, dx$$

to *dy dx dz*.

SOLUTION As seen in Figure 9.126(a), the region D is the solid in the first octant bounded by the three coordinate planes and the plane $2x + 3y + 4z = 12$. Referring to Figure 9.126(b) and the table, we conclude that

$$\int_0^6 \int_0^{4 - 2x/3} \int_0^{3 - x/2 - 3y/4} F(x, y, z) \, dz \, dy \, dx = \int_0^3 \int_0^{6 - 2z} \int_0^{4 - 2x/3 - 4z/3} F(x, y, z) \, dy \, dx \, dz.$$

Order of Integration	First Integration	Second Integration	Third Integration
dz dy dx	0 to $3 - x/2 - 3y/4$	0 to $4 - 2x/3$	0 to 6
dy dx dz	0 to $4 - 2x/3 - 4z/3$	0 to $6 - 2z$	0 to 3

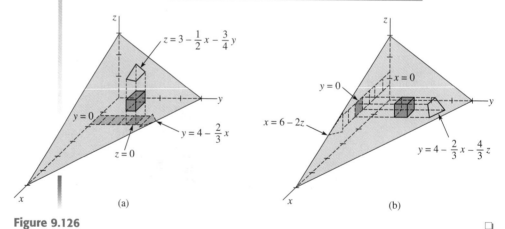

Figure 9.126 $\qquad \square$

Depending on the geometry of a region in 3-space, the evaluation of a triple integral over that region may be made easier by utilizing a new coordinate system.

■ Cylindrical Coordinates

The **cylindrical coordinate system** combines the polar description of a point in the plane with the rectangular description of the *z*-component of a point in space. As seen in Figure 9.127(a), the cylindrical coordinates of a point P are denoted by the ordered triple (r, θ, z). The word *cylindrical* arises from the fact that a point P in space is determined by the inter-

section of the planes $z =$ constant and $\theta =$ constant with a cylinder $r =$ constant. See Figure 9.127(b).

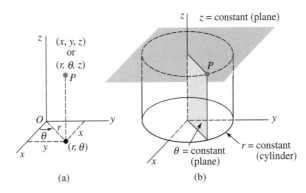

Figure 9.127

■ **Conversion of Cylindrical Coordinates to Rectangular Coordinates** From Figure 9.127(a) we also see that the rectangular coordinates (x, y, z) of a point can be obtained from the cylindrical coordinates (r, θ, z) by means of

$$x = r \cos \theta, \quad y = r \sin \theta, \quad z = z. \tag{3}$$

Example 3 **Cylindrical to Rectangular Coordinates**
Convert $(8, \pi/3, 7)$ in cylindrical coordinates to rectangular coordinates.
SOLUTION From (3),

$$x = 8 \cos \frac{\pi}{3} = 4, \qquad y = 8 \sin \frac{\pi}{3} = 4\sqrt{3}, \qquad z = 7.$$

Thus, $(8, \pi/3, 7)$ is equivalent to $(4, 4\sqrt{3}, 7)$ in rectangular coordinates. ❑

■ **Conversion of Rectangular Coordinates to Cylindrical Coordinates** To express rectangular coordinates (x, y, z) as cylindrical coordinates, we use

$$r^2 = x^2 + y^2, \qquad \tan \theta = \frac{y}{x}, \qquad z = z. \tag{4}$$

Example 4 **Rectangular to Cylindrical Coordinates**
Convert $(-\sqrt{2}, \sqrt{2}, 1)$ in rectangular coordinates to cylindrical coordinates.
SOLUTION From (4) we see that

$$r^2 = (-\sqrt{2})^2 + (\sqrt{2})^2 = 4, \qquad \tan \theta = \frac{\sqrt{2}}{-\sqrt{2}} = -1, \qquad z = 1.$$

If we take $r = 2$, then, consistent with the fact that $x < 0$ and $y > 0$, we take $\theta = 3\pi/4$.* Consequently, $(-\sqrt{2}, \sqrt{2}, 1)$ is equivalent to $(2, 3\pi/4, 1)$ in cylindrical coordinates. See Figure 9.128. ❑

Figure 9.128

———————
* If we use $\theta = \tan^{-1}(-1) = -\pi/4$, then we can use $r = -2$. Notice that the combinations $r = 2$, $\theta = -\pi/4$ and $r = -2$, $\theta = 3\pi/4$ are inconsistent.

■ **Triple Integrals in Cylindrical Coordinates** Recall from Section 9.11 that the area of a "polar rectangle" is $\Delta A = r^* \, \Delta r \, \Delta \theta$, where r^* is the average radius. From Figure 9.129(a) we see that the volume of a "cylindrical wedge" is simply $\Delta V = $ (area of base)(height) $= r^* \, \Delta r \, \Delta \theta \, \Delta z$. Thus, if $F(r, \theta, z)$ is a continuous function over the region D, as shown in Figure 9.129(b), then the triple integral of F over D is given by

$$\iiint_D F(r, \theta, z)\, dV = \iint_R \left[\int_{f_1(r,\theta)}^{f_2(r,\theta)} F(r, \theta, z)\, dz \right] dA = \int_\alpha^\beta \int_{g_1(\theta)}^{g_2(\theta)} \int_{f_1(r,\theta)}^{f_2(r,\theta)} F(r, \theta, z)\, r \, dz \, dr \, d\theta.$$

(a) (b)

Figure 9.129

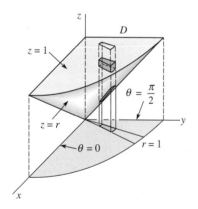

Figure 9.130

Example 5 Center of Mass

A solid in the first octant has the shape determined by the graph of the cone $z = \sqrt{x^2 + y^2}$ and the planes $z = 1$, $x = 0$, and $y = 0$. Find the center of mass if the density is given by $\rho(r, \theta, z) = r$.

SOLUTION In view of (4), the equation of the cone is $z = r$. Hence, we see from Figure 9.130 that

$$m = \iiint_D r \, dV = \int_0^{\pi/2} \int_0^1 \int_r^1 r(r \, dz \, dr \, d\theta)$$

$$= \int_0^{\pi/2} \int_0^1 r^2 z \Big]_r^1 \, dr \, d\theta$$

$$= \int_0^{\pi/2} \int_0^1 (r^2 - r^3) \, dr \, d\theta = \frac{\pi}{24}$$

$$M_{xy} = \iiint_D zr \, dV = \int_0^{\pi/2} \int_0^1 \int_r^1 zr^2 \, dz \, dr \, d\theta$$

$$= \int_0^{\pi/2} \int_0^1 \frac{z^2}{2} r^2 \Big]_r^1 \, dr \, d\theta$$

$$= \frac{1}{2} \int_0^{\pi/2} \int_0^1 (r^2 - r^4) \, dr \, d\theta = \frac{\pi}{30}.$$

In the integrals for M_{xz} and M_{yz} we substitute $y = r \sin \theta$ and $x = r \cos \theta$:

$$M_{xz} = \iiint_D r^2 \sin \theta \, dV = \int_0^{\pi/2} \int_0^1 \int_r^1 r^3 \sin \theta \, dz \, dr \, d\theta$$

$$= \int_0^{\pi/2} \int_0^1 r^3 z \sin \theta \Big]_r^1 dr \, d\theta$$

$$= \int_0^{\pi/2} \int_0^1 (r^3 - r^4) \sin \theta \, dr \, d\theta = \frac{1}{20}$$

$$M_{yz} = \iiint_D r^2 \cos \theta \, dV = \int_0^{\pi/2} \int_0^1 \int_r^1 r^3 \cos \theta \, dz \, dr \, d\theta = \frac{1}{20}.$$

Hence,

$$\bar{x} = \frac{M_{yz}}{m} = \frac{1/20}{\pi/24} \approx 0.38, \qquad \bar{y} = \frac{M_{xz}}{m} = \frac{1/20}{\pi/24} \approx 0.38, \qquad \bar{z} = \frac{M_{xy}}{m} = \frac{\pi/30}{\pi/24} \approx 0.8.$$

The center of mass has the approximate coordinates $(0.38, 0.38, 0.8)$. ❏

■ **Spherical Coordinates** As seen in Figure 9.131(a), the **spherical coordinates** of a point P are given by the ordered triple (ρ, ϕ, θ), where ρ is the distance from the origin to P, ϕ is the angle between the positive z-axis and the vector \overrightarrow{OP}, and θ is the angle measured from the positive x-axis to the vector projection \overrightarrow{OQ} of \overrightarrow{OP}.* Figure 9.131(b) shows that a point P in space is determined by the intersection of a cone $\phi = $ constant, a plane $\theta = $ constant, and a sphere $\rho = $ constant; whence arises the name "spherical" coordinates.

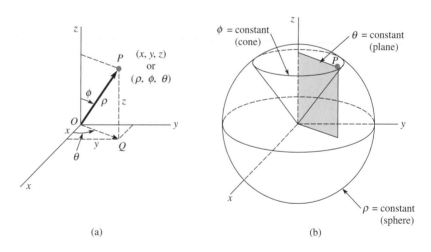

(a) (b)

Figure 9.131

■ **Conversion of Spherical Coordinates to Rectangular and Cylindrical Coordinates** To transform from spherical coordinates (ρ, ϕ, θ) to rectangular coordinates (x, y, z), we observe from Figure 9.131(a) that

$$x = \left| \overrightarrow{OQ} \right| \cos \theta, \qquad y = \left| \overrightarrow{OQ} \right| \sin \theta, \qquad z = \left| \overrightarrow{OP} \right| \cos \phi.$$

* θ is the same angle as in polar and cylindrical coordinates.

Since $|\overrightarrow{OQ}| = \rho \sin \phi$ and $|\overrightarrow{OP}| = \rho$, the foregoing equations become

$$x = \rho \sin \phi \cos \theta, \qquad y = \rho \sin \phi \sin \theta, \qquad z = \rho \cos \phi. \qquad (5)$$

It is customary to take $\rho \geq 0$ and $0 \leq \phi \leq \pi$. Also, since $|OQ| = \rho \sin \phi = r$, the formulas

$$r = \rho \sin \phi, \qquad \theta = \theta, \qquad z = \rho \cos \phi, \qquad (6)$$

enable us to transform from spherical coordinates (ρ, ϕ, θ) to cylindrical coordinates (r, θ, z).

Example 6 Spherical to Rectangular and Cylindrical Coordinates

Convert $(6, \pi/4, \pi/3)$ in spherical coordinates to rectangular coordinates and cylindrical coordinates.

SOLUTION Identifying $\rho = 6$, $\phi = \pi/4$, and $\theta = \pi/3$, we find from (5) that the rectangular coordinates of the point are given by

$$x = 6 \sin \frac{\pi}{4} \cos \frac{\pi}{3} = \frac{3\sqrt{2}}{2}, \qquad y = 6 \sin \frac{\pi}{4} \sin \frac{\pi}{3} = \frac{3\sqrt{6}}{2}, \qquad z = 6 \cos \frac{\pi}{4} = 3\sqrt{2}.$$

From (6) we obtain

$$r = 6 \sin \frac{\pi}{4} = 3\sqrt{2}, \qquad \theta = \frac{\pi}{3}, \qquad z = 6 \cos \frac{\pi}{4} = 3\sqrt{2}.$$

Thus, the cylindrical coordinates of the point are $(3\sqrt{2}, \pi/3, 3\sqrt{2})$. ❑

■ Conversion of Rectangular Coordinates to Spherical Coordinates

To transform from rectangular coordinates to spherical coordinates, we use

$$\rho^2 = x^2 + y^2 + z^2, \qquad \tan \theta = \frac{y}{x}, \qquad \cos \phi = \frac{z}{\sqrt{x^2 + y^2 + z^2}}. \qquad (7)$$

■ Triple Integrals in Spherical Coordinates

As seen in Figure 9.132, the volume of a "spherical wedge" is given by the approximation

$$\Delta V \approx \rho^2 \sin \phi \, \Delta\rho \, \Delta\phi \, \Delta\theta.$$

Thus, in a triple integral of a continuous spherical coordinate function $F(\rho, \phi, \theta)$, the differential of volume dV is given by

$$dV = \rho^2 \sin \phi \, d\rho \, d\phi \, d\theta.$$

A typical triple integral in spherical coordinates has the form

$$\iiint_D F(\rho, \phi, \theta) \, dV = \int_\alpha^\beta \int_{g_1(\theta)}^{g_2(\theta)} \int_{f_1(\phi, \theta)}^{f_2(\phi, \theta)} F(\rho, \phi, \theta) \rho^2 \sin \phi \, d\rho \, d\phi \, d\theta.$$

Example 7 Moment of Inertia

Find the moment of inertia about the z-axis of the homogeneous solid bounded between the spheres

$$x^2 + y^2 + z^2 = a^2 \quad \text{and} \quad x^2 + y^2 + z^2 = b^2, \qquad a < b.$$

SOLUTION If $\delta(\rho, \phi, \theta) = k$ is the density,* then

$$I_z = \iiint_D (x^2 + y^2) k \, dV.$$

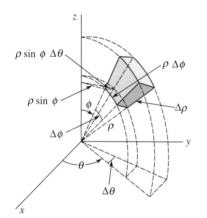

Figure 9.132

* We must use a different symbol to denote density to avoid confusion with the symbol ρ of spherical coordinates.

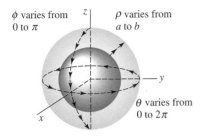

ϕ varies from 0 to π

ρ varies from a to b

θ varies from 0 to 2π

Figure 9.133

From (5) we find $x^2 + y^2 = \rho^2 \sin^2 \phi$ and $x^2 + y^2 + z^2 = \rho^2$. Thus the equations of the spheres are simply $\rho = a$ and $\rho = b$. See Figure 9.133. Consequently, in spherical coordinates the foregoing integral becomes

$$I_z = k \int_0^{2\pi} \int_0^\pi \int_a^b \rho^2 \sin^2\phi(\rho^2 \sin \phi \, d\rho \, d\phi \, d\theta)$$

$$= k \int_0^{2\pi} \int_0^\pi \int_a^b \rho^4 \sin^3\phi \, d\rho \, d\phi \, d\theta$$

$$= k \int_0^{2\pi} \int_0^\pi \frac{\rho^5}{5} \sin^3\phi \Big]_a^b d\phi \, d\theta$$

$$= \frac{k}{5} (b^5 - a^5) \int_0^{2\pi} \int_0^\pi (1 - \cos^2 \phi) \sin \phi \, d\phi \, d\theta$$

$$= \frac{k}{5} (b^5 - a^5) \int_0^{2\pi} \left[- \cos \phi + \frac{1}{3} \cos^3 \phi \right]_0^\pi d\theta$$

$$= \frac{4k}{15} (b^5 - a^5) \int_0^{2\pi} d\theta = \frac{8\pi k}{15} (b^5 - a^5). \qquad \Box$$

Remarks

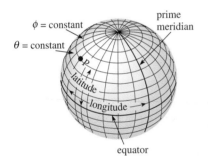

ϕ = constant

θ = constant

prime meridian

latitude

longitude

equator

Figure 9.134

Spherical coordinates are used in navigation. If we think of the earth as a sphere of fixed radius centered at the origin, then a point P can be located by specifying two angles θ and ϕ. As shown in Figure 9.134, when ϕ is held constant, the resulting curve is called a **parallel.** Fixed values of θ result in curves called **great circles.** Half of one of these great circles joining the north and south poles is called a **meridian.** The intersection of a parallel and a meridian gives the position of a point P. If $0° \le \phi \le 180°$ and $-180° \le \theta \le 180°$, the angles $90° - \phi$ and θ are said to be the **latitude** and **longitude** of P, respectively. The **prime meridian** corresponds to a longitude of $0°$. The latitude of the equator is $0°$; the latitudes of the north and south poles are, in turn, $+90°$ (or $90°$ North) and $-90°$ (or $90°$ South).

EXERCISES 9.15

Answers to odd-numbered problems begin on page A-51.

In Problems 1–8, evaluate the given iterated integral.

1. $\int_2^4 \int_{-2}^2 \int_{-1}^1 (x + y + z) \, dx \, dy \, dz$

2. $\int_1^3 \int_1^x \int_2^{xy} 24xy \, dz \, dy \, dx$

3. $\int_0^6 \int_0^{6-x} \int_0^{6-x-z} dy \, dz \, dx$

4. $\int_0^1 \int_0^{1-x} \int_0^{\sqrt{y}} 4x^2 z^3 \, dz \, dy \, dx$

5. $\int_0^{\pi/2} \int_0^{y^2} \int_0^y \cos \left(\frac{x}{y}\right) dz \, dx \, dy$

6. $\int_0^{\sqrt{2}} \int_{\sqrt{y}}^2 \int_0^{e^{x^2}} x \, dz \, dx \, dy$

7. $\int_0^1 \int_0^1 \int_0^{2-x^2-y^2} xye^z \, dz \, dx \, dy$

8. $\int_0^4 \int_0^{1/2} \int_0^{x^2} \frac{1}{\sqrt{x^2 - y^2}} \, dy \, dx \, dz$

9. Evaluate $\iiint_D z \, dV$, where D is the region in the first octant bounded by the graphs of $y = x$, $y = x - 2$, $y = 1$, $y = 3$, $z = 0$, and $z = 5$.

10. Evaluate $\iiint_D (x^2 + y^2) \, dV$, where D is the region bounded by the graphs of $y = x^2$, $z = 4 - y$, and $z = 0$.

In Problems 11 and 12, change the indicated order of integration to each of the other five orders.

11. $\int_0^2 \int_0^{4-2y} \int_{x+2y}^4 F(x, y, z) \, dz \, dx \, dy$

12. $\int_{0}^{2} \int_{0}^{\sqrt{36-9x^2}/2} \int_{1}^{3} F(x, y, z)\, dz\, dy\, dx$

In Problems 13 and 14, consider the solid given in the figure. Set up, but do not evaluate, the integrals giving the volume V of the solid using the indicated orders of integration.

13.

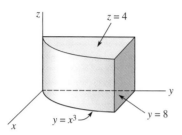

Figure 9.135

 (a) $dz\, dy\, dx$ **(b)** $dx\, dz\, dy$ **(c)** $dy\, dx\, dz$

14.

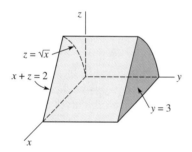

Figure 9.136

 (a) $dx\, dz\, dy$ **(b)** $dy\, dx\, dz$ **(c)** $dz\, dx\, dy$

[*Hint:* Part **(c)** will require two integrals.]

In Problems 15–20, sketch the region D whose volume V is given by the iterated integral.

15. $\int_{0}^{4} \int_{0}^{3} \int_{0}^{2-2z/3} dx\, dz\, dy$

16. $4\int_{0}^{3} \int_{0}^{\sqrt{9-y^2}} \int_{4}^{\sqrt{25-x^2-y^2}} dz\, dx\, dy$

17. $\int_{-1}^{1} \int_{-\sqrt{1-x^2}}^{\sqrt{1-x^2}} \int_{0}^{5} dz\, dy\, dx$

18. $\int_{0}^{2} \int_{0}^{\sqrt{4-x^2}} \int_{x^2+y^2}^{4} dz\, dy\, dx$

19. $\int_{0}^{2} \int_{0}^{2-y} \int_{-\sqrt{y}}^{\sqrt{y}} dx\, dz\, dy$

20. $\int_{1}^{3} \int_{0}^{1/x} \int_{0}^{3} dy\, dz\, dx$

In Problems 21–24, find the volume of the solid bounded by the graphs of the given equations.

21. $x = y^2$; $4 - x = y^2$, $z = 0$, $z = 3$

22. $x^2 + y^2 = 4$, $z = x + y$, the coordinate planes, first octant

23. $y = x^2 + z^2$, $y = 8 - x^2 - z^2$

24. $x = 2$, $y = x$, $y = 0$, $z = x^2 + y^2$, $z = 0$

25. Find the center of mass of the solid given in Figure 9.135 if the density at a point P is directly proportional to the distance from the xy-plane.

26. Find the centroid of the solid in Figure 9.136 if the density is constant.

27. Find the center of mass of the solid bounded by the graphs of $x^2 + z^2 = 4$, $y = 0$, and $y = 3$ if the density at a point P is directly proportional to the distance from the xz-plane.

28. Find the center of mass of the solid bounded by the graphs of $y = x^2$, $y = x$, $z = y + 2$, and $z = 0$ if the density at a point P is directly proportional to the distance from the xy-plane.

In Problems 29 and 30, set up, but do not evaluate, the iterated integrals giving the mass of the solid that has the given shape and density.

29. $x^2 + y^2 = 1$, $z + y = 8$, $z - 2y = 2$; $\rho(x, y, z) = x + y + 4$

30. $x^2 + y^2 - z^2 = 1$, $z = -1$, $z = 2$; $\rho(x, y, z) = z^2$ [*Hint:* Do not use $dz\, dy\, dx$.]

31. Find the moment of inertia of the solid in Figure 9.135 about the y-axis if the density is as given in Problem 25. Find the radius of gyration.

32. Find the moment of inertia of the solid in Figure 9.136 about the x-axis if the density is constant. Find the radius of gyration.

33. Find the moment of inertia about the z-axis of the solid in the first octant that is bounded by the coordinate planes and the graph of $x + y + z = 1$ if the density is constant.

34. Find the moment of inertia about the y-axis of the solid bounded by the graphs of $z = y$, $z = 4 - y$, $z = 1$, $z = 0$, $x = 2$, and $x = 0$ if the density at a point P is directly proportional to the distance from the yz-plane.

In Problems 35–38, convert the point given in cylindrical coordinates to rectangular coordinates.

35. $\left(10, \dfrac{3\pi}{4}, 5\right)$ **36.** $\left(2, \dfrac{5\pi}{6}, -3\right)$

37. $\left(\sqrt{3}, \dfrac{\pi}{3}, -4\right)$ **38.** $\left(4, \dfrac{7\pi}{4}, 0\right)$

In Problems 39–42, convert the point given in rectangular coordinates to cylindrical coordinates.

39. $(1, -1, -9)$

40. $(2\sqrt{3}, 2, 17)$

41. $(-\sqrt{2}, \sqrt{6}, 2)$

42. $(1, 2, 7)$

In Problems 43–46, convert the given equation to cylindrical coordinates.

43. $x^2 + y^2 + z^2 = 25$

44. $x + y - z = 1$

45. $x^2 + y^2 - z^2 = 1$

46. $x^2 + z^2 = 16$

In Problems 47–50, convert the given equation to rectangular coordinates.

47. $z = r^2$

48. $z = 2r \sin \theta$

49. $r = 5 \sec \theta$

50. $\theta = \pi/6$

In Problems 51–58, use triple integrals and cylindrical coordinates. In Problems 51–54, find the volume of the solid that is bounded by the graphs of the given equations.

51. $x^2 + y^2 = 4$, $x^2 + y^2 + z^2 = 16$, $z = 0$

52. $z = 10 - x^2 - y^2$, $z = 1$

53. $z = x^2 + y^2$, $x^2 + y^2 = 25$, $z = 0$

54. $y = x^2 + z^2$, $2y = x^2 + z^2 + 4$

55. Find the centroid of the homogeneous solid that is bounded by the hemisphere $z = \sqrt{a^2 - x^2 - y^2}$ and the plane $z = 0$.

56. Find the center of mass of the solid that is bounded by the graphs of $y^2 + z^2 = 16$, $x = 0$, and $x = 5$ if the density at a point P is directly proportional to distance from the yz-plane.

57. Find the moment of inertia about the z-axis of the solid that is bounded above by the hemisphere $z = \sqrt{9 - x^2 - y^2}$ and below by the plane $z = 2$ if the density at a point P is inversely proportional to the square of the distance from the z-axis.

58. Find the moment of inertia about the x-axis of the solid that is bounded by the cone $z = \sqrt{x^2 + y^2}$ and the plane $z = 1$ if the density at a point P is directly proportional to the distance from the z-axis.

In Problems 59–62, convert the point given in spherical coordinates to (**a**) rectangular coordinates and (**b**) cylindrical coordinates.

59. $\left(\dfrac{2}{3}, \dfrac{\pi}{2}, \dfrac{\pi}{6}\right)$

60. $\left(5, \dfrac{5\pi}{4}, \dfrac{2\pi}{3}\right)$

61. $\left(8, \dfrac{\pi}{4}, \dfrac{3\pi}{4}\right)$

62. $\left(\dfrac{1}{3}, \dfrac{5\pi}{3}, \dfrac{\pi}{6}\right)$

In Problems 63–66, convert the points given in rectangular coordinates to spherical coordinates.

63. $(-5, -5, 0)$

64. $(1, -\sqrt{3}, 1)$

65. $\left(\dfrac{\sqrt{3}}{2}, \dfrac{1}{2}, 1\right)$

66. $\left(-\dfrac{\sqrt{3}}{2}, 0, -\dfrac{1}{2}\right)$

In Problems 67–70, convert the given equation to spherical coordinates.

67. $x^2 + y^2 + z^2 = 64$

68. $x^2 + y^2 + z^2 = 4z$

69. $z^2 = 3x^2 + 3y^2$

70. $-x^2 - y^2 + z^2 = 1$

In Problems 71–74, convert the given equation to rectangular coordinates.

71. $\rho = 10$

72. $\phi = \pi/3$

73. $\rho = 2 \sec \phi$

74. $\rho \sin^2 \phi = \cos \phi$

In Problems 75–82, use triple integrals and spherical coordinates. In Problems 75–78, find the volume of the solid that is bounded by the graphs of the given equations.

75. $z = \sqrt{x^2 + y^2}$, $x^2 + y^2 + z^2 = 9$

76. $x^2 + y^2 + z^2 = 4$, $y = x$, $y = \sqrt{3}x$, $z = 0$, first octant

77. $z^2 = 3x^2 + 3y^2$, $x = 0$, $y = 0$, $z = 2$, first octant

78. Inside $x^2 + y^2 + z^2 = 1$ and outside $z^2 = x^2 + y^2$

79. Find the centroid of the homogeneous solid that is bounded by the cone $z = \sqrt{x^2 + y^2}$ and the sphere $x^2 + y^2 + z^2 = 2z$.

80. Find the center of mass of the solid that is bounded by the hemisphere $z = \sqrt{1 - x^2 - y^2}$ and the plane $z = 0$ if the density at a point P is directly proportional to the distance from the xy-plane.

81. Find the mass of the solid that is bounded above by the hemisphere $z = \sqrt{25 - x^2 - y^2}$ and below by the plane $z = 4$ if the density at a point P is inversely proportional to the distance from the origin. [*Hint:* Express the upper ϕ limit of integration as an inverse cosine.]

82. Find the moment of inertia about the z-axis of the solid that is bounded by the sphere $x^2 + y^2 + z^2 = a^2$ if the density at a point P is directly proportional to the distance from the origin.

9.16 Divergence Theorem

■ **Another Vector Form of Green's Theorem** Let $\mathbf{F} = P\mathbf{i} + Q\mathbf{j}$ and let $\mathbf{T} = (dx/ds)\mathbf{i} + (dy/ds)\mathbf{j}$ be a *unit tangent* to a simple closed plane curve C. In (1) of Section 9.14 we saw that $\oint_C (\mathbf{F} \cdot \mathbf{T})\, ds$ can be evaluated by a double integral involving curl \mathbf{F}. Similarly, if $\mathbf{n} = (dy/ds)\mathbf{i} - (dx/ds)\mathbf{j}$ is a *unit normal* to C (check $\mathbf{T} \cdot \mathbf{n}$), then $\oint_C (\mathbf{F} \cdot \mathbf{n})\, ds$ can be expressed in terms of a double integral of div \mathbf{F}. From Green's theorem,

$$\oint_C (\mathbf{F} \cdot \mathbf{n})\, ds = \oint_C P\, dy - Q\, dx = \iint_R \left[\frac{\partial P}{\partial x} - \left(-\frac{\partial Q}{\partial y} \right) \right] dA = \iint_R \left[\frac{\partial P}{\partial x} + \frac{\partial Q}{\partial y} \right] dA,$$

that is,

$$\oint_C (\mathbf{F} \cdot \mathbf{n})\, ds = \iint_R \operatorname{div} \mathbf{F}\, dA. \tag{1}$$

The result in (1) is a special case of the **divergence** or **Gauss' theorem.** The following is a generalization of (1) to 3-space:

THEOREM 9.15 **Divergence Theorem**

Let D be a closed and bounded region in 3-space with a piecewise smooth boundary S that is oriented outward. Let $\mathbf{F}(x, y, z) = P(x, y, z)\mathbf{i} + Q(x, y, z)\mathbf{j} + R(x, y, z)\mathbf{k}$ be a vector field for which P, Q, and R are continuous and have continuous first partial derivatives in a region of 3-space containing D. Then

$$\iint_S (\mathbf{F} \cdot \mathbf{n})\, dS = \iiint_D \operatorname{div} \mathbf{F}\, dV. \tag{2}$$

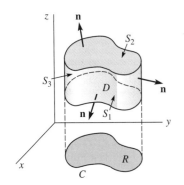

Figure 9.137

PARTIAL PROOF We will prove (2) for the special region D shown in Figure 9.137 whose surface S consists of three pieces:

(bottom) S_1: $z = f_1(x, y)$, (x, y) in R

(top) S_2: $z = f_2(x, y)$, (x, y) in R

(side) S_3: $f_1(x, y) \le z \le f_2(x, y)$, (x, y) on C,

where R is the projection of D onto the xy-plane and C is the boundary of R. Since

$$\operatorname{div} \mathbf{F} = \frac{\partial P}{\partial x} + \frac{\partial Q}{\partial y} + \frac{\partial R}{\partial z} \quad \text{and} \quad \mathbf{F} \cdot \mathbf{n} = P(\mathbf{i} \cdot \mathbf{n}) + Q(\mathbf{j} \cdot \mathbf{n}) + R(\mathbf{k} \cdot \mathbf{n}),$$

we can write

$$\iint_S (\mathbf{F} \cdot \mathbf{n})\, dS = \iint_S P(\mathbf{i} \cdot \mathbf{n})\, dS + \iint_S Q(\mathbf{j} \cdot \mathbf{n})\, dS + \iint_S R(\mathbf{k} \cdot \mathbf{n})\, dS$$

and

$$\iiint_D \operatorname{div} \mathbf{F}\, dV = \iiint_D \frac{\partial P}{\partial x}\, dV + \iiint_D \frac{\partial Q}{\partial y}\, dV + \iiint_D \frac{\partial R}{\partial z}\, dV.$$

To prove (2) we need only establish that

$$\iint_S P(\mathbf{i} \cdot \mathbf{n})\, dS = \iiint_D \frac{\partial P}{\partial x}\, dV \tag{3}$$

$$\iint_S Q(\mathbf{j} \cdot \mathbf{n}) \, dS = \iiint_D \frac{\partial Q}{\partial y} \, dV \qquad (4)$$

$$\iint_S R(\mathbf{k} \cdot \mathbf{n}) \, dS = \iiint_D \frac{\partial R}{\partial z} \, dV. \qquad (5)$$

Indeed, we shall prove only (5), since the proofs of (3) and (4) follow in a similar manner. Now,

$$\iiint_D \frac{\partial R}{\partial z} \, dV = \iint_R \left[\int_{f_1(x,\,y)}^{f_2(x,\,y)} \frac{\partial R}{\partial z} \, dz \right] dA = \iint_R [R(x, y, f_2(x, y)) - R(x, y, f_1(x, y))] \, dA. \qquad (6)$$

Next we write

$$\iint_S R(\mathbf{k} \cdot \mathbf{n}) \, dS = \iint_{S_1} R(\mathbf{k} \cdot \mathbf{n}) \, dS + \iint_{S_2} R(\mathbf{k} \cdot \mathbf{n}) \, dS + \iint_{S_3} R(\mathbf{k} \cdot \mathbf{n}) \, dS.$$

On S_1:

Since the outward normal points downward, we describe the surface as $g(x, y, z) = f_1(x, y) - z = 0$. Thus,

$$\mathbf{n} = \frac{\dfrac{\partial f_1}{\partial x} \mathbf{i} + \dfrac{\partial f_1}{\partial y} \mathbf{j} - \mathbf{k}}{\sqrt{1 + \left(\dfrac{\partial f_1}{\partial x}\right)^2 + \left(\dfrac{\partial f_1}{\partial y}\right)^2}} \qquad \text{so that} \qquad \mathbf{k} \cdot \mathbf{n} = \frac{-1}{\sqrt{1 + \left(\dfrac{\partial f_1}{\partial x}\right)^2 + \left(\dfrac{\partial f_1}{\partial y}\right)^2}}.$$

From the definition of dS we then have

$$\iint_{S_1} R(\mathbf{k} \cdot \mathbf{n}) \, dS = -\iint_R R(x, y, f_1(x, y)) \, dA. \qquad (7)$$

On S_2:

The outward normal points upward, so

$$\mathbf{n} = \frac{-\dfrac{\partial f_2}{\partial x} \mathbf{i} - \dfrac{\partial f_2}{\partial y} \mathbf{j} + \mathbf{k}}{\sqrt{1 + \left(\dfrac{\partial f_2}{\partial x}\right)^2 + \left(\dfrac{\partial f_2}{\partial y}\right)^2}} \qquad \text{so that} \qquad \mathbf{k} \cdot \mathbf{n} = \frac{1}{\sqrt{1 + \left(\dfrac{\partial f_2}{\partial x}\right)^2 + \left(\dfrac{\partial f_2}{\partial y}\right)^2}}$$

from which we find

$$\iint_{S_2} R(\mathbf{k} \cdot \mathbf{n}) \, dS = \iint_R R(x, y, f_2(x, y)) \, dA. \qquad (8)$$

On S_3:

Since this side is vertical, \mathbf{k} is perpendicular to \mathbf{n}. Consequently, $\mathbf{k} \cdot \mathbf{n} = 0$ and

$$\iint_{S_3} R(\mathbf{k} \cdot \mathbf{n}) \, dS = 0. \qquad (9)$$

Finally, adding (7), (8), and (9), we get

$$\iint_R [R(x, y, f_2(x, y)) - R(x, y, f_1(x, y))] \, dA,$$

which is the same as (6). ❏

Figure 9.138

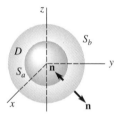

Figure 9.139

Although we proved (2) for a special region D that has a vertical side, we note that this type of region is not required in Theorem 9.15. A region D with no vertical side is illustrated in Figure 9.138; a region bounded by a sphere or an ellipsoid also does not have a vertical side. The divergence theorem also holds for the region D bounded between two closed surfaces, such as the concentric spheres S_a and S_b shown in Figure 9.139; the boundary surface S of D is the union of S_a and S_b. In this case $\iint_S (\mathbf{F} \cdot \mathbf{n})\, dS = \iiint_D \operatorname{div} \mathbf{F}\, dV$ becomes

$$\iint_{S_b} (\mathbf{F} \cdot \mathbf{n})\, dS + \iint_{S_a} (\mathbf{F} \cdot \mathbf{n})\, dS = \iiint_D \operatorname{div} \mathbf{F}\, dV,$$

where \mathbf{n} points outward from D; that is, on S_b \mathbf{n} points away from the origin and on S_a \mathbf{n} points toward the origin.

Example 1 Verifying Divergence Theorem

Let D be the region bounded by the hemisphere $x^2 + y^2 + (z - 1)^2 = 9$, $1 \le z \le 4$, and the plane $z = 1$. Verify the divergence theorem if $\mathbf{F} = x\mathbf{i} + y\mathbf{j} + (z - 1)\mathbf{k}$.

SOLUTION The closed region is shown in Figure 9.140.

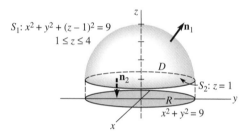

Figure 9.140

Triple Integral:

Since $\mathbf{F} = x\mathbf{i} + y\mathbf{j} + (z - 1)\mathbf{k}$, we see div $\mathbf{F} = 3$. Hence,

$$\iiint_D \operatorname{div} \mathbf{F}\, dV = \iiint_D 3\, dV = 3\iiint_D dV = 54\pi. \qquad (10)$$

In the last calculation, we used the fact that $\iiint_D dV$ gives the volume of the hemisphere ($\tfrac{2}{3}\pi 3^3$).

Surface Integral:

We write $\iint_S = \iint_{S_1} + \iint_{S_2}$, where S_1 is the hemisphere and S_2 is the plane $z = 1$. If S_1 is a level surface of $g(x, y, z) = x^2 + y^2 + (z - 1)^2$, then a unit outer normal is

$$\mathbf{n} = \frac{\nabla g}{\|\nabla g\|} = \frac{x\mathbf{i} + y\mathbf{j} + (z - 1)\mathbf{k}}{\sqrt{x^2 + y^2 + (z - 1)^2}} = \frac{x}{3}\mathbf{i} + \frac{y}{3}\mathbf{j} + \frac{z - 1}{3}\mathbf{k}.$$

Now

$$\mathbf{F} \cdot \mathbf{n} = \frac{x^2}{3} + \frac{y^2}{3} + \frac{(z - 1)^2}{3} = 3$$

and so

$$\iint_{S_1} (\mathbf{F} \cdot \mathbf{n})\, dS = \iint_R (3) \left(\frac{3}{\sqrt{9 - x^2 - y^2}}\, dA \right) \quad \leftarrow \text{polar coordinates}$$

$$= 9\int_0^{2\pi}\int_0^3 (9 - r^2)^{-1/2} r\, dr\, d\theta = 54\pi.$$

On S_2, we take $\mathbf{n} = -\mathbf{k}$ so that $\mathbf{F} \cdot \mathbf{n} = -z + 1$. But, since $z = 1$, $\iint_{S_2} (-z + 1)\, dS = 0$. Hence, we see that $\iint_S (\mathbf{F} \cdot \mathbf{n})\, dS = 54\pi + 0 = 54\pi$ agrees with (10). ❑

Example 2 Using Divergence Theorem

If $\mathbf{F} = xy\mathbf{i} + y^2 z\mathbf{j} + z^3\mathbf{k}$, evaluate $\iint_S (\mathbf{F} \cdot \mathbf{n})\, dS$, where S is the unit cube defined by $0 \le x \le 1, 0 \le y \le 1, 0 \le z \le 1$.

SOLUTION See Figure 9.114 and Problem 38 in Exercises 9.13. Rather than evaluate six surface integrals, we apply the divergence theorem. See div $\mathbf{F} = \nabla \cdot \mathbf{F} = y + 2yz + 3z^2$, we have from (2)

$$\iint_S (\mathbf{F} \cdot \mathbf{n})\, dS = \iiint_D (y + 2yz + 3z^2)\, dV$$

$$= \int_0^1 \int_0^1 \int_0^1 (y + 2yz + 3z^2)\, dx\, dy\, dz$$

$$= \int_0^1 \int_0^1 (y + 2yz + 3z^2)\, dy\, dz$$

$$= \int_0^1 \left(\frac{y^2}{2} + y^2 z + 3yz^2 \right) \Big]_0^1 dz$$

$$= \int_0^1 \left(\frac{1}{2} + z + 3z^2 \right) dz = \left(\frac{1}{2}z + \frac{1}{2}z^2 + z^3 \right) \Big]_0^1 = 2. \qquad \square$$

■ **Physical Interpretation of Divergence** In Section 9.14 we saw that we could express the normal component of the curl of a vector field \mathbf{F} at a point as a limit involving the circulation of \mathbf{F}. In view of (2), it is possible to interpret the divergence of \mathbf{F} at a point as a limit involving the flux of \mathbf{F}. Recall from Section 9.7 that the flux of the velocity field \mathbf{F} of a fluid is the rate of fluid flow—that is, the volume of fluid flowing through a surface per unit time. In Section 9.7 we saw that the divergence of \mathbf{F} is the flux per unit volume. To reinforce this last idea let us suppose $P_0(x_0, y_0, z_0)$ is any point in the fluid and S_r is a small sphere of radius r centered at P_0. See Figure 9.141. If D_r is the sphere S_r and its interior, then the divergence theorem gives

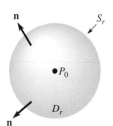

Figure 9.141

$$\iint_{S_r} (\mathbf{F} \cdot \mathbf{n})\, dS = \iint_{D_r} \text{div } \mathbf{F}\, dV. \qquad (11)$$

If we take the approximation div $\mathbf{F}(P) \approx$ div $\mathbf{F}(P_0)$ at all points $P(x, y, z)$ within the small sphere, then (11) gives

$$\iint_{S_r} (\mathbf{F} \cdot \mathbf{n})\, dS \approx \iiint_{D_r} \text{div } \mathbf{F}(P_0)\, dV$$

$$= \text{div } \mathbf{F}(P_0) \iiint_{D_r} dV \qquad (12)$$

$$= \text{div } \mathbf{F}(P_0)\, V_r,$$

where V_r is the volume ($\frac{4}{3}\pi r^3$) of the spherical region D_r. By letting $r \to 0$, we see from (12) that the divergence of \mathbf{F} is the limiting value of the ratio of the flux of \mathbf{F} to the volume of the spherical region:

$$\text{div } \mathbf{F}(P_0) = \lim_{r \to 0} \frac{1}{V_r} \iint_{S_r} (\mathbf{F} \cdot \mathbf{n})\, dS.$$

Hence, divergence \mathbf{F} is flux per unit volume.

The divergence theorem is extremely useful in the derivation of some of the famous equations in electricity and magnetism and hydrodynamics. In the discussion that follows we shall consider an example from the study of fluids.

■ **Continuity Equation** At the end of Section 9.7 we mentioned that one interpretation of div **F** was a measure of the rate of change of the density of a fluid at a point. To see why this is so, let us suppose that **F** is a velocity field of a fluid and that $\rho(x, y, z, t)$ is the density of the fluid at a point $P(x, y, z)$ at a time t. Let D be the closed region consisting of a sphere S and its interior. We know from Section 9.15 that the total mass m of the fluid in D is given by $m = \iiint_D \rho(x, y, z, t)\, dV$. The rate at which the mass increases in D is given by

$$\frac{dm}{dt} = \frac{d}{dt} \iiint_D \rho(x, y, z, t)\, dV = \iiint_D \frac{\partial \rho}{\partial t}\, dV. \tag{13}$$

Now from Figure 9.39 we saw that the volume of fluid flowing through an element of surface area ΔS per unit time is approximated by $(\mathbf{F} \cdot \mathbf{n})\, \Delta S$. The mass of the fluid flowing through an element of surface area ΔS per unit time is then $(\rho \mathbf{F} \cdot \mathbf{n})\, \Delta S$. If we assume that the change in mass in D is due only to the flow in and out of D, then the *volume of fluid* flowing out of D per unit time is given by (10) of Section 9.13, $\iint_S (\mathbf{F} \cdot \mathbf{n})\, dS$, whereas the *mass of the fluid* flowing out of D per unit time is $\iint_S (\rho \mathbf{F} \cdot \mathbf{n})\, dS$. Hence, an alternative expression for the rate at which the mass increases in D is

$$-\iint_S (\rho \mathbf{F} \cdot \mathbf{n})\, dS. \tag{14}$$

By the divergence theorem, (14) is the same as

$$-\iiint_D \operatorname{div}(\rho \mathbf{F})\, dV. \tag{15}$$

Equating (13) and (15) then yields

$$\iiint_D \frac{\partial \rho}{\partial t}\, dV = -\iiint_D \operatorname{div}(\rho \mathbf{F})\, dV \quad \text{or} \quad \iiint_D \left(\frac{\partial \rho}{\partial t} + \operatorname{div}(\rho \mathbf{F}) \right) dV = 0.$$

Since this last result is to hold for every sphere, we obtain the **equation of continuity** for fluid flows:

$$\frac{\partial \rho}{\partial t} + \operatorname{div}(\rho \mathbf{F}) = 0. \tag{16}$$

On page 484 we stated that if div $\mathbf{F} = \nabla \cdot \mathbf{F} = 0$, then a fluid is incompressible. This fact follows immediately from (16). If a fluid is incompressible (such as water), then ρ is constant, so consequently $\nabla \cdot (\rho \mathbf{F}) = \rho \nabla \cdot \mathbf{F}$. But in addition $\partial \rho / \partial t = 0$ and so (16) implies $\nabla \cdot \mathbf{F} = 0$.

E X E R C I S E S 9 . 1 6

Answers to odd-numbered problems begin on page A-52.

In Problems 1 and 2, verify the divergence theorem.

1. $\mathbf{F} = xy\mathbf{i} + yz\mathbf{j} + xz\mathbf{k}$; D the region bounded by the unit cube defined by $0 \le x \le 1$, $0 \le y \le 1$, $0 \le z \le 1$

2. $\mathbf{F} = 6xy\mathbf{i} + 4yz\mathbf{j} + xe^{-y}\mathbf{k}$; D the region bounded by the three coordinate planes and the plane $x + y + z = 1$

In Problems 3–14, use the divergence theorem to find the outward flux $\iint_S (\mathbf{F} \cdot \mathbf{n})\, dS$ of the given vector field **F**.

3. $\mathbf{F} = x^3\mathbf{i} + y^3\mathbf{j} + z^3\mathbf{k}$; D the region bounded by the sphere $x^2 + y^2 + z^2 = a^2$

4. $\mathbf{F} = 4x\mathbf{i} + y\mathbf{j} + 4z\mathbf{k}$; D the region bounded by the sphere $x^2 + y^2 + z^2 = 4$

5. $\mathbf{F} = y^2\mathbf{i} + xz^3\mathbf{j} + (z - 1)^2\mathbf{k}$; D the region bounded by the cylinder $x^2 + y^2 = 16$ and the planes $z = 1$, $z = 5$

6. $\mathbf{F} = x^2\mathbf{i} + 2yz\mathbf{j} + 4z^3\mathbf{k}$; D the region bounded by the parallelepiped defined by $0 \le x \le 1$, $0 \le y \le 2$, $0 \le z \le 3$

7. $\mathbf{F} = y^3\mathbf{i} + x^3\mathbf{j} + z^3\mathbf{k}$; D the region bounded within by $z = \sqrt{4 - x^2 - y^2}$, $x^2 + y^2 = 3$, $z = 0$

8. $\mathbf{F} = (x^2 + \sin y)\mathbf{i} + z^2\mathbf{j} + xy^3\mathbf{k}$; D the region bounded by $y = x^2$, $z = 9 - y$, $z = 0$

9. $\mathbf{F} = (x\mathbf{i} + y\mathbf{j} + z\mathbf{k})/(x^2 + y^2 + z^2)$; D the region bounded by the concentric spheres $x^2 + y^2 + z^2 = a^2$, $x^2 + y^2 + z^2 = b^2$, $b > a$

10. $\mathbf{F} = 2yz\mathbf{i} + x^3\mathbf{j} + xy^2\mathbf{k}$; D the region bounded by the ellipsoid $x^2/a^2 + y^2/b^2 + z^2/c^2 = 1$

11. $\mathbf{F} = 2xz\mathbf{i} + 5y^2\mathbf{j} - z^2\mathbf{k}$; D the region bounded by $z = y$, $z = 4 - y$, $z = 2 - \frac{1}{2}x^2$, $x = 0$, $z = 0$. See Figure 9.142.

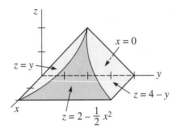

Figure 9.142

12. $\mathbf{F} = 15x^2y\mathbf{i} + x^2z\mathbf{j} + y^4\mathbf{k}$; D the region bounded by $x + y = 2$, $z = x + y$, $z = 3$, $x = 0$, $y = 0$

13. $\mathbf{F} = 3x^2y^2\mathbf{i} + y\mathbf{j} - 6zxy^2\mathbf{k}$; D the region bounded by the paraboloid $z = x^2 + y^2$ and the plane $z = 2y$

14. $\mathbf{F} = xy^2\mathbf{i} + x^2y\mathbf{j} + 6\sin x\mathbf{k}$; D the region bounded by the cone $z = \sqrt{x^2 + y^2}$ and the planes $z = 2$, $z = 4$

15. The electric field at a point $P(x, y, z)$ due to a point charge q located at the origin is given by the inverse square field

$$\mathbf{E} = q\frac{\mathbf{r}}{\|\mathbf{r}\|^3},$$

where $\mathbf{r} = x\mathbf{i} + y\mathbf{j} + z\mathbf{k}$.

(a) Suppose S is a closed surface, S_a is a sphere $x^2 + y^2 + z^2 = a^2$ lying completely within S, and D is the region bounded between S and S_a. See Figure 9.143. Show that the outward flux of \mathbf{E} for the region D is zero.

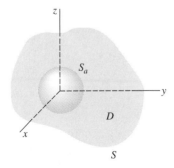

Figure 9.143

(b) Use the result of part **(a)** to prove **Gauss' law**:

$$\iint_S (\mathbf{E} \cdot \mathbf{n})\, dS = 4\pi q,$$

that is, the outward flux of the electric field \mathbf{E} through *any* closed surface (for which the divergence theorem applies) containing the origin is $4\pi q$.

16. Suppose there is a continuous distribution of charge throughout a closed and bounded region D enclosed by a surface S. Then the natural extension of Gauss' law is given by

$$\iint_S (\mathbf{E} \cdot \mathbf{n})\, dS = \iiint_D 4\pi\rho\, dV,$$

where $\rho(x, y, z)$ is the charge density or charge per unit volume.

(a) Proceed as in the derivation of the continuity equation (16) to show that div $\mathbf{E} = 4\pi\rho$.

(b) Given that \mathbf{E} is an irrotational vector field, show that the potential ϕ for \mathbf{E} satisfies Poisson's equation $\nabla^2\phi = 4\pi\rho$.

In Problems 17–21, assume that S forms the boundary of a closed and bounded region D.

17. If \mathbf{a} is a constant vector, show that $\iint_S (\mathbf{a} \cdot \mathbf{n})\, dS = 0$.

18. If $\mathbf{F} = P\mathbf{i} + Q\mathbf{j} + R\mathbf{k}$ and P, Q, and R have continuous second partial derivatives, prove that $\iint_S (\text{curl } \mathbf{F} \cdot \mathbf{n})\, dS = 0$.

In Problems 19 and 20, assume that f and g are scalar functions with continuous second partial derivatives. Use the divergence theorem to establish **Green's identities.**

19. $\displaystyle\iint_S (f\nabla g) \cdot \mathbf{n}\, dS = \iiint_D (f\nabla^2 g + \nabla f \cdot \nabla g)\, dV$

20. $\displaystyle\iint_S (f\nabla g - g\nabla f) \cdot \mathbf{n}\, dS = \iiint_D (f\nabla^2 g - g\nabla^2 f)\, dV$

21. If f is a scalar function with continuous first partial derivatives, prove that

$$\iint_S f\mathbf{n}\, dS = \iiint_D \nabla f\, dV$$

[*Hint:* Use (2) on $f\mathbf{a}$, where \mathbf{a} is a constant vector, and Problem 27 in Exercises 9.7.]

22. The buoyancy force on a floating object is $\mathbf{B} = -\iint_S p\mathbf{n}\, dS$, where p is the fluid pressure. The pressure p is related to the density of the fluid $\rho(x, y, z)$ by a law of hydrostatics: $\nabla p = \rho(x, y, z)\mathbf{g}$, where \mathbf{g} is the constant acceleration of gravity. If the weight of the object is $\mathbf{W} = m\mathbf{g}$, use the result of Problem 21 to prove Archimedes' principle, $\mathbf{B} + \mathbf{W} = \mathbf{0}$. See Figure 9.144.

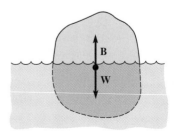

Figure 9.144

9.17 Change of Variables in Multiple Integrals

■ **Single Integral** In many instances it is convenient to make a substitution, or change of variable, in an integral in order to evaluate it. If f is continuous, $x = g(u)$ has a continuous derivative, and $dx = g'(u)\,du$, then

$$\int_a^b f(x)\,dx = \int_c^d f(g(u))g'(u)\,du, \tag{1}$$

where $c = g(a)$ and $d = g(b)$. There are three things that bear emphasizing in (1). To change the variable in a definite integral we replace x where it appears in the integrand by $g(u)$, we change the interval of integration $[a, b]$ on the x-axis to the corresponding interval $[c, d]$ on the u-axis, and we replace dx by a function multiple (namely, the derivative of g) of du. If we write $J(u) = dx/du$, then (1) has the form

$$\int_a^b f(x)\,dx = \int_c^d f(g(u))J(u)\,du. \tag{2}$$

For example, using $x = 2\sin\theta$, $-\pi/2 \le \theta \le \pi/2$, we get

$$\int_0^2 \overbrace{\sqrt{4 - x^2}}^{\substack{x\text{-limits}\\ \downarrow\ f(x)}}\,dx = \int_0^{\pi/2} \overbrace{2\cos\theta}^{\substack{\theta\text{-limits}\\ \downarrow\ f(2\sin\theta)}}\overbrace{(2\cos\theta)}^{J(\theta)}\,d\theta = 4\int_0^{\pi/2}\cos^2\theta\,d\theta = \pi.$$

■ **Double Integrals** Although changing variables in a multiple integral is not as straightforward as the procedure in (1), the basic idea illustrated in (2) carries over. To change variables in a double integral we need two equations such as

$$x = f(u, v), \qquad y = g(u, v). \tag{3}$$

To be analogous with (2), we expect that a change of variables in a double integral would take the form

$$\iint_R F(x, y)\,dA = \iint_S F(f(u, v),\, g(u, v))J(u, v)\,dA', \tag{4}$$

where S is the region in the uv-plane corresponding to the region R in the xy-plane and $J(u, v)$ is some function that depends on the partial derivatives of the equations in (3). The symbol dA' on the right side of (4) represents either $du\,dv$ or $dv\,du$.

In Section 9.11 we briefly discussed how to change a double integral $\iint_R F(x, y)\,dA$ from rectangular coordinates to polar coordinates. Recall that in Example 2 of that section the substitutions

$$x = r\cos\theta, \qquad y = r\sin\theta \tag{5}$$

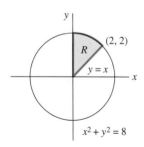

(a) Region R in xy-plane

(b) Region S in $r\theta$-plane

Figure 9.145

led to

$$\int_0^2 \int_x^{\sqrt{8 - x^2}} \frac{1}{5 + x^2 + y^2}\,dy\,dx = \int_{\pi/4}^{\pi/2}\int_0^{\sqrt{8}} \frac{1}{5 + r^2}\,r\,dr\,d\theta. \tag{6}$$

As we see in Figure 9.145, the introduction of polar coordinates changes the original region of integration R in the xy-plane to the more convenient rectangular region of integration S in the $r\theta$-plane. We note, too, that by comparing (4) with (6), we can identify $J(r, \theta) = r$ and $dA' = dr\,d\theta$.

The change-of-variable equations in (3) define a **transformation** or **mapping** T from the uv-plane to the xy-plane. A point (x_0, y_0) in the xy-plane determined from $x_0 = f(u_0, v_0)$, $y_0 = g(u_0, v_0)$ is said to be an **image** of (u_0, v_0).

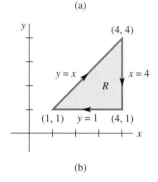

Figure 9.146

Example 1 Image of a Region

Find the image of the region S shown in Figure 9.146(a) under the transformation $x = u^2 + v^2$, $y = u^2 - v^2$.

SOLUTION We begin by finding the images of the sides of S that we have indicated by S_1, S_2, and S_3.

S_1: On this side $v = 0$ so that $x = u^2$, $y = u^2$. Eliminating u then gives $y = x$. Now imagine moving along the boundary from $(1, 0)$ to $(2, 0)$ (that is, $1 \leq u \leq 2$). The equations $x = u^2$, $y = u^2$ then indicate that x ranges from $x = 1$ to $x = 4$ and y ranges simultaneously from $y = 1$ to $y = 4$. In other words, in the xy-plane the image of S_1 is the line segment $y = x$ from $(1, 1)$ to $(4, 4)$.

S_2: On this boundary $u^2 + v^2 = 4$ and so $x = 4$. Now as we move from the point $(2, 0)$ to $(\sqrt{5/2}, \sqrt{3/2})$, the remaining equation $y = u^2 - v^2$ indicates that y ranges from $y = 2^2 - 0^2 = 4$ to $y = (\sqrt{5/2})^2 - (\sqrt{3/2})^2 = 1$. In this case the image of S_2 is the vertical line segment $x = 4$ starting at $(4, 4)$ and going down to $(4, 1)$.

S_3: Since $u^2 - v^2 = 1$, we get $y = 1$. But as we move on this boundary from $(\sqrt{5/2}, \sqrt{3/2})$ to $(1, 0)$, the equation $x = u^2 + v^2$ indicates that x ranges from $x = 4$ to $x = 1$. The image of S_3 is the horizontal line segment $y = 1$ starting at $(4, 1)$ and ending at $(1, 1)$.

The image of S is the region R given in Figure 9.146(b). ❏

Observe in Example 1 that as we traverse the boundary of S in the counterclockwise direction, the boundary of R is traversed in a clockwise manner. We say that the transformation of the boundary of S has *induced* an orientation on the boundary of R.

While a proof of the formula for changing variables in a multiple integral is beyond the level of this text, we will give *some* of the underlying assumptions that are made about the equations (3) and the region R and S. We assume that:

- The functions f and g have continuous first partial derivatives on S.
- The transformation is one-to-one.
- Each of the regions R and S consists of a piecewise smooth simple closed curve and its interior.
- The determinant

$$\begin{vmatrix} \dfrac{\partial x}{\partial u} & \dfrac{\partial x}{\partial v} \\[2mm] \dfrac{\partial y}{\partial u} & \dfrac{\partial y}{\partial v} \end{vmatrix} = \dfrac{\partial x}{\partial u}\dfrac{\partial y}{\partial v} - \dfrac{\partial x}{\partial v}\dfrac{\partial y}{\partial u} \tag{7}$$

is not zero on S.

A transformation T is said to be **one-to-one** if each point (x_0, y_0) in R is the image under T of a unique point (u_0, v_0) in S. Put another way, no two points in S have the same image in R. With the restrictions that $r \geq 0$ and $0 \leq \theta \leq 2\pi$, the equations in (5) define a one-to-one transformation from the $r\theta$-plane to the xy-plane. The determinant in (7) is called the **Jacobian** of the transformation T and is the key to changing variables in a multiple integral. The Jacobian of the transformation defined by the equations in (3) is denoted by the symbol

$$\frac{\partial(x, y)}{\partial(u, v)}.$$

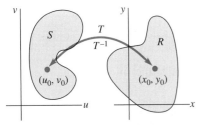

Figure 9.147

Similar to the notion of a one-to-one function, a one-to-one transformation T has an **inverse transformation** T^{-1} such that (u_0, v_0) is the image under T^{-1} of (x_0, y_0). See Figure 9.147. If it is possible to solve (3) for u and v in terms of x and y, then the inverse transformation is defined by a pair of equations

$$u = h(x, y), \qquad v = k(x, y). \tag{8}$$

The Jacobian of the inverse transformation T^{-1} is

$$\frac{\partial(u, v)}{\partial(x, y)} = \begin{vmatrix} \dfrac{\partial u}{\partial x} & \dfrac{\partial u}{\partial y} \\[2mm] \dfrac{\partial v}{\partial x} & \dfrac{\partial v}{\partial y} \end{vmatrix} \tag{9}$$

and is related to the Jacobian of the transformation T by

$$\frac{\partial(x, y)}{\partial(u, v)} \frac{\partial(u, v)}{\partial(x, y)} = 1. \tag{10}$$

Example 2 Jacobian
The Jacobian of the transformation $x = r \cos \theta$, $y = r \sin \theta$ is

$$\frac{\partial(x, y)}{\partial(r, \theta)} = \begin{vmatrix} \dfrac{\partial x}{\partial r} & \dfrac{\partial x}{\partial \theta} \\[2mm] \dfrac{\partial y}{\partial r} & \dfrac{\partial y}{\partial \theta} \end{vmatrix} = \begin{vmatrix} \cos \theta & -r \sin \theta \\ \sin \theta & r \cos \theta \end{vmatrix} = r(\cos^2\theta + \sin^2\theta) = r. \quad \square$$

We now turn our attention to the main point of this discussion: how to change variables in a multiple integral. The idea expressed in (4) is valid; the function $J(u, v)$ turns out to be $|\partial(x, y)/\partial(u, v)|$. Under the assumptions made above, we have the following result:

THEOREM 9.16 Change of Variables in a Double Integral

If F is continuous on R, then

$$\iint_R F(x, y) \, dA = \iint_S F(f(u, v), g(u, v)) \left| \frac{\partial(x, y)}{\partial(u, v)} \right| dA'. \tag{11}$$

Formula (3) of Section 9.11 for changing a double integral to polar coordinates is just a special case of (11) with

$$\left| \frac{\partial(x, y)}{\partial(r, \theta)} \right| = |r| = r,$$

since $r \geq 0$. In (6) then we have $J(r, \theta) = |\partial(x, y)/\partial(r, \theta)| = r$.

A change of variables in a multiple integral can be used for either a simplification of the integrand or a simplification of the region of integration. The actual change of variables used is often inspired by the structure of the integrand $F(x, y)$ or by equations that define the region R. As a consequence, the transformation is then defined by equations of the form given in (8); that is, we are dealing with the inverse transformation. The next two examples will illustrate these ideas.

Figure 9.148

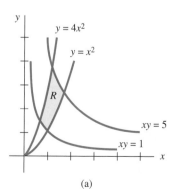

Figure 9.149

Example 3 **Changing Variables in a Double Integral**

Evaluate $\iint_R \sin(x + 2y) \cos(x - 2y)\, dA$ over the region R shown in Figure 9.148(a).

SOLUTION The difficulty in evaluating this double integral is clearly the integrand. The presence of the terms $x + 2y$ and $x - 2y$ prompts us to define the change of variables $u = x + 2y$, $v = x - 2y$. These equations will map R onto a region S in the uv-plane. As in Example 1, we transform the sides of the region.

S_1: $y = 0$ implies $u = x$ and $v = x$ or $v = u$. As we move from $(2\pi, 0)$ to $(0, 0)$, we see that the corresponding image points in the uv-plane lie on the line segment $v = u$ from $(2\pi, 2\pi)$ to $(0, 0)$.

S_2: $x = 0$ implies $u = 2y$ and $v = -2y$, or $v = -u$. As we move from $(0, 0)$ to $(0, \pi)$, the corresponding image points in the uv-plane lie on the line segment $v = -u$ from $(0, 0)$ to $(2\pi, -2\pi)$.

S_3: $x + 2y = 2\pi$ implies $u = 2\pi$. As we move from $(0, \pi)$ to $(2\pi, 0)$, the equation $v = x - 2y$ shows that v ranges from $v = -2\pi$ to $v = 2\pi$. Thus, the image of S_3 is the vertical line segment $u = 2\pi$ starting at $(-2\pi, -2\pi)$ and going up to $(2\pi, 2\pi)$. See Figure 9.148(b).

Now, solving for x and y in terms of u and v gives

$$x = \frac{1}{2}(u + v), \qquad y = \frac{1}{4}(u - v).$$

Therefore,
$$\frac{\partial(x, y)}{\partial(u, v)} = \begin{vmatrix} \dfrac{\partial x}{\partial u} & \dfrac{\partial x}{\partial v} \\[2mm] \dfrac{\partial y}{\partial u} & \dfrac{\partial y}{\partial v} \end{vmatrix} = \begin{vmatrix} \dfrac{1}{2} & \dfrac{1}{2} \\[2mm] \dfrac{1}{4} & -\dfrac{1}{4} \end{vmatrix} = -\frac{1}{4}.$$

Hence, from (11) we find that

$$\iint_R \sin(x + 2y) \cos(x - 2y)\, dA = \iint_S \sin u \cos v \left| -\frac{1}{4} \right| dA'$$

$$= \frac{1}{4} \int_0^{2\pi} \int_{-u}^{u} \sin u \cos v\, dv\, du$$

$$= \frac{1}{4} \int_0^{2\pi} \sin u \sin v \Big]_{-u}^{u} du$$

$$= \frac{1}{2} \int_0^{2\pi} \sin^2 u\, du$$

$$= \frac{1}{4} \int_0^{2\pi} (1 - \cos 2u)\, du$$

$$= \frac{1}{4} \left[u - \frac{1}{2} \sin 2u \right]_0^{2\pi} = \frac{\pi}{2}. \qquad \Box$$

Example 4 **Changing Variables in a Double Integral**

Evaluate $\iint_R xy\, dA$ over the region R shown in Figure 9.149(a).

SOLUTION In this case the integrand is fairly simple, but integration over the region R would be tedious since we would have to express $\iint_R xy\, dA$ as the sum of three integrals. (Verify this.)

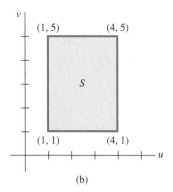

(b)

Figure 9.149

The equations of the boundaries of R suggest the change of variables

$$u = \frac{y}{x^2}, \qquad v = xy. \tag{12}$$

Obtaining the image of R is a straightforward matter in this case, since the images of the curves that make up the four boundaries are simply $u = 1$, $u = 4$, $v = 1$, and $v = 5$. In other words, the image of the region R is the rectangular region S: $1 \le u \le 4$, $1 \le v \le 5$. See Figure 9.149(b).

Now instead of trying to solve the equations in (12) for x and y in terms of u and v, we can compute the Jacobian $\partial(x, y)/\partial(u, v)$ by computing $\partial(u, v)/\partial(x, y)$ and using (10). We have

$$\frac{\partial(u, v)}{\partial(x, y)} = \begin{vmatrix} \dfrac{\partial u}{\partial x} & \dfrac{\partial u}{\partial y} \\[2mm] \dfrac{\partial v}{\partial x} & \dfrac{\partial v}{\partial y} \end{vmatrix} = \begin{vmatrix} -\dfrac{2y}{x^3} & \dfrac{1}{x^2} \\[2mm] y & x \end{vmatrix} = -\frac{3y}{x^2},$$

and so from (10),

$$\frac{\partial(x, y)}{\partial(u, v)} = \frac{1}{\dfrac{\partial(u, v)}{\partial(x, y)}} = -\frac{x^2}{3y} = -\frac{1}{3u}.$$

Hence,

$$\iint_R xy \, dA = \iint_S v \left| -\frac{1}{3u} \right| dA'$$

$$= \frac{1}{3} \int_1^4 \int_1^5 \frac{v}{u} \, dv \, du$$

$$= \frac{1}{3} \int_1^4 \frac{v^2}{2u} \Big]_1^5 du$$

$$= 4 \int_1^4 \frac{1}{u} \, du = 4 \ln u \Big]_1^4 = 4 \ln 4. \qquad \square$$

■ **Triple Integrals** To change variables in a triple integral, let

$$x = f(u, v, w), \qquad y = g(u, v, w), \qquad z = h(u, v, w),$$

be a one-to-one transformation T from a region E in uvw-space to a region D in xyz-space. If F is continuous on D, then

$$\iiint_D F(x, y, z) \, dV = \iiint_E F(f(u, v, w), g(u, v, w), h(u, v, w)) \left| \frac{\partial(x, y, z)}{\partial(u, v, w)} \right| dV'$$

where

$$\frac{\partial(x, y, z)}{\partial(u, v, w)} = \begin{vmatrix} \dfrac{\partial x}{\partial u} & \dfrac{\partial x}{\partial v} & \dfrac{\partial x}{\partial w} \\[2mm] \dfrac{\partial y}{\partial u} & \dfrac{\partial y}{\partial v} & \dfrac{\partial y}{\partial w} \\[2mm] \dfrac{\partial z}{\partial u} & \dfrac{\partial z}{\partial v} & \dfrac{\partial z}{\partial w} \end{vmatrix}.$$

We leave it as an exercise for the reader to show that if T is the transformation from spherical to rectangular coordinates defined by

$$x = \rho \sin \phi \cos \theta, \quad y = \rho \sin \phi \sin \theta, \quad z = \rho \cos \phi, \tag{13}$$

then

$$\frac{\partial(x, y, z)}{\partial(\rho, \phi, \theta)} = \rho^2 \sin \phi.$$

Answers to odd-numbered problems begin on page A-52.

1. Consider a transformation T defined by $x = 4u - v$, $y = 5u + 4v$. Find the images of the points $(0, 0)$, $(0, 2)$, $(4, 0)$, and $(4, 2)$ in the uv-plane under T.

2. Consider a transformation T defined by $x = \sqrt{v - u}$, $y = v + u$. Find the images of the points $(1, 1)$, $(1, 3)$, and $(\sqrt{2}, 2)$ in the xy-plane under T^{-1}.

In Problems 3–6, find the image of the set S under the given transformation.

3. S: $0 \le u \le 2$, $0 \le v \le u$; $x = 2u + v$, $y = u - 3v$

4. S: $-1 \le u \le 4$, $1 \le v \le 5$; $u = x - y$, $v = x + 2y$

5. S: $0 \le u \le 1$, $0 \le v \le 2$; $x = u^2 - v^2$, $y = uv$

6. S: $1 \le u \le 2$, $1 \le v \le 2$; $x = uv$, $y = v^2$

In Problems 7–10, find the Jacobian of the transformation T from the uv-plane to the xy-plane.

7. $x = ve^{-u}$, $y = ve^{u}$

8. $x = e^{3u} \sin v$, $y = e^{3u} \cos v$

9. $u = \dfrac{y}{x^2}$, $v = \dfrac{y^2}{x}$

10. $u = \dfrac{2x}{x^2 + y^2}$, $v = \dfrac{-2y}{x^2 + y^2}$

11. (a) Find the image of the region S: $0 \le u \le 1$, $0 \le v \le 1$ under the transformation $x = u - uv$, $y = uv$.

(b) Explain why the transformation is not one-to-one on the boundary of S.

12. Determine where the Jacobian $\partial(x, y)/\partial(u, v)$ of the transformation in Problem 11 is zero.

In Problems 13–22, evaluate the given integral by means of the indicated change of variables.

13. $\iint_R (x + y)\, dA$, where R is the region bounded by the graphs of $x - 2y = -6$, $x - 2y = 6$, $x + y = -1$, $x + y = 3$; $u = x - 2y$, $v = x + y$

14. $\iint_R \dfrac{\cos \frac{1}{2}(x - y)}{3x + y}\, dA$, where R is the region bounded by the graphs of $y = x$, $y = x - \pi$, $y = -3x + 3$, $y = -3x + 6$; $u = x - y$, $v = 3x + y$

15. $\iint_R \dfrac{y^2}{x}\, dA$, where R is the region bounded by the graphs of $y = x^2$, $y = \frac{1}{2}x^2$, $x = y^2$, $x = \frac{1}{2}y^2$; $u = \dfrac{x^2}{y}$, $v = \dfrac{y^2}{x}$

16. $\iint_R (x^2 + y^2)^{-3}\, dA$, where R is the region bounded by the circles $x^2 + y^2 = 2x$, $x^2 + y^2 = 4x$, $x^2 + y^2 = 2y$, $x^2 + y^2 = 6y$; $u = \dfrac{2x}{x^2 + y^2}$, $v = \dfrac{2y}{x^2 + y^2}$ [*Hint:* Form $u^2 + v^2$.]

17. $\iint_R (x^2 + y^2)\, dA$, where R is the region in the first quadrant bounded by the graphs of $x^2 - y^2 = a$, $x^2 - y^2 = b$, $2xy = c$, $2xy = d$, $0 < a < b$, $0 < c < d$; $u = x^2 - y^2$, $v = 2xy$

18. $\iint_R (x^2 + y^2) \sin xy\, dA$, where R is the region bounded by the graphs of $x^2 - y^2 = 1$, $x^2 - y^2 = 9$, $xy = 2$, $xy = -2$; $u = x^2 - y^2$, $v = xy$

19. $\iint_R \dfrac{x}{y + x^2}\, dA$, where R is the region in the first quadrant bounded by the graphs of $x = 1$, $y = x^2$, $y = 4 - x^2$; $x = \sqrt{v - u}$, $y = v + u$

20. $\iint_R y\, dA$, where R is the triangular region with vertices $(0, 0)$, $(2, 3)$ and $(-4, 1)$; $x = 2u - 4v$, $y = 3u + v$

21. $\iint_R y^4\, dA$, where R is the region in the first quadrant bounded by the graphs of $xy = 1$, $xy = 4$, $y = x$, $y = 4x$; $u = xy$, $v = y/x$

22. $\iiint_D (4z + 2x - 2y)\, dV$, where D is the parallelepiped $1 \le y + z \le 3$, $-1 \le -y + z \le 1$, $0 \le x - y \le 3$; $u = y + z$, $v = -y + z$, $w = x - y$

In Problems 23–26, evaluate the given double integral by means of an appropriate change of variables.

23. $\displaystyle\int_0^1 \int_0^{1-x} e^{(y-x)/(y+x)}\, dy\, dx$

24. $\displaystyle\int_{-2}^0 \int_0^{x+2} e^{y^2 - 2xy + x^2}\, dy\, dx$

25. $\iint_R (6x + 3y)\, dA$, where R is the trapezoidal region in the first quadrant with vertices $(1, 0)$, $(4, 0)$, $(2, 4)$, and $(\frac{1}{2}, 1)$

26. $\iint_R (x + y)^4 e^{x-y}\, dA$, where R is the square region with vertices $(1, 0)$, $(0, 1)$, $(1, 2)$, and $(2, 1)$

27. A problem in thermodynamics is to find the work done by an ideal Carnot engine. This work is defined to be the area

of the region R in the first quadrant bounded by the isothermals $xy = a$, $xy = b$, $0 < a < b$, and the adiabatics $xy^{1.4} = c$, $xy^{1.4} = d$, $0 < c < d$. Use $A = \iint_R dA$ and an appropriate substitution to find the area shown in Figure 9.150.

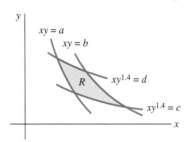

Figure 9.150

28. Use $V = \iiint_D dV$ and the substitutions $u = x/a$, $v = y/b$, $w = z/c$ to show that the volume of the ellipsoid $x^2/a^2 + y^2/b^2 + z^2/c^2 = 1$ is $V = \tfrac{4}{3}\pi abc$.

29. Evaluate the double integral $\iint_R \left(\dfrac{x^2}{25} + \dfrac{y^2}{9} \right) dA$, where R is the elliptical region whose boundary is the graph of $x^2/25 + y^2/9 = 1$. Use the substitutions $u = x/5$, $v = y/3$ and polar coordinates.

30. Verify that the Jacobian of the transformation given in (13) is $\partial(x, y, z)/\partial(\rho, \phi, \theta) = \rho^2 \sin \phi$.

CHAPTER 9
REVIEW EXERCISES

Answers to odd-numbered problems begin on page A-52.

Answer Problems 1–20 without referring back to the text. Fill in the blank or answer true/false. Where appropriate, assume continuity of P, Q, and their first partial derivatives.

1. A particle whose position vector is $\mathbf{r}(t) = \cos t\mathbf{i} + \cos t\mathbf{j} + \sqrt{2} \sin t\mathbf{k}$ moves with constant speed. _____

2. The path of a moving particle whose position vector is $\mathbf{r}(t) = (t^2 + 1)\mathbf{i} + 4\mathbf{j} + t^4\mathbf{k}$ lies in a plane. _____

3. The binormal vector is perpendicular to the osculating plane. _____

4. If $\mathbf{r}(t)$ is the position vector of a moving particle, then the velocity vector $\mathbf{v}(t) = \mathbf{r}'(t)$ and the acceleration vector $\mathbf{a}(t) = \mathbf{r}''(t)$ are orthogonal. _____

5. ∇z is perpendicular to the graph of $z = f(x, y)$. _____

6. If $\nabla f = 0$, then $f = $ constant. _____

7. The integral $\int_C (x^2 + y^2) \, dx + 2xy \, dy$, where C is given by $y = x^3$ from $(0, 0)$ to $(1, 1)$, has the same value on the curve $y = x^6$ from $(0, 0)$ to $(1, 1)$. _____

8. The value of the integral $\int_C 2xy \, dx - x^2 \, dy$ between two points A and B depends on the path C. _____

9. If C_1 and C_2 are two smooth curves such that $\int_{C_1} P \, dx + Q \, dy = \int_{C_2} P \, dx + Q \, dy$, then $\int_C P \, dx + Q \, dy$ is independent of the path. _____

10. If the work $\int_C \mathbf{F} \cdot d\mathbf{r}$ depends on the curve C, then \mathbf{F} is nonconservative. _____

11. If $\partial P/\partial x = \partial Q/\partial y$, then $\int_C P \, dx + Q \, dy$ is independent of the path. _____

12. In a conservative force field \mathbf{F}, the work done by \mathbf{F} around a simple closed curve is zero. _____

13. Assuming continuity of all partial derivatives, $\nabla \times \nabla f = \mathbf{0}$. _____

14. The surface integral of the normal component of the curl of a conservative vector field \mathbf{F} over a surface S is equal to zero. _____

15. Work done by a force \mathbf{F} along a curve C is due entirely to the tangential component of \mathbf{F}. _____

16. For a two-dimensional vector field \mathbf{F} in the plane $z = 0$, Stokes' theorem is the same as Green's theorem. _____

17. If \mathbf{F} is a conservative force field, then the sum of the potential and kinetic energies of an object is constant. _____

18. If $\int_C P \, dx + Q \, dy$ is independent of the path, then $\mathbf{F} = P\mathbf{i} + Q\mathbf{j}$ is the gradient of some function ϕ. _____

19. If $\phi = 1/\sqrt{x^2 + y^2}$ is a potential function for a conservative force field \mathbf{F}, then $\mathbf{F} = $ _____.

20. If $\mathbf{F} = f(x)\mathbf{i} + g(y)\mathbf{j} + h(z)\mathbf{k}$, then curl $\mathbf{F} = $ _____.

21. Find the velocity and acceleration of a particle whose position vector is $\mathbf{r}(t) = 6t\mathbf{i} + t\mathbf{j} + t^2\mathbf{k}$ as it passes through the plane $-x + y + z = -4$.

22. The velocity of a moving particle is $\mathbf{v}(t) = -10t\mathbf{i} + (3t^2 - 4t)\mathbf{j} + \mathbf{k}$. If the particle starts at $t = 0$ at $(1, 2, 3)$, what is its position at $t = 2$?

23. The acceleration of a moving particle is $\mathbf{a}(t) = \sqrt{2} \sin t\mathbf{i} + \sqrt{2} \cos t\mathbf{j}$. Given that the velocity and position of the particle

at $t = \pi/4$ are $\mathbf{v}(\pi/4) = -\mathbf{i} + \mathbf{j} + \mathbf{k}$ and $\mathbf{r}(\pi/4) = \mathbf{i} + 2\mathbf{j} + (\pi/4)\mathbf{k}$, respectively, what was the position of the particle at $t = 3\pi/4$?

24. Given that $\mathbf{r}(t) = \tfrac{1}{2}t^2\mathbf{i} + \tfrac{1}{3}t^3\mathbf{j} - \tfrac{1}{2}t^2\mathbf{k}$ is the position vector of a moving particle, find the tangential and normal components of the acceleration at any t. Find the curvature.

25. Sketch the curve traced by $\mathbf{r}(t) = \cosh t\,\mathbf{i} + \sinh t\,\mathbf{j} + t\mathbf{k}$.

26. Suppose that the vector function of Problem 25 is the position vector of a moving particle. Find the vectors \mathbf{T}, \mathbf{N}, and \mathbf{B} at $t = 1$. Find the curvature at that point.

In Problems 27 and 28, find the directional derivative of the given function in the indicated direction.

27. $f(x, y) = x^2y - y^2x$; $D_{\mathbf{u}}f$ in the direction of $2\mathbf{i} + 6\mathbf{j}$

28. $F(x, y, z) = \ln(x^2 + y^2 + z^2)$; $D_{\mathbf{u}}F$ in the direction of $-2\mathbf{i} + \mathbf{j} + 2\mathbf{k}$

29. Consider the function $f(x, y) = x^2y^4$. At $(1, 1)$ what is:

 (a) The rate of change of f in the direction of \mathbf{i}?

 (b) The rate of change of f in the direction of $\mathbf{i} - \mathbf{j}$?

 (c) The rate of change of f in the direction of \mathbf{j}?

30. Let $w = \sqrt{x^2 + y^2 + z^2}$.

 (a) If $x = 3 \sin 2t$, $y = 4 \cos 2t$, and $z = 5t^3$, find dw/dt.

 (b) If $x = 3 \sin 2\,\dfrac{t}{r}$, $y = 4 \cos 2\,\dfrac{r}{t}$, and $z = 5t^3r^3$, find $\partial w/\partial t$.

31. Find an equation of the tangent plane to the graph of $z = \sin xy$ at $\left(\dfrac{1}{2}, \dfrac{2\pi}{3}, \dfrac{\sqrt{3}}{2}\right)$.

32. Determine whether there are any points on the surface $z^2 + xy - 2x - y^2 = 1$ at which the tangent plane is parallel to $z = 2$.

33. Express the volume of the solid shown in Figure 9.151 as one or more iterated integrals using the order of integration **(a)** $dy\,dx$ and **(b)** $dx\,dy$. Choose either part **(a)** or part **(b)** to find the volume.

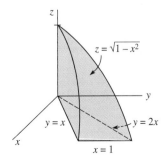

Figure 9.151

34. A lamina has the shape of the region in the first quadrant bounded by the graphs of $y = x^2$ and $y = x^3$. Find the center of mass if the density at a point P is directly proportional to the square of the distance from the origin.

35. Find the moment of inertia of the lamina described in Problem 34 about the y-axis.

36. Find the volume of the sphere $x^2 + y^2 + z^2 = a^2$ using a triple integral in **(a)** rectangular coordinates, **(b)** cylindrical coordinates, and **(c)** spherical coordinates.

37. Find the volume of the solid that is bounded between the cones $z = \sqrt{x^2 + y^2}$ and $z = \sqrt{9x^2 + 9y^2}$ and the plane $z = 3$.

38. Find the volume of the solid shown in Figure 9.152.

Figure 9.152

In Problems 39–42, find the indicated expression for the vector field $\mathbf{F} = x^2\,y\mathbf{i} + xy^2\mathbf{j} + 2xyz\mathbf{k}$.

39. $\nabla \cdot \mathbf{F}$ **40.** $\nabla \times \mathbf{F}$

41. $\nabla \cdot (\nabla \times \mathbf{F})$ **42.** $\nabla(\nabla \cdot \mathbf{F})$

43. Evaluate $\int_C \dfrac{z^2}{x^2 + y^2}\,ds$, where C is given by
$$x = \cos 2t, \quad y = \sin 2t, \quad z = 2t, \quad \pi \le t \le 2\pi$$

44. Evaluate $\int_C (xy + 4x)\,ds$, where C is given by $2x + y = 2$ from $(1, 0)$ to $(0, 2)$.

45. Evaluate $\int_C 3x^2y^2dx + (2x^3y - 3y^2)\,dy$, where C is given by $y = 5x^4 + 7x^2 - 14x$ from $(0, 0)$ to $(1, -2)$.

46. Show that $\displaystyle\oint_C \dfrac{-y\,dx + x\,dy}{x^2 + y^2} = 2\pi$, where C is the circle $x^2 + y^2 = a^2$.

47. Evaluate $\int_C y \sin \pi z\,dx + x^2e^y\,dy + 3xyz\,dz$, where C is given by $x = t$, $y = t^2$, $z = t^3$ from $(0, 0, 0)$ to $(1, 1, 1)$.

48. If $\mathbf{F} = 4y\mathbf{i} + 6x\mathbf{j}$ and C is given by $x^2 + y^2 = 1$, evaluate $\oint_C \mathbf{F} \cdot d\mathbf{r}$ in two different ways.

49. Find the work done by the force $\mathbf{F} = x \sin y\mathbf{i} + y \sin x\mathbf{j}$ acting along the line segments from $(0, 0)$ to $(\pi/2, 0)$ and from $(\pi/2, 0)$ to $(\pi/2, \pi)$.

50. Find the work done by $\mathbf{F} = \dfrac{2}{x^2 + y^2}\mathbf{i} + \dfrac{1}{x^2 + y^2}\mathbf{j}$ from $(-\frac{1}{2}, \frac{1}{2})$ to $(1, \sqrt{3})$ acting on the path shown in Figure 9.153.

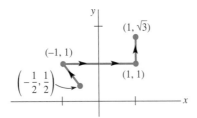

Figure 9.153

51. Evaluate $\iint_S (z/xy)\, dS$, where S is that portion of the cylinder $z = x^2$ in the first octant that is bounded by $y = 1$, $y = 3$, $z = 1$, $z = 4$.

52. If $\mathbf{F} = \mathbf{i} + 2\mathbf{j} + 3\mathbf{k}$, find the flux of \mathbf{F} through the square $0 \leq x \leq 1$, $0 \leq y \leq 1$, $z = 2$.

53. If $\mathbf{F} = c\nabla(1/r)$, where c is constant and $\|\mathbf{r}\| = r$, $\mathbf{r} = x\mathbf{i} + y\mathbf{j} + z\mathbf{k}$, find the flux of \mathbf{F} through the sphere $x^2 + y^2 + z^2 = a^2$.

54. Explain why the divergence theorem is not applicable in Problem 53.

55. Find the flux of $\mathbf{F} = c\nabla(1/r)$ through any surface S that forms the boundary of a closed bounded region of space not containing the origin.

56. If $\mathbf{F} = 6x\mathbf{i} + 7z\mathbf{j} + 8y\mathbf{k}$, use Stokes' theorem to evaluate $\iint_S (\text{curl } \mathbf{F} \cdot \mathbf{n})\, dS$, where S is that portion of the paraboloid $z = 9 - x^2 - y^2$ within the cylinder $x^2 + y^2 = 4$.

57. Use Stokes' theorem to evaluate $\oint_C - 2y\, dx + 3x\, dy + 10z\, dz$, where C is the circle $(x - 1)^2 + (y - 3)^2 = 25$, $z = 3$.

58. Find the work $\oint_C \mathbf{F} \cdot d\mathbf{r}$ done by the force $\mathbf{F} = x^2\mathbf{i} + y^2\mathbf{j} + z^2\mathbf{k}$ around the curve C that is formed by the intersection of the plane $z = 2 - y$ and the sphere $x^2 + y^2 + z^2 = 4z$.

59. If $\mathbf{F} = x\mathbf{i} + y\mathbf{j} + z\mathbf{k}$, use the divergence theorem to evaluate $\iint_S (\mathbf{F} \cdot \mathbf{n})\, dS$, where S is the surface of the region bounded by $x^2 + y^2 = 1$, $z = 0$, $z = 1$.

60. Repeat problem 59 for $\mathbf{F} = \frac{1}{3}x^3\mathbf{i} + \frac{1}{3}y^3\mathbf{j} + \frac{1}{3}z^3\mathbf{k}$.

61. If $\mathbf{F} = (x^2 - e^y \tan^{-1}z)\mathbf{i} + (x + y)^2\mathbf{j} - (2yz + x^{10})\mathbf{k}$, use the divergence theorem to evaluate $\iint_S (\mathbf{F} \cdot \mathbf{n})\, dS$, where S is

the surface of the region in the first octant bounded by $z = 1 - x^2$, $z = 0$, $z = 2 - y$, $y = 0$.

62. Suppose $\mathbf{F} = x\mathbf{i} + y\mathbf{j} + (z^2 + 1)\mathbf{k}$ and S is the surface of the region bounded by $x^2 + y^2 = a^2$, $z = 0$, $z = c$. Evaluate $\iint_S (\mathbf{F} \cdot \mathbf{n})\, dS$ without the aid of the divergence theorem. [*Hint:* The lateral surface area of the cylinder is $2\pi ac$.]

63. Evaluate the integral $\iint_R (x^2 + y^2)\sqrt[3]{x^2 - y^2}\, dA$, where R is the region bounded by the graphs of $x = 0$, $x = 1$, $y = 0$, and $y = 1$ by means of the change of variables $u = 2xy$, $v = x^2 - y^2$.

64. Evaluate the integral $\iint_R (1/\sqrt{(x - y)^2 + 2(x + y) + 1})\, dA$, where R is the region bounded by the graphs of $y = x$, $x = 2$, and $y = 0$ by means of the change of variables $x = u + uv$, $y = v + uv$.

65. As shown in Figure 9.154, a sphere of radius 1 has its center on the surface of a sphere of radius $a > 1$. Find the surface area of that portion of the larger sphere cut out by the smaller sphere.

Figure 9.154

66. On the surface of a globe or, more precisely, on the surface of the earth, the boundaries of the states of Colorado and Wyoming are both "spherical rectangles." (In this problem we assume that the earth is a perfect sphere.) Colorado is bounded by the lines of longitude 102°W and 109°W and the lines of latitude 37°N and 41°N. Wyoming is bounded by longitudes 104°W and 111°W and latitudes 41°N and 45°N. See Figure 9.155.

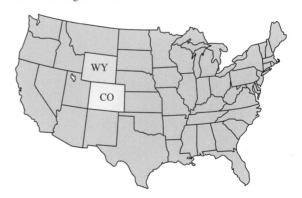

Figure 9.155

(a) Without explicitly computing their areas, determine which state is larger and explain why.

(b) By what percentage is Wyoming larger (or smaller) than Colorado? [*Hint:* Suppose the radius of the earth is R. Project a spherical rectangle in the Northern Hemisphere that is determined by latitudes θ_1 and θ_2 and longitudes ϕ_1 and ϕ_2 onto the xy-plane.]

(c) One reference book gives the areas of the two states as $104{,}247$ and $97{,}914$ mi^2. How does this answer compare with the answer in part **(b)**?

3

Systems of Differential Equations

Application *Essay*

Atmospheric Drag and the Decay of Satellite Orbits

John Ellison
Grove City College

Ever since the first Sputnik was launched into space, there have been artificial satellites circling the earth. These range from small pieces of space junk to large objects like the Hubble telescope and manned space stations. For most of their lives the main force acting on the objects is the earth's gravitational field. However, drag caused by the atmosphere will cause the orbit of any satellite to decay slowly, and, left alone, the satellite will ultimately fall to earth.

Small objects burn up in the earth's atmosphere and do not reach the earth's surface. Large objects have internal propulsion systems that can be used to maintain their orbits. However, in 1979 the internal propulsion system of one large object, Skylab, failed, and it entered the earth's atmosphere. Skylab was large enough that some small pieces survived and landed. They were very hot.

An exact mathematical model of the final few revolutions of the decaying orbit of a satellite is very complicated and difficult to obtain. In order to get an approximation we can solve, we must make many simplifying assumptions. Two common assumptions are that the earth is a perfect sphere and that the motion of the object is essentially two-dimensional. We will now derive a model for the motion of a decaying satellite.

The motion of the object is governed by Newton's second law: $F = ma$. In this analysis, we will assume that the only forces acting on the object are the gravitational attraction of the earth (ignoring the effects of the moon and sun and other bodies) and drag due to the atmosphere. Let the center of the earth be the origin, t be time, and $(x(t), y(t))$ be the position of the satellite. Then $r = \sqrt{x^2 + y^2}$ is the distance of the object from the center of the earth, $(x'(t), yx'(t))$ is the velocity, $v = \sqrt{(x')^2 + (y')^2}$ is the speed, and $(x''(t), y''(t))$ is the acceleration; see Figure 1. For a more complete discussion of these quantities, see Chapter 9. We now proceed to derive formulas for the two forces acting on the satellite.

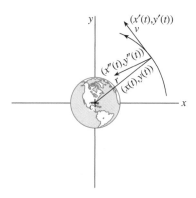

Figure 1 Position, velocity, speed, and acceleration of an object in earth orbit

By Newton's universal law of gravitation, the magnitude of the force due to gravity is $F = \dfrac{m_e m_s g}{r^2}$, where m_e is the mass of the earth, m_s is the mass of the satellite, and g is the gravitational constant. This force is directed toward the center of the earth, and so we find that in terms of components, the vector force due to gravity is $-\dfrac{m_e m_s g}{r^2} \dfrac{(x(t), y(t))}{r}$. This is obtained by multiplying the magnitude by a unit vector from the satellite to the center of the earth.

We will assume that the drag force has magnitude proportional to the product of the surface area A of the object facing the atmosphere, the density of the atmosphere ρ, and the square of velocity v^2. Specifically the magnitude takes the form $CA\rho v^2$, where C is a proportionality constant. One of the most significant assumptions we must make is in the model we take for the atmospheric density.

This density can vary considerably over the surface of the earth, depending upon time of day and year, weather conditions, and other factors. The reasons for variations in density are not particularly well understood, and for our simple model we will assume that atmospheric density depends only upon altitude. Even then, finding a formula for density is not easy. One method is to measure the density at different altitudes experimentally and then find a curve (such as a spline) through the

data. This is the approach taken in the example discussed below. Atmospheric scientists have also devised formulas that model the density. One such is

$$\rho = \begin{cases} (1.225 \times 10^{-3})e^{-0.1385(r-r_e)}, & r_e < r < r_e + 90, \\ 10^{1.274 - 4.41[\log_{10}(10.01(r-r_e) - 751.44)]}, & r > r_e + 90 \end{cases}$$

where r_e is the radius of the earth in kilometers.

As we did with the gravitational force, we must write the drag force in terms of components. The action of the force is perpendicular to the forward motion of the object, that is, perpendicular to the velocity. So we obtain the vector force due to drag by multiplying the magnitude of drag by a unit vector tangent to the path, $-CA\rho v^2 \dfrac{(x'(t), y'(t))}{v}$. The negative is due to the drag acting against the direction of the satellite.

Combining these two forces, we get from Newton's second law

$$m_s(x''(t), y''(t)) = -\frac{m_e m_s g}{r^2} \frac{(x(t), y(t))}{r} - CA\rho v^2 \frac{(x'(t), y'(t))}{v}.$$

This reduces to

$$(x''(t), y''(t)) = -\frac{m_e g}{r^3}(x(t), y(t)) - \frac{CA\rho v}{m_s}(x'(t), y'(t)).$$

In terms of components, we get the coupled system of second-order autonomous differential equations

$$x'' = -\frac{m_e g}{r^3}x - \frac{CA\rho v}{m_s}x'$$

$$y'' = -\frac{m_e g}{r^3}y - \frac{CA\rho v}{m_s}y'.$$

To solve this system we transform it to a set of four first-order autonomous differential equations, similar to what was done in Chapter 3. Let $y_1 = x$, $y_2 = x'$, $y_3 = y$, $y_4 = y'$, and $r = \sqrt{y_1^2 + y_3^2}$, $v = \sqrt{y_2^2 + y_4^2}$; then the system to be solved is

$$y_1' = y_2$$

$$y_2' = -\frac{m_e g}{r^3}y_1 - \frac{CA\rho v}{m_s}y_2$$

$$y_3' = y_4$$

$$y_4' = -\frac{m_e g}{r^3}y_3 - \frac{CA\rho v}{m_s}y_4.$$

This system of differential equations is highly nonlinear and cannot be solved analytically. Numerical methods and a computer must be used to get a numerical solution. Figure 2 shows the final one and a half revolutions of a decaying orbit of a satellite closely approximating Skylab. For our solution we used the numerical routines in *Mathematica,* which are similar to, but more sophisticated than, those presented in Chapter 6. The path of the satellite was tracked from a height of 100 kilometers with a velocity of just under 8 km/sec. We see that the satellite follows an elliptical path, spending most of the first revolution at a height greater than 100 kilometers. At the end of the first revolution, the height is approximately 97 km. The rate of decay then increases, and at the end of the next half orbit the height is slightly more than 80 km. Then very quickly the satellite (or what remains of it) strikes the earth. This is reasonable since the density of the atmosphere increases very rapidly as the altitude decreases.

Figure 2 Final orbits of satellite entering earth's atmosphere

Again, it must be stressed that this analysis is based on many assumptions and simplifications. To get a more accurate path, we must improve upon these assumptions—which only makes the solution more difficult to obtain. For these reasons the point of impact of a falling satellite on the earth's surface is hard to predict. We can only hope that it is in the ocean or on uninhabited land.

References

[1] Boulet, Dan, *Methods of Orbit Determination for the Microcomputer*. Richmond, VA: Willman-Bell. 1991.

[2] Danby, J. M. A., *Computing Applications to Differential Equations*. Reston, VA: Reston Publ. Co. 1985.

[3] Heicklen, G., *Atmospheric Chemistry*. New York: Academic Press. 1976.

[4] *Mathematica,* v 3.0.1 for Windows. Champaign, IL: Wolfram Research. 1998.

...

John H. Ellison received a B.A. from Whitman College, an M.A. from the University of Colorado, and a Ph.D. in Mathematics from the University of Pittsburgh. He has been a Professor of Mathematics at Grove City College for the past 25 years and is currently serving as chair of the department.

10

Systems of Linear Differential Equations

INTRODUCTION

We have seen systems of differential equations in Section 2.9, and we were able to solve some of these systems by means of either systematic elimination or the Laplace transform. In the present chapter we are going to concentrate only on systems of linear first-order equations. While most of the systems considered could be solved using elimination (Section 3.11) or the Laplace transform (Section 4.6), we are going to develop a general theory for these kinds of systems and, in the case of systems with constant coefficients, a method of solution that utilizes some basic concepts from the algebra of matrices. We shall see that this general theory and solution procedure is similar to that of linear higher-order DEs considered in Sections 3.3–3.5. This material is also fundamental to the analysis of systems of nonlinear first-order equations (Chapter 11).

10.1 Preliminary Theory

 Note to the student.

Matrix notation and properties are used extensively throughout this chapter. You should review Chapter 8 if you are unfamiliar with these concepts.

■ **Systems of DEs** Recall that in Section 3.11 we illustrated how to use systematic elimination to solve systems of n linear differential equations in n unknowns of the form

$$P_{11}(D)x_1 + P_{12}(D)x_2 + \cdots + P_{1n}(D)x_n = b_1(t)$$
$$P_{21}(D)x_1 + P_{22}(D)x_2 + \cdots + P_{2n}(D)x_n = b_2(t)$$
$$\vdots$$
$$P_{n1}(D)x_1 + P_{n2}(D)x_2 + \cdots + P_{nn}(D)x_n = b_n(t), \tag{1}$$

where the P_{ij} were polynomials of various degrees in the differential operator D. In this chapter we confine our study to systems of first-order differential equations that are special cases of the general form

$$\frac{dx_1}{dt} = g_1(t, x_1, x_2, ..., x_n)$$

$$\frac{dx_2}{dt} = g_2(t, x_1, x_2, ..., x_n) \tag{2}$$

$$\vdots$$

$$\frac{dx_n}{dt} = g_n(t, x_1, x_2, ..., x_n).$$

A system such as (2) of n first-order equations is called an ***nth-order system.***

■ **Linear Systems** When each of the functions g_1, g_2, \ldots, g_n in (2) is linear in the dependent variables x_1, x_2, \ldots, x_n, we get the **normal form** of an nth-order system of linear equations:

$$\frac{dx_1}{dt} = a_{11}(t)x_1 + a_{12}(t)x_2 + \cdots + a_{1n}(t)x_n + f_1(t)$$

$$\frac{dx_2}{dt} = a_{21}(t)x_1 + a_{22}(t)x_2 + \cdots + a_{2n}(t)x_n + f_2(t) \tag{3}$$

$$\vdots$$

$$\frac{dx_n}{dt} = a_{n1}(t)x_1 + a_{n2}(t)x_2 + \cdots + a_{nn}(t)x_n + f_n(t).$$

We refer to a system in (2) simply as a **linear system.** We assume that the coefficients a_{ij} and the functions f_i are continuous on a common interval I. When $f_i(t) = 0$, $i = 1, 2, \ldots, n$, the linear system is said to be **homogeneous;** otherwise it is **nonhomogeneous.**

■ **Matrix Form of a Linear System** If \mathbf{X}, $\mathbf{A}(t)$, and $\mathbf{F}(t)$ denote the respective matrices

$$\mathbf{X} = \begin{pmatrix} x_1(t) \\ x_2(t) \\ \vdots \\ x_n(t) \end{pmatrix}, \qquad \mathbf{A}(t) = \begin{pmatrix} a_{11}(t) & a_{12}(t) & \cdots & a_{1n}(t) \\ a_{21}(t) & a_{22}(t) & \cdots & a_{2n}(t) \\ \vdots & & & \vdots \\ a_{n1}(t) & a_{n2}(t) & \cdots & a_{nn}(t) \end{pmatrix}, \qquad \mathbf{F}(t) = \begin{pmatrix} f_1(t) \\ f_2(t) \\ \vdots \\ f_n(t) \end{pmatrix},$$

then the system of linear first-order differential equations (3) can be written as

$$\frac{d}{dt}\begin{pmatrix} x_1 \\ x_2 \\ \vdots \\ x_n \end{pmatrix} = \begin{pmatrix} a_{11}(t) & a_{12}(t) & \cdots & a_{1n}(t) \\ a_{21}(t) & a_{22}(t) & \cdots & a_{2n}(t) \\ \vdots & & & \vdots \\ a_{n1}(t) & a_{n2}(t) & \cdots & a_{nn}(t) \end{pmatrix}\begin{pmatrix} x_1 \\ x_2 \\ \vdots \\ x_n \end{pmatrix} + \begin{pmatrix} f_1(t) \\ f_2(t) \\ \vdots \\ f_n(t) \end{pmatrix}$$

or simply
$$\mathbf{X}' = \mathbf{AX} + \mathbf{F}. \tag{4}$$

If the system is homogeneous, its matrix form is then

$$\mathbf{X}' = \mathbf{AX}. \tag{5}$$

Example 1 Systems Written in Matrix Notation

(a) If $\mathbf{X} = \begin{pmatrix} x \\ y \end{pmatrix}$, then the matrix form of the homogeneous system

$$\frac{dx}{dt} = 3x + 4y$$

$$\frac{dy}{dt} = 5x - 7y$$

is $\quad \mathbf{X}' = \begin{pmatrix} 3 & 4 \\ 5 & -7 \end{pmatrix}\mathbf{X}.$

(b) If $\mathbf{X} = \begin{pmatrix} x \\ y \\ z \end{pmatrix}$, then the matrix form of the nonhomogeneous system

$$\frac{dx}{dt} = 6x + \ y + z + \ \ t$$

$$\frac{dy}{dt} = 8x + 7y - z + 10t$$

is $\quad \mathbf{X}' = \begin{pmatrix} 6 & 1 & 1 \\ 8 & 7 & -1 \\ 2 & 9 & -1 \end{pmatrix}\mathbf{X} + \begin{pmatrix} t \\ 10t \\ 6t \end{pmatrix}.$

$$\frac{dz}{dt} = 2x + 9y - z + \ \ 6t$$

❑

DEFINITION 10.1 Solution Vector

A solution vector on an interval I is any column matrix

$$\mathbf{X} = \begin{pmatrix} x_1(t) \\ x_2(t) \\ \vdots \\ x_n(t) \end{pmatrix}$$

whose entries are differentiable functions satisfying the system (4) on the interval.

A solution vector of (4), of course, is equivalent to n scalar equations $x_1 = \phi_1(t)$, $x_2 = \phi_2(t), \ldots, x_n = \phi_n(t)$, and can be interpreted geometrically as a set of parametric equations of a space curve. In the cases $n = 2$ and $n = 3$ the equations $x_1 = \phi_1(t)$, $x_2 = \phi_2(t)$, and $x_1 = \phi_1(t)$, $x_2 = \phi_2(t)$, $x_3 = \phi_3(t)$ represent curves in 2-space and 3-space, respectively. It is common practice to call such a solution curve a **trajectory.** The plane is also called the **phase plane.** We will illustrate these concepts in the section that follows, as well as in Chapter 11.

Example 2 Verification of Solutions

Verify that on the interval $(-\infty, \infty)$

$$\mathbf{X}_1 = \begin{pmatrix} 1 \\ -1 \end{pmatrix} e^{-2t} = \begin{pmatrix} e^{-2t} \\ -e^{-2t} \end{pmatrix} \quad \text{and} \quad \mathbf{X}_2 = \begin{pmatrix} 3 \\ 5 \end{pmatrix} e^{6t} = \begin{pmatrix} 3e^{6t} \\ 5e^{6t} \end{pmatrix}$$

are solutions of

$$\mathbf{X}' = \begin{pmatrix} 1 & 3 \\ 5 & 3 \end{pmatrix} \mathbf{X}. \tag{6}$$

SOLUTION From $\mathbf{X}_1' = \begin{pmatrix} -2e^{-2t} \\ 2e^{-2t} \end{pmatrix}$ and $\mathbf{X}_2' = \begin{pmatrix} 18e^{6t} \\ 30e^{6t} \end{pmatrix}$ we see that

$$\mathbf{AX}_1 = \begin{pmatrix} 1 & 3 \\ 5 & 3 \end{pmatrix} \begin{pmatrix} e^{-2t} \\ -e^{-2t} \end{pmatrix} = \begin{pmatrix} e^{-2t} - 3e^{-2t} \\ 5e^{-2t} - 3e^{-2t} \end{pmatrix} = \begin{pmatrix} -2e^{-2t} \\ 2e^{-2t} \end{pmatrix} = \mathbf{X}_1'$$

and

$$\mathbf{AX}_2 = \begin{pmatrix} 1 & 3 \\ 5 & 3 \end{pmatrix} \begin{pmatrix} 3e^{6t} \\ 5e^{6t} \end{pmatrix} = \begin{pmatrix} 3e^{6t} + 15e^{6t} \\ 15e^{6t} + 15e^{6t} \end{pmatrix} = \begin{pmatrix} 18e^{6t} \\ 30e^{6t} \end{pmatrix} = \mathbf{X}_2'. \quad \square$$

Much of the theory of systems of n linear first-order differential equations is similar to that of linear nth-order differential equations.

■ **Initial-Value Problem** Let t_0 denote a point on an interval I and

$$\mathbf{X}(t_0) = \begin{pmatrix} x_1(t_0) \\ x_2(t_0) \\ \vdots \\ x_n(t_0) \end{pmatrix} \quad \text{and} \quad \mathbf{X}_0 = \begin{pmatrix} \gamma_1 \\ \gamma_2 \\ \vdots \\ \gamma_n \end{pmatrix},$$

where the γ_i, $i = 1, 2, \ldots, n$ are given constants. Then the problem

$$\begin{aligned} \textit{Solve:} \quad & \mathbf{X}' = \mathbf{A}(t)\mathbf{X} + \mathbf{F}(t) \\ \textit{Subject to:} \quad & \mathbf{X}(t_0) = \mathbf{X}_0 \end{aligned} \tag{7}$$

is an **initial-value problem** on the interval.

> ### THEOREM 10.1 Existence of a Unique Solution
>
> Let the entries of the matrices $\mathbf{A}(t)$ and $\mathbf{F}(t)$ be functions continuous on a common interval I that contains the point t_0. Then there exists a unique solution of the initial-value problem (7) on the interval.

■ **Homogeneous Systems** In the next several definitions and theorems we are concerned only with homogeneous systems. Without stating it, we shall always assume that the a_{ij} and the f_i are continuous functions of t on some common interval I.

■ **Superposition Principle** The following result is a **superposition principle** for solutions of linear systems.

> ### THEOREM 10.2 Superposition Principle
>
> Let $\mathbf{X}_1, \mathbf{X}_2, \ldots, \mathbf{X}_k$ be a set of solution vectors of the homogeneous system (5) on an interval I. Then the linear combination
>
> $$\mathbf{X} = c_1\mathbf{X}_1 + c_2\mathbf{X}_2 + \cdots + c_k\mathbf{X}_k,$$
>
> where the c_i, $i = 1, 2, \ldots, k$ are arbitrary constants, is also a solution on the interval.

It follows from Theorem 10.2 that a constant multiple of any solution vector of a homogeneous system of linear first-order differential equations is also a solution.

Example 3 Using the Superposition Principle
You should practice by verifying that the two vectors

$$\mathbf{X}_1 = \begin{pmatrix} \cos t \\ -\tfrac{1}{2}\cos t + \tfrac{1}{2}\sin t \\ -\cos t - \sin t \end{pmatrix} \quad \text{and} \quad \mathbf{X}_2 = \begin{pmatrix} 0 \\ e^t \\ 0 \end{pmatrix}$$

are solutions of the system

$$\mathbf{X}' = \begin{pmatrix} 1 & 0 & 1 \\ 1 & 1 & 0 \\ -2 & 0 & -1 \end{pmatrix} \mathbf{X}. \tag{8}$$

By the superposition principle the linear combination

$$\mathbf{X} = c_1 \mathbf{X}_1 + c_2 \mathbf{X}_2 = c_1 \begin{pmatrix} \cos t \\ -\tfrac{1}{2}\cos t + \tfrac{1}{2}\sin t \\ -\cos t - \sin t \end{pmatrix} + c_2 \begin{pmatrix} 0 \\ e^t \\ 0 \end{pmatrix}$$

is yet another solution of the system. ❏

■ **Linear Dependence and Linear Independence** We are primarily interested in linearly independent solutions of the homogeneous system (5).

DEFINITION 10.2 Linear Dependence/Independence

Let $\mathbf{X}_1, \mathbf{X}_2, \ldots, \mathbf{X}_k$ be a set of solution vectors of the homogeneous system (5) on an interval I. We say that the set is **linearly dependent** on the interval if there exist constants c_1, c_2, \ldots, c_k, not all zero, such that

$$c_1 \mathbf{X}_1 + c_2 \mathbf{X}_2 + \cdots + c_k \mathbf{X}_k = \mathbf{0}$$

for every t in the interval. If the set of vectors is not linearly dependent on the interval, it is said to be **linearly independent.**

The case when $k = 2$ should be clear; two solution vectors \mathbf{X}_1 and \mathbf{X}_2 are linearly dependent if one is a constant multiple of the other, and conversely. For $k > 2$ a set of solution vectors is linearly dependent if we can express at least one solution vector as a linear combination of the remaining vectors.

■ **Wronskian** As in our earlier consideration of the theory of a single ordinary differential equation, we can introduce the concept of the **Wronskian** determinant as a test for linear independence. We state the following theorem without proof.

THEOREM 10.3 Criterion for Linearly Independent Solutions

Let $\qquad \mathbf{X}_1 = \begin{pmatrix} x_{11} \\ x_{21} \\ \vdots \\ x_{n1} \end{pmatrix}, \quad \mathbf{X}_2 = \begin{pmatrix} x_{12} \\ x_{22} \\ \vdots \\ x_{n2} \end{pmatrix}, \quad \ldots, \quad \mathbf{X}_n = \begin{pmatrix} x_{1n} \\ x_{2n} \\ \vdots \\ x_{nn} \end{pmatrix}$

continued

continued

be n solution vectors of the homogeneous system (5) on an interval I. Then the set of solution vectors is linearly independent on I if and only if the **Wronskian**

$$W(\mathbf{X}_1, \mathbf{X}_2, ..., \mathbf{X}_n) = \begin{vmatrix} x_{11} & x_{12} & \cdots & x_{1n} \\ x_{21} & x_{22} & \cdots & x_{2n} \\ \vdots & & & \vdots \\ x_{n1} & x_{n2} & \cdots & x_{nn} \end{vmatrix} \neq 0 \tag{9}$$

for every t in the interval.

It can be shown that if $\mathbf{X}_1, \mathbf{X}_2, \ldots, \mathbf{X}_n$ are solution vectors of (5), then for every t in I, either $W(\mathbf{X}_1, \mathbf{X}_2, \ldots, \mathbf{X}_n) \neq 0$ or $W(\mathbf{X}_1, \mathbf{X}_2, \ldots, \mathbf{X}_n) = 0$. Thus if we can show that $W \neq 0$ for some t_0 in I, then $W \neq 0$ for every t, hence the set of solutions is linearly independent on the interval.

Notice that, unlike our definition of the Wronskian in Section 3.1, here the definition of the determinant (9) does not involve differentiation.

Example 4 Linearly Independent Solutions

In Example 2 we saw that $\mathbf{X}_1 = \begin{pmatrix} 1 \\ -1 \end{pmatrix} e^{-2t}$ and $\mathbf{X}_2 = \begin{pmatrix} 3 \\ 5 \end{pmatrix} e^{6t}$ are solutions of system (6). Clearly \mathbf{X}_1 and \mathbf{X}_2 are linearly independent on the interval $(-\infty, \infty)$ since neither vector is a constant multiple of the other. In addition, we have

$$W(\mathbf{X}_1, \mathbf{X}_2) = \begin{vmatrix} e^{-2t} & 3e^{6t} \\ -e^{-2t} & 5e^{6t} \end{vmatrix} = 8e^{4t} \neq 0$$

for all real values of t. ❑

DEFINITION 10.3 Fundamental Set of Solutions

Any set $\mathbf{X}_1, \mathbf{X}_2, \ldots, \mathbf{X}_n$ of n linearly independent solution vectors of the homogeneous system (5) on an interval I is said to be a **fundamental set of solutions** on the interval.

THEOREM 10.4 Existence of a Fundamental Set

There exists a fundamental set of solutions for the homogeneous system (5) on an interval I.

The next two theorems are the linear system equivalents of Theorems 3.5 and 3.6.

THEOREM 10.5 General Solution—Homogeneous Systems

Let $\mathbf{X}_1, \mathbf{X}_2, \ldots, \mathbf{X}_n$ be a fundamental set of solutions of the homogeneous system (5) on an interval I. Then the **general solution** of the system on the interval is

$$\mathbf{X} = c_1 \mathbf{X}_1 + c_2 \mathbf{X}_2 + \cdots + c_n \mathbf{X}_n,$$

where the c_i, $i = 1, 2, \ldots, n$ are arbitrary constants.

Example 5 General Solution of System (6)

From Example 2 we know that $\mathbf{X}_1 = \begin{pmatrix} 1 \\ -1 \end{pmatrix} e^{-2t}$ and $\mathbf{X}_2 = \begin{pmatrix} 3 \\ 5 \end{pmatrix} e^{6t}$ are linearly independent solutions of (6) on $(-\infty, \infty)$. Hence \mathbf{X}_1 and \mathbf{X}_2 form a fundamental set of solutions on the interval. The general solution of the system on the interval is then

$$\mathbf{X} = c_1 \mathbf{X}_1 + c_2 \mathbf{X}_2 = c_1 \begin{pmatrix} 1 \\ -1 \end{pmatrix} e^{-2t} + c_2 \begin{pmatrix} 3 \\ 5 \end{pmatrix} e^{6t}. \qquad (10)$$

Example 6 General Solution of System (8)

The vectors

$$\mathbf{X}_1 = \begin{pmatrix} \cos t \\ -\tfrac{1}{2} \cos t + \tfrac{1}{2} \sin t \\ -\cos t - \sin t \end{pmatrix}, \qquad \mathbf{X}_2 = \begin{pmatrix} 0 \\ 1 \\ 0 \end{pmatrix} e^t, \qquad \mathbf{X}_3 = \begin{pmatrix} \sin t \\ -\tfrac{1}{2} \sin t - \tfrac{1}{2} \cos t \\ -\sin t + \cos t \end{pmatrix}$$

are solutions of the system (8) in Example 3 (see Problem 16 in Exercises 10.1). Now

$$W(\mathbf{X}_1, \mathbf{X}_2, \mathbf{X}_3) = \begin{vmatrix} \cos t & 0 & \sin t \\ -\tfrac{1}{2} \cos t + \tfrac{1}{2} \sin t & e^t & -\tfrac{1}{2} \sin t - \tfrac{1}{2} \cos t \\ -\cos t - \sin t & 0 & -\sin t + \cos t \end{vmatrix} = e^t \neq 0$$

for all real values of t. We conclude that \mathbf{X}_1, \mathbf{X}_2, and \mathbf{X}_3 form a fundamental set of solutions on $(-\infty, \infty)$. Thus the general solution of the system on the interval is the linear combination $\mathbf{X} = c\mathbf{X}_1 + c_2\mathbf{X}_2 + c_3\mathbf{X}_3$, that is,

$$\mathbf{X} = c_1 \begin{pmatrix} \cos t \\ -\tfrac{1}{2} \cos t + \tfrac{1}{2} \sin t \\ -\cos t - \sin t \end{pmatrix} + c_2 \begin{pmatrix} 0 \\ 1 \\ 0 \end{pmatrix} e^t + c_3 \begin{pmatrix} \sin t \\ -\tfrac{1}{2} \sin t - \tfrac{1}{2} \cos t \\ -\sin t + \cos t \end{pmatrix}.$$

■ **Nonhomogeneous Systems** For nonhomogeneous systems a **particular solution** \mathbf{X}_p on an interval I is any vector, free of arbitrary parameters, whose entries are functions that satisfy the system (4).

> **THEOREM 10.6** *General Solution—Nonhomogeneous Systems*
>
> Let \mathbf{X}_p be a given solution of the nonhomogeneous system (4) on an interval I, and let
>
> $$\mathbf{X}_c = c_1 \mathbf{X}_1 + c_2 \mathbf{X}_2 + \cdots + c_n \mathbf{X}_n$$
>
> denote the general solution on the same interval of the associated homogeneous system (5). Then the **general solution** of the nonhomogeneous system on the interval is
>
> $$\mathbf{X} = \mathbf{X}_c + \mathbf{X}_p.$$
>
> The general solution \mathbf{X}_c of the homogeneous system (5) is called the **complementary function** of the nonhomogeneous system (4).

Example 7 General Solution—Nonhomogeneous System

The vector $\mathbf{X}_p = \begin{pmatrix} 3t - 4 \\ -5t + 6 \end{pmatrix}$ is a particular solution of the nonhomogeneous system

$$\mathbf{X}' = \begin{pmatrix} 1 & 3 \\ 5 & 3 \end{pmatrix} \mathbf{X} + \begin{pmatrix} 12t - 11 \\ -3 \end{pmatrix} \qquad (11)$$

on the interval $(-\infty, \infty)$. (Verify this.) The complementary function of (11) on the same interval, or the general solution of $\mathbf{X}' = \begin{pmatrix} 1 & 3 \\ 5 & 3 \end{pmatrix}\mathbf{X}$, was seen in (10) of Example 5 to be $\mathbf{X}_c = c_1\begin{pmatrix} 1 \\ -1 \end{pmatrix}e^{-2t} + c_2\begin{pmatrix} 3 \\ 5 \end{pmatrix}e^{6t}$. Hence by Theorem 10.6

$$\mathbf{X} = \mathbf{X}_c + \mathbf{X}_p = c_1\begin{pmatrix} 1 \\ -1 \end{pmatrix}e^{-2t} + c_2\begin{pmatrix} 3 \\ 5 \end{pmatrix}e^{6t} + \begin{pmatrix} 3t - 4 \\ -5t + 6 \end{pmatrix}$$

is the general solution of (11) on $(-\infty, \infty)$. ❑

E X E R C I S E S 1 0 . 1

Answers to odd-numbered problems begin on page A-52.

In Problems 1–6, write the linear system in matrix form.

1. $\dfrac{dx}{dt} = 3x - 5y$

$\dfrac{dy}{dt} = 4x + 8y$

2. $\dfrac{dx}{dt} = 4x - 7y$

$\dfrac{dy}{dt} = 5x$

3. $\dfrac{dx}{dt} = -3x + 4y - 9z$

$\dfrac{dy}{dt} = 6x - y$

$\dfrac{dz}{dt} = 10x + 4y + 3z$

4. $\dfrac{dx}{dt} = x - y$

$\dfrac{dy}{dt} = x + 2z$

$\dfrac{dz}{dt} = -x + z$

5. $\dfrac{dx}{dt} = x - y + z + t - 1$

$\dfrac{dy}{dt} = 2x + y - z - 3t^2$

$\dfrac{dz}{dt} = x + y + z + t^2 - t + 2$

6. $\dfrac{dx}{dt} = -3x + 4y + e^{-t}\sin 2t$

$\dfrac{dy}{dt} = 5x + 9z + 4e^{-t}\cos 2t$

$\dfrac{dz}{dt} = y + 6z - e^{-t}$

In Problems 7–10, write the given system without the use of matrices.

7. $\mathbf{X}' = \begin{pmatrix} 4 & 2 \\ -1 & 3 \end{pmatrix}\mathbf{X} + \begin{pmatrix} 1 \\ -1 \end{pmatrix}e^t$

8. $\mathbf{X}' = \begin{pmatrix} 7 & 5 & -9 \\ 4 & 1 & 1 \\ 0 & -2 & 3 \end{pmatrix}\mathbf{X} + \begin{pmatrix} 0 \\ 2 \\ 1 \end{pmatrix}e^{5t} - \begin{pmatrix} 8 \\ 0 \\ 3 \end{pmatrix}e^{-2t}$

9. $\dfrac{d}{dt}\begin{pmatrix} x \\ y \\ z \end{pmatrix} = \begin{pmatrix} 1 & -1 & 2 \\ 3 & -4 & 1 \\ -2 & 5 & 6 \end{pmatrix}\begin{pmatrix} x \\ y \\ z \end{pmatrix} + \begin{pmatrix} 1 \\ 2 \\ 2 \end{pmatrix}e^{-t} - \begin{pmatrix} 3 \\ -1 \\ 1 \end{pmatrix}t$

10. $\dfrac{d}{dt}\begin{pmatrix} x \\ y \end{pmatrix} = \begin{pmatrix} 3 & -7 \\ 1 & 1 \end{pmatrix}\begin{pmatrix} x \\ y \end{pmatrix} + \begin{pmatrix} 4 \\ 8 \end{pmatrix}\sin t + \begin{pmatrix} t - 4 \\ 2t + 1 \end{pmatrix}e^{4t}$

In Problems 11–16, verify that the vector \mathbf{X} is a solution of the given system.

11. $\dfrac{dx}{dt} = 3x - 4y$

$\dfrac{dy}{dt} = 4x - 7y;\quad \mathbf{X} = \begin{pmatrix} 1 \\ 2 \end{pmatrix}e^{-5t}$

12. $\dfrac{dx}{dt} = -2x + 5y$

$\dfrac{dy}{dt} = -2x + 4y;\quad \mathbf{X} = \begin{pmatrix} 5\cos t \\ 3\cos t - \sin t \end{pmatrix}e^t$

13. $\mathbf{X}' = \begin{pmatrix} -1 & \frac{1}{4} \\ 1 & -1 \end{pmatrix}\mathbf{X};\quad \mathbf{X} = \begin{pmatrix} -1 \\ 2 \end{pmatrix}e^{-3t/2}$

14. $\mathbf{X}' = \begin{pmatrix} 2 & 1 \\ -1 & 0 \end{pmatrix}\mathbf{X};\quad \mathbf{X} = \begin{pmatrix} 1 \\ 3 \end{pmatrix}e^t + \begin{pmatrix} 4 \\ -4 \end{pmatrix}te^t$

15. $\mathbf{X}' = \begin{pmatrix} 1 & 2 & 1 \\ 6 & -1 & 0 \\ -1 & -2 & -1 \end{pmatrix}\mathbf{X};\quad \mathbf{X} = \begin{pmatrix} 1 \\ 6 \\ -13 \end{pmatrix}$

16. $\mathbf{X}' = \begin{pmatrix} 1 & 0 & 1 \\ 1 & 1 & 0 \\ -2 & 0 & -1 \end{pmatrix}\mathbf{X};\quad \mathbf{X} = \begin{pmatrix} \sin t \\ -\frac{1}{2}\sin t - \frac{1}{2}\cos t \\ -\sin t + \cos t \end{pmatrix}$

In Problems 17–20, the given vectors are solutions of a system $\mathbf{X}' = \mathbf{AX}$. Determine whether the vectors form a fundamental set on $(-\infty, \infty)$.

17. $\mathbf{X}_1 = \begin{pmatrix} 1 \\ 1 \end{pmatrix}e^{-2t},\quad \mathbf{X}_2 = \begin{pmatrix} 1 \\ -1 \end{pmatrix}e^{-6t}$

18. $\mathbf{X}_1 = \begin{pmatrix} 1 \\ -1 \end{pmatrix}e^t,\quad \mathbf{X}_2 = \begin{pmatrix} 2 \\ 6 \end{pmatrix}e^t + \begin{pmatrix} 8 \\ -8 \end{pmatrix}te^t$

19. $\mathbf{X}_1 = \begin{pmatrix} 1 \\ -2 \\ 4 \end{pmatrix} + t\begin{pmatrix} 1 \\ 2 \\ 2 \end{pmatrix},\quad \mathbf{X}_2 = \begin{pmatrix} 1 \\ -2 \\ 4 \end{pmatrix},\quad \mathbf{X}_3 = \begin{pmatrix} 3 \\ -6 \\ 12 \end{pmatrix} + t\begin{pmatrix} 2 \\ 4 \\ 4 \end{pmatrix}$

20. $\mathbf{X}_1 = \begin{pmatrix} 1 \\ 6 \\ -13 \end{pmatrix},\quad \mathbf{X}_2 = \begin{pmatrix} 1 \\ -2 \\ -1 \end{pmatrix}e^{-4t},\quad \mathbf{X}_3 = \begin{pmatrix} 2 \\ 3 \\ -2 \end{pmatrix}e^{3t}$

In Problems 21–24, verify that the vector \mathbf{X}_p is a particular solution of the given system.

21. $\dfrac{dx}{dt} = x + 4y + 2t - 7$

$\dfrac{dy}{dt} = 3x + 2y - 4t - 18; \quad \mathbf{X}_p = \begin{pmatrix} 2 \\ -1 \end{pmatrix} t + \begin{pmatrix} 5 \\ 1 \end{pmatrix}$

22. $\mathbf{X}' = \begin{pmatrix} 2 & 1 \\ 1 & -1 \end{pmatrix} \mathbf{X} + \begin{pmatrix} -5 \\ 2 \end{pmatrix}; \quad \mathbf{X}_p = \begin{pmatrix} 1 \\ 3 \end{pmatrix}$

23. $\mathbf{X}' = \begin{pmatrix} 2 & 1 \\ 3 & 4 \end{pmatrix} \mathbf{X} - \begin{pmatrix} 1 \\ 7 \end{pmatrix} e^t; \quad \mathbf{X}_p = \begin{pmatrix} 1 \\ 1 \end{pmatrix} e^t + \begin{pmatrix} 1 \\ -1 \end{pmatrix} t e^t$

24. $\mathbf{X}' = \begin{pmatrix} 1 & 2 & 3 \\ -4 & 2 & 0 \\ -6 & 1 & 0 \end{pmatrix} \mathbf{X} + \begin{pmatrix} -1 \\ 4 \\ 3 \end{pmatrix} \sin 3t; \quad \mathbf{X}_p = \begin{pmatrix} \sin 3t \\ 0 \\ \cos 3t \end{pmatrix}$

25. Prove that the general solution of

$$\mathbf{X}' = \begin{pmatrix} 0 & 6 & 0 \\ 1 & 0 & 1 \\ 1 & 1 & 0 \end{pmatrix} \mathbf{X}$$

on the interval $(-\infty, \infty)$ is

$$\mathbf{X} = c_1 \begin{pmatrix} 6 \\ -1 \\ -5 \end{pmatrix} e^{-t} + c_2 \begin{pmatrix} -3 \\ 1 \\ 1 \end{pmatrix} e^{-2t} + c_3 \begin{pmatrix} 2 \\ 1 \\ 1 \end{pmatrix} e^{3t}.$$

26. Prove that the general solution of

$$\mathbf{X}' = \begin{pmatrix} -1 & -1 \\ -1 & 1 \end{pmatrix} \mathbf{X} + \begin{pmatrix} 1 \\ 1 \end{pmatrix} t^2 + \begin{pmatrix} 4 \\ -6 \end{pmatrix} t + \begin{pmatrix} -1 \\ 5 \end{pmatrix}$$

on the interval $(-\infty, \infty)$ is

$$\mathbf{X} = c_1 \begin{pmatrix} 1 \\ -1 - \sqrt{2} \end{pmatrix} e^{\sqrt{2}\,t} + c_2 \begin{pmatrix} 1 \\ -1 + \sqrt{2} \end{pmatrix} e^{-\sqrt{2}\,t}$$

$$+ \begin{pmatrix} 1 \\ 0 \end{pmatrix} t^2 + \begin{pmatrix} -2 \\ 4 \end{pmatrix} t + \begin{pmatrix} 1 \\ 0 \end{pmatrix}.$$

10.2 Homogeneous Linear Systems

■ **Systems with Constant Coefficients** We saw in Example 5 of Section 10.1 that the general solution of the homogeneous system $\mathbf{X}' = \begin{pmatrix} 1 & 3 \\ 5 & 3 \end{pmatrix} \mathbf{X}$ is $\mathbf{X} = c_1 \mathbf{X}_1 + c_2 \mathbf{X}_2 = c_1 \begin{pmatrix} 1 \\ -1 \end{pmatrix} e^{-2t} + c_2 \begin{pmatrix} 3 \\ 5 \end{pmatrix} e^{6t}$. Since both solution vectors have the form $\mathbf{X}_i = \begin{pmatrix} k_1 \\ k_2 \end{pmatrix} e^{\lambda_i t}$, $i = 1, 2$, where k_1 and k_2 are constants, we are prompted to ask whether we can always find a solution of the form

$$\mathbf{X} = \begin{pmatrix} k_1 \\ k_2 \\ \vdots \\ k_n \end{pmatrix} e^{\lambda t} = \mathbf{K} e^{\lambda t} \tag{1}$$

for the general homogeneous linear first-order system

$$\mathbf{X}' = \mathbf{A}\mathbf{X}, \tag{2}$$

We will be dealing only with 👉 *linear systems with constant coefficients.*

where the coefficient matrix \mathbf{A} is an $n \times n$ matrix of constants.

■ **Eigenvalues and Eigenvectors** If (1) is to be a solution vector of the system, then $\mathbf{X}' = \mathbf{K}\lambda e^{\lambda t}$ so that (2) becomes $\mathbf{K}\lambda e^{\lambda t} = \mathbf{A}\mathbf{K} e^{\lambda t}$. After dividing out $e^{\lambda t}$ and rearranging, we obtain $\mathbf{A}\mathbf{K} = \lambda \mathbf{K}$ or $\mathbf{A}\mathbf{K} - \lambda \mathbf{K} = \mathbf{0}$. Since $\mathbf{K} = \mathbf{I}\mathbf{K}$, the last equation is the same as

$$(\mathbf{A} - \lambda \mathbf{I})\mathbf{K} = \mathbf{0}. \tag{3}$$

The matrix equation (3) is equivalent to the simultaneous algebraic equations

$$\begin{aligned} (a_{11} - \lambda)k_1 + \qquad a_{12}k_2 + \cdots + \qquad a_{1n}k_n &= 0 \\ a_{21}k_1 + (a_{22} - \lambda)k_2 + \cdots + \qquad a_{2n}k_n &= 0 \\ \vdots \qquad\qquad\qquad \vdots \\ a_{n1}k_1 + \qquad a_{n2}k_2 + \cdots + (a_{nn} - \lambda)k_n &= 0. \end{aligned}$$

Thus to find a nontrivial solution \mathbf{X} of (2) we must first find a nontrivial solution of the foregoing system; in other words, we must find a nontrivial vector \mathbf{K} that satisfies (3). But in order for (3) to have solutions other than the obvious solution $k_1 = k_2 = \ldots = k_n = 0$, we must have

$$\det(\mathbf{A} - \lambda \mathbf{I}) = 0.$$

This polynomial equation in λ is called the **characteristic equation** of the matrix \mathbf{A}; its solutions are the **eigenvalues** of \mathbf{A}. A solution $\mathbf{K} \neq \mathbf{0}$ of (3) corresponding to an eigenvalue λ is called an **eigenvector** of \mathbf{A}. A solution of the homogeneous system (2) is then $\mathbf{X} = \mathbf{K}e^{\lambda t}$.

In the following discussion we examine three cases: all the eigenvalues are real and distinct (that is, no eigenvalues are equal), repeated eigenvalues, and finally, complex eigenvalues.

10.2.1 Distinct Real Eigenvalues

When the $n \times n$ matrix \mathbf{A} possesses n distinct real eigenvalues $\lambda_1, \lambda_2, \ldots, \lambda_n$, then a set of n linearly independent eigenvectors $\mathbf{K}_1, \mathbf{K}_2, \ldots, \mathbf{K}_n$ can always be found and

$$\mathbf{X}_1 = \mathbf{K}_1 e^{\lambda_1 t}, \quad \mathbf{X}_2 = \mathbf{K}_2 e^{\lambda_2 t}, \quad \ldots, \quad \mathbf{X}_n = \mathbf{K}_n e^{\lambda_n t}$$

is a fundamental set of solutions of (2) on $(-\infty, \infty)$.

> **THEOREM 10.7 General Solution—Homogeneous Systems**
>
> Let $\lambda_1, \lambda_2, \ldots, \lambda_n$ be n distinct real eigenvalues of the coefficient matrix \mathbf{A} of the homogeneous system (2), and let $\mathbf{K}_1, \mathbf{K}_2, \ldots, \mathbf{K}_n$ be the corresponding eigenvectors. Then the **general solution** of (2) on the interval $(-\infty, \infty)$ is given by
>
> $$\mathbf{X} = c_1 \mathbf{K}_1 e^{\lambda_1 t} + c_2 \mathbf{K}_2 e^{\lambda_2 t} + \cdots + c_n \mathbf{K}_n e^{\lambda_n t}.$$

Example 1 Distinct Eigenvalues

Solve
$$\frac{dx}{dt} = 2x + 3y$$

$$\frac{dy}{dt} = 2x + y. \tag{4}$$

SOLUTION We first find the eigenvalues and eigenvectors of the matrix of coefficients.

From the characteristic equation

$$\det(\mathbf{A} - \lambda \mathbf{I}) = \begin{vmatrix} 2 - \lambda & 3 \\ 2 & 1 - \lambda \end{vmatrix} = \lambda^2 - 3\lambda - 4 = (\lambda + 1)(\lambda - 4) = 0$$

we see that the eigenvalues are $\lambda_1 = -1$ and $\lambda_2 = 4$.

Now for $\lambda_1 = -1$, (3) is equivalent to

$$3k_1 + 3k_2 = 0$$

$$2k_1 + 2k_2 = 0.$$

Thus $k_1 = -k_2$. When $k_2 = -1$, the related eigenvector is

$$\mathbf{K}_1 = \begin{pmatrix} 1 \\ -1 \end{pmatrix}.$$

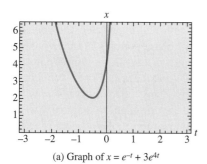

(a) Graph of $x = e^{-t} + 3e^{4t}$

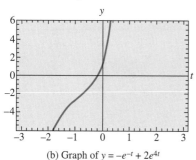

(b) Graph of $y = -e^{-t} + 2e^{4t}$

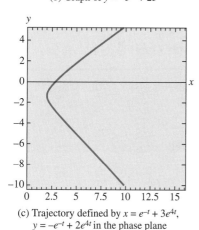

(c) Trajectory defined by $x = e^{-t} + 3e^{4t}$, $y = -e^{-t} + 2e^{4t}$ in the phase plane

Figure 10.1

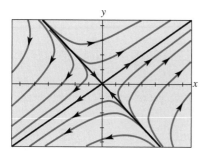

Figure 10.2 A phase portrait of the system (4)

For $\lambda_2 = 4$, we have
$$-2k_1 + 3k_2 = 0$$
$$2k_1 - 3k_2 = 0$$

so that $k_1 = 3k_2/2$, and therefore with $k_2 = 2$, the corresponding eigenvector is

$$\mathbf{K}_2 = \begin{pmatrix} 3 \\ 2 \end{pmatrix}.$$

Since the matrix of coefficients \mathbf{A} is a 2×2 matrix and since we have found two linearly independent solutions of (4),

$$\mathbf{X}_1 = \begin{pmatrix} 1 \\ -1 \end{pmatrix} e^{-t} \quad \text{and} \quad \mathbf{X}_2 = \begin{pmatrix} 3 \\ 2 \end{pmatrix} e^{4t},$$

we conclude that the general solution of the system is

$$\mathbf{X} = c_1\mathbf{X}_1 + c_2\mathbf{X}_2 = c_1 \begin{pmatrix} 1 \\ -1 \end{pmatrix} e^{-t} + c_2 \begin{pmatrix} 3 \\ 2 \end{pmatrix} e^{4t}. \qquad (5)\square$$

For the sake of review, you should keep firmly in mind that a solution of a system of linear first-order differential equations, when written in terms of matrices, is simply an alternative to the method that we employed in Section 3.11—namely, listing the individual functions and the relationship between the constants. If we add the vectors on the right side of (5) and then equate the entries with the corresponding entries in the vector of the left, we obtain the more familiar statement

$$x = c_1e^{-t} + 3c_2e^{4t}, \qquad y = -c_1e^{-t} + 2c_2e^{4t}.$$

As pointed out in Section 10.1 we can interpret these equations as parametric equations of a curve or **trajectory** in the xy-plane or **phase plane.** The three graphs shown in Figure 10.1, $x(t)$ in the tx-plane, $y(t)$ in the ty-plane, and the trajectory in the phase plane, correspond to the choice of constants $c_1 = c_2 = 1$ in the solution. A collection of trajectories in the phase plane as shown in Figure 10.2 is said to be a **phase portrait** of the given linear system. What appears to be two black lines in Figure 10.2 are actually four half-lines defined parametrically in the first, second, third, and fourth quadrants by the solutions \mathbf{X}_2, $-\mathbf{X}_1$, $-\mathbf{X}_2$, and \mathbf{X}_1, respectively. For example, the Cartesian equations $y = \frac{2}{3}x$, $x > 0$, and $y = -x$, $x > 0$, of the half-lines in the first and fourth quadrants were obtained by eliminating the parameter t in the solutions $x = 3e^{4t}$, $y = 2e^{4t}$, and $x = e^{-t}$, $y = -e^{-t}$, respectively. Moreover, each eigenvector can be visualized as a two-dimensional vector lying along one of these half-lines. The eigenvector $\mathbf{K}_2 = \begin{pmatrix} 3 \\ 2 \end{pmatrix}$ lies along $y = \frac{2}{3}x$ in the first quadrant and $\mathbf{K}_1 = \begin{pmatrix} 1 \\ -1 \end{pmatrix}$ lies along $y = -x$ in the fourth quadrant; each vector starts at the origin with \mathbf{K}_2 terminating at the point $(2, 3)$ and \mathbf{K}_1 terminating at $(1, -1)$.

The origin is not only a constant solution, $x = 0$, $y = 0$, of every 2×2 homogeneous linear system $\mathbf{X}' = \mathbf{AX}$ but is as well an important point in the qualitative study of such systems. If we think in physical terms, the arrowheads on a trajectory in Figure 10.2 indicate the direction that a particle with coordinates $(x(t), y(t))$ on a trajectory at time t would move as time increases. Observe that the arrowheads, with only those on the half-lines in the second and fourth quadrants being the exception, indicate that a particle would move away from the origin as time t increases. If we imagine time ranging from $-\infty$ to ∞, then inspection of the solution $x = c_1e^{-t} + 3c_2e^{4t}$, $y = -c_1e^{-t} + 2c_2e^{4t}$, $c_1 \neq 0$, $c_2 \neq 0$, shows that a trajectory, or moving particle, "starts" asymptotic to one of the half-lines defined by \mathbf{X}_1 or $-\mathbf{X}_1$ (since e^{4t} is negligible for $t \to -\infty$) and "finishes" asymptotic to one of the half-lines defined by \mathbf{X}_2 and $-\mathbf{X}_2$ (since e^{-t} is negligible for $t \to \infty$).

We note in passing that Figure 10.2 represents a phase portrait that is typical of *all* 2×2 homogeneous linear systems $\mathbf{X}' = \mathbf{AX}$ with real eigenvalues of opposite signs. See Problem 17 in Exercises 10.2. Moreover, phase portraits in the two cases when distinct real eigenvalues have the same algebraic sign would be typical portraits of all such 2×2 linear systems; the only difference is that the arrowheads would indicate that a particle would move away from the origin on any trajectory as $t \to \infty$ when both λ_1 and λ_2 are positive and would move toward the origin on any trajectory when both λ_1 and λ_2 are negative. Consequently, it is common to call the origin a **repeller** in the case $\lambda_1 > 0$, $\lambda_2 > 0$, and an **attractor** in the case $\lambda_1 < 0$, $\lambda_2 < 0$. See Problem 18 in Exercises 10.2. The origin in Figure 10.2 is neither a repeller nor an attractor. Investigation of the remaining case when $\lambda = 0$ is an eigenvalue of a 2×2 homogeneous linear system is left as an exercise. See Problem 48 in Exercises 10.2.

Example 2 Distinct Eigenvalues

Solve

$$\frac{dx}{dt} = -4x + y + z$$

$$\frac{dy}{dt} = x + 5y - z \qquad (6)$$

$$\frac{dz}{dt} = y - 3z.$$

SOLUTION Using the cofactors of the third row, we find

$$\det (\mathbf{A} - \lambda \mathbf{I}) = \begin{vmatrix} -4 - \lambda & 1 & 1 \\ 1 & 5 - \lambda & -1 \\ 0 & 1 & -3 - \lambda \end{vmatrix} = -(\lambda + 3)(\lambda + 4)(\lambda - 5) = 0,$$

and so the eigenvalues are $\lambda_1 = -3$, $\lambda_2 = -4$, $\lambda_3 = 5$.

For $\lambda_1 = -3$, Gauss–Jordan elimination gives

$$(\mathbf{A} + 3\mathbf{I}|\mathbf{0}) = \begin{pmatrix} -1 & 1 & 1 & | & 0 \\ 1 & 8 & -1 & | & 0 \\ 0 & 1 & 0 & | & 0 \end{pmatrix} \overset{\substack{\text{row} \\ \text{operations}}}{\Longrightarrow} \begin{pmatrix} 1 & 0 & -1 & | & 0 \\ 0 & 1 & 0 & | & 0 \\ 0 & 0 & 0 & | & 0 \end{pmatrix}.$$

Therefore $k_1 = k_3$ and $k_2 = 0$. The choice $k_3 = 1$ gives an eigenvector and corresponding solution vector

$$\mathbf{K}_1 = \begin{pmatrix} 1 \\ 0 \\ 1 \end{pmatrix}, \qquad \mathbf{X}_1 = \begin{pmatrix} 1 \\ 0 \\ 1 \end{pmatrix} e^{-3t}. \qquad (7)$$

Similarly, for $\lambda_2 = -4$,

$$(\mathbf{A} + 4\mathbf{I}|\mathbf{0}) = \begin{pmatrix} 0 & 1 & 1 & | & 0 \\ 1 & 9 & -1 & | & 0 \\ 0 & 1 & 1 & | & 0 \end{pmatrix} \overset{\substack{\text{row} \\ \text{operations}}}{\Longrightarrow} \begin{pmatrix} 1 & 0 & -10 & | & 0 \\ 0 & 1 & 1 & | & 0 \\ 0 & 0 & 0 & | & 0 \end{pmatrix}$$

implies $k_1 = 10k_3$ and $k_2 = -k_3$. Choosing $k_3 = 1$, we get a second eigenvector and solution vector

$$\mathbf{K}_2 = \begin{pmatrix} 10 \\ -1 \\ 1 \end{pmatrix}, \qquad \mathbf{X}_2 = \begin{pmatrix} 10 \\ -1 \\ 1 \end{pmatrix} e^{-4t}. \qquad (8)$$

Finally, when $\lambda_3 = 5$, the augmented matrices

$$(\mathbf{A} - 5\mathbf{I}|\mathbf{0}) = \begin{pmatrix} -9 & 1 & 1 & 0 \\ 1 & 0 & -1 & 0 \\ 0 & 1 & -8 & 0 \end{pmatrix} \overset{\text{row}}{\underset{\text{operations}}{\Longrightarrow}} \begin{pmatrix} 1 & 0 & -1 & 0 \\ 0 & 1 & -8 & 0 \\ 0 & 0 & 0 & 0 \end{pmatrix}$$

yield
$$\mathbf{K}_3 = \begin{pmatrix} 1 \\ 8 \\ 1 \end{pmatrix}, \qquad \mathbf{X}_3 = \begin{pmatrix} 1 \\ 8 \\ 1 \end{pmatrix} e^{5t}. \tag{9}$$

The general solution of (6) is a linear combination of the solution vectors in (7), (8), and (9):

$$\mathbf{X} = c_1 \begin{pmatrix} 1 \\ 0 \\ 1 \end{pmatrix} e^{-3t} + c_2 \begin{pmatrix} 10 \\ -1 \\ 1 \end{pmatrix} e^{-4t} + c_3 \begin{pmatrix} 1 \\ 8 \\ 1 \end{pmatrix} e^{5t}. \qquad\qquad \square$$

■ **Use of Computers** Software packages such as MATLAB, *Mathematica, Maple,* and DERIVE can be real time savers in finding eigenvalues and eigenvectors of a matrix. For example, to find the eigenvalues and eigenvectors of the matrix of coefficients in (6) using *Mathematica* we first input the definition of the matrix by rows:

$$\mathbf{m} = \{\{-4, 1, 1\}, \{1, 5, -1\}, \{0, 1, -3\}\}.$$

The commands

$$\textbf{Eigenvalues[m]} \qquad \text{and} \qquad \textbf{Eigenvectors[m]}$$

given in sequence yield

$$\{-4, -3, 5\} \qquad \text{and} \qquad \{\{10, -1, 1\}, \{1, 0, 1\}, \{1, 8, 1\}\},$$

respectively. In *Mathematica* eigenvalues and eigenvectors can also be obtained at the same time by using **Eigensystem[m]**.

10.2.2 Repeated Eigenvalues

Of course, not all of the n eigenvalues $\lambda_1, \lambda_2, \ldots, \lambda_n$ of an $n \times n$ matrix \mathbf{A} need be distinct, that is, some of the eigenvalues may be repeated. For example, the characteristic equation of the coefficient matrix in the system

$$\mathbf{X}' = \begin{pmatrix} 3 & -18 \\ 2 & -9 \end{pmatrix} \mathbf{X} \tag{10}$$

is readily shown to be $(\lambda + 3)^2 = 0$, and therefore $\lambda_1 = \lambda_2 = -3$ is a root of *multiplicity two.* For this value we find the single eigenvector

$$\mathbf{K}_1 = \begin{pmatrix} 3 \\ 1 \end{pmatrix}, \qquad \text{so} \qquad \mathbf{X}_1 = \begin{pmatrix} 3 \\ 1 \end{pmatrix} e^{-3t} \tag{11}$$

is one solution of (10). But since we are obviously interested in forming the general solution of the system, we need to pursue the question of finding a second solution.

In general, if m is a positive integer and $(\lambda - \lambda_1)^m$ is a factor of the characteristic equation while $(\lambda - \lambda_1)^{m+1}$ is not a factor, then λ_1 is said to be an **eigenvalue of multiplicity** m. The next three examples illustrate the following cases:

(*i*) For some $n \times n$ matrices \mathbf{A} it may be possible to find m linearly independent eigenvectors $\mathbf{K}_1, \mathbf{K}_2, \ldots, \mathbf{K}_m$ corresponding to an eigenvalue λ_1 of multiplicity $m \le n$. In this case the general solution of the system contains the linear combination

$$c_1 \mathbf{K}_1 e^{\lambda_1 t} + c_2 \mathbf{K}_2 e^{\lambda_1 t} + \cdots + c_m \mathbf{K}_m e^{\lambda_1 t}.$$

(*ii*) If there is only one eigenvector corresponding to the eigenvalue λ_1 of multiplicity m, then m linearly independent solutions of the form

$$\mathbf{X}_1 = \mathbf{K}_{11} e^{\lambda_1 t}$$
$$\mathbf{X}_2 = \mathbf{K}_{21} t e^{\lambda_1 t} + \mathbf{K}_{22} e^{\lambda_1 t}$$
$$\vdots$$
$$\mathbf{X}_m = \mathbf{K}_{m1} \frac{t^{m-1}}{(m-1)!} e^{\lambda_1 t} + \mathbf{K}_{m2} \frac{t^{m-2}}{(m-2)!} e^{\lambda_1 t} + \cdots + \mathbf{K}_{mm} e^{\lambda_1 t},$$

where \mathbf{K}_{ij} are column vectors, can always be found.

■ **Eigenvalue of Multiplicity Two** We begin by considering eigenvalues of multiplicity two. In the first example we illustrate a matrix for which we can find two distinct eigenvectors corresponding to a double eigenvalue.

Example 3 Repeated Eigenvalues

Solve $\mathbf{X}' = \begin{pmatrix} 1 & -2 & 2 \\ -2 & 1 & -2 \\ 2 & -2 & 1 \end{pmatrix} \mathbf{X}.$

SOLUTION Expanding the determinant in the characteristic equation

$$\det(\mathbf{A} - \lambda \mathbf{I}) = \begin{vmatrix} 1 - \lambda & -2 & 2 \\ -2 & 1 - \lambda & -2 \\ 2 & -2 & 1 - \lambda \end{vmatrix} = 0$$

yields $-(\lambda + 1)^2(\lambda - 5) = 0$. We see that $\lambda_1 = \lambda_2 = -1$ and $\lambda_3 = 5$.

For $\lambda_1 = -1$, Gauss–Jordan elimination immediately gives

$$(\mathbf{A} + \mathbf{I} | \mathbf{0}) = \begin{pmatrix} 2 & -2 & 2 & | & 0 \\ -2 & 2 & -2 & | & 0 \\ 2 & -2 & 2 & | & 0 \end{pmatrix} \overset{\text{row operations}}{\Longrightarrow} \begin{pmatrix} 1 & -1 & 1 & | & 0 \\ 0 & 0 & 0 & | & 0 \\ 0 & 0 & 0 & | & 0 \end{pmatrix}.$$

The first row of the last matrix means $k_1 - k_2 + k_3 = 0$ or $k_1 = k_2 - k_3$. The choices $k_2 = 1$, $k_3 = 0$ and $k_2 = 1$, $k_3 = 1$ yield, in turn, $k_1 = 1$ and $k_1 = 0$. Thus two eigenvectors corresponding to $\lambda_1 = -1$ are

$$\mathbf{K}_1 = \begin{pmatrix} 1 \\ 1 \\ 0 \end{pmatrix} \qquad \text{and} \qquad \mathbf{K}_2 = \begin{pmatrix} 0 \\ 1 \\ 1 \end{pmatrix}.$$

Since neither eigenvector is a constant multiple of the other, we have found, corresponding to the same eigenvalue, two linearly independent solutions

$$\mathbf{X}_1 = \begin{pmatrix} 1 \\ 1 \\ 0 \end{pmatrix} e^{-t} \qquad \text{and} \qquad \mathbf{X}_2 = \begin{pmatrix} 0 \\ 1 \\ 1 \end{pmatrix} e^{-t}.$$

Last, for $\lambda_3 = 5$, the reduction

$$(\mathbf{A} - 5\mathbf{I} | \mathbf{0}) = \begin{pmatrix} -4 & -2 & 2 & | & 0 \\ -2 & -4 & -2 & | & 0 \\ 2 & -2 & -4 & | & 0 \end{pmatrix} \overset{\text{row operations}}{\Longrightarrow} \begin{pmatrix} 1 & 0 & -1 & | & 0 \\ 0 & 1 & 1 & | & 0 \\ 0 & 0 & 0 & | & 0 \end{pmatrix}$$

implies $k_1 = k_3$ and $k_2 = -k_3$. Picking $k_3 = 1$ gives $k_1 = 1$, $k_2 = -1$, and thus a third eigenvector is

$$\mathbf{K}_3 = \begin{pmatrix} 1 \\ -1 \\ 1 \end{pmatrix}.$$

We conclude that the general solution of the system is

$$\mathbf{X} = c_1 \begin{pmatrix} 1 \\ 1 \\ 0 \end{pmatrix} e^{-t} + c_2 \begin{pmatrix} 0 \\ 1 \\ 1 \end{pmatrix} e^{-t} + c_3 \begin{pmatrix} 1 \\ -1 \\ 1 \end{pmatrix} e^{5t}.$$ ❑

The matrix of coefficients \mathbf{A} in Example 3 is a special kind of matrix known as a symmetric matrix. An $n \times n$ matrix \mathbf{A} is said to be **symmetric** if its transpose \mathbf{A}^T (where the rows and columns are interchanged) is the same as \mathbf{A}, that is, if $\mathbf{A}^T = \mathbf{A}$. It can be proved that if the matrix \mathbf{A} in the system $\mathbf{X}' = \mathbf{AX}$ is symmetric and has real entries, then we can always find n linearly independent eigenvectors $\mathbf{K}_1, \mathbf{K}_2, \ldots, \mathbf{K}_n$, and the general solution of such a system is as given in Theorem 10.7. As illustrated in Example 3, this result holds even when some of the eigenvalues are repeated.

■ **Second Solution** Now suppose that λ_1 is an eigenvalue of multiplicity two and that there is only one eigenvector associated with this value. A second solution can be found of the form

$$\mathbf{X}_2 = \mathbf{K}te^{\lambda_1 t} + \mathbf{P}e^{\lambda_1 t}, \tag{12}$$

where $$\mathbf{K} = \begin{pmatrix} k_1 \\ k_2 \\ \vdots \\ k_n \end{pmatrix} \quad \text{and} \quad \mathbf{P} = \begin{pmatrix} p_1 \\ p_2 \\ \vdots \\ p_n \end{pmatrix}.$$

To see this we substitute (12) into the system $\mathbf{X}' = \mathbf{AX}$ and simplify:

$$(\mathbf{AK} - \lambda_1 \mathbf{K})te^{\lambda_1 t} + (\mathbf{AP} - \lambda_1 \mathbf{P} - \mathbf{K})e^{\lambda_1 t} = \mathbf{0}.$$

Since this last equation is to hold for all values of t, we must have

$$(\mathbf{A} - \lambda_1 \mathbf{I})\mathbf{K} = \mathbf{0} \tag{13}$$

and $$(\mathbf{A} - \lambda_1 \mathbf{I})\mathbf{P} = \mathbf{K}. \tag{14}$$

Equation (13) simply states that \mathbf{K} must be an eigenvector of \mathbf{A} associated with λ_1. By solving (13), we find one solution $\mathbf{X}_1 = \mathbf{K}e^{\lambda_1 t}$. To find the second solution \mathbf{X}_2 we need only solve the additional system (14) for the vector \mathbf{P}.

Example 4 **Repeated Eigenvalues**

Find the general solution of the system given in (10).

SOLUTION From (11) we know that $\lambda_1 = -3$ and that one solution is $\mathbf{X}_1 = \begin{pmatrix} 3 \\ 1 \end{pmatrix} e^{-3t}$. Identifying $\mathbf{K} = \begin{pmatrix} 3 \\ 1 \end{pmatrix}$ and $\mathbf{P} = \begin{pmatrix} p_1 \\ p_2 \end{pmatrix}$, we find from (14) that we must now solve

$$(\mathbf{A} + 3\mathbf{I})\mathbf{P} = \mathbf{K} \quad \text{or} \quad \begin{array}{l} 6p_1 - 18p_2 = 3 \\ 2p_1 - 6p_2 = 1. \end{array}$$

Since this system is obviously equivalent to one equation, we have an infinite number of choices for p_1 and p_2. For example, by choosing $p_1 = 1$ we find $p_2 = \frac{1}{6}$. However, for simplicity, we shall choose $p_1 = \frac{1}{2}$ so that $p_2 = 0$. Hence $\mathbf{P} = \begin{pmatrix} \frac{1}{2} \\ 0 \end{pmatrix}$. Thus from (12) we find

$$\mathbf{X}_2 = \begin{pmatrix} 3 \\ 1 \end{pmatrix} te^{-3t} + \begin{pmatrix} \frac{1}{2} \\ 0 \end{pmatrix} e^{-3t}.$$

The general solution of (10) is then

$$\mathbf{X} = c_1 \begin{pmatrix} 3 \\ 1 \end{pmatrix} e^{-3t} + c_2 \left[\begin{pmatrix} 3 \\ 1 \end{pmatrix} te^{-3t} + \begin{pmatrix} \frac{1}{2} \\ 0 \end{pmatrix} e^{-3t} \right]. \qquad \square$$

By assigning various values to c_1 and c_2 in the solution in Example 4, we can plot trajectories of the system in (10). A phase portrait of (10) is given in Figure 10.3. The solutions \mathbf{X}_1 and $-\mathbf{X}_1$ determine two half-lines $y = \frac{1}{3}x$, $x > 0$, and $y = \frac{1}{3}x$, $x < 0$, respectively, that are shown in black in Figure 10.3. Because the single eigenvalue is negative and $e^{-3t} \to 0$ as $t \to \infty$ on *every* trajectory we have $(x(t), y(t)) \to (0, 0)$ as $t \to \infty$. This is why the arrowheads in Figure 10.3 indicate that a particle on any trajectory would move toward the origin as time increases and why the origin is an attractor in this case. Moreover, a moving particle on a trajectory $x = 3c_1 e^{-3t} + c_2(te^{-3t} + \frac{1}{2}e^{-3t})$, $y = c_1 e^{-3t} + c_2 te^{-3t}$, $c_2 \neq 0$, approaches $(0, 0)$ tangentially to one of the half-lines as $t \to \infty$. In contrast, when the repeated eigenvalue is positive the situation is reversed, and the origin is a repeller. See Problem 21 in Exercises 10.2. Analogous to Figure 10.2, Figure 10.3 is typical of all 2×2 homogeneous linear systems $\mathbf{X}' = \mathbf{AX}$ that have two repeated negative eigenvalues. See Problem 32 in Exercises 10.2.

Figure 10.3 A phase portrait of the system (10)

■ **Eigenvalue of Multiplicity Three** When the coefficient matrix \mathbf{A} has only one eigenvector associated with an eigenvalues λ_1 of multiplicity three, we can find a solution of the form (12) and a third solution of the form

$$\mathbf{X}_3 = \mathbf{K}\frac{t^2}{2}e^{\lambda_1 t} + \mathbf{P}te^{\lambda_1 t} + \mathbf{Q}e^{\lambda_1 t}, \tag{15}$$

where
$$\mathbf{K} = \begin{pmatrix} k_1 \\ k_2 \\ \vdots \\ k_n \end{pmatrix}, \qquad \mathbf{P} = \begin{pmatrix} p_1 \\ p_2 \\ \vdots \\ p_n \end{pmatrix}, \qquad \text{and} \qquad \mathbf{Q} = \begin{pmatrix} q_1 \\ q_2 \\ \vdots \\ q_n \end{pmatrix}.$$

By substituting (15) into the system $\mathbf{X}' = \mathbf{AX}$ we find that the column vectors \mathbf{K}, \mathbf{P}, and \mathbf{Q} must satisfy

$$(\mathbf{A} - \lambda_1 \mathbf{I})\mathbf{K} = \mathbf{0} \tag{16}$$

$$(\mathbf{A} - \lambda_1 \mathbf{I})\mathbf{P} = \mathbf{K} \tag{17}$$

and
$$(\mathbf{A} - \lambda_1 \mathbf{I})\mathbf{Q} = \mathbf{P}. \tag{18}$$

Of course, the solutions of (16) and (17) can be used in forming the solutions \mathbf{X}_1 and \mathbf{X}_2.

Example 5 **Repeated Eigenvalues**

Solve $\mathbf{X}' = \begin{pmatrix} 2 & 1 & 6 \\ 0 & 2 & 5 \\ 0 & 0 & 2 \end{pmatrix} \mathbf{X}.$

SOLUTION The characteristic equation $(\lambda - 2)^3 = 0$ shows that $\lambda_1 = 2$ is an eigenvalue of multiplicity three. By solving $(\mathbf{A} - 2\mathbf{I})\mathbf{K} = \mathbf{0}$ we find the single eigenvector

$$\mathbf{K} = \begin{pmatrix} 1 \\ 0 \\ 0 \end{pmatrix}.$$

We next solve the systems $(\mathbf{A} - 2\mathbf{I})\mathbf{P} = \mathbf{K}$ and $(\mathbf{A} - 2\mathbf{I})\mathbf{Q} = \mathbf{P}$ in succession and find that

$$\mathbf{P} = \begin{pmatrix} 0 \\ 1 \\ 0 \end{pmatrix} \quad \text{and} \quad \mathbf{Q} = \begin{pmatrix} 0 \\ -6/5 \\ 1/5 \end{pmatrix}.$$

Using (12) and (15), we see that the general solution of the system is

$$\mathbf{X} = c_1 \begin{pmatrix} 1 \\ 0 \\ 0 \end{pmatrix} e^{2t} + c_2 \left[\begin{pmatrix} 1 \\ 0 \\ 0 \end{pmatrix} te^{2t} + \begin{pmatrix} 0 \\ 1 \\ 0 \end{pmatrix} e^{2t} \right] + c_3 \left[\begin{pmatrix} 1 \\ 0 \\ 0 \end{pmatrix} \frac{t^2}{2} e^{2t} + \begin{pmatrix} 0 \\ 1 \\ 0 \end{pmatrix} te^{2t} + \begin{pmatrix} 0 \\ -6/5 \\ 1/5 \end{pmatrix} e^{2t} \right]. \quad \square$$

Remarks

When an eigenvalue λ_1 has multiplicity m, then we can either find m linearly independent eigenvectors or the number of corresponding eigenvectors is less than m. Hence the two cases listed on page 580 are not all the possibilities under which a repeated eigenvalue can occur. It could happen, say, that a 5×5 matrix has an eigenvalue of multiplicity 5 and there exist three corresponding linearly independent eigenvectors. See Problems 31 and 49 in Exercise 10.2.

10.2.3 Complex Eigenvalues

If $\lambda_1 = \alpha + \beta i$ and $\lambda_2 = \alpha - \beta i$, $\beta > 0$, $i^2 = -1$, are complex eigenvalues of the coefficient matrix \mathbf{A}, we can then certainly expect their corresponding eigenvectors to also have complex entries.*

For example, the characteristic equation of the system

$$\frac{dx}{dt} = 6x - y$$

$$\frac{dy}{dt} = 5x + 4y \tag{19}$$

is

$$\det (\mathbf{A} - \lambda\mathbf{I}) = \begin{vmatrix} 6 - \lambda & -1 \\ 5 & 4 - \lambda \end{vmatrix} = \lambda^2 - 10\lambda + 29 = 0.$$

From the quadratic formula we find $\lambda_1 = 5 + 2i$, $\lambda_2 = 5 - 2i$.

Now for $\lambda_1 = 5 + 2i$ we must solve

$$(1 - 2i)k_1 - k_2 = 0$$

$$5k_1 - (1 + 2i)k_2 = 0.$$

*When the characteristic equation has real coefficients, complex eigenvalues always appear in conjugate pairs.

Since $k_2 = (1 - 2i)k_1$,* the choice $k_1 = 1$ gives the following eigenvector and a solution vector:

$$\mathbf{K}_1 = \begin{pmatrix} 1 \\ 1 - 2i \end{pmatrix}, \qquad \mathbf{X}_1 = \begin{pmatrix} 1 \\ 1 - 2i \end{pmatrix} e^{(5 + 2i)t}.$$

In like manner, for $\lambda_2 = 5 - 2i$ we find

$$\mathbf{K}_2 = \begin{pmatrix} 1 \\ 1 + 2i \end{pmatrix}, \qquad \mathbf{X}_2 = \begin{pmatrix} 1 \\ 1 + 2i \end{pmatrix} e^{(5 - 2i)t}.$$

We can verify by means of the Wronskian that these solution vectors are linearly independent, and so the general solution of (19) is

$$\mathbf{X} = c_1 \begin{pmatrix} 1 \\ 1 - 2i \end{pmatrix} e^{(5 + 2i)t} + c_2 \begin{pmatrix} 1 \\ 1 + 2i \end{pmatrix} e^{(5 - 2i)t}. \tag{20}$$

Note that the entries in \mathbf{K}_2 corresponding to λ_2 are the conjugates of the entries in \mathbf{K}_1 corresponding to λ_1. The conjugate of λ_1 is, of course, λ_2. We write this as $\lambda_2 = \bar{\lambda}_1$ and $\mathbf{K}_2 = \bar{\mathbf{K}}_1$. We have illustrated the following general result.

> **THEOREM 10.8** Solutions Corresponding to a Complex Eigenvalue
>
> Let \mathbf{A} be the coefficient matrix having real entries of the homogeneous system (2), and let \mathbf{K}_1 be an eigenvector corresponding to the complex eigenvalue $\lambda_1 = \alpha + i\beta$, α and β real. Then
>
> $$\mathbf{K}_1 e^{\lambda_1 t} \qquad \text{and} \qquad \bar{\mathbf{K}}_1 e^{\bar{\lambda}_1 t}$$
>
> are solutions of (2).

It is desirable and relatively easy to rewrite a solution such as (20) in terms of real functions. To this end we first use Euler's formula to write

$$e^{(5 + 2i)t} = e^{5t} e^{2ti} = e^{5t}(\cos 2t + i \sin 2t)$$

$$e^{(5 - 2i)t} = e^{5t} e^{-2ti} = e^{5t}(\cos 2t - i \sin 2t).$$

Then, after we multiply complex numbers, collect terms, and replace $c_1 + c_2$ by C_1 and $(c_1 - c_2)i$ by C_2, (20) becomes

$$\mathbf{X} = C_1 \mathbf{X}_1 + C_2 \mathbf{X}_2, \tag{21}$$

where

$$\mathbf{X}_1 = \left[\begin{pmatrix} 1 \\ 1 \end{pmatrix} \cos 2t - \begin{pmatrix} 0 \\ -2 \end{pmatrix} \sin 2t \right] e^{5t}$$

and

$$\mathbf{X}_2 = \left[\begin{pmatrix} 0 \\ -2 \end{pmatrix} \cos 2t + \begin{pmatrix} 1 \\ 1 \end{pmatrix} \sin 2t \right] e^{5t}.$$

It is now important to realize that the two vectors \mathbf{X}_1 and \mathbf{X}_2 in (21) are themselves linearly independent *real* solutions of the original system. Consequently, we are justified in ignoring the relationship between C_1, C_2 and c_1, c_2, and we can regard C_1 and C_2 as completely arbitrary and real. In other words, the linear combination (21) is an alternative general solution of (19).

* Note that the second equation is simply $(1 + 2i)$ times the first.

The foregoing process can be generalized. Let \mathbf{K}_1 be an eigenvector of the coefficient matrix \mathbf{A} (with real entries) corresponding to the complex eigenvalue $\lambda_1 = \alpha + i\beta$. Then the two solution vectors in Theorem 10.8 can be written as

$$\mathbf{K}_1 e^{\lambda_1 t} = \mathbf{K}_1 e^{\alpha t} e^{i\beta t} = \mathbf{K}_1 e^{\alpha t}(\cos \beta t + i \sin \beta t)$$

$$\overline{\mathbf{K}}_1 e^{\overline{\lambda}_1 t} = \overline{\mathbf{K}}_1 e^{\alpha t} e^{-i\beta t} = \overline{\mathbf{K}}_1 e^{\alpha t}(\cos \beta t - i \sin \beta t).$$

By the superposition principle, Theorem 10.2, the following vectors are also solutions:

$$\mathbf{X}_1 = \frac{1}{2}(\mathbf{K}_1 e^{\lambda_1 t} + \overline{\mathbf{K}}_1 e^{\overline{\lambda}_1 t}) = \frac{1}{2}(\mathbf{K}_1 + \overline{\mathbf{K}}_1)e^{\alpha t} \cos \beta t - \frac{i}{2}(-\mathbf{K}_1 + \overline{\mathbf{K}}_1)e^{\alpha t} \sin \beta t$$

$$\mathbf{X}_2 = \frac{i}{2}(-\mathbf{K}_1 e^{\lambda_1 t} + \overline{\mathbf{K}}_1 e^{\overline{\lambda}_1 t}) = \frac{i}{2}(-\mathbf{K}_1 + \overline{\mathbf{K}}_1)e^{\alpha t} \cos \beta t + \frac{1}{2}(\mathbf{K}_1 + \overline{\mathbf{K}}_1)e^{\alpha t} \sin \beta t.$$

For *any* complex number $z = a + ib$, both $\frac{1}{2}(z + \overline{z}) = a$ and $\frac{i}{2}(-z + \overline{z}) = b$ are *real* numbers. Therefore, the entries in the column vectors $\frac{1}{2}(\mathbf{K}_1 + \overline{\mathbf{K}}_1)$ and $\frac{i}{2}(-\mathbf{K}_1 + \overline{\mathbf{K}}_1)$ are real numbers. By defining

$$\mathbf{B}_1 = \frac{1}{2}(\mathbf{K}_1 + \overline{\mathbf{K}}_1) \quad \text{and} \quad \mathbf{B}_2 = \frac{i}{2}(-\mathbf{K}_1 + \overline{\mathbf{K}}_1), \tag{22}$$

we are led to the following theorem.

> **THEOREM 10.9 Real Solutions Corresponding to a Complex Eigenvalue**
>
> Let $\lambda_1 = \alpha + i\beta$ be a complex eigenvalue of the coefficient matrix \mathbf{A} in the homogeneous system (2), and let \mathbf{B}_1 and \mathbf{B}_2 denote the column vectors defined in (22). Then
>
> $$\mathbf{X}_1 = [\mathbf{B}_1 \cos \beta t - \mathbf{B}_2 \sin \beta t]e^{\alpha t}$$
> $$\mathbf{X}_2 = [\mathbf{B}_2 \cos \beta t + \mathbf{B}_1 \sin \beta t]e^{\alpha t} \tag{23}$$
>
> are linearly independent solutions of (2) on $(-\infty, \infty)$.

The matrices \mathbf{B}_1 and \mathbf{B}_2 in (22) are often denoted by

$$\mathbf{B}_1 = \text{Re}(\mathbf{K}_1) \quad \text{and} \quad \mathbf{B}_2 = \text{Im}(\mathbf{K}_1) \tag{24}$$

since these vectors are, respectively, the *real* and *imaginary* parts of the eigenvector \mathbf{K}_1. For example, (21) follows from (23) with

$$\mathbf{K}_1 = \begin{pmatrix} 1 \\ 1 - 2i \end{pmatrix} = \begin{pmatrix} 1 \\ 1 \end{pmatrix} + i\begin{pmatrix} 0 \\ -2 \end{pmatrix}$$

$$\mathbf{B}_1 = \text{Re}(\mathbf{K}_1) = \begin{pmatrix} 1 \\ 1 \end{pmatrix} \quad \text{and} \quad \mathbf{B}_2 = \text{Im}(\mathbf{K}_1) = \begin{pmatrix} 0 \\ -2 \end{pmatrix}.$$

Example 6 Complex Eigenvalues

Solve the initial-value problem

$$\mathbf{X}' = \begin{pmatrix} 2 & 8 \\ -1 & -2 \end{pmatrix}\mathbf{X}, \qquad \mathbf{X}(0) = \begin{pmatrix} 2 \\ -1 \end{pmatrix}. \tag{25}$$

SOLUTION First we obtain the eigenvalues from

$$\det(\mathbf{A} - \lambda\mathbf{I}) = \begin{vmatrix} 2 - \lambda & 8 \\ -1 & -2 - \lambda \end{vmatrix} = \lambda^2 + 4 = 0.$$

The eigenvalues are $\lambda_1 = 2i$ and $\lambda_2 = \bar{\lambda}_1 = -2i$. For λ_1 the system

$$(2 - 2i)k_1 + 8k_2 = 0$$
$$-k_1 + (-2 - 2i)k_2 = 0$$

gives $k_1 = -(2 + 2i)k_2$. By choosing $k_2 = -1$ we get

$$\mathbf{K}_1 = \begin{pmatrix} 2 + 2i \\ -1 \end{pmatrix} = \begin{pmatrix} 2 \\ -1 \end{pmatrix} + i\begin{pmatrix} 2 \\ 0 \end{pmatrix}.$$

Now from (24) we form

$$\mathbf{B}_1 = \mathrm{Re}(\mathbf{K}_1) = \begin{pmatrix} 2 \\ -1 \end{pmatrix} \qquad \text{and} \qquad \mathbf{B}_2 = \mathrm{Im}(\mathbf{K}_1) = \begin{pmatrix} 2 \\ 0 \end{pmatrix}.$$

Since $\alpha = 0$, it follows from (23) that the general solution of the system is

$$\mathbf{X} = c_1\left[\begin{pmatrix} 2 \\ -1 \end{pmatrix} \cos 2t - \begin{pmatrix} 2 \\ 0 \end{pmatrix} \sin 2t \right] + c_2\left[\begin{pmatrix} 2 \\ 0 \end{pmatrix} \cos 2t + \begin{pmatrix} 2 \\ -1 \end{pmatrix} \sin 2t \right]$$

$$= c_1\begin{pmatrix} 2\cos 2t - 2\sin 2t \\ -\cos 2t \end{pmatrix} + c_2\begin{pmatrix} 2\cos 2t + 2\sin 2t \\ -\sin 2t \end{pmatrix}. \tag{26}$$

Some graphs of the curves or trajectories defined by the solution (26) of the system are illustrated in the phase portrait in Figure 10.4. Now the initial condition $\mathbf{X}(0) = \begin{pmatrix} 2 \\ -1 \end{pmatrix}$, or equivalently $x(0) = 2$, and $y(0) = -1$, yields the algebraic system $2c_1 + 2c_2 = 2$, $-c_1 = -1$ whose solution is $c_1 = 1$, $c_2 = 0$. Thus the solution to the problem is $\mathbf{X} = \begin{pmatrix} 2\cos 2t - 2\sin 2t \\ -\cos 2t \end{pmatrix}$. The specific trajectory defined parametrically by the particular solution $x = 2\cos 2t - 2\sin 2t$, $y = -\cos 2t$ is the black curve in Figure 10.4. Note that this curve passes through $(2, -1)$. ❑

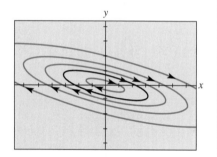

Figure 10.4 A phase portrait of the system in (26)

Answers to odd-numbered problems begin on page A-53.

10.2.1 Distinct Real Eigenvalues

In Problems 1–12, find the general solution of the given system.

1. $\dfrac{dx}{dt} = x + 2y$

$\dfrac{dy}{dt} = 4x + 3y$

2. $\dfrac{dx}{dt} = 2x + 2y$

$\dfrac{dy}{dt} = x + 3y$

3. $\dfrac{dx}{dt} = -4x + 2y$

$\dfrac{dy}{dt} = -\tfrac{5}{2}x + 2y$

4. $\dfrac{dx}{dt} = -\tfrac{5}{2}x + 2y$

$\dfrac{dy}{dt} = \tfrac{3}{4}x - 2y$

5. $\mathbf{X}' = \begin{pmatrix} 10 & -5 \\ 8 & -12 \end{pmatrix}\mathbf{X}$

6. $\mathbf{X}' = \begin{pmatrix} -6 & 2 \\ -3 & 1 \end{pmatrix}\mathbf{X}$

7. $\dfrac{dx}{dt} = x + y - z$

$\dfrac{dy}{dt} = 2y$

$\dfrac{dz}{dt} = y - z$

8. $\dfrac{dx}{dt} = 2x - 7y$

$\dfrac{dy}{dt} = 5x + 10y + 4z$

$\dfrac{dz}{dt} = 5y + 2z$

9. $\mathbf{X}' = \begin{pmatrix} -1 & 1 & 0 \\ 1 & 2 & 1 \\ 0 & 3 & -1 \end{pmatrix}\mathbf{X}$

10. $\mathbf{X}' = \begin{pmatrix} 1 & 0 & 1 \\ 0 & 1 & 0 \\ 1 & 0 & 1 \end{pmatrix}\mathbf{X}$

11. $\mathbf{X}' = \begin{pmatrix} -1 & -1 & 0 \\ \tfrac{3}{4} & -\tfrac{3}{2} & 3 \\ \tfrac{1}{8} & \tfrac{1}{4} & -\tfrac{1}{2} \end{pmatrix}\mathbf{X}$

12. $\mathbf{X}' = \begin{pmatrix} -1 & 4 & 2 \\ 4 & -1 & -2 \\ 0 & 0 & 6 \end{pmatrix}\mathbf{X}$

In Problems 13 and 14, solve the given initial-value problem.

13. $\mathbf{X}' = \begin{pmatrix} \tfrac{1}{2} & 0 \\ 1 & -\tfrac{1}{2} \end{pmatrix}\mathbf{X}$, $\mathbf{X}(0) = \begin{pmatrix} 3 \\ 5 \end{pmatrix}$

14. $\mathbf{X}' = \begin{pmatrix} 1 & 1 & 4 \\ 0 & 2 & 0 \\ 1 & 1 & 1 \end{pmatrix}\mathbf{X}$, $\mathbf{X}(0) = \begin{pmatrix} 1 \\ 3 \\ 0 \end{pmatrix}$

Computer Lab Assignment In Problems 15 and 16, use a CAS or linear algebra software as an aid in finding the general solution of the given system.

15. $\mathbf{X}' = \begin{pmatrix} 0.9 & 2.1 & 3.2 \\ 0.7 & 6.5 & 4.2 \\ 1.1 & 1.7 & 3.4 \end{pmatrix}\mathbf{X}$

16. $\mathbf{X}' = \begin{pmatrix} 1 & 0 & 2 & -1.8 & 0 \\ 0 & 5.1 & 0 & -1 & 3 \\ 1 & 2 & -3 & 0 & 0 \\ 0 & 1 & -3.1 & 4 & 0 \\ -2.8 & 0 & 0 & 1.5 & 1 \end{pmatrix} \mathbf{X}$

17. Computer Lab Assignment

(a) Use computer software to obtain the phase portrait of the system in Problem 5. If possible include the arrowheads as in Figure 10.2. Also, include four half-lines in your phase portrait.

(b) Obtain the Cartesian equations of each of the four half-lines in part **(a)**.

(c) Draw the eigenvectors on your phase portrait of the system.

18. Computer Lab Assignment Find phase portraits for the systems in Problems 2 and 4. For each system find any half-line trajectories and include these lines in your phase portrait.

10.2.2 Repeated Eigenvalues

In Problems 19–28, find the general solution of the given system.

19. $\dfrac{dx}{dt} = 3x - y$

$\dfrac{dy}{dt} = 9x - 3y$

20. $\dfrac{dx}{dt} = -6x + 5y$

$\dfrac{dy}{dt} = -5x + 4y$

21. $\mathbf{X}' = \begin{pmatrix} -1 & 3 \\ -3 & 5 \end{pmatrix} \mathbf{X}$

22. $\mathbf{X}' = \begin{pmatrix} 12 & -9 \\ 4 & 0 \end{pmatrix} \mathbf{X}$

23. $\dfrac{dx}{dt} = 3x - y - z$

$\dfrac{dy}{dt} = x + y - z$

$\dfrac{dz}{dt} = x - y + z$

24. $\dfrac{dx}{dt} = 3x + 2y + 4z$

$\dfrac{dy}{dt} = 2x + 2z$

$\dfrac{dz}{dt} = 4x + 2y + 3z$

25. $\mathbf{X}' = \begin{pmatrix} 5 & -4 & 0 \\ 1 & 0 & 2 \\ 0 & 2 & 5 \end{pmatrix} \mathbf{X}$

26. $\mathbf{X}' = \begin{pmatrix} 1 & 0 & 0 \\ 0 & 3 & 1 \\ 0 & -1 & 1 \end{pmatrix} \mathbf{X}$

27. $\mathbf{X}' = \begin{pmatrix} 1 & 0 & 0 \\ 2 & 2 & -1 \\ 0 & 1 & 0 \end{pmatrix} \mathbf{X}$

28. $\mathbf{X}' = \begin{pmatrix} 4 & 1 & 0 \\ 0 & 4 & 1 \\ 0 & 0 & 4 \end{pmatrix} \mathbf{X}$

In Problems 29 and 30, solve the given initial-value problem.

29. $\mathbf{X}' = \begin{pmatrix} 2 & 4 \\ -1 & 6 \end{pmatrix} \mathbf{X}, \quad \mathbf{X}(0) = \begin{pmatrix} -1 \\ 6 \end{pmatrix}$

30. $\mathbf{X}' = \begin{pmatrix} 0 & 0 & 1 \\ 0 & 1 & 0 \\ 1 & 0 & 0 \end{pmatrix} \mathbf{X}, \quad \mathbf{X}(0) = \begin{pmatrix} 1 \\ 2 \\ 5 \end{pmatrix}$

31. Show that the 5×5 matrix

$$\mathbf{A} = \begin{pmatrix} 2 & 1 & 0 & 0 & 0 \\ 0 & 2 & 0 & 0 & 0 \\ 0 & 0 & 2 & 0 & 0 \\ 0 & 0 & 0 & 2 & 1 \\ 0 & 0 & 0 & 0 & 2 \end{pmatrix}$$

has an eigenvalue λ_1 of multiplicity 5. Show that three linearly independent eigenvectors corresponding to λ_1 can be found.

32. Computer Lab Assignment Find phase portraits for the systems in Problems 20 and 21. For each system find any half-line trajectories and include these lines in your phase portrait.

10.2.3 Complex Eigenvalues

In Problems 33–44, find the general solution of the given system.

33. $\dfrac{dx}{dt} = 6x - y$

$\dfrac{dy}{dt} = 5x + 2y$

34. $\dfrac{dx}{dt} = x + y$

$\dfrac{dy}{dt} = -2x - y$

35. $\dfrac{dx}{dt} = 5x + y$

$\dfrac{dy}{dt} = -2x + 3y$

36. $\dfrac{dx}{dt} = 4x + 5y$

$\dfrac{dy}{dt} = -2x + 6y$

37. $\mathbf{X}' = \begin{pmatrix} 4 & -5 \\ 5 & -4 \end{pmatrix} \mathbf{X}$

38. $\mathbf{X}' = \begin{pmatrix} 1 & -8 \\ 1 & -3 \end{pmatrix} \mathbf{X}$

39. $\dfrac{dx}{dt} = z$

$\dfrac{dy}{dt} = -z$

$\dfrac{dz}{dt} = y$

40. $\dfrac{dx}{dt} = 2x + y + 2z$

$\dfrac{dy}{dt} = 3x + 6z$

$\dfrac{dz}{dt} = -4x - 3z$

41. $\mathbf{X}' = \begin{pmatrix} 1 & -1 & 2 \\ -1 & 1 & 0 \\ -1 & 0 & 1 \end{pmatrix} \mathbf{X}$

42. $\mathbf{X}' = \begin{pmatrix} 4 & 0 & 1 \\ 0 & 6 & 0 \\ -4 & 0 & 4 \end{pmatrix} \mathbf{X}$

43. $\mathbf{X}' = \begin{pmatrix} 2 & 5 & 1 \\ -5 & -6 & 4 \\ 0 & 0 & 2 \end{pmatrix} \mathbf{X}$

44. $\mathbf{X}' = \begin{pmatrix} 2 & 4 & 4 \\ -1 & -2 & 0 \\ -1 & 0 & -2 \end{pmatrix} \mathbf{X}$

In Problems 45 and 46, solve the given initial-value problem.

45. $\mathbf{X}' = \begin{pmatrix} 1 & -12 & -14 \\ 1 & 2 & -3 \\ 1 & 1 & -2 \end{pmatrix} \mathbf{X}, \quad \mathbf{X}(0) = \begin{pmatrix} 4 \\ 6 \\ -7 \end{pmatrix}$

46. $\mathbf{X}' = \begin{pmatrix} 6 & -1 \\ 5 & 4 \end{pmatrix} \mathbf{X}, \quad \mathbf{X}(0) = \begin{pmatrix} -2 \\ 8 \end{pmatrix}$

47. Computer Lab Assignment Find phase portraits for the systems in Problems 36, 37, and 38.

Miscellaneous Problems

48. Computer Lab Assignment Solve each of the following linear systems.

(a) $\mathbf{X}' = \begin{pmatrix} 1 & 1 \\ 1 & 1 \end{pmatrix} \mathbf{X}$ **(b)** $\mathbf{X}' = \begin{pmatrix} 1 & 1 \\ -1 & -1 \end{pmatrix} \mathbf{X}$.

Find a phase portrait of each system. What is the geometric significance of the line $y = -x$ in each portrait?

49. Consider the 5×5 matrix given in Problem 31. Solve the system $\mathbf{X}' = \mathbf{A}\mathbf{X}$ without the aid of matrix methods, but write the general solution using the matrix notation. Use the general solution as a basis for a discussion on how the system can be solved using the matrix methods of this section. Carry out your ideas.

50. Obtain a Cartesian equation of the curve defined parametrically by the solution of the linear system in Example 6. Identify the curve passing through $(2, -1)$ in Figure 10.4. [*Hint:* Compute x^2, y^2, and xy.]

51. Examine your phase portraits in Problem 47. Under what conditions will the phase portrait of a 2×2 homogeneous linear system with complex eigenvalues consist of a family of closed curves? Consist of a family of spirals? Under what conditions is the origin $(0, 0)$ a repeller? An attractor?

52. The system of linear second-order differential equations

$$m_1 x_1'' = -k_1 x_1 + k_2(x_2 - x_1)$$
$$m_2 x_2'' = -k_2(x_2 - x_1) \tag{27}$$

describes the motion of two coupled spring/mass systems (see Figure 3.59). We have already solved a special case of this system in Sections 3.11 and 4.6. In this problem we describe yet another method for solving the system.

(a) Show that (27) can be written as the matrix equation $\mathbf{X}'' = \mathbf{A}\mathbf{X}$, where

$$\mathbf{X} = \begin{pmatrix} x_1 \\ x_2 \end{pmatrix} \quad \text{and} \quad \mathbf{A} = \begin{pmatrix} -\dfrac{k_1 + k_2}{m_1} & \dfrac{k_2}{m_1} \\ \dfrac{k_2}{m_2} & -\dfrac{k_2}{m_2} \end{pmatrix}.$$

(b) If a solution is assumed of the form $\mathbf{X} = \mathbf{K}e^{\omega t}$, show that $\mathbf{X}'' = \mathbf{A}\mathbf{X}$ yields

$$(\mathbf{A} - \lambda \mathbf{I})\mathbf{K} = \mathbf{0} \quad \text{where } \lambda = \omega^2.$$

(c) Show that if $m_1 = 1$, $m_2 = 1$, $k_1 = 3$, and $k_2 = 2$, a solution of the system is

$$\mathbf{X} = c_1 \begin{pmatrix} 1 \\ 2 \end{pmatrix} e^{it} + c_2 \begin{pmatrix} 1 \\ 2 \end{pmatrix} e^{-it} + c_3 \begin{pmatrix} -2 \\ 1 \end{pmatrix} e^{\sqrt{6}it} + c_4 \begin{pmatrix} -2 \\ 1 \end{pmatrix} e^{-\sqrt{6}it}.$$

(d) Show that the solution in part **(c)** can be written as

$$\mathbf{X} = b_1 \begin{pmatrix} 1 \\ 2 \end{pmatrix} \cos t + b_2 \begin{pmatrix} 1 \\ 2 \end{pmatrix} \sin t$$
$$+ b_3 \begin{pmatrix} -2 \\ 1 \end{pmatrix} \cos \sqrt{6}t + b_4 \begin{pmatrix} -2 \\ 1 \end{pmatrix} \sin \sqrt{6}t.$$

10.3 Solution by Diagonalization

In this section we are going to consider an alternative method for solving a homogeneous system $\mathbf{X}' = \mathbf{A}\mathbf{X}$ of linear first-order differential equations. This method is applicable to such a system whenever the coefficient matrix \mathbf{A} is *diagonalizable*.

■ **Coupled Systems** A homogeneous linear system $\mathbf{X}' = \mathbf{A}\mathbf{X}$,

$$\begin{pmatrix} x_1' \\ x_2' \\ \vdots \\ x_n' \end{pmatrix} = \begin{pmatrix} a_{11} & a_{12} & \cdots & a_{1n} \\ a_{21} & a_{22} & \cdots & a_{2n} \\ \vdots & & & \vdots \\ a_{n1} & a_{n2} & \cdots & a_{nn} \end{pmatrix} \begin{pmatrix} x_1 \\ x_2 \\ \vdots \\ x_n \end{pmatrix}, \tag{1}$$

in which each x_i' is expressed as a linear combination of x_1, x_2, \ldots, x_n is said to be **coupled.** If the coefficient matrix \mathbf{A} is diagonalizable, then the system can be **uncoupled** in that each x_i' can be expressed solely in terms of x_i.

If the matrix \mathbf{A} has n linearly independent eigenvectors then we know from Theorem 8.27 that we can find a matrix \mathbf{P} such that $\mathbf{P}^{-1}\mathbf{A}\mathbf{P} = \mathbf{D}$, where \mathbf{D} is a diagonal matrix. If we make the substitution $\mathbf{X} = \mathbf{P}\mathbf{Y}$ in the system $\mathbf{X}' = \mathbf{A}\mathbf{X}$, then

$$\mathbf{P}\mathbf{Y}' = \mathbf{A}\mathbf{P}\mathbf{Y} \quad \text{or} \quad \mathbf{Y}' = \mathbf{P}^{-1}\mathbf{A}\mathbf{P}\mathbf{Y} \quad \text{or} \quad \mathbf{Y}' = \mathbf{D}\mathbf{Y}. \tag{2}$$

The last equation in (2) is the same as

$$
\begin{pmatrix} y_1' \\ y_2' \\ \vdots \\ y_n' \end{pmatrix} = \begin{pmatrix} \lambda_1 & 0 & 0 & \cdots & 0 \\ 0 & \lambda_1 & 0 & \cdots & 0 \\ \vdots & & & & \vdots \\ 0 & 0 & 0 & \cdots & \lambda_n \end{pmatrix} \begin{pmatrix} y_1 \\ y_2 \\ \vdots \\ y_n \end{pmatrix}. \tag{3}
$$

Since \mathbf{D} is diagonal, an inspection of (3) reveals that this new system is uncoupled: each differential equation in the system is of the form $y_i' = \lambda_i y_i$, $i = 1, 2, \ldots, n$. The solution of each of these linear equations is $y_i = c_i e^{\lambda_i t}$, $i = 1, 2, \ldots, n$. Hence the general solution of (3) can be written as the column vector

$$
\mathbf{Y} = \begin{pmatrix} c_1 e^{\lambda_1 t} \\ c_2 e^{\lambda_2 t} \\ \vdots \\ c_n e^{\lambda_n t} \end{pmatrix}. \tag{4}
$$

Since we now know \mathbf{Y} and since the matrix \mathbf{P} can be constructed from the eigenvectors of \mathbf{A}, the general solution of the original system $\mathbf{X}' = \mathbf{AX}$ is obtained from $\mathbf{X} = \mathbf{PY}$.

Example 1 Uncoupling a Linear System

Solve $\mathbf{X}' = \begin{pmatrix} -2 & -1 & 8 \\ 0 & -3 & 8 \\ 0 & -4 & 9 \end{pmatrix} \mathbf{X}$ by diagonalization.

SOLUTION We begin by finding the eigenvalues and corresponding eigenvectors of the coefficient matrix.

From $\det (\mathbf{A} - \lambda \mathbf{I}) = -(\lambda + 2)(\lambda - 1)(\lambda - 5)$ we get $\lambda_1 = -2$, $\lambda_2 = 1$, and $\lambda_3 = 5$. Since the eigenvalues are distinct, the eigenvectors are linearly independent. Solving $(\mathbf{A} - \lambda_i \mathbf{I})\mathbf{K} = \mathbf{0}$ for $i = 1, 2,$ and 3 gives, respectively,

$$
\mathbf{K}_1 = \begin{pmatrix} 1 \\ 0 \\ 0 \end{pmatrix}, \quad \mathbf{K}_2 = \begin{pmatrix} 2 \\ 2 \\ 1 \end{pmatrix}, \quad \mathbf{K}_3 = \begin{pmatrix} 1 \\ 1 \\ 1 \end{pmatrix}. \tag{5}
$$

Thus a matrix that diagonalizes the coefficient matrix is

$$
\mathbf{P} = \begin{pmatrix} 1 & 2 & 1 \\ 0 & 2 & 1 \\ 0 & 1 & 1 \end{pmatrix}.
$$

The entries on the main diagonal of \mathbf{D} are the eigenvalues of \mathbf{A} corresponding to the order in which the eigenvectors appear in \mathbf{P}:

$$
\mathbf{D} = \begin{pmatrix} -2 & 0 & 0 \\ 0 & 1 & 0 \\ 0 & 0 & 5 \end{pmatrix}.
$$

As we have shown above, the substitution $\mathbf{X} = \mathbf{PY}$ in $\mathbf{X}' = \mathbf{AX}$ gives the uncoupled system $\mathbf{Y}' = \mathbf{DY}$. The general solution of this last system is immediate:

$$
\mathbf{Y} = \begin{pmatrix} c_1 e^{-2t} \\ c_2 e^{t} \\ c_3 e^{5t} \end{pmatrix}.
$$

Hence the solution of the given system is

$$\mathbf{X} = \mathbf{PY} = \begin{pmatrix} 1 & 2 & 1 \\ 0 & 2 & 1 \\ 0 & 1 & 1 \end{pmatrix} \begin{pmatrix} c_1 e^{-2t} \\ c_2 e^{t} \\ c_3 e^{5t} \end{pmatrix} = \begin{pmatrix} c_1 e^{-2t} + 2c_2 e^{t} + c_3 e^{5t} \\ 2c_2 e^{t} + c_3 e^{5t} \\ c_2 e^{t} + c_3 e^{5t} \end{pmatrix}. \qquad (6)\ \square$$

Note that (6) can be written in the usual manner by expressing the last matrix as a sum of column matrices:

$$\mathbf{X} = c_1 \begin{pmatrix} 1 \\ 0 \\ 0 \end{pmatrix} e^{-2t} + c_2 \begin{pmatrix} 2 \\ 2 \\ 1 \end{pmatrix} e^{t} + c_3 \begin{pmatrix} 1 \\ 1 \\ 1 \end{pmatrix} e^{5t}.$$

Solution by diagonalization will always work provided we can find n linearly independent eigenvectors of the $n \times n$ matrix \mathbf{A}; the eigenvalues of \mathbf{A} could be real and distinct, complex, or repeated. The method fails when \mathbf{A} has repeated eigenvalues and n linearly independent eigenvectors cannot be found. Of course, in this last situation \mathbf{A} is not diagonalizable.

Since we have to find eigenvalues and eigenvectors of \mathbf{A}, this method is essentially equivalent to the procedure presented in the last section.

In the next section we shall see that diagonalization can also be used to solve nonhomogeneous linear systems $\mathbf{X}' = \mathbf{AX} + \mathbf{F}(t)$.

Answers to odd-numbered problems begin on page A-53.

In Problems 1–10, use diagonalization to solve the given system.

1. $\mathbf{X}' = \begin{pmatrix} 5 & 6 \\ 3 & -2 \end{pmatrix} \mathbf{X}$

2. $\mathbf{X}' = \begin{pmatrix} \tfrac{1}{2} & \tfrac{1}{2} \\ \tfrac{1}{2} & \tfrac{1}{2} \end{pmatrix} \mathbf{X}$

3. $\mathbf{X}' = \begin{pmatrix} 1 & \tfrac{1}{4} \\ 1 & 1 \end{pmatrix} \mathbf{X}$

4. $\mathbf{X}' = \begin{pmatrix} 1 & 1 \\ 1 & -1 \end{pmatrix} \mathbf{X}$

5. $\mathbf{X}' = \begin{pmatrix} -1 & 3 & 0 \\ 3 & -1 & 0 \\ -2 & -2 & 6 \end{pmatrix} \mathbf{X}$

6. $\mathbf{X}' = \begin{pmatrix} 1 & 1 & 2 \\ 1 & 2 & 1 \\ 2 & 1 & 1 \end{pmatrix} \mathbf{X}$

7. $\mathbf{X}' = \begin{pmatrix} 1 & -1 & -1 \\ -1 & 1 & -1 \\ -1 & -1 & 1 \end{pmatrix} \mathbf{X}$

8. $\mathbf{X}' = \begin{pmatrix} 1 & 1 & 1 & 1 \\ 1 & 1 & 1 & 1 \\ 1 & 1 & 1 & 1 \\ 1 & 1 & 1 & 1 \end{pmatrix} \mathbf{X}$

9. $\mathbf{X}' = \begin{pmatrix} -3 & 2 & 2 \\ -6 & 5 & 2 \\ -7 & 4 & 4 \end{pmatrix} \mathbf{X}$

10. $\mathbf{X}' = \begin{pmatrix} 0 & 2 & 0 \\ 2 & 0 & 2 \\ 0 & 2 & 0 \end{pmatrix} \mathbf{X}$

10.4 Nonhomogeneous Linear Systems

Before developing a matrix version of variation of parameters for nonhomogeneous linear systems $\mathbf{X}' = \mathbf{AX} + \mathbf{F}(t)$, we need to examine a special matrix that is formed out of the solution vectors of the corresponding homogeneous system $\mathbf{X}' = \mathbf{AX}$.

■ **A Fundamental Matrix** If $\mathbf{X}_1, \mathbf{X}_2, \ldots, \mathbf{X}_n$ is a fundamental set of solutions of the homogeneous system $\mathbf{X}' = \mathbf{AX}$ on an interval I, then its general solution on the interval is $\mathbf{X} = c_1 \mathbf{X}_1 + c_2 \mathbf{X}_2 + \ldots + c_n \mathbf{X}_n$, or

$$\mathbf{X} = c_1 \begin{pmatrix} x_{11} \\ x_{21} \\ \vdots \\ x_{n1} \end{pmatrix} + c_2 \begin{pmatrix} x_{12} \\ x_{22} \\ \vdots \\ x_{n2} \end{pmatrix} + \ldots + c_n \begin{pmatrix} x_{1n} \\ x_{2n} \\ \vdots \\ x_{nn} \end{pmatrix} = \begin{pmatrix} c_1 x_{11} + c_2 x_{12} + \cdots + c_n x_{1n} \\ c_1 x_{21} + c_2 x_{22} + \cdots + c_n x_{2n} \\ \vdots \\ c_1 x_{n1} + c_2 x_{n2} + \cdots + c_n x_{nn} \end{pmatrix}. \qquad (1)$$

The last matrix in (1) is recognized as the product of an $n \times n$ matrix with an $n \times 1$ matrix. In other words, the general solution (1) can be written as the product

$$\mathbf{X} = \mathbf{\Phi}(t)\mathbf{C}, \qquad (2)$$

where \mathbf{C} is the $n \times 1$ column vector of arbitrary constants c_1, c_2, \ldots, c_n, and the $n \times n$ matrix, whose columns consist of the entries of the solution vectors of the system $\mathbf{X}' = \mathbf{AX}$,

$$\boldsymbol{\Phi}(t) = \begin{pmatrix} x_{11} & x_{12} & \cdots & x_{1n} \\ x_{21} & x_{22} & \cdots & x_{2n} \\ \vdots & & & \vdots \\ x_{n1} & x_{n2} & \cdots & x_{nn} \end{pmatrix},$$

is called a **fundamental matrix** of the system on the interval.

In the discussion that follows we need to use two properties of a fundamental matrix:

- A fundamental matrix $\boldsymbol{\Phi}(t)$ is nonsingular.
- If $\boldsymbol{\Phi}(t)$ is a fundamental matrix of the system $\mathbf{X}' = \mathbf{AX,}$ then

$$\boldsymbol{\Phi}'(t) = \mathbf{A}\boldsymbol{\Phi}(t). \tag{3}$$

A reexamination of (9) of Theorem 10.3 shows that det $\boldsymbol{\Phi}(t)$ is the same as the Wronskian $W(\mathbf{X}_1, \mathbf{X}_2, \ldots, \mathbf{X}_n)$. Hence the linear independence of the columns of $\boldsymbol{\Phi}(t)$ on the interval I guarantees that det $\boldsymbol{\Phi}(t) \neq 0$ for every t in the interval. Since $\boldsymbol{\Phi}(t)$ is nonsingular, the multiplicative inverse $\boldsymbol{\Phi}^{-1}(t)$ exists for every t in the interval. The result given in (3) follows immediately from the fact that every column of $\boldsymbol{\Phi}(t)$ is a solution vector of $\mathbf{X}' = \mathbf{AX}$.

■ **Variation of Parameters** Analogous to the procedure in Section 3.5, we ask whether it is possible to replace the matrix of constants \mathbf{C} in (2) by a column matrix of functions

$$\mathbf{U}(t) = \begin{pmatrix} u_1(t) \\ u_2(t) \\ \vdots \\ u_n(t) \end{pmatrix} \qquad \text{so that} \qquad \mathbf{X}_p = \boldsymbol{\Phi}(t)\mathbf{U}(t) \tag{4}$$

is a particular solution of the nonhomogeneous system

$$\mathbf{X}' = \mathbf{AX} + \mathbf{F}(t). \tag{5}$$

By the product rule, the derivative of the last expression in (4) is

$$\mathbf{X}'_p = \boldsymbol{\Phi}(t)\mathbf{U}'(t) + \boldsymbol{\Phi}'(t)\mathbf{U}(t). \tag{6}$$

Note that the order of the products in (6) is very important. Since $\mathbf{U}(t)$ is a column matrix, the products $\mathbf{U}'(t)\boldsymbol{\Phi}(t)$ and $\mathbf{U}(t)\boldsymbol{\Phi}'(t)$ are not defined. Substituting (4) and (6) into (5) gives

$$\boldsymbol{\Phi}(t)\mathbf{U}'(t) + \boldsymbol{\Phi}'(t)\mathbf{U}(t) = \mathbf{A}\boldsymbol{\Phi}(t)\mathbf{U}(t) + \mathbf{F}(t). \tag{7}$$

Now if we use (3) to replace $\boldsymbol{\Phi}'(t)$, (7) becomes

$$\boldsymbol{\Phi}(t)\mathbf{U}'(t) + \mathbf{A}\boldsymbol{\Phi}(t)\mathbf{U}(t) = \mathbf{A}\boldsymbol{\Phi}(t)\mathbf{U}(t) + \mathbf{F}(t)$$

or

$$\boldsymbol{\Phi}(t)\mathbf{U}'(t) = \mathbf{F}(t). \tag{8}$$

Multiplying both sides of equation (8) by $\boldsymbol{\Phi}^{-1}(t)$ gives

$$\mathbf{U}'(t) = \boldsymbol{\Phi}^{-1}(t)\mathbf{F}(t) \qquad \text{and so} \qquad \mathbf{U}(t) = \int \boldsymbol{\Phi}^{-1}(t)\mathbf{F}(t)\, dt.$$

Since $\mathbf{X}_p = \boldsymbol{\Phi}(t)\mathbf{U}(t)$, we conclude that a particular solution of (5) is

$$\mathbf{X}_p = \boldsymbol{\Phi}(t) \int \boldsymbol{\Phi}^{-1}(t)\mathbf{F}(t)\, dt. \tag{9}$$

To calculate the indefinite integral of the column matrix $\mathbf{\Phi}^{-1}(t)\mathbf{F}(t)$ in (9) we integrate each entry. Thus the general solution of the system (5) is $\mathbf{X} = \mathbf{X}_c + \mathbf{X}_p$ or

$$\mathbf{X} = \mathbf{\Phi}(t)\mathbf{C} + \mathbf{\Phi}(t) \int \mathbf{\Phi}^{-1}(t)\mathbf{F}(t) \, dt. \tag{10}$$

Example 1 Variation of Parameters

Find the general solution of the nonhomogeneous system

$$\mathbf{X}' = \begin{pmatrix} -3 & 1 \\ 2 & -4 \end{pmatrix}\mathbf{X} + \begin{pmatrix} 3t \\ e^{-t} \end{pmatrix} \tag{11}$$

on the interval $(-\infty, \infty)$.

SOLUTION We first solve the homogeneous system

$$\mathbf{X}' = \begin{pmatrix} -3 & 1 \\ 2 & -4 \end{pmatrix}\mathbf{X}. \tag{12}$$

The characteristic equation of the coefficient matrix is

$$\det (\mathbf{A} - \lambda\mathbf{I}) = \begin{vmatrix} -3 - \lambda & 1 \\ 2 & -4 - \lambda \end{vmatrix} = (\lambda + 2)(\lambda + 5) = 0,$$

so the eigenvalues are $\lambda_1 = -2$ and $\lambda_2 = -5$. By the usual method we find that the eigenvectors corresponding to λ_1 and λ_2 are, respectively,

$$\begin{pmatrix} 1 \\ 1 \end{pmatrix} \quad \text{and} \quad \begin{pmatrix} 1 \\ -2 \end{pmatrix}.$$

The solution vectors of the system (11) are then

$$\mathbf{X}_1 = \begin{pmatrix} 1 \\ 1 \end{pmatrix}e^{-2t} = \begin{pmatrix} e^{-2t} \\ e^{-2t} \end{pmatrix} \quad \text{and} \quad \mathbf{X}_2 = \begin{pmatrix} 1 \\ -2 \end{pmatrix}e^{-5t} = \begin{pmatrix} e^{-5t} \\ -2e^{-5t} \end{pmatrix}.$$

The entries in \mathbf{X}_1 form the first column of $\mathbf{\Phi}(t)$, and the entries in \mathbf{X}_2 form the second column of $\mathbf{\Phi}(t)$. Hence

$$\mathbf{\Phi}(t) = \begin{pmatrix} e^{-2t} & e^{-5t} \\ e^{-2t} & -2e^{-5t} \end{pmatrix} \quad \text{and} \quad \mathbf{\Phi}^{-1}(t) = \begin{pmatrix} \tfrac{2}{3}e^{2t} & \tfrac{1}{3}e^{2t} \\ \tfrac{1}{3}e^{5t} & -\tfrac{1}{3}e^{5t} \end{pmatrix}.$$

From (9) we obtain

$$\mathbf{X}_p = \mathbf{\Phi}(t) \int \mathbf{\Phi}^{-1}(t)\mathbf{F}(t) \, dt = \begin{pmatrix} e^{-2t} & e^{-5t} \\ e^{-2t} & -2e^{-5t} \end{pmatrix} \int \begin{pmatrix} \tfrac{2}{3}e^{2t} & \tfrac{1}{3}e^{2t} \\ \tfrac{1}{3}e^{5t} & -\tfrac{1}{3}e^{5t} \end{pmatrix}\begin{pmatrix} 3t \\ e^{-t} \end{pmatrix} dt$$

$$= \begin{pmatrix} e^{-2t} & e^{-5t} \\ e^{-2t} & -2e^{-5t} \end{pmatrix} \int \begin{pmatrix} 2te^{2t} + \tfrac{1}{3}e^{t} \\ te^{5t} - \tfrac{1}{3}e^{4t} \end{pmatrix} dt$$

$$= \begin{pmatrix} e^{-2t} & e^{-5t} \\ e^{-2t} & -2e^{-5t} \end{pmatrix} \begin{pmatrix} te^{2t} - \tfrac{1}{2}e^{2t} + \tfrac{1}{3}e^{t} \\ \tfrac{1}{5}te^{5t} - \tfrac{1}{25}e^{5t} - \tfrac{1}{12}e^{4t} \end{pmatrix}$$

$$= \begin{pmatrix} \tfrac{6}{5}t - \tfrac{27}{50} + \tfrac{1}{4}e^{-t} \\ \tfrac{3}{5}t - \tfrac{21}{50} + \tfrac{1}{2}e^{-t} \end{pmatrix}.$$

Hence from (10) the general solution of (11) on the interval is

$$\mathbf{X} = \begin{pmatrix} e^{-2t} & e^{-5t} \\ e^{-2t} & -2e^{-5t} \end{pmatrix}\begin{pmatrix} c_1 \\ c_2 \end{pmatrix} + \begin{pmatrix} \tfrac{6}{5}t - \tfrac{27}{50} + \tfrac{1}{4}e^{-t} \\ \tfrac{3}{5}t - \tfrac{21}{50} + \tfrac{1}{2}e^{-t} \end{pmatrix}$$

$$= c_1\begin{pmatrix} 1 \\ 1 \end{pmatrix}e^{-2t} + c_2\begin{pmatrix} 1 \\ -2 \end{pmatrix}e^{-5t} + \begin{pmatrix} \tfrac{6}{5} \\ \tfrac{3}{5} \end{pmatrix}t - \begin{pmatrix} \tfrac{27}{50} \\ \tfrac{21}{50} \end{pmatrix} + \begin{pmatrix} \tfrac{1}{4} \\ \tfrac{1}{2} \end{pmatrix}e^{-t}. \qquad \square$$

■ **Initial-Value Problem** The general solution of the nonhomogeneous system (5) on an interval can be written in an alternative manner

$$X = \Phi(t)C + \Phi(t) \int_{t_0}^{t} \Phi^{-1}(s)F(s) \, ds, \tag{13}$$

where t and t_0 are points in the interval. The last form is useful in solving (5) subject to an initial condition $X(t_0) = X_0$, because the limits of integration are chosen so that the particular solution vanishes at $t = t_0$. Substituting $t = t_0$ in (13) yields $X_0 = \Phi(t_0)C$, from which we get $C = \Phi^{-1}(t_0)X_0$. Substituting this last result in (13) gives the following solution of the initial-value problem:

$$X = \Phi(t)\Phi^{-1}(t_0)X_0 + \Phi(t) \int_{t_0}^{t} \Phi^{-1}(s)F(s) \, ds. \tag{14}$$

■ **Diagonalization** As in Section 10.3, if the coefficient matrix A possesses n linearly independent eigenvectors, then we can use diagonalization to uncouple the system $X' = AX + F(t)$. Suppose P is the matrix such that $P^{-1}AP = D$, where D is a diagonal matrix. Substituting $X = PY$ into the nonhomogeneous system $X' = AX + F(t)$ gives

$$PY' = APY + F \quad \text{or} \quad Y' = P^{-1}APY + P^{-1}F \quad \text{or} \quad Y' = DY + G. \tag{15}$$

In the last equation in (15), $G = P^{-1}F$ is a column vector. So each differential equation in this new system has the form $y_i' = \lambda_i y_i + g_i(t)$, $i = 1, 2, \ldots, n$. But notice that, unlike the procedure for solving a homogeneous system $X' = AX$, we now are required to compute the inverse of the matrix P.

Example 2 **Diagonalization**

Solve $X' = \begin{pmatrix} 4 & 2 \\ 2 & 1 \end{pmatrix} X + \begin{pmatrix} 3e^t \\ e^t \end{pmatrix}$ by diagonalization.

SOLUTION The eigenvalues and corresponding eigenvectors of the coefficient matrix are found to be $\lambda_1 = 0$, $\lambda_2 = 5$, $K_1 = \begin{pmatrix} 1 \\ -2 \end{pmatrix}$, $K_2 = \begin{pmatrix} 2 \\ 1 \end{pmatrix}$. Thus we find $P = \begin{pmatrix} 1 & 2 \\ -2 & 1 \end{pmatrix}$ and $P^{-1} = \begin{pmatrix} 1/5 & -2/5 \\ 2/5 & 1/5 \end{pmatrix}$. Using the substitution $X = PY$ and

$$P^{-1}F = \begin{pmatrix} 1/5 & -2/5 \\ 2/5 & 1/5 \end{pmatrix} \begin{pmatrix} 3e^t \\ e^t \end{pmatrix} = \begin{pmatrix} 1/5 e^t \\ 7/5 e^t \end{pmatrix}$$

the uncoupled system is

$$Y' = \begin{pmatrix} 0 & 0 \\ 0 & 5 \end{pmatrix} Y + \begin{pmatrix} 1/5 e^t \\ 7/5 e^t \end{pmatrix}.$$

The solutions of the two differential equations

$$y_1' = 1/5 e^t \quad \text{and} \quad y_2' = 5y_2 + 7/5 e^t$$

are, respectively, $y_1 = 1/5 e^t + c_1$ and $y_2 = -7/20 e^t + c_2 e^{5t}$. Hence, the solution of the original system is

$$X = PY = \begin{pmatrix} 1 & 2 \\ -2 & 1 \end{pmatrix} \begin{pmatrix} 1/5 e^t + c_1 \\ -7/20 e^t + c_2 e^{5t} \end{pmatrix} = \begin{pmatrix} -1/2 e^t + c_1 + 2c_2 e^{5t} \\ -3/4 e^t - 2c_1 + c_2 e^{5t} \end{pmatrix}. \tag{16}$$

Written in the usual manner using column vectors, (16) is

$$X = c_1 \begin{pmatrix} 1 \\ -2 \end{pmatrix} + c_2 \begin{pmatrix} 2 \\ 1 \end{pmatrix} e^{5t} - \begin{pmatrix} 1/2 \\ 3/4 \end{pmatrix} e^t. \qquad ❑$$

Answers to odd-numbered problems begin on page A-54.

In Problems 1–20, use variation of parameters to solve the given system.

1. $\dfrac{dx}{dt} = 3x - 3y + 4$

$\dfrac{dy}{dt} = 2x - 2y - 1$

2. $\dfrac{dx}{dt} = 2x - y$

$\dfrac{dy}{dt} = 3x - 2y + 4t$

3. $\mathbf{X}' = \begin{pmatrix} 3 & -5 \\ \frac{3}{4} & -1 \end{pmatrix} \mathbf{X} + \begin{pmatrix} 1 \\ -1 \end{pmatrix} e^{t/2}$

4. $\mathbf{X}' = \begin{pmatrix} 2 & -1 \\ 4 & 2 \end{pmatrix} \mathbf{X} + \begin{pmatrix} \sin 2t \\ 2 \cos 2t \end{pmatrix} e^{2t}$

5. $\mathbf{X}' = \begin{pmatrix} 0 & 2 \\ -1 & 3 \end{pmatrix} \mathbf{X} + \begin{pmatrix} 1 \\ -1 \end{pmatrix} e^{t}$

6. $\mathbf{X}' = \begin{pmatrix} 0 & 2 \\ -1 & 3 \end{pmatrix} \mathbf{X} + \begin{pmatrix} 2 \\ e^{-3t} \end{pmatrix}$

7. $\mathbf{X}' = \begin{pmatrix} 1 & 8 \\ 1 & -1 \end{pmatrix} \mathbf{X} + \begin{pmatrix} 12 \\ 12 \end{pmatrix} t$

8. $\mathbf{X}' = \begin{pmatrix} 1 & 8 \\ 1 & -1 \end{pmatrix} \mathbf{X} + \begin{pmatrix} e^{-t} \\ te^{t} \end{pmatrix}$

9. $\mathbf{X}' = \begin{pmatrix} 3 & 2 \\ -2 & -1 \end{pmatrix} \mathbf{X} + \begin{pmatrix} 2e^{-t} \\ e^{-t} \end{pmatrix}$

10. $\mathbf{X}' = \begin{pmatrix} 3 & 2 \\ -2 & -1 \end{pmatrix} \mathbf{X} + \begin{pmatrix} 1 \\ 1 \end{pmatrix}$

11. $\mathbf{X}' = \begin{pmatrix} 0 & -1 \\ 1 & 0 \end{pmatrix} \mathbf{X} + \begin{pmatrix} \sec t \\ 0 \end{pmatrix}$

12. $\mathbf{X}' = \begin{pmatrix} 1 & -1 \\ 1 & 1 \end{pmatrix} \mathbf{X} + \begin{pmatrix} 3 \\ 3 \end{pmatrix} e^{t}$

13. $\mathbf{X}' = \begin{pmatrix} 1 & -1 \\ 1 & 1 \end{pmatrix} \mathbf{X} + \begin{pmatrix} \cos t \\ \sin t \end{pmatrix} e^{t}$

14. $\mathbf{X}' = \begin{pmatrix} 2 & -2 \\ 8 & -6 \end{pmatrix} \mathbf{X} + \begin{pmatrix} 1 \\ 3 \end{pmatrix} \dfrac{e^{-2t}}{t}$

15. $\mathbf{X}' = \begin{pmatrix} 0 & 1 \\ -1 & 0 \end{pmatrix} \mathbf{X} + \begin{pmatrix} 0 \\ \sec t \tan t \end{pmatrix}$

16. $\mathbf{X}' = \begin{pmatrix} 0 & 1 \\ -1 & 0 \end{pmatrix} \mathbf{X} + \begin{pmatrix} 1 \\ \cot t \end{pmatrix}$

17. $\mathbf{X}' = \begin{pmatrix} 1 & 2 \\ -\frac{1}{2} & 1 \end{pmatrix} \mathbf{X} + \begin{pmatrix} \csc t \\ \sec t \end{pmatrix} e^{t}$

18. $\mathbf{X}' = \begin{pmatrix} 1 & -2 \\ 1 & -1 \end{pmatrix} \mathbf{X} + \begin{pmatrix} \tan t \\ 1 \end{pmatrix}$

19. $\mathbf{X}' = \begin{pmatrix} 1 & 1 & 0 \\ 1 & 1 & 0 \\ 0 & 0 & 3 \end{pmatrix} \mathbf{X} + \begin{pmatrix} e^{t} \\ e^{2t} \\ te^{3t} \end{pmatrix}$

20. $\mathbf{X}' = \begin{pmatrix} 3 & -1 & -1 \\ 1 & 1 & -1 \\ 1 & -1 & 1 \end{pmatrix} \mathbf{X} + \begin{pmatrix} 0 \\ t \\ 2e^{t} \end{pmatrix}$

In Problems 21 and 22, use (14) to solve the given initial-value problem.

21. $\mathbf{X}' = \begin{pmatrix} 3 & -1 \\ -1 & 3 \end{pmatrix} \mathbf{X} + \begin{pmatrix} 4e^{2t} \\ 4e^{4t} \end{pmatrix}, \quad \mathbf{X}(0) = \begin{pmatrix} 1 \\ 1 \end{pmatrix}$

22. $\mathbf{X}' = \begin{pmatrix} 1 & -1 \\ 1 & -1 \end{pmatrix} \mathbf{X} + \begin{pmatrix} 1/t \\ 1/t \end{pmatrix}, \quad \mathbf{X}(1) = \begin{pmatrix} 2 \\ -1 \end{pmatrix}$

In Problems 23–26, use diagonalization to solve the given system.

23. $\mathbf{X}' = \begin{pmatrix} 5 & -2 \\ 21 & -8 \end{pmatrix} \mathbf{X} + \begin{pmatrix} 6 \\ 4 \end{pmatrix}$

24. $\mathbf{X}' = \begin{pmatrix} 1 & 3 \\ 2 & 2 \end{pmatrix} \mathbf{X} + \begin{pmatrix} e^{t} \\ e^{t} \end{pmatrix}$

25. $\mathbf{X}' = \begin{pmatrix} 5 & 5 \\ 5 & 5 \end{pmatrix} \mathbf{X} + \begin{pmatrix} 2t \\ 8 \end{pmatrix}$

26. $\mathbf{X}' = \begin{pmatrix} 0 & 1 \\ 1 & 0 \end{pmatrix} \mathbf{X} + \begin{pmatrix} 4 \\ 8e^{-2t} \end{pmatrix}$

27. The system of differential equations for the currents $i_1(t)$ and $i_2(t)$ in the electrical network shown in Figure 10.5 is

$$\frac{d}{dt}\begin{pmatrix} i_1 \\ i_2 \end{pmatrix} = \begin{pmatrix} -(R_1 + R_2)/L_2 & R_2/L_2 \\ R_2/L_1 & -R_2/L_1 \end{pmatrix}\begin{pmatrix} i_1 \\ i_2 \end{pmatrix} + \begin{pmatrix} E/L_2 \\ 0 \end{pmatrix}.$$

Figure 10.5

Use variation of parameters to solve the system if $R_1 = 8\ \Omega$, $R_2 = 3\ \Omega$, $L_1 = 1$ h, $L_2 = 1$ h, $E(t) = 100 \sin t$ V, $i_1(0) = 0$ and $i_2(0) = 0$.

28. The system of differential equations for the current $i_1(t)$ and charge $q(t)$ in the electrical network shown in Figure 10.6 is

$$\frac{d}{dt}\begin{pmatrix} i_1 \\ q \end{pmatrix} = \begin{pmatrix} -\dfrac{R_1 R_2}{L(R_1 + R_2)} & -\dfrac{R_1}{CL(R_1 + R_2)} \\ \dfrac{R_1}{R_1 + R_2} & -\dfrac{1}{C(R_1 + R_2)} \end{pmatrix}\begin{pmatrix} i_1 \\ q \end{pmatrix} + \begin{pmatrix} E/L \\ 0 \end{pmatrix}.$$

Figure 10.6

Use variation of parameters to solve the system if $R_1 = 50\ \Omega$, $R_2 = 100\ \Omega$, $L = 5$ h, $C = 0.004$ f, and $E(t) = 10$ V.

29. The system of differential equations for the currents $i_2(t)$ and $i_3(t)$ in the electrical network shown in Figure 10.7 is

$$\frac{d}{dt}\begin{pmatrix} i_2 \\ i_3 \end{pmatrix} = \begin{pmatrix} -R_1/L_1 & -R_1/L_1 \\ -R_1/L_2 & -(R_1 + R_2)/L_2 \end{pmatrix}\begin{pmatrix} i_2 \\ i_3 \end{pmatrix} + \begin{pmatrix} E/L_1 \\ E/L_2 \end{pmatrix}.$$

Figure 10.7

Use variation of parameters to solve the system if $R_1 = 2\ \Omega$, $R_2 = 3\ \Omega$, $L_1 = 1$ h, $L_2 = 1$ h, $E(t) = 60$ V, $i_2(0) = 0$ and $i_3(0) = 0$. Determine the current $i_1(t)$.

30. Computer Lab Assignment Solving a nonhomogeneous linear system $\mathbf{X}' = \mathbf{AX} + \mathbf{F}(t)$ by variation of parameters when \mathbf{A} is a 3×3 (or larger) matrix is almost an impossible task to do by hand. Consider the system

$$\mathbf{X}' = \begin{pmatrix} 2 & -2 & 2 & 1 \\ -1 & 3 & 0 & 3 \\ 0 & 0 & 4 & -2 \\ 0 & 0 & 2 & -1 \end{pmatrix} \mathbf{X} + \begin{pmatrix} te^t \\ e^{-t} \\ e^{2t} \\ 1 \end{pmatrix}.$$

(a) Use a CAS or linear algebra software to find the eigenvalues and eigenvectors of the coefficient matrix.

(b) Form a fundamental matrix $\mathbf{\Phi}(t)$ and use the computer to find $\mathbf{\Phi}^{-1}(t)$.

(c) Use the computer to carry out the computations of $\mathbf{\Phi}^{-1}(t)\mathbf{F}(t)$, $\int \mathbf{\Phi}^{-1}(t)\mathbf{F}(t)\ dt$, $\mathbf{\Phi}(t)\int \mathbf{\Phi}^{-1}(t)\mathbf{F}(t)\ dt$, $\mathbf{\Phi}(t)\mathbf{C}$, and $\mathbf{\Phi}(t)\mathbf{C} + \mathbf{\Phi}(t)\int \mathbf{\Phi}^{-1}(t)\mathbf{F}(t)\ dt$, where \mathbf{C} is a column matrix of constants c_1, c_2, c_3, and c_4.

(d) Rewrite the computer output for the general solution of the system in the form $\mathbf{X} = \mathbf{X}_c + \mathbf{X}_p$, where $\mathbf{X}_c = c_1\mathbf{X}_1 + c_2\mathbf{X}_2 + c_3\mathbf{X}_3 + c_4\mathbf{X}_4$.

10.5 Matrix Exponential

Matrices can be utilized in an entirely different manner to solve a system of linear first-order differential equations.

■ **Homogeneous Systems** Recall that the simple linear first-order differential equation $x' = ax$, where a is a constant, has the general solution $x = ce^{at}$. It seems natural, then, to ask whether we can define a **matrix exponential** $e^{\mathbf{A}t}$ so that the homogeneous system $\mathbf{X}' = \mathbf{AX}$, where \mathbf{A} is an $n \times n$ matrix of constants, has a solution

$$\mathbf{X} = e^{\mathbf{A}t}\mathbf{C}. \qquad (1)$$

Since \mathbf{C} is to be an $n \times 1$ column matrix of arbitrary constants, we want $e^{\mathbf{A}t}$ to be an $n \times n$ matrix. While the complete development of the meaning and theory of the matrix exponential would require a thorough knowledge of matrix algebra, one way of defining $e^{\mathbf{A}t}$ is inspired by the power series representation of the scalar exponential function e^{at}:

$$e^{at} = 1 + at + a^2\frac{t^2}{2!} + \cdots + a^k\frac{t^k}{k!} + \cdots = \sum_{k=0}^{\infty} a^k\frac{t^k}{k!}. \qquad (2)$$

The series in (2) converges for all t. Using this series, with 1 replaced by the identity \mathbf{I} and the constant a replaced with an $n \times n$ matrix \mathbf{A} of constants, we arrive at a definition for the $n \times n$ matrix $e^{\mathbf{A}t}$.

> **DEFINITION 10.4 Matrix Exponential**
>
> For any $n \times n$ matrix \mathbf{A},
>
> $$e^{\mathbf{A}t} = \mathbf{I} + \mathbf{A}t + \mathbf{A}^2\frac{t^2}{2!} + \cdots + \mathbf{A}^k\frac{t^k}{k!} + \cdots = \sum_{k=0}^{\infty} \mathbf{A}^k\frac{t^k}{k!}. \qquad (3)$$

It can be shown that the series given in (3) converges to an $n \times n$ matrix for every value of t. Also, in (3) $\mathbf{A}^0 = \mathbf{I}$, $\mathbf{A}^2 = \mathbf{AA}$, $\mathbf{A}^3 = \mathbf{A}(\mathbf{A}^2)$, and so on.

■ **Derivative of $e^{\mathbf{A}t}$** The derivative of the matrix exponential $e^{\mathbf{A}t}$ is analogous to that of the scalar exponential, that is, $d/dt\ e^{at} = ae^{at}$. To justify

$$\frac{d}{dt}e^{\mathbf{A}t} = \mathbf{A}e^{\mathbf{A}t} \qquad (4)$$

we differentiate (3) term-by-term:

$$\frac{d}{dt}\, e^{\mathbf{A}t} = \frac{d}{dt}\left[\mathbf{I} + \mathbf{A}t + \mathbf{A}^2\frac{t^2}{2!} + \cdots + \mathbf{A}^k\frac{t^k}{k!} + \cdots\right] = \mathbf{A} + \mathbf{A}^2t + \frac{1}{2!}\mathbf{A}^3t^2 + \cdots$$

$$= \mathbf{A}\left[\mathbf{I} + \mathbf{A}t + \mathbf{A}^2\frac{t^2}{2!} + \cdots\right] = \mathbf{A}e^{\mathbf{A}t}.$$

Because of (4), we can now prove that (1) is a solution of $\mathbf{X}' = \mathbf{A}\mathbf{X}$ for every $n \times 1$ vector \mathbf{C} of constants:

$$\mathbf{X}' = \frac{d}{dt}\, e^{\mathbf{A}t}\mathbf{C} = \mathbf{A}e^{\mathbf{A}t}\mathbf{C} = \mathbf{A}(e^{\mathbf{A}t}\mathbf{C}) = \mathbf{A}\mathbf{X}. \tag{5}$$

■ **$e^{\mathbf{A}t}$ Is a Fundamental Matrix** If we denote the matrix exponential $e^{\mathbf{A}t}$ by the symbol $\mathbf{\Psi}(t)$, then (4) is equivalent to the matrix differential equation $\mathbf{\Psi}'(t) = \mathbf{A}\mathbf{\Psi}$ (see (3) of Section 10.4). In addition, it follows immediately from Definition 10.4 that $\mathbf{\Psi}(0) = e^{\mathbf{A}0} = \mathbf{I}$ and so det $\mathbf{\Psi}(0) \neq 0$. It turns out that these two properties are sufficient for us to conclude that $\mathbf{\Psi}(t)$ is a fundamental matrix of the system $\mathbf{X}' = \mathbf{A}\mathbf{X}$.

■ **Nonhomogeneous Systems** We have seen in (4) of Section 2.3 that the general solution of the single linear first-order differential equation $x' = ax + f(t)$, where a is a constant, can be expressed as

$$x = x_c + x_p = ce^{at} + e^{at}\int_{t_0}^{t} e^{-as} f(s)\, ds.$$

For a nonhomogeneous system of linear first-order differential equations, it can be shown that the general solution of $\mathbf{X}' = \mathbf{A}\mathbf{X} + \mathbf{F}(t)$, where \mathbf{A} is an $n \times n$ matrix of constants, is

$$\mathbf{X} = \mathbf{X}_c + \mathbf{X}_p = e^{\mathbf{A}t}\mathbf{C} + e^{\mathbf{A}t}\int_{t_0}^{t} e^{-\mathbf{A}s}\mathbf{F}(s)\, ds. \tag{6}$$

Since the matrix exponential $e^{\mathbf{A}t}$ is a fundamental matrix, it is always nonsingular and $e^{-\mathbf{A}s} = (e^{\mathbf{A}s})^{-1}$. Note that $e^{-\mathbf{A}s}$ can be obtained from $e^{\mathbf{A}t}$ by replacing t by $-s$.

■ **Computation of $e^{\mathbf{A}t}$** The definition of $e^{\mathbf{A}t}$ given in (3) can, of course, always be used to compute $e^{\mathbf{A}t}$. However, the practical utility of (3) is limited by the fact that the entries in $e^{\mathbf{A}t}$ are power series in t. With one's natural desire to work with simple and familiar things, we then try to recognize whether these series define a closed form function. See Problems 1–4 in Exercises 10.5. Fortunately there are many alternative ways of computing $e^{\mathbf{A}t}$. We sketch two of these methods in the discussion that follows.

■ **Using the Laplace Transform** We saw in (5) that $\mathbf{X} = e^{\mathbf{A}t}$ is a solution of $\mathbf{X}' = \mathbf{A}\mathbf{X}$. Indeed, since $e^{\mathbf{A}0} = \mathbf{I}$, $\mathbf{X} = e^{\mathbf{A}t}$ is a solution of the initial-value problem

$$\mathbf{X}' = \mathbf{A}\mathbf{X}, \qquad \mathbf{X}(0) = \mathbf{I}. \tag{7}$$

If $\mathbf{x}(s) = \mathscr{L}\{\mathbf{X}(t)\} = \mathscr{L}\{e^{\mathbf{A}t}\}$, then the Laplace transform of (7) is

$$s\mathbf{x}(s) - \mathbf{X}(0) = \mathbf{A}\mathbf{x}(s) \quad \text{or} \quad (s\mathbf{I} - \mathbf{A})\mathbf{x}(s) = \mathbf{I}.$$

Multiplying the last equation by $(s\mathbf{I} - \mathbf{A})^{-1}$ implies $\mathbf{x}(s) = (s\mathbf{I} - \mathbf{A})^{-1}\mathbf{I} = (s\mathbf{I} - \mathbf{A})^{-1}$. In other words,

$$\mathscr{L}\{e^{\mathbf{A}t}\} = (s\mathbf{I} - \mathbf{A})^{-1}. \tag{8}$$

Example 1 **Matrix Exponential**

Use the Laplace transform to compute $e^{\mathbf{A}t}$ for $\mathbf{A} = \begin{pmatrix} 1 & -1 \\ 2 & -2 \end{pmatrix}$.

SOLUTION First we compute the matrix $s\mathbf{I} - \mathbf{A}$ and then find its inverse:

$$s\mathbf{I} - \mathbf{A} = \begin{pmatrix} s-1 & 1 \\ -2 & s+2 \end{pmatrix}$$

$$(s\mathbf{I} - \mathbf{A})^{-1} = \begin{pmatrix} s-1 & 1 \\ -2 & s+2 \end{pmatrix}^{-1} = \begin{pmatrix} \dfrac{s+2}{s(s+1)} & \dfrac{-1}{s(s+1)} \\ \dfrac{2}{s(s+1)} & \dfrac{s-1}{s(s+1)} \end{pmatrix}.$$

Then we decompose the entries of the last matrix into partial fractions:

$$(s\mathbf{I} - \mathbf{A})^{-1} = \begin{pmatrix} \dfrac{2}{s} - \dfrac{1}{s+1} & -\dfrac{1}{s} + \dfrac{1}{s+1} \\ \dfrac{2}{s} - \dfrac{2}{s+1} & \dfrac{-1}{s} + \dfrac{2}{s+1} \end{pmatrix}. \qquad (9)$$

Taking the inverse Laplace transform of (9) gives the desired result,

$$e^{\mathbf{A}t} = \begin{pmatrix} 2 - e^{-t} & -1 + e^{-t} \\ 2 - 2e^{-t} & -1 + 2e^{-t} \end{pmatrix}. \qquad \square$$

■ **Using Powers \mathbf{A}^m** In Section 8.8 we developed a method for computing an arbitrary power \mathbf{A}^k, k a nonnegative integer, of an $n \times n$ matrix \mathbf{A}. Recall from Section 8.8 that we can write

$$\mathbf{A}^k = \sum_{j=0}^{n-1} c_j \mathbf{A}^j \quad \text{and} \quad \lambda^k = \sum_{j=0}^{n-1} c_j \lambda^j, \qquad (10)$$

where the coefficients c_j are the same in each and the last expression is valid for the eigenvalues $\lambda_1, \lambda_2, \ldots, \lambda_n$ of \mathbf{A}. We assume here that the eigenvalues of \mathbf{A} are distinct. By setting $\lambda = \lambda_1, \lambda_2, \ldots, \lambda_n$ in the second expression in (10) we were able to find the c_j in the first expression by solving n equations in n unknowns. It will be convenient in the development that follows to emphasize the fact that the coefficients c_j in (10) depend on the power k by replacing c_j by $c_j(k)$. From (3) and (2), we have

$$e^{\mathbf{A}t} = \sum_{k=0}^{\infty} \mathbf{A}^k \frac{t^k}{k!} \quad \text{and} \quad e^{\lambda t} = \sum_{k=0}^{\infty} \lambda^k \frac{t^k}{k!}. \qquad (11)$$

We next use (10) in (11) to replace \mathbf{A}^k and λ^k as finite sums followed by an interchange of the order of summations

$$e^{\mathbf{A}t} = \sum_{k=0}^{\infty} \frac{t^k}{k!} \left(\sum_{j=0}^{n-1} c_j(k)\mathbf{A}^j \right) = \sum_{j=0}^{n-1} \mathbf{A}^j \left(\sum_{k=0}^{\infty} \frac{t^k}{k!} c_j(k) \right) = \sum_{j=0}^{n-1} \mathbf{A}^j b_j(t) \qquad (12)$$

$$e^{\lambda t} = \sum_{k=0}^{\infty} \frac{t^k}{k!} \left(\sum_{j=0}^{n-1} c_j(k)\lambda^j \right) = \sum_{j=0}^{n-1} \lambda^j \left(\sum_{k=0}^{\infty} \frac{t^k}{k!} c_j(k) \right) = \sum_{j=0}^{n-1} \lambda^j b_j(t), \qquad (13)$$

where $b_j(t) = \sum_{k=0}^{\infty} (t^k/k!)c_j(k)$. Analogous to how we used the eigenvalues of \mathbf{A} in (10) to determine the c_j, we again use the eigenvalues, but this time in the finite sum (13) to determine a system of equations to determine the b_j; these coefficients, in turn, are used in (12) to determine $e^{\mathbf{A}t}$.

Example 2 **Matrix Exponential**

Compute $e^{\mathbf{A}t}$ for $\mathbf{A} = \begin{pmatrix} -2 & 4 \\ -1 & 3 \end{pmatrix}$.

SOLUTION We have already seen the matrix \mathbf{A} in Section 8.8, and there we found its eigenvalues to be $\lambda_1 = -1$ and $\lambda_2 = 2$. Now since \mathbf{A} is a 2×2 matrix, we have from (12) and (13)

$$e^{\mathbf{A}t} = b_0\mathbf{I} + b_1\mathbf{A} \quad \text{and} \quad e^{\lambda t} = b_0 + b_1\lambda. \qquad (14)$$

Setting $\lambda = -1$ and $\lambda = 2$ in the second equation of (14) gives two equations in the two unknowns b_0 and b_1. Solving the system

$$e^{-t} = b_0 - b_1$$

$$e^{2t} = b_0 + 2b_1$$

yields $b_0 = \frac{1}{3}[e^{2t} + 2e^{-t}]$, $b_1 = \frac{1}{3}[e^{2t} - e^{-t}]$. Substituting these values in the first equation of (14), simplifying the entries, yields

$$e^{\mathbf{A}t} = \begin{pmatrix} -\frac{1}{3}e^{2t} + \frac{4}{3}e^{-t} & \frac{4}{3}e^{2t} - \frac{4}{3}e^{-t} \\ -\frac{1}{3}e^{2t} + \frac{1}{3}e^{-t} & \frac{4}{3}e^{2t} - \frac{1}{3}e^{-t} \end{pmatrix}. \qquad (15)\square$$

In Problems 25–28 in Exercises 10.5 we show how to compute the matrix exponential $e^{\mathbf{A}t}$ when the matrix \mathbf{A} is diagonalizable (see Section 8.11).

■ **Use of Computers** For those willing to momentarily trade understanding for speed of solution, $e^{\mathbf{A}t}$ can be computed in a mechanistic manner with the aid of computer software; for example, in *Mathematica* the function **MatrixExp[A t]** computes the matrix exponential for a square matrix **At**; in *Maple* the command is **exponential(A, t)**; in MATLAB the function is `expm(At)`. See Problems 27 and 28 in Exercises 10.5.

EXERCISES 10.5 Answers to odd-numbered problems begin on page A-54.

In Problems 1 and 2, use (3) to compute $e^{\mathbf{A}t}$ and $e^{-\mathbf{A}t}$.

1. $\mathbf{A} = \begin{pmatrix} 1 & 0 \\ 0 & 2 \end{pmatrix}$ **2.** $\mathbf{A} = \begin{pmatrix} 0 & 1 \\ 1 & 0 \end{pmatrix}$

In Problems 3 and 4, use (3) to compute $e^{\mathbf{A}t}$.

3. $\mathbf{A} = \begin{pmatrix} 1 & 1 & 1 \\ 1 & 1 & 1 \\ -2 & -2 & -2 \end{pmatrix}$ **4.** $\mathbf{A} = \begin{pmatrix} 0 & 0 & 0 \\ 3 & 0 & 0 \\ 5 & 1 & 0 \end{pmatrix}$

In Problems 5–8, use (1) and the results in Problems 1–4 to find the general solution of the given system.

5. $\mathbf{X}' = \begin{pmatrix} 1 & 0 \\ 0 & 2 \end{pmatrix}\mathbf{X}$ **6.** $\mathbf{X}' = \begin{pmatrix} 0 & 1 \\ 1 & 0 \end{pmatrix}\mathbf{X}$

7. $\mathbf{X}' = \begin{pmatrix} 1 & 1 & 1 \\ 1 & 1 & 1 \\ -2 & -2 & -2 \end{pmatrix}\mathbf{X}$ **8.** $\mathbf{X}' = \begin{pmatrix} 0 & 0 & 0 \\ 3 & 0 & 0 \\ 5 & 1 & 0 \end{pmatrix}\mathbf{X}$

In Problems 9–12, use (6) to find the general solution of the given system.

9. $\mathbf{X}' = \begin{pmatrix} 1 & 0 \\ 0 & 2 \end{pmatrix}\mathbf{X} + \begin{pmatrix} 3 \\ -1 \end{pmatrix}$ **10.** $\mathbf{X}' = \begin{pmatrix} 1 & 0 \\ 0 & 2 \end{pmatrix}\mathbf{X} + \begin{pmatrix} t \\ e^{4t} \end{pmatrix}$

11. $\mathbf{X}' = \begin{pmatrix} 0 & 1 \\ 1 & 0 \end{pmatrix}\mathbf{X} + \begin{pmatrix} 1 \\ 1 \end{pmatrix}$ **12.** $\mathbf{X}' = \begin{pmatrix} 0 & 1 \\ 1 & 0 \end{pmatrix}\mathbf{X} + \begin{pmatrix} \cosh t \\ \sinh t \end{pmatrix}$

13. Solve the system in Problem 7 subject to the initial condition $\mathbf{X}(0) = \begin{pmatrix} 1 \\ -4 \\ 6 \end{pmatrix}$.

14. Solve the system in Problem 9 subject to the initial condition $\mathbf{X}(0) = \begin{pmatrix} 4 \\ 3 \end{pmatrix}$.

In Problems 15–18, use the method of Example 1 to compute $e^{\mathbf{A}t}$ for the coefficient matrix. Use (1) to find the general solution of the given system.

15. $\mathbf{X}' = \begin{pmatrix} 4 & 3 \\ -4 & -4 \end{pmatrix}\mathbf{X}$ **16.** $\mathbf{X}' = \begin{pmatrix} 4 & -2 \\ 1 & 1 \end{pmatrix}\mathbf{X}$

17. $\mathbf{X}' = \begin{pmatrix} 5 & -9 \\ 1 & -1 \end{pmatrix}\mathbf{X}$ **18.** $\mathbf{X}' = \begin{pmatrix} 0 & 1 \\ -2 & -2 \end{pmatrix}\mathbf{X}$

In Problems 19–22, use the method of Example 2 to compute e^{At} for the coefficient matrix. Use (1) to find the general solution of the given system.

19. $\mathbf{X}' = \begin{pmatrix} 2 & -2 \\ -2 & 5 \end{pmatrix}\mathbf{X}$ **20.** $\mathbf{X}' = \begin{pmatrix} 1 & 2 \\ -1 & 4 \end{pmatrix}\mathbf{X}$

21. $\mathbf{X}' = \begin{pmatrix} 3 & 8 \\ 0 & -1 \end{pmatrix}\mathbf{X}$ **22.** $\mathbf{X}' = \begin{pmatrix} 1 & -\frac{3}{2} \\ \frac{1}{4} & -\frac{1}{4} \end{pmatrix}\mathbf{X}$

23. If the matrix \mathbf{A} can be diagonalized, then $\mathbf{P}^{-1}\mathbf{A}\mathbf{P} = \mathbf{D}$ or $\mathbf{A} = \mathbf{PDP}^{-1}$. Use this last result and (3) to show that $e^{At} = \mathbf{P}e^{\mathbf{D}t}\mathbf{P}^{-1}$.

24. Use $\mathbf{D} = \begin{pmatrix} \lambda_1 & 0 & \cdots & 0 \\ 0 & \lambda_2 & \cdots & 0 \\ \vdots & & & \vdots \\ 0 & 0 & \cdots & \lambda_n \end{pmatrix}$ and (3) to show that

$$e^{\mathbf{D}t} = \begin{pmatrix} e^{\lambda_1 t} & 0 & \cdots & 0 \\ 0 & e^{\lambda_2 t} & \cdots & 0 \\ \vdots & & & \vdots \\ 0 & 0 & \cdots & e^{\lambda_n t} \end{pmatrix}.$$

In Problems 25 and 26, use the results of Problems 23 and 24 to solve the given system.

25. $\mathbf{X}' = \begin{pmatrix} 2 & 1 \\ -3 & 6 \end{pmatrix}\mathbf{X}$ **26.** $\mathbf{X}' = \begin{pmatrix} 2 & 1 \\ 1 & 2 \end{pmatrix}\mathbf{X}$

27. Computer Lab Assignment
 (a) Use (1) to find the general solution of $\mathbf{X}' = \begin{pmatrix} 4 & 2 \\ 3 & 3 \end{pmatrix}\mathbf{X}$.
Use a CAS to find e^{At}. Then use the computer to find eigenvalues and eigenvectors of the coefficient matrix $\mathbf{A} = \begin{pmatrix} 4 & 2 \\ 3 & 3 \end{pmatrix}$ and form the general solution in the manner of Section 10.2. Finally, reconcile the two forms of the general solution of the system.

 (b) Use (1) to find the general solution of $\mathbf{X}' = \begin{pmatrix} -3 & -1 \\ 2 & -1 \end{pmatrix}\mathbf{X}$.
Use a CAS to find e^{At}. In the case of complex output, utilize the software to do the simplification; for example, in *Mathematica,* if $\mathbf{m} = \mathbf{MatrixExp[A\ t]}$ has complex entries, then try the command **Simplify[ComplexExpand[m]]**.

28. Computer Lab Assignment Use (1) to find the general solution of

$$\mathbf{X}' = \begin{pmatrix} -4 & 0 & 6 & 0 \\ 0 & -5 & 0 & -4 \\ -1 & 0 & 1 & 0 \\ 0 & 3 & 0 & 2 \end{pmatrix}\mathbf{X}.$$

Use a CAS to find e^{At}.

CHAPTER 10 REVIEW EXERCISES

Answers to odd-numbered problems begin on page A-55.

In Problems 1 and 2, fill in the blanks.

1. The vector $\mathbf{X} = k\begin{pmatrix} 4 \\ 5 \end{pmatrix}$ is a solution of $\mathbf{X}' = \begin{pmatrix} 1 & 4 \\ 2 & -1 \end{pmatrix}\mathbf{X} - \begin{pmatrix} 8 \\ 1 \end{pmatrix}$ for $k =$ ____ .

2. The vector $\mathbf{X} = c_1\begin{pmatrix} -1 \\ 1 \end{pmatrix}e^{-9t} + c_2\begin{pmatrix} 5 \\ 3 \end{pmatrix}e^{7t}$ is solution of the initial-value problem $\mathbf{X}' = \begin{pmatrix} 1 & 10 \\ 6 & -3 \end{pmatrix}\mathbf{X}$, $\mathbf{X}(0) = \begin{pmatrix} 2 \\ 0 \end{pmatrix}$ for $c_1 =$ ____ and $c_2 =$ ____ .

3. Consider the linear system $\mathbf{X}' = \begin{pmatrix} 4 & 6 & 6 \\ 1 & 3 & 2 \\ -1 & -4 & -3 \end{pmatrix}\mathbf{X}$. Without attempting to solve the system, which one of the following vectors,

$$\mathbf{K}_1 = \begin{pmatrix} 0 \\ 1 \\ 1 \end{pmatrix}, \quad \mathbf{K}_2 = \begin{pmatrix} 1 \\ 1 \\ -1 \end{pmatrix}, \quad \mathbf{K}_3 = \begin{pmatrix} 3 \\ 1 \\ -1 \end{pmatrix}, \quad \mathbf{K}_4 = \begin{pmatrix} 6 \\ 2 \\ -5 \end{pmatrix},$$

is an eigenvector of the coefficient matrix? What is the solution of the system corresponding to this eigenvector?

4. Consider the linear system $\mathbf{X}' = \mathbf{AX}$ of two differential equations where \mathbf{A} is a real coefficient matrix. What is the general solution of the system if it is known that $\lambda_1 = 1 + 2i$ is an eigenvalue and $\mathbf{K}_1 = \begin{pmatrix} 1 \\ i \end{pmatrix}$ is a corresponding eigenvector?

In Problems 5–14, solve the given linear system by the methods of this chapter.

5. $\dfrac{dx}{dt} = 2x + y$
$\dfrac{dy}{dt} = -x$

6. $\dfrac{dx}{dt} = -4x + 2y$
$\dfrac{dy}{dt} = 2x - 4y$

7. $\mathbf{X}' = \begin{pmatrix} 1 & 2 \\ -2 & 1 \end{pmatrix}\mathbf{X}$ **8.** $\mathbf{X}' = \begin{pmatrix} -2 & 5 \\ -2 & 4 \end{pmatrix}\mathbf{X}$

9. $\mathbf{X}' = \begin{pmatrix} 1 & -1 & 1 \\ 0 & 1 & 3 \\ 4 & 3 & 1 \end{pmatrix}\mathbf{X}$ **10.** $\mathbf{X}' = \begin{pmatrix} 0 & 2 & 1 \\ 1 & 1 & -2 \\ 2 & 2 & -1 \end{pmatrix}\mathbf{X}$

11. $\mathbf{X}' = \begin{pmatrix} 2 & 8 \\ 0 & 4 \end{pmatrix}\mathbf{X} + \begin{pmatrix} 2 \\ 16t \end{pmatrix}$ **12.** $\mathbf{X}' = \begin{pmatrix} 1 & 2 \\ -\frac{1}{2} & 1 \end{pmatrix}\mathbf{X} + \begin{pmatrix} 0 \\ e^t \tan t \end{pmatrix}$

13. $\mathbf{X}' = \begin{pmatrix} -1 & 1 \\ -2 & 1 \end{pmatrix} \mathbf{X} + \begin{pmatrix} 1 \\ \cot t \end{pmatrix}$ **14.** $\mathbf{X}' = \begin{pmatrix} 3 & 1 \\ -1 & 1 \end{pmatrix} \mathbf{X} + \begin{pmatrix} -2 \\ 1 \end{pmatrix} e^{2t}$

15. (a) Consider the linear system $\mathbf{X}' = \mathbf{AX}$ of three first-order differential equations where the coefficient matrix is

$$\mathbf{A} = \begin{pmatrix} 5 & 3 & 3 \\ 3 & 5 & 3 \\ -5 & -5 & -3 \end{pmatrix},$$

and $\lambda = 2$ is known to be an eigenvalue of multiplicity two. Find two different solutions of the system corresponding to

this eigenvalue without using any special formula (such as (12) of Section 10.2).

(b) Use the procedure in part **(a)** to solve $\mathbf{X}' = \begin{pmatrix} 1 & 1 & 1 \\ 1 & 1 & 1 \\ 1 & 1 & 1 \end{pmatrix} \mathbf{X}$.

16. Verify that $\mathbf{X} = \begin{pmatrix} c_1 \\ c_2 \end{pmatrix} e^t$ is a solution of the linear system $\mathbf{X}' = \begin{pmatrix} 1 & 0 \\ 0 & 1 \end{pmatrix} \mathbf{X}$ for arbitrary constants c_1 and c_2. By hand, draw a phase portrait of the system.

11

Systems of Nonlinear Differential Equations

INTRODUCTION

In Chapter 10 we concentrated on techniques for solving systems of linear DEs of the form $\mathbf{X}' = \mathbf{AX} + \mathbf{F}(t)$. When the system of differential equations is not linear, it is usually not possible to find solutions in terms of elementary functions. In this chapter we will see that valuable information about the geometric nature of solutions can be obtained by first analyzing special constant solutions called **critical points** and then searching for periodic solutions called **limit cycles.** The important concept of **stability** will be introduced and illustrated with examples from physics and biology.

11.1 Autonomous Systems, Critical Points, and Periodic Solutions

■ **Terminology and Notation** A system of first-order differential equations is called **autonomous** when the system can be written in the form

$$\frac{dx_1}{dt} = g_1(x_1, x_2, \ldots, x_n)$$

$$\frac{dx_2}{dt} = g_2(x_1, x_2, \ldots, x_n) \tag{1}$$

$$\vdots$$

$$\frac{dx_n}{dt} = g_n(x_1, x_2, \ldots, x_n).$$

Thus the variable t does *not* appear explicitly on the right-hand side of each differential equation.

> ### Example 1 A Nonautonomous System
>
> The system of differential equations
>
> $$\frac{dx_1}{dt} = x_1 - 3x_2 + t^2$$
>
> $$\frac{dx_2}{dt} = x_1 \sin(x_2 t)$$
>
> is *not* autonomous because of the presence of t^2 and $\sin(x_2 t)$ on the right-hand sides. ❏

Note that when $n = 1$ in (1), an autonomous differential equation takes the simple form $dx/dt = g(x)$. Explicit solutions can be constructed since this differential equation is separable, and we will make use of this fact to give illustrations of the concepts in this chapter.

Any nonlinear second-order differential equation of the form $x'' = g(x, x')$ can be written as an autonomous system. With the introduction of $y = x'$ and $y' = x'' = g(x, x')$, then the differential equation becomes

$$x' = y$$

$$y' = g(x, y).$$

> ### Example 2 The Pendulum DE as an Autonomous System
>
> In (6) of Section 3.10 we showed that the displacement angle θ for a pendulum satisfies the nonlinear second-order differential equation
>
> $$\frac{d^2\theta}{dt^2} + \frac{g}{l} \sin \theta = 0.$$
>
> If we let $x = \theta$ and $y = \theta'$, this second-order differential equation may be rewritten as the autonomous system
>
> $$x' = y$$
>
> $$y' = -\frac{g}{l} \sin x.$$ ❏

If $\mathbf{X}(t)$ and $\mathbf{g}(\mathbf{X})$ denote the respective column vectors

$$\mathbf{X}(t) = \begin{pmatrix} x_1(t) \\ x_2(t) \\ \vdots \\ x_n(t) \end{pmatrix}, \qquad \mathbf{g}(\mathbf{X}) = \begin{pmatrix} g_1(x_1, x_2, \ldots, x_n) \\ g_2(x_1, x_2, \ldots, x_n) \\ \vdots \\ g_n(x_1, x_2, \ldots, x_n) \end{pmatrix},$$

then the autonomous system (1) may be written in the compact **column vector form** $\mathbf{X}' = \mathbf{g}(\mathbf{X})$. The homogeneous linear system $\mathbf{X}' = \mathbf{A}\mathbf{X}$ studied in Section 10.2 is an important special case.

In this chapter it is also convenient to write (1) using row vectors. If we let

$$\mathbf{X}(t) = (x_1(t), x_2(t), \ldots, x_n(t))$$

and

$$\mathbf{g}(\mathbf{X}) = (g_1(x_1, x_2, \ldots, x_n), g_2(x_1, x_2, \ldots, x_n), \ldots, g_n(x_1, x_2, \ldots, x_n)),$$

then the autonomous system (1) may also be written in the compact **row vector form** $\mathbf{X}' = \mathbf{g}(\mathbf{X})$. *It should be clear from the context whether we are using column or row vector form, and therefore we will not distinguish between \mathbf{X} and \mathbf{X}^T, the transpose of \mathbf{X}.* In particular, when $n = 2$ it is convenient to use row vector form and write an initial condition as $\mathbf{X}(0) = (x_0, y_0)$.

When the variable t is interpreted as time, we can refer to the system of differential equations in (1) as a **dynamical system** and a solution $\mathbf{X}(t)$ as the **state of the system** or the **response of the system** at time t. With this terminology, a dynamical system is autonomous when the rate $\mathbf{X}'(t)$ at which the system changes depends only on the system's present state $\mathbf{X}(t)$. The linear system $\mathbf{X}' = \mathbf{A}\mathbf{X} + \mathbf{F}(t)$ studied in Chapter 10 is then autonomous when $\mathbf{F}(t)$ is constant. In the case $n = 2$ or 3 we call a solution a **path** or **trajectory** since we may think of $x = x_1(t)$, $y = x_2(t)$, $z = x_3(t)$ as the parametric equations of a curve.

■ **Vector Field Interpretation** When $n = 2$ the system in (1) is called a **plane autonomous system,** and we write the system as

$$\frac{dx}{dt} = P(x, y)$$

$$\frac{dy}{dt} = Q(x, y).$$

The vector $\mathbf{V}(x, y) = (P(x, y), Q(x, y))$ defines a **vector field** in a region of the plane, and a solution to the system may be interpreted as the resulting path of a particle as it moves through the region. To be more specific, let $\mathbf{V}(x, y) = (P(x, y), Q(x, y))$ denote the velocity of a stream at position (x, y), and suppose that a small particle (such as a cork) is released at a position (x_0, y_0) in the stream. If $\mathbf{X}(t) = (x(t), y(t))$ denotes the position of the particle at time t, then $\mathbf{X}'(t) = (x'(t), y'(t))$ is the velocity vector \mathbf{v}. When external forces are not present and frictional forces are neglected, the velocity of the particle at time t is the velocity of the stream at position $\mathbf{X}(t)$; that is, $\mathbf{X}'(t) = \mathbf{V}(x(t), y(t))$,

or

$$\frac{dx}{dt} = P(x(t), y(t))$$

$$\frac{dy}{dt} = Q(x(t), y(t)).$$

Thus the path of the particle is a solution to the system that satisfies the initial condition $\mathbf{X}(0) = (x_0, y_0)$. We will frequently call on this simple interpretation of a plane autonomous system to illustrate new concepts.

Example 3 Plane Autonomous System of a Vector Field

A vector field for the steady-state flow of a fluid around a cylinder of radius 1 is given by

$$\mathbf{V}(x, y) = V_0\left(1 - \frac{x^2 - y^2}{(x^2 + y^2)^2}, \frac{-2xy}{(x^2 + y^2)^2}\right),$$

where V_0 is the speed of the fluid far from the cylinder. If a small cork is released at $(-3, 1)$, the path $\mathbf{X}(t) = (x(t), y(t))$ of the cork satisfies the plane autonomous system

$$\frac{dx}{dt} = V_0\left(1 - \frac{x^2 - y^2}{(x^2 + y^2)^2}\right)$$

$$\frac{dy}{dt} = V_0\left(\frac{-2xy}{(x^2 + y^2)^2}\right)$$

subject to the initial condition $\mathbf{X}(0) = (-3, 1)$. See Figure 11.1.

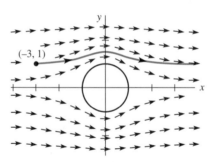

Figure 11.1 ❏

■ **Types of Solutions** If $P(x, y)$, $Q(x, y)$, and the first-order partial derivatives $\partial P/\partial x$, $\partial P/\partial y$, $\partial Q/\partial x$, and $\partial Q/\partial y$ are continuous in a region R of the plane, then a solution to the plane autonomous system

$$\frac{dx}{dt} = P(x, y)$$

$$\frac{dy}{dt} = Q(x, y)$$

that satisfies $\mathbf{X}(0) = \mathbf{X}_0$ is unique and of one of three basic types:

(*i*) A **constant solution** $x(t) = x_0$, $y(t) = y_0$ (or $\mathbf{X}(t) = \mathbf{X}_0$ for all t). A constant solution is called a **critical** or **stationary point**. When the particle is placed at a critical point \mathbf{X}_0 (that is, $\mathbf{X}(0) = \mathbf{X}_0$), it remains there indefinitely. For this reason a constant solution is also called an **equilibrium solution**. Note that since $\mathbf{X}'(t) = \mathbf{0}$, a critical point is a solution of the system of algebraic equations

$$P(x, y) = 0$$

$$Q(x, y) = 0.$$

(a)

(b)

Figure 11.2

Figure 11.3

(*ii*) A solution $x = x(t)$, $y = y(t)$ that defines an **arc**—a plane curve that does *not* cross itself. Thus the curve in Figure 11.2(a) can be a solution to a plane autonomous system, whereas the curve in Figure 11.2(b) cannot be a solution. There would be *two solutions* that start from the point **P** of intersection.

(*iii*) A **periodic solution** $x = x(t)$, $y = y(t)$. A periodic solution is called a **cycle.** If p is the period of the solution, then $\mathbf{X}(t + p) = \mathbf{X}(t)$ and a particle placed on the curve at \mathbf{X}_0 will cycle around the curve and return to \mathbf{X}_0 in p units of time. See Figure 11.3.

Example 4 Finding Critical Points

Find all critical points of each of the following plane autonomous systems.

(a) $x' = -x + y$	(b) $x' = x^2 + y^2 - 6$	(c) $x' = 0.01x(100 - x - y)$
$y' = x - y$	$y' = x^2 - y$	$y' = 0.05y(60 - y - 0.2x)$

SOLUTION We find the critical points by setting the right-hand sides of the differential equations equal to zero.

(a) The solution to the system

$$-x + y = 0$$
$$x - y = 0$$

consists of all points on the line $y = x$. Thus there are infinitely many critical points.

(b) To solve the system

$$x^2 + y^2 - 6 = 0$$
$$x^2 - y = 0$$

we substitute the second equation, $x^2 = y$, into the first equation to obtain $y^2 + y - 6 = (y + 3)(y - 2) = 0$. If $y = -3$, then $x^2 = -3$, and so there are no real solutions. If $y = 2$, then $x = \pm\sqrt{2}$, and so the critical points are $(\sqrt{2}, 2)$ and $(-\sqrt{2}, 2)$.

(c) Finding the critical points requires a careful consideration of cases. The equation $0.01x(100 - x - y) = 0$ implies $x = 0$ or $x + y = 100$.
 If $x = 0$, then substituting in $0.05y(60 - y - 0.2x) = 0$ we have $y(60 - y) = 0$. Thus $y = 0$ or 60, and so $(0, 0)$ and $(0, 60)$ are critical points.
 If $x + y = 100$, then $0 = y(60 - y - 0.2(100 - y)) = y(40 - 0.8y)$. It follows that $y = 0$ or 50, and so $(100, 0)$ and $(50, 50)$ are critical points. ❏

When the plane autonomous system is linear, we can use the methods in Chapter 10 to investigate solutions.

Example 5 Discovering Periodic Solutions

Determine whether the given linear dynamical system possesses a periodic solution.

(a) $x' = 2x + 8y$	(b) $x' = x + 2y$
$y' = -x - 2y$	$y' = -\frac{1}{2}x + y$

In each case sketch the graph of the solution that satisfies $\mathbf{X}(0) = (2, 0)$.

SOLUTION

(a) In Example 6 of Section 10.2 we used the eigenvalue-eigenvector method to show that

$$x = c_1(2 \cos 2t - 2 \sin 2t) + c_2(2 \cos 2t + 2 \sin 2t)$$

$$y = c_1(-\cos 2t) - c_2 \sin 2t.$$

(a)

(b)

Figure 11.4

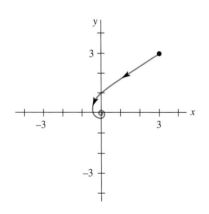

Figure 11.5

Thus every solution is periodic with period $p = \pi$. The solution satisfying $\mathbf{X}(0) = (2, 0)$ is

$$x = 2 \cos 2t + 2 \sin 2t, \qquad y = -\sin 2t.$$

This solution generates the ellipse shown in Figure 11.4(a).

(b) Using the eigenvalue-eigenvector method, we can show that

$$x = c_1(2e^t \cos t) + c_2(2e^t \sin t), \qquad y = c_1(-e^t \sin t) + c_2(e^t \cos t).$$

Because of the presence of e^t in the general solution, there are no periodic solutions (that is, cycles). The solution satisfying $\mathbf{X}(0) = (2, 0)$ is

$$x = 2e^t \cos t, \qquad y = -e^t \sin t,$$

and the resulting curve is shown in Figure 11.4(b). ❑

■ **Changing to Polar Coordinates** Except for the case of constant solutions, it is usually not possible to find explicit expressions for the solutions of a *nonlinear* autonomous system. We can solve some nonlinear systems, however, by changing to polar coordinates. From the formulas $r^2 = x^2 + y^2$ and $\theta = \tan^{-1}(y/x)$ we obtain

$$\frac{dr}{dt} = \frac{1}{r}\left(x\frac{dx}{dt} + y\frac{dy}{dt}\right), \qquad \frac{d\theta}{dt} = \frac{1}{r^2}\left(-y\frac{dx}{dt} + x\frac{dy}{dt}\right). \qquad (2)$$

We can sometimes use (2) to convert a plane autonomous system in rectangular coordinates to a simpler system in polar coordinates.

Example 6 Changing to Polar Coordinates

Find the solution of the nonlinear plane autonomous system

$$x' = -y - x\sqrt{x^2 + y^2}$$
$$y' = x - y\sqrt{x^2 + y^2}$$

satisfying the initial condition $\mathbf{X}(0) = (3, 3)$.

SOLUTION Substituting for dx/dt and dy/dt in the expressions for dr/dt and $d\theta/dt$ in (2), we obtain

$$\frac{dr}{dt} = \frac{1}{r}\,[x(-y - xr) + y(x - yr)] = -r^2$$

$$\frac{d\theta}{dt} = \frac{1}{r^2}\,[(-y(-y - xr) + x(x - yr)] = 1.$$

Since $(3, 3)$ is $(3\sqrt{2}, \pi/4)$ in polar coordinates, the initial condition $\mathbf{X}(0) = (3, 3)$ becomes $r(0) = 3\sqrt{2}$ and $\theta(0) = \pi/4$. Using separation of variables, we see that the solution of the system is

$$r = \frac{1}{t + c_1}, \qquad \theta = t + c_2,$$

for $r \neq 0$. (Check this!) Applying the initial conditions then gives

$$r = \frac{1}{t + \sqrt{2}/6}, \qquad \theta = t + \frac{\pi}{4}.$$

The spiral $r = \dfrac{1}{\theta + \sqrt{2}/6 - \pi/4}$ is sketched in Figure 11.5. ❑

Example 7 Solutions in Polar Coordinates

When expressed in polar coordinates, a plane autonomous system takes the form

$$\frac{dr}{dt} = 0.5(3 - r)$$

$$\frac{d\theta}{dt} = 1.$$

Find and sketch the solutions satisfying $\mathbf{X}(0) = (0, 1)$ and $\mathbf{X}(0) = (3, 0)$ in rectangular coordinates.

SOLUTION Applying separation of variables to $dr/dt = 0.5(3 - r)$ and integrating $d\theta/dt$ leads to the solution

$$r = 3 + c_1 e^{-0.5t}, \qquad \theta = t + c_2.$$

If $\mathbf{X}(0) = (0, 1)$, then $r(0) = 1$ and $\theta(0) = \pi/2$, and so $c_1 = -2$ and $c_2 = \pi/2$. The solution curve is the spiral $r = 3 - 2e^{-0.5(\theta - \pi/2)}$. Note that as $t \to \infty$, θ increases without bound and r approaches 3.

If $\mathbf{X}(0) = (3, 0)$, then $r(0) = 3$ and $\theta(0) = 0$. It follows that $c_1 = c_2 = 0$, and so $r = 3$ and $\theta = t$. Hence $x = r \cos \theta = 3 \cos t$ and $y = r \sin \theta = 3 \sin t$, and so the solution is periodic. The solution generates a circle of radius 3 about $(0, 0)$. Both solutions are shown in Figure 11.6. ❑

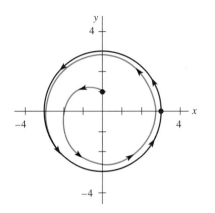

Figure 11.6

E X E R C I S E S 1 1 . 1

Answers to odd-numbered problems begin on page A-55.

In Problems 1–6, write the given nonlinear second-order differential equation as a plane autonomous system. Find all critical points of the resulting system.

1. $x'' + 9 \sin x = 0$

2. $x'' + (x')^2 + 2x = 0$

3. $x'' + x'(1 - x^3) - x^2 = 0$

4. $x'' + 4\dfrac{x}{1 + x^2} + 2x' = 0$

5. $x'' + x = \varepsilon x^3$ for $\varepsilon > 0$

6. $x'' + x - \varepsilon x|x| = 0$ for $\varepsilon > 0$.

In Problems 7–16, find all critical points of the given plane autonomous system.

7. $x' = x + xy$
$y' = -y - xy$

8. $x' = y^2 - x$
$y' = x^2 - y$

9. $x' = 3x^2 - 4y$
$y' = x - y$

10. $x' = x^3 - y$
$y' = x - y^3$

11. $x' = x(10 - x - \tfrac{1}{2}y)$
$y' = y(16 - y - x)$

12. $x' = -2x + y + 10$
$y' = 2x - y - 15\dfrac{y}{y + 5}$

13. $x' = x^2 e^y$
$y' = y(e^x - 1)$

14. $x' = \sin y$
$y' = e^{x - y} - 1$

15. $x' = x(1 - x^2 - 3y^2)$
$y' = y(3 - x^2 - 3y^2)$

16. $x' = -x(4 - y^2)$
$y' = 4y(1 - x^2)$

In Problems 17–22, for the given linear dynamical system (taken from Exercises 10.2)

(a) find the general solution and determine whether there are periodic solutions,

(b) find the solution satisfying the given initial condition, and

(c) with the aid of a graphing utility, plot the solution in part **(b)** and indicate the direction in which the curve is traversed.

17. $x' = x + 2y$
$y' = 4x + 3y$, $\mathbf{X}(0) = (2, -2)$ (Problem 1, Exercises 10.2)

18. $x' = -6x + 2y$
$y' = -3x + y$, $\mathbf{X}(0) = (3, 4)$ (Problem 6, Exercises 10.2)

19. $x' = 4x - 5y$
$y' = 5x - 4y$, $\mathbf{X}(0) = (4, 5)$ (Problem 37, Exercises 10.2)

20. $x' = x + y$
$y' = -2x - y$, $\mathbf{X}(0) = (-2, 2)$ (Problem 34, Exercises 10.2)

21. $x' = 5x + y$
$y' = -2x + 3y$, $\mathbf{X}(0) = (-1, 2)$ (Problem 35, Exercises 10.2)

22. $x' = x - 8y$
$y' = x - 3y$, $\mathbf{X}(0) = (2, 1)$ (Problem 38, Exercises 10.2)

In Problems 23–26, solve the given nonlinear plane autonomous system by changing to polar coordinates. Describe the geometric behavior of the solution that satisfies the given initial condition(s).

23. $x' = -y - x(x^2 + y^2)^2$
$y' = x - y(x^2 + y^2)^2$, $\mathbf{X}(0) = (4, 0)$

24. $x' = y + x(x^2 + y^2)$
$y' = -x + y(x^2 + y^2)$, $\quad \mathbf{X}(0) = (4, 0)$

25. $x' = -y + x(1 - x^2 - y^2)$
$y' = x + y(1 - x^2 - y^2)$, $\quad \mathbf{X}(0) = (1, 0)$, $\quad \mathbf{X}(0) = (2, 0)$

[*Hint:* The resulting differential equation for r is a Bernoulli differential equation. See Section 2.5.]

26. $x' = y - \dfrac{x}{\sqrt{x^2 + y^2}}(4 - x^2 - y^2)$

$y' = -x - \dfrac{y}{\sqrt{x^2 + y^2}}(4 - x^2 - y^2)$, $\mathbf{X}(0) = (1, 0)$, $\mathbf{X}(0) = (2, 0)$

[*Hint:* See Example 3 in Section 2.2.]

If a plane autonomous system has a periodic solution, then there must be at least one critical point inside the curve generated by the solution. In Problems 27–30, use this fact together with a numerical solver to investigate the possibility of periodic solutions.

27. $x' = -x + 6y$
$y' = xy + 12$

28. $x' = -x + 6xy$
$y' = -8xy + 2y$

29. $x' = y$
$y' = y(1 - 3x^2 - 2y^2) - x$

30. $x' = xy$
$y' = -1 - x^2 - y^2$

31. If $z = f(x, y)$ is a function with continuous first partial derivatives in a region R, then a flow $\mathbf{V}(x, y) = (P(x, y), Q(x, y))$ in R may be defined by letting $P(x, y) = -\dfrac{\partial f}{\partial y}(x, y)$ and $Q(x, y) = \dfrac{\partial f}{\partial x}(x, y)$. Show that if $\mathbf{X}(t) = (x(t), y(t))$ is a solution of the plane autonomous system

$$x' = P(x, y)$$
$$y' = Q(x, y),$$

then $f(x(t), y(t)) = c$ for some constant c. Thus a solution curve lies on the level curves of f. [*Hint:* Use the Chain Rule to compute $\dfrac{d}{dt} f(x(t), y(t))$.]

11.2 Stability of Linear Systems

In Section 11.1 we noted that the plane autonomous system

$$\frac{dx}{dt} = P(x, y)$$

$$\frac{dy}{dt} = Q(x, y)$$

gives rise to a vector field $\mathbf{V}(x, y) = (P(x, y), Q(x, y))$, and a solution $\mathbf{X} = \mathbf{X}(t)$ may be interpreted as the resulting path of a particle that is initially placed at position $\mathbf{X}(0) = \mathbf{X}_0$. If \mathbf{X}_0 is a critical point, the particle remains stationary. If \mathbf{X}_0 is placed near a critical point \mathbf{X}_1, however, we ask the following questions:

(*i*) Will the particle return to the critical point? More precisely, if $\mathbf{X} = \mathbf{X}(t)$ is the solution that satisfies $\mathbf{X}(0) = \mathbf{X}_0$, is $\lim_{t \to \infty} \mathbf{X}(t) = \mathbf{X}_1$?

(*ii*) If the particle does *not* return to the critical point, does it remain close to the critical point or move away from the critical point? It is conceivable, for example, that the particle may simply circle the critical point, or it may even return to a different critical point or to no critical point at all. See Figure 11.7.

If in some neighborhood of the critical point case (a) or (b) in Figure 11.7 *always* occurs, we call the critical point **locally stable.** If, however, an initial value \mathbf{X}_0 that results in behavior similar to (c) can be found in *any* given neighborhood, we call the critical point **unstable.** These concepts will be made more precise in Section 11.3, where questions (*i*) and (*ii*) will be investigated for nonlinear systems.

■ **Stability Analysis** We will first investigate these two stability questions for *linear* plane autonomous systems and lay the foundation for Section 11.3. The solution methods of Chapter 10 enable us to give a careful geometric analysis of the solutions to

\mathbf{X}_0

critical point

(a)

\mathbf{X}_0

critical point

(b)

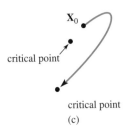

\mathbf{X}_0

critical point

critical point

(c)

Figure 11.7

$$x' = ax + by$$
$$y' = cx + dy \tag{1}$$

in terms of the eigenvalues and eigenvectors of the coefficient matrix

$$\mathbf{A} = \begin{pmatrix} a & b \\ c & d \end{pmatrix}.$$

Here a, b, c, and d are constants. To ensure that $\mathbf{X}_0 = (0, 0)$ is the only critical point, we will assume that the determinant $\Delta = ad - bc \neq 0$. If $\tau = a + d$ is the trace* of matrix \mathbf{A}, then the characteristic equation $\det (\mathbf{A} - \lambda\mathbf{I}) = 0$ may be rewritten as

$$\lambda^2 - \tau\lambda + \Delta = 0.$$

Therefore the eigenvalues of \mathbf{A} are $\lambda = (\tau \pm \sqrt{\tau^2 - 4\Delta})/2$, and the usual three cases for these roots occur according to whether $\tau^2 - 4\Delta$ is positive, negative, or zero. In the next example we use a numerical solver to discover the nature of the solutions corresponding to these cases.

Example 1 Eigenvalues and the Shape of Solutions

Find the eigenvalues of the linear system

$$x' = -x + y$$
$$y' = cx - y$$

in terms of c, and use a numerical solver to discover the shapes of solutions corresponding to the cases $c = \frac{1}{4}$, 4, 0, and −9.

SOLUTION The coefficient matrix $\begin{pmatrix} -1 & 1 \\ c & -1 \end{pmatrix}$ has trace $\tau = -2$ and determinant $\Delta = 1 - c$, and so the eigenvalues are

$$\lambda = \frac{\tau \pm \sqrt{\tau^2 - 4\Delta}}{2} = \frac{-2 \pm \sqrt{4 - 4(1 - c)}}{2} = -1 \pm \sqrt{c}.$$

The nature of the eigenvalues is therefore determined by the sign of c.

If $c = \frac{1}{4}$, then the eigenvalues are negative and distinct, $\lambda = -\frac{1}{2}$ and $-\frac{3}{2}$. In Figure 11.8(a) we have used a numerical solver to generate solution curves or trajectories that correspond to various initial conditions. Note that, except for the trajectories drawn in black in the figure, the trajectories all appear to approach **0** from a fixed direction. Recall from Chapter 10 that a collection of trajectories in the xy-plane or **phase plane** is called a **phase portrait** of the system.

When $c = 4$ the eigenvalues have opposite signs, $\lambda = 1$ and -3, and an interesting phenomenon occurs. All trajectories move away from the origin in a fixed direction except for solutions that start along the single line drawn in black in Figure 11.8(b). We have already seen behavior like this in the phase portrait given in Figure 10.2. Experiment with your numerical solver and verify these observations.

The selection $c = 0$ leads to a single real eigenvalue $\lambda = -1$. This case is very similar to the case $c = \frac{1}{4}$ with one notable exception. All solution curves in Figure 11.8(c) appear to approach **0** from a fixed direction as t increases.

*In general, if \mathbf{A} is an $n \times n$ matrix, the **trace** of \mathbf{A} is the sum of the main diagonal entries.

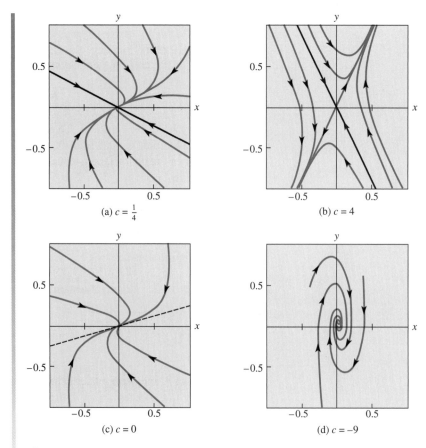

(a) $c = \frac{1}{4}$

(b) $c = 4$

(c) $c = 0$

(d) $c = -9$

Figure 11.8

Finally, when $c = -9$, $1 = \lambda = -1 \pm \sqrt{-9} = -1 \pm 3i$. Thus the eigenvalues are conjugate complex numbers with negative real part -1. Figure 11.8(d) shows that solution curves spiral in toward the origin **0** as t increases. ❑

The behaviors of the trajectories observed in the four phase portraits in Figure 11.8 in Example 1 can be explained using the eigenvalue-eigenvector solution results from Chapter 10.

Case I: Real Distinct Eigenvalues ($\tau^2 - 4\Delta > 0$)

According to Theorem 10.7 in Section 10.2, the general solution of (1) is given by

$$\mathbf{X}(t) = c_1 \mathbf{K}_1 e^{\lambda_1 t} + c_2 \mathbf{K}_2 e^{\lambda_2 t}, \tag{2}$$

where λ_1 and λ_2 are the eigenvalues and \mathbf{K}_1 and \mathbf{K}_2 are the corresponding eigenvectors. Note that $\mathbf{X}(t)$ can also be written as

$$\mathbf{X}(t) = e^{\lambda_1 t}[c_1 \mathbf{K}_1 + c_2 \mathbf{K}_2 e^{(\lambda_2 - \lambda_1)t}]. \tag{3}$$

(a) Both eigenvalues negative ($\tau^2 - 4\Delta > 0$, $\tau < 0$, and $\Delta > 0$)
Stable Node ($\lambda_2 < \lambda_1 < 0$): Since both eigenvalues are negative, it follows from (2) that $\lim_{t \to \infty} \mathbf{X}(t) = \mathbf{0}$. If we assume that $\lambda_2 < \lambda_1$, then

Figure 11.9 Stable node

Figure 11.10 Unstable node

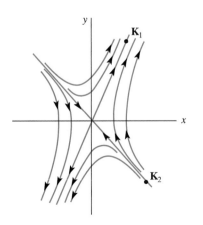

Figure 11.11 Saddle point

$\lambda_2 - \lambda_1 < 0$ and so $e^{(\lambda_2 - \lambda_1)t}$ is an exponential decay function. We may therefore conclude from (3) that $\mathbf{X}(t) \approx c_1\mathbf{K}_1e^{\lambda_1 t}$ for large values of t. When $c_1 \neq 0$, $\mathbf{X}(t)$ will approach $\mathbf{0}$ from one of the two directions determined by the eigenvector \mathbf{K}_1 corresponding to λ_1. If $c_1 = 0$, $\mathbf{X}(t) = c_2\mathbf{K}_2e^{\lambda_2 t}$ and $\mathbf{X}(t)$ approaches $\mathbf{0}$ along the line determined by the eigenvector \mathbf{K}_2. Figure 11.9 shows a collection of solution curves around the origin. A critical point is called a **stable node** when both eigenvalues are negative.

(b) Both eigenvalues positive ($\tau^2 - 4\Delta > 0$, $\tau > 0$, and $\Delta > 0$)
Unstable Node ($0 < \lambda_2 < \lambda_1$): The analysis for this case is similar to *(a)*. Again from (2), $\mathbf{X}(t)$ becomes unbounded as t increases. Moreover, again assuming $\lambda_2 < \lambda_1$ and using (3), we see that $\mathbf{X}(t)$ becomes unbounded in one of the directions determined by the eigenvector \mathbf{K}_1 (when $c_1 \neq 0$) or along the line determined by the eigenvector \mathbf{K}_2 (when $c_1 = 0$). Figure 11.10 shows a typical collection of solution curves. This type of critical point, corresponding to the case when both eigenvalues are positive, is called an **unstable node.**

(c) Eigenvalues have opposite signs ($\tau^2 - 4\Delta > 0$ and $\Delta < 0$)
Saddle Point ($\lambda_2 < 0 < \lambda_1$): The analysis of the solutions is identical to *(b)* with one exception. When $c_1 = 0$, $\mathbf{X}(t) = c_2\mathbf{K}_2e^{\lambda_2 t}$ and, since $\lambda_2 < 0$, $\mathbf{X}(t)$ will approach $\mathbf{0}$ along the line determined by the eigenvector \mathbf{K}_2. If $\mathbf{X}(0)$ does not lie on the line determined by \mathbf{K}_2, the line determined by \mathbf{K}_1 serves as an asymptote for $\mathbf{X}(t)$. Thus the critical point is unstable even though some solutions approach $\mathbf{0}$ as t increases. This unstable critical point is called a **saddle point.** See Figure 11.11.

Example 2 Real Distinct Eigenvalues

Classify the critical point $(0, 0)$ of each of the following linear systems $\mathbf{X}' = \mathbf{A}\mathbf{X}$ as either a stable node, an unstable node, or a saddle point.

$$\text{(a) } \mathbf{A} = \begin{pmatrix} 2 & 3 \\ 2 & 1 \end{pmatrix} \qquad \text{(b) } \mathbf{A} = \begin{pmatrix} -10 & 6 \\ 15 & -19 \end{pmatrix}$$

In each case discuss the nature of the solutions in a neighborhood of $(0, 0)$.

SOLUTION (a) Since the trace $\tau = 3$ and the determinant $\Delta = -4$, the eigenvalues are

$$\lambda = \frac{\tau \pm \sqrt{\tau^2 - 4\Delta}}{2} = \frac{3 \pm \sqrt{3^2 - 4(-4)}}{2} = \frac{3 \pm 5}{2} = 4, -1.$$

The eigenvalues have opposite signs, and so $(0, 0)$ is a saddle point. It is not hard to show (see Example 1, Section 10.2) that eigenvectors corresponding to $\lambda_1 = 4$ and $\lambda_2 = -1$ are

$$\mathbf{K}_1 = \begin{pmatrix} 3 \\ 2 \end{pmatrix} \qquad \text{and} \qquad \mathbf{K}_2 = \begin{pmatrix} 1 \\ -1 \end{pmatrix},$$

respectively. If $\mathbf{X}(0) = \mathbf{X}_0$ lies on the line $y = -x$, then $\mathbf{X}(t)$ approaches $\mathbf{0}$. For any other initial condition $\mathbf{X}(t)$ will become unbounded in the directions determined by \mathbf{K}_1. In other words, the line $y = \frac{2}{3}x$ serves as an asymptote for all these solution curves. See Figure 11.12.

Figure 11.12

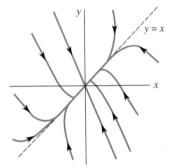

Figure 11.13

(b) From $\tau = -29$ and $\Delta = 100$ it follows that the eigenvalues of \mathbf{A} are $\lambda_1 = -4$ and $\lambda_2 = -25$. Both eigenvalues are negative, and so $(0, 0)$ is in this case a stable node. Since eigenvectors corresponding to $\lambda_1 = -4$ and $\lambda_2 = -25$ are

$$\mathbf{K}_1 = \begin{pmatrix} 1 \\ 1 \end{pmatrix} \quad \text{and} \quad \mathbf{K}_2 = \begin{pmatrix} 2 \\ -5 \end{pmatrix},$$

respectively, it follows that all solutions approach $\mathbf{0}$ from the direction defined by \mathbf{K}_1 except those solutions for which $\mathbf{X}(0) = \mathbf{X}_0$ lies on the line $y = -\frac{5}{2}x$ determined by \mathbf{K}_2. These solutions approach $\mathbf{0}$ along $y = -\frac{5}{2}x$. See Figure 11.13. ❑

Case II: A Repeated Real Eigenvalue ($\tau^2 - 4\Delta = 0$)

Degenerate Nodes: Recall from Section 10.2 that the general solution takes on one of two different forms depending on whether one or two linearly independent eigenvectors can be found for the repeated eigenvalue λ_1.

(a) Two linearly independent eigenvectors
If \mathbf{K}_1 and \mathbf{K}_2 are two linearly independent eigenvectors corresponding to λ_1, then the general solution is given by

$$\mathbf{X}(t) = c_1\mathbf{K}_1 e^{\lambda_1 t} + c_2\mathbf{K}_2 e^{\lambda_1 t} = (c_1\mathbf{K}_1 + c_2\mathbf{K}_2)e^{\lambda_1 t}.$$

If $\lambda_1 < 0$, then $\mathbf{X}(t)$ approaches $\mathbf{0}$ along the line determined by the vector $c_1\mathbf{K}_1 + c_2\mathbf{K}_2$ and the critical point is called a **degenerate stable node** (see Figure 11.14(a)). The arrows in Figure 11.14(a) are reversed when $\lambda_1 > 0$, and we have a **degenerate unstable node.**

(b) A single linearly independent eigenvector
When only a single linearly independent eigenvector \mathbf{K}_1 exists, the general solution is given by

$$\mathbf{X}(t) = c_1\mathbf{K}_1 e^{\lambda_1 t} + c_2(\mathbf{K}_1 t e^{\lambda_1 t} + \mathbf{P}e^{\lambda_1 t}),$$

where $(\mathbf{A} - \lambda_1\mathbf{I})\mathbf{P} = \mathbf{K}_1$ (see Section 10.2, (12)–(14)), and the solution may be rewritten as

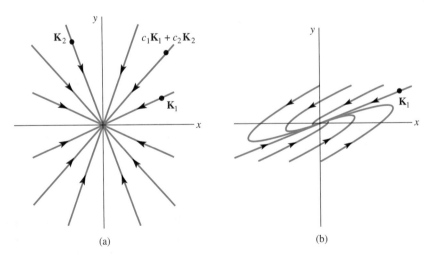

(a) (b)

Figure 11.14 Degenerate stable nodes

Figure 11.15 Center

(a)

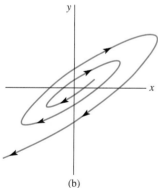

(b)

Figure 11.16 Stable and unstable spiral points

$$\mathbf{X}(t) = te^{\lambda_1 t}\left[c_2 \mathbf{K}_1 + \frac{c_1}{t}\,\mathbf{K}_1 + \frac{c_2}{t}\,\mathbf{P} \right].$$

If $\lambda_1 < 0$, then $\lim_{t \to \infty} te^{\lambda_1 t} = 0$ and it follows that $\mathbf{X}(t)$ approaches $\mathbf{0}$ in one of the directions determined by the vector \mathbf{K}_1 (see Figure 11.14(b)). The critical point is again called a **degenerate stable node.** When $\lambda_1 > 0$, the solutions look like those in Figure 11.14(b) with the arrows reversed. The line determined by \mathbf{K}_1 is an asymptote for *all* solutions. The critical point is again called a **degenerate unstable node.**

Case III: Complex Eigenvalues ($\tau^2 - 4\Delta < 0$)

If $\lambda_1 = \alpha + i\beta$ and $\bar{\lambda}_1 = \alpha - i\beta$ are the complex eigenvalues and $\mathbf{K}_1 = \mathbf{B}_1 + i\mathbf{B}_2$ is a complex eigenvector corresponding to λ_1, the general solution can be written as $\mathbf{X}(t) = c_1 \mathbf{X}_1(t) + c_2 \mathbf{X}_2(t)$, where

$$\mathbf{X}_1(t) = (\mathbf{B}_1 \cos \beta t - \mathbf{B}_2 \sin \beta t)e^{\alpha t} \text{ and } \mathbf{X}_2(t) = (\mathbf{B}_2 \cos \beta t + \mathbf{B}_1 \sin \beta t)e^{\alpha t}.$$

See equations (23) and (24) in Section 10.2. A solution can therefore be written in the form

$$x(t) = e^{\alpha t}(c_{11} \cos \beta t + c_{12} \sin \beta t), \quad y(t) = e^{\alpha t}(c_{21} \cos \beta t + c_{22} \sin \beta t), \quad (4)$$

and when $\alpha = 0$ we have

$$x(t) = c_{11} \cos \beta t + c_{12} \sin \beta t, \qquad y(t) = c_{21} \cos \beta t + c_{22} \sin \beta t. \quad (5)$$

(a) Pure imaginary roots ($\tau^2 - 4\Delta < 0$, $\tau = 0$)
Center: When $\alpha = 0$, the eigenvalues are pure imaginary and, from (5), all solutions are periodic with period $p = 2\pi/\beta$. Notice that if both c_{12} and c_{21} happened to be 0, then (5) would reduce to

$$x(t) = c_{11} \cos \beta t, \qquad y(t) = c_{22} \sin \beta t,$$

which is a standard parametric representation for the ellipse $x^2/c_{11}^2 + y^2/c_{22}^2 = 1$. By solving the system of equations in (4) for $\cos \beta t$ and $\sin \beta t$ and using the identity $\sin^2 \beta t + \cos^2 \beta t = 1$, it is possible to show that *all solutions are ellipses* with center at the origin. The critical point $(0, 0)$ is called a **center,** and Figure 11.15 shows a typical collection of solution curves. The ellipses are either *all* traversed in the clockwise direction or *all* traversed in the counterclockwise direction.

(b) Nonzero real part ($\tau^2 - 4\Delta < 0$, $\tau \neq 0$)
Spiral Points: When $\alpha \neq 0$, the effect of the term $e^{\alpha t}$ in (4) is similar to the effect of the exponential term in the analysis of damped motion given in Section 3.8. When $\alpha < 0$, $e^{\alpha t} \to 0$, and the elliptical-like solution spirals closer and closer to the origin. The critical point is called a **stable spiral point.** When $\alpha > 0$, the effect is the opposite. An elliptical-like solution is driven farther and farther from the origin, and the critical point is now called an **unstable spiral point.** See Figure 11.16.

Example 3 Repeated and Complex Eigenvalues
Classify the critical point $(0, 0)$ of each of the following linear systems $\mathbf{X}' = \mathbf{A}\mathbf{X}$.

(a) $\mathbf{A} = \begin{pmatrix} 3 & -18 \\ 2 & -9 \end{pmatrix}$ (b) $\mathbf{A} = \begin{pmatrix} -1 & 2 \\ -1 & 1 \end{pmatrix}$

In each case discuss the nature of the solution that satisfies $\mathbf{X}(0) = (1, 0)$. Determine parametric equations for each solution.

SOLUTION (a) Since $\tau = -6$ and $\Delta = 9$, the characteristic polynomial is $\lambda^2 + 6\lambda + 9 = (\lambda + 3)^2$, and so $(0, 0)$ is a degenerate stable node. For the repeated eigenvalue $\lambda = -3$ we find a single eigenvector $\mathbf{K}_1 = \begin{pmatrix} 3 \\ 1 \end{pmatrix}$, and so the solution $\mathbf{X}(t)$ that satisfies $\mathbf{X}(0) = (1, 0)$ approaches $(0, 0)$ from the direction specified by the line $y = x/3$.

(b) Since $\tau = 0$ and $\Delta = 1$, the eigenvalues are $\lambda = \pm i$, and so $(0, 0)$ is a center. The solution $\mathbf{X}(t)$ that satisfies $\mathbf{X}(0) = (1, 0)$ is an ellipse that circles the origin every 2π units of time.

From Example 4 of Section 10.2 the general solution of the system in (a) is

$$\mathbf{X}(t) = c_1 \begin{pmatrix} 3 \\ 1 \end{pmatrix} e^{-3t} + c_2 \left[\begin{pmatrix} 3 \\ 1 \end{pmatrix} t e^{-3t} + \begin{pmatrix} 1/2 \\ 0 \end{pmatrix} e^{-3t} \right].$$

The initial condition gives $c_1 = 0$ and $c_2 = 2$, and so

$$x = (6t + 1)e^{-3t}, \qquad y = 2te^{-3t}$$

are parametric equations for the solution.

The general solution of the system in (b) is

$$\mathbf{X}(t) = c_1 \begin{pmatrix} \cos t + \sin t \\ \cos t \end{pmatrix} + c_2 \begin{pmatrix} \cos t - \sin t \\ -\sin t \end{pmatrix}.$$

The initial condition gives $c_1 = 0$ and $c_2 = 1$, and so

$$x = \cos t - \sin t, \qquad y = -\sin t$$

are parametric equations for the ellipse. Note that $y < 0$ for small positive values of t, and therefore the ellipse is traversed in the clockwise direction.

The solutions of (a) and (b) are shown in Figures 11.17(a) and (b), respectively. ❑

(a)

(b)

Figure 11.17

Figure 11.18 conveniently summarizes the results of this section. The general geometric nature of the solutions can be determined by computing the trace and determinant of \mathbf{A}. In practice, graphs of the solutions are most easily obtained *not* by constructing explicit eigenvalue-eigenvector solutions but rather by generating the solutions numerically using a numerical solver and a method such as the Runge–Kutta method (Section 6.2) for first-order systems.

Example 4 Classifying Critical Points

Classify the critical point $(0, 0)$ of each of the following linear systems $\mathbf{X}' = \mathbf{A}\mathbf{X}$

$$\text{(a) } \mathbf{A} = \begin{pmatrix} 1.01 & 3.10 \\ -1.10 & -1.02 \end{pmatrix} \qquad \text{(b) } \mathbf{A} = \begin{pmatrix} -a\hat{x} & -ab\hat{x} \\ -cd\hat{y} & -d\hat{y} \end{pmatrix}$$

for positive constants a, b, c, d, \hat{x}, and \hat{y}.

SOLUTION (a) For this matrix $\tau = -0.01$, $\Delta = 2.3798$, and so $\tau^2 - 4\Delta < 0$. Using Figure 11.18, we see that $(0, 0)$ is a stable spiral point.

(b) This matrix arises from the Lotka–Volterra competition model, which we will study in Section 11.4. Since $\tau = -(a\hat{x} + d\hat{y})$ and all constants in the matrix are positive, $\tau < 0$. The determinant may be written as $\Delta = ad\hat{x}\hat{y}(1 - bc)$. If $bc > 1$, then $\Delta < 0$ and the critical point is a saddle point. If $bc < 1$, then $\Delta > 0$ and the critical point is either a stable node, a degenerate stable node, or a stable spiral point. In all three of these cases $\lim_{t \to \infty} \mathbf{X}(t) = 0$. ❑

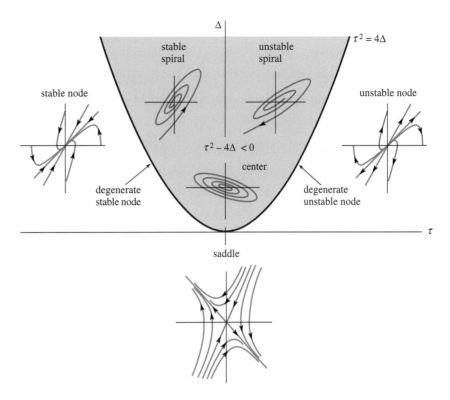

Figure 11.18

We can now give answers to each of the questions posed at the beginning of this section for the linear plane autonomous system

$$x' = ax + by$$
$$y' = cx + dy$$

with $ad - bc \neq 0$. The answers are summarized in the theorem below.

THEOREM 11.1 Stability Criteria for Linear Systems

For a linear plane autonomous system $\mathbf{X}' = \mathbf{AX}$ with $\det \mathbf{A} \neq 0$, let $\mathbf{X} = \mathbf{X}(t)$ denote the solution that satisfies the initial condition $\mathbf{X}(0) = \mathbf{X}_0$, where $\mathbf{X}_0 \neq \mathbf{0}$.

(a) $\lim_{t \to \infty} \mathbf{X}(t) = \mathbf{0}$ if and only if the eigenvalues of \mathbf{A} have negative real parts. This will occur when $\Delta > 0$ and $\tau < 0$.

(b) $\mathbf{X}(t)$ is periodic if and only if the eigenvalues of \mathbf{A} are pure imaginary. This will occur when $\Delta > 0$ and $\tau = 0$.

(c) In all other cases, given any neighborhood of the origin, there is at least one \mathbf{X}_0 in the neighborhood for which $\mathbf{X}(t)$ becomes unbounded as t increases.

Remarks

The terminology used to describe the types of critical points varies from text to text. The following table lists many of the alternative terms that you may encounter in your reading.

Term	Alternative Terms
critical point	equilibrium point, singular point, stationary point, rest point
spiral point	focus, focal point, vortex point
stable node or spiral point	attractor, sink
unstable node or spiral point	repeller, source

EXERCISES 11.2

Answers to odd-numbered problems begin on page A-55.

In Problems 1–8, the general solution of the linear system $\mathbf{X}' = \mathbf{AX}$ is given.

(a) In each case discuss the nature of the solutions in a neighborhood of (0, 0).

(b) With the aid of a graphing utility plot the solution that satisfies $\mathbf{X}(0) = (1, 1)$.

1. $\mathbf{A} = \begin{pmatrix} -2 & -2 \\ -2 & -5 \end{pmatrix}$, $\mathbf{X}(t) = c_1 \begin{pmatrix} 2 \\ -1 \end{pmatrix} e^{-t} + c_2 \begin{pmatrix} 1 \\ 2 \end{pmatrix} e^{-6t}$

2. $\mathbf{A} = \begin{pmatrix} -1 & -2 \\ 3 & 4 \end{pmatrix}$, $\mathbf{X}(t) = c_1 \begin{pmatrix} 1 \\ -1 \end{pmatrix} e^{t} + c_2 \begin{pmatrix} -4 \\ 6 \end{pmatrix} e^{2t}$

3. $\mathbf{A} = \begin{pmatrix} 1 & -1 \\ 1 & 1 \end{pmatrix}$, $\mathbf{X}(t) = e^{t} \left[c_1 \begin{pmatrix} -\sin t \\ \cos t \end{pmatrix} + c_2 \begin{pmatrix} \cos t \\ \sin t \end{pmatrix} \right]$

4. $\mathbf{A} = \begin{pmatrix} -1 & -4 \\ 1 & -1 \end{pmatrix}$, $\mathbf{X}(t) = e^{-t} \left[c_1 \begin{pmatrix} 2 \cos 2t \\ \sin 2t \end{pmatrix} + c_2 \begin{pmatrix} -2 \sin 2t \\ \cos 2t \end{pmatrix} \right]$

5. $\mathbf{A} = \begin{pmatrix} -6 & 5 \\ -5 & 4 \end{pmatrix}$, $\mathbf{X}(t) = c_1 \begin{pmatrix} 1 \\ 1 \end{pmatrix} e^{-t} + c_2 \left[\begin{pmatrix} 1 \\ 1 \end{pmatrix} t e^{-t} + \begin{pmatrix} 0 \\ \tfrac{1}{5} \end{pmatrix} e^{-t} \right]$

6. $\mathbf{A} = \begin{pmatrix} 2 & 4 \\ -1 & 6 \end{pmatrix}$, $\mathbf{X}(t) = c_1 \begin{pmatrix} 2 \\ 1 \end{pmatrix} e^{4t} + c_2 \left[\begin{pmatrix} 2 \\ 1 \end{pmatrix} t e^{4t} + \begin{pmatrix} 1 \\ 1 \end{pmatrix} e^{4t} \right]$

7. $\mathbf{A} = \begin{pmatrix} 2 & -1 \\ 3 & -2 \end{pmatrix}$, $\mathbf{X}(t) = c_1 \begin{pmatrix} 1 \\ 1 \end{pmatrix} e^{t} + c_2 \begin{pmatrix} 1 \\ 3 \end{pmatrix} e^{-t}$

8. $\mathbf{A} = \begin{pmatrix} -1 & 5 \\ -1 & 1 \end{pmatrix}$,

$\mathbf{X}(t) = c_1 \begin{pmatrix} 5 \cos 2t \\ \cos 2t - 2 \sin 2t \end{pmatrix} + c_2 \begin{pmatrix} 5 \sin 2t \\ 2 \cos 2t + \sin 2t \end{pmatrix}$

In Problems 9–16, classify the critical point (0, 0) of the given linear system by computing the trace τ and determinant Δ and using Figure 11.18.

9. $x' = -5x + 3y$
$y' = 2x + 7y$

10. $x' = -5x + 3y$
$y' = 2x - 7y$

11. $x' = -5x + 3y$
$y' = -2x + 5y$

12. $x' = -5x + 3y$
$y' = -7x + 4y$

13. $x' = -\dfrac{3}{2}x + \dfrac{1}{4}y$
$y' = -x - \dfrac{1}{2}y$

14. $x' = \dfrac{3}{2}x + \dfrac{1}{4}y$
$y' = -x + \dfrac{1}{2}y$

15. $x' = 0.02x - 0.11y$
$y' = 0.10x - 0.05y$

16. $x' = 0.03x + 0.01y$
$y' = -0.01x + 0.05y$

17. Determine conditions on the real constant μ so that (0, 0) is a center for the linear system

$x' = -\mu x + y$
$y' = -x + \mu y.$

18. Determine a condition on the real constant μ so that (0, 0) is a stable spiral point of the linear system

$x' = y$
$y' = -x + \mu y.$

19. Show that (0, 0) is always an unstable critical point of the linear system

$x' = \mu x + y$
$y' = -x + y,$

where μ is a real constant and $\mu \neq -1$. When is (0, 0) an unstable saddle point? When is (0, 0) an unstable spiral point?

20. Let $\mathbf{X} = \mathbf{X}(t)$ be the response of the linear dynamical system

$x' = \alpha x - \beta y$
$y' = \beta x + \alpha y$

that satisfies the initial condition $\mathbf{X}(0) = \mathbf{X}_0$. Determine conditions on the real constants α and β that will ensure $\lim_{t \to \infty} \mathbf{X}(t) = (0, 0)$. Can (0, 0) be a node or saddle point?

21. Show that the nonhomogeneous linear system $\mathbf{X}' = \mathbf{AX} + \mathbf{F}$ has a unique critical point \mathbf{X}_1 when $\Delta = \det \mathbf{A} \neq 0$. Conclude that if $\mathbf{X} = \mathbf{X}(t)$ is a solution to the nonhomogeneous system, $\tau < 0$ and $\Delta > 0$, then $\lim_{t \to \infty} \mathbf{X}(t) = \mathbf{X}_1$. [*Hint:* $\mathbf{X}(t) = \mathbf{X}_c(t) + \mathbf{X}_1$.]

22. In Example 4(b) show that (0, 0) is a stable node when $bc < 1$.

In Problems 23–26, a nonhomogeneous linear system $\mathbf{X}' = \mathbf{AX} + \mathbf{F}$ is given.

(a) In each case determine the unique critical point \mathbf{X}_1.

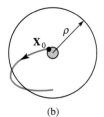

(b) Use a numerical solver to determine the nature of the critical point in **(a)**.

(c) Investigate the relationship between \mathbf{X}_1 and the critical point $(0, 0)$ of the homogeneous linear system $\mathbf{X}' = \mathbf{AX}$.

23. $x' = 2x + 3y - 6$
$\quad\ y' = -x - 2y + 5$

24. $x' = -5x + 9y + 13$
$\quad\ y' = -x - 11y - 23$

25. $x' = 0.1x - 0.2y + 0.35$
$\quad\ y' = 0.1x + 0.1y - 0.25$

26. $x' = 3x - 2y - 1$
$\quad\ y' = 5x - 3y - 2$

11.3 Linearization and Local Stability

We start this section by refining the stability concepts introduced in Section 11.2 in such a way that they will apply to nonlinear autonomous systems as well. Although the linear system $\mathbf{X}' = \mathbf{AX}$ had only one critical point when $\det \mathbf{A} \neq 0$, we saw in Section 11.1 that a nonlinear system may have many critical points. We therefore cannot expect that a particle placed initially at \mathbf{X}_0 will remain near a given critical point \mathbf{X}_1 unless \mathbf{X}_0 has been placed sufficiently close to \mathbf{X}_1 to begin with. The particle might well be driven to a second critical point \mathbf{X}_2.

To emphasize this idea we will consider the physical system shown in Figure 11.19 in which a bead slides along the curve $z = f(x)$ under the influence of gravity alone. We will show in Section 11.4 that the x-coordinate of the bead satisfies a nonlinear second-order differential equation $x'' = g(x, x')$, and therefore letting $y = x'$ satisfies the nonlinear autonomous system

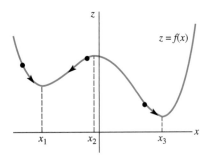

Figure 11.19

$$x' = y$$
$$y' = g(x, y).$$

If the bead is positioned at $P = (x, f(x))$ and given zero initial velocity, the bead will remain at P provided $f'(x) = 0$. If the bead is placed near the critical point located at $x = x_1$, it will remain near $x = x_1$ only if its initial velocity does not drive it over the "hump" at $x = x_2$ toward the critical point located at $x = x_3$. Therefore $\mathbf{X}(0) = (x(0), x'(0))$ must be near $(x_1, 0)$.

In the next definition we will denote the distance between two points \mathbf{X} and \mathbf{Y} by $|\mathbf{X} - \mathbf{Y}|$. Recall that if $\mathbf{X} = (x_1, x_2, ..., x_n)$ and $\mathbf{Y} = (y_1, y_2, ..., y_n)$, then

$$|\mathbf{X} - \mathbf{Y}| = \sqrt{(x_1 - y_1)^2 + (x_2 - y_2)^2 + \cdots + (x_n - y_n)^2}.$$

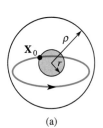

(a)

(b)

Figure 11.20

> **DEFINITION 11.1** Stable Critical Points
>
> Let \mathbf{X}_1 be a critical point of an autonomous system, and let $\mathbf{X} = \mathbf{X}(t)$ denote the solution that satisfies the initial condition $\mathbf{X}(0) = \mathbf{X}_0$, where $\mathbf{X}_0 \neq \mathbf{X}_1$. We say that \mathbf{X}_1 is a **stable critical point** when, given any radius $\rho > 0$, there is a corresponding radius $r > 0$ such that if the initial position \mathbf{X}_0 satisfies $|\mathbf{X}_0 - \mathbf{X}_1| < r$, then the corresponding solution $\mathbf{X}(t)$ satisfies $|\mathbf{X}(t) - \mathbf{X}_1| < \rho$ for all $t > 0$. If, in addition, $\lim_{t \to \infty} \mathbf{X}(t) = \mathbf{X}_1$ whenever $|\mathbf{X}_0 - \mathbf{X}_1| < r$, we call \mathbf{X}_1 an **asymptotically stable critical point.**

This definition is illustrated in Figure 11.20(a). Given any disk of radius ρ about the critical point \mathbf{X}_1, a solution will remain inside this disk provided $\mathbf{X}(0) = \mathbf{X}_0$ is selected sufficiently close to \mathbf{X}_1. It is *not* necessary that a solution approach the critical point in order for \mathbf{X}_1 to be stable. Stable nodes, stable spiral points, and centers are all examples of stable critical points for linear systems. To emphasize that \mathbf{X}_0 must be selected close to \mathbf{X}_1 the terminology **locally stable critical point** is also used.

By negating Definition 11.2 we obtain the definition of an unstable critical point.

> **DEFINITION 11.2** Unstable Critical Point
>
> Let X_1 be a critical point of an autonomous system, and let $X = X(t)$ denote the solution that satisfies the initial condition $X(0) = X_0$, where $X_0 \neq X_1$. We say that X_1 is an **unstable critical point** if there is a disk of radius $\rho > 0$ with the property that, for any $r > 0$, there is at least one initial position X_0 that satisfies $|X_0 - X_1| < r$, yet the corresponding solution $X(t)$ satisfies $|X(t) - X_1| \geq \rho$ for at least one $t > 0$.

If a critical point X_1 is unstable, no matter how small the neighborhood about X_1, an initial position X_0 can always be found that results in the solution leaving some disk of radius ρ at some future time t. See Figure 11.20(b). Therefore unstable nodes, unstable spiral points, and saddle points are all examples of unstable critical points for linear systems. In Figure 11.19 the critical point $(x_2, 0)$ is unstable. The slightest displacement or initial velocity results in the bead sliding away from the point $(x_2, f(x_2))$.

Example 1 A Stable Critical Point

Show that $(0, 0)$ is a stable critical point of the nonlinear plane autonomous system

$$x' = -y - x\sqrt{x^2 + y^2}$$
$$y' = x - y\sqrt{x^2 + y^2}$$

considered in Example 6 of Section 11.1.

SOLUTION In Example 6 of Section 11.1 we showed that in polar coordinates

$$r = \frac{1}{t + c_1}, \qquad \theta = t + c_2,$$

is the solution of the system. If $X(0) = (r_0, \theta_0)$ is the initial condition in polar coordinates, then

$$r = \frac{r_0}{r_0 t + 1}, \qquad \theta = t + \theta_0.$$

Note that $r \leq r_0$ for $t \geq 0$, and r approaches $(0, 0)$ as t increases. Therefore, given $\rho > 0$, a solution that starts less than ρ units from $(0, 0)$ remains within ρ units of the origin for all $t \geq 0$. Hence the critical point $(0, 0)$ is stable and is in fact asymptotically stable. A typical solution is shown in Figure 11.21. ❏

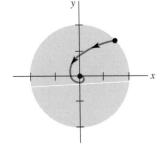

Figure 11.21

Example 2 An Unstable Critical Point

When expressed in polar coordinates, a plane autonomous system takes the form

$$\frac{dr}{dt} = 0.05r(3 - r)$$

$$\frac{d\theta}{dt} = -1.$$

Show that $(x, y) = (0, 0)$ is an unstable critical point.

SOLUTION Since $x = r\cos\theta$ and $y = r\sin\theta$, we have

$$\frac{dx}{dt} = -r\sin\theta \frac{d\theta}{dt} + \frac{dr}{dt}\cos\theta$$

$$\frac{dy}{dt} = r \cos \theta \frac{d\theta}{dt} + \frac{dr}{dt} \sin \theta.$$

From $dr/dt = 0.05r(3 - r)$ we see that $dr/dt = 0$ when $r = 0$ and can conclude that $(x, y) = (0, 0)$ is a critical point by substituting $r = 0$ into the new system.

The differential equation $dr/dt = 0.05r(3 - r)$ is a logistic equation that can be solved using separation of variables. If $r(0) = r_0$ and $r_0 \neq 0$, then

$$r = \frac{3}{1 + c_0 e^{-0.15t}},$$

where $c_0 = (3 - r_0)/r_0$. Since

$$\lim_{t \to \infty} \frac{3}{1 + c_0 e^{-0.15t}} = 3,$$

it follows that no matter how close to $(0, 0)$ a solution starts, the solution will leave a disk of radius 1 about the origin. Therefore $(0, 0)$ is an unstable critical point. A typical solution that starts near $(0, 0)$ is shown in Figure 11.22. ❑

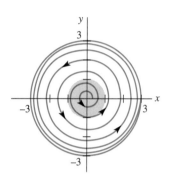

Figure 11.22

■ **Linearization** It is rarely possible to determine the stability of a critical point of a nonlinear system by finding explicit solutions, as in Examples 1 and 2. Instead we replace the term $\mathbf{g}(\mathbf{X})$ in the original autonomous system $\mathbf{X}' = \mathbf{g}(\mathbf{X})$ by a linear term $\mathbf{A}(\mathbf{X} - \mathbf{X}_1)$ that most closely approximates $\mathbf{g}(\mathbf{X})$ in a neighborhood of \mathbf{X}_1. This replacement process, called **linearization,** will be illustrated first for the first-order differential equation $x' = g(x)$.

An equation of the tangent line to the curve $y = g(x)$ at $x = x_1$ is $y = g(x_1) + g'(x_1)(x - x_1)$, and if x_1 is a critical point of $x' = g(x)$, we have

$$x' = g(x) \approx g'(x_1)(x - x_1).$$

The general solution to the linear differential equation

$$x' = g'(x_1)(x - x_1)$$

is $x = x_1 + ce^{\lambda_1 t}$, where $\lambda_1 = g'(x_1)$. Thus if $g'(x_1) < 0$, then $x(t)$ approaches x_1. Theorem 11.2 asserts that the same behavior occurs in the original differential equation provided $x(0) = x_0$ is selected close enough to x_1.

Autonomous first-order DEs were discussed in Section 2.1. ☞

> **THEOREM 11.2** **Stability Criteria for $x' = g(x)$**
>
> Let x_1 be a critical point of the autonomous differential equation $x' = g(x)$, where g is differentiable at x_1.
> (*a*) If $g'(x_1) < 0$, then x_1 is an asymptotically stable critical point.
> (*b*) If $g'(x_1) > 0$, then x_1 is an unstable critical point.

Example 3 Stability in a Nonlinear First-Order DE

Both $x = \pi/4$ and $x = 5\pi/4$ are critical points of the autonomous differential equation $x' = \cos x - \sin x$. This differential equation is difficult to solve explicitly, but we can use Theorem 11.2 to predict the behavior of solutions near these two critical points.

Since $g'(x) = -\sin x - \cos x$, $g'(\pi/4) = -\sqrt{2} < 0$ and $g'(5\pi/4) = \sqrt{2} > 0$. Therefore $x = \pi/4$ is an asymptotically stable critical point, but $x = 5\pi/4$ is

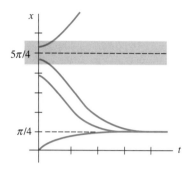

Figure 11.23

unstable. In Figure 11.23 we used a numerical solver to investigate solutions that start near $(0, \pi/4)$ and $(0, 5\pi/4)$. Observe that solution curves that start close to $(0, 5\pi/4)$ quickly move away from the line $x = 5\pi/4$, as predicted. ❏

Example 4 Stability Analysis of the Logistic DE

Without solving explicitly, analyze the critical points of the logistic differential equation (see Section 2.8) $x' = \dfrac{r}{K}x(K - x)$, where r and K are positive constants.

SOLUTION The two critical points are $x = 0$ and $x = K$. Since

$$g'(x) = \frac{r}{K}(K - 2x),$$

$g'(0) = r$ and $g'(K) = -r$. We may therefore conclude that $x = 0$ is an unstable critical point and $x = K$ is an asymptotically stable critical point. ❏

■ **Jacobian Matrix** A similar analysis may be carried out for a plane autonomous system. An equation of the tangent plane to the surface $z = g(x, y)$ at $\mathbf{X}_1 = (x_1, y_1)$ is

$$z = g(x_1, y_1) + \frac{\partial g}{\partial x}\bigg|_{(x_1, y_1)} (x - x_1) + \frac{\partial g}{\partial y}\bigg|_{(x_1, y_1)} (y - y_1),$$

and $g(x, y)$ may be approximated by its tangent plane in a neighborhood of \mathbf{X}_1.

When \mathbf{X}_1 is a critical point of a plane autonomous system, $P(x_1, y_1) = Q(x_1, y_1) = 0$ and we have

$$x' = P(x, y) \approx \frac{\partial P}{\partial x}\bigg|_{(x_1, y_1)} (x - x_1) + \frac{\partial P}{\partial y}\bigg|_{(x_1, y_1)} (y - y_1)$$

$$y' = Q(x, y) \approx \frac{\partial Q}{\partial x}\bigg|_{(x_1, y_1)} (x - x_1) + \frac{\partial Q}{\partial y}\bigg|_{(x_1, y_1)} (y - y_1).$$

The original system $\mathbf{X}' = \mathbf{g}(\mathbf{X})$ may be approximated in a neighborhood of the critical point \mathbf{X}_1 by the linear system $\mathbf{X}' = \mathbf{A}(\mathbf{X} - \mathbf{X}_1)$, where

$$\mathbf{A} = \begin{pmatrix} \dfrac{\partial P}{\partial x}\bigg|_{(x_1, y_1)} & \dfrac{\partial P}{\partial y}\bigg|_{(x_1, y_1)} \\ \dfrac{\partial Q}{\partial x}\bigg|_{(x_1, y_1)} & \dfrac{\partial Q}{\partial y}\bigg|_{(x_1, y_1)} \end{pmatrix}.$$

This matrix is called the **Jacobian matrix** at \mathbf{X}_1 and is denoted by $\mathbf{g}'(\mathbf{X}_1)$. If we let $\mathbf{H} = \mathbf{X} - \mathbf{X}_1$, then the linear system $\mathbf{X}' = \mathbf{A}(\mathbf{X} - \mathbf{X}_1)$ becomes $\mathbf{H}' = \mathbf{A}\mathbf{H}$, which is the form of the linear system analyzed in Section 11.2. The critical point $\mathbf{X} = \mathbf{X}_1$ for $\mathbf{X}' = \mathbf{A}(\mathbf{X} - \mathbf{X}_1)$ now corresponds to the critical point $\mathbf{H} = \mathbf{0}$ for $\mathbf{H}' = \mathbf{A}\mathbf{H}$. If the eigenvalues of \mathbf{A} have negative real parts, then, by Theorem 11.1, $\mathbf{0}$ is an asymptotically stable critical point for $\mathbf{H}' = \mathbf{A}\mathbf{H}$. If there is an eigenvalue with positive real part, $\mathbf{H} = \mathbf{0}$ is an unstable critical point. Theorem 11.3 asserts that the same conclusions can be made for the critical point \mathbf{X}_1 of the original system.

> **THEOREM 11.3** Stability Criteria for
> Plane Autonomous Systems
>
> Let \mathbf{X}_1 be a critical point of the plane autonomous system $\mathbf{X}' = \mathbf{g}(\mathbf{X})$, where $P(x, y)$ and $Q(x, y)$ have continuous first partials in a neighborhood of \mathbf{X}_1.
> (a) If the eigenvalues of $\mathbf{A} = \mathbf{g}'(\mathbf{X}_1)$ have negative real part, then \mathbf{X}_1 is an asymptotically stable critical point.
> (b) If $\mathbf{A} = \mathbf{g}'(\mathbf{X}_1)$ has an eigenvalue with positive real part, then \mathbf{X}_1 is an unstable critical point.

Example 5 Stability Analysis of Nonlinear Systems

Classify (if possible) the critical points of each of the following plane autonomous systems as stable or unstable.

$$\text{(a) } x' = x^2 + y^2 - 6 \qquad \text{(b) } x' = 0.01x(100 - x - y)$$
$$y' = x^2 - y \qquad\qquad\quad y' = 0.05y(60 - y - 0.2x)$$

SOLUTION The critical points of each system were determined in Example 4 of Section 11.1.

(a) The critical points are $(\sqrt{2}, 2)$ and $(-\sqrt{2}, 2)$, the Jacobian matrix is

$$\mathbf{g}'(\mathbf{X}) = \begin{pmatrix} 2x & 2y \\ 2x & -1 \end{pmatrix},$$

and so

$$\mathbf{A}_1 = \mathbf{g}'((\sqrt{2}, 2)) = \begin{pmatrix} 2\sqrt{2} & 4 \\ 2\sqrt{2} & -1 \end{pmatrix} \qquad \text{and} \qquad \mathbf{A}_2 = \mathbf{g}'((-\sqrt{2}, 2)) = \begin{pmatrix} -2\sqrt{2} & 4 \\ -2\sqrt{2} & -1 \end{pmatrix}.$$

Since the determinant of \mathbf{A}_1 is negative, \mathbf{A}_1 has a positive real eigenvalue. Therefore $(\sqrt{2}, 2)$ is an unstable critical point. Matrix \mathbf{A}_2 has a positive determinant and a negative trace, and so both eigenvalues have negative real parts. It follows that $(-\sqrt{2}, 2)$ is a stable critical point.

(b) The critical points are $(0, 0)$, $(0, 60)$, $(100, 0)$, and $(50, 50)$, the Jacobian matrix is

$$\mathbf{g}'(\mathbf{X}) = \begin{pmatrix} 0.01(100 - 2x - y) & -0.01x \\ -0.01y & 0.05(60 - 2y - 0.2x) \end{pmatrix},$$

and so

$$\mathbf{A}_1 = \mathbf{g}'((0, 0)) = \begin{pmatrix} 1 & 0 \\ 0 & 3 \end{pmatrix} \qquad\qquad \mathbf{A}_2 = \mathbf{g}'((0, 60)) = \begin{pmatrix} 0.4 & 0 \\ -0.6 & -3 \end{pmatrix}$$

$$\mathbf{A}_3 = \mathbf{g}'((100, 0)) = \begin{pmatrix} -1 & -1 \\ 0 & 2 \end{pmatrix} \qquad\qquad \mathbf{A}_4 = \mathbf{g}'((50, 50)) = \begin{pmatrix} -0.5 & -0.5 \\ -0.5 & -2.5 \end{pmatrix}.$$

Since the matrix \mathbf{A}_1 has a positive determinant and a positive trace, both eigenvalues have positive real parts. Therefore $(0, 0)$ is an unstable critical point. The determinants of matrices \mathbf{A}_2 and \mathbf{A}_3 are negative, and so in each case one of the eigenvalues is positive. Therefore both $(0, 60)$ and $(100, 0)$ are unstable critical points. Since the matrix \mathbf{A}_4 has a positive determinant and a negative trace, $(50, 50)$ is a stable critical point. ❑

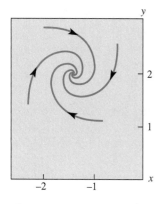

Figure 11.24

In Example 5 we did not compute $\tau^2 - 4\Delta$ (as in Section 11.2) and attempt to further classify the critical points as stable nodes, stable spiral points, saddle points, and so on. For example, for $\mathbf{X}_1 = (-\sqrt{2}, 2)$ in Example 4(a), $\tau^2 - 4\Delta < 0$, and if the system were linear, we would be able to conclude that \mathbf{X}_1 was a stable spiral point. Figure 11.24 shows several solution curves near \mathbf{X}_1 that were obtained with a numerical solver, and each solution does *appear* to spiral in toward the critical point.

■ **Classifying Critical Points** It is natural to ask whether we can infer more geometric information about the solutions near a critical point \mathbf{X}_1 of a nonlinear autonomous system from an analysis of the critical point of the corresponding linear system. The answer is summarized in Figure 11.25, but you should note the following comments.

(i) In five separate cases (stable node, stable spiral point, unstable spiral point, unstable node, and saddle) the critical point may be categorized like the critical point in the corresponding linear system. The solutions have *the same general geometric features* as the solutions to the linear system, and the smaller the neighborhood about \mathbf{X}_1, the closer the resemblance.

(ii) If $\tau^2 = 4\Delta$ and $\tau > 0$, the critical point \mathbf{X}_1 is unstable, but in this borderline case *we are not yet able to decide whether \mathbf{X}_1 is an unstable spiral, unstable node, or degenerate unstable node*. Likewise, if $\tau^2 = 4\Delta$ and $\tau < 0$, the critical point \mathbf{X}_1 is stable but may be either a stable spiral, a stable node, or a degenerate stable node.

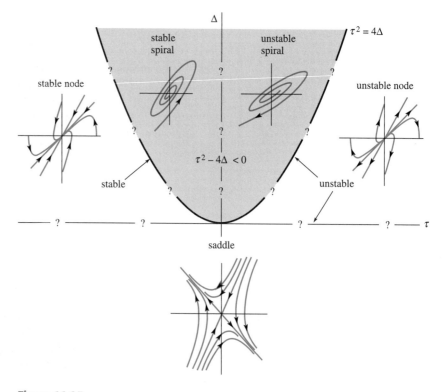

Figure 11.25

(*iii*) If $\tau = 0$ and $\Delta > 0$, the eigenvalues of $\mathbf{A} = \mathbf{g}'(\mathbf{X})$ are pure imaginary, and in this borderline case \mathbf{X}_1 may be either a stable spiral, an unstable spiral, or a center. *It is therefore not yet possible to determine whether \mathbf{X}_1 is stable or unstable.*

Example 6 Classifying Critical Points of a Nonlinear System

Classify each critical point of the plane autonomous system in Example 5(b) as a stable node, a stable spiral point, an unstable spiral point, an unstable node, or a saddle point.

SOLUTION For the matrix \mathbf{A}_1 corresponding to $(0, 0)$, $\Delta = 3$, $\tau = 4$, and so $\tau^2 - 4\Delta = 4$. Therefore $(0, 0)$ is an unstable node. The critical points $(0, 60)$ and $(100, 0)$ are saddles since $\Delta < 0$ in both cases. For matrix \mathbf{A}_4, $\Delta > 0$, $\tau < 0$, and $\tau^2 - 4\Delta > 0$. It follows that $(50, 50)$ is a stable node. Experiment with a numerical solver to verify these conclusions. ❑

Example 7 Stability Analysis for a Soft Spring

Recall from Section 3.10 that the second-order differential equation $mx'' + kx + k_1 x^3 = 0$, for $k > 0$, represents a general model for the free, undamped oscillations of a mass m attached to a nonlinear spring. If $k = 1$ and $k_1 = -1$, the spring is called *soft* and the plane autonomous system corresponding to the nonlinear second-order differential equation $x'' + x - x^3 = 0$ is

$$x' = y$$
$$y' = x^3 - x.$$

Find and classify (if possible) the critical points.

SOLUTION Since $x^3 - x = x(x^2 - 1)$, the critical points are $(0, 0)$, $(1, 0)$, and $(-1, 0)$. The corresponding Jacobian matrices are

$$\mathbf{A}_1 = \mathbf{g}'((0, 0)) = \begin{pmatrix} 0 & 1 \\ -1 & 0 \end{pmatrix}, \qquad \mathbf{A}_2 = \mathbf{g}'((1, 0)) = \mathbf{g}'((-1, 0)) = \begin{pmatrix} 0 & 1 \\ 2 & 0 \end{pmatrix}.$$

Since $\det \mathbf{A}_2 < 0$, critical points $(1, 0)$ and $(-1, 0)$ are both saddle points. The eigenvalues of matrix \mathbf{A}_1 are $\pm i$, and according to comment (*iii*), the status of the critical point at $(0, 0)$ remains in doubt. It may be either a stable spiral, an unstable spiral, or a center. ❑

■ **The Phase-Plane Method** The linearization method, when successful, can provide useful information on the local behavior of solutions near critical points. It is of little help if we are interested in solutions whose initial position $\mathbf{X}(0) = \mathbf{X}_0$ is not close to a critical point or if we wish to obtain a global view of the family of solution curves. The **phase-plane method** is based on the fact that

$$\frac{dy}{dx} = \frac{dy/dt}{dx/dt} = \frac{Q(x, y)}{P(x, y)}$$

and it attempts to find y as a function of x using one of the methods available for solving first-order differential equations (Chapter 2). As we show in Examples 8 and 9, the method can sometimes be used to decide whether a critical point such as $(0, 0)$ in Example 7 is a stable spiral, an unstable spiral, or a center.

Example 8 Phase-Plane Method

Use the phase-plane method to classify the sole critical point (0, 0) of the plane autonomous system

$$x' = y^2$$
$$y' = x^2.$$

SOLUTION The determinant of the Jacobian matrix

$$\mathbf{g}'(\mathbf{X}) = \begin{pmatrix} 0 & 2y \\ 2x & 0 \end{pmatrix}$$

is 0 at (0, 0), and so the nature of the critical point (0, 0) remains in doubt. Using the phase-plane method, we obtain the first-order differential equation

$$\frac{dy}{dx} = \frac{dy/dt}{dx/dt} = \frac{x^2}{y^2},$$

which can be easily solved by separation of variables:

$$\int y^2 \, dy = \int x^2 \, dx \qquad \text{or} \qquad y^3 = x^3 + c.$$

If $\mathbf{X}(0) = (0, y_0)$, it follows that $y^3 = x^3 + y_0^3$ or $y = \sqrt[3]{x^3 + y_0^3}$. Figure 11.26 shows a collection of solution curves corresponding to various choices for y_0, and the nature of the critical point is clear. No matter how close to (0, 0) the solution starts, $\mathbf{X}(t)$ moves away from the origin as t increases. The critical point at (0, 0) is therefore unstable. ❑

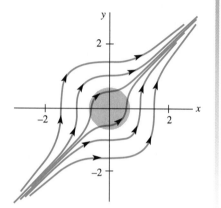

Figure 11.26

Example 9 Phase-Plane Analysis of a Soft Spring

Use the phase-plane method to determine the nature of the solutions to $x'' + x - x^3 = 0$ in a neighborhood of (0, 0).

SOLUTION If we let $dx/dt = y$, then $dy/dt = x^3 - x$. From this we obtain the first-order differential equation

$$\frac{dy}{dx} = \frac{dy/dt}{dx/dt} = \frac{x^3 - x}{y},$$

which can be solved by separation of variables. Integrating

$$\int y \, dy = \int (x^3 - x) \, dx \quad \text{gives} \quad \frac{y^2}{2} = \frac{x^4}{4} - \frac{x^2}{2} + c.$$

After completing the square, we can write the solution as $y^2 = (x^2 - 1)^2/2 + c_0$. If $\mathbf{X}(0) = (x_0, 0)$, where $0 < x_0 < 1$, then $c_0 = -(x_0^2 - 1)^2/2$, and so

$$y^2 = \frac{(x^2 - 1)^2}{2} - \frac{(x_0^2 - 1)^2}{2} = \frac{(2 - x^2 - x_0^2)(x_0^2 - x^2)}{2}.$$

Note that $y = 0$ when $x = -x_0$. In addition, the right-hand side is positive when $-x_0 < x < x_0$, and so each x has *two* corresponding values of y. The solution $\mathbf{X} = \mathbf{X}(t)$ that satisfies $\mathbf{X}(0) = (x_0, 0)$ is therefore periodic, and so (0, 0) is a center.

Figure 11.27 shows a family of solution curves or phase portrait of the original system. We used the original plane autonomous system to determine the directions indicated on each trajectory. ❑

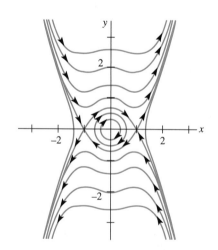

Figure 11.27

1. Show that $(0, 0)$ is an asymptotically stable critical point of the nonlinear autonomous system

$$x' = \alpha x - \beta y + y^2$$
$$y' = \beta x + \alpha y - xy$$

when $\alpha < 0$ and an unstable critical point when $\alpha > 0$. [*Hint:* Switch to polar coordinates.]

2. When expressed in polar coordinates, a plane autonomous system takes the form

$$\frac{dr}{dt} = \alpha r(5 - r)$$
$$\frac{d\theta}{dt} = -1.$$

Show that $(0, 0)$ is an asymptotically stable critical point if and only if $\alpha < 0$.

In Problems 3–10, without solving explicitly, classify the critical points of the given first-order autonomous differential equation as either asymptotically stable or unstable. All constants are assumed to be positive.

3. $\dfrac{dx}{dt} = kx(n + 1 - x)$ **4.** $\dfrac{dx}{dt} = -kx \ln \dfrac{x}{K}, \quad x > 0$

5. $\dfrac{dT}{dt} = k(T - T_0)$ **6.** $m\dfrac{dv}{dt} = mg - kv$

7. $\dfrac{dx}{dt} = k(\alpha - x)(\beta - x), \quad \alpha > \beta$

8. $\dfrac{dx}{dt} = k(\alpha - x)(\beta - x)(\gamma - x), \quad \alpha > \beta > \gamma$

9. $\dfrac{dP}{dt} = P(a - bP)(1 - cP^{-1}), \quad P > 0, \quad a < bc$

10. $\dfrac{dA}{dt} = k\sqrt{A}(K - \sqrt{A}), \quad A > 0$

In Problems 11–20, classify (if possible) each critical point of the given plane autonomous system as a stable node, a stable spiral point, an unstable spiral point, an unstable node, or a saddle point.

11. $x' = 1 - 2xy$ **12.** $x' = x^2 - y^2 - 1$
 $y' = 2xy - y$ $y' = 2y$

13. $x' = y - x^2 + 2$ **14.** $x' = 2x - y^2$
 $y' = x^2 - xy$ $y' = -y + xy$

15. $x' = -3x + y^2 + 2$ **16.** $x' = xy - 3y - 4$
 $y' = x^2 - y^2$ $y' = y^2 - x^2$

17. $x' = -2xy$ **18.** $x' = x(1 - x^2 - 3y^2)$
 $y' = y - x + xy - y^3$ $y' = y(3 - x^2 - 3y^2)$

19. $x' = x(10 - x - \frac{1}{2} y)$ **20.** $x' = -2x + y + 10$
 $y' = y(16 - y - x)$ $y' = 2x - y - 15\dfrac{y}{y + 5}$

In Problems 21–26, classify (if possible) each critical point of the given second-order differential equation as a stable node, a stable spiral point, an unstable spiral point, an unstable node, or a saddle point.

21. $\theta'' = (\cos \theta - 0.5) \sin \theta, \quad |\theta| < \pi$

22. $x'' + x = (\frac{1}{2} - 3(x')^2)x' - x^2$

23. $x'' + x'(1 - x^3) - x^2 = 0$

24. $x'' + 4\dfrac{x}{1 + x^2} + 2x' = 0$

25. $x'' + x = \varepsilon x^3$ for $\varepsilon > 0$

26. $x'' + x - \varepsilon x|x| = 0$ for $\varepsilon > 0 \left[\textit{Hint:} \dfrac{d}{dx} x|x| = 2|x|.\right]$

27. Show that the nonlinear second-order differential equation

$$(1 + \alpha^2 x^2)x'' + (\beta + \alpha^2(x')^2)x = 0$$

has a saddle point at $(0, 0)$ when $\beta < 0$.

28. Show that the dynamical system

$$x' = -\alpha x + xy$$
$$y' = 1 - \beta y - x^2$$

has a unique critical point when $\alpha\beta > 1$ and that this critical point is stable when $\beta > 0$.

29. (a) Show that the plane autonomous system

$$x' = -x + y - x^3$$
$$y' = -x - y + y^2$$

has two critical points by sketching the graphs of $-x + y - x^3 = 0$ and $-x - y + y^2 = 0$. Classify the critical point at $(0, 0)$.

(b) Show that the second critical point $\mathbf{X}_1 = (0.88054, 1.56327)$ is a saddle point.

30. (a) Show that $(0, 0)$ is the only critical point of the Raleigh differential equation

$$x'' + \varepsilon(\frac{1}{3}(x')^3 - x') + x = 0.$$

(b) Show that $(0, 0)$ is unstable when $\varepsilon > 0$. When is $(0, 0)$ an unstable spiral point?

(c) Show that $(0, 0)$ is stable when $\varepsilon < 0$. When is $(0, 0)$ a stable spiral point?

(d) Show that $(0, 0)$ is a center when $\varepsilon = 0$.

31. Use the phase-plane method to show that $(0, 0)$ is a center of the nonlinear second-order differential equation $x'' + 2x^3 = 0$.

32. Use the phase-plane method to show that the solution to the nonlinear second-order differential equation $x'' + 2x - x^2 = 0$ that satisfies $x(0) = 1$ and $x'(0) = 0$ is periodic.

33. (a) Find the critical points of the plane autonomous system

$$x' = 2xy$$
$$y' = 1 - x^2 + y^2,$$

and show that linearization gives no information about the nature of these critical points.

(b) Use the phase-plane method to show that the critical points in **(a)** are both centers. [*Hint:* Let $u = y^2/x$, and show that $(x - c)^2 + y^2 = c^2 - 1$.]

34. The origin is the only critical point of the nonlinear second-order differential equation $x'' + (x')^2 + x = 0$.

(a) Show that the phase-plane method leads to the Bernoulli differential equation $dy/dx = -y - xy^{-1}$.

(b) Show that the solution satisfying $x(0) = \frac{1}{2}$ and $x'(0) = 0$ is not periodic.

35. A solution of the nonlinear second-order differential equation $x'' + x - x^3 = 0$ satisfies $x(0) = 0$ and $x'(0) = v_0$. Use the phase-plane method to determine when the resulting solution is periodic. [*Hint:* See Example 9.]

36. The nonlinear differential equation $x'' + x = 1 + \varepsilon x^2$ arises in the analysis of planetary motion using relativity theory. Classify (if possible) all critical points of the corresponding plane autonomous system.

37. When a nonlinear capacitor is present in an *LCR* circuit, the voltage drop is no longer given by q/C but is more accurately described by $\alpha q + \beta q^3$, where α and β are constants and $\alpha > 0$. Differential equation (34) of Section 3.8 for the free circuit is then replaced by

$$L\frac{d^2q}{dt^2} + R\frac{dq}{dt} + \alpha q + \beta q^3 = 0.$$

Find and classify all critical points of this nonlinear differential equation. [*Hint:* Divide into the two cases $\beta > 0$ and $\beta < 0$.]

38. The nonlinear second-order differential equation

$$mx'' + kx + k_1x^3 = 0,$$

for $k > 0$, represents a general model for the free, undamped oscillations of a mass m attached to a spring. If $k_1 > 0$, the spring is called *hard* (see Example 1 in Section 3.10). Determine the nature of the solutions to $x'' + x + x^3 = 0$ in a neighborhood of $(0, 0)$.

39. The nonlinear differential equation

$$\frac{d^2\theta}{dt^2} + \sin\ \theta = \frac{1}{2}$$

can be interpreted as a model for a certain pendulum with a constant driving function.

(a) Show that $(\pi/6, 0)$ and $(5\pi/6, 0)$ are critical points of the corresponding plane autonomous system.

(b) Classify the critical point $(5\pi/6, 0)$ using linearization.

(c) Use the phase-plane method to classify the critical point $(\pi/6, 0)$.

40. (a) Show that $(0, 0)$ is an isolated critical point of the plane autonomous system

$$x' = x^4 - 2xy^3$$
$$y' = 2x^3y - y^4$$

but that linearization gives no useful information about the nature of this critical point.

(b) Use the phase-plane method to show that $x^3 + y^3 = 3cxy$. This classic curve is called a *folium of Descartes*. Parametric equations for a folium are

$$x = \frac{3ct}{1 + t^3}, \qquad y = \frac{3ct^2}{1 + t^3}.$$

[*Hint:* The differential equation in x and y is homogeneous.]

(c) Use a graphing utility or a numerical solver to obtain solution curves. Based on your phase portrait, would you classify the critical point as stable or unstable? Would you classify the critical point as a node, saddle point, center, or spiral point? Explain.

11.4 Modeling Using Autonomous Systems

A plane autonomous system can serve as a mathematical model for diverse physical phenomena. These include problems in steady-state fluid dynamics, mechanical oscillations, circuits containing nonlinear capacitors, predator-prey and competition interactions in ecology, and the kinetics of medical drugs.

Many applications from physics give rise to nonlinear second-order differential equations of the form $x'' = g(x, x')$. For example, in the analysis of free, damped motion in Section 3.8 we assumed that the damping force was proportional to the velocity x'. Frequently the magnitude of the damping force is proportional to the square of the velocity, and the new differential equation becomes

$$x'' = -\frac{\beta}{m}x'|x'| - \frac{k}{m}x.$$

The corresponding plane autonomous system is therefore

$$x' = y$$

$$y' = -\frac{\beta}{m}y|y| - \frac{k}{m}x.$$

In this section we will use the results of Section 11.3 to analyze the nonlinear pendulum, the motion of a bead on a curve, the Lotka–Volterra predator-prey model, and the Lotka–Volterra competition model. Additional models are presented in the exercises.

■ **The Nonlinear Pendulum** In (6) of Section 3.10 we showed that the displacement angle θ for a simple pendulum satisfies the nonlinear second-order differential equation

$$\frac{d^2\theta}{dt^2} + \frac{g}{l}\sin\theta = 0.$$

When we let $x = \theta$ and $y = \theta'$, this second-order differential equation may be rewritten as the dynamical system

$$x' = y$$

$$y' = -\frac{g}{l}\sin x.$$

The critical points are $(\pm k\pi, 0)$, and the Jacobian matrix is easily shown to be

$$\mathbf{g}'((\pm k\pi, 0)) = \begin{pmatrix} 0 & 1 \\ (-1)^{k+1}\dfrac{g}{l} & 0 \end{pmatrix}.$$

If $k = 2n + 1$, $\Delta < 0$, and so all critical points $(\pm(2n + 1)\pi, 0)$ are saddle points. In particular, the critical point at $(\pi, 0)$ is unstable as expected. See Figure 11.28. When $k = 2n$, the eigenvalues are pure imaginary, and so the nature of these critical points remains in doubt. Since we have assumed that there are no damping forces acting on the pendulum, we expect that all of the critical points $(\pm 2n\pi, 0)$ are centers. This can be verified using the phase-plane method. From

$$\frac{dy}{dx} = \frac{dy/dt}{dx/dt} = -\frac{g}{l}\frac{\sin x}{y}$$

(a) $\theta = 0,\ \theta' = 0$

(b) $\theta = \pi,\ \theta' = 0$

Figure 11.28

it follows that

$$y^2 = \frac{2g}{l}\cos x + c.$$

If $\mathbf{X}(0) = (x_0, 0)$, then

$$y^2 = \frac{2g}{l}(\cos x - \cos x_0).$$

Note that $y = 0$ when $x = -x_0$, and that $(2g/l)(\cos x - \cos x_0) > 0$ for $|x| < |x_0| < \pi$. Thus each such x has two corresponding values of y, and so the solution $\mathbf{X} = \mathbf{X}(t)$ that

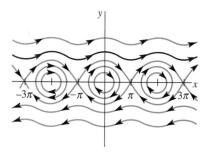

Figure 11.29

satisfies $\mathbf{X}(0) = (x_0, 0)$ is periodic. We may conclude that $(0, 0)$ is a center. Observe that $x = \theta$ increases for solutions, such as the one drawn in black in Figure 11.29, that correspond to large initial velocities. In this case the pendulum spins in complete circles about its pivot.

Example 1 Periodic Solutions of the Pendulum DE

A pendulum in an equilibrium position with $\theta = 0$ is given an initial angular velocity of ω_0 rad/s. Determine under what conditions the resulting motion is periodic.

SOLUTION We are asked to examine the solution of the plane autonomous system that satisfies $\mathbf{X}(0) = (0, \omega_0)$. From $y^2 = (2g/l) \cos x + c$ it follows that

$$y^2 = \frac{2g}{l}\left(\cos x - 1 + \frac{l}{2g}\omega_0^2\right).$$

To establish that the solution $\mathbf{X}(t)$ is periodic it is sufficient to show that there are two x-intercepts $x = \pm x_0$ between $-\pi$ and π and that the right-hand side is positive for $|x| < |x_0|$. Each such x then has two corresponding values of y.

If $y = 0$, $\cos x = 1 - (l/2g)\omega_0^2$, and this equation has two solutions $x = \pm x_0$ between $-\pi$ and π, provided $1 - (l/2g)\omega_0^2 > -1$. Note that $(2g/l)(\cos x - \cos x_0)$ is then positive for $|x| < |x_0|$. This restriction on the initial angular velocity may be written as

$$|\omega_0| < 2\sqrt{\frac{g}{l}}.$$

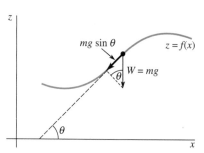

Figure 11.30

■ **Nonlinear Oscillations: The Sliding Bead** Suppose, as shown in Figure 11.30, a bead with mass m slides along a thin wire whose shape is described by the function $z = f(x)$. A wide variety of nonlinear oscillations can be obtained by changing the shape of the wire and by making different assumptions about the forces acting on the bead.

The tangential force \mathbf{F} due to the weight $W = mg$ has magnitude $mg \sin \theta$, and therefore the x-component of \mathbf{F} is $F_x = -mg \sin \theta \cos \theta$. Since $\tan \theta = f'(x)$, we may use the identities $1 + \tan^2\theta = \sec^2\theta$ and $\sin^2\theta = 1 - \cos^2\theta$ to conclude that

$$F_x = -mg \sin \theta \cos \theta = -mg\frac{f'(x)}{1 + [f'(x)]^2}.$$

We assume (as in Section 3.8) that a damping force \mathbf{D}, acting in the direction opposite to the motion, is a constant multiple of the velocity of the bead. The x-component of \mathbf{D} is therefore

$$D_x = -\beta\frac{dx}{dt}.$$

If we ignore the frictional force between the wire and the bead and assume that no other external forces are impressed on the system, it follows from Newton's second law that

$$mx'' = -mg\frac{f'(x)}{1 + [f'(x)]^2} - \beta x',$$

and the corresponding plane autonomous system is

$$x' = y$$

$$y' = -g\frac{f'(x)}{1 + [f'(x)]^2} - \frac{\beta}{m}y.$$

If $\mathbf{X}_1 = (x_1, y_1)$ is a critical point of the system, $y_1 = 0$ and therefore $f'(x_1) = 0$. The bead must therefore be at rest at a point on the wire where the tangent line is horizontal. When f is twice differentiable, the Jacobian matrix at \mathbf{X}_1 is

$$\mathbf{g}'(\mathbf{X}_1) = \begin{pmatrix} 0 & 1 \\ -gf''(x_1) & -\beta/m \end{pmatrix},$$

and so $\tau = -\beta/m$, $\Delta = gf''(x_1)$, and $\tau^2 - 4\Delta = \beta^2/m^2 - 4gf''(x_1)$. Using the results of Section 11.3, we can make the following conclusions:

(i) $f''(x_1) < 0$

A relative maximum therefore occurs at $x = x_1$, and since $\Delta < 0$, an *unstable saddle point* occurs at $\mathbf{X}_1 = (x_1, 0)$.

(ii) $f''(x_1) > 0$ and $\beta > 0$

A relative minimum therefore occurs at $x = x_1$, and since $\tau < 0$ and $\Delta > 0$, $\mathbf{X}_1 = (x_1, 0)$ is a *stable critical point*. If $\beta^2 > 4gm^2f''(x_1)$, the system is **overdamped** and the critical point is a *stable node*. If $\beta^2 < 4gm^2f''(x_1)$, the system is **underdamped** and the critical point is a *stable spiral point*. The exact nature of the stable critical point is still in doubt if $\beta^2 = 4gm^2f''(x_1)$.

(iii) $f''(x_1) > 0$ and the system is undamped ($\beta = 0$)

In this case the eigenvalues are pure imaginary, but the phase-plane method can be used to show that the critical point is a *center*. Therefore solutions with $\mathbf{X}(0) = (x(0), x'(0))$ near $\mathbf{X}_1 = (x_1, 0)$ are periodic.

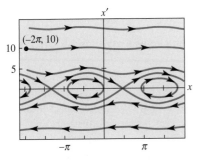

Figure 11.31

Example 2 Bead Sliding along a Sine Wave

A 10-gram bead slides along the graph of $z = \sin x$. According to conclusion (*ii*), the relative minima at $x_1 = -\pi/2$ and $3\pi/2$ give rise to stable critical points (see Figure 11.31). Since $f''(-\pi/2) = f''(3\pi/2) = 1$, the system will be underdamped provided $\beta^2 < 4gm^2$. If we use SI units, $m = 0.01$ kg and $g = 9.8$ m/s^2, and so the condition for an underdamped system becomes $\beta^2 < 3.92 \times 10^{-3}$.

If $\beta = 0.01$ is the damping constant, both of these critical points are stable spiral points. The two solutions corresponding to initial conditions $\mathbf{X}(0) = (x(0), x'(0)) = (-2\pi, 10)$ and $\mathbf{X}(0) = (-2\pi, 15)$, respectively, were obtained from a numerical solver and are shown in Figure 11.32. When $x'(0) = 10$, the bead has enough momentum to make it over the hill at $x = -3\pi/2$ but not over the hill at $x = \pi/2$. The bead then approaches the relative minimum based at $x = -\pi/2$. If $x'(0) = 15$, the bead has the momentum to make it over both hills, but then it rocks back and forth in the valley based at $x = 3\pi/2$ and approaches the point $(3\pi/2, -1)$ on the wire. Experiment with other initial conditions using your numerical solver.

Figure 11.33 shows a collection of solution curves obtained from a numerical solver for the undamped case. Since $\beta = 0$, the critical points corresponding to $x_1 = -\pi/2$ and $3\pi/2$ are now centers. When $\mathbf{X}(0) = (-2\pi, 10)$, the bead has sufficient momentum to move over *all* hills. The figure also indicates that when the bead is released from rest at a position on the wire between $x = -3\pi/2$ and $x = \pi/2$, the resulting motion is periodic. ❏

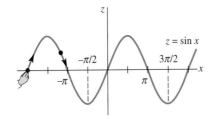

Figure 11.32 $\beta = 0.01$

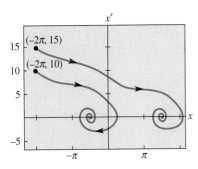

Figure 11.33 $\beta = 0$

■ Lotka–Volterra Predator-Prey Model A **predator-prey interaction** between two species occurs when one species (the predator) feeds on a second species (the prey). For example, the snowy owl feeds almost exclusively on a common arctic rodent called a lemming, while a lemming uses arctic tundra

plants as its food supply. Interest in using mathematics to help explain predator-prey interactions has been stimulated by the observation of population cycles in many arctic mammals. In the MacKenzie River district of Canada, for example, the principal prey of the lynx is the snowshoe hare, and both populations cycle with a period of about 10 years.

There are many predator-prey models that lead to plane autonomous systems with at least one periodic solution. The first such model was constructed independently by pioneer biomathematicians A. Lotka (1925) and V. Volterra (1926). If x denotes the number of predators and y denotes the number of prey, then the Lotka–Volterra model takes the form

$$x' = -ax + bxy = x(-a + by)$$
$$y' = -cxy + dy = y(-cx + d),$$

where a, b, c, and d are positive constants.

Note that in the absence of predators ($x = 0$), $y' = dy$, and so the number of prey grows exponentially. In the absence of prey, $x' = -ax$, and so the predator population becomes extinct. The term $-cxy$ represents the death rate due to predation. The model therefore assumes that this death rate is directly proportional to the number of *possible encounters* xy between predator and prey at a particular time t, and the term bxy represents the resulting positive contribution to the predator population.

The critical points of this plane autonomous system are $(0, 0)$ and $(d/c, a/b)$, and the corresponding Jacobian matrices are

$$\mathbf{A}_1 = \mathbf{g}'((0, 0)) = \begin{pmatrix} -a & 0 \\ 0 & d \end{pmatrix} \quad \text{and} \quad \mathbf{A}_2 = \mathbf{g}'((d/c, a/b)) = \begin{pmatrix} 0 & bd/c \\ -ac/b & 0 \end{pmatrix}.$$

The critical point at $(0, 0)$ is a saddle point, and Figure 11.34 shows a typical profile of solutions that are in the first quadrant and near $(0, 0)$.

Since the matrix \mathbf{A}_2 has pure imaginary eigenvalues $\lambda = \pm\sqrt{ad}\,i$, the critical point $(d/c, a/b)$ *may* be a center. This possibility can be investigated using the phase-plane method. Since

$$\frac{dy}{dx} = \frac{y(-cx + d)}{x(-a + by)},$$

we may separate variables and obtain

$$\int \frac{-a + by}{y}\, dy = \int \frac{-cx + d}{x}\, dx$$

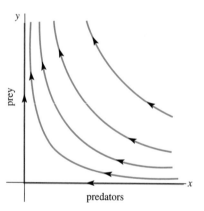

Figure 11.34 Solutions near $(0, 0)$

so that

$$-a \ln y + by = -cx + d \ln x + c_1 \quad \text{or} \quad (x^d e^{-cx})(y^a e^{-by}) = c_0.$$

The following argument establishes that all solution curves that originate in the first quadrant are periodic.

Typical graphs of the nonnegative functions $F(x) = x^d e^{-cx}$ and $G(y) = y^a e^{-by}$ are shown in Figure 11.35. It is not hard to show that $F(x)$ has an absolute maximum at $x = d/c$, whereas $G(y)$ has an absolute maximum at $y = a/b$. Note that, with the exception of 0 and the absolute maximum, F and G each take on all values in their range precisely twice.

These graphs can be used to establish the following properties of a solution curve that originates at a noncritical point (x_0, y_0) in the first quadrant.

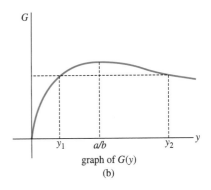

graph of $F(x)$ (a) graph of $G(y)$ (b)

Figure 11.35

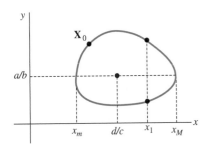

Figure 11.36

1. If $y = a/b$, the equation $F(x)G(y) = c_0$ has exactly two solutions x_m and x_M that satisfy $x_m < d/c < x_M$.
2. If $x_m < x_1 < x_M$ and $x = x_1$, then $F(x)G(y) = c_0$ has exactly two solutions y_1 and y_2 that satisfy $y_1 < a/b < y_2$.
3. If x is outside the interval $[x_m, x_M]$, then $F(x)G(y) = c_0$ has no solutions.

We will give the demonstration of (1) and outline parts (2) and (3) in the exercises. Since $(x_0, y_0) \neq (d/c, a/b)$, $F(x_0)G(y_0) < F(d/c)G(a/b)$. If $y = a/b$, then

$$0 < \frac{c_0}{G(a/b)} = \frac{F(x_0)G(y_0)}{G(a/b)} < \frac{F(d/c)G(a/b)}{G(a/b)} = F(d/c).$$

Therefore $F(x) = c_0/G(a/b)$ has precisely two solutions x_m and x_M that satisfy $x_m < d/c < x_M$. The graph of a typical periodic solution is shown in Figure 11.36.

> ### *Example 3* **Predator-Prey Population Cycles**
> If we let $a = 0.1$, $b = 0.002$, $c = 0.0025$, and $d = 0.2$ in the Lotka–Volterra predator-prey model, the critical point in the first quadrant is $(d/c, a/b) = (80, 50)$, and we know that this critical point is a center. See Figure 11.37 in which we have used a numerical solver to generate these cycles. The closer the initial condition \mathbf{X}_0 is to $(80, 50)$, the more the periodic solutions resemble the elliptical solutions to the corresponding linear system. The eigenvalues of $\mathbf{g}'((80, 50))$ are $\lambda = \pm\sqrt{ad}\,i = \pm(\sqrt{2}/10)i$, and so the solutions near the critical point have period $p \approx 10\sqrt{2}\,\pi$, or about 44.4. ❑

Figure 11.37

■ **Lotka–Volterra Competition Model** A **competitive interaction** occurs when two or more species compete for the food, water, light, and space resources of an ecosystem. The use of one of these resources by one population therefore inhibits the ability of another population to survive and grow. Under what conditions can two competing species coexist? A number of mathematical models have been constructed that offer insights into conditions that permit *coexistence*. If x denotes the number in species I and y denotes the number in species II, then the Lotka–Volterra model takes the form

$$x' = \frac{r_1}{K_1}x(K_1 - x - \alpha_{12}y)$$

$$y' = \frac{r_2}{K_2}y(K_2 - y - \alpha_{21}x). \tag{1}$$

(a) $\alpha_{12}\alpha_{21} < 1$

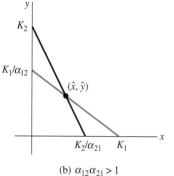

(b) $\alpha_{12}\alpha_{21} > 1$

Figure 11.38

Note that in the absence of species II ($y = 0$), $x' = (r_1/K_1)x(K_1 - x)$, and so the first population grows logistically and approaches the steady-state population K_1 (see Section 2.8 and Example 4 in Section 11.3). A similar statement holds for species II growing in the absence of species I. The term $-\alpha_{21}xy$ in the second equation stems from the competitive effect of species I on species II. The model therefore assumes that this rate of inhibition is directly proportional to the number of possible competitive pairs xy at a particular time t.

This plane autonomous system has critical points at $(0, 0)$, $(K_1, 0)$, and $(0, K_2)$. When $\alpha_{12}\alpha_{21} \neq 0$, the lines $K_1 - x - \alpha_{12}y = 0$ and $K_2 - y - \alpha_{21}x = 0$ intersect to produce a fourth critical point $\hat{\mathbf{X}} = (\hat{x}, \hat{y})$. Figure 11.38 shows the two conditions under which (\hat{x}, \hat{y}) is in the first quadrant.

The trace and determinant of the Jacobian matrix at (\hat{x}, \hat{y}) are, respectively,

$$\tau = -\hat{x}\frac{r_1}{K_1} - \hat{y}\frac{r_2}{K_2} \quad \text{and} \quad \Delta = (1 - \alpha_{12}\alpha_{21})\hat{x}\hat{y}\frac{r_1r_2}{K_1K_2}.$$

In case (a), $K_1/\alpha_{12} > K_2$ and $K_2/\alpha_{21} > K_1$. It follows that $\alpha_{12}\alpha_{21} < 1$, $\tau < 0$, and $\Delta > 0$. Since

$$\tau^2 - 4\Delta = \left(\hat{x}\frac{r_1}{K_1} + \hat{y}\frac{r_2}{K_2}\right)^2 + 4(\alpha_{12}\alpha_{21} - 1)\hat{x}\hat{y}\frac{r_1r_2}{K_1K_2}$$

$$= \left[\left(\hat{x}\frac{r_1}{K_1} - \hat{y}\frac{r_2}{K_2}\right)^2 - 4\hat{x}\hat{y}\frac{r_1r_2}{K_1K_2}\right] + 4\alpha_{12}\alpha_{21}\,\hat{x}\hat{y}\frac{r_1r_2}{K_1K_2}$$

$$= \left(\hat{x}\frac{r_1}{K_1} - \hat{y}\frac{r_2}{K_2}\right)^2 + 4\alpha_{12}\alpha_{21}\,\hat{x}\hat{y}\frac{r_1r_2}{K_1K_2},$$

$\tau^2 - 4\Delta > 0$, and so (\hat{x}, \hat{y}) is a stable node. Therefore if $\mathbf{X}(0) = \mathbf{X}_0$ is sufficiently close to $\hat{\mathbf{X}} = (\hat{x}, \hat{y})$, $\lim_{t \to \infty} \mathbf{X}(t) = \hat{\mathbf{X}}$, and we may conclude that coexistence is possible. The demonstration that case (b) leads to a saddle point and the investigation of the nature of critical points at $(0, 0)$, $(K_1, 0)$, and $(0, K_2)$ are left to the exercises.

When the competitive interactions between two species are weak, both of the coefficients α_{12} and α_{21} will be small, and so the conditions $K_1/\alpha_{12} > K_2$ and $K_2/\alpha_{21} > K_1$ may be satisfied. This might occur when there is a small overlap in the ranges of two predator species that hunt for a common prey.

Example 4 A Lotka–Volterra Competition Model

A competitive interaction is described by the Lotka–Volterra competition model

$$x' = 0.004x(50 - x - 0.75y)$$
$$y' = 0.001y(100 - y - 3.0x).$$

Find and classify all critical points of the system.

SOLUTION Critical points occur at $(0, 0)$, $(50, 0)$, $(0, 100)$ and at the solution $(20, 40)$ of the system

$$x + 0.75y = 50$$
$$3.0x + y = 100.$$

Since $\alpha_{12}\alpha_{21} = 2.25 > 1$, we have case (b) in Figure 11.38, and so the critical point at $(20, 40)$ is a saddle point. The Jacobian matrix is

$$\mathbf{g}'(\mathbf{X}) = \begin{pmatrix} 0.2 - 0.008x - 0.003y & -0.003x \\ -0.003y & 0.1 - 0.002y - 0.003x \end{pmatrix},$$

and we obtain

$$\mathbf{A}_1 = \mathbf{g}'((0, 0)) = \begin{pmatrix} 0.2 & 0 \\ 0 & 0.1 \end{pmatrix} \qquad\qquad \mathbf{A}_2 = \mathbf{g}'((50, 0)) = \begin{pmatrix} -0.2 & -0.15 \\ 0 & -0.05 \end{pmatrix}$$

$$\mathbf{A}_3 = \mathbf{g}'((20, 40)) = \begin{pmatrix} -0.08 & -0.12 \\ -0.06 & -0.04 \end{pmatrix} \qquad \mathbf{A}_4 = \mathbf{g}'((0, 100)) = \begin{pmatrix} -0.1 & 0 \\ -0.3 & -0.1 \end{pmatrix}.$$

Therefore $(0, 0)$ is an unstable node, whereas both $(50, 0)$ and $(0, 100)$ are stable nodes. (Check this!) Since $\det \mathbf{A}_3 < 0$, we have a second demonstration that $(20, 40)$ is a saddle point. ❑

Coexistence can also occur in the Lotka–Volterra competition model if there is at least one periodic solution lying entirely in the first quadrant. It is possible to show, however, that this model has no periodic solutions.

EXERCISES 11.4

Answers to odd-numbered problems begin on page A-56.

The Nonlinear Pendulum

1. A pendulum is released at $\theta = \pi/3$ and is given an initial angular velocity of ω_0 rad/s. Determine under what conditions the resulting motion is periodic.

2. (a) If a pendulum is released from rest at $\theta = \theta_0$, show that the angular velocity is again 0 when $\theta = -\theta_0$.

(b) The period T of the pendulum is the amount of time needed for θ to change from θ_0 to $-\theta_0$ and back to θ_0. Show that

$$T = \sqrt{\frac{2L}{g}} \int_{-\theta_0}^{\theta_0} \frac{1}{\sqrt{\cos\theta - \cos\theta_0}} \, d\theta.$$

The Sliding Bead

3. A bead with mass m slides along a thin wire whose shape is described by the function $z = f(x)$. If $\mathbf{X}_1 = (x_1, y_1)$ is a critical point of the plane autonomous system associated with the sliding bead, verify that the Jacobian matrix at \mathbf{X}_1 is

$$\mathbf{g}'(\mathbf{X}_1) = \begin{pmatrix} 0 & 1 \\ -gf''(x_1) & -\beta/m \end{pmatrix}.$$

4. A bead with mass m slides along a thin wire whose shape is described by the function $z = f(x)$. When $f'(x_1) = 0$, $f''(x_1) > 0$, and the system is undamped, the critical point $\mathbf{X}_1 = (x_1, 0)$ is a center. Estimate the period of the bead when $x(0)$ is near x_1 and $x'(0) = 0$.

5. A bead is released from the position $x(0) = x_0$ on the curve $z = x^2/2$ with initial velocity $x'(0) = v_0$ cm/s.

(a) Use the phase-plane method to show that the resulting solution is periodic when the system is undamped.

(b) Show that the maximum height z_{max} to which the bead rises is given by

$$z_{max} = \frac{1}{2} \left[e^{v_0^2/g}(1 + x_0^2) - 1 \right].$$

6. Rework Problem 5 with $z = \cosh x$.

Interaction Models

7. (Refer to Figure 11.36.) If $x_m < x_1 < x_M$ and $x = x_1$, show that $F(x)G(y) = c_0$ has exactly two solutions y_1 and y_2 that satisfy $y_1 < a/b < y_2$. [*Hint:* First show that $G(y) = c_0/F(x_1) < G(a/b)$.]

8. From (1) and (3) on page 631, conclude that the maximum number of predators occurs when $y = a/b$.

9. In many fishery science models the rate at which a species is caught is assumed to be directly proportional to its abundance. If both predator and prey are being exploited in this manner, the Lotka–Volterra differential equations take the form

$$x' = -ax + bxy - \varepsilon_1 x$$
$$y' = -cxy + dy - \varepsilon_2 y,$$

where ε_1 and ε_2 are positive constants.

(a) When $\varepsilon_2 < d$, show that there is a new critical point in the first quadrant that is a center.

(b) Volterra's principle states that a moderate amount of exploitation increases the average number of prey and decreases the average number of predators. Is this fisheries model consistent with Volterra's principle?

10. A predator-prey interaction is described by the Lotka–Volterra model

$$x' = -0.1x + 0.02xy$$
$$y' = 0.2y - 0.025xy.$$

(a) Find the critical point in the first quadrant, and use a numerical solver to sketch some population cycles.

(b) Estimate the period of the periodic solutions that are close to the critical point in (a).

11. A competitive interaction is described by the Lotka–Volterra competition model

$$x' = 0.08x(20 - 0.4x - 0.3y)$$
$$y' = 0.06y(10 - 0.1y - 0.3x).$$

Find and classify all critical points of the system.

12. In (1) show that $(0, 0)$ is always an unstable node.

13. In (1) show that $(K_1, 0)$ is a stable node when $K_1 > K_2/\alpha_{21}$ and a saddle point when $K_1 < K_2/\alpha_{21}$.

14. Use Problems 12 and 13 to establish that $(0, 0)$, $(K_1, 0)$, and $(0, K_2)$ are unstable when $\hat{\mathbf{X}} = (\hat{x}, \hat{y})$ is a stable node.

15. In (1) show that $\hat{\mathbf{X}} = (\hat{x}, \hat{y})$ is a saddle point when

$$\frac{K_1}{\alpha_{12}} < K_2 \quad \text{and} \quad \frac{K_2}{\alpha_{21}} < K_1.$$

Miscellaneous Nonlinear Models

16. If we assume that a damping force acts in a direction opposite to the motion of a pendulum and with a magnitude directly proportional to the angular velocity $d\theta/dt$, the displacement angle θ for the pendulum satisfies the nonlinear second-order differential equation

$$ml\frac{d^2\theta}{dt^2} = -mg \sin \theta - \beta \frac{d\theta}{dt}.$$

(a) Write the second-order differential equation as a plane autonomous system, and find all critical points.

(b) Find a condition on m, l, and β that will make $(0, 0)$ a stable spiral point.

17. In the analysis of free, damped motion in Section 3.8 we assumed that the damping force was proportional to the velocity x'. Frequently the magnitude of this damping force is proportional to the square of the velocity, and the new differential equation becomes

$$x'' = -\frac{\beta}{m}x'|x'| - \frac{k}{m}x.$$

(a) Write the second-order differential equation as a plane autonomous system, and find all critical points.

(b) The system is called *overdamped* when $(0, 0)$ is a stable node and is called *underdamped* when $(0, 0)$ is a stable spiral point. Physical considerations suggest that $(0, 0)$ must be an asymptotically stable critical point. Show that the system is necessarily underdamped.

$$\left[\text{Hint: } \frac{d}{dy}(y|y|) = 2|y|. \right]$$

18. A bead with mass m slides along a thin wire whose shape may be described by the function $z = f(x)$. Small stretches of the wire act like an inclined plane, and in mechanics it is assumed that the magnitude of the frictional force between

the bead and wire is directly proportional to $mg \cos \theta$. See Figure 11.30.

(a) Explain why the new differential equation for the x-coordinate of the bead is

$$x'' = g\frac{\mu - f'(x)}{1 + [f'(x)]^2} - \frac{\beta}{m}x'$$

for some positive constant μ.

(b) Investigate the critical points of the corresponding plane autonomous system. Under what conditions is a critical point a saddle point? A stable spiral point?

19. An undamped oscillation satisfies a nonlinear second-order differential equation of the form $x'' + f(x) = 0$, where $f(0) = 0$ and $xf(x) > 0$ for $x \neq 0$ and $-d < x < d$. Use the phase-plane method to investigate whether it is possible for the critical point $(0, 0)$ to be a stable spiral point. [*Hint:* Let $F(x) = \int_0^x f(u)\, du$, and show that $y^2 + 2F(x) = c$.]

20. The Lotka–Volterra predator-prey model assumes that, in the absence of predators, the number of prey grows exponentially. If we make the alternative assumption that the prey population grows logistically, the new system is

$$x' = -ax + bxy$$
$$y' = -cxy + \frac{r}{K}y(K - y),$$

where a, b, c, r, and K are positive and $K > a/b$.

(a) Show that the system has critical points at $(0, 0)$, $(0, K)$, and (\hat{x}, \hat{y}), where $\hat{y} = a/b$ and $c\hat{x} = \frac{r}{K}(K - \hat{y})$.

(b) Show that the critical points at $(0, 0)$ and $(0, K)$ are saddle points, whereas the critical point at (\hat{x}, \hat{y}) is either a stable node or a stable spiral point.

(c) Show that (\hat{x}, \hat{y}) is a stable spiral point if $\hat{y} < \frac{4bK^2}{r + 4bK}$. Explain why this case will occur when the carrying capacity K of the prey is large.

21. The nonlinear system

$$x' = \alpha\frac{y}{1 + y}x - x$$

$$y' = -\frac{y}{1 + y}x - y + \beta$$

arises in a model for the growth of microorganisms in a *chemostat*, a simple laboratory device in which a nutrient from a supply source flows into a growth chamber. In the system, x denotes the concentration of the microorganisms in the growth chamber, y denotes the concentration of

nutrients, and $\alpha > 1$ and $\beta > 0$ are constants that can be adjusted by the experimenter. Find conditions on α and β that ensure that the system has a single critical point (\hat{x}, \hat{y}) in the first quadrant, and investigate the stability of this critical point.

22. Use the methods of this chapter together with a numerical solver to investigate stability in the nonlinear spring/mass system modeled by

$$x'' + 8x - 6x^3 + x^5 = 0.$$

See Problem 8 in Exercises 3.10.

11.5 Periodic Solutions, Limit Cycles, and Global Stability

In this section we will investigate the existence of periodic solutions of nonlinear plane autonomous systems and introduce special periodic solutions called limit cycles.

We saw in Sections 11.3 and 11.4 that an analysis of critical points using linearization can provide valuable information on the behavior of solutions near critical points and insight into a variety of biological and physical phenomena. There are, however, some inherent limitations to this approach. When the eigenvalues of the Jacobian matrix are pure imaginary, we cannot conclude that there are periodic solutions near the critical point. In some cases we were able to solve $dy/dx = Q(x, y)/P(x, y)$, obtain an implicit representation $f(x, y) = c$ of the solution curves, and investigate whether any of these solutions formed closed curves. More often than not, this differential equation will not possess closed form solutions. For example, the Lotka–Volterra competition model cannot be handled by this procedure. The first goal of this section is *to determine conditions under which we can either exclude the possibility of periodic solutions or assert their existence.*

We encountered an additional problem in studying the models in Section 11.3. Figure 11.39 illustrates the common situation in which a region R contains a *single* asymptotically stable critical point \mathbf{X}_1. We can assert that $\lim_{t \to \infty} \mathbf{X}(t) = \mathbf{X}_1$ when the initial position $\mathbf{X}(0) = \mathbf{X}_0$ is "near" \mathbf{X}_1, but under what conditions is $\lim_{t \to \infty} \mathbf{X}(t) = \mathbf{X}_1$ for *all initial positions* in R? Such a critical point is called **globally stable** in R. A second goal is *to determine conditions under which an asymptotically stable critical point is globally stable.*

In motivating and discussing the methods in this section, we will use the fact that the vector field $\mathbf{V}(x, y) = (P(x, y), Q(x, y))$ can be interpreted as defining a fluid flow in a region of the plane, and a solution to the autonomous system may be interpreted as the resulting path of a particle as it moves through the region.

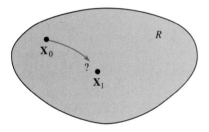

Figure 11.39

■ **Negative Criteria** A number of results can sometimes be used to establish that there are no periodic solutions in a given region R of the plane. We will assume that $P(x, y)$ and $Q(x, y)$ have continuous first partial derivatives in R and that R is simply connected. Recall that in a simply connected region, any simple closed curve C in R encloses only points in R. Therefore, if there is a periodic solution $\mathbf{X} = \mathbf{X}(t)$ in R, then R will contain all points in the interior of the resulting curve.

THEOREM 11.4 **Cycles and Critical Points**

If a plane autonomous system has a periodic solution $\mathbf{X} = \mathbf{X}(t)$ in a simply connected region R, then the system has at least one critical point inside the corresponding simple closed curve C. If there is a single critical point inside C, then that critical point cannot be a saddle point.

> ### COROLLARY
>
> If a simply connected region R either contains no critical points of a plane autonomous system or contains a single saddle point, then there are no periodic solutions in R.

Example 1 No Periodic Solutions

Show that the plane autonomous system

$$x' = xy$$
$$y' = -1 - x^2 - y^2$$

has no periodic solutions.

SOLUTION If (x, y) is a critical point, then, from the first equation, either $x = 0$ or $y = 0$. If $x = 0$, then $-1 - y^2 = 0$ or $y^2 = -1$. Likewise, $y = 0$ implies $x^2 = -1$. Therefore, this plane autonomous system has no critical points, and by the corollary possesses no periodic solutions in the plane. ❏

Example 2 Lotka–Volterra Competition Model

Show that the Lotka–Volterra competition model

$$x' = 0.004x(50 - x - 0.75y)$$
$$y' = 0.001y(100 - y - 3.0x)$$

has no periodic solutions in the first quadrant.

SOLUTION In Example 4 in Section 11.4 we showed that this system has critical points at $(0, 0)$, $(50, 0)$, $(0, 100)$, and $(20, 40)$, and that $(20, 40)$ is a saddle point. Since only $(20, 40)$ lies in the first quadrant, by the corollary there are no periodic solutions in the first quadrant. ❏

Another sometimes useful result can be formulated in terms of the divergence of the vector field $\mathbf{V}(x, y) = (P(x, y), Q(x, y))$:

> ### THEOREM 11.5 Bendixson Negative Criterion
>
> If div $\mathbf{V} = \partial P/\partial x + \partial Q/\partial y$ does not change sign in a simply connected region R, then the plane autonomous system has no periodic solutions in R.

PROOF Suppose, to the contrary, that there is a periodic solution $\mathbf{X} = \mathbf{X}(t)$ lying in R, and let C be the resulting simple closed curve and R_1 the region bounded by C. Green's theorem states that

$$\int_C M(x, y)\, dx + N(x, y)\, dy = \iint_{R_1} \left(\frac{\partial N}{\partial x} - \frac{\partial M}{\partial y} \right) dx\, dy$$

whenever $M(x, y)$ and $N(x, y)$ have continuous first partials in R. If we let $N = P$ and $M = -Q$, we obtain

$$\int_C -Q(x, y)\, dx + P(x, y)\, dy = \iint_{R_1} \left(\frac{\partial P}{\partial x} + \frac{\partial Q}{\partial y} \right) dx\, dy.$$

Since $\mathbf{X} = \mathbf{X}(t)$ is a solution with period p, we have $x'(t) = P(x(t), y(t))$ and $y'(t) = Q(x(t), y(t))$, and so

$$\int_C - Q(x, y)\, dx + P(x, y)\, dy = \int_0^p [-Q(x(t), y(t))x'(t) + P(x(t), y(t))y'(t)]\, dt$$

$$= \int_0^p [-QP + PQ]\, dt = 0.$$

Since div $\mathbf{V} = \partial P/\partial x + \partial Q/\partial y$ is continuous and does not change sign in R, it follows that either div $\mathbf{V} \geq 0$ in R or div $\mathbf{V} \leq 0$ in R, and so

$$\iint_{R_1} \left(\frac{\partial P}{\partial x} + \frac{\partial Q}{\partial y} \right) dx\, dy \neq 0.$$

This contradiction establishes that there are no periodic solutions in R. ❑

Example 3 Bendixson Negative Criterion

Investigate possible periodic solutions of each system.

(a) $x' = x + 2y + 4x^3 - y^2$ (b) $x' = y + x(2 - x^2 - y^2)$

 $y' = -x + 2y + yx^2 + y^3$ $y' = -x + y(2 - x^2 - y^2)$

SOLUTION (a) We have div $\mathbf{V} = \partial P/\partial x + \partial Q/\partial y = 1 + 12x^2 + 2 + x^2 + 3y^2 \geq 3$, and so there are no periodic solutions in the plane.

(b) For this system div $\mathbf{V} = (2 - 3x^2 - y^2) + (2 - x^2 - 3y^2) = 4 - 4(x^2 + y^2)$. Therefore, if R is the interior of the circle $x^2 + y^2 = 1$, div $\mathbf{V} > 0$, and so there are no periodic solutions inside this disk. Note that div $\mathbf{V} < 0$ on the exterior of the circle. If R is any simply connected subset of the exterior, then there are no periodic solutions in R. It follows that if there is a periodic solution in the exterior, it must enclose the circle $x^2 + y^2 = 1$. In fact, the reader can verify that $\mathbf{X}(t) = (\sqrt{2} \sin t, \sqrt{2} \cos t)$ is a periodic solution that generates the circle $x^2 + y^2 = 2$. ❑

Example 4 Sliding Bead and Periodic Solutions

The sliding bead discussed in Section 11.4 satisfies the differential equation

$$mx'' = -mg \frac{f'(x)}{1 + [f'(x)]^2} - \beta x'.$$

Show that there are no periodic solutions.

SOLUTION The corresponding plane autonomous system is

$$x' = y$$

$$y' = -g \frac{f'(x)}{1 + [f'(x)]^2} - \frac{\beta}{m} y$$

and so div $\mathbf{V} = \frac{\partial P}{\partial x} + \frac{\partial Q}{\partial y} = -\frac{\beta}{m} < 0.$ ❑

The following theorem is a generalization of the Bendixson negative criterion, which leaves it to the reader to construct an appropriate function $\delta(x, y)$.

> ### THEOREM 11.6 Dulac Negative Criterion
>
> If $\delta(x, y)$ has continuous first partial derivatives in a simply connected region R and $\dfrac{\partial(\delta P)}{\partial x} + \dfrac{\partial(\delta Q)}{\partial y}$ does not change sign in R, then the plane autonomous system has no periodic solutions in R.

There are no general techniques for constructing an appropriate function $\delta(x, y)$. Instead, we experiment with simple functions of the form $ax^2 + by^2$, $e^{ax + by}$, $x^a y^b$, and so on, and try to determine constants for which $\partial(\delta P)/\partial x + \partial(\delta Q)/\partial y$ is nonzero in a given region.

Example 5 Dulac Negative Criterion

Show that the second-order nonlinear differential equation

$$x'' = x^2 + (x')^2 - x - x'$$

has no periodic solutions.

SOLUTION The corresponding plane autonomous system is

$$x' = y$$
$$y' = x^2 + y^2 - x - y.$$

If we let $\delta(x, y) = e^{ax + by}$, then

$$\frac{\partial(\delta P)}{\partial x} + \frac{\partial(\delta Q)}{\partial y} = e^{ax + by}(ay + 2y - 1) + e^{ax + by}b(x^2 + y^2 - x - y).$$

If we set $a = -2$ and $b = 0$, then $\partial(\delta P)/\partial x + \partial(\delta Q)/\partial y = -e^{ax + by}$, which is always negative. Therefore, by the Dulac negative criterion, the second-order differential equation has no periodic solutions. ❏

Example 6 Dulac Negative Criterion

Use $\delta(x, y) = 1/(xy)$ to show that the Lotka–Volterra competition equations

$$x' = \frac{r_1}{K_1}x(K_1 - x - \alpha_{12}y)$$

$$y' = \frac{r_2}{K_2}y(K_2 - y - \alpha_{21}x)$$

have no periodic solutions in the first quadrant.

SOLUTION If $\delta(x, y) = 1/(xy)$, then

$$\delta P = \frac{r_1}{K_1}\left[\frac{K_1}{y} - \frac{x}{y} - \alpha_{12}\right] \quad \text{and} \quad \delta Q = \frac{r_2}{K_2}\left[\frac{K_2}{x} - \frac{y}{x} - \alpha_{21}\right]$$

and so

$$\frac{\partial(\delta P)}{\partial x} + \frac{\partial(\delta Q)}{\partial y} = \frac{r_1}{K_1}\left(-\frac{1}{y}\right) + \frac{r_2}{K_2}\left(-\frac{1}{x}\right).$$

For (x, y) in the first quadrant, the latter expression is always negative. Therefore, there are no periodic solutions. ❏

■ **Positive Criteria: Poincaré–Bendixson Theory** The Poincaré–Bendixson theorem is an advanced result that describes the long-range behavior of a *bounded* solution to a plane autonomous system. Rather than present the result in its full generality, we will concentrate on a number of special cases that occur frequently in applications. One of these cases will lead to a new type of periodic solution called a *limit cycle*.

> **DEFINITION 11.3** Invariant Region
>
> A region R is called an **invariant region** for a plane autonomous system if, whenever \mathbf{X}_0 is in R, the solution $\mathbf{X} = \mathbf{X}(t)$ satisfying $\mathbf{X}(0) = \mathbf{X}_0$ remains in R.

Figure 11.40 shows two standard types of invariant regions. A **Type I invariant region** R is bounded by a simple closed curve C, and the flow at the boundary defined by the vector field $\mathbf{V}(x, y) = (P(x, y), Q(x, y))$ is always directed into the region. This prevents a particle from crossing the boundary. A **Type II invariant region** is an annular region bounded by simple closed curves C_1 and C_2, and the flow at the boundary is again directed toward the interior of R. The following theorem provides a method for verifying that a given region is invariant.

(a) Type I invariant region

(b) Type II invariant region

Figure 11.40

> **THEOREM 11.7** Normal Vectors and Invariant Regions
>
> If $\mathbf{n}(x, y)$ denotes a normal vector on the boundary that points inside the region, then R will be an invariant region for the plane autonomous system provided $\mathbf{V}(x, y) \cdot \mathbf{n}(x, y) \geq 0$ for all points (x, y) on the boundary.

PROOF If θ is the angle between $\mathbf{V}(x, y)$ and $\mathbf{n}(x, y)$, then from $\mathbf{V} \cdot \mathbf{n} = \|\mathbf{V}\| \|\mathbf{n}\| \cos \theta$, we may conclude that $\cos \theta \geq 0$ and so θ is between $0°$ and $90°$. The flow is therefore directed into the region (or at worst along the boundary) for any boundary point (x, y). This prevents a solution that starts in R from leaving R. Therefore, R is an invariant region for the plane autonomous system. ❑

The problem of finding an invariant region for a given nonlinear system is an extremely difficult one. An excellent first step is to use software that plots the vector field $\mathbf{V}(x, y) = (P(x, y), Q(x, y))$ together with the curves $P(x, y) = 0$ (along which the vectors are vertical) and $Q(x, y) = 0$ (along which the vectors are horizontal). This can lead to choices for R. In the following examples, we will construct invariant regions bounded by lines and circles. In more complicated cases, we will be content to offer empirical evidence that an invariant region exists.

Example 7 A Circular Invariant Region

Find a circular region with center at $(0, 0)$ that serves as an invariant region for the plane autonomous system

$$x' = -y - x^3$$
$$y' = x - y^3.$$

SOLUTION For the circle $x^2 + y^2 = r^2$, $\mathbf{n} = (-2x, -2y)$ is a normal vector that points toward the interior of the circle. Since

$$\mathbf{V} \cdot \mathbf{n} = (-y - x^3, x - y^3) \cdot (-2x, -2y) = 2(x^4 + y^4),$$

we may conclude that $\mathbf{V} \cdot \mathbf{n} \geq 0$ on the circle $x^2 + y^2 = r^2$. Therefore, by Theorem 11.7 the circular region defined by $x^2 + y^2 \leq r^2$ serves as an invariant region for the system for any $r > 0$. ❑

Example 8 An Annular Invariant Region

Find an annular region bounded by circles that serves as an invariant region for the plane autonomous system

$$x' = x - y - 5x(x^2 + y^2) + x^5$$
$$y' = x + y - 5y(x^2 + y^2) + y^5.$$

SOLUTION As in Example 7, the normal vector $\mathbf{n}_1 = (-2x, -2y)$ points inside the circle $x^2 + y^2 = r^2$, while the normal vector $\mathbf{n}_2 = -\mathbf{n}_1$ is directed toward the exterior. Computing $\mathbf{V} \cdot \mathbf{n}_1$ and simplifying, we obtain

$$\mathbf{V} \cdot \mathbf{n}_1 = -2(r^2 - 5r^4 + x^6 + y^6).$$

Note that $r^2 - 5r^4 = r^2(1 - 5r^2)$ takes on both positive and negative values.

If $r = 1$, $\mathbf{V} \cdot \mathbf{n}_1 = 8 - 2(x^6 + y^6) \geq 0$, since the maximum value of $x^6 + y^6$ on the circle $x^2 + y^2 = 1$ is 1. The flow is therefore directed toward the interior of the circular region $x^2 + y^2 \leq 1$.

If $r = \frac{1}{4}$, $\mathbf{V} \cdot \mathbf{n}_1 \leq -2(r^2 - 5r^4) < 0$, and so $\mathbf{V} \cdot \mathbf{n}_2 = -\mathbf{V} \cdot \mathbf{n}_1 > 0$. The flow is therefore directed toward the exterior of the circle $x^2 + y^2 = \frac{1}{16}$, and so the annular region R defined by $\frac{1}{16} \leq x^2 + y^2 \leq 1$ is an invariant region for the system. ❑

Example 9 The Van der Pol Equation

The Van der Pol equation is a nonlinear second-order differential equation that arises in electronics, and as a plane autonomous system it takes the form

$$x' = y$$
$$y' = -\mu(x^2 - 1)y - x.$$

Figure 11.41 shows the corresponding vector field for $\mu = 1$, together with the curves $y = 0$ and $(x^2 - 1)y = -x$ along which the vectors are vertical and horizontal, respectively. (For convenience we have sketched the normalized vector field $\mathbf{V}/\|\mathbf{V}\|$). It is not possible to find a simple invariant region whose boundary consists of lines or circles. The figure does offer empirical evidence that an invariant region R, with $(0, 0)$ in its interior, does exist. Advanced methods are required to demonstrate this mathematically.* ❑

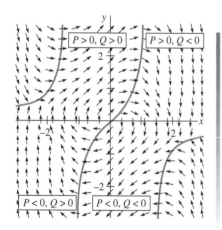

Figure 11.41

We next present two important special cases of the Poincaré–Bendixson theorem that guarantee the existence of periodic solutions.

THEOREM 11.8 Poincaré–Bendixson I

Let R be an invariant region for a plane autonomous system and suppose that R has no critical points on its boundary.

(i) If R is a Type I region that has a single unstable node or an unstable spiral point in its interior, then there is at least one periodic solution in R.

continued

* See M. Hirsch and S. Smale, *Differential Equations, Dynamical Systems, and Linear Algebra* (New York: Academic Press, 1974).

> (*ii*) If R is a Type II region that contains no critical points of the system, then there is at least one periodic solution in R.
>
> In either of the two cases, if $\mathbf{X} = \mathbf{X}(t)$ is a nonperiodic solution in R, then $\mathbf{X}(t)$ spirals toward a cycle that is a solution to the system. This periodic solution is called a **limit cycle.**

The flow interpretation portrayed in Figure 11.40 can be used to make the result plausible. If a particle is released at a point \mathbf{X}_0 in a Type II invariant region R, then, with no escape from the region and with no resting points, the particle will begin to rotate around the boundary C_2 and settle into a periodic orbit. It is not possible for the particle to return to an earlier position unless the solution is itself periodic.

Example 10 **Existence of a Periodic Solution**

Use Theorem 11.8 to show that the system

$$x' = -y + x(1 - x^2 - y^2) - y(x^2 + y^2)$$
$$y' = x + y(1 - x^2 - y^2) + x(x^2 + y^2)$$

has at least one periodic solution.

SOLUTION We first construct an invariant region that is bounded by circles. If $\mathbf{n}_1 = (-2x, -2y)$, then $\mathbf{V} \cdot \mathbf{n}_1 = -2r^2(1 - r^2)$. If we let $r = 2$ and then $r = \frac{1}{2}$, we may conclude that the annular region R defined by $\frac{1}{4} \leq x^2 + y^2 \leq 4$ is invariant. If (x_1, y_1) is a critical point of the system, then $\mathbf{V} \cdot \mathbf{n}_1 = (0, 0) \cdot \mathbf{n}_1 = 0$. Therefore, $r = 0$ or $r = 1$. If $r = 0$, then $(x_1, y_1) = (0, 0)$ is a critical point. If $r = 1$, the system reduces to $-2y = 0$, $2x = 0$ and we have reached a contradiction. Therefore, $(0, 0)$ is the only critical point, and this critical point is not in R. By part (*ii*) of Theorem 11.8, the system has at least one periodic solution in R.

The reader can verify that $\mathbf{X}(t) = (\cos 2t, \sin 2t)$ is a periodic solution. ❑

Example 11 **Limit Cycle in the Van der Pol Equation**

Show that the Van der Pol differential equation

$$x'' + \mu(x^2 - 1)x' + x = 0$$

has a periodic solution when $\mu > 0$.

SOLUTION We will assume that there is a Type I invariant region R for the corresponding plane autonomous system and that this region contains $(0, 0)$ in its interior (see Example 9 and Figure 11.41). The only critical point is $(0, 0)$, and the Jacobian matrix is given by

$$\mathbf{g}'((0, 0)) = \begin{pmatrix} 0 & 1 \\ -1 & \mu \end{pmatrix}.$$

Therefore, $\tau = \mu$, $\Delta = 1$, and $\tau^2 - 4\Delta = \mu^2 - 4$. Since $\mu > 0$, the critical point is either an unstable spiral point or an unstable node. By part (*i*) of Theorem 11.8, the system has at least one periodic solution in R. Figure 11.42 shows solutions corresponding to $\mathbf{X}(0) = (0.5, 0.5)$ and $\mathbf{X}(0) = (3, 3)$ for $\mu = 1$. Each of these solutions spirals around the origin and approaches a limit cycle. It can be shown that the Van der Pol differential equation has a unique limit cycle for all values of the parameter μ. ❑

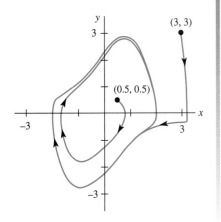

Figure 11.42

■ **Global Stability** Another version of the Poincaré–Bendixson theorem can be used to show that a locally stable critical point is globally stable:

> **THEOREM 11.9** Poincaré–Bendixson II
>
> Let R be a Type I invariant region for a plane autonomous system that has *no* periodic solutions in R.
>
> (*i*) If R has a finite number of nodes or spiral points, then given any initial position \mathbf{X}_0 in R, $\lim_{t \to \infty} \mathbf{X}(t) = \mathbf{X}_1$ for some critical point \mathbf{X}_1.
>
> (*ii*) If R has a *single* stable node or stable spiral point \mathbf{X}_1 in its interior and no critical points on its boundary, then $\lim_{t \to \infty} \mathbf{X}(t) = \mathbf{X}_1$ for all initial positions \mathbf{X}_0 in R.

In Theorem 11.9 the particle cannot escape from R, cannot return to any of its prior positions, and therefore, in the absence of cycles, must be attracted to some stable critical point \mathbf{X}_1.

Example 12 A Globally Stable Critical Point

Investigate global stability for the system from Example 7:

$$x' = -y - x^3$$
$$y' = x - y^3.$$

SOLUTION In Example 7 we showed that the circular region defined by $x^2 + y^2 \le r^2$ serves as an invariant region for the system for any $r > 0$. Since $\partial P/\partial x + \partial Q/\partial y = -3x^2 - 3y^2$ does not change sign, there are no periodic solutions by the Bendixson negative criterion. It is not hard to show that $(0, 0)$ is the only critical point and that the Jacobian matrix is

$$\mathbf{g}'((0, 0)) = \begin{pmatrix} 0 & -1 \\ 1 & 0 \end{pmatrix}.$$

Since $\tau = 0$ and $\Delta = 1$, $(0, 0)$ may be either a stable or an unstable spiral (it cannot be a center). Theorem 10.9, however, guarantees that $\lim_{t \to \infty} \mathbf{X}(t) = \mathbf{X}_1$ for some critical point \mathbf{X}_1. Since $(0, 0)$ is the only critical point, we must have

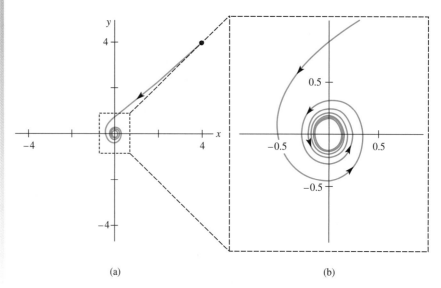

(a) (b)

Figure 11.43

$\lim_{t \to \infty} \mathbf{X}(t) = (0, 0)$ for *any* initial position \mathbf{X}_0 in the plane. The critical point is therefore a globally stable spiral point. Figure 11.43 shows two views of the solution satisfying $\mathbf{X}(0) = (4, 4)$. Part (b) is an enlarged view of the curve around $(0, 0)$. Notice how slowly the solution spirals toward $(0, 0)$. ❏

In Problems 1–8, show that the given plane autonomous system (or second-order differential equation) has no periodic solutions.

1. $x' = 2 + xy$
 $y' = x - y$

2. $x' = 2x - xy$
 $y' = -1 - x^2 + 2x - y^2$

3. $x' = -x + y^2$
 $y' = x - y$

4. $x' = xy^2 - x^2y$
 $y' = x^2y - 1$

5. $x' = -\mu x - y$ for $\mu < 0$
 $y' = x + y^3$

6. $x' = 2x + y^2$
 $y' = xy - y$

7. $x'' - 2x + (x')^4 = 0$

8. $x'' + x = [\frac{1}{2} + 3(x')^2]x' - x^2$

In Problems 9 and 10, use the Dulac negative criterion to show that the given plane autonomous system has no periodic solutions. Experiment with simple functions of the form $\delta(x, y) = ax^2 + by^2$, e^{ax+by}, or x^ay^b.

9. $x' = -2x + xy$
 $y' = 2y - x^2$

10. $x' = -x^3 + 4xy$
 $y' = -5x^2 - y^2$

11. Show that the plane autonomous system

$$x' = x(1 - x^2 - 3y^2)$$
$$y' = y(3 - x^2 - 3y^2)$$

has no periodic solutions in an elliptical region about the origin.

12. If $\partial g/\partial x' \neq 0$ in a region R, prove that $x'' = g(x, x')$ has no periodic solutions in R.

13. Show that the predator-prey model

$$x' = -ax + bxy$$
$$y' = -cxy + \frac{r}{K}y(K - y)$$

presented in Problem 20 of Exercises 11.4 has no periodic solutions in the first quadrant.

In Problems 14 and 15, find a circular invariant region for the given plane autonomous system.

14. $x' = -y - xe^{x+y}$
 $y' = x - ye^{x+y}$

15. $x' = -x + y + xy$
 $y' = x - y - x^2 - y^3$

16. Verify that the region bounded by the closed curve $x^6 + 3y^2 = 1$ is an invariant region for the second-order nonlinear differential equation $x'' + x' = -(x')^3 - x^5$. See Figure 11.44.

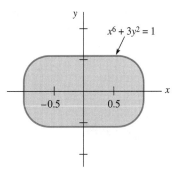

Figure 11.44

17. The plane autonomous system in Example 8 has only one critical point. Can we conclude that this system has at least one periodic solution?

18. Use the Poincaré–Bendixson theorem to show that the second-order nonlinear differential equation $x'' = x'[1 - 3x^2 - 2(x')^2] - x$ has at least one periodic solution. [*Hint:* Find an invariant annular region for the corresponding plane autonomous system.]

19. Let $\mathbf{X} = \mathbf{X}(t)$ be the solution of the plane autonomous system

$$x' = y$$
$$y' = -x - (1 - x^2)y$$

that satisfies $\mathbf{X}(0) = (x_0, y_0)$. Show that if $x_0^2 + y_0^2 < 1$, then $\lim_{t \to \infty} \mathbf{X}(t) = (0, 0)$. [*Hint:* Select $r < 1$ with $x_0^2 + y_0^2 < r^2$ and first show that the circular region R defined by $x^2 + y^2 \leq r^2$ is an invariant region.]

20. Investigate global stability for the system

$$x' = y - x$$
$$y' = -x - y^3.$$

21. Empirical evidence suggests that the plane autonomous system

$$x' = x^2y - x + 1$$
$$y' = -x^2y + \frac{1}{2}$$

has a Type I invariant region R that lies inside the rectangle $0 \leq x \leq 2, 0 \leq y \leq 1$.

(a) Use the Bendixson negative criterion to show that there are no periodic solutions in R.

(b) If \mathbf{X}_0 is in R and $\mathbf{X} = \mathbf{X}(t)$ is the solution satisfying $\mathbf{X}(t) = \mathbf{X}_0$, use Theorem 11.9 to find $\lim_{t \to \infty} \mathbf{X}(t)$.

22. (a) Find and classify all critical points of the plane autonomous system

$$x' = x\left(\frac{2y}{y+2} - 1\right)$$

$$y' = y\left(1 - \frac{2x}{y+2} - \frac{y}{8}\right).$$

(b) Figure 11.45 shows the vector field $\mathbf{V}/\|\mathbf{V}\|$ and offers empirical evidence that there is an invariant region R in the first quadrant with a critical point in its interior. Assuming that such a region exists, prove that there is at least one periodic solution.

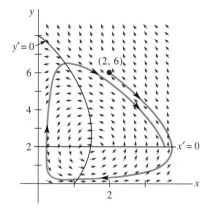

Figure 11.45

CHAPTER 11
REVIEW EXERCISES

Answers to odd-numbered problems begin on page A-56.

Answer Problems 1–10 without referring back to the text. Fill in the blank or answer true/false.

1. The second-order differential equation $x'' + f(x') + g(x) = 0$ can be written as a plane autonomous system. _____

2. If $\mathbf{X} = \mathbf{X}(t)$ is a solution of a plane autonomous system and $\mathbf{X}(t_1) = \mathbf{X}(t_2)$ for $t_1 \neq t_2$, then $\mathbf{X}(t)$ is a periodic solution. _____

3. If the trace of the matrix \mathbf{A} is 0 and det $\mathbf{A} \neq 0$, then the critical point $(0, 0)$ of the linear system $\mathbf{X}' = \mathbf{A}\mathbf{X}$ may be classified as _____.

4. If the critical point $(0, 0)$ of the linear system $\mathbf{X}' = \mathbf{A}\mathbf{X}$ is a stable spiral point, then the eigenvalues of \mathbf{A} are _____.

5. If the critical point $(0, 0)$ of the linear system $\mathbf{X}' = \mathbf{A}\mathbf{X}$ is a saddle point and $\mathbf{X} = \mathbf{X}(t)$ is a solution, then $\lim_{t \to \infty} \mathbf{X}(t)$ does not exist. _____

6. If the Jacobian matrix $\mathbf{A} = \mathbf{g}'(\mathbf{X}_1)$ at a critical point of a plane autonomous system has positive trace and determinant, then the critical point \mathbf{X}_1 is unstable. _____

7. It is possible to show that a nonlinear plane autonomous system has periodic solutions using linearization. _____

8. All solutions to the pendulum equation $\dfrac{d^2\theta}{dt^2} + \dfrac{g}{l} \sin \theta = 0$ are periodic. _____

9. If a simply connected region R contains no critical points of a plane autonomous system, then there are no periodic solutions in R. _____

10. If a plane autonomous system has no critical points in an annular invariant region R, then there is at least one periodic solution in R. _____

11. Solve the following nonlinear plane autonomous system by switching to polar coordinates, and describe the geometric behavior of the solution that satisfies the given initial condition.

$$x' = -y - x(\sqrt{x^2 + y^2})^3$$
$$y' = x - y(\sqrt{x^2 + y^2})^3, \qquad \mathbf{X}(0) = (1, 0)$$

12. Discuss the geometric nature of the solutions to the linear system $\mathbf{X}' = \mathbf{A}\mathbf{X}$ given the general solution.

(a) $\mathbf{X}(t) = c_1 \begin{pmatrix} 1 \\ 1 \end{pmatrix} e^{-t} + c_2 \begin{pmatrix} 1 \\ -2 \end{pmatrix} e^{-2t}$

(b) $\mathbf{X}(t) = c_1 \begin{pmatrix} 1 \\ -1 \end{pmatrix} e^{-t} + c_2 \begin{pmatrix} 1 \\ 2 \end{pmatrix} e^{2t}$

13. Classify the critical point $(0, 0)$ of the given linear system by computing the trace τ and determinant Δ.

(a) $\begin{aligned} x' &= -3x + 4y \\ y' &= -5x + 3y \end{aligned}$ **(b)** $\begin{aligned} x' &= -3x + 2y \\ y' &= -2x + y \end{aligned}$

14. Find and classify (if possible) the critical points of the plane autonomous system

$$x' = x + xy - 3x^2$$
$$y' = 4y - 2xy - y^2.$$

Does this system have any periodic solutions in the first quadrant?

15. Classify the critical point $(0, 0)$ of the plane autonomous system corresponding to the nonlinear second-order differential equation

$$x'' + \mu(x^2 - 1)x' + x = 0$$

where μ is a real constant.

16. Without solving explicitly, classify (if possible) the critical points of the autonomous first-order differential equation $x' = (x^2 - 1)e^{-x/2}$ as asymptotically stable or unstable.

17. Use the phase-plane method to show that the solutions of the nonlinear second-order differential equation

$$x'' = -2x\sqrt{(x')^2 + 1}$$

that satisfy $x(0) = x_0$ and $x'(0) = 0$ are periodic.

18. In Section 3.8 we assumed that the restoring force F of the spring satisfied Hookes' law $F = ks$, where s is the elongation of the spring and k is a positive constant of proportionality. If we replace this assumption with the nonlinear law $F = ks^3$, then the new differential equation for damped motion becomes $mx'' = -\beta x' - k(s + x)^3 + mg$, where $ks^3 = mg$. The system is called overdamped when $(0, 0)$ is a stable node and is called underdamped when $(0, 0)$ is a stable spiral point. Find new conditions on m, k, and β that will lead to overdamping and underdamping.

19. Show that the plane autonomous system

$$x' = 4x + 2y - 2x^2$$
$$y' = 4x - 3y + 4xy$$

has no periodic solutions.

20. Use the Poincaré–Bendixson theorem to show that the plane autonomous system

$$x' = \varepsilon x + y - x(x^2 + y^2)$$
$$y' = -x + \varepsilon y - y(x^2 + y^2)$$

has at least one periodic solution when $\varepsilon > 0$. What occurs when $\varepsilon < 0$?

21. The rod of a pendulum is attached to a movable joint at point P and rotates at an angular speed of ω (radians/s) in the plane perpendicular to the rod (see Figure 11.46). As a result, the bob of the pendulum experiences an additional centripetal force and the new differential equation for θ becomes

$$ml\,\frac{d^2\theta}{dt^2} = \omega^2 ml \sin\theta \cos\theta - mg\sin\theta - \beta\,\frac{d\theta}{dt}.$$

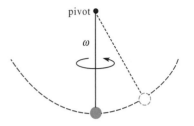

Figure 11.46

(a) Establish that there are no periodic solutions.

(b) If $\omega^2 < g/l$, show that $(0, 0)$ is a stable critical point and is the only critical point in the domain $-\pi < \theta < \pi$. Describe what occurs physically when $\theta(0) = \theta_0$, $\theta'(0) = 0$, and θ_0 is small.

(c) If $\omega^2 > g/l$, show that $(0, 0)$ is unstable and there are two additional stable critical points $(\pm\hat{\theta}, 0)$ in the domain $-\pi < \theta < \pi$. Describe what occurs physically when $\theta(0) = \theta_0$, $\theta'(0) = 0$, and θ_0 is small.

(d) Determine under what conditions the critical points in parts **(a)** and **(b)** are stable spiral points.

22. The nonlinear second-order differential equation

$$x'' - 2kx' + c(x')^3 + \omega^2 x = 0$$

arises in modeling the motion of an electrically driven tuning fork. See Figure 11.47, where $k = c = 0.1$ and $\omega = 1$. Assume that this differential equation possesses a Type I invariant region that contains $(0, 0)$. Show that there is at least one periodic solution.

Figure 11.47

PART
4

Fourier Series and Partial Differential Equations

Steady Transonic Flow Past Thin Airfoils

Sherif F. Ali
University of Arizona

When air is moving around a body, the speed of the air close to the body surface is different from that far away from the body. The main affecting parameter in this phenomenon is the geometry of the body. When the free stream speed (speed of air far away from the body) is close to the speed of sound (around 340 m/s), the flow is named transonic flow. In transonic flow, the local speed of air (speed on the surface of the body) is smaller than the speed of sound in the subsonic zone, greater than the speed of sound in the supersonic zone, and equal to the speed of sound in the sonic zone. For example, the flow on a thin circular arc airfoil, as shown in Figure 1, is accelerated near the leading edge of the airfoil until it reaches the speed of sound (sonic line). The flow continues to accelerate to supersonic speeds on the middle part of the airfoil. Then the supersonic zone is terminated by a shock wave near the trailing edge of the airfoil.

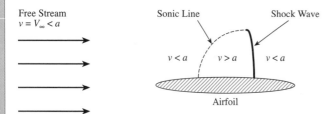

Figure 1 Transonic Flow Past an Airfoil

The basic principles of this problem are mass conservation, energy conservation, and momentum conservation. Starting from the basic principles and utilizing simplifying assumptions and geometrical transformations, one can simulate the problem mathematically by the following partial differential equation (PDE):

$$[K - (\gamma + 1)M_\infty^2 \phi_x]\phi_{xx} + \phi_{yy} = 0$$

where $K = (1 - M_\infty^2)/\delta^{2/3}$ is the transonic similarity parameter, δ is the airfoil thickness ratio, $M_\infty = v/a$ is the

Mach number, γ is the specific heat ratio (1.4 for air), and ϕ is the perturbation velocity potential (the differentiation of it at any direction gives the change of speed of air due to the existence of the body at this direction).

This is a second-order nonlinear PDE, and in order for the problem to be well posted, we need two boundary conditions:

(1) *The far-field boundary condition:* The flow far away from the body does not feel the existence of the body. Accordingly, the perturbations are zero in all directions $\phi_x = \phi_y = 0$ when x or y tends to infinity.

(2) *The no-penetration boundary condition:* The flow cannot penetrate the surface of the body. Accordingly, the perturbation normal to the surface ϕ_n is zero on the surface. For thin airfoils, this condition can be approximated by $\phi_x(x, 0) = F'(x)$, where $y = F(x)$ on the surface of the body.

Defining $A = K - (\gamma + 1)M_\infty^2 \phi_x$, the governing equation will be an elliptic, parabolic, or hyperbolic PDE according to the sign of A. When $A < 0$ the equation is hyperbolic, when $A > 0$ the equation is elliptic, and when $A = 0$ the equation is parabolic. It happens in this problem that the sign of A changes at different zones. That is, the characteristics of the governing equation change according to the value of the local flow speed ϕ_x.

One more simplifying transformation $\phi = (\Phi + Kx)/(\gamma + 1)$ can be used to put the governing equation in a form suitable for computations:

$$-\Phi_x\Phi_{xx} + \Phi_{yy} = 0.$$

Numerical Solution

The finite-differencing technique is used to solve the above problem. Since the governing equation is a mixed-type PDE, different differencing schemes were chosen at each zone to pass the information in the right direction as shown in Figure 2. The resulting algebraic system of equations is solved iteratively because of the nonlinear nature of the governing equation.

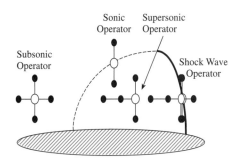

Figure 2 Finite-Differencing Scheme

■ Finite-Differencing Scheme

For the elliptic PDE there are no real characteristic lines. Accordingly, a disturbance at any point will affect the rest of the elliptic domain. For the hyperbolic PDE there are two real characteristic lines. Accordingly, a disturbance at a point will affect the part of the domain impeded between the characteristic lines (zone of action). For the parabolic PDE there is only one real characteristic line. Accordingly, a disturbance at any point will propagate along that line.

Subsonic Operator

Using central differencing in both the x and the y direction:

$$-\{(\Phi_{i+1,j} - \Phi_{i-1,j})/2\Delta x\}[(\Phi_{i+1,j} - 2\Phi_{i,j} + \Phi_{i-1,j})/\Delta x^2] + [(\Phi_{i,j+1} - 2\Phi_{i,j} + \Phi_{i,j-1})/\Delta y^2] = 0.$$

This is a stable elliptic operator when the first term $\Phi_c = \{(\Phi_{i+1,j} - \Phi_{i-1,j})/2\Delta x\}$ is negative.

Supersonic Operator

Using backward differencing in the x direction and central differencing in the y direction:

$$-\{(\Phi_{i,j} - \Phi_{i-2,j})/2\Delta x\}[(\Phi_{i,j} - 2\Phi_{i-1,j} + \Phi_{i-2,j})/\Delta x^2] + [(\Phi_{i,j+1} - 2\Phi_{i,j} + \Phi_{i,j-1})/\Delta y^2] = 0.$$

This is a stable hyperbolic operator when the first term $\Phi_b = \{(\Phi_{i,j} - \Phi_{i-2,j})/2\Delta x\}$ is positive.

Sonic Operator

There will be one point on j = constant line on which neither of the previous operators is locally stable ($\Phi_b < 0$, $\Phi_c > 0$). Thus a parabolic operator is introduced by setting $\Phi_x = 0$:

$$[(\Phi_{i,j+1} - 2\Phi_{i,j} + \Phi_{i,j-1})/\Delta y^2] = 0.$$

Shock Wave Operator

At any point on the shock wave surface there is a strong discontinuity in the flow parameters. The flow near its inner surface is supersonic ($\Phi_b > 0$), and on the outer surface the flow is subsonic ($\Phi_c < 0$). Accordingly, the shock wave operator is obtained by adding the subsonic and the supersonic operators.

■ Algebraic System of Equations

Two points should be elaborated here. First, the governing equation is nonlinear. Usually, that requires iterations to acquire a solution. Second, the chosen operators use three points at most (e.g., i, $i + 1$, $i - 1$). So the system of equations is expected to be tridiagonal.

Line Successive Over Relaxation (LSOR)

Let $\Phi = \{\Phi_1, \Phi_2, \ldots, \Phi_{\text{Max}}\}^T$ be a vector of i = constant line. The following LSOR formula was used to accelerate the convergence:

$$\Phi_i^{n+1} = \omega \Phi_i^{n+1} + (1 - \omega)\Phi_i^n,$$

where ω is a relaxation parameter. For the present problem, it is chosen to be over relaxation for the subsonic point $\omega = 1.2$ and under relaxation for the supersonic point $\omega = 0.4$, and n is the iteration number for the entire flow field. Iterations are performed on successive vertical lines and by sweeping the flow field along the flow direction which is the maximum gradient direction.

Tridiagonal System

For each line i = constant the equation is formed as:

$$[A_{i,j}]\{\Phi_i\} = \{f_i\},$$

where A is a tridiagonal matrix. This equation is solved using a tridiagonal solve from the United States Internet Mathematical Library called SGTSL. This subroutine requires the three diagonals and the constant vector f_i to be sent in four vectors (c,d,e,b) and returns with the solution in vector b.

Sample Results

Sample results for a circular arc airfoil with thickness ratio $\delta = 0.6$ at different values of the free stream Mach number are presented in this section. Results are presented in the form of coefficient of pressure $Cp \cup -2\Phi_x$ against

the distance from the leading edge of the airfoil. As shown in Figure 3, the supersonic pocket starts to build up at $M = 0.8$, and the shock wave discontinuity becomes very clear at $M = 0.84$.

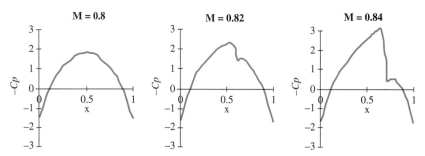

Figure 3 Variation of Pressure Coefficient with Distance from the Airfoil Leading Edge

Reference

[1] Ali, S. F., *"Steady Transonic Flow Past a Thin Airfoil,"* Eighth International Conference on Applied Mechanics and Mechanical Engineering, Military Technical College, Cairo, Egypt, May 12–14, 1998.

Sherif F. Ali received his B.S. and M.S. degrees, both in aeronautical engineering, from the Military Technical College in Cairo, Egypt, and his Ph.D. in aerospace technology from Arizona State University. Dr. Ali has over 10 years of research, teaching, and hands-on wind tunnel experience. His specialties include aerodynamics, aeroacoustics, fluid dynamics, and dynamics of flight, complemented by an extensive computer programming and numerical analysis background. He currently teaches at the University of Arizona and in Cairo, Egypt.

12

Orthogonal Functions and Fourier Series

TOPICS TO REVIEW FOR THIS CHAPTER

integration by parts
solution of the parametric Bessel DE (5.3)
solution of Legendre's DE (5.3)

IMPORTANT CONCEPTS IN THIS CHAPTER

orthogonal functions
orthogonal set
norm
square norm
orthogonality with respect to a weight function
orthogonal series expansions
Fourier series
even function
odd function
cosine series
sine series
half-range expansions
eigenvalues
eigenfunctions
regular Sturm–Liouville problem
singular Sturm–Liouville problem
Fourier–Bessel series
Fourier–Legendre series

INTRODUCTION

In this chapter we are setting the stage for the material in Chapters 13 and 14. Fundamental to the entire discussion are the two concepts of orthogonal functions and the expansion of a given function in a series of orthogonal functions.

12.1 Orthogonal Functions

■ **Inner Product** In advanced mathematics a function is considered to be a generalization of a vector. In this section we shall see how the two vector concepts of inner, or dot, product and orthogonality can be extended to functions. Recall that if **u** and **v** are two vectors in 3-space, then the inner product (**u**, **v**) (also denoted by **u** · **v**) possesses the following properties:

 (*i*) (**u**, **v**) = (**v**, **u**)
 (*ii*) (*k***u**, **v**) = *k*(**u**, **v**), *k* a scalar
 (*iii*) (**u**, **u**) = 0 if **u** = **0** and (**u**, **u**) > 0 if **u** ≠ **0**
 (*iv*) (**u** + **v**, **w**) = (**u**, **w**) + (**v**, **w**).

We expect that a generalization of the inner product concept should have these same properties.

Suppose that f_1 and f_2 are functions defined on an interval $[a, b]$.* Since a definite integral on the interval of the product $f_1(x)f_2(x)$ possesses properties (*i*)–(*iv*) of the inner product of vectors, whenever the integral exists we are prompted to make the following definition.

DEFINITION 12.1 **Inner Product of Functions**

The **inner product** of two functions f_1 and f_2 on an interval $[a, b]$ is the number

$$(f_1, f_2) = \int_a^b f_1(x)f_2(x)\,dx.$$

■ **Orthogonal Functions** Motivated by the fact that two vectors **u** and **v** are orthogonal whenever their inner product is zero, we define **orthogonal functions** in a similar manner.

DEFINITION 12.2 **Orthogonal Functions**

Two functions f_1 and f_2 are said to be **orthogonal** on an interval $[a, b]$ if

$$(f_1, f_2) = \int_a^b f_1(x)f_2(x)\,dx = 0. \tag{1}$$

For example, the functions $f_1(x) = x^2$ and $f_2(x) = x^3$ are orthogonal on the interval $[-1, 1]$ since

$$(f_1, f_2) = \int_{-1}^1 x^2 \cdot x^3\,dx = \tfrac{1}{6}x^6 \Big|_{-1}^{1} = 0.$$

Unlike vector analysis, where the word *orthogonal* is a synonym for *perpendicular,* in this present context the term *orthogonal* and condition (1) have no geometric significance.

* The interval could also be $(-\infty, \infty)$, $[0, \infty)$, and so on.

■ **Orthogonal Sets** We are primarily interested in infinite sets of orthogonal functions.

> **DEFINITION 12.3 Orthogonal Set**
>
> A set of real-valued functions $\{\phi_0(x), \phi_1(x), \phi_2(x), \ldots\}$ is said to be **orthogonal** on an interval $[a, b]$ if
>
> $$(\phi_m, \phi_n) = \int_a^b \phi_m(x)\phi_n(x)\, dx = 0, \quad m \neq n. \tag{2}$$

■ **Orthonormal Sets** The norm, or length $\|\mathbf{u}\|$, of a vector \mathbf{u} can be expressed in terms of the inner product. The expression $(\mathbf{u}, \mathbf{u}) = \|\mathbf{u}\|^2$ is called the square norm, and so the norm is $\|\mathbf{u}\| = \sqrt{(\mathbf{u}, \mathbf{u})}$. Similarly, the **square norm** of a function ϕ_n is $\|\phi_n(x)\|^2 = (\phi_n, \phi_n)$, and so the **norm,** or its generalized length, is $\|\phi_n(x)\| = \sqrt{(\phi_n, \phi_n)}$. In other words, the square norm and norm of a function ϕ_n in an orthogonal set $\{\phi_n(x)\}$ are, respectively,

$$\|\phi_n(x)\|^2 = \int_a^b \phi_n^2(x)\, dx \qquad \text{and} \qquad \|\phi_n(x)\| = \sqrt{\int_a^b \phi_n^2(x)\, dx}. \tag{3}$$

If $\{\phi_n(x)\}$ is an orthogonal set of functions on the interval $[a, b]$ with the property that $\|\phi_n(x)\| = 1$ for $n = 0, 1, 2, \ldots$, then $\{\phi_n(x)\}$ is said to be an **orthonormal set** on the interval.

Example 1 Orthogonal Set of Functions

Show that the set $\{1, \cos x, \cos 2x, \ldots\}$ is orthogonal on the interval $[-\pi, \pi]$.

SOLUTION If we make the identification $\phi_0(x) = 1$ and $\phi_n(x) = \cos nx$, we must then show that $\int_{-\pi}^{\pi} \phi_0(x)\phi_n(x)\, dx = 0$, $n \neq 0$, and $\int_{-\pi}^{\pi} \phi_m(x)\phi_n(x)\, dx = 0$, $m \neq n$. We have, in the first case,

$$(\phi_0, \phi_n) = \int_{-\pi}^{\pi} \phi_0(x)\phi_n(x)\, dx = \int_{-\pi}^{\pi} \cos nx\, dx$$

$$= \frac{1}{n} \sin nx \Big|_{-\pi}^{\pi}$$

$$= \frac{1}{n} [\sin n\pi - \sin(-n\pi)] = 0, \qquad n \neq 0,$$

and in the second,

$$(\phi_m, \phi_n) = \int_{-\pi}^{\pi} \phi_m(x)\phi_n(x)\, dx$$

$$= \int_{-\pi}^{\pi} \cos mx \cos nx\, dx$$

$$= \frac{1}{2} \int_{-\pi}^{\pi} [\cos(m+n)x + \cos(m-n)x]\, dx \qquad \leftarrow \text{trig identity}$$

$$= \frac{1}{2} \left[\frac{\sin(m+n)x}{m+n} + \frac{\sin(m-n)x}{m-n} \right]_{-\pi}^{\pi} = 0, \qquad m \neq n. \qquad \square$$

Example 2 Norms

Find the norms of each function in the orthogonal set given in Example 1.

SOLUTION For $\phi_0(x) = 1$ we have from (3)

$$\left\| \phi_0(x) \right\|^2 = \int_{-\pi}^{\pi} dx = 2\pi$$

so that $\left\| \phi_0(x) \right\| = \sqrt{2\pi}$. For $\phi_n(x) = \cos nx$, $n > 0$, it follows that

$$\left\| \phi_n(x) \right\|^2 = \int_{-\pi}^{\pi} \cos^2 nx \, dx = \frac{1}{2} \int_{-\pi}^{\pi} [1 + \cos 2nx] \, dx = \pi.$$

Thus for $n > 0$, $\left\| \phi_n(x) \right\| = \sqrt{\pi}$. ❑

An orthogonal set can be 👉
made into an orthonormal set.

Any orthogonal set of nonzero functions $\{\phi_n(x)\}$, $n = 0, 1, 2, \ldots$, can be *normalized*—that is, made into an orthonormal set—by dividing each function by its norm. It follows from Examples 1 and 2 that the set

$$\left\{ \frac{1}{\sqrt{2\pi}}, \frac{\cos x}{\sqrt{\pi}}, \frac{\cos 2x}{\sqrt{\pi}}, \ldots \right\}$$

is orthonormal on the interval $[-\pi, \pi]$.

We shall make one more analogy between vectors and functions. Suppose \mathbf{v}_1, \mathbf{v}_2, and \mathbf{v}_3 are three mutually orthogonal nonzero vectors in 3-space. Such an orthogonal set can be used as a basis for 3-space; that is, any three-dimensional vector can be written as a linear combination

$$\mathbf{u} = c_1 \mathbf{v}_1 + c_2 \mathbf{v}_2 + c_3 \mathbf{v}_3, \tag{4}$$

where the c_i, $i = 1, 2, 3$, are scalars called the components of the vector. Each component c_i can be expressed in terms of \mathbf{u} and the corresponding vector \mathbf{v}_i. To see this we take the inner product of (4) with \mathbf{v}_1:

$$(\mathbf{u}, \mathbf{v}_1) = c_1(\mathbf{v}_1, \mathbf{v}_1) + c_2(\mathbf{v}_2, \mathbf{v}_1) + c_3(\mathbf{v}_3, \mathbf{v}_1) = c_1 \left\| \mathbf{v}_1 \right\|^2 + c_2 \cdot 0 + c_3 \cdot 0.$$

Hence
$$c_1 = \frac{(\mathbf{u}, \mathbf{v}_1)}{\left\| \mathbf{v}_1 \right\|^2}.$$

In like manner we find that the components c_2 and c_3 are given by

$$c_2 = \frac{(\mathbf{u}, \mathbf{v}_2)}{\left\| \mathbf{v}_2 \right\|^2} \quad \text{and} \quad c_3 = \frac{(\mathbf{u}, \mathbf{v}_3)}{\left\| \mathbf{v}_3 \right\|^2}.$$

Hence (4) can be expressed as

$$\mathbf{u} = \frac{(\mathbf{u}, \mathbf{v}_1)}{\left\| \mathbf{v}_1 \right\|^2} \mathbf{v}_1 + \frac{(\mathbf{u}, \mathbf{v}_2)}{\left\| \mathbf{v}_2 \right\|^2} \mathbf{v}_2 + \frac{(\mathbf{u}, \mathbf{v}_3)}{\left\| \mathbf{v}_3 \right\|^2} \mathbf{v}_3 = \sum_{n=1}^{3} \frac{(\mathbf{u}, \mathbf{v}_n)}{\left\| \mathbf{v}_n \right\|^2} \mathbf{v}_n. \tag{5}$$

■ **Orthogonal Series Expansion** Suppose $\{\phi_n(x)\}$ is an infinite orthogonal set of functions on an interval $[a, b]$. We ask: If $y = f(x)$ is a function defined on the interval $[a, b]$, is it possible to determine a set of coefficients c_n, $n = 0, 1, 2, \ldots$, for which

$$f(x) = c_0 \phi_0(x) + c_1 \phi_1(x) + \cdots + c_n \phi_n(x) + \cdots? \tag{6}$$

As in the foregoing discussion on finding components of a vector, we can find the coefficients c_n by utilizing the inner product. Multiplying (6) by $\phi_m(x)$ and integrating over the interval $[a, b]$ gives

$$\int_a^b f(x)\phi_m(x)\,dx = c_0\int_a^b \phi_0(x)\phi_m(x)\,dx + c_1\int_a^b \phi_1(x)\phi_m(x)\,dx + \ldots + c_n\int_a^b \phi_n(x)\phi_m(x)\,dx + \cdots$$

$$= c_0(\phi_0,\ \phi_m) + c_1(\phi_1,\ \phi_m) + \cdots + c_n(\phi_n,\ \phi_m) + \cdots.$$

By orthogonality each term on the right-hand side of the last equation is zero *except* when $m = n$. In this case we have

$$\int_a^b f(x)\phi_n(x)\,dx = c_n \int_a^b \phi_n^2(x)\,dx.$$

It follows that the required coefficients are

$$c_n = \frac{\int_a^b f(x)\phi_n(x)\,dx}{\int_a^b \phi_n^2(x)\,dx}, \qquad n = 0, 1, 2,\ldots.$$

In other words,
$$f(x) = \sum_{n=0}^{\infty} c_n\phi_n(x), \tag{7}$$

where
$$c_n = \frac{\int_a^b f(x)\phi_n(x)\,dx}{\|\phi_n(x)\|^2}. \tag{8}$$

With inner product notation, (7) becomes

$$f(x) = \sum_{n=0}^{\infty} \frac{(f,\ \phi_n)}{\|\phi_n(x)\|^2}\ \phi_n(x). \tag{9}$$

Thus (9) is seen to be the functional analogue of the vector result given in (5).

DEFINITION 12.4 Orthogonal Set/Weight Function

A set of real-valued functions $\{\phi_0(x),\ \phi_1(x),\ \phi_2(x),\ \ldots\}$ is said to be **orthogonal with respect to a weight function** $w(x)$ on an interval $[a, b]$ if

$$\int_a^b w(x)\phi_m(x)\phi_n(x)\,dx = 0, \qquad m \neq n.$$

The usual assumption is that $w(x) > 0$ on the interval of orthogonality $[a, b]$. The set $\{1,\ \cos x,\ \cos 2x,\ \ldots\}$ in Example 1 is orthogonal with respect to the weight function $w(x) = 1$ on the interval $[-\pi, \pi]$.

If $\{\phi_n(x)\}$ is orthogonal with respect to a weight function $w(x)$ on the interval $[a, b]$, then multiplying (6) by $w(x)\phi_n(x)$ and integrating yields

$$c_n = \frac{\int_a^b f(x)w(x)\phi_n(x)\,dx}{\|\phi_n(x)\|^2}, \tag{10}$$

where
$$\|\phi_n(x)\|^2 = \int_a^b w(x)\phi_n^2(x)\,dx. \tag{11}$$

The series (7) with coefficients given by either (8) and (10) is said to be an **orthogonal series expansion** of f or a **generalized Fourier series**.

■ **Complete Sets** The procedure outlined for determining the coefficients c_n was *formal;* that is, basic questions on whether or not an orthogonal series expansion

such as (7) is actually possible were ignored. Also, to expand f in a series of orthogonal functions, it is certainly necessary that f not be orthogonal to each ϕ_n of the orthogonal set $\{\phi_n(x)\}$. (If f were orthogonal to every ϕ_n, then $c_n = 0$, $n = 0, 1, 2, \ldots$.) To avoid the latter problem we shall assume, for the remainder of the discussion, that an orthogonal set is **complete**. This means that the only function orthogonal to each member of the set is the zero function.

 Answers to odd-numbered problems begin on page A-56.

In Problems 1–6, show that the given functions are orthogonal on the indicated interval.

1. $f_1(x) = x$, $f_2(x) = x^2$; $[-2, 2]$

2. $f_1(x) = x^3$, $f_2(x) = x^2 + 1$; $[-1, 1]$

3. $f_1(x) = e^x$, $f_2(x) = xe^{-x} - e^{-x}$; $[0, 2]$

4. $f_1(x) = \cos x$, $f_2(x) = \sin^2 x$; $[0, \pi]$

5. $f_1(x) = x$, $f_2(x) = \cos 2x$; $[-\pi/2, \pi/2]$

6. $f_1(x) = e^x$, $f_2(x) = \sin x$; $[\pi/4, 5\pi/4]$

In Problems 7–12, show that the given set of functions is orthogonal on the indicated interval. Find the norm of each function in the set.

7. $\{\sin x, \sin 3x, \sin 5x, \ldots\}$; $[0, \pi/2]$

8. $\{\cos x, \cos 3x, \cos 5x, \ldots\}$; $[0, \pi/2]$

9. $\{\sin nx\}$, $n = 1, 2, 3, \ldots$; $[0, \pi]$

10. $\left\{\sin \dfrac{n\pi}{p} x\right\}$, $n = 1, 2, 3, \ldots$; $[0, p]$

11. $\left\{1, \cos \dfrac{n\pi}{p} x\right\}$, $n = 1, 2, 3, \ldots$; $[0, p]$

12. $\left\{1, \cos \dfrac{n\pi}{p} x, \sin \dfrac{m\pi}{p} x\right\}$, $n = 1, 2, 3, \ldots$,
$m = 1, 2, 3, \ldots$; $[-p, p]$

In Problems 13 and 14, verify by direct integration that the functions are orthogonal with respect to the indicated weight function on the given interval.

13. $H_0(x) = 1$, $H_1(x) = 2x$, $H_2(x) = 4x^2 - 2$; $w(x) = e^{-x^2}$, $(-\infty, \infty)$

14. $L_0(x) = 1$, $L_1(x) = -x + 1$, $L_2(x) = \dfrac{1}{2}x^2 - 2x + 1$;
$w(x) = e^{-x}$, $[0, \infty)$

15. Let $\{\phi_n(x)\}$ be an orthogonal set of functions on $[a, b]$ such that $\phi_0(x) = 1$. Show that $\int_a^b \phi_n(x)\, dx = 0$ for $n = 1, 2, \ldots$.

16. Let $\{\phi_n(x)\}$ be an orthogonal set of functions on $[a, b]$ such that $\phi_0(x) = 1$ and $\phi_1(x) = x$. Show that $\int_a^b (\alpha x + \beta)\phi_n(x)\, dx = 0$ for $n = 2, 3, \ldots$ and any constants α and β.

17. Let $\{\phi_n(x)\}$ be an orthogonal set of functions on $[a, b]$. Show that $\|\phi_m(x) + \phi_n(x)\|^2 = \|\phi_m(x)\|^2 + \|\phi_n(x)\|^2$, $m \neq n$.

18. From Problem 1 we know that $f_1(x) = x$ and $f_2(x) = x^2$ are orthogonal on $[-2, 2]$. Find constants c_1 and c_2 such that $f_3(x) = x + c_1 x^2 + c_2 x^3$ is orthogonal to both f_1 and f_2 on the same interval.

19. The set of functions $\{\sin nx\}$, $n = 1, 2, 3, \ldots$, is orthogonal on the interval $[-\pi, \pi]$. Show that the set is not complete.

20. Suppose f_1, f_2, and f_3 are functions continuous on the interval $[a, b]$. Show that $(f_1 + f_2, f_3) = (f_1, f_3) + (f_2, f_3)$.

21. A real-valued function f is said to be **periodic** with period T if $f(x + T) = f(x)$. For example, 4π is a period of $\sin x$ since $\sin(x + 4\pi) = \sin x$. The smallest value of T for which $f(x + T) = f(x)$ holds is called the **fundamental period** of f. For example, the fundamental period of $f(x) = \sin x$ is $T = 2\pi$. What is the fundamental period of each of the following functions?

(a) $f(x) = \cos 2\pi x$ **(b)** $f(x) = \sin \dfrac{4}{L} x$

(c) $f(x) = \sin x + \sin 2x$ **(d)** $f(x) = \sin 2x + \cos 4x$

(e) $f(x) = \sin 3x + \cos 2x$

(f) $f(x) = A_0 + \sum_{n=1}^{\infty} \left(A_n \cos \dfrac{n\pi}{p} x + B_n \sin \dfrac{n\pi}{p} x \right)$, A_n and B_n depend only on n

12.2 Fourier Series

We have just seen in the preceding section that if $\{\phi_0(x), \phi_1(x), \phi_2(x), \ldots\}$ is an orthogonal set on an interval $[a, b]$ and if f is a function defined on the same interval, then we formally expand f in an orthogonal series $c_0\phi_0(x) + c_1\phi_1(x) + c_2\phi_2(x) + \ldots$,

where the coefficients c_n can be determined using the inner product concept. The orthogonal set of trigonometric functions

$$\left\{ 1, \cos\frac{\pi}{p}x, \cos\frac{2\pi}{p}x, \cos\frac{3\pi}{p}x,..., \sin\frac{\pi}{p}x, \sin\frac{2\pi}{p}x, \sin\frac{3\pi}{p}x,... \right\} \quad (1)$$

will be of particular importance later on in the solution of certain kinds of boundary-value problems involving linear partial differential equations. The set (1) is orthogonal on the interval $[-p, p]$. See Problem 12 in Exercises 12.1.

■ **A Trigonometric Series** Suppose that f is a function defined on the interval $[-p, p]$ that can be expanded in an orthogonal series consisting of the trigonometric functions in the orthogonal set (1), that is,

$$f(x) = \frac{a_0}{2} + \sum_{n=1}^{\infty} \left(a_n \cos\frac{n\pi}{p}x + b_n \sin\frac{n\pi}{p}x \right). \quad (2)$$

The coefficients $a_0, a_1, a_2, \ldots, b_1, b_2, \ldots$, can be determined in exactly the same manner as in the general discussion of orthogonal series expansions on page 655. Before proceeding, note that we have chosen to write the coefficient of 1 in the set (1) as $a_0/2$ rather than a_0; this is for convenience only because the formula of a_n will then reduce to a_0 for $n = 0$.

This is why $a_0/2$ is used instead ☞
of a_0.

Now integrating both sides of (2) from $-p$ to p gives

$$\int_{-p}^{p} f(x)\, dx = \frac{a_0}{2} \int_{-p}^{p} dx + \sum_{n=1}^{\infty} \left(a_n \int_{-p}^{p} \cos\frac{n\pi}{p}x\, dx + b_n \int_{-p}^{p} \sin\frac{n\pi}{p}x\, dx \right). \quad (3)$$

Since $\cos(n\pi x/p)$ and $\sin(n\pi x/p)$, $n \geq 1$, are orthogonal to 1 on the interval, the right side of (3) reduces to a single term:

$$\int_{-p}^{p} f(x)\, dx = \frac{a_0}{2} \int_{-p}^{p} dx = \frac{a_0}{2} x \Big|_{-p}^{p} = pa_0.$$

Solving for a_0 yields

$$a_0 = \frac{1}{p} \int_{-p}^{p} f(x)\, dx. \quad (4)$$

Now we multiply (2) by $\cos(m\pi x/p)$ and integrate:

$$\int_{-p}^{p} f(x) \cos\frac{m\pi}{p}x\, dx = \frac{a_0}{2} \int_{-p}^{p} \cos\frac{m\pi}{p}x\, dx$$

$$+ \sum_{n=1}^{\infty} \left(a_n \int_{-p}^{p} \cos\frac{m\pi}{p}x \cos\frac{m\pi}{p}x\, dx + b_n \int_{-p}^{p} \cos\frac{m\pi}{p}x \sin\frac{n\pi}{p}x\, dx \right). \quad (5)$$

By orthogonality we have

$$\int_{-p}^{p} \cos\frac{m\pi}{p}x\, dx = 0, \quad m > 0, \qquad \int_{-p}^{p} \cos\frac{m\pi}{p}x \sin\frac{n\pi}{p}x\, dx = 0$$

and

$$\int_{-p}^{p} \cos\frac{m\pi}{p}x \cos\frac{n\pi}{p}x\, dx = \begin{cases} 0, & m \neq n \\ p, & m \neq n. \end{cases}$$

Thus (5) reduces to

$$\int_{-p}^{p} f(x) \cos\frac{n\pi}{p}x\, dx = a_n p,$$

and so
$$a_n = \frac{1}{p} \int_{-p}^{p} f(x) \cos \frac{n\pi}{p} x \, dx. \tag{6}$$

Finally, if we multiply (2) by $\sin(m\pi x/p)$, integrate, and make use of the results

$$\int_{-p}^{p} \sin \frac{m\pi}{p} x \, dx = 0, \quad m > 0, \qquad \int_{-p}^{p} \sin \frac{m\pi}{p} x \sin \frac{n\pi}{p} x \, dx = 0$$

and
$$\int_{-p}^{p} \sin \frac{m\pi}{p} x \sin \frac{n\pi}{p} x \, dx = \begin{cases} 0, & m \neq n \\ p, & m \neq n, \end{cases}$$

we find that
$$b_n = \frac{1}{p} \int_{-p}^{p} f(x) \sin \frac{n\pi}{p} x \, dx. \tag{7}$$

The trigonometric series (2) with coefficients a_0, a_n, and b_n defined by (4), (6), and (7), respectively, is said to be the **Fourier series** of the function f. The coefficients obtained from (4), (6), and (7) are referred to as **Fourier coefficients** of f.

In finding the coefficients a_0, a_n, and b_n we assumed that f was integrable on the interval and that (2), as well as the series obtained by multiplying (2) by $\cos(m\pi x/p)$, converged in such a manner as to permit term-by-term integration. Until (2) is shown to be convergent for a given function f, the equality sign is not to be taken in a strict or literal sense. Some texts use the symbol ~ in place of =. In view of the fact that most functions in applications are of a type that guarantees convergence of the series, we shall use the equality symbol. We summarize the results:

DEFINITION 12.5 Fourier Series

The **Fourier series** of a function f defined on the interval $(-p, p)$ is given by

$$f(x) = \frac{a_0}{2} + \sum_{n=1}^{\infty} \left(a_n \cos \frac{n\pi}{p} x + b_n \sin \frac{n\pi}{p} x \right), \tag{8}$$

where
$$a_0 = \frac{1}{p} \int_{-p}^{p} f(x) \, dx \tag{9}$$

$$a_n = \frac{1}{p} \int_{-p}^{p} f(x) \cos \frac{n\pi}{p} x \, dx \tag{10}$$

$$b_n = \frac{1}{p} \int_{-p}^{p} f(x) \sin \frac{n\pi}{p} x \, dx. \tag{11}$$

Example 1 Expansion in a Fourier Series

Expand
$$f(x) = \begin{cases} 0, & -\pi < x < 0 \\ \pi - x, & 0 \leq x < \pi \end{cases} \tag{12}$$

in a Fourier series.

SOLUTION The graph of f is given in Figure 12.1. With $p = \pi$ we have from (9) and (10) that

$$a_0 = \frac{1}{\pi} \int_{-\pi}^{\pi} f(x) \, dx = \frac{1}{\pi} \left[\int_{-\pi}^{0} 0 \, dx + \int_{0}^{\pi} (\pi - x) \, dx \right] = \frac{1}{\pi} \left[\pi x - \frac{x^2}{2} \right]_{0}^{\pi} = \frac{\pi}{2}$$

$$a_n = \frac{1}{\pi} \int_{-\pi}^{\pi} f(x) \cos nx \, dx = \frac{1}{\pi} \left[\int_{-\pi}^{0} 0 \, dx + \int_{0}^{\pi} (\pi - x) \cos nx \, dx \right]$$

Figure 12.1

$$= \frac{1}{\pi} \left[(\pi - x) \frac{\sin nx}{n} \Big|_0^\pi + \frac{1}{n} \int_0^\pi \sin nx \, dx \right]$$

$$= -\frac{1}{n\pi} \frac{\cos nx}{n} \Big|_0^\pi$$

$$= \frac{-\cos n\pi + 1}{n^2 \pi} \qquad\qquad \leftarrow \cos n\pi = (-1)^n$$

$$= \frac{1 - (-1)^n}{n^2 \pi}.$$

In like manner we find from (11) that

$$b_n = \frac{1}{\pi} \int_0^\pi (\pi - x) \sin nx \, dx = \frac{1}{n}.$$

Therefore $$f(x) = \frac{\pi}{4} + \sum_{n=1}^\infty \left\{ \frac{1 - (-1)^n}{n^2 \pi} \cos nx + \frac{1}{n} \sin nx \right\}. \qquad (13)\square$$

Note that a_n defined by (10) reduces to a_0 given by (9) when we set $n = 0$. But as Example 1 shows, this may not be the case *after* the integral for a_n is evaluated.

■ **Convergence of a Fourier Series** The following theorem gives sufficient conditions for convergence of a Fourier series at a point.

THEOREM 12.1 **Conditions for Convergence**

Let f and f' be piecewise continuous on the interval $(-p, p)$; that is, let f and f' be continuous except at a finite number of points in the interval and have only finite discontinuities at these points. Then the Fourier series of f on the interval converges to $f(x)$ at a point of continuity. At a point of discontinuity the Fourier series converges to the average

$$\frac{f(x+) + f(x-)}{2},$$

where $f(x+)$ and $f(x-)$ denote the limit of f at x from the right and from the left, respectively.*

For a proof of this theorem you are referred to the classic text by Churchill and Brown.[†]

Example 2 **Convergence of a Point of Discontinuity**

The function (12) in Example 1 satisfies the conditions of Theorem 12.1. Thus for every x in the interval $(-\pi, \pi)$, except at $x = 0$, the series (13) will converge to $f(x)$. At $x = 0$ the function is discontinuous, and so the series (13) will converge to

$$\frac{f(0+) + f(0-)}{2} = \frac{\pi + 0}{2} = \frac{\pi}{2}. \qquad\qquad \square$$

* In other words, for x a point in the interval and $h > 0$,

$$f(x+) = \lim_{h \to 0} f(x + h), \qquad f(x-) = \lim_{h \to 0} f(x - h).$$

[†] Ruel V. Churchill and James Ward Brown, *Fourier Series and Boundary Value Problems* (New York: McGraw-Hill).

■ **Periodic Extension** Observe that each of the functions in the basic set (1) has a different fundamental period,* namely, $2p/n$, $n \geq 1$, but since a positive integer multiple of a period is also a period we see that all of the functions have in common the period $2p$ (verify). Hence the right-hand side of (2) is $2p$-periodic; indeed, $2p$ is the fundamental period of the sum. We conclude that a Fourier series not only represents the function on the interval $(-p, p)$ but also gives the **periodic extension** of f outside this interval. We can now apply Theorem 12.1 to the periodic extension of f, or we may assume from the outset that the given function is periodic with period $T = 2p$; that is, $f(x + T) = f(x)$. When f is piecewise continuous and the right- and left-hand derivatives exist at $x = -p$ and $x = p$, respectively, then the series (8) converges to the average $[f(p-) + f(-p+)]/2$ at these endpoints and to this value extended periodically to $\pm 3p$, $\pm 5p$, $\pm 7p$, and so on. The Fourier series in (13) converges to the periodic extension of (12) on the entire x-axis. At 0, $\pm 2\pi$, $\pm 4\pi$, ..., and at $\pm \pi$, $\pm 3\pi$, $\pm 5\pi$, ..., the series converges to the values

We may assume that the given function f is periodic.

$$\frac{f(0+) + f(0-)}{2} = \frac{\pi}{2} \quad \text{and} \quad \frac{f(\pi+) + f(\pi-)}{2} = 0,$$

respectively. The solid dots in Figure 12.2 represent the value $\pi/2$.

Figure 12.2

■ **Sequence of Partial Sums** It is interesting to see how the sequence of partial sums $\{S_N(x)\}$ of a Fourier series approximates a function. For example, the first three partial sums of (13) are

$$S_1(x) = \frac{\pi}{4}, \quad S_2(x) = \frac{\pi}{4} + \frac{2}{\pi} \cos x + \sin x, \quad S_3(x) = \frac{\pi}{4} + \frac{2}{\pi} \cos x + \sin x + \frac{1}{2} \sin 2x.$$

In Figure 12.3 we have used a CAS to graph the partial sums $S_3(x)$, $S_5(x)$, $S_8(x)$, and $S_{15}(x)$ of (13) on the interval $(-\pi, \pi)$. Figure 12.3(e) shows the periodic extension using $S_{15}(x)$ on $(-4\pi, 4\pi)$.

(a) $S_3(x)$ on $(-\pi, \pi)$

(b) $S_5(x)$ on $(-\pi, \pi)$

Figure 12.3 Partial sums of a Fourier series

continued

* See Problem 21 in Exercises 12.1.

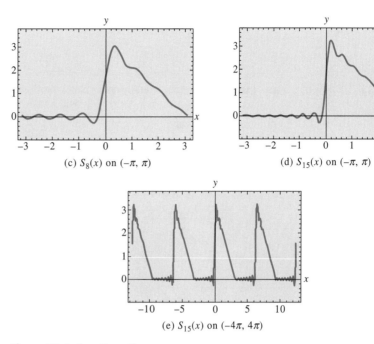

(c) $S_8(x)$ on $(-\pi, \pi)$ (d) $S_{15}(x)$ on $(-\pi, \pi)$

(e) $S_{15}(x)$ on $(-4\pi, 4\pi)$

Figure 12.3 (continued)

E X E R C I S E S 1 2 . 2 Answers to odd-numbered problems begin on page A-57.

In Problems 1–16, find the Fourier series of f on the given interval.

1. $f(x) = \begin{cases} 0, & -\pi < x < 0 \\ 1, & 0 \le x < \pi \end{cases}$

2. $f(x) = \begin{cases} -1, & -\pi < x < 0 \\ 2, & 0 \le x < \pi \end{cases}$

3. $f(x) = \begin{cases} 1, & -1 < x < 0 \\ x, & 0 \le x < 1 \end{cases}$

4. $f(x) = \begin{cases} 0, & -1 < x < 0 \\ x, & 0 \le x < 1 \end{cases}$

5. $f(x) = \begin{cases} 0, & -\pi < x < 0 \\ x^2, & 0 \le x < \pi \end{cases}$

6. $f(x) = \begin{cases} \pi^2, & -\pi < x < 0 \\ \pi^2 - x^2, & 0 \le x < \pi \end{cases}$

7. $f(x) = x + \pi, \quad -\pi < x < \pi$

8. $f(x) = 3 - 2x, \quad -\pi < x < \pi$

9. $f(x) = \begin{cases} 0, & -\pi < x < 0 \\ \sin x, & 0 \le x < \pi \end{cases}$

10. $f(x) = \begin{cases} 0, & -\pi/2 < x < 0 \\ \cos x, & 0 \le x < \pi/2 \end{cases}$

11. $f(x) = \begin{cases} 0, & -2 < x < -1 \\ -2, & -1 \le x < 0 \\ 1, & 0 \le x < 1 \\ 0, & 1 \le x < 2 \end{cases}$

12. $f(x) = \begin{cases} 0, & -2 < x < 0 \\ x, & 0 \le x < 1 \\ 1, & 1 \le x < 2 \end{cases}$

13. $f(x) = \begin{cases} 1, & -5 < x < 0 \\ 1 + x, & 0 \le x < 5 \end{cases}$

14. $f(x) = \begin{cases} 2 + x, & -2 < x < 0 \\ 2, & 0 \le x < 2 \end{cases}$

15. $f(x) = e^x, \quad -\pi < x < \pi$

16. $f(x) = \begin{cases} 0, & -\pi < x < 0 \\ e^x - 1, & 0 \le x < \pi \end{cases}$

17. Use the result of Problem 5 to show
$$\frac{\pi^2}{6} = 1 + \frac{1}{2^2} + \frac{1}{3^2} + \frac{1}{4^2} + \cdots$$
and
$$\frac{\pi^2}{12} = 1 - \frac{1}{2^2} + \frac{1}{3^2} - \frac{1}{4^2} + \cdots.$$

18. Use Problem 17 to find a series that gives the numerical value of $\pi^2/8$.

19. Use the result of Problem 7 to show
$$\frac{\pi}{4} = 1 - \frac{1}{3} + \frac{1}{5} - \frac{1}{7} + \cdots.$$

20. Use the result of Problem 9 to show
$$\frac{\pi}{4} = \frac{1}{2} + \frac{1}{1 \cdot 3} - \frac{1}{3 \cdot 5} + \frac{1}{5 \cdot 7} - \frac{1}{7 \cdot 9} + \cdots.$$

21. The **root-mean-square value** of a function $f(x)$ defined over an interval (a, b) is given by
$$\text{RMS}(f) = \sqrt{\frac{\int_a^b f^2(x)\, dx}{b - a}}.$$

If the Fourier series expansion of f is given by (8), show that the RMS value of f over the interval $(-p, p)$ is given by
$$\text{RMS}(f) = \sqrt{\tfrac{1}{4}a_0^2 + \tfrac{1}{2}\sum_{n=1}^{\infty}(a_n^2 + b_n^2)},$$

where a_0, a_n, and b_n are the Fourier coefficients in (9), (10), and (11).

12.3 Fourier Cosine and Sine Series

Figure 12.4

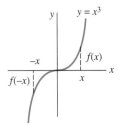

Figure 12.5

The effort expended in the evaluation of coefficients a_0, a_n, and b_n in expanding a function f in a Fourier series is reduced significantly when f is either an even or an odd function. A function f is said to be:

$$\textbf{even if } f(-x) = f(x) \qquad \text{and} \qquad \textbf{odd if } f(-x) = -f(x).$$

On a symmetric interval such as $(-p, p)$, the graph of an even function possesses symmetry with respect to the y-axis, whereas the graph of an odd function possesses symmetry with respect to the origin.

■ **Some Examples** It is likely the origin of the words *even* and *odd* derives from the fact that the graphs of polynomial functions that consist of all even powers of x are symmetric with respect to the y-axis whereas graphs of polynomials that consist of all odd powers of x are symmetric with respect to the origin. For example,

$$\downarrow \text{ even integer}$$
$$f(x) = x^2 \text{ is even since } f(-x) = (-x)^2 = x^2 = f(x)$$

$$\downarrow \text{ odd integer}$$
$$f(x) = x^3 \text{ is odd since } f(-x) = (-x)^3 = -x^3 = -f(x).$$

See Figures 12.4 and 12.5. The trigonometric cosine and sine functions are even and odd functions, respectively, since $\cos(-x) = \cos x$ and $\sin(-x) = -\sin x$. The exponential functions $f(x) = e^x$ and $f(x) = e^{-x}$ are neither odd nor even.

■ **Properties of Even and Odd Functions** The following theorem lists some properties of even and odd functions.

THEOREM 12.2 Properties of Even/Odd Functions

(*a*) The product of two even functions is even.
(*b*) The product of two odd functions is even.
(*c*) The product of an even function and an odd function is odd.
(*d*) The sum (difference) of two even functions is even.
(*e*) The sum (difference) of two odd functions is odd.

(*f*) If f is even, then $\displaystyle\int_{-a}^{a} f(x)\, dx = 2\int_{0}^{a} f(x)\, dx.$

(*g*) If f is odd, then $\displaystyle\int_{-a}^{a} f(x)\, dx = 0.$

PROOF OF (b) Let us suppose that f and g are odd functions. Then we have $f(-x) = -f(x)$ and $g(-x) = -g(x)$. If we define the product of f and g as $F(x) = f(x)g(x)$, then

$$F(-x) = f(-x)g(-x) = (-f(x))(-g(x)) = f(x)g(x) = F(x).$$

This shows that the product F of two odd functions is an even function. The proofs of the remaining properties are left as exercises. See Problem 49 in Exercises 12.3. ❏

■ **Cosine and Sine Series** If f is an even function on $(-p, p)$, then in view of the foregoing properties the coefficients (9), (10), and (11) of Section 12.2 become

$$a_0 = \frac{1}{p} \int_{-p}^{p} f(x)\, dx = \frac{2}{p} \int_{0}^{p} f(x)\, dx$$

$$a_n = \frac{1}{p} \int_{-p}^{p} \underbrace{f(x) \cos \frac{n\pi}{p} x}_{\text{even}}\, dx = \frac{2}{p} \int_{0}^{p} f(x) \cos \frac{n\pi}{p} x\, dx$$

$$b_n = \frac{1}{p} \int_{-p}^{p} \underbrace{f(x) \sin \frac{n\pi}{p} x}_{\text{odd}}\, dx = 0.$$

Similarly, when f is odd on the interval $(-p, p)$,

$$a_n = 0, \quad n = 0, 1, 2, \ldots, \qquad b_n = \frac{2}{p} \int_{0}^{p} f(x) \sin \frac{n\pi}{p} x\, dx.$$

We summarize the results in the following definition.

DEFINITION 12.6 Fourier Cosine and Sine Series

(*i*) The Fourier series of an even function on the interval $(-p, p)$ is the **cosine series**

$$f(x) = \frac{a_0}{2} + \sum_{n=1}^{\infty} a_n \cos \frac{n\pi}{p} x, \tag{1}$$

where

$$a_0 = \frac{2}{p} \int_{0}^{p} f(x)\, dx \tag{2}$$

$$a_n = \frac{2}{p} \int_{0}^{p} f(x) \cos \frac{n\pi}{p} x\, dx. \tag{3}$$

(*ii*) The Fourier series of an odd function on the interval $(-p, p)$ is the **sine series**

$$f(x) = \sum_{n=1}^{\infty} b_n \sin \frac{n\pi}{p} x, \tag{4}$$

where

$$b_n = \frac{2}{p} \int_{0}^{p} f(x) \sin \frac{n\pi}{p} x\, dx. \tag{5}$$

Example 2 **Expansion in a Sine Series**

Expand $f(x) = x$, $-2 < x < 2$, in a Fourier series.

SOLUTION Inspection of Figure 12.6 shows that the given function is odd on the interval $(-2, 2)$, and so we expand f in a sine series. With the identification $2p = 4$, we have $p = 2$. Thus (5), after integration by parts, is

$$b_n = \int_{0}^{2} x \sin \frac{n\pi}{2} x\, dx = \frac{4(-1)^{n+1}}{n\pi}.$$

Therefore

$$f(x) = \frac{4}{\pi} \sum_{n=1}^{\infty} \frac{(-1)^{n+1}}{n} \sin \frac{n\pi}{2} x. \tag{6} \ \square$$

$y = x,\ -2 < x < 2$

Figure 12.6

The function in Example 3 satisfies the conditions of Theorem 12.1. Hence the series (6) converges to the function on $(-2, 2)$ and the periodic extension (of period 4) given in Figure 12.7.

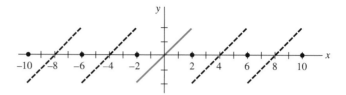

Figure 12.7

Example 3 Expansion in a Sine Series

Figure 12.8

The function $f(x) = \begin{cases} -1, & -\pi < x < 0 \\ 1, & 0 \le x < \pi \end{cases}$ shown in Figure 12.8 is odd on the interval $(-\pi, \pi)$. With $p = \pi$ we have from (5)

$$b_n = \frac{2}{\pi} \int_0^\pi (1) \sin nx \, dx = \frac{2}{\pi} \frac{1 - (-1)^n}{n},$$

and so

$$f(x) = \frac{2}{\pi} \sum_{n=1}^\infty \frac{1 - (-1)^n}{n} \sin nx. \tag{7}$$

■ **Gibbs Phenomenon** With the aid of a CAS we have plotted in Figure 12.9 the graphs $S_1(x)$, $S_2(x)$, $S_3(x)$, $S_{15}(x)$ of the partial sums of nonzero terms of (7). As seen in Figure 12.9(d) the graph of $S_{15}(x)$ has pronounced spikes near the discontinuities at $x = 0$, $x = \pi$, $x = -\pi$, and so on. This "overshooting" by the partial sums S_N from the functional values near a point of discontinuity does not smooth out but remains fairly

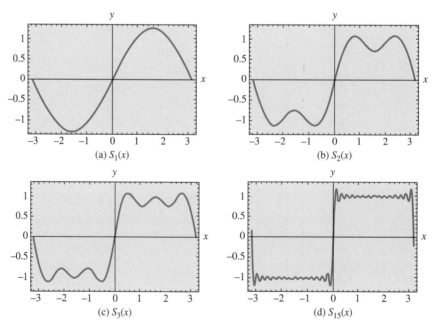

Figure 12.9 Partial sums of sine series (7) on $(-\pi, \pi)$

Figure 12.10

Figure 12.11

$f(x) = f(x + L)$

Figure 12.12

constant, even when the value N is taken to be large. This behavior of a Fourier series near a point at which f is discontinuous is known as the **Gibbs phenomenon.**

The periodic extension of f in Example 3 onto the entire x-axis is a meander function (see page 223).

■ **Half-Range Expansions** Throughout the preceding discussion it was understood that a function f was defined on an interval with the origin as midpoint, that is, $-p < x < p$. However, in many instances we are interested in representing a function that is defined only for $0 < x < L$ by a trigonometric series. This can be done in many different ways by supplying an arbitrary *definition* of the function on the interval $-L < x < 0$. For brevity we consider the three most important cases. If $y = f(x)$ is defined on the interval $0 < x < L$,

(*i*) reflect the graph of the function about the y-axis onto $-L < x < 0$; the function is now even on $-L < x < L$ (see Figure 12.10); or

(*ii*) reflect the graph of the function through the origin onto $-L < x < 0$; the function is now odd on $-L < x < L$ (see Figure 12.11); or

(*iii*) define f on $-L < x < 0$ by $f(x) = f(x + L)$ (see Figure 12.12).

Note that the coefficients of the series (1) and (4) utilize only the definition of the function on $0 < x < p$ (that is, half of the interval $-p < x < p$). Hence in practice there is no actual need to make the reflections described in (*i*) and (*ii*). If f is defined on $0 < x < L$, we simply identify the half-period as the length of the interval $p = L$. The coefficient formulas (2), (3), and (5) and the corresponding series yield either an even or an odd periodic extension of period $2L$ of the original function. The cosine and sine series obtained in this manner are known as **half-range expansions.** Lastly, in case (*iii*) we are defining the functional values on the interval $-L < x < 0$ to be the same as the values on $0 < x < L$. As in the previous two cases, there is no real need to do this. It can be shown that the set of functions in (1) of Section 12.2 is orthogonal on $a \leq x \leq a + 2p$ for any real number a. Choosing $a = -p$, we obtain the limits of integration in (9), (10), and (11) of that section. But for $a = 0$ the limits of integration are from $x = 0$ to $x = 2p$. Thus if f is defined over the interval $0 < x < L$, we identify $2p = L$ or $p = L/2$. The resulting Fourier series will give the periodic extension of f with period L. In this manner the values to which the series converges will be the same on $-L < x < 0$ as on $0 < x < L$.

Example 4 **Expansion in Three Series**

Expand $f(x) = x^2$, $0 < x < L$, (a) in a cosine series, (b) in a sine series, (c) in a Fourier series.

SOLUTION The graph of the function is given in Figure 12.13.

(a) We have

$$a_0 = \frac{2}{L} \int_0^L x^2 \, dx = \frac{2}{3} L^2, \qquad a_n = \frac{2}{L} \int_0^L x^2 \cos \frac{n\pi}{L} x \, dx = \frac{4L^2(-1)^n}{n^2 \pi^2},$$

where integration by parts was used twice in the evaluation of a_n.

$y = x^2, 0 < x < L$

Figure 12.13

Thus $$f(x) = \frac{L^2}{3} + \frac{4L^2}{\pi^2} \sum_{n=1}^{\infty} \frac{(-1)^n}{n^2} \cos \frac{n\pi}{L} x. \tag{8}$$

(b) In this case we must again integrate by parts twice:

$$b_n = \frac{2}{L} \int_0^L x^2 \sin \frac{n\pi}{L} x \, dx = \frac{2L^2(-1)^{n+1}}{n\pi} + \frac{4L^2}{n^3 \pi^3} [(-1)^n - 1].$$

Hence
$$f(x) = \frac{2L^2}{\pi} \sum_{n=1}^{\infty} \left\{ \frac{(-1)^{n+1}}{n} + \frac{2}{n^3 \pi^2} [(-1)^n - 1] \right\} \sin \frac{n\pi}{L} x. \tag{9}$$

(c) With $p = L/2$, $1/p = 2/L$, and $n\pi/p = 2n\pi/L$, we have

$$a_0 = \frac{2}{L} \int_0^L x^2 \, dx = \frac{2}{3} L^2, \qquad a_n = \frac{2}{L} \int_0^L x^2 \cos \frac{2n\pi}{L} x \, dx = \frac{L^2}{n^2 \pi^2}$$

and
$$b_n = \frac{2}{L} \int_0^L x^2 \sin \frac{2n\pi}{L} x \, dx = -\frac{L^2}{n\pi}.$$

Therefore
$$f(x) = \frac{L^2}{3} + \frac{L^2}{\pi} \sum_{n=1}^{\infty} \left\{ \frac{1}{n^2 \pi} \cos \frac{2n\pi}{L} x - \frac{1}{n} \sin \frac{2n\pi}{L} x \right\}. \tag{10}$$

The series (8), (9), (10) converge to the $2L$-periodic even extension of f, the $2L$-periodic odd extension of f, and the L-periodic extension of f, respectively. The graphs of these periodic extensions are shown in Figure 12.14.

(a) Cosine series

(b) Sine series

(c) Fourier series

Figure 12.14 ❏

■ **Periodic Driving Force** Fourier series are sometimes useful in determining a particular solution of a differential equation describing a physical system in which the input or driving force $f(t)$ is periodic. In the next example we find a particular solution of the differential equation

$$m \frac{d^2 x}{dt^2} + kx = f(t) \tag{11}$$

by first representing f by a half-range sine expansion and then assuming a particular solution of the form

$$x_p(t) = \sum_{n=1}^{\infty} B_n \sin \frac{n\pi}{p} t. \qquad (12)$$

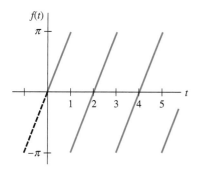

Figure 12.15

Example 5 Particular Solution of a DE

An undamped spring-mass system, in which the mass $m = \frac{1}{16}$ slug and the spring constant $k = 4$ lb/ft, is driven by the 2-periodic external force $f(t)$ shown in Figure 12.15. Although the force $f(t)$ acts on the system for $t > 0$, note that if we extend the graph of the function in a 2-periodic manner to the negative t-axis we obtain an odd function. In practical terms this means that we need only find the half-range sine expansion of $f(t) = \pi t$, $0 < t < 1$. With $p = 1$ it follows from (5) and integration by parts that

$$b_n = 2 \int_0^1 \pi t \sin n\pi t \, dt = \frac{2(-1)^{n+1}}{n}.$$

From (11) the differential equation of motion is seen to be

$$\frac{1}{16} \frac{d^2x}{dt^2} + 4x = \sum_{n=1}^{\infty} \frac{2(-1)^{n+1}}{n} \sin n\pi t. \qquad (13)$$

To find a particular solution $x_p(t)$ of (13), we substitute (12) into the equation and equate coefficients of $\sin n\pi t$. This yields

$$\left(-\frac{1}{16} n^2\pi^2 + 4 \right) B_n = \frac{2(-1)^{n+1}}{n} \qquad \text{or} \qquad B_n = \frac{32(-1)^{n+1}}{n(64 - n^2\pi^2)}.$$

Thus
$$x_p(t) = \sum_{n=1}^{\infty} \frac{32(-1)^{n+1}}{n(64 - n^2\pi^2)} \sin n\pi t. \qquad (14) \ \square$$

Observe in the solution (14) that there is no integer $n \geq 1$ for which the denominator $64 - n^2\pi^2$ of B_n is zero. In general, if there *is* a value of n, say N, for which $N\pi/p = \omega$, where $\omega = \sqrt{k/m}$, then the system described by (11) is in a state of pure resonance. In other words, we have pure resonance if the Fourier series expansion of the driving force $f(t)$ contains a term $\sin(N\pi/L)t$ (or $\cos(N\pi/L)t$) that has the same frequency as the free vibrations.

Of course, if the $2p$-periodic extension of the driving force f onto the negative t-axis yields an even function, then we expand f in a cosine series.

Answers to odd-numbered problems begin on page A-57.

In Problems 1–10, determine whether the function is even, odd, or neither.

1. $f(x) = \sin 3x$

2. $f(x) = x \cos x$

3. $f(x) = x^2 + x$

4. $f(x) = x^3 - 4x$

5. $f(x) = e^{|x|}$

6. $f(x) = e^x - e^{-x}$

7. $f(x) = \begin{cases} x^2, & -1 < x < 0 \\ -x^2, & 0 \leq x < 1 \end{cases}$

8. $f(x) = \begin{cases} x + 5, & -2 < x < 0 \\ -x + 5, & 0 \leq x < 2 \end{cases}$

9. $f(x) = x^3, \quad 0 \leq x \leq 2$

10. $f(x) = |x^5|$

In Problems 11–24, expand the given function in an appropriate cosine or sine series.

11. $f(x) = \begin{cases} -1, & -\pi < x < 0 \\ 1, & 0 \leq x < \pi \end{cases}$

12. $f(x) = \begin{cases} 1, & -2 < x < -1 \\ 0, & -1 < x < 1 \\ 1, & 1 < x < 2 \end{cases}$

13. $f(x) = |x|, \quad -\pi < x < \pi$

14. $f(x) = x, \quad -\pi < x < \pi$

15. $f(x) = x^2, \quad -1 < x < 1$

16. $f(x) = x|x|, \quad -1 < x < 1$

17. $f(x) = \pi^2 - x^2, \quad -\pi < x < \pi$

18. $f(x) = x^3, \quad -\pi < x < \pi$

19. $f(x) = \begin{cases} x - 1, & -\pi < x < 0 \\ x + 1, & 0 \leq x < \pi \end{cases}$

20. $f(x) = \begin{cases} x + 1, & -1 < x < 0 \\ x - 1, & 0 \leq x < 1 \end{cases}$

21. $f(x) = \begin{cases} 1, & -2 < x < -1 \\ -x, & -1 \le x < 0 \\ x, & 0 \le x < 1 \\ 1, & 1 \le x < 2 \end{cases}$

22. $f(x) = \begin{cases} -\pi, & -2\pi < x < -\pi \\ x, & -\pi \le x < \pi \\ \pi, & \pi \le x < 2\pi \end{cases}$

23. $f(x) = |\sin x|, \quad -\pi < x < \pi$

24. $f(x) = \cos x, \quad -\pi/2 < x < \pi/2$

In Problems 25–34, find the half-range cosine and sine expansions of the given function.

25. $f(x) = \begin{cases} 1, & 0 < x < \frac{1}{2} \\ 0, & \frac{1}{2} \le x < 1 \end{cases}$

26. $f(x) = \begin{cases} 0, & 0 < x < \frac{1}{2} \\ 1, & \frac{1}{2} \le x < 1 \end{cases}$

27. $f(x) = \cos x, \quad 0 < x < \pi/2$

28. $f(x) = \sin x, \quad 0 < x < \pi$

29. $f(x) = \begin{cases} x, & 0 < x < \pi/2 \\ \pi - x, & \pi/2 \le x < \pi \end{cases}$

30. $f(x) = \begin{cases} 0, & 0 < x < \pi \\ x - \pi, & \pi \le x < 2\pi \end{cases}$

31. $f(x) = \begin{cases} x, & 0 < x < 1 \\ 1, & 1 \le x < 2 \end{cases}$

32. $f(x) = \begin{cases} 1, & 0 < x < 1 \\ 2 - x, & 1 \le x < 2 \end{cases}$

33. $f(x) = x^2 + x, \quad 0 < x < 1$

34. $f(x) = x(2 - x), \quad 0 < x < 2$

In Problems 35–38, expand the given function in a Fourier series.

35. $f(x) = x^2, \quad 0 < x < 2\pi$

36. $f(x) = x, \quad 0 < x < \pi$

37. $f(x) = x + 1, \quad 0 < x < 1$

38. $f(x) = 2 - x, \quad 0 < x < 2$

In Problems 39 and 40, proceed as in Example 5 to find a particular solution $x_p(t)$ of equation (11) when $m = 1$, $k = 10$, and the driving force $f(t)$ is as given. Assume that when $f(t)$ is extended to the negative t-axis in a periodic manner, the resulting function is odd.

39. $f(t) = \begin{cases} 5, & 0 < t < \pi \\ -5, & \pi < t < 2\pi \end{cases}; \quad f(t + 2\pi) = f(t)$

40. $f(t) = 1 - t, \ 0 < t < 2; \quad f(t + 2) = f(t)$

In Problems 41 and 42, proceed as in Example 5 to find a particular solution $x_p(t)$ of equation (11) when $m = \frac{1}{4}$, $k = 12$, and the driving force $f(t)$ is as given. Assume that when $f(t)$ is extended to the negative t-axis in a periodic manner, the resulting function is even.

41. $f(t) = 2\pi t - t^2, \ 0 < t < 2\pi; \quad f(t + 2\pi) = f(t)$

42. $f(t) = \begin{cases} t, & 0 < t < \frac{1}{2} \\ 1 - t, & \frac{1}{2} < t < 1 \end{cases}; \quad f(t + 1) = f(t)$

43. (a) Solve the differential equation in Problem 39, $x'' + 10x = f(t)$, subject to the initial conditions $x(0) = 0$, $x'(0) = 0$.

(b) Use a CAS to plot the graph of the solution $x(t)$ in part **(a)**.

44. (a) Solve the differential equation in Problem 41, $\frac{1}{4} x'' + 12x = f(t)$, subject to the initial conditions $x(0) = 1$, $x'(0) = 0$.

(b) Use a CAS to plot the graph of the solution $x(t)$ in part **(a)**.

45. Suppose a uniform beam of length L is simply supported at $x = 0$ and at $x = L$. If the load per unit length is given by $w(x) = w_0 x/L$, $0 < x < L$, then the differential equation for the deflection $y(x)$ is

$$EI \frac{d^4 y}{dx^4} = \frac{w_0 x}{L},$$

where E, I, and w_0 are constants. (See (4) in Section 3.9.)

(a) Expand $w(x)$ in a half-range sine series.

(b) Use the method of Example 5 to find a particular solution $y(x)$ of the differential equation.

46. Proceed as in Problem 45 to find a particular solution $y(x)$

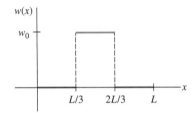

Figure 12.16

when the load per unit length is as given in Figure 12.16.

Computer Lab Assignment In Problems 47 and 48, use a CAS to plot graphs of partial sums $\{S_N(x)\}$ of the given trigonometric series. Experiment with different values of N and graphs on different intervals of the x-axis. Use your graphs to conjecture a closed-form expression for a function f defined for $0 < x < L$ that is represented by the series.

47. $f(x) = -\frac{\pi}{4} + \sum_{n=1}^{\infty} \left[\frac{(-1)^n - 1}{n^2 \pi} \cos nx + \frac{1 - 2(-1)^n}{n} \sin nx \right]$

48. $f(x) = \frac{1}{4} + \frac{4}{\pi^2} \sum_{n=1}^{\infty} \frac{1}{n^2} \left(1 - \cos \frac{n\pi}{2} \right) \cos \frac{n\pi}{2} x$

49. Prove properties (a), (c), (d), (f), and (g) in Theorem 12.2.

50. Is your answer in Problem 47 or in Problem 48 unique? Give a function f defined on a symmetric interval about the origin $-a < x < a$ that has the same trigonometric series as in Problem 47; as in Problem 48.

12.4 Complex Fourier Series and Frequency Spectrum

■ **Complex Fourier Series** The Fourier series (8) of Section 12.2 possesses a complex form. To obtain this form we convert cos x and sin x into complex exponential functions. From Euler's formula,

$$e^{ix} = \cos x + i \sin x \quad \text{and} \quad e^{-ix} = \cos x - i \sin x. \tag{1}$$

If we first add the two expressions in (1) and solve for cos x and then subtract the two expressions and solve for sin x, we arrive at

$$\cos x = \frac{e^{ix} + e^{-ix}}{2} \quad \text{and} \quad \sin x = \frac{e^{ix} - e^{-ix}}{2i}. \tag{2}$$

Using (2) to replace cos $(n\pi x/p)$ and sin $(n\pi x/p)$ in (8) of Section 12.2, the Fourier series of a function f can be written

$$\frac{a_0}{2} + \sum_{n=1}^{\infty} \left[a_n \frac{e^{in\pi x/p} + e^{-in\pi x/p}}{2} + b_n \frac{e^{in\pi x/p} - e^{-in\pi x/p}}{2i} \right]$$

$$= \frac{a_0}{2} + \sum_{n=1}^{\infty} \left[\frac{1}{2}(a_n - ib_n)e^{in\pi x/p} + \frac{1}{2}(a_n + ib_n)e^{-in\pi x/p} \right]$$

$$= c_0 + \sum_{n=1}^{\infty} c_n e^{in\pi x/p} + \sum_{n=1}^{\infty} c_{-n} e^{-in\pi x/p}, \tag{3}$$

where $c_0 = \frac{1}{2}a_0$, $c_n = \frac{1}{2}(a_n - ib_n)$, and $c_{-n} = \frac{1}{2}(a_n + ib_n)$. The symbols a_0, a_n, and b_n are the coefficients (9), (10), and (11) in Definition 12.5. When the function f is real, c_n and c_{-n} are complex conjugates and can also be written in terms of complex exponential functions:

$$c_0 = \frac{1}{2} \cdot \frac{1}{p} \int_{-p}^{p} f(x)\, dx, \tag{4}$$

$$c_n = \frac{1}{2}(a_n - ib_n) = \frac{1}{2} \left(\frac{1}{p} \int_{-p}^{p} f(x) \cos \frac{n\pi}{p} x\, dx - i \frac{1}{p} \int_{-p}^{p} f(x) \sin \frac{n\pi}{p} x\, dx \right)$$

$$= \frac{1}{2p} \int_{-p}^{p} f(x) \left[\cos \frac{n\pi}{p} x - i \sin \frac{n\pi}{p} x \right] dx$$

$$= \frac{1}{2p} \int_{-p}^{p} f(x)e^{-in\pi x/p}\, dx, \tag{5}$$

$$c_{-n} = \frac{1}{2}(a_n + ib_n) = \frac{1}{2} \left(\frac{1}{p} \int_{-p}^{p} f(x) \cos \frac{n\pi}{p} x\, dx + i \frac{1}{p} \int_{-p}^{p} f(x) \sin \frac{n\pi}{p} x\, dx \right)$$

$$= \frac{1}{2p} \int_{-p}^{p} f(x) \left[\cos \frac{n\pi}{p} x + i \sin \frac{n\pi}{p} x \right] dx$$

$$= \frac{1}{2p} \int_{-p}^{p} f(x)e^{in\pi x/p}\, dx. \tag{6}$$

Since the subscripts of the coefficients and exponents range over the entire set of nonnegative integers . . . −3, −2, −1, 0, 1, 2, 3, . . . , we can write the results in (3), (4), (5), and (6) in a more compact manner by summing over both the negative and nonnegative integers. In other words, we can use *one* summation and *one* integral that defines all three coefficients c_0, c_n, and c_{-n}.

DEFINITION 12.7 Complex Fourier Series

The **complex Fourier series** of functions f defined on an interval $(-p, p)$ is given by

$$f(x) = \sum_{n=-\infty}^{\infty} c_n e^{in\pi x/p},\tag{7}$$

where

$$c_n = \frac{1}{2p}\int_{-p}^{p} f(x)e^{-in\pi x/p}\,dx, \quad n = 0, \pm 1, \pm 2,\ldots.\tag{8}$$

If f satisfies the hypotheses of Theorem 12.1, a complex Fourier series converges to $f(x)$ at a point of continuity and to the average

$$\frac{f(x+) + f(x-)}{2}$$

at a point of discontinuity.

Example 1 Complex Fourier Series

Expand $f(x) = e^{-x}$, $-\pi < x < \pi$, in a complex Fourier series.

SOLUTION With $p = \pi$, (8) gives

$$c_n = \frac{1}{2\pi}\int_{-\pi}^{\pi} e^{-x}e^{-inx}\,dx = \frac{1}{2\pi}\int_{-\pi}^{\pi} e^{-(in+1)x}\,dx$$

$$= -\frac{1}{2\pi(in+1)}\left[e^{-(in+1)\pi} - e^{(in+1)\pi}\right].$$

We can simplify the coefficients c_n somewhat using Euler's formula:

$$e^{-(in+1)\pi} = e^{-\pi}(\cos n\pi - i\sin n\pi) = (-1)^n e^{-\pi}$$

and

$$e^{(in+1)\pi} = e^{\pi}(\cos n\pi + i\sin n\pi) = (-1)^n e^{\pi},$$

since $\cos n\pi = (-1)^n$ and $\sin n\pi = 0$. Hence

$$c_n = (-1)^n \frac{(e^{\pi} - e^{-\pi})}{2(in+1)\pi} = (-1)^n \frac{\sinh \pi}{\pi}\frac{1-in}{n^2+1}.\tag{9}$$

The complex Fourier series is then

$$f(x) = \frac{\sinh \pi}{\pi}\sum_{n=-\infty}^{\infty}(-1)^n\frac{1-in}{n^2+1}e^{inx}.\tag{10}$$

The series (10) converges to the 2π-periodic extension of f.

You may get the impression that we have just made life more complicated by introducing a complex version of a Fourier series. The reality of the situation is: in areas of engineering the form (7) given in Definition 12.7 is sometimes more useful than that given in (8) of Definition 12.5.

■ **Fundamental Frequency** The Fourier series in Definitions 12.5 and 12.7 define a periodic function and the **fundamental period** of that function (that is, the periodic extension of f) is $T = 2p$. Since $p = T/2$, (8) of Section 12.2 and (7) become, respectively,

$$\frac{a_0}{2} + \sum_{n=1}^{\infty}(a_n \cos n\omega x + b_n \sin n\omega x) \qquad \text{and} \qquad \sum_{n=-\infty}^{\infty} c_n e^{in\omega x},\tag{11}$$

where number $\omega = 2\pi/T$ is called the **fundamental angular frequency.** In Example 1 the periodic extension of the function has period $T = 2\pi$; the fundamental angular frequency is $\omega = 2\pi/2\pi = 1$.

■ **Frequency Spectrum** In the study of time-periodic signals, electrical engineers find it informative to examine various spectra of a wave form. If f is periodic and has fundamental period T, the plot of the points $(n\omega, |c_n|)$, where ω is the fundamental angular frequency and the c_n are the coefficients defined in (8), is called the **frequency spectrum** of f.

Example 2 Frequency Spectrum

In Example 1, $\omega = 1$ so that $n\omega$ takes on the values $0, \pm 1, \pm 2, \ldots$. Using $|\alpha + i\beta| = \sqrt{\alpha^2 + \beta^2}$, we see from (9) that

$$|c_n| = \frac{\sinh \pi}{\pi} \frac{1}{\sqrt{n^2 + 1}}.$$

The following table shows some values of n and corresponding values of c_n.

n	-3	-2	-1	0	1	2	3		
$	c_n	$	1.162	1.644	2.599	3.676	2.599	1.644	1.162

The graph in Figure 12.17, lines with arrowheads terminating at the points, is a portion of the frequency spectrum of f. ❑

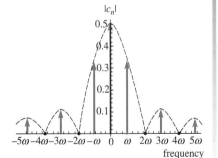

Figure 12.17 Frequency spectrum of f in Example 1

Figure 12.18 Periodic pulse

Figure 12.19 Frequency spectrum of f

Example 3 Frequency Spectrum

Find the frequency spectrum of the periodic square wave or periodic pulse shown in Figure 12.18. The wave is the periodic extension of the function f:

$$f(x) = \begin{cases} 0, & -\frac{1}{2} < x < -\frac{1}{4} \\ 1, & -\frac{1}{4} < x < \frac{1}{4} \\ 0, & \frac{1}{4} < x < \frac{1}{2} \end{cases}.$$

SOLUTION Here $T = 1 = 2p$ so $p = \frac{1}{2}$. Since f is 0 on the intervals $(-\frac{1}{2}, -\frac{1}{4})$ and $(\frac{1}{2}, \frac{1}{4})$, (8) becomes

$$c_n = \int_{-1/2}^{1/2} f(x)e^{2in\pi x}\, dx = \int_{-1/4}^{1/4} 1 \cdot e^{2in\pi x}\, dx$$

$$= \frac{e^{2in\pi x}}{2in\pi}\Big|_{-1/4}^{1/4}$$

$$= \frac{1}{n\pi} \frac{e^{in\pi/2} - e^{-in\pi/2}}{2i}.$$

That is, $c_n = \dfrac{1}{n\pi} \sin \dfrac{n\pi}{2}.$ ← by (2)

Since the last result is not valid at $n = 0$, we compute that term separately:

$$c_0 = \int_{-1/4}^{1/4} dx = \frac{1}{2}.$$

The following table shows some of the values of $|c_n|$, and Figure 12.19 shows the frequency spectrum of f. Since the fundamental frequency is $\omega = 2\pi/T = 2\pi$,

n	-5	-4	-3	-2	-1	0	1	2	3	4	5
$\lvert c_n\rvert$	$\dfrac{1}{5\pi}$	0	$\dfrac{1}{3\pi}$	0	$\dfrac{1}{\pi}$	$\dfrac{1}{2}$	$\dfrac{1}{\pi}$	0	$\dfrac{1}{3\pi}$	0	$\dfrac{1}{5\pi}$

the units $n\omega$ on the horizontal scale are $\pm 2\pi$, $\pm 4\pi$, $\pm 6\pi$, The curved dashed lines were added in Figure 12.19 to emphasize the presence of the zero values of $\lvert c_n\rvert$ when n is an even nonzero integer. ❑

In Problems 1–6, find the complex Fourier series of f on the given interval.

1. $f(x) = \begin{cases} -1, & -2 < x < 0 \\ 1, & 0 < x < 2 \end{cases}$

2. $f(x) = \begin{cases} 0, & 0 < x < 1 \\ 1, & 1 < x < 2 \end{cases}$

3. $f(x) = \begin{cases} 0, & -\frac{1}{2} < x < 0 \\ 1, & 0 < x < \frac{1}{4} \\ 0, & \frac{1}{4} < x < \frac{1}{2} \end{cases}$

4. $f(x) = \begin{cases} 0, & -\pi < x < 0 \\ x, & 0 < x < \pi \end{cases}$

5. $f(x) = x$, $\quad 0 < x < 2\pi$

6. $f(x) = e^{-|x|}$, $\quad -1 < x < 1$

7. Find the frequency spectrum of the periodic wave that is the periodic extension of the function f in Problem 1.

8. Find the frequency spectrum of the periodic wave that is the periodic extension of the function f in Problem 3.

In Problems 9 and 10, sketch the given periodic wave. Find the frequency spectrum of f.

9. $f(x) = 4 \sin x$, $\quad 0 < x < \pi$; $\quad f(x + \pi) = f(x)$
[*Hint:* Use (2).]

10. $f(x) = \begin{cases} \cos x, & 0 < x < \pi/2 \\ 0, & \pi/2 < x < \pi \end{cases}$; $\quad f(x + \pi) = f(x)$

11. (a) Show that $a_n = c_n + c_{-n}$ and $b_n = i(c_n - c_{-n})$.

(b) Use the results in part **(a)** and the complex Fourier series in Example 1 to obtain the Fourier series expansion of f.

12. The function f in Problem 1 is odd. Use the complex Fourier series to obtain the Fourier sine series expansion of f.

12.5 Sturm–Liouville Problem

■ **Review** For convenience we present here a brief review of some of the ordinary differential equations that will be of importance in the sections and chapters that follow.

Linear Equations	**General solutions**
$y' + \lambda y = 0$	$y = c_1 e^{\lambda x}$
$y'' + \lambda y = 0$, $\quad \lambda > 0$	$y = c_1 \cos \sqrt{\lambda}\, x + c_2 \sin \sqrt{\lambda}\, x$
$y'' - \lambda y = 0$, $\quad \lambda > 0$	$y = c_1 e^{-\sqrt{\lambda} x} + c_2 e^{\sqrt{\lambda} x}$, \quad or $y = c_1 \cosh \sqrt{\lambda}\, x + c_2 \sinh \sqrt{\lambda}\, x$

Cauchy–Euler equation	**General solutions, $x > 0$**
$x^2 y'' + xy' - \lambda^2 y = 0$	$y = c_1 x^{-\lambda} + c_2 x^{\lambda}$, $\quad \lambda \neq 0$ $y = c_1 + c_2 \ln x$, $\quad \lambda = 0$

Parametric Bessel equation ($\nu = 0$)	**General solution, $x > 0$**
$xy'' + y' + \lambda^2 xy = 0$	$y = c_1 J_0(\lambda x) + c_2 Y_0(\lambda x)$

Legendre's equation ($n = 0, 1, 2, \ldots$)	**Particular solutions are polynomials**
$(1 - x^2)y'' - 2xy' + n(n+1)y = 0$	$y = P_0(x) = 1$, $y = P_1(x) = x$, $y = P_2(x) = \frac{1}{2}(3x^2 - 1), \ldots$

This rule will be useful in ☞
Chapters 13–15.

As a rule, the exponential form $y = c_1 e^{-\sqrt{\lambda}\,x} + c_2 e^{\sqrt{\lambda}\,x}$ of the general solution of $y'' - \lambda y = 0$, $\lambda > 0$, is used when the domain of x is an infinite or semi-infinite interval; the hyperbolic form $y = c_1 \cosh \sqrt{\lambda}\,x + c_2 \sinh \sqrt{\lambda}\,x$ is used when the domain of x is a finite interval.

■ **Eigenvalues and Eigenfunctions** Orthogonal functions arise in the solution of differential equations. More to the point, an orthogonal set of functions can be generated by solving a two-point boundary-value problem involving a linear second-order differential equation containing a parameter λ. In Example 2 of Section 3.9 we saw that the boundary-value problem

$$y'' + \lambda y = 0, \qquad y(0) = 0, \quad y(L) = 0, \tag{1}$$

possessed nontrivial solutions only when the parameter λ took on the values $\lambda = n^2\pi^2/L^2$, $n = 1, 2, 3, \ldots$, called **eigenvalues.** The corresponding nontrivial solutions $y = c_2 \sin(n\pi x/L)$ or simply $y = \sin(n\pi x/L)$ are called the **eigenfunctions** of the problem. For example, for the boundary-value problem (1), $\lambda = -2$ *is not an eigenvalue,* whereas $\lambda = 9\pi^2/L^2$ $(n = 3)$ *is an eigenvalue,* and so for

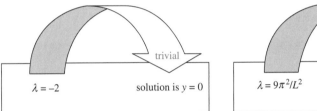

$\lambda = -2$ trivial solution is $y = 0$

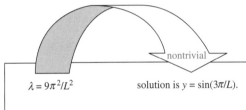

$\lambda = 9\pi^2/L^2$ nontrivial solution is $y = \sin(3\pi/L)$.

For our purposes in this chapter it is important to recognize that the set $\{\sin(n\pi x/L)\}$, $n = 1, 2, 3, \ldots$, is the orthogonal set of functions on the interval $[0, L]$ used as the basis for the Fourier sine series.

Example 1 Eigenvalues and Eigenfunctions

It is left as an exercise to verify that the eigenvalues and eigenfunctions for the boundary-value problem

$$y'' + \lambda y = 0, \qquad y'(0) = 0, \quad y'(L) = 0 \tag{2}$$

are, respectively, $\lambda = n^2\pi^2/L^2$, $n = 0, 1, 2, \ldots$, and $y = c_1 \cos(n\pi x/L)$, $c_1 \neq 0$. In contrast to (1), $\lambda = 0$ is an eigenvalue for this problem and $y = 1$ is the corresponding eigenfunction. The latter comes from solving $y'' = 0$ subject to the same boundary conditions $y'(0) = 0$, $y'(L) = 0$, but can be incorporated in $y = \cos(n\pi x/L)$ by permitting $n = 0$ and $c_1 = 1$. The set $\{\cos(n\pi x/L)\}$, $n = 0, 1, 2, 3, \ldots$, is orthogonal on the interval $[0, L]$. See Problem 3 in Exercises 12.5. ❏

■ **Regular Sturm–Liouville Problem** The problems in (1) and (2) are special cases of an important two-point boundary value problem. Let p, q, r, and r' be real-valued functions continuous on an interval $[a, b]$, and let $r(x) > 0$ and $p(x) > 0$ for every x in the interval. Then

Solve: $$\frac{d}{dx}[r(x)y'] + (q(x) + \lambda p(x))y = 0 \tag{3}$$

Subject to: $$\alpha_1 y(a) + \beta_1 y'(a) = 0 \tag{4}$$

$$\alpha_2 y(a) + \beta_2 y'(a) = 0 \tag{5}$$

is said to be a **regular Sturm–Liouville problem.** The coefficients in (4) and (5) are assumed to be real and independent of λ. In addition, α_1 and β_1 are not both zero, and α_2 and β_2 are not both zero.

Since the differential equation (3) is homogeneous, the Sturm–Liouville problem always has the trivial solution $y = 0$. However, this solution is of no interest to us. As in Example 1, in solving such a problem we seek to find numbers λ (eigenvalues) and nontrivial solutions y that depend on λ (eigenfunctions).

The boundary conditions in (4) and (5) (a linear combination of y and y' equal to zero at a point) are also **homogeneous.** The two boundary conditions in Example 1 and in Example 2, which follows, are homogeneous; a boundary condition such as $\alpha_1 y(a) + \beta_1 y'(a) = c$, where the constant c is nonzero, is **nonhomogeneous.** A boundary-value problem that consists of a homogeneous linear differential equation and homogeneous boundary conditions is, of course, said to be homogeneous; otherwise it is nonhomogeneous. The Sturm–Liouville problem given in (3)–(5) is a homogeneous boundary-value problem.

■ **Properties** Theorem 12.3 is a list of some of the more important of the many properties of the regular Sturm–Liouville problem. We shall prove only the last property.

THEOREM 12.3 Properties of the Regular Sturm–Liouville Problem

(a) There exist an infinite number of real eigenvalues that can be arranged in increasing order $\lambda_1 < \lambda_2 < \lambda_3 < \cdots \lambda_n < \cdots$ such that $\lambda_n \to \infty$ as $n \to \infty$.

(b) For each eigenvalue there is only one eigenfunction (except for nonzero multiples).

(c) Eigenfunctions corresponding to different eigenvalues are linearly independent.

(d) The set of eigenfunctions corresponding to the set of eigenvalues is orthogonal with respect to the weight function $p(x)$ on the interval $[a, b]$.

PROOF OF (d) Let y_m and y_n be eigenfunctions corresponding to eigenvalues λ_m and λ_n, respectively. Then

$$\frac{d}{dx}[r(x)y'_m] + (q(x) + \lambda_m p(x))y_m = 0 \tag{6}$$

$$\frac{d}{dx}[r(x)y'_n] + (q(x) + \lambda_n p(x))y_n = 0. \tag{7}$$

Multiplying (6) by y_n and (7) by y_m and subtracting the two equations gives

$$(\lambda_m - \lambda_n)p(x)y_m y_n = y_m \frac{d}{dx}[r(x)y'_n] - y_n \frac{d}{dx}[r(x)y'_m].$$

Integrating this last result by parts from $x = a$ to $x = b$ then yields

$$(\lambda_m - \lambda_n)\int_a^b p(x)y_m y_n\, dx = r(b)[y_m(b)y'_n(b) - y_n(b)y'_m(b)]$$
$$- r(a)[y_m(a)y'_n(a) - y_n(a)y'_m(a)]. \tag{8}$$

Now the eigenfunctions y_m and y_n must both satisfy the boundary conditions (4) and (5). In particular, from (4) we have

$$\alpha_1 y_m(a) + \beta_1 y_m'(a) = 0$$
$$\alpha_1 y_n(a) + \beta_1 y_n'(a) = 0.$$

In order for this system to be satisfied by α_1 and β_1, not both zero, the determinant of the coefficients must be zero:

$$y_m(a)y_n'(a) - y_n(a)y_m'(a) = 0.$$

A similar argument applied to (5) also gives

$$y_m(b)y_n'(b) - y_n(b)y_m'(b) = 0.$$

Since both members of the right side of (8) are zero, we have established the orthogonality relation

$$\int_a^b p(x)y_m(x)y_n(x)\,dx = 0, \qquad \lambda_m \neq \lambda_n. \tag{9} \; \square$$

Example 2 A Regular Sturm–Liouville Problem

Solve the boundary-value problem

$$y'' + \lambda y = 0, \qquad y(0) = 0, \quad y(1) + y'(1) = 0. \tag{10}$$

SOLUTION You should verify that for $\lambda < 0$ and for $\lambda = 0$ the problem in (10) possesses only the trivial solution $y = 0$. For $\lambda > 0$ the general solution of the differential equation is $y = c_1 \cos \sqrt{\lambda}x + c_2 \sin \sqrt{\lambda}x$. Now the condition $y(0) = 0$ implies $c_1 = 0$ in this solution. When $y = c_2 \sin \sqrt{\lambda}x$, the second boundary condition $y(1) + y'(1) = 0$ is satisfied if

$$c_2 \sin \sqrt{\lambda} + c_2 \sqrt{\lambda} \cos \sqrt{\lambda} = 0.$$

Choosing $c_2 \neq 0$, we see that the last equation is equivalent to

$$\tan \sqrt{\lambda} = -\sqrt{\lambda}. \tag{11}$$

If we let $x = \sqrt{\lambda}$, then Figure 12.20 shows the plausibility that there exist an infinite number of roots of the equation $\tan x = -x$. The eigenvalues of problem (10) are then $\lambda_n = x_n^2$, where x_n, $n = 1, 2, 3, \ldots$ are the consecutive *positive* roots. With the aid of a CAS it is easily shown that, to four rounded decimal places, $x_1 = 2.0288$, $x_2 = 4.9132$, $x_3 = 7.9787$, and $x_4 = 11.0855$, and the corresponding solutions are $y_1 = \sin 2.0288x$, $y_2 = \sin 4.9132x$, $y_3 = \sin 7.9787x$, and $y_4 = \sin 11.0855x$. In general, the eigenfunctions of the problem are $\{\sin \sqrt{\lambda_n}\,x\}$, $n = 1, 2, 3, \ldots$.

With identifications $r(x) = 1$, $q(x) = 0$, $p(x) = 1$, $\alpha_1 = 1$, $\beta_1 = 0$, $\alpha_2 = 1$, $\beta_2 = 1$ we see that (10) is a regular Sturm–Liouville problem. Thus $\{\sin \sqrt{\lambda_n}x\}$, $n = 1, 2, 3, \ldots$ is an orthogonal set with respect to the weight function $p(x) = 1$ on the interval $[0, 1]$. \square

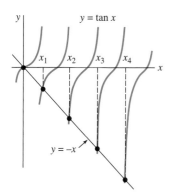

Figure 12.20

In some circumstances we can prove the orthogonality of the solutions of (3) without the necessity of specifying a boundary condition at $x = a$ and at $x = b$.

■ **Singular Sturm–Liouville Problem** If $r(a) = 0$, or $r(b) = 0$, or $r(a) = r(b) = 0$, the Sturm–Liouville problem consisting of the differential equation (3) and boundary conditions (4) and (5) is said to be **singular**. In the first case $x = a$ may be

a singular point of the differential equation, and consequently a solution of (3) may become unbounded as $x \to a$. However, we see from (8) that if $r(a) = 0$ then no boundary condition is required at $x = a$ to prove orthogonality of the eigenfunctions provided these solutions are bounded at that point. This latter requirement guarantees the existence of the integrals involved. By assuming the solutions of (3) are bounded on the closed interval $[a, b]$ we can say that

 (*i*) if $r(a) = 0$, then the orthogonality relation (9) holds with no boundary condition at $x = a$;

 (*ii*) if $r(b) = 0$, then the orthogonality relation (9) holds with no boundary condition at $x = b$;* and

 (*iii*) if $r(a) = r(b) = 0$, then the orthogonality relation (9) holds with no boundary condition at either $x = a$ or $x = b$.

We note in passing that a Sturm–Liouville problem is singular when the interval under consideration is infinite. See Problems 11 and 12 in Exercises 12.5.

■ **Self-Adjoint Form** If the coefficients are continuous and $a(x) \neq 0$ for all x in some interval, then any second-order differential equation

$$a(x)y'' + b(x)y' + (c(x) + \lambda d(x))y = 0$$

can be put into the so-called **self-adjoint form** (3) by multiplying the equation by the integrating factor

$$\frac{1}{a(x)}\, e^{\int (b(x)/a(x))\, dx}. \tag{12}$$

To see this we observe that the differential equation

$$e^{\int (b/a)\, dx}y'' + \frac{b(x)}{a(x)}\, e^{\int (b/a)\, dx}y' + \left(\frac{c(x)}{a(x)}\, e^{\int (b/a)\, dx} + \lambda \frac{d(x)}{a(x)}\, e^{\int (b/a)\, dx} \right) y = 0$$

is the same as

$$\frac{d}{dx}\, \underbrace{[e^{\int (b/a)\, dx}y']}_{r(x)} + \left(\underbrace{\frac{c(x)}{a(x)}\, e^{\int (b/a)\, dx}}_{q(x)} + \lambda \underbrace{\frac{d(x)}{a(x)}\, e^{\int (b/a)\, dx}}_{p(x)} \right) y = 0.$$

Example 3 Parametric Bessel Equation

In Section 5.3 we saw that the general solution of the parametric Bessel equation $x^2 y'' + xy' + (\lambda^2 x^2 - n^2)y = 0$, $n = 0, 1, 2, \ldots$, was $y = c_1 J_n(\lambda x) + c_2 Y_n(\lambda x)$. From (12) we find that $1/x$ is an integrating factor. Multiplying Bessel's equation by this factor yields the self-adjoint form

$$\frac{d}{dx}\, [xy'] + \left(\lambda^2 x - \frac{n^2}{x} \right) y = 0,$$

where we identify $r(x) = x$, $q(x) = -n^2/x$, and $p(x) = x$ and λ is replaced by λ^2. Now $r(0) = 0$, and of the two solutions $J_n(\lambda x)$ and $Y_n(\lambda x)$ only $J_n(\lambda x)$ is bounded at $x = 0$. Thus in view of (*i*) above, the set $\{J_n(\lambda_i x)\}$, $i = 1, 2, 3, \ldots$, is orthogonal with respect to the weight function $p(x) = x$ on an interval $[0, b]$. The orthogonality relation is

* Conditions (*i*) and (*ii*) are equivalent to choosing $\alpha_1 = 0$, $\beta_1 = 0$, and $\alpha_2 = 0$, $\beta_2 = 0$, respectively.

$$\int_0^b x J_n(\lambda_i x) J_n(\lambda_j x)\, dx = 0, \quad \lambda_i \neq \lambda_j,$$

provided the eigenvalues λ_i, $i = 1, 2, 3, \ldots$, are defined by means of a boundary condition at $x = b$ of the type

$$\alpha_2 J_n(\lambda b) + \beta_2 \lambda J_n'(\lambda b) = 0.^* \tag{13}$$

For any choice of α_2 and β_2, not both zero, it is known that (13) has an infinite number of roots $x_i = \lambda_i b$ from which we get $\lambda_i = x_i/b$. More will be said about the eigenvalues in the next chapter. ❑

Example 4 Legendre's Equation

The Legendre polynomials $P_n(x)$ are bounded solutions of Legendre's differential equation $(1 - x^2)y'' - 2xy' + n(n + 1)y = 0$, $n = 0, 1, 2, \ldots$, on the interval $[-1, 1]$. From the self-adjoint form

$$\frac{d}{dx}[(1 - x^2)y'] + n(n + 1)y = 0,$$

we read off $r(x) = 1 - x^2$, $q(x) = 0$, $p(x) = 1$, and $\lambda = n(n + 1)$. Observing that $r(-1) = r(1) = 0$, we find from (*iii*) above that the set $\{P_n(x)\}$, $n = 0, 1, 2, \ldots$, is orthogonal with respect to weight function $p(x) = 1$ on $[-1, 1]$. The orthogonality relation is

$$\int_{-1}^1 P_m(x)P_n(x)\, dx = 0, \qquad m \neq n.$$ ❑

* The extra factor of λ comes from the Chain Rule: $(d/dx)J_n(\lambda x) = \lambda J_n'(\lambda x)$.

E X E R C I S E S 1 2 . 5

Answers to odd-numbered problems begin on page A-58.

In Problems 1 and 2, find the eigenfunctions and the equation that defines the eigenvalues for the given boundary-value problem. Use a CAS to approximate the first four eigenvalues λ_1, λ_2, λ_3, and λ_4. Give the eigenfunctions corresponding to these approximations.

1. $y'' + \lambda y = 0$, $\quad y'(0) = 0$, $\quad y(1) + y'(1) = 0$

2. $y'' + \lambda y = 0$, $\quad y(0) + y'(0) = 0$, $\quad y(1) = 0$

3. Consider $y'' + \lambda y = 0$ subject to $y'(0) = 0$, $y'(L) = 0$. Show that the eigenfunctions are $\left\{ 1, \cos \dfrac{\pi}{L} x, \cos \dfrac{2\pi}{L} x, \ldots \right\}$. This set, which is orthogonal on $[0, L]$, is the basis for the Fourier cosine series.

4. Consider $y'' + \lambda y = 0$ subject to the periodic boundary conditions $y(-L) = y(L)$, $y'(-L) = y'(L)$. Show that the eigenfunctions are

$$\left\{ 1, \cos \frac{\pi}{L} x, \cos \frac{2\pi}{L} x, \ldots, \sin \frac{\pi}{L} x, \sin \frac{2\pi}{L} x, \sin \frac{3\pi}{L} x, \ldots \right\}.$$

This set, which is orthogonal on $[-L, L]$, is the basis for the Fourier series.

5. Find the square norm of each eigenfunction in Problem 1.

6. Show that for the eigenfunctions in Example 2,

$$\|\sin \sqrt{\lambda_n}\, x\|^2 = \frac{1}{2}[1 + \cos^2 \sqrt{\lambda_n}].$$

7. (a) Find the eigenvalues and eigenfunctions of the boundary-value problem

$$x^2 y'' + xy' + \lambda y = 0, \quad y(1) = 0, \quad y(5) = 0.$$

(b) Put the differential equation in self-adjoint form.

(c) Give an orthogonality relation.

8. (a) Find the eigenvalues and eigenfunctions of the boundary-value problem

$$y'' + y' + \lambda y = 0, \quad y(0) = 0, \quad y(2) = 0.$$

(b) Put the differential equation in self-adjoint form.

(c) Give an orthogonality relation.

9. (a) Give an orthogonality relation for the Sturm–Liouville problem in Problem 1.

(b) Use a CAS as an aid in verifying the orthogonality relation for the eigenfunctions y_1 and y_2 that correspond to the first two eigenvalues λ_1 and λ_2, respectively.

10. (a) Give an orthogonality relation for the Sturm–Liouville problem in Problem 2.

(b) Use a CAS as an aid in verifying the orthogonality relation for the eigenfunctions y_1 and y_2 that correspond to the first two eigenvalues λ_1 and λ_2, respectively.

11. Laguerre's differential equation $xy'' + (1-x)y' + ny = 0$, $n = 0, 1, 2, \ldots$, has polynomial solutions $L_n(x)$. Put the equation in self-adjoint form and give an orthogonality relation.

12. Hermite's differential equation $y'' - 2xy' + 2ny = 0$, $n = 0, 1, 2, \ldots$, has polynomial solutions $H_n(x)$. Put the

equation in self-adjoint form and give an orthogonality relation.

13. Consider the regular Sturm–Liouville problem

$$\frac{d}{dx}[(1+x^2)y'] + \frac{\lambda}{1+x^2}y = 0, \quad y(0) = 0, \quad y(1) = 0.$$

(a) Find the eigenvalues and eigenfunctions of the boundary-value problem. [*Hint:* Let $x = \tan\theta$ and then use the Chain Rule.]

(b) Give an orthogonality relation.

14. (a) Find the eigenfunctions and the equation that defines the eigenvalues for the boundary-value problem

$$x^2y'' + xy' + (\lambda^2x^2 - 1)y = 0,$$
$$y \text{ is bounded at } x = 0, \quad y(3) = 0.$$

(b) Use Table 5.1 to find the approximate values of the first four eigenvalues $\lambda_1, \lambda_2, \lambda_3,$ and λ_4.

12.6 Bessel and Legendre Series

Fourier series, Fourier cosine series, and Fourier sine series are three ways of expanding a function in terms of an orthogonal set of functions. But such expansions are by no means limited to orthogonal sets of trigonometric functions. We saw in Section 12.1 that a function f defined on an interval (a, b) could be expanded, at least formally, in terms of *any* set of functions $\{\phi_n(x)\}$ that is orthogonal with respect to a weight function on $[a, b]$. Many of these orthogonal series expansions or generalized Fourier series stem from Sturm–Liouville problems that, in turn, arise from attempts to solve linear partial differential equations serving as models for physical systems. Fourier series and orthogonal series expansions, as well as the two series considered in this section, will appear in the subsequent consideration of these applications in Chapters 13 and 14.

12.6.1 Fourier–Bessel Series

We saw in Example 3 of Section 12.5 that the set of Bessel functions $\{J_n(\lambda_i x)\}$, $i = 1, 2, 3, \ldots$, is orthogonal with respect to the weight function $p(x) = x$ on an interval $[0, b]$ when the eigenvalues λ_i are defined by means of a boundary condition of the form

$$\alpha_2 J_n(\lambda b) + \beta_2 \lambda J_n'(\lambda b) = 0. \tag{1}$$

From (7) and (8) of Section 12.1 the generalized Fourier expansion of a function f defined on $(0, b)$ in terms of this orthogonal set is

$$f(x) = \sum_{i=1}^{\infty} c_i J_n(\lambda_i x), \tag{2}$$

where

$$c_i = \frac{\int_0^b xJ_n(\lambda_i x) f(x)\, dx}{\int_0^b xJ_n^2(\lambda_i x)\, dx}. \tag{3}$$

The series (2) with coefficients (3) is called a **Fourier–Bessel series.**

■ **Differential Recurrence Relations** The **differential recurrence relations** that were given in (15) and (14) of Section 5.3,

$$\frac{d}{dx}[x^n J_n(x)] = x^n J_{n-1}(x) \tag{4}$$

$$\frac{d}{dx}[x^{-n} J_n(x)] = -x^{-n} J_{n+1}(x), \tag{5}$$

are often useful in the evaluation of the coefficients in (3).

■ **Square Norm** The value of the square norm $\|J_n(\lambda_i x)\|^2 = \int_0^b x J_n^2(\lambda_i x)\, dx$ depends on how the eigenvalues λ_i are defined. If $y = J_n(\lambda x)$, then we know from Example 3 of Section 12.5 that

$$\frac{d}{dx}[xy'] + \left(\lambda^2 x - \frac{n^2}{x}\right) y = 0.$$

After we multiply by $2xy'$, this equation can be written as

$$\frac{d}{dx}[xy']^2 + (\lambda^2 x^2 - n^2)\frac{d}{dx}[y]^2 = 0.$$

Integrating the last result by parts on $[0, b]$ then gives

$$2\lambda^2 \int_0^b xy^2\, dx = \left([xy']^2 + (\lambda^2 x^2 - n^2)y^2\right)\Big|_0^b.$$

Since $y = J_n(\lambda x)$, the lower limit is zero for $n > 0$ because $J_n(0) = 0$. For $n = 0$ the quantity $[xy']^2 + \lambda^2 x^2 y^2$ is zero at $x = 0$. Thus

$$2\lambda^2 \int_0^b x J_n^2(\lambda x)\, dx = \lambda^2 b^2 [J_n'(\lambda b)]^2 + (\lambda^2 b^2 - n^2)[J_n(\lambda b)]^2, \tag{6}$$

where we have used the Chain Rule to write $y' = \lambda J_n'(\lambda x)$.

We now consider three cases of (1).

Case I: If we choose $\alpha_2 = 1$ and $\beta_2 = 0$, then (1) is

$$J_n(\lambda b) = 0. \tag{7}$$

There are an infinite number of positive roots x_i of (7) (see Figure 5.3), and so the eigenvalues are positive and are given by $\lambda_i = x_i/b$. No new eigenvalues result from the negative roots of (7) since $J_n(-x) = (-1)^n J_n(x)$. (See page 259.) The number 0 is not an eigenvalue for any n since $J_n(0) = 0$ for $n = 1, 2, 3, \ldots$ and $J_0(0) = 1$. In other words, if $\lambda = 0$, we get the trivial function (which is never an eigenfunction) for $n = 1, 2, 3, \ldots$, and for $n = 0$, $\lambda = 0$ does not satisfy equation (7). When (5) is written in the form $x J_n'(x) = n J_n(x) - x J_{n+1}(x)$, it follows from (6) and (7) that the square norm of $J_n(\lambda_i x)$ is

$$\|J_n(\lambda_i x)\|^2 = \frac{b^2}{2} J_{n+1}^2(\lambda_i b). \tag{8}$$

Case II: If we choose $\alpha_2 = h \geq 0$, $\beta_2 = b$, then (1) is

$$h J_n(\lambda b) + \lambda b J_n'(\lambda b) = 0. \tag{9}$$

Equation (9) has an infinite number of positive roots x_i for each positive integer $n = 1, 2, 3, \ldots$. As before, the eigenvalues are obtained from $\lambda_i = x_i/b$. $\lambda = 0$ is not an eigenvalue for $n = 1, 2, 3, \ldots$. Substituting $\lambda_i b J_n'(\lambda_i b) = -h J_n(\lambda_i b)$ into (6), we find that the square norm of $J_n(\lambda_i x)$ is now

$$\left\| J_n(\lambda_i x) \right\|^2 = \frac{\lambda_i^2 b^2 - n^2 + h^2}{2\lambda_i^2} J_n^2(\lambda_i b). \tag{10}$$

Case III: If $h = 0$ and $n = 0$ in (9), the eigenvalues λ_i are defined from the roots of

$$J_0'(\lambda b) = 0. \tag{11}$$

Even though (11) is just a special case of (9), it is the only situation for which $\lambda = 0$ is an eigenvalue. To see this, observe that for $n = 0$ the result in (5) implies that

$$J_0'(\lambda b) = 0 \quad \text{is equivalent to} \quad J_1(\lambda b) = 0. \tag{12}$$

Since $x_1 = 0$ is a root of this equation and since $J_0(0) = 1$ is nontrivial, we conclude that $\lambda_1 = 0$ is an eigenvalue. But obviously we cannot use (10) when $h = 0$, $n = 0$, and $\lambda_1 = 0$. However, by definition of the square norm we have

$$\left\| 1 \right\|^2 = \int_0^b x\, dx = \frac{b^2}{2}. \tag{13}$$

For $\lambda_i > 0$ we can use (10) with $h = 0$ and $n = 0$:

$$\left\| J_0(\lambda_i x) \right\|^2 = \frac{b^2}{2} J_0^2(\lambda_i b). \tag{14}$$

The following summary gives the corresponding three forms of the series (2).

DEFINITION 12.8 Fourier–Bessel Series

The **Fourier–Bessel series** of a function f defined on the interval $(0, b)$ is given by

(*i*)
$$f(x) = \sum_{i=1}^{\infty} c_i J_n(\lambda_i x) \tag{15}$$

$$c_i = \frac{2}{b^2 J_{n+1}^2(\lambda_i b)} \int_0^b x J_n(\lambda_i x) f(x)\, dx, \tag{16}$$

where the λ_i are defined by $J_n(\lambda b) = 0$.

(*ii*)
$$f(x) = \sum_{i=1}^{\infty} c_i J_n(\lambda_i x) \tag{17}$$

$$c_i = \frac{2\lambda_i^2}{(\lambda_i^2 b^2 - n^2 + h^2) J_n^2(\lambda_i b)} \int_0^b x J_n(\lambda_i x) f(x)\, dx, \tag{18}$$

where the λ_i are defined by $h J_n(\lambda b) + \lambda b J_n'(\lambda b) = 0$.

(*iii*)
$$f(x) = c_1 + \sum_{i=2}^{\infty} c_i J_0(\lambda_i x) \tag{19}$$

continued

$$c_1 = \frac{2}{b^2} \int_0^b xf(x)\, dx$$

$$c_i = \frac{2}{b^2 J_0^2(\lambda_i b)} \int_0^b xJ_0(\lambda_i x)\, f(x)\, dx, \tag{20}$$

where the λ_i are defined by $J_0'(\lambda b) = 0$.

■ **Convergence of a Fourier–Bessel Series** Sufficient conditions for the convergence of a Fourier–Bessel series are not particularly restrictive.

> **THEOREM 12.4** **Conditions for Convergence**
>
> If f and f' are piecewise continuous on the open interval $(0, b)$, then a Fourier–Bessel expansion of f converges to $f(x)$ at any point where f is continuous and to the average $[f(x-) + f(x+)]/2$ at a point where f is discontinuous.

Example 1 **Expansion in a Fourier–Bessel Series**

Expand $f(x) = x$, $0 < x < 3$, in a Fourier–Bessel series, using Bessel functions of order one that satisfy the boundary condition $J_1(3\lambda) = 0$.

SOLUTION We use (15) where the coefficients c_i are given by (16):

$$c_i = \frac{2}{3^2 J_2^2(3\lambda_i)} \int_0^3 x^2 J_1(\lambda_i x)\, dx.$$

To evaluate this integral we let $t = \lambda_i x$ and use (4) in the form $(d/dt)[t^2 J_2(t)] = t^2 J_1(t)$:

$$c_i = \frac{2}{9\lambda_i^3 J_2^2(3\lambda_i)} \int_0^{3\lambda_i} \frac{d}{dt}[t^2 J_2(t)]\, dt = \frac{2}{\lambda_i J_2(3\lambda_i)}.$$

Therefore the desired expansion is

$$f(x) = 2 \sum_{i=1}^{\infty} \frac{1}{\lambda_i J_2(3\lambda_i)}\, J_1(\lambda_i x). \qquad \square$$

Example 2 **Expansion in a Fourier–Bessel Series**

If eigenvalues λ_i in Example 1 are defined by $J_1(3\lambda) + \lambda J_1'(3\lambda) = 0$, then the only thing that changes in the expansion is the value of the square norm. Multiplying the boundary condition by 3 gives $3J_1(3\lambda) + 3\lambda J_1'(3\lambda) = 0$, which now matches (9) when $h = 3$, $b = 3$, and $n = 1$. Thus (18) and (17) yield, in turn,

$$c_i = \frac{18\lambda_i J_2(3\lambda_i)}{(9\lambda_i^2 + 8)J_1^2(3\lambda_i)}$$

and

$$f(x) = 18 \sum_{i=1}^{\infty} \frac{\lambda_i J_2(3\lambda_i)}{(9\lambda_i^2 + 8)J_1^2(3\lambda_i)} J_1(\lambda_i x). \qquad \square$$

■ **Use of Computers** Since Bessel functions are "built-in functions" in a CAS, it is a straightforward task to find the approximate values of the eigenvalues λ_i and the coefficients c_i in a Fourier–Bessel series. For example, in (9) we can think of

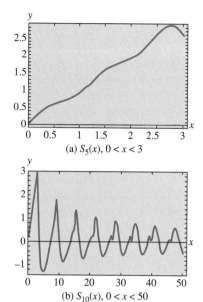

(a) $S_5(x)$, $0 < x < 3$

(b) $S_{10}(x)$, $0 < x < 50$

Figure 12.21 Partial sums of a Fourier–Bessel series

$x_i = \lambda_i b$ as a positive root of the equation $hJ_n(x) + xJ'_n(x) = 0$. Thus in Example 2 we have used a CAS to find the first five positive roots x_i of $3J_1(x) + xJ'_1(x) = 0$ and from these roots we obtain the first five eigenvalues: $\lambda_1 = x_1/3 = 0.98320$, $\lambda_2 = x_2/3 = 1.94704$, $\lambda_3 = x_3/3 = 2.95758$, $\lambda_4 = x_4/3 = 3.98538$, and $\lambda_5 = x_5/3 = 5.02078$. Knowing the roots $x_i = 3\lambda_i$ and the eigenvalues, we again use a CAS to calculate the numerical values of $J_2(3\lambda_i)$, $J_1^2(3\lambda_i)$, and finally the coefficients c_i. In this manner we find that the fifth partial sum $S_5(x)$ for the Fourier–Bessel series representation of $f(x) = x$, $0 < x < 3$ in Example 2 is

$$S_5(x) = 4.01844\, J_1(0.98320x) - 1.86937\, J_1(1.94704x)$$
$$+ 1.07106\, J_1(2.95758x) - 0.70306\, J_1(3.98538x) + 0.50343\, J_1(5.02078x).$$

The graph of $S_5(x)$ on the interval $0 < x < 3$ is shown in Figure 12.21(a). In Figure 12.21(b) we have graphed $S_{10}(x)$ on the interval $0 < x < 50$. Notice that outside the interval of definition $0 < x < 3$ the series does not converge to a periodic extension of f because Bessel functions are not periodic functions. See Problems 11 and 12 in Exercises 12.6.

12.6.2 Fourier–Legendre Series

From Example 4 of Section 12.5 we know that the set of Legendre polynomials $\{P_n(x)\}$, $n = 0, 1, 2, \ldots$, is orthogonal with respect to the weight function $p(x) = 1$ on the interval $[-1, 1]$. Furthermore, it can be proved the square norm of a polynomial $P_n(x)$ depends on n in the following manner:

$$\|P_n(x)\|^2 = \int_{-1}^{1} P_n^2(x)\, dx = \frac{2}{2n + 1}.$$

The orthogonal series expansion of a function in terms of the Legendre polynomials is summarized in the next definition.

DEFINITION 12.9 Fourier–Legendre Series

The **Fourier–Legendre series** of a function f and f_2 on an interval $(-1, 1)$ is given by

$$f(x) = \sum_{n=0}^{\infty} c_n P_n(x), \tag{21}$$

where

$$c_n = \frac{2n + 1}{2} \int_{-1}^{1} f(x) P_n(x)\, dx. \tag{22}$$

■ **Convergence of a Fourier–Legendre Series** Sufficient conditions for convergence of a Fourier–Legendre series are given in the next theorem.

THEOREM 12.5 Conditions for Convergence

If f and f' are piecewise continuous on $(-1, 1)$, then the Fourier–Legendre series (21) converges to $f(x)$ at a point of continuity and to $[f(x+) + f(x-)]/2$ at a point of discontinuity.

Example 3 Expansion in a Fourier–Legendre Series

Write out the first four nonzero terms in the Fourier–Legendre expansion of

$$f(x) = \begin{cases} 0, & -1 < x < 0 \\ 1, & 0 \leq x < 1. \end{cases}$$

SOLUTION The first several Legendre polynomials are listed on page 263. From these and (22) we find

$$c_0 = \frac{1}{2} \int_{-1}^{1} f(x)P_0(x)\, dx = \frac{1}{2} \int_{0}^{1} 1 \cdot 1\, dx = \frac{1}{2}$$

$$c_1 = \frac{3}{2} \int_{-1}^{1} f(x)P_1(x)\, dx = \frac{3}{2} \int_{0}^{1} 1 \cdot x\, dx = \frac{3}{4}$$

$$c_2 = \frac{5}{2} \int_{-1}^{1} f(x)P_2(x)\, dx = \frac{5}{2} \int_{0}^{1} 1 \cdot \frac{1}{2}(3x^2 - 1)\, dx = 0$$

$$c_3 = \frac{7}{2} \int_{-1}^{1} f(x)P_3(x)\, dx = \frac{7}{2} \int_{0}^{1} 1 \cdot \frac{1}{2}(5x^3 - 3x)\, dx = -\frac{7}{16}$$

$$c_4 = \frac{9}{2} \int_{-1}^{1} f(x)P_4(x)\, dx = \frac{9}{2} \int_{0}^{1} 1 \cdot \frac{1}{8}(35x^4 - 30x^2 + 3)\, dx = 0$$

$$c_5 = \frac{11}{2} \int_{-1}^{1} f(x)P_5(x)\, dx = \frac{11}{2} \int_{0}^{1} 1 \cdot \frac{1}{8}(63x^5 - 70x^3 + 15x)\, dx = \frac{11}{32}.$$

Hence $$f(x) = \frac{1}{2}P_0(x) + \frac{3}{4}P_1(x) - \frac{7}{16}P_3(x) + \frac{11}{32}P_5(x) + \cdots.$$ ❑

Like the Bessel functions, Legendre polynomials are built-in functions in computer algebra systems such as *Maple* and *Mathematica,* and so each of the coefficients just listed can be found using the integration application of such a program. Indeed, using a CAS, we further find that $c_6 = 0$ and $c_7 = -\,^{65}\!/_{256}$. The fifth partial sum of the Fourier–Legendre series representation of the function f defined in Example 3 is then

$$S_5(x) = \frac{1}{2}P_0(x) + \frac{3}{4}P_1(x) - \frac{7}{16}P_3(x) + \frac{11}{32}P_5(x) - \frac{65}{256}P_7(x).$$

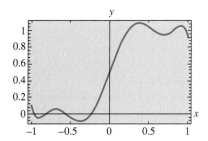

Figure 12.22 Partial sum $S_5(x)$ of Fourier–Legendre series

The graph of $S_5(x)$ on the interval $-1 < x < 1$ is given in Figure 12.22.

■ **Alternative Form of Series** In applications the Fourier–Legendre series appears in an alternative form. If we let $x = \cos\,\theta$, then $x = 1$ implies $\theta = 0$, whereas $x = -1$ implies $\theta = \pi$. Since $dx = -\sin\,\theta\, d\theta$, (21) and (22) become, respectively,

$$F(\theta) = \sum_{n=0}^{\infty} c_n P_n(\cos\,\theta) \qquad\qquad (23)$$

$$c_n = \frac{2n+1}{2} \int_{0}^{\pi} F(\theta)P_n(\cos\,\theta)\sin\,\theta\, d\theta, \qquad\qquad (24)$$

where $f(\cos\,\theta)$ has been replaced by $F(\theta)$.

Answers to odd-numbered problems begin on page A-58.

12.6.1 Fourier–Bessel Series

1. Find the first four eigenvalues λ_k defined by $J_1(3\lambda) = 0$.
[*Hint:* See Table 5.1 in Section 5.3.]

2. Find the first four eigenvalues λ_k defined by $J_0'(2\lambda) = 0$.

In Problems 3–6, expand $f(x) = 1$, $0 < x < 2$, in a Fourier–Bessel series using Bessel functions of order zero that satisfy the given boundary condition.

3. $J_0(2\lambda) = 0$ **4.** $J_0'(2\lambda) = 0$

5. $J_0(2\lambda) + 2\lambda J_0'(2\lambda) = 0$ **6.** $J_0(2\lambda) + \lambda J_0'(2\lambda) = 0$

In Problems 7–10, expand the given function in a Fourier–Bessel series using Bessel functions of the same order as in the indicated boundary condition.

7. $f(x) = 5x$, $\quad 0 < x < 4$

$3J_1(4\lambda) + 4\lambda J_1'(4\lambda) = 0$

8. $f(x) = x^2$, $\quad 0 < x < 1$

$J_2(\lambda) = 0$

9. $f(x) = x^2$, $\quad 0 < x < 3$

$J_0'(3\lambda) = 0$ [*Hint:* $t^3 = t^2 \cdot t$.]

10. $f(x) = 1 - x^2$, $\quad 0 < x < 1$

$J_0(\lambda) = 0$

11. Computer Lab Assignment

(a) Use a CAS to plot the graph of $y = 3J_1(x) + xJ_1'(x)$ on an interval so that the first five positive x-intercepts of the graph are shown.

(b) Use the root-finding capability of your CAS to approximate the first five roots x_i of the equation $3J_1(x) + xJ_1'(x) = 0$.

(c) Use the data obtained in part **(b)** to find the first five positive eigenvalues λ_i that satisfy $3J_1(4\lambda) + 4\lambda J_1'(4\lambda) = 0$. (See Problem 7.)

(d) If instructed, find the first ten positive eigenvalues.

12. Computer Lab Assignment

(a) Use the eigenvalues in part **(c)** of Problem 11 and a CAS to approximate the values of the first five coefficients c_i of the Fourier–Bessel series obtained in Problem 7.

(b) Use a CAS to plot the graphs of the partial sums $S_N(x)$, $N = 1, 2, 3, 4, 5$, of the Fourier–Bessel series in Problem 7.

(c) If instructed, plot the graph of the partial sum $S_{10}(x)$ on $0 < x < 4$ and on $0 < x < 50$.

13. If the partial sums in Problem 12 were plotted on a symmetric interval such as $-30 < x < 30$, would the graphs possess any symmetry? Explain.

14. (a) Sketch, by hand, a graph of what you think the Fourier–Bessel series in Problem 3 converges to on the interval $-2 < x < 2$.

(b) Sketch, by hand, a graph of what you think the Fourier–Bessel series would converge to on the interval $-1 < x < 1$ if the eigenvalues in Problem 7 were defined by $3J_2(2\lambda) + 4\lambda J_2'(2\lambda) = 0$.

12.6.2 Fourier–Legendre Series

In Problems 15 and 16, write out the first five nonzero terms in the Fourier–Legendre expansion of the given function. If instructed, use a CAS as an aid in evaluating the coefficients. Use a CAS to plot the graph of the partial sum $S_5(x)$.

15. $f(x) = \begin{cases} 0, & -1 < x < 0 \\ x, & 0 < x < 1 \end{cases}$ **16.** $f(x) = e^x$, $\quad -1 < x < 1$

17. The first three Legendre polynomials are $P_0(x) = 1$, $P_1(x) = x$, and $P_1(x) = \frac{1}{2}(3x^2 - 1)$. If $x = \cos\theta$, then $P_0(\cos\theta) = 1$ and $P_1(\cos\theta) = \cos\theta$. Show that $P_2(\cos\theta) = \frac{1}{4}(3\cos 2\theta + 1)$.

18. Use the results of Problem 17 to find a Fourier–Legendre expansion (23) of $F(\theta) = 1 - \cos 2\theta$.

19. A Legendre polynomial $P_n(x)$ is an even or odd function, depending on whether n is even or odd. Show that if f is an even function on $(-1, 1)$, then (21) and (22) become, respectively,

$$f(x) = \sum_{n=0}^{\infty} c_{2n} P_{2n}(x) \tag{25}$$

$$c_{2n} = (4n + 1) \int_0^1 f(x) P_{2n}(x)\, dx. \tag{26}$$

20. Show that if f is an odd function on $(-1, 1)$, then (21) and (22) become, respectively,

$$f(x) = \sum_{n=0}^{\infty} c_{2n+1} P_{2n+1}(x) \tag{27}$$

$$c_{2n+1} = (4n + 3) \int_0^1 f(x) P_{2n+1}(x)\, dx. \tag{28}$$

The series (25) and (27) can also be used when f is defined on only the interval $(0, 1)$. Both series represent f on $(0, 1)$; but on the interval $(-1, 0)$, (25) represents an even extension, whereas (27) represents an odd extension. In Problems 21 and 22, write out the first four nonzero terms in the indicated expansion of the given function. What function does the series represent on the interval $(-1, 1)$? Use a graphing utility to plot the graph of the partial sum $S_4(x)$.

21. $f(x) = x$, $\quad 0 < x < 1$; (25) **22.** $f(x) = 1$, $\quad 0 < x < 1$; (27)

Answers to odd-numbered problems begin on page A-58.

In Problems 1–6, fill in the blank or answer true/false without referring back to the text.

1. The functions $f(x) = x^2 - 1$ and $g(x) = x^5$ are orthogonal on $[-\pi, \pi]$._____

2. The product of an odd function f with an odd function g is _____.

3. To expand $f(x) = |x| + 1$, $-\pi < x < \pi$, in an appropriate trigonometric series we would use a _____ series.

4. $y = 0$ is never an eigenfunction of a Sturm–Liouville problem._____

5. $\lambda = 0$ is never an eigenvalue of a Sturm–Liouville problem._____

6. If the function $f(x) = \begin{cases} 3, & -\pi < x < 0 \\ 2, & x = 0 \\ 0, & 0 < x < \pi \end{cases}$ is expanded in a Fourier series, the series will converge to _____ at $x = 0$.

7. Suppose the function $f(x) = x^2 + 1$, $0 < x < 3$, is expanded in a Fourier series, a cosine series, and a sine series. Give the value to which each series will converge at $x = 0$.

8. What is the corresponding eigenfunction for the boundary-value problem $y'' + \lambda y = 0$, $y'(0) = 0$, $y(\pi/2) = 0$ for $\lambda = 25$?

9. Chebyshev's differential equation

$$(1 - x^2)y'' - xy' + n^2 y = 0$$

has a polynomial solution $y = T_n(x)$ for $n = 0, 1, 2, \dots$. Specify the weight function $w(x)$ and the interval over which the set of Chebyshev polynomials $\{T_n(x)\}$ is orthogonal. Give an orthogonality relation.

10. The set of Legendre polynomials $\{P_n(x)\}$, where $P_0(x) = 1$, $P_1(x) = x, \dots$, is orthogonal with respect to the weight function $w(x) = 1$ on the interval $[-1, 1]$. Explain why $\int_{-1}^{1} P_n(x)\, dx = 0$ for $n > 0$.

11. Without doing any work, explain why the cosine series of $f(x) = \cos^2 x$, $0 < x < \pi$, is the finite series

$$f(x) = \tfrac{1}{2} + \tfrac{1}{2} \cos 2x.$$

12. (a) Show that the set

$$\left\{ \sin \frac{\pi}{2L} x,\ \sin \frac{3\pi}{2L} x,\ \sin \frac{5\pi}{2L} x, \dots \right\}$$

is orthogonal on the interval $0 \le x \le L$.

(b) Find the norm of each function in part **(a)**. Construct an orthonormal set.

13. Expand $f(x) = |x| - x$, $-1 < x < 1$, in a Fourier series.

14. Expand $f(x) = 2x^2 - 1$, $-1 < x < 1$, in a Fourier series.

15. Expand $f(x) = e^{-x}$, $0 < x < 1$, in a cosine series; in a sine series.

16. In Problems 13, 14, and 15, sketch the periodic extension of f to which each series converges.

17. Find the eigenvalues and eigenfunctions of the boundary-value problem

$$x^2 y'' + xy' + 9\lambda y = 0, \quad y'(1) = 0, \quad y(e) = 0.$$

18. Give an orthogonality relation for the eigenfunctions in Problem 17.

19. Expand $f(x) = \begin{cases} 1, & 0 < x < 2 \\ 0, & 2 < x < 4 \end{cases}$ in a Fourier–Bessel series using Bessel functions of order zero that satisfy the boundary condition $J_0(4\lambda) = 0$.

20. Expand $f(x) = x^4$, $-1 < x < 1$, in a Fourier–Legendre series.

13 *Boundary-Value Problems in Rectangular Coordinates*

IMPORTANT CONCEPTS IN THIS CHAPTER

linear partial differential equation
separation of variables
separation constant
heat equation
wave equation
Laplace's equation
boundary-value problem
eigenvalues
eigenfunction
homogeneous boundary conditions
nonhomogeneous boundary conditions
initial conditions

INTRODUCTION

In this and the next two chapters, the emphasis will be on two procedures that are used in solving partial differential equations that occur frequently in problems involving temperature distributions, vibrations, and potentials. These problems, called boundary-value problems, are described by relatively simple linear second-order partial differential equations. The thrust of both of these procedures is to find particular solutions of a partial differential equation by reducing it to one or more ordinary differential equations.

We begin with the method of separation of variables for linear partial differential equations. The application of this method leads back to the important concepts in Chapter 12, namely, to eigenvalues, eigenfunctions, and the expansion of a function in an infinite series of orthogonal functions.

13.1 Separable Partial Differential Equations

■ **Linear Partial Differential Equation** Partial differential equations (PDEs), like ordinary differential equations, are classified as either linear or nonlinear. Analogous to a linear ODE (see (5) of Section 1.1), the dependent variable and its partial derivatives appear only to the first power in a linear PDE. Specifically, we are interested in **linear second-order partial differential equations.** If we let u denote the dependent variable and x and y the independent variables, then the general form of a linear second-order partial differential equation is given by

$$A \frac{\partial^2 u}{\partial x^2} + B \frac{\partial^2 u}{\partial x \partial y} + C \frac{\partial^2 u}{\partial y^2} + D \frac{\partial u}{\partial x} + E \frac{\partial u}{\partial y} + Fu = G, \tag{1}$$

where A, B, C, \ldots, G are functions of x and y. When $G(x, y) = 0$, equation (1) is said to be **homogeneous;** otherwise, it is **nonhomogeneous.** For example, the equations

$$\frac{\partial^2 u}{\partial x^2} + \frac{\partial^2 u}{\partial y^2} = 0 \qquad \text{and} \qquad \frac{\partial^2 u}{\partial x^2} - \frac{\partial u}{\partial y} = xy$$

are homogeneous and nonhomogeneous, respectively.

■ **Solution of a PDE** A **solution** of a linear partial differential equation (1) is a function $u(x, y)$ of two independent variables that possesses all partial derivatives occurring in the equation and that satisfies the equation in some region of the xy-plane.

 Note: We are interested only in particular solutions.

 It is not our intention to examine procedures for finding *general solutions* of linear partial differential equations. Not only is it often difficult to obtain a general solution of a linear second-order PDE, but a general solution is usually not all that useful in applications. Thus our focus will be on finding *particular solutions* of some of the important linear PDEs, that is, equations that appear in many applications.

■ **Separation of Variables** Although there are several methods that can be tried to find particular solutions of a linear PDE, in the **method of separation of variables** we seek to find a particular solution of the form of a *product* of a function of x and a function of y,

$$u(x, y) = X(x)Y(y).$$

With this assumption, it is sometimes possible to reduce a linear PDE in two variables to two ODEs. To this end we note

$$\frac{\partial u}{\partial x} = X'Y, \qquad \frac{\partial u}{\partial y} = XY'$$

and

$$\frac{\partial^2 u}{\partial x^2} = X''Y, \qquad \frac{\partial^2 u}{\partial y^2} = XY'',$$

where the primes denote ordinary differentiation.

> ### *Example 1* Using Separation of Variables
>
> Find product solutions of $\dfrac{\partial^2 u}{\partial x^2} = 4 \dfrac{\partial u}{\partial y}$.

SOLUTION If $u(x, y) = X(x)Y(y)$, the equation becomes

$$X''Y = 4XY'.$$

After dividing both sides by $4XY$, we have separated the variables:

$$\frac{X''}{4X} = \frac{Y'}{Y}.$$

Since the left-hand side of the last equation is independent of y and is equal to the right-hand side, which is independent of x, we conclude that both sides of the equation are independent of x *and* y. In other words, each side of the equation must be a constant. In practice it is convenient to write this real *separation constant* as either λ^2 or $-\lambda^2$. We distinguish the following three cases.

Case I: If $\lambda^2 > 0$, the two equalities

$$\frac{X''}{4X} = \frac{Y'}{Y} = \lambda^2$$

lead to $X'' - 4\lambda^2 X = 0$ and $Y' - \lambda^2 Y = 0.$

These latter equations have the solutions

$$X = c_1 \cosh 2\lambda x + c_2 \sinh 2\lambda x \qquad \text{and} \qquad Y = c_3 e^{\lambda^2 y},$$

respectively. Thus a particular solution of the partial differential equation is

$$\begin{aligned} u = XY &= (c_1 \cosh 2\lambda x + c_2 \sinh 2\lambda x)(c_3 e^{\lambda^2 y}) \\ &= A_1 e^{\lambda^2 y} \cosh 2\lambda x + B_1 e^{\lambda^2 y} \sinh 2\lambda x, \end{aligned} \qquad (2)$$

where $A_1 = c_1 c_3$ and $B_1 = c_2 c_3$.

Case II: If $-\lambda^2 < 0$, the two equalities

$$\frac{X''}{4X} = \frac{Y'}{Y} = -\lambda^2$$

give $X'' + 4\lambda^2 X = 0$ and $Y' + \lambda^2 Y = 0.$

Since the solutions of these equations are

$$X = c_4 \cos 2\lambda x + c_5 \sin 2\lambda x \qquad \text{and} \qquad Y = c_6 e^{-\lambda^2 y},$$

respectively, another particular solution is

$$u = A_2 e^{-\lambda^2 y} \cos 2\lambda x + B_2 e^{-\lambda^2 y} \sin 2\lambda x, \qquad (3)$$

where $A_2 = c_4 c_6$ and $B_2 = c_5 c_6$.

Case III: If $\lambda^2 = 0$, it follows that

$$X'' = 0 \qquad \text{and} \qquad Y' = 0.$$

In this case $X = c_7 x + c_8$ and $Y = c_9$ so that

$$u = A_3 x + B_3, \qquad (4)$$

where $A_3 = c_7 c_9$ and $B_3 = c_8 c_9$. ❑

It is left as an exercise to verify that (2), (3), and (4) satisfy the given equation. See Problem 29 in Exercises 13.1.

Separation of variables is not a general method for finding particular solutions; some linear partial differential equations are simply not separable. You should verify that the assumption $u = XY$ does not lead to a solution for $\partial^2 u/\partial x^2 - \partial u/\partial y = x$.

■ **Superposition Principle** The following theorem is analogous to Theorem 3.2 and is known as the **superposition principle.**

THEOREM 13.1 Superposition Principle

If u_1, u_2, \ldots, u_k are solutions of a homogeneous linear partial differential equation, then the linear combination

$$u = c_1 u_1 + c_2 u_2 + \cdots + c_k u_k,$$

where the c_i, $i = 1, 2, \ldots, k$ are constants, is also a solution.

Throughout the remainder of the chapter we shall assume that whenever we have an infinite set u_1, u_2, u_3, \ldots of solutions of a homogeneous linear equation, we can construct yet another solution u by forming the infinite series

$$u = \sum_{k=1}^{\infty} c_k u_k$$

where the c_i, $i = 1, 2, \ldots$, are constants.

■ **Classification of Equations** A linear second-order partial differential equation in two independent variables with constant coefficients can be classified as one of three types. This classification depends only on the coefficients of the second-order derivatives. Of course, we assume that at least one of the coefficients A, B, and C is not zero.

DEFINITION 13.1 Classification of Equations

The linear second-order partial differential equation

$$A \frac{\partial^2 u}{\partial x^2} + B \frac{\partial^2 u}{\partial x \partial y} + C \frac{\partial^2 u}{\partial y^2} + D \frac{\partial u}{\partial x} + E \frac{\partial u}{\partial y} + Fu = 0,$$

where A, B, C, D, E, and F are real constants, is said to be

$$\begin{aligned}
&\textbf{hyperbolic if} \quad B^2 - 4AC > 0, \\
&\textbf{parabolic if} \quad B^2 - 4AC = 0, \\
&\textbf{elliptic if} \quad B^2 - 4AC < 0.
\end{aligned}$$

Example 2 **Classifying Linear Second-Order PDEs**

Classify the following equations:

$$\text{(a) } 3 \frac{\partial^2 u}{\partial x^2} = \frac{\partial u}{\partial y} \qquad \text{(b) } \frac{\partial^2 u}{\partial x^2} = \frac{\partial^2 u}{\partial y^2} \qquad \text{(c) } \frac{\partial^2 u}{\partial x^2} + \frac{\partial^2 u}{\partial y^2} = 0.$$

SOLUTION (a) By rewriting the given equation as

$$3\frac{\partial^2 u}{\partial x^2} - \frac{\partial u}{\partial y} = 0$$

we can make the identifications $A = 3$, $B = 0$, and $C = 0$. Since $B^2 - 4AC = 0$, the equation is parabolic.

(b) By rewriting the equation as

$$\frac{\partial^2 u}{\partial x^2} - \frac{\partial^2 u}{\partial y^2} = 0,$$

we see that $A = 1$, $B = 0$, $C = -1$, and $B^2 - 4AC = -4(1)(-1) > 0$. The equation is hyperbolic.

(c) With $A = 1$, $B = 0$, $C = 1$, and $B^2 - 4AC = -4(1)(1) < 0$, the equation is elliptic. ❑

A detailed explanation of why we would want to classify a second-order partial differential equation is beyond the scope of this text. But the answer lies in the fact that we wish to solve partial differential equations subject to certain side conditions known as boundary and initial conditions. The kinds of side conditions appropriate for a given equation depend on whether the equation is hyperbolic, parabolic, or elliptic.

EXERCISES 13.1

Answers to odd-numbered problems begin on page A-58.

In Problems 1–6, use separation of variables to find, if possible, product solutions of the given partial differential equation.

1. $\dfrac{\partial u}{\partial x} = \dfrac{\partial u}{\partial y}$

2. $\dfrac{\partial u}{\partial x} + 3\dfrac{\partial u}{\partial y} = 0$

3. $u_x + u_y = u$

4. $u_x = u_y + u$

5. $x\dfrac{\partial u}{\partial x} = y\dfrac{\partial u}{\partial y}$

6. $y\dfrac{\partial u}{\partial x} + x\dfrac{\partial u}{\partial y} = 0$

7. $\dfrac{\partial^2 u}{\partial x^2} + \dfrac{\partial^2 u}{\partial x\partial y} + \dfrac{\partial^2 u}{\partial y^2} = 0$

8. $y\dfrac{\partial^2 u}{\partial x\partial y} + u = 0$

9. $k\dfrac{\partial^2 u}{\partial x^2} - u = \dfrac{\partial u}{\partial t}$, $k > 0$

10. $k\dfrac{\partial^2 u}{\partial x^2} = \dfrac{\partial u}{\partial t}$, $k > 0$

11. $a^2\dfrac{\partial^2 u}{\partial x^2} = \dfrac{\partial^2 u}{\partial t^2}$

12. $a^2\dfrac{\partial^2 u}{\partial x^2} = \dfrac{\partial^2 u}{\partial t^2} + 2k\dfrac{\partial u}{\partial t}$, $k > 0$

13. $\dfrac{\partial^2 u}{\partial x^2} + \dfrac{\partial^2 u}{\partial y^2} = 0$

14. $x^2\dfrac{\partial^2 u}{\partial x^2} + \dfrac{\partial^2 u}{\partial y^2} = 0$

15. $u_{xx} + u_{yy} = u$

16. $a^2 u_{xx} - g = u_{tt}$, g a constant

In Problems 17–26, classify the given partial differential equation as hyperbolic, parabolic, or elliptic.

17. $\dfrac{\partial^2 u}{\partial x^2} + \dfrac{\partial^2 u}{\partial x\partial y} + \dfrac{\partial^2 u}{\partial y^2} = 0$

18. $3\dfrac{\partial^2 u}{\partial x^2} + 5\dfrac{\partial^2 u}{\partial x\partial y} + \dfrac{\partial^2 u}{\partial y^2} = 0$

19. $\dfrac{\partial^2 u}{\partial x^2} + 6\dfrac{\partial^2 u}{\partial x\partial y} + 9\dfrac{\partial^2 u}{\partial y^2} = 0$

20. $\dfrac{\partial^2 u}{\partial x^2} - \dfrac{\partial^2 u}{\partial x\partial y} - 3\dfrac{\partial^2 u}{\partial y^2} = 0$

21. $\dfrac{\partial^2 u}{\partial x^2} = 9\dfrac{\partial^2 u}{\partial x\partial y}$

22. $\dfrac{\partial^2 u}{\partial x\partial y} - \dfrac{\partial^2 u}{\partial y^2} + 2\dfrac{\partial u}{\partial x} = 0$

23. $\dfrac{\partial^2 u}{\partial x^2} + 2\dfrac{\partial^2 u}{\partial x\partial y} + \dfrac{\partial^2 u}{\partial y^2} + \dfrac{\partial u}{\partial x} - 6\dfrac{\partial u}{\partial y} = 0$

24. $\dfrac{\partial^2 u}{\partial x^2} + \dfrac{\partial^2 u}{\partial y^2} = u$

25. $a^2\dfrac{\partial^2 u}{\partial x^2} = \dfrac{\partial^2 u}{\partial t^2}$

26. $k\dfrac{\partial^2 u}{\partial x^2} = \dfrac{\partial u}{\partial t}$, $k > 0$

27. Show that the equation

$$k\left(\frac{\partial^2 u}{\partial r^2} + \frac{1}{r}\frac{\partial u}{\partial r}\right) = \frac{\partial u}{\partial t}$$

possesses the product solution

$$u = e^{-k\lambda^2 t}(AJ_0(\lambda r) + BY_0(\lambda r)).$$

28. Show that the equation

$$\frac{\partial^2 u}{\partial r^2} + \frac{1}{r}\frac{\partial u}{\partial r} + \frac{1}{r^2}\frac{\partial^2 u}{\partial\theta^2} = 0$$

possesses the product solution

$$u = (c_1\cos\lambda\theta + c_2\sin\lambda\theta)(c_3 r^\lambda + c_4 r^{-\lambda}).$$

29. Verify that the products (2), (3), and (4) satisfy the first-order equation in Example 1.

30. Definition 13.1 generalizes to linear PDEs with coefficients that are functions of x and y. Determine the regions in the xy-plane for which the equation

$$(xy + 1)\frac{\partial^2 u}{\partial x^2} + (x + 2y)\frac{\partial^2 u}{\partial x\partial y} + \frac{\partial^2 u}{\partial y^2} + xy^2 u = 0$$

is hyperbolic, parabolic, or elliptic.

13.2 Classical Equations and Boundary-Value Problems

For the remainder of this and the next chapter we shall be concerned principally with finding product solutions of the partial differential equations

$$k \frac{\partial^2 u}{\partial x^2} = \frac{\partial u}{\partial t}, \qquad k > 0 \tag{1}$$

$$a^2 \frac{\partial^2 u}{\partial x^2} = \frac{\partial^2 u}{\partial t^2} \tag{2}$$

$$\frac{\partial^2 u}{\partial x^2} + \frac{\partial^2 u}{\partial y^2} = 0 \tag{3}$$

or slight variations of these equations. These classical equations of mathematical physics are known, respectively, as the **one-dimensional heat equation,** the **one-dimensional wave equation,** and **Laplace's equation in two dimensions.** "One-dimensional" refers to the fact that x denotes a spatial dimension whereas t represents time. Laplace's equation is abbreviated as $\nabla^2 u = 0$, where

$$\nabla^2 u = \frac{\partial^2 u}{\partial x^2} + \frac{\partial^2 u}{\partial y^2}$$

is called the **two-dimensional Laplacian** of the function u. In three dimensions the **Laplacian** of u is

$$\nabla^2 u = \frac{\partial^2 u}{\partial x^2} + \frac{\partial^2 u}{\partial y^2} + \frac{\partial^2 u}{\partial z^2}.$$

Note that the heat equation (1) is parabolic, the wave equation (2) is hyperbolic, and Laplace's equation (3) is elliptic.

■ **Heat Equation** Equation (1) occurs in the theory of heat flow—that is, heat transferred by conduction in a rod or thin wire. The function $u(x, t)$ is temperature. Problems in mechanical vibrations often lead to the wave equation (2). For purposes of discussion, a solution $u(x, t)$ of (2) will represent the displacement of an idealized string. Finally, a solution $u(x, y)$ of Laplace's equation (3) can be interpreted as the steady-state (that is, time-independent) temperature distribution throughout a thin, two-dimensional plate.

Even though we have to make many simplifying assumptions, it is worthwhile to see how equations such as (1) and (2) arise.

Suppose a thin circular rod of length L has a cross-sectional area A and coincides with the x-axis on the interval $[0, L]$. See Figure 13.1. Let us suppose:

cross-section of area A

Figure 13.1

- The flow of heat within the rod takes place only in the x-direction.
- The lateral, or curved, surface of the rod is insulated; that is, no heat escapes from this surface.
- No heat is being generated within the rod.
- The rod is homogeneous; that is, its mass per unit volume ρ is a constant.
- The specific heat γ and thermal conductivity K of the material of the rod are constants.

To derive the partial differential equation satisfied by the temperature $u(x, t)$, we need two empirical laws of heat conduction:

(i) *The quantity of heat Q in an element of mass m is*

$$Q = \gamma m u, \tag{4}$$

where u is the temperature of the element.

(ii) *The rate of heat flow Q_t through the cross-section indicated in Figure 13.1 is proportional to the area A of the cross-section and the partial derivative with respect to x of the temperature:*

$$Q_t = -K\,Au_x. \tag{5}$$

Since heat flows in the direction of decreasing temperature, the minus sign in (5) is used to ensure that Q_t is positive for $u_x < 0$ (heat flow to the right) and negative for $u_x > 0$ (heat flow to the left). If the circular slice of the rod shown in Figure 13.1 between x and $x + \Delta x$ is very thin, then $u(x, t)$ can be taken as the approximate temperature at each point in the interval. Now the mass of the slice is $m = \rho(A\,\Delta x)$, and so it follows from (4) that the quantity of heat in it is

$$Q = \gamma\rho A\,\Delta x\,u. \tag{6}$$

Furthermore, when heat flows in the positive x-direction, we see from (5) that heat builds up in the slice at the net rate

$$-K\,Au_x(x, t) - [-K\,Au_x(x + \Delta x, t)] = K\,A[u_x(x + \Delta x, t) - u_x(x, t)]. \tag{7}$$

By differentiating (6) with respect to t we see that this net rate is also given by

$$Q_t = \gamma\rho A\,\Delta x\,u_t. \tag{8}$$

Equating (7) and (8) gives

$$\frac{K}{\gamma\rho}\,\frac{u_x(x + \Delta x, t) - u_x(x, t)}{\Delta x} = u_t. \tag{9}$$

Taking the limit of (9) as $\Delta x \to 0$ finally yields (1) in the form*

$$\frac{K}{\gamma\rho}\,u_{xx} = u_t.$$

It is customary to let $k = K/\gamma\rho$ and call this positive constant the **thermal diffusivity.**

■ **Wave Equation** Consider a string of length L, such as a guitar string, stretched taut between two points on the x-axis—say, $x = 0$ and $x = L$. When the string starts to vibrate, assume that the motion takes place in the xy-plane in such a manner that each point on the string moves in a direction perpendicular to the x-axis (transverse vibrations). As shown in Figure 13.2(a), let $u(x, t)$ denote the vertical displacement of any point on the string measured from the x-axis for $t > 0$. We further assume:

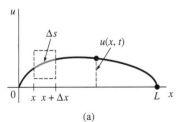

• The string is perfectly flexible.
• The string is homogeneous; that is, its mass per unit length ρ is a constant.
• The displacements u are small compared to the length of the string.
• The slope of the curve is small at all points.
• The tension **T** acts tangent to the string, and its magnitude T is the same at all points.
• The tension is large compared with the force of gravity.
• No other external forces act on the string.

 Now in Figure 13.2(b) the tensions \mathbf{T}_1 and \mathbf{T}_2 are tangent to the ends of the curve on the interval $[x, x + \Delta x]$. For small θ_1 and θ_2 the net vertical force acting on the corresponding element Δs of the string is then

$$T \sin \theta_2 - T \sin \theta_1 \approx T \tan \theta_2 - T \tan \theta_1$$
$$= T[u_x(x + \Delta x, t) - u_x(x, t)],^\dagger$$

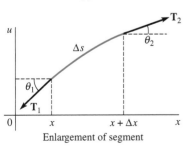

(a)

(b)

Figure 13.2

* Recall from calculus that $u_{xx} = \displaystyle\lim_{\Delta x \to 0} \frac{u_x(x + \Delta x, t) - u_x(x, t)}{\Delta x}$.

† $\tan \theta_2 = u_x(x + \Delta x, t)$ and $\tan \theta_1 = u_x(x, t)$ are equivalent expressions for slope.

where $T = |\mathbf{T}_1| = |\mathbf{T}_2|$. Now $\rho\,\Delta s \approx \rho\,\Delta x$ is the mass of the string on $[x, x + \Delta x]$, and so Newton's second law gives

$$T[u_x(x + \Delta x, t) - u_x(x, t)] = \rho\,\Delta x\, u_{tt}$$

or

$$\frac{u_x(x + \Delta x, t) - u_x(x, t)}{\Delta x} = \frac{\rho}{T}\, u_{tt}.$$

If the limit is taken as $\Delta x \to 0$, the last equation becomes $u_{xx} = (\rho/T)u_{tt}$. This of course is (2) with $a^2 = T/\rho$.

temperature as a
function of position
on the hot plate

thermometer

(x, y)

Figure 13.3

■ Laplace's Equation Although we shall not present its derivation, Laplace's equation in two and three dimensions occurs in time-independent problems involving potentials such as electrostatic, gravitational, and velocity in fluid mechanics. Moreover, a solution of Laplace's equation can also be interpreted as a steady-state temperature distribution. As illustrated in Figure 13.3, a solution $u(x, y)$ of (3) could represent the temperature that varies from point to point—but not with time—of a rectangular plate.

We often wish to find solutions of equations (1), (2), and (3) that satisfy certain side conditions.

■ Initial Conditions Since solutions of (1) and (2) depend on time t, we can prescribe what happens at $t = 0$; that is, we can give **initial conditions** (**IC**). If $f(x)$ denotes the initial temperature distribution throughout the rod in Figure 13.1, then a solution $u(x, t)$ of (1) must satisfy the single initial condition $u(x, 0) = f(x)$, $0 < x < L$. On the other hand, for a vibrating string, we can specify its initial displacement (or shape) $f(x)$ as well as its initial velocity $g(x)$. In mathematical terms we seek a function $u(x, t)$ satisfying (2) and the two initial conditions:

$$u(x, 0) = f(x), \qquad \frac{\partial u}{\partial t}\bigg|_{t=0} = g(x), \quad 0 < x < L. \tag{10}$$

For example, the string could be plucked, as shown in Figure 13.4, and released from rest ($g(x) = 0$).

$u = 0$
at $x = 0$

$u = 0$
at $x = L$

Figure 13.4

■ Boundary Conditions The string in Figure 13.4 is secured to the x-axis at $x = 0$ and $x = L$ for all time. We interpret this by the two **boundary conditions** (**BC**):

$$u(0, t) = 0, \qquad u(L, t) = 0, \quad t > 0.$$

Note that in this context the function f in (10) is continuous, and consequently $f(0) = 0$ and $f(L) = 0$. In general, there are three types of boundary conditions associated with equations (1), (2), and (3). On a boundary we can specify the values of *one* of the following:

$$(i)\; u, \qquad (ii)\; \frac{\partial u}{\partial n}, \qquad \text{or} \qquad (iii)\; \frac{\partial u}{\partial n} + hu, \quad h \text{ a constant.}$$

Here $\partial u/\partial n$ denotes the normal derivative of u (the directional derivative of u in the direction perpendicular to the boundary). A boundary condition of the first type (i) is called a **Dirichlet condition;** a boundary condition of the second type (ii) is called a **Neumann condition;** and a boundary condition of the third type (iii) is known as a **Robin condition.** For example, for $t > 0$ a typical condition at the right-hand end of the rod in Figure 13.1 can be

$(i)'$ $u(L,\ t) = u_0,$ u_0 a constant,

$(ii)'$ $\left.\dfrac{\partial u}{\partial x}\right|_{x=L} = 0,$ or

$(iii)'$ $\left.\dfrac{\partial u}{\partial x}\right|_{x=L} = -h(u(L,\ t) - u_m),$ $h > 0$ and u_m constants.

Condition $(i)'$ simply states that the boundary $x = L$ is held by some means at a constant *temperature* u_0 for all time $t > 0$. Condition $(ii)'$ indicates that the boundary $x = L$ is *insulated*. From the empirical law of heat transfer, the flux of heat across a boundary (that is, the amount of heat per unit area per unit time conducted across the boundary) is proportional to the value of the normal derivative $\partial u / \partial n$ of the temperature u. Thus when the boundary $x = L$ is thermally insulated, no heat flows into or out of the rod and so

$$\left.\frac{\partial u}{\partial x}\right|_{x=L} = 0.$$

We can interpret $(iii)'$ to mean that *heat is lost* from the right-hand end of the rod by being in contact with a medium, such as air or water, that is held at a constant temperature. From Newton's law of cooling, the outward flux of heat from the rod is proportional to the difference between the temperature $u(L,\ t)$ at the boundary and the temperature u_m of the surrounding medium. We note that if heat is lost from the left-hand end of the rod, the boundary condition is

$$\left.\frac{\partial u}{\partial x}\right|_{x=0} = h(u(0,\ t) - u_m).$$

The change in algebraic sign is consistent with the assumption that the rod is at a higher temperature than the medium surrounding the ends so that $u(0,\ t) > u_m$ and $u(L,\ t) > u_m$. At $x = 0$ and $x = L$, the slopes $u_x(0,\ t)$ and $u_x(L,\ t)$ must be positive and negative, respectively.

Of course, at the ends of the rod we can specify different conditions at the same time. For example, we could have

$$\left.\frac{\partial u}{\partial x}\right|_{x=0} = 0 \quad \text{and} \quad u(L,\ t) = u_0, \quad t > 0.$$

We note that the boundary condition in $(i)'$ is homogeneous if $u_0 = 0$; if $u_0 \neq 0$, the boundary condition is nonhomogeneous. The boundary condition $(ii)'$ is homogeneous; $(iii)'$ is homogeneous if $u_m = 0$ and nonhomogeneous if $u_m \neq 0$.

■ **Boundary-Value Problems** Problems such as

Solve: $a^2 \dfrac{\partial^2 u}{\partial x^2} = \dfrac{\partial^2 u}{\partial t^2},$ $0 < x < L,$ $t > 0$

Subject to: (BC) $u(0,\ t) = 0,$ $u(L,\ t) = 0,$ $t > 0$ (11)

 (IC) $u(x,\ 0) = f(x),$ $\left.\dfrac{\partial u}{\partial t}\right|_{t=0} = g(x),$ $0 < x < L$

and

Solve: $\dfrac{\partial^2 u}{\partial x^2} + \dfrac{\partial^2 u}{\partial y^2} = 0,$ $0 < x < a,$ $0 < y < b$

Subject to: (BC) $\begin{cases} \left.\dfrac{\partial u}{\partial x}\right|_{x=0} = 0, & \left.\dfrac{\partial u}{\partial x}\right|_{x=a} = 0, \quad 0 < y < b \\[2mm] u(x,\ 0) = 0, & u(x,\ b) = f(x), \quad 0 < x < a \end{cases}$ (12)

are called **boundary-value problems.**

■ **Modifications** The partial differential equations (1), (2), and (3) must be modified to take into consideration internal or external influences acting on the physical system. More general forms of the one-dimensional heat and wave equations are, respectively,

$$k \frac{\partial^2 u}{\partial x^2} + G(x, t, u) = \frac{\partial u}{\partial t} \tag{13}$$

and

$$a^2 \frac{\partial^2 u}{\partial x^2} + F(x, t, u, u_t) = \frac{\partial^2 u}{\partial t^2}. \tag{14}$$

For example, if there is heat transfer from the lateral surface of a rod into a surrounding medium that is held at a constant temperature u_m, then the heat equation (13) is

$$k \frac{\partial^2 u}{\partial x^2} - h(u - u_m) = \frac{\partial u}{\partial t}.$$

In (14) the function F could represent the various forces acting on the string. For example, when external, damping, and elastic restoring forces are taken into account, (14) assumes the form

$$a^2 \frac{\partial^2 u}{\partial x^2} + f(x, t) = \frac{\partial^2 u}{\partial t^2} + c \frac{\partial u}{\partial t} + ku. \tag{15}$$

$$\uparrow \qquad\qquad\qquad \uparrow \qquad \uparrow$$

external damping restoring
force force force

Remarks

The analysis of a wide variety of diverse phenomena yields the mathematical models (1), (2), or (3) or their generalizations involving a greater number of spatial variables. For example, (1) is sometimes called the **diffusion equation** since the diffusion of dissolved substances in solution is analogous to the flow of heat in a solid. The function $u(x, t)$ satisfying the partial differential equation in this case represents the concentration of the dissolved substance. Similarly, equation (1) arises in the study of the flow of electricity in a long cable or transmission line. In this setting (1) is known as the **telegraph equation.** It can be shown that under certain assumptions the current and the voltage in the line are functions satisfying two equations identical with (1). The wave equation (2) also appears in the theory of high-frequency transmission lines, fluid mechanics, acoustics, and elasticity. Laplace's equation (2) is encountered in the static displacement of membranes.

In Problems 1–4, a rod of length L coincides with the interval $[0, L]$ on the x-axis. Set up the boundary-value problem for the temperature $u(x, t)$.

1. The left end is held at temperature zero, and the right end is insulated. The initial temperature is $f(x)$ throughout.

2. The left end is held at temperature u_0, and the right end is held at temperature u_1. The initial temperature is zero throughout.

3. The left end is held at temperature 100, and there is heat transfer from the right end into the surrounding medium

at temperature zero. The initial temperature is $f(x)$ throughout.

4. The ends are insulated, and there is heat transfer from the lateral surface into the surrounding medium at temperature 50. The initial temperature is 100 throughout.

In Problems 5–8, a string of length L coincides with the interval $[0, L]$ on the x-axis. Set up the boundary-value problem for the displacement $u(x, t)$.

5. The ends are secured to the x-axis. The string is released from rest from the initial displacement $x(L - x)$.

6. The ends are secured to the x-axis. Initially the string is undisplaced but has the initial velocity $\sin(\pi x/L)$.

7. The left end is secured to the x-axis, but the right end moves in a transverse manner according to $\sin \pi t$. The string is released from rest from the initial displacement $f(x)$. For $t > 0$ the transverse vibrations are damped with a force proportional to the instantaneous velocity.

8. The ends are secured to the x-axis, and the string is initially at rest on that axis. An external vertical force proportional to the horizontal distance from the left end acts on the string for $t > 0$.

In Problems 9 and 10, set up the boundary-value problem for the steady-state temperature $u(x, y)$.

9. A thin rectangular plate coincides with the region defined by $0 \le x \le 4$, $0 \le y \le 2$. The left end and the bottom of the plate are insulated. The top of the plate is held at temperature zero, and the right end of the plate is held at temperature $f(y)$.

10. A semi-infinite plate coincides with the region defined by $0 \le x \le \pi$, $y \ge 0$. The left end is held at temperature e^{-y}, and the right end is held at temperature 100 for $0 < y \le 1$ and temperature zero for $y > 1$. The bottom of the plate is held at temperature $f(x)$.

13.3 Heat Equation

$u = 0$ $u = 0$

0 L x

Figure 13.5

Consider a thin rod of length L with an initial temperature $f(x)$ throughout and whose ends are held at temperature zero for all time $t > 0$. If the rod shown in Figure 13.5 satisfies the assumptions given on page 692, then the temperature $u(x, t)$ in the rod is determined from the boundary-value problem

$$k \frac{\partial^2 u}{\partial x^2} = \frac{\partial u}{\partial t}, \qquad 0 < x < L, \quad t > 0 \tag{1}$$

$$u(0, t) = 0, \qquad u(L, t) = 0, \quad t > 0 \tag{2}$$

$$u(x, 0) = f(x), \qquad 0 < x < L. \tag{3}$$

■ **Solution of the BVP** Using the product $u(x, t) = X(x)T(t)$, and $-\lambda^2$ as a separation constant, leads to

$$\frac{X''}{X} = \frac{T'}{kT} = -\lambda^2 \tag{4}$$

and

$$X'' + \lambda^2 X = 0 \tag{5}$$

$$T' + k\lambda^2 T = 0$$

$$X = c_1 \cos \lambda x + c_2 \sin \lambda x \tag{6}$$

$$T = c_3 e^{-k\lambda^2 t}. \tag{7}$$

Now since

$$u(0, t) = X(0)T(t) = 0$$

$$u(L, t) = X(L)T(t) = 0,$$

we must have $X(0) = 0$ and $X(L) = 0$. These homogeneous boundary conditions together with the homogeneous equation (5) constitute a regular Sturm–Liouville problem. Applying the first of these conditions in (6) immediately gives $c_1 = 0$. Therefore

$$X = c_2 \sin \lambda x.$$

The second boundary condition now implies

$$X(L) = c_2 \sin \lambda L = 0.$$

If $c_2 = 0$, then $X = 0$ so that $u = 0$. To obtain a nontrivial solution u we must have $c_2 \neq 0$, and so the last equation is satisfied when

$$\sin \lambda L = 0.$$

This implies that $\lambda L = n\pi$ or $\lambda = n\pi/L$, where $n = 1, 2, 3, \ldots$. The values

$$\lambda = \frac{n\pi}{L}, \qquad n = 1, 2, 3, \ldots, \tag{8}$$

and the corresponding solutions

$$X = c_2 \sin \frac{n\pi}{L} x, \qquad n = 1, 2, 3, \ldots, \tag{9}$$

are the **eigenvalues** and **eigenfunctions,** respectively, of the problem.

From (7) we have $T = c_3 e^{-k(n^2\pi^2/L^2)t}$, and so

$$u_n = XT = A_n e^{-k(n^2\pi^2/L^2)t} \sin \frac{n\pi}{L} x, \tag{10}$$

where we have replaced the constant $c_2 c_3$ by A_n. The products $u_n(x, t)$ satisfy the partial differential equation (1) and the boundary conditions (2) for each value of the positive integer n.* However, in order for the functions given in (10) to satisfy the initial condition (3), we would have to choose the coefficient A_n in such a manner that

$$u_n(x, 0) = f(x) = A_n \sin \frac{n\pi}{L} x. \tag{11}$$

In general, we would not expect condition (11) to be satisfied for an arbitrary, but reasonable, choice of f. Therefore we are forced to admit that $u_n(x, t)$ is *not a solution of the given problem.* Now by the superposition principle the function

$$u(x, t) = \sum_{n=1}^{\infty} u_n = \sum_{n=1}^{\infty} A_n e^{-k(n^2\pi^2/L^2)t} \sin \frac{n\pi}{L} x \tag{12}$$

must also, though formally, satisfy (1) and (2). Substituting $t = 0$ into (12) implies

$$u(x, 0) = f(x) = \sum_{n=1}^{\infty} A_n \sin \frac{n\pi}{L} x.$$

This last expression is recognized as the half-range expansion of f in a sine series. Thus if we make the identification $A_n = b_n$, $n = 1, 2, 3, \ldots$, it follows from (5) of Section 12.3 that

$$A_n = \frac{2}{L} \int_0^L f(x) \sin \frac{n\pi}{L} x \, dx.$$

* You are urged to verify that if the separation constant is chosen as either $\lambda^2 = 0$ or $\lambda^2 > 0$, then the only solution of (1) satisfying (2) is $u = 0$.

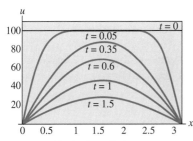

(a) $u(x, t)$ graphed as a
function of x for
various fixed times

(b) $u(x, t)$ graphed as a
function of t for
various fixed positions

Figure 13.6

We conclude that a solution of the boundary-value problem described in (1), (2), and (3) is given by the infinite series

$$u(x, t) = \frac{2}{L} \sum_{n=1}^{\infty} \left(\int_0^L f(x) \sin \frac{n\pi}{L} x\, dx \right) e^{-k(n^2\pi^2/L^2)t} \sin \frac{n\pi}{L} x.$$

In the special case when $u(x, 0) = 100$, $L = \pi$, and $k = 1$, you should verify that the coefficients A_n are given by

$$A_n = \frac{200}{\pi} \left[\frac{1 - (-1)^n}{n} \right],$$

and so

$$u(x, t) = \frac{200}{\pi} \sum_{n=1}^{\infty} \left[\frac{1 - (-1)^n}{n} \right] e^{-n^2 t} \sin nx. \qquad (13)$$

■ **Use of Computers** The graph of the solution (13) is a surface in 3-space. We could use the 3D-plot application of a computer algebra system to approximate this surface by graphing partial sums $S_n(x, t)$ over a rectangular region defined by $0 \le x \le \pi, 0 \le t \le T$. Alternatively, with the aid of the 2D-plot application of a CAS, we plot the solution $u(x, t)$ on the x-interval $[0, \pi]$ for increasing values of time t. See Figure 13.6(a). In Figure 13.6(b) the solution $u(x, t)$ is graphed on the t-interval $[0, 6]$ for increasing values of x ($x = 0$ is the left end and $x = \pi/2$ is the midpoint of the rod of length $L = \pi$.) Both sets of graphs verify that which is apparent in (13)—namely, $u(x, t) \to 0$ as $t \to \infty$.

In Problems 1 and 2, solve the heat equation (1) subject to the given conditions. Assume a rod of length L.

1. $u(0, t) = 0$, $u(L, t) = 0$

$$u(x, 0) = \begin{cases} 1, & 0 < x < L/2 \\ 0, & L/2 < x < L \end{cases}$$

2. $u(0, t) = 0$, $u(L, t) = 0$

$$u(x, 0) = x(L - x)$$

3. Find the temperature $u(x, t)$ in a rod of length L if the initial temperature is $f(x)$ throughout and if the ends $x = 0$ and $x = L$ are insulated.

4. Solve Problem 3 if $L = 2$ and $f(x) = \begin{cases} x, & 0 < x < 1 \\ 0, & 1 < x < 2. \end{cases}$

5. Suppose heat is lost from the lateral surface of a thin rod of length L into a surrounding medium at temperature zero. If the linear law of heat transfer applies, then the heat equation takes on the form

$$k \frac{\partial^2 u}{\partial x^2} - hu = \frac{\partial u}{\partial t}, \qquad 0 < x < L, \quad t > 0,$$

h a constant. Find the temperature $u(x, t)$ if the initial temperature is $f(x)$ throughout and the ends $x = 0$ and $x = L$ are insulated. See Figure 13.7.

6. Solve Problem 5 if the ends $x = 0$ and $x = L$ are held at temperature zero.

7. In Figure 13.6(b) we have the graphs of $u(x, t)$ on the interval $0 \le t \le 6$ for $x = 0$, $x = \pi/12$, $x = \pi/6$, $x = \pi/4$,

and $x = \pi/2$. Describe or sketch the graphs of $u(x, t)$ on the same time interval but for the fixed values $x = 3\pi/4$, $x = 5\pi/6$, $x = 11\pi/12$, and $x = \pi$.

8. Computer Lab Assignment

(a) Solve the heat equation (1) subject to

$$u(0, t) = 0, \qquad u(100, t) = 0, \quad t > 0$$

$$u(x, 0) = \begin{cases} 0.8x, & 0 \le x \le 50 \\ 0.8(100 - x), & 50 < x \le 100. \end{cases}$$

(b) Use the 3D-plot application of your CAS to graph the partial sum $S_5(x, t)$ consisting of the first five nonzero terms of the solution in part **(a)** for $0 \le x \le 100, 0 \le t \le 200$. Assume that $k = 1.6352$. Experiment with various three-dimensional viewing perspectives of the surface (called the ViewPoint option in *Mathematica*).

heat transfer from
lateral surface of
the rod

Figure 13.7

13.4 Wave Equation

We are now in a position to solve the boundary-value problem (11) discussed in Section 13.2. The vertical displacement $u(x, t)$ of the vibrating string of length L shown in Figure 13.2(a) is determined from

$$a^2 \frac{\partial^2 u}{\partial x^2} = \frac{\partial^2 u}{\partial t^2}, \qquad 0 < x < L, \quad t > 0 \tag{1}$$

$$u(0, t) = 0, \qquad u(L, t) = 0, \quad t > 0 \tag{2}$$

$$u(x, 0) = f(x), \qquad \frac{\partial u}{\partial t}\bigg|_{t=0} = g(x). \tag{3}$$

■ **Solution of the BVP** With the usual assumption that $u(x, t) = X(x)T(t)$, separating variables in (1) gives

$$\frac{X''}{X} = \frac{T''}{a^2 T} = -\lambda^2$$

so that

$$X'' + \lambda^2 X = 0 \quad \text{and} \quad T'' + \lambda^2 a^2 T = 0,$$

and therefore

$$X = c_1 \cos \lambda x + c_2 \sin \lambda x$$

$$T = c_3 \cos \lambda a t + c_4 \sin \lambda a t.$$

As before, the boundary conditions (2) translate into $X(0) = 0$ and $X(L) = 0$. In turn we find

$$c_1 = 0 \quad \text{and} \quad c_2 \sin \lambda L = 0.$$

This last equation yields the eigenvalues $\lambda = n\pi/L$, where $n = 1, 2, 3, \ldots$. The corresponding eigenfunctions are

$$X = c_2 \sin \frac{n\pi}{L} x, \qquad n = 1, 2, 3, \ldots.$$

Thus solutions of equation (1) satisfying the boundary conditions (2) are

$$u_n = \left(A_n \cos \frac{n\pi a}{L} t + B_n \sin \frac{n\pi a}{L} t \right) \sin \frac{n\pi}{L} x \tag{4}$$

and

$$u(x, t) = \sum_{n=1}^{\infty} \left(A_n \cos \frac{n\pi a}{L} t + B_n \sin \frac{n\pi a}{L} t \right) \sin \frac{n\pi}{L} x. \tag{5}$$

Setting $t = 0$ in (5) gives

$$u(x, 0) = f(x) = \sum_{n=1}^{\infty} A_n \sin \frac{n\pi}{L} x,$$

which is a half-range expansion for f in a sine series. As in the discussion of the heat equation, we can write $A_n = b_n$:

$$A_n = \frac{2}{L} \int_0^L f(x) \sin \frac{n\pi}{L} x \, dx. \tag{6}$$

To determine B_n we differentiate (5) with respect to t and then set $t = 0$:

$$\frac{\partial u}{\partial t} = \sum_{n=1}^{\infty} \left(-A_n \frac{n\pi a}{L} \sin \frac{n\pi a}{L} t + B_n \frac{n\pi a}{L} \cos \frac{n\pi a}{L} t \right) \sin \frac{n\pi}{L} x$$

$$\frac{\partial u}{\partial t}\bigg|_{t=0} = g(x) = \sum_{n=1}^{\infty} \left(B_n \frac{n\pi a}{L} \right) \sin \frac{n\pi}{L} x.$$

In order for this last series to be the half-range sine expansion of g on the interval, the *total* coefficient $B_n n\pi a/L$ must be given by the form of (5) in Section 12.3—that is,

$$B_n \frac{n\pi a}{L} = \frac{2}{L} \int_0^L g(x) \sin \frac{n\pi}{L} x \, dx,$$

from which we obtain

$$B_n = \frac{2}{n\pi a} \int_0^L g(x) \sin \frac{n\pi}{L} x \, dx. \qquad (7)$$

The solution of the problem consists of the series (5) with A_n and B_n defined by (6) and (7), respectively.

We note that when the string is released from *rest*, then $g(x) = 0$ for every x in $0 \le x \le L$ and consequently $B_n = 0$.

■ **Plucked String** A special case of the boundary-value problem (1)–(3) is a model of the **plucked string.** We can see the motion of the string by plotting the solution or displacement $u(x, t)$ for increasing values of time t and then using the animation feature of a CAS. Some frames of a movie generated in this manner are given in Figure 13.8. You are asked to emulate the results given in the figure by plotting a sequence of partial sums of (5). See Problems 7 and 21 in Exercises 13.4.

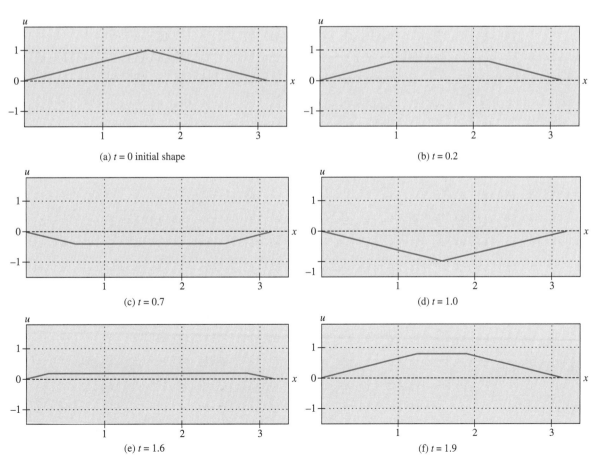

(a) $t = 0$ initial shape

(b) $t = 0.2$

(c) $t = 0.7$

(d) $t = 1.0$

(e) $t = 1.6$

(f) $t = 1.9$

Figure 13.8 Frames of plucked-string movie

■ **Standing Waves** Recall from the derivation of the wave equation in Section 13.2 that the constant a appearing in the solution of the boundary-value problem in (1), (2), and (3) is given by $\sqrt{T/\rho}$, where ρ is mass per unit length and T is the magnitude of the tension in the string. When T is large enough, the vibrating string produces a musical sound. This sound is the result of standing waves. The solution (5) is a superposition of product solutions called **standing waves** or **normal modes:**

$$u(x,\, t) = u_1(x,\, t) + u_2(x,\, t) + u_3(x,\, t) + \cdots.$$

In view of (6) and (7) of Section 3.8, the product solutions (4) can be written as

$$u_n(x,\, t) = C_n \sin\left(\frac{n\pi a}{L}\, t + \phi_n\right) \sin \frac{n\pi}{L}\, x, \tag{8}$$

where $C_n = \sqrt{A_n^2 + B_n^2}$ and ϕ_n is defined by $\sin \phi_n = A_n/C_n$ and $\cos \phi_n = B_n/C_n$. For $n = 1, 2, 3, \ldots$ the standing waves are essentially the graphs of $\sin(n\pi x/L)$, with a time-varying amplitude given by

$$C_n \sin\left(\frac{n\pi a}{L}\, t + \phi_n\right).$$

(a) First standing wave

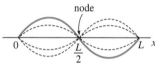

(b) Second standing wave

Alternatively, we see from (8) that at a fixed value of x each product function $u_n(x,\, t)$ represents simple harmonic motion with amplitude $C_n\left|\sin(n\pi x/L)\right|$ and frequency $f_n = na/2L$. In other words, each point on a standing wave vibrates with a different amplitude but with the same frequency. When $n = 1$,

$$u_1(x,\, t) = C_1 \sin\left(\frac{\pi a}{L}\, t + \phi_1\right) \sin \frac{\pi}{L}\, x$$

is called the **first standing wave,** the **first normal mode,** or the **fundamental mode of vibration.** The first three standing waves, or normal modes, are shown in Figure 13.9. The dashed graphs represent the standing waves at various values of time. The points in the interval $(0, L)$, for which $\sin(n\pi/L)x = 0$, correspond to points on a standing wave where there is no motion. These points are called **nodes.** For example, in Figures 13.9(b) and (c) we see that the second standing wave has one node at $L/2$ and the third standing wave has two nodes at $L/3$ and $2L/3$. In general, the nth normal mode of vibration has $n - 1$ nodes.

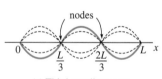

(c) Third standing wave

Figure 13.9

The frequency

$$f_1 = \frac{a}{2L} = \frac{1}{2L}\sqrt{\frac{T}{\rho}}$$

of the first normal mode is called the **fundamental frequency** or **first harmonic** and is directly related to the pitch produced by a stringed instrument. It is apparent that the greater the tension on the string, the higher the pitch of the sound. The frequencies f_n of the other normal modes, which are integer multiples of the fundamental frequency, are called **overtones.** The second harmonic is the first overtone, and so on.

In Problems 1–8, solve the wave equation (1) subject to the given conditions.

1. $u(0,\, t) = 0, \quad u(L,\, t) = 0$

$u(x,\, 0) = \dfrac{1}{4}x(L - x), \quad \left.\dfrac{\partial u}{\partial t}\right|_{t=0} = 0$

2. $u(0,\, t) = 0, \quad u(L,\, t) = 0$

$u(x,\, 0) = 0, \quad \left.\dfrac{\partial u}{\partial t}\right|_{t=0} = x(L - x)$

3. $u(0,\, t) = 0, \quad u(L,\, t) = 0$

$u(x,\, 0)$, as specified in Figure 13.10, $\left.\dfrac{\partial u}{\partial t}\right|_{t=0} = 0$

Figure 13.10

4. $u(0, t) = 0$, $u(\pi, t) = 0$

$u(x, 0) = \frac{1}{6}x(\pi^2 - x^2)$,

$\left.\dfrac{\partial u}{\partial t}\right|_{t=0} = 0$

5. $u(0, t) = 0$, $u(\pi, t) = 0$

$u(x, 0) = 0$, $\left.\dfrac{\partial u}{\partial t}\right|_{t=0} = \sin x$

6. $u(0, t) = 0$, $u(1, t) = 0$

$u(x, 0) = 0.01 \sin 3\pi x$,

$\left.\dfrac{\partial u}{\partial t}\right|_{t=0} = 0$

7. $u(0, t) = 0$, $u(L, t) = 0$

$u(x, 0) = \begin{cases} \dfrac{2hx}{L}, & 0 < x < L/2, \\ 2h\left(1 - \dfrac{x}{L}\right), & \dfrac{L}{2} \le x < L \end{cases}$

$\left.\dfrac{\partial u}{\partial t}\right|_{t=0} = 0$

8. $\left.\dfrac{\partial u}{\partial x}\right|_{x=0} = 0$, $\left.\dfrac{\partial u}{\partial x}\right|_{x=L} = 0$

$u(x, 0) = x$, $\left.\dfrac{\partial u}{\partial t}\right|_{t=0} = 0$

This problem could describe the longitudinal displacement $u(x, t)$ of a vibrating elastic bar. The boundary conditions at $x = 0$ and $x = L$ are called **free-end conditions.** See Figure 13.11.

Figure 13.11

9. A string is stretched and secured on the x-axis at $x = 0$ and $x = \pi$ for $t > 0$. If the transverse vibrations take place in a medium that imparts a resistance proportional to the instantaneous velocity, then the wave equation takes on the form

$$\frac{\partial^2 u}{\partial x^2} = \frac{\partial^2 u}{\partial t^2} + 2\beta \frac{\partial u}{\partial t}, \qquad 0 < \beta < 1, \quad t > 0.$$

Find the displacement $u(x, t)$ if the string starts from rest from the initial displacement $f(x)$.

10. Show that a solution of the boundary-value problem

$$\frac{\partial^2 u}{\partial x^2} = \frac{\partial^2 u}{\partial t^2} + u, \qquad 0 < x < \pi, \quad t > 0$$

$u(0, t) = 0$, $u(\pi, t) = 0$, $t > 0$

$u(x, 0) = \begin{cases} x, & 0 < x < \pi/2 \\ \pi - x, & \pi/2 \le x < \pi \end{cases}$

$\left.\dfrac{\partial u}{\partial t}\right|_{t=0} = 0$, $0 < x < \pi$

is

$$u(x, t) = \frac{4}{\pi} \sum_{k=1}^{\infty} \frac{(-1)^{k+1}}{(2k-1)^2} \sin (2k-1)x \cos \sqrt{(2k-1)^2 + 1}\, t.$$

11. The transverse displacement $u(x, t)$ of a vibrating beam of length L is determined from a fourth-order partial differential equation

$$a^2 \frac{\partial^4 u}{\partial x^4} + \frac{\partial^2 u}{\partial t^2} = 0, \qquad 0 < x < L, \quad t > 0.$$

If the beam is **simply supported,** as shown in Figure 13.12, the boundary and initial conditions are

$u(0, t) = 0$, $u(L, t) = 0$, $t > 0$

$\left.\dfrac{\partial^2 u}{\partial x^2}\right|_{x=0} = 0$, $\left.\dfrac{\partial^2 u}{\partial x^2}\right|_{x=L} = 0$, $t > 0$

$u(x, 0) = f(x)$, $\left.\dfrac{\partial u}{\partial t}\right|_{t=0} = g(x)$, $0 < x < L$.

Solve for $u(x, t)$. [*Hint:* For convenience use λ^4 instead of λ^2 when separating variables.]

Figure 13.12 Simply supported beam

12. If the ends of the beam in Problem 11 are **embedded** at $x = 0$ and $x = L$, the boundary conditions become, for $t > 0$,

$u(0, t) = 0$, $u(L, t) = 0$

$\left.\dfrac{\partial u}{\partial x}\right|_{x=0} = 0$, $\left.\dfrac{\partial u}{\partial x}\right|_{x=L} = 0$.

(a) Show that the eigenvalues of the problem are $\lambda = x_n/L$ where x_n, $n = 1, 2, 3, \ldots$, are the positive roots of the equation $\cosh x \cos x = 1$.

(b) Show graphically that the equation in part **(a)** has an infinite number of roots.

(c) Use a calculator or a computer to find approximations to the first four eigenvalues. Use four decimal places.

13. Consider the boundary-value problem given in (1), (2), and (3) of this section. If $g(x) = 0$ on $0 < x < L$, show that the solution of the problem can be written as

$$u(x, t) = \frac{1}{2} [f(x + at) + f(x - at)].$$

[*Hint:* Use the identity

$$2 \sin \theta_1 \cos \theta_2 = \sin(\theta_1 + \theta_2) + \sin(\theta_1 - \theta_2).]$$

14. The vertical displacement $u(x, t)$ of an infinitely long string is determined from the initial-value problem

$$a^2 \frac{\partial^2 u}{\partial x^2} = \frac{\partial^2 u}{\partial t^2}, \qquad -\infty < x < \infty, \quad t > 0 \qquad (9)$$

$u(x, 0) = f(x)$, $\left.\dfrac{\partial u}{\partial t}\right|_{t=0} = g(x)$.

This problem can be solved without separating variables.

(a) Show that the wave equation can be put into the form $\partial^2 u/\partial \eta \partial \xi = 0$ by means of the substitutions $\xi = x + at$ and $\eta = x - at$.

(b) Integrate the partial differential equation in part **(a)**, first respect to η and then with respect to ξ, to show that $u(x, t) = F(x + at) + G(x - at)$, where F and G are arbitrary twice differentiable functions, is a solution of the wave equation. Use this solution and the given initial conditions to show that

$$F(x) = \frac{1}{2}f(x) + \frac{1}{2a}\int_{x_0}^{x} g(s)\, ds + c$$

and

$$G(x) = \frac{1}{2}f(x) - \frac{1}{2a}\int_{x_0}^{x} g(s)\, ds - c,$$

where x_0 is arbitrary and c is a constant of integration.

(c) Use the results in part **(b)** to show that

$$u(x, t) = \frac{1}{2}\,[f(x + at) + f(x - at)] + \frac{1}{2a}\int_{x-at}^{x+at} g(s)\, ds. \quad (10)$$

Note that when the initial velocity $g(x) = 0$ we obtain

$$u(x, t) = \frac{1}{2}\,[f(x + at) + f(x - at)], \quad -\infty < x < \infty.$$

The last solution can be interpreted as a superposition of two **traveling waves,** one moving to the right (that is, $\frac{1}{2}f(x - at)$) and one moving to the left ($\frac{1}{2}f(x + at)$). Both waves travel with speed a and have the same basic shape as the initial displacement $f(x)$. The form of $u(x, t)$ given in (10) is called **d'Alembert's solution.**

In Problems 15–17, use d'Alembert's solution (10) to solve the initial-value problem in Problem 14 subject to the given initial conditions.

15. $f(x) = \sin x$, $g(x) = 1$

16. $f(x) = \sin x$, $g(x) = \cos x$

17. $f(x) = 0$, $g(x) = \sin 2x$

18. Suppose $f(x) = 1/(1 + x^2)$, $g(x) = 0$, and $a = 1$ for the initial-value problem stated in Problem 14. Graph d'Alembert's solution in this case at the times $t = 0$, $t = 1$, and $t = 3$.

19. Computer Lab Assignment A model for an infinitely long string that is initially held at the three points $(-1, 0)$, $(1, 0)$, and $(0, 1)$ and then simultaneously released at all three points at time $t = 0$ is given by (9) with

$$f(x) = \begin{cases} 1 - |x|, & |x| \le 1 \\ 0, & |x| > 1 \end{cases} \quad \text{and} \quad g(x) = 0.$$

(a) Plot the initial position of the string on the interval $[-6, 6]$.

(b) Use a CAS to plot d'Alembert's solution (10) on $[-6, 6]$ for $t = 0.2k$, $k = 0, 1, 2, \ldots, 25$.

(c) Use the animation feature of your computer algebra system to make a movie of the solution. Describe the motion of the string over time.

20. Computer Lab Assignment An infinitely long string coinciding with the x-axis is struck at the origin with a hammer whose head is 0.2 inch in diameter. A model for the motion of the string is given by (9) with

$$f(x) = 0 \quad \text{and} \quad g(x) = \begin{cases} 1, & |x| \le 0.1 \\ 0, & |x| \ge 0.1. \end{cases}$$

(a) Use a CAS to plot d'Alembert's solution (10) on $[-6, 6]$ for $t = 0.2k$, $k = 0, 1, 2, \ldots, 25$.

(b) Use the animation feature of your computer algebra system to make a movie of the solution. Describe the motion of the string over time.

21. Computer Lab Assignment The model of the vibrating string in Problem 7 is called the **plucked string.** The string is tied to the x-axis at $x = 0$ and $x = L$ and is held at $x = L/2$ at h units above the x-axis. See Figure 13.4. Starting at $t = 0$, the string is released from rest.

(a) Use a CAS to plot the partial sum $S_6(x, t)$, that is, the first six nonzero terms of your solution, for $t = 0.1k$, $k = 0, 1, 2, \ldots, 20$. Assume that $a = 1$, $h = 1$, and $L = \pi$.

(b) Use the animation feature of your computer algebra system to make a movie of the solution to Problem 7.

13.5 Laplace's Equation

Suppose we wish to find the steady-state temperature $u(x, y)$ in a rectangular plate whose vertical edges are insulated, as shown in Figure 13.13. When no heat escapes from the lateral faces of the plate, we solve the following boundary-value problem:

$$\frac{\partial^2 u}{\partial x^2} + \frac{\partial^2 u}{\partial y^2} = 0, \qquad 0 < x < a, \quad 0 < y < b \quad (1)$$

$$\left.\frac{\partial u}{\partial x}\right|_{x=0} = 0, \qquad \left.\frac{\partial u}{\partial x}\right|_{x=a} = 0, \quad 0 < y < b \quad (2)$$

$$u(x, 0) = 0, \qquad u(x, b) = f(x), \quad 0 < x < a. \quad (3)$$

Figure 13.13

■ **Solution of the BVP** With $u(x, y) = X(x)Y(y)$, separation of variables in (1) leads to

$$\frac{X''}{X} = -\frac{Y''}{Y} = -\lambda^2$$

$$X'' + \lambda^2 X = 0 \tag{4}$$

$$Y'' - \lambda^2 Y = 0 \tag{5}$$

$$X = c_1 \cos \lambda x + c_2 \sin \lambda x, \tag{6}$$

and since $0 < y < b$ is a finite interval, we use the solution

$$Y = c_3 \cosh \lambda y + c_4 \sinh \lambda y. \tag{7}$$

Now the first three boundary conditions translate into $X'(0) = 0$, $X'(a) = 0$, and $Y(0) = 0$. Differentiating X and setting $x = 0$ implies $c_2 = 0$ and, therefore, $X = c_1 \cos \lambda x$. Differentiating this last expression and then setting $x = a$ gives $-c_1 \lambda \sin \lambda a = 0$. This last condition is satisfied when $\lambda = 0$ or when $\lambda a = n\pi$ or $\lambda = n\pi/a$, $n = 1, 2, \ldots$. Observe that $\lambda = 0$ implies that (4) is $X'' = 0$. The general solution of this equation is given by the linear function $X = c_1 + c_2 x$ and *not* by (6). In this case the boundary conditions $X'(0) = 0$, $X'(a) = 0$ demand that $X = c_1$. In this example, unlike the previous two examples, we are forced to conclude that $\lambda = 0$ is an eigenvalue. Corresponding $\lambda = 0$ with $n = 0$, we get the eigenfunctions

$$X = c_1, \quad n = 0, \qquad \text{and} \qquad X = c_1 \cos \frac{n\pi}{a} x, \quad n = 1, 2, \ldots.$$

Finally, the condition $Y(0) = 0$ dictates that $c_3 = 0$ in (4) when $\lambda > 0$. However, when $\lambda = 0$, equation (5) becomes $Y'' = 0$, and thus the solution is given by $Y = c_3 + c_4 y$ rather than by (4). But $Y(0) = 0$ implies again that $c_3 = 0$, and so $Y = c_4 y$. Thus product solutions of the equation satisfying the first three boundary conditions are

$$A_0 y, \quad n = 0, \qquad \text{and} \qquad A_n \sinh \frac{n\pi}{a} y \cos \frac{n\pi}{a} x, \quad n = 1, 2, \ldots.$$

The superposition principle yields another solution

$$u(x, y) = A_0 y + \sum_{n=1}^{\infty} A_n \sinh \frac{n\pi}{a} y \cos \frac{n\pi}{a} x. \tag{8}$$

Substituting $y = b$ in (5) gives

$$u(x, b) = f(x) = A_0 b + \sum_{n=1}^{\infty} \left(A_n \sinh \frac{n\pi}{a} b \right) \cos \frac{n\pi}{a} x,$$

which in this case is a half-range expansion of f in a cosine series. If we make the identifications $A_0 b = a_0/2$ and $A_n \sinh(n\pi b/a) = a_n$, $n = 1, 2, 3, \ldots$, it follows from (2) and (3) of Section 12.3 that

$$2A_0 b = \frac{2}{a} \int_0^a f(x)\, dx$$

$$A_0 = \frac{1}{ab} \int_0^a f(x)\, dx \tag{9}$$

and

$$A_n \sinh \frac{n\pi}{a} b = \frac{2}{a} \int_0^a f(x) \cos \frac{n\pi}{a} x\, dx$$

$$A_n = \frac{2}{a \sinh \dfrac{n\pi}{a} b} \int_0^a f(x) \cos \frac{n\pi}{a} x\, dx. \tag{10}$$

The solution of this problem consists of the series given in (8), where A_0 and A_n are defined by (9) and (10), respectively.

■ **Dirichlet Problem** A boundary-value problem in which we seek a solution to an elliptic partial differential equation such as Laplace's equation $\nabla^2 u = 0$ within a region R (in the plane or 3-space) such that u takes on prescribed values on the entire boundary of the region is called a **Dirichlet problem.** In Problem 1 in Exercises 13.5 you are asked to show that the solution of the Dirichlet problem for a rectangular region

$$\frac{\partial^2 u}{\partial x^2} + \frac{\partial^2 u}{\partial y^2} = 0, \qquad 0 < x < a, \quad 0 < y < b$$

$$u(0, y) = 0, \qquad u(a, y) = 0$$

$$u(x, 0) = 0, \qquad u(x, b) = f(x)$$

is

$$u(x, y) = \sum_{n=1}^{\infty} A_n \sinh \frac{n\pi}{a} y \sin \frac{n\pi}{a} x \quad \text{where} \quad A_n = \frac{2}{a \sinh \dfrac{n\pi b}{a}} \int_0^a f(x) \sin \frac{n\pi}{a} x \, dx.$$

In the special case when $f(x) = 100$, $a = 1$, $b = 1$, the coefficients A_n are given by $A_n = 200 \dfrac{1 - (-1)^n}{n\pi \sinh n\pi}$. With the help of a CAS the plot of the surface defined by $u(x, y)$ over the region R: $0 \le x \le 1$, $0 \le y \le 1$ is given in Figure 13.14(a). You can see in the figure that boundary conditions are satisfied; especially note that along $y = 1$, $u = 100$ for $0 \le x \le 1$. The isotherms, or curves, in the rectangular region along which the temperature $u(x, y)$ is constant can be obtained using the contour plotting capabilities of a CAS and are illustrated in Figure 13.14(b). The isotherms can also be visualized as the curves of intersection (projected into the xy-plane) of horizontal planes $u = 80$, $u = 60$, and so on, with the surface in Figure 13.14(a). Notice that throughout the region the maximum temperature is $u = 100$ and occurs on the portion of the boundary corresponding to $y = 1$. This is no coincidence. There is a **maximum principle** that states a solution u of Laplace's equation within a bounded region R with boundary B (such as a rectangle, circle, sphere, and so on) takes on its maximum and minimum values on B. In addition, it can be proved that u can have no relative extrema (maxima or minima) in the interior of R. This last statement is clearly borne out by the surface shown in Figure 13.14(a).

Maximum of u is on the ☞
boundary of the region R.

$u(x, y)$

(a) Surface

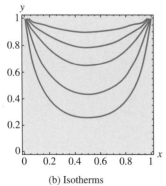

(b) Isotherms

Figure 13.14

■ **Superposition Principle** A Dirichlet problem for a rectangle can be readily solved by separation of variables when homogeneous boundary conditions are specified on two *parallel* boundaries. However, the method of separation of variables is not applicable to a Dirichlet problem when the boundary conditions on all four sides of the rectangle are nonhomogeneous. To get around this difficulty we break the problem

$$\frac{\partial^2 u}{\partial x^2} + \frac{\partial^2 u}{\partial y^2} = 0, \qquad 0 < x < a, \quad 0 < y < b$$

$$u(0, y) = F(y), \qquad u(a, y) = G(y), \quad 0 < y < b \qquad (11)$$

$$u(x, 0) = f(x), \qquad u(x, b) = g(x), \quad 0 < x < a$$

into two problems, each of which has homogeneous boundary conditions on parallel boundaries, as shown.

Problem 1

$$\frac{\partial^2 u_1}{\partial x^2} + \frac{\partial^2 u_1}{\partial y^2} = 0, \qquad 0 < x < a, \qquad 0 < y < b$$

$$u_1(0, y) = 0, \qquad u_1(a, y) = 0, \qquad 0 < y < b$$

$$u_1(x, 0) = f(x), \qquad u_1(x, b) = g(x), \qquad 0 < x < a$$

Problem 2

$$\frac{\partial^2 u_2}{\partial x^2} + \frac{\partial^2 u_2}{\partial y^2} = 0, \qquad 0 < x < a, \qquad 0 < y < b$$

$$u_2(0, y) = F(y), \qquad u_2(a, y) = G(y), \quad 0 < y < b$$

$$u_2(x, 0) = 0, \qquad u_2(x, b) = 0, \qquad 0 < x < a$$

Suppose u_1 and u_2 are the solutions of Problems 1 and 2, respectively. If we define $u(x, y) = u_1(x, y) + u_2(x, y)$, it is seen that u satisfies all boundary conditions in the original problem (11). For example,

$$u(0, y) = u_1(0, y) + u_2(0, y) = 0 + F(y) = F(y)$$

$$u(x, b) = u_1(x, b) + u_2(x, b) = g(x) + 0 = g(x)$$

and so on. Furthermore, u is a solution of Laplace's equation by Theorem 13.1. In other words, by solving Problems 1 and 2 and adding their solutions we have solved the original problem. This additive property of solutions is known as the superposition principle. See Figure 13.15.

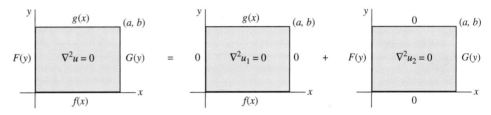

Figure 13.15 Solution $u =$ Solution u_1 of Problem 1 + Solution u_2 of Problem 2

We leave as exercises (see Problems 13 and 14 in Exercises 13.5) to show that a solution of Problem 1 is

$$u_1(x, y) = \sum_{n=1}^{\infty} \left\{ A_n \cosh \frac{n\pi}{a} y + B_n \sinh \frac{n\pi}{a} y \right\} \sin \frac{n\pi}{a} x,$$

where $\qquad A_n = \frac{2}{a} \int_0^a f(x) \sin \frac{n\pi}{a} x \, dx$

$$B_n = \frac{1}{\sinh \dfrac{n\pi}{a} b} \left(\frac{2}{a} \int_0^a g(x) \sin \frac{n\pi}{a} x \, dx - A_n \cosh \frac{n\pi}{a} b \right),$$

and that a solution of Problem 2 is

$$u_2(x, y) = \sum_{n=1}^{\infty} \left\{ A_n \cosh \frac{n\pi}{b} x + B_n \sinh \frac{n\pi}{b} x \right\} \sin \frac{n\pi}{b} y,$$

where $\qquad A_n = \frac{2}{b} \int_0^b F(y) \sin \frac{n\pi}{b} y \, dy$

$$B_n = \frac{1}{\sinh \dfrac{n\pi}{b} a} \left(\frac{2}{b} \int_0^b G(y) \sin \frac{n\pi}{b} y \, dy - A_n \cosh \frac{n\pi}{b} a \right).$$

In Problems 1–10, solve Laplace's equation (1) for a rectangular plate subject to the given boundary conditions.

1. $u(0, y) = 0$, $u(a, y) = 0$

 $u(x, 0) = 0$, $u(x, b) = f(x)$

2. $u(0, y) = 0$, $u(a, y) = 0$

 $\left.\dfrac{\partial u}{\partial y}\right|_{y=0} = 0$, $u(x, b) = f(x)$

3. $u(0, y) = 0$, $u(a, y) = 0$

 $u(x, 0) = f(x)$, $u(x, b) = 0$

4. $\left.\dfrac{\partial u}{\partial x}\right|_{x=0} = 0$, $\left.\dfrac{\partial u}{\partial x}\right|_{x=a} = 0$

 $u(x, 0) = x$, $u(x, b) = 0$

5. $u(0, y) = 0$, $u(1, y) = 1 - y$

 $\left.\dfrac{\partial u}{\partial y}\right|_{y=0} = 0$, $\left.\dfrac{\partial u}{\partial y}\right|_{y=1} = 0$

6. $u(0, y) = g(y)$, $\left.\dfrac{\partial u}{\partial x}\right|_{x=1} = 0$

 $\left.\dfrac{\partial u}{\partial y}\right|_{y=0} = 0$, $\left.\dfrac{\partial u}{\partial y}\right|_{y=\pi} = 0$

7. $\left.\dfrac{\partial u}{\partial x}\right|_{x=0} = u(0, y)$, $u(\pi, y) = 1$

 $u(x, 0) = 0$, $u(x, \pi) = 0$

8. $u(0, y) = 0$, $u(1, y) = 0$

 $\left.\dfrac{\partial u}{\partial y}\right|_{y=0} = u(x, 0)$, $u(x, 1) = f(x)$

9. $u(0, y) = 0$, $u(1, y) = 0$

 $u(x, 0) = 100$, $u(x, 1) = 200$

10. $u(0, y) = 10y$, $\left.\dfrac{\partial u}{\partial x}\right|_{x=1} = -1$

 $u(x, 0) = 0$, $u(x, 1) = 0$

In Problems 11 and 12, solve Laplace's equation (1) for the given semi-infinite plate extending in the positive y-direction. In each case assume $u(x, y)$ is bounded as $y \to \infty$.

11.

$u = 0$ $u = 0$

0 π

$u = f(x)$

Figure 13.16

12.

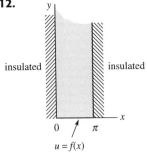

insulated insulated

0 π

$u = f(x)$

Figure 13.17

In Problems 13 and 14, solve Laplace's equation (1) for a rectangular plate subject to the given boundary conditions.

13. $u(0, y) = 0$, $u(a, y) = 0$

 $u(x, 0) = f(x)$, $u(x, b) = g(x)$

14. $u(0, y) = F(y)$, $u(a, y) = G(y)$

 $u(x, 0) = 0$, $u(x, b) = 0$

In Problems 15 and 16, use the superposition principle to solve Laplace's equation (1) for a square plate subject to the given boundary conditions.

15. $u(0, y) = 1$, $u(\pi, y) = 1$

 $u(x, 0) = 0$, $u(x, \pi) = 1$

16. $u(0, y) = 0$, $u(2, y) = y(2 - y)$

 $u(x, 0) = 0$, $u(x, 2) = \begin{cases} x, & 0 < x < 1 \\ 2 - x, & 1 \le x < 2 \end{cases}$

17. In Problem 16, what is the maximum value of the temperature u for $0 \le x \le 2, 0 \le y \le 2$?

18. Computer Lab Assignment

 (a) In Problem 1 suppose $a = b = \pi$ and $f(x) = 100x(\pi - x)$. Without using the solution $u(x, y)$ sketch, by hand, what the surface would look like over the rectangular region defined by $0 \le x \le \pi, 0 \le y \le \pi$.

 (b) What is the maximum value of the temperature u for $0 \le x \le \pi, 0 \le y \le \pi$?

 (c) Use the information in part **(a)** to compute the coefficients for your answer in Problem 1. Then use the 3D-plot application of your CAS to graph the partial sum $S_5(x, t)$ consisting of the first five nonzero terms of the solution in part **(a)** for $0 \le x \le \pi, 0 \le y \le \pi$. Use different perspectives and then compare with part **(a)**.

19. Computer Lab Assignment

 (a) Use the contour-plot application of your CAS to graph the isotherms $u = 170, 140, 110, 80, 60, 30$ for the solution of Problem 9. Use the partial sum $S_5(x, y)$ consisting of the first five nonzero terms of the solution.

 (b) Use the 3D-plot application of your CAS to graph the partial sum $S_5(x, y)$.

20. Computer Lab Assignment Use the contour-plot application of your CAS to graph the isotherms $u = 2, 1, 0.5, 0.2, 0.1, 0.05, 0, -0.05$ for the solution of Problem 10. Use the partial sum $S_5(x, y)$ consisting of the first five nonzero terms of the solution.

13.6 Nonhomogeneous Equations and Boundary Conditions

The method of separation of variables may not be applicable to a boundary-value problem when the partial differential equation or boundary conditions are nonhomogeneous. For example, when heat is generated at a constant rate r within a rod of finite length, the heat equation takes on the form

$$k\frac{\partial^2 u}{\partial x^2} + r = \frac{\partial u}{\partial t}. \tag{1}$$

Equation (1) is nonhomogeneous and is readily shown not to be separable. On the other hand, suppose we wish to solve the usual heat equation $ku_{xx} = u_t$ when the boundaries $x = 0$ and $x = L$ are held at nonzero temperatures k_1 and k_2. Even though the substitution $u(x, t) = X(x)T(t)$ separates the PDE, we quickly find ourselves at an impasse in determining eigenvalues and eigenfunctions since no conclusion can be drawn from $u(0, t) = X(0)T(t) = k_1$ and $u(L, t) = X(L)T(t) = k_2$.

■ **Change of Dependent Variable** A few problems involving nonhomogeneous equations and nonhomogeneous boundary conditions can be solved by changing the dependent variable u to a new dependent variable v by means of the substitution $u = v + \psi$. The basic idea is to determine ψ, a function of *one* variable, in such a manner that v, a function of *two* variables, is made to simultaneously satisfy a homogeneous partial differential equation and homogeneous boundary conditions. The following example illustrates the procedure.

> **Example 1** **Nonhomogeneous Equation and Boundary Condition**
>
> Solve equation (1) subject to
>
> $$u(0, t) = 0, \qquad u(1, t) = u_0, \quad t > 0$$
> $$u(x, 0) = f(x), \qquad 0 < x < 1.$$
>
> **SOLUTION** Both the equation and the boundary condition at $x = 1$ are nonhomogeneous. If we let $u(x, t) = v(x, t) + \psi(x)$, then
>
> $$\frac{\partial^2 u}{\partial x^2} = \frac{\partial^2 v}{\partial x^2} + \psi'' \qquad \text{and} \qquad \frac{\partial u}{\partial t} = \frac{\partial v}{\partial t}.$$
>
> Substituting these results into (1) gives
>
> $$k\frac{\partial^2 v}{\partial x^2} + k\psi'' + r = \frac{\partial v}{\partial t}. \tag{2}$$
>
> Equation (2) reduces to a homogeneous equation if we demand that ψ satisfy
>
> $$k\psi'' + r = 0 \qquad \text{or} \qquad \psi'' = -\frac{r}{k}.$$
>
> Integrating the last equation twice reveals that
>
> $$\psi(x) = -\frac{r}{2k}x^2 + c_1 x + c_2. \tag{3}$$
>
> Furthermore, $u(0, t) = v(0, t) + \psi(0) = 0$
>
> $$u(1, t) = v(1, t) + \psi(1) = u_0.$$

We have $v(0, t) = 0$ and $v(1, t) = 0$, provided

$$\psi(0) = 0 \quad \text{and} \quad \psi(1) = u_0.$$

Applying the latter two conditions to (3) gives, in turn, $c_2 = 0$ and $c_1 = r/2k + u_0$. Consequently

$$\psi(x) = -\frac{r}{2k} x^2 + \left(\frac{r}{2k} + u_0\right) x.$$

Last, the initial condition $u(x, 0) = v(x, 0) + \psi(x)$ implies $v(x, 0) = u(x, 0) - \psi(x) = f(x) - \psi(x)$. Thus to determine $v(x, t)$ we solve the *new* boundary-value problem

$$k \frac{\partial^2 v}{\partial x^2} = \frac{\partial v}{\partial t}, \quad 0 < x < 1, \quad t > 0$$

$$v(0, t) = 0, \quad v(1, t) = 0, \quad t > 0$$

$$v(x, 0) = f(x) + \frac{r}{2k} x^2 - \left(\frac{r}{2k} + u_0\right) x$$

by separation of variables. In the usual manner we find

$$v(x, t) = \sum_{n=1}^{\infty} A_n e^{-kn^2\pi^2 t} \sin n\pi x,$$

where

$$A_n = 2 \int_0^1 \left[f(x) + \frac{r}{2k} x^2 - \left(\frac{r}{2k} + u_0\right) x \right] \sin n\pi x \, dx. \tag{4}$$

Finally, a solution of the original problem is obtained by adding $\psi(x)$ and $v(x, t)$:

$$u(x, t) = -\frac{r}{2k} x^2 + \left(\frac{r}{2k} + u_0\right) x + \sum_{n=1}^{\infty} A_n e^{-kn^2\pi^2 t} \sin n\pi x, \tag{5}$$

where the A_n are defined in (4). ❑

Observe in (5) that $u(x, t) \to \psi(x)$ as $t \to \infty$. In the context of solving forms of the heat equation, ψ is called a **steady-state solution.** Since $v(x, t) \to 0$ as $t \to \infty$, v is called a **transient solution.**

The substitution $u = v + \psi$ can also be used on problems involving forms of the wave equation as well as Laplace's equation.

E X E R C I S E S 1 3 . 6

Answers to odd-numbered problems begin on page A-60.

In Problems 1 and 2, solve the heat equation $ku_{xx} = u_t$, $0 < x < 1$, $t > 0$ subject to the given conditions.

1. $u(0, t) = 100, \quad u(1, t) = 100$
$u(x, 0) = 0$

2. $u(0, t) = u_0, \quad u(1, t) = 0$
$u(x, 0) = f(x)$

In Problems 3 and 4, solve the partial differential equation (1) subject to the given conditions.

3. $u(0, t) = u_0, \quad u(1, t) = u_0$
$u(x, 0) = 0$

4. $u(0, t) = u_0, \quad u(1, t) = u_1$
$u(x, 0) = f(x)$

5. Solve the boundary-value problem

$$k \frac{\partial^2 u}{\partial x^2} + Ae^{-\beta x} = \frac{\partial u}{\partial t}, \quad \beta > 0, \quad 0 < x < 1, \quad t > 0$$

$$u(0, t) = 0, \quad u(1, t) = 0, \quad t > 0$$

$$u(x, 0) = f(x), \quad 0 < x < 1.$$

The partial differential equation is a form of the heat equation when heat is generated within a thin rod from radioactive decay of the material.

6. Solve the boundary-value problem

$$k \frac{\partial^2 u}{\partial x^2} - hu = \frac{\partial u}{\partial t}, \quad 0 < x < \pi, \quad t > 0$$

$$u(0, t) = 0, \quad u(\pi, t) = u_0, \quad t > 0$$

$$u(x, 0) = 0, \quad 0 < x < \pi.$$

7. Find a steady-state solution $\psi(x)$ of the boundary-value problem

$$k\frac{\partial^2 u}{\partial x^2} - h(u - u_0) = \frac{\partial u}{\partial t}, \qquad 0 < x < 1, \quad t > 0$$

$$u(0, t) = u_0, \qquad u(1, t) = 0, \quad t > 0$$

$$u(x, 0) = f(x), \qquad 0 < x < 1.$$

8. Find a steady-state solution $\psi(x)$ if the rod in Problem 7 is semi-infinite extending in the positive x-direction, radiates from its lateral surface into a medium at temperature zero, and

$$u(0, t) = u_0, \qquad \lim_{x \to \infty} u(x, t) = 0, \quad t > 0$$

$$u(x, 0) = f(x), \qquad x > 0.$$

9. When a vibrating string is subjected to an external vertical force that varies with the horizontal distance from the left end, the wave equation takes on the form

$$a^2 \frac{\partial^2 u}{\partial x^2} + Ax = \frac{\partial^2 u}{\partial t^2}.$$

Solve this partial differential equation subject to

$$u(0, t) = 0, \qquad u(1, t) = 0, \quad t > 0$$

$$u(x, 0) = 0, \qquad \frac{\partial u}{\partial t}\bigg|_{t=0} = 0, \quad 0 < x < 1.$$

10. A string initially at rest on the x-axis is secured on the x-axis at $x = 0$ and $x = 1$. If the string is allowed to fall under its own weight for $t > 0$, the displacement $u(x, t)$ satisfies

$$a^2 \frac{\partial^2 u}{\partial x^2} - g = \frac{\partial^2 u}{\partial t^2}, \qquad 0 < x < 1, \quad t > 0,$$

where g is the acceleration of gravity. Solve for $u(x, t)$.

11. Find the steady-state temperature $u(x, y)$ in the semi-infinite plate shown in Figure 13.18. Assume that the temperature is bounded as $x \to \infty$. [*Hint:* Try $u(x, y) = v(x, y) + \psi(y)$.]

Figure 13.18

12. Poisson's equation

$$\frac{\partial^2 u}{\partial x^2} + \frac{\partial^2 u}{\partial y^2} = -h, \qquad h > 0,$$

occurs in many problems involving electric potential. Solve the above equation subject to the conditions

$$u(0, y) = 0, \qquad u(\pi, y) = 1, \quad y > 0$$

$$u(x, 0) = 0, \qquad 0 < x < \pi.$$

13.7 Orthogonal Series Expansions

For certain of types of boundary conditions, the method of separation of variables and the superposition principle lead to an expansion of a function in a trigonometric series that is *not* a Fourier series. To solve the problems in this section we shall utilize the concept of orthogonal series expansions or generalized Fourier series developed in Section 12.1.

> ### *Example 1* Using Orthogonal Series Expansions
>
> The temperature in a rod of unit length in which there is heat transfer from its right boundary into a surrounding medium kept at a constant temperature zero is determined from
>
> $$k\frac{\partial^2 u}{\partial x^2} = \frac{\partial u}{\partial t}, \qquad 0 < x < 1, \quad t > 0$$
>
> $$u(0, t) = 0, \qquad \frac{\partial u}{\partial x}\bigg|_{x=1} = -hu(1, t), \quad h > 0, \quad t > 0$$
>
> $$u(x, 0) = 1, \qquad 0 < x < 1.$$
>
> Solve for $u(x, t)$.
>
> **SOLUTION** The method of separation of variables gives
>
> $$X'' + \lambda^2 X = 0, \qquad T' + k\lambda^2 T = 0 \tag{1}$$
>
> $$X(x) = c_1 \cos \lambda x + c_2 \sin \lambda x \qquad \text{and} \qquad T(t) = c_3 e^{-k\lambda^2 t}.$$

Since $u = XT$, we find that the boundary conditions become

$$X(0) = 0 \qquad \text{and} \qquad X'(1) = -hX(1). \tag{2}$$

The first condition in (2) immediately gives $c_1 = 0$. Applying the second condition in (2) to $X(x) = c_2 \sin \lambda x$ yields

$$\lambda \cos \lambda = -h \sin \lambda \qquad \text{or} \qquad \tan \lambda = -\frac{\lambda}{h}. \tag{3}$$

From the analysis in Example 2 of Section 12.4 we know that the last equation in (3) has an infinite number of roots. The consecutive positive roots λ_n, $n = 1, 2, 3, \ldots$, are the eigenvalues of the problem, and the corresponding eigenfunctions are $X(x) = c_2 \sin \lambda_n x$, $n = 1, 2, 3, \ldots$. Thus

$$u_n = XT = A_n e^{-k\lambda_n^2 t} \sin \lambda_n x \qquad \text{and} \qquad u(x, t) = \sum_{n=1}^{\infty} A_n e^{-k\lambda_n^2 t} \sin \lambda_n x.$$

Now at $t = 0$, $u(x, 0) = 1$, $0 < x < 1$, so that

$$1 = \sum_{n=1}^{\infty} A_n \sin \lambda_n x. \tag{4}$$

The series in (4) is not a Fourier sine series; it is an expansion in terms of the orthogonal functions arising from the regular Sturm–Liouville problem consisting of the first differential equation in (1) and boundary conditions (2). It follows that the set of eigenfunctions $\{\sin \lambda_n x\}$, $n = 1, 2, 3, \ldots$, where the λ's are defined by $\tan \lambda = -\lambda/h$, is orthogonal with respect to the weight function $p(x) = 1$ on the interval $[0, 1]$. With $f(x) = 1$ in (8) of Section 12.1 we can write

$$A_n = \frac{\int_0^1 \sin \lambda_n x \, dx}{\int_0^1 \sin^2 \lambda_n x \, dx}. \tag{5}$$

To evaluate the square norm of each of the eigenfunctions we use a trigonometric identity:

$$\int_0^1 \sin^2 \lambda_n x \, dx = \frac{1}{2} \int_0^1 [1 - \cos 2\lambda_n x] \, dx = \frac{1}{2} \left[1 - \frac{1}{2\lambda_n} \sin 2\lambda_n \right]. \tag{6}$$

Using $\sin 2\lambda_n = 2 \sin \lambda_n \cos \lambda_n$ and $\lambda_n \cos \lambda_n = -h \sin \lambda_n$, we simplify (6) to

$$\int_0^1 \sin^2 \lambda_n x \, dx = \frac{1}{2h} [h + \cos^2 \lambda_n].$$

Also

$$\int_0^1 \sin \lambda_n x \, dx = -\frac{1}{\lambda_n} \cos \lambda_n x \Big|_0^1 = \frac{1}{\lambda_n} [1 - \cos \lambda_n].$$

Consequently (5) becomes

$$A_n = \frac{2h(1 - \cos \lambda_n)}{\lambda_n(h + \cos^2 \lambda_n)}.$$

Finally, a solution of the boundary-value problem is

$$u(x, t) = 2h \sum_{n=1}^{\infty} \frac{1 - \cos \lambda_n}{\lambda_n(h + \cos^2 \lambda_n)} e^{-k\lambda_n^2 t} \sin \lambda_n x. \qquad \square$$

Example 2 **Using Orthogonal Series Expansions**

The twist angle $\theta(x, t)$ of a torsionally vibrating shaft of unit length is determined from

$$a^2 \frac{\partial^2 \theta}{\partial x^2} = \frac{\partial^2 \theta}{\partial t^2}, \qquad 0 < x < 1, \quad t > 0$$

$$\theta(0, t) = 0, \qquad \frac{\partial \theta}{\partial x}\bigg|_{x=1} = 0, \quad t > 0$$

$$\theta(x, 0) = x, \qquad \frac{\partial \theta}{\partial t}\bigg|_{t=0} = 0, \quad 0 < x < 1.$$

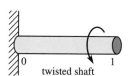

0 twisted shaft 1

Figure 13.19

See Figure 13.19. The boundary condition at $x = 1$ is called a free-end condition. Solve for $\theta(x, t)$.

SOLUTION Using $\theta = XT$, we find

$$X'' + \lambda^2 X = 0, \qquad T'' + a^2 \lambda^2 T = 0$$

$$X(x) = c_1 \cos \lambda x + c_2 \sin \lambda x \qquad \text{and} \qquad T(t) = c_3 \cos a\lambda t + c_4 \sin a\lambda t.$$

The boundary conditions $X(0) = 0$ and $X'(1) = 0$ give $c_1 = 0$ and $c_2 \cos \lambda = 0$, respectively. Since the cosine function is zero at odd multiples of $\pi/2$, the eigenvalues of the problem are $\lambda = (2n - 1)(\pi/2)$, $n = 1, 2, 3, \ldots$. The initial condition $T'(0) = 0$ gives $c_4 = 0$, so

$$\theta_n = XT = A_n \cos a \left(\frac{2n - 1}{2} \right) \pi t \sin \left(\frac{2n - 1}{2} \right) \pi x.$$

In order to satisfy the remaining initial condition we form

$$\theta(x, t) = \sum_{n=1}^{\infty} A_n \cos a \left(\frac{2n - 1}{2} \right) \pi t \sin \left(\frac{2n - 1}{2} \right) \pi x. \qquad (7)$$

When $t = 0$, we must have, for $0 < x < 1$,

$$\theta(x, 0) = x = \sum_{n=1}^{\infty} A_n \sin \left(\frac{2n - 1}{2} \right) \pi x. \qquad (8)$$

As in Example 1, the set of eigenfunctions $\left\{ \sin \left(\frac{2n - 1}{2} \right) \pi x \right\}$, $n = 1, 2, 3, \ldots$, is orthogonal with respect to the weight function $p(x) = 1$ on the interval $[0, 1]$. The series $\sum_{n=1}^{\infty} A_n \sin \left(\frac{2n - 1}{2} \right) \pi x$ is not a Fourier sine series since the argument of the sine is not an integer multiple of $\pi x/L$ ($L = 1$ in this case). The series is again a generalized Fourier series. Hence from (8) of Section 12.1 the coefficients in (7) are

$$A_n = \frac{\int_0^1 x \sin \left(\frac{2n - 1}{2} \right) \pi x \, dx}{\int_0^1 \sin^2 \left(\frac{2n - 1}{2} \right) \pi x \, dx}.$$

Carrying out the two integrations, we arrive at

$$A_n = \frac{8(-1)^{n+1}}{(2n - 1)^2 \pi^2}.$$

The twist angle is then

$$\theta(x, t) = \frac{8}{\pi^2} \sum_{n=1}^{\infty} \frac{(-1)^{n+1}}{(2n - 1)^2} \cos a \left(\frac{2n - 1}{2} \right) \pi t \sin \left(\frac{2n - 1}{2} \right) \pi x. \qquad \square$$

1. In Example 1 find the temperature $u(x, t)$ when the left end of the rod is insulated.

2. Solve the boundary-value problem

$$k \frac{\partial^2 u}{\partial x^2} = \frac{\partial u}{\partial t}, \qquad 0 < x < 1, \quad t > 0$$

$$u(0, t) = 0, \qquad \frac{\partial u}{\partial x}\bigg|_{x=1} = -h(u(1, t) - u_0), \quad h > 0, \quad t > 0$$

$$u(x, 0) = f(x), \qquad 0 < x < 1.$$

3. Find the steady-state temperature for a rectangular plate for which the boundary conditions are

$$u(0, y) = 0, \qquad \frac{\partial u}{\partial x}\bigg|_{x=a} = -hu(a, y), \quad 0 < y < b$$

$$u(x, 0) = 0, \qquad u(x, b) = f(x), \quad 0 < x < a.$$

4. Solve the boundary-value problem

$$\frac{\partial^2 u}{\partial x^2} + \frac{\partial^2 u}{\partial y^2} = 0, \qquad 0 < y < 1, \quad x > 0$$

$$u(0, y) = u_0, \qquad \lim_{x \to \infty} u(x, y) = 0, \quad 0 < y < 1$$

$$\frac{\partial u}{\partial y}\bigg|_{y=0} = 0, \qquad \frac{\partial u}{\partial y}\bigg|_{y=1} = -hu(x, 1), \quad h > 0, \quad x > 0.$$

5. Find the temperature $u(x, t)$ in a rod of length L if the initial temperature is $f(x)$ throughout and if the end $x = 0$ is kept at temperature zero and the end $x = L$ is insulated.

6. Solve the boundary-value problem

$$a^2 \frac{\partial^2 u}{\partial x^2} = \frac{\partial^2 u}{\partial t^2}, \qquad 0 < x < L, \quad t > 0$$

$$u(0, t) = 0, \qquad E \frac{\partial u}{\partial x}\bigg|_{x=L} = F_0, \quad t > 0$$

$$u(x, 0) = 0, \qquad \frac{\partial u}{\partial t}\bigg|_{t=0} = 0, \quad 0 < x < L.$$

The solution $u(x, t)$ represents the longitudinal displacement of a vibrating elastic bar that is anchored at its left end and is subjected to a constant force of magnitude F_0 at its right end. See Figure 13.11 on page 703. E is a constant called the modulus of elasticity.

7. Solve the boundary-value problem

$$\frac{\partial^2 u}{\partial x^2} + \frac{\partial^2 u}{\partial y^2} = 0, \qquad 0 < x < 1, \quad 0 < y < 1$$

$$\frac{\partial u}{\partial x}\bigg|_{x=0} = 0, \qquad u(1, y) = u_0, \quad 0 < y < 1$$

$$u(x, 0) = 0, \qquad \frac{\partial u}{\partial y}\bigg|_{y=1} = 0, \quad 0 < x < 1.$$

8. The initial temperature in a rod of unit length is $f(x)$ throughout. There is heat transfer from both ends, $x = 0$ and $x = 1$, into a surrounding medium kept at a constant temperature zero. Show that

$$u(x, t) = \sum_{n=1}^{\infty} A_n e^{-k\lambda_n^2 t}(\lambda_n \cos \lambda_n x + h \sin \lambda_n x),$$

where

$$A_n = \frac{2}{(\lambda_n^2 + 2h + h^2)} \int_0^1 f(x)(\lambda_n \cos \lambda_n x + h \sin \lambda_n x)\, dx$$

and the λ_n, $n = 1, 2, 3, \ldots$, are the consecutive positive roots of $\tan \lambda = 2\lambda h/(\lambda^2 - h^2)$.

9. A vibrating cantilever beam is embedded at its left end $(x = 0)$ and free at its right end $(x = 1)$. See Figure 13.20. The transverse displacement $u(x, t)$ of the beam is determined from the boundary-value problem

$$\frac{\partial^4 u}{\partial x^4} + \frac{\partial^2 u}{\partial t^2} = 0, \qquad 0 < x < 1, \quad t > 0$$

$$u(0, t) = 0, \qquad \frac{\partial u}{\partial x}\bigg|_{x=0} = 0, \quad t > 0$$

$$\frac{\partial^2 u}{\partial x^2}\bigg|_{x=1} = 0, \qquad \frac{\partial^3 u}{\partial x^3}\bigg|_{x=1} = 0, \quad t > 0$$

$$u(x, 0) = f(x), \qquad \frac{\partial u}{\partial t}\bigg|_{t=0} = g(x), \quad x > 0.$$

(a) Using the separation constant λ^4, show that the eigenvalues of the problem are determined from the equation $\cos \lambda \cosh \lambda = -1$.

(b) Use a calculator or a computer to find approximations to the first two positive eigenvalues.

Figure 13.20

10. (a) Find an equation that defines the eigenvalues when the ends of the beam in Problem 9 are embedded at $x = 0$ and $x = 1$.

(b) Use a calculator or a computer to find approximations to the first two positive eigenvalues.

13.8 Fourier Series in Two Variables

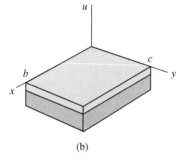

Figure 13.21

Suppose the rectangular region in Figure 13.21(a) is a thin plate in which the temperature u is a function of time t and position (x, y). Then, under suitable conditions, $u(x, y, t)$ can be shown to satisfy the **two-dimensional heat equation**

$$k\left(\frac{\partial^2 u}{\partial x^2} + \frac{\partial^2 u}{\partial y^2}\right) = \frac{\partial u}{\partial t}. \tag{1}$$

On the other hand, suppose Figure 13.21(b) represents a rectangular frame over which a thin flexible membrane has been stretched. If the membrane is set in motion, then its displacement u, measured from the xy-plane (transverse vibrations), is also a function of time t and position (x, y). When the vibrations are small, free, and undamped, $u(x, y, t)$ satisfies the **two-dimensional wave equation**

$$a^2\left(\frac{\partial^2 u}{\partial x^2} + \frac{\partial^2 u}{\partial y^2}\right) = \frac{\partial^2 u}{\partial t^2}. \tag{2}$$

As the next example will show, solutions of boundary-value problems involving (1) and (2) lead to the concept of a Fourier series in two variables.

Example 1 **Temperatures in a Plate**

Find the temperature $u(x, y, t)$ in the plate shown in Figure 13.21(a) if the initial temperature is $f(x, y)$ throughout and if the boundaries are held at temperature zero.

SOLUTION We must solve

$$k\left(\frac{\partial^2 u}{\partial x^2} + \frac{\partial^2 u}{\partial y^2}\right) = \frac{\partial u}{\partial t}, \qquad 0 < x < b, \quad 0 < y < c, \quad t > 0$$

subject to

$$u(0, y, t) = 0, \qquad u(b, y, t) = 0, \quad 0 < y < c, \quad t > 0$$

$$u(x, 0, t) = 0, \qquad u(x, c, t) = 0, \quad 0 < x < b, \quad t > 0$$

$$u(x, y, 0) = f(x, y), \qquad 0 < x < b, \quad 0 < y < c.$$

To separate variables for the partial differential equation in three independent variables we try to find a product solution $u(x, y, t) = X(x)Y(y)T(t)$. Substituting, we get

$$k(X''YT + XY''T) = XYT'$$

or

$$\frac{X''}{X} = -\frac{Y''}{Y} + \frac{T'}{kT}. \tag{3}$$

Since the left side of (3) depends only on x and the right side depends only on y and t, we must have both sides equal to a constant $-\lambda^2$:

$$\frac{X''}{X} = -\frac{Y''}{Y} + \frac{T'}{kT} = -\lambda^2,$$

and so

$$X'' + \lambda^2 X = 0 \tag{4}$$

$$\frac{Y''}{Y} = \frac{T'}{kT} + \lambda^2. \tag{5}$$

By the same reasoning, if we introduce another separation constant $-\mu^2$ in (5), then

$$\frac{Y''}{Y} = -\mu^2 \quad \text{and} \quad \frac{T'}{kT} + \lambda^2 = -\mu^2$$

$$Y'' + \mu^2 Y = 0 \quad \text{and} \quad T' + k(\lambda^2 + \mu^2)T = 0. \tag{6}$$

Now the solutions of the equations in (4) and (6) are, respectively,

$$X(x) = c_1 \cos \lambda x + c_2 \sin \lambda x \tag{7}$$

$$Y(y) = c_3 \cos \mu y + c_4 \sin \mu y \tag{8}$$

$$T(t) = c_5 e^{-k(\lambda^2 + \mu^2)t}. \tag{9}$$

But the boundary conditions

$$\left.\begin{array}{ll} u(0, y, t) = 0, & u(b, y, t) = 0 \\ u(x, 0, t) = 0, & u(x, c, t) = 0 \end{array}\right\} \text{ imply } \left\{\begin{array}{ll} X(0) = 0, & X(b) = 0 \\ Y(0) = 0, & Y(c) = 0. \end{array}\right.$$

Applying these conditions to (7) and (8) gives $c_1 = 0$, $c_3 = 0$, and $c_2 \sin \lambda b = 0$, $c_4 \sin \mu c = 0$. The latter equations in turn imply

$$\lambda = \frac{m\pi}{b}, \quad m = 1, 2, 3, \dots \qquad \mu = \frac{n\pi}{c}, \quad n = 1, 2, 3, \dots .$$

Thus a product solution of the two-dimensional heat equation that satisfies the boundary conditions is

$$u_{mn}(x, y, t) = A_{mn} e^{-k[(m\pi/b)^2 + (n\pi/c)^2]t} \sin \frac{m\pi}{b} x \sin \frac{n\pi}{c} y,$$

where A_{mn} is an arbitrary constant. Because we have two independent sets of eigenvalues, we are prompted to try the superposition principle in the form of a double sum

$$u(x, y, t) = \sum_{m=1}^{\infty} \sum_{n=1}^{\infty} A_{mn} e^{-k[(m\pi/b)^2 + (n\pi/c)^2]t} \sin \frac{m\pi}{b} x \sin \frac{m\pi}{c} y. \tag{10}$$

Now at $t = 0$ we must have

$$u(x, y, 0) = f(x, y) = \sum_{m=1}^{\infty} \sum_{n=1}^{\infty} A_{mn} \sin \frac{m\pi}{b} x \sin \frac{n\pi}{c} y. \tag{11}$$

We can find the coefficients A_{mn} by multiplying the double sum (11) by the product $\sin (m\pi x/b) \sin (n\pi y/c)$ and integrating over the rectangle $0 \le x \le b$, $0 \le y \le c$. It follows that

$$A_{mn} = \frac{4}{bc} \int_0^c \int_0^b f(x, y) \sin \frac{m\pi}{b} x \sin \frac{n\pi}{c} y \, dx \, dy. \tag{12}$$

Thus, the solution of the boundary-value problem consists of (1) with the A_{mn} defined by (12). ❏

The series (11) with coefficients (12) is called a **sine series in two variables,** or a **double sine series.**

E X E R C I S E S 1 3 . 8

In Problems 1 and 2, solve the heat equation (1) subject to the given conditions.

1. $u(0, y, t) = 0, \quad u(\pi, y, t) = 0$
$u(x, 0, t) = 0, \quad u(x, \pi, t) = 0$
$u(x, y, 0) = u_0$

2. $u(0, y, t) = 0, \quad u(1, y, t) = 0$
$u(x, 0, t) = 0, \quad u(x, 1, t) = 0$
$u(x, y, 0) = xy(x - 1)(y - 1)$

In Problems 3 and 4, solve the wave equation (2) subject to the given conditions.

3. $u(0, y, t) = 0, \quad u(\pi, y, t) = 0$
$u(x, 0, t) = 0, \quad u(x, \pi, t) = 0$
$u(x, y, 0) = xy(\pi - x)(\pi - y)$
$\left.\dfrac{\partial u}{\partial t}\right|_{t=0} = 0$

4. $u(0, y, t) = 0, \quad u(b, y, t) = 0$
$u(x, 0, t) = 0, \quad u(x, c, t) = 0$
$u(x, y, 0) = f(x, y)$
$\left.\dfrac{\partial u}{\partial t}\right|_{t=0} = g(x, y)$

The steady-state temperature $u(x, y, z)$ in the rectangular parallelepiped shown in Figure 13.22 satisfies Laplace's equation in three dimensions

$$\frac{\partial^2 u}{\partial x^2} + \frac{\partial^2 u}{\partial y^2} + \frac{\partial^2 u}{\partial z^2} = 0. \qquad (13)$$

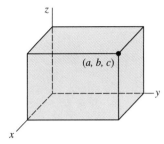

Figure 13.22

5. Solve Laplace's equation (13) if the top ($z = c$) of the parallelepiped is kept at temperature $f(x, y)$ and the remaining sides are kept at temperature zero.

6. Solve Laplace's equation (13) if the bottom ($z = 0$) of the parallelepiped is kept at temperature $f(x, y)$ and the remaining sides are kept at temperature zero.

7. Solve Laplace's equation (13) if the parallelepiped is a unit cube ($a = b = c = 1$) with the top $z = 1$ and bottom $z = 0$ kept at constant temperatures u_0 and $-u_0$, respectively, and the remaining sides kept at temperature zero.

C H A P T E R 1 3
R E V I E W E X E R C I S E S

1. Use separation of variables to find product solutions of

$$\frac{\partial^2 u}{\partial x \, \partial y} = u.$$

2. Use separation of variables to find product solutions of

$$\frac{\partial^2 u}{\partial x^2} + \frac{\partial^2 u}{\partial y^2} + 2\frac{\partial u}{\partial x} + 2\frac{\partial u}{\partial y} = 0.$$

Is it possible to choose a separation constant so that both X and Y are oscillatory functions?

3. Find a steady-state solution $\psi(x)$ of the boundary-value problem

$$k\frac{\partial^2 u}{\partial x^2} = \frac{\partial u}{\partial t}, \qquad 0 < x < \pi, \quad t > 0$$

$$u(0, t) = u_0, \qquad -\left.\frac{\partial u}{\partial x}\right|_{x=\pi} = u(\pi, t) - u_1, \quad t > 0$$

$$u(x, 0) = 0, \qquad 0 < x < \pi.$$

4. Give a physical interpretation for the boundary conditions in Problem 3.

5. At $t = 0$ a string of unit length is stretched on the positive x-axis. The ends of the string $x = 0$ and $x = 1$ are secured on the x-axis for $t > 0$. Find the displacement $u(x, t)$ if the initial velocity $g(x)$ is as given in Figure 13.23.

Figure 13.23

6. The partial differential equation

$$\frac{\partial^2 u}{\partial x^2} + x^2 = \frac{\partial^2 u}{\partial t^2}$$

is a form of the wave equation when an external vertical force proportional to the square of the horizontal distance from the left end is applied to the string. The string is secured at $x = 0$ one unit above the x-axis and on the x-axis at $x = 1$ for $t > 0$. Find the displacement $u(x, t)$ if the string starts from rest from the initial displacement $f(x)$.

7. Find the steady-state temperature $u(x, y)$ in the square plate shown in Figure 13.24.

Figure 13.24

8. Find the steady-state temperature $u(x, y)$ in the semi-infinite plate shown in Figure 13.25.

Figure 13.25

9. Solve Problem 8 if the boundaries $y = 0$ and $y = \pi$ are held at temperature zero for all time.

10. Find the temperature $u(x, t)$ in the infinite plate of width $2L$ shown in Figure 13.26 if the initial temperature is u_0 throughout. [*Hint:* $u(x, 0) = u_0$, $-L < x < L$ is an even function of x.]

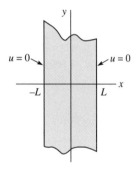

Figure 13.26

11. Solve the boundary-value problem

$$\frac{\partial^2 u}{\partial x^2} = \frac{\partial u}{\partial t}, \qquad 0 < x < \pi, \quad t > 0$$

$$u(0, t) = 0, \qquad u(\pi, t) = 0, \quad t > 0$$

$$u(x, 0) = \sin x, \qquad 0 < x < \pi.$$

12. Solve the boundary-value problem

$$k \frac{\partial^2 u}{\partial x^2} + \sin 2\pi x = \frac{\partial u}{\partial t}, \qquad 0 < x < 1, \quad t > 0$$

$$u(0, t) = 0, \qquad u(1, t) = 0, \quad t > 0$$

$$u(x, 0) = \sin \pi x, \qquad 0 < x < 1.$$

13. Find a formal series solution of the problem

$$\frac{\partial^2 u}{\partial x^2} + 2 \frac{\partial u}{\partial x} = \frac{\partial^2 u}{\partial t^2} + 2 \frac{\partial u}{\partial t} + u, \qquad 0 < x < \pi, \quad t > 0$$

$$u(0, t) = 0, \qquad u(\pi, t) = 0, \quad t > 0$$

$$\left. \frac{\partial u}{\partial t} \right|_{t=0} = 0, \qquad 0 < x < \pi.$$

Do not attempt to evaluate the coefficients in the series.

14. Project The concentration $c(x, t)$ of a substance that both diffuses in a medium and is convected by the currents in the medium satisfies the partial differential equation

$$k \frac{\partial^2 c}{\partial x^2} - h \frac{\partial c}{\partial x} = \frac{\partial c}{\partial t}, \qquad k \text{ and } h \text{ constants.}$$

Solve the PDE subject to

$$c(0, t) = 0, \qquad c(1, t) = 0, \quad t > 0$$

$$c(x, 0) = c_0, \qquad 0 < x < 1,$$

where c_0 is a constant. Solve for $c(x, t)$.

14

Boundary-Value Problems in Other Coordinate Systems

TOPICS TO REVIEW FOR THIS CHAPTER

polar coordinates
cylindrical coordinates
spherical coordinates
solution of Cauchy–Euler DEs (3.6)
Fourier–Bessel series (12.5)
Fourier–Legendre series (12.5)

IMPORTANT CONCEPTS IN THIS CHAPTER

Laplacian in
 polar coordinates
 cylindrical coordinates
 spherical coordinates
radial symmetry
radial vibrations
standing waves
nodal lines
steady-state temperatures

INTRODUCTION

All the boundary-value problems that have been considered so far have been expressed in terms of rectangular coordinates. If, however, we wish to find temperatures in a *circular disk,* a *circular cylinder,* or in a *sphere,* we would naturally try to describe the problems in polar coordinates, cylindrical coordinates, or spherical coordinates, respectively. It is important, then, that we express the Laplacian $\nabla^2 u$ in terms of these different coordinate systems. We shall now put to practical use the theory of Fourier–Bessel series (Section 14.2) and Fourier–Legendre series (Section 14.3).

14.1 Problems Involving Laplace's Equation in Polar Coordinates

■ **Laplacian in Polar Coordinates** The relationships between polar coordinates in the plane and rectangular coordinates are given by

$$x = r \cos \theta, \quad y = r \sin \theta \quad \text{and} \quad r^2 = x^2 + y^2, \quad \tan \theta = \frac{y}{x}.$$

Figure 14.1

See Figure 14.1. The first pair of equations transform polar coordinates (r, θ) into rectangular coordinates (x, y); the second pair of equations enable us to transform rectangular coordinates into polar coordinates. These equations also make it possible to convert the two-dimensional Laplacian of the function u, $\nabla^2 u = \partial^2 u/\partial x^2 + \partial^2 u/\partial y^2$, into polar coordinates. You are encouraged to work through the details of the Chain Rule and show that

$$\frac{\partial u}{\partial x} = \frac{\partial u}{\partial r}\frac{\partial r}{\partial x} + \frac{\partial u}{\partial \theta}\frac{\partial \theta}{\partial x} = \cos \theta \, \frac{\partial u}{\partial r} - \frac{\sin \theta}{r}\frac{\partial u}{\partial \theta}$$

$$\frac{\partial u}{\partial y} = \frac{\partial u}{\partial r}\frac{\partial r}{\partial y} + \frac{\partial u}{\partial \theta}\frac{\partial \theta}{\partial y} = \sin \theta \, \frac{\partial u}{\partial r} + \frac{\cos \theta}{r}\frac{\partial u}{\partial \theta}$$

$$\frac{\partial^2 u}{\partial x^2} = \cos^2 \theta \, \frac{\partial^2 u}{\partial r^2} - \frac{2 \sin \theta \cos \theta}{r}\frac{\partial^2 u}{\partial r \partial \theta} + \frac{\sin^2 \theta}{r^2}\frac{\partial^2 u}{\partial \theta^2} + \frac{\sin^2 \theta}{r}\frac{\partial u}{\partial r} + \frac{2 \sin \theta \cos \theta}{r}\frac{\partial u}{\partial \theta} \quad (1)$$

$$\frac{\partial^2 u}{\partial y^2} = \sin^2 \theta \, \frac{\partial^2 u}{\partial r^2} + \frac{2 \sin \theta \cos \theta}{r}\frac{\partial^2 u}{\partial r \partial \theta} + \frac{\cos^2 \theta}{r^2}\frac{\partial^2 u}{\partial \theta^2} + \frac{\cos^2 \theta}{r}\frac{\partial u}{\partial r} - \frac{2 \sin \theta \cos \theta}{r}\frac{\partial u}{\partial \theta}. \quad (2)$$

Adding (1) and (2) and simplifying yields the Laplacian of u in polar coordinates:

$$\nabla^2 u = \frac{\partial^2 u}{\partial x^2} + \frac{\partial^2 u}{\partial y^2} = \frac{\partial^2 u}{\partial r^2} + \frac{1}{r}\frac{\partial u}{\partial r} + \frac{1}{r^2}\frac{\partial^2 u}{\partial \theta^2}.$$

In this section we shall focus only on problems involving Laplace's equation in polar coordinates:

$$\frac{\partial^2 u}{\partial r^2} + \frac{1}{r}\frac{\partial u}{\partial r} + \frac{1}{r^2}\frac{\partial^2 u}{\partial \theta^2} = 0.$$

Our first example is the Dirichlet problem for a circular disk.

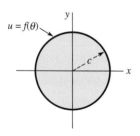

Figure 14.2

Example 1 Steady Temperatures in a Circular Plate

Find the steady-state temperature $u(r, \theta)$ in a circular plate of radius c if the temperature of the circumference is $u(c, \theta) = f(\theta)$, $0 < \theta < 2\pi$. See Figure 14.2.

SOLUTION We must solve Laplace's equation

$$\frac{\partial^2 u}{\partial r^2} + \frac{1}{r}\frac{\partial u}{\partial r} + \frac{1}{r^2}\frac{\partial^2 u}{\partial \theta^2} = 0, \quad 0 < \theta < 2\pi, \quad 0 < r < c$$

subject to $u(c, \theta) = f(\theta)$, $0 < \theta < 2\pi$. If we define $u = R(r)\Theta(\theta)$, then separation of variables gives

$$\frac{r^2 R'' + rR'}{R} = -\frac{\Theta''}{\Theta} = \lambda^2$$

and

$$r^2 R'' + rR' - \lambda^2 R = 0 \quad (3)$$

$$\Theta'' + \lambda^2 \Theta = 0. \quad (4)$$

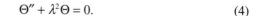

The solution of (4) is

$$\Theta = c_1 \cos \lambda\theta + c_2 \sin \lambda\theta. \tag{5}$$

Equation (3) is recognized as a Cauchy–Euler equation. Its general solution is

$$R = c_3 r^{\lambda} + c_4 r^{-\lambda}. \tag{6}$$

Now there are no explicit conditions in the statement of this problem that enable us to determine any of the coefficients or the eigenvalues. However, there are some implicit conditions.

First, our physical intuition leads us to expect that the temperature $u(r, \theta)$ should be bounded inside the circle $r = c$. Moreover, the temperature u should be the same at a specified point in the circle regardless of the polar description of that point. Since $(r, \theta + 2\pi)$ is an equivalent description of the point (r, θ), we must have $u(r, \theta) = u(r, \theta + 2\pi)$. In other words, the temperature $u(r, \theta)$ must be periodic with period 2π. But the only way the solutions of (4) can be 2π-periodic is if we take $\lambda = n$, where $n = 0, 1, 2, \ldots$. Thus for $n > 0$, (5) and (6) become

$$\Theta = c_1 \cos n\theta + c_2 \sin n\theta \tag{7}$$

$$R = c_3 r^n + c_4 r^{-n}. \tag{8}$$

Now observe in (8) that $r^{-n} = 1/r^n$. In order for the solution to be bounded at the center of the plate (that is, at $r = 0$) we must define $c_4 = 0$. Hence for $\lambda = n > 0$,

$$u_n = R\Theta = r^n(A_n \cos n\theta + B_n \sin n\theta), \tag{9}$$

where $c_3 c_1$ and $c_3 c_2$ have been replaced by A_n and B_n, respectively.

There is one last item of consideration: namely, $n = 0$ is an eigenvalue. For this value the solutions of (3) and (4) are not given by (5) and (6). The solutions of the differential equations $\Theta'' = 0$ and $r^2 R'' + rR' = 0$ are

$$\Theta = c_5\theta + c_6 \quad \text{and} \quad R = c_7 + c_8 \ln r.$$

Periodicity demands that we take $c_5 = 0$, whereas the expectation that the temperature is finite at $r = 0$ demands that we take $c_8 = 0$. In this case a product solution of the partial differential equation is simply

$$u_0 = A_0, \tag{10}$$

where A_0 represents $c_6 c_7$.

The superposition principle then gives $u(r, \theta) = \sum_{n=0}^{\infty} u_n$ or

$$u(r, \theta) = A_0 + \sum_{n=1}^{\infty} r^n(A_n \cos n\theta + B_n \sin n\theta). \tag{11}$$

Finally, by applying the boundary condition at $r = c$ to the result in (11) we recognize

$$f(\theta) = A_0 + \sum_{n=1}^{\infty} c^n(A_n \cos n\theta + B_n \sin n\theta)$$

as an expansion of f in a full Fourier series. Consequently we can make the identifications

$$A_0 = \frac{a_0}{2}, \quad c^n A_n = a_n, \quad \text{and} \quad c^n B_n = b_n.$$

That is,
$$A_0 = \frac{1}{2\pi} \int_0^{2\pi} f(\theta)\, d\theta \tag{12}$$

$$A_n = \frac{1}{c^n \pi} \int_0^{2\pi} f(\theta) \cos n\theta\, d\theta \tag{13}$$

$$B_n = \frac{1}{c^n \pi} \int_0^{2\pi} f(\theta) \sin n\theta\, d\theta. \tag{14}$$

The solution of the problem consists of the series given in (11), where A_0, A_n, and B_n are defined in (12), (13), and (14). ❑

Observe in Example 1 that corresponding to each positive eigenvalue, $n = 1, 2, 3, \ldots$, there are two different eigenfunctions—namely, $\cos n\theta$ and $\sin n\theta$. In this situation the eigenvalues are sometimes called **double eigenvalues.**

Example 2 Steady Temperatures in a Semicircular Plate

Find the steady-state temperature $u(r, \theta)$ in the semicircular plate shown in Figure 14.3.

SOLUTION The boundary-value problem is

$$\frac{\partial^2 u}{\partial r^2} + \frac{1}{r} \frac{\partial u}{\partial r} + \frac{1}{r^2} \frac{\partial^2 u}{\partial \theta^2} = 0, \qquad 0 < \theta < \pi, \quad 0 < r < c$$

$$u(c, \theta) = u_0, \qquad 0 < \theta < \pi$$

$$u(r, 0) = 0, \qquad u(r, \pi) = 0, \quad 0 < r < c.$$

Defining $u = R(r)\Theta(\theta)$ and separating variables gives

$$\frac{r^2 R'' + rR'}{R} = -\frac{\Theta''}{\Theta} = \lambda^2$$

and

$$r^2 R'' + rR' - \lambda^2 R = 0 \tag{15}$$

$$\Theta'' + \lambda^2 \Theta = 0. \tag{16}$$

Applying the boundary conditions $\Theta(0) = 0$ and $\Theta(\pi) = 0$ to the solution $\Theta = c_1 \cos \lambda\theta + c_2 \sin \lambda\theta$ of (16) gives, in turn, $c_1 = 0$ and $\lambda = n$, $n = 1, 2, 3, \ldots$. Hence $\Theta = c_2 \sin n\theta$. In this problem, unlike the problem in Example 1, $n = 0$ is not an eigenvalue. With $\lambda = n$ the solution of (15) is $R = c_3 r^n + c_4 r^{-n}$. But the assumption that $u(r, \theta)$ is bounded at $r = 0$ prompts us to define $c_4 = 0$. Therefore $u_n = R(r)\Theta(\theta) = A_n r^n \sin n\theta$ and

$$u(r, \theta) = \sum_{n=1}^{\infty} A_n r^n \sin n\theta.$$

The remaining boundary condition at $r = c$ gives the sine series

$$u_0 = \sum_{n=1}^{\infty} A_n c^n \sin n\theta.$$

Consequently
$$A_n c^n = \frac{2}{\pi} \int_0^{\pi} u_0 \sin n\theta\, d\theta,$$

and so
$$A_n = \frac{2u_0}{\pi c^n} \frac{1 - (-1)^n}{n}.$$

Figure 14.3

(figure labels: y; $u = u_0$; $\theta = \pi$; c; x; $u = 0$ at $\theta = \pi$; θ; $u = 0$ at $\theta = 0$)

Hence the solution of the problem is given by

$$u(r, \theta) = \frac{2u_0}{\pi} \sum_{n=1}^{\infty} \frac{1 - (-1)^n}{n} \left(\frac{r}{c}\right)^n \sin n\theta. \qquad \square$$

Answers to odd-numbered problems begin on page A-61.

In Problems 1–4, find the steady-state temperature $u(r, \theta)$ in a circular plate of radius $r = 1$ if the temperature on the circumference is as given.

1. $u(1, \theta) = \begin{cases} u_0, & 0 < \theta < \pi \\ 0, & \pi < \theta < 2\pi \end{cases}$

2. $u(1, \theta) = \begin{cases} \theta, & 0 < \theta < \pi \\ \pi - \theta, & \pi < \theta < 2\pi \end{cases}$

3. $u(1, \theta) = 2\pi\theta - \theta^2, \quad 0 < \theta < 2\pi$

4. $u(1, \theta) = \theta, \quad 0 < \theta < 2\pi$

5. Solve the exterior Dirichlet problem for a circular disk of radius c if $u(c, \theta) = f(\theta), 0 < \theta < 2\pi$. In other words, find the steady-state temperature $u(r, \theta)$ in a plate that coincides with the entire xy-plane in which a circular hole of radius c has been cut out around the origin and the temperature on the circumference of the hole is $f(\theta)$. [*Hint:* Assume that the temperature is bounded as $r \to \infty$.]

6. Find the steady-state temperature in the quarter-circular plate shown in Figure 14.4.

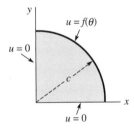

Figure 14.4

7. If the boundaries $\theta = 0$ and $\theta = \pi/2$ in Figure 14.4 are insulated, we then have, respectively,

$$\left.\frac{\partial u}{\partial \theta}\right|_{\theta=0} = 0, \qquad \left.\frac{\partial u}{\partial \theta}\right|_{\theta=\pi/2} = 0.$$

Find the steady-state temperature if

$$u(c, \theta) = \begin{cases} 1, & 0 < \theta < \pi/4 \\ 0, & \pi/4 < \theta < \pi/2. \end{cases}$$

8. Find the steady-state temperature in the infinite wedge-shaped plate shown in Figure 14.5. [*Hint:* Assume that the temperature is bounded as $r \to 0$ and as $r \to \infty$.]

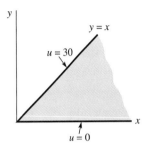

Figure 14.5

9. Find the steady-state temperature $u(r, \theta)$ in the circular ring shown in Figure 14.6. [*Hint:* Proceed as in Example 1.]

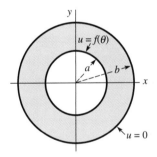

Figure 14.6

10. If the boundary conditions for the circular ring in Figure 14.6 are $u(a, \theta) = u_0, u(b, \theta) = u_1, 0 < \theta < 2\pi, u_0$ and u_1 constants, show that the steady-state temperature is given by

$$u(r, \theta) = \frac{u_0 \ln(r/b) - u_1 \ln(r/a)}{\ln(a/b)}.$$

[*Hint:* Try a solution of the form $u(r, \theta) = v(r, \theta) + \psi(r)$.]

11. Find the steady-state temperature $u(r, \theta)$ in a semicircular ring if

$$u(a, \theta) = \theta(\pi - \theta), \qquad u(b, \theta) = 0, \quad 0 < \theta < \pi$$

$$u(r, 0) = 0, \qquad u(r, \pi) = 0, \quad a < r < b.$$

12. Find the steady-state temperature $u(r, \theta)$ in a semicircular plate of radius $r = 1$ if

$$u(1, \theta) = u_0, \qquad 0 < \theta < \pi$$

$$u(r, 0) = 0, \qquad u(r, \pi) = u_0, \quad 0 < r < 1,$$

u_0 a constant.

13. Find the steady-state temperature $u(r, \theta)$ in a semicircular plate of radius $r = 2$ if

$$u(2, \theta) = \begin{cases} u_0, & 0 < \theta < \pi/2 \\ 0, & \pi/2 < \theta < \pi, \end{cases}$$

u_0 a constant, and the edges $\theta = 0$ and $\theta = \pi$ are insulated.

14. Consider the circular ring shown in Figure 14.6. Explain how the steady-state temperature $u(r, \theta)$ can be found when the boundary conditions are $u(a, \theta) = f(\theta)$, $u(b, \theta) = g(\theta)$, $0 \leq \theta \leq 2\pi$.

15. Computer Lab Assignment

(a) Find the series solution for $u(r, \theta)$ in Example 1 when

$$u(1, \theta) = \begin{cases} 100, & 0 < \theta < \pi \\ 0, & \pi < \theta < 2\pi. \end{cases}$$

(See Problem 1.)

(b) Use a CAS or a graphing utility to plot the partial sum $S_5(r, \theta)$ consisting of the first five nonzero terms of the solution in part **(a)** for $r = 0.9$, $r = 0.7$, $r = 0.5$, $r = 0.3$, and $r = 0.1$. Superimpose the graphs on the same coordinate axes.

(c) Approximate the temperatures $u(0.9, 1.3)$, $u(0.7, 2)$, $u(0.5, 3.5)$, $u(0.3, 4)$, $u(0.1, 5.5)$. Then approximate $u(0.9, 2\pi - 1.3)$, $u(0.7, 2\pi - 2)$, $u(0.5, 2\pi - 3.5)$, $u(0.3, 2\pi - 4)$, $u(0.1, 2\pi - 5.5)$.

(d) What is the temperature at the center of the circular plate? Why is it appropriate to call this value the average temperature in the plate? [*Hint:* Look at the graphs in part **(b)** and look at the numbers in part **(c)**.]

14.2 Problems in Polar and Cylindrical Coordinates: Bessel Functions

The two-dimensional heat and wave equations

$$k\left(\frac{\partial^2 u}{\partial x^2} + \frac{\partial^2 u}{\partial y^2}\right) = \frac{\partial u}{\partial t} \qquad \text{and} \qquad a^2\left(\frac{\partial^2 u}{\partial x^2} + \frac{\partial^2 u}{\partial y^2}\right) = \frac{\partial^2 u}{\partial t^2}$$

expressed in polar coordinates are, respectively,

$$k\left(\frac{\partial^2 u}{\partial r^2} + \frac{1}{r}\frac{\partial u}{\partial r} + \frac{1}{r^2}\frac{\partial^2 u}{\partial \theta^2}\right) = \frac{\partial u}{\partial t} \qquad \text{and} \qquad a^2\left(\frac{\partial^2 u}{\partial r^2} + \frac{1}{r}\frac{\partial u}{\partial r} + \frac{1}{r^2}\frac{\partial^2 u}{\partial \theta^2}\right) = \frac{\partial^2 u}{\partial t^2},$$

where $u = u(r, \theta, t)$. To solve a boundary-value problem involving either of these equations by separation of variables we define $u = R(r)\Theta(\theta)T(t)$. As in Section 13.8, this assumption leads to multiple infinite series. See Problem 15 in Exercises 14.2. In the following discussion we shall consider the simpler, but still important, problems that possess **radial symmetry**—that is, problems that are independent of the angular coordinate θ. In this case the heat and wave equations take, in turn, the forms

$$k\left(\frac{\partial^2 u}{\partial r^2} + \frac{1}{r}\frac{\partial u}{\partial r}\right) = \frac{\partial u}{\partial t} \qquad \text{and} \qquad a^2\left(\frac{\partial^2 u}{\partial r^2} + \frac{1}{r}\frac{\partial u}{\partial r}\right) = \frac{\partial^2 u}{\partial t^2}, \qquad (1)$$

where $u = u(r, t)$. Vibrations described by the second equation in (1) are said to be **radial vibrations.**

The first example deals with the free radial vibrations of a thin circular membrane. We assume that the displacements are small and that the motion is such that each point on the membrane moves in a direction perpendicular to the xy-plane (transverse vibrations)—that is, the u-axis is perpendicular to the xy-plane. A physical model to keep in mind while reading this example is a vibrating drumhead.

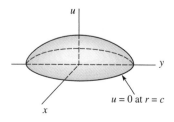

$u = 0$ at $r = c$

Figure 14.7

Example 1 Radial Vibrations of a Circular Membrane

Find the displacement $u(r, t)$ of a circular membrane of radius c clamped along its circumference if its initial displacement is $f(r)$ and its initial velocity is $g(r)$. See Figure 14.7.

SOLUTION The boundary-value problem to be solved is

$$a^2 \left(\frac{\partial^2 u}{\partial r^2} + \frac{1}{r} \frac{\partial u}{\partial r} \right) = \frac{\partial^2 u}{\partial t^2}, \qquad 0 < r < c, \quad t > 0$$

$$u(c, t) = 0, \qquad t > 0$$

$$u(r, 0) = f(r), \qquad \left. \frac{\partial u}{\partial t} \right|_{t=0} = g(r), \quad 0 < r < c.$$

Substituting $u = R(r)T(t)$ into the partial differential equation and separating variables gives

$$\frac{R'' + \frac{1}{r} R'}{R} = \frac{T''}{a^2 T} = -\lambda^2$$

and

$$rR'' + R' + \lambda^2 rR = 0 \tag{2}$$

$$T'' + a^2 \lambda^2 T = 0. \tag{3}$$

Now (2) is not a Cauchy–Euler equation but the parametric Bessel differential equation with order $\nu = 0$. Its general solution is

$$R = c_1 J_0(\lambda r) + c_2 Y_0(\lambda r).$$

The general solution of the familiar equation (3) is

$$T = c_3 \cos a\lambda t + c_4 \sin a\lambda t.$$

Now recall $Y_0(\lambda r) \to -\infty$ as $r \to 0^+$, and so the implicit assumption that the displacement $u(r, t)$ should be bounded at $r = 0$ forces us to define $c_2 = 0$. Thus $R = c_1 J_0(\lambda r)$.

 Since the boundary condition $u(c, t) = 0$ is equivalent to $R(c) = 0$, we must have $c_1 J_0(\lambda c) = 0$. We rule out $c_1 = 0$ (this would lead to a trivial solution of the partial differential equation), so consequently

$$J_0(\lambda c) = 0. \tag{4}$$

This equation defines the positive eigenvalues λ_n of the problem: If x_n are the positive roots of (4), then $\lambda_n = x_n/c$. Product solutions that satisfy the partial differential equation and the boundary condition are

$$u_n = RT = (A_n \cos a\lambda_n t + B_n \sin a\lambda_n t)J_0(\lambda_n r), \tag{5}$$

where we have done the usual relabeling of the constants. The superposition principle then gives

$$u(r, t) = \sum_{n=1}^{\infty} (A_n \cos a\lambda_n t + B_n \sin a\lambda_n t)J_0(\lambda_n r). \tag{6}$$

The given initial conditions determine the coefficients A_n and B_n.

 Setting $t = 0$ in (6) and using $u(r, 0) = f(r)$ gives

$$f(r) = \sum_{n=1}^{\infty} A_n J_0(\lambda_n r). \tag{7}$$

This last result is recognized as the Fourier–Bessel expansion of the function f on the interval $(0, c)$. Hence by a direct comparison of (4) and (7) with (7) and (15) of Section 12.5 we can identify the coefficients A_n with those given in (16) of Section 12.5:

$$A_n = \frac{2}{c^2 J_1^2(\lambda_n c)} \int_0^c r J_0(\lambda_n r) f(r) \, dr. \tag{8}$$

Next, we differentiate (6) with respect to t, set $t = 0$, and use $u_t(r, 0) = g(r)$:

$$g(r) = \sum_{n=1}^{\infty} a\lambda_n B_n J_0(\lambda_n r).$$

This is now a Fourier–Bessel expansion of the function g. By identifying the total coefficient $a\lambda_n B_n$ with (16) of Section 12.5 we can write

$$B_n = \frac{2}{a\lambda_n c^2 J_1^2(\lambda_n c)} \int_0^c r J_0(\lambda_n r) g(r) \, dr. \qquad (9)$$

Finally, the solution of the given boundary-value problem is the series (6) with A_n and B_n defined in (8) and (9). ❑

Analogous to (8) of Section 13.4, the product solutions (5) are called **standing waves**. For $n = 1, 2, 3, \ldots$, the standing waves are basically the graph of $J_0(\lambda_n r)$ with the time-varying amplitude

$$A_n \cos a\lambda_n t + B_n \sin a\lambda_n t.$$

$n = 1$
(a)

The standing waves at different values of time are represented by the dashed graphs in Figure 14.8. The zeros of each standing wave in the interval $(0, c)$ are the roots of $J_0(\lambda_n r) = 0$ and correspond to the set of points on a standing wave where there is no motion. This set of points is called a **nodal line**. If (as in Example 1) the positive roots of $J_0(\lambda_n c) = 0$ are denoted by x_n, then $\lambda_n c = x_n$ implies $\lambda_n = x_n/c$ and consequently the zeros of the standing waves are determined from

$$J_0(\lambda_n r) = J_0\left(\frac{x_n}{c} r\right) = 0.$$

Now the first three positive zeros of J_0 are (approximately) $x_1 = 2.4$, $x_2 = 5.5$, and $x_3 = 8.7$. Thus for $n = 1$, the first positive root of

$$J_0\left(\frac{x_1}{c} r\right) = 0 \qquad \text{is} \qquad \frac{2.4}{c} r = 2.4 \qquad \text{or} \qquad r = c.$$

$n = 2$
(b)

Since we are seeking zeros of the standing waves in the open interval $(0, c)$, the last result means that the first standing wave has no nodal line. For $n = 2$, the first two positive roots of

$$J_0\left(\frac{x_2}{c} r\right) = 0 \qquad \text{are determined from} \qquad \frac{5.5}{c} r = 2.4 \qquad \text{and} \qquad \frac{5.5}{c} r = 5.5.$$

Thus the second standing wave has one nodal line defined by $r = x_1 c/x_2 = 2.4c/5.5$. Note that $r \approx 0.44c < c$. For $n = 3$, a similar analysis shows that there are two nodal lines defined by $r = x_1 c/x_3 = 2.4c/8.7$ and $r = x_2 c/x_3 = 5.5c/8.7$. In general, the nth standing wave has $n - 1$ nodal lines $r = x_1 c/x_n, r = x_2 c/x_n, \ldots, r = x_{n-1} c/x_n$. Since $r = constant$ is an equation of a circle in polar coordinates, we see in Figure 14.8 that the nodal lines of a standing wave are concentric circles.

$n = 3$
(c)

Figure 14.8

■ **Use of Computers** It is possible to see the effect of a single drumbeat for the model solved in Example 1 by means of the animation capabilities of a computer algebra system. In Problem 14 in Exercises 14.2 you are asked to find the solution given in (6) when

$$c = 1, \qquad f(r) = 0, \qquad \text{and} \qquad g(r) = \begin{cases} -v_0, & 0 \le r < b \\ 0, & b \le r < 1. \end{cases}$$

Some frames of a "movie" of the vibrating drumhead are given in Figure 14.9.

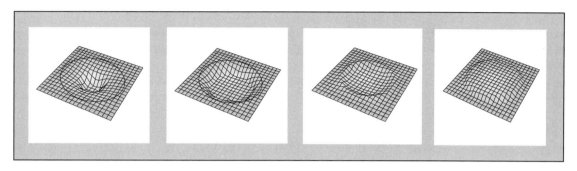

Figure 14.9 The shape of a drumbeat

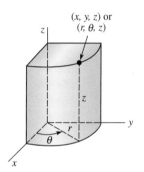

Figure 14.10

Laplacian in Cylindrical Coordinates From Figure 14.10 we see that the relationship between the cylindrical coordinates of a point in space and its rectangular coordinates is given by

$$x = r \cos \theta, \qquad y = r \sin \theta, \qquad z = z.$$

It follows immediately from the derivation of the Laplacian in polar coordinates (see Section 14.1) that the Laplacian of a function u in cylindrical coordinates is

$$\nabla^2 u = \frac{\partial^2 u}{\partial x^2} + \frac{\partial^2 u}{\partial y^2} + \frac{\partial^2 u}{\partial z^2} = \frac{\partial^2 u}{\partial r^2} + \frac{1}{r}\frac{\partial u}{\partial r} + \frac{1}{r^2}\frac{\partial^2 u}{\partial \theta^2} + \frac{\partial^2 u}{\partial z^2}.$$

Figure 14.11

Example 2 **Steady Temperatures in a Circular Cylinder**

Find the steady-state temperature in the circular cylinder shown in Figure 14.11.

SOLUTION The prescribed boundary conditions suggest that the temperature u has radial symmetry. Accordingly, $u(r, z)$ is determined from

$$\frac{\partial^2 u}{\partial r^2} + \frac{1}{r}\frac{\partial u}{\partial r} + \frac{\partial^2 u}{\partial z^2} = 0, \qquad 0 < r < 2, \quad 0 < z < 4$$

$$u(2, z) = 0, \qquad 0 < z < 4$$

$$u(r, 0) = 0, \qquad u(r, 4) = u_0, \quad 0 < r < 2.$$

Using $u = R(r)Z(z)$ and separating variables gives

$$\frac{R'' + \dfrac{1}{r}R'}{R} = -\frac{Z''}{Z} = -\lambda^2$$

and

$$rR'' + R' + \lambda^2 rR = 0 \tag{10}$$

$$Z'' - \lambda^2 Z = 0. \tag{11}$$

A negative separation constant is used here since there is no reason to expect the solution $u(r, z)$ to be periodic in z. The solution of (10) is

$$R = c_1 J_0(\lambda r) + c_2 Y_0(\lambda r),$$

and since the solution of (11) is defined on the finite interval $[0, 2]$, we write its general solution as

$$Z = c_3 \cosh \lambda z + c_4 \sinh \lambda z.$$

As in Example 1, the assumption that the function u is bounded at $r = 0$ demands that $c_2 = 0$. The condition $u(2, z) = 0$ implies $R(2) = 0$. This equation,

$$J_0(2\lambda) = 0, \tag{12}$$

defines the positive eigenvalues λ_n of the problem. Last, $Z(0) = 0$ implies $c_3 = 0$. Hence we have $R = c_1 J_0(\lambda_n r)$, $Z = c_4 \sinh \lambda_n z$, and

$$u_n = RZ = A_n \sinh \lambda_n z J_0(\lambda_n r)$$

$$u(r, z) = \sum_{n=1}^{\infty} A_n \sinh \lambda_n z J_0(\lambda_n r).$$

The remaining boundary condition at $z = 4$ then gives the Fourier–Bessel series

$$u_0 = \sum_{n=1}^{\infty} A_n \sinh 4\lambda_n J_0(\lambda_n r),$$

so that in view of (12) the coefficients are defined by (16) of Section 12.5,

$$A_n \sinh 4\lambda_n = \frac{2u_0}{2^2 J_1^2(2\lambda_n)} \int_0^2 r J_0(\lambda_n r) \, dr.$$

To evaluate the last integral we first use the substitution $t = \lambda_n r$, followed by $\dfrac{d}{dt}[tJ_1(t)] = tJ_0(t)$:

$$A_n \sinh 4\lambda_n = \frac{u_0}{2\lambda_n^2 J_1^2(2\lambda_n)} \int_0^{2\lambda_n} \frac{d}{dt}[tJ_1(t)] \, dt = \frac{u_0}{\lambda_n J_1(2\lambda_n)}.$$

Finally we arrive at $A_n = \dfrac{u_0}{\lambda_n \sinh 4\lambda_n J_1(2\lambda_n)}.$

Thus the temperature in the cylinder is given by

$$u(r, z) = u_0 \sum_{n=1}^{\infty} \frac{\sinh \lambda_n z J_0(\lambda_n r)}{\lambda_n \sinh 4\lambda_n J_1(2\lambda_n)}. \qquad \square$$

EXERCISES 14.2

Answers to odd-numbered problems begin on page A-61.

1. Find the displacement $u(r, t)$ in Example 1 if $f(r) = 0$ and the circular membrane is given an initial unit velocity in the upward direction.

2. A circular membrane of unit radius 1 is clamped along its circumference. Find the displacement $u(r, t)$ if the membrane starts from rest from the initial displacement $f(r) = 1 - r^2$, $0 < r < 1$. [*Hint:* See Problem 10 in Exercises 12.5.]

3. Find steady-state temperature $u(r, z)$ in the cylinder in Example 2 if the boundary conditions are $u(2, z) = 0$, $0 < z < 4$, $u(r, 0) = u_0$, $u(r, 4) = 0$, $0 < r < 2$.

4. If the lateral side of the cylinder in Example 2 is insulated, then

$$\left. \frac{\partial u}{\partial r} \right|_{r=2} = 0, \qquad 0 < z < 4.$$

(a) Find the steady-state temperature $u(r, z)$ when $u(r, 4) = f(r)$, $0 < r < 2$.

(b) Show that the steady-state temperature in part **(a)** reduces to $u(r, z) = u_0 z/4$ when $f(r) = u_0$. [*Hint:* Use (11) of Section 12.5.]

5. The temperature in a circular plate of radius c is determined from the boundary-value problem

$$k\left(\frac{\partial^2 u}{\partial r^2} + \frac{1}{r} \frac{\partial u}{\partial r} \right) = \frac{\partial u}{\partial t}, \qquad 0 < r < c, \quad t > 0$$

$$u(c, t) = 0, \qquad t > 0$$

$$u(r, 0) = f(r), \qquad 0 < r < c.$$

Solve for $u(r, t)$.

6. Solve Problem 5 if the edge $r = c$ of the plate is insulated.

7. When there is heat transfer from the lateral side of an infinite circular cylinder of unit radius (see Figure 14.12) into a surrounding medium at temperature zero, the temperature inside the cylinder is determined from

$$k\left(\frac{\partial^2 u}{\partial r^2} + \frac{1}{r}\frac{\partial u}{\partial r}\right) = \frac{\partial u}{\partial t}, \quad 0 < r < 1, \quad t > 0$$

$$\frac{\partial u}{\partial r}\bigg|_{r=1} = -hu(1, t), \quad h > 0, \quad t > 0$$

$$u(r, 0) = f(r), \quad 0 < r < 1.$$

Solve for $u(r, t)$.

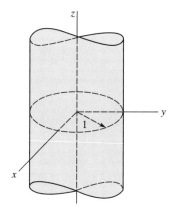

Figure 14.12

8. Find the steady-state temperature $u(r, z)$ in a semi-infinite cylinder of unit radius ($z \geq 0$) if there is heat transfer from its lateral side into a surrounding medium at temperature zero and if the temperature of the base $z = 0$ is held at a constant temperature u_0.

9. A circular plate is a composite of two different materials in the form of concentric circles. See Figure 14.13. The temperature in the plate is determined from the boundary-value problem

$$\frac{\partial^2 u}{\partial r^2} + \frac{1}{r}\frac{\partial u}{\partial r} = \frac{\partial u}{\partial t}, \quad 0 < r < 2, \quad t > 0$$

$$u(2, t) = 100, \quad t > 0$$

$$u(r, 0) = \begin{cases} 200, & 0 < r < 1 \\ 100, & 1 < r < 2. \end{cases}$$

Solve for $u(r, t)$. [*Hint:* Let $u(r, t) = v(r, t) + \psi(r)$.]

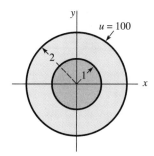

Figure 14.13

10. Solve the boundary-value problem

$$\frac{\partial^2 u}{\partial r^2} + \frac{1}{r}\frac{\partial u}{\partial r} + \beta = \frac{\partial u}{\partial t}, \quad 0 < r < 1, \quad t > 0, \quad \beta \text{ a constant}$$

$$u(1, t) = 0, \quad t > 0$$

$$u(r, 0) = 0, \quad 0 < r < 1.$$

11. The horizontal displacement $u(x, t)$ of a heavy chain of length L oscillating in a vertical plane satisfies the partial differential equation

$$g\frac{\partial}{\partial x}\left(x\frac{\partial u}{\partial x}\right) = \frac{\partial^2 u}{\partial t^2}, \quad 0 < x < L, \quad t > 0.$$

See Figure 14.14.

(a) Using $-\lambda^2$ as a separation constant, show that the ordinary differential equation in the spatial variable x is $xX'' + X' + \lambda^2 X = 0$. Solve this equation by means of the substitution $x = \tau^2/4$.

(b) Use the result of part **(a)** to solve the given partial differential equation subject to

$$u(L, t) = 0, \quad t > 0$$

$$u(x, 0) = f(x), \quad \frac{\partial u}{\partial t}\bigg|_{t=0} = 0, \quad 0 < x < L.$$

[*Hint:* Assume the oscillations at the free end $x = 0$ are finite.]

Figure 14.14

12. Computer Lab Assignment

(a) Consider Example 1 with $a = 1$, $c = 10$, $g(r) = 0$, and $f(r) = 1 - r/10$, $0 < r < 10$. Use a CAS as an aid in finding the numerical values of the first three eigenvalues $\lambda_1, \lambda_2, \lambda_3$ of the boundary-value problem and the first three coefficients A_1, A_2, A_3 of the solution $u(r, t)$ given in (6). Write the third partial sum $S_3(r, t)$ of the series solution.

(b) Use a CAS to plot the graph of $S_3(r, t)$ for $t = 0, 4,$ 10, 12, 20.

13. Computer Lab Assignment Solve Problem 5 with boundary conditions $u(c, t) = 200$, $u(r, 0) = 0$. With these imposed conditions, one would expect intuitively that at any interior point of the plate, $u(r, t) \to 200$ as $t \to \infty$. Assume that $c = 10$ and that the plate is cast iron so that $k = 0.1$ (approximately). Use a CAS as an aid in finding the numerical values of the first five eigenvalues $\lambda_1, \lambda_2, \lambda_3, \lambda_4, \lambda_5$ of

the boundary-value problem and the five coefficients A_1, A_2, A_3, A_4, A_5 in the solution $u(r, t)$. Let the corresponding approximate solution be denoted by $S_5(r, t)$. Plot $S_5(5, t)$ and $S_5(0, t)$ on a sufficiently large time interval $0 \le t \le T$. Use the plots of $S_5(5, t)$ and $S_5(0, t)$ to estimate the times (in seconds) for which $u(5, t) \approx 100$ and $u(0, t) \approx 100$. Repeat for $u(5, t) \approx 200$ and $u(0, t) \approx 200$.

14. Computer Lab Assignment Consider an idealized drum consisting of a thin membrane stretched over a circular frame of unit radius. When such a drum is struck at its center, one hears a sound that is frequently described as a dull thud rather than a melodic tone. We can model a single drumbeat using the boundary-value problem solved in Example 1.

(a) Find the solution $u(r, t)$ given in (6) when $c = 1$, $f(r) = 0$, and

$$g(r) = \begin{cases} -v_0, & 0 \le r < b \\ 0, & b \le r < 1. \end{cases}$$

(b) Show that the frequency of the standing wave $u_n(r, t)$ is $f_n = a\lambda_n/2\pi$, where λ_n is the nth positive zero of $J_0(x)$. Unlike the solution of the one-dimensional wave equation in Section 13.4, the frequencies are not integer multiples of the fundamental frequency f_1. Show that $f_2 \approx 2.295f_1$ and $f_3 \approx 3.598f_1$. We say that the drumbeat produces **anharmonic overtones.** As a result the displacement function $u(r, t)$ is not periodic, and so our ideal drum cannot produce a sustained tone.

(c) Let $a = 1$, $b = \frac{1}{4}$, and $v_0 = 1$ in your solution in part **(a)**. Use a CAS to graph the fifth partial sum $S_5(r, t)$ for the times $t = 0, 0.1, 0.2, 0.3, \ldots, 5.9, 6.0$ on the interval $-1 \le r \le 1$. Use the animation capabilities of your CAS to produce a movie of these vibrations.

(d) For a greater challenge, use the 3D plotting capabilities of your CAS to make a movie of the motion of the circular drumhead that is shown in cross-section in part **(c)**. [*Hint:* There are several ways of proceeding. For a fixed time either graph u as a function of x and y using $r = \sqrt{x^2 + y^2}$ or use the equivalent of *Mathematica*'s CylindricalPlot3D.]

15. Project Before attempting this problem reread Section 13.8.

Consider the general case of the vibrating circular membrane of radius $r = c$.

$$a^2\left(\frac{\partial^2 u}{\partial r^2} + \frac{1}{r}\frac{\partial u}{\partial r} + \frac{1}{r^2}\frac{\partial^2 u}{\partial \theta^2}\right) = \frac{\partial^2 u}{\partial t^2}, \quad 0 < r < c, \quad t > 0$$

$$u(c, \theta, t) = 0, \quad 0 < \theta < 2\pi, \quad t > 0$$

$$u(r, \theta, 0) = f(r, \theta), \quad 0 < r < c, \quad 0 < \theta < 2\pi$$

$$\left.\frac{\partial u}{\partial t}\right|_{t=0} = g(r, \theta), \quad 0 < r < c, \quad 0 < \theta < 2\pi.$$

(a) Assume that $u = R(r)\Theta(\theta)T(t)$ and that the separation constants are $-\lambda^2$ and $-v^2$. Show that the separated differential equations are

$$T'' + a^2\lambda^2 T = 0, \quad \Theta'' + v^2\Theta = 0$$

$$r^2 R'' + rR' + (\lambda^2 r^2 - v^2)R = 0.$$

(b) Solve the separated equations.

(c) Show that the eigenvalues and eigenfunctions of the problem are as follows:

Eigenvalues: $v = n$, $n = 0, 1, 2, \ldots$; eigenfunctions: $1, \cos n\theta, \sin n\theta$.

Eigenvalues: $\lambda_{ni} = x_{ni}/c$, $i = 1, 2, \ldots$, where, for each n, x_{ni} are the positive roots of $J_n(\lambda c) = 0$; eigenfunctions: $J_n(\lambda_{ni}r)$.

(d) Use the superposition principle to determine a multiple series solution. Do not attempt to evaluate the coefficients.

14.3 Problems in Spherical Coordinates: Legendre Polynomials

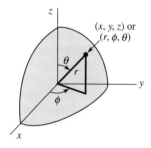

Figure 14.15

■ **Laplacian in Spherical Coordinates** In the spherical coordinate system shown in Figure 14.15 we have $x = r \sin\theta \cos\phi$, $y = r \sin\theta \sin\phi$, $z = r \cos\theta$. The Laplacian $\nabla^2 u = \dfrac{\partial^2 u}{\partial x^2} + \dfrac{\partial^2 u}{\partial y^2} + \dfrac{\partial^2 u}{\partial z^2}$ then becomes

$$\nabla^2 u = \frac{\partial^2 u}{\partial r^2} + \frac{2}{r}\frac{\partial u}{\partial r} + \frac{1}{r^2 \sin^2\theta}\frac{\partial^2 u}{\partial \phi^2} + \frac{1}{r^2}\frac{\partial^2 u}{\partial \theta^2} + \frac{\cot\theta}{r^2}\frac{\partial u}{\partial \theta}. \tag{1}$$

As you might imagine, problems involving (1) can be quite formidable. Consequently we shall consider only a few of the simpler problems that are independent of the azimuthal angle ϕ.

Example 1 **Steady Temperatures in a Sphere**

Find the steady-state temperature $u(r, \theta)$ in the sphere shown in Figure 14.16.

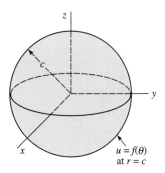

Figure 14.16

SOLUTION The temperature is determined from

$$\frac{\partial^2 u}{\partial r^2} + \frac{2}{r}\frac{\partial u}{\partial r} + \frac{1}{r^2}\frac{\partial^2 u}{\partial \theta^2} + \frac{\cot\theta}{r^2}\frac{\partial u}{\partial \theta} = 0, \qquad 0 < r < c, \quad 0 < \theta < \pi$$

$$u(c, \theta) = f(\theta), \qquad 0 < \theta < \pi.$$

If $u = R(r)\Theta(\theta)$, the partial differential equation separates as

$$\frac{r^2 R'' + 2rR'}{R} = -\frac{\Theta'' + \cot\theta\,\Theta'}{\Theta} = \lambda^2,$$

and so

$$r^2 R'' + 2rR' - \lambda^2 R = 0 \tag{2}$$

$$\sin\theta\,\Theta'' + \cos\theta\,\Theta' + \lambda^2 \sin\theta\,\Theta = 0. \tag{3}$$

After we substitute $x = \cos\theta$, $0 \le \theta \le \pi$, (3) becomes

$$(1 - x^2)\frac{d^2\Theta}{dx^2} - 2x\frac{d\Theta}{dx} + \lambda^2\Theta = 0, \qquad -1 \le x \le 1. \tag{4}$$

The latter equation is a form of Legendre's equation (see Problem 36 in Exercises 5.3). Now the only solutions of (4) that are continuous and have continuous derivatives on the closed interval $[-1, 1]$ are the Legendre polynomials $P_n(x)$ corresponding to $\lambda^2 = n(n + 1)$, $n = 0, 1, 2, \ldots$. Thus we take the solutions of (3) to be

$$\Theta = P_n(\cos\theta).$$

Furthermore, when $\lambda^2 = n(n + 1)$, the general solution of the Cauchy–Euler equation (2) is

$$R = c_1 r^n + c_2 r^{-(n+1)}.$$

Since we again expect $u(r, \theta)$ to be bounded at $r = 0$, we define $c_2 = 0$. Hence $u_n = A_n r^n P_n(\cos\theta)$, and

$$u(r, \theta) = \sum_{n=0}^{\infty} A_n r^n P_n(\cos\theta).$$

At $r = c$,

$$f(\theta) = \sum_{n=0}^{\infty} A_n c^n P_n(\cos\theta).$$

Therefore $A_n c^n$ are the coefficients of the Fourier–Legendre series (23) of Section 12.5:

$$A_n = \frac{2n + 1}{2c^n}\int_0^\pi f(\theta)P_n(\cos\theta)\sin\theta\,d\theta.$$

It follows that the solution is

$$u(r, \theta) = \sum_{n=0}^{\infty}\left(\frac{2n+1}{2}\int_0^\pi f(\theta)P_n(\cos\theta)\sin\theta\,d\theta\right)\left(\frac{r}{c}\right)^n P_n(\cos\theta). \qquad \square$$

E X E R C I S E S 1 4 . 3 Answers to odd-numbered problems begin on page A-61.

1. Solve the problem in Example 1 if

$$f(\theta) = \begin{cases} 50, & 0 < \theta < \pi/2 \\ 0, & \pi/2 < \theta < \pi. \end{cases}$$

Write out the first four nonzero terms of the series solution. [*Hint:* See Example 3, Section 12.5.]

2. The solution $u(r, \theta)$ in Example 1 could also be interpreted as the potential inside the sphere due to a charge distribution $f(\theta)$ on its surface. Find the potential outside the sphere.

3. Find the solution of the problem in Example 1 if $f(\theta) = \cos\theta$, $0 < \theta < \pi$. [*Hint:* $P_1(\cos\theta) = \cos\theta$. Use orthogonality.]

4. Find the solution of the problem in Example 1 if $f(\theta) = 1 - \cos 2\theta$, $0 < \theta < \pi$. [*Hint:* See Problem 16, Exercises 12.5.]

5. Find the steady-state temperature $u(r, \theta)$ within a hollow sphere $a < r < b$ if its inner surface $r = a$ is kept at temperature $f(\theta)$ and its outer surface $r = b$ is kept at temperature zero. The sphere in the first octant is shown in Figure 14.17.

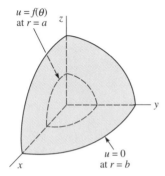

$u = f(\theta)$
at $r = a$

$u = 0$
at $r = b$

Figure 14.17

6. The steady-state temperature in a hemisphere of radius $r = c$ is determined from

$$\frac{\partial^2 u}{\partial r^2} + \frac{2}{r}\frac{\partial u}{\partial r} + \frac{1}{r^2}\frac{\partial^2 u}{\partial\theta^2} + \frac{\cot\theta}{r^2}\frac{\partial u}{\partial\theta} = 0, \quad 0 < r < c,\ 0 < \theta < \frac{\pi}{2}$$

$$u\left(r, \frac{\pi}{2}\right) = 0, \quad 0 < r < c$$

$$u(c, \theta) = f(\theta), \quad 0 < \theta < \frac{\pi}{2}.$$

Solve for $u(r, \theta)$. [*Hint:* $P_n(0) = 0$ only if n is odd. Also see Problem 18, Exercises 12.5.]

7. Solve Problem 6 when the base of the hemisphere is insulated; that is,

$$\frac{\partial u}{\partial\theta}\bigg|_{\theta=\pi/2} = 0, \quad 0 < r < c.$$

8. Solve Problem 6 for $r > c$.

9. The time-dependent temperature within a sphere of unit radius is determined from

$$\frac{\partial^2 u}{\partial r^2} + \frac{2}{r}\frac{\partial u}{\partial r} = \frac{\partial u}{\partial t}, \quad 0 < r < 1,\ t > 0$$

$$u(1, t) = 100, \quad t > 0$$

$$u(r, 0) = 0, \quad 0 < r < 1.$$

Solve for $u(r, t)$. [*Hint:* Verify that the left side of the partial differential equation can be written as $\dfrac{1}{r}\dfrac{\partial^2}{\partial r^2}(ru)$. Let $ru(r, t) = v(r, t) + \psi(r)$. Use only functions that are bounded as $r \to 0$.]

10. A uniform solid sphere of radius 1 at an initial constant temperature u_0 throughout is dropped into a large container of fluid that is kept at a constant temperature u_1 ($u_1 > u_0$) for all time. See Figure 14.18. Since there is heat transfer across the boundary $r = 1$, the temperature $u(r, t)$ in the sphere is determined from the boundary-value problem

$$\frac{\partial^2 u}{\partial r^2} + \frac{2}{r}\frac{\partial u}{\partial r} = \frac{\partial u}{\partial t}, \quad 0 < r < 1,\ t > 0$$

$$\frac{\partial u}{\partial r}\bigg|_{r=1} = -h(u(1, t) - u_1), \quad 0 < h < 1$$

$$u(r, 0) = u_0, \quad 0 < r < 1.$$

Solve for $u(r, t)$. [*Hint:* Proceed as in Problem 9.]

u_1

Figure 14.18

11. Solve the boundary-value problem involving spherical vibrations:

$$a^2\left(\frac{\partial^2 u}{\partial r^2} + \frac{2}{r}\frac{\partial u}{\partial r}\right) = \frac{\partial^2 u}{\partial t^2}, \quad 0 < r < c,\ t > 0$$

$$u(c, t) = 0, \quad t > 0$$

$$u(r, 0) = f(r), \quad \frac{\partial u}{\partial t}\bigg|_{t=0} = g(r), \quad 0 < r < c.$$

[*Hint:* Write the left side of the partial differential equation as $a^2\dfrac{1}{r}\dfrac{\partial^2}{\partial r^2}(ru)$. Let $v(r, t) = ru(r, t)$.]

12. A conducting sphere of radius $r = c$ is grounded and placed in a uniform electric field that has intensity E in the z-direction. The potential $u(r, \theta)$ outside the sphere is determined from the boundary-value problem

$$\frac{\partial^2 u}{\partial r^2} + \frac{2}{r}\frac{\partial u}{\partial r} + \frac{1}{r^2}\frac{\partial^2 u}{\partial\theta^2} + \frac{\cot\theta}{r^2}\frac{\partial u}{\partial\theta} = 0, \quad r > c,\ 0 < \theta < \pi$$

$$u(c, \theta) = 0, \quad 0 < \theta < \pi$$

$$\lim_{r\to\infty} u(r, \theta) = -Ez = -Er\cos\theta.$$

Show that $u(r, \theta) = -Er\cos\theta + E\dfrac{c^3}{r^2}\cos\theta$.

[*Hint:* Explain why $\int_0^\pi \cos\theta\, P_n(\cos\theta)\sin\theta\, d\theta = 0$ for all nonnegative integers except $n = 1$. See (24) of Section 12.5.]

1. Find the steady-state temperature $u(r, \theta)$ in a circular plate of radius c if the temperature on the circumference is given by

$$u(c, \theta) = \begin{cases} u_0, & 0 < \theta < \pi \\ -u_0, & \pi < \theta < 2\pi. \end{cases}$$

2. Find the steady-state temperature in the circular plate in Problem 1 if

$$u(c, \theta) = \begin{cases} 1, & 0 < \theta < \pi/2 \\ 0, & \pi/2 < \theta < 3\pi/2 \\ 1, & 3\pi/2 < \theta < 2\pi. \end{cases}$$

3. Find the steady-state temperature $u(r, \theta)$ in a semicircular plate of radius 1 if

$$u(1, \theta) = u_0(\pi\theta - \theta^2), \quad 0 < \theta < \pi$$

$$u(r, 0) = 0, \quad u(r, \pi) = 0, \quad 0 < r < 1.$$

4. Find the steady-state temperature $u(r, \theta)$ in the semicircular plate in Problem 3 if $u(1, \theta) = \sin \theta, 0 < \theta < \pi$.

5. Find the steady-state temperature $u(r, \theta)$ in the plate shown in Figure 14.19.

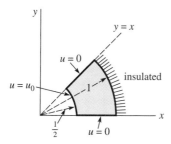

Figure 14.19

6. Find the steady-state temperature $u(r, \theta)$ in the infinite plate shown in Figure 14.20.

Figure 14.20

7. Suppose heat is lost from the flat surfaces of a very thin circular unit disk into a surrounding medium at temperature zero. If the linear law of heat transfer applies, the heat equation assumes the form

$$\frac{\partial^2 u}{\partial r^2} + \frac{1}{r}\frac{\partial u}{\partial r} - hu = \frac{\partial u}{\partial t}, \quad h > 0, \quad 0 < r < 1, \quad t > 0.$$

See Figure 14.21. Find the temperature $u(r, t)$ if the edge $r = 1$ is kept at temperature zero and if initially the temperature of the plate is unity throughout.

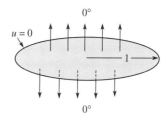

Figure 14.21

8. Suppose x_k is a positive zero of J_0. Show that a solution of the boundary-value problem

$$a^2\left(\frac{\partial^2 u}{\partial r^2} + \frac{1}{r}\frac{\partial u}{\partial r}\right) = \frac{\partial^2 u}{\partial t^2}, \quad 0 < r < 1, \quad t > 0$$

$$u(1, t) = 0, \quad t > 0$$

$$u(r, 0) = u_0 J_0(x_k r), \quad \frac{\partial u}{\partial t}\bigg|_{t=0} = 0, \quad 0 < r < 1$$

is $u(r, t) = u_0 J_0(x_k r) \cos a x_k t$.

9. Find the steady-state temperature $u(r, z)$ in the cylinder in Figure 14.11 if the lateral side is kept at temperature zero, the top $z = 4$ is kept at temperature 50, and the base $z = 0$ is insulated.

10. Solve the boundary-value problem

$$\frac{\partial^2 u}{\partial r^2} + \frac{1}{r}\frac{\partial u}{\partial r} + \frac{\partial^2 u}{\partial z^2} = 0, \quad 0 < r < 1, \quad 0 < z < 1$$

$$\frac{\partial u}{\partial r}\bigg|_{r=1} = 0, \quad 0 < z < 1$$

$$u(r, 0) = f(r), \quad u(r, 1) = g(r), \quad 0 < r < 1.$$

11. Find the steady-state temperature $u(r, \theta)$ in a sphere of unit radius if the surface is kept at

$$u(1, \theta) = \begin{cases} 100, & 0 < \theta < \pi/2 \\ -100, & \pi/2 < \theta < \pi. \end{cases}$$

[*Hint:* See Problem 20 in Exercises 12.5.]

12. Solve the boundary-value problem

$$\frac{\partial^2 u}{\partial r^2} + \frac{2}{r}\frac{\partial u}{\partial r} = \frac{\partial^2 u}{\partial t^2}, \quad 0 < r < 1, \quad t > 0$$

$$\frac{\partial u}{\partial r}\bigg|_{r=1} = 0, \quad t > 0$$

$$u(r, 0) = f(r), \quad \frac{\partial u}{\partial t}\bigg|_{t=0} = g(r), \quad 0 < r < 1.$$

[*Hint:* Proceed as in Problems 9 and 10 in Exercises 14.3, but let $v(r, t) = ru(r, t)$. See Section 13.7.]

13. The function $u(x) = Y_0(\lambda a)J_0(\lambda x) - J_0(\lambda a)Y_0(\lambda x)$, $a > 0$ is a solution of the parametric Bessel equation

$$x^2 \frac{d^2u}{dx^2} + x \frac{du}{dx} + \lambda^2 x^2 u = 0$$

on the interval $a \leq x \leq b$. If the eigenvalues λ_n are defined by the positive roots of the equation

$$Y_0(\lambda a)J_0(\lambda b) - J_0(\lambda a)Y_0(\lambda b) = 0,$$

show that the functions

$$u_m(x) = Y_0(\lambda_m a)J_0(\lambda_m x) - J_0(\lambda_m a)Y_0(\lambda_m x)$$

$$u_n(x) = Y_0(\lambda_n a)J_0(\lambda_n x) - J_0(\lambda_n a)Y_0(\lambda_n x)$$

are orthogonal with respect to the weight function $p(x) = x$ on the interval $[a, b]$; that is,

$$\int_a^b x\, u_m(x)u_n(x)\, dx = 0, \; m \neq n.$$

[*Hint:* Follow the procedure on page 674.]

14. Use the results of Problem 13 to solve the following boundary-value problem for the temperature $u(r, t)$ in a circular ring:

$$\frac{\partial^2 u}{\partial r^2} + \frac{1}{r}\frac{\partial u}{\partial r} = \frac{\partial u}{\partial t}, \qquad a < r < b, \quad t > 0$$

$$u(a, t) = 0, \qquad u(b, t) = 0, \quad t > 0$$

$$u(r, 0) = f(r), \qquad a < r < b.$$

15

Integral Transform Method

INTRODUCTION

The method of separation of variables that we employed in Chapters 13 and 14 is a powerful, but not universally applicable, method for solving boundary-value problems. If the partial differential equation in question is nonhomogeneous, or if the boundary conditions are time-dependent, or if the domain of the spatial variable is infinite: $(-\infty, \infty)$, or semi-infinite: (a, ∞), we may be able to use an integral transform to solve the problem. In Section 15.2 we shall solve problems that involve the heat and wave equations by means of the familiar Laplace transform. In Section 15.4 three new integral transforms, Fourier transforms, will be introduced and used.

15.1 Error Function

■ **Functions Defined by Integrals** There are many functions in mathematics that are defined by means of an integral. For example, in many traditional calculus texts the natural logarithm is defined in the following manner: $\ln x = \int_1^x \frac{1}{t}\, dt$, $x > 0$. In preceding sections we have already seen, albeit briefly, the error function erf(x), the complementary error function erfc(x), the sine integral function Si(x), and the Fresnel sine integral $S(x)$ and the gamma function $\Gamma(\alpha)$; all of these functions are defined in terms of an integral. Before applying the Laplace transform to boundary-value problems, we need to examine the graphs and a few of the more obvious properties of erf(x) and erfc(x).

■ **Properties and Graphs** Recall from (14) of Section 2.3 that the definitions of the **error function** erf(x) and **complementary error function** erfc(x) are, respectively,

$$\text{erf}(x) = \frac{2}{\sqrt{\pi}} \int_0^x e^{-u^2}\, du \qquad \text{and} \qquad \text{erfc}(x) = \frac{2}{\sqrt{\pi}} \int_x^\infty e^{-u^2}\, du. \qquad (1)$$

With the aid of polar coordinates it can be demonstrated that

$$\int_0^\infty e^{-u^2}\, du = \frac{\sqrt{\pi}}{2} \qquad \text{or} \qquad \frac{2}{\sqrt{\pi}} \int_0^\infty e^{-u^2}\, du = 1.$$

Thus from the additive interval property of definite integrals the last result is the same as

$$\frac{2}{\sqrt{\pi}} \left[\int_0^x e^{-u^2}\, du + \int_x^\infty e^{-u^2}\, du \right] = 1.$$

This shows that the erf(x) and erfc(x) are related by the identity

$$\text{erf}(x) + \text{erfc}(x) = 1. \qquad (2)$$

The graphs of erf(x) and erfc(x) for $x \geq 0$ are given in Figure 15.1. Note that erf(0) = 0, erfc(0) = 1 and that erf(x) → 1, erfc(x) → 0 as $x \to \infty$.

Other numerical values of erf(x) and erfc(x) can be obtained from a CAS or tables. In tables the error function is often referred to as the **probability integral.** The domain of erf(x) and of erfc(x) is $(-\infty, \infty)$. In Problem 11 in Exercises 15.1 you are asked to obtain the graph of each function on this interval and to deduce a few additional properties.

Table 15.1, of Laplace transforms, will be useful in the exercises in the next section. The proofs of these results are complicated and will not be given.

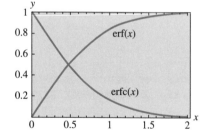

Figure 15.1 erf(x) and erfc(x), $x \geq 0$

Table 15.1

$f(t)$, $a > 0$	$\mathscr{L}\{f(t)\} = F(s)$	$f(t)$, $a > 0$	$\mathscr{L}\{f(t)\} = F(s)$
1. $\dfrac{1}{\sqrt{\pi t}}\, e^{-a^2/4t}$	$\dfrac{e^{-a\sqrt{s}}}{\sqrt{s}}$	4. $2\sqrt{\dfrac{t}{\pi}}\, e^{-a^2/4t} - a\, \text{erfc}\left(\dfrac{a}{2\sqrt{t}}\right)$	$\dfrac{e^{-a\sqrt{s}}}{s\sqrt{s}}$
2. $\dfrac{a}{2\sqrt{\pi t^3}}\, e^{-a^2/4t}$	$e^{-a\sqrt{s}}$	5. $e^{ab}e^{b^2 t}\, \text{erfc}\left(b\sqrt{t} + \dfrac{a}{2\sqrt{t}}\right)$	$\dfrac{e^{-a\sqrt{s}}}{\sqrt{s}(\sqrt{s}+b)}$
3. $\text{erfc}\left(\dfrac{a}{2\sqrt{t}}\right)$	$\dfrac{e^{-a\sqrt{s}}}{s}$	6. $-e^{ab}e^{b^2 t}\, \text{erfc}\left(b\sqrt{t} + \dfrac{a}{2\sqrt{t}}\right) + \text{erfc}\left(\dfrac{a}{2\sqrt{t}}\right)$	$\dfrac{be^{-a\sqrt{s}}}{s(\sqrt{s}+b)}$

Answers to odd-numbered problems begin on page A-62.

1. (a) Show that $\operatorname{erf}(\sqrt{t}) = \dfrac{1}{\sqrt{\pi}} \displaystyle\int_0^t \dfrac{e^{-\tau}}{\sqrt{\tau}}\, d\tau.$

(b) Use the convolution theorem and the results of Problems 39 and 40 in Exercises 4.1 to show that

$$\mathscr{L}\{\operatorname{erf}(\sqrt{t})\} = \dfrac{1}{s\sqrt{s+1}}.$$

2. Use the result of Problem 1 to show that

$$\mathscr{L}\{\operatorname{erfc}(\sqrt{t})\} = \dfrac{1}{s}\left[1 - \dfrac{1}{\sqrt{s+1}}\right].$$

3. Use the result of Problem 1 to show that

$$\mathscr{L}\{e^t \operatorname{erf}(\sqrt{t})\} = \dfrac{1}{\sqrt{s}(s-1)}.$$

4. Use the result of Problem 2 to show that

$$\mathscr{L}\{e^t \operatorname{erfc}(\sqrt{t})\} = \dfrac{1}{\sqrt{s}(\sqrt{s}+1)}.$$

5. Let C, G, R, and x be constants. Use Table 15.1 to show that

$$\mathscr{L}^{-1}\left\{\dfrac{C}{Cs+G}(1 - e^{-x\sqrt{RCs+RG}})\right\} = e^{-Gt/C}\operatorname{erf}\left(\dfrac{x}{2}\sqrt{\dfrac{RC}{t}}\right).$$

6. Let a be a constant. Show that

$$\mathscr{L}^{-1}\left\{\dfrac{\sinh a\sqrt{s}}{s\sinh\sqrt{s}}\right\} = \sum_{n=0}^{\infty}\left[\operatorname{erf}\left(\dfrac{2n+1+a}{2\sqrt{t}}\right) - \operatorname{erf}\left(\dfrac{2n+1-a}{2\sqrt{t}}\right)\right].$$

[*Hint:* Use the exponential definition of the hyperbolic sine. Expand $1/(1 - e^{-2\sqrt{s}})$ in a geometric series.]

7. Use the Laplace transform and Table 15.1 to solve the integral equation

$$y(t) = 1 - \int_0^t \dfrac{y(\tau)}{\sqrt{t-\tau}}\, d\tau.$$

8. Use the third and fifth entries in Table 15.1 to derive the sixth entry.

9. Show that $\displaystyle\int_a^b e^{-u^2}\, du = \dfrac{\sqrt{\pi}}{2}[\operatorname{erf}(b) - \operatorname{erf}(a)].$

10. Show that $\displaystyle\int_{-a}^a e^{-u^2}\, du = \sqrt{\pi}\operatorname{erf}(a).$

11. Computer Lab Assignment The functions $\operatorname{erf}(x)$ and $\operatorname{erfc}(x)$ are defined for $x < 0$. Use a CAS to superimpose the graphs of $\operatorname{erf}(x)$ and $\operatorname{erfc}(x)$ on the same axes for $-10 \le x \le 10$. Do the graphs possess any symmetry? What are $\lim_{x\to-\infty}\operatorname{erf}(x)$ and $\lim_{x\to-\infty}\operatorname{erfc}(x)$?

15.2 Applications of the Laplace Transform

We define the Laplace transform of a function of two variables $u(x, t)$ with respect to the variable t by

$$\mathscr{L}\{u(x, t)\} = \int_0^\infty e^{-st}u(x, t)\, dt = U(x, s),$$

where x is treated as a parameter. Throughout this section we shall assume that all the operational properties of Sections 4.3 and 4.4 apply to functions of two variables. For example, by Theorem 4.4 the transform of the partial derivative $\partial u/\partial t$ is

$$\mathscr{L}\left\{\dfrac{\partial u}{\partial t}\right\} = s\mathscr{L}\{u(x, t)\} - u(x, 0);$$

that is,
$$\mathscr{L}\left\{\dfrac{\partial u}{\partial t}\right\} = sU(x, s) - u(x, 0). \tag{1}$$

Similarly,
$$\mathscr{L}\left\{\dfrac{\partial^2 u}{\partial t^2}\right\} = s^2 U(x, s) - su(x, 0) - u_t(x, 0). \tag{2}$$

Since we are transforming with respect to t, we further suppose that it is legitimate to interchange integration and differentiation in the transform of $\partial^2 u/\partial x^2$:

$$\mathscr{L}\left\{\dfrac{\partial^2 u}{\partial x^2}\right\} = \int_0^\infty e^{-st}\dfrac{\partial^2 u}{\partial x^2}\, dt = \int_0^\infty \dfrac{\partial^2}{\partial x^2}[e^{-st}u(x, t)]\, dt$$

$$= \frac{d^2}{dx^2} \int_0^\infty e^{-st} u(x, t)\, dt = \frac{d^2}{dx^2}\, \mathcal{L}\{u(x, t)\};$$

that is,

$$\mathcal{L}\left\{\frac{\partial^2 u}{\partial x^2}\right\} = \frac{d^2 U}{dx^2}. \tag{3}$$

In view of (1) and (2) we see that the Laplace transform is suited to problems with initial conditions—namely, those problems associated with the heat equation or the wave equation.

Example 1 Laplace Transform of a PDE

Find the Laplace transform of the wave equation $a^2 \dfrac{\partial^2 u}{\partial x^2} = \dfrac{\partial^2 u}{\partial t^2}$, $t > 0$.

SOLUTION From (2) and (3),

$$\mathcal{L}\left\{a^2 \frac{\partial^2 u}{\partial x^2}\right\} = \mathcal{L}\left\{\frac{\partial^2 u}{\partial t^2}\right\}$$

becomes

$$a^2 \frac{d^2}{dx^2}\, \mathcal{L}\{u(x, t)\} = s^2 \mathcal{L}\{u(x, t)\} - su(x, 0) - u_t(x, 0)$$

or

$$a^2 \frac{d^2 U}{dx^2} - s^2 U = -su(x, 0) - u_t(x, 0). \tag{4} \square$$

The Laplace transform with respect to t of either the wave equation or the heat equation eliminates that variable, and for the one-dimensional equations the transformed equations are then *ordinary differential equations* in the spatial variable x. In solving a transformed equation we treat s as a parameter.

Example 2 Using the Laplace Transform to Solve a BVP

Solve:

$$\frac{\partial^2 u}{\partial x^2} = \frac{\partial^2 u}{\partial t^2}, \quad 0 < x < 1, \quad t > 0$$

Subject to:

$$u(0, t) = 0, \quad u(1, t) = 0, \quad t > 0$$

$$u(x, 0) = 0, \quad \left.\frac{\partial u}{\partial t}\right|_{t=0} = \sin \pi x, \quad 0 < x < 1.$$

SOLUTION The partial differential equation is recognized as the wave equation with $a = 1$. From (4) and the given initial conditions, the transformed equation is

$$\frac{d^2 U}{dx^2} - s^2 U = -\sin \pi x, \tag{5}$$

where $U(x, s) = \mathcal{L}\{u(x, t)\}$. Since the boundary conditions are functions of t, we must also find their Laplace transforms:

$$\mathcal{L}\{u(0, t)\} = U(0, s) = 0 \quad \text{and} \quad \mathcal{L}\{u(1, t)\} = U(1, s) = 0. \tag{6}$$

The results in (6) are boundary conditions for the ordinary differential equation (5). Since (5) is defined over a finite interval, its complementary function is

$$U_c(x, s) = c_1 \cosh sx + c_2 \sinh sx.$$

The method of undetermined coefficients yields a particular solution

$$U_p(x, s) = \frac{1}{s^2 + \pi^2} \sin \pi x.$$

Hence
$$U(x, s) = c_1 \cosh sx + c_2 \sinh sx + \frac{1}{s^2 + \pi^2} \sin \pi x.$$

But the conditions $U(0, s) = 0$ and $U(1, s) = 0$ yield, in turn, $c_1 = 0$ and $c_2 = 0$. We conclude that

$$U(x, s) = \frac{1}{s^2 + \pi^2} \sin \pi x$$

$$u(x, t) = \mathcal{L}^{-1}\left\{\frac{1}{s^2 + \pi^2} \sin \pi x\right\} = \frac{1}{\pi} \sin \pi x \mathcal{L}^{-1}\left\{\frac{\pi}{s^2 + \pi^2}\right\}.$$

Therefore
$$u(x, t) = \frac{1}{\pi} \sin \pi x \sin \pi t. \qquad \square$$

Example 3 **Using the Laplace Transform to Solve a BVP**

A very long string is initially at rest on the nonnegative x-axis. The string is secured at $x = 0$, and its distant right end slides down a frictionless vertical support. The string is set in motion by letting it fall under its own weight. Find the displacement $u(x, t)$.

SOLUTION Since the force of gravity is taken into consideration, it can be shown that the wave equation has the form

$$a^2 \frac{\partial^2 u}{\partial x^2} - g = \frac{\partial^2 u}{\partial t^2}, \qquad x > 0, \quad t > 0.$$

The boundary and initial conditions are, respectively,

$$u(0, t) = 0, \qquad \lim_{x \to \infty} \frac{\partial u}{\partial x} = 0, \quad t > 0$$

$$u(x, 0) = 0, \qquad \left.\frac{\partial u}{\partial t}\right|_{t=0} = 0, \quad x > 0.$$

The second boundary condition $\lim_{x \to \infty} \partial u/\partial x = 0$ indicates that the string is horizontal at a great distance from the left end. Now from (2) and (3),

$$\mathcal{L}\left\{a^2 \frac{\partial^2 u}{\partial x^2}\right\} - \mathcal{L}\{g\} = \mathcal{L}\left\{\frac{\partial^2 u}{\partial t^2}\right\}$$

becomes
$$a^2 \frac{d^2 U}{dx^2} - \frac{g}{s} = s^2 U - su(x, 0) - u_t(x, 0)$$

or, in view of the initial conditions,

$$\frac{d^2 U}{dx^2} - \frac{s^2}{a^2} U = \frac{g}{a^2 s}.$$

The transforms of the boundary conditions are

$$\mathcal{L}\{u(0, t)\} = U(0, s) = 0 \qquad \text{and} \qquad \mathcal{L}\left\{\lim_{x \to \infty} \frac{\partial u}{\partial x}\right\} = \lim_{x \to \infty} \frac{dU}{dx} = 0.$$

With the aid of undetermined coefficients, the general solution of the transformed equation is found to be

$$U(x, s) = c_1 e^{-(x/a)s} + c_2 e^{(x/a)s} - \frac{g}{s^3}.$$

The boundary condition $\lim_{x \to \infty} dU/dx = 0$ implies $c_2 = 0$, and $U(0, s) = 0$ gives $c_1 = g/s^3$. Therefore

$$U(x, s) = \frac{g}{s^3} e^{-(x/a)s} - \frac{g}{s^3}.$$

Now by the second translation theorem we have

$$u(x, t) = \mathcal{L}^{-1}\left\{\frac{g}{s^3} e^{-(x/a)s} - \frac{g}{s^3}\right\} = \frac{1}{2} g \left(t - \frac{x}{a}\right)^2 \mathcal{U}\left(t - \frac{x}{a}\right) - \frac{1}{2} gt^2$$

or

$$u(x, t) = \begin{cases} -\dfrac{1}{2} gt^2, & 0 \le t < \dfrac{x}{a} \\ -\dfrac{g}{2a^2}(2axt - x^2), & t \ge \dfrac{x}{a}. \end{cases}$$

To interpret the solution let us suppose $t > 0$ is fixed. For $0 \le x \le at$ the string is the shape of a parabola passing through $(0, 0)$ and $(at, -\frac{1}{2}gt^2)$. For $x > at$ the string is described by the horizontal line $u = -\frac{1}{2}gt^2$. See Figure 15.2. ❑

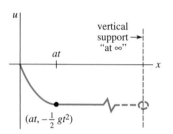

Figure 15.2

Observe that the problem in the next example could be solved by the procedure in Section 13.6. The Laplace transform provides an alternative solution.

Example 4 A Solution in Terms of erf(x)

Solve the heat equation

$$\frac{\partial^2 u}{\partial x^2} = \frac{\partial u}{\partial t}, \qquad 0 < x < 1, \quad t > 0$$

subject to

$$u(0, t) = 0, \qquad u(1, t) = u_0, \quad t > 0$$
$$u(x, 0) = 0, \qquad 0 < x < 1.$$

SOLUTION From (1) and (3) and the given initial condition,

$$\mathcal{L}\left\{\frac{\partial^2 u}{\partial x^2}\right\} = \mathcal{L}\left\{\frac{\partial u}{\partial t}\right\}$$

becomes

$$\frac{d^2 U}{dx^2} - sU = 0. \tag{7}$$

The transforms of the boundary conditions are

$$U(0, s) = 0 \qquad \text{and} \qquad U(1, s) = \frac{u_0}{s}. \tag{8}$$

Since we are concerned with a finite interval on the x-axis, we choose to write the general solution of (7) as

$$U(x, s) = c_1 \cosh(\sqrt{s}\,x) + c_2 \sinh(\sqrt{s}\,x).$$

Applying the two boundary conditions in (8) yields, respectively, $c_1 = 0$ and $c_2 = u_0/(s \sinh \sqrt{s})$. Thus

$$U(x, s) = u_0 \frac{\sinh (\sqrt{s}\,x)}{s \sinh \sqrt{s}}.$$

Now the inverse transform of the latter function cannot be found in most tables. However, by writing

$$\frac{\sinh (\sqrt{s}\,x)}{s \sinh \sqrt{s}} = \frac{e^{\sqrt{s}\,x} - e^{-\sqrt{s}\,x}}{s(e^{\sqrt{s}} - e^{-\sqrt{s}})} = \frac{e^{(x-1)\sqrt{s}} - e^{-(x+1)\sqrt{s}}}{s(1 - e^{-2\sqrt{s}})}$$

and using the geometric series

$$\frac{1}{1 - e^{-2\sqrt{s}}} = \sum_{n=0}^{\infty} e^{-2n\sqrt{s}}$$

we find

$$\frac{\sinh(\sqrt{s}\,x)}{s \sinh \sqrt{s}} = \sum_{n=0}^{\infty} \left[\frac{e^{-(2n+1-x)\sqrt{s}}}{s} - \frac{e^{-(2n+1+x)\sqrt{s}}}{s} \right].$$

If we assume that the inverse Laplace transform can be done term by term, it follows from entry 3 of Table 15.1 that

$$u(x, t) = u_0 \mathscr{L}^{-1} \left\{ \frac{\sinh(\sqrt{s}\,x)}{s \sinh \sqrt{s}} \right\}$$

$$= u_0 \sum_{n=0}^{\infty} \left[\mathscr{L}^{-1} \left\{ \frac{e^{-(2n+1-x)\sqrt{s}}}{s} \right\} - \mathscr{L}^{-1} \left\{ \frac{e^{-(2n+1+x)\sqrt{s}}}{s} \right\} \right]$$

$$= u_0 \sum_{n=0}^{\infty} \left[\text{erfc} \left(\frac{2n+1-x}{2\sqrt{t}} \right) - \text{erfc} \left(\frac{2n+1+x}{2\sqrt{t}} \right) \right]. \tag{9}$$

The solution (9) can be rewritten in terms of the error function using $\text{erfc}(x) = 1 - \text{erf}(x)$:

$$u(x, t) = u_0 \sum_{n=0}^{\infty} \left[\text{erf} \left(\frac{2n+1+x}{2\sqrt{t}} \right) - \text{erf} \left(\frac{2n+1-x}{2\sqrt{t}} \right) \right]. \tag{10} \ \square$$

Figure 15.3(a), obtained with the aid of the 3D plot function in a CAS, shows the surface over the rectangular region $0 \le x \le 1$, $0 \le t \le 6$ defined by the partial sum $S_{10}(x, t)$ of the solution (10). It is apparent from the surface and the accompanying two-dimensional graphs that at a fixed value of x (the curve of intersection of a plane slicing the surface perpendicular to the x-axis on the interval $0 \le x \le 1$) the temperature $u(x, t)$ increases rapidly to a constant value as time increases. See Figures 15.3(b) and (c). For a fixed time (the curve of intersection of a plane slicing the surface perpendicular to the t-axis) the temperature $u(x, t)$ naturally increases from 0 to 100. See Figures 15.3(d) and (e).

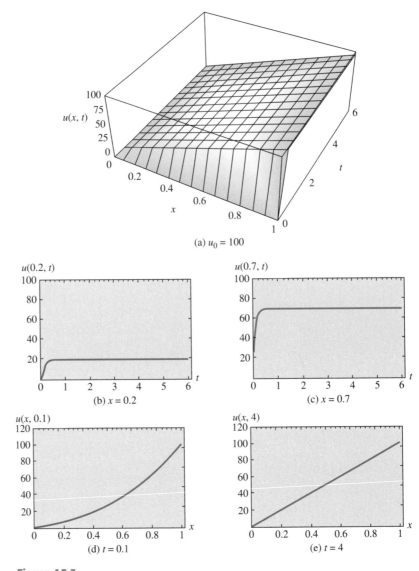

(a) $u_0 = 100$

(b) $x = 0.2$

(c) $x = 0.7$

(d) $t = 0.1$

(e) $t = 4$

Figure 15.3

EXERCISES 15.2

Answers to odd-numbered problems begin on page A-62.

In the following problems use tables as necessary.

1. A string is stretched along the x-axis between $(0, 0)$ and $(L, 0)$. Find the displacement $u(x, t)$ if the string starts from rest in the initial position $A \sin(\pi x/L)$.

2. Solve the boundary-value problem

$$\frac{\partial^2 u}{\partial x^2} = \frac{\partial^2 u}{\partial t^2}, \qquad 0 < x < 1, \quad t > 0$$

$$u(0, t) = 0, \qquad u(1, t) = 0$$

$$u(x, 0) = 0, \qquad \left.\frac{\partial u}{\partial t}\right|_{t=0} = 2 \sin \pi x + 4 \sin 3\pi x.$$

3. The displacement of a semi-infinite elastic string is determined from

$$a^2 \frac{\partial^2 u}{\partial x^2} = \frac{\partial^2 u}{\partial t^2}, \qquad x > 0, \quad t > 0$$

$$u(0, t) = f(t), \qquad \lim_{x \to \infty} u(x, t) = 0, \quad t > 0$$

$$u(x, 0) = 0, \qquad \left.\frac{\partial u}{\partial t}\right|_{t=0} = 0, \quad x > 0.$$

Solve for $u(x, t)$.

4. Solve the boundary-value problem in Problem 3 when

$$f(t) = \begin{cases} \sin \pi t, & 0 \le t \le 1 \\ 0, & t > 1. \end{cases}$$

Sketch the displacement $u(x, t)$ for $t > 1$.

5. In Example 3 find the displacement $u(x, t)$ when the left end of the string at $x = 0$ is given an oscillatory motion described by $f(t) = A \sin \omega t$.

6. The displacement $u(x, t)$ of a string that is driven by an external force is determined from

$$\frac{\partial^2 u}{\partial x^2} + \sin \pi x \sin \omega t = \frac{\partial^2 u}{\partial t^2}, \qquad 0 < x < 1, \quad t > 0$$

$$u(0, t) = 0, \qquad u(1, t) = 0, \quad t > 0$$

$$u(x, 0) = 0, \qquad \frac{\partial u}{\partial t}\bigg|_{t=0} = 0, \quad 0 < x < 1.$$

Solve for $u(x, t)$.

7. A uniform bar is clamped at $x = 0$ and is initially at rest. If a constant force F_0 is applied to the free end at $x = L$, the longitudinal displacement $u(x, t)$ of a cross-section of the bar is determined from

$$a^2 \frac{\partial^2 u}{\partial x^2} = \frac{\partial^2 u}{\partial t^2}, \qquad 0 < x < L, \quad t > 0$$

$$u(0, t) = 0, \qquad E \frac{\partial u}{\partial x}\bigg|_{x=L} = F_0, \quad E \text{ a constant}, \quad t > 0$$

$$u(x, 0) = 0, \qquad \frac{\partial u}{\partial t}\bigg|_{t=0} = 0, \quad 0 < x < L.$$

Solve for $u(x, t)$. [*Hint:* Expand $1/(1 + e^{-2sL/a})$ in a geometric series.]

8. A uniform semi-infinite elastic beam moving along the x-axis with a constant velocity $-v_0$ is brought to a stop by hitting a wall at time $t = 0$. See Figure 15.4. The longitudinal displacement $u(x, t)$ is determined from

$$a^2 \frac{\partial^2 u}{\partial x^2} = \frac{\partial^2 u}{\partial t^2}, \qquad x > 0, \quad t > 0$$

$$u(0, t) = 0, \qquad \lim_{x \to \infty} \frac{\partial u}{\partial x} = 0, \quad t > 0$$

$$u(x, 0) = 0, \qquad \frac{\partial u}{\partial t}\bigg|_{t=0} = -v_0, \quad x > 0.$$

Solve for $u(x, t)$.

Figure 15.4

9. Solve the boundary-value problem

$$\frac{\partial^2 u}{\partial x^2} = \frac{\partial^2 u}{\partial t^2}, \qquad x > 0, \quad t > 0$$

$$u(0, t) = 0, \qquad \lim_{x \to \infty} u(x, t) = 0, \quad t > 0$$

$$u(x, 0) = xe^{-x}, \qquad \frac{\partial u}{\partial t}\bigg|_{t=0} = 0, \quad x > 0.$$

10. Solve the boundary-value problem

$$\frac{\partial^2 u}{\partial x^2} = \frac{\partial^2 u}{\partial t^2}, \qquad x > 0, \quad t > 0$$

$$u(0, t) = 1, \qquad \lim_{x \to \infty} u(x, t) = 0, \quad t > 0$$

$$u(x, 0) = e^{-x}, \qquad \frac{\partial u}{\partial t}\bigg|_{t=0} = 0, \quad x > 0.$$

11. Computer Lab Assignment

(a) The temperature in a semi-infinite solid is modeled by the boundary-value problem

$$k \frac{\partial^2 u}{\partial x^2} = \frac{\partial u}{\partial t}, \qquad x > 0, \quad t > 0$$

$$u(0, t) = u_0, \qquad \lim_{x \to \infty} u(x, t) = 0, \quad t > 0$$

$$u(x, 0) = 0, \qquad x > 0.$$

Solve for $u(x, t)$. Use the solution to determine analytically the value of $\lim_{t \to \infty} u(x, t), x > 0$.

(b) Use a CAS to obtain the graph of $u(x, t)$ over the rectangular region $0 \le x \le 10, 0 \le t \le 15$. Assume $u_0 = 100$ and $k = 1$. Indicate the two boundary conditions and initial condition on your graph. Use 2D and 3D plots of $u(x, t)$ to verify your answer to part **(a)**.

12. Computer Lab Assignment

(a) In Problem 11 if there is a constant flux of heat into the solid at its left-hand boundary, then the boundary condition is

$$\frac{\partial u}{\partial x}\bigg|_{x=0} = -A, A > 0, t > 0.$$ Solve for $u(x, t)$. Use the solution to determine analytically the value of $\lim_{t \to \infty} u(x, t), x > 0$.

(b) Use a CAS to obtain the graph of $u(x, t)$ over the rectangular region $0 \le x \le 10, 0 \le t \le 15$. Assume $u_0 = 100$ and $k = 1$. Use 2D and 3D plots of $u(x, t)$ to verify your answer to part **(a)**.

In Problems 13–20, use the Laplace transform to solve the heat equation $u_{xx} = u_t$, $x > 0$, $t > 0$ subject to the given conditions.

13. $u(0, t) = u_0$, $\lim\limits_{x \to \infty} u(x, t) = u_1$

$u(x, 0) = u_1$

14. $u(0, t) = u_0$, $\lim\limits_{x \to \infty} \dfrac{u(x, t)}{x} = u_1$

$u(x, 0) = u_1 x$

15. $\dfrac{\partial u}{\partial x}\Big|_{x=0} = u(0, t)$, $\lim\limits_{x \to \infty} u(x, t) = u_0$

$u(x, 0) = u_0$

16. $\dfrac{\partial u}{\partial x}\Big|_{x=0} = u(0, t) - 50$, $\lim\limits_{x \to \infty} u(x, t) = 0$

$u(x, 0) = 0$

17. $u(0, t) = f(t)$, $\lim\limits_{x \to \infty} u(x, t) = 0$

$u(x, 0) = 0$

[*Hint:* Use the convolution theorem.]

18. $\dfrac{\partial u}{\partial x}\Big|_{x=0} = -f(t)$, $\lim\limits_{x \to \infty} u(x, t) = 0$

$u(x, 0) = 0$

19. $u(0, t) = 60 + 40\,\mathcal{U}(t - 2)$, $\lim\limits_{x \to \infty} u(x, t) = 60$

$u(x, 0) = 60$

20. $u(0, t) = \begin{cases} 20, & 0 < t < 1 \\ 0, & t \geq 1 \end{cases}$, $\lim\limits_{x \to \infty} u(x, t) = 100$

$u(x, 0) = 100$

21. Solve the boundary-value problem

$$\frac{\partial^2 u}{\partial x^2} = \frac{\partial u}{\partial t}, \qquad -\infty < x < 1, \quad t > 0$$

$$\frac{\partial u}{\partial x}\Big|_{x=1} = 100 - u(1, t), \qquad \lim\limits_{x \to -\infty} u(x, t) = 0, \quad t > 0$$

$$u(x, 0) = 0, \qquad -\infty < x < 1.$$

22. Show that a solution of the boundary-value problem

$$k\frac{\partial^2 u}{\partial x^2} + r = \frac{\partial u}{\partial t}, \qquad x > 0, \quad t > 0$$

$$u(0, t) = 0, \qquad \lim\limits_{x \to \infty} \frac{\partial u}{\partial x} = 0, \quad t > 0$$

$$u(x, 0) = 0, \qquad x > 0,$$

where r is a constant, is given by

$$u(x, t) = rt - r\int_0^t \operatorname{erfc}\left(\frac{x}{2\sqrt{k\tau}}\right) d\tau.$$

23. A rod of length L is held at a constant temperature u_0 at its ends $x = 0$ and $x = L$. If the rod's initial temperature is

$u_0 + u_0 \sin(x\pi/L)$, solve the heat equation $u_{xx} = u_t$, $0 < x < L$, $t > 0$ for the temperature $u(x, t)$.

24. If there is a heat transfer from the lateral surface of a thin wire of length L into a medium at constant temperature u_m, then the heat equation takes on the form

$$k\frac{\partial^2 u}{\partial x^2} - h(u - u_m) = \frac{\partial u}{\partial t}, \qquad 0 < x < L, \quad t > 0,$$

where h is a constant. Find the temperature $u(x, t)$ if the initial temperature is a constant u_0 throughout and the ends $x = 0$ and $x = L$ are insulated.

25. A rod of unit length is insulated at $x = 0$ and is kept at temperature zero at $x = 1$. If the initial temperature of the rod is a constant u_0, solve $ku_{xx} = u_t$, $0 < x < 1$, $t > 0$ for the temperature $u(x, t)$. [*Hint:* Expand $1/(1 + e^{-2\sqrt{s/k}})$ in a geometric series.]

26. An infinite porous slab of unit width is immersed in a solution of constant concentration c_0. A dissolved substance in the solution diffuses into the slab. The concentration $c(x, t)$ in the slab is determined from

$$D\frac{\partial^2 c}{\partial x^2} = \frac{\partial c}{\partial t}, \qquad 0 < x < 1, \quad t > 0$$

$$c(0, t) = c_0, \qquad c(1, t) = c_0, \quad t > 0$$

$$c(x, 0) = 0, \qquad 0 < x < 1,$$

where D is a constant. Solve for $c(x, t)$.

27. A very long telephone transmission line is initially at a constant potential u_0. If the line is grounded at $x = 0$ and insulated at the distant right end, then the potential $u(x, t)$ at a point x along the line at time t is determined from

$$\frac{\partial^2 u}{\partial x^2} - RC\frac{\partial u}{\partial t} - RGu = 0, \qquad x > 0, \quad t > 0$$

$$u(0, t) = 0, \qquad \lim\limits_{x \to \infty} \frac{\partial u}{\partial x} = 0, \quad t > 0$$

$$u(x, 0) = u_0, \qquad x > 0,$$

where R, C, and G are constants known as resistance, capacitance, and conductance, respectively. Solve for $u(x, t)$. [*Hint:* See Problem 5 in Exercises 15.1.]

28. Show that a solution of the boundary-value problem

$$\frac{\partial^2 u}{\partial x^2} - hu = \frac{\partial u}{\partial t}, \qquad x > 0, \quad t > 0$$

$$u(0, t) = u_0, \qquad \lim\limits_{x \to \infty} u(x, t) = 0, \quad t > 0$$

$$u(x, 0) = 0, \qquad x > 0$$

is

$$u(x, t) = \frac{u_0 x}{2\sqrt{\pi}}\int_0^t \frac{e^{-h\tau - x^2/4\tau}}{\tau^{3/2}}\, d\tau.$$

29. Computer Lab Assignment Humans gather most of their information of the outside world through sight and sound. But many creatures use chemical signals as their primary means of communication; for example, honeybees, when alarmed, emit a substance and fan their wings feverishly to relay the warning signal to the bees that attend to the queen. These molecular messages between members of the same species are called pheromones. The signals may be carried by moving air or water or by a *diffusion process* in which the random movement of gas molecules transports the chemical away from its source. Figure 15.5 shows an ant emitting an alarm chemical into the still air of a tunnel. If $c(x, t)$ denotes the concentration of the chemical x centimeters from the source at time t, then $c(x, t)$ satisfies

$$k \frac{\partial^2 c}{\partial x^2} = \frac{\partial c}{\partial t}, \qquad x > 0, \quad t > 0,$$

and k is a positive constant. The emission of pheromones as a discrete pulse gives rise to a boundary condition of the form

$$\frac{\partial c}{\partial x}\bigg|_{x=0} = -A\delta(t).$$

Figure 15.5

(a) Solve the boundary-value problem if it is further known that $c(x, 0) = 0$, $x > 0$, and $\lim_{x \to \infty} c(x, t) = 0$, $t > 0$.

(b) Use a CAS to plot the graph of the solution in part **(a)** for $x \geq 0$ at the fixed times $t = 0.1$, $t = 0.5$, $t = 1$, $t = 2$, $t = 5$.

(c) For any fixed time t, show that $\int_0^\infty c(x, t)\, dx = Ak$. Thus Ak represents the total amount of chemical discharged.

30. Starting at $t = 0$, a concentrated load of magnitude F_0 moves with a constant velocity v_0 along a semi-infinite string. In this case the wave equation becomes

$$a^2 \frac{\partial^2 u}{\partial x^2} = \frac{\partial^2 u}{\partial t^2} + F_0\delta\left(t - \frac{x}{v_0}\right).$$

Solve the above partial differential equation subject to

$$u(0, t) = 0, \qquad \lim_{x \to \infty} u(x, t) = 0, \quad t > 0$$

$$u(x, 0) = 0, \qquad \frac{\partial u}{\partial t}\bigg|_{t=0} = 0, \quad x > 0,$$

(a) when $v_0 \neq a$; **(b)** when $v_0 = a$.

15.3 Fourier Integral

We used Fourier series in Chapters 12 and 13 to represent a function f defined on a *finite* interval $(-p, p)$ or $(0, L)$. When f and f' are piecewise continuous on such an interval, a Fourier series represents the function on the interval and converges to the periodic extension of f outside the interval. In this way we are justified in saying that Fourier series are associated only with periodic functions. We shall now derive, in a nonrigorous fashion, a means of representing certain kinds of nonperiodic functions that are defined on either an *infinite* interval $(-\infty, \infty)$ or a *semi-infinite* interval $(0, \infty)$.

■ **From Fourier Series to Fourier Integral** Suppose a function f is defined on $(-p, p)$. If we use the integral definitions of the coefficients (9), (10), (11) of Section 12.2 in (8) of that section, then the Fourier series of f on the interval is

$$f(x) = \frac{1}{2p}\int_{-p}^{p} f(t)\, dt + \frac{1}{p}\sum_{n=1}^{\infty}\left[\left(\int_{-p}^{p} f(t)\cos\frac{n\pi}{p}t\, dt\right)\cos\frac{n\pi}{p}x + \left(\int_{-p}^{p} f(t)\sin\frac{n\pi}{p}t\, dt\right)\sin\frac{n\pi}{p}x\right]. \qquad (1)$$

If we let $\alpha_n = n\pi/p$, $\Delta\alpha = \alpha_{n+1} - \alpha_n = \pi/p$, then (1) becomes

$$f(x) = \frac{1}{2\pi}\left(\int_{-p}^{p} f(t)\, dt\right)\Delta\alpha + \frac{1}{\pi}\sum_{n=1}^{\infty}\left[\left(\int_{-p}^{p} f(t)\cos\alpha_n t\, dt\right)\cos\alpha_n x + \left(\int_{-p}^{p} f(t)\sin\alpha_n t\, dt\right)\sin\alpha_n x\right]\Delta\alpha. \qquad (2)$$

We now expand the interval $(-p, p)$ by letting $p \to \infty$. Since $p \to \infty$ implies that $\Delta\alpha \to 0$, the limit of (2) has the form $\lim_{\Delta\alpha \to 0} \sum_{n=1}^{\infty} F(\alpha_n)\Delta\alpha$, which is suggestive of the definition of the integral $\int_0^\infty F(\alpha)\, d\alpha$. Thus if $\int_{-\infty}^\infty f(t)\, dt$ exists, the limit of the first term in (2) is zero and the limit of the sum becomes

$$f(x) = \frac{1}{\pi} \int_0^\infty \left[\left(\int_{-\infty}^\infty f(t) \cos \alpha t \, dt \right) \cos \alpha x + \left(\int_{-\infty}^\infty f(t) \sin \alpha t \, dt \right) \sin \alpha x \right] d\alpha. \quad (3)$$

The result given in (3) is called the **Fourier integral** of f on $(-\infty, \infty)$. As the following summary shows, the basic structure of the Fourier integral is reminiscent of that of a Fourier series.

DEFINITION 15.1 Fourier Integral

The **Fourier integral** of a function f defined on the interval $(-\infty, \infty)$ is given by

$$f(x) = \frac{1}{\pi} \int_0^\infty [A(\alpha) \cos \alpha x + B(\alpha) \sin \alpha x]\, d\alpha, \quad (4)$$

where

$$A(\alpha) = \int_{-\infty}^\infty f(x) \cos \alpha x \, dx \quad (5)$$

$$B(\alpha) = \int_{-\infty}^\infty f(x) \sin \alpha x \, dx. \quad (6)$$

■ **Convergence of a Fourier Integral** Sufficient conditions under which a Fourier integral converges to $f(x)$ are similar to, but slightly more restrictive than, the conditions for a Fourier series.

THEOREM 15.1 Conditions for Convergence

Let f and f' be piecewise continuous on every finite interval, and let f be absolutely integrable on $(-\infty, \infty)$.* Then the Fourier integral of f on the interval converges to $f(x)$ at a point of continuity. At a point of discontinuity the Fourier integral will converge to the average

$$\frac{f(x+) + f(x-)}{2},$$

where $f(x+)$ and $f(x-)$ denote the limit of f at x from the right and from the left, respectively.

Example 1 **Fourier Integral Representation**

Find the Fourier integral representation of the function

$$f(x) = \begin{cases} 0, & x < 0 \\ 1, & 0 < x < 2 \\ 0, & x > 2. \end{cases}$$

* This means that the integral $\int_{-\infty}^\infty |f(x)|\, dx$ converges.

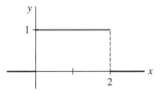

Figure 15.6

SOLUTION The function, whose graph is shown in Figure 15.6, satisfies the hypotheses of Theorem 15.1. Hence from (5) and (6) we have at once

$$A(\alpha) = \int_{-\infty}^{\infty} f(x) \cos \alpha x \, dx$$

$$= \int_{-\infty}^{0} f(x) \cos \alpha x \, dx + \int_{0}^{2} f(x) \cos \alpha x \, dx + \int_{2}^{\infty} f(x) \cos \alpha x \, dx$$

$$= \int_{0}^{2} \cos \alpha x \, dx = \frac{\sin 2\alpha}{\alpha}$$

$$B(\alpha) = \int_{-\infty}^{\infty} f(x) \sin \alpha x \, dx = \int_{0}^{2} \sin \alpha x \, dx = \frac{1 - \cos 2\alpha}{\alpha}.$$

Substituting these coefficients into (4) then gives

$$f(x) = \frac{1}{\pi} \int_{0}^{\infty} \left[\left(\frac{\sin 2\alpha}{\alpha} \right) \cos \alpha x + \left(\frac{1 - \cos 2\alpha}{\alpha} \right) \sin \alpha x \right] d\alpha.$$

When we use trigonometric identities, the last integral simplifies to

$$f(x) = \frac{2}{\pi} \int_{0}^{\infty} \frac{\sin \alpha \cos \alpha(x - 1)}{\alpha} \, d\alpha. \tag{7} \;\square$$

The Fourier integral can be used to evaluate integrals. For example, at $x = 1$ it follows from Theorem 15.1 that (7) converges to $f(1)$; that is,

$$\int_{0}^{\infty} \frac{\sin \alpha}{\alpha} \, d\alpha = \frac{\pi}{2}.$$

The latter result is worthy of special note since it cannot be obtained in the "usual" manner; the integrand $(\sin x)/x$ does not possess an antiderivative that is an elementary function.

■ **Cosine and Sine Integrals** When f is an even function on the interval $(-\infty, \infty)$, then the product $f(x) \cos \alpha x$ is also an even function whereas $f(x) \sin \alpha x$ is an odd function. As a consequence of property (g) of Section 12.3, $B(\alpha) = 0$, and so (4) becomes

$$f(x) = \frac{2}{\pi} \int_{0}^{\infty} \left(\int_{0}^{\infty} f(t) \cos \alpha t \, dt \right) \cos \alpha x \, d\alpha.$$

Here we have also used property (f) of Section 12.3 to write

$$\int_{-\infty}^{\infty} f(t) \cos \alpha t \, dt = 2 \int_{0}^{\infty} f(t) \cos \alpha t \, dt.$$

Similarly, when f is an odd function on $(-\infty, \infty)$, products $f(x) \cos \alpha x$ and $f(x) \sin \alpha x$ are odd and even functions, respectively. Therefore $A(\alpha) = 0$ and

$$f(x) = \frac{2}{\pi} \int_{0}^{\infty} \left(\int_{0}^{\infty} f(t) \sin \alpha t \, dt \right) \sin \alpha x \, d\alpha.$$

We summarize in the following definition.

DEFINITION 15.2 Fourier Cosine and Sine Integrals

(*i*) The Fourier integral of an even function on the interval $(-\infty, \infty)$ is the **cosine integral**

$$f(x) = \frac{2}{\pi} \int_0^\infty A(\alpha) \cos \alpha x \, d\alpha, \tag{8}$$

where

$$A(\alpha) = \int_0^\infty f(x) \cos \alpha x \, dx. \tag{9}$$

(*ii*) The Fourier integral of an odd function on the interval $(-\infty, \infty)$ is the **sine integral**

$$f(x) = \frac{2}{\pi} \int_0^\infty B(\alpha) \sin \alpha x \, d\alpha, \tag{10}$$

where

$$B(\alpha) = \int_0^\infty f(x) \sin \alpha x \, dx. \tag{11}$$

Example 2 Cosine Integral Representation

Find the Fourier integral representation of the function

$$f(x) = \begin{cases} 1, & |x| < a \\ 0, & |x| > a. \end{cases}$$

SOLUTION It is apparent from Figure 15.7 that f is an even function. Hence we represent f by the Fourier cosine integral (8). From (9) we obtain

$$A(\alpha) = \int_0^\infty f(x) \cos \alpha x \, dx = \int_0^a f(x) \cos \alpha x \, dx + \int_a^\infty f(x) \cos \alpha x \, dx$$

$$= \int_0^a \cos \alpha x \, dx = \frac{\sin a\alpha}{\alpha},$$

and so

$$f(x) = \frac{2}{\pi} \int_0^\infty \frac{\sin a\alpha \cos \alpha x}{\alpha} \, d\alpha. \qquad \Box$$

Figure 15.7

The integrals (8) and (10) can be used when f is neither odd nor even and defined only on the half-line $(0, \infty)$. In this case (8) represents f on the interval $(0, \infty)$ and its even (but not periodic) extension to $(-\infty, 0)$, whereas (10) represents f on $(0, \infty)$ and its odd extension to the interval $(-\infty, 0)$. The next example illustrates this concept.

Example 3 Cosine and Sine Integral Representations

Represent $f(x) = e^{-x}$, $x > 0$
(a) by a cosine integral; (b) by a sine integral.

SOLUTION The graph of the function is given in Figure 15.8.
(a) Using integration by parts, we find

$$A(\alpha) = \int_0^\infty e^{-x} \cos \alpha x \, dx = \frac{1}{1 + \alpha^2}.$$

Figure 15.8

(a) Cosine integral

(b) Sine integral

Figure 15.9

Therefore the cosine integral of f is

$$f(x) = \frac{2}{\pi} \int_0^\infty \frac{\cos \alpha x}{1 + \alpha^2} \, d\alpha.$$

(b) Similarly, we have

$$B(\alpha) = \int_0^\infty e^{-x} \sin \alpha x \, dx = \frac{\alpha}{1 + \alpha^2}.$$

The sine integral of f is then

$$f(x) = \frac{2}{\pi} \int_0^\infty \frac{\alpha \sin \alpha x}{1 + \alpha^2} \, d\alpha.$$

Figure 15.9 shows the graphs of the functions and their extensions represented by the two integrals. ❑

■ **Complex Form** The Fourier integral (4) also possesses an equivalent **complex form,** or **exponential form,** that is analogous to the complex form of a Fourier series (see Section 12.4). If (5) and (6) are substituted into (4), then

$$f(x) = \frac{1}{\pi} \int_0^\infty \int_{-\infty}^\infty f(t)[\cos \alpha t \cos \alpha x + \sin \alpha t \sin \alpha x] \, dt \, d\alpha$$

$$= \frac{1}{\pi} \int_0^\infty \int_{-\infty}^\infty f(t) \cos \alpha(t - x) \, dt \, d\alpha$$

$$= \frac{1}{2\pi} \int_{-\infty}^\infty \int_{-\infty}^\infty f(t) \cos \alpha(t - x) \, dt \, d\alpha \qquad (12)$$

$$= \frac{1}{2\pi} \int_{-\infty}^\infty \int_{-\infty}^\infty f(t)[\cos \alpha(t - x) + i \sin \alpha(t - x)] \, dt \, d\alpha \qquad (13)$$

$$= \frac{1}{2\pi} \int_{-\infty}^\infty \int_{-\infty}^\infty f(t)e^{i\alpha(t - x)} \, dt \, d\alpha$$

$$= \frac{1}{2\pi} \int_{-\infty}^\infty \left(\int_{-\infty}^\infty f(t)e^{i\alpha t} \, dt \right) e^{-i\alpha x} \, d\alpha. \qquad (14)$$

We note that (12) follows from the fact that the integrand is an even function of α. In (13) we have simply added zero to the integrand,

$$i \int_{-\infty}^\infty \int_{-\infty}^\infty f(t) \sin \alpha(t - x) \, dt \, d\alpha = 0,$$

because the integrand is an odd function of α. The integral in (14) can be expressed as

$$f(x) = \frac{1}{2\pi} \int_{-\infty}^\infty C(\alpha)e^{-i\alpha x} \, d\alpha, \qquad (15)$$

where

$$C(\alpha) = \int_{-\infty}^\infty f(x)e^{i\alpha x} \, dx. \qquad (16)$$

This latter form of the Fourier integral will be put to use in the next section when we return to the solution of boundary-value problems.

In Problems 1–6, find the Fourier integral representation of the given function.

1. $f(x) = \begin{cases} 0, & x < -1 \\ -1, & -1 < x < 0 \\ 2, & 0 < x < 1 \\ 0, & x > 1 \end{cases}$

2. $f(x) = \begin{cases} 0, & x < \pi \\ 4, & \pi < x < 2\pi \\ 0, & x > 2\pi \end{cases}$

3. $f(x) = \begin{cases} 0, & x < 0 \\ x, & 0 < x < 3 \\ 0, & x > 3 \end{cases}$

4. $f(x) = \begin{cases} 0, & x < 0 \\ \sin x, & 0 \le x \le \pi \\ 0, & x > \pi \end{cases}$

5. $f(x) = \begin{cases} 0, & x < 0 \\ e^{-x}, & x > 0 \end{cases}$

6. $f(x) = \begin{cases} e^x, & |x| < 1 \\ 0, & |x| > 1 \end{cases}$

In Problems 7–12, represent the given function by an appropriate cosine or sine integral.

7. $f(x) = \begin{cases} 0, & x < -1 \\ -5, & -1 < x < 0 \\ 5, & 0 < x < 1 \\ 0, & x > 1 \end{cases}$

8. $f(x) = \begin{cases} 0, & |x| < 1 \\ \pi, & 1 < |x| < 2 \\ 0, & |x| > 2 \end{cases}$

9. $f(x) = \begin{cases} |x|, & |x| < \pi \\ 0, & |x| > \pi \end{cases}$

10. $f(x) = \begin{cases} x, & |x| < \pi \\ 0, & |x| > \pi \end{cases}$

11. $f(x) = e^{-|x|} \sin x$

12. $f(x) = xe^{-|x|}$

In Problems 13–16, find the cosine and sine integral representations of the given function.

13. $f(x) = e^{-kx}, \quad k > 0, x > 0$

14. $f(x) = e^{-x} - e^{-3x}, \quad x > 0$

15. $f(x) = xe^{-2x}, \quad x > 0$

16. $f(x) = e^{-x} \cos x, \quad x > 0$

In Problems 17 and 18, solve the given integral equation for the function f.

17. $\displaystyle\int_0^\infty f(x) \cos \alpha x \, dx = e^{-\alpha}$

18. $\displaystyle\int_0^\infty f(x) \sin \alpha x \, dx = \begin{cases} 1, & 0 < \alpha < 1 \\ 0, & \alpha > 1 \end{cases}$

19. **(a)** Use (7) to show that $\displaystyle\int_0^\infty \frac{\sin 2x}{x} \, dx = \frac{\pi}{2}$.

[*Hint:* α is a dummy variable of integration.]

(b) Show in general that, for $k > 0$, $\displaystyle\int_0^\infty \frac{\sin kx}{x} \, dx = \frac{\pi}{2}$.

20. Use the complex form (15) to find the Fourier integral representation of $f(x) = e^{-|x|}$. Show that the result is the same as that obtained from (8).

15.4 Fourier Transforms

■ **Transform Pairs** The Laplace transform $F(s)$ of a function $f(t)$ is defined by an integral, but up to now we have been using the symbolic representation $f(t) = \mathcal{L}^{-1}\{F(s)\}$ to denote the inverse Laplace transform of $F(s)$. Actually, the inverse Laplace transform is also an integral. If

$$\mathcal{L}\{f(t)\} = \int_0^\infty e^{-st} f(t) \, dt = F(s), \tag{1}$$

then

$$\mathcal{L}^{-1}\{F(s)\} = \frac{1}{2\pi i} \int_{\gamma - i\infty}^{\gamma + i\infty} e^{st} F(s) \, ds = f(t). \tag{2}$$

The last integral is called a **contour integral**; its evaluation requires the use of complex variables (see Chapter 17). The point is this: integral transforms appear in **transform pairs**. If $f(x)$ is transformed into $F(\alpha)$ by an **integral transform**

$$F(\alpha) = \int_a^b f(x) K(\alpha, x) \, dx, \tag{3}$$

then the function f can be recovered by another integral transform

$$f(x) = \int_c^d F(\alpha) H(\alpha, x) \, d\alpha, \tag{4}$$

called the **inverse transform.** The functions K and H are called the **kernels** of their respective transforms. We identify $K(s,t) = e^{-st}$ as the kernel of the Laplace transform and $H(s,t) = e^{st}/2\pi i$ as the kernel of the inverse Laplace transform.

■ **Fourier Transform Pairs** The Fourier integral is the source of three new integral transforms. From (8) and (9), (10) and (11), and (15) and (16) of the preceding section we are prompted to define the following **Fourier transform pairs.**

DEFINITION 15.3 Fourier Transform Pairs

(*i*) Fourier transform:
$$\mathscr{F}\{f(x)\} = \int_{-\infty}^{\infty} f(x)e^{i\alpha x}\, dx = F(\alpha) \tag{5}$$

Inverse Fourier transform:
$$\mathscr{F}^{-1}\{F(\alpha)\} = \frac{1}{2\pi}\int_{-\infty}^{\infty} F(\alpha)e^{-i\alpha x}\, d\alpha = f(x) \tag{6}$$

(*ii*) Fourier sine transform:
$$\mathscr{F}_s\{f(x)\} = \int_{0}^{\infty} f(x)\sin \alpha x\, dx = F(\alpha) \tag{7}$$

Inverse Fourier sine transform:
$$\mathscr{F}_s^{-1}\{F(\alpha)\} = \frac{2}{\pi}\int_{0}^{\infty} F(\alpha)\sin \alpha x\, d\alpha = f(x) \tag{8}$$

(*iii*) Fourier cosine transform:
$$\mathscr{F}_c\{f(x)\} = \int_{0}^{\infty} f(x)\cos \alpha x\, dx = F(\alpha) \tag{9}$$

Inverse Fourier cosine transform:
$$\mathscr{F}_c^{-1}\{F(\alpha)\} = \frac{2}{\pi}\int_{0}^{\infty} F(\alpha)\cos \alpha x\, d\alpha = f(x) \tag{10}$$

■ **Existence** The conditions under which (5), (7), and (9) exist are more stringent than those for the Laplace transform. For example, you should verify that $\mathscr{F}\{1\}$, $\mathscr{F}_s\{1\}$, and $\mathscr{F}_c\{1\}$ do not exist. Sufficient conditions for existence are that f be absolutely integrable on the appropriate interval and that f and f' be piecewise continuous on every finite interval.

■ **Operational Properties** Since our immediate goal is to apply these new transforms to boundary-value problems, we need to examine the transforms of derivatives.

Fourier Transform

Suppose that f is continuous and absolutely integrable on the interval $(-\infty, \infty)$ and f' is piecewise continuous on every finite interval. If $f(x) \to 0$ as $x \to \pm\,\infty$, then integration by parts gives

$$\mathscr{F}\{f'(x)\} = \int_{-\infty}^{\infty} f'(x)e^{i\alpha x}\, dx$$

$$= f(x)e^{i\alpha x}\bigg|_{-\infty}^{\infty} - i\alpha \int_{-\infty}^{\infty} f(x)e^{i\alpha x}\, dx$$

$$= -i\alpha \int_{-\infty}^{\infty} f(x)e^{i\alpha x}\, dx;$$

that is,
$$\mathscr{F}\{f'(x)\} = -i\alpha F(\alpha). \tag{11}$$

Similarly, under the added assumptions that f' is continuous on $(-\infty, \infty)$, $f''(x)$ is piecewise continuous on every finite interval, and $f'(x) \to 0$ as $x \to \pm\,\infty$, we have

$$\mathscr{F}\{f''(x)\} = (-i\alpha)^2 \mathscr{F}\{f(x)\} = -\alpha^2 F(\alpha). \tag{12}$$

It is important to be aware that the sine and cosine transforms are not suitable for transforming the first derivative (or, for that matter, any derivative of *odd* order). It is readily shown that

$$\mathscr{F}_s\{f'(x)\} = -\alpha\,\mathscr{F}_c\{f(x)\} \qquad \text{and} \qquad \mathscr{F}_c\{f'(x)\} = \alpha\,\mathscr{F}_s\{f(x)\} - f(0).$$

The difficulty is apparent; the transform of $f'(x)$ is not expressed in terms of the original integral transform.

Fourier Sine Transform

Suppose that f and f' are continuous, f is absolutely integrable on the interval $[0, \infty)$, and f'' is piecewise continuous on every finite interval. If $f \to 0$ and $f' \to 0$ as $x \to \infty$, then

$$\mathscr{F}_s\{f''(x)\} = \int_0^\infty f''(x)\,\sin\,\alpha x\,dx$$

$$= f'(x)\,\sin\,\alpha x\,\Big|_0^\infty - \alpha\int_0^\infty f'(x)\,\cos\,\alpha x\,dx$$

$$= -\alpha\left[f(x)\,\cos\,\alpha x\,\Big|_0^\infty + \alpha\int_0^\infty f(x)\,\sin\,\alpha x\,dx \right]$$

$$= \alpha f(0) - \alpha^2\mathscr{F}_s\{f(x)\};$$

that is, $$\mathscr{F}_s\{f''(x)\} = -\alpha^2 F(\alpha) + \alpha f(0). \tag{13}$$

Fourier Cosine Transform

Under the same assumptions that lead to (9), we find the Fourier cosine transform of $f''(x)$ to be

$$\mathscr{F}_c\{f''(x)\} = -\alpha^2 F(\alpha) - f'(0). \tag{14}$$

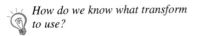

How do we know what transform to use?

A natural question is "How do we know which transform to use on a given boundary-value problem?" Clearly, to use a Fourier transform, the domain of the variable to be eliminated must be $(-\infty, \infty)$. To utilize a sine or cosine transform, the domain of at least one of the variables in the problem must be $[0, \infty)$. But the determining factor in choosing between the sine transform and the cosine transform is the type of boundary condition specified at zero.

In the examples that follow we shall assume without further mention that both u and $\partial u/\partial x$ (or $\partial u/\partial y$) approach zero as $x \to \pm\infty$. This is not a major restriction since these conditions hold in most applications.

Example 1 Using the Fourier Transform

Solve the heat equation $k\dfrac{\partial^2 u}{\partial x^2} = \dfrac{\partial u}{\partial t}$, $-\infty < x < \infty$, $t > 0$, subject to

$$u(x, 0) = f(x), \quad \text{where} \quad f(x) = \begin{cases} u_0, & |x| < 1 \\ 0, & |x| > 1. \end{cases}$$

SOLUTION The problem can be interpreted as finding the temperature $u(x, t)$ in an infinite rod. Since the domain of x is the infinite interval $(-\infty, \infty)$, we use the Fourier transform (5) and define

$$\mathscr{F}\{u(x,t)\} = \int_{-\infty}^{\infty} u(x,t)e^{i\alpha x}\,dx = U(\alpha, t).$$

Transforming the partial differential equation and using (12),

$$\mathscr{F}\left\{k\frac{\partial^2 u}{\partial x^2}\right\} = \mathscr{F}\left\{\frac{\partial u}{\partial t}\right\}$$

yields $\qquad -k\alpha^2 U(\alpha, t) = \dfrac{dU}{dt}$ or $\dfrac{dU}{dt} + k\alpha^2 U(\alpha, t) = 0.$

Solving the last equation gives $U(\alpha, t) = ce^{-k\alpha^2 t}$. Now the transform of the initial condition is

$$\mathscr{F}\{u(x,0)\} = \int_{-\infty}^{\infty} f(x)e^{i\alpha x}\, dx = \int_{-1}^{1} u_0 e^{i\alpha x}\, dx = u_0\,\frac{e^{i\alpha} - e^{-i\alpha}}{i\alpha}.$$

This result is the same as $U(\alpha, 0) = 2u_0\,\dfrac{\sin\alpha}{\alpha}$. Applying this condition to the solution $U(\alpha, t)$ gives $U(\alpha, 0) = c = (2u_0 \sin\alpha)/\alpha$, and so

$$U(\alpha, t) = 2u_0\,\frac{\sin\alpha}{\alpha}\,e^{-k\alpha^2 t}.$$

It then follows from the inversion integral (6) that

$$u(x, t) = \frac{u_0}{\pi}\int_{-\infty}^{\infty}\frac{\sin\alpha}{\alpha}e^{-k\alpha^2 t}e^{-i\alpha x}\, d\alpha.$$

The last expression can be simplified somewhat by Euler's formula $e^{-i\alpha x} = \cos\alpha x - i\sin\alpha x$ and noting that $\displaystyle\int_{-\infty}^{\infty}\frac{\sin\alpha}{\alpha}e^{-k\alpha^2 t}\sin\alpha x\, d\alpha = 0$ since the integrand is an odd function of α. Hence we finally have

$$u(x, t) = \frac{u_0}{\pi}\int_{-\infty}^{\infty}\frac{\sin\alpha\cos\alpha x}{\alpha}e^{-k\alpha^2 t}\, d\alpha. \qquad (15)\;\square$$

It is left to the reader to show that the solution (15) can be expressed in terms of the error function. See Problem 4 in Exercises 15.4.

Example 2 Using the Cosine Transform

The steady-state temperature in a semi-infinite plate is determined from

$$\frac{\partial^2 u}{\partial x^2} + \frac{\partial^2 u}{\partial y^2} = 0, \qquad 0 < x < \pi, \quad y > 0$$

$$u(0, y) = 0, \qquad u(\pi, u) = e^{-y}, \quad y > 0$$

$$\left.\frac{\partial u}{\partial y}\right|_{y=0} = 0, \qquad 0 < x < \pi.$$

Solve for $u(x, y)$.

SOLUTION The domain of the variable y and the prescribed condition at $y = 0$ indicate that the Fourier cosine transform is suitable for the problem. We define

$$\mathscr{F}_c\{u(x, y)\} = \int_{0}^{\infty} u(x, y)\cos\alpha y\, dy = U(x, \alpha).$$

In view of (14), $\qquad \mathscr{F}_c\left\{\dfrac{\partial^2 u}{\partial x^2}\right\} + \mathscr{F}_c\left\{\dfrac{\partial^2 u}{\partial y^2}\right\} = \mathscr{F}_c\{0\}$

becomes

$$\frac{d^2U}{dx^2} - \alpha^2 U(x, \alpha) - u_y(x, 0) = 0 \qquad \text{or} \qquad \frac{d^2U}{dx^2} - \alpha^2 U = 0.$$

Since the domain of x is a finite interval, we choose to write the solution of the ordinary differential equation as

$$U(x, \alpha) = c_1 \cosh \alpha x + c_2 \sinh \alpha x. \tag{16}$$

Now $\mathscr{F}_c\{u(0, y)\} = \mathscr{F}_c\{0\}$ and $\mathscr{F}_c\{u(\pi, y)\} = \mathscr{F}_c\{e^{-y}\}$ are in turn equivalent to

$$U(0, \alpha) = 0 \qquad \text{and} \qquad U(\pi, \alpha) = \frac{1}{1 + \alpha^2}.$$

When we apply these latter conditions, the solution (16) gives $c_1 = 0$ and $c_2 = 1/[(1 + \alpha^2) \sinh \alpha \pi]$. Therefore

$$U(x, \alpha) = \frac{\sinh \alpha x}{(1 + \alpha^2) \sinh \alpha \pi},$$

and so from (10) we arrive at

$$u(x, y) = \frac{2}{\pi} \int_0^\infty \frac{\sinh \alpha x}{(1 + \alpha^2) \sinh \alpha \pi} \cos \alpha y \, d\alpha. \tag{17} \square$$

Had $u(x, 0)$ been given in Example 2 rather than $u_y(x, 0)$, then the sine transform would have been appropriate.

Answers to odd-numbered problems begin on page A-63.

In Problems 1–21, use the Fourier integral transforms of this section to solve the given boundary-value problem. Make assumptions about boundedness where necessary.

1. $k\dfrac{\partial^2 u}{\partial x^2} = \dfrac{\partial u}{\partial t}, \quad -\infty < x < \infty, \quad t > 0$

$u(x, 0) = e^{-|x|}, \quad -\infty < x < \infty$

2. $k\dfrac{\partial^2 u}{\partial x^2} = \dfrac{\partial u}{\partial t}, \quad -\infty < x < \infty, \quad t > 0$

$$u(x, 0) = \begin{cases} 0, & x < -1 \\ -100, & -1 < x < 0 \\ 100, & 0 < x < 1 \\ 0, & x > 1 \end{cases}$$

3. Use the result $\mathscr{F}\{e^{-x^2/4p^2}\} = 2\sqrt{\pi} p e^{-p^2\alpha^2}$ to solve the boundary-value problem

$$k\frac{\partial^2 u}{\partial x^2} = \frac{\partial u}{\partial t}, \quad -\infty < x < \infty, \quad t > 0$$

$$u(x, 0) = e^{-x^2}, \quad -\infty < x < \infty.$$

4. (a) If $\mathscr{F}\{f(x)\} = F(\alpha)$ and $\mathscr{F}\{g(x)\} = G(\alpha)$, then the **convolution theorem** for the Fourier transform is given by

$$\int_{-\infty}^\infty f(\tau)g(x - \tau) \, d\tau = \mathscr{F}^{-1}\{F(\alpha)G(\alpha)\}.$$

Use this result and $\mathscr{F}\{e^{-x^2/4p^2}\} = 2\sqrt{\pi} p e^{-p^2\alpha^2}$ to show that a solution of the boundary-value problem

$$k\frac{\partial^2 u}{\partial x^2} = \frac{\partial u}{\partial t}, \quad -\infty < x < \infty, \quad t > 0$$

$$u(x, 0) = f(x), \quad -\infty < x < \infty$$

is given by

$$u(x, t) = \frac{1}{2\sqrt{k\pi t}} \int_{-\infty}^\infty f(\tau)e^{-(x - \tau)^2/4kt} \, d\tau.$$

(b) Use the change of variables $u = (x - \tau)/2\sqrt{kt}$, Problem 9 in Exercises 15.1, and part **(a)** to show that the solution of Example 1 is

$$u(x, t) = \frac{u_0}{2}\left[\text{erf}\left(\frac{x + 1}{2\sqrt{kt}}\right) - \text{erf}\left(\frac{x - 1}{2\sqrt{kt}}\right)\right].$$

5. Computer Lab Assignment Assume $u_0 = 100$ and $k = 1$ in the solution in part **(b)** of Problem 4. Use a CAS to obtain the graph of $u(x, t)$ over the rectangular region $-4 \leq x \leq 4$, $0 \leq t \leq 6$. Use a 2D plot to superimpose the graphs of $u(x, t)$ for $t = 0.05, 0.125, 0.5, 1, 2, 4, 6,$ and 15 on the interval $-4 \leq x \leq 4$. Use the graphs to conjecture the values of $\lim_{t \to \infty} u(x, t)$ and $\lim_{x \to \infty} u(x, t)$. Then prove these results analytically using the properties of erf(x).

6. (a) Find the temperature $u(x, t)$ in a semi-infinite rod if $u(0, t) = u_0$, $t > 0$, and $u(x, 0) = 0$, $x > 0$.

(b) Use the result $\int_0^\infty \frac{\sin \alpha x}{\alpha} \, d\alpha = \frac{\pi}{2}$, $x > 0$, to show that the solution in part **(a)** can be written as

$$u(x, t) = u_0 - \frac{2u_0}{\pi} \int_0^\infty \frac{\sin \alpha x}{\alpha} e^{-k\alpha^2 t} \, d\alpha.$$

7. Find the temperature $u(x, t)$ in a semi-infinite rod if $u(0, t) = 0$, $t > 0$, and

$$u(x, 0) = \begin{cases} 1, & 0 < x < 1 \\ 0, & x > 1. \end{cases}$$

8. Solve Problem 5 if the condition at the left boundary is

$$\frac{\partial u}{\partial x}\bigg|_{x=0} = -A, \qquad t > 0.$$

9. Solve Problem 7 if the end $x = 0$ is insulated.

10. Find the temperature $u(x, t)$ in a semi-infinite rod if $u(0, t) = 1$, $t > 0$, and $u(x, 0) = e^{-x}$, $x > 0$.

11. (a) $a^2 \dfrac{\partial^2 u}{\partial x^2} = \dfrac{\partial^2 u}{\partial t^2}$, $\quad -\infty < x < \infty$, $\quad t > 0$

$$u(x, 0) = f(x), \qquad \frac{\partial u}{\partial t}\bigg|_{t=0} = g(x), \quad -\infty < x < \infty$$

(b) If $g(x) = 0$, show that the solution of part **(a)** can be written as $u(x, t) = \frac{1}{2}[f(x + at) + f(x - at)]$.

12. Find the displacement $u(x, t)$ of a semi-infinite string if

$$u(0, t) = 0, \qquad t > 0$$

$$u(x, 0) = xe^{-x}, \qquad \frac{\partial u}{\partial t}\bigg|_{t=0} = 0, \quad x > 0.$$

13. Solve the problem in Example 2 if the boundary conditions at $x = 0$ and $x = \pi$ are reversed: $u(0, y) = e^{-y}$, $u(\pi, y) = 0$, $y > 0$.

14. Solve the problem in Example 2 if the boundary condition at $y = 0$ is $u(x, 0) = 1$, $0 < x < \pi$.

15. Find the steady-state temperature $u(x, y)$ in a plate defined by $x \geq 0$, $y \geq 0$ if the boundary $x = 0$ is insulated and, at $y = 0$,

$$u(x, 0) = \begin{cases} 50, & 0 < x < 1 \\ 0, & x > 1. \end{cases}$$

16. Solve Problem 15 if the boundary condition at $x = 0$ is $u(0, y) = 0$, $y > 0$.

17. $\dfrac{\partial^2 u}{\partial x^2} + \dfrac{\partial^2 u}{\partial y^2} = 0$, $\quad x > 0$, $\quad 0 < y < 2$

$$u(0, y) = 0, \quad 0 < y < 2$$

$$u(x, 0) = f(x), \quad u(x, 2) = 0, \quad x > 0$$

18. $\dfrac{\partial^2 u}{\partial x^2} + \dfrac{\partial^2 u}{\partial y^2} = 0$, $\quad 0 < x < \pi$, $\quad y > 0$

$$u(0, y) = f(y), \qquad \frac{\partial u}{\partial x}\bigg|_{x=\pi} = 0, \quad y > 0$$

$$\frac{\partial u}{\partial y}\bigg|_{y=0} = 0, \quad 0 < x < \pi$$

In Problems 19 and 20, find the steady-state temperature in the plate given in the figure. [*Hint:* One way of proceeding is to express Problems 19 and 20 as two- and three-boundary-value problems, respectively. Use the superposition principle (see Section 13.5).]

19.

Figure 15.10

20.

Figure 15.11

21. Use the transform $\mathscr{F}\{e^{-x^2/4p^2}\}$ given in Problem 3 to find the steady-state temperature in the infinite strip shown in Figure 15.12.

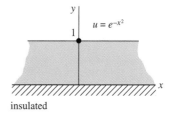

Figure 15.12

22. The solution of Problem 16 can be integrated. Use entries 42 and 43 of the table in Appendix III to show that

$$u(x, y) = \frac{100}{\pi}\left[\arctan \frac{x}{y} - \frac{1}{2}\arctan \frac{x+1}{y} - \frac{1}{2}\arctan \frac{x-1}{y}\right].$$

15.5 Fast Fourier Transform

■ **Discrete Fourier Transform** Consider a function f defined on the interval $[0, 2p]$. From (11) of Section 12.4 we saw that f can be written in a complex Fourier series,

$$f(x) = \sum_{n=-\infty}^{\infty} c_n e^{in\omega x} \quad \text{where} \quad c_n = \frac{1}{2p} \int_0^{2p} f(x) e^{-in\omega x} \, dx, \tag{1}$$

where the $\omega = 2\pi/2p = \pi/p$ is the fundamental angular frequency and $2p$ is the fundamental period. In the *discrete case,* however, the input is f_0, f_1, f_2, \ldots, which are the values of the function f at equally spaced points $x = nT$, $n = 0, 1, 2, \ldots$. The number T is called the **sampling rate** or the length of the sampling interval.* If f is continuous at T, then the **sample** of f at T is defined to be the product $f(x)\delta(x - T)$, where $\delta(x - T)$ is the Dirac delta function (see Section 4.5). We can then represent this discrete version of f, or **discrete signal,** as the sum of unit impulses acting on the function at $x = nT$:

$$\sum_{n=-\infty}^{\infty} f(x)\delta(x - nT). \tag{2}$$

If we apply the Fourier transform to the discrete signal (2), we have

$$\int_{-\infty}^{\infty} \sum_{n=-\infty}^{\infty} f(x)\delta(x - nT) e^{i\alpha x} \, dx. \tag{3}$$

By the sifting property of the Dirac delta function (see the Remarks at the end of Section 4.5), (3) is the same as

$$F(\alpha) = \sum_{n=-\infty}^{\infty} f(nT) e^{i\alpha nT}. \tag{4}$$

The expression $F(\alpha)$ in (4) is called the **discrete Fourier transform** (DFT) of the function f. We often write the coefficients $f(nT)$ in (4) as $f(n)$ or f_n. It is also worth noting that since $e^{i\alpha x}$ is periodic in α and $e^{i\alpha T} = e^{i(\alpha T + 2\pi)} = e^{i(\alpha + 2\pi/T)T}$, we only need to consider the function for α in $[0, 2\pi/T]$. Let $N = 2\pi/T$. This places x in the interval $[0, 2\pi]$. So, because we sample over one period, the sum in (4) is actually finite.

Now consider the functional values $f(x)$ at N equally spaced points, $x = nT$, $n = 0, 1, 2, \ldots, N - 1$, in the interval $[0, 2\pi]$, that is, $f_0, f_1, f_2, \ldots, f_{N-1}$. The (finite) discrete Fourier series $f(x) = \sum_{n=-\infty}^{\infty} c_n e^{inx}$ using these N terms gives us

$$
\begin{aligned}
f_0 &= c_0 + c_1 e^{i1 \cdot 0} &&+ c_2 e^{i2 \cdot 0} + \cdots &&+ c_{N-1} e^{i(N-1)\cdot 0} \\
f_1 &= c_0 + c_1 e^{i2\pi/N} &&+ c_2 e^{i4\pi/N} + \cdots &&+ c_{N-1} e^{i2(N-1)\pi/N} \\
f_2 &= c_0 + c_1 e^{i4\pi/N} &&+ c_2 e^{i8\pi/N} + \cdots &&+ c_{N-1} e^{i4(N-1)\pi/N} \\
&\vdots &&&& \vdots \\
f_{N-1} &= c_0 + c_1 e^{i2(N-1)\pi/N} &&+ c_2 e^{i4(N-1)\pi/N} + \cdots + c_{N-1} e^{i2(N-1)^2\pi/N}.
\end{aligned}
$$

* Note that the symbol T used here does not have the same meaning as in Section 12.4.

If we let $\omega_n = e^{i2\pi/n} = \cos\dfrac{2\pi}{n} + i \sin\dfrac{2\pi}{n}$ and use the usual laws of exponents, this system of equations is the same as

$$
\begin{aligned}
f_0 &= c_0 + c_1 & + c_2 + \cdots & & + c_{N-1} \\
f_1 &= c_0 + c_1\omega_N & + c_2\omega_N^2 + \cdots & & + c_{N-1}\omega_N^{N-1} \\
f_2 &= c_0 + c_1\omega_N^2 & + c_2\omega_N^4 + \cdots & & + c_{N-1}\omega_N^{2(N-1)} \\
&\ \ \vdots & & & \qquad\quad \vdots \\
f_{N-1} &= c_0 + c_1\omega_N^{N-1} + c_2\omega_N^{2(N-1)} + \cdots + c_{N-1}\omega_N^{(N-1)^2}.
\end{aligned}
\tag{5}
$$

If we use matrix notation (see Sections 8.1 and 8.2), then (5) is

$$
\begin{pmatrix} f_0 \\ f_1 \\ f_2 \\ \vdots \\ f_{N-1} \end{pmatrix}
=
\begin{pmatrix}
1 & 1 & 1 & \cdots & 1 \\
1 & \omega_N & \omega_N^2 & & \omega_N^{N-1} \\
1 & \omega_N^2 & \omega_N^4 & & \omega_N^{2(N-1)} \\
\vdots & & & & \vdots \\
1 & \omega_N^{N-1} & \omega_N^{2(N-1)} & \cdots & \omega_N^{(N-1)^2}
\end{pmatrix}
\begin{pmatrix} c_0 \\ c_1 \\ c_2 \\ \vdots \\ c_{N-1} \end{pmatrix}.
\tag{6}
$$

Let the $N \times N$ matrix in (6) by denoted by the symbol \mathbf{F}_N. Given the inputs f_0, f_1, f_2, \ldots, f_{N-1}, is there an easy way to find the Fourier coefficients $c_0, c_1, c_2, \ldots, c_{N-1}$? If $\overline{\mathbf{F}}_N$ is the matrix consisting of the complex conjugates of the entries of \mathbf{F}_N and if \mathbf{I} denotes the $N \times N$ identity matrix, then we have

$$
\mathbf{F}_N\overline{\mathbf{F}}_N = \overline{\mathbf{F}}_N\mathbf{F}_N = N\mathbf{I} \qquad \text{and so} \qquad \mathbf{F}_N^{-1} = \frac{1}{N}\overline{\mathbf{F}}_N.
$$

It follows from (6) and the last equation that

$$
\begin{pmatrix} c_0 \\ c_1 \\ c_2 \\ \vdots \\ c_{N-1} \end{pmatrix}
= \frac{1}{N}\overline{\mathbf{F}}_N
\begin{pmatrix} f_0 \\ f_1 \\ f_2 \\ \vdots \\ f_{N-1} \end{pmatrix}.
$$

■ **Discrete Transform Pair** Recall from Section 15.4 that in the Fourier transform pair we use a function $f(x)$ as input and compute the coefficients that give the amplitude for each frequency k (c_k in the case of periodic functions of period 2π) or we compute the coefficients that give the amplitude for each frequency $\alpha(F(\alpha)$ in the case of nonperiodic functions).

Also, given these frequencies and coefficients, we could reconstruct the original function $f(x)$. In the discrete case, we use a sample of N values of the function $f(x)$ as input and compute the coefficients that give the amplitude for each sampled frequency. Given these frequencies and coefficients, it is possible to reconstruct the n sampled values of $f(x)$. The transform pair, the **discrete Fourier transform pair,** is given by

$$
\mathbf{c} = \frac{1}{N}\overline{\mathbf{F}}_N\mathbf{f} \qquad \text{and} \qquad \mathbf{f} = \mathbf{F}_N\mathbf{c},
\tag{7}
$$

where
$$
\mathbf{c} = \begin{pmatrix} c_0 \\ c_1 \\ c_2 \\ \vdots \\ c_{N-1} \end{pmatrix}
\qquad \text{and} \qquad
\mathbf{f} = \begin{pmatrix} f_0 \\ f_1 \\ f_2 \\ \vdots \\ f_{N-1} \end{pmatrix}.
$$

Example 1 Discrete Fourier Transform

Let $N = 4$ so that the input is f_0, f_1, f_2, f_3 at the four points $x = 0, \pi/2, \pi, 3\pi/2$. Since $\omega_4 = e^{i\pi/2} = \cos\left(\dfrac{\pi}{2}\right) + i \sin\left(\dfrac{\pi}{2}\right) = i$, the matrix \mathbf{F}_4 is

$$\mathbf{F}_4 = \begin{pmatrix} 1 & 1 & 1 & 1 \\ 1 & i & -1 & -i \\ 1 & -1 & 1 & -1 \\ 1 & -i & -1 & i \end{pmatrix}.$$

Hence from (7) the Fourier coefficients are given by $\mathbf{c} = \frac{1}{4}\, \overline{\mathbf{F}}_4 \mathbf{f}$:

$$\begin{pmatrix} c_0 \\ c_1 \\ c_2 \\ c_3 \end{pmatrix} = \frac{1}{4} \begin{pmatrix} 1 & 1 & 1 & 1 \\ 1 & -i & -1 & i \\ 1 & -1 & 1 & -1 \\ 1 & i & -1 & -i \end{pmatrix} \begin{pmatrix} f_0 \\ f_1 \\ f_2 \\ f_3 \end{pmatrix}.$$

If we let f_0, f_1, f_2, f_3 be 0, 2, 4, 6, respectively, we find from the preceding matrix product that

$$\mathbf{c} = \begin{pmatrix} c_0 \\ c_1 \\ c_2 \\ c_3 \end{pmatrix} = \begin{pmatrix} 3 \\ -1 + i \\ -1 \\ -1 - i \end{pmatrix}.$$

Note that we obtain the same result using (4), that is, $F(\alpha) = \displaystyle\sum_{n=0}^{3} f(nT)e^{i\alpha nT}$, with $T = \pi/2$, $\alpha = 0, 1, 2, 3$. The graphs of $|c_n|$, $n = 0, 1, 2, 3$, or equivalently $|F(\alpha)|$ for $\alpha = 0, 1, 2, 3$, are given in Figure 15.13. ❑

Figure 15.13

Finding the coefficients involves multiplying by matrices \mathbf{F}_n and $\overline{\mathbf{F}}_n$. Because of the nature of these matrices, these multiplications can be done in a very computationally efficient manner, using the **Fast Fourier Transform** (FFT), which is discussed later in this section.

■ Heat Equation and Discrete Fourier Series

If the function f in the initial-value problem

$$k\frac{\partial^2 u}{\partial x^2} = \frac{\partial u}{\partial t}, \qquad -\infty < x < \infty, \quad t > 0 \tag{8}$$

$$u(x,\, 0) = f(x)$$

is periodic with period 2π, the solution can be written in terms of a Fourier series for $f(x)$. We can also approximate this solution with a finite sum

$$u(x,\, t) = \sum_{k=0}^{n-1} c_k(t)e^{ikx}.$$

If we examine both sides of the one-dimensional heat equation in (8), we see that

$$\frac{\partial u}{\partial t} = \sum_{j=0}^{n-1} \frac{dc_j}{dt}\, e^{ijx}$$

and

$$k\frac{\partial^2 u}{\partial x^2} = k\sum_{j=0}^{n-1} c_j(t)(ij)^2 e^{ijx},$$

since $\dfrac{d^2 e^{ijx}}{dx^2} = (ij)^2 e^{ijx}$.

Equating these last two expressions, we have the first-order DE

$$\frac{dc_j}{dt} = -kj^2 c_j(t) \qquad \text{with solution} \qquad c_j(t) = c_j(0)\, e^{-kj^2 t}.$$

The final task is to find the values $c_j(0)$. However, recall that $u(x,\ t) = \displaystyle\sum_{k=0}^{n-1} c_k(t) e^{ikx}$ and $u(x,\ 0) = f(x)$, so $c_j(0)$ are the coefficients of the discrete Fourier series of $f(x)$! Compare this with Section 13.3.

■ Heat Equation and Discrete Fourier Transform

The initial-value problem (8) can be interpreted as the mathematical model for the temperature $u(x,\ t)$ in an infinitely long bar. In Section 15.4 we saw that we can solve (8) using the Fourier transform and that the solution $u(x,\ t)$ depends on the Fourier transform $F(\alpha)$ of $f(x)$ (see page 752). We can approximate $F(\alpha)$ by taking a different look at the discrete Fourier transform.

First we approximate values of the transform by discretizing the integral $\mathscr{F}\{f(x)\} = F(\alpha) = \int_{-\infty}^{\infty} f(x) e^{i\alpha x}\, dx$. Consider an interval $[a,\ b]$. Let $f(x)$ be given at n equally spaced points

$$x_j = a + \frac{b-a}{n}\, j, \qquad j = 0,\ 1,\ 2, ...,\ n-1.$$

Now approximate:

$$F(\alpha) \approx \frac{b-a}{n} \sum_{j=0}^{n-1} f(x_j) e^{i\alpha x_j}$$

$$= \frac{b-a}{n} \sum_{j=0}^{n-1} f\!\left(a + \frac{b-a}{n}\, j\right) e^{i\alpha x_j}$$

$$= \frac{b-a}{n} \sum_{j=0}^{n-1} f\!\left(a + \frac{b-a}{n}\, j\right) e^{i\alpha a} e^{i\alpha \frac{b-a}{n} j}$$

$$= \frac{b-a}{n}\, e^{i\alpha a} \sum_{j=0}^{n-1} f\!\left(a + \frac{b-a}{n}\, j\right) e^{i\alpha \frac{b-a}{n} j}.$$

If we now choose a convenient value for α, say $\dfrac{2\pi M}{b-a}$ with M an integer, we have

$$F\!\left(\frac{2\pi M}{b-a}\right) \approx \frac{b-a}{n}\, e^{i\frac{2\pi M a}{b-a}} \sum_{j=0}^{n-1} f\!\left(a + \frac{b-a}{n}\, j\right) e^{i\frac{2\pi j M}{n}}$$

$$= \frac{b-a}{n}\, e^{i\frac{2\pi M a}{b-a}} \sum_{j=0}^{n-1} f\!\left(a + \frac{b-a}{n}\, j\right) \omega_n^{jM},$$

where recall that $\omega_n = e^{i2\pi/n}$. This is a numerical approximation to the Fourier transform of $f(x)$ evaluated at points $\dfrac{2\pi M}{b-a}$ with M an integer.

Example 2 Example 1, Section 15.4—Revisited

Recall from Example 1 in Section 15.4 (with $u_0 = 1$) that the Fourier transform of a rectangular pulse defined by

$$f(x) = \begin{cases} 1, & |x| < 1 \\ 0, & |x| > 1 \end{cases}$$

is

$$F(\alpha) = \frac{2 \sin \alpha}{\alpha}.$$

The frequency spectrum is the graph of $|F(\alpha)|$ versus α given in Figure 15.14(a). Using $n = 16$ equally spaced points between $a = -2$ and $b = 2$, and M running from -6 to 6, we get the discrete Fourier transform of $f(x)$, superimposed over the graph of $|F(\alpha)|$, in Figure 15.14(b).

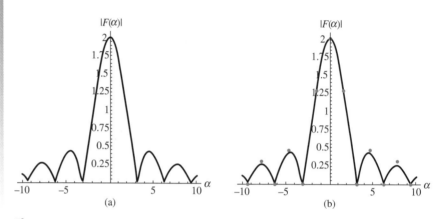

(a) (b)

Figure 15.14

■ **Aliasing** A problem known as aliasing may appear whenever one is sampling data at equally spaced intervals. If you have ever seen a motion picture where rotating wheels appear to be rotating slowly (or even backwards!), you have experienced aliasing. The wheels may rotate at a high rate, but because the frames in a motion picture are "sampled" at equally spaced intervals, we see a low rate of rotation.

Graphing calculators also suffer from aliasing due to the way they sample points to create graphs. For example, plot the trigonometric function $y = \sin 20\pi x$ with frequency 10 on a Texas Instrument TI-92 and you get the nice graph in Figure 15.15(a). At higher frequencies, say $y = \sin 100\pi x$ with frequency 50, you get the correct amount of cycles, but the amplitudes of the graph in Figure 15.15(b) are clearly not 1.

(a) $y = \sin 20\pi x$; x range: $[0, 1]$; y range: $[-1, 1]$

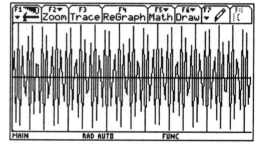

(b) $y = \sin 100\pi x$; x range: $[0, 1]$; y range: $[-1, 1]$

Figure 15.15 TI-92

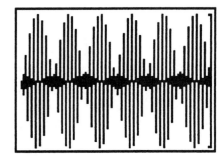

(a) $y = \sin 20\pi x$; x range: [0, 1]; y range: [−1, 1] (b) $y = \sin 100\pi x$; x range: [0, 1]; y range: [−1, 1]

Figure 15.16 TI-83

On a calculator such as the Texas Instruments TI-83 the graphs in Figure 15.16 show aliasing much more clearly.

The problem lies in the fact that $e^{2n\pi i} = \cos 2n\pi + i \sin 2n\pi = 1$ for all integer values of n. The discrete Fourier series cannot distinguish e^{inx} from 1 as these functions are equal at sampled points $x = \dfrac{2k\pi}{n}$. The higher frequency is seen as the lower one. Consider the functions $\cos \dfrac{\pi n}{2}$ and $\cos \dfrac{7\pi n}{2}$. If we sample at the points $n = 0, 1,$ 2, . . . , these two functions appear the same, the lower frequency is assumed, and the amplitudes (Fourier coefficients) associated with the higher frequencies are added in with the amplitude of the low frequency. If these Fourier coefficients at large frequencies are small, however, we do not have a big problem. In the Sampling Theorem below, we will see what can take care of this problem.

■ **Signal Processing** Beyond solving PDEs as we have done earlier, the ideas of this section are useful in **signal processing.** Consider the functions we have been dealing with as **signals** from a source. We would like to reconstruct a signal transmitted by sampling it at discrete points. The problem of calculating an infinite number of Fourier coefficients and summing an infinite series to reconstruct a signal (function) is not practical. A finite sum could be a decent approximation, but certain signals can be reconstructed by a finite number of samples.

> **SAMPLING THEOREM**
>
> If a signal $f(x)$ is **band-limited,** that is, if the range of frequencies of the signal lie in a band $-A < k < A$, then the signal can be reconstructed by sampling two times for every cycle of the highest frequency present; in fact,
>
> $$f(x) = \sum_{n=-\infty}^{\infty} f(\frac{n\pi}{A}) \frac{\sin (Ax - n\pi)}{Ax - n\pi}.$$

To justify this **Sampling Theorem,** consider the Fourier transform $F(\alpha)$ of $f(x)$ as a periodic extension so that $F(\alpha)$ is defined for all values of α, not just those in $-A < k < A$. Using the Fourier transform, we have

$$F(\alpha) = \int_{-\infty}^{\infty} f(x)e^{i\alpha x}\, dx \qquad (9)$$

$$f(x) = \frac{1}{2\pi} \int_{-\infty}^{\infty} F(\alpha)e^{-i\alpha x}\, d\alpha = \frac{1}{2\pi} \int_{-A}^{A} F(\alpha)e^{-i\alpha x}\, d\alpha. \tag{10}$$

Treating $F(\alpha)$ as a periodic extension, the Fourier series for $F(\alpha)$ is

$$F(\alpha) = \sum_{n=-\infty}^{\infty} c_n e^{in\pi\alpha/A} \tag{11}$$

$$c_n = \frac{1}{2A} \int_{-A}^{A} F(\alpha)e^{-in\pi\alpha/A}\, d\alpha. \tag{12}$$

Using (10), note that

$$\frac{\pi}{A} f\left(\frac{n\pi}{A}\right) = \frac{\pi}{A}\frac{1}{2\pi} \int_{-A}^{A} F(\alpha)e^{-i\alpha n\pi/A}\, d\alpha,$$

which by (12) is equal to c_n. Substituting $c_n = \frac{\pi}{A} f\left(\frac{n\pi}{A}\right)$ into (11) yields

$$F(\alpha) = \sum_{n=-\infty}^{\infty} \frac{\pi}{A} f\left(\frac{n\pi}{A}\right) e^{in\pi\alpha/A}.$$

Substituting this expression for $F(\alpha)$ back into (10), we have

$$f(x) = \frac{1}{2\pi} \int_{-A}^{A} \left(\sum_{n=-\infty}^{\infty} \frac{\pi}{A} f\left(\frac{n\pi}{A}\right) e^{in\pi\alpha/A} \right) e^{-i\alpha x}\, d\alpha$$

$$= \frac{1}{2A} \sum_{n=-\infty}^{\infty} f\left(\frac{n\pi}{A}\right) \int_{-A}^{A} e^{in\pi\alpha/A}\, e^{-i\alpha x}\, d\alpha$$

$$= \frac{1}{2A} \sum_{n=-\infty}^{\infty} f\left(\frac{n\pi}{A}\right) \int_{-A}^{A} e^{i\alpha\left(\frac{n\pi}{A} - x\right)}\, d\alpha$$

$$= \frac{1}{2A} \sum_{n=-\infty}^{\infty} f\left(\frac{n\pi}{A}\right) \frac{1}{i\left(\frac{n\pi}{A} - x\right)} \left\{ e^{iA\left(\frac{n\pi}{A} - x\right)} - e^{-iA\left(\frac{n\pi}{A} - x\right)} \right\}$$

$$= \frac{1}{2A} \sum_{n=-\infty}^{\infty} f\left(\frac{n\pi}{A}\right) \frac{1}{i\left(\frac{n\pi}{A} - x\right)}\, 2i\,\sin(n\pi - Ax)$$

$$= \sum_{n=-\infty}^{\infty} f\left(\frac{n\pi}{A}\right) \frac{\sin(n\pi - Ax)}{n\pi - Ax}$$

$$= \sum_{n=-\infty}^{\infty} f\left(\frac{n\pi}{A}\right) \frac{\sin(Ax - n\pi)}{Ax - n\pi}.$$

Note that we used, in succession, an interchange of summation and integration (not always allowed, but is OK here), integration of an exponential function, $\sin\theta = \dfrac{e^{i\theta} - e^{-i\theta}}{2i}$, and the fact that $\sin(-\theta) = -\sin\theta$.

So, from samples at intervals of $\dfrac{\pi}{A}$ all values of f can be reconstructed. Note that if we allow e^{iAx} (in other words, we allow $k = A$), then the Sampling Theorem will fail. If, for example, $f(x) = \sin Ax$, then all samples will be 0 and f cannot be reconstructed, as aliasing appears again.

■ **Band-Limited Signals**　A signal that contains many frequencies can be **filtered** so that only frequencies in an interval survive, and it becomes a band-limited signal. Consider the signal $f(x)$. Multiply the Fourier transform $F(\alpha)$ of f by a function $G(\alpha)$ that is 1 on the interval containing the frequencies α you wish to keep, and 0 elsewhere. This multiplication of two Fourier transforms in the frequency domain is a convolution of $f(x)$ and $g(x)$ in the time domain. Recall that Problem 4 in Exercises 15.4 states that

$$\mathscr{F}^{-1}\{F(\alpha)G(\alpha)\} = \int_{-\infty}^{\infty} f(\tau)g(x - \tau)\,d\tau.$$

The integral on the right-hand side is called the **convolution of f and g** and is written $f*g$. The last statement can be written more compactly as

$$\mathscr{F}\{f*g\} = F(\alpha)G(\alpha).$$

The analogous idea for Laplace transforms is in Section 4.4. The function $g(x) = \dfrac{\sin Ax}{\pi x}$ has as its Fourier transform the pulse function

$$G(\alpha) = \begin{cases} 1, & -A < \alpha < A \\ 0, & \text{elsewhere.} \end{cases}$$

This implies that the function $(f*g)(x)$ is band-limited, with frequencies between $-A$ and A.

■ **Computing with the Fast Fourier Transform**　Return to the discrete Fourier transform of $f(x)$, where we have f sampled at n equally spaced points a distance of T apart, namely $0, T, 2T, 3T, \ldots, (n - 1)T$. (We used $T = \pi/n$ at the beginning of this section.) Substituting this, the discrete Fourier transform

$$F\left(\frac{2\pi M}{b - a}\right) = \frac{b - a}{n}\,e^{i\frac{2\pi Ma}{b-a}}\sum_{j=0}^{n-1} f\left(a + \frac{b - a}{n}\,j\right)\omega_n^{jM}$$

becomes　　$$F\left(\frac{2\pi k}{nT}\right) = T\sum_{j=0}^{n-1} f(jT)\omega_n^{kj}, \qquad k = 0, 1, 2, \ldots, n - 1.$$

For simplicity of notation, write this instead as

$$c_k = \sum_{j=0}^{n-1} f_j\omega_n^{kj}, \qquad k = 0, 1, 2, \ldots, n - 1.$$

This should remind you of (6), where we had

$$\begin{pmatrix} f_0 \\ f_1 \\ f_2 \\ \vdots \\ f_{n-1} \end{pmatrix} = \begin{pmatrix} 1 & 1 & 1 & \cdots & 1 \\ 1 & \omega_n & \omega_n^2 & \cdots & \omega_n^{n-1} \\ 1 & \omega_n^2 & \omega_n^4 & \cdots & \omega_n^{2(n-1)} \\ \vdots & \vdots & \vdots & & \vdots \\ 1 & \omega_n^{n-1} & \omega_n^{2(n-1)} & \cdots & \omega_n^{(n-1)^2} \end{pmatrix} \begin{pmatrix} c_0 \\ c_1 \\ c_2 \\ \vdots \\ c_{n-1} \end{pmatrix},$$

or $\mathbf{f} = \mathbf{F}_n\mathbf{c}$. The key to the FFT is properties of ω_n and matrix factorization. If $n = 2^N$, we can write \mathbf{F}_n in the following way (which we will not prove):

$$\mathbf{F}_{2^N} = \begin{pmatrix} \mathbf{I}_{2^{N-1}} & \mathbf{D}_{2^{N-1}} \\ \mathbf{I}_{2^{N-1}} & -\mathbf{D}_{2^{N-1}} \end{pmatrix} \begin{pmatrix} \mathbf{F}_{2^{N-1}} & \mathbf{0} \\ \mathbf{0} & \mathbf{F}_{2^{N-1}} \end{pmatrix} \mathbf{P}, \tag{13}$$

where \mathbf{I}_k is the $k \times k$ identity matrix and \mathbf{P} is the permutation matrix that rearranges the matrix \mathbf{c} so that the even subscripts are ordered on the top and the odd ones are ordered on the bottom. The matrix \mathbf{D} is a diagonal matrix defined by

$$\mathbf{D}_{2^{N-1}} = \begin{pmatrix} 1 & & & & \\ & \omega_{2^N} & & & \\ & & (\omega_{2^N})^2 & & \\ & & & \ddots & \\ & & & & (\omega_{2^N})^{2^{N-1}-1} \end{pmatrix}.$$

Note that each of the $\mathbf{F}_{2^{N-1}}$ matrices can in turn be factored. In the end, the matrix \mathbf{F}_n with n^2 nonzero entries is factored into the product of n simpler matrices at a great savings to the number of computations needed on the computer.

Example 3 The FFT

Let $n = 2^2 = 4$ and let \mathbf{F}_4 be the matrix in Example 1:

$$\mathbf{F}_4 = \begin{pmatrix} 1 & 1 & 1 & 1 \\ 1 & i & -1 & -i \\ 1 & -1 & 1 & -1 \\ 1 & -i & -1 & i \end{pmatrix}.$$

From (13) the desired factorization of \mathbf{F}_4 is

$$\mathbf{F}_4 = \begin{pmatrix} 1 & 0 & 1 & 0 \\ 0 & 1 & 0 & i \\ 1 & 0 & -1 & 0 \\ 0 & 1 & 0 & -i \end{pmatrix} \begin{pmatrix} 1 & 1 & 0 & 0 \\ 1 & -1 & 0 & 0 \\ 0 & 0 & 1 & 1 \\ 0 & 0 & 1 & -1 \end{pmatrix} \begin{pmatrix} 1 & 0 & 0 & 0 \\ 0 & 0 & 1 & 0 \\ 0 & 1 & 0 & 0 \\ 0 & 0 & 0 & 1 \end{pmatrix}. \qquad (14)$$
$$\qquad\qquad \mathbf{A} \qquad\qquad\qquad \mathbf{B} \qquad\qquad\qquad \mathbf{P}$$

We have inserted dashed lines in the matrices marked \mathbf{A} and \mathbf{B} so that you can identify the submatrices \mathbf{I}_2, \mathbf{D}_2, $-\mathbf{D}_2$, and \mathbf{F}_2 by comparing (14) directly with (13). You are also encouraged to multiply out the right side of (14) and verify that you get \mathbf{F}_4. Now if $\mathbf{c} = \begin{pmatrix} 3 \\ 5 \\ 8 \\ 20 \end{pmatrix}$, then

$$\mathbf{F}_4\mathbf{c} = \begin{pmatrix} 1 & 0 & 1 & 0 \\ 0 & 1 & 0 & i \\ 1 & 0 & -1 & 0 \\ 0 & 1 & 0 & -i \end{pmatrix} \begin{pmatrix} 1 & 1 & 0 & 0 \\ 1 & -1 & 0 & 0 \\ 0 & 0 & 1 & 1 \\ 0 & 0 & 1 & -1 \end{pmatrix} \begin{pmatrix} 1 & 0 & 0 & 0 \\ 0 & 0 & 1 & 0 \\ 0 & 1 & 0 & 0 \\ 0 & 0 & 0 & 1 \end{pmatrix} \begin{pmatrix} 3 \\ 5 \\ 8 \\ 20 \end{pmatrix}$$

$$= \begin{pmatrix} 1 & 0 & 1 & 0 \\ 0 & 1 & 0 & i \\ 1 & 0 & -1 & 0 \\ 0 & 1 & 0 & -i \end{pmatrix} \begin{pmatrix} 1 & 1 & 0 & 0 \\ 1 & -1 & 0 & 0 \\ 0 & 0 & 1 & 1 \\ 0 & 0 & 1 & -1 \end{pmatrix} \begin{pmatrix} 3 \\ 8 \\ 5 \\ 20 \end{pmatrix}$$

$$= \begin{pmatrix} 1 & 0 & 1 & 0 \\ 0 & 1 & 0 & i \\ 1 & 0 & -1 & 0 \\ 0 & 1 & 0 & -i \end{pmatrix} \begin{pmatrix} 11 \\ -5 \\ 25 \\ -15 \end{pmatrix} = \begin{pmatrix} 36 \\ -5 - 15i \\ -14 \\ -5 + 15i \end{pmatrix} = \mathbf{f}.$$

Without going into details, a computation of \mathbf{F}_n requires n^2 computations, while using the matrix factorization (the FFT) means the number of computations is reduced to one proportional to $n \ln n$. Try a few larger values of n and you will see substantial savings.

1. Show that $\mathbf{F}_4^{-1} = \frac{1}{4}\,\overline{\mathbf{F}}_4$.

2. Prove the sifting property of the Dirac delta function:

$$\int_{-\infty}^{\infty} f(x)\delta(x-a)\,dx = f(a).$$

[*Hint:* Consider the function $\delta_\varepsilon(x-a) = \begin{cases} \dfrac{1}{2\varepsilon}, & |x-a| < \varepsilon \\ 0, & \text{elsewhere.} \end{cases}$

Use the mean value theorem for integrals and then let $\varepsilon \to 0$.]

3. Find the Fourier transform of the Dirac delta function $\delta(x)$.

4. Show that the Dirac delta function is the identity under the convolution operation, that is, show $f*\delta = \delta*f = f$. [*Hint:* Use Fourier transforms and Problem 3.]

5. Show that the derivative of the Dirac delta function $\delta'(x-a)$ has the property that it sifts out the derivative of a function f at a. [*Hint:* Use integration by parts.]

6. Use a CAS to show that the Fourier transform of the function $g(x) = \dfrac{\sin Ax}{\pi x}$ is the pulse function

$$G(\alpha) = \begin{cases} 1, & -A < \alpha < A \\ 0, & \text{elsewhere.} \end{cases}$$

7. Write the matrix \mathbf{F}_8 and then write it in factored form (13). Verify that the product of the factors is \mathbf{F}_8. If instructed, use a CAS to verify the result.

8. Let $\omega_n = e^{i2\pi/n} = \cos\dfrac{2\pi}{n} + i\sin\dfrac{2\pi}{n}$. Since $e^{i2\pi k} = 1$, the numbers ω_n^k, $k = 0, 1, 2, \ldots, n-1$, all have the property that $(\omega_n^k)^n = 1$. Because of this, ω_n^k, $k = 0, 1, 2, \ldots, n-1$, are called the ***n*th roots of unity** and are solutions of the equation $z^n - 1 = 0$. Find the 8th roots of unity and plot them in the xy-plane where a complex number is written $z = x + iy$. What do you notice?

9. Computer Lab Assignment Use a CAS to verify that the function $f*g$, where $f(x) = e^{-5x^2}$ and $g(x) = \dfrac{\sin 2x}{\pi x}$, is band-limited. If your CAS can handle it, plot the graphs of $\mathscr{F}\{f*g\}$ and $F(\alpha)G(\alpha)$ to verify the result.

10. Computer Lab Assignment If your CAS has a discrete Fourier transform command, choose any six points and compare the result obtained using this command with that obtained from $\mathbf{c} = \frac{1}{6}\,\overline{\mathbf{F}}_6\mathbf{f}$.

In Problems 1–13, solve the given boundary-value problem by an appropriate integral transform. Make assumptions about boundedness where necessary.

1. $\dfrac{\partial^2 u}{\partial x^2} + \dfrac{\partial^2 u}{\partial y^2} = 0, \quad x > 0, \quad 0 < y < \pi$

$\left.\dfrac{\partial u}{\partial x}\right|_{x=0} = 0, \quad 0 < y < \pi$

$u(x, 0) = 0, \quad \left.\dfrac{\partial u}{\partial y}\right|_{y=\pi} = e^{-x}, \quad x > 0$

2. $\dfrac{\partial^2 u}{\partial x^2} = \dfrac{\partial u}{\partial t}, \quad 0 < x < 1, \quad t > 0$

$u(0, t) = 0, \quad u(1, t) = 0, \quad t > 0$

$u(x, 0) = 50 \sin 2\pi x, \quad 0 < x < 1$

3. $\dfrac{\partial^2 u}{\partial x^2} - hu = \dfrac{\partial u}{\partial t}, \quad h > 0, \quad x > 0, \quad t > 0$

$u(0, t) = 0, \quad \lim_{x \to \infty} \dfrac{\partial u}{\partial x} = 0, \quad t > 0$

$u(x, 0) = u_0, \quad x > 0$

4. $\dfrac{\partial u}{\partial t} - \dfrac{\partial^2 u}{\partial x^2} = e^{-|x|}, \quad -\infty < x < \infty, \quad t > 0$

$u(x, 0) = 0, \quad -\infty < x < \infty$

5. $\dfrac{\partial^2 u}{\partial x^2} = \dfrac{\partial u}{\partial t}, \quad x > 0, \quad t > 0$

$u(0, t) = t, \quad \lim_{x \to \infty} u(x, t) = 0$

$u(x, 0) = 0, \quad x > 0$ [*Hint:* Use Theorem 4.9.]

6. $\dfrac{\partial^2 u}{\partial x^2} = \dfrac{\partial^2 u}{\partial t^2}, \quad 0 < x < 1, \quad t > 0$

$u(0, t) = 0, \quad u(1, t) = 0, \quad t > 0$

$u(x, 0) = \sin \pi x, \quad \left.\dfrac{\partial u}{\partial t}\right|_{t=0} = -\sin \pi x, \quad 0 < x < 1$

7. $k\dfrac{\partial^2 u}{\partial x^2} = \dfrac{\partial u}{\partial t}, \quad -\infty < x < \infty, \quad t > 0$

$u(x, 0) = \begin{cases} 0, & x < 0 \\ u_0, & 0 < x < \pi \\ 0, & x > \pi \end{cases}$

8. $\dfrac{\partial^2 u}{\partial x^2} + \dfrac{\partial^2 u}{\partial y^2} = 0, \quad 0 < x < \pi, \quad y > 0$

$u(0, y) = 0, \quad u(\pi, y) = \begin{cases} 0, & 0 < y < 1 \\ 1, & 1 < y < 2 \\ 0, & y > 2 \end{cases}$

$\left.\dfrac{\partial u}{\partial y}\right|_{y=0} = 0, \quad 0 < x < \pi$

9. $\dfrac{\partial^2 u}{\partial x^2} + \dfrac{\partial^2 u}{\partial y^2} = 0, \quad x > 0, \quad y > 0$

$u(0, y) = \begin{cases} 50, & 0 < y < 1 \\ 0, & y > 1 \end{cases}$

$u(x, 0) = \begin{cases} 100, & 0 < x < 1 \\ 0, & x > 1 \end{cases}$

10. $\dfrac{\partial^2 u}{\partial x^2} + r = \dfrac{\partial u}{\partial t}, \quad 0 < x < 1, \quad t > 0$

$\left.\dfrac{\partial u}{\partial x}\right|_{x=0} = 0, \quad u(1, t) = 0, \quad t > 0$

$u(x, 0) = 0, \quad 0 < x < 1$

11. $\dfrac{\partial^2 u}{\partial x^2} + \dfrac{\partial^2 u}{\partial y^2} = 0, \quad x > 0, \quad 0 < y < \pi$

$u(0, y) = A, \quad 0 < y < \pi$

$\left.\dfrac{\partial u}{\partial y}\right|_{y=0} = 0, \quad \left.\dfrac{\partial u}{\partial y}\right|_{y=\pi} = Be^{-x}, \quad x > 0$

12. $\dfrac{\partial^2 u}{\partial x^2} = \dfrac{\partial u}{\partial t}, \quad 0 < x < 1, \quad t > 0$

$u(0, t) = u_0, \quad u(1, t) = u_0, \quad t > 0$

$u(x, 0) = 0, \quad 0 < x < 1$

[*Hint:* Use the identity

$$\sinh(x - y) = \sinh x \cosh y - \cosh x \sinh y,$$

and then use Problem 6 in Exercises 15.1.]

13. $k\dfrac{\partial^2 u}{\partial x^2} = \dfrac{\partial u}{\partial t}, \quad -\infty < x < \infty, \quad t > 0$

$u(x, 0) = \begin{cases} 0, & x < 0 \\ e^{-x}, & x > 0 \end{cases}$

16

Numerical Solutions of Partial Differential Equations

INTRODUCTION

In Section 6.5 we saw that one way of approximating a solution of a second-order boundary-value problem was to work with a finite difference equation replacement of the ordinary differential equation. The same idea carries over to BVPs involving partial differential equations. In the respective sections of this chapter, we shall construct a difference equation replacement for Laplace's equation, the heat equation, and the wave equation by replacing the partial derivatives u_{xx}, u_{yy}, u_{tt}, and u_t by difference quotients.

16.1 Elliptic Equations

Using the difference quotients derived in Section 6.5, we can easily replace a linear second-order partial differential equation by a difference equation. In the discussion that follows we shall confine our attention to elliptic partial differential equations such as Laplace's equation.

■ **A Difference Equation Replacement for Laplace's Equation** If u is a function of two variables x and y, we can form two central differences

$$u(x + h, y) - 2u(x, y) + u(x - h, y) \quad \text{and} \quad u(x, y + h) - 2u(x, y) + u(x, y - h).$$

It then follows from (6) of Section 6.5 that difference quotient approximations for the second partial derivatives u_{xx} and u_{yy} are given by

$$\frac{\partial^2 u}{\partial x^2} \approx \frac{1}{h^2} [u(x + h, y) - 2u(x, y) + u(x - h, y)] \tag{1}$$

$$\frac{\partial^2 u}{\partial y^2} \approx \frac{1}{h^2} [u(x, y + h) - 2u(x, y) + u(x, y - h)]. \tag{2}$$

Now by adding (1) and (2) we obtain a **five-point approximation** to the Laplacian:

$$\frac{\partial^2 u}{\partial x^2} + \frac{\partial^2 u}{\partial y^2} \approx \frac{1}{h^2} [u(x + h, y) + u(x, y + h) + u(x - h, y) + u(x, y - h) - 4u(x, y)].$$

Hence we can replace Laplace's equation $\dfrac{\partial^2 u}{\partial x^2} + \dfrac{\partial^2 u}{\partial y^2} = 0$ by the difference equation

$$u(x + h, y) + u(x, y + h) + u(x - h, y) + u(x, y - h) - 4u(x, y) = 0. \tag{3}$$

If we adopt the notation $u(x, y) = u_{ij}$ and

$$u(x + h, y) = u_{i+1,j}, \qquad u(x, y + h) = u_{i,j+1}$$

$$u(x - h, y) = u_{i-1,j}, \qquad u(x, y - h) = u_{i,j-1},$$

then (3) becomes

$$u_{i+1,j} + u_{i,j+1} + u_{i-1,j} + u_{i,j-1} - 4u_{ij} = 0. \tag{4}$$

To understand (4) a little better, suppose a rectangular grid consisting of horizontal lines spaced h units apart and vertical lines spaced h units apart is placed over a region R bounded by a curve C in which we are seeking a solution of Laplace's equation. The number h is called the **mesh size**. See Figure 16.1(a). The points of intersection $P_{ij} = P(ih, jh)$, i and j integers, of the lines are called **mesh points** or **lattice points**. A mesh point is an **interior point** if its four nearest neighboring mesh points are points of R. Points in R or on C that are not interior points are called **boundary points**. For example, in Figure 16.1(a) we have

$$P_{20} = P(2h, 0), \qquad P_{11} = P(h, h), \qquad P_{21} = P(2h, h), \qquad P_{22} = P(2h, 2h),$$

and so on. Of the points listed, P_{21} and P_{22} are interior points, P_{20} and P_{11} are boundary points. In Figure 16.1(a) interior points are the colored dots and the boundary points are the black dots. Now from (4) we see that

$$u_{ij} = \frac{1}{4} [u_{i+1,j} + u_{i,j+1} + u_{i-1,j} + u_{i,j-1}], \tag{5}$$

(a)

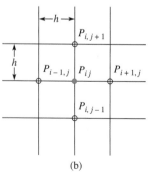

(b)

Figure 16.1

and so, as shown in Figure 16.1(b), the value of u_{ij} at an interior mesh point of R is the average of the values of u at four neighboring mesh points. The neighboring points $P_{i+1,j}$, $P_{i,j+1}$, $P_{i-1,j}$, and $P_{i,j-1}$ correspond, respectively, to the four points on a compass E, N, W, and S.

■ **Dirichlet Problem** Recall that in the **Dirichlet problem** for Laplace's equation $\nabla^2 u = 0$ the values of $u(x, y)$ are prescribed on the boundary C of a region R. The basic idea is to find an approximate solution to Laplace's equation at interior mesh points by replacing the partial differential equation at these points by the difference equation (4). Hence the approximate values of u at the mesh points— namely, the u_{ij}—are related to each other and, possibly, to known values of u if a mesh point lies on the boundary C. In this manner we obtain a system of linear algebraic equations that we solve for the unknown u_{ij}. The following example illustrates the method for a square region.

Example 1 A BVP Revisited

In Problem 14 of Exercises 13.5 you were asked to solve the boundary-value problem

$$\frac{\partial^2 u}{\partial x^2} + \frac{\partial^2 u}{\partial y^2} = 0, \qquad 0 < x < 2, \quad 0 < y < 2$$

$$u(0, y) = 0, \qquad u(2, y) = y(2 - y), \quad 0 < y < 2$$

$$u(x, 0) = 0, \qquad u(x, 2) = \begin{cases} x, & 0 < x < 1 \\ 2 - x, & 1 \le x < 2 \end{cases}$$

utilizing the superposition principle. To apply the present numerical method let us start with a mesh size of $h = \frac{2}{3}$. As we see in Figure 16.2, that choice yields four interior points and eight boundary points. The numbers listed next to the boundary points are the exact values of u obtained from the specified condition along that boundary. For example, at $P_{31} = P(3h, h) = P(2, \frac{2}{3})$ we have $x = 2$ and $y = \frac{2}{3}$, and so the condition $u(2, y)$ gives $u(2, \frac{2}{3}) = \frac{2}{3}(2 - \frac{2}{3}) = \frac{8}{9}$. Similarly, at $P_{13} = P(\frac{2}{3}, 2)$ the condition $u(x, 2)$ gives $u(\frac{2}{3}, 2) = \frac{2}{3}$. We now apply (4) at each interior point. For example, at P_{11} we have $i = 1$ and $j = 1$, so (4) becomes

$$u_{21} + u_{12} + u_{01} + u_{10} - 4u_{11} = 0.$$

Since $u_{01} = u(0, \frac{2}{3}) = 0$ and $u_{10} = u(\frac{2}{3}, 0) = 0$, the foregoing equation becomes $-4u_{11} + u_{21} + u_{12} = 0$. Repeating this, in turn, at P_{21}, P_{12}, and P_{22}, we get three additional equations:

$$\begin{array}{rcl} -4u_{11} + u_{21} + u_{12} & = & 0 \\[4pt] u_{11} - 4u_{21} + u_{22} & = & -\dfrac{8}{9} \\[4pt] u_{11} - 4u_{12} + u_{22} & = & -\dfrac{2}{3} \\[4pt] u_{21} + u_{12} - 4u_{22} & = & -\dfrac{14}{9}. \end{array} \qquad (6)$$

Using a computer algebra system to solve the system, we find the approximate temperatures at the four interior points to be

$$u_{11} = \frac{7}{36} = 0.1944, \qquad u_{21} = \frac{5}{12} = 0.4167,$$

$$u_{12} = \frac{13}{36} = 0.3611, \qquad u_{22} = \frac{7}{12} = 0.5833. \qquad \square$$

Figure 16.2

As in the discussion of ordinary differential equations, we expect that a smaller value of h will improve the accuracy of the approximation. However, using a smaller mesh size means, of course, that there are more interior mesh points, and correspondingly there is a much larger system of equations to be solved. For a *square* region whose length of side is L, a mesh size of $h = L/n$ will yield a total of $(n-1)^2$ interior mesh points. In Example 1, for $n = 8$, the mesh size is a reasonable $h = \frac{2}{8} = \frac{1}{4}$, but the number of interior points is $(8-1)^2 = 49$. Thus we have 49 equations in 49 unknowns. In the next example we use a mesh size of $h = \frac{1}{2}$.

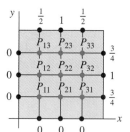

Figure 16.3

Example 2 Example 1 with More Mesh Points

As we see in Figure 16.3, with $n = 4$, a mesh size $h = \frac{2}{4} = \frac{1}{2}$ for the square in Example 1 gives $3^2 = 9$ interior mesh points. Applying (4) at these points and using the indicated boundary conditions, we get nine equations in nine unknowns. So that you can verify the results, we give the system in an unsimplified form:

$$u_{21} + u_{12} + \quad 0 + \quad 0 - 4u_{11} = 0$$
$$u_{31} + u_{22} + u_{11} + \quad 0 - 4u_{21} = 0$$
$$\frac{3}{4} + u_{32} + u_{21} + \quad 0 - 4u_{31} = 0$$
$$u_{22} + u_{13} + u_{11} + \quad 0 - 4u_{12} = 0$$
$$u_{32} + u_{23} + u_{12} + u_{21} - 4u_{22} = 0 \qquad (7)$$
$$1 + u_{33} + u_{22} + u_{31} - 4u_{32} = 0$$
$$u_{23} + \frac{1}{2} + \quad 0 + u_{12} - 4u_{13} = 0$$
$$u_{33} + \quad 1 + u_{13} + u_{22} - 4u_{23} = 0$$
$$\frac{3}{4} + \frac{1}{2} + u_{23} + u_{32} - 4u_{33} = 0.$$

In this case a CAS yields

$$u_{11} = \frac{7}{64} = 0.1094, \qquad u_{21} = \frac{51}{224} = 0.2277, \qquad u_{31} = \frac{177}{448} = 0.3951$$

$$u_{12} = \frac{47}{224} = 0.2098, \qquad u_{22} = \frac{13}{32} = 0.4063, \qquad u_{32} = \frac{135}{224} = 0.6027$$

$$u_{13} = \frac{145}{448} = 0.3237, \qquad u_{23} = \frac{131}{224} = 0.5848, \qquad u_{33} = \frac{39}{64} = 0.6094. \qquad \square$$

After we simplify (7), it is interesting to note that the 9×9 matrix of coefficients is

$$\begin{pmatrix} -4 & 1 & 0 & 1 & 0 & 0 & 0 & 0 & 0 \\ 1 & -4 & 1 & 0 & 1 & 0 & 0 & 0 & 0 \\ 0 & 1 & -4 & 0 & 0 & 1 & 0 & 0 & 0 \\ 1 & 0 & 0 & -4 & 1 & 0 & 1 & 0 & 0 \\ 0 & 1 & 0 & 1 & -4 & 1 & 0 & 1 & 0 \\ 0 & 0 & 1 & 0 & 1 & -4 & 0 & 0 & 1 \\ 0 & 0 & 0 & 1 & 0 & 0 & -4 & 1 & 0 \\ 0 & 0 & 0 & 0 & 1 & 0 & 1 & -4 & 1 \\ 0 & 0 & 0 & 0 & 0 & 1 & 0 & 1 & -4 \end{pmatrix}. \qquad (8)$$

This is an example of a **sparse matrix** in that a large percentage of the entries are zeros. The matrix (8) is also an example of a **banded matrix.** These kinds of matrices are characterized by the properties that the entries on the main diagonal and on diagonals (or bands) parallel to the main diagonal are all nonzero. The bands shown in color in (8) are separated by diagonals consisting of all zeros or not.

■ **Gauss–Seidel Iteration** Problems requiring approximations to solutions of partial differential equations invariably lead to large systems of linear algebraic equations. It is not uncommon to have to solve systems involving hundreds of equations. Although a direct method of solution such as Gaussian elimination leaves unchanged the zero entries outside the bands in a matrix such as (8), it does fill in the positions between the bands with nonzeros. Since storing very large matrices uses up a large portion of computer memory, it is usual practice to solve a large system in an indirect manner. One popular indirect method is called **Gauss–Seidel iteration.**

We shall illustrate this method for the system in (6). For the sake of simplicity we replace the double-subscripted variables u_{11}, u_{21}, u_{12}, and u_{22} by x_1, x_2, x_3, and x_4, respectively.

Example 3 Gauss–Seidel Iteration

STEP 1: *Solve each equation for the variables on the main diagonal of the system.* That is, in (6) solve the first equation for x_1, the second equation for x_2, and so on:

$$x_1 = 0.25x_2 + 0.25x_3$$
$$x_2 = 0.25x_1 + 0.25x_4 + 0.2222$$
$$x_3 = 0.25x_1 + 0.25x_4 + 0.1667 \tag{9}$$
$$x_4 = 0.25x_2 + 0.25x_3 + 0.3889.$$

These equations can be obtained directly by using (5) rather than (4) at the interior points.

STEP 2: *Iterations.* We start by making an initial guess for the values of x_1, x_2, x_3, and x_4. If this were simply a system of linear equations and we knew nothing about the solution, we could start with $x_1 = 0$, $x_2 = 0$, $x_3 = 0$, $x_4 = 0$. But since the solution of (9) represents approximations to a solution of a boundary-value problem, it would seem reasonable to use as the initial guess for the values of $x_1 = u_{11}$, $x_2 = u_{21}$, $x_3 = u_{12}$, and $x_4 = u_{22}$ the average of all the boundary conditions. In this case the average of the numbers at the eight boundary points shown in Figure 16.2 is approximately 0.4. Thus our initial guess is $x_1 = 0.4$, $x_2 = 0.4$, $x_3 = 0.4$, and $x_4 = 0.4$. Iterations of the Gauss–Seidel method use the x values as soon as they are computed. Note that the first equation in (9) depends only on x_2 and x_3; thus substituting $x_2 = 0.4$ and $x_3 = 0.4$ gives $x_1 = 0.2$. Since the second and third equations depend on x_1 and x_4, we use the newly calculated values $x_1 = 0.2$ and $x_4 = 0.4$ to obtain $x_2 = 0.3722$ and $x_3 = 0.3167$. The fourth equation depends on x_2 and x_3, so we use the new values $x_2 = 0.3722$ and $x_3 = 0.3167$ to get $x_4 = 0.5611$. In summary, the first iteration has given the values

$$x_1 = 0.2, \qquad x_2 = 0.3722, \qquad x_3 = 0.3167, \qquad x_4 = 0.5611.$$

Note how close these numbers are already to the actual values given at the end of Example 1.

The second iteration starts with substituting $x_2 = 0.3722$ and $x_3 = 0.3167$ into the first equation. This gives $x_1 = 0.1722$. From $x_1 = 0.1722$ and the last computed value of x_4 (namely, $x_4 = 0.5611$), the second and third equations give, in turn, $x_2 = 0.4055$ and $x_3 = 0.3500$. Using these two values, we find from the fourth equation that $x_4 = 0.5678$. At the end of the second iteration we have

$$x_1 = 0.1722, \qquad x_2 = 0.4055, \qquad x_3 = 0.3500, \qquad x_4 = 0.5678.$$

The third through seventh iterations are summarized in Table 16.1.

Table 16.1 Iteration

	3rd	4th	5th	6th	7th
x_1	0.1889	0.1931	0.1941	0.1944	0.1944
x_2	0.4139	0.4160	0.4165	0.4166	0.4166
x_3	0.3584	0.3605	0.3610	0.3611	0.3611
x_4	0.5820	0.5830	0.5833	0.5833	0.5833

Note

To apply Gauss–Seidel iteration to a general system of n linear equations in n unknowns, the variable x_i must actually appear in the ith equation of the system. Moreover, after each equation is solved for x_i, $i = 1, 2, \ldots, n$, the resulting system has the form $\mathbf{X} = \mathbf{AX} + \mathbf{B}$, where all the entries on the main diagonal of \mathbf{A} are zero.

Remarks

(*i*) In the examples given in this section the values of u_{ij} were determined using known values of u at boundary points. But what do we do if the region is such that boundary points do not coincide with the actual boundary C of the region R? In this case the required values can be obtained by interpolation.

(*ii*) It is sometimes possible to cut down the number of equations to solve by using symmetry. Consider the rectangular region shown in Figure 16.4. The boundary conditions are $u = 0$ along $x = 0$, $x = 2$, $y = 1$, and $u = 100$ along $y = 0$. The region is symmetric about the lines $x = 1$ and $y = \frac{1}{2}$, and the interior points P_{11} and P_{31} are equidistant from the neighboring boundary points at which the specified values of u are the same. Consequently we assume that $u_{11} = u_{31}$, and so the system of three equations in three unknowns reduces to two equations in two unknowns. See Problem 2 in Exercises 16.1.

(*iii*) Although on a computer this may not be noticeable, convergence of Gauss–Seidel iteration may not be particularly fast. Also, in a more general setting Gauss–Seidel iteration may not converge at all. If a system of n linear equations in n unknowns is written in the standard form $\mathbf{AX} = \mathbf{B}$, then a sufficient condition for convergence is that the coefficient matrix \mathbf{A} be **diagonally dominant**. This means that the absolute value of each entry on the main diagonal of \mathbf{A} is greater than the sum of the absolute values of the remaining entries in the same row. For example, the coefficient matrix (8) for the system (7) is clearly diagonally dominant, since in each row $|-4|$ is greater than the sum of the absolute values of the other entries. You should verify that the coefficient matrix for the system (6) has the same property. Diagonal dominance guarantees that Gauss–Seidel iteration converges for any initial values.

Figure 16.4

In Problems 1–8, use a computer as a computation aid.

In Problems 1–4, use (4) to approximate the solution of Laplace's equation at the interior points of the given region. Use symmetry when possible.

1. $u(0, y) = 0$, $u(3, y) = y(2 - y)$, $0 < y < 2$
$u(x, 0) = 0$, $u(x, 2) = x(3 - x)$, $0 < x < 3$
mesh size: $h = 1$

2. $u(0, y) = 0$, $u(2, y) = 0$, $0 < y < 1$
$u(x, 0) = 100$, $u(x, 1) = 0$, $0 < x < 2$
mesh size: $h = \frac{1}{2}$

3. $u(0, y) = 0$, $u(1, y) = 0$, $0 < y < 1$
$u(x, 0) = 0$, $u(x, 1) = \sin \pi x$, $0 < x < 1$
mesh size: $h = \frac{1}{3}$

4. $u(0, y) = 108y^2(1 - y)$, $u(1, y) = 0$, $0 < y < 1$
$u(x, 0) = 0$, $u(x, 1) = 0$, $0 < x < 1$
mesh size: $h = \frac{1}{3}$

In Problems 5 and 6, use (5) and Gauss–Seidel iteration to approximate the solution of Laplace's equation at the interior points of a unit square. Use the mesh size $h = \frac{1}{4}$. In Problem 5 the boundary conditions are given; in Problem 6 the values of u at boundary points are given in Figure 16.5.

5. $u(0, y) = 0$, $u(1, y) = 100y$, $0 < y < 1$
$u(x, 0) = 0$, $u(x, 1) = 100x$, $0 < x < 1$

6.

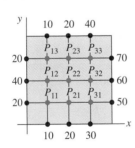

Figure 16.5

7. (a) In Problem 12 of Exercises 13.6, you solved a potential problem using a special form of Poisson's equation $\frac{\partial^2 u}{\partial x^2} + \frac{\partial^2 u}{\partial y^2} = f(x, y)$. Show that the difference equation replacement for Poisson's equation is

$$u_{i+1,j} + u_{i,j+1} + u_{i-1,j} + u_{i,j-1} - 4u_{ij} = h^2 f(x, y).$$

(b) Use the result in part **(a)** to approximate the solution of the Poisson equation $\frac{\partial^2 u}{\partial x^2} + \frac{\partial^2 u}{\partial y^2} = -2$ at the interior points of the region in Figure 16.6. The mesh size is $h = \frac{1}{2}$, $u = 1$ at every point along $ABCD$, and $u = 0$ at every point along $DEFGA$. Use symmetry and, if necessary, Gauss–Seidel iteration.

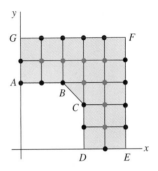

Figure 16.6

8. Use the result in part **(a)** of Problem 7 to approximate the solution of the Poisson equation $\frac{\partial^2 u}{\partial x^2} + \frac{\partial^2 u}{\partial y^2} = -64$ at the interior points of the region in Figure 16.7. The mesh is $h = \frac{1}{8}$, and $u = 0$ at every point on the boundary of the region. If necessary, use Gauss–Seidel iteration.

Figure 16.7

16.2 Parabolic Equations

In this section we examine two methods for approximating the solution of a boundary-value problem of the type

$$c\frac{\partial^2 u}{\partial x^2} = \frac{\partial u}{\partial t}, \qquad 0 < x < a, \quad t > 0$$

$$u(0, t) = T_1, \qquad u(a, t) = T_2, \quad t > 0 \tag{1}$$

$$u(x, 0) = f(x), \qquad 0 < x < a.$$

The equation in this problem is the one-dimensional heat equation and is, as know from Section 13.1, a parabolic PDE. Recall that the function f can be interpreted as the initial temperature distribution in a homogeneous rod extending from $x = 0$ to $x = a$. The constants T_1 and T_2 represent temperatures at the endpoints of the rod. Although we shall not prove it, the boundary-value problem has a unique solution when f is continuous on the closed interval $[0, a]$. This latter condition will be assumed, and so we replace the initial condition in (1) by $u(x, 0) = f(x)$, $0 \le x \le a$.

■ **A Difference Equation Replacement for the Heat Equation** To approximate the solution of (1) we first replace the heat equation by a difference equation. Using the central difference approximation (1) of Section 6.5,

$$\frac{\partial^2 u}{\partial x^2} \approx \frac{1}{h^2}\,[u(x + h, t) - 2u(x, t) + u(x - h, t)],$$

and the forward difference approximation (3) of Section 6.5,

$$\frac{\partial u}{\partial t} \approx \frac{1}{k}\,[u(x, t + k) - u(x, t)],$$

we can replace the heat equation $c\dfrac{\partial^2 u}{\partial x^2} = \dfrac{\partial u}{\partial t}$ by the following difference equation:

$$\frac{c}{h^2}[u(x + h, t) - 2u(x, t) + u(x - h, t)] = \frac{1}{k}[u(x, t + k) - u(x, t)]. \qquad (2)$$

If we let $\lambda = ck/h^2$ and

$$u(x, t) = u_{ij}, \quad u(x + h, t) = u_{i+1,j}, \quad u(x - h, t) = u_{i-1,j}, \quad u(x, t + k) = u_{i,j+1},$$

then (2) becomes, after simplifying,

$$u_{i,j+1} = \lambda u_{i+1,j} + (1 - 2\lambda)u_{ij} + \lambda u_{i-1,j}. \qquad (3)$$

We wish to approximate a solution of a problem such as (1) on a rectangular region in the xt-plane defined by the inequalities $0 \le x \le a$, $0 \le t \le T$ for some value of time T. Over this region we place a rectangular grid consisting of vertical lines h units apart and horizontal lines k units apart. If we choose two positive integers n and m and define

$$h = \frac{a}{n} \qquad \text{and} \qquad k = \frac{T}{m},$$

then the vertical and horizontal grid lines are defined by

$$x_i = ih, \quad i = 0, 1, 2, \dots, n \qquad \text{and} \qquad t_j = jk, \quad j = 0, 1, 2, \dots, m.$$

See Figure 16.8

As illustrated in Figure 16.9, the idea here is to use formula (3) to estimate the values of the solution $u(x, t)$ at the points on the $(j + 1)$st time line using only values from the jth time line. For example, the values on the first time line $(j = 1)$ depend on the initial condition $u_{i,0} = u(x_i, 0) = f(x_i)$ given on the zeroth time line $(j = 0)$. This kind of numerical procedure is called an **explicit finite difference method.**

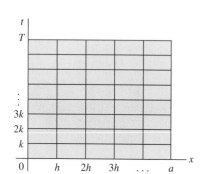

Figure 16.8

Figure 16.9

(j + 1)st time line

jth time line

Example 1 Using the Finite Difference Method

Consider the boundary-value problem

$$\frac{\partial^2 u}{\partial x^2} = \frac{\partial u}{\partial t}, \qquad\qquad 0 < x < 1, \quad 0 < t < 0.5$$

$$u(0, t) = 0, \qquad\qquad u(1, t) = 0, \quad 0 \le t \le 0.5$$
$$u(x, 0) = \sin \pi x, \qquad\quad 0 \le x \le 1.$$

First we identify $c = 1$, $a = 1$, and $T = 0.5$. If we choose, say, $n = 5$ and $m = 50$, then $h = 1/5 = 0.2$, $k = 0.5/50 = 0.01$, $\lambda = 0.25$,

$$x_i = i\frac{1}{5}, \quad i = 0, 1, 2, 3, 4, 5, \qquad \text{and} \qquad t_j = j\frac{1}{100}, \quad j = 0, 1, 2, \ldots, 50.$$

Thus (3) becomes

$$u_{i,j+1} = 0.25(u_{i+1,j} + 2u_{ij} + u_{i-1,j}).$$

By setting $j = 0$ in this formula we get a formula for the approximations to the temperature u on the first time line:

$$u_{i,1} = 0.25(u_{i+1,0} + 2u_{i,0} + u_{i-1,0}).$$

If we then let $i = 1, \ldots, 4$ in the last equation, we obtain, in turn,

$$u_{11} = 0.25(u_{20} + 2u_{10} + u_{00})$$

$$u_{21} = 0.25(u_{30} + 2u_{20} + u_{10})$$

$$u_{31} = 0.25(u_{40} + 2u_{30} + u_{20})$$

$$u_{41} = 0.25(u_{50} + 2u_{40} + u_{30}).$$

The first equation in this list is interpreted as

$$u_{11} = 0.25(u(x_2, 0) + 2u(x_1, 0) + u(0, 0))$$

$$= 0.25(u(0.4, 0) + 2u(0.2, 0) + u(0, 0)).$$

From the initial condition $u(x, 0) = \sin \pi x$ the last line becomes

$$u_{11} = 0.25(0.951056516 + 2(0.587785252) + 0) = 0.531656755.$$

This number represents an approximation to the temperature $u(0.2, 0.01)$. The remaining calculations are summarized in Table 16.2.

Table 16.2 Explicit Difference Equation Approximation with $h = 0.2$, $k = 0.01$, $\lambda = 0.25$

Time	x = 0.20	x = 0.40	x = 0.60	x = 0.80	Time	x = 0.20	x = 0.40	x = 0.60	x = 0.80
0.00	0.5878	0.9511	0.9511	0.5878	0.26	0.0432	0.0700	0.0700	0.0432
0.01	0.5317	0.8602	0.8602	0.5317	0.27	0.0391	0.0633	0.0633	0.0391
0.02	0.4809	0.7781	0.7781	0.4809	0.28	0.0354	0.0572	0.0572	0.0354
0.03	0.4350	0.7038	0.7038	0.4350	0.29	0.0320	0.0518	0.0518	0.0320
0.04	0.3934	0.6366	0.6366	0.3934	0.30	0.0289	0.0468	0.0468	0.0289
0.05	0.3559	0.5758	0.5758	0.3559	0.31	0.0262	0.0424	0.0424	0.0262
0.06	0.3219	0.5208	0.5208	0.3219	0.32	0.0237	0.0383	0.0383	0.0237
0.07	0.2911	0.4711	0.4711	0.2911	0.33	0.0214	0.0347	0.0347	0.0214
0.08	0.2633	0.4261	0.4261	0.2633	0.34	0.0194	0.0313	0.0313	0.0194
0.09	0.2382	0.3854	0.3854	0.2382	0.35	0.0175	0.0284	0.0284	0.0175
0.10	0.2154	0.3486	0.3486	0.2154	0.36	0.0159	0.0256	0.0256	0.0159
0.11	0.1949	0.3153	0.3153	0.1949	0.37	0.0143	0.0232	0.0232	0.0143
0.12	0.1763	0.2852	0.2852	0.1763	0.38	0.0130	0.0210	0.0210	0.0130
0.13	0.1594	0.2580	0.2580	0.1594	0.39	0.0117	0.0190	0.0190	0.0117
0.14	0.1442	0.2333	0.2333	0.1442	0.40	0.0106	0.0172	0.0172	0.0106
0.15	0.1304	0.2111	0.2111	0.1304	0.41	0.0096	0.0155	0.0155	0.0096
0.16	0.1180	0.1909	0.1909	0.1180	0.42	0.0087	0.0140	0.0140	0.0087
0.17	0.1067	0.1727	0.1727	0.1067	0.43	0.0079	0.0127	0.0127	0.0079
0.18	0.0965	0.1562	0.1562	0.0965	0.44	0.0071	0.0115	0.0115	0.0071
0.19	0.0873	0.1413	0.1413	0.0873	0.45	0.0064	0.0104	0.0104	0.0064

continued

Table 16.2 Continued

Time	$x = 0.20$	$x = 0.40$	$x = 0.60$	$x = 0.80$	Time	$x = 0.20$	$x = 0.40$	$x = 0.60$	$x = 0.80$
0.20	0.0790	0.1278	0.1278	0.0790	0.46	0.0058	0.0094	0.0094	0.0058
0.21	0.0714	0.1156	0.1156	0.0714	0.47	0.0053	0.0085	0.0085	0.0053
0.22	0.0646	0.1045	0.1045	0.0646	0.48	0.0048	0.0077	0.0077	0.0048
0.23	0.0584	0.0946	0.0946	0.0584	0.49	0.0043	0.0070	0.0070	0.0043
0.24	0.0529	0.0855	0.0855	0.0529	0.50	0.0039	0.0063	0.0063	0.0039
0.25	0.0478	0.0774	0.0774	0.0478					

Exact	Approximation
$u(0.4, 0.05) = 0.5806$	$u_{25} = 0.5758$
$u(0.6, 0.06) = 0.5261$	$u_{36} = 0.5208$
$u(0.2, 0.10) = 0.2191$	$u_{1,10} = 0.2154$
$u(0.8, 0.14) = 0.1476$	$u_{4,14} = 0.1442$

Computer programs for the method illustrated in Example 1 are provided in the software supplement for this text.

You should verify, using the methods of Chapter 13, that the exact solution of the boundary-value problem in Example 1 is given by $u(x, t) = e^{-\pi^2 t} \sin \pi x$. Some sample values are given in the margin table.

■ **Stability** These approximations are comparable to the exact values and accurate enough for some purposes. But there is a difficulty with the foregoing method. Recall that a numerical method is **unstable** if round-off errors or any other errors grow too rapidly as the computations proceed. The numerical procedure in Example 1 can exhibit this kind of behavior. It can be proved that the procedure is stable if λ is less than or equal to 0.5 but unstable otherwise. To obtain $\lambda = 0.25 \le 0.5$ in Example 1, we had to choose the value $k = 0.01$; the necessity of using very small step sizes in the time direction is the principal fault of this method. You are urged to work Problem 12 in Exercises 16.2 and witness the predictable instability when $\lambda = 1$.

■ **Crank–Nicholson Method** There are **implicit finite difference methods** for solving parabolic partial differential equations. These methods require that we solve a system of equations to determine the approximate values of u on the $(j + 1)$st time line. However, implicit methods do not suffer from instability problems.

The algorithm introduced by J. Crank and P. Nicholson in 1947 is used mostly for solving the heat equation. The algorithm consists of replacing the second partial derivative in $c\dfrac{\partial^2 u}{\partial x^2} = \dfrac{\partial u}{\partial t}$ by an average of two central difference quotients, one evaluated at t and the other at $t + k$:

$$\frac{c}{2}\left[\frac{u(x + h, t) - 2u(x, t) + u(x - h, t)}{h^2} + \frac{u(x + h, t + k) - 2u(x, t + k) + u(x - h, t + k)}{h^2} \right] = \frac{1}{k}[u(x, t + k) - u(x, t)]. \quad (4)$$

If we again define $\lambda = ck/h^2$, then after rearranging terms we can write (4) as

$$-u_{i-1,j+1} + \alpha u_{i,j+1} - u_{i+1,j+1} = u_{i+1,j} - \beta u_{ij} + u_{i-1,j}, \quad (5)$$

where $\alpha = 2(1 + 1/\lambda)$ and $\beta = 2(1 - 1/\lambda)$, $j = 0, 1, \ldots, m - 1$, and $i = 1, 2, \ldots, n - 1$.

For each choice of j the difference equation (5) for $i = 1, 2, \ldots, n - 1$ gives $n - 1$ equations in $n - 1$ unknowns $u_{i,j+1}$. Because of the prescribed boundary conditions, the values of $u_{i,j+1}$ are known for $i = 0$ and for $i = n$. For example, in the case $n = 4$ the system of equations for determining the approximate values of u on the $(j + 1)$st time line is

$$-u_{0,j+1} + \alpha u_{1,j+1} - u_{2,j+1} = u_{2,j} - \beta u_{1,j} + u_{0,j}$$

$$-u_{1,j+1} + \alpha u_{2,j+1} - u_{3,j+1} = u_{3,j} - \beta u_{2,j} + u_{1,j}$$

$$-u_{2,j+1} + \alpha u_{3,j+1} - u_{4,j+1} = u_{4,j} - \beta u_{3,j} + u_{2,j}$$

or

$$\alpha u_{1,j+1} - u_{2,j+1} \qquad\qquad = b_1$$
$$-u_{1,j+1} + \alpha u_{2,j+1} - u_{3,j+1} = b_2 \qquad (6)$$
$$- u_{2,j+1} + \alpha u_{3,j+1} = b_3,$$

where

$$b_1 = u_{2,j} - \beta u_{1,j} + u_{0,j} + u_{0,j+1}$$
$$b_2 = u_{3,j} - \beta u_{2,j} + u_{1,j}$$
$$b_3 = u_{4,j} - \beta u_{3,j} + u_{2,j} + u_{4,j+1}.$$

In general, if we use the difference equation (5) to determine values of u on the $(j+1)$st time line, we need to solve a linear system $\mathbf{AX} = \mathbf{B}$, where the coefficient matrix \mathbf{A} is a **tridiagonal matrix,**

$$\mathbf{A} = \begin{vmatrix} \alpha & -1 & 0 & 0 & 0 & \cdots & & 0 \\ -1 & \alpha & -1 & 0 & 0 & & & 0 \\ 0 & -1 & \alpha & -1 & 0 & & & 0 \\ 0 & 0 & -1 & \alpha & -1 & & & 0 \\ \vdots & & & & & \ddots & & \vdots \\ 0 & 0 & 0 & 0 & 0 & & \alpha & -1 \\ 0 & 0 & 0 & 0 & 0 & \cdots & -1 & \alpha \end{vmatrix},$$

and the entries of the column matrix \mathbf{B} are

$$b_1 = u_{2,j} - \beta u_{1,j} + u_{0,j} + u_{0,j+1}$$
$$b_2 = u_{3,j} - \beta u_{2,j} + u_{1,j}$$
$$b_3 = u_{4,j} - \beta u_{3,j} + u_{2,j}$$
$$\vdots$$
$$b_{n-1} = u_{n,j} - \beta u_{n-1,j} + u_{n-2,j} + u_{n,j+1}.$$

Computer programs for the Crank–Nicholson method are provided in the software supplement for this text.

Example 2 Using the Crank–Nicholson Method

Use the Crank–Nicholson method to approximate the solution of the boundary-value problem

$$0.25 \frac{\partial^2 u}{\partial x^2} = \frac{\partial u}{\partial t}, \qquad 0 < x < 2, \quad 0 < t < 0.3$$

$$u(0, t) = 0, \qquad\qquad u(2, t) = 0, \quad 0 \le t \le 0.3$$

$$u(x, 0) = \sin \pi x, \qquad 0 \le x \le 2,$$

using $n = 8$ and $m = 30$.

SOLUTION From the identifications $a = 2$, $T = 0.3$, $h = \frac{1}{4} = 0.25$, $k = \frac{1}{100} = 0.01$, and $c = 0.25$ we get $\lambda = 0.04$. With the aid of a computer we get the results in Table 16.3.

Table 16.3 Crank–Nicholson Method with $h = 0.25$, $k = 0.01$, $\lambda = 0.25$

Time	$x = 0.25$	$x = 0.50$	$x = 0.75$	$x = 1.00$	$x = 1.25$	$x = 1.50$	$x = 1.75$
0.00	0.7071	1.0000	0.7071	0.0000	−0.7071	−1.0000	−0.7071
0.01	0.6907	0.9768	0.6907	0.0000	−0.6907	−0.9768	−0.6907
0.02	0.6747	0.9542	0.6747	0.0000	−0.6747	−0.9542	−0.6747
0.03	0.6591	0.9321	0.6591	0.0000	−0.6591	−0.9321	−0.6591

continued

Table 16.3 Continued

Time	x = 0.25	x = 0.50	x = 0.75	x = 1.00	x = 1.25	x = 1.50	x = 1.75
0.04	0.6438	0.9105	0.6438	0.0000	−0.6438	−0.9105	−0.6438
0.05	0.6289	0.8894	0.6289	0.0000	−0.6289	−0.8894	−0.6289
0.06	0.6144	0.8688	0.6144	0.0000	−0.6144	−0.8688	−0.6144
0.07	0.6001	0.8487	0.6001	0.0000	−0.6001	−0.8487	−0.6001
0.08	0.5862	0.8291	0.5862	0.0000	−0.5862	−0.8291	−0.5862
0.09	0.5727	0.8099	0.5727	0.0000	−0.5727	−0.8099	−0.5727
0.10	0.5594	0.7911	0.5594	0.0000	−0.5594	−0.7911	−0.5594
0.11	0.5464	0.7728	0.5464	0.0000	−0.5464	−0.7728	−0.5464
0.12	0.5338	0.7549	0.5338	0.0000	−0.5338	−0.7549	−0.5338
0.13	0.5214	0.7374	0.5214	0.0000	−0.5214	−0.7374	−0.5214
0.14	0.5093	0.7203	0.5093	0.0000	−0.5093	−0.7203	−0.5093
0.15	0.4975	0.7036	0.4975	0.0000	−0.4975	−0.7036	−0.4975
0.16	0.4860	0.6873	0.4860	0.0000	−0.4860	−0.6873	−0.4860
0.17	0.4748	0.6714	0.4748	0.0000	−0.4748	−0.6714	−0.4748
0.18	0.4638	0.6559	0.4638	0.0000	−0.4638	−0.6559	−0.4638
0.19	0.4530	0.6407	0.4530	0.0000	−0.4530	−0.6407	−0.4530
0.20	0.4425	0.6258	0.4425	0.0000	−0.4425	−0.6258	−0.4425
0.21	0.4323	0.6114	0.4323	0.0000	−0.4323	−0.6114	−0.4323
0.22	0.4223	0.5972	0.4223	0.0000	−0.4223	−0.5972	−0.4223
0.23	0.4125	0.5834	0.4125	0.0000	−0.4125	−0.5834	−0.4125
0.24	0.4029	0.5699	0.4029	0.0000	−0.4029	−0.5699	−0.4029
0.25	0.3936	0.5567	0.3936	0.0000	−0.3936	−0.5567	−0.3936
0.26	0.3845	0.5438	0.3845	0.0000	−0.3845	−0.5438	−0.3845
0.27	0.3756	0.5312	0.3756	0.0000	−0.3756	−0.5312	−0.3756
0.28	0.3669	0.5189	0.3669	0.0000	−0.3669	−0.5189	−0.3669
0.29	0.3584	0.5068	0.3584	0.0000	−0.3584	−0.5068	−0.3584
0.30	0.3501	0.4951	0.3501	0.0000	−0.3501	−0.4951	−0.3501

Exact	Approximation
$u(0.75, 0.05) = 0.6250$	$u_{35} = 0.6289$
$u(0.50, 0.20) = 0.6105$	$u_{2,20} = 0.6259$
$u(0.25, 0.10) = 0.5525$	$u_{1,10} = 0.5594$

You should verify that the function satisfying all the conditions in Example 2 is $u(x, t) = e^{-\pi^2 t/4} \sin \pi x$. The sample comparisons listed in the margin table show that the absolute errors are of the order 10^{-2} or 10^{-3}. Smaller errors can be obtained by decreasing either h or k.

EXERCISES 16.2

Answers to odd-numbered problems begin on page A-64.

In Problems 1–12, use a computer as a computation aid.

1. Use the difference equation (3) to approximate the solution of the boundary-value problem

$$\frac{\partial^2 u}{\partial x^2} = \frac{\partial u}{\partial t}, \qquad 0 < x < 2, \quad 0 < t < 1$$

$$u(0, t) = 0, \qquad u(2, t) = 0, \quad 0 \le t \le 1$$

$$u(x, 0) = \begin{cases} 1, & 0 \le x \le 1 \\ 0, & 1 < x \le 2. \end{cases}$$

Use $n = 8$ and $m = 40$.

2. Using the Fourier series solution obtained in Problem 1 of Exercises 13.3 with $L = 2$, one can sum the first 20 terms to estimate the values for $u(0.25, 0.1)$, $u(1, 0.5)$, and $u(1.5, 0.8)$ for the solution $u(x, t)$ of Problem 1 above. A student wrote a computer program to do this and obtained the results

$u(0.25, 0.1) = 0.3794$, $u(1, 0.5) = 0.1854$, and $u(1.5, 0.8) = 0.0623$. Assume these results are accurate for all digits given. Compare these values with the approximations obtained in Problem 1 above. Find the absolute errors in each case.

3. Solve Problem 1 by the Crank–Nicholson method with $n = 8$ and $m = 40$. Use the values for $u(0.25, 0.1)$, $u(1, 0.5)$, and $u(1.5, 0.8)$ given in Problem 2 to compute the absolute errors.

4. Repeat Problem 1 using $n = 8$ and $m = 20$. Use the values for $u(0.25, 0.1)$, $u(1, 0.5)$, and $u(1.5, 0.8)$ given in Problem 2 to compute the absolute errors. Why are the approximations so inaccurate in this case?

5. Solve Problem 1 by the Crank–Nicholson method with $n = 8$ and $m = 20$. Use the values for $u(0.25, 0.1)$, $u(1, 0.5)$, and $u(1.5, 0.8)$ given in Problem 2 to compute the absolute

errors. Compare the absolute errors with those obtained in Problem 4.

6. It was shown in Section 13.2 that if a rod of length L is made of a material with thermal conductivity K, specific heat γ, and density ρ, the temperature $u(x, t)$ satisfies the partial differential equation

$$\frac{K}{\gamma\rho}\frac{\partial^2 u}{\partial x^2} = \frac{\partial u}{\partial t}, \qquad 0 < x < L.$$

Consider the boundary-value problem consisting of the foregoing equation and conditions

$$u(0, t) = 0, \qquad u(L, t) = 0, \quad 0 \le t \le 10$$

$$u(x, 0) = f(x), \qquad 0 \le x \le L.$$

Use the difference equation (3) in this section with $n = 10$ and $m = 10$ to approximate the solution of the boundary-value problem when

(a) $L = 20, \ K = 0.15, \ \rho = 8.0, \ \gamma = 0.11, \ f(x) = 30$

(b) $L = 50, \ K = 0.15, \ \rho = 8.0, \ \gamma = 0.11, \ f(x) = 30$

(c) $L = 20, \ K = 1.10, \ \rho = 2.7, \ \gamma = 0.22, \ f(x) = 0.5x(20 - x)$

(d) $L = 100, \ K = 1.04, \ \rho = 10.6, \ \gamma = 0.06,$

$$f(x) = \begin{cases} 0.8x, & 0 \le x \le 50 \\ 0.8(100 - x), & 50 < x \le 100. \end{cases}$$

7. Solve Problem 6 by the Crank–Nicholson method with $n = 10$ and $m = 10$.

8. Repeat Problem 6 if the endpoint temperatures are $u(0, t) = 0, u(L, t) = 20, 0 \le t \le 10$.

9. Solve Problem 8 by the Crank–Nicholson method.

10. Consider the boundary-value problem in Example 2. Assume that $n = 4$.

(a) Find the new value of λ.

(b) Use the Crank–Nicholson difference equation (5) to find the system of equations for u_{11}, u_{21}, and u_{31}, that is, the approximate values of u on the first time line. [*Hint:* Set $j = 0$ in (5), and let i take on the values 1, 2, 3.]

(c) Solve the system of three equations without the aid of a computer program. Compare your results with the corresponding entries in Table 16.3.

11. Consider a rod whose length is $L = 20$ for which $K = 1.05$, $\rho = 10.6$, and $\gamma = 0.056$. Suppose

$$u(0, t) = 20, \qquad u(20, t) = 30$$

$$u(x, 0) = 50.$$

(a) Use the method outlined in Section 13.6 to find the steady-state solution $\psi(x)$.

(b) Use the Crank–Nicholson method to approximate the temperatures $u(x, t)$ for $0 \le t \le T_{max}$. Select T_{max} large enough to allow the temperatures to approach the steady-state values. Compare the approximations for $t = T_{max}$ with the values of $\psi(x)$ found in part **(a)**.

12. Use the difference equation (3) to approximate the solution of the boundary-value problem

$$\frac{\partial^2 u}{\partial x^2} = \frac{\partial u}{\partial t}, \qquad 0 < x < 1, \quad 0 < t < 1$$

$$u(0, t) = 0, \qquad u(1, t) = 0, \quad 0 \le t \le 1$$

$$u(x, 0) = \sin \pi x, \qquad 0 \le x \le 1.$$

Use $n = 5$ and $m = 25$.

16.3 Hyperbolic Equations

We turn now to a numerical method for approximating a solution of a boundary-value problem involving the one-dimensional wave equation:

$$c^2 \frac{\partial^2 u}{\partial x^2} = \frac{\partial^2 u}{\partial t^2} \qquad 0 < x < a, \quad t > 0$$

$$u(0, t) = 0, \qquad u(a, t) = 0, \quad t > 0 \tag{1}$$

$$u(x, 0) = f(x), \qquad \frac{\partial u}{\partial t}\bigg|_{t=0} = g(x), \quad 0 \le x \le a.$$

The wave equation is an example of a hyperbolic partial differential equation. This problem has a unique solution if the functions f and g have continuous second derivatives on the interval $(0, a)$ and $f(a) = f(0) = 0$.

■ **A Difference Equation Replacement for the Wave Equation** As in the preceding two sections, our first objective is to replace the partial differential equation in (1) by a difference equation. In this case we replace the two second-order

partial derivatives by the central difference quotients

$$\frac{\partial^2 u}{\partial x^2} \approx \frac{1}{h^2}[u(x + h, t) - 2u(x, t) + u(x - h, t)]$$

and

$$\frac{\partial^2 u}{\partial t^2} \approx \frac{1}{k^2}[u(x, t + k) - 2u(x, t) + u(x, t - k)].$$

Thus we replace $c^2 \dfrac{\partial^2 u}{\partial x^2} = \dfrac{\partial^2 u}{\partial t^2}$ by

$$\frac{c^2}{h^2}[u(x + h, t) - 2u(x, t) + u(x - h, t)] = \frac{1}{k^2}[u(x, t + k) - 2u(x, t) + u(x, t - k)]. \qquad (2)$$

If we let $\lambda = ck/h$, then (2) can be expressed as

$$u_{i,j+1} = \lambda^2 u_{i+1,j} + 2(1 - \lambda^2)u_{ij} + \lambda^2 u_{i-1,j} - u_{i,j-1} \qquad (3)$$

for $i = 1, 2, \ldots, n - 1$ and $j = 1, 2, \ldots, m - 1$.

The numerical method based on equation (3), like the first method considered in Section 16.2, is an explicit finite difference method. As before, we use the difference equation to approximate the solution $u(x, t)$ of (1) over a rectangular region in the xt-plane defined by the inequalities $0 \leq x \leq a$, $0 \leq t \leq T$. If n and m are positive integers and

$$h = \frac{a}{n} \qquad \text{and} \qquad k = \frac{T}{m},$$

the vertical and horizontal grid lines on this region are defined by

$$x_i = ih, \quad i = 0, 1, 2, \ldots, n \qquad \text{and} \qquad t_j = jk, \quad j = 0, 1, 2, \ldots, m.$$

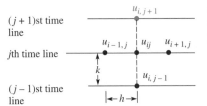

(j + 1)st time line

jth time line

(j − 1)st time line

Figure 16.10

As shown in Figure 16.10, (3) enables us to obtain the approximation $u_{i,j+1}$ on the $(j + 1)$st time line from the values indicated on the jth and $(j - 1)$st time lines. Moreover, we use

$$u_{0,j} = u(0, jk) = 0, \qquad u_{n,j} = u(a, jk) = 0 \qquad \leftarrow \text{boundary conditions}$$

and

$$u_{i,0} = u(x_i, 0) = f(x_i). \qquad \leftarrow \text{initial condition}$$

There is one minor problem in getting started. You can see from (3) that for $j = 1$ we need to know the values of $u_{i,1}$ (that is, the estimates of u on the first time line) in order to find $u_{i,2}$. But from Figure 16.10, with $j = 0$, we see that the values of $u_{i,1}$ on the first time line depend on the values of $u_{i,0}$ on the zeroth time line and on the values of $u_{i,-1}$. To compute these latter values we make use of the initial-velocity condition $u_t(x, 0) = g(x)$. At $t = 0$ it follows from (5) of Section 6.5 that

$$g(x_i) = u_t(x_i, 0) \approx \frac{u(x_i, k) - u(x_i, -k)}{2k}. \qquad (4)$$

In order to make sense of the term $u(x_i, -k) = u_{i,-1}$ in (4) we have to imagine $u(x, t)$ extended backward in time. It follows from (4) that

$$u(x_i, -k) \approx u(x_i, k) - 2kg(x_i).$$

This last result suggests that we define

$$u_{i,-1} = u_{i,1} - 2kg(x_i) \qquad (5)$$

in the iteration of (3). By substituting (5) into (3) when $j = 0$, we get the special case

$$u_{i,1} = \frac{\lambda^2}{2}(u_{i+1,0} + u_{i-1,0}) + (1 - \lambda^2)u_{i,0} + kg(x_i). \qquad (6)$$

Example 1 Using the Finite Difference Method

Approximate the solution of the boundary-value problem

$$4\frac{\partial^2 u}{\partial x^2} = \frac{\partial^2 u}{\partial t^2}, \qquad 0 < x < 1, \quad 0 < t < 1$$

$$u(0, t) = 0, \qquad u(1, t) = 0, \quad 0 \le t \le 1$$

$$u(x, 0) = \sin \pi x, \qquad \left.\frac{\partial u}{\partial t}\right|_{t=0} = 0, \quad 0 \le x \le 1,$$

using (3) with $n = 5$ and $m = 20$.

SOLUTION We make the identifications $c = 2$, $a = 1$, and $T = 1$. With $n = 5$ and $m = 20$ we get $h = \frac{1}{5} = 0.2$, $k = \frac{1}{20} = 0.05$, and $\lambda = 0.5$. Thus, with $g(x) = 0$, equations (6) and (3) become, respectively,

$$u_{i,1} = 0.125(u_{i+1,0} + u_{i-1,0}) + 0.75u_{i,0} \qquad (7)$$

$$u_{i,j+1} = 0.25u_{i+1,j} + 1.5u_{ij} + 0.25u_{i-1,j} - u_{i,j-1}. \qquad (8)$$

For $i = 1, 2, 3, 4$, equation (7) yields the following values for the $u_{i,1}$ on the first time line:

$$\begin{aligned}
u_{11} &= 0.125(u_{20} + u_{00}) + 0.75u_{10} = 0.55972100 \\
u_{21} &= 0.125(u_{30} + u_{10}) + 0.75u_{20} = 0.90564761 \\
u_{31} &= 0.125(u_{40} + u_{20}) + 0.75u_{30} = 0.90564761 \\
u_{41} &= 0.125(u_{50} + u_{30}) + 0.75u_{40} = 0.55972100.
\end{aligned} \qquad (9)$$

Note that the results given in (9) were obtained from the initial condition $u(x, 0) = \sin \pi x$. For example, $u_{20} = \sin(0.2\pi)$, and so on. Now $j = 1$ in (8) gives

$$u_{i,2} = 0.25u_{i+1,1} + 1.5u_{i,1} + 0.25u_{i-1,1} - u_{i,0},$$

and so for $i = 1, 2, 3, 4$ we get

$$\begin{aligned}
u_{12} &= 0.25u_{21} + 1.5u_{11} + 0.25u_{01} - u_{10} \\
u_{22} &= 0.25u_{31} + 1.5u_{21} + 0.25u_{11} - u_{20} \\
u_{32} &= 0.25u_{41} + 1.5u_{31} + 0.25u_{21} - u_{30} \\
u_{42} &= 0.25u_{51} + 1.5u_{41} + 0.25u_{31} - u_{40}.
\end{aligned}$$

Using the boundary conditions, the initial conditions, and the data obtained in (9), we get from these equations the approximations for u on the second time line. These results and the remaining calculations are summarized in Table 16.4.

Table 16.4 Explicit Difference Equation Approximation with $h = 0.2$, $k = 0.05$, $\lambda = 0.5$

Time	$x = 0.20$	$x = 0.40$	$x = 0.60$	$x = 0.80$	Time	$x = 0.20$	$x = 0.40$	$x = 0.60$	$x = 0.80$
0.00	0.5878	0.9511	0.9511	0.5878	0.55	−0.5663	−0.9163	−0.9163	−0.5663
0.05	0.5597	0.9056	0.9056	0.5597	0.60	−0.4912	−0.7947	−0.7947	−0.4912
0.10	0.4782	0.7738	0.7738	0.4782	0.65	−0.3691	−0.5973	−0.5973	−0.3691
0.15	0.3510	0.5680	0.5680	0.3510	0.70	−0.2119	−0.3428	−0.3428	−0.2119
0.20	0.1903	0.3080	0.3080	0.1903	0.75	−0.0344	−0.0556	−0.0556	−0.0344
0.25	0.0115	0.0185	0.0185	0.0115	0.80	0.1464	0.2369	0.2369	0.1464
0.30	−0.1685	−0.2727	−0.2727	−0.1685	0.85	0.3132	0.5068	0.5068	0.3132

continued

Table 16.5 Continued

Time	$x = 0.20$	$x = 0.40$	$x = 0.60$	$x = 0.80$	Time	$x = 0.20$	$x = 0.40$	$x = 0.60$	$x = 0.80$
0.35	−0.3324	−0.5378	−0.5378	−0.3324	0.90	0.4501	0.7283	0.7283	0.4501
0.40	−0.4645	−0.7516	−0.7516	−0.4645	0.95	0.5440	0.8803	0.8803	0.5440
0.45	−0.5523	−0.8936	−0.8936	−0.5523	1.00	0.5860	0.9482	0.9482	0.5860
0.50	−0.5873	−0.9503	−0.9503	−0.5873					

Exact	Approximation
$u(0.4, 0.25) = 0$	$u_{25} = 0.0185$
$u(0.6, 0.3) = -0.2939$	$u_{36} = -0.2727$
$u(0.2, 0.5) = -0.5878$	$u_{1,10} = -0.5873$
$u(0.8, 0.7) = -0.1816$	$u_{4,14} = -0.2119$

Exact	Approximation
$u(0.25, 0.3125) =$ -0.2706	$u_{25} = -0.2706$
$u(0.375, 0.375) =$ -0.6533	$u_{36} = -0.6533$
$u(0.125, 0.625) =$ -0.2706	$u_{1,10} = -0.2706$

It is readily verified that the exact solution of the problem in Example 1 is $u(x, t) = \sin \pi x \cos 2\pi t$. With this function we can compare the exact results with the approximations given in Table 16.4. For example, some sample values are given in the first margin table.

As you can see, the approximations are in the same "ball park" as the exact values, but the accuracy is not particularly impressive. We can, however, obtain more accurate results. The accuracy of this algorithm varies with the choice of λ. Of course, λ is determined by the choice of the integers n and m, which in turn determine the values of the step sizes h and k. It can be proved that the best accuracy is always obtained from this method when the ratio $\lambda = kc/h$ is equal to one—in other words, when the step in the time direction is $k = h/c$. For example, the choice $n = 8$ and $m = 16$ yields $h = ⅛$, $k = 1/16$, and $\lambda = 1$. The sample values listed in the second margin table clearly show the improved accuracy.

■ **Stability** We note in conclusion that this explicit finite difference method for the wave equation is stable when $\lambda \leq 1$ and unstable when $\lambda > 1$.

E X E R C I S E S 1 6 . 3

Answers to odd-numbered problems begin on page A-69.

In Problems 1, 3, 5, and 6, use a computer as a computation aid.

1. Use the difference equation (3) to approximate the solution of the boundary-value problem

$$c^2 \frac{\partial^2 u}{\partial x^2} = \frac{\partial^2 u}{\partial t^2}, \qquad 0 < x < a, \quad 0 < t < T$$

$$u(0, t) = 0, \qquad u(a, t) = 0, \quad 0 \leq t \leq T$$

$$u(x, 0) = f(x), \qquad \left. \frac{\partial u}{\partial t} \right|_{t=0} = 0, \quad 0 \leq x \leq a,$$

when

(a) $c = 1$, $a = 1$, $T = 1$, $f(x) = x(1 - x)$; $n = 4$ and $m = 10$

(b) $c = 1$, $a = 2$, $T = 1$, $f(x) = e^{-16(x-1)^2}$; $n = 5$ and $m = 10$

(c) $c = \sqrt{2}$, $a = 1$, $T = 1$, $f(x) = \begin{cases} 0, & 0 \leq x \leq 0.5 \\ 0.5, & 0.5 < x \leq 1; \end{cases}$

$n = 10$ and $m = 25$.

2. Consider the boundary-value problem

$$\frac{\partial^2 u}{\partial x^2} = \frac{\partial^2 u}{\partial t^2}, \qquad 0 < x < 1, \quad 0 < t < 0.5$$

$$u(0, t) = 0, \qquad u(1, t) = 0, \quad 0 \leq t \leq 0.5$$

$$u(x, 0) = \sin \pi x, \qquad \left. \frac{\partial u}{\partial t} \right|_{t=0} = 0, \quad 0 \leq x \leq 1.$$

(a) Use the methods of Chapter 13 to verify that the solution of the problem is $u(x, t) = \sin \pi x \cos \pi t$.

(b) Use the method of this section to approximate the solution of the problem without the aid of a computer program. Use $n = 4$ and $m = 5$.

(c) Compute the absolute error at each interior grid point.

3. Approximate the solution of the boundary-value problem in Problem 2 using a computer program with

(a) $n = 5$, $m = 10$ **(b)** $n = 5$, $m = 20$.

4. Given the boundary-value problem

$$\frac{\partial^2 u}{\partial x^2} = \frac{\partial^2 u}{\partial t^2}, \qquad 0 < x < 1, \quad 0 < t < 1$$

$$u(0, t) = 0, \qquad u(1, t) = 0, \quad 0 \leq t \leq 1$$

$$u(x, 0) = x(1 - x), \qquad \left.\frac{\partial u}{\partial t}\right|_{t=0} = 0, \quad 0 \leq x \leq 1,$$

use $h = k = \frac{1}{5}$ in equation (6) to compute the values of $u_{i,1}$ by hand.

5. It was shown in Section 13.2 that the equation of a vibrating string is

$$\frac{T}{\rho} \frac{\partial^2 u}{\partial x^2} = \frac{\partial^2 u}{\partial t^2},$$

where T is the constant magnitude of the tension in the string and ρ is its mass per unit length. Suppose a string of length 60 centimeters is secured to the x-axis at its ends and is released from rest from the initial displacement

$$f(x) = \begin{cases} 0.01x, & 0 \leq x \leq 30 \\ 0.30 - \dfrac{x - 30}{100}, & 30 < x \leq 60. \end{cases}$$

Use the difference equation (3) in this section to approximate the solution of the boundary-value problem when $h = 10$, $k = 5\sqrt{\rho/T}$ and where $\rho = 0.0225$ g/cm, $T = 1.4 \times 10^7$ dynes. Use $m = 50$.

6. Repeat Problem 5 using

$$f(x) = \begin{cases} 0.2x, & 0 \leq x \leq 15 \\ 0.30 - \dfrac{x - 15}{150}, & 15 < x \leq 60 \end{cases}$$

and $h = 10$, $k = 2.5\sqrt{\rho/T}$. Use $m = 50$.

CHAPTER 16 REVIEW EXERCISES

Answers to odd-numbered problems begin on page A-71.

1. Consider the boundary-value problem

$$\frac{\partial^2 u}{\partial x^2} + \frac{\partial^2 u}{\partial y^2} = 0, \qquad 0 < x < 2, \quad 0 < y < 1$$

$$u(0, y) = 0, \qquad u(2, y) = 50, \quad 0 < y < 1$$

$$u(x, 0) = 0, \qquad u(x, 1) = 0, \quad 0 < x < 2.$$

Approximate the solution of the differential equation at the interior points of the region with mesh size $h = \frac{1}{2}$. Use Gaussian elimination or Gauss–Seidel iteration.

2. Solve Problem 1 using mesh size $h = \frac{1}{4}$. Use Gauss–Seidel iteration.

3. Consider the boundary-value problem

$$\frac{\partial^2 u}{\partial x^2} = \frac{\partial u}{\partial t}, \qquad 0 < x < 1, \quad 0 < t < 0.05$$

$$u(0, t) = 0, \qquad u(1, t) = 0, \quad t > 0$$

$$u(x, 0) = x, \qquad 0 < x < 1.$$

(a) Note that the initial temperature $u(x, 0) = x$ indicates that the temperature at the right boundary $x = 1$ should be $u(1, 0) = 1$, whereas the boundary conditions imply that $u(1, 0) = 0$. The BASIC program in Appendix V for the explicit finite difference method was written so that the boundary conditions prevail for all times considered,

including $t = 0$. Use the program (or write your own) to complete the following table.

Time	x = 0.00	x = 0.20	x = 0.40	x = 0.60	x = 0.80	x = 1.00
0.00	0.0000	0.2000	0.4000	0.6000	0.8000	0.0000
0.01	0.0000					0.0000
0.02	0.0000					0.0000
0.03	0.0000					0.0000
0.04	0.0000					0.0000
0.05	0.0000					0.0000

(b) Modify the BASIC program (or your own) so that the initial condition prevails at the boundaries at $t = 0$. Use this program to complete the following table.

Time	x = 0.00	x = 0.20	x = 0.40	x = 0.60	x = 0.80	x = 1.00
0.00	0.0000	0.2000	0.4000	0.6000	0.8000	1.0000
0.01	0.0000					0.0000
0.02	0.0000					0.0000
0.03	0.0000					0.0000
0.04	0.0000					0.0000
0.05	0.0000					0.0000

(c) Are the tables in parts **(a)** and **(b)** related in any way? Use a larger time interval if necessary.

PART

5

Complex Analysis

Application *Essay*

Forebody Drag of Bluff Bodies

Robert E. Breidenthal
University of Washington

The shape of the front end of automobiles and especially trucks has markedly changed over the past two decades. Prior to this change, cars and trucks were boxy at the front, with sharp corners and prominent radiators. See [1] for vehicle aerodynamics and its history and [2] for the general subject of aerodynamic drag from a former Messerschmitt designer. However, modern vehicles are now much different, with rounded corners at the front. This has dramatically lowered the aerodynamic drag of these vehicles at highway speeds, with a corresponding increase in cruise fuel economy. In this example, taken from class notes of Professor Anatol Roshko at Caltech, we will explore the underlying physics of the flow around the front of these vehicles.

For simplicity, we consider first the flow of a frictionless fluid at speed U past a flat-sided bluff body. The flowfield is readily calculated using the **Schwarz–Christoffel transformation.** A forward-facing step is illustrated in Figure 1, which represents the upper half of a blunt-faced bluff body. Using the Schwarz–Christoffel transformation, the potential flow past the step is calculated.

Figure 1 z-plane

First, the actual flow in the physical z-plane is transformed to the ζ-plane (where the speed is U′) by the transformation

$$\frac{dz}{d\zeta} = k\left(\frac{\zeta-1}{\zeta+1}\right)^{1/2} = k\frac{\zeta-1}{\sqrt{\zeta^2-1}} = k\left(\frac{\zeta}{\sqrt{\zeta^2-1}} - \frac{1}{\sqrt{\zeta^2-1}}\right).$$

See Figure 2.

Figure 2 ζ-plane

Integrating, we find

$$z = k\left[\sqrt{\zeta^2-1} - \ln(\zeta+\sqrt{\zeta^2-1})\right] + \text{const.}$$

Evaluating the constants, at $z = 0$, $\zeta = -1$, so const. $= ki\pi$. At $z = ih$, $\zeta = 1$, so that $k = h/\pi$. Therefore

$$z = \frac{h}{\pi}\left[\sqrt{\zeta^2-1} - \ln(\zeta+\sqrt{\zeta^2-1})\right] + ih.$$

Now, the complex velocity is

$$w(z) = u - iv = \frac{dF}{dz} = \frac{dF}{d\zeta}\frac{d\zeta}{dz} = \frac{U'}{k\left(\dfrac{\zeta-1}{\zeta+1}\right)^{1/2}},$$

where the velocity is U' in the ζ-plane. Thus

$$w(z) = \left(\frac{\zeta+1}{\zeta-1}\right)^{1/2} U.$$

For ζ real going to infinity, z goes to infinity, and w goes to U. For ζ real going to minus infinity, z goes to minus infinity, and w goes to U. We conclude that $U = U'/k = \pi U'/h$. However, we require $U = U'$, so $h = \pi$.

The pressure distribution on the boundary is then determined from Bernoulli's equation. The pressure coefficient c_p is a dimensionless measure of the pressure,

$$c_p = \frac{p - p_\infty}{\frac{1}{2}\rho U^2} = 1 - \frac{\text{vel}^2}{U^2},$$

where the magnitude of the velocity is defined by

$$\text{vel}^2 = u^2 + v^2,$$

and the fluid density is ρ. In Figure 1, on the segments AB and CD, the vertical velocity v is zero, so

$$c_p = 1 - \frac{u^2}{U^2}.$$

On BC, the horizontal velocity component $u = 0$, so

$$c_p = 1 - \frac{v^2}{U^2}.$$

The total drag force is the integral of the pressure on the face of the step

$$D = \tfrac{1}{2}\rho U^2 \int_0^h c_p \, dy.$$

On BC, $\zeta = \xi$, where $-1 < \xi < 1$. The complex velocity

$$u - iv = U\left(\frac{\xi + 1}{\xi - 1}\right)^{1/2} = iU\left(\frac{1 + \xi}{1 - \xi}\right)^{1/2}.$$

Also on BC, $u = 0$ and

$$v = U\left(\frac{1 + \xi}{1 - \xi}\right)^{1/2}.$$

Consequently,

$$c_p = 1 - \frac{1 + \xi}{1 - \xi}.$$

From the transformation earlier,

$$\frac{dz}{d\zeta} = i\frac{dy}{d\xi} = k\left(\frac{\xi - 1}{\xi + 1}\right)^{1/2} = ik\left(\frac{1 - \xi}{1 + \xi}\right)^{1/2},$$

so that

$$\frac{dy}{d\xi} = k\left(\frac{1 - \xi}{1 + \xi}\right)^{1/2} = \left(\frac{1 - \xi}{1 + \xi}\right)^{1/2}.$$

The drag becomes

$$D = \tfrac{1}{2}\rho U^2 \int_{-1}^{1} c_p \frac{dy}{d\xi} \, d\xi$$

$$= \tfrac{1}{2}\rho U^2 \int_{-1}^{1} \left(1 - \frac{1 + \xi}{1 - \xi}\right)\left(\frac{1 - \xi}{1 + \xi}\right)^{1/2} d\xi.$$

Normalizing the drag by the dynamic pressure $q = \tfrac{1}{2}\rho U^2$,

$$\frac{D}{q} = \int_{-1}^{1}\left[\left(\frac{1 - \xi}{1 + \xi}\right)^{1/2} - \left(\frac{1 + \xi}{1 - \xi}\right)^{1/2}\right]d\xi =$$

$$\int_{-1}^{1}\left[\frac{-2\xi}{\sqrt{1 - \xi^2}}\right]d\xi = 0.$$

We see that the forebody drag force is zero. How can this be? High pressure exists in the stagnation region near the re-entrant corner, and infinitely negative pressure exists at the convex corner, where the fluid velocity is infinite.

The line $y = 0$ represents a plane of symmetry for the problem of flow past a bluff forebody. Figure 3 is a sketch of the pressure distribution over the front surface of a bluff body in frictionless flow. The low pressure at the convex corners is called "leading edge suction." The forces from these two pressures cancel each other; the average pressure on the front face of the step is found to be exactly equal to the freestream pressure! Thus the drag on a semi-infinite body in potential flow is zero (see also references [3] and [4]). This fact has important consequences for the engineering design of vehicles that you see every day.

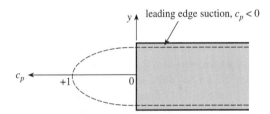

Figure 3 Ideal flow

If the fluid is not frictionless but instead has viscosity, a boundary layer forms next to any solid surface. The boundary layer may not always follow the contours of the body but may separate away from it in regions of rapidly rising pressure, which typically occur at the sharp corners. Thus the flow does not follow the sharp turn at the corner, and the low pressure there is lost. See Figure 4. However, if the surface contours are sufficiently gentle that the boundary layer can remain attached [5], then the flow outside the thin boundary layer is nearly that of a perfect fluid, with a corresponding pressure distribution. The leading edge suction is preserved, and the drag is nearly zero.

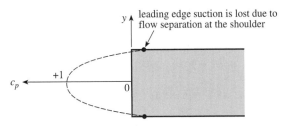

Figure 4 Real flow

Modern trucks (and cars) have a rounded shape to the front corners, so that the boundary layer can remain attached there, preserving the leading edge suction and dramatically reducing the drag. Without the rounding, the boundary layer would separate, the outer flow would see a much different effective shape, and the leading edge suction would be lost, corresponding to a much higher drag.

Airfoils have a rounded leading edge for the same reason. By maintaining attached flow there, the intense low pressure on the leading edge helps yield a very low drag. Turbine engine inlets likewise have a rounded inlet lip, so that at the start of a takeoff, the leading edge suction on the lip helps the aircraft accelerate. For a Boeing 747 at the start of the takeoff roll, about 40% of the entire thrust is due to leading edge suction pulling the aircraft forward! If you look closely, you can see the multiple rows of rivets securing the inlet lip to the rest of the engine nacelle.

Modern automobile and truck design exploits the leading edge suction to nearly eliminate the aerodynamic drag from their forebodies [1]. However, the drag from the base region on the backside of such bluff bodies is not so easily eliminated. Research continues on the aerodynamics of bluff bodies, making this subject much more interesting and complicated than that of streamlined objects [6].

Suggested Problems

1. Plot $c_p(y)$ for ideal flow past a two-dimensional step.
2. Estimate the drag if all leading edge suction is lost, such as from a sharp corner.
3. Estimate the energy saved per mile by one truck driving on level ground at 70 mph at sea level if it can preserve its leading edge suction by rounding off the upstream corners.

References

[1] *Aerodynamic Drag Mechanisms of Bluff Bodies and Road Vehicles* (ed. G. Sovran, T. Morel, & W. T. Mason). New York: Plenum, 1978.
[2] Hoerner, S. F., *Fluid Dynamic Drag*. Bricktown, N.J.: Hoerner Fluid Dynamics, 1965.
[3] Prandtl, L. and O. G. Tietjens, *Applied Hydro- and Aeromechanics* (translated by J. P. Hartog). London: Dover, 1934.
[4] Morel, T., "Theoretical lower limits of forebody drag." *Aero. J.* 83 (1979): 23–27.
[5] Polhamus, E. C., "Effect of nose shape on aerodynamic characteristics of a body of revolution having a fineness ratio of 10.94." *NACA RM L57F25* (1957).
[6] Koenig, K. and A. Roshko, "An experimental study of geometrical effects on the drag and flow field of two bluff bodies separated by a gap." *J. Fluid Mech.* 156 (1985): 167–204.

Robert E. Breidenthal, Jr., is currently a professor in the Department of Aeronautics and Astronautics, University of Washington, Seattle. After graduating from Wichita State University, Mr. Breidenthal obtained both his master's and doctoral degrees from the California Institute of Technology. He is the author of several publications in his field of expertise.

Functions of a Complex Variable

INTRODUCTION

In elementary algebra courses you learned about the existence, and some of the properties, of complex numbers. But in courses such as calculus, it is most likely that you did not even see a complex number. Introductory calculus is basically the study of functions of a real variable. In advanced courses you used complex numbers only occasionally (see Sections 3.3, 8.8, and 10.2). However, in the next four chapters we are going to introduce you to complex analysis. We are going to examine limits, derivatives, integrals, sequences, series, and applications of functions of a complex variable. Although there are many similarities between this and real analysis, there are many interesting differences and some surprises.

17.1 Complex Numbers

You have undoubtedly encountered complex numbers in your earlier mathematics classes. From the quadratic formula we know that the zeros of the quadratic function $f(x) = ax^2 + bx + c$ are complex whenever the discriminant $b^2 - 4ac$ is negative. Simple equations such as $x^2 = -5$ and $x^2 + x + 1 = 0$ have no real-number solutions.

> **DEFINITION 17.1** **Complex Number**
>
> A number of the form $z = x + iy$ where x and y are real numbers and i is a number such that $i^2 = -1$ is called a **complex number.**

■ **Terminology** The number i in Definition 17.1 is called the **imaginary unit.** The real number x in $z = x + iy$ is called the **real part** of z; the real number y is called the **imaginary part** of z. The real and imaginary parts of a complex number z are abbreviated $\text{Re}(z)$ and $\text{Im}(z)$, respectively. For example, if $z = 4 - 9i$, then $\text{Re}(z) = 4$ and $\text{Im}(z) = -9$. A real constant multiple of the imaginary unit is called a **pure imaginary number.** For example, $z = 6i$ is a pure imaginary number. Two complex numbers are **equal** if their real and imaginary parts are equal. Since this simple concept is sometimes useful, we formalize the last statement in the next definition.

> **DEFINITION 17.2** **Equality**
>
> Complex numbers $z_1 = x_1 + iy_1$ and $z_2 = x_2 + iy_2$ are **equal,** $z_1 = z_2$, if $\text{Re}(z_1) = \text{Re}(z_2)$ and $\text{Im}(z_1) = \text{Im}(z_2)$.

A complex number $x + iy = 0$ if $x = 0$ and $y = 0$.

■ **Arithmetic Operations** Complex numbers can be added, subtracted, multiplied, and divided. If $z_1 = x_1 + iy_1$ and $z_2 = x_2 + iy_2$, these operations are defined as follows.

Addition: $z_1 + z_2 = (x_1 + iy_1) + (x_2 + iy_2) = (x_1 + x_2) + i(y_1 + y_2)$

Subtraction: $z_1 - z_2 = (x_1 + iy_1) - (x_2 + iy_2) = (x_1 - x_2) + i(y_1 - y_2)$

Multiplication: $z_1 \cdot z_2 = (x_1 + iy_1)(x_2 + iy_2)$
$$= x_1 x_2 - y_1 y_2 + i(y_1 x_2 + x_1 y_2)$$

Division: $\dfrac{z_1}{z_2} = \dfrac{x_1 + iy_1}{x_2 + iy_2}$

$$= \frac{x_1 x_2 + y_1 y_2}{x_2^2 + y_2^2} + i\,\frac{y_1 x_2 - x_1 y_2}{x_2^2 + y_2^2}$$

The familiar commutative, associative, and distributive laws hold for complex numbers.

Commutative laws: $\begin{cases} z_1 + z_2 = z_2 + z_1 \\ \quad z_1 z_2 = z_2 z_1 \end{cases}$

Associative laws: $\begin{cases} z_1 + (z_2 + z_3) = (z_1 + z_2) + z_3 \\ \quad z_1(z_2 z_3) = (z_1 z_2)z_3 \end{cases}$

Distributive law: $z_1(z_2 + z_3) = z_1 z_2 + z_1 z_3$

In view of these laws, there is no need to memorize the definitions of addition, subtraction, and multiplication. To add (subtract) two complex numbers, we simply add (subtract) the corresponding real and imaginary parts. To multiply two complex numbers, we use the distributive law and the fact that $i^2 = -1$.

Example 1 Addition and Multiplication

If $z_1 = 2 + 4i$ and $z_2 = -3 + 8i$, find (a) $z_1 + z_2$ and (b) $z_1 z_2$.

SOLUTION (a) By adding the real and imaginary parts of the two numbers, we get

$$(2 + 4i) + (-3 + 8i) = (2 - 3) + (4 + 8)i = -1 + 12i.$$

(b) Using the distributive law, we have

$$(2 + 4i)(-3 + 8i) = (2 + 4i)(-3) + (2 + 4i)(8i)$$

$$= -6 - 12i + 16i + 32i^2$$

$$= (-6 - 32) + (16 - 12)i = -38 + 4i.$$ ❏

There is also no need to memorize the definition of division, but before discussing that we need to introduce another concept.

■ **Conjugate** If z is a complex number, then the number obtained by changing the sign of its imaginary part is called the **complex conjugate** or, simply, the **conjugate** of z. If $z = x + iy$, then its conjugate is

$$\bar{z} = x - iy.$$

For example, if $z = 6 + 3i$, then $\bar{z} = 6 - 3i$; if $z = -5 - i$, then $\bar{z} = -5 + i$. If z is a real number, say $z = 7$, then $\bar{z} = 7$. From the definition of addition it can be readily shown that the conjugate of a sum of two complex numbers is the sum of the conjugates:

$$\overline{z_1 + z_2} = \bar{z}_1 + \bar{z}_2.$$

Moreover, we have the additional three properties

$$\overline{z_1 - z_2} = \bar{z}_1 - \bar{z}_2, \qquad \overline{z_1 z_2} = \bar{z}_1 \bar{z}_2, \qquad \overline{\left(\frac{z_1}{z_2}\right)} = \frac{\bar{z}_1}{\bar{z}_2}.$$

The definitions of addition and multiplication show that the sum and product of a complex number z and its conjugate \bar{z} are also real numbers:

$$z + \bar{z} = (x + iy) + (x - iy) = 2x \tag{1}$$

$$z\bar{z} = (x + iy)(x - iy) = x^2 - i^2 y^2 = x^2 + y^2. \tag{2}$$

The difference between a complex number z and its conjugate \bar{z} is a pure imaginary number:

$$z - \bar{z} = (x + iy) - (x - iy) = 2iy. \tag{3}$$

Since $x = \text{Re}(z)$ and $y = \text{Im}(z)$, (1) and (3) yield two useful formulas:

$$\text{Re}(z) = \frac{z + \bar{z}}{2} \quad \text{and} \quad \text{Im}(z) = \frac{z - \bar{z}}{2i}.$$

However, (2) is the important relationship that enables us to approach division in a more practical manner: To divide z_1 by z_2, we multiply both numerator and denominator of z_1/z_2 by the conjugate of z_2. This will be illustrated in the next example.

Example 2 Division

If $z_1 = 2 - 3i$ and $z_2 = 4 + 6i$, find (a) $\dfrac{z_1}{z_2}$ and (b) $\dfrac{1}{z_1}$.

SOLUTION In both parts of this example we shall multiply both numerator and denominator by the conjugate of the denominator and then use (2).

(a)
$$\frac{2 - 3i}{4 + 6i} = \frac{2 - 3i}{4 + 6i}\,\frac{4 - 6i}{4 - 6i} = \frac{8 - 12i - 12i + 18i^2}{16 + 36}$$

$$= \frac{-10 - 24i}{52} = -\frac{5}{26} - \frac{6}{13}\,i.$$

(b)
$$\frac{1}{2 - 3i} = \frac{1}{2 - 3i}\,\frac{2 + 3i}{2 + 3i} = \frac{2 + 3i}{4 + 9} = \frac{2}{13} + \frac{3}{13}\,i. \qquad \square$$

■ **Geometric Interpretation** A complex number $z = x + iy$ is uniquely determined by an *ordered pair* of real numbers (x, y). The first and second entries of the ordered pairs correspond, in turn, with the real and imaginary parts of the complex number. For example, the ordered pair $(2, -3)$ corresponds to the complex number $z = 2 - 3i$. Conversely, $z = 2 - 3i$ determines the ordered pair $(2, -3)$. In this manner we are able to associate a complex number $z = x + iy$ with a point (x, y) in a coordinate plane. But, as we saw in Section 7.1, an ordered pair of real numbers can be interpreted as the components of a vector. Thus, a complex number $z = x + iy$ can also be viewed as a *vector* whose initial point is the origin and whose terminal point is (x, y). The coordinate plane illustrated in Figure 17.1 is called the **complex plane** or simply the **z-plane**. The horizontal or x-axis is called the **real axis** and the vertical or y-axis is called the **imaginary axis.** The length of a vector z, or the distance from the origin to the point (x, y), is clearly $\sqrt{x^2 + y^2}$. This real number is given a special name.

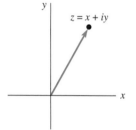

Figure 17.1

> **DEFINITION 17.3 Modulus or Absolute Value**
>
> The **modulus** or **absolute value** of $z = x + iy$, denoted by $|z|$, is the real number
> $$|z| = \sqrt{x^2 + y^2} = \sqrt{z\bar{z}}. \qquad (4)$$

Example 3 Modulus of a Complex Number

If $z = 2 - 3i$, then $|z| = \sqrt{2^2 + (-3)^2} = \sqrt{13}$. $\qquad \square$

As Figure 17.2 shows, the sum of the vectors z_1 and z_2 is the vector $z_1 + z_2$. For the triangle given in the figure we know that the length of the side of the triangle corresponding to the vector $z_1 + z_2$ cannot be longer than the sum of the remaining two sides. In symbols this is

$$|z_1 + z_2| \le |z_1| + |z_2|. \qquad (5)$$

The result in (5) is known as the **triangle inequality** and extends to any finite sum:

$$|z_1 + z_2 + z_3 + \cdots + z_n| \le |z_1| + |z_2| + |z_3| + \cdots + |z_n|. \qquad (6)$$

Using (5) on $z_1 + z_2 + (-z_2)$, we obtain another important inequality:

$$|z_1 + z_2| \ge |z_1| - |z_2|. \qquad (7)$$

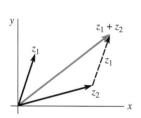

Figure 17.2

> ### Remarks
>
> Many of the properties of the real system hold in the complex number system, but there are some remarkable differences as well. For example, we cannot compare two complex numbers $z_1 = x_1 + iy_1$, $y_1 \neq 0$, and $z_2 = x_2 + iy_2$, $y_2 \neq 0$, by means of inequalities. In other words, statements such as $z_1 < z_2$ and $z_2 \geq z_1$ have no meaning except in the case when the two numbers z_1 and z_2 are real. We can, however, compare the absolute values of two complex numbers. Thus, if $z_1 = 3 + 4i$ and $z_2 = 5 - i$, then $|z_1| = 5$ and $|z_2| = \sqrt{26}$, and consequently $|z_1| < |z_2|$. This last inequality means that the point $(3, 4)$ is closer to the origin than is the point $(5, -1)$.

In Problems 1–26, write the given number in the form $a + ib$.

1. $2i^3 - 3i^2 + 5i$

2. $3i^5 - i^4 + 7i^3 - 10i^2 - 9$

3. i^8

4. i^{11}

5. $(5 - 9i) + (2 - 4i)$

6. $3(4 - i) - 3(5 + 2i)$

7. $i(5 + 7i)$

8. $i(4 - i) + 4i(1 + 2i)$

9. $(2 - 3i)(4 + i)$

10. $(\tfrac{1}{2} - \tfrac{1}{4}i)(\tfrac{2}{3} + \tfrac{5}{3}i)$

11. $(2 + 3i)^2$

12. $(1 - i)^3$

13. $\dfrac{2}{i}$

14. $\dfrac{i}{1 + i}$

15. $\dfrac{2 - 4i}{3 + 5i}$

16. $\dfrac{10 - 5i}{6 + 2i}$

17. $\dfrac{(3 - i)(2 + 3i)}{1 + i}$

18. $\dfrac{(1 + i)(1 - 2i)}{(2 + i)(4 - 3i)}$

19. $\dfrac{(5 - 4i) - (3 + 7i)}{(4 + 2i) + (2 - 3i)}$

20. $\dfrac{(4 + 5i) + 2i^3}{(2 + i)^2}$

21. $i(1 - i)(2 - i)(2 + 6i)$

22. $(1 + i)^2(1 - i)^3$

23. $(3 + 6i) + (4 - i)(3 + 5i) + \dfrac{1}{2 - i}$

24. $(2 + 3i)\left(\dfrac{2 - i}{1 + 2i}\right)^2$

25. $\left(\dfrac{i}{3 - i}\right)\left(\dfrac{1}{2 + 3i}\right)$

26. $\dfrac{1}{(1 + i)(1 - 2i)(1 + 3i)}$

In Problems 27–32, let $z = x + iy$. Find the indicated expression.

27. $\text{Re}(1/z)$

28. $\text{Re}(z^2)$

29. $\text{Im}(2z + 4\bar{z} - 4i)$

30. $\text{Im}(\bar{z}^2 + z^2)$

31. $|z - 1 - 3i|$

32. $|z + 5\bar{z}|$

In Problems 33–36, use Definition 17.2 to find a complex number z satisfying the given equation.

33. $2z = i(2 + 9i)$

34. $z - 2\bar{z} + 7 - 6i = 0$

35. $z^2 = i$

36. $\bar{z}^2 = 4z$

In Problems 37 and 38, determine which complex number is closer to the origin.

37. $10 + 8i$, $11 - 6i$

38. $\tfrac{1}{2} - \tfrac{1}{4}i$, $\tfrac{2}{3} + \tfrac{1}{6}i$

39. Prove that $|z_1 - z_2|$ is the distance between the points z_1 and z_2 in the complex plane.

40. Show for all complex numbers z on the circle $x^2 + y^2 = 4$ that $|z + 6 + 8i| \leq 12$.

17.2 Polar Form of Complex Numbers; Powers and Roots

■ **Polar Form** Recall from calculus that a point (x, y) in rectangular coordinates has the polar description (r, θ), where x, y, r, and θ are related by $x = r \cos \theta$ and $y = r \sin \theta$. Thus a nonzero complex number $z = x + iy$ can be written as $z = (r \cos \theta) + i(r \sin \theta)$ or

$$z = r(\cos \theta + i \sin \theta). \tag{1}$$

We say that (1) is the **polar form** of the complex number z. We see from Figure 17.3 that the coordinate r can be interpreted as the distance from the origin to the

Figure 17.3

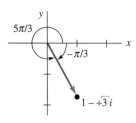

Figure 17.4

point (x, y). In other words, we adopt the convention that r is never negative so that we can take r to be the modulus of z, that is, $r = |z|$. The angle θ of inclination of the vector z measured in radians from the positive real axis is positive when measured counterclockwise and negative when measured clockwise. The angle θ is called an **argument** of z and is written $\theta = \arg z$. From Figure 17.3 we see that an argument of a complex number must satisfy the equation $\tan \theta = y/x$. The solutions of this equation are not unique, since if θ_0 is an argument of z, then necessarily the angles $\theta_0 \pm 2\pi, \theta_0 \pm 4\pi, \ldots$, are also arguments. The argument of a complex number in the interval $-\pi < \theta \le \pi$ is called the **principal argument** of z and is denoted by Arg z. For example, Arg $(i) = \pi/2$.

Example 1 A Complex Number in Polar Form

Express $1 - \sqrt{3}i$ in polar form.

SOLUTION With $x = 1$ and $y = -\sqrt{3}$, we obtain $r = |z| = \sqrt{(1)^2 + (-\sqrt{3})^2} = 2$. Now since the point $(1, -\sqrt{3})$ lies in the fourth quadrant, we can take the solution of $\tan \theta = -\sqrt{3}/1 = -\sqrt{3}$ to be $\theta = \arg z = 5\pi/3$. It follows from (1) that a polar form of the number is

$$z = 2\left(\cos \frac{5\pi}{3} + i \sin \frac{5\pi}{3}\right).$$

As we see in Figure 17.4, the argument of $1 - \sqrt{3}i$ that lies in the interval $(-\pi, \pi]$, the principal argument of z, is Arg $z = -\pi/3$. Thus, an alternative polar form of the complex number is

$$z = 2\left[\cos\left(-\frac{\pi}{3}\right) + i \sin\left(-\frac{\pi}{3}\right)\right].$$ ❏

■ **Multiplication and Division** The polar form of a complex number is especially convenient to use when multiplying or dividing two complex numbers. Suppose

$$z_1 = r_1(\cos \theta_1 + i \sin \theta_1) \quad \text{and} \quad z_2 = r_2(\cos \theta_2 + i \sin \theta_2),$$

where θ_1 and θ_2 are any arguments of z_1 and z_2, respectively. Then

$$z_1 z_2 = r_1 r_2 [(\cos \theta_1 \cos \theta_2 - \sin \theta_1 \sin \theta_2) + i(\sin \theta_1 \cos \theta_2 + \cos \theta_1 \sin \theta_2)] \quad (2)$$

and for $z_2 \ne 0$,

$$\frac{z_1}{z_2} = \frac{r_1}{r_2} [(\cos \theta_1 \cos \theta_2 + \sin \theta_1 \sin \theta_2) + i(\sin \theta_1 \cos \theta_2 - \cos \theta_1 \sin \theta_2)]. \quad (3)$$

From the addition formulas from trigonometry, (2) and (3) can be rewritten, in turn, as

$$z_1 z_2 = r_1 r_2 [\cos(\theta_1 + \theta_2) + i \sin(\theta_1 + \theta_2)] \quad (4)$$

and

$$\frac{z_1}{z_2} = \frac{r_1}{r_2} [\cos(\theta_1 - \theta_2) + i \sin(\theta_1 - \theta_2)]. \quad (5)$$

Inspection of (4) and (5) shows that

$$|z_1 z_2| = |z_1| \, |z_2|, \qquad \left|\frac{z_1}{z_2}\right| = \frac{|z_1|}{|z_2|}, \quad (6)$$

and

$$\arg (z_1 z_2) = \arg z_1 + \arg z_2, \qquad \arg\left(\frac{z_1}{z_2}\right) = \arg z_1 - \arg z_2. \quad (7)$$

Example 2 **Argument of a Product and of a Quotient**

We have seen that Arg $z_1 = \pi/2$ for $z_1 = i$. In Example 1 we saw that Arg $z_2 = -\pi/3$ for $z_2 = 1 - \sqrt{3}i$. Thus, for

$$z_1 z_2 = i(1 - \sqrt{3}i) = \sqrt{3} + i \quad \text{and} \quad \frac{z_1}{z_2} = \frac{1 - \sqrt{3}i}{i} = -\sqrt{3} - i$$

it follows from (7) that

$$\arg (z_1 z_2) = \frac{\pi}{2} - \frac{\pi}{3} = \frac{\pi}{6} \quad \text{and} \quad \arg \left(\frac{z_1}{z_2}\right) = \frac{\pi}{2} - \left(-\frac{\pi}{3}\right) = \frac{5\pi}{6}. \qquad \Box$$

In Example 2 we used the principal arguments of z_1 and z_2 and obtained arg $(z_1 z_2) =$ Arg $(z_1 z_2)$ and arg $(z_1/z_2) =$ Arg (z_1/z_2). It should be observed, however, that this was a coincidence. Although (7) is true for any arguments of z_1 and z_2, *it is not true,* in general, that Arg $(z_1 z_2) =$ Arg $z_1 +$ Arg z_2 and Arg $(z_1/z_2) =$ Arg $z_1 -$ Arg z_2. See Problem 39 in Exercises 17.2.

■ **Powers of z** We can find integer powers of the complex number z from the results in (4) and (5). For example, if $z = r(\cos \theta + i \sin \theta)$, then with $z_1 = z$ and $z_2 = z$, (4) gives

$$z^2 = r^2[\cos (\theta + \theta) + i \sin (\theta + \theta)] = r^2(\cos 2\theta + i \sin 2\theta).$$

Since $z^3 = z^2 z$, it follows that

$$z^3 = r^3(\cos 3\theta + i \sin 3\theta).$$

Moreover, since arg $(1) = 0$, it follows from (5) that

$$\frac{1}{z^2} = z^{-2} = r^{-2}[\cos(-2\theta) + i \sin(-2\theta)].$$

Continuing in this manner, we obtain a formula for the nth power of z for any integer n:

$$z^n = r^n(\cos n\theta + i \sin n\theta). \qquad (8)$$

Example 3 **Power of a Complex Number**

Compute z^3 for $z = 1 - \sqrt{3}i$.

SOLUTION In Example 1 we saw that

$$z = 2\left[\cos\left(-\frac{\pi}{3}\right) + i \sin\left(-\frac{\pi}{3}\right)\right].$$

Hence from (8) with $r = 2$, $\theta = -\pi/3$, and $n = 3$, we get

$$(1 - \sqrt{3}i)^3 = 2^3\left[\cos\left(3\left(-\frac{\pi}{3}\right)\right) + i \sin\left(3\left(-\frac{\pi}{3}\right)\right)\right]$$

$$= 8[\cos(-\pi) + i \sin(-\pi)] = -8. \qquad \Box$$

■ **DeMoivre's Formula** When $z = \cos \theta + i \sin \theta$, we have $|z| = r = 1$ and so (8) yields

$$(\cos \theta + i \sin \theta)^n = \cos n\theta + i \sin n\theta. \qquad (9)$$

This last result is known as **DeMoivre's formula** and is useful in deriving certain trigonometric identities.

■ **Roots** A number w is said to be an ***n*th root** of a nonzero complex number z if $w^n = z$. If we let $w = \rho(\cos \phi + i \sin \phi)$ and $z = r(\cos \theta + i \sin \theta)$ be the polar forms of w and z, then in view of (8) $w^n = z$ becomes

$$\rho^n(\cos n\phi + i \sin n\phi) = r(\cos \theta + i \sin \theta).$$

From this we conclude that $\rho^n = r$ or $\rho = r^{1/n}$ and

$$\cos n\phi + i \sin n\phi = \cos \theta + i \sin \theta.$$

By equating the real and imaginary parts, we get from this equation

$$\cos n\phi = \cos \theta \quad \text{and} \quad \sin n\phi = \sin \theta.$$

These equalities imply that $n\phi = \theta + 2k\pi$, where k is an integer. Thus,

$$\phi = \frac{\theta + 2k\pi}{n}.$$

As k takes on the successive integer values $k = 0, 1, 2, \ldots, n - 1$, we obtain n *distinct* roots with the same modulus but different arguments. But for $k \geq n$ we obtain the same roots because the sine and cosine are 2π-periodic. To see this, suppose $k = n + m$, where $m = 0, 1, 2, \ldots$. Then

$$\phi = \frac{\theta + 2(n + m)\pi}{n} = \frac{\theta + 2m\pi}{n} + 2\pi$$

and so $$\sin \phi = \sin\left(\frac{\theta + 2m\pi}{n}\right), \qquad \cos \phi = \cos\left(\frac{\theta + 2m\pi}{n}\right).$$

We summarize this result. The nth roots of a nonzero complex number $z = r(\cos \theta + i \sin \theta)$ are given by

$$w_k = r^{1/n}\left[\cos\left(\frac{\theta + 2k\pi}{n}\right) + i \sin\left(\frac{\theta + 2k\pi}{n}\right)\right], \tag{10}$$

where $k = 0, 1, 2, \ldots, n - 1$.

> ### *Example 4* Roots of a Complex Number
> Find the three cube roots of $z = i$.
>
> **SOLUTION** With $r = 1$, $\theta = \arg z = \pi/2$, the polar form of the given number is $z = \cos(\pi/2) + i \sin(\pi/2)$. From (10) with $n = 3$ we obtain
>
> $$w_k = (1)^{1/3}\left[\cos\left(\frac{\pi/2 + 2k\pi}{3}\right) + i \sin\left(\frac{\pi/2 + 2k\pi}{3}\right)\right], \qquad k = 0, 1, 2.$$
>
> Hence, the three roots are:
>
> $$k = 0, \qquad w_0 = \cos \frac{\pi}{6} + i \sin \frac{\pi}{6} = \frac{\sqrt{3}}{2} + \frac{1}{2} i$$
>
> $$k = 1, \qquad w_1 = \cos \frac{5\pi}{6} + i \sin \frac{5\pi}{6} = -\frac{\sqrt{3}}{2} + \frac{1}{2} i$$
>
> $$k = 2, \qquad w_2 = \cos \frac{3\pi}{2} + i \sin \frac{3\pi}{2} = -i. \qquad \square$$

Figure 17.5

The root w of a complex number z obtained by using the principal argument of z with $k = 0$ is sometimes called the **principal nth root** of z. In Example 4, since Arg $(i) = \pi/2$, $w_0 = (\sqrt{3}/2) + (1/2)i$ is the principal third root of i.

Since the roots given by (8) have the same modulus, the n roots of a nonzero complex number z lie on a circle of radius $r^{1/n}$ centered at the origin in the complex plane. Moreover, since the difference between the arguments of any two successive roots is $2\pi/n$, the nth roots of z are equally spaced on this circle. Figure 17.5 shows the three roots of i equally spaced on a unit circle; the angle between roots (vectors) w_k and w_{k+1} is $2\pi/3$.

As the next example will show, the roots of a complex number do not have to be "nice" numbers as in Example 3.

Example 5 **Roots of a Complex Number**

Find the four fourth roots of $z = 1 + i$.

SOLUTION In this case, $r = \sqrt{2}$ and $\theta = \arg z = \pi/4$. From (10) with $n = 4$ we obtain

$$w_k = (2)^{1/4}\left[\cos\left(\frac{\pi/4 + 2k\pi}{4}\right) + i\sin\left(\frac{\pi/4 + 2k\pi}{4}\right)\right], \qquad k = 0, 1, 2, 3.$$

Thus,

$$k = 0, \qquad w_0 = (2)^{1/4}\left[\cos\frac{\pi}{16} + i\sin\frac{\pi}{16}\right] = 1.1664 + 0.2320i$$

$$k = 1, \qquad w_1 = (2)^{1/4}\left[\cos\frac{9\pi}{16} + i\sin\frac{9\pi}{16}\right] = -0.2320 + 1.1664i$$

$$k = 2, \qquad w_2 = (2)^{1/4}\left[\cos\frac{17\pi}{16} + i\sin\frac{17\pi}{16}\right] = -1.1664 - 0.2320i$$

$$k = 3, \qquad w_3 = (2)^{1/4}\left[\cos\frac{25\pi}{16} + i\sin\frac{25\pi}{16}\right] = 0.2320 - 1.1664i. \qquad \square$$

EXERCISES 17.2 Answers to odd-numbered problems begin on page A-71.

In Problems 1–10, write the given complex number in polar form.

1. 2

2. -10

3. $-3i$

4. $6i$

5. $1 + i$

6. $5 - 5i$

7. $-\sqrt{3} + i$

8. $-2 - 2\sqrt{3}i$

9. $\dfrac{3}{-1 + i}$

10. $\dfrac{12}{\sqrt{3} + i}$

In Problems 11–14, write the number given in polar form in the form $a + ib$.

11. $z = 5\left(\cos\dfrac{7\pi}{6} + i\sin\dfrac{7\pi}{6}\right)$

12. $z = 8\sqrt{2}\left(\cos\dfrac{11\pi}{4} + i\sin\dfrac{11\pi}{4}\right)$

13. $z = 6\left(\cos\dfrac{\pi}{8} + i\sin\dfrac{\pi}{8}\right)$

14. $z = 10\left(\cos\dfrac{\pi}{5} + i\sin\dfrac{\pi}{5}\right)$

In Problems 15 and 16, find $z_1 z_2$ and z_1/z_2. Write the number in the form $a + ib$.

15. $z_1 = 2\left(\cos\dfrac{\pi}{8} + i\sin\dfrac{\pi}{8}\right)$, $z_2 = 4\left(\cos\dfrac{3\pi}{8} + i\sin\dfrac{3\pi}{8}\right)$

16. $z_1 = \sqrt{2}\left(\cos\dfrac{\pi}{4} + i\sin\dfrac{\pi}{4}\right)$, $z_2 = \sqrt{3}\left(\cos\dfrac{\pi}{12} + i\sin\dfrac{\pi}{12}\right)$

In Problems 17–20, write each complex number in polar form. Then use either (4) or (5) to obtain a polar form of the given number. Write the polar form in the form $a + ib$.

17. $(3 - 3i)(5 + 5\sqrt{3}i)$

18. $(4 + 4i)(-1 + i)$

19. $\dfrac{-i}{2-2i}$ **20.** $\dfrac{\sqrt{2}+\sqrt{6}i}{-1+\sqrt{3}i}$

In Problems 21–26, use (8) to compute the indicated power.

21. $(1+\sqrt{3}i)^9$ **22.** $(2-2i)^5$

23. $(\tfrac{1}{2}+\tfrac{1}{2}i)^{10}$ **24.** $(-\sqrt{2}+\sqrt{6}i)^4$

25. $\left(\cos\dfrac{\pi}{8}+i\sin\dfrac{\pi}{8}\right)^{12}$

26. $\left[\sqrt{3}\left(\cos\dfrac{2\pi}{9}+i\sin\dfrac{2\pi}{9}\right)\right]^6$

In Problems 27–32, use (10) to compute all roots. Sketch these roots on an appropriate circle centered at the origin.

27. $(8)^{1/3}$ **28.** $(1)^{1/8}$

29. $(i)^{1/2}$ **30.** $(-1+i)^{1/3}$

31. $(-1+\sqrt{3}i)^{1/2}$ **32.** $(-1-\sqrt{3}i)^{1/4}$

In Problems 33 and 34, find all solutions of the given equation.

33. $z^4+1=0$ **34.** $z^8-2z^4+1=0$

In Problems 35 and 36, express the given complex number first in polar form and then in the form $a+ib$.

35. $\left(\cos\dfrac{\pi}{9}+i\sin\dfrac{\pi}{9}\right)^{12}\left[2\left(\cos\dfrac{\pi}{6}+i\sin\dfrac{\pi}{6}\right)\right]^5$

36. $\dfrac{\left[8\left(\cos\dfrac{3\pi}{8}+i\sin\dfrac{3\pi}{8}\right)\right]^3}{\left[2\left(\cos\dfrac{\pi}{16}+i\sin\dfrac{\pi}{16}\right)\right]^{10}}$

37. Use the result $(\cos\theta+i\sin\theta)^2=\cos 2\theta+i\sin 2\theta$ to find trigonometric identities for $\cos 2\theta$ and $\sin 2\theta$.

38. Use the result $(\cos\theta+i\sin\theta)^3=\cos 3\theta+i\sin 3\theta$ to find trigonometric identities for $\cos 3\theta$ and $\sin 3\theta$.

39. (a) If $z_1=-1$ and $z_2=5i$, verify that

$$\text{Arg}(z_1 z_2)\neq \text{Arg}(z_1)+\text{Arg}(z_2).$$

(b) If $z_1=-1$ and $z_2=-5i$, verify that

$$\text{Arg}(z_1/z_2)\neq \text{Arg}(z_1)-\text{Arg}(z_2).$$

40. For the complex numbers given in Problem 39, verify in both parts **(a)** and **(b)** that

$$\arg(z_1 z_2)=\arg(z_1)+\arg(z_2)\quad\text{and}\quad \arg\left(\dfrac{z_1}{z_2}\right)=\arg(z_1)-\arg(z_2).$$

17.3 Sets of Points in the Complex Plane

$|z - z_0| = \rho$

Figure 17.6

■ **Terminology** Before discussing the concept of functions of a complex variable, we need to introduce some essential terminology about sets in the complex plane.
Suppose $z_0=x_0+iy_0$. Since $|z-z_0|=\sqrt{(x-x_0)^2+(y-y_0)^2}$ is the distance between the points $z=x+iy$ and $z_0=x_0+iy_0$, the points $z=x+iy$ that satisfy the equation

$$|z-z_0|=\rho,$$

$\rho>0$, lie on a **circle** of radius ρ centered at the point z_0. See Figure 17.6.

Figure 17.7

| **Example 1** **Circles**

(a) $|z|=1$ is the equation of a unit circle centered at the origin.
(b) $|z-1-2i|=5$ is the equation of a circle of radius 5 centered at $1+2i$. ❑

The points z satisfying the inequality $|z-z_0|<\rho$, $\rho>0$, lie within, but not on, a circle of radius ρ centered at the point z_0. This set is called a **neighborhood** of z_0 or an **open disk**. A point z_0 is said to be an **interior point** of a set S of the complex plane if there exists some neighborhoood of z_0 that lies entirely within S. If every point z of a set S is an interior point, then S is said to be an **open set**. See Figure 17.7. For example, the inequality $\text{Re}(z)>1$ defines a *right half-plane*, which is an open set. All complex numbers $z=x+iy$ for which $x>1$ are in this set. If we choose, for example, $z_0=1.1+2i$, then a neighborhood of z_0 lying entirely in the set is defined by $|z-(1.1+2i)|<0.05$. See Figure 17.8. On the other hand, the set S of points in the complex plane defined by $\text{Re}(z)\geq 1$ is *not* open, since every neighborhood of a point on the line $x=1$ must contain points in S and points not in S. See Figure 17.9.

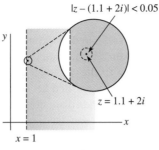

$|z - (1.1 + 2i)| < 0.05$

$z = 1.1 + 2i$

$x = 1$

open set
magnified view of a point near $x = 1$

Figure 17.8

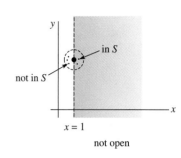

in S

not in S

$x = 1$

not open

Figure 17.9

Example 2 **Open Sets**

Figure 17.10 illustrates some additional open sets.

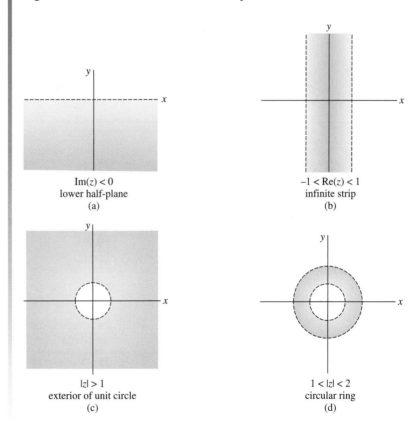

Im$(z) < 0$
lower half-plane
(a)

$-1 < $ Re$(z) < 1$
infinite strip
(b)

$|z| > 1$
exterior of unit circle
(c)

$1 < |z| < 2$
circular ring
(d)

Figure 17.10 ❏

The set of numbers satisfying the inequality

$$\rho_1 < |z - z_0| < \rho_2,$$

such as illustrated in Figure 17.10(d), is also called an open **annulus.**

If every neighborhood of a point z_0 contains at least one point that is in a set S and at least one point that is not in S, then z_0 is said to be a **boundary point** of S.

Figure 17.11

The **boundary** of a set S is the set of all boundary points of S. For the set of points defined by $\text{Re}(z) \geq 1$, the points on the line $x = 1$ are boundary points. The points on the circle $|z - i| = 2$ are boundary points for the disk $|z - i| \leq 2$.

If any pair of points z_1 and z_2 in an open set S can be connected by a polygonal line that lies entirely in the set, then the open set S is said to be **connected.** See Figure 17.11. An open connected set is called a **domain.** All the open sets in Figure 17.10 are connected and so are domains. The set of numbers satisfying $\text{Re}(z) \neq 4$ is an open set but is not connected, since it is not possible to join points on either side of the vertical line $x = 4$ by a polygonal line without leaving the set (bear in mind that the points on $x = 4$ are not in the set).

A **region** is a domain in the complex plane with all, some, or none of its boundary points. Since an open connected set does not contain any boundary points, it is automatically a region. A region containing all its boundary points is said to be **closed.** The disk defined by $|z - i| \leq 2$ is an example of a closed region and is referred to as a closed disk. A region may be neither open nor closed; the annular region defined by $1 \leq |z - 5| < 3$ contains only some of its boundary points and so is neither open nor closed.

Remarks

Often in mathematics the same word is used in entirely different contexts. Do not confuse the concept of "domain" defined above with the concept of the "domain of a function."

EXERCISES 17.3

Answers to odd-numbered problems begin on page A-71.

In Problems 1–8, sketch the graph of the given equation.

1. $\text{Re}(z) = 5$

2. $\text{Im}(z) = -2$

3. $\text{Im}(\bar{z} + 3i) = 6$

4. $\text{Im}(z - i) = \text{Re}(z + 4 - 3i)$

5. $|z - 3i| = 2$

6. $|2z + 1| = 4$

7. $|z - 4 + 3i| = 5$

8. $|z + 2 + 2i| = 2$

In Problems 9–22, sketch the set of points in the complex plane satisfying the given inequality. Determine whether the set is a domain.

9. $\text{Re}(z) < -1$

10. $|\text{Re}(z)| > 2$

11. $\text{Im}(z) > 3$

12. $\text{Im}(z - i) < 5$

13. $2 < \text{Re}(z - 1) < 4$

14. $-1 \leq \text{Im}(z) < 4$

15. $\text{Re}(z^2) > 0$

16. $\text{Im}(1/z) < \dfrac{1}{2}$

17. $0 \leq \arg(z) \leq 2\pi/3$

18. $|\arg(z)| < \pi/4$

19. $|z - i| > 1$

20. $|z - i| > 0$

21. $2 < |z - i| < 3$

22. $1 \leq |z - 1 - i| < 2$

23. Describe the set of points in the complex plane that satisfies $|z + 1| = |z - i|$.

24. Describe the set of points in the complex plane that satisfies $|\text{Re}(z)| \leq |z|$.

25. Describe the set of points in the complex plane that satisfies $z^2 + \bar{z}^2 = 2$.

26. Describe the set of points in the complex plane that satisfies $|z - i| + |z + i| = 1$.

17.4 Functions of a Complex Variable; Analyticity

■ **Function** Recall that a function f is a rule of correspondence that assigns to each element of a set A one and only one element of a set B. If b is the element in B assigned to the element a in A by f, we say that b is the **image** of a and write $b = f(a)$. The set A is called the **domain** of the function f but is not necessarily a domain in the

sense defined in Section 17.3. The set of all images in B is called the **range** of the function. For example, suppose the set A is a set of real numbers defined by $3 \leq x < \infty$ and the function is given by $f(x) = \sqrt{x - 3}$; then $f(3) = 0$, $f(4) = 1$, $f(8) = \sqrt{5}$, and so on. In other words, the range of f is the set given by $0 \leq y < \infty$. Since A is a set of real numbers, we say f is a function of a real variable x. When the domain A is the set of complex numbers z, we naturally say that f is a function of a complex variable z or a **complex function.** The image w of a complex number z will be some complex number $u + iv$, that is,

$$w = f(z) = u(x, y) + iv(x, y), \tag{1}$$

where u and v are the real and imaginary parts of w and are real-valued functions. Inherent in the mathematical statement (1) is the fact that *we cannot draw a graph of a complex function $w = f(z)$* since a graph would require four axes in a four-dimensional coordinate system.

Some examples of functions of a complex variable are

$$f(z) = z^2 - 4z, \qquad z \text{ any complex number}$$

$$f(z) = \frac{z}{z^2 + 1}, \qquad z \neq i \text{ and } z \neq -i$$

$$f(z) = z + \text{Re}(z), \qquad z \text{ any complex number.}$$

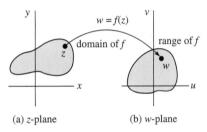

(a) z-plane (b) w-plane

Figure 17.12

Each of these functions can be expressed in form (1). For example,

$$f(z) = z^2 - 4z = (x + iy)^2 - 4(x + iy) = (x^2 - y^2 - 4x) + i(2xy - 4y).$$

Thus, $u(x, y) = x^2 - y^2 - 4x$, and $v(x, y) = 2xy - 4y$.

Although we cannot draw a graph, a complex function $w = f(z)$ can be interpreted as a **mapping** or **transformation** from the z-plane to the w-plane. See Figure 17.12.

Example 1 Image of a Vertical Line

Find the image of the line $\text{Re}(z) = 1$ under the mapping $f(z) = z^2$.

SOLUTION For the function $f(z) = z^2$ we have $u(x, y) = x^2 - y^2$ and $v(x, y) = 2xy$. Now, $\text{Re}(z) = x$ and so by substituting $x = 1$ into the functions u and v, we obtain $u = 1 - y^2$ and $v = 2y$. These are parametric equations of a curve in the w-plane. Substituting $y = v/2$ into the first equation eliminates the parameter y to give $u = 1 - v^2/4$. In other words, the image of the line in Figure 17.13(a) is the parabola shown in Figure 17.13(b). ❑

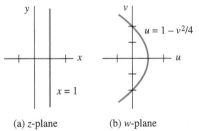

(a) z-plane (b) w-plane

Figure 17.13

We shall pursue the idea of $f(z)$ as a mapping in greater detail in Chapter 20.

It should be noted that a complex function is completely determined by the real-valued functions u and v. This means a complex function $w = f(z)$ can be defined by arbitrarily specifying $u(x, y)$ and $v(x, y)$, even though $u + iv$ may not be obtainable through the familiar operations on the symbol z alone. For example, if $u(x, y) = xy^2$ and $v(x, y) = x^2 - 4y^3$, then $f(z) = xy^2 + i(x^2 - 4y^3)$ is a function of a complex variable. To compute, say, $f(3 + 2i)$, we substitute $x = 3$ and $y = 2$ into u and v to obtain $f(3 + 2i) = 12 - 23i$.

■ **Complex Functions as Flows** We also may interpret a complex function $w = f(z)$ as a **two-dimensional fluid flow** by considering the complex number $f(z)$ as a vector based at the point z. The vector $f(z)$ specifies the speed and direction of the

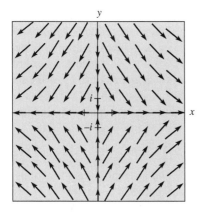

Figure 17.14 $f_1(z) = \bar{z}$
(normalized)

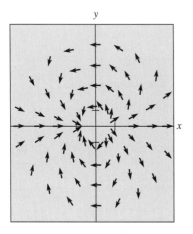

Figure 17.15 $f_2(z) = z^2$
(normalized)

flow at a given point z. Figures 17.14 and 17.15 show the flows corresponding to the complex functions $f_1(z) = \bar{z}$ and $f_2(z) = z^2$, respectively.

If $x(t) + iy(t)$ is a parametric representation for the path of a particle in the flow, the tangent vector $\mathbf{T} = x'(t) + iy'(t)$ must coincide with $f(x(t) + iy(t))$. When $f(z) = u(x, y) + iv(x, y)$, it follows that the path of the particle must satisfy the system of differential equations

$$\frac{dx}{dt} = u(x, y)$$

$$\frac{dy}{dt} = v(x, y).$$

We call the family of solutions of this system the *streamlines* of the flow associated with $f(z)$.

Example 2 Streamlines

Find the streamlines of the flows associated with the complex functions (a) $f_1(z) = \bar{z}$ and (b) $f_2(z) = z^2$.

SOLUTION (a) The streamlines corresponding to $f_1(z) = x - iy$ satisfy the system

$$\frac{dx}{dt} = x$$

$$\frac{dy}{dt} = -y$$

and so $x(t) = c_1 e^t$ and $y(t) = c_2 e^{-t}$. By multiplying these two parametric equations, we see that the point $x(t) + iy(t)$ lies on the hyperbola $xy = c_1 c_2$.

(b) To find the streamlines corresponding to $f_2(z) = (x^2 - y^2) + i2xy$, note that $dx/dt = x^2 - y^2$, $dy/dt = 2xy$, and so

$$\frac{dy}{dx} = \frac{2xy}{x^2 - y^2}.$$

This homogeneous differential equation has the solution $x^2 + y^2 = c_2 y$, a family of circles that have centers on the y-axis and pass through the origin. ❏

■ **Limits and Continuity** The definition of a limit of a complex function $f(z)$ as $z \to z_0$ has the same outward appearance as the limit in real variables.

DEFINITION 17.4 Limit of a Function

Suppose the function f is defined in some neighborhood of z_0, except possibly at z_0 itself. Then f is said to possess a **limit** at z_0, written

$$\lim_{z \to z_0} f(z) = L$$

if, for each $\varepsilon > 0$, there exists a $\delta > 0$ such that $\left| f(z) - L \right| < \varepsilon$ whenever $0 < \left| z - z_0 \right| < \delta$.

In words, $\lim_{z \to z_0} f(z) = L$ means that the points $f(z)$ can be made arbitrarily close to the point L if we choose the point z sufficiently close to, but not equal to,

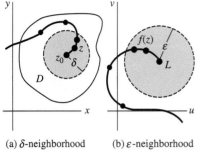

(a) δ-neighborhood (b) ε-neighborhood

Figure 17.16

the point z_0. As shown in Figure 17.16, for each ε-neighborhood of L (defined by $|f(z) - L| < \varepsilon$) there is a δ-neighborhood of z_0 (defined by $|z - z_0| < \delta$) so that the images of all points $z \neq z_0$ in this neighborhood lie in the ε-neighborhood of L.

The fundamental difference between this definition and the limit concept in real variables lies in the understanding of $z \to z_0$. For a function f of a single real variable x, $\lim_{x \to x_0} f(x) = L$ means $f(x)$ approaches L as x approaches x_0 either from the right of x_0 or from the left of x_0 on the real number line. But since z and z_0 are points in the complex plane, when we say that $\lim_{z \to z_0} f(z)$ exists, we mean that $f(z)$ approaches L as the point z approaches z_0 from *any* direction.

The following theorem will summarize some properties of limits:

THEOREM 17.1 Limit of Sum, Product, Quotient

Suppose $\lim_{z \to z_0} f(z) = L_1$ and $\lim_{z \to z_0} g(z) = L_2$. Then:

(*i*) $\displaystyle \lim_{z \to z_0} [f(z) + g(z)] = L_1 + L_2$

(*ii*) $\displaystyle \lim_{z \to z_0} f(z)g(z) = L_1 L_2$

(*iii*) $\displaystyle \lim_{z \to z_0} \frac{f(z)}{g(z)} = \frac{L_1}{L_2}, \qquad L_2 \neq 0.$

DEFINITION 17.5 Continuity at a Point

A function f is **continuous** at a point z_0 if $\displaystyle \lim_{z \to z_0} f(z) = f(z_0)$.

As a consequence of Theorem 17.1, it follows that if two functions f and g are continuous at point z_0, then their sum and product are continuous at z_0. The quotient of the two functions is continuous at z_0 provided $g(z_0) \neq 0$.

A function f defined by

$$f(z) = a_n z^n + a_{n-1} z^{n-1} + \cdots + a_2 z^2 + a_1 z + a_0, \qquad a_n \neq 0, \qquad (2)$$

where n is a nonnegative integer and the coefficients a_i, $i = 0, \ldots, n$, are complex constants, is called a **polynomial** of degree n. Although we shall not prove it, the limit result $\lim_{z \to z_0} z = z_0$ indicates that the simple polynomial function $f(z) = z$ is continuous everywhere—that is, on the entire z-plane. With this result in mind and with repeated applications of Theorems 17.1 (*ii*) and (*i*), it follows that a polynomial function (2) is continuous everywhere. A **rational function**

$$f(z) = \frac{g(z)}{h(z)},$$

where g and h are polynomial functions, is continuous except at those points at which $h(z)$ is zero.

■ **Derivative** The derivative of a complex function is defined in terms of a limit. The symbol Δz used in the following definition is the complex number $\Delta x + i\Delta y$.

DEFINITION 17.6 **Derivative**

Suppose the complex function f is defined in a neighborhood of a point z_0. The **derivative** of f at z_0 is

$$f'(z_0) = \lim_{\Delta z \to 0} \frac{f(z_0 + \Delta z) - f(z_0)}{\Delta z}, \tag{3}$$

provided this limit exists.

If the limit in (3) exists, the function f is said to be **differentiable** at z_0. The derivative of a function $w = f(z)$ is also written dw/dz.

As in real variables, differentiability implies continuity:

If f is differentiable at z_0, then f is continuous at z_0.

Moreover, the **rules of differentiation** are the same as in the calculus of real variables. If f and g are differentiable at a point z and c is a complex constant, then:

Constant Rules: $\dfrac{d}{dz} c = 0, \qquad \dfrac{d}{dz} cf(z) = cf'(z)$ (4)

Sum Rule: $\dfrac{d}{dz} [f(z) + g(z)] = f'(z) + g'(z)$ (5)

Product Rule: $\dfrac{d}{dz} [f(z)g(z)] = f(z)g'(z) + g(z)f'(z)$ (6)

Quotient Rule: $\dfrac{d}{dz} \left[\dfrac{f(z)}{g(z)} \right] = \dfrac{g(z)f'(z) - f(z)g'(z)}{[g(z)]^2}$ (7)

Chain Rule: $\dfrac{d}{dz} f(g(z)) = f'(g(z))g'(z).$ (8)

The usual rule for differentiation of powers of z is also valid:

$$\frac{d}{dz} z^n = nz^{n-1}, \qquad n \text{ an integer.} \tag{9}$$

Example 3 **Using the Rules of Differentiation**

Differentiate (a) $f(z) = 3z^4 - 5z^3 + 2z$ and (b) $f(z) = \dfrac{z^2}{4z + 1}$.

SOLUTION (a) Using the Power Rule (9) along with the Sum Rule (5), we obtain

$$f'(z) = 3 \cdot 4z^3 - 5 \cdot 3z^2 + 2 = 12z^3 - 15z^2 + 2.$$

(b) From the quotient rule (7),

$$f'(z) = \frac{(4z + 1) \cdot 2z - z^2 \cdot 4}{(4z + 1)^2} = \frac{4z^2 + 2z}{(4z + 1)^2}. \qquad \square$$

In order for a complex function f to be differentiable at a point z_0, $\displaystyle\lim_{\Delta z \to 0} \frac{f(z_0 + \Delta z) - f(z_0)}{\Delta z}$ must approach the same complex number from any direction.

Thus in the study of complex variables, to require the differentiability of a function is a greater demand than in real variables. If a complex function is made up, such as $f(z) = x + 4iy$, there is a good chance that it is not differentiable.

> ### *Example 4* A Function That Is Nowhere Differentiable
> Show that the function $f(z) = x + 4iy$ is nowhere differentiable.
>
> **SOLUTION** With $\Delta z = \Delta x + i\Delta y$, we have
>
> $$f(z + \Delta z) - f(z) = (x + \Delta x) + 4i(y + \Delta y) - x - 4iy = \Delta x + 4i\Delta y$$
>
> and so
> $$\lim_{\Delta z \to 0} \frac{f(z + \Delta z) - f(z)}{\Delta z} = \lim_{\Delta z \to 0} \frac{\Delta x + 4i\Delta y}{\Delta x + i\Delta y}. \tag{10}$$
>
> Now, if we let $\Delta z \to 0$ along a line parallel to the x-axis, then $\Delta y = 0$ and the value of (10) is 1. On the other hand, if we let $\Delta z \to 0$ along a line parallel to the y-axis, then $\Delta x = 0$ and the value of (10) is seen to be 4. Therefore, $f(z) = x + 4iy$ is not differentiable at any point z. ❑

■ **Analytic Functions** While the requirement of differentiability is a stringent demand, there is a class of functions that is of great importance whose members satisfy even more severe requirements. These functions are called **analytic functions.**

> **DEFINITION 17.7** Analyticity at a Point
>
> A complex function $w = f(z)$ is said to be **analytic at a point** z_0 if f is differentiable at z_0 and at every point in some neighborhood of z_0.

A function f is analytic in a domain D if it is analytic at every point in D.

The student should read Definition 17.7 carefully. Analyticity *at a point* is a neighborhood property. Analyticity at a point is, therefore, not the same as differentiability at a point. It is left as an exercise to show that the function $f(z) = |z|^2$ is differentiable at $z = 0$ but is differentiable nowhere else. Hence, $f(z) = |z|^2$ is nowhere analytic. In contrast, the simple polynomial $f(z) = z^2$ is differentiable at every point z in the complex plane. Hence, $f(z) = z^2$ is analytic everywhere. A function that is analytic at every point z is said to be an **entire function.** Polynomial functions are differentiable at every point z and so are entire functions.

> ### *Remarks*
>
> Recall from algebra that a number c is a zero of a polynomial function if and only if $x - c$ is a factor of $f(x)$. The same result holds in complex analysis. For example, since $f(z) = z^4 + 5z^2 + 4 = (z^2 + 1)(z^2 + 4)$, the zeros of f are $-i$, i, $-2i$, and $2i$. Hence, $f(z) = (z + i)(z - i)(z + 2i)(z - 2i)$. Moreover, the quadratic formula is also valid. For example, using this formula, we can write
>
> $$f(z) = z^2 - 2z + 2 = (z - (i + i))(z - (i - i)) = (z - 1 - i)(z - 1 + i).$$
>
> See Problems 21 and 22 in Exercises 17.4.

In Problems 1–6, find the image of the given line under the mapping $f(z) = z^2$.

1. $y = 2$ **2.** $x = -3$

3. $x = 0$ **4.** $y = 0$

5. $y = x$ **6.** $y = -x$

In Problems 7–14, express the given function in the form $f(z) = u + iv$.

7. $f(z) = 6z - 5 + 9i$ **8.** $f(z) = 7z - 9i\bar{z} - 3 + 2i$

9. $f(z) = z^2 - 3z + 4i$ **10.** $f(z) = 3\bar{z}^2 + 2z$

11. $f(z) = z^3 - 4z$ **12.** $f(z) = z^4$

13. $f(z) = z + 1/z$ **14.** $f(z) = \dfrac{z}{z+1}$

In Problems 15–18, evaluate the given function at the indicated points.

15. $f(z) = 2x - y^2 + i(xy^3 - 2x^2 + 1)$

 (a) $2i$ **(b)** $2 - i$ **(c)** $5 + 3i$

16. $f(z) = (x + 1 + 1/x) + i(4x^2 - 2y^2 - 4)$

 (a) $1 + i$ **(b)** $2 - i$ **(c)** $1 + 4i$

17. $f(z) = 4z + i\bar{z} + \operatorname{Re}(z)$

 (a) $4 - 6i$ **(b)** $-5 + 12i$ **(c)** $2 - 7i$

18. $f(z) = e^x \cos y + ie^x \sin y$

 (a) $\pi i/4$ **(b)** $-1 - \pi i$ **(c)** $3 + \pi i/3$

In Problems 19–22, the given limit exists. Find its value.

19. $\lim\limits_{z \to i} (4z^3 - 5z^2 + 4z + 1 - 5i)$

20. $\lim\limits_{z \to 1-i} \dfrac{5z^2 - 2z + 2}{z + 1}$

21. $\lim\limits_{z \to i} \dfrac{z^4 - 1}{z - i}$ **22.** $\lim\limits_{z \to 1+i} \dfrac{z^2 - 2z + 2}{z^2 - 2i}$

In Problems 23 and 24, show that the given limit does not exist.

23. $\lim\limits_{z \to 0} \dfrac{z}{\bar{z}}$ **24.** $\lim\limits_{z \to 1} \dfrac{x + y - 1}{z - 1}$

In Problems 25 and 26, use (3) to obtain the indicated derivative of the given function.

25. $f(z) = z^2, \quad f'(z) = 2z$

26. $f(z) = 1/z, \quad f'(z) = -1/z^2$

In Problems 27–34, use (4)–(8) to find the derivative $f'(z)$ for the given function.

27. $f(z) = 4z^3 - (3 + i)z^2 - 5z + 4$

28. $f(z) = 5z^4 - iz^3 + (8 - i)z^2 - 6i$

29. $f(z) = (2z + 1)(z^2 - 4z + 8i)$

30. $f(z) = (z^5 + 3iz^3)(z^4 + iz^3 + 2z^2 - 6iz)$

31. $f(z) = (z^2 - 4i)^3$ **32.** $f(z) = (2z - 1/z)^6$

33. $f(z) = \dfrac{3z - 4 + 8i}{2z + i}$ **34.** $f(z) = \dfrac{5z^2 - z}{z^3 + 1}$

In Problems 35–38, give the points at which the given function will not be analytic.

35. $f(z) = \dfrac{z}{z - 3i}$ **36.** $f(z) = \dfrac{2i}{z^2 - 2z + 5iz}$

37. $f(z) = \dfrac{z^3 + z}{z^2 + 4}$ **38.** $f(z) = \dfrac{z - 4 + 3i}{z^2 - 6z + 25}$

39. Show that the function $f(z) = \bar{z}$ is nowhere differentiable.

40. The function $f(z) = |z|^2$ is continuous throughout the entire complex plane. Show, however, that f is differentiable only at the point $z = 0$. [*Hint:* Use (3) and consider two cases, $z = 0$ and $z \neq 0$. In the second case let Δz approach zero along a line parallel to the x-axis and then let Δz approach zero along a line parallel to the y-axis.]

In Problems 41–44, find the streamlines of the flow associated with the given complex function.

41. $f(z) = 2z$ **42.** $f(z) = iz$

43. $f(z) = 1/\bar{z}$ **44.** $f(z) = x^2 - iy^2$

In Problems 45 and 46, use a graphics calculator or computer to obtain the image of the given parabola under the mapping $f(z) = z^2$.

45. $y = \tfrac{1}{2}x^2$ **46.** $y = (x - 1)^2$

17.5 Cauchy–Riemann Equations

In the preceding section we saw that a function f of a complex variable z is analytic in a domain D if f is differentiable at all points in D. We shall now develop a simple test for analyticity of a complex-valued function f based on partial derivatives of its real and imaginary parts. We first prove a necessary condition for analyticity.

> ### THEOREM 17.2 Cauchy–Riemann Equations
>
> Suppose $f(z) = u(x, y) + iv(x, y)$ is continuous in some neighborhood of the point $z = x + iy$ and is differentiable at z. Then at the point z the first-order partial derivatives of u and v exist and satisfy the **Cauchy–Riemann equations**
>
> $$\frac{\partial u}{\partial x} = \frac{\partial v}{\partial y} \quad \text{and} \quad \frac{\partial u}{\partial y} = -\frac{\partial v}{\partial x}. \tag{1}$$

PROOF Since $f'(z)$ exists, we know that

$$f'(z) = \lim_{\Delta z \to 0} \frac{f(z + \Delta z) - f(z)}{\Delta z}. \tag{2}$$

By writing $f(z) = u(x, y) + iv(x, y)$ and $\Delta z = \Delta x + i\Delta y$, we get from (2)

$$f'(z) = \lim_{\Delta z \to 0} \frac{u(x + \Delta x, y + \Delta y) + iv(x + \Delta x, y + \Delta y) - u(x, y) - iv(x, y)}{\Delta x + i\Delta y}. \tag{3}$$

Since this limit exists, Δz can approach zero from any convenient direction. In particular, if $\Delta z \to 0$ horizontally, then $\Delta z = \Delta x$ and so (3) becomes

$$f'(z) = \lim_{\Delta x \to 0} \frac{u(x + \Delta x, y) - u(x, y)}{\Delta x} + i \lim_{\Delta x \to 0} \frac{v(x + \Delta x, y) - v(x, y)}{\Delta x}. \tag{4}$$

Since $f'(z)$ exists, the two limits in (4) exist. But by definition the limits in (4) are the first partial derivatives of u and v with respect to x. Thus, we have shown that

$$f'(z) = \frac{\partial u}{\partial x} + i\frac{\partial v}{\partial x}. \tag{5}$$

Now if we let $\Delta z \to 0$ vertically, then $\Delta z = i\Delta y$ and (3) becomes

$$f'(z) = \lim_{\Delta y \to 0} \frac{u(x, y + \Delta y) - u(x, y)}{i\Delta y} + i \lim_{\Delta y \to 0} \frac{v(x, y + \Delta y) - v(x, y)}{i\Delta y}, \tag{6}$$

which is the same as

$$f'(z) = -i\frac{\partial u}{\partial y} + \frac{\partial v}{\partial y}. \tag{7}$$

Equating the real and imaginary parts of (5) and (7) yields the pair of equations in (1). $\qquad \square$

If a complex function $f(z) = u(x, y) + iv(x, y)$ is analytic throughout a domain D, then the real functions u and v must satisfy the Cauchy–Riemann equations (1) at every point in D.

Example 1 Using the Cauchy–Riemann Equations

The polynomial $f(z) = z^2 + z$ is analytic for all z and $f(z) = x^2 - y^2 + x + i(2xy + y)$. Thus, $u(x, y) = x^2 - y^2 + x$ and $v(x, y) = 2xy + y$. For any point (x, y) we see that the Cauchy–Riemann equations are satisfied:

$$\frac{\partial u}{\partial x} = 2x + 1 = \frac{\partial v}{\partial y} \quad \text{and} \quad \frac{\partial u}{\partial y} = -2y = -\frac{\partial v}{\partial x}. \qquad \square$$

Example 2 Using the Cauchy–Riemann Equations

Show that the function $f(z) = (2x^2 + y) + i(y^2 - x)$ is not analytic at any point.

SOLUTION We identify $u(x, y) = 2x^2 + y$ and $v(x, y) = y^2 - x$. Now from

$$\frac{\partial u}{\partial x} = 4x \quad \text{and} \quad \frac{\partial v}{\partial y} = 2y$$

$$\frac{\partial u}{\partial y} = 1 \quad \text{and} \quad \frac{\partial v}{\partial x} = -1$$

we see that $\partial u/\partial y = -\partial v/\partial x$ but that the equality $\partial u/\partial x = \partial v/\partial y$ is satisfied only on the line $y = 2x$. However, for any point z on the line there is no neighborhood or open disk about z in which f is differentiable. We conclude that f is nowhere analytic. ❑

By themselves, the Cauchy–Riemann equations are not sufficient to ensure analyticity. However, when we add the condition of continuity to u and v *and* the four partial derivatives, the Cauchy–Riemann equations can be shown to imply analyticity. The proof is long and complicated and so we state only the result.

> **THEOREM 17.3 Criterion for Analyticity**
>
> Suppose the real-valued functions $u(x, y)$ and $v(x, y)$ are continuous and have continuous first-order partial derivatives in a domain D. If u and v satisfy the Cauchy–Riemann equations at all points of D, then the complex function $f(z) = u(x, y) + iv(x, y)$ is analytic in D.

Example 3 Using Theorem 17.3

For the function $f(z) = \dfrac{x}{x^2 + y^2} - i\,\dfrac{y}{x^2 + y^2}$ we have

$$\frac{\partial u}{\partial x} = \frac{y^2 - x^2}{(x^2 + y^2)^2} = \frac{\partial v}{\partial y} \quad \text{and} \quad \frac{\partial u}{\partial y} = -\frac{2xy}{(x^2 + y^2)^2} = -\frac{\partial v}{\partial x}.$$

In other words, the Cauchy–Riemann equations are satisfied except at the point where $x^2 + y^2 = 0$, that is, at $z = 0$. We conclude from Theorem 17.3 that f is analytic in any domain not containing the point $z = 0$. ❑

The results in (5) and (7) were obtained under the basic assumption that f was differentiable at the point z. In other words, (5) and (7) give us a formula for computing $f'(z)$:

$$f'(z) = \frac{\partial u}{\partial x} + i\,\frac{\partial v}{\partial x} = \frac{\partial v}{\partial y} - i\,\frac{\partial u}{\partial y}. \tag{8}$$

For example, we know that $f(z) = z^2$ is differentiable for all z. With $u(x, y) = x^2 - y^2$, $\partial u/\partial x = 2x$, $v(x, y) = 2xy$, and $\partial v/\partial x = 2y$, we see that

$$f'(z) = 2x + i2y = 2(x + iy) = 2z.$$

Recall that analyticity implies differentiability but not vice versa. Theorem 17.3 has an analogue that gives sufficient conditions for differentiability:

> *If the real-valued functions $u(x, y)$ and $v(x, y)$ are continuous and have continuous first-order partial derivatives in a neighborhood of z, and if u and v satisfy the Cauchy–Riemann equations at the point z, then the complex function $f(z) = u(x, y) + iv(x, y)$ is differentiable at z and $f'(z)$ is given by (8).*

The function $f(z) = x^2 - y^2 i$ is nowhere analytic. With the identifications $u(x, y) = x^2$ and $v(x, y) = -y^2$, we see from

$$\frac{\partial u}{\partial x} = 2x, \quad \frac{\partial v}{\partial y} = -2y \quad \text{and} \quad \frac{\partial u}{\partial y} = 0, \quad \frac{\partial v}{\partial x} = 0$$

that the Cauchy–Riemann equations are satisfied only when $y = -x$. But since the functions u, $\partial u/\partial x$, $\partial u/\partial y$, v, $\partial v/\partial x$, and $\partial v/\partial y$ are continuous at every point, it follows that f is differentiable on the line $y = -x$ and on that line (8) gives the derivative $f'(z) = 2x = -2y$.

■ **Harmonic Functions** We saw in Chapter 13 that Laplace's equation $\partial^2 u/\partial x^2 + \partial^2 u/\partial y^2 = 0$ occurs in certain problems involving steady-state temperatures. This partial differential equation also plays an important role in many areas of applied mathematics. Indeed, as we now see, the real and imaginary parts of an analytic function cannot be chosen arbitrarily, since both u and v must satisfy Laplace's equation. It is this link between analytic functions and Laplace's equation that makes complex variables so essential in the serious study of applied mathematics.

DEFINITION 17.8 Harmonic Functions

A real-valued function $\phi(x, y)$ that has continuous second-order partial derivatives in a domain D and satisfies Laplace's equation is said to be **harmonic** in D.

THEOREM 17.4 A Source of Harmonic Functions

Suppose $f(z) = u(x, y) + iv(x, y)$ is analytic in a domain D. Then the functions $u(x, y)$ and $v(x, y)$ are harmonic functions.

PROOF In this proof we shall assume that u and v have continuous second-order partial derivatives. Since f is analytic, the Cauchy–Riemann equations are satisfied. Differentiating both sides of $\partial u/\partial x = \partial v/\partial y$ with respect to x and differentiating both sides of $\partial u/\partial y = -\partial v/\partial x$ with respect to y then give

$$\frac{\partial^2 u}{\partial x^2} = \frac{\partial^2 v}{\partial x\, \partial y} \quad \text{and} \quad \frac{\partial^2 u}{\partial y^2} = -\frac{\partial^2 v}{\partial y\, \partial x}.$$

With the assumption of continuity, the mixed partials are equal. Hence, adding these two equations gives

$$\frac{\partial^2 u}{\partial x^2} + \frac{\partial^2 u}{\partial y^2} = 0.$$

This shows that $u(x, y)$ is harmonic.

Now differentiating both sides of $\partial u/\partial x = \partial v/\partial y$ with respect to y and differentiating both sides of $\partial u/\partial y = -\partial v/\partial x$ with respect to x and subtracting yield

$$\frac{\partial^2 v}{\partial x^2} + \frac{\partial^2 v}{\partial y^2} = 0. \qquad ❑$$

■ **Conjugate Harmonic Functions** If $f(z) = u(x, y) + iv(x, y)$ is analytic in a domain D, then u and v are harmonic in D. Now suppose $u(x, y)$ is a given function that is harmonic in D. It is then sometimes possible to find another function $v(x, y)$ that is harmonic in D so that $u(x, y) + iv(x, y)$ is an analytic function in D. The function v is called a **conjugate harmonic function** of u.

Example 4 **Harmonic Function/Conjugate Harmonic Function**

(a) Verify that the function $u(x, y) = x^3 - 3xy^2 - 5y$ is harmonic in the entire complex plane. (b) Find the conjugate harmonic function of u.

SOLUTION (a) From the partial derivatives

$$\frac{\partial u}{\partial x} = 3x^2 - 3y^2, \quad \frac{\partial^2 u}{\partial x^2} = 6x, \quad \frac{\partial u}{\partial y} = -6xy - 5, \quad \frac{\partial^2 u}{\partial y^2} = -6x$$

we see that u satisfies Laplace's equation:

$$\frac{\partial^2 u}{\partial x^2} + \frac{\partial^2 u}{\partial y^2} = 6x - 6x = 0.$$

(b) Since the conjugate harmonic function v must satisfy the Cauchy–Riemann equations, we must have

$$\frac{\partial v}{\partial y} = \frac{\partial u}{\partial x} = 3x^2 - 3y^2 \quad \text{and} \quad \frac{\partial v}{\partial x} = -\frac{\partial u}{\partial y} = 6xy + 5. \qquad (9)$$

Partial integration of the first equation in (9) with respect to y gives $v(x, y) = 3x^2 y - y^3 + h(x)$. From this we get

$$\frac{\partial v}{\partial x} = 6xy + h'(x).$$

Substituting this result into the second equation in (9) gives $h'(x) = 5$, and so $h(x) = 5x + C$. Therefore, the conjugate harmonic function of u is $v(x, y) = 3x^2 y - y^3 + 5x + C$. The analytic function is

$$f(z) = x^3 - 3xy^2 - 5y + i(3x^2 y - y^3 + 5x + C). \qquad \square$$

Remarks

Suppose u and v are the harmonic functions that comprise the real and imaginary parts of an analytic function $f(z)$. The level curves $u(x, y) = c_1$ and $v(x, y) = c_2$ defined by these functions form two *orthogonal* families of curves. (See Problem 32 in Exercises 17.5.) For example, the level curves generated by the simple analytic function $f(z) = z = x + iy$ are $x = c_1$ and $y = c_2$. The family of vertical lines defined by $x = c_1$ is clearly orthogonal to the family of horizontal lines defined by $y = c_2$. In electrostatics, if $u(x, y) = c_1$ defines the equipotential curves, then the other, and orthogonal, family $v(x, y) = c_2$ defines the lines of force.

EXERCISES 17.5 Answers to odd-numbered problems begin on page A-72.

In Problems 1 and 2, the given function is analytic for all z. Show that the Cauchy–Riemann equations are satisfied at every point.

In Problems 3–8, show that the given function is not analytic at any point.

1. $f(z) = z^3$

2. $f(z) = 3z^2 + 5z - 6i$

3. $f(z) = \text{Re}(z)$

4. $f(z) = y + ix$

5. $f(z) = 4z - 6\bar{z} + 3$ **6.** $f(z) = \bar{z}^2$

7. $f(z) = x^2 + y^2$

8. $f(z) = \dfrac{x}{x^2 + y^2} + i\,\dfrac{y}{x^2 + y^2}$

In Problems 9–14, use Theorem 17.3 to show that the given function is analytic in an appropriate domain.

9. $f(z) = e^x \cos y + ie^x \sin y$

10. $f(z) = x + \sin x \cosh y + i(y + \cos x \sinh y)$

11. $f(z) = e^{x^2 - y^2} \cos 2xy + ie^{x^2 - y^2} \sin 2xy$

12. $f(z) = 4x^2 + 5x - 4y^2 + 9 + i(8xy + 5y - 1)$

13. $f(z) = \dfrac{x - 1}{(x - 1)^2 + y^2} - i\,\dfrac{y}{(x - 1)^2 + y^2}$

14. $f(z) = \dfrac{x^3 + xy^2 + x}{x^2 + y^2} + i\,\dfrac{x^2 y + y^3 - y}{x^2 + y^2}$

In Problems 15 and 16, find real constants a, b, c, and d so that the given function is analytic.

15. $f(z) = 3x - y + 5 + i(ax + by - 3)$

16. $f(z) = x^2 + axy + by^2 + i(cx^2 + dxy + y^2)$

In Problems 17–20, show that the given function is not analytic at any point, but is differentiable along the indicated curve(s).

17. $f(z) = x^2 + y^2 + 2xyi$; x-axis

18. $f(z) = 3x^2 y^2 - 6x^2 y^2 i$; coordinate axes

19. $f(z) = x^3 + 3xy^2 - x + i(y^3 + 3x^2 y - y)$; coordinate axes

20. $f(z) = x^2 - x + y + i(y^2 - 5y - x)$; $y = x + 2$

21. Use (8) to find the derivative of the function in Problem 9.

22. Use (8) to find the derivative of the function in Problem 11.

In Problems 23–28, verify that the given function u is harmonic. Find v, the conjugate harmonic function of u. Form the corresponding analytic function $f(z) = u + iv$.

23. $u(x, y) = x$ **24.** $u(x, y) = 2x - 2xy$

25. $u(x, y) = x^2 - y^2$

26. $u(x, y) = 4xy^3 - 4x^3 y + x$

27. $u(x, y) = \log_e(x^2 + y^2)$

28. $u(x, y) = e^x(x \cos y - y \sin y)$

29. Sketch the level curves $u(x, y) = c_1$ and $v(x, y) = c_2$ of the analytic function $f(z) = z^2$.

30. Consider the function $f(z) = 1/z$. Describe the level curves.

31. Consider the function $f(z) = z + 1/z$. Describe the level curve $v(x, y) = 0$.

32. Suppose u and v are the harmonic functions forming the real and imaginary parts of an analytic function. Show that the level curves $u(x, y) = c_1$ and $v(x, y) = c_2$ are orthogonal. [*Hint:* Consider the gradient of u and the gradient of v. Ignore the case where a gradient vector is the zero vector.]

17.6 Exponential and Logarithmic Functions

In this and the next section, we shall examine the exponential, logarithmic, trigonometric, and hyperbolic functions of a complex variable z. Although the definitions of these complex functions are motivated by their real variable analogues, the properties of these complex functions will yield some surprises.

■ **Exponential Function** Recall that in real variables the exponential function $f(x) = e^x$ has the properties

$$f'(x) = f(x) \quad \text{and} \quad f(x_1 + x_2) = f(x_1)f(x_2). \tag{1}$$

We certainly want the definition of the complex function $f(z) = e^z$, where $z = x + iy$, to reduce e^x for $y = 0$ and to possess the same properties as in (1).

We have already used an exponential function with a pure imaginary exponent. Euler's formula,

$$e^{iy} = \cos y + i \sin y, \qquad y \text{ a real number}, \tag{2}$$

played an important role in Section 3.3. We can formally establish the result in (2) by using the Maclaurin series for e^x and replacing x by iy and rearranging terms:

$$e^{iy} = \sum_{k=0}^{\infty} \frac{(iy)^k}{k!} = 1 + iy + \frac{(iy)^2}{2!} + \frac{(iy)^3}{3!} + \frac{(iy)^4}{4!} + \cdots$$

$$= \left(1 - \frac{y^2}{2!} + \frac{y^4}{4!} - \frac{y^6}{6!} + \cdots\right) + i\left(y - \frac{y^3}{3!} + \frac{y^5}{5!} - \frac{y^7}{7!} + \cdots\right)$$

$$= \cos y + i \sin y.$$

For $z = x + iy$ it is natural to expect that

$$e^{x+iy} = e^x e^{iy}$$

and so by (2), $e^{x+iy} = e^x(\cos y + i \sin y).$

Inspired by this formal result, we make the following definition:

DEFINITION 17.9 Exponential Function

$$e^z = e^{x+iy} = e^x(\cos y + i \sin y). \tag{3}$$

The exponential function e^z is also denoted by the symbol exp z. Note that (3) reduces to e^x when $y = 0$.

Example 1 Complex Value of the Exponential Function

Evaluate $e^{1.7 + 4.2i}$.

SOLUTION With the identifications $x = 1.7$ and $y = 4.2$ and the aid of a calculator, we have, to four rounded decimal places,

$$e^{1.7} \cos 4.2 = -2.6837 \quad \text{and} \quad e^{1.7} \sin 4.2 = -4.7710.$$

It follows from (3) that $e^{1.7 + 4.2i} = -2.6837 - 4.7710i$. ❑

The real and imaginary parts of e^z, $u(x, y) = e^x \cos y$ and $v(x, y) = e^x \sin y$, are continuous and have continuous first partial derivatives at every point z of the complex plane. Moreover, the Cauchy–Riemann equations are satisfied at all points of the complex plane:

$$\frac{\partial u}{\partial x} = e^x \cos y = \frac{\partial v}{\partial y} \quad \text{and} \quad \frac{\partial u}{\partial y} = -e^x \sin y = -\frac{\partial v}{\partial x}.$$

It follows from Theorem 17.3 that $f(z) = e^z$ is analytic for all z; in other words, f is an entire function.

We shall now demonstrate that e^z possesses the two desired properties given in (1). First, the derivative of f is given by (5) of Section 17.5:

$$f'(z) = e^x \cos y + i(e^x \sin y) = e^x(\cos y + i \sin y) = f(z).$$

As desired, we have established that

$$\frac{d}{dz} e^z = e^z.$$

Second, if $z_1 = x_1 + iy_1$ and $z_2 = x_2 + iy_2$, then by multiplication of complex numbers and the addition formulas of trigonometry, we obtain

Figure 17.17

Figure 17.18

$$f(z_1)f(z_2) = e^{x_1}(\cos y_1 + i \sin y_1)e^{x_2}(\cos y_2 + i \sin y_2)$$
$$= e^{x_1 + x_2}[(\cos y_1 \cos y_2 - \sin y_1 \sin y_2) + i(\sin y_1 \cos y_2 + \cos y_1 \sin y_2)]$$
$$= e^{x_1 + x_2}[\cos (y_1 + y_2) + i \sin (y_1 + y_2)] = f(z_1 + z_2).$$

In other words,
$$e^{z_1}e^{z_2} = e^{z_1 + z_2}. \tag{4}$$

It is left as an exercise to prove that

$$\frac{e^{z_1}}{e^{z_2}} = e^{z_1 - z_2}.$$

■ **Periodicity** Unlike the real function e^x, the complex function $f(z) = e^z$ is **periodic** with the complex period $2\pi i$. Since $e^{2\pi i} = \cos 2\pi + i \sin 2\pi = 1$ and, in view of (4), $e^{z + 2\pi i} = e^z e^{2\pi i} = e^z$ for all z, it follows that $f(z + 2\pi i) = f(z)$. Because of this complex periodicity, all possible functional values of $f(z) = e^z$ are assumed in any infinite horizontal strip of width 2π. Thus, if we divide the complex plane into horizontal strips defined by $(2n - 1)\pi < y \le (2n + 1)\pi$, $n = 0, \pm 1, \pm 2, \dots$, then, as shown in Figure 17.17, for any point z in the strip $-\pi < y \le \pi$, the values $f(z)$, $f(z + 2\pi i)$, $f(z - 2\pi i)$, $f(z + 4\pi i)$, and so on are the same. The strip $-\pi < y \le \pi$ is called the **fundamental region** for the exponential function $f(z) = e^z$. The corresponding flow over the fundamental region is shown in Figure 17.18.

■ **Polar Form of a Complex Number** In Section 17.2 we saw that the complex number z could be written in polar form as $z = r(\cos \theta + i \sin \theta)$. Since $e^{i\theta} = \cos \theta + i \sin \theta$, we can now write the polar form of a complex number as

$$z = re^{i\theta}.$$

For example, in polar form $z = 1 + i$ is $z = \sqrt{2}e^{\pi i/4}$.

■ **Remarks** In applying mathematics, mathematicians and engineers often approach the same problem in completely different ways. Consider, for example, the solution of Example 10 in Section 3.8. In this example we used strictly real analysis to find the steady-state current $i_p(t)$ in an *LRC* series circuit described by the differential equation

$$L \frac{d^2q}{dt^2} + R \frac{dq}{dt} + \frac{1}{C} q = E_0 \sin \gamma t.$$

Electrical engineers often solve circuit problems such as this using complex analysis. To illustrate, let us first denote the imaginary unit $\sqrt{-1}$ by the symbol j to avoid confusion with the current i. Since current i is related to charge q by $i = dq/dt$, the differential equation is the same as

$$L \frac{di}{dt} + Ri + \frac{1}{C} q = E_0 \sin \gamma t.$$

Moreover, the impressed voltage $E_0 \sin \gamma t$ can be replaced by $\text{Im}(E_0 e^{j\gamma t})$, where Im means the "imaginary part of." Because of this last form, the method of undetermined coefficients suggests that we assume a solution in the form of a constant multiple of complex exponential—that is, $i_p(t) = \text{Im}(Ae^{j\gamma t})$. We substitute this expression into the last differential equation, use the fact that q is an antiderivative of i, and equate coefficients of $e^{j\gamma t}$:

$$\left(jL\gamma + R + \frac{1}{jC\gamma}\right)A = E_0 \quad \text{gives} \quad A = \frac{E_0}{R + j\left(L\gamma - \dfrac{1}{C\gamma}\right)}.$$

The quantity $Z = R + j(L\gamma - 1/C\gamma)$ is called the **complex impedance** of the circuit. Note that the modulus of the complex impedance, $|Z| = \sqrt{R^2 + (L\gamma - 1/C\gamma)^2}$, was denoted in Example 10 of Section 3.8 by the letter Z and called the impedance.

Now, in polar form the complex impedance is

$$Z = |Z|e^{j\theta} \quad \text{where} \quad \tan\theta = \frac{L\gamma - \dfrac{1}{C\gamma}}{R}.$$

Hence, $A = E_0/Z = E_0/(|Z|e^{j\theta})$, and so the steady-state current can be written as

$$i_p(t) = \text{Im}\frac{E_0}{|Z|}e^{-j\theta}e^{j\gamma t}.$$

The reader is encouraged to verify that this last expression is the same as (35) in Section 3.8.

■ **Logarithmic Function** The logarithm of a complex number $z = x + iy$, $z \neq 0$, is defined as the inverse of the exponential function—that is,

$$w = \ln z \quad \text{if} \quad z = e^w. \tag{5}$$

In (5) we note that $\ln z$ is not defined for $z = 0$, since there is no value of w for which $e^w = 0$. To find the real and imaginary parts of $\ln z$ we write $w = u + iv$ and use (3) and (5):

$$x + iy = e^{u+iv} = e^u(\cos v + i\sin v) = e^u\cos v + ie^u\sin v.$$

The last equality implies $x = e^u\cos v$ and $y = e^u\sin v$. We can solve these two equations for u and v. First, by squaring and adding the equations, we find

$$e^{2u} = x^2 + y^2 = r^2 = |z|^2 \quad \text{and so} \quad u = \log_e|z|,$$

where $\log_e|z|$ denotes the *real natural logarithm* of the modulus of z. Second, to solve for v we divide the two equations to obtain

$$\tan v = \frac{y}{x}.$$

This last equation means that v is an argument of z, that is, $v = \theta = \arg z$. But since there is no unique argument of a given complex number $z = x + iy$, if θ is an argument of z, then so is $\theta + 2n\pi$, $n = 0, \pm1, \pm2, \ldots$.

DEFINITION 17.10 **Logarithm of a Complex Number**

For $z \neq 0$, and $\theta = \arg z$,

$$\ln z = \log_e|z| + i(\theta + 2n\pi), \qquad n = 0, \pm1, \pm2,\ldots. \tag{6}$$

As is clearly indicated in (6), there are infinitely many values of the logarithm of a complex number z. This should not be any great surprise since the exponential function is periodic.

In real calculus, logarithms of negative numbers are not defined. As the next example will show, this is not the case in complex calculus.

Example 2 Complex Values of the Logarithmic Function

Find the values of (a) ln (−2), (b) ln i, and (c) ln (−1 − i).

SOLUTION (a) With $\theta = \arg(-2) = \pi$ and $\log_e|-2| = 0.6932$, we have from (6)

$$\ln(-2) = 0.6932 + i(\pi + 2n\pi).$$

(b) With $\theta = \arg(i) = \pi/2$ and $\log_e|i| = \log_e 1 = 0$, we have from (6)

$$\ln i = i\left(\frac{\pi}{2} + 2n\pi\right).$$

In other words, ln $i = \pi i/2, -3\pi i/2, 5\pi i/2, -7\pi i/2$, and so on.

(c) With $\theta = \arg(-1 - i) = 5\pi/4$ and $\log_e|-1 - i| = \log_e \sqrt{2} = 0.3466$, we have from (6)

$$\ln(-1 - i) = 0.3466 + i\left(\frac{5\pi}{4} + 2n\pi\right). \qquad \square$$

Example 3 Solving an Exponential Equation

Find all values of z such that $e^z = \sqrt{3} + i$.

SOLUTION From (5), with the symbol w replaced by z, we have $z = \ln(\sqrt{3} + i)$. Now $|\sqrt{3} + i| = 2$ and $\tan\theta = 1/\sqrt{3}$ imply that $\arg(\sqrt{3} + i) = \pi/6$, and so (6) gives

$$z = \log_e 2 + i\left(\frac{\pi}{6} + 2n\pi\right) \quad \text{or} \quad z = 0.6931 + i\left(\frac{\pi}{6} + 2n\pi\right). \qquad \square$$

■ **Principal Value** It is interesting to note that as a consequence of (6) the logarithm of a positive real number has many values. For example, in real calculus $\log_e 5$ has only one value: $\log_e 5 = 1.6094$, whereas in complex calculus ln 5 = $1.6094 + 2n\pi i$. The value of ln 5 corresponding to $n = 0$ is the same as the real logarithm $\log_e 5$ and is called the **principal value** of ln 5. Recall that in Section 17.2 we stipulated that the principal argument of a complex number, written Arg z, lies in the interval $(-\pi, \pi]$. In general, we define the **principal value** of ln z as that complex logarithm corresponding to $n = 0$ and $\theta =$ Arg z. To emphasize the principal value of the logarithm we shall adopt the notation Ln z. In other words,

$$\text{Ln } z = \log_e|z| + i \text{ Arg } z. \tag{7}$$

Since Arg z is unique, there is only one value of Ln z for each $z \neq 0$.

Example 4 Principal Values

The principal values of the logarithms in Example 2 are as follows.

(a) Since Arg $(-2) = \pi$, we need only set $n = 0$ in the result given in part (a) of Example 2:

$$\text{Ln}(-2) = 0.6932 + \pi i.$$

(b) Similarly, since Arg $(i) = \pi/2$, we set $n = 0$ in the result in part (b) of Example 2 to obtain

$$\text{Ln } i = \frac{\pi}{2}i.$$

(c) In part (c) of Example 2, arg $(-1 - i) = 5\pi/4$ is not the principal argument of $z = -1 - i$. The argument of z that lies in the interval $(-\pi, \pi]$ is Arg $(-1 - i) = -3\pi/4$. Hence, it follows from (7) that

$$\text{Ln } (-1 - i) = 0.3466 - \frac{3\pi}{4}i.$$ ❑

Up to this point we have avoided the use of the word *function* for the obvious reason that ln z defined in (6) is not a function in the strictest interpretation of that word. Nonetheless, it is customary to write $f(z) = \ln z$ and to refer to $f(z) = \ln z$ by the seemingly contradictory phrase *multiple-valued function*. Although we shall not pursue the details, (6) can be interpreted as an infinite collection of logarithmic functions (standard meaning of the word). Each function in the collection is called a **branch** of ln z. The function $f(z) = \text{Ln } z$ is then called the **principal branch** of ln z, or the **principal logarithmic function.** To minimize the confusion, we shall hereafter simply use the words *logarithmic function* when referring to either $f(z) = \ln z$ or $f(z) = \text{Ln } z$.

Some familiar properties of the logarithmic function hold in the complex case:

$$\ln (z_1 z_2) = \ln z_1 + \ln z_2 \quad \text{and} \quad \ln \left(\frac{z_1}{z_2}\right) = \ln z_1 - \ln z_2. \tag{8}$$

Equations (8) and (9) are to be interpreted in the sense that if values are assigned to two of the terms, then a correct value is assigned to the third term.

Example 5 **Properties of Logarithms**

Suppose $z_1 = 1$ and $z_2 = -1$. Then if we take ln $z_1 = 2\pi i$ and ln $z_2 = \pi i$, we get

$$\ln (z_1 z_2) = \ln (-1) = \ln z_1 + \ln z_2 = 2\pi i + \pi i = 3\pi i$$

$$\ln \left(\frac{z_1}{z_2}\right) = \ln (-1) = \ln z_1 - \ln z_2 = 2\pi i - \pi i = \pi i.$$ ❑

Just as (7) of Section 17.2 was not valid when arg z was replaced with Arg z, so too (8) is not true, in general, when ln z is replaced by Ln z. See Problems 45 and 46 in Exercises 17.6.

■ **Analyticity** The logarithmic function $f(z) = \text{Ln } z$ is not continuous at $z = 0$ since $f(0)$ is not defined. Moreover, $f(z) = \text{Ln } z$ is discontinuous at all points of the negative real axis. This is because the imaginary part of the function, $v = \text{Arg } z$, is discontinuous only at these points. To see this, suppose x_0 is a point on the negative real axis. As $z \to x_0$ from the upper half-plane, Arg $z \to \pi$, whereas if $z \to x_0$ from the lower half-plane, then Arg $z \to -\pi$. This means that $f(z) = \text{Ln } z$ is not analytic on the nonpositive real axis. However, $f(z) = \text{Ln } z$ is analytic throughout the domain D consisting of all the points in the complex plane except those on the nonpositive real axis. It is convenient to think of D as the complex plane from which the nonpositive real axis has been cut out. Since $f(z) = \text{Ln } z$ is the principal branch of ln z, the nonpositive real axis is referred to as a **branch cut** for the function. See Figure 17.19. It is left as exercises to show that the Cauchy–Riemann equations are satisfied throughout this cut plane and that the derivative of Ln z is given by

$$\frac{d}{dz} \text{Ln } z = \frac{1}{z} \tag{9}$$

for all z in D.

Figure 17.19

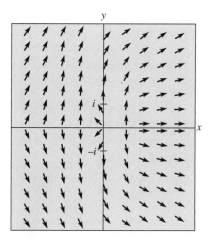

Figure 17.20

Figure 17.20 shows $w = \text{Ln } z$ as a flow. Note that the vector field is not continuous along the branch cut.

■ **Complex Powers** Inspired by the identity $x^a = e^{a \ln x}$ in real variables, we can define complex powers of a complex number. If α is a complex number and $z = x + iy$, then z^α is defined by

$$z^\alpha = e^{\alpha \ln z}, \qquad z \neq 0. \tag{10}$$

In general, z^α is multiple-valued since $\ln z$ is multiple-valued. However, in the special case when $\alpha = n$, $n = 0, \pm 1, \pm 2, \ldots$, (10) is single-valued since there is only one value for z^2, z^3, z^{-1}, and so on. To see that this is so, suppose $\alpha = 2$ and $z = re^{i\theta}$, where θ is any argument of z. Then

$$e^{2 \ln z} = e^{2(\log_e r + i\theta)} = e^{2 \log_e r + 2i\theta} = e^{2 \log_e r} e^{2i\theta} = r^2 e^{i\theta} e^{i\theta} = (re^{i\theta})(re^{i\theta}) = z^2.$$

If we use $\text{Ln } z$ in place of $\ln z$, then (10) gives the **principal value** of z^α.

Example 6 Complex Power
Find the value of i^{2i}.

SOLUTION With $z = i$, $\arg z = \pi/2$, and $\alpha = 2i$ it follows from (9) that

$$i^{2i} = e^{2i[\log_e 1 + i(\pi/2 + 2n\pi)]} = e^{-(1 + 4n)\pi}$$

where $n = 0, \pm 1, \pm 2, \ldots$. Inspection of the equation shows that i^{2i} is real for every value of n. Since $\pi/2$ is the principal argument of $z = i$, we obtain the principal value of i^{2i} for $n = 0$. To four rounded decimal places, this principal value is $i^{2i} = e^{-\pi} = 0.0043$. ❑

EXERCISES 17.6

Answers to odd-numbered problems begin on page A-73.

In Problems 1–10, express e^z in the form $a + ib$.

1. $z = \dfrac{\pi}{6} i$

2. $z = -\dfrac{\pi}{3} i$

3. $z = -1 + \dfrac{\pi}{4} i$

4. $z = 2 - \dfrac{\pi}{2} i$

5. $z = \pi + \pi i$

6. $z = -\pi + \dfrac{3\pi}{2} i$

7. $z = 1.5 + 2i$

8. $z = -0.3 + 0.5i$

9. $z = 5i$

10. $z = -0.23 - i$

In Problems 11 and 12, express the given number in the form $a + ib$.

11. $e^{1 + 5\pi i/4} e^{-1 - \pi i/3}$

12. $\dfrac{e^{2 + 3\pi i}}{e^{-3 + \pi i/2}}$

In Problems 13–16, use Definition 17.9 to express the given function in the form $f(z) = u + iv$.

13. $f(z) = e^{-iz}$

14. $f(z) = e^{2\bar{z}}$

15. $f(z) = e^{z^2}$

16. $f(z) = e^{1/z}$

In Problems 17–20, verify the given result.

17. $|e^z| = e^x$

18. $\dfrac{e^{z_1}}{e^{z_2}} = e^{z_1 - z_2}$

19. $e^{z + \pi i} = e^{z - \pi i}$

20. $(e^z)^n = e^{nz}$, n an integer

21. Show that $f(z) = e^{\bar{z}}$ is nowhere analytic.

22. (a) Use the result in Problem 15 to show that $f(z) = e^{z^2}$ is an entire function.

(b) Verify that $u(x, y) = \text{Re}(e^{z^2})$ is a harmonic function.

In Problems 23–28, express $\ln z$ in the form $a + ib$.

23. $z = -5$

24. $z = -ei$

25. $z = -2 + 2i$

26. $z = 1 + i$

27. $z = \sqrt{2} + \sqrt{6}i$

28. $z = -\sqrt{3} + i$

In Problems 29–34, express $\text{Ln } z$ in the form $a + ib$.

29. $z = 6 - 6i$

30. $z = -e^3$

31. $z = -12 + 5i$

32. $z = 3 - 4i$

33. $z = (1 + \sqrt{3}i)^5$ **34.** $z = (1 + i)^4$

In Problems 35–38, find all values of z satisfying the given equation.

35. $e^z = 4i$ **36.** $e^{1/z} = -1$

37. $e^{z-1} = -ie^2$ **38.** $e^{2z} + e^z + 1 = 0$

In Problems 39–42, find all values of the given quantity.

39. $(-i)^{4i}$ **40.** $3^{i/\pi}$

41. $(1 + i)^{(1+i)}$ **42.** $(1 + \sqrt{3}i)^{3i}$

In Problems 43 and 44, find the principal value of the given quantity. Express answers in the form $a + ib$.

43. $(-1)^{(-2i/\pi)}$ **44.** $(1 - i)^{2i}$

45. If $z_1 = i$ and $z_2 = -1 + i$, verify that $\text{Ln}\,(z_1 z_2) \neq \text{Ln}\,z_1 + \text{Ln}\,z_2$.

46. Find two complex numbers z_1 and z_2 such that

$$\text{Ln}\,(z_1/z_2) \neq \text{Ln}\,z_1 - \text{Ln}\,z_2.$$

47. Determine whether the given statement is true.

(a) $\text{Ln}\,(-1 + i)^2 = 2\,\text{Ln}\,(-1 + i)$

(b) $\text{Ln}\,i^3 = 3\,\text{Ln}\,i$

(c) $\ln i^3 = 3 \ln i$

48. The laws of exponents hold for complex numbers α and β:

$$z^\alpha z^\beta = z^{\alpha+\beta}, \qquad \frac{z^\alpha}{z^\beta} = z^{\alpha-\beta}, \qquad (z^\alpha)^n = z^{n\alpha}, \quad n \text{ an integer.}$$

However, the last law is not valid if n is a complex number. Verify that $(i^i)^2 = i^{2i}$, but $(i^2)^i \neq i^{2i}$.

49. For complex numbers z satisfying $\text{Re}(z) > 0$, show that (7) can be written as

$$\text{Ln}\,z = \frac{1}{2}\log_e(x^2 + y^2) + i\,\tan^{-1}\frac{y}{x}.$$

50. The function given in Problem 49 is analytic.

(a) Verify that $u(x, y) = \log_e(x^2 + y^2)$ is a harmonic function.

(b) Verify that $v(x, y) = \tan^{-1}(y/x)$ is a harmonic function.

17.7 Trigonometric and Hyperbolic Functions

■ **Trigonometric Functions** If x is a real variable, then Euler's formula gives

$$e^{ix} = \cos x + i \sin x \quad \text{and} \quad e^{-ix} = \cos x - i \sin x.$$

By subtracting and then adding these equations, we see that the real functions $\sin x$ and $\cos x$ can be expressed as a combination of exponential functions:

$$\sin x = \frac{e^{ix} - e^{-ix}}{2i} \qquad \cos x = \frac{e^{ix} + e^{-ix}}{2}. \tag{1}$$

Using (1) as a model, we now define the sine and cosine of a complex variable:

DEFINITION 17.11 Trigonometric Sine and Cosine

For any complex number $z = x + iy$,

$$\sin z = \frac{e^{iz} - e^{-iz}}{2i} \qquad \cos z = \frac{e^{iz} + e^{-iz}}{2}. \tag{2}$$

As in trigonometry, we define four additional trigonometric functions in terms of $\sin z$ and $\cos z$:

$$\tan z = \frac{\sin z}{\cos z}, \qquad \cot z = \frac{1}{\tan z}, \qquad \sec z = \frac{1}{\cos z}, \qquad \csc z = \frac{1}{\sin z}. \tag{3}$$

When $y = 0$, each function in (2) and (3) reduces to its real counterpart.

■ **Analyticity** Since the exponential functions e^{iz} and e^{-iz} are entire functions, it follows that $\sin z$ and $\cos z$ are entire functions. Now, as we shall see shortly, $\sin z = 0$ only for the real numbers $z = n\pi$, n an integer, and $\cos z = 0$ only for the real numbers $z = (2n + 1)\pi/2$, n an integer. Thus, $\tan z$ and $\sec z$ are analytic except at the points $z = (2n + 1)\pi/2$, and $\cot z$ and $\csc z$ are analytic except at the points $z = n\pi$.

■ **Derivatives** Since $(d/dz)e^z = e^z$, it follows from the Chain Rule that $(d/dz)e^{iz} = ie^{iz}$ and $(d/dz)e^{-iz} = -ie^{-iz}$. Hence,

$$\frac{d}{dz}\sin z = \frac{d}{dz}\frac{e^{iz} - e^{-iz}}{2i} = \frac{e^{iz} + e^{-iz}}{2} = \cos z.$$

In fact, it is readily shown that the forms of the derivatives of the complex trigonometric functions are the same as the real functions. We summarize the results.

$$\frac{d}{dz}\sin z = \cos z \qquad \frac{d}{dz}\cos z = -\sin z$$

$$\frac{d}{dz}\tan z = \sec^2 z \qquad \frac{d}{dz}\cot z = -\csc^2 z \qquad (4)$$

$$\frac{d}{dz}\sec z = \sec z \tan z \qquad \frac{d}{dz}\csc z = -\csc z \cot z$$

■ **Identities** The familiar trigonometric identities are also the same in the complex case.

$$\sin(-z) = -\sin z \qquad \cos(-z) = \cos z$$

$$\cos^2 z + \sin^2 z = 1$$

$$\sin(z_1 \pm z_2) = \sin z_1 \cos z_2 \pm \cos z_1 \sin z_2$$

$$\cos(z_1 \pm z_2) = \cos z_1 \cos z_2 \mp \sin z_1 \sin z_2$$

$$\sin 2z = 2\sin z \cos z \qquad \cos 2z = \cos^2 z - \sin^2 z$$

■ **Zeros** To find the zeros of $\sin z$ and $\cos z$ we need to express both functions in the form $u + iv$. Before proceeding, recall from calculus that if y is real, then the hyperbolic sine and hyperbolic cosine are defined in terms of the real exponential functions e^y and e^{-y}:

$$\sinh y = \frac{e^y - e^{-y}}{2} \quad \text{and} \quad \cosh y = \frac{e^y + e^{-y}}{2}. \qquad (5)$$

Now from Definition 17.11 and Euler's formula we find, after simplifying,

$$\sin z = \frac{e^{i(x + iy)} - e^{-i(x + iy)}}{2i} = \sin x\left(\frac{e^y + e^{-y}}{2}\right) + i\cos x\left(\frac{e^y - e^{-y}}{2}\right).$$

Thus from (5) we have

$$\sin z = \sin x \cosh y + i \cos x \sinh y. \qquad (6)$$

It is left as an exercise to show that

$$\cos z = \cos x \cosh y - i \sin x \sinh y. \qquad (7)$$

From (6), (7), and $\cosh^2 y = 1 + \sinh^2 y$, we find

$$|\sin z|^2 = \sin^2 x + \sinh^2 y \qquad (8)$$

and

$$|\cos z|^2 = \cos^2 x + \sinh^2 y. \qquad (9)$$

Now a complex number z is zero if and only if $|z|^2 = 0$. Thus, if $\sin z = 0$, then from (8) we must have $\sin^2 x + \sinh^2 y = 0$. This implies that $\sin x = 0$ and $\sinh y = 0$, and so $x = n\pi$ and $y = 0$. Thus the only **zeros** of $\sin z$ are the real numbers $z = n\pi + 0i = n\pi$, $n = 0, \pm 1, \pm 2, \ldots$. Similarly, it follows from (9) that $\cos z = 0$ only when $z = (2n + 1)\pi/2$, $n = 0, \pm 1, \pm 2, \ldots$.

Example 1 Complex Value of the Sine Function

From (6) we have, with the aid of a calculator,

$$\sin (2 + i) = \sin 2 \cosh 1 + i \cos 2 \sinh 1 = 1.4031 - 0.4891i. \qquad \square$$

In ordinary trigonometry we are accustomed to the fact that $|\sin x| \le 1$ and $|\cos x| \le 1$. Inspection of (8) and (9) shows that these inequalities do not hold for the complex sine and cosine, since $\sinh y$ can range from $-\infty$ to ∞. In other words, it is perfectly feasible to have solutions for equations such as $\cos z = 10$.

Example 2 Solving a Trigonometric Equation

Solve the equation $\cos z = 10$.

SOLUTION From (2), $\cos z = 10$ is equivalent to $(e^{iz} + e^{-iz})/2 = 10$. Multiplying the last equation by e^{iz} then gives the quadratic equation in e^{iz}:

$$e^{2iz} - 20e^{iz} + 1 = 0.$$

From the quadratic formula we find $e^{iz} = 10 \pm 3\sqrt{11}$. Thus, for $n = 0, \pm 1, \pm 2, \ldots$, we have $iz = \log_e(10 \pm 3\sqrt{11}) + 2n\pi i$. Dividing by i and utilizing $\log_e(10 - 3\sqrt{11}) = -\log_e(10 + 3\sqrt{11})$, we can express the solutions of the given equation as $z = 2n\pi \pm i \log_e(10 + 3\sqrt{11})$. $\qquad \square$

■ **Hyperbolic Functions** We define the complex hyperbolic sine and cosine in a manner analogous to the real definitions given in (5):

> **DEFINITION 17.12 Hyperbolic Sine and Cosine**
>
> For any complex number $z = x + iy$,
>
> $$\sinh z = \frac{e^z - e^{-z}}{2} \qquad \cosh z = \frac{e^z + e^{-z}}{2}. \qquad (10)$$

The hyperbolic tangent, cotangent, secant, and cosecant functions are defined in terms of $\sinh z$ and $\cosh z$:

$$\tanh z = \frac{\sinh z}{\cosh z} \qquad \coth z = \frac{1}{\tanh z} \qquad \operatorname{sech} z = \frac{1}{\cosh z} \qquad \operatorname{csch} z = \frac{1}{\sinh z}. \qquad (11)$$

The hyperbolic sine and cosine are entire functions, and the functions defined in (11) are analytic except at points where the denominators are zero. It is also easy to see from (10) that

$$\frac{d}{dz} \sinh z = \cosh z \quad \text{and} \quad \frac{d}{dz} \cosh z = \sinh z. \qquad (12)$$

It is interesting to observe that, in contrast to real calculus, the trigonometric and hyperbolic functions are related in complex calculus. If we replace z by iz everywhere in (10) and compare the results with (2), we see that $\sinh(iz) = i \sin z$ and $\cosh(iz) = \cos z$. These equations enable us to express $\sin z$ and $\cos z$ in terms of $\sinh(iz)$ and $\cosh(iz)$, respectively. Similarly, by replacing z by iz in (2) we can express, in turn, $\sinh z$ and $\cosh z$ in terms of $\sin(iz)$ and $\cos(iz)$. We summarize the results:

$$\sin z = -i \sinh(iz), \qquad \cos z = \cosh(iz) \tag{13}$$

$$\sinh z = -i \sin(iz), \qquad \cosh z = \cos(iz). \tag{14}$$

■ **Zeros** The relationships given in (14) enable us to derive identities for the hyperbolic functions utilizing results for the trigonometric functions. For example, to express $\sinh z$ in the form $u + iv$ we write $\sinh z = -i \sin(iz)$ in the form $\sinh z = -i \sin(-y + ix)$ and use (6):

$$\sinh z = -i[\sin(-y) \cosh x + i \cos(-y) \sinh x].$$

Since $\sin(-y) = -\sin y$ and $\cos(-y) = \cos y$, the foregoing simplifies to

$$\sinh z = \sinh x \cos y + i \cosh x \sin y. \tag{15}$$

Similarly, $\qquad \cosh z = \cosh x \cos y + i \sinh x \sin y. \tag{16}$

It also follows directly from (14) that the **zeros** of $\sinh z$ and $\cosh z$ are pure imaginary and are, respectively,

$$z = n\pi i \quad \text{and} \quad z = (2n + 1)\frac{\pi i}{2}, \qquad n = 0, \pm 1, \pm 2, \dots.$$

■ **Periodicity** Since $\sin x$ and $\cos x$ are 2π-periodic we can easily demonstrate that $\sin z$ and $\cos z$ are also periodic with the same real period 2π. For example, from (6) note that

$$\sin(z + 2\pi) = \sin(x + 2\pi + iy)$$
$$= \sin(x + 2\pi) \cosh y + i \cos(x + 2\pi) \sinh y$$
$$= \sin x \cosh y + i \cos x \sinh y,$$

that is, $\sin(z + 2\pi) = \sin z$. In exactly the same manner, it follows from (7) that $\cos(z + 2\pi) = \cos z$. In addition, the hyperbolic functions $\sinh z$ and $\cosh z$ have the imaginary period $2\pi i$. This last result follows from either Definition 17.12 and the fact e^z is periodic with period $2\pi i$, or from (15) and (16) and replacing z by $z + 2\pi i$.

Answers to odd-numbered problems begin on page A-73.

In Problems 1–12, express the given quantity in the form $a + ib$.

1. $\cos(3i)$

2. $\sin(-2i)$

3. $\sin\left(\dfrac{\pi}{4} + i\right)$

4. $\cos(2 - 4i)$

5. $\tan(i)$

6. $\cot\left(\dfrac{\pi}{2} + 3i\right)$

7. $\sec(\pi + i)$

8. $\csc(1 + i)$

9. $\cosh(\pi i)$

10. $\sinh\left(\dfrac{3\pi}{2} i\right)$

11. $\sinh\left(1 + \dfrac{\pi}{3} i\right)$

12. $\cosh(2 + 3i)$

In Problems 13 and 14, verify the given result.

13. $\sin\left(\dfrac{\pi}{2} + i \ln 2\right) = \dfrac{5}{4}$

14. $\cos\left(\dfrac{\pi}{2} + i \ln 2\right) = -\dfrac{3}{4} i$

In Problems 15–20, find all values of z satisfying the given equation.

15. $\sin z = 2$

16. $\cos z = -3i$

17. $\sinh z = -i$

18. $\sinh z = -1$

19. $\cos z = \sin z$

20. $\cos z = i \sin z$

In Problems 21 and 22, use the definition of equality of complex numbers to find all values of z satisfying the given equation.

21. $\cos z = \cosh 2$

22. $\sin z = i \sinh 2$

23. Prove that $\cos z = \cos x \cosh y - i \sin x \sinh y$.

24. Prove that $\sinh z = \sinh x \cos y + i \cosh x \sin y$.

25. Prove that $\cosh z = \cosh x \cos y + i \sinh x \sin y$.

26. Prove that $|\sinh z|^2 = \sin^2 y + \sinh^2 x$.

27. Prove that $|\cosh z|^2 = \cos^2 y + \sinh^2 x$.

28. Prove that $\cos^2 z + \sin^2 z = 1$.

29. Prove that $\cosh^2 z - \sinh^2 z = 1$.

30. Show that $\tan z = u + iv$, where

$$u = \frac{\sin 2x}{\cos 2x + \cosh 2y} \quad \text{and} \quad y = \frac{\sinh 2y}{\cos 2x + \cosh 2y}.$$

31. Prove that $\tanh z$ is periodic with period πi.

32. Prove that (a) $\overline{\sin z} = \sin \overline{z}$ and (b) $\overline{\cos z} = \cos \overline{z}$.

17.8 Inverse Trigonometric and Hyperbolic Functions

As functions of a complex variable z, we have seen that both the trigonometric and hyperbolic functions are periodic. Consequently, these functions do not possess inverses that are functions in the strictest interpretation of that word. The inverses of these analytic functions are multiple-valued functions. As we did in Section 17.6, in the discussion of the logarithmic function, we shall drop the adjective "multiple-valued."

■ **Inverse Sine** The **inverse sine** function, written as $\sin^{-1}z$ or arcsin z, is defined by

$$w = \sin^{-1}z \quad \text{if} \quad z = \sin w. \tag{1}$$

The inverse sine can be expressed in terms of the logarithmic function. To see this we use (1) and the definition of the sine function:

$$\frac{e^{iw} - e^{-iw}}{2i} = z \quad \text{or} \quad e^{2iw} - 2ize^{iw} - 1 = 0.$$

From the last equation and the quadratic formula, we then obtain

$$e^{iw} = iz + (1 - z^2)^{1/2}. \tag{2}$$

Note in (2) we did not use the customary symbolism $\pm\sqrt{1 - z^2}$, since we know from Section 17.2 that $(1 - z^2)^{1/2}$ is two-valued. Solving (2) for w then gives

$$\sin^{-1}z = -i \ln[iz + (1 - z^2)^{1/2}]. \tag{3}$$

Proceeding in a similar manner, we find the inverses of the cosine and tangent to be

$$\cos^{-1}z = -i \ln[z + i(1 - z^2)^{1/2}] \tag{4}$$

$$\tan^{-1}z = \frac{i}{2} \ln \frac{i + z}{i - z}. \tag{5}$$

│ ***Example 1*** **Values of an Inverse Sine**
Find all values of $\sin^{-1} \sqrt{5}$.

SOLUTION From (3) we have

$$\sin^{-1}\sqrt{5} = -i\,\ln[\sqrt{5}i + (1 - (\sqrt{5})^2)^{1/2}].$$

With $(1 - (\sqrt{5})^2)^{1/2} = (-4)^{1/2} = \pm 2i$, the preceding expression becomes

$$\sin^{-1}\sqrt{5} = -i\,\ln[(\sqrt{5} \pm 2)i]$$

$$= -i\left[\log_e(\sqrt{5} \pm 2) + \left(\frac{\pi}{2} + 2n\pi\right)i\right], \qquad n = 0, \pm 1, \pm 2, \dots.$$

The foregoing result can be simplified a little by noting that $\log_e(\sqrt{5} - 2) = \log_e(1/(\sqrt{5} + 2)) = -\log_e(\sqrt{5} + 2)$. Thus for $n = 0, \pm 1, \pm 2, \dots,$

$$\sin^{-1}\sqrt{5} = \frac{\pi}{2} + 2n\pi \pm i\,\log_e(\sqrt{5} + 2). \tag{6} \ \square$$

To obtain particular values of, say, $\sin^{-1}z$, we must choose a specific root of $1 - z^2$ and a specific branch of the logarithm. For example, if we choose $(1 - (\sqrt{5})^2)^{1/2} = (-4)^{1/2} = 2i$ and the principal branch of the logarithm, then (6) gives the single value

$$\sin^{-1}\sqrt{5} = \frac{\pi}{2} - i\,\log_e(\sqrt{5} + 2).$$

■ **Derivatives** The derivatives of the three inverse trigonometric functions considered above can be found by implicit differentiation. To find the derivative of the inverse sine function $w = \sin^{-1}z$, we begin by differentiating $z = \sin w$:

$$\frac{d}{dz}z = \frac{d}{dz}\sin w \quad \text{gives} \quad \frac{dw}{dz} = \frac{1}{\cos w}.$$

Using the trigonometric identity $\cos^2 w + \sin^2 w = 1$ (see Problem 28 in Exercises 17.7) in the form $\cos w = (1 - \sin^2 w)^{1/2} = (1 - z^2)^{1/2}$, we obtain

$$\frac{d}{dz}\sin^{-1}z = \frac{1}{(1 - z^2)^{1/2}}. \tag{7}$$

Similarly, we find that

$$\frac{d}{dz}\cos^{-1}z = \frac{-1}{(1 - z^2)^{1/2}} \tag{8}$$

$$\frac{d}{dz}\tan^{-1}z = \frac{1}{1 + z^2}. \tag{9}$$

It should be noted that the square roots used in (7) and (8) must be consistent with the square roots used in (3) and (4).

Example 2 **Evaluating a Derivative**

Find the derivative of $w = \sin^{-1}z$ at $z = \sqrt{5}$.

SOLUTION In Example 1 if we use $(1 - (\sqrt{5})^2)^{1/2} = (-4)^{1/2} = 2i$, then that same root must be used in (7). The value of the derivative consistent with this choice is given by

$$\left.\frac{dw}{dz}\right|_{z=\sqrt{5}} = \frac{1}{(1 - (\sqrt{5})^2)^{1/2}} = \frac{1}{(-4)^{1/2}} = \frac{1}{2i} = -\frac{1}{2}i. \qquad \square$$

■ **Inverse Hyperbolic Functions** The inverse hyperbolic functions can also be expressed in terms of the logarithm. We summarize these results for the inverse hyperbolic sine, cosine, and tangent along with their derivatives:

$$\sinh^{-1}z = \ln[z + (z^2 + 1)^{1/2}] \tag{10}$$

$$\cosh^{-1}z = \ln[z + (z^2 - 1)^{1/2}] \tag{11}$$

$$\tanh^{-1}z = \frac{1}{2}\ln\frac{1+z}{1-z} \tag{12}$$

$$\frac{d}{dz}\sinh^{-1}z = \frac{1}{(z^2+1)^{1/2}} \tag{13}$$

$$\frac{d}{dz}\cosh^{-1}z = \frac{1}{(z^2-1)^{1/2}} \tag{14}$$

$$\frac{d}{dz}\tanh^{-1}z = \frac{1}{1-z^2}. \tag{15}$$

Example 3 **Values of an Inverse Hyperbolic Cosine**
Find all values of $\cosh^{-1}(-1)$.

SOLUTION From (11) with $z = -1$, we get

$$\cosh^{-1}(-1) = \ln(-1) = \log_e 1 + (\pi + 2n\pi)i.$$

Since $\log_e 1 = 0$ we have for $n = 0, \pm 1, \pm 2, \dots,$

$$\cosh^{-1}(-1) = (2n + 1)\pi i.$$ ❑

E X E R C I S E S 1 7 . 8 Answers to odd-numbered problems begin on page A-73.

In Problems 1–14, find all values of the given quantity.

1. $\sin^{-1}(-i)$

2. $\sin^{-1}\sqrt{2}$

3. $\sin^{-1} 0$

4. $\sin^{-1} 13/5$

5. $\cos^{-1} 2$

6. $\cos^{-1} 2i$

7. $\cos^{-1} \frac{1}{2}$

8. $\cos^{-1} \frac{5}{3}$

9. $\tan^{-1} 1$

10. $\tan^{-1} 3i$

11. $\sinh^{-1} \frac{4}{3}$

12. $\cosh^{-1} i$

13. $\tanh^{-1}(1 + 2i)$

14. $\tanh^{-1}(-\sqrt{3}i)$

C H A P T E R 1 7
R E V I E W E X E R C I S E S Answers to odd-numbered problems begin on page A-73.

Answer Problems 1–16 without referring back to the text. Fill in the blank or answer true/false.

1. $\text{Re}((1 + i)^{10}) = $ _____ and $\text{Im}((1 + i)^{10}) = $ _____ .

2. If z is a point in the third quadrant, then $i\bar{z}$ is in the _____ quadrant.

3. If $z = 3 + 4i$, then $\text{Re}\left(\frac{z}{\bar{z}}\right) = $ _____ .

4. $i^{127} - 5i^9 + 2i^{-1} = $ _____

5. If $z = \frac{4i}{-3 - 4i}$, then $|z| = $ _____ .

6. Describe the region defined by $1 \le |z + 2| \le 3$. _____

7. $\text{Arg}(z + \bar{z}) = 0$ _____

8. If $z = \frac{5}{-\sqrt{3} + i}$, then $\text{Arg } z = $ _____ .

9. If $e^z = 2i$, then $z = $ _____ .

10. If $|e^z| = 1$, then z is a pure imaginary number. ____

11. The principal value of $(1 + i)^{(2+i)}$ is _____ .

12. If $f(z) = x^2 - 3xy - 5y^3 + i(4x^2y - 4x + 7y)$, then $f(-1 + 2i) = $ _____ .

13. If the Cauchy–Riemann equations are satisfied at a point, then the function is necessarily analytic there. ____

14. $f(z) = e^z$ is periodic with period _____.

15. Ln $(-ie^3) =$ _____

16. $f(z) = \sin(x - iy)$ is nowhere analytic. ____

In Problems 17–20, write the given number in the form $a + ib$.

17. $i(2 - 3i)^2(4 + 2i)$

18. $\dfrac{3 - i}{2 + 3i} + \dfrac{2 - 2i}{1 + 5i}$

19. $\dfrac{(1 - i)^{10}}{(1 + i)^3}$

20. $4e^{\pi i/3}e^{-\pi i/4}$

In Problems 21–24, sketch the set of points in the complex plane satisfying the given inequality.

21. $\operatorname{Im}(z^2) \le 2$

22. $\operatorname{Im}(z + 5i) > 3$

23. $\dfrac{1}{|z|} \le 1$

24. $\operatorname{Im}(z) < \operatorname{Re}(z)$

25. Look up the definitions of conic sections in a calculus text. Now describe the set of points in the complex plane that satisfy the equation $|z - 2i| + |z + 2i| = 5$.

26. Let z and w be complex numbers such that $|z| = 1$ and $|w| \ne 1$. Prove that
$$\left| \frac{z - w}{1 - z\overline{w}} \right| = 1.$$

In Problems 27 and 28, find all solutions of the given equation.

27. $z^4 = 1 - i$

28. $z^{3/2} = \dfrac{1}{2 - i}$

29. If $f(z) = z^{24} - 3z^{20} + 4z^{12} - 5z^6$, find $f\left(\dfrac{1 + i}{\sqrt{2}}\right)$.

30. Write $f(z) = \operatorname{Im}(z - 3\overline{z}) + z \operatorname{Re}(z^2) - 5z$ in the form $f(z) = u(x, y) + iv(x, y)$.

In Problems 31 and 32, find the image of the line $x = 1$ in the w-plane under the given mapping.

31. $f(z) = x^2 - y + i(y^2 - x)$

32. $f(z) = \dfrac{1}{z}$

In Problems 33–36, find all complex numbers for which the given statement is true.

33. $z = z^{-1}$

34. $\overline{z} = \dfrac{1}{z}$

35. $\overline{z} = -z$

36. $z^2 = (\overline{z})^2$

37. Show that the function $f(z) = -(2xy + 5x) + i(x^2 - 5y - y^2)$ is analytic for all z. Find $f'(z)$.

38. Determine whether the function
$$f(z) = x^3 + xy^2 - 4x + i(4y - y^3 - x^2 y)$$
is differentiable. Is it analytic?

In Problems 39 and 40, verify the given equality.

39. $\operatorname{Ln}(1 + i)(1 - i) = \operatorname{Ln}(1 + i) + \operatorname{Ln}(1 - i)$

40. $\operatorname{Ln} \dfrac{1 + i}{1 - i} = \operatorname{Ln}(1 + i) - \operatorname{Ln}(1 - i)$

INTRODUCTION

To define an integral of a complex function f, we start with f defined along some curve C or **contour** in the complex plane. We shall see in this section that the definition of a complex integral, its properties, and method of evaluation are quite similar to those of a real line integral in the plane.

18.1 Contour Integrals

Integration in the complex plane is defined in a manner similar to that of a line integral in the plane. In other words, we shall be dealing with an integral of a complex function $f(z)$ that is defined along a curve C in the complex plane. These curves are defined in terms of parametric equations $x = x(t)$, $y = y(t)$, $a \le t \le b$, where t is a real parameter. By using $x(t)$ and $y(t)$ as real and imaginary parts, we can also describe a curve C in the complex plane by means of a complex-valued function of a real variable t: $z(t) = x(t) + iy(t)$, $a \le t \le b$. For example, $x = \cos t$, $y = \sin t$, $0 \le t \le 2\pi$, describes a unit circle centered at the origin. This circle can also be described by $z(t) = \cos t + i \sin t$, or even more compactly by $z(t) = e^{it}$, $0 \le t \le 2\pi$. The same definitions of smooth curve, piecewise smooth curve, closed curve, and simple closed curve given in Section 9.8 carry over to this discussion. The student is encouraged to review these concepts. As before, we shall assume that the positive direction on C corresponds to increasing values of t. In complex variables, a piecewise smooth curve C is also called a **contour** or **path**. An integral of $f(z)$ on C is denoted by $\int_C f(z)\, dz$ or $\oint_C f(z)\, dz$ if the contour C is closed; it is referred to as a **contour integral** or a **complex line integral.**

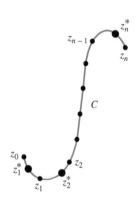

$f(z) = u(x, y) + iv(x, y)$

1. Let f be defined at all points on a smooth curve C defined by $x = x(t)$, $y = y(t)$, $a \le t \le b$.
2. Divide C into n subarcs according to the partition $a = t_0 < t_1 < \cdots < t_n = b$ of $[a, b]$. The corresponding points on the curve C are $z_0 = x_0 + iy_0 = x(t_0) + iy(t_0)$, $z_1 = x_1 + iy_1 = x(t_1) + iy(t_1)$, \ldots, $z_n = x_n + iy_n = x(t_n) + iy(t_n)$. Let $\Delta z_k = z_k - z_{k-1}$, $k = 1, 2, \ldots, n$.
3. Let $\|P\|$ be the **norm** of the partition—that is, the maximum value of $|\Delta z_k|$.
4. Choose a point $z_k^* = x_k^* + iy_k^*$ on each subarc.
5. Form the sum $\sum_{k=1}^{n} f(z_k^*)\, \Delta z_k$.

DEFINITION 18.1 Contour Integral

Let f be defined at points of a smooth curve C defined by $x = x(t)$, $y = y(t)$, $a \le t \le b$. The **contour integral** of f along C is

$$\int_C f(z)\, dz = \lim_{\|P\| \to 0} \sum_{k=1}^{n} f(z_k^*)\, \Delta z_k. \tag{1}$$

The limit in (1) exists if f is continuous at all points on C and C is either smooth or piecewise smooth. Consequently we shall, hereafter, assume these conditions as a matter of course.

■ **A Method of Evaluation** We shall turn now to the question of evaluating a contour integral. To facilitate the discussion let us suppress the subscripts and write (1) in the abbreviated form

$$\int_C f(z)\, dz = \lim \sum (u + iv)(\Delta x + i\, \Delta y) = \lim \left\{ \sum (u\, \Delta x - v\, \Delta y) + i \sum (v\, \Delta x + u\, \Delta y) \right\}.$$

This means

$$\int_C f(z)\,dz = \int_C u\,dx - v\,dy + i\int_C v\,dx + u\,dy. \tag{2}$$

In other words, a contour integral $\int_C f(z)\,dz$ is a combination of two real-line integrals $\int_C u\,dx - v\,dy$ and $\int_C v\,dx + u\,dy$. Now, since $x = x(t)$ and $y = y(t)$, $a \le t \le b$, the right side of (2) is the same as

$$\int_a^b [u(x(t),\,y(t))x'(t) - v(x(t),\,y(t))y'(t)]\,dt + i\int_a^b [v(x(t),\,y(t))x'(t) + u(x(t),\,y(t))y'(t)]\,dt.$$

But if we use $z(t) = x(t) + iy(t)$ to describe C, the last result is the same as $\int_a^b f(z(t))z'(t)\,dt$ when separated into two integrals. Thus we arrive at a practical means of evaluating a contour integral:

THEOREM 18.1 Evaluation of a Contour Integral

If f is continuous on a smooth curve C given by $z(t) = x(t) + iy(t)$, $a \le t \le b$, then

$$\int_C f(z)\,dz = \int_a^b f(z(t))z'(t)\,dt. \tag{3}$$

If f is expressed in terms of the symbol z, then to evaluate $f(z(t))$ we simply replace the symbol z by $z(t)$. If f is not expressed in terms of z, then to evaluate $f(z(t))$ we replace x and y wherever they appear by $x(t)$ and $y(t)$, respectively.

Example 1 Evaluating a Contour Integral

Evaluate $\displaystyle\int_C \bar{z}\,dz$, where C is given by $x = 3t$, $y = t^2$, $-1 \le t \le 4$.

SOLUTION We write $z(t) = 3t + it^2$ so that $z'(t) = 3 + 2it$ and $f(z(t)) = 3t + it^2 = 3t - it^2$. Thus,

$$\int_C \bar{z}\,dz = \int_{-1}^4 (3t - it^2)(3 + 2it)\,dt$$

$$= \int_{-1}^4 (2t^3 + 9t)\,dt + i\int_{-1}^4 3t^2\,dt = 195 + 65i. \qquad \square$$

Example 2 Evaluating a Contour Integral

Evaluate $\displaystyle\oint_C \frac{1}{z}\,dz$, where C is the circle $x = \cos t$, $y = \sin t$, $0 \le t \le 2\pi$.

SOLUTION In this case $z(t) = \cos t + i\sin t = e^{it}$, $z'(t) = ie^{it}$, and $f(z) = 1/z = e^{-it}$. Hence,

$$\oint_C \frac{1}{z}\,dz = \int_0^{2\pi} (e^{-it})ie^{it}\,dt = i\int_0^{2\pi} dt = 2\pi i. \qquad \square$$

For some curves, the real variable x itself can be used as the parameter. For example, to evaluate $\int_C (8x^2 - iy)\,dz$ on $y = 5x$, $0 \le x \le 2$, we write $\int_C (8x^2 - iy)\,dz = \int_0^2 (8x^2 - 5ix)(1 + 5i)\,dx$ and integrate in the usual manner.

■ **Properties** The following properties of contour integrals are analogous to the properties of line integrals:

> **THEOREM 18.2 Properties of Contour Integrals**
>
> Suppose f and g are continuous in a domain D and C is a smooth curve lying entirely in D. Then:
> - (*i*) $\int_C kf(z)\,dz = k\int_C f(z)\,dz$, k a constant
> - (*ii*) $\int_C [f(z) + g(z)]\,dz = \int_C f(z)\,dz + \int_C g(z)\,dz$
> - (*iii*) $\int_C f(z)\,dz = \int_{C_1} f(z)\,dz + \int_{C_2} f(z)\,dz$, where C is the union of the smooth curves C_1 and C_2
> - (*iv*) $\int_{-C} f(z)\,dz = -\int_C f(z)\,dz$, where $-C$ denotes the curve having the opposite orientation of C.

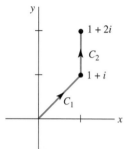

Figure 18.1

Example 3 Evaluating a Contour Integral

Evaluate $\int_C (x^2 + iy^2)\,dz$, where C is the contour shown in Figure 18.1.

SOLUTION In view of Theorem 18.2(*iii*) we write

$$\int_C (x^2 + iy^2)\,dz = \int_{C_1} (x^2 + iy^2)\,dz + \int_{C_2} (x^2 + iy^2)\,dz.$$

Since the curve C_1 is defined by $y = x$, it makes sense to use x as a parameter. Therefore, $z(x) = x + ix$, $z'(x) = 1 + i$, $f(z(x)) = x^2 + ix^2$, and

$$\int_{C_1} (x^2 + iy^2)\,dz = \int_0^1 (x^2 + ix^2)(1 + i)\,dx$$

$$= (1 + i)^2 \int_0^1 x^2\,dx = \frac{(1 + i)^2}{3} = \frac{2}{3}\,i.$$

The curve C_2 is defined by $x = 1$, $1 \le y \le 2$. Using y as a parameter, we have $z(y) = 1 + iy$, $z'(y) = i$, and $f(z(y)) = 1 + iy^2$. Thus,

$$\int_{C_2} (x^2 + iy^2)\,dz = \int_1^2 (1 + iy^2)\,i\,dy = -\int_1^2 y^2\,dy + i\int_1^2 dy = -\frac{7}{3} + i.$$

Finally, we have $\int_C (x^2 + iy^2)\,dz = \tfrac{2}{3}i + (-\tfrac{7}{3} + i) = -\tfrac{7}{3} + \tfrac{5}{3}\,i.$ ❑

There are times in the application of complex integration that it is useful to find an upper bound for the absolute value of a contour integral. In the next theorem we shall use the fact that the length of a plane curve is $s = \int_a^b \sqrt{[x'(t)]^2 + [y'(t)]^2}\,dt$. But if $z'(t) = x'(t) + iy'(t)$, then $|z'(t)| = \sqrt{[x'(t)]^2 + [y'(t)]^2}$ and consequently $s = \int_a^b |z'(t)|\,dt$.

> **THEOREM 18.3 A Bounding Theorem**
>
> If f is continuous on a smooth curve C and if $|f(z)| \le M$ for all z on C, then $\left|\int_C f(z)\,dz\right| \le ML$, where L is the length of C.

PROOF From the triangle inequality (6) of Section 17.1 we can write

$$\left| \sum_{k=1}^n f(z_k^*)\,\Delta z_k \right| \le \sum_{k=1}^n |f(z_k^*)|\,|\Delta z_k| \le M \sum_{k=1}^n |\Delta z_k|. \tag{4}$$

Now, $|\Delta z_k|$ can be interpreted as the length of the chord joining the points z_k and z_{k-1}. Since the sum of the lengths of the chords cannot be greater than the length

of C, (4) becomes $\left| \sum_{k=1}^{n} f(z_k^*) \, \Delta z_k \right| \le ML$. Hence, as $\|P\| \to 0$, the last inequality yields $\left| \int_C f(z) \, dz \right| \le ML$. ❑

Theorem 18.3 is used often in the theory of complex integration and is sometimes referred to as the **ML-inequality.**

> **Example 4** **A Bound for a Contour Integral**
>
> Find an upper bound for the absolute value of $\oint_C \dfrac{e^z}{z+1} \, dz$, where C is the circle $|z| = 4$.
>
> **SOLUTION** First, the length s of the circle of radius 4 is 8π. Next, from the inequality (7) of Section 17.1, it follows that $|z + 1| \ge |z| - 1 = 4 - 1 = 3$, and so
>
> $$\left| \frac{e^z}{z+1} \right| \le \frac{|e^z|}{|z| - 1} = \frac{|e^z|}{3}. \tag{5}$$
>
> In addition, $|e^z| = |e^x(\cos y + i \sin y)| = e^x$. For points on the circle $|z| = 4$ the maximum that x can be is 4, and so (5) becomes
>
> $$\left| \frac{e^z}{z+1} \right| \le \frac{e^4}{3}.$$
>
> Hence from Theorem 18.3 we have
>
> $$\left| \oint_C \frac{e^z}{z+1} \, dz \right| \le \frac{8\pi e^4}{3}. \tag{❑}$$

■ **Circulation and Net Flux** Let \mathbf{T} and \mathbf{N} denote the unit tangent vector and the unit normal vector to a positively oriented simple closed contour C. When we interpret the complex function $f(z) = u(x, y) + iv(x, y)$ as a vector, the line integrals

$$\oint_C f \cdot \mathbf{T} \, ds = \oint_C u \, dx + v \, dy \tag{6}$$

and

$$\oint_C f \cdot \mathbf{N} \, ds = \oint_C u \, dy - v \, dx \tag{7}$$

have special interpretations. The line integral in (6) is called the **circulation** around C and measures the tendency of the flow to rotate the curve C. See Section 9.8 for the derivation. The **net flux** across C is the difference between the rate at which fluid enters and the rate at which fluid leaves the region bounded by C. The net flux across C is given by the line integral in (7), and a nonzero value for $\oint_C f \cdot \mathbf{N} \, ds$ indicates the presence of sources or sinks for the fluid inside the curve C. Note that

$$\left(\oint_C f \cdot \mathbf{T} \, ds \right) + i \left(\oint_C f \cdot \mathbf{N} \, ds \right) = \oint_C (u - iv)(dx + i \, dy) = \oint_C \overline{f(z)} \, dz$$

and so

$$\text{circulation} = \text{Re}\left(\oint_C \overline{f(z)} \, dz \right) \tag{8}$$

$$\text{net flux} = \text{Im}\left(\oint_C \overline{f(z)} \, dz \right). \tag{9}$$

Thus, both of these key quantities may be found by computing a single complex integral.

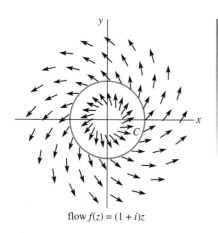

flow $f(z) = (1 + i)z$

Figure 18.2

Example 5 Net Flux

Given the flow $f(z) = (1 + i)z$, compute the circulation around, and the net flux across, the circle $C: |z| = 1$.

SOLUTION Since $\overline{f(z)} = (1 - i)\bar{z}$ and $z(t) = e^{it}$, $0 \le t \le 2\pi$, we have

$$\oint_C \overline{f(z)} \, dz = \int_0^{2\pi} (1 - i)e^{-it}ie^{it} \, dt = (1 + i) \int_0^{2\pi} dt = 2\pi(1 + i).$$

Thus the circulation around C is 2π and the net flux across C is 2π. See Figure 18.2. ❑

E X E R C I S E S 1 8 . 1

Answers to odd-numbered problems begin on page A-73.

In Problems 1–16, evaluate the given integral along the indicated contour.

1. $\int_C (z + 3) \, dz$, where C is $x = 2t$, $y = 4t - 1$, $1 \le t \le 3$

2. $\int_C (2\bar{z} - z) \, dz$, where C is $x = -t$, $y = t^2 + 2$, $0 \le t \le 2$

3. $\int_C z^2 \, dz$, where C is $z(t) = 3t + 2it$, $-2 \le t \le 2$

4. $\int_C (3z^2 - 2z) \, dz$, where C is $z(t) = t + it^2$, $0 \le t \le 1$

5. $\int_C \dfrac{1 + z}{z} \, dz$, where C is the right half of the circle $|z| = 1$ from $z = -i$ to $z = i$

6. $\int_C |z|^2 \, dz$, where C is $x = t^2$, $y = 1/t$, $1 \le t \le 2$

7. $\oint_C \text{Re}(z) \, dz$, where C is the circle $|z| = 1$

8. $\oint_C \left(\dfrac{1}{(z + i)^3} - \dfrac{5}{z + i} + 8 \right) dz$, where C is the circle $|z + i| = 1$, $0 \le t \le 2\pi$

9. $\int_C (x^2 + iy^3) \, dz$, where C is the straight line from $z = 1$ to $z = i$

10. $\int_C (x^3 - iy^3) \, dz$, where C is the lower half of the circle $|z| = 1$ from $z = -1$ to $z = 1$

11. $\int_C e^z \, dz$, where C is the polygonal path consisting of the line segments from $z = 0$ to $z = 2$ and from $z = 2$ to $z = 1 + \pi i$

12. $\int_C \sin z \, dz$, where C is the polygonal path consisting of the line segments from $z = 0$ to $z = 1$ and from $z = 1$ to $z = 1 + i$

13. $\int_C \text{Im}(z - i) \, dz$, where C is the polygonal path consisting of the circular arc along $|z| = 1$ from $z = 1$ to $z = i$ and the line segment from $z = i$ to $z = -1$

14. $\int_C dz$, where C is the left half of the ellipse $x^2/36 + y^2/4 = 1$ from $z = 2i$ to $z = -2i$

15. $\oint_C ze^z \, dz$, where C is the square with vertices $z = 0$, $z = 1$, $z = 1 + i$, and $z = i$

16. $\int_C f(z) \, dz$, where $f(z) = \begin{cases} 2, & x < 0 \\ 6x, & x > 0 \end{cases}$ and C is the parabola $y = x^2$ from $z = -1 + i$ to $z = 1 + i$

In Problems 17–20, evaluate the given integral along the contour C given in Figure 18.3.

17. $\oint_C x \, dz$ **18.** $\oint_C (2z - 1) \, dz$

19. $\oint_C z^2 \, dz$ **20.** $\oint_C \bar{z}^2 \, dz$

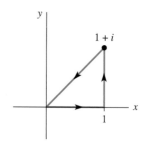

Figure 18.3

In Problems 21–24, evaluate $\int_C (z^2 - z + 2)$ from i to 1 along the indicated contours.

21. **22.**

 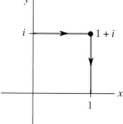

Figure 18.4 **Figure 18.5**

23.

Figure 18.6

24.

Figure 18.7

In Problems 25–28, find an upper bound for the absolute value of the given integral along the indicated contour.

25. $\oint_C \dfrac{e^z}{z^2+1}\, dz$, where C is the circle $|z| = 5$

26. $\displaystyle\int_C \dfrac{1}{z^2 - 2i}\, dz$, where C is the right half of the circle $|z| = 6$ from $z = -6i$ to $z = 6i$

27. $\displaystyle\int_C (z^2 + 4)\, dz$, where C is the line segment from $z = 0$ to $z = 1 + i$

28. $\displaystyle\int_C \dfrac{1}{z^3}\, dz$, where C is one quarter of the circle $|z| = 4$ from $z = 4i$ to $z = 4$

29. (a) Use Definition 18.1 to show for any smooth curve C between z_0 and z_n that $\int_C dz = z_n - z_0$.

(b) Use the result in part **(a)** to verify the answer to Problem 14.

30. Use Definition 18.1 to show for any smooth curve C between z_0 and z_n that $\int_C z\, dz = \frac{1}{2}(z_n^2 - z_0^2)$. [*Hint:* The integral exists, so choose $z_k^* = z_k$ and $z_k^* = z_{k-1}$.]

31. Use the results of Problems 29 and 30 to evaluate $\oint_C (6z + 4)\, dz$ where C is:

(a) The straight line from $1 + i$ to $2 + 3i$.

(b) The closed contour $x^4 + y^4 = 4$.

In Problems 32–35, compute the circulation and net flux for the given flow and the indicated closed contour.

32. $f(z) = 1/z$, where C is the circle $|z| = 2$

33. $f(z) = 2z$, where C is the circle $|z| = 1$

34. $f(z) = 1/(z - 1)$, where C is the circle $|z - 1| = 2$

35. $f(z) = \bar{z}$, where C is the square with vertices $z = 0$, $z = 1$, $z = 1 + i$, $z = i$

18.2 Cauchy–Goursat Theorem

(a) Simply connected domain

(b) Multiply connected domain

Figure 18.8

■ **Simply and Multiply Connected Domains** In the discussion that follows we shall concentrate on contour integrals where the contour C is a **simple closed curve** with a positive (counterclockwise) orientation. Before doing this, we need to distinguish two kinds of domains. A domain D is said to be **simply connected** if every simple closed contour C lying entirely in D can be shrunk to a point without leaving D. In other words, in a simply connected domain every simple closed contour C lying entirely within it encloses only points of the domain D. Expressed yet another way, a simply connected domain has no "holes" in it. The entire complex plane is an example of a simply connected domain. A domain that is not simply connected is called a **multiply connected domain;** that is, a multiply connected domain has "holes" in it. See Figure 18.8. As in Section 9.9, we call a domain with one "hole" **doubly connected,** a domain with two "holes" **triply connected,** and so on.

■ **Cauchy's Theorem** In 1825 the French mathematician Louis-Augustin Cauchy proved one of the most important theorems in complex analysis. **Cauchy's theorem** says:

> *Suppose that a function f is analytic in a simply connected domain D and that f′ is continuous in D. Then for every simple closed contour C in D,*
> $\oint_C f(z)\, dz = 0.$

The proof of this theorem is an immediate consequence of Green's theorem and the Cauchy–Riemann equations. Since f' is continuous throughout D, the real and

imaginary parts of $f(z) = u + iv$ and their first partial derivatives are continuous throughout D. By (2) of Section 18.1 we write $\oint_C f(z)\, dz$ in terms of real-line integrals and use Green's theorem on each line integral:

$$\oint_C f(z)\, dz = \oint_C u(x, y)\, dx - v(x, y)\, dy + i \oint_C v(x, y)\, dx + u(x, y)\, dy$$

$$= \iint_D \left(-\frac{\partial v}{\partial x} - \frac{\partial u}{\partial y} \right) dA + i \iint_D \left(\frac{\partial u}{\partial x} - \frac{\partial v}{\partial y} \right) dA. \tag{1}$$

Now since f is analytic, the Cauchy–Riemann equations, $\partial u/\partial x = \partial v/\partial y$ and $\partial u/\partial y = -\partial v/\partial x$, imply that the integrands in (1) are identically zero. Hence, we have $\oint_C f(z)\, dz = 0$.

In 1883 the French mathematician Edouard Goursat proved Cauchy's theorem without the assumption of continuity of f'. The resulting modified version of Cauchy's theorem is known as the **Cauchy–Goursat theorem:**

THEOREM 18.4 Cauchy–Goursat Theorem

Suppose a function f is analytic in a simply connected domain D. Then for every simple closed contour C in D, $\oint_C f(z)\, dz = 0$.

Since the interior of a simple closed contour is a simply connected domain, the Cauchy–Goursat theorem can be stated in the slightly more practical manner:

If f is analytic at all points within and on a simple closed contour C, then
$$\oint_C f(z)\, dz = 0. \tag{2}$$

Example 1 Applying The Cauchy–Goursat Theorem

Evaluate $\oint_C e^z\, dz$, where C is the curve shown in Figure 18.9.

SOLUTION The function $f(z) = e^z$ is entire and C is a simple closed contour. It follows from the form of the Cauchy–Goursat theorem given in (2) that $\oint_C e^z\, dz = 0$. ❑

Example 2 Applying The Cauchy–Goursat Theorem

Evaluate $\oint_C \dfrac{dz}{z^2}$, where C is the ellipse $(x - 2)^2 + \dfrac{(y - 5)^2}{4} = 1$.

SOLUTION The rational function $f(z) = 1/z^2$ is analytic everywhere except at $z = 0$. But $z = 0$ is not a point interior to or on the contour C. Thus, from (2) we have $\oint_C dz/z^2 = 0$. ❑

Example 3 Applying The Cauchy–Goursat Theorem

Given the flow $f(z) = \overline{\cos z}$, compute the circulation around and net flux across C, where C is the square with vertices $z = 1$, $z = i$, $z = -1$, and $z = -i$.

SOLUTION We must compute $\oint_C \overline{f(z)}\, dz = \oint_C \cos z\, dz$ and then take the real and imaginary parts of the integral to find the circulation and net flux, respectively. The function $\cos z$ is analytic everywhere, and so $\oint_C \overline{f(z)}\, dz = 0$ from (2). The circulation and net flux are therefore both zero. Figure 18.10 shows the flow $f(z) = \overline{\cos z}$ and the contour C. ❑

Figure 18.9

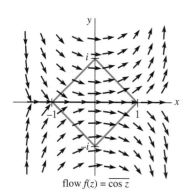

flow $f(z) = \overline{\cos z}$

Figure 18.10

(a)

(b)

Figure 18.11

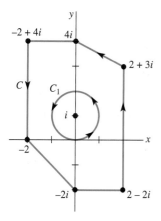

Figure 18.12

■ **Cauchy–Goursat Theorem for Multiply Connected Domains** If f is analytic in a multiply connected domain D, then we cannot conclude that $\oint_C f(z)\,dz = 0$ for every simple closed contour C in D. To begin, suppose D is a doubly connected domain and C and C_1 are simple closed contours such that C_1 surrounds the "hole" in the domain and is interior to C. See Figure 18.11(a). Suppose, also, that f is analytic on each contour and at each point interior to C but exterior to C_1. When we introduce the cut AB shown in Figure 18.11(b), the region bounded by the curves is simply connected. Now the integral from A to B has the opposite value of the integral from B to A, and so from (2) we have $\oint_C f(z)\,dz + \oint_{C_1} f(z)\,dz = 0$ or

$$\oint_C f(z)\,dz = \oint_{C_1} f(z)\,dz. \tag{3}$$

The last result is sometimes called the principle of **deformation of contours,** since we can think of the contour C_1 as a continuous deformation of the contour C. Under this deformation of contours the value of the integral does not change. Thus, on a practical level, (3) allows us to evaluate an integral over a complicated simple closed contour by replacing that contour with one that is more convenient.

Example 4 **Applying Deformation of Contours**

Evaluate $\oint_C \dfrac{dz}{z-i}$, where C is the outer contour shown in Figure 18.12.

SOLUTION In view of (3), we choose the more convenient circular contour C_1 in the figure. By taking the radius of the circle to be $r = 1$, we are guaranteed that C_1 lies within C. In other words, C_1 is the circle $|z - i| = 1$, which can be parameterized by $x = \cos t$, $y = 1 + \sin t$, $0 \le t \le 2\pi$, or equivalently by $z = i + e^{it}$, $0 \le t \le 2\pi$. From $z - i = e^{it}$ and $dz = ie^{it}\,dt$ we obtain

$$\oint_C \frac{dz}{z-i}\,dz = \oint_{C_1} \frac{dz}{z-i}\,dz = \int_0^{2\pi} \frac{ie^{it}}{e^{it}}\,dt = i\int_0^{2\pi} dt = 2\pi i. \qquad \square$$

The result in Example 4 can be generalized. Using the principle of deformation of contours (2) and proceeding as in the example, we can show that if z_0 is any constant complex number interior to *any* simple closed contour C, then

$$\oint_C \frac{dz}{(z - z_0)^n} = \begin{cases} 2\pi i, & n = 1 \\ 0, & n \text{ an integer} \ne 1. \end{cases} \tag{4}$$

The fact that the integral in (4) is zero when n is an integer $\ne 1$ follows only partially from the Cauchy–Goursat theorem. When n is zero or a negative integer, $1/(z - z_0)^n$ is a polynomial (for example, $n = -3$, $1/(z - z_0)^{-3} = (z - z_0)^3$) and therefore entire. Theorem 18.4 then implies $\oint_C dz/(z - z_0)^n = 0$. It is left as an exercise to show that the integral is still zero when n is a positive integer different from one. See Problem 22 in Exercises 18.2.

Example 5 **Applying Formula (4)**

Evaluate $\oint_C \dfrac{5z + 7}{z^2 + 2z - 3}\,dz$, where C is the circle $|z - 2| = 2$.

SOLUTION Since the denominator factors as $z^2 + 2z - 3 = (z - 1)(z + 3)$, the integrand fails to be analytic at $z = 1$ and $z = -3$. Of these two points, only $z = 1$

lies within the contour C, which is a circle centered at $z = 2$ of radius $r = 2$. Now by partial fractions,

$$\frac{5z+7}{z^2+2z-3} = \frac{3}{z-1} + \frac{2}{z+3}$$

and so
$$\oint_C \frac{5z+7}{z^2+2z-3}\, dz = 3\oint_C \frac{dz}{z-1} + 2\oint_C \frac{dz}{z+3}. \tag{5}$$

In view of the result given in (4), the first integral in (5) has the value $2\pi i$. By the Cauchy–Goursat theorem the value of the second integral is zero. Hence, (5) becomes

$$\oint_C \frac{5z+7}{z^2+2z-3}\, dz = 3(2\pi i) + 2(0) = 6\pi i. \qquad \square$$

If C, C_1, and C_2 are the simple closed contours shown in Figure 18.13 and if f is analytic on each of the three contours as well at each point interior to C but exterior to both C_1 and C_2, then by introducing cuts, we get from Theorem 18.4 that $\oint_C f(z)\,dz + \oint_{C_1} f(z)\,dz + \oint_{C_2} f(z)\,dz = 0$. Hence,

$$\oint_C f(z)\, dz = \oint_{C_1} f(z)\, dz + \oint_{C_2} f(z)\, dz.$$

The next theorem will summarize the general result for a multiply connected domain with n "holes":

Figure 18.13

THEOREM 18.5 Cauchy–Goursat Theorem for Multiply Connected Domains

Suppose C, C_1, \ldots, C_n are simple closed curves with a positive orientation such that C_1, C_2, \ldots, C_n are interior to C but the regions interior to each C_k, $k = 1, 2, \ldots, n$, have no points in common. If f is analytic on each contour and at each point interior to C but exterior to all the C_k, $k = 1, 2, \ldots, n$, then

$$\oint_C f(z)\, dz = \sum_{k=1}^{n} \oint_{C_k} f(z)\, dz. \tag{6}$$

Example 6 Applying Theorem 18.5

Evaluate $\oint_C \dfrac{dz}{z^2+1}$, where C is the circle $|z| = 3$.

SOLUTION In this case the denominator of the integrand factors as $z^2 + 1 = (z - i)(z + i)$. Consequently, the integrand $1/(z^2 + 1)$ is not analytic at $z = i$ and $z = -i$. Both of these points lie within the contour C. Using partial fraction decomposition once more, we have

$$\frac{1}{z^2+1} = \frac{1/2i}{z-i} - \frac{1/2i}{z+i}$$

and
$$\oint_C \frac{dz}{z^2+1} = \frac{1}{2i}\oint_C \left[\frac{1}{z-i} - \frac{1}{z+i}\right] dz.$$

We now surround the points $z = i$ and $z = -i$ by circular contours C_1 and C_2, respectively, that lie entirely within C. Specifically, the choice $|z - i| = \frac{1}{2}$ for C_1 and $|z + i| = \frac{1}{2}$ for C_2 will suffice. See Figure 18.14. From Theorem 18.5 we can then write

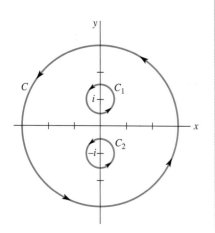

Figure 18.14

$$\oint_C \frac{dz}{z^2+1} = \frac{1}{2i} \oint_{C_1} \left[\frac{1}{z-i} - \frac{1}{z+i} \right] dz + \frac{1}{2i} \oint_{C_2} \left[\frac{1}{z-i} - \frac{1}{z+i} \right] dz$$

(7)

$$= \frac{1}{2i} \oint_{C_1} \frac{dz}{z-i} - \frac{1}{2i} \oint_{C_1} \frac{dz}{z+i} + \frac{1}{2i} \oint_{C_2} \frac{dz}{z-i} - \frac{1}{2i} \oint_{C_2} \frac{dz}{z+i}.$$

Because $1/(z+i)$ is analytic on C_1 and at each point in its interior and because $1/(z-i)$ is analytic on C_2 and at each point in its interior, it follows from (3) that the second and third integrals in (4) are zero. Moreover, it follows from (4), with $n = 1$, that

$$\oint_{C_1} \frac{dz}{z-i} = 2\pi i \quad \text{and} \quad \oint_{C_2} \frac{dz}{z+i} = 2\pi i.$$

Thus (7) becomes $\qquad \oint_C \frac{dz}{z^2+1} = \pi - \pi = 0.$ ❑

Figure 18.15

Remarks

Throughout the foregoing discussion we assumed that C was a simple closed contour; in other words, C did not intersect itself. Although we shall not give the proof, it can be shown that the Cauchy–Goursat theorem is valid for any closed contour C in a simply connected domain D. As shown in Figure 18.15, the contour C is closed but not simple. Nevertheless, if f is analytic in D, then $\oint_C f(z)\,dz = 0$.

EXERCISES 18.2

Answers to odd-numbered problems begin on page A-73.

In Problems 1–8, prove that $\oint_C f(z)\,dz = 0$, where f is the given function and C is the unit circle $|z| = 1$.

1. $f(z) = z^3 - 1 + 3i$

2. $f(z) = z^2 + \dfrac{1}{z-4}$

3. $f(z) = \dfrac{z}{2z+3}$

4. $f(z) = \dfrac{z-3}{z^2+2z+2}$

5. $f(z) = \dfrac{\sin z}{(z^2-25)(z^2+9)}$

6. $f(z) = \dfrac{e^z}{2z^2+11z+15}$

7. $f(z) = \tan z$

8. $f(z) = \dfrac{z^2-9}{\cosh z}$

9. Evaluate $\oint_C \dfrac{1}{z}\,dz$, where C is the contour shown in Figure 18.16.

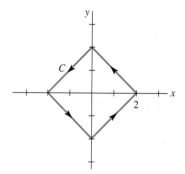

Figure 18.16

10. Evaluate $\oint_C \dfrac{5}{z+1+i}\,dz$, where C is the contour shown in Figure 18.17.

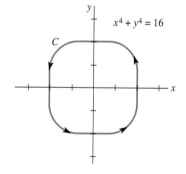

Figure 18.17

In Problems 11–20, use any of the results in this section to evaluate the given integral along the indicated closed contour(s).

11. $\oint_C \left(z + \dfrac{1}{z}\right)dz; \ |z| = 2$

12. $\oint_C \left(z + \dfrac{1}{z^2}\right)dz; \ |z| = 2$

13. $\oint_C \dfrac{z}{z^2-\pi^2}\,dz; \ |z| = 3$

14. $\oint_C \dfrac{10}{(z+i)^4}\, dz;\ |z+i|=1$

15. $\oint_C \dfrac{2z+1}{z^2+z}\, dz;$ **(a)** $|z|=\tfrac12,$ **(b)** $|z|=2,$ **(c)** $|z-3i|=1$

16. $\oint_C \dfrac{2z}{z^2+3}\, dz;$ **(a)** $|z|=1,$ **(b)** $|z-2i|=1,$ **(c)** $|z|=4$

17. $\oint_C \dfrac{-3z+2}{z^2-8z+12}\, dz;$ **(a)** $|z-5|=2,$ **(b)** $|z|=9$

18. $\oint_C \left(\dfrac{3}{z+2}-\dfrac{1}{z-2i}\right) dz;$ **(a)** $|z|=5,$ **(b)** $|z-2i|=\tfrac12$

19. $\oint_C \dfrac{z-1}{z(z-i)(z-3i)}\, dz;\ |z-i|=\tfrac12$

20. $\oint_C \dfrac{1}{z^3+2iz^2}\, dz;\ |z|=1$

21. Evaluate $\oint_C \dfrac{8z-3}{z^2-z}\, dz,$ where C is the closed contour shown in Figure 18.18. [*Hint:* Express C as the union of two closed curves C_1 and C_2.]

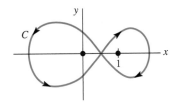

Figure 18.18

22. Suppose z_0 is any constant complex number interior to any simple closed contour C. Show that

$$\oint_C \dfrac{dz}{(z-z_0)^n} = \begin{cases} 2\pi i, & n=1 \\ 0, & n \text{ a positive integer} \neq 1. \end{cases}$$

In Problems 23 and 24, evaluate the given integral by any means.

23. $\oint_C \left(\dfrac{e^z}{z+3}-3\bar{z}\right) dz,$ C is the unit circle $|z|=1$

24. $\oint_C (z^3+z^2+\mathrm{Re}(z))\, dz,$ C is the triangle with vertices $z=0,$ $z=1+2i,$ $z=1$

18.3 Independence of Path

In real calculus, when a function f possesses an antiderivative F, a definite integral can be evaluated by the Fundamental Theorem of calculus:

$$\int_a^b f(x)\, dx = F(b) - F(a). \tag{1}$$

Moreover, in Section 9.9 we saw that when a real line integral $\int_C P\, dx + Q\, dy$ is independent of the path—that is, depends on only the endpoints of the curve C—it can be evaluated by the Fundamental Theorem of Line Integrals (Theorem 9.8). It seems natural then to ask:

> *Can a contour integral $\int_C f(z)\, dz$ be independent of the path? Is there a complex version of the Fundamental Theorem (1)?*

The answer to both of these questions is yes.

DEFINITION 18.2 Independence of the Path

Let z_0 and z_1 be points in a domain D. A contour integral $\int_C f(z)\, dz$ is said to be **independent of the path** if its value is the same for all contours C in D with an initial point z_0 and a terminal point z_1.

At the end of the preceding section we noted that the Cauchy–Goursat theorem also holds for closed contours, not just simple closed contours, in a simply connected domain D. Now suppose, as shown in Figure 18.19, that C and C_1 are two contours in a simply connected domain D, both with initial point z_0 and terminal

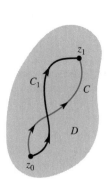

Figure 18.19

point z_1. Note that C and $-C_1$ form a closed contour. Thus, if f is analytic in D, it follows from the Cauchy–Goursat theorem that

$$\int_C f(z)\, dz + \int_{-C_1} f(z)\, dz = 0. \tag{2}$$

But (2) is equivalent to

$$\int_C f(z)\, dz = \int_{C_1} f(z)\, dz. \tag{3}$$

The result in (3) is also an example of the principle of deformation of contours introduced in (3) of Section 18.2. We shall summarize the last result as a theorem:

THEOREM 18.6 Analyticity Implies Path Independence

If f is an analytic function in a simply connected domain D, then $\int_C f(z)\, dz$ is independent of the path C.

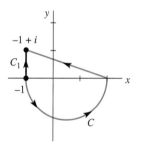

Figure 18.20

Example 1 **Choosing a Different Path**

Evaluate $\int_C 2z\, dz$, where C is the contour with initial point $z = -1$ and terminal point $z = -1 + i$ shown in Figure 18.20.

SOLUTION Since the function $f(z) = z$ is entire, we can replace the path C by any convenient contour C_1 joining $z = -1$ and $z = -1 + i$. In particular, by choosing C_1 to be the straight line segment $x = -1, 0 \le y \le 1$, shown in black in Figure 18.20, we have $z = -1 + iy, dz = i\, dy$. Therefore,

$$\int_C 2z\, dz = \int_{C_1} 2z\, dz = -2 \int_0^1 y\, dy - 2i \int_0^1 dy = -1 - 2i. \qquad \square$$

A contour integral $\int_C f(z)\, dz$ that is independent of the path C is usually written $\int_{z_0}^{z_1} f(z)\, dz$, where z_0 and z_1 are the initial and terminal points of C. Hence in Example 1 we can write $\int_{-1}^{-1+i} 2z\, dz$.

There is an easier way to evaluate the contour integral in Example 1, but before proceeding we need another definition.

DEFINITION 18.3 Antiderivative

Suppose f is continuous in a domain D. If there exists a function F such that $F'(z) = f(z)$ for each z in D, then F is called an **antiderivative** of f.

For example, the function $F(z) = -\cos z$ is an antiderivative of $f(z) = \sin z$, since $F'(z) = \sin z$. As in real calculus, the most general antiderivative, or **indefinite integral,** of a function $f(z)$ is written $\int f(z)\, dz = F(z) + C$, where $F'(z) = f(z)$ and C is some complex constant.

Since an antiderivative F of a function f has a derivative at each point in a domain D, it is necessarily analytic and hence continuous in D (recall that differentiability implies continuity).

We are now in a position to prove the complex analogue of (1).

THEOREM 18.7 **Fundamental Theorem for Contour Integrals**

Suppose f is continuous in a domain D and F is an antiderivative of f in D. Then for any contour C in D with initial point z_0 and terminal point z_1,

$$\int_C f(z)\,dz = F(z_1) - F(z_0). \tag{4}$$

PROOF We will prove (4) in the case when C is a smooth curve defined by $z = z(t)$, $a \le t \le b$. Using (3) of Section 18.1 and the fact that $F'(z) = f(z)$ for each z in D, we have

$$\int_C f(z)\,dz = \int_a^b f(z(t))z'(t)\,dt = \int_a^b F'(z(t))z'(t)\,dt$$

$$= \int_a^b \frac{d}{dt}\,F(z(t))\,dt \quad \leftarrow \text{Chain Rule}$$

$$= F(z(t))\Big|_a^b$$

$$= F(z(b)) - F(z(a)) = F(z_1) - F(z_0). \qquad \square$$

Example 2 Using an Antiderivative

In Example 1 we saw that the integral $\int_C 2z\,dz$, where C is shown in Figure 18.20, is independent of the path. Now since $f(z) = 2z$ is an entire function, it is continuous. Moreover, $F(z) = z^2$ is an antiderivative of f, since $F'(z) = 2z$. Hence by (4) we have

$$\int_{-1}^{-1+i} 2z\,dz = z^2\,\Big|_{-1}^{-1+i} = (-1+i)^2 - (-1)^2 = -1 - 2i. \qquad \square$$

Example 3 Using an Antiderivative

Evaluate $\int_C \cos z\,dz$, where C is any contour with initial point $z = 0$ and terminal point $z = 2 + i$.

SOLUTION $F(z) = \sin z$ is an antiderivative of $f(z) = \cos z$, since $F'(z) = \cos z$. Therefore from (4) we have

$$\int_C \cos z\,dz = \int_0^{2+i} \cos z\,dz = \sin z\,\Big|_0^{2+i} = \sin(2+i) - \sin 0 = \sin(2+i).$$

If we desired a complex number of the form $a + ib$ for an answer, we can use $\sin(2+i) = 1.4031 - 0.4891i$ (see Example 1 in Section 17.7). Hence,

$$\int_C \cos z\,dz = 1.4031 - 0.4891i. \qquad \square$$

We can draw several immediate conclusions from Theorem 18.7. First, observe that if the contour C is closed, then $z_0 = z_1$ and consequently

$$\oint_C f(z)\,dz = 0. \tag{5}$$

Next, since the value of $\int_C f(z)\,dz$ depends on only the points z_0 and z_1, this value is the same for any contour C in D connecting these points. In other words:

If a continuous function f has an antiderivative F in D, then $\int_C f(z)\,dz$ is independent of the path. (6)

In addition we have the following sufficient condition for the existence of an antiderivative:

If f is continuous and $\int_C f(z)\,dz$ is independent of the path in a domain D, then f has an antiderivative everywhere in D. (7)

The last statement is important and deserves a proof. Assume that f is continuous, $\int_C f(z)\,dz$ is independent of the path in a domain D, and F is a function defined by $F(z) = \int_{z_0}^{z} f(s)\,ds$ where s denotes a complex variable, z_0 is a fixed point in D, and z represents any point in D. We wish to show that $F'(z) = f(z)$; that is, F is an antiderivative of f in D. Now,

$$F(z + \Delta z) - F(z) = \int_{z_0}^{z + \Delta z} f(s)\,ds - \int_{z_0}^{z} f(s)\,ds = \int_{z}^{z + \Delta z} f(s)\,ds. \qquad (8)$$

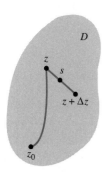

Figure 18.21

Because D is a domain we can choose Δz so that $z + \Delta z$ is in D. Moreover, z and $z + \Delta z$ can be joined by a straight segment lying in D, as shown in Figure 18.21. This is the contour we use in the last integral in (8). With z fixed, we can write*

$$f(z)\,\Delta z = f(z) \int_{z}^{z + \Delta z} ds = \int_{z}^{z + \Delta z} f(z)\,ds \quad \text{and} \quad f(z) = \frac{1}{\Delta z} \int_{z}^{z + \Delta z} f(z)\,ds. \qquad (9)$$

From (8) and (9) it follows that

$$\frac{F(z + \Delta z) - F(z)}{\Delta z} - f(z) = \frac{1}{\Delta z} \int_{z}^{z + \Delta z} [f(s) - f(z)]\,ds.$$

Now f is continuous at the point z. This means that for any $\varepsilon > 0$ there exists a $\delta > 0$ so that $|f(s) - f(z)| < \varepsilon$ whenever $|s - z| < \delta$. Consequently, if we choose Δz so that $|\Delta z| < \delta$, we have

$$\left| \frac{F(z + \Delta z) - F(z)}{\Delta z} - f(z) \right| = \left| \frac{1}{\Delta z} \int_{z}^{z + \Delta z} [f(s) - f(z)]\,ds \right|$$

$$= \left| \frac{1}{\Delta z} \right| \left| \int_{z}^{z + \Delta z} [f(s) - f(z)]\,ds \right| \leq \frac{1}{|\Delta z|}\,\varepsilon |\Delta z| = \varepsilon.$$

Hence, we have shown that

$$\lim_{\Delta z \to 0} \frac{F(z + \Delta z) - F(z)}{\Delta z} = f(z) \quad \text{or} \quad F'(z) = f(z).$$

If f is an analytic function in a simply connected domain D, it is necessarily continuous throughout D. This fact, when put together with the results in Theorem 18.6 and (7), leads to a theorem which states that an analytic function possesses an analytic antiderivative.

> **THEOREM 18.8** **Existence of an Antiderivative**
>
> If f is analytic in a simply connected domain D, then f has an antiderivative in D; that is, there exists a function F such that $F'(z) = f(z)$ for all z in D.

* See Problem 29 in Exercises 18.1.

In (9) of Section 17.6 we saw that $1/z$ is the derivative of Ln z. This means that under some circumstances Ln z is an antiderivative of $1/z$. Care must be exercised in using this result. For example, suppose D is the entire complex plane without the origin. The function $1/z$ is analytic in this *multiply connected* domain. If C is any simple closed contour containing the origin, it does *not* follow from (5) that $\oint_C dz/z = 0$. In fact, from (4) of Section 18.2 with the identification $z_0 = 0$, we see that

$$\oint_C \frac{1}{z}\, dz = 2\pi i.$$

In this case, Ln z is not an antiderivative of $1/z$ in D, since Ln z is not analytic in D. Recall that Ln z fails to be analytic on the nonpositive real axis (the branch cut of the principal branch of the logarithm).

Figure 18.22

Example 4 Using the Logarithmic Function

Evaluate $\displaystyle\int_C \frac{1}{z}\, dz$, where C is the contour shown in Figure 18.22.

SOLUTION Suppose that D is the simply connected domain defined by $x = \mathrm{Re}(z) > 0$, $y = \mathrm{Im}(z) > 0$. In this case, Ln z is an antiderivative of $1/z$, since both these functions are analytic in D. Hence by (4),

$$\int_3^{2i} \frac{1}{z}\, dz = \mathrm{Ln}\ z \Big|_3^{2i} = \mathrm{Ln}\ 2i - \mathrm{Ln}\ 3.$$

From (7) of Section 17.6, we have

$$\mathrm{Ln}\ 2i = \log_e 2 + \frac{\pi}{2}\, i \quad \text{and} \quad \mathrm{Ln}\ 3 = \log_e 3$$

and so $\displaystyle\int_3^{2i} \frac{1}{z}\, dz = \log_e \frac{2}{3} + \frac{\pi}{2}\, i = -0.4055 + 1.5708i.$ ❑

Remarks

Suppose f and g are analytic in a simply connected domain D that contains the contour C. If z_0 and z_1 are the initial and terminal points of C, then the **integration by parts** formula is valid in D:

$$\int_{z_0}^{z_1} f(z)g'(z)\, dz = f(z)g(z) \Big|_{z_0}^{z_1} - \int_{z_0}^{z_1} f'(z)g(z)\, dz.$$

This can be proved in a straightforward manner using Theorem 18.7 on the function $(d/dz)(fg)$. See Problems 21–24 in Exercises 18.3.

EXERCISES 18.3

Answers to odd-numbered problems begin on page A-73.

In Problems 1 and 2, evaluate the given integral, where C is the contour given in the figure, by **(a)** finding an alternative path of integration and **(b)** using Theorem 18.7.

1. $\displaystyle\int_C (4z - 1)\, dz$

Figure 18.23

2. $\displaystyle\int_C e^z\, dz$

Figure 18.24

In Problems 3 and 4, evaluate the given integral along the indicated contour C.

3. $\displaystyle\int_C 2z\, dz$, where C is $z(t) = 2t^3 + i(t^4 - 4t^3 + 2), -1 \le t \le 1$

4. $\displaystyle\int_C 6z^2\, dz$, where C is $z(t) = 2\cos^3 \pi t - i\sin^2\dfrac{\pi}{4}t, 0 \le t \le 2$

In Problems 5–24, use Theorem 18.7 to evaluate the given integral. Write each answer in the form $a + ib$.

5. $\displaystyle\int_0^{3+i} z^2\, dz$

6. $\displaystyle\int_{-2i}^{1} (3z^2 - 4z + 5i)\, dz$

7. $\displaystyle\int_{1-i}^{1+i} z^3\, dz$

8. $\displaystyle\int_{-3i}^{2i} (z^3 - z)\, dz$

9. $\displaystyle\int_{-i/2}^{1-i} (2z + 1)^2\, dz$

10. $\displaystyle\int_1^{i} (iz + 1)^3\, dz$

11. $\displaystyle\int_{i/2}^{i} e^{\pi z}\, dz$

12. $\displaystyle\int_{1-i}^{1+2i} ze^{z^2}\, dz$

13. $\displaystyle\int_{\pi}^{\pi+2i} \sin\dfrac{z}{2}\, dz$

14. $\displaystyle\int_{1-2i}^{\pi i} \cos z\, dz$

15. $\displaystyle\int_{\pi i}^{2\pi i} \cosh z\, dz$

16. $\displaystyle\int_{i}^{1+(\pi/2)i} \sinh 3z\, dz$

17. $\displaystyle\int_C \dfrac{1}{z}\, dz$, C is the arc of the circle $z = 4e^{it}, -\pi/2 \le t \le \pi/2$

18. $\displaystyle\int_C \dfrac{1}{z}\, dz$, C is the straight line segment between $z = 1 + i$ and $z = 4 + 4i$

19. $\displaystyle\int_{-4i}^{4i} \dfrac{1}{z^2}\, dz$, C is any contour not passing through the origin

20. $\displaystyle\int_{1-i}^{1+\sqrt{3}i} \left(\dfrac{1}{z} + \dfrac{1}{z^2}\right)\, dz$, C is any contour in the right half-plane $\operatorname{Re}(z) > 0$

21. $\displaystyle\int_{\pi}^{i} e^z \cos z\, dz$

22. $\displaystyle\int_0^{i} z \sin z\, dz$

23. $\displaystyle\int_{i}^{1+i} ze^z\, dz$

24. $\displaystyle\int_0^{\pi i} z^2 e^z\, dz$

18.4 Cauchy's Integral Formula

In the discussion that follows we shall consider another important consequence of the Cauchy–Goursat theorem:

> The value of an analytic function f at any point z_0 in a simply connected domain can be represented by a contour integral.

As a direct consequence of this last result, we shall then show:

> An analytic function in a simply connected domain possesses derivatives of all orders.

We begin with the Cauchy integral formula. The idea in the next theorem is this: If f is analytic in a simply connected domain and z_0 is any point D, then the quotient $f(z)/(z - z_0)$ is *not* analytic in D. As a consequence, the integral of $f(z)/(z - z_0)$ around a simple closed contour C that contains z_0 is not necessarily zero but has, as we shall now see, the value $2\pi i f(z_0)$. This remarkable result indicates that the values of an analytic function f at points *inside* a simple closed contour C are determined by the values of f on the contour C.

THEOREM 18.9 Cauchy's Integral Formula

Let f be analytic in a simply connected domain D, and let C be a simple closed contour lying entirely within D. If z_0 is any point within C, then

$$f(z_0) = \dfrac{1}{2\pi i} \oint_C \dfrac{f(z)}{z - z_0}\, dz. \qquad (1)$$

PROOF Let D be a simply connected domain, C a simple closed contour in D, and z_0 an interior point of C. In addition, let C_1 be a circle centered at z_0 with radius small enough that it is interior to C. By the principle of deformation of contours, we can write

$$\oint_C \frac{f(z)}{z - z_0}\, dz = \oint_{C_1} \frac{f(z)}{z - z_0}\, dz. \tag{2}$$

We wish to show that the value of the integral on the right is $2\pi i f(z_0)$. To this end we add and subtract the constant $f(z_0)$ in the numerator:

$$\oint_{C_1} \frac{f(z)}{z - z_0}\, dz = \oint_{C_1} \frac{f(z_0) - f(z_0) + f(z)}{z - z_0}\, dz$$

$$= f(z_0) \oint_{C_1} \frac{dz}{z - z_0} + \oint_{C_1} \frac{f(z) - f(z_0)}{z - z_0}\, dz. \tag{3}$$

Now from (4) of Section 18.2 we know that

$$\oint_{C_1} \frac{dz}{z - z_0} = 2\pi i.$$

Thus, (3) becomes

$$\oint_{C_1} \frac{f(z)}{z - z_0}\, dz = 2\pi i f(z_0) + \oint_{C_1} \frac{f(z) - f(z_0)}{z - z_0}\, dz. \tag{4}$$

Since f is continuous at z_0 for any arbitrarily small $\varepsilon > 0$, there exists a $\delta > 0$ such that $|f(z) - f(z_0)| < \varepsilon$ whenever $|z - z_0| < \delta$. In particular, if we choose the circle C_1 to be $|z - z_0| = \delta/2 < \delta$, then by the *ML*-inequality (Theorem 18.3) the absolute value of the integral on the right side of (4) satisfies

$$\left| \oint_{C_1} \frac{f(z) - f(z_0)}{z - z_0}\, dz \right| \leq \frac{\varepsilon}{\delta/2} 2\pi \left(\frac{\delta}{2} \right) = 2\pi\varepsilon.$$

In other words, the absolute value of the integral can be made arbitrarily small by taking the radius of the circle C_1 to be sufficiently small. This can happen only if the integral is zero. The Cauchy integral formula (1) follows from (4) by dividing both sides by $2\pi i$. ❏

The Cauchy integral formula (1) can be used to evaluate contour integrals. Since we often work problems without a simply connected domain explicitly defined, a more practical restatement of Theorem 18.9 is:

If f is analytic at all points within and on a simple closed contour C, and z_0 is any point interior to C, then $f(z_0) = \dfrac{1}{2\pi i} \displaystyle\oint_C \frac{f(z)}{z - z_0}\, dz.$ $\tag{5}$

Example 1 Using Cauchy's Integral Formula

Evaluate $\displaystyle\oint_C \frac{z^2 - 4z + 4}{z + i}\, dz$, where C is the circle $|z| = 2$.

SOLUTION First, we identify $f(z) = z^2 - 4z + 4$ and $z_0 = -i$ as a point within the circle C. Next, we observe that f is analytic at all points within and on the contour C. Thus by the Cauchy integral formula we obtain

$$\oint_C \frac{z^2 - 4z + 4}{z + i}\, dz = 2\pi i f(-i) = 2\pi i(3 + 4i) = 2\pi(-4 + 3i).$$ ❏

Figure 18.25

(a) Source: $k > 0$

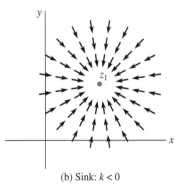

(b) Sink: $k < 0$

Figure 18.26

Example 2 **Using Cauchy's Integral Formula**

Evaluate $\oint_C \dfrac{z}{z^2 + 9}\, dz$, where C is the circle $|z - 2i| = 4$.

SOLUTION By factoring the denominator as $z^2 + 9 = (z - 3i)(z + 3i)$, we see that $3i$ is the only point within the closed contour at which the integrand fails to be analytic. See Figure 18.25. Now by writing

$$\frac{z}{z^2 + 9} = \frac{\dfrac{z}{z + 3i}}{z - 3i},$$

we can identify $f(z) = z/(z + 3i)$. This function is analytic at all points within and on the contour C. From the Cauchy integral formula we then have

$$\oint_C \frac{z}{z^2 + 9}\, dz = \oint_C \frac{\dfrac{z}{z + 3i}}{z - 3i}\, dz = 2\pi i f(3i) = 2\pi i \,\frac{3i}{6i} = \pi i. \qquad \square$$

Example 3 **Flux and Cauchy's Integral Formula**

The complex function $f(z) = k/(\bar{z} - \bar{z}_1)$, where $k = a + ib$ and z_1 are complex numbers, gives rise to a flow in the domain $z \neq z_1$. If C is a simple closed contour containing $z = z_1$ in its interior, then from the Cauchy integral formula we have

$$\oint_C \overline{f(z)}\, dz = \oint_C \frac{a - ib}{z - z_1}\, dz = 2\pi i(a - ib).$$

Thus the circulation around C is $2\pi b$ and the net flux across C is $2\pi a$. If z_1 were in the exterior of C, both the circulation and net flux would be zero by Cauchy's theorem.

Note that when k is real, the circulation around C is zero but the net flux across C is $2\pi k$. The complex number z_1 is called a **source** for the flow when $k > 0$ and a **sink** when $k < 0$. Vector fields corresponding to these two cases are shown in Figure 18.26(a) and (b). $\qquad \square$

■ **Derivatives of Analytic Functions** We can now use Theorem 18.9 to prove that an analytic function possesses derivatives of all orders; that is, if f is analytic at a point z_0, then f', f'', f''', and so on are also analytic at z_0. Moreover, the values of the derivatives $f^{(n)}(z_0)$, $n = 1, 2, 3, \ldots$, are given by a formula similar to (1).

> **THEOREM 18.10** Cauchy's Integral Formula for Derivatives
>
> Let f be analytic in a simply connected domain D, and let C be a simple closed contour lying entirely within D. If z_0 is any point interior to C, then
>
> $$f^{(n)}(z_0) = \frac{n!}{2\pi i} \oint_C \frac{f(z)}{(z - z_0)^{n+1}}\, dz. \qquad (6)$$

PARTIAL PROOF We will prove (6) only for the case $n = 1$. The remainder of the proof can be completed using the principle of mathematical induction. We begin with the definition of the derivative and (1):

$$f'(z_0) = \lim_{\Delta z \to 0} \frac{f(z_0 + \Delta z) - f(z_0)}{\Delta z}$$

$$= \lim_{\Delta z \to 0} \frac{1}{2\pi i \, \Delta z} \left[\oint_C \frac{f(z)}{z - (z_0 + \Delta z)} \, dz - \oint_C \frac{f(z)}{z - z_0} \, dz \right]$$

$$= \lim_{\Delta z \to 0} \frac{1}{2\pi i} \oint_C \frac{f(z)}{(z - z_0 - \Delta z)(z - z_0)} \, dz.$$

Before proceeding, let us set up some preliminaries. Since f is continuous on C, it is bounded; that is, there exists a real number M such that $|f(z)| \le M$ for all points z on C. In addition, let L be the length of C and let δ denote the shortest distance between points on C and the point z_0. Thus for all points z on C, we have

$$|z - z_0| \ge \delta \quad \text{or} \quad \frac{1}{|z - z_0|^2} \le \frac{1}{\delta^2}.$$

Furthermore, if we choose $|\Delta z| \le \delta/2$, then

$$|z - z_0 - \Delta z| \ge \big| |z - z_0| - |\Delta z| \big| \ge \delta - |\Delta z| \ge \frac{\delta}{2} \quad \text{and so} \quad \frac{1}{|z - z_0 - \Delta z|} \le \frac{2}{\delta}.$$

Now,

$$\left| \oint_C \frac{f(z)}{(z - z_0)^2} \, dz - \oint_C \frac{f(z)}{(z - z_0 - \Delta z)(z - z_0)} \, dz \right| = \left| \oint_C \frac{-\Delta z \, f(z)}{(z - z_0)^2(z - z_0 - \Delta z)} \, dz \right| \le \frac{2ML|\Delta z|}{\delta^3}.$$

Because the last expression approaches zero as $\Delta z \to 0$, we have shown that

$$f'(z_0) = \lim_{\Delta z \to 0} \frac{f(z_0 + \Delta z) - f(z_0)}{\Delta z} = \frac{1}{2\pi i} \oint_C \frac{f(z)}{(z - z_0)^2} \, dz. \qquad \Box$$

If $f(z) = u(x, y) + iv(x, y)$ is analytic at a point, then its derivatives of all orders exist at that point and are continuous. Consequently, from

$$f'(z) = \frac{\partial u}{\partial x} + i \frac{\partial v}{\partial x} = \frac{\partial v}{\partial y} - i \frac{\partial u}{\partial y}$$

$$f''(z) = \frac{\partial^2 u}{\partial x^2} + i \frac{\partial^2 v}{\partial x^2} = \frac{\partial^2 v}{\partial y \, \partial x} - i \frac{\partial^2 u}{\partial y \, \partial x}$$

$$\vdots$$

we can conclude that the real functions u and v have continuous partial derivatives of all orders at a point of analyticity.

Like (1), (6) can sometimes be used to evaluate integrals.

Example 4 Using Cauchy's Integral Formula for Derivatives

Evaluate $\displaystyle \oint_C \frac{z + 1}{z^4 + 4z^3} \, dz$, where C is the circle $|z| = 1$.

SOLUTION Inspection of the integrand shows that it is not analytic at $z = 0$ and $z = -4$, but only $z = 0$ lies within the closed contour. By writing the integrand as

$$\frac{z + 1}{z^4 + 4z^3} = \frac{\dfrac{z+1}{z+4}}{z^3},$$

we can identify $z_0 = 0$, $n = 2$, and $f(z) = (z + 1)/(z + 4)$. By the quotient rule, $f''(z) = -6/(z + 4)^3$ and so by (6) we have

$$\oint_C \frac{z + 1}{z^4 + 4z^3}\, dz = \frac{2\pi i}{2!} f''(0) = -\frac{3\pi}{32} i. \qquad \square$$

Example 5 Using Cauchy's Integral Formula for Derivatives

Evaluate $\displaystyle\oint_C \frac{z^2 + 3}{z(z - i)^2}\, dz$, where C is the contour shown in Figure 18.27.

SOLUTION Although C is not a simple closed contour, we can think of it as the union of two simple closed contours C_1 and C_2 as indicated in Figure 18.27. By writing

$$\oint_C \frac{z^3 + 3}{z(z - i)^2}\, dz = \oint_{C_1} \frac{z^3 + 3}{z(z - i)^2}\, dz + \oint_{C_2} \frac{z^3 + 3}{z(z - i)^2}\, dz$$

$$= -\oint_{C_1} \frac{\dfrac{z^3 + 3}{(z - i)^2}}{z}\, dz + \oint_{C_2} \frac{\dfrac{z^3 + 3}{z}}{(z - i)^2}\, dz = -I_1 + I_2,$$

we are in a position to use both (1) and (6).

To evaluate I_1, we identify $z_0 = 0$ and $f(z) = (z^3 + 3)/(z - i)^2$. By (1) it follows that

$$I_1 = \oint_{C_1} \frac{\dfrac{z^3 + 3}{(z - i)^2}}{z}\, dz = 2\pi i f(0) = -6\pi i.$$

To evaluate I_2 we identify $z_0 = i$, $n = 1$, $f(z) = (z^3 + 3)/z$, and $f'(z) = (2z^3 - 3)/z^2$. From (6) we obtain

$$I_2 = \oint_{C_2} \frac{\dfrac{z^3 + 3}{z}}{(z - i)^2}\, dz = \frac{2\pi i}{1!} f'(i) = 2\pi i(3 + 2i) = 2\pi(-2 + 3i).$$

Finally we get

$$\oint_C \frac{z^3 + 3}{z(z - i)^2}\, dz = -I_1 + I_2 = 6\pi i + 2\pi(-2 + 3i) = 4\pi(-1 + 3i). \qquad \square$$

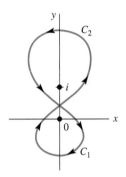

Figure 18.27

■ **Liouville's Theorem** If we take the contour C to be the circle $|z - z_0| = r$, it follows from (6) and the *ML*-inequality that

$$\left| f^{(n)}(z_0) \right| = \frac{n!}{2\pi} \left| \oint_C \frac{f(z)}{(z - z_0)^{n+1}}\, dz \right| \le \frac{n!}{2\pi} M \frac{1}{r^{n+1}} 2\pi r = \frac{n!M}{r^n}, \qquad (7)$$

where M is a real number such that $|f(z)| \le M$ for all points z on C. The result in (7), called **Cauchy's inequality,** is used to prove the next result.

> ### THEOREM 18.11 Liouville's Theorem
>
> The only bounded entire functions are constants.

PROOF Suppose f is an entire function and is bounded, that is, $|f(z)| \le M$ for all z. Then for any point z_0, (7) gives $|f'(z_0)| \le M/r$. By taking r arbitrarily large we can make $|f'(z_0)|$ as small as we wish. This means $f'(z_0) = 0$ for all points z_0 in the complex plane. Hence f must be a constant. ❏

■ **Fundamental Theorem of Algebra** Liouville's theorem enables us to prove, in turn, a result that is learned in elementary algebra: If $P(z)$ is a nonconstant polynomial, then the equation $P(z) = 0$ has at least one root. This result is known as the **Fundamental Theorem of Algebra.** To prove it, let us suppose that $P(z) \ne 0$ for all z. This implies that the reciprocal of P, $f(z) = 1/P(z)$, is an entire function. Now since $|f(z)| \to 0$ as $|z| \to \infty$, the function f must be bounded for all finite z. It follows from Liouville's theorem that f is a constant and therefore P is a constant. But this is a contradiction to our underlying assumption that P was *not* a constant polynomial. We conclude that there must exist at least one number z for which $P(z) = 0$.

EXERCISES 18.4

Answers to odd-numbered problems begin on page A-73.

In Problems 1–24, use Theorems 18.9 and 18.10, when appropriate, to evaluate the given integral along the indicated closed contour(s).

1. $\oint_C \dfrac{4}{z - 3i}\, dz$; $|z| = 5$

2. $\oint_C \dfrac{z^2}{(z - 3i)^2}\, dz$; $|z| = 5$

3. $\oint_C \dfrac{e^z}{z - \pi i}\, dz$; $|z| = 4$

4. $\oint_C \dfrac{1 + 2e^z}{z}\, dz$; $|z| = 1$

5. $\oint_C \dfrac{z^2 - 3z + 4i}{z + 2i}\, dz$; $|z| = 3$

6. $\oint_C \dfrac{\cos z}{3z - \pi}\, dz$; $|z| = 1.1$

7. $\oint_C \dfrac{z^2}{z^2 + 4}\, dz$; **(a)** $|z - i| = 2$, **(b)** $|z + 2i| = 1$

8. $\oint_C \dfrac{z^2 + 3z + 2i}{z^2 + 3z - 4}\, dz$; **(a)** $|z| = 2$, **(b)** $|z + 5| = \frac{3}{2}$

9. $\oint_C \dfrac{z^2 + 4}{z^2 - 5iz - 4}\, dz$; $|z - 3i| = 1.3$

10. $\oint_C \dfrac{\sin z}{z^2 + \pi^2}\, dz$; $|z - 2i| = 2$

11. $\oint_C \dfrac{e^{z^2}}{(z - i)^3}\, dz$; $|z - i| = 1$

12. $\oint_C \dfrac{z}{(z + i)^4}\, dz$; $|z| = 2$

13. $\oint_C \dfrac{\cos 2z}{z^5}\, dz$; $|z| = 1$

14. $\oint_C \dfrac{e^{-z} \sin z}{z^3}\, dz$; $|z - 1| = 3$

15. $\oint_C \dfrac{2z + 5}{z^2 - 2z}\, dz$; **(a)** $|z| = \frac{1}{2}$, **(b)** $|z + 1| = 2$, **(c)** $|z - 3| = 2$, **(d)** $|z + 2i| = 1$

16. $\oint_C \dfrac{z}{(z - 1)(z - 2)}\, dz$; **(a)** $|z| = \frac{1}{2}$, **(b)** $|z + 1| = 1$, **(c)** $|z - 1| = \frac{1}{2}$, **(d)** $|z| = 4$

17. $\oint_C \dfrac{z + 2}{z^2(z - 1 - i)}\, dz$; **(a)** $|z| = 1$, **(b)** $|z - 1 - i| = 1$

18. $\oint_C \dfrac{1}{z^3(z - 4)}\, dz$; **(a)** $|z| = 1$, **(b)** $|z - 2| = 1$

19. $\oint_C \left(\dfrac{e^{2iz}}{z^4} - \dfrac{z^4}{(z - i)^3} \right) dz$; $|z| = 6$

20. $\oint_C \left(\dfrac{\cosh z}{(z - \pi)^3} - \dfrac{\sin^2 z}{(2z - \pi)^3} \right) dz$; $|z| = 3$

21. $\oint_C \dfrac{1}{z^3(z - 1)^2}\, dz$; $|z - 2| = 5$

22. $\oint_C \dfrac{1}{z^2(z^2 + 1)}\, dz$; $|z - i| = \frac{3}{2}$

23. $\displaystyle\oint_C \frac{3z+1}{z(z-2)^2}\, dz$; C is given in Figure 18.28

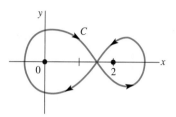

Figure 18.28

24. $\displaystyle\oint_C \frac{e^{iz}}{(z^2+1)^2}\, dz$; C is given in Figure 18.29

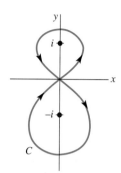

Figure 18.29

CHAPTER 18
REVIEW EXERCISES

Answers to odd-numbered problems begin on page A-74.

Answer Problems 1–12, without referring back to the text. Fill in the blank or answer true/false.

1. The sector defined by $-\pi/6 < \arg z < \pi/6$ is a simply connected domain. _____

2. If $\displaystyle\oint_C f(z)\, dz = 0$ for every simple closed contour C, then f is analytic within and on C. _____

3. The value of $\displaystyle\int_C \frac{z-2}{z}\, dz$ is the same for any path C in the right half-plane $\mathrm{Re}(z) > 0$ between $z = 1 + i$ and $z = 10 + 8i$. _____

4. If g is entire, then $\displaystyle\oint_C \frac{g(z)}{z-i}\, dz = \oint_{C_1} \frac{g(z)}{z-i}\, dz$, where C is the circle $|z| = 3$ and C_1 is the ellipse $x^2 + y^2/9 = 1$. _____

5. If f is a polynomial and C is a simple closed curve, then $\displaystyle\oint_C f(z)\, dz = $ _____.

6. If $f(z) = \displaystyle\oint_C \frac{\xi^2 + 6\xi - 2}{\xi - z}\, d\xi$, where C is $|z| = 3$, then $f(1 + i) = $ _____.

7. If $f(z) = z^3 + e^z$ and C is the contour $z = 8e^{it}$, $0 \le t \le 2\pi$, then $\displaystyle\oint_C \frac{f(z)}{(z + \pi i)^3}\, dz = $ _____.

8. If f is entire and $|f(z)| \le 10$ for all z, then $f(z) = $ _____.

9. $\displaystyle\oint_C \frac{1}{(z - z_0)(z - z_1)}\, dz = 0$ for every simple closed contour C that encloses the points z_0 and z_1. _____

10. If f is analytic within and on the simple closed contour C and z_0 is a point within C, then
$$\oint_C \frac{f'(z)}{z - z_0}\, dz = \oint_C \frac{f(z)}{(z - z_0)^2}\, dz. \ ___$$

11. $\displaystyle\oint_C z^n\, dz = \begin{cases} 0, & \text{if } n \ ___ \\ 2\pi i, & \text{if } n \ ___ \end{cases}$
where n is an integer and C is $|z| = 1$.

12. If $|f(z)| \le 2$ on $|z| = 3$, then $\left|\displaystyle\oint_C f(z)\, dz\right| \le$ _____.

In Problems 13–28, evaluate the given integral using the techniques considered in this chapter.

13. $\displaystyle\int_C (x + iy)\, dz$; C is the contour shown in Figure 18.30

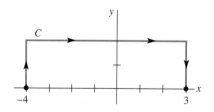

Figure 18.30

14. $\displaystyle\int_C (x - iy)\, dz$; C is the contour shown in Figure 18.30

15. $\displaystyle\int_C |z^2|\, dz$; C is $z(t) = t + it^2$, $0 \le t \le 2$

16. $\displaystyle\int_C e^{\pi z}\, dz$; C is the line segment from $z = i$ to $z = 1 + i$

17. $\displaystyle\oint_C e^{\pi z}\, dz$; C is the ellipse $x^2/100 + y^2/64 = 1$

18. $\displaystyle\int_{3i}^{1-i} (4z - 6)\, dz$

19. $\int_C \sin z \, dz$; C is $z(t) = t^4 + i(1 + t^3)^2$, $-1 \le t < 1$

20. $\int_C (4z^3 + 3z^2 + 2z + 1) \, dz$; C is the line segment from 0 to $2i$

21. $\oint_C (z^{-2} + z^{-1} + z + z^2) \, dz$; C is the circle $|z| = 1$

22. $\oint_C \dfrac{3z + 4}{z^2 - 1} \, dz$; C is the circle $|z| = 2$

23. $\oint_C \dfrac{e^{-2z}}{z^4} \, dz$; C is the circle $|z - 1| = 3$

24. $\oint_C \dfrac{\cos z}{z^3 - z^2} \, dz$; C is the circle $|z| = \frac{1}{2}$

25. $\oint_C \dfrac{1}{2z^2 + 7z + 3} \, dz$; C is the ellipse $x^2/4 + y^2 = 1$

26. $\oint_C z \csc z \, dz$; C is the rectangle with vertices $1 + i$, $1 - i$, $2 + i$, $2 - i$

27. $\oint_C \dfrac{z}{z + i} \, dz$; C is the contour shown in Figure 18.31

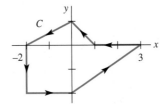

Figure 18.31

28. $\oint_C \dfrac{e^{i\pi z}}{2z^2 - 5z + 2} \, dz$; C is **(a)** $|z| = 1$, **(b)** $|z - 3| = 2$,

(c) $|z + 3| = 2$

29. Let $f(z) = z^n g(z)$, where n is a positive integer, $g(z)$ is entire, and $g(z) \ne 0$ for all z. Let C be a circle with center at the origin. Evaluate $\oint_C \dfrac{f'(z)}{f(z)} \, dz$.

30. Let C be the straight line segment from i to $2 + i$. Show that

$$\left| \int_C \mathrm{Ln}(z + 1) \, dz \right| \le \log_e 10 + \frac{\pi}{2}.$$

19

Series and Residues

INTRODUCTION

Cauchy's integral formula for derivatives indicates that if a function f is analytic at a point z_0, then it possesses derivatives of all orders at that point. As a consequence of this result we shall see that f can always be expanded in a power series centered at that point. On the other hand, if f fails to be analytic at a point z_0, we may still be able to expand it in a different kind of series known as a **Laurent series.** The notion of Laurent series leads to the concept of a **residue,** and this, in turn, leads to yet another way of evaluating complex integrals.

19.1 Sequences and Series

Much of the theory of complex sequences and series is analogous to that encountered in real calculus.

■ **Sequences** A **sequence** $\{z_n\}$ is a function whose domain is the set of positive integers; in other words, to each integer $n = 1, 2, 3, \ldots$, we assign a complex number z_n. For example, the sequence $\{1 + i^n\}$ is

$$1 + i, \quad 0, \quad 1 - i, \quad 2, \quad 1 + i, \ldots$$
$$\uparrow \qquad \uparrow \qquad \uparrow \qquad \uparrow \qquad \uparrow$$
$$n = 1, \, n = 2, \, n = 3, \, n = 4, \, n = 5, \ldots. \tag{1}$$

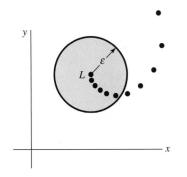

Figure 19.1

If $\lim_{n \to \infty} z_n = L$ we say the sequence $\{z_n\}$ is **convergent.** In other words, $\{z_n\}$ **converges** to the number L if, for each positive number ε, an N can be found such that $|z_n - L| < \varepsilon$ whenever $n > N$. As shown in Figure 19.1, when a sequence $\{z_n\}$ converges to L, all but a finite number of the terms of the sequence are within every ε-neighborhood of L. The sequence $\{1 + i^n\}$ illustrated in (1) is divergent, since the general term $z_n = 1 + i^n$ does not approach a fixed complex number as $n \to \infty$. Indeed, the first four terms of this sequence repeat endlessly as n increases.

Example 1 A Convergent Sequence

The sequence $\left\{\dfrac{i^{n+1}}{n}\right\}$ converges, since

$$\lim_{n \to \infty} \frac{i^{n+1}}{n} = 0.$$

As we see from

$$-1, \quad -\frac{i}{2}, \quad \frac{1}{3}, \quad \frac{i}{4}, \quad -\frac{1}{5}, \ldots,$$

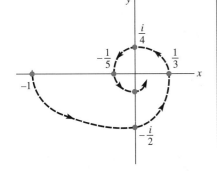

Figure 19.2

and Figure 19.2, the terms of the sequence spiral in toward the point $z = 0$. ❏

The following theorem should make intuitive sense:

> **THEOREM 19.1 Criterion for Convergence**
>
> A sequence $\{z_n\}$ converges to a complex number L if and only if $\mathrm{Re}(z_n)$ converges to $\mathrm{Re}(L)$ and $\mathrm{Im}(z_n)$ converges to $\mathrm{Im}(L)$.

Example 2 Illustrating Theorem 19.1

The sequence $\left\{\dfrac{ni}{n + 2i}\right\}$ converges to i. Note that $\mathrm{Re}(i) = 0$ and $\mathrm{Im}(i) = 1$. Then from

$$z_n = \frac{ni}{n + 2i} = \frac{2n}{n^2 + 4} + i\,\frac{n^2}{n^2 + 4},$$

we see that $\mathrm{Re}(z_n) = 2n/(n^2 + 4) \to 0$ and $\mathrm{Im}(z_n) = n^2/(n^2 + 4) \to 1$ as $n \to \infty$. ❏

■ **Series** An **infinite series** of complex numbers

$$\sum_{k=1}^{\infty} z_k = z_1 + z_2 + z_3 + \cdots + z_n + \cdots$$

is **convergent** if the sequence of partial sums $\{S_n\}$, where

$$S_n = z_1 + z_2 + z_3 + \cdots + z_n,$$

converges. If $S_n \to L$ as $n \to \infty$, we say that the **sum** of the series is L.

■ **Geometric Series** For the geometric series

$$\sum_{k=1}^{\infty} az^{k-1} = a + az + az^2 + \cdots + az^{n-1} + \cdots \tag{2}$$

the nth term of the sequence of partial sums is

$$S_n = a + az + az^2 + \cdots + az^{n-1}. \tag{3}$$

By multiplying S_n by z and subtracting this result from S_n, we obtain $S_n - zS_n = a - az^n$. Solving for S_n gives

$$S_n = \frac{a(1 - z^n)}{1 - z}. \tag{4}$$

Since $z^n \to 0$ as $n \to \infty$, whenever $|z| < 1$ we conclude from (4) that (2) converges to

$$\frac{a}{1 - z}$$

when $|z| < 1$; the series diverges when $|z| \geq 1$. The special geometric series

$$\frac{1}{1 - z} = 1 + z + z^2 + z^3 + \cdots \tag{5}$$

and

$$\frac{1}{1 + z} = 1 - z + z^2 - z^3 + \cdots \tag{6}$$

valid for $|z| < 1$ will be of particular usefulness in the next two sections. In addition, we shall use

$$\frac{1 - z^n}{1 - z} = 1 + z + z^2 + z^3 + \cdots + z^{n-1} \tag{7}$$

in the alternative form

$$\frac{1}{1 - z} = 1 + z + z^2 + z^3 + \cdots + z^{n-1} + \frac{z^n}{1 - z} \tag{8}$$

in the proofs of the two principal theorems of this chapter.

> **Example 3** **Convergent Geometric Series**
>
> The series
>
> $$\sum_{k=1}^{\infty} \frac{(1 + 2i)^k}{5^k} = \frac{1 + 2i}{5} + \frac{(1 + 2i)^2}{5^2} + \frac{(1 + 2i)^3}{5^3} + \cdots$$
>
> is a geometric series with $a = (1 + 2i)/5$ and $z = (1 + 2i)/5$. Since $|z| = \sqrt{5}/5 < 1$, the series converges and we can write

$$\sum_{k=1}^{\infty} \frac{(1+2i)^k}{5^k} = \frac{\dfrac{1+2i}{5}}{1 - \dfrac{1+2i}{5}} = \frac{i}{2}.$$ ❑

THEOREM 19.2 Necessary Condition for Convergence

If $\sum_{k=1}^{\infty} z_k$ converges, then $\lim_{n \to \infty} z_n = 0$.

An equivalent form of Theorem 19.2 is the familiar nth term test for divergence of an infinite series.

THEOREM 19.3 The nth Term Test for Divergence

If $\lim_{n \to \infty} z_n \neq 0$, then the series $\sum_{k=1}^{\infty} z_k$ diverges.

For example, the series $\sum_{k=1}^{\infty} (k + 5i)/k$ diverges since $a_n = (n + 5i)/n \to 1$ as $n \to \infty$. The geometric series (2) diverges when $|z| \geq 1$, since, in this case, $\lim_{n \to \infty} |z^n|$ does not exist.

DEFINITION 19.1 Absolute Convergence

An infinite series $\sum_{k=1}^{\infty} z_k$ is said be **absolutely convergent** if $\sum_{k=1}^{\infty} |z_k|$ converges.

Example 4 **Absolute Convergence**

The series $\sum_{k=1}^{\infty} (i^k/k^2)$ is absolutely convergent since $|i^k/k^2| = 1/k^2$ and the real series $\sum_{k=1}^{\infty} (1/k^2)$ converges. Recall from calculus that a real series of the form $\sum_{k=1}^{\infty} (1/k^p)$ is called a p-series and converges for $p > 1$ and diverges for $p \leq 1$. ❑

As in real calculus,

Absolute convergence implies convergence.

Thus in Example 4, the series

$$\sum_{k=1}^{\infty} \frac{i^k}{k^2} = i - \frac{1}{2^2} - \frac{i}{3^2} + \cdots$$

converges.

The following two tests are the complex versions of the ratio and root tests that are encountered in calculus:

THEOREM 19.4 Ratio Test

Suppose $\sum_{k=1}^{\infty} z_k$ is a series of nonzero complex terms such that

$$\lim_{n \to \infty} \left| \frac{z_{n+1}}{z_n} \right| = L. \tag{9}$$

continued

(*i*) If $L < 1$, then the series converges absolutely.
(*ii*) If $L > 1$ or $L = \infty$, then the series diverges.
(*iii*) If $L = 1$, the test is inconclusive.

THEOREM 19.5　Root Test

Suppose $\sum_{k=1}^{\infty} z_k$ is a series of complex terms such that

$$\lim_{n \to \infty} \sqrt[n]{|z_n|} = L. \qquad (10)$$

(*i*) If $L < 1$, then the series converges absolutely.
(*ii*) If $L > 1$ or $L = \infty$, then the series diverges.
(*iii*) If $L = 1$, the test is inconclusive.

We are interested primarily in applying these tests to power series.

■ **Power Series**　The notion of a power series is important in the study of analytic functions. An infinite series of the form

$$\sum_{k=0}^{\infty} a_k(z - z_0)^k = a_0 + a_1(z - z_0) + a_2(z - z_0)^2 + \cdots, \qquad (11)$$

where the coefficients a_k are complex constants, is called a **power series** in $z - z_0$. The power series (11) is said to be **centered at z_0**, and the complex point z_0 is referred to as the **center** of the series. In (11) it is also convenient to define $(z - z_0)^0 = 1$ even when $z = z_0$.

■ **Circle of Convergence**　Every complex power series has **radius of convergence** R. Analogous to the concept of an *interval* of convergence in real calculus, when $0 < R < \infty$, a complex power series (11) has a **circle of convergence** defined by $|z - z_0| = R$. The power series converges absolutely for all z satisfying $|z - z_0| < R$ and diverges for $|z - z_0| > R$. See Figure 19.3. The radius R of convergence can be:

(*i*) zero (in which case (11) converges at only $z = z_0$),
(*ii*) a finite number (in which case (11) converges at all interior points of the circle $|z - z_0| = R$), or
(*iii*) ∞ (in which case (11) converges for all z).

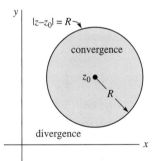

Figure 19.3

A power series may converge at some, all, or none of the points *on* the actual circle of convergence.

Example 5　Circle of Convergence

Consider the power series $\sum_{k=1}^{\infty} (z^{k+1}/k)$. By the ratio test (9),

$$\lim_{n \to \infty} \left| \frac{\dfrac{z^{n+2}}{n+1}}{\dfrac{z^{n+1}}{n}} \right| = \lim_{n \to \infty} \frac{n}{n+1} |z| = |z|.$$

continued

Thus the series converges absolutely for $|z| < 1$. The circle of convergence is $|z| = 1$ and the radius of convergence is $R = 1$. Note that on the circle of convergence, the series does not converge absolutely, since the series of absolute values is the well-known divergent harmonic series $\sum_{k=1}^{\infty} (1/k)$. Bear in mind this does not say, however, that the series diverges on the circle of convergence. In fact, at $z = -1$, $\sum_{k=1}^{\infty} ((-1)^{k+1}/k)$ is the convergent alternating harmonic series. Indeed, it can be shown that the series converges at all points on the circle $|z| = 1$ except at $z = 1$. ❏

It should be clear from Theorem 19.4 and Example 5 that for a power series $\sum_{k=0}^{\infty} a_k(z - z_0)^k$, the limit (9) depends on only the coefficients a_k. Thus, if

(i) $\lim\limits_{n \to \infty} \left| \dfrac{a_{n+1}}{a_n} \right| = L \neq 0$, the radius of convergence is $R = 1/L$;

(ii) $\lim\limits_{n \to \infty} \left| \dfrac{a_{n+1}}{a_n} \right| = 0$, the radius of convergence is ∞;

(iii) $\lim\limits_{n \to \infty} \left| \dfrac{a_{n+1}}{a_n} \right| = \infty$, the radius of convergence is $R = 0$.

Similar remarks can be made for the root test (10) by utilizing $\lim_{n \to \infty} \sqrt[n]{|a_n|}$.

Example 6 Radius of Convergence

Consider the power series $\displaystyle\sum_{k=1}^{\infty} \dfrac{(-1)^{k+1}(z - 1 - i)^k}{k!}$. Identifying $a_n = (-1)^{n+1}/n!$, we have

$$\lim_{n \to \infty} \left| \dfrac{\dfrac{(-1)^{n+2}}{(n+1)!}}{\dfrac{(-1)^{n+1}}{n!}} \right| = \lim_{n \to \infty} \dfrac{1}{n+1} = 0.$$

Thus the radius of convergence is ∞; the power series with center $1 + i$ converges absolutely for all z. ❏

Example 7 Radius of Convergence

Consider the power series $\displaystyle\sum_{k=1}^{\infty} \left(\dfrac{6k+1}{2k+5} \right)^k (z - 2i)^k$. With $a_n = \left(\dfrac{6n+1}{2n+5} \right)^n$, the root test in the form

$$\lim_{n \to \infty} \sqrt[n]{|a_n|} = \lim_{n \to \infty} \dfrac{6n+1}{2n+5} = 3$$

shows that the radius of convergence of the series is $R = \frac{1}{3}$. The circle of convergence is $|z - 2i| = \frac{1}{3}$; the series converges absolutely for $|z - 2i| < \frac{1}{3}$. ❏

In Problems 1–4, write out the first five terms of the given sequence.

1. $\{5i^n\}$

2. $\{2 + (-i)^n\}$

3. $\{1 + e^{n\pi i}\}$

4. $\{(1 + i)^n\}$ [*Hint:* Write in polar form.]

In Problems 5–10, determine whether the given sequence converges or diverges.

5. $\left\{\dfrac{3ni+2}{n+ni}\right\}$

6. $\left\{\dfrac{ni+2^n}{3ni+5^n}\right\}$

7. $\left\{\dfrac{(ni+2)^2}{n^2i}\right\}$

8. $\left\{\dfrac{n(1+i^n)}{n+1}\right\}$

9. $\left\{\dfrac{n+i^n}{\sqrt{n}}\right\}$

10. $\{e^{1/n}+2(\tan^{-1}n)i\}$

In Problems 11 and 12, show that the given sequence $\{z_n\}$ converges to a complex number L by computing $\lim_{n\to\infty}\operatorname{Re}(z_n)$ and $\lim_{n\to\infty}\operatorname{Im}(z_n)$.

11. $\left\{\dfrac{4n+3ni}{2n+i}\right\}$

12. $\left\{\left(\dfrac{1+i}{4}\right)^n\right\}$

In Problems 13 and 14, use the sequence of partial sums to show that the given series is convergent.

13. $\displaystyle\sum_{k=1}^{\infty}\left[\dfrac{1}{k+2i}-\dfrac{1}{k+1+2i}\right]$

14. $\displaystyle\sum_{k=2}^{\infty}\dfrac{i}{k(k+1)}$

In Problems 15–20, determine whether the given geometric series is convergent or divergent. If convergent, find its sum.

15. $\displaystyle\sum_{k=0}^{\infty}(1-i)^k$

16. $\displaystyle\sum_{k=1}^{\infty}4i\left(\dfrac{1}{3}\right)^{k-1}$

17. $\displaystyle\sum_{k=1}^{\infty}\left(\dfrac{i}{2}\right)^k$

18. $\displaystyle\sum_{k=0}^{\infty}\dfrac{1}{2}i^k$

19. $\displaystyle\sum_{k=0}^{\infty}3\left(\dfrac{2}{1+2i}\right)^k$

20. $\displaystyle\sum_{k=2}^{\infty}\dfrac{i^k}{(1+i)^{k-1}}$

In Problems 21–28, find the circle and radius of convergence of the given power series.

21. $\displaystyle\sum_{k=0}^{\infty}\dfrac{1}{(1-2i)^{k+1}}(z-2i)^k$

22. $\displaystyle\sum_{k=1}^{\infty}\dfrac{1}{k}\left(\dfrac{i}{1+i}\right)^k z^k$

23. $\displaystyle\sum_{k=1}^{\infty}\dfrac{(-1)^k}{k2^k}(z-1-i)^k$

24. $\displaystyle\sum_{k=1}^{\infty}\dfrac{1}{k^2(3+4i)^k}(z+3i)^k$

25. $\displaystyle\sum_{k=0}^{\infty}(1+3i)^k(z-i)^k$

26. $\displaystyle\sum_{k=1}^{\infty}\dfrac{z^k}{k^k}$

27. $\displaystyle\sum_{k=0}^{\infty}\dfrac{(z-4-3i)^k}{5^{2k}}$

28. $\displaystyle\sum_{k=0}^{\infty}(-1)^k\left(\dfrac{1+2i}{2}\right)^k(z+2i)^k$

29. Show that the power series $\displaystyle\sum_{k=1}^{\infty}\dfrac{(z-i)^k}{k2^k}$ is not absolutely convergent on its circle of convergence. Determine at least one point on the circle of convergence at which the power series converges.

30. (a) Show that the power series $\displaystyle\sum_{k=1}^{\infty}\dfrac{z^k}{k^2}$ converges at every point on its circle of convergence.

(b) Show that the power series $\displaystyle\sum_{k=1}^{\infty}kz^k$ diverges at every point on its circle of convergence.

19.2 Taylor Series

We saw in the last section that every power series has a radius of convergence R. The next three theorems will give some important facts about the nature of a power series within its circle of convergence $|z-z_0|=R$, $R\ne 0$.

THEOREM 19.6 Continuity

A power series $\sum_{k=0}^{\infty}a_k(z-z_0)^k$ represents a continuous function f within its circle of convergence $|z-z_0|=R$, $R\ne 0$.

THEOREM 19.7 Term-by-Term Integration

A power series $\sum_{k=0}^{\infty}a_k(z-z_0)^k$ can be integrated term by term within its circle of convergence $|z-z_0|=R$, $R\ne 0$, for every contour C lying entirely within the circle of convergence.

> **THEOREM 19.8 Term-by-Term Differentiation**
>
> A power series $\sum_{k=0}^{\infty} a_k(z - z_0)^k$ can be differentiated term by term within its circle of convergence $|z - z_0| = R$, $R \neq 0$.

■ **Taylor Series** Suppose a power series represents a function f for $|z - z_0| < R$, $R \neq 0$; that is,

$$f(z) = \sum_{k=0}^{\infty} a_k(z - z_0)^k = a_0 + a_1(z - z_0) + a_2(z - z_0)^2 + a_3(z - z_0)^3 + \cdots. \tag{1}$$

It follows from Theorem 19.8 that the derivatives of f are

$$f'(z) = \sum_{k=1}^{\infty} k a_k(z - z_0)^{k-1} = a_1 + 2a_2(z - z_0) + 3a_3(z - z_0)^2 + \cdots \tag{2}$$

$$f''(z) = \sum_{k=2}^{\infty} k(k-1) a_k(z - z_0)^{k-2} = 2 \cdot 1 a_2 + 3 \cdot 2 a_3(z - z_0) + \cdots \tag{3}$$

$$f'''(z) = \sum_{k=3}^{\infty} k(k-1)(k-2) a_k(z - z_0)^{k-3} = 3 \cdot 2 \cdot 1 a_3 + \cdots \tag{4}$$

and so on. Each of the differentiated series has the same radius of convergence as the original series. Moreover, since the original power series represents a differentiable function f within its circle of convergence, we conclude that when $R \neq 0$, *a power series represents an analytic function within its circle of convergence.*

There is a relationship between the coefficients a_k and the derivatives of f. Evaluating (1), (2), (3), and (4) at $z = z_0$ gives

$$f(z_0) = a_0, \quad f'(z_0) = 1!a_1, \quad f''(z_0) = 2!a_2, \quad \text{and} \quad f'''(z_0) = 3!a_3,$$

respectively. In general, $f^{(n)}(z_0) = n!a_n$ or

$$a_n = \frac{f^{(n)}(z_0)}{n!}, \qquad n \geq 0. \tag{5}$$

When $n = 0$ we interpret the zeroth derivative as $f(z_0)$ and $0! = 1$. Substituting (5) into (1) yields

$$f(z) = \sum_{k=0}^{\infty} \frac{f^{(k)}(z_0)}{k!} (z - z_0)^k. \tag{6}$$

This series is called the **Taylor series** for f centered at z_0. A Taylor series with center $z_0 = 0$,

$$f(z) = \sum_{k=0}^{\infty} \frac{f^{(k)}(0)}{k!} z^k, \tag{7}$$

is referred to as a **Maclaurin series.**

We have just seen that a power series with a nonzero radius of convergence represents an analytic function. On the other hand, if we are given a function f that is analytic in some domain D, can we represent it by a power series of the form (6) and (7)? Since a power series converges in a circular domain, and a domain D is generally not circular, the question becomes: Can we expand f in *one or more* power series that are valid in circular domains that are all contained in D? The question will be answered in the affirmative by the next theorem.

THEOREM 19.9 **Taylor's Theorem**

Let f be analytic within a domain D and let z_0 be a point in D. Then f has the series representation

$$f(z) = \sum_{k=0}^{\infty} \frac{f^{(k)}(z_0)}{k!} (z - z_0)^k \qquad (8)$$

valid for the largest circle C with center at z_0 and radius R that lies entirely within D.

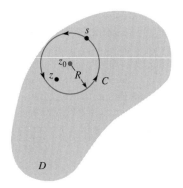

Figure 19.4

PROOF Let z be a fixed point within the circle C and let s denote the variable of integration. The circle C is then described by $|s - z_0| = R$. See Figure 19.4. To begin, we use the Cauchy integral formula to obtain the value of f at z:

$$f(z) = \frac{1}{2\pi i} \oint_C \frac{f(s)}{s - z}\, dz = \frac{1}{2\pi i} \oint_C \frac{f(s)}{(s - z_0) - (z - z_0)}\, ds$$

$$= \frac{1}{2\pi i} \oint_C \frac{f(s)}{s - z_0} \left[\frac{1}{1 - \dfrac{z - z_0}{s - z_0}} \right] ds. \qquad (9)$$

By replacing z by $(z - z_0)(s - z_0)$ in (8) of Section 19.1, we have

$$\frac{1}{1 - \dfrac{z - z_0}{s - z_0}} = 1 + \frac{z - z_0}{s - z_0} + \left(\frac{z - z_0}{s - z_0}\right)^2 + \cdots + \left(\frac{z - z_0}{s - z_0}\right)^{n-1} + \frac{(z - z_0)^n}{(s - z)(s - z_0)^{n-1}},$$

and so (9) becomes

$$f(z) = \frac{1}{2\pi i} \oint_C \frac{f(s)}{s - z_0}\, ds + \frac{z - z_0}{2\pi i} \oint_C \frac{f(s)}{(s - z_0)^2}\, ds + \frac{(z - z_0)^2}{2\pi i} \oint_C \frac{f(s)}{(s - z_0)^3}\, ds + \cdots$$

$$+ \frac{(z - z_0)^{n-1}}{2\pi i} \oint_C \frac{f(s)}{(s - z_0)^n}\, ds + \frac{(z - z_0)^n}{2\pi i} \oint_C \frac{f(s)}{(s - z)(s - z_0)^n}\, ds. \qquad (10)$$

Utilizing Cauchy's integral formula for derivatives, we can write (10) as

$$f(z) = f(z_0) + \frac{f'(z_0)}{1!} (z - z_0) + \frac{f''(z_0)}{2!} (z - z_0)^2 + \cdots +$$

$$+ \frac{f^{(n-1)}(z_0)}{(n-1)!} (z - z_0)^{n-1} + R_n(z), \qquad (11)$$

where

$$R_n(z) = \frac{(z - z_0)^n}{2\pi i} \oint_C \frac{f(s)}{(s - z)(s - z_0)^n}\, ds.$$

Equation (11) is called Taylor's formula with remainder R_n. We now wish to show that $R_n(z) \to 0$ as $n \to \infty$. Since f is analytic in D, $|f(z)|$ has a maximum value M on the contour C. In addition, since z is inside C, we have $|z - z_0| < R$, and, consequently,

$$|s - z| = |s - z_0 - (z - z_0)| \geq |s - z_0| - |z - z_0| = R - d,$$

where $d = |z - z_0|$ is the distance from z and z_0. The ML-inequality then gives

$$\left|R_n(z)\right| = \left|\frac{(z-z_0)^n}{2\pi i}\oint_C \frac{f(s)}{(s-z)(s-z_0)^n}\,ds\right| \leq \frac{d^n}{2\pi}\cdot\frac{M}{(R-d)R^n}\cdot 2\pi R = \frac{MR}{R-d}\left(\frac{d}{R}\right)^n.$$

Because $d < R$, $(d/R)^n \to 0$ as $n \to \infty$, we conclude that $\left|R_n(z)\right| \to 0$ as $n \to \infty$. It follows that the infinite series

$$f(z_0) + \frac{f'(z_0)}{1!}(z-z_0) + \frac{f''(z_0)}{2!}(z-z_0)^2 + \cdots$$

converges to $f(z)$. In other words, the result in (8) is valid for any point z interior to C. ❑

We can find the radius of convergence of a Taylor series in exactly the same manner illustrated in Examples 5–7 of Section 19.1. However, we can simplify matters even further by noting that the radius of convergence is the *distance* from the center z_0 of the series to the nearest *isolated singularity* of f. We shall elaborate more on this concept in the next section, but an isolated singularity is a point at which f fails to be analytic but is, nonetheless, analytic at all other points throughout *some* neighborhood of the point. For example, $z = 5i$ is an isolated singularity of $f(z) = 1/(z - 5i)$. If the function f is entire, then the radius of convergence of a Taylor series centered at any point z_0 is necessarily infinite. Using (8) and the last fact, we can say that the Maclaurin series representations

$$e^z = 1 + \frac{z}{1!} + \frac{z^2}{2!} + \cdots = \sum_{k=0}^{\infty}\frac{z^k}{k!} \tag{12}$$

$$\sin z = z - \frac{z^3}{3!} + \frac{z^5}{5!} - \cdots = \sum_{k=0}^{\infty}(-1)^k\frac{z^{2k+1}}{(2k+1)!} \tag{13}$$

$$\cos z = 1 - \frac{z^2}{2!} + \frac{z^4}{4!} - \cdots = \sum_{k=0}^{\infty}(-1)^k\frac{z^{2k}}{(2k)!} \tag{14}$$

are valid for all z.

If two power series with center z_0:

$$\sum_{k=0}^{\infty}a_k(z-z_0)^k \quad\text{and}\quad \sum_{k=0}^{\infty}b_k(z-z_0)^k$$

represent the same function and have the same nonzero radius of convergence, then $a_k = b_k$, $k = 0, 1, 2, \dots$. Stated in another way, the power series expansion of a function with center z_0 is *unique*. On a practical level this means that a power series expansion of an analytic function f centered at z_0, irrespective of the method used to obtain it, is the Taylor series expansion of the function. For example, we can obtain (14) by simply differentiating (13) term by term. The Maclaurin series for e^{z^2} can be obtained by replacing the symbol z in (12) by z^2.

Example 1 Maclaurin Series

Find the Maclaurin expansion of $f(z) = \dfrac{1}{(1-z)^2}$.

SOLUTION We could, of course, begin by computing the coefficients using (8). However, we know from (5) of Section 19.1 that for $|z| < 1$,

$$\frac{1}{1-z} = 1 + z + z^2 + z^3 + \cdots. \tag{15}$$

Differentiating both sides of the last result with respect to z then yields

$$\frac{1}{(1-z)^2} = 1 + 2z + 3z^2 + \cdots = \sum_{k=1}^{\infty} kz^{k-1}.$$

Since we are using Theorem 19.8, the radius of convergence of this last series is the same as the original series, $R = 1$. ❏

Example 2 **Taylor Series**

Expand $f(z) = \dfrac{1}{1-z}$ in a Taylor series with center $z_0 = 2i$.

SOLUTION We shall solve this problem in two ways. We begin by using (8). From the first several derivatives,

$$f'(z) = \frac{1}{(1-z)^2}, \quad f''(z) = \frac{2 \cdot 1}{(1-z)^3}, \quad f'''(z) = \frac{3 \cdot 2}{(1-z)^4},$$

we conclude that $f^{(n)}(z) = n!/(1-z)^{n+1}$ and so $f^{(n)}(2i) = n!/(1-2i)^{n+1}$. Thus from (8) we obtain the Taylor series

$$\frac{1}{1-z} = \sum_{k=0}^{\infty} \frac{1}{(1-2i)^{k+1}} (z-2i)^k. \tag{16}$$

Since the distance from the center $z_0 = 2i$ to the nearest singularity $z = 1$ is $\sqrt{5}$, we conclude that the circle of convergence for the power series in (16) is $|z - 2i| = \sqrt{5}$. This can be verified by the ratio test of the preceding section.

ALTERNATIVE SOLUTION In this solution we again use the geometric series (15). By adding and subtracting $2i$ in the denominator of $1/(1-z)$, we can write

$$\frac{1}{1-z} = \frac{1}{1-z+2i-2i} = \frac{1}{1-2i-(z-2i)} = \frac{1}{1-2i} \frac{1}{1 - \dfrac{z-2i}{1-2i}}.$$

We now write $\dfrac{1}{1 - \dfrac{z-2i}{1-2i}}$ as a power series by using (15) with the symbol z

replaced by $(z-2i)/(1-2i)$:

$$\frac{1}{1-z} = \frac{1}{1-2i} \left[1 + \frac{z-2i}{1-2i} + \left(\frac{z-2i}{1-2i}\right)^2 + \left(\frac{z-2i}{1-2i}\right)^3 + \cdots \right]$$

$$= \frac{1}{1-2i} + \frac{1}{(1-2i)^2} (z-2i) + \frac{1}{(1-2i)^3} (z-2i)^2 + \frac{1}{(1-2i)^4} (z-2i)^3 + \cdots.$$

The reader should verify that this last series is exactly the same as that in (16). ❏

In (15) and (16) we represented the same function $1/(1-z)$ by two different power series. The first series

$$\frac{1}{1-z} = 1 + z + z^2 + z^3 + \cdots$$

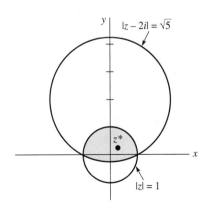

$|z - 2i| = \sqrt{5}$

z^*

$|z| = 1$

Figure 19.5

has center zero and radius of convergence one. The second series

$$\frac{1}{1-z} = \frac{1}{1-2i} + \frac{1}{(1-2i)^2} (z-2i) + \frac{1}{(1-2i)^3} (z-2i)^2 + \frac{1}{(1-2i)^4} (z-2i)^3 + \cdots$$

has center $2i$ and radius of convergence $\sqrt{5}$. The two different circles of convergence are illustrated in Figure 19.5. The interior of the intersection of the two circles (shaded) is the region where *both* series converge; in other words, at a specified point z^* in this region both series converge to the same value $f(z^*) = 1/(1 - z^*)$. Outside the shaded region at least one of the two series must diverge.

Answers to odd-numbered problems begin on page A-74.

In Problems 1–12, expand the given function in a Maclaurin series. Give the radius of convergence of each series.

1. $f(z) = \dfrac{z}{1+z}$

2. $f(z) = \dfrac{1}{4-2z}$

3. $f(z) = \dfrac{1}{(1+2z)^2}$

4. $f(z) = \dfrac{z}{(1-z)^3}$

5. $f(z) = e^{-2z}$

6. $f(z) = ze^{-z^2}$

7. $f(z) = \sinh z$

8. $f(z) = \cosh z$

9. $f(z) = \cos \dfrac{z}{2}$

10. $f(z) = \sin 3z$

11. $f(z) = \sin z^2$

12. $f(z) = \cos^2 z$ [*Hint:* Use a trigonometric identity.]

In Problems 13–22, expand the given function in a Taylor series centered at the indicated point. Give the radius of convergence of each series.

13. $f(z) = 1/z$, $z_0 = 1$

14. $f(z) = 1/z$, $z_0 = 1 + i$

15. $f(z) = \dfrac{1}{3-z}$, $z_0 = 2i$

16. $f(z) = \dfrac{1}{1+z}$, $z_0 = -i$

17. $f(z) = \dfrac{z-1}{3-z}$, $z_0 = 1$

18. $f(z) = \dfrac{1+z}{1-z}$, $z_0 = i$

19. $f(z) = \cos z$, $z_0 = \pi/4$

20. $f(z) = \sin z$, $z_0 = \pi/2$

21. $f(z) = e^z$, $z = 3i$

22. $f(z) = (z-1)e^{-2z}$, $z = 1$

In Problems 23 and 24, use (7) to find the first three nonzero terms of the Maclaurin series of the given function.

23. $f(z) = \tan z$

24. $f(z) = e^{1/(1+z)}$

In Problems 25 and 26, use partial fractions as an aid in obtaining the Maclaurin series for the given function. Give the radius of convergence of the series.

25. $f(z) = \dfrac{i}{(z-i)(z-2i)}$

26. $f(z) = \dfrac{z-7}{z^2 - 2z - 3}$

In Problems 27 and 28, without actually expanding, determine the radius of convergence of the Taylor series of the given function centered at the indicated point.

27. $f(z) = \dfrac{4+5z}{1+z^2}$, $z_0 = 2 + 5i$

28. $f(z) = \cot z$, $z_0 = \pi i$

In Problems 29 and 30, expand the given function in the Taylor series centered at the indicated points. Give the radius of convergence of each series. Sketch the region within which both series converge.

29. $f(z) = \dfrac{1}{2+z}$, $z_0 = -1$, $z_0 = i$

30. $f(z) = \dfrac{1}{z}$, $z_0 = 1 + i$, $z_0 = 3$

31. (a) Suppose the principal branch of the logarithm $f(z) = \text{Ln } z = \log_e |z| + i \text{ Arg } z$ is expanded in a Taylor series with center $z_0 = -1 + i$. Explain why $R = 1$ is the radius of the largest circle centered at $z_0 = -1 + i$ within which f is analytic.

(b) Show that within the circle $|z - (-1 + i)| = 1$ the Taylor series for f is

$$\text{Ln } z = \frac{1}{2}\log_e 2 + \frac{3\pi}{4}i - \sum_{k=1}^{\infty} \frac{1}{k}\left(\frac{1+i}{2}\right)^k (z + 1 - i)^k.$$

(c) Show that the radius of convergence for the power series in part **(b)** is $R = \sqrt{2}$. Explain why this does not contradict the result in part **(a)**.

32. (a) Consider the function $f(z) = \text{Ln}(1 + z)$. What is the radius of the largest circle centered at the origin within which f is analytic?

(b) Expand f in a Maclaurin series. What is the radius of convergence of this series?

(c) Use the result in part (b) to find a Maclaurin series for Ln(1 − z).

(d) Find a Maclaurin series for $Ln\left(\dfrac{1+z}{1-z}\right)$.

In Problems 33 and 34, approximate the value of the given expression using the indicated number of terms of a Maclaurin series.

33. $e^{(1+i)/10}$, three terms **34.** $\sin\left(\dfrac{1+i}{10}\right)$, two terms

35. In Section 15.1 we defined the error function as

$$\text{erf}(z) = \frac{2}{\sqrt{\pi}} \int_0^z e^{-t^2}\, dt.$$

Find a Maclaurin series for erf(z).

36. Use the Maclaurin series for e^{iz} to prove Euler's formula for complex z:

$$e^{iz} = \cos z + i \sin z.$$

19.3 Laurent Series

■ **Singularities** If a complex function f fails to be analytic at $z = z_0$, then this point is said to be a **singularity** or a **singular point** of the function. For example, the complex numbers $z = 2i$ and $z = -2i$ are singularities of the function $f(z) = z/(z^2 + 4)$ since f is not defined at either of these points. Recall from Section 17.6 that the principal branch of the logarithm, $f(z) = \text{Ln } z$, is analytic at all points except points on the nonpositive x-axis. In other words, $z = 0$ as well as all negative real numbers are singular points of $f(z) = \text{Ln } z$. Recall from Section 19.2 that a point $z = z_0$ is said to be an **isolated singularity** of a function f if f is analytic throughout some neighborhood of z_0 but not at z_0 itself. For example, $z = 2i$ (as well as $z = -2i$) is an isolated singularity of $f(z) = z/(z^2 + 4)$ since f is analytic at every point in the neighborhood defined by $|z - 2i| < 1$ except at $z = 2i$. In other words, if f is analytic in the deleted neighborhood $0 < |z - 2i| < 1$. The point $z = 0$ is not an isolated singularity of $f(z) = \text{Ln } z$ since every neighborhood of this point must contain points on the negative x-axis. We say that a singular point $z = z_0$ of a function f is **nonisolated** if every neighborhood of z_0 contains at least one singularity of f other than z_0.

If $z = z_0$ is a singularity of a function f, then f cannot be expanded in a power series with z_0 as its center. (Why?) However, about an isolated singularity $z = z_0$ it is possible to represent f by a new kind of series involving negative and nonnegative integer powers of $z - z_0$; that is,

$$f(z) = \cdots + \frac{a_{-2}}{(z - z_0)^2} + \frac{a_{-1}}{z - z_0} + a_0 + a_1(z - z_0) + a_2(z - z_0)^2 + \cdots.$$

Using summation notation, we can write the last expression as the sum of two series

$$f(z) = \sum_{k=1}^{\infty} \frac{a_{-k}}{(z - z_0)^k} + \sum_{k=0}^{\infty} a_k(z - z_0)^k. \tag{1}$$

The first series in (1) will converge for $\left|1/(z - z_0)\right| < r^*$ or equivalently for $|z - z_0| > 1/r^* = r$. The second series in (1) converges for $|z - z_0| < R$. Hence, the sum of the two series converges when z satisfies both these conditions—that is, when z is a point in an annular domain defined by $r < |z - z_0| < R$, $r < R$.

By summing over negative and nonnegative integers, we can write (1) compactly as

$$f(z) = \sum_{k=-\infty}^{\infty} a_k(z - z_0)^k.$$

Example 1 A New Kind of Series

The function $f(z) = (\sin z)/z^3$ is not analytic at $z = 0$ and hence cannot be expanded in a Maclaurin series. However, $\sin z$ is an entire function, and from (13) of Section 19.2 we know that its Maclaurin series,

$$\sin z = z - \frac{z^3}{3!} + \frac{z^5}{5!} - \frac{z^7}{7!} + \cdots,$$

converges for all z. Dividing this power series by z^3 gives the following series with negative and nonnegative integer powers of z:

$$f(z) = \frac{\sin z}{z^3} = \frac{1}{z^2} - \frac{1}{3!} + \frac{z^2}{5!} - \frac{z^4}{7!} + \cdots. \tag{2}$$

This series converges for all z except $z = 0$, that is, for $0 < |z|$. ❑

A series representation of a function f that has the form given in (1), and (2) is such an example, is called a **Laurent series** or a **Laurent expansion** of f.

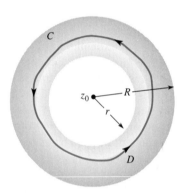

Figure 19.6

THEOREM 19.10 Laurent's Theorem

Let f be analytic within the annular domain D defined by $r < |z - z_0| < R$. Then f has the series representation

$$f(z) = \sum_{k=-\infty}^{\infty} a_k(z - z_0)^k \tag{3}$$

valid for $r < |z - z_0| < R$. The coefficients a_k are given by

$$a_k = \frac{1}{2\pi i} \oint_c \frac{f(s)}{(s - z_0)^{k+1}} \, ds, \qquad k = 0, \pm 1, \pm 2, \ldots, \tag{4}$$

where C is a simple closed curve that lies entirely within D and has z_0 in its interior (see Figure 19.6).

Figure 19.7

PROOF Let C_1 and C_2 be concentric circles with center z_0 and radii r_1 and R_2, where $r < r_1 < R_2 < R$. Let z be a fixed point in D that also satisfies $r_1 < |z - z_0| < R_2$. See Figure 19.7. By introducing a cross cut between C_2 and C_1, we find from Cauchy's integral formula that

$$f(z) = \frac{1}{2\pi i} \oint_{C_2} \frac{f(s)}{s - z} \, ds - \frac{1}{2\pi i} \oint_{C_1} \frac{f(s)}{s - z} \, ds. \tag{5}$$

Proceeding as in the proof of Theorem 19.9, we can write

$$\frac{1}{2\pi i} \oint_{C_2} \frac{f(s)}{s - z} \, ds = \sum_{k=0}^{\infty} a_k(z - z_0)^k, \tag{6}$$

where

$$a_k = \frac{1}{2\pi i} \oint_{C_2} \frac{f(s)}{(s - z_0)^{k+1}} \, ds, \qquad k = 0, 1, 2, \ldots. \tag{7}$$

Now using (5) and (8) of Section 19.1, we have

$$-\frac{1}{2\pi i}\oint_{C_1}\frac{f(s)}{s-z}\,ds = \frac{1}{2\pi i}\oint_{C_1}\frac{f(s)}{(z-z_0)-(s-z_0)}\,ds$$

$$= \frac{1}{2\pi i}\oint_{C_1}\frac{f(s)}{z-z_0}\left\{\frac{1}{1-\dfrac{s-z_0}{z-z_0}}\right\}\,ds$$

$$= \frac{1}{2\pi i}\oint_{C_1}\frac{f(s)}{z-z_0}\left\{1+\frac{s-z_0}{z-z_0}+\left(\frac{s-z_0}{z-z_0}\right)^2+\cdots\right.$$

$$\left.+\left(\frac{s-z_0}{z-z_0}\right)^{n-1}+\frac{(s-z_0)^n}{(z-s)(z-z_0)^{n-1}}\right\}\,ds \qquad (8)$$

$$= \sum_{k=1}^{n}\frac{a_{-k}}{(z-z_0)^k}+R_n(z),$$

where

$$a_{-k}=\frac{1}{2\pi i}\oint_{C_1}\frac{f(s)}{(s-z_0)^{-k+1}}\,ds, \qquad k=1,2,3,\ldots, \qquad (9)$$

and

$$R_n(z)=\frac{1}{2\pi i(z-z_0)^n}\oint_{C_1}\frac{f(s)(s-z_0)^n}{z-s}\,ds.$$

Now let d denote the distance from z to z_0, that is, $|z-z_0|=d$, and let M denote the maximum value of $|f(z)|$ on the contour C_1. Since $|s-z_0|=r_1$,

$$|z-s|=|z-z_0-(s-z_0)|\geq|z-z_0|-|s-z_0|=d-r_1.$$

The *ML*-inequality then gives

$$|R_n(z)|=\left|\frac{1}{2\pi i(z-z_0)^n}\oint_{C_1}\frac{f(s)(s-z_0)^n}{z-s}\,ds\right|\leq\frac{1}{2\pi d^n}\cdot\frac{Mr_1^n}{d-r_1}\cdot 2\pi r_1=\frac{Mr_1}{d-r_1}\left(\frac{r_1}{d}\right)^n.$$

Because $r_1<d$, $(r_1/d)^n\to 0$ as $n\to\infty$ and so $|R_n(z)|\to 0$ as $n\to\infty$. Thus we have shown that

$$-\frac{1}{2\pi i}\oint_{C_1}\frac{f(s)}{s-z}\,ds=\sum_{k=1}^{\infty}\frac{a_{-k}}{(z-z_0)^k}, \qquad (10)$$

where the coefficients a_{-k} are given in (9). Combining (6) and (10), we see that (5) yields

$$f(z)=\sum_{k=0}^{\infty}a_k(z-z_0)^k+\sum_{k=1}^{\infty}\frac{a_{-k}}{(z-z_0)^k}. \qquad (11)$$

Finally, by summing over nonnegative and negative integers, we can write (11) as $f(z)=\sum_{k=-\infty}^{\infty}a_k(z-z_0)^k$. However, (7) and (9) can be written as a single integral:

$$a_k=\frac{1}{2\pi i}\oint_C\frac{f(z)}{(z-z_0)^{k+1}}, \qquad k=0,\pm 1,\pm 2,\ldots,$$

where, in view of (3) of Section 18.2, we have replaced the contours C_1 and C_2 by any simple closed contour C in D with z_0 in its interior. ❏

In the case when $a_{-k}=0$ for $k=1,2,3,\ldots$, the Laurent series (3) is a Taylor series. Because of this, a Laurent expansion is a generalization of a Taylor series.

The annular domain in Theorem 19.10 defined by $r < |z - z_0| < R$ need not have the "ring" shape illustrated in Figure 19.7. Some other possible annular domains are: (i) $r = 0$, R finite; (ii) $r \neq 0$, $R \to \infty$; and (iii) $r = 0$, $R \to \infty$. In the first case, the series converges in the annular domain defined by $0 < |z - z_0| < R$. This is the interior of the circle $|z - z_0| = R$ except the point z_0. In the second case, the annular domain is defined by $r < |z - z_0|$; in other words, the domain consists of all points exterior to the circle $|z - z_0| = r$. In the third case, the domain is defined by $0 < |z - z_0|$. This represents the entire complex plane except the point z_0. The series we obtained in (2) is valid on this last type of domain.

In actual practice the formula in (4) for the coefficients of a Laurent series is seldom used. As a consequence, finding the Laurent series of a function in a specified annular domain is generally not an easy task. We often use the geometric series (5) and (6) of Section 19.1 or, as we did in Example 1, other known series. But regardless of how a Laurent expansion of a function f is obtained in a specified annular domain, it is *the* Laurent series; that is, the series we obtain is unique.

Example 2 Laurent Expansions

Expand $f(z) = \dfrac{1}{z(z-1)}$ in a Laurent series valid for (a) $0 < |z| < 1$, (b) $1 < |z|$,

(c) $0 < |z - 1| < 1$, and (d) $1 < |z - 1|$.

SOLUTION The four specified annular domains are shown in Figure 19.8. In parts (a) and (b) we want to represent f in a series involving only negative and nonnegative integer powers of z, whereas in parts (c) and (d) we want to represent f in a series involving negative and nonnegative integer powers of $z - 1$.

(a) By writing
$$f(z) = -\frac{1}{z}\frac{1}{1-z},$$
we can use (5) of Section 19.1:
$$f(z) = -\frac{1}{z}[1 + z + z^2 + z^3 + \cdots].$$
The series in the brackets converges for $|z| < 1$, but after this expression is multiplied by $1/z$, the resulting series
$$f(z) = -\frac{1}{z} - 1 - z - z^2 - \cdots$$
converges for $0 < |z| < 1$.

(b) To obtain a series that converges for $1 < |z|$ we start by constructing a series that converges for $|1/z| < 1$. To this end we write the given function f as
$$f(z) = \frac{1}{z^2}\frac{1}{1 - \dfrac{1}{z}}$$
and again use (5) of Section 19.1 with z replaced by $1/z$:
$$f(z) = \frac{1}{z^2}\left[1 + \frac{1}{z} + \frac{1}{z^2} + \frac{1}{z^3} + \cdots\right].$$
The series in the brackets converges for $|1/z| < 1$ or equivalently for $1 < |z|$. Thus the required Laurent series is
$$f(z) = \frac{1}{z^2} + \frac{1}{z^3} + \frac{1}{z^4} + \frac{1}{z^5} + \cdots.$$

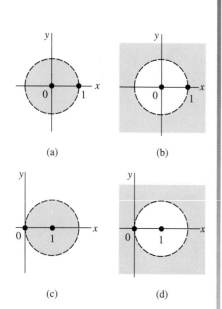

(a) (b)

(c) (d)

Figure 19.8

(c) This is basically the same problem as part (a) except that we want all powers of $z - 1$. To that end we add and subtract 1 in the denominator and use (6) of Section 19.1 with z replaced by $z - 1$:

$$f(z) = \frac{1}{(1 - 1 + z)(z - 1)}$$

$$= \frac{1}{z - 1} \frac{1}{1 + (z - 1)}$$

$$= \frac{1}{z - 1} [1 - (z - 1) + (z - 1)^2 - (z - 1)^3 + \cdots]$$

$$= \frac{1}{z - 1} - 1 + (z - 1) - (z - 1)^2 + \cdots.$$

The series in brackets converges for $|z - 1| < 1$, and so the last series converges for $0 < |z - 1| < 1$.

(d) Proceeding as in part (b), we write

$$f(z) = \frac{1}{z - 1} \frac{1}{1 + (z - 1)} = \frac{1}{(z - 1)^2} \frac{1}{1 + \dfrac{1}{z - 1}}$$

$$= \frac{1}{(z - 1)^2} \left[1 - \frac{1}{z - 1} + \frac{1}{(z - 1)^2} - \frac{1}{(z - 1)^3} + \cdots \right]$$

$$= \frac{1}{(z - 1)^2} - \frac{1}{(z - 1)^3} + \frac{1}{(z - 1)^4} - \frac{1}{(z - 1)^5} + \cdots.$$

Because the series within the brackets converges for $|1/(z - 1)| < 1$, the final series converges for $1 < |z - 1|$. ❏

Example 3 Laurent Expansions

Expand $f(z) = \dfrac{1}{(z - 1)^2(z - 3)}$ in a Laurent series valid for (a) $0 < |z - 1| < 2$ and (b) $0 < |z - 3| < 2$.

SOLUTION (a) As in parts (c) and (d) of Example 2, we want only powers of $z - 1$, and so we need to express $z - 3$ in terms of $z - 1$. This can be done by writing

$$f(z) = \frac{1}{(z - 1)^2(z - 3)} = \frac{1}{(z - 1)^2} \frac{1}{-2 + (z - 1)}$$

$$= \frac{-1}{2(z - 1)^2} \frac{1}{1 - \dfrac{z - 1}{2}}$$

and then using (5) of Section 19.1 with z replaced by $(z - 1)/2$:

$$f(z) = \frac{-1}{2(z - 1)^2} \left[1 + \frac{z - 1}{2} + \frac{(z - 1)^2}{2^2} + \frac{(z - 1)^3}{2^3} + \cdots \right]$$

$$= - \frac{1}{2(z - 1)^2} - \frac{1}{4(z - 1)} - \frac{1}{8} - \frac{1}{16}(z - 1) - \cdots. \tag{12}$$

(b) To obtain powers of $z - 3$ we write $z - 1 = 2 + (z - 3)$ and

$$f(z) = \frac{1}{(z - 1)^2(z - 3)} = \frac{1}{z - 3} [2 + (z - 3)]^{-2} = \frac{1}{4(z - 3)} \left[1 + \frac{z - 3}{2} \right]^{-2}.$$

At this point we can expand $\left[1 + \dfrac{z-3}{2} \right]^{-2}$ in a power series using the general binomial theorem:

$$f(z) = \frac{1}{4(z-3)} \left[1 + \frac{(-2)}{1!}\left(\frac{z-3}{2} \right) + \frac{(-2)(-3)}{2!}\left(\frac{z-3}{2} \right)^2 + \frac{(-2)(-3)(-4)}{3!}\left(\frac{z-3}{2} \right)^3 + \cdots \right].$$

The binomial series is valid for $|(z-3)/2| < 1$ or $|z-3| < 2$. Multiplying this series by $1/4(z-3)$ gives a Laurent series that is valid for $0 < |z-3| < 2$:

$$f(z) = \frac{1}{4(z-3)} - \frac{1}{4} + \frac{3}{16}(z-3) - \frac{1}{8}(z-3)^2 + \cdots. \qquad \square$$

Example 4 A Laurent Expansion

Expand $f(z) = \dfrac{8z+1}{z(1-z)}$ in a Laurent series valid for $0 < |z| < 1$.

SOLUTION By (5) of Section 19.1 we can write

$$f(z) = \frac{8z+1}{z(1-z)} = \frac{8z+1}{z}\,\frac{1}{1-z} = \left(8 + \frac{1}{z} \right)(1 + z + z^2 + z^3 + \cdots).$$

We then multiply the series by $8 + 1/z$ and collect like terms:

$$f(z) = \frac{1}{z} + 9 + 9z + 9z^2 + \cdots.$$

The geometric series converges for $|z| < 1$. After multiplying by $8 + 1/z$, the resulting Laurent series is valid for $0 < |z| < 1$. $\qquad \square$

In the preceding examples the point at the center of the annular domain of validity for each Laurent series was an isolated singularity of the function f. A reexamination of Theorem 19.10 shows that this need not be the case.

Example 5 A Laurent Expansion

Expand $f(z) = \dfrac{1}{z(z-1)}$ in a Laurent series valid for $1 < |z-2| < 2$.

SOLUTION The specified annular domain is shown in Figure 19.9. The center of this domain, $z = 2$, is a point of analyticity of the function f. Our goal now is to find two series involving integer powers of $z-2$: one converging for $1 < |z-2|$ and the other converging for $|z-2| < 2$. To accomplish this we start with the decomposition of f into partial fractions:

$$f(z) = -\frac{1}{z} + \frac{1}{z-1} = f_1(z) + f_2(z). \qquad (13)$$

Now, $\quad f_1(z) = -\dfrac{1}{z} = -\dfrac{1}{2+z-2}$

$$= -\frac{1}{2}\,\frac{1}{1 + \dfrac{z-2}{2}}$$

$$= -\frac{1}{2}\left[1 - \frac{z-2}{2} + \frac{(z-2)^2}{2^2} - \frac{(z-2)^3}{2^3} + \cdots \right]$$

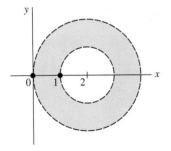

Figure 19.9

$$= -\frac{1}{2} + \frac{z-2}{2^2} - \frac{(z-2)^2}{2^3} + \frac{(z-2)^3}{2^4} - \cdots.$$

This series converges for $|(z-2)/2| < 1$ or $|z-2| < 2$. Furthermore,

$$f_2(z) = \frac{1}{z-1} = \frac{1}{1+z-2} = \frac{1}{z-2}\,\frac{1}{1+\dfrac{1}{z-2}}$$

$$= \frac{1}{z-2}\left[1 - \frac{1}{z-2} + \frac{1}{(z-2)^2} - \frac{1}{(z-2)^3} + \cdots\right]$$

$$= \frac{1}{z-2} - \frac{1}{(z-2)^2} + \frac{1}{(z-2)^3} - \frac{1}{(z-2)^4} + \cdots$$

converges for $|1/(z-2)| < 1$ or $1 < |z-2|$. Substituting these two results in (13) then gives

$$f(z) = \cdots - \frac{1}{(z-2)^4} + \frac{1}{(z-2)^3} - \frac{1}{(z-2)^2} + \frac{1}{z-2} - \frac{1}{2} + \frac{z-2}{2^2} - \frac{(z-2)^2}{2^3} + \frac{(z-2)^3}{2^4} - \cdots.$$

This representation is valid for $1 < |z-2| < 2$. ❑

Example 6 A Laurent Expansion

Expand $f(z) = e^{3/z}$ in a Laurent series valid for $0 < |z|$.

SOLUTION From (12) of Section 19.2 we know that for all finite z,

$$e^z = 1 + z + \frac{z^2}{2!} + \frac{z^3}{3!} + \cdots. \tag{14}$$

By replacing z in (14) by $3/z$, $z \neq 0$, we obtain the Laurent series

$$e^{3/z} = 1 + \frac{3}{z} + \frac{3^2}{2!z^2} + \frac{3^3}{3!z^3} + \cdots.$$

This series is valid for $0 < |z|$. ❑

In conclusion, we point out a result that will be of special interest to us in Sections 19.5 and 19.6. Replacing the complex variable s with the usual symbol z, we see that when $k = -1$, (4) for the Laurent series coefficients yields

$$a_{-1} = \frac{1}{2\pi i} \oint_C f(z)\, dz$$

or, more important,

$$\oint_C f(z)\, dz = 2\pi i a_{-1}. \tag{15}$$

In Problems 1–6, expand the given function in a Laurent series valid for the indicated annular domain.

1. $f(z) = \dfrac{\cos z}{z}, \quad 0 < |z|$

2. $f(z) = \dfrac{z - \sin z}{z^5}, \quad 0 < |z|$

3. $f(z) = e^{-1/z^2}, \quad 0 < |z|$

4. $f(z) = \dfrac{1 - e^z}{z^2}, \quad 0 < |z|$

5. $f(z) = \dfrac{e^z}{z-1}, \quad 0 < |z-1|$

6. $f(z) = z \cos \dfrac{1}{z}, \quad 0 < |z|$

In Problems 7–12, expand $f(z) = \dfrac{1}{z(z-3)}$ in a Laurent series valid for the indicated annular domain.

7. $0 < |z| < 3$ **8.** $|z| > 3$

9. $0 < |z-3| < 3$ **10.** $|z-3| > 3$

11. $1 < |z-4| < 4$ **12.** $1 < |z+1| < 4$

In Problems 13–16, expand $f(z) = \dfrac{1}{(z-1)(z-2)}$ in a Laurent series valid for the indicated annular domain.

13. $1 < |z| < 2$ **14.** $|z| > 2$

15. $0 < |z-1| < 1$ **16.** $0 < |z-2| < 1$

In Problems 17–20, expand $f(z) = \dfrac{z}{(z+1)(z-2)}$ in a Laurent series valid for the indicated annular domain.

17. $0 < |z+1| < 3$ **18.** $|z+1| > 3$

19. $1 < |z| < 2$ **20.** $0 < |z-2| < 3$

In Problems 21 and 22, expand $f(z) = \dfrac{1}{z(1-z)^2}$ in a Laurent series valid for the indicated annular domain.

21. $0 < |z| < 1$ **22.** $|z| > 1$

In Problems 23 and 24, expand $f(z) = \dfrac{1}{(z-2)(z-1)^3}$ in a Laurent series valid for the indicated annular domain.

23. $0 < |z-2| < 1$ **24.** $0 < |z-1| < 1$

In Problems 25 and 26, expand $f(z) = \dfrac{7z-3}{z(z-1)}$ in a Laurent series valid for the indicated annular domain.

25. $0 < |z| < 1$ **26.** $0 < |z-1| < 1$

In Problems 27 and 28, expand $f(z) = \dfrac{z^2-2z+2}{z-2}$ in a Laurent series valid for the indicated annular domain.

27. $1 < |z-1|$ **28.** $0 < |z-2|$

19.4 Zeros and Poles

■ **Principal Part** Suppose that $z = z_0$ is an isolated singularity of a function f and that

$$f(z) = \sum_{k=-\infty}^{\infty} a_k(z-z_0)^k = \sum_{k=1}^{\infty} \frac{a_{-k}}{(z-z_0)^k} + \sum_{k=0}^{\infty} a_k(z-z_0)^k \qquad (1)$$

is the Laurent series representation of f valid for $0 < |z - z_0| < R$. The part of the series involving negative powers of $z - z_0$, that is, $\sum_{k=1}^{\infty} a_{-k}/(z-z_0)^k$, is called the **principal part** of f at $z = z_0$. Isolated singular points are further classified according to whether the principal part contains zero, a finite number, or an infinite number of terms.

(*i*) If the principal part is zero—that is, all the coefficients a_{-k} are zero—then $z = z_0$ is called a **removable singularity.**

(*ii*) If the principal part contains a finite number of nonzero terms, then $z = z_0$ is called a **pole.** If in this case a_{-n} is the last nonzero coefficient in the principal part, then $z = z_0$ is called a **pole of order n.** A pole of order 1 is called a **simple pole.**

(*iii*) If the principal part contains an infinite number of nonzero terms, then $z = z_0$ is called an **essential singularity.**

The following table summarizes the form of the Laurent series for a function f when $z = z_0$ is one of the above singularities.

$z = z_0$	**Laurent Series**
Removable singularity	$a_0 + a_1(z - z_0) + a_2(z - z_0)^2 + \cdots$
Pole of order n	$\dfrac{a_{-n}}{(z - z_0)^n} + \dfrac{a_{-(n-1)}}{(z - z_0)^{n-1}} + \cdots + \dfrac{a_{-1}}{z - z_0} + a_0 + a_1(z - z_0) + \cdots$
Simple pole	$\dfrac{a_{-1}}{z - z_0} + a_0 + a_1(z - z_0) + a_2(z - z_0)^2 + \cdots$
Essential singularity	$\cdots + \dfrac{a_{-2}}{(z - z_0)^2} + \dfrac{a_{-1}}{z - z_0} + a_0 + a_1(z - z_0) + a_2(z - z_0)^2 + \cdots$

Example 1 Removable Discontinuity

Proceeding as we did in (2) of Section 19.3, we see from

$$\frac{\sin z}{z} = 1 - \frac{z^2}{3!} + \frac{z^4}{5!} - \cdots \qquad (2)$$

that $z = 0$ is a removable singularity of the function $f(z) = (\sin z)/z$. ❏

If a function f has a removable singularity at the point $z = z_0$, then we can always supply an appropriate definition for the value of $f(z_0)$ so that f becomes analytic at the point. For instance, since the right side of (2) is 1 at $z = 0$, it makes sense to define $f(0) = 1$. With this definition, the function $f(z) = (\sin z)/z$ in Example 1 is now analytic at $z = 0$.

Example 2 Poles and Essential Singularity

(a) From

$$\frac{\sin z}{z^2} = \underset{\underset{\text{principal part}}{\uparrow}}{\frac{1}{z}} - \frac{z}{3!} + \frac{z^3}{5!} - \cdots,$$

$0 < |z|$, we see that $a_{-1} \neq 0$, and so $z = 0$ is a simple pole of the function $f(z) = (\sin z)/z^2$. The function $f(z) = (\sin z)/z^3$ represented by the series in (2) of Section 19.3 has a pole of order 2 at $z = 0$.

(b) In Example 3 of Section 19.3 we showed that the Laurent expansion of $f(z) = 1/(z - 1)^2(z - 3)$ valid for $0 < |z - 1| < 2$ was

$$f(z) = \overbrace{-\frac{1}{2(z - 1)^2} - \frac{1}{4(z - 1)}}^{\text{principal part}} - \frac{1}{8} - \frac{z - 1}{16} - \cdots.$$

Since $a_{-2} \neq 0$, we conclude that $z = 1$ is pole of order 2.

(c) From Example 6 of Section 19.3 we see from the Laurent series that the principal part of the function $f(z) = e^{3/z}$ contains an infinite number of terms. Thus $z = 0$ is an essential singularity. ❏

In part (b) of Example 2 in Section 19.3 we showed that the Laurent series representation of $f(z) = 1/z(z - 1)$ valid for $1 < |z|$ is

$$f(z) = \frac{1}{z^2} + \frac{1}{z^3} + \frac{1}{z^4} + \frac{1}{z^5} + \cdots.$$

The point $z = 0$ is an isolated singularity of f and the Laurent series contains an infinite number of terms involving negative integer powers of z. Does this mean that $z = 0$ is an essential singularity of f? The answer is no. For this particular function, a reexamination of (1) shows that the Laurent series we are interested in is the one with the annular domain $0 < |z| < 1$. From part (a) of that same example we saw that

$$f(z) = -\frac{1}{z} - 1 - z - z^2 - \cdots$$

was valid for $0 < |z| < 1$. Thus we see that $z = 0$ is a simple pole.

■ **Zeros** Recall that z_0 is a **zero** of a function f if $f(z_0) = 0$. An analytic function f has a **zero of order** n at $z = z_0$ if

$$f(z_0) = 0, \quad f'(z_0) = 0, \quad f''(z_0) = 0, \quad \ldots, \quad f^{(n-1)}(z_0) = 0, \quad \text{but} \quad f^{(n)}(z_0) \neq 0. \quad (3)$$

For example, for $f(z) = (z - 5)^3$ we see that $f(5) = 0, f'(5) = 0, f''(5) = 0$, but $f'''(5) = 6$. Thus $z = 5$ is a zero of order 3. If an analytic function f has a zero of order n at $z = z_0$, it follows from (3) that the Taylor series expansion of f centered at z_0 must have the form

$$f(z) = a_n(z - z_0)^n + a_{n+1}(z - z_0)^{n+1} + a_{n+2}(z - z_0)^{n+2} + \cdots$$
$$= (z - z_0)^n[a_n + a_{n+1}(z - z_0) + a_{n+2}(z - z_0)^2 + \cdots],$$

where $a_n \neq 0$.

Example 3 Order of a Zero

The analytic function $f(z) = z \sin z^2$ has a zero at $z = 0$. By replacing z by z^2 in (13) of Section 19.2 we obtain

$$\sin z^2 = z^2 - \frac{z^6}{3!} + \frac{z^{10}}{5!} - \cdots$$

and so

$$f(z) = z \sin z^2 = z^3 \left[1 - \frac{z^4}{3!} + \frac{z^8}{5!} - \cdots \right].$$

Hence, $z = 0$ is a zero of order 3. ❏

A zero z_0 of a nontrivial analytic function f is *isolated* in the sense that there exists some neighborhood of z_0 for which $f(z) \neq 0$ at every point z in that neighborhood except at $z = z_0$. As a consequence, if z_0 is a zero of a nontrivial analytic function f, then the function $1/f(z)$ has an isolated singularity at the point $z = z_0$. The following result enables us, in some circumstances, to determine the poles of a function by inspection.

THEOREM 19.11 Pole of Order n

If the functions f and g are analytic at $z = z_0$ and f has a zero of order n at $z = z_0$ and $g(z_0) \neq 0$, then the function $F(z) = g(z)/f(z)$ has a pole of order n at $z = z_0$.

Example 4 Order of Poles

(a) Inspection of the rational function

$$F(z) = \frac{2z+5}{(z-1)(z+5)(z-2)^4}$$

shows that the denominator has zeros of order 1 at $z = 1$ and $z = -5$, and a zero of order 4 at $z = 2$. Since the numerator is not zero at these points, it follows from Theorem 19.11 that f has simple poles at $z = 1$ and $z = -5$, and a pole of order 4 at $z = 2$.

(b) In Example 3 we saw that $z = 0$ is a zero of order 3 of $f(z) = z \sin z^2$. From Theorem 19.11 we conclude that the function $F(z) = 1/(z \sin z^2)$ has a pole of order 3 at $z = 0$. ❑

From the preceding discussion, it should be intuitively clear that if a function has a pole at $z = z_0$, then $|f(z)| \to \infty$ as $z \to z_0$ from any direction.

Answers to odd-numbered problems begin on page A-75.

In Problems 1 and 2, show that $z = 0$ is a removable singularity of the given function. Supply a definition of $f(0)$ so that f is analytic at $z = 0$.

1. $f(z) = \dfrac{e^{2z} - 1}{z}$ **2.** $f(z) = \dfrac{\sin 4z - 4z}{z^2}$

In Problems 3–8, determine the zeros and their orders for the given function.

3. $f(z) = (z + 2 - i)^2$ **4.** $f(z) = z^4 - 16$

5. $f(z) = z^4 + z^2$ **6.** $f(z) = z + \dfrac{9}{z}$

7. $f(z) = e^{2z} - e^z$ **8.** $f(z) = \sin^2 z$

In Problems 9–12, the indicated number is a zero of the given function. Use a Maclaurin or Taylor series to determine the order of the zero.

9. $f(z) = z(1 - \cos z^2)$; $z = 0$

10. $f(z) = z - \sin z$; $z = 0$

11. $f(z) = 1 - e^{z-1}$; $z = 1$

12. $f(z) = 1 - \pi i + z + e^z$; $z = \pi i$

In Problems 13–22, determine the order of the poles for the given function.

13. $f(z) = \dfrac{3z - 1}{z^2 + 2z + 5}$ **14.** $f(z) = 5 - \dfrac{6}{z^2}$

15. $f(z) = \dfrac{1 + 4i}{(z + 2)(z + i)^4}$ **16.** $f(z) = \dfrac{z - 1}{(z + 1)(z^3 + 1)}$

17. $f(z) = \tan z$ **18.** $f(z) = \dfrac{\cot \pi z}{z^2}$

19. $f(z) = \dfrac{1 - \cosh z}{z^4}$ **20.** $f(z) = \dfrac{e^z}{z^2}$

21. $f(z) = \dfrac{1}{1 - e^z}$ **22.** $f(z) = \dfrac{\sin z}{z^2 - z}$

23. Determine whether $z = 0$ is an isolated or nonisolated singularity of $f(z) = \tan (1/z)$.

24. Show that $z = 0$ is an essential singularity of $f(z) = z^3 \sin(1/z)$.

19.5 Residues and Residue Theorem

■ **Residue** We saw in the last section that if the complex function f has an isolated singularity at the point z_0, then f has a Laurent series representation

$$f(z) = \sum_{k=-\infty}^{\infty} a_k(z - z_0)^k = \cdots + \frac{a_{-2}}{(z - z_0)^2} + \frac{a_{-1}}{z - z_0} + a_0 + a_1(z - z_0) + \cdots,$$

which is valid in some deleted neighborhood $0 < |z - z_0| < R$. The coefficient a_{-1} of $1/(z - z_0)$ is of special interest to us; a_{-1} is called the **residue** of f at z_0 and is denoted by Res$(f(z), z_0)$.

Example 1 Residues

(a) In Example 2 of Section 19.4 we saw that $z = 1$ is a pole of order 2 of the function $f(z) = 1/(z-1)^2(z-3)$. From the Laurent series given in that example we see that the coefficient of $1/(z-1)$ is $a_{-1} = \text{Res}(f(z), 1) = -\frac{1}{4}$.

(b) Example 6 of Section 19.3 showed that $z = 0$ is an essential singularity of $f(z) = e^{3/z}$. From the Laurent series given in that example we see that the coefficient of $1/z$ is $a_{-1} = \text{Res}(f(z), 0) = 3$. ❏

Later on in this section we will see why the coefficient a_{-1} is so important. In the meantime we are going to examine ways of obtaining this complex number when z_0 is a *pole* of a function f without the necessity of expanding f in a Laurent series at z_0. We begin with the residue at a simple pole.

THEOREM 19.12 Residue at a Simple Pole

If f has a simple pole at $z = z_0$, then

$$\text{Res}(f(z), z_0) = \lim_{z \to z_0} (z - z_0)f(z). \tag{1}$$

PROOF Since $z = z_0$ is a simple pole, the Laurent expansion of f about that point has the form

$$f(z) = \frac{a_{-1}}{z - z_0} + a_0 + a_1(z - z_0) + a_2(z - z_0)^2 + \cdots.$$

By multiplying both sides by $z - z_0$ and then taking the limit as $z \to z_0$, we obtain

$$\lim_{z \to z_0} (z - z_0)f(z) = \lim_{z \to z_0} [a_{-1} + a_0(z - z_0) + a_1(z - z_0)^2 + \cdots] = a_{-1} = \text{Res}(f(z), z_0). \quad ❏$$

THEOREM 19.13 Residue at a Pole of Order n

If f has a pole of order n at $z = z_0$, then

$$\text{Res}(f(z), z_0) = \frac{1}{(n-1)!} \lim_{z \to z_0} \frac{d^{n-1}}{dz^{n-1}} (z - z_0)^n f(z). \tag{2}$$

PROOF Since f is assumed to have a pole of order n, its Laurent expansion for $0 < |z - z_0| < R$ must have the form

$$f(z) = \frac{a_{-n}}{(z - z_0)^n} + \cdots + \frac{a_{-2}}{(z - z_0)^2} + \frac{a_{-1}}{z - z_0} + a_0 + a_1(z - z_0) + \cdots.$$

We multiply the last expression by $(z - z_0)^n$:

$$(z - z_0)^n f(z) = a_{-n} + \cdots + a_{-2}(z - z_0)^{n-2} + a_{-1}(z - z_0)^{n-1} + a_0(z - z_0)^n + a_1(z - z_0)^{n+1} + \cdots$$

and then differentiate $n - 1$ times:

$$\frac{d^{n-1}}{dz^{n-1}} (z - z_0)^n f(z) = (n-1)!a_{-1} + n!a_0(z - z_0) + \cdots. \tag{3}$$

Since all the terms on the right side after the first involve positive integer powers of $z - z_0$, the limit of (3) as $z \to z_0$ is

$$\lim_{z \to z_0} \frac{d^{n-1}}{dz^{n-1}} (z - z_0)^n f(z) = (n - 1)! a_{-1}.$$

Solving the last equation for a_{-1} gives (2). ❏

Note that (2) reduces to (1) when $n = 1$.

Example 2 Residue at a Pole

The function $f(z) = \dfrac{1}{(z - 1)^2(z - 3)}$ has a simple pole at $z = 3$ and a pole of order 2 at $z = 1$. Use Theorems 19.12 and 19.13 to find the residues.

SOLUTION Since $z = 3$ is a simple pole, we use (1):

$$\text{Res}(f(z),3) = \lim_{z \to 3}(z - 3)f(z) = \lim_{z \to 3} \frac{1}{(z - 1)^2} = \frac{1}{4}.$$

Now at the pole of order 2 it follows from (2) that

$$\text{Res}(f(z), 1) = \frac{1}{1!} \lim_{z \to 1} \frac{d}{dz} (z - 1)^2 f(z)$$

$$= \lim_{z \to 1} \frac{d}{dz} \frac{1}{z - 3}$$

$$= \lim_{z \to 1} \frac{-1}{(z - 3)^2} = -\frac{1}{4}.$$ ❏

An alternative method for computing a residue at a simple pole. ☞

When f is not a rational function, calculating residues by means of (1) can sometimes be tedious. It is possible to devise alternative residue formulas. In particular, suppose a function f can be written as a quotient $f(z) = g(z)/h(z)$, where g and h are analytic at $z = z_0$. If $g(z_0) \neq 0$ and if the function h has a zero of order 1 at z_0, then f has a simple pole at $z = z_0$ and

$$\text{Res}(f(z), z_0) = \frac{g(z_0)}{h'(z_0)}. \tag{4}$$

To see this last result, we use (1) and the facts that $h(z_0) = 0$ and $\lim_{z \to z_0}(h(z) - h(z_0))/(z - z_0)$ is a definition of the derivative $h'(z_0)$:

$$\text{Res}(f(z), z_0) = \lim_{z \to z_0} (z - z_0) \frac{g(z)}{h(z)} = \lim_{z \to z_0} \frac{g(z)}{\dfrac{h(z) - h(z_0)}{z - z_0}} = \frac{g(z_0)}{h'(z_0)}.$$

Analogous formulas for residues at poles of order greater than 1 are complicated and will not be given.

Example 3 Using (4) to Compute a Residue

The polynomial $z^4 + 1$ can be factored as $(z - z_1)(z - z_2)(z - z_3)(z - z_4)$, where z_1, z_2, z_3, and z_4 are the four distinct roots of the equation $z^4 + 1 = 0$. It follows from Theorem 19.11 that the function

$$f(z) = \frac{1}{z^4 + 1}$$

has four simple poles. Now from (10) of Section 17.2 we have $z_1 = e^{\pi i/4}$, $z_2 = e^{3\pi i/4}$, $z_3 = e^{5\pi i/4}$, $z_4 = e^{7\pi i/4}$. To compute the residues we use (4) and Euler's formula:

$$\text{Res}(f(z), z_1) = \frac{1}{4z_1^3} = \frac{1}{4} e^{-3\pi i/4} = -\frac{1}{4\sqrt{2}} - \frac{1}{4\sqrt{2}} i$$

$$\text{Res}(f(z), z_2) = \frac{1}{4z_2^3} = \frac{1}{4} e^{-9\pi i/4} = \frac{1}{4\sqrt{2}} - \frac{1}{4\sqrt{2}} i$$

$$\text{Res}(f(z), z_3) = \frac{1}{4z_3^3} = \frac{1}{4} e^{-15\pi i/4} = \frac{1}{4\sqrt{2}} + \frac{1}{4\sqrt{2}} i$$

$$\text{Res}(f(z), z_4) = \frac{1}{4z_4^3} = \frac{1}{4} e^{-21\pi i/4} = -\frac{1}{4\sqrt{2}} + \frac{1}{4\sqrt{2}} i.$$ ❑

■ **Residue Theorem** We come now to the reason for the importance of the residue concept. The next theorem states that under some circumstances, we can evaluate complex integrals $\oint_C f(z)\, dz$ by summing the residues at the isolated singularities of f within the closed contour C.

THEOREM 19.14 **Cauchy's Residue Theorem**

Let D be a simply connected domain and C a simple closed contour lying entirely within D. If a function f is analytic on and within C, except at a finite number of singular points z_1, z_2, \ldots, z_n within C, then

$$\oint_C f(z)\, dz = 2\pi i \sum_{k=1}^{n} \text{Res}(f(z), z_k). \qquad (5)$$

Figure 19.10

PROOF Suppose C_1, C_2, \ldots, C_n are circles centered at z_1, z_2, \ldots, z_n, respectively. Suppose further that each circle C_k has a radius r_k small enough so that C_1, C_2, \ldots, C_n are mutually disjoint and are interior to the simple closed curve C. See Figure 19.10. Recall that (15) of Section 19.3 implies $\oint_{C_k} f(z)\, dz = 2\pi i\, \text{Res}(f(z), z_k)$, and so Theorem 18.5 gives

$$\oint_C f(z)\, dz = \sum_{k=1}^{n} \oint_{C_k} f(z)\, dz = 2\pi i \sum_{k=1}^{n} \text{Res}(f(z), z_k). \qquad ❑$$

Example 4 **Evaluation by the Residue Theorem**

Evaluate $\oint_C \dfrac{1}{(z-1)^2(z-3)}\, dz$, where

(a) the contour C is the rectangle defined by $x = 0$, $x = 4$, $y = -1$, $y = 1$, and
(b) the contour C is the circle $|z| = 2$.

SOLUTION (a) Since both poles $z = 1$ and $z = 3$ lie within the square, we have from (5) that

$$\oint_C \frac{1}{(z-1)^2(z-3)}\, dz = 2\pi i[\text{Res}(f(z), 1) + \text{Res}(f(z), 3)].$$

We found these residues in Examples 2 and 3, and so

$$\oint_c \frac{1}{(z-1)^2(z-3)}\, dz = 2\pi i \left[-\frac{1}{4} + \frac{1}{4} \right] = 0.$$

(b) Since only the pole $z = 1$ lies within the circle $|z| = 2$, we have from (5) that

$$\oint_c \frac{1}{(z-1)^2(z-3)}\, dz = 2\pi i \operatorname{Res}(f(z),\, 1) = 2\pi i \left(-\frac{1}{4} \right) = -\frac{\pi}{2} i. \qquad \square$$

Example 5 Evaluation by the Residue Theorem

Evaluate $\oint_c \dfrac{2z+6}{z^2+4}\, dz$, where the contour C is the circle $|z - i| = 2$.

SOLUTION By writing $z^2 + 4 = (z - 2i)(z + 2i)$, we see that the integrand has simple poles at $-2i$ and $2i$. Now since only $2i$ lies within the contour C, it follows from (5) that

$$\oint_c \frac{2z+6}{z^2+4}\, dz = 2\pi i \operatorname{Res}(f(z),\, 2i).$$

But

$$\operatorname{Res}(f(z),\, 2i) = \lim_{z \to 2i} (z - 2i)\, \frac{2z+6}{(z-2i)(z+2i)}$$

$$= \frac{6+4i}{4i} = \frac{3+2i}{2i}.$$

Hence,

$$\oint_c \frac{2z+6}{z^2+4}\, dz = 2\pi i \left(\frac{3+2i}{2i} \right) = \pi(3 + 2i). \qquad \square$$

Example 6 Evaluation by the Residue Theorem

Evaluate $\oint_c \dfrac{e^z}{z^4 + 5z^3}\, dz$, where the contour C is the circle $|z| = 2$.

SOLUTION Since $z^4 + 5z^3 = z^3(z+5)$ we see that the integrand has a pole of order 3 at $z = 0$ and a simple pole at $z = -5$. Since only $z = 0$ lies within the given contour, we have from (5) and (2)

$$\oint_c \frac{e^z}{z^4 + 5z^3}\, dz = 2\pi i \operatorname{Res}(f(z),\, 0)$$

$$= 2\pi i\, \frac{1}{2!} \lim_{z \to 0} \frac{d^2}{dz^2}\, z^3\, \frac{e^z}{z^3(z+5)}$$

$$= \pi i \lim_{z \to 0} \frac{(z^2 + 8z + 17)e^z}{(z+5)^3} = \frac{17\pi}{125} i. \qquad \square$$

Example 7 Evaluation by the Residue Theorem

Evaluate $\oint_c \tan z\, dz$, where the contour C is the circle $|z| = 2$.

SOLUTION The integrand $\tan z = \sin z / \cos z$ has simple poles at the points where $\cos z = 0$. We saw in Section 17.7 that the only zeros for $\cos z$ are the real numbers $z = (2n + 1)\pi/2$, $n = 0,\, \pm1,\, \pm2,\, \ldots$. Since only $-\pi/2$ and $\pi/2$ are within the circle $|z| = 2$, we have

$$\oint_c \tan z\, dz = 2\pi i \left[\operatorname{Res}\left(f(z),\, -\frac{\pi}{2}\right) + \operatorname{Res}\left(f(z),\, \frac{\pi}{2}\right) \right].$$

Now from (4) with $g(z) = \sin z$, $h(z) = \cos z$, and $h'(z) = -\sin z$, we see that

$$\text{Res}\left(f(z), -\frac{\pi}{2}\right) = \frac{\sin(-\pi/2)}{-\sin(-\pi/2)} = -1 \quad \text{and} \quad \text{Res}\left(f(z), \frac{\pi}{2}\right) = \frac{\sin(\pi/2)}{-\sin(\pi/2)} = -1.$$

Therefore,
$$\oint_C \tan z \, dz = 2\pi i[-1-1] = -4\pi i.$$
□

Example 8 Evaluation by the Residue Theorem

Evaluate $\oint_C e^{3/z} \, dz$, where the contour C is the circle $|z| = 1$.

SOLUTION As we have seen, $z = 0$ is an essential singularity of the integrand $f(z) = e^{3/z}$ and so neither formula (1) nor (2) is applicable to find the residue of f at that point. Nevertheless, we saw in Example 1 that the Laurent series of f at $z = 0$ gives $\text{Res}(f(z), 0) = 3$. Hence from (5) we have

$$\oint_C e^{3/z} \, dz = 2\pi i \, \text{Res}(f(z), 0) = 6\pi i.$$
□

R e m a r k s

In the application of the limit formulas (1) and (2) for computing residues, the indeterminate form 0/0 may result. Although we are not going to prove it, it should be pointed out that **L'Hôpital's rule** is valid in complex analysis. If $f(z) = g(z)/h(z)$, where g and h are analytic at $z = z_0$, $g(z_0) = 0$, $h(z_0) = 0$, and $h'(z_0) \neq 0$, then

$$\lim_{z \to z_0} \frac{g(z)}{h(z)} = \frac{g'(z_0)}{h'(z_0)}.$$

E X E R C I S E S 1 9 . 5 Answers to odd-numbered problems begin on page A-75.

In Problems 1–6, use a Laurent series to find the indicated residue.

1. $f(z) = \dfrac{2}{(z-1)(z+4)}$; $\text{Res}(f(z), 1)$

2. $f(z) = \dfrac{1}{z^3(1-z)^3}$; $\text{Res}(f(z), 0)$

3. $f(z) = \dfrac{4z-6}{z(2-z)}$; $\text{Res}(f(z), 0)$

4. $f(z) = (z+3)^2 \sin \dfrac{2}{z+3}$; $\text{Res}(f(z), -3)$

5. $f(z) = e^{-2/z^2}$; $\text{Res}(f(z), 0)$

6. $f(z) = \dfrac{e^{-z}}{(z-2)^2}$; $\text{Res}(f(z), 2)$

In Problems 7–16, use (1), (2), or (4) to find the residue at each pole of the given function.

7. $f(z) = \dfrac{z}{z^2 + 16}$

8. $f(z) = \dfrac{4z+8}{2z-1}$

9. $f(z) = \dfrac{1}{z^4 + z^3 - 2z^2}$

10. $f(z) = \dfrac{1}{(z^2 - 2z + 2)^2}$

11. $f(z) = \dfrac{5z^2 - 4z + 3}{(z+1)(z+2)(z+3)}$

12. $f(z) = \dfrac{2z-1}{(z-1)^4(z+3)}$

13. $f(z) = \dfrac{\cos z}{z^2(z-\pi)^3}$

14. $f(z) = \dfrac{e^z}{e^z - 1}$

15. $f(z) = \sec z$

16. $f(z) = \dfrac{1}{z \sin z}$

In Problems 17–20, use Cauchy's residue theorem, where appropriate, to evaluate the given integral along the indicated contours.

17. $\oint_C \dfrac{1}{(z-1)(z+2)^2} \, dz$

(a) $|z| = \frac{1}{2}$ **(b)** $|z| = \frac{3}{2}$ **(c)** $|z| = 3$

18. $\displaystyle\oint_C \frac{z+1}{z^2(z-2i)}\,dz$

(a) $|z| = 1$　　　**(b)** $|z - 2i| = 1$　　　**(c)** $|z - 2i| = 4$

19. $\displaystyle\oint_C z^3 e^{-1/z^2}\,dz$

(a) $|z| = 5$　　　**(b)** $|z + i| = 2$　　　**(c)** $|z - 3| = 1$

20. $\displaystyle\oint_C \frac{1}{z \sin z}\,dz$

(a) $|z - 2i| = 1$　　　**(b)** $|z - 2i| = 3$　　　**(c)** $|z| = 5$

In Problems 21–32, use Cauchy's residue theorem to evaluate the given integral along the indicated contour.

21. $\displaystyle\oint_C \frac{1}{z^2 + 4z + 13}\,dz, \quad C: |z - 3i| = 3$

22. $\displaystyle\oint_C \frac{1}{z^3(z-1)^4}\,dz, \quad C: |z - 2| = \tfrac{3}{2}$

23. $\displaystyle\oint_C \frac{z}{z^4 - 1}\,dz, \quad C: |z| = 2$

24. $\displaystyle\oint_C \frac{z}{(z+1)(z^2+1)}\,dz, \quad C$ is the ellipse $16x^2 + y^2 = 4$

25. $\displaystyle\oint_C \frac{ze^z}{z^2 - 1}\,dz, \quad C: |z| = 2$

26. $\displaystyle\oint_C \frac{e^z}{z^3 + 2z^2}\,dz, \quad C: |z| = 3$

27. $\displaystyle\oint_C \frac{\tan z}{z}\,dz, \quad C: |z - 1| = 2$

28. $\displaystyle\oint_C \frac{\cot \pi z}{z^2}\,dz, \quad C: |z| = \tfrac{1}{2}$

29. $\displaystyle\oint_C \cot \pi z \,dz, \quad C$ is the rectangle defined by $x = \tfrac{1}{2}, x = \pi,$ $y = -1, y = 1$

30. $\displaystyle\oint_C \frac{2z - 1}{z^2(z^3 + 1)}\,dz, \quad C$ is the rectangle defined by $x = -2, x = 1,$ $y = -\tfrac{1}{2}, y = 1$

31. $\displaystyle\oint_C \frac{e^{iz} + \sin z}{(z - \pi)^4}\,dz, \quad C: |z - 3| = 1$

32. $\displaystyle\oint_C \frac{\cos z}{(z - 1)^2(z^2 + 9)}\,dz, \quad C: |z - 1| = 1$

19.6 Evaluation of Real Integrals

Residue theory is useful in evaluating real integrals of the forms

$$\int_0^{2\pi} F(\cos\theta, \sin\theta)\,d\theta \tag{1}$$

$$\int_{-\infty}^{\infty} f(x)\,dx \tag{2}$$

$$\int_{-\infty}^{\infty} f(x)\cos x\,dx \quad \text{and} \quad \int_{-\infty}^{\infty} f(x)\sin x\,dx, \tag{3}$$

where F in (1) is a rational function of $\cos\theta$ and $\sin\theta$, and in (2) and (3) $f(x) = P(x)/Q(x)$ is a rational function such that the polynomials P and Q have no common factors.

■ **Integrals of the Form** $\int_0^{2\pi} F(\cos\theta, \sin\theta)\,d\theta$　　The basic idea here is to convert an integral of form (1) into a complex integral where the contour C is the unit circle centered at the origin. As we saw on page 853, this contour can be parameterized by $z = \cos\theta + i\sin\theta = e^{i\theta}, 0 \le \theta \le 2\pi$. Using

$$dz = ie^{i\theta}\,d\theta, \qquad \cos\theta = \frac{e^{i\theta} + e^{-i\theta}}{2}, \qquad \sin\theta = \frac{e^{i\theta} - e^{-i\theta}}{2i},$$

we replace, in turn, $d\theta$, $\cos\theta$, and $\sin\theta$ by

$$d\theta = \frac{dz}{iz}, \qquad \cos\theta = \frac{1}{2}(z + z^{-1}), \qquad \sin\theta = \frac{1}{2i}(z - z^{-1}). \tag{4}$$

The integral in (1) then becomes

$$\oint_C F\left(\frac{1}{2}(z+z^{-1}), \frac{1}{2i}(z-z^{-1})\right) \frac{dz}{iz},$$

where C is $|z| = 1$.

Example 1 A Real Trigonometric Integral

Evaluate $\displaystyle\int_0^{2\pi} \frac{1}{(2+\cos\theta)^2}\, d\theta.$

SOLUTION Using the substitutions in (4) and simplifying yield the contour integral

$$\frac{4}{i}\oint_C \frac{z}{(z^2+4z+1)^2}\, dz.$$

With the aid of the quadratic formula we can write

$$f(z) = \frac{z}{(z^2+4z+1)^2} = \frac{z}{(z-z_0)^2(z-z_1)^2},$$

where $z_0 = -2 - \sqrt{3}$ and $z_1 = -2 + \sqrt{3}$. Since only z_1 is inside the unit circle C, we have

$$\oint_C \frac{z}{(z^2+4z+1)^2}\, dz = 2\pi i\, \text{Res}(f(z), z_1).$$

Now z_1 is a pole of order 2 and so from (2) of Section 19.5,

$$\text{Res}(f(z), z_1) = \lim_{z\to z_1} \frac{d}{dz}(z-z_1)^2 f(z) = \lim_{z\to z_1} \frac{d}{dz}\frac{z}{(z-z_0)^2}$$

$$= \lim_{z\to z_1} -\frac{z+z_0}{(z-z_0)^3} = \frac{1}{6\sqrt{3}}.$$

Hence,

$$\frac{4}{i}\oint_C \frac{z}{(z^2+4z+1)^2}\, dz = \frac{4}{i}\cdot 2\pi i\, \text{Res}(f(z), z_1) = \frac{4}{i}\cdot 2\pi i \cdot \frac{1}{6\sqrt{3}}$$

and finally

$$\int_0^{2\pi} \frac{1}{(2+\cos\theta)^2}\, d\theta = \frac{4\pi}{3\sqrt{3}}. \qquad \square$$

■ Integrals of the Form $\int_{-\infty}^{\infty} f(x)\, dx$
When f is continuous on $(-\infty, \infty)$, recall from calculus that the improper integral $\int_{-\infty}^{\infty} f(x)\, dx$ is defined in terms of two distinct limits:

$$\int_{-\infty}^{\infty} f(x)\, dx = \lim_{r\to\infty}\int_{-r}^0 f(x)\, dx + \lim_{R\to\infty}\int_0^R f(x)\, dx. \tag{5}$$

If both limits in (5) exist, the integral is said to be **convergent;** if one or both of the limits fail to exist, the integral is **divergent.** In the event that we know (a priori) that an integral $\int_{-\infty}^{\infty} f(x)\, dx$ converges, we can evaluate it by means of a single limiting process:

$$\int_{-\infty}^{\infty} f(x)\, dx = \lim_{R\to\infty}\int_{-R}^R f(x)\, dx. \tag{6}$$

It is important to note that the symmetric limit in (6) may exist even though the improper integral is divergent. For example, the integral $\int_{-\infty}^{\infty} x \, dx$ is divergent since $\lim_{R \to \infty} \int_{0}^{R} x \, dx = \lim_{R \to \infty} \frac{1}{2} R^2 = \infty$. However, using (6), we obtain

$$\lim_{R \to \infty} \int_{-R}^{R} x \, dx = \lim_{R \to \infty} \left[\frac{R^2}{2} - \frac{(-R)^2}{2} \right] = 0. \tag{7}$$

The limit in (6) is called the **Cauchy principal value** of the integral and is written

$$\text{P.V.} \int_{-\infty}^{\infty} f(x) \, dx = \lim_{R \to \infty} \int_{-R}^{R} f(x) \, dx.$$

In (7) we have shown that P.V. $\int_{-\infty}^{\infty} x \, dx = 0$. To summarize, when an integral of form (2) converges, its Cauchy principal value is the same as the value of the integral. If the integral diverges, it may still possess a Cauchy principal value.

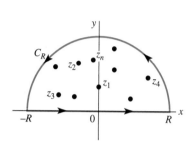

Figure 19.11

To evaluate an integral $\int_{-\infty}^{\infty} f(x) \, dx$, where $f(x) = P(x)/Q(x)$ is continuous on $(-\infty, \infty)$, by residue theory we replace x by the complex variable z and integrate the complex function f over a closed contour C that consists of the interval $[-R, R]$ on the real axis and a semicircle C_R of radius large enough to enclose all the poles of $f(z) = P(z)/Q(z)$ in the upper half-plane $\text{Re}(z) > 0$. See Figure 19.11. By Theorem 19.14 we have

$$\oint_C f(z) \, dz = \int_{C_R} f(z) \, dz + \int_{-R}^{R} f(x) \, dx = 2\pi i \sum_{k=1}^{n} \text{Res}(f(z), z_k),$$

where z_k, $k = 1, 2, \ldots, n$, denotes poles in the upper half-plane. If we can show that the integral $\int_{C_R} f(z) \, dz \to 0$ as $R \to \infty$, then we have

$$\text{P.V.} \int_{-\infty}^{\infty} f(x) \, dx = \lim_{R \to \infty} \int_{-R}^{R} f(x) \, dx = 2\pi i \sum_{k=1}^{n} \text{Res}(f(z), z_k). \tag{8}$$

Example 2 **Cauchy P.V. of an Improper Integral**

Evaluate the Cauchy principal value of

$$\int_{-\infty}^{\infty} \frac{1}{(x^2 + 1)(x^2 + 9)} \, dx.$$

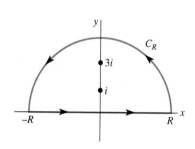

Figure 19.12

SOLUTION Let $f(z) = 1/(z^2 + 1)(z^2 + 9)$. Since

$$(z^2 + 1)(z^2 + 9) = (z - i)(z + i)(z - 3i)(z + 3i),$$

we let C be the closed contour consisting of the interval $[-R, R]$ on the x-axis and the semicircle C_R of radius $R > 3$. As seen from Figure 19.12,

$$\oint_C \frac{1}{(z^2 + 1)(z^2 + 9)} \, dz = \int_{-R}^{R} \frac{1}{(x^2 + 1)(x^2 + 9)} \, dx + \int_{C_R} \frac{1}{(z^2 + 1)(z^2 + 9)} \, dz$$

$$= I_1 + I_2$$

and

$$I_1 + I_2 = 2\pi i [\text{Res}(f(z), i) + \text{Res}(f(z), 3i)].$$

At the simple poles $z = i$ and $z = 3i$ we find, respectively,

$$\text{Res}(f(z), i) = \frac{1}{16i} \quad \text{and} \quad \text{Res}(f(z), 3i) = -\frac{1}{48i},$$

so that

$$I_1 + I_2 = 2\pi i \left[\frac{1}{16i} - \frac{1}{48i} \right] = \frac{\pi}{12}. \tag{9}$$

We now want to let $R \to \infty$ in (9). Before doing this, we note that on C_R,

$$\left|(z^2 + 1)(z^2 + 9)\right| = \left|z^2 + 1\right|\left|z^2 + 9\right| \geq \left|\,|z|^2 - 1\,\right|\left|\,|z|^2 - 9\,\right| = (R^2 - 1)(R^2 - 9),$$

and so from the *ML*-inequality of Section 18.1 we can write

$$|I_2| = \left|\int_{C_R} \frac{1}{(z^2 + 1)(z^2 + 9)}\, dz\right| \leq \frac{\pi R}{(R^2 - 1)(R^2 - 9)}.$$

This last result shows that $|I_2| \to 0$ as $R \to \infty$, and so we conclude that $\lim_{R \to \infty} I_2 = 0$. It follows from (9) that $\lim_{R \to \infty} I_1 = \pi/12$; in other words,

$$\lim_{R \to \infty} \int_{-R}^{R} \frac{1}{(x^2 + 1)(x^2 + 9)}\, dx = \frac{\pi}{12} \quad \text{or} \quad \text{P.V.} \int_{-\infty}^{\infty} \frac{1}{(x^2 + 1)(x^2 + 9)}\, dx = \frac{\pi}{12}. \quad \square$$

It is often tedious to have to show that the contour integral along C_R approaches zero as $R \to \infty$. Sufficient conditions under which this is always true are as follows:

THEOREM 19.15 Behavior of Integral as $R \to \infty$

Suppose $f(z) = P(z)/Q(z)$, where the degree of $P(z)$ is n and the degree of $Q(z)$ is $m \geq n + 2$. If C_R is a semicircular contour $z = Re^{i\theta}$, $0 \leq \theta \leq \pi$, then $\int_{C_R} f(z)\, dz \to 0$ as $R \to \infty$.

In other words, the integral along C_R approaches zero as $R \to \infty$ when the denominator of f is of a power at least 2 more than its numerator. The proof of this fact follows in the same manner as in Example 2. Notice in that example that the conditions stipulated in Theorem 19.15 are satisfied, since the degree of $P(z) = 1$ is 0 and the degree of $Q(z) = (z^2 + 1)(z^2 + 9)$ is 4.

Example 3 Cauchy P.V. of an Improper Integral

Evaluate the Cauchy principal value of $\int_{-\infty}^{\infty} \frac{1}{x^4 + 1}\, dx$.

SOLUTION By inspection of the integrand, we see that the conditions given in Theorem 19.15 are satisfied. Moreover, we know from Example 3 of Section 19.5 that f has simple poles in the upper half-plane at $z_1 = e^{\pi i/4}$ and $z_2 = e^{3\pi i/4}$. We also saw in that example that the residues at these poles are

$$\text{Res}(f(z), z_1) = -\frac{1}{4\sqrt{2}} - \frac{1}{4\sqrt{2}}\, i \quad \text{and} \quad \text{Res}(f(z), z_2) = \frac{1}{4\sqrt{2}} - \frac{1}{4\sqrt{2}}\, i.$$

Thus, by (8),

$$\text{P.V.} \int_{-\infty}^{\infty} \frac{1}{x^4 + 1}\, dx = 2\pi i[\text{Res}(f(z), z_1) + \text{Res}(f(z), z_2)] = \frac{\pi}{\sqrt{2}}. \quad \square$$

■ **Integrals of the Forms $\int_{-\infty}^{\infty} f(x) \cos \alpha x\, dx$ or $\int_{-\infty}^{\infty} f(x) \sin \alpha x\, dx$**
We encountered integrals of this type in Section 15.4 in the study of Fourier transforms. Accordingly, $\int_{-\infty}^{\infty} f(x) \cos \alpha x\, dx$ and $\int_{-\infty}^{\infty} f(x) \sin \alpha x\, dx$, $\alpha > 0$, are referred to as **Fourier integrals.** Fourier integrals appear as the real and imagi-

nary parts in the improper integral $\int_{-\infty}^{\infty} f(x)e^{i\alpha x}\,dx$. Using Euler's formula $e^{i\alpha x} = \cos \alpha x + i \sin \alpha x$, we get

$$\int_{-\infty}^{\infty} f(x)e^{i\alpha x}\,dx = \int_{-\infty}^{\infty} f(x) \cos \alpha x\,dx + i \int_{-\infty}^{\infty} f(x) \sin \alpha x\,dx \tag{10}$$

whenever both integrals on the right side converge. When $f(x) = P(x)/Q(x)$ is continuous on $(-\infty, \infty)$ we can evaluate both Fourier integrals at the same time by considering the integral $\int_C f(z)e^{i\alpha z}\,dz$, where $\alpha > 0$ and the contour C again consists of the interval $[-R, R]$ on the real axis and a semicircular contour C_R with radius large enough to enclose the poles of $f(z)$ in the upper half-plane.

Before proceeding we give, without proof, sufficient conditions under which the contour integral along C_R approaches zero as $R \to \infty$:

THEOREM 19.16 Behavior of Integral as $R \to \infty$

Suppose $f(z) = P(z)/Q(z)$, where the degree of $P(z)$ is n and the degree of $Q(z)$ is $m \geq n + 1$. If C_R is a semicircular contour $z = Re^{i\theta}$, $0 \leq \theta \leq \pi$, and $\alpha > 0$, then $\int_{C_R} (P(z)/Q(z))e^{i\alpha z}\,dz \to 0$ as $R \to \infty$.

Example 4 **Using Symmetry**

Evaluate the Cauchy principal value of $\displaystyle\int_0^{\infty} \frac{x \sin x}{x^2 + 9}\,dx$.

SOLUTION First note that the limits of integration are not from $-\infty$ to ∞ as required by the method. This can be rectified by observing that since the integrand is an even function of x, we can write

$$\int_0^{\infty} \frac{x \sin x}{x^2 + 9}\,dx = \frac{1}{2} \int_{-\infty}^{\infty} \frac{x \sin x}{x^2 + 9}\,dx. \tag{11}$$

With $\alpha = 1$ we now form the contour integral

$$\oint_C \frac{z}{z^2 + 9} e^{iz}\,dz,$$

where C is the same contour shown in Figure 19.12. By Theorem 19.14,

$$\int_{C_R} \frac{z}{z^2 + 9} e^{iz}\,dz + \int_{-R}^{R} \frac{x}{x^2 + 9} e^{ix}\,dx = 2\pi i \operatorname{Res}(f(z)e^{iz}, 3i),$$

where $f(z) = z/(z^2 + 9)$. From (4) of Section 19.5,

$$\operatorname{Res}(f(z)e^{iz}, 3i) = \frac{ze^{iz}}{2z}\bigg|_{z=3i} = \frac{e^{-3}}{2}.$$

Hence, in view of Theorem 19.16 we conclude $\int_{C_R} f(z)e^{iz}\,dz \to 0$ as $R \to \infty$ and so

$$\text{P.V.} \int_{-\infty}^{\infty} \frac{x}{x^2 + 9} e^{ix}\,dx = 2\pi i\left(\frac{e^{-3}}{2}\right) = \frac{\pi}{e^3}\,i.$$

But by (10),

$$\int_{-\infty}^{\infty} \frac{x}{x^2 + 9} e^{ix}\,dx = \int_{-\infty}^{\infty} \frac{x \cos x}{x^2 + 9}\,dx + i\int_{-\infty}^{\infty} \frac{x \sin x}{x^2 + 9}\,dx = \frac{\pi}{e^3}\,i.$$

Equating real and imaginary parts in the last line gives the bonus result

$$\text{P.V.} \int_{-\infty}^{\infty} \frac{x \cos x}{x^2 + 9} \, dx = 0 \quad \text{along with} \quad \text{P.V.} \int_{-\infty}^{\infty} \frac{x \sin x}{x^2 + 9} \, dx = \frac{\pi}{e^3}.$$

Finally, in view of (11) we obtain the value of the prescribed integral:

$$\int_{0}^{\infty} \frac{x \sin x}{x^2 + 9} \, dx = \frac{1}{2} \text{ P.V.} \int_{-\infty}^{\infty} \frac{x \sin x}{x^2 + 9} \, dx = \frac{\pi}{2e^3}. \qquad \square$$

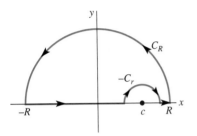

Figure 19.13

■ **Indented Contours** The improper integrals of form (2) and (3) that we have considered up to this point were continuous on the interval $[-\infty, \infty]$. In other words, the complex function $f(z) = P(z)/Q(z)$ did not have poles on the real axis. In the event f has poles on the real axis, we must modify the procedure used in Examples 2–4. For example, to evaluate $\int_{-\infty}^{\infty} f(x) \, dx$ by residues when $f(z)$ has a pole at $z = c$, where c is a real number, we use an **indented contour** as illustrated in Figure 19.13. The symbol C_r denotes a semicircular contour centered at $z = c$ oriented in the *positive* direction. The next theorem is important to this discussion.

THEOREM 19.17 Behavior of Integral as $r \to 0$

Suppose f has a simple pole $z = c$ on the real axis. If C_r is the contour defined by $z = c + re^{i\theta}$, $0 \le \theta \le \pi$, then

$$\lim_{r \to 0} \int_{C_r} f(z) \, dz = \pi i \, \text{Res}(f(z), c).$$

PROOF Since f has a simple pole at $z = c$, its Laurent series is

$$f(z) = \frac{a_{-1}}{z - c} + g(z),$$

where $a_{-1} = \text{Res}(f(z), c)$ and g is analytic at c. Using the Laurent series and the parameterization of C_r, we have

$$\int_{C_r} f(z) \, dz = a_{-1} \int_{0}^{\pi} \frac{ire^{i\theta}}{re^{i\theta}} \, d\theta + ir \int_{0}^{\pi} g(c + re^{i\theta})e^{i\theta} \, d\theta = I_1 + I_2. \tag{12}$$

First, we see that

$$I_1 = a_{-1} \int_{0}^{\pi} \frac{ire^{i\theta}}{re^{i\theta}} \, d\theta = a_{-1} \int_{0}^{\pi} i \, d\theta = \pi i a_{-1} = \pi i \, \text{Res}(f(z), c).$$

Next, g is analytic at c and so it is continuous at this point and bounded in a neighborhood of the point; that is, there exists an $M > 0$ for which $\left| g(c + re^{i\theta}) \right| \le M$. Hence,

$$|I_2| = \left| ir \int_{0}^{\pi} g(c + re^{i\theta})e^{i\theta} \, d\theta \right| \le r \int_{0}^{\pi} M \, d\theta = \pi r M.$$

It follows from this last inequality that $\lim_{r \to 0} |I_2| = 0$ and consequently $\lim_{r \to 0} I_2 = 0$. By taking the limit of (12) as $r \to 0$, we have proved the theorem. \square

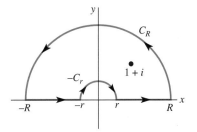

Figure 19.14

Example 5 Using an Indented Contour

Evaluate the Cauchy principal value of $\displaystyle\int_{-\infty}^{\infty} \frac{\sin x}{x(x^2 - 2x + 2)}\, dx.$

SOLUTION Since the integral is of form (3), we consider the contour integral $\oint_C e^{iz}\, dz/z(z^2 - 2z + 2)$. The function $f(z) = 1/z(z^2 - 2z + 2)$ has simple poles at $z = 0$ and at $z = 1 + i$ in the upper half-plane. The contour C shown in Figure 19.14 is indented at the origin. Adopting an obvious notation, we have

$$\oint_C = \int_{C_R} + \int_{-R}^{-r} + \int_{-C_r} + \int_{r}^{R} = 2\pi i\, \mathrm{Res}(f(z)e^{iz}, 1 + i), \tag{13}$$

where $\int_{-C_r} = -\int_{C_r}$. Taking the limits of (13) as $R \to \infty$ and as $r \to 0$, we find from Theorems 19.16 and 19.17 that

$$\mathrm{P.V.} \int_{-\infty}^{\infty} \frac{e^{ix}}{x(x^2 - 2x + 2)}\, dx - \pi i\, \mathrm{Res}(f(z)e^{iz}, 0) = 2\pi i\, \mathrm{Res}(f(z)e^{iz}, 1 + i).$$

Now,

$$\mathrm{Res}(f(z)e^{iz}, 0) = \frac{1}{2} \quad \text{and} \quad \mathrm{Res}(f(z)e^{iz}, 1 + i) = -\frac{e^{-1+i}}{4}(1 + i).$$

Therefore,

$$\mathrm{P.V.} \int_{-\infty}^{\infty} \frac{e^{ix}}{x(x^2 - 2x + 2)}\, dx = \pi i\left(\frac{1}{2}\right) + 2\pi i\left(-\frac{e^{-1+i}}{4}(1 + i)\right).$$

Using $e^{-1+i} = e^{-1}(\cos 1 + i \sin 1)$, simplifying, and then equating real and imaginary parts, we get from the last equality

$$\mathrm{P.V.} \int_{-\infty}^{\infty} \frac{\cos x}{x(x^2 - 2x + 2)}\, dx = \frac{\pi}{2}\, e^{-1}(\sin 1 + \cos 1)$$

and $\quad\displaystyle \mathrm{P.V.} \int_{-\infty}^{\infty} \frac{\sin x}{x(x^2 - 2x + 2)}\, dx = \frac{\pi}{2}\, [1 + e^{-1}(\sin 1 - \cos 1)].$ ❑

E X E R C I S E S 1 9 . 6 Answers to odd-numbered problems begin on page A-75.

In Problems 1–10, evaluate the given trigonometric integral.

1. $\displaystyle\int_0^{2\pi} \frac{1}{1 + 0.5 \sin \theta}\, d\theta$

2. $\displaystyle\int_0^{2\pi} \frac{1}{10 - 6 \cos \theta}\, d\theta$

3. $\displaystyle\int_0^{2\pi} \frac{\cos \theta}{3 + \sin \theta}\, d\theta$

4. $\displaystyle\int_0^{2\pi} \frac{1}{1 + 3 \cos^2 \theta}\, d\theta$

5. $\displaystyle\int_0^{\pi} \frac{1}{2 - \cos \theta}\, d\theta$ [*Hint:* Let $t = 2\pi - \theta$.]

6. $\displaystyle\int_0^{\pi} \frac{1}{1 + \sin^2 \theta}\, d\theta$

7. $\displaystyle\int_0^{2\pi} \frac{\sin^2 \theta}{5 + 4 \cos \theta}\, d\theta$

8. $\displaystyle\int_0^{2\pi} \frac{\cos^2 \theta}{3 - \sin \theta}\, d\theta$

9. $\displaystyle\int_0^{2\pi} \frac{\cos 2\theta}{5 - 4 \cos \theta}\, d\theta$

10. $\displaystyle\int_0^{2\pi} \frac{1}{\cos \theta + 2 \sin \theta + 3}\, d\theta$

In Problems 11–30, evaluate the Cauchy principal value of the given improper integral.

11. $\displaystyle\int_{-\infty}^{\infty} \frac{1}{x^2 - 2x + 2}\, dx$

12. $\displaystyle\int_{-\infty}^{\infty} \frac{1}{x^2 - 6x + 25}\, dx$

13. $\displaystyle\int_{-\infty}^{\infty} \frac{1}{(x^2 + 4)^2}\, dx$

14. $\displaystyle\int_{-\infty}^{\infty} \frac{x^2}{(x^2 + 1)^2}\, dx$

15. $\displaystyle\int_{-\infty}^{\infty} \frac{1}{(x^2 + 1)^3}\, dx$

16. $\displaystyle\int_{-\infty}^{\infty} \frac{x}{(x^2 + 4)^3}\, dx$

17. $\displaystyle\int_{-\infty}^{\infty} \frac{2x^2 - 1}{x^4 + 5x^2 + 4}\, dx$

18. $\displaystyle\int_{-\infty}^{\infty} \frac{dx}{(x^2 + 1)^2(x^2 + 9)}$

19. $\displaystyle\int_0^\infty \frac{x^2+1}{x^4+1}\,dx$

20. $\displaystyle\int_0^\infty \frac{1}{x^6+1}\,dx$

21. $\displaystyle\int_{-\infty}^\infty \frac{\cos x}{x^2+1}\,dx$

22. $\displaystyle\int_{-\infty}^\infty \frac{\cos 2x}{x^2+1}\,dx$

23. $\displaystyle\int_{-\infty}^\infty \frac{x \sin x}{x^2+1}\,dx$

24. $\displaystyle\int_0^\infty \frac{\cos x}{(x^2+4)^2}\,dx$

25. $\displaystyle\int_0^\infty \frac{\cos 3x}{(x^2+1)^2}\,dx$

26. $\displaystyle\int_{-\infty}^\infty \frac{\sin x}{x^2+4x+5}\,dx$

27. $\displaystyle\int_0^\infty \frac{\cos 2x}{x^4+1}\,dx$

28. $\displaystyle\int_0^\infty \frac{x \sin x}{x^4+1}\,dx$

29. $\displaystyle\int_{-\infty}^\infty \frac{\cos x}{(x^2+1)(x^2+9)}\,dx$

30. $\displaystyle\int_0^\infty \frac{x \sin x}{(x^2+1)(x^2+4)}\,dx$

In Problems 31 and 32, use an indented contour and residues to establish the given result.

31. P.V. $\displaystyle\int_{-\infty}^\infty \frac{\sin x}{x}\,dx = \pi$

32. P.V. $\displaystyle\int_{-\infty}^\infty \frac{\sin x}{x(x^2+1)}\,dx = \pi(1-e^{-1})$

33. Establish the general result

$$\int_0^\pi \frac{d\theta}{(a+\cos\theta)^2} = \frac{a\pi}{(\sqrt{a^2-1})^3}, \qquad a>1$$

and use this formula to verify the answer in Example 1.

34. Establish the general result

$$\int_0^{2\pi} \frac{\sin^2\theta}{a+b\cos\theta}\,d\theta = \frac{2\pi}{b^2}(a-\sqrt{a^2-b^2}), \qquad a>b>0$$

and use this formula to verify the answer to Problem 7.

35. Use the contour shown in Figure 19.15 to show that

$$\text{P.V.} \int_{-\infty}^\infty \frac{e^{ax}}{1+e^x}\,dx = \frac{\pi}{\sin a\pi}, \qquad 0<a<1.$$

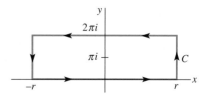

Figure 19.15

36. The steady-state temperature $u(x,y)$ in a semi-infinite plate is determined from

$$\frac{\partial^2 u}{\partial x^2} + \frac{\partial^2 u}{\partial y^2} = 0, \qquad 0<x<\pi, \quad y>0$$

$$u(0,y)=0, \qquad u(\pi,y)=\frac{2y}{y^4+4}, \qquad y>0$$

$$u(x,0)=0, \qquad 0<x<\pi.$$

Use a Fourier transform and the residue method to show that

$$u(x,y)=\int_0^\infty \frac{e^{-\alpha}\sin\alpha\sinh\alpha x}{\sinh\alpha\pi}\sin\alpha y\,d\alpha.$$

CHAPTER 19 REVIEW EXERCISES

Answers to odd-numbered problems begin on page A-75.

Answer Problems 1–12 without referring back to the text. Fill in the blank or answer true/false.

1. A function f is analytic at a point z_0 if f can be expanded in a convergent power series centered at z_0. _____

2. A power series represents a continuous function at every point within and on its circle of convergence. _____

3. For $f(z)=1/(z-3)$, the Laurent series valid for $|z|>3$ is $z^{-1}+3z^{-2}+9z^{-3}+\cdots$. Since there are an infinite number of negative powers of $z=z-0$, $z=0$ is an essential singularity. _____

4. The only possible singularities of a rational function are poles. _____

5. The function $f(z)=e^{1/(z-1)}$ has an essential singularity at $z=1$. _____

6. The function $f(z)=z/(e^z-1)$ has a removable singularity at $z=0$. _____

7. The function $f(z)=z(e^z-1)$ possesses a zero of order 2 at $z=0$. _____

8. The function $f(z)=(z+5)/(z^3\sin^2 z)$ has a pole of order _____ at $z=0$.

9. If $f(z)=\cot\pi z$, then $\text{Res}(f(z),0)=$ _____.

10. The Laurent series of f valid for $0<|z-1|$ is given by

$$(z-1)^{-3}-\frac{(z-1)^{-1}}{3!}+\frac{(z-1)}{5!}-\frac{(z-1)^3}{7!}+\cdots.$$

From this series we see that f has a pole of order _____ at $z=1$ and $\text{Res}(f(z),1)=$ _____.

11. The circle of convergence of the power series $\displaystyle\sum_{k=1}^\infty \frac{(z-i)^k}{(2+i)^{k+1}}$ is _____.

12. The power series $\sum_{k=1}^{\infty} \dfrac{z^k}{2^{k+1}}$ converges at $z = 2i$. ____

13. Find a Maclaurin expansion of $f(z) = e^z \cos z$. [*Hint:* Use the identity $\cos z = (e^{iz} + e^{-iz})/2$.]

14. Show that the function $f(z) = 1/\sin(\pi/z)$ has an infinite number of singular points. Are any of these isolated singular points?

In Problems 15–18, use known results as an aid in expanding the given function in a Laurent series valid for the indicated annular region.

15. $f(z) = \dfrac{1 - e^{iz}}{z^4}, \quad 0 < |z|$

16. $f(z) = e^{z/(z-2)}, \quad 0 < |z - 2|$

17. $f(z) = (z - i)^2 \sin \dfrac{1}{z - i}, \quad 0 < |z - i|$

18. $f(z) = \dfrac{1 - \cos z^2}{z^5}, \quad 0 < |z|$

19. Expand $f(z) = \dfrac{2}{z^2 - 4z + 3}$ in an appropriate series valid for

(a) $|z| < 1$ **(b)** $1 < |z| < 3$ **(c)** $|z| > 3$

(d) $0 < |z - 1| < 2$.

20. Expand $f(z) = \dfrac{1}{(z - 5)^2}$ in an appropriate series valid for

(a) $|z| < 5$ **(b)** $|z| > 5$ **(c)** $0 < |z - 5|$.

In Problems 21–30, use Cauchy's residue theorem to evaluate the given integral along the indicated contour.

21. $\displaystyle\oint_C \dfrac{2z + 5}{z(z + 2)(z - 1)^4} \, dz, \quad C: |z + 2| = \tfrac{1}{2}$

22. $\displaystyle\oint_C \dfrac{z^2}{(z - 1)^3(z^2 + 4)} \, dz, \quad C$ is the ellipse $x^2/4 + y^2 = 1$

23. $\displaystyle\oint_C \dfrac{1}{2 \sin z - 1} \, dz, \quad C: |z - \tfrac{1}{2}| = \tfrac{1}{3}$

24. $\displaystyle\oint_C \dfrac{z + 1}{\sinh z} \, dz, \quad C$ is the rectangle defined by $x = -1, x = 1,$ $y = 4, \quad y = -1$

25. $\displaystyle\oint_C \dfrac{e^{2z}}{z^4 + 2z^3 + 2z^2} \, dz, \quad C: |z| = 4$

26. $\displaystyle\oint_C \dfrac{1}{z^4 - 2z^2 + 4} \, dz, \quad C$ is the square defined by $x = -2,$ $x = 2, \quad y = 0, \quad y = 1$

27. $\displaystyle\oint_C \dfrac{1}{z(e^z - 1)} \, dz, \quad C: |z| = 1$ [*Hint:* Use the Maclaurin series for $z(e^z - 1)$.]

28. $\displaystyle\oint_C \dfrac{z}{(z + 1)(z - 1)^{10}} \, dz, \quad C: |z - 1| = 3$

29. $\displaystyle\oint_C \left[ze^{3/z} + \dfrac{\sin z}{z^2(z - \pi)^3} \right] dz, \quad C: |z| = 6$

30. $\displaystyle\oint_C \csc \pi z \, dz, \quad C$ is the rectangle defined by $x = -\tfrac{1}{2}, x = \tfrac{5}{2},$ $y = -1, y = 1$

In Problems 31 and 32, evaluate the Cauchy principal value of the given improper integral.

31. $\displaystyle\int_{-\infty}^{\infty} \dfrac{x^2}{(x^2 + 2x + 2)(x^2 + 1)^2} \, dx$

32. $\displaystyle\int_{-\infty}^{\infty} \dfrac{a \cos x + x \sin x}{x^2 + a^2} \, dx, a > 0$ [*Hint:* Consider $e^{iz}/(z - ai)$.]

In Problems 33 and 34, evaluate the given trigonometric integral.

33. $\displaystyle\int_0^{2\pi} \dfrac{\cos^2\theta}{2 + \sin \theta} \, d\theta$ **34.** $\displaystyle\int_0^{2\pi} \dfrac{\cos 3\theta}{5 - 4 \cos \theta} \, d\theta$

35. Use an indented contour to show that

$$\text{P.V.} \int_0^{\infty} \dfrac{1 - \cos x}{x^2} \, dx = \dfrac{\pi}{2}.$$

36. Show that $\int_0^{\infty} e^{-a^2 x^2} \cos bx \, dx = e^{-b^2/4a^2} \sqrt{\pi}/2a$ by considering the complex integral $\oint_C e^{-a^2 z^2} e^{ibz} \, dz$ along the contour C shown in Figure 19.16. Use the known result $\int_{-\infty}^{\infty} e^{-a^2 x^2} \, dx = \sqrt{\pi}/a$.

Figure 19.16

37. The Laurent expansion of $f(z) = e^{(u/2)(z - 1/z)}$ valid for $0 < |z|$ can be shown to be $f(z) = \sum_{k=-\infty}^{\infty} J_k(u) z^k$, where $J_k(u)$ is the Bessel function of the first kind of order k. Use (4) of Section 19.3 and the contour $C: |z| = 1$ to show that the coefficients $J_k(u)$ are given by

$$J_k(u) = \dfrac{1}{2\pi} \int_0^{2\pi} \cos (kt - u \sin t) \, dt.$$

20 *Conformal Mappings and Applications*

INTRODUCTION

A **conformal mapping** is an analytic function that maps a given region in the z-plane to another region in the w-plane so that the angles between curves in the z-plane are preserved. In this chapter we shall study the mapping properties of the elementary functions introduced in Chapter 17 and develop two new classes of special mappings called the **linear fractional transformations** and the **Schwarz–Christoffel transformations.**

In earlier chapters we used Fourier series and integral transforms to solve boundary-value problems involving Laplace's equation. Conformal mapping methods can be used to transfer known solutions to Laplace's equation from one region to another. In addition, fluid flows around obstacles and through channels can be determined using conformal mappings.

20.1 Complex Functions as Mappings

■ **Mappings Between Planes** In Chapter 17 we emphasized the algebraic definitions and properties of complex functions. In order to give a geometric interpretation of a complex function $w = f(z)$, we place a z-plane and a w-plane side by side and imagine that a point $z = x + iy$ in the domain of the definition of f has mapped (or transformed) to the point $w = f(z)$ in the second plane. Thus the complex function $w = f(z) = u(x, y) + iv(x, y)$ may be considered as the **planar transformation**

$$u = u(x, y) \qquad v = v(x, y)$$

and $w = f(z)$ is called the **image** of z under f.

Figure 20.1 indicates the images of a finite number of complex numbers in the region R. More useful information is obtained by finding the image of the region R together with the images of a family of curves lying inside R. Common choices for the curves are families of lines, families of circles, and the system of level curves for the real and imaginary parts of f.

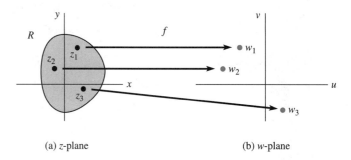

(a) z-plane (b) w-plane

Figure 20.1

Note that if $z(t) = x(t) + iy(t)$, $a \leq t \leq b$, describes a curve C in the region, then $w = f(z(t))$, $a \leq t \leq b$, is a parametric representation of the corresponding curve C' in the w-plane. In addition, a point z on the level curve $u(x, y) = a$ will be mapped to a point w that lies on the vertical line $u = a$, and a point z on the level curve $v(x, y) = b$ will be mapped to a point w that lies on the horizontal line $v = b$.

Example 1 The Mapping $f(z) = e^z$

The horizontal strip $0 \leq y \leq \pi$ lies in the fundamental region of the exponential function $f(z) = e^z$. A vertical line segment $x = a$ in this region can be described by $z(t) = a + it$, $0 \leq t \leq \pi$, and so $w = f(z(t)) = e^a e^{it}$. Thus the image is a semicircle with center at $w = 0$ and with radius $r = e^a$. Similarly, a horizontal line $y = b$ can be parametrized by $z(t) = t + ib$, $-\infty < t < \infty$, and so $w = f(z(t)) = e^t e^{ib}$. Since Arg $w = b$ and $|w| = e^t$, the image is a ray emanating from the origin, and since $0 \leq$ Arg $w \leq \pi$, the image of the entire horizontal strip is the upper half-plane $v \geq 0$. Note that the horizontal lines $y = 0$ and $y = \pi$ are mapped onto the positive and negative u-axis, respectively. See Figure 20.2 for the mapping by $f(z) = e^z$.

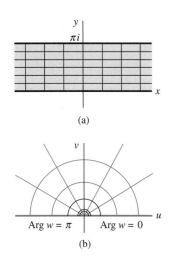

(a)

(b)

Arg $w = \pi$ Arg $w = 0$

Figure 20.2

From $w = e^x e^{iy}$, we can conclude that $|w| = e^x$ and $y = \text{Arg } w$. Hence, $z = x + iy = \log_e |w| + i \text{ Arg } w = \text{Ln } w$. The inverse function $f^{-1}(w) = \text{Ln } w$ therefore maps the upper half-plane $v \geq 0$ back to the horizontal strip $0 \leq y \leq \pi$. ❏

Example 2 The Mapping $f(z) = 1/z$

The complex function $f(z) = 1/z$ has domain $z \neq 0$ and real and imaginary parts $u(x, y) = x/(x^2 + y^2)$ and $v(x, y) = -y/(x^2 + y^2)$, respectively. When $a \neq 0$, a level curve $u(x, y) = a$ can be written as

$$x^2 - \frac{1}{a} x + y^2 = 0 \quad \text{or} \quad \left(x - \frac{1}{2a}\right)^2 + y^2 = \left(\frac{1}{2a}\right)^2.$$

The level curve is therefore a circle with its center on the x-axis and passing through the origin. A point z on this circle other than zero is mapped to a point w on the line $u = a$. Likewise, the level curve $v(x, y) = b$, $b \neq 0$, can be written as

$$x^2 + \left(y + \frac{1}{2b}\right)^2 = \left(\frac{1}{2b}\right)^2,$$

and a point z on this circle is mapped to a point w on the line $v = b$. Figure 20.3 shows the mapping by $f(z) = 1/z$. Figure 20.3(a) shows the two collections of circular level curves, and Figure 20.3(b) shows their corresponding images in the w-plane.

Since $w = 1/z$, we have $z = 1/w$. Thus $f^{-1}(w) = 1/w$, and so $f = f^{-1}$. We can therefore conclude that f maps the horizontal line $y = b$ to the circle $u^2 + (v + 1/2b)^2 = (1/2b)^2$, and f maps the vertical line $x = a$ to the circle $(u - 1/2a)^2 + v^2 = (1/2a)^2$. ❏

■ **Translation and Rotation** The elementary function $f(z) = z + z_0$ may be interpreted as a **translation** in the z-plane. To see this, we let $z = x + iy$ and $z_0 = h + ik$. Since $w = f(z) = (x + h) + i(y + k)$, the point (x, y) has been translated h units in the horizontal direction and k units in the vertical direction to the new position at $(x + h, y + k)$. In particular, the origin O has been mapped to $z_0 = h + ik$.

The elementary function $g(z) = e^{i\theta_0} z$ may be interpreted as a **rotation** through θ_0 degrees, for if $z = re^{i\theta}$, then $w = g(z) = re^{i(\theta + \theta_0)}$. Note that if the complex mapping $h(z) = e^{i\theta_0} z + z_0$ is applied to a region R that is centered at the origin, then image region R' may be obtained by first rotating R through θ_0 degrees and then translating the center to the new position z_0. See Figure 20.4 for the mapping by $h(z) = e^{i\theta_0} z + z_0$.

Example 3 Rotation and Translation

Find a complex function that maps the horizontal strip $-1 \leq y \leq 1$ onto the vertical strip $2 \leq x \leq 4$.

SOLUTION If the horizontal strip $-1 \leq y \leq 1$ is rotated through 90°, the vertical strip $-1 \leq x \leq 1$ results, and the vertical strip $2 \leq x \leq 4$ can be obtained by shifting this vertical strip 3 units to the right. See Figure 20.5. Since $e^{i\pi/2} = i$, we obtain $h(z) = iz + 3$ as the desired complex mapping.

(a)

(b)

Figure 20.3

(a)

(b)

Figure 20.4

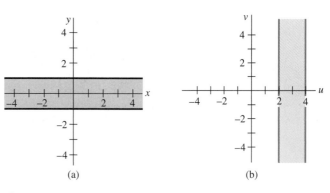

Figure 20.5 ❑

■ **Magnification** A **magnification** is a complex function of the form $f(z) = \alpha z$, where α is a fixed positive real number. Note that $|w| = |\alpha z| = \alpha|z|$, and so f changes the length (but not the direction) of the complex number z by a fixed factor α. If $g(z) = az + b$ and $a = r_0 e^{i\theta_0}$, then the vector z is rotated through θ_0 degrees, magnified by a factor of r_0, and then translated using b.

> *Example 4* **Contraction and Translation**
>
> Find a complex function that maps the disk $|z| \leq 1$ onto the disk $|w - (1 + i)| \leq \frac{1}{2}$.
>
> **SOLUTION** We must first contract the radius of the disk by a factor of $\frac{1}{2}$ and then translate its center to the point $1 + i$. Therefore, $w = f(z) = \frac{1}{2}z + (1 + i)$ maps $|z| \leq 1$ to the disk $|w - (1 + i)| \leq \frac{1}{2}$. ❑

■ **Power Functions** A complex function of the form $f(z) = z^\alpha$, where α is a fixed positive real number, is called a **real power function.** Figure 20.6 shows the effect of the complex function $f(z) = z^\alpha$ on the angular wedge $0 \leq \text{Arg } z \leq \theta_0$. If $z = re^{i\theta}$, then $w = f(z) = r^\alpha e^{i\alpha\theta}$. Hence, $0 \leq \text{Arg } w \leq \alpha\theta_0$ and the opening of the wedge is changed by a factor of α. It is not hard to show that a circular arc with center at the origin is mapped to a similar circular arc, and rays emanating from the origin are mapped to similar rays.

(a)

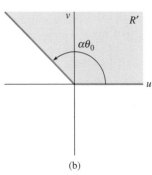

(b)

Figure 20.6

> *Example 5* **The Power Function $f(z) = z^{1/4}$**
>
> Find a complex function that maps the upper half-plane $y \geq 0$ onto the wedge $0 \leq \text{Arg } w \leq \pi/4$.
>
> **SOLUTION** The upper half-plane $y \geq 0$ can also be described by the inequality $0 \leq \text{Arg } z \leq \pi$. We must therefore find a complex mapping that reduces the angle $\theta_0 = \pi$ by a factor of $\alpha = \frac{1}{4}$. Hence, $f(z) = z^{1/4}$. ❑

■ **Successive Mappings** To find a complex mapping between two regions R and R', it is often convenient to first map R onto a third region R'' and then find a complex mapping from R'' onto R'. More precisely, if $\zeta = f(z)$ maps R onto R'', and $w = g(\zeta)$ maps R'' onto R', then the composite function $w = g(f(z))$ maps R onto R'. See Figure 20.7 for a diagram of successive mappings.

> *Example 6* **Successive Mappings**
>
> Find a complex function that maps the horizontal strip $0 \leq y \leq \pi$ onto the wedge $0 \leq \text{Arg } w \leq \pi/4$.

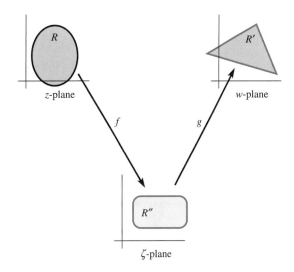

Figure 20.7

SOLUTION We saw in Example 1 that the complex function $f(z) = e^z$ mapped the horizontal strip $0 \le y \le \pi$ onto the upper half-plane $0 \le \text{Arg } \zeta \le \pi$. From Example 5, the upper half-plane $0 \le \text{Arg } \zeta \le \pi$ is mapped onto the wedge $0 \le \text{Arg } w \le \pi/4$ by $g(\zeta) = \zeta^{1/4}$. It therefore follows that the composite function $w = g(f(z)) = g(e^z) = e^{z/4}$ maps the horizontal strip $0 \le y \le \pi$ onto the wedge $0 \le \text{Arg } w \le \pi/4$. ❑

Example 7 **Successive Mappings**

Find a complex function that maps the wedge $\pi/4 \le \text{Arg } z \le 3\pi/4$ onto the upper half-plane $v \ge 0$.

SOLUTION We first rotate the wedge $\pi/4 \le \text{Arg } z \le 3\pi/4$ so that it is in the standard position shown in Figure 20.6. If $\zeta = f(z) = e^{-i\pi/4}z$, then the image of this wedge is the wedge R'' defined by $0 \le \text{Arg } \zeta \le \pi/2$. The real power function $w = g(\zeta) = \zeta^2$ expands the opening of R'' by a factor of 2 to give the upper half-plane $0 \le \text{Arg } w \le \pi$ as its image. Therefore, $w = g(f(z)) = (e^{-i\pi/4}z)^2 = -iz^2$ is the desired mapping. ❑

In Sections 20.2–20.4 we will expand our knowledge of complex mappings and show how they can be used to solve Laplace's equation in the plane.

E X E R C I S E S 2 0 . 1 Answers to odd-numbered problems begin on page A-76.

In Problems 1–10, a curve in the z-plane and a complex mapping $w = f(z)$ are given. In each case, find the image curve in the w-plane.

1. $y = x$ under $w = 1/z$ **2.** $y = 1$ under $w = 1/z$

3. Hyperbola $xy = 1$ under $w = z^2$

4. Hyperbola $x^2 - y^2 = 4$ under $w = z^2$

5. Semicircle $|z| = 1$, $y > 0$, under $w = \text{Ln } z$

6. Ray $\theta = \pi/4$ under $w = \text{Ln } z$

7. Ray $\theta = \theta_0$ under $w = z^{1/2}$

8. Circular arc $r = 2$, $0 \le \theta \le \pi/2$, under $w = z^{1/2}$

9. Curve $e^x \cos y = 1$ under $w = e^z$

10. Circle $|z| = 1$ under $w = z + 1/z$

In Problems 11–20, a region in the z-plane and a complex mapping $w = f(z)$ are given. In each case, find the image region in the w-plane.

11. First quadrant under $w = 1/z$

12. Strip $0 \le y \le 1$ under $w = 1/z$

13. Strip $\pi/4 \le y \le \pi/2$ under $w = e^z$

14. Rectangle $0 \le x \le 1$, $0 \le y \le \pi$ under $w = e^z$

15. Circle $|z| = 1$ under $w = z + 4i$

16. Circle $|z| = 1$ under $w = 2z - 1$

17. Strip $0 \le y \le 1$ under $w = iz$

18. First quadrant under $w = (1 + i)z$

19. Wedge $0 \le \text{Arg } z \le \pi/4$ under $w = z^3$

20. Wedge $0 \le \text{Arg } z \le \pi/4$ under $w = z^{1/2}$

In Problems 21–30, find a complex mapping from the given region R in the z-plane to the image region R' in the w-plane.

21. Strip $1 \le y \le 4$ to the strip $0 \le u \le 3$

22. Strip $1 \le y \le 4$ to the strip $0 \le v \le 3$

23. Disk $|z - 1| \le 1$ to the disk $|w| \le 2$

24. Strip $-1 \le x \le 1$ to the strip $-1 \le v \le 1$

25. Wedge $\pi/4 \le \text{Arg } z \le \pi/2$ to the upper half-plane $v \ge 0$

26. Strip $0 \le y \le 4$ to the upper half-plane $v \ge 0$

27. Strip $0 \le y \le \pi$ to the wedge $0 \le \text{Arg } w \le 3\pi/2$

28. Wedge $0 \le \text{Arg } z \le 3\pi/2$ to the half-plane $u \ge 2$

29.

Figure 20.8

30.

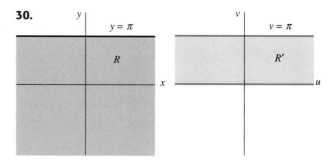

Figure 20.9

31. Project The mapping in Problem 10 is a special case of the mapping $w = z + k^2/z$, where k is a positive constant, called the **Joukowski transformation.**

(a) Show that the Joukowski transformation maps any circle $x^2 + y^2 = R^2$ into the ellipse

$$\frac{u^2}{\left(1 + \dfrac{k^2}{R^2}\right)^2} + \frac{v^2}{\left(1 - \dfrac{k^2}{R^2}\right)^2} = R^2, \qquad k \ne R.$$

(b) What is the image of the circle when $R = k$?

(c) The importance of the transformation $w = z + k^2/z$ does not lie in its effect on circles $|z| = R$ centered at the origin but on off-centered circles with center on the real axis. Show that the Joukowski transformation can be written

$$\frac{w - 2k}{w + 2k} = \left(\frac{z - k}{z + k}\right)^2.$$

With $k = 1$, this particular transformation maps a circle passing through $z = -1$ and containing the point $z = 1$ into a closed curve with a sharply pointed trailing edge. This kind of curve, which resembles a cross-section of an airplane wing, is known as a Joukowski airfoil.

 Write a report on the use of the Joukowski transformation in the study of the flow of air around an airfoil. There is a lot of information on this topic on the Internet; for example, you might start with the URL:

www.ae.su.oz.au/aero/jouk/jouk.html

20.2 Conformal Mapping and the Dirichlet Problem

■ **Angle-Preserving Mappings** A complex mapping $w = f(z)$ defined on a domain D is called **conformal at** $z = z_0$ in D when f preserves the angles between any two curves in D that intersect at z_0. More precisely, if C_1 and C_2 intersect in D at z_0, and C_1' and C_2' are the corresponding images in the w-plane, we require that the angle θ between C_1 and C_2 equal the angle ϕ between C_1' and C_2'. See Figure 20.10.

 These angles can be computed in terms of tangent vectors to the curves. If z_1' and z_2' denote tangent vectors to curves C_1 and C_2, respectively, then, applying the law of cosines to the triangle determined by z_1' and z_2', we have

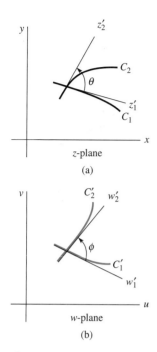

Figure 20.10

$$|z_1' - z_2'|^2 = |z_1'|^2 + |z_2'|^2 - 2|z_1'||z_2'| \cos \theta$$

or

$$\theta = \cos^{-1}\left(\frac{|z_1'|^2 + |z_2'|^2 - |z_1' - z_2'|^2}{2|z_1'||z_2'|}\right). \tag{1}$$

Likewise, if w_1' and w_2' denote tangent vectors to curves C_1' and C_2', respectively, then

$$\phi = \cos^{-1}\left(\frac{|w_1'|^2 + |w_2'|^2 - |w_1' - w_2'|^2}{2|w_1'||w_2'|}\right). \tag{2}$$

The next theorem will give a simple condition which guarantees that $\theta = \phi$.

THEOREM 20.1* Conformal Mapping

If $f(z)$ is analytic in the domain D and $f'(z_0) \neq 0$, then f is conformal at $z = z_0$.

PROOF If a curve C in D is parameterized by $z = z(t)$, then $w = f(z(t))$ describes the image curve in the w-plane. Applying the Chain Rule to $w = f(z(t))$ gives $w' = f'(z(t))z'(t)$. If curves C_1 and C_2 intersect in D at z_0, then $w_1' = f'(z_0)z_1'$ and $w_2' = f'(z_0)z_2'$. Since $f'(z_0) \neq 0$, we can use (2) to obtain

$$\phi = \cos^{-1}\left(\frac{|f'(z_0)z_1'|^2 + |f'(z_0)z_2'|^2 - |f'(z_0)z_1' - f'(z_0)z_2'|^2}{2|f'(z_0)z_1'||f'(z_0)z_2'|}\right).$$

We can apply the laws of absolute value to factor out $|f'(z_0)|^2$ in the numerator and denominator and obtain

$$\phi = \cos^{-1}\left(\frac{|z_1'|^2 + |z_2'|^2 - |z_1' - z_2'|^2}{2|z_1'||z_2'|}\right).$$

Therefore, from (1), $\phi = \theta$. ❑

Example 1 Conformal Mappings

(a) The analytic function $f(z) = e^z$ is conformal at all points in the z-plane, since $f'(z) = e^z$ is never zero.
(b) The analytic function $g(z) = z^2$ is conformal at all points except $z = 0$ since $g'(z) = 2z \neq 0$ for $z \neq 0$. From Figure 20.6 we see that $g(z)$ doubles the angle formed by the two rays at the origin. ❑

If $f'(z_0) = 0$ but $f''(z_0) \neq 0$, it is possible to show that f *doubles* the angle between any two curves in D that intersect at $z = z_0$. The next two examples will introduce two important complex mappings that are conformal at all but a finite number of points in their domains.

Example 2 $f(z) = \sin z$ as a Conformal Mapping

The vertical strip $-\pi/2 \leq x \leq \pi/2$ is called the fundamental region of the trigonometric function $w = \sin z$. A vertical line $x = a$ in the interior of this region can be described by $z(t) = a + it$, $-\infty < t < \infty$. From (6) in Section 17.7, we have

$$\sin z = \sin x \cosh y + i \cos x \sinh y$$

* It is also possible to prove that f preserves the *sense of direction* between the tangent vectors.

(a)

(b)

Figure 20.11

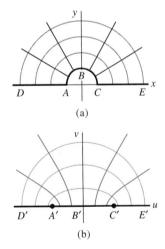

(a)

(b)

Figure 20.12

and so

$$u + iv = \sin(a + it) = \sin a \cosh t + i \cos a \sinh t.$$

From the identity $\cosh^2 t - \sinh^2 t = 1$, it follows that

$$\frac{u^2}{\sin^2 a} - \frac{v^2}{\cos^2 a} = 1.$$

The image of the vertical line $x = a$ is therefore a hyperbola with $\pm \sin a$ as u-intercepts, and since $-\pi/2 < a < \pi/2$, the hyperbola crosses the u-axis between $u = -1$ and $u = 1$. Note that if $a = -\pi/2$, then $w = -\cosh t$, and so the line $x = -\pi/2$ is mapped onto the interval $(-\infty, -1]$ on the negative u-axis. Likewise, the line $x = \pi/2$ is mapped onto the interval $[1, \infty)$ on the positive u-axis.

A similar argument establishes that the horizontal line segment described by $z(t) = t + ib$, $-\pi/2 < t < \pi/2$, is mapped onto either the upper portion or the lower portion of the ellipse

$$\frac{u^2}{\cosh^2 b} + \frac{v^2}{\sinh^2 b} = 1$$

according to whether $b > 0$ or $b < 0$. These results are summarized in Figure 20.11, which shows the mapping by $f(z) = \sin z$. Note that we have carefully used capital letters to indicate where portions of the boundary are mapped. Thus, for example, boundary segment AB is transformed to $A'B'$.

Since $f'(z) = \cos z$, f is conformal at all points in the region except $z = \pm \pi/2$. The hyperbolas and ellipses are therefore orthogonal since they are images of the orthogonal families of horizontal segments and vertical lines. Note that the 180° angle at $z = -\pi/2$ formed by segments AB and AC is doubled to form a single line segment at $w = -1$. ❏

Example 3 $f(z) = z + 1/z$ as a Conformal Mapping

The complex mapping $f(z) = z + 1/z$ is conformal at all values of z except $z = \pm 1$ and $z = 0$. In particular the function is conformal for all values of z in the upper half-plane that satisfy $|z| > 1$. If $z = re^{i\theta}$, then $w = re^{i\theta} + (1/r)e^{-i\theta}$, and so

$$u = \left(r + \frac{1}{r}\right) \cos \theta, \qquad v = \left(r - \frac{1}{r}\right) \sin \theta. \qquad (3)$$

Note that if $r = 1$, then $v = 0$ and $u = 2 \cos \theta$. Therefore, the semicircle $z = e^{it}$, $0 \le t \le \pi$, is mapped to the segment $[-2, 2]$ on the u-axis. It follows from (3) that if $r > 1$, then the semicircle $z = re^{it}$, $0 \le t \le \pi$, is mapped onto the upper half of the ellipse $u^2/a^2 + v^2/b^2 = 1$, where $a = r + 1/r$ and $b = r - 1/r$. See Figure 20.12 for the mapping by $f(z) = z + 1/z$.

For a fixed value of θ, the ray $z = te^{i\theta}$, for $t \ge 1$, is mapped to the portion of the hyperbola $u^2/\cos^2 \theta - v^2/\sin^2 \theta = 4$ in the upper half-plane $v \ge 0$. This follows from (3), since

$$\frac{u^2}{\cos^2 \theta} - \frac{v^2}{\sin^2 \theta} = \left(t + \frac{1}{t}\right)^2 - \left(t - \frac{1}{t}\right)^2 = 4.$$

Since f is conformal for $|z| > 1$ and a ray $\theta = \theta_0$ intersects a circle $|z| = r$ at a right angle, the hyperbolas and ellipses in the w-plane are orthogonal. ❏

■ **Conformal Mappings Using Tables** Conformal mappings are given in Appendix IV. The mappings have been categorized as elementary mappings (E-1 to E-9), mappings to half-planes (H-1 to H-6), mappings to circular regions (C-1 to

C-5), and miscellaneous mappings (M-1 to M-10). Some of these complex mappings will be derived in Sections 20.3 and 20.4.

The entries indicate not only the images of the region R but also the images of various portions of the boundary of R. This will be especially useful when we attempt to solve boundary-value problems using conformal maps. You should use the appendix much like you use integral tables to find antiderivatives. In some cases a single entry can be used to find a conformal mapping between two given regions R and R'. In other cases successive transformations may be required to map R to R'.

Example 4 Using a Table of Conformal Mappings

Use the conformal mappings in Appendix IV to find a conformal mapping between the strip $0 \le y \le 2$ and the upper half-plane $v \ge 0$. What is the image of the negative x-axis?

SOLUTION An appropriate mapping may be obtained directly from entry H-2. Letting $a = 2$ then $f(z) = e^{\pi z/2}$ and noting the positions of E, D, E', and D' in the figure, we can map the negative x-axis onto the interval (0, 1) on the u-axis. ❑

Example 5 Using a Table of Conformal Mappings

Use the conformal mappings in Appendix IV to find a conformal mapping between the strip $0 \le y \le 2$ and the disk $|w| \le 1$. What is the image of the negative x-axis?

SOLUTION Appendix IV does not have an entry that maps the strip $0 \le y \le 2$ directly onto the disk. In Example 5 the strip was mapped by $f(z) = e^{\pi z/2}$ onto the upper half-plane and, from entry C-4, the complex mapping $w = \dfrac{i - \zeta}{i + \zeta}$ maps the upper half-plane to the disk $|w| \le 1$. Therefore, $w = g(f(z)) = \dfrac{i - e^{\pi z/2}}{i + e^{\pi z/2}}$ maps the strip $0 \le y \le 2$ onto the disk $|w| \le 1$.

The negative x-axis is first mapped to the interval (0, 1) in the ζ-plane, and from the position of points C and C' in C-4, the interval (0, 1) is mapped to the circular arc $w = e^{i\theta}$, $0 < \theta < \pi/2$, in the w-plane. ❑

■ **Harmonic Functions and the Dirichlet Problem** A bounded harmonic function $u = u(x, y)$ that takes on prescribed values on the entire boundary of a region R is called a solution to a **Dirichlet problem** on R. In Chapters 13–15 we introduced a number of techniques for solving Laplace's equation in the plane, and we interpreted the solution to a Dirichlet problem as the steady-state temperature distribution in the interior of R that results from the fixed temperatures on the boundary.

There are at least two disadvantages to the Fourier series and integral transform methods presented in Chapters 13–15. The methods work only for simple regions in the plane and the solutions typically take the form of either infinite series or improper integrals. As such, they are difficult to evaluate. In Section 17.5 we saw that the real and imaginary parts of an analytic function are both harmonic. Since we have a large stockpile of analytic functions, we can find closed-form solutions to many Dirichlet problems and use these solutions to sketch the isotherms and lines of flow of the temperature distribution.

We will next show how conformal mappings can be used to solve a Dirichlet problem in a region R once the solution to the corresponding Dirichlet problem in the image region R' is known. The method depends on the following theorem:

> **THEOREM 20.2** **Transformation Theorem for Harmonic Functions**
>
> Let f be an analytic function that maps a domain D onto a domain D'. If U is harmonic in D', then the real-valued function $u(x, y) = U(f(z))$ is harmonic in D.

PROOF We will give a proof for the special case in which D' is simply connected. If U has a harmonic conjugate V in D', then $H = U + iV$ is analytic in D', and so the composite function $H(f(z)) = U(f(z)) + iV(f(z))$ is analytic in D. By Theorem 17.4, it follows that the real part $U(f(z))$ is harmonic in D, and the proof is complete.

To establish that U has a harmonic conjugate, let $g(z) = \partial U/\partial x - i\,\partial U/\partial y$. The first Cauchy–Riemann equation $(\partial/\partial x)(\partial U/\partial x) = (\partial/\partial y)(-\partial U/\partial y)$ is equivalent to Laplace's equation $\partial^2 U/\partial x^2 + \partial^2 U/\partial y^2 = 0$, which is satisfied because U is harmonic in D'. The second Cauchy–Riemann equation $(\partial/\partial y)(\partial U/\partial x) = -(\partial/\partial x)(-\partial U/\partial y)$ is equivalent to the equality of the second-order mixed partial derivatives. Therefore, $g(z)$ is analytic in the simply connected domain D' and so, by Theorem 18.8, has an antiderivative $G(z)$. If $G(z) = U_1 + iV_1$, then $g(z) = G'(z) = \partial U_1/\partial x - i\,\partial U_1/\partial y$. Since $g(z) = \partial U/\partial x - i\,\partial U/\partial y$, it follows that U and U_1 have equal first partial derivatives. Therefore, $H = U + iV_1$ is analytic in D', and so U has a harmonic conjugate in D'. ❏

Theorem 20.2 can be used to solve a Dirichlet problem in a region R by transforming the problem to a region R' in which the solution U either is apparent or has been found by prior methods (including the Fourier series and integral transform methods of Chapters 13–15). The key steps are summarized next.

■ Solving Dirichlet Problems Using Conformal Mapping

1. Find a conformal mapping $w = f(z)$ that transforms the original region R onto the image region R'. The region R' may be a region for which many explicit solutions to Dirichlet problems are known.
2. Transfer the boundary conditions from the boundary of R to the boundary of R'. The value of u at a boundary point ξ of R is assigned as the value of U at the corresponding boundary point $f(\xi)$. See Figure 20.13 for an illustration of transferring boundary conditions.

Figure 20.13

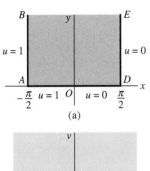

Figure 20.14

3. Solve the corresponding Dirichlet problem in R'. The solution U may be apparent from the simplicity of the problem in R' or may be found using Fourier or integral transform methods. (Additional methods will be presented in Sections 20.3 and 20.5.).

4. The solution to the original Dirichlet problem is $u(x, y) = U(f(z))$.

Example 6 **Solving a Dirichlet Problem**

The function $U(u, v) = (1/\pi)$ Arg w is harmonic in the upper half-plane $v > 0$ since it is the imaginary part of the analytic function $g(w) = (1/\pi)$ Ln w. Use this function to solve the Dirichlet problem in Figure 20.14(a).

SOLUTION The analytic function $f(z) = \sin z$ maps the original region to the upper half-plane $v \geq 0$ and maps the boundary segments to the segments shown in Figure 20.14(b). The harmonic function $U(u, v) = (1/\pi)$ Arg w satisfies the transferred boundary conditions $U(u, 0) = 0$ for $u > 0$ and $U(u, 0) = 1$ for $u < 0$. Therefore, $u(x, y) = U(\sin z) = (1/\pi)$ Arg$(\sin z)$ is the solution to the original problem. If $\tan^{-1}(v/u)$ is chosen to lie between 0 and π, the solution can also be written as

$$u(x, y) = \frac{1}{\pi} \tan^{-1}\left(\frac{\cos x \sinh y}{\sin x \cosh y}\right).$$

❑

Example 7 **Solving a Dirichlet Problem**

From C-1 in Appendix IV of conformal mappings, the analytic function $f(z) = (z - a)/(az - 1)$, where $a = (7 + 2\sqrt{6})/5$, maps the region outside the two open disks $|z| < 1$ and $|z - \frac{3}{2}| < \frac{1}{2}$ onto the circular region $r_0 \leq |w| \leq 1$, where $r_0 = 5 - 2\sqrt{6}$. Figure 20.15(a) shows the original Dirichlet problem, and Figure 20.15(b) shows the transferred boundary conditions.

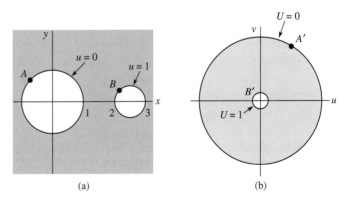

(a) (b)

Figure 20.15

In Problem 10 in Exercises 14.1 we discovered that $U(r, \theta) = (\log_e r)/(\log_e r_0)$ is the solution to the new Dirichlet problem. From Theorem 20.2 we can conclude that the solution to the original boundary-value problem is

$$u(x, y) = \frac{1}{\log_e(5 - 2\sqrt{6})} \log_e \left|\frac{z - (7 + 2\sqrt{6})/5}{(7 + 2\sqrt{6})z/5 - 1}\right|$$

❑

A favorite image region R' for a simply connected region R is the upper half-plane $y \geq 0$. For any real number a, the complex function $\text{Ln}(z - a) = \log_e|z - a| + i \, \text{Arg}(z - a)$

Figure 20.16

is analytic in R'. Therefore, $\text{Arg}(z - a)$ is harmonic in R' and is a solution to the Dirichlet problem shown in Figure 20.16.

It follows that the solution in R' to the Dirichlet problem with

$$U(x, 0) = \begin{cases} c_0, & a < x < b \\ 0, & \text{otherwise} \end{cases}$$

is the harmonic function $U(x, y) = (c_0/\pi)(\text{Arg}(z - b) - \text{Arg}(z - a))$. A large number of Dirichlet problems in the upper half-plane $y \geq 0$ can be solved by adding together harmonic functions of this form.

E X E R C I S E S 2 0 . 2 Answers to odd-numbered problems begin on page A-76.

In Problems 1–6, determine where the given complex mapping is conformal.

1. $f(z) = z^3 - 3z + 1$

2. $f(z) = \cos z$

3. $f(z) = z + e^z + 1$

4. $f(z) = z + \text{Ln } z + 1$

5. $f(z) = (z^2 - 1)^{1/2}$

6. $f(z) = \pi i - \frac{1}{2}[\text{Ln}(z + 1) + \text{Ln}(z - 1)]$

In Problems 7–10, use the results in Examples 3 and 4.

7. Use the identity $\cos z = \sin(\pi/2 - z)$ to find the image of the strip $0 \leq x \leq \pi$ under the complex mapping $w = \cos z$. What is the image of a horizontal line in the strip?

8. Use the identity $\sinh z = -i \sin(iz)$ to find the image of the strip $-\pi/2 \leq y \leq \pi/2$, $x \geq 0$, under the complex mapping $w = \sinh z$. What is the image of a vertical line segment in the strip?

9. Find the image of the region defined by $-\pi/2 \leq x \leq \pi/2$, $y \geq 0$, under the complex mapping $w = (\sin z)^{1/4}$. What is the image of the line segment $[-\pi/2, \pi/2]$ on the x-axis?

10. Find the image of the region $|z| \leq 1$ in the upper half-plane under the complex mapping $w = z + 1/z$. What is the image of the line segment $[-1, 1]$ on the x-axis?

In Problems 11–18, use the conformal mappings in Appendix IV to find a conformal mapping from the given region R in the z-plane onto the target region R' in the w-plane, and find the image of the given boundary curve.

11.

Figure 20.17

12.

Figure 20.18

13.

Figure 20.19

14.

Figure 20.20

15.

Figure 20.21

16.

Figure 20.22

17.

Figure 20.23

18.

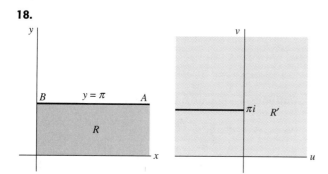

Figure 20.24

In Problems 19–22, use an appropriate conformal mapping and the harmonic function $U = (1/\pi) \operatorname{Arg} w$ to solve the given Dirichlet problem.

19.

Figure 20.25

20.

Figure 20.26

21.

Figure 20.27

22.

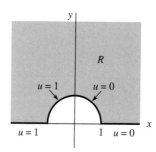

Figure 20.28

In Problems 23–26, use an appropriate conformal mapping and the harmonic function $U = (c_0/\pi)[\operatorname{Arg}(w - 1) - \operatorname{Arg}(w + 1)]$ to solve the given Dirichlet problem.

23.

Figure 20.29

24.

Figure 20.30

25.

Figure 20.31

26.

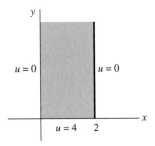

Figure 20.32

27. A real-valued function $\phi(x, y)$ is called **biharmonic** in a domain D when the fourth-order differential equation

$$\frac{\partial^4 \phi}{\partial x^4} + 2 \frac{\partial^4 \phi}{\partial x^2 \partial y^2} + \frac{\partial^4 \phi}{\partial y^4} = 0$$

at all points in D. Examples of biharmonic functions are the Airy stress function in the mechanics of solids and velocity potentials in the analysis of viscous fluid flow.

(a) Show that if ϕ is biharmonic in D, then $u = \partial^2 \phi / \partial x^2 + \partial^2 \phi / \partial y^2$ is harmonic in D.

(b) If $g(z)$ is analytic in D and $\phi(x, y) = \text{Re}(\bar{z} g(z))$, show that ϕ is biharmonic in D.

20.3 Linear Fractional Transformations

A **linear fractional transformation** is a complex function of the form

$$T(z) = \frac{az + b}{cz + d},$$

where a, b, c, and d are complex constants. Since

$$T'(z) = \frac{ad - bc}{(cz + d)^2},$$

T is conformal at z provided $\Delta = ad - bc \neq 0$ and $z \neq -d/c$. (If $\Delta = 0$, then $T'(z) = 0$ and $T(z)$ would be a constant function.) Linear fractional transformations are circle preserving in a sense that we will make precise in this section, and, as we saw in Example 8 of Section 20.2, they can be useful in solving Dirichlet problems in regions bounded by circles.

Note that when $c \neq 0$, $T(z)$ has a simple pole at $z_0 = -d/c$ and so

$$\lim_{z \to z_0} |T(z)| = \infty.$$

We will write $T(z_0) = \infty$ as shorthand for this limit. In addition, if $c \neq 0$, then

$$\lim_{|z| \to \infty} T(z) = \lim_{|z| \to \infty} \frac{a + b/z}{c + d/z} = \frac{a}{c},$$

and we write $T(\infty) = a/c$.

Example 1 A Linear Fractional Transformation

If $T(z) = (2z + 1)/(z - i)$, compute $T(0)$, $T(\infty)$, and $T(i)$.

SOLUTION Note that $T(0) = 1/(-i) = i$ and $T(\infty) = \lim_{|z| \to \infty} T(z) = 2$. Since $z = i$ is a simple pole for $T(z)$, we have $\lim_{z \to i} |T(z)| = \infty$ and we write $T(i) = \infty$. ❏

■ **Circle-Preserving Property** If $c = 0$, the linear fractional transformation reduces to a linear function $T(z) = Az + B$. We saw in Section 20.1 that such a complex mapping can be considered as the composite of a rotation, magnification, and translation. As such, a linear function will map a circle in the z-plane to a circle in the w-plane. When $c \neq 0$, we can divide $cz + d$ into $az + b$ to obtain

$$w = \frac{az + b}{cz + d} = \frac{bc - ad}{c} \frac{1}{cz + d} + \frac{a}{c}. \tag{1}$$

If we let $A = (bc - ad)/c$ and $B = a/c$, $T(z)$ can be written as the composite of transformations:

$$z_1 = cz + d, \qquad z_2 = \frac{1}{z_1}, \qquad w = Az_2 + B. \tag{2}$$

A general linear fractional transformation can therefore be written as the composite of two linear functions and the inversion $w = 1/z$. Note that if $|z - z_1| = r$ and $w = 1/z$, then

$$\left| \frac{1}{w} - \frac{1}{w_1} \right| = \frac{|w - w_1|}{|w| \, |w_1|} = r \qquad \text{or} \qquad |w - w_1| = (r|w_1|)|w - 0|. \tag{3}$$

It is not hard to show that the set of all points w that satisfy

$$|w - w_1| = \lambda |w - w_2| \tag{4}$$

is a line when $\lambda = 1$ and is a circle when $\lambda > 0$ and $\lambda \neq 1$. It follows from (3) that the image of the circle $|z - z_1| = r$ under the inversion $w = 1/z$ is a circle except when $r = 1/|w_1| = |z_1|$. In the latter case, the original circle passes through the origin and the image is a line. See Figure 20.3. From (2), we can deduce the following theorem:

THEOREM 20.3 Circle-Preserving Property

A linear fractional transformation maps a circle in the z-plane to either a line or a circle in the w-plane. The image is a line if and only if the original circle passes through a pole of the linear fractional transformation.

PROOF We have shown that a linear function maps a circle to a circle, whereas an inversion maps a circle to a circle or a line. It follows from (2) that a circle in the z-plane will be mapped to either a circle or a line in the w-plane. If the original circle passes through a pole z_0, then $T(z_0) = \infty$, and so the image is unbounded. Therefore, the image of such a circle must be a line. If the original circle does not pass through z_0, then the image is bounded and must be a circle. ❏

Example 2 Images of Circles

Find the images of the circles $|z| = 1$ and $|z| = 2$ under $T(z) = (z + 2)/(z - 1)$. What are the images of the interiors of these circles?

SOLUTION The circle $|z| = 1$ passes through the pole $z_0 = 1$ of the linear fractional transformation and so the image is a line. Since $T(-1) = -\frac{1}{2}$ and $T(i) = -\frac{1}{2} - \frac{3}{2}i$, we can conclude that the image is the line $u = -\frac{1}{2}$. The image of the interior $|z| < 1$ is either the half-plane $u < -\frac{1}{2}$ or the half-plane $u > -\frac{1}{2}$. Using $z = 0$ as a *test point*, $T(0) = -2$, and so the image is the half-plane $u < -\frac{1}{2}$.

The circle $|z| = 2$ does not pass through the pole and so the image is a circle. For $|z| = 2$,

$$|\bar{z}| = 2 \qquad \text{and} \qquad \overline{T(z)} = \overline{\frac{z+2}{z-1}} = \frac{\bar{z}+2}{\bar{z}-1} = T(\bar{z}).$$

Therefore, $\overline{T(z)}$ is a point on the image circle and so the image circle is symmetric with respect to the u-axis. Since $T(-2) = 0$ and $T(2) = 4$, the center of the circle is $w = 2$ and the image is the circle $|w - 2| = 2$ (see Figure 20.33). The image of the interior $|z| < 2$ is either the interior or the exterior of the image circle $|w - 2| = 2$. Since $T(0) = -2$, we can conclude that the image is $|w - 2| > 2$.

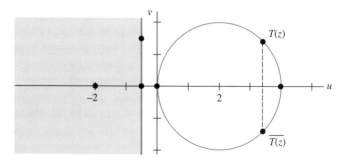

Figure 20.33

■ Constructing Special Mappings

In order to use linear fractional transformations to solve Dirichlet problems, we must construct special functions that map a given circular region R to a target region R' in which the corresponding Dirichlet problem is solvable. Since a circular boundary is determined by three of its points, we must find a linear fractional transformation $w = T(z)$ that maps three given points z_1, z_2, and z_3 on the boundary of R to three points w_1, w_2, and w_3 on the boundary of R'. In addition, the interior of R' must be the image of the interior of R. See Figure 20.34.

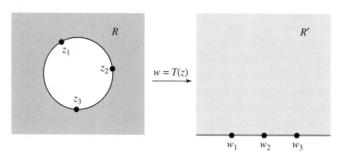

Figure 20.34

■ **Matrix Methods** Matrix methods can be used to simplify many of the computations. We can associate the matrix

$$\mathbf{A} = \begin{pmatrix} a & b \\ c & d \end{pmatrix}$$

with $T(z) = (az + b)/(cz + d)$.* If $T_1(z) = (a_1z + b_1)/(c_1z + d_1)$ and $T_2(z) = (a_2z + b_2)/(c_2z + d_2)$, then the composite function $T(z) = T_2(T_1(z))$ is given by $T(z) = (az + b)/(cz + d)$, where

$$\begin{pmatrix} a & b \\ c & d \end{pmatrix} = \begin{pmatrix} a_2 & b_2 \\ c_2 & d_2 \end{pmatrix}\begin{pmatrix} a_1 & b_1 \\ c_1 & d_1 \end{pmatrix}. \tag{5}$$

If $w = T(z) = (az + b)/(cz + d)$, we can solve for z to obtain $z = (dw - b)/(-cw + a)$. Therefore the inverse of the linear fractional transformation T is $T^{-1}(w) = (dw - b)/(-cw + a)$ and we associate the matrix

$$\text{adj } \mathbf{A} = \begin{pmatrix} d & -b \\ -c & a \end{pmatrix} \tag{6}$$

with T^{-1}. The matrix adj \mathbf{A} is the adjoint matrix of \mathbf{A} (see Section 8.6), the matrix for T.

Example 3 Using Matrices to Find an Inverse Transform

If $T(z) = \dfrac{2z - 1}{z + 2}$ and $S(z) = \dfrac{z - i}{iz - 1}$, find $S^{-1}(T(z))$.

SOLUTION From (5) and (6), we have $S^{-1}(T(z)) = (az + b)/(cz + d)$, where

$$\begin{pmatrix} a & b \\ c & d \end{pmatrix} = \text{adj}\begin{pmatrix} 1 & -i \\ i & -1 \end{pmatrix}\begin{pmatrix} 2 & -1 \\ 1 & 2 \end{pmatrix}$$

$$= \begin{pmatrix} -1 & i \\ -i & 1 \end{pmatrix}\begin{pmatrix} 2 & -1 \\ 1 & 2 \end{pmatrix} = \begin{pmatrix} -2 + i & 1 + 2i \\ 1 - 2i & 2 + i \end{pmatrix}.$$

Therefore, $S^{-1}(T(z)) = \dfrac{(-2 + i)z + 1 + 2i}{(1 - 2i)z + 2 + i}$. ❑

■ **Triples to Triples** The linear fractional transformation

$$T(z) = \frac{z - z_1}{z - z_3}\frac{z_2 - z_3}{z_2 - z_1}$$

has a zero at $z = z_1$, a pole at $z = z_3$, and $T(z_2) = 1$. Therefore, $T(z)$ maps three *distinct* complex numbers z_1, z_2, and z_3 to 0, 1, and ∞, respectively. The term $\dfrac{z - z_1}{z - z_3}\dfrac{z_2 - z_3}{z_2 - z_1}$ is called the **cross-ratio** of the complex numbers z, z_1, z_2, and z_3.

Likewise, the complex mapping $S(w) = \dfrac{w - w_1}{w - w_3}\dfrac{w_2 - w_3}{w_2 - w_1}$ sends w_1, w_2, and w_3 to 0, 1, and ∞, and so S^{-1} maps 0, 1, and ∞ to w_1, w_2, and w_3. It follows that the linear fractional transformation $w = S^{-1}(T(z))$ maps the triple z_1, z_2, and z_3 to w_1, w_2, and w_3. From $w = S^{-1}(T(z))$, we have $S(w) = T(z)$ and we can conclude that

$$\frac{w - w_1}{w - w_3}\frac{w_2 - w_3}{w_2 - w_1} = \frac{z - z_1}{z - z_3}\frac{z_2 - z_3}{z_2 - z_1}. \tag{7}$$

*The matrix \mathbf{A} is not unique since the numerator and denominator in $T(z)$ can be multiplied by a nonzero constant.

In constructing a linear fractional transformation that maps the triple z_1, z_2, and z_3 to w_1, w_2, and w_3, we can use matrix methods to compute $w = S^{-1}(T(z))$. Alternatively, we can substitute into (7) and solve the resulting equation for w.

Example 4 Constructing a Linear Fractional Transformation

Construct a linear fractional transformation that maps the points 1, i, and -1 on the circle $|z| = 1$ to the points -1, 0, and 1 on the real axis.

SOLUTION Substituting into (7), we have

$$\frac{w+1}{w-1}\frac{0-1}{0-(-1)} = \frac{z-1}{z+1}\frac{i+1}{i-1}$$

or

$$-\frac{w+1}{w-1} = -i\frac{z-1}{z+1}.$$

Solving for w, we obtain $w = -i(z-i)/(z+i)$. Alternatively, we could use the matrix method to compute $w = S^{-1}(T(z))$. ❑

When $z_k = \infty$ plays the role of one of the points in a triple, the definition of the cross-ratio is changed by replacing each factor that contains z_k by 1. For example, if $z_2 = \infty$, both $z_2 - z_3$ and $z_2 - z_1$ are replaced by 1, giving $(z - z_1)/(z - z_3)$ as the cross-ratio.

Example 5 Constructing a Linear Fractional Transformation

Construct a linear fractional transformation that maps the points ∞, 0, and 1 on the real axis to the points 1, i, and -1 on the circle $|w| = 1$.

SOLUTION Since $z_1 = \infty$, the terms $z - z_1$ and $z_2 - z_1$ in the cross product are replaced by 1. It follows that

$$\frac{w-1}{w+1}\frac{i+1}{i-1} = \frac{1}{z-1}\frac{0-1}{1}$$

or

$$S(w) = -i\frac{w-1}{w+1} = \frac{-1}{z-1} = T(z).$$

If we use the matrix method to find $w = S^{-1}(T(z))$, then

$$\begin{pmatrix} a & b \\ c & d \end{pmatrix} = \mathrm{adj}\begin{pmatrix} -i & i \\ 1 & 1 \end{pmatrix}\begin{pmatrix} 0 & -1 \\ 1 & -1 \end{pmatrix} = \begin{pmatrix} -i & -1+i \\ -i & 1+i \end{pmatrix}$$

and so $w = \dfrac{-iz - 1 + i}{-iz + 1 + i} = \dfrac{z - 1 - i}{z - 1 + i}$. ❑

Example 6 Solving a Dirichlet Problem

Solve the Dirichlet problem in Figure 20.35(a) using conformal mapping by constructing a linear fractional transformation that maps the given region into the upper half-plane.

SOLUTION The boundary circles $|z| = 1$ and $|z - \frac{1}{2}| = \frac{1}{2}$ each pass through $z = 1$. We can therefore map each boundary circle to a line by selecting a linear fractional transformation that has $z = 1$ as a pole. If we further require that $T(i) = 0$ and $T(-1) = 1$, then

$$T(z) = \frac{z-i}{z-1}\frac{-1-1}{-1-i} = (1-i)\frac{z-i}{z-1}.$$

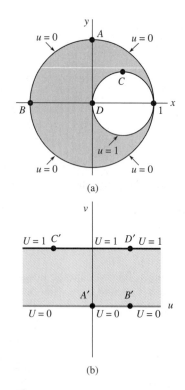

(a)

(b)

Figure 20.35

Since $T(0) = 1 + i$ and $T(\frac{1}{2} + \frac{1}{2}i) = -1 + i$, T maps the interior of the circle $|z| = 1$ onto the upper half-plane and maps the circle $|z - \frac{1}{2}| = \frac{1}{2}$ onto the line $v = 1$. Figure 20.35(b) shows the transferred boundary conditions.

The harmonic function $U(u, v) = v$ is the solution to the simplified Dirichlet problem in the w-plane, and so, by Theorem 20.2, $u(x, y) = U(T(z))$ is the solution to the original Dirichlet problem in the z-plane.

Since the imaginary part of $T(z) = (1 - i)\dfrac{z - i}{z - 1}$ is $\dfrac{1 - x^2 - y^2}{(x - 1)^2 + y^2}$, the solution is given by

$$u(x, y) = \frac{1 - x^2 - y^2}{(x - 1)^2 + y^2}.$$

The level curves $u(x, y) = c$ can be written as

$$\left(x - \frac{c}{1 + c}\right)^2 + y^2 = \left(\frac{1}{1 + c}\right)^2$$

and are therefore circles that pass through $z = 1$. See Figure 20.36. These level curves may be interpreted as the isotherms of the steady-state temperature distribution induced by the boundary temperatures. ❏

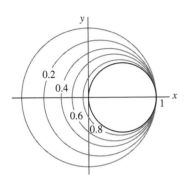

Figure 20.36

Answers to odd-numbered problems begin on page A-76.

In Problems 1–4, a linear fractional transformation is given.

(a) Compute $T(0)$, $T(1)$, and $T(\infty)$.

(b) Find the images of the circles $|z| = 1$ and $|z - 1| = 1$.

(c) Find the image of the disk $|z| \leq 1$.

1. $T(z) = \dfrac{i}{z}$

2. $T(z) = \dfrac{1}{z - 1}$

3. $T(z) = \dfrac{z + 1}{z - 1}$

4. $T(z) = \dfrac{z - i}{z}$

In Problems 5–8, use the matrix method to compute $S^{-1}(w)$ and $S^{-1}(T(z))$ for each pair of linear fractional transformations.

5. $T(z) = \dfrac{z}{iz - 1}$ and $S(z) = \dfrac{iz + 1}{z - 1}$

6. $T(z) = \dfrac{iz}{z - 2i}$ and $S(z) = \dfrac{2z + 1}{z + 1}$

7. $T(z) = \dfrac{2z - 3}{z - 3}$ and $S(z) = \dfrac{z - 2}{z - 1}$

8. $T(z) = \dfrac{z - 1 + i}{iz - 2}$ and $S(z) = \dfrac{(2 - i)z}{z - 1 - i}$

In Problems 9–16, construct a linear fractional transformation that maps the given triple z_1, z_2, and z_3 to the triple w_1, w_2, and w_3.

9. -1, 0, 2 to 0, 1, ∞

10. i, 0, $-i$ to 0, 1, ∞

11. 0, 1, ∞ to 0, i, 2

12. 0, 1, ∞ to $1 + i$, 0, $1 - i$

13. -1, 0, 1 to i, ∞, 0

14. -1, 0, 1 to ∞, $-i$, 1

15. 1, i, $-i$ to -1, 0, 3

16. 1, i, $-i$ to i, $-i$, 1

17. Use the results in Example 2 and the harmonic function $U = (\log_e r)/(\log_e r_0)$ to solve the Dirichlet problem in Figure 20.37. Explain why the level curves must be circles.

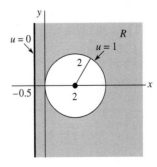

Figure 20.37

18. Use the linear fractional transformation that maps -1, 1, 0 to 0, 1, ∞ to solve the Dirichlet problem in Figure 20.38. Explain why, with one exception, all level curves must be circles. Which level curve is a line?

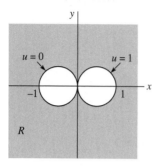

Figure 20.38

19. Derive the conformal mapping H-1 in the conformal mappings in Appendix IV.

20. Derive the conformal mapping H-5 in the conformal mappings in Appendix IV by first mapping 1, i, -1 to ∞, i, 0.

21. Show that the composite of two linear fractional transformations is a linear fractional transformation and verify (5).

22. If $w_1 \neq w_2$ and $\lambda > 0$, show that the set of all points w that satisfy $|w - w_1| = \lambda |w - w_2|$ is a line when $\lambda = 1$ and is a circle when $\lambda \neq 1$. [*Hint:* Write as $|w - w_1|^2 = \lambda^2 |w - w_2|^2$ and expand.]

20.4 Schwarz–Christoffel Transformations

(a) Bounded polygonal region

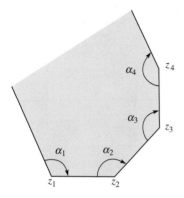

(b) Unbounded polygonal region

Figure 20.39

■ **Mapping to Polygonal Regions** If D' is a simply connected domain with at least one boundary point, then the famous **Riemann mapping theorem** asserts the existence of an analytic function g that conformally maps the unit open disk $|z| < 1$ onto D'. The Riemann mapping theorem is a pure existence theorem that does not specify a formula for the conformal mapping. Since the upper half-plane $y > 0$ can be conformally mapped onto this disk using a linear fractional transformation, it follows that there exists a conformal mapping f between the upper half-plane and D'. In particular, there are analytic functions that map the upper half-plane onto polygonal regions of the types shown in Figure 20.39. Unlike the Riemann mapping theorem, the Schwarz–Christoffel formula specifies a form for the derivative $f'(z)$ of a conformal mapping from the upper half-plane to a bounded or unbounded polygonal region.

■ **Special Cases** To motivate the general Schwarz–Christoffel formula, we first examine the effect of the mapping $f(z) = (z - x_1)^{\alpha/\pi}$, $0 < \alpha < 2\pi$, on the upper half-plane $y \geq 0$ shown in Figure 20.40(a). This mapping is the composite of the translation $\zeta = z - x_1$ and the real power function $w = \zeta^{\alpha/\pi}$. Since $w = \zeta^{\alpha/\pi}$ changes the angle in a wedge by a factor of α/π, the interior angle in the image region is $(\alpha/\pi)\pi = \alpha$. See Figure 20.40(b).

Note that $f'(z) = A(z - x_1)^{(\alpha/\pi) - 1}$ for $A = \alpha/\pi$. Next assume that $f(z)$ is a function that is analytic in the upper half-plane and that has the derivative

$$f'(z) = A(z - x_1)^{(\alpha_1/\pi) - 1}(z - x_2)^{(\alpha_2/\pi) - 1}, \tag{1}$$

where $x_1 < x_2$. In determining the images of line segments on the x-axis, we will use the fact that a curve $w = w(t)$ in the w-plane is a line segment when the argument of its tangent vector $w'(t)$ is constant. From (1), an argument of $f'(t)$ is given by

$$\arg f'(t) = \text{Arg } A + \left(\frac{\alpha_1}{\pi} - 1\right)\text{Arg }(t - x_1) + \left(\frac{\alpha_2}{\pi} - 1\right)\text{Arg }(t - x_2). \tag{2}$$

Since $\text{Arg}(t - x) = \pi$ for $t < x$, we can find the variation of $\arg f'(t)$ along the x-axis. The results are shown in the following table.

Figure 20.40

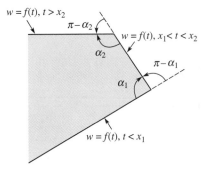

Figure 20.41

Interval	$\arg f'(t)$	Change in argument
$(-\infty, x_1)$	$\text{Arg } A + (\alpha_1 - \pi) + (\alpha_2 - \pi)$	0
(x_1, x_2)	$\text{Arg } A + (\alpha_2 - \pi)$	$\pi - \alpha_1$
(x_2, ∞)	$\text{Arg } A$	$\pi - \alpha_2$

Since $\arg f'(t)$ is constant on the intervals in the table, the images are line segments, and Figure 20.41 shows the image of the upper half-plane. Note that the interior angles of the polygonal image region are α_1 and α_2. This discussion generalizes to produce the Schwarz–Christoffel formula.

THEOREM 20.4 Schwarz–Christoffel Formula

Let $f(z)$ be a function that is analytic in the upper half-plane $y > 0$ and that has the derivative

$$f'(z) = A(z - x_1)^{(\alpha_1/\pi) - 1}(z - x_2)^{(\alpha_2/\pi) - 1} \cdots (z - x_n)^{(\alpha_n/\pi) - 1}, \qquad (3)$$

where $x_1 < x_2 < \cdots < x_n$ and each α_i satisfies $0 < \alpha_i < 2\pi$. Then $f(z)$ maps the upper half-plane $y \geq 0$ to a polygonal region with interior angles $\alpha_1, \alpha_2, \ldots, \alpha_n$.

In applying this formula to a particular polygonal target region, the reader should carefully note the following comments:

i.　One can select the location of three of the points x_k on the x-axis. A judicious choice can simplify the computation of $f(z)$. The selection of the remaining points depends on the shape of the target polygon.

ii.　A general formula for $f(z)$ is

$$f(z) = A\left(\int (z - x_1)^{(\alpha_1/\pi) - 1}(z - x_2)^{(\alpha_2/\pi) - 1} \cdots (z - x_n)^{(\alpha_n/\pi) - 1} \, dz \right) + B,$$

and therefore $f(z)$ may be considered as the composite of the conformal mapping

$$g(z) = \int (z - x_1)^{(\alpha_1/\pi) - 1}(z - x_2)^{(\alpha_2/\pi) - 1} \cdots (z - x_n)^{(\alpha_n/\pi) - 1} \, dz$$

and the linear function $w = Az + B$. The linear function $w = Az + B$ allows us to magnify, rotate, and translate the image polygon produced by $g(z)$. (See Section 20.1.)

iii.　If the polygonal region is bounded, only $n - 1$ of the n interior angles should be included in the Schwarz–Christoffel formula. As an illustration, the interior angles α_1, α_2, α_3, and α_4 are sufficient to determine the Schwarz–Christoffel formula for the pentagon shown in Figure 20.39(a).

Example 1　Constructing a Conformal Mapping

Use the Schwarz–Christoffel formula to construct a conformal mapping from the upper half-plane to the strip $|v| \leq 1$, $u \geq 0$.

(a)

(b)

Figure 20.42

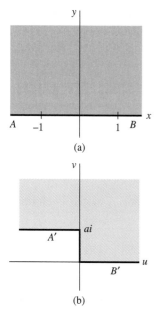

(a)

(b)

Figure 20.43

SOLUTION We may select $x_1 = -1$ and $x_2 = 1$ on the x-axis, and we will construct a conformal mapping f with $f(-1) = -i$ and $f(1) = i$. See Figure 20.42. Since $\alpha_1 = \alpha_2 = \pi/2$, the Schwarz–Christoffel formula (3) gives

$$f'(z) = A(z+1)^{-1/2}(z-1)^{-1/2} = A\,\frac{1}{(z^2-1)^{1/2}} = \frac{A}{i}\,\frac{1}{(1-z^2)^{1/2}}.$$

Therefore, $f(z) = -Ai\,\sin^{-1}z + B$. Since $f(-1) = -i$ and $f(1) = i$, we obtain, respectively,

$$-i = Ai\,\frac{\pi}{2} + B \qquad \text{and} \qquad i = -Ai\,\frac{\pi}{2} + B$$

and conclude that $B = 0$ and $A = -2/\pi$. Thus, $f(z) = (2/\pi)i\,\sin^{-1}z$. ❏

Example 2 Constructing a Conformal Mapping

Use the Schwarz–Christoffel formula to construct a conformal mapping from the upper half-plane to the region shown in Figure 20.43(b).

SOLUTION We again select $x_1 = -1$ and $x_2 = 1$, and we will require that $f(-1) = ai$ and $f(1) = 0$. Since $\alpha_1 = 3\pi/2$ and $\alpha_2 = \pi/2$, the Schwarz–Christoffel formula (3) gives

$$f'(z) = A(z+1)^{1/2}(z-1)^{-1/2}.$$

If we write $f'(z)$ as $A(z/(z^2-1)^{1/2} + 1/(z^2-1)^{1/2})$, it follows that

$$f(z) = A[(z^2-1)^{1/2} + \cosh^{-1}z] + B.$$

Note that $\cosh^{-1}(-1) = \pi i$ and $\cosh^{-1}1 = 0$, and so $ai = f(-1) = A(\pi i) + B$ and $0 = f(1) = B$. Therefore, $A = a/\pi$ and $f(z) = (a/\pi)[(z^2-1)^{1/2} + \cosh^{-1}z]$. ❏

The next example will show that it may not always be possible to find $f(z)$ in terms of elementary functions.

Example 3 Constructing a Conformal Mapping

Use the Schwarz–Christoffel formula to construct a conformal mapping from the upper half-plane to the interior of the equilateral triangle shown in Figure 20.44(b).

SOLUTION Since the polygonal region is bounded, only two of the three 60° interior angles should be included in the Schwarz–Christoffel formula. If $x_1 = 0$ and $x_2 = 1$, we obtain $f'(z) = Az^{-2/3}(z-1)^{-2/3}$. It is not possible to evaluate $f(z)$ in terms of elementary functions; however, we can use Theorem 18.8 to construct the antiderivative

$$f(z) = A \int_0^z \frac{1}{s^{2/3}(s-1)^{2/3}}\,ds + B.$$

If we require that $f(0) = 0$ and $f(1) = 1$, it follows that $B = 0$ and

$$1 = A \int_0^1 \frac{1}{x^{2/3}(x-1)^{2/3}}\,dx.$$

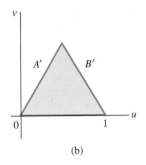

(a) (b)

Figure 20.44

It can be shown that this last integral is $\Gamma(\frac{1}{3})$, where Γ denotes the gamma function. Therefore, the required conformal mapping is

$$f(z) = \frac{1}{\Gamma(\frac{1}{3})} \int_0^z \frac{1}{s^{2/3}(s-1)^{2/3}}\, ds.$$

The Schwarz–Christoffel formula can sometimes be used to suggest a possible conformal mapping from the upper half-plane onto a nonpolygonal region R'. A key first step is to approximate R' by polygonal regions. This will be illustrated in the final example.

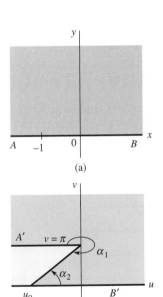

Figure 20.45

Example 4 **Constructing a Conformal Mapping**

Use the Schwarz–Christoffel formula to construct a conformal mapping from the upper half-plane to the upper half-plane with the horizontal line $v = \pi$, $u \le 0$, deleted.

SOLUTION The nonpolygonal target region can be approximated by a polygonal region by adjoining a line segment from $w = \pi i$ to a point u_0 on the negative u-axis. See Figure 20.45(b). If we require that $f(-1) = \pi i$ and $f(0) = u_0$, the Schwarz–Christoffel transformation satisfies

$$f'(z) = A(z+1)^{(\alpha_1/\pi)-1} z^{(\alpha_2/\pi)-1}.$$

Note that as u_0 approaches $-\infty$, the interior angles α_1 and α_2 approach 2π and 0, respectively. This suggests we examine conformal mappings that satisfy $w' = A(z+1)^1 z^{-1} = A(1 + 1/z)$ or $w = A(z + \text{Ln } z) + B$.

We will first determine the image of the upper half-plane under $g(z) = z + \text{Ln } z$ and then translate the image region if needed. For t real,

$$g(t) = t + \log_e|t| + i \text{ Arg } t.$$

If $t < 0$, Arg $t = \pi$ and $u(t) = t + \log_e|t|$ varies from $-\infty$ to -1. It follows that $w = g(t)$ moves along the line $v = \pi$ from $-\infty$ to -1. When $t > 0$, Arg $t = 0$ and $u(t)$ varies from $-\infty$ to ∞. Therefore, g maps the positive x-axis onto the u-axis. We can conclude that $g(z) = z + \text{Ln } z$ maps the upper half-plane onto the upper half-plane with the horizontal line $v = \pi$, $u \le -1$, deleted. Therefore, $w = z + \text{Ln } z + 1$ maps the upper half-plane onto the original target region.

Many of the conformal mappings in Appendix IV can be derived using the Schwarz–Christoffel formula, and we will show in Section 20.6 that these mappings are especially useful in analyzing two-dimensional fluid flows.

In Problems 1–4, use (2) to describe the image of the upper half-plane $y \geq 0$ under the conformal mapping $w = f(z)$ that satisfies the given conditions. Do *not* attempt to find $f(z)$.

1. $f'(z) = (z-1)^{-1/2}, \quad f(1) = 0$

2. $f'(z) = (z+1)^{-1/3}, \quad f(-1) = 0$

3. $f'(z) = (z+1)^{-1/2}(z-1)^{1/2}, \quad f(-1) = 0$

4. $f'(z) = (z+1)^{-1/2}(z-1)^{-3/4}, \quad f(-1) = 0$

In Problems 5–8, find $f'(z)$ for the given polygonal region using $x_1 = -1$, $x_2 = 0$, $x_3 = 1$, $x_4 = 2$, and so on. Do *not* attempt to find $f(z)$.

5. $f(-1) = 0, \quad f(0) = 1$

Figure 20.46

6. $f(-1) = -1, \quad f(0) = 0$

Figure 20.47

7. $f(-1) = -1, \quad f(0) = 1$

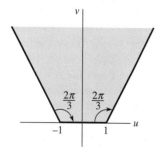

Figure 20.48

8. $f(-1) = i, \quad f(0) = 0$

Figure 20.49

9. Use the Schwarz–Christoffel formula to construct a conformal mapping from the upper half-plane $y \geq 0$ to the region in Figure 20.50. Require that $f(-1) = \pi i$ and $f(1) = 0$.

Figure 20.50

10. Use the Schwarz–Christoffel formula to construct a conformal mapping from the upper half-plane $y \geq 0$ to the region in Figure 20.51. Require that $f(-1) = -ai$ and $f(1) = ai$.

Figure 20.51

11. Use the Schwarz–Christoffel formula to construct a conformal mapping from the upper half-plane $y \geq 0$ to the horizontal strip $0 \leq v \leq \pi$ by first approximating the strip by the polygonal region shown in Figure 20.52. Require that $f(-1) = \pi i$, $f(0) = w_2 = -\overline{w_1}$, and $f(1) = 0$, and let $w_1 \to \infty$ in the horizontal direction.

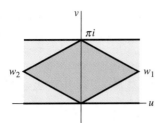

Figure 20.52

12. Use the Schwarz–Christoffel formula to construct a conformal mapping from the upper half-plane $y \geq 0$ to the wedge $0 \leq \text{Arg } w \leq \pi/4$ by first approximating the wedge by the region shown in Figure 20.53. Require that $f(0) = 0$ and $f(1) = 1$ and let $\theta \to 0$.

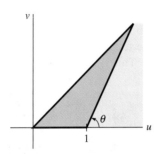

Figure 20.53

13. Verify M-4 in Appendix IV by first approximating the region R' by the polygonal region shown in Figure 20.54. Require that $f(-1) = -u_1, f(0) = ai,$ and $f(1) = u_1$ and let $u_1 \to 0$ along the u-axis.

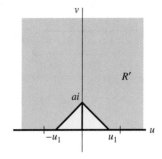

Figure 20.54

14. Show that if a curve in the w-plane is parametrized by $w = w(t), a \leq t \leq b,$ and arg $w'(t)$ is constant, then the curve is a line segment. [*Hint:* If $w(t) = u(t) + iv(t)$, then $\tan(\arg w'(t)) = dv/du$.]

20.5 Poisson Integral Formulas

The success of the conformal mapping method depends on the recognition of the solution to the new Dirichlet problem in the image region R'. It would therefore be helpful if a general solution could be found for Dirichlet problems in either the upper half-plane $y \geq 0$ or the unit disk $|z| \leq 1$. The **Poisson integral formula** for the upper half-plane provides such a solution by expressing the value of a harmonic function at a point in the interior of the upper half-plane in terms of its values on the boundary $y = 0$.

■ **Formulas for the Upper Half-Plane** To develop the formula, we first assume that the boundary function is given by $u(x, 0) = f(x)$, where $f(x)$ is the step function indicated in Figure 20.55. The solution of the corresponding Dirichlet problem in the upper half-plane is

$$u(x, y) = \frac{u_i}{\pi} [\text{Arg}(z - b) - \text{Arg}(z - a)]. \tag{1}$$

Since $\text{Arg}(z - b)$ is an exterior angle in the triangle formed by z, a, and b, $\text{Arg}(z - b) = \theta(z) + \text{Arg}(z - a)$, where $0 < \theta(z) < \pi$, and we can write

$$u(x, y) = \frac{u_i}{\pi} \theta(z) = \frac{u_i}{\pi} \text{Arg}\left(\frac{z - b}{z - a}\right). \tag{2}$$

Figure 20.55

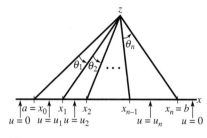

Figure 20.56

The superposition principle can be used to solve the more general Dirichlet problem in Figure 20.56. If $u(x, 0) = u_i$ for $x_{i-1} \le x \le x_i$ and $u(x, 0) = 0$ outside of the interval $[a, b]$, then from (1),

$$u(x, y) = \sum_{i=1}^{n} \frac{u_i}{\pi} [\mathrm{Arg}(z - x_i) - \mathrm{Arg}(z - x_{i-1})] = \frac{1}{\pi} \sum_{i=1}^{n} u_i \theta_i(z). \tag{3}$$

Note that $\mathrm{Arg}(z - t) = \tan^{-1}(y/(x - t))$, where \tan^{-1} is selected between 0 and π, and therefore $(d/dt) \mathrm{Arg}(z - t) = y/((x - t)^2 + y^2)$. From (3),

$$u(x, y) = \frac{1}{\pi} \sum_{i=1}^{n} \int_{x_{i-1}}^{x_i} u_i \frac{d}{dt} \mathrm{Arg}(z - t)\, dt = \frac{1}{\pi} \sum_{i=1}^{n} \int_{x_{i-1}}^{x_i} \frac{u_i y}{(x - t)^2 + y^2}\, dt.$$

Since $u(x, 0) = 0$ outside of the interval $[a, b]$, we have

$$u(x, y) = \frac{y}{\pi} \int_{-\infty}^{\infty} \frac{u(t, 0)}{(x - t)^2 + y^2}\, dt. \tag{4}$$

A bounded piecewise continuous function can be approximated by step functions, and therefore our discussion suggests that (4) is the general solution to the Dirichlet problem in the upper half-plane. This is the content of Theorem 20.5.

THEOREM 20.5 Poisson Integral Formula for the Upper Half-Plane

Let $u(x, 0)$ be a piecewise continuous function on every finite interval and bounded on $-\infty < x < \infty$. Then the function defined by

$$u(x, y) = \frac{y}{\pi} \int_{-\infty}^{\infty} \frac{u(t, 0)}{(x - t)^2 + y^2}\, dt$$

is the solution of the corresponding Dirichlet problem on the upper half-plane $y \ge 0$.

There are a few functions for which it is possible to evaluate the integral in (4), but in general numerical methods are required to evaluate the integral.

Example 1 Solving a Dirichlet Problem

Find the solution of the Dirichlet problem in the upper half-plane that satisfies the boundary condition $u(x, 0) = x$ when $|x| < 1$, and $u(x, 0) = 0$ otherwise.

SOLUTION By the Poisson integral formula,

$$u(x, y) = \frac{y}{\pi} \int_{-1}^{1} \frac{t}{(x - t)^2 + y^2}\, dt.$$

Using the substitution $s = x - t$, we can show that

$$u(x, y) = \frac{1}{\pi} \left[\frac{y}{2} \log_e ((x - t)^2 + y^2) - x \tan^{-1}\left(\frac{x - t}{y}\right) \right]\Bigg|_{t=-1}^{t=1},$$

which can be simplified to

$$u(x, y) = \frac{y}{2\pi} \log_e \left[\frac{(x - 1)^2 + y^2}{(x + 1)^2 + y^2} \right] + \frac{x}{\pi} \left[\tan^{-1}\left(\frac{x + 1}{y}\right) - \tan^{-1}\left(\frac{x - 1}{y}\right) \right]. \quad \square$$

In most of the examples and exercises $u(x, 0)$ is a step function, and we will use the integrated solution (3) rather than (4). If the first interval is $(-\infty, x_1)$, then the term $\text{Arg}(z - x_1) - \text{Arg}(z - a)$ in the sum should be replaced by $\text{Arg}(z - x_1)$. Likewise, if the last interval is (x_{n-1}, ∞), then $\text{Arg}(z - b) - \text{Arg}(z - x_{n-1})$ should be replaced by $\pi - \text{Arg}(z - x_{n-1})$.

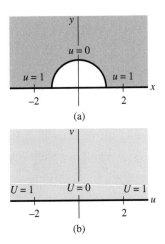

(a)

(b)

Figure 20.57

Example 2 Solving a Dirichlet Problem

The conformal mapping $f(z) = z + 1/z$ maps the region in the upper half-plane and outside the circle $|z| = 1$ onto the upper half-plane $v \geq 0$. Use this mapping and the Poisson integral formula to solve the Dirichlet problem shown in Figure 20.57(a).

SOLUTION Using the results of Example 4 in Section 20.2, we can transfer the boundary conditions to the w-plane. See Figure 20.57(b). Since $U(u, 0)$ is a step function, we will use the integrated solution (3) rather than the Poisson integral. The solution to the new Dirichlet problem is

$$U(u, v) = \frac{1}{\pi} \text{Arg}(w + 2) + \frac{1}{\pi}[\pi - \text{Arg}(w - 2)] = 1 + \frac{1}{\pi}\text{Arg}\!\left(\frac{w + 2}{w - 2}\right),$$

and therefore

$$u(x, y) = U\!\left(z + \frac{1}{z}\right) = 1 + \frac{1}{\pi}\text{Arg}\!\left(\frac{z + 1/z + 2}{z + 1/z - 2}\right),$$

which can be simplified to $u(x, y) = 1 + \dfrac{1}{\pi}\text{Arg}\!\left(\dfrac{z + 1}{z - 1}\right)^2$. ❏

■ **Formula for the Unit Disk** A Poisson integral formula can also be developed to solve the general Dirichlet problem for the unit disk:

THEOREM 20.6 **Poisson Integral Formula for the Unit Disk**

Let $u(e^{i\theta})$ be bounded and piecewise continuous for $-\pi \leq \theta \leq \pi$. Then the solution to the corresponding Dirichlet problem on the open unit disk $|z| < 1$ is given by

$$u(x, y) = \frac{1}{2\pi} \int_{-\pi}^{\pi} u(e^{it}) \frac{1 - |z|^2}{|e^{it} - z|^2}\, dt. \tag{5}$$

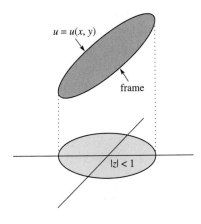

Figure 20.58

■ **Geometric Interpretation** Figure 20.58 shows a thin membrane (such as a soap film) that has been stretched across a frame defined by $u = u(e^{i\theta})$. The displacement u in the direction perpendicular to the z-plane satisfies the **two-dimensional wave equation**

$$a^2\!\left(\frac{\partial^2 u}{\partial x^2} + \frac{\partial^2 u}{\partial y^2}\right) = \frac{\partial^2 u}{\partial t^2},$$

and so at equilibrium, the displacement function $u = u(x, y)$ is harmonic. Formula (5) provides an explicit solution for the displacement u and has the advantage that the integral is over the finite interval $[-\pi, \pi]$. When the integral cannot be evaluated, standard numerical integration procedures can be used to estimate $u(x, y)$ at a fixed point $z = x + iy$ with $|z| < 1$.

Example 3 **Displacement of a Membrane**

A frame for a membrane is defined by $u(e^{i\theta}) = |\theta|$ for $-\pi \le \theta \le \pi$. Estimate the equilibrium displacement of the membrane at $(-0.5, 0)$, $(0, 0)$, and $(0.5, 0)$.

SOLUTION Using (5), we get $u(x, y) = \dfrac{1}{2\pi} \displaystyle\int_{-\pi}^{\pi} |t| \dfrac{1 - |z|^2}{|e^{it} - z|^2} \, dt$. When $(x, y) = (0, 0)$, we get

$$u(x, y) = \frac{1}{2\pi} \int_{-\pi}^{\pi} |t| \, dt = \frac{\pi}{2}.$$

For the other two values of (x, y), the integral is not elementary and must be estimated using a numerical integration procedure. Using Simpson's rule, we obtain (to four decimal places) $u(-0.5, 0) = 2.2269$ and $u(0.5, 0) = 0.9147$. ❏

■ **Fourier Series Form** The Poisson integral formula for the unit disk is actually a compact way of writing the Fourier series solution to Laplace's equation that we developed in Chapter 14. To see this, first note that $u_n(r, \theta) = r^n \cos n\theta$ and $v_n(r, \theta) = r^n \sin n\theta$ are each harmonic, since these functions are the real and imaginary parts of z^n. If a_0, a_n, and b_n are chosen to be the Fourier coefficients of $u(e^{i\theta})$ for $-\pi < \theta < \pi$, then, by the superposition principle,

$$u(r, \theta) = \frac{a_0}{2} + \sum_{n=1}^{\infty} (a_n r^n \cos n\theta + b_n r^n \sin n\theta) \tag{6}$$

is harmonic and $u(1, \theta) = (a_0/2) + \sum_{n=1}^{\infty} (a_n \cos n\theta + b_n \sin n\theta) = u(e^{i\theta})$. Since the solution of the Dirichlet problem is also given by (5), we have

$$u(r, \theta) = \frac{1}{2\pi} \int_{-\pi}^{\pi} u(e^{it}) \frac{1 - r^2}{|e^{it} - re^{i\theta}|^2} \, dt = \frac{a_0}{2} + \sum_{n=1}^{\infty} (a_n r^n \cos n\theta + b_n r^n \sin n\theta).$$

Example 4 **Solving a Dirichlet Problem**

Find the solution of the Dirichlet problem in the unit disk satisfying the boundary condition $u(e^{i\theta}) = \sin 4\theta$. Sketch the level curve $u = 0$.

SOLUTION Rather than working with the Poisson integral (5), we will use the Fourier series solution (6), which reduces to $u(r, \theta) = r^4 \sin 4\theta$. Therefore, $u = 0$ if and only if $\sin 4\theta = 0$. This implies $u = 0$ on the lines $x = 0$, $y = 0$, and $y = \pm x$.

If we switch to rectangular coordinates, $u(x, y) = 4xy(x^2 - y^2)$. The surface $u(x, y) = 4xy(x^2 - y^2)$, the frame $u(e^{i\theta}) = \sin 4\theta$, and the system of level curves were sketched using graphics software and are shown in Figure 20.59. ❏

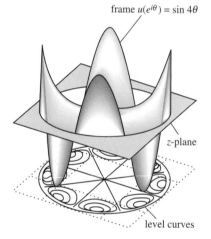

frame $u(e^{i\theta}) = \sin 4\theta$

z-plane

level curves

Figure 20.59

Answers to odd-numbered problems begin on page A-77.

In Problems 1–4, use the integrated solution (3) to the Poisson integral formula to solve the given Dirichlet problem in the upper half-plane.

1.

$u = 0 \quad u = -1 \quad u = 1 \quad u = 0$

Figure 20.60

2.

$u = 0 \quad u = 5 \quad u = 1 \quad u = 0$

Figure 20.61

3.

Figure 20.62

4.

Figure 20.63

5. Find the solution of the Dirichlet problem in the upper half-plane that satisfies the boundary condition $u(x, 0) = x^2$ when $0 < x < 1$, and $u(x, 0) = 0$ otherwise.

6. Find the solution of the Dirichlet problem in the upper half-plane that satisfies the boundary condition $u(x, 0) = \cos x$. [*Hint:* Let $s = t - x$ and use the Section 19.6 formulas

$$\int_{-\infty}^{\infty} \frac{\cos s}{s^2 + a^2}\, ds = \frac{\pi e^{-a}}{a}, \qquad \int_{-\infty}^{\infty} \frac{\sin s}{s^2 + a^2}\, ds = 0$$

for $a > 0$.]

In Problems 7–10, solve the given Dirichlet problem by finding a conformal mapping from the given region onto the upper half-plane $v \geq 0$.

7.

Figure 20.64

8.

Figure 20.65

9.

Figure 20.66

10.

Figure 20.67

11. A frame for a membrane is defined by $u(e^{i\theta}) = \theta^2/\pi^2$ for $-\pi \leq \theta \leq \pi$. Use the Poisson integral formula for the unit disk to estimate the equilibrium displacement of the membrane at $(-0.5, 0)$, $(0, 0)$, and $(0.5, 0)$.

12. A frame for a membrane is defined by $u(e^{i\theta}) = e^{-|\theta|}$ for $-\pi \leq \theta \leq \pi$. Use the Poisson integral formula for the unit disk to estimate the equilibrium displacement of the membrane at $(-0.5, 0)$, $(0, 0)$, and $(0.5, 0)$.

13. Use the Poisson integral formula for the unit disk to show that $u(0, 0)$ is the average value of the function $u = u(e^{i\theta})$ on the boundary $|z| = 1$.

In Problems 14 and 15, solve the given Dirichlet problem for the unit disk using the Fourier series form of the Poisson integral formula, and sketch the system of level curves.

14. $u(e^{i\theta}) = \cos 2\theta$ **15.** $u(e^{i\theta}) = \sin\theta + \cos\theta$

20.6 Applications

In Sections 20.2, 20.3, and 20.5 we demonstrated how Laplace's equation can be solved with conformal mapping methods, and we interpreted a solution $u = u(x, y)$ of the Dirichlet problem as either the steady-state temperature at the point (x, y) or the equilibrium displacement of a membrane at the point (x, y). Laplace's equation is a fundamental partial differential equation that arises in a variety of contexts. In this section we will establish a general relationship between vector fields and analytic functions and use our conformal mapping techniques to solve problems involving electrostatic force fields and two-dimensional fluid flows.

■ **Vector Fields** A vector field $\mathbf{F}(x, y) = P(x, y)\mathbf{i} + Q(x, y)\mathbf{j}$ in a domain D can also be expressed in the complex form

$$\mathbf{F}(x, y) = P(x, y) + iQ(x, y)$$

and thought of as a complex function. Recall from Chapter 9 that div $\mathbf{F} = \partial P/\partial x + \partial Q/\partial y$ and curl $\mathbf{F} = (\partial Q/\partial x - \partial P/\partial y)\mathbf{k}$. If we require that both div $\mathbf{F} = 0$ and curl $\mathbf{F} = \mathbf{0}$, then

$$\frac{\partial P}{\partial x} = -\frac{\partial Q}{\partial y} \qquad \text{and} \qquad \frac{\partial P}{\partial y} = \frac{\partial Q}{\partial x}. \tag{1}$$

This set of equations is reminiscent of the Cauchy–Riemann criterion for analyticity presented in Theorem 17.3 and suggests that we examine the complex function $g(z) = P(x, y) - iQ(x, y)$.

THEOREM 20.7 Vector Fields and Analyticity

(*i*) Suppose that $\mathbf{F}(x, y) = P(x, y) + iQ(x, y)$ is a vector field in a domain D and $P(x, y)$ and $Q(x, y)$ are continuous and have continuous first partial derivatives in D. If div $\mathbf{F} = 0$ and curl $\mathbf{F} = \mathbf{0}$, then the complex function

$$g(z) = P(x, y) - iQ(x, y)$$

is analytic in D.

(*ii*) Conversely, if $g(z)$ is analytic in D, then $\mathbf{F}(x, y) = \overline{g(z)}$ defines a vector field in D for which div $\mathbf{F} = 0$ and curl $\mathbf{F} = \mathbf{0}$.

PROOF If $u(x, y)$ and $v(x, y)$ denote the real and imaginary parts of $g(z)$, then $u = P$ and $v = -Q$. Therefore the equations in (1) are equivalent to the equations

$$\frac{\partial u}{\partial x} = -\frac{\partial(-v)}{\partial y} \qquad \text{and} \qquad \frac{\partial u}{\partial y} = \frac{\partial(-v)}{\partial x};$$

that is,

$$\frac{\partial u}{\partial x} = \frac{\partial v}{\partial y} \qquad \text{and} \qquad \frac{\partial u}{\partial y} = -\frac{\partial v}{\partial x}. \tag{2}$$

The equations in (2) are the Cauchy–Riemann equations for analyticity. ❏

Example 1 Vector Field Gives an Analytic Function

The vector field defined by $\mathbf{F}(x, y) = (-kq/|z - z_0|^2)(z - z_0)$ may be interpreted as the electric field produced by a wire that is perpendicular to the z-plane at $z = z_0$ and carries a charge of q coulombs per unit length. The corresponding complex function is

$$g(z) = \frac{-kq}{|z - z_0|^2}\,\overline{(z - z_0)} = \frac{-kq}{z - z_0}.$$

Since $g(z)$ is analytic for $z \neq z_0$, div $\mathbf{F} = 0$ and curl $\mathbf{F} = \mathbf{0}$. ❏

Example 2 Analytic Function Gives a Vector Field

The complex function $g(z) = Az$, $A > 0$, is analytic in the first quadrant and therefore gives rise to the vector field $\mathbf{V}(x, y) = \overline{g(z)} = Ax - iAy$, which satisfies div $\mathbf{V} = 0$ and curl $\mathbf{V} = \mathbf{0}$. We will show toward the end of this section that $\mathbf{V}(x, y)$ may be interpreted as the velocity of a fluid that moves around the corner produced by the boundary of the first quadrant. ❏

The physical interpretation of the conditions div $\mathbf{F} = 0$ and curl $\mathbf{F} = \mathbf{0}$ depends on the setting. If $\mathbf{F}(x, y)$ represents the force in an electric field that acts on a unit test charge placed at (x, y), then, by Theorem 9.9, curl $\mathbf{F} = \mathbf{0}$ if and only if the field is **conservative.** The work done in transporting a test charge between two points in D must be independent of the path.

If C is a simple closed contour that lies in D, *Gauss's law* asserts that the line integral $\oint_C (\mathbf{F} \cdot \mathbf{n})\, ds$ is proportional to the total charge enclosed by the curve C. If D is simply connected and all the electric charge is distributed on the boundary of D, then $\oint_C (\mathbf{F} \cdot \mathbf{n})\, ds = 0$ for any simple closed contour in D. By the divergence theorem in the form (1) of Section 9.16,

$$\oint_C (\mathbf{F} \cdot \mathbf{n})\, ds = \iint_R \text{div } \mathbf{F}\, dA, \tag{3}$$

where R is the region enclosed by C, and we can conclude that div $\mathbf{F} = 0$ in D. Conversely, if div $\mathbf{F} = 0$ in D, the double integral is zero and therefore the domain D contains no charge.

■ **Potential Functions** Suppose that $\mathbf{F}(x, y)$ is a vector field in a *simply connected* domain D with both div $\mathbf{F} = 0$ and curl $\mathbf{F} = \mathbf{0}$. By Theorem 18.8, the analytic function $g(z) = P(x, y) - iQ(x, y)$ has an antiderivative

$$G(z) = \phi(x, y) + i\psi(x, y) \tag{4}$$

in D, which is called a **complex potential** for the vector field \mathbf{F}. Note that

$$g(z) = G'(z) = \frac{\partial \phi}{\partial x}(x, y) + i\frac{\partial \psi}{\partial x}(x, y) = \frac{\partial \phi}{\partial x}(x, y) - i\frac{\partial \phi}{\partial y}(x, y) \text{ and so}$$

$$\frac{\partial \phi}{\partial x} = P \qquad \text{and} \qquad \frac{\partial \phi}{\partial y} = Q. \tag{5}$$

Therefore, $\mathbf{F} = \nabla\phi$ and, as in Section 9.9, the harmonic function ϕ is called a (real) **potential function** for \mathbf{F}.* When the potential ϕ is specified on the boundary of a region R, we can use conformal mapping techniques to solve the resulting Dirichlet problem. The equipotential lines $\phi(x, y) = c$ can be sketched and the vector field \mathbf{F} can be determined using (5).

Example 3 Complex Potential

The potential ϕ in the half-plane $x \geq 0$ satisfies the boundary conditions $\phi(0, y) = 0$ and $\phi(x, 0) = 1$ for $x \geq 1$ (see Figure 20.68(a)). Determine a complex potential, the equipotential lines, and the force field \mathbf{F}.

SOLUTION We saw in Example 3 of Section 20.2 that the analytic function $z = \sin w$ maps the strip $0 \leq u \leq \pi/2$ in the w-plane to the region R in question. Therefore, $f(z) = \sin^{-1} z$ maps R onto the strip, and Figure 20.68(b) shows the transferred boundary conditions. The simplified Dirichlet problem has the solution $U(u, v) = \frac{2}{\pi}u$, and so $\phi(x, y) = U(\sin^{-1} z) = \text{Re}(\frac{2}{\pi} \sin^{-1} z)$ is the potential function on D, and $G(z) = \frac{2}{\pi} \sin^{-1} z$ is a complex potential for the force field \mathbf{F}.

Note that the equipotential lines $\phi = c$ are the images of the equipotential lines $U = c$ in the w-plane under the inverse mapping $z = \sin w$. In Example 3

(a)

(b)

Figure 20.68

* If \mathbf{F} is an electric field, the electric potential function Φ is defined to be $-\phi$ and $\mathbf{F} = -\nabla\Phi$.

of Section 20.2 we showed that the vertical line $u = a$ is mapped onto a branch of the hyperbola

$$\frac{x^2}{\sin^2 a} - \frac{y^2}{\cos^2 a} = 1.$$

Since the equipotential line $U = c$, $0 < c < 1$, is the vertical line $u = \frac{\pi}{2}c$, it follows that the equipotential line $\phi = c$ is the right branch of the hyperbola

$$\frac{x^2}{\sin^2(\pi c/2)} - \frac{y^2}{\cos^2(\pi c/2)} = 1.$$

Since $\mathbf{F} = \overline{G'(z)}$ and $(d/dz) \sin^{-1} z = 1/(1 - z^2)^{1/2}$, the force field is given by

$$\mathbf{F} = \frac{2}{\pi} \overline{\frac{1}{(1 - z^2)^{1/2}}} = \frac{2}{\pi} \frac{1}{(1 - \bar{z}^2)^{1/2}}. \qquad \square$$

■ **Steady-State Fluid Flow** The vector $\mathbf{V}(x, y) = P(x, y) + iQ(x, y)$ may also be interpreted as the velocity vector of a two-dimensional steady-state fluid flow at a point (x, y) in a domain D. The velocity at all points in the domain is therefore independent of time, and all movement takes place in planes that are parallel to a z-plane.

The physical interpretation of the conditions div $\mathbf{V} = 0$ and curl $\mathbf{V} = \mathbf{0}$ was discussed in Section 9.7. Recall that if curl $\mathbf{V} = \mathbf{0}$ in D, the flow is called **irrotational.** If a small circular paddle wheel is placed in the fluid, the net angular velocity on the boundary of the wheel is zero, and so the wheel will not rotate. If div $\mathbf{V} = 0$ in D, the flow is called **incompressible.** In a simply connected domain D, an incompressible flow has the special property that the amount of fluid in the interior of any simple closed contour C is independent of time. The rate at which fluid enters the interior of C matches the rate at which it leaves, and consequently there can be no fluid sources or sinks at points in D.

If div $\mathbf{V} = 0$ and curl $\mathbf{V} = \mathbf{0}$, \mathbf{V} has a **complex velocity potential**

$$G(z) = \phi(x, y) + i\psi(x, y)$$

that satisfies $\overline{G'(z)} = \mathbf{V}$. In this setting, special importance is placed on the level curves $\psi(x, y) = c$. If $z(t) = x(t) + iy(t)$ is the path of a particle (such as a small cork) that has been placed in the fluid, then

$$\frac{dx}{dt} = P(x, y)$$

$$\frac{dy}{dt} = Q(x, y). \tag{6}$$

Hence, $dy/dx = Q(x, y)/P(x, y)$ or $-Q(x, y)\, dx + P(x, y)\, dy = 0$. This differential equation is exact, since div $\mathbf{V} = 0$ implies $\partial(-Q)/\partial y = \partial P/\partial x$. By the Cauchy–Riemann equations, $\partial\psi/\partial x = -\partial\phi/\partial y = -Q$ and $\partial\psi/\partial y = \partial\phi/\partial x = P$, and therefore all solutions of (6) satisfy $\psi(x, y) = c$. The function $\psi(x, y)$ is therefore called a **stream function** and the level curves $\psi(x, y) = c$ are **streamlines** for the flow.

Example 4 Uniform Flow

The uniform flow in the upper half-plane is defined by $\mathbf{V}(x, y) = A(1, 0)$, where A is a fixed positive constant. Note that $|\mathbf{V}| = A$ and so a particle in the fluid moves at a constant speed. A complex potential for the vector field is $G(z) = Az = Ax + iAy$, and so the streamlines are the horizontal lines $Ay = c$. See Figure 20.69(a). Note that the boundary $y = 0$ of the region is itself a streamline. $\qquad \square$

Figure 20.69

Figure 20.70

Example 5 Flow Around a Corner

The analytic function $G(z) = z^2$ gives rise to the vector field $\mathbf{V}(x,\ y) = \overline{G'(z)} = (2x,\ -2y)$ in the first quadrant. Since $z^2 = x^2 - y^2 + i(2xy)$, the stream function is $\psi(x,\ y) = 2xy$ and the streamlines are the hyperbolas $2xy = c$. This flow, called flow around a corner, is depicted in Figure 20.69(b). As in Example 4, the boundary lines $x = 0$ and $y = 0$ in the first quadrant are themselves streamlines. ❏

■ **Constructing Special Flows** The process of constructing an irrotational and incompressible flow that remains inside a given region R is called **streamlining.** Since the streamlines are described by $\psi(x,\ y) = c$, two distinct streamlines do not intersect. Therefore, if the boundary is itself a streamline, a particle that starts inside R cannot leave R. This is the content of the following theorem:

> **THEOREM 20.8 Streamlining**
>
> Suppose that $G(z) = \phi(x,\ y) + i\psi(x,\ y)$ is analytic in a region R and $\psi(x,\ y)$ is constant on the boundary of R. Then $\mathbf{V}(x,\ y) = \overline{G'(z)}$ defines an irrotational and incompressible fluid flow in R. Moreover, if a particle is placed inside R, its path $z = z(t)$ remains in R.

Example 6 Flow Around a Cylinder

The analytic function $G(z) = z + 1/z$ maps the region R in the upper half-plane and outside the circle $|z| = 1$ onto the upper half-plane $v \geq 0$. The boundary of R is mapped onto the u-axis, and so $v = \psi(x,\ y) = y - y/(x^2 + y^2)$ is zero on the boundary of R. Figure 20.70 shows the streamlines of the resulting flow. The velocity field is given by $\overline{G'(z)} = 1 - 1/\overline{z}^2$, and so

$$\overline{G'(re^{i\theta})} = 1 - \frac{1}{r^2}\,e^{2i\theta}.$$

It follows that $\mathbf{V} \approx (1,\ 0)$ for large values of r, and so the flow is approximately uniform at large distances from the circle $|z| = 1$. The resulting flow in the region R is called flow around a cylinder. The mirror image of the flow can be adjoined to give a flow around a complete cylinder. ❏

If R is a polygonal region, we can use the Schwarz–Christoffel formula to find a conformal mapping $z = f(w)$ from the upper half-plane R' onto R. The inverse function $G(z) = f^{-1}(z)$ maps the boundary of R onto the u-axis. Therefore, if $G(z) = \phi(x,\ y) + i\psi(x,\ y)$, then $\psi(x,\ y) = 0$ on the boundary of R. Note that the streamlines $\psi(x,\ y) = c$ in the z-plane are the images of the horizontal lines $v = c$ in the w-plane under $z = f(w)$.

Example 7 Streamlines Defined Parametrically

The analytic function $f(w) = w + \text{Ln } w + 1$ maps the upper half-plane $v \geq 0$ to the upper half-plane $y \geq 0$ with the horizontal line $y = \pi$, $x \leq 0$, deleted. See Example 4 in Section 20.4. If $G(z) = f^{-1}(z) = \phi(x,\ y) + i\psi(x,\ y)$, then $G(z)$ maps R onto the upper half-plane and maps the boundary of R onto the u-axis. Therefore, $\psi(x,\ y) = 0$ on the boundary of R.

It is not possible to find an explicit formula for the stream function $\psi(x,\ y)$. The streamlines, however, are the images of the horizontal lines $v = c$ under

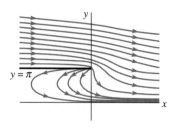

Figure 20.71

$z = f(w)$. If we write $w = t + ic$, $c > 0$, then the streamlines can be represented in the parametric form

$$z = f(t + ic) = t + ic + \text{Ln}(t + ic) + 1,$$

that is, $x = t + 1 + \dfrac{1}{2}\log_e(t^2 + c^2)$, $y = c + \text{Arg}(t + ic)$.

Graphing software was used to generate the streamlines in Figure 20.71. ❏

A stream function $\psi(x, y)$ is harmonic but, unlike a solution to a Dirichlet problem, we do *not* require $\psi(x, y)$ to be bounded (see Examples 4–6) or to assume a fixed set of constants on the boundary. Therefore, there may be many different stream functions for a given region that satisfy Theorem 20.8. This will be illustrated in the final example.

Example 8 Streamlines Defined Parametrically

The analytic function $f(w) = w + e^w + 1$ maps the horizontal strip $0 \le v \le \pi$ onto the region R shown in Figure 20.71. Therefore, $G(z) = f^{-1}(z) = \phi(x, y) + i\psi(x, y)$ maps R back to the strip and, from M-1 in the conformal mappings in Appendix IV, maps the boundary line $y = 0$ onto the u-axis and maps the boundary line $y = \pi$, $x \le 0$, onto the horizontal line $v = \pi$. Therefore, $\psi(x, y)$ is constant on the boundary of R.

The streamlines are the images of the horizontal lines $v = c$, $0 < c < \pi$, under $z = f(w)$. As in Example 7, a parametric representation of the streamlines is

$$z = f(t + ic) = t + ic + e^{t + ic} + 1$$

or $x = t + 1 + e^t \cos c$, $y = c + e^t \sin c$.

The streamlines are shown in Figure 20.72. Unlike the flow in Example 7, the fluid appears to emerge from the strip $0 \le y \le \pi$, $x \le 0$. ❏

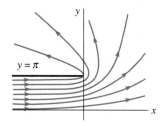

Figure 20.72

In Problems 1–4, verify that div $\mathbf{F} = 0$ and curl $\mathbf{F} = \mathbf{0}$ for the given vector field $\mathbf{F}(x, y)$ by examining the corresponding complex function $g(z) = P(x, y) - iQ(x, y)$. Find a complex potential for the vector field and sketch the equipotential lines.

1. $\mathbf{F}(x, y) = (\cos\theta_0)\mathbf{i} + (\sin\theta_0)\mathbf{j}$

2. $\mathbf{F}(x, y) = -y\mathbf{i} - x\mathbf{j}$

3. $\mathbf{F}(x, y) = \dfrac{x}{x^2 + y^2}\mathbf{i} + \dfrac{y}{x^2 + y^2}\mathbf{j}$

4. $\mathbf{F}(x, y) = \dfrac{x^2 - y^2}{(x^2 + y^2)^2}\mathbf{i} + \dfrac{2xy}{(x^2 + y^2)^2}\mathbf{j}$

5. The potential ϕ on the wedge $0 \le \text{Arg } z \le \pi/4$ satisfies the boundary conditions $\phi(x, 0) = 0$ and $\phi(x, x) = 1$ for $x > 0$. Determine a complex potential, the equipotential lines, and the corresponding force field \mathbf{F}.

6. Use the conformal mapping $f(z) = 1/z$ to determine a complex potential, the equipotential lines, and the corre-

sponding force field \mathbf{F} for the potential ϕ that satisfies the boundary conditions shown in Figure 20.73.

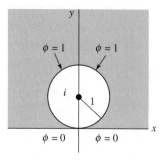

Figure 20.73

7. The potential ϕ on the semicircle $|z| \le 1$, $y \ge 0$, satisfies the boundary conditions $\phi(x, 0) = 0$, $-1 < x < 1$, and $\phi(e^{i\theta}) = 1$, $0 < \theta < \pi$. Show that

$$\phi(x, y) = \frac{1}{\pi} \operatorname{Arg}\left(\frac{z-1}{z+1}\right)^2$$

and use the mapping properties of linear fractional transformations to explain why the equipotential lines are arcs of circles.

8. Use the conformal mapping C-1 in Appendix IV to find the potential ϕ in the region outside the two circles $|z| = 1$ and $|z - 3| = 1$ if the potential is kept at zero on $|z| = 1$ and one on $|z - 3| = 1$. Use the mapping properties of linear fractional transformations to explain why the equipotential lines are, with one exception, circles.

In Problems 9–14, a complex velocity potential $G(z)$ is defined on a region R.

(a) Find the stream function and verify that the boundary of R is a streamline.

(b) Find the corresponding velocity vector field $\mathbf{V}(x, y)$.

(c) Use a graphing utility to sketch the streamlines of the flow.

9. $G(z) = z^4$

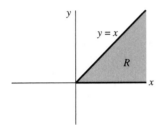

Figure 20.74

10. $G(z) = z^{2/3}$

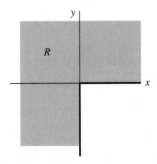

Figure 20.75

11. $G(z) = \sin z$

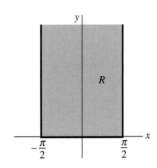

Figure 20.76

12. $G(z) = i \sin^{-1} z$

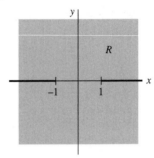

Figure 20.77

13. $G(z) = z^2 + 1/z^2$

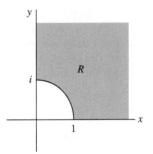

Figure 20.78

14. $G(z) = e^z$

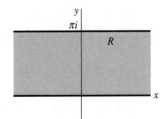

Figure 20.79

In Problems 15–18, a conformal mapping $z = f(w)$ from the upper half-plane $v \geq 0$ to a region R in the z-plane is given and the flow in R with complex potential $G(z) = f^{-1}(z)$ is constructed.

(a) Verify that the boundary of R is a streamline for the flow.

(b) Find a parametric representation for the streamlines of the flow.

(c) Use a graphing utility to sketch the streamlines of the flow.

15. M-9 in Appendix IV

16. M-4 in Appendix IV; use $a = 1$

17. M-2 in Appendix IV; use $a = 1$

18. M-5 in Appendix IV

19. A stagnation point in a flow is a point at which $\mathbf{V} = \mathbf{0}$. Find all stagnation points for the flows in Examples 5 and 6.

20. For any two real numbers k and x_1, the function $G(z) = k \, \text{Ln}(z - x_1)$ is analytic in the upper half-plane and therefore is a complex potential for a flow. The real number x_1 is called a *sink* when $k < 0$ and a *source* for the flow when $k > 0$.

(a) Show that the streamlines are rays emanating from x_1.

(b) Show that $\mathbf{V} = (k/|z - x_1|^2)(z - x_1)$ and conclude that the flow is directed toward x_1 precisely when $k < 0$.

21. If $f(z)$ is a conformal mapping from a domain D onto the upper half-plane, a flow with a source at a point ξ_0 on the boundary of D is defined by the complex potential $G(z) = k \, \text{Ln}(f(z) - f(\xi_0))$, where $k > 0$. Determine the streamlines for a flow in the first quadrant with a source at $\xi_0 = 1$ and $k = 1$.

22. (a) Construct a flow on the horizontal strip $0 < y < \pi$ with a sink at the boundary point $\xi_0 = 0$. [*Hint:* See Problem 21.]

(b) Use a graphing utility to sketch the streamlines of the flow.

23. The complex potential $G(z) = k \, \text{Ln}(z - 1) - k \, \text{Ln}(z + 1)$ with $k > 0$ gives rise to a flow on the upper half-plane with a single source at $z = 1$ and a single sink at $z = -1$. Show that the streamlines are the family of circles $x^2 + (y - c)^2 = 1 + c^2$. See Figure 20.80.

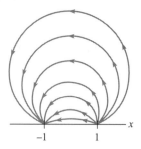

Figure 20.80

24. The flow with velocity vector $\mathbf{V} = (a + ib)/\bar{z}$ is called a *vortex* at $z = 0$, and the geometric nature of the streamlines depends on the choice of a and b.

(a) Show that if $z = x(t) + iy(t)$ is the path of a particle, then
$$\frac{dx}{dt} = \frac{ax - by}{x^2 + y^2}$$
$$\frac{dy}{dt} = \frac{bx + ay}{x^2 + y^2}.$$

(b) Change to polar coordinates to establish that $dr/dt = a/r$ and $d\theta/dt = b/r^2$, and conclude that $r = ce^{a\theta/b}$ for $b \neq 0$. [*Hint:* See (2) of Section 11.1.]

(c) Conclude that the logarithmic spirals in part **(b)** spiral inward if and only if $a < 0$, and the curves are traversed clockwise if and only if $b < 0$. See Figure 20.81.

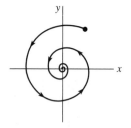

Figure 20.81

Answer Problems 1–10 without referring back to the text. Fill in the blank or answer true/false.

1. Under the complex mapping $f(z) = z^2$, the curve $xy = 2$ is mapped onto the line _____.

2. The complex mapping $f(z) = -iz$ is a rotation through _____ degrees.

3. The image of the upper half-plane $y \geq 0$ under the complex mapping $f(z) = z^{2/3}$ is _____.

4. The analytic function $f(z) = \cosh z$ is conformal except at $z =$ _____.

5. If $w = f(z)$ is an analytic function that maps a domain D onto the upper half-plane $v > 0$, then the function $u = \text{Arg}(f(z))$ is harmonic in D. ____

6. Is the image of the circle $|z - 1| = 1$ under the complex mapping $T(z) = (z - 1)/(z - 2)$ a circle or a line? ____

7. The linear fractional transformation $T(z) = \dfrac{z - z_1}{z - z_3} \dfrac{z_2 - z_3}{z_2 - z_1}$ maps the triple z_1, z_2, and z_3 to _____.

8. If $f'(z) = z^{-1/2}(z + 1)^{-1/2}(z - 1)^{-1/2}$, then $f(z)$ maps the upper half-plane $y > 0$ onto the interior of a rectangle. ____

9. If $\mathbf{F}(x, y) = P(x, y)\mathbf{i} + Q(x, y)\mathbf{j}$ is a vector field in a domain D with div $\mathbf{F} = 0$ and curl $\mathbf{F} = \mathbf{0}$, then the complex function $g(z) = P(x, y) + iQ(x, y)$ is analytic in D. ____

10. If $G(z) = \phi(x, y) + i\psi(x, y)$ is analytic in a region R and $\mathbf{V}(x, y) = \overline{iG'(z)}$, then the streamlines of the corresponding flow are described by $\phi(x, y) = c$. ____

11. Find the image of the first quadrant under the complex mapping $w = \text{Ln } z = \log_e|z| + i \text{ Arg } z$. What are images of the rays $\theta = \theta_0$ that lie in the first quadrant?

In Problems 12 and 13, use the conformal mappings in Appendix IV to find a conformal mapping from the given region R in the z-plane onto the target region R' in the w-plane, and find the image of the given boundary curve.

12.

Figure 20.82

13.

Figure 20.83

In Problems 14 and 15, use an appropriate conformal mapping to solve the given Dirichlet problem.

14.

Figure 20.84

15.

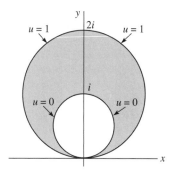

Figure 20.85

16. Derive conformal mapping C-4 in Appendix IV by constructing the linear fractional transformation that maps $1, -1, \infty$ to $i, -i, -1$.

17. (a) Approximate the region R' in M-9 in Appendix IV by the polygonal region shown in Figure 20.86. Require that $f(-1) = u_1$, $f(0) = \pi i/2$, and $f(1) = u_1 + \pi i$.

(b) Show that when $u_1 \to \infty$,

$$f'(z) = Az(z + 1)^{-1}(z - 1)^{-1} = \frac{1}{2} A\left[\frac{1}{z + 1} + \frac{1}{z - 1}\right].$$

(c) If we require that $\text{Im}(f(t)) = 0$ for $t < -1$, $\text{Im}(f(t)) = \pi$ for $t > 1$, and $f(0) = \pi i/2$, conclude that

$$f(z) = \pi i - \frac{1}{2}[\text{Ln}(z + 1) + \text{Ln}(z - 1)].$$

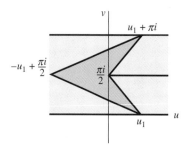

Figure 20.86

18. (a) Find the solution $u(x, y)$ of the Dirichlet problem in the upper half-plane $y \geq 0$ that satisfies the boundary condition $u(x, 0) = \sin x$. [*Hint:* See Problem 6 in Exercises 20.5.]

(b) Find the solution $u(x, y)$ of the Dirichlet problem in the unit disk $|z| \leq 1$ that satisfies the boundary condition $u(e^{i\theta}) = \sin \theta$.

19. Explain why the streamlines in Figure 20.72 may also be interpreted as the equipotential lines of the potential ϕ that satisfies $\phi(x, 0) = 0$ for $-\infty < x < \infty$ and $\phi(x, \pi) = 1$ for $x < 0$.

20. Verify that the boundary of the region R defined by $y^2 \geq 4(1 - x)$ is a streamline for the fluid flow with complex potential $G(z) = i(z^{1/2} - 1)$. Sketch the streamlines of the flow.

Appendices

APPENDIX I

Some Derivative and Integral Formulas

■ **Derivative Formulas** u and v are functions of x; c_1 and c_2 are constants.

1. $\dfrac{d}{dx}(c_1 u + c_2 v) = c_1 u' + c_2 v'$ (Sum Rule)

2. $\dfrac{d}{dx} uv = uv' + vu'$ (Product Rule)

3. $\dfrac{d}{dx}\dfrac{u}{v} = \dfrac{vu' - uv'}{v^2}$ (Quotient Rule)

4. $\dfrac{dy}{dx} = \dfrac{dy}{du}\cdot\dfrac{du}{dx}$ (Chain Rule)

5. $\dfrac{d}{dx} u^n = nu^{n-1}\dfrac{du}{dx}$

6. $\dfrac{d}{dx}\ln u = \dfrac{1}{u}\dfrac{du}{dx}$

7. $\dfrac{d}{dx} e^u = e^u\dfrac{du}{dx}$

8. $\dfrac{d}{dx} b^u = b^u \ln b\dfrac{du}{dx}$

9. $\dfrac{d}{dx}\sin u = \cos u\dfrac{du}{dx}$

10. $\dfrac{d}{dx}\cos u = -\sin u\dfrac{du}{dx}$

11. $\dfrac{d}{dx}\tan u = \sec^2 u\dfrac{du}{dx}$

12. $\dfrac{d}{dx}\cot u = -\csc^2 u\dfrac{du}{dx}$

13. $\dfrac{d}{dx}\sec u = \sec u \tan u\dfrac{du}{dx}$

14. $\dfrac{d}{dx}\csc u = -\csc u \cot u\dfrac{du}{dx}$

15. $\dfrac{d}{dx}\sin^{-1} u = \dfrac{1}{\sqrt{1 - u^2}}\dfrac{du}{dx}$

16. $\dfrac{d}{dx}\cos^{-1} u = \dfrac{-1}{\sqrt{1 - u^2}}\dfrac{du}{dx}$

17. $\dfrac{d}{dx}\tan^{-1} u = \dfrac{1}{1 + u^2}\dfrac{du}{dx}$

18. $\dfrac{d}{dx}\sec^{-1} u = \dfrac{1}{u\sqrt{u^2 - 1}}\dfrac{du}{dx}$

19. $\dfrac{d}{dx}\sinh u = \cosh u\dfrac{du}{dx}$

20. $\dfrac{d}{dx}\cosh u = \sinh u\dfrac{du}{dx}$

21. $\dfrac{d}{dx}\sinh^{-1} u = \dfrac{1}{\sqrt{u^2 + 1}}\dfrac{du}{dx}$

22. $\dfrac{d}{dx}\cosh^{-1} u = \dfrac{1}{\sqrt{u^2 - 1}}\dfrac{du}{dx}$

23. $\dfrac{d}{dx}\tanh^{-1} u = \dfrac{1}{1 - u^2}\dfrac{du}{dx}, \quad |u| < 1$

24. $\dfrac{d}{dx}\coth^{-1} u = \dfrac{1}{1 - u^2}\dfrac{du}{dx}, \quad |u| > 1$

■ Integral Formulas u and v are functions of x; c_1 and c_2 are constants.

1. $\displaystyle\int (c_1 u + c_2 v)\,dx = c_1 \int u\,dx + c_2 \int v\,dx$

2. $\displaystyle\int u\,dv = uv - \int v\,du$ (integration by parts)

3. $\displaystyle\int u^n\,du = \frac{u^{n+1}}{n+1} + C,\quad n \neq -1$

4. $\displaystyle\int \frac{1}{u}\,du = \ln |u| + C$

5. $\displaystyle\int b^u\,du = \frac{b^u}{\ln b} + C$

6. $\displaystyle\int e^u\,du = e^u + C$

7. $\displaystyle\int \cos u\,du = \sin u + C$

8. $\displaystyle\int \sin u\,du = -\cos u + C$

9. $\displaystyle\int \sec^2 u\,du = \tan u + C$

10. $\displaystyle\int \csc^2 u\,du = -\cot u + C$

11. $\displaystyle\int \sec u \tan u\,du = \sec u + C$

12. $\displaystyle\int \csc u \cot u\,du = -\csc u + C$

13. $\displaystyle\int \frac{1}{\sqrt{a^2 - u^2}}\,du = \sin^{-1} \frac{u}{a} + C$

14. $\displaystyle\int \frac{1}{a^2 + u^2}\,du = \frac{1}{a} \tan^{-1} \frac{u}{a} + C$

15. $\displaystyle\int \cosh u\,du = \sinh u + C$

16. $\displaystyle\int \sinh u\,du = \cosh u + C$

17. $\displaystyle\int \frac{1}{\sqrt{u^2 + a^2}}\,du = \begin{cases} \sinh^{-1} \dfrac{u}{a} + C \\[2mm] \ln(u + \sqrt{u^2 + a^2}) + C \end{cases}$

18. $\displaystyle\int \frac{1}{\sqrt{u^2 - a^2}}\,du = \begin{cases} \cosh^{-1} \dfrac{u}{a} + C \\[2mm] \ln(u + \sqrt{u^2 - a^2}) + C \end{cases}$

19. $\displaystyle\int \frac{1}{a^2 - u^2}\,du = \begin{cases} \dfrac{1}{a} \tanh^{-1} \dfrac{u}{a} + C,\ |u| < a \\[2mm] \dfrac{1}{a} \coth^{-1} \dfrac{u}{a} + C,\ |u| > a \end{cases}$

20. $\displaystyle\int \frac{1}{a^2 - u^2}\,du = \frac{1}{2a} \ln \left| \frac{a + u}{a - u} \right| + C$

21. $\displaystyle\int \tan u\,du = -\ln |\cos u| + C$

22. $\displaystyle\int \cot u\,du = \ln |\sin u| + C$

23. $\displaystyle\int \sec u\,du = \ln |\sec u + \tan u| + C$

24. $\displaystyle\int \csc u\,du = \ln |\csc u - \cot u| + C$

Gamma Function

Euler's integral definition of the **gamma function*** is

$$\Gamma(x) = \int_0^\infty t^{x-1} e^{-t} \, dt. \tag{1}$$

Convergence of the integral requires that $x - 1 > -1$, or $x > 0$. The recurrence relation

$$\Gamma(x + 1) = x\Gamma(x), \tag{2}$$

that we saw in Section 5.3, can be obtained from (1) by employing integration by parts. Now when $x = 1$,

$$\Gamma(1) = \int_0^\infty e^{-t} \, dt = 1,$$

and thus (2) gives

$$\Gamma(2) = 1\Gamma(1) = 1$$
$$\Gamma(3) = 2\Gamma(2) = 2 \cdot 1$$
$$\Gamma(4) = 3\Gamma(3) = 3 \cdot 2 \cdot 1$$

and so on. In this manner it is seen that when n is a positive integer,

$$\Gamma(n + 1) = n!.$$

For this reason the gamma function is often called the **generalized factorial function.**

Although the integral form (1) does not converge for $x < 0$, it can be shown by means of alternative definitions that the gamma function is defined for all real and complex numbers *except* $x = -n$, $n = 0, 1, 2, \ldots$. As a consequence, (2) is actually valid for $x \neq -n$. Considered as a function of a real variable x, the graph of $\Gamma(x)$ is as given in Figure A.1. Observe that the nonpositive integers correspond to the vertical asymptotes of the graph.

In Problems 23 and 24 in Exercises 5.3, we utilized the fact that $\Gamma(\frac{1}{2}) = \sqrt{\pi}$. This result can be derived from (1) by setting $x = \frac{1}{2}$:

$$\Gamma(\tfrac{1}{2}) = \int_0^\infty t^{-1/2} e^{-t} \, dt. \tag{3}$$

*This function was first defined by Leonhard Euler in his text *Institutiones calculi integralis* published in 1768.

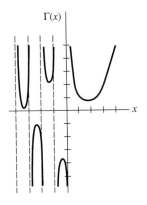

$\Gamma(x)$

Figure A.1

By letting $t = u^2$, we can write (3) as

$$\Gamma(\tfrac{1}{2}) = 2 \int_0^\infty e^{-u^2}\, du.$$

But

$$\int_0^\infty e^{-u^2}\, du = \int_0^\infty e^{-v^2}\, dv$$

and so

$$[\Gamma(\tfrac{1}{2})]^2 = \left(2 \int_0^\infty e^{-u^2}\, du\right)\left(2 \int_0^\infty e^{-v^2}\, dv\right) = 4 \int_0^\infty \int_0^\infty e^{-(u^2 + v^2)}\, du\, dv.$$

Switching to polar coordinates $u = r \cos\theta$, $v = r \sin\theta$ enables us to evaluate the double integral:

$$4 \int_0^\infty \int_0^\infty e^{-(u^2 + v^2)}\, du\, dv = 4 \int_0^{\pi/2} \int_0^\infty e^{-r^2}\, r\, dr\, d\theta = \pi.$$

Hence,

$$[\Gamma(\tfrac{1}{2})]^2 = \pi \quad \text{or} \quad \Gamma(\tfrac{1}{2}) = \sqrt{\pi}. \tag{4}$$

In view of (2) and (4) we can find additional values of the gamma function. For example, when $x = -\tfrac{1}{2}$, it follows from (2) that $\Gamma(\tfrac{1}{2}) = -\tfrac{1}{2}\Gamma(-\tfrac{1}{2})$. Therefore, $\Gamma(-\tfrac{1}{2}) = -2\Gamma(\tfrac{1}{2}) = -2\sqrt{\pi}$.

APPENDIX II EXERCISES

Answers to odd-numbered problems begin on page A-79.

1. Evaluate the following.

(a) $\Gamma(5)$ (b) $\Gamma(7)$

(c) $\Gamma(-\tfrac{3}{2})$ (d) $\Gamma(-\tfrac{5}{2})$

2. Use (1) and the fact that $\Gamma(\tfrac{6}{5}) = 0.92$ to evaluate $\displaystyle\int_0^\infty x^5 e^{-x^5}\, dx$. [*Hint:* Let $t = x^5$.]

3. Use (1) and the fact that $\Gamma(\tfrac{5}{3}) = 0.89$ to evaluate $\displaystyle\int_0^\infty x^4 e^{-x^3}\, dx$.

4. Evaluate $\displaystyle\int_0^1 x^3 \left(\ln \frac{1}{x}\right)^3 dx$. [*Hint:* Let $t = -\ln x$.]

5. Use the fact that $\Gamma(x) > \displaystyle\int_0^1 t^{x-1} e^{-t}\, dt$ to show that $\Gamma(x)$ is unbounded as $x \to 0^+$.

6. Use (1) to derive (2) for $x > 0$.

Table of Laplace Transforms

$f(t)$	$\mathscr{L}\{f(t)\} = F(s)$
1. 1	$\dfrac{1}{s}$
2. t	$\dfrac{1}{s^2}$
3. t^n	$\dfrac{n!}{s^{n+1}}$, $\quad n$ a positive integer
4. $t^{-1/2}$	$\sqrt{\dfrac{\pi}{s}}$
5. $t^{1/2}$	$\dfrac{\sqrt{\pi}}{2s^{3/2}}$
6. t^{α}	$\dfrac{\Gamma(\alpha+1)}{s^{\alpha+1}}$, $\quad \alpha > -1$
7. $\sin kt$	$\dfrac{k}{s^2+k^2}$
8. $\cos kt$	$\dfrac{s}{s^2+k^2}$
9. $\sin^2 kt$	$\dfrac{2k^2}{s(s^2+4k^2)}$
10. $\cos^2 kt$	$\dfrac{s^2+2k^2}{s(s^2+4k^2)}$
11. e^{at}	$\dfrac{1}{s-a}$
12. $\sinh kt$	$\dfrac{k}{s^2-k^2}$
13. $\cosh kt$	$\dfrac{s}{s^2-k^2}$

14. $\sinh^2 kt$ $\dfrac{2k^2}{s(s^2 - 4k^2)}$

15. $\cosh^2 kt$ $\dfrac{s^2 - 2k^2}{s(s^2 - 4k^2)}$

16. $e^{at}t$ $\dfrac{1}{(s - a)^2}$

17. $e^{at}t^n$ $\dfrac{n!}{(s - a)^{n+1}}$, n a positive integer

18. $e^{at} \sin kt$ $\dfrac{k}{(s - a)^2 + k^2}$

19. $e^{at} \cos kt$ $\dfrac{s - a}{(s - a)^2 + k^2}$

20. $e^{at} \sinh kt$ $\dfrac{k}{(s - a)^2 - k^2}$

21. $e^{at} \cosh kt$ $\dfrac{s - a}{(s - a)^2 - k^2}$

22. $t \sin kt$ $\dfrac{2ks}{(s^2 + k^2)^2}$

23. $t \cos kt$ $\dfrac{s^2 - k^2}{(s^2 + k^2)^2}$

24. $\sin kt + kt \cos kt$ $\dfrac{2ks^2}{(s^2 + k^2)^2}$

25. $\sin kt - kt \cos kt$ $\dfrac{2k^3}{(s^2 + k^2)^2}$

26. $t \sinh kt$ $\dfrac{2ks}{(s^2 - k^2)^2}$

27. $t \cosh kt$ $\dfrac{s^2 + k^2}{(s^2 - k^2)^2}$

28. $\dfrac{e^{at} - e^{bt}}{a - b}$ $\dfrac{1}{(s - a)(s - b)}$

29. $\dfrac{ae^{at} - be^{bt}}{a - b}$ $\dfrac{s}{(s - a)(s - b)}$

30. $1 - \cos kt$ $\dfrac{k^2}{s(s^2 + k^2)}$

31. $kt - \sin kt$ $\dfrac{k^3}{s^2(s^2 + k^2)}$

32. $\cos at - \cos bt$ $\dfrac{s(b^2 - a^2)}{(s^2 + a^2)(s^2 + b^2)}$

33. $\sin kt \sinh kt$ $\dfrac{2k^2 s}{s^4 + 4k^4}$

34. $\sin kt \cosh kt$ $\dfrac{k(s^2 + 2k^2)}{s^4 + 4k^4}$

35. $\cos kt \sinh kt$ $\qquad\qquad\qquad\qquad\qquad\qquad \dfrac{k(s^2 - 2k^2)}{s^4 + 4k^4}$

36. $\cos kt \cosh kt$ $\qquad\qquad\qquad\qquad\qquad\qquad \dfrac{s^3}{s^4 + 4k^4}$

37. $\delta(t)$ $\qquad\qquad\qquad\qquad\qquad\qquad\qquad\quad 1$

38. $\delta(t - a)$ $\qquad\qquad\qquad\qquad\qquad\qquad\quad e^{-as}$

39. $\mathcal{U}(t - a)$ $\qquad\qquad\qquad\qquad\qquad\qquad\quad \dfrac{e^{-as}}{s}$

40. $J_0(kt)$ $\qquad\qquad\qquad\qquad\qquad\qquad\quad \dfrac{1}{\sqrt{s^2 + k^2}}$

41. $\dfrac{e^{bt} - e^{at}}{t}$ $\qquad\qquad\qquad\qquad\qquad\qquad \ln\dfrac{s - a}{s - b}$

42. $\dfrac{2(1 - \cos at)}{t}$ $\qquad\qquad\qquad\qquad\qquad \ln\dfrac{s^2 + a^2}{s^2}$

43. $\dfrac{2(1 - \cosh at)}{t}$ $\qquad\qquad\qquad\qquad\qquad \ln\dfrac{s^2 - a^2}{s^2}$

44. $\dfrac{\sin at}{t}$ $\qquad\qquad\qquad\qquad\qquad\qquad \arctan\left(\dfrac{a}{s}\right)$

45. $\dfrac{\sin at \cos bt}{t}$ $\qquad\qquad\qquad \dfrac{1}{2}\arctan\dfrac{a + b}{s} + \dfrac{1}{2}\arctan\dfrac{a - b}{s}$

46. $\dfrac{1}{\sqrt{\pi t}}\, e^{-a^2/4t}$ $\qquad\qquad\qquad\qquad\qquad \dfrac{e^{-a\sqrt{s}}}{\sqrt{s}}$

47. $\dfrac{a}{2\sqrt{\pi t^3}}\, e^{-a^2/4t}$ $\qquad\qquad\qquad\qquad\qquad e^{-a\sqrt{s}}$

48. $\operatorname{erfc}\left(\dfrac{a}{2\sqrt{t}}\right)$ $\qquad\qquad\qquad\qquad\qquad \dfrac{e^{-a\sqrt{s}}}{s}$

49. $2\sqrt{\dfrac{t}{\pi}}\, e^{-a^2/4t} - a\operatorname{erfc}\left(\dfrac{a}{2\sqrt{t}}\right)$ $\qquad\qquad \dfrac{e^{-a\sqrt{s}}}{s\sqrt{s}}$

50. $e^{ab}e^{b^2t}\operatorname{erfc}\left(b\sqrt{t} + \dfrac{a}{2\sqrt{t}}\right)$ $\qquad\qquad \dfrac{e^{-a\sqrt{s}}}{\sqrt{s}(\sqrt{s} + b)}$

51. $-e^{ab}e^{b^2t}\operatorname{erfc}\left(b\sqrt{t} + \dfrac{a}{2\sqrt{t}}\right) + \operatorname{erfc}\left(\dfrac{a}{2\sqrt{t}}\right)$ $\qquad \dfrac{be^{-a\sqrt{s}}}{s(\sqrt{s} + b)}$

52. $e^{at}f(t)$ $\qquad\qquad\qquad\qquad\qquad\qquad F(s - a)$

53. $f(t - a)\mathcal{U}(t - a)$ $\qquad\qquad\qquad\qquad\quad e^{-as}F(s)$

54. $g(t)\mathcal{U}(t - a)$ $\qquad\qquad\qquad\qquad\quad e^{-as}\mathcal{L}\{g(t + a)\}$

55. $f^{(n)}(t)$ $\qquad\qquad\qquad s^nF(s) - s^{(n-1)}f(0) - \cdots - f^{(n-1)}(0)$

56. $t^nf(t)$ $\qquad\qquad\qquad\qquad\qquad\qquad (-1)^n\dfrac{d^n}{ds^n}\,F(s)$

57. $\displaystyle\int_0^t f(\tau)g(t - \tau)\, d\tau$ $\qquad\qquad\qquad\qquad F(s)G(s)$

APPENDIX IV

Conformal Mappings

■ Elementary Mappings

E-1

$$w = z + z_0$$

E-2

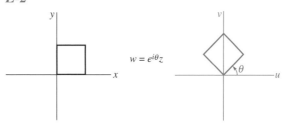

$$w = e^{i\theta}z$$

E-3

$$w = \alpha z, \ \alpha > 0$$

E-4

$$w = z^{\alpha},\ \alpha > 0$$

E-5

$$w = e^{z}$$
$$z = \operatorname{Ln} w$$

E-6

$$w = \sin z$$
$$z = \sin^{-1} w$$

E-7

$$w = \frac{1}{z}$$

E-8

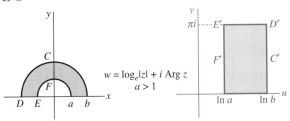

$$w = \log_{e} |z| + i \operatorname{Arg} z$$
$$a > 1$$

E-9

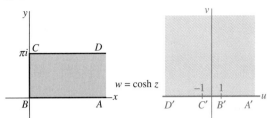

$w = \cosh z$

■ Mappings to Half-Planes

H-1

$w = i\dfrac{1-z}{1+z}$

H-2

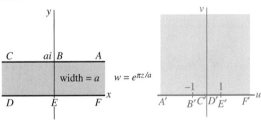

$w = e^{\pi z/a}$

H-3

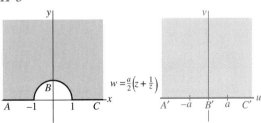

$w = \dfrac{a}{2}\left(z + \dfrac{1}{z}\right)$

H-4

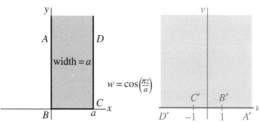

$$w = \cos\left(\frac{\pi z}{a}\right)$$

H-5

$$w = \left(\frac{1+z}{1-z}\right)^2$$

H-6

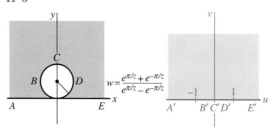

$$w = \frac{e^{\pi/z} + e^{-\pi/z}}{e^{\pi/z} - e^{-\pi/z}}$$

■ Mappings to Circular Regions

C-1

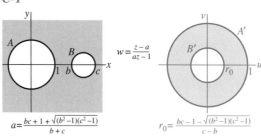

$$a = \frac{bc + 1 + \sqrt{(b^2-1)(c^2-1)}}{b+c}$$

$$r_0 = \frac{bc - 1 - \sqrt{(b^2-1)(c^2-1)}}{c-b}$$

C-2

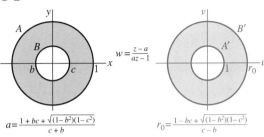

$$w = \frac{z-a}{az-1}$$

$$a = \frac{1+bc+\sqrt{(1-b^2)(1-c^2)}}{c+b}$$

$$r_0 = \frac{1-bc+\sqrt{(1-b^2)(1-c^2)}}{c-b}$$

C-3

$$w = e^z$$

C-4

$$w = \frac{i-z}{i+z}$$

C-5

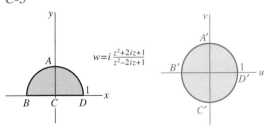

$$w = i\,\frac{z^2+2iz+1}{z^2-2iz+1}$$

■ Miscellaneous Mappings

M-1

$$w = z + e^z + 1$$

M-2

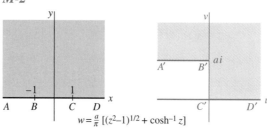

$$w = \frac{a}{\pi}\left[(z^2-1)^{1/2} + \cosh^{-1} z\right]$$

M-3

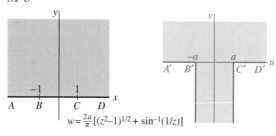

$$w = \frac{2a}{\pi}\left[(z^2-1)^{1/2} + \sin^{-1}(1/z)\right]$$

M-4

$$w = a(z^2-1)^{1/2}$$

M-5

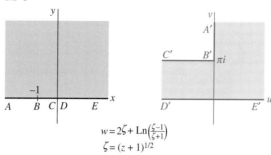

$$w = 2\zeta + \text{Ln}\left(\frac{\zeta-1}{\zeta+1}\right)$$
$$\zeta = (z+1)^{1/2}$$

M-6

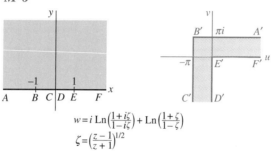

$$w = i\,\text{Ln}\left(\frac{1+i\zeta}{1-i\zeta}\right) + \text{Ln}\left(\frac{1+\zeta}{1-\zeta}\right)$$
$$\zeta = \left(\frac{z-1}{z+1}\right)^{1/2}$$

M-7

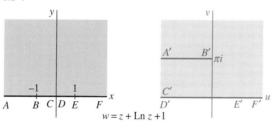

$$w = z + \text{Ln}\,z + 1$$

M-8

$$w = \frac{e^z + 1}{e^z - 1}$$

M-9

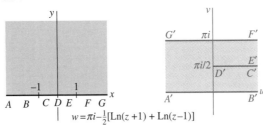

$$w = \pi i - \tfrac{1}{2}[\mathrm{Ln}(z+1) + \mathrm{Ln}(z-1)]$$

M-10

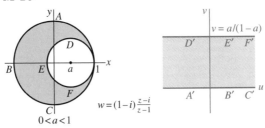

$$w = (1-i)\frac{z-i}{z-1}$$

$0 < a < 1$

APPENDIX V

Some BASIC Programs for Numerical Methods

■ **1. Euler's Method**

```
100   REM EULER'S METHOD TO SOLVE Y' = FNF(X, Y)
110   REM DEFINE FNF(X, Y) HERE
120   REM GET INPUTS
130   PRINT
140   INPUT "STEP SIZE =",H
150   INPUT "NUMBER OF STEPS =",N
160   INPUT "XO =",X
170   INPUT "YO =",Y
180   PRINT
190   REM SET UP TABLE
200   PRINT "X", "Y"
210   PRINT
220   PRINT X, Y
230   REM COMPUTE X AND Y VALUES
240   FOR I = 1 TO N
250   Y = Y + H * FNF(X, Y)
260   X = X + H
270   PRINT X, Y
280   NEXT I
290   END
```

■ **2. Improved Euler's Method**

```
100   REM IMPROVED EULER'S METHOD TO SOLVE Y' = FNF(X, Y)
110   REM DEFINE FNF(X, Y) HERE
120   REM GET INPUTS
130   PRINT
140   INPUT "STEP SIZE =",H
150   INPUT "NUMBER OF STEPS =",N
160   INPUT "XO =",X
170   INPUT "YO =",Y
180   PRINT
190   REM SET UP TABLE
200   PRINT "X", "Y"
210   PRINT
```

```
220   PRINT X, Y
230   REM COMPUTE X AND Y VALUES
240   FOR I = 1 TO N
250   FVAL = FNF(X, Y)
260   Y = Y + H * (FVAL + FNF(X + H, Y + H * FVAL))/2
270   X = X + H
280   PRINT X, Y
290   NEXT I
300   END
```

■ 3. Runge–Kutta Method

```
100   REM RUNGE-KUTTA METHOD TO SOLVE Y' = FNF(X, Y)
110   REM DEFINE FNF(X, Y) HERE
120   REM GET INPUTS
130   PRINT
140   INPUT "STEP SIZE =",H
150   INPUT "NUMBER OF STEPS =",N
160   INPUT "XO =",X
170   INPUT "YO =",Y
180   PRINT
190   REM SET UP TABLE
200   PRINT "X", "Y"
210   PRINT
220   PRINT X, Y
230   REM COMPUTE X AND Y VALUES
240   FOR I = 1 TO N
250   K1 = H * FNF(X, Y)
260   K2 = H * FNF(X + H/2, Y + K1/2)
270   K3 = H * FNF(X + H/2, Y + K2/2)
280   K4 = H * FNF(X + H, Y + K3)
290   Y = Y + (K1 + 2 * K2 + 2 * K3 + K4)/6
300   X = X + H
310   PRINT X, Y
320   NEXT I
330   END
```

■ 4. Gauss–Seidel Iteration

```
 10   REM GAUSS-SEIDEL ITERATION TO SOLVE THE SYSTEM
      X = AX + B
 20   REM WHERE THE MATRIX A HAS ZEROS ON THE MAIN
      DIAGONAL
 30   INPUT "THE NUMBER OF UNKNOWNS IS",N
 40   INPUT "THE TOLERANCE IS",TOL
 50   DIM A(N,N)
 60   DIM B(N)
 70   DIM X(N)
 80   DIM Y(N)
 90   PRINT "INPUT THE COEFFICIENTS OF THE MATRIX A"
100   FOR I = 1 TO N
110   FOR J = 1 TO N
```

```
120  PRINT "A("I","J") =";
130  INPUT " ", A(I,J)
140  NEXT J
150  NEXT I
160  REM PRINT "INPUT THE CONSTANT COEFFICIENTS"
170  FOR I = 1 TO N
180  PRINT "B("I") =";
190  INPUT " ", B(I)
200  NEXT I
210  PRINT "INPUT THE INITIAL GUESS"
220  FOR I = 1 TO N
230  PRINT "X("I") =";
240  INPUT " ", X(I)
250  NEXT I
260  REM COMPUTE THE NEXT ITERATE
270  FOR I = 1 TO N
280  Y(I) = X(I)
290  X(I) = 0
300  FOR J = 1 TO N
310  X(I) = X(I) + A(I,J)*X(J)
320  NEXT J
330  X(I) = X(I) + B(I)
340  NEXT I
350  REM CHECK THE TOLERANCE
360  FOR I = 1 TO N
370  IF ABS(X(I)−Y(I)) > TOL GOTO 270
380  NEXT I
390  REM DISPLAY THE SOLUTION
400  PRINT "THE SOLUTION WITH TOLERANCE" TOL "IS"
410  FOR I = 1 TO N
420  PRINT "X("I") = "X(I)
430  NEXT I
440  END
```

▌ 5. Numerical Solution of the One-Dimensional Heat Equation

```
 10  REM NUMERICAL SOLUTION OF THE ONE-DIMENSIONAL
     HEAT EQUATION
 20  REM DEFINE THE INITIAL TEMPERATURE FUNCTION FNF(X)
     BELOW
 30  DEF FNF(X) =
 40  REM GET INPUTS
 50  INPUT "THE COEFFICIENT IN THE PDE IS C =",C
 60  INPUT "THE LENGTH OF THE ROD IS A =",A
 70  INPUT "THE ENDING TIME IS T =",T
 80  INPUT "THE CONSTANT LEFT-HAND TEMPERATURE IS T1 =",T1
 90  INPUT "THE CONSTANT RIGHT-HAND TEMPERATURE IS T2 =",T2
100  INPUT "THE NUMBER OF X SUBINTERVALS IS N =",N
110  INPUT "THE NUMBER OF T SUBINTERVALS IS M =",M
120  REM COMPUTE H, K, AND L
130  H = A/N: K = T/M: L = C*K/H^2
```

```
140   REM FORMAT THE TABLE HEADER AND PRINT THE FIRST LINE
150   PRINT "TIME";
160   FOR I = 1 TO N − 1
170   PRINT TAB(2 + 9*I);"X =";
180   PRINT USING "#.##";I*H;
190   NEXT I
200   PRINT: PRINT STRING$(10*N − 1,45)
210   PRINT USING "##.##";0;
220   FOR I = 1 TO N − 1
230   PRINT TAB(9*I);
240   PRINT USING "###.####";FNF(I*H);
250   NEXT I
260   PRINT
270   DIM U(N + 1,M + 1)
280   REM INITIALIZE THE U(I,J) VALUES
290   FOR I = 1 TO N − 1: U(I,0) = FNF(I*H): NEXT I
300   FOR J = 0 TO M: U(0,J) = T1: U(N,J) = T2: NEXT J
310   REM COMPUTE THE U(I,J) VALUES
320   FOR J = 0 TO M − 1
330   PRINT USING "##.##";(J + 1)*K;
340   FOR I = 1 TO N − 1
350   U(I,J + 1) = L*U(I + 1,J) + (1 − 2*L)*U(I,J) + L*U(I − 1,J)
360   PRINT TAB(9*I);
370   PRINT USING "###.####";U(I,J + 1);
380   NEXT I
390   PRINT
400   NEXT J
410   END
```

■ 6. Crank–Nicholson Method

```
 10   REM CRANK-NICHOLSON METHOD FOR THE ONE-DIMENSIONAL
      HEAT EQUATION
 20   REM DEFINE THE INITIAL TEMPERATURE FUNCTION FNF(X)
      BELOW
 30   DEF FNF(X) =
 40   REM GET INPUTS
 50   INPUT "THE COEFFICIENT IN THE PDE IS C =",C
 60   INPUT "THE LENGTH OF THE ROD IS A =",A
 70   INPUT "THE ENDING TIME IS T =",T
 80   INPUT "THE CONSTANT LEFT-HAND TEMPERATURE IS T1 =",T1
 90   INPUT "THE CONSTANT RIGHT-HAND TEMPERATURE IS T2 =",T2
100   INPUT "THE NUMBER OF X SUBINTERVALS IS N =",N
110   INPUT "THE NUMBER OF T SUBINTERVALS IS M =",M
120   REM COMPUTE H, K, AND L
130   H = A/N: K = T/M: L = C*K/H^2
140   REM FORMAT THE TABLE HEADER AND PRINT THE FIRST LINE
150   PRINT "TIME";
160   FOR I = 1 TO N − 1
170   PRINT TAB(2 + 9*I);"X =";
180   PRINT USING "#.##";I*H;
```

```
190   NEXT I
200   PRINT: PRINT STRING$(10*N − 1,45)
210   PRINT USING "##.##";0;
220   FOR I = 1 TO N − 1
230   PRINT TAB(9*I)
240   PRINT USING "###.####"; FNF(I*H);
250   NEXT I
260   PRINT
270   DIM U(N + 1,M + 1): DIM B(N − 1): DIM R(N − 1): DIM S(N − 1)
280   REM INITIALIZE THE U(I,J) VALUES
290   FOR I = 1 TO N − 1: U(I,0) = FNF(I*H): NEXT I
300   FOR J = 0 TO M: U(0,J) = T1: U(N,J) = T2: NEXT J
310   REM COMPUTE THE U(I,J) VALUES
320   REM THE SYSTEM OF EQUATIONS IS SOLVED USING
      GAUSS ELIMINATION
330   REM ADAPTED TO THE APPROPRIATE TRIDIAGONAL MATRIX
340   ALPHA = 2*(1 + 1/L): BETA = 2*(1 − 1/L)
350   R(1) = ALPHA
360   FOR I = 2 TO N − 1: R(I) = ALPHA − 1/R(I − 1): NEXT I
370   FOR J = 0 TO M − 1
380   B(1) = U(2,J) − BETA*U(1,J) + U(0,J) + U(0,J + 1)
390   FOR I = 2 TO N − 2: B(I) = U(I + 1,J) − BETA*U(I,J) + U(I − 1,J): NEXT I
400   B(N − 1) = U(N,J) − BETA*U(N − 1,J) + U(N − 2,J) + U(N,J + 1)
410   S(1) = B(1)
420   FOR I = 2 TO N − 1: S(I) = B(I) + (1/R(I − 1))*S(I − 1): NEXT I
430   U(N − 1,J + 1) = S(N − 1)/R(N − 1)
440   FOR I = N − 2 TO 1 STEP − 1: U(I,J + 1) = (1/R(I))*(S(I) + U(I + 1,J + 1)):
      NEXT I
450   PRINT USING "##.##"; (J + 1)*K;
460   FOR I = 1 TO N − 1
470   PRINT TAB(9*I);
480   PRINT USING "###.####"; U(I,J + 1);
490   NEXT I
500   PRINT
510   NEXT J
520   END
```

■ 7. Numerical Solution of the One-Dimensional Wave Equation

```
 10   REM NUMERICAL SOLUTION OF THE ONE-DIMENSIONAL
      WAVE EQUATION
 20   REM WITH 0 < X < A AND U(0,T) = U(A,T) = 0 FOR T > 0
 30   REM DEFINE THE INITIAL POSITION FUNCTION FNF(X) BELOW
 40   DEF FNF(X) =
 50   REM DEFINE THE INITIAL VELOCITY FUNCTION FNG(X) BELOW
 60   DEF FNG(X) =
 70   REM GET INPUTS
 80   INPUT "THE COEFFICIENT IN THE PDE IS THE SQUARE OF C =",C
 90   INPUT "THE LENGTH OF THE X-INTERVAL IS A =",A
100   INPUT "THE ENDING TIME IS T =",T
110   INPUT "THE NUMBER OF X SUBINTERVALS IS N =",N
```

```
120  INPUT "THE NUMBER OF T SUBINTERVALS IS M =",M
130  REM COMPUTE H, K, AND L
140  H = A/N: K = T/M: L = C*K/H
150  REM FORMAT THE TABLE HEADER AND PRINT THE FIRST LINE
160  PRINT " TIME";
170  FOR I = 1 TO N – 1
180  PRINT TAB(2 + 9*I);"X =";
190  PRINT USING "#.##";I*H;
200  NEXT I
210  PRINT: PRINT STRING$(10*N – 1,45)
220  PRINT USING "##.##"; 0;
230  FOR I = 1 TO N – 1
240  PRINT TAB(9*I)
250  PRINT USING "###.####"; FNF(I*H);
260  NEXT I
270  PRINT
280  DIM U(N + 1,M + 1)
290  REM INITIALIZE THE U(0,J) AND U(A,J) VALUES
300  FOR J = 0 TO M: U(0,J) = 0: U(A,J) = 0: NEXT J
310  REM INITIALIZE THE U(I,0) VALUES
320  FOR I = 1 TO N – 1: U(I,0) = FNF(I*H): NEXT I
330  REM COMPUTE THE U(I,1) VALUES
340  PRINT USING "##.##";K;
350  FOR I = 1 TO N – 1
360  U(I,1) = (L^2/2)*(U(I + 1,0) + U(I – 1,0)) + (1 – L^2)*U(I,0) + K*FNG(I*H)
370  PRINT TAB(9*I);
380  PRINT USING "###.####"; U(I,1)
390  NEXT I
400  PRINT
410  REM COMPUTE THE U(I,J) K VALUES
420  FOR J = 1 TO M – 1
430  PRINT USING "##.##"; (J + 1)*K;
440  FOR I = 1 TO N – 1
450  U(I,J + 1) = L^2*U(I + 1,J) + 2*(1 – L^2)*U(I,J) + L^2*U(I – 1,J) – U(I,J – 1)
460  PRINT TAB(9*I);
470  PRINT USING "###.####"; U(I,J + 1):
480  NEXT I
490  PRINT
500  NEXT J
510  END
```

Selected Answers
for Odd-Numbered
Problems

Exercises 1.1, Page 13

1. linear, second order

3. linear in x but nonlinear in y, first order

5. linear, fourth order

7. nonlinear, second order

9. linear, third order

15. $X = \dfrac{e^t - 1}{e^t - 2}$ defined on $(-\infty, \ln 2)$ or on $(\ln 2, \infty)$

25. one interval is $-\pi/10 < x < \pi/10$

29. $y' = y$; $y = ce^x$; $y' = ky$, $y = ce^{kx}$, where c is an arbitrary constant

37. (a) $m = -2$ **(b)** $m = 2, m = 3$ **(c)** $m = 0, m = -1$ **(d)** $m = 3, m = 5$

39. (a) $y = 2$ **(b)** $y = -3, y = 1$ **(c)** no constant solutions **(d)** $y = \frac{5}{3}$

47. (a) $y = 0, y = a/b$

 (b) $y = \phi(x)$ is increasing on $0 < y < a/b$, decreasing for $y < 0$ and for $y > a/b$

Exercises 1.2, Page 19

1. $y = 1/(1 - 4e^{-x})$ **3.** $x = -\cos t + 8 \sin t$ **5.** $x = \dfrac{\sqrt{3}}{4} \cos t + \dfrac{1}{4} \sin t$

7. $y = \frac{3}{2} e^x - \frac{1}{2} e^{-x}$ **9.** $y = 5e^{-x-1}$ **11.** $y = 0, y = x^3$

13. half-planes defined by either $y > 0$ or $y < 0$

15. half-planes defined by either $x > 0$ or $x < 0$

17. the regions defined by $y > 2$, $y < -2$, or $-2 < y < 2$

19. any region not containing $(0, 0)$

21. yes **23.** no

25. (a) $y = cx$ **(b)** any rectangular region not touching the y-axis

 (c) No, the function is not differentiable at $x = 0$.

27. (b) $y = 1/(1 - x)$ on $(-\infty, 1)$; $y = -1/(x + 1)$ on $(-1, \infty)$

 (c) $y = y_0/(1 - y_0 x)$, $y_0 \neq 0$. Consider the cases $y_0 < 0$ and $y_0 > 0$.

29. the black curve

33. $y = 4e^{2x} + 3x^2 + 5$

Exercises 1.3, Page 28

1. $\dfrac{dP}{dt} = kP + r;$ $\dfrac{dP}{dt} = kP - r$ **3.** $\dfrac{dP}{dt} = k_1 P - k_2 P^2$

7. $\dfrac{dx}{dt} = kx(1000 - x)$ **9.** $\dfrac{dA}{dt} = -\dfrac{A}{100}$

11. $\dfrac{dh}{dt} = -\dfrac{c\pi}{450}\sqrt{h}$ **13.** $L\dfrac{di}{dt} + Ri = E(t)$

15. $m\dfrac{dv}{dt} = mg - kv^2$ **17.** $m\dfrac{d^2x}{dt^2} = -kx$

19. $x\dfrac{d^2x}{dt^2} + \left(\dfrac{dx}{dt}\right)^2 + 32x = 160$ **21.** $\dfrac{d^2r}{dt^2} + \dfrac{gR^2}{r^2} = 0$

23. $\dfrac{dA}{dt} = k(M - A), k > 0$ **25.** $\dfrac{dx}{dt} + kx = r, k > 0$

27. $\dfrac{dy}{dx} = \dfrac{-x + \sqrt{x^2 + y^2}}{y}$ **33.** $\dfrac{d^2s}{dt^2} = -g$

Chapter 1 Review Exercises, Page 33

1. $\dfrac{dy}{dx} = -\dfrac{y}{x}$ **3.** $y'' + k^2 y = 0$

5. $y'' - 2y' + y = 0$ **7.** a, d

9. b **11.** b

13. $y = c_1$ and $y = c_2 e^x$, c_1 and c_2 constants

15. $y' = x^2 + y^2$

17. (a) The domain is all real numbers.

 (b) $(-\infty, 0)$ or $(0, \infty)$

19. (c) $y = \begin{cases} -x^2, & x < 0 \\ x^2, & x \geq 0 \end{cases}$ **21.** $(0, \infty)$

25. $\dfrac{dv}{dt} = 16(1 - \sqrt{3}\,\mu)$

Exercises 2.1, Page 42

17. 0 is asymptotically stable (attractor); 3 is unstable (repeller).

19. 2 is semistable.

21. -2 is unstable (repeller); 0 is semistable; 2 is asymptotically stable (attractor).

23. -1 is asymptotically stable (attractor); 0 is unstable (repeller).

25. (a) mg/k **(b)** $\sqrt{mg/k}$

29. Lineal elements drawn through points (x, y) on the circle $x^2 + y^2 = c^2$ all have the same slope c^2.

33. (b) The y-coordinate of the point of inflection is $\frac{1}{2}$.

Exercises 2.2, Page 49

1. $y = -\frac{1}{5}\cos 5x + c$ **3.** $y = \frac{1}{3}e^{-3x} + c$

5. $y = cx^4$ **7.** $-3e^{-2y} = 2e^{3x} + c$

9. $\frac{1}{3}x^3 \ln x - \frac{1}{9}x^3 = \frac{1}{2}y^2 + 2y + \ln|y| + c$

11. $4\cos y = 2x + \sin 2x + c$ **13.** $(e^x + 1)^{-2} + 2(e^y + 1)^{-1} = c$

15. $S = ce^{kr}$ **17.** $P = \dfrac{ce^t}{1 + ce^t}$

19. $(y + 3)^5 e^x = c(x + 4)^5 e^y$ **21.** $y = \sin(\frac{1}{2}x^2 + c)$

23. $x = \tan\left(4y - \dfrac{3\pi}{4}\right)$ **25.** $xy = e^{-(1 + 1/x)}$

27. $y = \dfrac{\sqrt{3}}{2}\sqrt{1 - x^2} + \frac{1}{2}x$

29. (a) $y = 2, y = -2, y = 2\dfrac{3 - e^{4x-1}}{3 + e^{4x-1}}$

33. $y = 1$ **35.** $y = 1 + \frac{1}{10}\tan(\frac{1}{10}x)$

39. (a) $y = -\sqrt{x^2 + x - 1}$ **(c)** $\left(-\infty, -\dfrac{1}{2} - \dfrac{\sqrt{5}}{2}\right)$

Exercises 2.3, Page 57

1. $y = ce^{5x}, -\infty < x < \infty$ **3.** $y = \frac{1}{4}e^{3x} + ce^{-x}, -\infty < x < \infty$

5. $y = \frac{1}{3} + ce^{-x^3}, -\infty < x < \infty$ **7.** $y = x^{-1}\ln x + cx^{-1}, x > 0$

9. $y = cx - x\cos x, x > 0$ **11.** $y = \frac{1}{7}x^3 - \frac{1}{5}x + cx^{-4}, x > 0$

13. $y = \dfrac{1}{2x^2}e^x + \dfrac{c}{x^2}e^{-x}, x > 0$ **15.** $x = 2y^6 + cy^4, y > 0$

17. $y = \sin x + c\cos x, -\pi/2 < x < \pi/2$

19. $(x + 1)e^x y = x^2 + c, x > -1$

21. $(\sec\theta + \tan\theta)r = \theta - \cos\theta + c, -\pi/2 < \theta < \pi/2$

23. $y = e^{-3x} + \dfrac{c}{x}e^{-3x}, 0 < x < \infty$

25. $y = \dfrac{e^x}{x} + \dfrac{2 - e}{x}, x > 0$

27. $i = \dfrac{E}{R} + \left(i_0 - \dfrac{E}{R}\right)e^{-Rt/L}, -\infty < t < \infty$

29. $(x + 1)y = x\ln x - x + 21, x > 0$

31. $y = \begin{cases} \frac{1}{2}(1 - e^{-2x}), & 0 \leq x \leq 3 \\ \frac{1}{2}(e^6 - 1)e^{-2x}, & x > 3 \end{cases}$

33. $y = \begin{cases} \frac{1}{2} + \frac{3}{2}e^{-x^2}, & 0 \leq x < 1 \\ (\frac{1}{2}e + \frac{3}{2})e^{-x^2}, & x \geq 1 \end{cases}$

35. $y = \begin{cases} 2x - 1 + 4e^{-2x}, & 0 \leq x \leq 1 \\ 4x^2 \ln x + (1 + 4e^{-2})x^2, & x > 1 \end{cases}$

37. $y = e^{x^2-1} + \dfrac{\sqrt{\pi}}{2} e^{x^2} [\text{erf}(x) - \text{erf}(1)]$

41. $y = \dfrac{c}{\sqrt{9-x^2}}$

Exercises 2.4, Page 63

1. $x^2 - x + \frac{3}{2} y^2 + 7 y = c$ **3.** $\frac{5}{2} x^2 + 4xy - 2y^4 = c$

5. $x^2 y^2 - 3x + 4y = c$ **7.** not exact

9. $xy^3 + y^2 \cos x - \frac{1}{2}x^2 = c$ **11.** not exact

13. $xy - 2xe^x + 2e^x - 2x^3 = c$ **15.** $x^3 y^3 - \tan^{-1} 3x = c$

17. $-\ln|\cos x| + \cos x \sin y = c$

19. $t^4 y - 5t^3 - ty + y^3 = c$ **21.** $\frac{1}{3}x^3 + x^2 y + xy^2 - y = \frac{4}{3}$

23. $4ty + t^2 - 5t + 3y^2 - y = 8$

25. $y^2 \sin x - x^3 y - x^2 + y \ln y - y = 0$

27. $k = 10$ **29.** $x^2 y^2 \cos x = c$

31. $x^2 y^2 + x^3 = c$ **33.** $3x^2 y^3 + y^4 = c$

35. (b) $\cos y = \frac{1}{2} + \sin x$

 (c) $y = \arccos(\frac{1}{2} + \sin x), 5\pi/6 \le x \le 13\pi/6$

39. (a) $M(x, y) = ye^{xy} + y^2 - x^{-2}y + h(x)$

 (b) $N(x, y) = x^{1/2}y^{-1/2} + \frac{1}{2}(x^2 + y)^{-1} + g(y)$

Exercises 2.5, Page 67

1. $x \ln|x| + y = cx$

3. $(x - y) \ln|x - y| = y + c(x - y)$

5. $x + y \ln|x| = cy$

7. $\ln(x^2 + y^2) + 2 \tan^{-1}\left(\dfrac{y}{x}\right) = c$

9. $4x = y(\ln|y| - c)^2$ **11.** $y^3 + 3x^3 \ln|x| = 8x^3$

13. $\ln|x| = e^{y/x} - 1$ **15.** $y^3 = 1 + cx^{-3}$

17. $y^{-3} = x + \frac{1}{3} + ce^{3x}$ **19.** $e^{t/y} = ct$

21. $y^{-3} = -\frac{9}{5}x^{-1} + \frac{49}{5}x^{-6}$ **23.** $y = -x - 1 + \tan(x + c)$

25. $2y - 2x + \sin 2(x + y) = c$ **27.** $4(y - 2x + 3) = (x + c)^2$

29. $-\cot(x + y) + \csc(x + y) = x + \sqrt{2} - 1$

33. (b) $y = \dfrac{2}{x} + (-\frac{1}{4}x + cx^{-3})^{-1}$

Exercises 2.6, Page 72

1. $y_2 = 2.9800, y_4 = 3.1151$

3. $y_{10} = 2.5937, y_{20} = 2.6533; y = e^x$

5. $y_5 = 0.4198, y_{10} = 0.4124$

7. $y_5 = 0.5639, y_{10} = 0.5565$

9. $y_5 = 1.2194, y_{10} = 1.2696$

13. Euler: $y_{10} = 3.8191, y_{20} = 5.9363$
 Runge–Kutta: $y_{10} = 42.9931, y_{20} = 84.0132$

Exercises 2.7, Page 79

1. 7.9 years; 10 years

3. 760

5. $I(15) = 0.00098I_0$ or approximately 0.1% of I_0

7. 11 hours

9. 136.5 hours

11. 15,600 years

13. $T(1) = 36.76°F$; approximately 3.06 minutes

15. approximately 82.1 seconds; approximately 145.7 seconds

17. approximately 2.57 hours prior to the discovery of the body

19. $A(t) = 200 - 170e^{-t/50}$

21. $A(t) = 1000 - 1000e^{-t/100}$; 0.0975 lb/gal

23. 64.38 lb

25. $i(t) = \frac{3}{5} - \frac{3}{5}e^{-500t}; i \to \frac{3}{5}$ as $t \to \infty$

27. $q(t) = \frac{1}{100} - \frac{1}{100} e^{-50t}; i(t) = \frac{1}{2} e^{-50t}$

29. $i(t) = \begin{cases} 60 - 60e^{-t/10}, 0 \le t \le 20 \\ 60(e^2 - 1)e^{-t/10}, \quad t > 20 \end{cases}$

31. (a) $v(t) = \dfrac{mg}{k} + \left(v_0 - \dfrac{mg}{k}\right)e^{-kt/m}$

 (b) $v \to \dfrac{mg}{k}$ as $t \to \infty$

 (c) $s(t) = \dfrac{mg}{k} t - \dfrac{m}{k}\left(v_0 - \dfrac{mg}{k}\right)e^{-kt/m} + \dfrac{m}{k}\left(v_0 - \dfrac{mg}{k}\right) + s_0$

35. (a) $v(t) = \dfrac{\rho g}{4k}\left(\dfrac{k}{\rho} t + r_0\right) - \dfrac{\rho g r_0}{4k}\left(\dfrac{r_0}{\frac{k}{\rho} t + r_0}\right)^3$

 (b) $33\frac{1}{3}$ minutes

37. (a) $P(t) = P_0 e^{(k_1 - k_2)t}$

39. (a) As $t \to \infty, A \to \dfrac{k_1 M}{k_1 + k_2}$. Observe that if $k_1 > 0$ and $k_2 > 0$, then $\dfrac{k_1 M}{k_1 + k_2} < M$.

 (b) $A(t) = \dfrac{k_1 M}{k_1 + k_2} (1 - e^{-(k_1 + k_2)t})$

Exercises 2.8, Page 87

1. (a) $N = 2000$

(b) $N(t) = \dfrac{1.0005e^t}{1 + 0.0005e^t}$; $N(10) = 1834$

3. 1,000,000; 5.29 mo

5. (a) When $P_0 > 4$ or $1 < P_0 < 4$, $P(t) \to 4$ as $t \to \infty$.

When $0 < P_0 < 1$, $P(t) \to -\infty$.

(b) $P(t) = \dfrac{4(P_0 - 1) - (P_0 - 4)e^{-3t}}{(P_0 - 1) - (P_0 - 4)e^{-3t}}$

(c) $t = \frac{1}{3} \ln\left(\dfrac{P_0 - 4}{4P_0 - 4}\right)$

7. (c) $P(t) = e^{a/b}e^{-ce^{-bt}}$, where $c = \dfrac{a}{b} - \ln P_0$

9. 29.3 g; $X \to 60$ as $t \to \infty$; 0 g of A and 30 g of B

11. (a) $h(t) = \left(\sqrt{H} - \dfrac{4A_h}{A_w} t\right)^2$; I is $0 \le t \le \sqrt{H}\, A_w/4A_h$

(b) $576\sqrt{10}$ s or 30.36 min

13. (a) approximately 856.65 s or 14.31 min

(b) 243 s or 4.05 min

15. (a) $v(t) = \sqrt{\dfrac{mg}{k}} \tanh\left(\sqrt{\dfrac{kg}{m}}\, t + c_1\right)$,

where $c_1 = \tanh^{-1}\left(\sqrt{\dfrac{k}{mg}}\, v_0\right)$

(b) $\sqrt{\dfrac{mg}{k}}$

(c) $s(t) = \dfrac{m}{k} \ln \cosh\left(\sqrt{\dfrac{kg}{m}}\, t + c_1\right) + c_2$, where
$c_2 = -(m/k) \ln \cosh c_1$

21. (a) $y^2 = 2c_1 x + c_1^2$

23. (b) $h(t) = \frac{1}{200}(-t + \sqrt{t^2 + 4000t})$

(c) 2000/3 min or approximately 11.11 hr

(d) approximately 5.49 ft

Exercises 2.9, Page 96

1. $x(t) = x_0 e^{-\lambda_1 t}$

$y(t) = \dfrac{x_0 \lambda_1}{\lambda_2 - \lambda_1}(e^{-\lambda_1 t} - e^{-\lambda_2 t})$

$z(t) = x_0\left(1 - \dfrac{\lambda_2}{\lambda_2 - \lambda_1} e^{-\lambda_1 t} + \dfrac{\lambda_1}{\lambda_2 - \lambda_1} e^{-\lambda_2 t}\right)$

3. 5, 20, 147 days. The time when $y(t)$ and $z(t)$ are the same makes sense because most of A and half of B are gone, so half of C should have been formed.

5. $\dfrac{dx_1}{dt} = 6 - \frac{2}{25} x_1 + \frac{1}{50} x_2$

$\dfrac{dx_2}{dt} = \frac{2}{25} x_1 - \frac{2}{25} x_2$

7. (a) $\dfrac{dx_1}{dt} = 3\,\dfrac{x_2}{100 - t} - 2\,\dfrac{x_1}{100 + t}$

$\dfrac{dx_2}{dt} = 2\,\dfrac{x_1}{100 + t} - 3\,\dfrac{x_1}{100 - t}$

(b) $x_1(t) + x_2(t) = 150$; $x_2(30) \approx 47.4$ lb

13. $L_1\dfrac{di_2}{dt} + (R_1 + R_2)i_2 + R_1 i_3 = E(t)$

$L_2\dfrac{di_3}{dt} + R_1 i_2 + (R_1 + R_3)i_3 = E(t)$

15. $i(0) = i_0$, $s(0) = n - i_0$, $r(0) = 0$

Chapter 2 Review Exercises, Page 98

1. $-A/k$, a repeller for $k > 0$, an attractor for $k < 0$

3. $\dfrac{dy}{dx} = (y - 1)^2(y - 3)^3$

5. semistable for n even and unstable for n odd; semistable for n even and asymptotically stable for n odd

9. $2x + \sin 2x = 2 \ln(y^2 + 1) + c$

11. $(6x + 1)y^3 = -3x^3 + c$

13. $Q = \dfrac{c}{t} + \dfrac{t^4}{25}(-1 + 5 \ln t)$ **15.** $y = \frac{1}{4} + c(x^2 + 4)^{-4}$

17. $y = \csc x$, $\pi < x < 2\pi$

19. $y = \frac{1}{4}(x + 2\sqrt{y_0} - x_0)^2$, $x \ge x_0 - 2\sqrt{y_0}$

23. $P(45) = 8.99$ billion **25.** $E(t) = E_0 e^{-(t - t_1)/RC}$

27. (a) $T(t) = \dfrac{T_2 + BT_1}{1 + B} + \dfrac{T_1 - T_2}{1 + B} e^{k(1 + B)t}$

(b) $\dfrac{T_2 + BT_1}{1 + B}$

(c) $\dfrac{T_2 + BT_1}{1 + B}$

29. $x(\theta) = k\theta - \dfrac{k}{2}\sin 2\theta + c$, $y(\theta) = k \sin^2 \theta$

31. $x(t) = \dfrac{\alpha c_1 e^{\alpha k_1 t}}{1 + c_1 e^{\alpha k_1 t}}$, $y(t) = c_2(1 + c_1 e^{\alpha k_1 t})^{k_2/k_1}$

Exercises 3.1, Page 113

1. $y = \frac{1}{2} e^x - \frac{1}{2} e^{-x}$ **3.** $y = 3x - 4x \ln x$

9. $(-\infty, 2)$ **11.** $y = \dfrac{e}{e^2 - 1}\left(e^x - e^{-x}\right)$

13. (a) $y = e^x \cos x - e^x \sin x$

(b) no solution

(c) $y = e^x \cos x + e^{-\pi/2} e^x \sin x$

(d) $y = c_2 e^x \sin x$, where c_2 is arbitrary

15. dependent **17.** dependent

19. dependent **21.** independent

23. The functions satisfy the DE and are linearly independent on the interval since $W(e^{-3x}, e^{4x}) = 7e^x \neq 0$; $y = c_1 e^{-3x} + c_2 e^{4x}$.

25. The functions satisfy the DE and are linearly independent on the interval since $W(e^x \cos 2x, e^x \sin 2x) = 2e^{2x} \neq 0$; $y = c_1 e^x \cos 2x + c_2 e^x \sin 2x$.

27. The functions satisfy the DE and are linearly independent on the interval since $W(x^3, x^4) = x^6 \neq 0$; $y = c_1 x^3 + c_2 x^4$.

29. The functions satisfy the DE and are linearly independent on the interval since $W(x, x^{-2}, x^{-2} \ln x) = 9x^{-6} \neq 0$; $y = c_1 x + c_2 x^{-2} + c_3 x^{-2} \ln x$.

31. (d) y_1, y_2, and Y_2

33. dependent

39. (b) $y_p = x^2 + 3x + 3e^{4x}$; $y_p = -2x^2 - 6x - \frac{1}{3}e^{2x}$

Exercises 3.2, Page 116

1. $y_2 = xe^{2x}$ **3.** $y_2 = \sin 4x$

5. $y_2 = \sinh x$ **7.** $y_2 = xe^{2x/3}$

9. $y_2 = x^4 \ln|x|$ **11.** $y_2 = 1$

13. $y_2 = x \cos(\ln x)$ **15.** $y_2 = x^2 + x + 2$

17. $y_2 = e^{2x}$, $y_p = -\frac{1}{2}$ **19.** $y_2 = e^{2x}$, $y_p = \frac{5}{2}e^{3x}$

Exercises 3.3, Page 122

1. $y = c_1 + c_2 e^{-x/4}$ **3.** $y = c_1 e^{3x} + c_2 e^{-2x}$

5. $y = c_1 e^{-4x} + c_2 xe^{-4x}$ **7.** $y = c_1 e^{2x/3} + c_2 xe^{-x/4}$

9. $y = c_1 \cos 3x + c_2 \sin 3x$ **11.** $y = e^{2x}(c_1 \cos x + c_2 \sin x)$

13. $y = e^{-x/3}\left(c_1 \cos \frac{\sqrt{2}}{3}x + c_2 \sin \frac{\sqrt{2}}{3}x\right)$

15. $y = c_1 + c_2 e^{-x} + c_3 e^{5x}$ **17.** $y = c_1 e^{-x} + c_2 e^{3x} + c_3 xe^{3x}$

19. $u = c_1 e^t + e^{-t}(c_2 \cos t + c_3 \sin t)$

21. $y = c_1 e^{-x} + c_2 xe^{-x} + c_3 x^2 e^{-x}$

23. $y = c_1 + c_2 x + e^{-x/2}\left(c_3 \cos \frac{\sqrt{3}}{2}x + c_4 \sin \frac{\sqrt{3}}{2}x\right)$

25. $y = c_1 \cos \frac{\sqrt{3}}{2}x + c_2 \sin \frac{\sqrt{3}}{2}x + c_3 x \cos \frac{\sqrt{3}}{2}x + c_4 x \sin \frac{\sqrt{3}}{2}x$

27. $u = c_1 e^r + c_2 re^r + c_3 e^{-r} + c_4 re^{-r} + c_5 e^{-5r}$

29. $y = 2 \cos 4x - \frac{1}{2} \sin 4x$ **31.** $y = -\frac{1}{3}e^{-(t-1)} + \frac{1}{3}e^{5(t-1)}$

33. $y = 0$ **35.** $y = \frac{5}{36} - \frac{5}{36}e^{-6x} + \frac{1}{6}xe^{-6x}$

37. $y = 2 - 2e^x + 2xe^x - \frac{1}{2}x^2 e^x$ **39.** $y = e^{5x} - xe^{5x}$

41. $y = -2 \cos x$

43. $y = \frac{1}{2}\left(1 - \frac{5}{\sqrt{3}}\right)e^{-\sqrt{3}x} - \frac{1}{2}\left(1 + \frac{5}{\sqrt{3}}\right)e^{\sqrt{3}x}$;

$\qquad y = \cosh \sqrt{3}x + \frac{5}{\sqrt{3}} \sinh \sqrt{3}x$

55. (a) $x(t) = c_1 \cosh \frac{8}{\sqrt{L}}t + c_2 \sinh \frac{8}{\sqrt{L}}t$

(b) $x(t) = x_0 \cosh \frac{8}{\sqrt{L}}t$ **(c)** 17.69 ft/s

57. $y''' + 6y'' - 15y' - 100y = 0$

59. $y = c_1 e^{-4x} \cos x + c_2 e^{-4x} \sin x + c_3 e^{2x}$

Exercises 3.4, Page 131

1. $y = c_1 e^{-x} + c_2 e^{-2x} + 3$

3. $y = c_1 e^{5x} + c_2 xe^{5x} + \frac{6}{5}x + \frac{3}{5}$

5. $y = c_1 e^{-2x} + c_2 xe^{-2x} + x^2 - 4x + \frac{7}{2}$

7. $y = c_1 \cos \sqrt{3}x + c_2 \sin \sqrt{3}x + (-4x^2 + 4x - \frac{4}{3})e^{3x}$

9. $y = c_1 + c_2 e^x + 3x$

11. $y = c_1 e^{x/2} + c_2 xe^{x/2} + 12 + \frac{1}{2}x^2 e^{x/2}$

13. $y = c_1 \cos 2x + c_2 \sin 2x - \frac{3}{4}x \cos 2x$

15. $y = c_1 \cos x + c_2 \sin x - \frac{1}{2}x^2 \cos x + \frac{1}{2}x \sin x$

17. $y = c_1 e^x \cos 2x + c_2 e^x \sin 2x + \frac{1}{4}xe^x \sin 2x$

19. $y = c_1 e^{-x} + c_2 xe^{-x} - \frac{1}{2} \cos x + \frac{12}{25} \sin 2x - \frac{9}{25} \cos 2x$

21. $y = c_1 + c_2 x + c_3 e^{6x} - \frac{1}{4}x^2 - \frac{6}{37} \cos x + \frac{1}{37} \sin x$

23. $y = c_1 e^x + c_2 xe^x + c_3 x^2 e^x - x - 3 - \frac{2}{3}x^3 e^x$

25. $y = c_1 \cos x + c_2 \sin x + c_3 x \cos x + c_4 x \sin x + x^2 - 2x - 3$

27. $y = \sqrt{2} \sin 2x - \frac{1}{2}$

29. $y = -200 + 200e^{-x/5} - 3x^2 + 30x$

31. $y = -10e^{-2x} \cos x + 9e^{-2x} \sin x + 7e^{-4x}$

33. $x = \frac{F_0}{2\omega^2} \sin \omega t - \frac{F_0}{2\omega} t \cos \omega t$

35. $y = 11 - 11e^x + 9xe^x + 2x - 12x^2 e^x + \frac{1}{2}e^{5x}$

37. $y = 6 \cos x - 6(\cot 1) \sin x + x^2 - 1$

39. $y = \begin{cases} \cos 2x + \frac{5}{6} \sin 2x + \frac{1}{3} \sin x, & 0 \le x \le \pi/2 \\ \frac{2}{3} \cos 2x + \frac{5}{6} \sin 2x, & x > \pi/2 \end{cases}$

Exercises 3.5, Page 136

1. $y = c_1 \cos x + c_2 \sin x + x \sin x + \cos x \ln |\cos x|$

3. $y = c_1 \cos x + c_2 \sin x - \frac{1}{2} x \cos x$

5. $y = c_1 \cos x + c_2 \sin x + \frac{1}{2} - \frac{1}{6} \cos 2x$

7. $y = c_1 e^x + c_2 e^{-x} + \frac{1}{2} x \sinh x$

9. $y = c_1 e^{2x} + c_2 e^{-2x} + \frac{1}{4} \left(e^{2x} \ln |x| - e^{-2x} \int_{x_0}^x \frac{e^{4t}}{t} dt \right), x_0 > 0$

11. $y = c_1 e^{-x} + c_2 e^{-2x} + (e^{-x} + e^{-2x}) \ln (1 + e^x)$

13. $y = c_1 e^{-2x} + c_2 e^{-x} - e^{-2x} \sin e^x$

15. $y = c_1 e^{-t} + c_2 t e^{-t} + \frac{1}{2} t^2 e^{-t} \ln t - \frac{3}{4} t^2 e^{-t}$

17. $y = c_1 e^x \sin x + c_2 e^x \cos x + \frac{1}{3} x e^x \sin x + \frac{1}{3} e^x \cos x \ln |\cos x|$

19. $y = \frac{1}{4} e^{-x/2} + \frac{3}{4} e^{x/2} + \frac{1}{8} x^2 e^{x/2} - \frac{1}{4} x e^{x/2}$

21. $y = \frac{4}{9} e^{-4x} + \frac{25}{36} e^{2x} - \frac{1}{4} e^{-2x} + \frac{1}{9} e^{-x}$

23. $y = c_1 x^{-1/2} \cos x + c_2 x^{-1/2} \sin x + x^{-1/2}$

25. $y = c_1 + c_2 \cos x + c_3 \sin x - \ln |\cos x| - \sin x \ln |\sec x + \tan x|$

27. $y = y_c + y_p$ where $y_c = c_1 e^x \cos 3x + c_2 e^x \sin 3x$, $y_p = y_{p_1} + y_{p_2} = \frac{9}{17} \sin x + \frac{2}{17} \cos x - \frac{1}{27} e^x \cos 3x \ln |\sec 3x + \tan 3x|$

29. $-\pi/2 < x < \pi/2, -\infty < x < \infty, 0 < x < \infty, -1 \le x \le 1$

Exercises 3.6, Page 141

1. $y = c_1 x^{-1} + c_2 x^2$ **3.** $y = c_1 + c_2 \ln x$

5. $y = c_1 \cos(2 \ln x) + c_2 \sin(2 \ln x)$

7. $y = c_1 x^{(2-\sqrt{6})} + c_2 x^{(2+\sqrt{6})}$

9. $y = c_1 \cos(\frac{1}{5} \ln x) + c_2 \sin(\frac{1}{5} \ln x)$

11. $y = c_1 x^{-2} + c_2 x^{-2} \ln x$

13. $y = x^{-1/2} \left[c_1 \cos\left(\frac{\sqrt{3}}{6} \ln x\right) + c_2 \sin\left(\frac{\sqrt{3}}{6} \ln x\right) \right]$

15. $y = c_1 x^3 + c_2 \cos(\sqrt{2} \ln x) + c_3 \sin(\sqrt{2} \ln x)$

17. $y = c_1 + c_2 x + c_3 x^2 + c_4 x^{-3}$ **19.** $y = c_1 + c_2 x^5 + \frac{1}{5} x^5 \ln x$

21. $y = c_1 x + c_2 x \ln x + x(\ln x)^2$ **23.** $y = 2 - 2x^{-2}$

25. $y = \cos(\ln x) + 2 \sin(\ln x)$ **27.** $y = \frac{3}{4} - \ln x + \frac{1}{4} x^2$

29. $y = c_1 x^{-1} + c_2 x^{-8} + \frac{1}{30} x^2$

31. $y = x^2 [c_1 \cos(3 \ln x) + c_2 \sin(3 \ln x)] + \frac{4}{13} + \frac{3}{10} x$

37. $y = 2(-x)^{1/2} - 5(-x)^{1/2} \ln(-x), x < 0$

39. $y = c_1 \cos[\ln(x + 2)] + c_2 \sin[\ln(x + 2)]$

41. y_0 and y_1 arbitrary; $y_0 = 0$ and y_1 arbitrary; $y_0 = 0$ and $y_1 = 0$

Exercises 3.7, Page 146

3. $y = \ln |\cos(c_1 - x)| + c_2$

5. $y = \frac{1}{c_1^2} \ln |c_1 x + 1| - \frac{1}{c_1} x + c_2$

7. $\frac{1}{3} y^3 - c_1 y = x + c_2$

9. $y = \tan\left(\frac{\pi}{4} - \frac{x}{2}\right), -\frac{\pi}{2} < x < \frac{3\pi}{2}$

11. $y = -\frac{1}{c_1} \sqrt{1 - c_1^2 x^2} + c_2$

13. $y = 1 + x + \frac{1}{2} x^2 + \frac{1}{2} x^3 + \frac{1}{6} x^4 + \frac{1}{10} x^5 + \cdots$

15. $y = 1 + x - \frac{1}{2} x^2 + \frac{2}{3} x^3 - \frac{1}{4} x^4 + \frac{7}{60} x^5 + \cdots$

17. $y = -\sqrt{1 - x^2}$

21. $y = \sinh(x + c_1) + c_2 x + c_3$

Exercises 3.8, Page 158

1. $\dfrac{\sqrt{2}\pi}{8}$ **3.** $x(t) = -\frac{1}{4} \cos 4\sqrt{6} t$

5. (a) $x\left(\frac{\pi}{12}\right) = -\frac{1}{4}; x\left(\frac{\pi}{8}\right) = -\frac{1}{2}; x\left(\frac{\pi}{6}\right) = -\frac{1}{4};$
$x\left(\frac{\pi}{4}\right) = \frac{1}{2}; x\left(\frac{9\pi}{32}\right) = \frac{\sqrt{2}}{4}$

(b) 4 ft/s; downward

(c) $t = \dfrac{(2n + 1)\pi}{16}, n = 0, 1, 2, \ldots$

7. (a) the 20-kg mass

(b) the 20-kg mass; the 50-kg mass

(c) $t = n\pi, n = 0, 1, 2, \ldots$; at the equilibrium position; the 50-kg mass is moving upward whereas the 20-kg mass is moving upward when n is even and downward when n is odd.

9. $x(t) = \frac{1}{2} \cos 2t + \frac{3}{4} \sin 2t = \frac{\sqrt{13}}{4} \sin(2t + 0.5880)$

11. (a) $x(t) = -\frac{2}{3} \cos 10t + \frac{1}{2} \sin 10t = \frac{5}{6} \sin(10t - 0.927)$

(b) $\frac{5}{6}$ ft; $\frac{\pi}{5}$ **(c)** 15 cycles **(d)** 0.721 s

(e) $\dfrac{(2n + 1)\pi}{20} + 0.0927, n = 0, 1, 2, \ldots$

(f) $x(3) = -0.597$ ft

(g) $x'(3) = -5.814$ ft/s

(h) $x''(3) = 59.702$ ft/s^2

(i) $\pm 8\frac{1}{3}$ ft/s

(j) $0.1451 + \dfrac{n\pi}{5}$; $0.3545 + \dfrac{n\pi}{5}$, $n = 0, 1, 2, \ldots$

(k) $0.3545 + \dfrac{n\pi}{5}$, $n = 0, 1, 2, \ldots$

13. 120 lb/ft; $x(t) = \dfrac{\sqrt{3}}{12} \sin 8\sqrt{3}\, t$

17. (a) above **(b)** heading upward

19. (a) below **(b)** heading upward

21. $\frac{1}{4}s$; $\frac{1}{2}s$, $x(\frac{1}{2}) = e^{-2}$; that is, the weight is approximately 0.14 ft below the equilibrium position.

23. (a) $x(t) = \frac{4}{3}e^{-2t} - \frac{1}{3}e^{-8t}$

(b) $x(t) = -\frac{2}{3}e^{-2t} + \frac{5}{3}e^{-8t}$

25. (a) $x(t) = e^{-2t}(-\cos 4t - \frac{1}{2}\sin 4t)$

(b) $x(t) = \dfrac{\sqrt{5}}{2}\, e^{-2t} \sin(4t + 4.249)$

(c) $t = 1.294$ s

27. (a) $\beta > \frac{5}{2}$ **(b)** $\beta = \frac{5}{2}$ **(c)** $0 < \beta < \frac{5}{2}$

29. $x(t) = e^{-t/2}\left(-\dfrac{4}{3}\cos\dfrac{\sqrt{47}}{2}t - \dfrac{64}{3\sqrt{47}}\sin\dfrac{\sqrt{47}}{2}t\right)$

$+ \dfrac{10}{3}(\cos 3t + \sin 3t)$

31. $x(t) = \frac{1}{4}e^{-4t} + te^{-4t} - \frac{1}{4}\cos 4t$

33. $x(t) = -\frac{1}{2}\cos 4t + \frac{9}{4}\sin 4t + \frac{1}{2}e^{-2t}\cos 4t$

$-2e^{-2t}\sin 4t$

35. (a) $m\dfrac{d^2x}{dt^2} = -k(x - h) - \beta\dfrac{dx}{dt}$ or

$\dfrac{d^2x}{dt^2} + 2\lambda\dfrac{dx}{dt} + \omega^2 x = \omega^2 h(t)$,

where $2\lambda = \beta/m$ and $\omega^2 = k/m$

(b) $x(t) = e^{-2t}(-\frac{56}{13}\cos 2t - \frac{72}{13}\sin 2t)$
$+ \frac{56}{13}\cos t + \frac{32}{13}\sin t$

37. $x(t) = -\cos 2t - \frac{1}{8}\sin 2t + \frac{3}{4}t\sin 2t + \frac{5}{4}t\cos 2t$

39. (b) $\dfrac{F_0}{2\omega}t\sin\omega t$ **45.** 4.568 C; 0.0509 s

47. $q(t) = 10 - 10e^{-3t}(\cos 3t + \sin 3t)$
$i(t) = 60e^{-3t}\sin 3t$; 10.432 C

49. $q_p = \frac{100}{13}\sin t + \frac{150}{13}\cos t$
$i_p = \frac{100}{13}\cos t - \frac{150}{13}\sin t$

53. $q(t) = -\frac{1}{2}e^{-10t}(\cos 10t + \sin 10t) + \frac{3}{2}$; $\frac{3}{2}$ C

57. $q(t) = \left(q_0 - \dfrac{E_0 C}{1 - \gamma^2 LC}\right)\cos\dfrac{t}{\sqrt{LC}}$

$+ \sqrt{LC}\,i_0\sin\dfrac{t}{\sqrt{LC}} + \dfrac{E_0 C}{1 - \gamma^2 LC}\cos\gamma t$

$i(t) = i_0\cos\dfrac{t}{\sqrt{LC}}$

$- \dfrac{1}{\sqrt{LC}}\left(q_0 - \dfrac{E_0 C}{1 - \gamma^2 LC}\right)\sin\dfrac{t}{\sqrt{LC}}$

$- \dfrac{E_0 C\gamma}{1 - \gamma^2 LC}\sin\gamma t$

Exercises 3.9, Page 168

1. (a) $y_2 = \dfrac{w_0}{24EI}(6L^2 x^2 - 4Lx^3 + x^4)$

3. (a) $y_2 = \dfrac{w_0}{48EI}(3L^2 x^2 - 5Lx^3 + 2x^4)$

5. (a) $y_{\max} = \dfrac{w_0 L^4}{8EI}$

(b) $\frac{1}{16}$th of the maximum deflection in part **(a)**

7. $y(x) = -\dfrac{w_0 EI}{P^2}\cosh\sqrt{\dfrac{P}{EI}}x$

$+ \left(\dfrac{w_0 EI}{P^2}\sinh\sqrt{\dfrac{P}{EI}}L - \dfrac{w_0 L\sqrt{EI}}{P\sqrt{P}}\right)\dfrac{\sinh\sqrt{\dfrac{P}{EI}}x}{\cosh\sqrt{\dfrac{P}{EI}}L}$

$+ \dfrac{w_0}{2P}x^2 + \dfrac{w_0 EI}{P^2}$

9. $\lambda = n^2$, $n = 1, 2, 3, \ldots$; $y = \sin nx$

11. $\lambda = \dfrac{(2n-1)^2\pi^2}{4L^2}$, $n = 1, 2, 3, \ldots$; $y = \cos\dfrac{(2n-1)\pi x}{2L}$

13. $\lambda = n^2$, $n = 0, 1, 2, \ldots$; $y = \cos nx$

15. $\lambda = \dfrac{n^2\pi^2}{25}$, $n = 1, 2, 3, \ldots$; $y = e^{-x}\sin\dfrac{n\pi x}{5}$

17. $\lambda = \dfrac{n\pi}{L}$, $n = 1, 2, 3, \ldots$; $y = \sin\dfrac{n\pi x}{L}$

19. $\lambda = n^2$, $n = 1, 2, 3, \ldots$; $y = \sin(n\ln x)$

21. $\lambda = 0$; $y = 1$

$\lambda = \dfrac{n^2\pi^2}{4}$, $n = 1, 2, 3, \ldots$; $y = \cos\left(\dfrac{n\pi}{2}\ln x\right)$

25. $x = L/4$, $x = L/2$, $x = 3L/4$

27. $\omega_n = \dfrac{n\pi\sqrt{T}}{L\sqrt{\rho}}$, $n = 1, 2, 3, \ldots$; $y = \sin\dfrac{n\pi x}{L}$

29. $u(r) = \left(\dfrac{u_0 - u_1}{b - a}\right)\dfrac{ab}{r} + \dfrac{u_1 b - u_0 a}{b - a}$

Exercises 3.10, Page 176

7. $\dfrac{d^2x}{dt^2} + x = 0$

19. (a) 5 ft **(b)** $4\sqrt{10}$ ft/s

(c) $0 \le t \le 15/(4\sqrt{10})$; 7.5 ft

21. (a) $xv\dfrac{dv}{dx} + v^2 = 32x$ **(b)** $\frac{1}{2}v^2x^2 - \frac{32}{3}x^3 = -288$

23. (a) $xy'' = r\sqrt{1 + (y')^2}$. When $t = 0$, $x = a$, $y = 0$, $dy/dx = 0$.

(b) When $r \ne 1$,

$$y(x) = \frac{a}{2}\left[\frac{1}{1+r}\left(\frac{x}{a}\right)^{1+r} - \frac{1}{1-r}\left(\frac{x}{a}\right)^{1-r}\right] + \frac{ar}{1-r^2}.$$

When $r = 1$,

$$y(x) = \frac{1}{2}\left[\frac{1}{2a}(x^2 - a^2) + \frac{1}{a}\ln\frac{a}{x}\right].$$

(c) The paths intersect when $r < 1$.

Exercises 3.11, Page 185

1. $x = c_1 e^t + c_2 t e^t$
$y = (c_1 - c_2)e^t + c_2 t e^t$

3. $x = c_1 \cos t + c_2 \sin t + t + 1$
$y = c_1 \sin t - c_2 \cos t + t - 1$

5. $x = \frac{1}{2}c_1 \sin t + \frac{1}{2}c_2 \cos t - 2c_3 \sin \sqrt{6}t - 2c_4 \cos \sqrt{6}t$
$y = c_1 \sin t + c_2 \cos t + c_3 \sin \sqrt{6}t + c_4 \cos \sqrt{6}t$

7. $x = c_1 e^{2t} + c_2 e^{-2t} + c_3 \sin 2t + c_4 \cos 2t + \frac{1}{5}e^t$
$y = c_1 e^{2t} + c_2 e^{-2t} - c_3 \sin 2t - c_4 \cos 2t - \frac{1}{5}e^t$

9. $x = c_1 - c_2 \cos t + c_3 \sin t + \frac{17}{15}e^{3t}$
$y = c_1 + c_2 \sin t + c_3 \cos t - \frac{4}{15}e^{3t}$

11. $x = c_1 e^t + c_2 e^{-t/2} \cos \dfrac{\sqrt{3}}{2} t + c_3 e^{-t/2} \sin \dfrac{\sqrt{3}}{2} t$

$y = \left(-\dfrac{3}{2}c_2 - \dfrac{\sqrt{3}}{2}c_3\right)e^{-t/2} \cos \dfrac{\sqrt{3}}{2} t$

$\quad + \left(\dfrac{\sqrt{3}}{2}c_2 - \dfrac{3}{2}c_3\right)e^{-t/2} \sin \dfrac{\sqrt{3}}{2} t$

13. $x = c_1 e^{4t} + \frac{4}{3}e^t$
$y = -\frac{3}{4}c_1 e^{4t} + c_2 + 5e^t$

15. $x = c_1 + c_2 t + c_3 e^t + c_4 e^{-t} - \frac{1}{2}t^2$
$y = (c_1 - c_2 + 2) + (c_2 + 1)t + c_4 e^{-t} - \frac{1}{2}t^2$

17. $x = c_1 e^t + c_2 e^{-t/2} \sin \dfrac{\sqrt{3}}{2} t + c_3 e^{-t/2} \cos \dfrac{\sqrt{3}}{2} t$

$y = c_1 e^t + \left(-\dfrac{1}{2}c_2 - \dfrac{\sqrt{3}}{2}c_3\right)e^{-t/2} \sin \dfrac{\sqrt{3}}{2} t$

$\quad + \left(\dfrac{\sqrt{3}}{2}c_2 - \dfrac{1}{2}c_3\right)e^{-t/2} \cos \dfrac{\sqrt{3}}{2} t$

$z = c_1 e^t + \left(-\dfrac{1}{2}c_2 + \dfrac{\sqrt{3}}{2}c_3\right)e^{-t/2} \sin \dfrac{\sqrt{3}}{2} t$

$\quad + \left(-\dfrac{\sqrt{3}}{2}c_2 - \dfrac{1}{2}c_3\right)e^{-t/2} \cos \dfrac{\sqrt{3}}{2} t$

19. $x = -6c_1 e^{-t} - 3c_2 e^{-2t} + 2c_3 e^{3t}$
$y = c_1 e^{-t} + c_2 e^{-2t} + c_3 e^{3t}$
$z = 5c_1 e^{-t} + c_2 e^{-2t} + c_3 e^{3t}$

21. $x = e^{-3t+3} - te^{-3t+3}$
$y = -e^{-3t+3} + 2te^{-3t+3}$

25. $m\dfrac{d^2x}{dt^2} = 0$

$m\dfrac{d^2y}{dt^2} = -mg$

$x = c_1 t + c_2$

$y = -\frac{1}{2}gt^2 + c_3 t + c_4$

Chapter 3 Review Exercises, Page 186

1. $y = 0$ **3.** false **5.** 8 ft

7. $\frac{3}{4}$ **9.** $(-\infty, 0); (0, \infty)$

11. $y = c_1 e^{3x} + c_2 e^{-5x} + c_3 x e^{-5x} + c_4 e^x + c_5 x e^x + c_6 x^2 e^x$,
$y = c_1 x^3 + c_2 x^{-5} + c_3 x^{-5} \ln x + c_4 x + c_5 x \ln x + c_6 x(\ln x)^2$

13. $y = c_1 e^{(1 + \sqrt{3})x} + c_2 e^{(1 - \sqrt{3})x}$ **15.** $y = c_1 + c_2 e^{-5x} + c_3 x e^{-5x}$

17. $y = c_1 e^{-x/3} + e^{-3x/2}\left(c_2 \cos \dfrac{\sqrt{7}}{2}x + c_3 \sin \dfrac{\sqrt{7}}{2}x\right)$

19. $y = e^{3x/2}\left(c_2 \cos \dfrac{\sqrt{11}}{2}x + c_3 \sin \dfrac{\sqrt{11}}{2}x\right) + \frac{4}{5}x^3 + \frac{36}{25}x^2$
$\quad + \frac{46}{125}x - \frac{222}{625}$

21. $y = c_1 + c_2 e^{2x} + c_3 e^{3x} + \frac{1}{5}\sin x - \frac{1}{5}\cos x + \frac{4}{3}x$

23. $y = e^x(c_1 \cos x + c_2 \sin x) - e^x \cos x \ln |\sec x + \tan x|$

25. $y = c_1 x^{-1/3} + c_2 x^{1/2}$ **27.** $y = c_1 x^2 + c_2 x^3 + x^4 - x^2 \ln x$

29. (a) $y = c_1 \cos \omega x + c_2 \sin \omega x + A \cos \alpha x + B \sin \alpha x$, $\omega \ne \alpha$,
$\quad y = c_1 \cos \omega x + c_2 \sin \omega x + Ax \cos \omega x + Bx \sin \omega x$, $\omega = \alpha$

(b) $y = c_1 e^{-\omega x} + c_2 e^{\omega x} + Ae^{\alpha x}$, $\omega \ne \alpha$,
$\quad y = c_1 e^{-\omega x} + c_2 e^{\omega x} + Axe^{\omega x}$, $\omega = \alpha$

31. (a) $y = c_1 \cosh x + c_2 \sinh x + c_3 x \cosh x + c_4 x \sinh x$

(b) $y_p = Ax^2 \cosh x + Bx^2 \sinh x$

33. $y = e^{x - \pi} \cos x$

35. $y = \frac{13}{4}e^x - \frac{5}{4}e^{-x} - x - \frac{1}{2}\sin x$

37. $y = x^2 + 4$

41. $x = -c_1 e^t - \frac{3}{2}c_2 e^{2t} + \frac{5}{2}$ **43.** $x = c_1 e^t + c_2 e^{5t} + te^t$
$\quad y = c_1 e^t + c_2 e^{2t} - 3$ $\quad y = -c_1 e^t + 3c_2 e^{5t} - te^t + 2e^t$

45. 14.4 lb **47.** $0 < m \leq 2$

49. (a) $q(t) = -\frac{1}{150} \sin 100t + \frac{1}{75} \sin 50t$

 (b) $i(t) = -\frac{2}{3} \cos 100t + \frac{2}{3} \cos 50t$

 (c) $t = \dfrac{n\pi}{50}$, $n = 0, 1, 2, \ldots$

Exercises 4.1, Page 195

1. $\dfrac{2}{s} e^{-s} - \dfrac{1}{s}$ **3.** $\dfrac{1}{s^2} - \dfrac{1}{s^2} e^{-s}$ **5.** $\dfrac{1 + e^{-s\pi}}{s^2 + 1}$

7. $\dfrac{e^{-s}}{s} + \dfrac{e^{-s}}{s^2}$ **9.** $\dfrac{1}{s} - \dfrac{1}{s^2} + \dfrac{e^{-s}}{s^2}$ **11.** $\dfrac{e^7}{s-1}$

13. $\dfrac{1}{(s-4)^2}$ **15.** $\dfrac{1}{s^2 + 2s + 2}$ **17.** $\dfrac{s^2 - 1}{(s^2 + 1)^2}$

19. $\dfrac{48}{s^5}$ **21.** $\dfrac{4}{s^2} - \dfrac{10}{s}$ **23.** $\dfrac{2}{s^3} + \dfrac{6}{s^2} - \dfrac{3}{s}$

25. $\dfrac{6}{s^4} + \dfrac{6}{s^3} + \dfrac{3}{s^2} + \dfrac{1}{s}$ **27.** $\dfrac{1}{s} + \dfrac{1}{s-4}$

29. $\dfrac{1}{s} + \dfrac{2}{s-2} + \dfrac{1}{s-4}$ **31.** $\dfrac{8}{s^3} - \dfrac{15}{s^2 + 9}$

33. Use $\sinh kt = \dfrac{e^{kt} - e^{-kt}}{2}$ to show that
$\mathscr{L}\{\sinh kt\} = \dfrac{k}{s^2 - k^2}$.

35. $\dfrac{1}{2(s-2)} - \dfrac{1}{2s}$ **37.** $\dfrac{2}{s^2 + 16}$

43. For $s > $ maximum of $\{s_1, s_2\}$

Exercises 4.2, Page 203

1. $\frac{1}{2} t^2$ **3.** $t - 2t^4$

5. $1 + 3t + \frac{3}{2} t^2 + \frac{1}{6} t^3$ **7.** $t - 1 + e^{2t}$

9. $\frac{1}{4} e^{-t/4}$ **11.** $\frac{5}{7} \sin 7t$

13. $\cos \dfrac{t}{2}$ **15.** $2 \cos 3t - 2 \sin 3t$

17. $\frac{1}{3} - \frac{1}{3} e^{-3t}$ **19.** $\frac{3}{4} e^{-3t} + \frac{1}{4} e^t$

21. $0.3e^{0.1t} + 0.6e^{-0.2t}$ **23.** $\frac{1}{2} e^{2t} - e^{3t} + \frac{1}{2} e^{6t}$

25. $\frac{1}{5} - \frac{1}{5} \cos \sqrt{5}t$ **27.** $-4 + 3e^{-t} + \cos t + 3 \sin t$

29. $\frac{1}{3} \sin t - \frac{1}{6} \sin 2t$ **31.** $y = -1 + e^t$

33. $y = \frac{1}{10} e^{4t} + \frac{19}{10} e^{-6t}$ **35.** $y = \frac{4}{3} e^{-t} - \frac{1}{3} e^{-4t}$

37. $y = 10 \cos t + 2 \sin t - \sqrt{2} \sin \sqrt{2}t$

39. $y = -\frac{8}{9} e^{-t/2} + \frac{1}{9} e^{-2t} + \frac{5}{18} e^t + \frac{1}{2} e^{-t}$

41. $y_0(t) = 2e^{-6t}$, $y_1(t) = \frac{1}{10} e^{4t} - \frac{1}{10} e^{-6t}$

Exercises 4.3, Page 212

1. $\dfrac{1}{(s-10)^2}$ **3.** $\dfrac{6}{(s+2)^4}$

5. $\dfrac{1}{(s-2)^2} + \dfrac{2}{(s-3)^2} + \dfrac{1}{(s-4)^2}$ **7.** $\dfrac{3}{(s-1)^2 + 9}$

9. $\dfrac{s}{s^2 + 25} - \dfrac{s-1}{(s-1)^2 + 25} + 3 \dfrac{s+4}{(s-1)^2 + 25}$

11. $\frac{1}{2} t^2 e^{-t}$ **13.** $e^{3t} \sin t$

15. $e^{-2t} \cos t - 2e^{-2t} \sin t$ **17.** $e^{-t} - te^{-t}$

19. $5 - t - 5e^{-t} - 4te^{-t} - \frac{3}{2}t^2 e^{-t}$

21. $y = te^{-4t} + 2e^{-4t}$ **23.** $y = e^{-t} + 2te^{-t}$

25. $y = \frac{1}{9} t + \frac{2}{27} - \frac{2}{27} e^{3t} + \frac{10}{9} te^{3t}$

27. $y = -\frac{3}{2} e^{3t} \sin 2t$

29. $y = \frac{1}{2} - \frac{1}{2} e^t \cos t + \frac{1}{2} e^t \sin t$

31. $y = (e + 1)te^{-t} + (e - 1)e^{-t}$

33. $x(t) = -\frac{3}{2} e^{-7t/2} \cos \dfrac{\sqrt{15}}{2} t - \dfrac{7\sqrt{15}}{10} e^{-7t/2} \sin \dfrac{\sqrt{15}}{2} t$

37. $\dfrac{e^{-s}}{s^2}$ **39.** $\dfrac{e^{-2s}}{s^2} + 2 \dfrac{e^{-2s}}{s}$ **41.** $\dfrac{s}{s^2 + 4} e^{-\pi s}$

43. $\frac{1}{2} (t - 2)^2 \, \mathcal{U}(t - 2)$ **45.** $-\sin t \, \mathcal{U}(t - \pi)$

47. $\mathcal{U}(t - 1) - e^{-(t-1)} \, \mathcal{U}(t - 1)$ **49.** c

51. f **53.** a

55. $f(t) = 2 - 4 \, \mathcal{U}(t - 3)$; $\mathscr{L}\{f(t)\} = \dfrac{2}{s} - \dfrac{4}{s} e^{-3s}$

57. $f(t) = t^2 \, \mathcal{U}(t - 1)$; $\mathscr{L}\{f(t)\} = 2\dfrac{e^{-s}}{s^3} + 2\dfrac{e^{-s}}{s^2} + \dfrac{e^{-s}}{s}$

59. $f(t) = t - t \, \mathcal{U}(t - 2)$; $\mathscr{L}\{f(t)\} = \dfrac{1}{s^2} - \dfrac{e^{-2s}}{s^2} - 2\dfrac{e^{-2s}}{s}$

61. $f(t) = \mathcal{U}(t - a) - \mathcal{U}(t - b)$; $\mathscr{L}\{f(t)\} = \dfrac{e^{-as}}{s} - \dfrac{e^{-bs}}{s}$

63. $y = [5 - 5e^{-(t-1)}] \, \mathcal{U}(t - 1)$

65. $y = -\frac{1}{4} + \frac{1}{2} t + \frac{1}{4} e^{-2t} - \frac{1}{4} \, \mathcal{U}(t - 1) - \frac{1}{2} (t - 1) \, \mathcal{U}(t - 1)$
 $+ \frac{1}{4} e^{-2(t-1)} \, \mathcal{U}(t - 1)$

67. $y = \cos 2t - \frac{1}{6} \sin 2(t - 2\pi) \, \mathcal{U}(t - 2\pi)$
 $+ \frac{1}{3} \sin(t - 2\pi) \, \mathcal{U}(t - 2\pi)$

69. $y = \sin t + [1 - \cos(t - \pi)] \, \mathcal{U}(t - \pi)$
 $-[1 - \cos(t - 2\pi)] \, \mathcal{U}(t - 2\pi)$

71. $x(t) = \frac{5}{4} t - \frac{5}{16} \sin 4t - \frac{5}{4}(t - 5) \, \mathcal{U}(t - 5)$
 $+ \frac{5}{16} \sin 4(t - 5) \, \mathcal{U}(t - 5) - \frac{25}{4} \, \mathcal{U}(t - 5)$
 $+ \frac{25}{4} \cos 4(t - 5) \, \mathcal{U}(t - 5)$

73. $q(t) = \frac{2}{5}\,\mathcal{U}(t-3) - \frac{2}{5}\,e^{-5(t-3)}\,\mathcal{U}(t-3)$

75. (a) $i(t) = \frac{1}{101}e^{-10t} - \frac{1}{101}\cos t + \frac{10}{101}\sin t$

$$- \frac{10}{101}e^{-10(t-3\pi/2)}\,\mathcal{U}\!\left(t-\frac{3\pi}{2}\right)$$

$$+ \frac{10}{101}\cos\!\left(t-\frac{3\pi}{2}\right)\mathcal{U}\!\left(t-\frac{3\pi}{2}\right) + \frac{1}{101}\sin\!\left(t-\frac{3\pi}{2}\right)$$

(b) $i_{\max} \approx 0.1$ at $t \approx 1.6$, $i_{\min} \approx -0.1$ at $t \approx 4.7$

77. $y(x) = \dfrac{w_0 L^2}{16EI}\,x^2 - \dfrac{w_0 L}{12EI}\,x^3 + \dfrac{w_0}{24EI}\,x^4 -$

$$\dfrac{w_0}{24EI}\left(x-\dfrac{L}{2}\right)^4 \mathcal{U}\!\left(x-\dfrac{L}{2}\right)$$

79. $y(x) = \dfrac{w_0 L^2}{48EI}\,x^2 - \dfrac{w_0 L}{24EI}\,x^3 + \dfrac{w_0}{60EI}\left[\dfrac{5L}{2}\,x^4 - x^5 +\right.$

$$\left.\left(x-\dfrac{L}{2}\right)^5 \mathcal{U}\!\left(x-\dfrac{L}{2}\right)\right]$$

Exercises 4.4, Page 222

1. $\dfrac{s^2-4}{(s^2+4)^2}$ **3.** $\dfrac{6s^2+2}{(s^2-1)^3}$

5. $\dfrac{12s-24}{[(s-2)^2+36]^2}$ **7.** $\dfrac{6}{s^5}$

9. $\dfrac{s-1}{(s+1)[(s-1)^2+1]}$ **11.** $\dfrac{1}{s(s-1)}$

13. $\dfrac{s+1}{s[(s+1)^2+1]}$ **15.** $\dfrac{1}{s^2(s-1)}$ **17.** $\dfrac{3s^2+1}{s^2(s^2+1)^2}$

19. (a) e^t-1 **(b)** e^t-t-1 **(c)** $e^t-\frac{1}{2}t^2-t-1$

21. $\dfrac{e^{-t}-e^{3t}}{t}$ **23.** $\dfrac{1-e^{-as}}{s(1+e^{-as})}$

25. $\dfrac{a}{s}\left(\dfrac{1}{bs}-\dfrac{1}{e^{bs}-1}\right)$ **27.** $\dfrac{\coth(\pi s/2)}{s^2+1}$

29. $y = -\frac{1}{2}e^{-t} + \frac{1}{2}\cos t - \frac{1}{2}t\cos t + \frac{1}{2}t\sin t$

31. $y = 2\cos 3t + \frac{5}{3}\sin 3t + \frac{1}{6}t\sin 3t$

33. $y = \frac{1}{4}\sin 4t + \frac{1}{8}t\sin 4t - \frac{1}{8}(t-\pi)\sin 4(t-\pi)\,\mathcal{U}(t-\pi)$

35. $y = \frac{1}{2}\sin t - \frac{1}{2}t\cos t + \frac{1}{4}t\sin t - \frac{1}{4}t^2\cos t$

37. $f(t) = \sin t$ **39.** $f(t) = -\frac{1}{8}e^{-t} + \frac{1}{8}e^t + \frac{3}{4}te^t + \frac{1}{4}t^2 e^t$

41. $f(t) = e^{-t}$

43. $f(t) = \frac{3}{8}e^{2t} + \frac{1}{8}e^{-2t} + \frac{1}{2}\cos 2t + \frac{1}{4}\sin 2t$

45. $y = \sin t - \frac{1}{2}t\sin t$

47. $i(t) = 100[e^{-10(t-1)} - e^{-20(t-1)}]\,\mathcal{U}(t-1)$
$$- 100[e^{-10(t-2)} - e^{-20(t-2)}]\,\mathcal{U}(t-2)$$

49. $i(t) = \dfrac{1}{R}(1-e^{-Rt/L}) + \dfrac{2}{R}\displaystyle\sum_{n=1}^{\infty}(-1)^n(1-e^{-R(t-n)/L})\,\mathcal{U}(t-n)$

51. $x(t) = 2(1 - e^{-t}\cos 3t - \frac{1}{3}e^{-t}\sin 3t)$

$$+ 4\sum_{n=1}^{\infty}(-1)^n\,[1 - e^{-(t-n\pi)}\cos 3(t-n\pi)$$

$$- \tfrac{1}{3}e^{-(t-n\pi)}\sin 3(t-n\pi)]\,\mathcal{U}(t-n\pi)$$

Exercises 4.5, Page 227

1. $y = e^{3(t-2)}\,\mathcal{U}(t-2)$ **3.** $y = \sin t + \sin t\,\mathcal{U}(t-2\pi)$

5. $y = -\cos t\,\mathcal{U}\!\left(t-\dfrac{\pi}{2}\right) + \cos t\,\mathcal{U}\!\left(t-\dfrac{3\pi}{2}\right)$

7. $y = \frac{1}{2} - \frac{1}{2}e^{-2t} + [\frac{1}{2} - \frac{1}{2}e^{-2(t-1)}]\mathcal{U}(t-1)$

9. $y = e^{-2(t-2\pi)}\sin t\,\mathcal{U}(t-2\pi)$

11. $y = e^{-2t}\cos 3t + \frac{2}{3}e^{-2t}\sin 3t$
$$+ \tfrac{1}{3}e^{-2(t-\pi)}\sin 3(t-\pi)\mathcal{U}(t-\pi)$$
$$+ \tfrac{1}{3}e^{-2(t-3\pi)}\sin 3(t-3\pi)\mathcal{U}(t-3\pi)$$

13. $y(x) = \begin{cases} \dfrac{P_0}{EI}\left(\dfrac{L}{4}x^2 - \dfrac{1}{6}x^3\right), & 0 \le x < \dfrac{L}{2} \\[2mm] \dfrac{P_0 L^2}{4EI}\left(\dfrac{1}{2}x - \dfrac{L}{12}\right), & \dfrac{L}{2} \le x \le L \end{cases}$

15. $y(t) = \dfrac{1}{\omega}\sin \omega t$

Exercises 4.6, Page 230

1. $x = -\frac{1}{3}e^{-2t} + \frac{1}{3}e^t$ **3.** $x = -\cos 3t - \frac{5}{3}\sin 3t$
 $y = \frac{1}{3}e^{-2t} + \frac{2}{3}e^t$ $y = 2\cos 3t - \frac{7}{3}\sin 3t$

5. $x = -2e^{3t} + \frac{5}{2}e^{2t} - \frac{1}{2}$ **7.** $x = -\frac{1}{2}t - \frac{3}{4}\sqrt{2}\sin\sqrt{2}t$
 $y = \frac{8}{3}e^{3t} - \frac{5}{2}e^{2t} - \frac{1}{6}$ $y = -\frac{1}{2}t + \frac{3}{4}\sqrt{2}\sin\sqrt{2}t$

9. $x = 8 + \dfrac{2}{3!}t^3 + \dfrac{1}{4!}t^4$

 $y = -\dfrac{2}{3!}t^3 + \dfrac{1}{4!}t^4$

11. $x = \frac{1}{2}t^2 + t + 1 - e^{-t}$
 $y = -\frac{1}{3} + \frac{1}{3}e^{-t} + \frac{1}{3}te^{-t}$

13. $x_1 = \dfrac{1}{5}\sin t + \dfrac{2\sqrt{6}}{15}\sin\sqrt{6}t + \dfrac{2}{5}\cos t - \dfrac{2}{5}\cos\sqrt{6}t$

 $x_2 = \dfrac{2}{5}\sin t - \dfrac{\sqrt{6}}{15}\sin\sqrt{6}t + \dfrac{4}{5}\cos t + \dfrac{1}{5}\cos\sqrt{6}t$

15. (b) $i_2 = \frac{100}{9} - \frac{100}{9}e^{-900t}$
 $i_3 = \frac{80}{9} - \frac{80}{9}e^{-900t}$

 (c) $i_1 = 20 - 20e^{-900t}$

17. $i_2 = -\frac{20}{13}e^{-2t} + \frac{375}{1469}e^{-15t} + \frac{145}{113}\cos t + \frac{85}{113}\sin t$

 $i_3 = \frac{30}{13}e^{-2t} + \frac{250}{1469}e^{-15t} - \frac{280}{113}\cos t + \frac{810}{113}\sin t$

19. $i_1 = \dfrac{6}{5} - \dfrac{6}{5}e^{-100t}\cosh 50\sqrt{2}t - \dfrac{9\sqrt{2}}{10}e^{-100t}\sinh 50\sqrt{2}t$

 $i_2 = \dfrac{6}{5} - \dfrac{6}{5}e^{-100t}\cosh 50\sqrt{2}t - \dfrac{6\sqrt{2}}{5}e^{-100t}\sinh 50\sqrt{2}t$

Chapter 4 Review Exercises, Page 232

1. $\dfrac{1}{s^2} - \dfrac{2}{s^2} e^{-s}$ **3.** false **5.** true

7. $\dfrac{1}{s+7}$ **9.** $\dfrac{2}{s^2+4}$ **11.** $\dfrac{4s}{(s^2+4)^2}$

13. $\tfrac{1}{6} t^5$ **15.** $\tfrac{1}{2} t^2 e^{5t}$

17. $e^{5t} \cos 2t + \tfrac{5}{2} e^{5t} \sin 2t$

19. $\cos \pi(t-1)\,\mathcal{U}(t-1) + \sin \pi(t-1)\,\mathcal{U}(t-1)$

21. -5 **23.** $e^{-k(s-a)} F(s-a)$

25. $f(t)\,\mathcal{U}(t-t_0)$ **27.** $f(t-t_0)\,\mathcal{U}(t-t_0)$

29. $f(t) = t - (t-1)\mathcal{U}(t-1) - \mathcal{U}(t-4)$;

$$\mathcal{L}\{f(t)\} = \frac{1}{s^2} - \frac{1}{s^2} e^{-s} - \frac{1}{s} e^{-4s};$$

$$\mathcal{L}\{e^t f(t)\} = \frac{1}{(s-1)^2} - \frac{1}{(s-1)^2} e^{-(s-1)}$$

$$- \frac{1}{s-1} e^{-4(s-1)}$$

31. $f(t) = 2 + (t-2)\mathcal{U}(t-2)$;

$$\mathcal{L}\{f(t)\} = \frac{2}{s} + \frac{1}{s^2} e^{-2s};$$

$$\mathcal{L}\{e^t f(t)\} = \frac{2}{s-1} + \frac{1}{(s-1)^2} e^{-2(s-1)}$$

33. $y = 5te^t + \tfrac{1}{2} t^2 e^t$

35. $y = -\tfrac{6}{25} + \tfrac{1}{5} t^2 + \tfrac{3}{2} e^{-t} - \tfrac{13}{50} e^{-5t}$
$\quad - \tfrac{4}{25}\,\mathcal{U}(t-2) - \tfrac{1}{5}(t-2)^2\,\mathcal{U}(t-2)$
$\quad + \tfrac{1}{4} e^{-(t-2)}\,\mathcal{U}(t-2) - \tfrac{9}{100} e^{-5(t-2)}\,\mathcal{U}(t-2)$

37. $y = 1 + t + \tfrac{1}{2} t^2$

39. $x = -\tfrac{1}{4} + \tfrac{9}{8} e^{-2t} + \tfrac{1}{8} e^{2t}$
$\quad y = t + \tfrac{9}{4} e^{-2t} - \tfrac{1}{4} e^{2t}$

41. $i(t) = -9 + 2t + 9e^{-t/5}$

43. $y(x) = \dfrac{w_0}{12EIL}\left[-\dfrac{1}{5} x^5 + \dfrac{L}{2} x^4 - \dfrac{L^2}{2} x^3 + \dfrac{L^3}{4} x^2 \right.$

$$\left. + \frac{1}{5}\left(x - \frac{L}{2}\right)^5 \mathcal{U}\left(x - \frac{L}{2}\right)\right]$$

Exercises 5.1, Page 245

1. $[-\tfrac{1}{2}, \tfrac{1}{2})$ **3.** $(-5, 15)$

5. $x - \tfrac{2}{3} x^3 + \tfrac{2}{15} x^5 - \tfrac{4}{315} x^7 + \cdots$

7. $1 + \tfrac{1}{2} x^2 + \tfrac{5}{24} x^4 + \tfrac{61}{720} x^6 + \cdots,\ (-\pi/2, \pi/2)$

9. $2c_1 + \displaystyle\sum_{k=1}^{\infty}[2(k+1)c_{k+1} + 6c_{k-1}]x^k$

13. $y_1(x) = c_0\left[1 + \dfrac{1}{3\cdot 2} x^3 + \dfrac{1}{6\cdot 5\cdot 3\cdot 2} x^6 \right.$

$$\left. + \frac{1}{9\cdot 8\cdot 6\cdot 5\cdot 3\cdot 2} x^9 + \cdots\right]$$

$$y_2(x) = c_1\left[x + \frac{1}{4\cdot 3} x^4 + \frac{1}{7\cdot 6\cdot 4\cdot 3} x^7 \right.$$

$$\left. + \frac{1}{10\cdot 9\cdot 7\cdot 6\cdot 4\cdot 3} x^{10} + \cdots\right]$$

15. $y_1(x) = c_0\left[1 - \dfrac{1}{2!} x^2 - \dfrac{3}{4!} x^4 - \dfrac{21}{6!} x^6 - \cdots\right]$

$$y_2(x) = c_1\left[x + \frac{1}{3!} x^3 + \frac{5}{5!} x^5 + \frac{45}{7!} x^7 + \cdots\right]$$

17. $y_1(x) = c_0\left[1 - \dfrac{1}{3!} x^3 + \dfrac{4^2}{6!} x^6 - \dfrac{7^2\cdot 4^2}{9!} x^9 + \cdots\right]$

$$y_2(x) = c_1\left[x - \frac{2^2}{4!} x^4 + \frac{5^2\cdot 2^2}{7!} x^7 - \frac{8^2\cdot 5^2\cdot 2^2}{10!} x^{10} + \cdots\right]$$

19. $y_1(x) = c_0;\ y_2(x) = c_1 \displaystyle\sum_{n=1}^{\infty} \dfrac{1}{n} x^n$

21. $y_1(x) = c_0\,[1 + \tfrac{1}{2} x^2 + \tfrac{1}{6} x^3 + \tfrac{1}{6} x^4 + \cdots]$

$$y_2(x) = c_1\,[x + \tfrac{1}{2} x^2 + \tfrac{1}{2} x^3 + \tfrac{1}{4} x^4 + \cdots]$$

23. $y_1(x) = c_0\left[1 + \dfrac{1}{4} x^2 - \dfrac{7}{4\cdot 4!} x^4 + \dfrac{23\cdot 7}{8\cdot 6!} x^6 - \cdots\right]$

$$y_2(x) = c_1\left[x - \frac{1}{6} x^3 + \frac{14}{2\cdot 5!} x^5 - \frac{34\cdot 14}{4\cdot 7!} x^7 - \cdots\right]$$

25. $y(x) = -2\left[1 + \dfrac{1}{2!} x^2 + \dfrac{1}{3!} x^3 + \dfrac{1}{4!} x^4 + \cdots\right] + 6x$

$$= 8x - 2e^x$$

27. $y(x) = 3 - 12x^2 + 4x^4$

29. $y_1(x) = c_0\,[1 - \tfrac{1}{6} x^3 + \tfrac{1}{120} x^5 + \cdots]$

$$y_2(x) = c_1\,[x - \tfrac{1}{12} x^4 + \tfrac{1}{180} x^6 + \cdots]$$

31. $y(x) = c_0\,(1 + \tfrac{1}{6} x^3 + \tfrac{1}{180} x^6 + \cdots) + c_1(x + \tfrac{1}{12} x^4$
$\quad + \tfrac{1}{504} x^7 + \cdots) + \tfrac{1}{2} x^2 + \tfrac{1}{40} x^5 + \tfrac{1}{2240}x^8 \cdots$;

$$y(x) = c_0\,(1 + 2x^2 + 2x^4 + \cdots) + c_1\,(x + \tfrac{4}{3} x^3$$
$$+ \tfrac{16}{15} x^5 + \cdots) + \tfrac{1}{2} x^2 + \tfrac{1}{6} x^3 + \tfrac{13}{24} x^4 + \cdots$$

33. (a) $y_1(x) = \displaystyle\sum_{n=0}^{\infty} \dfrac{(-1)^n}{2^n n!} x^{2n}$;

$$y_2(x) = \sum_{n=0}^{\infty} \frac{(-1)^n}{1\cdot 3\cdot 5 \cdots (2n+1)} x^{2n+1}$$

35. $y_1(x) = e^{-x^2/2}$

Exercises 5.2, Page 254

1. $x = 0$, irregular singular point

3. $x = -3$, regular singular point; $x = 3$, irregular singular point

5. $x = 0, 2i, -2i$, regular singular points

7. $x = -3, 2$, regular singular points

9. $x = 0$, irregular singular point; $x = -5, 5, 2$, regular singular points

11. for $x = 1$: $p(x) = 5$, $q(x) = \dfrac{x(x-1)^2}{x+1}$

for $x = -1$: $p(x) = \dfrac{5(x+1)}{x-1}$, $q(x) = x^2 + x$

13. $r_1 = \frac{1}{3}$, $r_2 = -1$

15. $r_1 = \frac{3}{2}$, $r_2 = 0$

$y(x) = C_1 x^{3/2}\left[1 - \dfrac{2}{5}x + \dfrac{2^2}{7\cdot 5\cdot 2}x^2 - \dfrac{2^3}{9\cdot 7\cdot 5\cdot 3!}x^3 + \cdots\right]$

$\quad + C_2\left[1 + 2x - 2x^2 + \dfrac{2^3}{3\cdot 3!}x^3 - \cdots\right]$

17. $r_1 = \frac{7}{8}$, $r_2 = 0$

$y(x) = C_1\, x^{7/8}\left[1 - \dfrac{2}{15}x + \dfrac{2^2}{23\cdot 15\cdot 2}x^2\right.$

$\qquad\qquad \left. - \dfrac{2^3}{31\cdot 23\cdot 15\cdot 3!}x^3 + \cdots\right]$

$\quad + C_2\left[1 - 2x + \dfrac{2^2}{9\cdot 2}x^2 - \dfrac{2^3}{17\cdot 9\cdot 3!}x^3 + \cdots\right]$

19. $r_1 = \frac{1}{3}$, $r_2 = 0$

$y(x) = C_1 x^{1/3}\left[1 + \dfrac{1}{3}x + \dfrac{1}{3^2\cdot 2}x^2 + \dfrac{1}{3^3\cdot 3!}x^3 + \cdots\right]$

$\quad + C_2\left[1 + \dfrac{1}{2}x + \dfrac{1}{5\cdot 2}x^2 + \dfrac{1}{8\cdot 5\cdot 2}x^3 + \cdots\right]$

21. $r_1 = \frac{5}{2}$, $r_2 = 0$

$y(x) = C_1 x^{5/2}\left[1 + \dfrac{2\cdot 2}{7}x + \dfrac{2^2\cdot 3}{9\cdot 7}x^2 + \dfrac{2^3\cdot 4}{11\cdot 9\cdot 7}x^3 + \cdots\right]$

$\quad + C_2\left[1 + \dfrac{1}{3}x - \dfrac{1}{6}x^2 - \dfrac{1}{6}x^3 - \cdots\right]$

23. $r_1 = \frac{2}{3}$, $r_2 = \frac{1}{3}$

$y(x) = C_1 x^{2/3}\,[1 - \frac{1}{2}x + \frac{5}{28}x^2 - \frac{1}{21}x^3 + \cdots]$

$\quad + C_2 x^{1/3}\,[1 - \frac{1}{2}x + \frac{1}{5}x^2 - \frac{7}{120}x^3 + \cdots]$

25. $r_1 = 0$, $r_2 = -1$

$y(x) = C_1\displaystyle\sum_{n=0}^{\infty}\dfrac{1}{(2n+1)!}x^{2n} + C_2 x^{-1}\sum_{n=0}^{\infty}\dfrac{1}{(2n)!}x^{2n}$

$\quad = C_1 x^{-1}\displaystyle\sum_{n=0}^{\infty}\dfrac{1}{(2n+1)!}x^{2n+1} + C_2 x^{-1}\sum_{n=0}^{\infty}\dfrac{1}{(2n)!}x^{2n}$

$\quad = \dfrac{1}{x}\,[C_1 \sinh x + C_2 \cosh x]$

27. $r_1 = 1$, $r_2 = 0$

$y(x) = C_1 x + C_2\,[x \ln x - 1 + \frac{1}{2}x^2 + \frac{1}{12}x^3 + \frac{1}{72}x^4 + \cdots]$

29. $r_1 = r_2 = 0$

$y(x) = C_1 y(x) + C_2\left[y_1(x)\ln x + y_1(x)(-x + \frac{1}{4}x^2\right.$

$\qquad\qquad \left. - \dfrac{1}{3\cdot 3!}x^3 + \dfrac{1}{4\cdot 4!}x^4 - \cdots)\right]$

where $y_1(x) = \displaystyle\sum_{n=0}^{\infty}\dfrac{1}{n!}x^n = e^x$

33. **(b)** $y_1(t) = \displaystyle\sum_{n=0}^{\infty}\dfrac{(-1)^n}{(2n+1)!}\,(\sqrt{\lambda}t)^{2n} = \dfrac{\sin(\sqrt{\lambda}t)}{\sqrt{\lambda}t}$

$y_2(t) = t^{-1}\displaystyle\sum_{n=0}^{\infty}\dfrac{(-1)^n}{(2n)!}\,(\sqrt{\lambda}t)^{2n} = \dfrac{\cos(\sqrt{\lambda}t)}{t}$

(c) $y = c_1 x \sin\!\left(\dfrac{\sqrt{\lambda}}{x}\right) + c_2 x \cos\!\left(\dfrac{\sqrt{\lambda}}{x}\right)$

Exercises 5.3, Page 264

1. $y = c_1 J_{1/3}(x) + c_2 J_{-1/3}(x)$

3. $y = c_1 J_{5/2}(x) + c_2 J_{-5/2}(x)$

5. $y = c_1 J_0(x) + c_2 Y_0(x)$

7. $y = c_1 J_2(3x) + c_2 Y_2(3x)$

9. $y = c_1 x^{-1/2} J_{1/2}(\lambda x) + c_2 x^{-1/2} J_{-1/2}(\lambda x)$

13. From Problem 10, $y = x^{1/2} J_{1/2}(x)$; from Problem 11, $y = x^{1/2} J_{-1/2}(x)$.

15. From Problem 10, $y = x^{-1} J_{-1}(x)$; from Problem 11, $y = x^{-1} J_1(x)$. Since $J_{-1}(x) = -J_1(x)$, no new solution results.

17. From Problem 12 and $\lambda = 1$ and $\nu = \pm\frac{3}{2}$, $y = \sqrt{x}J_{3/2}(x)$ and $y = \sqrt{x}J_{-3/2}(x)$.

23. $J_{-1/2}(x) = \sqrt{\dfrac{2}{\pi x}}\,\cos x$

35. **(a)** $P_6(x) = \frac{1}{16}\,(231x^6 - 315x^4 + 105x^2 - 5)$
$P_7(x) = \frac{1}{16}\,(429x^7 - 693x^5 + 315x^3 - 35x)$

(b) $P_6(x)$ satisfies $(1 - x^2)y'' - 2xy' + 42y = 0$.
$P_7(x)$ satisfies $(1 - x^2)y'' - 2xy' + 56y = 0$.

Chapter 5 Review Exercises, Page 266

3. $R = \sqrt{10}$

7. $r_1 = \frac{1}{2}$, $r_2 = 0$

$y_1(x) = C_1 x^{1/2}\,[1 - \frac{1}{3}x + \frac{1}{30}x^2 - \frac{1}{630}x^3 + \cdots]$

$y_2(x) = C_2\,[1 - x + \frac{1}{6}x^2 - \frac{1}{90}x^3 + \cdots]$

9. $y_1(x) = c_0\,[1 + \frac{3}{2}x^2 + \frac{1}{2}x^3 + \frac{5}{8}x^4 + \cdots]$

$y_2(x) = c_1\,[x + \frac{1}{2}x^3 + \frac{1}{4}x^4 + \cdots]$

11. $r_1 = 7$, $r_2 = 0$

$y_1(x) = C_1 x^7\left[1 + \displaystyle\sum_{k=1}^{\infty}\dfrac{(-1)^k\, 4\cdot 5\cdot 6\cdots(k+3)}{k!\, 8\cdot 9\cdot 10\cdots(k+7)}x^k\right]$

$y_2(x) = C_2\,[1 - \frac{1}{2}x + \frac{1}{10}x^2 - \frac{1}{120}x^3]$

13. $y(x) = 3[1 - x^2 + \frac{1}{3} x^4 - \frac{1}{15} x^6 + \cdots] - 2[x - \frac{1}{2} x^3 + \frac{1}{8} x^5 - \frac{1}{48} x^7 + \cdots]$

15. $x = 0$ is an ordinary point

17. $y(x) = c_0 \left[1 - \frac{1}{3} x^3 + \frac{1}{3^2 \cdot 2!} x^6 - \frac{1}{3^3 \cdot 3!} x^9 + \cdots \right]$

$\qquad + c_1 \left[x - \frac{1}{4} x^4 + \frac{1}{4 \cdot 7} x^7 - \frac{1}{4 \cdot 7 \cdot 10} x^{10} + \cdots \right]$

$\qquad + \left[\frac{5}{2} x^2 - \frac{1}{3} x^3 + \frac{1}{3^2 \cdot 2!} x^6 - \frac{1}{3^3 \cdot 3!} x^9 + \cdots \right]$

Exercises 6.1, Page 272

1. $h = 0.1$

x_n	y_n
1.00	5.0000
1.10	3.9900
1.20	3.2545
1.30	2.7236
1.40	2.3451
1.50	2.0801

$h = 0.05$

x_n	y_n
1.00	5.0000
1.05	4.4475
1.10	3.9763
1.15	3.5751
1.20	3.2342
1.25	2.9452
1.30	2.7009
1.35	2.4952
1.40	2.3226
1.45	2.1786
1.50	2.0592

3. $h = 0.1$

x_n	y_n
0.00	0.0000
0.10	0.1005
0.20	0.2030
0.30	0.3098
0.40	0.4234
0.50	0.5470

$h = 0.05$

x_n	y_n
0.00	0.0000
0.05	0.0501
0.10	0.1004
0.15	0.1512
0.20	0.2028
0.25	0.2554
0.30	0.3095
0.35	0.3652
0.40	0.4230
0.45	0.4832
0.50	0.5465

5. $h = 0.1$

x_n	y_n
0.00	0.0000
0.10	0.0952
0.20	0.1822
0.30	0.2622
0.40	0.3363
0.50	0.4053

$h = 0.05$

x_n	y_n
0.00	0.0000
0.05	0.0488
0.10	0.0953
0.15	0.1397
0.20	0.1823
0.25	0.2231
0.30	0.2623
0.35	0.3001
0.40	0.3364
0.45	0.3715
0.50	0.4054

7. $h = 0.1$

x_n	y_n
0.00	5.0000
0.10	0.5215
0.20	0.5362
0.30	0.5449
0.40	0.5490
0.50	0.5503

$h = 0.05$

x_n	y_n
0.00	0.5000
0.05	0.5116
0.10	0.5214
0.15	0.5294
0.20	0.5359
0.25	0.5408
0.30	0.5444
0.35	0.5469
0.40	0.5484
0.45	0.5492
0.50	0.5495

9. $h = 0.1$

x_n	y_n
1.00	1.0000
1.10	1.0095
1.20	1.0404
1.30	1.0967
1.40	1.1866
1.50	1.3260

$h = 0.05$

x_n	y_n
1.00	1.0000
1.05	1.0024
1.10	1.0100
1.15	1.0228
1.20	1.0414
1.25	1.0663
1.30	1.0984
1.35	1.1389
1.40	1.1895
1.45	1.2526
1.50	1.3315

11. $h = 0.1$

x_n	y_n	exact
0.00	2.0000	2.0000
0.10	2.1220	2.1230
0.20	2.3049	2.3085
0.30	2.5858	2.5958
0.40	3.0378	3.0650
0.50	3.8254	3.9082

$h = 0.05$

x_n	y_n	exact
0.00	2.0000	2.0000
0.05	2.0553	2.0554
0.10	2.1228	2.1230
0.15	2.2056	2.2061
0.20	2.3075	2.3085
0.25	2.4342	2.4358
0.30	2.5931	2.5958
0.35	2.7953	2.7997
0.40	3.0574	3.0650
0.45	3.4057	3.4189
0.50	3.8840	3.9082

13. (a) $y_1 = 1.2$

(b) $y''(c) \dfrac{h^2}{2} = 4e^{2c} \dfrac{(0.1)^2}{2} = 0.02e^{2c} \le 0.02e^{0.2} = 0.0244$

(c) Actual value is $y(0.1) = 1.2214$. Error is 0.0214.

(d) If $h = 0.05$, $y_2 = 1.21$.

(e) Error with $h = 0.1$ is 0.0214. Error with $h = 0.05$ is 0.0114.

15. (a) $y_1 = 0.8$

(b) $y''(c) \dfrac{h^2}{2} = 5e^{-2c} \dfrac{(0.1)^2}{2} = 0.025e^{-2c} \le 0.025$ for

$0 \le c \le 0.1$

(c) Actual value is $y(0.1) = 0.8234$. Error is 0.0234.

(d) If $h = 0.05$, $y_2 = 0.8125$.

(e) Error with $h = 0.1$ is 0.0234. Error with $h = 0.05$ is 0.0109.

17. (a) Error is $19h^2e^{-3(c-1)}$.

(b) $y''(c) \dfrac{h^2}{2} \le 19(0.1)^2(1) = 0.19$

(c) If $h = 0.1$, $y_5 = 1.8207$. If $h = 0.05$, $y_{10} = 1.9424$.

(d) Error with $h = 0.1$ is 0.2325. Error with $h = 0.05$ is 0.1109.

19. (a) Error is $\dfrac{1}{(c+1)^2} \dfrac{h^2}{2}$.

(b) $\left| y''(c) \dfrac{h^2}{2} \right| \le (1)\dfrac{(0.1)^2}{2} = 0.005$

(c) If $h = 0.1$, $y_5 = 0.4198$. If $h = 0.05$, $y_{10} = 0.4124$.

(d) Error with $h = 0.1$ is 0.0143. Error with $h = 0.05$ is 0.0069.

Exercises 6.2, Page 277

1.

x_n	y_n	True value
0.00	2.0000	2.0000
0.10	2.1230	2.1230
0.20	2.3085	2.3085
0.30	2.5958	2.5958
0.40	3.0649	3.0650
0.50	3.9078	3.9082

3.

x_n	y_n
1.00	5.0000
1.10	3.9724
1.20	3.2284
1.30	2.6945
1.40	2.3163
1.50	2.0533

5.

x_n	y_n
0.00	0.0000
0.10	0.1003
0.20	0.2027
0.30	0.3093
0.40	0.4228
0.50	0.5463

7.

x_n	y_n
0.00	0.0000
0.10	0.0953
0.20	0.1823
0.30	0.2624
0.40	0.3365
0.50	0.4055

9.

x_n	y_n
0.00	0.5000
0.10	0.5213
0.20	0.5358
0.30	0.5443
0.40	0.5482
0.50	0.5493

11.

x_n	y_n
1.00	1.0000
1.10	1.0101
1.20	1.0417
1.30	1.0989
1.40	1.1905
1.50	1.3333

13. (a) $v(5) = 35.7678$

(c) $v(t) = \sqrt{\dfrac{mg}{k}} \tanh \sqrt{\dfrac{kg}{m}}\, t;\ v(5) = 35.7678$

15. (a) $h = 0.1$

x_n	y_n
1.00	1.0000
1.10	1.2511
1.20	1.6934
1.30	2.9425
1.40	903.0282

$h = 0.05$

x_n	y_n
1.00	1.0000
1.05	1.1112
1.10	1.2511
1.15	1.4348
1.20	1.6934
1.25	2.1047
1.30	2.9560
1.35	7.8981
1.40	1.1E + 15

17. (a) $y_1 = 0.82341667$

(b) $y^{(5)}(c) \dfrac{h^5}{5!} = 40e^{-2c} \dfrac{h^5}{5!} \leq 40e^{2(0)} \dfrac{(0.1)^5}{5!} = 3.333 \times 10^{-6}$

(c) Actual value is $y(0.1) = 0.8234134413$. Error is $3.225 \times 10^{-6} \leq 3.333 \times 10^{-6}$.

(d) If $h = 0.05$, $y_2 = 0.82341363$.

(e) Error with $h = 0.1$ is 3.225×10^{-6}. Error with $h = 0.05$ is 1.854×10^{-7}.

19. (a) $y^{(5)}(c) \dfrac{h^5}{5!} = \dfrac{24}{(c+1)^5} \dfrac{h^5}{5!}$

(b) $\dfrac{24}{(c+1)^5} \dfrac{h^5}{5!} \leq 24 \dfrac{(0.1)^5}{5!} = 2.0000 \times 10^{-6}$

(c) From calculation with $h = 0.1$, $y_5 = 0.40546517$. From calculation with $h = 0.05$, $y_{10} = 0.40546511$.

Exercises 6.3, Page 280

1. $y(x) = -x + e^x$; $y(0.2) = 1.0214$, $y(0.4) = 1.0918$,

$y(0.6) = 1.2221$, $y(0.8) = 1.4255$

3.

x_n	y_n
0.00	1.0000
0.20	0.7328
0.40	0.6461
0.60	0.6585
0.80	0.7232

5.

x_n	y_n
0.00	0.0000
0.20	0.2027
0.40	0.4228
0.60	0.6841
0.80	1.0297
1.00	1.5569

x_n	y_n
0.00	0.0000
0.10	0.1003
0.20	0.2027
0.30	0.3093
0.40	0.4228
0.50	0.5463
0.60	0.6842
0.70	0.8423
0.80	1.0297
0.90	1.2603
1.00	1.5576

7.

x_n	y_n
0.00	0.0000
0.20	0.0026
0.40	0.0201
0.60	0.0630
0.80	0.1360
1.00	0.2385

x_n	y_n
0.00	0.0000
0.10	0.0003
0.20	0.0026
0.30	0.0087
0.40	0.0200
0.50	0.0379
0.60	0.0629
0.70	0.0956
0.80	0.1360
0.90	0.1837
1.00	0.2384

Exercises 6.4, Page 285

1. $y(x) = -2e^{2x} + 5xe^{2x}$; $y(0.2) = -1.4918$, $y_2 = -1.6800$

3. $y_1 = -1.4928$, $y_2 = -1.4919$

5. $y_1 = 1.4640$, $y_2 = 1.4640$

7. $x_1 = 8.3055$, $y_1 = 3.4199$; $x_2 = 8.3055$, $y_2 = 3.4199$

9. $x_1 = -3.9123$, $y_1 = 4.2857$; $x_2 = -3.9123$, $y_2 = 4.2857$

11. $x_1 = 0.4179$, $y_1 = -2.1824$; $x_2 = 0.4173$, $y_2 = -2.1821$

Exercises 6.5, Page 290

1. $y_1 = -5.6774$, $y_2 = -2.5807$, $y_3 = 6.3226$

3. $y_1 = -0.2259$, $y_2 = -0.3356$, $y_3 = -0.3308$, $y_4 = -0.2167$

5. $y_1 = 3.3751$, $y_2 = 3.6306$, $y_3 = 3.6448$, $y_4 = 3.2355$, $y_5 = 2.1411$

7. $y_1 = 3.8842$, $y_2 = 2.9640$, $y_3 = 2.2064$, $y_4 = 1.5826$, $y_5 = 1.0681$, $y_6 = 0.6430$, $y_7 = 0.2913$

9. $y_1 = 0.2660$, $y_2 = 0.5097$, $y_3 = 0.7357$, $y_4 = 0.9471$, $y_5 = 1.1465$, $y_6 = 1.3353$, $y_7 = 1.5149$, $y_8 = 1.6855$, $y_9 = 1.8474$

11. $y_1 = 0.3492$, $y_2 = 0.7202$, $y_3 = 1.1363$, $y_4 = 1.6233$, $y_5 = 2.2118$, $y_6 = 2.9386$, $y_7 = 3.8490$

13. (c) $y_0 = -2.2755$, $y_1 = -2.0755$, $y_2 = -1.8589$, $y_3 = -1.6126$, $y_4 = -1.3275$

Chapter 6 Review Exercises, Page 290

1. Comparison of Numerical Methods with $h = 0.1$

x_n	Euler	Improved Euler	Runge–Kutta
1.00	2.0000	2.0000	2.0000
1.10	2.1386	2.1549	2.1556
1.20	2.3097	2.3439	2.3454
1.30	2.5136	2.5672	2.5695
1.40	2.7504	2.8246	2.8278
1.50	3.0201	3.1157	3.1197

Comparison of Numerical Methods with $h = 0.05$

x_n	Euler	Improved Euler	Runge–Kutta
1.00	2.0000	2.0000	2.0000
1.05	2.0693	2.0735	2.0736
1.10	2.1469	2.1554	2.1556
1.15	2.2329	2.2459	2.2462
1.20	2.3272	2.3450	2.3454
1.25	2.4299	3.4527	2.4532
1.30	2.5410	2.5689	2.5695
1.35	2.6604	2.6937	2.6944
1.40	2.7883	2.8269	2.8278
1.45	2.9245	2.9686	2.9696
1.50	3.0690	3.1187	3.1197

3. Comparison of Numerical Methods with $h = 0.1$

x_n	Euler	Improved Euler	Runge–Kutta
0.50	0.5000	0.5000	0.5000
0.60	0.6000	0.6048	0.6049
0.70	0.7095	0.7191	0.7194
0.80	0.8283	0.8427	0.8431
0.90	0.9559	0.9752	0.9757
1.00	1.0921	1.1163	1.1169

Comparison of Numerical Methods with $h = 0.05$

x_n	Euler	Improved Euler	Runge–Kutta
0.50	0.5000	0.5000	0.5000
0.55	0.5500	0.5512	0.5512
0.60	0.6024	0.6049	0.6049
0.65	0.6573	0.6610	0.6610
0.70	0.7144	0.7194	0.7194
0.75	0.7739	0.7802	0.7801
0.80	0.8356	0.8431	0.8431
0.85	0.8996	0.9083	0.9083
0.90	0.9657	0.9757	0.9757
0.95	1.0340	1.0453	1.0452
1.00	1.1044	1.1170	1.1169

5. $h = 0.2$: $y(0.2) \approx 3.2$; $h = 0.1$: $y(0.2) \approx 3.23$

7. $x(0.2) \approx 1.62$, $y(0.2) \approx 1.84$

Exercises 7.1, Page 302

1. $6\mathbf{i} + 12\mathbf{j}$; $\mathbf{i} + 8\mathbf{j}$; $3\mathbf{i}$; $\sqrt{65}$; 3

3. $\langle 12, 0 \rangle$; $\langle 4, -5 \rangle$; $\langle 4, 5 \rangle$; $\sqrt{41}$; $\sqrt{41}$

5. $-9\mathbf{i} + 6\mathbf{j}$; $-3\mathbf{i} + 9\mathbf{j}$; $-3\mathbf{i} - 5\mathbf{j}$; $3\sqrt{10}$; $\sqrt{34}$

7. $-6\mathbf{i} + 27\mathbf{j}$; $\mathbf{0}$; $-4\mathbf{i} + 18\mathbf{j}$; 0; $2\sqrt{85}$

9. $\langle 6, -14 \rangle$; $\langle 2, 4 \rangle$ **11.** $10\mathbf{i} - 12\mathbf{j}$; $12\mathbf{i} - 17\mathbf{j}$

13. $\langle 20, 52 \rangle$; $\langle -2, 0 \rangle$

15. $2\mathbf{i} + 5\mathbf{j}$ **17.** $2\mathbf{i} + 2\mathbf{j}$

19. $(1, 18)$ **21.** a, b, c, e, f

23. $\langle 6, 15 \rangle$

25. $\left\langle \dfrac{1}{\sqrt{2}}, \dfrac{1}{\sqrt{2}} \right\rangle$; $\left\langle -\dfrac{1}{\sqrt{2}}, -\dfrac{1}{\sqrt{2}} \right\rangle$

27. $\langle 0, -1 \rangle$; $\langle 0, 1 \rangle$ **29.** $\left\langle \dfrac{5}{13}, \dfrac{12}{13} \right\rangle$

31. $\dfrac{6}{\sqrt{58}}\mathbf{i} + \dfrac{14}{\sqrt{58}}\mathbf{j}$ **33.** $\left\langle -3, -\dfrac{15}{2} \right\rangle$

35. **37.** $-(\mathbf{a} + \mathbf{b})$

41. $\mathbf{a} = \dfrac{5}{2}\mathbf{b} - \dfrac{1}{2}\mathbf{c}$ **43.** $\pm\dfrac{1}{\sqrt{2}}(\mathbf{i} + \mathbf{j})$

45. (b) approximately 31° **47.** $\mathbf{F} = \dfrac{qQ}{4\pi\varepsilon_0}\dfrac{1}{L\sqrt{L^2 + a^2}}\mathbf{i}$

Exercises 7.2, Page 308

1.–5.

7. The set $\{(x, y, 5)\,|\,x, y \text{ real numbers}\}$ is a plane perpendicular to the z-axis, 5 units above the xy-plane.

9. The set $\{(2, 3, z)\,|\,z \text{ a real number}\}$ is a line perpendicular to the xy-plane at $(2, 3, 0)$.

11. $(0, 0, 0)$, $(2, 0, 0)$, $(2, 5, 0)$, $(0, 5, 0)$, $(0, 0, 8)$, $(2, 0, 8)$, $(2, 5, 8)$, $(0, 5, 8)$

13. $(-2, 5, 0)$, $(-2, 0, 4)$, $(0, 5, 4)$; $(-2, 5, -2)$; $(3, 5, 4)$

15. the union of the coordinate planes

17. the point $(-1, 2, -3)$

19. the union of the planes $z = -5$ and $z = 5$

21. $\sqrt{70}$ **23.** 7; 5 **25.** right triangle

27. isosceles **29.** $d(P_1, P_2) + d(P_1, P_3) = d(P_2, P_3)$

31. 6 or -2 **33.** $(4, \frac{1}{2}, \frac{3}{2})$

35. $P_1(-4, -11, 10)$ **37.** $\langle -3, -6, 1\rangle$ **39.** $\langle 2, 1, 1\rangle$

41. $\langle 2, 4, 12\rangle$ **43.** $\langle -11, -41, -49\rangle$ **45.** $\sqrt{139}$

47. 6 **49.** $\langle -\frac{2}{3}, \frac{1}{3}, -\frac{2}{3}\rangle$ **51.** $4\mathbf{i} - 4\mathbf{j} + 4\mathbf{k}$

53.

Exercises 7.3, Page 315

1. $25\sqrt{2}$ **3.** 12 **5.** -16 **7.** 48

9. 29 **11.** 25 **13.** $\langle -\frac{2}{5}, \frac{4}{5}, 2\rangle$

15. a and f, c and d, b and e

17. $\langle \frac{4}{9}, -\frac{1}{3}, 1\rangle$ **21.** 1.11 radians or 63.43°

23. 1.89 radians or 108.43°

25. $\cos \alpha = 1/\sqrt{14}$, $\cos \beta = 2/\sqrt{14}$, $\cos \gamma = 3/\sqrt{14}$; $\alpha = 74.5°$, $\beta = 57.69°$, $\gamma = 36.7°$

27. $\cos \alpha = \frac{1}{2}$, $\cos \beta = 0$, $\cos \gamma = -\sqrt{3}/2$; $\alpha = 60°$, $\beta = 90°$, $\gamma = 150°$

29. 0.955 radian or 54.74°; 0.616 radian or 35.26°

31. $\alpha = 58.19°$, $\beta = 42.45°$, $\gamma = 65.06°$ **33.** $\frac{5}{7}$

35. $-6\sqrt{11}/11$ **37.** $72\sqrt{109}/109$

39. $\langle -\frac{21}{5}, \frac{28}{5}\rangle$; $\langle -\frac{4}{5}, -\frac{3}{5}\rangle$

41. $\langle -\frac{12}{7}, \frac{6}{7}, \frac{4}{7}\rangle$; $\langle \frac{5}{7}, -\frac{20}{7}, \frac{45}{7}\rangle$ **43.** $\langle \frac{72}{25}, \frac{96}{25}\rangle$

45. 1000 ft-lb **47.** 0; 150 N-m

49. approximately 1.80 angstroms

Exercises 7.4, Page 321

1. $-5\mathbf{i} - 5\mathbf{j} + 3\mathbf{k}$ **3.** $\langle -12, -2, 6\rangle$

5. $-5\mathbf{i} + 5\mathbf{k}$ **7.** $\langle -3, 2, 3\rangle$ **9.** 0

11. $6\mathbf{i} + 14\mathbf{j} + 4\mathbf{k}$ **13.** $-3\mathbf{i} - 2\mathbf{j} - 5\mathbf{k}$

17. $-\mathbf{i} + \mathbf{j} + \mathbf{k}$ **19.** $2\mathbf{k}$ **21.** $\mathbf{i} + 2\mathbf{j}$

23. $-24\mathbf{k}$ **25.** $5\mathbf{i} - 5\mathbf{j} - \mathbf{k}$ **27.** 0

29. $\sqrt{41}$ **31.** $-\mathbf{j}$ **33.** 0 **35.** 6

37. $12\mathbf{i} - 9\mathbf{j} + 18\mathbf{k}$ **39.** $-4\mathbf{i} + 3\mathbf{j} - 6\mathbf{k}$

41. $-21\mathbf{i} + 16\mathbf{j} + 22\mathbf{k}$ **43.** -10

45. 14 square units **47.** $\frac{1}{2}$ square unit

49. $\frac{7}{2}$ square units **51.** 10 cubic units

53. coplanar

55. 32; in the xy-plane, 30° from the positive x-axis in the direction of the negative y-axis; $16\sqrt{3}\mathbf{i} - 16\mathbf{j}$

57. $\mathbf{A} = \mathbf{i} - \mathbf{k}$, $\mathbf{B} = \mathbf{j} - \mathbf{k}$, $\mathbf{C} = 2\mathbf{k}$

Exercises 7.5, Page 328

1. $\langle x, y, z\rangle = \langle 1, 2, 1\rangle + t\langle 2, 3, -3\rangle$

3. $\langle x, y, z\rangle = \langle \frac{1}{2}, -\frac{1}{2}, 1\rangle + t\langle -2, 3, -\frac{3}{2}\rangle$

5. $\langle x, y, z\rangle = \langle 1, 1, -1\rangle + t\langle 5, 0, 0\rangle$

7. $x = 2 + 4t$, $y = 3 - 4t$, $z = 5 + 3t$

9. $x = 1 + 2t$, $y = -2t$, $z = -7t$

11. $x = 4 + 10t$, $y = \frac{1}{2} + \frac{3}{4}t$, $z = \frac{1}{3} + \frac{1}{6}t$

13. $\dfrac{x - 1}{9} = \dfrac{y - 4}{10} = \dfrac{z + 9}{7}$

15. $\dfrac{x+7}{11} = \dfrac{z-5}{-4},\ y = 2$

17. $x = 5,\ \dfrac{y-10}{9} = \dfrac{z+2}{12}$

19. $x = 4 + 3t,\ y = 6 + \frac{1}{2}t,\ z = -7 - \frac{3}{2}t;$

$\dfrac{x-4}{3} = 2y - 12 = \dfrac{2z+14}{-3}$

21. $x = 5t,\ y = 9t,\ z = 4t;\ \dfrac{x}{5} = \dfrac{y}{9} = \dfrac{z}{4}$

23. $x = 6 + 2t,\ y = 4 - 3t,\ z = -2 + 6t$

25. $x = 2 + t,\ y = -2,\ z = 15$

27. Both lines pass through the origin and have parallel direction vectors.

29. $(0, 5, 15),\ (5, 0, \frac{15}{2}),\ (10, -5, 0)$

31. $(2, 3, -5)$ **33.** Lines do not intersect.

35. $40.37°$ **37.** $x = 4 - 6t,\ y = 1 + 3t,\ z = 6 + 3t$

39. $2x - 3y + 4z = 19$ **41.** $5x - 3z = 51$

43. $6x + 8y - 4z = 11$ **45.** $5x - 3y + z = 2$

47. $3x - 4y + z = 0$ **49.** The points are collinear.

51. $x + y - 4z = 25$ **53.** $z = 12$

55. $-3x + y + 10z = 18$ **57.** $9x - 7y + 5z = 17$

59. $6x - 2y + z = 12$

61. orthogonal: a and d, b and c, d and f, b and e; parallel: a and f, c and e

63. c and d **65.** $x = 2 + t,\ y = \frac{1}{2} - t,\ z = t$

67. $x = \frac{1}{2} - \frac{1}{2}t,\ y = \frac{1}{2} - \frac{3}{2}t,\ z = t$

69. $(-5, 5, 9)$ **71.** $(1, 2, -5)$

73. $x = 5 + t,\ y = 6 + 3t,\ z = -12 + t$

75. $3x - y - 2z = 10$

77. **79.**

81.

Exercises 7.6, Page 337

1. not a vector space, axiom (*vi*) is not satisfied

3. not a vector space, axiom (*x*) is not satisfied

5. vector space

7. not a vector space, axiom (*ii*) is not satisfied

9. vector space **11.** a subspace

13. not a subspace **15.** a subspace

17. a subspace **19.** not a subspace

23. (b) $a = 7u_1 - 12u_2 + 8u_3$ **25.** linearly dependent

27. linearly independent

29. f is discontinuous at $x = -1$ and at $x = -3$.

31. $2\sqrt{\frac{2}{3}}\pi^{2/3},\ \sqrt{\pi}$

Chapter 7 Review Exercises, Page 338

1. true **3.** false **5.** true **7.** true

9. true **11.** $9\mathbf{i} + 2\mathbf{j} + 2\mathbf{k}$ **13.** $5\mathbf{i}$

15. 14 **17.** $-6\mathbf{i} + \mathbf{j} - 7\mathbf{k}$ **19.** $(4, 7, 5)$

21. $(5, 6, 3)$ **23.** $-36\sqrt{2}$ **25.** 12, -8, and 6

27. $3\sqrt{10}/2$ **29.** 2 units

31. $(\mathbf{i} - \mathbf{j} - 3\mathbf{k})/\sqrt{11}$ **33.** 2

35. $\frac{26}{9}\mathbf{i} + \frac{7}{9}\mathbf{j} + \frac{20}{9}\mathbf{k}$ **37.** sphere; plane

39. $\dfrac{x-7}{4} = \dfrac{y-3}{-2} = \dfrac{z+5}{6}$

41. The direction vectors are orthogonal and the point of intersection is $(3, -3, 0)$.

43. $14x - 5y - 3z = 0$ **45.** $30\sqrt{2}$ N-m

47. approximately 153 lb

49. not a vector space **51.** a subspace; 1, x

Exercises 8.1, Page 348

1. 2×4 **3.** 3×3 **5.** 3×4

7. not equal **9.** not equal

11. $x = 2$, $y = 4$ **13.** $c_{23} = 9$, $c_{12} = 12$

15. $\begin{pmatrix} 2 & 11 \\ 2 & -1 \end{pmatrix}$; $\begin{pmatrix} -6 & 1 \\ 14 & -19 \end{pmatrix}$; $\begin{pmatrix} 2 & 28 \\ 12 & -12 \end{pmatrix}$

17. $\begin{pmatrix} -11 & 6 \\ 17 & -22 \end{pmatrix}$; $\begin{pmatrix} -32 & 27 \\ -4 & -1 \end{pmatrix}$;

$\begin{pmatrix} 19 & -18 \\ -30 & 31 \end{pmatrix}$; $\begin{pmatrix} 19 & 6 \\ 3 & 22 \end{pmatrix}$

19. $\begin{pmatrix} 9 & 24 \\ 3 & 8 \end{pmatrix}$; $\begin{pmatrix} 3 & 8 \\ -6 & -16 \end{pmatrix}$; $\begin{pmatrix} 0 & 0 \\ 0 & 0 \end{pmatrix}$; $\begin{pmatrix} -4 & -5 \\ 8 & 10 \end{pmatrix}$

21. 180; $\begin{pmatrix} 4 & 8 & 10 \\ 8 & 16 & 20 \\ 10 & 20 & 25 \end{pmatrix}$; $\begin{pmatrix} 6 \\ 12 \\ -5 \end{pmatrix}$

23. $\begin{pmatrix} 7 & 38 \\ 10 & 75 \end{pmatrix}$; $\begin{pmatrix} 7 & 38 \\ 10 & 75 \end{pmatrix}$ **25.** $\begin{pmatrix} -14 \\ 1 \end{pmatrix}$

27. $\begin{pmatrix} -38 \\ -2 \end{pmatrix}$ **29.** 4×5

37. $\mathbf{A} = \begin{pmatrix} 1 & 2 \\ 0 & 0 \end{pmatrix}$, $\mathbf{B} = \begin{pmatrix} 2 & -4 \\ -1 & 2 \end{pmatrix}$

39. \mathbf{AB} is not necessarily the same as \mathbf{BA}.

41. $a_{11}x_1 + a_{12}x_2 = b_1$
$a_{21}x_1 + a_{22}x_2 = b_2$

45. (b) $M_R = \begin{pmatrix} \cos\beta & 0 & -\sin\beta \\ 0 & 1 & 0 \\ \sin\beta & 0 & \cos\beta \end{pmatrix}$,

$M_P = \begin{pmatrix} 1 & 0 & 0 \\ 0 & \cos\alpha & \sin\alpha \\ 0 & -\sin\alpha & \cos\alpha \end{pmatrix}$

(c) $x_S = \dfrac{3\sqrt{2} + 2\sqrt{3} - \sqrt{6} + 6}{8} \approx 1.4072$

$y_S = \dfrac{3\sqrt{2} + 2\sqrt{3} - 3\sqrt{6} + 2}{8} \approx 0.2948$

$z_S = \dfrac{\sqrt{2} + \sqrt{6}}{8} \approx 0.9659$

Exercises 8.2, Page 361

1. $x_1 = 4$, $x_2 = -7$ **3.** $x_1 = -\frac{2}{3}$, $x_2 = \frac{1}{3}$

5. $x_1 = 0$, $x_2 = 4$, $x_3 = -1$ **7.** $x_1 = -t$, $x_2 = t$, $x_3 = 0$

9. inconsistent **11.** $x_1 = 0$, $x_2 = 0$, $x_3 = 0$

13. $x_1 = -2$, $x_2 = -2$, $x_3 = 4$ **15.** $x_1 = 1$, $x_2 = 2 - t$, $x_3 = t$

17. $x_1 = 0$, $x_2 = 1$, $x_3 = 1$, $x_4 = 0$

19. inconsistent

21. $x_1 = 0.3$, $x_2 = -0.12$, $x_3 = 4.1$

23. $2Na + 2H_2O \rightarrow 2NaOH + H_2$

25. $Fe_3O_4 + 4C \rightarrow 3Fe + 4CO$

27. $3Cu + 8NHO_3 \rightarrow 3Cu(NO_3)_2 + 4H_2O + 2NO$

29. $i_1 = \frac{35}{9}$, $i_2 = \frac{38}{9}$, $i_3 = \frac{1}{3}$

31. Interchange row 1 and row 2 in \mathbf{I}_3.

33. Multiply the second row of \mathbf{I}_3 by c and add to the third row.

35. $\mathbf{EA} = \begin{pmatrix} a_{21} & a_{22} & a_{23} \\ a_{11} & a_{12} & a_{13} \\ a_{31} & a_{32} & a_{33} \end{pmatrix}$

37. $\mathbf{EA} = \begin{pmatrix} a_{11} & a_{12} & a_{13} \\ a_{21} & a_{22} & a_{23} \\ ca_{21} + a_{31} & ca_{22} + a_{32} & ca_{23} + a_{33} \end{pmatrix}$

39. $x_1 = \frac{8}{3}$, $x_2 = \frac{5}{3}$ **41.** $x_1 = \frac{5}{3}$, $x_2 = -\frac{1}{3}$, $x_3 = -\frac{1}{3}$

Exercises 8.3, Page 367

1. 2 **3.** 1 **5.** 3

7. 2 **9.** 3 **11.** linearly independent

13. linearly independent **15.** 5

17. $\text{rank}(\mathbf{A}) = 2$

Exercises 8.4, Page 373

1. 9 **3.** 1 **5.** 2 **7.** 10 **9.** -7

11. 17 **13.** $\lambda^2 - 3\lambda - 4$ **15.** -48

17. 62 **19.** 0 **21.** -85

23. $-x + 2y - z$ **25.** -104 **27.** 48

29. $\lambda_1 = -5$, $\lambda_2 = 7$

Exercises 8.5, Page 379

1. Theorem 8.11 **3.** Theorem 8.14

5. Theorem 8.12 **7.** Theorem 8.10

9. Theorem 8.8 **11.** -5 **13.** -5

15. 80 **17.** -105 **23.** 0 **25.** -15

27. -9 **29.** 0 **31.** 16

Exercises 8.6, Page 389

3. $\begin{pmatrix} \frac{1}{9} & \frac{1}{9} \\ -\frac{4}{9} & \frac{5}{9} \end{pmatrix}$ **5.** $\begin{pmatrix} \frac{1}{6} & 0 \\ \frac{1}{4} & \frac{1}{2} \end{pmatrix}$

7. $\begin{pmatrix} -\frac{1}{2} & \frac{1}{2} & \frac{1}{2} \\ -\frac{1}{8} & \frac{1}{4} & -\frac{3}{8} \\ \frac{3}{8} & -\frac{1}{4} & \frac{1}{8} \end{pmatrix}$ **9.** $\begin{pmatrix} \frac{7}{15} & -\frac{13}{30} & -\frac{8}{15} \\ \frac{1}{15} & -\frac{2}{15} & \frac{1}{15} \\ \frac{2}{15} & \frac{7}{30} & \frac{2}{15} \end{pmatrix}$

11. $\begin{pmatrix} \frac{1}{3} & 0 & 0 \\ 0 & \frac{1}{6} & 0 \\ 0 & 0 & -\frac{1}{2} \end{pmatrix}$

13. $\begin{pmatrix} \frac{2}{9} & \frac{7}{9} & -\frac{1}{3} & -\frac{4}{3} \\ -\frac{1}{27} & \frac{1}{27} & \frac{2}{9} & -\frac{1}{9} \\ \frac{10}{27} & \frac{17}{27} & -\frac{2}{9} & -\frac{17}{9} \\ \frac{4}{27} & -\frac{4}{27} & \frac{1}{9} & \frac{4}{9} \end{pmatrix}$

15. $\begin{pmatrix} \frac{1}{6} & \frac{1}{12} \\ 0 & \frac{1}{4} \end{pmatrix}$

17. $\begin{pmatrix} -\frac{1}{4} & \frac{1}{4} \\ \frac{5}{12} & -\frac{1}{12} \end{pmatrix}$

19. Singular matrix

21. $\begin{pmatrix} 0 & \frac{2}{3} & \frac{1}{3} \\ 0 & -\frac{1}{3} & -\frac{2}{3} \\ \frac{1}{3} & -\frac{2}{3} & 0 \end{pmatrix}$

23. $\begin{pmatrix} 5 & 6 & -3 \\ 2 & 2 & -1 \\ -1 & -1 & 1 \end{pmatrix}$

25. $\begin{pmatrix} -\frac{1}{2} & & & \\ 1-\frac{2}{3} & -\frac{1}{6} & & \frac{7}{6} \\ 0 & \frac{1}{3} & \frac{1}{3} & -\frac{4}{3} \\ -\frac{1}{2}-\frac{1}{3} & -\frac{1}{3} & & \frac{1}{3} \\ 1 & \frac{1}{2} & & \frac{1}{2} \end{pmatrix}$

27. $\begin{pmatrix} -\frac{1}{3} & \frac{1}{3} \\ -1 & \frac{10}{3} \end{pmatrix}$ **29.** $\begin{pmatrix} -2 & 3 \\ 3 & -4 \end{pmatrix}$ **31.** $x = 5$

35. By Theorem 8.13, det \mathbf{AB} = det \mathbf{A} det \mathbf{B}. Since det $\mathbf{A} \neq 0$ and det $\mathbf{B} \neq 0$, it follows that det $\mathbf{AB} \neq 0$. By Theorem 8.19, \mathbf{AB} is nonsingular.

37. The result follows from det (\mathbf{AA}^{-1}) = det \mathbf{I}.

39. Multiply $\mathbf{AB} = \mathbf{0}$ by \mathbf{A}^{-1}. **41.** no

43. $x_1 = 6, x_2 = -2$ **45.** $x_1 = \frac{3}{4}, x_2 = -\frac{1}{2}$

47. $x_1 = 2, x_2 = 4, x_3 = -6$ **49.** $x_1 = 21, x_2 = 1, x_3 = -11$

51. $x_1 = \frac{9}{10}, x_2 = \frac{13}{20};$
$x_1 = 6, x_2 = 16;$
$x_1 = -2, x_2 = -7$

53. System has only the trivial solution.

55. System has nontrivial solutions.

57. $i_1 = \dfrac{-R_3E_2 + R_3E_1 + R_2E_1 - R_2E_3}{R_3R_1 + R_3R_2 + R_1R_2},$

$i_2 = \dfrac{R_3E_2 - R_3E_1 - R_1E_3 + R_1E_2}{R_3R_1 + R_3R_2 + R_1R_2},$

$i_3 = \dfrac{-R_2E_1 + R_1E_3 + R_2E_3 - R_1E_2}{R_3R_1 + R_3R_2 + R_1R_2}$

Exercises 8.7, Page 394

1. $x_1 = -\frac{3}{5}, x_2 = \frac{6}{5}$ **3.** $x_1 = 0.1, x_2 = -0.3$

5. $x = 4, y = -7$ **7.** $x_1 = -4, x_2 = 4, x_3 = -5$

9. $u = 4, v = \frac{3}{2}, w = 1$ **11.** $k = \frac{6}{5}$

13. $T_1 \approx 450.8$ lb, $T_2 \approx 423$ lb

Exercises 8.8, Page 399

1. $\mathbf{K}_3, \lambda = -1$ **3.** $\mathbf{K}_3, \lambda = 0$

5. $\mathbf{K}_2, \lambda = 3; \mathbf{K}_3, \lambda = 1$

7. $\lambda_1 = 6, \lambda_2 = 1, \mathbf{K}_1 = \begin{pmatrix} 2 \\ 7 \end{pmatrix}, \mathbf{K}_2 = \begin{pmatrix} 1 \\ 1 \end{pmatrix}$

9. $\lambda_1 = \lambda_2 = -4, \mathbf{K}_1 = \begin{pmatrix} 1 \\ -4 \end{pmatrix}$

11. $\lambda_1 = 3i, \lambda_2 = -3i$
$\mathbf{K}_1 = \begin{pmatrix} 1 - 3i \\ 5 \end{pmatrix}, \mathbf{K}_2 = \begin{pmatrix} 1 + 3i \\ 5 \end{pmatrix}$

13. $\lambda_1 = 4, \lambda_2 = -5,$
$\mathbf{K}_1 = \begin{pmatrix} 1 \\ 0 \end{pmatrix}, \mathbf{K}_2 = \begin{pmatrix} -8 \\ 9 \end{pmatrix}$

15. $\lambda_1 = 0, \lambda_2 = 4, \lambda_3 = -4,$
$\mathbf{K}_1 = \begin{pmatrix} 9 \\ 45 \\ 25 \end{pmatrix}, \mathbf{K}_2 = \begin{pmatrix} 1 \\ 1 \\ 1 \end{pmatrix}, \mathbf{K}_3 = \begin{pmatrix} 1 \\ 9 \\ 1 \end{pmatrix}$

17. $\lambda_1 = \lambda_2 = \lambda_3 = -2,$
$\mathbf{K}_1 = \begin{pmatrix} 2 \\ -1 \\ 0 \end{pmatrix}, \mathbf{K}_2 = \begin{pmatrix} 0 \\ 0 \\ 1 \end{pmatrix}$

19. $\lambda_1 = -1, \lambda_2 = i, \lambda_3 = -i,$
$\mathbf{K}_1 = \begin{pmatrix} 1 \\ -1 \\ 1 \end{pmatrix}, \mathbf{K}_2 = \begin{pmatrix} -1 \\ i \\ i \end{pmatrix}, \mathbf{K}_3 = \begin{pmatrix} -1 \\ -i \\ -i \end{pmatrix}$

21. $\lambda_1 = 1, \lambda_2 = 5, \lambda_3 = -7,$
$\mathbf{K}_1 = \begin{pmatrix} 1 \\ 0 \\ 0 \end{pmatrix}, \mathbf{K}_2 = \begin{pmatrix} 1 \\ 2 \\ 0 \end{pmatrix}, \mathbf{K}_3 = \begin{pmatrix} -1 \\ -2 \\ 4 \end{pmatrix}$

23. For $\mathbf{A}, \lambda_1 = 4, \lambda_2 = 6,$
$\mathbf{K}_1 = \begin{pmatrix} 1 \\ -1 \end{pmatrix}, \mathbf{K}_2 = \begin{pmatrix} 1 \\ 1 \end{pmatrix}$

25. $\lambda_1 = 0, \lambda_2 = 6$

Exercises 8.9, Page 403

3. $\begin{pmatrix} \frac{1}{7}[3(-1)^m 2^{m+1} + (5)^m] & -\frac{3}{7}[(-2)^m - 5^m] \\ -\frac{2}{7}[(-2)^m - 5^m] & \frac{1}{7}[(-2)^m - 6(5)^m] \end{pmatrix}; \begin{pmatrix} 11 & 57 \\ 38 & 106 \end{pmatrix}$

5. $\begin{pmatrix} \frac{1}{3}2^{m-1}[5^m + (-1)^m] & \frac{5}{3}2^{m-2}[5^m - (-1)^m] \\ \frac{1}{3}[10^m - (-2)^m] & \frac{1}{3}2^{m-1}[5^m + 5(-1)^m] \end{pmatrix};$

$\begin{pmatrix} 83328 & 41680 \\ 33344 & 16640 \end{pmatrix}$

7. $\begin{pmatrix} 1 & 2^m - 1 & 2^m - 1 \\ 0 & \frac{1}{3}[2^{m+1} + (-1)^m] & \frac{2}{3}[2^m - (-1)^m] \\ 0 & \frac{1}{3}[2^m - (-1)^m] & \frac{1}{3}[2^m + 2(-1)^m] \end{pmatrix};$

$\begin{pmatrix} 1 & 1023 & 1023 \\ 0 & 683 & 682 \\ 0 & 341 & 342 \end{pmatrix}$

9. $\begin{pmatrix} \frac{1}{3}[2^{2m+1}+(-2)^m] & \frac{1}{3}[4^m-(-2)^m] & 0 \\ \frac{1}{3}[2^{2m+1}-(-1)^m 2^{m+1}] & \frac{1}{3}[4^m+(-1)^m 2^{m+1}] & 0 \\ \frac{1}{3}[2^{2m+1}+(-2)^m-3] & \frac{1}{3}[4^m-(-2)^m] & 1 \end{pmatrix}$;

$\begin{pmatrix} 699392 & 349184 & 0 \\ 698368 & 350208 & 0 \\ 699391 & 349184 & 1 \end{pmatrix}$

11. $\begin{pmatrix} 7m4^{m-1}+(1-m)4^m & 3m4^{m-1} \\ -3m4^{m-1} & m4^{m-1}+(1-m)4^m \end{pmatrix}$;

$\begin{pmatrix} 22528 & 18432 \\ -18432 & -14336 \end{pmatrix}$

13. (a) $4^m \begin{pmatrix} 1 & 1 \\ 3 & 3 \end{pmatrix}$, $m > 1$

(b) $\mathbf{A}^m = \mathbf{0}$, $m > 1$

(c) $\begin{pmatrix} 2(3)^{m-1} & 3^{m-1} & 3^{m-1} \\ \frac{1}{6}[9(2)^m - 4(3)^m] & \frac{1}{6}[3(2)^m - 2(3)^m] & \frac{1}{6}[-3(2)^m - 2(3)^m] \\ \frac{1}{6}[-9(2)^m + 8(3)^m] & \frac{1}{6}[-3(2)^m + 4(3)^m] & \frac{1}{6}[3(2)^m + 4(3)^m] \end{pmatrix}$

15. $\begin{pmatrix} \frac{3}{10} & \frac{2}{5} \\ -\frac{1}{10} & \frac{1}{5} \end{pmatrix}$ **17.** b–f

Exercises 8.10, Page 410

1. (b) $\lambda_1 = -4, \lambda_2 = -1, \lambda_3 = 16$

3. (b) $\lambda_1 = 18, \lambda_2 = \lambda_3 = -8$ **5.** orthogonal

7. orthogonal **9.** not orthogonal

11. $\begin{pmatrix} \dfrac{1}{\sqrt{2}} & \dfrac{1}{\sqrt{2}} \\ -\dfrac{1}{\sqrt{2}} & \dfrac{1}{\sqrt{2}} \end{pmatrix}$ **13.** $\begin{pmatrix} \dfrac{3}{\sqrt{10}} & \dfrac{1}{\sqrt{10}} \\ -\dfrac{1}{\sqrt{10}} & -\dfrac{3}{\sqrt{10}} \end{pmatrix}$

15. $\begin{pmatrix} -\dfrac{1}{\sqrt{2}} & \dfrac{1}{\sqrt{2}} & 0 \\ 0 & 0 & 1 \\ \dfrac{1}{\sqrt{2}} & \dfrac{1}{\sqrt{2}} & 0 \end{pmatrix}$

17. $\begin{pmatrix} -\dfrac{3}{\sqrt{11}} & \dfrac{1}{\sqrt{66}} & \dfrac{1}{\sqrt{6}} \\ \dfrac{1}{\sqrt{11}} & -\dfrac{4}{\sqrt{66}} & \dfrac{2}{\sqrt{6}} \\ \dfrac{1}{\sqrt{11}} & \dfrac{7}{\sqrt{66}} & \dfrac{1}{\sqrt{6}} \end{pmatrix}$

19. $a = -\frac{4}{5}, b = \frac{3}{5}$

21. (b) $\begin{pmatrix} 4 \\ 1 \\ -1 \end{pmatrix}, \begin{pmatrix} 0 \\ 1 \\ 1 \end{pmatrix}, \begin{pmatrix} -1 \\ 2 \\ -2 \end{pmatrix}$

23. Use $(\mathbf{AB})^T = \mathbf{B}^T\mathbf{A}^T$.

Exercises 8.11, Page 417

1. $2, \begin{pmatrix} 1 \\ 1 \end{pmatrix}$ **3.** $14, \begin{pmatrix} 1 \\ 3 \end{pmatrix}$ **5.** $10, \begin{pmatrix} 1 \\ 1 \\ 0.5 \end{pmatrix}$

7. 7 and 2 **9.** 4, 3, and 1

11. approximately 0.2087

13. (c) $\dfrac{1}{4}\begin{pmatrix} 3 & 2 & 1 \\ 2 & 4 & 2 \\ 1 & 2 & 3 \end{pmatrix}$

(d) 0.59

(e) approximately $9.44EI/L^2$

Exercises 8.12, Page 426

1. $\mathbf{P} = \begin{pmatrix} -3 & 1 \\ 1 & 1 \end{pmatrix}, \mathbf{D} = \begin{pmatrix} 1 & 0 \\ 0 & 5 \end{pmatrix}$

3. not diagonalizable

5. $\mathbf{P} = \begin{pmatrix} 13 & 1 \\ 2 & 1 \end{pmatrix}, \mathbf{D} = \begin{pmatrix} -7 & 0 \\ 0 & 4 \end{pmatrix}$

7. $\mathbf{P} = \begin{pmatrix} 1 & 1 \\ -1 & 1 \end{pmatrix}, \mathbf{D} = \begin{pmatrix} \frac{1}{3} & 0 \\ 0 & \frac{2}{3} \end{pmatrix}$

9. $\mathbf{P} = \begin{pmatrix} 1 & 1 \\ -i & i \end{pmatrix}, \mathbf{D} = \begin{pmatrix} -i & 0 \\ 0 & i \end{pmatrix}$

11. $\mathbf{P} = \begin{pmatrix} 1 & 0 & 1 \\ 0 & 1 & 1 \\ 0 & 0 & 1 \end{pmatrix}, \mathbf{D} = \begin{pmatrix} 1 & 0 & 0 \\ 0 & -1 & 0 \\ 0 & 0 & 2 \end{pmatrix}$

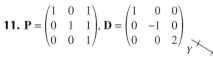

13. $\mathbf{P} = \begin{pmatrix} 1 & 1 & 1 \\ 0 & 1 & 0 \\ -1 & 1 & 1 \end{pmatrix}, \mathbf{D} = \begin{pmatrix} 0 & 0 & 0 \\ 0 & 1 & 0 \\ 0 & 0 & 2 \end{pmatrix}$

15. not diagonalizable

17. $\mathbf{P} = \begin{pmatrix} 0 & 1+\sqrt{5} & 1-\sqrt{5} \\ 0 & 2 & 2 \\ 1 & 0 & 0 \end{pmatrix}$,

$\mathbf{D} = \begin{pmatrix} 1 & 0 & 0 \\ 0 & \sqrt{5} & 0 \\ 0 & 0 & -\sqrt{5} \end{pmatrix}$

19. $\mathbf{P} = \begin{pmatrix} -3 & -1 & -1 & 1 \\ 0 & 1 & 0 & 0 \\ -3 & 0 & 0 & 1 \\ 1 & 0 & 1 & 0 \end{pmatrix}$,

$\mathbf{D} = \begin{pmatrix} 2 & 0 & 0 & 0 \\ 0 & 2 & 0 & 0 \\ 0 & 0 & 1 & 0 \\ 0 & 0 & 0 & -1 \end{pmatrix}$

21. $P = \begin{pmatrix} \frac{1}{\sqrt{2}} & \frac{1}{\sqrt{2}} \\ -\frac{1}{\sqrt{2}} & \frac{1}{\sqrt{2}} \end{pmatrix}$, $D = \begin{pmatrix} 0 & 0 \\ 0 & 2 \end{pmatrix}$

23. $P = \begin{pmatrix} -\frac{\sqrt{10}}{\sqrt{14}} & \frac{\sqrt{10}}{\sqrt{35}} \\ \frac{2}{\sqrt{14}} & \frac{5}{\sqrt{35}} \end{pmatrix}$, $D = \begin{pmatrix} 3 & 0 \\ 0 & 10 \end{pmatrix}$

25. $P = \begin{pmatrix} -\frac{1}{\sqrt{2}} & \frac{1}{\sqrt{2}} & 0 \\ \frac{1}{\sqrt{2}} & \frac{1}{\sqrt{2}} & 0 \\ 0 & 0 & 1 \end{pmatrix}$, $D = \begin{pmatrix} -1 & 0 & 0 \\ 0 & 1 & 0 \\ 0 & 0 & 1 \end{pmatrix}$

27. $P = \begin{pmatrix} \frac{2}{3} & \frac{2}{3} & \frac{1}{3} \\ \frac{2}{3} & -\frac{1}{3} & -\frac{2}{3} \\ \frac{1}{3} & -\frac{2}{3} & \frac{2}{3} \end{pmatrix}$, $D = \begin{pmatrix} 3 & 0 & 0 \\ 0 & 6 & 0 \\ 0 & 0 & 9 \end{pmatrix}$

29. $P = \begin{pmatrix} 0 & \frac{1}{\sqrt{2}} & \frac{1}{\sqrt{2}} \\ 1 & 0 & 0 \\ 0 & -\frac{1}{\sqrt{2}} & \frac{1}{\sqrt{2}} \end{pmatrix}$, $D = \begin{pmatrix} 1 & 0 & 0 \\ 0 & -6 & 0 \\ 0 & 0 & 8 \end{pmatrix}$

31. Ellipse; using

$$X = \begin{pmatrix} \frac{1}{\sqrt{2}} & \frac{1}{\sqrt{2}} \\ -\frac{1}{\sqrt{2}} & \frac{1}{\sqrt{2}} \end{pmatrix} X'$$

we get $X^2/4 + Y^2/6 = 1$.

33. Hyperbola; using

$$X = \begin{pmatrix} \frac{1}{\sqrt{5}} & -\frac{2}{\sqrt{5}} \\ \frac{2}{\sqrt{5}} & \frac{1}{\sqrt{5}} \end{pmatrix} X'$$

we get $X^2/4 - Y^2/4 = 1$.

35. $A = \begin{pmatrix} 4 & -1 \\ 2 & 1 \end{pmatrix}$ **39.** $A^5 = \begin{pmatrix} 21 & 11 \\ 22 & 10 \end{pmatrix}$

Exercises 8.13, Page 429

1. (a) $\begin{pmatrix} 35 & 15 & 38 & 36 & 0 \\ 27 & 10 & 26 & 20 & 0 \end{pmatrix}$

3. (a) $\begin{pmatrix} 48 & 64 & 120 & 107 & 40 \\ 32 & 40 & 75 & 67 & 25 \end{pmatrix}$

5. (a) $\begin{pmatrix} 31 & 44 & 15 & 61 & 50 & 49 & 41 \\ 24 & 29 & 15 & 47 & 35 & 31 & 21 \\ 1 & -15 & 15 & 0 & -15 & -5 & -19 \end{pmatrix}$

7. STUDY_HARD

9. MATH_IS_IMPORTANT_

11. DAD_I_NEED_MONEY_TODAY

13. (a) $B' = \begin{pmatrix} 15 & 22 & 20 & 8 & 23 & 6 & 21 & 22 \\ 10 & 22 & 18 & 23 & 25 & 2 & 23 & 25 \\ 3 & 26 & 26 & 14 & 23 & 16 & 26 & 12 \end{pmatrix}$

Exercises 8.14, Page 435

1. $(0\ 1\ 1\ 0)$ **3.** $(0\ 0\ 0\ 1\ 1)$

5. $(1\ 0\ 1\ 0\ 1\ 0\ 0\ 1)$ **7.** $(1\ 0\ 0)$

9. parity error **11.** $(1\ 0\ 0\ 1\ 1)$

13. $(0\ 0\ 1\ 0\ 1\ 1\ 0)$ **15.** $(0\ 1\ 0\ 0\ 1\ 0\ 1)$

17. $(1\ 1\ 0\ 0\ 1\ 1\ 0)$ **19.** code word; $(0\ 0\ 0\ 0)$

21. $(0\ 0\ 0\ 1)$ **23.** code word; $(1\ 1\ 1\ 1)$

25. $(1\ 0\ 0\ 1)$ **27.** $(1\ 0\ 1\ 0)$

29. $2^7 = 128$; $2^4 = 16$;
$(0\ 0\ 0\ 0\ 0\ 0\ 0)$, $(0\ 1\ 0\ 0\ 1\ 0\ 1)$,
$(0\ 1\ 1\ 0\ 0\ 1\ 1)$, $(0\ 1\ 0\ 1\ 0\ 1\ 0)$,
$(0\ 1\ 1\ 1\ 1\ 0\ 0)$, $(0\ 0\ 1\ 0\ 1\ 1\ 0)$,
$(0\ 0\ 1\ 1\ 0\ 0\ 1)$, $(0\ 0\ 0\ 1\ 1\ 1\ 1)$,
$(1\ 0\ 0\ 0\ 0\ 1\ 1)$, $(1\ 1\ 0\ 0\ 1\ 1\ 0)$,
$(1\ 0\ 1\ 0\ 1\ 0\ 1)$, $(1\ 0\ 0\ 1\ 1\ 0\ 0)$,
$(1\ 1\ 1\ 0\ 0\ 0\ 0)$, $(1\ 1\ 0\ 1\ 0\ 0\ 1)$,
$(1\ 0\ 1\ 1\ 0\ 1\ 0)$, $(1\ 1\ 1\ 1\ 1\ 1\ 1)$

Exercises 8.15, Page 438

1. $y = 0.4x + 0.6$ **3.** $y = 1.1x - 0.3$

5. $y = 1.3571x + 1.9286$

7. $v = -0.84T + 234$, 116.4, 99.6

Exercises 8.16, Page 442

1. (a) $T = \begin{pmatrix} 0.8 & 0.4 \\ 0.2 & 0.6 \end{pmatrix}$, $X_0 = \begin{pmatrix} 90 \\ 60 \end{pmatrix}$

(b) $\begin{pmatrix} 96 \\ 54 \end{pmatrix}$, $\begin{pmatrix} 98.4 \\ 51.6 \end{pmatrix}$ **(c)** $\hat{X} = \begin{pmatrix} 100 \\ 50 \end{pmatrix}$

3. (a) $T = \begin{pmatrix} 0.2 & 0.5 & 0 \\ 0.3 & 0.1 & 0 \\ 0.5 & 0.4 & 1 \end{pmatrix}$, $X_0 = \begin{pmatrix} 100 \\ 0 \\ 0 \end{pmatrix}$

(b) $\begin{pmatrix} 20 \\ 30 \\ 50 \end{pmatrix}$, $\begin{pmatrix} 19 \\ 9 \\ 72 \end{pmatrix}$ **(c)** $\hat{X} = \begin{pmatrix} 0 \\ 0 \\ 100 \end{pmatrix}$

Chapter 8 Review Exercises, Page 444

1. $\begin{pmatrix} 2 & 3 & 4 \\ 3 & 4 & 5 \\ 4 & 5 & 6 \\ 5 & 6 & 7 \end{pmatrix}$ **3.** $\begin{pmatrix} 3 & 4 \\ 6 & 8 \end{pmatrix}$, (11) **5.** false

7. $\frac{5}{8}$, -5 **9.** 0 **11.** false **13.** true

15. false **17.** true **19.** false

23. (a) $\begin{pmatrix} 1 & 1 \\ -1 & -1 \end{pmatrix}$ **25.** $x_1 = -\frac{1}{2}$, $x_2 = 7$, $x_3 = \frac{1}{2}$

29. 240 **31.** trivial solution only

33. $I_2 + 10\,HNO_3 \rightarrow 2\,HIO_3 + 10\,NO_2 + 4\,H_2O$

35. $x_1 = -\frac{1}{2}$, $x_2 = \frac{1}{4}$, $x_3 = \frac{2}{3}$

37. $x = X \cos\theta - Y \sin\theta$,
$y = X \sin\theta + Y \cos\theta$

39. $x_1 = 7$, $x_2 = 5$, $x_3 = 23$

41. $\lambda_1 = 5$, $\lambda_2 = -1$, $\mathbf{K}_1 = \begin{pmatrix} 1 \\ 2 \end{pmatrix}$, $\mathbf{K}_2 = \begin{pmatrix} 1 \\ -1 \end{pmatrix}$

43. $\lambda_1 = \lambda_2 = -1$, $\lambda_3 = 8$,

$\mathbf{K}_1 = \begin{pmatrix} 1 \\ -2 \\ 0 \end{pmatrix}$, $\mathbf{K}_2 = \begin{pmatrix} 1 \\ 0 \\ -1 \end{pmatrix}$, $\mathbf{K}_3 = \begin{pmatrix} 2 \\ 1 \\ 2 \end{pmatrix}$

45. $\lambda_1 = \lambda_2 = -3$, $\lambda_3 = 5$,

$\mathbf{K}_1 = \begin{pmatrix} -2 \\ 1 \\ 0 \end{pmatrix}$, $\mathbf{K}_2 = \begin{pmatrix} 3 \\ 0 \\ 1 \end{pmatrix}$, $\mathbf{K}_3 = \begin{pmatrix} 1 \\ 2 \\ -1 \end{pmatrix}$

47. $\begin{pmatrix} \dfrac{1}{\sqrt{6}} \\ -\dfrac{2}{\sqrt{6}} \\ \dfrac{1}{\sqrt{6}} \end{pmatrix}$ **49.** hyperbola

51. $\begin{pmatrix} 204 & 13 & 208 & 55 & 124 & 120 & 105 & 214 & 50 & 6 & 138 & 19 & 210 \\ 185 & 12 & 188 & 50 & 112 & 108 & 96 & 194 & 45 & 6 & 126 & 18 & 189 \end{pmatrix}$

53. HELP_IS_ON_THE_WAY

55. (1 1 0 0 1); parity error

Exercises 9.1, Page 453

1.

3.

5.

7.

9.

11.

$\mathbf{r}(t) = t\mathbf{i} + t\mathbf{j} + 2t^2\mathbf{k}$

13.

$\mathbf{r}(t) = 3\cos t\,\mathbf{i} + 3\sin t\,\mathbf{j} + 9\sin^2 t\,\mathbf{k}$

15. $2\mathbf{i} - 32\mathbf{j}$

17. $(1/t)\mathbf{i} - (1/t^2)\mathbf{j}$; $-(1/t^2)\mathbf{i} + (2/t^3)\mathbf{j}$

19. $\langle e^{2t}(2t+1), 3t^2, 8t-1 \rangle$; $\langle 4e^{2t}(t+1), 6t, 8 \rangle$

21.

23.

25. $x = 2 + t$, $y = 2 + 2t$, $z = \frac{8}{3} + 4t$

27. $\mathbf{r}(t) \times \mathbf{r}''(t)$ **29.** $\mathbf{r}(t) \cdot [\mathbf{r}'(t) \times \mathbf{r}'''(t)]$

31. $2\mathbf{r}_1'(2t) - (1/t^2)\mathbf{r}_2'(1/t)$ **33.** $\frac{3}{2}\mathbf{i} + 9\mathbf{j} + 15\mathbf{k}$

35. $e^t(t - 1)\mathbf{i} + \frac{1}{2}e^{-2t}\mathbf{j} + \frac{1}{2}e^{t^2}\mathbf{k} + \mathbf{c}$

37. $(6t + 1)\mathbf{i} + (3t^2 - 2)\mathbf{j} + (t^3 + 1)\mathbf{k}$

39. $(2t^3 - 6t + 6)\mathbf{i} + (7t - 4t^{3/2} - 3)\mathbf{j} + (t^2 - 2t)\mathbf{k}$

41. $2\sqrt{a^2 + c^2}\,\pi$ **43.** $6(e^{3\pi} - 1)$

45. $a \cos(s/a)\mathbf{i} + a \sin(s/a)\mathbf{j}$

47. Differentiate $\mathbf{r}(t) \cdot \mathbf{r}(t) = c^2$.

Exercises 9.2, Page 457

1. Speed is $\sqrt{5}$. **3.** Speed is 2.

5. Speed is $\sqrt{5}$. **7.** Speed is $\sqrt{14}$.

9. $(0, 0, 0)$ and $(25, 115, 0)$;

 $\mathbf{v}(0) = -2\mathbf{j} - 5\mathbf{k}$, $\mathbf{a}(0) = 2\mathbf{i} + 2\mathbf{k}$,

 $\mathbf{v}(5) = 10\mathbf{i} + 73\mathbf{j} + 5\mathbf{k}$, $\mathbf{a}(5) = 2\mathbf{i} + 30\mathbf{j} + 2\mathbf{k}$

11. $\mathbf{r}(t) = (-16t^2 + 240t)\mathbf{j} + 240\sqrt{3}\,t\mathbf{i}$ and $x(t) = 240\sqrt{3}\,t$, $y(t) = -16t^2 + 240t$; 900 ft; approximately 6235 ft; 480 ft/s

13. 72.11 ft/s

15. 97.98 ft/s

17. Assume that (x_0, y_0) are the coordinates of the center of the target at $t = 0$. Then $\mathbf{r}_p = \mathbf{r}_t$ when $t = x_0/(v_0 \cos \theta) = y_0/(v_0 \sin \theta)$. This implies $\tan \theta = y_0/x_0$. In other words, aim directly at the target at $t = 0$.

21. 191.33 lb

23. $y = -\dfrac{g}{2v_0^2 \cos^2\theta}\,x^2 + (\tan \theta)x + s_0$ is an equation of a parabola.

25. $\mathbf{r}(t) = k_1\, e^{2t^3}\mathbf{i} + \dfrac{1}{2t^2 + k_2}\,\mathbf{j} + (k_3 e^{t^2} - 1)\mathbf{k}$

27. Since \mathbf{F} is directed along \mathbf{r}, we must have $\mathbf{F} = c\mathbf{r}$ for some constant c. Hence $\boldsymbol{\tau} = \mathbf{r} \times (c\mathbf{r}) = c(\mathbf{r} \times \mathbf{r}) = \mathbf{0}$. If $\boldsymbol{\tau} = \mathbf{0}$, then $d\mathbf{L}/dt = \mathbf{0}$. This implies that \mathbf{L} is a constant.

Exercises 9.3, Page 463

1. $\mathbf{T} = (\sqrt{5}/5)(-\sin t\mathbf{i} + \cos t\mathbf{j} + 2\mathbf{k})$

3. $\mathbf{T} = (a^2 + c^2)^{-1/2}(-a \sin t\mathbf{i} + a \cos t\mathbf{j} + c\mathbf{k})$,

 $\mathbf{N} = -\cos t\mathbf{i} - \sin t\mathbf{j}$,

 $\mathbf{B} = (a^2 + c^2)^{-1/2}(c \sin t\mathbf{i} - c \cos t\mathbf{j} + a\mathbf{k})$,

 $\kappa = a/(a^2 + c^2)$

5. $3\sqrt{2}x - 3\sqrt{2}y + 4z = 3\pi$

7. $4t/\sqrt{1 + 4t^2}$, $2/\sqrt{1 + 4t^2}$

9. $2\sqrt{6}$, 0, $t > 0$ **11.** $2t/\sqrt{1 + t^2}$, $2/\sqrt{1 + t^2}$

13. $0, 5$ **15.** $-\sqrt{3}e^{-t}$, 0

17. $\kappa = \dfrac{\sqrt{b^2c^2 \sin^2 t + a^2c^2 \cos^2 t + a^2b^2}}{(a^2 \sin^2 t + b^2 \cos^2 t + c^2)^{3/2}}$

23. $\kappa = 2$, $\rho = \frac{1}{2}$; $\kappa = 2/\sqrt{125} \approx 0.18$, $\rho = \sqrt{125}/2 \approx 5.59$; the curve is sharper at $(0, 0)$.

25. κ is close to zero.

Exercises 9.4, Page 469

1. **3.**

5.

7. elliptical cylinders **9.** ellipsoids

11.

$c = 0$ $c > 0$ $c < 0$

13. $\partial z/\partial x = 2x - y^2$, $\partial z/\partial y = -2xy + 20y^4$

15. $\partial z/\partial x = 20x^3y^3 - 2xy^6 + 30x^4$,
$\partial z/\partial y = 15x^4y^2 - 6x^2y^5 - 4$

17. $\partial z/\partial x = 2x^{-1/2}/(3y^2 + 1)$,
$\partial z/\partial y = -24\sqrt{x}y/(3y^2 + 1)^2$

19. $\partial z/\partial x = -3x^2(x^3 - y^2)^{-2}$,
$\partial z/\partial y = 2y(x^3 - y^2)^{-2}$

21. $\partial z/\partial x = -10 \cos 5x \sin 5x$,
$\partial z/\partial y = 10 \sin 5y \cos 5y$

23. $f_x = e^{x^3y}(3x^3y + 1)$, $f_y = x^4e^{x^3y}$

25. $f_x = 7y/(x + 2y)^2$, $f_y = -7x/(x + 2y)^2$

27. $g_u = 8u/(4u^2 + 5v^3)$, $g_v = 15v^2/(4u^2 + 5v^3)$

29. $w_x = x^{-1/2}y$, $w_y = 2\sqrt{x} - (y/z)e^{y/z} - e^{y/z}$,
$w_z = (y^2/z^2)e^{y/z}$

31. $F_u = 2uw^2 - v^3 - vwt^2 \sin(ut^2)$,
$F_v = -3uv^2 + w \cos(ut^2)$, $F_x = 128x^7t^4$,
$F_t = -2uvwt \sin(ut^2) + 64x^8t^3$

39. $\partial z/\partial x = 3x^2v^2e^{uv^2} + 2uve^{uv^2}$, $\partial z/\partial y = -4yuve^{uv^2}$

41. $\partial z/\partial u = 16u^3 - 40y(2u - v)$,
$\partial z/\partial v = -96v^2 + 20y(2u - v)$

43. $\partial w/\partial t = -3u(u^2 + v^2)^{1/2}e^{-t} \sin \theta$
$\qquad -3v(u^2 + v^2)^{1/2}e^{-t} \cos \theta$,
$\partial w/\partial\theta = 3u(u^2 + v^2)^{1/2}e^{-t} \cos \theta$
$\qquad -3v(u^2 + v^2)^{1/2}e^{-t} \sin \theta$

45. $\partial R/\partial u = s^2t^4e^{v^2} - 4rst^4uve^{-u^2} + 8rs^2t^3uv^2e^{u^2v^2}$,
$\partial R/\partial v = 2s^2t^4uve^{v^2} + 2rst^4e^{-u^2} + 8rs^2t^3u^2ve^{u^2v^2}$

47. $\dfrac{\partial w}{\partial t} = \dfrac{xu}{(x^2 + y^2)^{1/2}(rs + tu)} + \dfrac{y \cosh rs}{u(x^2 + y^2)^{1/2}}$,

$\dfrac{\partial w}{\partial r} = \dfrac{xs}{(x^2 + y^2)^{1/2}(rs + tu)} + \dfrac{sty \sinh rs}{u(x^2 + y^2)^{1/2}}$,

$\dfrac{\partial w}{\partial u} = \dfrac{xt}{(x^2 + y^2)^{1/2}(rs + tu)} - \dfrac{ty \cosh rs}{u^2(x^2 + y^2)^{1/2}}$

49. $dz/dt = (4ut - 4vt^{-3})/(u^2 + v^2)$

51. $dw/dt\big|_{t - \pi} = -2$ **57.** 5.31 cm²/s

Exercises 9.5, Page 475

1. $(2x - 3x^2y^2)\mathbf{i} + (-2x^3y + 4y^3)\mathbf{j}$

3. $(y^2/z^3)\mathbf{i} + (2xy/z^3)\mathbf{j} - (3xy^2/z^4)\mathbf{k}$

5. $4\mathbf{i} - 32\mathbf{j}$ **7.** $2\sqrt{3}\mathbf{i} - 8\mathbf{j} - 4\sqrt{3}\mathbf{k}$

9. $\sqrt{3}x + y$ **11.** $^{15}\!/_2(\sqrt{3} - 2)$

13. $-1/2\sqrt{10}$ **15.** $98/\sqrt{5}$ **17.** $-3\sqrt{2}$

19. -1 **21.** $-12/\sqrt{17}$

23. $\sqrt{2}\mathbf{i} + (\sqrt{2}/2)\mathbf{j}$, $\sqrt{5/2}$

25. $-2\mathbf{i} + 2\mathbf{j} - 4\mathbf{k}$, $2\sqrt{6}$

27. $-8\sqrt{\pi/6}\mathbf{i} - 8\sqrt{\pi/6}\mathbf{j}$, $- 8\sqrt{\pi/3}$

29. $-\tfrac{3}{8}\mathbf{i} - 12\mathbf{j} - \tfrac{2}{3}\mathbf{k}$, $-\sqrt{83,281}/24$

31. $\pm 31/\sqrt{17}$

33. $\mathbf{u} = \tfrac{3}{5}\mathbf{i} - \tfrac{4}{5}\mathbf{j}$; $\mathbf{u} = \tfrac{4}{5}\mathbf{i} + \tfrac{3}{5}\mathbf{j}$; $\mathbf{u} = -\tfrac{4}{5}\mathbf{i} - \tfrac{3}{5}\mathbf{j}$

35. $D_\mathbf{u} f = (9x^2 + 3y^2 - 18xy^2 - 6x^2y)/\sqrt{10}$;
$D_\mathbf{u} F = (-6x^2 - 54y^2 + 54x + 6y - 72xy)/10$

37. $(2, 5), (-2, 5)$

39. $-16\mathbf{i} - 4\mathbf{j}$

41. $x = 3e^{-4t}$, $y = 4e^{-2t}$

43. One possible function is
$$f(x, y) = x^3 - \tfrac{2}{3}y^3 + xy^3 + e^{xy}.$$

Exercises 9.6, Page 479

1.

3.

5.

7.

9.

11.
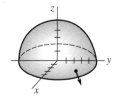

13. $(-4, -1, 17)$

15. $-2x + 2y + z = 9$

17. $6x - 2y - 9z = 5$

19. $6x - 8y + z = 50$

21. $2x + y - \sqrt{2}z = (4 + 5\pi)/4$

23. $\sqrt{2}x + \sqrt{2}y - z = 2$

25. $(1/\sqrt{2}, \sqrt{2}, 3/\sqrt{2}), (-1/\sqrt{2}, -\sqrt{2}, -3/\sqrt{2})$

27. $(-2, 0, 5), (-2, 0, -3)$

33. $x = 1 + 2t, y = -1 - 4t, z = 1 + 2t$

35. $(x - \frac{1}{2})/4 = (y - \frac{1}{3})/6 = -(z - 3)$

Exercises 9.7, Page 484

1.

3.

5.

7. $(x - y)\mathbf{i} + (x - y)\mathbf{j}; 2z$ **9.** $0; 4y + 8z$

11. $(4y^3 - 6xz^2)\mathbf{i} + (2z^3 - 3x^2)\mathbf{k}; 6xy$

13. $(3e^{-z} - 8yz)\mathbf{i} - xe^{-z}\mathbf{j}; e^{-z} + 4z^2 - 3ye^{-z}$

15. $(xy^2e^y + 2xye^y + x^3yze^z + x^3ye^z)\mathbf{i} - y^2e^y\mathbf{j} + (-3x^2yze^z - xe^x)\mathbf{k};$
$xye^x + ye^x - x^3ze^z$

35. $2\mathbf{i} + (1 - 8y)\mathbf{j} + 8z\mathbf{k}$

45. div $\mathbf{F} = 1 \neq 0$. If there existed a vector field \mathbf{G} such that $\mathbf{F} = \text{curl } \mathbf{G}$, then necessarily div $\mathbf{F} = \text{div curl } \mathbf{G} = 0$.

Exercises 9.8, Page 493

1. $-125/3\sqrt{2}; -250(\sqrt{2} - 4)/12; {}^{125}/_2$

3. $3; 6; 3\sqrt{5}$ **5.** $-1; (\pi - 2)/2; \pi^2/8; \sqrt{2}\pi^2/8$

7. 21 **9.** 30 **11.** 1 **13.** 1

15. 460 **17.** ${}^{26}/_9$ **19.** $-{}^{64}/_3$ **21.** $-{}^8/_3$

23. 0 **25.** ${}^{123}/_2$ **27.** 70 **29.** $-{}^{19}/_8$

31. e **33.** -4 **35.** 0

37. On each curve the line integral has the value ${}^{208}/_3$.

41. $\bar{x} = \frac{3}{2}, \bar{y} = 2/\pi$

Exercises 9.9, Page 501

1. ${}^{16}/_3$ **3.** 14 **5.** 3 **7.** 330

9. 1096 **11.** $\phi = x^4y^3 + 3x + y$

13. not a gradient field **15.** $\phi = \frac{1}{4}x^4 + xy + \frac{1}{4}y^4$

17. $3 + e^{-1}$ **19.** 63 **21.** $8 + 2e^3$

23. 16 **25.** $\pi - 4$ **27.** $\phi = (Gm_1m_2)/|\mathbf{r}|$

Exercises 9.10, Page 509

1. $24y - 20e^y$ **3.** $x^2e^{3x^2} - x^2e^x$ **5.** $\dfrac{x}{2} \ln 5$

7. $2 - \sin y$

9.

11.

13. $\frac{1}{21}$ **15.** ${}^{25}/_{84}$ **17.** 96

19. $2 \ln 2 - 1$ **21.** ${}^{14}/_3$ **23.** c, 16π

25. 18 **27.** 2π **29.** 4 **31.** $30 \ln 6$

33. $15\pi/4$ **35.** $(2^{3/2} - 1)/18$ **37.** $\frac{2}{3} \sin 8$

39. $\pi/8$ **41.** $\bar{x} = \frac{8}{3}, \bar{y} = 2$ **43.** $\bar{x} = 3, \bar{y} = \frac{3}{2}$

45. $\bar{x} = {}^{17}/_{21}, \bar{y} = {}^{55}/_{147}$ **47.** $\bar{x} = 0, \bar{y} = \frac{4}{7}$

49. $\bar{x} = (3e^4 + 1)/[4(e^4 - 1)],$
$\bar{y} = 16(e^5 - 1)/[25(e^4 - 1)]$

51. $\frac{1}{105}$ **53.** $4k/9$ **55.** ${}^{256}/_{21}$ **57.** ${}^{941}/_{10}$

59. $a\sqrt{10}/5$ **61.** $ab^3\pi/4$; $a^3b\pi/4$; $b/2$; $a/2$

63. $ka^4/6$ **65.** $16\sqrt{2}k/3$ **67.** $a\sqrt{3}/3$

Exercises 9.11, Page 514

1. $27\pi/2$ **3.** $(4\pi - 3\sqrt{3})/6$ **5.** $25\pi/3$

7. $(2\pi/3)(15^{3/2} - 7^{3/2})$ **9.** $\frac{5}{4}$

11. $\bar{x} = 13/3\pi$, $\bar{y} = 13/3\pi$ **13.** $\bar{x} = \frac{12}{5}$, $\bar{y} = 3\sqrt{3}/2$

15. $\bar{x} = (4 + 3\pi)/6$, $\bar{y} = \frac{4}{3}$ **17.** $\pi a^4 k/4$

19. $(ka/12)(15\sqrt{3} - 4\pi)$ **21.** $\pi a^4 k/2$

23. $4k$ **25.** 9π **27.** $(\pi/4)(e - 1)$

29. $3\pi/8$ **31.** 250

33. approximately 1450 m^3 **35.** $\sqrt{\pi}/2$

Exercises 9.12, Page 520

1. 3 **3.** 0 **5.** 75π **7.** 48π

9. $\frac{56}{3}$ **11.** $\frac{2}{3}$ **13.** $\frac{1}{8}$

15. $(b \times a) \times$ (area of region bounded by C)

19. $3a^2\pi/8$ **23.** $45\pi/2$ **25.** π

27. $27\pi/2$ **29.** $3\pi/2$ **33.** 3π

Exercises 9.13, Page 527

1. $3\sqrt{29}$ **3.** $10\pi/3$ **5.** $(\pi/6)(17^{3/2} - 1)$

7. $25\pi/6$ **9.** $2a^2(\pi - 2)$ **11.** $8a^2$

13. $2\pi a(c_2 - c_1)$ **15.** $\frac{26}{3}$ **17.** 0

19. 972π **21.** $(3^{5/2} - 2^{7/2} + 1)/15$

23. $9(17^{3/2} - 1)$ **25.** $12\sqrt{14}$ **27.** $k\sqrt{3}/12$

29. 18 **31.** 28π **33.** 8π **35.** $5\pi/2$

37. $-8\pi a^3$ **39.** $4\pi kq$ **41.** $(1, \frac{2}{3}, 2)$

Exercises 9.14, Page 533

1. -40π **3.** $\frac{45}{2}$ **5.** $\frac{3}{2}$ **7.** -3

9. $-3\pi/2$ **11.** π **13.** -152π

15. 112 **17.** Take the surface to be $z = 0$; $81\pi/4$.

Exercises 9.15, Page 543

1. 48 **3.** 36 **5.** $\pi - 2$

7. $\frac{1}{4}e^2 - \frac{1}{2}e$ **9.** 50

11. $\displaystyle\int_0^4 \int_0^{2-(x/2)} \int_{x+2y}^4 F(x, y, z) \, dz \, dy \, dx;$

$\displaystyle\int_0^2 \int_{2y}^4 \int_0^{z-2y} F(x, y, z) \, dx \, dz \, dy;$

$\displaystyle\int_0^4 \int_0^{z/2} \int_0^{z-2y} F(x, y, z) \, dx \, dy \, dz;$

$\displaystyle\int_0^4 \int_x^4 \int_0^{(z-x)/2} F(x, y, z) \, dy \, dz \, dx;$

$\displaystyle\int_0^4 \int_0^z \int_0^{(z-x)/2} F(x, y, z) \, dy \, dx \, dz$

13. $\displaystyle\int_0^2 \int_{x^3}^8 \int_0^4 dz \, dy \, dx;$

$\displaystyle\int_0^8 \int_0^4 \int_0^{\sqrt[3]{y}} dx \, dz \, dy;$

$\displaystyle\int_0^4 \int_0^2 \int_{x^3}^8 dy \, dx \, dz$

15.

17. **19.**

21. $16\sqrt{2}$ **23.** 16π

25. $\bar{x} = \frac{4}{5}$, $\bar{y} = \frac{32}{7}$, $\bar{z} = \frac{8}{3}$ **27.** $\bar{x} = 0$, $\bar{y} = 2$, $\bar{z} = 0$

29. $\displaystyle\int_{-1}^1 \int_{-\sqrt{1-x^2}}^{\sqrt{1-x^2}} \int_{2y+2}^{8-y} (x + y + 4) \, dz \, dy \, dx$

31. $2560k/3$; $\sqrt{80/9}$ **33.** $k/30$

35. $(-10/\sqrt{2}, 10/\sqrt{2}, 5)$ **37.** $(\sqrt{3}/2, \frac{3}{2}, -4)$

39. $(\sqrt{2}, -\pi/4, -9)$ **41.** $(2\sqrt{2}, 2\pi/3, 2)$

43. $r^2 + z^2 = 25$ **45.** $r^2 - z^2 = 1$

47. $z = x^2 + y^2$ **49.** $x = 5$

51. $(2\pi/3)(64 - 12^{3/2})$ **53.** $625\pi/2$

55. $(0, 0, 3a/8)$ **57.** $8\pi k/3$

59. $(\sqrt{3}/3, \frac{1}{3}, 0)$; $(\frac{2}{3}, \pi/6, 0)$

61. $(-4, 4, 4\sqrt{2})$; $(4\sqrt{2}, 3\pi/4, 4\sqrt{2})$

63. $(5\sqrt{2}, \pi/2, 5\pi/4)$ **65.** $(\sqrt{2}, \pi/4, \pi/6)$

67. $\rho = 8$

69. $\phi = \pi/6,\ \phi = 5\pi/6$

71. $x^2 + y^2 + z^2 = 100$

73. $z = 2$

75. $9\pi(2 - \sqrt{2})$

77. $2\pi/9$

79. $(0, 0, \frac{7}{6})$

81. πk

Exercises 9.16, Page 550

1. $\frac{3}{2}$ **3.** $12a^5\pi/5$ **5.** 256π **7.** $62\pi/5$

9. $4\pi(b - a)$ **11.** 128 **13.** $\pi/2$

Exercises 9.17, Page 557

1. $(0, 0),\ (-2, 8),\ (16, 20),\ (14, 28)$

3.

5.

7. $-2v$

9. $-1/3u^2$

11.

$(0, 0)$ is the image of every point on the boundary $u = 0$.

13. 16 **15.** $\frac{1}{2}$ **17.** $\frac{1}{4}(b - a)(d - c)$

19. $\frac{1}{2}(1 - \ln 2)$ **21.** $^{315}\!/_4$ **23.** $\frac{1}{4}(e - e^{-1})$

25. 126 **27.** $\frac{5}{2}(b - a) \ln \dfrac{d}{c}$ **29.** $15\pi/2$

Chapter 9 Review Exercises, Page 558

1. true **3.** true **5.** false **7.** true

9. false **11.** false **13.** true

15. true **17.** true

19. $\nabla\phi = -\dfrac{x}{(x^2 + y^2)^{3/2}}\mathbf{i} - \dfrac{y}{(x^2 + y^2)^{3/2}}\mathbf{j}$

21. $\mathbf{v}(1) = 6\mathbf{i} + \mathbf{j} + 2\mathbf{k},\ \mathbf{v}(4) = 6\mathbf{i} + \mathbf{j} + 8\mathbf{k},\ \mathbf{a}(t) = 2\mathbf{k}$ for all t

23. $\mathbf{i} + 4\mathbf{j} + (3\pi/4)\mathbf{k}$

25.

27. $(6x^2 - 2y^2 - 8xy)/\sqrt{40}$ **29.** $2;\ -2/\sqrt{2};\ 4$

31. $4\pi x + 3y - 12z = 4\pi - 6\sqrt{3}$

33. $\displaystyle\int_0^1\int_x^{2x} \sqrt{1 - x^2}\ dy\ dx;$

$\displaystyle\int_0^1\int_{y/2}^{y} \sqrt{1 - x^2}\ dx\ dy + \int_1^2\int_{y/2}^{1} \sqrt{1 - x^2}\ dx\ dy;\ \frac{1}{3}$

35. $41k/1512$ **37.** 8π **39.** $6xy$ **41.** 0

43. $56\sqrt{2}\pi^3/3$ **45.** 12 **47.** $2 + 2/3\pi$

49. $\pi^2/2$ **51.** $(\ln 3)(17^{3/2} - 5^{3/2})/12$

53. $-4\pi c$ **55.** 0 **57.** 125π **59.** 3π

61. $\frac{5}{3}$ **63.** 0 **65.** π

Exercises 10.1, Page 574

1. $\mathbf{X}' = \begin{pmatrix} 3 & -5 \\ 4 & 8 \end{pmatrix}\mathbf{X}$, where $\mathbf{X} = \begin{pmatrix} x \\ y \end{pmatrix}$

3. $\mathbf{X}' = \begin{pmatrix} -3 & 4 & -9 \\ 6 & -1 & 0 \\ 10 & 4 & 3 \end{pmatrix}\mathbf{X}$, where $\mathbf{X} = \begin{pmatrix} x \\ y \\ z \end{pmatrix}$

5. $\mathbf{X}' = \begin{pmatrix} 1 & -1 & 1 \\ 2 & 1 & -1 \\ 1 & 1 & 1 \end{pmatrix}\mathbf{X}$, where $\mathbf{X} = \begin{pmatrix} x \\ y \\ z \end{pmatrix}$

7. $\dfrac{dx}{dt} = 4x + 2y + e^t$

$\dfrac{dy}{dt} = -x + 3y - e^t$

9. $\dfrac{dx}{dt} = x - y + 2z + e^{-t} - 3t$

$\dfrac{dy}{dt} = 3x - 4y + z + 2e^{-t} + t$

$\dfrac{dz}{dt} = -2x + 5y + 6z + 2e^{-t} - t$

17. Yes; $W(\mathbf{X}_1, \mathbf{X}_2) = -2e^{-8t} \neq 0$ implies that \mathbf{X}_1 and \mathbf{X}_2 are linearly independent on $(-\infty, \infty)$.

19. No; $W(\mathbf{X}_1, \mathbf{X}_2, \mathbf{X}_3) = 0$ for every t. The solution vectors are linearly dependent on $(-\infty, \infty)$. Note that $\mathbf{X}_3 = 2\mathbf{X}_1 + \mathbf{X}_2$.

Exercises 10.2, Page 586

1. $\mathbf{X} = c_1 \begin{pmatrix} 1 \\ 2 \end{pmatrix} e^{5t} + c_2 \begin{pmatrix} 1 \\ -1 \end{pmatrix} e^{-t}$

3. $\mathbf{X} = c_1 \begin{pmatrix} 2 \\ 1 \end{pmatrix} e^{-3t} + c_2 \begin{pmatrix} 2 \\ 5 \end{pmatrix} e^{t}$

5. $\mathbf{X} = c_1 \begin{pmatrix} 5 \\ 2 \end{pmatrix} e^{8t} + c_2 \begin{pmatrix} 1 \\ 4 \end{pmatrix} e^{-10t}$

7. $\mathbf{X} = c_1 \begin{pmatrix} 1 \\ 0 \\ 0 \end{pmatrix} e^{t} + c_2 \begin{pmatrix} 2 \\ 3 \\ 1 \end{pmatrix} e^{2t} + c_3 \begin{pmatrix} 1 \\ 0 \\ 2 \end{pmatrix} e^{-t}$

9. $\mathbf{X} = c_1 \begin{pmatrix} -1 \\ 0 \\ 1 \end{pmatrix} e^{-t} + c_2 \begin{pmatrix} 1 \\ 4 \\ 3 \end{pmatrix} e^{3t} + c_3 \begin{pmatrix} 1 \\ -1 \\ 3 \end{pmatrix} e^{-2t}$

11. $\mathbf{X} = c_1 \begin{pmatrix} 4 \\ 0 \\ -1 \end{pmatrix} e^{-t} + c_2 \begin{pmatrix} -12 \\ 6 \\ 5 \end{pmatrix} e^{-t/2} + c_3 \begin{pmatrix} 4 \\ 2 \\ -1 \end{pmatrix} e^{-3t/2}$

13. $\mathbf{X} = 3 \begin{pmatrix} 1 \\ 1 \end{pmatrix} e^{t/2} + 2 \begin{pmatrix} 0 \\ 1 \end{pmatrix} e^{-t/2}$

19. $\mathbf{X} = c_1 \begin{pmatrix} 1 \\ 3 \end{pmatrix} + c_2 \left[\begin{pmatrix} 1 \\ 3 \end{pmatrix} t + \begin{pmatrix} \frac{1}{4} \\ -\frac{1}{4} \end{pmatrix} \right]$

21. $\mathbf{X} = c_1 \begin{pmatrix} 1 \\ 1 \end{pmatrix} e^{2t} + c_2 \left[\begin{pmatrix} 1 \\ 1 \end{pmatrix} te^{2t} + \begin{pmatrix} -\frac{1}{3} \\ 0 \end{pmatrix} e^{2t} \right]$

23. $\mathbf{X} = c_1 \begin{pmatrix} 1 \\ 1 \\ 1 \end{pmatrix} e^{t} + c_2 \begin{pmatrix} 1 \\ 1 \\ 0 \end{pmatrix} e^{2t} + c_3 \begin{pmatrix} 1 \\ 0 \\ 1 \end{pmatrix} e^{2t}$

25. $\mathbf{X} = c_1 \begin{pmatrix} -4 \\ -5 \\ 2 \end{pmatrix} + c_2 \begin{pmatrix} 2 \\ 0 \\ -1 \end{pmatrix} e^{5t}$

$+ c_3 \left[\begin{pmatrix} 2 \\ 0 \\ -1 \end{pmatrix} te^{5t} + \begin{pmatrix} -\frac{1}{2} \\ -\frac{1}{2} \\ -1 \end{pmatrix} e^{5t} \right]$

27. $\mathbf{X} = c_1 \begin{pmatrix} 0 \\ 1 \\ 1 \end{pmatrix} e^{t} + c_2 \left[\begin{pmatrix} 0 \\ 1 \\ 1 \end{pmatrix} te^{t} + \begin{pmatrix} 0 \\ 1 \\ 0 \end{pmatrix} e^{t} \right]$

$+ c_3 \left[\begin{pmatrix} 0 \\ 1 \\ 1 \end{pmatrix} \frac{t^2}{2} e^{t} + \begin{pmatrix} 0 \\ 1 \\ 0 \end{pmatrix} te^{t} + \begin{pmatrix} \frac{1}{2} \\ 0 \\ 0 \end{pmatrix} e^{t} \right]$

29. $\mathbf{X} = -7 \begin{pmatrix} 2 \\ 1 \end{pmatrix} e^{4t} + 13 \begin{pmatrix} 2t+1 \\ t+1 \end{pmatrix} e^{4t}$

31. Corresponding to the eigenvalue $\lambda_1 = 2$ of multiplicity five, eigenvectors are

$$\mathbf{K}_1 = \begin{pmatrix} 1 \\ 0 \\ 0 \\ 0 \\ 0 \end{pmatrix}, \mathbf{K}_2 = \begin{pmatrix} 0 \\ 0 \\ 1 \\ 0 \\ 0 \end{pmatrix}, \mathbf{K}_3 = \begin{pmatrix} 0 \\ 0 \\ 0 \\ 1 \\ 0 \end{pmatrix}.$$

33. $\mathbf{X} = c_1 \begin{pmatrix} \cos t \\ 2 \cos t + \sin t \end{pmatrix} e^{4t} + c_2 \begin{pmatrix} \sin t \\ 2 \sin t - \cos t \end{pmatrix} e^{4t}$

35. $\mathbf{X} = c_1 \begin{pmatrix} \cos t \\ -\cos t - \sin t \end{pmatrix} e^{4t} + c_2 \begin{pmatrix} \sin t \\ -\sin t + \cos t \end{pmatrix} e^{4t}$

37. $\mathbf{X} = c_1 \begin{pmatrix} 5 \cos 3t \\ 4 \cos 3t + 3 \sin 3t \end{pmatrix} + c_2 \begin{pmatrix} 5 \sin 3t \\ 4 \sin 3t - 3 \cos 3t \end{pmatrix}$

39. $\mathbf{X} = c_1 \begin{pmatrix} 1 \\ 0 \\ 0 \end{pmatrix} + c_2 \begin{pmatrix} -\cos t \\ \cos t \\ \sin t \end{pmatrix} + c_3 \begin{pmatrix} \sin t \\ -\sin t \\ \cos t \end{pmatrix}$

41. $\mathbf{X} = c_1 \begin{pmatrix} 0 \\ 2 \\ 1 \end{pmatrix} e^{t} + c_2 \begin{pmatrix} \sin t \\ \cos t \\ \cos t \end{pmatrix} e^{t} + c_3 \begin{pmatrix} \cos t \\ -\sin t \\ -\sin t \end{pmatrix} e^{t}$

43. $\mathbf{X} = \begin{pmatrix} 28 \\ -5 \\ 25 \end{pmatrix} e^{2t} + c_2 \begin{pmatrix} 5 \cos 3t \\ -4 \cos 3t - 3 \sin 3t \\ 0 \end{pmatrix} e^{-2t}$

$+ c_3 \begin{pmatrix} 5 \sin 3t \\ -4 \sin 3t + 3 \cos 3t \\ 0 \end{pmatrix} e^{-2t}$

45. $\mathbf{X} = - \begin{pmatrix} 25 \\ -7 \\ 6 \end{pmatrix} e^{t} - \begin{pmatrix} \cos 5t - 5 \sin 5t \\ \cos 5t \\ \cos 5t \end{pmatrix}$

$+ 6 \begin{pmatrix} 5 \cos 5t + \sin 5t \\ \sin 5t \\ \sin 5t \end{pmatrix}$

Exercises 10.3, Page 590

1. $\mathbf{X} = \begin{pmatrix} 3c_1 e^{7t} - 2c_2 e^{-4t} \\ c_1 e^{7t} + 3c_2 e^{-4t} \end{pmatrix}$

3. $\mathbf{X} = \begin{pmatrix} c_1 e^{t/2} + c_2 e^{3t/2} \\ -2c_1 e^{t/2} + 2c_2 e^{3t/2} \end{pmatrix}$

5. $\mathbf{X} = \begin{pmatrix} -c_1 e^{-4t} + c_2 e^{2t} \\ c_1 e^{-4t} + c_2 e^{2t} \\ c_2 e^{2t} + c_3 e^{6t} \end{pmatrix}$

7. $\mathbf{X} = \begin{pmatrix} c_1 e^{-t} - c_2 e^{2t} - c_3 e^{2t} \\ c_1 e^{-t} + c_2 e^{2t} \\ c_1 e^{-t} + c_3 e^{2t} \end{pmatrix}$

9. $\mathbf{X} = \begin{pmatrix} c_1 e^{t} + 2c_2 e^{2t} + 3c_3 e^{3t} \\ c_1 e^{t} + 2c_2 e^{2t} + 4c_3 e^{3t} \\ c_1 e^{t} + 2c_2 e^{2t} + 5c_3 e^{3t} \end{pmatrix}$

Exercises 10.4, Page 594

1. $X = c_1 \begin{pmatrix} 1 \\ 1 \end{pmatrix} + c_2 \begin{pmatrix} 3 \\ 2 \end{pmatrix} e^t - \begin{pmatrix} 11 \\ 11 \end{pmatrix} t - \begin{pmatrix} 15 \\ 10 \end{pmatrix}$

3. $X = c_1 \begin{pmatrix} 2 \\ 1 \end{pmatrix} e^{t/2} + c_2 \begin{pmatrix} 10 \\ 3 \end{pmatrix} e^{3t/2}$

$- \begin{pmatrix} 13/2 \\ 13/4 \end{pmatrix} te^{t/2} - \begin{pmatrix} 15/2 \\ 9/4 \end{pmatrix} e^{t/2}$

5. $X = c_1 \begin{pmatrix} 2 \\ 1 \end{pmatrix} e^t + c_2 \begin{pmatrix} 1 \\ 1 \end{pmatrix} e^{2t} + \begin{pmatrix} 3 \\ 3 \end{pmatrix} e^t + \begin{pmatrix} 4 \\ 2 \end{pmatrix} te^t$

7. $X = c_1 \begin{pmatrix} 4 \\ 1 \end{pmatrix} e^{3t} + c_2 \begin{pmatrix} -2 \\ 1 \end{pmatrix} e^{-3t} + \begin{pmatrix} -12 \\ 0 \end{pmatrix} t - \begin{pmatrix} 4/3 \\ 4/3 \end{pmatrix}$

9. $X = c_1 \begin{pmatrix} 1 \\ -1 \end{pmatrix} e^t + c_2 \begin{pmatrix} -t \\ 1/2 - t \end{pmatrix} e^t + \begin{pmatrix} 1/2 \\ -2 \end{pmatrix} e^{-t}$

11. $X = c_1 \begin{pmatrix} \cos t \\ \sin t \end{pmatrix} + c_2 \begin{pmatrix} \sin t \\ -\cos t \end{pmatrix}$

$+ \begin{pmatrix} \cos t \\ \sin t \end{pmatrix} t + \begin{pmatrix} -\sin t \\ \cos t \end{pmatrix} \ln|\cos t|$

13. $X = c_1 \begin{pmatrix} \cos t \\ \sin t \end{pmatrix} e^t + c_2 \begin{pmatrix} \sin t \\ -\cos t \end{pmatrix} e^t + \begin{pmatrix} \cos t \\ \sin t \end{pmatrix} te^t$

15. $X = c_1 \begin{pmatrix} \cos t \\ -\sin t \end{pmatrix} + c_2 \begin{pmatrix} \sin t \\ \cos t \end{pmatrix} + \begin{pmatrix} \cos t \\ -\sin t \end{pmatrix} t$

$+ \begin{pmatrix} -\sin t \\ \sin t \tan t \end{pmatrix} - \begin{pmatrix} \sin t \\ \cos t \end{pmatrix} \ln|\cos t|$

17. $X = c_1 \begin{pmatrix} 2\sin t \\ \cos t \end{pmatrix} e^t + c_2 \begin{pmatrix} 2\cos t \\ -\sin t \end{pmatrix} e^t + \begin{pmatrix} 3\sin t \\ 3/2 \cos t \end{pmatrix} te^t$

$+ \begin{pmatrix} \cos t \\ -1/2 \sin t \end{pmatrix} e^t \ln|\sin t| + \begin{pmatrix} 2\cos t \\ -\sin t \end{pmatrix} e^t \ln|\cos t|$

19. $X = c_1 \begin{pmatrix} 1 \\ -1 \\ 0 \end{pmatrix} + c_2 \begin{pmatrix} 1 \\ 1 \\ 0 \end{pmatrix} e^{2t} + c_3 \begin{pmatrix} 0 \\ 0 \\ 1 \end{pmatrix} e^{3t}$

$+ \begin{pmatrix} -1/4 e^{2t} + 1/2 te^{2t} \\ -e^t + 1/4 e^{2t} + 1/2 te^{2t} \\ 1/2 t^2 e^{3t} \end{pmatrix}$

21. $X = \begin{pmatrix} 2 \\ 2 \end{pmatrix} te^{2t} + \begin{pmatrix} -1 \\ 1 \end{pmatrix} e^{2t} + \begin{pmatrix} -2 \\ 2 \end{pmatrix} te^{4t} + \begin{pmatrix} 2 \\ 0 \end{pmatrix} e^{4t}$

23. $X = \begin{pmatrix} 20 + c_1 e^{-t} + 2c_2 e^{-2t} \\ 53 + 3c_1 e^{-t} + 7c_2 e^{-2t} \end{pmatrix}$

25. $X = \begin{pmatrix} 1/2 t^2 - 41/10 t - 41/110 + c_1 + c_2 e^{10t} \\ -1/2 t^2 + 39/10 t - 41/110 - c_1 + c_2 e^{10t} \end{pmatrix}$

27. $\begin{pmatrix} i_1 \\ i_2 \end{pmatrix} = 2 \begin{pmatrix} 1 \\ 3 \end{pmatrix} e^{-2t} + \frac{6}{29} \begin{pmatrix} 3 \\ -1 \end{pmatrix} e^{-12t} + \begin{pmatrix} 332/29 \\ 276/29 \end{pmatrix} \sin t - \begin{pmatrix} 76/29 \\ 168/29 \end{pmatrix} \cos t$

29. $\begin{pmatrix} i_2 \\ i_3 \end{pmatrix} = -12 \begin{pmatrix} 2 \\ -1 \end{pmatrix} e^{-t} - 6 \begin{pmatrix} 1 \\ 2 \end{pmatrix} e^{-6t} + \begin{pmatrix} 30 \\ 0 \end{pmatrix};$

$i_1(t) = -12e^{-t} - 18e^{-6t} + 30$

Exercises 10.5, Page 598

1. $e^{At} = \begin{pmatrix} e^t & 0 \\ 0 & e^{2t} \end{pmatrix};\ e^{-At} = \begin{pmatrix} e^{-t} & 0 \\ 0 & e^{-2t} \end{pmatrix}$

3. $e^{At} = \begin{pmatrix} t+1 & t & t \\ t & t+1 & t \\ -2t & -2t & -2t+1 \end{pmatrix}$

5. $X = c_1 \begin{pmatrix} 1 \\ 0 \end{pmatrix} e^t + c_2 \begin{pmatrix} 0 \\ 1 \end{pmatrix} e^{2t}$

7. $X = c_1 \begin{pmatrix} t+1 \\ t \\ -2t \end{pmatrix} + c_2 \begin{pmatrix} t \\ t+1 \\ -2t \end{pmatrix} + c_3 \begin{pmatrix} t \\ t \\ -2t+1 \end{pmatrix}$

9. $X = c_3 \begin{pmatrix} 1 \\ 0 \end{pmatrix} e^t + c_4 \begin{pmatrix} 0 \\ 1 \end{pmatrix} e^{2t} + \begin{pmatrix} -3 \\ 1/2 \end{pmatrix}$

11. $X = c_1 \begin{pmatrix} \cosh t \\ \sinh t \end{pmatrix} + c_2 \begin{pmatrix} \sinh t \\ \cosh t \end{pmatrix} e^{2t} - \begin{pmatrix} 1 \\ 1 \end{pmatrix}$

13. $X = \begin{pmatrix} t+1 \\ t \\ -2t \end{pmatrix} - 4 \begin{pmatrix} t \\ t+1 \\ -2t \end{pmatrix} + 6 \begin{pmatrix} t \\ t \\ -2t+1 \end{pmatrix}$

15. $e^{At} = \begin{pmatrix} 3/2 e^{2t} - 1/2 e^{-2t} & 3/4 e^{2t} - 3/4 e^{-2t} \\ -e^{2t} + e^{-2t} & -1/2 e^{2t} + 3/2 e^{-2t} \end{pmatrix};$

$X = c_1 \begin{pmatrix} 3/2 e^{2t} - 1/2 e^{-2t} \\ -e^{2t} + e^{-2t} \end{pmatrix} + c_2 \begin{pmatrix} 3/4 e^{2t} - 3/4 e^{-2t} \\ -1/2 e^{2t} + 3/2 e^{-2t} \end{pmatrix}$ or

$X = c_3 \begin{pmatrix} 3 \\ -2 \end{pmatrix} e^{2t} + c_4 \begin{pmatrix} 1 \\ -2 \end{pmatrix} e^{-2t}$

17. $e^{At} = \begin{pmatrix} e^{2t} + 3te^{2t} & -9te^{2t} \\ te^{2t} & e^{2t} - 3te^{2t} \end{pmatrix};$

$X = c_1 \begin{pmatrix} 1 + 3t \\ t \end{pmatrix} e^{2t} + c_2 \begin{pmatrix} -9t \\ 1 - 3t \end{pmatrix} e^{2t}$

19. $e^{At} = \begin{pmatrix} 4/5 e^t + 1/5 e^{6t} & 2/5 e^t - 2/5 e^{6t} \\ 2/5 e^t - 2/5 e^{6t} & 1/5 e^t + 4/5 e^{6t} \end{pmatrix};$

$X = c_1 \begin{pmatrix} 4/5 e^t + 1/5 e^{6t} \\ 2/5 e^t - 2/5 e^{6t} \end{pmatrix} + c_2 \begin{pmatrix} 2/5 e^t - 2/5 e^{6t} \\ 1/5 e^t + 4/5 e^{6t} \end{pmatrix}$ or

$X = c_3 \begin{pmatrix} 2 \\ 1 \end{pmatrix} e^t + c_4 \begin{pmatrix} 1 \\ -2 \end{pmatrix} e^{6t}$

21. $e^{At} = \begin{pmatrix} e^{3t} & -2e^{-t} + 2e^{3t} \\ 0 & e^{-t} \end{pmatrix};$

$X = c_1 \begin{pmatrix} 1 \\ 0 \end{pmatrix} e^{3t} + c_2 \begin{pmatrix} -2e^{-t} + 2e^{3t} \\ e^{-t} \end{pmatrix}$ or $X = c_2 \begin{pmatrix} -2 \\ 1 \end{pmatrix} e^{-t} + c_4 \begin{pmatrix} 1 \\ 0 \end{pmatrix} e^{3t}$

25. $X = c_1 \begin{pmatrix} 3/2 e^{3t} - 1/2 e^{5t} \\ 3/2 e^{3t} - 3/2 e^{5t} \end{pmatrix} + c_2 \begin{pmatrix} -1/2 e^{3t} + 1/2 e^{5t} \\ -1/2 e^{3t} + 3/2 e^{5t} \end{pmatrix}$ or

$X = c_3 \begin{pmatrix} 1 \\ 1 \end{pmatrix} e^{3t} + c_4 \begin{pmatrix} 1 \\ 3 \end{pmatrix} e^{5t}$

Chapter 10 Review Exercises, Page 599

1. $k = \frac{1}{3}$

5. $\mathbf{X} = c_1 \begin{pmatrix} 1 \\ -1 \end{pmatrix} e^t + c_2 \left[\begin{pmatrix} 1 \\ -1 \end{pmatrix} t e^t + \begin{pmatrix} 0 \\ 1 \end{pmatrix} e^t \right]$

7. $\mathbf{X} = c_1 \begin{pmatrix} \cos 2t \\ -\sin 2t \end{pmatrix} e^t + c_2 \begin{pmatrix} \sin 2t \\ \cos 2t \end{pmatrix} e^t$

9. $\mathbf{X} = c_1 \begin{pmatrix} -2 \\ 3 \\ 1 \end{pmatrix} e^{2t} + c_2 \begin{pmatrix} 0 \\ 1 \\ 1 \end{pmatrix} e^{4t} + c_3 \begin{pmatrix} 7 \\ 12 \\ -16 \end{pmatrix} e^{-3t}$

11. $\mathbf{X} = c_1 \begin{pmatrix} 1 \\ 0 \end{pmatrix} e^{2t} + c_2 \begin{pmatrix} 4 \\ 1 \end{pmatrix} e^{4t} + \begin{pmatrix} 16 \\ -4 \end{pmatrix} t + \begin{pmatrix} 11 \\ -1 \end{pmatrix}$

13. $\mathbf{X} = c_1 \begin{pmatrix} \cos t \\ \cos t - \sin t \end{pmatrix} + c_2 \begin{pmatrix} \sin t \\ \sin t + \cos t \end{pmatrix}$

$\qquad - \begin{pmatrix} 1 \\ 1 \end{pmatrix} + \begin{pmatrix} \sin t \\ \sin t + \cos t \end{pmatrix} \ln|\csc t - \cot t|$

15. (b) $\mathbf{X} = c_1 \begin{pmatrix} -1 \\ 1 \\ 0 \end{pmatrix} + c_2 \begin{pmatrix} -1 \\ 0 \\ 1 \end{pmatrix} + c_3 \begin{pmatrix} 1 \\ 1 \\ 1 \end{pmatrix} e^{3t}$

Exercises 11.1, Page 607

1. $x' = y$

$\quad y' = -9 \sin x$; critical points at $(\pm n\pi, 0)$

3. $x' = y$

$\quad y' = x^2 + y(x^3 - 1)$; critical point at $(0, 0)$

5. $x' = y$

$\quad y' = \varepsilon x^3 - x$;

\quad critical points at $(0, 0)$, $\left(\dfrac{1}{\sqrt{\varepsilon}}, 0 \right)$, $\left(-\dfrac{1}{\sqrt{\varepsilon}}, 0 \right)$

7. $(0, 0)$ and $(-1, -1)$

9. $(0, 0)$ and $(\frac{4}{3}, \frac{4}{3})$

11. $(0, 0)$, $(10, 0)$, $(0, 16)$, and $(4, 12)$

13. $(0, y)$, y arbitrary

15. $(0, 0)$, $(0, 1)$, $(0, -1)$, $(1, 0)$, $(-1, 0)$

17. (a) $x = c_1 e^{5t} - c_2 e^{-t}$ **(b)** $x = -2e^{-t}$

$\qquad y = 2c_1 e^{5t} + c_2 e^{-t}$ $y = 2e^{-t}$

19. (a) $x = c_1(4 \cos 3t - 3 \sin 3t) + c_2(4 \sin 3t + 3 \cos 3t)$

$\qquad y = c_1(5 \cos 3t) + c_2(5 \sin 3t)$

(b) $x = 4 \cos 3t - 3 \sin 3t$

$\qquad y = 5 \cos 3t$

21. (a) $x = c_1(\sin t - \cos t)e^{4t} + c_2(-\sin t - \cos t)e^{4t}$

$\qquad y = 2c_1(\cos t)e^{4t} + 2c_2(\sin t)e^{4t}$

(b) $x = (\sin t - \cos t)e^{4t}$

$\qquad y = 2(\cos t)e^{4t}$

23. $r = \dfrac{1}{\sqrt[4]{4t + c_1}}$, $\theta = t + c_2$; $r = 4\dfrac{1}{\sqrt[4]{1024t + 1}}$, $\theta = t$;

the solution spirals toward the origin as t increases.

25. $r = \dfrac{1}{\sqrt{1 + c_1 e^{-2t}}}$, $\theta = t + c_2$; $r = 1$, $\theta = t$ (or $x = \cos t$ and $y = \sin t$) is the solution that satisfies $\mathbf{X}(0) = (1, 0)$;

$r = \dfrac{1}{\sqrt{1 - \frac{3}{4}e^{-2t}}}$, $\theta = t$ is the solution that satisfies $\mathbf{X}(0) = (2, 0)$. This solution spirals toward the circle $r = 1$ as t increases.

27. There are no critical points and therefore no periodic solutions.

29. There appears to be a periodic solution enclosing the critical point $(0, 0)$.

Exercises 11.2, Page 616

1. (a) If $\mathbf{X}(0) = \mathbf{X}_0$ lies on the line $y = 2x$, then $\mathbf{X}(t)$ approaches $(0, 0)$ along this line. For all other initial conditions, $\mathbf{X}(t)$ approaches $(0, 0)$ from the direction determined by the line $y = -x/2$.

3. (a) All solutions are unstable spirals that become unbounded as t increases.

5. (a) All solutions approach $(0, 0)$ from the direction specified by the line $y = x$.

7. (a) If $\mathbf{X}(0) = \mathbf{X}_0$ lies on the line $y = 3x$, then $\mathbf{X}(t)$ approaches $(0, 0)$ along this line. For all other initial conditions, $\mathbf{X}(t)$ becomes unbounded and $y = x$ serves as the asymptote.

9. saddle point **11.** saddle point

13. degenerate stable node **15.** stable spiral

17. $|\mu| < 1$

19. $\mu < -1$ for a saddle point; $-1 < \mu < 3$ for an unstable spiral point

23. (a) $(-3, 4)$

(b) unstable node or saddle point

(c) $(0, 0)$ is a saddle point.

25. (a) $(\frac{1}{2}, 2)$

(b) unstable spiral point

(c) $(0, 0)$ is an unstable spiral point.

Exercises 11.3, Page 625

1. $r = r_0 e^{\alpha t}$

3. $x = 0$ is unstable; $x = n + 1$ is asymptotically stable.

5. $T = T_0$ is unstable.

7. $x = \alpha$ is unstable; $x = \beta$ is asymptotically stable.

9. $P = a/b$ is asymptotically stable; $P = c$ is unstable.

11. $(\frac{1}{2}, 1)$ is a stable spiral point.

13. $(\sqrt{2}, 0)$ and $(-\sqrt{2}, 0)$ are saddle points; $(\frac{1}{2}, -\frac{7}{4})$ is a stable spiral point.

15. $(1, 1)$ is a stable node; $(1, -1)$ is a saddle point; $(2, 2)$ is a saddle point; $(2, -2)$ is an unstable spiral point.

17. $(0, -1)$ is a saddle point; $(0, 0)$ is unclassified; $(0, 1)$ is stable but we are unable to classify further.

19. $(0, 0)$ is an unstable node; $(10, 0)$ is a saddle point; $(0, 16)$ is a saddle point; $(4, 12)$ is a stable node.

21. $\theta = 0$ is a saddle point. It is not possible to classify either $\theta = \pi/3$ or $\theta = -\pi/3$.

23. It is not possible to classify $x = 0$.

25. It is not possible to classify $x = 0$, but $x = 1/\sqrt{\varepsilon}$ and $x = -1/\sqrt{\varepsilon}$ are each saddle points.

29. (a) $(0, 0)$ is a stable spiral point.

33. (a) $(1, 0), (-1, 0)$

35. $|v_0| < \frac{1}{2}\sqrt{2}$

37. If $\beta > 0$, $(0, 0)$ is the only critical point and is stable. If $\beta < 0$, $(0, 0)$, $(\hat{x}, 0)$, and $(-\hat{x}, 0)$, where $\hat{x}^2 = -\alpha/\beta$, are critical points. $(0, 0)$ is stable, while $(\hat{x}, 0)$ and $(-\hat{x}, 0)$ are each saddle points.

39. (b) $(5\pi/6, 0)$ is a saddle point.

(c) $(\pi/6, 0)$ is a center.

Exercises 11.4, Page 633

1. $|\omega_0| < \sqrt{\dfrac{3g}{L}}$

5. (a) First show that $y^2 = v_0^2 + g \ln\left(\dfrac{1 + x^2}{1 + x_0^2}\right)$.

9. (a) The new critical point is $(d/c - \varepsilon_2/c, a/b + \varepsilon_1/b)$.

(b) yes

11. $(0, 0)$ is an unstable node, $(0, 100)$ is a stable node, $(50, 0)$ is a stable node, and $(20, 40)$ is a saddle point.

17. (a) $(0, 0)$ is the only critical point.

Exercises 11.5, Page 643

1. The system has no critical points.

3. $\dfrac{\partial P}{\partial x} + \dfrac{\partial Q}{\partial y} = -2 < 0$

5. $\dfrac{\partial P}{\partial x} + \dfrac{\partial Q}{\partial y} = -\mu + 9y^2 > 0$ if $\mu < 0$

7. The single critical point $(0, 0)$ is a saddle point.

9. $\delta(x, y) = e^{-y/2}$

11. $\dfrac{\partial P}{\partial x} + \dfrac{\partial Q}{\partial y} = 4(1 - x^2 - 3y^2) > 0$ for $x^2 + 3y^2 < 1$

13. Use $\delta(x, y) = 1/(xy)$ and show that $\dfrac{\partial(\delta P)}{\partial x} + \dfrac{\partial(\delta Q)}{\partial y} = -\dfrac{r}{Kx}$.

15. If $\mathbf{n} = (-2x, -2y)$, show that $\mathbf{V} \cdot \mathbf{n} = 2(x - y)^2 + 2y^4$.

17. Yes; the sole critical point $(0, 0)$ lies outside the invariant region $\frac{1}{16} \le x^2 + y^2 \le 1$, and so Theorem 11.8(*ii*) applies.

19. $\mathbf{V} \cdot \mathbf{n} = 2y^2(1 - x^2) \ge 2y^2(1 - r^2)$ and $\partial P/\partial x + \partial Q/\partial y = x^2 - 1 < 0$. The sole critical point is $(0, 0)$ and this critical point is a stable spiral point. Therefore, Theorem 11.9(*ii*) applies.

21. (a) $\dfrac{\partial P}{\partial x} + \dfrac{\partial Q}{\partial y} = 2xy - 1 - x^2 \le 2x - 1 - x^2 = -(x - 1)^2 \le 0$

(b) $\lim_{t \to \infty} \mathbf{X}(t) = (\frac{3}{2}, \frac{2}{9})$, a stable spiral point

Chapter 11 Review Exercises, Page 644

1. true **3.** a center or a saddle point

5. false **7.** false **9.** true

11. $r = \dfrac{1}{\sqrt[3]{3t + 1}}$, $\theta = t$; the solution curve spirals toward the origin.

13. center; degenerate stable node

15. stable node for $\mu < -2$; stable spiral point for $-2 < \mu < 0$; unstable spiral point for $0 < \mu < 2$; unstable node for $\mu > 2$

17. Show that $y^2 = (1 + x_0^2 - x^2)^2 - 1$.

19. $\dfrac{\partial P}{\partial x} + \dfrac{\partial Q}{\partial y} = 1$

21. (a) *Hint:* Use the Bendixson negative criterion.

(d) In b, $(0, 0)$ is a stable spiral point when $\beta < 2ml\sqrt{g/l - \omega^2}$. In c, $(\hat{x}, 0)$ and $(-\hat{x}, 0)$ are stable spiral points when $\beta < 2ml\sqrt{\omega^2 - g^2/(\omega^2l^2)}$.

Exercises 12.1, Page 656

7. $\dfrac{\sqrt{\pi}}{2}$ **9.** $\sqrt{\dfrac{\pi}{2}}$

11. $\|1\| = \sqrt{p}$; $\left\|\cos\dfrac{n\pi}{p}x\right\| = \sqrt{\dfrac{p}{2}}$

21. (a) $T = 1$ **(b)** $T = \pi L/2$ **(c)** $T = 2\pi$

(d) $T = \pi$ **(e)** $T = 2\pi$ **(f)** $T = 2p$

Exercises 12.2, Page 661

1. $f(x) = \dfrac{1}{2} + \dfrac{1}{\pi} \sum\limits_{n=1}^{\infty} \dfrac{1-(-1)^n}{n} \sin nx$

3. $f(x) = \dfrac{3}{4} + \sum\limits_{n=1}^{\infty} \left\{ \dfrac{(-1)^n - 1}{n^2 \pi^2} \cos n\pi x - \dfrac{1}{n\pi} \sin n\pi x \right\}$

5. $f(x) = \dfrac{\pi^2}{6} + \sum\limits_{n=1}^{\infty} \left\{ \dfrac{2(-1)^n}{n^2} \cos nx \right.$
$\left. + \left(\dfrac{(-1)^{n+1}\pi}{n} + \dfrac{2}{\pi n^3}[(-1)^n - 1] \right) \sin nx \right\}$

7. $f(x) = \pi + 2 \sum\limits_{n=1}^{\infty} \dfrac{(-1)^{n+1}}{n} \sin nx$

9. $f(x) = \dfrac{1}{\pi} + \dfrac{1}{2} \sin x + \dfrac{1}{\pi} \sum\limits_{n=2}^{\infty} \dfrac{(-1)^n + 1}{1 - n^2} \cos nx$

11. $f(x) = -\dfrac{1}{4} + \dfrac{1}{\pi} \sum\limits_{n=1}^{\infty} \left\{ -\dfrac{1}{n} \sin \dfrac{n\pi}{2} \cos \dfrac{n\pi}{2} x \right.$
$\left. + \dfrac{3}{n}\left(1 - \cos \dfrac{n\pi}{2}\right) \sin \dfrac{n\pi}{2} x \right\}$

13. $f(x) = \dfrac{9}{4} + 5 \sum\limits_{n=1}^{\infty} \left\{ \dfrac{(-1)^n - 1}{n^2\pi^2} \cos \dfrac{n\pi}{5} x + \dfrac{(-1)^{n+1}}{n\pi} \sin \dfrac{n\pi}{5} x \right\}$

15. $f(x) = \dfrac{2 \sinh \pi}{\pi} \left[\dfrac{1}{2} + \sum\limits_{n=1}^{\infty} \dfrac{(-1)^n}{1+n^2}(\cos nx - n \sin nx) \right]$

19. Set $x = \pi/2$.

Exercises 12.3, Page 667

1. odd **3.** neither even nor odd **5.** even

7. odd **9.** neither even nor odd

11. $f(x) = \dfrac{2}{\pi} \sum\limits_{n=1}^{\infty} \dfrac{1-(-1)^n}{n} \sin nx$

13. $f(x) = \dfrac{\pi}{2} + \dfrac{2}{\pi} \sum\limits_{n=1}^{\infty} \dfrac{(-1)^n - 1}{n^2} \cos nx$

15. $f(x) = \dfrac{1}{3} + \dfrac{4}{\pi^2} \sum\limits_{n=1}^{\infty} \dfrac{(-1)^n}{n^2} \cos n\pi x$

17. $f(x) = \dfrac{2\pi^2}{3} + 4 \sum\limits_{n=1}^{\infty} \dfrac{(-1)^{n+1}}{n^2} \cos nx$

19. $f(x) = \dfrac{2}{\pi} \sum\limits_{n=1}^{\infty} \dfrac{1-(-1)^n(1+\pi)}{n} \sin nx$

21. $f(x) = \dfrac{3}{4} + \dfrac{4}{\pi^2} \sum\limits_{n=1}^{\infty} \dfrac{\cos \dfrac{n\pi}{2} - 1}{n^2} \cos \dfrac{n\pi}{2} x$

23. $f(x) = \dfrac{2}{\pi} + \dfrac{2}{\pi} \sum\limits_{n=2}^{\infty} \dfrac{1+(-1)^n}{1-n^2} \cos nx$

25. $f(x) = \dfrac{1}{2} + \dfrac{2}{\pi} \sum\limits_{n=1}^{\infty} \dfrac{\sin \dfrac{n\pi}{2}}{n} \cos n\pi x$

$f(x) = \dfrac{2}{\pi} \sum\limits_{n=1}^{\infty} \dfrac{1 - \cos \dfrac{n\pi}{2}}{n} \sin n\pi x$

27. $f(x) = \dfrac{2}{\pi} + \dfrac{4}{\pi} \sum\limits_{n=1}^{\infty} \dfrac{(-1)^n}{1-4n^2} \cos 2nx$

$f(x) = \dfrac{8}{\pi} \sum\limits_{n=1}^{\infty} \dfrac{n}{4n^2 - 1} \sin 2nx$

29. $f(x) = \dfrac{\pi}{4} + \dfrac{2}{\pi} \sum\limits_{n=1}^{\infty} \dfrac{2 \cos \dfrac{n\pi}{2} - (-1)^n - 1}{n^2} \cos nx$

$f(x) = \dfrac{4}{\pi} \sum\limits_{n=1}^{\infty} \dfrac{\sin \dfrac{n\pi}{2}}{n^2} \sin nx$

31. $f(x) = \dfrac{3}{4} + \dfrac{4}{\pi^2} \sum\limits_{n=1}^{\infty} \dfrac{\cos \dfrac{n\pi}{2} - 1}{n^2} \cos \dfrac{n\pi}{2} x$

$f(x) = \sum\limits_{n=1}^{\infty} \left\{ \dfrac{4}{n^2\pi^2} \sin \dfrac{n\pi}{2} - \dfrac{2}{n\pi}(-1)^n \right\} \sin \dfrac{n\pi}{2} x$

33. $f(x) = \dfrac{5}{6} + \dfrac{2}{\pi^2} \sum\limits_{n=1}^{\infty} \dfrac{3(-1)^n - 1}{n^2} \cos n\pi x$

$f(x) = 4 \sum\limits_{n=1}^{\infty} \left\{ \dfrac{(-1)^{n+1}}{n\pi} + \dfrac{(-1)^n - 1}{n^3\pi^3} \right\} \sin n\pi x$

35. $f(x) = \dfrac{4\pi^2}{3} + 4 \sum\limits_{n=1}^{\infty} \left\{ \dfrac{1}{n^2} \cos nx - \dfrac{\pi}{n} \sin nx \right\}$

37. $f(x) = \dfrac{3}{2} - \dfrac{1}{\pi} \sum\limits_{n=1}^{\infty} \dfrac{1}{n} \sin 2n\pi x$

39. $x_p(t) = \dfrac{10}{\pi} \sum\limits_{n=1}^{\infty} \dfrac{1-(-1)^n}{n(10-n^2)} \sin nt$

41. $x_p(t) = \dfrac{\pi^2}{18} + 16 \sum\limits_{n=1}^{\infty} \dfrac{1}{n^2(n^2 - 48)} \cos nt$

43. $x(t) = \dfrac{10}{\pi} \sum\limits_{n=1}^{\infty} \dfrac{1-(-1)^n}{10-n^2} \left\{ \dfrac{1}{n} \sin nt - \dfrac{1}{\sqrt{10}} \sin \sqrt{10}\,t \right\}$

45. (b) $y(x) = \dfrac{2w_0 L^4}{EI\pi^5} \sum\limits_{n=1}^{\infty} \dfrac{(-1)^{n+1}}{n^5} \sin \dfrac{n\pi}{L} x$

Exercises 12.4, Page 672

1. $f(x) = \sum\limits_{n=-\infty,\, n \neq 0}^{\infty} \dfrac{1-(-1)^n}{n\pi i} e^{in\pi x/2}$

3. $f(x) = \frac{1}{4} + \sum\limits_{n=-\infty,\, n \neq 0}^{\infty} \dfrac{1 - e^{-in\pi/2}}{2n\pi i} e^{2in\pi x}$

5. $f(x) = \pi + \sum\limits_{n=-\infty,\, n \neq 0}^{\infty} \dfrac{i}{n} e^{inx}$

Exercises 12.5, Page 677

1. $y = \cos\sqrt{\lambda_n}\,x$; $\cot\sqrt{\lambda} = \sqrt{\lambda}$; 0.7402, 11.7349, 41.4388, 90.8082; $\cos 0.8603x$, $\cos 3.4256x$, $\cos 6.4373x$, $\cos 9.5293x$

5. $\frac{1}{2}(1 + \sin^2\sqrt{\lambda_n})$

7. (a) $\lambda = \left(\dfrac{n\pi}{\ln 5}\right)^2$, $y = \sin\left(\dfrac{n\pi}{\ln 5}\ln x\right)$, $n = 1, 2, 3, \ldots$

(b) $\dfrac{d}{dx}[xy'] + \dfrac{\lambda}{x}\,y = 0$

(c) $\displaystyle\int_1^5 \frac{1}{x}\sin\left(\frac{m\pi}{\ln 5}\ln x\right)\sin\left(\frac{n\pi}{\ln 5}\ln x\right)dx = 0$, $m \neq n$

9. (a) $\displaystyle\int_0^1 \cos(x_m x)\cos(x_n x)\,dx = 0$, $m \neq n$, where x_m and x_n are positive roots of $\cot x = x$.

11. $\dfrac{d}{dx}[xe^{-x}y'] + ne^{-x}y = 0$; $\displaystyle\int_0^\infty e^{-x}L_m(x)L_n(x)\,dx = 0$, $m \neq n$

13. (a) $\lambda = 16n^2$, $y = \sin(4n\tan^{-1}x)$, $n = 1, 2, 3, \ldots$

(b) $\displaystyle\int_0^1 \frac{1}{1+x^2}\sin(4m\tan^{-1}x)\sin(4n\tan^{-1}x)\,dx = 0$, $m \neq n$

Exercises 12.6, Page 684

1. 1.277, 2.339, 3.391, 4.441

3. $f(x) = \displaystyle\sum_{i=1}^\infty \frac{1}{\lambda_i J_1(2\lambda_i)}J_0(\lambda_i x)$

5. $f(x) = 4\displaystyle\sum_{i=1}^\infty \frac{\lambda_i J_1(2\lambda_i)}{(4\lambda_i^2 + 1)J_0^2(2\lambda_i)}J_0(\lambda_i x)$

7. $f(x) = 20\displaystyle\sum_{i=1}^\infty \frac{\lambda_i J_2(4\lambda_i)}{(2\lambda_i^2 + 1)J_1^2(4\lambda_i)}J_1(\lambda_i x)$

9. $f(x) = \frac{9}{2} - 4\displaystyle\sum_{i=1}^\infty \frac{J_2(3\lambda_i)}{\lambda_i^2 J_0^2(3\lambda_i)}J_0(\lambda_i x)$

13. The graphs would be symmetric with respect to the origin.

15. $f(x) = \frac{1}{4}P_0(x) + \frac{1}{2}P_1(x) + \frac{5}{16}P_2(x) - \frac{3}{32}P_4(x) + \cdots$

21. $f(x) = \frac{1}{2}P_0(x) + \frac{5}{8}P_2(x) - \frac{3}{16}P_4(x) + \cdots$, $f(x) = |x|$ on $(-1, 1)$

Chapter 12 Review Exercises, Page 685

1. true **3.** cosine

5. false **7.** 5.5, 1, 0

9. $\dfrac{1}{\sqrt{1-x^2}}$, $-1 \le x \le 1$, $\displaystyle\int_{-1}^1 \frac{1}{\sqrt{1-x^2}}T_m(x)T_n(x)\,dx = 0$, $m \neq n$

13. $f(x) = \frac{1}{2} + \dfrac{2}{\pi}\displaystyle\sum_{n=1}^\infty \left\{\frac{1}{n^2\pi}[(-1)^n - 1]\cos n\pi x + \frac{2}{n}(-1)^n \sin n\pi x\right\}$

15. $f(x) = 1 - e^{-1} + 2\displaystyle\sum_{n=1}^\infty \frac{1-(-1)^n e^{-1}}{1 + n^2\pi^2}\cos n\pi x$,

$f(x) = \displaystyle\sum_{n=1}^\infty \frac{2n\pi[1 - (-1)^n e^{-1}]}{1 + n^2\pi^2}\sin n\pi x$

17. $\lambda = \dfrac{(2n-1)^2\pi^2}{36}$, $n = 1, 2, 3, \ldots$,

$y = \cos\left(\dfrac{2n-1}{2}\pi\ln x\right)$

19. $f(x) = \frac{1}{4}\displaystyle\sum_{i=1}^\infty \frac{J_1(2\lambda_i)}{\lambda_i J_1^2(4\lambda_i)}J_0(\lambda_i x)$

Exercises 13.1, Page 691

1. The possible cases can be summarized in one form $u = c_1 e^{c_2(x+y)}$, where c_1 and c_2 are constants.

3. $u = c_1 e^{y + c_2(x-y)}$

5. $u = c_1(xy)^{c_2}$

7. not separable

9. $u = e^{-t}(A_1 e^{k\lambda^2 t}\cosh \lambda x + B_1 e^{k\lambda^2 t}\sinh \lambda x)$
$u = e^{-t}(A_2 e^{-k\lambda^2 t}\cos \lambda x + B_2 e^{-k\lambda^2 t}\sin \lambda x)$
$u = (c_7 x + c_8)c_9 e^{-t}$

11. $u = (c_1\cosh \lambda x + c_2\sinh \lambda x)(c_3\cosh \lambda at + c_4\sinh \lambda at)$
$u = (c_5\cos \lambda x + c_6\sin \lambda x)(c_7\cos \lambda at + c_8\sin \lambda at)$
$u = (c_9 x + c_{10})(c_{11}t + c_{12})$

13. $u = (c_1\cosh \lambda x + c_2\sinh \lambda x)(c_3\cos \lambda y + c_4\sin \lambda y)$
$u = (c_5\cos \lambda x + c_6\sin \lambda x)(c_7\cosh \lambda y + c_8\sinh \lambda y)$
$u = (c_9 x + c_{10})(c_{11}y + c_{12})$

15. For $\lambda^2 > 0$ there are three possibilities:

$u = (c_1\cosh \lambda x$
$\quad + c_2\sinh \lambda x)(c_3\cosh \sqrt{1-\lambda^2}\,y$
$\quad + c_4\sinh \sqrt{1-\lambda^2}\,y)$, $\lambda^2 < 1$

$u = (c_1\cosh \lambda x$
$\quad + c_2\sinh \lambda x)(c_3\cos \sqrt{\lambda^2-1}\,y$
$\quad + c_4\sin \sqrt{\lambda^2-1}\,y)$, $\lambda^2 > 1$

$u = (c_1\cosh x + c_2\sinh x)$
$\quad \times (c_3 y + c_4)$, $\lambda^2 = 1$.

The results for the case $-\lambda^2 < 0$ are similar. For $\lambda^2 = 0$ we have

$u = (c_1 x + c_2)(c_3\cosh y + c_4\sinh y)$.

17. elliptic **19.** parabolic **21.** hyperbolic

23. parabolic **25.** hyperbolic

Exercises 13.2, Page 696

1. $k\dfrac{\partial^2 u}{\partial x^2} = \dfrac{\partial u}{\partial t}$, $0 < x < L$, $t > 0$

$u(0, t) = 0, \dfrac{\partial u}{\partial x}\Big|_{x=L} = 0, t > 0$

$u(x, 0) = f(x), 0 < x < L$

3. $k \dfrac{\partial^2 u}{\partial x^2} = \dfrac{\partial u}{\partial t}, 0 < x < L, t > 0$

$u(0, t) = 100, \dfrac{\partial u}{\partial x}\Big|_{x=L} = -hu(L, t), t > 0$

$u(x, 0) = f(x), 0 < x < L$

5. $a^2 \dfrac{\partial^2 u}{\partial x^2} = \dfrac{\partial^2 u}{\partial t^2}, 0 < x < L, t > 0$

$u(0, t) = 0, u(L, t) = 0, t > 0$

$u(x, 0) = x(L - x), \dfrac{\partial u}{\partial t}\Big|_{t=0} = 0, 0 < x < L$

7. $a^2 \dfrac{\partial^2 u}{\partial x^2} - 2\beta \dfrac{\partial u}{\partial t} = \dfrac{\partial^2 u}{\partial t^2}, 0 < x < L, t > 0$

$u(0, t) = 0, u(L, t) = \sin \pi t, t > 0$

$u(x, 0) = f(x), \dfrac{\partial u}{\partial t}\Big|_{t=0} = 0, 0 < x < L$

9. $\dfrac{\partial^2 u}{\partial x^2} + \dfrac{\partial^2 u}{\partial y^2} = 0, 0 < x < 4, 0 < y < 2$

$\dfrac{\partial u}{\partial x}\Big|_{x=0} = 0, u(4, y) = f(y), 0 < y < 2$

$\dfrac{\partial u}{\partial y}\Big|_{y=0} = 0, u(x, 2) = 0, 0 < x < 4$

Exercises 13.3, Page 699

1. $u(x, t) = \dfrac{2}{\pi} \displaystyle\sum_{n=1}^{\infty} \left(\dfrac{-\cos \frac{n\pi}{2} + 1}{n} \right) e^{-k(n^2 \pi^2/L^2)t} \sin \dfrac{n\pi}{L} x$

3. $u(x, t) = \dfrac{1}{L} \displaystyle\int_0^L f(x)\, dx$

$+ \dfrac{2}{L} \displaystyle\sum_{n=1}^{\infty} \left(\int_0^L f(x) \cos \dfrac{n\pi}{L} x\, dx \right)$

$\times e^{-k(n^2 \pi^2/L^2)t} \cos \dfrac{n\pi}{L} x$

5. $u(x, t) = e^{-ht} \left[\dfrac{1}{L} \displaystyle\int_0^L f(x)\, dx \right.$

$+ \dfrac{2}{L} \displaystyle\sum_{n=1}^{\infty} \left(\int_0^L f(x) \cos \dfrac{n\pi}{L} x\, dx \right)$

$\left. \times e^{-k(n^2 \pi^2/L^2)t} \cos \dfrac{n\pi}{L} x \right]$

Exercises 13.4, Page 702

1. $u(x, t) = \dfrac{L^2}{\pi^3} \displaystyle\sum_{n=1}^{\infty} \dfrac{1 - (-1)^n}{n^3} \cos \dfrac{n\pi a}{L} t \sin \dfrac{n\pi}{L} x$

3. $u(x, t) = \dfrac{6\sqrt{3}}{\pi^2} \left(\cos \dfrac{\pi a}{L} t \sin \dfrac{\pi}{L} x - \dfrac{1}{5^2} \cos \dfrac{5\pi a}{L} t \sin \dfrac{5\pi}{L} x \right.$

$\left. + \dfrac{1}{7^2} \cos \dfrac{7\pi a}{L} t \sin \dfrac{7\pi}{L} x - \cdots \right)$

5. $u(x, t) = \dfrac{1}{a} \sin at \sin x$

7. $u(x, t) = \dfrac{8h}{\pi^2} \displaystyle\sum_{n=1}^{\infty} \dfrac{\sin \frac{n\pi}{2}}{n^2} \cos \dfrac{n\pi a}{L} t \sin \dfrac{n\pi}{L} x$

9. $u(x, t) = e^{-\beta t} \displaystyle\sum_{n=1}^{\infty} A_n \left\{ \cos q_n t + \dfrac{\beta}{q_n} \sin q_n t \right\} \sin nx$,

where $A_n = \dfrac{2}{\pi} \displaystyle\int_0^{\pi} f(x) \sin nx\, dx$ and $q_n = \sqrt{n^2 - \beta^2}$

11. $u(x, t) = \displaystyle\sum_{n=1}^{\infty} \left(A_n \cos \dfrac{n^2 \pi^2}{L^2} at + B_n \sin \dfrac{n^2 \pi^2}{L^2} at \right)$

$\times \sin \dfrac{n\pi}{L} x$,

where $A_n = \dfrac{2}{L} \displaystyle\int_0^L f(x) \sin \dfrac{n\pi}{L} x\, dx$

$B_n = \dfrac{2L}{n^2 \pi^2 a} \displaystyle\int_0^L g(x) \sin \dfrac{n\pi}{L} x\, dx$

15. $u(x, t) = t + \sin x \cos 2at$

17. $u(x, t) = \dfrac{1}{2a} \sin 2x \sin 2at$

Exercises 13.5, Page 708

1. $u(x, y) = \dfrac{2}{a} \displaystyle\sum_{n=1}^{\infty} \left(\dfrac{1}{\sinh \frac{n\pi}{a} b} \int_0^a f(x) \sin \dfrac{n\pi}{a} x\, dx \right)$

$\times \sinh \dfrac{n\pi}{a} y \sin \dfrac{n\pi}{a} x$

3. $u(x, y) = \dfrac{2}{a} \displaystyle\sum_{n=1}^{\infty} \left(\dfrac{1}{\sinh \frac{n\pi}{a} b} \int_0^a f(x) \sin \dfrac{n\pi}{a} x\, dx \right)$

$\times \sinh \dfrac{n\pi}{a} (b - y) \sin \dfrac{n\pi}{a} x$

5. $u(x, y) = \dfrac{1}{2} x + \dfrac{2}{\pi^2} \displaystyle\sum_{n=1}^{\infty} \dfrac{1 - (-1)^n}{n^2 \sinh n\pi} \sinh n\pi x \cos n\pi y$

7. $u(x, y) = \dfrac{2}{\pi} \displaystyle\sum_{n=1}^{\infty} \dfrac{[1 - (-1)^n]}{n}$

$\times \dfrac{n \cosh nx + \sinh nx}{n \cosh n\pi + \sinh n\pi} \sin ny$

9. $u(x, y) = \displaystyle\sum_{n=1}^{\infty} (A_n \cosh n\pi y + B_n \sinh n\pi y) \sin n\pi x,$

where $A_n = 200 \dfrac{[1 - (-1)^n]}{n\pi}$

$B_n = 200 \dfrac{[1 - (-1)^n]}{n\pi} \dfrac{[2 - \sinh n\pi]}{\sinh n\pi}$

11. $u(x, y) = \dfrac{2}{\pi} \displaystyle\sum_{n=1}^{\infty} \left(\int_0^{\pi} f(x) \sin nx \, dx \right) e^{-ny} \sin nx$

13. $u(x, y) = \displaystyle\sum_{n=1}^{\infty} \left(A_n \cosh \dfrac{n\pi}{a} y + B_n \sinh \dfrac{n\pi}{a} y \right) \sin \dfrac{n\pi}{a} x,$

where $A_n = \dfrac{2}{a} \displaystyle\int_0^a f(x) \sin \dfrac{n\pi}{a} x \, dx$

$B_n = \dfrac{1}{\sinh \dfrac{n\pi}{a} b} \left(\dfrac{2}{a} \displaystyle\int_0^a g(x) \sin \dfrac{n\pi}{a} x \, dx \right.$

$\left. - A_n \cosh \dfrac{n\pi}{a} b \right)$

15. $u = u_1 + u_2$ where

$u_1(x, y) = \dfrac{2}{\pi} \displaystyle\sum_{n=1}^{\infty} \dfrac{1 - (-1)^n}{n \sinh n\pi} \sinh ny \sin nx$

$u_2(x, y) = \dfrac{2}{\pi} \displaystyle\sum_{n=1}^{\infty} \dfrac{[1 - (-1)^n]}{n}$

$\times \dfrac{\sinh nx + \sinh n(\pi - x)}{\sinh n\pi} \sin ny$

17. max temperature is $u = 1$

Exercises 13.6, Page 710

1. $u(x, t) = 100 + \dfrac{200}{\pi} \displaystyle\sum_{n=1}^{\infty} \dfrac{(-1)^n - 1}{n} e^{-kn^2\pi^2 t} \sin n\pi x$

3. $u(x, t) = u_0 - \dfrac{r}{2k} x(x - 1) + 2 \displaystyle\sum_{n=1}^{\infty} \left[\dfrac{u_0}{n\pi} + \dfrac{r}{kn^3\pi^3} \right]$

$\times [(-1)^n - 1] e^{-kn^2\pi^2 t} \sin n\pi x$

5. $u(x, t) = \psi(x) + \displaystyle\sum_{n=1}^{\infty} A_n e^{-kn^2\pi^2 t} \sin n\pi x,$

where $\psi(x) = \dfrac{A}{k\beta^2} [-e^{-\beta x} + (e^{-\beta} - 1)x + 1]$

$A_n = 2 \displaystyle\int_0^1 [f(x) - \psi(x)] \sin n\pi x \, dx$

7. $\psi(x) = u_0 \left[1 - \dfrac{\sinh \sqrt{h/k}\, x}{\sinh \sqrt{h/k}} \right]$

9. $u(x, t) = \dfrac{A}{6a^2}(x - x^3)$

$+ \dfrac{2A}{a^2\pi^3} \displaystyle\sum_{n=1}^{\infty} \dfrac{(-1)^n}{n^3} \cos n\pi at \sin n\pi x$

11. $u(x, y) = (u_0 - u_1)\, y + u_1$

$+ \dfrac{2}{\pi} \displaystyle\sum_{n=1}^{\infty} \dfrac{u_0(-1)^n - u_1}{n} e^{-n\pi x} \sin n\pi y$

Exercises 13.7, Page 714

1. $u(x, t) = 2h \displaystyle\sum_{n=1}^{\infty} \dfrac{\sin \lambda_n}{\lambda_n[h + \sin^2\lambda_n]} e^{-k\lambda_n^2 t} \cos \lambda_n x,$ where the λ_n

are the consecutive positive roots of $\cot \lambda = \lambda/h$

3. $u(x, y) = \displaystyle\sum_{n=1}^{\infty} A_n \sinh \lambda_n y \sin \lambda_n x,$ where

$A_n = \dfrac{2h}{\sinh \lambda_n b[ah + \cos^2\lambda_n a]} \displaystyle\int_0^a f(x) \sin \lambda_n x \, dx$ and the λ_n are

the consecutive positive roots of $\tan \lambda a = -\lambda/h$

5. $u(x, t) = \displaystyle\sum_{n=1}^{\infty} A_n e^{-k(2n-1)^2\pi^2 t/4L^2} \sin \left(\dfrac{2n-1}{2L} \right) \pi x,$

where $A_n = \dfrac{2}{L} \displaystyle\int_0^L f(x) \sin \left(\dfrac{2n-1}{2L} \right) \pi x \, dx$

7. $u(x, y) = \dfrac{4u_0}{\pi} \displaystyle\sum_{n=1}^{\infty} \dfrac{1}{(2n-1) \cosh \left(\dfrac{2n-1}{2} \right) \pi}$

$\times \cosh \left(\dfrac{2n-1}{2} \right) \pi x \sin \left(\dfrac{2n-1}{2} \right) \pi y$

9. (b) 1.8751, 4.6941

Exercises 13.8, Page 717

1. $u(x, y, t) = \displaystyle\sum_{m=1}^{\infty} \sum_{n=1}^{\infty} A_{mn} e^{-k(m^2 + n^2)t} \sin mx \sin ny,$

where $A_{mn} = \dfrac{4u_0}{mn\pi^2} [1 - (-1)^m][1 - (-1)^n]$

3. $u(x, y, t) = \displaystyle\sum_{m=1}^{\infty} \sum_{n=1}^{\infty} A_{mn} \sin mx \sin ny \cos a\sqrt{m^2 + n^2}\, t,$

where $A_{mn} = \dfrac{16}{m^3 n^3 \pi^2} [(-1)^m - 1][(-1)^n - 1]$

5. $u(x, y, z) = \displaystyle\sum_{m=1}^{\infty} \sum_{n=1}^{\infty} A_{mn} \sinh \omega_{mn} z \sin \dfrac{m\pi}{a} x \sin \dfrac{n\pi}{b} y,$

where $\omega_{mn} = \sqrt{(m\pi/a)^2 + (n\pi/b)^2}$

$A_{mn} = \dfrac{4}{ab \sinh(c\omega_{mn})} \displaystyle\int_0^b \int_0^a f(x, y)$

$\times \sin \dfrac{m\pi}{a} x \sin \dfrac{n\pi}{b} y \, dx \, dy$

7. Use $a = b = c = 1$ with $f(x, y) = u_0$ in Problem 5 and $f(x, y) = -u_0$ in Problem 6. Add the two solutions.

Chapter 13 Review Exercises, Page 717

1. $u = c_1 e^{(c_2 x + y/c_2)}$

3. $\psi(x) = u_0 + \dfrac{(u_1 - u_0)}{1 + \pi} x$

5. $u(x, t) = \dfrac{2h}{\pi^2 a} \sum\limits_{n=1}^{\infty} \dfrac{\left(\cos \dfrac{n\pi}{4} - \cos \dfrac{3n\pi}{4} \right)}{n^2}$

$\times \sin n\pi at \sin n\pi x$

7. $u(x, y) = \dfrac{100}{\pi} \sum\limits_{n=1}^{\infty} \dfrac{1 - (-1)^n}{n \sinh n\pi} \sinh nx \sin ny$

9. $u(x, y) = \dfrac{100}{\pi} \sum\limits_{n=1}^{\infty} \dfrac{1 - (-1)^n}{n} e^{-nx} \sin ny$

11. $u(x, t) = e^{-t} \sin x$

13. $u(x, t) = e^{-(x+t)} \sum\limits_{n=1}^{\infty} A_n \left[\sqrt{n^2 + 1} \cos \sqrt{n^2 + 1} \, t \right.$

$\left. + \sin \sqrt{n^2 + 1} \, t \right] \sin nx$

Exercises 14.1, Page 723

1. $u(r, \theta) = \dfrac{u_0}{2} + \dfrac{u_0}{\pi} \sum\limits_{n=1}^{\infty} \dfrac{1 - (-1)^n}{n} r^n \sin n\theta$

3. $u(r, \theta) = \dfrac{2\pi^2}{3} - 4 \sum\limits_{n=1}^{\infty} \dfrac{r^n}{n^2} \cos n\theta$

5. $u(r, \theta) = A_0 + \sum\limits_{n=1}^{\infty} r^{-n}(A_n \cos n\theta + B_n \sin n\theta)$,

where $A_0 = \dfrac{1}{2\pi} \displaystyle\int_0^{2\pi} f(\theta) \, d\theta$

$A_n = \dfrac{c^n}{\pi} \displaystyle\int_0^{2\pi} f(\theta) \cos n\theta \, d\theta$

$B_n = \dfrac{c^n}{\pi} \displaystyle\int_0^{2\pi} f(\theta) \sin n\theta \, d\theta$

7. $u(r, \theta) = \dfrac{1}{2} + \dfrac{2}{\pi} \sum\limits_{n=1}^{\infty} \dfrac{\sin \dfrac{n\pi}{2}}{n} \left(\dfrac{r}{c} \right)^{2n} \cos 2n\theta$

9. $u(r, \theta) = A_0 \ln \left(\dfrac{r}{b} \right) + \sum\limits_{n=1}^{\infty} \left[\left(\dfrac{b}{r} \right)^n - \left(\dfrac{r}{b} \right)^n \right]$

$\times [A_n \cos n\theta + B_n \sin n\theta]$,

where $A_0 \ln \left(\dfrac{a}{b} \right) = \dfrac{1}{2\pi} \displaystyle\int_0^{2\pi} f(\theta) \, d\theta$

$\left[\left(\dfrac{b}{a} \right)^n - \left(\dfrac{a}{b} \right)^n \right] A_n = \dfrac{1}{\pi} \displaystyle\int_0^{2\pi} f(\theta) \cos n\theta \, d\theta$

$\left[\left(\dfrac{b}{a} \right)^n - \left(\dfrac{a}{b} \right)^n \right] B_n = \dfrac{1}{\pi} \displaystyle\int_0^{2\pi} f(\theta) \sin n\theta \, d\theta$

11. $u(r, \theta) = \dfrac{4}{\pi} \sum\limits_{n=1}^{\infty} \dfrac{1 - (-1)^n}{n^3} \dfrac{r^{2n} - b^{2n}}{a^{2n} - b^{2n}} \left(\dfrac{a}{r} \right)^n \sin n\theta$

13. $u(r, \theta) = \dfrac{u_0}{2} + \dfrac{2u_0}{\pi} \sum\limits_{n=1}^{\infty} \dfrac{\sin \dfrac{n\pi}{2}}{n} \left(\dfrac{r}{2} \right)^n \cos n\theta$

Exercises 14.2, Page 728

1. $u(r, t) = \dfrac{2}{ac} \sum\limits_{n=1}^{\infty} \dfrac{\sin \lambda_n at \, J_0(\lambda_n r)}{\lambda_n^2 J_1(\lambda_n c)}$

3. $u(r, z) = u_0 \sum\limits_{n=1}^{\infty} \dfrac{\sinh \lambda_n(4 - z) J_0(\lambda_n r)}{\lambda_n \sinh 4\lambda_n J_1(2\lambda_n)}$

5. $u(r, t) = \sum\limits_{n=1}^{\infty} A_n J_0(\lambda_n r) e^{-k\lambda_n^2 t}$,

where $A_n = \dfrac{2}{c^2 J_1^2(\lambda_n c)} \displaystyle\int_0^c r J_0(\lambda_n r) f(r) \, dr$

7. $u(r, t) = \sum\limits_{n=1}^{\infty} A_n J_0(\lambda_n r) e^{-k\lambda_n^2 t}$,

where $A_n = \dfrac{2\lambda_n^2}{(\lambda_n^2 + h^2) J_0^2(\lambda_n)} \displaystyle\int_0^1 r J_0(\lambda_n r) \, f(r) \, dr$

9. $u(r, t) = 100 + 50 \sum\limits_{n=1}^{\infty} \dfrac{J_1(\lambda_n) J_0(\lambda_n r)}{\lambda_n J_1^2(2\lambda_n)} e^{-\lambda_n^2 t}$

11. (b) $u(x, t) = \sum\limits_{n=1}^{\infty} A_n \cos(\lambda_n \sqrt{gt}) J_0(2\lambda_n \sqrt{x})$,

where $A_n = \dfrac{2}{LJ_1^2(2\lambda_n \sqrt{L})} \displaystyle\int_0^{\sqrt{L}} v J_0(2\lambda_n v) \, f(v^2) \, dv$

Exercises 14.3, Page 731

1. $u(r, \theta) = 50 \left[\dfrac{1}{2} P_0(\cos \theta) + \dfrac{3}{4} \left(\dfrac{r}{c} \right) P_1(\cos \theta) \right.$

$\left. - \dfrac{7}{16} \left(\dfrac{r}{c} \right)^3 P_3(\cos \theta) + \dfrac{11}{32} \left(\dfrac{r}{c} \right)^5 P_5(\cos \theta) + \cdots \right]$

3. $u(r, \theta) = \dfrac{r}{c} \cos \theta$

5. $u(r, \theta) = \sum\limits_{n=1}^{\infty} A_n \dfrac{b^{2n+1} - r^{2n+1}}{b^{2n+1} r^{n+1}} P_n(\cos \theta)$, where

$\dfrac{b^{2n+1} - a^{2n+1}}{b^{2n+1} a^{n+1}} A_n = \dfrac{2n + 1}{2} \displaystyle\int_0^\pi f(\theta) P_n(\cos \theta) \sin \theta \, d\theta$

7. $u(r, \theta) = \sum\limits_{n=0}^{\infty} A_n r^{2n} P_{2n}(\cos \theta)$,

where $A_n = \dfrac{(4n + 1)}{c^{2n}} \displaystyle\int_0^{\pi/2} f(\theta) P_{2n}(\cos \theta) \sin \theta \, d\theta$

9. $u(r, t) = 100 + \dfrac{200}{\pi r} \sum\limits_{n=1}^{\infty} \dfrac{(-1)^n}{n} e^{-n^2 \pi^2 t} \sin n\pi r$

11. $u(r, t) = \sum_{n=1}^{\infty} \left(A_n \cos \frac{n\pi a}{c} t + B_n \sin \frac{n\pi a}{c} t \right) \sin \frac{n\pi}{c} r,$

where $A_n = \frac{2}{c} \int_0^c rf(r) \sin \frac{n\pi}{c} r \, dr$

$B_n = \frac{2}{n\pi a} \int_0^c rg(r) \sin \frac{n\pi}{c} r \, dr$

Chapter 14 Review Exercises, Page 733

1. $u(r, \theta) = \frac{2u_0}{\pi} \sum_{n=1}^{\infty} \frac{1-(-1)^n}{n} \left(\frac{r}{c} \right)^n \sin n\theta$

3. $u(r, \theta) = \frac{4u_0}{\pi} \sum_{n=1}^{\infty} \frac{1-(-1)^n}{n^3} \sin n\theta$

5. $u(r, \theta) = \frac{2u_0}{\pi} \sum_{n=1}^{\infty} \frac{r^{4n} + r^{-4n}}{2^{4n} + 2^{-4n}} \frac{1-(-1)^n}{n} \sin 4n\theta$

7. $u(r, t) = 2e^{-ht} \sum_{n=1}^{\infty} \frac{J_0(\lambda_n r)}{\lambda_n J_1(\lambda_n)} e^{-\lambda_n^2 t}$

9. $u(r, z) = 50 \sum_{n=1}^{\infty} \frac{\cosh \lambda_n z J_0(\lambda_n r)}{\lambda_n \cosh 4\lambda_n J_1(2\lambda_n)}$

11. $u(r, \theta) = 100 \left[\frac{3}{2} rP_1(\cos \theta) - \frac{7}{8} r^3 P_3(\cos \theta) \right.$

$\left. + \frac{11}{16} r^5 P_5(\cos \theta) + \cdots \right]$

Exercises 15.1, Page 737

1. (a) Let $\tau = u^2$ in the integral erf (\sqrt{t}).

7. $y(t) = e^{\pi t} \, \text{erf}(\sqrt{\pi t})$

9. Use the property $\int_0^b - \int_0^a = \int_0^b + \int_a^0.$

Exercises 15.2, Page 742

1. $u(x, t) = A \cos \frac{a\pi t}{L} \sin \frac{\pi x}{L}$

3. $u(x, t) = f\left(t - \frac{x}{a}\right) \mathcal{U}\left(t - \frac{x}{a}\right)$

5. $u(x, t) = \left[\frac{1}{2} g\left(t - \frac{x}{a}\right)^2 + A \sin \omega\left(t - \frac{x}{a}\right) \right]$

$\times \mathcal{U}\left(t - \frac{x}{a}\right) - \frac{1}{2} gt^2$

7. $u(x, t) = a \frac{F_0}{E} \sum_{n=0}^{\infty} (-1)^n \left\{ \left(t - \frac{2nL + L - x}{a}\right) \right.$

$\times \mathcal{U}\left(t - \frac{2nL + L - x}{a}\right)$

$\left. - \left(t - \frac{2nL + L + x}{a}\right) \times \mathcal{U}\left(t - \frac{2nL + L + x}{a}\right) \right\}$

9. $u(x, t) = (t - x) \sinh(t - x) \, \mathcal{U}(t - x)$

$+ xe^{-x} \cosh t - e^{-x}t \sinh t$

13. $u(x, t) = u_1 + (u_0 - u_1) \, \text{erfc}\left(\frac{x}{2\sqrt{t}}\right)$

15. $u(x, t) = u_0 \left[1 - \left\{ \text{erfc}\left(\frac{x}{2\sqrt{t}}\right) \right.\right.$

$\left.\left. - e^{x+1} \, \text{erfc}\left(\sqrt{t} + \frac{x}{2\sqrt{t}}\right) \right\} \right]$

17. $u(x, t) = \frac{x}{2\sqrt{\pi}} \int_0^t \frac{f(t - \tau)}{\tau^{3/2}} e^{-x^2/4\tau} \, d\tau$

19. $u(x, t) = 60 + 40 \, \text{erfc}\left(\frac{x}{2\sqrt{t-2}}\right) \mathcal{U}(t - 2)$

21. $u(x, t) = 100 \left[-e^{1-x+t} \, \text{erfc}\left(\sqrt{t} + \frac{1-x}{2\sqrt{t}}\right) \right.$

$\left. + \text{erfc}\left(\frac{1-x}{2\sqrt{t}}\right) \right]$

23. $u(x, t) = u_0 + u_0 e^{-(\pi^2/L^2)t} \sin\left(\frac{\pi}{L} x\right)$

25. $u(x, t) = u_0 - u_0 \sum_{n=0}^{\infty} (-1)^n \left[\text{erfc}\left(\frac{2n + 1 - x}{2\sqrt{kt}}\right) \right.$

$\left. + \text{erfc}\left(\frac{2n + 1 + x}{2\sqrt{kt}}\right) \right]$

27. $u(x, t) = u_0 e^{-Gt/C} \, \text{erf}\left(\frac{x}{2} \sqrt{\frac{RC}{t}}\right)$

Exercises 15.3, Page 750

1. $f(x) = \frac{1}{\pi} \int_0^{\infty} \frac{\sin \alpha \cos \alpha x + 3(1 - \cos \alpha) \sin \alpha x}{\alpha} \, d\alpha$

3. $f(x) = \frac{1}{\pi} \int_0^{\infty} [A(\alpha) \cos \alpha x + B(\alpha) \sin \alpha x] \, d\alpha,$

where $A(\alpha) = \frac{3\alpha \sin 3\alpha + \cos 3\alpha - 1}{\alpha^2}$

$B(\alpha) = \frac{\sin 3\alpha - 3\alpha \cos 3\alpha}{\alpha^2}$

5. $f(x) = \frac{1}{\pi} \int_0^{\infty} \frac{\cos \alpha x + \alpha \sin \alpha x}{1 + \alpha^2} \, d\alpha$

7. $f(x) = \frac{10}{\pi} \int_0^{\infty} \frac{(1 - \cos \alpha) \sin \alpha x}{\alpha} \, d\alpha$

9. $f(x) = \frac{2}{\pi} \int_0^{\infty} \frac{(\pi\alpha \sin \pi\alpha + \cos \pi\alpha - 1) \cos \alpha x}{\alpha^2} \, d\alpha$

11. $f(x) = \frac{4}{\pi} \int_0^{\infty} \frac{\alpha \sin \alpha x}{4 + \alpha^4} \, d\alpha$

13. $f(x) = \dfrac{2k}{\pi} \displaystyle\int_0^\infty \dfrac{\cos \alpha x}{k^2 + \alpha^2}\, d\alpha$

$f(x) = \dfrac{2}{\pi} \displaystyle\int_0^\infty \dfrac{\alpha \sin \alpha x}{k^2 + \alpha^2}\, d\alpha$

15. $f(x) = \dfrac{2}{\pi} \displaystyle\int_0^\infty \dfrac{(4 - \alpha^2) \cos \alpha x}{(4 + \alpha^2)^2}\, d\alpha$

$f(x) = \dfrac{8}{\pi} \displaystyle\int_0^\infty \dfrac{\alpha \sin \alpha x}{(4 + \alpha^2)^2}\, d\alpha$

17. $f(x) = \dfrac{2}{\pi}\, \dfrac{1}{1 + x^2},\ x > 0$

19. Let $x = 2$ in (7). Use a trigonometric identity and replace α by x. In part b make the change of variable $2x = kt$.

Exercises 15.4, Page 754

1. $u(x, t) = \dfrac{1}{\pi} \displaystyle\int_{-\infty}^\infty \dfrac{e^{-k\alpha^2 t}}{1 + \alpha^2}\, e^{-i\alpha x}\, d\alpha$

$= \dfrac{1}{\pi} \displaystyle\int_{-\infty}^\infty \dfrac{\cos \alpha x}{1 + \alpha^2}\, e^{-k\alpha^2 t}\, d\alpha$

3. $u(x, t) = \dfrac{1}{\sqrt{1 + 4kt}}\, e^{-x^2/(1 + 4kt)}$

7. $u(x, t) = \dfrac{2}{\pi} \displaystyle\int_0^\infty \dfrac{1 - \cos \alpha}{\alpha}\, e^{-k\alpha^2 t} \sin \alpha x\, d\alpha$

9. $u(x, t) = \dfrac{2}{\pi} \displaystyle\int_0^\infty \dfrac{\sin \alpha}{\alpha}\, e^{-k\alpha^2 t} \cos \alpha x\, d\alpha$

11. (a) $u(x, t) = \dfrac{1}{2\pi} \displaystyle\int_{-\infty}^\infty \left(F(\alpha) \cos \alpha a t + G(\alpha)\, \dfrac{\sin \alpha a t}{\alpha a}\right) e^{-i\alpha x}\, d\alpha$

13. $u(x, y) = \dfrac{2}{\pi} \displaystyle\int_0^\infty \dfrac{\sinh \alpha(\pi - x)}{(1 + \alpha^2) \sinh \alpha \pi}\, \cos \alpha y\, d\alpha$

15. $u(x, y) = \dfrac{100}{\pi} \displaystyle\int_0^\infty \dfrac{\sin \alpha}{\alpha}\, e^{-\alpha y} \cos \alpha x\, d\alpha$

17. $u(x, y) = \dfrac{2}{\pi} \displaystyle\int_0^\infty F(\alpha)\, \dfrac{\sinh \alpha(2 - y)}{\sinh 2\alpha}\, \sin \alpha x\, d\alpha$

19. $u(x, y) = \dfrac{2}{\pi} \displaystyle\int_0^\infty \dfrac{\alpha}{1 + \alpha^2}\, [e^{-\alpha x} \sin \alpha y + e^{-\alpha y} \sin \alpha x]\, d\alpha$

21. $u(x, y) = \dfrac{1}{2\sqrt{\pi}} \displaystyle\int_{-\infty}^\infty \dfrac{e^{-\alpha^2/4} \cosh \alpha y}{\cosh \alpha}\, e^{-i\alpha x}\, d\alpha$

$= \dfrac{1}{2\sqrt{\pi}} \displaystyle\int_{-\infty}^\infty \dfrac{e^{-\alpha^2/4} \cosh \alpha y}{\cosh \alpha}\, \cos \alpha x\, d\alpha$

Exercises 15.5, Page 765 (See bottom of page)

Chapter 15 Review Exercises, Page 765

1. $u(x, y) = \dfrac{2}{\pi} \displaystyle\int_0^\infty \dfrac{\sinh \alpha y}{\alpha(1 + \alpha^2) \cosh \alpha \pi}\, \cos \alpha x\, d\alpha$

3. $u(x, t) = u_0 e^{-ht}\, \text{erf}\left(\dfrac{x}{2\sqrt{t}}\right)$

5. $u(x, t) = \displaystyle\int_0^t \text{erfc}\left(\dfrac{x}{2\sqrt{\tau}}\right) d\tau$

7. $u(x, t) = \dfrac{u_0}{2\pi} \displaystyle\int_{-\infty}^\infty \dfrac{\sin \alpha(\pi - x) + \sin \alpha x}{\alpha}\, e^{-k\alpha^2 t}\, d\alpha$

9. $u(x, y) = \dfrac{100}{\pi} \displaystyle\int_0^\infty \left(\dfrac{1 - \cos \alpha}{\alpha}\right)$

$\qquad \times [e^{-\alpha x} \sin \alpha y + 2e^{-\alpha y} \sin \alpha x]\, d\alpha$

11. $u(x, y) = \dfrac{2}{\pi} \displaystyle\int_0^\infty \left(\dfrac{B \cosh \alpha y}{(1 + \alpha^2) \sinh \alpha \pi} + \dfrac{A}{\alpha}\right) \sin \alpha x\, d\alpha$

13. $u(x, y) = \dfrac{1}{2\pi} \displaystyle\int_{-\infty}^\infty \dfrac{\cos \alpha x + \alpha \sin \alpha x}{1 + \alpha^2}\, e^{-k\alpha^2 t}\, d\alpha$

Exercises 16.1, Page 773

1. $u_{11} = {}^{11}\!/_{15},\ u_{21} = {}^{14}\!/_{15}$

3. $u_{11} = u_{21} = \sqrt{3}/16,$
$\quad u_{22} = u_{12} = 3\sqrt{3}/16$

5. $u_{21} = u_{12} = 12.50,$
$\quad u_{31} = u_{13} = 18.75,$
$\quad u_{32} = u_{23} = 37.50,$
$\quad u_{11} = 6.25,\ u_{22} = 25.00,\ u_{33} = 56.25$

3. 1

7.

$F_8 = \begin{pmatrix} 1 & 1 & 1 & 1 & 1 & 1 & 1 & 1 \\ 1 & \sqrt{2}/2 + i\sqrt{2}/2 & i & -\sqrt{2}/2 + i\sqrt{2}/2 & -1 & -\sqrt{2}/2 - i\sqrt{2}/2 & -i & \sqrt{2}/2 - i\sqrt{2}/2 \\ 1 & i & -1 & -i & 1 & i & -1 & -i \\ 1 & -\sqrt{2}/2 + i\sqrt{2}/2 & -i & \sqrt{2}/2 + i\sqrt{2}/2 & -1 & \sqrt{2}/2 - i\sqrt{2}/2 & i & -\sqrt{2}/2 - i\sqrt{2}/2 \\ 1 & -1 & 1 & -1 & 1 & -1 & 1 & -1 \\ 1 & -\sqrt{2}/2 - i\sqrt{2}/2 & i & \sqrt{2}/2 - i\sqrt{2}/2 & -1 & \sqrt{2}/2 + i\sqrt{2}/2 & -i & -\sqrt{2}/2 + i\sqrt{2}/2 \\ 1 & -i & -1 & i & 1 & -i & -1 & i \\ 1 & \sqrt{2}/2 - i\sqrt{2}/2 & -i & -\sqrt{2}/2 - i\sqrt{2}/2 & -1 & -\sqrt{2}/2 + i\sqrt{2}/2 & i & \sqrt{2}/2 + i\sqrt{2}/2 \end{pmatrix}$

7. (b) $u_{14} = u_{41} = 0.5427$,
$\quad u_{24} = u_{42} = 0.6707$,
$\quad u_{34} = u_{43} = 0.6402$,
$\quad u_{33} = 0.4451$, $u_{44} = 0.9451$

Exercises 16.2, Page 778

1.

Time	$x = 0.25$	$x = 0.50$	$x = 0.75$	$x = 1.00$	$x = 1.25$	$x = 1.50$	$x = 1.75$
0.000	1.0000	1.0000	1.0000	1.0000	0.0000	0.0000	0.0000
0.025	0.6000	1.0000	1.0000	0.6000	0.4000	0.0000	0.0000
0.050	0.5200	0.8400	0.8400	0.6800	0.3200	0.1600	0.0000
0.075	0.4400	0.7120	0.7760	0.6000	0.4000	0.1600	0.0640
0.100	0.3728	0.6288	0.6800	0.5904	0.3840	0.2176	0.0768
0.125	0.3261	0.5469	0.6237	0.5437	0.4000	0.2278	0.1024
0.150	0.2840	0.4893	0.5610	0.5182	0.3886	0.2465	0.1116
0.175	0.2525	0.4358	0.5152	0.4835	0.3836	0.2494	0.1209
0.200	0.2248	0.3942	0.4708	0.4562	0.3699	0.2517	0.1239
0.225	0.2027	0.3571	0.4343	0.4275	0.3571	0.2479	0.1255
0.250	0.1834	0.3262	0.4007	0.4021	0.3416	0.2426	0.1242
0.275	0.1672	0.2989	0.3715	0.3773	0.3262	0.2348	0.1219
0.300	0.1530	0.2752	0.3448	0.3545	0.3101	0.2262	0.1183
0.325	0.1407	0.2541	0.3209	0.3329	0.2943	0.2166	0.1141
0.350	0.1298	0.2354	0.2990	0.3126	0.2787	0.2067	0.1095
0.375	0.1201	0.2186	0.2790	0.2936	0.2635	0.1966	0.1046
0.400	0.1115	0.2034	0.2607	0.2757	0.2488	0.1865	0.0996
0.425	0.1036	0.1895	0.2438	0.2589	0.2347	0.1766	0.0945
0.450	0.0965	0.1769	0.2281	0.2432	0.2211	0.1670	0.0896
0.475	0.0901	0.1652	0.2136	0.2283	0.2083	0.1577	0.0847
0.500	0.0841	0.1545	0.2002	0.2144	0.1961	0.1487	0.0800
0.525	0.0786	0.1446	0.1876	0.2014	0.1845	0.1402	0.0755
0.550	0.0736	0.1354	0.1759	0.1891	0.1735	0.1320	0.0712
0.575	0.0689	0.1269	0.1650	0.1776	0.1632	0.1243	0.0670
0.600	0.0645	0.1189	0.1548	0.1668	0.1534	0.1169	0.0631
0.625	0.0605	0.1115	0.1452	0.1566	0.1442	0.1100	0.0594
0.650	0.0567	0.1046	0.1363	0.1471	0.1355	0.1034	0.0559
0.675	0.0532	0.0981	0.1279	0.1381	0.1273	0.0972	0.0525
0.700	0.0499	0.0921	0.1201	0.1297	0.1196	0.0914	0.0494
0.725	0.0468	0.0864	0.1127	0.1218	0.1124	0.0859	0.0464
0.750	0.0439	0.0811	0.1058	0.1144	0.1056	0.0807	0.0436
0.775	0.0412	0.0761	0.0994	0.1074	0.0992	0.0758	0.0410
0.800	0.0387	0.0715	0.0933	0.1009	0.0931	0.0712	0.0385
0.825	0.0363	0.0671	0.0876	0.0948	0.0875	0.0669	0.0362
0.850	0.0341	0.0630	0.0823	0.0890	0.0822	0.0628	0.0340
0.875	0.0320	0.0591	0.0772	0.0836	0.0772	0.0590	0.0319
0.900	0.0301	0.0555	0.0725	0.0785	0.0725	0.0554	0.0300
0.925	0.0282	0.0521	0.0681	0.0737	0.0681	0.0521	0.0282
0.950	0.0265	0.0490	0.0640	0.0692	0.0639	0.0489	0.0265
0.975	0.0249	0.0460	0.0601	0.0650	0.0600	0.0459	0.0249
1.000	0.0234	0.0432	0.0564	0.0610	0.0564	0.0431	0.0233

3.

Time	$x = 0.25$	$x = 0.50$	$x = 0.75$	$x = 1.00$	$x = 1.25$	$x = 1.50$	$x = 1.75$
0.000	1.0000	1.0000	1.0000	1.0000	0.0000	0.0000	0.0000
0.025	0.7074	0.9520	0.9566	0.7444	0.2545	0.0371	0.0053
0.050	0.5606	0.8499	0.8685	0.6633	0.3303	0.1034	0.0223
0.075	0.4684	0.7473	0.7836	0.6191	0.3614	0.1529	0.0462
0.100	0.4015	0.6577	0.7084	0.5837	0.3753	0.1871	0.0684
0.125	0.3492	0.5821	0.6428	0.5510	0.3797	0.2101	0.0861
0.150	0.3069	0.5187	0.5857	0.5199	0.3778	0.2247	0.0990
0.175	0.2721	0.4652	0.5359	0.4901	0.3716	0.2329	0.1078
0.200	0.2430	0.4198	0.4921	0.4617	0.3622	0.2362	0.1132
0.225	0.2186	0.3809	0.4533	0.4348	0.3507	0.2358	0.1160
0.250	0.1977	0.3473	0.4189	0.4093	0.3378	0.2327	0.1166
0.275	0.1798	0.3181	0.3881	0.3853	0.3240	0.2275	0.1157
0.300	0.1643	0.2924	0.3604	0.3626	0.3097	0.2208	0.1136
0.325	0.1507	0.2697	0.3353	0.3412	0.2953	0.2131	0.1107
0.350	0.1387	0.2495	0.3125	0.3211	0.2808	0.2047	0.1071
0.375	0.1281	0.2313	0.2916	0.3021	0.2666	0.1960	0.1032
0.400	0.1187	0.2150	0.2725	0.2843	0.2528	0.1871	0.0989
0.425	0.1102	0.2002	0.2549	0.2675	0.2393	0.1781	0.0946
0.450	0.1025	0.1867	0.2387	0.2517	0.2263	0.1692	0.0902
0.475	0.0955	0.1743	0.2236	0.2368	0.2139	0.1606	0.0858
0.500	0.0891	0.1630	0.2097	0.2228	0.2020	0.1521	0.0814
0.525	0.0833	0.1525	0.1967	0.2096	0.1906	0.1439	0.0772
0.550	0.0779	0.1429	0.1846	0.1973	0.1798	0.1361	0.0731
0.575	0.0729	0.1339	0.1734	0.1856	0.1696	0.1285	0.0691
0.600	0.0683	0.1256	0.1628	0.1746	0.1598	0.1214	0.0653
0.625	0.0641	0.1179	0.1530	0.1643	0.1506	0.1145	0.0617
0.650	0.0601	0.1106	0.1438	0.1546	0.1419	0.1080	0.0582
0.675	0.0564	0.1039	0.1351	0.1455	0.1336	0.1018	0.0549
0.700	0.0530	0.0976	0.1270	0.1369	0.1259	0.0959	0.0518
0.725	0.0497	0.0917	0.1194	0.1288	0.1185	0.0904	0.0488
0.750	0.0467	0.0862	0.1123	0.1212	0.1116	0.0852	0.0460
0.775	0.0439	0.0810	0.1056	0.1140	0.1050	0.0802	0.0433
0.800	0.0413	0.0762	0.0993	0.1073	0.0989	0.0755	0.0408
0.825	0.0388	0.0716	0.0934	0.1009	0.0931	0.0711	0.0384
0.850	0.0365	0.0674	0.0879	0.0950	0.0876	0.0669	0.0362
0.875	0.0343	0.0633	0.0827	0.0894	0.0824	0.0630	0.0341
0.900	0.0323	0.0596	0.0778	0.0841	0.0776	0.0593	0.0321
0.925	0.0303	0.0560	0.0732	0.0791	0.0730	0.0558	0.0302
0.950	0.0285	0.0527	0.0688	0.0744	0.0687	0.0526	0.0284
0.975	0.0268	0.0496	0.0647	0.0700	0.0647_	0.0495	0.0268
1.000	0.0253	0.0466	0.0609	0.0659	0.0608	0.0465	0.0252

Absolute errors are approximately 2.2×10^{-2}, 3.7×10^{-2}, 1.3×10^{-2}.

5.

Time	$x = 0.25$	$x = 0.50$	$x = 0.75$	$x = 1.00$	$x = 1.25$	$x = 1.50$	$x = 1.75$
0.00	1.0000	1.0000	1.0000	1.0000	0.0000	0.0000	0.0000
0.05	0.5265	0.8693	0.8852	0.6141	0.3783	0.0884	0.0197
0.10	0.3972	0.6551	0.7043	0.5883	0.3723	0.1955	0.0653
0.15	0.3042	0.5150	0.5844	0.5192	0.3812	0.2261	0.1010
0.20	0.2409	0.4171	0.4901	0.4620	0.3636	0.2385	0.1145
0.25	0.1962	0.3452	0.4174	0.4092	0.3391	0.2343	0.1178
0.30	0.1631	0.2908	0.3592	0.3624	0.3105	0.2220	0.1145
0.35	0.1379	0.2482	0.3115	0.3208	0.2813	0.2056	0.1077
0.40	0.1181	0.2141	0.2718	0.2840	0.2530	0.1876	0.0993
0.45	0.1020	0.1860	0.2381	0.2514	0.2265	0.1696	0.0904
0.50	0.0888	0.1625	0.2092	0.2226	0.2020	0.1523	0.0816
0.55	0.0776	0.1425	0.1842	0.1970	0.1798	0.1361	0.0732
0.60	0.0681	0.1253	0.1625	0.1744	0.1597	0.1214	0.0654
0.65	0.0599	0.1104	0.1435	0.1544	0.1418	0.1079	0.0582
0.70	0.0528	0.0974	0.1268	0.1366	0.1257	0.0959	0.0518
0.75	0.0466	0.0860	0.1121	0.1210	0.1114	0.0851	0.0460
0.80	0.0412	0.0760	0.0991	0.1071	0.0987	0.0754	0.0408
0.85	0.0364	0.0672	0.0877	0.0948	0.0874	0.0668	0.0361
0.90	0.0322	0.0594	0.0776	0.0839	0.0774	0.0592	0.0320
0.95	0.0285	0.0526	0.0687	0.0743	0.0686	0.0524	0.0284
1.00	0.0252	0.0465	0.0608	0.0657	0.0607	0.0464	0.0251

Absolute errors are approximately 1.8×10^{-2}, 3.7×10^{-2}, 1.3×10^{-2}.

7.

(a) Time	$x = 2.00$	$x = 4.00$	$x = 6.00$	$x = 8.00$	$x = 10.00$	$x = 12.00$	$x = 14.00$	$x = 16.00$	$x = 18.00$
0.00	30.0000	30.0000	30.0000	30.0000	30.0000	30.0000	30.0000	30.0000	30.0000
1.00	28.7733	29.9749	29.9995	30.0000	30.0000	30.0000	29.9995	29.9749	28.7733
2.00	27.6450	29.9037	29.9970	29.9999	30.0000	29.9999	29.9970	29.9037	27.6450
3.00	26.6051	29.7938	29.9911	29.9997	30.0000	29.9997	29.9911	29.7938	26.6051
4.00	25.6452	29.6517	29.9805	29.9991	29.9999	29.9991	29.9805	29.6517	25.6452
5.00	24.7573	29.4829	29.9643	29.9981	29.9998	29.9981	29.9643	29.4829	24.7573
6.00	23.9347	29.2922	29.9421	29.9963	29.9996	29.9963	29.9421	29.2922	23.9347
7.00	23.1711	29.0836	29.9134	29.9936	29.9992	29.9936	29.9134	29.0836	23.1711
8.00	22.4612	28.8606	29.8782	29.9898	29.9986	29.9898	29.8782	28.8606	22.4612
9.00	21.7999	28.6263	29.8362	29.9848	29.9977	29.9848	29.8362	28.6263	21.7999
10.00	21.1829	28.3831	29.7878	29.9782	29.9964	29.9782	29.7878	28.3831	21.1829

(b)

Time	$x = 5.00$	$x = 10.00$	$x = 15.00$	$x = 20.00$	$x = 25.00$	$x = 30.00$	$x = 35.00$	$x = 40.00$	$x = 45.00$
0.00	30.0000	30.0000	30.0000	30.0000	30.0000	30.0000	30.0000	30.0000	30.0000
1.00	29.7968	29.9993	30.0000	30.0000	30.0000	30.0000	30.0000	29.9993	29.7968
2.00	29.5964	29.9973	30.0000	30.0000	30.0000	30.0000	30.0000	29.9973	29.5964
3.00	29.3987	29.9939	30.0000	30.0000	30.0000	30.0000	30.0000	29.9939	29.3987
4.00	29.2036	29.9893	29.9999	30.0000	30.0000	30.0000	29.9999	29.9893	29.2036
5.00	29.0112	29.9834	29.9998	30.0000	30.0000	30.0000	29.9998	29.9834	29.0112
6.00	28.8212	29.9762	29.9997	30.0000	30.0000	30.0000	29.9997	29.9762	28.8213
7.00	28.6339	29.9679	29.9995	30.0000	30.0000	30.0000	29.9995	29.9679	28.6339
8.00	28.4490	29.9585	29.9992	30.0000	30.0000	30.0000	29.9993	29.9585	28.4490
9.00	29.2665	29.9479	29.9989	30.0000	30.0000	30.0000	29.9989	29.9479	28.2665
10.00	28.0864	29.9363	29.9986	30.0000	30.0000	30.0000	29.9986	29.9363	28.0864

(c)

Time	$x = 2.00$	$x = 4.00$	$x = 6.00$	$x = 8.00$	$x = 10.00$	$x = 12.00$	$x = 14.00$	$x = 16.00$	$x = 18.00$
0.00	18.0000	32.0000	42.0000	48.0000	50.0000	48.0000	42.0000	32.0000	18.0000
1.00	16.4489	30.1970	40.1561	46.1495	48.1486	46.1495	40.1561	30.1970	16.4489
2.00	15.3312	28.5348	38.3465	44.3067	46.3001	44.3067	38.3465	28.5348	15.3312
3.00	14.4216	27.0416	36.6031	42.4847	44.4619	42.4847	36.6031	27.0416	14.4216
4.00	13.6371	25.6867	34.9416	40.6988	42.6453	40.6988	34.9416	25.6867	13.6371
5.00	12.9378	24.4419	33.3628	38.9611	40.8634	38.9611	33.3628	24.4419	12.9378
6.00	12.3012	23.2863	31.8624	37.2794	39.1273	37.2794	31.8624	23.2863	12.3012
7.00	11.7137	22.2051	30.4350	35.6578	37.4446	35.6578	30.4350	22.2051	11.7137
8.00	11.1659	21.1877	29.0757	34.0984	35.8202	34.0984	29.0757	21.1877	11.1659
9.00	10.6517	20.2261	27.7799	32.6014	34.2567	32.6014	27.7799	20.2261	10.6517
10.00	10.1665	19.3143	26.5439	31.1662	32.7549	31.1662	26.5439	19.3143	10.1665

(d)

Time	$x = 10.00$	$x = 20.00$	$x = 30.00$	$x = 40.00$	$x = 50.00$	$x = 60.00$	$x = 70.00$	$x = 80.00$	$x = 90.00$
0.00	8.0000	16.0000	24.0000	32.0000	40.0000	32.0000	24.0000	16.0000	8.0000
1.00	8.0000	16.0000	24.0000	31.9979	39.7425	31.9979	24.0000	16.0000	8.0000
2.00	8.0000	16.0000	23.9999	31.9918	39.4932	31.9918	23.9999	16.0000	8.0000
3.00	8.0000	16.0000	23.9997	31.9820	39.2517	31.9820	23.9997	16.0000	8.0000
4.00	8.0000	16.0000	23.9993	31.9686	39.0175	31.9686	23.9993	16.0000	8.0000
5.00	8.0000	16.0000	23.9987	31.9520	38.7905	31.9520	23.9987	16.0000	8.0000
6.00	8.0000	15.9999	23.9978	31.9323	38.5701	31.9323	23.9978	15.9999	8.0000
7.00	8.0000	15.9999	23.9966	31.9097	38.3561	31.9097	23.9966	15.9999	8.0000
8.00	8.0000	15.9998	23.9950	31.8844	38.1483	31.8844	23.9950	15.9998	8.0000
9.00	8.0000	15.9997	23.9931	31.8566	37.9463	31.8566	23.9931	15.9997	8.0000
10.00	8.0000	15.9996	23.9908	31.8265	37.7498	31.8265	23.9908	15.9996	8.0000

9.

(a)

Time	x = 2.00	x = 4.00	x = 6.00	x = 8.00	x = 10.00	x = 12.00	x = 14.00	x = 16.00	x = 18.00
0.00	30.0000	30.0000	30.0000	30.0000	30.0000	30.0000	30.0000	30.0000	30.0000
1.00	28.7733	29.9749	29.9995	30.0000	30.0000	30.0000	29.9998	29.9916	29.5911
2.00	27.6450	29.9037	29.9970	29.9999	30.0000	30.0000	29.9990	29.9679	29.2150
3.00	26.6051	29.7938	29.9911	29.9997	30.0000	29.9999	29.9970	29.9313	28.8684
4.00	25.6452	29.6517	29.9805	29.9991	30.0000	29.9997	29.9935	29.8839	28.5484
5.00	24.7573	29.4829	29.9643	29.9981	29.9999	29.9994	29.9881	29.8276	28.2524
6.00	23.9347	29.2922	29.9421	29.9963	29.9997	29.9988	29.9807	29.7641	27.9782
7.00	23.1711	29.0836	29.9134	29.9936	29.9995	29.9979	29.9711	29.6945	27.7237
8.00	22.4612	28.8606	29.8782	29.9899	29.9991	29.9966	29.9594	29.6202	27.4870
9.00	21.7999	28.6263	29.8362	29.9848	29.9985	29.9949	29.9454	29.5421	27.2666
10.00	21.1829	28.3831	29.7878	29.9783	29.9976	29.9927	29.9293	29.4610	27.0610

(b)

Time	x = 5.00	x = 10.00	x = 15.00	x = 20.00	x = 25.00	x = 30.00	x = 35.00	x = 40.00	x = 45.00
0.00	30.0000	30.0000	30.0000	30.0000	30.0000	30.0000	30.0000	30.0000	30.0000
1.00	29.7968	29.9993	30.0000	30.0000	30.0000	30.0000	30.0000	29.9998	29.9323
2.00	29.5964	29.9973	30.0000	30.0000	30.0000	30.0000	30.0000	29.9991	29.8655
3.00	29.3987	29.9939	30.0000	30.0000	30.0000	30.0000	30.0000	29.9980	29.7996
4.00	29.2036	29.9893	29.9999	30.0000	30.0000	30.0000	30.0000	29.9964	29.7345
5.00	29.0112	29.9834	29.9998	30.0000	30.0000	30.0000	29.9999	29.9945	29.6704
6.00	28.8212	29.9762	29.9997	30.0000	30.0000	30.0000	29.9999	29.9921	29.6071
7.00	28.6339	29.9679	29.9995	30.0000	30.0000	30.0000	29.9998	29.9893	29.5446
8.00	28.4490	29.9585	29.9992	30.0000	30.0000	30.0000	29.9997	29.9862	29.4830
9.00	28.2665	29.9479	29.9989	30.0000	30.0000	30.0000	29.9996	29.9827	29.4222
10.00	28.0864	29.9363	29.9986	30.0000	30.0000	30.0000	29.9995	29.9788	29.3621

(c)

Time	x = 2.00	x = 4.00	x = 6.00	x = 8.00	x = 10.00	x = 12.00	x = 14.00	x = 16.00	x = 18.00
0.00	18.0000	32.0000	42.0000	48.0000	50.0000	48.0000	42.0000	32.0000	18.0000
1.00	16.4489	30.1970	40.1562	46.1502	48.1531	46.1773	40.3274	31.2520	22.9449
2.00	15.3312	28.5350	38.3477	44.3130	46.3327	44.4671	39.0872	31.5755	24.6930
3.00	14.4219	27.0429	36.6090	42.5113	44.5759	42.9362	38.1976	31.7478	25.4131
4.00	13.6381	25.6913	34.9606	40.7728	42.9127	41.5716	37.4340	31.7086	25.6986
5.00	12.9409	24.4545	33.4091	39.1182	41.3519	40.3240	36.7033	31.5136	25.7663
6.00	12.3088	23.3146	31.9546	37.5566	39.8880	39.1565	36.9745	31.2134	25.7128
7.00	11.7294	22.2589	30.5939	36.0884	38.5109	38.0470	35.2407	30.8434	25.5871
8.00	11.1946	21.2785	29.3217	34.7092	37.2109	36.9834	34.5032	30.4279	25.4167
9.00	10.6987	20.3660	28.1318	33.4130	35.9801	35.9591	33.7660	29.9836	25.2181
10.00	10.2377	19.5150	27.0178	32.1929	34.8117	34.9710	33.0338	29.5224	25.0019

(d)

Time	$x = 10.00$	$x = 20.00$	$x = 30.00$	$x = 40.00$	$x = 50.00$	$x = 60.00$	$x = 70.00$	$x = 80.00$	$x = 90.00$
0.00	8.0000	16.0000	24.0000	32.0000	40.0000	32.0000	24.0000	16.0000	8.0000
1.00	8.0000	16.0000	24.0000	31.9979	39.7425	31.9979	24.0000	16.0026	8.3218
2.00	8.0000	16.0000	23.9999	31.9918	39.4932	31.9918	24.0000	16.0102	8.6333
3.00	8.0000	16.0000	23.9997	31.9820	39.2517	31.9820	24.0001	16.0225	8.9350
4.00	8.0000	16.0000	23.9993	31.9686	39.0175	31.9687	24.0002	16.0391	9.2272
5.00	8.0000	16.0000	23.9987	31.9520	38.7905	31.9520	24.0003	16.0599	9.5103
6.00	8.0000	15.9999	23.9978	31.9323	38.5701	31.9324	24.0005	16.0845	9.7846
7.00	8.0000	15.9999	23.9966	31.9097	38.3561	31.9098	24.0008	16.1126	10.0506
8.00	8.0000	15.9998	23.9950	31.8844	38.1483	31.8846	24.0012	16.1441	10.3084
9.00	8.0000	15.9997	23.9931	31.8566	37.9463	31.8569	24.0017	16.1786	10.5585
10.00	8.0000	15.9996	23.9908	31.8265	37.7499	31.8269	24.0023	16.2160	10.8012

11. (a) $\psi(x) = \frac{1}{2}x + 20$

(b)

Time	$x = 4.00$	$x = 8.00$	$x = 12.00$	$x = 16.00$
0.00	50.0000	50.0000	50.0000	50.0000
10.00	32.7433	44.2679	45.4228	38.2971
20.00	29.9946	36.2354	38.3148	35.8160
30.00	26.9487	32.1409	34.0874	32.9644
40.00	25.2691	29.2562	31.2704	31.2580
50.00	24.1178	27.4348	29.4296	30.1207
60.00	23.3821	26.2339	28.2356	29.3810
70.00	22.8995	25.4560	27.4554	28.8998
80.00	22.5861	24.9481	26.9482	28.5859
90.00	22.3817	24.6176	26.6175	28.3817
100.00	22.2486	24.4022	26.4023	28.2486
110.00	22.1619	24.2620	26.2620	28.1619
120.00	22.1055	24.1707	26.1707	28.1055
130.00	22.0687	24.1112	26.1112	28.0687
140.00	22.0447	24.0724	26.0724	28.0447
150.00	22.0291	24.0472	26.0472	28.0291
160.00	22.0190	24.0307	26.0307	28.0190
170.00	22.0124	24.0200	26.0200	28.0124
180.00	22.0081	24.0130	26.0130	28.0081
190.00	22.0052	24.0085	26.0085	28.0052
200.00	22.0034	24.0055	26.0055	28.0034
210.00	22.0022	24.0036	26.0036	28.0022
220.00	22.0015	24.0023	26.0023	28.0015
230.00	22.0009	24.0015	26.0015	28.0009
240.00	22.0006	24.0010	26.0010	28.0006
250.00	22.0004	24.0007	26.0007	28.0004
260.00	22.0003	24.0004	26.0004	28.0003
270.00	22.0002	24.0003	26.0003	28.0002
280.00	22.0001	24.0002	26.0002	28.0001
290.00	22.0001	24.0001	26.0001	28.0001
300.00	22.0000	24.0001	26.0001	28.0000
310.00	22.0000	24.0001	26.0001	28.0000
320.00	22.0000	24.0000	26.0000	28.0000
330.00	22.0000	24.0000	26.0000	28.0000
340.00	22.0000	24.0000	26.0000	28.0000
350.00	22.0000	24.0000	26.0000	28.0000

Exercises 16.3, Page 782

1.

(a)

Time	x = 0.25	x = 0.50	x = 0.75
0.00	0.1875	0.2500	0.1875
0.10	0.1775	0.2400	0.1775
0.20	0.1491	0.2100	0.1491
0.30	0.1066	0.1605	0.1066
0.40	0.0556	0.0938	0.0556
0.50	0.0019	0.0148	0.0019
0.60	−0.0501	−0.0682	−0.0501
0.70	−0.0970	−0.1455	−0.0970
0.80	−0.1361	−0.2072	−0.1361
0.90	−0.1648	−0.2462	−0.1648
1.00	−0.1802	−0.2591	−0.1802

(b)

Time	x = 0.4	x = 0.8	x = 1.2	x = 1.6
0.00	0.0032	0.5273	0.5273	0.0032
0.10	0.0194	0.5109	0.5109	0.0194
0.20	0.0652	0.4638	0.4638	0.0652
0.30	0.1318	0.3918	0.3918	0.1318
0.40	0.2065	0.3035	0.3035	0.2065
0.50	0.2743	0.2092	0.2092	0.2743
0.60	0.3208	0.1190	0.1190	0.3208
0.70	0.3348	0.0413	0.0413	0.3348
0.80	0.3094	−0.0180	−0.0180	0.3094
0.90	0.2443	−0.0568	−0.0568	0.2443
1.00	0.1450	−0.0768	−0.0768	0.1450

(c)

Time	x = 0.1	x = 0.2	x = 0.3	x = 0.4	x = 0.5	x = 0.6	x = 0.7	x = 0.8	x = 0.9
0.00	0.0000	0.0000	0.0000	0.0000	0.0000	0.5000	0.5000	0.5000	0.5000
0.04	0.0000	0.0000	0.0000	0.0000	0.0800	0.4200	0.5000	0.5000	0.4200
0.08	0.0000	0.0000	0.0000	0.0256	0.2432	0.2568	0.4744	0.4744	0.2312
0.12	0.0000	0.0000	0.0082	0.1126	0.3411	0.1589	0.3792	0.3710	0.0462
0.16	0.0000	0.0026	0.0472	0.2394	0.3076	0.1898	0.2108	0.1663	−0.0496
0.20	0.0008	0.0187	0.1334	0.3264	0.2146	0.2651	0.0215	−0.0933	−0.0605
0.24	0.0071	0.0657	0.2447	0.3159	0.1735	0.2463	−0.1266	−0.3056	−0.0625
0.28	0.0299	0.1513	0.3215	0.2371	0.2013	0.0849	−0.2127	−0.3829	−0.1223
0.32	0.0819	0.2525	0.3168	0.1737	0.2033	−0.1345	−0.2580	−0.3223	−0.2264
0.36	0.1623	0.3197	0.2458	0.1657	0.0877	−0.2853	−0.2843	−0.2104	−0.2887
0.40	0.2412	0.3129	0.1727	0.1583	−0.1223	−0.3164	−0.2874	−0.1473	−0.2336
0.44	0.2657	0.2383	0.1399	0.0658	−0.3046	−0.2761	−0.2549	−0.1565	−0.0761
0.48	0.1965	0.1410	0.1149	−0.1216	−0.3593	−0.2381	−0.1977	−0.1715	0.0800
0.52	0.0466	0.0531	0.0225	−0.3093	−0.2992	−0.2260	−0.1451	−0.1144	0.1300
0.56	−0.1161	−0.0466	−0.1662	−0.3876	−0.2188	−0.2114	−0.1085	0.0111	0.0602
0.60	−0.2194	−0.2069	−0.3875	−0.3411	−0.1901	−0.1662	−0.0666	0.1140	−0.0446
0.64	−0.2485	−0.4290	−0.5362	−0.2611	−0.2021	−0.0969	0.0012	0.1084	−0.0843
0.68	−0.2559	−0.6276	−0.5625	−0.2503	−0.1993	−0.0298	0.0720	0.0068	−0.0354
0.72	−0.3003	−0.6865	−0.5097	−0.3230	−0.1585	0.0156	0.0893	−0.0874	0.0384
0.76	−0.3722	−0.5652	−0.4538	−0.4029	−0.1147	0.0289	0.0265	−0.0849	0.0596
0.80	−0.3867	−0.3464	−0.4172	−0.4068	−0.1172	−0.0046	−0.0712	−0.0005	0.0155
0.84	−0.2647	−0.1633	−0.3546	−0.3214	−0.1763	−0.0954	−0.1249	0.0665	−0.0386
0.88	−0.0254	−0.0738	−0.2202	−0.2002	−0.2559	−0.2215	−0.1079	0.0385	−0.0468
0.92	0.2064	−0.0157	−0.0325	−0.1032	−0.3067	−0.3223	−0.0804	−0.0636	−0.0127
0.96	0.3012	0.1081	0.1380	−0.0487	−0.2974	−0.3407	−0.1250	−0.1548	0.0092
1.00	0.2378	0.3032	0.2392	−0.0141	−0.2223	−0.2762	−0.2481	−0.1840	−0.0244

3.

(a)

Time	x = 0.2	x = 0.4	x = 0.6	x = 0.8
0.00	0.5878	0.9511	0.9511	0.5878
0.05	0.5808	0.9397	0.9397	0.5808
0.10	0.5599	0.9059	0.9059	0.5599
0.15	0.5256	0.8505	0.8505	0.5256
0.20	0.4788	0.7748	0.7748	0.4788
0.25	0.4206	0.6806	0.6806	0.4206
0.30	0.3524	0.5701	0.5701	0.3524
0.35	0.2757	0.4460	0.4460	0.2757
0.40	0.1924	0.3113	0.3113	0.1924
0.45	0.1046	0.1692	0.1692	0.1046
0.50	0.0142	0.0230	0.0230	0.0142

(b)

Time	x = 0.2	x = 0.4	x = 0.6	x = 0.8
0.00	0.5878	0.9511	0.9511	0.5878
0.03	0.5860	0.9482	0.9482	0.5860
0.05	0.5808	0.9397	0.9397	0.5808
0.08	0.5721	0.9256	0.9256	0.5721
0.10	0.5599	0.9060	0.9060	0.5599
0.13	0.5445	0.8809	0.8809	0.5445
0.15	0.5257	0.8507	0.8507	0.5257
0.18	0.5039	0.8153	0.8153	0.5039
0.20	0.4790	0.7750	0.7750	0.4790
0.23	0.4513	0.7302	0.7302	0.4513
0.25	0.4209	0.6810	0.6810	0.4209
0.28	0.3879	0.6277	0.6277	0.3879
0.30	0.3527	0.5706	0.5706	0.3527
0.33	0.3153	0.5102	0.5102	0.3153
0.35	0.2761	0.4467	0.4467	0.2761
0.38	0.2352	0.3806	0.3806	0.2352
0.40	0.1929	0.3122	0.3122	0.1929
0.43	0.1495	0.2419	0.2419	0.1495
0.45	0.1052	0.1701	0.1701	0.1052
0.48	0.0602	0.0974	0.0974	0.0602
0.50	0.0149	0.0241	0.0241	0.0149

5.

Time	x = 10	x = 20	x = 30	x = 40	x = 50
0.00000	0.1000	0.2000	0.3000	0.2000	0.1000
0.20045	0.1000	0.2000	0.2750	0.2000	0.1000
0.40089	0.1000	0.1938	0.2125	0.1938	0.1000
0.60134	0.0984	0.1688	0.1406	0.1688	0.0984
0.80178	0.0898	0.1191	0.0828	0.1191	0.0898
1.00223	0.0661	0.0531	0.0432	0.0531	0.0661
1.20268	0.0226	−0.0121	0.0085	−0.0121	0.0226
1.40312	−0.0352	−0.0635	−0.0365	−0.0635	−0.0352
1.60357	−0.0913	−0.1011	−0.0950	−0.1011	−0.0913
1.80401	−0.1271	−0.1347	−0.1566	−0.1347	−0.1271
2.00446	−0.1329	−0.1791	−0.2072	−0.1719	−0.1329
2.20491	−0.1153	−0.2081	−0.2402	−0.2081	−0.1153
2.40535	−0.0920	−0.2292	−0.2571	−0.2292	−0.0920
2.60580	−0.0801	−0.2230	−0.2601	−0.2230	−0.0801
2.80624	−0.0838	−0.1903	−0.2445	−0.1903	−0.0838
3.00669	−0.0932	−0.1445	−0.2018	−0.1445	−0.0932
3.20713	−0.0921	−0.1003	−0.1305	−0.1003	−0.0921
3.40758	−0.0701	−0.0615	−0.0440	−0.0615	−0.0701
3.60803	−0.0284	−0.0205	0.0336	−0.0205	−0.0284
3.80847	0.0224	0.0321	0.0842	0.0321	0.0224
4.00892	0.0700	0.0953	0.1087	0.0953	0.0700
4.20936	0.1064	0.1555	0.1265	0.1555	0.1064
4.40981	0.1285	0.1962	0.1588	0.1962	0.1285
4.61026	0.1354	0.2106	0.2098	0.2106	0.1354
4.81070	0.1273	0.2060	0.2612	0.2060	0.1273
5.01115	0.1070	0.1955	0.2851	0.1955	0.1070
5.21159	0.0821	0.1853	0.2641	0.1853	0.0821
5.41204	0.0625	0.1689	0.2038	0.1689	0.0625
5.61249	0.0539	0.1347	0.1260	0.1347	0.0539
5.81293	0.0520	0.0781	0.0526	0.0781	0.0520
6.01338	0.0436	0.0086	−0.0080	0.0086	0.0436
6.21382	0.0156	−0.0564	−0.0604	−0.0564	0.0156
6.41427	−0.0343	−0.1043	−0.1107	−0.1043	−0.0343
6.61472	−0.0931	−0.1364	−0.1578	−0.1364	−0.0931
6.81516	−0.1395	−0.1630	−0.1942	−0.1630	−0.1395
7.01561	−0.1568	−0.1915	−0.2150	−0.1915	−0.1568
7.21605	−0.1436	−0.2173	−0.2240	−0.2173	−0.1436
7.41650	−0.1129	−0.2263	−0.2297	−0.2263	−0.1129
7.61695	−0.0824	−0.2078	−0.2336	−0.2078	−0.0824
7.81739	−0.0625	−0.1644	−0.2247	−0.1644	−0.0625
8.01784	−0.0526	−0.1106	−0.1856	−0.1106	−0.0526
8.21828	−0.0440	−0.0611	−0.1091	−0.0611	−0.0440
8.41873	−0.0287	−0.0192	−0.0085	−0.0192	−0.0287
8.61918	−0.0038	0.0229	0.0867	0.0229	−0.0038
8.81962	0.0287	0.0743	0.1500	0.0743	0.0287
9.02007	0.0654	0.1332	0.1755	0.1332	0.0654
9.22051	0.1027	0.1858	0.1799	0.1858	0.1027
9.42096	0.1352	0.2160	0.1872	0.2160	0.1352
9.62140	0.1540	0.2189	0.2089	0.2189	0.1540
9.82185	0.1506	0.2030	0.2356	0.2030	0.1506
10.02230	0.1226	0.1822	0.2461	0.1822	0.1226

Note: Time is expressed in milliseconds.

Chapter 16 Review Exercises, Page 783

1. $u_{11} = 0.8929$, $u_{21} = 3.5714$, $u_{31} = 13.3929$

3. (a)

Time	$x = 0.00$	$x = 0.20$	$x = 0.40$	$x = 0.60$	$x = 0.80$	$x = 1.00$
0.00	0.0000	0.2000	0.4000	0.6000	0.8000	0.0000
0.01	0.0000	0.2000	0.4000	0.6000	0.5500	0.0000
0.02	0.0000	0.2000	0.4000	0.5375	0.4250	0.0000
0.03	0.0000	0.2000	0.3844	0.4750	0.3469	0.0000
0.04	0.0000	0.1961	0.3609	0.4203	0.2922	0.0000
0.05	0.0000	0.1883	0.3346	0.3734	0.2512	0.0000

(b)

Time	$x = 0.00$	$x = 0.20$	$x = 0.40$	$x = 0.60$	$x = 0.80$	$x = 1.00$
0.00	0.0000	0.2000	0.4000	0.6000	0.8000	1.0000
0.01	0.0000	0.2000	0.4000	0.6000	0.8000	0.0000
0.02	0.0000	0.2000	0.4000	0.6000	0.5500	0.0000
0.03	0.0000	0.2000	0.4000	0.5375	0.4250	0.0000
0.04	0.0000	0.2000	0.3844	0.4750	0.3469	0.0000
0.05	0.0000	0.1961	0.3609	0.4203	0.2922	0.0000

(c) Yes; the table in part b is the table in part a shifted downward.

Exercises 17.1, Page 793

1. $3 + 3i$

3. 1

5. $7 - 13i$

7. $-7 + 5i$

9. $11 - 10i$

11. $-5 + 12i$

13. $-2i$

15. $-\frac{7}{17} - \frac{11}{17}i$

17. $8 - i$

19. $\frac{23}{37} - \frac{64}{37}i$

21. $20i$

23. $\frac{102}{5} + \frac{116}{5}i$

25. $\frac{7}{130} + \frac{9}{130}i$

27. $x/(x^2 + y^2)$

29. $-2y - 4$

31. $\sqrt{(x-1)^2 + (y-3)^2}$

33. $x = -\frac{9}{2}$, $y = 1$

35. $\frac{\sqrt{2}}{2} + \frac{\sqrt{2}}{2}i$, $-\frac{\sqrt{2}}{2} - \frac{\sqrt{2}}{2}i$

37. $11 - 6i$

Exercises 17.2, Page 797

1. $2(\cos 0 + i \sin 0)$ or $2(\cos 2\pi + i \sin 2\pi)$

3. $3\left(\cos \dfrac{3\pi}{2} + i \sin \dfrac{3\pi}{2}\right)$

5. $\sqrt{2}\left(\cos \dfrac{\pi}{4} + i \sin \dfrac{\pi}{4}\right)$

7. $2\left(\cos \dfrac{5\pi}{6} + i \sin \dfrac{5\pi}{6}\right)$

9. $\dfrac{3\sqrt{2}}{2}\left(\cos \dfrac{5\pi}{4} + i \sin \dfrac{5\pi}{4}\right)$

11. $-\frac{53}{2} - \frac{5}{2}i$

13. $5.5433 + 2.2961i$

15. $8i$; $\dfrac{\sqrt{2}}{4} - \dfrac{\sqrt{2}}{4}i$

17. $30\sqrt{2}\left(\cos \dfrac{25\pi}{12} + i \sin \dfrac{25\pi}{12}\right)$; $40.9808 + 10.9808i$

19. $\dfrac{1}{2\sqrt{2}}\left[\cos\left(-\dfrac{\pi}{4}\right) + i \sin\left(-\dfrac{\pi}{4}\right)\right]$; $\dfrac{1}{4} - \dfrac{1}{4}i$

21. -512

23. $\frac{1}{32}i$

25. $-i$

27. $w_0 = 2$, $w_1 = -1 + \sqrt{3}i$, $w_2 = -1 - \sqrt{3}i$

29. $w_0 = \dfrac{\sqrt{2}}{2} + \dfrac{\sqrt{2}}{2}i$, $w_1 = -\dfrac{\sqrt{2}}{2} - \dfrac{\sqrt{2}}{2}i$

31. $w_0 = \dfrac{\sqrt{2}}{2} + \dfrac{\sqrt{6}}{2}i$, $w_1 = -\dfrac{\sqrt{2}}{2} - \dfrac{\sqrt{6}}{2}i$

33. $\pm \dfrac{\sqrt{2}}{2}(1 + i)$, $\pm \dfrac{\sqrt{2}}{2}(1 - i)$

35. $32\left(\cos \dfrac{13\pi}{6} + i \sin \dfrac{13\pi}{6}\right)$, $16\sqrt{3} + 16i$

37. $\cos 2\theta = \cos^2\theta - \sin^2\theta$, $\sin 2\theta = 2 \sin \theta \cos \theta$

Exercises 17.3, Page 800

1.

3.

5.

7.

9. domain

11. domain

13. domain

15. not a domain

17. not a domain

19. domain

21. domain

23. the line $y = -x$

25. the hyperbola $x^2 - y^2 = 1$

Exercises 17.4, Page 806

1.

$u = \dfrac{v^2}{16} - 4$

3.

$u \leq 0$
$v = 0$

5.

$v \geq 0$
$u = 0$

7. $f(z) = (6x - 5) + i(6y + 9)$

9. $f(z) = (x^2 - y^2 - 3x) + i(2xy - 3y + 4)$

11. $f(z) = (x^3 - 3xy^2 - 4x) + i(3x^2 y - y^3 - 4y)$

13. $f(z) = \left(x + \dfrac{x}{x^2 + y^2}\right) + i\left(y - \dfrac{y}{x^2 + y^2}\right)$

15. $-4 + i;\ 3 - 9i;\ 1 + 86i$

17. $14 - 20i;\ -13 + 43i;\ 3 - 26i$ **19.** $6 - 5i$

21. $-4i$ **27.** $12z^2 - (6 + 2i)z - 5$

29. $6z^2 - 14z - 4 + 16i$ **31.** $6z(z^2 - 4i)^2$

33. $\dfrac{8 - 13i}{(2z + i)^2}$ **35.** $3i$ **37.** $2i, -2i$

41. $x(t) = c_1 e^{2t}$ and $y(t) = c_2 e^{2t}$; the streamlines lie on lines through the origin.

43. $y = cx$; the streamlines are lines through the origin.

45.

Exercises 17.5, Page 810

15. $a = 1, b = 3$ **21.** $f'(z) = e^x \cos y + ie^x \sin y$

23. $f(z) = x + i(y + C)$

25. $f(z) = x^2 - y^2 + i(2xy + C)$

27. $f(z) = \log_e(x^2 + y^2) + i\left(2 \tan^{-1} \dfrac{y}{x} + C\right)$

29.

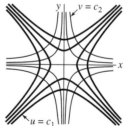

31. the x-axis and the circle $|z| = 1$

Exercises 17.6, Page 817

1. $\dfrac{\sqrt{3}}{2} + \dfrac{1}{2}i$

3. $e^{-1}\left(\dfrac{\sqrt{2}}{2} + \dfrac{\sqrt{2}}{2}i\right)$

5. $-e^{\pi}$

7. $-1.8650 + 4.0752i$

9. $0.2837 - 0.9589i$

11. $-0.9659 + 0.2588i$

13. $e^{y}(\cos x - i \sin x)$

15. $e^{x^2 - y^2}(\cos 2xy + i \sin 2xy)$

23. $1.6094 + i(\pi + 2n\pi)$

25. $1.0397 + i\left(\dfrac{3\pi}{4} + 2n\pi\right)$

27. $1.0397 + i\left(\dfrac{\pi}{3} + 2n\pi\right)$

29. $2.1383 - \dfrac{\pi}{4}i$

31. $2.5649 + 2.7468i$

33. $3.4657 - \dfrac{\pi}{3}i$

35. $1.3863 + i\left(\dfrac{\pi}{2} + 2n\pi\right)$

37. $3 + i\left(-\dfrac{\pi}{2} + 2n\pi\right)$

39. $e^{(2 - 8n)\pi}$

41. $e^{-2n\pi}(0.2740 + 0.5837i)$

43. e^2

47. no; no; yes

Exercises 17.7, Page 821

1. 10.0677

3. $1.0911 + 0.8310i$

5. $0.7616i$

7. -0.6481

9. -1

11. $0.5876 + 1.3363i$

15. $\dfrac{\pi}{2} + 2n\pi - i \log_e (2 \pm \sqrt{3})$

17. $\left(-\dfrac{\pi}{2} + 2n\pi\right)i$

19. $\dfrac{\pi}{4} + n\pi$

21. $2n\pi \pm 2i$

Exercises 17.8, Page 824

1. $n\pi + (-1)^{n+1} i \log_e (1 + \sqrt{2})$　**3.** $n\pi$

5. $2n\pi \pm i \log_e (2 + \sqrt{3})$

7. $\pm \dfrac{\pi}{3} + 2n\pi$

9. $\dfrac{\pi}{4} + n\pi$

11. $(-1)^n \log_e 3 + n\pi i$

Chapter 17 Review Exercises, Page 824

1. $0; 32$　　**3.** $-\dfrac{7}{25}$　　**5.** $\dfrac{4}{5}$　　　**7.** false

9. $0.6931 + i\left(\dfrac{\pi}{2} + 2n\pi\right)$　　**11.** $-0.3097 + 0.8577i$

13. false

15. $3 - \dfrac{\pi}{2}i$

17. $58 - 4i$

19. $-8 + 8i$

21.

23.

25. an ellipse with foci $(0, -2)$ and $(0, 2)$

27. $1.0696 - 0.2127i, 0.2127 + 1.0696i, -1.0696 + 0.2127i,$
$- 0.2127 - 1.0696i$

29. $5i$

31. the parabola $v = u^2 - 2u$

33. $1, -1$

35. pure imaginary numbers

37. $f'(z) = (-2y - 5) + 2xi$

Exercises 18.1, Page 832

1. $-28 + 84i$

3. $-48 + \dfrac{736}{3}i$

5. $(2 + \pi)i$

7. πi

9. $-\dfrac{7}{12} + \dfrac{1}{12}i$

11. $-e - 1$

13. $\dfrac{3}{2} - \dfrac{\pi}{4}$　**15.** 0

17. $\dfrac{1}{2}i$

19. 0

21. $\dfrac{4}{3} - \dfrac{5}{3}i$

23. $\dfrac{4}{3} - \dfrac{5}{3}i$

25. $\dfrac{5\pi e^5}{12}$

27. $6\sqrt{2}$

31. $-11 + 38i; 0$

33. circulation $= 0$, net flux $= 4\pi$

35. circulation $= 0$, net flux $= 0$

Exercises 18.2, Page 837

9. $2\pi i$

11. $2\pi i$　　**13.** 0

15. $2\pi i; 4\pi i; 0$

17. $-8\pi i; -6\pi i$

19. $-\pi(1 + i)$

21. $-4\pi i$　　**23.** $-6\pi i$

Exercises 18.3, Page 842

1. $-2i$　　**3.** $48 + 24i$　　**5.** $6 + \dfrac{26}{3}i$　　**7.** 0

9. $-\dfrac{7}{16} - \dfrac{22}{3}i$　　**11.** $-\dfrac{1}{\pi} - \dfrac{1}{\pi}i$　　**13.** $2.3504i$

15. 0　　　**17.** πi　　　　**19.** $\dfrac{1}{2}i$

21. $11.4928 + 0.9667i$　　**23.** $-0.9056 + 1.7699i$

Exercises 18.4, Page 848

1. $8\pi i$　　**3.** $-2\pi i$　　**5.** $-\pi(20 + 8i)$

7. $-2\pi; 2\pi$　　**9.** -8π　　**11.** $-2\pi e^{-1}i$

13. $\frac{4}{3}\pi i$

15. $-5\pi i; -5\pi i; 9\pi i; 0$

17. $-\pi(3 + i); \pi(3 + i)$

19. $\pi(\frac{8}{3} + 12i)$

21. 0

23. $-\pi i$

Chapter 18 Review Exercises, Page 849

1. true **3.** true **5.** 0 **7.** $\pi(6\pi - i)$

9. true **11.** 0 if $n \neq -1$, $2\pi i$ if $n = -1$

13. $-\frac{1}{2}$ **15.** $\frac{136}{15} + \frac{88}{3}i$ **17.** 0

19. $-14.2144 + 22.9637i$ **21.** $2\pi i$

23. $-\frac{8}{3}\pi i$ **25.** $\frac{2}{5}\pi i$ **27.** 2π **29.** $2n\pi i$

Exercises 19.1, Page 856

1. $5i, -5, -5i, 5, 5i$ **3.** $0, 2, 0, 2, 0$

5. converges **7.** converges **9.** diverges

11. $\lim_{n \to \infty} \text{Re}(z_n) = 2$ and $\lim_{n \to \infty} \text{Im}(z_n) = \frac{3}{2}$

13. The series converges to $1/(1 + 2i)$. **15.** divergent

17. convergent, $-\frac{1}{5} + \frac{2}{5}i$

19. convergent, $\frac{9}{5} - \frac{12}{5}i$

21. $|z - 2i| = \sqrt{5}$, $R = \sqrt{5}$

23. $|z - 1 - i| = 2$, $R = 2$

25. $|z - i| = 1/\sqrt{10}$, $R = 1/\sqrt{10}$

27. $|z - 4 - 3i| = 25$, $R = 25$

29. The series converges at $z = -2 + i$.

Exercises 19.2, Page 862

1. $\sum_{k=1}^{\infty} (-1)^{k+1} z^k$, $R = 1$

3. $\sum_{k=1}^{\infty} (-1)^{k-1} k(2z)^{k-1}$, $R = \frac{1}{2}$

5. $\sum_{k=0}^{\infty} \frac{(-1)^k}{k!} (2z)^k$, $R = \infty$

7. $\sum_{k=0}^{\infty} \frac{z^{2k+1}}{(2k+1)!}$, $R = \infty$

9. $\sum_{k=0}^{\infty} \frac{(-1)^k}{(2k)!} \left(\frac{z}{2}\right)^{2k}$, $R = \infty$

11. $\sum_{k=0}^{\infty} \frac{(-1)^k}{(2k+1)!} z^{4k+2}$, $R = \infty$

13. $\sum_{k=0}^{\infty} (-1)^k (z - 1)^k$, $R = 1$

15. $\sum_{k=0}^{\infty} \frac{(z - 2i)^k}{(3 - 2i)^{k+1}}$, $R = \sqrt{13}$

17. $\sum_{k=1}^{\infty} \frac{(z - 1)^k}{2^k}$, $R = 2$

19. $\frac{\sqrt{2}}{2} - \frac{\sqrt{2}}{2 \cdot 1!} \left(z - \frac{\pi}{4}\right) - \frac{\sqrt{2}}{2 \cdot 2!} \left(z - \frac{\pi}{4}\right)^2$

$\quad + \frac{\sqrt{2}}{2 \cdot 3!} \left(z - \frac{\pi}{4}\right)^3 + \cdots$, $R = \infty$

21. $e^{3i} \sum_{k=0}^{\infty} \frac{(z - 3i)^k}{k!}$, $R = \infty$

23. $z + \frac{1}{3} z^3 + \frac{2}{15} z^5 + \cdots$

25. $\frac{1}{2i} + \frac{3}{(2i)^2} z + \frac{7}{(2i)^3} z^2 + \frac{15}{(2i)^4} z^3 + \cdots$, $R = 1$

27. $2\sqrt{5}$

29. $\sum_{k=0}^{\infty} (-1)^k (z + 1)^k$, $R = 1$;

$\quad \sum_{k=0}^{\infty} \frac{(-1)^k}{(2 + i)^{k+1}} (z - i)^k$, $R = \sqrt{5}$

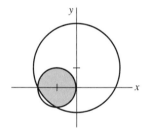

31. (a) The distance from z_0 to the branch cut is one unit.

 (c) The series converges within the circle $|z + 1 - i| = \sqrt{2}$. Although the series converges in the shaded region, it does not converge to (or represent) Ln z in this region.

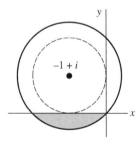

33. $1.1 + 0.12i$ **35.** $\frac{2}{\sqrt{\pi}} \sum_{k=0}^{\infty} \frac{(-1)^k}{(2k+1)k!} z^{2k+1}$

Exercises 19.3, Page 869

1. $\dfrac{1}{z} - \dfrac{z}{2!} + \dfrac{z^3}{4!} - \dfrac{z^5}{6!} + \cdots$

3. $1 - \dfrac{1}{1! \cdot z^2} + \dfrac{1}{2! \cdot z^4} - \dfrac{1}{3! \cdot z^6} + \cdots$

5. $\dfrac{e}{z-1} + e + \dfrac{e(z-1)}{2!} + \dfrac{e(z-1)^2}{3!} + \cdots$

7. $-\dfrac{1}{3z} - \dfrac{1}{3^2} - \dfrac{z}{3^3} - \dfrac{z^2}{3^4} - \cdots$

9. $\dfrac{1}{3(z-3)} - \dfrac{1}{3^2} + \dfrac{z-3}{3^3} - \dfrac{(z-3)^2}{3^4} + \cdots$

11. $\cdots - \dfrac{1}{3(z-4)^2} + \dfrac{1}{3(z-4)} - \dfrac{1}{12} + \dfrac{z-4}{3 \cdot 4^2}$
$- \dfrac{(z-4)^2}{3 \cdot 4^3} + \cdots$

13. $\cdots - \dfrac{1}{z^2} - \dfrac{1}{z} - \dfrac{1}{2} - \dfrac{z}{2^2} - \dfrac{z^2}{2^3} - \cdots$

15. $\dfrac{-1}{z-1} - 1 - (z-1) - (z-1)^2 - \cdots$

17. $\dfrac{1}{3(z+1)} - \dfrac{2}{3^2} - \dfrac{2(z+1)}{3^3} - \dfrac{2(z+1)^2}{3^4} - \cdots$

19. $\cdots - \dfrac{1}{3z^2} + \dfrac{1}{3z} - \dfrac{1}{3} - \dfrac{z}{3 \cdot 2} - \dfrac{z^2}{3 \cdot 2^2} - \cdots$

21. $\dfrac{1}{z} + 2 + 3z + 4z^2 + \cdots$

23. $\dfrac{1}{z-2} - 3 + 6(z-2) - 10(z-2)^2 + \cdots$

25. $\dfrac{3}{z} - 4 - 4z - 4z^2 - \cdots$

27. $\cdots + \dfrac{2}{(z-1)^3} + \dfrac{2}{(z-1)^2} + \dfrac{2}{z-1} + 1 + (z-1)$

Exercises 19.4, Page 873

1. Define $f(0) = 2$.

3. $-2 + i$ is a zero of order 2.

5. $-i$ and i are zeros of order 1; 0 is a zero of order 2.

7. $2n\pi i$, $n = 0, \pm 1, \ldots$, are zeros of order 1.

9. order 5 **11.** order 1

13. $-1 \pm 2i$ are simple poles.

15. -2 is a simple pole; $-i$ is a pole of order 4.

17. $(2n+1)\pi/2$, $n = 0, \pm 1, \ldots$, are simple poles.

19. 0 is a pole of order 2.

21. $2n\pi i$, $n = 0, \pm 1, \ldots$, are simple poles.

23. nonisolated

Exercises 19.5, Page 878

1. $\frac{2}{5}$ **3.** -3 **5.** 0

7. $\mathrm{Res}(f(z), -4i) = \frac{1}{2}$, $\mathrm{Res}(f(z), 4i) = \frac{1}{2}$

9. $\mathrm{Res}(f(z), 1) = \frac{1}{3}$, $\mathrm{Res}(f(z), -2) = -\frac{1}{12}$,
$\mathrm{Res}(f(z), 0) = -\frac{1}{4}$

11. $\mathrm{Res}(f(z), -1) = 6$, $\mathrm{Res}(f(z), -2) = -31$,
$\mathrm{Res}(f(z), -3) = 30$

13. $\mathrm{Res}(f(z), 0) = -3/\pi^4$, $\mathrm{Res}(f(z), \pi) = (\pi^2 - 6)/2\pi^4$

15. $\mathrm{Res}(f(z), (2n+1)\pi/2) = (-1)^{n+1}$, $n = 0, \pm 1, \pm 2, \cdots$

17. 0; $2\pi i/9$; 0 **19.** πi; πi; 0 **21.** $\dfrac{\pi}{3}$

23. 0 **25.** $2\pi i \cosh 1$ **27.** $-4i$ **29.** $6i$

31. $-\dfrac{\pi}{3} + \dfrac{\pi}{3} i$

Exercises 19.6, Page 885

1. $\dfrac{4\pi}{\sqrt{3}}$ **3.** 0 **5.** $\dfrac{\pi}{\sqrt{3}}$ **7.** $\dfrac{\pi}{4}$

9. $\dfrac{\pi}{6}$ **11.** π **13.** $\dfrac{\pi}{16}$ **15.** $\dfrac{3\pi}{8}$

17. $\dfrac{\pi}{2}$ **19.** $\dfrac{\pi}{\sqrt{2}}$ **21.** πe^{-1} **23.** πe^{-1}

25. πe^{-3} **27.** $\dfrac{\pi e^{-\sqrt{2}}}{2\sqrt{2}} (\cos \sqrt{2} + \sin \sqrt{2})$

29. $-\dfrac{\pi}{8} \left(\dfrac{e^{-3}}{3} - e^{-1} \right)$

Chapter 19 Review Exercises, Page 886

1. true **3.** false **5.** true **7.** true

9. $\dfrac{1}{\pi}$ **11.** $|z - i| = \sqrt{5}$

13. $1 + \displaystyle\sum_{k=1}^{\infty} \dfrac{(\sqrt{2})^k \cos \dfrac{k\pi}{4}}{k!} z^k$

15. $-\dfrac{i}{z^3} + \dfrac{1}{2!z^2} + \dfrac{i}{3!z} - \dfrac{1}{4!} - \dfrac{i}{5!} z + \cdots$

17. $\cdots + \dfrac{1}{5!(z-i)^3} - \dfrac{1}{3!(z-i)} + (z-i)$

19. $\dfrac{2}{3} + \dfrac{8}{9} z + \dfrac{26}{27} z^2 + \cdots$;

$\cdots - \dfrac{1}{z^3} - \dfrac{1}{z^2} - \dfrac{1}{z} + \dfrac{1}{3} - \dfrac{z}{3^2} - \dfrac{z^2}{3^3} - \cdots$;

$\dfrac{2}{z^2} + \dfrac{8}{z^3} + \dfrac{26}{z^4} + \cdots$;

$-\dfrac{1}{z-1} - \dfrac{1}{2} - \dfrac{z-1}{2^2} - \dfrac{(z-1)^2}{2^3} - \cdots$

21. $\dfrac{404\pi}{81} i$ **23.** $\dfrac{2\pi}{\sqrt{3}} i$

25. $(\pi + \pi e^{-2} \cos 2)i$ **27.** $-\pi i$

29. $\dfrac{9\pi^3 + 2}{\pi^2} i$ **31.** $\dfrac{7\pi}{50}$

33. $\pi\left(\dfrac{90 - 52\sqrt{3}}{12 - 7\sqrt{3}}\right)$

Exercises 20.1, Page 893

1. the line $v = -u$ **3.** the line $v = 2$

5. open line segment from 0 to πi

7. the ray $\theta = \frac{1}{2}\theta_0$ **9.** the line $u = 1$

11. the fourth quadrant

13. the wedge $\pi/4 \le \text{Arg } w \le \pi/2$

15. the circle with center $w = 4i$ and radius $r = 1$

17. the strip $-1 \le u \le 0$

19. the wedge $0 \le \text{Arg } w \le 3\pi/4$

21. $w = -i(z - i) = -iz - 1$

23. $w = 2(z - 1)$ **25.** $w = -z^4$

27. $w = e^{3z/2}$ **29.** $w = -z + i$

Exercises 20.2, Page 900

1. conformal at all points except $z = \pm 1$

3. conformal at all points except $z = \pi i \pm 2n\pi i$

5. conformal at all points outside the interval $[-1, 1]$ on the x-axis

7. The image is the region shown in Figure 20.11(b). A horizontal segment $z(t) = t + ib$, $0 < t < \pi$, is mapped onto the lower or upper portion of the ellipse

$$\frac{u^2}{\cosh^2 b} + \frac{v^2}{\sinh^2 b} = 1$$

according to whether $b > 0$ or $b < 0$.

9. The image of the region is the wedge $0 \le \text{Arg } w \le \pi/4$. The image of the line segment $[-\pi/2, \pi/2]$ is the union of the line segments joining $e^{i\pi/4}$ to 0 and 0 to 1.

11. $w = \cos(\pi z/2)$ using H-4

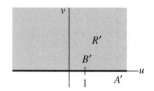

13. $w = \left(\dfrac{1+z}{1-z}\right)^{1/2}$ using H-5 and $w = z^{1/4}$

15. $w = \left(\dfrac{e^{\pi/z} + e^{-\pi/z}}{e^{\pi/z} - e^{-\pi/z}}\right)^{1/2}$ using H-6 and $w = z^{1/2}$

17. $w = \sin(-i \text{ Ln } z - \pi/2)$; $A'B'$ is the real interval $(-\infty, -1]$.

19. $u = \dfrac{1}{\pi} \text{Arg}(z^4)$ or $u(r, \theta) = \dfrac{4}{\pi}\theta$

21. $u = \dfrac{1}{\pi} \text{Arg}\left(i\,\dfrac{1-z}{1+z}\right) = \dfrac{1}{\pi} \tan^{-1}\left(\dfrac{1 - x^2 - y^2}{2y}\right)$

23. $u = \dfrac{1}{\pi} [\text{Arg}(z^2 - 1) - \text{Arg}(z^2 + 1)]$

25. $u = \dfrac{10}{\pi} [\text{Arg}(e^{\pi z} - 1) - \text{Arg}(e^{\pi z} + 1)]$

Exercises 20.3, Page 907

1. $T(0) = \infty$, $T(1) = i$, $T(\infty) = 0$; $|w| = 1$ and the line $v = \frac{1}{2}$; $|w| \ge 1$

3. $T(0) = -1$, $T(1) = \infty$, $T(\infty) = 1$; the line $u = 0$ and the circle $|w - 1| = 2$; the half-plane $u \le 0$

5. $S^{-1}(w) = \dfrac{-w - 1}{-w + i} = \dfrac{w + 1}{w - i}$, $S^{-1}(T(z)) = \dfrac{(1 + i)z - 1}{2z + i}$

7. $S^{-1}(w) = \dfrac{-w + 2}{-w + 1} = \dfrac{w - 2}{w - 1}$, $S^{-1}(T(z)) = \dfrac{3}{z}$

9. $w = -2\dfrac{z+1}{z-2}$

11. $w = \dfrac{2z}{z-1-2i}$

13. $w = \dfrac{i}{2}\dfrac{z-1}{z}$

15. $w = 3\dfrac{(1+i)z+1-i}{(-3+5i)z-3-5i}$

17. $u = \dfrac{1}{\log_e 2}\log_e\left|\dfrac{z+2}{z-1}\right|$. The level curves are the images of

the circles $|w| = r$, $1 < r < 2$, under the linear fractional transformation $T(w) = (w + 2)/(w - 1)$. Since the circles do not pass through the pole at $w = 1$, the images are circles.

19. Construct the linear fractional transformation that sends $1, i, -i$ to $0, 1, -1$.

21. Simplify $T_2(T_1(z)) = \dfrac{a_2\left(\dfrac{a_1 z + b_1}{c_1 z + d_1}\right) + b_2}{c_2\left(\dfrac{a_1 z + b_1}{c_1 z + d_1}\right) + d_2}$.

Exercises 20.4, Page 912

1. first quadrant

3.

5. $f'(z) = A(z + 1)^{-1/2}z^{-1/2}(z - 1)^{-1/2}$ for some constant A

7. $f'(z) = A(z + 1)^{-1/3}z^{-1/3}$ for some constant A

9. Show that $f'(z) = \dfrac{A}{(z^2 - 1)^{1/2}}$ and conclude that $f(z) = \cosh^{-1}z$.

11. Show that $f'(z) \to A/z$ as $w_1 \to \infty$ and conclude that $f(z) = \text{Ln } z$.

13. Show that $f'(z) \to A(z + 1)^{-1/2}z(z - 1)^{-1/2} = Az/(z^2 - 1)^{1/2}$ as $u_1 \to 0$.

Exercises 20.5, Page 916

1. $u = \dfrac{1}{\pi}\text{Arg}\left(\dfrac{z-1}{z}\right) - \dfrac{1}{\pi}\text{Arg}\left(\dfrac{z}{z+1}\right)$

3. $u = \dfrac{5}{\pi}[\pi - \text{Arg}(z - 1)] + \dfrac{1}{\pi}\text{Arg}\left(\dfrac{z}{z+1}\right)$

$\quad - \dfrac{1}{\pi}\text{Arg}\left(\dfrac{z+1}{z+2}\right)$

5. $u = \dfrac{y}{\pi}\left\{1 + \dfrac{y^2 - x^2}{y}\left[\tan^{-1}\left(\dfrac{x-1}{y}\right) - \tan^{-1}\left(\dfrac{x}{y}\right)\right]\right.$

$\quad \left. + x\log_e\left[\dfrac{(x-1)^2 + y^2}{x^2 + y^2}\right]\right\}$

7. $u = \dfrac{1}{\pi}\text{Arg}\left(\dfrac{z^2 - 1}{z^2}\right) + \dfrac{5}{\pi}\text{Arg}(z^2 + 1)$

9. $u = \dfrac{1}{\pi}\text{Arg}\left(\dfrac{(1 - i)z - (1 + i)}{1 - z}\right)$

$\quad - \dfrac{1}{\pi}\text{Arg}\left(\dfrac{1 - z}{-(1 + i)z + 1 - i}\right)$

11. $u(0, 0) = \frac{1}{3}$, $u(-0.5, 0) = 0.5693$, $u(0.5, 0) = 0.1516$

13. Show that $u(0, 0) = \dfrac{1}{2\pi}\displaystyle\int_{-\pi}^{\pi} u(e^{it})\, dt$.

15. $u(r, \theta) = r\sin\theta + r\cos\theta$ or $u(x, y) = x + y$

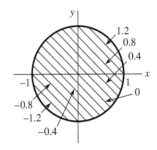

Exercises 20.6, Page 922

1. $g(z) = e^{-i\theta_0}$ is analytic everywhere and $G(z) = e^{-i\theta_0}z$ is a complex potential. The equipotential lines are the lines $x\cos\theta_0 + y\sin\theta_0 = c$.

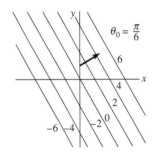

3. $g(z) = 1/z$ is analytic for $z \neq 0$ and $G(z) = \text{Ln } z$ is analytic except for $z = x \leq 0$. The equipotential lines are the circles $x^2 + y^2 = e^{2c}$.

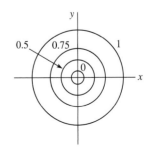

5. $\phi = \dfrac{4}{\pi}$ Arg z or $\phi(r, \theta) = \dfrac{4}{\pi}\theta$, and $G(z) = \dfrac{4}{\pi}$ Ln z is a complex potential. The equipotential lines are the rays $\theta = \dfrac{\pi}{4}c$ and $\mathbf{F} = \dfrac{4}{\pi}\left(\dfrac{x}{x^2+y^2}, \dfrac{y}{x^2+y^2}\right)$.

7. The equipotential lines are the images of the rays $\theta = \theta_0$ under the successive transformations $\zeta = w^{1/2}$ and $z = (\zeta + 1)/(-\zeta + 1)$. The transformation $\zeta = w^{1/2}$ maps the ray $\theta = \theta_0$ to the ray $\theta = \theta_0/2$ in the ζ-plane, and $z = (\zeta + 1)/(-\zeta + 1)$ maps this ray onto an arc of a circle that passes through $z = -1$ and $z = 1$.

9. (a) $\psi(x, y) = 4xy(x^2 - y^2)$ or, in polar coordinates, $\psi(r, \theta) = r^4 \sin 4\theta$. Note that $\psi = 0$ on the boundary of R.

(b) $\mathbf{V} = \overline{4z^3} = 4(x^3 - 3xy^2, y^3 - 3x^2y)$

(c)

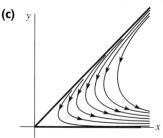

11. (a) $\psi(x, y) = \cos x \sinh y$ and $\psi = 0$ on the boundary of R.

(b) $\mathbf{V} = \overline{\cos z} = (\cos x \cosh y, \sin x \sinh y)$

(c)

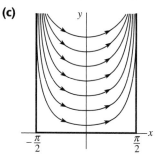

13. (a) $\psi(x, y) = 2xy - \dfrac{2xy}{(x^2+y^2)^2}$ or, in polar coordinates, $\psi(r, \theta) = (r^2 - 1/r^2) \sin 2\theta$. Note that $\psi = 0$ on the boundary of R.

(b) $\mathbf{V} = 2\bar{z} - 2/\bar{z}^3$

(c)

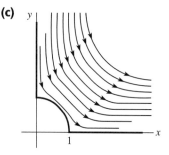

15. (a) $f(t) = \pi i - \dfrac{1}{2}$ $[\log_e|t + 1| + \log_e|t - 1|$
$+ i$ Arg$(t + 1) + i$ Arg$(t - 1)]$ and so

$$\mathrm{Im}(f(t)) = \begin{cases} 0, & t < -1 \\ \pi/2, & -1 < t < 1 \\ \pi, & t > 1. \end{cases}$$

Hence, $\mathrm{Im}(G(z)) = \psi(x, y) = 0$ on the boundary of R.

(b) $x = -\frac{1}{2}[\log_e|t + 1 + ic| + \log_e|t - 1 + ic|]$

$y = \pi - \frac{1}{2}[\mathrm{Arg}(t + 1 + ic) + \mathrm{Arg}(t - 1 + ic)]$, for $c > 0$

(c)

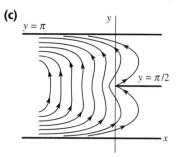

17. (a) $f(t) = \dfrac{1}{\pi}((t^2 - 1)^{1/2} + \cosh^{-1} t) = \dfrac{1}{\pi}((t^2 - 1)^{1/2} +$

$\mathrm{Ln}(t + (t^2 - 1)^{1/2}))$ and so $\mathrm{Im}(f(t)) = \begin{cases} 1, & t < -1 \\ 0, & t > 1 \end{cases}$ and

$\mathrm{Re}(f(t)) = 0$ for $-1 < t < 1$. Hence, $\mathrm{Im}(G(z)) = \psi(x, y) = 0$ on the boundary of R.

(b) $x = \mathrm{Re}\left[\dfrac{1}{\pi}\left(((t + ic)^2 - 1)^{1/2} + \cosh^{-1}(t + ic)\right)\right]$

$y = \mathrm{Im}\left[\dfrac{1}{\pi}\left(((t + ic)^2 - 1)^{1/2} + \cosh^{-1}(t + ic)\right)\right]$

for $c > 0$

(c)

19. $z = 0$ in Example 5; $z = 1, z = -1$ in Example 6

21. The streamlines are the branches of the family of hyperbolas $x^2 + Bxy - y^2 - 1 = 0$ that lie in the first quadrant. Each member of the family passes through $(1, 0)$.

23. *Hint:* For z in the upper half-plane,

$$k[\mathrm{Arg}(z - 1) - \mathrm{Arg}(z + 1)] = k\,\mathrm{Arg}\left(\dfrac{z - 1}{z + 1}\right).$$

Chapter 20 Review Exercises, Page 924

1. $v = 4$　　　**3.** the wedge $0 \leq \text{Arg } w \leq 2\pi/3$

5. true　　　**7.** $0, 1, \infty$　　　**9.** false

11. The image of the first quadrant is the strip $0 < v < \pi/2$. Rays $\theta = \theta_0$ are mapped onto horizontal lines $v = \theta_0$ in the w-plane.

13. $w = \dfrac{i - \cos \pi z}{i + \cos \pi z}$

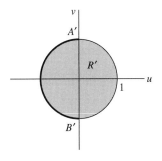

15. $u = 2 - 2y/(x^2 + y^2)$

17. (a) Note that $\alpha_1 \to 0$, $\alpha_2 \to 2\pi$, and $\alpha_3 \to 0$ as $u_1 \to \infty$.

(b) *Hint:* Write $f(t) = \frac{1}{2} A[\log_e |t + 1| + \log_e |t - 1| + i \text{ Arg}(t + 1) + i \text{ Arg}(t - 1)] + B$.

19. $G(z) = f^{-1}(z)$ maps R to the strip $0 \leq v \leq \pi$, and $U(u, v) = v/\pi$ is the solution to the transferred boundary problem. Hence, $\phi(x, y) = (1/\pi)\text{Im}(G(z)) = (1/\pi)\psi(x, y)$, and so the equipotential lines $\phi(x, y) = c$ are the streamlines $\psi(x, y) = c\pi$.

Appendix II Exercises, Page 932

1. $24; 720; 4\sqrt{\pi}/3; -8\sqrt{\pi}/15$　　　**3.** 0.297

5. $\Gamma(x) > \displaystyle\int_0^1 t^{x-1} e^{-t} \, dt > e^{-1} \int_0^1 t^{x-1} \, dt = \dfrac{1}{xe}$ for $x > 0$.

As $x \to 0^+$, $1/x \to +\infty$.

Index